COMPREHENSIVE BIOTECHNOLOGY

IN 4 VOLUMES

COMPREHENSIVE BIOTECHNOLOGY

The Principles, Applications and Regulations of Biotechnology in Industry, Agriculture and Medicine

EDITOR-IN-CHIEF

MURRAY MOO-YOUNG

University of Waterloo, Ontario, Canada

Volume 3

The Practice of Biotechnology: Current Commodity Products

VOLUME EDITORS

HARVEY W. BLANCH

University of California, Berkeley, CA, USA

STEPHEN DREW

Merck, Sharp & Dohme, Rahway, NJ, USA

and

DANIEL I. C. WANG

Massachusetts Institute of Technology, Cambridge, MA, USA

PERGAMON PRESS

OXFORD · NEW YORK · TORONTO · SYDNEY · FRANKFURT

U.K.	Pergamon Press Ltd., Headington Hill Hall, Oxford OX3 0BW, England
U.S.A.	Pergamon Press Inc., Maxwell House, Fairview Park, Elmsford, New York 10523, U.S.A.
CANADA	Pergamon Press Canada Ltd., Suite 104, 150 Consumers Road, Willowdale, Ontario M2J 1P9, Canada
AUSTRALIA	Pergamon Press (Aust.) Pty. Ltd., P.O. Box 544, Potts Point, N.S.W. 2011, Australia
FEDERAL REPUBLIC OF GERMANY	Pergamon Press GmbH, Hammerweg 6, D-6242 Kronberg-Taunus, Federal Republic of Germany

First edition 1985

Library of Congress Cataloging in Publication Data

Main entry under title:
Comprehensive biotechnology.
Includes bibliographies and index.
Contents: v. 1. The principles of biotechnology – scientific fundamentals / volume editors, Alan T. Bull, Howard Dalton –
v. 2. The principles of biotechnology – engineering considerations / volume editors, A.E. Humphrey, Charles L. Cooney – [etc.] –
v. 4. The practice of biotechnology – speciality products and service activities / volume editors, C.W. Robinson, John A. Howell.
1. Biotechnology.   I. Moo-Young, Murray.
TP248.2.C66   1985   660′.6   85–6509

British Library Cataloguing in Publication Data

Comprehensive biotechnology: the principles, applications and regulations of biotechnology in industry, agriculture and medicine.
1. Biotechnology
I. Title   II. Moo-Young, Murray
660′.6   TP248.3

ISBN 0–08–032511–4 (vol. 3)
ISBN 0–08–026204–X (4–vol. set)

Printed in Great Britain by A. Wheaton & Co. Ltd., Exeter

Contents

Preface

In his recent book, entitled 'Megatrends', internationally-celebrated futurist John Naisbitt observed that recent history has taken industrialized civilizations through a series of technology-based eras: from the chemical age (plastics) to an atomic age (nuclear energy) and a microelectronics age (computers) and now we are at the beginning of an age based on biotechnology. Biotechnology deals with the use of microbial, plant or animal cells to produce a wide variety of goods and services. As such, it has ancient roots in the agricultural and brewing arts. However, recent developments in genetic manipulative techniques and remarkable advances in bioreactor design and computer-aided process control have founded a 'new biotechnology' which considerably extends the present range of technical possibilities and is expected to revolutionize many facets of industrial, agricultural and medical practices.

Biotechnology has evolved as an ill-defined field from inter-related activities in the biological, chemical and engineering sciences. Inevitably, its literature is widely scattered among many specialist publications. There is an obvious need for a comprehensive treatment of the basic principles, methods and applications of biotechnology as an integrated multidisciplinary subject. *Comprehensive Biotechnology* fulfils this need. It delineates and collates all aspects of the subject and is intended to be the standard reference work in the field.

In the preparation of this work, the following conditions were imposed. (1) Because of the rapid advances in the field, it was decided that the work would be comprehensive but concise enough to enable completion within a set of four volumes published simultaneously rather than a more encyclopedic series covering a period of years to complete. In addition, supplementary volumes will be published as appropriate and the work will be updated regularly via *Biotechnology Advances*, a review journal, also published by Pergamon Press with the same executive editor. (2) Because of the multidisciplinary nature of biotechnology, a multi-authored work having an international team of experts was required. In addition, a distinguished group of editors was established to handle specific sections of the four volumes. As a result, this work has 10 editors and over 250 authors representing 15 countries. (3) Again, because of the multidisciplinary nature of the work, it was virtually impossible to use a completely uniform system of nomenclature for symbols. However, provisional guidelines on a more unified nomenclature of certain key variables, as provided by IUPAC, was recommended. (4) According to our definition, aspects of biomedical engineering (such as biomechanics in the development of prosthetic devices) and food engineering (such as product formulations) are not included in this work. (5) Since the work is intended to be useful to both beginners as well as veterans in the field, basic elementary material as well as advanced specialist aspects are covered. For convenience, a glossary of terms is supplied. (6) Since each of the four volumes is expected to be fairly self-contained, a certain degree of duplication of material, especially of basic principles, is inevitable. (7) Because of space constraints, a value judgement was made on the relative importance of topics in terms of their actual rather than potential commercial significance. For example, 'agricultural biotechnology' is given relatively less space compared to 'industrial biotechnology', the current raison d'être of biotechnology as a major force in the manufacture of goods and services. (8) Finally, a delicate balance of material was required in order to meet the objective of providing a comprehensive and stimulating coverage of important practical aspects as well as the intellectual appeal of the field. Readers may wish to use this work for initial information before possibly delving deeper into the literature as a result of the critical discussions and wide range of references provided in it.

Comprehensive Biotechnology is aimed at a wide range of user needs. Students, teachers, researchers, administrators and others in academia, industry and government are addressed. The requirements of the following groups have been given particular consideration: (1) chemists, especially biochemists, who require information on the chemical characteristics of enzymes, metabolic processes, products and raw materials, and on the basic mechanisms and analytical techniques involved in biotechnological transformations; (2) biologists, especially microbiologists and molecular biologists, who require information on the biological characteristics of living organisms involved in biotechnology and the development of new life forms by genetic engineering techniques; (3) health scientists, especially nutritionists and toxicologists, who require information on biohazards and containment techniques, and on the quality of products and by-products of biotechnological processes, including the pharmaceutical, food and beverage industries; (4) chemical engineers, especially biochemical engineers, who require information on mass and energy balances and rates of processes, including fermentations, product recovery and feedstock pretreatment, and the equipment for carrying out these processes; (5) civil engineers, especially environmental engineers, who require information on biological waste treatment methods and equipment, and on contamination potentials of the air, water and land within the ecosystem, by industrial and domestic effluents; (6) other engineers, especially agricultural and biomedical engineers, who require information on advances in the relevant sciences that could significantly affect the future practice of their professions; (7) administrators, particularly executives and legal advisors, who require information on national and international governmental regulations and guidelines on patents, environmental pollution, external aid programs and the control of raw materials and the marketing of products.

No work of this magnitude could have been accomplished without suitable assistance. For guidance on the master plan, I am indebted to the International Advisory Board (J. D. Bu'Lock, T. K. Ghose, G. Hamer, J. M. Lebault, P. Linko, C. Rolz, H. Sahm, B. Sikyta and H. Taguchi). For structuring details of the various sections, the invaluable assistance of the section editors is gratefully acknowledged, especially Alan Bull, Charles Cooney, Harvey Blanch and Campbell Robinson, who also acted as coordinators for each of the four volumes. For the individual chapters, the 250 authors are to be commended for their hard work and patience during the two years of preparation of the work. For checking the hundreds of literature references cited in the various chapters, the many graduate students are thanked for a tedious but important task well done. A special note of thanks is due to Jonathan and Arlene Lamptey, who acted as editorial assistants in many diverse ways. At Pergamon Press, I wish to thank Don Crawley for originally suggesting this project and Colin Drayton for managing it. Finally, I am pleased to note the favourable evaluations of the work by two distinguished authorities, Sir William Henderson and Nobel Laureate Donald Glaser, who provided a foreword and a guest editorial, respectively, to the treatise.

MURRAY MOO-YOUNG
Waterloo, Canada
December 1984

Foreword

This very comprehensive reference work on biotechnology is published ten years after the call by the National Academy of Sciences of the United States of America for a voluntary worldwide moratorium to be placed on certain areas of genetic engineering research thought to be of potential hazard. The first priority then became the evaluation of the conjectural risks and the development of guidelines for the continuation of the research within a degree of containment. There had hardly been a more rapid response to this type of situation than that of the British Advisory Board for the Research Councils. The expression of concern by Professor Paul Berg and the committee under his chairmanship, and the call for the moratorium, was published in *Nature* on 19 July 1974. The Advisory Board agreed at their meeting on the 26 July to establish a Working Party with the following terms of reference:

'To assess the potential benefits and potential hazards of the techniques which allow the experimental manipulation of the genetic composition of micro-organisms, and to report to the Advisory Board for the Research Councils.'

Because of the conviction of those concerned that recombinant DNA techniques could lead to great benefits, the word order used throughout the report of the Working Party (Chairman, Lord Ashby) always put 'benefits' before 'hazards'. The implementation of the recommendations led to the development of codes of practice. This was followed by the establishment of the Genetic Manipulation Advisory Group as a standing central advisory authority operating within the framework of the Health and Safety at Work, *etc.* Act 1974 and, later, more specifically within the framework of the Health and Safety (Genetic Manipulation) Regulations 1978. Similar moves took place in many other countries but the other most prominent and important activity was that of the US National Institutes of Health. This resulted in the adoption by most countries of the NIH or the UK guidelines, or the use of practices based on both.

The significant consequence of the debates, the discussions and of the recommendations that emerged during these early years of this decade (1974–1984) was that research continued, expanded and progressed under increasingly less restriction at such a pace that now makes it possible and necessary to devote the first Section of Volume 1 of this work to genetic engineering. Many chapters of the subsequent Sections and Volumes are of direct relevance to the application of genetic engineering.

The reason for identifying today's genetic engineering for first mention in this foreword is its novelty. It was being conceived barely more than ten years ago. Ten years by most standards is a short time. Although in the biological context it represents at least 10^4 generation times of the most vigorous viruses, it is less than one of man even for the most precocious. The current developments in biotechnology, whether they be in recombinant DNA, monoclonal antibodies, immobilized enzymes, *etc.* are mostly directed towards producing a better product, or a better process. This is commendable and is supportable by the ensuing potential commercial benefits. The newer challenge is the application of the new biotechnology to achieve what previously could scarcely have been contemplated. Limited biological sources of hormones, growth regulators, *etc.* are being, and will be increasingly, replaced by the use of transformed microorganisms, providing a vastly increased scale of production. Complete safety of vaccines by the absence of ineffectively inactivated virus is one of the great advantages of the genetically engineered antigen. This is quite apart from the ability to prepare products for which, at present, there is a technical difficulty or which is economically not feasible by standard methods.

A combination of advances in recombinant DNA research, molecular biology and in blastomere manipulation has provided the technology to insert genetic material into the totipotent animal cell. The restriction on the application of this technology for improved animal production is the lack of knowledge on the genetic control of desirable biological characteristics for transfer from one breed line to another.

There are probably greater potential benefits to be won in the cultivation of the domesticated plants than in the production of the domesticated animals. In both cases, the objectives are to increase the plant's or the animal's resistance to the prejudicial components of its environment and to increase the yield, quality and desired composition of the marketable commodities. These include the leaf, the tuber, the grain, the berry, the fruit or the milk, meat and other products of animal origin. This is not taking into account the other valuable products of horticulture, of oil or wax palms, rubber trees and forestry in general. Genetic engineering should be able to provide short-cuts to reach objectives attainable by traditional procedures, for example by by-passing the sequential stages of a traditional plant breeding programme by the transfer of the genetic material in one step. Examples of desirable objectives are better to meet user specifications with regard to yield, quality, biochemical composition, disease and pest resistance, cold tolerance, drought resistance, nitrogen fixation, *etc.* One of the constraints in this work in plants is the scarcity of vectors compared with the many available for the transformation of microorganisms. The highest research priority on the plant side is to determine by one means or another how to increase the efficiency of photosynthesis. The photosynthetic efficiency of temperate crop plants is no more than 2–2.5% in terms of conversion of intercepted solar energy. These plants possess the C_3 metabolic pathway with the energy loss of photorespiration. Tropical species of plants with the C_4 metabolic pathway have a higher efficiency of photosynthesis in that they do not photorespire. One approach for the breeder of C_3 plants is to endow them with a C_4 metabolism. If this transformation is ever to be achieved, it is most likely to be by genetic engineering. Such an advance has obvious advantages with regard, say, to increased wheat production for the ever-increasing human population. Nitrogen fixation as an agricultural application of biotechnology is given prominence in Section 1 of Volume 4. Much knowledge has been acquired about the chemistry and the biology of the fixation of atmospheric nitrogen. This provides a solid foundation from which to attempt to exploit the potential for transfer of nitrogen-fixing genes to crop plants or to the symbiotic organisms in their root systems. If plants could be provided with their own capability for nitrogen fixation, the energy equation might not be too favourable in the case of high yielding varieties. Without an increase in the efficiency of photosynthesis, any new property so harnessed would have to be at the expense of the energy requirements of existing characteristics such as yield.

Enzymes have been used for centuries in the processing of food and in the making of beverages. The increasing availability of enzymes for research, development and industrial use combined with systems for their immobilization, or for the immobilization of cells for the utilization of their enzymes, is greatly expanding the possibilities for their exploitation. Such is the power of the new biotechnology that it will be possible to produce the most suitable enzymes for the required reaction with the specific substrate. An increasing understanding at the molecular level of enzyme degradation will make it possible for custom-built enzymes to have greater stability than those isolated from natural sources.

The final section of Volume 4 deals with waste and its management. This increasingly voluminous by-product of our society can no longer be effectively dealt with by the largely empirical means that continue to be practised. Biological processes are indispensable components in the treatment of many wastes. The new biotechnology provides the opportunity for moving from empiricism to processes dependent upon the use of complex biological reactions based on the selection or the construction of the most appropriate cells or their enzymes.

The very comprehensive coverage of biotechnology provided by this four-volume work of reference reflects that biotechnology is the integration of molecular biology, microbiology, biochemistry, cell biology, chemical engineering and environmental engineering for application to manufacturing and servicing industries. Viruses, bacteria, yeasts, fungi, algae, the cells and tissues of higher plants and animals, or their enzymes, can provide the means for the improvement of existing industrial processes and can provide the starting points for new industries, for the manufacture of novel products and for improved processes for management of the environment.

SIR WILLIAM HENDERSON, FRS
Formerly of the *Agricultural Research Council*
and *Celltech Ltd.*, *London, UK*

Guest Editorial

Since 1950, the new science of molecular biology has produced a remarkable outpouring of new ideas and powerful techniques. From this revolution has sprung a new discipline called genetic engineering, which gives us the power to alter living organisms for important purposes in medicine, agriculture and industry. The resulting biotechnologies span the range from the ancient arts of fermentation to the most esoteric use of gene splicing and monoclonal antibodies. With unprecedented speed, new scientific findings are translated into industrial processes, sometimes even before the scientific findings have been published. In earlier times there was a more or less one-way flow of new discoveries and techniques from scientific institutions to industrial organizations where they were exploited to make useful products. In the burgeoning biotechnology industry, however, developments are so rapid that there is a close intimacy between science and technology which blurs the boundaries between them. Modern industrial laboratories are staffed with sophisticated scientists and equipped with modern facilities so that they frequently produce new scientific discoveries in areas that were previously the exclusive province of universities and research institutes, and universities not infrequently develop inventions and processes of industrial value in biotechnology and other fields as well.

Even the traditional flow of new ideas from science to application is no longer so clear. In many applications, process engineers may find that the most economical and efficient process design requires an organism with new properties or an enzyme of previously unknown stability. These requirements often motivate scientists to try to find in nature, or to produce through genetic engineering or other techniques of molecular biology, novel organisms or molecules particularly suited for the requirements of production. A recent study done for the United States Congress* concluded that "in the next decade, competitive advantage in areas related to biotechnology may depend as much on developments in bioprocess engineering as on innovations in genetics, immunology, and other areas of basic science."

These volumes bring together for the first time in one unified publication the scientific and engineering principles on which the multidisciplinary field of biotechnology is based. Following accounts of the scientific principles is a large set of illustrations of the diverse applications of these principles in the practice of biotechnology. Finally, there are sections dealing with important regulatory aspects of the potential hazards of the growing field and of the need for promoting biotechnology in developing countries.

Comprehensive Biotechnology has been produced by a team of some of the world's foremost experts in various aspects of biotechnology and will be an invaluable resource for those wishing to build bridges between 'academic' and 'commercial' biotechnology, the ultimate form of any technology.

DONALD A. GLASER
University of California, Berkeley
and *Cetus Corp., Palo Alto, CA, USA*

*"Commercial Biotechnology: An International Analysis," Office of Technology Assessment Report, U.S. Congress, Pergamon Press, Oxford, 1984.

Executive Summary

In this work, biotechnology is interpreted in a fairly broad context: the evaluation and use of biological agents and materials in the production of goods and services for industry, trade and commerce. The underlying scientific fundamentals, engineering considerations and governmental regulations dealing with the development and applications of biotechnological processes and products for industrial, agricultural and medical uses are addressed. In short, a comprehensive but concise treatment of the principles and practice of biotechnology as it is currently viewed is presented. An outline of the main topics in the four volumes is given in Figure 1.

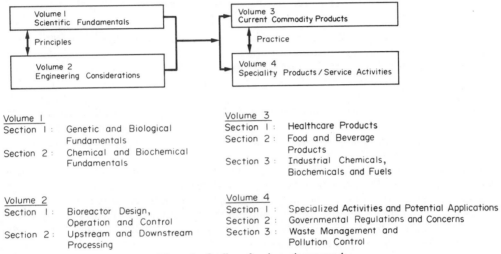

Volume 1
Section 1 : Genetic and Biological Fundamentals
Section 2 : Chemical and Biochemical Fundamentals

Volume 2
Section 1 : Bioreactor Design, Operation and Control
Section 2 : Upstream and Downstream Processing

Volume 3
Section 1 : Healthcare Products
Section 2 : Food and Beverage Products
Section 3 : Industrial Chemicals, Biochemicals and Fuels

Volume 4
Section 1 : Specialized Activities and Potential Applications
Section 2 : Governmental Regulations and Concerns
Section 3 : Waste Management and Pollution Control

Figure 1 Outline of main topics covered

As depicted in Figure 2, it is first recognized that biotechnology is a multidisciplinary field having its roots in the biological, chemical and engineering sciences leading to a host of specialities, *e.g.* molecular genetics, microbial physiology, biochemical engineering. As shown in Figure 3, this is followed by a description of technical developments and commercial implementation,

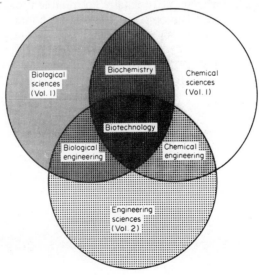

Figure 2 Multidisciplinary nature of biotechnology

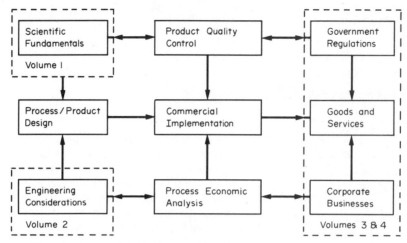

Figure 3 Interrelationships between biotechnology principles and applications

the ultimate form of any technology, which takes into account other important factors such as socio-economic and geopolitical constraints in the marketplace.

There are two main divisions of the subject matter: a pedagogical academic coverage of the disciplinary underpinnings of the field (Volumes 1 and 2) followed by a utilitarian practical view of the various commercial processes and products (Volumes 3 and 4). In the integration of these two areas, other common factors dealing with product quality, process economics and government policies are introduced at appropriate points throughout all four volumes. Since biotechnological advances are often ahead of theoretical understanding, some process descriptions are primarily based on empirical knowledge.

The four volumes are relatively self-contained according to the following criteria. Volume 1 delineates and integrates the unifying multidisciplinary principles in terms of relevant scientific fundamentals. Volume 2 delineates and integrates the unifying multidisciplinary principles of biotechnology in terms of relevant engineering fundamentals. Volume 3 describes the various biotechnological processes which are involved in the manufacture of bulk commodity products. Volume 4 describes various specialized services, potential applications of biotechnology and related government concerns. In each volume, a glossary of terms and nomenclature guideline are included to assist the beginner and the non-specialist.

This work takes into account the relative importance of the various topics, primarily in terms of current practice. Thus, bulk commodity products of the manufacturing industries (Volume 3) are accorded more space compared to less major ones and for potential applications (part of Volume 4). This proportional space distribution may be contrasted with the expectations generated by the recent news media 'biohype'. For example, virtually no treatment of 'biochips' is presented. In addition, since the vast majority of commercial ventures involve microbial cells and cell-derived enzymes, relatively little coverage is given to the possible use of whole plant or animal cells in the manufacturing industries. As future significant areas of biotechnology develop, supplementary volumes of this work are planned to cover them. In the meantime, on-going progress and trends will be covered in Pergamon's complementary review journal, *Biotechnology Advances*.

M. Moo-Young
University of Waterloo, Canada

Contributors to Volume 3

Dr R. G. Aldi
Hiram Walker and Sons Ltd, Walkerville, Box 2518, Windsor, Ontario N8Y 4S5, Canada

Dr J. T. Baker
Sir George Fisher Centre for Tropical Marine Science, James Cook University, Queensland 4811, Australia

Dr A. Bhadra, Department of Chemical Engineering, University of Waterloo, Waterloo, Ontario N2L 3G1, Canada

Professor H. W. Blanch
Department of Chemical Engineering, University of California, Berkeley, CA 94720, USA

Dr G. Boguslawski
Miles Laboratories, PO Box 932, Elkhart, IN 46515, USA

Dr D. M. Booth
Hiram Walker and Sons Ltd, Walkerville, Box 2518, Windsor, Ontario N8Y 4S5, Canada

Dr C. A. Boulton
Bass PLC, 137 High Street, Burton-on-Trent DE14 1JZ, UK

Dr B. C. Buckland
Department of Biochemical Engineering, Merck, Sharp and Dohme Research Laboratories, Rahway, NJ 07065, USA

Dr S. L. Chen
Universal Foods Corporation, 6143 North 60th Street, Milwaukee, WI 53218, USA

Dr I. Chibata
Research and Development Headquarters, Tanabe Seiyaku Co Ltd, 16–89 Kashima-3-chome, Yodogawa-ku, Osaka 532, Japan

Dr M. Chiger
Universal Foods Corporation, 6143 North 60th Street, Milwaukee, WI 53218, USA

Dr L. E. Coker
Hubinger Co, One Progress Street, Keokuk, IA 52632, USA

Dr S. Drew
Merck, Sharp & Dohme, PO Box 2000, Rahway, NJ 07065, USA

Dr H. Enei
Central Research Laboratories, Ajinomoto Co Inc, 1-1 Suzuki-cho, Kawasaki-ku, Kawasaki-shi 210, Japan

Dr M. C. Flickinger
Institute for Advanced Studies in Biological Process Technology, 240 Gortner Laboratory, 1479 Gortner Avenue, University of Minnesota, St Paul, MN 55108, USA

Dr J. Florent
Département Biochimie, Rhône-Poulenc Santé, Centre de Recherches de Vitry-sur-Seine, 13 quai Jules Guesde, 94407 Vitry-sur-Seine Cedex, France

Dr B. A. Friend
Ralston Purina Company, Checkerboard Square, St Louis, MO 63164, USA

Dr P. A. Franks
Mauri Dairy Laboratories, Moorebank Avenue, Moorebank, Sydney, NSW 2170, Australia

Professor T. K. Ghose
Biochemical Engineering Research Centre, Indian Institute of Technology, Hauz Khas, New Delhi 110016, India

Dr J. E. Gonzalez
Fermentation Research and Development, The Upjohn Company, Kalamazoo, MI 49001, USA

Dr P. P. Gray
School of Biotechnology, University of New South Wales, Kensington, Sydney, NSW 2033, Australia

Dr R. J. Hall
Department of Bioprocess Engineering, School of Chemical Engineering and Industrial Chemistry, University of New South Wales, Kensington, NSW 2033, Sydney, Australia

Dr A. R. Hill
Department of Food Science, University of Guelph, Guelph, Ontario N1G 2W1, Canada

Dr Y. Hirose
Central Research Laboratories, Ajinomoto Co Inc, 1-1 Suzuki-cho, Kawasaki-ku, Kawasaki-shi 210, Japan

Dr G. H. Hunt
Department of Biochemical Engineering, Merck, Sharp & Dohme Research Laboratories, Rahway NJ 07065, USA

Professor D. M. Irvine
Department of Food Science, University of Guelph, Guelph, Ontario N1G 2W1, Canada

Dr N. B. Jansen
Laboratory of Renewable Resources Engineering, Potter Center, Purdue University, West Lafayette, IN 47907, USA

Dr Y. C. Jao
Miles Laboratories, PO Box 932, Elkhart, IN 46515, USA

Dr H. Kase
Kyowa Hakko Kogyo Co Ltd, Tokyo Research Laboratory, 3-6-6 Asahicho, Machida-shi 194, Tokyo, Japan

Dr B. Khosrovi
Cetus Corporation, 1400 Fifty-third Street, Emeryville, CA 94608, USA

Professor A. Kilara
Department of Food Science, Pennsylvania State University, University Park, PA 16802, USA

Professor H. Kleinkauf
Institut für Biochemie und Molekulare Biologie, Technische Universität Berlin, Franklinstrasse 29, D-1000 Berlin 10, Federal Republic of Germany

Professor M. R. Ladisch
Laboratory of Renewable Resources Engineering, Potter Center, Purdue University, West Lafayette, IN 47907, USA

Dr O. J. Lantero
Miles Laboratories, PO Box 932, Elkhart, IN 46515, USA

Dr T. G. Lenz
Department of Agricultural and Chemical Engineering, Colorado State University, Fort Collins, CO 80523, USA

Professor J. C. Linden
Department of Agricultural and Chemical Engineering, Colorado State University, Fort Collins, CO 80523, USA

Dr J. H. Litchfield
Battelle Memorial Institute, Columbus Laboratories, 505 King Avenue, Columbus, OH 43201, USA

Professor B. L. Maiorella
Cetus Corporation, 1400 Fifty-third Street, Emeryville, CA 94608, USA

Professor A. Margaritis
Faculty of Engineering Science, University of Western Ontario, London, Ontario N6A 5BG, Canada

Dr J. L. Meers
John & E Sturge Ltd, Denison Road, Selby, North Yorkshire YO8 8EF, UK

Dr A. J. M. Messenger
Department of Biochemistry, University of Hull, Hull HU6 7RX, UK

Dr T. L. Miller
Fermentation Products Production, The Upjohn Company, Kalamazoo, MI 49001, USA

Mr P. E. Milsom
John & E Sturge Ltd, Denison Road, Selby, North Yorkshire YO8 8EF, UK

Dr A. R. Moreira
Department of Agricultural and Chemical Engineering, Colorado State University, Fort Collins, CO 80523, USA

Dr K. Nakayama
Zama Research Laboratories, Bior Inc, 5-5410 Hibarigaoka, Zama-shi, Kanagawa-ken 228, Japan

Dr R. Narayan
Laboratory of Renewable Resources Engineering, Potter Center, Purdue University, West Lafayette, IN 47907, USA

Dr D. R. Omstead
Department of Biochemical Engineering, Merck, Sharp & Dohme Research Laboratories, Rahway, NJ 07065, USA

Dr G. W. Pace
Biotechnology Australia Pty Ltd, 28 Barcoo Street, East Roseville, NSW 2069, Australia

Dr M. J. Playne
Biotechnology Section, CSIRO Division of Chemical and Wood Technology, Private Bag 10, Clayton, Victoria 3168, Australia

Professor C. Ratledge
Department of Biochemistry, University of Hull, Hull HU6 7RX, UK

Dr J. L. Reichelt
Sir George Fisher Centre for Tropical Marine Science, James Cook University, Queensland 4811, Australia

Dr V. W. Rodwell
Laboratory of Renewable Resources Engineering, Potter Center, Purdue University, West Lafayette, IN 47907, USA

Dr I. Russell
Labatt Brewing Co Ltd, 150 Simcoe Street, London, Ontario N6A 4M3, Canada

Dr V. Santamarina
Department of Biochemical Engineering, Merck, Sharp & Dohme Research Laboratories, Rahway, NJ 07065, USA

Dr T. Sato
Research and Development Headquarters, Tanabe Seiyaku Co Ltd, 16-89 Kashima-3-chome, Yodogawa-ku, Osaka 532, Japan

Dr S. T. Schlager
Miles Laboratories, PO Box 932, Elkhart, IN 46515, USA

Dr K. M. Shahani
Department of Food Science and Technology, University of Nebraska, Lincoln, NE 68583, USA

Dr F. H. Sharpell, Jr
Givaudan Corporation, 125 Delawanna Avenue, Clifton, NJ 07015, USA

Dr H. Shibai
Central Research Laboratories, Ajinomoto Co Inc, 1-1 Suzuki-cho, Kawasaki-ku, Kawasaki-shi 210, Japan

Professor K. Shimura
Department of Agricultural Chemistry, Tohoku University, Sendai 980, Japan

Mr A. Smith
Biotechnology K681–145, Ciba-Geigy AG, CH-4002 Basle, Switzerland

Dr M. Sobolov
Hiram Walker and Sons Ltd, Walkerville, Box 2518, Windsor, Ontario N8Y 4S5, Canada

Dr G. L. Solomons
RHM Research Ltd, The Lord Rank Research Centre, Lincoln Road, High Wycombe, Bucks HP12 3QR, UK

Dr G. G. Stewart
Labatt Brewing Co Ltd, 150 Simcoe Street, London, Ontario N6A 4M3, Canada

Dr H. Stockdale
Shell Research Ltd, Sittingbourne Research Centre, Sittingbourne, Kent ME9 8AG, UK

Dr D. C. Sutton
Sir George Fisher Centre for Tropical Marine Science, James Cook University, Queensland 4811, Australia

Dr R. W. Swartz
Department of Chemical Engineering, Tufts University, Medford, MA 02155, USA

Dr T. Tosa
Research and Development Headquarters, Tanabe Seiyaku Co Ltd, 16-89 Kashima-3-chome, Yodogawa-ku, Osaka 532, Japan

Dr G. T. Tsao
Laboratory of Renewable Resources Engineering, Potter Center, Purdue University, West Lafayette, IN 47907, USA

Dr K. Venkatasubramanian
H J Heinz Co, World Headquarters, PO Box 57, Pittsburgh, PA 15230, USA

Dr F. H. Verhoff
Miles Laboratories, PO Box 932, Elkhart, IN 46515, USA

Dr T. B. Vickroy
W R Grace & Co, Washington Research Center, 7379 Route 32, Columbia, MD 21044, USA

Dr M. Voloch
Laboratory of Renewable Resources Engineering, Potter Center, Purdue University, West Lafayette, IN 47907, USA

Dr H. von Döhren
Institut für Biochemie und Molekulare Biologie, Technische Universität Berlin, Franklinstrasse 29, D-1000 Berlin 10, Federal Republic of Germany

Professor D. I. C. Wang
Department of Nutrition and Food Science, Massachusetts Institute of Technology, Room 16-114, Cambridge, MA 02139, USA

Dr O. P. Ward
School of Biological Sciences, National Institute for Higher Education, Dublin 9, Republic of Ireland

Dr T. Yokotsuka
Kikkoman Corporation, 339 Noda, Noda-shi, Chiba-ken 278, Japan

Contents of All Volumes

SECTION 1

HEALTHCARE PRODUCTS

1
Introduction

S. W. DREW
Merck & Co. Inc., Rahway, NJ, USA

1.1 INTRODUCTION

Biotechnology has played an essential role in the development of the healthcare chemical industries. The range of products includes diagnostic, prophylactic and therapeutic agents. The pharmaceutical industries are the largest manufacturers of these agents with chemicals in the latter two categories accounting for major markets. The search for and discovery of new drugs almost always involves a biological model of action related to disease. The discovery of a potentially active compound starts a sequence of exhaustive chemical and biological testing that may culminate in manufacture of the agent or an improved analog. The role of biotechnology in this complex path to regulatory approval and marketing is diverse, often starting with preparation of the agents to be screened or the biological targets for action. Scale-up to commercial manufacture of the compound or its precursors often involves biological processes as the route of choice to complex chemistries.

1.2 THE PHARMACEUTICAL INDUSTRIES

The world market for healthcare drugs in 1981 was nearly $25 billion and rose to approximately $35 billion by 1983. Over 110 000 tonnes of bulk medicinals were produced in the United States in 1980 relative to research and development expenditures of nearly $1.9 billion (Austin, 1984). Research and development expenditures for ethical products produced by the major US pharmaceutical manufacturers could exceed $2.7 billion in 1985.

Products in the pharmaceutical industry are typically high potency, low volume and high value. The amount of product in an effective dose may range from less than a microgram per dose for live-virus vaccines to tens of grams per patient-day for antibiotic therapy. Annual manufacturing level for significant market penetration ranges from a few grams of live-virus vaccine to about 100 000 kg of an antibiotic such as tetracycline. The value of a drug, an article intended for use in the diagnosis, prevention, treatment, mitigation or cure of disease in humans or animals (Osol, 1980), may range to over $1 million per gram for a new biologic (*e.g.* a vaccine).

The path to product approval and licensure by the regulatory authority is arduous and long, typically requiring seven to ten years to reach licensure of human healthcare products. This process is research intensive with major efforts spent on extensive clinical testing and development of commercial process technology. Establishment of proprietary rights, protected by patent or license (usually with an exclusivity clause) is often a prerequisite for corporate commitment to develop a new drug. A primary goal of drug discovery programs is to identify new, unique drugs that are safe, effective and broadly patentable.

1.2.1 Marketing Strategy

The general strategy in marketing human and animal healthcare drugs is to achieve market entry first with a new drug of high quality, priced relative to the cost/benefit value of the treated disease. Market share in the human drug category can usually be taken from an established product by one with qualities that are clearly recognized as improved. Product differentiation and the manufacturer's reputation are major factors in establishing market share between similar, competing therapeutics. While the manufacturer's reputation may be the largest factor in establishing the market share for similar, competing biologics, first entry in the marketplace usually dominates the impact of early product differentiation. Although product characteristics (absence of side effects, supplemental label claim, color and form, purity, *etc.*) seem to control the human healthcare drug market, the animal healthcare market for food production seems to be clearly more sensitive to the price for benefit received.

1.3 COMMON FEATURES IN THE EVOLUTION OF PRODUCTS AND PROCESSES

Table 1 lists 11 classes of drugs of major commercial importance in human healthcare and five categories of importance in the animal healthcare market. Biological synthesis is the preferred route to most products of complex chemistry because of its efficiency, selectivity and remarkable propensity for continued yield improvement. While biology and biotechnology have played important roles in discovering and developing drugs in virtually all of these classes, biotechnology for commercial production is currently limited to relatively few of the marketed compounds (roughly 23% of the total sales for the human healthcare markets shown in Table 1). The largest single class of drugs made entirely or in part through biological processes are the antibiotics, comprising nearly 90% of the systemic antiinfectives sales. Antibiotics are the largest selling group of drugs in the human healthcare market with cardiovascular agents and antiinflammatories running a distant second and third.

Table 1　Major Classes of Human and Animal Healthcare Drugs

Human Healthcare Drug Targets	
Antiinflammatories	Ophthalmologicals
Cardiovascular/renal system drugs	Respiratory system drugs
Dermatologicals	Systemic antiinfectives
Diabetics	Urologicals
Gastrointestinals	Vaccines and biologicals
Nervous system/mental health drugs	
Animal Healthcare Drug Targets	
Antiparasitics	Growth promotion (feed efficiency)
Gastrointestinals	Respiratory system drugs
Growth Permitants	

Most of the biological processes for the manufacture of human and animal healthcare drugs involve common steps in their evolution and commercial practice.

1.4 PROCESS TECHNOLOGY

1.4.1 Fermentation

Most of the biological routes to pharmaceuticals involve aerobic submerged cultivation of microorganisms (fermentation). (The term fermentation refers in this context not to anaerobic cultivation as in the alcoholic beverage industry, but to the general cultivation of microorganisms in liquid media.) Discussions of the principles of microbial cultivation can be found in the references by Drew (1982) and Wang and coworkers (1979). Commercial fermentations for manufacture of therapeutics are most often executed at large scale (45 to 150 m^3) while fermentations supporting manufacture of vaccines may be no more than 1000 l at full scale.

Fermenters in the pharmaceutical industry are low pressure vessels designed for high horsepower agitation to facilitate dissolution of oxygen in the growth medium. Although many of the production fermenters in the industry are well over 20 years old, the continual refinement of these vessels has kept them quite serviceable. The relatively short commercial period of patent protection for healthcare products combined with the high capital cost of building new fermenters has led the industry toward multiple use facilities rather than construction of new, dedicated factories. The continual drive toward higher volumetric productivity has led to high oxygen demand. While most of the larger fermenters were designed on the basis of oxygen transfer, modification of existing fermenters may be facing new constraints. Higher volumetric productivity has often meant increased broth viscosity (with increased cell mass) and increased heat load (increased power transfer from higher horsepower agitators and increased metabolic rates). Fluid bulk mixing and heat transfer are becoming increasingly more important parameters for achieving high fermenter productivity, particularly in older units. Wang and colleagues (1979) provide a detailed analysis of fermenter design. See Section 1 in Volume 2 of *Comprehensive Biotechnology* for additional detailed information on fermenter design and process control.

1.4.2 Product Recovery

Product characteristics such as purity and form are controlled by purification and final isolation procedures. Human healthcare products are generally of high purity (usually greater than 95% and often over 98% purity). Where possible, crystalline products are sought as a means of achieving high purity and desirable form (color, stability, dissolution rate). The ultimate use of these products contraindicates use of toxic separation agents in the final steps of product isolation. Fermentation products often have limited chemical and thermal stability. Generally, the larger the molecular weight of the product, the more gentle must be the recovery.

Belter (1979) defines four general steps in conventional isolation of fermentation products; with little modification they apply to recovery of non-fermentation biological products. Belter lists (1) removal of insolubles by filtration, centrifugation and decantation; (2) primary isolation of product through sorption, solvent extraction, precipitation and ultrafiltration; (3) purification by fractional precipitation, chromatography, chemical derivatization and decolorization; and (4) final product isolation by crystallization (or chromatography for larger molecular weight products). A sound fundamental approach to high product purity is to choose multiple purification procedures, each based on a different mechanistic principle. Additional insight into recovery of biological products can be found in Section 2, Volume 2 of this work.

1.5 NEW TRENDS IN BIOTECHNOLOGY

The defining of complex environmental conditions in the fermenter (that support still more complex process biochemistries) constrains the processes for recovery of active product. Impurities in the biological product may strongly influence subsequent chemical modification. The need for interdisciplinary coordination to facilitate product recovery has long existed, but historically, processes have been defined from the top down, starting with the genetics of the producing microorganisms but involving little feedback from downstream processing.

The famous admonition to the recovery engineer, 'It is the microbiologist's responsibility to put the product in the fermenter and your responsibility to get it out', speaks to the importance of high titers but shorts the opportunities for cost savings during recovery. Product recovery may contribute as much as 50% to the cost of bulk therapeutic drug manufacture and as much as 90% to the cost of manufacture of bulk biologicals and recombinant DNA products. As biologist, chemist and chemical engineer learn more about their respective fields, the opportunities for interaction become more apparent. The development of DNA recombination techniques in the commercial environment has accelerated the pace of communication between the biologist and the chemical engineer. This new technology enables the use of genetic modification as a tool for whole-process design. Fish and Lilly (1984) have addressed the potential for this integration between microbiology and engineering. Michaels (1984a, 1984b) has written two very entertaining and thought provoking articles on this topic.

Recombinant DNA technology has already had significant impact on new drug discovery and development. The products have been mostly biologics in the pharmaceutical industry. The future will see the use of recombinant DNA technology to improve the performance of other fermentations (*e.g.* antibiotics, cardiovascular agents, *etc.*) and to extend the process options now available for synthetic chemistry (biotransformations, complex chemistries reconstructed in live microorganisms). The rate at which these new developments come about will depend on interdisciplinary communication, a much improved understanding of process biochemistry, and the particular details of product age, value, state of development and competitive pressure.

The following contributions focus predominantly on commercially important drugs. The emphasis on antibiotics recognizes the importance of this class of drug. The concepts and techniques developed are, however, broadly applicable to other drugs that are or can be produced through process biochemistry. The final contribution deals with products that currently have very specific function of a biological nature. Many of the process concepts developed for these recombinant DNA products are generally applicable to the entire class of biologics.

1.6 REFERENCES

Austin, G. T. (1984). In *Shreve's Chemical Process Industries*, 5th edn., pp. 796–797. McGraw-Hill, New York.

Belter, P. A. (1979). Isolation of fermentation products. In *Microbial Technology*, ed. H. J. Peppler and D. Perlman, 2nd ed., vol. 2, pp. 403–432. Academic, New York.

Drew, S. W. (1982). Cultivation of microorganisms. In *Manual of Methods for General Bacteriology*, ed. P. Gerhardt, chap. 12, American Society for Microbiology, New York.

Fish, N. M. and M. D. Lilly (1984). The interactions between fermentations and protein recovery. *Bio/Technology*, **2**, 623–627.

Michaels, A. S. (1984a). The impact of genetic engineering. *Chem. Eng. Prog.*, April, 9–15.

Michaels, A. S. (1984b). Adapting modern biology to industrial practice. *Chem. Eng. Prog.*, June, 19–25.

Osol, A. (1980). *Remington's Pharmaceutical Sciences*, 16th edn., Mack Publishing Co., Easton, PA.

Wang, D. I. C., C. L. Cooney, A. L. Demain, P. Dunnill, A. E. Humphrey and M. D. Lilly (1979). Fermentation and enzyme technology. In *Techniques in Pure and Applied Microbiology*, ed. C.-G. Heden. Wiley, New York.

2

Penicillins

R. W. SWARTZ
Swartz Associates, Winchester, MA, USA

2.1 HISTORICAL PERSPECTIVE

2.1.1 History

In 1928, Alexander Fleming's curiosity was aroused by zones of inhibition surrounding mold colonies contaminating a used Petri dish. On further investigation he concluded that the mold had secreted a compound which further investigation showed to have activity against several pathogenic bacteria (Fleming, 1929). The lytic filtrate of the mold *Penicillium notatum* he called penicillin. Further development of this compound has revolutionized the treatment of bacterial infection and has led to the discovery and served as a model for the development of numerous pharmacologically active compounds produced by fermentation (Perlman, 1975). In 1941, when penicillin manufacture began in the United States, the process was to grow the *P. notatum* on the surface of a simple medium for 5–10 days and use the liquid underlying the culture, which contained the penicillin in concentrations of 10–20 Oxford units ml^{-1} (an activity equivalent to 0.006–0.012 mg pure benzyl penicillin, Na salt). A solvent extraction procedure was used to isolate material for clinical, pharmacological and chemical characterization.

Experiments at the USDA Northern Regional Research Laboratory in Peoria, IL, led to the development of the submerged culture method using a medium containing starch, lactose and corn-steep liquor and also discovered a new strain, *P. chrysogenum* NRRL 1951 (from a moldy cantaloupe), which led to all modern production strains. Moyer (1948) discovered that the addition of phenylacetic acid to penicillin-producing cultures increased the antibiotic yield and changed the side chain from a mixture to predominantly penicillin G. The basic understanding of process kinetics and essential nutritional requirements was established in defined medium studies of Hosler and Johnson (1953).

2.1.2 Biosynthetic Penicillins

Many different penicillins are produced by *P. chrysogenum* provided the appropriate carboxylic side chain is added to the medium in sufficient quantity. The original stimulatory effect of corn-steep liquor was in part due to its providing additional quantities of a mixture of carboxylic acids which had the free methylene adjacent to the carboxyl which is necessary for incorporation by the organism (Behrens *et al.*, 1948a,b, 1949). Many biosynthesized penicillins, all having the structure shown in (**1**), have been identified, but only *penicillin G* (R = Ph) and *penicillin V* (R = PhO) are important therapeutically.

Both have a similar spectrum (Gram positive bacteria) and activity (Table 1) but whereas peni-

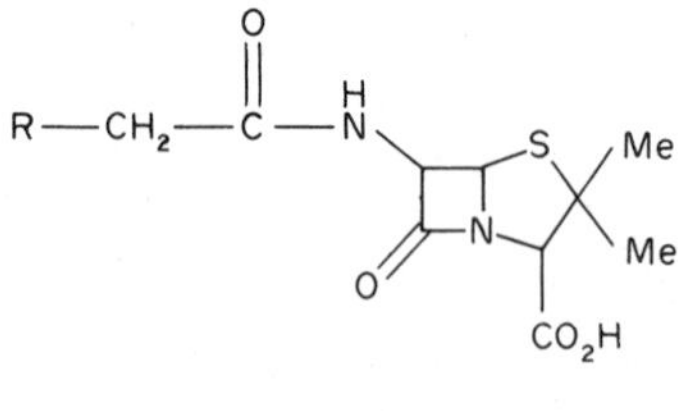

(1)

Table 1 Molecular Weights and Activities of Biosynthetic Penicillins

	M.W.	*Activity* (mg^{-1})
Penicillin G, Na salt	356.38	1667
Penicillin G, K salt	372.47	1595
Penicillin V, Na salt	372.38	1595
Penicillin V, K salt	388.47	1529
6-Aminopenicillanic acid	216.28	—

cillin G is degraded by stomach acid and must be administered parenterally, penicillin V is more acid stable and is orally administered. Penicillin V is relatively inactive against gonococci and haemophili and both have poor activity against Gram negative rods and are rapidly degraded by penicillinase-producing bacteria. See Table 1 for molecular weights and relative activities.

2.1.3 Process Overview

The technology of the manufacturing process is still regarded as highly proprietary by the major commercial firms even though excellent fermentation technology and organisms have been publicly available for purchase for many years from Panlab Genetics, and others. Similarly, purification of these compounds is thoroughly described in the literature. The distinguishing aspects of commercial processes are of an operational character and cost advantages result from which of the well understood alternatives are implemented and with what degree of optimization and attention to detail.

The process, involving culture maintenance, fermentation and isolation is presented in Figures 1 and 2. Nutrient utilization and accumulation of microbial mass and penicillin are presented in Figure 3. In analysing data of the type presented in Figure 3 a quantitive approach, which clearly presents rates of utilization and production, is essential to understanding process behavior. Penicillin fermentation is such that even substantial apparent improvements may result in little or no decrease in unit manufacturing cost and it is not always obvious where process improvement dollars should be expended. In this author's experience it is difficult to maintain a competitive position without some effort in each area of the process: strain, fermentation and separation.

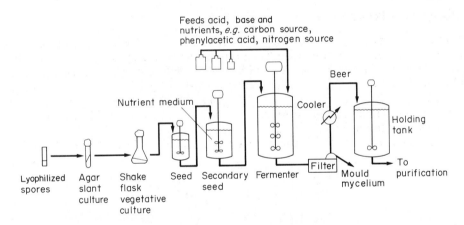

Figure 1 Penicillin production: fermentation

Basically the process involves cultivation of *P. chrysogenum* in vessels of increasing size up to 200 000 liters, beginning with a slant or a vial of frozen vegetative mycelium. At harvest, the batch is filtered to remove the mycelium and the filtrate is extracted and crystallized or acid precipitated.

2.2 FERMENTATION TECHNOLOGY

2.2.1 The Culture: Strain Development

2.2.1.1 *Strain maintenance and initial seed*

As shown in Figure 1, the process begins with lyophilized spores of a production strain. Alternatives are to store the master or secondary cell bank in the form of spore suspensions in liquid nitrogen or as frozen vegetative material with glycerol and/or lactose as suspending agents at either −70 °C or under liquid nitrogen. Additional details regarding storage and maintenance of a master cell bank are given in Queener and Swartz (1979). Frozen spores (a standard number) are inoculated on slants, and allowed to sporulate, then are transferred to vegetative cultures in

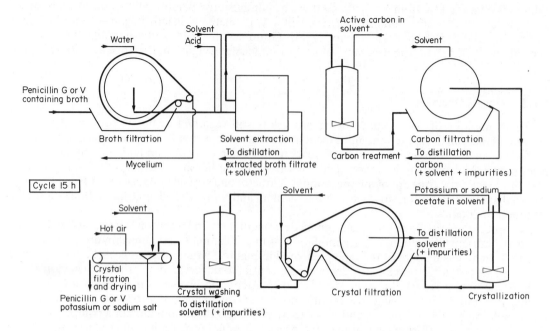

Figure 2   Penicillin purification process of Gist-Brocades (From Hersbach, 1984)

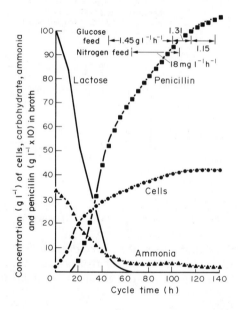

Figure 3   The time-course of carbohydrate, nitrogen, penicillin and biomass concentrations during a penicillin fermentation (From Queener and Swartz, 1978)

Erlenmeyer flasks containing a vegetative medium generally similar to the production medium (see Section 2.2.2.2). Cultures stored as frozen mycelium avoid the introduction of variability which may result from sporulation.

Regardless of the culture maintenance details, eventually the master lot will be exhausted and so, before this occurs, a secondary lot is prepared. It is common for a sporulation, cloning and natural selection procedure to be carried out in which colonies are evaluated for their production capability and similarity to the parent in other criteria, often including the degree of sporulation and some morphological traits. These evaluations include shake flask and tank fermentations and new lots are often run side by side against the parent in production trials before full approval is given by those responsible for production results.

It is not uncommon for some natural selectants to exhibit slightly greater product formation than the parent, but such increases are generally small. Because of the many factors which may affect the process and because small changes often seem to be ameliorated by a selection procedure, a minimal strain maintenance/improvement effort is deemed essential to the long term success of a production effort.

Lein (1985) presents various culture preservation, reisolation, assay procedures and both seed and fermentation media used in a program in the early 1970s.

2.2.1.2 Mutation

Lein diagramatically represented the pathway to secondary metabolites as DNA → RNA → enzyme → primary metabolites + enzymes → secondary metabolites, calling attention to the complex multigene pathway which is strongly influenced by genetic and physiological factors (Lein, 1983). In practice, antibiotic synthesis has been correlated with various complex physiological factors such as sporulation and growth rate and levels of various intermediates within the cell, all of which are regulated in complex ways by many genes. The consistent success of the empirical and various semi-empirical approaches to mutation and screening must result in part from these characteristics and from the fact that even as we work out the details of the metabolic pathways, isolate the enzymes involved and clone most of them (a task that will likely be complete within one or two years), we remain ignorant of which are rate limiting in a particular industrial strain under the current physiological conditions enforced in the production fermenters. Until the middle 1960s, UV irradiation and nitrogen mustard were the preferred mutagens in penicillin strain improvement studies, as seen in Figures 4 and 5 (Elander, 1967; Elander *et al.*, 1973). Table 2 lists various mutagens which have been used. In recent years, NTG, a mutagen providing extremely high mutation rate relative to its killing effect, has become the preferred compound for this reason, and simply because its mechanism differs from the techniques previously emphasized, NTG has the potential thereby of inducing different groups of mutational events. Details are given by Ball (1973a).

Screening (*e.g.* examining all members of a population which has been mutagenized) has been the most significant technique in raising penicillin yields to date. Table 3 lists media used for screening in the Panlabs program. One approach to improving the efficiency of this procedure is the potency index agar plate technique used by Ball and McGonagle (1978) in the improvement of high yielding cultures. In this case the zone size was reduced by incorporating penicillinase in the agar to reduce the sensitivity to a useful range. Additional details on this and other approaches to strain improvement are presented by Ball (1978, 1983).

2.2.1.3 Selection

The proven success of the empirical approach (mutation and screening for increased potency) notwithstanding, this is nevertheless a tedious procedure. A specific selection procedure designed to identify those organisms possessing a desired phenotype (generally by allowing only that group to survive) can greatly increase the effectiveness of the program provided the trait selected for is related as expected to improved yield. One company, Panlab Genetics Inc., established a multi-client penicillin strain development program in 1973. Participants paid a fee for access to strains developed by Panlab. Tables 4 and 5 present the results of the Panlab strain improvement program to 1980 in Oxford units ml^{-1} (1667 Ou pen G-Na mg^{-1} or 1595 Ou pen V-Na mg^{-1}). The strains were provided with protocols for their use in flasks and in pilot fermenters of up to 3000 liters. Lein (1983, 1985) has presented the program in considerable detail. Several different and clearly quite successful selective techniques are described. The improved yields presented in the tables were achieved largely through the sequential application of selective procedures involving high concentrations of amino acids, biosynthetic intermediates, and amino acid analogs. These selective techniques presumably lead to organisms with higher levels of amino acids and intermediates. Specific selections leading to the improvements in Tables 4 and 5 are presented by Lein (1985). Changes in colony morphology during the program are presented in Figure 6. Note the more clumpy character of later strains, presumably due to rapid hyphal branching.

An important contaminant of the high potency broths turned out to be the oxidized *para*-hydroxy form of phenylacetic acid, which when incorporated interferes with the semisynthetic chemistries. A clever selection technique used by Panlab involved choosing for further evaluation

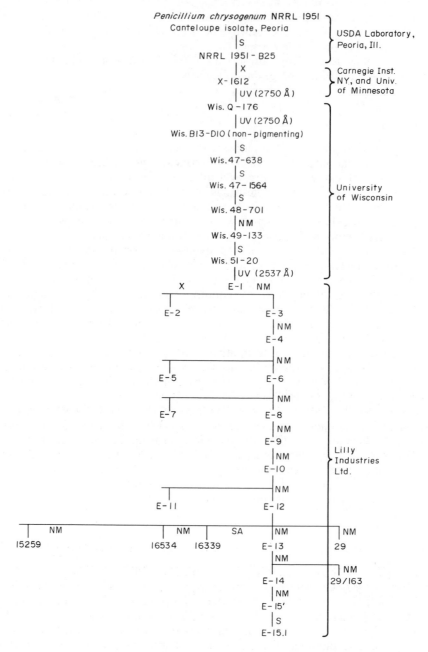

Figure 4 Strain lineage in the penicillin-producing mould *Penicillium chrysogenum*. S indicates natural selection; UV, ultraviolet radiation; X, X-ray radiation; NM, nitrogen mustard; and SA, sarcolysine (*p*-di(2-chloroethyl)aminophenyl-alanine) (From Elander, 1967)

small colonies from plates with phenylacetic acid as sole carbon source, except for just enough glucose to produce a small colony. Thus the small colonies selected were those impaired in the ability to oxidize the side chain precursor. The technique was successful and several commercial users of P14B-4 and its progeny no longer are concerned about interference by *p*-hydroxy penicillin G. Also this loss of precursor is eliminated (Lein, 1985).

Substantial experimental detail on operation of the strain improvement and culture maintenance programs is presented in Ball (1979), Lein (1985) and Queener and Swartz (1979). Elander (1967) notes the importance of making the shake flask mimic plant conditions. Emphasis on other critical parameters such as design of the screen or selection to identify cultures producing at high specific rates (product/cell mass/time) and at high overall yield from the carbon source (reducing

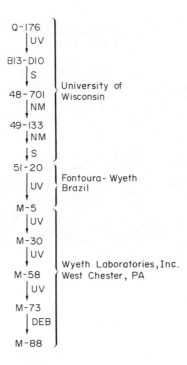

Q-176
│ UV
▼
B13-D10
│ S
▼
48-701 } University of
│ NM   Wisconsin
▼
49-133
│ NM
▼
│ S
▼
51-20 } Fontoura-Wyeth
│ UV   Brazil
▼
M-5
│ UV
▼
M-30
│ UV   Wyeth Laboratories,Inc.
▼   West Chester, PA
M-58
│ UV
▼
M-73
│ DEB
▼
M-88

Figure 5 Lineage of improved Wyeth strain of *Penicillium chrysogenum*. UV indicates ultraviolet radiation; S, spontaneous; NM, methylbis(2-chloroethyl)amine; DEB, 1,2,3,4-diepoxybutane (From Elander *et al.*, 1973)

Table 2 A List of Mutagens Used in the Improvement of Penicillin Yields by *Penicillium chrysogenum*

Methylbis(2-chloroethyl)amine	Ethyl methanesulfonate
Ultraviolet light irradiation (275 and 253 nm)	X-rays
N-Methyl-*N'*-nitro-*N*-nitrosoguanidine	γ-Rays
Nitrous acid	Ethylenimine
Diepoxybutane	

Table 3 Fermentation Screening Media Used in the Panlabs P-2 Program[a]

Medium	% (w/v)	Medium	% (w/v)
Lactose	12.0	Lard oil	1.0
Pharmamedia	2.5	KH_2PO_4	0.05
$(NH_4)_2SO_4$	1.0	K_2SO_4	0.5
$CaCO_3$	1.0	Na-phenoxyacetate	1.25

[a] Details are for a 500 ml shake flask containing 35 ml.

cost and increasing total yield (Lein, 1985; Swartz, 1979) have been discussed. Increasing sugar efficiency also tends to improve efficiency relative to heat evolution and oxygen use. Other techniques which have been employed are screening for cultures which are lower in viscosity due to pelleting or fragmentation. One company raised the oxygen transport capability of their underpowered (0.61 HP/100 gal in relatively short tanks) fermenters using this approach (Bartholomew *et al.*, 1977).

Recognition of the importance of screening really large numbers of isolates in conjunction with selective techniques has led to the development of automated procedures and miniaturization of the flasks (Cape, 1976; Bylinsky, 1974) and to the plate procedures described by Ditchburn *et al.* (1974) and by Ball and McGonagle (1978), who adapted an early plate technique of Elander's to high-yield penicillin screening.

Demain (1974) and Ball (1978) have reviewed a substantial body of work and have presented the biochemical rationalizations for the selective techniques described by Lein (1983, 1985) relat-

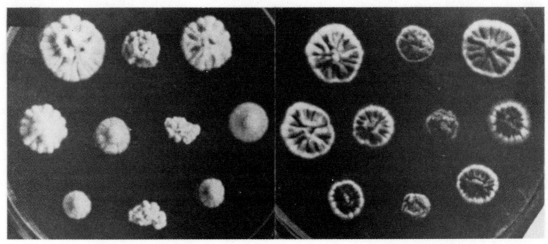

P-1, P-2, P-3
P-4, P-5, P-6, P-7
P-8, P-9, P-10

P-1, P-2, P-3
P-4, P-5, P-6, P-7
P-8, P-9, P-10

Figure 6

ing to amino acid and metabolism of biochemical intermediates. Careful examination of the literature would make it possible for someone skilled in the techniques to duplicate this work but since entirely satisfactory strains are available for license, few are likely to do so. The literature is mainly useful as a source of information on approaches which may prove useful in the development of other strains such as those producing cephalosporin C and in helping the physiologist understand the organism for which the nutritional environment must be optimized.

2.2.1.4 *Genetics*

Genetics of penicillin producing organisms has been reviewed by Ball (1983), O'Sullivan and Ball (1983) and MacDonald (1983). The biology of *Penicillium* has been reviewed by Peberdy (1985). The biosynthetic pathway to the penicillins has been largely worked out. Genetic studies have been utilized in two ways to improve penicillin production. First, studies with blocked mutants have led to an understanding of the pathway and suggested appropriate selective techniques described previously. Second, two techniques have allowed the merging of potentially desirable traits present in different lines. These are parasexual breeding and protoplast fusion. The parasexual breeding technique was discussed in some detail in Queener and Swartz (1979) and in MacDonald *et al.* (1972) and MacDonald and Holt (1976) and basically involves the construction of complementary auxotrophic blocks or different resistance mutations in two strains having traits deemed desirable to incorporate in a single strain. The parents are grown together on complete medium, then selected for presence of both markers, allowed to sporulate and stable diploid spores are isolated. After various treatments to allow recombination, haploid colonies are derived and evaluated for production and other characteristics. Recombinants with superior yields over the parents have been reported by Ball (1973a) and certain diploids have been used industrially (Calam *et al.*, 1976; Elander *et al.*, 1973).

In *Commercial Biotechnology: An International Analysis* (OTA, 1984) Bristol-Myers reported that protoplast fusion followed by selection techniques had allowed it to improve penicillin productivity (*sic*) by 8% per year over the four previous years, a quite respectable improvement. Lein (1985) presents details of a program utilizing this technique to introduce desirable characteristics such as stable sporulation and good growth of seed cultures. Protoplast fusion protocols are presented in appendixes. The literature on protoplast fusion was reviewed by Peberdy (1985), who also mentioned reports of success in crosses between *P. chrysogenum* and *C. acremonium* from Hoechst and from Bristol-Myers.

No discussion of modern genetic approaches would be complete without mention of recombinant DNA techniques. Basically the goal of any program is to increase the rate of the rate limiting

Table 4 Panlabs' Penicillin Strain-development Program[a]

Year	Culture	Penicillin	Shaker yield				Pilot plant yield				Production yield			
			Min.	Max.	Av.	Time (d)	Min.	Max.	Av.	Time (h)	Min.	Max.	Av.	Time (h)
1973	P-1	V	6 100	7 500	7 100	5	14 500	17 600	15 900	140	11 700	14 400	13 560	130
	P-3	V	11 000	13 100	12 000	8	13 900	20 700	18 000	180	14 200	19 100	16 500	180
		V	17 400	21 300	19 800	9	15 400	21 800	18 200	182	—	—	—	132
		G	—	—	—	—	—	—	—	—	17 600	17 900	17 700	132
	P-4	V	23 300	25 400	24 400	10	—	—	22 100	182	19 700	20 700	20 120	182
		G	—	—	—	—	—	—	—	—	15 800	20 600	17 900	182
1974	P-5	V	27 500	31 200	29 800	10	22 600	23 700	23 150	182	23 600	25 900	24 700	182
		G	—	—	—	—	—	—	—	—	21 300	24 600	23 150	182
	P-7	V	29 200	33 900	32 100	10	25 100	33 400	30 200	210	23 400	29 600	27 300	190
		G	—	—	—	—	25 200	29 000	27 700	185	23 700	28 200	25 600	190
1975	P-8	V	34 700	36 900	36 000	10	31 500	33 000	32 250	185	29 200	34 800	32 400	203
		G	—	—	—	—	—	—	—	—	27 900	30 800	29 600	203
1976	P-10	V	36 800	42 600	38 700	—	32 500	36 500	33 700	185	—	—	—	—
	P-10	V	—	—	—	—	—	—	—	—	—	—	—	—
	P-12	V	40 911	42 800	41 900	9	32 300	38 000	35 000	185	—	—	—	—
		G	—	—	—	—	—	—	—	—	—	—	—	—
1978	P-13	V	33 700	39 000	37 200	7	—	—	41 500	185	—	—	—	—
		G	—	—	—	—	—	—	42 250	185	—	—	—	—
1979	P-15	G	45 200	49 200	46 200	9	—	47 800	46 900	185	—	—	—	—

[a] P-1 line; yields in units ml^{-1}. Data supplied by Panlabs Inc.

Table 5 Panlabs' Penicillin Strain-development Program[a]

Year	Culture	Penicillin	Shaker yield				Pilot plant yield				Production yield			
			Min.	*Max.*	*Av.*	*Time (d)*	*Min.*	*Max.*	*Av.*	*Time (h)*	*Min.*	*Max.*	*Av.*	*Time (h)*
1973	P-2	G	9 600	11 700	10 500	10	12 300	16 070	14 270	160	12 350	14 150	13 189	160
		G	—	—	20 000	10		—		—	—	—	—	—
1974	P-6	V	26 800	29 000	28 000	10	(medium 1)	—	—		—	—	—	—
		V	—	—	25 000	7	(medium 2)	—	—		—	—	—	—
1975	P-9	V	31 600	35 100	33 100	10	32 000	35 300	33 600	185	—	—	—	—
1976	P-11	V	36 000	39 800	38 500	9	32 500	40 100	36 200	185	—	—	—	—
		G	—	—	—	—	33 200	36 000	34 500	185	—	—	—	—

[a] P-2 line; yields in units ml^{-1}. Data supplied by Panlabs Inc.

step (with each success a new opportunity is created). The problem with a deterministic approach here is the great difficulty in determining which step is rate limiting. A major research effort would be needed and at the end one would identify a particular enzyme, isolate it, clone it and reintroduce multiple copies into the organism. At this point one is faced with a new, equally complex problem for the new rate limiting step. According to Ball of Panlab Genetics, 'any practical approach one takes is likely to be empirical, but there are different degrees of empiricism. If one has a suitable vector and a properly sized gene bank of a high producing strain, then in principle, screening 3000 clones should allow one to obtain increased copies of each gene in the organism in turn. Provided that transcriptional, translational or substrate level regulation did not overcome the effect of increased gene dose, one would have created an improved strain. If one could amplify and integrate the gene responsible into the chromosome, a more stable strain might result. Some rate limiting steps will involve regulation making this difficult. An alternative would be to take a fairly simplistic view of synthesis, obtain blocked mutants at key points in the pathway, identify regions of DNA from the gene bank that relieve the blocks and introduce these on plasmids or into the genome of a high producing strain. Techniques required for these approaches are becoming available.'

In recent developments, a group at Eli Lilly have developed a transformation system for *cephalosporium* and have cloned the gene for the *C. acremonium* cyclase (Queener, 1985, personal communication). With the enzymes for the steps from the tripeptide already isolated, cloning of the remaining genes in the *C. acremonium* and *P. chrysogenum* pathway cannot be far behind. This should allow substantial deregulation of the terminal steps in the pathway but the overall path to penicillin is complex and the semi-empirical approach of Ball described above may well have a significant role. Combination of all the above techniques should continue to lead to improvements in penicillin production efficiency for several years, further reducing today's costs of $15–18 kg^{-1}. It should be recognized that another application of this work, its potential application to the engineering of novel β-lactams through modification of substrate specificities of key enzymes, and other techniques, may be quite important also.

2.2.2 Fermentation Process: Flow Sheet

2.2.2.1 *Facilities*

Production of penicillin G or V by fermentation is carried out in liquid cultures. In modern practice the culture volume is typically 40 000–200 000 litres although one or two plants may be successfully operating somewhat larger air lift vessels. The process is aerobic, having a volumetric oxygen uptake rate in the range of 0.4–1.0 mmol l^{-1} min^{-1} and an RQ of approximately 0.95 (Nelligan and Calam, 1983; Sylvester and Coghill, 1954; Perlman, 1970). Oxygen is supplied by passing air through the culture at a rate of 0.5–1.0 volumes of air (volume of fluid)$^{-1}$ min^{-1} and the air is vigorously contacted with the fluid using turbine agitators of various designs. Power introduced to the culture is generally of the order of 1–4 W l^{-1} including that introduced by the air stream. Other fermenter designs, including Waldhof fermenters and air-blown columns, are used, some of which are more energy efficient than the conventional designs, or offer other advantages in terms of capital costs or higher oxygen-transport capacity.

All of these vessels are fitted with coils, jackets or cascade systems for heat removal, and control systems for the maintenance of the desired temperature, air-flow rates, agitator speed, pH value and various nutrient feed rates. Air filtration and aseptic-vessel design ensure culture sterility. The fundamentals of equipment design, asepsis and the control of environmental conditions are discussed by Aiba *et al.* (1973); however, there is no source in the literature which fully details the design and construction of large fermenters. Real success here depends on the shared experience of the engineers and scientists actually operating a facility. The time-course of penicillin production is depicted in Table 6 and presented graphically in Figure 3. A flow sheet of the typical production unit is presented in Figures 1 and 2.

2.2.2.2 *Inoculum development*

The main purpose of vegetative cultures and subsequent inoculum development steps is to increase the biomass to give a population which can be added to the next stage, such that each step will be reasonably short and large-scale equipment is used efficiently.

Table 6   Changes during Fed-batch or Semicontinuous Penicillin Fermentation

Time	0–40 h	40–?	?–Harvest
Penicillin production	Slight	High but the rate peaks then declines over most of the period	Some production occurs but at a low and declining rate
Specific growth rate	8–20 h; 0.07 h^{-1} (McCann and Calam, 1972) 20–40 h; 0.01–0.02 h^{-1} (McCann and Calam, 1972)	40–75 h; ~0.009 h^{-1} (McCann and Calam, 1972)	115–209 h; ~0 h^{-1} (McCann and Calam, 1972) 90–186 h; ~0.0005 h^{-1} (Lur'e *et al.*, 1976)
Carbon and energy source used	Organic nitrogen (*e.g.* corn-steep liquor and/or carbohydrates, oils)	Carbohydrates, triglyceride oils, ethanol, fed or hydrolyzed slowly and generally thought to regulate growth	Various energy sources as available, sometimes insufficient for maintenance. Growth limitation may shift to other nutrients
Nitrogen source used	Mainly organic nitrogen	Organic nitrogen and/or ammonium and/or nitrate salts depending on the process. Used for cell growth and product formation	Nitrogen source may become growth-rate limiting. Cell lysis may occur (Lur'e *et al.*, 1976)
Sulfur source used	In crude organic nitrogen and inorganic salts, *e.g.* $(NH_4)_2SO_4$. Used rapidly for cell growth	Often fed as an inorganic salt. Used for penicillin synthesis and slow cell growth	—
Phosphorus source used	In crude organic nitrogen and as inorganic salts	Very little if any is fed as quantitative needs are very low. Growth-rate limitation may be harmful (Mason and Righelato, 1976)	—
Other nutrients	Required for cell growth	Some, like iron, may be regulatory (Pan *et al.*, 1975)	—

Inoculum development stages are typically conducted at 25 °C in shake cultures and agitated vessels. A typical vegetative or seed-stage medium contains an organic nitrogen source, such as corn-steep liquor, and a sufficient concentration of a fermentable carbohydrate, such as 2% (w/v) sucrose or glucose. Calcium carbonate is often included as a buffer at 0.5–1.0% (w/v). Other inorganic salts may be required for proper growth (Pan *et al.*, 1975). Log-phase growth is usually desirable in these stages, and a mass-doubling of time of about six hours (minimum for typical production strains) is achieved. Problems with growth rate or inoculum quality are usually associated with raw material variability, mishandling of earlier inoculum steps or culture storage.

Details of culture conditions are controlled in industrial practice to ensure uniformity. Criteria for monitoring the inoculum stages include cycle time, changes in pH value, residual carbon source concentration, packed cell volume, and respiration rate. When lactose is the carbon source in the fermentation stage, it is often included with glucose or sucrose in the final stage, and that stage is allowed to proceed until the readily used sugar is exhausted and β-galactosidase is induced. This allows for rapid initiation of growth in the fermenter.

2.2.2.3   Fermentation stage: medium

(i)   Carbon and energy sources other than glucose and lactose

Penicillium chrysogenum can utilize a variety of carbon and energy sources. Among those that are used for penicillin production are sucrose, dextrins and starches (Moyer, 1948). A variety of crude carbohydrate sources of lower purity, such as molasses, has been mentioned. Animal and vegetable oils (Pan *et al.*, 1959; Abu Shady *et al.*, 1976) and ethanol (Matelova, 1976) have also been used. Pan *et al.* (1959) compared the use of fatty oils and lactose as the carbon-energy source for benzylpenicillin synthesis. Under the same conditions, one gram of soybean oil produced as much benzylpenicillin as 2.5 g lactose. Under their optimized batch conditions, fed corn oil produced penicillin at a faster rate and for longer duration than lactose (added at make-up). Vegetable oils are effective carbon sources when fed.

Giona *et al.* (1976a) state that the stoichiometric number of moles of oxygen consumed to kilograms of sugar consumed is constant at 33 regardless of growth rate, and that the concentration of glucose after the rapid growth phase is constant within a range of glucose feed rates to the fermenter. These data indicate that, under certain conditions, the carbon in the fed glucose is completely combusted for energy production and is not incorporated into cell mass.

(ii) Nutrients other than carbon-energy sources

In general, the media used in the early years of penicillin production, and in in 1977, are quite similar. Perlman (1970) listed typical formulations for 1945 and 1967 (Table 7). Sylvester and Coghill (1954) described a similar medium but also mentioned sodium nitrate at 0–0.5%. The ingredients of Table 3 are more typical of media utilized today. Hersbach *et al.* (1984) list additional ingredients often used.

Table 7 Typical Fermentation Media for Production of Penicillin G or V in 1945 and 1967[a]

1945		1967	
Lactose	3–4%	Glucose or molasses (by continuous feed)	10% total
Corn-steep liquor solids	3.5%	Corn-steep liquor solids	4–5%
CaCO$_3$	1.0%	Phenylacetic acid (by continuous feed)	0.5–0.8% total
KH$_2$PO$_4$	0.4%	Lard oil (or vegetable oil) antifoam (by continuous	
Lard oil antifoam	0.25%	addition)	0.5% total

[a] Adapted from Perlman (1970).

(iii) Nitrogen source

Until 1975, most industrial processes included corn-steep liquor as the organic nitrogen source in the medium. This was included because it gave a marked improvement in the yield of penicillin (Sylvester and Coghill, 1954) due to its content of side-chain precursors. Smith and Bide (1948) and Mead and Stack (1948) found that compounds in corn-steep liquor which could be broken down to phenylacetic acid (*e.g.* phenylalanine) explained the corn-steep liquor effects.

The introduction of specific side-chain precursors allowed producers to substitute for corn-steep liquor because of its variability and occasionally short supply. Foster *et al.* (1946) showed cotton-seed meal to be a satisfactory substitute. Lein (personal communication, 1977) noted that a first step in their improvement program for penicillin yield was to replace corn-steep liquor. The relatively high yields reported for the Pan Laboratories processes (Tables 4 and 5) show that corn-steep liquor can be successfully replaced with Pharmedia, a cottonseed flour. Hockenhull and Mackenzie (1968) of Glaxo Laboratories reported synthetic media which equalled their corn-steep liquor medium in performance. All of the crude protein sources mentioned above are somewhat more expensive per kilogram of crude protein than corn-steep liquor. Other commonly available meals such as soy or peanut are also satisfactory.

Lur'e *et al.* (1976) demonstrated the importance of a continued supply of ammonia nitrogen. They found it essential to maintain a concentration of ammonia (by feeding ammonium sulfate) between 250 and 340 μg ml^{-1} for continued synthesis of penicillin. A lack of available ammonia nitrogen caused lysis of the mycelium and a drastic decrease in respiration relative to controls with a suitable nitrogen feed. Nitrogen is, of course, required for penicillin synthesis.

(iv) Precursors of the penicillin side-chain

Early work on side-chain precursors is reviewed by Behrens (1949), Perlman (1970), Sylvester and Coghill (1954) and Demain (1967). Later work showed that side-chain attachment was relatively nonspecific, and that relatively high concentrations of phenylacetic acid or phenoxyacetic acid had to be included in the medium, or in feeds to the fermenter, if the desired single-component penicillin was to be obtained (Demain, 1974). These compounds represent a significant portion of the unit cost of penicillin G and V, respectively. The theoretical requirement for sodium phenylacetate is 0.47 g g^{-1} penicillin G acid. For sodium phenoxyacetate, it is 0.50 g g^{-1} penicillin V acid.

To avoid the toxicity of sodium phenylacetate to the producing organism, and to minimize hydroxylation of these compounds by the producing organism, they are fed continuously. Perlman (1970) noted that greater than 90% conversion of phenylacetic acid into the penicillin G side-chain has been obtained through continuous feeding of one of its salts to maintain a low residual concentration as determined by gas chromatography. This was common practice by 1967. The penicillin V precursor, phenoxyacetic acid may be fed in a few larger doses. More than 80% of the material can be incorporated into the penicillin V side-chain. Lein has noted (personal communication, 1977) that, in his experience with two separately derived lineages of production strains, each strain with an improved yield of penicillin V invariably showed improved yield for penicillin G. Representative data (Table 8) show that, for the six strains in one lineage for which comparable results are available, the average ratio of yields of G to V by weight is 0.95, the same as the ratio of the molecular weights of the free acids.

Table 8   Relative Production of Penicillins V and G by Representative Pan Laboratories' Strains at Production Scale

Strain	Year	Average titre G^a (mg ml^{-1})	Cycle (h)	Ratio G/V
P-3	1973	9.9	130	1.21
P-4	1973	10.0	182	0.85
P-5	1974	13.0	182	0.89
P-7	1974	14.4	190	0.89
P-8	1975	16.6	203	0.87
P-13	1978	23.7	185	0.97
Average				0.95

[a] Single experiments were conducted with harvest volumes of about 110% of the initial volumes. Production media and feeds have been changed along with cultures (Queener and Swartz, 1979).

(v)   Sulfur source

Pan *et al.* (1975) mentioned ammonium sulfate, a common ingredient in production media, as providing nitrogen and sulfur in the proportions necessary for penicillin synthesis. Sulfur is also provided in significant quantity by the organic nitrogen source.

(vi)   Phosphorus source

Phytic acid in corn-steep liquor has been reported as an important source of phosphorus (El-Saied *et al.*, 1977). Phosphorus may also be added in the form of inorganic salts and can be maintained feeding inorganic phosphorus salts. A level of 250–500 μg ml^{-1} of phosphate phosphorus should be maintained.

(vii)   Other nutrients

Inorganic salts are commonly included as noted by Hersbach *et al.* (1984). These provide potassium and magnesium as well as additional nitrogen, phosphorus and sulfur for growth and product formation. A Romanian patent (number 55 510, issued June 10, 1973) details a batch medium containing 21 ingredients, 10 of which are inorganic salts. Chemically defined media (Jarvis and Johnson, 1947; Hosler and Johnson, 1953; Hockenhull and Mackenzie, 1968; El-Saied *et al.*, 1976) list requirements for phosphorus, sulfur, potassium, magnesium, manganese, zinc, cobalt and copper. Al-Haffar (1970) described optimum cation concentrations for penicillin synthesis and response surface techniques for determination of optimum concentrations of inorganic salts for batch fermentation.

2.2.2.4   Process control

(i)   Physical factors affecting q_{pen} values

A pH value near 6.5 must be maintained (Lur'e and Levitov, 1967; Pan *et al.*, 1972). Andreyeva *et al.* (1973) discussed the effects of pH on growth and product formation. They also discussed the determination of optimum pH profiles which should not be constant, as a rule, since optimal values are likely to be different for growth and product formation. They observed different pH values for a maximum specific oxygen-uptake rate for 66-hour (pH = 6.7) and 90-hour (pH = 7.0) samples removed from fermenters and placed in synthetic media. This is not really surprising in light of the profound shift in sugar metabolism reported by Righelato *et al.* (1968) in the range of specific growth rates through which a batch process passes during that period.

When a constant temperature is chosen for a batch penicillin fermentation, it is generally in the range 23–28 °C. McCann and Calam (1972) reported an optimal temperature between 25 and 27 °C for a fed batch process. Unfortunately, the penicillin concentration was quite low in all cases, a common problem with such reports in the literature. Pan *et al.* (1972), working with three industrial strains, selected 25 °C as the temperature for their experiments. Constantinides *et al.*

(1970a,b) illustrate the use of Pontryagin's continuous maximum principle to predict an optimum temperature profile for a fed batch process. This work was discussed further by Constantinides and Rai (1974). The procedure appears to have merit. Unfortunately, the optimum temperature profile determined was based on limited data collected in the Squibb pilot plant, and a test of the proposed 'optimum' profile was not reported.

Carbon dioxide at a partial pressure of 0.08 atm caused a 50% decrease in penicillin production in continuous culture. No inhibition was observed at 0.0006 atm (Pirt and Mancini, 1975). Lengyel and Nyiri (1966) showed that, with a decreased air flow in an industrial fermenter, penicillin synthesis was impaired. The respiration rate also dropped when the carbon dioxide concentration in the exit gas rose over 4%. Iron, if present at too high a concentration, specifically inhibits penicillin production but not cell growth (Pan *et al.*, 1972).

(ii)   Optimization criteria

Penicillin per unit volume in a fermenter is the product of three parameters: X, the concentration of cells; q_{pen}, the specific rate of penicillin synthesis; and t, the duration of the fermentation. Obviously the object is to achieve high X and q_{pen} values as quickly as possible, and to maintain maximal X and q_{pen} (both of which vary as the fermentation proceeds) as long as possible. That is, one optimizes the integral

$$\int_{t_1 \to t_2} [q_{pen}(t)\, X(t)] \mathrm{d}t$$

where t_1 is the time of inoculation and t_2 the time of harvest, within the constraints of equipment and overall economics.

(iii)   Rapidly achieving the desired cell mass

In a typical penicillin fermentation, most of the cell mass necessary for high penicillin yields is obtained during the first 40 hours of the fermentation, starting from an inoculum consisting of, for example, 10% (v/v) of a vegetative culture containing 20 g dry cell weight l^{-1}. This means that, during the first portion of the process, the organism is growing near μ_{max}, its maximum specific growth rate, which corresponds to a mass doubling time of about six hours.

Once a cell concentration sufficient to support a satisfactory volumetric yield of penicillin has been obtained, data obtained with some strains suggests that growth must continue at a certain minimum rate if a high q_{pen} value is to be maintained (Ryu and Hospodka, 1977). Hence, in the typical industrial penicillin fermentation, the concentration of cells achieved at the end of the rapid growth phase (X_{rgp}) is limited to allow for the additional cell mass to be added during growth in the 'slow growth' or 'production' phase of the fermentation.

A certain minimum supply of oxygen is required to support maximum q_{pen} values. Since oxygen demand and viscosity increase with increasing cell concentration, and since oxygen transfer decreases with increasing viscosity, beyond a certain cell concentration X_1, the oxygen supply in a particular fermenter will be insufficient to support maximal q_{pen} for sufficient duration (Fig. 7). There is a range beyond X_1 in which satisfactory results are obtained because the final penicillin yield is a product of X and q_{pen}. However, beyond a certain concentration X_2, the average overall volumetric rate of penicillin synthesis will be less.

An obvious means of improving penicillin yield, therefore, has been to increase the maximum oxygen-transport rate of fermenters and thus increase the mycelial mass which may be supported at a particular physiological state. This approach has been used extensively in industry. Several companies have doubled the agitator power on existing fermenters, and some have selected strains and culture conditions which decrease broth viscosity (Bartholomew *et al.*, 1977).

Heat transfer can limit the mycelial concentration in fermenters where cooling capacity is low and a strain generates excessive heat. The temperature for obtaining maximal q_{pen} values can be maintained only by removal of heat generated by the organism and the mechanical heat generated by the agitator and aeration. The cell concentration obtained at the end of the rapid growth phase, X_{rgp}, is limited by restricting the amount of one or more nutrients supplied to the culture during this period. Typically the source of carbon and energy, in the form of sugar, has been restricted.

(iv)   Yield of penicillin from sugar

A key element in economic evaluation is the overall yield of penicillin G (potassium salt) from carbon and energy sources and from other major raw materials such as the precursor (sodium phenylacetate) or nitrogen source. A summary of the factors which influence conversion yield is provided by Cooney and Acevedo (1977).

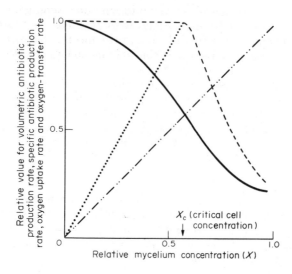

Figure 7 Relationships between volumetric antibiotic production rate (...) and specific antibiotic production rate (– – –), oxygen-transfer rate (———), oxygen uptake rate (–··–··–), and mycelium concentration (X) in a typical penicillin fermentation. All values are normalized with respect to their maximum values (Adapted from Ryu and Humphrey, 1972)

A major factor in the economics of penicillin production is the yield of penicillin from carbon and energy sources (glucose equivalent) (Swartz, 1979). Over the course of the Panlabs program this yield doubled and since yield from oxygen and heat can be expected to follow the yield from carbon and energy sources, considerable economies can result from the yield improvement. (This effect results because only a portion of the cell's energy goes to make penicillin, so as each cell makes more penicillin, the yield per cell increases.)

In much of the literature it is assumed that a single substrate (lactose, sucrose, glucose, *etc.*) provides the carbon and energy for cell growth, maintenance and penicillin formation. In industrial practice a combination of carbon sources, primarily sugar, starch and triglyceride oils, are used. Table 17 (p. 39) presents the overall yield of penicillin from carbon and energy sources (sugars, oils, *etc.*). These calculations include materials used in inoculum development but exclude that amount of organic nitrogen (protein) which provides 2 g nitrogen l^{-1}, approximately the amount of nitrogen required to produce 25–30 g l^{-1} of cell mass.

McCann and Calam (1972) published results of a 'high yielding strain' of *P. chrysogenum*. They referred to a combination of sucrose and peanut oil as their major carbon and energy sources. Their results show a productivity of 1.0 g (1000 l)$^{-1}$ d^{-1} and a yield of penicillin G–K from glucose equivalent to 0.046 g g^{-1}. This is in the range of the oldest Panlabs culture. In this calculation one must account for the sucrose feed and the peanut oil used as 'antifoam'.

This author uses a factor of 2.5 to convert triglyceride oil to glucose equivalents based on the literature (Pan *et al.*, 1959). It is very close to the ratio of the energy densities of triglyceride oils and glucose on a weight basis (2.6). This approach allows us to relate the sugar use and productivity for a penicillin plant. There are of course many other cost elements to be considered. This sort of analysis is also useful for changing from one carbon source to another. The standard feed is glucose, given at various, preset rates at different times. Lactose (a 'slowly utilized' and non 'glucose repressing' sugar) has been replaced in today's processes by feeds in the early hours of the process which increase stepwise, initially approximately a logarithmic increase, and thereby supporting a rapidly growing cell mass under limiting conditions.

2.2.2.5 *Physiological variables and their effect on product formation*

(i) *Transition from logarithmic growth to slow growth*

The successful shift from a culture with a high q_{pen} value growing at an exponential rate to one with a high q_{pen} growing at a low rate is a function of many variables. Volumetric rates of penicillin synthesis, obtained in continuous culture by Wright and Calam (1968), were not competitive with batch fermentation because the high q_{pen} values attained at the end of logarithmic growth declined before the growth rate was controlled in continuous culture (Fig. 8). In general, the same parameters which are important in the development of high q_{pen} values are probably equally important in the maintenance of a high q_{pen} value as the culture decreases in growth rate; however, it may not be possible to maintain these parameters in continuous culture.

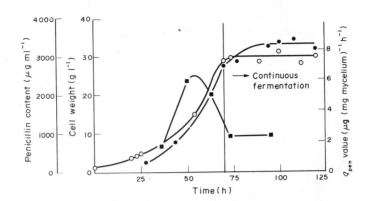

Figure 8 Relationship between rate of penicillin production and time. ● indicates penicillin content of the fermentation; ■, values for q_{pen}; ○, mycelia dry weight (Adapted from Wright and Calam, 1968)

(ii) Maintaining high q*pen* values for long durations (the 'production phase')

Once a large cell mass with a high q_{pen} value is growing at a rate which can be maintained for 120–160 hours, the industrial penicillin fermentation is said to be in its 'production phase' and it functions much like a continuous culture. In fact the physiological and nutritional environment during this phase can be simulated by studies similar to continuous culture. The key difference between continuous culture and the productive phase in the fed-batch or repeated fed-batch culture is that a constant cell concentration is maintained in the former, whereas the cell mass slowly increases in the latter two and batch cultures have a rapid growth phase and shift down in μ immediately preceding the productive phase. With some strains, q_{pen} and q_{O_2} values increase as μ decreases. Growth kinetics in fed-batch culture are discussed by Pirt (1975). Because quantitative measurements, such as rate of penicillin production, specific rate of penicillin production, specific growth rate, and specific yield from a particular nutrient, can be made most easily during this last segment of the penicillin fermentation, it is probably the best understood.

(iii) Optimal growth rate for the production phase

The desired specific growth rate $\mu(t)$ for the producton phase, $\mu_{prod}(t)$, has an upper limit, $U(t)$, which is dictated by the same constraints which dictate the limit of the concentration of cells obtained at the end of the rapid growth phase, X_{rgp}. If, for a given X_{rgp} value, $\mu(t)$ exceeds $U(t)$, then a cell mass which requires oxygen in excess of supply results prematurely during the production phase, and the q_{pen} value decreases dramatically and too soon. For example, Ryu and Hospodka (1977) reported that a specific oxygen-uptake rate of 1.2 mmol oxygen (g cell)$^{-1}$ h^{-1} was required to maintain one *P. chrysogenum* strain at a μ value of 0.015 h^{-1}. Thus, a cell concentration of 40 g l^{-1} would have an oxygen uptake rate of 48 mmol l^{-1} h^{-1}, which exceeds the transport capacity of some conventional fermenters for mycelial cultures.

The lower limit to $\mu_{prod}(t)$ will vary with strain, and may be determined experimentally. Using continuous-culture techniques, Ryu and Hospodka (1977) determined specific growth rates and specific rates of penicillin synthesis as a function of rates of uptake of different nutrients. The availability of any required nutrient can limit the growth rate according to the equation:

$$\mu = \mu_{max} [S/(K_S + S)]$$

where S is the concentration of a particular required nutrient and K_S is the concentration in the fermenter when the growth rate is one-half of maximum with all other required nutrients in excess. Excess is defined for each required substrate as the condition where S is much greater than K_S (typically 10 times greater; Cooney and Mateles, 1971). In the experiments of Ryu and Hospodka (1977), in which the availability of glucose limited specific growth rate, q_{pen} values varied linearly with μ up to a glucose feed rate of 0.33 mmol glucose $(\text{g cell})^{-1} \text{h}^{-1}$ at which point the μ value was 0.015 h^{-1} (Fig. 9). At higher feed rates the value for μ increased linearly but that for q_{pen} remained constant. With the experiments in which oxygen limited growth rate, q_{pen} values varied linearly with increasing oxygen supply up to an oxygen-uptake rate of 1.6 mmol oxygen $(\text{g cell})^{-1} \text{h}^{-1}$ and a μ value of 0.03 h^{-1}. With oxygen limiting, at a μ value of 0.015 h^{-1}, the specific oxygen-uptake rate was 1.2 mmol oxygen $(\text{g cell})^{-1} \text{h}^{-1}$ (Fig. 10). From this it was deduced that, for the strain tested, a specific growth rate of 0.015 h^{-1} and a specific oxygen-transfer rate of 1.6 mmol oxygen $(\text{g cell})^{-1} \text{h}^{-1}$ should be maintained during the 'slow growth' or 'production' phase of a fed-batch industrial fermentation in which glucose limited growth rate.

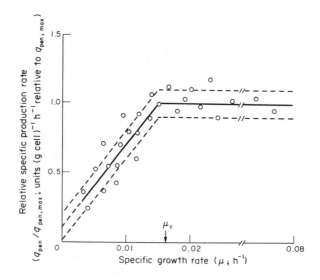

Figure 9   Relationship between specific cell growth rate and specific rate of penicillin synthesis with glucose limiting growth rate (Adapted from Ryu and Hospodka, 1977)

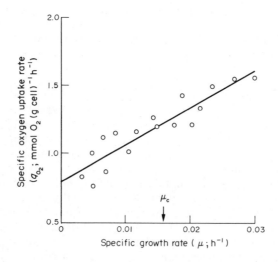

Figure 10   Relationship between oxygen uptake rate and specific growth rate with oxygen limiting growth rate (Adapted from Ryu and Hospodka, 1977)

Although an attempt is made in most industrial fermentations to limit growth rate by means of the availability of the carbon-energy source in the form of glucose or lactose, in actual fact, towards the end of some batch fermentations, oxygen becomes colimiting with sugar. That is, the concentration of oxygen is no longer excessively greater than K_{O_2} and the sugar concentration is not substantially greater than K_{sugar} (see discussion of colimitation by Cooney and Mateles,

1971). To maintain [sugar]/K_{sugar} less than $[O_2]/K_{O_2}$, *i.e.* to keep sugar the primary colimiting substrate, the sugar feed rate must be reduced. Specific growth rates thus fall as both of these ratios decrease and eventually μ falls below that required to maintain the optimal q_{pen}.

Ryu and Humphrey (1972) have measured specific oxygen-uptake rates (q_{O_2}) as a function of μ for an industrial production strain under study at Squibb. They discuss a method by which the balance between the available supply and the culture demand for oxygen can be managed. Righelato *et al.* (1968) also determined q_{O_2} values as a function of μ. They distinguished the q_{O_2} value required just minimally to sustain cells ($\mu = 0$), termed maintenance q_{O_2}, and q_{O_2} associated with growth.

The optimal μ_{prod} value determined by Ryu and Hospodka (1977) for their strain was 0.015 h^{-1}. However, the typical μ value for the production phase of most industrial fed-batch fermentations is less than 0.01 h^{-1}. Early results (Pirt and Righelato, 1967), which indicated that q_{pen} values decayed exponentially to zero if μ was less than 0.01 h^{-1}, do not appear to hold for many modern production strains and processes. Thus in the process reported by McCann and Calam (1972), μ was less than 0.01 h^{-1} from 41 to 200 hours. Oxygen-transfer rates may limit μ_{prod} to values below the strain optimum in some processes (Ryu and Humphrey, 1973).

(iv) The 'glucose effect'

During the production phase the specific rate of penicillin synthesis, q_{pen}, is sensitive to excess glucose, not only because excess glucose can stimulate growth to rates above the upper limit of μ_{prod}, but also because glucose, in the absence of growth, may directly inhibit the activity and/or synthesis of enzymes involved in penicillin synthesis as hypothesized by Demain (1968). Thus, *P. chrysogenum* grown in a glucose–lactose medium, sampled in the production phase (also termed 'idiophase'), washed, and transferred to a phenylacetate–salts solution produced penicillin as a suspension of non-growing or 'resting' cells. Addition of glucose to the phenylacetate–salts solution depressed penicillin production, whereas lactose stimulated production (Demain, 1968).

(v) Principle of minimum sufficient nutrient supply

In industrial penicillin fermentation, the carbohydrate feed rate in the production phase is regulated so that a specific growth rate profile $\mu(t)$, consistent with a maximal $\int[q_{pen}X]dt$, is achieved. With growth rate thus controlled by available sugar, in principle, it is desirable that other nutrients be present in concentrations minimally in excess of the μ and q_{pen} dictated demand for oxygen, nitrogen, phosphorus, sulfur, trace elements and side-chain precursor. These demands have been precisely determined in a quasi-steady state following a precisely defined growth phase for some strains (Ryu and Hospodka, 1977; Table 9). Deviations from these ideal conditions are sometimes dictated by the limitations of existing production equipment.

Table 9 Optimal Feed Rates for Penicillin Production by *Penicillium chrysogenum* Determined in Continuous-culture Experiments[a]

Feed	Specific uptake rates (optimal rate at μ = 0.015 h^{-1} unless specified otherwise) (mmol (g cell)$^{-1}$ h^{-1})	Feed rates (optimal rates at μ = 0.015 h^{-1} and X = 40 g l^{-1}) (mmol l^{-1} h^{-1})
Carbon	0.33	13.2
Oxygen	1.6 (optimal) 1.2 (at μ = 0.015 h^{-1})	64.0
Nitrogen (ammonia nitrogen basis)	2.0	80.0
Phosphate (phosphate basis)	0.6	24.0
Sulfate (sulfate basis)	2.8	112.0
Precursor (phenylacetic acid basis)	1.8	72.0

[a] Adapted from Ryu and Hospodka (1977).

(vi) Means other than carbohydrate limitation for control of cell mass and growth rate

Although limitation of hexose availability is essential to the success of a typical industrial penicillin fermentation, conceivably both X_{rgp} and μ could be controlled by other means.

(a) *Limitation by other nutrients.* Processes in which a slowly hydrolyzed carbohydrate is added to the medium in sufficient concentration to generate X_{rgp} and then is fed small batches of the same carbohydrate may, as previously mentioned, limit growth rate *via* another nutrient. Even in processes which are glucose-fed throughout, there may be advantage to deliberate colimitation of growth by a second nutrient. For example, in processes which rely on acid production from sugar to maintain an appropriate pH value in the early hours of the fermentation, phosphorus colimitation might be used to prevent excessive cell growth when extra sugar was required to lower the pH value. Certainly the choice of growth rate-limiting nutrient can significantly affect the metabolism of eukaryotes in continuous culture (Tempest *et al.*, 1965).

Mason and Righelato (1976) examined the effect of several growth-limiting nutrients (sucrose, oxygen, phosphorus, nitrate, magnesium, ammonia and sulfate) on the specific rate of sucrose and oxygen utilization. Although the specific rate of penicillin synthesis was not reported, many measurements were made at a μ value of 0.02 h^{-1} (near the optimum design point reported by Ryu and Hospodka, 1977) and the choice of limiting nutrient was seen to have profound effects on energy metabolism. When phosphate was growth limiting, a range of sucrose concentrations in the feed could be completely metabolized. The respiration rate did not change, and excess sucrose was metabolized to gluconic acid and malic acid. With oxygen limiting growth (at $\mu = 0.02$ h^{-1} and 0.039 h^{-1}) and sucrose near the growth-limiting rate of supply, the steady-state culture showed a two-to-three-fold increase in the efficiency of oxygen utilization for cell formation. The respiratory quotient was close to 1.0, and most of the carbon was recovered as cell mass and carbon dioxide.

With commercial fermenters being pushed to the limit of available $K_L a$ and with clumping and non-Newtonian rheological behavior of penicillin, it is possible that this situation is close to that which is obtained in commercial fermenters at least part of the time.

(*b*) *Temperature as a means of controlling specific growth rate.* Giona *et al.* (1976a) described growth rate as a function of temperature. They noted that temperatures above or below 26 °C decreased growth rate and prevented the biomass from becoming too great for the oxygen-transfer capacity of the fermenter. Commercial fermenters often operate at or near their limit of heat removal capacity. One might find that raising the temperature was the only way to salvage a runaway growth. This is an unconventional idea, but has been claimed by Giona.

(*vii*) *Basis for relationship between the specific growth rate, μ, and the specific rate of penicillin synthesis, q_{pen}*

The relationship between q_{pen} and μ observed in continuous culture and in the production phase of an industrial fermentation may reflect both the necessity of replacing damaged or old cells which have irreversibly lost their capacity to synthesize penicillin and the necessity for prolonging a particular state of metabolism in healthy cells. That the value for μ correlates with metabolic state can be easily observed at the level of morphology. Righelato *et al.* (1968) studied the morphological states of *P. chrysogenum* in continuous culture as a function of growth rate. When the value for μ was greater than 0.023 h^{-1}, only vegetative growth occurred, and, as growth rate increased to 0.075 h^{-1}, hyphae became more swollen. Below 0.014 h^{-1}, conidiation began to occur and was maximal at 0.009 h^{-1}.

Correlation of the value for μ with metabolic state is also reflected in respiratory quotient. In the study by Righelato *et al.* (1968), the respiratory quotient for a growth rate of 0.014–0.075 h^{-1} was 0.97, but the value fell to 0.72 at zero growth rate, and acid from the incomplete catabolism of carbohydrate accumulated (0.1–0.2 mmol H^+ (g cell)$^{-1}$ h^{-1}). Such profound effects of μ on physiological state are common.

2.2.2.6 *Duration of the fermentation*

The volumetric product formation rate eventually declines to the point where prolonging the fermentation will lower overall productivity (average volumetric rate of penicillin synthesis is equal to the total penicillin produced/volume of fermenter/hours of fermentation). The fermentation is terminated near to this time. Because preparation and harvesting of fermenters are cost intensive, the typical penicillin fermentation may be allowed to proceed somewhat beyond this time to lower unit costs. In fact, there is a choice here: whether to optimize unit cost or productivity.

(*i*) *Factors limiting duration of the fed-batch fermentation*

The duration of penicillin fermentation can be limited by the oxygen-transfer capacity of the fermenter. Since a certain μ value is required to maintain high q_{pen} values, at some time in the fermentation the cell concentration will exceed the oxygen supply necessary to maintain high q_{pen} values and synthesis will decay. An equation for the time (t) necessary for the oxygen concentration to decrease to a particular value (C_L) has been derived for a constant growth rate (μ) in a fermentation with no withdrawal (Giona *et al.*, 1976b):

$$t' = \frac{1}{(n + 1)}\ qn\ \frac{bV^{n+1}\ (C^* - C_L)}{Kx_0^{n+1}[M - (B-a)m]}$$

where n is an exponent relating oxygen-transport capacity to cell concentration, b is a constant relating oxygen-transport capacity to cell concentration, m is the specific growth rate, V is the volume of the tank, K is the product of the stoichiometric coefficient of the carbohydrate combustion reaction multiplied by the conversion factor for weight units to mole units, x_0 is the cell concentration just after inoculation of the tank at time zero, M is the rate of glucose utilization required to maintain the basic function of the cells at zero growth rate, B is the slope of the experimental plot of rate of glucose utilization as a function of growth rate, a is a constant relating the instantaneous amount of increase in cell concentration, dx, to the amount of carbohydrate consumed by dx, C^* is the equilibrium concentration of oxygen in the broth obtained by Henry's Law, and C_L is the concentration of oxygen measured in the culture.

In some processes, build-up of toxicants and inhibitors or other unknown factors may be a cause for declining q_{pen} values late in the production phase. Penicillin itself may serve as an inhibitor under certain conditions (Gordee and Day, 1972).

(ii) Factors limiting duration of the continuous or semicontinuous fermentation

In fed fermentations, the volume of the fermenter may be exceeded before q_{pen} values begin to decline significantly. In order to maximize the value of t during the duration of the fermentation, it becomes necessary to withdraw, either continuously or in batches, part of the culture, thereby making room for additional feeds and continued growth. The disadvantage of the withdrawals is loss of unused nutrient in the withdrawal and the discarding of cells which required a substantial expenditure of raw materials to produce. This causes lower yields of penicillin per kilogram of raw material. When periodic dilution and withdrawal are used to allow maintenance of high values for μ (and thus a high q_{pen}), the potency is lower due to dilution, and the volume of broth to be purified is higher. These characteristics make purification more difficult.

A novel approach to circumventing the waste of raw materials and the decreased potencies in continuous or semicontinuous fermentations has been patented (Young and Koplove, 1975). In this process a portion of the cells is destroyed and returned to the fermenter as nutrient, thereby allowing values for μ and X to be controlled at optimum levels and decreasing or eliminating the need for feeds and withdrawals. Disruption of the chitinous wall of *P. chrysogenum* and its conversion of conserved nutrients to utilizable forms represent major obstacles to the practical application of this approach in the industrial penicillin fermentation.

When the fermentation is prolonged by withdrawals and feeds, the duration of the fermentation may be limited by factors other than oxygen supply. Toxicants can be produced faster than they are withdrawn. Nutrients that are unidentified, or which cannot be fed conveniently or economically, may become exhausted. The lifespan of the cell's active penicillin-producing state may not be sufficiently long to allow for complete replacement of old or damaged cells leaving that state by young cells entering that state. Mutants of lower productivity can be produced and selected for. The latter limitation has been documented for carbon (sucrose)-limited continuous fermentation (Righelato, 1976).

Finally, though some commercial strains show growth associated kinetics of product formation, this behavior is not universal. With some, the repeated fed-batch approach may be of no value because their kinetic behavior differs from the strain of Ryu and Hospodka (1977). Thus, when yield optimization is required, an experimental approach is appropriate.

Where process design for highly mutated strains is concerned, the literature is only a guide and a source of ideas. Hersbach *et al.* (1984) have reviewed the recent literature on process control.

2.3 RECOVERY OF PENICILLIN

2.3.1 Carbon Process (Obsolete)

The original commercial process for the recovery of penicillin from fermentation broths was based upon the adsorption of the product on activated carbon. The carbon was collected, washed with water, and the carbon adsorbate was eluted with 80% acetone. The penicillin was concentrated by distillation or evaporation under vacuum at 15–32 °C. The remaining aqueous solution was cooled to 2 °C, acidified to pH 2.0–3.0 and the penicillin extracted into amyl acetate from which it was crystallized with excess mineral salt or buffer near neutral pH under vacuum. As higher quantities of penicillin were produced, this approach became impractical because of increased carbon requirements. There are other disadvantages to this process when compared to direct solvent extraction, which is described below. The carbon process has been discussed by Sylvester and Coghill (1954) and by Whitmore *et al.* (1946).

2.3.2   Solvent Extraction Process (Industry Standard)

2.3.2.1   *Process overview*

An outline of the current standard process is presented in Table 10 and in Figures 2 and 12.

Solvent extraction forms the basis for isolation and purification of penicillin today. The outline of the process was reviewed by Sylvester and Coghill (1954), and the theoretical basis for the separations involved was discussed by Rowley *et al.* (1946) and Whitmore *et al.* (1946).

The basic process has changed very little since the late 1940s except for the addition of automatic controls and simplifications permitted by broths of higher potency and purity, by increased scale of operation and by improved design of specific equipment. Typical step yields for the solvent extraction process are presented in Table 11.

Table 10   Outline of the Penicillin Purification Process

Step	*Filtration* →	*Extraction*	→ *Carbon treatment* →	*Crystallization* →	*Drying*
Purpose:	Separate mycelia from penicillin	Remove soluble contaminants	Remove soluble contaminants	Further purification and stabilization	Stabilization
Equipment:	Rotary drum vacuum filter	Countercurrent extractors	Mixer and drum filter	Tank and drum filter, settlers, or basket centrifuge	Horizontal belt filter/warm, dry air
Basis:	Size of particles	Differential extraction based on pH change	Adsorption of impurities	*Via* addition of Na or K salts	Dry solvent, vacuum, or air

Table 11   Typical Yields in Penicillin Purification

Step	*Yield (approx.)*
Holding	95% but losses can be large if broth is not rapidly chilled or if microbial contamination occurs
Filtration	90–95% based on 'filtrate' assay. 5% of the 'loss' is accounted for by insoluble solids in the broth. Other losses are due to degradation and leakage to drain
Solvent extraction single stage	80–90%
lead-trail	92–96%
Aqueous (back) extraction	95–97%
Crystallization	95%
Drying	95%
Overall	~78%

It is noteworthy that Hersbach *et al.* (1984) list a purification yield of 90% overall, even with their carbon treatment step. This requires intensive optimization of each step on an ongoing basis and is most impressive.

2.3.2.2 *Filtration*

Filtration is usually accomplished on high-capacity, rotary vacuum drum filters. At harvest the culture has the consistency of dilute sludge and is light tan to dark brown in color. The mycelium is separated from the penicillin-containing broth and is washed on the filter. Filter aids or pre-coats are sometimes needed and can represent a significant raw material cost. The 'rich' penicillin-containing filtrate is cooled in a heat exchanger to 0–4 °C in order to minimize chemical and enzymic degradation during solvent extraction. In some early processes the filtrate was further clarified by a second filtration with 1–1.5% Hyflo, occasionally with the addition of aluminum sulfate or tannic acid (Lovens *et al.*, 1949) to precipitate dissolved proteinaceous material. It might be possible to use such agents or even a pH change prior to initial filtration or to remove these precipitants prior to solvent extraction. New extractors or higher potency and purity broths may permit this second filtration clarification to be omitted. If this step is omitted, emulsion difficulties encountered during extraction may be more severe, particularly with older designs of extractor. Any material left in the product may cause severe problems with later chemical modifications. As an alternative to filtration, direct extraction of penicillin-containing whole broth can be accomplished with acceptable recovery and favorable economics (Podbielniak *et al.*, 1970). Most large-scale penicillin-recovery facilities of recent design, such as the Glaxo plant at Cambois (Berkovitch, 1976) which was built in 1969, incorporate the filtration step.

In recent years, membrane technology has evolved to the point that macroporous filters, produced by several manufacturers including Dorr-Oliver, Millipore, Romicon and others, can be considered in this application. In principle it can be cost effective in both installed cost and avoiding the use of filter aids. At least one antibiotic production facility (the Merck cephamycin C plant) is believed to use these filters; however, the organism involved is a streptomycete. Such filters may be less flexible than the standard approach so pilot scale performance over extended periods and savings should be considered carefully.

It is important to recognize that one step, that of broth holding prior to filtration or extraction, has not been mentioned. Losses due to degradation can be minimized with clean, refrigerated holding tanks, filter hoppers, and surge. Contamination with penicillinase, *etc.* must be avoided. Capital is often well spent on improved broth holding and increased throughput.

2.3.2.3 *Solvent extraction*

Penicillins G and V are strong acids (pK_a values in the range 2.5–3.1) with moderate molecular weights (334 and 350, respectively). The acid forms are soluble in many organic solvents and can be extracted with high efficiency into amyl acetate or butyl acetate at pH 2.5–3.0.

Experimental and analytical aspects of the solvent-extraction process have been thoroughly discussed by Rowley *et al.* (1946) who recommended several solvents, including cyclic ketones, which are more selective than amyl acetate and which allow efficient extraction at pH 4.0, thus minimizing acid inactivation. These authors also discuss the design parameters for separation of penicillins from acidic impurities based on selective extraction. Whitmore *et al.* (1946) also measured distribution coefficients in various solvents at 0 °C and pH 2.5, and discussed continuous countercurrent extraction in a bench-scale column as well as subsequent back extraction into the aqueous phase at pH 7.4 with phosphate buffer. Souders *et al.* (1970) discussed the history of penicillin extraction, particularly the early work at the Shell Development laboratories. The conceptual basis for the early process, which required separation of penicillins from other acids which were both stronger and weaker, was discussed. Figure 11, adapted from Souders *et al.* (1970), presents the basis for the separation. Today's broths contain between 20 and 35 g of penicillin G or V l^{-1} and minimal concentrations of other penicillins and 6-aminopenicillanic acid. Thus the problems of purification are much simpler than in the mid 1940s and 1950s as discussed by Sylvester and Coghill (1954), Whitmore *et al.* (1946) and Podbielniak *et al.* (1970). They are summarized in Table 12, which is adapted from Whitmore *et al.* (1946) and which follows the product path. In the modern process the penicillin-containing filtered broth is extracted in continuous, countercurrent, multistage centrifugal extractors into butyl acetate or amyl acetate at 0–3 °C and pH 2.5–3.0. The major examples are the Podbielniak D-36 extractors and the Alfa Laval ABE 216, used extensively in Europe. The ABE 216 allows the process engineer substantially greater flexibility in altering operating parameters; however, operation of the 'Pods' in many facilities seems hampered by the failure of process engineers to take advantage of the substantial literature on their operation. Both manufacturers make short-term support available to potential users. Operation of the continuous, multistage, countercurrent centrifugal extractor is discussed by Podbielniak *et al.* (1970), Todd and Davies (1973) and Todd (1972), and in literature available from the manufacturer (Baker-Perkins, Saginaw, Michigan, USA). Alfa Laval has leased the ABE 216 on favorable terms in order to assist prospective users in their evaluation. The author, in his only use of the machine, was favorably impressed with its performance and quality and would definitely specify it in the future where the multi-plate efficiency is required.

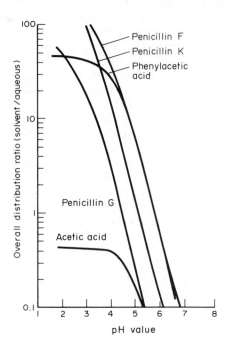

Figure 11 Distribution ratio of penicillin as a function of pH value of extraction solvent (Adapted from Souders *et al.*, 1970)

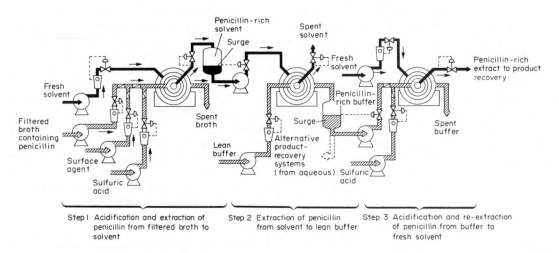

Figure 12 Diagram of three-stage extraction/purification process used in penicillin production (Adapted from Baker-Perkins Literature for 1976)

Other approaches such as static mixers and decanters in 'lead trail' mode, or in combination with more efficient extractors, offer capital cost reduction in new or expanded facilities and should be considered.

Penicillins G and V degrade under acid conditions. The kinetics of degradation are first order,

Table 12 Multiple Extraction Process for Penicillin Used in the 1940s and 1950s[a]

Step:	Broth → Amyl acetate	→ Aqueous buffer	→ Chloroform →	Sodium salt solution
Degree of concentration	0.4	0.2	0.4	0.4
pH Value:	2.0	7.4	2.0	6.0–7.0
Reagent (%, w/v)	H_3PO_4 (10)	Phosphate (2)	H_3SO_4 (10)	NaOH (2)

[a] Queener and Swartz (1979).

and the rate is proportional to temperature and to the reciprocal of the pH value. The degradation kinetics are described by Brodersen (1946). The half-life of penicillin under conditions necessary for efficient extraction made short-residence times (measured in seconds) a constraint on solvent extraction at low pH values. Penicillin V is much more stable than penicillin G under acid conditions, offering potential yield advantages during purification. This, combined with the lower toxicity of the precursor in fermentation, makes penicillin V the best choice for 6-APA production.

The theoretical basis for extraction of penicillin in butyl acetate using a Podbielniak or other extractor is discussed by Zhukovskaya (1972), who relates the general distribution coefficient (K) for benzylpenicillin between butyl acetate and water to pH value and pK_a value as follows:

$$K = K_0 \frac{1}{1 + \dfrac{K_a}{C_{H+}}} = 47 \times \frac{1}{1 + 10(\text{pH} - 2.75)}$$

where K_0 is the distribution coefficient for undissolved penicillin, K_a is the dissociation constant for penicillin and C_{H+} is the hydrogen ion concentration. Zhukovaskaya developed a series of nomographs relating the percentage of penicillin extracted, the degree of concentration, the pH value and the number of theoretical stages provided by the equipment. In general, the degree of concentration achieved experimentally with 2% residual penicillin in the spent beer was approximately 10–40% less than theory with five theoretical stages. The discrepancy was said to result from emulsion formation, and further treatment of rich beer to remove emulsifying impurities or the use of demulsifiers was suggested. Zhukovskaya suggests that extraction of penicillin G into butyl acetate be performed at pH 3.0 with a 5–7-fold concentration.

Basically, rich beer is mixed with dilute (10%, w/v) sulfuric or phosphoric acid and a demulsifier (0.003–0.1%, w/w, in solvent). The use of demulsifiers and wetting agents is discussed by Podbielniak *et al.* (1970), Todd and Davies (1973), Todd (1972) and Sylvester and Coghill (1954). The mixture is fed to the heavy liquid in-port of the extractor. Butyl acetate, amyl acetate or other solvent is then fed to the light liquid in-port. The back pressure of light liquid coming out is used to control the position of the interface, hopefully such that the solvent (lowest flow) is the continuous phase for the extraction, thus maximizing efficiency. In modern practice, control of extractor operation would include automatic pressure controls to optimize operation based on relationships described by Todd (1972). Potencies of the solvent could be as high as 125–200 g l^{-1} depending on the details of operating parameters.

Concentration and extraction are accomplished in one step. Many producers link two extractors in a lead-trail (or series) arrangement as shown in Figure 13, which is taken from Todd and Davies (1973). When only a single extractor is used, as much as 5–10% of the recoverable penicillin may be lost with the spent broth. With lead-trail operation, the overall recovery of extractable penicillin is increased to as high as 98–99%, and the solvent flow rate can sometimes be decreased to as low as 10–20% of the rich-beer flow rate. Rich solvent potency could reach 100–200 g l^{-1}, making short contact time crucial.

Podbielniak *et al.* (1970) note that whole-broth extraction in the Podbielniak extractor is possible. When this is done, the lead-trail mode is usually justified (Podbielniak *et al.*, 1970), and the lead extractor is larger and of more open design to avoid plugging (Todd and Davies, 1973). Modification of concentric elements for increased turbulence might also help. At least one major manufacturer uses whole-broth extraction in production, and Baker Perkins has used this to interest others in the approach. Though never explicitly identified, it is generally thought that the facility referred to is the Pfizer Plant at Groton, Connecticut. Choice of wetting and demulsifying agents is probably a key to satisfactory operation. The system may be more sensitive to the vagaries of variable broths, but the entire filtration step is eliminated.

Whole broth extraction could be used to produce lower quality material for animal feed or for

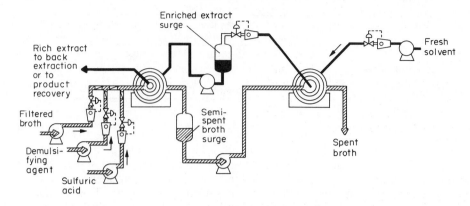

Figure 13 Centrifugal extractors operating in series to extract penicillin from filtered broth

sales as intermediate to other companies whose product specifications are inadequate. Here, the purchaser may be forced to recrystallize the product, especially for semisynthetic uses. For semisynthetic uses the author believes that contract specifications should prohibit whole-broth extraction and require initial precipitation of protein.

The use of whole-broth extraction may have implications in waste treatment, since residual solvent could make extracted mycelial cakes difficult to dispose of. The introduction of large-scale ultrafiltration and macroporous filters may provide potentially attractive alternatives to precoat filters in terms of capital cost and cost of filter aids and may reduce the advantages of whole broth extraction.

Todd and Davies (1973) and Todd (1972) discuss optimization of operational parameters for countercurrent centrifugal extractors. Spent-broth and spent-solvent streams are processed for solvent recovery and re-used throughout the process.

2.3.2.4   Carbon treatment

Depending on the final specifications or end use, the penicillin-containing solvent may be treated with carbon (0.25–0.5% according to Sylvester and Coghill, 1954) to remove pigments and other impurities. This could be achieved with several columns in series. With today's high broth potencies and proportionately lower impurities, together with efficient extraction using solvent-to-broth ratios of 0.1, one may proceed directly to crystallization through addition of potassium or sodium acetate. This may be especially attractive if the product is to be further processed to provide 6-APA for semisynthetic products or sold as commodity-crude penicillin. Simplification of the process scheme may result in products of lower quality, and purchasers of these materials should exercise caution, particularly as regards precursor content, color and purity. Hersbach *et al.* (1984) list carbon absorption as a routine step in the Gist Brocades commercial process.

2.3.2.5   Further extraction

Penicillin may be back extracted into water by addition of sufficient alkali (potassium or sodium hydroxide) or buffer at pH 5.0–7.5. Choice of the pH value to be used is dependent upon the impurities present. This step is generally eliminated from modern processes.

The volume ratio of water-to-solvent could be as low as 0.1–0.2, if necessary, and is adjusted along with the ratio in the primary extraction to give the desired potency during crystallization. A continuous, multistage, countercurrent extractor may be used (Todd and Davies, 1973). The role of the two phases is reversed, and the rotor is kept full of the heavy liquid by maintaining low back pressure on the light liquid out-port. Separation is quite clean, no emulsion forms, no additives are needed (Todd and Davies, 1973), and the phases separate readily even without centrifugation (Sylvester and Coghill, 1954). This back extraction could be performed using an agitated tank or static mixer followed by a centrifuge or preferably a gravity separation, particularly if little concentration is required. The position of the solvent–aqueous interface should be monitored and used to control the flow of rich aqueous and spent solvent streams.

Crystallization may be performed from the aqueous phase at this step if desired. The critical parameters are potassium or sodium concentration, pH value, penicillin concentration and temperature. Todd and Davies (1973) note that some older plants were designed for crystallization of penicillin from solvent and, therefore, may perform a third extraction into solvent and then crystallize. The final solvent and pH value could be different from the initial one to effect further purification, but this is generally unnecessary. A three-stage extraction is illustrated in Fig. 2. In most cases, however, crystallization follows either the initial solvent extraction or the back extraction depending upon the intended use.

2.3.2.6 *Crystallization*

As already discussed, an excess of potassium or sodium is added as the alkali or acetate to an aqueous or solvent stream containing a high concentration of penicillin. This step is performed in an agitated tank, and the crystals are collected in a bottom unloading centrifuge or on a filter of the rotary vacuum or horizontal belt type.

2.3.2.7 *Drying*

The crystals may be washed and predried with anhydrous isopropyl alcohol, butyl alcohol or other volatile solvent which removes some impurities. Removal of color is achieved here also, and butyl alcohol is particularly effective. Drying may be accomplished with warm air, vacuum or radiant heat. In modern plants, large horizontal belt filters are used on which crystals are collected, washed and dried.

2.3.3 Further Processing

Crystalline penicillin G or V is sold as an intermediate, converted to 6-APA or is further processed to pharmaceutical grade. Detailed discussion of final procedures is beyond the scope of this chapter. Briefly, the dried crystals may be further purified by recrystallization. They may be sterile-filtered in water or solvent using standard procedures. Aqueous penicillin G could be diluted with butanol and azeotropically distilled under vacuum (aseptically) to produce the injectable crystalline product (Sylvester and Coghill, 1954). Penicillin V is processed into the oral dosage forms. Either penicillin V or G may be used for the production of semisynthetic penicillins or cephalosporins.

2.3.4 Penicillin Acid Process (State of the Art)

A US patent issued recently to Allanson and Edmundowicz (1983) of Eli Lilly describes a novel approach to the isolation and purification of penicillin from culture fluids. In this process, penicillin acid is isolated as a crystal from filtered, acid polished broth. A crystalline potassium salt is formed through the addition of potassium 2-ethylhexanoate to an acetone solution of the free acid. The process yields penicillin in a pharmaceutical intermediate form. Capital and raw material cost components should be significantly below those for the conventional process. Organizations manufacturing large quantities of penicillin or contemplating construction of purification facilities should assess the potential for licensing this technology.

2.4 SEMISYNTHETIC PENICILLINS

Penicillins (2) with substituents on the methylene α to the 6-amido moiety of the penicillin molecule are produced by enzymic or chemical cleavage to give 6-aminopenicillanic acid (6-APA; 3) which is then re-acylated with the desired acyl side chain at the 6^{α}-amino position. Alternatively, diacylpenicillins (4) may be produced chemically from penicillin G or V and the RCO group hydrolyzed to give the desired semisynthetic penicillin. Hersbach *et al.* (1984) have recently reviewed semisynthetic processes.

(2)   (3)   (4)

2.4.1 6-Aminopenicillanic Acid

6-Aminopenicillanic acid (6-APA), used to produce semisynthetic penicillins, can be prepared enzymatically or chemically (Carrington, 1971). Both processes are used commercially. For example, both Beecham and Squibb have at some time split penicillins to produce 6-APA using enzymes, while Gist-Brocades, Bayer, Bristol-Meyers and Lilly have used chemical cleavage. Details of the procedures used at any given time by a particular company are, however, proprietary information which is not generally available for publication.

6-APA may be converted to 7-ADCA by a chemical ring expansion. This material is then further processed to yield oral cephalosporins. Cloning and expression of the expandase enzyme from *C. acremonium* or *Streptomyces*, combined with protein engineering, could represent an alternative to chemical ring expansion.

2.4.1.1 *Enzymic cleavage of penicillins to yield 6-aminopenicillanic acid*

Use of penicillin acylase to prepare 6-APA has been reviewed by Huber *et al.* (1972). More recently, penicillin acylases have been reviewed by Moss (1977). The source of enzyme used in production of 6-APA from penicillin will vary with the manufacturer, and details are trade secrets. However, the general procedures employed may be surmised from patents and, to a lesser extent, from journal publications.

There is a great variety of known sources of penicillin acylase. Furthermore, active screening for new acylases continues (Romanova *et al.*, 1976). Two general classes of penicillin acylase exist, namely those that cleave penicillin G more rapidly than V, and those that split V more rapidly than G (Lemke and Brannon, 1972). Most bacteria produce the former type, while fungi produce the latter. Acylase from *Escherichia coli* has been actively studied, and some processes probably utilize an industrially improved strain of this species (Savidge and Cole, 1976). Numerous recent patents and publications on acylases from other bacteria suggest that some producers, particularly in Japan, may use bacterial sources other than *E. coli. Bacillus megaterium, Kluyvera citrophilia* and *Pseudomonas melanogenum* are probably employed industrially to produce 6-APA from penicillin G. Studies on acylases from fungi have continued and some may be presently employed in industrial preparations of 6-APA from penicillin V. For a detailed discussion see Queener and Swartz (1978).

Only a few studies have dealt with other penicillin substrates, *e.g.* penicillin G sulfoxide (**5**), penicillin V sulfoxide (**6**) (Kondo *et al.*, 1975), benzyl- or phenoxymethyl-penicillin tetrazoles (**7**) (Hamsher, 1975), and the dimethylsilane esters of penicillin G (**8**) or V (Aukowski and Skrzypkowska, 1974).

Industrially developed acylase-producing strains and details of fermentation processes for these strains are the property of producing companies; however, studies on media (Dogariu *et al.*, 1973; Vojtisek and Slezak, 1975; Nara *et al.*, 1975; Savidge and Cole, 1976), aeration (Kleiner and Lopatnev, 1972, 1973, 1974; Vojisek and Slezak, 1975) and regulation (Thadhani *et al.*, 1972; Acevedo and Cooney, 1973; Savidge and Cole, 1976) have been published, which suggest import-

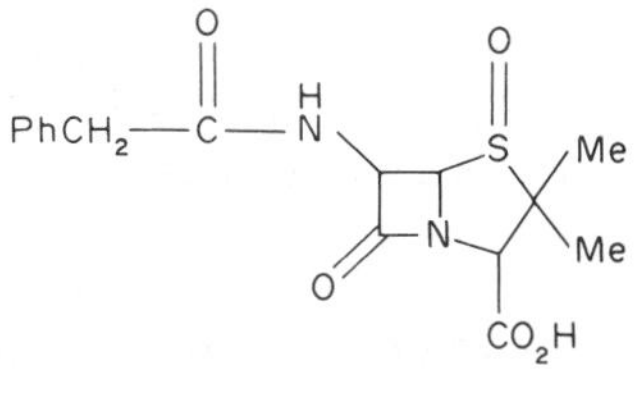

(5)

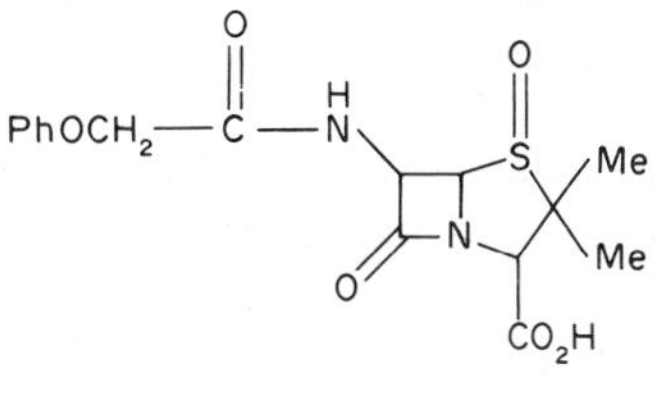

(6)

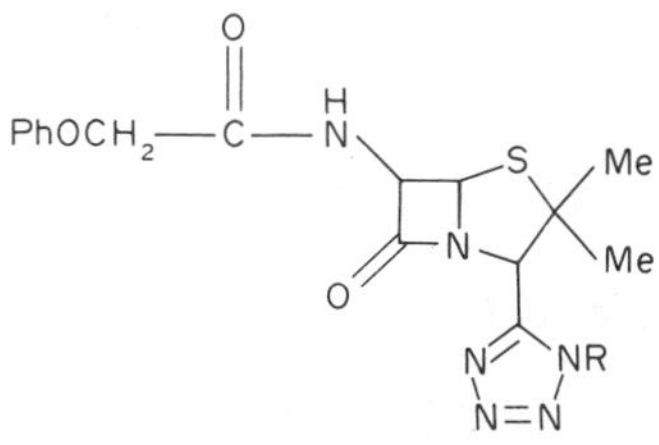

(7)

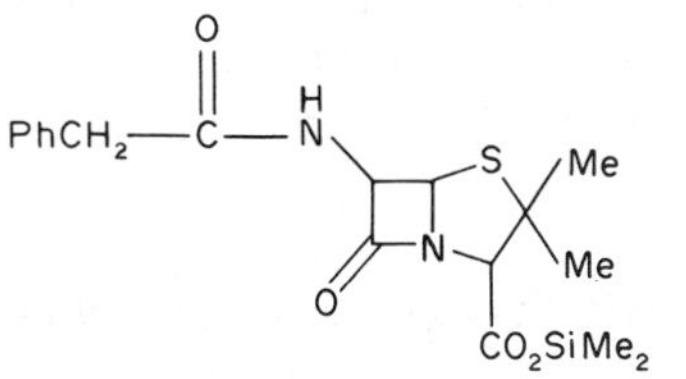

(8)

ant general requirements for efficient preparation of many penicillin acylases. For example, coproduction of penicillin acylase and penicillinase in certain microorganisms has been recognized (Takasawa *et al.*, 1972; Tochenaya and Chaikovskaya, 1973), and the advantages of penicillinase-deficient mutants have been demonstrated (Okachi *et al.*, 1973). The importance of the presence of phenylacetic acid for stimulation of penicillin acylase synthesis (Savidge and Cole, 1976) by many microorganisms is known. Fermentations with certain strains of *E. coli* and with *Bacillus megaterium* ATCC 14 945, *Penicillium chrysogenum* Wis 49.408 and *Fusarium semitectum* BC 805 have been summarized (Savidge and Cole, 1976; Vanderhueghe, 1976).

Four basic procedures exist for enzymic preparation of 6-APA. Contact with the penicillin may be with cells of penicillin acylase-producing microorganisms which are in suspension or immobilized on a support. Examples of each method are given below.

(*a*) *Contact with free cells of penicillin acylase-producing organisms. Aspergillus repens* (ATCC 20 149) is grown aerobically at 30 °C for 90 h in kerosene (7.5%), corn-steep liquor (0.1%), peptone (0.25%), CaCO$_3$ (1%), KH$_2$PO$_4$ (0.08%), MgSO$_4$ (0.05%), ZnSO$_4$ (0.003%) and FeSO$_4$ (0.005%). Cells are removed, washed and dried. Dried cells (5 g l^{-1}) are suspended in phosphate buffer (pH 7.5) and potassium penicillin V added (20 g l^{-1}). The mixture is incubated at 37 °C for 5 h, and potassium 6-aminopenicillanate (12.6 g l^{-1}) is isolated (92% yield; Nara and Misawa, 1974). Because of its simplicity, variations of this method are probably the most widely used for enzymic preparation of 6-APA.

(*b*) *Contact with cell-free penicillin acylase.* Cells of *E. coli* from a 24 h aerobic culture are passed through a narrow slit at high pressure, and the cells removed by filtration to give an enzyme solution containing 48 units g^{-1} dry weight. Aqueous potassium penicillin G is added to achieve a solution with 34 units of enzyme g^{-1} penicillin G and 2.5 N NaOH added to adjust the pH value of 7.8. The mixture is incubated 6 h at 37 °C, cooled to 10 °C, and extracted with methyl isobutyrate at pH 2 (Delin *et al.*, 1972a).

(*c*) *Contact with immobilized penicillin acylase.* Purified enzyme is obtained from *Bovista plumbea* NRRL 3501 by extraction with ground glass, filtration, acid precipitation, absorption and elution from bentonite, and concentration under vacuum. To obtain fibre-bound enzyme, penicillin acylase in 20% glycerol is added dropwise to a solution of cellulose acetate in methylene chloride, and the resulting emulsion is extruded through an orifice into toluene for coagulation. Fibres are dried at room temperature under vacuum and placed in a column. An aqueous solution of penicillin V is recycled over the column at 32 °C with 10% NH$_4$OH added continuously to maintain pH 7.5. In three hours, 97% conversion of 6-APA was obtained. 6-APA is obtained from the solution withdrawn from the column by isoelectric precipitation (91% yield; Brandle and Knauseder, 1975). There is an extensive literature on immobilized penicillin acylase. See Queener and Swartz (1978) for references dealing with immobilization of penicillin acylase from *E.coli*.

(*d*) *Contact with immobilized cells.* Immobilization of cells producing penicillin acylase (Chibata *et al.*, 1974; Sato *et al.*, 1976) is desirable, since the method increases enzyme stability, expedites separation of the catalyst from the product, and allows for convenient re-use of the catalyst. Unlike immobilized cell-free penicillin acylase, the trouble and expense of purifying the enzyme (Ramachandran *et al.*, 1970; Delin *et al.*, 1972a,b; Kutzbach, 1973a,b,c; Shimizu *et al.*, 1975; Schneider and Roehr, 1976) are avoided. For example, *E.coli* ATCC 9637, containing 6.5 units penicillin acylase per gram dry weight, is suspended in saline (48 M) to which is added acrylamide (9 g), *N,N'*-methylenebisacrylamide (0.48 g), 6 ml of 5% dimethylaminopropionitrile and 6 ml of 2.5% potassium persulfate. After 30 min at 37 °C the resulting gel is granulated by passing through a sieve to produce particles of diameter 3 mm. A volume of 120 ml of immobilized cells is added to a column (2.1 cm × 34.8 cm). One percent potassium penicillin G is passed at a rate of 16 ml h^{-1} over the column buffered at pH 9.0. After acidification to pH 2.0, unreacted penicillin is removed by extraction into methyl isobutyl ketone, and the 6-APA is concentrated at pH 7.0 (NaOH) under decreased pressure. The 6-APA is crystallized at pH 4.3 (HCl; yield is about 75%; Chibata *et al.*, 1974).

2.4.1.2   *Chemical preparation of 6-aminopenicillanic acid*

Chemical production of 6-APA from penicillins involves protection of the C-3 carboxyl in penicillin with groups which are readily hydrolyzed in the presence of water. Such groups include silyl (Weissenburger and Van der Hoeven, 1970) and analogous groups (Pereira and Luz, 1975). Mixed (Chauvette *et al.*, 1972; Stojanovic and Jovanovic, 1971) or symmetrical anhydrides and various phosphate esters (Sellstedt, 1975) can also be used. Esters such as benzyl or benzhydryl, which require hydrogenolysis for removal, are not used because hydrogenolysis extensively opens the β-lactam ring and lowers the yield of 6-APA. The esters of penicillin G and V are not isolated. Phosphorus pentachloride in the same solvent used to prepare the esters (*e.g.* methylene chloride or chloroform) is added to solutions of the penicillin esters at low temperature (*e.g.* −50 °C). Formation of the iminohalide of the esterified penicillin is allowed to proceed for about 3 h, and a large excess of alcohol is then added at low temperature (*e.g.* −60 °C). Alcoholysis is allowed to proceed for several hours, and water is then added and the pH value immediately adjusted to 5.4 with a bicarbonate salt. A seed of 6-APA is added and the mixture stirred at 0–5 °C to precipitate 6-APA. Precipitated 6-APA is washed with an aqueous solvent (*e.g.* 50% acetone). Yields of 6-APA as high as 95% with purities up to 95% can be attained (Huber *et al.*, 1972).

2.4.2   Synthesis of Clinically Useful Penicillins and Closely Related Congeners

Acylation of 6-APA to produce semisynthetic penicillins has been reviewed (Kaiser and Kukolja, 1972; Moll and Kastenmeier, 1971; Queener and Swartz, 1979). In Queener and Swartz are described examples from patents illustrating typical processes for each of the semisynthetic penicillins used routinely in clinical practice. In addition, examples of other semisynthetic penicillins which are similar in structure to and preparation of their clinically accepted counterparts are given. Only compounds which have at least achieved clinical trial status were dealt with. The interested reader is referred there for details. The discussion is lengthy and beyond the scope of this chapter.

2.5   AUTOMATION

In the Glaxo plant at Cambois in England, which was built in 1969 and uses a three-stage extraction with crystallization from solvent, a computer control system included 100 analog control loops, 300 digital inputs and 600 digital outputs (Berkovitch, 1976). Automation of the extraction and other steps is certainly extensive, and no doubt other steps, including valve control of liquid flow, transfer of penicillin by fluidization, blending and packaging by weight, could be automated (Podbielniak *et al.*, 1970). Such operational controls should not be confused with physiological control of the fermentation process, which is an entirely separate issue. The new Wyeth (now Fermenta) penicillin plant, which has four 40 000 gallon fermenters, is also automated in an operational sense but does not employ extensive, interactive, multiparameter control of the fermentation. Thus, this facility is probably not particularly unique except in that it is one of very few penicillin plants built in the late 1970s when the economics of penicillin supply clearly argued against new construction. The basis for a practical control scheme for control of cell growth and volumetric oxygen uptake and CO_2 evolution rates is provided in a paper by Nelligan and Calam (1983). The control is anticipatory and is based on cell mass and growth rate predictions, combined with kinetic constants determined experimentally. The authors note that an alternative would be to feed to a particular predetermined curve and adjust to avoid exceeding the $K_L a$ limits of the vessel at a particular cell mass. The article presents the approach in adequate detail for this proposed application. One would, of course, want to build in automatic safety measures, perhaps including that proposed by Giona of heating the vessel to slow a runaway growth.

The limiting factor for some vessels may be the cooling capacity, in which case the concept remains the same, but the limiting value changes. In some situations it may be useful to use DO_2 or heat evolution measurements as consensus data to avoid problems due to failures in the exit gas analysis system (used for CO_2 analysis) or air mass flow meaurements.

2.6 PROCESS ECONOMICS

Because of the need for confidentiality to maintain a competitive position, actual producers of penicillin will not disclose their manufacturing costs. Little information on the economics of antibiotic production is available. Fortuitously, Arthur G. McKee and Company has developed a manufacturing cost breakdown of production costs for a typical modern non-sterile bulk penicillin G production facility. The design criteria for the model plant developed by McKee based on their experience are presented in Table 13.

Table 13 Design Criteria and Assumptions for Model Penicillin Plant[a]

1.	Five 50 000 gallon vessels (40 000–45 000 gallon working volume, 304 stainless steel) with a 700 horsepower agitator (~1.5 HP/100 gallons delivered).
2.	Three compressors (10 500 CFM each), one of which is on standby (*eg.* 0.82 v/v min^{-1}).
3.	Three inoculum tanks plus tanks for sterile feeding.
4.	205 h cycle plus 10 h turnaround with annual production of 11.36×10^6 gallons of broth/year or 11.74×10^5 gallons/year with continuous sterilization (an increase of 3.34%).
5.	Product recovery equipment and labor sized for $1.6 \times$ fermentation volume. Process: belt filtration, two-step extraction (Podbielniak) (two solvents), crystallization and vacuum drying. For complete discussion see Queener and Swartz (1979). We have assumed further that the purification cost per 1000 gallons of broth is independent of the potency. This simplification is reasonable in the light of the purification scheme.
6.	Solid waste ~1 000 tons dry substance/year. Equipment provided for primary treatment of liquid waste. Solids disposal by landfill. No odor removal for off-gas is provided.
7.	Steam generation by gas or oil or purchased. Electrical rate assumed is $0.0375 kWh^{-1}.
8.	Manpower excludes 'front office staff.'
9.	Additional costs for plant are $2 000 000–$3 000 000 for land, client engineering process royalty, working capital, contingency, maintenance equipment, office equipment and vehicles.
10.	The process is assumed to be fed batch with intermediate withdrawals (semi-continuous). The starting volume is 40 000 gallons and a total of 55 000 gallons are harvested. Later we adjust and modify this assumption to make the Panlabs and McKee data consistent. We assumed a fed batch process without withdrawal, having a harvest volume of 45 000 gallons.

[a] Adapted from the study by A. G. McKee & Co. (Swartz, 1979).

Further results from the McKee study are presented in Tables 14 and 15. Table 14 shows the details of the plant construction budget estimate. Table 15 develops the overall fixed charges per 1000 gallons of broth produced. The author has adjusted for the costs omitted in the McKee study. Table 16 presents the direct production costs per the McKee study. The raw material costs will be approached independently.

Table 17 presents the total glucose used per litre of harvested broth (45 000 gallons are assumed later as a harvest volume; this is consistent with actual results for fed batch operations in a 50 000 gallon vessel). The 205 hour cycle is segmented into a 25 hour 'growth phase' and a 180 hour synthetic phase for ease of analysis (after Cooney and Acevedo, 1977).

In subsequent analysis the average glucose use of 240.3 g l^{-1} is used as a basis. The overall yield then is 0.125 g pen G-K/g glucose equivalent. Fermentation medium formulations are trade secrets. For our purposes we have developed a cost based on material balance considerations using very few assumptions (see Table 18). The organic nitrogen source was assumed to supply nitrogen for cell mass only. Though somewhat arbitrary in light of the wide belief in the special significance of corn-steep liquor (CSL) in modern penicillin fermentation, reference is made to the work of Hockenhull and MacKenzie (1968) of Glaxo, who reported equivalent results with either the production medium containing CSL or a synthetic medium with inorganic nitrogen. Elemental cell and product compositions (Tables 19 and 20), costs (Table 21) and the assumption that 20% (considered a worst case assumption) of the precursor was wasted (oxidized as carbon source but not credited to glucose use) are the other information required (see Swartz, 1979, for details).

Except for glucose, the ingredients in our model medium are among those described by McCann and Calam (1972). The fermentation raw material cost was approximately $454/1000 gallons of harvest broth.

Table 14   Capital Investment[a]

	Cost ($1 000)		Cost ($1 000)
Process equipment	6 082	Spare parts	121
Installation	1 338		
Insulation	486	DIRECT INSTALLED COST	25 724
Instruments	608	Engineering	2 572
Pipe	3 041	Construction	1 543
Electrical	4 073	Fee	895
Building	2 912		
Utilities	5 474	FIXED INVESTMENT	30 734
Site	608	EXCLUDED COSTS[b]	3 000
Laboratory equipment	981		
		TOTAL	33 734

[a] Adapted from the study by A. G. McKee & Co. Mold fermentation, long cycle.   [b] Allowance for costs excluded from McKee study (Table 13, item 9). $1 000 000 depreciated at 0.03%, $2 000 000 other depreciated at 0.1% (Swartz, 1979).

Table 15   Fixed Charges per 1000 Gallons Broth Produced[a]

	Cost ($)
Depreciation (10% on equipment and 3% on buildings)	268
Taxes 3% on fixed investment	89
Insurance 0.7% on fixed investment	21
TOTAL	378

[a] Mold fermentation, long cycle (Swartz, 1979).

2.6.1   Costs

The total cost per 1000 gallons is summarized in Tables 22 and 23. Modifications from the McKee study include the raw material calculations shown, increased labor charges and correction to a fed batch from a semicontinuous operation. The latter was done to make the McKee economics consistent with the Panlabs results.

The unit cost calculated from the information described above is $18.92 kg^{-1} bulk crystals (see Table 24). The theoretical yield is 124.7 kg/1000 gallons. A real production unit, with contamination and non-optimal conditions due to equipment shortcomings and the variability of biological processes, might actually average 80–85% of this value. This factor is here termed the 'fermenter plant efficiency'.

Table 16 Direct Production Cost per 1000 Gallons of Broth[a]

	Cost per 1000 gallons ($)		
Item	Fermentation	Purification	Total
1. Raw materials	454.00	32.99	486.99
2. Operating labor @ $10 h^{-1}	27.45	52.45	79.90
3. Direct supervision @ $15 h^{-1}	15.00	15.00	30.00
4. Maintenance	116.26	35.91	152.17
5. Laboratory	1.27	2.02	3.29
6. Utility			
A. Steam ($3.00 Mlb^{-1})	17.75	28.83	46.58
B. Electricity ($0.0375 kWh^{-1})	127.77	18.18	145.95
C. Water ($0.47 Mgal^{-1})	2.43	0.92	3.35
D. Waste ($250 000 year^{-1})	11.00	11.00	22.00
7. Supplies	9.66	9.66	19.32
Total	782.59	206.96	989.55

[a] Adapted from the study by A. G. McKee & Co. to increase labor charges and modified raw materials calculations. This is based on the McKee semicontinuous process with 55 000 gallons harvested (Swartz, 1979).

Table 17 Total Glucose Use Per Liter[a]

	High cost	Low cost	Average
Dry cell weight	35 g l^{-1}	25 g l^{-1}	
Penicillin G-K	30 g l^{-1}	30 g l^{-1}	
$Y_{cells/glucose} = 0.45$ g g^{-1}	78 g	56 g	
$M = \dfrac{0.027 \text{ g glucose}}{\text{g cells h}^{-1}}$	170 g	122 g	
$Y^{b}_{penicillin \text{ } G\text{-}K/glucose}$			
case I: 1.13 g g^{-1}	33.9 g	(33.9 g)	
case II: 0.69 g g^{-1}	(20.7 g)	20.7 g	
	281.9 g	198.7 g	240.3 g
Overall yield			
$Y \dfrac{\text{g pen G-K}}{\text{g glucose}}$	0.106	0.15	0.128

[a] Analysis is per Cooney and Acevedo (1977). Case I and case II represent with and without α-aminoadipate recycle. [b] Y is a theoretical requirement for synthesis only (Swartz, 1979).

Table 18 Raw Material Requirements and Their Costs Per Liter[a]

	Amount (g l^{-1})	Cost ($ l^{-1})	Amount (g l^{-1})	Cost ($ l^{-1})
Corn-steep liquor (dry basis)	21.88	0.0048	15.63	0.0034
$(NH_4)_2SO_4$	10.7	0.0027	10.7	0.0027
KH_2PO_4	2.74	0.0027	1.96	0.0020
Sodium phenylacetate (80% efficient) ($2.95 kg^{-1})	15.9	0.0469	15.9	0.0469
Glucose (from Table 17)	281.9	0.062	198.7	0.0437
Other (10%)		0.0119		0.0099
Total		0.131		0.1086
		or		or
		$496 per 1000 gal		$411 per 1000 gal
Average			$454 per 1000 gal	

[a] Swartz (1979).

Table 19 Elemental Composition of
P. chrysogenum during Penicillin
Fermentation (% of Dry Cell Pellet)[a]

Element	60 h	113 h
C	46	44
H	6.3	6
N	8	4.8
S	<0.6	<0.6
P	3	1.8

[a] A residue amounting to 20% of the dry cell weight and containing ~50% calcium was subtracted from the total and the results were corrected to a 100% basis. Water and acetone washed filtered mycelia (Swartz, 1979).

Table 20 Elemental Composition of
Penicillin G[a]

Element	Potassium salt	Free acid
C	51.6	57.6
H	4.6	5.2
K	10.5	—
N	7.5	8.4
O	17.2	19.2
S	8.6	9.6
MW	372.47	333.38

[a] Swartz (1979).

Table 21 Raw Material Prices, Composition, *etc.*

Item	US cost[a] ($ kg^{-1})	% Nitrogen	% Sulfur
Carbon sources			
glucose (70% w/v)	0.154		
glucose from cellulose (Purdue) (60% w/v)	0.057		
enzose (72% w/v)	0.099		
sucrose (92% w/v)	0.44		
beet molasses (53% w/v sucrose)	0.081		
starch (pearl, 90% w/v)	0.165		
lactose (fermentation, 95% w/v)	0.44		
soybean oil (refined)	0.606		
prime burning lard oil	0.584		
ethanol (95%)	0.366		
Nitrogen sources			
soybean flour	0.265	8.32	—
cottonseed flour (Pharmamedia) (62% protein)	0.43	9.92	—
corn-steep liquor (24% protein)	0.11	3.84	—
$(NH_4)_2SO_4$ (tech.) (by product)	0.254	21	24
NH_4OH	0.187	23	—

[a] Costs are predictive of 1979 based on futures or inflated from actual costs in 3rd quarter, 1978, and are as delivered to Indianapolis, Indiana, in bulk (Swartz, 1979).

An overall purification yield of 85% is used, resulting in the unit cost above. The unit cost is quite sensitive to fermenter plant efficiency and purification yield. If both drop to 80%, the unit cost becomes $20.72 kg^{-1}. Actual unit costs probably fall within this range, the lower end representing either a very efficient producer in fermentation and particularly product recovery or one achieving unusually high recovery yields by relaxing quality (purity) standards. Producers of feed grade penicillin using simpler purification procedures and with relaxed quality requirements may well reduce unit costs further. This illustrates the importance of comprehensive purchasing specifications on purchased bulk penicillin.

Table 22 Production Costs per 1000 Gallons of Broth[a]

Item	Cost per 1000 gallons broth ($)	Cost as % of fermentation cost	Cost as % of total cost
Direct costs			
Glucose	200.06	15	12
Sodium phenylacetate	177.53	13.5	11
Corn-steep liquor	15.52		
$(NH_4)_2SO_4$	10.22		
KH_2PO_4	8.9		
Other	41.26		
Total raw materials	(454.00)	35	27
Operating labor	33.55	4	3
Direct supervision	18.33		
Maintenance	142.10		
Laboratory	1.55		
Utility			
A. Steam	21.69	2	1
B. Electric	156.16	12	9
C. Water	2.97		
D. Waste	13.44		
Supplies	11.81		
Fixed charges	351.80	27	21
Plant overhead	103.89		
	1 311.29		

[a] Corrected to 45 000 gallons harvest per 40 000 gallons initial volume (Swartz, 1979).

Table 23 Manufacturing Cost per 1000 Gallons Broth Produced[a]

		Fermentation ($)	Purification ($)	Total ($)
A.	Direct production cost	855.60	206.96	989.55
B.	Fixed charges	351.80	90.16	378.00
C.	Plant overhead	103.89	46.00	131.00
Total		1 311.29	343.12	1 654.41

[a] Adapted from study by A. G. McKee & Co. (Swartz, 1979).

Table 24 Unit Cost Calculation in $ kg^{-1} for Penicillin G-K[a]

(a) *kg in broth per 1000 gallons broth harvested*

$$\frac{29.4 \text{ g}}{1} \times \frac{205 \text{ h}}{183 \text{ h}} \times \frac{3.785 \text{ l}}{\text{gal}} \times \frac{1000 \text{ gal}}{1000 \text{ gal}} = \frac{125 \text{ kg}}{1000 \text{ gal}}$$

Potency, Cycle correction

(b) *kg bulk crystals per 1000 gallons broth harvested*

$$\frac{0.83 \text{ kg actual}}{\text{kg optimal}} \times \frac{0.85 \text{ kg bulk crystals}}{\text{kg in broth}} \times \frac{125 \text{ kg}}{1000 \text{ gal}} = \frac{87.5 \text{ kg}}{1000 \text{ gal}}$$

Fermenter plant efficiency, Purification yield

(c) *Unit cost*

$$\frac{\$1654.41 \times 1000 \text{ gal}}{1000 \text{ gal} \times 87.45 \text{ kg}} = \frac{\$18.92}{\text{kg bulk crystals}}$$

[a] Swartz (1979).

2.6.2   Alternatives—The Name of the Game

Table 25 (from Swartz, 1979) is presented here. It summarizes the expectations from several approaches. The most desirable is the genetic approach, but experience shows it to be less predictable. There is no basis on which to expect q_{pen} to increase by 18% except that it has worked before. Both the increased cell mass and the alternative raw material approaches are much more predictable. Continuous processes could be operated at large scale. The analysis suggests that little incentive exists to do so.

Table 25   Approaches to Process Improvement Summarized[a]

Approach	Unit cost ($ kg^{-1})	Savings per 10^6 kg produced ($)	Fermenter productivity turnaround (kg $(1000\,l)^{-1}\,d^{-1}$)
Fed batch	18.5	—	3.3
Genetics or nutritional control	16.1	2.46×10^6	3.9
Cell mass	17.3	1.28×10^6	3.9
Cheaper raw materials			
Ethanol		No advantage	
Enzose	17.7	0.84×10^6	—
Fermentable sugar	16.9	1.7×10^6	—
Continuous operation			
$\mu = 0.0068$ h^{-1}	32.0	No advantage	1.3
$\mu = 0.015$ h^{-1}, $q_{pen} = 5.56$	21.0	No advantage	2.1
$\mu = 0.015$ h^{-1}, $q_{pen} = 6.56$	18.0	0.40×10^6	2.5
Recycle of Killes cells ($\mu = 0.015$ h^{-1})			
$q_{pen} = 5.56$	19.3	No advantage	2.1
$q_{pen} = 6.56$	16.4	2.19×10^6	2.5

[a] 30 g l^{-1} harvest potency used in this example (Swartz, 1979).

Many factors influence the cost of penicillin production. Even the age of the facility can have a significant influence. Many large-scale facilities are 15–20 years old, and the fixed charges are greatly reduced. If the fixed charges drop by $200 (out of $378 total) due to operation in a fully depreciated facility, the unit cost becomes $1454.41/89.23 kg^{-1} = $16.30 kg^{-1}, substantially less than our standard of $18.54 kg^{-1}. Of course, maintenance would increase and labor tends to be higher in older plants which often have smaller fermenters, greater contamination and less efficient process strategies.

In Section 2.5 the economic judgment exercised by Wyeth in the construction of a new facility in the late 1970s was raised. This issue is also pertinent to the question of investment in new penicillin facilities generally. Wyeth's new facility was sold in the fourth quarter of 1983 to AB Fermenta (Sweden). Most of the process development group was discharged, and the new owners elected to abandon Wyeth's proprietary penicillin G culture and install a process for penicillin V which is based on a culture developed by Panlabs. Outsiders can only speculate on the decisions made by the management in first electing to source penicillin internally, then selecting new facility construction as the best alternative at a time when antibiotic capacity was in excess and penicillin was abundantly available.

The choice among development and facility expansion alternatives depends upon the scale of production and the importance of a given product among a group of similar products which the process improvement team might work on. Economic analysis can provide a guide to the potential of competing approaches to process improvement.

The decision to develop a cellulose to sugar conversion installation would require a substantial technical and capital expenditure in an unproven technology and would probably not be undertaken by a smaller producer. The decision to implement a major culture improvement program for penicillin would probably not be made by a large company with many newer products unless they were semisynthetic cephalosporins or penicillins made from 6-APA. In some cases this alone will determine the priorities. Usually though there are other much more subjective factors relating to the level of development effort and time required to accomplish each project and the probability of success. Attempts can be made to quantify these concepts and apply factors to the annual savings expected from alternative approaches. These are not usually of much help, but the basic economic analysis is useful in spite of its subjectivity. Maintenance of such a facility to bring it to a reasonably high standard of performance will not add back nearly $200/100 gallons.

Other major factors in determining unit cost are scale of operation and such issues as waste treatment and energy cost. A really low cost operation would utilize (a) 50 000 gallon fermenters (a trade-off between reduced labor costs and the saving of a purification facility for an unstable product), (b) land application of mycelial waste, (c) cogeneration of steam and electricity from coal and steam turbine air compressors (and perhaps agitators), (d) continuous sterilization of media and (e) extensive automation in both process control and of support functions such as media make-up. Several of these factors are of significant benefit only if the scale of operation is large. Thus they will be used only in the facility of a really large-scale producer of penicillin such as Giste Brocades or one producing a number of products in very large, multiproduct facilities such as might be feasible for Pfizer or Lilly. Large-scale producers are also those who have most to gain from really optimal processes and thereby should be most willing to invest in process improvement work.

2.7 SUMMARY

Penicillin remains a major pharmaceutical compound after more than 40 years of use, and there is no reason why this will not be so after another 40 years. It and its derivatives are effective antibacterials with few side effects.

Penicillin has been very important in modern medicine, and its wide use created an economic motive for substantial process improvement research by several competing organizations striving to remain competitive. Because so much work has been done on this process, and because it was the first fermentation process to benefit from the efforts of the discipline now called biochemical engineering, development work on penicillin formed the basis first for work on other antibiotics such as tetracycline and later for improved processes for such products as amino acids, vitamins, enzymes and more recently products of rDNA technology. An article by Perlman (1975) reviews this aspect of the significance of penicillin process development.

Referring to Figures 1, 2 and 3, and to Tables 6 and 10, provides a review of the process involving mycelial growth under stringently controlled conditions, maintenance of culture physiology to optimize product formation, separation and purification using filtration and solvent extraction followed by crystallization, drying, chemical modification if appropriate, and conversion to final dosage forms such as sterile powder for injection or tablets for oral use.

The extensive technology of penicillin production is practiced throughout the world. Large quantities, tens and hundreds of tons per annum, are bought and sold in many forms with the largest quantities being as bulk non-sterile pharmaceutical intermediates, either as penicillin or as 6-APA. For these reasons penicillin process technology is worthy of study not only by specialists in its production but also by those interested in biotechnology from a broader perspective.

2.8 ACKNOWLEDGEMENTS

Many figures and tables and other material are adapted from the work of numerous other authors whose invaluable contributions to this area of technology are gratefully recognized. Many of the figures have appeared previously in two earlier works as follows: *Secondary Products of Metabolism* (ed. A. H. Rose), *Economic Microbiology*, volume 3 (1979) and *Annual Reports on Fermentation Processes*, volume 3 (1979), both published by Academic Press. The material is included for completeness of this contribution in order to make it comprehensive, especially with respect to optimization criteria and economic issues. The original publication in these works is recognized.

2.9 REFERENCES

Abu-Shady, M. R., F. M. El-Beih and S. S. Radwan (1976). Effects of pure lipids and natural oils on the production of antibiotics by microorganisms. *Fette Seifen Anstrichm.*, **78**, 478–480.
Acevedo, F. and C. L. Cooney (1973). Penicillin amidose production by *Bacillus megaterium. Biotechnol. Bioeng.*, **15**, 493–503.
Aiba, S., A. E. Humphrey and N. F. Millis (1973). *Biochemical Engineering*, 2nd edn. Academic, New York.
Al-Haffar, S. (1970). *Ann. Physiol. Veg. Univ. Bruxelles*, **15**, 101.
Alikanian, S. I. (1962). Induced mutagenesis in the selection of microorganisms. *Adv. Appl. Microbiol.*, **4**, 1–48.
Allanson, E. E. and J. M. Edmundowicz (1983). *US Pat.* 4 354 971.
Andreyeva, L. N., V. V. Biryukov, B. Siktya, A. Prokop and M. Novik (1973). Analysis of mathematical models of the

effect of pH on fermentation processes and their use for calculating optimal fermentation conditions. *Biotechnol. Bioeng. Symp.*, **4**, 61–76.

Ball, C. (1971). Haploidization analysis in *Penicillium chrysogenum*. *J. Gen. Microbiol.*, **66**, 63–69.

Ball, C. (1973a). In *Genetics of Industrial Microorganisms*, ed. Z. Vanek, Z. Hostalek and J. Cudlin, vol. 2, pp. 227–238. Elsevier, Amsterdam.

Ball, C. (1973b). The genetics of *Penicillium chrysogenum*. *Prog. Ind. Microbiol.*, **12**, 47–72.

Ball, C. (1978). Genetics in the development of the penicillin process. In *Antibiotics and Other Secondary Metabolites*, ed. R. Huter, pp. 165–175. Academic, New York.

Ball, C. (1983). The genetics of beta-lactam producing fungi. In *Antibiotics Containing the Beta Lactam Structure*, ed. A. L. Demain and N. A. Solomon, pp. 147–162. Springer Verlag, New York.

Ball, C. and M. P. MaGonagle (1978). Development and evaluation of a potency index screen for detecting mutants of *P. chrysogenum* having increased penicillin yield. *J. Appl. Bacteriol.*, **45**, 67–74.

Bartholomew, W. H., B. T. Sheehan, P. Shu and R. W. Squires (1977). Potential effects of the energy shortage on the fermentation industries. *Abstracts of the 174th National Meeting of the American Chemical Society, Chicago*, Abstr. 37. ACS Division of Microbial and Biochemical Technology, Washington, DC.

Behrens, O. K. (1949). Biosynthesis of penicillins. In *The Chemistry of Penicillins*, ed. H. T. Clarke, J. R. Johnson and R. Robinson, pp. 657–679. Princeton University Press, NJ.

Behrens, O. K., J. Corse, D. E. Huff, R. G. Jones, Q. F. Soper and C. W. Whitehead (1948a). Biosynthesis of penicillins. III. Preparation and evaluation of precursors for new penicillins. *J. Biol. Chem.*, **175**, 771–792.

Behrens, O. K., J. Corse, J. P. Edwards, L. Garrison, R. G. Jones, Q. F. Soper, F. R. Van Abeele and C. W. Whitehead (1948b). Biosynthesis of penicillins. IV. New crystalline biosynthetic penicillins. *J. Biol. Chem.*, **175**, 793–809.

Berkovitch, I. (1976). Painless computer control of penicillin plant. *Process Eng. (London)*, Oct., 95.

Brandle, E. and F. Knauseder (1975). *Ger. Pat.* 2 503 584.

Brodersen, R. (1946). Three antibacterial components in commercial penicillin. *Acta Pharmacol. Toxicol.*, **2**, 1–8.

Bungay, H. R. (1963). Economic definition of continuous fermentation goals. *Biotechnol. Bioeng.*, **5**, 1–7.

Bylinsky, G. (1974). *Fortune*, **2**, 93.

Calam, C. T., L. B. Daglish and E. P. McCann (1976). In *Proceedings of the Second International Symposium on the Genetics of Industrial Microorganisms*, ed. K. D. Macdonald, pp. 273–287. Academic, New York.

Cape, R. E. (1976). Microbial genetics and the antibiotics industry. *Abstracts of the Annual Meeting of the American Chemical Society, New York*, Abstr. 3. ACS Division of Microbial Chemistry and Technology, Washington, DC.

Carrington, T. R. (1971). The development of commercial processes for the production of 6-aminopenicillanic acid (6-APA). *Proc. R. Soc. London, Ser. B*, **179**, 321–333.

Chauvette, R. R., H. B. Hayes, G. L. Huff and P. A. Pennington (1972). Preparation of 7-aminocephalosporanic acid and 6-aminopenicillanic acid. *J. Antibiot.*, **25**, 248–250.

Chibata, J., T. Tosa and T. Sato (1974). *Ger. Pat.* 2 414 128.

Cole, D. S., G. Holt and K. D. Macdonald (1976). Relationship of the genetic determination of impaired penicillin production in naturally occurring strains to that in induced mutants of *Aspergillus nidulans*. *J. Gen. Microbiol.*, **96**, 423–426.

Constantinides, A. and V. R. Rai (1974). Application of the continuous minimum principle to fermentation processes. *Biotechnol. Bioeng. Symp.*, **4**, 663–680.

Constantinides, A., J. L. Spencer and E. L. Gaden, Jr. (1970a). Optimization of batch fermentation processes. I. Development of mathematical models for batch penicillin fermentations. *Biotechnol. Bioeng.*, **12**, 803–830.

Constantinides, A., J. L. Spencer and E. L. Gaden, Jr. (1970b). Optimization of batch fermentation processes. II. Optimum temperature profiles for batch penicillin fermentations. *Biotechnol. Bioeng.*, **12**, 1081–1098.

Cooney, C. L. and F. Acevedo (1977). Theoretical conversion yields for penicillin synthesis. *Biotechnol. Bioeng.*, **19**, 1449–1462.

Cooney, C. L. and R. I. Mateles (1971). In *Recent Advances in Microbiology*, ed. A. Perez-Hiravete and D. Palaez, pp. 441–449.

Delin, P. S., B. A. Ekström, B. O. H. Sjöberg, K. H. Thelin and L. S. Narthorst-Westfeld (1972a). *Ger. Pat.* 1 966 427.

Delin, P. S., B. A. Ekström, B. O. H. Sjöberg, K. H. Thelin and L. S. Narthorst-Westfeld (1972b). *Ger. Pat.* 1 988 428.

Demain, A. L. (1967). Biosynthesis of penicillins and cephalosporins. In *Biosynthesis of Antibiotics*, ed. J. F. Snell, pp. 29–94, Academic, New York.

Demain, A. L. (1968). Regulatory mechanisms and the industrial production of microbial metabolites, bacteria, fungi, enzymes, antibiotics, vitamins, steroids. *Lloydia*, **31**, 395–418.

Demain, A. L. (1974). Biochemistry of penicillin and cephalosporin fermentation. *Lloydia*, **37**, 147–167.

Ditchburn, P., B. Giddings and K. D. Macdonald (1974). Rapid screening for the isolation of mutants of *Aspergillus nidulans* with increased penicillin yields. *J. Appl. Bacteriol.*, **37**, 515–523.

Ditchburn, P., G. Holt and K. D. Macdonald (1976). In *Proceedings of The Second International Symposium on the Genetics of Industrial Microorganisms*, ed. K. D. Macdonald, pp. 213–227. Academic, New York.

Dogariun, M., G. Alupei, E. Grumeza, M. Verdes, T. Cobzariu, G. Friedman and Ionescu (1973). *Ger. Pat.* 2 205 735.

Elander, R. P. (1967). *Abh. Dtsch. Akad. Wiss. Berlin*, **2**, 403.

Elander, R. P., M. A. Espenshade, S. G. Pathak and C. H. Pan (1973). The use of parasexual genetics in an industrial-strain improvement program with *Penicillium chrysogenum*. In *Genetics of Industrial Microorganisms*, ed. Z. Vanek, Hostalek and J. Cudlin, vol. 2, pp. 239–253. Elsevier, Amsterdam.

El-Saied, H. M., M. K. El-Marsafy and M. M. Darwish (1976). Chemically defined medium for penicillin production. *Staerke*, **28**, 282–284.

El-Saied, H. M., S. B. El-Din and M. A. Akher (1977). Replacement of corn steep liquor in penicillin fermentation. *Process Biochem.*, **12**, Oct., 31–32.

Fleming, A. (1929). On the antibacterial action of cultures of a *Penicillium*, with special reference to their use in the isolation of *B. influenza*. *Br. J. Exp. Pathol.*, **10**, 226–236.

Foster, J. W., H. B. Woodruff, D. Perlman, I. E. McDaniel, B. I. Wilker and D. Hendlin (1946). Microbiological aspects of penicillin. IX. Cottonseed meal as a substitute for corn steep liquor in penicillin production. *J. Bacteriol.*, **51**, 695–698.

Giona, A. R., L. Marrelli, L. Toro and R. DeSantis (1976a). The influence of oxygen concentration and of specific rate of growth on the kinetics of penicillin production. *Biotechnol. Bioeng.*, **18**, 493–512.

Giona, A. R., L. Marrelli, L. Toro and R. DeSantis (1976b). Kinetic analysis of penicillin production by semicontinuous fermenters. *Biotechnol. Bioeng.*, **18**, 473–492.

Gordee, E. Z. and L. E. Day (1972). Effect of exogenous penicillin on penicillin biosynthesis. *Antimicrob. Agents Chemother.*, **1**, 315–322.

Hamsher, J. J. (1975). *US Pat.* 3 905 868.

Hersbach, G. J. M., C. P. Van Der Beek and P. W. M. Van Dijck (1984). The penicillins: properties, biosynthesis and fermentation. In *Biotechnology of Industrial Antibiotics*, ed. E. J. Vandamme, pp. 45–139.

Hockenhull, D. J. D. and R. M. MacKenzie (1968). Preset nutrient feeds for penicillin fermentation on defined media. *Chem. Ind. (London)*, 607–610.

Hosler, P. and M. J. Johnson (1953). Penicillin from chemically defined media. *Ind. Eng. Chem.*, **45**, 871–874.

Huber, F. M., R. R. Chauvette and B. G. Jackson (1972). Preparative methods for 7-aminocephalosporanic acid and 6-aminopenicillanic acid. In *Cephalosporin and Penicillins: Chemistry and Biology*, ed. E. H. Flynn, pp. 27–73. Academic, New York.

Jarvis, B. G. and M. J. Johnson (1947). The role of the constituents of synthetic media for penicillin production. *J. Am. Chem. Soc.*, **69**, 3010–3017.

Kaiser, G. V. and S. Kukolja (1972). Modifications of the β-lactam system. In *Cephalosporins and Penicillins: Chemistry and Biology*, ed. E. H. Flynn, pp. 74–133. Academic, New York.

Kleiner, G. and S. V. Lopatnev (1972). *Prikl. Biokhim. Mikrobiol.*, **8**, 559.

Kleiner, G. and S. V. Lopatnev (1974). *Latv. PSR Zinat. Akad. Vestis*, **7**, 101.

Kleiner, G., S. V. Lopatnev, B. Sikyta, A. Prokop and M. Novak (1973). Effect of aeration and agitation on production of penicillin acylase in submerged culture of *Escherichia coli*. *Biotechnol. Bioeng. Symp.: Adv. Microb. Eng.*, *Part I*, *Symp. 4*, pp. 241–243. Wiley, New York.

Kondo, E., T. Mitsugi and R. Muneyuki (1975). *Ger. Pat.* 2 441 637.

Kutzbach, K. (1973a). *Ger. Pat.* 2 151 236.

Kutzbach, K. (1973b). *Ger. Pat.* 2 217 745.

Kutzbach, K. (1973c). *Neth. Pat.* 7 213 899.

Lein, J. (1983). Strain development with non-recombinant techniques. *ASM News*, **49**, 576–579.

Lein, J. (1985). The Panlabs penicillin strain development program. In *Excess Metabolites*. Benjamin-Cummings, Menlo Park, CA.

Lemke, P. A. and D. R. Brannon (1972). Microbial synthesis of cephalosporins and penicillin compounds. In *Cephalosporins and Penicillins: Chemistry and Biology*, ed. E. H. Flynn, pp. 370–437. Academic, New York.

Lengyel, Z. L. and L. Nyiri (1966). The inhibitory effect of CO_2 on the penicillin biosynthesis. In *Antibiotics—Advances in Research, Production and Clinical Use*, ed. M. Harold and Z. Gabriel, pp. 733–735. Butterworths, London.

Lovens, Kemeske and Fabrek (1949). *Danish Pat.* 294.

Lur'e, L. M. and M. M. Levitov (1967). Biosynthesis of penicillin and pH values. *Antibiotiki*, **12**, 395–400.

Lur'e, L. M., T. P. Verkhovtseva, A. I. Orlova and M. M. Levitov (1976). *Pharm. Chem. J. (Engl. Transl.)*, **10**, 218.

MacDonald, K. D. (1983). Fungal genetics and antibiotic production. In *Biochemistry and Genetics of Commercially Important Antibiotics*, ed. L. C. Vining, pp. 25–47. Addison-Wesley.

Macdonald, K. D. and P. Ditchburn (1973). The genetics of penicillin production in *Aspergillus nidulans*. *Heredity*, **31**, 131–132.

Macdonald, K. D. and G. Holt (1976). Genetics of biosynthesis and overproduction of penicillin. *Sci. Prog. (Oxford)*, **63**, 547–573.

Macdonald, K. D., G. Holt and P. Ditchburn (1972). The genetics of penicillin production. In *Fermentation Technology Today: Proceedings of the Fourth International Fermentation Symposium*, ed. G. Terui, pp. 251–257. Society of Fermentation Technology, Tokyo.

Malek, I. (1975). In *Continuous Culture 6, Applications and New Fields*, ed. A. C. R. Dean, D. C. Ellwood, C. G. T. Evans and J. Melling, p. 31. Ellis Horwood, Chichester.

Mason, H. R. S. and R. C. Righelato (1976). Energetics of fungal growth: the effect of growth-limiting substrate on respiration of *Penicillium chrysogenum*. *J. Appl. Chem. Biotechnol.*, **26**, 145–152.

Matelova, V. (1976). Utilization of carbon sources during penicillin biosynthesis. *Folia Microbiol.*, **21**, 208–209.

McCann, E. P. and C. T. Calam (1972). The metabolism of *Penicillium chrysogenum* and the product of penicillin using a high yielding strain at different temperatures. *J. Appl. Chem. Biotechnol.*, **22**, 1201–1208.

McDaniel, L. E. and E. G. Bailey (1969). Effect of shaking speed and type of closure on shake flask cultures. *Appl. Microbiol.*, **17**, 286–290.

Mead, T. H. and M. V. Stack (1948). Penicillin precursors in corn-steep liquor. *Biochem. J.*, **47**, 18.

Merrick, M. J. (1975). The inheritance of penicillin titre in crosses between lines of *Aspergillus nidulans* selected for increased productivity. *J. Gen. Microbiol.*, **91**, 287–294.

Merrick, M. J. and C. E. Catean (1975). The design of fermentation and biological assay procedures for assessment of penicillin production in populations of *Aspergillus nidulans*. *J. Appl. Bacteriol.*, **38**, 121–131.

Moll, F. and P. Kastenmeier (1971). *Pharm. Ztg.*, **116**, 1345.

Moss, M. O. (1977). *Top. Enzyme Ferment. Biotechnol.*, **1**, 111.

Moyer, A. J. (1948). *US Pat.* 2 442 141.

Nara, T. and M. Misawa (1974). *Jpn. Pat.* 46 080.

Nara, T., M. Misawa and M. Okaji (1975). *Jpn. Pat.* 37 755.

Nelligan, I. and C. T. Calam (1983). Optimal control of penicillin production using a mini-computer. *Biotechnol. Lett.*, **5**, 561–566.

Okachi, R., I. Kawamoto, M. Yamamoto, S. Takasawa and T. Nara (1973). Isolation of penicillinase-deficient mutants of *Kluyvera citrophila* KY 3641. *Agric. Biol. Chem.*, **37**, 335–339.

O'Sullivan, J. and C. Ball (1983). Beta-lactams. In *Biochemistry and Genetics of Commercially Important Antibiotics*, ed. L. C. Vining, pp. 25–47. Addison-Wesley.

Pan, S. C., S. Bonanno and G. H. Wagman (1959). Efficient utilization of fatty oils as energy sources in penicillin fermentation. *Appl. Microbiol.*, **7**, 176–180.

Pan, C. H., L. Hepler and R. P. Elander (1972). Control of pH and carbohydrate addition in the penicillin fermentation. *Dev. Ind. Microbiol.*, **13**, 103–112.

Pan, C. H., L. Hepler and R. P. Elander (1975). The effect of iron on a high yielding industrial strain of *Penicillium chrysogenum* and production levels of penicillin G. *J. Ferment. Technol.*, **53**, 854–861.

Peberdy, J. F. (1985). The biology of penicillium. In *Biology of Industrial Microorganisms*, ed. A. L. Demain and N. A. Solomon, pp. 407–431. Benjamin-Cummings, Menlo Park, CA.

Pereira, G. and A. Luz (1975). 6-Aminopenicillanic acid derivatives. *Austrian Pat.* 322 101.

Perlman, D. (1970). The evolution of penicillin manufacturing processes. In *The History of Penicillin Production*, ed. A. L. Elder, *Chem. Eng. Prog. Symp. Ser. 100*, vol. 66, pp. 24–30. American Institute of Chemical Engineers, New York.

Perlman, D. (1975). The influence of the development of penicillin fermentation on the technology used for other antibiotics. *Process Biochem.*, **10**, 23–32.

Pirt, S. J. (1971). The diffusion capsule, a novel device for the addition of a solute at a constant rate to a liquid medium: its application to metabolic regulation. *Biochem. J.*, **121**, 293–297.

Pirt, S. J. (1975). Batch cultures with substrate feeds. In *Principles of Microbe and Cell Cultivation*, pp. 211–222. Wiley, New York.

Pirt, S. J. and B. Mancini (1975). Inhibition of penicillin production by carbon dioxide. *J. Appl. Chem. Biotechnol.*, **25**, 781–783.

Pirt, S. J. and R. C. Righelato (1967). Effect of growth rate on the synthesis of penicillin by *Penicillium chrysogenum* in batch and chemostat cultures. *Appl. Microbiol.*, **15**, 1284–1290.

Podbielniak, W. J., H. R. Kaiser and G. J. Ziegenhorn (1970). In *The History of Penicillin Production*, ed. A. L. Elder, *Chem. Eng. Prog. Symp. Ser. 100*, vol. 66, pp. 44–50. American Institute of Chemical Engineers, New York.

Pontecorvo, G. P. A. and J. A. Roper (1958). *US Pat.* 2 820 742.

Queener, S. W. and R. W. Swartz (1979). Penicillins: biosynthetic and semisynthetic. In *Economic Microbiology*, ed. A. H. Rose, vol. 3. pp. 35–123. Academic, New York.

Queener, S. W., O. K. Sebek and G. Vezina (1978). Mutants blocked in antibiotic synthesis. *Annu. Rev. Microbiol.*, **32**, 593–636.

Ramachandran, S., P. S. Borkar, S. B. Tradani and V. K. Singh (1970). *S. Afr. Pat.* 69 08 045.

Righelato, R. C. (1976). Selections of strains of *Penicillium chrysogenum* with reduced penicillin yields in continuous cultures. *J. Appl. Chem. Biotechnol.*, **26**, 153–159.

Righelato, R. C., A. P. J. Trinci, S. J. Pirt and A. Peat (1968). The influence of maintenance energy and growth rate on the metabolic activity, morphology and conidiation of *Penicillium chrysogenum*. *J. Gen. Microbiol.*, **50**, 399–412.

Romanova, N. B., E. I. Golub and A. M. Belkind (1976). *USSR Pat.* 511 344.

Rowley, D., H. Steiner and E. Zimkin (1946). *J. Soc. Chem. Ind.*, **65**, 237.

Ryu, D. Y. and J. Hospodka (1977). Quantitative physiology in penicillin fermentation as a directed fermentation. *Abstr. 174th Nat. Meet. Am. Chem. Soc.*, *Chicago*, Abstr. 8. ACS Division of Microbial and Biochemical Technology, Washington, DC.

Sato, T., T. Tosa and I. Chibata (1976). Continuous production of 6-aminopenicillanic acid from penicillin by immobilized microbial cells. *Eur. J. Appl. Microbiol.*, **2**, 153–160.

Savidge, T. A. and M. Cole (1976). β-Lactam antibiotics/penicillin acylase (bacterial). *Methods Enzymol.*, **43**, 705.

Schneider, W. J. and M. Roehr (1976). Purification and properties of penicillin acylase of *Bovista plumbea*. *Biochim. Biophys. Acta*, **452**, 177–185.

Sellstedt, J. H. (1975). *US Pat.* 3 896 110.

Shimizu, M., R. Okachi, K. Kimura and T. Nara (1975). Purification and properties of penicillin acylase from *Khuyvera citrophila*. *Agric. Biol. Chem.*, **39**, 1655–1661.

Shultz, J. S. and P. I. Gerhardt (1969). Dialysis culture of microorganisms: design, theory and results. *Bacteriol. Rev.*, **33**, 1–47.

Smith, E. L. and A. E. Bide (1948). Biosynthesis of the penicillins. *Biochem. J.*, **42**, 17–18.

Souders, M., G. J. Pierotti and C. L. Dunn (1970). The recovery of penicillin by extraction with a pH gradient. In *The History of Penicillin Production*, ed. A. L. Elder, *Chem. Eng. Prog. Symp. Ser. 100*, vol. 66, pp. 37–42. American Institute of Chemical Engineers, New York.

Stojanovic, O. and R. Jovanovic (1971). *Ger. Pat.* 2 126 037.

Stouthamer, A. H. (1977). *Natuurkunde*, **86**, 134.

Swartz, R. W. (1979). The use of economic analysis of penicillin by manufacturing costs in establishing priorities for fermentation process improvement. *Annu. Rep. Ferment. Process.*, **3**, 75–110.

Sylvester, J. C. and R. D. Coghill (1954). The penicillin fermentation. In *Industrial Fermentations*, ed. L. A. Underkofler and R. J. Hickey, vol. 2, pp. 219–263. Chemical Publishing Co., New York.

Takasawa, S., R. Okachi, I. Kawamoto, M. Yamamoto and T. Nara (1972). Some problems involved in ampicillin formation by Khuyvera's penicillin acylase. *Agric. Biol. Chem.*, **36**, 1701–1706.

Tempest, D. W., T. R. Hunter and J. Sykes (1965). Magnesium-limited growth of *Aerobacter aerogenes* in a chemostat. *J. Gen. Microbiol.*, **39**, 355–366.

Thadhani, S. B., P. S. Borkar and S. Ramachandran (1972). Structural requirements of inducer for the formation of penicillin acylase in *Fusarium* sp. 7.5-5. *Biochem. J.*, **128**, 49–50.

Tochenaya, N. P. and S. M. Chaikovskaya (1973). In *Enzymes in Laboratory Diagnosis*, ed. A. S. Petrova, vol. 3, pp. 144–145. Materials of the All Union Congress of Physicians—Analysts, Moscow.

Todd, D. B. (1972). *Chem. Eng.*, **79**, 153.

Todd, D. B. and G. R. Davies (1973). Centrifugal pharmaceutical extraction. *Filtr. Sep.*, **10**, 663–666.

Vanderhueghe, H. (1975). Penicillin acylase (fungal). *Methods Enzymol.*, **43**, 721–728.

Vojtisek, V. and J. Slezak (1975). Penicillin aminohydrolase in *Escherichia coli*. II. Synthesis of the enzyme, kinetics and specificity of its introduction and the effect of O_2. *Folia Microbiol. (Prague)*, **20**, 289.

Weissenburger, H. W. O. and M. G. Van der Hoeven (1970). *US Pat.* 3 499 909.

Whitmore, F. C., R. B. Wagner, C. Noll, *et al.* (1946). Processing penicillin. *Ind. Eng. Chem.*, **38**, 942–948.

Wright, D. G. and C. T. Calam (1968). Importance of the introductory phase in penicillin production using continuous flow culture. *Chem. Ind. (London)*, 1274–1275.

Young, T. B. and H. M. Koplove (1975). *US Pat.* 3 886 046.

Zhukovskaya, S. A. (1972). *Khim. Farm. Zh.*, **6**, 393.

3

Novel β-Lactam Antibiotics

B. C. BUCKLAND, D. R. OMSTEAD and V. SANTAMARINA
Merck & Co. Inc., Rahway, NJ, USA

3.1 INTRODUCTION

From 1952 to 1975, novel β-lactam antibiotics came almost exclusively from modifications to either the penicillin molecule or the cephalosporin molecule and, more recently, the cephamycin C molecule. Since that time, the situation has changed. Several important new classes of naturally occurring β-lactam antibiotics have recently been discovered which include thienamycin, clavulanic acid, norcardicin, olivanic acid and the monobactams.

A combination of reasons lies behind this sudden change in events. The nocardicins were discovered (Aoki *et al.*, 1976) only after a test strain was developed which was hypersensitive to β-lactam antibiotics. Both clavulanic acid (Brown *et al.*, 1976) and some of the olivanic acids (Butterworth *et al.*, 1978) were discovered using an enzymatic screen to detect inhibitors of β-lactamase. Thienamycin (Kahan *et al.*, 1979) was discovered using a screen for detecting inhibitors of peptidoglycan synthesis. Only two cultures are known to produce thienamycin and these occur very infrequently in the soil. Coupling its rarity with the chemical instability of thienamycin served to prolong the purification and recognition of this important new antibiotic. Modern analytical techniques for determining structures from small amounts of materials have also played a crucial role in the discovery of some of these compounds because broth titers have often been very low in the original culture.

These discoveries stand as a testimony to the remarkable versatility of microorganisms to elaborate novel types of compound. The naturally occurring bicyclic β-lactam antibiotics have only the β-lactam ring in common. To this is fused a thiazolidine ring in the penicillins, a dihydrothiazine ring in the cephalosporins and cephamycins, an oxazolidine ring in clavulanic acid and a pyrroline ring in the carbapenems. The carbapenems have neither a sulfur nor an oxygen atom in their ring structure (Figure 1).

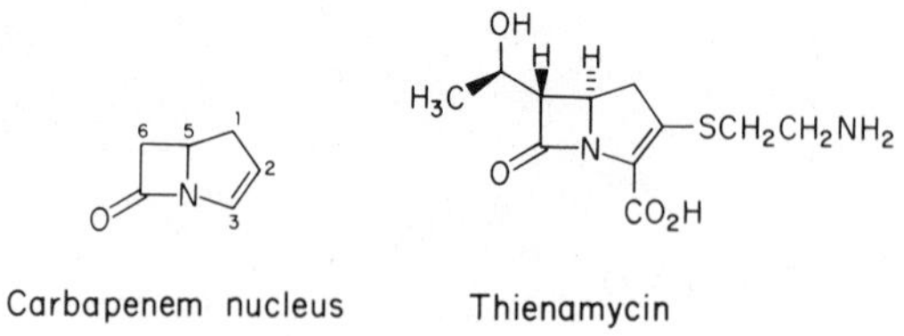

Carbapenem nucleus   Thienamycin

Figure 1   Structure of thienamycin molecule.

3.2   THIENAMYCIN

3.2.1   Introduction

3.2.1.1   *Discovery*

Thienamycin was discovered by Kahan *et al.* (1976) as a result of a screening procedure in which soil microorganisms were tested for their production of inhibitors of peptidoglycan synthesis. It is an antibiotic with an unusual and highly desirable antibiotic spectrum (Weaver *et al.*, 1979) against both Gram-positive and Gram-negative bacteria. Its activity is undiminished when tested against organisms which produce β-lactamases and are consequently resistant to many penicillins and cephalosporins.

It is produced in the culture filtrate of *Streptomyces cattleya* as a component of a complex of β-lactam antibiotics including penicillin N, cephamycin C and the *N*-acetyl derivative of thienamycin. This species is rare in Nature and has been found only once in more than 10 000 isolates from soil. Until recently, *S. cattleya* was the only organism known to produce thienamycin. A recent patent (Shionogi, 1981) claimed that *S. perenifacins* also produces this antibiotic.

3.2.1.2   *Chemistry*

Thienamycin is a zwitterionic compound with an acidic dissociation constant of *ca.* 3.1. The first derivatives were prepared when only small amounts of partially purified antibiotic were available. Using field-desorption mass spectra of the antibiotic (MH^+, 273) and high-resolution mass spectra of certain derivatives, the elemental composition was determined to be $C_{11}H_{16}N_2O_4S$ with a molecular weight of 272. The structure and absolute configuration of thienamycin has been described by Albers-Schonberg *et al.* (1978).

In contrast to well-known penam and cephem antibiotics, it contains no sulfur atom in the ring system and the two β-lactam ring protons are *trans* to one another. The carbapenem nucleus was unknown prior to the discovery of thienamycin (see Figure 1).

Thienamycin decomposes in dilute aqueous solution by apparent first-order reaction and has greatest stability at neutral pH. However, the decomposition accelerates as the antibiotic concentration is increased. This behavior is believed to be due to aminolysis of the β-lactam by the primary amine of a second thienamycin molecule. The semisynthetic derivative of *N*-formimidoylthienamycin (Imipemide) overcomes this stability problem while retaining the full biological activity of thienamycin (Smith and Schoenewaldt, 1981).

3.2.1.3 Pharmacological activity

Imipemide appears to be the most potent and broad spectrum β-lactam antibiotic reported. Minimum inhibitory concentrations (MICs) of 10 μg ml^{-1} and less were reported (Weaver *et al.*, 1979) for both Gram-positive and Gram-negative bacteria including *Pseudomonas*. It is fully effective against β-lactamase-producing strains at the same levels. Imipemide is not absorbed orally, but is effective subcutaneously at levels of 0.005–0.2 mg kg^{-1} for Gram-positive and 2–10 mg kg^{-1} for Gram-negative bacteria in mice.

3.2.1.4 Chemical synthesis

There are a number of potential chemical synthesis routes for this product. Although the molecule is small, it has not been easy to make chemically because of two main problems: the construction of the bicyclic nucleus and the introduction of substituents at the 2- and 6-positions. Despite these difficulties, remarkable progress has been made on this front (Johnston *et al.*, 1978; Reider and Grabowski, 1982) and chemical synthesis may become the preferred route for production (Pines, 1981).

3.2.2 Biosynthesis and Regulation of Thienamycin

3.2.2.1 Biosynthesis

The thienamycin molecule is illustrated in Figure 1. The pyrrolinecarboxylic acid portion was found to be derived from glutamate. Other precursors of the molecule have been determined by E. Inamine and J. Williamson (personal communication).

3.2.2.2 Regulation

In a typical fermentation, after the rate of accumulation of thienamycin has fallen to zero (Figure 2), resuspension of washed cells in buffer alone again permits antibiotic synthesis to resume. When antibiotic synthesis ceases in 36–48 h, the cells can again be made productive by resuspension in fresh buffer. This phenomenon indicates that thienamycin may regulate its own synthesis by feedback inhibition. Experiments to prove this beyond doubt have not been clear-cut. The natural tendency of the molecule to degrade creates a practical difficulty and contradictory evidence has been found in a resting cell system and in a thienamycin-synthesizing protoplast system. Inamine and co-workers at Merck have studied this phenomenon and this information will be the subject of some future publications (personal communication).

Lilley *et al.* (1981) in chemostat studies found that thienamycin production by *S. cattleya* was highly 'growth dissociated' and was only produced in detectable amounts during phosphate-limited growth. Evidence of the 'growth-dissociated' synthesis can be seen in Figure 2 in which no thienamycin was produced during the growth phase (as measured by carbon dioxide evolution). Thienamycin synthesis only starts when rapid growth ceases and when the readily assimilable carbon source (in this case glycerol) is completely utilized. There appears to be both growth-regulated synthesis and carbon catabolite repression of synthesis. The cessation of thienamycin synthesis coincides with a large increase in ammonium ion. Therefore, there may be a correlation between ammonium concentration and thienamycin regulation in the range of ammonium concentrations observed here ($>$500 mg l^{-1} as ammonium). Alternatively, this could be a result of chemical degradation or an indication of cell lysis.

Thienamycin and cephamycin C are β-lactam antibiotics produced by *S. cattleya*. Co^{2+} was found to have no effect on the growth of the culture but was found to be essential for thienamycin synthesis (Foor *et al.*, 1982) and had little effect on cephamycin synthesis. Foor *et al.* also found that $FeCl_3$ and Na_2SO_4 were essential for thienamycin synthesis.

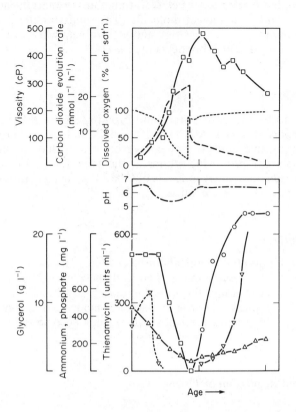

Figure 2 Thienamycin production in a 50 000 liter fermenter. Bottom: ○, thienamycin; □, glycerol; ▽, ammonium; △, phosphate; — · —, pH. Top: – – – – –, dissolved oxygen; — — — —, carbon dioxide evolution rate; □ viscosity

Foor *et al.* (1982) demonstrated, using studies on agar, that cellular differentiation and antibiotic synthesis were not associated with one another. Aerial mycelial formation, pigment formation and sporulation occurred at pH < 6.2, whereas antibiotic synthesis occurred in the pH range 6.5–7.5. In liquid fermentation studies, if the pH is allowed to rise above 7.0, then very little antibiotic is accumulated. This is probably due to very high rates of thienamycin chemical degradation which occur in an alkaline environment.

Lilley *et al.* (1981) have reported that synthesis of the antibiotic is controlled by inorganic phosphate. Other reported results are less definitive (Foor *et al.*, 1982), perhaps because the high rate of degradation of thienamycin is accelerated by inorganic phosphate or perhaps because different mutants were used for the studies. Foor *et al.* found that using phosphate at the concentration which just satisfied growth requirements (about 5–10 mM) resulted in maximum production of thienamycin. Further increases in phosphate concentration resulted in a decrease in production. This decrease, however, correlated quite well with a decrease in half-life of the antibiotic. More recent data (Williamson, personal communication) have provided a different interpretation of the role of phosphate in this fermentation and will be described in a future publication.

Foor *et al.* (1982) determined that glucose supported growth equal to that seen with glycerol while ribose, galactose, D-xylose, mannitol, gluconate and arabinase were increasingly less effective as carbon sources. No growth occurred with glucose 6-phosphate, fructose, sorbitol or the disaccharides lactose, maltose, sucrose, cellobiose or melibiose. Glutamate was fair as a carbon source while pyruvate, alanine, aspartate and ketoglutarate were poor. Acetate, ethanol, lactate, citrate, succinate, fumarate and malate failed to support growth.

3.2.3 Classical Fermentation Process

3.2.3.1 Introduction

Imipemide is not yet commercially available (1983) and so is not being manufactured on a production scale. However, information can be given for those procedures which were used to make thienamycin product for use in clinical trials and this involved running fermentations at the 50 000 liter scale.

3.2.3.2 Seed stages

These are shown in Figure 3.

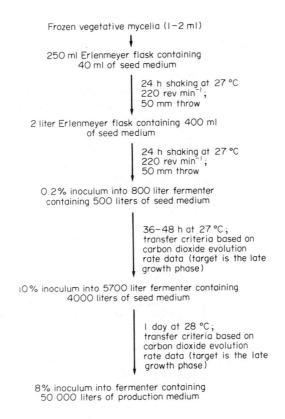

Figure 3   Seed stages.

3.2.3.3 Production stage

The thienamycin fermentation has proved to be unusually difficult to optimize for a variety of reasons. Under most conditions of growth, absolutely no product is formed, presumably as a result of carbon catabolite repression. Thienamycin synthesis also appears to be inhibited toward the latter part of the fermentation cycle by a yet to be determined mechanism. This limited the amount of carbon which could be added to the fermentation and, while feeding of glycerol had a stimulatory effect in the synthetic medium, this was never obtained in the complex medium fermentation.

The media developed to date fall into two extreme categories. The complex medium of choice is particularly variable and difficult to work with, whereas the synthetic medium, while more elegant and invaluable for biochemical regulation studies, is vastly more expensive because of a requirement for isoleucine. The complex medium was used for making product and primarily consisted of solulac, proflo, corn-steep liquor and glycerol. A comparison is made between the two media in Table 1.

A detailed analysis of a typical fermentation batch at the 50 000 liter scale is provided in Figure 2. The kinetics of the fermentation are very unusual and, under most conditions, product formation is completely non-growth associated. The batch cycle can be divided almost exactly into two parts. During the first half of the cycle, rapid cell growth occurs and the viscosity of the broth increases in parallel with increase in cell mass of *S. cattleya*. The phosphate concentration

Table 1   Comparison of Complex Medium and Synthetic Medium

	Complex medium	*Synthetic medium*
Carbon (g l^{-1})	18.3	6.6
Nitrogen (g l^{-1})	1.5	0.77
Phosphorus (g l^{-1})	0.43	0.18
Glycerol carbon (g l^{-1})	7.8	3.9
Non-glycerol carbon (g l^{-1})	10.5	3.7
NH$_4^+$ (mg l^{-1})	60.0	252.0
NH$_4^+$/total N (%)	3.0	26.0
PO$_4^{3-}$ (mg l^{-1})	500.0	525.0
PO$_4^{3-}$/total P (%)	37.0	96.0

decreases to 80 mg l^{-1} and the ammonium concentration decreases to zero. As the viscosity increases, the $K_L a$ decreases to a minimum by the time of maximum oxygen transfer requirements. At this point the $K_L a$ is 60% of that available in the beginning of the cycle (Figure 4). At the beginning of the cycle, cell growth presumably occurs using proteinaceous carbon and the culture switches to glycerol. Growth continues until the glycerol is depleted. The whole fermentation radically changes at this point. The carbon dioxide evolution rate immediately drops by 16 mmol l^{-1} h^{-1}, the pH starts to increase and thienamycin synthesis begins. Synthesis occurs for a relatively short time and then abruptly stops. This coincides with a rapid increase in ammonium concentration. The phosphate level is above 100 mg l^{-1} during the period of rapid synthesis.

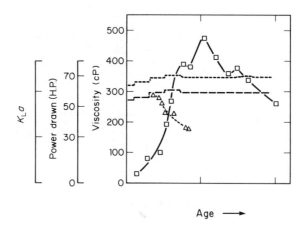

Figure 4   Effect of viscosity on $K_L a$ and power inputs in a 50 000 liter fermentation. □, viscosity; △, $K_L a$; — — —, shaft power input; ------, power consumed

Two curves for power consumption are shown during the cycle (Figure 4). The lower level corresponds to the shaft horsepower and is calculated based on a motor efficiency of 90% and a power factor of 89%.

Data at the 500 l scale indicated that the culture was relatively shear sensitive. Once the period of peak oxygen demand was over, it was found that a significant increase in titer could be obtained by decreasing the agitator speed and airflow rate. The difference in dissolved oxygen concentration was not that great between the two conditions and the effect can best be explained by a decrease in culture shear damage when the r.p.m. is reduced. Further demonstration of this phenomenon was provided by the growth of *S. cattleya* in the presence of celite (see Section 3.2.5.2). This improvement could not be tested at the 50 000 l scale because the large fermenters do not have variable speed drive.

3.2.4   Fermentation Process Development

3.2.4.1   *Strain improvement*

For a process such as this one, which is at an early stage of development, the most fruitful avenue to pursue is generally that of culture improvement in parallel with media development.

Classical techniques were used for culture mutation followed by testing of individual colonies in shaker flask fermentations. Promising mutants were then evaluated at the 500 l scale in the pilot plant. A particularly successful case study of a new culture introduction is described below.

A promising mutant (M1) was confirmed in shaker flasks and was immediately evaluated in 500 l tanks using a scaled-down version of the 50 000 l fermentation. From vent gas analysis using a Perkin-Elmer mass spectrometer (Buckland and Fastert, 1982) and from dissolved oxygen data, it became immediately apparent that the M1 mutant grew more rapidly than the parent culture. In fact, under the conditions used, the dissolved oxygen dropped to 0% for 20 h and a titer no higher than that obtained with the parent culture was achieved. Based on historical experience with this fermentation, it was clear that the period of oxygen limitation was excessive. However, the oxygen transfer limitations of the existing 50 000 l fermenter provided a boundary within which it was necessary to operate. Further experimentation at the pilot plant scale using lower concentrations of the same medium constitutents and consideration of the oxygen uptake rate data led to the first 50 000 l trial with this culture to be run using a 0.9× medium formulation. A further complication resulted from a change in the quality of the corn-steep liquor being used. Analysis of the new lot had revealed that the percentage of solids in the liquor was considerably lower than previous lots, and so an adjustment was made to the amount added to the fermenter based on adding the same amount of solids as previously used. The new culture performed in a radically different way to the parent. Initial growth was much more rapid and the broth became considerably more viscous than previously observed (Figure 5). The titer in the broth almost doubled and the cycle time was reduced by 25%. After further rebalancing of the medium (the corn-steep liquor concentration was increased while keeping the other ingredients at the 0.9× level), the peak volumetric synthesis rate was at least three-fold higher than for the parent culture (Figure 5).

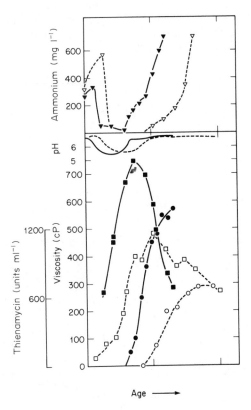

Figure 5 Difference in performance between superior mutant culture and its parent in a classical fermentation. Dashed lines represent data from parent culture and continuous lines represent data from the derived mutant culture. ∇ and ▼, ammonium; □ and ■, viscosity; ○ and ●, thienamycin; ——— and — — —, pH

The purity of thienamycin in the broth was considerably higher with this new mutant and this provided important processing benefits. The overall percentage recovery from thienamycin in the broth to purified derivatized thienamycin was increased (unpublished data by K. Shultis). The yield obtained per fermenter batch was therefore tripled when the new culture was introduced.

3.2.4.2 Fed-batch techniques

Preliminary trials with fed-batch techniques, while successful with synthetic media (see later), did not provide an immediate titer increase with the complex medium. Attempts to extend the productive cycle length in most media by slow feeding of lactic acid or glycerol had no effect on the fermentation. This may be a result of the rapid increase in ammonium concentration which is possibly inhibitory to biosynthesis of thienamycin. This putative effect of ammonium inhibition can possibly be reversed using an ammonium scavenger.

3.2.4.3 Synthetic media

A titer was achieved in the synthetic medium at the 500 l scale similar to that obtained in the complex medium with the earlier cultures. Although of great interest, this had little practical value because of the high cost of the isoleucine used in the synthetic medium. In fact, for many mutants, the specific rate of thienamycin synthesis was considerably higher for cells grown on the synthetic medium than for cells grown in the complex medium. The more recent cultures have performed much better in the complex medium than in the synthetic medium, which is not surprising considering that selection for better mutants at the shaker flask scale was made using the complex medium. This illustrates the point that conditions used today in a mutant screening program determine the fermentation process of the future.

The profile from a typical synthetic medium based fermentation at the 500 l scale is described in Figure 6. Cell growth is moderate, reaching a maximum value of 6 g l^{-1}. The maximum carbon dioxide evolution rate is also low at 16 mmol l^{-1} h^{-1}. Synthesis of antibiotic was increased by use of a slow glycerol feed during the latter part of the cycle (Figure 6). The glycerol feed was started as soon as the carbon dioxide evolution rate decreased.

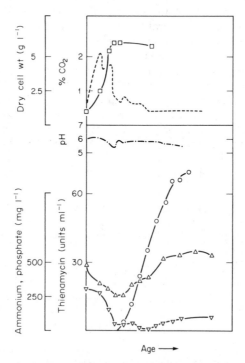

Figure 6 Thienamycin production in a 500 liter fermenter using a synthetic medium. ○, thienamycin; ▽, ammonium; △, phosphate; —— · ——, pH; — — —, carbon dioxide evolution rate; □, dry cell weight

Fermentations were run using different levels of inorganic phosphate and this appeared to have only a minor effect on the fermentation. The best results were obtained with a starting concentration in the range of 500–600 mg l^{-1} as phosphate. Under these conditions, phosphate is in excess throughout the fermentation. This result is in contrast to results obtained by Lilley *et al.* (1981), who found that under their conditions and with a different culture, phosphate limitation was necessary for thienamycin synthesis. Results obtained by Foor *et al.* (1982), although not in liquid culture, also indicated that inorganic phosphate does not play a major regulatory role in this fermentation.

3.2.5   Novel Fermentation Processes

3.2.5.1   *Ultrafiltration coupled fermenter*

Because of the unusual growth-dissociated kinetics of this fermentation (Figure 6), the suspicion of feedback inhibition effects, and high rates of thienamycin degradation (20% d^{-1}), ultrafiltration coupled fermentation was investigated. Broth was continuously recycled through an ultrafiltration membrane and a clear liquid stream containing product continuously withdrawn. If medium conditions could be correctly adjusted then, in theory, product could be made at relatively high rates for extended periods. Preliminary results were disappointing with regard to product formation but this may be a result of the extreme sensitivity of *S. cattleya* to small changes in environmental conditions. The fermenter was operated aseptically in this manner for over a week and good flux rates of filtrate removal were obtained.

3.2.5.2   *Immobilized cells*

The immobilized cell approach potentially offers a number of advantages to this fermentation which would result in reduced capital, raw material, energy, waste treatment and product recovery costs. (1) Long-term continuous or semicontinuous operation with high volumetric productivities. (2) Reduced end-product and substrate inhibition. (3) Decrease in medium viscosity, thus improving bulk mixing as well as oxygen transfer coefficients. (4) Separate optimization of the growth and production phases to a degree not possible in a classical fermentation. (5) A decrease in the loss of chemically unstable thienamycin (20% d^{-1} in the classical fermentation) when the process mode is changed from batch to continuous or semicontinuous. This loss is further decreased the more the continuous operation tends toward plug flow. (6) Purer broth which improves the efficiency of the downstream processing.

Initial trials with this approach have been encouraging and production of thienamycin has been maintained for extended periods using either polyacrylamide-entrapped cells or celite-attached cells. For celite, a simple *in situ* procedure for cell attachment was used in both shaker flasks and in a bubble column reactor (Baker *et al.*, 1983; Arcuri *et al.*, 1983). The attached cells have then been maintained in a productive state for extended periods of time (Buckland *et al.*, 1982; Baker *et al.*, 1983a)—up to 6 months in the shaker flask (see Figure 7). Moreover, it was shown by Baker *et al.* (1983b) that the volumetric synthesis rate was proportional to immobilized cell concentration over a wide range (Figure 8).

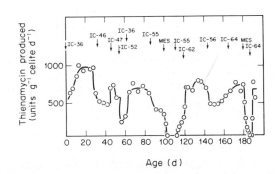

Figure 7   Thienamycin production rate by *Streptomyces cattleya* immobilized on to celite particles. IC-XX refer to different media used (Baker *et al.*, 1983a). MES refers to the use of MES buffer without nutrients

Attachment of the cells *per se* does not increase productive stability. Productivities of free cells were strikingly similar to the immobilized cells over a 30 day period in shaker flasks (Figure 9)

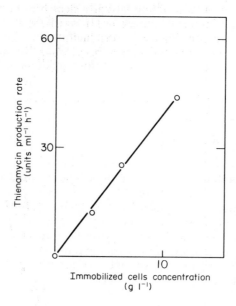

Figure 8 Relationship between thienamycin production rate and concentration of immobilized cells

using a daily regime of decanting spent broth (containing thienamycin) and replacing with fresh nutrients (Baker *et al.*, 1983). Attachment to celite, however, does greatly facilitate solid–liquid separation in a reactor system and also significantly reduces the viscosity which in turn improves oxygen transfer in the reactor (see Figure 2).

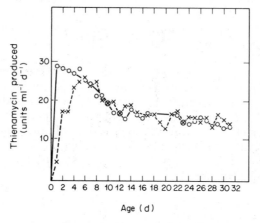

Figure 9 Comparison of thienamycin production rate by resting cells and immobilized cells of *Streptomyces cattleya*: ×, immobilized; ○, non-immobilized

The importance of viscosity effects was vividly illustrated during initial studies using celite-attached cells in a 10 l bubble column reactor (Arcuri *et al.*, 1983). A major objective from the beginning in this program was to attach the cells in the celite *in situ*. This eliminates costly separate immobilized steps, as well as greatly increasing the probability of avoiding microbial contamination. During the initial trials, only partial attachment of the *S. cattleya* was achieved and the resulting viscous broth made it impossible to obtain rapid settling of the celite-attached cells. This problem was overcome by making use of the washout phenomenon observed in classical chemostat operation. As soon as the bubble column reactor was inoculated, the nutrient feed was increased to a dilution rate which was greater than the growth rate of the *S. cattleya*. In this way, free cells were washed out and only attached cells remained (Figure 10). Most interestingly, the attachment of the *S. cattleya* to the celite was enhanced by the formation of fibrils by the culture (Figure 11). After 72 h of operation, very few unattached cells remained, the viscosity of the broth was low, and rapid settling of the celite-attached cells occurred. This allowed relatively high oxygen transfer to occur, as well as enabling virtually clear broth containing thienamycin of a relatively high purity to be continuously pumped out of a vigorously aerated broth using a modified bubble column reactor shown in Figure 12. Production of thienamycin over extended time periods has been possible using this reactor (data to be published by Arcuri and co-workers).

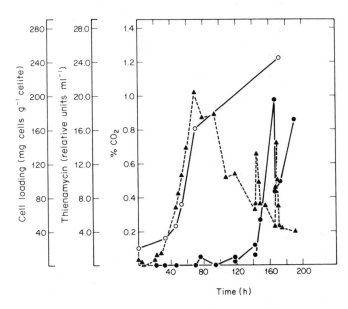

Figure 10 CO$_2$, thienamycin and cell loading *versus* time. ▲, CO$_2$; ●, thienamycin; ○, cell loading

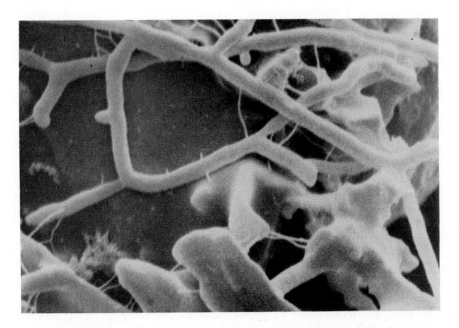

Figure 11 SEM of celite with attached *S. cattleya*; magnification × 10 000

An attempt was made to attach *S. cattleya* to celite in a 500 l stirred tank fermenter using high aeration with either a turbine impeller at low speed or a propeller at low speed (Arcuri *et al.*, 1983). All attempts with the turbine resulted in no growth of the culture, indicating a high degree of shear sensitivity. In the case of the propeller, good growth was obtained but there was a very low degree of attachment of cells to celite.

Potentially the immobilized cell approach is very attractive and this potential can be illustrated

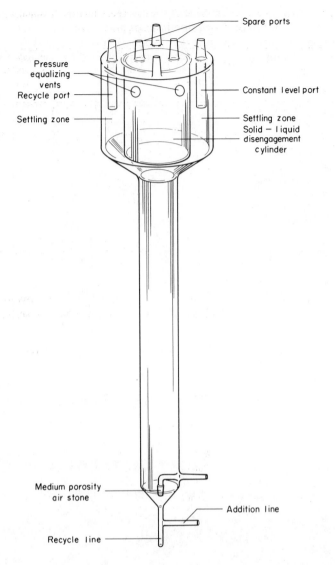

Spare ports

Pressure
equalizing
vents
Recycle port

Constant level port

Settling zone

Settling zone
Solid — liquid
disengagement
cylinder

Medium porosity
air stone

Addition line

Recycle line

Figure 12 Experimental apparatus employed for fermentations with immobilized cells of *S. cattleya*

in Table 2. The key component in capital cost for this example is volumetric productivity and potentially the introduction of immobilized cell technology for thienamycin production could increase overall volumetric productivity by a factor of ten (Table 2). Also an immobilized cell reactor would cost about one third less to build than a classical fermenter. Therefore, the cost of constructing a plant could be very much less if immobilized cell technology were used. In addition, operating costs would be less for the immobilized cell reactor because the percentage conversion of substrate to product is approximately three times higher than that in the classical fermentation, thus reducing raw material costs as well as downstream processing costs.

Balanced against these potential advantages remains the fact that immobilized cell technology has yet to be applied for the large-scale biosynthesis of secondary metabolites and so the scale-up risks associated with this technology are greater than for the classical fermentation process or chemical synthesis process. Certain problems, such as build-up of cell film thickness, would have to be overcome.

3.2.6 Thienamycin Purification

The purification scheme was developed sufficiently so that product could be made for clinical trials. A number of modifications would be made to this process before it was implemented at a

Table 2 Comparison between Use of Classical Fermentation and Immobilized Cell Technology for Thienamycin Production

	Classical fermentation (Actual)	Shaker flask immobilized cell (Actual)	Bubble column immobilized cell reactor	
			Actual	Potential
Cell concentration (g dcw l^{-1})	12	11.5	25	30
Maximum volumetric productivity (arbitrary units ml^{-1} h^{-1})	100	63	35	249[a]
Period for maintaining peak rate (h)	10	96	48	1000
Overall volumetric productivity (arbitrary units ml^{-1} h^{-1}) (include downtime)	24	52	24	240
Maximum specific productivity (arbitrary units mg^{-1} h^{-1})	8.3	5.5	1.4	8.3
Conversion of substrate to product (arbitrary units)	0.144	0.42	0.22	0.42

[a] Calculated from maximum specific productivity observed in the classical fermentation.

true production scale. This interim process for a classical fermentation, developed by Molnar and Davidson, based on pilot-plant scale studies by Treiber *et al.* (1981), is described in Figure 13.

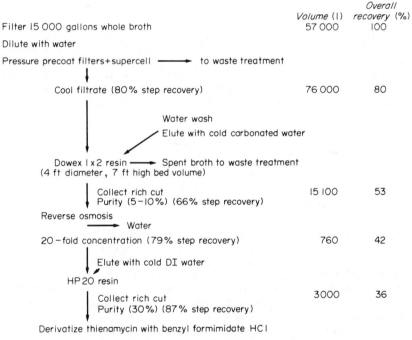

Figure 13 Interim process scheme for thienamycin production

The fermentation broth is chilled to 2–5 °C and the mycelia are removed by pressure rotary filters. Thienamycin is then adsorbed from the filtrate on to Dowex 1 × 2 (HCO$_3^-$) resin. The Dowex 1 × 2 resin has a low degree of cross-linking (2%) which results in practical operational difficulties because it is soft and easily deforms. It also changes size in response to osmotic changes and tends to form clumps with resultant channelling. The resin is regenerated in an upflow mode after every batch fermentation. The thienamycin could be unloaded from the column using carbonated water. This provides an elegant solution to the problem of loading additional metal ions on to the resin. This step provided a five-fold increase in concentration as well as great purification. Four columns were used in a sequential mode of loading and unloading. Typically, the resin could be used for 50 cycles. The Dowex 1 × 2 rich cuts are then further concentrated by reverse osmosis. It is essential to keep the broth cold during the whole process because of the inherent chemical instability of the thienamycin molecule.

Introduction of new cultures had a dramatic impact on yields. For example, when the culture described in Figure 5 was introduced, the overall yield to derivatized thienamycin-formamidine (Imipemide) jumped from 7.8% to 10.5%. Presumably, this is because of a combination of both higher concentration and better purity of thienamycin in broth with the new culture.

Purification of the product-containing stream from an immobilized reactor system would be rather different to this. Presumably, conditions in the reactor could be adjusted so that even if the thienamycin would be somewhat lower in concentration than from a classical fermentation process, it would be much higher in purity. The first step in a proposed process would involve a polish filter to remove minor cell debris. The broth would then be collected in a surge tank and fed continuously to a reverse osmosis system which would increase concentration of the thienamycin without affecting the purity. Further purification and concentration could be done using the Dowex 1×2 resin and the HP20 resin as in the classical fermentation process.

3.2.7 Future Prospects

Chemical stability problems have been overcome by chemically modifying the parent molecule to form *N*-formimidoylthienamycin (Imipemide), and the market prospects are excellent. It is planned for introduction in 1984 and will, at that time, be the most broad spectrum antibiotic available. Use of an enzyme inhibitor has made the antibiotic more effective *in vivo* by preventing degradation of the antibiotic.

3.2.8 Market Projections

Imipemide is the most broad spectrum antibiotic yet discovered and has few side effects. Although production costs will inevitably remain high in the foreseeable future, it has a high probability of becoming a very successful pharmaceutical agent. Most importantly, it will undoubtedly save many lives which would otherwise have been lost as a result of serious infections which are not treatable in a timely fashion using existing antibiotics.

3.3 CLAVULANIC ACID

3.3.1 Introduction

Clavulanic acid is a novel major β-lactam antibiotic which was initially discovered as a lactamase inhibiting agent. This compound was discovered independently by Brown *et al.* (1976) and Napier *et al.* (1981). Like thienamycin, it was found to inhibit the activity of β-lactamase enzymes which are produced by many penicillin- and cephalosporin-resistant clinical pathogens. However, the apparent mechanism by which clavulanic acid acts to inhibit β-lactamases is unusual. Apparently, the antibiotic irreversibly binds to and inhibits lactamase enzymes. Most lactamase-resistant β-lactams are merely competitive inhibitors.

Clavulanic acid appears to be active against a wide spectrum of Gram-positive and Gram-negative bacteria; however, the level of observed activity is quite low, relative to other broad spectrum antibiotics such as thienamycin. As a result, it has not been possible to use clavulanic acid as a solely administered antibacterial product. Rather, clavulanic acid has been coformulated with broad spectrum antibiotics which are susceptible to lactamase. Coadministration of clavulanic acid with other antibiotics has dramatically enhanced their antibacterial activity (Brown, 1981). For example, *in vitro* studies indicate that the minimum inhibitory concentration (MIC) of ampicillin alone exceeds 500 μg ml^{-1} when measured against a lactamase-producing strain of *S. aureus* (Brown *et al.*, 1981). Upon supplementation with 5 μg ml^{-1} of clavulanic acid, the MIC for combined addition is reduced to less than 0.1 μg ml^{-1}. Although the combined effects of clavulanic acid and other drugs have not been evaluated, extensive clinical studies have been performed with clavulanic acid and amoxicillin.

Following its discovery, the chemical structure of clavulanic acid was identified by Howarth *et al.* (1976). This compound was found to be an analog of the basic penicillin structure (Figure 14). In the clavulanic acid molecule, an oxygen atom is substituted for the sulfur, characteristic of penicillin. As such, the clavulanic acid derivatives all contain an oxazolidine ring structure.

Synthesis of clavulanic acid has been observed in several species of *Streptomyces*. While the

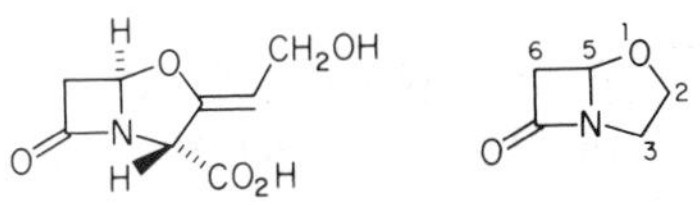

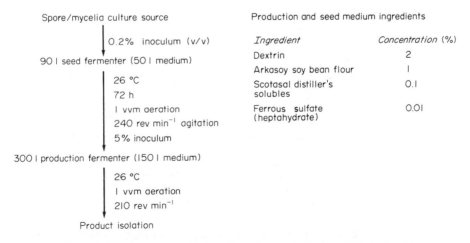

Figure 14 Structure of clavulanic acid molecule

originally isolated producing culture is *S. clavuligerus*, *S. jumonjinensis* (Box, 1978) and *S. kat-surahamanus* (Kitano *et al.*, 1978) are also known to produce clavulanic acid. Little has been reported regarding the biosynthetic pathways which are utilized by these organisms to produce clavulanic acid. However, initial studies (Elson *et al.*, 1982; Elson and Oliver, 1978) suggest that synthesis of this compound by *S. clavuligeras* does not follow the classical tripeptide theory. The absence of a 6-amino group or an α-aminoadipoyl side chain in clavulanic acid, coupled with glutamic acid incorporation into the oxazolidine carbon backbone, indicates that the traditional Arstein tripeptide is not formed. Apparently, another assembly mechanism is in place. This is surprising as *S. clavuligerus* also produces penicillin N, cephamycin C and other β-lactam compounds (Elson and Oliver, 1978).

3.3.2 Production

Reported processes for the preparation of clavulanic acid indicate that it is produced in conventional fermentation processes, which employ a series of stirred, aerated fermenter vessels. A pilot plant scale version of a clavulanic acid fermentation process has been reported by Reading and Cole (1977) and is illustrated in Figure 15. While commercial production of clavulanic acid undoubtedly occurs in much larger vessels, utilizing additional seed development stages, Figure 15 depicts a conventional batch fermentation process. The *S. clavuligerus* culture source is a suspension of mycelia and spores. This material is aseptically inoculated into a 90 l seed development fermenter containing 50 l of the culture medium detailed in Figure 15. The culture is then incubated under the indicated conditions for 72 h, whereupon a fraction of the fermenter contents is aseptically transferred into the production fermenter. This vessel is then operated at the conditions illustrated. Presumably, incubation of the culture in this production stage continues until product titer is maximized. Upon attainment of the proper clavulanic acid potency, the production fermenter contents would be harvested and subjected to subsequent product isolation techniques.

Spore/mycelia culture source

| 0.2% inoculum (v/v)

90 l seed fermenter (50 l medium)

| 26 °C
| 72 h
| 1 vvm aeration
| 240 rev min^{-1} agitation
| 5% inoculum

300 l production fermenter (150 l medium)

| 26 °C
| 1 vvm aeration
| 210 rev min^{-1}

Product isolation

Production and seed medium ingredients

Ingredient	Concentration (%)
Dextrin	2
Arkasoy soy bean flour	1
Scotasal distiller's solubles	0.1
Ferrous sulfate (heptahydrate)	0.01

Figure 15 Downstream processing scheme for clavulanic acid (pilot plant scale)

This production process is not meant to delineate specifically the exact means by which clavulanic acid is currently commercially produced. Rather, it is included principally to demonstrate the general approach to antibiotic production. As mentioned, a commercial process would no doubt occur in larger vessels. As such, several flask and fermenter inoculum development stages would be required.

The yield of product obtained in the illustrated clavulanic acid production process has not been reported. However, product titers as high as 1100 μg ml^{-1} clavulanic acid have been reported (Kitano *et al.*, 1978).

3.3.3 Market

Clavulanic acid in combination with amoxycillin or ampicillin has pronounced synergistic activity against a wide variety of β-lactamase-producing strains. It is being manufactured by Beecham Pharmaceuticals under the trade name 'Augmentin' and has every prospect of becoming an important member of the arsenal of antibiotics used to fight infectious diseases.

3.4 OLIVANIC ACIDS AND EPITHIENAMYCINS

Certain strains of *Streptomyces olivaceus* were discovered to be producers of a complex mixture of antibiotic β-lactamase inhibitors (Brown *et al.*, 1977). Three compounds were isolated, MM4550, MM13902 and MM17880, and determined to be closely related members of a new family of fused β-lactams based on the structure of olivanic acid (Figure 16).

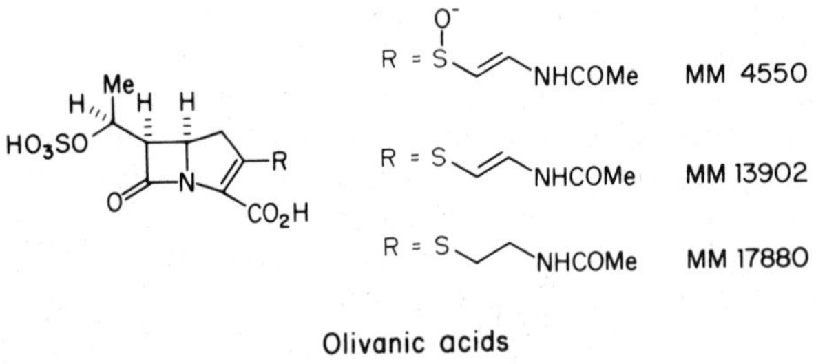

Olivanic acids

Figure 16 Structure of olivanic acids

Four further olivanic acid derivatives were discovered as products made by different strains of *S. olivaceus*. These were discovered at Beecham Pharmaceuticals (Brown *et al.*, 1979). The same compounds were discovered independently at Merck (Stapley *et al.*, 1977) as products of *S. flavogriseus* and were called the epithienamycins (see Figure 17).

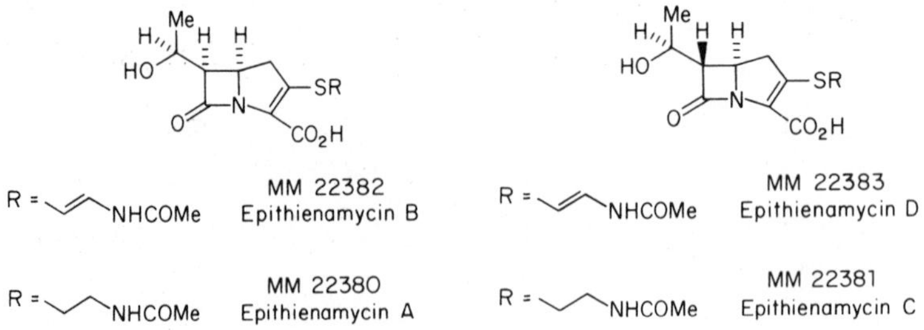

Figure 17 Structure of epithienamycins

A number of the olivanic acids are potent antibiotics. For example, MM13902 and MM17880 have activity similar to thienamycin against most Gram-negative organisms although are less active against *Ps. aeruginosa* and *Straph. aureus*.

MM4550, MM13902 and MM17880 are potent inhibitors of a wide range of β-lactamase types and are more effective than clavulanic acid. Synergism with penicillins and cephalosporins is difficult to evaluate because of the intrinsic activity of the olivanic acid as an antibiotic.

3.5 NOCARDICINS

3.5.1 Introduction

The nocardicins were discovered from a strain of *Nocardia* (Aoki *et al.*, 1976) by using a strain of *E. coli* as a detector which was supersensitive to penicillin G. Only one strain of Actinomycetes has been found to produce nocardicin A among tens of thousands that have been tested. Antibiotics detected in this screen were also examined for their sensitivity to β-lactamase and their effect on an *in vitro* peptidoglycan system.

Seven closely related compounds, nocardicins A to G, are monocyclic β-lactams with features resembling the fused β-lactams (Hashimoto *et al.*, 1976). Nocardicin A (Figure 18) is the most active antibiotic in this group. It has weak to moderate activity against Gram-negative organisms, low activity against Gram-positive bacteria, and is ineffective against fungi. It appears to be non-toxic (LD_{50} is 2 g kg^{-1}), is stable in solution, is not broken down by β-lactamases, and is inactive orally (Aoki *et al.*, 1976).

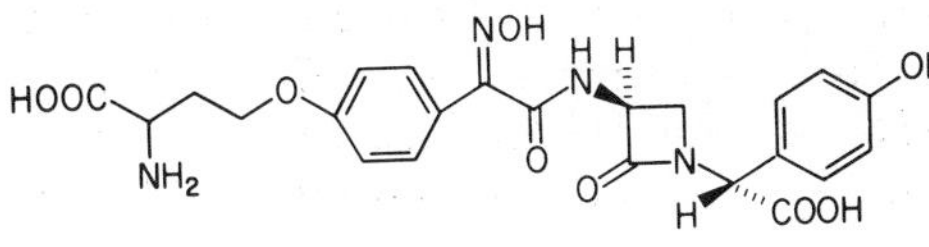

$C_{23}H_{24}N_4O_9$

Figure 18 Structure of nocardicin A

3.5.2 Production of Nocardicin A

The producing organism (*Nocardia uniformis* subsp. *rsuyamanensis*) was found to be sensitive to β-lactam antibiotics including nocardicin A itself. Addition of 2 mg l^{-1} of nocardicin at various times during the fermentation completely inhibited growth of the culture and the cells became fragmented and lysed. A strain selection procedure was developed based on the ability of mutated cultures to form colonies on agar plates containing varying concentrations of nocardicin A. As soon as ability to make β-lactamase was mutated out of the culture, this was proved to be a practical technique for the selection of superior mutants (Elander and Aoki, 1982).

Aoki *et al.* (1976) have described a pilot plant scale procedure for production of the antibiotic using 20 l of medium in a 30 l fermenter. Seed flasks (500 ml) containing 100 ml of the seed medium (sucrose, 20 g l^{-1}; cotton seed meal, 20 g l^{-1}; dried yeast 10 g l^{-1}) were inoculated with spores from the slant culture and incubated at 30 °C on a shaker with a 3 inch throw at 180 rev min^{-1} for 2–3 days to obtain good growth. Fermentations were carried out at 30 °C for 4 days under aeration of 20 l min^{-1} and agitation of 300 rev min^{-1} using a medium which has subsequently been greatly modified (Table 3).

An isolation procedure for nocordicin A is described in Figure 19 (Kamiya *et al.*, 1982).

3.5.3 Market Projections

Nocardicin A appears to be a very safe antibiotic but has relatively low activity as an antimicrobial agent. It is difficult to estimate at this time whether or not it will become an important therapeutic agent in view of the large number of interesting new β-lactams which will soon be available.

3.6 MONOBACTAMS

The monobactams represent a new class of monocyclic, bacterially produced, β-lactam antibiotics which were recently discovered at the Squibb Institute for Medical Research in Princeton, New Jersey, USA. The natural monobactams only have moderate activity. However, a synthetic analog called SQ26776 or azthreonam (Figure 20) has excellent activity against Gram-negative

Table 3 Nocardicin A Production

	Medium I	*Medium II*	*Medium III*	*Supplemented medium III*[a]
Titer ($\mu g\ ml^{-1}$)	110	400	500	1000
Soluble starch ($g\ l^{-1}$)	20	10	10	10
Yeast extract ($g\ l^{-1}$)	4	2	—	—
K_2HPO_4 ($g\ l^{-1}$)	3.5	—	—	—
$Na_2HPO_4 \cdot 12H_2O$ ($g\ l^{-1}$)	1.5	12	14.3	14.3
$MgSO_4 \cdot 7H_2O$ ($g\ l^{-1}$)	1.0	5	5	5
Glucose ($g\ l^{-1}$)	—	5	—	—
Peptone ($g\ l^{-1}$)	—	10	—	—
Ca partothenate ($g\ l^{-1}$)	—	0.2	—	—
Cotton seed meal ($g\ l^{-1}$)	—	—	20	20
Dried yeast ($g\ l^{-1}$)	—	—	20	20
KH_2PO_4 ($g\ l^{-1}$)	—	18	21.8	21.8

[a] Addition of 1 $g\ l^{-1}$ L-tyrosine and 1 $g\ l^{-1}$ of glycine increased the nocardicin A titer to 1000 $\mu g\ ml^{-1}$.

Fermentation broth
Adjust pH to
4.0 with IN HCl
Filter
Discard mycelia
Diaion HP-20 adsorption
Elute with
30% methanol
Vacuum concentration
pH to 2.5 with
IN HCl
Crystallization
Recrystallization from ethanol-water
to form nocardicin A crystals (95%)

Figure 19 Isolation procedure for nocordicin A

bacteria and appears to have few side effects. It can be chemically synthesized relatively easily from the amino acid threonine.

S
O
H
N
NH3
N
CH3
N
O
O
N
SO3
CH3
CO2H
CH3

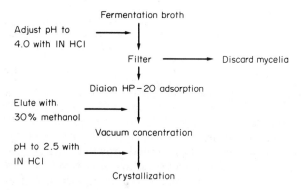

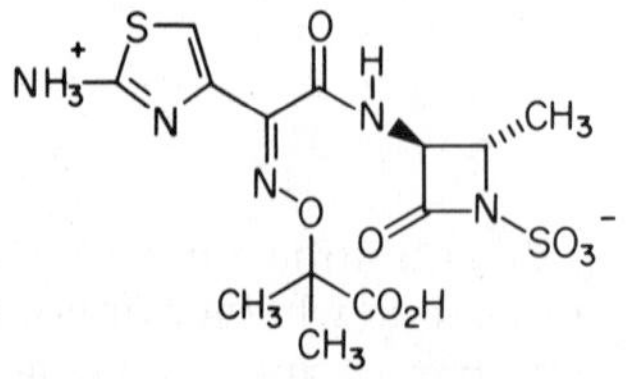

Figure 20 Structure of a monobactam (SQ26776)

ACKNOWLEDGEMENTS

The following people employed at Merck & Co., USA, have made important contributions to this manuscript, either through their comments or through their experimental contributions: E. Arcuri, E. Baker, S. Builder, S. Drew, O. Davidson, M. Milder, E. Inamine, M. Nallin,

T. Molnar, K. Shultis, L. Treiber and J. Williamson. Our special thanks go to Ms. S. Sodomora for typing the manuscript.

3.7 REFERENCES

Albers-Schonberg, G., B. H. Arison, O. D. Hensens, J. Hirshfield, K. Hoogsteen, E. A. Kaczka, R. E. Rhodes, J. S. Kahan, F. M. Kahan, R. W. Ratcliffe, E. Walton, L. J. Ruswinkle, R. B. Morin and B. G. Christensen (1978). Structure and absolute configuration of thienamycin. *J. Am. Chem. Soc.*, **100**, 6491–6499.

Aoki, H., H. Sakai, M. Kohsaka, T. Konomi, J. Hosoda, Y. Kubochi, E. Iguchi and H. Imanaka (1976). Nocardicin A, a new monocyclic β-lactam antibiotic. *J. Antibiot.*, **29**, 492–500.

Arcuri, E. J., J. R. Nichols, T. S. Brix, V. G. Santamarina, B. C. Buckland and S. W. Drew (1983). Thienamycin production by immobilized cells of *Streptomyces cattleya* in a bubble column. *Biotechnol. Bioeng.*, **25**, 2399–2411.

Baker, E. E., R. J. Prevoznak, S. W. Drew and B. C. Buckland (1983). Thienamycin production by *Streptomyces cattleya* cells immobilized in celite beads. *Dev. Ind. Microbiol.*, **24**, 467–474.

Box, S. J. (1978). Preparation of clavulanic acid using *Streptomyces jumonjinesis*. *US Pat.* 4 072 569.

Brown, A. G., D. Butterworth, M. Cole, G. Hanscomb, J. D. Hood, C. Reading and G. N. Rolinson (1976). Naturally occurring β-lactamase inhibitors with antibacterial activity. *J. Antibiot.*, **29**, 668–669.

Brown, A. G., D. F. Corbett, A. J. Eglington and T. T. Howarth (1977). Structures of olivanic acid derivatives MM4550 and MM13902, two new fused β-lactams isolated from *Streptomyces olivaceus*. *J. Chem. Soc., Chem. Commun.*, 523–525.

Brown, A. G., D. F. Corbett, A. J. Eglington and T. T. Howarth (1979). Structures of olivanic acid derivatives MM22380, MM22381, MM22382 and MM22383: four new antibiotics isolated from *Streptomyces olivaceus*. *J. Antibiot.*, **32**, 961–963.

Brown, A. G. (1981). New naturally occurring β-lactam antibiotics and related compounds. *J. Antimicrob. Chemother.*, **7**, 15–48.

Buckland, B. C. and H. Fastert (1982). Analysis of fermentation exhaust gas using a mass spectrometer. In *Computer Applications in Fermentation Technology*, pp. 119–126. Society of Chemical Industry, London.

Buckland, B. C., E. E. Baker and R. J. Prevoznak (1982). Process for producing thienamycin employing *Streptomyces cattleya* cells immobilized to celite beads. *US Pat. Appl.*

Butterworth, D., M. Cole, G. Hanscomb and G. N. Rolinson (1978). Olivanic acids, a family of β-lactam antibiotics with β-lactamase inhibitory properties produced by *Streptomyces* species. *J. Antibiot.*, **32**, 287–294.

Elander, R. P. and H. Aoki (1982). β-Lactam-producing microorganisms: their biology and fermentation behavior. In *The Chemistry and Biology of β-Lactam Antibiotics*, vol. 3, pp. 88–153. Academic, New York.

Elson, S. W. and R. S. Oliver (1978). Studies on biosynthesis of clavulanic acid. I. Incorporation of ^{13}C-labelled precursors. *J. Antibiot.*, **31**, 586–592.

Elson, S. W., R. S. Oliver, B. W. Bycroft and E. A. Faruk (1982). Studies on the biosynthesis of clavulanic acid. III. Incorporation of DL-[3,4-^{13}C$_2$] glutamic acid. *J. Antibiot.*, **35**, 81–86.

Foor, F., B. Tyler and N. Morin (1982). Effect of glycerol, phosphate and pH on cellular differentiation and production of the beta-lactam antibiotic thienamycin by *Streptomyces cattleya*. *Dev. Ind. Microbiol.*, **23**, 305.

Hashinoto, M., T. Konori and T. Kaniya (1976). Nocardicin A, a new monocyclic β-lactam antibiotic. II. Structure determination of nocardicin A and B. *J. Antibiot.*, **29**, 890–891.

Hosoda, J., T. Konomi, N. Tani, H. Aoki and H. Iminaka (1977). Isolation of new nocardicins from *Nocardia uniforms* subsp. *tsuyamanensis*. *Agric. Biol. Chem.*, **41**, 2013–2020.

Howarth, T. T., A. G. Brown and T. J. King (1976). Clavulanic acid, a novel β-lactam isolated from *Streptomyces clavuligerus*. *J. Chem. Soc., Chem. Commun.*, 266–267.

Johnston, D. B. R., S. M. Schmitt, F. A. Bouffard and B. G. Christensen (1978). Total synthesis of thienamycin. *J. Am. Chem. Soc.*, **100**, 313–315.

Kahan, J. S., F. M. Kahan, E. O. Stapley, R. T. Goegelman and S. Hernandez (1976). *US Pat.* 3 950 357.

Kahan, J. S., F. M. Kahan, R. Goegelman, S. A. Currie, M. Jackson, E. O. Stapley, T. W. Miller, A. K. Miller, D. Hendlin, S. Mochales, S. Hernandez, H. B. Woodruff and J. Birnbaum (1979). Thienamycin, a new β-lactam antibiotic. 1. Discovery, taxonomy, isolation, and physical properties. *J. Antibiot.*, **32**, 1–12.

Kamiya, T., H. Aoki and Y. Mine (1982). Nocardicins. In *Chemistry and Biology of β-lactam Antibiotics*, ed. R. B. Morin and M. Gorman, vol. 2, p. 5. Academic, New York.

Kitano, K., K. Kintaka and K. Katamoto (1978). Clavulanic acid production by *Streptomyces katsurahamanus*. *Jpn. Pat.* 78 104 796, *Appl.* 78 20 080.

Lilley, G., A. E. Clark, and G. C. Lawrence (1981). Control of the production of cephamycin C and thienamycin by *Streptomyces cattleya* NRRL 8057. *J. Chem. Technol. Biotechnol.*, **31**, 127–134.

Napier, E. J., J. R. Evans, D. Noble, M. E. Bushell, G. Webb and D. Brown (1981). Clavam derivatives. *Br. Pat.* 1 585 661.

Pines, S. H. (1981). A practicable synthesis of thienamycin. In *Organic Synthesis Today and Tomorrow*, ed. B. M. Trost and C. R. Hutchinson, pp. 327–334. Pergamon, Oxford.

Reading, C. and M. Cole (1977). Clavulanic acid: a beta-lactamase inhibiting beta-lactam from *Streptomyces clavuligerus*. *Antimicrob. Agents Chemother.*, **11**, 852–857.

Reider, P. J. and F. J. J. Grabowski (1982). Total synthesis of thienamycin — a new approach from aspartic acid. *Tetrahedron Lett.*, **23**, 2293–2296.

Shionogi and Co. Ltd. (1981). Process for preparing thienamycin. *Eur. Pat. Appl.* 0038–534.

Smith, G. B. and E. F. Schoenewaldt (1981). Stability of N-formimidoylthienamycin in aqueous solution. *J. Pharm. Sci.*, **70**, 272–276.

Stapley, E. O., P. Cassidy, S. A. Currie, D. Daoust, R. Goegelman, S. Hernandez, M. Jackson, M. S. Manta, A. K. Miller, R. L. Monaghan, J. B. Tunac, S. B. Zimmerman and D. Hendlin (1977). Epithienamycins, biological studies of a new family of β-lactam antibiotics. In *17th Intersci. Conf. Antimicrob. Agents Chemother.*, *New York*, Abstr. 80.

Treiber, L. R., V. P. Gullo and I. Putter (1981). Procedure for isolation of thienamycin from fermentation broths. *Biotechnol. Bioeng.*, **23**, 1255–1265.
Weaver, S., G. P. Bodey and B. M. Leblance (1979). Thienamycin: new beta-lactam antibiotic with potent broad spectrum activity. *Antimicrob. Agents Chemother.*, **15**, 518–521.

4

Aminoglycoside Antibiotics

H. KASE
Kyowa Hakko Kogyo Co. Ltd., Tokyo, Japan

4.1 INTRODUCTION

Streptomycin was the first member of an entirely new group of antibiotics called the aminoglycoside (aminocyclitol) antibiotics. The antibiotic was isolated from a species of *Streptomyces* by Waksman's group in 1944 (Schatz *et al.*, 1946). Since that time, many other aminoglycoside antibiotics have been isolated as fermentation products of soil microorganisms: neomycin, kanamycin, paromomycin, gentamicin, ribostamycin, tobramycin, spectinomycin, micronomicin (formerly sagamicin), astromicin (formerly fortimicin A), validamycin, *etc.* The number of antibiotics belonging to this class has been increased further by addition of semisynthetic derivatives which have biological activity. Many aminoglycoside antibiotics have been used successfully as therapeutic agents. Table 1 shows 13 aminoglycoside antibiotics available for therapeutic use at present. Among others, astromicin is now in clinical trials. In 1981 the estimated worldwide sales were approximately 500 million dollars. Gentamicin sales comprise about one-half of this total (Daniels, 1982). Figs. 1–6 show the chemical structures of clinically important aminoglycoside antibiotics.

Aminoglycoside antibiotics are among the most potent antibiotics known. They are broad-spectrum, active against both Gram-positive and Gram-negative bacteria as well as mycobacteria, although they have no useful activity against anaerobic bacteria or fungi. The aminoglycosides are limited to parenteral routes of administration for the treatment of systemic infection. They are not absorbed when given by mouth, but are excreted unchanged in the faeces. To varying degrees, all of the aminoglycosides have some potential for toxicity to the kidney (nephrotoxicity) and the inner ear (eighth nerve or ototoxicity).

All aminoglycoside antibiotics experience bacterial resistance, usually following several years

Table 1   Aminoglycoside Antibiotics Used in Therapy

Antibiotic	Preparation	Producer	Discovery by the group of
Streptomycin	Fermentation	*Streptomyces griseus*	S. Waksman (1944)
Dihydrostreptomycin	Fermentation	*S. fumidus*	R. Peck (1946)
Fradiomycin (neomycin)	Fermentation	*S. fradiae*	H. Umezawa (1948)
Kanamycin (kanamycin A)	Fermentation	*S. kanamyceticus*	H. Umezawa (1957)
Paromomycin	Fermentation	*S. rimosus* forma *paromomycinus*	R. Forhardt (1958)
Aminodeoxykanamycin (kanamycin B)	Fermentation	*S. kanamyceticus*	H. Umezawa (1961)
Gentamicin	Fermentation	*Micromonospora purpurea*	M. Weinstein (1963)
Tobramycin	Fermentation	*S. tenebrarius*	W. Stark (1967)
Ribostamycin	Fermentation	*S. ribosidificus*	N. Niida (1970)
Sisomicin	Fermentation	*M. inyoensis*	W. Weinstein (1970)
Dibekacin	Semisynthetic		H. Umezawa (1971)
Amikacin	Semisynthetic		H. Kawaguchi (1972)
Micronomicin (sagamicin)	Fermentation	*M. sagamiensis*	T. Nara (1975)

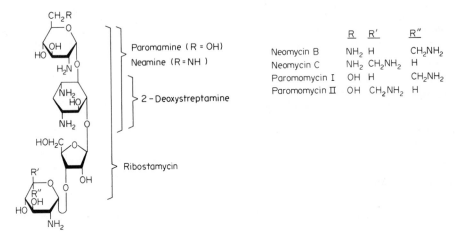

	R	R′	R″
Neomycin B	NH₂	H	CH₂NH₂
Neomycin C	NH₂	CH₂NH₂	H
Paromomycin I	OH	H	CH₂NH₂
Paromomycin II	OH	CH₂NH₂	H

Figure 1   Neomycin and paromomycin

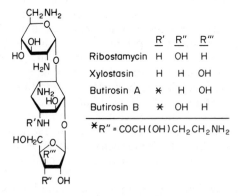

	R′	R″	R‴
Ribostamycin	H	OH	H
Xylostasin	H	H	OH
Butirosin A	*	H	OH
Butirosin B	*	OH	H

*R″ = COCH(OH)CH₂CH₂NH₂

Figure 2   Ribostamycin, xylostasin and butirosin

of extensive use. In many species of bacteria this resistance is determined by the presence of R factors which direct the synthesis of antibiotic-inactivating enzymes. These enzymes inactivate aminoglycosides by three separate mechanisms: acetylation, adenylation, and phosphorylation. An understanding of these resistant mechanisms at a molecular level has led to the preparation of some semisynthetic derivatives resistant to enzyme inactivation. This approach has yielded two

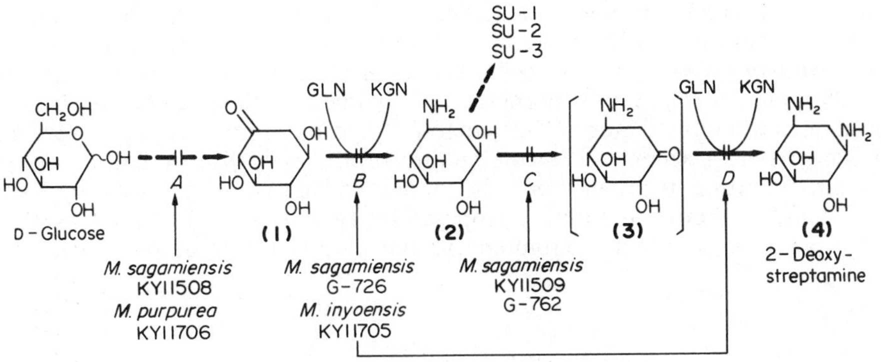

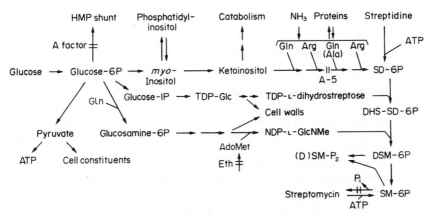

Figure 7 Proposed biosynthetic pathway for 2-deoxystreptamine and blocked steps of the D⁻ mutants of *M. sagamiensis*, *M. purpurea* and *M. inyoensis* (Kase *et al.*, 1982a)

4.2.2 Streptomycin

The enzymic sequences involved in streptomycin biosynthesis have been extensively studied by cell-free systems, and now the biosynthetic sequence of streptidine and streptose has been established in detail (Walker and Walker, 1982; Grisebach *et al.*, 1981). Though enzymes involved in the synthesis of *N*-methylglucosamine have not yet been identified, it can be estimated that about 28 enzymes take part in the conversion of glucose into streptomycin (Figure 8).

Figure 8 General scheme of streptomycin biosynthesis, showing branch points between idiabolic and trophabolic pathway and sites of experimental interventions. DSM, dihydrostreptomycin; SD, streptidine; Eth, ethionine; A-5, Demain's mutant; NDP, nucleoside-diphosphate (Walker and Walker, 1982)

4.2.3 Neomycin, Paromomycin, Ribostamycin and Butirosin

The extensive work of Rinehart and his group has shown that the individual subunits of neomycin and paromomycin are first synthesized and then subsequently joined together to form the complete molecule (Rinehart and Stroshane, 1976). The mode of assembly of the subunits is still a matter for conjecture. Pearce *et al.* (1976) have demonstrated that neamine (which consists of neosamine coupled to a deoxystreptamine unit) can be incorporated into the neomycin molecule in *S. rimosus* forma *paromomycinus*. From the postulated precursor neamine, two possible routes to neomycin are available: one by addition of a ribose unit, thus forming ribostamycin prior to conversion to neomycin, the other by addition of a molecule of neobiosamine to produce neomycin directly. Baud *et al.* (1977) have provided evidence in favor of the route *via* ribostamycin, although the ^{13}C studies by Stroshane (1976) have demonstrated the concurrent synthesis of neamine and neobiosamine during fermentation, thus indicating the viability of the alternative route. In the biosynthesis of ribostamycin and butirosin in *S. ribosidificus* (Kojima and Satoh, 1973) and *B. circulans* (Claridge *et al.*, 1974; Takeda *et al.*, 1978), ribose seems to be added last. More recently, Autisser *et al.* (1981) proposed a biosynthetic pathway of neomycin and paromomycin including new intermediates, 6'''-deamino-6'''-hydroxy derivatives of neomycin and paromomycin, which were accumulated in the culture broths of *S. fradiae* and *S. rimosus* forma *paromomycinus*. Putting together both the neomycin biosynthetic pathway postulated by Autisser *et al.* and the butirosin pathway proposed by Takeda *et al.* (1978), a biosynthetic pathway for ribostamycin, neomycin, paromomycin and butirosin may be postulated as shown in Figure 9.

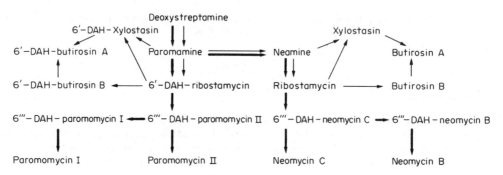

Figure 9   Postulated pathway for biosynthesis of neomycin, paromomycin, ribostamycin, butirosins and related compounds. DAH, deaminohydroxy; ➤, *Streptomyces* sp.; →, *Bacillus circulans*

4.2.4   Gentamicin, Micronomicin and Sisomicin

Biosynthetic pathways of gentamicin and sisomicin have been proposed from the biotransformation experiments using idiotrophic mutants of *M. purpurea* and *M. inyoensis*, respectively (Testa and Tilley, 1975, 1976). The results with *M. purpurea* suggest that gentamicin biosynthesis involves a branched point at gentamicin X_2 with one branch leading to gentamicin C_{1a} and C_{2b} (micronomicin) and the other leading to gentamicin C_1 and C_2. The identity of the biotransformation products was indicated by chromatographic evidence alone and should be confirmed by other techniques. The authors (Kase *et al.*, 1981a,b, 1982a) investigated the biosynthesis of micronomicin and gentamicin by using mutants blocked in various steps in the biosynthetic pathway. A variety of compounds was isolated from the culture broths or biotransformation mixtures with the mutants and identified from the spectral data. Based on the analyses of biotransformation and biochemical blocks in the mutants, a biosynthetic pathway of micronomicin and gentamicin was proposed (Figure 10). In this scheme, sisomicin group antibiotics (sisomicin, verdamicin and antibiotic G-52), which are usually not produced by gentamicin- and micronomicin-producing microorganisms, are involved in the biosynthesis of micronomicin and gentamicin. Actually, a mutant which appeared to be blocked at step L in Fig. 10 (4',5'-dehydrogenation) was isolated from *M. sagamiensis* and found to produce antibiotic G-52 and sisomicin. Additionally, a novel aminoglycoside antibiotic, 6'-*N*-methylverdamicin, was isolated from the transformation products of JI-20B or verdamicin with D⁻ mutants of *M. sagamiensis*.

4.2.5   Fortamine and Fortimicins

The biosynthetic pathway of fortimicins and fortamine, containing the 1,4-diaminocyclitol unit of the fortimicin class of compounds, has recently been elucidated by use of techniques similar to those used in the experiments on micronomicin biosynthesis. Odakura *et al.* (1983) isolated a number of mutants which were blocked in fortimicin biosynthesis from astromicin (formerly fortimicin A) producing *M. olivasterospora*. Biosynthetic intermediates produced by the mutants

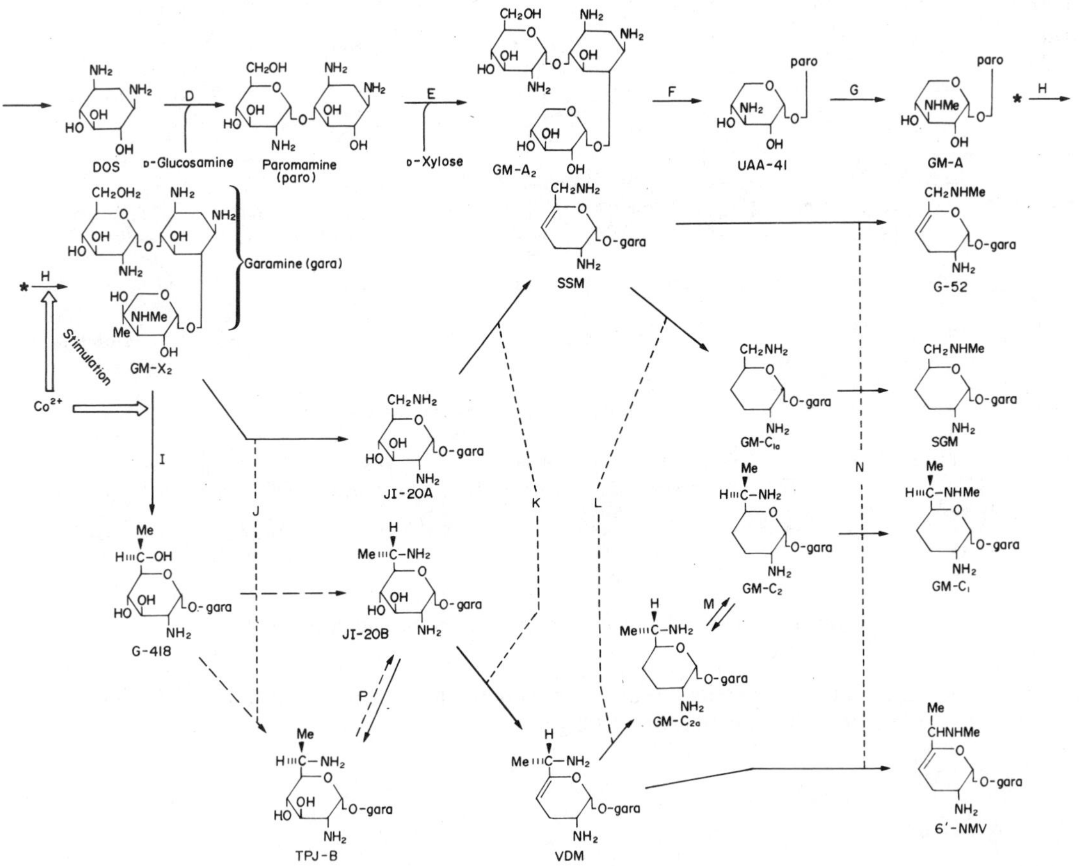

Figure 10   Proposed biosynthetic pathway for micronomicin and gentamicin

were isolated and identified: KY11554, KY11581, KY11583, KY11556, KY11560, KY11555, KY11557 and KY11582 produced fortimicin FU-10, AO, KL$_1$, KK$_1$, AP, KH, KR and B as a main product, respectively. Based on cosynthesis and biotransformation analyses by use of the mutants and the isolated compounds, the biosynthetic sequence was proposed as follows (Ito *et al.*, 1983) ; FU-10 → AO → KL$_1$ → KK$_1$ → AP → KH → KR → B - - → astromicin.

Two other mutants, KY11559 and KY11558, which appeared to be blocked in the earlier parts in the pathway, biotransformed all of the intermediates described above into astromicin. KY11559, but not KY11558, converted *scyllo*-inosose and *syllo*-inosamine to astromocin. Fortamine KH or B, the aminocyclitol unit of fortimicin KH or B, respectively, was not utilized by the mutants. These results strongly suggest that *scyllo*-inosose and *scyllo*-inosamine are intermediates of fortimicin biosynthesis and that fortamine itself is not utilized as a precursor. Fortamine is apparently formed from *scyllo*-inosamine (derived from *scyllo*-inosose) which is already attached to D-glucosamine in FU-10 and is then aminated at position 4. Consequently, a biosynthetic pathway of fortamine and fortimicins is proposed as shown in Figure 11. The fortimicin pathway may have a unique feature in that the aminocyclitol itself is not incorporated as a unit into the antibiotics. In nearly every case studied thus far the aminocyclitol is, in fact, incorporated into the antibiotics (Rinehart, 1979). Only the 1-*N*-(aminohydroxybutyryl)deoxystreptamine unit of butirosin is not the case (Fig. 9), but deoxystreptamine itself is incorporated.

4.2.6   Mutasynthesis

D$^-$ mutants were used not only to study biosynthesis but also to prepare new antibiotics by incorporation of related or analogous aminocyclitols ('mutasynthesis' or 'mutational biosynthesis': Rinehart and Stroshane, 1976; Nagaoka and Demain, 1975). Mutasynthesis has now been used to produce novel antibiotics related to nearly all the clinically useful aminoglycoside antibio-

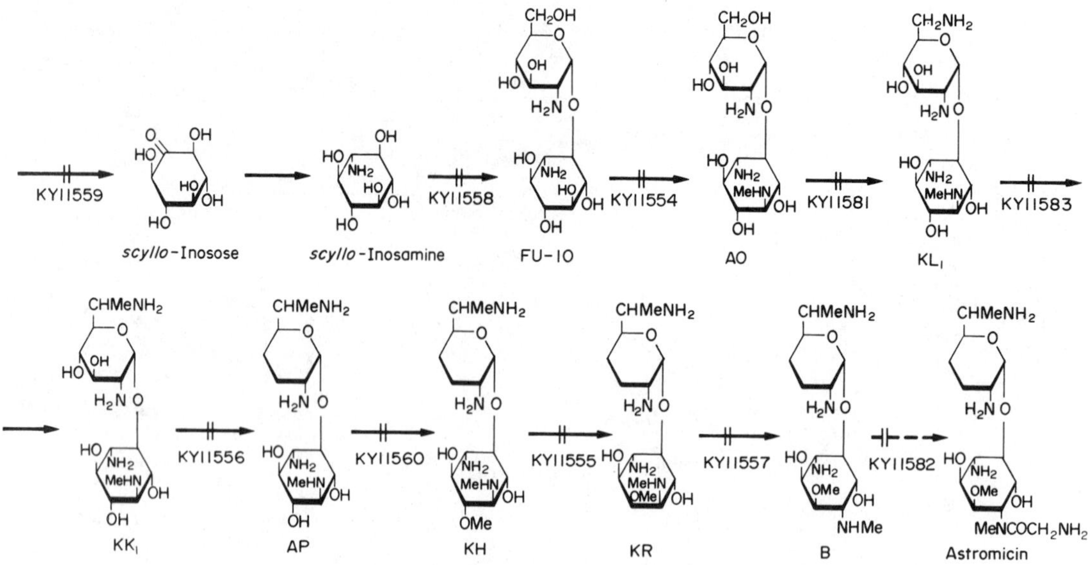

Figure 11   Proposed biosynthetic pathway for fortamine and fortimicin

tics, some of which have properties greatly superior to the parent compounds (Rinehart, 1979; Daum and Lemke, 1979; Kitamura *et al.*, 1982).

4.2.7   A-Factor

Induction, a well known regulatory mechanism in primary metabolism, appears to be involved in the control of the biosynthesis of some antibiotics. A-factor, a compound which appears to induce streptomycin production and spore-forming ability, was found by Khokhlov *et al.* (1982) and its structure determined as 3S-isocapryloyl-4S-hydroxymethyl-γ-butyrolactone. Hara and Beppu (1982) reported that A-factor not only regulated streptomycin biosynthesis but also induced resistance of the producing organisms to streptomycin. They demonstrated that A-factor-induced resistance was due to the inactivation by streptomycin-6-phosphotransferase, and that the enzyme synthesis was completely dependent on the presence of A-factor. A-factor seems to regulate streptomycin biosynthesis, not through an indirect metabolic sequence proposed by Voronina *et al.* (1978) (Fig. 8), but by directly stimulating synthesis of enzyme(s) in the biosynthetic pathway.

4.2.8   Metabolic Grid

Antibiotics are produced typically as members of a particular chemical family. For example, *M. sagamiensis* and *M. olivasterospora* were found to produce no fewer than 30 aminocyclitols (Kase *et al.*, 1982a; Nara, 1978; Shirahata *et al.*, 1980; McAlpine *et al.*, 1980). Depending on genetic and environmental factors, the proportion of each component was affected, probably because of the low specificity of the enzymes involved in the biosynthesis. Figure 12 shows the 'metabolic grid' proposed in *M. sagamiensis* (Kase *et al.*, 1982a).

4.3   MANUFACTURE

At present all of the aminoglycosides are derived from fermentation processes. Two semisynthetic aminoglycosides in Table 1, amikacin and dibekacin, are produced by a combination of fermentation and subsequent chemical manipulation of the fermentation product. Reviews covering the semisynthetic aminoglycoside work have appeared (Price *et al.*, 1974; Umezawa, S., 1975).

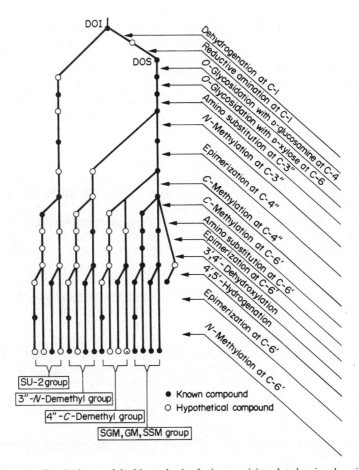

Figure 12 Total scheme of the biosynthesis of micronomicin-related aminoglycosides

4.3.1 Fermentation

The aminoglycoside antibiotics are produced by submerged fermentation under optimal physical and nutritional conditions to provide the desired product. The specific characteristics of industrial microbial strains, media and cultural conditions cannot be described in detail since these facts are considered trade secrets. Here, the outline of the aminoglycoside fermentation process is discussed, referring to recently published literature.

4.3.2 Microorganisms

The majority of aminoglycoside antibiotics are produced by actinomycetes of the *Streptomyces* genus, although the important antibiotics such as gentamicin, micronomicin and astromycin are products of *Micromonospora* species (Table 1). Additionally, a few aminoglycoside antibiotics such as butirosin have now been isolated from bacterial genera. The original strains isolated from soils usually show too low productivity of the antibiotics to apply industrially and are subjected to mutation to isolate higher-yielding strains (see Section 4.4).

4.3.3 Equipment

Aminoglycoside antibiotics are produced by large-scale submerged fermentation in tanks providing high levels of aeration and agitation. These fermenters vary in size, often in excess of 40 liters, and are designed to provide conditions that are optimal for the specific strains used.

4.3.4 Inoculum Development

For the fermentation set up in the manufacturing plant, an inoculum must be developed from the original spore or mycelial preparation of producing organism in sufficient quantity to provide a proper rate of growth and antibiotic production in the final tank. The size of production fermenters usually requires that the inoculum be developed in several stages of increasing volume until a 5–10% inoculum is reached. The early stages are usually grown in shake flasks or bottles; the later stages in small replicas of the production fermenters.

4.3.5 Media

Media for production of aminoglycoside antibiotics are individually designed for the specific requirements of the producing strains and fermentation equipment used. The development of these media has been largely empirical. It should be noted that the antibiotics are produced by a metabolic route which differs from that responsible for the growth of microorganisms. They are usually produced only at low specific growth rates of the producing cultures. This type of regulation affects a whole range of biosynthetic processes (Martin and Demain, 1980). Complex media containing natural nutrients are usually used for the production of aminoglycosides, not only because these nutrients are inexpensive but also because they are utilized slowly by the producing organisms and provide a C-N-P balance optimal for the antibiotic production. Glucose is usually an excellent carbon source for growth, but interferes with the biosynthesis of many antibiotics. During the studies on fermentation medium development, slowly utilized polysaccharides or oligosaccharides are often found to be better than glucose as carbon sources, probably because of deregulation from 'carbon catabolite regulation' by use of these carbon sources. In some cases, glucose directly represses biosynthetic enzymes. For instance, glucose represses N-acetylkanamycin amidohydrase, apparently the last enzyme in kanamycin biosynthesis (Satoh *et al.*, 1976). In some industrial processes the carbohydrate source is continuously added to the fermentation.

High concentrations of readily utilized nitrogen sources such as inorganic ammonium salts usually inhibit antibiotic production (nitrogen metabolite regulation). Screening of nitrogen sources for antibiotic production in chemically defined media has frequently resulted in the selection of slowly metabolized amino acids. A classic case is the superiority of proline in the streptomycin fermentation (Dulaney, 1948).

Although phosphate ions are essential for both growth and antibiotic production, addition of excess phosphate to the media resulted in the inhibition of antibiotic production. In streptomycin fermentation, excess inorganic phosphate not only increases sugar consumption by activating the further metabolism of phosphoglyceraldehyde, but also inhibits the formation of streptomycin from glucose 6-phosphate (Hockenhull, 1978). In addition, inorganic phosphate inhibits streptomycin phosphatase by which streptomycin 6-phosphate, a postulated terminal intermediate in streptomycin biosynthesis, is cleaved to form streptomycin (Fig. 8). Actually, the phosphorylated derivative of streptomycin accumulates in cultures of *S. griseus* growing in excess inorganic phosphate (Miller and Walker, 1970). A similar involvement of the phosphatase control has been described in the fermentation of neomycin by *S. fradiae* (Majumdar and Majumdar, 1970), lividomycin in *S. lividus* (Kimizuka *et al.*, 1977) and butirosin in *B. vitellinus* (Shirafuji *et al.*, 1982a,b).

Some metal ions (Fe^{2+}, Fe^{3+}, Mg^{2+}, Mn^{2+}, Co^{2+}, Zn^{2+}, Ni^{2+}) affect antibiotic production to a great extent (Weinberg, 1982). Here, three examples reported recently are cited. In nebramycin fermentation with *S. tenebrarius*, the addition of zinc sulfate dramatically changed the whole fermentation process; growth was accelerated, soybean oil used as the nitrogen source was metabolized more rapidly, glucose disappeared earlier, and nebramycin titer was increased (Stark *et al.*, 1976).

Cobalt or vitamin B_{12} stimulated the production of gentamicin, astromicin and seldomycin factor 5 (Shimizu *et al.*, 1978; Yamamoto *et al.* 1977; Testa and Tilly, 1979). The effect of cobalt on the production of gentamicin and micronomicin is rather complicated. A gentamicin producing strain, *M. purpurea*, fermented in the soybean–dextrin medium with added cobalt resulted in the normal production of gentamicin C complex (C_{1a}, C_2 and C_1), while in the absence of added cobalt an approximate 10-fold decrease in antibiotic titer was observed (Testa and Tilley, 1979). In a micronomicin producing strain, *M. sagamiensis* KY11510, cobalt chloride significantly reduced micronomicin production and stimulated gentamicin C_1 production (Kase *et al.*, 1982a; Odakura *et al.*, 1983). Since micronomicin was chosen as a candidate for development, attempts were made by mutation to increase the amount of micronomicin and to decrease that of gentamicin C_1 in the fermentation. Consequently, mutants blocked in 6'-C-methylation (step I in Fig. 10) were isolated to produce no gentamicin C_1 and higher levels of micronomicin than the parent. Moreover, micronomicin production was stimulated by the addition of cobalt, the maxima' yield of such a mutant KY11562 being approximately two-fold that of the parent. Cobalt was shown to stimulate two C-methylation steps (steps H and I) in the biosynthesis (Testa and Tilley, 1979; Kase *et al.*, 1982b). In the parent strain, the stimulatory effect of cobalt on gentamicin C_1 production may be due to the activation of both C-methylation steps by cobalt. In the mutants deficient in the 6'-C-methylation, cobalt may activate only 4''-C-methylation (step H) and hence has stimulated micronomicin production.

Mg^{2+} at concentrations from 5 to 20 mM stimulated significantly the production of kanamycin and neomycin in a complex medium by *S. kanamyceticus* K-2j and *S. fradiae* 117, respectively. This effect is partly ascribed to stimulation by Mg^{2+} of the release of antibiotic into the medium from bound cells (Hotta and Okami, 1976).

Addition of some amino acids often stimulates the production of aminoglycoside antibiotics. For example, Yoshikawa and Taniguchi (1976) demonstrated that alanine increased butirosin formation by *B. circulans* NRRL-B-3313. Addition of 0.5–1% of DL-, D- or L-alanine enhanced butirosin yields from 590 up to 1200–1400 μg ml^{-1}. This alanine effect was also noticed with gentamicin, micronomicin and sisomicin fermentation (Nara, 1977). The stimulatory effect of glutamine was reported in nebramycin fermentation (Stark *et al.*, 1976).

Several other additives significantly stimulate the production of aminoglycoside antibiotics. The stimulatory effect of barbital on streptomycin biosynthesis in *S. griseus* was reported by Ferguson *et al.* in 1957. Barbital retards cell lysis and may exhibit a positive and economically feasible effect on the streptomycin fermentation (Heding and Gurtu, 1977).

Screlin, a physiologically active substance stimulating the mitochondrial oxidative phosphorylation in plants, microorganisms and animals, was found to enhance the production of kanamycin, ribostamycin and streptomycin (Satoh *et al.*, 1975). For example, addition of 5 p.p.m. of sclerin initially increased the production of kanamycin from 1300 to 2000 μg ml^{-1}.

Various inhibitors of cell wall synthesis were shown to stimulate the production of some aminoglycoside antibiotics. Barabás and Szabó (1977) demonstrated that addition of 3 μg ml^{-1} of penicillin to 48 h culture broth gave a 90% increase in streptomycin production. Similar observations were reported in the production of neomycin (Barabás *et al.*, 1978), kanamycin and micronomicin (Nakayama, 1977). Aminoglycoside antibiotics appear to be shunt products diverting from a pathway leading to the cell wall. Actually, biosynthetic correlation between streptomycin and mucopeptide in utilization of D-glucosamine as a common precursor has recently been demonstrated by Nimi *et al.* (1976, 1981).

4.3.6 Procedures

In general, antibiotic fermentation can be divided into two interrelated phases: trophophase (growth phase) and idiophase (production phase). For example, a typical time course of streptomycin fermentation was described as follows: rapid growth occurs during the first phase which lasts approximately 40 h; then the antibiotic is produced at a rapid rate during the second phase which lasts 160 h. The growth rate is markedly reduced during this period (Hockenhull, 1978). Streptomycin can be produced in high yield (10 g l^{-1}) on a simple glucose–ammonium salt medium (Walker and Walker, 1982).

In many antibiotic fermentations, typical trophophase–idiophase dynamics occur in complex media capable of supporting rapid growth, but the two phases overlap in defined media supporting slow growth (Martin and Demain, 1980). Moreover, growth associated production of antibiotic can be observed in the nutrient limited continuous culture. For example, Inoue *et al.* (1982) demonstrated that streptomycin formation by *S. griseus* could be associated with cellular growth in a glucose- or phosphate-limited chemostat. In a process of this type, deregulation of antibiotic synthesis can occur from the very start of cultivation by limiting some nutritional factors which control the onset of antibiotic biosynthesis. In this respect, processes for continuous production may become increasingly popular for aminoglycoside antibiotics.

4.3.7 Isolation

The isolation procedures rely primarily on solubility, adsorption and ionic characteristics of the antibiotic to separate it from the larger number of other substances present in the fermentation mixture. The aminoglycoside antibiotics are relatively small, basic, water-soluble molecules which form stable acid addition salts. A general isolation procedure for aminoglycoside antibiotics is as follows. Fermentation broth is usually acidified, filtered to remove the cells and other solids, and then absorbed with an acidic ion exchange resin. Elution with acid or base gives the crude material, which almost always contains a complex of closely related compounds. Further purification is carried out by any of the known techniques of chromatography, crystallization or precipitation of insoluble salts.

4.4 STRAIN IMPROVEMENT

Routine strain improvement programs in industry have successfully induced high-yielding mutants of antibiotic producers. Most efforts have been made by random screening: testing for productivity of survivors of mutagenesis on agar plate or in liquid culture and selecting high-yielding mutants. Several attempts toward more rational screening procedures have recently been made on the basis of knowledge of regulation mechanisms in biosynthesis (Martin and Demain, 1980). Although most of them are still indirect methods, the use of rational screening techniques for strain improvement will increase, accompanied with research progress in the biochemical genetics of antibiotic biosynthetic pathways. Moreover, the antibiotic industry will increase the use of genetic approaches such as recombination, protoplast fusion and genetic engineering of antibiotic producing strains, including the manipulation of plasmids containing genes involved in antibiotic production (Hopwood, 1978, 1982; Queener and Baltz, 1979; Davies, 1982).

4.5 REFERENCES

Autissier, D., P. Barthelemy, N. Mazieres, M. Peyre and L. Penasse (1981). 6‴-Deamino-6‴-hydroxy derivatives as intermediates in the biosynthesis of neomycin and paromomycin. *J. Antibiot.*, **34**, 536–543.

Barabás, G., A. Ottenberger, I. Szabó, J. Erdei and G. Szabó (1978). The biological role of aminoglycoside antibiotics in *Streptomycetes*. In *Nocardia and Streptomyces*, ed. M. Mordarski, W. Kurytowicz and J. Jeljaszewicz, pp. 353–361. Gustav Fischer Verlag, Stuttgart.

Barabás, G. and G. Szabó (1977). Effect of penicillin on streptomycin production by *Streptomyces griseus*. *Antimicrob. Agents Chemother.*, **11**, 392–395.

Baud, H., A. Betencourt, M. Peyre and L. Penasse (1977). Ribostamycin as an intermediate in the biosynthesis of neomycin. *J. Antibiot.*, **30**, 720–723.

Claridge, C. A., J. A. Bush, M. D. DeFuria and K. E. Price (1974). Fermentation and mutation studies with a butirosin-producing strain of *Bacillus circulans*. *Dev. Ind. Microbiol.*, **15**, 101–113.

Cox, D. A., K. Richardson and B. C. Ross (1977). The aminoglycosides. In *Topics in Antibiotic Chemistry*, ed. P. G. Sammes, vol. 1, pp. 1–90. Ellis Horwood, Chichester.

Daniels, P. J. L. (1982). Aminoglycosides. In *Antibiotics, Chemotherapeutics, and Antibacterial Agents for Disease Control*, ed. M. Grayson, pp. 38–71. Wiley, New York.

Daum, S. J., D. Rosi and W. A. Goss (1977). Mutational biosynthesis by idiotrophs of *Micromonospora purpurea*. II. Conversion of non-amino containing cyclitols to aminoglycoside antibiotics. *J. Antibiot.*, **30**, 98–105.

Daum, S. J. and J. R. Lemke (1979). Mutational biosynthesis of new antibiotics. *Annu. Rev. Microbiol.*, **33**, 241–265.

Davies, J. E. (1982). Plasmids in antibiotic-producing organisms: their roles in biosynthesis and resistance. In *Genetics of Industrial Microorganisms*, ed. Y. Ikeda and T. Beppu, pp. 103–111. Kodansha, Tokyo.

Dulaney, E. L. (1948). Observation on *Streptomyces griseus*. II. Nitrogen sources for growth and streptomycin production, *J. Bacteriol.*, **56**, 305–313.

Elseviers, D. and L. Gorini (1975). Misreading and mode of action of streptomycin. In *Drug Action and Drug Resistance in Bacteria*, ed. S. Mitsuhashi, vol. 2, pp. 147–175. University Park Press, Tokyo.

Ferguson, J. H., H. T. Huang and J. W. Davisson (1957). Stimulation of streptomycin production by a series of synthetic organic compounds. *Appl. Microbiol.*, **5**, 339–343.

Fujiwara, T., Y. Takahashi, K. Matsumoto and E. Kondo (1980). Isolation of an intermediate of 2-deoxystreptamine biosynthesis from a mutant of *Bacillus circulans*. *J. Antibiot.*, **33**, 824–829.

Fujiwara, T. and E. Kondo (1981). Biosynthetic pathway of 2-deoxystreptamine. *J. Antibiot.*, **34**, 13–15.

Furumai, T., K. Takeda, A. Kinumaki, Y. Ito and T. Okuda (1979). Biosynthesis of butirosin. II. Biosynthetic pathway of butirosins elucidated from co-synthesis and feeding experiments. *J. Antibiot.*, **32**, 891–899.

Grisebach, H (1978). Biosynthesis of sugar components of antibiotic substances. *Adv. Carbohydr. Chem. Biochem.*, **35**, 81–126.

Grisebach, H., B. Kniep and H. Wahl (1981). Biosynthesis of streptomycins. In *Advances in Biotechnology*, ed. C. Vezina and K. Singh, vol. 3, pp. 95–99. Pergamon, Oxford.

Hara, O. and T. Beppu (1982). Induction of streptomycin-inactivating enzyme by A-factor in *Streptomyces griseus*. *J. Antibiot.*, **35**, 1208–1215.

Heding, H. and A. K. Gurtu (1977). Effect of barbital on the biosynthesis of streptomycin in *Streptomyces griseus*. *J. Antibiot.*, **30**, 879–880.

Hockenhull (1978). Production of antibiotics by fermentation. In *Essay in Applied Microbiology*, ed. J. R. Norris and N. H. Richmond, pp. 4/1–4/31. Wiley, New York.

Hopwood, D. A. (1978). Extrachromosomally determined antibiotic production. *Annu. Rev. Microbiol.*, **32**, 373–392.

Hopwood, D. A. (1982). Genetic manipulation in *Streptomyces*. In *Genetics of Industrial Microorganisms*, ed. Y. Ikeda and T. Beppu, pp. 3–8. Kodansha, Tokyo.

Hotta, K. and Y. Okami (1976). Resistance of producers to their own antibiotics. I. Effect of magnesium(2+) on binding of aminoglycoside antibiotics to their producers. *J. Ferment. Technol.*, **54**, 563–571.

Igarashi, K., T. Honma, T. Fujiwara and E. Kondo (1980). Structure elucidation of an intermediate of 2-deoxystreptamine biosynthesis. *J. Antibiot.*, **33**, 830–835.

Inoue, S., Y. Nishizawa and S. Nagai (1982). Stimulation of streptomycin formation by *Streptomyces griseus* grown in a phosphate deficient culture. *J. Ferment. Technol.*, **60**, 417–422.

Ito, S., Y. Odakura, H. Kase, S. Sato, K. Takahashi, T. Iida, K. Shirahata and K. Nakayama (1983). Biosynthesis of astromicin and related antibiotics. I. Biosynthetic studies by conversion experiments. *J. Antibiot.*, **37**, 1664–1669.

Kakinuma, K., Y. Ogawa, T. Sasaki, H. Seto and N. Otake (1981). Stereochemistry of ribostamycin biosynthesis—an application of ^{2}H NMR spectroscopy. *J. Am. Chem. Soc.*, **103**, 5614–5616.

Kase, H., T. Iida, Y. Odakura, K. Shirahata and K. Nakayama (1980). Accumulation of 2-deoxy-*scyllo*-inosamine by a 2-deoxystreptamine-requiring idiotroph of *Micromonospora sagamiensis*. *J. Antibiot.*, **33**, 1210–1212.

Kase, H., G. Shimura, T. Iida and K. Nakayama (1981a). Biotransformation of sisomicin and verdamicin by *Micromonospora sagamiensis*. *Agric. Biol. Chem.*, **46**, 515–522.

Kase, H., Y. Odakura, Y. Takazawa, S. Kitamura and Y. Nakayama (1981b). Fermentation and biosynthesis of sagamicin and related aminoglycosides in *Micromonospora sagamiensis*. In *Advances in Biotechnology*, ed. C. Vezina and K. Singh, vol. 3, pp. 123–128. Pergamon, Oxford.

Kase, H., Y. Odakura, Y. Takazawa, S. Kitamura and K. Nakayama (1982a). Biosynthesis of sagamicin and related aminoglycosides. In *Trends in Antibiotic Research*, ed. H. Umezawa, A. L. Demain, T. Hata and C. R. Hutchinson, pp. 195–212. Japan Antibiotics Res. Assoc., Tokyo.

Kase, H., Y. Odakura and K. Nakayama (1982b). Sagamicin and the related aminoglycosides: fermentation and biosynthesis. I. Biosynthetic studies with the blocked mutants of *Micromonospora sagamiensis*. *J. Antibiot.*, **35**, 1–9.

Khokhlov, A. S. (1982). Low molecular weight microbiol bioregulators of secondary metabolism. In *Overproduction of Microbial Products*, ed. V. Krumphanzl, B. Sikyta and Z. Vanek, pp. 97–109. Academic, New York.

Kimizuka, F., N. Naito, K. Kumagaya, A. Obayashi and O. Tanabe (1977). Accumulation of lividomycin 6'-phosphate by lividomycin-producing strains. *Abstr. Annu. Meet. of the Agric. Chem. Soc. Jpn.*, *Yokohama, April 1–4*, No. 1L-21.

Kitamura, S., H. Kase, Y. Odakura, T. Iida, K. Shirahata and K. Nakayama (1982). 2-Hydroxysagamicin: a new antibiotic produced by mutational biosynthesis of *Micromonospora sagamiensis*. *J. Antibiot.*, **35**, 94–97.

Kojima, M. and Satoh, A. (1973). Microbial semi-synthesis of aminoglycoside antibiotics by mutants of *S. ribosidificus* and *S. kanamyceticus*. *J. Antibiot.*, **26**, 784–786.

Loewus, M. W., F. A. Leowus, G.-U. Brillinger, H. Otsuka and H. G. Floss (1980). Stereochemistry of the *myo*-inositol-1-phosphate synthase reaction. *J. Biol. Chem.*, **255**, 11710–11712.

Majumdar, M. K. and S. K. Majumdar (1970). Isolation and characterization of three phosphoamido-neomycins and their conversion into neomycin by *Streptomyces fradiae*, *Biochem. J.*, **120**, 271–278.

Martin, J. F. and A. L. Demain (1980). Control of antibiotic biosynthesis. *Microbiol. Rev.*, **44**, 230–251.

McAlpine, J. B., R. S. Egan, R. S. Stanaszek, M. Cirovic, S. L. Mueller, R. E. Carney, P. Collum, E. E. Fager, A. W. Goldstein, J. Grampovnik, P. Kurath, J. R. Martin, G. C. Post, J. H. Sedy and J. Tadanier (1980). The structure of minor components of the fortimicin complex. In *Aminocyclitol Antibiotics*, ed. K. L. Rinehart, Jr. and T. Suami, pp. 295–308. American Chemical Society, Washington, DC.

Miller, A. L. and J. B. Walker (1970). Accumulation of streptomycin phosphate in cultures of streptomycin producers grown on a high-phosphate medium. *J. Bacteriol.*, **104**, 8–12.

Nagaoka, K. and A. L. Demain (1975), Mutational biosynthesis of a new antibiotic, streptomutin-A, by an idiotroph of *Streptomyces griseus*. *J. Antibiot.*, **28**, 627–635.

Nakayama, K. (1977). Regulation of microbial secondary metabolism. *Amino Acid, Nucleic Acid*, **36**, 1–33.

Nara, T. (1977). Aminoglycoside antibiotics. *Annu. Rep. Ferment. Process.*, **1**, 299–326.

Nara, T. (1978). Aminoglycoside antibiotics. *Annu. Rep. Ferment. Process.*, **2**, 223–266.

Nimi, O., A. Kokan, K. Manabe, K. Maehara and R. Nomi (1976). Correlation between streptomycin formation and mucopeptide biosynthesis. *J. Ferment. Technol.*, **54**, 587–595.

Nimi, O., H. Kawashima, A. Ikeda, M. Sugiyama and R. Nomi (1981). Biosynthetic correlation between streptomycin and mucopeptide in utilization of D-glucosamine as a common precursor. *J. Ferment. Technol.*, **59**, 91–96.

Odakura, Y., H. Kase, S. Ito, S. Sato, S. Takasawa, K. Takahashi, K. Shirahata and K. Nakayama (1983). Biosynthesis of astromicin and related antibiotics. II. Biosynthetic studies with blocked mutants of *Micromonospora olivasterospora*. *J. Antibiot.*, **37**, 1670–1680.

Odakura, Y., H. Kase and K. Nakayama (1983). Sagamicin and the related aminoglycosides: fermentation and biosynthesis. III. Isolation and characterization of *Micromonospora sagamiensis* mutants blocked in gentamicin C_1 pathway. *J. Antibiot.*, **36**, 125–130.

Otsuka, H., O. A. Mascaretti, L. H. Hurley and H. G. Floss (1980). Stereochemical aspects of the biosynthesis of spectinomycin. *J. Am. Chem. Soc.*, **102**, 6817–6820.

Pearce, C. J., J. E. G. Barnett, C. Anthony, M. Achtar and S. D. Gero (1976). The role of the pseudo-disaccharide neamine as an intermediate in the biosynthesis of neomycin. *Biochem. J.*, **159**, 601–606.

Pearce, C. J. and K. L. Rinehart, Jr. (1981). Biosynthesis of aminocyclitol antibiotics. In *Antibiotics Biosynthesis*, ed. J. W. Corcoran, vol. 4, pp. 74–100. Springer-Verlag, Berlin.

Price, K. E., J. C. Godfrey and H. Kawaguchi (1974). Effect of structural modification on the biological properties of aminoglycoside antibiotics containing 2-deoxystreptamine. *Adv. Appl. Microbiol.*, **18**, 191–307.

Queener, S. W. and R. H. Baltz (1979). Genetics of industrial microorganisms. *Annu. Rep. Ferment. Process.*, **3**, 5–45.

Rinehart, K. L., Jr. and R. M. Stroshane (1976). Biosynthesis of aminocyclitol antibiotics. *J. Antibiot.*, **26**, 319–353.

Rinehart, K. L., Jr. (1979). Biosynthesis and mutasynthesis of aminocyclitol antibiotics. *Jpn. J. Antibiot.*, **32**, suppl. s-32–s-44.

Satoh, A., H. Ogawa and Y. Satomura (1975). Effect of sclerin on production of the aminoglycoside antibiotics accompanied by salvage function in *Streptomyces*. *Agric. Biol. Chem.*, **39**, 1593–1598.

Satoh, A., H. Ogawa and Y. Satomura (1976). Regulation of *N*-acetylkanamycin amidohydrolase in the idiophase in kanamycin fermentation. *Agric. Biol. Chem.*, **40**, 191–196.

Schatz, A., E. Bugie and S. Waksman (1946). Streptomycin, a substance exhibiting antibiotic activity against Gram-positive and Gram-negative bacteria. *Proc. Soc. Exp. Biol. Med.*, **55**, 66–69.

Schlessinger, D. and G. Medoff (1975). Streptomycin, dihydrostreptomycin, and the gentamicins. In *Antibiotics*, ed. J. W. Corcoran and F. E. Hahn, vol. 3, pp. 535–550. Springer-Verlag, Berlin.

Shimizu, M., I. Takahashi and T. Nara (1978). Some problems involved in seldomycin fermentation. *Agric. Biol. Chem.*, **42**, 653–658.

Shirafuji, H., K. Nakahama, I. Nogami, M. Kida and M. Yoneda (1982a). Accumulation of diphosphorylated butirosin A derivatives by phosphatase-negative mutants of *Bacillus vitellinus*. *Agric. Biol. Chem.*, **46**, 1599–1611.

Shirafuji, H., I. Nogami, M. Kida and M. Yoneda (1982b). Two alkaline phosphatases from a butirosin A producer *Bacillus vitellinus*. *Agric. Biol. Chem.*, **46**, 2465–2476.

Shirahata, K., G. Shimura, S. Takasawa, T. Iida and K. Takahashi (1980). The structure of new fortimicins having double bonds in their purpurosamine moieties. In *Aminocyclitol Antibiotics*, ed. K. L. Rinehart, Jr. and T. Suami, pp. 308–320. American Chemical Society, Washington, DC.

Stark, W. M., N. G. Knox, R. Wilgus and R. DuBus (1976). The nebramycin fermentation: culture and fermentation development. *Dev. Ind. Microbiol.*, **17**, 61–77.

Stroshane, R. M. (1976). *Biosynthetic Studies on the Aminocyclitol Antibiotics Neomycin and Spectinomycin*. PhD Thesis, University of Illinois, Xerox Univ. Microfilm, 76–16202.

Takeda, K., K. Aihara, T. Furumai and Y. Ito (1978). An approach to the biosynthetic pathway of butirosins and the

Tanaka, N. (1975). Aminoglycoside antibiotics. In *Antibiotics*, ed. J. W. Corcoran and F. E. Hahn, vol. 3, pp. 340–364. Springer-Verlag, Berlin.

Testa, R. T. and B. C. Tilley (1975). Biotransformation, a new approach to aminoglycoside biosynthesis. I. Sisomicin. *J. Antibiot.*, **28**, 573–579.

Testa, R. T. and B. C. Tilley (1976). Biotransformation, a new approach to aminoglycoside biosynthesis. II. Gentamicin. *J. Antibiot.*, **29**, 140–146.

Testa, R. T. and B. C. Tilley (1979). Biosynthesis of sisomicin and gentamicin. *Jpn. J. Antibiot.*, **32**, suppl. s-47–s-59.

Umezawa, H. (1975). Biochemical mechanism of resistance to aminoglycoside antibiotics. In *Drug Action and Drug Resistance in Bacteria*, ed. S. Mitsuhashi, vol. 2, pp. 211–248. University Park Press, Tokyo.

Umezawa, S. (1975). The chemistry and conformation of aminoglycoside antibiotics. In *Drug Action and Drug Resistance in Bacteria*, ed. S. Mitsuhashi, vol. 2, pp. 3–43. University Park Press, Tokyo.

Voronina, O. I., I. I. Tovarova and A. S. Khokhlov (1978). Studies on the A-factor induced inhibition of glucose-6-phosphate dehydrogenase in *Actinomyces streptomycini*. *Bioorg. Khim.*, **4**, 1538–1546.

Walker, J. B. (1980). Biosynthesis of aminoglycoside antibiotics. *Dev. Ind. Microbiol.*, **21**, 105–113.

Walker, J. B. and M. S. Walker (1982). Enzymic synthesis of streptomycin as a model system for study of the regulation and evolution of antibiotic biosynthetic pathways. In *Overproduction of Microbial Products*, ed. V. Krumphanzl, B. Sikyta and Z. Vaněk, pp. 271–281. Academic, New York.

Weinberg, E. C. (1982). Biosynthesis of microbial metabolites — Regulation by mineral elements and temperature. In *Overproduction of Microbial Products*, ed. V. Krumphanzl, B. Sikyta and Z. Vaněk, pp. 181–194. Academic, New York.

Yamanoto, M., R. Okachi, I. Kawamoto and T. Nara (1977). Fortimicin A production by *Micromonospora olivoasterospora* in a chemically defined medium. *J. Antibiot.*, **30**, 1064–1072.

Yoshikawa, H. and Y. Taniguchi (1976). Effect of alanine on the fermentative production of butirosin. *Abstr.*, *Meet. Ferment. Technol. Soc. Jpn.*, *Osaka, October 25–27*, No. 324.

5

Tylosin

P. P. GRAY
University of New South Wales, Sydney, NSW, Australia

5.1 INTRODUCTION

Tylosin is a 16-membered antibiotic produced commercially by strains of *Streptomyces fradiae* (McGuire *et al.*, 1961). Production of tylosin by *S. rimosus* (Pape and Brillinger, 1973) and by *S. hygroscopius* (Jensen *et al.*, 1963) has also been reported. The initial isolation of two tylosin-producing strains of *S. fradiae* from soil samples obtained from Nongkhai in the north-eastern part of Thailand has been described by Hamill and co-workers (1965). Tylosin is composed of a branched lactone (tylonolide) and three sugars: mycarose, mycaminose and mycinose.

Tylosin is used exclusively for animal nutrition and veterinary medicine. Its use is permitted as an improver of feed efficiency, as a growth promoter, for medicinal use in chickens and swine, and for medicinal use in cattle (Burg, 1982). There have not been any reports detailing volume of sales of tylosin, but Perlman (1978) lists it amongst products with sales volumes greater than $50 million per annum. Perlman also lists Eli Lilly and Co., Indianapolis, USA and Dista Products Ltd., Liverpool, UK as two producers.

5.2 PRODUCTION TECHNOLOGY

5.2.1 Structure of Tylosin and Related Compounds

The structure of tylosin is shown in Figures 1 and 2. Morin and Gorman (1964) were the first to describe the structure of the antibiotic as consisting of a central 16-membered lactone ring with the three sugars, mycarose, mycaminose and mycinose attached. Mycinose is projected from C-14 of the lactone ring while mycarose is attached to the basic sugar mycaminose, the resulting disaccharide being attached to the oxygen at the C-5 position of the lactone ring (Morin *et al.*, 1970). Tylosin has been classified in the tylosin chalcomycin group (Omura and Nakagawa, 1975), other members of the group including cirramycin A, rosamycin, angolamycin and B-58941.

During tylosin production by *Streptomyces fradiae* other structurally related compounds were found to accumulate, in particular macrocin, relomycin and desmycosin (Seno *et al.*, 1977). The

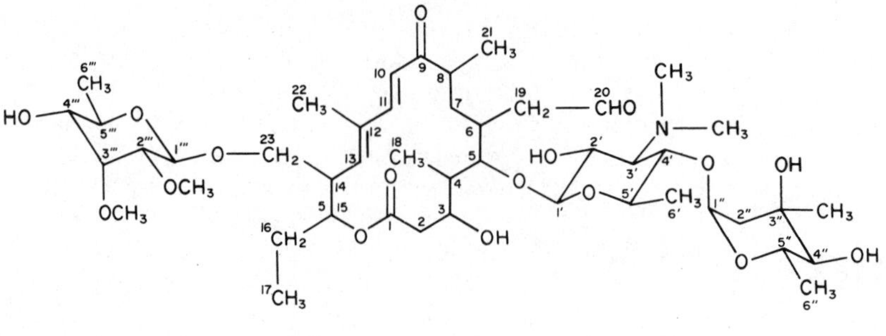

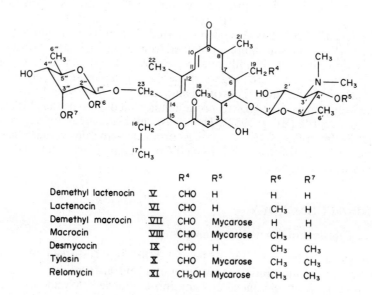

Figure 1 Structure of tylosin

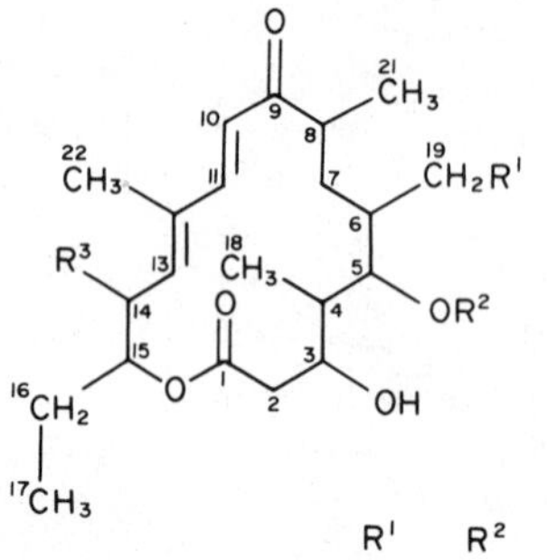

		R¹	R²	R³
Tylactone	I	CH_3	H	CH_3
O-Mycaminosyl tylactone	II	CH_3	Mycaminose	CH_3
O-Mycaminosyl tylonolide	III	CHO	Mycaminose	CH_2OH

		R⁴	R⁵	R⁶	R⁷
Demethyl lactenocin	V	CHO	H	H	H
Lactenocin	VI	CHO	H	CH_3	H
Demethyl macrocin	VII	CHO	Mycarose	H	H
Macrocin	VIII	CHO	Mycarose	CH_3	H
Desmycocin	IX	CHO	H	CH_3	CH_3
Tylosin	X	CHO	Mycarose	CH_3	CH_3
Relomycin	XI	CH_2OH	Mycarose	CH_3	CH_3

Figure 2 Structure of tylactone and related compounds (top) and of tylosin and related compounds (bottom)

structures of these compounds and other related compounds found on the biosynthetic pathway are shown in Figure 2. Proof of the absolute configuration of tylosin has recently been provided by Jones *et al.* (1982).

5.2.2 Biosynthetic Pathway

The tylosin biosynthetic pathway can be divided into three main sections, *viz.* synthesis of tylactone, synthesis of the sugars mycarose, mycaminose and mycinose, and the terminal stages in the pathway involving the conversion of tylactone to tylosin.

Synthesis of the tylactone is the initial stage in the pathway, followed by addition of the sugars and modification of the lactone ring. Studies on the incorporation of ^{13}C precursors into tylosin indicated that the carbon skeleton of tylactone (Figure 2) is derived from five propionates, two acetates and one butyrate (Omura *et al.*, 1975, 1977). Corcoran (1974) and Rossi and Corcoran (1973) studied the biosynthesis of the lactone ring of the 14-membered macrolide erythromycin and found that it was synthesized by a multienzyme synthase complex. Based on the evidence for the erythromycin lactone ring, Masamune *et al.* (1977) and Martin (1977) proposed that tylactone is formed by a mechanism similar to the synthesis of saturated long-chain fatty acids. The work of Omura *et al.* (1978) on the inhibition of lactone formation by the antibiotic cerulenin adds further weight to this hypothesis.

The formation of mycarose is best documented of the three sugars occurring in tylosin. Pape and Brillinger (1973) showed that TDP-mycarose is synthesized from TDP-D-glucose and *S*-adenosyl-l-methionine in *Streptomyces rimosus* fermentations producing tylosin. The reaction required NADPH and has TDP-4-keto-6-deoxy-D-glucose as an intermediate together with a second methylated TDP-sugar, the structure of which is unknown. Formation of TDP-4-keto-6-deoxy-D-glucose is catalyzed by the enzyme TDP-D-glucose oxidoreductase. Pape *et al.* (1969) showed that the methyl group at C-3 of mycarose is transferred from methionine. The other two sugars, mycaminose and mycinose, are also derived from glucose and the methyl groups are transferred from methionine (Achenbach and Grisebach, 1964; Grisebach, 1978). Martin (1979) has suggested that the synthesis of mycaminose is similar to mycarose, and has proposed a pathway.

Work by Seno and co-workers in recent years has elucidated the terminal stages in the tylosin biosynthetic pathway. These workers made extensive use of mutants blocked in various steps in the pathway, cofermentations using the blocked mutants, bioconversion efficiencies for tylosin intermediates using mutants blocked in tylactone formation but with the remainder of the pathway either partially or wholly intact, and analysis of the *O*-methylation reactions in wild-type and mutant extracts (Seno *et al.*, 1977; Baltz and Seno, 1981; Baltz *et al.*, 1981, 1982, 1983). The preferred biosynthetic pathway from tylactone to tylosin is shown in Figure 3. Following the formation of the tylactone, the first step is the addition of mycinose to C-5 followed by oxidation reactions of C-20 and C-23. Mycinose is then added to produce demethyllactenocin which is then glycosylated with mycarose to form demethylmacrocin. Demethylmacrocin is then methylated to macrocin which is subsequently methylated to tylosin. Omura *et al.* (1982a) have proposed a pathway where the route was from tylactone to *O*-mycaminosyl tylactone to either 23-deoxy-*O*-mycaminosyl tylonolide or 20-deoxy-20-dehydrodemycinosyl tylosin, both of which can be converted to *O*-mycaminosyl tylonolide which is converted to desmycosin and then to tylosin. In a subsequent paper it was concluded that *O*-mycaminosyl tylactone is first hydroxylated at C-20 which is oxidized to formyl, followed by the hydroxylation of C-23 (Omura *et al.*, 1982b).

5.2.3 Growth of Producer Microorganisms

A scheme along the lines of the following is usually adopted for the production of tylosin by *Streptomyces fradiae*:

$$\begin{array}{ccccccccc} \text{Lyophilized} & \rightarrow & \text{Liquid} & \rightarrow & \text{Agar} & \rightarrow & \text{Liquid} & \rightarrow & \text{Tylosin} \\ \text{pellet} & & \text{vegetative} & & \text{slants} & & \text{vegetative} & & \text{production} \\ & & \text{medium} & & & & \text{medium} & & \text{medium} \end{array}$$

Several different formulations for agar slant media have been described (McGuire *et al.*, 1961; Stark *et al.*, 1961; Seno *et al.*, 1977). The slant medium described by Seno *et al.* (1977) had the following composition (g l^{-1}): glucose (10.0), phytone (10.0), agar (25.0), biotin (0.001), sodium thiosulfate (1.0). Slants of this medium were incubated for 10 days at 28 °C and then stored at 4 °C until used. It has been reported that slant cultures held under refrigeration for longer than six weeks before use showed decreases in antibiotic yields (Stark *et al.*, 1961). Spore suspensions obtained from agar slants are used to inoculate liquid vegetative medium, which is formulated to provide consistent amounts of mycelial growth, a terminal pH near neutrality and good yields of the antibiotic in the resulting tylosin production medium. Of the various vegetative media which have been described, the most commonly described consists of (g l^{-1}): glucose (15.0), cornsteep liquor (10.0), yeast extract (5.0–6.25), calcium carbonate (3.0–3.8) (Stark *et al.*, 1961; McGuire *et al.*, 1961; Seno *et al.*, 1977). Aerobic growth on the vegetative medium was carried out for 48 hours and the resulting suspension of vegetative mycelia was used to inoculate the tylosin production medium.

Both complex and defined media have been described for the production media. Initial reports on tylosin production described a complex medium consisting of molasses, nutrisoy flour, distillers' solubles and calcium carbonate (McGuire *et al.*, 1961). A more recent publication describes a complex medium consisting of (g l^{-1}): beet molasses (20.0), corn meal (15.0), fish meal (9.0), corn gluten (9.0), NaCl (1.0), $(NH_4)_2HPO_4$ (0.4), calcium carbonate (2.0), crude soybean oil (30.0) (Baltz and Seno, 1981). Despite the lack of any detailed reports on the optimization of complex media, it would seem from the limited information available that the following ingredients need to be present in a complex medium: a source of readily assimilable carbohydrate and a source of carbohydrate in the form of starch; an insoluble protein source; a source of mineral salts; and a source of lipid to supply energy and precursors during tylosin synthesis.

The production of tylosin on synthetic media has also been reported. Stark and co-workers (1961) published a detailed study on the effects of carbon source, amino acids, methylated fatty acids and inorganic components on mycelial growth and tylosin biosynthesis. The optimum medium developed by these workers contained the following (g l^{-1} except where stated): NaCl (2.0), $MgSO_4$ (5.0), $CoCl_2 \cdot 6H_2O$ (0.001), iron (III) ammonium citrate (1.0), $ZnSO_4 \cdot 7H_2O$ (0.01), $CaCO_3$ (3.0), glycine (7.0), L-alanine (2.0), L-valine (1.0), betaine (5.0), glucose (35.0), methyl oleate (25.0 ml l^{-1}), K_2HPO_4 (2.3). Gray and Bhuwapathanapun (1980) modified the above medium by the substitution of calcium chloride for the calcium carbonate and sodium glutamate instead of glycine, valine and alanine, resulting in a soluble medium suitable for use in continuous culture if the methyl oleate is fed separately. Madry and Pape (1982) have described a synthetic medium for tylosin production consisting of glucose, betaine, glycine, potassium nitrate, magnesium sulfate, calcium carbonate, methyl oleate, potassium phosphate buffer and trace elements.

There have been no reports on the kinetics of growth and tylosin production in high yielding fermentations on optimized complex media. Most reports on tylosin-producing fermentations incubate the cultures at 28–30 °C under aerobic conditions for 7–10 days in batch culture. Initial pH is usually in the range 7.0–7.8 with final pH values when quoted in a similar range.

5.2.4 Product Recovery and Purification

Tylosin can be recovered from fermentation broth by applying either adsorption or extraction techniques. Extractants which can be used include water-immiscible polar organic solvents such as ethyl acetate and amyl acetate, chlorinated hydrocarbons such as chloroform, and water-immiscible alcohols, ketones and ethers (Hamill *et al.*, 1965). For recovery of tylosin by adsorption, a range of adsorbents and ion exchange resins can be used and then the tylosin can be eluted with an organic solvent. The organic solvent extract can then be either evaporated to dryness to provide a crude tylosin or concentrated *in vacuo* and a precipitant added. The formation of various salts of tylosin has also been described (Hamill *et al.*, 1965). The isolation of compounds after 23-deoxy-20-dehydro-*O*-mycaminosyl tylonolide on the tylosin synthetic pathway (Figure 3) by methods suitable for organic solvent extractable basic compounds has been described by Kirst *et al.* (1982). The fermentation broths were filtered, extracted using amyl acetate or ethyl acetate at pH 9.0–9.5 and then back extracted into water at about pH 4.0 to separate the products from organic solvent soluble neutral compounds. The pH of the aqueous phase was carefully controlled so as to prevent the hydrolysis of mycarose, which is cleaved under acidic conditions, from the products. The products were then either crystallized directly from the aqueous solution, or extracted into a volatile solvent such as methylene chloride or ethyl acetate at pH 9.0–9.5. Evaporation of the solvent yielded a dry product which could be used directly or further purified by crystallization. Further details on the separation by multistage countercurrent extraction of closely related compounds produced by mutant strains of *S. fradiae* are described by Kirst *et al.* (1982). Vasileva-Lukanova *et al.* (1980a, 1980b) described the recovery of tylosin from filtered *S. fradiae* fermentation liquor by extraction into either chloroform or dichloromethane at pH 9.5 and 2–5 °C with a 95% efficiency. The same workers studied the effect of different extracting solvents and temperature on the efficiency of extraction of tylosin and examined the reextraction from butyl acetate back into water at pH 4.0.

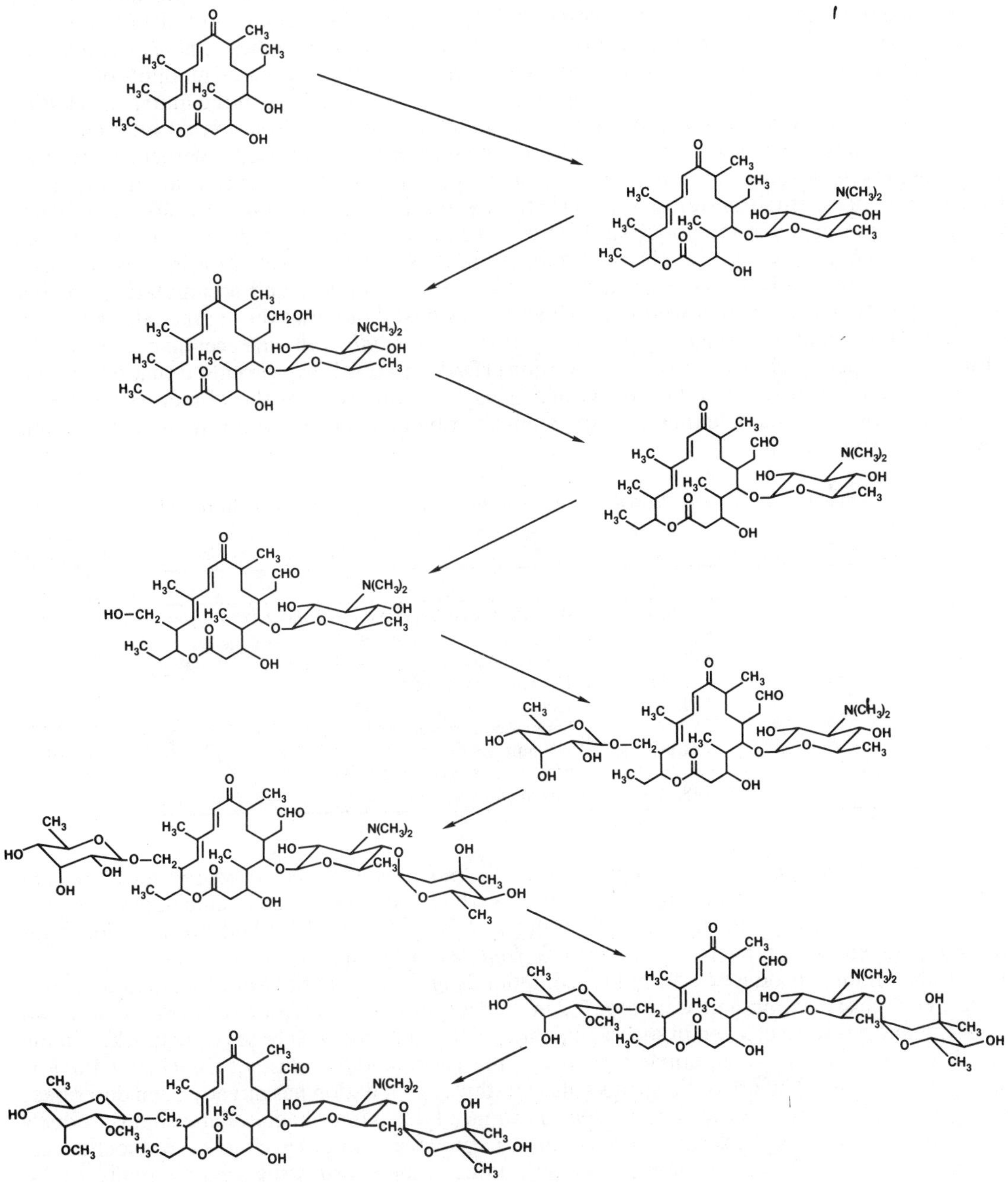

Figure 3 The preferred biosynthetic pathway from tylactone to tylosin. The compounds listed in sequence are tylactone → O-mycaminosyl tylactone → 23-deoxy-20-dehydro-O-mycaminosyl tylonolide → 23-deoxy-O-mycaminosyl tylonide − −→ O-mycaminosyl tylonolide − −→ demethyllactenocin − −→ demethylmacrocin − −→ macrocin − −→ tylosin (from Baltz *et al.*, 1983)

The isolation of tylactone and 5-O-mycarosyltylactone has been described by Jones *et al.* (1982). A mutant strain of *Streptomyces fradiae* which accumulated the two compounds was grown on a complex medium. The broth was extracted with petroleum ether and the extract concentrated to an oil. The oil was dissolved in ethyl acetate, heptane was added and the ethyl acetate allowed to evaporate slowly to permit crystallization. The crystals consisted of a mixture of the two compounds which were separated by silica gel chromatography.

5.3 PRODUCT DEVELOPMENT

5.3.1 Developments in the Genetic Improvement of Producing Strains

There have been no detailed reports on strain improvement programs which presumably have been carried out since tylosin became a commercial product in 1962. Seno and Baltz (1982) compared the S-adenosyl-L-methionine:macrocin O-methyltransferase activities of nine *Streptomyces fradiae* mutants which had been selected for their ability to produce increased amounts of tylosin. These authors described the relationship of the nine mutants and the use of the mutagens HNO_2, UV and MNNG (*N*-methyl-*N'*-nitro-*N*-nitrosoguanidine) in their isolation (Table 1). They showed that increased tylosin production was accompanied by increased macrocin O-methyltransferase levels in some of the mutants. Comparisons of final tylosin concentrations and averaged macrocin O-methyltransferase specific activities in the S. *fradiae* strains are shown in Figure 4. The results obtained in the above study confirmed earlier results by these workers (Seno and Baltz, 1981) where it was shown that a mutant strain of S. *fradiae* selected for increased tylosin production synthesized macrocin O-methyltransferase more rapidly and accumulated a higher enzyme specific activity than a wild-type strain. It was postulated that during the later stages of the tylosin fermentation in highly productive strains of S. *fradiae*, the intracellular levels of the tylosin macrolides reach inhibitory concentrations (Baltz, 1982). It was proposed that this apparent rate limitation might be relieved by increasing concentrations of the O-methyltransferase enzyme by cloning and amplifying the copy number or by inducing mutations to alter the kinetic properties of the enzyme.

Table 1 *Streptomyces fradiae* Strains Selected for Ability to Produce Increased Tylosin Levels (Seno and Baltz, 1982)

Strain no.	Designation	Source
1	T59235	Recloned from original soil isolate (ATCC 19609)
2	NA171	HNO_2 mutagenesis of T59235
3	NC199	HNO_2 mutagenesis of NA171
4	OD198	UV mutagenesis of NC199
5	TPH72	HNO_2 mutagenesis of OD198
6	TPQ96	Spontaneous mutant of TPH72
7	TTM85	UV mutagenesis of TPQ96
8	C4	MNNG mutagenesis of TTM85[a]
9	T482	MNNG mutagenesis of C4[a]

[a] MNNG, *N*-methyl-*N'*-nitro-*N*-nitrosoguanidine.

Other reports on the genetic manipulation of S. *fradiae* which have appeared have concentrated on the work with blocked and other mutants which have been used in elucidating the tylosin biosynthetic pathway (Section 5.2.2). Baltz and Seno (1981) described the isolation after MNNG mutagenesis of numerous mutants of S. *fradiae* blocked in tylosin biosynthesis. Less than 20% of the mutants produced tylosin-like compounds and the mutants were classified into nine classes. Four classes of mutants were blocked in the biosynthesis or addition of tylosin sugars; two classes of mutants were blocked in specific oxidations of tylactone; two classes were blocked in the specific O-methylations of demethylmacrocin and macrocin; and one class was blocked in the formation of tylactone. The use of these mutants in pathway elucidation has already been described; in addition the mutants have also been used to produce tylosin intermediates and shunt products with antibiotic activities which can be used to produce novel antibiotics in chemical or biochemical conversions. Baltz (1982b) described the modifications of the tylosin structure potentially attainable by such methods and concluded that well over 1000 novel structures were theoretically possible. The production of some of these compounds has been described (*e.g.* Sadakane *et al.*, 1982; Okamoto *et al.*, 1980), and the study of these compounds will provide information on the structure–activity relationships of the antibiotics. There has been an initial report on the structure–activity relationships for compounds with tylosin-related structures by Kirst *et al.* (1982), who concluded that none of the macrolides they evaluated appeared to be superior to tylosin in treating bacterial or mycoplasmal infections caused by sensitive microorganisms.

Before recombinant DNA techniques can be used either for antibiotic yield improvement or for the production of novel antibiotics with tylosin-related structures, it will be necessary to obtain more information on the location and organization of tylosin genes. Work by Baltz and coworkers (Baltz, 1978, 1980; Baltz and Matsushima, 1981) showed that expression of tylosin structural genes was lost at a very high frequency after fusion of protoplasts, genetic recombination

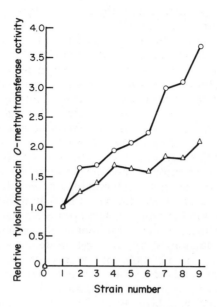

Figure 4 Comparison of final tylosin productivities and average macrocin *O*-methyltransferase specific activities in *S. fradiae* strains. Macrocin *O*-methyltransferase specific activities were averaged between 2 and 6 days and normalized to the average specific activity of strain 1 (△). Final tylosin yields were determined by the UV absorbance assay (○) and normalized to that of strain 1 (Seno and Baltz, 1982)

and cell regeneration. These workers concluded that the tylosin structural genes are clearly self-transmissable and may be plasmid borne, although the presence of plasmids in *S. fradiae* was not demonstrated. Other fundamental areas where work is necessary to allow the application of gene cloning techniques have been described (Baltz, 1982a, 1982b) as the need to develop suitable vectors and problems associated with restriction endonucleases in *Streptomyces*.

5.3.2 Developments in Fermentation Technology

Development of the fermentation conditions which allow maximization of tylosin production is dependent on knowledge of the factors regulating biosynthesis of the antibiotic. There have been a number of studies on the role of key medium ingredients on tylosin biosynthesis.

High levels of inorganic phosphate have been reported to inhibit many antibiotics. Stark *et al.* (1961) were the first to report that high phosphate levels inhibited tylosin formation. Subsequent work by Madry *et al.* (1979) also demonstrated inhibition of tylosin synthesis by high phosphate levels and showed a parallel decrease in methyl oleate uptake. Vu-Trong *et al.* (1980) showed that in batch cultures when elevated levels of phosphate were added in the idiophase (120 h) there was no metabolic response or inhibition of tylosin synthesis. When the same amount of phosphate was added at the start of the batch culture, tylosin synthesis was inhibited and there was a marked increase in the level of the adenylates (ATP, ADP, AMP) over control, although the energy charge was unchanged. Also, high initial levels of phosphate suppressed the activities of the enzymes methylmalonylcoenzyme A carboxyltransferase (EC 2.1.3.1) and propionylcoenzyme A carboxylase (EC 6.4.1.3). Gray and Bhuwapathanapun (1980) showed that in steady state chemostat culture at a fixed dilution rate, increasing the specific uptake rate of inorganic phosphate resulted in a decrease in the specific rate of tylosin production, although the ratio of relomycin to tylosin increased. Madry and Pape (1981) concluded that the phosphate effect can be considered as an effect on primary metabolism leading to altered secondary metabolite formation, probably *via* a limitation of precursors necessary for the synthesis of the secondary metabolite. These authors also presented evidence that phosphate or a metabolic signal controlled by the phosphate concentration could be involved in the regulation of transcription of the genes associated with tylosin biosynthesis.

Initial work by Stark and co-workers (1961) had shown that both a source of carbohydrate (preferably glucose) and a source of lipid are necessary for good growth and tylosin production. Subsequent work has failed to find a carbohydrate which can successfully replace the lipid source, which is usually either soybean oil or a methylated fatty acid, *e.g.* methyl oleate. Sprinkmeyer and Pape (1978) showed that oleyl-CoA inhibits the enzyme citrate synthase and that long-chain fatty acids, as well as supplying precursors for tylosin synthesis, also led to increased levels of acetyl-CoA by reducing its final oxidation through inhibition of citrate synthase. The presence of an insoluble lipid phase during the fermentation can be thought of as a feeding system, providing positive lipid levels are maintained. Although several workers have shown that the uptake rate of the lipid source is correlated to the rate of synthesis of tylosin and is dependent on the metabolic state of the microorganism, there have been no reports of such factors as the size of oil droplets or the presence of emulsifiers affecting tylosin synthesis rates. Although glucose has been shown to be a good source of carbohydrate for growth and tylosin synthesis and to be involved in the synthesis of the tylosin sugars (Section 5.2.2), excessively high uptake rates of glucose result in suppression of tylosin synthesis. Sprinkmeyer and Pape (1978) showed that increased initial levels of glucose inhibited both methyl oleate uptake and tylosin synthesis in batch culture. Madry *et al.* (1979) showed that 2-deoxyglucose but not 3-*O*-methyl-D-glucose inhibited tylosin formation and methyl oleate uptake. Vu-Trong *et al.* (1980) demonstrated that the addition of 30 g l^{-1} glucose to idiophase batch cultures resulted in the rapid cessation of tylosin synthesis, rapid increase in the total adenylate levels although the energy charge remained unchanged, and a decrease in the activities of the enzymes methylmalonyl-CoA carboxyltransferase and propionyl-CoA carboxylase. Once the additional glucose was metabolized, tylosin synthesis resumed at the same rate as before the addition of the glucose and the level of total adenylates decreased. Gray and Bhuwapathanapun (1980) showed that in steady state chemostat cultures at a constant dilution rate, increasing the specific rate of glucose uptake above 10 mg glucose per g dry weight of cells per h resulted in suppression of the specific rate of tylosin synthesis.

The importance of various amino acids in the nutritional requirements for tylosin production and the differences in requirements of different strains of *S. fradiae* were first reported by Stark *et al.* (1961). Although this is an area of major importance in maximizing tylosin synthesis, there have only been a few relevant reports (Dotzlaf *et al.*, 1983; Gray and Bhuwapathanapun, 1980; Vu-Trong and Gray, 1982).

In spite of the information which is accumulating on the regulation of tylosin biosynthesis, all reports on the production of the antibiotic on complex media have been straight batch fermentations with no additional feeds, where presumably the medium composition had been optimized by empirical methods. Under such conditions, antibiotic production follows a classical tropophase–idiophase kinetic pattern with a rapid decrease in the specific rate of tylosin production towards the end of the fermentation (*e.g.* Seno and Baltz, 1982). Most studies on tylosin production on synthetic media have also been straight batch fermentations with similar kinetic patterns. The only report on fed-batch fermentations on a synthetic medium has been by Bhuwapathanapun and Gray (1981), who showed that with linear feeds of complete medium it was possible to increase the final concentration of tylosin in the fermentation by increasing biomass levels, the specific rate of tylosin production being only marginally increased. Chemostat studies on a defined medium showed that the activities of methylmalonyl-CoA carboxyltransferase and propionyl-CoA carboxylase (two enzymes involved in the synthesis of one of the tylactone precursors) and the activity of macrocin methyltransferase (the enzyme involved in the final step on the pathway) all changed in concert with each other and with the specific production rate of tylosin, *i.e.* the pathway seemed to be tightly regulated in a concerted fashion.

Using a synthetic medium derived in chemostat cultures, Gray and Bhuwapathanapun (1980) described a batch fermentation where there were high initial rates of specific tylosin production during the period of active growth of the culture. These high rates rapidly decreased after 48 h and it was not possible to maintain the high initial rates by linear feeds of either the complete medium or medium components. Vu-Trong and Gray (1982) showed that it was possible to maintain high values of specific tylosin production throughout the fermentation if the correct levels of glucose and sodium glutamate were fed in square wave profiles to the fermentation (Figure 5). It was shown that only this transient feeding profile could maintain linear tylosin accretion throughout the fermentation and that both the period and amplitude of the profiles were of importance in maximizing tylosin productivity. It appeared that the rapid uptake of glucose/glutamate followed by periods of zero uptake allowed the cell to maintain the correct balance between supply of precursors, maintaining high levels of the enzymes on the tylosin biosynthetic pathway, minimal catabolite inhibition/repression problems to allow long term tylosin synthesis at high rates. The findings by Madry and Pape (1981), that one or more of the proteins involved in tylosin biosynthesis appears to be quite unstable and that the corresponding mRNAs are either relatively stable or transcribed continuously, are of particular interest in this context.

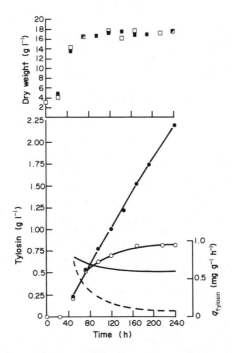

Figure 5 Time course of a fed batch fermentation where square-wave feeding of glucose and sodium glutamate was started at 48 hours and continued until the end of the fermentation. Feeds were controlled by Apple microprocessor with the period of the square waves being 36 hours. Control fermentation had no feeds. Open symbols, control; closed symbols, fed fermentation: ●, ○, tylosin concentration; ■, □ cell concentration; ————, $q_{tylosin}$ for fed fermentation; ———, $q_{tylosin}$ for control (all other conditions as per Vu-Trong and Gray, 1982b)

The production of tylosin by immobilized *S. fradiae* cells has been described by Veelken and Pape (1982).

5.4 CONCLUSIONS

The last few years have seen a rapid expansion in the information available on tylosin, with the elucidation of the biosynthetic pathway, information on structure–activity relationships between tylosin-related compounds, and more detailed information on the factors regulating biosynthesis. It will be necessary to obtain further information on the key factors regulating tylosin biosynthesis such as enzyme levels and controls, the supply of precursors and the role of primary metabolism in regulating biosynthesis in order that the developing techniques of recombinant DNA in *Streptomyces* can be rationally applied to obtaining improved strains.

The application of more sophisticated control strategies to the fermentation coupled with these improved strains should see further improvements in tylosin productivities and control of production of related factors.

5.5 REFERENCES

Achenbach, H. and H. Grisebach (1964). Zur Biogenese der Macrolide XII— Mitt Weitzf untersuchungen zer Biogenese des Magnamycin. *Z. Naturforsch., Teil B*, **193**, 561–568.
Baltz, R. H. (1978). Genetic recombination in *Streptomyces fradiae* by protoplast fusion and cell regeneration. *J. Gen. Microbiol.*, **107**, 93–102.
Baltz, R. H. (1980). Genetic recombination by protoplast fusion in *Streptomyces. Dev. Indust. Microbiol.*, **21**, 43–54.
Baltz, R. H. (1982a). Genetics and biochemistry of tylosin production; a model for genetic engineering in antibiotic-producing *Streptomyces. Basic Life Sci.*, **19**, 431–444.
Baltz, R. H. (1982b). Genetics and biochemistry of tylosin production: a model for genetic engineering in antibiotic-producing *Streptomyces*. In *Genetic Engineering of Microorganisms for Chemicals*, ed. A. Hollaender, R. D. De Moss, S. Kaplin, J. Konisky, D. Savage and R. S. Wolfe. Plenum, New York.
Baltz, R. H. and P. Matsushima (1981). Protoplast fusion in *Streptomyces*: Conditions for efficient genetic recombination and cell regeneration. *J. Gen. Microbiol.*, **127**, 137–146.

Baltz, R. H. and E. T. Seno (1981). Properties of *Streptomyces fradiae* mutants blocked in biosynthesis of the macrolide antibiotic tylosin. *Antimicrob. Agents Chemother.*, **20** (2), 214–225.

Baltz, R. H., E. T. Seno, J. Stonesifer and G. M. Wild (1983). Biosynthesis of the macrolide antibiotic tylosin: a preferred pathway from tylactone to tylosin. *J. Antibiot.*, **36** (2), 131–141.

Baltz, R. H., E. T. Seno, J. Stonesifer, P. Matsushima and G. M. Wild (1981). Genetics and biochemistry of tylosin production by *Streptomyces fradiae*. In *Microbiology—1981*, ed. D. Schlessinger, pp. 371–375. American Society for Microbiology, Washington, DC.

Baltz, R. H., E. T. Seno, J. Stonesifer, P. Matsushima and G. M. Wild (1982). Genetics and biochemistry of tylosin production. *Trends Antibiot. Res.*, 65–72.

Bhuwapathanapun, S. and P. P. Gray (1981). Production of the macrolide antibiotic tylosin in fed-batch cultures. *J. Ferment. Technol.*, **95** (5), 419–421.

Burg, R. W. (1982). Fermentation products in animal health. *ASM News*, **48** (10), 460–463.

Corcoran, J. W. (1974). Lipid and macrolide lactone biosynthesis in *Streptomyces erythreus*. *Dev. Ind. Microbiol.*, **15**, 93–100.

Dotzlaf, J. E., L. S. Metzger and M. A. Foglesong (1983). Incorporation of amino acid derived carbon into tylosin aglycone by *Streptomyces fradiae*. Abstract 0 63, ASM Annual Meeting.

Gray, P. P. and S. Bhuwapathanapun (1980). Production of the macrolide antibiotic tylosin in batch and continuous cultures. *Biotechnol. Bioeng.*, **22**, 1785–1804.

Grisebach, H. (1978). Biosynthesis of macrolide antibiotics. In *Antibiotics and Other Secondary Metabolites: Biosynthesis and Production*, ed. R. Hunter, T. Leisinger, J. Neusch and W. Wehrli, pp. 113–127. Academic, London.

Hamill, R. L., M. E. Harvey, Jr., M. Stamper and P. F. Wiley (1961). Tylosin, a new antibiotic: II. Isolation, properties and preparation of desmycosin, a microbiologically active degradation product. *Antibiot. Chemother.*, **11** (5), 328–334.

Hamill, R. L., M. E. Harvey, Jr., G. M. McGuire and M. C. Stamper (1965). *US Pat.* 3 178 341.

Jensen, A. L., M. A. Darken, J. S. Schultz and A. J. Shay (1963). Relomycin: flask and tank fermentation studies. *Antimicrob. Agents Chemother.*, 49–53.

Jones, N. D., M. O. Chaney, H. A. Kirst, G. M. Wild, R. H. Baltz, R. L. Hamill and J. W. Paschal (1982). Novel fermentation products from *Streptomyces fradiae*: X-ray crystal structure of 5-*O*-mycarosyltylactone and proof of the absolute configuration of tylosin. *J. Antibiot.*, **35** (4), 420–425.

Kirst, H. A., G. M. Wild, R. H. Baltz, R. L. Hamill, J. L. Ott, F. T. Counter and E. E. Ose (1982). Structure–activity studies among 16-membered macrolide antibiotics related to tylosin. *J. Antibiot.*, **35** (12), 1675–1682.

Madry, N. and H. Pape (1981). Regulation of tylosin biosynthesis by phosphate—possible involvement of transcriptional control. In *Actinomyces, Zbl. Bakt. Suppl. II*, ed. Schaal and Pulverer. Fischer, Stuttgart.

Madry, N. and H. Pape (1982). Formation of secondary metabolism enzymes in the tylosin producer *Streptomyces* T59-235. *Arch. Microbiol.*, **131**, 170–173.

Madry, N., R. Sprinkmeyer and H. Pape (1979). Regulation of tylosin synthesis in *Streptomyces*: Effects of glucose analogs and inorganic phosphate. *Eur. J. Appl. Microbiol.*, **7**, 365–370.

Martin, J. F. (1977). Biosynthesis of polyene macrolide antibiotics. *Annu. Rev. Microbiol.*, **31**, 13–38.

Martin, J. F. (1979). Nonpolyene macrolide antibiotics. In *Economic Microbiogy III—Secondary Products of Metabolism*, ed. A. H. Rose, pp. 239–289. Academic, New York.

Masamune, S., G. S. Bates and J. W. Corcoran (1977). Macrolides—recent progress in chemistry and biochemistry. *Angew. Chem., Int. Ed. Engl.*, **16**, 585–607.

McGuire, J. M., W. S. Boniece, C. E. Higgins, M. M. Hoehn, W. M. Stark, J. Westhead and R. N. Wolfe (1961). Tylosin, a new antibiotic: I. Microbiological studies. *Antibiot. Chemother.*, **11** (5), 320–327.

Morin, R. B. and M. Gorman (1964). The partial structure of tylosin, a macrolide antbiotic. *Tetrahedron Lett.*, 2339–2345.

Morin, R. B., M. Gorman, R. L. Hamill and P. V. Demarco (1970). The structure of tylosin. *Tetrahedron Lett.*, 4737–4740.

Okamata, R., M. Tsuchiya, H. Nomura, H. Iguchi, K. Kiyoshim, S. Hori, T. Inui, T. Sawa, T. Takeuchi and H. Umezawa (1980). Biological properties of new acyl derivatives of tylosin. *J. Antibiot.*, **33**, 1309–1315.

Omura, S. and A. Nakagawa (1975). Chemical and biological studies on 16-membered macrolide antibiotics. *J. Antibiot.*, **28**, 401–433.

Omura, S., A. Nakagawa, H. Takeshima, J. Miyazawa, C. Kitao, F. Piriou and G. Lukacs (1975). A ^{13}C nuclear magnetic resonance study of the biosynthesis of the 16-membered macrolide antibiotic tylosin. *Tetrahedron Lett.*, 4503–4506.

Omura, S., A. Takeshima, A. Nakagawa, J. Miyazawa, F. Piriou and G. Lukacs (1977). Studies on the biosynthesis of 16-membered macrolide antibiotics using carbon-13 nuclear magnetic resonance spectroscopy. *Biochemistry*, **16**, 2860–2866.

Omura, S., C. Kitao, J. Miyazawa, H. Imai and H. Takeshima (1978). Bioconversion and biosynthesis of 16-membered macrolide antibiotic, tylosin, using enzyme inhibitor: cerulenin. *J. Antibiot.*, **31**, 254–256.

Omura, S., H. Tanaka and M. Tsuki (1982a). Biosynthesis of tylosin: oxidations of 5-*O*-mycaminosylprotylonolide at C-20 and C-23 with a cell free extract from *Streptomyces fradiae*. *Biochem. Biophys. Res. Commun.*, **107** (2), 554–560.

Omura, S., N. Sadakane and H. Matsubara (1982b). Bioconversion and biosynthesis of 16-membered macrolide antibiotics. Biosynthesis of tylosin after protylonolide formation. *Chem. Pharm. Bull.*, **30** (1), 223–229.

Pape, H. and G. U. Brillinger (1973). Metabolic products of microorganisms. 113. Biosynthesis of thymidine di-phospho mycarose in a cell-free system from *Streptomyces rimosus*. *Arch. Microbiol.*, 25–35.

Pape, H., R. Schmid and H. Grisebach (1969). Ubertragung der intaken Methylgruppe des Methionins bei der Biosynthese der L-Mycarose. *Eur. J. Biochem.*, **11**, 479–483.

Perlman, D. (1978) Fermentation. In *Kirk–Othmer Encyclopedia of Chemical Technology*, 3rd edn., vol. 9, pp. 861–880, Wiley, New York.

Rossi, A. and J. W. Corcoran (1973). Identification of a multi-enzyme complex synthesising fatty acids in the actinomycete *Streptomyces erythreus*. *Biochem. Biophys. Res. Commun.*, **50**, 597–602.

Sadakane, N., Y. Tanaka and S. Omura (1982). Hybrid biosynthesis of derivatives of protylonolide and M-4365 by macrolide-producing microorganisms. *J. Antibiot.*, **35** (6), 680–687.

Seno, E. T. and R. H. Baltz (1981). Properties of *S*-adenosyl-L-methionine: macrocin *O*-methyltransferase in extracts of *Streptomyces fradiae* strains which produce normal or elevated levels of tylosin and in mutants blocked in specific *O*-methylations. *Antimicrob. Agents Chemother.*, **20** (3), 370–377.

Seno, E. T. and R. H. Baltz (1982). *S*-Adenosyl-L-methionine: macrocin *O*-methyltransferase activities in a series of *Streptomyces fradiae* mutants that produce different levels of the macrolide antibiotic tylosin. *Antimicrob. Agents Chemother.*, **21** (5), 758–763.

Seno, E. T., R. T. Pieper and F. M. Huber (1977). Terminal stages in the biosynthesis of tylosin. *Antimicrob. Agents Chemother.*, **11** (3), 455–461.

Sprinkmeyer, R. and H. Pape (1978). Effects of glucose and fatty acids on the formation of the macrolide antibiotic tylosin by *Streptomyces*. In *Genetics of the Actinomycetales*, ed. E. Freerksen, I. Tarnok and J. H. Thumim. Fischer, Stuttgart.

Stark, W. M., W. A. Daily and J. M. McGuire (1961). A fermentation study of the biosynthesis of tylosin in synthetic media. *Sci. Rep. Ist. Super. Sanita*, **1**, 340–354.

Vasileva-Lukanova, B., T. Atanasova and E. Khlebarova (1980a). Studies on the isolation of tylosin from the culture liquid of *Streptomyces fradiae*. I. Method for tylosin production. *Khim. Ind. (Sofia)*, 200–202.

Vasileva-Lukanova, B., T. Atanasova, V. Gancheva and E. Khlebarova (1980b). Studies on the isolation of tylosin from the culture fluid of *Streptomyces fradiae*. II. Extraction of tylosin from a native solution. *Khim. Ind. (Sofia)*, 300–302.

Veelken, M. and H. Pape (1982). Production of tylosin and nikkomycin by immobilised *Streptomyces* cells. *Eur. J. Appl. Microbiol. Biotechnol.*, **15**, 206–210.

Vu-Trong, K. and P. P. Gray (1982a). Continuous culture studies on the regulation of tylosin biosynthesis. *Biotechnol. Bioeng.*, **24**, 1093–1103.

Vu-Trong, K. and P. P. Gray (1982b). Simulation of tylosin productivity resulting from cyclic feeding profiles in fed batch cultures. *Biotechnol. Lett.*, **4** (11), 725–728.

Vu-Trong, K., S. Bhuwapatanapun and P. P. Gray (1980). Metabolic regulation in tylosin producing *Streptomyces fradiae*: regulatory role of adenylate nucleotide pool and enzymes involved in biosynthesis of tylonolide precursors. *Antimicrob. Agents Chemother.*, **17** (4), 519–525.

Vu-Trong, K., S. Bhuwapathanapun and P. P. Gray (1981). Metabolic regulation in tylosin producing *Streptomyces fradiae*: phosphate control of tylosin biosynthesis. *Antimicrob. Agents Chemother.*, **19** (2), 209–212.

6

Peptide Antibiotics

H. KLEINKAUF and H. VON DÖHREN
Technische Universität Berlin, Federal Republic of Germany

6.1 INTRODUCTION

The present 'modern era' (Bérdy, 1980b) of antibiotic research is characterized by advanced screening procedures aiming specifically not only at antibacterials or antifungals but at compounds like enzyme inhibitors, immunomodulators and antitumor drugs. New sources like rare microorganisms (actinomycetes, pseudomonales) or various marine organisms are currently at the beginning of exploitation. Significant improvements have been obtained by semisynthetic derivatives of natural compounds. Concerning peptide antibiotics, β-lactams still have the major part of the market. Since these have been treated earlier in this volume, we focus on recent trends of the extended concept of antibiotics and biosynthetic implications for the production of modified compounds. Less attention will be given to chemical aspects of drug improvement.

6.1.1 Current Applications of Peptides

Peptides that are currently produced on an industrial scale are listed together with their applications in Table 1. To illustrate a few general approaches in product evaluation some notes on blasticidin S (agricultural antibiotic, improvement by antagonists), bleomycin (antitumor agent, improvement by semisynthesis), bestatin (immunoenhancer, used in combination with bleomycin) and cyclosporin (isolation from rare microorganisms) have been included. Structural formulae not given in the text have been added in an appendix (Section 6.5).

Table 1   Current Applications of Peptide Antibiotics

No.	G⁺ antibacterial	G⁻ antibacterial	Mycobacterial	Antifungal	Antitumor	Antiviral	Immunomodulating	Topical antibiotic	Veterinary drug	Feed additive	Food preservative	Chemical reagent	Notes	Molecular mechanisms, targets and actions	Related or identical compounds
1 Actinomycins	+				+	+						+	Most potent antitumor drug	DNA intercalating, inhibition of DNA and RNA synthesis	Cactinomycin, dactinomycin, cosmagen
2 Albomycins		+											Penicillin resistant cocci	Membrane transport, Fe^{3+} complex	Sideromycins
3 Amphomycin	+							+	+				Poor gastrointestinal absorption, only in ointments		Parvulin
4 Bacitracins	+							+		+			Topical antibiotic	Cell wall synthesis (lipid carrier, PPase)	Ayfivin
5 Bestatin					+		+						Immunoenhancing, in cancer therapy	Aminopeptidase inhibitor	Amastatin, pepstatins
6 Bicyclomycin		+								+			On clinical trial	Affects biosynthesis of lipoproteins (G⁻)	
7 Blasticidin S	+	+		+	+								Agricultural antibiotic (rice blast)	rRNA and protein synthesis	
8 Bleomycins	+	+	+	+	+		+						Broad spectrum antibiotic, antitumor drug	DNA strand scission; inhibits DNA, RNA and protein synthesis	Phleomycins, tallysomycins
9 Capreomycins			+										Limited use as antitubercular drug, long term nephrotoxic effects	Inhibits protein synthesis	Capstat
10 Cycloserine		+	+										Broad spectrum, used for resistant mycobacteria, neurotoxic side effects	Cell wall synthesis (D-Ala analogue)	Cyclomycin, orientmycin, oxymycin
11 Cyclosporins				+			+						Immunosuppressor in transplantation		
12 Distamycins	+	+	+	+	+	+							Topical antiviral	Inhibits T 2 multiplication and R-factor transfer in E. coli; inducer of enzyme synthesis in E. coli	Netropsin, stallimycin
13 Duazomycins					+		+		+				Gastrointestinal carcinomas, soft tissue carcinomas	Purine synthesis, glutamine antagonist	Azaserine, DON

No.	Name	Remarks / clinical	Mechanism of action	Other names
14	Enduracidin A	Long lasting activity, limited clinical application		Enramycin, enradine
15	Gramicidin	Topical use, hemolysis of red blood cells		
16	Gramicidin S	Topical use, hemolysis of erythrocytes, surfactant		Tyrocidines
17	Hadacidin	Antileukemic, on clinical trial antiprotozoal	Nucleotide biosynthesis	
18	Iturin A	Dermatomycose treatment, on clinical trial		Bacillomycin, fungimycin
19	Negamycin	Clinical for G⁻ (including *Pseudomonas*-carrying R-factor)	Translational termination, misreading	
20	Neocarcinostatin	Tumor treatment	DNA degradation, DNA synthesis	Zinostatin
21	Netropsin	Antiphage, antiviral, larvicide and molluscicide	DNA complex, inhibits DNA, RNA and protein synthesis	Sinanomycin, distamycin
22	Nisin	Antiprotozoal, antimalarial		Subtilin
23	Nosiheptid			Thiopeptin, thiostrepton
24	Ostreogrycin A + B	Synergistic compounds	Inhibition of cell wall, protein (ribosome binding) and nucleic acid synthesis	Staphylomycins, pristinamycins, mikamycins, vernamycins, synergistins, amphomycin
25	Parvulin			
26	Piperazinedione	Clinical trial (human lymphomas)		
27	Polymyxins	Urinary tract infections, slight nephrotoxic effects	Membrane active	Colistins
28	Saramycetin	Effective antifungal (*Cryptococcus, Blastomyces, Histoplasma*), on clinical trial		
29	Siomycin			
30	Staphylomycin S			Virginiamycin S, ostreogrycin

Table 1 (continued)

No.	G+ antibacterial	G− antibacterial	Mycobacterial	Antifungal	Antitumor	Antiviral	Immunomodulating	Topical antibiotic	Veterinary drug	Feed additive	Food preservative	Chemical reagent	Notes	Molecular mechanisms, targets and actions	Related or identical compounds
31 Subtilin	+		+			+					+				Nisin
32 Thiopeptin	+								+	+					Thiostrepton, nusoheptid
33 Thiostrepton	+								+				Poor absorption, no toxic effects in clinical studies, veterinary drug (bovine mastitis)	Inhibition of protein synthesis	Thiopeptin, nosiheptid
34 Tuberactinomycins			+										Limited use as antitubercular drug, less toxic than 9 and 38	Inhibition of protein synthesis	Enviomycin
35 Tyrocidines	+							+				+	Hemolysis of erythrocytes, only topical		
36 U-42126				+									Antileukemic, clinical trial	Glutamic antagonist	
37 Valinomycin												+		K$^+$ carrier, uncouples oxidative phosphorylation	Aminomycin, amidomycin
38 Viomycin			+										Limited use as antitubercular drug, long term nephrotoxic effects	Inhibition of protein synthesis	Viomycin, vinactane

6.1.1.1 Blasticidin S: an agricultural antibiotic

Blasticidin S, the first successful Japanese agricultural antibiotic, is effective against the rice blast pathogen *Pyricularia oryzae*. Its properties have been summarized in a review on agricultural antibiotics by Misato *et al.* (1977). The benzylaminobenzenesulfonate, being less phytotoxic to the host plant, has been produced industrially.

Blasticidin S is produced by several strains of *Streptomyces*, first isolated from *griseochromogenes* in 1958. It is certainly not a peptide, but a related product, leucylblasticidin S, can be considered a modified dipeptide. Another biogenetically related compound, cytomycin, also exhibits antitumor properties like blasticidin S (Figure 1). The obvious precursors are cytosine, glucose and arginine, as has been shown by *in vivo* incorporation studies. Biochemical studies in cell-free translation systems indicate an interaction with the peptidyl-transferase center. Resistance to blasticidin S in *P. oryzae* may arise from reduced mycelial permeability.

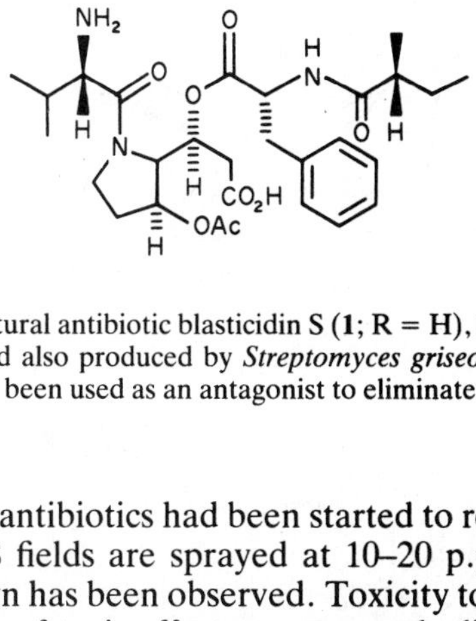

Figure 1 Structures of the agricultural antibiotic blasticidin S (**1**; R = H), leucylblasticidin S (**2**; R = Leu) and cytomycin (**3**), a related antitumor compound also produced by *Streptomyces griseochromogenes*. Detoxin (**4**) from *Streptomyces caespitosus* has been used as an antagonist to eliminate side effects of blasticidin S

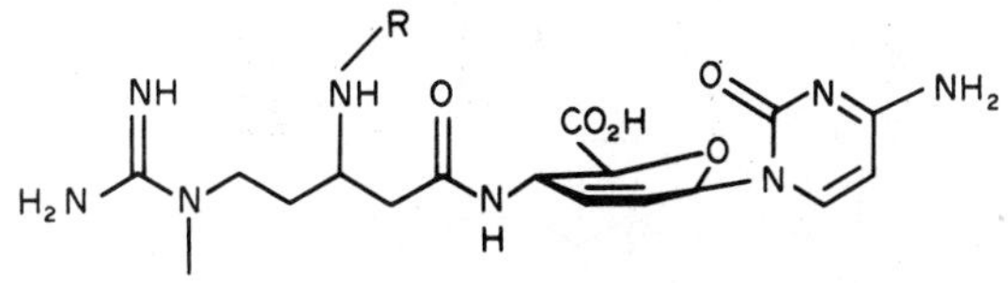

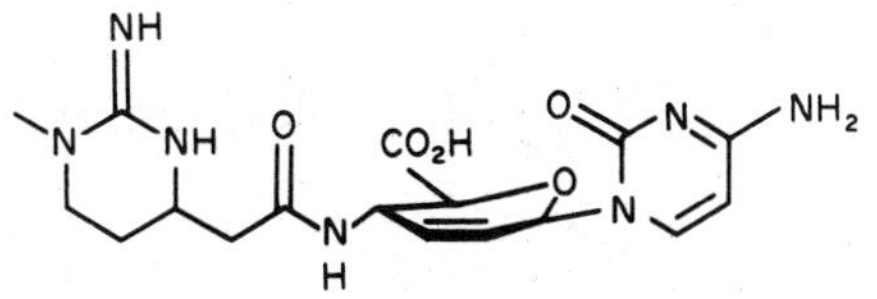

Research for agricultural antibiotics had been started to replace organic mercurials as rice blast controls. With blasticidin S fields are sprayed at 10–20 p.p.m. From studies of environmental metabolism rapid breakdown has been observed. Toxicity to fish is rather low, so it can be used in the paddy field. Elimination of toxic effects on mammals, like conjunctivitis upon accidental eye contact or severe inflammation of mucous membrane or injured skin, has been attempted many times by chemical or enzymatic modifications. The discovery of selective antagonists, the detoxins from *Streptomyces caespitosus*, led to combinations with reduced phytotoxicity and eye irritation properties. At the same time it was found that simple addition of calcium acetate selectively reduced the eye effect, so that this combination is now in use.

6.1.1.2 Bleomycin and bestatin: peptides used in anticancer therapy

Since many antimicrobials are able to suppress or retard the growth of tumors, screening for cytostatics had already started in the 1950s and has led in the meantime to more than 1000 compounds (for a summary up to 1975 see Fuska and Proksa, 1976). As has been pointed out by Oki (1980) the establishment of chemotherapy of cancer and, more recently, applications in combination with surgery, radiotherapy and immunotherapy have led to a remarkable expanding market. Currently the most promising directions are modification of established compounds as well as the application of various *biochemical* screens. Such screens, as has recently been discussed by Aoyagi (1980) for enzyme inhibitors acting on proteinases, esterases or phosphatases, have led to compounds with immunomodulating properties, that may also find application in cancer treatment (Umezawa 1980). To illustrate these concepts the examples of bleomycin and bestatin have been selected.

The structures of several bleomycin-type glycopeptides are summarized in Figure 2. Phleomycins had already been discovered as Cu^{2+}-containing antibiotics by Umezawa *et al.* in the 1950s, and inhibition of Ehrlich carcinoma with a high therapeutic index was found soon afterwards. However, it was not clinically tested since it showed irreversible nephrotoxicity in the dog. The search for phleomycin analogs led to the isolation of bleomycin (BLM) in 1966. It was also a blue Cu^{2+} complex, and Cu^{2+} was essential for its fermentative production from *Streptomyces verticillus*. BLM caused reversible hepatotoxicity, but was not nephrotoxic. Its current main clinical use is in the palliative treatment of squamous cell carcinoma of the head and neck as a mixture of natural BLMs, a 'first generation agent'. In the 1970s hundreds of semisynthetic analogs were prepared by directed biosynthesis and chemical transformation. In order to modify the terminal amine moiety the preparation of bleomycinic acid has been accomplished first enzymatically, and then chemically.

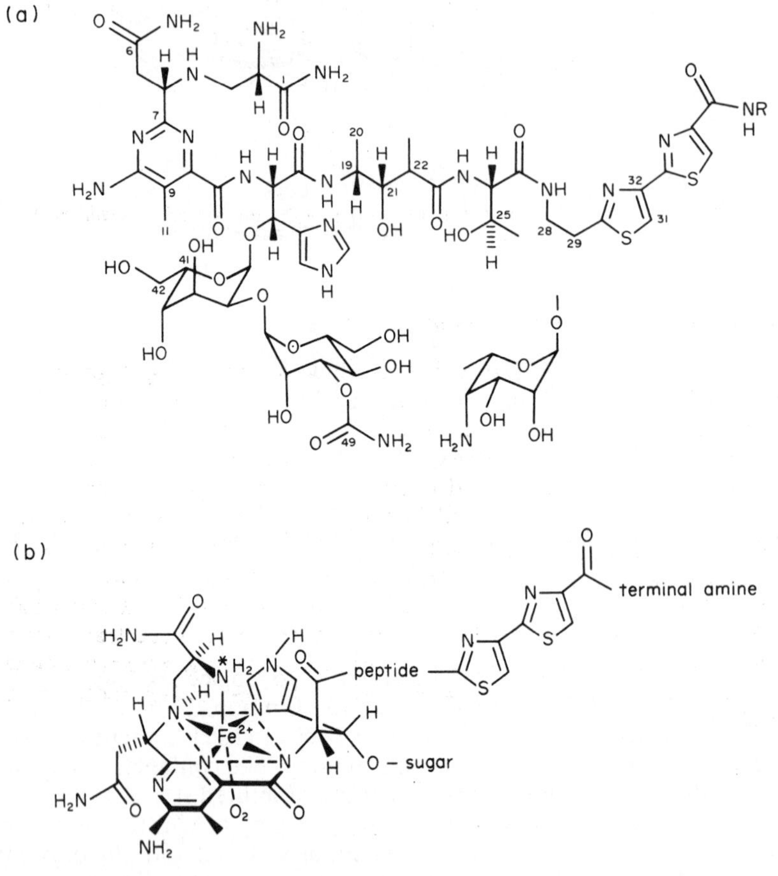

Figure 2 (a) Structures of some bleomycin-type compounds. The structure shown, with R representing various amine compounds like $NH(CH_2)_3SOCH_3$, $NH(CH_2)_3NH(CH_2)_4NH_2$ *etc.*, is bleomycin. In phleomycins the C(31)—C(32) double bond is missing. Zorbamycins (YA-56 x) contain an extra CH_2OH at methyl group 20, an extra methyl group at 25, no double bond 31–32, and no hydroxy group at 42. In YA-56 Y the hydroxy group at 21 and the methyl group at 22 are missing, while an extra methyl is found at 25. In tallysomycin an extra amino sugar is found at 28, a hydroxy group at 29, and the methyl group at 22 is missing. Obviously variations are restricted to certain regions of the compound. (b) Structure of the active Fe^{2+}/O_2 complex involved in DNA double strand scission. The indicated amino group is essential for this reaction, and its removal by BLM-hydrolase leads to inactivation. The hydrolase concentration is low in tumor cells

Since all available proteases did not remove agmatine from BLM B2, a screening was initiated based on the low antimycobacterial activity of bleomycinic acid (5% of BLM B2). An acylagmatine amidohydrolase was then detected in a strain of *Fusarium anguioides*, and has been used in the preparation of semisynthetic BLMs. The first 'second-generation' BLM, peplomycine (with ethyl-3-diaminopropane as terminal amine), introduced clinically in 1981, has several advantages such as high distribution in the stomach, inhibition of gastric carcinoma, and lower lung toxicity; it is now used in the treatment of prostatic cancer.

During their study of mammalian cell membrane enzymes, Umezawa *et al.* investigated the mechanism of influenza virus. They found activities of aminopeptidase, alkaline phosphatase and esterase located on the host cell membrane and transferred to the virus envelope. In a screening for aminopeptidase B, bestatin was obtained from the broth of *Streptomyces olivoreticuli* (Figure 3), a competitive inhibitor also of leucine aminopeptidase, but not of aminopeptidase A or endopeptidases. Labelled bestatin has been used to detect aminopeptidases on the surface of various mammalian cells including lymphocytes and tumor cells. It has been found to enhance the immune response in mice, and to exhibit a suppressive effect on some slowly growing tumors in mice. First clinical studies were reported at the Bestatin Conference in Tokyo in 1979. A metabolite of orally given bestatin, *p*-hydroxybestatin, has been found to be a more effective immunostimulator. These compounds should be useful in cancer treatment, in the annihilation of minimal residual tumors, and in the treatment of infections of immunodeficient patients.

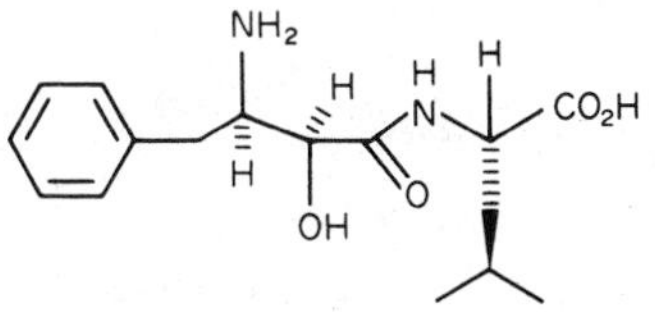

Figure 3 Bestatin, an aminopeptidase inhibitor produced by *Streptomyces olivoreticuli*, with immunoenhancing properties

6.1.1.3 Cyclosporin: an immunosuppressor

Another immunomodulator, not with enhancing but with promising suppressive properties, is cyclosporin. This cyclic 11-peptide was isolated in 1970 from two new strains of Fungi Imperfecti, *Cyclindrocarpon lucidum* and *Tolypocladium inflatum*. Since then, the unusual structures of nine analogs have been established (Figure 4). These and other analogs have been prepared by directed biosynthesis, transformation and chemical synthesis. Since screens for cytostatic and immunosuppressive drugs were developed in the 1970s, immunosuppressive effects of cyclosporin were recognized together with its antifungal properties (narrow spectrum, *e.g. Aspergillus* and *Neurospora*). Extensive clinical studies have been carried out, and conferences on cyclosporin were held in 1981 and 1982 (White, 1982). The peptide has a remarkable affinity for lymphocytes. An interference with an early event during lymphocyte transformation has been suggested. There have been numerous encouraging results in clinical transplantations. However, long-term effects and reversible nephrotoxicity have not been studied in sufficient detail to permit general applications.

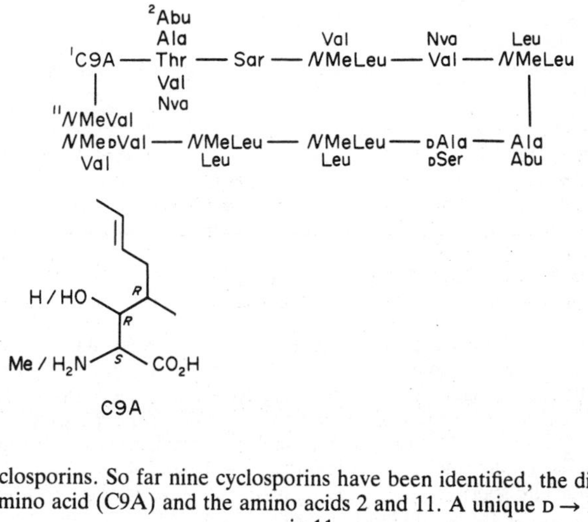

Figure 4 Structures of cyclosporins. So far nine cyclosporins have been identified, the differences being located in the unusual unsaturated C-9 amino acid (C9A) and the amino acids 2 and 11. A unique D → L replacement has been found in 11

6.1.2 Structural Types of Peptides

Peptides and compounds containing peptide structures have been classified by Perlman (1978) into linear and cyclic types, the latter being subdivided according to the mode of ring closure with either peptide or ester bonds. In a systematic approach Bérdy (1980a) lists 77 groups of amino acid derivatives, peptides and peptolides, together with protein and proteide types of antibiotic. This classification is based on structural peculiarities of well-known compounds and contains 780 entries, analogs and double citations included. It has the advantage of being a quite comprehensive compilation, the data being available in computerized form. The user has to be familiar with

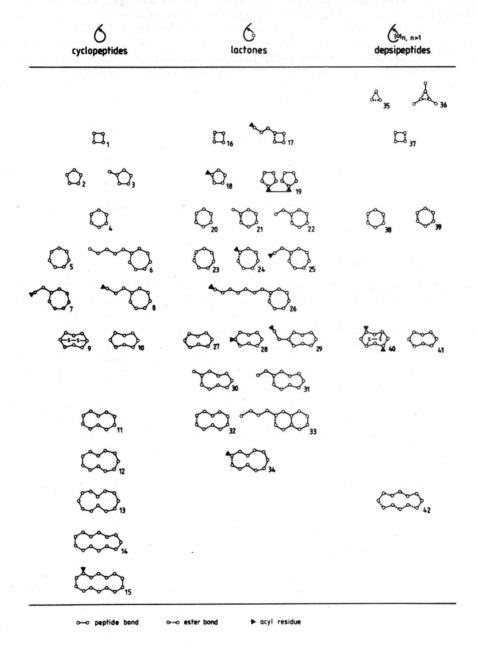

Figure 5 The major structural types of cyclic peptides: 1, tentoxin, cyl-2; 2, cyclochlorotine; 3, capreomycine, tuberactinomycin; 4, destruxin, echinocandin, ferrichrome; 5, ilamycin, ulicylamid; 6, bacitracin; 7, octapeptins; 8, polymyxins; 9, ulithiacylamide; 10, mycosubtilin; 11, tyrocidin, gramicidin S; 12, cyclosporin; 13, gratisin; 14, mycobacillin; 15, enduracin; 16, AM-toxin; 17, ostreogrycin A; 18, brevigellin; 19, actinomycin; 20, isariin, destruxin, globomycin; 21, ostreogrycin B; 22, esperin; 23, A-43-F; 24, etamycin; 25, viscosin; 26, stendomycin; 27, surfactin; 28, lipopeptins; 29, griselimycin, mycoplanecin; 30, LL-AO-341; 31, telomycin; 32, polypeptins; 33, A-21978. 34, brevistin; 35, pyridomycin; 36, enterobactin; 37, angolid, serratamolide; 38, sporidesmolides; 39, enniatins, beauvericin; 40, triostin, echinomycin; 41, bassianolide; 42, valinomycin

the structures of individual groups, but does not have the possibility of searching for similar constituents, ring sizes or modifications. A more general approach has been suggested by us (Kleinkauf and von Döhren, 1981) from the biogenetic viewpoint. A scheme based on the terms linear, cyclic or lactone appears to be of limited use since many compounds carry two or all of these characteristics. Any cyclic structure, however, can be linearized to a precursor structure, which is then modified by enzymatic catalysis. Thus sequence data and constituents could be compared and localized within families or groups. This approach has not yet been realized, and many types of modification reaction have not yet been characterized. The limited number of approximately 300 peptide structures evaluated so far still shows surprisingly little structural variance compared to the synthetic organic chemist's possibilities, but also surprisingly few structural homologies regarding sequence data. To illustrate the limited structural variance the principal types of structure are shown in Figure 5. As far as sequence data are concerned, convenient short notations are not available for all nonprotein amino acids identified so far. For example, more than 10 differently substituted prolines have been described. Considering two different substitutions together with stereochemical arrangement and a possible D-configuration, it is quite difficult to design, say, a three-letter notation compared to the one-letter characterization of protein amino acids. Considering the biochemistry of enzyme systems involved, many more homologies seem to evolve than would be expected from the peptide structure (see Section 6.2.2).

6.2 BIOSYNTHESIS OF PEPTIDE ANTIBIOTICS

6.2.1 Ribosomal and Nonribosomal Mechanisms

From structural considerations the nonribosomal origin of the majority of known peptide antibiotics is obvious. Nonprotein components like D-amino acids and fatty acids, cyclic or modified structures are linked to antibiotic properties. It should be realized, however, that conventional screening procedures using compounds isolated from culture broth and detection of growth inhibitory properties might well miss large peptides not being transported through the membrane in either way. A good reason to select low molecular weight compounds is their possible lack of antigenicity.

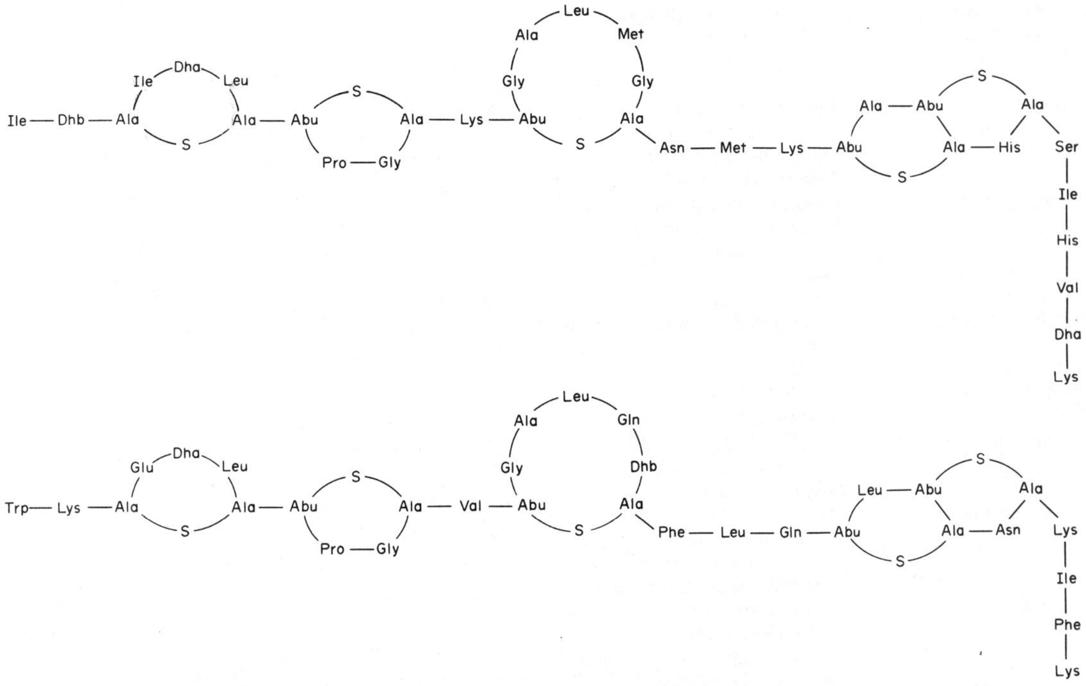

Figure 6 Structures of the food preservatives subtilin (bottom) (*B. subtilis*) and nisin (*Streptococcus lactis*), which are presumably of ribosomal origin

Knowledge of the biogenetic origin of a compound is essential to direct efforts towards cultural improvement and possible directed biosynthesis. Structural genetic manipulations should rely on a detailed knowledge of biosynthetic events. Proposals for manipulations presented so far consider multistep enzymic procedures involving a large number of discrete functions. A single function changed or knocked out could then lead to a modified product. Biosynthetic studies of several peptide antibiotics within the past decade have led to new types of enzymic organization, the multifunctional enzyme type. Although not yet definitely proved, a single gene cluster may provide the message for a single polypeptide catalyzing a multistep sequence of reactions (Kleinkauf and von Döhren, 1982a, 1983a, 1983b). Changes within a single function could then deteriorate the functioning of the multienzyme.

The famous examples of presumably ribosomal origin are nisin and subtilin (Figure 6), two modified peptides of 34 and 32 amino acid residues produced by *Streptococcus lactis* and *Bacillus subtilis* ATCC 6633. The structures have been postulated to originate from an all-L precursor of protein amino acids. Biosynthesis *in vivo* of both nisin and subtilin has been found to be sensitive to chloramphenicol. Ring-closure reactions leading to a compact peptide structure may proceed by inversion with a final D-configuration in the peptide (for references see Kleinkauf and von Döhren, 1981). Both antibiotics have antibacterial (also *Streptococcus*), antifungal, antiprotozoal and antimalarial activity and are used as preservatives in the milk and canned food industries (Lipińska, 1977).

6.2.2 Reactions Involved in Enzymatic Peptide Formation

The topic of enzymatic peptide formation has been reviewed quite extensively (Kleinkauf and Koischwitz, 1978; Kleinkauf and von Döhren, 1981, 1982a). So here, after a brief discussion of functional organization of enzyme systems, we just focus on the types of reaction involved, from the point of view of possible applications in manipulated biosynthesis.

Table 2 Current State of Research in Peptide-forming Enzyme Systems (Kleinkauf and von Döhren, 1981, 1982)

Compound	Organism/source	Activation of precursors		Total enzymatic synthesis	Enzyme studies	
		Adenylate	Phosphate		Preliminary	Advanced
Actinomycin	*Streptomyces antibioticus*	+		−	+	
Alamethicin	*Trichoderma viride*	+		+	+	
Bacitracin	*Bacillus licheniformis* ATCC 10716	+		+		+
Beauvericin	*Beauveria bassiana*	+		+	+	
Bleomycin	*Streptomyces verticillus*			−		
Carnosine	Chick muscle/rat brain		+	+		+
Edeine	*Bacillus brevis* Vm4	+		(+)	+	
Enniatins	*Fusarium oxysporum*	+		+		+
Enterochelin	*E. coli* K-12/*Salmonella typhimurium*	+		+		+
Ergot peptides	*Claviceps purpurea*	+		−		
Ferrichrome	*Aspergillus quadricintus*			+	+	
Folyl-poly-γ-Glu	*Corynebacterium* sp./*E. coli*/rat liver/chinese hamster ovary cells			+		+
Glutathione	*E. coli* K-12, bovine erythrocytes		+	+		+
Gramicidin (linear)	*Bacillus brevis* ATCC 8185	+		(+)	+	+
Gramicidin S	*Bacillus brevis* ATCC 9999	+		+		+
Leupeptin	*Streptomyces roseus*	+		+		+
Mycobacillin	*Bacillus subtilis* B 3		+	+	+	
Penicillin	*Cephalosporium acremonium, Penicillium chrysogenum, Streptomyces clavuligenus*			+	+	
4′-Phosphopantetheine	*E. coli* B/rat liver/*Brevibacterium ammoniagenes*		+	+		+
Poly-γ-D-Glu	*Bacillus licheniformis*			+	+	
Polymyxin	*Aerobacillus polyaerogenes Bacillus polymyxa*	+		+	+	
Tyrocidine	*Bacillus brevis* ATCC 8185	+		+		+

Table 3 Survey of Multienzyme Systems in Peptide Biosynthesis

Peptide structure	Organism	Enzyme[a]	Mol. size (kDa)	Multi-functional	Pantetheine	Activated compounds (*epimerization)	Other functions
Alamethicin (AL) AcAib-Pro-Aib-Ala-Aib-Ala-Gln-Aib-Val-Aib-Gly-Leu-Aib-Pro-Val-Aib-Aib-Glu-Gln-Pheol	Trichoderma viride	AL 1	320	nD	nD	Aib, Pro, Ala, Gln, Val, Glu,	AcCoA + Aib pheol add.
		AL 2	150	nD	nD	Leu, Pro, Aib, Val, Gln, (Glu?)	
Bacitracin (BA) Ile-Cys-Leu-DGlu-Ile-Lys-DOrn-Ile-DPhe Asn-DAsp-His	Bacillus licheniformis ATCC 10716	BA 1	330	+	+	Ile, Cys, Leu, Glu	Cyclization
		BA 2	210	+	+	Lys, Orn*, (Ile ?)	
		BA 3	380	+	+	Phe*, His, Asp*, Asn	
Beauvericin (BE) cyclo (MePhe-DHiv)3	Beauveria bassiana	BE	250	+	nD	Phe, DHiv	Methylation of Phe (trimerization)
Edeine (ED) βTyr-Ise-Dpr-Dha-Gly-spermidine	Bacillus brevis Vm4	ED 1	210	nD	+	βTyr	Spermidine
		ED 2	180	nD	+	Ise, Dpr, Dha, Gly	
		ED 3	100	nD	-	?	
		ED 1A	210	nD	+	βTyr, Dpr	
		ED 1B	160	nD	+	Dpr, Ise, Dha, Gly	
		ED 1C	10	nD	-		
Enniatin B (EN) cyclo (MeVal-DHiv)3	Fusarium oxysporum ETH 1566/9	EN	250	+	+	Val, DHiv	Methylation of Val (trimerization)
Gramicidin A (LG) fVal-Gly-Ala-DLeu-Ala-DVal-Val-DVal-Trp-(DLeu-Trp)3-ethanolamine	Bacillus brevis ATCC 8185	LG 1	160	+	+	Gly, Val	
		LG 2	350	+	+	Ala, Leu*, Val*, Val	
Gramicidin S (GS) cyclo (DPhe-Pro-Val-Orn-Leu)2	Bacillus brevis ATCC 9999 or Nagano	GS 1	100	+	-	Phe*	(Dimerization)
		GS 2	280	+	+	Pro, Val, Orn, Leu	Cyclization
Leupeptic acid (LP) AcLeu-Arg-Arginal	Streptomyces roseus MA 839-A1	LP	260	nD	nD	AcLeu, AcLeuLeu	
Polymyxin (PO) OctanoylDab-Thr-Dab-Dab-Dab-Leu-Leu Thr-Dab-Dab	Aerobacillus polyaerogenes	POx	300	nD	nD	Dab, Leu, Thr	
Tyrocidine A (TY) DPhe-Pro-Phe-DPhe-Asn Leu-Orn-Val-Tyr-Glu	Bacillus brevis ATCC 8185	TY 1	100	+	-	Phe*	
		TY 2	230	+	+	Pro, Phe, Phe*	
		TY 3	450	+	+	Asn, Gln, Tyr, Val, Orn, Leu	Cyclization

[a] Proposed abbreviation.

The current state of research on enzyme systems is summarized in Table 2. Contrary to earlier expectations the multifunctional type of enzyme organization dominates. Amino acid activating units have been found associated together with a transport function 4'-phosphopantetheine that functionally resembles CoA. The tight association is required for proper functioning; if broken up by proteinase, no peptide bond formation is observed, although amino acid activation still proceeds. From characteristics of amino acid activation with regard to specificity towards analogs of ATP as well as amino acids, the bound cofactor pantetheine, and the high molecular weight retained under denaturing conditions, multienzymes forming linear, cyclic, branched cyclic or depsipeptides are very similar, as summarized in Table 3. The main conclusion arrived at for this type of mechanism is that only the direct precursor amino or hydroxy acids can be fed to the system. Only enzyme systems not utilizing stable covalent intermediates accept intermediate peptides.

6.2.2.1 *Carboxyl activation*

Activated carboxyls may be stabilized in aqueous solutions as active esters of CoA, tRNA or enzyme thiol groups. Enzymatic activation reactions characterized to date are summarized in Table 4. Each of the three types of synthetase may be involved directly in nonribosomal peptide formation: acyl-CoA derivatives in *N*-terminal modifications, aminoacyl tRNAs as substrates for interpeptide bridges of bacterial cell walls or aminoacyl transferases, and peptide synthetases as the protein templates directing reaction sequences. The carrier function of CoA or tRNA is fulfilled by enzyme-bound 4'-phosphopantetheine.

Table 4 Types of Carboxyl Activation Reactions (Kleinkauf and Koischwitz, 1982a)

	Type I (AMP)[a]	*Type II (NDP)*[b]
Acid-CoA ligases	Acetate, medium chains (C_4–C_{12}), intermediate chains (C_3–C_7), long chains (C_{12}–C_{18}), benzoate, cholate, luciferin, lysergate, phenylacetate, oxalate, biotin	Succinate, succinate (GDP), medium chains (C_4–C_{12}) (GDP), glutarate
Aminoacyl-tRNA	Ala, Arg, Asn, Asp, Cys, Glu, Gln, Gly, His, Ile, Leu, Lys, Met, Phe, Pro, Ser, Thr, Trp, Tyr, Val	
Peptide synthetases	AcLeu	Pro, DAla
(1) amino acids α-carboxyl	Aib, Ala, DAla, Asn, Cys, Dab, Dap, Glu, Gln, Gly, His, Ile, DLeu, Leu, Lys, Orn, DPhe, Phe, Pro, Thr, Trp, Tyr, Val	
(2) amino acids other carboxyls	Dha, Ise, βTyr	βAla, Gly (γ), DGlu (γ), DAsp (β)
(3) peptides α-carboxyl	AcLeu-Leu	Glu-Cys, UDP-Mur-*N*AcGly (Ala), UDP-Mur*N*Ac-A1-DGlu-A3 (Ay = Lys, Dpm, Dbu, Orn, Hsr)
(4) other carboxyls		UDP-Mur*N*Ac-A1-DGlu (γ)
(5) carboxylic acids	Dhb, pantoate, DHiv	UDP-G*N*Ac-lactate

[a] Carboxyl activation by cleavage of ATP to AMP and PP_i.
[b] Carboxyl activation by cleavage of NTP to NDP and P_i (ATP if not indicated differently).

Activation reactions generally proceed by cleavage of the α,β- or β,γ-bond of an ATP-M^{2+} complex (M = Mg, Mn, *etc.*). As is evident from Table 3, amino acids are predominantly activated by adenylate formation, an intermediate that can easily be detected by the reverse reaction with labelled pyrophosphate to give labelled ATP. Amino acid β- or γ-carboxyls are preferably activated by β,γ-phosphate bond cleavage with formation of intermediate phosphates. Specificities of activation reaction are considerably lower than in protein synthesizing systems. This is already evident from numerous peptide analogs frequently isolated in minor amounts. Thus the branched chain hydrophobic amino acids valine, leucine and isoleucine have been found exchanged in many examples. Quite often different strains show a phenylalanine–leucine exchange. This could indicate that mutational alteration easily leads to this change of specificity.

Several *in vivo* and *in vitro* studies have shown that the product spectrum or a single product may be directed by precursor selection and concentration. Numerous amino acid substitutions have been demonstrated with the gramicidin S enzyme system. Products have been obtained with tyrosines, tryptophan and ring-substituted aromatic amino acids replacing phenylalanine, or arginine and lysine replacing ornithine (Kleinkauf and von Döhren, 1982b). Such product analogs generally have properties similar to those of the main component; however, some analogs could be used with advantage in later modification reactions.

6.2.2.2 Peptide bond formation

Peptide bond formation proceeds by enzymatic catalysis from the activated carboxyl intermediate by addition of an amino group of either amino acid or peptide origin. According to a definition proposed by Lipmann (1968) the activated carboxyl *donor* carrying the leaving group adenylate, phosphate, tRNA or enzyme-thiol undergoes an electrophilic addition with the *acceptor* amino (imino) group. Types of reaction described so far have been compiled in Table 5. The most important types are the elongation reactions of the ribosomal or multienzymic systems.

Table 5 Peptide Bond Formation — Reaction Types

			'Activated carboxyl' donor	*'Amino' acceptor*	*Examples*
'Free' soluble products	1.1	Acyl transfer	Acyl-CoA	Amino acid peptide	Acyl transferases (E.C. 2.3.1)
	1.2	Synthetases	Acyl-P	Amino acid	Ala-adding enzyme (cell wall)
		Noncovalent	Acyl-AMP	Amino acid	Pantothenate synthetase
		intermediates	Aminoacyl-P	Amino acid	DAla-DAla-ligase
		and products	Aminoacyl-AMP	Amino acid	Leupeptine acid synthetase
			Peptidyl-P	Amino acid	Glutathione synthetase
			Peptidyl-AMP	Amino acid	Leupeptin acid synthetase
			Peptidyl-P (?)	Peptide	DAla-DAla adding enzyme (cell wall)
Covalently bound products	1.3	Aminoacyl transfer	Aminoacyl-tRNA	Peptide	Aminoacyl transferases (E.C. 2.3.1)
	1.4	Cyclization reaction	Peptidyl-S-enzyme	Peptide	Gramicidin S, termination, piperazinedione formation
	2.2	Synthetases Ribosomal	Aminoacyl-tRNA	Aminoacyl-tRNA	Ribosomal initiation
		Covalent	f-met-tRNA	Aminoacyl-tRNA	Ribosomal initiation
		intermediates	Peptidyl-tRNA	Aminoacyl-tRNA	Ribosomal elongation
		Nonribosomal	Aminoacyl-S-enzyme[1]	Aminoacyl-S-enzyme[1]	Bacitracin, initiation
			Aminoacyl-S-enzyme[1]	Aminoacyl-S-enzyme[2]	Gramicidin S, initiation
			Formylaminoacyl-S-enzyme	Aminoacyl-S-enzyme	Gramicidin, initiation
			Acylaminoacyl-S-enzyme	Aminoacyl-S-enzyme	Polymyxin, initiation
			Peptidyl-S-enzyme	Aminoacyl-S-enzyme[1]	Peptide synthetases, elongation
			Peptidyl-S-enzyme	Peptidyl-S-enzyme	Gramicidin S, termination

Activated amino acids are acceptors of activated peptide donors. Thus the product remains activated, ribosome-bound or enzyme-bound. When the acceptor is not activated, the product is free or soluble, and may be activated for further elongation (glutathione, leupeptin). Obviously the reaction of enzymic intermediates with an amine compound is a termination with product release (see Section 6.2.2.3).

6.2.2.3 *Modification reactions*

In analogy to the formylmethionyl starter of ribosomal translation, formyl- or acyl-amino acids frequently serve as starter in enzymatic peptide formation. The compounds are then activated and transferred to the acceptor amino acid. The amino-terminal acylation is the major modification reaction, and is often of surprisingly low specificity. Thus pepstatin analogs have been isolated with acyl moieties ranging from C_2 to C_{20} (Aoyagi, 1978). The acylation site is a candidate for enzymatic deacylation and modification by acylases.

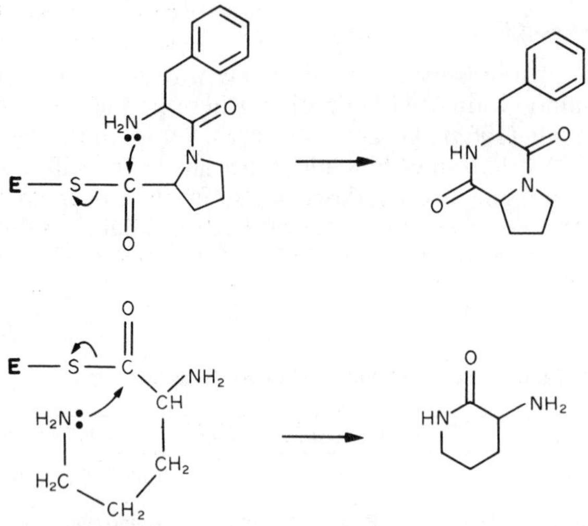

Figure 7   Stereochemically favored side reactions of gramicidin S-synthetase. *Top*: Cyclization of enzyme-bound D-phenylalanylproline to the 2,5-piperazinedione. Bottom: Cyclization of thiol-ester bound L-ornithine to the β-amino-α-piperidone

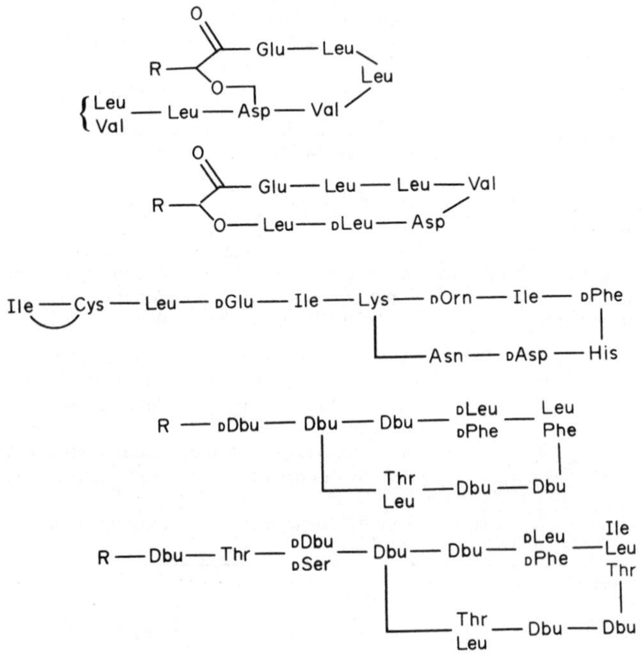

Figure 8   Cyclic peptide structures that indicate enzymatically directed cyclization of linear peptide; esperin (**1**) and surfactin (**2**) have an identical peptide sequence, although optical configuration of esperin amino acids is not yet determined. Esperin (**1**), produced by a strain of *B. mesentericus*, is a hemolytic antibiotic, while (**2**), produced by *B. subtilis* strains, has been described as a clotting inhibitor in the thrombin–fibrinogen reaction, and has recently been redetected as an inhibitor of cyclo-AMP diesterase. (**3**)–(**5**) are the seven-membered ring structures of bacitracin (*B. licheniformis*), octapeptins (*B. circulans*) and polymyxins (*B. polymyxa, B. colistinus*)

The origin of D-amino acids still remains largely a matter of speculation. In the *Bacillus brevis* systems producing gramicidin S and tyrocidines both L- and D-phenylalanine are activated. The L-form is enzyme-bound as D,L-amino acid (D/L as 2/1), and only the D-form serves as donor for amino (imino) groups. The epimerization seems to involve an exchange of the α-H with an enzymic proton. Some of the known activation sites of multienzyme systems leading to D-amino acids activate both isomers while others accept only the L-form. On a few occasions, the L- as well as the D-configuration has been found (cyclosporins).

Methylation of amino groups is a major modification, the methyl group being derived from *S*-adenosylmethionine (SAM). In the only available enzymic study of enniatin synthetase (Zocher *et al.*, 1982) activated valine is methylated by SAM and then incorporated into the depsipeptide. Omission of SAM *in vitro* leads with a tenfold reduced rate to presumably unmethylated compounds.

Most carboxy-terminal modifications can be considered termination reactions releasing enzyme-bound activated carboxyls, and thus include several types of cyclization reaction. Obviously, amine compounds like spermidine or ethanolamine (edeine or gramicidin) could act as chain terminators. Stereochemically favored cyclization reactions are ornithyl-release or piperazinedione formation (Figure 7, gramicidin S-system). Generally cyclization reactions appear to be strictly enzymatically controlled, since only one specific product is formed, although several reactive groups are present (Figure 8).

Table 6 Directed Biosynthesis of Peptides

Compound	Organism	Amino acid/ compound replaced	Amino acid/ compound incorporated	Ref.
Actinomycin	*Streptomyces antibioticus* *Streptomyces chrysomallus*	2′-DVal 3′-Pro	Ile Sarcosine Pipecolic acid Azetidine-2-carboxylic acid 4-Methyl/fluoro/chloro/ bromo-Pro Thioproline	Meienhoffer and Atherton (1973), Katz (1974), Kleinkauf and von Döhren (1981)
		5′-Val	Ile	
Bleomycin	*Streptomyces verticillus* ATCC-15003	Terminal amine	Various amine compounds	Takita and Maeda (1980)
Cyclosporins	*Tolypocladium inflatum Gams*	Abu/Thr	Ala Val norVal	Kobel and Traber (1982)
Enniatins	*Fusarium oxysporum*	Val	Ile/Leu	Madry *et al.* (1983)
Ergot peptides	*Claviceps purpurea*	Phe Val/Ile/Leu	*p*-Fluoro/chloro-Phe norVal/norLeu, α-Abu 5,5,5-Fluoro-Leu β-OH-Leu	Kleinkauf and von Döhren (1981)
Gliotoxin	*Trichoderma viride*	Ser	Ala	Kleinkauf and von Döhren (1981)
Gramicidin S	*Bacillus brevis* ATCC 9999	Phe	*p*-Fluoro-Phe Thienylalanine	Katz and Demain (1977)
		Orn	Lys	Kleinkauf *et al.* (unpublished)
Tyrocidine	*Bacillus brevis* ATCC 8185	Phe/Tyr	Trp Thienylalanine	Katz and Demain (1977)
		Leu/Val	Ile	
Viridogrisein	*Streptomyces griseoviridus*	OH-Pro	Pro	Kleinkauf and von Döhren (1981)

Some carboxyl modifications are reductions to aldehydes (leupeptin) or alcohols (peptaibophols).

Cyclization reactions to small four- or five-membered rings have not been studied in detail. For β-lactam-type compounds some details can be found in the preceding chapter. Most prominent are cysteine-derived thiazole-type structures (thiostrepton, bacitracin).

6.3 PRODUCTION OF PEPTIDES

6.3.1 Screening Methods

As has been summarized by Bérdy (1980b), microbiological selection methods (culture preservation, pretreatment of soil samples with inhibitors) and new biochemical assay procedures have been successfully introduced. The classical antimicrobial screens can be supplemented with specific procedures for cell-free synthesis of protein, nucleic acid or cell wall, membrane permeability, or enzyme inhibition. Antitumor drugs should be more easily detected by biochemical screens, such as inhibition of enzyme systems involved in tumor growth (proteinases, phosphatases, esterases), antimetabolite screens, immunostimulants or replication inhibitors.

6.3.2 Biotechnological Production Methods

Since most compounds of interest here are microbial secondary metabolites their production is carried out by conventional batch and continuous fermentation procedures. Most important is the medium composition and controlled growth conditions (*e.g.* by oxygen supply, pH or redox control). An efficient immobilized cell procedure has been devised for bacitracin production from *Bacillus licheniformis* (*e.g.* Morikawa *et al.*, 1980). Peptides are not always excreted into the medium, but may be isolated from the cell-mass. However, changes in medium composition or mutational alteration could afford peptide release.

It has been pointed out by Campbell (1983) that several fungal metabolites are not produced in submerged culture, but need surface cultivation.

A few promising examples of enzymatic peptide production have been carried out (von Döhren, 1982). If a simple multienzyme system is involved, an enzyme system could be used with advantage in the production of peptide analogs. Genetic and/or fermentation improvement of enzyme levels are generally required (Kleinkauf and von Döhren, 1983a).

6.3.3 Improvements and Modification Procedures

As has been pointed out by Demain (1973) the production of secondary metabolites appears to be affected by the same regulatory mechanisms that control primary metabolism: induction, feed back and catabolite regulation. Thus conventional mutational improvements have led to 100- to 1000-fold increases of productivity of a metabolite. The manipulation of fermentational procedures (selection of carbon or nitrogen source, restriction of growth rate by control of pH or oxygen uptake) may also lead to significant improvements (see *e.g.* Vandamme, 1981: Flickinger and Perlman, 1979). No detailed molecular genetics are available, and a possible extrachromosomal location of genes involved in antibiotic production has been studied. At least in peptide antibiotics (actinomycin, gramicidin S) evidence favors a chromosomal location, although DNA intercalators cause the formation of nonproducer mutants with a high rate (Marahiel *et al.*, 1979).

It is evident from biogenetic studies that amino acids as direct precursors are important tools directing product concentration and composition. No detailed studies have been carried out on amino acid pools, uptake or toxicity of certain amino acid analogs. The potential of the method has by no means been fully exploited, especially perhaps in the case of the amine modification of bleomycins. Obviously the limitation is the availability of structural analogs as well as the unclear prediction of biological activities of the peptide analogs. The most well documented cases have been summarized in Table 6.

6.4 COMPILATION OF PEPTIDES

This table of bioactive peptides is similar to a compilation by Perlman (1978), which has been updated and extended. Data on modified amino acids, proteins and largely unknown structures, and from outdated references have been omitted. Besides a list of alternative names to avoid repeated citations, the gross peptide structures have been assigned if available, and references have focused on most recent work and patents (most patent numbers have been obtained from Bérdy (1980)). Since penicillins and cephalosporins are treated separately in this volume data on these modified tripeptides have not been entered in the table.

Abbreviations Used in the Table

Activities. G$^+$/G$^-$, Gram positive/negative; MY, mycobacterial; AF, antifungal; AT/AV, antitumor/antiviral.

Producers. Systematic names: *S.*, *Streptomyces*; *Stv.*, *Streptoverticillum*; *B.*, *Bacillus*; *Noc.*, *Nocardia*; *Act.*, *Actinomyces*.

Structural types. If no structure is available a given name refers to a type of compound or structure with similar properties. To describe peptide structures, the following system is used: P = peptide, 2-P = dipeptide, *etc.*; C = cyclopeptide, where the given number refers to the number of amino acids in the cyclic structure; thus 10-P-7-C means a decapeptide structure containing a heptapeptide ring; L = lactone (a cyclic structure closed with an ester bond); D = depsipeptide (a peptide structure also containing ester bonds); the type of bonds other than α-peptide bonds is given as β, γ, δ or ε, while O refers to ester bond, and U refers to a urea-type bond (—NHCONH—) between two amino groups. All types of modifications are not differentiated and marked M (cyclizations, methylations, glycosylations, *etc.*).

Alternative Names and Synonyms of Compounds Listed in the Table

A-128 (telomycin), A-3302-B (TL-119), A-19009 (fumarylcarboxamido-L-2,3-diaminopropionyl-L-alanine), A-30641 (aspirochlorine), acinoleukin (echinomycin), aerosporin (polymyxin), aizomycin (bicyclomycin), alazopeptin (duazomycin), allomycin (amicetin), amanitin (amatoxins), amidomycin (valinomycin), AMPBA (phosphinothricyl-alanylalanine), AM-toxin (alternarolide), APD (surfactin), aspartocin (amphomycin), avenacein (nikkomycin), ayfivin (bacitracin), bacillin (bacilysin), bialaphos (phosphinothricyl-alanylalanine), blastmycin (antimycins), bleomycin S (phleomycins), bryamycin (thiostrepton), BU-2470 (octapeptins), circulins (polymyxins), cirratiomycin (antrimycin), cleomycin (bleomycin), colistins (polymyxins), cyclohexenyl-1-glycine (nikkomycin), E-49 (octapeptins), enterochelins (enterobactins), enviomycin (tuberactinomycin), eryzomycin (pyridomycin), F-1370-A (etamycin), F-1370-B (griseoviridin), florimycin (viomycin), fructigenin (enniatins), furanomycin (threomycin), gliovirin (FA-2097), glumamycin (amphomycin), grisein (albomycin), hydroxyviomycin (tuberactinomycin), K-16 (malonomycin), KM-208 (bacilysin), koluopthisin (viomycin), lanosulin (fumitremorgin), lateritin (enniatins), levomycin (echinomycin), matamycin (althiomycin), melinacidin (11,11-dihydroxychaetocin), mikamycin A (ostreogrycin A), mikamycin B (ostreogrycin B), mutabicillin (siomycin), multhiomycin (nosiheptide), myxoviromycin (amidinomycin), neopeptin (lipopeptin), neotelomycin (telomycin), ostreogrycin (pristinamycin), P-168 (1907), PA-114-A (ostreogrycin A), PA-114-B (ostreogrycin B), phtiomycin (viomycin), pilosomycin (ferrimycin), pristinamycin (ostreogrycin), polymycin (streptothricin), quinomycin (echinomycin), racemomycin (streptothricin), ristocetin (ristomycin), RP-9671 (nosiheptide), RP-11072 (grisellimycin), rufomycin (ilamycin), S-300 (viomycin), S.A. (ostreogrycin), sacromycin (amicetin), sambucin (enniatin), SF-1293 (phosphinothricyl-alanylalanine), SF-1902 (globomycin), sinanomycin (netropsin), siomycin A (A-59-A), sporangiomycin (siomycin), staphylomycin M (ostreogrycin A), stilbellin (antiamoebin), stravidin (MSD-235), streptogramin A (ostreogrycin A), streptolin (streptothricin), streptolysin (surfactin), subtilysin (surfactin), synergistin A (matchamycin), synergistin (ostreogrycin), tatemycin (thiopeptin A), tetaine (bacilysin), thermomycetin (theiomycetin), thiactin (thiostrepton), tuberactin (tuberactinomycin), tuberactinomycin (viomycin), U-27810 (berninamycin), U-43946 (CC-1014), vernamycin (ostreogrycin), virginiamycin (ostreogrycin), viridogrisein (etamycin), vivicil (fluvomycin), WF-3161 (FR 900261), wildfire toxin (tabtoxin), WS-4545 (bicyclomycin), X-5079-C (saramycetin), YA-56-X (zorbamycin), 333-25 (octapeptin), 657-A-2 (echinomycin), 899 (ostreogrycin), 1491 (echinomycin), 2928 (5590), 5520 (echinomycin).

Table 7 Compilation of Peptides

Name	Producers	Structural type/ characterization	G⁺	G⁻	My	AF	AT AV	Others/uses	Ref.
A 43 F	Fungus ATCC 20529	7-L (δ)				+			*US Pat.* 4 254 224
A-59-A	*S. sp.*	Thiostrepton	+		+				*J. Antibiot.,* **14A**, 194 (1961)
A 287	*Actinoplanes utahensis*	Cyclopeptide	+						*US Pat.* 3 824 305, *Ger. Pat.* 2 402 956
A 1787	*S. sp.*	Albomycin	+	+					*Science,* **125**, 357 (1957)
A 2315	*Actinoplanes philippinensis, Actinoplanes flava*	C-2-P-M	+						*J. Antiobiot.,* **30**, 199 (1977), *J. Chem. Soc., Perkin Trans. 1,* 2464 (1977)
A-3302-A	*B. subtilis*	R-7-P-4-L	+		+				*Jpn. Pat.* 75 106 492 (*Chem. Abstr.,* **85**, 3872)
A 4992	*S. kentuckensis*	2-P-M	+	+					*US Pat.* 3 629 405
A-7413-A	*Actinoplanes sp.*	Thiostrepton	+			+			*Ger. Pat.* 2 103 938, *Belg. Pat.* 850 899
A-16686	*Actinoplanes sp.* ATCC 33076	Glycopeptide-M	+						*J. Antibiot.,* **37**, 309 (1984), *Jpn. Kokai* 82 54 592
A 19009	*S. collinus*	R-2-P	+			+			*J. Antibiot.,* **36**, 784 (1983), *US Pat.* 3 832 287
A 21978	*S. roseosporus* NRRL 11379	R-13-P-10-L	+					Growth promoter	*US Pat.* 4 331 594, 4 208 403 *Jpn. Kokai* 80 92 353
A 22765	*S. aureofaciens*	Succinimycin	+						*Arch. Mikrobiol.,* **68**, 107 (1969), *US Pat.* 3 155 578, 3 211 246, *Ger. Pat.* 1 129 252
A 30912	*Asp. nidulans, Asp. rugulosus,* var. *roseus, A.* 42355, NRRL 8113	R-C-6 (δ)				+			*US Pat.* 4 074 246, 4 074 245, 4 288 549, 4 293 490, 4 293 491, *Ger. Pat.* 2 643 485
AB-1	*B. circulans*	Octapeptin	+	+					*Jpn. Pat.* 76 118 828
AB 97	*Actinomadura helvata*	Thiostrepton	+						*Jpn. Pat.* 78 101 301
Acetylleucyl- argininal	Bacterium, unidentified	M-2-P						Inhibitor dipeptidyl aminopeptidase	*J. Antibiot.,* **37**, 680 (1984)
Actaplanin	*Actinoplanes missouriensis*	7-P-M	+					Growth promoter	*J. Antibiot.,* **37**, 85 (1984), *Jpn. Kokai* 82 128 689
Actinobolin	*S. griseoviridus*	M-2-P	+				+		*J. Am. Chem. Soc.,* **90**, 1087 (1968)
Actinoidin	*Nocardia actinoides*	7-P-M	+						*Antibiotiki,* **6**, 609 (1961), *Tetrahedron Lett.,* 2861 (1979)
Actinomycins	*S. antibioticus, S. chrysomallus, S. parvulus* and various *S. sp.*	R-(5-L)₂	+				+		*Adv. Appl. Microbiol.,* **17**, 203 (1974)
Actinonin	*S. felis, S. roseopallidus*	M-3-P	+	+					*US Pat.* 3 240 787, *J. Chem. Soc., Perkin Trans. 1,* 819 (1975)

Name	Code / synonym	Producer organism						Remarks	References
Actinothiocin	Althiomycin	*Actinomadura pusilla*	+						*J. Antibiot.*, **26**, 343 (1973), Jpn. Pat. 73 78 692
Aculeacins	Echinocandin	*Asp. aculeatus*		+				Fungal cell wall formation (glucan)	*J. Antibiot.*, **30**, 297, 303, 308 (1977), **35**, 203 (1982), US Pat. 3 978 210, Ger. Pat. 2 509 820, Belg. Pat. 826 393
Alamethicin	R-19-P-M	*Trichoderma viride*	+	+					*J. Am. Chem. Soc.*, **99**, 8469 (1977)
(S)-Alanyl-3-(α-(S)-hydroxy-2-oxo-3-azetidinylmethyl)-(S)-alanine	2-P	*S. sp.*	+	+					*J. Antibiot.*, **28**, 1 (1975), US Pat. 3 901 880, 3 956 067
AL-Antibiotic		*B. subtilis*	+				+		*Kuchne Archiv.*, **74**, 19 (1960)
Aboleutin	Bacillomycin	*B. subtilis* AF 8	+		+			Plant phyt. fungi	*J. Antibiot.*, **33**, 758 (1980)
Albomycin	Sideromycin	*S. griseus, S. subtropicus*	+	+	+				*J. Antibiot.*, **24**, 830 (1971), *Eur. J. Appl. Microbiol.* **5**, 51 (1978), US Pat. 2 505 053
Albonoursin	C-2	*S. noursei, Act. albus*	+		+				*Antibiotiki*, **8**, 201 (1963), *Tetrahedron Lett.*, 1881 (1963), US Pat. 2 516 267
Alboverticillin		*S. alboverticillatus*	+			+			*J. Antibiot.*, **11**, 30 (1958), Jpn. Pat. 60 10 997
Almarcetin		*S. albus*	+		+	+			*Antimicrob. Agents Chemother.*, 53 (1964)
Alternarolide	C-4-L	*Act. mali*	+		+				Jpn. Pat. 75 123 881, *Chem. Lett.*, 1411 (1977)
Althiomycin	M-4-P	*S. althioticus, S. matensis, S. bellus, Cystobacter fuscus*	+						*J. Antibiot.*, **27**, 897 (1974), **35**, 635 (1982)
Alvein		*B. alvei*	+						*Br. J. Exp. Path.*, **30**, 209, 214 (1949)
Alveomycin	Sideromycin	*S. sp.*	+	+					*Med. Chem.*, **7**, 528 (1963)
AM-630	Basic peptide	*S. lavendofoliae*							Jpn. Kokai 80 118 500
AM 2504		*S. sp.*							*Agric. Biol. Chem.*, **41**, 1827 (1977)
Amatoxins	C-8-M	*Amanita* sp., *Galerina* sp.						Toxins	*Crit. Rev. Biochem.*, **5**, 185 (1978)
Amicetin	Glycopeptide M-2-P	*S. vinaceus-drappus, S. fasciculatus, S. sacromyceticus*	+						*J. Am. Chem. Soc.*, **80**, 743 (1958)
Amidinomycin	M-aa	*S. sp., S. kasugaensis, S. flavochromogenes*	+						*J. Antibiot.*, **14**, 103, 163, 251 (1961), Jpn. Pat. 61 4598
Amphomycin	R-11-P-C-2	*S. canus, S. violaceus, S. albogriseolus*	+				+	Peptidoglycan synth. inhib. feed additive	*J. Am. Chem. Soc.*, **95**, 2352 (1973), Br. Pat. 736 325, Ger. Pat. 1 021 538, Can. Pat. 494 191, *Biochemistry*, **20**, 1561 (1981)

Table 7 *(continued)*

Name	Producers	Structural type/characterization	G⁺	G⁻	My	AF	AT AV	Others/uses	Ref.
AN-1	S. albus AJ 9003	Basic poly-peptide						Highly mutagenic	J. Antibiot., **37**, 27 (1984)
AN-7	S. griseoincarnatus	Acidic poly-peptides 16-P							J. Antibiot., **37**, 20 (1984)
Ancovenin	S. sp.							Inh. angiotensin converting enzyme	J. Antibiot., **36**, 1295 (1983)
Angolide	Hitomyces sacchari	C-4						Protection of spore surface	Biochem. J., **98**, 8 (1966), Tetrahedron Lett., 3313 (1964)
Anthelvencin	S. venezuelae	Netropsin M-3-P	+	+			+		Antimicrob. Agents Chemother., 789, (1965), US Pat. 3 467 750
Anthramycin	S. refuineus var. thermotolerans		+	+		+	+		J. Am. Chem. Soc., **90**, 5641 (1969)
Antiamoebin	Emericellopsis salmosynnemata, E. poonensis, Cephalosporium, pimprina, Stilbella sp.		+	+				AP	J. Am. Chem. Soc., **99**, 5203 (1977), Ger. Pat. 1 467 945, US Pat. 3 657 419, J. Antibiot., **31**, 241 (1978)
Antimycins	S. kitasawaensis, S. griseus, S. blastmyceticus	M-2-P				+			J. Antibiot., **25**, 373 (1972), Adv. Biotechnol., **3**, 37 (1980)
Antipain	S. michiganensis, S. yokosukaensis	4-P(U)-M						Proteinase inhibitory	J. Antibiot., **25**, 263, 267 (1972)
Antrimycin	S. xanthocidicus, S. cirratus	7-P			+				J. Antibiot., **34**, 1615 (1981), Agric. Biol. Chem., **46**, 865, 1861, 1885 (1982), Jpn. Pat. 81 144 306
AO 341	S. candidus	M-C-2	+		+				US Pat. 3 377 244
Aranotin	Arachniotus aureus	3-P				+	+		Chem. Commun., 359 (1969)
Arginyl-d-allothreon-yl-L-phenyl-alanine	Keratinophyton terreum, Trichophyton indicum								Ber., **106**, 816 (1973)
Arphamenines	Chromobacterium violaceum	2-P-M						Aminopeptidase B inhibitor	J. Antibiot., **37**, 518 (1984), **36**, 1572 (1983)
Arsimycin	S. arsitiensis, S. roseus	Leucinamycin C-8	+	+		+			Ger. Pat. 1 090 380
Ascidiacyclamide	Unidentified ascidian							Cytotoxic	J. Chem. Soc., Chem. Commun., 323 (1983)
ASK-753	S. sp.	Pseudosidero-mycin	+	+		+			J. Antibiot., **27**, 874 (1974), Ind. J. Biochem., **8**, 723 (1977)
Aspartyl-N-hydroxy-aspartyl-D-cyclo-serine	Corynebacterium kutscheri	3-P-M							J. Bacteriol., **137**, 243 (1979)

Name	Producing organism	Code					Property/Use	References
Asperchromes / Aspirochlorine	*Asp. ochraceous* / *Asp. oxyzae*	C-6 / C-2-M					Iron transport	*J. Bacteriol.*, **158**, 683 (1984); *Agric. Biol. Chem.*, **47**, 2673 (1983)
Aspochracin	*Asp. ochraceus*	R-C-3 (γ)					Insecticide	*Jpn. Pat.* 71 21 789; *Agric. Biol. Chem.*, **33**, 1491, 1501 (1969)
Athlestain / Austamid	*Asp. niger* / *Asp. ustus*	Echinocandin / M-C-2					Toxic	*Jpn. Pat.* 66 12 688; *Tetrahedron Lett.*, 3331 (1971)
Avoparcin	*S. candidus*	7-P-M	+				Feed additive	*Annu. Rev. Microbiol.*, **38**, 339 (1984)
Azomultin / B-7	*S. noboritoensis* / *B. subtilis*	Kikumycin / Bacillomycin	+	+				*Jpn. Pat.* 70 6073; *J. Appl. Bacteriol.*, **23**, 114 (1960)
B-43 / B-344 / B-52653	*B. circulans* / *B. subtilis* / *S. albulus* ATCC 31713	Polypeptin / Cyclopeptide / 2-P-M	+ +	+ +	+ +		Phyt. fungi, inh. collagen-Pro-hydroxylase	*J. Antibiot.*, **29**, 814 (1976); *J. Chem.*, **1**, 135 (1963); *Jpn. Kokai* 82 54 151
Bacillomycins	*B. subtilis*	C-8 (β)	+	+				*Eur. J. Biochem.*, **77**, 61 (1977), **118**, 323 (1981); *J. Antibiot.*, **35**, 306 (1982); *Fr. Pat.* 2 508 766
Bacillocin / Bacilysin	*B. subtilis antiblasti* / *B. subtilis*, *B. pumilus*	Bacillomycin / 2-P	+					*Jpn. Pat.* 61 1149; *J. Gen. Microbiol.*, **94**, 23, 46 (1976); *J. Antibiot.*, **34**, 1608 (1981)
Bacitracin	*B. licheniformis*, *B. subtilis*	12-P-C-7 (ε)	+				Feed additive	*Pharmacol. Ther.*, **16**, 199 (1978)
Bassianolide	*Beauveria bassiana*, *Verticillium lecanii*	C-8-D	+				Insecticide	*Tetrahedron Lett.*, 2167, 4049 (1977); *Agric. Biol. Chem.*, **42**, 629 (1978)
BBM 928-A-D	*Actinomyces* G 455-101	Quinoline depsipeptide, R(4-L)$_2$, 6-P	+			+	DNA binding	*J. Antibiot.*, **33**, 1087 (1980); *US Pat.* 4 360 458
BE-4	*Microcystis aeruginosa*						Animal toxin	*J. Chem. Soc., Chem. Commun.*, 652 (1983); *Aust. J. Chem.*, **31**, 1397 (1978)
Beauvericin	*Beauveria bassiana*, *Paecylomyces fumoroseus*, *Polyporus sulphureus*	C-6-D	+	+				*J. Chem. Soc., Perkin Trans.* 1, 2878 (1980); *Tetrahedron Lett.*, 2791, (1978); *J. Gen. Microbiol.*, **128**, 875 (1982); *US Pat.* 4 001 090, *Jpn. Pat.* 70 17 599
Beauverolide	*Beauveria bassiana*	4-L						
Berninamycin	*S. berniensis*, *S. verticillus*	13-P-C-12-M	+					

Table 7 (*continued*)

Name	Producers	Structural type/ characterization	G+	G−	My	AF	AT AV	Others/uses	Ref.
Bestatin	S. olivoreticuli	2-P (β)					+	Proteinase inhibitor, immunoenhancer	J. Antibiot., **29**, 100, 600 (1976), **36**, 695 (1983), Ger. Pat. 2 628 354, Jpn. Pat. 77 116 435
Bicyclomycin	S. sapporoensis, S. trebensis	C-2-M	+	+				Feed additive, inhibitor of lipoprotein synthesis (cell wall)	J. Antibiot., **26**, 155 (1976), US Pat. 3 784 447, Ger. Pat. 2 150 593, 2 501 958, Jpn. Pat. 77 108 092, 77 108 093
Blasticidin S	S. griseochromogenes, S. globifer, S. morvokaensis	2(3)-P-M (β)				+			Tetrahedron Lett., 3785 (1966)
Bleomycin	S. verticillus	R-8-P-M (γ, β)	+	+	+	+	+		J. Antibiot., **36**, 92 (1983), Jpn. Pat. 65 8117, Ger. Pat. 2 307 986
BN-240	Erwinia sp. BN-240	Basic peptide	+						Jpn. Kokai 81 5495
BN-7	B. circulans	Polypeptin	+	+					Jpn. Pat. 73 56 895
BN-109	B. polymyxa	Tridecaptin	+	+		+			Ger. Pat. 2 344 932, US Pat. 3 940 479
Botrycidin Botryticidin Bottromycins	B. subtilis NRRL B 12231 B. subtilis AS 1316 S. bottropensis itasatoa kanaensis	62-P 7-P-C-4-M	+	+		+ +		Antimycoplasma	Jpn. Pat. 79 130 168 J. Antibiot., **36**, 1076 (1983), Br. Pat. 762 736
Bouvardin	Bouvardia terniflora	C-6-M					+	Antitumor	Chem. Abstr., **93**, 145 132; **92**, 191 361
Bresein	B. brevis	Tyrocidin	+						Dokl. Akad. Nauk SSSR, **204**, 465 (1972)
Brevigellin Brevistin	Penicillium brevicompactum B. brevis	5-L (β, γ) R-C-11-L	+					Gelatinous subst.	Tetrahedron, **37**, 1795 (1981) J. Antibiot., **29**, 375, 380 (1976). Ger. Pat. 2 526 250, Neth. Pat. 75 7346
Bryamycin	S. hawaiiensis	Thiostrepton	+			+			J. Antibiot., **16**, 76 (1963), Br. Pat. 790 521
BSA	B. subtilis								Jpn. Pat. 75 132 188, 76 11 198
BU-2743 E	B. circulans	R-2-P						Inh. leucine aminopeptidase	J. Antibiot., **36**, 1396 (1983)
C-22-4 Cairomycin	S. viridochromogenes S. sp.	Basic peptide CC-2-P	+	+					Jpn. Kokai 82 91 993 Antimicrob. Agents Chemother., **11**, 373 (1977), **19**, 941 (1981)
Capreomycin	S. capreolys, Dactylosporangium variesporum	6-P-C-5 (β)-M	+	+		+			Tetrahedron, **39**, 921 (1978), US Pat. 3 143 468, 4 026 766, Br. Pat. 920 563

Name	Source	Code	1	2	3	4	5	Property	Reference
CC-1014	*Paecilomyces abruptus* VC 7200	Similar to 1907	+					Toxic	*J. Antibiot.*, **35**, 1231 (1982), US Pat. 4 123 521, 4 282 327
CC-1065	*S. zelensis*	2-P-M							*J. Antibiot.*, **36**, 383 (1983)
Celenamides	*Cliona celata*	4-P							*J. Org. Chem.*, **45**, 3687 (1980)
Cerexins	*B. cereus*	R-10-P	+						*J. Antibiot.*, **29**, 1281 (1976), US Pat. 3 590 436, Ger. Pat. 2 420 103/4
Chaetocin	*Chaetomium minutum*	C-2-P-M	+	+				Cytostatic	*Helv. Chim. Acta*, **53**, 1061 (1970)
Chaetomin	*Chaetomium cochliodes, C. globosum, C. umbonatum*	C-2-P-M	+	+					*J. Am. Chem. Soc.*, **98**, 6741 (1976)
Chalcidine	*Micromonospora chalcea*								*Antibioiki*, 483 (1970)
Chlamydocin	*Diheterospora chlamydospora*	C-4-P	+	+		+		Cytostatic	*Helv. Chim. Acta*, **57**, 533 (1974), *Tetrahedron Lett.* **24**, 5305 (1983), Belg. Pat. 747 441, Br. Pat. 1 300 137
Chymostatin		4-P(U)-M	+					Proteinase inhibitor	*J. Antibiot.*, **26**, 625 (1973)
Cinnamycin	*S. hygroscopicus, S. lavendulae*							Nisin/subtilin analog	*Antibiot. Chemother.*, **4**, 1135, 1242 (1954)
	S. cinnamoneus	+	+				+		*Antibiotiki*, 323 (1981)
Cinropeptin	*Act. cineraceus*	+	+						*Agric. Biol. Chem.*, **46**, 865, 1885, 1861 (1982)
Cirratiomycin	*S. cirratus*	7-P-M	+				+		
Congocidin	*S. ambofaciens*	M-3-P, netropsin C-10	+		+				*Antimicrob. Agents Chemother.*, **1**, 483 (1972)
Cortinarins	*Cortinarius speciosissimus*	Amphomycin	+					Toxic	*Experientia*, **40**, 441 (1984)
CP-41012	*Actinoplanes nipponensis*	Thiostrepton	+						US Pat. 4 001 397
CP-41043/41494	*Pseudonocardia fastidiosa*	Thiostrepton	+						US Pat. 4 031 206
CP-46192	*Streptosporangium roseum-incarnatum*		+						US Pat. 4 083 963
Cryomycin A	*S. griseus psychrophylus*		+	+					*J. Antibiot.*, **27**, 138 (1974)
Cryomycin B	*S.* sp. + alkane		+	+					*J. Antibiot.*, **27**, 138 (1974)
Crystallomycin	*S. violaceae-niger*	Amphomycin	+						*Biokhimiya*, **24**, 399, 425 (1960)
Cyclochlorotine	*Penicillium islandicum* Sopp	C-5 (β) M-7-6-L	+				+	Toxic	*Chem. Lett.*, 1319 (1973)
Cycloheptamycine	*S.* sp.							Plant growth retardant, insecticidal	*Tetrahedron*, **26**, 4931 (1970)
Cyclo-Pro-Phe	*Oospora destructer*	C-2-P				+		Plant growth retardant, insecticidal	*Agric. Biol. Chem.*, **25**, 216 (1961)
Cyclo-Pro-Val	*S. tanashiensis*	C-2-P							Jpn. Pat. 75 116 689
Cyclo-Tyr-Pro	*S. gelaticus*	C-2						Cytotoxic	*Agric. Biol. Chem.*, **45**, 2613 (1981)
Cyclosporins	*Trichoderma polysporium, Cylindrocarpon lucidum, Tolypocladium inflatum*	C-11-P-M	+			+		Immunsuppr.	*Helv. Chim. Acta*, **60**, 1568 (1977), **65**, 1655 (1982), Belg. Pat. 823 008, Ger. Pat. 2 455 859

Table 7 (*continued*)

Name	Producers	Structural type/ characterization	G+	G−	My	AF	AV	AT	Others/uses	Ref.
Cyl-2	*Cylindrocladium scoparium*	C-4-P	−						Plant growth inhibitory	*Agric. Biol. Chem.*, **37**, 643, 1185 (1973)
Cystamycin	*S. sp.* ANS-734	Acidic peptide								*Jpn. Kokai* 82 118 595
Cytomycin	*S. griseochromogenes*	2-P-M					+		Phyt. fungi	*Agric. Biol. Chem.*, **30**, 132 (1966)
Danomycin	*S. albaduncus*	Succinimycin	+	+						*Belg. Pat.* 634 041, *Br. Pat.* 975 492. *Jpn. Pat.* 65 13 796
Desferrioxamin	*S. sp.*	Sideromycins		+						*Antimicrob. Agents Chemother.*, **7**, 377 (1975)
Destruxins	*Metarrhizium anisopliae, Asp. ochraceus, Oospora destructer*	6-L(β)							Insecticidal	*Phytochemistry*, **20**, 715 (1981)
Detoxins	*S. caespitosus, S. sp.*	R-3-D (O)							Effects uptake of blasticidins (antagonist)	*J. Antibiot.*, **27**, 484 (1974), *Experientia*, **37**, 365 (1981)
N-(2,6-Diamino-6-hydroxymethyl-pimelyl)alanine	*Micromonospora chalcea*	2-P-M								*J. Antibiot.*, **34**, 374 (1981)
11,11-Dihydroxy-chaetocin	*Verticillium tererum, Acrostalagmus cinnabarinus*	C-2-P-M	+				+			*J. Antibiot.*, **30**, 468 (1977), *US Pat.* 3 857 936, *Br. Pat.* 1 229 297
Diprotins	*B. cereus*	3-P							Dipeptidyl-amino peptidase-inhibitor	*J. Antibiot.*, **37**, 422 (1984)
Distamycin	*S. distallicus*	Netropsin, M-3-P	+	+	+	+			DNA binding, inh. RNA polymerase	*FEBS Lett.* **19**, 154, 327, **21**, 154 (1972), *US Pat.* 3 190 801
Doricin	*S. loidensis*	Virginiamycin, R-6-L-M	+							*Tetrahedron Lett.* 4231 (1966), *US Pat.* 3 299 047
Duazomycin	*S. ambofaciens, S. griseoplanus, S. candidus var. azaticus*	M-3-P	+			+	+			*Antimicrob. Agents Chemother.*, 115 (1965), *Br. Pat.* 935 321, *Jpn. Pat.* 79 50 193
Duramycin	*S. cinnamoneus azacoluta*	R-C-6	+						Nisin/subtilin analog inh. Na+/K+ ATPase	*J. Am. Chem. Soc.*, **80**, 3912 (1958), *Biochemistry*, **23**, 385 (1984)
Echinocandin	*Aspergillus rugulosus, Asp. nidulans-echinulatus*				+	+				*Helv. Chim. Acta*, **62**, 1252 (1979), *Swiss Pat.* 568 386
Echinomycin	*S. echinatus, S. lavendulae, S. griseolus, S. aureus, S. flaveolus, S. flacochromogenes, S. sp.*	M-(R-4-P)$_2$	+	+	+		+			*J. Antibiot.*, **28**, 332 (1975), *J. Am. Chem. Soc.*, **97**, 2497 (1975), *Swiss Pat.* 34 665

Antibiotic	Organism	Designation	1	2	3	4	5	Function	References
Edeines	*B. brevis* Vm 4	M-5-P	+	+			+	Inhibits DNA synthesis	*J. Antibiot.*, **36**, 793, 1001 (1983), *Biochemistry*, **9**, 1224 (1970)
Elastatinal	*S. griseoruber*	4-P(U)-M	+	+				Proteinase inhibitory	*J. Antibiot.*, **28**, 337 (1975)
EM-49	*B. circulans*	R-8-P-C-7 (γ)	+	+					*J. Antibiot.*, **30**, 756 (1977)
Emericins	*Emericellopsis microspora*	R-14-P-M	+	+					*J. Am. Chem. Soc.*, **99**, 5205 (1977), *US Pat.* 3 821 367, *Ger. Pat.* 1 809 875
Enduracidins	*S. fungicidicus*	R-17-P-16-L	+	+	+			Feed additive, murein inhibitor	*Jpn. Pat.* 70 17 158, 70 114, *Ger. Pat.* 1 809 875, *US Pat.* 3 786 142
Enniatins	*Fus. oxysporum, Fus. orthoceras, Fus. sambucinum, Fus. latentium, Fus. roseum-acumination*	C-6-D (3 × 2)	+	+		+	+	Cation binding (Li$^+$, K$^+$)	*Chem. Pharm. Bull.*, **21**, 1171, 1175, 1184 (1973)
Enterobactins	*E. coli, Aerobacter aerogenes, Salmonella typhimurium*	Siderochrome, R-3-L						Ion transport	
3-Epideoxynega-mycins	*S. goshikiensis*	3-P-M (ε, β)	+	+					*J. Antibiot.*, **30**, 1137 (1977)
Epidermidins	*Staphylococcus epidermidis*	C-11	+	+					*Can. J. Microbiol.*, **18**, 121 (1972)
Esein	*B. brevis* GB	Gramicidin	+						*Dokl. Akad. Nauk SSSR*, **204**, 405 (1972)
Esperin	*B. mesentericus*	8-P-6-L	+		+				*Tetrahedron*, **25**, 1985 (1969), *Jpn. Pat.* 51 1145
Etamycin	*S. griseus, S. griseovirdus, S. lavendulus, S. sp.*	R-7-L-M	+	+	+			Membrane active, ion carrier, broad spectrum	*J. Am. Chem. Soc.*, **95**, 875 (1973), *Ger. Pat.* 1 072 773, 2 725 163, *Belg. Pat.* 855 591
FA-2097	*Eupenicillium abidjanum, Gliocladium virens*	C-2-M	+					Anaerobic bact.	*J. Antibiot.*, **35**, 374 (1982)
Feldamycin	*S. ficellus*	2-P-M	+	+			+	Bact. DNA repl.	*US Pat.* 3 965 515, *Biochemistry*, **16**, 3406 (1977)
Ferramidochloro-mycin	*S. sp.*	Pseudosidero-mycin	+	+			+		*J. Antibiot.*, **19**, 110, 250 (1966)
Ferrichrocin	*Asp. sp.*	Sideramine, C-6	+					Iron transport	*Eur. J. Appl. Microbiol.*, **5**, 51 (1978)
Ferrichrome	*Ustilago sphaerogena*	Sideramine, C-6	+					Iron transport	*Bull. Chem. Soc. Jpn.*, **47**, 215 (1974)
Ferrichrysin	*Asp. sp., S. griseoflavus, S. galilaeus, S. lavendulae, S. sp.*	Sideramine, C-6	+	+				Iron transport	*Belg. Pat.* 625 235, *Helv. Chim. Acta*, **43**, 901, 1868, 2105 (1960), *US Pat.* 3 093 550, 3 033 760, *Ger. Pat.* 1 058 216, 1 089 122, 1 157 735, *Can. Pat.* 798 312, *Jpn. Pat.* 63 6595
Ferrimycins			+	+			+		

Table 7 (continued)

Name	Producers	Structural type/characterization	G+	G−	My	AF	AT AV	Others/uses	Ref.
Ficellomycin	S. ficellus		+			+		Inh. DNA repl.	J. Antibiot., **29**, 1001 (1976), Biochemistry, **16**, 3406 (1977)
FK-156	S. olivaceogriseus, sp. nov., S. violaceus	R-4-P (γ)					+	Immunostimulatory	J. Antibiot., **35**, 1280, 1286, 1293 (1982), **36**, 1045, 1051, 1059 (1983)
Fluvomycin	B. subtilis		+	+		+			Appl. Microbiol., **4**, 13 (1956), Br. Pat. 722 433, Ger. Pat. 915 852
FM-1001 FMPI	Ampulariella regulans S. rishiriensis	Neutral peptide M-2-P							Jpn. Kokai 82 88 128 Agric. Biol. Chem., **46**, 855, 2697, 2979 (1983)
FR 900261 Fumarylalanine	Petriella guttulata P. resticulosum, Asp. indicus	C-4 M-AA	+			+	+	Inh. metalloproteinase	J. Antibiot., **36**, 478 (1983) Aust. J. Chem., **21**, 2775 (1968)
Fumarylcarbox-amido-1-2,3-diaminopro-pionyl-l-alanine	S. collinus	M-2-P	+			+			J. Antibiot., **25**, 137 (1972), US Pat. 3 832 287
Fumitremorgin B	Asp. fumigatus, Asp. caespitosus, Penicillium lanosum	M-C-2						Tremorganic toxin	J. Chem. Soc., Chem. Commun., 408 (1974)
Fusafungin	Fus. latentium	Valinomycin	+						Br. Pat. 944 131, 1 018 626, Fr. Pat. 1 164 181
Fusarinine Galantins	Fus. roseum, Fus. arbense B. pulvifaciens		+	+	+				Biochemistry, **7**, 184 (1969) J. Antibiot., **28**, 122 (1975), Tetrahedron Lett., **25**, 1587 (1984)
Garlandosin Gatavalin	S. althioticus B. polymyxa-colistinus, Aerobacillus colistinus	Althiomycin	+ +	+ +	+	+			US Pat. 3 642 984 J. Antibiot., **25**, 243 (1972), US Pat. 3 923 978, Br. Pat. 1 346 972, Ger. Pat. 2 251 916
Gliotoxin	Gliocladium fimbriatum, Asp. fumigatus, Penicillium sp., Trichoderma viride	M-C-2	+	+		+			J. Med. Chem., **21**, 796, 799 (1978)
Globomycin	Streptomyces halstedii, Streptoverticillium cinnamoneum	R-5-L		+				Inhibitory of cell wall synth., prolipoprotein-processing	J. Antibiot., **31**, 421, 426 (1978)
Glucomycin Gluconimycin	S. sp. A S 9 S. sp.	Actinomycin	+ +	+ +		+ +			Jpn. Kokai 60 21 395 Arch. Mikrobiol., **54**, 246 (1966)
Gougerotin	S. gougerotti, S. kitakiensis	P-2-M	+	+					Tetrahedron Lett., 6029 (1968)

Name	Producing organism	Structure	1	2	3	4	5	Action	References
Gramicidins	*B. brevis*	15-P-M	+	+					*J. Sci. Ind. Res.*, **34**, 249 (1975)
Gramicidin S	*B. brevis*, *B.* sp.	C-10 (2×5)	+	+	+				*Topics in Enzyme Fermentation and Biotechnology* (ed. A. Wiseman), p. 187 (1981). *J. Antibiot.*, **34**, 1227 (1981), **36**, 751 (1983)
Gratisin	*B. brevis*	C-12 (2×6)	+	+					*Bull. Soc. Chim. Fr.*, 2363 (1971), Ger. Pat. 1 805 280, 1 180 893
Grisellimycins	*S. griseus*, *S. coelicus*	R-10-P-8-L-M	+		+				*J. Antibiot.*, **15**, 141 (1962)
Griseococcin	*S. griseus*	Netropsin-like, M-3-P	+						*J. Am. Chem. Soc.*, **98**, 1926 (1976), US Pat. 3 174 902, 3 023 204, Ger. Pat. 1 072 773
Griseoviridin	*S. griseoviridus*, *S. griseus*	C-2-P-M	+	+			+		*Experientia*, **16**, 129 (1960), US Pat. 3 147 184, Ger. Pat. 1 070 782, Swiss Pat. 366 123, Br. Pat. 876 096
Grisonomycin	*S. griseus*	Ferrimycin	+	+					
HC-Toxin	*Helmintosporium carbonum*	C-4						Plant pathogen	*Tetrahedron*, **38**, 45 (1982), *Tetrahedron Lett.*, **24**, 5309 (1983)
Herbicolin	*Erwinia herbicola* A 111	Acylated peptide				+			*J. Antibiot.*, **33**, 353 (1980)
Hodydamycin HOE 467 A	*S.* sp. AS-Y-400, *S. tendae* ATCC 31210	74-P (two disulfides)	+					α-Amylase inhibitor	*J. Antibiot.*, **23**, 388 (1970), US Pat. 4 282 318, Jpn. Kokai 82 93 916
Hyalodendrin	*Hyalodendron* sp.	C-2-P-M	+			+			*J. Chem. Soc., Perkin Trans. 1*, 2600 (1973), US Pat. 3 715 352
1-(S)-Hydroxy-2-(S,S)-valyl-amidocyclobutane-1-acetic acid	*S.* sp. X-1092	2-P							*J. Antibiot.*, **27**, 754 (1974)
Hypelcin	*Hypetrea peltata*	R-19-P-M	+					Uncoupler ox. phosphorylation	*J. Chem. Soc., Chem. Commun.*, 413 (1979)
I-1/2	*Saccharomyces cerevisiae*	Acidic cyclopeptide	+	+					*Cereal Chem.*, **39**, 183 (1962), *Nature (London)*, **182**, 415 (1958)
ICI-13595	*Paecylomyces* sp.	Basic peptide						Antitrypanosomal	*Nature (London)*, **181**, 48 (1957)
Ikarugamycin	*S. paeochromogenes*	Cyclic polyketide + Orn	+					Antitrichomonas	*Bull. Chem. Soc. Jpn.*, **50**, 1813 (1977), Jpn. Pat. 71 88 833

Table 7 (*continued*)

Name	Producers	Structural type/characterization	G$^+$	G$^-$	Activities My	AF	AT AV	Others/uses	Ref.
Ilamycin	S. islandicus, S. insubtus	C-7-P-M			+				FEBS Lett., **17**, 145 (1971), Jpn. Pat. 64 1657
Imacidine	S. olivaceus	6-P-5L						Inh. cell wall formation (Actinomyces)	Liebigs Ann. Chem., 28 (1982)
Isariins	Isaria cretacea, I. felina	S-L (β)	+					Insecticidal	J. Antibiot., **34**, 1261, 1266 (1981)
Iturins	B. subtilis-ituriensiens	C-7-(β)	+	+		+			Biochemistry, **17**, 3992 (1978), Biochem. Biophys. Res. Commun., **81**, 297 (1978), Tetrahedron Lett., **30**, 3065 (1982)
Janiemycin	S. macrosporeus	Enduracidin	+		+				J. Antibiot., **23**, 502 (1976), Ger. Pat. 2 035 655
Jolipeptin	B. polymyxa-colistinus, Aerobacillus colistinus	Polypeptin	+	+					J. Antibiot., **25**, 147, 304, 309 (1972), Jpn. Pat. 72 43 291, Br. Pat. 1 346 973, US Pat. 38 833 649
K-582	Metarhizium anisopliae	7-P			+	+		Interferon induction	J. Antibiot., **33**, 533 (1980), **36**, 335 (1983)
Kikumycins	S. phaeochromogenes	2-P-M	+	+			+		Org. Mass Spectrom., **9**, 635 (1974)
Kobenomycin	S. kobenensis	Leucinamycin	+						Jpn. Pat. 74 35 085
Komamycins	S. pyridomyceticus, S. mitakaensis				+				J. Antibiot., **20**, 194 (1967), Jpn. Pat. 70 8636
Kuwaitimycin	S. kuwaitinensis	Lipopeptide	+						J. Antibiot., **30**, 749 (1977), Belg. Pat. 856 273, Ger. Pat. 2 724 090, US Pat. 4 100 273
L-13365	Actinoplanes sarveparensis		+						
L-681, 176	S. sp.	Marasmin-type						Inh. angiotensin converting enzyme	J. Antibiot., **37**, 462 (1984)
LA-5253	S. sp.	Albomycin	+						Antibiot. Chemother., **9**, 160 (1959)
LA-5937	S. bobilae-sporificans	Albomycin	+						Antibiot. Chemother., **9**, 160 (1959), Br. Pat. 920 799
Laspartomycin	S. viridochromogenes	Amphomycin	+						J. Antibiot., **23**, 423 (1970), Ann. Inst. Pasteur, **120**, 609 (1971)
Laterosporin	B. laterosporus		+			+			
Lathumycin	S. lathumensis	Virginiamycin	+						Neth. Pat. 106 644, J. Antibiot., **20**, 194 (1967), Jpn. Pat. 74 135 085
Leucinamycin	S. cinnamoneus		+			+			
Leucinostatins	Paecilomyces lilacinus A-267	R-9-P-M	+			+		ATPase inh.	J. Chem. Soc., Chem. Commun., 94 (1982), J. Antibiot., **36**, 1084 (1983)

Name	Source	Designation	Activity	Remarks	Reference
Leucopeptin	*S. hachijoensis-takahaziensis*	Leucinamycin	+		*J. Antibiot.*, **17**, 262 (1964)
Leucylnegamycin	*S. purpeofuscus*	3-P-M (ε, β)	+	Proteinase inhibitory	*J. Antibiot.*, **24**, 732 (1971)
Leupeptins	*S. roseus*	R-3-P-M	+		*J. Antibiot.*, **25**, 515 (1972), *Biochim. Biophys. Acta,* **661**, 175 (1981) *Biochem. J.*, **51**, 538, 558 (1982)
Licheniformins	*B. licheniformis*	Basic polypeptide	+		*Agric. Biol. Chem.*, **45**, 895 (1981), **46**, 2621 (1982)
Lipopeptin	*S. violachromogenes*	R-C-8-L (α)	+	Phyt. bact./cell wall formation inh. *in vitro*	—
LL-AO-341	*S. candidus*	10-P-9-L-M	+ / —		*Antimicrob. Agents Chemother.*, 1966, 587, 591, *US Pat.* 3 377 244
LL-BM-547	*Noc.* sp.	6-P-C-5(β)-M	+		*J. Org. Chem.*, **42**, 1282 (1977)
Longicatenamycin	*S. diastaticus*	C-6	+		*J. Antibiot.*, **28**, 561 (1975). *Jpn. Pat.* 70 27 800
Lydicmycin	*S. lydicus*	Actithiazic acid	+		*Antimicrob. Agents Chemother.*, **1**, 135 (1972), *US Pat.* 3 395 220
M-81	*S. globisporus, S. griseus-psychrophylus*	Basic polypeptide	+		*J. Antibiot.*, **27**, 128, 138 (1974)
M-5093/4	*Gnomoniella* sp., *Nodulisporium*	Acidic large peptide	+	Inh. glucosyl transferase	*Jpn. Kokai* 82 98 215, 82 99 518
Majusculamide C	*Lyngbya majuscula*	8-L-M (complex)		Phyt. fungi	*US Pat.* 4 342 751
Malformins	*Asp. niger*	C-5-M	+		*J. Am. Chem. Soc.*, **98**, 3366 (1976), *Int. J. Pept. Prot. Res.*, **20**, 16 (1982)
Malonomycin	*S. rimosus-paromomyceticus*	2-P-M	+	Antitrypanosomal	*Tetrahedron*, **34**, 223 (1978). **38**, 1775 (1982), *Neth. Pat.* 67 1356, *Br. Pat.* 1 178 783
MAPI	*S. nigrescens* WT-27	4-P (U)	+	Proteinase inhibitor	*Appl. Microbiol.*, **21**, 837 (1971)
Marcescin	*Serratia marcescens*	Basic polypeptide	+		*J. Antibiot.*, **23**, 461 (1970). *Jpn. Pat.* 70 14 879
Matchamycin	*S. amagasakaensis*	Chelate-forming	+		*Jpn. Pat.* 76 76 492, 77 10 2201
MC-902	*S. platensis*		+	Phyt. bact. insecticide	*J. Antibiot.*, **8**, 164 (1955)
Mesenterin	*Noc. mesenterica*	Ostreogrycin	+		*J. Chem. Soc. Chem. Commun.*, 265 (1978)
Micrococcins	*Micrococcus* sp., *B. pumilus*		+		*Antibiotiki*, 587 (1966)
Micropolysporin	*Micropolyspora caesia, Thermonospora viridus*		+		
Minosaminomycin	*S.* sp., *A. aureo monopodiales*	M-2-P (U)	+		*J. Antibiot.*, **28**, 613 (1975)

Table 7 (continued)

Name	Producers	Structural type/characterization	Activities G⁺	G⁻	My	AF	AT AV	Others/uses	Ref.
Monamycins	S. jamaicensis	6-L-M	+						J. Chem. Soc., Perkin Trans. I, 2369 (1977); Jpn. Pat. 70 955
Morimycin	S. diastatochromogenas-inteus							Phyt. bact.	Jpn. Pat. 70 955
MSD-235	S. avidinii, S. lavendulae	2-P	+					Phyt. bact.	Chem. Commun., 101, (1969), Ger. Pat. 1 232 315, Br. Pat. 1 077 999
Muraceins	Nocardia orientalis	R-3-P-M	+					Inhibits angiotensin converting enzyme	J. Antibiot., 37, 336 (1984)
MX-A	B. vitelinus, B. circulans	Octapeptin	+						Jpn. Pat. 76 15 692, 76 29 294, 77 83 803
Mycobacillin Mycobactins	B. subtilis V 3 Mycobact. sp.	C-13 (two γ) Lipodepsi-peptide (hydroxamate)				+		Fe³⁺ chelator	J. Antibiot., 31, 147 (1978); Biochem. J., 111, 785 (1969)
Mycoplanecins	Actinoplanes sp. No 41042	R-10-P-8-L-M			+				Jpn. Kokai 81 140 957, 81 142 250, 82 176 942
Mycosubtilin	B. subtilis, B. subtilis niger	C-8 (β)				+			Eur. J. Biochem., 63, 391 (1976), US Pat. 2 602 767
Myroridin	Myrothecium	Basic peptide	+			+	+		Tohoku J. Exp. Med. 122, 403 (1977), Jpn. Pat. 70 12 276
Myxothiazol	Myxococcus fulvus	R-2-P-M	+			+			J. Antibiot., 33, 1474, 1480 (1980), Antimicrob. Agents Chemother., 19, 504 (1981)
Myxovalargin	Myxococcus fulvus	R-8-P-M	+	+					J. Antibiot., 36, 6 (1983)
Negamycin	S. purpeofuscus, S. sp.	2-P-M	+	+				Translational term. misreading	J. Antibiot., 29, 937 (1976), US Pat. 3 743 580, Ger. Pat. 2 022 311, Belg. Pat. 750 096
Neginamycin Neoantimycin Netropsin	Stv. cinnamoneum Stv. orinoco S. netropsis, S. reticuli, S. chromogenes, S. sp.	Leucinamycin R-C-4-ester M-3-P	+ +	+	+ +	+	+		Chem. Abstr. 83, 41 414 Tetrahedron, 25, 2193 (1969) J. Antibiot., 17, 220 (1964), US Pat. 2 586 762
Neurosporin	Neurospora crassa	C-6-D-(γ,δ)						Siderophore	J. Am. Chem. Soc., 106, 1285 (1984)
Nikkomycin	S. antibioticus	M-aa	+						Naturwissenschaften, 58, 603 (1971), Ger. Pat. 1 230 525, Belg. Pat. 644 682
Nikkomycin	S. tendae	M-2-P				+		Chitin formation inh.	Arch. Mikrobiol., 107, 143 (1976), J. Antibiot., 37, 80 (1984)

Name	Producing organism	Code					Function	References
Nisin	*Streptococcus lactis*	34-P-M	+		+		Food preservative, murein inh.	*Ger. Pat.* 200 818, *Sov. Pat.* 410 081
Nocardamin	*Act. buchanan, Noc.* sp., *S.* sp., *S. hygroscopicus-geldanus*	Sideramine		+	+			*Z. Naturforsch., Teil B*, **32**, 937 (1977)
Nocobactin Noformicin	*Nocardia asteroides Noc. formica*	M-4-P (e) M-aa				+	Fe³⁺ chelator	*Biochem. J.*, **138**, 407 (1974) *J. Med. Chem.*, **16**, 857 (1973)
Nosiheptide	*S. antibioticus, S. glauco-griseus* 12514	2-P-M	+				Growth factor food add.	*J. Antibiot.*, **31**, 623 (1978), *Experientia*, **36**, 414 (1980), *US Pat.* 3 155 581. *Belg. Pat.* 847 684. *Jpn. Pat.* 78 101 593
Nourseimycin NRCS-15	*S. noursei* 804 *S.* sp.	2-P-M Pseudosideromycin	+	+	+		Pro-antagonist	*J. Antibiot.*, **21**, 1 (1974)
Octapeptins	*B. circulans* ATCC 31805	R-8-P-C-7 (γ)	+	+	+	+	Antiprotozoal feed suppl.	*J. Antibiot.*, **30**, 756 (1977), **36**, 625, 634 (1983), *Neth. Pat.* 72 5751, *Ger. Pat.* 2 219 993, *US Pat.* 4 341 768
Ostreogrycins A	*S. ostreogriseus, S. mitakaensis, S. pristinae, S. piralis, S. virginiae, S. olivaceus, S. loidensis, S. graminofaciens*	C-2-D-M	+			+	Ribosome binding, food suppl., membrane active	*J. Chem. Soc., Perkin Trans.* 1, 2464 (1977), *Tetrahedron Lett.*, 707 (1982) *US Pat.* 1 301 857, 2 990 325, 3 137 640, 3 299 047, 3 311 538, *Br. Pat.* 776 035, 848 195
Ostreogrycins B	As for ostreogrycin A	R-6-L-M	+				Ribosome binding	*J. Antibiot.*, **30**, 665 (1977), *US Pat.* 2 787 580, 2 990 325, 3 137 640, 3 154 475, 3 299 047, 3 311 538, *Br. Pat.* 819 872, *Fr. Pat.* 1 301 837
Parvulin	*S. parvulus, S. pseudogriseolus*	Amphomycin	+				Feed additive	*US Pat.* 3 798 129, *Hung. Pat.* 157 984, *Can. Pat.* 956 254, *Belg. Pat.* 827 051 *Chem. Abstr.*, **82**, 153 748 *J. Org. Chem.*, **47**, 1807 (1982)
Patellamides Pepstatins	*Lissoclinum patella S. testaceus*, many other species	C-8-M R-5-P-M	+			+	Proteinase inhibitory	*J. Antibiot.*, **27**, 267 (1974) *J. Antibiot.*, **21**, 429 (1968)
Pepthiomycins	*S. roseo spinus*	Peptide	+				Phyt. bact.	*Biochim. Biophys. Acta*, **125**, 75 (1966)
Peptidolipin NA	*Nocardia asteroides*	C-8 (β)		+				*J. Antibiot.*, **32**, 115, 121 (1979)
Permetin Phalamycin Phallotoxins	*B. circulans S. noursei Amanita* sp.	Lipopeptide Berninamycin C-7-M	+				Actin binding	*Antibiot. Chemother.*, **3**, 815 (1953), *Br. Pat.* 790 521 *Crit. Rev. Biochem.*, **5**, 185 (1978)

Table 7 (*continued*)

Name	Producers	Structural type/ characterization	G+	G−	My	AF	AT AV	Others/uses	Ref.
Phleomycins	*S. verticillus*	R-8-P-M (γ, β)	+	+	+		+		*J. Antibiot.*, **25**, 752, 755, (1972), *Jpn. Pat.* 59 2595, 61 10 697
Phomopsin A	*Phomopsis leptostromiformis*	C-6						Mycotoxin	*J. Chem. Soc., Chem. Commun.*, 1259 (1983)
Phosphinothricyl-alanylalanine	*S. viridochromogenes, S. hygroscopicus*	3-P-M	+	+		+		Herbicide	*J. Antibiot.*, **36**, 96, 1040 (1983), *Neth. Pat.* 72 10 308
L-N⁵-Phosphono-methionine-S-sulfoximinyl-alanylalanine	*S. sp.*	3-P-M	+	+					*J. Antibiot.*, **26**, 261 (1973)
Phosphoramidin	*S. tanashiensis*	2-P-M						Proteinase inhibitory	*Biochem. Biophys. Res. Commun.*, **65**, 352 (1975)
Phytoactin	*S. hygroscopicus*	Acidic polypeptide	+					Phyt. fungi	*Mycologia*, **61**, 136 (1969), *US Pat.* 3 032 471, 3 155 520
Phytostreptin	*S. hygroscopicus*	Acidic polypeptide	+	+		+		Phyt. fungi	*Phytopathology*, **55**, 1366 (1965), *US Pat.* 3 032 470, 3 155 520
Piperazinedione	*S. griseoluteus*	C-2					+		*J. Am. Chem. Soc.*, **98**, 6742 (1976), *US Pat.* 3 718 651, 3 987 046, *Ger. Pat.* 2 029 708
Piperazinomycin	*Stv. olivoreticuli*	C-2-M							*J. Antibiot.*, **35**, 1130, 1137 (1982)
Planothiocins	*Actinoplanes* sp.	Peptide						Growth promoter	*Br. Pat.* 2 055 794, *Jpn. Kokai* 81 26 899
Platomycins	*Streptosporangium violaceochromogenes-globophilum*	Bleomycin	+	+			+		*J. Antibiot.*, **28**, 656, 662 (1975), *Ger. Pat.* 2 408 121, *Belg. Pat.* 811 311
Plumbemycins	*S. plumbeus*	M-3-P		+					*Agric. Biol. Chem.*, **41**, 573 (1977)
Polcillin	*B. subtilis*	Bacillomycin				+			*Agric. Biol. Chem.*, **29**, 548 (1968), *Swiss Pat.* 446 615, *Jpn. Pat.* 61 1149
Polymyxins	*B. polymyxa, B. colistinus, B. polymyxacolistinus*	R-10-P-C-7 (γ)	+	+				Membrane active	*Annu. Rev. Biochem.*, **46**, 723 (1977), *J. Sci. Ind. Res.*, **38**, 695 (1979), *Br. Pat.* 646 258, 645 750, 658 766, 742 589, *Jpn. Pat.* 72 27 038
Polyoxins	*S. cacaoi var. asoensis*	M-2-P (nucleoside)				+		Agric. fungicide, feed add.	*J. Am. Chem. Soc.*, **91**, 7490 (1969)
Polypeptins	*B. circulans, B. krzemieniewskii*	10-L (β)	+	+					*J. Med. Chem.*, **19**, 1228 (1976)
Pteroyl-poly-γ-glutamic acid	*Corynebacterium*	γ-Glu-chain						Growth factor	*J. Biol. Chem.*, **255**, 5649 (1980)

Name	Producing organism	Code	Activity	Reference
Pumilin	*B. pumilus*			*Nature (London)*, **175**, 816 (1955)
Pyridomycin	*S. pyridomyceticus, S. albidofuscus*	R-C-3-D		*Bull. Chem. Soc. Jpn.*, **48**, 2081 (1975), *Jpn. Pat.* 54 1349, 54 1048, 56 9566, *US Pat.* 3 367 833
RA	*Rubiae radix*	C-6		*Chem. Pharm. Bull.*, **31**, 1424 (1983), **32**, 284 (1984)
Radicicolin	*Cylindrocarpon radicicola*	Cyclosporin		*Trans. Br. Mycol. Soc.*, **49**, 563 (1966), *Br. Pat.* 1 006 724
Reumycin	*Act. sp., S. sp.*	Cyclopeptide	Phyt. bact.	*Antibiotiki*, 1014 (1968)
Rhizomycin	*S. novoverticillus*	C-7-M	Mycotoxin	*Jpn. Pat.* 70 17 155
Rhizoins	*Rhizopus microsporus*			*J. Chem. Soc., Chem. Commun.*, 47 (1983)
Ristomycin	*Nocardia lurida*	6-P-M	Cell wall formation inh.	*J. Chem. Soc., Perkin Trans.* 1, 1483 (1981)
S-685	*S. sp.*	Netropsin		*Jpn. Pat.* 70 05 435
S-31794-F	*Acrophialophora limonispora*	Echinocandin		*Ger. Pat.* 2 628 965
S-41062-F	*Cryptosporiopsis* ATCC 20594	R-C-6 (γ)		*Eur. Pat.* 57 724 (1982)
S-53210-A	*Microellobosporia brunea*	Thiostrepton		*Br. Pat.* 2 000 120, *Ger. Pat.* 2 825 618
Saihochins	*Ps. sp.*	Lipopeptide		*Jpn. Med. Gaz.*, **4**, 7 (1972), *Jpn. Pat.* 74 45 596
Samarosporin	*Amarospora sp.*	Peptaibophol		*J. Antibiot.*, **29**, 618 (1976)
Saramycetin	*S. saraceticus*	Thiazolyl peptide		*J. Colloid Interface Sci.*, **33**, 439 (1970), *US Pat.* 3 118 813
SCA-18640-B	*Mic. arborensis*	Thiostrepton		*J. Am. Chem. Soc.*, **103**, 5231 (1981), *US Pat.* 4 078 056
Serratamolide	*Serratia marcescens, Bacterium prodigiosum*	C-4-D		*Tetrahedron Lett.*, 47 (1969)
SF-1691	*S. filamentosus*	Bleomycin		*Jpn. Pat.* 78 127 895, *Chem. Abstr.*, **90**, 184 892
SH-50 Silk-worm sugar peptide	*S. griseolus* AH-SH-50 *Bombyx mori*	51-Glycopeptide		*Egypt. J. Bot.*, **22**, 215 (1979)
Siomycin	*S. sioyaensis, Planomonospora parontospora, Act. mutabilis*		Feed additive	*J. Antibiot.*, **33**, 1563 (1980), **34**, 124 (1981), *Antibiotiki*, 1971, 204, *US Pat.* 3 082 153, *Ger. Pat.* 1 178 172
Sirodesmins	*Sirodesmium diversum*	C-2-P-M		*J. Chem. Soc., Perkin Trans.* 1, 180 (1977), *Ger. Pat.* 2 346 389, *Br. Pat.* 1 387 504
Spergualin	*B. laterosporus*	2-P-M		*J. Antibiot.*, **34**, 1625 (1981)

Table 7 (continued)

Name	Producers	Structural type/ characterization	G⁺	G⁻	My	AF	AV	AT	Others/uses	Ref.
Sporidesmolides	Pithomyces dartamm	C-2-P-M	+				+			J. Chem. Soc., Perkin Trans. 1, 1476 (1978)
SS-70-A	S. olivogriseus	Bleomycin	+	+						Jpn. Pat. 76 15 694
Stendomycins	S. endus, S. sp.	R-14-P-7-L	+	+		+			Phyt. fungi	Biochemistry, 11, 4132 (1972)
Stenothricin	S. sp.	R-8-P-M	+							Liebigs Ann. Chem., 2011 (1976)
Stravidin S	S. avidinii, S. lavendulae, S. sp.	2-P		+						Chem. Commun., 101 (1969), Ger. Pat. 1 232 315, Br. Pat. 1 077 999
Streptothricins	S. lavendulae, S. griseus, S. candidus, S. hygroscopicus and many others	Nucleoside-β-Lys-chains	+	+						Bull. Chem. Soc. Jpn., 49, 3611 (1976), J. Antibiot., 35, 925 (1982), 36, 1638 (1983)
Subsporin	B. subtilis	Bacillomycin	+			+				J. Antibiot., 22, 467 (1969), Jpn. Pat. 71 40 195
Subtilin	B. subtilis	M-32-P	+				+		Food preservative	Hoppe Seyler's Z. Physiol. Chem., 354, 810 (1973), Biochim. Biophys. Res. Commun., 50, 559 (1973)
Succinimycin	S. olivochromogenes	Sideromycin	+							J. Antibiot., 26, 67 (1963)
Sulfactin	Act. roseus, S. roseus	Thiostrepton	+							J. Biol. Chem., 168, 765 (1947)
Sulfomycin	S. viridochromogenes, S. cineroviridis	Berninamycin	+							Tetrahedron Lett., 2791 (1978), US Pat. 4 007 090, Jpn. Pat. 70 17 599
Surfactin	B. subtilis, B. natto	8-L(β)			+				cAMP diesterase inhibitor	J. Antibiot., 36, 667, 674, 679 (1983), Chem. Pharm. Bull., 22, 938 (1974), US Pat. 3 687 926, Neth. Pat. 68 15 030
Suzukacillin	Trichoderma viridae	R-23-P-M	+						Phyt. bact.	J. Am. Chem. Soc., 99, 8469 (1977), Jpn. Pat. 65 13 795
Syriamycin	S. violaceoniger S-303		+				+			Drugs Future, 5, 414 (1980), Chem. Abstr., 91, 191 294, 92, 15 368
Syringomycin	Ps. syringae		+			+			Phyt. bact.	J. Appl. Bacteriol., 43, 453 (1977)
Tabtoxin	Ps. tabaci, Ps. coronafaciens, Ps. garcae	2-P							Phyt. Bact.	J. Antibiot., 28, 1 (1975)
Takaomycin	S. sp. AC-1978		+	+						J. Antibiot., 37, 700 (1984)
Tallysomycin	Streptoalloteichus hindustanus	R-8-P-M (γ, β)	+	+		+	+			J. Antibiot., 30, 779, 789 (1977), Belg. Pat. 845 513, US Pat. 4 051 237
Tatumine	B. brevis Vm4	Edeine				+				J. Antibiot., 33, 359 (1980)

Name	Source	Code						Remarks	Reference
Teichomycin A2	*Actinoplanes teichomyceticus*	Glycopeptide	+						J. Antibiot., **31**, 170 (1978)
Teicoplanin	*Actinoplanes teichomyceticus*	Glycopeptide	+						J. Antibiot., **37**, 621 (1984)
Telomycin	*S. canus, S. sp.*	11-P-9-L	+						Biochemistry, **12**, 3811 (1973), Bioorg. Khim., **3**, 422 (1977)
Tentoxine	*Alternaria tenuis, A. mali*	C-4-P	+					Chlorosis inducing toxin	Tetrahedron Lett. 5263 (1982)
Teprotide	*Bothrops jararaca*	9-P				+		Inh. angiotensin converting enzyme	Br. J. Pharmacol., **24**, 163 (1965)
Theiomycetin	*S. sp.*	Berninamycin	+				+		Jpn. Pat. 60 11 599, Chem. Abstr., **62**, 9481, US Pat. 3 697 646
Thermothiocin	*Thermoactinopolyspora coremialis, Noc. madurae B. thiaminolyticus*		+						Jpn. Pat. 71 42 959, Chem. Abstr., **76**, 32 827, 84 518
Thianosine		Polymyxin			+				J. Antibiot., **29**, 366 (1976), **34**, 1126 (1981) Jpn. Pat. 76 79 789
Thiocillines	*B. cereus G-15, B. megaterium I-13, B. badius AR-91*	Micrococcin	+						J. Antibiot., **36**, 832 (1983), Ger. Pat. 1 929 355
Thiopeptin A	*S. tateyamensis*	Thiostrepton	+					Feed additive	J. Antibiot., **36**, 799 (1983), US Pat. 2 982 689, 2 982 698, 4 064 013, Br. Pat. 790 521
Thiostrepton	*S. azureus, S. laurenti, S. sp.*	R-16-P-M	+					Veterinary drug, ribosome binding	J. Med. Chem., **10**, 1149 (1967), Jpn. Pat. 67 3076
Threomycin	*S. sp.*	AA	+	+		+			J. Antibiot., **28**, 126, 1004 (1975), Jpn. Pat. 75 106 492
TL-119	*B. subtilis*	R-7-P-4-L	+		+	+			Jpn. Pat. 76 9789, Fr. Pat. 2 277 593, Chem. Abstr., **85**, 31 609
TM-743	*B. circulans*	Octapeptin	+	+	+				
Toximycin	*B. subtilis*	Bacillomycin	+				+	Phyt. fungi	Phytopathology, **42**, 23 (1953)
Triacetyldesferri-fusigen	*Asp. deflectus*	Sideramine	+		+				J. Antibiot., **30**, 125 (1977)
Trichopolins	*Trichoderma polysporum*	R-P-9-M	+		+			Phyt. fungi	Experientia, **34**, 238 (1978), J. Chem. Soc. Chem. Commun., 585 (1981)
Trichorzianine	*Trichoderma harzianum*	R-P-2-M						Fe^{3+} chelator	Biochem. Biophys. Res. Commun., **116**, 1 (1983)
Trichostatins	*S. hygrosopicus*	Chelate-forming							J. Antibiot., **29**, 1 (1976), Jpn. Pat. 74 14 691
Trichotoxin A-40	*Trichoderma viride NRAL 5242*	R-18-P-M	+				+	Membrane modifying	Biochim. Biophys. Acta, **507**, 470, 485 (1978)

Table 7 (continued)

Name	Producers	Structural type/ characterization	Activities						Others/uses	Ref.
			G+	G−	My	AF	AT	AV		
Triculamin	S. triculaminicus		+		+					Agric. Biol. Chem., **33**, 1737 (1969)
Tridecaptins	B. polymyxa	R-13-P		+						Jpn. Pat. 76 144 796, Chem. Abstr., **86**, 16 933
Triostins	S. sp., S. aureus	(R-4-P)$_2$-L	+							Tetrahedron Lett., 1613 (1978), Bull. Chem. Soc. Jpn., **51**, 1501 (1978)
Tryphtophan-dehydrobutyrine-piperazinedione	S. spectabilis	C-2-P							RLV reverse transcriptase (weak)	J. Antibiot., **27**, 733 (1974)
Tsushimycin	S. pseudogriseolus	Amphomycin	+	+						J. Antibiot., **23**, 473 (1969), US Pat. 3 639 582, Jpn. Pat. 72 5717
Tuberactinomycin	S. griseo verticillatus-tuberacticus, S. sp.	6-P-C-5-(β)-M			+				Antibacterial drug	J. Antibiot., **30**, 1008, 1073 (1977), Neth. Pat. 71 9029, Jpn. Pat. 73 30 396, US Pat. 3 892 732
Tyrocidines	B. brevis ATCC 8185, 10068	C-10-P	+							Experientia, **26**, 476, 587 (1970)
Ulicylamide	Lissoclinum patella	C-7-M								J. Am. Chem. Soc., **102**, 5688 (1980)
Ulithiacylamide	Lissoclinum patella	C-(4-P)$_2$								J. Am. Chem. Soc., **102**, 5688 (1980)
Ussamycin	S. lavendulae		+							Ger. Pat. 1 208 499
Valinomycin	S. fulvissimus, S. tsusimaensis, S. sp.	C-(4-D)$_3$	+	+	+	+		+		J. Am. Chem. Soc., **97**, 7242 (1975), US Pat. 3 520 973
Vancomycin	S. orientalis	7-P-M	+		+				Inh. formation of murein	Topics Antibiot. Chem., **5**, 119 (1980), J. Am. Chem. Soc., **104**, 4293 (1982), J. Antibiot., **37**, 446 (1984)
Vanoxinin	Saccharopolyspora hirsuta	R-2-P-M						(+)	Inh. of thymidylate synthetase	J. Antibiot., **36**, 656 (1983), Jpn. Kokai 83 134 064
Verruculogen	Penicillium verruculosum	C-2-P-M							Tumor producing	J. Am. Chem. Soc., **96**, 6785 (1974)
Victomycin	Streptosporangium violaceachromogenes	Bleomycin	+					+		Belg. Pat. 804 529, Ger. Pat. 2 344 780
Violacetin	S. purpeochromogenes	Netropsin	+	+	+			+		J. Antibiot., **9**, 226 (1956)
Viomycin	S. vinaceus, S. puniceus, S. floridae, S. abikoensis, S. olivoreticuli, S. californicus, S. griseus-purpureus	6-P-C-5 (β) M	+	+	+				Antibacterial drug	Hung. Pat. 153 683, US Pat. 2 828 245, Br. Pat. 651 269, Jpn. Pat. 55 3096, J. Antibiot., **26**, 528 (1973)

Name	Source	Type	Reference
Viridomycin A	S. viridans, S. roseoviridus, S. olivoviridis	Chelate-forming	Bioorg. Khim., **2**, 365 (1976)
Viscosin	Ps. viscosa	R-9-P-7-L	Tetrahedron Lett., 1087 (1970)
X-73	Asp. rugulosis	Echinocandin	Ind. J. Biochem., **7**, 81 (1970)
XK-33-F2	S. olivoreticuliccellulophylus	Viomycin	Neth. Pat. 71 18 132, Ger. Pat. 2 165 644, Br. Pat. 1 368 153, Jpn. Pat. 73 30 400, Jpn. Kokai 82 105 194
XK-210	Actinomadura sp. MK-210	Acidic large peptide	
Ya-56-Y	S. humidus-antitumoris	Bleomycin	J. Antibiot., **26**, 77, 83 (1973); Ger. Pat. 2 206 637, Jpn. Pat. 72 2557
Yakusimycin A	S. sp. 61-26	C-2-D-M	Jpn. Pat. 73 30 398, 73 10 294, Chem. Abstr., **78**, 134 498, **80**, 106 851
Yakusimycin B/C	S. antibioticus	R-6-L-M	Jpn. Pat. 73 30 398, 73 10 294, Chem. Abstr., **78**, 134 498, **80**, 106 851
Yemenimycin	S. albus		J. Antibiot., **24**, 283 (1971), US Pat. 3 839 560
Zaomycin	S. zaomyceticus	Amphomycin	J. Antibiot., **7**, 134 (1954), Jpn. Pat. 55 8150
Zervamycin	Emericellopsis salmosynnemata		J. Antibiot., **27**, 321 (1974), J. Am. Chem. Soc., **103**, 6517 (1981), US Pat. 3 907 990
Zizyphines	Zizyphus oenoplia (root bark)	R-4-P-M	Tetrahedron Lett., 2577 (1973)
Zorbamycin	S. bikiniensis	R-8-P-M (γ, β)	J. Antibiot., **24**, 543 (1971), Br. Pat. 1 277 150
61-26	B. sp.	Peptolide	J. Antibiot., **28**, 129 (1975)
339-29	B. pumilus	Lipopeptide	J. Antibiot., **29**, 810 (1976)
583	S. orientalis	M-3-P	Jpn. Pat. 70 17 596
681-17	B. brevis	Polymyxin	Jpn. Pat. 78 121 703
1037	Trichoderma viridae	Peptaibophol	Jpn. Pat. 77 72 891, Chem. Abstr., **87**, 166 000 (Antiprotozoal)
1456	S. sp.	Thiostrepton	Med. Chem., **6**, 276 (1958)
1907	Paecilomyces lilacinus 1907	R-9-9-P-M	Agric. Biol. Chem., **44**, 3033, 3037 (1980), Jpn. Pat. 80 142 019
4205	B. sp.	Polypeptin	Appl. Microbiol., **14**, 79 (1966)
5590	S. sp. 5590	Basic chromopeptide	Antibiotiki, **26**, 483 (1981)
102804	B. cereus		J. Antibiot., **30**, 283 (1977)

6.5 APPENDIX

Abbreviations Used

a	amide
AA	amino acid
Ac	acetyl
Aib	aminoisobutyric acid
Abu	aminobutyric acid
αAbu	α-aminobutyric acid
AHOrn	N-δ-aceylhydroxy-Orn
βA1a	β-alanine
ΔAla	dehydroalanine
Δ But	α, β-dehydrobutyric acid
C8A	C_8-β-amino acid (iturin)
C13A	C_{13}-acid (amphomycin)
Cap	capreomycidine
Cit	citrulline
D	D-configuration (capital letter following!)
Dbu	diaminobutyric acid
Dha	2,6-diamino-7-hydroxyazaleic acid
Dhb	2,3-dihydroxybenzoic acid
Dhi	2,3-dihydroxyisoleucine
Dpr	diaminopropionic acid
ΔDprU	dehydrodiaminopropionic acid (β-amino $\leftarrow$ amide)
f	formyl
FA	fatty acid
FG	phenylglycine
FG1	4-hydroxyphenylglycine
FG2	3,5-dichloro-4-hydroxyphenylglycine
GMβLys	N-(ε)-guanidino-N-(β)-methyl-βLys
Hiv	hydroxyisovaleric acid
HyPic	3-hydroxypyridine-2-carboxylic acid
4HyPip	4-hydroxypipecolic acid
3Hy4OP	3-hydroxy-4-oxopipecolic acid
HyPro	hydroxy-Pro
Ise	isoserine
βLys	β-lysine
Me	methyl
Mur$N$Ac	N-acetylmuramic acid
OHβLys	γ-hydroxy-β-Lys
4OPPip	4-oxopipecolic acid
Orn	ornithine
OxPro	oxoproline
Pip	pipecolic acid
Pheol	phenylalaninol
Sar	sarcosine
βTyr	β-tyrosine
U	urea type of bond (—NH—CO—NH—)

Structural Formulae of Compounds of Table 1

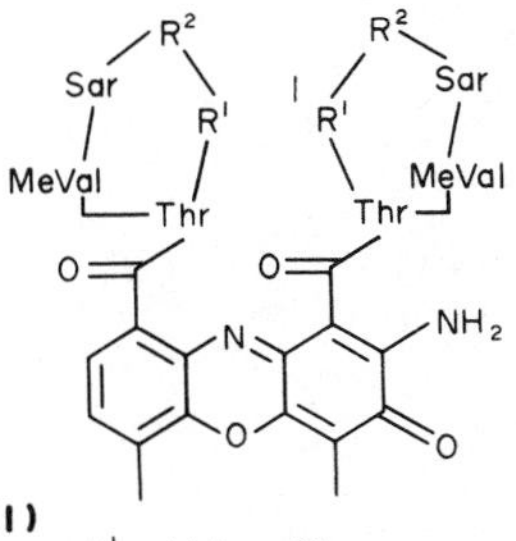

(1)

R¹ = ᴅVal, ᴅaIle
R² = Sar, Pro, HyPro, OxPro

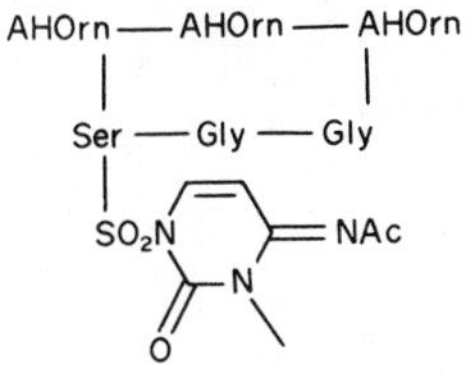

(2)

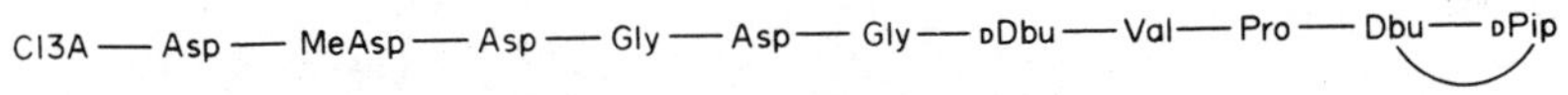

CI3A — Asp — MeAsp — Asp — Gly — Asp — Gly — ᴅDbu — Val — Pro — Dbu — ᴅPip

(3)

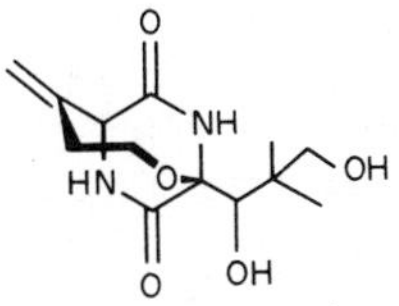

(6)

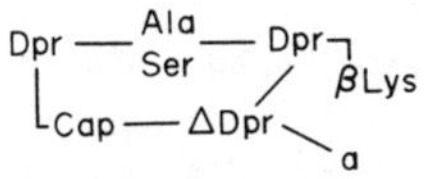

(9)

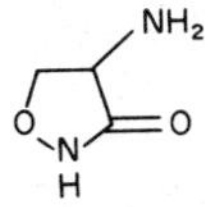

(10)

(12)

(21)

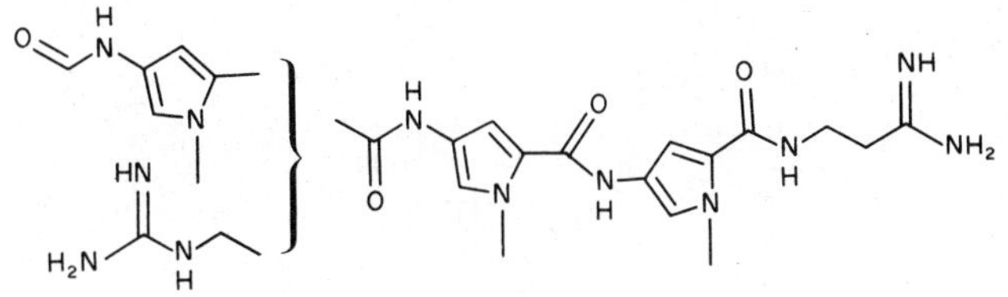

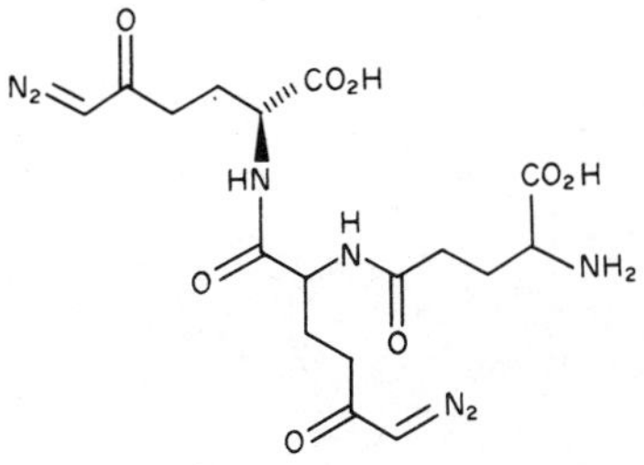

(13)

FA — Asp — Thr — DFG — ᴅOrn — ᴅaThr — FGl — ᴅFGl — aThr — Cit
      └ FG — ᴅAla — ᴅEnd — Gly — FG2 — ᴅSer — FG — End

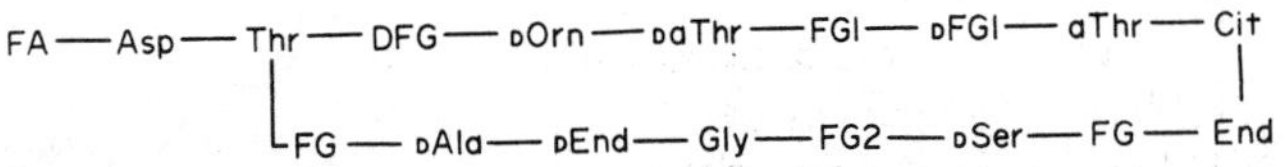

(14)

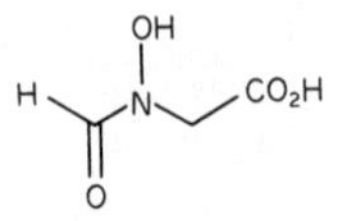

(17)

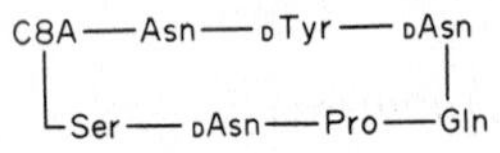

(18)

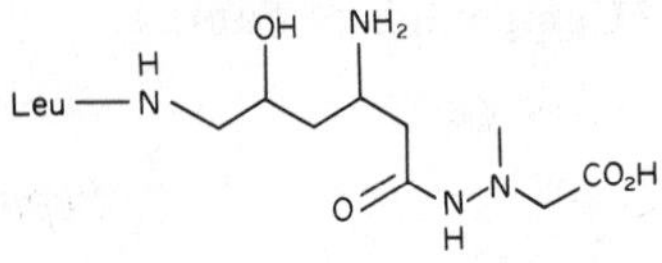

CBA—Asn—$_D$Tyr—$_D$Asn
└Ser—$_D$Asn—Pro—Gln

(19)

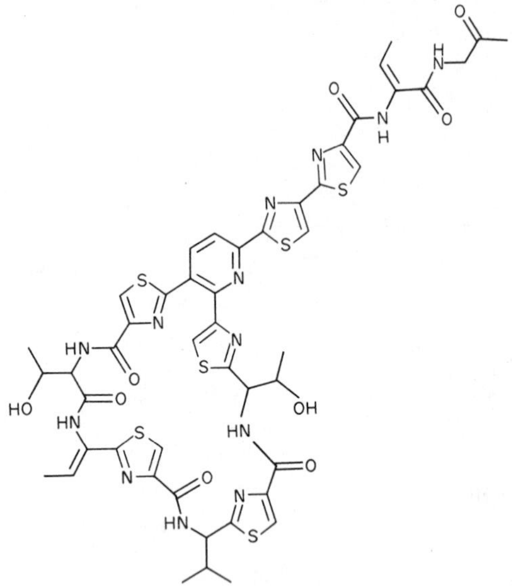

(23)

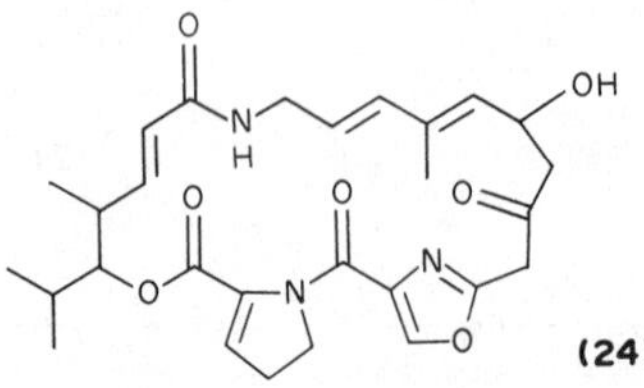

(24)

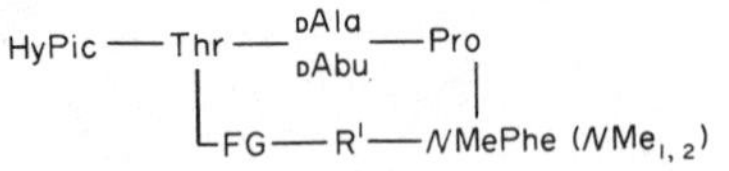

HyPic—Thr—$_D$Ala/$_D$Abu—Pro
        └FG—R^1—NMePhe (NMe$_{1,2}$)

R^1 = Pro, Pip, 4HyPip, 4OPip, 3Hy4OP, Asp

(25)

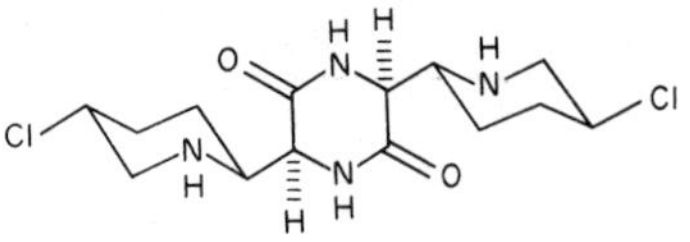

R^1 =

NH$_2$ **(29, 33)**

OH **(32)**

(26)

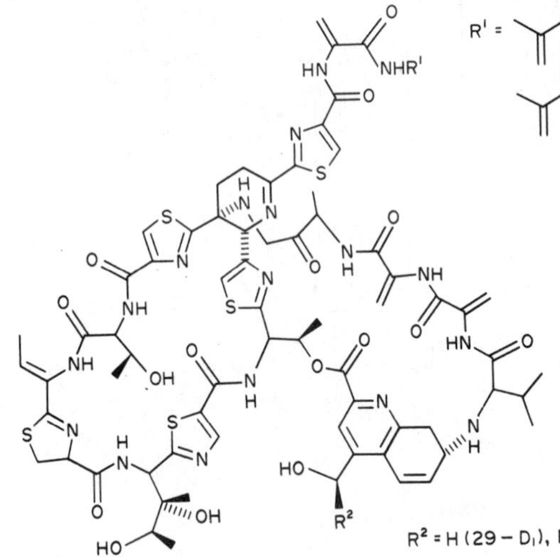

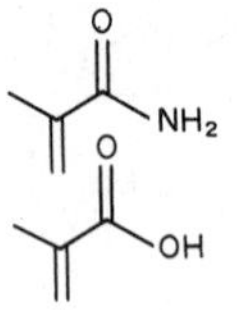

R^2 = H (29 – D$_1$), Me

βLys
OHβLys—Dpr—Ser—Ser
GMβLys—Ala Dpr Dpr
        └Cap—ΔDprU

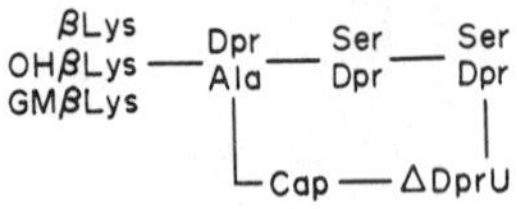

(34, 38)

6.6 REFERENCES

Bérdy, J. (1980a). *Handbook of Antibiotic Compounds*, vol. IV, parts 1 and 2. CRC Press, Boca Raton, Florida.
Bérdy, J. (1980b). Recent advances in prospects of antibiotic research. *Process Biochem.*, 28–35.
Campbell, I. M. (1983). Fungal secondary metabolism research: past, present and future. *J. Nat. Prod.*, **46**, 60–70.
Demain, A. L. (1973). Mutation and the production of secondary metabolites. *Adv. Appl. Microbiol.*, **16**, 177–202.
Flickinger, M. C. and D. Perlman (1979). Application of oxygen-enriched aeration on the production of bacitracin by *Bacillus licheniformis*. *Antimicrob. Agents Chemother.*, **15**, 653–661.
Fuska, J. and B. Proksa (1976). Cytotoxic and antitumor antibiotics produced by microorganisms. *Adv. Appl. Microbiol.*, **20**, 259–370.
Hecht, S. M. (ed.) (1979). *Bleomycin*. Springer, New York.
Katz, E. (1974). Controlled biosynthesis of actinomycins. *Cancer Chemother. Rep.*, **58**, 83–91.
Katz, E. and A. L. Demain (1977). The peptide antibiotics of *Bacillus*: biogenesis and possible functions. *Bacteriol. Rev.*, **41**, 449–474.
Kleinkauf, H. and H. Koischwitz (1978). Peptide bond formation in non-ribosomal systems. *Prog. Mol. Subcell. Biol.*, **6**, 59–112.
Kleinkauf, H. and H. von Döhren (1981). Nucleic acid independent synthesis of peptides. *Curr. Top. Microbiol. Immunol.*, **91**, 129–177.
Kleinkauf, H. and H. von Döhren (eds.) (1982a). *Peptide Antibiotics—Biosynthesis and Functions*. de Gruyter, Berlin.
Kleinkauf, H. and H. von Döhren (1982b). A survey of enzymatic biosynthesis of peptide antibiotics. In *Trends in Antibiotic Research*, pp. 220–232. Japan Antibiotics Research Association, Tokyo.
Kleinkauf, H. and H. von Döhren (1983a). Peptides. In *Biochemistry and Genetic Regulation of Commercially Important Antibiotics*, ed. L. C. Vining, pp. 95–146. Addison-Wesley, Reading.
Kleinkauf, H. and H. von Döhren (1983b). Non-ribosomal peptide formation on multifunctional proteins. *Trends Biochem. Sci.*, **8**, 281–283.
Kobel, H. and R. Traber (1982). Directed biosynthesis of cyclosporins. *Eur. J. Appl. Microbiol. Biotechnol.*, **14**, 237–240.
Lipińska, E. (1977). Nisin and its application. In *Antibiotics and Antibiosis in Agriculture*, ed. M. Woodbine, pp. 103–130. Butterworth, London.
Lipmann, F. (1968). The relation between direction and mechanism of polymerisation. *Essays Biochem.*, **4**, 1–23.
Madry, N., R. Zocher and H. Kleinkauf (1983). Enniatin production by *Fusarium oxysporum* in chemically defined media. *Eur. J. Appl. Microbiol. Biotechnol.*, **17**, 75–79.
Marahiel, M., W. Danders, M. Krause and H. Kleinkauf (1979). Biological role of Gramicidin S in spore functions. *Eur. J. Biochem.*, **99**, 49–55.
Meienhofer, J. and E. Atherton (1973). Structure–activity relationships in the actinomycins. *Adv. Appl. Microbiol.*, **16**, 203–300.
Misato, T., K. Keido and I. Yamaguchi (1977). Use of antibiotics in agriculture. *Adv. Appl. Microbiol.*, **21**, 53–88.
Morikawa, Y., I. Karube and S. Suzuki (1980). Continuous production of bacitracin by immobilized living whole cells of *Bacillus* sp. *Biotechnol. Bioeng.*, **22**, 1015–1023.
Oki, T. (1980). Cytotoxic and antitumor antibiotics produced by microorganisms. *Biotechnol. Bioeng.*, **22**, Suppl. 1, 83–97.
Perlman, D. (1978). Antibiotics (Peptides). In *Kirk-Othmer Encyclopedia of Chemical Technology*, pp. 991–1036. Wiley, New York.
Takita, T. and K. Maeda (1980). Chemical and biological modification of bleomycin, an antitumor antibiotic. *J. Heterocycl. Chem.*, **17**, 1799–1802.
Umezawa, H. (1980). Low-molecular-weight immunomodulators produced by microorganisms. *Biotechnol. Bioeng.*, **22**, Suppl. 1, 99–110.
Umezawa, H. (1981). *Small Molecular Immunomodifiers of Microbial Origin*. Pergamon, Oxford.
Umezawa, H. (1982). Low-molecular-weight enzyme inhibitors of microbial origin. *Annu. Rev. Microbiol.*, **36**, 75–99.
Umezawa, H., T. Takita and T. Shiba (eds.) (1978). *Bioactive Peptides Produced by Microorganisms*. Kodansha, Tokyo, and Wiley, New York.
Vandamme, E. J. (1981). Environmental influences on the dynamics of the Gramicidin S fermentation. In *Topics in Enzyme and Fermentation Biotechnology*, ed. A. Wiseman, pp. 187–261. Ellis Horwood, Chichester.
von Döhren, H. (1982). Applications of multienzyme systems in the production of antibiotics. In *Peptide Antibiotics—Biosynthesis and Functions*, ed. H. Kleinkauf and H. von Döhren, pp. 169–182. de Gruyter, Berlin.
White, D. J. G. (ed.) (1982). *Cyclosporin A*. Elsevier, Cambridge.
Zocher, R., U. Keller and H. Kleinkauf (1982). Enniatin synthetase, a novel type of multifunctional enzyme catalyzing depsipeptide synthesis in *Fusarium oxysporum*. *Biochemistry*, **21**, 43–48.

7

Streptomycin and Commercially Important Aminoglycoside Antibiotics

J. FLORENT
Rhône-Poulenc Santé, Vitry-sur-Seine, France

7.1 GENERALITIES ON AMINOGLYCOSIDE ANTIBIOTICS

The antibiotics of clinical importance which are commonly designated as aminoglycosides form an extremely large group. As their name indicates, all these products possess one or more amino sugars. However, this characteristic alone is not sufficient to define them accurately, as a considerable number of antibiotics would also be included, resulting in a group whose structural and therapeutic features are very different from one another. In fact, the products designated by the term aminoglycosides (or aminosides) possess an aminocyclitol, linked to one or several sugars, which are, in most cases, aminated. This definition was accepted in particular by Claridge (1979), Daniels (1978), Davies and Yagisawa (1983) and Rinehart and Stroshane (1976), in recent reviews. So defined, the aminosides constitute a really homogeneous group of antibiotics, not only because of their structure but also because of their type of activity, of their mechanism of action and, unfortunately, of some undesirable secondary effects.

7.1.1 Historical Background

The story of aminoglycosides started with two important discoveries by Waksman and coworkers in the USA: the discovery of streptomycin, isolated from the fermentation broths of *Streptomyces griseus* (Schatz *et al.*, 1944) and that of neomycin, produced by *Streptomyces fradiae* (Waksman and Lechevalier, 1949). The outstanding properties of these two products prompted manufacturers to undertake research in the same field in order to isolate more active aminoglycosides, with broader spectrum and less noticeable secondary effects. The last 30 years have been marked by regular announcements of discoveries of new aminosides of natural origin endowed with great therapeutic value: kanamycin (1957), paromomycin (1958), spectinomycin (1961), gentamicin (1963), tobramycin (1967), lividomycin (1968), ribostamycin (1969), sisomicin (1969) and sagamicin (1973), to mention only the main ones which have been commercialized. The aminosides which have been isolated in recent years differ little from the point of view of their biological properties from the best of the above-listed products, although some of them (fortimicins, istamycins, *etc.*) possess original structures.

Although far from exhaustive, a fuller list of natural glycosidic aminocyclitols is shown in Table 1. There are, in total, over a hundred of them, mostly obtained from *Streptomyces* cultures, but also from other Actinomycetales (*Micromonospora*, *Dactylosporangium*, *Saccharopolyspora*) and even from *Bacillus*. This family of antibiotics has also been gradually enriched with semisynthetic and synthetic products, whose remarkable biological activity has enabled them to supplant in certain cases, in clinical use, the original natural products; for instance, amikacin and dibekacin which derive from kanamycins or netilmicin derived from sisomicin.

Table 1   Principal Aminoglycosides from Natural Origins

Aminoglycoside	Producer strain	Ref.
Streptomycin	*Streptomyces griseus*	Schatz *et al.* (1944)
Neomycin	*Streptomyces fradiae*	Waksman and Lechevalier (1949)
Kanamycin	*Streptomyces kanamyceticus*	Umezawa *et al.* (1957)
Paromomycin	*Streptomyces rimosus forma paromomycinus*	Parke-Davis (1958)
Hygromycin B	*Streptomyces hygroscopicus*	Mann and Brower (1958)
Spectinomycin (actinospectacin)	*Streptomyces spectabilis*	Mason (1961)
Gentamicin	*Micromonospora purpurea*	Schering (1963)
Destomycin	*Streptomyces rimofaciens*	Kondo (1965)
Tobramycin (nebramycin factor 6) Apramycin (nebramycin factor 2)	*Streptomyces tenebrarius*	Stark *et al.* (1968)
Lividomycin	*Streptomyces lividus*	Kowa (1968)
Ribostamycin	*Streptomyces ribosidoficus*	Meiji (1969)
Sisomicin	*Micromonospora inyoensis*	Schering (1969)
Butirosin	*Bacillus circulans*	Parke-Davis (1969)
Verdamicin	*Micromonospora grisea*	Schering (1973)
Sagamicin (micronomicin)	*Micromonospora sagamiensis* var. *nonreductans*	Kyowa (1973)
Minosaminomycin	*Streptomyces* sp.	Hameda (1974)
Fortimicin	*Micromonospora olivoasterospora*	Kyowa (1974)
Seldomycin	*Streptomyces hofuensis*	Kyowa (1975)
G 52	*Micromonospora zionensis*	Schering (1976)
Sporaricin	*Saccharopolyspora hirsuta* subsp. *kobensis*	Kowa (1978)
Istamycin	*Streptomyces tenjimariensis*	Okami (1979)
Dactimicin (SF 2052)	*Dactylosporangium matsuzakiensis*	Inouye (1979)
G 367 S 1	*Dactylosporangium thaïlandense*	Toyo Jozo (1980)
Saccharocin	*Saccharopolyspora* sp.	Awata (1983)

7.1.2 Structure of Different Classes of Aminoglycoside Antibiotics

The aminosides are usually classified according to the structure of their aminocyclitol unit (Figure 1). They fall into three main groups. (1) The aminosides whose aminocyclitol is a streptamine derivative; the only clinically important representatives are streptomycin and dihydrostreptomycin which possess the streptidine ring (streptamine with two guanidine functions; see Section 7.2.1, Figure 3). Bluensomycin (Bannister and Argoudelis, 1963) also belongs to this category. Its aminocyclitol, bluensidine, differs from the streptidine in possessing a carbamoyl group instead of a guanidino group. (2) These are the most numerous aminosides, possessing a 2-deoxystreptamine ring which is sometimes N-substituted. With most of them, two of the hydroxy groups of the ring are linked to amino sugar moieties either in positions 4 and 5 (neomycins, paromomycins, lividomycins, ribostamycin, butirosins) (see Section 7.3, Figure 10) or in positions 4 and 6; the latter include the clinically most important aminoglycoside antibiotics (kanamycins, tobramycin, gentamicins and sisomicin, and their semisynthetic derivatives, amikacin, dibekacin and netilmicin) (see Section 7.3, Figures 11 and 12). With others, the 2-deoxystreptamine ring is only linked to one amino sugar residue, either in position 4 (apramycin) or in position 5 (destomycins, hygromycin B). (3) The other aminosides (fortimicins, *etc.*) and the antibiotics which, strictly speaking, are not aminosides but which are related by similarities of structures or very close biological properties (spectinomycin, kasugamycin, hygromycin A, validamycins). Among those products, only spectinomycin has had a commercial outlet owing to its being particularly active against gonorrhoea.

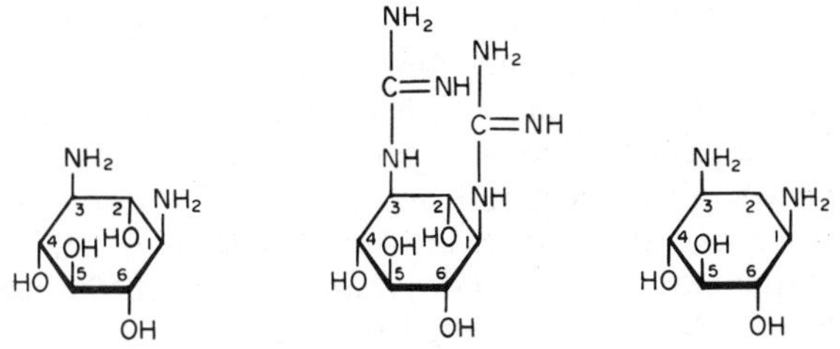

Figure 1   Structure of streptamine, streptidine and 2-deoxystreptamine

7.1.3 Microbiological Activity and Clinical Use

Aminoglycosides are among the most potent antibiotics known so far. Having a broad spectrum, they are active both against Gram-positive and Gram-negative bacteria and against mycobacteria. They are therefore clinically widely used despite their tendency to provoke serious secondary effects (see Section 7.1.5). Their use finds its justification in that they inhibit a great number of pathogenic microorganisms which cannot be effectively treated with other antibacterial agents, despite the recent discoveries in the β-lactam field.

The antibacterial action of aminoglycosides is mostly of a bactericidal type. Their activity is particularly noticeable against aerobic Gram-negative bacteria such as *Escherichia coli*, *Klebsiella*, *Proteus* and *Enterobacter*. Some of them can inhibit *Pseudomonas aeruginosa*. Although a lot of them are highly active against *Staphylococcus aureus in vitro*, their clinical efficacy, when dealing with severe *Staphylococcus* infections, is not guaranteed. They are ineffective against streptococci. Finally their activities against *Mycobacterium tuberculosis* are variable, streptomycin being still the product most widely used against that germ.

No aminoside is active against anaerobes or against aerobic bacteria growing in anaerobiosis. Their activity is null against enterococci, as well as against fungi, viruses and protozoa.

Aminoglycoside activity *in vitro* can be only roughly estimated since the answers vary considerably according to whether the strains examined have already been in previous contact with the antibiotic, that is to say, according to whether they contain one or several plasmids carrying resistance factors (inactivation enzymes, see Section 7.1.5); this particularly emerges from results presented by Daniels (1978), Korzybski *et al.* (1978) and Price *et al.* (1977). However the gradual improvement in potentialities of six major aminosides discovered from 1944 to 1970 clearly appears from the results in Table 2.

Table 2 Susceptibility of Various Sensitive Bacteria to some Major Aminoglycoside Antibiotics[a]

| Organism | Streptomycin | Mean minimum inhibitory concentration (μg ml^{-1}) | | | | |
		Neomycin	Kanamycin	Gentamicin	Tobramycin	Sisomicin
Staphylococcus aureus	2	0.5	1.2	0.3	0.2	0.2
Enterobacter aerogenes	—	—	2.6	0.7	1.2	0.7
Escherichia coli	8	8	4	0.6	0.7	0.4
Klebsiella pneumoniae	4	2	2.6	0.8	0.7	0.4
Proteus vulgaris	4	4	1.1	0.4	0.4	0.2
Proteus mirabilis	8	8	1.4	0.5	0.4	0.4
Salmonella spp.	16	2	4	0.7	0.8	0.4
Pseudomonas aeruginosa	32	32	>32	2.3	0.4	0.9

[a] Adapted from Moellering, 1983.

Sensitivity of clinical Gram-negative isolates to the same antibiotics varies in the same positive way. Moellering (1983) reports the particularly telling results of a survey of several thousands of strains (Massachusetts General Hospital, 1978–1980): in that example, the respective sensitivities to streptomycin, kanamycin, gentamicin and tobramycin are 64, 85, 99 and 98%, respectively, in the case of *E. coli*, 52, 91, 100 and 100% in the case of *Salmonella* and 5, 4, 78 and 94% with *Pseudomonas aeruginosa*.

Aminoglycosides are generally administered by means of intramuscular injections. They are rapidly absorbed. In case of very serious infections, it can be desirable to give, with precaution, intravenous injections. The drug is eliminated through the kidneys without having undergone any modifications. In case of renal failure, blood concentration can reach toxic levels.

No aminoside is active when administered orally to treat systemic infections. There is no absorption through the digestive tract. In case of oral ingestion, only the sensitive intestinal microbial flora are, to a large extent, sterilized. Aminosides are often used locally in the treatment of injuries or to prevent infections resulting from burns.

A certain amount of information concerning pharmacology, pharmacokinetics and toxicity of the main aminoglycoside antibiotics can be found in a succinct article by Lazowski and Lypka (1983).

7.1.4 Mode of Action

All aminoglycosides inhibit protein synthesis in sensitive bacteria. However, that action does not seem sufficient to account for the death of the cells in cases of bactericidal effect, which is the most frequent. That phenomenon could bear a relation to the interaction of that type of antibiotic with the function and integrity of the bacterial membrane (Dalhoff, 1983).

The intracellular points of impact of aminosides are the ribosomes. At this level, the effect on biochemical mechanisms varies somewhat according to the aminoglycosides being considered. Complete reviews on this question have been written by Franklin and Snow (1981), Gale *et al.* (1981), Schlessinger and Medoff (1975), Tanaka (1975), Vazquez *et al.* (1979) and Wallace *et al.* (1979).

The most advanced works deal with the mode of action of streptomycin. The specific antibacterial activity of this antibiotic is based on its ability to inhibit protein synthesis on the 70S ribosomes while having no effect on the 80S ribosomes of eukaryotes. It has been shown that the target is the 30S subunit of the 70S ribosome. Streptomycin strongly inhibits initiation of peptidic chains and slows down the elongation of the chains in the process of formation. These effects on initiation and elongation are attributed to a distortion, both of the acceptor and donor sites of the ribosome, by the antibiotic. Moreover, streptomycin causes misreading of mRNA, so that erroneous amino acids are chosen. Their incorporation into any peptidic chain in the process of formation results in a defective polypeptide (a nonsense polypeptide). Finally, under streptomycin influence, some molecules of nucleic acids such as denatured DNA, ribosomal RNA and transfer RNA can act as messengers in the ribosomes although they ordinarily do not have this property.

Important progress has been made in attempts at identifying the proteins of the 30S subunit of the *E. coli* ribosome which determine streptomycin action. Protein S_{12} (initially called P_{10}) seems to be the ultimate target of the antibiotic. Nevertheless the precise site of its binding is not well determined; besides, the mechanism involved is not clear and brings into play at least four other ribosomal proteins (S_3, S_5, S_9 and S_{14}).

All aminoglycoside antibiotics have an effect on the protein biosynthesis, yet some of them have a very different behaviour from that of streptomycin; for instance, spectinomycin has no effect on the decoding of messenger RNA and is not bactericidal. Then again, gentamicin, kanamycin and neomycin show effects on isolated ribosomes which vary according to the concentrations of antibiotics brought into play and there seem to be at least two binding sites on the ribosome for each of these antibiotics instead of a single one for streptomycin.

7.1.5 Problems with Toxicity and Bacterial Resistance

The keen interest roused by the outstanding clinical properties of aminoglycoside antibiotics must unfortunately be moderated owing to two major problems found, to various degrees, in all the representatives of this family. That accounts for the fact that many of their applications were given up when other antibiotics with broad spectrum, which did not have these drawbacks, appeared on the market. The problems concerned are those of toxicity and of bacterial resistance (inactivation).

7.1.5.1 *Toxicity*

One of the problems linked to the use of aminoglycosides, particularly in case of prolonged use (as with for instance, tuberculosis being treated with streptomycin), concerns the toxicity of this type of molecule. It appears in the kidney and inner ear, with variable intensities according to the product used.

(i) *Nephrotoxicity*

Nephrotoxicity of aminoglycoside antibiotics is due to their accumulation in the renal tissue, where they provoke morphological and functional alterations of the nephron both at glomerular and tubular levels. There seems to be a relation between the molecular structure of aminosides, in particular the number of amino groups per molecule, and their nephrotoxicity: neomycin is very toxic, streptomycin very little, and kanamycin, gentamicin and tobramycin stand halfway between those extremes. Cojocel and Hook (1983) as well as Lietman and Smith (1983) have recently restated the questions of both the clinical and physiological aspects of nephrotoxicity and the potential mechanisms involved.

In most cases with aminosides, nephrotoxicity is relatively slight and reversible. Moreover, as it is easily detected, it is possible, if it appears, to reduce its importance by shortening the length of treatment and reducing the doses of antibiotics prescribed.

(ii) *Ototoxicity*

The prolonged use of streptomycin and of a few other aminoglycosides can cause permanent damage to the eighth pair of cranial nerves. It can affect either one or the other, or both of the branches of the nerve. Damage to the auditory branch (cochlear function) is associated with a permanent weakening of hearing power, which can end up in irreversible definitive deafness. Damage to the vestibular branch (labyrinthic disturbances) shows itself in particular through a loss of balance.

Streptomycin in prolonged treatment against tuberculosis is especially toxic to the vestibular branch whereas dihydrostreptomycin affects the auditory branch. Neomycin shows high cochlear toxicity; it is therefore used only for local applications or as intestinal antiseptic. Kanamycin is less toxic than neomycin; unlike streptomycin, it causes lesions which are cochlear rather than vestibular. With gentamicin and tobramycin, phenomena of ototoxicity which are more vestibular than cochlear have also been reported, but as these products are highly active, even in small doses, it usually allows quite a large safety margin in the treatment of *Pseudomonas* infections (Fee, 1983).

No aminoside is completely devoid of ototoxicity but, as regards certain semisynthetic products, noticeable improvements have been made on this point.

7.1.5.2 *Bacterial resistance*

As is the case, to various degrees, with all antibiotics, inactivation phenomena are observed with aminoglycosides. Such bacterial resistance can have three origins: (1) the reduction in, or suppression of, transport of the antibiotic into the bacterial cell through the modification of membrane permeability, which leads to a concentration at the ribosomal site of action which is insufficient for inhibition; (2) the alteration of the target of the drug through spontaneous mutation affecting one of the proteins of the 30S subunit of the bacterial ribosome. This transformation, genetically controlled by the chromosome (chromosomal resistance), occurs only with aminoglycosides of the streptomycin type; (3) the modification of the antibiotic through bacterial enzymes, leading to a product which is inactive or at least ineffective because it is unable to spot the ribosome or because it is expelled outside the cell (Le Goffic, 1979). The enzymes responsible for this inactivation are coded by plasmids (R factors), hence the term 'extrachromosomal' attributed to this type of resistance; it is very common with aminosides and is prevalent owing to the frequent dissemination of R factors from donor resistant bacteria to recipient sensitive bacteria, through transduction (transmission by a bacteriophage, as with staphylococci) or conjugation (as with Gram-negative microorganisms, in particular with enterobacteria).

This third mechanism, which has recently been the subject of a restatement by Davies (1983), is clinically the most significant. At least 12 types of enzymes can play a role in the inactivation of the various aminoglycosides by carrying out, on some specific amino or hydroxy groups of those antibiotics, some N-acetylations, some O-adenylations or O-nucleotidylations and some O-phosphorylations (Claridge, 1979; Davies and Yagisawa, 1983) (Table 3). The enzymes of clinical importance that are mostly found are APH (3'), ANT (2''), AAC (6') and AAC (3) (Davies, 1983).

Table 3   Different Classes of Aminoglycoside-modifying Enzymes

Mechanism	Enzyme Trivial name	Abbreviation	Subtype[a]
N-acetylation	3-acetyltransferase	AAC (3)	I, II, III, IV
	2'-acetyltransferase	AAC (2')	—
	6'-acetyltransferase	AAC (6')	I, II
O-nucleotidylation[b]	6-adenylyltransferase	AAD (6)	—
	4'-nucleotidyltransferase	ANT (4')	I, II
	2''-nucleotidyltransferase	ANT (2'')	—
	3''-adenylyltransferase	AAD (3'')	—
O-phosphorylation	6-phosphotransferase	APH (6)	—
	3'-phosphotransferase	APH (3')	I, II, III
	2''-phosphotransferase	APH (2'')	—
	3''-phosphotransferase	APH (3'')	—
	5''-phosphotransferase	APH (5'')	—

[a] There are several aminoglycoside substrates for every subtype; some substrates may be modified by several subtypes of the same enzyme.   [b] AAD = ATP: aminoglycoside adenylyltransferase; ANT = NTP: aminoglycoside nucleotidyltransferase.

Some of the enzymatic sites of inactivation are common to a lot of aminoside antibiotics. For each site likely to be modified in an aminoglycoside, one or several enzymes can be involved. One given enzyme can have the ability to inactivate more than one antibiotic, sometimes on different sites. For instance, APH (3') I and APH (3') III, normally phosphorylating in the 3' position, carry out phosphorylation of the 5''-hydroxy group of lividomycins (Daniels, 1978). Finally, a given antibiotic can be inactivated by more than one enzyme.

The nature of the structural modifications brought to the various aminosides have been reviewed by Davies and Smith (1978). As an example, the modes of inactivation of streptomycin and kanamycin B known at present are represented in Figure 2.

It is easy to understand that, as soon as this mode of bacterial resistance was known, one of the researchers' major aims was to bring about chemical modifications to the enzymatic sites of inactivation, so as to make them insensitive to enzymes.

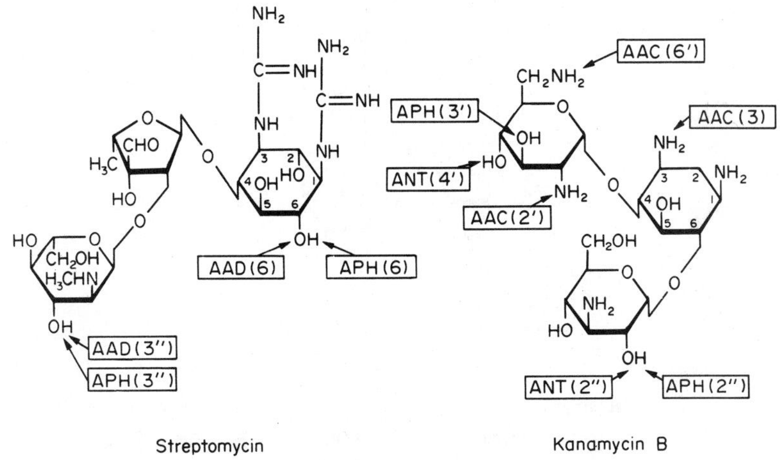

Figure 2 Enzymatic modifications of streptomycin and kanamycin B: AAC(3), AAC(2′), AAC(6′), AAD(6), ANT(4′), ANT(2″), AAD(3″), APH(6), APH(3′), APH(2″), APH(3″) (see Table 3)

This study on resistance mechanisms cannot be concluded without a reminder that genes coding for aminoglycoside-inactivating enzymes are also found in the microorganisms producing these antibiotics (Davies *et al.*, 1979). The precise role of these enzymes is not known, but it can be assumed that they constitute, for those microorganisms, one of the means of protecting themselves against the antibiotics produced (Davies and Yagisawa, 1983).

7.2 STREPTOMYCIN

7.2.1 Generalities

The first aminoglycoside antibiotic which was discovered, streptomycin, was isolated from the culture broths of an Actinomycetale, *Streptomyces griseus*, by Waksman and coworkers (Schatz *et al.*, 1944). Three years later, its structure was elucidated in its broad outlines by Kuehl *et al.* (1947), who showed that the molecule, looking like a trisaccharide, was formed from three subunits: an aminocyclitol, the streptidine, linked to a disaccharide, the streptobiosamine, composed of L-streptose and of *N*-methyl-L-glucosamine. That structure was definitively established, as regards the accurate configuration of glucosidic bonds, thanks to the works of Dyer and Todd (1963). The complete synthesis of the molecule has been carried out by Umezawa *et al.* (1974), but presents no interest on an industrial scale.

Numerous species of *Streptomyces* produce streptomycin or products of the same family, *e.g.* mannosidostreptomycin (or streptomycin B), dihydrostreptomycin, mannosidodihydrostreptomycin, *N*-demethylstreptomycin, hydroxystreptomycin (see Figure 3), whose activity is of the same degree as, or inferior to, that of streptomycin. Among those, only streptomycin and its derivative, dihydrostreptomycin, produced industrially through catalytic reduction of streptomycin, are clinically used and commercialized. Another semisynthetic derivative, deoxydihydrostreptomycin, as active as streptomycin, but, it seems, slightly less toxic, has been clinically used in the past, particularly in Japan.

Streptomycin and its two semisynthetic derivatives are used, on the one hand, in the treatment of tuberculosis and, on the other hand, to treat systemic infections with Gram-negative bacteria, and generally speaking, bacterial infections of the urinary tracts. Comparing aminoglycosides with 2-deoxystreptamine rings, they are usually far less active but their cost price and their nephrotoxicity are noticeably lower.

A non-exhaustive list of the main microorganisms producing streptomycin and its derivatives is given in Table 4. All of them are *Streptomyces*. The most famous is *Streptomyces griseus* (first described by Krainsky in 1914), whose subcultures, described by Waksman and Curtis in 1916, led much later to the discovery of streptomycin. In 1949, Krassilnikov, in the USSR, contested that Krainsky's strain and that from which streptomycin was isolated, and which was again described in 1948 by Waksman and Henrici, were strictly the same ones, and he named the latter *Actino-*

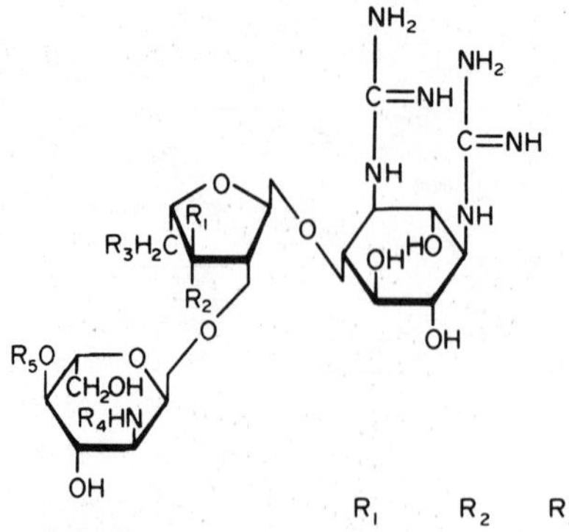

	R$_1$	R$_2$	R$_3$	R$_4$	R$_5$
Streptomycin	CHO	OH	H	CH$_3$	H
Dihydrostreptomycin	CH$_2$OH	OH	H	CH$_3$	H
Hydroxystreptomycin	CHO	OH	OH	CH$_3$	H
N−Demethylstreptomycin	CHO	OH	H	H	H
Deoxydihydrostreptomycin	CH$_2$OH	H	H	CH$_3$	H
Mannosidostreptomycin	CHO	OH	H	CH$_3$	M*
Mannosidodihydrostreptomycin	CH$_2$OH	OH	H	CH$_3$	M*

With M* =

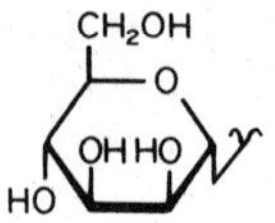

Figure 3 Structure of different members of the streptomycin family

myces globisporus streptomycini. Whatever its name, it is the mutants of this strain that have been used industrially (see Section 7.2.5.1).

Table 4 Different Producers of Streptomycin and Derivatives

Antibiotic	Producer strain	Ref.
Streptomycin	*Streptomyces griseus*	Schatz *et al.* (1944)
	Streptomyces bikiniensis	Johnstone and Waksman (1948)
	Streptomyces olivaceus	Kurosawa (1951)
	Streptomyces mashuensis	Sawazaki *et al.* (1955)
	Streptomyces rameus	Shibata (1959)
	Streptomyces galbus	Murase *et al.* (1959)
	Streptomyces erythrochromogenes var. *narutoensis*	Kondo *et al.* (1962)
Dihydrostreptomycin	*Streptomyces humidus*	Tatsuoka *et al.* (1957)
Hydroxystreptomycin	*Streptomyces griseocarneus*	Benedict *et al.* (1951)
	Streptomyces subrutilus	Arai *et al.* (1964)
	Streptomyces glaucescens	Hutter (1967)

7.2.2 Physicochemical Properties

Streptomycin C$_{21}$H$_{39}$N$_7$O$_{12}$ (PM 581,58) comes in the form of a white crystalline powder. It is very soluble in water and strong base. It is usually used in the form of a salt: sesquisulfate, trichlorhydrate, trichlorhydrate and calcium chloride double salt, whose aqueous solutions are levorotatory and have respectively for rotary power ($[\alpha]_D^{25}$, $c = 1$) : $-79°$, $-84°$ and $-76°$. These salts, which are very hygroscopic, turn deliquescent when exposed to the air. They are very soluble in water (> 20 g l^{-1}), slightly soluble in ethanol, acetone and chloroform. The chlorhydrate is very soluble in methanol (> 20 g l^{-1}), unlike the sulfate (0.85 g l^{-1}) (Weiss *et al.*, 1957).

Streptomycin is very stable; its aqueous solutions stand up well to pH modifications, and can be kept for several days in ambient temperature and sterilized through boiling without loss of activity.

The UV spectrum of streptomycin is not very characteristic (end adsorption). Its IR spectrum is shown in Figure 4.

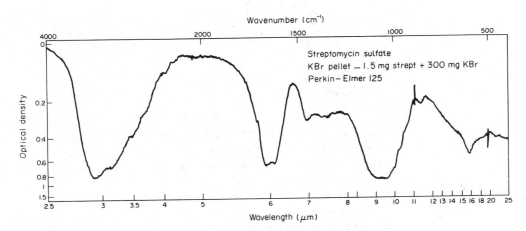

Figure 4   IR spectrum of streptomycin sulfate

7.2.3   Assay and Identification Methods

7.2.3.1   *Assay methods*

Numerous physical, chemical and microbiological methods have been described relating to the analysis of streptomycin and products of the same family, in particular mannosidostreptomycin (streptomycin B), often present in fermentation broths, and dihydrostreptomycin. Those which have been mostly used formerly have been described in a book by Grove and Randall (1955). Recently, Hughes *et al.* (1978) have presented a comprehensive survey of the many modern techniques at present applied to the quality control of nine aminoglycosides, among them streptomycin and dihydrostreptomycin. Among all those methods, the microbiological and colorimetric ones are the most widely used.

The classic microbiological methods are the method of diffusion in agar with *B. subtilis* ATCC 6633 as test organism, and the turbidimetric method using *Klebsiella pneumoniae* ATCC 10 031. It should be noted that mannosidostreptomycin microbiological activity is four to five times lower than that of streptomycin. Colorimetric methods are very numerous. The major types are as follows. (1) The method with maltol (Grove and Randall, 1955): when streptomycin is heated in dilute alkaline solution, maltol (2-methyl-3-hydroxy-γ-pyrone) is formed. It is a product of streptose degradation. With iron(III) chloride, maltol produces a red colouring, stable to acid pH, measured at 550 nm. This method is however not suitable for dihydrostreptomycin. (2) The method with nitroprussiate (Grove and Randall, 1955): the guanidino groups of streptomycins react with sodium nitroprussiate and with potassium ferricyanide in alkaline solution, producing a red orange colour, measured at 490 nm. Isolated streptidine, often present in broths although in small quantity, also reacts to this test. (3) Halliday's method (1952) or the method with naphthol. (4) The more recent method of Duda (1973), based on the oxidation of the deoxy sugars, such as streptose; the derivative formed, coupled with thiobarbituric acid, gives a chromogen with maximum absorption at 532 nm. (5) The method of specific assay of mannosyl derivatives, streptomycin B and dihydrostreptomycin B (Grove and Randall, 1955): the methanolysis of these products releases methylmannoside which is then isolated, and the quantity of mannose is colorimetrically determined with anthrone reagent. This method is used particularly to estimate the quantity of streptomycin B in fermentation broths, in presence of streptomycin, and in the crude or purified preparations of that antibiotic.

Among the other methods used to measure streptomycins, polarography (Siegerman, 1975) can also be mentioned as well as titrimetry, which turns the highly basic characteristic of these molecules to account, and which is perfectly suitable for use with pure solutions. Recently an immunological method with fluorescein-labelled streptomycin used as tracer has been developed by Schwenzer and Anhalt (1983).

7.2.3.2   Identification methods

The most common methods used to separate, identify and possibly determine the streptomycins present in mixtures are, and have been for a long time, paper and thin-layer chromatography. The numerous protocols (composition of the solvent mixtures, *etc.*) used in both methods cannot be explained in detail here. Aszalos and Frost (1975), Betina (1975) and Wagman and Weinstein (1973) have collected a great number of them. Electrophoresis (Umezawa and Kondo, 1975) has its followers too, either on paper (Kniep and Grisebach, 1980b) or in polyacrylamide gel (Kniep and Grisebach, 1980a). Recently, high performance liquid chromatography (HPLC) has been introduced (Whall, 1981). Combined with UV detection at 195 nm for the measurement of separate constituents, it is at present the easiest, fastest, most sensitive and most accurate technique available to the analyst.

7.2.4   Biosynthesis

The biosynthesis of streptomycin in *S. griseus* (and that of dihydrostreptomycin which has the same pathway) has been the object of several reviews in recent years (Claridge, 1979; Grisebach, 1978; Pearce and Rinehart, 1981; Rinehart and Stroshane, 1976; Walker, 1980). They bring some modifications to, and supplement, the data collected earlier by Demain and Inamine (1970).

Studies on the subject have used three main approaches: the techniques of isotopic competition with assumed precursors, labelled with ^{14}C, ^{13}C or ^{15}N; the use of idiotrophic mutants; and the detection of enzymatic reactions likely to be concerned.

Without going into the detail of the works that have made it possible to establish the pathway of biosynthesis which is the most generally accepted (see Figure 5), the following points can be made. (a) D-Glucose is the precursor common to the three subunits of the streptomycin molecule. (b) Streptomycin biosynthesis from D-glucose leads to a series of reactions bringing into play about thirty enzymes (Grisebach *et al.*, 1981). (c) Carbons 1, 2, 3, 4, 5 and 6 of D-glucose provide respectively: carbons 5, 4, 3, 2, 1 and 6 of streptidine, carbons 1, 2, 3, 4, 5 and 6 of N-methyl-L-glucosamine and carbons 1, 2, 3', 3, 4 and 5 of L-streptose. (d) Biosynthetic pathways of streptidine and L-streptose subunits are now clear. (i) As regards streptidine, glucose is first converted to *myo*-inositol, which is then transformed in two sequential sets of five analogous enzymatic steps: oxidation, amination, phosphorylation in *para* position to the aminated group, carbamidinylation, dephosphorylation. The first amination is carried out from glutamine, the second one from alanine, both operations bringing into play different enzymes. For the two carbamidinylations, arginine is the carbamidinyl donor, the enzymes involved being different in both cases. On the other hand, phosphorylation of *scyllo*-inosamine and that of N-amidinostreptamine utilize the same enzyme. It is generally accepted that the second dephosphorylation step happens in the final stage of the streptomycin biosynthesis. (ii) As for streptose, the conversion is carried out *via* a carbon–carbon intramolecular rearrangement bringing into play deoxythymidine-5'-diphosphoglucose (D-glucose-1-dTDP), which is converted to 4-keto-6-deoxy-D-glucose, then to 4-keto-L-rhamnose and finally to L-dihydrostreptose (Wahl and Grisebach, 1979). (e) The formation of N-methyl-L-glucosamine from D-glucose is not so clear. At the onset, a phosphorylation undoubtedly occurs, followed by a transamination leading to D-glucosamine-6-phosphate. Then, several epimerizations probably take place. The methyl group comes from methionine. UDP-N-methyl-D-glucosamine phosphate intermediate has been mentioned in recent works (Hirose-Kumagai *et al.*, 1982). Grisebach *et al.* (1981) report the presence of N-methyl-L-glucosamine linked to a nucleotide whose identity is yet to be established. (f) The coming together of the various elements of streptomycin molecule takes place in two stages: (i) formation of 4-O-(α-L-dihydrostreptosyl)streptidine phosphate from streptidine phosphate and dTDP-L-dihydrostreptose (Kniep and Grisebach, 1980a); (ii) formation of dihydrostreptomycin phosphate from dihydrostreptosylstreptidine phosphate and XDP-N-methyl-L-glucosamine (Kniep and Grisebach, 1980b). (g) The final stages include oxidation of dihydrostreptomycin phosphate to streptomycin phosphate (Maier and Grisebach, 1979), then dephosphorylation of the latter which leads eventually to streptomycin.

It should be noted that *myo*-inositol is a specific precursor for the aminoglycosides having the

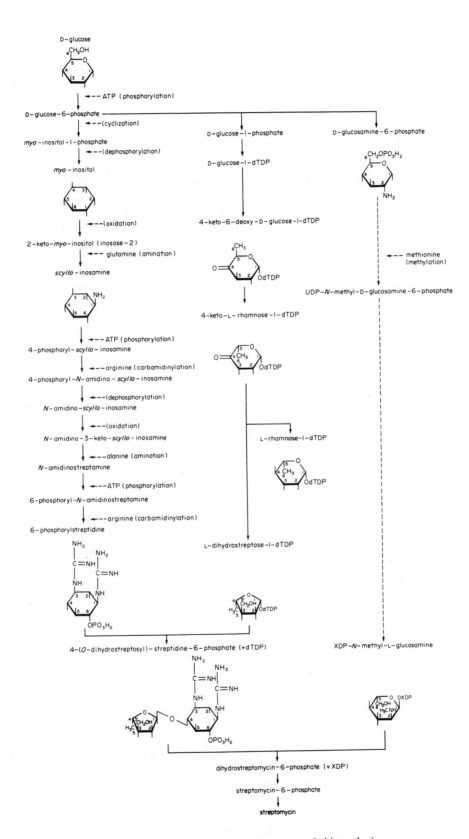

Figure 5   Probable pathway of streptomycin biosynthesis

streptidine ring (or its related product, bluensidine); it has no part in the biosynthesis of desoxys-treptamine aminosides, of which aminocyclitol would be formed from glucose *via* viburnitol, the latter being a deoxyinositol (Daum *et al.*, 1977 ; Walker, 1980).

7.2.5 Production Technology

For industrial production, streptomycin is prepared by fermentation tanks of 100 to 200 m³; it is then isolated and purified through various methods.

7.2.5.1 Fermentation

(i) Strains

From Waksman's discovery up to the present day, the productivity of *S. griseus* strains used in industry has been increased over 100 fold by using classical programmes of mutation involving physical agents (UV, X-rays, *etc.*; Dulaney, 1953), chemical ones (*N*-methyl-*N'*-nitro-*N*-nitroso-guanidine, ethyleneimine, chloroethyleneimine, diethyl sulfate, nitrosomethylurea, nitrogen mustard; Alikhanyan and Shishkina, 1967; Borisova *et al.*, 1970) and biological ones (actino-phages; Alikhanyan and Teteryatnik, 1962). With these programmes, various techniques are used for the screening of the strains, such as determination of streptomycin production, of secondary products occurring in its biosynthesis or of a particular enzymatic activity (Alikhanyan and Shishkina, 1967; Borisova *et al.*, 1970; Shishkina, 1967). An attempt is then made to find out the best culture conditions, especially medium composition, for the selected mutants. The increase in production, which is at present over 10 g l⁻¹, is shown in Table 5. The industrial yields are probably about 15 g l⁻¹ at the very least (Daniels, 1978), but the exact figure is kept secret.

Table 5 Improvement of Streptomycin Production by *Streptomyces griseus* (1944–1979)

Culture medium	Yield (g l⁻¹)	Ref.
Glucose–peptone–meat extract	0.1	Schatz *et al.* (1944)
Glucose–proline	1	Woodruff and Ruger (1948)
Glucose–yeast–ammonium sulfate	1	Savage (1949)
Glucose–soybean–distillers' solubles	2	Dulaney (1953)
Glucose–soybean–yeast	3.4	Shirato and Motoyama (1966)
Glucose–soybean	5–10	Horner (1967)
Glucose–soybean–ammonium sulfate	10.3	Singh *et al.* (1975)
Glucose–ammonium sulfate	10.5	Bormann (1979)

In recent years, thanks to the increasingly accurate knowledge of streptomycin biosynthetic pathways (see Section 7.2.4) and to the use of genetic recombination methods (protoplast fusion, Hopwood *et al.*, 1977; recombination *in vitro*, Zhuang *et al.*, 1982), new possibilities have been offered. They have certainly been swiftly put into use by industrial researchers to improve the producing strains. However, if interesting results have been obtained, they have not been revealed.

(ii) Culture media

In 1970, Demain and Inamine, using numerous publications written in the course of the pre-ceding 25 years, presented a complete review of the various carbon and nitrogen sources and minerals used for streptomycin production by *S. griseus*. Their advantages and drawbacks were also examined.

Among carbohydrates, glucose, the starting-point in streptomycin biosynthesis, has always been the product of choice for its production. Other simple or complex sugars (fructose, maltose, lactose, dextrin, starch, *etc.*) can also be used, but with less satisfactory results. Various lipids have been successfully tested in the past, as a replacement for glucose; a small quantity of soy-bean oil is often introduced in glucose media.

As regards nitrogenous nutrients, most industrial strains require a complex source for optimum production, although *Streptomyces griseus* can grow very well too on certain chemically defined media. To begin with, Waksman used peptone and meat extract, but Rake and Donovick (1946) showed that soybean flour was much more favourable; besides it is more economical. No better nitrogen sources have been found since then. The value of this material is probably due to its rather slow breakdown and to the very gradual release of ammonia over the course of fermentation. The stimulating effect of ammonium ions on production has been known for a long time and has been recently shown by Inoue *et al.* (1983). Other complex substances, with a high nitrogen content (corn-steep liquor, distillers' solubles), are sometimes added to the medium to complement soybean flour. Naturally, the optimal concentrations of nitrogen nutrients and of glucose in the media must be carefully adjusted according to the requirements of the increasing performances of the *S. griseus* strains that are available. A few examples of such media taken from the literature are shown in Table 6.

Table 6 Composition of Some Culture Media used for Streptomycin Production

Nutrient	Schatz et al., 1944	Rake and Donovick, 1946	Merck, 1951	Hockenhull et al., 1954	Singh et al., 1976	Bormann, 1979
Glucose	10	10	10	25	$60 (+10)^a$	$70 (+10)^a$
Peptone	5	—	—	—	—	—
Meat extract	3	—	—	—	—	—
Soybean meal	—	10	20	40	30	—
Distillers' solubles	—	—	2.5	5	—	—
Corn-steep (100% solids)	—	—	—	—	4	—
Ammonium sulfate	—	—	—	—	$9 (+1.5)^a$	$60(+10)^a$
Sodium chloride	5	5	10	2.5	2.5	—
Potassium dihydrogen phosphate	—	—	—	—	0.025	—
Calcium carbonate	—	—	—	—	0.5	—
Soybean oil	—	—	—	—	7	—
Lard oil	—	—	—	—	—	2

Heading: *Nutrient concentration according to the medium* $(g\,l^{-1})$

[a] Complementary addition during the fermentation.

Mineral salts have an important effect both on the growth of the microorganism and on streptomycin biosynthesis: the favourable effect of Na^+, Mg^{2+}, Fe^{2+}, Zn^{2+}, Cu^{2+}, K^+, Ca^{2+} and PO_4^{3-} ions has been shown by using synthetic media. Usually, complex media are automatically inclusive of these elements, but a complement of sodium chloride is always thought necessary. Moreover, with strains having high potentialities, it is common to add calcium carbonate and phosphate to the medium. Their optimal doses are carefully determined. Both play an essential part in regulation phenomena in the course of fermentation. Phosphate concentration in particular is very critical; a slight excess, although it stimulates growth, can inhibit the production (Shirato and Motoyama, 1965). As for Ca^{2+} ions, they might show their favourable effect in various ways: by allowing the concentration of PO_4^{3-} ions available in the medium to be modulated, by retarding the mycelium lysis and by promoting the enzymatic activity of mannosidostreptomycinase at the end of the culture (see Section 7.2.5.1(i)).

Various other substances can stimulate production, but their beneficial effect is particularly noticeable when low-producing strains and synthetic media are used. These are in particular: (i) barbital (Ferguson *et al.*, 1957) which, by delaying the mycelium lysis at the end of the fermentation, prolongs the secretion of the antibiotic; (ii) factor A (Khokhlov *et al.*, 1967; Figure 6), a component that is secreted by the streptomycin-producing strains; isolated and reintroduced into the culture medium, it stimulates both the production of streptomycin and the formation of spores. The effect of this factor is particularly noticeable in non-streptomycin-producing mutants. It can restore their ability to produce the antibiotic (Hara and Beppu, 1982); (iii) sclerine, whose favourable effect on production has been shown by Satoh *et al.* (1975); and (iv) *myo*-inositol, the precursor of the streptidine subunit of the streptomycin molecule (Demain and Inamine, 1970).

(iii) Description of the fermentation

The precise fermentation conditions vary, on small points, according to the strain and nutrients used, to the scale considered and to the 'know-how' proper to each manufacturer. But the general outline is based on the small-scale production method described by Singh *et al.* (1976; see Figure 7).

The operations start in the laboratory: spores, kept lyophilized and usually mixed with soil, are spread on a soybean flour agar medium, in Petri dishes or Roux bottles. After 2–3 weeks of incubation at 27 °C, the culture obtained is well sporulated. The spores are harvested and used to prepare the inoculum intended to seed the fermenter used for production.

The inoculum is usually prepared in two stages, the first one in the laboratory, in 0.5–2 l flasks

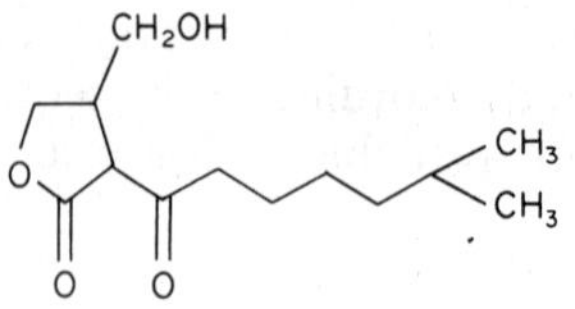

Figure 6 Structure of factor A

***Streptomyces griseus* DTH 2 stock culture** ↓	Lyophilized with soil
Maintenance culture ↓	Roux bottle with soybean agar medium, incubated for 2 weeks at 27 °C
First-stage seed culture ↓	0.5 l baffled Erlenmeyer flask with 50 ml of medium (a), incubated for 48 h at 26 °C on a rotary shaker (250 r.p.m.)
Second-stage seed culture ↓	0.5 l baffled Erlenmeyer flask with 100 ml of medium (a), incubated for 24 h at 26 °C on a rotary shaker (250 r.p.m.)
Main culture	5 l baffled glass fermenter with 4 l of medium (b) sterilized for 90 min at 120 °C (glucose separately), inoculated with 100 ml of the second-stage seed culture and incubated at 26 °C for 185 h under strong aeration (0.25 m^3 h^{-1}) and strong agitation with two impellers (600 r.p.m.); pH held above 6.4 by 4 N KOH addition along the whole fermentation; foaming controlled by automatic addition of sterilized soybean oil
Medium (a):	glucose, 40 g; dextrin, 2 g; soybean flour, 30 g; ammonium sulfate, 5.5 g; calcium carbonate, 6.5 g; sodium chloride, 2 g; potassium dihydrogen phosphate, 0.05 g; with tap water to 1 l
Medium (b):	glucose, 60 g; soybean flour, 30 g; ammonium sulfate, 9 g; calcium carbonate, 9 g; sodium chloride, 2.5 g; corn-steep liquor (100% solids), 4 g; potassium dihydrogen phosphate, 0.025 g; soybean oil, 7 ml; with tap water to 1 l and pH adjusted to 7.4

Figure 7 Streptomycin from *Streptomyces griseus*: laboratory scale fermentation process
(Adapted from Singh *et al.*, 1976)

and agitated on a shaker, the second one in a fermenter. In both cases, the composition of the medium is relatively close to that of the medium of production; temperature is maintained at about 26 °C, the duration of culture varying from 2 to 4 days. The production fermenter is then inoculated at a rate of 5–10% (v/v). The pH, which is around 7.0 at the start, is mostly in the range of 6.5–7.5 in the course of the fermentation, which can last up to 8 days. Temperature is maintained at 26–27 °C. Vigorous agitation and much aeration are necessary to obtain good streptomycin production. In case of restricted aeration, glucose is rapidly consumed, with formation of lactate and pyruvate, and streptomycin production is lower. Excess of phosphate in the medium (see Section 7.2.5.1(ii)) leads to a similar phenomenon.

The fermentation has a biphasic characteristic, as is the case with the production of numerous antibiotics (Demain and Inamine, 1970). During the first stage (trophophase), the growth of *Streptomyces* takes place. At the end of this phase and at the beginning of the second stage (idiophase), the enzymes involved in streptomycin biosynthesis are suddenly derepressed, initiating the production, which is exocellular although part of the streptomycin usually remains bound to the mycelium.

Numerous phenomena regulating the biosynthesis occur during the idiophase. For instance, excess phosphate, in addition to its effect on glucose consumption, might inhibit and repress the numerous phosphatases coming into play in streptomycin biosynthesis.

Mannosidostreptomycin, an undesirable product which is not an obligatory intermediate in streptomycin biosynthesis, but which is produced concurrently, is hydrolyzed with release of streptomycin only at the end of the fermentation. The main reason for the late appearance of mannosidostreptomycinase is the catabolite repression by glucose. The regulation of biosynthesis of this enzyme and its properties have been thoroughly examined by Demain and Inamine (1970) and Inamine *et al.* (1969). Mannosidostreptomycin can account for up to 40% of the total amount of streptomycins produced in a fermentation, but this percentage seems to have been brought down to a much lower level with the mutant strains used at present in industry.

Streptomycin production comes to an end when the culture autolysis finally takes place. The improvement of the production capacities of the strains, the adjustment for optimized media and the addition of supplementary nutrients during the fermentation have allowed the increasing delay of the autolysis phenomenon and prolonging the streptomycin production phase. The evolution of such an operation is shown in Figure 8.

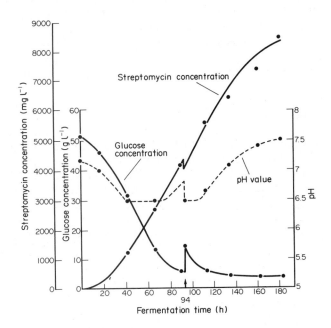

Figure 8 Time-course of streptomycin production in 5 l fermenter (adapted from Singh *et al.*, 1976). Culture medium and fermentation conditions as in Figure 7, plus a complementary addition of glucose (10 g l^{-1}) and ammonium sulfate (1.5 g l^{-1}) at 94 h

Culture degeneration is a major problem in streptomycin manufacture. Repeated transfer of the strain causes a progressive change in the population to 'nocardial' types of morphology, with mycelium fragmentation and loss of conidia formation. In fact, asporogenous cultures do not produce streptomycin. That is why the strains in the laboratory need to be controlled, so as to maintain high productivity in the fermenters of important capacities.

7.2.5.2 *Product recovery*

Numerous processes, often rather complex, have been used to recover the streptomycin present in fermentation broths, and to purify it in the form of various salts (sulfate, chlorhydrate, calcium complex, *etc.*). They always involve several steps requiring some of the following techniques: (i) adsorption on a support, followed by elution: the support can be activated carbon (Norite), with elution with a lower alcohol in aqueous acid solution (such as methanol–water–formic acid, for instance) or a non-ionic resin (such as Amberlite XAD 1 or XAD 2) with elution with methanol or methyl ethyl ketone (Upjohn, 1970); (ii) fixation by ion exchange on a carboxylic type of a weak cationic resin (Amberlite IRC 50, XE 89, XE 222, Wofatit CP 300), with elution with a diluted mineral acid, HCl or H$_2$SO$_4$; (iii) chromatography on alumina, followed by elution with methanol; (iv) extraction with a water-immiscible solvent in the presence of a carrier (for instance a 5% mono-2-ethylhexyl phthalate solution in chloroform; Novo, 1967) followed by a re-isolation with water having acid pH; (v) selective precipitation as helianthate, picrate, silicotungstate, reineckate, various sulfonates, *etc.*, which are then converted into chlorhydrate or sulfate; (vi) formation of a Schiff base by reaction with an amine (benzylamine, β-phenethylamine, dibenzylmethylamine, *etc.*); the crystallized product obtained is recovered and later subjected to the action of an acid, sulfuric acid, for instance, which leads to streptomycin sulfate (Olin-Mathieson, 1963).

In addition to those operations, in which streptomycin is primarily involved, there are a certain number of other processing steps which are intended to eliminate impurities without directly interfering with the antibiotic: (i) broth filtration and discarding insolubles; (ii) precipitation of the calcium ions by oxalic acid; (iii) washing of the columns of IRC 50 resin with aqueous solutions of a complexing agent (Complexon III, EDTA) to eliminate the heavy metals and alkaline earth metals, or with aqueous solutions of carbon dioxide under pressure, to eliminate sodium ions (Squibb, 1969); (iv) passage through strong cationic resin (Dowex 50 × 16; Pfizer, 1967) or through a mixture of anionic and strong cationic resins (Amberlite IR 45 + IR 124, Fabrica de Antibiotice, 1974; or IR 4 B + IR 120, Olin-Mathieson, 1956) to eliminate certain impurities, in particular heavy metals and alkaline and alkaline earth metals; (v) addition of carbon DARCO G 60 or passage through a strong alkaline resin (Amberlite IR 401 S) to take the colour off; (vi) filtration through dextrane beds (Sephadex) or polyacrylamide gels (Biogel) or dialysis through cellulose acetate membrane, to eliminate antigens from the purified batches of streptomycin (Beecham, 1968).

Among all the above-mentioned processing steps, a certain number of them have been analyzed in a review by Perlman and Ogawa (1978). Those chosen in industry have to combine, at best, simplicity, economy, yield and quality of the final product, according to the broth treated (composition of the medium, streptomycin concentration). Most of the time, in production plants, after preliminary treatment of the broth (acidification, filtration, neutralization), the antibiotic is successively fixed and eluted in one or several ion-exchange columns; it goes as a solution through other different ion-exchange columns (demineralization, neutralization columns, *etc.*) to get rid of various impurities; the operation ends with decolorization with carbon, followed by concentration under vacuum and drying or lyophilization. A patent was taken out by Squibb (1969) (see Figure 9) for a process based on this principle, but much simplified and therefore very attractive.

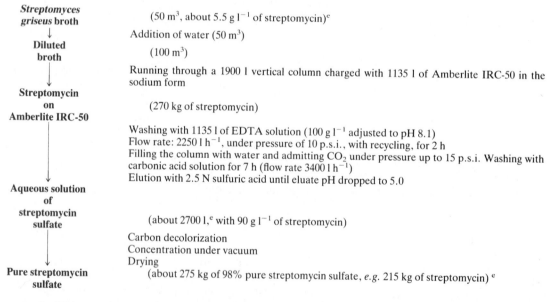

Figure 9   Industrial process of streptomycin recovery. Adapted from Squibb (1969). [e]Estimation

Streptomycin is usually obtained, industrially, as a sulfate salt of approximately 99% purity (790 mg of streptomycin base per gramme of sulfate). The global yield of the preparation, as streptomycin base in the final product in comparison with the broth, is between 75 and 85%.

7.3   OTHER MAJOR AMINOGLYCOSIDE ANTIBIOTICS

The three major aminoside antibiotics discovered before 1960 (streptomycin, neomycin, kanamycin) are, commercially speaking, quite important. However, owing to their weak points (toxicity, particularly marked for neomycin; appearance of a great number of resistances; lack of activity against some very potent pathogens), their expansion on the market has been considerably limited.

In order to remedy this situation, studies have been made for 25 years within this class of antibiotics to find intrinsically more active products, having broader spectrum (active in particular against *Pseudomonas*), lower sensitivity to inactivating enzymes (active against the germs that had become resistant to the aminosides of the first generation) and with lower toxicity. To arrive at this result, several kinds of approaches have been used, and still are; they have led to the aminosides known as the second and the third generations (Davies, 1983): (i) screening and genetic engineering of strains for new aminosides; (ii) modification of the structure of known aminosides; and (iii) chemical synthesis of new aminosides.

Some spectacular results have been obtained, mostly with aminosides having a 2-deoxystreptamine cycle (Nara, 1977, 1978).

7.3.1 Screening and Genetic Engineering of Strains for New Aminosides

7.3.1.1 Screening of new strains

The systematic exploration of new producing strains (in particular by going beyond the traditional field of *Streptomyces* alone, see Table 1) led between 1960 and 1970 to the discovery of a great number of new products, six of them being commercialized: spectinomycin, gentamicin (a mixture of gentamicins C_1, C_{1a} and C_2), tobramycin, lividomycin, ribostamycin and sisomicin (see Figures 10, 11 and 12). Then, other products appeared, such as sagamicin (or micronomicin) and seldomycin both close to gentamicin; verdamicin and G.52, both close to sisomicin; fortimicin and istamycin, all products with original structures. However, at present, none of those more recent products, except sagamicin (gentamicin C_{2b}), are clinically used because, as far as biological properties are concerned, they do not sufficiently differ from the products already found on the market; some of them are even less interesting (Moellering, 1983).

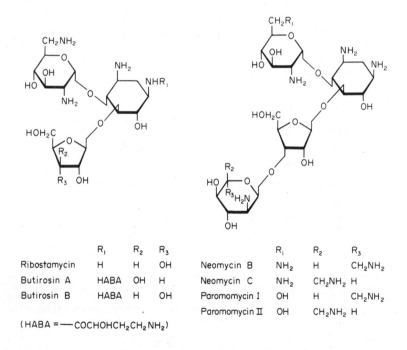

	R_1	R_2	R_3
Ribostamycin	H	H	OH
Butirosin A	HABA	OH	H
Butirosin B	HABA	H	OH

(HABA = —COCHOHCH$_2$CH$_2$NH$_2$)

	R_1	R_2	R_3
Neomycin B	NH$_2$	H	CH$_2$NH$_2$
Neomycin C	NH$_2$	CH$_2$NH$_2$	H
Paromomycin I	OH	H	CH$_2$NH$_2$
Paromomycin II	OH	CH$_2$NH$_2$	H

Figure 10   Structure of aminoglycosides containing 4,5-disubstituted 2-deoxystreptamine

7.3.1.2 Genetic engineering for new strains

The present knowledge in the fields of genetics and genetic engineering allows great hope about the possibilities to obtain original aminosides by using recombinant DNA techniques:

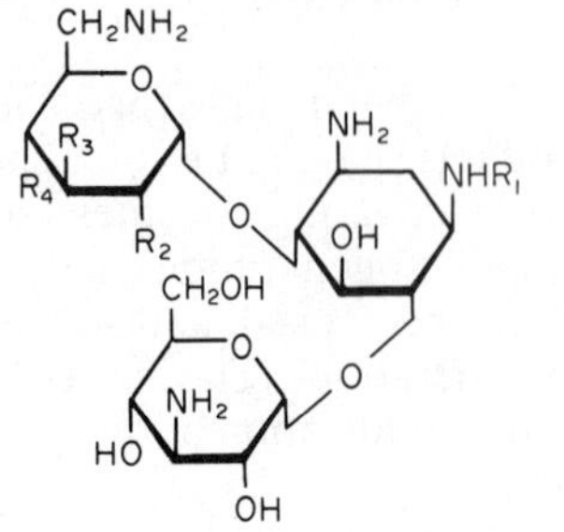

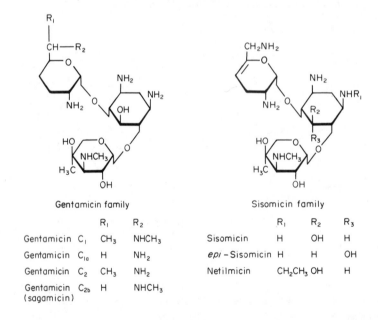

Figure 11 Structure of aminoglycosides containing 4,6-disubstituted 2-deoxystreptamine (kanamycin family)

	R_1	R_2	R_3	R_4
Kanamycin A	H	OH	OH	OH
Kanamycin B	H	NH_2	OH	OH
Tobramycin	H	NH_2	H	OH
Amikacin	HABA	OH	OH	OH
Dibekacin	H	NH_2	H	H
Butikacin	HAB	OH	OH	OH
Habekacin	HABA	NH_2	H	H

$(HABA = \text{—} CO \text{——} CHOH \text{——} CH_2 \text{——} CH_2 \text{——} NH_2)$

$(HAB\ \ = \text{——} CH_2 \text{——} CHOH \text{——} CH_2 \text{——} CH_2 \text{——} NH_2)$

Figure 12 Structure of aminoglycosides containing 4,6-disubstituted 2-deoxystreptamine (gentamicin and sisomicin families)

Gentamicin family	R_1	R_2		Sisomicin family	R_1	R_2	R_3
Gentamicin C_1	CH_3	$NHCH_3$		Sisomicin	H	OH	H
Gentamicin C_{1a}	H	NH_2		*epi*–Sisomicin	H	H	OH
Gentamicin C_2	CH_3	NH_2		Netilmicin	CH_2CH_3	OH	H
Gentamicin C_{2b} (sagamicin)	H	$NHCH_3$					

interspecific recombination *in vivo* (Mazieres *et al.*, 1981) and *in vitro* (Vournakis and Elander, 1983). However, it would seem that, up to now, no interesting products have yet been obtained in that way.

7.3.1.3 *Use of idiotrophic mutants*

A great number of idiotrophic mutants (or blocked mutants) of *Streptomyces* and *Micromonospora*, which normally produce aminosides, have been obtained through classical mutagenesis techniques. Most of them cannot produce the antibiotic normally synthesized if they are not provided with the corresponding aminocyclitol. By providing such mutants with a whole series of aminocyclitols prepared by synthesis, a great number of new aminosides have been discovered (Daum and Lemke, 1979; Demain, 1981; Sepulchre *et al.*, 1980). For instance, a blocked mutant of *Micromonospora inyoensis*, usually a sisomicin producer, has produced the mutabilicins 1 and 2 (Mu 1 and Mu 2) once provided respectively with streptamine and 2,5-dideoxystreptamine; unfortunately these mutabilicins do not show biological properties distinctly superior to those of sisomicin. On the other hand, by providing the same mutant with 5-*epi*-2-deoxystreptamine, it has been possible to obtain the 5-*epi*-sisomicin (Testa *et al.*, 1976) (Figure 12). This product is very active against the sisomicin-resistant germs; the *epi* configuration of the molecule makes it less easily accessible to inactivating enzymes. *epi*-Sisomicin is at present the only aminoside, obtained through this process, known as mutational synthesis (or mutasynthesis), which justifies clinical experiments. It is now commonly produced through the chemical modification of sisomicin, a less expensive process.

7.3.2 Structural Modification of Known Aminosides

It is possible to carry out in a rational way the modification of known aminosides, through hemisynthesis or bioconversion, in order to increase their potentialities, so long as we know: (i) the structure of the products; (ii) the structure–activity relationships (Davies *et al.*, 1969; Philippe *et al.*, 1983; Price *et al.*, 1977); and (iii) the mechanisms of inactivation by the enzymes of the resistant microorganisms. Methods have been described concerning the rapid determination of the toxicity of the products obtained, auditory toxicity in particular (Sato *et al.*, 1983).

The main aims of the work undertaken have been (Le Goffic, 1979): (i) to modify the chemical structure of the target ordinarily hit by a given inactivating enzyme, so that it should stop being the enzyme substrate; (ii) to modify the spatial configuration of the target, which, owing to the high specificity of the enzymes, may not be recognized any more; and (iii) to prevent the formation of the substrate–enzyme complex by modifying the bulk of the molecule.

It is to be noted that a hemisynthesis molecule less sensitive to inactivating enzymes will be interesting only if its biological properties are retained or, even better, increased.

7.3.2.1 Hemisynthesis

In the past 10 to 15 years, the modifications, brought about through chemical means, to existing aminosides on precise points of their molecules have been the object of important and fruitful studies. The question was summed up in a well-documented review by Price *et al.* (1977) and the most notable recent works have been reported by Moellering (1983). It has been possible to direct the experiments using comparisons with natural molecules: *e.g.* (i) the absence of a 3′-hydroxy group in tobramycin makes it insensitive to the enzyme APH (3′), which inactivates kanamycins A and B; (ii) the absence of 3′- and 4′-hydroxy groups in gentamicins makes them insensitive to the enzymes APH (3′) and ANT (4′); and (iii) butirosin B differs from ribostamycin only in the presence of the *N*-hydroxy-γ-aminobutyryl (HABA) substituent in position 1 which gives it anti-*Pseudomonas aeruginosa* activity, not to be found in ribostamycin, and which makes it insensitive to the enzyme APH (3′) I.

The main chemical modifications resulting in very interesting molecules have concerned: (a) the 3′, 4′ and 6′ positions: the activity of the 3′-deoxykanamycin A is close to that of tobramycin (natural 3′-deoxykanamycin B). The 6′-*N*-methylkanamycin A is active against the bacteria containing AAC (6′). The most interesting result was obtained with 3′,4′-dideoxykanamycin B (dibekacin; Umezawa, 1982; Umezawa *et al.*, 1971) active against the bacteria containing APH (3′) and ANT (4′): it is commonly used clinically (see Figure 11); (b) the 1 position: the preparation of 1-*N*-substituted derivatives has led to two very potent semisynthetic aminosides which are at present commercialized: the 1-*N*-HABA-kanamycin A (amikacin, BBK-8; Kawaguchi *et al.*, 1972) and the 1-*N*-ethylsisomicin (netilmicin; Wright, 1976). Amikacin is quite active against *Pseudomonas aeruginosa* and inhibits the bacteria harbouring APH (3′), AAC (3) and ANT (2′). Netilmicin is hardly toxic and is very active.

Another derivative of kanamycin A, the 1-*N*-(2-hydroxy-4-aminobutyl)kanamycin A (butikacin, UK 18892; Richardson *et al.*, 1977) is also very active but its spectrum is virtually the same as that of amikacin.

By associating the transformations in both directions, Tsuchiya *et al.* (1979) obtained the 3'-deoxyamikacin and Iwasawa *et al.* (1982) the 3',6"-dideoxyamikacin. These products are more active than amikacin and have low toxicity; they are therefore potentially very interesting and probably justify clinical experiments. It is on the same basis that the 1-*N*-HABA-dibekacin (habekacin; Kondo *et al.*, 1973) was prepared earlier on.

7.3.2.2 *Bioconversion*

So far, the biotransformation of the aminosides does not seem to have given spectacular results. However, one can mention: (i) the transformation of kanamycin A into amikacin with a mutant strain of *Bacillus circulans* (Cappelletti *et al.*, 1983). That approach can offer an alternative to the costly chemical synthesis of amikacin; (ii) the transformation, carried out in two stages, of gentamicin into 3-*N*-ethylgentamicin: in an enzymatic stage, the acetyltransferase AAC (3) of a gentamicin-resistant *E. coli* strain is used; it gives the 3-acetylgentamicin, which can then be very easily chemically reduced, producing, in a highly specific way, the 3-*N*-ethylgentamicin (Le Goffic *et al.*, 1977). In a similar way, Okutani *et al.* (1977) have prepared the 3'-deoxy derivatives of several aminoglycosides (such as tobramycin from kanamycin B) by successively carrying out enzymatic phosphorylation with a phosphorylase APH (3') of *Pseudomonas aeruginosa*, and catalytic hydrogenation on the same site.

7.3.3 Chemical Synthesis of New Aminosides

Aminosides can be synthesized from the rings constituting them. That is how Umezawa carried out the chemical synthesis of kanamycins A, B and C. It is also possible to operate more easily by linking, for instance, a third ring to a pre-existing bicyclic component.

Several very active aminosides have a neamine residue. The preparation of neamine is very easy; it is obtained through acid hydrolysis of neomycin, a low-cost aminoside. By linking it to a third chemically synthesized ring, then, by preparing the 1-*N*-HABA derivative of the product obtained, Gasc (1979) produced the antibiotic RU 25434, whose level of activity is superior to those of amikacin, tobramycin and gentamicin against a number of resistant strains.

7.4 STREPTOTHRICINS, AMINOGLYCOSIDE-LIKE ANTIBIOTICS

In 1942, Waksman and Woodruff isolated an antibiotic from some culture broths of *Streptomyces lavendulae* which was very active against Gram-positive and Gram-negative bacteria, mycobacteria and fungi. This antibiotic, which they named streptothricin, was one of the very first discovered among Actinomycetales. Very shortly after the discovery, numerous antibiotics very close to streptothricin were described. As the techniques of analysis improved, it appeared that there existed not one but several streptothricins, as well as a great number of closely related products.

Streptothricins are not aminoglycosides in the strict sense of the word. They do contain an amino sugar, but their global structures do not correspond to the definition given in Section 7.1. Only a certain kinship at the chemical level (strongly basic character) and the biological one (broad spectrum of activity and toxicity) relates them to aminosides. A confusion was made possible by the fact that neomycin A or neamine, an aminoglycoside antibiotic from *Streptomyces fradiae*, had been initially named streptothricin. Likewise, at the time, neomycins B and C were named respectively streptothricins B_{II} and B_I.

7.4.1 Structure

The family of genuine streptothricins is made up of seven members. The most simple from the structural point of view is probably the one discovered by Waksman and Woodruff (streptothricin F); it contains three subunits: streptolidine, D-gulosamine carrying a carbamoyl group and L-β-lysine linked to gulosamine by an amide link. The other different streptothricins are produced by varying the length of the peptidic chain (poly-β-lysine): they have been called streptothricins A, B, C, D, E and X (see Figure 13).

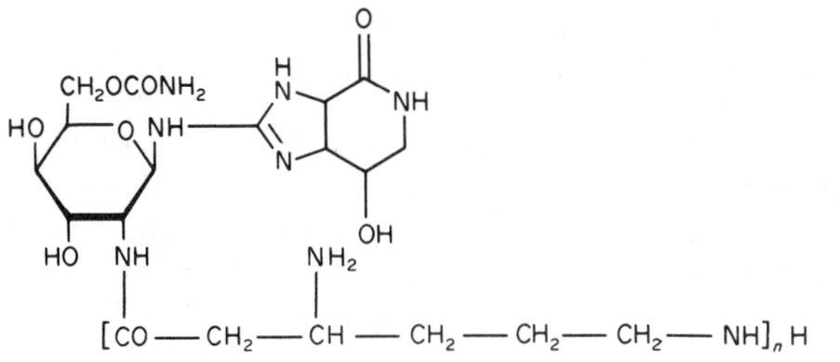

Streptothricins

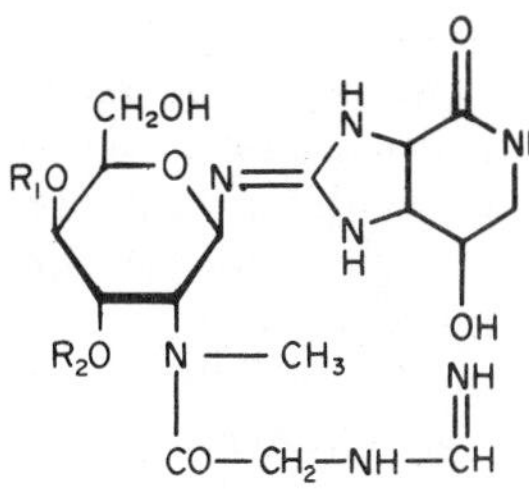

$(R_1, R_2 = H, CONH_2)$

Citromycin

n	1	2	3	4	5	6	7
Streptothricin	F	E	D	C	B	A	X
Racemomycin	A	C	B	D	E		
Boseimycin	I	II	III				
Grasseriomycin	*						
Akimycin	*						

Figure 13 Structure of streptothricins and citromycin

Numerous antibiotics have been identified with one or the other of these seven streptothricins: racemomycins A, B, C, D, E, grasseriomycin, boseimycins I, II, III, akimycin. Khokhlov (1978) has noted about 70 antibiotics, which are in fact mixtures, in fairly well-defined variable proportions, of several streptothricins (roseothricin, streptolin, polymycin, *etc*.). He has also given a review of about ten analogous products differing from streptothricins in some structural modifications on one or several parts of the molecule, citromycin for instance (see Figure 13).

7.4.2 Physicochemical and Biological Properties

Streptothricins are basic substances, very soluble in water, less soluble in lower alcohols and insoluble in non-polar organic solvents. They are very often produced in mixtures; their alkaline strength, varying according to the number of β-lysine units of their peptidic chains, allows them to be easily analyzed by electrophoresis or circular chromatography on paper (solvent; 1-propanol:pyridine:acetic acid:water 15:10:31:12) (Khokhlov, 1978).

Streptothricins are usually prepared as salts (chlorhydrate, sulfate, acetate). They are very active against Gram-negative and Gram-positive bacteria, mycobacteria and numerous clinical isolates resistant to other antibiotics. Unlike aminosides, they are also active against fungi (*Aspergillus*, *Penicillium*, *Trichophyton*) and against some viruses (influenza). Finally, several of them show good insecticidal activity (Kato *et al*., 1983; Kubo *et al*., 1981, 1983). These activities, which are conditioned by the integrity of the lactam nucleus of streptolidin, increase with the number of β-lysine units of the peptidic chain.

Streptothricins are active *in vivo*. The acute toxicity of streptothricin F is low; it is important with all the others. They all show very high chronic toxicity, which prevents their clinical use.

7.4.3 Production by Fermentation and Isolation

Only streptothricin F can be produced independently by fermentation. The others are obtained as mixtures. Streptothricin F (racemomycin A) can be prepared by using, for instance, the process described by Sawada *et al*. (1976). A *Streptomyces lavendulae* strain is grown in a culture medium containing, in g l^{-1}: polypeptone 10, yeast extract 5 and sodium chloride 3. About 0.2 g l^{-1} of racemomycin A is obtained within 35 h. It is better not to incorporate glucose in the medium as sugar represses the production. At the end of the fermentation, the broth, whose pH is near 9, is neutralized and filtered. The filtrate is passed through an Amberlite IRC 50 column in acid cycle and the antibiotic is fixed there. It is eluted with 0.3 N hydrochloric acid. The eluate is neutralized, concentrated and passed through a Sephadex G 10 column, where the antibiotic is adsorbed and then eluted with water. A subsequent purification can be carried out by chromatography on cellulose powder column, the streptothricin is then eluted with a solvent mixture of *n*-butanol, pyridine, acetic acid, water and *t*-butanol and is then lyophilized.

Other methods can be used for the isolation and purification of streptothricin, such as adsorption on activated carbon (Norite) followed by elution with aqueous methanol acid solution and precipitation with acetone, precipitation in the form of picrate or helianthate followed by conversion into chlorhydrate, or chromatography on an alumina column, elution with methanol and then precipitation with acetone or ether. To separate the streptothricins contained in a mixture, an excellent method has been developed by Khokhlov and Reshetov (1965) which consists of chromatography on a carboxymethylcellulose column followed by elution with NaCl solution of 0.19 M, 0.22 M, 0.25 M, 0.30 M, 0.34 M and 0.36 M respectively, to elute streptothricins F, E, D, C, B and A. The separation is fast, with good yields. Later on, Borders (1975) has described an ion-exchange chromatography method which can be applied to streptothricins and analogous products.

7.4.4   Uses

At the time of its discovery, streptothricin was thought to have good prospects in the medical field, yet this antibiotic has never really been used in medicine, particularly because of its toxicity. On the other hand, in the agricultural field it may well be used some day to fight insects or phytopathogenic bacteria, or as an additive to animal feed, provided that its cost does not constitute an insuperable obstacle.

7.5   MARKETING PROSPECTS

Notwithstanding the great importance of the semisynthetic penicillins and cephalosporins, the place of the aminoglycosides on the antibiotic market is far from negligible. However, the market varies according to the products. Streptomycin and neomycin come first by far, as far as quantity is concerned. On the other hand, the global turnover of these two antibiotics was, in 1977, lower than that of gentamicin, according to the results of a survey published in 1980 by the Stanford Research Institute (Table 7).

Table 7   World-wide Market of the Most Commercially Important Aminoglycosides[a]

| Product | World production (metric tons) | Market value ($ 10^6$) | Price ($ kg^{-1}$) | World market (%) | | | | |
				North America	Latin America	Western Europe	Asia	Africa + Oceania
Streptomycins	785	38.5	49	8	21	14	40	17
Neomycin	600	30	50	52	13	21	10	4
Kanamycin	42[b]	8.4	200	20	4	24	38	14
Gentamicin	7.2	82.8	11 500[c]	30	21	26	19	4
Amikacin	5.2	2.1	400	21	13	25	33	8
Tobramycin	2.4	3.6	1 500	21	13	25	33	8

[a] Stanford Research Institute, 1977.   [b] Excludes quantity for making amikacin.   [c] $800–1000, in 1982, patent having lapsed.

One half of the streptomycin market is constituted by streptomycin proper and the other half by dihydrostreptomycin. Two thirds of the whole are intended for medical use, the rest for veterinary use. These products are widely used in Asia and in Latin America. In 1983, the market can be estimated at about 1200 tons per year, two thirds of which is manufactured by three firms: Pfizer (USA), Rhône-Poulenc (France) and Meiji (Japan); the rest is produced by India, the Eastern European countries including the USSR, Brazil and Spain. Neomycin has an important place on the market despite its toxicity which prevents its parenteral use for human therapy. 50% of the production is used for animal therapy. The main manufacturers are Upjohn (USA) and Roussel-UCLAF (France) which, together, make up 75% of this developing market. Kanamycin comes third in quantity. It is manufactured mainly by Bristol-Myers (USA) and Meiji (Japan). It is used exclusively for human therapy, like its hemisynthesis derivative, amikacin, whose market is held by Bristol-Myers. In 1977, gentamicin ranked first among the aminosides in its turnover. Since then, its price has been considerably reduced, but remains high. It is manufactured nearly exclusively by Schering-Plough (USA). It is beginning to be seriously threatened by tobramycin, produced by Eli Lilly (USA), and recently, it has also been challenged by sisomicin, manufactured by Schering-Plough. The kanamycin B (bekanamycin, kanendomycin), spectinomycin, ribostamycin and lividomycin markets are not very important. On the contrary, the dibekacin one is rather active. The netilmicin and *epi*-sisomicin markets will very likely grow in the future. As there are still many research opportunities in this family of antibiotics, it is anticipated that, in spite of the important place of semisynthetic penicillins and cephalosporins on the market, the clinical use of aminoglycosides is not anywhere near to disappearing.

7.6 REFERENCES

Alikhanyan, S. I. and T. A. Shishkina (1967). Induced variation in *Actinomyces streptomycini* in relation to antibiotic formation and proteolytic activity. *Genetika*, 3, 159–164.

Alikhanyan, S. I. and A. F. Teteryatnik (1962). Formation of streptomycin-producing variants from mutants of *Actinomyces streptomycini* strain LS 1 through the action of actinophages. *Mikrobiologiya*, 31, 54–60.

Arai, T., S. Kuroda, S. Yamagishi and Y. Katoh (1964). A new hydroxystreptomycin source, *Streptomyces subrutilus*. *J. Antibiot.*, *Ser. A*, 17, 23–28.

Aszalos, A. and D. Frost (1975). Thin-layer chromatography of antibiotics. *Methods Enzymol*, 43, 172–213.

Awata, M., S. Satoi, N. Muto, M. Hayashi, H. Sagai and H. Sakakibara (1983). Saccharocin, a new aminoglycoside antibiotic: fermentation, isolation, characterization and structural study. *J. Antibiot.*, 36, 651–655.

Bannister, B. and A. D. Argoudelis (1963). The chemistry of bluensomycin. II. The structure of bluensomycin. *J. Am. Chem. Soc.*, 85, 234–235.

Beecham Group Ltd (1968). *Neth. Pat.* 68 06 832.

Benedict, R. G., L. A. Lindenfelser, F. H. Stodola and D. H. Traufler (1951). Studies on *Streptomyces griseocarneus* and the production of hydroxystreptomycin. *J. Bacteriol.*, 62, 487–497.

Betina, V. (1975). Paper chromatography of antibiotics. *Methods Enzymol.*, 43, 100–172.

Borisova, L. N., N. S. Ivkina, L. A. Gomenyuk and I. A. Rapoport (1970). Mutations and expression of characters in *Actinomyces streptomycini*. *Mikrobiologiya*, 39, 675–679.

Borders, D. B. (1975). Ion-exchange chromatography of streptothricin-like antibiotics. *Methods Enzymol.*, 43, 256–263.

Bormann, E. J. (1979). *Ger. (East) Pat.* 137 945.

Cappelletti, L. M. and R. Spagnoli (1983). Biological transformation of kanamycin to amikacin (BBK-8). *J. Antibiot.*, 36, 328–330.

Claridge, C. A. (1979). Aminoglycoside antibiotics. In *Economic Microbiology*, ed. A. H. Rose, vol. 3, pp.151–238. Academic, London.

Cojocel, C. and J. B. Hook (1983). Aminoglycoside nephrotoxicity. *Trends Pharmacol. Sci.*, 4, 174–179.

Dalhoff, A. (1983). Studies on the action of aminoglycosides on bacterial membranes *Zbl. Bakteriol. 1 Abt. Orig. A*, 253, 427.

Daniels, P. J. L. (1978). Aminoglycosides. In *Kirk-Othmer Encyclopedia of Chemical Technology*. vol. 2, 3rd edn., pp. 819–852. Wiley-Interscience, New York.

Daum, S. J. and P. A. Lemke (1979). Mutational biosynthesis of new antibiotics. *Annu. Rev. Microbiol.*, 33, 241–265.

Daum, S. J., D. Rosi and W. A. Goss (1977). Mutational biosynthesis by idiotrophs of *Micromonospora purpurea*. II. Conversion of non-amino containing cyclitols to aminoglycoside antibiotics. *J. Antibiot.*, 30, 98–105.

Davies, J. E. (1983). Resistance to aminoglycosides: mechanisms and frequency. *Rev. Infect. Dis.*, 5 (Suppl. 2), S261–S267.

Davies, J. E. and D. I. Smith (1978). Plasmid-determined resistance to antimicrobial agents. *Annu. Rev. Microbiol.*, 32, 469–518.

Davies, J. E. and M. Yagisawa (1983). The aminocyclitol glycosides (aminoglycosides). In *Biochemistry and Genetic Regulation of Commercially Important Antibiotics*, ed. L. C. Vining, pp. 329–354. Addison-Wesley, Reading, MA.

Davies, J. E., C. Houk, M. Yagisawa and T. J. White (1979). Occurrence and function of aminoglycoside-modifying enzymes. In *Proceedings of the 3rd International Symposium on Genetics of Industrial Microorganisms*, ed. O. K. Sebek and A. J. Laskin, pp. 166–169. American Society for Microbiology, Washington, DC.

Davies, J. E., R. Benveniste, K. Kviteck, B. Ozanne and T. Yamada (1969). Aminoglycosides: biological effects of molecular manipulation. *J. Infect. Dis.*, 119, 351–354.

Demain, A. L. (1981). Directed and mutational biosynthesis. In *The Future of Antibiotherapy and Antibiotic Research*, ed. L. Ninet, P. E. Bost, D. H. Bouanchaud and J. Florent, pp. 417–435. Academic, London.

Demain, A. L. and E. Inamine (1970). Biochemistry and regulation of streptomycin and mannosidostreptomycinase (α-D-mannosidase) formation. *Bacteriol. Rev.*, 34, 1–19.

Duda, E. (1973). A new sensitive method for the determination of streptomycin. *Anal. Biochem.*, 51, 651–653.

Dulaney, E. L. (1953). Observations on *Streptomyces griseus*. VI. Further studies on strain selection for improved streptomycin production. *Mycologia*, 45, 481–487.

Dyer, J. R. and A. W. Todd (1963). The absolute configuration of streptidine in streptomycin. *J. Am. Chem. Soc.*, 85, 3896–3897.

Fabrica de Antibiotice (1974). *Br. Pat.* 1 377 293.

Fee, W. E., Jr. (1983). Gentamicin and tobramycin: comparison of ototoxicity. *Rev. Infect. Dis.*, 5, S304–S313.

Ferguson, J. H., H. T. Huang, J. W. Huang and J. W. Davisson (1957). Stimulation of streptomycin production by a series of synthetic organic compounds. *Appl. Microbiol.*, 5, 339–343.

Franklin, T. J. and G. A. Snow (1981). Suppression of gene function 2. Interference with the translation of the genetic message: inhibitors of protein synthesis. In *Biochemistry of Antimicrobial Action*. 3rd edn., pp. 110–137. Chapman and Hall, London.

Gale, E. F., E. Cundliffe, P. E. Reynolds, M. H. Richmond and M. J. Waring (1981). Antibiotic inhibitors of ribosome function. In *The Molecular Basis of Antibiotic Action*, 2nd edn., pp. 402–547.

Gasc, J. C. (1979). Aminosides de synthèse: recherches et problèmes industriels. *Actualités de Chimie Thérapeutique*, 6è série, pp. 95–120. Société de Chimie Thérapeutique, Chatenay-Malabry (France).

Grisebach, H. (1978). Biosynthesis of sugar components of antibiotic substances. *Adv. Carbohydr. Chem. Biochem.*, **35**, 81–126.

Grisebach, H., B. Kniep and H. Wahl (1981). Biosynthesis of streptomycins. In *Advances in Biotechnology*, ed. M. Moo-Young, C. Vezina and K. Singh, vol. III, pp. 95–99. Pergamon, Toronto.

Grove, D. C. and W. A. Randall (1955). Streptomycin and dihydrostreptomycin. In *Antibiotics Monographs*, ed. H. Welch and F. Marti-Ibanez, vol. 2, pp. 34–43 and 211–213. Medical Encyclopedia, New York.

Halliday, W. J. (1952). A new colour reaction of streptomycin. *Nature (London)*, **169**, 335–336.

Hamada, M., S. Kondo, T. Yokoyama, K. Miura, K. Iinuma, H. Yamamoto, K. Maeda, T. Takeuchi and H. Umezawa (1974). Minosaminomycin, a new antibiotic containing *myo*-inosamine. *J. Antibiot.*, **27**, 81–83.

Hara, O. and T. Beppu (1982). Mutants blocked in streptomycin production in *S. griseus*. The role of A-factor. *J. Antibiot.*, **35**, 349–358.

Hirose-Kumagai, A., A. Yagita and N. Akamatsu (1982). UDP-*N*-methyl-D-glucosamine phosphate, a possible intermediate of *N*-methyl-L-glucosamine moiety of streptomycin. *J. Antibiot.*, **35**, 1571–1577.

Hockenhull, D. J. D., K. H. Fantes, M. Herbert and B. Whitehead (1954). Glucose utilization by *Streptomyces griseus*. *J. Gen. Microbiol.*, **10**, 353–370.

Hopwood, D. A., H. M. Wright, M. J. Bibb and S. N. Cohen (1977). Genetic recombination through protoplast fusion in *Streptomyces*. *Nature (London)*, **268**, 171–174.

Horner, W. H. (1967). Streptomycin. In *Antibiotics*, ed. D. Gottlieb and P. D. Shaw, vol. II, pp. 373–399. Springer Verlag, Berlin.

Hugues, D. W., A. Vilim and W. L. Wilson (1978). Chemical and physical analysis of antibiotics. III. Aminoglycosides. *Can. J. Pharm. Sci.*, **13**, 21–30

Hutter, R. (1967). *Systematik der Streptomyceten*. Karger Verlag, Basel.

Inamine, E., B. D. Lago and A. L. Demain (1969). Regulation of mannosidase, an enzyme of streptomycin biosynthesis. In *Fermentation Advances*, ed. D. Perlman, pp.199–209. Academic, New York.

Inoue, S., Y. Nishizawa and S. Nagai (1983). Stimulatory effect of ammonium on streptomycin formation by *Streptomyces griseus* growing on a glucose minimal medium: study on the role of glutamic acid and glutamine in streptomycin formation. *J. Ferment. Technol.*, **61**, 7–12.

Inoue, S., K. Ohba, T. Shomura, M. Kojima, T. Tsuruoka, J. Yoshida, N. Kato, M. Ito, S. Amano, S. Omoto, N. Ezaki, T. Ito and T. Niida (1979). A novel aminoglycoside antibiotic, substance SF-2052. *J. Antibiot.*, **32**, 1354–1356.

Iwasawa, H., D. Ikeda, S. Kondo, U. Umezawa (1982). Chemical modification of 3′-deoxyamikacin. *J. Antibiot.*, **35**, 1715–1718.

Johnstone, D. B. and S. A. Waksman (1948). The production of streptomycin by *Streptomyces bikiniensis*. *J. Bacteriol.*, **55**, 317–328.

Kato, Y., M. Kubo, K. Morisaka, Y. Waku, K. Hayashiya and Y. Inamori (1983). The mechanisms of delayed insecticidal action of streptothricin antibiotics. II. Effect of racemomycin-D on excretion function of the 5th instar larvae of silkworm, *Bombyx mori* Linne. *Chem. Pharm. Bull.*, **31**, 305–311.

Kawaguchi, H., T. Naito, S. Nakagawa and K. Fujisawa (1972). BB K-8, a new semisynthetic aminoglycoside antibiotic. *J. Antibiot.*, **25**, 695–708.

Khokhlov, A. S. (1978). Streptothricins and related antibiotics. *J. Chromatogr. Lib.*, **15**, 617–713.

Khokhlov, A. S. and P. D. Reshetov (1964). Chromatography of streptothricins on carboxymethylcellulose. *J. Chromatogr.*, **14**, 495–496.

Khokhlov, A. S., I. I. Tovarova, L. N. Borisova, S. A. Pliner, L. A. Shevchenko, E. Y. Kornitskaia, N. S. Ivkina and I. A. Rapoport (1967). A-Factor securing the biosynthesis of streptomycin by a mutant strain of *Actinomyces streptomycini*. *Dokl. Akad. Nauk SSSR*, **177**, 232–235.

Kniep, B. and H. Grisebach (1980a). Purification and properties of a dTDP-L-dihydrostreptose: streptidine-6-phosphate dihydrostreptosyltransferase from *Streptomyces griseus*. *Eur. J. Biochem.*, **105**, 139–144.

Kniep, B. and H. Grisebach (1980b). Biosynthesis of streptomycin. Enzymatic formation of dihydrostreptomycin-6-phosphate from dihydrostreptosylstreptidine-6-phosphate. *J. Antibiot.*, **33**, 416–419.

Kondo, S., K. Iinuma, H. Yamamoto, K. Maeda and H. Umezawa (1973). Synthesis of 1-*N*-[(S)-4-amino-2-hydroxybutyryl]-kanamycin B and 3′,4′-dideoxykanamycin B active against kanamycin-resistant bacteria. *J. Antibiot.*, **26**, 412–415.

Kondo, S., M. Sezaki, M. Koike, M. Shimura, E. Akita, K. Satoh and T. Hara (1965). Destomycins A and B, two new antibiotics produced by a *Streptomyces*. *J. Antibiot.*, *Ser. A*, **18**, 38–42.

Kondo, S., H. Yumoto, T. Miyakawa, K. Hamamoto, M. Sezaki, K. Sato and T. Niida (1962). *Meiji Seika Kenkyu Nempo*, **5**, 9–13.

Korzybski, T., Z. Kowszyk-Gindifer and W. Kurilowicz (1978). Streptomycin. In *Antibiotics — Origin, Nature and Properties*, vol. I, pp. 570–580. American Society for Microbiology. Washington, DC.

Kowa (1968). *Neth. Pat.* 68 08 802.

Kowa (1978). *Ger. Pat.* 2 813 021.

Kubo, M., Y. Kato, K. Morisaka, Y. Inamori, K. Nomoto, T. Takemoto, M. Sakai, Y. Sawada and H. Taniyama (1981). Insecticidal activity of streptothricin antibiotics. *Chem. Pharm. Bull.*, **29**, 3727–3730.

Kubo, M., Y. Kato, K. Morisaka, K. Nomoto and Y. Inamori (1983). The mechanisms of delayed insecticidal action of streptothricin antibiotics. I. Toxic symptoms and distribution of racemomycin-D into the tissues of the 5th instar larvae of silkworm, *Bombyx mori* Linne. *Chem. Pharm. Bull.*, **31**, 325–329.

Kuehl, F. A., Jr., R. L. Peck, C. E. Hoffhine, Jr., E. W. Peel and K. Folkers (1947). *Streptomyces* antibiotics. XIV. The position of the linkage of streptobiosamine to streptidine in streptomycin. *J. Am. Chem. Soc.*, **69**, 1234.

Kurosawa, H. (1951). Mycological characters of antagonistic *Streptomyces*. I. On the correlation between Pridham's classification method and antibiotic character. *J. Antibiot.*, *Ser. A*, **4**, 183–193.

Kyowa Hakko Kogyo (1973). *Ger. Pat.* 2 326 781.

Kyowa Hakko Kogyo (1974). *Ger. Pat.* 2 418 349.

Kyowa Hakko Kogyo (1975). *Ger. Pat.* 2 450 411.

Lazowski, J. and A. Lypka (1983). Aminoglycoside antibiotics. Clinical pharmacology and pharmacokinetics. *Farm. Pol.*, **39**, 7–13.

Le Goffic, F. (1979). Antibiotiques aminoglycosidiques et leurs récepteurs. In *Actualités de Chimie Thérapeutique*, 6è série, pp. 61–78. Société de Chimie Thérapeutique, Chatenay-Malabry (France).

Le Goffic, F., S. Sicsic and C. Vincent (1977). Hémisynthèse d'antibiotiques aminoglycosidiques. I. Mise au point d'un réacteur enzymatique à cofacteur. *Biochimie*, **59**, 927–932.

Lietman, P. S. and C. R. Smith (1983). Aminoglycoside nephrotoxicity in humans. *Rev. Infect. Dis.*, **5**, S284–S291.

Maier, S. and H. Grisebach (1979). Biosynthesis of streptomycin. Enzymatic oxidation of dihydrostreptomycin-6-phosphate to streptomycin-6-phosphate with a particulate fraction of *Streptomyces griseus*. *Biochim. Biophys. Acta*, **586**, 231–241.

Mann, R. L. and W. W. Bromer (1958). The isolation of a second antibiotic from *Streptomyces hygroscopicus*. *J. Am. Chem. Soc.*, **80**, 2714–2716.

Mason, D. J., A. Dietz and R. M. Smith (1961). Actinospectacin, a new antibiotic. I. Discovery and biological properties. *Antibiot. Chemother.*, **11**, 118–122.

Mazieres, N., M. Peyre and L. Penasse (1981). Interspecific recombination among aminoglycoside-producing streptomycetes. *J. Antibiot.*, **34**, 544–550.

Meiji (1969). *Neth. Pat.* 68 18 105.

Merck and Co. (1951). *US Pat.* 2 538 942.

Moellering, R. C., Jr. (1983). *In vitro* antibacterial activity of the aminoglycoside antibiotics. *Rev. Infect. Dis.*, **5**, S212–S232.

Murase, M., T. Takita, K. Ohi, H. Kondo, Y. Okami and H. Umezawa (1959). Identification of an antibiotic produced by *S. galbus* n. sp. with streptomycin. *J. Antibiot.*, *Ser. A*, **12**, 126–132.

Nara, T. (1977). Aminoglycoside antibiotics. In *Annual Reports on Fermentation Processes*, ed. D. Perlman, vol. 1, pp. 299–326. Academic, New York.

Nara, T. (1978). Aminoglycoside antibiotics. In *Annual Reports on Fermentation Processes*, ed. D. Perlman, vol. 2, pp. 223–266. Academic, New York.

Novo Terapeutisk Laboratorium (1967). *Danish Pat.* 106 272.

Okami, Y., K. Hotta, M. Yoshida, D. Ikeda, S. Kondo and H. Umezawa (1979). New aminoglycoside antibiotics, istamycins A and B. *J. Antibiot.*, **32**, 964–966.

Okutani, T., T. Asako, K. Yoshioka, K. Hiraga and M. Kida (1977). Conversion of aminoglycosidic antibiotics: Novel and efficient approaches to 3'-deoxyaminoglycosides *via* 3'-phosphoryl esters. *J. Am. Chem. Soc.*, **99**, 1278–1279.

Olin-Mathieson (1956). *US Pat.* 2 765 302.

Olin-Mathieson (1963). *Br. Pat.* 927 486.

Parke-Davis (1958). *Br. Pat.* 797 568.

Parke-Davis (1969). *Neth. Pat.* 69 04 408.

Perlman, D. and Y. Ogawa (1978). Streptamine-containing antibiotics. *J. Chromatogr. Lib.*, **15**, 587–616.

Pfizer (1967). *US Pat.* 3 313 694.

Philippe, M., B. Quiclet Sire, A. M. Sepulchre and S. D. Gero (1983). Role of the 1-amino group in aminocyclitol antibiotics: synthesis of 1-deaminogentamicin C 2. *J. Antibiot.*, **36**, 250–255.

Pierce, C. J. and K. L. Rinehart, Jr. (1981). Biosynthesis of aminocyclitol antibiotics. In *Antibiotics*, ed. J. W. Corcoran, vol. IV, pp. 74–100. Springer Verlag, Berlin.

Price, K. E., J. C. Godfrey, H. Kawaguchi (1977). Effect of structural modifications on the biological properties of aminoglycoside antibiotics containing 2-deoxystreptamine. In *Structure–Activity Relationships among the Semisynthetic Antibiotics*, ed. D. Perlman, pp. 239–395. Academic, New York.

Rake, G. and R. Donovick (1946). Studies on the nutritional requirements of *S. griseus* for the formation of streptomycin. *J. Bacteriol.*, **52**, 222–226.

Richardson, K., S. Jevons, J. W. Moore, B. C. Ross and J. R. Wright (1977). Synthesis and antibacterial activities of 1-*N*-[(S)-ω-amino-2-hydroxyalkyl]-kanamycin A derivatives. *J. Antibiot.*, **30**, 843–846.

Rinehart, K. L. Jr. and R. M. Stroshane (1976). Biosynthesis of aminocyclitol antibiotics. *J. Antibiot.*, **29**, 319–353.

Sato, K., T. Saito and T. Matsuhira (1983). Comparative study by scanning electron microscopy on vestibular toxicities of dibekacin, ribostamycin and other aminoglycoside antibiotics in guinea pigs. *Int. J. Clin. Pharmacol. Ther. Toxicol.*, **21**, 109–114.

Satoh, A., H. Ogawa and Y. Satomura (1975). Effect of sclerin on production of the aminoglycoside antibiotics accompanied by salvage function in *Streptomyces*. *Agric. Biol. Chem.*, **39**, 1593–1598.

Savage, G. M. (1949). Improvement in streptomycin-producing strains of *Streptomyces griseus* by ultraviolet and X-ray energy. *J. Bacteriol.*, **57**, 429–441.

Sawada, Y., H. Sakamoto, T. Kubo and H. Taniyama (1976). Biosynthesis of streptothricin antibiotics. II. Catabolite inhibition of glucose on racemomycin-A production. *Chem. Pharm. Bull.*, **24**, 2480–2485.

Sawazaki, T., S. Suzuki, G. Nakamura, M. Kawazaki, S. Yamashita, K. Isono, K. Anzai, Y. Serizawa and Y. Sekiyama (1955). Streptomycin production by a new strain, *Streptomyces mashuensis*. *J. Antibiot.*, *Ser. A*, **8**, 44–47.

Schatz, A., E. Bugie and S. A. Waksman (1944). Streptomycin, a new substance exhibiting antibiotic activity against Gram+ and Gram− bacteria. *Proc. Soc. Exp. Biol. Med.*, **55**, 66–69.

Schering (1963). *US Pat.* 3 091 572.

Schering (1969). *Neth. Pat.* 69 09 642.

Schering (1973). *Belg. Pat.* 787 758.

Schering (1976). *US Pat.* 3 956 068.

Schlessinger, D. and G. Medoff (1975). Streptomycin, dihydrostreptomycin and the gentamicins. In *Antibiotics*, ed. J. W. Corcoran and F. E. Hahn, vol. III, pp. 535–550. Springer Verlag, Berlin.

Schwenzer, K. S. and J. P. Anhalt (1983). Automated fluorescence polarization immunoassay for monitoring streptomycin. *Antimicrob. Agents Chemother.*, **23**, 683–687.

Sepulchre, A.-M., B. Quiclet and S. D. Gero (1980). Bioconversion dans le domaine des antibiotiques aminocyclitolglycosidiques. *Bull. Soc. Chim. Fr.*, **1–2**, II 56–II 65.

Shibata, M. (1959). *J. Antibiot.*, *Ser. B.*, **12**, 398–400.

Shirato, S. and H. Motoyama (1965). Fermentation studies with *Streptomyces griseus*. I. Carbohydrate sources for the production of protease and streptomycin. *Appl. Microbiol.*, **13**, 669–672.

Shirato, S. and H. Motoyama (1966). Fermentation studies with *Streptomyces griseus*. II. Synthetic media for the production of streptomycin. *Appl. Microbiol.*, **14**, 706–710.

Shishkina, T. A. (1967). Comparative investigation of the proteolytic activity of different strains of *Actinomyces streptomycini*. *Prikl. Biokhim. Mikrobiol.*, **3**, 64–69.

Siegerman, H. (1975). Differential pulse polarography of antibiotics. *Methods Enzymol.*, **43**, 373–388.

Singh, A., E. Bruzelius and H. Heding (1976). Streptomycin. A fermentation study. *Eur. J. Appl. Microbiol.*, **3**, 97–101.

Squibb (1969). *US Pat.* 3 451 992.

Stark, W. M., M. M. Hoehn and N. G. Knox (1968). Nebramycin, a new broad-spectrum antibiotic complex. I. Detection and biosynthesis. *Antimicrob. Agents Chemother.*, 314–323.

Tanaka, N. (1975). Aminoglycoside antibiotics. In *Antibiotics*, ed. J. W. Corcoran and F. E. Hahn, vol. III, pp. 340–364. Springer Verlag, Berlin.

Tatsuoka, S., T. Kusaka, A. Miyake, M. Inoue, H. Hitomi, Y. Shiraishi, H. Iwasaki and M. Imanishi (1957). Antibiotics. XVI. Isolation and identification of dihydrostreptomycin produced by a new *Streptomyces: Streptomyces humidus* nov. sp. *Chem. Pharm. Bull.*, **5**, 343–349.

Testa, T. R., G. H. Wagman, P. J. L. Daniels and M. J. Weinstein (1976). *Abst. 76th Annu. Meeting of Am. Soc. Microbiol.*, Atlantic City, NY, No. 223.

Toyo Jozo (1980). *Ger. Pat.* 3 013 210.

Tsuchiya, T., T. Jikihara, T. Miyake, S. Umezawa, M. Hamada and H. Umezawa (1979). 3′-Deoxyamikacin and 3′,4′-dideoxyamikacin and their antibacterial activities. *J. Antibiot.*, **32**, 1351–1353.

Umezawa, H. (1982). Découverte de la dibékacine et de ses aspects chimiques. *Nouv. Presse Méd.*, **11**, 3379–3384.

Umezawa, H. and S. Kondo (1975). Electrophoresis of antibiotics. *Methods Enzymol.*, **43**, 279–290.

Umezawa, S., Y. Takahashi, T. Usui and T. Tsuchiya (1974). Total synthesis of streptomycin. *J. Antibiot.*, **27**, 997–999.

Umezawa, H., S. Umezawa, T. Tsuchiya and Y. Okasaki (1971). 3′,4′-Dideoxykanamycin B action against kanamycin-resistant *Escherichia coli* and *Pseudomonas fluorescens*. *J. Antibiot.*, **24**, 485–487.

Umezawa, H., M. Ueda, K. Maeda, K. Yagishita, S. Kondo, Y. Okami, R. Utahara, Y. Osato, K. Nitta and T. Takeuchi (1957). Production and isolation of a new antibiotic, kanamycin. *J. Antibiot.*, *Ser. A*, **10**, 181–188.

Upjohn (1970). *US Pat.* 3 515 717-S.

Vazquez, D., M. J. Cabanas, A. Gonzalez, A. Jimenez and J. Modolell (1979). Effects of the aminoglycoside antibiotics in translocation by bacterial and eukaryotic ribosomes. In *Actualités de Chimie Thérapeutique*, 6è série, pp. 81–91. Société de Chimie Thérapeutique, Chatenay-Malabry (France).

Vournakis, J. N. and R. P. Elander (1983). Genetic manipulation of antibiotic-producing microorganisms. *Science*, **219**, 703–709.

Wagman, G. H. and M. J. Weinstein (1973). Streptomycin. *J. Chromatogr. Lib.*, **1**, 174–175.

Wahl, H. P. and H. Grisebach (1979). Biosynthesis of streptomycin. dTDP-dihydrostreptose synthase from *Streptomyces griseus* and dTDP-4-keto-L-rhamnose 3,5 epimerase from *S. griseus* and *Escherichia coli* Y10. *Biochim. Biophys. Acta*, **568**, 243–252.

Waksman, S. A. and H. B. Woodruff (1942). Streptothricin, a new selective bacteriostatic and bactericidal agent, particularly active against Gram-negative bacteria. *Proc. Soc. Exp. Biol. Med.*, **49**, 207–210.

Waksman, S. A. and H. A. Lechevalier (1949). A Neomycin, a new antibiotic against streptomycin-resistant bacteria, including tuberculosis organisms. *Science*, **109**, 305–307.

Walker, J. B. (1980). Biosynthesis of aminoglycoside antibiotics. *Dev. Ind. Microbiol.*, **21**, 105–113.

Wallace, B. J., P.-C. Tai and B. D. Davis (1979). Streptomycin and related antibiotics. In *Antibiotics*, ed. F. E. Hahn, vol. 5, pp. 272–303. Springer Verlag, Berlin.

Weiss, P. J., M. L. Andrew and W.W. Wright (1957). Solubility of antibiotics in twenty four solvents: use in analysis. *Antibiot. Chemother.*, **7**, 374–377.

Whall, T. J. (1981). Determination of streptomycin sulfate and dihydrostreptomycin sulfate by high-performance liquid chromatography. *J. Chromatogr.*, **219**, 89–100.

Wright, J. J. (1976). Synthesis of 1-*N*-ethylsisomicin: a broad-spectrum semisynthetic aminoglycoside antibiotic. *J. Chem. Soc.*, *Chem. Commun.*, 206–208.

Woodruff, H. B. and M. Ruger (1948). Studies on the physiology of a streptomycin-producing strain of *Streptomyces griseus* on proline medium. *J. Bacteriol.*, **56**, 315–321.

Zhuang, Z., Y. Zhu, H. Tan and Y. Xue (1982). Plasmid in relation to streptomycin biosynthesis: isolation from *S. griseus* (conference abstract). *Proceedings of the 4th International Symposium on the Genetics of Industrial Microorganisms*, p. 76.

8
Cephalosporins

A. SMITH
Ciba-Geigy AG, Basle, Switzerland

8.1 INTRODUCTION

Thirty years have passed since the discovery at Oxford, England of the first cephalosporin β-lactam, cephalosporin C. From this compound, present in such trace amounts in the original cultures as to be below the limits of detection then available, has developed a multi-million dollar industry. Scientists from many different disciplines ranging from geneticists to chemical engineers and working in Britain, USA, Europe and Japan have, by their combined efforts, produced an impressive array of drugs to combat infectious disease and unravelled many of the mysteries surrounding the biosynthesis of the cephalosporin antibiotics.

The cephalosporins are just part of a large group of β-lactams produced by microorganisms. The story of their discovery and development would be incomplete without reference to the earlier discovery of penicillin. Alexander Fleming made his rather fortuitous discovery in 1929 with the observation that growth of bacteria on an agar plate was inhibited by a mould contaminant which turned out to be *Penicillium notatum*. Years elapsed before this discovery was finally exploited by Florey and Chain who isolated the active compound and initiated a new era in the treatment of infectious disease. Penicillin G however appeared to be limited to use against Gram-positive bacteria; its relatively low acid stability precluded oral administration; a small percentage of patients developed allergic reactions and, perhaps more important, there was an increasing number of penicillin-resistant isolates appearing.

Like penicillin before it, the cephalosporin C story had a rather curious beginning. Giuseppe Brotzu, working at an institute at Cagliari on the island of Sardinia, isolated from seawater near a sewage outlet an organism which had antibiotic activity against both Gram-positive and Gram-negative bacteria. The organism was identified as a *Cephalosporium* species and was subsequently sent to Oxford in 1948. The discovery of cephalosporin C was still five years into the future. The early work with the Brotzu isolate revealed two compounds which probably account for the observations made in Sardinia. Cephalosporin P had activity against Gram-positive bacteria and subsequently proved to be less interesting than cephalosporin N which had Gram-negative activity. Cephalosporin N was found to be a β-lactam with the fused β-lactam and thiazolidine rings of a penicillin and was therefore renamed penicillin N. The compound was first purified by Abraham *et al.* (1954) and its structure elucidated by Newton and Abraham (1954). During the purification of crude preparations of penicillin N a similar type of compound was detected which had strong UV absorption, was relatively acid stable, and was resistant to the enzyme penicillinase which destroys penicillin N. Thus, in the autumn of 1953 a new chapter in the β-lactam story began. The cephalosporins industry has been founded largely on derivatives of the cephalosporin C produced by *Cephalosporium acremonium*. A mutant strain M8650 isolated at Clevedon Antibiotics Research Station, England permitted work on the isolation of the new antibiotic to proceed and led to the publication of its structure by Abraham and Newton (1961) (see Figure 1.).

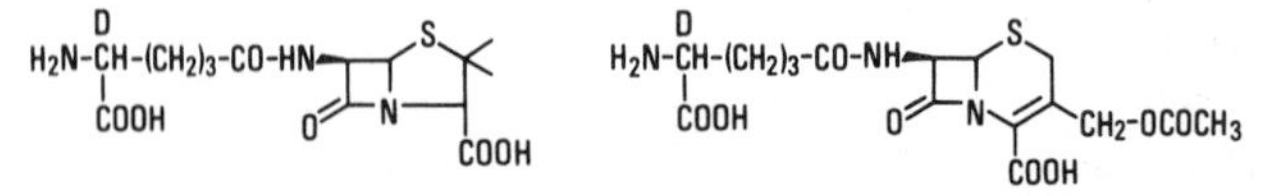

Figure 1   Structural formulae of penicillin N and cephalosporin C

For many years it appeared that the β-lactam antibiotics were confined to just a few species of organisms, all of them fungi. However, a new class of cephalosporins, the cephamycins, were isolated from streptomycetes by two groups working independently (Nagarajan *et al.*, 1971; Stapley *et al.*, 1972). The cephamycins are characterized by the presence of a 7-α-methoxy group attached to the β-lactam ring. More recently novel β-lactam structures have been isolated from a range of microorganisms including bacteria. *Streptomyces clavuligerus* produces a β-lactam which is a potent β-lactamase inhibitor, clavulanic acid. Other streptomycetes produce novel carbapenems such as the olivanic acids and thienamycin. A *Nocardia* species produces a monocyclic β-lactam, nocardicin, and certain bacteria produce monocyclic β-lactams now known as the monobactams. Many of these compounds are not cephalosporins and belong to the broader category of β-lactams. However, their structures are of interest, in particular with respect to the mode of action and structure–activity relationships of the cephalosporin antibiotics. The relationship between the cephalosporins and other important classes of β-lactams is shown in Table 1. Nomenclature and the physical and chemical properties of the cephalosporins have been reviewed by Hoover and Nash (1978).

8.2   MODE OF ACTION OF CEPHALOSPORINS

8.2.1   Structure and Biosynthesis of Bacterial Cell Wall

The successful application of the cephalosporins, and the β-lactam antibiotics in general, for the treatment of infectious disease is due in no small way to their unique mode of action. It was recognized at an early stage that these antibiotics had as their target the biosynthesis of the bacterial cell wall. As inhibitors of a part of metabolism unique to bacteria they are consequently relatively non-toxic to the animal host.

The Gram-positive bacterial cell wall comprises layers of peptidoglycan and teichoic acid polymer outside the cell membrane. The peptidoglycan accounts for about 50% of the wall dry

Table 1 Structural Formulae and Producers of Important β-Lactams

Formula	Name	Producing organism	Ref.
	Penicillin	*Penicillium* sp.	Florey *et al.* (1949)
	Cephalosporin C	*Cephalosporium* sp.	Abraham and Newton (1961)
	Cephamycin C	*Streptomyces* sp.	Nagarajan *et al.* (1971) Stapley *et al.* (1972)
	Clavulanic acid	*Streptomyces clavuligerus*	Howarth *et al.* (1976)
	Thienamycin	*Streptomyces cattleya*	Kahan *et al.* (1979)
	Monobactam	*Pseudomonas* sp. *Acetobacter* sp *Chromobacterium* sp.	Imada *et al.* (1981) Sykes *et al.* (1981)

weight. The outer envelope of the Gram-negative bacteria is more complex. Again peptidoglycan is the important structural polymer in the wall but there is an additional outer membrane composed of proteins, lipoproteins, lipopolysaccharides and phospholipids. This layer presents a barrier to the uptake of small molecules including the antibiotics. The common feature in all bacterial cell walls is the peptidoglycan polymer. The glycan polymer consists of a repeating disaccharide unit comprising *N*-acetylglucosamine and the related sugar, *N*-acetylmuramic acid. Short peptides can be linked to the muramic acid residues. These oligopeptide chains usually contain one diamino-amino acid (*e.g.* lysine or diaminopimelic acid) which is the basis for cross-link formation between the peptide chains of adjacent glycan strands. The cross-linked peptidoglycan is the structural polymer which protects the osmotically fragile bacterial membrane. In *Staphylococcus aureus* the cross-linked peptide has five glycine residues whilst in *Escherichia coli* the terminal amino acid of one peptide chain is directly cross-linked to the free amino group of diaminopimelic acid in the adjacent peptide chain. The presence of D-amino acids, especially D-alanine, is characteristic of the bacterial cell wall peptidoglycan.

Cell wall biosynthesis occurs by a series of enzyme-catalyzed reactions, some occurring inside and some on the surface of the cell. The first building block is a UDP-*N*-acetylglucosamine sugar nucleotide which is also the precursor of the second component of the disaccharide, *N*-acetylmuramic acid. In *E. coli* attachment of a pentapeptide unit (L-ala-D-glu-DAP-D-ala-D-ala) through the amino group of L-alanine to the muramic acid derivative occurs within the cell. A transglycosylase combines the two sugar derivatives to yield a disaccharide with its pentapeptide substituent which becomes the repeating sequence in the peptidoglycan polymer. These peptidoglycan precursors are transported across the lipophilic membrane by combination with a lipid phosphate carrier. The incorporation of newly synthesized units into the pre-existing or nascent polymer occurs at the outer surface of the cell membrane. The lipid phosphate carrier is released to capture and transport a new subunit.

The final step in wall biosynthesis and the step implicated in β-lactam inhibition is the cross-linking of the oligopeptide residues to yield the final polymer matrix. This occurs by the reaction between the penultimate D-alanine residue of one peptide chain and the free amino group of a glycine (*S. aureus*) or DAP residue (*E. coli*) in an adjacent peptide chain with concomitant release of the terminal D-alanine. The reaction is driven by the energy in the D-ala-D-ala peptide linkage and overcomes the lack of an energy source as ATP outside the cell. It is the inhibition of this final step catalyzed by transpeptidase enzymes which forms the basis of the lethal action of β-lactam antibiotics. Other enzymes such as carboxypeptidase which removes terminal D-alanine residues, endopeptidase which cleaves the cross-linked peptides and muramidase which splits the disaccharide bond are also present in the bacterial membrane. Their detailed function is not yet completely understood but they certainly play an important role in the expansion of the cell wall surface as the bacterium increases in size, allowing insertion of new wall material, elongation and septum formation. For recent reviews see Baddiley (1982) and Mirelman (1982). Cell wall biosynthesis is illustrated in Figure 2.

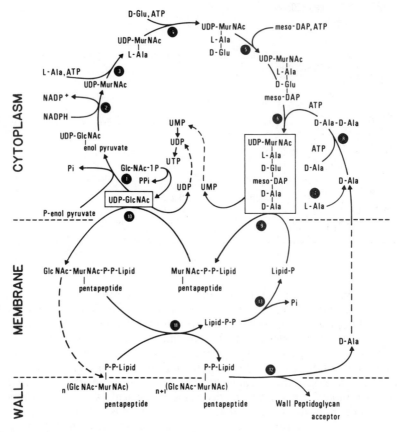

Figure 2   Biosynthesis of *E. coli* peptidoglycan (reproduced from Mirelman (1982) with permission from the author and Academic Press)

8.2.2   Sensitivity and Resistance

The action of cephalosporins and other β-lactam antibiotics on bacteria such as *E. coli* results in various morphological responses including inhibition of cell division, formation of filaments and spherical cells. By using radioactively labelled benzylpenicillin it was shown that a series of proteins within the cell membrane reacts covalently with the β-lactam antibiotics. In some cases these so-called penicillin binding proteins (PBPs) have been allocated specific enzyme activities in cell wall biosynthesis such as transpeptidase and carboxypeptidase activity. Inhibition of these enzymes by the β-lactam involves the formation of a stable, covalent enzyme inhibitor complex which can subsequently break down to regenerate the active enzyme with concomitant release of inactive β-lactam breakdown products. The efficacy of a particular antibiotic often depends upon the speed of formation of the complex and its stability. Changes in these reaction rates can result in resistance to the antibiotic.

Tipper and Strominger (1965) proposed that penicillin inhibited the transpeptidase reaction by virtue of its structural analogy with the terminal D-ala-D-ala in the peptidoglycan peptide chain. Modification of the side chain substituent appears to change the reactivity of the antibiotic. Inhibition of the transpeptidase or carboxypeptidase alone does not account for the bacteriocidal effect of the β-lactams and resulting cell lysis. It appears that other autolytic reactions are triggered which bring about the final demise of the cell (see Tomasz, 1979).

Bacteria vary in their sensitivity to β-lactam antibiotics in general and in their response to individual penicillin or cephalosporin derivatives. Three main factors have been recognized to account for this variable response. First, the antibiotic must penetrate the cell wall in order to reach the target in the cytoplasmic membrane. In Gram-positive bacteria the cell wall seems not to be a very effective barrier to the penetration of the low molecular weight penicillins and cephalosporins. However, in Gram-negative bacteria the outer membrane is sometimes responsible for exclusion of the antibiotic and therefore resistance to it. This seems to be the prime cause of resistance to cephalosporins in the pseudomonads where most derivatives fail to penetrate easily *via* the porins, channels for diffusion of these antibiotics through the outer membrane. The second important barrier to effective action of the cephalosporins is the presence of β-lactamase enzymes. These enzymes are secreted into the medium by Gram-positive bacteria and are found in the periplasmic space of Gram-negative bacteria (Sykes and Matthew, 1976). The β-lactamases are widely distributed and were observed when resistant organisms were first encountered. The evolutionary origin of these enzymes is not yet clear although there is often some structural homology with PBPs. Much of the effort in synthesizing the bewildering range of cephalosporin derivatives has been directed at finding compounds resistant to the β-lactamases. The occurrence of new resistant strains, especially in the hospital environment, and the interspecific transfer of resistance factors on plasmid vectors have presented a never ending challenge to the organic chemist to synthesize new potent derivatives. The third factor determining the efficacy of the cephalosporin is the sensitivity of the target itself, that is the transpeptidase enzyme or other PBP in the cell membrane. Modification of the relative rates of formation and breakdown of the enzyme inhibitor complex can lead to changes in susceptibility of the bacterium.

Although penetration, the presence of β-lactamases and target specificity are the important factors determining *in vitro* activity of the cephalosporin, other factors are important when the antibiotic is used for the clinical treatment of infectious disease. These come under the heading of pharmacokinetic properties. The ideal antibiotic has a very high antibacterial activity, a long half-life in the bloodstream and is subsequently recovered unaltered in the urine. These properties can be influenced by the various substituents of the cephem nucleus. Cephalosporins retaining the 3-acetoxymethyl substituent of the parent compound cephalosporin C, *e.g.* cephalothin, are attacked by acetyl esterases to yield the less active desacetyl derivative. In the first generation of cephalosporins, the replacement of the 3-acetoxy group by pyridine to give cephaloridine overcame this problem but led to another, its nephrotoxicity.

8.2.3 Structure/Activity Relationships

The classical cephalosporins can be modified at three positions on the cephem nucleus: 7β, 7α and 3.

It was assumed, correctly, that replacement of the α-aminoadipyl substituent of cephalosporin C by a hydrophobic group would increase the antibacterial activity. This came by straight analogy with the penicillins, benzylpenicillin being much more active than penicillin N. A thienylacetyl group proved to be more active than the phenylacetyl and led to the first generation compounds cephalothin and cephaloridine which had the advantages over penicillins of Gram-negative activity and resistance to β-lactamases from Gram-positive organisms. As mentioned earlier, the replacement of the 3-acetoxy by pyridine prevented deactivation by mammalian esterases.

The spread of β-lactamase producers and the regular occurrence of resistant pathogenic strains was the impetus behind the search for new cephalosporin derivatives with β-lactamase resistance. The penalty for having increased resistance to β-lactamase is often that the antibacterial activity is reduced (see O'Callaghan, 1980). Of the three possible sites for modification of the classical β-lactams those at positions 7β and 7α have the main impact on β-lactamase resistance. Modification at position 3 can affect antibacterial activity and pharmacokinetic properties of the cephalosporin. A major breakthrough occurred with the discovery in the streptomycetes of the cephamycins, cephalosporins having a 7α-methoxy substituent. The methoxy substituent confers nearly complete resistance to β-lactamases. Other bulkier 7α substituents, *e.g.* 7-ethoxy, also confer resistance to β-lactamases but at the expense of drastically reduced antibacterial activity. The first cephalosporin derivative incorporating the 7α-methoxy group was cefoxitin, which combines resistance to β-lactamase and broad spectrum activity.

Changes in the 7β group also affect the antibacterial activity and β-lactamase sensitivity. Steric effects are important here in regulating the reactivity of the β-lactam ring. By substitution in the α-carbon of the side-chain significant improvements were obtained over the straight substituted acetic acid derivatives such as cephalothin. Compounds of this second generation of cephalosporin derivatives include cephamandole and cefuroxime. These compounds, although having a broad spectrum and improved β-lactamase stability, are still relatively inactive against pathogenic pseudomonads and must be administered parenterally. Despite their acid stability in comparison with the penicillins, the cephalosporins in general are not readily absorbed from the gut. The exception is cephalexin. Here the 7β substituent is the D-phenylglycine residue. This, or a slightly modified analogue, is common to all orally absorbed cephalosporins. Cephalexin has a methyl substituent at position 3 on the cephem nucleus. Although this group occurs naturally in desacetoxycephalosporin C, an intermediate in the biosynthesis, it is produced in relatively low yields, even in blocked mutants.

Cephalexin is therefore normally produced by chemical ring expansion from pencillin V followed by deacylation to 7-aminodesacetoxycephalosporanic acid (7-ADCA) and re-acylation with the phenylglycine substituent. Cefaclor with a 3-chloro substituent and cefroxadine having a 3-methoxy group are new orally active derivatives which have improved antibacterial activity compared with cephalexin.

A third generation of cephalosporin derivatives is already working its way through the clinics and challenging the existing market leaders. Worthy of mention are cefotaxime, ceftizoxime and ceftazidime, the latter having activity against the pseudomonads. One of the most interesting derivatives at present in the clinic is moxalactam. This cephalosporin has an oxygen atom to replace the sulfur atom present in the cephem ring of all the other derivatives. This confers increased antibacterial activity. The 7α-methoxy substituent and the substituted malonic acid side-chain result in resistance to many β-lactamases.

Table 2 shows the structures of some of the important cephalosporin derivatives in current therapeutic use and some new compounds undergoing clinical trials.

8.2.4 Cephalosporin Market

An estimate of the worldwide systemic antibiotic market for 1980 showed that the cephalosporins occupied first place with a market value of approximately US $1750 million (equivalent to 29%). The ratio of oral to parenteral forms was 44:56.

Amongst the oral products, cephalexin had an 83% market share and was the biggest single best-seller of the cephalosporins. Amongst the parenteral products, well-established derivatives such as cephalothin and cephazolin had a large share of the market (over 20% each). However the trend for these products is downwards as the second and third generation products take a larger bite of the cake. In the immediate future it is probable that products such as cephamandole, cefoxitin and cefaclor will take an increasing share of the market.

8.3 BIOSYNTHESIS OF CEPHALOSPORINS

8.3.1 Biosynthetic Pathway

During the 1950s Arnstein and co-workers had shown that the penicillin nucleus was formed from the amino acids cysteine and valine. These workers also isolated sulfur-containing peptides from the mycelium of *P. chrysogenum*, in particular a linear tripeptide comprising α-aminoadipic acid, cysteine and valine. Several pieces of evidence implicated this tripeptide in the biosynthesis of penicillin. The inhibitory effect of lysine could be rationalized by feedback regulation of its own synthesis resulting in a reduced supply of the pathway intermediate, α-aminoadipic acid. The discovery that penicillin N from *C. acremonium* contained an α-aminoadipyl side-chain and

Table 2   The Structures of Important Cephalosporin Antibiotics

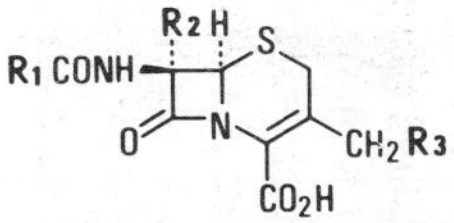

Core structure: R_1CONH, R_2, H, S, cephem ring with CH_2R_3, CO_2H, O.

Name	R^1	R^2	R^3		Name	R^1	R^2	R^3
Cephalothin	thienyl–CH_2–	H	$OCOCH_3$		Cefotiam	aminothiazolyl–CH_2–	H	tetrazolyl, CH_2-CH_2-$N(CH_3)_2$
Cephaloridine	thienyl–CH_2–	H	pyridinium		Ceftazidime	aminothiazolyl–$C(=N$-$C(CH_3)_2$-$COOH)$–	H	pyridinium
Cefazolin	tetrazolyl $N=N$, N-CH_2–	H	thiadiazolyl–CH_3, $OCONH_2$		Moxalactam (Oxygen in cephem ring)	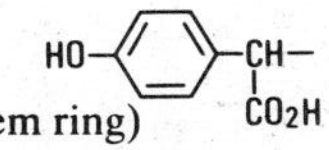 HO-phenyl–CH-CO_2H–	OCH_3	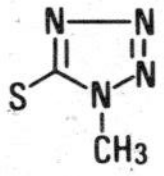 thiadiazolyl–CH_3
Cefoxitin	thienyl–CH_2–	OCH_3	$OCONH_2$		Cephalexin	phenyl–CH-NH_2–	H	H
Cefamandole	phenyl–CH-OH–	H	tetrazolyl–CH_3		Cefaclor	phenyl–CH-NH_2–	H	Cl
Cefuroxime	furyl–$C(=N$-$OCH_3)$–	H	$OCONH_2$		Cephradine	cyclohexadienyl–CH-NH_2–	H	H
Cefotaxime	H_2N-thiazolyl–$C(=N$-$OCH_3)$–	H	$OCOCH_3$		Cefroxadine	cyclohexadienyl–CH-NH_2–	H	O-CH_3
Ceftizoxime	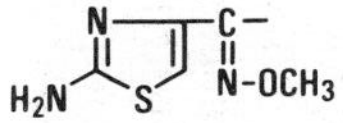 H_2N-thiazolyl–$C(=N$-$OCH_3)$–	H	H					

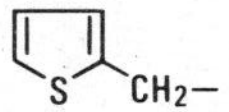

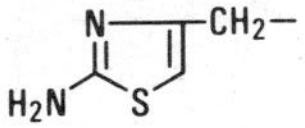

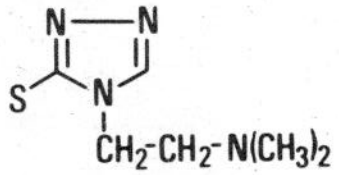

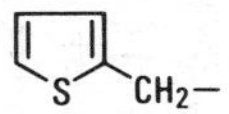

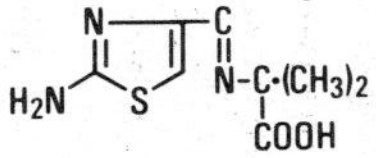

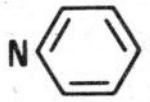

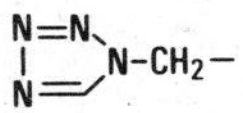

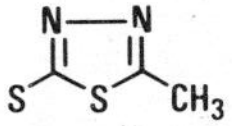

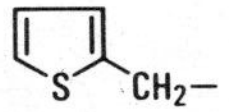

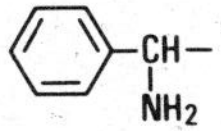

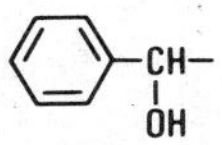

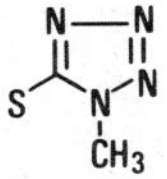

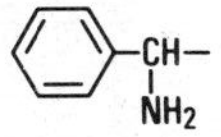

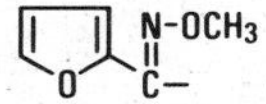

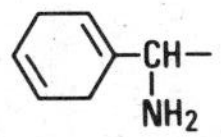

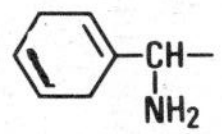

The last four listed compounds are oral cephalosporins whilst the others are parenteral

finally the isolation by Flynn *et al.* (1962) of isopenicillin N from *P. chrysogenum* provided additional evidence that the tripeptide was an important intermediate. With the elucidation of the structure of cephalosporin C by Abraham and Newton (1961) it was clear that the pathways to penicillins and cephalosporins had much in common.

Investigations with radioactively labelled precursors by the Oxford group showed that the three amino acids were incorporated into penicillin N and cephalosporin C (Trown *et al.*, 1963; Warren *et al.*, 1967a, 1967b). A peptide isolated from a high yielding strain of *C. acremonium* was found to be δ-(L-α-aminoadipyl)-L-cysteinyl-D-valine (Loder and Abraham, 1971a). A labelled tripeptide with this configuration was subsequently shown to be incorporated into a penicillin by a protoplast lysate (Fawcett *et al.*, 1976). Previous work had shown that the tripeptide was synthesized from α-aminoadipylcysteine rather than from cysteinylvaline (Loder and Abraham, 1971b). The tripeptide has also been found to have the same configuration (LLD) when isolated from *Penicillium* and *Streptomyces* sp.

The conversion of valine from the L- to the D-epimer appears to take place during the synthesis of the tripeptide.

The apparently obligatory role of α-aminoadipic acid in the biosynthesis not only of penicillin N and cephalosporins but also in the biosynthesis of the hydrophobic penicillins such as penicillin G is interesting. The discovery of β-lactam antibiotics in the streptomycetes provides additional evidence of a special role for α-aminoadipic acid, since unlike the eukaryotes, where it is a normal metabolic intermediate of the lysine pathway, in the streptomycetes it is produced only by catabolism of lysine produced by an alternative route. It is also significant that tripeptide synthesis begins from the α-aminoadipyl residue. The amphoteric group of α-aminoadipic acid may be important for attachment to the peptide synthesizing enzyme and the cyclization enzyme.

The development of cell-free systems was the next crucial step in biosynthetic studies. Early work with ultrasonic or mechanically ruptured cells was not successful. However, the technique of protoplast formation by use of cell-wall lytic enzymes and the subsequent gentle lysis of the osmotically fragile protoplasts allowed further progress with labelled tripeptide precursors. The earliest demonstration by Fawcett *et al.* (1976) of the conversion of the tripeptide to a penicillin had not distinguished between penicillin N and isopenicillin N. Subsequent work by O'Sullivan *et al.* (1979) and Konomi *et al.* (1979) showed that the product of cyclization of the tripeptide was in fact isopenicillin N. A similar investigation by Meesschaert *et al.* (1980) using *P. chrysogenum* showed that isopenicillin N was also the product of cyclization of the LLD-tripeptide in this organism. Isopenicillin N is apparently synthesized by a pathway common to both the penicillin and cephalosporin producing organisms and is a pivotal compound/intermediate from which the biosyntheses of these two classes of β-lactam diverge. The enzyme responsible for cyclization of the tripeptide to isopenicillin N requires iron(II) ions for activity. The reaction mechanism still remains to be elucidated. It now seems very unlikely that a dehydrovaline- or β-hydroxyvaline-containing intermediate is involved in formation of the thiazolidine ring of isopenicillin N (Baldwin *et al.*, 1981). A monocyclic β-lactam intermediate may be the first heterocyclic compound formed but this is probably enzyme bound throughout the cyclization reaction.

Although isopenicillin N is now well established as an intermediate in cephalosporin biosynthesis, it does not accumulate either intra- or extra-cellularly. Abraham *et al.* (1954) had identified the antibiotic secreted by *C. acremonium* as penicillin N and Jayatilake *et al.* (1981) demonstrated a labile epimerase which converts the L-α-aminoadipyl side-chain of isopenicillin N to the D-configuration and thus to penicillin N. This is the key step in the divergence of the pathways for hydrophobic penicillins and hydrophilic cephalosporins in *P. chrysogenum* and *C. acremonium* respectively.

In the synthesis of hydrophobic penicillins, *e.g.* penicillin G, isopenicillin N is the substrate for an acyltransferase enzyme (absent in *C. acremonium*) which exchanges the L-α-aminoadipyl side-chain for phenylacetyl, the acyltransferase enzyme being first acylated by phenylacetyl-coenzyme A. Under certain fermentation conditions penicillin N can be secreted by *C. acremonium* as an end-product in high yield by strains normally producing cephalosporins or can be produced by mutant strains blocked at some later stage in the biosynthetic pathway.

For many years it was not clear whether penicillin N was a product of a branched pathway or whether it was indeed a direct precursor of the cephalosporins. Kohsaka and Demain (1976) first demonstrated the synthesis of a cephalosporin from penicillin N by the remarkable 'ring expansion' enzyme using a cell-free extract of *C. acremonium*. This enzyme has since been studied by several groups of workers including those at MIT (Yoshida *et al.*, 1978; Kupka *et al.*, 1983), Oxford (Baldwin *et al.*, 1980) and industrial groups at Bristol-Myers (Hook *et al.*, 1979) and Ciba-Geigy (Felix *et al.*, 1981). It is an oxygenase requiring iron(II) ions and ascorbate for activity.

The product of the reaction is desacetoxycephalosporin C which therefore appears to be the first true cephalosporin to be synthesized.

Studies of a complementary kind had already provided strong evidence that desacetoxycephalosporin C was a precursor of the main fermentation product, cephalosporin C. Higgens *et al.* (1974) identified desacetoxycephalosporin C (DAOC) as a product of both fungi and streptomycetes and Queener *et al.* (1974) showed that a mutant blocked in the biosynthesis of cephalosporin C produced DAOC. Fujisawa *et al.* (1975a) first found a cephalosporin C negative mutant which accumulated desacetylcephalosporin C (DAC) and then introduced a second mutation which resulted in the accumulation of DAOC (Fujisawa *et al.*, 1975b). Thus the first evidence was found to suggest that the later stages of cephalosporin biosynthesis involved the conversion of DAOC first to DAC and then finally to cephalosporin C itself.

Enzymological proof of the step from DAOC to DAC was reported by Brewer *et al.* (1977a). They demonstrated the oxygenation of 3-methylcephalosporin (DAOC) to the 3-hydroxymethyl-cephalosporin (DAC) using a cell-free extract of *C. acremonium*. The same group (Turner *et al.*, 1978) showed that this reaction was catalyzed by a 2-oxoglutarate-linked dioxygenase in both *Cephalosporium* and *Streptomyces*. Scheidegger *et al.* (1984) have recently published evidence that the ring expansion of penicillin N and the oxygenation of DAOC are catalyzed by a single bifunctional dioxygenase in *C. acremonium*.

Enzymological evidence for the final reaction, the acetylation of DAC, was found by Fujisawa and Kanzaki (1975). They detected an acetyl-coenzyme A dependent transferase which catalyzes the acetylation of DAC to form cephalosporin C. Blocked mutants accumulating DAC lacked this activity which was restored in a revertant. Felix *et al.* (1980) were able to demonstrate both the hydroxylation and acetylation steps using ether-permeabilized cells of *C. acremonium*.

The presence of DAC in cultures of *C. acremonium* can be the result of a partial block in the last stage in cephalosporin C biosynthesis or that of enzymic degradation of the cephalosporin C itself. Fujisawa *et al.* (1973) isolated mutants of both types. In one case a mutant synthesized DAC as the sole cephalosporin. A second mutant type synthesized first cephalosporin C which was then quantitatively deacetylated by an acetyl esterase expressed towards the end of the fermentation. The extracellular esterase investigated by Hinnen and Nüesch (1976) was apparently catabolite repressed. The source of DAC which accumulates during the cephalosporin C fermentation with high yielding strains is not absolutely clear and may be due partly to *de novo* synthesis and partly to deacetylation. Smith *et al.* (unpublished) detected a constitutive intracellular esterase with low activity in a high yielding strain. A mutant producing larger amounts of DAC had a greatly increased esterase activity.

The production of cephalosporin C represents the normal endpoint of the biosynthesis in *C. acremonium*. In the β-lactam-producing streptomycetes two further reactions are of special significance. The cephamycins are characterized by the 7α-methoxy substituent. This group is introduced in a two step reaction involving molecular oxygen and *S*-adenosylmethionine as methyl donor (O'Sullivan *et al.*, 1979; O'Sullivan and Abraham, 1980).

The streptomycetes also produce cephalosporins with a carbamoyl substituent in the 3-position. Brewer *et al.* (1977b, 1980) showed the carbamoylation of 7α-methoxydesacetylcephalosporin C and desacetylcephalosporin C by a carbamoyl phosphate utilizing enzyme in *S. clavuligerus*. The biosynthetic pathway is summarized by Figure 3.

8.3.2 Regulation of Cephalosporin Biosynthesis

It will already be apparent that a large volume of literature has been published concerning the biosynthetic pathway for the cephalosporins. Scientists, and especially those working in the competitive industrial environment, are also interested in the factors which regulate the activity and flux through the pathway. Through identification of a rate-limiting step a rational approach to mutant selection can be undertaken with the hope of selecting strains with increased productivity or other favourable characteristics. Primary and secondary metabolism are intimately joined sharing common substrates and interlinked pathways of metabolism. In the case of the cephalosporins which are synthesized from the three amino acid precursors it is logical to ask whether their availability limits the rate of antibiotic synthesis. Conclusive published data are difficult to find and undoubtedly there are strain to strain variations. Amino acids, particularly those at the ends of long biosynthetic pathways, regulate their own synthesis by two mechanisms. Feedback inhibition is a fine control operating at the level of enzyme activity. The first enzyme unique to the pathway is usually inhibited non-competitively by the end-product combining with an allos-

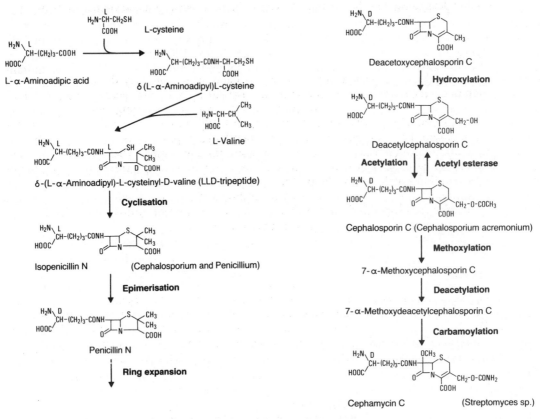

Figure 3   Cephalosporin biosynthetic pathway

teric site to cause a conformational change. Enzyme repression is a coarser control operating at the level of gene transcription and requiring growth to dilute out the already existing enzymes of the pathway.

8.3.2.1   α-Aminoadipic acid

Demain (1957) showed that penicillin production by *P. chrysogenum* was inhibited by lysine and reversed by α-aminoadipic acid. This has since been shown to be caused by lysine feedback on the enzyme homocitrate synthetase, this leading to depletion of the pathway intermediate, α-aminoadipic acid, and consequent reduction in the rate of antibiotic synthesis. The regulation of this pathway and the effect of lysine on cephalosporin synthesis is certainly different in *C. acremonium*. Mehta *et al.* (1979) reported two effects depending upon the lysine concentration. At a low concentration lysine stimulated cephalosporin C production but inhibited at higher concentrations. Inhibition might be explained according to the penicillin model where the first enzyme is regulated. A second regulatory site beyond the α-aminoadipic acid step which was inhibited by low concentrations of lysine might account for the stimulatory effect by raising the α-aminoadipate pool. Enzymological proof is not yet published. Using a steady state perturbation methodology in phosphate-limited chemostat culture the author also observed a stimulation of cephalosporin production by lysine. After wash-out of the residual lysine the rate of antibiotic production returned to its original value and could be stimulated again by a further perturbation with lysine. The mechanism of this stimulation was not apparent. A related high yielding strain had high cell pool levels of both α-aminoadipic acid and lysine in fed-batch cultures suggesting that the former was unlikely to be limiting and that the pathway was deregulated. In the cephalosporin-producing streptomycetes the situation could be quite different. Lysine is synthesized in prokaryotes *via* the diaminopimelic acid pathway and α-aminoadipic acid is produced by lysine catabolism. Mendelovitz and Aharonowitz (1982) showed that lysine stimulated cephamycin synthesis in *S. clavuligerus* and that the biosynthesis of lysine was regulated by a concerted feedback inhibition by lysine and threonine together. The site of regulation was the enzyme aspartate kinase. Mutants deregulated in their aspartate kinase activity overproduced cephalosporins.

8.3.2.2 Valine

End product regulation of valine biosynthesis has been investigated by the author in a high yielding strain of *C. acremonium*. Regulation of the enzyme acetohydroxy acid synthetase, which catalyzes the first step in valine synthesis from pyruvate, was studied using benzene-permeabilized cells. The enzyme was specifically and non-competitively inhibited by L-valine. In a complex fermentation medium there was also evidence of a four-fold derepression of enzyme activity during the course of the fermentation. These investigations did not indicate whether or not the supply of valine was rate-limiting for antibiotic synthesis. Steady state perturbation experiments using a phosphate-limited chemostat helped to answer this question. It was shown that the cell pool concentration of valine could be inflated by addition of a pulse of valine to a steady state culture producing cephalosporin C at a high rate. No stimulation of antibiotic production was observed and it was concluded that the valine supply was not limiting.

8.3.2.3 Cysteine

The carbon skeleton of cysteine is derived from serine. Studies in the chemostat like those described above showed that in the strain under investigation the supply of serine was not limiting.

The sulfur of cysteine is derived either from inorganic sulfate or by reverse trans-sulfurylation from exogenously added methionine (see Figure 4). Methionine was recognized as a preferred, and in some cases an obligatory, sulfur source for cephalosporin C biosynthesis.

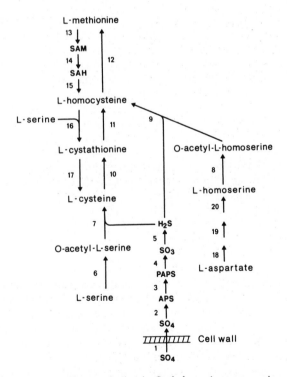

Figure 4 Sulfur metabolism in *Cephalosporium acremonium*

The reverse trans-sulfurylation pathway (reactions 13–17 in Figure 4), demonstrated by Caltrider and Niss (1966), was believed to be the preferred pathway to the antibiotic precursor, cysteine, and to overcome a limitation in the sulfate reduction pathway (reactions 1–5). An alternative explanation for the stimulation of cephalosporin C synthesis by methionine was proposed by

Demain's group. Using a mutant blocked in reactions 1–5 and reaction 11 Drew and Demain (1975) showed that when only sufficient methionine to support growth was added, the mutant failed to produce cephalosporin C even in the presence of the direct precursor cysteine. Addition of methionine stimulated production. A role for methionine independent from its role as a sulfur donor was postulated. This was supported by the evidence that norleucine, a non-sulfur-containing analogue of methionine, could mimic the stimulatory effect. Treichler *et al*. (1979) isolated a range of mutants blocked in various parts of the sulfur pathway. Their evidence suggested that the preferred route of cysteine synthesis from sulfate was *via* the *O*-acetyl-L-homoserine pathway (reactions 9, 16 and 17). Mutants unable to synthesize L-cystathionine had reduced ability to synthesize cephalosporin C. A mutant blocked in the normal sulfur assimilation pathway (reaction 7) and therefore obliged to use the cystathionine loop produced increased amounts of cephalosporin C from sulfate without the need to add methionine. Queener *et al*. (1984) have recently reviewed the regulatory aspects of sulfur metabolism in relation to cephalosporin C production. The author used the chemostat with phosphate as the growth limiting nutrient in order to separate the effects of added amino acids on growth and cephalosporin production. High rates of cephalosporin C synthesis could be established and the steady state perturbed by addition of a continuous feed or a single pulse of the amino acid under investigation. Inflation of the steady state cell pool was proof of uptake. It was found that L-methionine had no effect on antibiotic production but that D-methionine stimulated the specific rate by over 40%. At a low dilution rate $(0.016\ \mathrm{h}^{-1})$ the methionine analogue norleucine also stimulated cephalosporin production. The methionine effect was accompanied by a pronounced increase in the degree of fragmentation of the mycelium into unicells. Similar effects of methionine on cephalosporin production and morphological differentiation were published by Matsumura *et al*. (1980a, 1980b). This and a recent report by Demain (1983) that the activities of the cyclization and ring expansion enzymes were higher in methionine-supplemented cultures support a regulatory role for methionine.

8.3.2.4 *Effect of oxygen tension*

The cyclization of the tripeptide, the ring expansion of penicillin N and the hydroxylation to produce desacetylcephalosporin C are oxidation reactions. A high oxygen transfer rate is necessary for the realization of maximum yields of cephalosporin C. Depletion of the dissolved oxygen tension leads to the accumulation of the intermediate, penicillin N. It therefore appears that the ring expansion reaction is most sensitive to oxygen depletion. If the oxygen deprivation is not too severe the rate of penicillin N accumulation is similar to the previous rate of cephalosporin C synthesis. Under more severely limiting conditions the overall molar rate of β-lactam production is reduced, perhaps due to secondary effects on the cyclization enzyme. An oxygen-limited chemostat culture produced cephalosporin C and penicillin N at about equal rates. Restoration of carbon limitation caused a doubling of the rate of cephalosporin C synthesis, whilst the penicillin N synthesis was virtually zero. Some high yielding strains produce significant amounts of penicillin N even under conditions of oxygen abundance suggesting that for these strains the rate limiting step may lie in the terminal stages of the biosynthesis. Mutants blocked in a late step in the biosynthetic pathway have been shown to accumulate penicillin N.

8.3.2.5 *Catabolite repression*

Recent findings by Martin (1983) suggest a temporal separation of penicillin N and cephalosporin C production. Catabolite repression of β-lactam synthesis by glucose was evident with a sequential derepression of the cyclization and ring expansion reactions. The catabolite repression/inhibition effects in *Cephalosporium* are probably not as pronounced as those in *Penicillium* sp. The reduction in cephalosporin synthesis by high levels of phosphate is a related phenomenon. In batch culture Küenzi (1980) observed a reduction of cephalosporin production at high phosphate concentrations. By controlling the rate of glucose consumption in a glucose-limited chemostat he obtained high rates of cephalosporin production even at high phosphate levels. In batch and fed-batch culture, repression by glucose can be overcome by using vegetable oils as the carbon source.

In some strains catabolite repression of acetyl esterase activity (Hinnen and Nüesch, 1976) may be an important determinant of the desacetylcephalosporin C (DAC) concentration. Cephalosporin C itself is efficiently exported from the cell. When large quantities of cephalosporin C were added to shake flask cultures before inoculation, no evidence of either growth inhibition or product inhibition was found. It is necessary in this type of experiment to take account of the non-enzymic decomposition of the added cephalosporin which occurs at a rate of 5×10^{-3} h^{-1} (Konecny *et al.*, 1973).

8.3.2.6 *Specific growth rate*

The effect of specific growth rate on β-lactam antibiotic synthesis is best studied using the chemostat. Using the isolate M8650, a low yielder, Matsumura *et al.* (1978) found that in glucose-limited chemostats there was an inverse relationship between specific growth rate and specific production rate. Similar observations were made by Lilley *et al.* (1981) for *S. cattleya* producing cephamycin. In both cases the strains employed had not been exposed to an extensive strain improvement programme. For high yielding mutants of *C. acremonium* a different picture has been observed. The author found a linear relationship between specific growth rate and cephalosporin C specific production rate in chemostats where soya bean oil was the growth limiting nutrient. Thus, the specific production rate (q_{ceph}) increased by a factor of 2 between growth rates of 0.01–0.04 h^{-1}. Küenzi has made similar observations using glucose-limited chemostats.

8.4 FERMENTATION PROCESS

8.4.1 The Fermenter — Its Design and Instrumentation

A large variety of bioreactor types have become available in recent years, each offering some advantage for the culture of aerobic microorganisms. With certain remarkable exceptions, such as the ICI air-lift fermenter for the production of single cell protein, most of the non-conventional bioreactors have been restricted to the experimental pilot plant in their application. Large scale producers of antibiotics have large capital sums invested in conventional stirred tank reactors and are understandably reluctant to introduce alternative reactor designs unless a substantial economic advantage would accrue.

The basic design of the production fermenter has therefore changed very little since the first deep culture fermenters were installed for the production of penicillin G in the late 1940s. The early fermenters were constructed in mild steel but modern practice is to use stainless steel. Fermenters have a height to diameter ratio normally in the range 3:1 to 4:1. A typical fermenter for the production of cephalosporin C has a working volume of 50 m^3 although fermenters of double this size are not uncommon. The cephalosporin fermentation is a highly aerobic fermentation and consequently it is usual to find a relatively high installed power of more than 4 kW m^{-3}. The drive is mounted above the production fermenter. The agitation speed can be changed either by means of a continuously variable speed controller or by stepwise changes in pulley size. The impracticality of frequent interruptions to change the pulley size effectively removes the variation of power input as a variable for fermentation control in these fermenters. Power is transmitted to the fermentation broth by means of agitators mounted at intervals on the stirrer shaft. Various designs of agitator have been investigated. In general the ratio of tank diameter to agitator diameter is around 3:1. The number of agitators depends largely on the fermenter size. Thus a single six-bladed disc turbine might be sufficient for a 0.5 m^3 pilot plant fermenter whilst several such might be required at the 50 m^3 level. Agitators with swept-back blades are sometimes used as an alternative to the disc turbine. The function of the agitation system is three-fold. First, to create conditions in the region of the agitator which allow a high rate of gas exchange. Vigorous agitation has the dual effect of reducing the depth of the gas–liquid interfacial film and also increasing the interfacial area by producing small gas bubbles. The combined effect is to increase $k_L a$. The rate of oxygen transfer is given by the relationship $dc/dt = k_L a(C_s - C)$, where C_s and C are the concentrations of oxygen at the interface and in the bulk liquid respectively.

The second function of the agitation system is the distribution of the culture and added nutrients uniformly throughout the fermenter. Here the difficulties tend to become greater as the size increases. Poor bulk-mixing of the culture results in pockets of stagnant broth which are deprived of essential nutrients and results also in excessive air hold-up in the culture.

The third function of the agitation system is to promote the dissipation of heat. Heat is generated both mechanically by the agitator itself and metabolically from the oxidation of nutrients during the fermentation. With increasing size of fermenter the surface area to volume ratio decreases and it is often necessary to have internally mounted cooling coils. In some cases the coils are mounted vertically in the fermenter and so serve an additional role as baffles.

The air supply is delivered to a position just below the bottom agitator to a sparger ring. Air flow rates of 1 volume per volume culture per minute are to be expected for cephalosporin C production. The air flow rate has a relatively small effect on the oxygen transfer rate compared with changes in power input. However, as can be seen from the above equation for conditions of constant $k_L a$ the rate of oxygen transfer depends on the value of (C_s-C), known as the driving force. In practice this can be increased by raising the overpressure in the fermenter although enrichment of the air with pure oxygen would be an alternative strategy.

In general, for a well-established production process the agitator speed is kept constant at a speed which varies with the fermenter size but normally gives an agitator tip velocity of around 5 m s^{-1}. The air flow is increased stepwise during the first 48 h to reach its maximum value and further regulation of the dissolved oxygen concentration is achieved by changes in overpressure and nutrient feed rate.

The modern production fermenter is equipped with sterilizable probes for on-line measurement of temperature, pH value and p_{O_2} and with gauges for the measurement of head pressure, batch weight, air flow and power consumption. A further important on-line facility for the control of the fermentation process is the analysis of exhaust gas. The rates of carbon dioxide production and oxygen consumption are critical diagnostic factors in fermentation control, analysis and mass balancing. The rate of antibiotic production often has a characteristic relationship to either the instantaneous oxygen uptake rate or the integrated value over the whole fermentation (Küenzi, 1978). These relationships will certainly be further exploited in the derivation of mathematical models and the application of computer control to fermentation processes including that for cephalosporin C.

Cephalosporin C is produced by a fed-batch process requiring the semicontinuous addition of nutrients to supply energy, carbon and to regulate the pH value. This requires a variety of addition systems. For the addition of shots of liquid nutrients pneumatically controlled kits operated from a variable timer are often used to deliver regular amounts of 1–2 l. Where the pH is controlled with aqueous ammonia solution the operation of a valve can be triggered from the pH controller. In an alternative system gaseous ammonia is admitted into the air supply by manual operation at the flow meter. Finally it is often necessary to control foaming in highly aerated cultures. This can also be achieved automatically by using a sensitive electrode which detects the foam head or by manual addition through the so-called antifoam pot, a vessel mounted alongside the fermenter in which antifoam agents or other desired additions can be conveniently sterilized before addition to the main culture.

8.4.2 Fermentation Microbiology

For the large scale production of cephalosporin C a complex nutrient medium is used which can be sterilized batch-wise in the fermenter or by means of a continuous sterilizer. Although media containing very high levels of vegetable meals, such as soya bean or peanut meal, can present problems with the continuous sterilizer, the media normally used for cephalosporin C production are not particularly difficult. Continuous sterilization is the method of choice on the grounds of energy saving and is also less damaging to labile components because of the reduced exposure time at high temperature. The continuous sterilizer itself is simply a system of heat exchangers across which the medium is pumped at a predetermined rate. Heat given up by the cooling sterile medium is used to heat up medium entering the circuit. Industrial equipment allowing flow rates of 40 m^3 h^{-1} is available. For conventional batch sterilization in the fermenter, steam is admitted to the jacket or internal coils and subsequently to the medium itself. A temperature of 121 °C at a steam pressure of 15 p.s.i. is usually considered essential. This is maintained for a period of time which varies from medium to medium and with fermenter size. Generally a rather shorter period at peak temperature is required in the larger fermenters because of the extended cooling time. A further problem generated by batch-wise sterilization is the variable amount of condensate which results in uncertainty about the start volume in the fermenter.

Part of the sterilization cycle involves the steam sterilization of ancillary equipment such as air filters, addition kits, sampling points, *etc.* The air filter is contained within a metal housing and is one of two types, either a depth or a membrane filter. Perkowski (1981) reviewed the methods of air filtration and recommended a filter cartridge system in which an auxiliary steam jacket is used to keep the air temperature in the filter above the dew-point. In more traditional types of filter a multi-layered tower of glass wool packing is used.

There is no single magic recipe for the medium to produce cephalosporin C. It is fairly certain that various manufacturers will have optimized on media with both quantitative and qualitative differences. It is often necessary to re-optimize for each new high yielding mutant which reaches the production scale. As earlier with the penicillin fermentation, corn-steep liquor has proved to be a valuable nitrogen source providing a rich variety of amino acids, peptides and proteins plus numerous trace elements. The starting level will normally be in the range from 0.2–0.6% N. Other nitrogen sources used in addition to corn-steep are peanut meal and ammonium sulfate. When a relatively dilute medium is used it is customary to supplement with inorganic salts to satisfy the requirements for Mg^{2+}, SO_4^{2-} and PO_4^{3-}. Early strains required the addition of methionine for maximum cephalosporin production but this can be avoided by the selection of mutants utilizing sulfate for cephalosporin C synthesis.

A range of carbon sources have also found application in the cephalosporin process. Thus, glucose, dextrin and vegetable oils either singly or in combination can be used. For some strains the inclusion of batched glucose and dextrin in the medium promotes more rapid growth than if vegetable oil alone were used. There may be a switch from carbohydrate to oil as the main component of the feed as the fermentation progresses. When media rich in corn-steep liquor are used the readily available amino acids provide an excellent carbon source and addition of other carbohydrates is sometimes unnecessary. Many vegetable oils and fish oils can be used for cephalosporin C production. The preferred one is soya bean oil on account of cost and availability.

The first prerequisite for a high yielding fermentation is a good strain. The mutant M8650 is the parent from which all the high yielding isolates have been derived by a process of repeated mutation and selection. High yielding strains are preserved by lyophilization or by storage in liquid nitrogen. Recovery of viability after preservation is often desperately poor since in most cases vegetative cells rather than conidia are preserved. Increased productivity of a strain line is often accompanied by a fall in conidiation of the strain. Optimization of the support medium can offset the poor recoveries to some extent. From the lyophilized ampoule the next step in the propagation sequence is the cultivation on a solid medium. Surface cultures in the form of slants or medicinal flats provide inoculum for shaken flask cultures in large conical or balloon-shaped flasks. These vegetative cultures can be used to provide inoculum for the first fermenter seed stage. It is normally considered necessary to have two fermenter seed stages in order to step up the inoculum volume to a size appropriate for a production fermenter. A flow diagram showing a possible scheme for inoculum build up is shown in Figure 5.

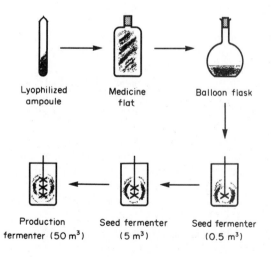

Figure 5 Flow diagram for cephalosporin C production

The inoculum level for the main production fermenter varies from one process to another. When the medium is rich in readily metabolized carbon sources such as glucose, a relatively low inoculum of 5–7.5% is adequate. When growth in the production medium is slow, this can be offset by increasing the inoculum level to 20%. If the seed stage fails for some reason, *e.g.* contamination, it may be possible to back-inoculate from another production fermenter.

The first 48 h of the production fermenter is the period of most rapid growth. The author determined a specific growth rate of 0.04–0.05 h^{-1} from measurements of carbon dioxide in the exhaust gas during the phase when growth was at the expense of corn-steep liquor in the medium. This phase was characterized by a high level of the catabolic NAD-linked glutamate dehydrogenase. The culture has a mainly filamentous morphology during this period. With the sequential exhaustion of rapidly catabolized nutrients the specific growth rate falls and the culture morphology changes. First the filamentous hyphae become shorter and fatter and then fragmentation into unicells occurs. The unicells can produce germination tubes giving rise to new hyphae and slowly reproducing the growth cycle. This morphological differentiation is a key factor in the onset of cephalosporin C production and may be promoted by the addition of methionine to the medium or the switch to non-repressing carbon sources such as soya bean oil. The anabolic NADPH-linked glutamate dehydrogenase is derepressed as complex nitrogen sources are exhausted and the culture begins to assimilate inorganic ammonia.

The culture pH value is not absolutely critical but a value in the range 6.0–7.0 is maintained by addition of ammonia. Temperature is a useful regulator of the rate of metabolism. Values in the range 24–28 °C can be used to speed up or slow down the fermentation as required. Very important is the dissolved oxygen tension. There is no absolute value which can be quoted since the measured value depends upon the positioning of the electrode in the fermenter relative to the agitator and effects of hydrostatic pressure in large fermenters. However, it is important to prevent a precipitous fall in p_{O_2} by arranging for the culture to become carbon-limited at the appropriate time. The period in the middle of the fermentation is important as the culture biomass level increases and the critical transition to carbon limitation is made. Off-line analyses of glucose or tests for excess oil by lipid stains are useful controls during this phase. The fermentation may last for about 120–160 h. The cycle time is determined by productivity, cost effectiveness and the influence of filtration difficulties. The culture broth becomes increasingly difficult to filter with age due to morphological changes and the increased number of autolyzed cells.

8.4.3 Production Kinetics

Cephalosporin C titres in excess of 15 g l^{-1} (whole broth, HPLC assay) can be realized by high yielding strains in production scale fermentations. Substantial quantities of desacetylcephalosporin C and significant amounts of desacetoxycephalosporin C and penicillin N are also present in the harvest broth. The actual biosynthetic potential of the organism is not fully revealed by the simple measurements of product formation. Non-enzymic decomposition of cephalosporin C by attack on the β-lactam and destruction of the cephalosporin ring system occurs as a first order reaction at a rate of 5×10^{-3} h^{-1} (Konecny *et al.*, 1973). This results in the disappearance of something like 25% of the cephalosporin C synthesized to products which are unaccounted for. Since the non-enzymic decomposition has first order kinetics it is concentration dependent. Therefore the rate of decomposition in absolute terms is greatest when the broth titre is highest towards the end of the fermentation. In practice a levelling off of titre accretion is observed as the rate of decomposition rises to equal the rate of synthesis. The halt in apparent cephalosporin C accumulation is therefore a consequence of kinetics of decomposition and has no microbiological explanation.

The proportion of desacetylcephalosporin C tends to increase towards the end of the fermentation. This is partly because of the relative stability of DAC compared with cephalosporin C and the increased production by enzymic deacetylation. It is also possible in some strains that the final acetylation reaction becomes limiting leading to *de novo* excretion of DAC.

An analysis of the fermentation carbon balance (Küenzi, personal communication) showed that about 60% was evolved as CO_2 and 30% fixed in biomass.

An analysis of the production costs is important for large scale manufacturers of cephalosporin C or 7-ACA. The total production costs are made up of components relating to the fermentation itself and those relating to the recovery process. It is normal to subdivide the costs into variable costs and allocated costs. The variable costs are those relating directly to raw materials used in the process. The allocated costs include factory costs allocated to the product which might be difficult to assign precisely, *e.g.* steam, electricity, engineering costs, air compression, *etc.* In general a unit cost is calculated for each component and used to calculate the cost to produce one kilogram of product in the harvest broth.

Since only about 70% of the fermented cephalosporin C is actually recovered, these values have to be adjusted to include the cost of producing the extra product needed to realize a kilogram of pure cephalosporin C salt. The variable costs and allocated costs for the recovery process are calculated in a similar way.

The recovery costs might well account for over half the total production costs for cephalosporin C salt.

8.4.4 Strain Development

Cephalosporin C was present only in trace amounts in cultures of the original Brotzu isolate. Current high yielding industrial strains produce greater than 15 g l^{-1} of cephalosporin C and more than 20 g l^{-1} of total β-lactam. This has been achieved by a combination of strain selection and fermentation development over a period of 25 years. The selection of higher yielding mutants has largely been an empirical process but nonetheless a very successful one. Traditionally the conidia or vegetative cells are exposed to a mutagen, normally ultraviolet light, and survivors randomly selected for yield testing. This process tends to be one of diminishing returns. As the yield increases it becomes increasingly difficult not only to generate improved mutants but also to identify these isolates against the background of the parental population. In order to detect a relatively rare event it is desirable to test a large number of isolates. This means that a model test system has to be developed in which the aspiring high yielding mutant will express some of its potential for production in the fermenter itself. In the interests of testing greater numbers of isolates it is often necessary to test under conditions increasingly remote from those in the stirred/aerated fermenter. Shake flask test systems are a good compromise allowing large numbers to be tested in liquid media. By careful choice of media the conditions can be selected to approach the oxygen abundant environment of the fermenter. Some attempts have been made to increase the numbers tested by using screening methods on solid media. Thus, high yielding colonies can be identified after bioassay by an overlayering technique. Problems associated with the effect of colony size and variation in the germination time make this a difficult method to quantify. Trilli *et al.* (1978) developed a procedure in which colonies of *Cephalosporium acremonium* were grown on the surface of small agar discs. After incubation the antibiotic content of the disc was assayed by bioassay by placing the discs on plates of a sensitive organism and determining the size of the inhibition zones. With improved methods of chemical assay the speed and precision could be increased by eluting the antibiotic from the agar piece. This type of methodology is open to the criticism that the selection conditions are remote from those in the fermenter.

Elander (1982) has advocated an alternative strategy to random screening. He recommends selective screens of various types on the basis that the frequency of improved mutants is thereby increased. Among the methods used are the selection of strains resistant to toxic precursors or products, strains resistant to analogues which overproduce biosynthetic intermediates, resistance to metal ions, reversion of auxotrophs and strains less sensitive to catabolite repression. In the particular case of cephalosporin-producing organisms toxic analogues of methionine, such as selenomethionine, and metal ions which react with pathway intermediates or the β-lactam ring itself, such as Hg^{2+} and Cu^{2+}, have been advocated as selective agents. Mendelovitz and Aharonowitz selected mutants of *S. clavuligerus* deregulated in aspartate kinase activity which overproduced cephalosporins by virtue of increased α-aminoadipate production from lysine.

A third strategy is the selection of recombinant strains following protoplast fusion. Hamlyn and Ball (1979) demonstrated that polyethylene glycol promoted the fusion and subsequent regeneration of recombinant cells. They selected recombinants having improved productivity, growth rate and sporulation compared with the parent strains. Classical genetical methods appear to be less readily used in *C. acremonium* than in *P. chrysogenum* where the parasexual cycle and the relative ease of producing stable diploids have been an advantage.

8.4.5 Fermentation Development

The development of the cephalosporin fermentation has followed that of other secondary metabolites. The key observation was that the growth rate had to be regulated in order to maintain an adequate oxygen transfer rate. This is achieved by operating a fed-batch fermentation in which the carbon source is the growth-limiting nutrient. The fermentation has essentially two phases. One under conditions of nutrient excess in which the growth rate is relatively high and the p_{O_2} falls quite sharply. This phase must be concluded before oxygen becomes limiting. This is achieved by a switch to carbon limiting conditions. The kinetics of product formation vary according to the strain and according to the influence of catabolite repression. In media rich in rapidly utilized carbon sources there is a pronounced temporal separation of growth and product formation. In media in which vegetable oil is the major or sole carbon source this temporal separation is not so apparent and production of cephalosporin follows the linear growth curve.

The best experimental system for studying the complex kinetic relationships in this and other fermentations is undoubtedly the chemostat where quantitative steady state data are readily obtained. The freedom to change experimentally the specific growth rate and the growth-limiting nutrient makes this culture method a most powerful tool. A further advantage is the relative independence from the size and quality of inoculum which affects the interpretation of results in batch and fed-batch culture. The chemostat can be used to develop quantitative mathematical models of the fermentation which might have application in design of a computer-controlled fermentation process. Because of the genetic instability of high yielding strains and the relative instability of the product itself, continuous production methods for cephalosporin C are unlikely to have any advantages. Present recovery methods are also more efficient when a high titre broth of the type produced in fed-batch culture is used.

8.4.6 Alternative Process — DAC process

Cephalosporin C is a relatively labile product and undergoes decomposition by non-enzymic attack on the β-lactam ring and enzymic deacetylation. The observed cephalosporin C titre is therefore some 40% below that which would be attained if no decomposition occurred. Observations in the author's laboratory that desacetylcephalosporin C is, in contrast, very stable led to the development of a fermentation process for production of DAC instead of cephalosporin C (Smith and Bailey, 1980). This was achieved by the addition of a potent acetyl esterase to the culture at the onset of antibiotic accretion. Synthesized cephalosporin C was deacetylated before any decomposition could occur. A suitable source of esterase was the yeast *Rhodosporidium toruloides* (Smith and Larner, 1977). The predicted increases in yield of cephalosporin nucleus obtained by fermenting to produce DAC were achieved. Indeed it was possible to extend the normal fermentation cycle by 48 h since accretion of DAC was maintained. This alternative process for the production of cephalosporin nucleus results in a large reduction in fermentation costs. However, until now its exploitation on an industrial scale has been limited by doubts about the efficiency of DAC recovery and further processing to the various cephalosporin derivatives.

8.5   RECOVERY PROCESS

8.5.1   Purification of Cephalosporin C

Hydrophobic compounds, such as penicillin G, can be recovered from the culture broth efficiently and relatively cheaply by extraction at low pH into an organic solvent. Unfortunately no such easy solution is available for the recovery of cephalosporin C from fermentation broths. The hydrophilic nature of the α-aminoadipyl side-chain of the cephalosporins renders them highly soluble in water and precludes direct solvent extraction methods. Four techniques used either separately or sometimes in combination have been used for the isolation of cephalosporin from broths. These can be briefly classified as adsorption techniques, ion exchange techniques, derivatization followed by solvent extraction, and enzymic modification. In each case the first step in the recovery process is the removal of the microorganism. This is achieved by filtration. The highly fragmented morphology of the culture broth also complicates this process. A rotary drum filter which may be 75 m^2 in size is precoated with a layer of filter-aid. The broth itself may also be mixed with filter-aid, the amount varying according to the difficulty of filtration, which varies from strain to strain and sometimes from batch to batch. In general the older the culture the more difficult the filtration. At the same time the broth is acidified to about pH 3.0 to precipitate protein from the culture medium and to improve the loading efficiency on to the adsorbent. Penicillin N is also destroyed at the reduced pH value. Sprinklers positioned above the filter provide a water-wash to ensure efficient separation of the antibiotic from the mycelium/filter-aid mixture. The culture filtrate collected from the filter may be diluted by 20–50% of wash water.

In many industrial processes for the recovery of cephalosporin C filtration is followed by an adsorption step. Two types of adsorbent have gained widespread use. The first processes used activated carbon following the early work of Abraham and Newton. Loading on to carbon is best achieved by upflow since this allows good flow rates to be maintained and facilitates the easy removal of fouled carbon from the bottom of the column. Cephalosporin C is adsorbed on to the carbon whilst many impurities and inorganic salts are lost in the percolate and column washings. The cephalosporin C is relatively vulnerable to decomposition whilst it is adsorbed on carbon and substantial losses can occur at this stage. Efficient loading of the column is achieved by monitoring the percolate by HPLC or by simpler methods such as polarimetry. The antibiotic is eluted from the carbon by an aqueous solvent mixture, such as a mixture of water and acetone (Nara *et al.*, 1975).

In 1965 Rohm and Haas commercialized the first synthetic organic macroreticular resins of the polystyrene type. Their use for the isolation of hydrophilic fermentation products has been reviewed by Voser (1982). The various macroreticular resins are characterized by their surface polarity, surface area, porosity and the distribution of pore size. This variety of surface properties results in a wide range of sorption behaviour. For the recovery of cephalosporin C, non-polar resins of the type Amberlite XAD-2, XAD-4 and ER-180 (Pirotta, 1982) have been found to be most useful. Factors to consider in choosing an adsorbent are its capacity, its selectivity and other physical properties such as swelling of the resin. For the adsorption of cephalosporin C pH is an important determinant of the resin capacity. Although XAD-4 resin has a higher capacity than XAD-2, the former has a relatively more difficult elution and regeneration cycle on an industrial scale. Elution of the resin with aqueous isopropanol yields a cephalosporin of higher purity than that from the carbon adsorption process.

In some cases, especially when working with high titre broths, it is possible simply to concentrate the eluate from the macroreticular resin prior to crystallization of cephalosporin C as a salt. For broths of more modest potency an intermediate ion-exchange resin is necessary if a product of reasonable purity is to be obtained. The preferred resin appears to be Amberlite IRA-68 or its equivalent. The resin is used in the acetate cycle and loaded directly with the eluate from the adsorption stage. A low pH value favours a high loading efficiency. Further impurities are lost in the column percolate and washings. The resin can be eluted with sodium or potassium acetate to give the appropriate salt of cephalosporin C. In practice a forerun and tail fraction are rejected since these are rich in impurities which bind to the resin more or less efficiently than cephalosporin C. Amongst the impurities, desacetylcephalosporin C is more strongly retained and is partially separated from cephalosporin C during elution.

A variety of processes have been described in the patent literature in which ion exchange resins are used without a prior adsorption step. Thus McCormick (1971) described a process in which impurities were first precipitated with acetone, the pH reduced with a cation exchange resin and then the cephalosporin adsorbed on an anion exchanger. Similarly, Stables and Briggs (1978) developed a three resin process in which impurities and salts are removed using an anion exchange resin followed by pH reduction using a cation exchanger to facilitate adsorption on to an anion exchanger. This was eluted with acetate as described above.

It is usually necessary to concentrate the resin eluate prior to crystallization of the cephalosporin C. Various methods are available, the important point being the avoidance of long contact times at elevated temperatures which destroy the labile cephalosporin. Crystallization of the cephalosporin C is brought about by addition of several volumes of solvent, for example acetone, methanol or isopropanol, to the aqueous concentrate. The resulting sodium or potassium salt is recovered by centrifugation or filtration and dried to give the intermediate product which is used to synthesize the wide range of cephalosporin derivatives. In earlier processes the zinc complex of cephalosporin C was often isolated. It had the advantage of reduced solubility in the aqueous/solvent mother liquors but led to problems in downstream processing. Modern processes for the isolation of cephalosporin C have an overall efficiency from whole broth to the isolated salt of about 70%. The salt can be between 70–80% pure. The process is costly because of the modest efficiency, the cost of raw materials for filtration, elution of columns and regeneration of resins, the large volumes of solvent required, the energy demanding concentration step and the large capital outlay for an industrial plant.

Alternative methods of cephalosporin recovery have been sought, among them derivatization and solvent extraction. Acylation of the α-amino group in the aminoadipyl side-chain destroys the amphoteric nature of the molecule and renders the derivative solvent extractable. A large number of *N*-acylcephalosporins have been synthesized. Among the more important are the *N*-isobutoxycarbonyl (Johnson *et al.*, 1970), *N*-chloroacetyl (Hayes and Lee, 1971), *N*-(2,2-dicarbethoxyl)vinyl (Ascher, 1981) and *N*-benzoyl derivatives (Sciavolino, 1972). The resulting *N*-blocked cephalosporin is extracted into an organic solvent such as methyl isobutyl ketone at acid pH values. The problem with this type of process is the lack of specificity of the acylating agent. Thus in complex fermentation broths other compounds having a free amino group compete with the cephalosporin for the acylating species. In some cases the acylating reagent may also be readily hydrolyzed by water. It may be necessary to use between 5–10 molar equivalents of an expensive reagent to achieve complete acylation of the cephalosporin C. A partial solution of the problem may be to carry out the derivatization after partial purification by adsorption and subsequent elution. The factors governing the derivatization/extraction process are the costs of reagents for acylation and solvent extraction. In the case where derivatization of the cephalosporin is carried out in a broth culture filtrate, substantial volumes of solvent may be required for the subsequent partition. The choice of solvent must therefore take into consideration the possibility of recovery by distillation and subsequent reuse.

The fate of recovered cephalosporin C is subsequent cleavage to 7-ACA. This process also requires that the amino and carboxyl functions of cephalosporin C be protected. The recovery of a ready protected cephalosporin was achieved by preparing the bis(dibenzhydryl) ester (Robinson and Walker, 1975), obtained by treating an *N*-blocked cephalosporin C with diphenyldiazomethane. The protected cephalosporin is extracted at low pH into a water-immiscible solvent such as methylene chloride where it is readily esterified. The resulting fully protected cephalosporin can be used directly in the cleavage process for production of 7-ACA without isolation of the intermediate cephalosporin salt. Again the cost and stoichiometry of the reaction are a potential problem in industrial application.

The fourth method investigated for the recovery of the cephalosporin nucleus is the use of enzymes. Many groups have searched for a cephalosporin amidase which would cleave the α-aminoadipyl residue to yield 7-ACA directly. Such a process operating on a broth culture filtrate might have economic attractions provided that the resulting 7-ACA could be efficiently and simply recovered. However, despite the occasional false alarm, no convincing evidence for the existence of such an enzyme has been published. The known penicillin-type acylases are inactive against the hydrophilic cephalosporins.

An alternative enzymic approach was described by Arnold *et al.* (1969). A D-amino acid oxidase from the yeast *Trigonopsis variabilis* was used to oxidize the amino group. The reaction proceeds through a keto acid intermediate to yield glutarimidocephalosporin C. Under appropriate conditions the reaction goes to completion. The product can be isolated by solvent extraction at acid pH or can be further subjected to enzymatic hydrolysis to give 7-ACA itself. A number of microbial sources have been described which hydrolyze the glutarimido derivative to 7-ACA (*e.g.* Banyu Pharmaceutical, 1977; Asahi Chemical, 1978). A further interesting possibility (Toyo Jozo, 1977) involved the synthesis of an *N*-blocked cephalosporin with phenylaminothiocyanate. The derivative could be cleaved directly to 7-ACA by an enzyme from *Bacillus megaterium*. Other *N*-blocked derivatives such as the *N*-benzoylcephalosporin C would be reconverted back to cephalosporin C by bacterial amidases.

It is not known whether enzymic processes of the type described above have reached the stage where they are employed on an industrial scale. Companies with a large capital investment in more conventional adsorption or ion-exchange processes are reluctant to introduce new technology unless the advantages are clear cut. The cost of enzyme production, perhaps the need to recover the enzyme itself and immobilize it and the uncertainty about scale-up of laboratory yields are factors to be considered.

8.5.2 Cleavage of Cephalosporin C to 7-ACA

The key to the synthesis of a wide range of cephalosporin derivatives was the chemical cleavage of cephalosporin C to 7-ACA. The nitrosyl chloride cleavage developed by Morin *et al.* (1962) opened the way to industrial production of 7-ACA on a large scale. This process has now been superseded by processes based on that developed by Ciba (Fechtig *et al.*, 1966). Cephalosporin C is cleaved using phosphorus pentachloride after protection of the amino and carboxyl functions. The reaction proceeds as far as an imino chloride intermediate in the presence of base, to an imino ether intermediate by addition of alcohol and finally to an ester of 7-ACA by hydrolysis with acid.

Further improvements (*Belg. Pat.* 718 824, 1969) have been made by using a silylyl protection which simultaneously blocks the amino and carboxyl functions of cephalosporin C and permits the cleavage with PCl_5 to be carried out in a common solvent, methylene chloride.

8.6 CONCLUDING REMARKS

Pharmaceutical companies will continue to invest substantial sums in the search for new cephalosporins with improved antibacterial activity, broad spectrum and resistance to β-lactamases, and for compounds which can be orally administered. The cephalosporins are likely to remain the most important weapon in the continued fight against infectious disease. However, they are likely to face renewed competition from the latest penams, carbapenems and monocyclic β-lactams. The cost of production might play an increasing part in the decisions about which new compounds to develop.

Much has already been achieved in establishing the basic biosynthetic pathway to the cephalosporins. Further progress can be expected in determining the mechanisms of the key enzymic steps such as the ring cyclization of the tripeptide and the ring expansion of penicillin N. More will undoubtedly be learned about the regulation of antibiotic synthesis and the expression of biosynthetic enzymes during the fermentation process. The technique of mutasynthesis might lead to the production of novel biosynthetic products.

The fermentation technologists will continue to improve yields by exploiting new strains and introducing better fermentation control algorithms. Computer systems can be expected to play an increased role both at pilot and production level. Advances in the recovery methodology might stem from the use of better adsorbents or the exploitation of enzymes such as amino acid oxidases, amidases and acetyl esterases to achieve cheaper, more efficient processes to the cephalosporin nucleus.

ACKNOWLEDGEMENT

The author would like to thank colleagues in the Department of Biotechnology, Ciba-Geigy, Basel for their help and advice in preparing this article. The following deserve special thanks: J. Auden, M. Küenzi, Prof. J. Nüesch, W. Voser and W. Zimmermann for their critical reading of the manuscript, and Mrs E. Schmutz and M. Klingler for typing the manuscript and preparing the figures.

8.7 REFERENCES

Abraham, E. P. and G. G. F. Newton (1961). The structure of cephalosporin C. *Biochem. J.*, **79**, 377.
Abraham, E. P., G. G. F. Newton and C. W. Hale (1954). *Biochem. J.*, **58**, 94.
Aoki, H., H. Sakai, M. Kohsaka, T. Konomi, J. Hosoda, Y. Kubochi, E. Iguchi and H. Imanaka (1976). Nocardicin A, a new monocyclic β-lactam antibiotic. 1. Discovery, isolation and characterisation. *J. Antibiot.*, **29**, 492.
Arnold, B. H., R. A. Fildes and D. A. Gilbert (1969). *Ger. Pat.* 1 939 341.
Asahi Chemical Industries KK (1978). Microbiological preparation of 7-aminocephem compounds using minimum reaction period which avoids loss of yield caused by decomposition. *Jpn. Pat.* 78 59 095.
Ascher, G. (1981). Cephalosporin derivatives. *US Pat.* 4 267 321.
Baddiley, J. (1982). Cell wall structure and biosynthesis. In *β-Lactam Antibiotics — Mode of Action, New Developments and Future Prospects*, ed. M. Salton and G. D. Shockman, pp. 13–30. Academic, New York.
Baldwin, J. E., M. Jung, P. Singh, T. Wan, S. Haber, S. Herchen, J. Kitchin, A. L. Demain, N. A. Hunt, M. Kohsaka, T. Konomi and M. Yoshida (1980). *Philos. Trans. R. Soc. London, Ser. B*, **289**, 169.
Baldwin, J. E., M. Jung, J. J. Uster, E. P. Abraham, J. A. Huddleston and R. L. White (1981). Penicillin biosynthesis: Conversion of deuteriated (L-α-amino-δ-adipyl)-L-cysteinyl-D-valine into isopenicillin N by a cell-free extract of *Cephalosporium acremonium. J. Chem. Soc., Chem. Commun.*, 246.
Banyu Pharmaceutical Co. Ltd (1977). 7-Aminocephalosporanic acid derivative preparation by treating cephalosporin with *Bacillus* or *Arthrobacter* microorganisms. *Jpn. Pat.* 77 128 293.
Brewer, S. J., T. T. Boyle and M. K. Turner (1977a). The carbamoylation of the 3-hydroxymethyl group of desacetyl-7-methoxycephalosporin C by homogenates of *Streptomyces clavuligerus. Biochem. Soc. Trans.*, **5**, 1026.
Brewer, S. J., J. E. Farthing and M. K. Turner (1977b). The oxygenation of the 3-methyl group of 7-β-(5-D-aminoadipamido)-3-methylceph-3-em-4-carboxylic acid (desacetoxycephalosporin C) by extracts of *Acremonium chrysogenum. Biochem. Soc. Trans.*, **5**, 1024.
Brewer, S. J., P. M. Taylor and M. K. Turner (1980). An ATP-dependent carbamoylphosphate-3-hydroxymethylcephem-*O*-carbamoyltransferase from *Streptomyces clavuligerus. Biochem. J.*, **185**, 555.
Caltrider, P. G. and H. F. Niss (1966). Role of methionine in cephalosporin synthesis. *Appl. Microbiol.*, **14** (5), 746.
Demain, A. L. (1957). Inhibition of penicillin formation by lysine. *Arch. Biochem. Biophys.*, **67**, 244.
Demain, A. L. (1983). Formation of β-lactam antibiotics. In *Proc. Eur. Conf. Ind. Microbiol. 2nd*, pp. 225–244.
Drew, S. W. and A. L. Demain (1975). Production of cephalosporin C by single and double sulphur auxotrophic mutants of *Cephalosporium acremonium. Antimicrob. Agents Chemother.*, **8** (1), 5.
Elander, R. P. (1982). In *Trends in Antibiotic Research*, ed. H. Umezawa, A. L. Demain, T. Hata and C. R. Hutchinson, pp. 16–31. Japan Antibiotics Research Association, Tokyo.
Fawcett, P. A., J. J. Usher, J. A. Huddleston, R. C. Bleaney, J. J. Nisbet and E. P. Abraham (1976). Synthesis of δ-(α-aminoadipyl)cysteinylvaline and its role in penicillin biosynthesis. *Biochem. J.*, **157**, 651.

Fechtig, B., H. H. Peter, H. Bickel and E. Vischer (1966). Process for the manufacture of 7-aminocephalosporanic acids. *Br. Pat.* 1 041 985. See also *Helv. Chim. Acta*, **51**, 1108 (1968).

Felix, H. R., J. Nüesch and W. Wehrli (1980). Investigation of the two final steps in the biosynthesis of cephalosporin C using permeabilized cells of *Cephalosporium acremonium*. *FEMS Microbiol. Lett.*, **8**, 55.

Felix, H. R., H. H. Peter and H. J. Treichler (1981). Microbiological ring expansion of penicillin N. *J. Antibiot.*, **34**, 567.

Florey, H. W., E. B. Chain, N. G. Heatley, M. A. Jennings, A. G. Sanders, E. P. Abraham and M. E. Florey (1949). *Antibiotics*, vol. 2. Oxford University Press, London.

Flynn, E. H., M. H. McCormick, M. C. Stamper, H. DeValeria and C. W. Godzeski (1962). A new natural penicillin from *Penicillium chrysogenum*. *J. Am. Chem. Soc.*, **84**, 4594.

Fujisawa, Y. and T. Kanzaki (1975a). Role of acetyl-CoA: desacetylcephalosporin C acetyltransferase in cephalosporin C biosynthesis by *Cephalosporium acremonium*. *Agric. Biol. Chem.*, **39**, 2043.

Fujisawa, Y., K. Kitano and T. Kanzaki (1975a). Accumulation of deacetoxycephalosporin C by a deacetylcephalosporin C negative mutant of *Cephalosporium acremonium*. *Agric. Biol. Chem.*, **39**, 2049.

Fujisawa, Y., H. Shirafuji, M. Kida, K. Nara, M. Yoneda and T. Kanzaki (1973). New findings on cephalosporin C biosynthesis. *Nature New Biol.*, **246**, 154.

Fujisawa, Y., H. Shirafuji, M. Kida, K. Nara, M. Yoneda and T. Kanzaki (1975b). Accumulation of deacetylcephalosporin C by cephalosporin C negative mutants of *Cephalosporium acremonium*. *Agric. Biol. Chem.*, **39**, 1295.

Hamlyn, P. F. and C. Ball (1979). Recombination studies with *Acremonium chrysogenum*. In *Genetics of Industrial Microorganisms*, ed. O. K. Sebek and A. I. Lasking, p. 185. American Society for Microbiology, Washington, DC.

Hayes, H. B. and G. L. Lee (1971). Verfahren zur Herstellung von 7-Aminocephalosporansäure (7-ACA) aus Cephalosporin C und neue Cephalosporin C-Derivate als Zwischenprodukte. *Ger. Pat.* 2 031 754.

Higgens, C. E., R. L. Hamill, T. H. Sands, M. M. Hoehen, N. E. Davies, R. Nagarajan and L. D. Boeck (1974). The occurrence of desacetoxycephalosporin C in fungi and streptomycetes. *J. Antibiot.*, **27**, 298.

Hinnen, A. and J. Nüesch (1976). Enzymatic hydrolysis of cephalosporin C by an extracellular acetylhydrolase of *Cephalosporium acremonium*. *Antimicrob. Agents Chemother.*, **9**, 824.

Hook, D. J., L. T. Chang, R. P. Elander and R. B. Morin (1979). Stimulation of the conversion of penicillin N to cephalosporin by ascorbic acid, α-ketoglutarate, and ferrous ions in cell-free extracts of strains of *Cephalosporium acremonium*. *Biochem. Biophys. Res. Commun.*, **87**, 258.

Hoover, J. R. E. and C. H. Nash (1978). In *Kirk-Othmer Encyclopedia of Chemical Technology*, vol. 2, 3rd edn., pp. 871–919. Wiley, New York.

Howarth, T. T., A. G. Brown and T. J. King (1976). Clavulanic acid, a novel β-lactam isolated from *Streptomyces clavuligerus*; X-ray crystal structure analysis. *J. Chem. Soc., Chem. Commun.*, 266.

Imada, A., K. Kitano, K. Kintaka, M. Muroi and M. Asai (1981). Sulfazecin and isosulfazecin, novel β-lactam antibiotics of bacterial origin. *Nature (London)*, **289**, 590.

Jayatilake, S., J. A. Huddleston and E. P. Abraham (1981). Conversion of isopenicillin N into penicillin N in cell-free extracts of *Cephalosporium acremonium*. *Biochem. J.*, **194**, 645.

Johnson, D. A., R. R. Smith, E. J. Richardson, J. M. Roubie and H. H. Silvestri (1970). Process for the preparation of 7-aminocephalosporanic acid. *Can. Pat.* 858 566.

Kahan, J. S., F. M. Kahan, R. Goegelman, S. A. Currie, M. Jackson, E. O. Stapley, T. W. Miller, A. K. Miller, D. Hendlin, S. Mochales, S. Hernandez, H. B. Woodruff and J. Birnbaum (1979). Thienamycin. A new β-lactam antibiotic. 1. Discovery, taxonomy, isolation and physical properties, *J. Antibiot.*, **32**, 1.

Kohsaka, M. and A. L. Demain (1976). Conversion of penicillin N to cephalosporin(s) by cell-free extracts of *Cephalosporium acremonium*. *Biochem. Biophys. Res. Commun.*, **70**, 465.

Konecny, J., E. Felber and J. Gruner (1973). Kinetics of the hydrolysis of cephalosporin C. *J. Antibiot.*, **26** (3), 135.

Konomi, T., S. Herchen, J. E. Baldwin, M. Yoshida, N. A. Hunt and A. L. Demain (1979). Cell-free conversion of δ-(L-α-aminoadipyl)-L-cysteinyl-D-valine to an antibiotic with the properties of isopenicillin N in *Cephalosporium acremonium*. *Biochem. J.*, **184**, 427.

Küenzi, M. T. (1978). Process design and control in antibiotic fermentation. In *Antibiotics and Other Secondary Metabolites, Biosynthesis and Production*, ed. R. Hütter, T. Leisinger, J. Nüesch and W. Wehrli, pp. 39–56. Academic, New York.

Küenzi, M. T. (1980). Regulation of cephalosporin synthesis in *Cephalosporium acremonium* by phosphate and glucose. *Arch. Microbiol.*, **128**, 78.

Kupka, J., Y. Shen, S. Wolfe and A. L. Demain (1983). Partial purification and properties of the α-ketoglutarate-linked ring-expansion enzyme of β-lactam biosynthesis of *Cephalosporium acremonium*. *FEMS Microbiol. Lett.*, **16**, 1.

Lilley, G., A. E. Clark and G. C. Lawrence (1981). Control of production of cephamycin C and thienamycin by *Streptomyces cattleya* NRRL 8057. *J. Chem. Technol. Biotechnol.*, **31**, 127.

Loder, P. B. and E. P. Abraham (1971a). Isolation and nature of intracellular peptides from a cephalosporin C-producing *Cephalosporium* sp. *Biochem. J.*, **123**, 471.

Loder, P. B. and E. P. Abraham (1971b). Biosynthesis of peptides containing α-aminoadipic acid and cysteine in extracts of a *Cephalosporium* sp. *Biochem. J.*, **123**, 477.

Martin, J. F. (1983). Carbon catabolite regulation of β-lactam antibiotic production. In *Proc. Eur. Conf. Ind. Microbiol, 2nd*, pp. 259–262.

Matsumura, M., T. Imanaka, T. Yoshida and H. Taguchi (1978). Effect of glucose and methionine consumption rates on cephalosporin C production by *Cephalosporium acremonium*. *J. Ferment. Technol.*, **56** (4), 345.

Matsumura, M., T. Imanaka, T. Yoshida and H. Taguchi (1980a). Morphological differentiation in relation to cephalosporin C synthesis by *Cephalosporium acremonium*. *J. Ferment. Technol.*, **58** (3), 197.

Matsumura, M., T. Imanaka, T. Yoshida and H. Taguchi (1980b). Regulation of cephalosporin C production by endogenous methionine in *Cephalosporium acremonium*. *J. Ferment. Technol.*, **58** (3), 205.

Matsumura, M., T. Imanaka, T. Yoshida and H. Taguchi (1981). Modeling of cephalosporin C production and its application to fed-batch culture. *J. Ferment. Technol.*, **59** (2), 115.

McCormick, M. H. (1971). *Ger. Pat.* 1 617 456.

Meesschaert, B., P. Adriaens and H. Eyssen (1980). Studies on the biosynthesis of isopenicillin N with a cell-free preparation of *Penicillium chrysogenum*. *J. Antibiot.*, **33**, 722.

Mehta, R. J., J. L. Speth and C. H. Nash (1979). Lysine stimulation of cephalosporin C synthesis in *Cephalosporium acremonium. Eur. J. Appl. Microbiol. Biotechnol.*, **8**, 177.

Mendelovitz, S. and Y. Aharonowitz (1982). Regulation of lysine synthesis, aspartokinase, dihydrodipicolinic acid synthetase and homoserine dehydrogenase by aspartic acid family amino acids in *Streptomyces. Antimicrob. Agents Chemother.*, **21** (1), 74.

Mirelman, D. (1982). Assembly of wall peptidoclycan polymers. In *β-Lactam Antibiotics — Mode of Action, New Developments and Future Prospects*, ed. M. Salton and G. D. Shockman, pp. 67–86. Academic, New York.

Morin, R. B., B. G. Jackson, E. H. Flynn and R. W. Roeske (1962). Chemistry of cephalosporin antibiotics. 1. 7-amino-cephalosporanic acid from cephalosporin C. *J. Am. Chem. Soc.*, **84**, 3400.

Nagarajan, R., L. D. Boeck, M. Gorman, R. L. Hamill, C. E. Higgens, M. M. Hoehn, W. M. Stark and J. G. Whitney (1971). *β*-Lactam antibiotics from *Streptomyces. J. Am. Chem. Soc.*, **93**, 2308.

Nara, K., K. Ohta, K. Katamoto, N. Mizokami and H. Fukuda (1975). Method for separating cephalosporin C. *US Pat.* 3 926 973.

Newton, G. G. F. and E. P. Abraham (1954). *Biochem. J.*, **58**, 103.

O'Callaghan, C. H. (1980). Structure–activity relations and *β*-lactamase resistance. *Philos. Trans. R. Soc. London, Ser. B*, **289**, 197.

O'Sullivan, J. and E. P. Abraham (1980). The conversion of cephalosporins to 7*α*-methoxycephalosporins by cell-free extracts of *Streptomyces clavuligerus. Biochem. J.*, **186**, 613.

O'Sullivan, J., R. T. Aplin, C. M. Stevens and E. P. Abraham (1979). Biosynthesis of 7*α*-methoxycephalosporin. Incorporation of molecular oxygen. *Biochem. J.*, **179**, 47.

O'Sullivan, J., R. C. Bleaney, J. A. Huddleston and E. P. Abraham (1979). Incorporation of ³H from δ-(L-*α*-amino[4,5-³H]adipyl)-L-cysteinyl-D-[4,4-³H]valine into isopenicillin N. *Biochem. J.*, **184**, 421.

Perkowski, C. A. (1981). Fermentation process air filtration via cartridge filters. Paper presented at Am. Chem. Soc. Ann. Meeting 1981, New York.

Pirotta, M. (1982). Amberlite ER-180 — a new styrene divinylbenzene adsorbent specifically designed for industrial chromatography and particularly for the extraction of cephalosporin C. *Angew. Makromol. Chem.*, **109/110**, 197.

Queener, S. W., J. J. Capone, A. B. Radue and R. Nagarajan (1974). Synthesis of deacetoxycephalosporin C by a mutant of *Cephalosporium acremonium. Antimicrob. Agents Chemother.*, **6**, 334.

Queener, S. W., S. Wilkerson, D. R. Tunin, J. P. McDermott, J. L. Chapman, C. Nash, C. Platt and J. Westpheling (1984). Cephalosporin C production: biochemical and regulatory aspects of sulfur metabolism. In *Biotechnology of Industrial Antibiotics*, ed. E. J. Vandamme. Dekker, New York.

Robinson, C. and D. Walker (1975). *Ger. Pat.* 2 436 772.

Scheidegger, A., M. T. Küenzi and J. Nüesch (1984). Partial purification and catalytic properties of a bifunctional enzyme in the biosynthetic pathway of *β*-lactams in *Cephalosporium acremonium. J. Antibiot.*, **37** (5), 522.

Sciavolino, F. (1972). Verfahren zur Isolierung von Cephalosporin C in Form eines mit einem Lösungsmittel extrahierbaren *N*-subsituierten Derivates. *Ger. Pat.* 2 157 693.

Smith, A. and P. J. Bailey (1980). Desacetylcephalosporin C production. *Br. Pat. Appl.* 2 060 610 A.

Smith, A. and R. W. Larner (1977). *Br. Pat.* 1 531 212.

Stables, H. C. and K. Briggs (1978). *Ger. Pat.* 2 852 596.

Stapley, E. O., M. Jackson, S. Hernandez, S. B. Zimmerman, S. A. Currie, S. Mochales, J. M. Mahta, H. B. Woodruff and D. Hendlin (1972). Cephamycins, a new family of *β*-lactam antibiotics. 1. Production by actinomycetes, including *Streptomyces lactamdurans* sp. *Antimicrob. Agents Chemother.*, **2**, 122.

Sykes, R. B., C. M. Cimarusti, D. P. Bonner, K. Bush, D. M. Floyd, N. H. Georgopapadakou, W. H. Koster, W. C. Liu, W. L. Parker, P. A. Principe, M. L. Rathnum, W. A. Slusarchyk, W. H. Trejo and J. S. Wells (1981). Monocyclic *β*-lactam antibiotics produced by bacteria. *Nature (London)*, **291**, 489.

Sykes, R. B. and M. Matthew (1976). The *β*-lactamases of gram negative bacteria and their role in resistance to *β*-lactam antibiotics. *J. Antimicrob. Chemother.*, **2** (2), 115.

Tipper, D. J. and J. L. Strominger (1965). Mechanism of action of penicillins: A proposal based on their structural similarity to acyl-D-alanyl-D-alanine. *Proc. Natl. Acad. Sci. USA*, **54**, 1133.

Tomasz, A. (1979). The mechanism of the irreversible antimicrobial effects of penicillins. *Annu. Rev. Microbiol.*, **33**, 113.

Toyo Jozo K. K. (1977). 7-Aminocephalosporins preparation by microbial transformation of the corresponding 7-acylamino derivative. *Jpn. Pat.* 77 82 791.

Treichler, H. J., M. Liersch, J. Nüesch and H. Döbeli (1979). Role of sulphur metabolism in cephalosporin C and penicillin biosynthesis. In *Genetics of Industrial Microorganisms*, ed. O. K. Sebek and A. I. Laskin, 97–104 American Society for Microbiology, Washington, DC.

Trilli, A., V. Michelini, V. Mantovani and S. J. Pirt (1978). Development of an agar disk method for the rapid selection of cephalosporin producers with improved yields. *Antimicrob. Agents Chemother.*, **13** (1), 7.

Trown, P. W., B. Smith and E. P. Abraham (1963). Biosynthesis of cephalosporin C from amino acids. *Biochem. J.*, **86**, 284.

Turner, M. K., J. E. Farthing and S. J. Brewer (1978). The oxygenation of [3-methyl-³H]desacetoxycephalosporin C to [3-hydroxymethyl-³H]desacetylcephalosporin C by 2-oxoglutarate-linked dioxogenases from *Acremonium chrysogenum* and *Streptomyces clavuligerus. Biochem. J.*, **173**, 839.

Voser, W. (1982). Isolation of hydrophilic fermentation products by adsorption chromatography. *J. Chem. Technol. Biotechnol.*, **32**, 109.

Warren, S. C., G. G. F. Newton and E. P. Abraham (1967a). Use of *α*-aminoadipic acid for the biosynthesis of penicillin N and cephalosporin C by a *Cephalosporium* sp. *Biochem. J.*, **103**, 891.

Warren, S. C., G. G. F. Newton and E. P. Abraham (1967b). The role of valine in the biosynthesis of penicillin N and cephalosporin C by a *Cephalosporium* sp. *Biochem. J.*, **103**, 902.

Yoshida, M., T. Konomi, M. Kohsaka, J. E. Baldwin, S. Herchen, P. Singh, N. A. Hunt and A. L. Demain (1978). Cell-free ring expansion of penicillin N to deacetoxycephalosporin C by *Cephalosporium acremonium* CW-19 and its mutants. *Proc. Natl. Acad. Sci. USA*, **75**, 6253.

9

Commercial Production of Cephamycin Antibiotics

D. R. OMSTEAD, G. R. HUNT and B. C. BUCKLAND
Merck and Co. Inc., Rahway, NJ, USA

9.1 INTRODUCTION

Penicillin and cephalosporin antibiotics are clinical products which were first marketed in recent decades for the treatment of bacterial infections in humans and animals. These products have generally been made *via* a fermentation process in which various fungal species are grown in large agitated vessels and used to synthesize the desired product. The penicillin or cephalosporin produced by these organisms has been isolated from the fermenter broth and then purified and modified to produce a clinical product.

Since the original introduction of cephalosporin derivatives and penicillin, a large number of new β-lactam products with superior antibacterial activity have been marketed. The great majority of new products have been chemical derivatives of the original penicillin and cephalosporin compounds. These new products are manufactured by first preparing penicillin or cephalosporin *via* an existing fermentation process. In this way, new antibiotics have been introduced without the need to develop wholly new fermentation processes. This chemical modification approach has resulted in the clinical availability of several 'generations' of antibiotic products exhibiting antibacterial spectra which far exceed those of microbially synthesized penicillin or cephalosporin compounds. More recently, however, new classes of broadly active β-lactam antibiotics have been discovered which are not chemical derivatives of penicillin or cephalosporin. The cephamycins, for example, represent such a class of microbially synthesized β-lactam antibiotics. The molecular structure of these compounds resembles that of the cephalosporins. However, two important characteristics, β-lactamase resistance and the nature of the producing species, distinguish the cephamycins from the cephalosporins. The cephamycins are resistant to the action of β-lactamase enzymes whereas cephalosporins are inactivated by lactamases. Since many clinical pathogens synthesize lactamase, they are resistant to both penicillin and cephalosporin. Frequently, such bacteria are susceptible to cephamycins. As a result, cephamycins can be used to treat a more broad spectrum of human infections. The taxonomic classification of synthesizing microorganisms also distinguishes cephamycins from cephalosporins. Cephamycins are synthesized only by the filamentous bacteria, actinomycetes. In contrast, the vast majority of observed penicillin- and cephalosporin-producing organisms are eukaryotes, principally fungi.

The differences in the metabolic requirements of fungi and actinomycetes required development of new fermentation processes for the cephamycins. Biochemical similarities do exist between the β-lactam biosynthetic pathways in fungi and actinomycetes and these similarities were used as the basis for development of the initial cephamycin fermentation processes. The intent of this chapter is to review the metabolic pathways in penicillin-, cephalosporin- and cephamycin-producing organisms which are relevant to cephamycin production and to discuss the means by which this information was used to formulate the current cephamycin production processes.

9.2 CEPHAMYCIN PRODUCT DESCRIPTION

9.2.1 Discovery

The penicillins were the first β-lactam antibiotics to become commercially available in the United States. Following their introduction into clinical use, this class of compounds was found to be capable of controlling many human infections for which there had previously been no effective treatment. Upon further investigation, it was observed that there were several naturally occurring penicillin derivatives which were lethal to a wide variety of Gram-positive bacteria. However, there were certain classes of Gram-positive bacteria which were resistant to the effect of penicillin. In addition, the spectrum of penicillin activity against Gram-negative organisms was not as broad.

A continuing effort was then made to find naturally occurring compounds whose bactericidal spectra complemented or surpassed those exhibited by the existing penicillins. This was accomplished by randomly screening marine and soil samples containing a wide variety of bacterial and fungal species. New compounds were identified by testing the antibacterial activity of these samples toward a group of indicator organisms. Among many other discoveries, Brotzu (1948) isolated a novel activity produced by a previously unknown fungal species. The producing culture was originally classified as *Cephalosporium acremonium*. This organism attracted great attention because it produced penicillin N, a compound which is more active against certain Gram-negative bacteria than were existing penicillins. Of even greater importance, however, was the finding (Abraham and Newton, 1954) that this organism produced another β-lactam product which was highly active against Gram-negative bacteria. This novel product was termed cephalosporin C. The chemical structure of cephalosporin was determined by Abraham and Newton (1961) and Hodgkin and Maslen (1961). Like penicillin, the basic cephalosporin C nucleus was found to be bicyclic, but differed from the penicillin nucleus by the presence of a six-membered ring in place of the five-membered thiazolidine ring.

Since the discovery of cephalosporin C, a number of species have been isolated which produce a variety of related antibiotic products. In all, over 1000 naturally occurring or chemically modified antibiotic products have been reported, of which over 100 are commercially produced (Aharonowitz and Cohen, 1981). One important class of microbially produced compounds, the cephamycins, was separately isolated in the early 1970s by Merck and Co., Inc. and by Eli Lilly and Co. The first member of this family to be isolated was cephamycin C. It is produced by both *Nocardia lactamdurans* (Stapley *et al.*, 1972) and *Streptomyces clavuligerus* (Nagajarian *et al.*, 1971) as well as several other actinomycetes (Stapley and Birnbaum, 1981). The basic structure for this compound is illustrated in Figure 1, as are the structures for other naturally occurring cephamycins. The cephamycin family of compounds is produced only by the filamentous bacteria, actinomycetes. A characteristic methoxy substitution at the 7α position of the bicyclic ring structure distinguishes between cephamycins and cephalosporins.

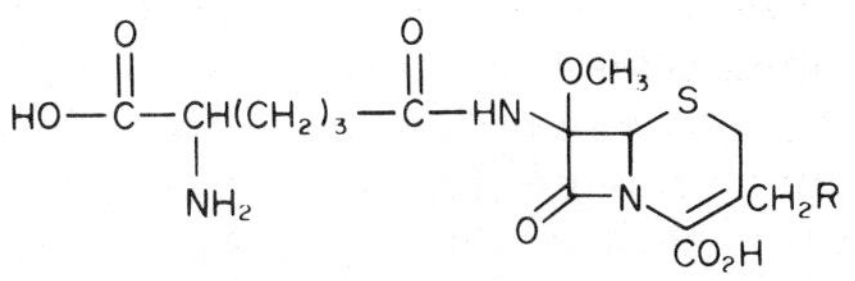

Compound	Substituent (R)
7—Methoxy cephalosporin C	$-O-\overset{\displaystyle O}{\overset{\|}{C}}-CH_3$
Cephamycin A	$-O-\overset{\displaystyle O}{\overset{\|}{C}}-\underset{\displaystyle OCH_3}{\overset{\|}{C}}=CH\!-\!\langle\text{ring}\rangle\!-\!OSO_3H$
Cephamycin B	$-O-\overset{\displaystyle O}{\overset{\|}{C}}-\underset{\displaystyle OCH_3}{\overset{\|}{C}}-CH\!-\!\langle\text{ring}\rangle\!-\!OH$
Cephamycin C	$-O-\overset{\displaystyle O}{\overset{\|}{C}}-NH_2$
Cephamycin C—2801X	$-O-\overset{\displaystyle O}{\overset{\|}{C}}-\underset{\displaystyle OCH_3}{\overset{\|}{C}}=CH\!-\!\langle\text{ring}\rangle\!-\!OH,\ OH$
7—Methoxydeacetoxy cephalosporin	—H
7—Methoxydeacetyl cephalosporin	—OH

Figure 1 Naturally occurring cephamycin antibiotics

9.2.2 Mode of Action

As with all β-lactam compounds, the antibacterial activity of the cephamycins derives from their ability to bind to a receptor protein in the susceptible organism and then to inhibit further cell wall synthesis. This inhibition leads to a loss of viability in the affected cell. The β-lactam structure specifically inhibits activity of the transpeptidase enzymes involved in peptidoglycan biosynthesis. As a result, assembly of the cell envelope in both Gram-positive and Gram-negative organisms is prevented.

The primary means by which microorganisms express resistance to β-lactam antibiotics is by synthesizing enzymes which have the ability to cleave the β-lactam ring. Occurrence of such lactamase activity is widespread. Therefore, in order to control proliferation of organisms which synthesize lactamase, it was necessary to identify structures which would resist lactamase mediated breakdown. Two approaches were taken. In one, chemical modifications of existing antibiotics were made with the intent of conferring resistance to lactamase. A second approach was to identify microbially synthesized products which are resistant to lactamase. Cephamycin C was first identified as an inhibitor of cell wall synthesis (Daust *et al.*, 1973). Subsequently, it was found that this compound is particularly resistant to lactamase activity. It was determined that the characteristic 7α methoxy group of the cephamycins was the substituent which confers lactamase resistance. This group acted to stablize the β-lactam ring structure, reducing lactamase sensitivity.

9.2.3 Cefoxitin

While cephamycin C is particularly active against Gram-negative organisms, it is not as active against Gram-positive organisms. As a result, this molecule was chemically modified to enhance its antibacterial spectrum. The intent was to alter the existing side-chains of the cephamycin C molecule such that a wide spectrum of Gram-positive antibacterial activity would be added without sacrificing the Gram-negative activity or *in vivo* stability of the parent molecule (Christensen, 1982).

Replacement of the aminoadipoyl side-chain at the 7β position of cephamycin with a thienyla-cetyl substitution similar to that in cephalothin had the desired effect. The resulting semi-synthetic β-lactam compound, cefoxitin (see Figure 2), has a particularly wide antibacterial spectrum. The substituents of the cefoxitin molecule which contribute various specific activities (as discussed by Christensen, 1982) toward the overall spectrum are also indicated in Figure 2. Cefoxitin is currently manufactured and marketed by Merck and Co., Inc. It has found a number of uses in prevention and treatment of human infection. This product is widely used in preventing post-operative infection by penicillin resistant bacteria. An additional attribute of the cefoxitin spectrum is its unusual activity against a variety of anaerobic bacteria.

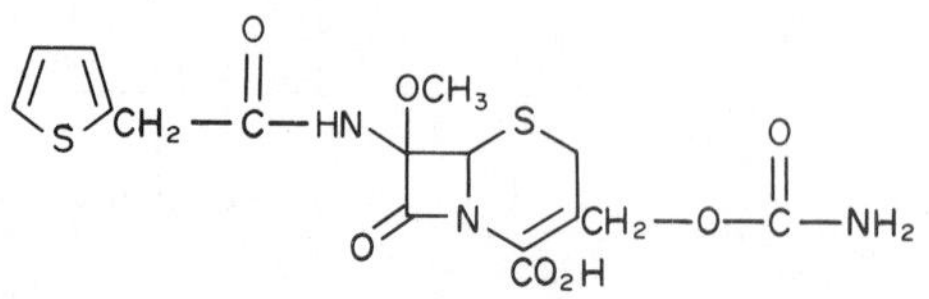

Substituent	Structure	Contribution toward activity
Thienylacetyl	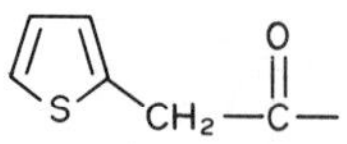	Good Gram-positive activity
Cephem nucleus	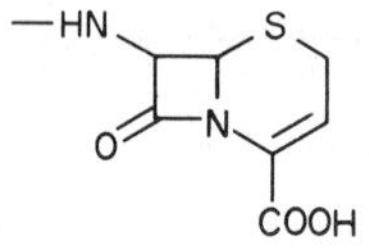	Overall activity, particularly Gram – negative
Methoxy	—OCH₃	Lactamase stability
Carbamoyl		*In vivo* metabolic stability

Figure 2 Structure–function relationships of cefoxitin

9.2.4 Physicochemical Characteristics

The chemical characteristics of cephamycin C have been summarized by Kamogashira *et al.* (1982). In purified form, this compound is a white powder which is quite soluble in water; it is slightly soluble in ethanol and dimethyl sulfoxide. Ultraviolet (UV), proton nuclear magnetic resonance (NMR) and infrared (IR) spectra of this material have been constructed, but are not reproduced here.

9.2.5 Cephamycin C Assay Techniques

As is the case with most antibiotics, quantitative analysis of microbially produced cephamycin C is normally accomplished by either high performance liquid chromatographic (HPLC) or biological assay techniques. HPLC analysis of filtered aqueous samples of cephamycin C has been demonstrated by Kamogashira *et al.* (1982). Concentrations of product are measured relative to those of freshly prepared standard solutions of cephamycin dissolved in distilled water or an appropriate buffer.

Bioassays of cephamycin concentration in aqueous solution can also be made utilizing disc plate techniques. Such assays (Stapley *et al.*, 1972) involve use of agar Petri plates containing an appropriate medium as well as a suitably grown indicator organism (*Vibrio percolans*) as the basis for assay. Filter paper discs saturated with cephamycin solutions are set on the assay plates and incubated. Resulting zones of inhibition of *Vibrio* growth are then measured. Inhibition zones can be compared with analogous zones produced by standard solutions of cephamycin and can be used for quantitative assessment of antibiotic concentration.

A third assay technique for quantitative measurement of cephamycin has been reported (Trieber, 1981). This method utilizes one dimensional thin-layer chromatography (TLC) applied to precoated silica gel plates. Spotted plates are developed in a solvent and scanned in a high speed TLC scanner at 273 nm. This technique permitted quantitative analysis of samples containing 1 μg of antibiotic.

9.3 FERMENTATION MICROBIOLOGY

9.3.1 Introduction

Most patented processes for the biological production of antibiotics and other secondary metabolic products consist of a batch biological reaction in which a producing organism is inoculated into a stirred, aerated fermenter vessel containing an aqueous mixture of organic raw materials. These medium ingredients are often quite complex and sometimes consist of nonhomogeneous by-products of other manufacturing processes. On first inspection, the complex mixtures of which most classical fermentation media are composed might seem ill-begotten. However, most frequently, medium constituents result from a rational attempt to satisfy the complicated metabolic requirements of the producing organism.

Any given microorganism has specific requirements for nutrient supplies of carbon, hydrogen, nitrogen, phosphorus and sulfur as well as trace requirements for a host of other elements. Traditionally, complex materials such as corn steep liquor and distillers' solubles, which are rich in a variety of nutrients, have been used to satisfy these requirements. Use of such complex raw materials has been considered desirable for several reasons. Since the cost of these materials is usually quite low, they provide an inexpensive supply of nutrients. In addition, when precise minimum nutrient requirements are not known, levels in excess of predicted minima can be added to assure optimal product synthesis without a major cost effect on the final product. Further, corn steep liquor or yeast extract often contain trace levels of components which stimulate product synthesis. The stimulatory component is often an amino acid or trace metal which constitutes only a small fraction of the raw material. Corn steep liquor may also be a source of phosphate. However, until the active ingredient is identified, the entire complex product is added to the medium to insure inclusion of the stimulant.

Unfortunately, the use of complex raw materials can have deleterious effects as well. These inexpensive products are usually quite variable in consistency on a lot to lot basis. This may not present a problem in the laboratory, but frequently translates into subsequent variability in process yield at the commercial scale. As a result, it is preferable to minimize the use of complex raw materials. A possible alternative is to use only well defined, purified medium ingredients such as amino acids, sugars and inorganic salts. However, utilization of such products leads to increased expenditure and to a more complicated process in that nutrient feeding is often required. Cost considerations thus dictate that the minimum metabolic requirements be determined. In turn, this requires that there be a comprehensive understanding of the metabolic pathways and regulatory mechanisms which control biosynthesis of the desired product. For some well established antibiotic products such as penicillin and cephalosporin, this information is becoming available and has led to the development of sophisticated fermentation processes (Ryu and Hospodka, 1980).

The structural similarity which exists between the cephamycins, penicillins and cephalosporins suggests that the metabolic origins and regulatory mechanisms which control their biosynthesis might be similar. The existence of such a common biosynthetic pathway might then result in similar fermentation media for all these products. However, several metabolic differences in the organisms which produce these compounds are known. It is necessary to contrast the cephamycin producing actinomycete species with other antibiotic producers in order to determine where existing process development strategies can be utilized and where novel approaches must be invented.

Several studies which determine the metabolic origins and regulatory mechanisms of cephamycin synthesis have been reported. They will be discussed in light of historical observations made with other β-lactam products. Together this body of work constitutes an evolutionary path which leads to the cephamycin fermentation processes discussed in a later section.

9.3.2 Metabolic Origins

There are many interacting factors which determine the quantity of product which is synthesized in any natural product fermentation. In order to determine which medium components might potentially limit synthesis of a given compound, it is important to determine the precursors from which that compound is derived as well as the means by which those precursors are assembled into the final structure. In the case of both penicillin and cephalosporin fermentations, the basic structures are the penam and cephem rings, respectively. These compounds are derived from a tripeptide precursor molecule composed of valine, cysteine and α-aminoadipic acid moieties. The means by which this precursor is cyclized to form the basic cephalosporin and penicillin nuclei are documented (Queener and Neusch, 1982). Microbiologically directed substitutions at various locations on the resulting structures are known to occur.

In the case of the cephamycins, a similar tripeptide precursor occurs (Abraham, 1978). The structure is then assembled to form a β-lactam, dihydrothiazine double ring structure. The resulting cephamycin family of compounds are substituted at the C-3 and C-7 position as illustrated in Figure 3. A methoxy substitution is present at the 7α carbon of all cephamycins. It is this characteristic which imparts β-lactamase resistance of the cephamycins. The methyl group of this substitution is derived from the methyl group of methionine (Whitney *et al.*, 1972). The oxygen atom is derived from molecular O_2 (O'Sullivan *et al.*, 1979). A second substitution at the C-7 position of cephamycins is provided by the α-aminoadipic acid constituent and is derived from lysine. Substitution at the C-3 position varies among individual cephamycins. In the case of cephamycin C, the commercially significant compound, there is a carbamoyl substitution, which is derived from carbamoyl phosphate (Brewer *et al.*, 1977).

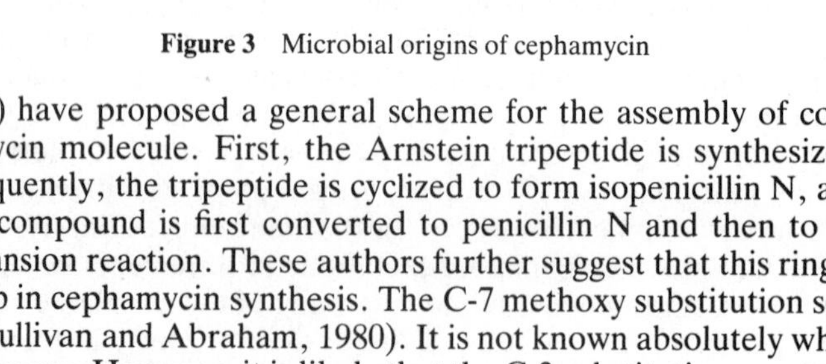

Figure 3 Microbial origins of cephamycin

Kern *et al.* (1980) have proposed a general scheme for the assembly of components into the completed cephamycin molecule. First, the Arnstein tripeptide is synthesized from individual amino acids. Subsequently, the tripeptide is cyclized to form isopenicillin N, a common β-lactam intermediate. This compound is first converted to penicillin N and then to deacetoxycephalosporin in a ring expansion reaction. These authors further suggest that this ring expansion may be the rate limiting step in cephamycin synthesis. The C-7 methoxy substitution seems to occur after ring expansion (O'Sullivan and Abraham, 1980). It is not known absolutely when the substitution at the C-3 position occurs. However, it is likely that the C-3 substitution occurs first. It is precisely the substitutions at this position which distinguish between naturally occurring cephamycins.

9.3.3 Carbon Metabolism

The importance of carbon metabolism on β-lactam biosynthesis was recognized even before there was a viable, submerged penicillin fermentation process. Moyer and Coghill (1946a) determined with surface cultivation of *Penicillium notatum* that carbon sources such as glucose and sucrose did not support penicillin yields as high as those observed with less rapidly used carbon sources (such as lactose and starch). As flask scale submerged fermentations for penicillin production were developed (Moyer and Coghill, 1946b; Koffler *et al.*, 1946; Foster *et al.*, 1946), the importance of the carbon source was confirmed. Lactose continued to support higher levels of antibiotic synthesis when compared to glucose. It was concluded (Stefaniak *et al.*, 1946) that the advantage of lactose arose from its characteristic slow utilization by the *Penicillium* organism. Subsequently, it was observed that glucose could in fact support high penicillin titers, but only when added at relatively low concentrations, either continuously or intermittently through the course of the fermentation.

One of the earliest demonstrations of a specific carbon source effect in the penicillin fermentation was made by Soltero and Johnson (1954). These authors evaluated a number of penicillin-producing cultures using both lactose (added prior to inoculation) and glucose (fed throughout the fermentation) as carbon sources. Several cultures which produced high levels of penicillin in the presence of lactose did not support high antibiotic production in glucose-fed fermentations. Conversely, two cultures which did not produce well when lactose was the carbon source elaborated very high levels of penicillin when glucose was fed throughout the fermentation. When the entire glucose charge was added to the medium prior to inoculation, very low antibiotic synthesis was observed.

These initial, empirical results strongly suggested the presence of a regulatory role for the carbon source. It was reasoned that carbon sources which support high specific growth rates lead to a suppression of antibiotic synthesis while with carbon sources which do not allow such rapid growth, higher levels of antibiotic synthesis occur. This has been demonstrated for several antibiotic fermentations (Drew and Demain, 1977). However, in some cases, sole use of a slowly used carbon source leads to wide variations in product yield. In the actinomycin fermentation, for example, inclusion of a rapidly used carbon source such as glucose in addition to the more slowly metabolized galactose (Gallo and Katz, 1972) has increased yields. Presumably, addition of glucose allows rapid growth of the organism. Once the glucose supply is exhausted and its regulatory role is relieved, biosynthesis of antibiotic can occur on the alternate carbon source. This concept of dual carbon source utilization appears to be the basis for published media which support commercial antibiotic synthesis. As another example, a patented medium for the production of cephamycin C by *Streptomyces clavuligerus* contains both glucose and glycerol (Gorman *et al.*, 1979). It is not unlikely that these compounds support growth and antibiotic production, respectively.

It has also been observed that increasing levels of a single carbon source can suppress antibiotic synthesis. Matsumura *et al.* (1978) demonstrated that increasing the rate of glucose addition to a *Cephalosporium acremonium* fermentation led to a substantial decrease in cephalosporin C production. These authors also evaluated cephalosporin production in continuous culture. As glucose utilization in the chemostat increased, specific production of cephalosporin decreased. Further, when both batch and continuous culture data were combined, specific growth rate (the reciprocal of dilution rate) was inversely proportional to specific yield over a wide range in specific growth rate. These data suggest that growth rate, whether altered by varying concentrations of a single carbon source or by varying carbon sources altogether, is a strong affector of antibiotic production.

Aharonowitz and Demain (1978) evaluated the effect of various carbon sources on antibiotic production in *Streptomyces clavuligerus*. This organism elaborates four β-lactam products: cephamycin C, penicillin N, clavulanic acid and a cephalosporin compound, 7-(5-amino-5-carboxyvalerimido)-3-carbamoyl-3-cephem-4-carboxylic acid. In order to isolate regulatory effects, Aharonowitz and Demain adjusted the assay and fermentation conditions respectively, such that neither clavulanic acid nor penicillin N was observed. Initially, the authors used a minimal medium supplemented with asparagine as the sole nitrogen source. They then added various carbon sources and identified those which supported the largest increases in biomass. The effect of varying these initial concentrations of rapidly used carbon sources on biomass yield, on volumetric antibiotic potency as well as on the specific β-lactam production rate was then evaluated. When two such carbon sources (glycerol and maltose) were independently studied, the maximum cell yield increased and specific productivity decreased as their initial concentration was increased from 0 to 10 g l^{-1}. Such an increase in cell yield would be expected if the carbon source represented the rate limiting nutrient. However, this inverse relationship between specific productivity and carbon source concentration suggested the presence of some type of carbon catabolite regulation. This is consistent with work described for cephalosporin and actinomycin.

This study clearly shows that to obtain maximum volumetric cephamycin potency, it is not sufficient to determine the conditions which independently maximize biomass yield or specific productivity. Rather, since it is likely that the conditions which optimize these two parameters will not coincide, maximizing broth titer (which usually leads to the best fermentation process) makes it necessary to modify the physicochemical environment so that neither of these factors is optimal. Hunt (unpublished data) has found this to be true for the commercial production of cephamycin C.

Aharonowitz and Demain also investigated the effect of varying the concentration of poorly used carbon sources on net β-lactam biosynthesis. It was observed that for such compounds as sodium α-ketoglutarate and sodium succinate, specific β-lactam production was high. Despite the fact that biomass concentration was significantly lower than for the preferred carbon sources (maltose and glycerol), the volumetric concentration of product in the fermentation broth was nearly equivalent, resulting in improved specific productivity. Furthermore, the production of antibiotic seemed to be growth associated. This suggests that in commercial fermentation process development, it might be possible to obtain improved product yield by coupling a relatively poorly utilized carbon source to increased biomass. If the poorly used carbon source supports high specific productivity, any obtainable increase in biomass should result in an increase in broth potency. It should be possible to take advantage of this in commercial fermentation process development.

Lilley *et al.* (1981) have also studied the effect of carbon metabolism on production of antibiotic products. *Streptomyces cattleya* was studied in both batch and chemostat culture. This organism produces cephamycin C, penicillin N and thienamycin (Kahan *et al.*, 1979). Under the experimental conditions imposed, only cephamycin C was observed; thienamycin production was below the level of detection. Penicillin N was not measured. Lilley *et al.* evaluated growth rate and yield, as well as volumetric and specific cephamycin yield in batch culture as a function of glucose concentration. Glucose broth concentration was varied between 10 and 60 g l^{-1}. Both cephamycin titer and cell yield were found to increase as carbon concentration increased. Specific growth rate varied inversely with glucose concentration. Specific productivity declined as the glucose concentration rose to 40 g l^{-1}. At higher concentrations, specific antibiotic yield increased. This study would appear to contradict the earlier work of Aharonowitz and Demain. However, Lilley *et al.* suggest that the increase in specific production above 40 g l^{-1} glucose indicates that at such high concentrations, the fermentation is not carbon limited. In such a case, carbon regulatory effects might not be observed.

To verify the apparent inverse relationship between specific growth rate and specific cephamycin in batch culture, Lilley *et al.* evaluated the specific productivity of *S. cattleya* as a function of dilution rate in a chemostat. Specific growth rate remained inversely proportional to specific productivity. The specific uptake of both glucose and oxygen varied directly with dilution rate. These data are consistent with the presence of a carbon regulatory effect on cephamycin biosynthesis similar to that observed with other fermentations. Whether this effect is due to catabolite repression, catabolite inhibition or some other phenomenon awaits elucidation. The individual cephamycin biosynthetic enzymes and their metabolic interactions with carbon-containing compounds must be determined before the exact nature of any carbon regulation in the cephamycin fermentation will be understood. Perhaps the use of resting cell systems similar to those developed for cephalosporin producers (Demain and Kennel, 1978) will lead to a more thorough understanding.

9.3.4 Nitrogen Metabolism

Nitrogenous compounds also play an important role in determining both the growth and product yield in antibiotic fermentations. In the production of β-lactams for example, a supply of nitrogen is required both for cell growth and for secondary metabolite production. As a result, the choice of compounds which provide nitrogen must be made based on the requirements imposed by these divergent metabolic processes. Nitrogen must be introduced to the culture in a manner which satisfies both growth and antibiotic synthesis, but which restricts neither. As a result, the choice of nitrogen source(s) is sometimes quite complicated, for example it is often found that the compound which allows the greatest cell yield in some way limits antibiotic accumulation. Alternatively, the preferred nitrogen source based on antibiotic production may limit cell yield. In such cases, a nitrogen source is chosen which is not optimal for either parameter, but which maximizes volumetric potency of the antibiotic. Alternatively, several compounds which independently optimize growth and product yield are added to the production medium to optimize both parameters simultaneously. Post inoculation feeding of nitrogenous compounds has allowed independent optimization of cell growth and secondary metabolite production. Striking the proper balance among nitrogen sources is frequently a major obstacle to fermentation process optimization.

The earliest efforts at developing a submerged fermentation for the production of penicillins centered about finding the proper combination of nitrogen sources. Evaluation of a large number of raw materials in various combinations led to a determination that the combined inclusion of corn steep liquor (a product of corn milling) and soybean meal supported the highest penicillin yields. Corn steep liquor is a fermentative by-product which contains a variety of soluble peptide oligomers, inorganic salts, lactic acid and other low molecular weight products. Soybean meal, on the other hand, is a polymeric, proteinaceous raw material. In combination, these two products provided the desired mix of nitrogen sources in early penicillin fermentation processes. Corn steep liquor provided readily assimilable organic amino nitrogen for uptake during growth of the *Penicillium* strains. Soybean meal was slowly degraded during the course of the fermentation and contributed a slowly released supply of organic nitrogen during antibiotic synthesis.

The combination of these two ingredients allowed the initial development of a submerged fermentation process for penicillin production. Prior to this, penicillin was produced on surface culture in a manner analogous to classical fungal food fermentations. However, use of corn steep liquor and soybean meal presented a number of practical problems when used in large, agitated production vessels. Corn steep is a liquid material which is the by-product of another manufacturing process. As such, it is variable in nature and is susceptible to microbial degradation prior to its use in penicillin production. As a result, use of corn steep liquor often led to variable penicillin yields. Soybean meal is highly particulate and therefore quite difficult to sterilize. In addition, this material contains a number of non-fermentable sugars which accumulate during the fermentation and which negatively impact on the isolation of penicillin from the fermenter broth. The operating problems associated with these products led to a search for alternative nitrogen sources. Specifically desirable would be non-particulate, homogeneous sources which would not significantly impact production costs. Indeed, to this day, such criteria for media components remain intact.

Attempts to replace corn steep and soybean meal in the penicillin production media were aimed at finding inexpensive, reproducible materials which would simulate the combined effect of these two products. The use of ammonium salts, which were known to be absorbed rapidly by many organisms during growth, was evaluated quite extensively. Ammonium could be used to obtain high cell yields but would not support acceptable levels of antibiotic synthesis when used as the sole nitrogen source. Rather, cottonseed products (such as Proflo), when used alone or in combination with low levels of ammonium salts were found to support the highest levels of penicillin biosynthesis. These products completely replaced corn steep and soybean meal as the primary sources of nitrogen. Proflo and similar raw materials have been used for many years in penicillin production and remain the basic nitrogen sources in modern antibiotic fermentation processes (Ryu and Hospodka, 1980).

Ammonia regulation of antibiotic synthesis, similar to that observed in penicillin fermentations, has been reported for several other secondary metabolite producing organisms (Martin and Demain, 1980). Aharonowitz and Demain (1979) have recently observed a strong suppressive effect of ammonia on antibiotic production in the cephamycin producer *S. clavuligerus*. The nature of this regulation appears to be not unlike that observed with rapidly metabolizable carbon sources: when the concentration of soluble, inorganic ammonium is kept low, β-lactam production is relatively high. As the residual concentration of ammonium is increased, antibiotic accumulation decreases rapidly. Aharonowitz and Demain suggest that the regulatory nature of rapidly metabolized carbohydrates and ammonium are tightly coupled. These authors claim that if either of these regulatory elements are limiting in the fermentation broth, antibiotic synthesis is stimulated in *S. clavuligerus*.

This hypothesis is supported by studies made with another cephamycin producing organism *N. lactamdurans*. Ginther (1979) has observed that conditions which lead to such secondary processes as sporulation and protease formation, also favor cephamycin C production. Onset of sporulation is associated with the presence of metabolic starvation conditions, typically loss of available assimilable nitrogen. Similarly, excretion of extracellular proteolytic enzymes would be consistent with a runout of soluble amino nitrogen in the fermentation medium. This study is consistent with the idea that starvation conditions imposed by a nitrogen or carbohydrate limitation could induce sporulation, protease excretion and antibiotic synthesis in *N. lactamdurans*. However, this hypothesis has by no means been quantitatively proved.

The nature of ammonia regulation in the cephamycin fermentation has several implications for fermentation process development. Nitrogen sources should be included in or added to the fermentation in a manner which precludes the transient or extended elevation of soluble ammonia concentration. In addition, since fermenter operating conditions such as batch temperature and mass transfer conditions can affect the degradation and uptake rates of various nutrients, such parameters should be programmed in such a way to prevent accumulation of ammonia.

A number of studies also have been undertaken to determine the means by which amino acids are incorporated directly into β-lactam antibiotics. The principal amino acids involved are those which comprise the Arnstein tripeptide: cysteine, valine and α-aminoadipic acid. Incorporation of α-aminoadipic acid depends on the type of producing organism. In general, however, the availability of α-aminoadipic acid (αAAA) is coupled to lysine metabolism. In fungi like *Penicillium* and *Cephalosporium*, αAAA is an intermediate in the biosynthesis of lysine (Metzler, 1977). As such, synthesis of β-lactams represents a branched pathway in lysine biosynthesis. Higher rates of lysine biosynthesis result in greater availability of αAAA and subsequent incorporation into and biosynthesis of antibiotic. Conversely, decreased rates of lysine biosynthesis lead to a decrease in both availability of αAAA and synthesis of β-lactams. Accumulation of lysine serves to regulate its own synthesis in fungi by feedback repressing a synthetic enzyme in the initial steps of its biosynthesis (Masurekar and Demain, 1972). Since the repressed enzyme precedes the αAAA intermediate, lysine induced regulation of lysine synthesis inhibits AAA formation as well. As a result, accumulation of lysine in fungi inhibits biosynthesis of β-lactams.

In bacteria like the cephamycin-producing *Streptomyces* and *Nocardia*, incorporation of αAAA into the Arnstein tripeptide is regulated in quite a different manner. In these organisms, αAAA is a catabolic product of lysine degradation. As a result, lysine accumulation leads to a greater lysine breakdown rate. Presumably the resultant higher levels of αAAA resulting from lysine accumulation would increase the rate of incorporation into β-lactam antibiotics and lead to higher levels of accumulation. If αAAA incorporation into the Arnstein tripeptide is the step which limits biosynthesis in a *Streptomyces* antibiotic fermentation, it should be possible to increase product yield by adding lysine to the production medium. Such stimulation has been observed with a number of cephamycin-producing organisms. Hallada *et al.* (1975) have demonstrated a 50% increase in total cephamycin A and B production by *Streptomyces griseus* (NRRL 3912) when 0.20% L-lysine hydrochloride was added to a complex fermentation. Direct addition of a similar level of DL-αAAA to the fermentation medium also increased product titer, but by 30%. Inamine and Birnbaum (1975a, 1975b) and Inamine *et al.* (1974) have observed stimulation of cephamycin C biosynthesis by strains of *N. lactamdurans* and *S. clavuligerus* upon addition of various levels of D-lysine and DL-lysine to fermentation media. It is interesting to note that L-lysine does not stimulate cephamycin production in *N. lactamdurans*. This suggests the possibility of a regulatory role of D- and DL-lysine for *N. lactamdurans* in addition to or in place of a precursor role.

Mendelowitz and Aharonowitz (1982) have also studied the effect of lysine addition to *S. clavuligerus*. Addition of lysine to cultures of this organism grown in a chemically defined medium increased specific antibiotic production by 75%. Addition of DL-*meso*-diaminopimelate, a precursor of lysine and of αAAA, had a similar stimulatory effect on β-lactam production. Further, combined inclusion of lysine and DL-*meso*-diaminopimelic acid resulted in a four-fold increase in antibiotic production. These studies suggest that α-aminoadipic acid incorporation into the precursor tripeptide as well as biosynthesis of the final product can be greatly enhanced by αAAA addition.

Valine is also incorporated into the Arnstein tripeptide. Mendelowitz and Aharonowitz observed that inclusion of this compound, in addition to lysine and DL-*meso*-diaminopimelate, stimulated antibiotic production by *S. clavuligerus*. The role of cysteine, the third component of the Arnstein tripeptide, will be discussed in the following section.

9.3.5 Sulfur Metabolism

There are two major pathways by which antibiotic-producing microorganisms incorporate sulfur into cysteine and ultimately into β-lactam products (Figure 4). One way is by reduction of inorganic compounds such as sulfate or thiosulfate. An alternative means is the conversion of methionine into cysteine prior to formation of the Arnstein tripeptide. In most commercial antibiotic producing strains, one of these two pathways predominates, for example in high yielding *Penicillium* species, an efficient system for sulfate reduction usually exists (Tardrew *et al.*, 1958; Segel and Johnson, 1961). As a result, commercial processes for the production of penicillin utilize ammonium sulfate as the principal source of sulfur. Although these *Penicillium* strains are capable of using methionine as a sulfur source, it is not routinely included in penicillin production media because methionine tends to inhibit sulfate uptake and, as such, may interfere with sulfate supported antibiotic biosynthesis.

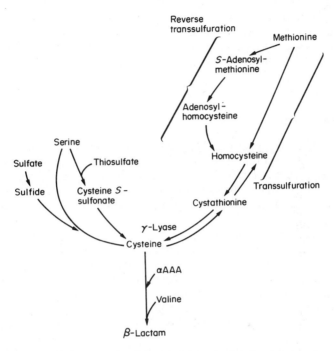

Figure 4 Incorporation of sulfur into β-lactam products

With other fungal species, however, sulfate is not the optimal source of sulfur. In *Cephalosporium* strains used for commercial production of cephalosporin antibiotics, inorganic sulfur does not support high product titers. Rather, methionine is effectively incorporated into β-lactam products *via* the reverse transsulfuration pathway. In these organisms, methionine can be used as a sole source of carbon or sulfur, and its presence markedly stimulates antibiotic synthesis (Caltrider and Niss, 1966). The role of methionine in these species is two-fold. First, it is an efficient precursor of β-lactams; methionine is converted to cysteine which is then incorporated into the final product. In addition, methionine has a second stimulatory role which is independent of its ability to donate sulfur. This role has been demonstrated in mutants of *Cephalosporium acremonium* which were not able to convert methionine to cysteine (Drew and Demain, 1975a, 1975b). As such, the reverse transsulfuration pathway is inactive and these mutants should not be able to incorporate methionine into cephalosporin. On the other hand, antibiotic synthesis from other sulfur sources such as cysteine and sulfate should continue. Despite this, Drew and Demain found that methionine in excess of growth demand (these mutants were also methionine auxotrophs) was necessary for β-lactam synthesis using any sulfur source. The authors concluded that methionine had a regulatory role in β-lactam synthesis, independent of its precursor role. This hypothesis was supported by the fact that norleucine, an analog of methionine which contains no sulfur, exhibited a stimulatory role similar to that observed with methionine. The exact nature of this stimulatory effect has not yet been elucidated. However, Komatsu and Kodaira (1977) have suggested that methionine and norleucine exert their effect by manipulating the intracellular cysteine pool in *Cephalosporium* organisms.

Sulfur metabolism in cephamycin-producing actinomycetes has been investigated by Inamine and coworkers. Cysteine, methionine and inorganic thiosulfate, the major sources of sulfur for antibiotic production, have all been found to inhibit growth of *Nocardia lactamdurans*. However, Inamine and Birnbaum (1973) have found that if thiosulfate is added to this culture, subsequent to growth in order not to interfere with cell growth, cephamycin production is dramatically stimulated. This concept has been used in commercial production of cephamycins and is discussed in Section 9.3.4.

In addition to the uptake system observed for thiosulfate, other sulfur incorporation systems have been observed in cephamycin-producing organisms. Whitney *et al.* (1972) reported that a radioactive sulfur label from methionine was incorporated into cephamycin C by the filamentous bacterium *S. clavuligerus*. This would be unexpected since bacteria have been thought to universally lack the enzyme cystathionine γ-lyase (Delavier-Klutchko and Flavin, 1965), which is necessary for catalyzing the final step in conversion of methionine to cysteine (Figure 4). However, more recently, Kern and Inamine (1981) have demonstrated that this enzyme is present in *N. lactamdurans*. These data are the first to suggest that reverse transsulfuration is present in these actinomycetes and that methionine may be a significant contributor of sulfur to the cephamycin molecule. At present, however, thiosulfate incorporation appears to be of greater significance to cephamycin biosynthesis (Inamine and Birnbaum, 1972).

9.3.6 Phosphate Metabolism

The role of phosphate in regulating β-lactam antibiotic synthesis is less well understood than that of sulfur. Despite the fact that it is not incorporated directly into the β-lactam nucleus, phosphate is known to effect secondary metabolism in β-lactam-producing organisms. Elucidation of the exact interaction between phosphate metabolism and antibiotic biosynthesis has been complicated by the fact that phosphate is involved in a host of related and non-related metabolic pathways.

In general, three separate regulatory roles have been postulated for phosphate. Martín (1977) suggests that phosphate can independently inhibit enzyme action, regulate protein expression and limit cell growth. Miller and Walker (1969) have demonstrated phosphate mediated enzyme inhibition of antibiotic synthesis in the streptomycin fermentation. Streptidine is a phosphorylated precursor in the biosynthesis of this aminoglycoside antibiotic. Walker and Walker (1971) observed that dephosphorylation of streptidine phosphate to form streptomycin was inhibited by phosphate. These authors suggested that this effect was due to suppression of phosphatase activity.

The role of phosphate in regulating protein expression is not fully understood, however, several laboratories are currently addressing this effect (Martin and Demain, 1980). High levels of exogenously added phosphate can repress synthesis of phosphatase enzymes (Demain and Inamine, 1970).

A general role of phosphate in effecting antibiotic expression has been observed in the synthesis of macrolide antibiotics such as candicidin (Martin and Demain, 1980). Typically, it is observed that phosphate concentrations of 1 to 10 mM are optimal for combined growth and antibiotic synthesis. Frequently, when phosphate concentrations are increased beyond 10 mM, growth yield increases while antibiotic accumulation decreases, suggesting a regulatory role for phosphate.

Analogous observations have been made with the cephamycin producing organism, *Streptomyces clavuligerus*. Aharonowitz and Demain (1977) have varied initial medium phosphate concentrations over a range of 1 to 100 mM. When phosphate concentration was less than 10 mM, both cell growth and antibiotic accumulation were limited. When phosphate concentration was between 10 and 25 mM, significant growth was observed and antibiotic synthesis was optimal. Above 25 mM phosphate concentrations, growth yield increased but antibiotic synthesis was depressed. A similar role for phosphate in effecting β-lactam biosynthesis by *S. cattleya* was observed by Lilley *et al.* (1981). In continuous culture studies, phosphate-induced growth limitation led to increased cephamycin C synthesis.

9.4 CEPHAMYCIN PRODUCTION TECHNOLOGY

Classical fermentation processes for the production of secondary metabolic products can be subdivided into three phases. The first stage is one of inoculum development and occurs sequentially in a series of mixed reaction vessels. The second phase usually takes place in a single reactor and is the stage in which antibiotics are microbiologically synthesized. The third phase is one of product recovery. Based on metabolic regulatory mechanisms discussed in previous sections, there are specific requirements imposed by the producing organism in the first two phases.

9.4.1 Inoculum Development Stage

The overall goal of the inoculum development stage is to rapidly produce a large quantity of viable, active biomass capable of synthesizing the desired product. This cellular material is ultimately used to inoculate the production stage fermenter. The source of microorganism is usually a few milliliters of frozen vegetative mycelium or spore suspension. This stock culture supply is aseptically inoculated into a laboratory flask containing 20–500 ml of growth or 'seed' medium and incubated at some optimal temperature with external mixing. After initiation of growth, the microorganism multiplies exponentially and eventually reaches a point at which growth slows. After an experimentally determined growth period, the culture is transferred to a larger flask for continued growth. Several flask stages usually are employed in sequentially larger flasks. Since it is difficult to measure growth parameters in such flask stages, the criteria (*i.e.* the required cell mass and inoculum volume) used for transferring inoculum from one flask stage to the next are often determined empirically.

At some point, the contents from a flask growth stage are used to inoculate an intermediate sized aerated, agitated 'seed' fermenter. This transfer step is usually the critical stage in inoculum development. Practical considerations (*i.e.* the maximum available flask size and smallest available fermenter) often dictate that the inoculum volume is small. This can lead to aberrant growth patterns in the initial stirred vessel stage. A protracted lag prior to the initiation of growth can occur. Occasionally, small inoculum volumes can lead to aggregated or 'pelleted' growth in a seed fermenter. Hunt (unpublished data) has observed such a phenomenon in growth of the cephamycin producer *N. lactamdurans*. In all cases, pelleted growth in seed fermenters led directly to depressed antibiotic yield in a subsequent production stage.

The volumes of typical β-lactam antibiotic production fermenters are 50–200 thousand l. Since required inocula volumes are usually 2 to 10% of the production stage medium volume, several agitated, fermenter seed stages are typically utilized. Transfer from one fermenter seed stage to a subsequent one or to a production stage vessel is usually based on quantitative assessment of critical indicators of growth. On-line measurements of pH and dissolved oxygen tension are now used routinely to measure these variables. In addition, it is possible to sample fermenter seed vessels repeatedly without jeopardizing aseptic operation. Off-line measurement of cell weight provides an accurate measurement of growth yield. The most important recent development in fermenter monitoring has been the introduction of an industrial mass spectrometer for vent gas analysis. With such an instrument instantaneous assessment of oxygen uptake, carbon dioxide evolution and respiratory quotient is possible. This respiratory data can be closely correlated with optimum transfer conditions. Mass spectrometry is now used widely for the monitoring of fermenter off-gas streams in both pilot plant and factory environments (Buckland and Fastert, 1982).

As previously stated, the primary requirement of inoculum or 'seed' development is to provide an environment for proliferation of the producing organism. This requires that a rapidly assimilable supply of the basic elemental nutrients be provided. Carbon, hydrogen, oxygen (as air) and nitrogen are usually added in excess to prevent any possible limitation in the growth stages. The seed media included in four separate processes for the production of cephamycin products are listed in Table 1, as are the required growth conditions and producing organisms. Processes patented by Eli Lilly and Co., Takeda Industries and Otsuka Pharmaceuticals are included as is a non-patented process used at Merck and Co., Inc. All the strains utilized are from the order Actinomycetales. The strain used by Merck and Co., Inc. (Currie, 1985) is a *Nocardia*; the other three organisms are *Streptomyces*.

In all cases, readily metabolizable sources of carbon, nitrogen and phosphorus are provided in complex form. Glycerol, sucrose and glucose are carbon sources well utilized by the respective organisms. Ardamine YEP (Yeast Products) supplies both carbon and nitrogen for growth in the Merck process. Ardamine YEP is approximately 40% carbon with the remainder consisting of soluble and insoluble nitrogenous compounds as well as a variety of inorganic salts, including sources of phosphate. In the other three processes, soluble and insoluble organic nitrogen are supplied separately from the source of carbon. Most of the raw materials used are highly complex, contributing a wide variety of nutrients. The cell yields (grams biomass attained per liter of broth) are not available. However, all but one of these seed development media are rich in the principal required nutrients. The Merck seed medium is relatively lean and would not be expected to support biomass levels which could be obtained in the other media.

Most of the processes illustrated utilize operating temperatures of 28 to 30 °C. Only the Otsuka process (37 °C) maintains temperature at a higher level. Higher temperatures result in higher

Table 1   Cephamycin C Seed Development Parameters

US Pat.	4 132 790 (example 21)	4 017 485	Not patented	4 332 891 (example 1)
Company (reference)	Eli Lilly (Gorman *et al.*, 1979)	Takeda Chemical Ind. (Hasegawa *et al.*, 1977)	Merck and Co. (Hunt, 1983)	Otsuka Pharmaceutical (Kamogashira *et al.*, 1982)
Product	Cephamycin C	C2801X	Cephamycin C	Cephamycin C
Producing organism	*S. clavuligerus* (NRRL 3585)	*S. heteromorphus* (ATTC 31054)	*N. lactamdurans* (MA2908)	*Streptomyces* sp. OFR 1022 (ATCC 31666)
Culture source	2.0 ml of frozen spore suspension	N/A[a]	Frozen vegetative mycelium	1% inoculum
Growth media				
Carbon sources (g l^{-1})	Glycerol (10) Sucrose (20)	Glucose (20) Soluble starch (30)	Ardamine YEP (10)	Starch (30) Sucrose (10)
Nitrogen sources (g l^{-1})	Nutrisoy grits (15) Amber BYF 300 (5) Tryptone (5)	Corn steep liquor (10) Soybean meal (10) Peptone (5)	Ardamine YEP (10)	Soybean flour (10) Dry yeast (3)
Phosphorus source (g l^{-1})	Potassium phosphate (0.20)	—	—	—
Other (g l^{-1})	None	Sodium chloride (3) Calcium carbonate (5)	None	None
Post sterile pH	6.5	7.0	6.9	7.0
Temperature (°C)	30	28	28	37
Cycle (h)	N/A[a]	40	14–18[b]	48

[a] Not available.   [b] Last seed stage.

intrinsic metabolic rates and usually lead to less reproducible, but shorter, fermenter cycle times. The fact that the Otsuka process reports use the longest cycle is somewhat surprising.

9.4.2   Antibiotic Production Stage

The antibiotic production stage is separated from inoculum development only for the sake of discussion. It is well known that inoculum conditions have a dramatic effect on net biosynthesis of secondary metabolites. Parameters such as inoculum volume and the so called 'physiological state' of the inoculating culture are critical to both initiation and maintenance of antibiotic synthesis.

It should be noted that in the initial phase of the production stage, substantial growth of the producing organism occurs. Since inoculum volume represents less than 10% of the production volume, added growth must occur in order for the culture to reach its final density. In the case of the four processes previously discussed, volume percent inoculum is summarized in Table 2. In the three cases where this information is available, production stage inoculum volume is 5% (v/v). In all cases described, glycerol or sucrose is provided in the production media. Presumably, these compounds support initial growth. Assimilable nitrogen is provided by NZ amine A (Humko Sheffield), Distiller's solubles (Nadrisol; Brown and Forman), Proflo (Trader's Protein) or cottonseed flour.

The production media detailed in Table 2 contain more ingredients than do the seed media previously described. This is due to the complex requirements of antibiotic synthesis. Cephamycin biosynthesis initiates during, or just subsequent to, the initial growth period. The precise factors which lead to initiation of synthesis have yet to be elucidated; some of the influencing factors have been discussed in Section 9.3.

The classes of compounds utilized for synthesis of the cephamycins include a slowly metabolizable carbon source as well as a slowly utilized nitrogen source. Continued uptake of these materials supports both maintenance requirements of the producing organism and biosynthesis of cephamycin. At high rates of synthesis, carbon and nitrogen requirements can be significant. The Lilly and Otsuka production media contain slowly used starch. In addition, soybean and cottonseed flour provide a slowly released source of nitrogen. Presumably, the Merck and Takeda processes utilize complex nitrogen sources as a slowly metabolized supply of both nitrogen and carbon.

Table 2 Cephamycin C Production Stage Parameters

US Pat.	4 132 790	4 017 790	Not patented	4 332 891
Company	Eli Lilly	Takeda Chemical Industries	Merck and Co.	Otsuka Pharmaceuticals
Inoculum volume (% v/v)	5	5	5	N/A[a]
Medium sterilization (min)	30	N/A[a]	30	N/A[a]
Initial pH	6.5	7.0	7.0	6.0
Operating temperature (°C)	30	28	28	37
Air flow rate (v/vm)	0.35	100% aeration	0.25	1.0
Fermenter volume (l)	24	N/A[a]	N/A[a]	20
Production medium (g l^{-1})				
Carbon sources	Starch (47) Glycerol (5.2) Soybean flour grits (21)	Sucrose (30) Proflo (20)	Glycerol (12) Glycine (0.5)	Starch (30) Sucrose (10) Cottonseed flour (20)
Nitrogen sources	NZ amine A (5.2) Nadrisol	Corn steep liquor (10)	Distillers' solubles (30) Primary dried yeast (10)	Dry yeast (10)
Phosphate source	—	Potassium phosphate (0.5)	—	Potassium phosphate (0.2)
Sulfur source	Ferrous sulfate (heptahydrate) (2.5)	Iron sulfate (0.5)	Sodium thiosulfate (1)	Magnesium sulfate (0.5)
Other	Antifoam A (5)	Calcium carbonate (5) Sodium chloride (3)	Mobil Par S (2.5) Polyethylene glycol (0.15) Dimethylformamide (9.4)	Silicone (5)
Product yield (g l^{-1})	N/A[a]	N/A[a]	0.5	2
Cycle time (h)	66	66	88	90

[a] Not available.

In all four cases described, sulfur, a constituent of the cephamycin molecule, is added to the production medium primarily as an inorganic salt. The value of an inorganic sulfur source relative to an organic source, at least with *N. lactamdurans*, has been discussed. In the Merck process described in Table 2, sodium thiosulfate is added to the production stage subsequent to cessation of growth. In the Takeda and Otsuka production media, phosphate is added at relatively low levels in the form of potassium salts. In all four media, phosphate is supplied with the complex nitrogen sources.

Oxygen is added to all antibiotic fermentations by the injection of compressed air. While the cephamycin molecule directly incorporates oxygen, the principal requirement for oxygen derives from growth and maintenance of the culture. Although some studies concerning oxygen consumption in other fermentations have been published (Feren and Squires, 1969), no studies on oxygen uptake in cephamycin fermentations have been reported. In the processes described in Table 2, airflow requirements vary from 0.25 to 1.0 volume air/volume fermenter/minute. The highest air flow rate is used in the elevated temperature Otsuka process. This is not unexpected since there would be a higher maintenance requirement and lower oxygen saturation concentration at elevated temperature.

Production stage operating temperatures for all four processes match those reported for seed growth. The Otsuka process uses both a higher temperature and longer cycle time than the other processes. However, the antibiotic yield in this process is 2.0 mg cephamycin C ml^{-1}. This is several times greater than the Merck process illustrated in Table 2. Antibiotic potencies obtained in the other processes are not indicated.

A protocol for seed development and production stage operating conditions in the Merck process is illustrated in Figure 5. This protocol varies with specific strains of cephamycin-producing organisms and with media. However, Figure 5 represents a typical scheme for production scale manufacturing of cephamycin products. Long term storage of *N. lactamdurans* culture stock is accomplished in lyophilized ampoules. These ampoules are periodically used as inoculum for the production of vegetative mycelium in flasks. After growth has occurred in these flasks, the entire broth contents are subdivided into small vials and frozen. Frozen vegetative mycelia produced in this way are used as a routine source of the producing organisms for factory scale cephamycin production lots. The cell matter contained in these vials is used to inoculate the first flask inoculum stage. After 24 h of incubation on a rotary shaker, the contents of the flask are used to inoculate a 2 l seed flask stage. After an additional 24 h, the contents of the 2 l flask are used to inoculate a seed fermenter. In this process, transfer from a given seed flask stage is based on attainment of maximum cell concentration. The conditions for transfer vary significantly with the strain and media utilized.

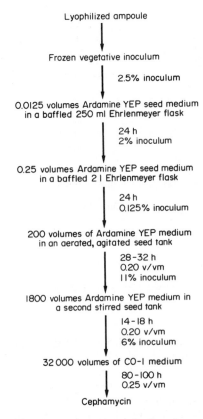

Figure 5 Cephamycin production scheme

Seed fermenters are incubated with appropriate mixing and aeration. Transfer to subsequent stages is made based on cellular respiration. Exponential growth of the organism is monitored; transfer to a subsequent stage is accomplished when carbon dioxide evolution deviates from exponential increase. Typically, this technique minimizes the lag which is observed prior to initiation of growth in the subsequent stage.

Production of cephamycin is accomplished in large commercial fermenters. The profiles of relevant parameters during such a production stage fermentation are illustrated in Figure 6. The medium (CO-1) utilized in this study was derived directly from a medium developed by Orgelfinger and Inamine (unpublished data). CO-1 medium components are detailed in the Merck process previously described (Table 2). Measured values of pH, dissolved solids, cell growth and cephamcyin C titer are presented. These data indicate that there is an initial lag period followed by rapid growth of the organism and finally by sustained synthesis of cephamycin. The initial lag period extends through the first 10 to 15 h of the production cycle. This is followed by a burst of carbon dioxide evolution, cell growth and dissolved solids depletion. The indicator of cell growth is packed cell volume.This value represents the total concentration of particulate solids and viable cells. Since the production medium used contains several insoluble ingredients, the initial value of packed cell volume is much greater than zero. When the packed cell volume increases, it is a result of cell replication. When it decreases, the combined effect of particulate solubilization and/or cell lysis is measured. As such, packed cell volume (pcv) is not a quantitative indicator of cell growth. However, the simultaneous increase in pcv and carbon dioxide evolution as well as a decrease in dissolved solids (a relative indicator of soluble carbohydrate) indicates that rapidly assimilable nutrients are being converted into cell mass.

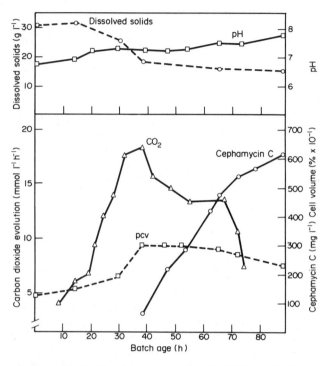

Figure 6   Cephamycin production profile (CO-1 medium)

At approximately 35 h after inoculation, net cell accumulation ceases and antibiotic synthesis begins. The rate of antibiotic accumulation is approximately 14 μg ml^{-1} h^{-1} and is maintained until 70 to 75 h after inoculation. Subsequently, cephamycin accumulation slows to 7.0 μg ml^{-1} h^{-1}. After 88 h, at which point the cephamycin titer reached 620 μg ml^{-1}, little additional product accumulates.

An effect of medium component selection is indicated in Figure 7. The illustrated data represent a production stage fermentation in which DSG medium was used. This medium was developed by Hallada and Inamine (unpublished data) and is composed of the following ingredients: dextrose monohydrate (40 g l^{-1}), glycerol (12 g l^{-1}), cottonseed flour (20 g l^{-1}), DL-lysine (1 g l^{-1}), dimethylformamide (9.2 g l^{-1}), corn steep liquor (20 g l^{-1}), sodium thiosulfate (0.5 g l^{-1}) and polyethylene glycol P2000 (0.15 g l^{-1}). Inoculum development proceeded in a manner identical to that described in Figure 5.

The basic patterns which are illustrated in Figure 6 are evident in the fermentation of the Pro-flo–corn steep liquor (DSG) medium. An initial lag period is followed by rapid growth and subsequent antibiotic production. The major difference between results obtained with the CO-1 and DSG medium is a sustained uptake of sugar and evolution of carbon dioxide throughout the fermentation cycle. This indicates that a higher level of primary metabolism is maintained after initial growth slows.

Initiation of antibiotic synthesis in DSG medium occurs at the same point in the fermentation as was observed in the CO-1 medium. However, the rate of synthesis in DSG medium is significantly higher. In this medium, cephamycin accumulates at a rate of 20 μg ml^{-1} h^{-1}, approximately 50% higher than that observed in CO-1 medium. Further, a decline in the rate of synthesis is not so strongly indicated. The net result is that 50 to 60% more cephamycin accumulates in the DSG medium.

It is interesting to speculate about the cause of increased antibiotic accumulation observed

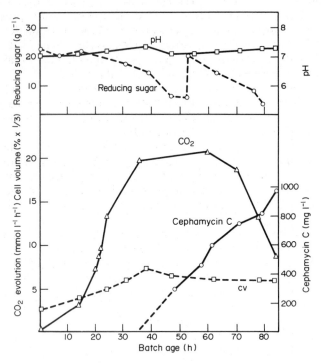

Figure 7   Cephamycin production profile (DSG medium)

between these two media. Such increases are usually due to either an increase in specific productivity of the producing organism or to an increase in the quantity of biomass accumulated. An increase in either of these factors can lead to higher antibiotic potencies. A quantitative assessment of specific productivity requires accurate measurement of cell concentration. It was not possible to measure dried cell weight in either of these media due to the presence of various particulate organic materials. However, it is possible to estimate relative cell concentrations based on carbon dioxide evolution. If it is assumed that the metabolic pathways utilized by *N. lactamdurans* are similar in both these media, then maximum carbon dioxide evolution should be proportional to cell yield. Based on this assumption, the DSG medium supports only slightly increased cell yields. The large difference in volumetric productivity observed in this medium can then be attributed primarily to a medium-directed increase in the specific productivity of *N. lactamdurans*.

9.4.3   Isolation and Purification Stage

Fermentation processes involve the production of an aqueous broth which contains the desired antibiotic or other microbially produced product. The compound of interest must be isolated from the fermenter broth, purified, and possibly derivatized before it can be used as a clinical product. Typically, the quantity of β-lactam antibiotic present in fermentation broths represents only a small fraction of the total fermenter contents. The desired product must be separated from a variety of suspended and soluble broth components. Suspended materials include remnants of the microorganisms which produced the product as well as a variety of non-degraded materials which were included in the original fermentation production medium. Soluble broth components usually include monomers and oligomers of organic medium components as well as a variety of inorganic solutes. Several purification steps must be used to isolate the desired antibiotic from broth. Conventional antibiotic isolation processes typically include a cell removal step followed by a number of product purification and concentration steps.

The majority of β-lactam antibiotics accumulate extracellularly. Typically, the concentration of antibiotic inside the cell at the conclusion of the fermentation stage is quite small. Since the cell mass contains little product, removal of the suspended cellular material from the broth supernatant is an important and effective purification step. Cell removal is usually the first stage of the purification of extracellular antibiotics. This step is normally accomplished using a conventional vacuum filter device. Filtration of the broth is often difficult due to the presence of organic species which reduce filtration rate. Chemical pretreatment of the broth can lessen this effect. However, the most effective means of maintaining filtration rate is to add diatomaceous earth or another suitable 'filter aid' prior to cell separation. Although filter aid usually increases filtration rate, it also creates an expensive and undesirable disposal problem.

Ultrafiltration is an alternative means to separate cellular material from fermenter broth. The recent availability of large, non-fouling ultrafiltration systems has eliminated problems historically associated with the use of this technique. Ultrafiltration has been used for fermentation-broth filtration with excellent results (Paul *et al.*, 1981).

A number of product purification steps follow cell removal in typical antibiotic purification processes. Anion- and cation-exchange chromatography, non-specific adsorption, solvent extraction, and crystallization are used in various combinations to purify β-lactam antibiotics. The specific order and optimal combination of isolation steps depend on the exact physicochemical characteristics of the target product. Typically, ion exchange and solvent extraction steps follow suspended solids removal.

Usually product concentration (*i.e.* water removal) is one of the final steps in antibiotic purification, although dewatering techniques have been used at all stages of product purification (Treiber *et al.*, 1981). Low temperature vacuum evaporation is the conventional means of concentrating antibiotics. However, a membrane process, reverse osmosis, has come into greater use in recent years. Datta *et al.* (1977) demonstrated that reverse osmosis could be used to concentrate antibiotics and that this technique was far more energy efficient than vacuum evaporation.

While a number of other processes have been reported (Gorman *et al.*, 1979; Kamogashira *et al.*, 1982), Merck and Co. has patented a series of cephamycin C purification processes. In one such process (Pines, 1976), the pH of a fermentation broth was adjusted to 2.5, after which it was filtered to remove suspended solids (see Figure 8). The resulting filtrate was then passed through a column containing acid washed activated carbon. Presumably, cephamycin and other organic material was adsorbed by the activated carbon, while non-cephamycin material passed through the column. The adsorbed material (including cephamycin) was then eluted from the column with one of a number of aqueous solutions of polar solvents. Preferred eluant solutions included mixtures of equal volumes of acetone and water as well as saturated solutions of butanol or ethyl acetate.

The eluate from the activated carbon column was then loaded onto a weakly basic anion exchange resin and eluted with a salt solution. This second step effected a further purification of the product. Cephamycin in the eluate from the anion exchange column accounted for 45–65% of the organic matter present in the product stream. Subsequent dewatering steps could be used to prepare a concentrated solution from which a relatively pure product could be crystallized.

In a second process for purification of cephamycin C (see Figure 9) from fermenter broth (Pines *et al.*, 1973), the carbon adsorption column previously described was replaced with a cation exchange step. Filtered broth, pH adjusted to 2.5, was passed through a column of strongly acidic ion exchange resin. Since the isoelectric point of cephamycin is 3.5, this compound is positively charged at the low pH and would remain with the resin as would other positively charged species. The remainder of organic and inorganic material would pass through the column. The cation exchange resin acts to effect a selective separation of charged species. Elution from the column with aqueous pyridine (or some other weak base) followed. The eluted fraction containing cephamycin, pyridine and other species was then concentrated by a vacuum technique to remove the pyridine solvent component.

The pH of the concentrate was then raised to a value of 5–7. This fraction was then passed through a column containing an anion exchange resin. The cephamycin absorbed on the resin was then washed with an aqueous solution of an alkanoic acid such as acetic. Pines *et al.* (1973) have found that washing the resin column in this manner results in a preferential release of as much as 90% of the adsorbed non-cephamycin species. This, in turn, results in an effective purification step upon subsequent elution of cephamycin with a weak organic base such as pyridine. Inorganic phosphate buffer can also be used to elute cephamycin from the 'washed' anion exchange resin. The cephamycin C in the eluate can be concentrated by evaporation or can be used as a substrate for subsequent derivatization steps.

This cation/anion exchange process can be used to prepare substantially purified cephamycin antibiotic. However, the effluent from the anion exchange column is relatively dilute and requires significant concentration. Schubert (1980) has recently patented an even more efficient recycle

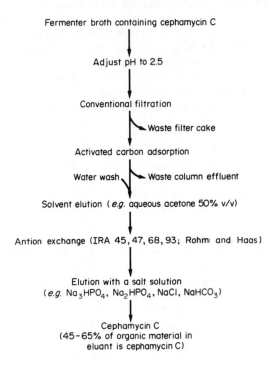

Figure 8 Purification of cephamycin C: carbon adsorption/anion exchange process

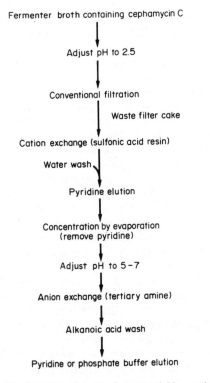

Figure 9 Purification of cephamycin C: sequential ion exchange process

process which produces a concentrated stream of purified cephamycin C (see Figure 10). In this process, fermenter broth was pH adjusted to a low value (1.8–2.5) and then subjected to conventional filtration. The filtrate from this step was then passed through a cation exchange resin column. The column was then washed with water or, alternatively, a dilute aqueous solution of cephamycin (containing only 5–6% of the cephamycin present in the fermentation broth). The cation column was then partially eluted with an aqueous solution of sodium chloride (10% w/v). The wash and eluate streams were combined and pH adjusted to 7.0. This combined fraction was then recycled through the column. The pH of the resulting effluent was then readjusted to 7.0. Increased chemical degradation at this high pH was compensated for by operating the process at low temperature. Recirculation and pH adjustment continued until the pH of the column eluate exceeded 5.5. Typically, 10–15 cycles were required. The final effluent was concentrated or derivatized as appropriate.

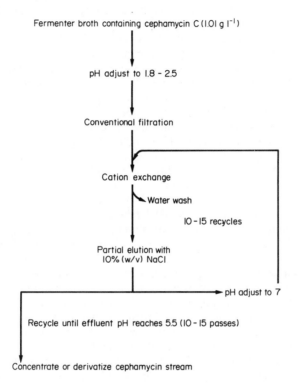

Figure 10   Purification of cephamycin C: cation exchange/recycle process

The efficiency of this recycle process was significantly greater than the sequential ion exchange or carbon adsorption processes. Schubert found that when 36 500 l of filtered fermenter broth containing 1.01 g cephamycin C l^{-1} was purified by this process, a 1500 l fraction containing 94% of the original product could be obtained.

9.5   COMMERCIAL ASPECTS OF CEPHAMYCIN PRODUCTION

Merck and Co. is the only current manufacturer of a cephamycin type antibiotic. This company produces cephamycin C by fermentation using an industrial strain of *N. lactamdurans*. Cephamycin C is isolated from fermenter broth and chemically modified to yield a derivative called cefoxitin. This derivative is marketed worldwide as Mefoxin®. Cefoxitin has become a major injectable antibiotic and is used mainly to treat infections caused by β-lactamase resistant bacteria. Worldwide sales of cefoxitin in 1982 have been estimated at $215 million (Jaffe, 1982). The principal competitors of cefoxitin are second generation cephalosporin derivatives (Neu, 1982). It is expected that cefoxitin will continue to be a major broad spectrum antibiotic.

Several third generation cephalosporins will become available in the United States in the next few years. These compounds typically exhibit better activity against Gram-negative bacteria and *Pseudomonas* species than does cefoxitin. However, the Gram-positive activity demonstrated by cefoxitin is generally superior. Future competition for cefoxitin may then derive from antibiotic formulations which combine third generation cephalosporins with other antibiotics exhibiting broad Gram-positive activity. Alternatively, the clinical availability of other broad spectrum antibiotics currently under development may affect cefoxitin sales. One recently discovered class of compounds, the thienamycins, appears to exhibit an even broader spectrum of antibacterial activity than cephamycin (Kahan *et al.*, 1977). Future availability of thienamycin antibiotics may compete with cefoxitin.

9.6 CONCLUSIONS AND IMPLICATIONS

Relative to penicillins and cephalosporins, the cephamycins represent a new family of broad spectrum β-lactam antibiotics. The fact that cephamycins are synthesized by a class of organisms not previously known to produce β-lactams required that new fermentation processes be developed. The classical literature regarding production of penicillin and cephalosporin was used as a basis for developing processes for cephamycin production. This body of information, modified to suit the requirements of cephamycin-producing organisms, has yielded several processes for the production of cephamycin antibiotics. These processes have been used to prepare cephamycin for clinical use. At present, cephamycin is used widely for treatment of infection in humans. However, with time, competition from other products will require that the cost of preparing cefoxitin be made more economical. Undoubtedly, all aspects of the manufacturing process will be improved. Current processes for the preparation of cephamycin C reflect improvements in media, process conditions and in productivity of the producing culture relative to the initial processes. Conventional mutation techniques have been successful in development of superior cultures. These techniques along with applications of recently developed genetic methods such as protoplast fusion and genetic recombination may also yield higher-producing factory strains. Utilizing genetic methods, Wesseling and Lago (1979) have already isolated several strains which produce increased levels of cephamycin C in flask fermentations.

Media development and process modification should also contribute significantly to improving the existing cephamycin processes. As the fundamental parameters which control biosynthesis of cephamycins are elucidated, the currently employed media can be modified accordingly. As the precise nature of the carbon and nitrogen regulatory mechanisms become known, high yielding fed batch and possibly continuous processes will result. Additional improvement will be derived from the use of advanced instrumentation for monitoring and control of the cephamycin fermentation. Elucidation of the controlling factors in this fermentation will allow application of the sophisticated process control systems which have recently come into availability. Finally, use of immobilized cell techniques to maximize volumetric productivity may be applied in the future. Such techniques have been used in other fermentations (Arcuri *et al.*, 1983). Production of cephamycins by immobilized cell preparations has also been observed (Freeman and Aharonowitz, 1981).

ACKNOWLEDGEMENTS

The authors are pleased to acknowledge the contributions of S. Budavari, S. Drew, S. Gatchalian, R. Greasham, M. Hunt, E. Inamine, J. Lago, A. Kaufman, R. Murawski, M. Nallin, D. Sloan and E. Stapley in the formulation and completion of this manuscript.

9.7 REFERENCES

Abraham, E. P. and G. G. F. Newton (1954). Synthesis of D-δ-Amino-δ-carboxyvalerylglycine (a degradation product of cephalosporin N) and of DL-δ-amino-δ-carboxyvaleramide. *Biochem. J.*, **58**, 266–8.
Abraham, E. P. and G. G. F. Newton (1961). The structure of cephalosporin C. *Biochem. J.*, **79**, 377–93.
Abraham, E. P. (1978). Developments in the chemistry and biochemistry of β-lactam antibiotics. In *Antibiotics and Other Secondary Metabolites*, ed. R. Hutter, T. Leisinger, J. Neusch and W. Wehrli, pp. 141–164. Academic, London.
Aharonowitz, Y. and A. L. Demain (1977). Influence of inorganic phosphate and organic buffers on cephalosporin production by *Streptomyces clavuligerus*. *Arch. Microbiol.*, **115**, 169–73.
Aharonowitz, Y. and A. L. Demain (1978). Carbon catabolite regulation of cephalosporin production in *Streptomyces clavuligerus*. *Antimicrob. Agents Chemother.*, **14**, 159–64.
Aharonowitz, Y. and A. L. Demain (1979). Nitrogen nutrition and regulation of cephalosporin production in *Streptomyces clavuligerus*. *Can. J. Microbiol.*, **25**, 61–7.
Aharonowitz, Y. and G. Cohen (1981). The microbiological production of pharmaceuticals. *Sci. Am.*, **245** (3), 140–52.
Arcuri, E. J., J. R. Nichols, T. S. Brix, V. G. Santamaniam, B. C. Buckland and S. W. Drew (1983). Threnamycin production by immobilized cells of *Streptomyces cattleya* in a bubble column. *Biotechnol. Bioeng.*, **25**, 2399–2411.
Brewer, S. J., T. T. Boyle and M. K. Turner (1977). The carbamoylation of the 3-hydroxymethyl group of 7α-methoxy-7β-(5-D-aminoadipamido)-3-hydroxymethylceph-3-em-4-carboxylic acid (desacetyl-7α-methoxycephalosporin C) by homogenates of *Streptomyces clavuligerus*. *Biochem. Soc. Trans.*, **5**, 1026–9.

Brotzu, G. (1948). *Lavori dell'instituto D'Igiene diCagliari.*

Buckland, B. C. and H. Fastert (1982). Analysis of fermentation exhaust gas using a mass spectrometer. In *Computer Applications in Fermentation Technology*, pp. 119–26. Society of Chemical Industries, London.

Caltrider, P. G. and H. F. Niss (1966). Role of methionine in cephalosporin synthesis. *Appl. Microbiol.*, **14**, 746–53.

Christensen, B. G. (1982). Cefoxitin. *Chronicles of Drug Discovery*, **1**, 223–38.

Currie, S. (1985). In preparation.

Daust, D. R., H. R. Onishi, H. Wallick, D. Hendlin and E. O. Stapley (1973). Cephamycins, a new family of beta-lactam antibiotics: antibacterial activity and resistance to beta-lactamase degradation. *Antimicrob. Agents Chemother.*, **3**, 254–61.

Datta, R., L. Fries and G. T. Wildman (1977). Concentration of antibiotics by reverse osmosis. *Biotechnol. Bioeng.*, **19**, 1419–29.

Delavier-Klutchko, C. and M. Flavin (1965). Enzymatic synthesis and cleavage of cystathionine in fungi and bacteria. *J. Biol. Chem.*, **240**, 2537–49.

Demain, A. L. and E. Inamine (1970). Biochemistry and regulation of streptomycin and mannosidostreptomycinase (α-D-mannosidase) formation. *Bacteriol. Rev.*, **34**, 1–19.

Demain, A. L. and Y. M. Kennel (1978). Resting cell studies of carbon source regulation of beta-lactam antibiotic biosynthesis. *J. Ferment., Technol.* **56**, 323–8.

Drew, S. W. and A. L. Demain (1975a). Stimulation of cephalosporin production by methionine peptides in a mutant blocked in reverse transsulfuration. *J. Antibiot.*, **28**, 889–95.

Drew, S. W. and A. L. Demain (1975b). The obligatory role of methionine in the conversion of sulfate to cephalosporin C. *Eur. J. Appl. Microbiol.*, **2**, 121–8.

Drew, S. W. and A. L. Demain (1977). Effect of primary metabolites on secondary metabolism. *Annu. Rev. Microbiol.*, **31**, 343–56.

Feren, C. J. and R. W. Squires (1969). The relationship between critical oxygen level and antibiotic synthesis of capreomycin and cephalosporin C. *Biotechnol. Bioeng.*, **11**, 583–92.

Foster, J. W., H. B. Woodruff and L. E. McDaniel (1946). Microbiological aspects of penicillin production in submerged cultures of *Penicillium notatum*. *J. Bacteriol.*, **51**, 465–78.

Freeman, A. and Y. Aharonowitz (1981). Immobilization of microbial cells in cross linked, prepolymerized, linear polyacrylamide gels: antibiotic production by immobilized *Streptomyces clavuligerus* cells. *Biotechnol. Bioeng.*, **23**, 2747–59.

Gallo, M. and E. Katz (1972). Regulation of secondary metabolite biosynthesis: catabolite repression of phenoxazinone synthase and actinomycin formation by glucose. *J. Bacteriol.*, **109**, 659–67.

Ginther, C. L. (1979). Sporulation and the production of serine protease and cephamycin C by *Streptomyces lactamdurans*. *Antimicrob. Agents Chemother.*, **15**, 522–6.

Gorman, M., C. E. Higgens and R. Nagajarian (1979). Antibiotic A16886II. *US Pat. 4 132 790.*

Hallada, T. C., E. Inamine and J. Birnbaum (1975). Fermentation process. *US Pat. 3 922 202.*

Hasegawa, T., K. Hatano, H. Fukase and H. Iwasaki (1977). 7-Methoxycephalosporin derivatives. *US Pat. 4 017 485.*

Hodgkin, D. C. and E. N. Maslen (1961). The X-ray analysis of the structure of cephalosporin C. *Biochem. J.*, **79**, 393–402.

Inamine, E. and J. Birnbaum (1972). Cephamycin biosynthesis: Isotope incorporation studies. *Abstr. 72nd Annu. Meeting ASM E68*, p. 12. American Society for Microbiology, Washington, DC.

Inamine, E. and J. Birnbaum (1973). Fermentation process. *US Pat. 3 770 590.*

Inamine, E. and J. Birnbaum (1975a). Fermentation process. *US Pat. 3 902 968.*

Inamine, E. and J. Birnbaum (1975b). Process of making cephamycin C by fermentation. *US Pat. 3 886 044.*

Jaffe, T. (1982). Companies: Merck and Co., Inc., a stitch in time. *Forbes*, Nov. 12, 114.

Kahan, J. S., F. M. Kahan, E. O. Stapley, R. T. Goegelman and S. Hernandez (1977). Thienamycin production. *US Pat. 4 006 060.*

Kahan, J. S., F. M. Kahan, R. Goegelman, S. A. Currie, M. Jackson, E. O. Stapley, T. W. Miller, D. Hendlin, S. Mochales, S. Hernandez, H. B. Woodruff and J. Birnbaum (1979). Thienamycin, a new β-lactam antibiotic. I. Discovery, taxonomy, isolation and physical properties. *J. Antibiot.*, **33**, 1–12.

Kamogashira, T., T. Nishida, M. Nihno and S. Takegata (1982). Process for production of antibiotic cephamycin C. *US Pat. 4 332 891.*

Kern, B. A., D. Hendlin and E. Inamine (1980). L-Lysine ε-aminotransferase in cephamycin C synthesis in *Streptomyces lactamdurans*. *Antimicrob. Agents Chemother.*, **17**, 674–85.

Kern, B. A. and E. Inamine (1981). Cystathionine γ-lyase activity in the cephamycin producer *Streptomyces lactamdurans*. *J. Antibiot.*, **34**, 583–9.

Koffler, H., S. G. Knight, W. C. Frazier and R. H. Burris (1946). Metabolic changes in submerged penicillin fermentations on synthetic media. *J. Bacteriol.*, **51**, 385–92.

Komatsu, K. and R. Kodaira (1977). Sulfur metabolism of a mutant of *Cephalosporium acremonium* with enhanced potential to utilize sulfate for cephalosporin C production. *J. Antibiot.*, **30**, 226–33.

Lemke, P. A. and D. R. Brannon (1972). Microbial synthesis of cephalosporin and penicillin compounds. In *Cephalosporins and Penicillins*, ed. G. Flynn, pp. 370–437. Academic, New York.

Lilley, G., A. E. Clark and G. C. Lawrence (1981). Control of the production of cephamycin C and thienamycin by *Streptomyces cattleya* NRRL 8057. *J. Chem. Technol. Biotechnol.*, **31**, 127–34.

Martin, J. F. (1977). Control of antibiotic synthesis by phosphate. In *Advances in Biochemical Engineering*, ed. L. T. K. Ghose, A. Fiechter and N. Blakeborough, vol. 6, pp. 105–27. Springer, New York.

Martin, J. F. and A. L. Demain (1980). Control of antibiotic synthesis. *Microbiol. Rev.*, **44**, 230–51.

Masurekar, P. and A. L. Demain (1972). Lysine control of penicillin biosynthesis. *Can. J. Microbiol.*, **18**, 1045–48.

Matsumura, M., T. Imanaka, T. Yoshida and H. Taguchi (1978). Effect of glucose and net consumption rates of cephalosporin C production by *Cephalosporium acremonium*. *J. Ferment. Technol.*, **56**, 345–53.

Mendelowitz, S. and Y. Aharonowitz (1982). Regulation of cephamycin C, synthesis aspartokinase, dihydrodipicolinic acid synthetase, and homoserine dehydrogenase by aspartic acid family of amino acids in *Streptomyces clavuligerus*. *Antimicrob. Agents Chemother.*, **21**, 74–84.

Metzler, D. (1977). *Biochemistry*, p. 813. Academic, New York.

Miller, A. L. and J. B. Walker (1969). Enzymatic phosphorylation of streptomycin by extracts of streptomycin producing strains of *Streptomyces. J. Bacteriol.*, **99**, 401–5.

Moyer, A. J. and R. D. Coghill (1946a). Penicillin VIII. Production of penicillin in surface cultures. *J. Bacteriol.*, **51**, 57–78.

Moyer, A. J. and R. D. Coghill (1946b). Penicillin—the laboratory scale production of penicillin in submerged culture by *Penicillium notatum Westling* (NRRL 832). *J. Bacteriol.*, **51**, 79–93.

Nagajarian, R., L. D. Boeck, M. Gorman, R. L. Hamill, C. E. Higgins, M. M. Hoehn, W. M. Stark and J. G. Whitney (1971). β-Lactams from *Streptomyces. J. Am. Chem. Soc.*, **93**, 2308–10.

Neu, H. (1982). The new beta-lactamase stable cephalosporins. *Ann. Internal Med.*, **97**, 408–19.

O'Sullivan, J. and E. P. Abraham (1980). The conversion of cephalosporins to 7α-methoxycephalosporins by cell-free extracts of *Streptomyces clavuligerus. Biochem. J.*, **186**, 613–16.

O'Sullivan, J., R. T. Aplin, C. M. Stevens and E. P. Abraham (1979). Biosynthesis of a 7-α-methoxycephalosporin incorporation of molecular oxygen. *Biochem. J.*, **179**, 47–52.

Paul, E. L., A. Kaufman and W. A. Sklarz (1981). An industrial approach to integrated fermentation/isolation process development. *Ann. N. Y. Acad. Sci.*, **369**, 181–6.

Pines, S. (1976). Antibiotic purification process. *US Pat.* 3 983 108.

Pines, S., N. C. Jamieson and M. A. Kozlowski (1973). Antibiotic purification process. *US Pat.* 3 733 320.

Queener, S. W. and N. Neusch (1982). The biosynthesis of β-lactam antibiotics. In *Biochemistry*, ed. R. B. Morin and M. Gorman, vol. 3, pp. 1–81. Academic, New York.

Ryu, D. D. Y. and J. Hospodka (1980). Quantitative physiology of *Penicillium chrysogenum* in penicillin fermentation. *Biotechnol. Bioeng.*, **22**, 289–98.

Schubert, P. F. (1980). Antibiotic purification process. *US Pat.* 4 196 285.

Segel, I. H. and M. J. Johnson (1961). Accumulation of intracellular inorganic sulfate by *Penicillium chrysogenum. J. Bacteriol.*, **81**, 91–98.

Soltero, F. V. and M. J. Johnson (1954). Continuous addition of glucose for evaluation of penicillin producing cultures. *Appl. Microbiol.*, **2**, 41–4.

Stapley, E. O. and J. Birnbaum (1981). Chemistry and microbiological properties of the cephamycins. In *Beta-Lactam Antibiotics*, pp. 327–51. Academic, New York.

Stapley, E. O., M. Jackson, S. Hernandez, S. B. Zimmerman, S. A. Currie, S. Mochale, J. M. Mata, H. B. Woodruff and D. Henlin (1972). Cephamycins, a new family of β-lactam antibiotics. *Antimicrob. Agents Chemother.*, **2**, 122–35.

Stefaniak, J. J., F. B. Gailey, F. G. Jarvis and M. J. Johnson (1946). The effect of environmental conditions of penicillin fermentations with *Penicillium chrysogenum*, X-1612. *J. Bacteriol.* **52**, 119–27.

Tardrew, P. L. and M. J. Johnson (1958). Sulfate utilization by penicillin producing mutants of *Penicillium chrysogenum. J. Bacteriol.*, **76**, 400–5.

Treiber, L. R. (1981). Quantitative analysis of cephamycin C in fermentation broths by means of thin layer spectrodensitometry. *J. Chromatogr.*, **213**, 129–36.

Treiber, L. R., V. P. Gullo and I. Putter (1981). Procedure for isolation of thienamycin from fermentation broths. *Biotechnol. Bioeng.*, **23**, 1255–65.

Walker, M. S. and J. B. Walker (1971). Streptomycin biosynthesis-separation and substrate specificities of phosphatases acting on guanidinodeoxy-scyllo-inositol phosphate and streptomycin-(streptidino) phosphate. *J. Biol. Chem.*, **246**, 7034–40.

Whitney, J. G., D. R. Brannon, J. A. Mabe and K. J. Wicker (1972). Incorporation of labeled precursor into A16886B, a novel beta-lactam antibiotic produced by *Streptomyces clavuligerus. Antimicrob. Agents Chemother.*, **1**, 247–51.

Wesseling, A. C. and B. D. Lago (1980). Strain improvement by genetic recombination of cephamycin producers, *Nocardia lactamdurans* and *Streptomyces griseus. Dev. Ind. Microbiol.*, **22**, 641–51.

10
Lincomycin

J. E. GONZALEZ and T. L. MILLER
The Upjohn Co., Kalamazoo, MI, USA

10.1 INTRODUCTION AND BACKGROUND

10.1.1 Discovery

A certain amount of background information is important for an understanding of how and why lincomycin is produced. Therefore, this section providing supplementary information is included.

Lincomycin, an antibiotic with mainly Gram-positive antibacterial activity, was first reported in 1962 (Mason *et al.*, 1963). The original patent was granted in 1963 to The Upjohn Company (Bergy *et al.*, 1963). This antibiotic is produced by a soil streptomycete that was designated as *Streptomyces lincolnensis* var. *lincolnensis*. However, it has also been reported that the antibiotic is produced by a number of other organisms to be described below.

10.1.2 Chemistry

The chemical structure of lincomycin as determined by Hoeksema *et al.* (1964) is shown in Figure 1.

Cleavage of lincomycin at the amide bond was found to give a methylthio amino sugar and an amino acid. The stereochemistry of the carbohydrate moiety was studied (Schroeder *et al.*, 1967; Slomp *et al.*, 1967) and assigned as methyl 6-amino-6,8-dideoxy-1-thio-D-erythro-α-D-galacto-octopyranoside, which was given the trivial name methylthiolincosaminide (MTL).

The amino acid portion of lincomycin was shown to be *trans-N*-methyl-4-*n*-propyl-L-proline

211

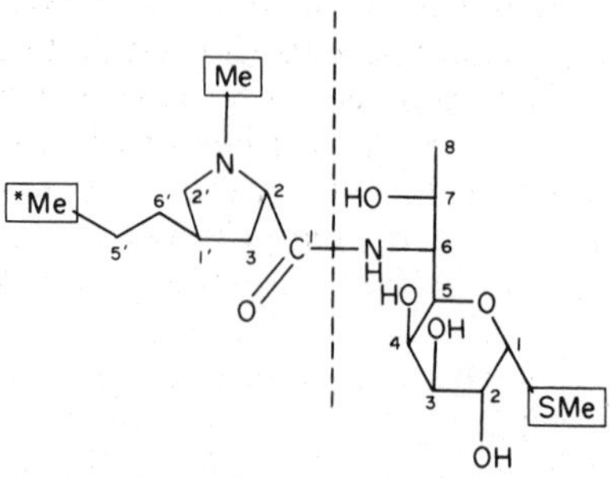

Figure 1   Lincomycin structure and biosynthetic origin

(Magerlein *et al.*, 1967); the trivial name for this compound is propylhygric acid (PHA). It was found to be of key importance to have the natural *trans* isomer of PHA, since the *cis* isomer of lincomycin was only about one-half as active as lincomycin.

10.1.3   Spectrum

The antimicrobial activities of lincomycin and several of its analogs have been described (Grady, 1968; Lewis *et al.*, 1963, 1965; Mason *et al.*, 1963; Mason and Lewis, 1965). Lincomycin was found to possess excellent activity against clinical strains of staphylococci, streptococci and diplococci both *in vitro* and *in vivo*. The antibiotic is effective when administered either subcutaneously or orally and has a low order of acute toxicity. In mice CD_{50} values of 0.8 mg kg^{-1} and 3.8 mg kg^{-1} were obtained with *Streptococcus hemolyticus* when the antibiotic was administered subcutaneously and orally, respectively. The antimicrobial spectrum of lincomycin is generally similar to that of erythromycin, but lincomycin is not cross-resistant with any other major antibiotic. However, a phenomenon designated as the macrolide effect or dissociated cross-resistance has been reported to occur between lincomycin and erythromycin (Grady, 1968); this effect will be discussed in more detail later in this report. Analogs of lincomycin have been described that possess altered and extended activities.

In one study, the incubation of *Streptomyces rochei* in the presence of lincomycin resulted in the formation of lincomycin-3-phosphate (Argoudelis *et al.*, 1969). This derivative was inactive *in vitro*, but was apparently hydrolyzed *in vivo* giving lincomycin. Other important chemical derivatives will be considered in more detail below.

10.1.4   Mode of Action

It was first reported by Josten and Allen (1964) that lincomycin inhibits the incorporation of ^{14}C-lysine into log phase cells of *Staphylococcus aureus* by 87%. Further analysis indicated that incorporation of labeled lysine into protein was inhibited by 92% while incorporation into cell walls did not appear to be inhibited. Under the test conditions protein synthesis was shown to cease immediately after lincomycin addition to the growing culture while RNA synthesis stopped 15 minutes later and DNA synthesis was unaffected for 60 minutes. These data suggested that lincomycin acts by inhibiting protein synthesis.

Since lincomycin was known to inhibit Gram-positive but not Gram-negative organisms, it was of interest to determine whether this difference is because of differences in protein synthesizing mechanisms or in cell permeabilities. For these studies the effect of lincomycin on the incorporation of ^{14}C-phenylalanine into protein by cell-free systems of *Escherichia coli* and *Bacillus stearothermophilus* was examined (Chang *et al.*, 1966). It was noted that lincomycin had little effect on the incorporation of ^{14}C-phenylalanine into protein by the Gram-negative organism, *E. coli*. On the other hand, incorporation into protein by the *B. stearothermophilus* was strongly inhibited by lincomycin. These data suggest that the protein synthesizing system of *E. coli* is insensitive to lincomycin. By mixing ribosomal and supernatant fractions of *E. coli* and *B. stearothermophilus* in various combinations, it was shown that the inhibition of protein synthesis by lincomycin (10^{-4} M) occurred only when ribosomes of *B. stearothermophilus* were used. Studies were then conducted with 30S and 50S ribosomal subunits from *E. coli* and *B. stearothermophilus* in an effort to determine which subunit was sensitive to lincomycin. The mixing of the ribosomal subunits of each organism in various combinations showed that the incorporation of ^{14}C-phenyl-alanine into protein was inhibited by lincomycin only in systems containing the *B. stearothermophilus* 50S subunit. Thus, inhibition of protein synthesis by lincomycin appeared to occur at the 50S subunit. Further studies on the effect of lincomycin on binding of ^{14}C-phenylalanine-tRNA to the ribosomal poly U complex were conducted. The results were consistent with the lincomycin mode of action being inhibition of binding of aminoacyl-tRNA to the mRNA-ribosome complex at the 50S subunit.

Dilution experiments indicated that the lincomycin–ribosome complex can dissociate at 25 °C suggesting that the binding is reversible. Erythromycin antagonism of lincomycin was reported by Griffith *et al.* (1965). It was found that erythromycin-resistant and lincomycin-sensitive strains of *S. aureus* have greatly reduced sensitivity to lincomycin in the presence of erythromycin. This phenomenon was referred to earlier in this report as dissociated cross-resistance.

The work on lincomycin mode of action has been reviewed by Chang and Weisblum (1967). Most of the data suggest that the mode of action of lincomycin is inhibition of aminoacyl-tRNA binding to ribosomes. More specifically, lincomycin stimulated the hydrolysis of formylmethionyl-tRNA while inhibiting esterification of formylmethionine and peptide bond formation (Campbell *et al.*, 1979).

10.1.5 Lincomycin Assays for Fermentation Development and Production

Both biological and chemical assays have been developed for lincomycin. The biological assay is accomplished by using the diffusion disc plate assay procedure. Penassay seed agar is inoculated with *Sarcina lutea* and added to Petri plates. Samples and standards are then spotted onto 12.7 mm assay discs placed on the agar. Incubation is at 30 °C for 16–18 h (Hanka *et al.*, 1963). This assay has a standard deviation of about 8%. The biological assay has been modified in order to determine the lincomycin content of animal feeds (Neff *et al.*, 1967). In addition, the disc plate assay provides the basis for an automated turbidometric bioassay (unpublished or Coded Federal Regulation 436.106).

The chemical assay is based on the liberation of methanethiol by the acid hydrolysis of lincomycin. The thiol is then distilled from the sample solution and reacted with the color reagent, a 0.01% solution of 5,5'-dithiobis(2-nitrobenzoic acid) in water. The color develops almost instantaneously at room temperature and is then read spectrophotometrically at 420 nm. This procedure has been adapted for use with an automatic analyzer (AutoAnalyzer) (Prescott, 1966). The standard deviation of the assay is about 5%.

An assay for lincomycin based on gas–liquid chromatography (GLC) has also been described (Houtman *et al.*, 1968). By the use of the GLC procedure, the ratio of lincomycin to lincomycin B may be determined since the two components are easily resolved. Quantitation was achieved by the use of an internal standard. More recently high performance liquid chromatography assays for lincomycin have been developed that are quite accurate and also resolve the two components.

10.2 PRODUCTION TECHNOLOGY

10.2.1 Lincomycin Biosynthesis

The structure of lincomycin is presented in Figure 1. The molecule is composed of two moieties: an amino acid (PHA) and a sugar (MTL) joined by a peptide bond. Seven of the nine carbons and the nitrogen in the amino acid moiety are derived from L-tyrosine. The *N*-methyl carbon and the third carbon in the side-chain have L-methionine as precursor. The atoms derived from methionine are enclosed in squares in Figure 1. L-3-Hydroxytyrosine or L-3,4-dihydroxyphenylalanine (L-DOPA) is a more immediate precursor to lincomycin than L-tyrosine. The carbons derived from L-DOPA are numbered in Figure 1 using the order in which they are found in the parent compound (see Figure 2).

Most of the biosynthetic work has been based on fermentation using chemically defined medium (CDM) (Argoudelis, 1969). Media composition and incubation conditions are summarized in Table 1.

Table 1 Media Used for Biosynthetic Studies with *S. lincolnensis*

| Fermentation | | Seed | |
Substance	Amount (g l^{-1})	Substance	Amount (g l^{-1})
Glucose	30.0	Cerelose	10
Sodium citrate	3.0	NZ-amine B	5
$ZnSO_4 \cdot 7H_2O$	0.001	Yeastolac	10
$FeSO_4 \cdot 7H_2O$	0.001		
$MgSO_4$	1.0		
K_2HPO_4	2.5		
NaCl	0.5		
NH_4NO_3	2.0		

Seed cultures were incubated at 28 °C for 48 hours on a rotary shaker. Generally the seed culture was centrifuged and the mycelium was washed three times with sterile deionized water prior to use as inoculum. The synthetic medium was inoculated at a rate of 2% (v/v) with washed mycelium, and was incubated at 28 °C on a rotary shaker for periods up to 6–8 days.

The more successful approach to the elucidation of the lincomycin biosynthetic pathway to date is the use of either radioactively labeled or stable isotopes of precursor candidates followed by isolation and appropriate analysis. Witz *et al.* (1971) fed L[1-^{14}C]-tyrosine to cultures of *S. lincolnensis* resulting in a 14.1% incorporation into lincomycin A. Lincomycin B, lacking the methionine-derived carbon in the side-chain of the amino acid moiety, also incorporated radioactivity. After degradation, 96% of the radioactivity of the amino acid moiety was assigned to the carboxyl carbon. Thus, this carbon is derived from C-1 of L-tyrosine. Feeding experiments using radioactively labeled L-DOPA also showed that this amino acid is a precursor to lincomycin.

A similar experiment using L[U-^{14}C]-tyrosine as a precursor showed a seven-fold increase in incorporation of radioactivity when compared to the use of L[1-^{14}C]-tyrosine. Essentially all the radioactivity was found in the amino acid moiety. These two experiments suggest that seven of the nine carbons in the amino acid moiety were derived from L-tyrosine.

Experiments using ^{15}N-labeled tyrosine demonstrated a higher level of incorporation into the nitrogen of the amino acid moiety than into the one found in the peptide bond when analyzed by mass spectrometry.

Another experimental avenue that confirms and extends these results involves the use of ^{13}C NMR studies using suitably labeled precursors of lincomycin. Deuterated carbons reduce the ^{13}C NMR signal and produce a multiplet corresponding to the resonance of the assigned carbon (J. P. Rolls, unpublished data). In this work [3′,5′-^{2}H]-L-tyrosine was fed to *S. lincolnensis* resulting in the formation of a monodeuterated lincomycin A suggesting that one of the carbon–deuterium bonds is lost in the course of biosynthesis. The remaining carbon–deuterium bond appeared in the β-position of the side-chain of the amino acid moiety of lincomycin (designated 5′ in Figure 1).

Feeding experiments using [2′,5′,6′-^{2}H]-L-DOPA further clarified the events in the biosynthetic pathway. A trideuterated lincomycin resulted from these experiments. The 2′-, 5′- and 6′-positions of L-DOPA corresponded identically to the identified positions in Figure 1. These findings suggest that carbons 3′ and 4′ of L-DOPA are lost during biosynthesis. These data are also consistent with the results of the feeding experiments with deuterated and L[1-^{14}C]-labeled tyrosine. The original location of the L-DOPA-derived carbons retained in the antibiotic can be inferred from this group of experiments. These results also suggest that L-tyrosine is converted to L-DOPA in an early step of biosynthesis. Hurley and his coworkers (1979) have found evidence for similar pathways in the biosynthesis of anthramycin, tomaymycin and sibiromycin that share the same precursors (L-tyrosine and L-DOPA) and obtain acid moieties related to PHA in the antibiotics.

The enzyme activity converting L-tyrosine to L-DOPA has been measured and partially characterized in *S. lincolnensis* (Michalik *et al.*, 1975). Figure 2 shows that this tyrosinase activity seems to be involved in both the lincomycin and melanin pathways in *S. lincolnensis*. However, L-DOPA ring cleavage precedes cyclization when L-DOPA becomes a precursor of lincomycin according to the labeling studies discussed above. The reverse order of events would result in an amino acid moiety with vastly different labeling patterns (compare structures A and B in Figure 2).

The origin of the methyl groups in the lincomycin molecule was studied using the same techniques. Methionine labeled with ^{14}C in the methyl group showed a 20% incorporation into lincomycin. Hydrazinolysis of the peptide bond and purification of the two resulting compounds

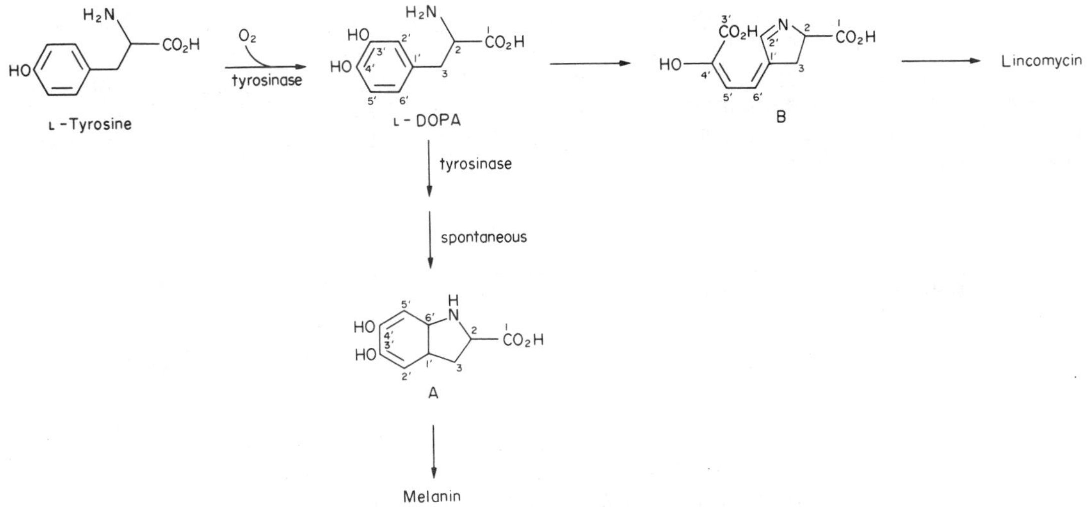

Figure 2 Fate of L-3,4-dihydroxyphenylalanine in *S. lincolnensis*

indicated that twice as much radioactivity was found in the amino acid moiety as in the sugar. Further degradation pointed to the radioactive carbons being located in the third position (methyl) of the propyl side-chain and on the *N*-methyl group in the amino acid moiety. Likewise, the thiomethyl group in the sugar moiety was identified as the other radioactive source (Argoudelis *et al.*, 1969). Mass spectral analysis of an analogous experiment using [^{13}CH$_3$]-L-methionine tended to corroborate the above findings regarding the origin of the carbons in the *N*-methyl and the third position of the side-chain in the amino acid moiety. An experiment using [^{35}S]-L-methionine also established that the sulfur in the thiomethyl group of the antibiotic was derived from methionine.

An attempt to answer the question whether this reaction occurred in one or two steps, transthiomethylation or sulfur transfer followed by transmethylation, was undertaken. Specifically, enriched [^{13}CH$_3$, ^{34}S]-D,L-methionine was fed to *S. lincolnensis* resulting in a lincomycin product of mixed molecular weights. Using mass spectral analysis transthiomethylation would be detected as a peak with an increased molecular weight over the base peak of three atomic units ($M + 3$). Stepwise addition of the thiomethyl group would be indicated by a peak at two atomic units above the base peak ($M + 2$) or by a peak one atomic unit above the base peak ($M + 1$). The $M + 3$ peak occurred at 30% the abundance of the $M + 2$ peak and at the same abundance as the $M + 1$ peak suggesting that transthiomethylation is an important mechanism in the biosynthesis of the sugar moiety (J. P. Rolls, unpublished data).

The anomeric carbon of the sugar moiety (MTL) is derived from the carbon equally positioned in glucose according to the results of feeding experiments using [1-^{13}C]-D-glucose. Only the C-8 carbon of the amino sugar skeleton is enriched by [6-^{13}C]-D-glucose suggesting that glucose specifically loses the carbon on the sixth position during conversion to lincomycin (see Figure 1) resulting in a pentose ring.

Likewise, [1-^{13}C]-pyruvate labels the C-6 carbon atom of the amino sugar skeleton. [2-^{13}C]-Glycerol shows an enrichment of the C-7 carbon of MTL. [2-^{13}C]-Acetate results in a dual label at C-7 and C-8 of the amino sugar. This can be explained by the scrambling of the acetate precursor label by anabolic reactions such as those encountered in the citric acid cycle into the C-2 and C-3 positions of phosphoenol pyruvate or an analogous immediate precursor of the antibiotic prior to incorporation of the ^{13}C label (Rolls *et al.*, 1977).

The work on the biosynthesis of lincomycin was described in some detail since it provides an illustration of how the information may be useful in process development. For example, it was pointed out above that lincomycin B is much less potent than lincomycin; therefore, it is an undesirable side-product. The unusual pathway from tyrosine to PHA provided the basis for reducing the level of lincomycin B in fermentation beers. It was demonstrated that the addition of tyrosine and some of its precursors or degradation products to the fermentation medium resulted in reduced levels of lincomycin B in the harvest beer (Witz *et al.*, 1971) although the mechanism is not yet understood.

10.2.2 Fermentation

The sequence of operations in the lincomycin fermentation is outlined in Figure 3. The sporu-lated culture is stored in liquid nitrogen or in lyophile tubes. These are used to prepare agar slants. Spore suspensions from these slants are inoculated into the primary seed medium. After growth has advanced to stationary phase in this stage, this seed is used to inoculate the next scale. Finally, the fermentation broth is inoculated with the secondary seed at a 5% ratio of seed volume to fermenter volume.

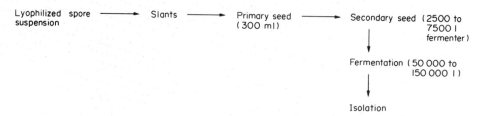

Figure 3   Operational flowsheet of the fermentation process

The two seed stages are used to minimize the growth time in the secondary seed and obtain more economical operation. The initial seed conditions used in this process were described by Bergy *et al.*, (1963). A soil slant of *Streptomyces lincolnensis* var. *lincolnensis*, NRRL 2936, was used to inoculate a series of 500 ml Erlenmeyer flasks each containing 100 ml of seed medium consisting of the following ingredients: Yeastolac (10 g), glucose monohydrate (10 g), NZ Amine B (5 g) and tap water brought to a volume of 1 l. The seed medium presterilization pH was 7.3. The seed was grown for 2 days at 28 °C on a rotary shaker with a two inch stroke at 250 r.p.m.

Flask fermentations were started with a 5% seed inoculum using 500 ml Erlenmeyer flasks containing 100 ml of the following fermentation medium: glucose monohydrate (20 g), molasses (20 g), cornsteep liquor (20 g), Wilson's peptone liquor #159 (10 g), calcium carbonate (4 g) brought to a 1 l volume with tap water. The pH of the medium after sterilization was 7.0. The shake flasks were harvested after four days of fermentation at 28 °C on a rotary shaker at 250 r.p.m. The harvested beer pH was 8.6.

The time course profile of some chemical changes in pilot scale fermentations is illustrated in Figure 4. The fermentation medium is similar to the one used for the flask fermentations above. Lincomycin production begins when the rapid growth phase is over. This also coincides with the depletion of glucose and sucrose from the medium. Once initiated, the rate of lincomycin production is nearly linear over the span of the experiment. Soluble nitrogen, excepting ammonia, remains constant after 20 hours of fermentation, whereas viscosity peaks at two days when the faster growth rate ends.

Over the years productivity improvements have been sought in the laboratory and pilot plants in both culture and process variables. Strain improvement using the classical techniques of mutation followed by selection for higher antibiotic producers on agar medium or shake-flask fermentations has been a successful approach to increased productivity of the process. Types of mutagenic treatments have been varied including chemicals such as nitrosoguanidine and radiation including UV and X-rays. Resistance to phage infections has also been an effective selection method for obtaining higher lincomycin producers. The mechanisms by which these mutations enhanced productivity remain largely unknown.

Lapchinskaya *et al.* (1968) reported that treatment of *S. lincolnensis* var. *lincolnensis* sp. nov. with various physical and chemical mutagens resulted in mutants with increased production capacity. From an initial antibiotic titer of 40 μg ml^{-1} with the parent strain, mutagenic treatment and selection for higher lincomycin producing strains led to cultures capable of producing 700 μg ml^{-1}. Mutagenic treatments included γ-rays, UV light, nitrosoethylurea and 1-methyl-3-nitro-1-nitrosoguanidine.

Another report, possibly involving the use of *S. lincolnensis* as the wild-type organism, described the use of UV light as a mutagenic agent resulting in a strain (UV-914) that produced 650 μg of lincomycin per ml. Optimal medium conditions for production were also investigated. The best yield was obtained in a medium composed of 0.5% soyflour, 5% sucrose, 0.15% ammonium sulfate, 1–2% corn extract, 0.3% sodium chloride and 0.5% calcium carbonate. Apparently, sul-

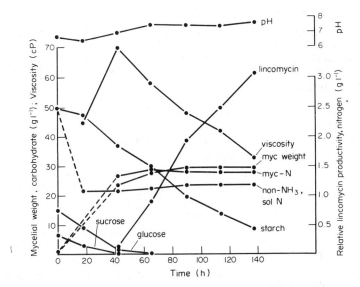

Figure 4 Chemical changes in a lincomycin fermentation with *S. lincolnensis*. Myc weight = dry weight of mycelium; myc-N = total mycelial nitrogen; non-NH₃, sol N = non-ammonia, soluble nitrogen

fate could be substituted by bisulfate or thiosulfate. The corn extract played an important role in yields since its omission reduced titers by 80%. The 5% sucrose ingredient could be replaced by galactose or by molasses or by a combination of 2% sucrose and 4% whale oil which gave both higher yields and increased pH (Gavrilina *et al.*, 1970).

Optimization of process variables comprised adjusting medium ingredients and operational parameters. Typically, a round of fine tuning of these variables was required following the introduction of a superior culture in production. The main tool for medium ingredient optimization in this fermentation has been shake flask fermentations. These were used to determine which ingredients were best suited for the fermentation and provided the starting point regarding the amounts in which these ingredients should be added to the fermentation. Shake flask fermentations were also used to optimize slant and seed composition.

Among the process variables optimized, pH, temperature, aeration, back-pressure and agitation were found to be important in the fermentation stage. The timing of the harvest of seeds to be used as inocula was also critical. Process variable optimization was studied most intensely in laboratory, pilot and production plant fermenters rather than in shake flask experiments.

10.2.2.1 *Lincomycin production by other* Actinomyces *species*

Besides *S. lincolnensis* several other actinomycetes have been identified as lincomycin producers. Table 2 summarizes the strains described in the U.S. patents and scientific literature.

Table 2 Lincomycin-producing Cultures

Name	Comment	Ref.
S. lincolnensis var. lincolnensis	Original culture	Bergy *et al.* (1963)
S. variabilis chemovar *liniabilis* Dietz, var. nova	Lower lincomycin B production	Argoudelis and Coats (1974)
S. espinosus Dietz, sp. n.	1. Lower lincomycin B production, melanin negative	1. Argoudelis *et al.* (1972)
	2. Thermophillic	2. Reusser and Argoudelis (1974)
S. pseudogriseoulus chemovar *linmyceticus* Dietz var. nova	Melanin negative	Argoudelis and Coats (1973)
S. vellosus Dietz, sp. n.	Lower lincomycin B production	Bergy *et al.* (1981)
Actinomyces roseolus	USSR	Kuznetsov *et al.* (1974)

An interesting lincomycin-producing microorganism is *S. espinosus* Dietz. The salient features of the strain include an increased selectivity, compared to *S. lincolnensis*, for the exclusive synthesis of lincomycin A, a melanin-negative phenotype and an ability to produce lincomycin at higher temperatures. Recently, a plasmid, pUC6, has been identified and isolated from *S. espinosus* opening the way for recombinant DNA experiments applied to lincomycin development (Manis, 1981). Recombinant plasmids, using pUC6 and *E. coli*-derived pBR322, have been obtained yielding DNA molecules which may function as vectors in both *E. coli* and *Streptomyces*.

Published work regarding *Actinomyces roseolus* emphasizes the effect of mutagenic treatment and conditions of culture storage on lincomycin production and culture survival. Chemical and radiation treatments were effective in eliciting mutations that resulted in a 40-fold stimulation of lincomycin-producing activity when compared to the parent strain. Mutagenic agents could be ranked according to their ability to produce stable mutations in the following order: bruneomycin > γ-ray > UV light > rubomycin (Lapchinskaya and Pogozheva, 1971, 1977).

Survival of spores of *Actinomyces roseolus* was enhanced when these were lyophilized on cotton, exceeding 83–89% of the original spore count. When the whole population was subjected to lyophilization, increased numbers of the asporogenous and proactinomycete variants and preservation of the natural ratio of these variants in the population was observed. Interestingly, when each variant (basic, oligosporogenous, asporogenous and proactinomycete-like) was lyophilized separately, almost all of them died (Kuznetsov and Bushueva, 1973, 1974). Optimal lyophilization conditions also included the prevention of exposing the spores to oxygen in the lyophilization medium and storage at 5 °C. Higher storage temperatures led to an increased death rate and accumulation of lower producing oligosporic and asporogenic variants (Kuznetsov and Bushueva, 1974; Kuznetsov *et al.*, 1975).

10.2.3 Fermentation Power Requirements

Engineering estimates of the fermentation power requirements for both the secondary seed and fermentation stages are summarized in Table 3 (unpublished data). The seed stage is typically performed in a 7500 l fermenter and the fermentation itself in a 150 000 l fermenter. The estimates for the seed stage refer to the typical length that the seed requires to reach maturity, while the estimates for the fermentation are reported on a daily basis (with the exception of the steam used in sterilization) since the fermentation length will vary according to scheduling constraints.

Table 3   Typical Fermentation Power Requirements

Item	Amount required for seed stage	Amount required for fermentation stage
Steam		
Sterilization	10 000 pounds	78 300 pounds
Bleeders	2880 pounds	1680 pounds d^{-1}
Tracers	—	1000 pounds d^{-1}
Electricity		
Agitation	442 kW h d^{-1}	6100 kW h d^{-1}
Aeration	1160 kW h d^{-1}	7900 kW h d^{-1}

10.2.4 Isolation

A number of processes for the isolation of lincomycin have been developed. The choice of method may depend on the intended use of the product since both pharmaceutical and agricultural grades are manufactured. Most processes take advantage of the fact that the antibiotic is basic, with a pK_a of 7.6.

One of the earliest production methods for isolating lincomycin was based on solvent extraction (Bergy *et al.*, 1963). Such a process will be described in some detail since it entails many aspects of classical isolation and purification. In a typical process the whole beer was adjusted to pH 6.7 with sulfuric acid and filtered after the addition of 4.0% filter aid. The filter cake was washed with 0.1 volume of water and the wash added to the filtered beer. Then the filtered beer was adjusted to pH 10 with sodium hydroxide and extracted twice with 0.33 volume of *n*-butanol. The combined butanol extracts were mixed with 0.5 volume of water and adjusted to pH 2 with concentrated sulfuric acid. The aqueous extract was separated, adjusted to pH 10.1 with sodium hydroxide and extracted twice with 0.33 volume of *n*-butanol. The combined butanol extracts were washed with 0.1 volume of water. The washed butanol extract could be freeze dried and further upgraded.

The upgrading could be accomplished by dissolving the freeze dried product in water. The solution could be adjusted to pH 2.0 with concentrated sulfuric acid and the precipitate removed by filtration. The filtrate was extracted once with an equal volume of methylene chloride to remove impurities. The aqueous layer was adjusted to pH 5 with sodium hydroxide and extracted with an equal volume of methylene chloride. The aqueous phase was then adjusted to pH 10.2 and extracted five times with equal volumes of methylene chloride. A small amount of water was added to the combined extracts and the methylene chloride removed by vacuum distillation. The resulting solution could be freeze dried to a solid.

Further purification could be accomplished by partition chromatography. A typical solvent system consisted of cyclohexane, methyl ethyl ketone (MEK) and pH 10 $NaHCO_3$ buffer (70:30:20). The upper phase could be slurried with diatomite and this mixture could be slurried and poured into a column. The column could be developed with the antibiotic dissolved in the lower phase, homogenized with the upper phase and placed on top of the column. Fractions from the column could be collected and concentrated by well-known classical procedures. Such a process was obviously very cumbersome and was replaced by the carbon adsorption process described below. A variation of this solvent extraction procedure served as the early production scale isolation process.

Ion exchange procedures can also be used (Bergy *et al.*, 1963). Cation exchange resins such as Amberlite IRC 50® or Dowex 50® are suitable. Either fixed bed or fluidized bed columns may be used. Elution can be accomplished with water at an acid pH. The eluate is neutralized and concentrated to get crystals.

An unusual process has been described that consists of contacting filtered beer with a water immiscible liquid cation exchanger such as dichloromethane containing a small amount of a suitable salt of an aromatic sulfonic acid (Jahnke, 1967). This operation may be carried out in a centrifugal countercurrent extractor. The loaded exchanger is concentrated followed by addition of octylamine and water added with agitation. The phases are separated and lincomycin extracted from the dichloromethane into water at pH 10. Concentration gives a residue that is again taken up in water, and crystallization is brought about by the addition of HCl.

Lincomycin may also be isolated by the use of a non-ionic macroporous copolymer styrene cross-linked with divinylbenzene, *e.g.* Amberlite® XAD 1 (Cha and Jahnke, 1970). This procedure may be carried out as a batch process using whole beer sorption. Elution may be done with an organic or aqueous organic solvent in which lincomycin is soluble. Crystallization may be brought about by methods described elsewhere in this review.

In a unique process, whole or screened beer is mixed with a nonvolatile relatively water-immiscible organic oil (Jariwala, 1978). The water is azeotroped off and the solids are separated from the oil by conventional means and dried. Further purification of the dried product may be carried out depending on the quality requirement of the final product.

In the traditional production method of isolating lincomycin, shown in Figure 5, the filtered beer is passed over activated charcoal (Bergy *et al.*, 1963). Then the adsorbed antibiotic is eluted with an organic solvent such as acetone or methyl ethyl ketone. The eluate is concentrated and the crystals are filtered. The product may be purified by transfer of the protonated and non-protonated forms between water and an organic solvent followed by crystallization. This product may be recrystallized by dissolving the salt in water followed by the addition of a water-miscible solvent such as acetone and then cooling. The final product is filtered and dried and can be obtained as the free base or as the acid salt (usually HCl). The literature does not contain information regarding yields and material balances for each of the above processes. However, yields would be expected to range between 50 and 90%. In this process the spent carbon is reactivated by heating it to a high temperature.

10.3 CHEMICAL DERIVATIVES OF LINCOMYCIN

Inspection of the lincomycin structure suggests a number of positions for possible chemical substitution or modification. Lincomycin analogs have been produced both by fermentation and chemical methods. One compound, lincomycin B (4′-depropyl-4′-ethyllincomycin), is produced normally during the fermentation as a minor component. As can be seen in Figure 1, lincomycin

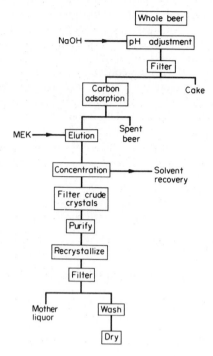

Figure 5 Simplified flow sheet for lincomycin isolation

B contains one less carbon than lincomycin, having an ethyl group in place of the propyl group in the PHA moiety (Argoudelis *et al.*, 1965). Lincomycin B has about 25% the bioactivity of lincomycin (Grady, 1968). It was noted that lincomycin analogs produced from the *cis* isomers of the alkyl hygric acid moiety were less active than those with the naturally occurring *trans* isomers. The antibacterial spectrum of lincomycin did not change with 4'-alkyl analogs.

A significant portion of the parent antibiotic goes into the manufacture of chemical derivatives with improved antibiotic properties. Indeed a very large number of derivatives with modification of both the amino acid and carbohydrate portions of the molecule have been synthesized and tested for efficacy (Magerlein, 1971).

To date, the most important lincomycin derivative is obtained when the 7(*R*)-hydroxy group is replaced by a 7(*S*)-chloro substituent giving 7-chlorolincomycin (Birkenmeyer, 1970a, 1970b); this new antibiotic is called clindamycin. Clindamycin is much more active both *in vitro* and *in vivo* than the parent compound, *e.g.* clindamycin has about 16 times the activity of lincomycin against *E. coli* (Birkenmeyer, 1970b). Furthermore, clindamycin is more lipid soluble than lincomycin and therefore much more rapidly absorbed into the blood stream when administered orally. Clindamycin has been found to be active against a large number of clinical strains of staphylococci, streptococci and pneumococci (Modde, 1971). In addition, it has been shown that 7-chlorolincomycin has antiplasmodial activity which the parent antibiotic lacks altogether. However, this derivative has not yet been approved for use in the treatment of diseases such as malaria. Other 7-halo lincomycin derivatives were also found to be more potent than the parent compound. Ester substitutions at the 7-position of lincomycin have produced a series of analogs with less *in vivo* activity when orally administered than the parent antibiotic (Sinkula *et al.*, 1969).

Two clinically important derivatives of clindamycin have been synthesized. These are clindamycin 2-phosphate (Morozowich and Lamb, 1969) and clindamycin 2-palmitate (Morozowich and Sinkula, 1971). Both of these antibiotics are useful for oral formulations since they lack the bitter taste of lincomycin. More detail on these compounds is provided in a later section of this review.

10.4 COMMERCIAL MARKETS

10.4.1 Current Manufacturers

The Upjohn Company produces most of the world's lincomycin at plants located in Kalamazoo, Michigan; Barceloneta, Puerto Rico; and Cuernavaca, Mexico. Clindamycin and its derivatives (clindamycin phosphate and clindamycin palmitate) are produced at the facility in Puerto Rico by processes protected by patents.

There is ample evidence that there is interest in lincomycin in the USSR. For example, there are many reports of clinical studies and comparisons of lincomycin with other antibiotics in the Soviet literature. In addition, there are also reports dealing with biosynthesis, mutation and process development as indicated in the fermentation section of this review. This information leads to the belief that the USSR is interested in producing lincomycin and may in fact be doing so.

It is known that lincomycin is being produced in the People's Republic of China at plants in Suzhou and Shijiazhuang (personal communication) and marketed as jiemycin (Pharmaceuticals for Export Catalogue, exported by China National Chemicals Import and Export (1980). Erh Li Kou, Beijing, China). However, no information is available regarding the process, titers, culture or other details, but total quantities produced are considered to be low.

10.4.2 Product Outlook

The product outlook for lincomycin and its derivatives continues to be good and the market is expanding. Lincomycin is well received by the medical community for the treatment of serious infections caused by streptococci, pneumococci and staphylococci. It is used in both oral and injectable forms.

Clindamycin is widely used in the treatment of serious aerobic Gram-positive infections and also Gram-negative and Gram-positive anaerobic infections. Clindamycin is the drug of choice in life-threatening infections by Gram-negative anaerobic infections of hospitalized patients where the therapy also includes gentamicin or a similar antibiotic. Both oral and injectable forms of this antibiotic are available.

The market for lincomycin-based products was greatly expanded by the introduction of clindamycin phosphate and clindamycin palmitate. The indications for these antibiotics are nearly identical to those for clindamycin and in fact both drugs are hydrolyzed to clindamycin *in vivo*. Clindamycin palmitate is used in oral pediatric formulations that are prepared by dissolving flavored granules in water. Clindamycin phosphate is an injectable formulation that comes as a sterile solution. In addition, a relatively new formulation of clindamycin phosphate (Cleocin T) has been approved for topical treatment of acne vulgaris. The lincomycin family of pharmaceutical drugs is growing and indications are that it will continue to do so.

Lincomycin is also used in agricultural products, either alone or in combination (Bergy *et al.*, 1965). Lincomycin-containing products are approved for use as a growth promoter in poultry. Further, the antibiotic has been approved for therapeutic uses such as the treatment of swine dysentery and infections in small animals of the same type as indicated for humans.

10.5 SUMMARY

Lincomycin is an important antibiotic that has been produced for 20 years, principally by The Upjohn Company. Lincomycin is effective against Gram-positive and anaerobic bacteria and finds use both in pharmaceutical and agricultural products. Its activity is based on inhibition of protein synthesis at the level of the 50S ribosomal subunit.

Extensive biosynthetic studies have been done providing information on the pathway to both the amino acid and carbohydrate portions of the molecule. This information has provided the basis for process improvement both by mutation of the culture and fermentation condition development techniques. The fermentation process is conducted using conventional techniques of culture storage, seed and fermentation stage scale-up. Power requirements have been determined and are in the range to be expected for a *Streptomyces* fermentation.

A number of different approaches are available for isolation of the antibiotic. They include solvent extraction, ion exchange resin and carbon adsorption procedures. Further processing is dictated by the grade of product required.

Much of the lincomycin produced is used in the manufacture of chemical derivatives: clindamycin, clindamycin 2-phosphate and clindamycin 2-palmitate. These represent more potent antibiotics than the parent with a somewhat expanded antimicrobial spectrum.

New formulations of lincomycins have been introduced in recent years and the market outlook remains good.

ACKNOWLEDGEMENTS

We wish to thank many of our colleagues at Upjohn whose work provided much of the information in this paper, especially J. Visser, J. L. Jost and D. F. Witz.

10.6 REFERENCES

Argoudelis, A. D., J. A. Fox and T. E. Eble (1965). A new lincomycin related antibiotic. *Biochemistry*, **4**, 698–703.
Argoudelis, A. D. and J. H. Coats (1969). Microbial transformation of antibiotics. II. Phosphorylation of lincomycin by *Streptomyces* species. *J. Antibiot.*, **22**, 341–343.
Argoudelis, A. D., T. E. Eble, J. A. Fox and D. J. Mason (1969). Studies on the biosynthesis of lincomycin. IV. The origin of the methyl groups. *Biochemistry*, **8**, 3408–3411.
Argoudelis, A. D., J. H. Coats and T. R. Pyke (1972). Lincomycin production. *US Pat.* 3 697 380.
Argoudelis, A. D. and J. H. Coats (1973). Process for producing lincomycin. *US Pat.* 3 726 766.
Argoudelis, A. D. and J. H. Coats (1974). Process for preparing lincomycin. *US Pat.* 3 812 014.
Bergy, E., R. R. Herr and D. J. Mason (1963). Antibiotic lincolnensin and method of production. *US Pat.* 3 086 912.
Bergy, M. E., R. R. Herr and D. J. Mason (1965). Animal feed containing lincomycin and spectinomycin. *US Pat.* 3 261 687.
Bergy, M. E., J. H. Coats and V. S. Malik (1981). Process for preparing lincomycin. *US Pat.* 4 271 266.
Birkenmeyer, R. D. and F. Kagan (1970a). Lincomycin XI. Synthesis and structure of clindamycin. A potent antibacterial agent. *J. Med. Chem.*, **13**, 616–619.
Birkenmeyer, R. D. and F. Kagan (1970b). 7-Halo-7-deoxylincomycins and process for preparing the same. *US Pat.* 3 496 163.
Campbell, J. M., F. Reusser and C. T. Caskey (1979). Specificity of lincomycin action on peptidyl transferase activity. *Biochem. Biophys. Res. Commun.*, **90**, 1032–1037.
Cha, D. Y. and H. K. Jahnke (1970). Process for isolating antibiotics. *US Pat.* 3 515 717.
Chang, F. N., C. J. Sih and B. Weisblum (1966). Lincomycin, an inhibitor of aminoacyl SRNA binding to ribosomes. *Proc. Natl. Acad. Sci. USA*, **55**, 431–438.
Chang, F. N. and B. Weisblum (1967). Lincomycin. In *Antibiotics*, ed. D. Gottlieb and P. D. Shaw, vol. 1, pp. 440–445. Springer-Verlag, New York.
Gavrilina, G. V., N. P. Nechaeva and R. S. Ukholina (1970). Studies on lincomycin production by strain UV-914. *Antibiotiki*, **15**, 595–599.
Grady, J. E. (1968). Recent developments in lincomycin research. *Int. J. Clin. Pharmacol.*, **16**, 533–538.
Griffith, L. J., W. E. Ostrander, C. G. Mullins and D. E. Beswick (1965). Drug antagonism between lincomycin and erythromycin. *Science*, **147**, 746–747.
Hanka, L. J., D. J. Mason, M. R. Burch and R. W. Treick (1963). Lincomycin, a new antibiotic. III. Microbiological assay. In *Antimicrobial Agents and Chemotherapy — 1962*, ed. J. C. Sylvester, pp. 565–569. American Society for Microbiology, Ann Arbor, Michigan.
Hoeksema, H., B. Bannister, R. D. Birkenmeyer, F. Kagan, B. J. Magerlein, F. A. MacKellar, W. Schroeder, G. Slomp and R. R. Herr (1964). Chemical studies on lincomycin. I. The structure of lincomycin. *J. Am. Chem. Soc.*, **86**, 4223–4224.
Houtman, R. L., D. G. Kaiser and A. J. Taraszka (1968). Gas–liquid chromatographic determination of lincomycin. *J. Pharm. Sci.*, **57**, 693–695.
Hurley, L. H., W. L. Laswell, J. M. Ostrander and R. Parry (1979). Pyrrolo[1,4]benzodiazepine antibiotics. Biosynthetic conversion of tyrosine to the C_2- and C_3-proline moieties of anthramycin, tomaymycin and sibiromycin. *Biochemistry*, **18**, 4230–4237.
Jahnke, H. K. (1967). Process for separating nitrogen-basic materials from aqueous solution. *US Pat.* 3 318 867.
Jariwala, S. L. (1978). Process for recovering lincomycin from fermentation beer. *US Pat.* 4 091 204.
Josten, J. J. and P. Allen (1964). The mode of action of lincomycin. *Biochem. Biophys. Res. Commun.*, **14**, 241–244.
Kuznetsov, V. D. and O. A. Bushueva (1973). Effect of lyophilization on survival of spores of actinomycetes producing streptomycin and lincomycin. *Antibiotiki*, **19**, 311–315.
Kuznetsov, V. D., O. A. Bushueva and L. S. Bryzgalova (1974). Variation of *Actinomyces reseolus*. A lincomycin-producing organism. *Antibiotiki*, **19**, 690–693.
Kuznetsov, V. D. and Bushueva, O. A. (1974). Effect of temperature of storage of lyophilized spores of streptomycin and lincomycin producing actinomycetes on their viability and variation. *Antibiotiki*, **19**, 690–693.
Kuznetsov, V. D., O. A. Bushueva and E. K. Ruuge (1975). Effect of lyophilization method on formation of free radicals, survival and population composition of lincomycin-producing organisms. *Antibiotiki*, **20**, 1065–1068.
Lapchinskaya, O. A., G. A. Trenina, T. P. Korabkova, N. P. Nechaeva and R. S. Ukholina (1968). Selection of lincomycin-producing strains and comparative studies on efficacy of some mutagens. *Antibiotiki*, **13**, 309–316.
Lapchinskaya, O. A. and V. V. Pogozheva (1971). Comparative efficiency of X-rays with different dose capacity in selection of lincomycin-producing organisms. *Antibiotiki*, **16**, 534–538.
Lapchinskaya, O. A. and V. V. Pogozheva (1977). Role of mutagenic factors in selection of *Actinomyces roseolus* producing lincomycin. *Antibiotiki*, **22**, 1100–1103.
Lewis, C., H. W. Clapp and J. E. Grady (1963). *In vitro* and *in vivo* evaluation of lincomycin, a new antibiotic. In *Antimicrobial Agents and Chemotherapy — 1962*, ed. J. C. Sylvester, pp. 570–582. American Society for Microbiology, Ann Arbor, Michigan.
Lewis, C., K. F. Stern and J. E. Grady (1965). Comparison in laboratory animals of the antibacterial activity and adsorption of lincomycin hexadecylsulfamate to lincomycin hydrochloride. In *Antimicrobial Agents and Chemotherapy — 1964*, ed. J. C. Sylvester, pp. 13–17. American Society for Microbiology, Ann Arbor, Michigan.
Magerlein, B. J. (1971). Modification of lincomycin. In *Advances in Applied Microbiology*, ed. D. Perlman, vol. 14, pp. 185–227. Academic, New York.

Magerlein, B. J., R. D. Birkenmeyer, R. R. Herr and F. Kagan (1967). Lincomycin. V. Amino acid fragment. *J. Am. Chem. Soc.*, **89**, 2459–2464.

Manis, J. J. (1981). Plasmid and process of isolating. *US Pat.* 4 273 875.

Mason, D. J., A. Dietz and C. DeBoer (1963). Lincomycin, a new antibiotic. III. Discovery and biological properties. In *Antimicrobial Agents and Chemotherapy — 1962*, ed. J. C. Sylvester, pp. 554–559. American Society for Microbiology, Ann Arbor, Michigan.

Mason, D. J. and C. Lewis (1965). Biological activity of the lincomycin related antibiotics. In *Antimicrobial Agents and Chemotherapy — 1964*, ed. J. C. Sylvester, pp. 7–12. American Society for Microbiology, Ann Arbor, Michigan.

Michalik, J., W. E. Czerska, L. Switalski and K. R. Bojanowska (1975). Monophenol monooxygenase and lincomycin biosynthesis in *S. lincolnensis. Antimicrob. Agents Chemother.*, **8**, 526–531.

Modde, H. (1971). Die bakteriostatische and bakterizide Wirkung des 7-chlor-7-desoxy-lincomycins gegen Staphlakokken, Streptokokken and Pneumokokken. *Schweiz. Med. Wochenschr.*, **101**, 1629–1631.

Morozowich, W. and D. J. Lamb (1969). Lincomycin-2-phosphate, 7-substituted compounds and salts thereof. *US Pat.* 3 487 068.

Morozowich, W. and A. A. Sinkula (1971). 7-Halo-7-deoxy-lincomycin derivatives. *US Pat.* 3 580 904.

Neff, A. W., A. R. Barbiers and J. I. Northam (1967). Microbiological methods for assaying lincomycin in animal feed. *J. Assoc. Off. Anal. Chem.* **50**, 442–446.

Prescott, G. C. (1966). Automated assay for the antibiotic lincomycin. *J. Pharm. Sci.*, **55**, 423–425.

Reusser, F. and A. D. Argoudelis (1974). Process for preparing lincomycin. *US Pat.* 3 833 475.

Rolls, J. P., B. D. Ruff, W. J. Haak and E. J. Hessler (1977). Biosynthesis of the amino sugar moiety of lincomycin. In Abstracts of the 77th Annual Meeting of the American Society for Microbiology, New Orleans, LA. p. 250.

Schroeder, W., B. Bannister and H. Hoeksema (1967). Lincomycin. III. The structure and stereochemistry of the carbohydrate moiety. *J. Am. Chem. Soc.*, **89**, 2448–2453.

Sinkula, A. A., W. Morozowich, C. Lewis and F. A. MacKellar (1969). Synthesis and bioactivity of lincomycin-7-monoesters. *J. Pharm. Sci.*, **58**, 1389–1392.

Slomp, G. and F. A. MacKellar (1967). Lincomycin. IV. Nuclear magnetic resonance studies on the structure of lincomycin, its degradation products, and some analogs. *J. Am. Chem. Soc.*, **89**, 2454–2459.

Witz, D. F., E. J. Hessler and T. L. Miller (1971). Bioconversion of tyrosine into the propylhygric acid moiety of lincomycin. *Biochemistry*, **10**, 1128–1133.

11

Pharmacologically Active and Related Marine Microbial Products

J. T. BAKER, J. L. REICHELT and D. C. SUTTON
James Cook University, Queensland, Australia

11.1 INTRODUCTION

Marine microorganisms have been little studied for their application to human needs. Early research into marine microorganisms was concerned largely with their physiology, photosynthetic capacities and potential therapeutic, antibiotic and toxicological properties. This area of research remains significant, particularly in view of the development of techniques enabling culture of microalgae and bacteria on scales which had previously been seen as impracticable.

Recent studies have demonstrated that, in addition to the useful products which can be obtained from marine microorganisms, direct use of cell biomass has potential in applications ranging from pure research to agriculture, and to wastewater treatment.

11.2 PHARMACOLOGICALLY ACTIVE COMPOUNDS FROM MARINE MICROORGANISMS

Most of the useful medical substances derived from microbial cultures have been antibiotics from three groups of microorganisms generally found in soils, namely the filamentous fungi, the spore-forming bacilli and the actinomycetes. However, the most common microorganisms in the marine environment are the Gram-negative eubacteria, the cyanobacteria and the myxobacteria, groups not generally thought to produce many medically useful substances (see Table 1).

The common groups of microorganisms in the ocean were investigated for the production of medically useful substances at the Roche Research Institute of Marine Pharmacology from 1976 to 1981. The significant results of these and related investigations are described in this section and the range of animals and the biological activity of compounds from them are shown in Table 2.

Marine Gram-negative eubacteria and microalgae were found to produce a high frequency of potentially useful pharmacological activities, but the marine cultures tested were not found to produce many significant antibiotic activities. *In vitro* antibiotic activity was observed but, in nearly all cases, no *in vivo* activity. A possible exception was the ability of extracts of the sponge *Dysidea herbacea* to cure mice of experimental, local infections of *Staphylococcus aureus*. This antibiotic activity was only observed in samples of the sponge which contained large concentrations of a cyanobacterial symbiont. Conventional testing methods for microbial cultures would

Table 1 Distribution and Production of Antibiotics of the Major
Microbial Groups

Major microbial group	Presently known frequency of antibiotics	Presence in the ocean
EUKARYOTIC		
Fungi		
filamentous	+++	+
yeasts	+	++
aquatic	−	++++
Microalgae	−	++++
Protozoa	−	+++
Slime moulds	+	+
PROKARYOTIC		
G +ve eubacteria		
actinomycetes	++++	+
bacilli	+++	+
others	+	+
G −ve eubacteria		
pseudomonads	+	++++
enterobacteria	+	++++
others	+	++++
Cyanobacteria	−	++++
Myxobacteria	+	+++
Spirochaetes	−	+

Table 2 Pharmacologically Important Effects of Extracts of Marine Microorganisms

	Type of extract	Biological activity
EUKARYOTIC		
Aquatic fungi	Cells	Nicotinic blocker
		Polysynaptic blocker
Diatoms		
Navicula spp.	Cells	Neuromuscular blocker
Phaeodactylum spp.	Cells	Nicotinic blocker
Dinoflagellates		
Amphidinium carterae	Cells	$\beta2$ blocker
Green microalgae		
Dunaliella tertiolecta	Culture fluids (a)	Antioedema
	Culture fluids (b)	Bronchodilator, polysynaptic blocker, antiwrithing, hypotensive
PROKARYOTIC		
Gram-Negative bacteria		
Alteromonas rubra		
(log)	Culture fluid (a)	Bronchodilator
	Culture fluid (b)	Inotropic
A. rubra (autolysed)	Culture fluid	Hypotensive
Alteromonas spp.	Cells	Neuromuscular blocker
	Culture fluid (a)	Antiserotonin, bronchodilator
	Culture fluid (b)	Antiamphetamine
A. luteoviolaceus	Culture fluid	Nicotinic blocker
Cyanobacteria		
Rivularia firma	Cells	Antiinflammatory, analgesic, antiallergic
Lyngbya lutea	Cells	Antiamphetamine

not have detected the pharmacological activities observed in this program because of the very low yields of biologically active compounds produced by marine microorganisms. The method which was successful in detecting the pharmacologically active compounds was to generate concentrated extracts from relatively large culture volumes (20–200 l) for all initial bioassays. It was also found to be important to test these initial extracts using *in vivo* bioassays for both antibiotic and pharmacological activities.

With few exceptions, the investigation of marine microorganisms by Roche was not continued long enough to determine the chemical structures of the biologically active compounds. One series of compounds which was investigated was the group of new aromatic acids responsible for the bronchodilator activity of extracts of the marine bacterium *Alteromonas rubra*. The proposed structures for the active compounds, called rubrenoic acids A, B and C, are shown in Figure 1.

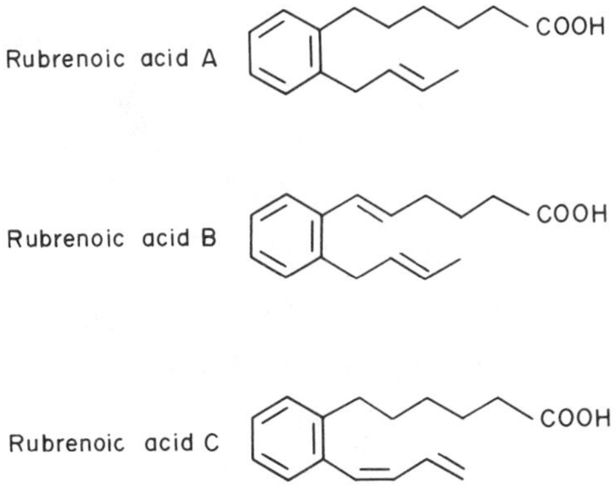

Rubrenoic acid A

Rubrenoic acid B

Rubrenoic acid C

Figure 1

Their general superficial similarity to prostaglandin-type structures attracted some interest, and syntheses were achieved by Hoffman-La Roche in New Jersey.

In an assay for bronchodilator activity using the isolated, intact, guinea-pig trachea the rubrenoic acids A, B and C each had a potency similar to the standard drug, theophylline, namely an ED_{50} of $1.5-7.5 \times 10^{-5}$ M. Preliminary *in vivo* tests of rubrenoic acids A and C were carried out using a modification of the Konzett–Rossler technique developed in the Roche Research Institute by Jamieson and Taylor (1979), with disappointing results. At doses up to 20 mg kg^{-1} *i.v.*, there was no bronchodilator effect as evidenced by abolition of histamine-induced bronchoconstriction, whereas at doses above 10 mg kg^{-1} *i.v.*, blood pressure and heart rate were reduced. However, the rubrenoic acids are a novel group of microbial metabolites and a full exploration of all three rubrenoic acids and selected analogues may yield compounds with a medically-useful pharmacological profile.

It is important to note that the active metabolite purified from the marine bacterium *Alteromonas rubra* was not one of the many known antibiotics produced by the soil microorganisms. The probability of discovering a range of new medically useful compounds from terrestrial microorganisms is comparatively low. This group of microorganisms has been studied extensively, as evidenced by the large number of previously described antibiotics from them.

In the case where the symbiosis between the sponge *Dysidea herbacea* and cyanobacteria was found to produce an antibacterial metabolite, hexabromodihydroxybiphenyl ether (Figure 2) was found to be the active compound. This is also a novel compound, not produced by any soil microorganisms studied to date. The purified compound was found to have activity against localized infections of mice caused by both Gram-positive bacteria such as *Staphylococcus aureus* and Gram-negative bacteria such as *Escherichia coli* at 30 to 100 μg ml^{-1}. In the same *in vivo* tests, hexachlorophene was only one third as active against *Staphylococcus aureus* and inactive against *Escherichia coli*. No systemic antimicrobial action was observed. Two very surprising properties of this novel compound were: (a) its relatively broad spectrum, since activity against Gram-negative bacteria is generally not observed for compounds with antimicrobial activity based on the phenyl moiety; and (b) its lack of observable toxicity when injected intravenously into mice, since related compounds have been found to have toxicity attributable to the phenyl moiety. These properties suggested that the hexabromodihydroxybiphenyl ether be considered as a skin cleanser to replace hexachlorophene.

The structure of the hexabromodihydroxybiphenyl ether was elucidated by Norton and Wells (1977), and although the substance showed the promising activity profile as above, Roche was reluctant to proceed with development because of the possible by-production, in synthesis or in use, of one or other of the dioxins shown in Figure 3, and the possible relation of certain by-products to the dioxin TCDD, accidentally generated and released at Seveso in Italy.

The problem of by-products has long been recognized in all branches of the production of biologically active substances for commercial utilization, and this negative aspect of the by-product

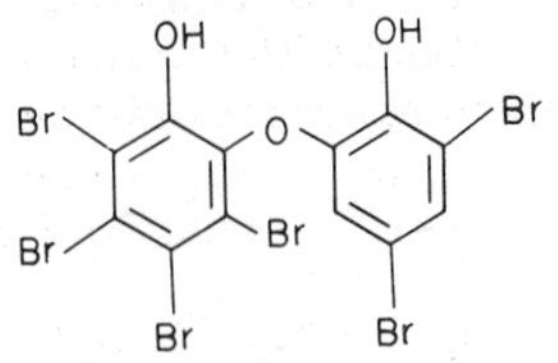

Figure 2

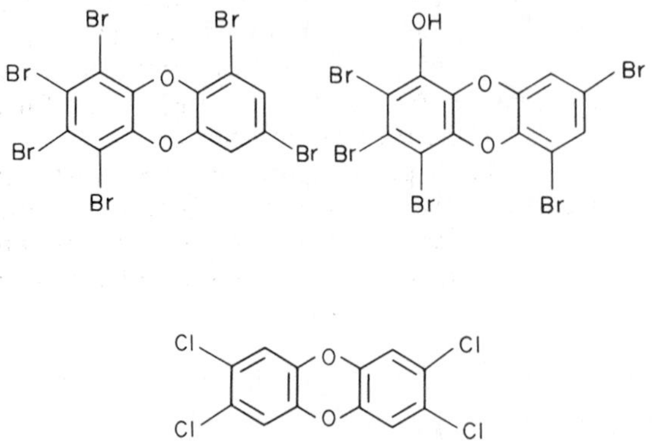

Figure 3

compared with the interesting activities of the naturally occurring hexabromodihydroxybiphenyl ether should not be over-accentuated. In fact, not all dioxins are toxic or teratogenic and the brominated dioxins above have not been fully screened for their biological activity (Figure 3).

11.3   PRODUCTS FROM THE CULTURE OF MICROALGAE IN COASTAL PONDS

Marine microbial technology is also being developed for the production of fine chemicals, vitamins and animal feeds, by the use of large coastal ponds to culture certain types of microalgae. Seawater is modified by evaporation and/or addition of inorganic nutrients to allow the growth of the desired species of microalgae in very large quantities.

Until the technology of algal pond culture is further developed, the cost of producing a tonne of algal biomass may be too high to allow production of cheap fuel or cheap food material. Initial priority is thus being given to higher value products. For example, the green microalga, *Dunaliella salina*, produces high concentrations of β-carotene. Because of the market price of chemically synthesized β-carotene, the goal of achieving a cheaper source by algal culture is more feasible than producing cheaper food. Algal pond culture is being successfully used in the specialized market of health foods. The cyanobacterium, *Spirulina*, is cultured in Mexico, California and Israel and sold as a relatively high-priced health food.

Not all species of algae can be grown in large open ponds because of the tendency for a succession of different types of algae to overgrow the species being cultivated. However, certain algae can be maintained successfully in open ponds because of their ability to grow under conditions which severely inhibit the growth of most other species. The green microalga, *Dunaliella salina*, and the cyanobacterium, *Aphanothece halophytica*, are favoured by high salinity and high temperatures; the cyanobacteria, *Spirulina platensis* and *Spirulina maxima*, are favoured by high alkalinity. For any other alga proposed as a source of algal products, it will be necessary either (a) to demonstrate conditions under which the alga is known to 'bloom' (*i.e.* consistently grow to high concentrations as the dominant species) or (b) to consider the possibility of genetic transfer of the proposed product into one of the algal species known to grow well in open ponds.

Recent research on the genetics of cyanobacteria has made it more feasible to consider the genetic transfer of a proposed product into microalgae. As a model system for genetic studies of microalgae, the cyanobacteria have several advantages. Mutant isolation is facilitated by the relative ease of handling prokaryotic cells. Most cyanobacteria can be grown in defined liquid media under rigorously-controlled conditions with generation times of 5–20 hours depending on the strain. Unicellular and some filamentous species can be grown on agar medium as discrete colonies. Many bacteriological techniques for mutant isolation have been successfully adapted to the cyanobacteria. Phenotypic segregation in prokaryotes is rapid and does not require a sexual life-cycle as in diploid organisms.

Of the known gene transfer mechanisms in bacteria, only transformation has been unequivocally demonstrated in cyanobacteria. Transformation is the term used to describe genetic recombination brought about by the uptake of DNA from the culture medium. Transformation with chemically-extracted DNA has been reported for a number of unicellular cyanobacterial strains. Recombinant DNA techniques have revolutionized the study of genetics by allowing the isolation, amplification and purification of specific genes from any organism. The use of these techniques with the cyanobacteria has gained momentum over the last five years in both the development of plasmid vectors for cloning of genes in cyanobacteria and the cloning of cyanobacterial genes in the bacterium *E. coli*. A number of shuttle vectors for cyanobacteria now exist which replicate stably in both *E. coli* and the cyanobacterial host.

These developments mean that algal pond culture methods may now be considered for a wider range of products. To exploit this possibility, the genetic research on the cyanobacteria will have to be focused on the *Spirulina* and *Alphanothece* strains which grow well in open ponds. Extension of the latest genetic techniques to the eukaryotic microalgae, particulary *Dunaliella* species, is also a priority.

11.4 AGRICULTURAL APPLICATIONS

One further area under study in marine microbial technology is in agriculture, where plant nutrition and soil improvement may be enhanced. This application stems firstly from the ability of microorganisms to absorb and mobilize nutrients which might otherwise be lost from the soil, and secondly through the effect of microbial biomass on improving soil structure.

Of particular significance is the ability of certain microorganisms, including many marine cyanobacteria, to fix nitrogen. Nitrogen is a major limiting nutrient to the growth of many crop species. Field trials have demonstrated that the addition of nitrogen-fixing cyanobacteria to rice fields results in a progressive increase in crop yields, reaching levels comparable to applications of nitrogenous fertilizers of more than 70 kg ha^{-1}. The potential for future use of microalgae as biological fertilizers is easily appreciated in light of the knowledge that most agricultural crops require the addition of nitrogen in order to achieve economic yields. The other potential use of algal biomass in soil improvement is through their effect on soil structure; the addition of a living biomass would be expected to increase water retention and improve aeration and availability of nutrient ions through chelation in addition to reducing free salts and erosion.

11.5 CONCLUSIONS

There is little doubt that the scope for utilization of marine microorganisms in technological applications is yet to be fully appreciated. Marine microorganisms lend themselves readily to large scale culture conditions. The medium (sterilized salt water) and the temperature range for growth are two factors which are important in suggesting that cost-effective processes can be developed.

Indications have been given of applications in the fields of pharmaceuticals, fine chemicals, biomass, aquaculture and agriculture, as well as in the purification of wastewaters.

It could well be that in a world where the necessity for recycling water for re-use by industry, agriculture or for human consumption is increasing, the potential of microorganism technology (particularly with those capable of tolerating and utilizing saline conditions) in water purification should be given the highest priority.

11.6 REFERENCES

Jamieson, D. and K. M. Taylor (1979). Roche Research Institute of Marine Pharmacology, Internal Research Report.
Norton, R. N. and R. J. Wells (1977). Roche Research Institute of Marine Pharmacology, Internal Research Report.

Steward, W. D. P. (ed.) (1975). *Nitrogen Fixation by Free-living Microorganisms*. Cambridge University Press, Cambridge.

Venkataramon, G. S. (1975). The role of blue-green algae in tropical rice cultivation. In *Nitrogen Fixation by Free-living Microorganisms*. ed. W. D. P. Steward, pp. 207–218. Cambridge University Press, Cambridge.

Wells, R. J. (1980). Roche Research Institute of Marine Pharmacology, Internal Research Report.

12

Anticancer Agents

M. C. FLICKINGER
National Cancer Institute, Frederick, MD, USA

12.1 INTRODUCTION

Few biochemical features unique to tumor cells have so far been selectively exploited for treatment. Cancer chemotherapy has primarily been directed at discovery of cytotoxic agents capable of inhibiting many aspects of mammalian cell division (Creasey, 1981). This approach to development of human clinical anticancer agents has resulted in investigation of a tremendous variety of naturally occurring compounds produced by microorganisms, plants and more recently by mammalian cells in culture.

Process development for production of this variety of anticancer agents may require more diversity of biotechnology than any other single field because of the many bioengineering problems associated with these potent agents (Table 1). The demand for this process technology is increasing as more antitumor antibiotics are scaled up for commercial production and as new treatment approaches using biologicals such as interferons and monoclonal antibodies begin to be evaluated for human therapy.

Table 1 Biotechnological Problems in Production of Anticancer Agents

(1)	General toxicity to the biochemistry of cell division often limits the final concentration of drugs produced by microbial or cell culture systems to very low levels
(2)	Recovery of agents from dilute medium in the presence of large quantities of contaminating proteins and subsequent purification for use as a human pharmaceutical
(3)	Containment of potent chemical and biological activity of the agent throughout processing and purification to protect workers and the environment
(4)	Reduction in the risks associated with production and purification of experimental biologicals using large-scale transformed mammalian cells in culture
(5)	Decontamination of process wastes for inactivation of biological and chemical activity

The majority of anticancer agents produced by microorganisms or by transformed mammalian cells in culture are in the early stages of drug evaluation. Few reports of biochemical engineering data, process scale-up or strain development exist other than in the medicinal chemistry and patent literature. The extensive information documenting the discovery of the naturally occurring anticancer agents is well summarized in recent reviews and monographs (Cassady and Douros, 1980; Glasby, 1976; Aszalos, 1981; Douros and Suffness, 1981a, 1981b; Aszalos and Berdy, 1978; Oki and Yoshimoto, 1979; Marquenz, 1982; Korzybski *et al.*, 1978). The appendix to this chapter (Tables A1–A12) briefly summarizes many of the major classes of anticancer agents and their microbial sources, with references to process patents and isolation procedures.

Anticancer drug process development and production processes are different from those for other antibiotics and biologicals in that they are primarily small scale, *i.e.* designed to produce small quantities of drugs for investigational use or clinical application for specific types of tumors. Anticancer antibiotics marketed for general medical practice are produced on a significantly reduced scale compared with antibacterial or antifungal antibiotics.

Some of this technology, however, does not differ significantly from that associated with the production and purification of antibacterial and antifungal drugs. Thus biotechnology associated with laboratory-scale production of many types of antibiotics with antitumor activity, large-scale production of hormones, alkylating agents, semisynthetic antimetabolites and extraction of plant materials important in cancer treatment is not included in this chapter. Process examples of cytotoxic drug production by actinomycetes, therapeutic enzyme production and biologicals production by mammalian cell culture are reviewed in this chapter in order to illustrate process scale-up, purification, containment and decontamination technologies unique to the anticancer field.

12.1.1 The Drug Development Process

The drug development scheme used by the National Cancer Institute (Figure 1) begins with a drug discovery program (prescreen and screening phase). Since ascites tumors are useful in identifying potentially active agents, a prescreening step using the P388 mouse leukemia *in vitro* has been evaluated along with other tumor clonogenic, microbial phage induction or enzyme inhibition screens (Douros and Suffness, 1981a; White, 1982a, 1982b; Garretson, 1981; Von Hoff, 1979). In the next *in vivo* evaluation step, the potentially active agents are tested against a number of model murine tumors such as mouse colon, breast, lung, B16 melanoma and L1210 leukemia and occasionally xenografts of human colon, breast and lung cancers in immunodeficient nude mice (Vendetti, 1982). If the compounds are active in the animal tumor screens, they are then formulated for intravenous or oral use, tested for toxicity in large animals (dogs, monkeys), and, if toxicities are reasonable, brought to Phase I clinical trials (toxicology and activity). This initial small-scale clinical trial is designed to find the maximally tolerated human dose. Often the Phase I trials are concurrent with Phase II trials (potential usefulness and dosage) designed to evaluate the maximally tolerated dose of a drug against a panel of several common solid tumors. Phase III and IV trials (safety and efficacy) explore the activity of a new drug against established agents. Finally, if the drug is found to be superior to other agents, it is brought into general medical practice. This overall development scheme may be as long as seven to twelve years, during which a process for large-scale production of the desired agent is developed. Extensive genetics, selection of a cell line or strain that is an efficient producer of the drug, or engineering optimization of fermentation, scale-up, recovery and purification often cannot be justified until after the new compound's potential clinical effectiveness has been determined (Phase I and II clinical trials). In order for a compound to progress through the stages of evaluation, however, increasing quantities of purified material must be available. This can require several grams (for tumor panel evaluation) to kilogram quantities of clinical grade material for analogue development or Phase I and II trials (Douros and Flickinger, 1983). The estimated development cost for an anticancer antibiotic is $7–15 million, excluding unsuccessful drugs.

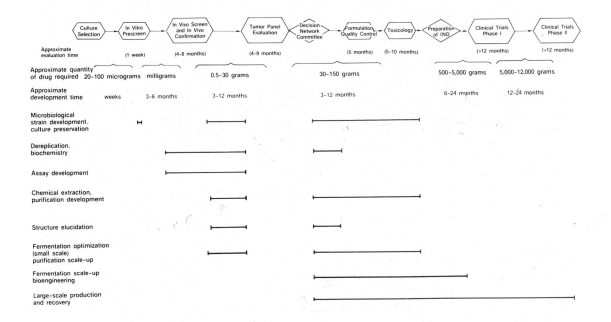

Figure 1 Natural product preclinical drug development (Douros and Flickinger, 1983)

12.2 MARKET INFORMATION

The commercial market for all anticancer agents has grown dramatically in the last five to seven years. Many new drugs discovered in the early 1970s are now completing clinical trials and will soon enter medical practice. The current annual world market for antineoplastic agents is estimated to be in excess of $1–5 billion (U.S.) with projections to increase to $27 billion by the end of the century. The rate of growth of sales of anticancer drugs in individual countries can best be seen in Japan. In 1970, the total sales of anticancer agents was Y500 million, 0.1% of total pharmaceutical sales. This figure has risen steadily: Y10 billion in 1975, Y60 billion in 1977, Y107 billion in 1980 and Y122 billion in 1981. Total sales of anticancer agents in Japan now comprise greater than 3.1% of all pharmaceuticals, an increase of over 30-fold in 10 years.

The total sales of anticancer agents in the U.S. in 1981 was $200 million, with 30–45% of this market resulting from sales of fermentation-derived drugs (Table 2). The largest selling antitumor antibiotic continues to be adriamycin ($40–60 million 1981, ≈ $32 million 1982). This single anthracycline antibiotic has led U.S. sales of all anticancer agents for the past several years but may soon be eclipsed by synthetic platinum compounds.

Anticancer biologicals such as interferon and monoclonal antibodies for treatment of breast, lung, colon and prostate cancers and certain types of leukemia and lymphoma will be the fastest growing area of cancer therapy in the next five years. Monoclonal antibodies for cancer treatment and diagnosis are currently in the early stages of preclinical and pilot clinical evaluation. Many previously scarce human biologicals may become commercially available by 1990 as the result of gene cloning and microbial production processes. The current (1982) U.S. and world markets for all diagnostic uses of monoclonal antibodies are $58 million and $870 million, respectively. This market is expected to expand dramatically during the next 10 years to greater than $8.4 billion

Table 2   Anticancer Fermentation Products in Commercial Production (1982)

Name	Trade name	Marketing firm	Manufacturer
Adriamycin (Doxorubicin)	Adriamycin	Adria Labs (Montedison)	Farmitalia Carlo Erba SpA, Italy
Daunorubicin HCl	Cerubidine	Ives Labs	Ives Laboratories, New York, NY Rhone-Poulenc, France
Dactinomycin (Actinomycin D)	Cosmegen	Merck	Merck Sharp and Dohme, West Point, PA Reanal, Budapest, Hungary
Mitomycin C	Mutamycin	Bristol	Bristol Laboratories, Syracuse, NY
	Mitomycin C-Kyowa	Kyowa Hakko	Kyowa Hakko Kogyo Co. Ltd, Japan
Mithramycin	Mithracin	Dome Division Miles Laboratories West Haven, CT	Pfizer Laboratories, New York, NY
Bleomycin sulfate	Blenoxane	Bristol	Bristol Laboratories, Syracuse, NY
Peplomycin (PEP-bleomycin)	Bleo	Nippon Kayaku	Nippon Kayaku, Japan
	Pepleo Injection	Nippon Kayaku	Nippon Kayaku, Japan
Chromomycin A$_3$	Toyomycin	Takeda	Takeda Chemical Industries Ltd, Japan
L-Asparaginase (*E. coli* EC2)	Elspar	Merck	Merck Sharp and Dohme, West Point, PA
	Leunase	Kyowa Hakko	Kyowa Hakko Kogyo Co. Ltd, Japan
	Crasnitin		Farbenfabriken Bayer AG, Wuppertal, Federal Republic of Germany
Aclarubicin	Aclacinon	Roger Ballon (Rhone-Poulenc)	Sanraku Ocean, Japan
	Aclacinomycine injectable	Yamanouchi	Yamanouchi Pharmaceutical Co., Japan
Zorubicin	Rubidazome injectable	Specia	Rhone-Poulenc, France
Bestatin (1984)		Nippon Kayaku	Nippon Kayaku, Japan
Streptozotocin		Upjohn	Upjohn Co., Kalamazoo, MI
Siomycin		Shionogi	Shionogi and Co. Ltd, Osaka, Japan

per year (U.S. market). Anticancer monoclonal antibodies will share a significant percentage of this market.

12.3   CONTAINMENT TECHNOLOGY FOR CYTOTOXIC AGENTS

Containment technology for large-scale production and purification of microbially produced cytotoxic agents and biologicals from transformed human cells in suspension culture has evolved from the National Cancer Institute guidelines for design of biomedical research facilities for investigation of oncogenic viruses (see the following: U.S. Department of Health, Education and Welfare, Public Health Service, National Institutes of Health publications: *National Cancer Institute Safety Standards for Research Involving Oncogenic Viruses*, DHEW No. NIH 75-790 [1974]; *National Institutes of Health Biohazards Safety Guide*, NIH 1740-00383 [1974]; *Laboratory Safety at the Center for Disease Control*, CDC 75-8118 [1974]; *Design Criteria for Viral Oncology Research Facilities*, DHEW 78-891 [1978]; *Design of Biomedical Research Facilities*, NIH 81-2305 [1979]; *Management of Hazardous Chemical Wastes in Research Institutions*, DHEW 82-2459 [1981]); and the technology developed by the U.S. Army Chemical Corps (Dick and Hanel, 1970).

The basic concepts for design of an effective biological barrier to contain spills and large-scale aerosols are summarized in Table 3 (Flickinger and Sansone, 1984). Aerosols can be generated from any liquids involved in fermentor inoculation procedures, sampling, vessel overpressure situations and drug recovery operations. Accidental spills often occur due to leaks in fermentation equipment (seals, head plates, quick-disconnect fittings, probe penetrations); during transfer of process liquids; and in recovery operations such as filtration, centrifugation and crystallization.

Table 3 Design Concepts for Containment of Cytotoxic Agents on the Pilot and Production Scale (Flickinger and Sansone, 1984)

- Controlled-access facility
- Negative air pressure at work site (class I, II, III biological cabinets; containment suites)
- Air exhausted to the environment only through HEPA filters
- Additional containment of aerosol-generating process equipment
- Appropriate personnel protection
- Biological and chemical decontamination of process wastes
- Medical monitoring of 'at risk' personnel
- Environmental monitoring

12.3.1 Containment of Process Equipment

Pilot- and production-scale fermentors can be isolated from the environment by locating them inside a biological barrier. The fermentor serves as the primary biological barrier, with the controlled-access, HEPA-filter-exhausted building containing the fermentors being the secondary biological barrier. For production-scale quantities of cytotoxic antitumor compounds, fermentation equipment is often modified by the addition of parallel exhaust filters, overpressure sensors and automatic sparge shut-off systems, with the rupture disc piped into a contained waste treatment system (see Section 12.3.3).

Production of large quantities of nonhuman primate or human transformed cells in suspension culture for isolation of cell paste or products produced by these cells presents contamination problems different from those associated with containment of cytotoxic antibiotics. However, the same design concepts are applied in that the cell culture incubators and fermentors (the primary biological barrier) are located within an HEPA-filter-exhausted secondary barrier equipped with air locks, change room, showers and double-door autoclaves, and connected to a sterilizing sewer or kill tank system (Runkle and Phillips, 1969).

Process recovery equipment for extraction, concentration and purification of large quantities of antitumor drugs must be protected from explosion hazards as well as biologically contained to prevent spills and generation of toxic aerosols. For these reasons, process equipment that generates significant quantities of aerosolized microorganisms or drug is further isolated within the secondary barrier structure. Continuous centrifuges can be isolated in a contained HEPA-filter-exhausted suite within the fermentation plant so that fermentation broth or crystal slurries can be processed. Large-scale filtration operations involving plate and frame press equipment can be completely enclosed in a hinged-lid box (Figure 2). This configuration can contain and direct filtered culture medium to various recovery vessels as well as divert contaminated press cake for further chemical decontamination or incineration. Transferable surface contamination by cytotoxic agents can result from hose couplings and leaks at couplings during recovery operations. To reduce this exposure, a completely enclosed fixed-transfer system can be installed along with local exhaust ventilation to control particulates and vapors (Flickinger and Sansone, 1984).

Recovery of transformed cells or biologicals from supernatants by centrifugation as well as concentration of cytotoxic proteins by large-scale ultrafiltration requires aerosol containment within isolated HEPA-filter-exhausted cubicles inside the secondary barrier (Hellman *et al.*, 1973; DHEW publication [NIH] 78-373 *Centrifuge Biohazards*, 1973).

12.3.2 Personnel Protection

Direct operator exposure to any liquids or aerosols containing cytotoxic, potentially mutagenic or oncogenic agents is avoided by operator safety training and the appropriate personnel protective devices.

Fermentor operators involved in sampling vessels should be protected by use of appropriate gloves, respirators and disposable lab coats. Recovery plant operators are further protected by the use of nitrile gloves (Sansone and Tewari, 1978), Tyvek coveralls, boots, face shield and full-face respirators when working on purification of cytotoxic drugs (Figure 3) (Flickinger and Sansone, 1984). Ventilated hoods are used for procedures such as centrifugation operations, recovery of spilled material, decontamination of process equipment and final packaging of purified

Figure 2

drug (Figure 2). Final lyophilization and packaging of potent cytotoxic agents are carried out in isolated HEPA-filter-exhausted suites, with fully suited workers manipulating the nonsterile bulk drug within HEPA-filter-exhausted Class I cabinets. Most process equipment capable of producing aerosols of cytotoxic materials may also be contained and modified for remote operation (Runkle and Phillips, 1969; Hellman *et al.*, 1973; Flickinger, 1979). However, when operators are required to enter areas where aerosols may be produced, they can be appropriately protected with disposable suits, respirators, shoe and head covers, or complete inflatable suits (Figure 4).

Personnel protective equipment, such as gloves, coveralls, lab coats, shoe covers and hair covers, that becomes contaminated with or is assumed to be contaminated with cytotoxic or mutagenic agents is disposed of by incineration.

12.3.3 Decontamination of Waste Streams

Spills can be contained and decontaminated with a kill tank or waste treatment system that is capable of sterilizing all process wastes before they proceed to sanitary sewage treatment. To chemically inactivate cytotoxic agents, both heat and extreme pH or hypochlorite are often required. An anhydrous ammonia addition system in conjunction with steam injection can be used to decontaminate (both biologically and chemically) aqueous process wastes (Castegnaro *et al.*, 1980).

Large quantities of organic solvents contaminated with a cytotoxic drug can be recycled by distillation, and the contaminated liquid residue can be incinerated at 850 °C (Shen *et al.*, 1978). Hydrocarbons containing cytotoxic residues are incinerated using a liquid injection incinerator with water and sodium hydroxide scrubbers (Berky *et al.*, 1981; Carnes and Whitmore, 1981). Aqueous process streams containing cytotoxic agents contaminated with organic solvents (*e.g.* from column washing) must often be decontaminated with heat and pH treatments. A modification of the biological kill tank or blow-case decontamination system can be used to process, hold and validate the decontamination of biological and chemical activity. Batch and continuous designs have been used (Figure 5) (Dick and Hanel, 1970). The chemically and biologically contaminated process wastes are injected with steam into the blow-case at high pH, held at sterilization temperature, sampled for validation of decontamination, and then dumped into a sanitary sewer system, where an additional sterilization may occur. Chemical and biological decontamina-

Figure 3

tion flow sheets for fermentation and recovery equipment are summarized in Figure 6 (Flickinger and Sansone, 1984). Equipment is held at each decontamination step until assay results are received and validation of decontamination is certified by analysis.

12.4 MICROBIAL PROCESS EXAMPLES

The following microbial process examples demonstrate many of the bioengineering techniques used for large-scale production of cytotoxic materials for human clinical trials. Containment of process equipment and waste decontamination technologies described in the previous section are often incorporated at the early stages of process design and continually modified and improved based on environmental monitoring (Flickinger and Sansone, 1984).

12.4.1 Fermentation Processes for Production of Anthracyclines

The anthracycline antibiotics are among the most widely used antitumor drugs. They are characterized by an anthraquinone chromophore that is substituted with one or more sugars. Daunorubicin (daunomycin) and adriamycin (doxorubicin) have the amino sugar daunosamine as a substituent (Appendix, Table A4). Daunorubicin, the first clinically effective anthracycline, was isolated in 1963 at Farmitalia Calo Erba SpA in Italy (Grein *et al.*, 1963) and an identical compound called 'rubidomycin' was isolated from *Streptomyces coeruleorubidus* (Dubost *et al.*, 1963). Adriamycin was produced later by a mutant strain of *S. peucetius*, referred to as var. *caesius*, in an attempt to improve daunorubicin (Arcamone *et al.*, 1969). Hundreds of related anthracyclines have since been isolated or prepared semisynthetically in efforts to find an anthracycline antibiotic with a wide spectrum of activity and an improved therapeutic index with reduced human cardiotoxicity (Arcamone, 1981; Oki and Yoshimoto, 1979).

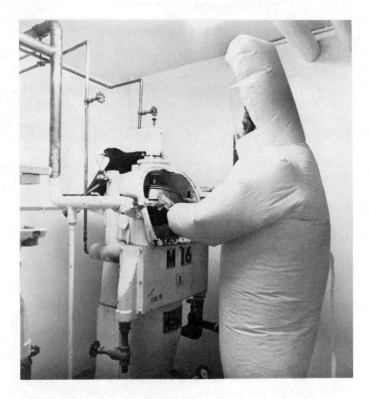

Figure 4

Adriamycin is one of the most used broad-spectrum cytotoxic antibiotics in treatment of a number of solid tumors (breast, lung) as well as acute leukemia (Davis and Davis, 1979). Daunorubicin has primarily been used for the treatment of acute leukemia.

12.4.1.1 Strain improvement

Daunorubicin, produced by *S. peucetius*, occurs in fermentations in the form of glycosides that liberate daunorubicin on simple acid hydrolysis (McGuire *et al.*, 1980). Fermentation titers for daunorubicin were originally reported in the range of 60–70 μg ml^{-1} (DiMarco *et al.*, 1977) and 5–15 μg ml^{-1} for adriamycin (Arcamone *et al.*, 1968). White and Stroshane (1983) have recently summarized the production technology for both daunorubicin and adriamycin with emphasis on the process developed at the National Cancer Institute using an unspeciated streptomycete V8. For this process, cerulenin, a specific inhibitor of fatty acid and polyketide biosynthesis (Omura, 1976), was used to select for pigmented (daunorubicin) producing colonies from UV irradiated spores (McGuire *et al.*, 1980). Low producers in the presence of cerulenin were found to accumulate high concentrations of the aglycone ε-rhodomycinone. A high-producing mutant (C5) produced three- to five-fold higher titers and was scaled up to the 7000 l scale (Hamilton *et al.*, 1981).

12.4.1.2 Batch fermentation processes

Seed culture and production media for two daunorubicin batch fermentation processes are summarized in the Appendix (Table A5) (White, 1983). In one batch process, inoculum preparation begins with inoculation of agar slants from frozen stock cultures (DiMarco *et al.*, 1977). After 10 days of incubation at 26–27 °C, mycelium is inoculated into shake flask cultures at 28 °C for two days. A 170 l seed fermentor is batched with 100 l of seed medium and inoculated with 200 ml (0.17% v/v) of shake flask culture. The seed culture is grown for 27 h (26–27 °C) with agitation and 0.69 vvm aeration. Fifty l of 27-hour seed culture are used to inoculate an 800 l fermentor containing 500 l of production medium. The production stage is agitated and aerated (0.5 vvm) at 28 °C for 67 h.

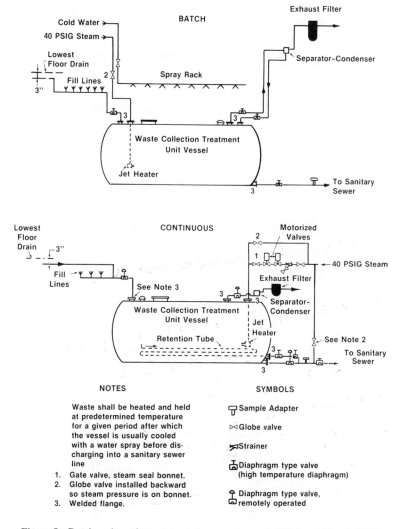

Figure 5 Batch and continuous waste treatment vessels (Dick and Hanel, 1970)

A second batch method (McGuire *et al.*, 1979) (Figure 7) also begins with a frozen stock inoculated into a shake flask culture incubated at 28 °C for 2.5 days. This culture is used to inoculate 75 l of seed medium (1.3% v/v, 30 °C, 0.5 vvm, 125 r.p.m., impeller tip velocity V_t=750 ft min^{-1}, 24 h). This seed serves as inoculum (5% v/v) for a 1000 l seed fermentor (same conditions) for 12 days. The production stage of this process (7000 l) is inoculated with 750 l from the 1000 l scale stage and is operated under the identical agitation, aeration and temperature conditions for 7–10 days (Figure 8) (Hamilton *et al.*, 1981).

Adriamycin production from *S. peucetius* var. *caesius* (Figure 7) begins with growth on agar slants at 26–27 °C for 10 days followed by homogenization of the mycelium, suspension in sterile distilled water and inoculation into 500 ml of growth medium in 2 l baffled round-bottomed flasks (Arcamone *et al.*, 1968, 1969). After 48 h of shaking at 28 °C (120 r.p.m., 70 mm stroke) the culture is transferred to a 170 l seed fermentor containing 120 l of medium and incubated at 26–27 °C for 27 h. Fifty l (10% v/v) are transferred into 500 l of production medium in an 800 l fermentor and incubated at 27 °C for 67–145 h. The semisynthesis of adriamycin from fermentation-derived daunorubicin may be commercially interesting if adriamycin fermentation titers are significantly lower than obtained for daunorubicin.

12.4.1.3 Isolation and purification

Recovery and purification processes for anthracyclines have been summarized by White (1983) (Figures 9,10). These agents are cardiotoxic and mutagenic and require special handling

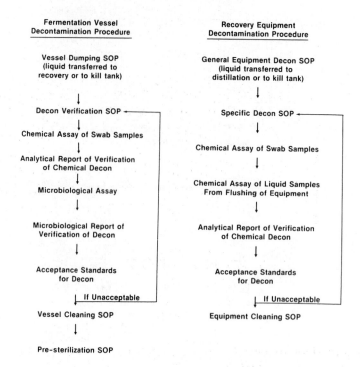

Figure 6

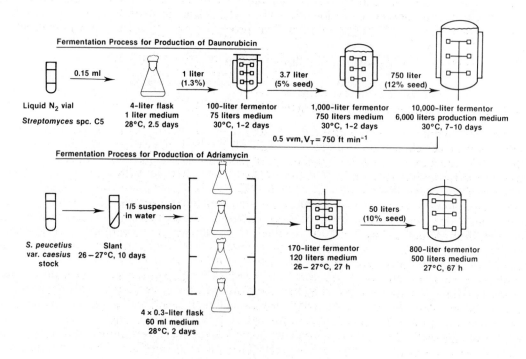

Figure 7

precautions during isolation and purification from fermentation broth (see Section 12.3). In addition, daunorubicin fermentation broths contain very little daunorubicin *per se* since the antibiotic is present in both the mycelium and broth as higher glycosides requiring an initial mild hydrolysis step to convert these glycosides to the desired product. Excess hydrolysis must be avoided to minimize generation of the aglycone.

The Rhone–Poulenc daunorubicin process (Figure 9) acidifies the whole fermentation broth with excess oxalic acid and with heat (50 °C) (Pinnent *et al.*, 1976). This lyses the cells and hydro-

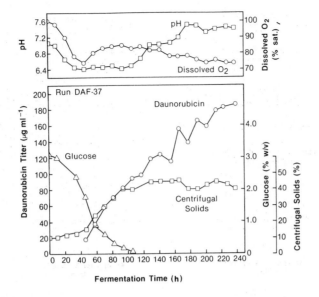

Figure 8

lyzes the higher glycosides (baumycins) to daunorubicin. This procedure is reported to yield ≈90% pure daunorubicin hydrochloride.

The National Cancer Institute process (Figure 9) involves acidification of the whole broth with sulfuric acid to pH 1.5, which solubilizes the derivable daunorubicin and lyses the cells. Following filtration, extraction and concentration, the higher glycosides are hydrolyzed to daunorubicin using ethanolic HCl at 45 °C for 10 minutes. This procedure produces daunorubicin hydrochloride at ≈95% purity.

Adriamycin is recovered from both the mycelium and the broth (Figure 10).

12.4.2 Fermentation Processes for Production of Nucleosides

Purine and pyrimidine antagonists such as 5-fluorouracil and Ara-A are important drugs in combination cancer chemotherapy and have stimulated interest in synthesis of many analogues (Ohno, 1980; Suhakolnik, 1970; Marquenz, 1982). Nucleosides and analogues of nucleosides are among the most potent cytotoxic agents known. They are inhibitors of RNA, protein and DNA syntheses. They exhibit antibiotic, antiviral and antineoplastic activity as well as cytotoxic and immunosuppressive effects. These drugs have strong activity against a wide variety of animal tumors, but toxicity and local inflammatory reactions have limited their widespread use in human therapy (Sortorelli and Johns, 1975). Even though the majority of anticancer nucleosides currently in development are synthetic, microbial processes are still used for production of starting material for chemical conversion to analogues (Appendix, Table A7). Antitumor nucleoside fermentations involve primarily *Streptomyces* and *Nocardia* species in contrast to nucleotide and nucleoside production for seasonings and drug applications (Hirose *et al.*, 1979; Nako, 1979).

Processing problems for nucleoside production are different from those for many other antitumor agents in that these compounds are colorless and easily soluble in aqueous media. In contrast, spills and surface contamination from many other classes of antitumor antibiotics are readily apparent because the drugs are strongly colored (red, blue, orange, yellow) even in dilute solutions. Surface decontamination of process equipment from nucleosides may require large volumes of acid–water, methanol or DMSO, in contrast to other antitumor compounds that are sparingly soluble in water and can be more easily removed from surfaces by organic solvents.

One example of a nucleoside fermentation process is the production of toyocamycin, a pyrrolopyrimidine antibiotic that serves as a precursor for chemical conversion to sangivamycin and a tricyclic nucleoside phosphate. Production of toyocamycin has been described using *Streptomyces toyocaensis* (Nishihara *et al.*, 1956) and *S. rimosus* (Rao *et al.*, 1963). The National Cancer Institute has developed a process utilizing strains of *S. chrestomyceticus* obtained by cloning and UV mutation (Flickinger *et al.*, 1982).

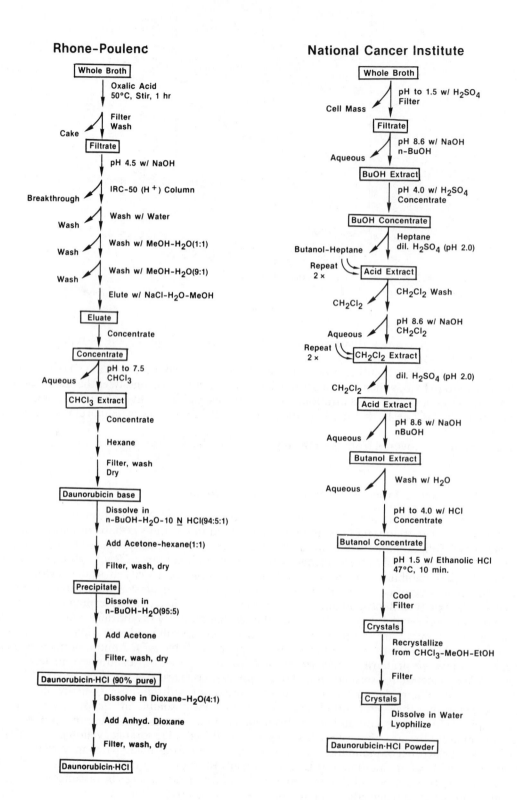

Figure 9   Recovery and purification of daunorubicin

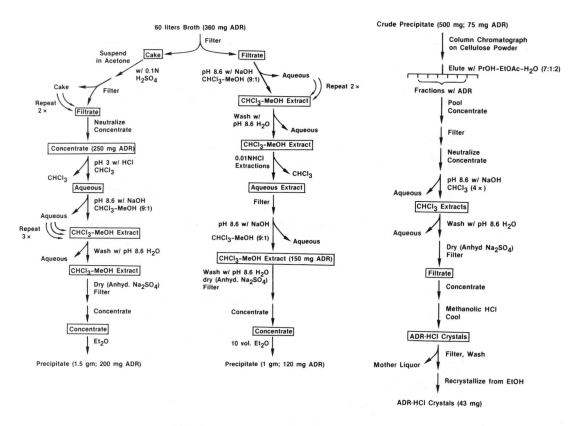

Figure 10 Recovery and purification of adriamycin

12.4.2.1 Strain improvement

Cloning of the parent organism FCRF 341 followed by UV mutagenesis produced survivors (such as strain U190) demonstrating titers of over twice the levels found for the parent. Feedback regulation mutants at the key biosynthetic enzymes IMP dehydrogenase and XMP aminase were sought among 8-azaguanine- and decoyinine-resistant populations of FCRF 341. However, these did not demonstrate higher levels of purine nucleotide production. Carbon catabolite repression was evaluated with 2-deoxyglucose (DOG) and found not to affect antibiotic production.

Production and seed media were developed based upon modification of tubercidin media (Ohkurma, 1961). A synthetic medium was also used, modified from that previously described for novobiocin production (Kominck, 1972) (Appendix, Table A8). Free phosphate levels were controlled by addition of magnesium chloride to reduce phosphate repression. The addition of oils to the production medium at the time of inoculation stimulated antibiotic production and controlled pH between 6.0 and 6.5.

12.4.2.2 Batch production process

A large-scale batch process for production of toyocamycin and chemical conversion to sangiva-mycin is summarized in Figures 11 and 12 (Flickinger *et al.*, 1982). Vegetative cells or spores of *S. chrestomyceticus* FCRF U190 were maintained at $-70\,^{\circ}C$ in liquid nitrogen. Inoculum tanks were transferred to the production vessel at the point of peak CO_2 evolution, corresponding to the end of exponential growth.

Microbial production of this antibiotic demonstrated an increase in specific antibiotic production rate with increasing incubation temperature. Optimal toyocamycin production rates were observed at $36\,^{\circ}C$. Fermentation of *S. chrestomyceticus* at $36\,^{\circ}C$ did not appear to adversely affect

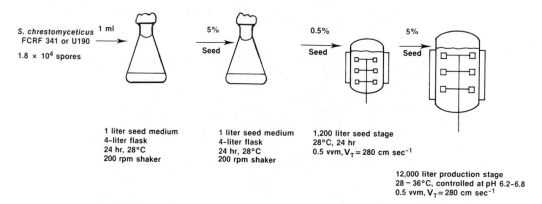

Figure 11 Fermentation process for production of tocoyamycin or sangivamycin (Flickinger *et al.*, 1982)

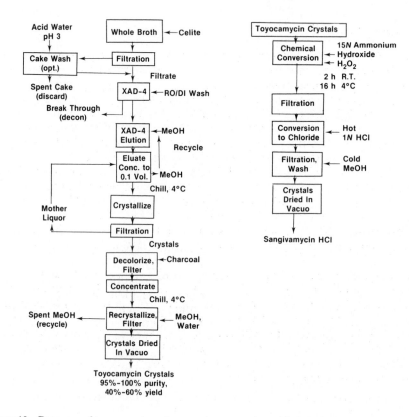

Figure 12 Recovery of tocoyamycin and conversion to sangivamycin HCl (Flickinger *et al.*, 1982)

the percentage of antibiotic retained within the mycelium and not released into the supernatant. The 20% of the total antibiotic associated with the cell cake following filtration could be removed by a pH 3 H_2SO_4 wash (Figure 12).

A low impeller tip velocity (V_t) of 275 cm s^{-1} significantly improved antibiotic productivity during 36 to 120 hours of fermentation, and a higher V_t of 612 cm s^{-1} resulted in increased initial rates of antibiotic accumulation. A V_t program was evaluated at the 7000 l scale for both the parent strain FCRF 341 as well as a mutant, U190. The parent culture was stimulated by a shear rate program while the mutant appeared to be more mechanically fragile during growth and antibiotic production (Figure 13). Titers of 2000–2500 μg ml^{-1} could be obtained using the U190 strain.

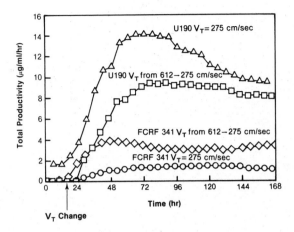

Figure 13 Comparison of the effect of V_T programming on FCRF 341 Clone 3 and U190 (28 °C, 12 000 l scale; Flickinger *et al.*, 1982)

12.4.3 Therapeutic Enzymes

Anticancer biotechnology includes large-scale bacterial enzyme production and isolation technology. The literature on the chemistry, biochemistry and pharmacological aspects of anticancer therapeutic enzymes has been summarized by Cooney and Handschumacher (1970), Wriston and Yellin (1973), Cooney and Rosenbluth (1975) and Abell and Uren (1981). Only a limited number of microbially produced enzymes that deplete nutritionally essential amino acids or nonessential amino acids have been produced commercially for treatment of human leukemias and solid tumors. The most successful antineoplastic therapeutic enzyme with respect to clinical utility has been L-asparaginase originally isolated from guinea pig serum (Kidd, 1953).

In neoplastic cells, L-asparaginase (EC 3.5.1.1) catalyzes the conversion of the amino acid asparagine to aspartic acid and ammonia, inhibiting asparagine-dependent protein synthesis and eventually the synthesis of DNA and RNA. Clinically, L-asparaginase EC-2 from *E. coli* has been used in the treatment of acute lymphocytic leukemia in children. However, the remissions have been fairly short because the malignant cells become resistant as a result of increased biosynthetic capacity of the tumor and the clearance rate of the enzyme from the body (Broome and Schenkein, 1971).

Because the enzymes used from bacterial sources are antigenic proteins, antibody appears during the course of treatment and effectively neutralizes later doses. Treatment may be prolonged by using a sequence of different enzymes from antigenically different species such as *Erwinia* (Roberts *et al.*, 1976; King, 1974). Attempts have been made to reduce the antigenicity and prolong the plasma half-life of *E. coli* and *Erwinia* L-asparaginases by chemical deamination, acetylation and carbodimide reactions with free amino groups to decrease the isoelectric point (Wagner *et al.*, 1969). Plasma half-lives have also been extended by attachment of the enzymes to dextran polymers and poly-DL-alanine peptides (Uren and Ragin, 1979).

12.4.3.1 *Batch and continuous fermentation processes*

The present commercial sources of L-asparaginase for clinical use are from *E. coli* B and *Erwinia carotovora*. Although continuous culture production of the enzyme at titers from 80 to 200 International Units (IU) per ml have been reported (Callow *et al.*, 1971, 1975; Liu and Zajic, 1973a), current production may be limited to a batch process.

A summary of an extensive screen of 200 strains of L-asparaginase-producing cultures from 78 species was prepared by Wade *et al.* (1971) in efforts to find a serologically distinct asparaginase from that of *E. coli* EC-2. From this study, high enzyme-producing cultures of *Erwinia* were evaluated and scaled up for experimental production (Buck *et al.*, 1971; Callow *et al.*, 1975). Numerous processes have been patented for L-asparaginase production (Table 4), primarily using *E. coli* B strains.

The effects of culture media, pH and oxygen transfer rate on asparaginase synthesis have been

Table 4   Process Patents for Production of L-Asparaginase

Organism/strain	Patent no.	Date	Equivalent patents	Assignee	Inventors
E. coli	Hung. 2414	July 21, 1971	—	Biogal Gyogyszergyar and Debreceni, Orvostudomanyi Egyetem Gyogyszertani Intezete	Nyiri, L. et al. (1971)
E. coli	US 3 440 142	Apr. 22, 1969	—	Worthington Biochemical Corp.	Teller, J.D. (1969)
E. coli	US 3 542 647	Nov. 24, 1970	UK 1 271 287 Ger. 1 941 539 Fr. 2 015 664	Eli Lilly and Co.	Boeck, L.V.D. and Ho, P.P.K. (1970)
E. coli	US 3 528 887	Sept. 15, 1970		E. R. Squibb & Sons, Inc.	Robison (1970)
E. coli	US 3 511 755	May 12, 1970	Fr. 1 562 365 Ger. 1 642 655	E. R. Squibb & Sons, Inc.	Berk et al. (1970a)
E. coli	US 3 511 754	May 12, 1970	Fr. 1 562 365 Ger. 1 642 655	E. R. Squibb & Sons, Inc.	Berk et al. (1970b)
Erwinia carotovora	Ger. 2 031 759	Jan. 7, 1971	—	UK Secretary of State for Social Services, London	Christie et al. (1971)
	UK 1 379 728	Jan. 8, 1975	—		Callow et al. (1975)
Erwinia aroideae, Hydrogenomonas eutropha	US 3 589 982	June 29, 1971	—	US Dept. of Agriculture	Peterson (1971)
E. coli	Jpn. 71 42 954	Dec. 18, 1971		Kyowa Fermentation Ind. Co. Ltd	Saito et al. (1971)
Candida utilis IAM 215	Jpn. 71 42 592	Dec. 16, 1971		Sanraka-Ocean Co. Ltd	Arima et al. (1971)
Bacillus spp.	Ger. 2 126 181	Dec. 9, 1971		Kyowa Fermentation Ind. Co. Ltd	Kato et al. (1971)
—	Jpn. 71 27 714	Aug. 11, 1971	Ger. 1 904 330	Kyowa Fermentation Ind. Co. Ltd	Tanaka et al. (1971a)
—	Jpn. 71 28 149	Aug. 16, 1971	Ger. 1 925 951	Kyowa Fermentation Ind. Co. Ltd	Tanaka et al. (1971b)
—	Jpn. 71 29 150	Aug. 16, 1971	Ger. 1 928 051	Kyowa Fermentation Ind. Co. Ltd	Mochizieki et al. (1971)
Serratia	Jpn. 71 28 148	Aug. 16, 1971	Ger. 1 920 955	Kyowa Fermentation Ind. Co. Ltd	Tanaka et al. (1971c)
Erwinia carotovora	US 3 686 072	Aug. 22, 1972	—	—	Herbert and Wade (1972)
Erwinia carotovora or Serratia marcescens	US 3 660 238	May 2, 1972	—	—	Wade (1972)
Pseudomonas aurontiaca	SU 782 363-E29 SU 739 096-D06	Jan. 23, 1982 June 5, 1980	— —	Moscow Lomonosov University	Berezov et al. (1980)

shown to vary for different organisms (Barnes *et al.*, 1977; Grossowicz and Rasooly, 1972; Liu and Zajic, 1973b). Production of L-asparaginase by *E. coli* B is greatest in media rich in amino acids, with a limited oxygen supply, over a pH range of 7.0 to 7.8. When the culture is grown aerobically, cell yields are high but enzyme levels are low. Anaerobically, cell yields are low but with high specific enzyme levels. Expression of enzyme in *E. coli* appears to be maximal after eight hours of rapid aerobic growth followed by no more than two hours of anaerobic incubation. The cells are then harvested and the enzyme is extracted. [Production of enzyme is measured in units, with one IU being that amount of enzyme that liberates 1 μmol NH_3 per minute at 37 °C in pH 8.5 borate buffer (Wade and Phillips, 1971)].

The temperature optimum and oxygen transfer rates for maximal enzyme production by *Erwinia* are different than for *E. coli* (Liu and Zajic, 1973a, 1973b; Peterson and Ciegler, 1969; Buck *et al.*, 1971; Grossowicz and Rasooly, 1972). Optimal $K_L a$ values for enzyme production (0.98 min^{-1}) appear to be slightly lower than for maximal cell growth (1.2–1.9 min^{-1}).

Lactic acid, glutamic acid, succinate and glutamine have a beneficial effect on enzyme yield. Readily fermented carbohydrates such as glucose, glycerol and mannitol cause catabolite repression at a level as low as 0.1%, which may not be relieved by addition of cAMP. Mutations that adversely affect growth depress L-asparaginase production (Jefferies, 1976). Culture media for enzyme production are summarized in the Appendix (Table A12).

12.4.3.2 *Isolation and purification*

Many intracellular enzyme extraction and crystallization schemes have been devised for isolation and purification of clinical grade L-asparaginase (Table 4). The extraction and crystallization procedures summarized by Rauenbusch *et al.* (1970) take advantage of the solvent stability of *E. coli* L-asparaginase (Figure 14). The overall yield of this process appears to be 6%, starting with a fermentation titer of 2 IU per mg protein. A different recovery and purification procedure for *Erwinia* enzyme claims 40% overall recovery yield, starting from 20 IU per ml of culture (Buck *et al.*, 1971). Grossowicz and Rasooly (1972) extract intracellular enzyme using a combined toluene and hypertonic urea method, with a reported yield of 85–90%. Nucleic acids are precipitated by the addition of 0.05 M manganese chloride. Ammonium sulfate precipitation, Bio Gel and DEAE-cellulose column chromatography are used for further purification to a final specific activity of >600 IU per mg protein with a 35–40% overall yield. A large-scale process for the production of *Erwinia* L-asparaginase is outlined in Figure 15.

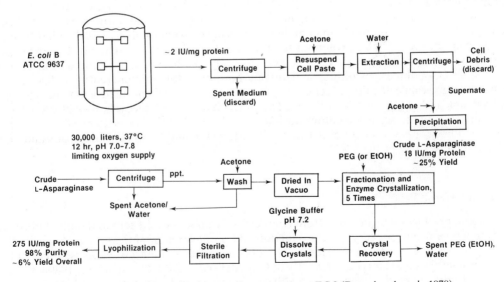

Figure 14 Process for recovery of *E. coli* L-asparaginase EC 2 (Rauenbusch *et al.*, 1970)

12.5 EXAMPLES OF PRODUCTS OF MAMMALIAN CELLS IN CULTURE

Previous biological approaches to cancer therapy have included isolation of immunopotentiating agents from bacterial and yeast cell wall material (Aszalos, 1981) and development of anti-

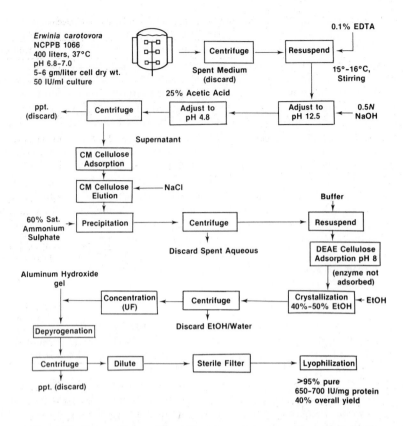

Figure 15   Process for recovery of *Erwinia* L-asparaginase (Buck *et al.*, 1971)

cancer vaccines for animals (Table 5). More recent biological approaches to cancer therapy are directed at *in vivo* manipulation of human immunological molecules produced by mammalian cells, which regulate production of tumor specific antibodies, proteins that interfere with virus and cell proliferation (interferons), natural killer cells, cytotoxic T lymphocytes and activated macrophages (Mitchison and Landy, 1978; Herberman, 1980; Goldstein and Chirigos, 1981).

Although gene cloning and recombinant DNA technology offer the future advantage of being able to produce many products of mammalian cells using conventional large-scale microbial technology, many molecules with protential for use in anticancer therapy may not be amenable to production by recombinant methods. The need to study the therapeutic potential of interferons, lymphokines, monokines, cytokines and thymic hormones for human treatment has renewed interest in large-scale cell culture technology for cultivation of cells in suspension and aggregates in stirred tank and perfusion reactors as well as attached to glass surfaces, hollow fibers and beads (microcarriers). Large-scale cell culture technologies have recently been extensively reviewed by Tolbert and Feder (1983), Feder and Tolbert (1983), Acton and Lynn (1977), Jakoby and Pastan (1979) and Fleischaker *et al.* (1981).

12.5.1   Interferon Production

Interferon production technology from leukocytes, lymphoblastoid cells and diploid fibroblasts are examples of recent processes for production and purification of anticancer biologicals. Many human interferons and hybrid interferons have been cloned, sequenced, expressed and purified from microbial systems (Derynck *et al.*, 1980; Goeddel *et al.*, 1980; Nagata *et al.*, 1980a; Strauli *et al.*, 1980; Taniguchi *et al.*, 1980a, 1980b, 1980c); however, little process-related information on these systems has been published (Weissman, 1981). In spite of this microbial recombinant DNA effort, large-scale leukocyte and lymphoblastoid interferon production facilities have been constructed for production of material for antiviral and anticancer evaluation. This technology literature has recently been reviewed by Peska (1981), White *et al.* (1980), Berg (1982) and Vilcek (1979).

Evaluation of the anticancer effect of exogenous interferon was initially limited by the supply

Table 5 Process Patents for Immunomodulators and Leukemia Vaccines

Description	Patent Number	Date	Author/Assignee
Preparation of feline leukemia vaccine from feline lymphoblastoid cells chronically infected with feline leukemia virus (FL 74 cells)	US 4 117 112 US 4 086 134 US 4 034 081 US 3 966 907	Sept. 26, 1978 April 25, 1978 July 5, 1977 June 29, 1976	Jarrett *et al.*/University of Glasgow, Scotland
Method for preparation of a vaccine for Marek's disease in poultry (virus strain ATCC VR-736)	US 3 981 771	Sept. 21, 1976	Seroian
Method for production of Marek's disease vaccine	US 3 965 258	June 22, 1976	McAleer *et al.*/Merck & Co., US
Method for isolating a cancer-associated polypeptide antigen useful as an immunizing agent	US 3 960 827	June 1, 1976	Bjorklund
Method for dissociating a culture of mycobacteria (BCG, *M. tuberculosis* H37Rv, *M. kansasii* P8, *M. intracellulare* P2) on liquid media for cancer treatment	US 4 038 142	July 26, 1977	Turcotte and Quevillon/Institute Armand Frappier, Canada
Preparation of a vaccine for stimulating the immune mechanism of mammals to species-specific tumors which comprises a viral oncolysate of the tumor and a virus having a lytic action on the cells of the tumor	US 4 108 983	Aug. 22, 1978	Wallack/The Wistar Institute
Immunopotentiating agent from yeast cell wall material	US 4 138 479	Feb. 6, 1979	Truscheit *et al.*/Bayer AG, Germany
Process for preparation of an antileukemia agent from the culture solution of *Staphylococcus epidermidis* (ATCC 31310)	US 4 159 321	June 26, 1979	Toyoshima *et al.*/Tobishi Pharmaceutical Co. Ltd, Japan

of material of high purity. However, even with relatively impure material, early studies of interferon as an adjuvant following surgery to evaluate the onset of metastases in osteogenic sarcoma patients were encouraging (Strander, 1977). Since this initial evaluation, interferon clinical trial results on a variety of human cancers using both 'natural' and recombinant interferons have been reported (Merigan, 1981; Scorticatti *et al.*, 1982; Rasmussen and Merigan, 1980) along with the ability of interferons to stimulate lymphocytes to lyse tumor cells (Gidlund *et al.*, 1978; Watlach, 1981).

12.5.1.1 *Fibroblast processes (HuIFN-β)*

Human fibroblast interferon (HuIFN-β) processes utilize a variety of diploid human fibroblasts (*e.g.* infant foreskin cells, FS-4) induced by exposure to paramyxovirus or poly I:C. These anchorage-dependent cells have a finite life span and frequently can only be cultivated for 35–60 population doublings (White *et al.*, 1980). The yield of HuIFN-β can be increased eightfold by pretreating (priming) the cells with moderate amounts (100 units ml^{-1}) of interferon and by superinduction. In this procedure, the fibroblasts are treated with poly I:C (500 μg ml^{-1}) and cycloheximide (10 μg ml^{-1}) to block transcription but not translation. As the translation block is released, actinomycin D (1 μg ml^{-1}) or another transcription inhibitor is added, resulting in a greater than 100-fold increase in HuIFN-β production (Havell and Vilcek, 1972). Fibroblast interferon titers of 4000 to >60 000 units ml^{-1} have been reported for various cell lines and induction systems (Berg, 1982). Examples of HuIFN-β processes using a continuous cell line are summarized in Figure 16.

The major obstacle to scale-up and production of HuIFN-β is to provide adequate surface area for the growth of anchorage-dependent cells. Technologies originally developed for vaccine production using stationary flasks, roller bottles, multisurface stack plate propagators, fibers (in spirals and horizontal beds) and microcarriers in stirred tank reactors have been used. Inoculum volumes for successive steps are generally large (20–25% v/v) and final cell densities obtained are 2–5 × 10^6 viable cells per milliliter (White *et al.*, 1980). Using superinduction methods, only 5–10 cells are required for production of one unit of HuIFN-β.

Many purification methods have evolved for large-scale recovery of material for clinical trials based on trichloroacetic acid precipitation, ethanol extraction and antibody affinity chromatography, methods originally developed for purification of biologicals such as tetanus and cholera toxoid (Beale, 1979). Fibroblast HuIFN-β purification methods have been summarized by Berg (1982) and are presented in Table 6.

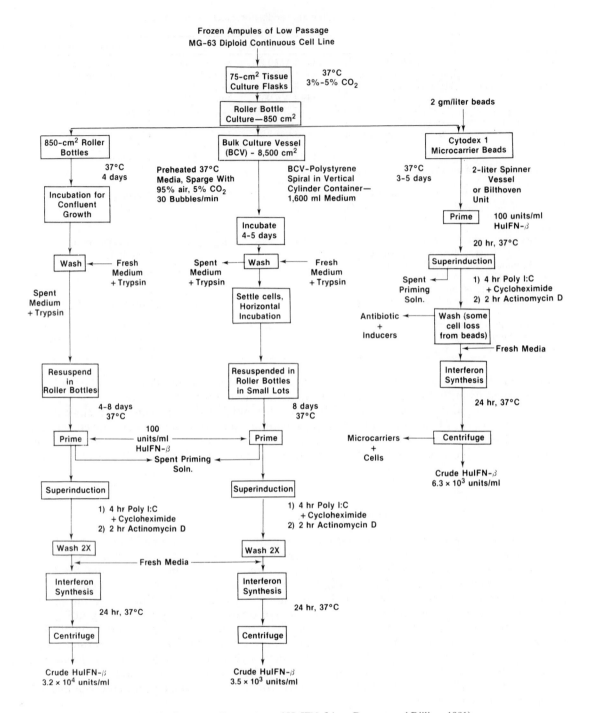

Figure 16   Processes for recovery of HuIFN-β (van Damme and Rillian, 1981)

12.5.1.2   *Leukocyte processes (HuIFN-α)*

Production technology for preparation of human leukocyte interferon (HuIFN-α) has recently been reviewed by Mogensen and Cantell (1977), Kauppinen *et al.* (1978) and Berg (1982). Leukocyte interferon is produced from small-scale suspensions of leukocytes with HuIFN-α priming and induction by Sendai or Newcastle disease virus (Cantell *et al.*, 1981; Waldman *et al.*, 1981). Large quantities of HuIFN-α can also be produced from human chronic myelogenous leukemia

Table 6 Purification of Human Fibroblast IFN (Berg, 1982)

Methods of purification	Inducer of IFN	Type of cells	Specific activity of purified interferon (units mg^{-1})	% yield of purified IFN	Purification factor	Amounts of IFN purified	Titer of anti-IFN	Type of immunogen
Antibody affinity chromatography	UV + Poly I:C	Diploid fibroblasts	10^8	90	200	3.6×10^6	2×10^5	PIF
ConA followed by phenylsepharose	Poly I:C	C-10	$2\text{--}10 \times 10^8$	4–20	~1000	$2\text{--}10 \times 10^7$	—	—
Controlled pore glass and ConA chromatography	Poly I:C	C-10	5×10^8	10	~1000	3×10^7	—	—
ConA and phenylsepharose in agarose (tandem affinity chromatography)	Poly I:C	BG-27	4×10^7	60	4000	$10\text{--}20 \times 10^6$	—	—
Antibody affinity chromatography and prep. SDS-PAGE	Poly I:C	C-10	$\sim 10^8$	70	2000	$1\text{--}10 \times 10^6$	10^5	Pure HuIFN-β
Blue Dextran and prep. SDS-PAGE	Poly I:C	FS-4	$2\text{--}10 \times 10^8$	5–20	2000	$5\text{--}7 \times 10^7$	—	—
Blue Dextran and HPLC	Poly I:C	FS-4	3×10^8	15	~2000	11×10^6	—	—

PIF, partially purified human interferon; Poly I:C, polyinosinic acid—polycytidylic acid.

(CML) donors (Hershberg *et al.*, 1981) and by myeloblast cultures such as the KG-1 line (Familletti *et al.*, 1981).

Venous blood is collected and stabilized with citrate; 10-30 ml of 'buffy coat' leukocytes are recovered from each unit of whole blood by centrifugation. The leukocytes are stored overnight at 4 °C followed by lysis of contaminating lymphocytes using two ammonium chloride treatments (0.83%, 4 °C, 10 min). Following centrifugation, the leukocytes are resuspended in tricine buffer with antibiotics plus 4% human agamma serum, primed with 100 units ml^{-1} of HuIFN-α for 1–2 h at 37 °C and induced with virus (100-150 HA units ml^{-1}). Yields of 20 000–60 000 units ml^{-1} have been reported (Berg, 1982). An example large-scale process is summarized in Figure 17.

Application of hollow fiber and plate and frame ultrafiltration for initial isolation of interferons

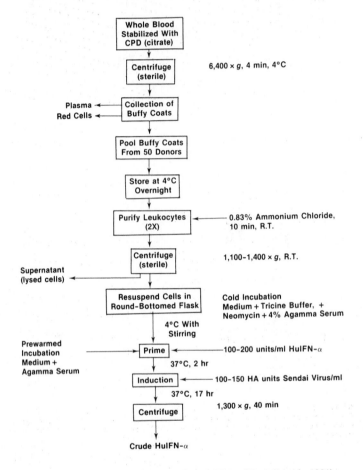

Figure 17   Process for production of HuIFN-α (Cantell *et al.*, 1981)

is increasing. A recent example of a cascade filtration system, including mass transfer and sieving coefficient data, has been reported by Van Reis *et al.* (1982) for production (Figure 18) and recovery (Figure 19) of HuIFN-α. HuIFN-α purification is accomplished by either ethanol fractionation and blue dextran chromatography or antibody affinity chromatography methods (Table 7) (Berg, 1982).

12.5.1.3   *Lymphoblastoid processes (Hu Ly IFN)*

Human lymphoblastoid interferon processes [Hu Ly IFN or HuIFN ($\alpha + \beta$)] utilize transformed cell lines capable of rapid growth in large-scale suspension cultures that have been cloned for optimum IFN production. The Namalva (Namalwa) human lymphoblastoid cell line has been scaled up to the 2000–4000 l scale (Figure 20) for production of human interferon for clinical trials. Almost all lymphoblastoid cell lines contain Epstein–Barr virus (EBV) DNA; however, several lines only produce portions of the viral genome and are suitable for production of material for human clinical use. The acceptability and safety of interferon prepared for human

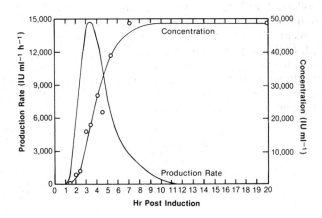

Figure 18 HuIFN-α production kinetics in a primed batch process (Van Reis *et al.*, 1982)

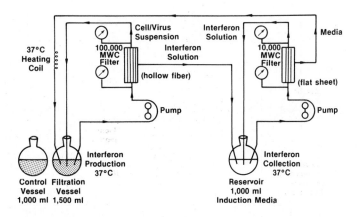

All vessels are 2–liter round–bottomed flasks with magnetic stirring bars kept at 37°C. Filtration membranes were saturated with serum proteins prior to use. The initial induction media in the collection reservoir contained 200 U crude HuIFN-α/ml + 150 HAu/ml Sendai virus. Filtration was started 0 to 4 hr post induction.

Figure 19 Cascade filtration system for production and recovery of HuIFN-α (Van Reis *et al.*, 1982)

treatment from lymphoblastoid cells have recently been reviewed by Finter and Fantes (1980), Petricciani (1979), Parks and Hubbell (1979), Hillman (1979) and Beale (1979).

Several large-scale Namalva interferon processes have been reported using Sendai or Newcastle disease virus (Strain B1) as the inducing agent (Bridgen *et al.*, 1979; Klein *et al.*, 1979; Mizrahi, 1981). The addition of butyric or propionic acids has recently been reported to stimulate interferon production in some systems up to levels of 30 000 units per 10^6 cells (Monser *et al.*, 1980). Two Namalva cell lymphoblastoid interferon process options which use either a batch (White *et al.*, 1980; Bodo, 1981; Klein and Ricketts, 1981) or semicontinuous cell growth mode (Mizrahi, 1981) are summarized in Figure 21. Purification of lymphoblastoid interferon has been extensively reviewed by Peska (1981).

12.5.1.4 *Immune interferon processes (HuIFN-γ)*

Human immune interferon (HuIFN-γ) is produced by human immunocompetent cells after stimulation with mitogens or antigens and is thus a lymphokine. HuIFN-γ may have the greatest

Healthcare Products

Table 7 Purification of Human Leukocyte IFN (Berg, 1982)

Method of purification	Starting material	Specific activity of purified IFN (units mg^{-1})	% yield of purified interferon	Purification factor	Amounts of interferon purified	Titer of anti-IFN	Adsorption of anti-IFN
Antibody affinity chromatography	CIF	N.D.	50–100	N.D.	$\sim10^6$	5 000–10 000	+
Antibody affinity chromatography	CIF	4–30×10^6	80–100	$\sim1\,000$	2–3×10^6	50 000	+
Sequential antibody affinity chromatographies	CIF	50×10^6 (2×10^8)	25	$5\,000$ $(20\,000)$	6×10^5	130 000	+
Antibody affinity chromatography	CIF	30–50×10^6	~100	$5\,000$	1–3×10^6	200 000	+
	PIF	10	50	100	8×10		
Gel filtration, Cu-chelate, blue dextran and antibody affinity chromatography	CIF	$\sim10^9$	50	$350\,000$	$\sim2 \times 10^6$	200 000	+
Ethanol fractionation or antibody affinity chromatography, gel filtration	CIF	5–10×10^6	~50	$4\,000$	1–5×10^6	$\sim100\,000$	+
Ethanol fractionation and blue dextran	CIF	2–5×10^6	~50	$1\,000$	1–5×10^6	—	—
Gel filtration HPLC	CIF	2–4×10^8	25–80	$60\,000$	12×10^6	—	—
Periodate gel filtration	PIF	10^9	25	$1\,000$	2.5×10^8	—	—

CIF, crude human leukocyte interferon; PIF, partially purified human leukocyte interferon.

Figure 20 1000, 2000 and 4000 l cell culture fermentors for commercial production of human lymphoblastoid interferon at Wellcome Research (courtesy of N. B. Finter)

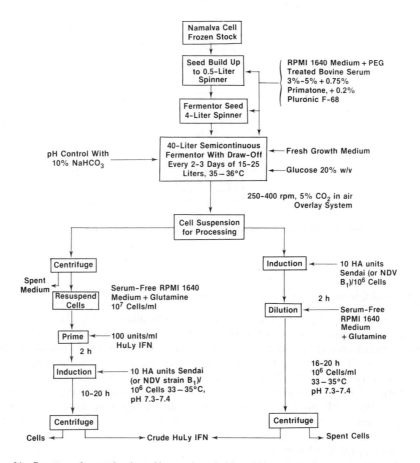

Figure 21 Processes for production of human lymphoblastoid interferon (Hu Ly IFN) (Mizrahi, 1981)

potential as an anticancer agent because it appears to be more efficient as an antiproliferative agent and may also actually be cytotoxic to certain tumor cells (Rubin and Gupta, 1980; Vilcek and Gresser, 1982).

Several different large-scale roller bottle procedures have been described for production of HuIFN-γ for clinical evaluation using induction by mitogens, galactose oxidase treatment or staphylococcal enterotoxin A (SEA), yielding titers in the 100–10 000 units ml^{-1} (Berg, 1982; Epstein, 1981a, 1981b; Johnson *et al.*, 1981). Two types of HuIFN-γ process are summarized in Figure 22. Purification procedures for this interferon using controlled pore glass (CPG) and Con A sepharose have been summarized by Epstein (1981a) (Table 8).

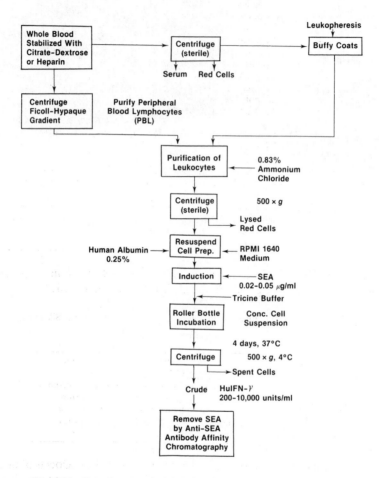

Figure 22   Processes for production of HuIFN-γ (Johnson *et al.*, 1981)

12.5.2   Future Technologies: Lymphokines and Monoclonal Antibodies

The application of bioengineering to the growth of mammalian cells and human hybridomas in serum-free large-scale suspension cultures for production of lymphokines, cytokines and monoclonal antibodies may become the dominant technology for production of anticancer agents in the next 10 years (Oldham 1983). Many lymphokines and thymic hormones are currently being produced using laboratory-scale cell culture technology for evaluation of the potential utilization in cancer therapeutics (Goldstein and Chirigos, 1981). Certainly some of these agents may eventually be produced efficiently on a large scale through cloning of the genes into microbial systems. Prior to this development, however, scale-up of mammalian cell culture systems using hybridomas, normal human immunocompetent and tumor cells for production of these agents will present a challenging problem for biotechnology.

Initially this future technology may be concerned with production of cell growth stimulating

Table 8 Purification of Human Immune IFN (Epstein, 1981)

Species	Cell source	Inducer	Method of purification	pI	Specific activity (units mg^{-1})	Mol. wt.
Human	Peripheral blood mononuclear cells	SEA	Chromatography on controlled pore glass (CPG) and Poly U agarose	8.3	5×10^7	—
Human	Peripheral blood mononuclear cells	SEA	PEG precipitation, chromatography on DEAE Sepharose; adsorption of SP-Sephadex, Ultragel Ac 44	—	10^7	50 000
Human	Peripheral blood mononuclear cells	TPA and PHA	Chromatography on CPG and ConA Sepharose and Biogel 200	8.6	10^7	58 000 (±3000)
Human	Peripheral blood lymphocytes	PHA or ConA	Chromatography on ConA Sepharose, gel filtration and anti BSA Sepharose	4.0–4.5	10^6	36 000
Human	Peripheral blood mononuclear cells	PHA	Chromatography on CPG and poly U agarose	—	10^6	—
Human	Peripheral blood mononuclear cells	ConA	Batch adsorption on CPG	—	$10^{5.3}$	45 000
Human	Peripheral blood mononuclear cells	SEA	Adsorbed with Matrex Blue; chromatography on CPG Ultragel AcA 54	4.0–4.3; 4.91–5.38	$10^{6.3}$	35 000–70 000
Mouse	Spleen cells	SEA	NH$_4$SO$_4$; chromatography on BSA-Affigel or hydroxylapatite, Ultragel AcA 34	—	—	40 000 70 000–90 000
Mouse	Spleen cells	PHA	Chromatography on ConA Sepharose, Affigel 202, Blue Sepharose CL 6B, Phenyl Sepharose CL 4B	—	—	

factors elaborated by human T cells (interleukins) (Klein *et al.*, 1983) and macrophages as well as anticancer monoclonal antibodies for pilot human clinical evaluation (Table 9).

Table 9 Uses of Monoclonal Antibodies in Cancer (Lennox and Sikora, 1982)

Tumor diagnosis	Antigen detection in serum
Tumor localization	Using labelled antibodies
Targeting toxic agents to tumors	Toxins, radioisotopes, liposomes
Monitoring of therapeutic agents	Detection of agent in serum
Monitoring of intrinsic agents	Detection of serum hormones, *etc.*
Monitoring of the state of the immune and haematopoietic systems	Assaying different cell types

There are few reported investigations of the biochemical engineering aspects of large-scale processes for production of clinical monoclonal antibodies for cancer treatment and diagnosis at this time. Recently, a small-scale cytostat has been described for continuous production of monoclonal antibodies and recovery using immunoaffinity chromatography (Fazekas de St. Groth, 1983). Many biological and technical obstacles remain in order to begin to develop efficient large-scale production procedures from the hybridoma technology originally reported by Kohler and Milstein (1975). The majority of investigational anticancer monoclonal antibodies have been produced from murine cells on the laboratory scale using the mouse ascites tumors as the antibody production stage (Kennett *et al.*, 1980; McMichael and Fabre, 1982; Galfre and Milstein, 1981; Maurer and Callahan, 1980; Hubbard, 1983). Scale-up of hybridoma growth and antibody production in suspension culture in the absence of serum may be essential to replace the current laboratory-scale method of harvesting small quantities of ascites fluid from large numbers of mice for antibody isolation and purification. The ability of hybridomas to secrete antibody in static and suspension culture currently results in accumulation of approximately 30–60 μg ml^{-1} levels (Maurer and Callahan, 1980), in contrast to antibody titers as high as 5–25 mg ml^{-1} from ascites fluids which allows for production of 100–200 mg of antibody from each mouse. Hybridoma microencapsulation using porous carbohydrate capsules and immobilization techniques (Klinman and McKearn, 1981) may be able to stimulate production of high specific activity monoclonal antibodies at levels comparable to ascites fluids (Marcipar *et al.*, 1983; Liu and Sun, 1980; Hubbard, 1983).

The possibility of using monoclonal antibody–toxin conjugates for human therapy combines microbial fermentation technology for production of chemotherapeutics and toxins with large-scale production, isolation and conjugation methods for human antibodies. A summary of the cytotoxic effects of antibodies conjugated to intact (A + B chain) toxins and toxin fragments (conjugation with A chains only) has recently been made by Thorpe *et al.* (1982). Only a few examples of techniques for production of antibody drug conjugates (Masuho *et al.*, 1983) exist. One example is a method for conjugating daunomycin (Sela *et al.*, 1978) (Figure 23). A more general procedure for conjugation of a number of cytotoxic agents to immunoglobulins using a polymer carrier has been summarized by Rowland (1977).

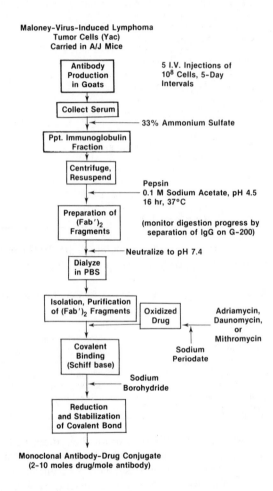

Figure 23   Procedure for production of an antibody–drug conjugate against a lymphoma (Sela *et al.*, 1978)

A small number of animal and pilot clinical trials have begun to evaluate both monoclonal antibodies and monoclonal antibody–chemotherapeutic drug conjugates for human therapy (Sobol *et al.* 1982). Antibodies to lung, rectal, colon, breast, prostate, neuroblastoma, lymphoma and nasopharyngeal carcinoma have been produced on the laboratory scale from mouse × mouse and mouse × human fusions. Further expansion of this field appears to be limited at present because of the lack of a suitable human myeloma cell line to be used as the malignant partner in cell fusions and a source of adequate numbers of human lymphocytes immunized to human tumor specific antigens (Yelton and Scharff, 1981; Lennox and Sikora, 1982). Once human hybridoma techniques become established, many of the methods developed for large-scale production of lymphoblastoid interferon and lymphokines may be applied to produce monoclonal antibodies of high specific activity for cancer treatment, diagnosis and use in large-scale affinity chromatography (Hubbard, 1983).

12.6 SUMMARY

A wide variety of biochemical production and purification technologies are required for development of anticancer agents, since both products of microbial and mammalian cellular metabolism are being applied to human therapy. Because many of these drugs and biologicals have potent biological activity or human toxicity, they are produced, isolated and purified under contained conditions to avoid worker exposure. Process wastes must be chemically and biologically decontaminated before disposal. Process equipment is often isolated within a biological barrier system designed to minimize aerosol generation and spills.

Cytotoxic drugs and therapeutic enzymes currently have a rapidly expanding market in established anticancer therapy. These agents are produced by microbial fermentation technology, however, on a reduced scale from that of antibacterial and antifungal agents. Anticancer biologicals such as interferons are currently being produced for clinical evaluation using suspension cultures of lymphoblastoid cells or by stimulation of human fibroblast or immunocompetent cells in small-scale suspensions.

The next 10 years, however, may require development of new large-scale biotechnology for production of anticancer monoclonal antibodies, antibody–drug conjugates and biologicals for *in vivo* manipulation of the human immunological response to tumors. Many of these future agents may be able to be genetically transferred and expressed in microbial systems for large-scale production. Others may be produced by large-scale cell culture techniques similar to the methods used for clinical production of lymphoblastoid and fibroblast interferons.

12.7 APPENDIX

Table A1 Actinomycins (Precursor Fed Fermentations)

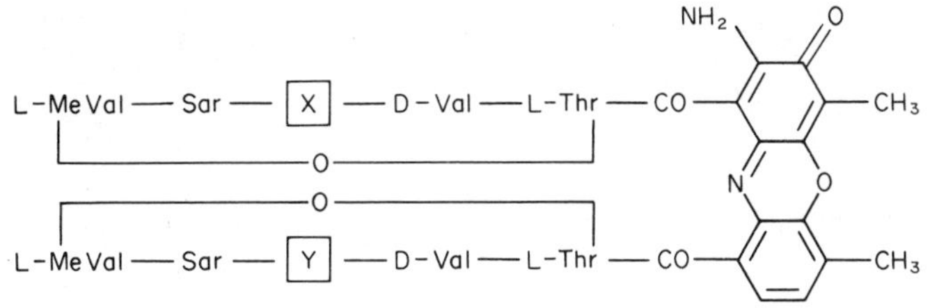

Member	Structure	Producing organism	Ref.
Actinomycin D (actinomycin, C_1, dactinomycin, DIV, IV, X)	X, Y = L-proline	*S. antibioticus* *S. chrysomallus*	Katz (1967), Meienhofer and Atherton (1973), Katz and Weissbach (1967)
Actinomycin Pip 1 α	X = pipecolic acid Y = 4-oxopipecolic acid	*S. antibioticus* *S. chrysomallus*	Formica and Katz (1973) Korzybski *et al.* (1978)
Actinomycin Pip 1 β	X = pipecolic acid Y = L-proline	*S. antibioticus* *S. chrysomallus*	
Actinomycin Pip 2	X, Y = pipecolic acid	*S. antibioticus* *S. chrysomallus*	
Azetomycin I	X = L-proline Y = L-azetidine-2-carboxylic acid (AZC)	*S. antibioticus*	Formica *et al.* (1968), Formica and Apple (1976)
Azetomycin II	X, Y = AZC	*S. antibioticus*	

Healthcare Products

Table A2 Ansamycins

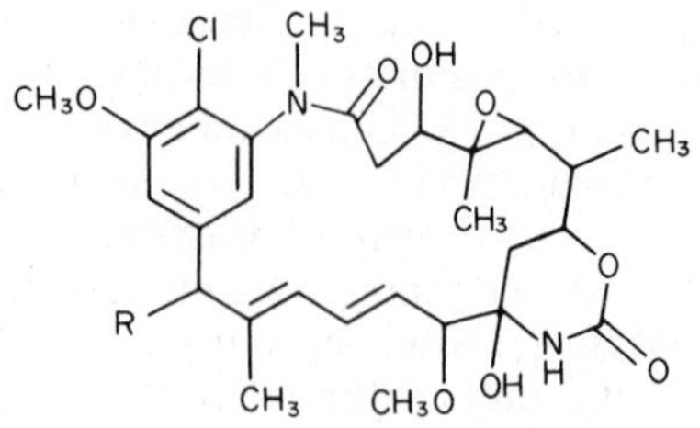

Member	Structure	Producing organism	Ref.
Ansamitocins[a]			
P-0	R = H	*Nocardia* sp.	Higashide *et al.* (1977)
P-1	R = COCH₃	C-15003	
P-2	R = COCH₂CH₃	*Nocardia*	Tanida *et al.* (1980)
P-3	R = COCH(CH₃)₂	*meditenarei*	
P-4	R = COCH₂(CH₃)₂	N813	

[a] Used as precursors for semisynthesis of matansine derivatives currently under development by Takeda Chemical Industries and Bristol Laboratories.

Table A3 Anthramycins

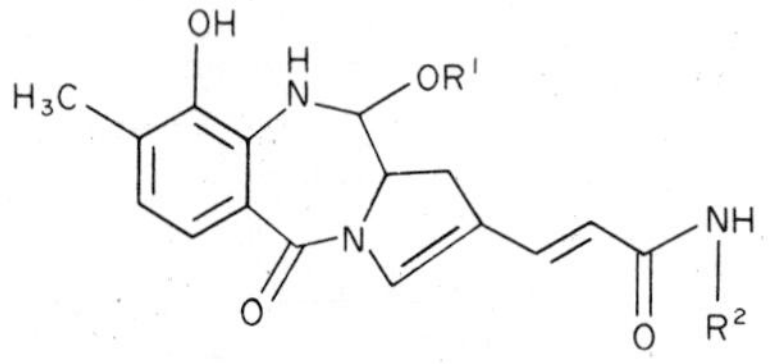

Anthramycins, mezethamycins

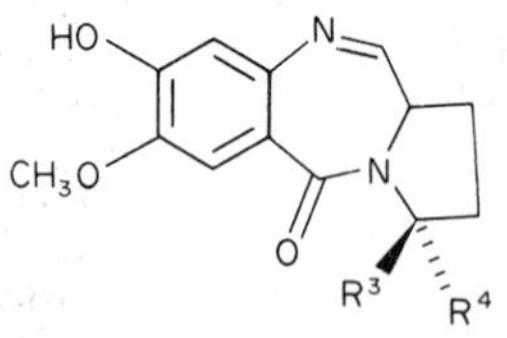

Neothamycins

Member	Structure	Producing organism	Ref.
Anthramycin	R¹, R² = H	*S. refuineus* var. *thermo-tolerans* (NRRL-3143)	Leimgruber *et al.* (1968), Hurley (1977)
Mezethramycin A	R¹ = H	*S. thioluteus* ME-561-4	Umezawa *et al.* (1978)
	R² = CH₃		
Mezethramycin B	R¹ = CH₃		
	R² = CH₃		
Mezethramycin C	R¹ = CH₂CH₃		
	R² = CH₃		
Neothramycin A	R³ = OH	*Streptomyces* MC 916-14	Miyamoto *et al.* (1977)
	R⁴ = H		
Neothramycin B	R³ = H	*Streptomyces* MC 916-14	
	R⁴ = OH		

Table A4 Anthracyclines

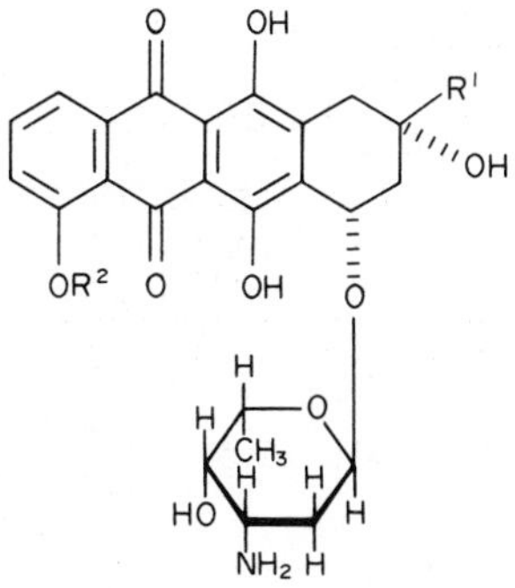

	R_1	R_2
Daunorubicin	CH_3CO	CH_3
Adriamycin	CH_2OHCO	CH_3
Carminomycin	CH_3CO	H

Member of class	Producing organism	Ref.
Adriamycin (14-hydroxy daunomycin, doxorubicin)	*S. peucetius* var. *caesius* or *carneus* *S. galilaeus* MA 144-M1	White (1983)
Daunomycin (daunorubicin)	*S. peucetius* *S. bifurcus* 23219 *S. coeruleorubidus* 8899, 31723 *S. coeruleorubidus* *S. griseoruber* *S. griseus* 32041 *S. griseus* IMETSA, 3933, 5570, 10086, 10431 *S. viridochromogenes*	White (1983)
Carminomycin (figaroic acid complex)	*Actinomadura carminata* *Streptosparangium* sp. C-31751	Gauze *et al.* (1974)
Aclacinomycin	*S. galilaeus*	Oki *et al.* (1977)
Nogalomycin	*S. nogalater* *S. glomeratus*	Bhuyan and Dietz (1965)
Steffimycins	*S. elegreteus* NRRL 5634 *S. steffisburgensis*	Brodasky and Reusser (1974)
Rhodirubins	*S.* sp. ME505	Oki and Yoshimoto (1979)
Rhodomycins	*S. purpurescens*	Brockman *et al.* (1951)
Setomimycin	*S. pseudovenezuelae*	Omura *et al.* (1978)
Roseorubicins	*Act. roseoviolaceus* *Act. violacens* *Act. violacechromogenes*	Matsuzawa *et al.* (1979)

Table A5   Media Used for Production of Daunorubicin (White, 1983)

Rhone-Poulenc		National Cancer Institute	
Medium	Concentration (% w/v)	*Medium*	Concentration (% w/v)
Medium for slant culture		*Seed medium*	
Saccharose	2	Corn starch	1.5
Dry yeast	0.1	Cottonseed flour	0.5
Potassium hydrogen phosphate	0.2	Defatted soy flour	0.5
Sodium nitrate	0.2	Autolyzed nutritional yeast	0.1
Magnesium sulfate	0.2	NaCl	0.25
Agar	2	$CaCO_3$	0.5
Tap water	q.s.	Prochem #51, antifoam	0.8[a]
		Tap water	q.s.
Medium for shake flask		*Production medium*	
Peptone	0.6	Glucose	5.0
Dry yeast	0.3	Herring meal	1.2
Calcium nitrate, hydrate	0.05	Bakers' nutrisoy	1.25
Tap water	q.s.	Autolyzed nutritional yeast	0.75
pH after sterilization: 7.2		NaCl	0.33
		$CaCO_3$	1.0
Vegetative medium		Prochem #51, antifoam	1.0[b]
Corn steep	2.4[c]	Tap water	q.s.
Sucrose	3.6		
Calcium carbonate	0.9		
Ammonium sulfate	0.24		
Water	q.s.		
pH after sterilization: 7.2			
Production medium			
Soy flour	4		
Distillers' solubles	0.5		
Starch	4		
Soy oil	0.5		
Sodium chloride	1		
Water	q.s.		
pH is adjusted to 7.2 with sodium hydroxide solution			

[a] Initial concentration; no antifoam addition during seed fermentation is necessary. [b] Initial concentration; antifoam addition during production fermentation is necessary to control foam. [c] As batched. Volume increases 20% on sterilization.

Messenger, A. J. M. and R. Barclay (1983). Bacteria, iron and pathogenicity. *Biochem. Educ.*, **11**, 54–64.

Neilands, J. B. (1981a). Iron absorption and transport in micro-organisms. *Annu. Rev. Nutr.*, **1**, 27–46.

Neilands, J. B. (1981b). Microbial iron transport compounds (siderophores) as chelating agents. In *Development of Iron Chelators for Clinical Use*, ed. A. E. Martell, W. J. Anderson and D. G. Badman, pp. 13–31. Elsevier/North Holland, Amsterdam.

Neilands, J. B. (1982). Microbial envelope proteins related to iron. *Annu. Rev. Microbiol.*, **36**, 285–309.

Neilands, J. B. and C. Ratledge (1982). Microbial iron transport compounds. In *CRC Handbook of Microbiology*, ed. A. J. Laskin and H. A. Lechevalier, 2nd edn., vol. 4, pp. 565–574. CRC Press, Cleveland, Ohio.

Neinhuis, A. W., D. T. Peterson and W. Henry (1977). Evaluation of endocrine and cardiac function in patients with iron overload on chelation therapy. In *Chelation Therapy in Chronic Iron Overload*, ed. E. C. Zaino and R. H. Roberts, pp. 1–15. Symposia Specialists, Miami.

Philson, S. B. and M. Llinas (1982). Siderochromes from *Pseudomonas fluorescens*. II. Structural homology as revealed by NMR spectroscopy. *J. Biol. Chem.*, **257**, 8081–8085.

Pitt, C. G. (1981). Structure and activity relationship of iron chelating drugs. In *Development of Iron Chelators for Clinical Use*, ed. A. E. Martell, W. F. Anderson and D. G. Badman, pp. 105–131. Elsevier/North Holland, Amsterdam.

Pogglitsch, H., W. Petek, O. Wawschinek and W. Holzer (1981). Treatment of early stages of dialysis encephalopathy by aluminium. *Lancet* **2**, 1344–1345.

Powell, P. E., G. R. Cline, C. P. P. Reid and P. J. Szaniszlo (1980). Occurrence of hydroxamate siderophore iron chelators in soils. *Nature (London)*, **287**, 833–834.

Ratledge, C. (1982a). Nutrition, growth and metabolism. In *Biology of the Mycobacteria*, ed. C. Ratledge and J. Stanford, vol. 1, pp. 187–271. Academic, London.

Ratledge, C. (1982b). Mycobactins and nocobactins. In *Handbook of Microbiology*, 2nd edn., CRC Press, Cleveland, OH. ed. A. J. Laskin and H. A. Lechavalier, 2nd edn., vol. 4, pp. 575–581. CRC Press, Cleveland, OH.

Ratledge, C. and M. J. Hall (1971). Influence of metal ions on the formation of mycobactin and salicylic acid in *Mycobacterium smegmatis* grown in static culture. *J. Bacteriol.*, **108**, 312–319.

Robotham, J. L. and P. S. Lietman (1980). Acute iron poisoning. *Am. J. Dis. Child.*, **134**, 875–879.

Rogers, H. J., V. E. Woods and C. Synge (1982). Antibacterial effect of the scandium and indium complexes of enterochelin on *Escherichia coli*. *J. Gen. Microbiol.*, **128**, 2389–2394.

Snow, G. A. (1970). Mycobactin: iron-chelating growth factors from mycobacteria. *Bacteriol. Rev.*, **34**, 99–125.

Stiefel, E. I. and G. D. Watt (1979). *Azotobacter* cytochrome b is a bacterioferritin. *Nature (London)*, **279**, 81–83.

Teintze, M., M. B. Hossain, C. L. Barnes, J. Leong and D. van der Helm (1981). Structure of ferric pseudobactin, a siderophore from a plant growth promoting *Pseudomonas*. *Biochemistry*, **20**, 6446–6457.

Volf, V. (1978). Treatment of Incorporated Transuranium Elements (*Tech. Rep. Ser. IAEA*, no. 184). International Atomic Energy Agency, Vienna.

Waring, W. S. and C. H. Werkman (1942). Growth of bacteria in an iron-free medium. *Arch. Biochem.*, **1**, 303–310.

Weinberg, E. D. (1978). Iron and infection. *Microbial Rev.*, **42**, 45–66.

White, G. P., R. Bailey-Wood and A. Jacobs (1976). The effect of chelating agents on cellular iron metabolism. *Clin. Sci. Mol. Med.*, **50**, 145–152.

Winder, F. G. (1982). Mode of action of the antimycobacterial agents and associated aspects of the molecular biology of mycobacteria. In *Biology of the Mycobacteria*, ed. C. Ratledge and J. Stanford, vol. 1, pp. 353–438. Academic, London.

Yancey, R. J. and R. A. Finklestein (1981). Siderophore production by pathogenic *Neisseria* spp. *Infect. Immun.*, **32**, 600–608.

Yariv, J., A. J. Kalb, R. Sperling, E. R. Bauminger, S. G. Cohen and S. Ofer (1981). The composition and the structure of bacterioferritin of *Escherichia coli*. *Biochem. J.*, **197**, 171–175.

Yokoyama, A., Y. Ohmomo, K. Horiuchi, H. Saji, H. Tanaka, K. Yamamoto, Y. Ishii and K. Torizuka (1982). Deferoxamine, a promising bifunctional chelating agent for labelling proteins with gallium: G-67 DF-HSA. *J. Nucl. Med.*, **23**, 909–914.

14

Steroid Fermentations

T. L. MILLER
The Upjohn Company, Kalamazoo, MI, USA

14.1 INTRODUCTION

In the late 1940s the second wave of modern wonder drugs began. Just as the early 1940s ushered in the era of antibiotics, the last part of that decade saw the demonstration of the use of corticosteroids for treatment of inflammatory diseases (*i.e.* rheumatoid arthritis) and injuries. The naturally occurring corticosteroids, cortisone and cortisol (hydrocortisone), were used in the successful treatment of arthritis and other acute inflammatory and allergic diseases. Immediately there was a great demand for synthetic sources of these compounds. Various plant and animal steroids were examined as potential sources of this new class of medicinal compounds. For example, cholesterol was examined as a possible steroid source, but it was found to be very difficult to convert the saturated eight-carbon side chain into the required two-carbon cortical side chain. However, a successful process starting with bile acids (deoxycholic acid) for the production of corticosteroids was developed in Europe. It soon became apparent that phytosterols, especially diosgenin and stigmasterol, offered excellent potential sources of pregnenolone and progesterone, respectively. Diosgenin was obtained from the root of the barbasco plant (*Dioscorea villosa*), which grows wild principally in Mexico and Central America, while stigmasterol is abundant in soya sterols obtained from soybean seed oil. Other plant sterols such as hecogenin and solasodine from *Agave* and *Solanum* species, respectively, have been used to a much lesser extent as steroid starting materials.

Pregnenolone from diosgenin is easily converted to progesterone (Oppenauer reaction), which has the proper 3-one-4-ene A-ring configuration of corticosteroids. Unfortunately, progesterone has female hormone activity influencing pregnancy, rather than corticosteroid activity. Chemical introduction of a 21-hydroxyl group into the progesterone gives deoxycorticosterone, a naturally occurring corticosteroid which is medically useful, but it does not have the desired activity of, for example, hydrocortisone. Figure 1 shows the structure of deoxycorticosterone as well as the carbon atom numbering system for steroids. What is still required to obtain cortisone or corticol (hydrocortisone) from deoxycorticosterone is a 17α-hydroxyl which could be introduced chemically or microbiologically and an 11-hydroxyl or 11-keto function. This proved much more difficult to obtain. The important bioconversion of steroids really got started with microbiological methods for introducing an 11-hydroxyl group, although the first steroid conversions were apparently reductions of 17-keto steroids that were reported by Mamoli and Vercellone (1937a, 1937b); somewhat later the 7-hydroxylation of cholesterol was reported by Krámli and Horváth (1948), but these bioconversions were not of commercial significance.

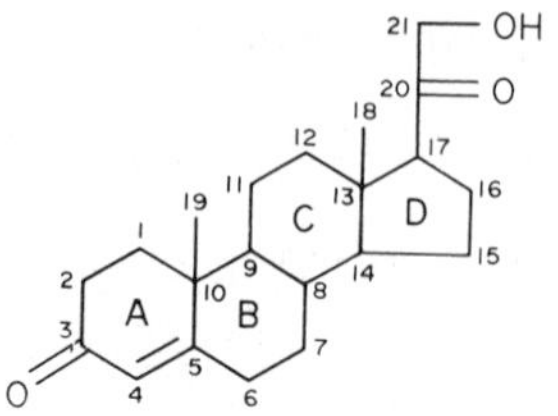

Figure 1   Structure of deoxycorticosterone

Thus, the stage was set for the first report of the very important 11α-hydroxylation by Peterson and Murray (1952). A patent covering a process employing *Rhizopus nigricans* was issued to Upjohn in that year (Murray and Peterson, 1952). Chemistry was subsequently developed for converting the 11α-hydroxyl into the desired 11-keto or 11β-hydroxyl group. It should be pointed out that in steroid stereochemistry the ring system is considered essentially planar and substituents on carbon atoms that project below the plane of the molecule as shown on the page are designated as α while those projected above the molecule are β, and are represented by dotted and solid bonds, respectively.

In the late 1950s and early 1960s steroid compounds with hormonal activity began to find wide use as oral contraceptives, but production of these compounds did not involve fermentation. Two problems that plague steroid bioconversion studies should be considered: firstly, the steroid presents ∼19 to 29 carbon atoms available for attack; secondly, most steroids are quite insoluble in the aqueous systems usually used for bioconversions.

Presently the first problem seems to have been adequately resolved since a wide range of reactions at various positions on steroid substrates have been well documented. Generally these reaction are stereospecific and a given culture usually produces mainly a single product and one or more minor by-products. Indeed the wide variety of bioconversions is detailed in several excellent books and reviews that have appeared during the last two decades (Čapek *et al.*, 1966; Charney and Herzog, 1967; Fonken and Johnson, 1972; Iizuka and Naito, 1967; Marsheck, 1971; Vézina *et al.*, 1971). Therefore no attempt will be made to review the steroid conversions in general. It is sufficient to point out the classes of reactions that have been reported (see Table 1). Also total steroid degradation, which encompasses several of these types of reactions, is known to occur with certain organisms. With regard to the second problem, that of steroid solubility, a more complete discussion will be given later in this review.

Table 1   Common Microbiological Reactions of Steroids

Dehydrogenation	Hydrolysis of esters
Epoxidation	Hydrolysis of epoxides
Esterification	Hydroxylation
Isomerization	Oxidation of alcohols and ketones
Hydrolysis of acetals	Reduction of ketones and double bonds

The steroid bioconversions that are of practical importance will be covered in some detail;

these are 11-hydroxylation (both α and β positions), 16α-hydroxylation and 1-dehydrogenation. The 17α- and 21-hydroxylations are also of potential practical importance. The important aspects covered in this discussion should apply equally well to other steroid conversions of the same type. Should information be desired on other steroid bioconversions, it would only be necessary to consult one or more of the available reference texts mentioned above to determine the organisms and conditions to be used. Other important topics such as steroid substrate availability including, for example, solubility, suspension, derivatization, solvent conversions, *etc.*, and current areas of greatest active research interest, *i.e.* sterol degradation, use of immobilized cells and enzymes and conversions at high substrate levels in organic solvents, will be discussed.

There is an area that will not be covered in depth, but is important and should at least be mentioned, namely the matter of choosing the proper medium for a given bioconversion. Generally any of a wide variety of media will work, but the products obtained may be dependent on medium composition. Therefore, it is usually necessary to test several media before selecting the one to be used for a given bioconversion. However, in general with steroid bioconversions, a well balanced medium in terms of carbon, nitrogen and cofactors will give reasonable results and fine tuning is not nearly as critical as in antibiotic fermentations.

14.2 BIOCONVERSIONS OF PRACTICAL IMPORTANCE

14.2.1 11-Hydroxylations

As mentioned above, 11α-hydroxylation was the first reported bioconversion of great practical significance. Peterson and Murray (1952) reported that molds of the order Mucorales, and especially *Rhizopus arrhizus*, could convert progesterone to 11α-hydroxyprogesterone. However, in practice today, the organisms most widely used for 11α-hydroxylation are *Rhizopus nigricans* and *Aspergillus ochraceus*. A typical reaction including the most common by-product, 6α,11β-dihydroxyprogesterone, is shown in Figure 2. Most of the discussion of 11α-hydroxylation will focus on the use of progesterone as the substrate since this is the reaction most often reported in the literature. The 11α-hydroxylase enzyme is inducible, with progesterone serving as an excellent inducer (Sallam *et al.*, 1971; Shibahara *et al.*, 1970). The 6β-hydroxylase is also inducible and 11α-hydroxyprogesterone was the best known inducer (Shibahara *et al.*, 1970). With *Aspergillus ochraceus*, Zn^{2+} appears to be required for the induction of the 6β-hydroxylase (Dulaney *et al.*, 1955). This finding is of obvious significance since reduction of Zn^{2+} in the fermentation medium would result in production of less of the undesirable 6β-hydroxylated side product. On the other hand, there is some evidence suggesting that the 11α-hydroxylase is an Fe^{2+}-requiring enzyme. Many studies of these enzymes have been done using cell-free systems from both *A. ochraceus* and *R. nigricans*.

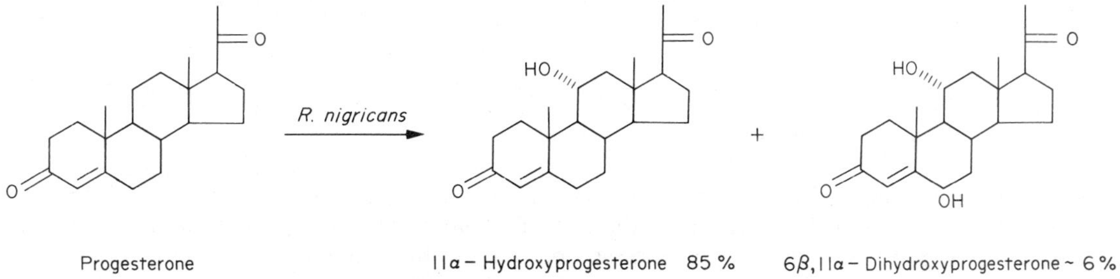

Figure 2 Conversion of progesterone by *Rhizopus nigricans*

A number of other minor products are formed from progesterone by *R. nigricans*; these include the 17α- and 21-hydroxy derivatives as well as 11α-hydroxyallopregnane-3,20-dione (El-Monem *et al.*, 1970; Sallam *et al.*, 1971). Thus, this organism must possess 17α- and 21-hydroxylases, as well as a 4,5 reductase. It should be pointed out that the spectrum of by-products may vary with different organisms or different strains of the same species, and also with different culture conditions.

The literature indicates that usually conversions of 70 to 90% to the desired 11α-hydroxylated product are obtained (Abd-Elsamie *et al.*, 1969; Hanson and Maxon, 1965). Substrate levels in excess of 20 g l⁻¹ progesterone may be used (Hanson and Maxon, 1965); the matter of substrate levels as well as the problem of mixed crystal formation will be covered in more detail below.

Much evidence suggests that mixed function oxygenases are responsible for 11α-hydroxylation by microorganisms. It has been found that the reaction requires molecular oxygen, a reduced pyridine nucleotide and may involve an iron-containing enzyme. It has also been demonstrated that the addition of Mg^{2+} and PO_4^{3-} to the medium stimulates the conversion of progesterone to 11α-hydroxyprogesterone by *R. nigricans* (El-Refai *et al.*, 1970), however, the side reactions were also stimulated by these ions and the significance of this observation is not clear.

In the case of steroid 11β-hydroxylation the first reported and most commonly used microorganisms are *Curvularia lunata* (Shull *et al.*, 1953) and *Cunninghamella blakesleeana* (Hanson *et al.*, 1953). The most frequently used substrate for 11β-hydroxylation is compound S (17α,21-dihydroxypregn-4-ene-3,20-dione), and the product of the reaction is hydrocortisone. The literature indicates that this reaction is not as efficient as 11α-hydroxylation. The substrate concentrations reported are lower and higher levels of side products are formed (Dulaney and Stapley, 1959). However, process improvements have been introduced recently (Dahl *et al.*, 1979; DeFlines and Vander Waard, 1970).

The conversion of compound S by cell-free extracts of *C. lunata* has been studied, although apparently not extensively. This reaction along with the most common by-product is shown in Figure 3. 14α-Hydroxylation seemingly always accompanies 11β-hydroxylation in this fungus. In fact, there is speculation that both reactions are catalyzed by the same enzyme. Like the 11α-hydroxylase, this enzyme requires molecular oxygen and NADPH, therefore, it is probably a mixed-function oxygenase. The enzyme is inducible with compound S and other substituted pregnanes lacking functionality at the C-11 or C-14 position serving as inducers. The principal position of hydroxylation can vary depending on the substrate, for example C-19 (androstane series) steroids are 14α-hydroxylated, 19-nortestosterone is 10β- and 11β-hydroxylated and C-27 sterols are not hydroxylated. Without exception, the inducers also serve as substrates for the hydroxylating enzyme. The purified enzyme has been found to be unstable at temperatures much above 2 °C, but it is quite stable at −30 °C. The pH optimum is about pH 8.0. Minor products formed by *C. lunata* usually include the 7α-hydroxy, 11α-hydroxy or the 20-hydroxy derivative; obviously the latter derivative arises *via* the reduction of the 20-oxo group of compound S (Zuidweg, 1968; Zuidweg *et al.*, 1962).

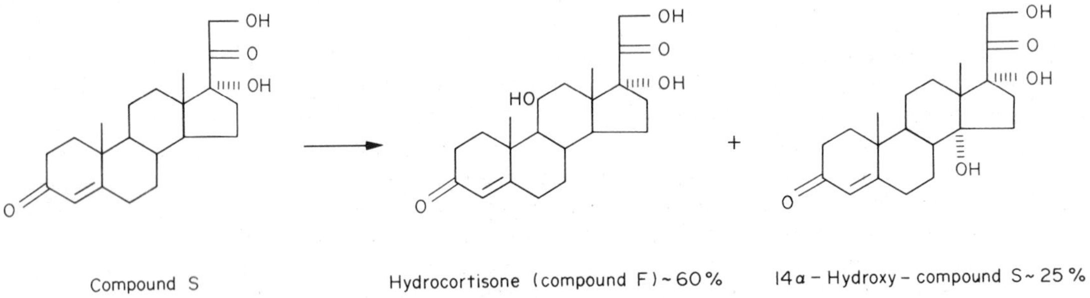

Compound S   Hydrocortisone (compound F) ~ 60%   14α – Hydroxy – compound S ~ 25%

Figure 3   Conversion of compound S by *Curvularia lunata*

Many other organisms have been mentioned in the literature as being capable of bringing about 11β-hydroxylation. However, in addition to *C. lunata* only *Cunninghamella blakesleeana* appears to give good enough conversion of compound S to be of potential commercial interest. The latter organism gives bioconversions of compound S commonly in the proportion of about 70% hydrocortisone and 20% cortisone, that is, the 11-keto compound is also formed (Mann *et al.*, 1955; O'Connell *et al.*, 1955). With both organisms conversions to hydrocortisone generally appear to be below 65% as compared to conversions of 85% or slightly higher for 11α-hydroxylation of progesterone. Moreover, the substrate levels achieved with 11β-hydroxylation appear lower than with 11α-hydroxylation as mentioned above. Conversions with organisms such as *Streptomyces fradiae* and *Absidia orchidis* have given much poorer results (Colingsworth *et al.*, 1952; Čapek and Hanč, 1961).

One last point of possible interest is the report of a *Pseudomonas* mutant that converted compound S directly to prednisolone (Takeda *et al.*, 1959). Thus, this organism was apparently capable of both 1-dehydrogenation and 11β-hydroxylation. However, follow-ups of this old report do not appear in the literature, and it is, therefore, unlikely that an efficient combined 11β-hydroxylation–1-dehydrogenation process was ever developed.

14.2.2 1-Dehydrogenations

The second most important and common practical conversion of steroids brought about by microorganisms is the 1-dehydrogenation reaction. It was found quite early (1954) during the period of intensive research on steroids that the introduction of a 1,2 double bond into steroids such as cortisone or hydrocortisone formed products (prednisone and prednisolone, respectively) that had 3 to 5 times the potency of their precursors. In addition, side effects such as salt retention in humans were greatly diminished.

There are many microorganisms that reportedly 1-dehydrogenate steroids. Examples include the bacteria *Arthrobacter (Corynebacterium) simplex*, *Bacillus sphaericus* and *Nocardia restrictus* and the molds *Septomyxa affinis* and *Fusarium solani*. However, only a few representative examples will be discussed. A typical 1-dehydrogenation, the conversion of hydrocortisone to prednisolone, is shown in Figure 4. Some details will be provided on what is known about the mechanism and some other characteristics of this reaction.

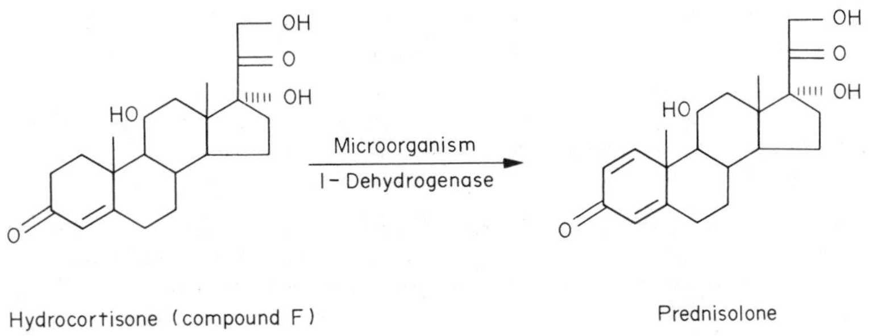

Hydrocortisone (compound F)   Prednisolone

Figure 4   Steroid 1-dehydrogenation by microorganisms

It is significant that in each of the organisms that have been studied 1-dehydrogenase appears to be an inducible enzyme (Sih and Bennet, 1962; Thoma *et al.*, 1957; Koepsell, 1962). For example, with *Septomyxa affinis*, the enzyme is not formed in the absence of an inducer. Table 2 shows the effectiveness of several steroids for enzyme induction in *S. affinis*. It is noted that BNA (3-ketobisnor-4-cholen-22-al) and progesterone are excellent inducers while dienediol (11β,21-dihydroxy-4,17(20)-pregnadien-3-one) is not as good, and cortisone, hydrocortisone, *etc.*, do not induce at all. Therefore, for certain steroids such as hydrocortisone, an enzyme inducer must be added to the fermentation beer before the bioconversion can occur. Also, with the *S. affinis* enzyme, induction appears to occur only after glucose has been exhausted in the medium. The pH optimum of the enzyme is 7.0–8.0. Kinetic data using dienediol as substrate suggest that product inhibition occurs. A mathematical model was developed by Chen *et al.* (1965) to optimize the process for this bioconversion. The model indicated that incremental additions of substrate resulted in greater conversion to the product and resulted in lower residual substrate levels. As with 11-hydroxylation, conversions have been demonstrated using both mycelia and spores. No enzyme induction was noted with *S. affinis* spores, but it is likely that the substrate itself, compound S, was acting as the inducer (Singh *et al.*, 1965).

The mechanism of steroid 1-dehydrogenation by *Bacillus sphaericus* has been studied using isotopically labeled substrates (Ringold *et al.*, 1963; Jerussi and Ringold, 1965). It has been concluded that the reaction involves trans diaxial elimination of the 1α,2β-hydrogens. The proposed mechanism involves first enolization and then hydride abstraction as shown in Figure 5. This enzymatic mechanism is identical to that proposed for the chemical 1-dehydrogenation by dichlorodicyanoquinone (DDQ). There are numerous indications that 1-dehydrogenation is mediated by a flavoprotein. It should also be mentioned that in cell-free systems, artificial electron acceptors, such as phenazine methosulfate, 2,6-dichlorophenol, indophenol and menadione, have been used successfully (Sih and Bennett, 1962; Jerussi and Ringold, 1965). The pH optimum of this enzyme was $\sim$7.0 and the K_m was 1.2×10^{-4} M for androstenedione (Jerussi and

Table 2 Induction of 1-Dehydrogenase in *Septomyxa affinis* by Various Steroids[a]

Steroid tested[b]	1-Dehydrogenase potency of beer (DU ml^{-1})	Steroid tested[b]	1-Dehydrogenase potency of beer (DU ml^{-1})
None	0	Cortisone	0
BNA[c]	2330	Hydrocortisone	0
Dienediol[d]	480	6α-Methylhydrocortisone	0
Progesterone	870	Prednisolone	0

[a] Koepsell (1962).
[b] The steroids at 50 μg ml^{-1} were added to shake flasks at 18 h after inoculation, and the beer was assayed for enzyme 16 h later.
[c] BNA = 3-Ketobisnor-4-cholen-22-al.
[d] Dienediol = 11β,21-dihydroxy-4,17(20)-pregnadien-3-one.

Ringold, 1965), while the *Nocardia restrictus* 1-dehydrogenase has an optimum pH of 9.0–9.5 and a K_m of 1.4×10^{-5} M for testosterone (Sih and Bennett, 1962).

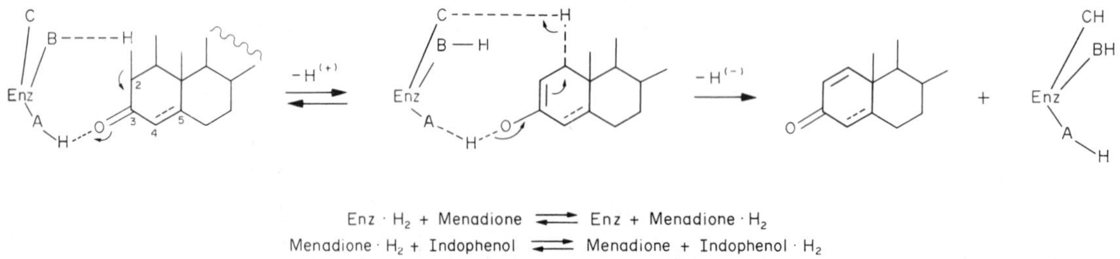

$$Enz \cdot H_2 + Menadione \rightleftharpoons Enz + Menadione \cdot H_2$$
$$Menadione \cdot H_2 + Indophenol \rightleftharpoons Menadione + Indophenol \cdot H_2$$

Figure 5 Proposed mechanism of enzymatic steroid dehydrogenation in the presence of menadione and 2,6-dichloroindophenol. A=proton donor; B=proton acceptor; C=hydride acceptor (Reprinted with permission from Jerussi and Ringold, 1965. Copyright 1965, American Chemical Society)

More recent reports suggest that *Mycobacterium phlei* and *Septomyxa affinis* bring about 1-dehydrogenation by a mechanism that is either identical to, or very similar to, that of *B. sphaericus* (Phillips and Ross, 1974; Abul-Hajj, 1972). However, with bacteria there are indications that there is a certain endogenous enzyme level, but inducers may increase it 8–10 fold. It may be assumed that most other microbial 1-dehydrogenating enzymes have the same or a very similar mechanism.

The literature suggests that in conversions of practical importance substrate levels are generally 0.5 to 2.0 g l^{-1} and conversions reach 80–90% completion. It seems clear that many industrial 1-dehydrogenation fermentations must be running at considerably higher substrate levels in order to achieve acceptable economics. Indeed, there is a report of the conversion of hydrocortisone to prednisolone by *Arthrobacter simplex* at the remarkable substrate level of 500 g l^{-1}. The authors coined the term 'pseudo-crystallofermentation' to describe this bioconversion (Kondo and Masuo, 1961). Typical side-reactions accompanying 1-dehydrogenations include steroid side-chain cleavage, reductions, and the complete degradation of some steroids.

14.2.3 16α-Hydroxylations

Another reaction of considerable commercial importance is microbial 16α-hydroxylation. In the mid-1950s it was discovered that the addition of a 9α-fluorine atom to some corticosteroids greatly enhanced their antiinflammatory activity. Unfortunately, this substitution also increased salt retention in humans. Introduction of the 1,2 double bond was not sufficient to decrease this undesirable side effect; however, the introduction of a 16α-hydroxyl group gave a product, 9α-fluoro-16α-hydroxyprednisolone, that had the desired properties (Thoma *et al.*, 1957). This reaction is shown in Figure 6. This reaction had been described earlier using an organism later classified as *S. argenteolus* and with progesterone as the substrate (Perlman *et al.*, 1952). The literature indicates that substrate levels up to 0.5 g l^{-1} may be used and conversions of about 50% are reported.

It should be noted that in some cases the 2β-hydroxy derivative either with or without the 16α-

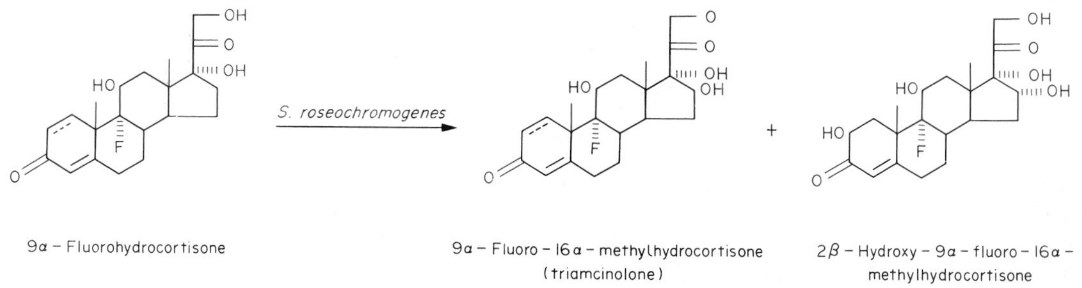

9α – Fluorohydrocortisone   9α – Fluoro – 16α – methylhydrocortisone   2β – Hydroxy – 9α – fluoro – 16α –
(triamcinolone)   methylhydrocortisone

Figure 6   16α-Hydroxylation by *Streptomyces roseochromogenes*

hydroxyl group may be formed as a side product. However, this side reaction can reportedly be eliminated or minimized by selection of the proper strain or mutant. Furthermore, it has also been found that if the 1,2 double bond is introduced before 16α-hydroxylation is carried out, *i.e* if the substrate is 9α-fluoroprednisolone instead of 9α-fluorohydrocortisone, then the 2β-hydroxylation is eliminated (Goodman and Smith, 1961).

One non-enzymatic side reaction that may be a problem should be mentioned. It has been found that if the concentration of iron in the medium is about 100 mg 1^{-1} or higher, the D-ring homoannulated product is formed (9α-fluoro-11β,16α,17aα-trihydroxy-17aα-hydroxymethyl-4-D-homoandrostene-3,17-dione). Its structure is show in Figure 7. This latter product can be eliminated by either reducing the iron concentration in the medium or by adding phosphate to complex the iron (Goodman and Smith, 1960). Apparently the mechanism of 16-hydroxylation has not yet been elucidated, but it is not unreasonable to assume that this reaction, like the 11-hydroxylations, occurs *via* a mixed function oxygenase. The enzyme appears to be constitutive and it is not very substrate specific since a rather wide variety of steroids are 16-hydroxylated.

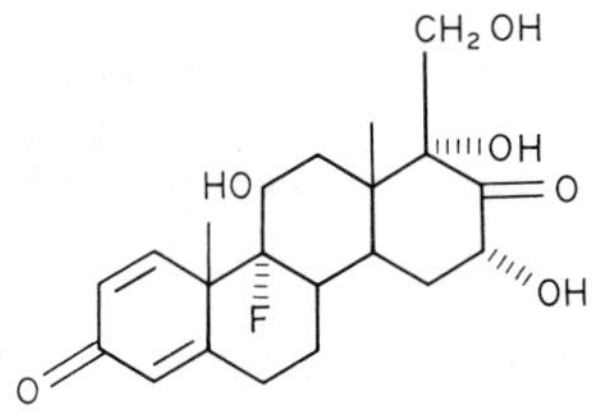

Figure 7   D Homo isomer of triamcinolone

14.3   BIOCONVERSIONS OF LIMITED OR POTENTIAL PRACTICAL IMPORTANCE

14.3.1   Progesterone Side Chain Cleavage

There are several products that are made from the microbiological cleavage of the progesterone side chain that are of limited practical importance. For example, a number of molds cleave the two-carbon side chains of progesterone (Miller, 1972). Figure 8 shows the usual sequence. The first reaction is the formation of the 17-acetate by the introduction of oxygen between C-17 and C-20. Next an esterase cleaves the acetate leaving the 17-hydroxy steroid, testosterone. The 17-hydroxyl group is then oxidized to the 17-ketone, which in some cases is the final product. However, most organisms insert an oxygen molecule between C-17 and C-13 forming a ring D lactone which is sometimes further enzymatically hydrolyzed to give the acid as shown in Figure 8. There are methods available for inhibiting the last two reactions so as to accumulate the 17-keto product (Miller, 1972). In addition, some of the organisms form the 1-dehydrogenated derivatives. The intermediates, *e.g.* androsta-1,4-diene-3,17-dione, in this sequence may be useful for the production of androgens, estrogens, *etc.*, by further chemical modification. Presently this type of bioconversion does not represent a major source of these compounds. However, the product of the fermentation can be 1-dehydrotestololactone, which is approved for the treatment of mammary cancer.

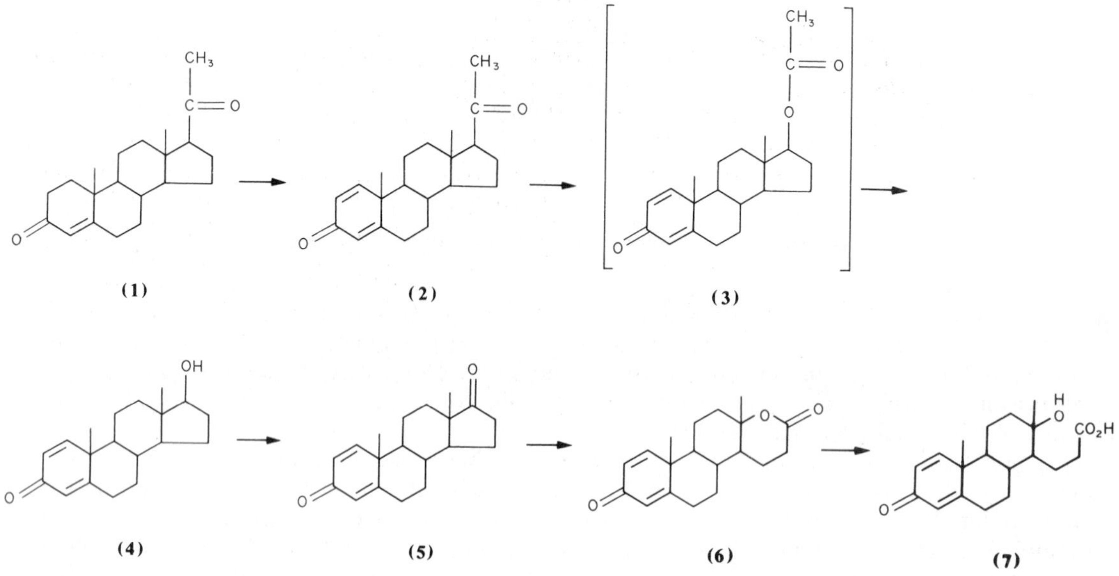

Figure 8 Proposed sequence for the conversion of progesterone by some fungi: progesterone (**1**); Δ^1-progesterone (**2**); Δ^1-testosterone acetate (**3**); Δ^1-testosterone (**4**); androstadienedione (**5**); Δ^1-testololactone (**6**); Δ^1-testolic acid (**7**)

14.3.2 Ring A Aromatization

It is also possible to produce ring A aromatic steroids microbiologically. Dodson and Muir (1961) found that 19-hydroxyandrostenedione was converted to estrone by a pseudomonad. They proposed that the organism 1-dehydrogenated the steroid and then the hydroxymethyl group was removed by reverse aldol cleavage. Later Sih *et al.* (1965b) expanded on this finding by showing that some sterol-degrading organisms convert 19-hydroxy sterols to estrone. It has also been well established that 1-dehydrogenating organisms will convert 19-nor steroids into the corresponding steroids with an aromatic ring A. However it is unlikely that such processes are presently used for the production of estrogens.

14.3.3 17α and 21-Hydroxylations

There are two additional microbial hydroxylations of steroids that could be of practical importance, but neither is thought to be in use today because superior chemical routes are available. As mentioned earlier, the 21-hydroxylation of a steroid such as progesterone is necessary in order to obtain a product with corticosteroid activity. There are few reports in the literature of organisms that carry out this reaction. The organism most frequently mentioned for 21-hydroxylation is *Ophiobolus herpotrichus* (Meystre *et al.*, 1954). There have been very few studies done with this reaction. A typical reaction would be the conversion of progesterone to 21-hydroxyprogesterone (deoxycortisone) as shown in Figure 9.

The situation is very similar for 17-hydroxylation. In this case the organism most often mentioned is *Trichothecium roseum* (Meystre *et al.*, 1954; El-Refai *et al.*, 1970). Some strains of this organism are also capable of carrying out 11α-hydroxylation and 6β-hydroxylation giving a product mix that is dependent on the composition of the medium. The conversion of progesterone to 17α-hydroxyprogesterone is low-yielding and the other hydroxylated products are generally present. The use of high substrate levels (greater than 0.5 g l^{-1}) has apparently not been successful. Typical of this reaction is the conversion of deoxycorticosterone (21-hydroxyprogesterone) to deoxycortisol (compound S), as shown in Figure 9.

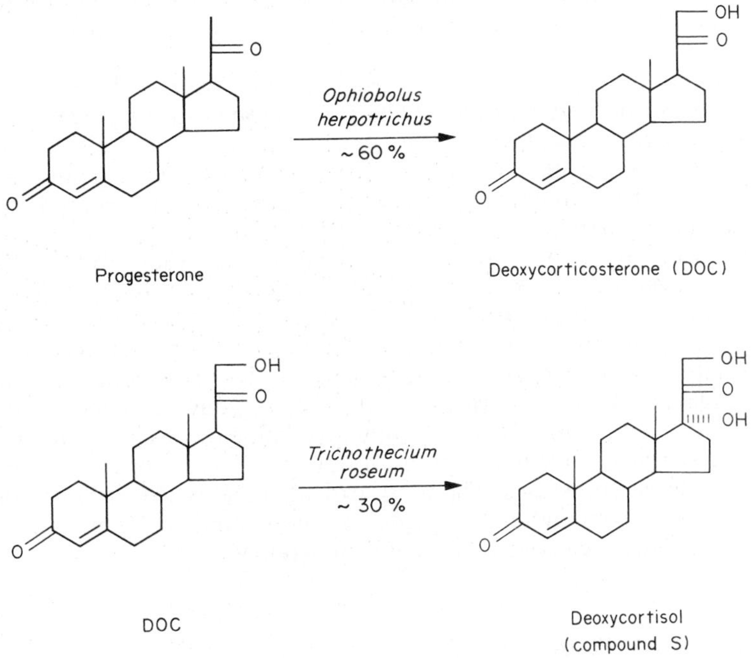

Figure 9 Microbiological 21- and 17α-hydroxylations

14.3.4 Alternative Bioconversion Methods

Generally the bioconversions discussed up to this point have centered around conventional fermentation technology, that is proper fermentation medium in a stirred tank reactor is inoculated with the desired microorganism. After a suitable period (10 to 30 h) of growth and sometimes enzyme induction, the steroid substrate is added to the fermentaion beer either as a slurried powdered solid or dissolved in an organic solvent. In this case bioconversion is brought about by living cells or in some cases mixtures of mycelia and spores. The fermentation is then harvested when the bioconversion is complete.

There are, however, other means of conducting bioconversions. For example, 1-dehydrogenations have been accomplished by several other procedures. Erickson *et al.* (1967) described a procedure for 1-dehydrogenation of 9α-fluoro-16α-hydroxyhydrocortisone by the use of acetone-dried cells of *Arthrobacter simplex*. Treating the cells in this way greatly decreased the levels of 20-keto reductase, an enzyme producing an undesirable side product. Also the cells could presumably be stored for longer periods of time in this form allowing greater freedom in scheduling bioconversions. The addition to the reaction mixture of an artificial electron acceptor such as 2-methyl-1,4-naphthoquinone (menadione) was required.

It has also been demonstrated that spores of *Septomyxa affinis* can be used for the 1-dehydrogenation of various steroids of the pregnane series (Singh *et al.*, 1963; Singh *et al.*, 1968). In this case the culture was grown on a solid or liquid medium and then the spores were harvested, concentrated and resuspended in phosphate buffer for the actual bioconversion. It was also reported that substitution at the 17α-position of the steroid by an alkyl group greatly reduced side chain cleavage.

The recovery and reuse of spores has apparently not been demonstrated, but does offer an interesting opportunity for study. Another possibility for the reuse of 1-dehydrogenase is provided by immobilization of the enzyme by entrapping it in a crosslinked polyacrylamide gel (Mosbach and Larsson, 1970). However, much of the enzyme activity was lost by the immobilization procedure. In any case, the conversion of hydrocortisone to prednisolone was demonstrated when the substrate dissolved in ethanol along with phenazine methosulfate was passed through a column packed with the immobilized enzyme. It has been shown more recently that it is much easier to simply entrap whole cells of *Corynebacterium simplex* instead of the enzyme into the polyacrylamide gel (Larsson *et al.*, 1976). In this case it frequently is not necessary to add an artificial electron acceptor and continuous operation is possible. Furthermore, much less enzyme activity was lost during the whole cell immobilization procedure than when a cell-free enzyme preparation was used.

11-Hydroxylations have also been carried out using spores. For example, *Aspergillus ochraceus* spores were prepared, suspended in phosphate buffer and used to 11α-hydroxylate progesterone (Vézina *et al.*, 1970; Singh *et al.*, 1968). The spores could be prepared in either liquid or on solid media and stored frozen until use. The best conversion was obtained when spores were resuspended in phosphate buffer containing glucose at pH 5.2 and the progesterone was added as a micronized powder at a level of 5 g l^{-1}. A number of other steroid substrates were tested under the above conditions and found to be efficiently 11α-hydroxylated (Sehgal *et al.*, 1968).

The use of immobilized cells of *Rhizopus nigricans* was also tested for 11α-hydroxylation of progesterone (Maddox *et al.*, 1981). It was found that cells immobilized in alginate or agar gels, but not polyacrylamine gel, could carry out the bioconversion. The presence of glucose in the reaction mixture improved the bioconversion. In addition, it was demonstrated that *Curvularia lunata* immobilized in a crosslinked polyacrylamide gel could convert compound S to hydrocortisone (Mosbach and Larsson, 1970). That is, 11β-hydroxylation by this procedure is also possible.

The procedures described above using spores or immobilized cells or enzymes offer intriguing possibilities for commercial processes. The most attractive features of such processes would be the reuse of the bioconversion system coupled with the continuous operation of the process. However, it is believed that such systems are not in use commercially.

14.4 STEROL DEGRADATION

The steroid bioconversions discussed up to this point could be considered more or less classical since most have been known for several decades. Recently, however, some bioconversions have been developed that have opened a new vista for the synthesis of important products. This area of increasing practical importance is sterol fermentation. The first reports of microbiological sterol degradation were by Söhngen in 1913 (Charney and Herzog, 1967). As the more common sources of steroids such as diosgenin have become more scarce and expensive, new starting materials have been sought. Sterols are common both in plants and animals as pointed out earlier and indeed stigmasterol is already an important steroid starting material for the Upjohn Company. However, sterols such as cholesterol and sitosterol had not until recently been used as starting materials because of the great difficulty in removing the side chain by chemical or microbial means. Figure 10 shows the structure of several common sterols. Sitosterol is even more abundant in soybean oil than is stigmasterol and large quantities of cholesterol are available from wool grease. It has been thought for a number of years that sterols such as these might be subjected to microbial degradation to give useful steroids. In the past few years this area of investigation has received a great deal of attention. A brief review will be given covering what is known about microbial sterol degradation. This degradation of sterols appears to be restricted to a few bacterial genera; those most commonly mentioned in the literature are *Arthrobacter (Corynebacterium)*, *Bacillus*, *Brevibacterium*, *Mycobacterium*, *Microbacteria*, *Nocardia*, *Protaminobacter*, *Serratia* and *Streptomyces*.

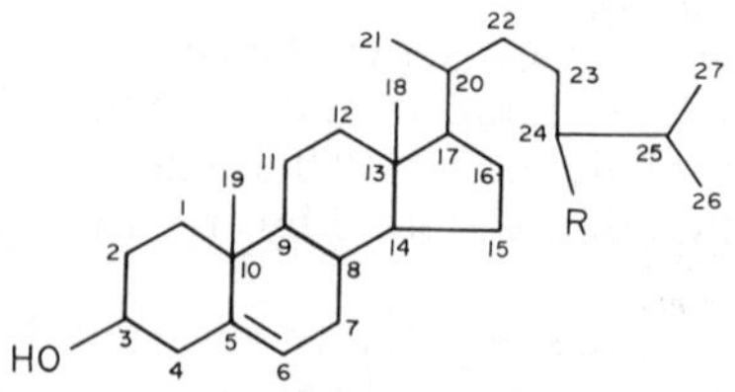

Figure 10   Structure of several sterols: R= H, cholesterol; R=CH$_3$ (28), campesterol; R=CH$_2$CH$_3$ (28,29) sitosterol

Currently it is not known exactly at which end of the sterol molecule the organism acts first or whether in some cases degradation begins at both ends simultaneously. However, it has been reported that when C^{14}-4-cholesterol or C^{14}-26-cholesterol were oxidized by *Mycobacterium* spp. carbon-4 was oxidized to CO$_2$ about four times more rapidly than carbon-26 (Stadtman *et al.*, 1954). This suggests that the steroid rings are degraded more rapidly than the side chain.

This review will begin by considering degradation at the 'tail end', *i.e.* the side chain. Side chain degradation has been studied by first blocking ring degradation. It has been found in several studies using mycobacteria that hydroxylation of the terminal carbon is the first step (Zaretskaya *et al.*, 1968; Schubert *et al.*, 1969). This reaction is probably catalyzed by a mixed-function oxygenase just as is the case with alkanes. The evidence with cholesterol indicates that the hydroxylated carbon is next converted to a carboxyl group probably with the aldehyde as an intermediate. Figure 11 shows how the cleavage of the cholesterol side chain is thought to occur as worked out by Sih and his coworkers (1967a) using *Nocardia restrictus*. Following the formation of the carboxyl group, cleavage occurs between carbons 24 and 25 and results eventually in a carboxyl group at carbon-24. Next a two-carbon cleavage occurs leaving a carboxyl group at carbon-22. This C_{22}-acid is clearly a key intermediate and has been isolated from bioconversion beers on several occasions. Cleavage of another three-carbon fragment leaves the well established 17-keto intermediate. The final carboxylated three-carbon side chain may be cleaved by dehydrogenation followed by hydration and aldolytic fission as shown in Figure 12 (Sih *et al.*, 1967b). The exact mechanisms of the reactions involved in side chain cleavage are not known; however, they could be similar, but not identical, to the classical β-oxidation of fatty acids. Of course, when the campesterol or sitosterol side chains are cleaved the pattern must differ somewhat because of the extra methyl or ethyl function, respectively, at carbon-24. Recently, studies using cell free preparations of *Mycobacterium* spp. have shown that side chain cleavage with the sterols occurs in a fashion similar to that of cholesterol, that is the first reaction is hydroxylation at carbon-26 followed by further oxidation to give the 26-carboxyl derivative. The carbon–carbon fission reactions lead to the same C-24 carboxyl intermediate (3-oxochol-4-ene-24-oic acid) as formed from cholesterol (Fujimoto *et al.*, 1982a).

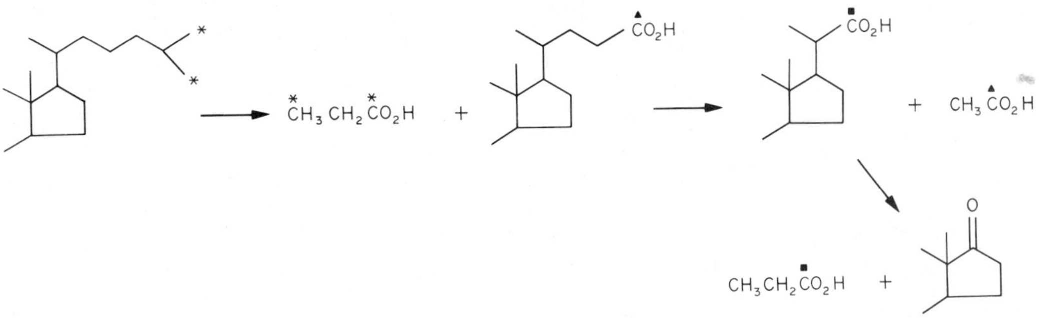

Figure 11 Degradation of the cholesterol side chain by *Nocardia restrictus* (Reprinted with permission from Sih *et al.*, 1967a. Copyright 1967, American Chemical Society)

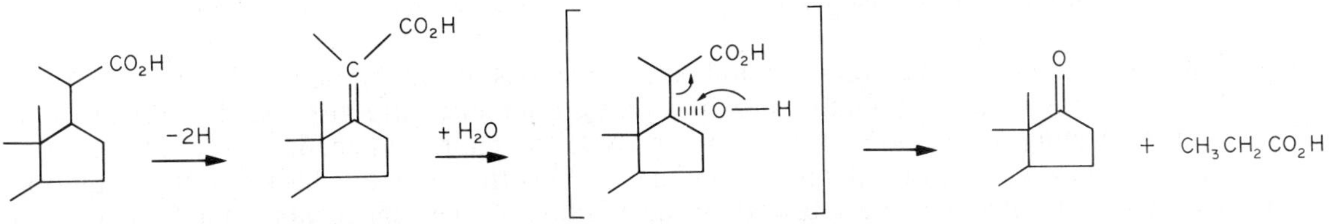

Figure 12 Possible mechanisms of degradation of the three-carbon side chain during sterol degradation (Reprinted with permission from Sih *et al.*, 1967. Copyright 1967, American Chemical Society)

Further experiments with identical cell free systems provided evidence for the mechanism of side-chain cleavage for sitosterol and campesterol. With both substrates HCO_3^- was incorporated into the C-28 position of the sterol. The resulting C-24 propionate or acetate group was then cleaved to give the common C-24-carboxyl intermediate (Fujimoto *et al.*, 1982b). Thus, for cholesterol, campesterol and sitosterol, the side-chain cleavage pattern has been determined.

The degradation of the steroid ring system will be considered next. Again some of this pathway was worked out by Sih's group using *Nocardia restrictus* (Sih, 1963; Wang and Sih, 1963; Sih and Wang, 1963; Sih *et al.*, 1965a; Gibson *et al.*, 1966; Sih *et al.*, 1966). However, work in other laboratories, including our own, has confirmed that at least in general detail, the pathway is also followed by other genera of sterol-degrading bacteria. The mechanisms and steps in this degradation are too involved to cover in depth, so only a general outline of steroid ring degradation will be given.

Androstenedione will be used as the starting substrate in this discussion although as mentioned earlier the sterol side chain, or a portion of it, could still be present during ring degradation. It seems clear that the 3-keto-$\triangle^4$- ring A configuration must be present for ring degradation to occur. As shown in Figure 13, a 1,2 double bond and a 9-hydroxy group are introduced in an order that is not definitely known, to form an unstable intermediate; this intermediate undergoes non-enzymatic reverse aldol cleavage yielding the 9,10 seco steroid. This 9,10 seco compound is hydroxylated at the 4 position and then split between carbons 4 and 5 by a dioxygenase as shown. The remainder of the ring A carbons are split off between carbons 5 and 10 to give 2-oxo-*cis*-4-hexenoic acid and a substituted indanedione with the cumbersome chemical name 3aα-*H*-4α [3'-propionic acid]-7aβ-methylhexahydro-1,5-indanedione. This compound has been given the trivial name of diketo acid (DKA). The former compound is hydrated and then split to propionaldehyde and pyruvate, while the degradation of the latter compound is uncertain, but it is probably split into two four-carbon fragments and one five-carbon fragment that would feed into the TCA cycle.

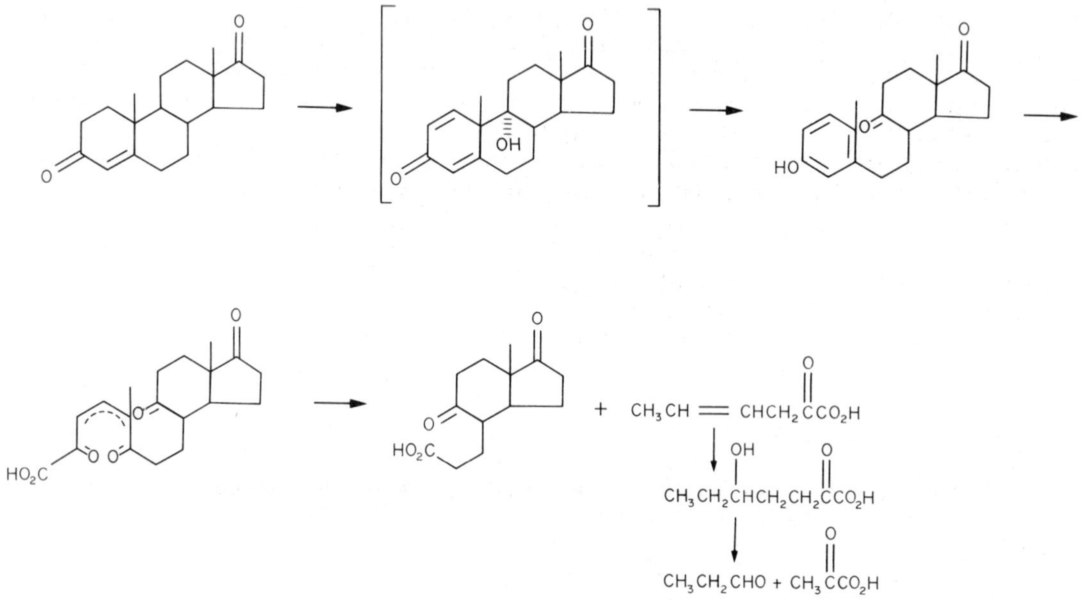

Figure 13   Steroid ring degradation by *Nocardia restrictus* (From Gibson *et al.*, 1966)

Table 3 shows some of the literature and patent methods for inhibiting microbial degradation of the steroid rings; the degradative enzymes that are inhibited are also listed. Most of these methods yield steroids such as androstenedione (AD) and androstadienedione (ADD). Each of the methods shown, except for the use of mutants, suffers from the drawback of either requiring the use of a chemically modified substrate or the addition of chemical agents to the fermentation medium. On a commercial scale this causes obvious complications and resultant increased costs.

The basic studies described above laid the groundwork for some very important process developments. Obviously, it was desirable to find microorganisms that would degrade the sterol side chain while leaving the ring system intact. Available information suggested that four enzymatic reactions were necessary for ring degradation. They are: (1) 1-dehydrogenation, (2) 9α-hydroxylation, (3) 3-dehydrogenation, and (4) 4,5 isomerization. Elimination of the enzyme for catalyzing any of these reactions should leave the ring system intact while allowing side chain degradation. Indeed, Kraychy and Marsheck described processes for producing androsta-1,4-diene-3,17-dione and androst-4-ene-3,17-dione with mutants of a *Mycobacterium* spp. (Kraychy and Marsheck, 1972; Marsheck and Kraychy, 1973). These and other key compounds are shown in Figure 14. The mutants used for these processes lacked 9α-hydroxylase and 1-dehydrogenase.

Table 3 Methods of Inhibiting Sterol Ring Degradation

Method	Inhibition	Major products[a]	Ref.
Chelating agents	9-Hydroxylase	AD and ADD	Arima *et al.* (1968)
Inorganic ions: Ni^{2+}, Pb^{3+}, Cd^{2+}, Co^{2+}, *etc.*	9-Hydroxylase	AD and ADD	Vander Waard (1965)
19-OH sterols	9-Hydroxylase	Estrone	Sih *et al.* (1965b)
6,19-Oxido sterols	1-Dehydrogenase and 9-hydroxylase	6,19-Oxido AD	Sih *et al.* (1965a)
I-Sterols (3,5 cyclo)	1-Dehydrogenase and 9-hydroxylase	17-Keto I-steroid	Vander Waard (1966)
Mutation	9-Hydroxylase	AD and ADD	Kraychy and Marschek (1972), Marschek and Kraychy (1973), Wovcha and Biggs (1981), Wovcha and Brooks (1981), Wovcha *et al.* (1980)
Mutation	1-Dehyrogenase	9HAD	Wovcha (1977)
Mutation	Uncertain	DKA	Biggs *et al.* (1977)

[a] AD = androstenedione; ADD = androstadienedione; 9HAD = 9α-hydroxyandrostenedione; DKA = diketo acid

Wovcha *et al.* (1978) described procedures for systematically selecting cultures of the desired type. Important compounds formed by mutants selected in this manner are also shown in Figure 14. The most important compound was 9α-hydroxyandrost-4-ene-3,17-dione. New chemistry was developed for adding the two-carbon cortical side chain and the 11β-hydroxyl group to this compound (Barton, 1969; Shephard and VanRheenen, 1977; Beaton *et al.*, 1978; Hessler and VanRheenen, 1980). Thus, whole new routes to many important commercial steroid products were recently developed. These combined microbiological and chemical processes represent the most significant advances in steroid chemistry in the past two decades. The other three compounds shown in Figure 14, androsta-1,4-diene-3,17-dione, androst-4-ene-3,17-dione and diketo acid, can be converted into important estrogens and androgens by known chemical means.

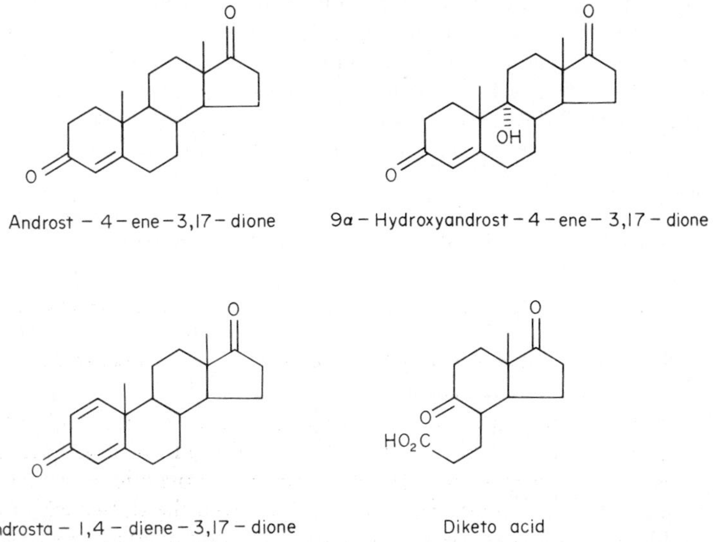

Androst − 4 − ene − 3,17 − dione   9α − Hydroxyandrost − 4 − ene − 3,17 − dione

Androsta − 1,4 − diene − 3,17 − dione   Diketo acid

Figure 14 Several important compounds provided by the degradation of sterols by mutants of mycobacteria

An improved process for the conversion of androsta-1,4-diene-3,17-dione to estrone is reportedly being used in Japan (personal communication). In addition androst-4-ene-3,17-dione is chemically converted to spironolactone which is an important drug in the treatment of hypertension. Androst-4-ene-3,17-dione may also be chemically reduced to give testosterone and some derivatives that have important medicinal uses.

In addition, a number of processes have been described for the bioconversion of sterols into steroids with side chains of varying lengths (Antosz *et al.*, 1977; Wovcha *et al.*, 1979; Wovcha *et al.*, 1980; Knight and Wovcha, 1980). All of these processes employ mutants of sterol-degrading organisms. To date, the most interesting processes make products with three-carbon side chains.

Some of these compounds can be chemically converted into steroids with the two-carbon cortico side chain. Unfortunately, no mutant has yet been described that converts sterols directly into steroids with the important two-carbon side chain.

The products with longer side chains offer interesting possibilities as starting materials for making products such as chenodeoxycholic acid and the newly discovered vitamin D_3 metabolites, *e.g.* 25-hydroxycholecalciferol.

14.5 STEROID SOLUBILITY

14.5.1 Methods of Steroid Addition

Various methods have been employed for the addition of steroid substrates to bioconversion systems. It is safe to say that there is no set rule and the steroid may be added to suit a particular bioconversion. In some cases the steroid is added to the medium before sterilization and inoculation. Of course, in this way the steroid substrate is sterilized along with the medium. However, it is frequently more desirable to add the substrate following a growth phase. In some cases an enzyme inducer must be added prior to or at the same time as the steroid substrate, as mentioned previously. If the steroid is added after the growth phase it is usually unnecessary to maintain sterility during the bioconversion phase. That is, the desired bioconversion usually occurs even in the presence of low levels of the usual microbial contaminants.

Steroids may be added to the medium after they have been dissolved in suitable water-miscible organic solvents. Frequently used solvents include *N*,*N*-dimethylformamide (DMF), acetone, dimethyl sulfoxide (DMSO), ethylene glycol, ethanol and others. Upon addition to an aqueous medium at levels above their solubilities the steroids usually precipitate as very fine particles or crystals. The limitation to this procedure is imposed by the fact that the solvents used are usually toxic to the converting organisms at levels between 2 and 10%. Under such conditions of solvent toxicity little or no conversion would be expected.

In order to obtain high steroid substrate levels and to eliminate solvent toxicity effects, finely powdered substrates have been used. In this event, the substrate may first be passed through a suiable grinder or mill and then added to the medium. Progesterone at levels of 15 to 40 g l^{-1} has been converted by *Rhizopus nigricans* in this way (Hanson and Maxon, 1965). The results of one such conversion are shown in Table 4 (Weaver *et al.*, 1960). It is noted that powdered progesterone at 50 g l^{-1} was 65% converted to 11α-hydroxyprogesterone while untreated progesterone was only 40% converted, and the recovery was also much less in the latter case. An interesting observation was made that high substrate levels tended to inhibit the formation of the by-product 6β,11α-dihydroxyprogesterone.

Table 4 Conversion of Progesterone to 11α-Hydroxyprogesterone by *Aspergillus ochraceus*[a]

Steroid treatment	Progesterone[b] (g l^{-1})	Time (h)	11α-Hydroxyprogesterone Found (g l^{-1})	Conversion[c] (%)
Ground	20.0	72	17.4	87
None	50.0	72	20.0	40
Ground	50.0	72	32.5	65

[a] From Weaver *et al.* (1960).
[b] All samples have volume of 50 ml.
[c] Based on initial progesterone content.

In fermentations conducted with steroid levels above their solubilities problems with mixed crystal formation are sometimes encountered. Maxon *et al.* (1966) reported that the conversion of progesterone to 11α-hydroxyprogesterone at 20 g l^{-1} ended with a residual substrate level of about 14%. These workers isolated mixed crystals from the harvest beer with a progesterone:11α-hydroxyprogesterone ratio of 1:6. Thus, the substrate was effectively tied up and not available for further bioconversion. It was found that a continuous slow feed of the substrate helped to minimize this problem, but did not eliminate it entirely.

One way to get around the problems of converting relatively insoluble steroids at high substrate levels (*i.e.* >2.0 g l^{-1}) is to form water-soluble steroid derivatives to serve as substrates. A process has been described for the 1-dehydrogenation of water soluble cycloborate esters of steroids (*Belg. Pat.* 63 0877). Kominek (1973) has patented a process using steroid 21-hemisuccinates as substrates for bioconversions. In the salt form these steroid derivatives are water-soluble and may be bioconverted at much higher than normal substrate levels. It should be mentioned that with this process the hemisuccinate group is usually hydrolyzed during the bioconversion, but this does not interfere with the bioconversion.

14.5.2 Steroid Conversion in Organic Solvents

Still another way of carrying out a steroid conversion is by using cells or enzymes and the substrate in an organic solvent. In this way the problems of substrate solubility, mixed crystal formation, *etc.*, are minimized or eliminated entirely. However, from a biochemical standpoint one might suspect that enzymes would be inactive in a non-aqueous environment. Indeed this may be the case with most enzyme systems, but Buckland *et al.* (1975) have successfully converted cholesterol to cholest-4-en-3-one in such a system. This conversion was carried out by a strain of *Nocardia* which reportedly contained high levels of cholesterol oxidase. The culture was grown in a 1000 l fermenter in a medium containing cholesterol to induce cholesterol oxidase synthesis. The growth conditions were regulated so that the organism did not degrade cholestenone (the desired product). The harvested cells (wet paste) were tested for their ability to convert cholesterol to cholestenone in a number of organic solvents. It was found that several organic solvents unexpectedly gave much higher conversion rates than were obtained in an aqueous system. Frozen and thawed cells (rehydrated) also were capable of carrying out the reaction, but freeze-dried cells were not. Carbon tetracholoride was a very good, but not the best, organic solvent tested and was chosen as the solvent for most of the work described. The results of the conversion of cholesterol in carbon tetrachloride in a stirred tank reactor are shown in Figure 15. The aeration rate was 0.5 vvm. It can be seen that the conversion in carbon tetrachloride is much more rapid than in the aqueous system. The conversion rate increased slightly when the agitator speed was changed from 550 to 900 r.p.m.; a further increase did not result in a bioconversion rate increase. In this experiment the substrate level (cholesterol) was 80 g l^{-1}. A small amount of unknown product was also formed at 20 °C; much more of the by-product was reportedly formed at 37 °C.

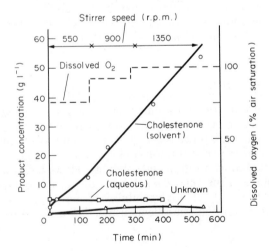

Figure 15 Conversion of cholesterol to cholestenone in carbon tetrachloride by *Nocardia* cells (From Buckland *et al.*, 1975)

A similar conversion was run at an initial cholesterol level of 160 g l^{-1}. In this experiment the reaction rate was 35 g l^{-1} h^{-1} and the reaction (in 200 ml) was complete in 5.5 h. It was observed that in this case the conversion of cholesterol in an organic solvent was very rapid and almost complete. The system consisted of an organic solvent and the only requirement was that the cells be hydrated. These investigators also found that the cells could be separated and reused. It should be noted that the solubilities of both cholesterol and oxygen are much greater in organic solvents than in water. Studies with cell-free systems were also carried out, but not reported in detail.

This work would seem to represent a major step forward in the conversion of steroids at high substrate levels by microorganisms. However, further reports on this type of process surprisingly have not appeared in the literature.

A similar study was made by Cremonesi *et al.* (1975) using a two-phase system. They proposed using a two-phase system in order to get around the problem of low steroid solubility in aqueous systems. A crystalline preparation of 20β-hydroxysteroid dehydrogenase was used in these studies that employed a 50:50 mixture or an organic solvent (butyl acetate) and an aqueous buffer. As with the work of Buckland *et al.* (1975) described above, the enzymatic activity decreased with the increasing polarity of the organic solvent. The reduction of cortisone was studied, but at much lower substrate levels (2.0 g l^{-1}) than those reported by Buckland *et al.* Nevertheless, these studies also demonstrated that very good conversion of steroids is possible in a system containing a very large amount of an organic solvent. However, this system required the use of serum albumin as an enzyme stabilizer. Under these conditions nearly 100% reduction of the 20-keto group of cortisone was obtained.

Both of these studies demonstrate the potential of converting steroids in systems employing organic solvents at high substrate levels. The recycle of cells or immobilized enzymes and the possibility of using high substrate levels offers very exciting possibilities.

These two fairly recent studies demonstrate that both whole cells and purified enzymes may be used in systems containing large proportions of organic solvents. However, in systems using cell-free enzymes, it appears likely that an aqueous phase may always be required, while with whole cells it may only be necessary to use hydrated cells.

14.6 FUTURE TRENDS IN STEROID BIOCONVERSIONS

A few comments about the areas of steroid bioconversion that might be expected to receive more attention in the near future seem appropriate. First, is the area of sterol degradation. The literature indicates that there is presently a great deal of activity in this area. Of course, the ultimate goal is to obtain new intermediate steroid sources. Breakthroughs in this area are already being made according to literature reports cited above. The second area of intensified investigation will be the conversion of steroids by immobilized cells or enzymes. Again some encouraging reports are beginning to appear. The impetus for this is that the side reactions can be reduced or eliminated and that the cells or enzymes may be reused. The third area that should gain increasing attention is the conversion of steroids at very high substrate levels using cells or enzymes in organic solvents. As mentioned above, work in this area is presently under way.

14.7 RECOVERY OF STEROIDS

A long discussion of steroid isolation seems unnecessary since most processes are based on simple variations of several primary solvent extraction procedures. Of course, with each fermentation fine tuning of the extraction process is required, which is dictated by the nature of the fermentation beer, *i.e.* pH, viscosity, distribution coefficient of the product between the solvent and beer, *etc.*, as well as product concentration and solubility. However, conceptually the usual isolation process takes advantage of the very low solubility of steroids in aqueous systems, generally less than 100 mg l^{-1}, and the relatively high solubility of steroids in organic solvents. Therefore, the isolation of steroids from fermentation beers usually utilizes one of three schemes described below.

14.7.1 Split Process

In the early steroid fermentations the substrate levels were low, *i.e.* about 0.1 to 0.5 g l^{-1}. In this type of bioconversion the product is approximately equally distributed between the soluble and insoluble form in the fermentation beer. This situation makes it imperative to recover the product that is in solution as well as that which is not. Such a process usually starts with a filtration step. Both the filter cake and the filtrate are then extracted with organic solvents as shown in Figure 16 (Colingsworth *et al.*, 1953). The cake and the filtrate extracts are then combined in later steps, concentrated, and the crystals are collected. Purification can be accomplished by recrystallization to give the final product.

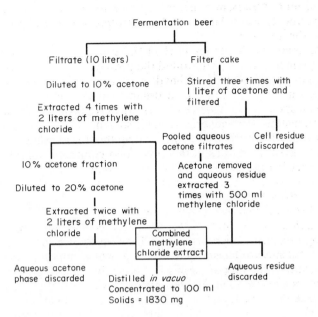

Figure 16 Typical steroid extraction flow chart: total starting medium volume=10 l; a total of 1 g of compound S was bioconverted (From Colingsworth *et al.*, 1953)

14.7.2 Whole-beer Process

With fermentation processes at either relatively high or low substrate concentrations, the product can be recovered by extracting the fermentation beer with a water-immiscible solvent. The procedure can be carried out by contacting the aqueous fermentation beer with water-immiscible solvent such as methylene chloride or butyl acetate. The phases can be separated by settling and decantation or by continuous countercurrent extraction, *e.g.* in a Podbielniak extractor. With this type of process, the steroid product remains in the organic solvent phase. With such processes a frequent problem is caused by the emulsions formed between the two phases. Media containing high levels of protein combined with the protein and nucleic acids released from lysed cells are the cause of the problem. There are various methods for breaking the emulsions, but operations are usually complex and therefore whole beer extraction processes are not generally used today.

14.7.3 Cake-extraction Process

In the more highly developed fermentations that are conducted at high steroid substrate levels, such as 11α-hydroxylation of progesterone, most of the steroid product is out of solution. In this case the usual process starts with filtration of the whole beer, *e.g.* with a rotary vacuum precoat filter. The great majority of the steroid material remains in the filter cake which is then extracted with an organic solvent such as acetone. Since at high substrate levels most of the product will be in the filter cake, the filtrate is usually discarded without further extraction. The solvent is then concentrated by distillation to give an aqueous slurry of steroid crystals. This slurry is extracted with a second solvent and the solvent concentrated to give crystals that are harvested by filtration.

Each of the above approaches involves extraction of the product into an organic solvent. This is a partially selective procedure since aqueous soluble materials are eliminated. However, other organic soluble materials, such as triglycerides, fatty acids, some antifoam agents, *etc.*, are retained in the organic solvent phase along with the desired steroid product. Therefore, further purification procedures including decolorization and recrystallization are usually required. Of course, the nearer the steroid is to the final product, the greater is the requirement for purity. For example, 11α-hydroxyprogesterone is an early intermediate that requires many subsequent chemical steps and sometimes bioconversions to reach final products; consequently, the demands for high purity of this product are not overly stringent. On the other hand, in the bioconversion of hydrocortisone to prednisolone, the bioconversion product is the final product and the demands for purity are high.

14.8 PRODUCTS OF COMMERCIAL IMPORTANCE

A large number of very important steroid compounds are produced by routes that require one or more bioconversion steps. The generic names and structures of some of the more important ones are given in Table 5. They are used in the treatment of a wide variety of serious diseases and injuries and represent combined sales in the hundreds of millions of dollars for the many companies that produce and market them.

14.9 SUMMARY

The conversion of steroids by microorganisms has been known for many years. However, the first bioconversion by a microorganism that was of practical importance was the 11α-hydroxylation of progesterone which was reported in 1952. Since that time a great many other bioconversions of steroids have been described. However, only a very few of the reactions are of practical importance. These include 11α-, 11β-, 16α-, 17α-, and 21-hydroxylations and 1-dehydrogenation. A good deal of biochemical information is available about the most important of these reactions.

Side chain cleavage, both of steroids such as progesterone and of sterols such as cholesterol and β-sitosterol, is also important for the production of useful products. The conversion of sterols is especially important since it provides new sources of steroid starting materials. The products of these bioconversions include androst-4-ene-3,17-dione, androsta-1,4-diene-3,17-dione, 9α-hydroxyandrost-4-ene-3,17-dione, substituted indanediones and estrone. These compounds serve as intermediates in the synthesis of corticosteroids, hormones (estrogens and androgens) and other medicinals such as antihypertensive and chemotherapeutic agents.

Traditionally these steroid conversions have been brought about by the use of conventional fermentation technology, that is a properly prepared fermenter was inoculated with the desired microorganism and the steroid substrate to be converted was added to the fermentation beer. After a suitable bioconversion period the product was isolated, usually by extraction into an

Table 5 Some Steroid Products of Commercial Importance that have One or More Bioconversion Steps as Part of their Synthesis

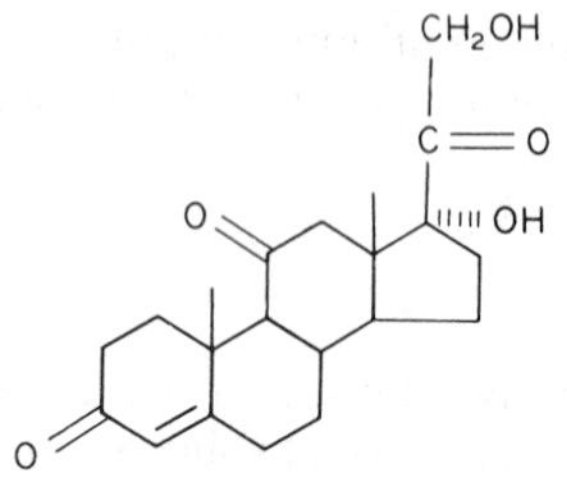

Cortisone

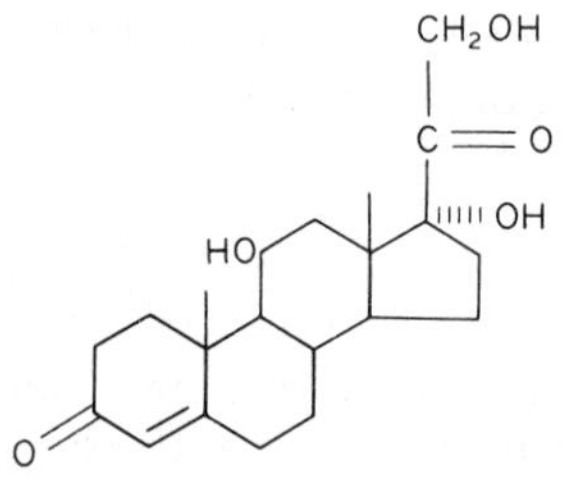

Hydrocortisone

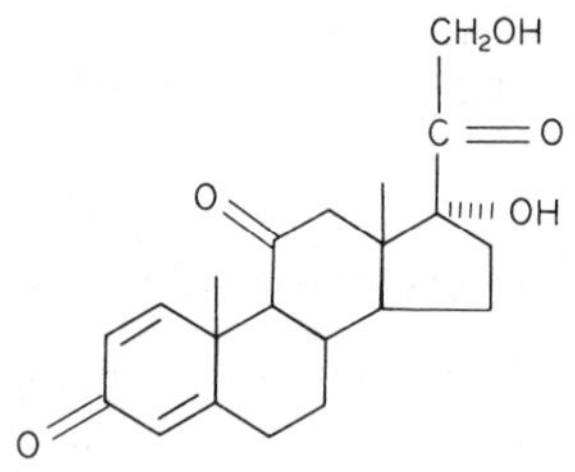

Prednisone

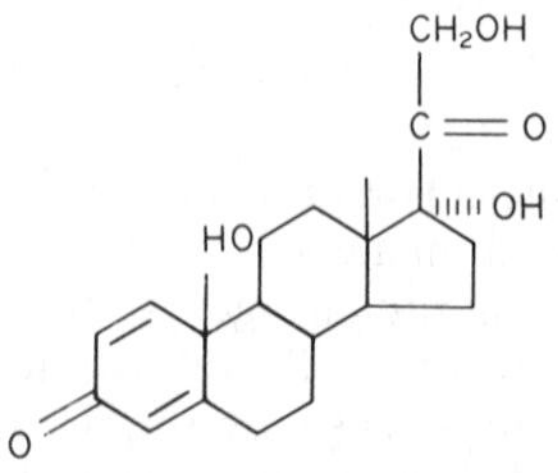

Prednisolone

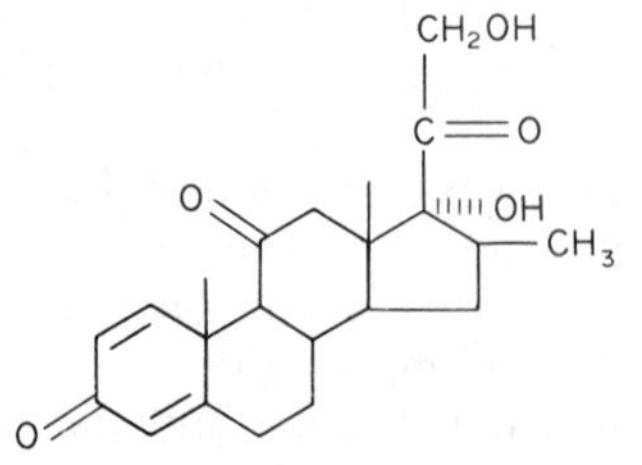

Meprednisone

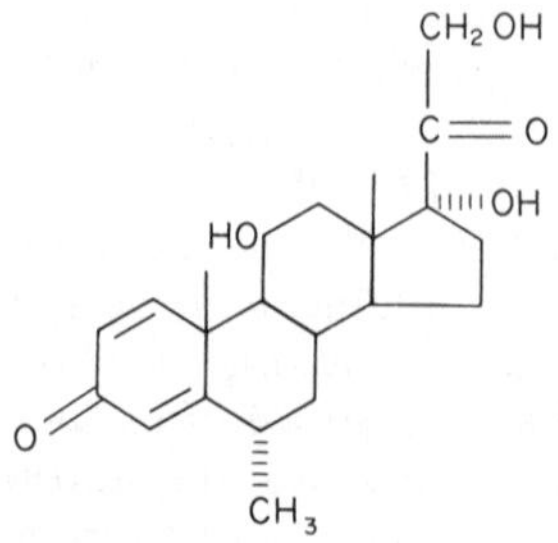

6α – Methylprednisolone

Table 5 *(continued)*

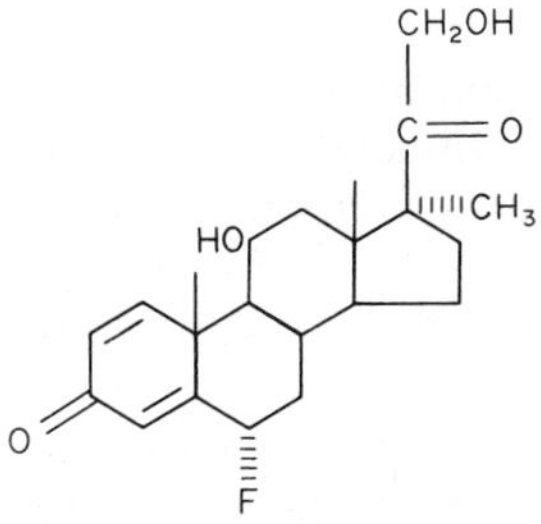

Flucortolone

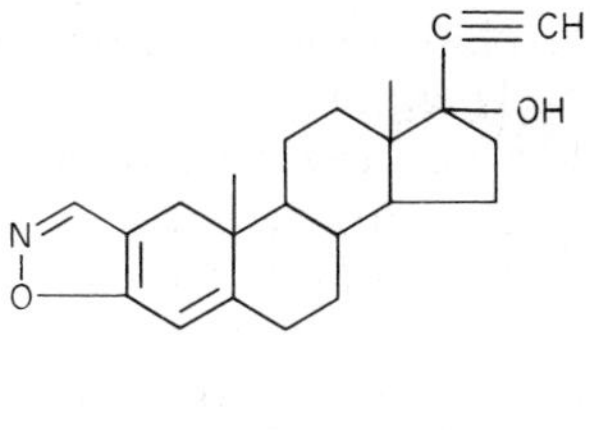

Danazol

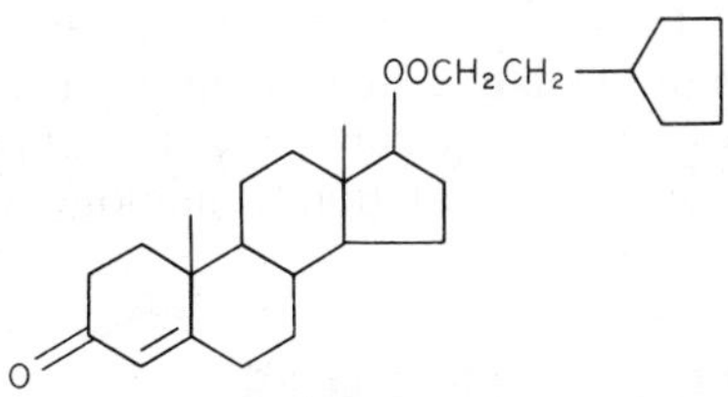

Testosterone cypionate

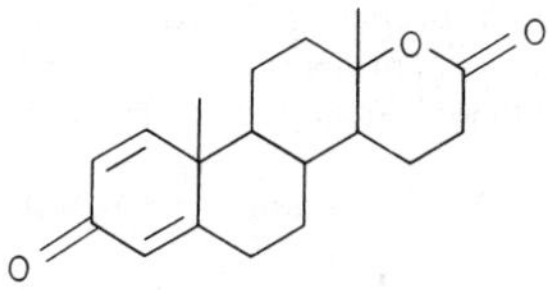

Testolactone

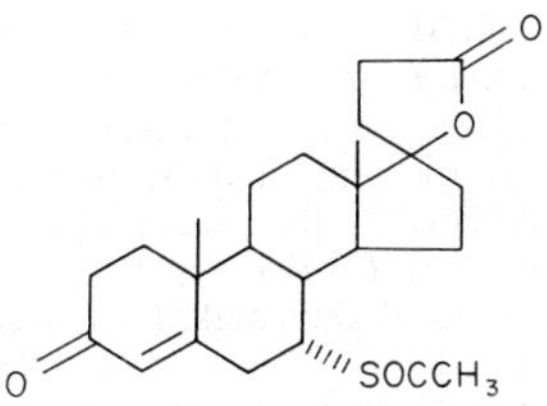

Spironolactone

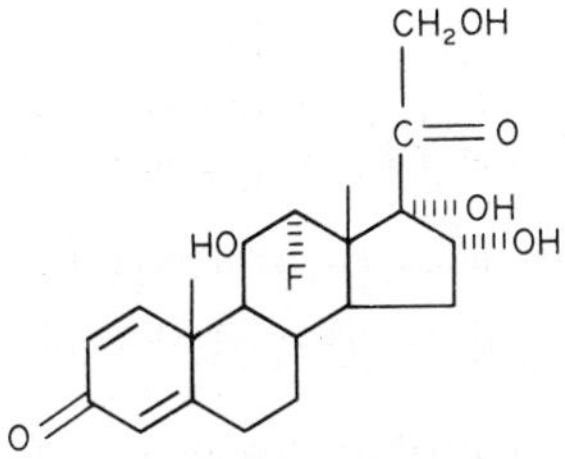

Triamcinolone

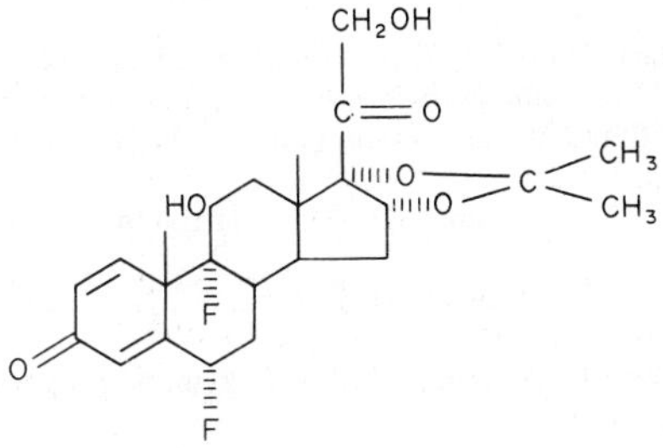

Fluocinolone acetonide

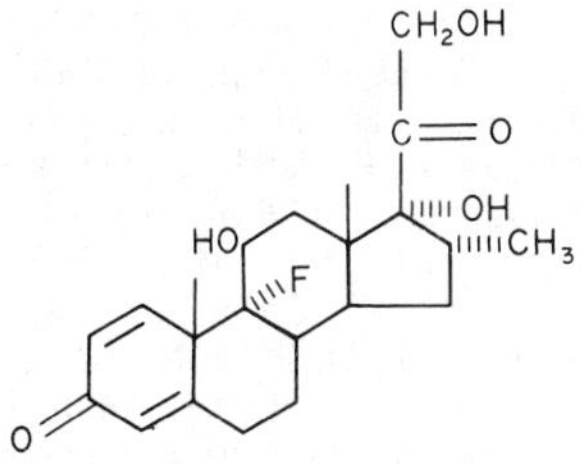

Dexamethasone

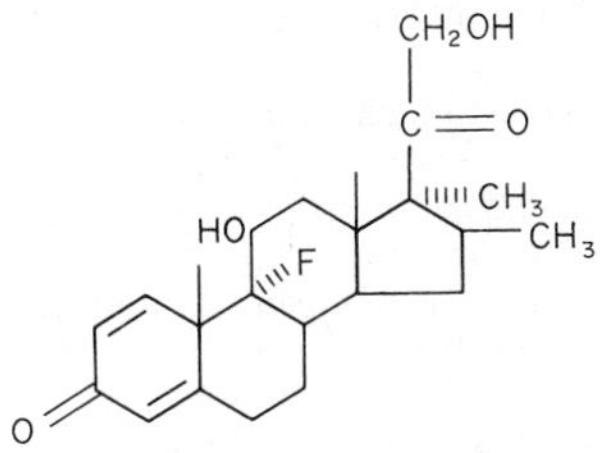

Betamethasone

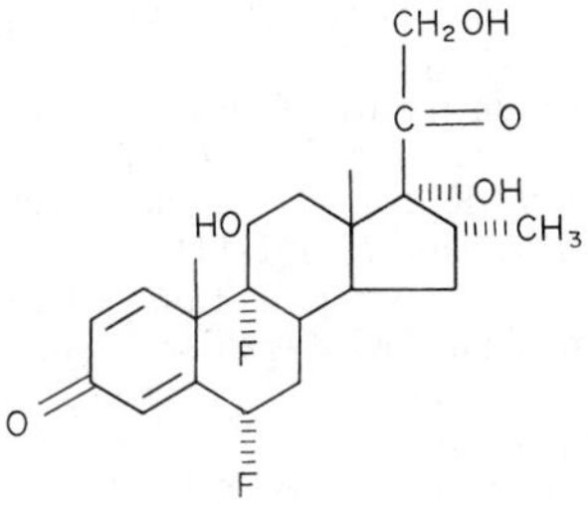

Flumethasone

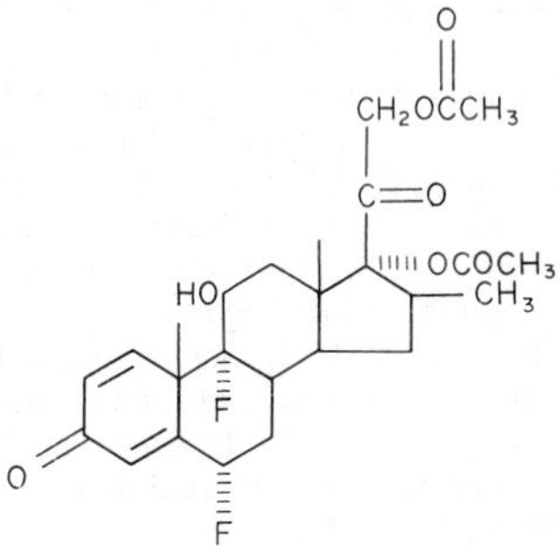

Diflorasone diacetate

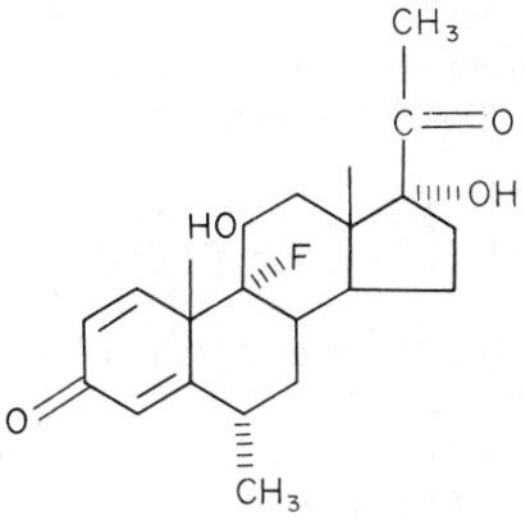

Fluorometholone

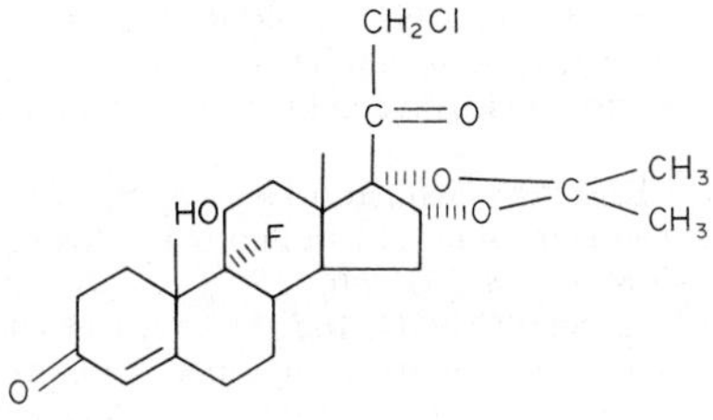

Halcinonide

organic solvent. However, other bioconversion methods have been described using fungal spores and more recently immobilized cells and enzymes. This, of course, offers an interesting area for future study.

Currently there is a great deal of interest in, and work on, the conversion of sterols to useful intermediates. New chemistry is being developed for converting these intermediates into a wide variety of very important steroid products. Therefore, after a short lull in the development of new steroid fermentation technology, this field is again active and fertile.

14.10 REFERENCES

Abd-Elsamie, M. E., M. B. Fayez, H. G. Osman and L. A. R. Sallam (1969). Microbiological transformation. I. Hydroxylation of progesterone at C-11 by *Rhizopus nigricans* and *Penicillium oxalicum*. *Z. Allg. Mikrobiol.*, **9**, 173–182.

Abul-Hajj, Y. J. (1972). Stereochemistry of C-1,2-dehydrogenation of 5β-pregnane-3,11,20-trione by *Septomyxa affinis*. *J. Biol. Chem.*, **247**, 686–691.

Antosz, F. J., W. J. Haak and M. G. Wovcha (1977). Process of producing 9α-hydroxy-3-ketobisnor-chol-4-en-22-oic with *Mycobacterium fortuitum*. *US Pat.* 4 029 549.

Arima, K., G. Tamura, M. Nazasawa and M. Bae (1968). Process for preparing androst-4-ene-3,17-dione and androsta-1,4-diene-3,17-dione from sterol compounds by the action of microorganisms. *US Pat.* 3 388 042.

Barton, D. H. R. (1969). Process for the preparation of debrominated analogues of bromo-steroids of the pregnane, androstane and cholestane series. *US Pat.* 3 480 622.

Beaton, J. M., J. E. Huber, A. G. Padilla and M. E. Breuer (1978). Non-aromatic oxygenated strong acid dehydrogenation of 9α-hydroxyandrostenediones. *US Pat.* 4 127 596.

Biggs, C. B., T. R. Pyke and M. G. Wovcha (1977). Microbial transformation of steroids. *US Pat.* 4 062 729.

Buckland, B. C., P. Dunhill and M. D. Lilly (1975). The enzymatic transformation of water insoluble reactants in non-aqueous solvents. Conversion of cholesterol to cholest-4-ene-3-one by a *Nocardia* spp. *Biotechnol. Bioeng.*, **17**, 815–826.

Čapek, A. and O. Hanč (1961). Microbiological transformation of steroids. XIV. Microbial 11β-hydroxylation of steroids. *Fol. Microbiol.*, **6**, 237–242.

Čapek, A., O. Hanč and M. Tadra (1966). *Microbial Transformations of Steroids*. Academia, Prague.

Charney, W. and H. L. Herzog (1967). *Microbial Transformations of Steroids*. Academic, New York.

Chen, J. W., F. G. Hills, H. J. Koepsell and W. D. Maxon (1965). 1-Dehydrogenation of steroids at levels above their solubilities. *IEC Proc. Des. Dev.*, **4**, 421–425.

Colingsworth, D. R., M. P. Brunner and W. J. Haines (1952). A partial microbiological synthesis of adrenal cortex hormones. *J. Am. Chem. Soc.*, **74**, 2381.

Colingsworth, D. R., J. N. Karnemaat, F. R. Hanson, M. P. Brunner, K. M. Mann and W. J. Haines (1953). A partial microbiological synthesis of hydrocortisone. *J. Biol. Chem.*, **203**, 807–813.

Cremonesi, P., G. Carrea, L. Ferrara and E. Antonini (1975). Enzymatic preparation of 20β-hydroxysteroids in a two-phase system. *Biotechnol. Bioeng.*, **17**, 1101–1108.

Dahl, H., E. Schöttle, R. Wieske, A. Weber and M. Kennecke (1979). Steroids in the pregnane series esterified in the 17α-position, and preparation and use thereof. *Br. Pat.* 2 013 685.

DeFlines, J. and W. F. Vander Waard (1970). Process for preparation of trihydroxy steroids. *US Pat.* 3 530 038.

Dodson, R. M. and R. D. Muir (1961). Microbiological transformations. I. The microbiological aromatization of steroids. *J. Am. Chem. Soc.*, **83**, 4627–4635.

Dulaney, E. L., E. O. Stapley and C. Hlavac (1955). Hydroxylation of steroids, principally progesterone, by a strain of *Aspergillus ochraceus*. *Mycologia*, **47**, 464–474.

Dulaney, E. L. and E. O. Stapley (1959). Studies on the transformation of 11-deoxy-17α-hydroxy corticosterone to hydrocortisone with a strain of *Curvularia lunata*. *Appl. Microbiol.*, **7**, 276–284.

El-Monem, A., H. El-Refai, L. A. Sallam and I. A. El-Kady (1970). Studies on the mechanism of progesterone transformation with *Rhizopus nigricans*. *J. Gen. Appl. Microbiol.*, **16**, 137–144.

El-Refai, A-M., L. Sallam and I. El-Kady (1970). Transformation of progesterone by *Rhizopus nigricans* REF 129 as influenced by modification of the fermentation medium. *Bull. Chem. Soc. Jpn.*, **43**, 2878–2884.

Erickson, R. C., W. E. Brown and R. W. Thoma (1967). Process for preparing 1-dehydro steroids. *US Pat.* 3 360 439.

Fonken, G. S. and R. A. Johnson (1972). *Chemical Oxidations with Microorganisms*. Marcel Dekker, New York.

Fujimoto, Y., C. S. Chen, Z. Szeleczky, D. DiTullio and C. J. Sih (1982a). Microbial degradation of the phytosterol side chain. 1. Enzymatic conversion of 3-oxo-24-ethylcholest-4-en-26-oic acid into 3-oxochol-4-ene-24-oic acid and androst-4-ene-3,17-dione. *J. Am. Chem. Soc.*, **104**, 4718–4720.

Fujimoto, Y., C. S. Chen, A. S. Gopalan and C. J. Sih (1982b). Microbial degradation of the phytosterol side chain. 2. Incorporation of NaH^{14}CO$_3$ onto the C-28 position. *J. Am. Chem. Soc.*, **104**, 4720–4722.

Gibson, D. T., K. C. Wang, C. J. Sih and H. Whitlock (1966). Mechanisms of steroid oxidation by microorganisms. IX. On the mechanism of ring A cleavage in the degradation of 9,10-seco steroids by microorganisms. *J. Biol. Chem.*, **241**, 551–559.

Goodman, J. J. and L. L. Smith (1960). 16α-Hydroxy steroids. IX. Effect of medium composition on isomerization of 9α-fluoro-16α-hydroxyhydrocortisone and 9α-fluoro-16α-hydroxyprednisolone (triamcinolone) during microbiological fermentation. *Appl. Microbiol.*, **8**, 363–366.

Goodman, J. J. and L. L. Smith (1961). 16β-Hydroxy steroids. XI. 2α- and 16α-hydroxylation and 9α-fluorohydrocortisone by strains of *Streptomyces roseochromogenes*. *Appl. Microbiol.*, **9**, 372–375.

Hanson, F. R., K. M. Mann, E. D. Neilson, H. V. Anderson, M. P. Brunner, J. N. Karnemaat, D. R. Colingsworth and W. J. Haines (1953). Microbiological transformation of steroids. VIII. Preparation of 17α-hydroxycortisone. *J. Am. Chem. Soc.*, **75**, 5369–5370.

Hanson, F. R. and W. D. Maxon (1965). Process for the microbiological oxygenation of progesterone. *US Pat.* 3 201 324.

Hessler, E. J. and V. H. VanRheenen (1980). Synthesis of 16-unsaturated pregnanes from 17-keto steroids. *US Pat.* 4 216 159.

Iizuka, H. and A. Naito (1967). *Microbial Transformation of Steroids and Alkaloids.* University Park Press, State College, PA.

Jerussi, R. and H. J. Ringold (1965). The mechanism of the bacterial C-1,2-dehydrogenation of steroids. III. Kinetics and isotope effects. *Biochemistry*, **4**, 2113–2126.

Knight, J. C. and M. G. Wovcha (1980). Microbial degradation of the phytosterol side-chain to 24-oxo products. *Steroids*, **36**, 723–730.

Koepsell, H. J. (1962). 1-Dehydrogenation of steroid by *Septomyxa affinis*. *Biotechnol. Bioeng.*, **4**, 57–63.

Kominek, L. A. (1973). Process for the microbiological 1-dehydrogenation of certain 4,9-(11)-pregnadienes. *US Pat.* 3 770 586.

Kondo, E. and E. Masuo (1961). 'Pseudo-crystallofermentations' of steroid: a new process for preparing prednisolone by a microorganism. *J. Gen. Appl. Microbiol.*, **7**, 113–117.

Krámli, A. and J. Horváth (1948). Microbial oxidation of sterols. *Nature (London)*, **162**, 619.

Kraychy, S. and W. J. Marsheck (1972). Selective microbiological degradation of steroidal 17-alkyls. *US Pat.* 3 684 657.

Larsson, P. O., S. Ohlson and K. Mosbach (1976). New approach to steroid conversion using activated immobilized microorganisms. *Nature (London)*, **263**, 796–797.

Maddox, I. S., P. Dunnill and M. D. Lilly (1981). Use of immobilized cells of *Rhizopus nigricans* for the 11α-hydroxylation of progesterone. *Biotechnol. Bioeng.*, **23**, 345–354.

Mamoli, L. and A. Vercellone (1937a). Biochemische Umwandlung von Δ⁴-Androstenedion in Δ⁴-Testosteron. Ein Beitrag zur Genese des Keimdrüsenhormons. *Ber.*, **70**, 470–471.

Mamoli, L. and A. Vercellone (1937b). Biochemische Umwandlung von Δ⁴-Androstendion in Isoandrostandiol und Δ⁴-Testosteron. Weiterer Beitrag zur Genese der Keimdrüsenhormone. *Ber.*, **70**, 2079–2082.

Mann, K. M., F. R. Hanson, P. W. O'Connell, H. V. Anderson, M. P. Brunner and J. N. Karmemaat (1955). Studies of the microbiological oxidation of steroids by *Cunninghamella blakesleeana* H-334. I. The effect of alcohol and phenol. *Appl. Microbiol.*, **3**, 14–16.

Marsheck, W. J. and S. Kraychy (1973). Selection microbiological preparation of androst-4-ene-3,17-dione. *US Pat.* 3 759 791.

Marsheck, W. J. (1971). Current trends in the microbiological transformation of steroids. In *Progress in Industrial Microbiology*, ed. D. J. D. Hockenhull, vol. 10, pp. 49–103. Churchill Livingstone, London.

Maxon, W. D., J. W. Chen and F. R. Hanson (1966). Simulation of a steroid bioconversion with a mathematical model. *IEC Proc. Des. Dev.*, **5**, 285–289.

Meystre, C., E. Vischer and A. Wettstein (1954). Mikrobiologische Hydroxylierung von Steroiden in der 17α- und 21-Stellung. *Helv. Chim. Acta*, **37**, 1548–1553.

Miller, T. L. (1972). The inhibition of microbial steroid D-ring lactonization by high levels of progesterone. *Biochim. Biophys. Acta*, **270**, 167–180.

Mosbach, K. and P. O. Larsson (1970). Preparation and application of polymer-entrapped enzymes and microorganisms in microbial transformation processes with special reference to steroid 11β-hydroxylation and Δ¹-dehydrogenation. *Biotechnol. Bioeng.*, **12**, 19–27.

Murray, H. C. and D. H. Peterson (1952). Oxygenation of steroids by mucorales fungi. *US Pat.* 2 602 769.

Nagasawa, M., N. Watanabe, H. Hishiba, M. Murakami, M. Bae, G. Tamura and K. Arima (1980). Microbial transformation of steroids. Part V. Inhibitors of microbial degradation of cholesterol. *Agric. Biol. Chem.*, **34**, 838–844.

O'Connell, P. W., K. M. Mann, E. D. Neilson and F. R. Hanson (1955). Studies on the microbiological oxidation of steroids by *Cunninghamella blakesleeana* H-344. II. Medium design. *Appl. Microbiol.*, **3**, 16–20.

Perlman, D., E. Titus and J. Fried (1952). Microbiological hydroxylation of progesterone. *J. Am Chem. Soc.*, **74**, 2126.

Peterson, D. H. and H. C. Murray (1952). Microbial oxygenation of steroids at carbon 11. *J. Am. Chem. Soc.*, **74**, 1871–1872.

Phillips, G. T. and F. P. Ross (1974). Stereochemistry of C-1,2-dehydrogenation during cholesterol degradation by *Mycobacterium phlei*. *Eur. J. Biochem.*, **44**, 603–610.

Ringold, H. J., M. Hayano and V. Stefanovic (1963). Concerning the stereochemistry and mechanism of the bacterial C-1,2-dehydrogenation of steroids. *J. Biol. Chem.*, **238**, 1960–1965.

Sallam, L. A. R., A. H. El-Refai and I. A. El-Kady (1971). The *in vitro* transformation of progesterone by *Rhizopus nigricans* REF 129. *Z. Allg. Mikrobiol.*, **11**, 325–330.

Schubert, K., G. Kaufmann and H. Budzikiewicz (1969). Cholest-4-en-26-ol-3-on und cholesta-1,4-dien-26-ol-3-on als Komponenten eines neuen mikrobiell gebildeten Estertyps. *Biochim. Biophys. Acta*, **176**, 197–177.

Sehgal, S. N., K. Singh and C. Vézina (1968). 11α-Hydroxylation of steroids by spores of *Asperigullus ochraceus*. *Can. J. Microbiol.*, **14**, 529–532.

Shephard, K. P. and V. H. VanRheenen (1977). Process for the preparation of 17α-hydroxyprogesterone and corticoids from androstanes. *US Pat.* 4 041 055.

Shibahara, M., J. A. Moody and L. L. Smith (1970). Microbial hydroxylations V. 11α-Hydroxylation of progesterone by cell-free preparations of *Aspergillus ochraceus*. *Biochem. Biophys. Acta*, **202**, 172–179.

Shull, G. M., D. A. Kita, and J. W. Davidsson (1953). Oxygenation of steroids. *US Pat.* 2 658 023.

Singh, K., S. N. Sehgal and C. Vézina (1963). C-1 dehydrogenation of steroids by spores of *Septomyxa affinis*. *Steroids*, **2**, 513–520.

Singh, K., S. N. Sehgal and C. Vézina (1965). Transformation of Reichstein's compound S and oxidation of carbohydrates by spores of *Septomyxa affinis*. *Can. J. Microbiol.*, **11**, 351–365.

Singh, K., S. N. Sehgal and C. Vézina (1968). Large-scale transformation of steroids by fungal spores. *Appl. Microbiol.*, **16**, 393–400.

Sih, C. J. and R. E. Bennett (1962). Steroid 1-dehydrogenase of *Nocarbia restrictus*. *Biochim. Biophys. Acta*, **56**, 584–592.

Sih, C. J. (1963). Mechanisms of steroid oxidation by microorganisms. *Biochim. Biophys. Acta*, **62**, 541–547.

Sih, C. J. and K. C. Wang (1963). Mechanisms of steroid oxidation by microorganisms. II. Isolation and characterization of 3aα-*H*-4α[3'-propionic acid]-7aβ-methylhexahydro-1,5-indanedione. *J. Am. Chem. Soc.*, **85**, 2135–2137.

Sih, C. J., S. S. Lee, Y. Y. Tsong and K. C. Wang (1965a). 3,4-Dihydroxy-9,10-secoandrosta-1,3,5(10)-triene-9,17-dione. An intermediate in the microbiological degradation of ring A of androst-4-ene-3,17-dione. *J. Am. Chem. Soc.*, **76**, 1385–1386.

Sih, C. J., S. S. Lee, Y. Y. Tsong, K. C. Wang and F. N. Chang (1965b). An efficient synthesis of estrone and 19-norsteroids from cholesterol. *J. Am. Chem. Soc.*, **87**, 2765–2766.

Sih, C. J., S. S. Lee, Y. Y. Tsong and K. C. Wang (1966). Mechanisms of steroid oxidation by microorganisms. VIII. 3,4-Dihydroxy-9,10-secoandrosta-1,3,5-(10)-triene-9,17-dione. An intermediate in the microbiological degradation of ring A of androst-4-ene-3,17-dione. *J. Biol. Chem.*, **241**, 450–550.

Sih, C. J., H. H. Tai and Y. Y. Tsong (1967a). The mechanism of microbial conversion of cholesterol into 17-keto steroids. *J. Am. Chem. Soc.*, **89**, 1957–1958.

Sih, C. J., K. C. Wang and H. H. Tai (1967b). C$_{22}$-Acid intermediates in the microbiological cleavage of the cholesterol side chain. *J. Am. Chem. Soc.*, **89**, 1956–1957.

Stadtman, T. C., A. Cherkes and C. B. Anfinsen (1954). Studies on the microbiological degradation of cholesterol. *J. Biol. Chem.*, **206**, 511–523.

Takeda, R., I. Nakanishi, J. Terumichi, M. Uchida, M. Katsumata, M. Uchibayashi and H. Nawa (1959). Transformation of Reichstein's substance S to prednisolone by *Pseudomonas*. *Tetrahedron Lett.*, No. 18. 17–19.

Thoma, R. W., J. Fried, S. Bonanno and P. Grabowich (1957). Oxidation of steroids by microorganisms. IV. 16α-Hydroxylation of 9α-fluorohydrocortisone and 9α-fluoroprednisolone by *Streptomyces roseochromogenes*. *J. Am. Chem. Soc.*, **79**, 4818.

Vander Waard, W. F. (1965). Process for the microbiological preparation of steroids. *Neth. Pat.* 65 13 718.

Vander Waard, W. F. (1966). Improved methods for removal of 17-alkyl steroidal side chains. *Neth. Pat.* 66 05 738.

Vézina, C., S. N. Sehgal, K. Singh and D. Kluepfel (1971). Microbial aromatization of steroids. In *Progress in Industrial Microbiology*, ed. D. J. D. Hockenhull, vol. 10. pp. 1–47. Churchill Livingstone, London.

Wang, K. C. and C. J. Sih (1963). Mechanisms of steroid oxidation by microorganisms. IV. Seco intermediates. *Biochemistry*, **2**, 1238–1243.

Weaver, E. A., H. E. Kinney and M. E. Wall (1960). Effect of concentration of the microbiological hydroxylation of progesterone. *Appl. Microbiol.*, **8**, 345–348.

Wovcha, M. G. (1977). Process for preparing 9α-hydroxyandrostenedione. *US Pat.* 4 035 236.

Wovcha, M. G., J. F. Antosz, J. C. Knight, L. A. Kominek and T. R. Pyke (1978). Bioconversion of sitosterol to useful steroidal intermediates by mutants of *Mycobacterium fortuitum*. *Biochim. Biophys. Acta*, **531**, 308–321.

Wovcha, M. G., F. J. Antosz, J. M. Beaton, A. B. Garcia, and L. A. Kominek (1979). Composition of matter and process. *US Pat.* 4 175 006.

Wovcha, M. G., F. J. Antosz, J. M. Beaton, A. B. Garcia and L. A. Kominek (1980). Process for preparing 9α-OH BN acid methyl ester. *US Pat.* 4 214 051.

Wovcha, M. G. and C. B. Biggs (1981). Process for preparing androst-4-ene-3,17-dione. *US Pat.* 4 293 644.

Wovcha, M. G. and K. E. Brooks (1981). Composition of matter and process. *US Pat.* 4 293 646.

Zaretskaya, I. I., L. M. Kogan, O. B. Tekhomirova, J. D. Sis, N. S. Wolfson, V. I. Zaretskii, V. G. Zaikin, G. K. Skryabin and I. V. Torgov (1968). Microbial hydroxylation of the cholesterol side chain. *Tetrahedron*, **24**, 1595–1600.

Zuidweg, M. H. J., W. F. Vander Waard and J. DeFlines (1962). Formation of hydrocortisone by hydroxylation of Reichstein's compound S with an enzyme preparation from *Curvularia lunata*. *Biochim. Biophys. Acta*, **58**, 131–133.

Zuidweg, M. H. J. (1968). Hydroxylation of compound S with cell-free preparations from *Curvularia lunata*. *Biochim. Biophys. Acta*, **152**, 144–158.

15

Products from Recombinant DNA

B. KHOSROVI
Cetus Corporation, Emeryville, CA, USA
and
P. P. GRAY
University of New South Wales, Sydney, NSW, Australia

15.1 INTRODUCTION

The use of recombinant DNA technology for the manufacture of useful products is one of the most exciting and rapidly developing areas in the field of biotechnology. In a very short time, recombinant DNA technology has progressed from a powerful tool available to molecular biologists to a technology available for the production of proteins of interest. The rapid pace in the development of the technology is evidenced by the fact that only five years elapsed from the first reports on the cloning of human hormones (Shine *et al.*, 1977; Itakura *et al.*, 1977), and FDA approval in 1982 for use of rDNA insulin as a licensed product. Recombinant techniques have allowed the production of protein products which were previously either not available or were only available in short supply at prohibitively high costs. Examples of such compounds include insulin, various growth hormones and growth factors, interferons and other lymphokines, vaccines, blood factors, enzymes and various other biologically active compounds. A second class of product exists where recombinant DNA is being used to construct strains which will then overproduce the product of interest. Examples of products in this category include amino acids and antibiotics.

In this chapter the emphasis will be on compounds in the first category, *i.e.* where the gene coding for the protein of interest has been inserted into a microorganism in such a fashion as to allow expression of the gene at a high level. There have been many reviews of the genetic developments underlying the technology (*e.g.* Volume 1, Chapter 6). The aim of this review is to concentrate on aspects relevant to the use of the technology for the production of useful products.

For the production of polypeptides with a small number of amino acid residues, there will be competition between recombinant methods and organic synthesis. The smaller the protein, the more favorable economically will be the chemical synthesis. O'Neill (1981) describes the principal factors influencing the choice of chemical *vs.* biological synthesis and quotes an estimate for production costs by chemical synthesis of a 34 amino acid parathyroid hormone of $5 per milligram. Many factors will influence the number of residues of the protein above which production costs will swing from favouring chemical synthesis to favouring biological synthesis. In this review an arbitrary change-over at 50 residues is taken and polypeptides with less residues than this, such as calcitonin, β-endorphin, somatostatin, *etc.*, will not be discussed in detail.

15.2 PRODUCTION TECHNOLOGY

15.2.1 Methods for Cloning and Expression

Cloning and expression involve the isolation and identification of the gene coding for the protein of interest and its insertion into the chosen host cell in a manner which allows synthesis of the desired protein. The amino acid sequence for a given protein is encoded at the genetic level by the corresponding sequence of nucleic acid triplets, or codons, in an oligonucleotide molecule. By way of example, the structural genetic information for β-interferon, which is made up of 165 amino acids, is contained in an RNA sequence, or its complementary DNA sequence, with 495 bases. Prior knowledge of the amino acid sequence of a protein allows a corresponding DNA sequence to be predicted. Although it is possible to synthesize chemically long DNA sequences — and this has been done for α-interferon (Edge *et al.*, 1981) — in general it is more practical to obtain the DNA from a donor cell.

Each human cell contains the genetic information coding for the production of all the proteins of the body. The process of synthesizing a particular protein starts at the nucleus, where genomic DNA is transcribed to messenger RNA. This RNA, containing the structural gene, acts as the template from which the protein is synthesized in the cytoplasm of the cell. The task of finding the gene of interest is difficult because of its rarity within the cell.

Numerous methods of cloning and expression have been used (Friesen and An, 1983) and are constantly being refined and developed. In order to illustrate the various steps involved a general method for making human proteins in *E.coli* will be described. The difficulty of finding the gene of interest can be appreciated by considering that a typical cell synthesizes a large number and wide variety of protein molecules together with their RNA precursors.

In order to facilitate the isolation of the structural gene, a donor cell is chosen which either already makes significant amounts of the desired protein or can be induced to make the protein. Correspondingly, these cells are enriched in the messenger RNA of interest. For example, fibroblast cells are known to produce small but significant amounts of β-interferon. Culturing these cells in the presence of certain inducers, chemical or viral, further increases synthesis of this protein and its corresponding messenger RNA. RNA is extracted from such cells and complementary DNA prepared using an enzymatic process. This DNA is randomly inserted into a cloning vehicle to prepare a gene 'library'. Also known as vectors, these cloning vehicles may be plasmids or bacteriophages which naturally populate or infect *E.coli*. The vectors carrying the recombinant DNA are used to transform *E.coli* in which they will then self-replicate. Grown on agar plates the bacteria will form colonies each containing faithful copies of the cloned gene. Screening of these colonies will determine which contains the gene of interest.

For example, plasmid DNA can be extracted from a colony and tested with a synthetic length of DNA, or probe, corresponding to a portion of the sequence of the protein of interest. If hybridization occurs between the probe and a complementary region of the plasmid DNA, then that plasmid will be known to contain the gene of interest. If the sequence of the protein is unknown, colonies must be screened indirectly for presence of the gene, for example by detection of the synthesis of the protein using antibodies. Once identified and its DNA sequence determined, the gene may be inserted into an expression vector which is then transformed into the production organism. In the case of *E. coli* a plasmid expression vector is commonly used.

An idealized expression plasmid is shown in Figure 1. By using a battery of enzymes that can cut and splice DNA, the human gene is inserted into the plasmid together with sequences of bacterial DNA responsible for control of protein synthesis. The choice of promoter used will determine the rate of protein synthesis and whether or not its synthesis is constitutive or regulated. It is usual to include in the design of the plasmid a selectable marker such as resistance to a specific antibiotic. This simplifies monitoring of the presence of the plasmid and allows the use of antibiotics to select against bacteria which have lost the plasmid.

Although expression systems are well developed for *E. coli*, systems for other host organisms, including *Bacillus* spp. (Dubnau, 1982; Ganesan *et al.*, 1982), yeasts (Botstein *et al.*, 1982) and mammalian cells (Gluzman, 1982; Hamer and Khoury, 1983) have received much attention and are developing rapidly.

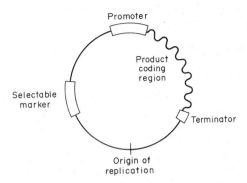

Figure 1 Idealized plasmid expresson vector. The diagram depicts circular, double-stranded plasmid DNA with the major features important for its use. The origin of replication is a region of DNA required for the plasmid to replicate in a given host. The selectable marker, usually coding for resistance to an antibiotic, provides a means to select for cells retaining the plasmid. The region of DNA coding for the product of interest is bounded by promoter and terminator control elements which affect the rate of product synthesis.

15.2.2 Range and Relative Advantages of Host Microorganisms

A range of microorganisms have been used as the host organisms for recombinant DNA techniques. By far the most common host is *Escherichia coli*, owing to the well-developed genetic systems and vectors for this organism. Compounds which have been cloned into *E. coli* include the A and B chains of human insulin, human proinsulin, several α-, β-, and γ-human interferons, various animal interferons, human and various animal growth hormones, epidermal and other growth factors, lymphokines, human serum albumin, a range of enzymes and plasminogen acltivators. The only recombinant product with FDA approval for use as a licensed product, insulin, uses *E. coli* as the host organism, as do several other products currently in clinical trials.

An advantage of using *E. coli* is that it has the most developed systems for genetic manipulation. This has resulted in the achievement of higher levels of expression than for other hosts. In addition, more is known about the stability of constructed plasmids and the stability of these plasmids in the host than is known for other systems. Additional advantages of *E. coli* include the fact that it can be grown easily on defined media at high growth rates, and techniques for collecting and disrupting the cells have been described. Disadvantages associated with *E. coli* as a host include the fact that the products are, with a few exceptions, intracellular, and the cells contain endotoxins and pyrogenic lipopolysaccharides. The presence of these compounds places additional pressure on the purification process.

After *E. coli*, the most commonly used hosts are probably various strains of yeast. The advantages of using a eukaryotic host, such as yeast, include the absence of pyrogenic lipopolysaccharides, the use of a GRAS organism and having the potential for protein secretion and protein glycosylation. However, genetic systems for yeast such as *Saccharomyces* are not as well developed as for *E. coli*, and expression levels which have been quoted for proteins cloned into yeast are approximately an order of magnitude lower than those quoted for *E. coli*.

Another host which has been used for the production of recombinant DNA products is *Bacillus subtilis*. *B.subtilis* is currently widely used for the commercial production of enzymes such as α-amylase and proteases. However, it is only in the last few years that several *B.subtilis* cloning systems have been developed, and the genetic systems in this host are much less developed than for *E.coli*. McLaughlin and Chang (in press) list the following advantages for *B.subtilis* as host: *B.subtilis* is a non-pathogenic soil microbe which grows strictly under aerobic conditions and does not contain pyrogenic lipopolysaccharides; it is a Gram-positive microorganism in which secreted proteins are transported across the cytoplasmic membrane and can be released directly into the culture medium.

High expression levels have been observed when genes have been cloned from one *Bacillus* species to another, *e.g.* Palva (1982) reported that when the gene coding for α-amylase was cloned from *Bacillus amyloliquefaciens* into *B.subtilis*, 1.6 mg ml^{-1} of the enzyme was secreted after 45 h of fermentation. However, when heterologous genes of the type successfully cloned into *E.coli* are cloned in *B.subtilis*, expression levels fall by several orders of magnitude (*e.g.* Palva *et al.*, 1983). Other microbial hosts are described in the article by Hofschneider and Goebel (1982).

An area of rapid development in recent years is the use of cultured mammalian cells as hosts for recombinant products. The development of high-level genetic systems has enabled genes of interest to be inserted into established cell lines to allow the production of such compounds as human β-interferon (McCormick *et al.*, 1984), human growth hormone (Pavlakis and Hamer, 1983) and human plasma factor VIII (Wood *et al.*, 1984). The genetic constructions being used with mammalian cells have been summarized by Wilson (1984). The advantage of mammalian cells as host is that the cells can process the protein exactly as occurs in the body (including glycosylation if necessary), then secrete the protein in the correct conformation into the culture medium, simplifying product recovery.

The traditional disadvantages of tissue culture, *viz.* high media costs due to the need to supply sera and difficulties in growing shear sensitive cells, have been largely overcome by the developments of defined media and of bioreactors providing good environmental control whilst maintaining a low shear environment suitable for the growth of such cells.

15.2.3 Stability of Strains and Plasmids

The stability of constructed strains is of major concern in running industrial fermentations. In the case of recombinant strains, both the stability of the plasmid in the cell and stability within the plasmid will be of importance. There is increasing knowledge of the genetic and environmental factors which influence plasmid retention and stability (*e.g.* McLaughlin and Chang, 1982; Helling *et al.*, 1981; Shogman *et al.*, 1983; Edlin *et al.*, 1984; Dwivedi *et al.*, 1982; Lee and Bailey, 1984). Edlin and coworkers showed (1984) that pBR322 is maintained in *E.coli* RR1 for about 60 generations. With this stability, one cell is able to divide sufficient times to produce 11 500 l of culture of cell density 10 g l^{-1}. In batch fermentations lasting 24 hours or less, stabilities of this order or even considerably less are then sufficient to allow production of proteins such as insulin, growth hormone, interferons, *etc*. With these types of products it may not be desirable to maintain stability by use of antibiotic resistance; however, the fermentations are usually operated in such a fashion that a very high level of expression is only maintained for a relatively short period of several hours at the end of the fermentation. An example of such a fermentation is shown in Figure 2.

The question of stability will become of increasing importance as the cost of products made by recombinant techniques decreases, particularly if economics dictate maintaining high volumetric productivities by the use of multicopy plasmids expressing at high levels for long time periods, *e.g.* in continuous culture.

The final concentration in the fermentation of the desired products, whether intra- or extracellular, will result from the difference between the integral of the rate of synthesis minus the integral of the rate of proteolytic breakdown, both with respect to time. Early workers showed that when small peptides were cloned into *E.coli*, instability was a problem. For this reason fused proteins were produced where the protein of choice was linked to a larger protein. These chimeric proteins together with larger molecular weight proteins occur as insoluble cytoplasmic inclusion bodies when they are expressed in high levels in *E.coli* (*e.g.* Williams *et al.*, 1982).

The importance of the effect of cellular location on protein stability has been reported by Talmadge *et al.* (1980) who showed that proinsulin molecules transported to the periplasm of *E.coli* had ten times the half life of molecules which remained in the cytoplasm ($t_{1/2} = 2$ minutes for cytoplasm). The use of protease deficient mutants to minimize proteolytic breakdown has been reported (Zehnbauer and Markovitz, 1980).

15.2.4 Product Recovery and Purification

For many of the first generation recombinant products produced in *E.coli*, the major contribution to the manufacturing cost will be from the product recovery and purification, often referred to as downstream processing. The ease of product recovery is influenced by

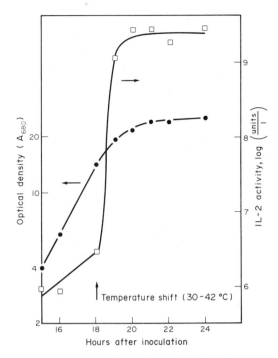

Figure 2 Growth curve for *E.coli* producing recombinant interleukin-2. *E.coli* K12 harboring a plasmid coding for human interleukin-2 (IL-2) under the control of a temperature-sensitive promoter was grown in a ten l fermenter. The bacteria are propagated at 30 °C at which temperature synthesis of IL-2 is repressed. A shift of temperature to 42 °C induces rapid product synthesis (figure courtesy of K. Bauer, Cetus Corporation)

the level of expression of the desired protein by the cell at the end of the fermentation. Thus, a cloned protein which is only being expressed at a level of 0.1% of the cell's protein is going to take considerably more stages to purify than a protein present as 20% of the cell protein. Low expression levels are going to have a magnifying effect on the final cost of the product, in that not only are more stages required together with their concomitant costs but also losses associated with the additional stages will decrease the overall yield.

For *E.coli* expression levels for cloned proteins of 20–30% of the final protein, with some reports of higher levels, are not uncommon. At these high expression levels the protein forms as insoluble inclusion bodies in the cell, which offers the possibility of substantial initial purification in a single centrifugation stage following cell disruption.

A sample flowsheet for the recovery of an intracellular product from *E.coli* is shown in Figure 3. In Figure 4 the magnitude of the purification problem for such a process is demonstrated by showing SDS Page gels for an extract of *E.coli* before and after purification.

A similar overall scheme to that shown in Figure 3 could be used for the recovery of an intracellular product from yeast. In the case of proteins secreted extracellularly or those which can be released from the periplasmic space by relatively mild treatments which do not result in cell lysis, the purification is going to be simplified since cell disruption is not required. However, the comments made above regarding expression levels and their influence on ease of purification will also apply to an extracellular product as other proteins are also usually present in the medium.

As containment is of prime concern in recombinant DNA processes (Table 1), it is common for the culture to be killed in the fermentation stage if the cells are to be harvested by centrifugation (Ross, 1981). Another approach which can be used to concentrate the cells while maintaining containment is the use of cross-flow filtration (Hanisch *et al.*, 1982). For products made by *E.coli*, it would appear that the most common method of cell disruption is a high-pressure homogenizer (*e.g.* Staehelin *et al.*, 1981). Protein release from *E.coli* using such a machine has been described by Gray *et al.* (1973).

Initial reports on the purification of recombinant proteins followed relatively traditional protocols. The recovery of human insulin A and B chains from *E.coli* has been described in some detail (Goeddel *et al.*, 1979; Riggs, 1981). In these constructions, synthetic genes for human insulin A

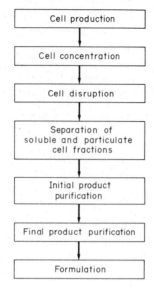

Figure 3   Sample flowsheet for the recovery of an intracellular product from *E.coli*

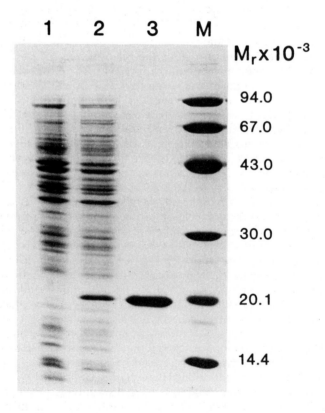

Figure 4   Production of recombinant interferon-beta in bacteria. Separation of proteins is by gel electrophoresis. Lane 1: whole extract of *E.coli* cells not coding for interferon. Lane 2: whole extract of *E.coli* cells expressing beta-interferon. Lane 3: purified protein. Lane M: molecular weight markers (figure courtesy of L. Lin, Cetus Corporation)

Table 1　Summary of National Institutes of Health Guidelines for the Physical Containment of Organisms Containing Recombinant DNA Molecules when Cultured at Scales of Greater than 10 liters[a]

Requirement	Level of containment[b]		
	P1-LS	P2-LS	P3-LS
Recombinant organism must be inactivated before removal from primary containment equipment using procedures validated to be effective on the host organism	√	√	√
Sample collection or culture transfer must minimize aerosols or contamination of exposed surfaces	√		
Sample collection or culture transfer must prevent aerosols or contamination of exposed surfaces		√	√
Exhaust gases from primary containment equipment to be filtered to HEPA efficiency or incinerated	√	√	√
Primary containment equipment can only be opened for maintenance after sterilization using procedures validated to be effective on the host organism	√	√	√
Plans for handling large losses of culture on an emergency basis	√	√	√
Rotating seals and mechanical devices associated with containment equipment must be designed to prevent leakage or be fully enclosed		√	√
Monitoring of containment integrity during operations. Prior testing of the containment features using the host organism. Revalidation following modification or equipment replacement		√	√
Permanent identification of equipment. Documentation of testing, maintenance and operation		√	√
Posting of universal biohazard sign during operation		√	√
Operation at pressures as low as possible, consistent with equipment design			√
Primary containment equipment to be located within a controlled area designed to prevent further release of organisms in the event of a spill from the primary system and to facilitate decontamination of personnel and surfaces			√

[a] *Federal Register*, vol 48, no. 106, pp. 24 577–24 580.
[b] Level of containment depends on nature of host organism and recombinant DNA in question.

and B genes were fused to an *E.coli* β-galactosidase gene. Approximately 20% of the bacterial protein produced by the insulin strains was found to be β-galactosidase-insulin fused protein. The fused protein is insoluble in the cell and can be enriched to more than 50% purity by centrifugation following cell disruption. Williams *et al.* (1982) have described the formation in insulin-producing *E.coli* of intracellular inclusion bodies whose accumulation corresponds to the formation of chimeric protein. Similar inclusion bodies were present in cells bearing plasmids containing gene sequences coding for insulin A chain, insulin B chain, or proinsulin fused to portions of either the lactose or tryptophan operons. At peak production, the inclusion bodies were observed to occupy as much as 20% of the *E.coli* cellular volume. Following collection by precipitation, the insulin chains were then cleaved from the fused protein by treatment with cyanogen bromide in 70% formic acid overnight at room temperature. The free insulin chains were converted to *S*-sulfonated derivatives and purified by ion exchange chromatography and gel filtration. After separate purification, the insulin A and B chains were joined by air oxidation.

A similar purification procedure has been described for the recovery of human growth hormone (hGH). In this *E.coli* construction, the hGH is present at greater than 1% of the soluble protein (Ross, 1981). A multistep purification is required to separate the hGH from all the other *E.coli* proteins. The initial lysate was fractionated by precipitations to enrich hGH relative to the other proteins and to remove nucleic acids and intact ribosomes. The first chromatography stage is a DEAE cellulose anion exchange resin, resulting in a preparation of greater than 50% purity. The next chromatography stage is a cation exchange CM cellulose column resulting in material of greater than 98% purity. The final stage in the purification process is a gel filtration, on the basis of molecular size, to yield the final product.

There have been rapid developments in a range of purification techniques which are being increasingly applied to the purification of recombinant products. These techniques include: affinity chromatography, high performance liquid chromatography (HPLC) and fast protein liquid chromatography (FPLC).

Affinity chromatography utilizing a number of ligands as well as immunoaffinity utilizing monoclonal antibodies has been described for a range of systems. Human fibroblast interferon has been purified using concanavalin A-sepharose (Davey *et al.*, 1976) and blue dextran-sepharose (De Maeyer, 1981).

Staehelin *et al.* (1981) described a process for the purification of recombinant human leukocyte interferon consisting of affinity chromatography by immobilized monoclonal antibodies followed

by final purification on reverse phase HPLC. Monoclonal antibodies to human leukocyte interferon were produced in mouse ascites. Ascitic fluid containing the equivalent of 4–7 mg ml^{-1} of
antibody was collected and cells and debris removed by centrifugation. Following purification
involving ammonium sulfate precipitation and ion exchange chromatography, the purified antibodies were coupled to a preactivated support and packed in a column. The immunoadsorbent
column, of bed volume 17 ml, contained 408 mg of purified monoclonal antibody. Cells of *E.coli*
containing recombinant DNA producing human leukocyte interferon were disrupted in a high
pressure homogenizer and the intracellular protein released was subjected to fractional precipitation with ammonium sulfate.

Protein precipitating between 30 and 65% saturation with ammonium sulphate was collected,
redissolved and applied to the immunoadsorbent column. The protein added to the column
totalled 37 100 mg. After washing the column, interferon totalling 30 mg was eluted off. A purification factor of 1150 and a 95% recovery were claimed. Following further purification on a CM52
column, the purified interferon exhibited a single band on sodium dodecyl sulfate polyacrylamide
gel electrophoresis.

The use of both reverse phase and normal partition HPLC in the purification of various interferons has been described (Pestka *et al.*, 1981; Pestka, 1983). It was claimed (Pestka *et al.*, 1981)
that HPLC using *n*-propanol as eluent with octyl silica columns was a major factor in the success
of the purification. Another major chromatographic advance has been the development of separation media designed for fast, high resolution preparative chromatography of proteins. The fast
separations possible with these systems minimize the chances of proteolytic attack on the product
as well as increasing the throughput of separation techniques which previously had long retention
times. These fast separations coupled with the high specificity of immunoaffinity chromatography
will be of increasing importance in minimizing the problems associated with the purification of
recombinant products.

In addition there are several other novel separation techniques such as the biostream continuous free flow electrophoresis system (John Brown, UK, 1984), purifications utilizing extraction in
aqueous two-phase systems (Kula *et al.*, 1982) and novel applications of affinity techniques
(Lowe, 1984), all of which could assume increasing importance in product recovery.

There has been relatively little work on some of the basic problems associated with the
recovery of protein products. The work of Dunnill and coworkers stands out in this area, particularly their studies on shear effects on proteins (Thomas and Dunnill, 1979; Virkar *et al.*,
1981) and their studies on the influence of precipitation conditions and ageing on the properties
of protein precipitates (Bell and Dunnill, 1982; Salt *et al.*, 1982). Problems associated with protein denaturation/renaturation during the recovery process are also of major importance (*e.g.*
Marston *et al.*, 1984).

15.3 COMMERCIAL MARKETS

15.3.1 Markets for Recombinant Products

In this chapter we have discussed the use of recombinant DNA to produce human proteins
which are either currently in use as therapeutic products or are expected to show therapeutic
value. In the former case the recombinant products will compete with the existing methods of
production on basis of cost and quality. In the latter case the newer products are generally rare
products and production using recombinant DNA is expected to dominate supply.

Those proteins in current use are usually obtained from glands, for example pancreatic insulin
and pituitary growth hormone; from blood, for example human serum albumin, immunoglobins
and antihemophilic factor VIII; or from urine such as the thrombolytic enzyme urokinase. Additionally, proteins produced by culture of human cells are in limited use, for example certain
native interferons have been licensed in Europe and in Israel (Rich, 1984). A 1983 analysis by
Arthur D. Little Inc. estimates worldwide markets for albumin at $400 million, factor VIII at
$100 million, insulin at $400 million, growth hormone at $20 million and urokinase at $200 million.

For the newer, recently studied, biologically active proteins it is more difficult to predict their
market value since these are still under investigation and the extent of their therapeutic activity is
as yet unknown. Current emphasis on these future products is on the interferons and lymphokines. Interferons are expected to find both antiviral and anticancer applications whilst the lymphokines (for example IL-2, TNF and lymphotoxin), which are a family of immune regulators, are

expected to find applications in fighting cancer, immune-deficiency diseases and infectious diseases. Although these products may radically affect the practice of medicine, their market predictions should be seen in the context of the existing markets for cancer chemotherapeutic agents and antiinfective agents, which are expected to grow in annual sales from $500 to $800 million and from $2500 to $4000 million respectively between 1985 and 1990.

In order to gain a perspective of the progress of these products towards commercial reality it is informative to consider the process of new drug registration in the USA and the actual number of products in clinical trials. Initiation of clinical trials requires an IND (Notice of Investigational New Drug Exemption) to be filed with the FDA which will grant approval provided satisfactory preclinical data on the safety, characterization and potential efficacy of the product are provided by the manufacturer. Clinical trials are generally conducted in three phases: the first phase establishes safety and optimum route of administration, the second establishes efficacy and the third proves therapeutic value. The clinical data are then submitted to the FDA as part of an NDA (New Drug Application). The complete process from the pre-clinical stage to an approved product may take five to ten years. Table 2 provides a partial listing of human therapeutic products currently being addressed by recombinant DNA technology. In the USA only one recombinant product, human insulin, has been approved. Six recombinant products (human growth hormone, α-, β- and γ-interferon, interleukin-2 and tissue plasminogen activator) are in various stages of clinical trials.

Table 2 A Partial Listing of Human Therapeutic Products Being Addressed by Recombinant DNA Technology

Product	Disease/use
Insulin	Diabetes
Growth hormone	Short stature
Serum albumin	Surgery, shock, burns
Factor VIII	Hemophilia
Urokinase	
Tissue plasminogen activator	Heart attack, stroke, pulmonary embolism
Interferons	Cancer, viral infection
Lymphokines	Cancer, auto-immune disease, infectious disease

15.3.2 Patent Situation

Today the application of recombinant DNA technology to biotechnology and medicine is marked by aggressive policies of both corporations and universities to patent techniques, products and processes. Within the US, patents began to be issued in increasing numbers in 1981. A publication of the US Department of Commerce (Baldridge *et al.*, 1983) summarizes patent activity in the biotechnology area under six classifications. Total US Patents under a mutation/genetic engineering category were 10, 38 and 122 in 1980, 1981 and 1982 respectively. The total of 170 patents for these three years compares with a total of 38 for the preceding five years. It is noteworthy that prior to 1980 patents in this category were dominated by those of non-US origin.

Although this technology was discovered in the early part of the 1970s the apparent delay in the corresponding build up of patent activity may be attributed to several factors. First, this period should be viewed in the context of significant uncertainty surrounding the patentability of genetically modified microorganisms. In 1972, Chakrabarty, a scientist working for the General Electric Company, filed a patent application claiming a bacterium in which he had inserted plasmids providing it with the capacity to break down multiple components of crude oil. This patent was initially rejected by the US Patent and Trademark Office (PTO) and remained in the courts until the 1980 decision of the US Supreme Court, in the now celebrated case of Diamond *vs.* Chakrabarty (447 US 303, 1980) which held that a live, man-made microorganism is patentable. Although this case did not involve recombinant DNA it did pave the way for the current policy of the PTO which allows claims for organisms. It is noteworthy that Chakrabarty deposited his strains in a recognized culture collection prior to the date of filing of his application and promised to make them available to the public subsequent to the issue of his patent. A second factor is that outside the universities this field has been pioneered by fledgling genetic engineering companies whose early energies have been directed towards establishing a strong technical base. An additional factor is that since the field involves a new and highly complex technology, patent offices have been obliged to train staff and to acquire specialists, which has inevitably caused some delay in examining cases. Lastly, the granting of the patent to Cohen and Boyer in 1980 which covers certain basic cloning processes and its subsequent publicity alerted the business community to the importance of this area.

In an extensive review Halluin (1982) discusses the US PTO and foreign policy towards patenting the results of genetic engineering and US legislative developments, as well as selected inventions. He describes the distinction between process patents and composition of matter or product patents as they relate to the field of recombinant DNA and hybridomas.

A key composition of matter patent (assigned to Stanford University) which makes a general claim to recombinant plasmids containing foreign genes functioning biologically in cells was granted to Cohen and Boyer in 1984 after eight years of review by the PTO. However, the claim has been limited to prokaryotes and although more than sixty companies are already paying license fees to Stanford, the strength of the patent is still in question (Budiansky, 1984). In general, composition of matter patents may prove the most valuable, since they are less readily circumvented, especially where a specific protein product or derivative is claimed.

The trend of rapidly expanding patent activity in the field of recombinant DNA technology and products as well as in the allied hybridoma field will continue. As products of recombinant DNA become commercial reality it can be expected that many patents will be challenged in the courts. How broadly the courts will interpret the scope of composition of matter claims on recombinant proteins remains to be seen.

15.4 CONCLUSIONS

In this chapter the use of recombinant DNA technology for the production of compounds of medical importance has been discussed. The technology is being used to produce proteins of medical interest which were previously either not available or only available at prohibitively high cost and scarce supply.

The implications of recombinant technology for medical science are far reaching, particularly when taken in conjunction with related technologies using DNA probes for diagnosis of disease and monoclonal antibodies for diagnosis and therapy. In spite of massive research activity, probably less than 0.2% of the proteins in the body have been cloned and only a handful have reached the stage of clinical trials. Not only is it possible to make the exact protein as occurs in the body, but it is also possible to introduce subtle differences which are desirable from the viewpoint of using the proteins clinically, *e.g.* using the technique of site specific mutagenesis to substitute serine for cysteine at position 17 in human β-interferon to obtain a product with a good specific activity and stability (Khosrovi, 1984). The possibility of producing novel bioactive protein analogues capable of acting as agonists or antagonists is another area of major potential.

The rate at which it is now possible to develop recombinant processes to make proteins of interest, means that clinical evaluation will be the rate limiting step for the new compounds.

Developments in genetic manipulations are expanding the range of host organisms available for recombinant processes. The use of *E.coli* as host is widespread and it will continue to be the host of choice where the high expression levels and high volumetric productivities obtainable with this microorganism are necessary to produce a low cost product. Other hosts such as yeast and *B.subtilis* will be used in specialized applications, *e.g.* the use of yeast for the cloning of the hepatitis B surface antigen and for producing products of interest to the food industry and allied areas. The use of mammalian cells as hosts will assume increasing importance, particularly in the production of large complex proteins, *e.g.* plasma factor VIII, and where glycosylation or conformational considerations are of importance.

The continuing rapid developments in genetics, bioreactor design and operation, and in new high speed and high resolution product recovery stages will ensure that an increasing range of products become available at a decreasing cost of manufacture.

15.5 REFERENCES

Baldrige, M., G. J. Mossinghoff and J. F. Terapane (1983). *Patent Profiles; Biotechnology 1983 Update*. US Department of Commerce.

Bell, D. J. and P. Dunnill (1982). Shear disruption of soya protein precipitate particles and the effect of ageing in a stirred tank. *Biotechnol. Bioeng.*, **24**, 1271–1285.

Botstein, D. and P. W. Davis (1982). In *The Molecular Biology of the Yeast* Saccharomyces: *Metabolism and Gene Expression*, ed. J. N. Strathern, E. W. Jones and J. R. Broach, pp. 607–636. Cold Spring Harbor Laboratory, Cold Spring Harbor, NY.

Budiansky, S. (1984). Cohen–Boyer patent: licensees ponder weaknesses. *Nature (London)*, **312**, 487.

Cohen, S. N. and H. W. Boyer (1980). Process for producing biologically functional molecular chimeras. *US Pat.* 4 237 224.

Cohen, S. N. and H. W. Boyer (1984). Biologically functional molecular chimeras. *US Pat.* 4 468 464.

Davey, M. W., E. Sulkowski and W. A. Carter (1976). Hydrophobic interactions of human mouse and rabbit interferons with immobilised hydrocarbons. *J. Biol. Chem.*, **251**, 7620–7625.

De Maeyer, J. (1981). *US Pat.* 4 172 071.

Dubnau, D. A. (ed.) (1982). *The Molecular Biology of the Bacilli*, vol. 1. Academic, New York.

Dwivedi, C. P., T. Imanaka and S. Aiba. (1982). Instability of plasmid-harbouring strain of *E.coli* in continuous culture. *Biotechnol. Bioeng.*, **24**, 1465–1468.

Edge, M. D. *et al.* (1981). Total synthesis of a human leukocyte interferon gene. *Nature (London)*, **292**, 756–762.

Edlin, G., R. C. Tait and R. L. Rodriguez (1984). A bacteriophage cohesive ends (cos) DNA fragment enhances the fitness of plasmid-containing bacteria growing in energy limited chemostats. *Bio/Technology*, **2**, 251–259.

Friesen, J. D. and G. An (1983). Expression vehicles used in recombinant DNA technology. *Biotechnol. Adv.* **1**, 205–227.

Ganesan, A. T., S. Chang and J. A. Hoch (eds.) (1982). *Molecular Cloning and Gene Regulation in Bacilli*. Academic, New York.

Gluzman, Y. (1982). In *Eukaryotic Viral Vectors*, ed. Y. Gluzman, p. 1. Cold Spring Harbor Laboratory, Cold Spring Harbor, NY.

Goeddel, D. V., D. G. Kleid, F. Bolivar, H. L. Heyneker, D. G. Yansura, R. Crea, T. Hose, A. Kaszewski, K. Itakura and A. D. Riggs (1979). Expression in *Escherichia coli* of chemically synthesized genes for human insulin. *Proc. Natl. Acad. Sci. USA*, **76**, 106–110.

Gray, P. P., P. Dunnill and M. D. Lilly (1973). The continuous flow isolation of enzymes. In *Fermentation Technology Today*, ed. Gg. Terui, pp. 347–351. Society of Fermentation Technology, Osaka.

Halluin, A. P. (1982). In *Banbury Report 10: Patenting Life Forms*. p. 67–126. Cold Spring Harbor Laboratory, Cold Spring Harbor, NY.

Hamer, D. H. and G. Khoury (1983). In *Enhancers and Eukaryotic Gene Expression*, ed. Y. Gluzman and T. Shenk, p. 1. Cold Spring Harbor Laboratory, Cold Spring Harbor, New York.

Hanisch, W. H., S. Fuhrman, D. Harano and M. Pemberton (1982). Separation and purification techniques applicable to the biological processes. In *Proceedings of the Fifth Australian Biotechnology Conference*, pp. 153–170. University of New South Wales, Sydney.

Helling, R. B., T. Kinney and J. Adams (1981). The maintenance of plasmid-containing organisms in populations of *Escherichia coli*. *J. Gen. Microbiol.*, **123**, 129–141.

Hofschneider, P. H. and W. Goebel (1982). Gene cloning in organisms other than *E.coli*. *Curr. Top. Microbiol. Immunol.*, **96**, 254.

Itakura, K., T. Hirose, R. Crea and A. D. Riggs (1977). Expression in *Escherichia coli* of a chemically synthesized gene for the hormone somatostatin. *Science*, **198**, 1056–1063.

Khosrovi, B. (1984). The production, characterization, and testing of a modified recombinant human interferon beta. In *Interferon: Research, Clinical Application and Regulatory Consideration*, eds. K. C. Zoon, P. D. Nogucki and T.-Y. Liu, pp. 89–99. Elsevier, New York.

Kula, M-R., K. H. Kroner and H. Hustedt (1982). Purification of enzymes by liquid–liquid extraction. *Adv. Biochem. Eng.*, **24**, 73–118.

Lowe, C. R. (1984). New developments in downstream processing. *J. Biotechnol*, **1**, 3–12.

Lee, S. B. and J. E. Bailey (1984). Analysis of growth rate effects on productivity of recombinant *E.coli* populations using molecular mechanism models. *Biotechnol. Bioeng.*, **26**, 66–73.

Arthur D. Little, Inc. (1983). Biotechnology: opportunities in the biomedical industry. Acorn Park, Cambridge, MA.

McCormick, F., M. Trahey, M. Innis, B. Dieckmann and G. Ringold (1984). Inducible expression of amplified human beta interferon genes in CHO cells. *Mol. Cell. Biol.*, **4** (1), 166–172.

McLaughlin, J. R. and S. Chang (in press). Molecular cloning and expression of heterologous genes in *Bacillus subtilis*. In *DNA Recombinant Technology*, ed. S. Woo, vol. 1. CRC Press, Boca Raton, Florida.

Marston, F. A. O., P. A. Lowe, M. T. Doel, J. M. Schoemaker, S. White and S. Angal (1984). Purification of calf prochymosin (prorennin) synthesised in *E.coli*. *Bio/Technology*, **2**, 800–804.

O'Neill, W. P. (1981). Medical impacts of molecular genetics. In *Genetic Engineering Applications for Industry*, ed. J. K. Paul, pp. 144–487. Noyes Data Corp., NJ.

Palva, I. (1982). Molecular cloning of alpha-amylase gene from *Bacillus amyloliquefaciens* and its expression in *B.subtilis*. *Gene*, **19**, 81–87.

Palva, I., P. Lehtovaara, L. Kaariainen, M. Sibakov, K. Cantell, C. H. Schein, K. Kashiwagi and C. Weissmann (1983). Secretion of interferon by *Bacillus subtilis*. *Gene*, **23**, 229–235.

Pavlakis, G. N. and D. H. Hamer (1983). Regulation of a metallothionein-growth hormone hybrid gene in bovine papilloma virus. *Proc. Natl. Acad. Sci. USA.*, **80**, 397–401.

Pestka, S., S. Maeda, D. S. Hobbs, T. R. C. Chiang, L. L. Costello, E. Rehberg, W. P. Levy, N. T. Chang, N. R. Wainwright, J. B. Hiscott, R. McCandliss, S. Stein, J. A. Moschera and T. Staehelin (1981). The human interferons: the proteins and their expression in bacteria. *Recombinant DNA, Proceedings of the Third Cleveland Symposium on Macromolecules*, Cleveland, Ohio, ed. A. G. Walton, pp. 51–74. Elsevier, New York.

Pestka, S. (1983). The human interferons — from protein purification and sequence to cloning and expression in bacteria: before, between and beyond. *Arch. Biochem. Biophys.*, **221**, 1–37.

Riggs, A. D. (1981). Bacterial production of human insulin. *Diabetes Care*, **4**, 64–68.

Rich, V. (1984). First commercial interferon. *Biotechnology*, **2**, 672.

Ross, M. J. (1981). Production of medically important polypeptides using recombinant DNA technology. In *Insulins, Growth Hormone and Recombinant DNA Technology*, ed. J. L. Gueriguian, pp. 33–48. Raven Press, New York.

Salt, D. J., R. B. Leslie, P. J. Lillford and P. Dunnill (1982). Factors influencing protein structure during and after precipitation: a study of soya proteins. *Eur. J. Appl. Microbiol. Biotechnol.*, **14**, 144–148.

Shine, J., P. H. Seeburg, J. A. Martial, J. D. Baxter and H. M. Goodman (1977). Construction and analysis of recombinant DNA for human chorionic somatomammotropin. *Nature* (*London*), **270**, 494–499.

Shogman, G., J. Nilsson and P. Gustafsson (1983). The use of a partition locus to increase stability of tryptophan-operon bearing plasmids in *E.coli*. *Gene*, **23**, 105–115.

Staehelin, T., D. S. Hobbs, H.-F. Kung, C.-Y. Lai, and S. Pestka (1981). Purification and characterization of recombinant human leukocyte interferon (IFLrA) with monoclonal antibodies. *J. Biol. Chem.*, **256**, 9750–9754.

Talmadge, K., J. Kaufman and W. Gilbert (1980). Bacteria mature preproinsulin to proinsulin. *Proc. Natl. Acad. Sci. USA*, **77**, 3988–3992.

Thomas, C. R. and P. Dunnill (1979). Action of shear on enzymes: studies with catalase and urease. *Biotechnol. Bioeng.*, **21**, 2279–2302.

Virkar, P. D., T. J. Narendranathan, M. Hoare and P. Dunnill (1981). Studies on the effects of shear on globular proteins: extension to high shear fields and pumps. *Biotechnol. Bioeng.*, **23**, 425–426.

Williams, D. C., R. M. Van Frank, W. L. Muth and J. P. Burnett (1982). Cytoplasmic inclusion bodies in *Escherichia coli* producing biosynthetic human insulin proteins. *Science*, **215**, 687–688.

Wilson, T. (1984). More protein from mammalian cells. *Bio/Technology*, **2**, 753–755.

Wood, W. I., D. J. Capon, C. C. Simonsen, D. L. Eaton, J. Gitschier, B. Keyt, P. H. Seeburg, D. H. Smith, P. Hollingshead, K. L. Wion, E. Delwart, E. G. D. Tuddenham, G. A. Vehar and R. M. Lawn (1984). Expression of active human factor VIII from recombinant DNA clones. *Nature* (*London*), **312**, 330–337.

Zehnbauer, B. A. and A. Markovitz (1980). Cloning of gene lon (cap R) of *E.coli* K-12 and identification of polypeptides specified by the cloned deoxyribonucleic acid fragment. *J. Bacteriol.*, **143**, 852–863.

SECTION 2

FOOD AND BEVERAGE PRODUCTS

16
Introduction

D. I. C. WANG
Massachusetts Institute of Technology, Cambridge, MA, USA

Some of the products and processes included in this section of *Comprehensive Biotechnology* can be considered as being well established and have to some degree reached maturity. For example, the bakers' yeast industry has been in existence for a large number of years and further growth of this industry is not anticipated to be monumental. Furthermore, one can also consider bakers' yeast to be a commodity product of the biotechnology industry. In general, most commodity products will be dependent on the economics of production where technological improvements will play a vital role with respect to its future. One might address the technological developments in biotechnology which might in the future have an impact on the production of bakers' yeast.

One of the major economic factors for the production of bakers' yeast is the cost of the raw materials. Molasses has traditionally been the carbon source for yeast production. The cost of molasses, a by-product of sugar (sucrose) manufacturing, has often been volatile due to the demand and fluctuation of sugar prices. This fact is of special importance in recent years due to the market success of sugar substitutes such as high fructose corn syrup and artificial sweeteners. One is therefore led to believe that future prices for sugar will increase due to lower volumes and thus the cost of molasses will increase proportionately.

The displacement of traditional thinking in the light of other technological developments might be worthy of consideration. For example, alternate carbon sources with lower costs and stable prices for the growth of *Saccharomyces cerevisiae* might be important in overall future planning. As illustrative examples, perhaps molecular biology and genetic engineering concepts might be explored so that carbon sources such as starch or methanol might one day serve as the raw material for bakers' yeast production. One might also ask how modern biology might affect the production of bakers' yeast from a more fundamental point of view. Specifically, among the properties of bakers' yeast are its ability to produce flavor components and its functional roles in baking. The biochemical machinery in the form of enzymes is the key for achieving the functional properties of *S. cerevisiae* for the baker's needs. Here, modern advances in molecular biology, especially the recent developments in eukaryotic organisms, must not be neglected. One can imagine using the present day bakers' yeast manufacturing process to produce greater levels of the key enzymes for functionality in baking per unit of product. Thus one is able to maintain the same unit manufacturing cost but at a higher level of functionality, leading to a greater economic margin.

Let us now address the second major area within this section of *Comprehensive Biotechnology* — the amino acids. It has been reported that the growth of this industry will be moderate, in the 5 to 7% range per year. Two of the amino acids are worthy of special emphasis: aspartic acid and phenylalanine. Due to the acceptance of the artificial sweetener aspartylmethylphenylalanine, the sales of these two amino acids have been truly phenomenal, reaching nearly $600 million in 1984. Although new products from amino acids might not be as dramatic as this dipeptide sweetener, this success has certainly attracted attention to many non-traditional industries and companies in the amino acid market. Much of this attention has been healthy in that new concepts and technologies are beginning to be considered and introduced to a well established manufacturing industry. For example, through molecular biology, it has been shown that polypeptides of many amino acids can be biosynthetically produced through the use of synthetic genes. Alternate technologies to fermentation have even been conceived for the production of amino acids. These include the use of immobilized whole cells for bioconversions such as that reported in this section for aspartic acid production. In addition, the production of phenylalanine *via* enzymatic routes in conjunction with organic synthesis is another example: this process uses cinnamic acid plus ammonia. Lastly, the specificity of enzymes in non-aqueous peptide synthesis represents a further example for the production of dipeptides from DL-mixtures. These technological advancements certainly must be considered as to the future of this specific industry.

There is no doubt that the achievements in modern biology will also play a predominant role in the manufacturing processes for amino acid production. For example, the traditionally random mutation and selection for enhanced microbial producers of desired amino acids will undoubtedly be displaced by directed mutagenesis. The alteration of critical enzymes in the biochemical regulatory pathways will enable increased product formation to be achieved once our understanding of the genetic systems for amino acid producing microorganisms is more clear. One might even ask whether the traditional organisms such as *Corynebacterium* or *Brevibacterium* for the production of amino acids will not one day be replaced by more efficient genetically engineered hybrids. However, as shown by past experience, it will be the finding of new uses of amino acids that will drive the advancements of the more traditional technologies.

Let us now address the third category in this section, which is the single-cell proteins. The excitement surrounding the use of microbial proteins diminished due to the increased oil prices in the 1970s, and microbial proteins have met with difficulty in penetrating the marketplace. This scenario will undoubtedly not undergo any drastic changes, especially in the light of the potential of biotechnological concepts in agricultural products, which are the primary competitors of microbial proteins. Although modern biological concepts (increasing cell yield through genetic engineering) have been introduced toward the production of single-cell proteins, the marginal economic gains have not offset the incremental cost differential between the competitors. However, there is still the hope of achieving higher value-addition to microbial proteins with respect to their functional properties, such as those of the filamentous fungi. This will tell whether the economic viability which is needed can be attained.

What one can derive from the production of microbial proteins are the technological spin-offs of this technology. For example, new concepts in fermenter (bioreactor) design as a result of the single-cell protein production from methanol by Imperial Chemical Industries of the UK certainly represent examples of technological breakthrough. The demonstration that a non-mechanically agitated fermenter at an unheard of scale (1500 m^3) can be designed and operated aerobically for long periods of asepsis is proof of spin-off to other biotechnology industries.

To conclude this introduction, the processes and products presented in this section must be considered in the light of future process innovations and other technological advancements. These factors will surely offer opportunities to the existing industries. Of particular significance are the following concepts and developments: computer technology for on-line control and optimization; cell harvesting *via* technologies such as cross-flow membrane filtration and flocculation; improved separation technologies for desired end-products, which include affinity and immuno chromatography, liquid-membrane ion exchange technology, and others.

These illustrate but a few examples where the more traditional processes could benefit economically in the future.

17

Modern Brewing Biotechnology

G. G. STEWART and I. RUSSELL
Labatt Brewing Company Ltd., London, Ontario, Canada

17.1 INTRODUCTION

For the purpose of this chapter, beer refers to a beverage which involves in its production (i) extracting malted barley with water; other carbohydrate-rich materials may also be employed, (ii) boiling this extract, usually with hops, and (iii) cooling the extract and fermenting it using yeast. The fermented beverage is then normally clarified and dispensed in an effervescent condition.

The word beer stems from the Latin infinitive 'bibere' meaning to drink. Medieval monasteries brewed beer and were responsible for spreading their name for it throughout Central Europe. The old High German word for beer was 'boer' or 'bior' corresponding to the old English word 'beer'.

Monasteries and other establishments which brewed beer during medieval times brewed beer much as we know it today, using malted barley and hops. Beverages produced as a result of the fermentation of grain or grain extracts, however, did not begin in medieval times. They reach far back into the prehistory of man. Artifacts taken from the mines of ancient cities give evidence that a formal practice of brewing existed over 5000 years ago. A drawing packed into a pottery piece excavated in Mesopotamia shows two brewery workers using two long poles to stir the contents of a brewery vat. This artifact was estimated to have been made in the 37th century BC. Numerous writings and drawings have been recovered from ancient Egypt that give details of commercial brewing and distribution of such beers in commerce. There is evidence that the Chinese also produced a kind of beer called 'Kin' over 4000 years ago. These beers were brewed from barley, wheat, spelt (a hard grained variety of wheat) and millet, as well as from rice, but in ancient Egypt, Mesopotamia and in Southern Europe, barley was the grain of preference. There is indirect evidence that in prehistoric times man fermented natural sugar-containing substances long before he had learned to bake bread and to brew beer from grain. Presumably man fermented grapes for hundreds of years before he fermented grain, while his use of fermented honey and the fermented saps from plants, such as the moor plant found in ancient India, must go far back beyond man's earliest use of grapes. No one can estimate with accuracy just when man first began to consume fermented beverages, but estimates have been made that it was at least 30 000 years ago. Man's discovery of beverage alcohol doubtlessly took place as did his discovering of wine, *i.e.* by chance; they were both in existence long before man!

Beer is a dilute solution of ethanol which also contains a unique mixture of flavouring substances. The characteristic flavour of beer arises from the use of malt as the predominant source of fermentable carbohydrate and other yeast nutrients, and of hops as a source of bitter components. These characteristics immediately distance beer in flavour terms from products derived from vegetable and fruit products. It also separates beer from distilled liquids which are flavoured either with added vegetable products, for example gin, or which contain only the volatile constituents of fermented broths such as with whisky.

Classically, beer is produced from malted cereals, water, hops and yeast. Only one country, West Germany, still abides by this strict definition. In that country, the Bavarian Purity Law that was established in the 16th century is still enforced and only the previously mentioned four ingredients are allowed, if the product is to be called beer and sold as such in West Germany (Narziss, 1984). The German Beer Law, as it is now known, applies to the more than 1200 breweries that produce 5000 brands and deals with the amount of the beer tax, the gravity of the individual beer classes and their taxation, the time when the beer tax is due as well as defining the permitted raw materials. Furthermore, the Beer Law describes the product methods and the composition of some traditional beers, like 'Berliner Weissbier'.

The Purity Law clearly states that for the production of bottom fermented beers only barley malt, hops, yeast and water are permitted. Top fermented beers follow the same regulations, but additionally wheat malt is allowed and furthermore, for special beers, pure beet, cane or invert sugar and colouring substances derived from sugar are allowed. Beers which are explicitly brewed for export can deviate within allowable limits. Auxiliary definitions for malt production, special malts, the treatment of brewery liquor and the stabilization of beer are laid down in the regulations.

Brewing was one of the earliest processes to be undertaken on the commercial scale. Of necessity, it became one of the first processes to be developed from an art into a technology. In return, science in general and technologies other than brewing have derived much greater benefit than is commonly appreciated. For example, as discussed by Hudson (1983), few scientists are aware that Joule, who devised the mathematical control which was so important in establishing the physics of heat and the calculations of electrical energy, was a practising brewer. The great debates between Liebig, Wohler, Pasteur and others as to the nature of fermentation are somewhat more familiar. The results of Pasteur's research reflect in many aspects of daily life throughout the civilized world. It is appropriate to recognize, therefore, that Pasteur accomplished much of his research with the brewing process as a tool for study and as a reward, the pasteurization technique has allowed trade in beer to develop from its localized form to regional, national and international scales. From the Carlsberg Institute in Denmark, Sørensen's concept of the pH scale has been adopted throughout science and technology, while agricultural biochemistry has depended heavily on the Kjeldahl technique of measuring nitrogen.

Despite contrary superficial appearance, the brewing industry has been quite ready to adopt techniques from other industries. A very early example was the commission of Linde, who later established the technology of liquefaction and separation of gases, to install a refrigeration plant in the storage cellars of the Munich breweries. Eventually, cooling equipment for use in fermentation vessels was developed and when, much later, plants became available which could dissipate the heat generated in large bulks of germinating grain, the brewing and malting processes could be operated throughout the year and not seasonally, as had formerly been the case.

The beer industry is a large one and beer is the largest product by value of all the biotechnological industries. It is produced by an industry which has made a rapid transition from a craft organized in small units, to an industry making its products in large complexes. In a craft industry the usual way to produce a constant product is to change nothing. The modern brewing technologist must produce a constant product even though raw materials, types of plant and scale of operation are changing radically.

The current interest in biotechnology and the belief that it will expand considerably are based upon three factors: (i) the raw materials can be obtained from renewable sources; (ii) biotechnological processes appear likely to be economical against the chemical processing of vegetable materials; (iii) a wide range of possibly valuable products is being defined through traditional biological methods and through genetic manipulation.

Biotechnology has many technical handicaps to overcome and as Atkinson (1983) has pointed out, it is appropriate to acknowledge that brewing has, in part at least, overcome many of these handicaps. Brewing is operated very largely non-aseptically, it avoids separation processes other than solid/liquid and has come to terms with a range of raw materials of an exceedingly complex nature. Medium to large scale biotechnology will have to overcome many problems for which the brewing industry has devised partial answers, although in several cases, complete technical solutions have to date been prevented by the narrowness of the economic margins. Nevertheless, the brewing industry can instead instruct emerging new biotechnology industries in the following areas: (i) by-product recovery of materials rejected raw material preparation; (ii) use of low value carbon dioxide; (iii) sale and/or disposal of biomass; (iv) downstream processing, for example, the use of inexpensive flocculation and cold storage for clarification; (v) concentration liquid processing — high gravity brewing; (vi) non-aseptic fermentations, achieved in brewing by a semi-anaerobic operation at low pH and by attenuating fermentations as completely as possible; and (vii) recycling operations will have to be established for water and unconsumed nutrients; brewing yeast is recycled, last runnings from the lauter tun are re-used in mashing and water usage is kept to a minimum (8–12 hl water/hl beer); this latter aspect also positively influences liquid waste disposal.

As Atkinson (1983) has very distinctly remarked, 'the establishment of a flow-sheet for a large scale biotechnological process is a salutary experience, highlighting as it inevitably does, low overall process conversion efficiencies, large losses of raw materials, large quantities of water, the extent of recycle necessary, and the large number of preparatory and product recovery stages required.' Obviously significant improvements are necessary and the experiences of the brewing industry will be invaluable during the optimization process.

17.2 THE TRADITIONAL BREWING PROCESS

Beer production is divided into three quite clear-cut processes: malting, mashing and fermentation. The first two processes together produce a medium known as wort; to this yeast is added, and fermentation is allowed to proceed. Wort is essentially an aqueous extract of malted barley, the primary raw material in the manufacture of beer.

Beer brewing is a biotechnological process whereby agricultural products, such as barley and hops, are converted into beer by control of the biochemical reactions in malting, mashing and fermentation. It has been suggested that malting and mashing may not always be a part of the manufacture of beer (Lewis, 1968); however, fermentation has, at the present time, and for the foreseeable future, no prospect of being replaced. Indeed, Slater (1977) has pointed out that although the 12 enzymes that are necessary to convert glucose to ethanol and CO_2 have all been highly purified, and most have been crystallized, it would be unthinkable to make beer from crystallized enzymes. Why go to the enormous expense of isolating and purifying these enzymes and then adding them individually to the wort when they are nicely packaged in the yeast cell, which moreover reproduces itself!

In principle, the production of beer is a simple process. Yeast cells are added to a nutrient medium (wort) and the cells take up nutrients and utilize them so as to increase the yeast population; in so doing they excrete into the medium major end-products such as ethanol and carbon dioxide, together with a host of other minor metabolites. The final medium, after yeast removal, is the product, beer. It contains some wort constituents, as well as all the non-volatile and many of the volatile substances produced by the metabolism of the yeast. The objective of individual breweries is to bring about the appropriate degree of metabolism so that the product contains the required mixture of by-products and to achieve this in a reasonable time.

17.3   RAW MATERIALS

It has been stated previously that the basic raw materials employed in brewing are barley malt, hops, water and yeast. In addition non-malt cereals are often used as an adjunct with barley malt. These non-malted cereals are employed both for economic reasons and to produce a lighter product that currently has consumer appeal in the market place. There are two major barleys currently in use: six-row and two-row. Barley heads on the cereal may have either six rows or two rows. In six-row barley, there are three kernels at each node on alternate sides of the head, resulting in six rows of kernels. In two-row barley, only one kernel develops at each node on alternate sides of the head and results in two rows of barley. There is significant debate and discussion about the merits and demerits of these two types of barley for malting. Analytically there is a notable difference between two-row and six-row barleys, in that six-row barleys contain proportionately more hemicellulose and insoluble carbohydrate and less potential extract-forming substances. Since extract as determined on barley is a measure of starch content and soluble enzymes, it follows that grain containing a high proportion of total carbohydrate and hence a low proportion of nitrogenous materials, will be rich in starch.

Over the years efforts to improve barley by classical breeding and selection programmes have resulted in new varieties with (i) much greater yields per acre, (ii) shorter, stiffer straw that is resistant to lodging even after the application of nitrogenous fertilizers, (iii) early ripening, (iv) disease resistance, and (v) generally greater uniformity. The modern plant breeders have at their disposal many hundreds of strains of barley and related grasses from which to breed new desirable varieties and with the advent of novel genetic manipulation techniques such as protoplast fusion and recombinant DNA/transformation the potential has been significantly increased.

17.3.1   Malt

Malt is cereal, usually barley (sometimes wheat and oats are malted), which has been allowed to germinate for a limited period of time, growth of the embryo being terminated by drying. Germination of the grain occurs until the starchy store (endosperm) available for the development of the germ of the grain has suffered some degradation by enzymes. The maltster is concerned with both the degradation of the endosperm and the accumulation of the enzymes in the grain that effect the degradation. Growth of the germ or embryo is an unwanted incidental to the making of malt, because it leads to depletion of the endosperm material through respiration of the embryo together with rootlet and acrospire formation. When the degradation of the endosperm has progressed to only a limited extent, the maltster terminates the growth of the embryo by drying the grain. In order that he can store the malt for long periods in a stable state, it has been customary for the maltster to continue this drying, beyond that required to arrest growth, by kilning. In summary, therefore, the malting process involves (i) the collection of stocks of suitable barley, (ii) the storage of cereal until it is required, (iii) steeping the grain in water, (iv) germination of the grain, and finally (v) drying and curing in a kiln. The complete malting process usually takes 6–9 days.

At its most basic, malting involves providing the correct environment within the grain for the synthesis of hydrolytic enzymes and the controlled action of these enzymes to hydrolyse the cell wall and the reserve proteins of the endosperm. The correct environment is a hydrated grain (44–47% moisture) and adequate available oxygen in the aleurone and in the embryo. The best malting barleys are those which can synthesize the relevant enzymes most effectively and complete the necessary hydrolyses within the endosperm most rapidly.

The art of malting originated in prehistoric times and it is reasonable to assume that, by processes of trial and error, the early maltsters developed their craft to a modest level of efficiency (MacLeod, 1977). Certainly, adequate malt was being manufactured long before there was any real understanding of the biochemical mechanisms involved in transforming the tough barley to the desirably friable malt kernel which carries within it all the necessary enzymes to convert its starch to fermentable sugars. To exploit a natural phenomenon such as germination, so as to secure the optimum end-product for brewing, requires the application of scientific method — the ability to observe, to change a procedure, to record any differences from the control and then to make a considered judgement based on the results of the experiment. There can be little doubt that over the past 4000 years maltsters have conducted such an experimental programme and in so doing, have accommodated their technology within the constraints imposed by the physiological mechanisms evolved by the grain which enable it to compete successfully in its natural environment.

Over the last 20 years, there has been an accelerated understanding of the physiology and biochemistry of cereal seed germination, and much of this knowledge has been translated into malting practice. Not all scientific innovations require a profound understanding of the fine details of all known biochemical pathways. A very simple observation, carefully interpreted, can lead to significant improvements in technology. An excellent example of this scientific approach to malting was the demonstration (Pollock *et al.*, 1955a, 1955b) that in estimating the germination energy of a sample of barley, very different results could be obtained depending on whether optimal or slightly supra-optimal amounts of water were provided. From this essentially simple observation there has arisen the concept of water sensitivity, a terminology misnomer because the excess water is not detrimental *per se* but excess water restricts access of oxygen to the embryo. This finding has led to a critical appraisal of steeping systems. Though it appears that embryos of 'water sensitive' barley may require higher than normal oxygen tensions before they can grow, there is as yet no accepted biochemical explanation of this requirement.

For several years the alterations in steeping, complete with superior temperature control and mechanical handling of the grain, which was also introduced, seemed to be sufficient change and improvement for the industry. However, research on malting continued and the next development represents a reversal of the usual order of discovery and exploitation. The plant hormone gibberellic acid was discovered in Japan, evaluated but not used in Sweden, produced in the UK and employed in the malting industry of that country and several others. Extraneous treatments of malting barley with gibberellic acid were employed to regulate and accelerate the germination process (Hough and Rudin, 1958).

Due to the fact that commercial application of gibberellic acid had been achieved prior to complete elucidation of it function and effects, further laboratory investigations were undertaken in the late 1960s. It was soon realized that in the time available for malting, the barley embryo passes gibberellin to only a small proportion of all aleurone cells (Palmer, 1969). This led to the development of the process of abrasion whereby the husk and underlying layers are just sufficiently damaged for extraneous gibberellic acid, applied in sprays or in steep-water, to pass directly to the aleurone layer. In this way the enzymes which modify the grain, both chemically and mechanically, can be distributed throughout the barley corn much more homogeneously. The abrasion process is notable in that it passed from laboratory discovery to application in as little as three years.

During malting the term modification is frequently employed. In this context modification refers to 'an alteration in structure'. There are three major categories of polymer in the barley endosperm which undergo modification during malting. These are the cell wall polysaccharides (essentially β-glucan and pentosan), the reserve protein hordein and the starch granules. Thus a term frequently used in malting circles is index of modification. This has traditionally been used to quantify protein solubilization; equally well a well-modified malt is one in which all the cell walls have been eliminated and degraded to oligosaccharides of low molecular weight. Further, over-modification of malt may be one in which extensive corrosion of the large starch granules has occurred with a resultant overproduction of products of starch hydrolysis and a high cold water extract. This confusion in terminology is unfortunate and care should be taken in writing about modification in order to state which of the possible meanings is intended.

As previously stated, the cell walls of barley endosperm contain primarily β-glucans and pentosans. The amount of the water-insoluble hemicelluloses as well as the water soluble gums depends upon barley variety and agronomic conditions. β-Glucans are technologically far more important than the pentosans, as they are highly viscous and thus can cause a great deal of problems if their degradation during germination is not satisfactory. β-Glucans complicate the recovery of the malt extract during the brewing process; their high specific viscosity impedes the run-off of the wort away from the spent grains and adversely affects the filtration of the beer.

When considering proteolysis during barley germination, the structure of the barley grain and the location of the enzymes and their substrates must be borne in mind. At least two thirds of the grain's reserve protein is located in the starchy endosperm and in order to act upon it the enzymes must also be present in this location. The pH is also of importance for the activity of these enzymes. The starchy endosperm contains high activities of proteinases especially sulfhydryl enzymes with a pH optimum around 5, the pH of the endosperm. The endosperm also contains high activities of carboxypeptidases, but both neutral aminopeptidases and alkaline peptidases are absent. The break down of the reserve proteins in the endosperm is obviously carried out by the proteinases and carboxypeptidases present. The resulting amino acids and small peptides are taken up by the scutellum where the peptides are further hydrolysed by the peptidases before they are transported to the growing tissues of the embryo. The sulfhydryl proteinases and carboxypeptidases have a central role in the production of amino acids during germination.

The moisture content of malt is reduced from *ca.* 45% to *ca.* 5% during kilning. To reach this dryness economically, malt has to be subjected to rather high temperatures. During drying malt loses its gherkin-like taste and obtains the typical malt aroma. By employing the kilning conditions of humidity, time and temperature, it is possible to produce malt with a greater or lesser degree of colour and aroma (Runkel, 1975). Moreover, the greatest portion of the rootlets are removed during kilning. The purposes of kilning can be summarized as follows: (i) removal of water to obtain keeping qualities of the highly perishable unkilned 'green' malt; (ii) interruption of germination, *i.e.* inhibition of metabolism (drying); and (iii) formation of taste and aroma compounds (curing).

Attention has to be paid to the preservation in the malt of enzymes necessary for brewing. For several of these enzymes it is initially necessary to predry malt at *ca.* 50–60 °C to a moisture content of approximately 23%. The second stage of kilning reflects the slower removal of moisture. The surface and easily removed moisture has already been eliminated and, moreover, kernel shrinkage occurs with interior moisture reaching the surface with increasing difficulty. Total surface area has also decreased. At this stage air flow is usually reduced and the temperature is increased in stages to *ca.* 71 °C. This stage continues until approximately 12% moisture remains in the malt.

In the final drying stage the moisture is reduced to 4.0–5.0% by the use of decreased air flow and higher temperatures of 71–92 °C. The final kilning (or curing) at the highest temperature may be from two to four hours long. During this third stage moisture removal is difficult and slow as all the water is physically bound. This stage eliminates the green, grainy taste and supplies most of the malty flavour. After kilning, the dried malt is cooled to 38 °C or lower by passage of air and then dropped into collection hoppers prior to transfer to the elevator. Sprouts are generally severed by passing over a malt cleaner and the grain is stored for at least three weeks prior to shipment.

Runkel (1975) has studied the influence of kilning on malt enzymes. The activities of α- and β-amylase are only slightly decreased during the drying phase. During curing β-amylase is inactivated to a greater extent than is α-amylase. Proteolytic enzymes are noticeably damaged only during curing over 100 °C. Catalase activity is strongly diminished during drying, and curing temperatures completely inactivate the enzyme. The activity of polyphenoloxidase is hardly affected during kilning. Depending upon the temperature level, lipase is partially inactivated during curing.

The degradation of protein and starch has been found to be dependent upon moisture content and temperature. The lower the moisture content, the higher the temperature should be to guarantee degradation. Proteolytic enzymes are still active at lower moisture contents then the amylases. With increasing curing temperatures and moisture contents, protein coagulation increases. The more sugars and amino acids that are formed during germination and drying, and the higher the curing temperature, the greater the concentration of melanoidins formed during kilning. The acidity of malt also increases as does beer colour. For the taste of light lager beers a curing temperature of 80 °C is favourable. In some experiments concerning different curing temperatures, superior foam keeping properties of beer derived from higher cured malt have been established. Due to damage of the amylase, the attenuation limit (degree of fermentation) of the beer falls as a result of an increase in curing temperature. Low molecular nitrogen decreases as a result of a rise in curing temperature, due to melanoidin formation. It has also been shown that a higher curing temperature decreases the control of dimethyl sulfide (DMS) of malt and beer. DMS is an important beer flavouring compound and will be discussed in greater depth later in this chapter.

In 1978 it was reported that volatile nitrosamines, particularly nitrosodimethlyamine (NDMA), were present in minute amounts in many beers (Spiegelhalder *et al.*, 1979). It was soon found that the principle source of NDMA was malted barley. As a result of close collaboration between brewers and maltsters all over the globe, much is now known about the formation of NDMA in malt and beer (Wainwright, 1981). The major precursor of NDMA is hordenine, which is produced in the developing seedling during barley germination. It can be synthesized from pre-existing tyrosine but can presumably be formed *de novo*. Any conditions which restrict growth decrease NDMA formation by decreasing hordenine biosynthesis. Shortly thereafter, it was determined that malt NDMA was formed during the malt kilning operation, especially when combustion gases from sulfur-free fuels (*e.g.* natural gas) are used in the drying medium. This method of kilning is called 'direct', as distinguished from 'indirect' where combustion gases do not directly contact malt on the kiln. Where indirect kilning is employed the malt NDMA levels are negligible. The formation of NDMA in malt is now thought to occur according to the scheme depicted in Figure 1. The oxides of nitrogen (NO_x) found in the drying air nitrosate the amines which are formed during germination. NDMA formation can be controlled by burning elemental sulfur if the sulfur is applied continuously at a prescribed time during the first hour of drying. There is still some dispute as to the ideal conditions of sulfuring (Wainwright, 1981). Addition of SO_2 to the kiln gas should convert NO_2 to NO. Although air from burners using sulfur-containing fuels often has high NO_x levels, most of it is NO. Sulfur in the fuel, or added as carbon disulfide, will be beneficial because it competes for O_2 thus decreasing NO_x formation and partly by reducing the NO_2 which is formed to NO. Conversion of all the NO_2 to NO gives malt with very little NDMA.

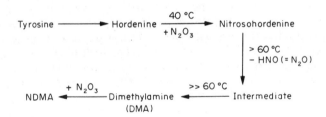

Figure 1 Nitrosodimethylamine production

Although part of the effect of sulfuring is to convert NO_x to a less reactive form (NO) it also 'inhibits' nitrosamine formation in other ways. Trials have been conducted in which a selection of burners were tested in the same kiln. With the oil-fired burner the drying air contained high concentrations of NO and NO_2 as it was shown that NO_x was removed by the malt bed but low NDMA values were obtained. It is clear that part of the benefit of sulfuring is due to the formation of sulfuric acid. This converts hordenine into its protonated form which cannot be nitrosated. NDMA formation can be prevented if the green malt is sprayed with sulfuric acid prior to kilning and indeed all other acids tested have been found to be effective. Further, a process for controlling NDMA formation by spraying green malt with sugars has been patented (Aalbers *et al.*, 1980). Experiments show that this causes the production of acids but the sugar will also produce reducing substances which reduce NO_2.

Gushing frequently occurs after malting barley has been harvested during periods of rainy weather and it is probable that this barley has been contaminated by a large number of fungal species. Gushing is the violent uncontrolled ejection of product when the package is first opened and involves the loss of a significant portion of the contents. Gushing is completely unacceptable to the consumer, and if the problem is not corrected this condition could prove catastrophic to the brewing company concerned. Although fungal contamination of the malting barley is not the only cause of gushing, it is certainly one of the primary factors (Prentice and Sloey, 1960). The action of the contaminating fungi can moreover continue during storage, if the barley is not sufficiently dried. During malting, conditions prevail which are favourable for the multiplication of the microorganisms present in barley. A heavy fungal contamination of the barley can give rise to an active and abundant mycelium, which especially during germination of the barley produces meta-bolites which induce gushing. Gushing factors can also be formed by moulds which originate not from the barley but from the malting plant (Gyllang and Martinson, 1976a, 1976b). The gushing inducing substances have been shown to be peptides or compounds containing peptides produced mainly by species of the genera *Fusarium*, *Aspergillus*, *Penicillium*, *Nigrospora* and *Stemphylum* (Haikara, 1980).

Haikara (1980) has studied gushing problems that occurred in Finnish beer in the late 1970s. It has been concluded that there is a good correlation between *Fusarium* contamination in barley and malt and the gushing of beer, both on the brewery and laboratory scale. Great differences in gushing activity have been found between species belonging to the genus *Fusarium* and, indeed, even beween strains of the same species. On the basis of the material investigated, direct conclu-sions could not be drawn about the 'safe' level of *Fusarium* moulds on barley and malt. It is how-ever probable, that if less than 50% of malt grown is contaminated with *Fusarium*, gushing of beer is rare.

17.3.2 Adjuncts

Many brewers today employ some form of brewing adjunct to supplement malt. Adjuncts are defined as non-malted carbohydrate materials of suitable composition and properties that ben-eficially complement or supplement barley malt. These adjuncts vary considerably in their carbo-hydrate, nitrogenous, lipid and mineral composition and are used firstly for the influence they have on regulating the composition of the resulting wort and secondly as a cost saving initiative due to the fact that most non-malted carbohydrate sources are less expensive than barley malt. Wort properties are thus deliberately modified by incorporating brewing adjunct materials into the grist and this action in turn influences beer properties.

The genesis of adjuncts in brewing was in North America, where it was realized that the indige-nous barleys differed from those of Europe in that the former contained higher nitrogen levels and had thicker husks than the latter. Because of these differences, often (but not always) their use in all-malt brewing produced beers with poor physical stability. However, the thicker husk provided a more efficient filter bed for lautering and the high nitrogen levels provided higher enzymatic activity. This higher diastatic activity was sufficient to convert a much larger concen-tration of starch than is contained in the malt itself. Consequently, a variety of other starch-con-taining materials could be employed in order to obtain more extract at a lower cost per unit weight than is obtained from an all-malt grist.

Although adjuncts are employed primarily because they provide extract at a lower cost than that available from malt, other advantages can be achieved. Beers produced with adjuncts are lighter in colour with a less satiating thinner taste, greater brilliancy, enhanced physical stability and superior chill-proof qualities. In addition, the use of certain adjuncts makes it possible to increase products in cases of limited brewhouse capacity. This is especially true with the use of syrups and sugar adjuncts that can be added directly to the kettle and thus by-pass the mashing operation.

The non-malted brewing materials used in greatest quantity today are those derived from corn and rice, although barley, wheat and sorghum grain are also employed. Also, in some parts of the world, particularly countries in the southern hemisphere (for example Australia and South Africa), sucrose from both cane and beet are employed as are cassava and potatoes. Unlike the distilled spirits industry which uses whole kernel cereal grains as basic ingredients, malt beverage products utilize cereal grains which have been preprocessed to some degree before delivery to the brewery. In this sense, all of the adjuncts used in brewing are manufactured products that have been obtained from a variety of cereal grains.

The starch of these adjunct products is in its native form and is not readily attacked by the malt diastase during mashing. Consequently, these adjuncts must be processed by boiling in a cereal cooker mash to bring about solubilization and gelatinization of the starch granules to render them susceptible to diastatic enzyme attack. As the starch granules gelatinize, they swell and become viscous. A portion of the malt mash (approximately 10% by weight) is generally added to the cooker mash. The α-amylase of the malt, under suitable conditions, hydrolyses the gelatinized starch leaving it partially degraded and fluid prior to addition of the cooker mash to the malt mash.

In a refinement of brewing adjuncts, the wet milling industry now provides pure dried starch (*i.e.* refined grits) and so-called liquid adjuncts, which are clean, non-crystallizing mixtures of fermentable sugars and dextrins in syrup form. In these cases, the wet miller has processed whole grain (usually, but not always, corn) to obtain pure starch with essentially all of the protein and lipid materials removed. The wet miller can further process the pure starch slurry with acid and/ or diastatic enzymes to produce syrup (a liquid adjunct).

Today's liquid adjuncts (syrups) are clear, colourless, non-crystallizing liquids consisting of carefully controlled mixtures of glucose, maltose, maltotriose and higher saccharides (dextrins). Uniformity of raw material is of great importance in brewing. Variations in product, which sometimes occur when a brewer adjusts his cooker operation for differences in corn grits, are eliminated by the use of liquid adjuncts. Liquid adjunct use results in a better run-off and clearer wort from the all-malt mash than a grits mash, and because syrups are added to the kettle, the mashing procedure can be controlled for maximum benefits from the malt. The faster mash run-off and resultant rapid filling of the kettle means greater control over the kettle operation. The use of liquid adjuncts makes it possible to use higher wort concentration (°Plato) in the kettle and to reduce the concentration before fermentation by the addition of water. In this way more hectolitres per brew can be achieved.

Brewing syrups have a spectrum of fermentable sugars and unfermentable dextrins closely approximating the carbohydrate spectrum formed during mashing when malt and dry cereal adjuncts are employed. Indeed, Pfisterer *et al.* (1978) have shown that the sugar spectrum of brewing syrups must closely approximate the sugar spectrum of wort with a high maltose concentration (50–60%), a glucose and maltotriose level of significantly less (10–15%) and a fermentability of approximately 70% of the total carbohydrate material in the syrup. Syrups with elevated glucose levels (40–60%) have presented fermentation problems in some brewing situations. Some brewing yeast strains have been found to be sensitive to elevated glucose levels in wort and appear to exhibit a repression of the maltose uptake system. This has resulted in incomplete fermentations ('hanging fermentations') with unacceptably high amounts of yeast fermentable extracts in the beers at the end of fermentation. When syrups with high maltose concentrations were employed, hanging fermentations did not occur. These high maltose syrups are produced by using a maltose-releasing enzyme (β-amylase) rather than glucoamylase and an acid conversion of starch, thus achieving a distribution of fermentable carbohydrates in the syrup which closely matches the pattern found in wort. In terms of the technology of syrup production, overall efficiencies have been improved by the use of immobilized enzyme systems and a plethora of syrups both for fermentation and their organoleptic properties are beginning to be employed in brewing.

Although cost is obviously an important factor in the selection and use of adjuncts, the economic advantage available is limited by the amount of adjunct used. Opinions vary concerning the extent of adjunct use and each brewing operation must determine its own ratio for the type of beer it intends to produce. Due to reasons of economy or availability, it may be necessary to substitute one adjunct for another. When it is not desired to modify wort properties, if these are already satisfactory, it might well be necessary to alter the character of the malt in use in order to accommodate the adjunct change. Usually such changes involve malt because its properties can be varied so much more conveniently than those of the adjuncts. Consequently, the flexibility offered by the maltster permits the best use of the many brewing adjuncts now available.

17.3.3 Hops

Hops as used in brewing are the dried blossoms of the female hop plant (*Humulus lupulus*). The earliest regular use of hops is generally believed to have been in Germany during the twelfth century. Although many other flavourings had been employed in fermented beverages as far back as 10 000 BC, the use of hops, to produce what is now regarded as beer, gained wide popularity only in the past few hundred years. The description of the characteristic flavour of hops in beer is subject to a great deal of debate, but all agree that this flavour is an essential part of the total organoleptic impact of beer. Hops also contribute to flavour stability and foam retention. In most countries there are legal restrictions which dictate the minimum level of hops if a product is to be labelled beer. It is also possible that hops were first added to beer for their known bacteriostatic effect. Considering the high hopping ratios required for bacteriostatic activity (*ca.* 400 g hl^{-1}) and the reduced hopping ratios employed in many countries, particularly in North America, this effect is probably no longer important.

Hops are employed in the brewing process to impart a bitter taste to beer. They can also provide hop aroma and flavour, or hop character. The chemistry of the bittering process is largely understood: the α-acids present in hops are converted during kettle boil into the more soluble iso-α-acids, which account for most of the bitterness of beer. This understanding has led to large benefits for the brewing industry in terms of hop breeding, hop growing, hop processing and the control of hop usage in the brewery (Seaton *et al.*, 1981). Surprisingly, the flavour contribution of hops to beer (as distinct to bitterness) is less well understood, and it is not possible to measure it either quantitatively or qualitatively. This presents problems for the hop breeder because relevant guidance is unavailable for assessing new varieties for aroma and flavour. Indeed, at this time, good 'aroma hops' are associated with low α-acid varieties.

Hop varieties can be roughly divided into two classes, bittering hops and aroma hops. As their name implies, bittering hop varieties are those that impart the bitter flavour to beer and have an α-acid composition of approximately 5–8%, but high α-acid varieties with α-acid contents of 12–14% are now available. Aroma hops impart characteristic hop aromas to beer and include such varieties as Hallertau, Brewers Gold and Cascade.

The composition of hops is shown in Table 1 and is affected by variety, area grown, climate, time of harvest, processing, storage conditions, *etc.* Of the compounds in Table 1, only the α-acids, β-acids, essential oils and polyphenols have been shown to have a significant effect on beer bitterness and beer hop flavour. Beer bittering substances derived from hops arise predominately from the α- and β-acids, also known as humulones and lupulones (Verzele, 1979). Although this fraction as extracted from hops has no significant bitterness, when boiled in water it yields a solution that possesses a sharp, intense bitterness. The major source of bitterness in the resulting solution is a group of compounds named iso-α-acids or isohumulones, two of which arise from the base-catalysed isomerization of each α-acid. Because these compounds are the most bitter and abundant of the hop-derived bittering compounds, the hop industry has sought to maximize the α-acid content of its hops and the brewing industry uses α-acid content as a criterion for purchase. The iso-α-acid content of beer is employed as a quality control guide in tandem with taste testing.

Table 1   Chemical Composition of Hops

Component	Percentage
α–Acids	2–12
β–Acids	2–10
Essential oils	0.5–1
Polyphenols	2–4
Oil and fatty acids	Traces to 25%
Wax and steroids	Trace
Protein	15
Cellulose	40–50
Water	8–12
Chlorophyll	Trace
Pectins	2
Ash	10

At harvesting, the hop vines are cut off at ground level and the hop cones are stripped off and separated from the leaves and other plant material. The freshly picked cones, containing 80% moisture, are immediately loaded into a drying kiln where they are dried and moved to a cooling room where they are then packed into rectangular bales (covered with jute burlap or plastic cloth) about 50 × 80 × 14 cm with a net weight of approximately 90 kg.

Hop pellets have become available over the past 10–15 years as a more convenient alternative to hop cones. Pellets are produced by shredding the baled hops, removing woody stems and other foreign material, hammermilling the hops to a fine powder and then running the powder through a high pressure pelletizing die. The lupulin glands in the hops are ruptured and the released resin binds the other vegetable matter together to form a dense pellet 4.5–6 mm in diameter by 4–6 mm long. These pellets are then cooled and vacuum packed in plastic or foil laminated bags and packed into boxes or polyethylene lined drums.

Hops are maintained in storage rooms at the brewery at *ca.* 2–4 °C. Nevertheless even in this environment they lose their bittering potential and α-acids during storage. The decrease in α-acid content is larger than the decrease in bittering potential because of the ability of the α-acid oxidation products to produce bitterness in beer. The rate of loss of α-acid is dependent on variety of hops and storage conditions, and is much slower at low temperatures and in the absence of oxygen. Similar results with hop pellets have been noted. Although it has been claimed that properly packaged pellets are stable at ambient temperatures or at least do not lose their bittering potential, many brewers have found that pellets can lose substantial amounts of α-acid and bitter potential if stored at ambient temperatures. Consequently, it is recommended that hop pellets be stored at cold temperatures in a similar fashion to leaf hops.

Hop extracts, which are used as a direct replacement for hops in the kettle, and isomerized extracts, which are used to best advantage by addition to beer in the conditioning or aging tank, are both available from hop processors (Laws, 1981). Commercial quantities of hop extracts have been available since the turn of the century, however, it was only during the last 20 years that significant quantities have been used by the brewing industry. On a world wide basis, about 20% of the 1978 crop was converted into hop extracts and a further 6% was processed into isomerized hop extracts (Hudson, 1979). The United States and West Germany together produced over 50% of the world hop crop and it is not surprising that extensive hop extraction industries have developed in these countries.

Hop extracts are concentrations of the α-acids from which beer bitter substances are produced as a result of isomerization during wort boiling. The extracts are normally obtained by treating hops with organic solvents followed by evaporation. The α-acid content of such extracts will depend on: (i) the level of α-acid in the hops, (ii) the age of the hop and conditions of storage, and (iii) the organic solvent employed.

The main reasons for using hop extracts as a replacement for kettle hops can be summarized as follows: (i) more consistent bitterness between successive brews, (ii) improved hop utilization, (iii) improved stability on long term storage, and (iv) reduced transport, storage and handling costs. The improvement in bitterness consistency of successive brews is usually most marked when brewing on the small or pilot scale. In such cases it is often difficult to select and add to the kettle portions of hops containing similar amounts of α-acids. Extracts are more homogeneous than hops and it is easier to maintain a constant level of α-acid in wort when using extracts to brew similar beers. The extent of the improvement in hop utilization when using hop extract varies from brewery to brewery. In most plants the increase is usually about 10%, but occasionally the increase in utilization can exceed 20%. The reasons for such variations in hop utilization are still not fully understood. Pure resin extracts have been stored at ambient temperatures for up to six years without losing significant amounts of α-acids. However, extracts containing added water soluble material are often very unstable and display considerable loss of α-acid after storage for 12 weeks. Hence pure resin extracts are preferable as a brewing raw material because of their superior storage characteristics.

Significant losses of potential bittering substances normally occur when hops are boiled with wort in the kettle and also during fermentation but only small losses occur during aging and subsequent packaging. As a result, the utilization of the bitter substances rarely exceeds 40% in commercial breweries and is often as low as 25%. In contrast when isomerized extracts are added to beer after fermentation the utilization of the bitter substances is often in excess of 80%. Substantial savings can be made by employing isomerized extracts to bitter beer and the United States, Australia and British brewing industries all use significant amounts of such extracts.

α-Acids can be isomerized in aqueous sodium carbonate and under these conditions the formation of degradation products such as humulinic acid is minimized. This reaction forms the basis of a number of processes which have been described for producing iso-α-acids by treating α-acids with aqueous sodium or potassium carbonate. The products of such reactions are either sodium or potassium salts of iso-α-acids which can then be readily metered into beer. Potassium salts of iso-α-acids are more soluble than the sodium salts. α-Acids can also be isomerized at neutral pH in aqueous methanolic solution containing divalent cations such as magnesium or calcium. A very high yield ($\sim$95%) of pure iso-α-acid can be obtained from this reaction on the laboratory scale. However, attempts to scale up the reaction gave variable yields of iso-α-acids together with substantial quantities of degradation products.

The United States brewing industry uses substantial quantities of reduced isomerized extracts to bitter beer. This product is obtained by reduction of the carbonyl group on one of the side chains of isohumulone, to a secondary alcohol group using sodium borohydride as a reducing agent. The reduced extract is claimed to give beer protection from the characteristic 'light struck' aroma which occurs when beer is exposed to sunlight as a result of the formation of 3-methyl-2-butene-1-thiol. Reduced iso-α-acids are less bitter on a weight-for-weight basis than are iso-α-acids.

Methylene chloride is the most common solvent used to extract hops but hexane and methanol are also employed. Using these solvents it is normally possible to extract at least 95% of the available α-acids from fresh hops. Hop extracts usually contain a small proportion of the available essential oil present in the hops because most is lost during the recovery of the organic solvent. Hence hop extracts are not suitable for imparting hop character to beer because they are also devoid of essential oils. Some hop extract manufacturers treat the residual hops, after extraction with organic solvents, with hot water to obtain tannins, amino acids and water soluble resins. This water soluble material is then blended with the solvent extract to standardize the α-acid content of the product. These water soluble fractions are expensive to produce, and, moreover, they accelerate the deterioration of the α-acids. Hence the use of water soluble hop components is an expensive and unsatisfactory way to standardize α-acids.

In the 1960s it was demonstrated in both the USA and the USSR that liquid carbon dioxide was a selective solvent for extracting fruits, spices and vegetables. A pilot scale extraction plant was constructed in the USSR in 1966 and during the following six years over 1000 tonnes of a wide range of raw materials was extracted with liquid carbon dioxide. A USSR patent filed in 1968 (Pechov *et al.*, 1968) was the first report of extracting hops with liquid carbon dioxide. The extract was a light brown coloured viscous mass which dissolved completely in ethanol and had a strong odour of hops. Using liquid carbon dioxide it was possible to remove over 90% of the total extractables from hops after two hours although precise details of the extraction conditions have not been published.

A Japanese patent was filed in 1970 (Kato *et al.*, 1970) which described the extraction of hops with liquid carbon dioxide over a temperature range of -15 to $+25$ °C. Yields were very variable and some extracts were contaminated with high levels of tannins. Since 1970 there has been renewed interest in extracting hops with carbon dioxide. Two main processes have been described: (i) the extraction of hops with carbon dioxide under supercritical conditions (*i.e.* at temperatures of 45 to 50 °C and at pressures of 300–400 atm) and (ii) the extraction with liquid carbon dioxide at temperatures of -10 to $+15$ °C and pressures of 40 to 60 atm (Laws, 1981).

When hops are extracted with liquid carbon dioxide at 7 to 10 °C, selective extraction of certain hop components occurs. The resulting yellow extracts consist mainly of α-acids, β-acids, essential oils and water and are free of hard resins, tannins and green pigments. The extraction time required for removal of a high percentage of available α-acids from hops when employing liquid carbon dioxide depends on: (i) the temperature of extraction, (ii) the physical state of the hops, and (iii) the rate of flow of liquid carbon dioxide. Hop α-acids have a maximum solubility in liquid carbon dioxide of 8 g kg^{-1} at a temperature of 7 °C. Surprisingly the solubility decreases at higher temperatures and at 75 °C is only approximately 5 g kg^{-1}. Hence the optimum temperature of extraction is 7 °C and pilot-scale studies have shown that the efficiency of extracting α-acids is satisfactory when hops are treated with liquid carbon dioxide over a temperature range of 5 to 10 °C. The efficiency of extraction of both oils and resins is poor when dried cone hops are extracted with liquid carbon dioxide. Carbon dioxide extracts can be stored for prolonged periods without showing significant deterioration of either oils or resins. Pilot and production scale brewing trials have shown that carbon dioxide extracts can be used as a direct replacement for kettle hops when brewing both ales and lagers. The resulting beers were of sound quality and were indistinguishable in flavour from controls using hop pellets or hop extracts produced using organic solvents.

Liquid carbon dioxide extracts are ideal starting materials for preparing isomerized extracts because of their freedom from tannins, hard resins and plant pigments. Initial laboratory scale isomerized extracts encountered problems because the β-acids did not always form a granular precipitate after pH adjustment (Laws *et al.*, 1977). However, Harold and Clarke (1979) have described a commercial process for removal of α-acids from carbon dioxide prior to isomerization. The extract is converted into a finely divided form and the α-acids are selectively removed by treatment with aqueous alkali. Isomerization can be carried out either in alkaline solution or by precipitating the α-acids as soluble magnesium salts and then converting them into iso-α-acids by heating. A pale yellow powder containing approximately 85% of iso-α-acids can be obtained by the latter process.

The use of carbon dioxide to extract hops provides a means of producing extracts which are of higher quality than those obtained using organic solvents. It is also possible to obtain extracts rich in essential oils which are paricularly suitable for imparting hop character and aroma to beer. High quality isomerized extracts can be prepared from liquid carbon dioxide extracts. Liquid carbon dioxide extracts are free from traces of residual organic solvents and are stable on storage at ambient temperature.

17.3.4 Brewing Water

For centuries brewing water quality has been recognized as a major factor in determining beer quality. Breweries were located where water quality was consistent and very often this water was drawn from underground sources where its composition remained relatively constant and it was protected from pollution by the geological structure. Older established brewing centres that have become famous for a particular type of beer were often associated with water supplies of a special composition that was presumably beneficial to the type of beer being produced, for example, Burton-on-Trent, Dortmund, Munich, Dublin and Pilsen.

Unfortunately, water supplies everywhere are becoming polluted at an ever increasing rate. Small amounts of toxic metals, detergents, herbicides, pesticides, petroleum hydrocarbons and trihalomethanes are finding their way not only into surface waters but also into underground supplies. Brewers can no longer take the quality of water for granted. From whatever source it may come, strict water quality control procedures must be instituted.

Brewing water must not only meet the general requirements for potable water (*i.e.* free from disease-producing organisms and from substances producing adverse physiological effects) but must also comply with specific requirements in order to ensure proper mash pH, efficient hop extraction, good kettle break, sound fermentation and acceptable flavour and colour development in the finished beer. In many breweries all incoming water is subjected to activated carbon filtration and in some cases it is filtered through ion exchange resins. A sufficient amount of calcium is essential in brewing water, particularly during mashing. It protects α-amylase against heat destruction and thus assists in the liquefaction of the mash. Calcium also stimulates the activity of proteases and amylases. Calcium assists in monitoring proper mash pH and aids in the flocculation of proteinaceous material in the kettle. In later stages of the brewing process calcium is required to assist in yeast flocculation and in oxalate removal. In summary, a proper calcium level in the brewing water is required in order to obtain a stable and palatable beer. In addition to calcium, the correct levels of magnesium, sodium, potassium, sulfate, chloride and nitrate are all necessary in the correct proportion in brewing water. Most brewers consider it advantageous to control the composition of their brewing water by the addition of a mixture of salts. This process is often referred to as 'Burtonizing the brewing water' (after Burton-on-Trent). The principal component of such a salt mixture is calcium sulfate or gypsum.

17.4 WORT PRODUCTION

17.4.1 Mashing

The first step in wort production is the process of mashing whereby soluble materials are extracted from ground malt (Figure 2). Initially, mashing was simply the mixing of warm water with ground malt, and, after a suitable standing period, as much liquid was recovered as was possible with the primitive equipment on hand. Sparging or washing of the grains was probably not universal. Breweries must soon have realized that hotter water yields a different result from cooler water but they were unable to control the temperature accurately because thermometers were either not invented or unavailable.

Historically, brewers have been very inventive in adapting their methods and equipment to utilize to best advantage the ingredients available to them to produce beers to the taste of their customers. This inventiveness has been particularly apparent in mashing (Dougherty, 1977). There are many methods of mashing practised in the world today. Most mashing systems in current use can be classified as either infusion systems, or decoction sytems, or a combination of both, or a modification of one or the other. The single most widely employed mashing system in North America is described as a double-mash, upward infusion system. This system utilizes a cereal cooker in which the adjuncts are prepared by boiling and a mash mixer (or 'tun' or 'tub') in which malt mash is prepared and in which the two mashes are ultimately combined. This system was developed to utilize solid adjuncts, such as corn grits and rice, as a means of producing lighter and more drinkable beers. The extensive use of solid adjuncts has been facilitated by the relatively high diastatic power of North American barleys.

The objectives of mashing can be summarized as follows: (i) to dissolve the substances that are in the ingredients which are immediately soluble; this fraction constitutes only 10 to 15% of the total weight of the ingredients; (ii) to render soluble through enzymatic action substances which are insoluble in their natural state; and (iii) to change the chemical structures, through

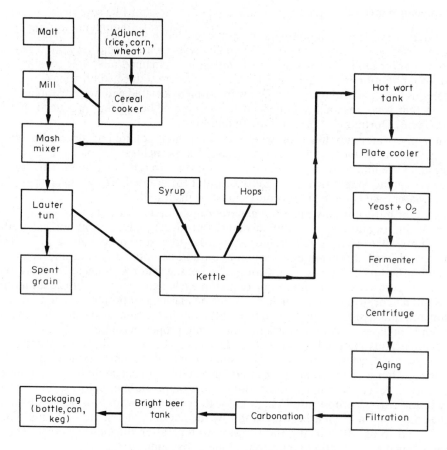

Figure 2   Schematic diagram of the brewing process

simultaneous enzymatic action, of some of the constituent substances in a planned or predictable manner.

Variations in: (i) the enzyme complement contributed to the mash by the malt, (ii) the mashing temperature, and (iii) the duration of mashing, all alter the composition of the first wort. Fortunately, mashes show many self-regulatory properties and worts vary less with alterations in mashing conditions, with certain exceptions, than might be expected. Wort contains simple sugars, more complex polysaccharides (dextrins), amino acids, peptides, proteins, other nitrogenous materials, vitamins, organic and inorganic phosphates, mineral salts, polyphenols, tannin precursors and tannins, small quantities of lipids and numerous other components, many of which have not been identified. Wort is unstable microbiologically. It also throws a heavy precipitate which carries down with it a variable amount of its constituents. Surface active materials such as lipids, hop bitter substances and proteins are removed preferentially in the precipitate.

The most notable change brought about in mashing is the dissolution of starch to yield the greater part of the wort carbohydrates: The major factor that influences starch dissolution as it occurs in an infusion mash may be explained by the joint action of the α- and β-amylases, perhaps with small contributions from R-enzyme and limit dextrinase. In model experiments acting on starch, α-amylase alone will ultimately produce a dextrinous wort containing a mixture of sugars, of which only 16–20% are fermentable by normal brewing yeast strains. In contrast, β-amylase will produce only maltose and β-limit dextrins sufficiently large to give a blue colour with iodine. However, acting in concert, an excess of a mixture of pure amylases (α- and β-) will theoretically break down starch to yield a wort that is approximatly 70% fermentable. This figure increases to approximately 80% if limit dextrinase is present. On mashing it is the liquefying action of α-amylase that is chiefly responsible for the dissolution of the starch granules, which do not begin to gelatinize below 65 °C. The subsequent formation of the characteristic spectrum of sugars is due to the concerted action of all the relevant enzymes on the products of liquefaction.

From the point of view of unit operations, mashing entails mixing, enzyme reactions, liquid–solid separation, elution and liquid–solid separation. Consequently, when the enzyme reactions discussed above have reached a suitable point the liquid must now be separated from the grains (spent grains). Wort separation is essentially a physical process, although not a completely simple one. At the completion of mash conversion the temperature of the mash is raised to 75–77 °C in preparation for the separation of the wort from the solids. The 'mashing-off' temperature is to some extent a compromise. At this temperature, the wort viscosity is favourable for quick and complete separation from the spent grains, enzymes are inactivated although there is some residual α-amylase activity, bacterial action is precluded and extraction of the previously insoluble substances from the grains, principally malt husk tannins, is not a problem.

17.4.2 Wort Filtration

The method of separating the wort from the mash solids and thus the equipment employed for the purpose is mainly a matter of choice and sometimes even tradition on the part of the individual brewing operation. In some instances, different methods of wort separation are employed in plants of the same brewing company and in some cases even within the same brewery. Essentially there are two types of equipment for wort separation, the lauter tun (or tub) and the mash filter.

The lauter tun is the most widely employed wort separation vessel system in North America and Europe at the present time. A lauter tun is a vertical cylinder of large diameter to depth ratio. It is usually constructed of stainless steel although copper lauter tuns are still in operation. The top of the tun is usually spherical or conical. The top of the tun may be flat or sloped or constructed with several concentric valleys with intervening ridges. Fitted at the bottom of the tun is a wort collecting system of manifolded pipes, usually arranged in concentric rings leading through a system of valves through which the wort is delivered to a collecting vessel called a grant. There is often a system of water jets built into the bottom of the tun by which the floor and wort collecting system can be flushed. Suspended above the true bottom of the tun is a false bottom of precisely slotted stainless steel or brass plates. The plates are carefully fitted together to form a very flat and level floor upon which the mash comes to rest. The void between the false bottom and true bottom is typically 8–12 cm in a flat or slotted bottom tun and 10–15 cm in a large diameter valley bottom tun.

The lauter tun is equipped with a lautering machine (the rakes). The lautering machine consists of a heavily constructed shaft positioned precisely in the centre of the tun and connected either from above or below to a multispeed drive by which it can be rotated, raised and lowered. Connected radially from the main shaft and extending out about to the vertical wall of the tun are several (usually 2 or 3) rigid arms or booms. Connected vertically downward from the arms are lautering blades or rakes which can be positioned to almost touch the false bottom of the tun. The blades are of stainless steel or brass, designed with a thin knife-like cross section, interspaced with small elliptical flights perpendicular to the axis of the blade. The blades are approximately 15 cm apart on the arms. The blades on each arm are interconnected by an operator bar so they may be partially rotated on the vertical axis. In the lautering position, they are tangential to the wall of the tun. In the 'grains out' position, they are adjusted to approximately 45 °.

The lauter tun is also equipped with a sparging (hot water delivery) system. This is usually a system of concentric pipes suspended just below the ceiling of the tun and well above the highest position of the lautering machine and manifolded to the sparge water supply. The sparging rings are either drilled with numerous small holes on the lower radius or are equipped with spray nozzles directed downward to deliver the sparge water in a uniform pattern over the entire surface area of the mash bed in the tun.

In preparation for receiving a mash, the lauter tun is thoroughly rinsed and heated to mash temperature, or slightly above, by sparging or underletting hot water through the wort collection system. On the first brew of the week, (or if the tun has been allowed to cool for some time between brews), it would first be sterilized by boiling. It is common practice to provide a cushion for the mash by filling the tun with hot sparge water up to slightly above the level of the false bottom. This prevents the incoming mash from impacting directly onto the false bottom with resultant partial blinding of the slots by compacted mash. The cushion water dilutes this wort only slightly but must, just as mashing water and sparge water must, be of 'ingredient quality'.

The mash is delivered into the lauter tun as gently as possible consistent with a reasonable tun allowance for the transfer. The mash must be distributed uniformly over the floor of the tun and is usually levelled with the lautering machine. In some breweries, the mash is allowed to settle after being levelled and wort circulation (run-off) is not initiated until 15 to 30 minutes after all of the mash has been transferred into the lauter tun. In other operations, wort circulation is initiated almost immediately after transfer of the mash into the lauter tun has begun. The wort can be virtually classified as ready to go into the kettle by the time all of the mash is in the lauter tun and the grain bed levelled. These represent the operational extremes and most procedures would be intermediate between the two. During wort circulation the rakes may be employed to assist in the classification by particle size and stratification of the filter bed. If the malt is properly modified, has been correctly ground and has not been too harshly treated during mashing, the grist fractions should be at least partly recognizable in the mash in the lauter tun.

After the filter bed is established and the wort has achieved satisfactory clarity, circulation is stopped and the 'first wort' is diverted to the kettle. The flow of the wort must be controlled at a rate that will maintain hydraulic equilibrium with the system. As the first wort is driven off, sparging is initiated shortly before the wort level reaches the top of the grain bed. The top surface of the grain should not be allowed to become dry until sparging is complete. Sparging of the grain bed fulfils several functions during wort separation. It dilutes the first wort and reduces viscosity, thus encouraging the rapid flow down through the filter bed. As the concentration of the sparged water passing down through the bed decreases, it dilutes and elutes the wort which is intermixed and entrained upon and within the insoluble particles of the mash. The sparge water should be maintained at all times at a level approximately 210 cm above the surface of the grain bed until sparging is completed. This assumes an uninterrupted flow down through the bed without excessive weight and resulting compaction of the bed. If the entire lauter tun operating cycle is properly computed and managed, simultaneously the kettle will be filled, the sparge completed, the grain bed left relatively dry and the extract remaining in the grains reduced to a satisfactory level. During run-off of the wort, the lautering machine is operated at slow speeds, the rakes being gradually lowered into the grain bed in order to keep the bed permeable. If the grain bed becomes compacted or the hydraulic equilibrium is upset due to a too rapid wort withdrawal, stopping the wort flow for a few minutes usually helps correct the situation. This tends to restore balance to the fluid pressures and helps restore porosity and permeability. A brief recirculation of the wort may be necessary to recover clarity before running it into the kettle. If a severe compaction of the grain bed occurs, it may be necessary to halt the wort flow and underlet through the wort collecting system with hot (sparging temperature) water. The pressure of the water is thus utilized to 'lift' the bed. In a well operated system underletting of the grain bed is a procedure employed only infrequently. Recirculation of the wort for clarification prior to running into the kettle following such an underletting procedure is essential.

Upon completion of the sparge, the dilute wort is collected until the kettle has been filled to the predetermined level or until the clarity of the wort becomes impaired. Collection of the wort is discontinued and it is either diverted to the sewer, to by-product recovery or this material, called the post run-off (PRO), is recycled into the foundation water in the lauter tun of the subsequent flow. After the last wort and drainings have left the grain bed, the spent grains are removed from the lauter tun and the tun is cleaned for the next brew.

Although lauter tuns are widely employed pieces of equipment for wort separation in both North America and Europe, many large volume brewers favour mash filters. The use of filter presses for wort separation was developed in Europe in the late nineteenth century and introduced into North America about the turn of the century. A modern mash filter has advantages over lauter tuns but a comparison of the merits and demerits of the two systems is a controversial topic and one likely to generate considerable discussion and argument between brewers. Amongst the advantages of mash filters most certainly are a minimal space requirement and high extraction efficiency in terms of both wort recovery and brewhouse cycle time.

A mash filter consists of a series of alternating plates and hollow frames suspended on side rails in a very heavily constructed frame. Channels through which the mash and sparge are admitted to the filter and the wort is removed are formed out of eyes cast into the plates and frames. The frames are gasketed on both sides. The filter medium is a rectangular cloth, slightly larger than the cross section of the filter, which is draped over the top of the filter frame and down both sides. Modern mash filter cloths are generally of polyethylene or polypropylene fibre.

To initiate the wort separation cycle, the mash filter (with filter cloths in place and hydraulically closed so that it is fluid tight) is flushed and preheated with hot water. The mash is then pumped into the filter through the top channel, completely filling the filter frames. It is important that the volume of the mash be very precisely controlled so that the filter is exactly full. Overfilling causes a mash cake of excessive density which results in loss of filtration efficiency. Underfilling results in voids at the top of the frame which permit surpassing of sparge water. The total mash volume of a filter can be changed to precisely accommodate the mash by removing or adding plates and frames. As the mash fills the voids in the frames, the displaced air and vapour are vented off through the opposite topside channel. As soon as the filter is full, the vent is closed, and the small amount of mash remaining is pumped in at reduced flow to avoid excessive pressure build up. When the filter is full of mash, the wort collecting system is opened, and the first wort is drawn from the mash horizontally through the filter cloths. It flows downward along the plates out through cocks or valves into the wort collecting trough or pipe.

The wort is recirculated through the filter until satisfactory clarity is achieved. After this first wort is partially drained, but before the filter core becomes dry, sparging is initiated. After the sparge is complete, the filter core is drained to relative dryness and wort separation is concluded. The filter is then opened and the plates are separated to permit the spent grains to drop from the frames and into the grain collecting system. Following removal of the spent grains, the filter press is closed and tightened to a predetermined point necessary to assume a fluid-tight joint between the gasketed channel elements of the plates and frames. The filter is then prepared to receive mash, and the next cycle begins.

17.4.3 Wort Boiling

As previously discussed, once the wort has been separated from the spent grains it is collected in a vessel usually termed a kettle (or copper in the UK). As its name implies it is employed to boil the wort. Wort boiling is in itself a relatively simple, although energy intensive, operation. However, the complexities of the interactions which affect the wort constituents during boiling tend to frustrate attempts to refine the process. The commonly quoted reasons for wort boiling are: (i) to extract bitter and other substances from hops and promote necessary chemical changes, (ii) to precipitate unwanted nitrogenous material, (iii) to terminate enzymatic action, (iv) to remove undesirable volatile compounds, (v) to sterilize the wort, and (vi) to evaporate excess water. The objectives to be achieved by wort boiling can be simply defined as (i) stabilization, (ii) flavour development and hop isomerization, (iii) concentration, and (iv) spent hop removal.

Many complex reactions take place during wort boiling. As the wort is heated, the residual amylases and other enzymes are inactivated, which terminates the mashing process and fixes the carbohydrate composition and spectrum of the wort. At the boiling point the wort is sterilized and the microflora of the malt, hops and adjuncts employed are destroyed. As the boil continues some proteins are coagulated and some, together with simple nitrogenous constituents, interact with carbohydrates and/or polyphenolic components (tannins). The visible precipitate resulting from these interactions is known as the 'break'. Part of the break separates from the boiling wort (the hot break) but a further amount precipitates when the wort is cooled (cold break). At the same time the hop principles are extracted into solution and undergo transformation or interact with other wort constituents. In open kettles ideally between 4 and 10% of the volume is evaporated during one hour's boil so that some of the wort dilution caused by sparging is compensated. At the same time, most of the steam-volatile essential hop oils are lost. It is necessary that the bulk of these oils are eliminated in this way or the flavour of the hop oil would be too pronounced in the resulting beer.

In medieval times wort was boiled in iron cauldrons over open fires but as the scale of production increased, the vessels were covered over and fitted with a chimney to carry the steam out of the building. To construct these more complex vessels, copper became the material of choice due to its malleability, superior conductivity and resistance to corrosion. In the past 25 years stainless steel has become a popular construction material. Kettles were originally heated over open coal or wood fires housed in iron furnaces but the control of such fires required considerable skill and energy and were difficult to maintain in efficient operation. Such difficulties were overcome with the introduction of superheated steam for heating kettles and the majority of breweries employ this method today.

To ensure complete dissolution of the hop principles and the formation of a satisfactory break it is necessary that there should be vigorous ebullition during wort boiling. Such a boil was a feature of open-fired kettles and there was a marked deterioration in the vigour of the kettle boil when steam jackets were employed. Kettles are often fitted with fountains or calandria which work on the principle of a coffee percolator and increase the circulation of the boiling wort. It has also been found that internal steam coils are more successful in promoting a vigorous boil then a steam jacket. Indeed on many kettles both a steam jacket and internal steam coils can be found.

The question of the necessity of vigorous wort boiling in order to produce a stable beer of sound flavour has been the subject of much debate and discussion and is still an unresolved issue today. Rennie (1972) has suggested that wort boiling is indeed unnecessary and has taken the position that boiling can be replaced by a 'hold' at 85 °C if the wort is vigorously agitated. Although the non-boiled wort was very cloudy they produced beers of good shelf-life. This study has suggested that there are possibilities of energy cost savings if a simmered wort boil is employed instead of a full scale boil. However, to the authors' knowledge few breweries employ this method of wort preparation and the merits and demerits of wort boiling will be argued for many years.

17.4.4   Wort Cooling

At the end of the prescribed boil period, which may be based on time, evaporation rate or more usually a combination of both factors the wort is cooled in order that it may be aerated/oxygenated and the yeast added and fermentation commence. Wort cooling has a number of objectives (i) reduction of wort temperature from approximately 100 °C to the pitching temperatures of 8–16 °C, (ii) removal of haze and denatured protein (trub), and (iii) adequate aeration/oxygenation of the wort to permit the yeast to ferment efficiently. Wort cooling can be accomplished employing one or more of the following techniques (i) cooling with atmospheric air, (ii) cooling with water, and (iii) cooling by a refrigerant such as glycol, brine, ice, direct expression ammonia or freon.

Wort cooling would be a simple operation were it not for the important changes in the physical state of some wort constitutents involved (*e.g.* hops and wort proteins) and the danger of microbiological infection which is ever present. The desire to minimize the threat of infection has been largely responsible for the substitution of closed hot wort tanks for open cooling vessels called coolships and closed coolers for open coolers. The cooled sterilized wort that has had a significant amount of its solid material removed and is consequently in a more stable condition is now ready to be fermented by yeast in the central stage of the brewing process, namely fermentation.

17.5   YEAST AND FERMENTATION

In the production of fermentation ethanol (be it for beer, wine, potable spirit or industrial ethanol), the microorganisms being employed should possess a number of important characteristics: (i) rapid and relevant carbohydrate fermentation ability, (ii) appropriate flocculation and sedimentation characteristics, (iii) genetic stability, (iv) osmotolerance (*i.e.* the ability to ferment concentrated carbohydrate solutions), (v) ethanol tolerance and the ability to produce elevated concentrations of ethanol, (vi) high cell viability for repeated recycling, and (vii) temperature tolerance.

17.5.1   Brewer's Yeast Taxonomy

As previously discussed in this chapter, the two main types of beer are lager and ale. These are fermented with strains of *Saccharomyces uvarum (carlsbergensis)* and *Saccharomyces cerevisiae*, respectively. Traditionally, lager is produced by bottom-fermenting yeasts at fermentation temperatures between 7 and 15 °C. This means that at the end of the fermentation, bottom yeasts flocculate and collect at the bottom of the fermenter. Top-fermenting yeasts, used in the production of ale at fermentation temperatures between 18 and 22 °C, tend to be somewhat less flocculent and loose clumps of cells are carried to the fermenting wort surface adsorbed to carbon dioxide bubbles. Consequently, top yeasts are collected for reuse from the surface of the fermenting wort (a process called skimming), whereas bottom yeasts are collected (or cropped) from the fermenter bottom. The differentiation of lagers and ales on the basis of bottom and top cropping is becoming less and less distinct with the advent of vertical bottom fermenters and centrifuges. With centrifuges, non-flocculent yeast strains are required for both lagers and ales, where, as soon as fermentation is complete and before the yeast has had the opportunity to sediment, the fermented medium is passed through a centrifuge in order to separate the yeast from the 'young' beer. With conical bottom fermenters, a more sedimentary yeast (lager or ale) settles into the cone of the fermenter at the completion of fermentation. It is then removed and a portion reused as an inoculum for a subsequent fermentation.

Taxonomically the two species *Saccharomyces cerevisiae* and *Saccharomyces uvarum (carlsbergensis)* have been distinguished on the basis of their ability to ferment the disaccharide melibiose

(Lodder, 1970). Strains of *Saccharomyces uvarum (carlsbergensis)* possess the *MEL* gene(s) and consequently produce the extracellular enzyme α-galactosidase (melibiase) and are, therefore, able to utilize melibiose whereas strains of *Saccharomyces cerevisiae* are unable to produce α-galactosidase and are, therefore, unable to utilize melibiose. However, in two recent texts on yeast taxomony (Barnett *et al.*, 1983; Krejer-van Rij, 1984) these two groups of yeasts have been consolidated into one species, *Saccharomyces cerevisiae*. Indeed the texts have proposed that the number of species with the genus *Saccharomyces* be reduced from 41, as described by Lodder (1970), to either six (Barnett *et al.*, 1983) or seven (Krejer-van Rij, 1984). Whether or not these proposals will receive acceptance by scientists active in the field of yeast research and by fermentation technologists employing yeasts at the industrial level will have to await the test of time. Indeed, it is at the strain level that interest in brewing yeast centres. At the last count there were at least 1000 separate strains of *Saccharomyces cerevisiae*; these strains may be brewing, baking, wine, distilling or laboratory cultures. There is a problem classifying such strains in the brewing context; the minor differences between strains that the taxonomist dismisses are vitally important to brewers.

The behaviour, performance and quality of a yeast strain are influenced by two sets of determining factors; these two factors have been collectively called nature–nurture effects. The nurture effects are all the environmental factors (*i.e.* the phenotype) to which the yeast is subjected from inoculation onwards as distinguished from the nature or heredity. On the other hand the nature influence is the genetic make-up (*i.e.* the genotype) of a particular yeast strain.

The requirements of an acceptable brewer's yeast strain can be defined as follows: 'In order to achieve a beer of high quality, the yeast culture must be effective in removing the desired nutrients from the growth medium (*i.e.* the wort), it must impart the required flavour to the beer and finally, the microorganisms themselves must be effectively removed from the fermented wort after they have fulfilled their metabolic role.' (Stewart, 1977). This definition although wide ranging and very general allows the major activities of a brewing yeast strain during the fermentation of wort to be divided into a number of stages: (i) nutrition, *i.e.* sugar and amino acid uptake, *etc.*, (ii) by-product formation, *i.e.* the excretion of compounds that contribute to the flavour of the product, and (iii) removal of the culture from the fermented wort by flocculation or centrifugation.

In terms of the development and selection of yeast strains for use in a brewery there are a number of parameters that are important as scale-up criteria: (i) fermentation rate, (ii) decrease in specific gravity (°Plato of the wort), (iii) taste and flavour match of the final product, (iv) fusel oil, ester and organosulfur production, (v) consistent high cell viability, (vi) ethanol production and tolerance, and (vii) inoculation (pitching) rate. More specifically, there are a number of factors that will affect fermentation rate *per se*, and these include: (i) inoculation (pitching) rate, (ii) yeast cell viability, (iii) fermentation temperature, (iv) wort dissolved oxygen concentration at pitching, (v) wort soluble nitrogen concentration, (vi) wort fermentable carbohydrate concentration, and (vii) yeast storage conditions, *e.g.* the influence of intracellular glycogen levels.

17.5.2 Wort Oxygenation

Although wort fermentations in the production of beer are very largely anaerobic this is not the case when the yeast is pitched into the wort. If conventional beer fermentations are to be successful some oxygen must be made available to the yeast. The need for oxygen arises because: (i) brewing yeasts are unable to synthesize sterols and unsaturated fatty acids, which are essential membrane components, in the absence of molecular oxygen, (ii) sterols and unsaturated fatty acids are normally present in wort in suboptimal quantities; although they are abundant in malt, normal manufacturing procedures prevent them passing into wort and oxygen must be supplied to allow their synthesis by yeast. If ergosterol and an unsaturated fatty acid such as oleic acid are added to wort, the requirement for oxygen disappears. If a brewer's yeast strain is grown anaerobically it accumulates sterols and unsaturated lipids within the cells in excess of its minimum requirements and the lipids can be 'diluted' to a degree by subsequent growth without untoward effects. Accordingly, cells prepared aerobically can grow to some extent anaerobically. If, however, yeast is harvested at the end of fermentation and used to inoculate a second batch of wort, oxygen is required because the new inoculum contains no reserves of the necessary lipids. Such an inoculum can be used to determine how much oxygen is required and it has been suggested that this requirement varies with circumstance and yeast strain (Kirsop, 1974).

Kirsop (1978) has examined the oxygen requirements of a large number of brewer's yeast strains. The oxygen requirement has been defined as, 'the amount of oxygen needed to make a yeast ferment as rapidly as the control non-oxygen requiring sample of the same yeast strain'. Examination of a large number of strains has shown that the requirement for oxygen varies greatly from strain to strain and it has proven convenient to group the strains into four classes, on the basis of their need for oxygen: Class 01 — need satisfied by wort which is half saturated with air; Class 02 — need satisfied by air-saturated wort; Class 03 — need satisfied by oxygen-saturated wort; Class 04 — need not satisfied by oxygen-saturated wort.

More than 50% of the strains examined were classified as 03 or 04 types having a requirement for oxygen in excess of the quantity in air-saturated and oxygen-saturated wort respectively. The reason for this high oxygen requirement appeared to be that most of the strains studied were ale yeasts that had previously been used for production purposes in relatively small vessels which were either open or had a large head-space in comparison to the wort volume. It could be speculated that lager yeast strains employed in large vessels would be classified as either 01 or 02 types.

17.5.3　Wort Sugar Uptake

When yeast is pitched into wort, it is introduced into an extremely complex environment due to the fact that the wort is a medium consisting of simple sugars, dextrins, amino acids, peptides, proteins, vitamins, ions, nucleic acids and other constituents too numerous to mention. One of the major advances in brewing science during the past 25 years has been the elucidation of the mechanisms by which the yeast cell, under normal circumstances, utilizes, in a very orderly manner, the plethora of wort nutrients.

Wort contains the sugars sucrose, fructose, glucose, maltose and maltotriose together with dextrin material. In the normal situation, brewing yeast strains are capable of utilizing sucrose, glucose, fructose, maltose and maltotriose in this approximate sequence, although some degree of overlap does occur, and the majority of brewing strains leave the maltotetraose and other dextrins unfermented. The fermentability of wort can be measured by two general methods. The first method involves fermentation of wort in the presence of excess brewer's yeast with shaking for 24 hours at 20 °C. In most cases, maltotriose is completely fermented under these conditions. The fermentability is defined in the following equation:

$$\text{Fermentability (\%)} = \frac{\text{original gravity} - \text{final gravity}}{\text{original gravity}} \times 100 \times 0.8192$$

The factor 0.8192 converts apparent fermentability into real fermentability. In the second method, wort fermentability is determined by the ratio of fermentable sugars, estimated by gas chromatography, to the total carbohydrate content. It has been found that for most worts, the results given by the latter procedure are approximately 5% higher than those obtained by the former method. This discrepancy has been the subject of much debate in the industry (Hargitt and Buckee, 1978).

Saccharomyces cerevisiae and related species have the ability to take up and ferment a wide range of sugars, for example sucrose, glucose, fructose, galactose, mannose, maltose and maltotriose. In addition, *Saccharomyces diastaticus* is able to utilize dextrin material. The initial step in the utilization of any sugar by yeast is usually either its passage intact across the cell membrane, or its hydrolysis outside the cell membrane followed by entry into the cell by some or all of the hydrolysis products. Maltose, and maltotriose are examples of sugars that pass intact across the cell membrane whereas sucrose (and dextrin with *Saccharomyces diastaticus*) is hydrolysed by extracellular enzymes and the hydrolysis products are taken up into the cell.

Maltose and maltotriose are the major sugars in brewer's wort and as a consequence, a brewer's yeast's ability to use these two sugars is vital and depends upon the correct genetic complement. Brewer's yeast possesses independent uptake mechanisms (maltose and maltotriose permease), to transport the two sugars across the cell membrane into the cell. Once inside the cell, both sugars are hydrolysed to glucose units by the α-glucosidase system. The transport, hydrolysis and fermentation of maltose are particularly important in brewing, distilling and baker's yeast strains since maltose is the major sugar component of brewing wort, spirit mash and wheat dough. There are (at least) five unlinked polymeric genes that control the ability of yeast to produce α-glucoside permease in response to maltose, *MAL1*, *MAL2*, *MAL3*, *MAL4* and *MAL6*. Strains carrying an active allele at any one of these loci are inducible. The role of the *MAL* genes is still not fully understood. It has been suggested that the *MAL* loci are either: (i) structural genes for α-glucosidase, (ii) regulatory genes controlling both α-glucosidase and α-glucoside permease, or (iii) complex loci containing both regulatory and structural elements. Indeed recent evidence would appear to suggest that the *MAL* loci include the structural genes for α-glucosidase and maltose permease.

Isogenic triploid yeast strains (strains genetically identical with the exception of the particular gene) constructed by hybridization have been employed to study the influence of *MAL* gene dosage on wort fermentation rate (Stewart *et al.*, 1983). The triploids *MAL2/malo/malo*, *MAL2/MAL2/malo* and *MAL2/MAL2/MAL2* were compared to production brewing strains and their rates of fermentation in wort under agitated conditions at 21 °C monitored. In 12 °P and 16 °P worts, the strains containing the *MAL2/MAL2/malo* or *MAL2/MAL2/MAL2* genotype fermented at a much faster rate than either production ale or lager yeast strains. It would thus appear that strains with multiple *MAL2* genes would be invaluable strains for increased fermentation rates in wort.

In the normal situation the presence of glucose in the fermentation will inhibit the uptake of maltose and maltotriose. Thus the presence of glucose in the fermenting wort will exert a major repressing influence upon wort fermentation rate. However, with the use of the glucose analogue 2-deoxyglucose (2-DOG), mutants of brewing strains have been selected in which the maltose uptake is not repressed by glucose and as a consequence they have significantly increased fermentation rates when compared to repressed wild-type strains (Stewart *et al.*, 1985b).

Although sucrose is not employed extensively in brewing in North America it is used as an adjunct in such countries as Australia and South Africa. It has already been discussed that *Saccharomyces* species metabolize sucrose by virtue of the excretion of the extracellular enzyme invertase. Invertase hydrolyses the sucrose to glucose and fructose which are taken up into the cell. The production of invertase, both internal and external, is controlled by the *SUC* genes. A number of polymeric *SUC* genes have been identified and these are dispersed in the yeast genome. Each *SUC* gene confers upon the strain carrying it the ability to produce and excrete invertase.

17.5.4 Wort Nitrogen Metabolism

Active yeast growth requires the uptake of nitrogen, mainly in the form of amino acids, for the synthesis of protein and other nitrogenous components of the cells. Nitrogen uptake slows or ceases later in the fermentation as yeast multiplication stops; the cessation of growth is rarely due to lack of oxygen and consequently a deficiency of membrane components. Fermentation of sugars continues, although more slowly than before, until inhibition by ethanol or settling of the yeast from the fermenting wort brings the fermentation to a gradual end.

In wort, the main source of nitrogen for synthesis of protein, nucleic acids and other nitrogenous cell components is the variety of amino acids formed by proteolysis of barley protein. Wort contains 19 amino acids and, under brewery fermentation conditions, brewer's yeast takes them up in an orderly manner, different amino acids being removed at various points in the fermentation cycle (Jones and Pierce, 1967). Short chains of amino acids in the form of di- or tri-peptides may also be taken up by yeast cells. Amino acids, like many sugars, do not permeate freely into the cells by simple diffusion; there is a regulated uptake by a limited number of transport enzymes. Therefore, at the start of fermentation only eight amino acids (arginine, aspartic acid, asparagine, glutamic acid, glutamine, lysine, serine and threonine) are absorbed rapidly. The other amino acids are absorbed only slowly, or not at all until later in the fermentation. Under strictly anaerobic conditions (those encountered late in a brewery fermentation) proline, the most plentiful amino acid present in wort, is scarcely assimilated. Whilst 95% of the other amino acids have disappeared by the end of the fermentation, there is a considerable amount of proline in the finished product (approximately 200–300 μg ml^{-1}). Under aerobic laboratory conditions however, proline is assimilated after exhaustion of the other amino acids.

The inability of *Saccharomyces* spp. to assimilate proline under brewery conditions is the result of several phenomena. As long as other amino acids or ammonium ions are present in the wort, the activity of proline permease (the enzyme that catalyses the transport of proline across the cell membrane) is repressed; as a consequence of this permease repression, proline absorption is slight. The first catabolic reaction of proline, once inside the cell, involves proline oxidase which requires the participation of cytochrome *c* and molecular oxygen. By the time all the other amino acids have been assimilated, however, thus removing the repression of the proline permease system, conditions are strongly anaerobic and, as a consequence, the activity of proline oxidase is inhibited and proline uptake cannot occur.

17.5.5 Yeast Excretion Products

A major excretion product produced during wort fermentation by yeast is ethanol, however, this primary alcohol has little or no impact on the flavour of the final beer. It is the type and concentration of the plethora of the other yeast excretion products during wort fermentation that will determine, to a significant extent, the flavour of the product; their formation will depend upon the overall metabolic balance of the yeast culture. There are many factors that can alter the balance and hence the flavour of the product; these include yeast strain, incubation temperature, adjunct level, wort pH, buffering capacity, wort gravity, *etc.* Research on beer flavour has been intensively studied in a number of laboratories over the past 25 years.

A great many volatiles, most but not all are fermentation by-products, have been identified in beer, and different substances may influence the aroma and flavour of the product to a different degree. Some volatiles are of great importance and may contribute significantly to beer flavour, whilst others are of importance merely in building the background flavour of the product. The composition and concentration of beer volatiles will depend upon the raw materials used, brewery procedures in mashing, fermentation, *etc.*, and the yeast strain employed. The following groups of substances are to be found in beer: alcohols, esters, carbonyls, organic acids, sulfur compounds, amines, phenols and a number of miscellaneous compounds.

17.5.5.1 *Alcohols*

In addition to ethanol a great number of other alcohols are found in beer, and higher alcohols or fusel oils constitute an important part of the by-products formed during wort fermentation. Their formation is linked to yeast protein synthesis and they are formed from β-acids, which in turn may be formed by transamination and deamination of the amino acids in wort, or synthesized from wort carbohydrates. The yeast strain used for fermentation is of great significance in determining the level of higher alcohols in beer. With other conditions constant, some brewery yeasts have been reported to produce five times as much higher alcohols as others. Laboratory and pilot plant fermentations, as well as full scale trials in different breweries, have demonstrated the importance of the yeast strain (for a review see Engan, 1981). The formation of higher alcohols is also very dependent upon the fermentation temperature with an increase in temperature resulting in increased concentrations of higher alcohols in beer. It would appear that temperature changes have a more significant influence upon the formation of the aromatic alcohol 2-phenylethanol than on the aliphatic alcohols such as propanol, butanol and hexanol.

17.5.5.2 *Esters*

The esters constitute an important group amongst the beer volatiles due to their strong, penetrating fruity flavours. Most of the esters found in beer are formed during fermentation and their formation is linked to the lipid metabolism of the yeast (ester formation is also considered in the section on high gravity brewing). The direct, enzyme-free formation of esters is an equilibrium reaction between an alcohol and an acid, and this reaction is a possible route to ester formation in beer. As wort and beer contain a large number of alcohols and acids, and they may all react to form esters, the theoretical number of esters in beer is large, however, direct formation of esters would be too slow to account for the concentration of some of the esters found in beer. It is now well documented (for review see MacDonald *et al.*, 1984), that the formation of ethyl acetate during fermentation proceeds according to the following reaction:

$$CH_3COSCoA + C_2H_5OH \rightleftharpoons CH_3CO_2C_2H_5 + CoASH$$
(acetyl coenzyme A) + (ethanol) $\rightleftharpoons$ (ethyl acetate) + (coenzyme A)

A number of factors (Nordstrom, 1964) have been found to influence the amount of esters formed during fermentation and these include yeast strain, fermentation temperature (where an increase in temperature from 10 to 25 °C has been found to increase the concentration of ethyl acetate from 12.5 to 21.5 $\mu g\ ml^{-1}$), fermentation method (continuous fermentation appears to result in higher levels of esters than conventional batch fermentation), pitching rate (higher rates have been found to result in a marked reduction in the formation of ethyl acetate) and wort aeration (which seems to influence ester production in such a manner that low levels of oxygen may enhance ester formation).

17.5.5.3 Carbonyl compounds

Many carbonyls have a high flavour potential and, as will be discussed in a later section of this chapter, have a significant influence on the flavour stability of beer thus making them an important group of beer volatiles. In this section the carbonyls that are produced during fermentation will be discussed and the staling carbonyls will be subsequently discussed. The carbonyl found in highest concentration in beer is acetaldehyde. It is formed during fermentation and is a metabolic branch point in the pathway leading from carbohydrate to ethanol. The acetaldehyde formed may either by reduced to ethanol or oxidized to acetic acid, and in the final step of alcoholic fermentation, acetaldehyde is reduced to ethanol by an enzymatic reaction. The concentration of acetaldehyde varies during fermentation and aging/conditioning and reaches a maximum during the main fermentation and then decreases. The question of the influence of yeast strain upon acetaldehyde formation is very much open to question but greater variations in acetaldehyde seem to occur from brewery to brewery than from yeast to yeast. An increase in fermentation temperature does not increase beer acetaldehyde levels but yeast pitching rate does have a significant influence with increasing yeast concentrations resulting in higher acetaldehyde levels.

Excess levels of acetaldehyde in beer can be the result of bacterial spoilage, especially by strains of *Zymomonas anaeroba*. High levels of acetaldehyde can also be caused by high air levels during fermentation, and acetaldehyde levels in bottled beer have been observed to increase during pasteurization and storage, especially if there is a high air content in the bottle head-space.

17.5.5.4 Diacetyl and pentane-2,3-dione

Diacetyl and pentane-2,3-dione both impart a characteristic aroma and taste to beer; this is variously described as 'buttery', 'honey- or toffee-like' or as 'butterscotch'. The flavour is detectable more readily in lager, where it is almost universally regarded as a defect, than in typical heavily hopped British ales (in the light Canadian ales diacetyl is regarded as much a defect as in lager). The taste threshold concentration for diacetyl in lager is of the order of 0.1–0.14 μg ml^{-1} and is somewhat higher in ale. In recent years there has been a great deal of interest in the factors that influence the concentration of diacetyl in beer (the topic has been well reviewed by Wainwright, 1973).

It is now accepted that diacetyl and pentane-2,3-dione are formed outside the yeast cell, by the oxidative decarboxylation of α-acetolactate and α-acetohydroxybutyrate, respectively. These α-acetohydroxy acids are intermediates in the biosynthesis of leucine and valine (acetolactate) and isoleucine (acetobutyrate) and are leaked into the wort by yeast during fermentation. Once diacetyl and pentane-2,3-dione have been formed in the fermenting wort, they are normally converted to acetoin or pentane-2,3-diol respectively by the action of yeast reductases. Thus, the final concentration of diacetyl in beer is the net result of three separate steps: (i) synthesis and excretion of α-acetohydroxy acids by yeast, (ii) oxidative decarboxylation of α-acetohydroxy acids to their respective diketones, and (iii) reduction of diacetyl and pentane-2,3-dione by yeast.

The presence of diacetyl in beer at above threshold levels occurs when α-acetolactate has decomposed to give diacetyl at a time when the yeast cells are either absent or have lost their ability to reduce diacetyl to acetoin. Commonly the fault arises because α-acetolactate breakdown has been curtailed by the use of temperatures conducive to yeast settling when the potential to produce diacetyl remains. When the beer becomes warm, which is usually when it is packaged and pasteurized but may not be until the beer is disposed at the point of sale, diacetyl is produced and in the absence of yeast cannot be converted to acetoin and therefore accumulates. Diacetyl levels can thus be controlled by ensuring that there is sufficient active yeast in contact with the beer at the end of fermentation to reduce diacetyl to acetoin. Diacetyl formation from α-acetolactate has been shown to be dependent upon pH, the concentration of α-acetolactate, temperature, the presence of oxygen, the vigour of the fermentation, and certain metal ions. Vigorous fermentations produce more acetohydroxy acids but the decomposition of acetohydroxy acids to vicinal diketones is also more rapid. In addition, since diacetyl is formed earlier in the fermentation, there is more time for diacetyl removal by the yeast. The production of beer containing undesirable concentrations of diacetyl, as a result of yeast metabolism, should today be rare in normal brewing conditions. However, excessive levels of diacetyl can be the result of beer spoilage by certain strains of bacteria such as *Pediococcus* and *Lactobacillus*.

Enzymatic procedures have been reported, either for the rapid conversion of acetohydroxy acids to their respective diols or for the reduction of diacetyl. As the rate limiting step in the removal of the diacetyl precursor, α-acetolactate, from beer is its conversion to diacetyl, the use of diacetyl reductases is impractical whereas α-acetolactate converting enzymes might be employed (Godtfredsen *et al.*, 1983). It has been found possible to reduce diacetyl to acceptable levels in 24 hours at 10 °C by 'shunting' α-acetolactate in maturing beer directly to acetoin with the aid of acetolactate decarboxylase obtained from a strain of *Enterobacter aerogenes*:

$$CH_3C(OH)(COCH_3)COSH \rightleftharpoons CH_3CH(OH)COCH_3 + CO_2$$
$$\alpha\text{-acetolactate} \rightleftharpoons \text{acetoin}$$

This finding has important implications in brewing. Provided suitable microbial sources of acetolactate decarboxylase can be identified it may be possible to benefit from such enzymes in commercial brewing.

Attempts have also been made to obtain yeast mutants incapable of producing vicinal diketones (Ramos-Jeunehomme and Masschelein, 1977). Such mutants have been isolated from haploid laboratory strains of *Saccharomyces cerevisiae* after treatment with ethyl methanesulfonate. The initial isolates produced excessive amounts of propanol but this was avoided by using recombinants which lacked both acetolactate synthetase as well as threonine deaminase. Such mutants, from one of the haploid strains, have been able to produce beers which, regardless of their quality and their composition of higher alcohols and esters, are reported to resemble beers obtained with brewer's yeast. However, the industrial use of such mutants will involve improvement of their stability and fermentation characteristics, particularly those relating to the use of maltotriose.

17.5.6 Sulfur Compounds

The biological importance of sulfur has come to be recognized over the past 25 years (Greenberg, 1975). The sulfur cycle in nature is just as indispensable for the existence of life as are the carbon and nitrogen cycles. In recent years, considerable progress has been made in understanding the sulfur biochemistry of animals, plants and microorganisms but many facets of the subject are incomplete and obscure. From the standpoint of brewing, sulfur is of additional importance because traces of volatile sulfur compounds such as hydrogen sulfide, dimethyl sulfide, sulfur dioxide and thiols significantly contribute to the flavour of the beer. As early as 1890 it was shown that the stench occurring during primary fermentation of beer was due to the evolution of H_2S caused by the reduction of sulfur compounds in wort. Since that time numerous studies have been published concerning volatile compounds in beer.

Although small amounts of sulfur compounds may be acceptable or even desirable in beer, in excess they give rise to unpleasant off-flavours and special measures such as purging with CO_2 or prolonged maturation times are necessary to remove them. Although volatile organic sulfur compounds are contributed to the wort and beer by hops, adjuncts and malt, a significant proportion of those present in finished beer are formed during or after fermentation. During fermentation, yeasts usually excrete significant amounts of hydrogen sulfide and sulfur dioxide. Indeed, recent analytical studies have suggested that inorganic volatile compounds containing sulfur do not make as important a contribution to beer flavour as was originally thought. Sulfur dioxide is usually present at concentrations below its taste threshold and normal beers when free of infection contain low levels of free H_2S. This latter compound is removed during processing and even beer described as 'sulfury' does not usually contain free H_2S (Wainwright, 1972). However, as will be discussed in a later section of this chapter, sulfur dioxide is important in beer as an antioxidant and as a complexer of carbonyl compounds; some of these latter compounds in their free state will generate a stale flavour in beer.

17.5.6.1 *Hydrogen sulfide*

Factors that influence the formation and final concentration of hydrogen sulfide in beer have been the subject of a large number of publications over the past 25 years. Nevertheless, the biochemical pathways and the metabolic control are not completely understood. Yeast strains requiring the B vitamin pantothenate for growth form sulfide from sulfite or sulfate if the wort is deficient in this vitamin. It has also been demonstrated that amino acids such as threonine and glycine when added to wort will stimulate H_2S production whereas methionine will retard it. It is therefore important to maintain a high level of methionine in the wort. As previously discussed, methionine is removed from the wort early in the fermentation leaving a relative excess of amino acids which promotes H₂S production. Consequently, there is usually a delay in H_2S production early in the fermentation which corresponds to methionine utilization before the maximum rate of H_2S evolution is observed.

A relationship has been noted (Nagami *et al.*, 1980 and Takahashi *et al.*, 1980) between H_2S production during a wort fermentation and the bud index of the yeast cultures (the bud index being defined as: 'the percentage of budded cells of yeast in the suspension', that is 'the ratio of the number of daughter cells to the number of mother cells of yeast in suspension'; 100% budding index means that all yeast cells in suspension have buds). When the budding index was nearly equal to zero, the maximum appearance of H_2S was always observed, which means that the H_2S peaks appeared when almost all of the cells did not possess a bud. Further, when the bud index increased, the H_2S content always decreased, which means that the H_2S content decreased when the cells were budding.

17.5.6.2 *Dimethyl sulfide*

Dimethyl sulfide (DMS) is a volatile thioether which makes a significant contribution to the flavour and aroma of lager beers. It is the major organosulfur volatile of most beers and has received much attention in recent years. It is now fairly well established that most of the DMS present in beer originates from *S*-methylmethionine (SMM) which is synthesized during the germination of barley or from small peptides containing SMM. This component (SMM) breaks down on heating, such as occurs during malt kilning or wort boiling, to give DMS. However, the free DMS is not inert and can be oxidized to dimethyl sulfoxide (DMSO), a heat-stable non-volatile compound, which may in turn be reduced back to DMS by yeast and bacteria. Hence DMS in beer arises from SMM either by chemical decomposition of SMM or *via* the metabolism of DMSO by microorganisms (Anness, 1981).

17.5.7 Flocculation

The flocculation properties, or conversely lack of flocculation, of a particular brewing yeast culture is the other major factor when considering the important characteristics during wort fermentation. Flocculation involves the formation of an open agglomeration of cells, the mechanisms of which depend upon molecules acting as bridges between the cells; this open structure is indeed implicit in the word flocculation. Unfortunately, a certain degree of confusion has arisen by the use of the term 'flocculation' in the scientific literature to describe different phenomena in yeast cell behaviour. The term flocculation, as used most commonly within the brewing industry in particular, is defined here as 'the reversible aggregation of dispersed yeast cells into flocs, generally toward the end of fermentation, and the subsequent segregation of the flocs from the suspending liquid'. This definition excludes other forms of yeast aggregation, particularly those of 'clumpy growth' and 'chain formation'. This non-separation of daughter and mother cells during growth has sometimes been referred to as flocculation. The term 'non-flocculent' thereby implies the lack of cell aggregation and consequently a much slower separation of (dispersed) yeast cells from the liquid medium. Flocculation occurs in the absence of cell division and only under rather circumscribed environmental conditions and involves the cross bridging of divalent ions, usually calcium, bridging anionic groups at the cell surface.

Although yeast separation very often occurs by sedimentation, it may also be by flotation because of cell aggregates entrapping bubbles of carbon dioxide, as in the case of 'top-fermenting' ale brewing yeasts.

The most important aspect of the flocculation characteristics of a brewing yeast strain is the period during the fermentation cycle that the yeast culture flocculates. As previously discussed, one of the most troublesome problems encountered by the brewer is 'hung' or 'stuck' fermentations, *i.e.* incomplete attenuation of the wort. 'Hung' fermentations are invariably caused by one of two factors: (i) premature flocculation of the yeast cultures in the fermenting wort, and (ii) failure of the yeast, although still in suspension, to utilize all of the fermentable sugars; this is usually due to the inability of the yeast to take up and metabolize maltotriose. It is also possible that the yeast may fail to flocculate, thus making its removal from the fermented wort very troublesome, if a centrifuge is not available. This in turn will cause difficulty in obtaining a bright sparkling beer and under such circumstances off-flavours, due to yeast autolysis, can often result. However, with the advent of centrifuges into the brewing process, at the end of the fermentation the yeast is centrifuged out of the fermented wort and the centrifuged fermented wort is cooled and placed in an aging tank. A non-flocculent yeast is now required because prior to centrifugation the absolute minimum of yeast sedimentation is essential. Indisputably, knowledge of the mechanisms that control flocculation at the biochemical, molecular and genetic level are of paramount importance to both the brewer and the brewing microbiologist (for reviews see: Calleja, 1984; Johnston and Reader, 1983; Rose, 1984; Stewart and Russell, 1981).

There is little doubt that differences in the flocculation characteristics of various yeast cultures are primarily manifestations of the yeast culture's cell wall structure. Studies in many laboratories have failed to reveal any meaningful differences in gross composition between the walls of the two culture types that could be directly correlated to the phenomenon of flocculation. Consequently, it has been stated on many occasions that only when the yeast cell surface is examined by rather subtle means, for instance with an electron microscope, will any meaningful differences in the microstructure of flocculent and non-flocculent cultures be revealed. When ether-washed cells were shadowed with tungsten oxide, differences between flocculent and non-flocculent cultures became immediately apparent. Whereas the cells of a non-flocculent culture appeared to possess an extracellular projection, cells from flocculent cultures were covered with an extensive layer of fibrous or hair-like protuberances. A number of flocculent and non-flocculent strains were examined and all of the flocculent cultures were found to possess a 'hairy' outer surface, whereas non-flocculent cells possessed a smooth one (Day *et al.*, 1975). It has also been shown that the anionic groups involved in floc formation are the carboxyl groups of wall proteins (Beavan *et al.*, 1979); floc formation in yeast involves bridging between calcium, magnesium or manganese ions and anionic polymers on the cell surface.

Genetic studies on yeast flocculation date from the early 1950s (Gilliland, 1951; Thorne, 1952). However, because of the polyploid/aneuploid nature of brewing strains most, but not all (Johnston and Reader, 1983), of the research on flocculation genetics has been conducted on haploid genetically defined laboratory yeast strains. Using a flocculent haploid yeast strain, a single dominant gene for flocculation has been identified, *FLO1*, and found to be allelic with *FLO2* and *FLO4* which had previously been identified (Russell *et al.*, 1980). Mapping studies have revealed that *FLO1* is linked to *ade1* and thereby located on chromosome I, 45 centimorgans (cM) from the centromere and 40 cM from *ade1*. A further dominant flocculation gene (*FLO5*) has been identified. To date attempts to map this *FLO* gene have been to no avail (Johnston and Reader, 1983).

Although the technological importance of the flocculation characteristic in yeast has been associated predominantly with brewing, it is becoming increasingly evident that the ability to flocculate is a characteristic to be considered in the choice of yeast strains for other alcoholic fermentations. The ability to flocculate strongly is included amongst criteria for selection of improved rum yeasts, yeasts for the Scotch whisky industry and for the production of fuel ethanol from sugarcane juice in Brazil.

17.5.8 Antifoams in Fermentation

During a brewery fermentation, because of the complexity of the wort and the vigour of the fermentation, the evolution of carbon dioxide will inevitably result in the formation of foam at the liquid surface. (Foam stability on the finished beer is also of importance and this will be considered in a later section of this chapter.) In a brewing operation there are a number of advantages to controlling foam, these include: (i) higher production through increased fermentation capacity — up to 20% increased fermenter utilization, (ii) the use of closed fermentations resulting in reduced oxidation and improved sanitation, (iii) reduced foam results in less protein denaturation thus improving the physical, foam and flavour stability of the beer, (iv) less yeast build-up on the sides of the fermenter, thus less cleaning, and (v) improved hop utilization as less bitter material is lost. A method of resolving the problems associated with foam formation requires the deliberate addition of a foam control additive. A number of antifoam types have been employed in the fermentation industry and these include: (i) natural oils and fats, (ii) alcohol, (iii) sorbitan derivatives, (iv) polyethers, and (v) silicones. Most of these antifoam agents possess both hydrophilic and hydrophobic groups in their molecular structure, which ensures appropriate spreading at the gas–water interface. An alternative approach is to employ a hydrophobic antifoam in an aqueous emulsion with a hydrophilic dispersing agent. Among the more effective hydrophobic antifoams are the polydimethylsiloxanes, which are usually used in an emulsion containing one or more surface-active agents, such as glyceryl stearate plus a preservative such as potassium metabisulfate. Polydimethylsiloxanes (PDS) have the general structural formula $(CH_3)_3SiO[Si(CH_3)_2O]_nSi(CH_3)_3$ where n has a value of approximately 300. An additional advantage in using PDS as an antifoam is that, unlike most other antifoams, it cannot be metabolized by micro-organisms including yeast. However, the use of PDS-containing antifoams in brewery fermentations could cause problems, not the least of which is the possible retention of the antifoam in the finished beer, thus having an adverse effect on the beer's foam stability. However, it has been reported (Evans, 1972; Hall *et al.*, 1973) that the residual PDS attached to the carbonyl groups of cell wall proteins (Vernon and Rose, 1976; Jayatissa and Rose, 1976) is almost completely removed from the fermentation by the yeast. Consequently, the use of PDS-containing antifoams has been successful in employing fermenters with a variety of geometries (*e.g.* cylinder-conical and rectangular vessels). The resulting beer does not have a reduced foam when compared to a control beer produced without the use of antifoam.

17.5.9 Yeast Management and Propagation

An area of particular concern to the brewer and the brewing microbiologist is that of yeast management and propagation. In the long history of the brewing industry the introduction of contaminant free yeast and of pure yeast cultures were major advances in brewing technology. Interestingly enough they both occurred in the 1870s. In his book *Etudes sur la biere*, Pasteur (1876) presented persuasive evidence that the yeast cultures employed at the time to produce beer, wine, alcohol and bread contained contaminants in the form of acetic bacteria (*Acetobacter aceti* and *Acetobacter pasteurianus*), lactic acid bacteria, *Penicillium glaucum*, *Aspergillus glaucus*, *Mucor racemosus* and *Mucor mucedo*. Also a variety of wild yeasts including *Saccharomyces bayanus* and *Hanseniospora valbyensis* were present in the yeast population. Pasteur devised cleaning procedures which favoured the growth of the desired yeast. Thus, for instance, treatment of the cultures with potassium bitartrate and ethanol will reduce the bacterial flora which prefer alkaline conditions. Yeast washing with acid solutions is still practised in many breweries today but a solution of ammonium persulfate and phosphoric acid is usually employed.

In his copy of Pasteur's (1876) book, Hansen of the Carlsberg Institute in Copenhagen annotated in the margin: 'How do you get that absolute pure culture?' and 'This question is much more complicated, since not one but several yeast species will survive the mentioned treatment!'. Hansen recognized that Pasteur's cleaning procedures could, in the short term, improve a yeast culture by removing bacteria, but that this does not lead to an absolute pure culture top fermenting yeast, bottom fermenting yeast or wine yeast (von Wettstein, 1983).

The experimentation which led to the discovery of how one obtains colonies consisting of the progeny from a single cell, a pure strain or line (also called a clone) is described by Hansen in his 1883 paper *Recherches sur la physiologie et la morphologie des ferments alcooliques II–IV*. To begin with he carried out a fluctuation test in which he suspended yeast cells in water, counted the number of cells per unit with a haemocytometer, and diluted the suspension so that it only contained 1 cell in every 2 ml. Several hundred plates containing molten wort–gelatin were inoculated with 1 ml of the yeast dilution. After approximately one week, he observed a single colony appearing on approximately half of the plates and no colonies on the other half. A few plates contained two colonies and very rarely more than two colonies of yeast. Hansen concluded that it was possible to obtain a single colony from a single isolated cell (this work has been reviewed by von Wettstein, 1983).

Thereafter, Hansen produced a number of pure cultures, notably of a top fermenting yeast from a brewery in Edinburgh, a bottom fermenting yeast from the Carlsberg brewery, a bottom fermenting yeast from the Tuborg brewery as well as several strains of *Saccharomyces pastorianus* and *Saccharomyces ellipsoideus*. Besides differences in fermentation characteristics, these strains were different in fermentation gravity, morphology and temperature optima for ascospore formation. The Carlsberg brewery in 1882 and 1883 experienced another problem, frequently the beer had an unpleasant bitter taste and a bad smell. Hansen was of the opinion that the problem could be due to contamination of the brewing yeast with a strain of *Saccharomyces pastorianus* and he recommended to the owner of the brewery, Jacobsen, that pure cultures of brewing yeasts should be established and employed in production. Hansen isolated four different *Saccharomyces* strains from the yeast culture used at the Old Carlsberg Brewery. Each of the strains was allowed to ferment wort as a separate pure culture as well as in mixtures. One strain, namely Carlsberg Uterhefe No. 1, gave consistent beer with good taste and flavour and high stability. A second strain, Carlsberg Uterhefe No. 2, was likewise a useful brewing strain with rapid clearing of the wort, providing a fuller taste than No. 1, but prolonged shelf-life could not be achieved with this yeast. The third strain was identified as *Saccharomyces pastorianus* and it was the culprit that caused the bitter taste and unacceptable flavour.

As a result of these experiments, Jacobsen was convinced that a single pure strain of bottom fermenting yeast, *Saccharomyces uvarum (carlsbergensis)*, could carry out the main and secondary fermentation successfully and alone. On November 12, 1883 the Old Carlsberg Brewery began to employ the strain Carlsberg Uterhefe No. 1 in production and by the middle of 1884 the entire production of 200 000 hl beer was based on this pure strain of yeast. By 1892 Pabst, Schlitz and Anheuser-Busch in North America alone produced 2.3 million hl of beer with pure yeast strains as did an additional 50 breweries on the North American continent.

As to ale production by top fermenting yeast strains of the species *Saccharomyces cerevisiae*, the situation proved to be much more complicated. Indeed in 1959 it was estimated that a large proportion of ale cultures employed for ale production in the United Kingdom consisted of two or more strains (Hough, 1959). Indeed, in some specialized ale products, a separate yeast such as *Brettanomyces* was formerly used to produce the desired flavour in the post-fermentation period, whilst the main fermentation was conducted by a strain(s) of *Saccharomyces cerevisiae*. Similarly, the Berliner Weissbier is made with a mixture of five parts yeast and one part *Lactobacillus brevis* or *Lactobacillus delbrueckii*.

In order to translate Hansen's findings onto the industrial scene he and a colleague, Kuhle, invented an apparatus for culturing yeast. This was brought into use at Old Carlsberg in 1884. So successful was it that a similar plant was installed and operated in Heineken's brewery in Rotterdam in 1885. This plant also has historical importance, not just because it was the second, but because it was customary for quite a number of German breweries to obtain their change of yeast from Heineken. Thus, they too became quickly aware of the beneficial properties of culture yeast. The third 'Hansen' plant was installed in the Tuborg Brewery in Copenhagen in 1887. The essential features of the Hansen plant (which is still the basis of the pure yeast culture plants of today) was a wort holding vessel into which was run hot wort from the brewery hop separation. The wort was collected at a temperature sufficiently high to avoid reheating for sterilization. It was considered that reheating would cause greater precipitation of protein–tannin material (trub) and this would alter the nutritional properties of the wort. The wort was cooled in the wort cylinder and then run into the yeast culture vessel. When the plant was started up, the cooled wort in the culture vessel would be inoculated with actively fermenting wort from a Carlsberg flask which originated from the laboratory, but this would not occur at subsequent cultivations unless, due to infection or some other fault, the plant had to be restarted. Instead a sufficient quantity of yeast would be left behind in the culture vessel for the next cultivation after the bulk of the contents had been run into the brewery.

As well as yeast propagation and the maintenance of pure cultures another important factor in yeast management is storage of the yeast between fermentations and the condition under which the yeast is stored, because this will greatly influence both the rate of subsequent fermentations and the degree of yeast autolysis, which will affect the flavour of the final product. Of particular importance in this regard is the influence of these storage conditions on the yeast cell's intracellular glycogen level. Glycogen is the major reserve carbohydrate stored within the yeast cell and is similar in function and structure to plant amylopectin. It serves as a store of biochemical energy and carbon to sustain the yeast cell during periods of starvation and also to provide the cell with an immediate supply of energy during the lag phase of fermentation, when the demand for energy is intense for the synthesis of such compounds as sterols and fatty acids (*i.e.* lipids). Thus an intracellular source of glucose is required to fuel the synthesis of lipid at the same time that oxygen is available to the cell. As previously discussed, brewery fermentations are unique in this regard in that oxygen is supplied in limited amounts and on a one time basis, usually with the incoming wort. The uptake of oxygen by the yeast cell is very rapid and at the same time there is a delay in the passive diffusion of wort glucose into the cell. For example, with a 16 °Plato wort, there is no appreciable wort glucose uptake until 6 hours or even later after pitching, whilst the wort dissolved oxygen is almost completely depleted in this same time period (Murray *et al.*, 1984; Quain and Tubb, 1983).

In order to synthesize lipid, the yeast immediately mobilizes its reserve of glycogen in order to fulfil the cell's requirement for glucose. The high levels of ATP resulting from respiration activate the phosphorylase system which is necessary for the hydrolysis of glycogen to glucose. Phosphorylase activity during wort fermentation peaks coincidentally with glycogen hydrolysis which is within the first 10 hours after pitching; dissimilation of glycogen and the synthesis of lipid are both rapid. The hydrolysis of glycogen from approximately 27% to 5% and the corresponding production of lipid from 5% to 11.5% of the cell dry weight occurs within the first 6 hours after pitching. Towards the later stages of fermentation, the yeast restores its reserve of glycogen. The actual maximum glycogen content is a function of yeast strain, fermentation temperature, wort gravity and a plethora of other factors. However, the concentration of glycogen stored and the degree of depletion and the end of fermentation will, to a great extent, determine the yeast culture's ability to survive extended storage periods and still ferment at an acceptable rate when pitched into wort.

Storage in most brewery yeast handling systems usually involves maintaining the yeast in beer or water, both of which will vary in concentration from collection to collection. These conditions are far from ideal for growth or even maintenance, having limited assimilable carbon and soluble nitrogen present and also relatively high concentrations of ethanol. Under these conditions the yeast must survive for an indeterminate period of time and to do so requires a basal level of metabolic energy. Glycogen, to a great extent, must provide the cell with these requirements. Storage temperature has a direct influence on the rate of glycogen dissimilation, as might be expected considering the effect which temperature has on metabolic rates in general. The conditions under which yeast is stored and collected and the time of storage can result in detrimental changes to the yeast which will result in sluggish fermentation rates and modification to the flavour and stability of the final beer. Good yeast handling practices should include collection and storage procedures which avoid inclusion of oxygen in the slurry, cooling of the yeast slurry to 4–6 °C as soon as possible after collection and perhaps most importantly, recognition prior to pitching of a yeast that contains low glycogen in order that appropriate corrections in the pitching rate can be made.

17.6 AGING, FILTRATION AND FINAL PROCESSING

17.6.1 Aging and Flavour Maturation

The terms 'aging', 'storage' and 'lagering' are used interchangeably to describe the process of holding beer in a tank at refrigerated temperatures for a period of time (up to four months) following fermentation. A storage period is not essential, but is desirable, for beer production. In the UK, ale beers are produced and put in kegs without a storage period. However, a storage period (usually at low temperatures, 2–6 °C) is necessary for lagers (and for Canadian ales) with the following process functions: (i) flavour maturation, (ii) chillproofing and stabilization, (iii) clarification and filtration, and (iv) carbonation. Storage periods today are usually two to four weeks.

The most important function in a traditional brewing process is flavour maturation, but of the four functions listed above, it is the least understood. Three reactions have a major influence on flavour maturation: reduction in the concentrations of hydrogen sulfide, acetaldehyde and diacetyl. However, these three reactions only occur in the presence of residual amounts of yeast fermentation. In this way a minimal amount of diacetyl, acetaldehyde and other undesirable flavours remain at the end of fermentation. However, at elevated temperatures, some lager yeast strains can be susceptible to autolysis. In many 'modern' operations immediately following fermentation, the beer is cooled to 0 °C and the yeast separated with a centrifuge; in some operations cooling occurs immediately following centrifugation. At this point, flavour maturation is virtually complete. The beer is then stored at 0–2 °C for a few more days to enhance physical stability, usually with the aid of an adsorbent such as silica gel or polyvinylpolypyrrolidone, and the beer is then carbonated and final polish filtered. This 'modern' process requires less capital equipment than traditional processes, *i.e.* less tankage, but operating costs can be higher.

The stability of beer, physical, flavour and foam, has assumed increasing importance in the past 10–20 years and will be discussed in a later section of this chapter. Measures to increase stability are usually taken during aging and this is particularly true of physical stability (haze). Consequently, several techniques are employed such as the addition of proteolytic enzymes, adsorbents and tannic acid to remove the constituents that contribute to beer physical instability. In addition, initiatives can be taken at this point to remove molecular oxygen from beer which if allowed to remain can cause flavour instability. These activities usually involve the use of antioxidants such as sodium/potassium metabisulfate or ascorbic acid.

After fermentation, beer is extremely turbid due to the presence of significant amounts of yeast still in suspension and protein/tannin materials which precipitate out of solution due to the cold temperatures, lower pH (approximately 4.2) and lower insolubility in alcohol solutions. This turbidity must be significantly reduced in order to render the beer marketable, and with the advent of centrifuges, little or no yeast is present during aging. A modification of flavour maturation is the krausen storage process. The term krausen is a term applied to the most active stage of fermentation and during which foaming is most prevalent. Krausen storage is a process in which wort (usually 10 to 20% of the total liquid volume) is added to nearly or completely fermented wort. The krausen creates a secondary fermentation which produces a beer with a characteristic estery flavour. Krausening can also be employed to reduce the diacetyl level of a fermentation if it is unacceptably high.

Since flavour maturation is a prime reason for storage, many brewers have adopted novel technology to reduce processing time. In some brewing processes, the beer is fermented beginning at 10–15 °C and allowed to warm up by the heat of fermentation to 16–18 °C and is held at that temperature until it is brilliant enough to market. There are a variety of processing techniques employed to clarify beer and these include gravity sedimentation, fining and centrifugation.

During storage at temperatures from 0 to 5 °C, the vast majority of the suspended yeast and turbidity will settle to the bottom of the storage vessel if fermentation has stopped and if the tank is under counter-pressure. At least a ten-fold reduction in turbidity can be expected by this approach. Although this is a simple technique for reducing turbidity and is a traditional technique employed by many brewers, it has some drawbacks. The precipitated material at the tank bottom can be self-insulating and can warm up, which will permit yeast autolysis to occur and this will impart an unclean sulfury aroma to the beer. Also cleaning of the tank can be a problem. Fining agents can be added at the onset of aging in an attempt to accelerate clarification and to effect more complete sedimentation. Some fining agents employed are bentonite, tannic acid, isinglass, Irish moss (carrageenan) and silica gel. Obviously the use of one or more of these agents will add to overall operating costs.

17.6.2 Yeast and Particle Removal

As discussed in the previous section of this chapter, yeast removal can be accelerated by the use of a centrifuge, thereby eliminating the rather lengthy sedimentation time required by gravity sedimentation in a storage tank. Centrifuges used for beer clarification are normally the disk self-opening type. Turbid beer enters the top of the centrifuge and is distributed between each of the disks. Clarified beer works toward the centre and is discharged by means of a centripetal pump. Sludge and yeast collect in the bowl, which periodically separates to discharge the sludge. Timing of the 'shoot' or sludge discharge is controlled such that its frequency coincides with an increase in turbidity of the clarified beer and its duration minimizes beer lost to the sludge. Since the concentration of solids in the feed stream is usually variable, most centrifuges are now controlled by a turbidimeter on the clarified beer side of the centrifuge which activates the shoot when the turbidity limit is reached.

The advantages of using a centrifuge include (i) yeast and other suspended solids are quickly removed from fermented beer and if the conditions are appropriate the yeast can be repitched into a subsequent fermentation, (ii) beer losses are minimized, (iii) capital cost is reduced when compared to tankage costs for gravity clarification, (iv) clarified beer can be controlled to a consistent turbidity level, and (v) cleaning costs of tankage are reduced. The disadvantages of using a centrifuge for beer clarification include (i) there is an increase in beer temperature during the centrifugation which must be carefully monitored in order to prevent yeast autolysis and the development of undesirable flavours, (ii) there can be an increase in the concentration of fine haze particles which will render final filtration difficult, (iii) there is the potential for oxygen pick-up, and (iv) as yeast is usually completely removed from the aging beer the influence of yeast on flavour maturation is eliminated. It should be noted that the technology is not far enough advanced to clarify beer to the final clarity stage (*i.e.* bright beer) and a polishing filter is still absolutely necessary. Also centrifuges are not recommended for clarifying beer after lengthy storage during which yeast is present. Yeast cell walls become fragile after lengthy storage, particularly in an ethanolic medium, and degrade during centrifugation causing a troublesome unfilterable haze in the clarified beer as well as autolysed yeast flavour problems.

Gravity sedimentation, fining and clarification are employed to clarify beer to a certain degree but do not give the brilliance required by the consumer. It is normal to filter beer at least once at the end of the aging process, prior to packaging, in order to obtain brilliant clarity. Various techniques are employed, including Kieselguhr (diatomaceous earth, D.E.) filtration, sheet filtration and pulp filtration.

17.6.2.1 Diatomaceous earth filtration

Diatomaceous earth is the skeletal remains of microscopic plants which were deposited on ocean and lake bottoms some 20 million years ago. Today it is mined from high chalk-like deposits, ground to a powder, sterilized and calcined at 800–900 °C, and classified into various size grades. As a filter medium, the D.E. is deposited upon a filter septum usually made of fine stainless steel wire. The small diatoms of infinite configuration form a rigid, but porous filter cake which sieves out the particulate matter in beer as it passes through the filter. In order to prevent 'blinding' of the filter and to achieve extended filter runs, D.E. is continuously metered into the unfiltered beer as 'body feed', thereby constantly building up the depth of the filter cake. A filtration run begins by rapidly recirculating a slurry of D.E. precoat through the filter and back to the precoat tank. The precoat filter cake becomes a thin, 1.5 mm protective coating on the filter septum. When the recirculating liquid (usually beer, but sometimes water) becomes clear, the precoat has been established and beer filtration is begun. Beer is pumped through the filter whilst D.E. is constantly metered into the beer. A normal length of run would be 8 to 12 hours. After this, either the filter cake begins to exceed the physical space limitations within the filter or the pressure drop across the filter becomes excessive. Filtration is then stopped and the filter cake removed; the filter is cleaned and sterilized as necessary and a precoat re-established for the next run.

17.6.2.2 Sheet filtration

The technique of sheet filtration was pioneered in Germany in the 1930s and has since gained global popularity. The sheets are made from cellulose, diatomaceous earth and other ingredients in varying proportions to achieve various degrees of adsorptivity and retentivity. The sheets are positioned between stainless steel or plastic plates which direct the beer through the sheet and collect it on the other side. The appropriate number of plates and sheets are sandwiched together, in a filter frame, in order to obtain volume capacity. The total quantity of beer passed through a filter sheet during a run is inversely proportional to the flow rate and quantity of turbidity. For some types of sheet filter it is possible to backwash the filter, resterilize and run several times before pressure drop limits of 1.5 to 2.0 atm are exceeded.

Sheet filters do not have adequate capacity unless the beer is prefiltered to nearly final beer clarity prior to the sheet filters. Some brewers are of the opinion that sheet filters will also produce microbiologically-free beer, thereby eliminating pasteurization. However, great care should be exercised in this approach.

17.6.2.3 Pulp filtration

The first filtration employed for beer was pulp filtration. It was introduced by Enzinger in the latter part of the 19th century and it soon became popular as a means of improving shelf stability and marketability of beer. Although pulp filtration ushered a new unit operation into brewing, with the exception of some developing countries, it has been almost entirely replaced by D.E. and sheet filtration technology. Pulp filters utilize a pad made from cotton and cellulose fibres. The pulp can be rewashed using cold caustic and chlorine bleach, thereby reducing the environmental problems that are present with discarded D.E. or sheets. Pulp filters are less retentive than sheet filters and rely more on the depth of pad and adsorption. Therefore, they can be used directly after storage for prefiltration of beer with considerable turbidity.

Once the beer has been filtered to acceptable clarity, it is ready for packaging and this can either be into a keg or barrel (for draught beer), bottle or can. Draught beer is either usually unpasteurized or subjected to a flash pasteurization procedure, however, some British keg beer is subjected to a modified tunnel pasteurization. Bottles and cans are usually pasteurized through a tunnel pasteurizer, although some brewing companies do not pasteurize their beer but subject it to membrane filtration and sterile filling of either the can or bottle. Developments in packaging technology and the plethora of packaging modes are beyond the scope of this document, suffice to say that in the marketing oriented societies of North America and Europe, it has become an important tool for merchandizing beer! Also in the age of the environmental lobby, the question of returnable, non-returnable and recyclable packaging is also an important consideration to be borne in mind by the management of brewing companies.

17.7　BEER STABILITY

17.7.1　Physical Stability

Once beer has been final filtered to acceptable clarity and packaged either with or without pasteurization and distributed into the marketplace, it is important that it retains the characteristics of the fresh product for as long as possible (*i.e.* it possess acceptable stability). This is an oversimplification of the problem because stability can be considered as a number of different phenomena, namely, physical stability (non-biological haze formation), flavour stability, foam stability and microbiological stability.

Beers infected with bacteria or wild yeast will rapidly develop a turbidity producing a biological haze but with the widespread use of pasteurization and sterile filtration such infections are fairly rare. However, uninfected beers when stored for any length of time, usually in bottles or cans, also become cloudy and deposit a haze. Such beers are usually unacceptable and the rate of development of this non-biological haze is one of the major factors that determines the shelf-life of bottled or canned beer. Prior to the appearance of a permanent haze at room temperature beer will usually form a chill haze if suddenly cooled to 0 °C or below. Such hazes usually redissolve when beer is warmed up again to room temperature (20 °C).

Colloidal material is of importance in the formation of haze and such compounds include polypeptides and polysaccharides, which form aggregates with each other and with polyphenols. Colloids containing polyphenols often form hydrophobic sols and the extent of the hydrophobicity largely depends on the amount, nature and distribution of the polyphenols in the particles. The formation of hazes in beer usually reflects changes in the composition of the sol particles which make them less soluble, although in some cases (β-glucan precipitation, for example) the particle composition is fairly constant but the particle grows in size. When the size of haze particles is sufficiently large, the haze is not stable in suspension and a sediment is formed. There is no fixed size at which this happens; it depends upon the stage of the particle and its density relative to the density of the suspending solution. It also depends on forces such as agitation or thermal motion which maintain the particle in suspension.

Although the presence of polyphenolic material (derived from barley) associated with protein in beer hazes was first postulated early in this century, it was only in 1955 that phenolic acids were positively identified in haze hydrolysates from haze and beer (Bengough and Harris, 1955). In the same year, McFarlane *et al.* (1955) observed cyanidin and delphinidin in acidified extracts of beers and beer sediments and reported that anthocyanogens contributed by malt and hops could be readily precipitated or adsorbed from beer with polyvinylpolypyrrolidone (PVPP). These findings stimulated considerable research into the phenolic components of beers and brewing materials in the expectation that these materials would hold the key to the formation mechanisms of non-biological hazes. The realization that anthocyanogens are constituents of haze has resulted in several methods being proposed for their extraction in beer. Most of the methods depend upon adsorbing the anthocyanogens onto polyamide resins (insoluble PVP (polyclar AT), Nylon 66 or Perlon) followed by treatment with acid to generate the anthocyanidin pigments which are estimated spectroscopically. These methods suffer the disadvantage that the efficiency of anthocyanidin formation is highly dependent on the structure and the degree of polymerization of individual anthocyanogens. Also, employing these methods, no direct correlation has been found between anthocyanogen values for beers and the beer's potential for producing haze.

In order to gain greater insight into the role of beer polyphenols in haze formation, new analytical methods have been developed that yield quantitative information on individual polyphenol components and the fate of these components in beers held in storage for prolonged periods in order to induce haze formation (Gracey and Barker, 1976; McGuinness *et al.*, 1975). These methods have involved gas chromatography, with its capability of yielding both resolution and quantitative data, coupled with mass spectrometry for identification studies. Although beer phenolics are non-volatile compounds and, therefore, their direct analysis by gas chromatography is precluded, this problem has been overcome by preparation of their volatile trimethylsilyl (TMS) or ether derivatives.

Nevertheless, even in the wake of significant study on the mechanism of haze formation in beer, the process is far from understood. The most widely held theory is that of Gramshaw (1970) which proposes that simple phenolics in beer polymerize to yield 'active' polymers (tannins) which complex with beer protein to form first a reversible chill haze that precipitates from solution at 0 °C but redissolves at approximately 20 °C, then upon further polymerization and complexing, a non-reversible permanent haze forms. However, this theory has not been convincingly defined, nor is knowledge, other than theoretical, available on the binding between the tannins and proteins. An alternate theory of haze formation (Gardner and McGuinness, 1977) discounts polyphenol polymerization and invokes 'activation' of raw material derived dimeric flavanoids prior to their direct interaction with proteins. Again, this theory presents the difficulties of a lack of knowledge on the mode of activation or the links developed to protein.

Gracey and Barker (1981) uncovered a new facet of haze structure and have proposed a novel mechanism in an attempt, at least, to explain part of the mechanism by which haze forms in beer over time. An analysis of naturally formed haze in commercial beers has revealed hydrogen bonded components (2–3% of total) comprised of mainly carbohydrates and glycerol. The 'back-bone' of the haze material was found to be protein (65–70%) which possessed an amino acid composition rich in glutamic and aspartic acid residues. Only traces of anthocyanidins were detected and phenolic fragments were not detected amongst hydrolysis products of the haze material which included carbohydrates, phosphoric acid and an organophosphate residue. Total haze phosphate was found to be 2.2% of the total haze and the most significant metal ion was calcium (0.12%). When in solution, the haze was cleaved into protein and phosphate enriched fractions by pH manipulation. The phosphate fraction was macromolecular and when this material was added back to bright beer, haze formation was immediately induced.

As a result of these findings, a novel structure has been proposed where the primary contact between the macromolecules is calcium ion bridges between phosphate and protein carboxyl residues, with polyphenols attached to the haze by weaker secondary hydrogen bonds. Analysis of haze hydrolysis products has indicated that the phosphate macromolecules include a glycerol phosphate polymer and phosphopolysaccharides. Glycerol phosphate polymers suggest teichoic acid, a component of the cell wall of Gram-positive bacteria that are malt contaminants, and these may come into wort during mashing and wort boiling.

It would thus appear that there are a number of conflicting theories available for haze formation but, in all likelihood, all of the proposed mechanisms are valid to a lesser or greater degree. Obviously from the point of view of the brewer, prevention of haze formation is the priority and over the years a number of such measures have been proposed, many of which are employed in production. These include (i) The use of cold cellar temperatures for prolonged periods and good tight filtration techniques in order to remove all precipitated material. Because of the long storage period required, this is an expensive technique. (ii) The use of proteolytic (chillproofing) enzymes in the aging beer in order to hydrolyse protein material thus removing the 'backbone' material of haze and, therefore, preventing its formation. The use of chillproofing enzymes leads to foreign material being present in the beer and has a negative effect on foam stability. This aspect will be discussed further in the section on foam stability. (iii) Removal of proteins with adsorbants such as tannic acid, bentonite and silica gel. Tannic acid is a traditional method of removing haze material and is still employed in West Germany and by at least one large brewing company in the United States. The addition of tannic acid to beer in storage produces an unusual amount of precipitate which settles to the bottom of the tank, and is removed by decantation and subsequent filtration. By forcing the haze forming reaction to completion, the components which can create the haze are eliminated from the beer. The use of tannic acid is a very effective technique to obtain colloidal stability but on the negative side, the cost of handling the precipitate, the beer lost in the precipitate and the presence of foreign material, however minute a concentration, are factors that should not be overlooked.

The use of an insoluble protein adsorbant is an alternative method of physically stabilizing beer. One type of protein adsorbant, the silica hydrogels, has found considerable application for chill and permanent haze protection in beer over the past ten years (Hough, 1976). Silica hydrogels have been developed which will selectively take up into their internal space the proteins involved in haze production. At the same time, they leave within the beer, and in an unchanged form, those proteins concerned with the development of beer foam. Similarly hop bittering material, colouring material and compounds important in taste and aroma are not adsorbed.

The use of PVPP as an effective adsorbant of polyphenol/tannic material has already been discussed (McFarlane, 1961). The tannin molecules are electrostatically attracted to the adsorbant to form a floc which precipitates from the beer. By elimination of the tannin from solution, the protein cannot form an insoluble complex and haze formation is reduced, if not eliminated. PVPP is most effective when added to prefiltered beer and the rate of adsorption is quite rapid with only a few hours' contact time being required. There is no doubt that PVPP treatment is a very effective method of preventing haze formation but of all the methods discussed, with the possible exception of long term storage at cold temperatures, it is the most expensive. However, this cost can be reduced if the PVPP is reused following regeneration with caustic solution.

The alternative, instead of removing haze or haze precursors, is to reduce the potential for haze development. This can be achieved in a number of ways which include reduced proportion of malt (increased adjunct rate), use of well modified malt (low soluble nitrogen), formaldehyde in the mash water (not permitted in many countries), use of low mash pH to precipitate wort tannins and the use of a protein rest in the mash cycle. Another novel method is to employ a malting barley that is free of anthocyanogens, thereby eliminating a great proportion of the tannins from wort and beer. Anthocyanogens are secondary plant metabolites and are thus not required for the normal growth and development of the barley plant and, as a consequence of this, scientists at the Carlsberg Research Laboratory in Copenhagen have bred anthocyanogen-free barley varieties (von Wettstein *et al.*, 1980). Some 50 anthocyanogen-free barley mutants have been isolated in barley varieties after mutagen treatments with ethyl methanesulfonate and sodium azide. Malt, wort and beer produced from a few of these mutants were free of catechins and anthocyanogens. The bottled beer had excellent physical stability without any stabilizing treatment, and was palatable with a good all-round flavour. Large scale agronomic trials are currently ongoing in both North America and Europe, the main objective being to study yield per acre and susceptibility to infection by moulds, *etc.*

17.7.2 Flavour Stability

By comparison with other biologically stable food products, packaged beer has a rather short shelf-life. As discussed above, the clarity can be preserved by enzyme or adsorbent treatment but stale flavours also soon develop. In milk, butter, vegetables, vegetable oils and many other foods and beverages, staling is caused by the appearance of various unsaturated carbonyl compounds. It is now becoming increasingly clear that the same is true of beer staling. This phenomenon of beer staling has been intensively investigated by the brewing industry with a view to understanding and controlling it. Despite these studies, the mechanism of staling is still not fully understood. The actual compounds responsible for stale flavour vary during prolonged storage as evidenced by changes in the flavour profile of beer. Although the compounds causing the sweetish, leathery character of very old beers have not been identified, there is evidence that the papery, cardboard character of 2–4 month old beer is due to unsaturated aldehydes. The most flavour-active aldehyde which has been conclusively proven to rise beyond threshold is *trans*-2-nonenal. Others such as nonadienal, decadienal and undecadienal may also exceed threshold.

The adverse effects of oxidation on the flavour of finished beer have been known for a considerable time and brewers add bisulfites or other antioxidants, such as ascorbic acid, to beer prior to packaging to provide protection against oxygen pick-up, and this can improve flavour stability. The effectiveness of bisulfite, besides its antioxidant properties, is also its ability to bind carbonyl compounds into flavour neutral complexes (Barker *et al.*, 1983). Its addition to fresh beer reduces increases in free aldehyde concentration during aging. In addition, when added to stale beer, bisulfite lowers the concentration of free aldehydes and effects the removal of the cardboard flavour.

The above observations supported the involvement of aldehydes in flavour staling and indicated a positive role for bisulfite in flavour stability. The next question to be answered is 'what is the source of the aldehydes?' Several aldehydes, including unsaturated aldehydes, have been found in staled beer and have also been observed in malt and wort, but until recently, theories for aldehydes appearing in staled beer have only considered *de novo* synthesis in beer. Although aldehydes are produced in finished beer, the relative importance of these as contributors to flavour staling under normal beer storage conditions is questionable. The amounts of simple aldehydes produced, for example, by Strecker degradation of amino acids or by oxidative degradation of isohumulones are considered too small for them to contribute to staling in any significant manner. It would appear that the greater proportion of staling aldehydes is already present in wort at the time that the yeast is pitched. The sulfur dioxide produced by the yeast during fermentation, together with the bisulfite added as an antioxidant, forms flavour neutral bisulfite complexes with the aldehydes. The aldehydes, their bisulfite adducts and free bisulfite participate in a dynamic equilibrium in beer, and flavour stability depends on the concentration of the bisulfite and its rate of decline (usually by oxidation to sulfate) during storage of the packaged product. Flavour staling is thus the result of the equilibrium system shifting in favour of free aldehydes caused by diminished bisulfite concentrations. It would, therefore, appear that the flavour staling aldehydes of malt and wort probably participate in the flavour staling of beer. Consequently, improved flavour stability will have its genesis in preventing the staling aldehydes from occurring in wort, and also ensuring that the aldehyde–bisulfite complex is maintained in pack-aged beer as long as possible. This can be achieved either by increasing the level of free bisulfite in beer or preventing the oxidation of bisulfite to sulfate by rigidly controlling the level of oxygen in the packaged product, thus ensuring that as much bisulfite as possible is employed as a flavour neutralizer of the staling aldehydes.

17.7.3 Foam Stability

Foam stability has also been discussed in this chapter with reference to the use of antifoams in the fermenter. Beer foam has been a topic for research and discussion since the early part of this century. When beer is sold, the stability of the foam in a glass of beer is usually considered by the consumer to reflect the quality of the product. Also, because of this importance, foam or foaming potential, often expressed as head retention value, is one of the process control tests by which beer is judged before sale. Many researchers have endeavoured to identify the ingredients or pro-cess conditions that contribute to a good head retention value. The chemical engineer, on the other hand, has to design plants to minimize the tendency to foam because each time the beer foams during production, a little of its foaming potential is permanently lost.

Pure liquids will not give stable foams and, for any reasonable stability, the presence of a foam-ing agent is required. The foaming agent must be surface active, that is it should position itself at, and thus modify the properties of, the liquid–gas interface. It is the properties of this interface or, more correctly, of the film formed at the interface by the surface-active agent, that determine the properties of the foam. The proteinaceous material in the beer is the major foaming agent. Although differences in the quantity and composition of this protein fraction will influence foam quality, they are never as great as can be caused by the interaction of other materials with the protein. It is unlikely that any native proteins occur in beer since, during the brewing process, there has been extensive proteolysis and heat denaturation of the proteins present in the cereals employed. However, the final beer does contain denatured and degraded fragments of cereal protein with molecular weights varying from over 100 000 Daltons down to small peptides and simple amino acids. There is a continuous range of molecular weights present and, likewise, the protein fraction from beer is relatively heterogeneous in terms of other characteristics normally used in protein fractionation.

This situation is ideal for foam formation because the relative consistency of the heterogeneous protein mixture found in the final beer makes it likely that there will be sufficient of the right kind of proteins in the beer, irrespective of a reasonable degree of variation in malt and other adjuncts employed in the mash tun. However, the increasing concentration of adjuncts and the associated decrease in malt being employed today is having a negative effect on foam values in many beers. Indeed, it is appropriate to ask how much protein is required to produce a reasonable foam. Roberts (1977) has shown that an all-malt 10 °Plato original gravity beer brewed by a constant temperature infusion process has approximately four times as much protein as is necessary to pro-duce a reasonable foam, provided that foam inhibitors are not present in significant amounts.

The period of wort boiling affects foam stability. A wort boiled for only half an hour showed a 7% higher head retention value than a wort boiled for two hours. This is probably due to more heat-coagulable protein being precipitated during the long boil and this fact was reflected in the figures for total soluble nitrogen. However, it has often been stated that brewing is a series of compromises because the 7% increase in head retention on reducing the wort boil from two hours to half an hour was accompanied by a 40% reduction in physical stability. Any attempt to improve head retention by simply increasing the amount of lager proteins in the beer can be associated with the danger of increased haze formation.

Lipids are another broad class of surface active materials found in beer. Lipids in this context are taken to be those substances containing a paraffin hydrocarbon moiety, such as fatty acids, glycerides and phospholipids. When dissolved in water, the molecules form a surface film up to a concentration at which the surface becomes saturated, and above this concentration they will form association structures or micelles in solution. Thus, together with the protein, lipids cause beer to be a mixed surfactant system.

In beer, the lipids present are derived from three main sources: cereals, hops and as by-pro-ducts of the fermentation process. The principle lipids found in unhopped wort are fatty acids, especially C_{12} to C_{18-3}, the mono-, di- and tri-glycerides and phospholipids. Most of these materials are removed as protein breaks (trub) during the brewing process but some remain in the final beer and can prove troublesome. Two adjunct materials commonly employed, corn and rice, are rich in lipid material in their native state and consequently are not suitable for use as adjuncts without prior treatment, such as degermination. The essential oils of hops could be considered as lipid-like and can impair foam quality. These materials have antigushing properties which are probably related to their lipophilic nature.

During fermentation, short chain-length fatty acids (C_6–C_{12}) are liberated by the yeast; the actual amounts being liberated are very strain dependent. Although these smaller fatty acids can be an important feature of beer flavour, they can prove harmful to foam, however, their effect is much less than that of the longer chain fatty acids found in the wort.

Knowledge of the absolute amount of any particular lipid in a beer gives little information on its effect on foam unless the state of the lipid in the beer is known. For example, lipid present in the dispersed state (complexed as lipid–protein) often affects only the quality of the foam (producing an open, coarse foam as opposed to a thick, creamy foam) but lipid present in a non-dispersed state leads to poor foaming ability and stability.

Substances also exist in beer which affect the stability of the foam without themselves being surface active or interacting with material in the surface film. These molecules affect foam stability by causing an increase in the viscosity of the beer. Such molecules that will secure good foam quality are large polysaccharides such as β-linked glucans and dextrans. Linear dextrans with molecular weights in the range 10 000 to 2 000 000 Daltons have been added to beer and it was found that the increase they caused in head retention value was directly proportional to the increase in beer viscosity. The larger dextrans gave greater increases in viscosity for the same concentration, so less was required for a given increase in head retention. Therefore, in order to secure good foam quality, it is advantageous to maintain the β-glucans and large dextrans at the highest practical levels in the finished beer. However, problems are encountered with flow rates and filtration when the β-glucans are present in too large a concentration.

Charged polysaccharides such as the propylene glycol ester of alginic acid give increases in head retention greater than that predicted from the increase in viscosity which these polymers cause. This disproportional increase in head retention is probably due to the interaction of the propylene glycol alginate with polypeptides in the surface film so that protein–polysaccharide complexes are formed at the surface. Such complexes would be similar to glycoproteins and it has been shown that glycoproteins are more effective than are free polypeptides in stabilizing foam.

In order to enhance foam stability, it is common to employ stabilizing agents in many breweries. This is particularly so when proteolytic enzymes are used to improve physical stability due to the fact that foam-enhancing proteins will be destroyed by such treatment. The most commonly used foam stabilizer is propylene glycol alginate but great care should be exercised in its use in order to avoid the formation of an unfilterable haze. Other foam-stabilizing agents include glucans, partially hydrolyzed protein and trace amounts of iron, but all these agents may have the penalty of physical instability in the beer.

17.7.4 Microbiological Stability

Microbiological stability of beer is a subject that could warrant a chapter all to itself but it will not be discussed in any detail in this document. A microbiologically stable beer can be produced by employing good housekeeping practices in the brewhouse, fermentation and aging cellars, and by ensuring that the culture yeast is free of infection from either bacteria or wild yeasts. It has often been stated that brewing is a very forgiving process microbiologically due to: (i) the wort boiling step, (ii) the advent of stainless steel tanks and equipment and therefore increased ease of cleaning, (iii) use of cold fermentation temperatures and even colder aging temperatures, (iv) vigorous filtration procedures in order to ensure a bright product, as well as to remove microorganisms, and (v) the pasteurization of bottled and canned beer. Nevertheless, the microbiological state of all parts of the brewery should be constantly monitored and every attention to cleaning procedures taken.

17.8 DEVELOPMENTS IN BREWING TECHNOLOGY AND PRODUCTS

17.8.1 Continuous Fermentation

Continuous fermentation for the production of beer was first attempted before 1900. Indeed by 1906 at least five separate systems had been proposed including simple stirred tanks, multiple arrangements of such vessels and towers packed with supporting materials upon which a culture of yeast was maintained. The reasons why these systems at that time failed to gain a foothold in commercial operations are obscure, but it is likely that inability to guard adequately against contamination and also resistance to change were major factors.

A re-awakening of interest was stimulated in the late 1950s when multivessel systems were in operation in Canada and in New Zealand. This was shortly followed by a new system in the UK which exploited the ability of flocculent strains of yeast to sediment, thus enabling a high concentration of cells to be held within the system (Portno, 1978). This opened up the possibility of much more rapid fermentation than had hitherto been possible. In the decade between 1960 and 1970 substantial interest arose in the brewing industry in the field of continuous fermentation. Increases in knowledge of brewing science, together with the relatively advanced engineering and electronic control equipment which was becoming available, offered real hope that continuous fermentation could be developed into a viable process. It was anticipated that the following advantages would result from the use of continuous fermentation for beer products: (i) reduced capital costs as a result of faster throughput; (ii) less beer tied up in process also as a result of faster throughput; (iii) reduced labour costs due to less downtime and therefore less cleaning, and automatic control at steady state; and (iv) lower product cost resulting from the production of more ethanol and less yeast, reduced beer losses, improved hop utilization and reduced detergent usage. The major economic gains were therefore in respect of capital investment, labour costs and value of the product in process. Today, this view has substantially changed. With the exception of two brewers in New Zealand, no company is dependent on continuous fermentation. An increase in its use in the UK for ale products in the late 1960s has proved transitory; continuous fermentation never proved acceptable for lager production.

Why did the brewing industry fail to make a commercial success of continuous fermentation? Essentially batch fermentation is simple; a vessel is cleaned, sterilized and rinsed; it is filled with wort and the required quantity of yeast is inoculated. The temperature cycle can be pre-programmed and little further attention is required until it is necessary for further processing, 3 or 4 days later for an ale, 7 to 10 days later for a lager. Operation by trained but not highly qualified staff is straightforward. On the other hand, continuous fermentation requires constant laboratory monitoring and complex automatic control of flow rates, temperature gradients, yeast recycle and oxygen levels. Cell morphology and fermentation gravity need regular checking. Engineering support to correct possible faults in control systems, pumps, heat exchangers and pasteurizers are required. All these must be available 24 hours a day, 7 days a week.

The much more rapid flow rate from continuous fermentation is, in part, an illusion. It is necessary to have a reservoir of wort to feed the fermenter. Because other types of beer are likely to be produced in the same plant, a beer reservoir to accumulate the output into suitable batches for further processing is required. Although the residence time within the fermenter may be very short, this is not the economic factor that should be considered. It is the residence time in the plant that matters and this may be in excess of 24–36 hours. The use of continuous fermentation significantly reduces the flexibility of a brewery. Not all consumers drink the same beer, they drink more in summer than winter, they drink more on a hot dry weekend than on a cold wet one. An ability to provide the required diversity of products in varying and unforeseeable amounts is a prerequisite of a successful brewing operation. Batch fermentation can meet this need for flexibility far better than a continuous process which is best suited to the production of a high volume product at an unvarying rate.

Portno (1978) is of the view that 'any change in the process which results in it becoming increasingly complex and demanding must offer corresponding advantages before it will be accepted. Continuous fermentation in the brewing industry tends to create more problems than it offers solutions.' Continuous fermentation is scientifically and technically viable and can give rise to products of excellent quality, although not always a match of existing brands. However, for the reasons discussed above it is most unlikely to be the fermentation method of choice in the brewing industry.

17.8.2 High Temperature Wort Boiling

The greatest energy consumer in a brewery is the wort boiling process which accounts for 38–40% of the total heat requirement of the brewing process. For this reason, the significant increase in the price of fossil fuel, especially fuel oil and natural gas, that has occurred since 1973 has forced the brewing industry to seek ways of reducing energy costs. Obviously, wort boiling has received particular attention and continuous boiling is considered to have advantages over conventional batch systems. Continuous wort boiling at high temperatures has the advantages of energy saving and compactness of plant in comparison to batch boiling (Seaton *et al.*, 1981; Chantrell, 1983).

A continuous wort boiling plant has been installed in a brewhouse in the UK (Chantrell, 1983) in such a way that the contents of two of the existing four wort kettles can be boiled conventionally and two through the high temperature wort boiling plant (HTWB). In this way, precise comparisons of the attributes of the two operations have been made. The HTWB plant has a capacity of 450 hl h^{-1} with a boiling time of 3 min at 140 °C. The evaporation rate is 8%, identical with that obtained in the conventional system. The wort is collected in the existing kettle at approximately 72 °C from the lauter tun. Hop addition and colour and gravity adjustments are made at this step. The wort is then piped through the HTWB to the whirlpool separator (hot wort tank where the solid particles suspended in a rotating mass of liquid will migrate to the centre and bottom of the vessel). In passing through the plant the wort temperature is raised in three successive heat exchanges to 160 °C. The wort is held at this temperature for 3 min during its passage through an insulated holding tube. The wort then passes into the first of two expansion vessels, where the pressure is reduced to a predetermined level. The wort continues into the second expansion vessel where the pressure is reduced to atmospheric pressure. The flash vapours from these two expansion steps are employed to drive the first two wort heat exchangers. Only the third heat exchanger requires an external steam supply.

The beer originally produced from the wort brewed in the HTWB showed a significant difference to the beer produced from wort brewed in a conventional kettle boil but, after adjustment of the hop rate, no significant flavour differences could be detected. Detailed examination of the beers by a trained taste panel did identify some areas of consistent differences. Prior to the adjustment of hop addition rate, increased bitterness was detected in the HTWB beers. After adjustment to the hop rate, this difference was eliminated as were sulfidic and oxidized flavour notes.

Brewers that have installed the HTWB system have concluded that it is an energy efficient technique, capable of satisfying both brewer and customer. The rate of return on investment is quite attractive and if energy costs resume an upward trend, a continued pressure on energy costs will favour the installation of this type of plant.

17.8.3 High Gravity Brewing

In recent years an increasing number of breweries have adopted the 'high gravity brewing' procedure and at this time certainly more beer is produced in North America according to this production method than by conventional means. High gravity brewing is a procedure which employs wort of higher than normal concentration and hence requires dilution with water at a later stage in processing. By reducing the amount of water, increasing production demands can be met without expanding existing brewing, fermenting and storage facilities.

Beers produced according to the high gravity brewing procedure (i) have improved colloidal haze and flavour stability, (ii) provide a more efficient use of existing plant facilities, (iii) reduce energy costs, (iv) yield more alcohol per unit of fermentable extract, (v) may contain high adjunct ratios, and (vi) are rated smoother in taste. On the other hand, beers produced according to this procedure suffer from the fact that they (i) decrease brewhouse material efficiency, and (ii) are sometimes difficult to flavour match with an existing normal gravity product (Pfisterer and Stewart, 1976).

The dilution step in the production process is the major innovation in the procedure and it may be carried out before fermentation, or at some point during the aging process. Each system offers advantages over the regular brewing procedure. In West Germany, for instance, as a part of the Purity Law, it is not permitted to change the concentration of wort and beer after the inoculation (pitching) of the yeast. In Canada, in the early days of high gravity brewing, dilution was quite frequently carried out in the fermenter (at approximately 5 °P prior to the excise dip), since the former excise regulations made it economically unattractive to add water at a later stage. In this context it should be pointed out that Canadian excise duties used to be based on the beer volume in the fermenter; the volume being determined prior to 'dropping' the fermented wort out of the fermenter by means of a calibrated dip-stick. A shrinkage allowance of 5% for losses in subsequent processes was permitted. In 1975 the excise regulations were modified and today all excise duties are based on the quantity of packaged beer that enters the shipping room. Nevertheless, dilution in the fermenter still improves fermentation cellar capacity as less head space is required. Further, when water is added during fermentation rather than immediately prior to final filtration, the oxygen in the dilution water will be removed by the yeast and the requirement for expensive oxygen deaeration equipment is circumvented. However, the longer the beer is maintained undiluted, the greater is the capacity efficiency and further investigations have indicated that beers are more stable when they are processed in the undiluted stage. Consequently most breweries add the water to the concentrated beer immediately after the final polishing filter but prior to the trap filter.

The water for dilution at this point in the process requires special treatment in order to ensure the quality of the finished beer. Such treatment is to secure biological purity and chemical consistency and encompasses filtration, pH adjustment and, occasionally, ozonization or pasteurization. In addition to these conventional measures, the dissolved oxygen content of the water must be reduced to a level of approximately 0.1 p.p.m. This can be achieved by a vacuum deaeration using either a hot or cold process. The hot system flashes water at 77 °C, the cold system flashes water at a temperature of 3–24 °C through the vacuum deaeration.

The blending of this deaerated water with the concentrated beer is usually based on the concept of the primary flow of beer. The dilution rate is set according to the alcohol level and may reach as high as 40% using conventional brewing techniques with lauter tuns. Indeed, wort concentrates above 16 °P may be achieved by employing alternative procedures such as using syrup in the kettle or applying mash filters to separate the spent grains from the extractable matter.

Two factors have a major impact on the properties of high gravity worts. One is the grist-to-liquor ratio, *i.e.* the ratio of cereal grain (both malt and adjunct) to brewing water, in the mash mixer and the other is the run-off characteristics during the lautering process. By maintaining sparging and boiling constant, only an increase in the grist-to-liquor ratio will inevitably result in a higher wort concentration. Additionally, such a change in grist-to-liquor ratio has been found to influence the pattern of fermentable and non-fermentable carbohydrates. Worts produced from high grist-to-liquor ratios in the mash mixer have undergone further degradation than worts from low grist-to-liquor ratios (Pfisterer, 1971). This is evidenced by a significant increase in glucose and maltotriose at the expense of dextrins. However, little change in the maltose concentration has been observed. It appears that amylolytic enzymes in mashes with high grist-to-liquor ratios are more resistant to thermal inactivation and, therefore, prolonged enzymatic action occurs.

Differences between the run-off characteristics of varying gravity worts have been noted (Pfisterer and Stewart, 1975). It has been found that the grain beds of high gravity worts are sparged less intensively than conventional worts. Consequently, high gravity worts do not contain those substances which are extracted during the final stages of the lautering process of normal gravity worts. The composition of the extract from first wort differs significantly from the composition of the extract of the last runnings. The latter, for instance, contains more nitrogenous matter, ash and tannins. Further, marked differences in oleic acid and linoleic acid levels are found when comparing the amounts in high gravity worts to the amounts in a normal gravity wort, or a first wort *versus* a last wort. Most of the above mentioned compounds have been implicated in beer taste stability and ester formation (Ayrapaa and Lindstrom, 1973) and as some of these compounds are modified in high gravity worts, it is not unreasonable to assume that the characteristics of high gravity beers are directly attributable to the reduced levels of these substances.

In a detailed analysis of a large-scale brewery trial of three worts, 11.5, 13.7 and 16.5 °P, it was readily apparent that the brewhouse material efficiency is reduced as the gravity of the wort is increased. This fact is reflected in the increase in last wort concentration and by the decrease in hop utilization. However, the portion of the total extract representing the α-amino nitrogen remained constant. This latter observation was somewhat surprising since the fermentability of high corn adjuncts has been found to be increased when compared to all malt worts of similarly elevated gravity and also to low-gravity/high-adjunct worts. The improved fermentability of high-adjunct/high gravity worts despite the constant percentage of α-amino nitrogen in wort solids has been correlated to the observation that a doubling of wort concentration did not double the total yeast crop (Anderson and Kirsop, 1974). Therefore, since high gravity worts do not produce more yeast on a proportional basis, it becomes clear that they would not require the same amount of assimilable nitrogen on a proportional basis. It is also clear that the concentration of assimilable nitrogen in high-gravity worts does not become limiting as rapidly as in normal-gravity worts.

High gravity worts affect yeast performance and in particular their flocculation characteristics. In many instances it has been found that the higher the wort gravity the more flocculent and sedimentary the yeast will become. Beer produced under high gravity conditions has a marked improvement in both flavour and physical stability. It was obvious, however, that beers brewed by the high-gravity process were different in flavour from their conventional counterparts. Studies by Anderson and Kirsop (1974) have revealed that high-gravity fermentations produced significantly increased amounts of acetate esters over their normal gravity counterparts. It has been demonstrated that the excretion of these esters by yeast was amenable to control by oxygenation during fermentation. Such treatment involved the introduction of oxygen, in the form of oxen saturated water, into the fermenter (Anderson *et al.*, 1975, Kirsop, 1978). The addition of up to 9% of the wort volume as oxygenated water was necessary for the control of acetate esters. The rationale behind this approach was to channel cell energy away from ester production and into cell growth.

Pfisterer and Stewart (1976) have taken a different approach. Their research efforts have been directed to finding fermentation conditions which allow for control of the acetate esters in high gravity worts by means other than stimulating growth. The investigation into the mechanism of ester formation was based on the following reaction:

$$CH_3CONSCoA + C_2H_5OH \rightarrow CH_3CO_2C_2H_5 + CoASH$$

Any means to reduce the pool of acetyl-SCoA in the yeast cell, either by depletion or by inhibition of its synthesis, might also reduce the level of ethyl acetate. In the presence of high levels of fermentable sugar in the wort medium, acetyl-SCoA molecules are directed, in large numbers, towards lipid synthesis and away from the formation of acetate esters. It has been observed in an 18 °P wort that the amounts of ethyl acetate were reduced when increasing portions of brewing syrup were used as adjunct material. In addition, analysis for total lipids revealed a simultaneous accumulation of these compounds in yeast. It would, therefore, appear that in the early aerobic stages of fermentation, the yeast cell develops the metabolic mechanism for the production of fatty acids and lipids which are required for membrane structural purposes. Brewery worts, which contain higher than normal levels of fermentable sugars (brewing syrup adjunct), may prolong this condition and thus enable the yeast to build up a substantial 'lipid credit'. Later in the fermentation, under anaerobic conditions, the yeast can use this potential by drawing upon the pool of acetyl-SCoA for synthesis, thus reducing its availability for ethyl acetate formation.

17.8.4   Brewing With Unmalted Cereals and Microbial Enzymes

The use of adjuncts such as corn, wheat and rice has already been discussed in some detail in this chapter, however, unmalted barley is also being employed in a number of countries, particularly Australia and South Africa. Malted barley has a fully developed enzyme system by virtue of the germination steps and consequently does not require the addition of further enzymes during mashing. However, when employing regular barley, it is necessary to add enzymes to efficiently extract and convert all the starch to fermentable sugars. It was forecast by Quittenton (1969) that malt would largely be replaced by unmalted barley and that cane sugar would be employed in place of corn adjuncts. Although this development has not occurred on the scale envisaged, the use of unmalted barley is a development worthy of consideration (Weig *et al.*, 1969; Nielsen, 1971).

The advent of industrial enzymes, mainly of microbial origin, has resulted in considerable interest in their application within brewing. Although the primary motive has been economic in nature, due to the possible use of less expensive raw materials, there has also been a growing appreciation of the interesting scientific and technical features offered by the new processes. As a contrast to traditional brewing, the enzymes are separated from their substrate prior to enzymatic action and, consequently, may be selected in respect to nature and quantity by the brewer. Furthermore, the time of enzyme addition is not necessarily that of the start of mashing but may, for a part of the enzymes, be postponed to a later stage of the mashing temperature program. These possibilities yield a number of degrees of freedom in selecting mashing conditions and consequently in control of wort composition.

The enzymes in question are mainly bacterial or fungal α-amylases, amyloglucosidases and proteases but also less familiar enzymes, such as pullulanase, may be considered. The optimal enzymatic action is likely to be achieved through a combination of enzymes and substrate. The selection of a commercial enzyme preparation should be accompanied by an awareness of the possible inhibitive effects of barley components. Also β-glucanases will be required in order to degrade the barley β-glucans, which if present will give rise to filtration problems. Although interest in 'barley brewing' is not as intense as it was, the technical possibilities of brewing with barley and enzymes are being pursued commercially by a number of companies offering enzymes, know-how or enzyme converted barley as a concentrated wort.

17.8.5   Low Carbohydrate (Low Calorie) 'Lite' Beer

Beers which have a caloric value lower than conventional beers, usually by a third or more, are enjoying increasing popularity. These light (lite) beers usually have a reduced alcohol content, 20–50% generally, but this is often not sufficient to give the desired reduction in caloric value. It is, therefore, necessary to reduce the amount of non-fermentable sugars which emanate from the malt and adjunct, and which generally carry through the brewing process unchanged. It should be noted that the non-fermentables, such as the dextrins, and the alcohol content are largely responsible for the caloric content of beer. Low calorie beer in North America represents a significant share of the beer volume; in 1982 it constituted 14% of the market or 24 million hl per annum in the USA and 9% or 1.9 million hl per annum in Canada. In addition to low calorie, low alcohol beers, there are low carbohydrate beers in which the alcohol level may be the same or even higher than in normal beers but in which the level of residual carbohydrates (*i.e.* the real extract) has been reduced. Such beers are targeted at people, such as diabetics, who desire a low carbohydrate diet. It should also be noted that there are a number of low alcohol beers that cannot be considered to be low calorie due to the fact that they possess a high level of residual carbohydrates.

There have been numerous attempts to solve the problem of reducing the amount of non-fermentable sugars. Most of these attempts centre on one of the following techniques: (i) dilution of regular strength beer with water, (ii) addition of fungal α-amylase or glucoamylase and bacterial pullulanase to the wort during either mashing or fermentation, (iii) use of barley or malt enzyme preparations during mashing or fermentation, (iv) use of a totally fermentable sugar such as glucose, fructose or sucrose as an adjunct, and (v) use of a brewing yeast strain with amylolytic activity.

One solution that is employed by a number of brewers is to use a process wherein the amylolytic enzyme glucoamylase (amyloglucosidase) is added to the mash and/or during fermentation (Marshall *et al.*, 1982). The resulting beers have a lower calorific value than normally produced beers due to the reduction in dextrin content and it is also reported that the beers, surprisingly, remain free from haze for periods longer than previously possible. Due to the fact that the amount of debranching enzymes present is usually low in normal brewers' mashing materials, glucoamylase is necessary to effect the additional dextrin hydrolysis necessary to produce a reduced carbohydrate beer. However, the glucoamylase normally employed for this purpose is of fungal origin (usually *Aspergillus niger* or *Aspergillus oryzae*) and is thermostable and when employed in fermentation will result, unless specific steps to the contrary are taken, in an active residue of the enzyme in the final beer, which may continue to react, and this can lead to flavour instability. Moreover, the introduction of additives into food products is presently causing increased concern because of possible health-related effects.

Another means of producing a low carbohydrate beer is to employ an aqueous extract of either barley (which will be high in β-amylase, thereby producing maltose as the sole hydrolysis product from starch dextrins) or malt (which will be high in α-amylase, thereby producing a spectrum of sugars from starch dextrins). The aqueous extraction procedure is employed under conditions favourable to protein extraction whilst unfavourable to the extraction of carbohydrates, hence the amount of non-fermentable sugars extracted is reduced. A highly fermentable sugar (*e.g.* glucose) is then added to the resulting extract to produce, effectively, a reconstituted brewer's wort low in unfermentable carbohydrates. The final step involves fermenting the wort. The main advantage of this process is the fact that there will be no residual enzymes, such as glucoamylase, in the final beer and that the product is claimed to be analytically and organoleptically comparable to conventional beers. However, it is well documented that high concentrations of glucose in a wort (in excess of 20%) will effect a different response from the yeast. It has previously been discussed in this chapter that in the presence of high concentrations of glucose the yeast does not adequately develop its capability to metabolize the more complex fermentable sugars such as maltotriose and as a result is unable to effect the latter part of the fermentation. Complete generation of the components (particularly organoleptic components) which provide the beer with its fundamental characteristics is not achieved. In addition, such systems are proven to result in hanging fermentations. Therefore, in processes which vary the barley/malt extract:glucose ratio in the wort, the products will in all probability have widely variable characteristics.

A brewing process has been devised that seeks to circumvent some of the problems discussed above (Geiger, 1983). The process involves the use of a highly fermentable sugar such as glucose, fructose or sucrose but instead of adding this material to wort abruptly at the start of a fermentation, it is added into a partially fermented wort over a period of time. The addition does not commence until the yeast has developed the ability to utilize both maltose and maltotriose and, as a consequence, the likelihood of a 'hung' fermentation is minimized. During this controlled addition, the rate of addition of the sugar is such that the °Plato value of the fermenting wort is not significantly increased from the value when addition of the sugar commenced. This method of introducing the sugar into the fermenting wort over a period of time, which is obviously different from the alternate case where the total amount of sugar is added to the fermenting wort in one batch over a relatively short period of time, is termed 'infusion'. It is believed that the presentation of a highly fermentable sugar to the yeast in this fashion approximates the natural generation of fermentable sugars by any amylolytic enzyme that may be present in or added to the wort during the fermentation stage, thus removing the possible adverse effects of osmotic shock and glucose repression, which would affect the performance of the yeast and thus render product quality control difficult and product consistency very uncertain.

Yeasts of the species *Saccharomyces diastaticus* (Lodder, 1970) have been classified as a distinct species from that of *Saccharomyces cerevisiae* due to the fact that the former produces the extracellular enzyme glucoamylase. Three genes have been identified that are associated with glucoamylase production of *Saccharomyces diastaticus*, *DEX1*, *DEX2* and *STA3* (Stewart *et al.*, 1984). Using classical hybridization techniques, a diploid strain containing the *DEX* and *STA* genes in the homozygous condition has been constructed and its fermentation rate studied in brewer's wort under static fermentation conditions. The initial fermentation rate of this strain was slower than a production ale brewing strain, however, the *DEX*-containing strain fermented the wort to a greater extent than the brewing strain due to the partial hydrolysis of the dextrins by the action of glucoamylase. Thus, *Saccharomyces diastaticus* strains possess the capacity to produce beer which has been fermented to a high degree, which is desirable in the production of low carbohydrate beer; however, the beer produced by these strains had a characteristic phenolic-off-flavour. Phenolic-off-flavours in beer are due on many occasions to the presence of 4-vinylguaiacol (4-VG) which arises by the enzymatic decarboxylation of ferulic acid, a wort constituent. It has been found that a single dominant nuclear gene designated *POF* (phenolic-off-flavour) codes for the ferulic acid decarboxylation enzyme. Therefore, strains possessing the *POF* gene can produce the enzyme capable of decarboxylating ferulic acid. Whereas brewing *Saccharomyces* strains normally cannot decarboxylate ferulic acid, all the *Saccharomyces diastaticus* strains initially studied produced 4-VG in the presence of ferulic acid. Assuming that the *POF* and *DEX* genes are independent characteristics, it could be possible to construct a strain containing the *DEX* but not the *POF* gene by means of hybridization. Thus, a haploid that was *DEX* positive and that carried the *POF* characteristic was mated with a dextrin negative phenolic-off-flavour negative haploid. The resultant diploid fermented dextrin and decarboxylated ferulic acid. When tetrad dissection was carried out a 2:2 segregation for dextrin fermentation and a 2:2 segregation for phenolic-off-flavour was obtained. The *DEX* and *POF* genes segregated independently of each other, therefore, it was possible to select haploids that were *DEX* positive and *POF* negative. Subsequently, a diploid with the genotype *DEX2/DEX2 pofo/pofo* was constructed and a fermentation of an 11.3 °Plato wort conducted. Although the initial wort attenuation rate was found to be slower than that of a polyploid ale yeast strain, the yeast was capable of superattenuating the wort, *i.e.* it was able to hydrolyze part of the dextrins into glucose, which is readily fermentable, whereas the brewing strain was unable to utilize the dextrins. 'Expert' taste panel assessment has deemed the beer produced from this dextrin positive diploid to be rather winey and to have a slightly sulfury character, however, the characteristic phenolic-off-flavour associated with the *POF* gene (4-VG) could not be detected.

A further problem with the use of strains of *Saccharomyces diastaticus* (*POF* or *pof*) for the production of beer is the fact that the extracellular glucoamylase is heat stable and remains active in the packaged beer after pasteurization. As a consequence, upon storage at 21 °C an increasing concentration of glucose can be found in the beer.

Although *Saccharomyces diastaticus* produces a thermostable glucoamylase, no traces of α-amylase or debranching ability could be detected. As starch is a polysaccharide composed of two polymers, 20–25% in the form of amylose (linear chains of α-1,4 linked glucose residues) and 75–80% in the form of amylopectin (a highly branched polymer occurring by α-1,6 linkages) debranching activity is essential for complete hydrolysis of the polysaccharide. Two *Endomycopsis fibuligera* strains studied were found to possess α-amylase and glucoamylase activity but no debranching activity. A strain of *Pichia burtonii* produced very low levels of debranching activity, however, the yeasts *Schwanniomyces castellii* and *Schwanniomyces occidentalis* produced significant amounts of α-amylase, glucoamylase and debranching activity. These amylolytic systems have been isolated, purified and characterized and it has been found that *Schwanniomyces castellii* possesses a glucoamylase with debranching activity.

It has already been discussed in this chapter that one of the techniques for producing low carbohydrate (low calorie) 'lite' beers is to add fungal amylases to the wort during fermentation. Since 70–75% of the dextrins in wort are of the branched type, a debranching enzyme is essential for total hydrolysis of wort dextrins to fermentable sugars. The fungal glucoamylase used extensively in the production of light beer possesses debranching activity, therefore it can hydrolyze the dextrins. However, normal pasteurization of the final product employing a conventional temperature/time cycle, does not completely inactivate this enzyme. An important characteristic of amylases from *Schwanniomyces castellii* is their sensitivity to the normal pasteurization cycle employed in brewing. A representative curve of pasteurization temperature *versus* time indicates that there is an eight minute period during which the temperature is maintained at 60–62 °C. It has been reported that 15 minutes at 60 °C is the time required to inactivate the *Schwanniomyces castellii* glucoamylase, however, this study was conducted at pH 5.5 (this enzyme's optimal pH) and in the absence of ethanol. When a commercial fungal glucoamylase preparation derived from *Aspergillus niger* and the glucoamylase from *Schwanniomyces castellii* were compared for their sensitivity to pasteurization at pH 4.0 (normal beer pH) and pH 6.0 in the presence and absence of ethanol, the ethanol enhanced the inactivation effect of the pasteurization. In addition, at pH 4.0, the pasteurization cycle inactivated glucoamylase as well as α-amylase, from *Schwanniomyces castellii* with or without ethanol, but at pH 6.0, the presence of ethanol was necessary for enzyme inactivation.

A two stage fermentation system has been devised for the production of low carbohydrate beers (Sills *et al.*, 1983). Amylases (gluco- and α-amylase) from *Schwanniomyces castellii* are produced in a highly inducing medium containing maltose. Subsequently, the cells are removed, the culture filtrate is concentrated and added to wort previously inoculated with a genetically manipulated strain of *Saccharomyces diastaticus* or with a brewing production strain of *Saccharomyces uvarum (carlsbergensis)*. The *Saccharomyces diastaticus* strain is a diploid containing both the *DEX1* and *DEX2* genes, which code for glucoamylase production and, more importantly, this strain lacks the capability to decarboxylate ferulic acid to 4-vinylguaiacol.

In an attempt to establish the optimal amount of enzyme culture filtrate to be added to the fermenting wort, several concentrations of enzymes were added. As anticipated, increasing concentrations of enzyme had a direct correlation with apparent attenuation, *i.e.* with the maximal amounts of enzymes added, the degree of fermentation increased to 100% and up to 99.4% with 70% less added enzyme. Fermentations conducted with the diploid *Saccharomyces diastaticus* strains were more sluggish because this particular strain lacked *MAL* genes, therefore, maltose was only hydrolyzed by the extracellular glucoamylase and a longer time of fermentation was required to achieve 100% of apparent attenuation (Sills *et al.*, 1983).

A direct correlation has been found between ethanol production and enzyme concentration employed, *i.e.* the greater the enzyme concentration, the higher the level of ethanol obtained. Thus, with fermentations employing the lager yeast strain, a maximal increase of 19% ethanol (compared to the control) could be obtained in the final product, whereas with *Saccharomyces diastaticus*, an improvement of 15.1% could be obtained with the maximal enzyme addition.

17.8.6 Genetic Manipulation of Brewing Yeast Strains

There are a number of methods that are employed in genetic research and development of brewer's yeast strains. These include hybridization, mutation and selection, rare mating, spheroplast fusion and transformation. Transformation can be carried out using native DNA, recombinant DNA and by liposome-mediated DNA transfer. Hybridization, which employs the haploid/diploid life cycle of a yeast strain, cannot be employed directly as a means to manipulate brewer's yeast strains because, due to their polyploid/aneuploid nature, they do not possess a mating type, they sporulate poorly, and the few spores that do form often fail to germinate. Nevertheless, hybridization is a technique that has made an invaluable contribution to the field of yeast genetics and is today by no means obsolete. Hybridization has been used in conjunction with more novel genetic techniques to verify the genetic composition of recombinants. It can also be employed to provide a great deal of relevant information about traits that are germane to brewing fermentation systems. As has already been discussed in this chapter, hybridization has been used to study the genetic control of flocculation, phenolic-off-flavour production and the uptake of wort sugars and dextrins.

Techniques that have the greatest potential and promise as aids in the genetic manipulation of industrial (including brewing) yeast strains are rare mating, mutation and selection, spheroplast/ protoplast fusion (also called somatic fusion) and transformation (usually associated with recombinant DNA techniques). All of these methods have total disregard for ploidy and mating type and thus have great applicability to brewing strains. Rare mating has been used in conjunction with the *kar* (karyogamy defective) strains to introduce zymocidal (killer) activity into brewing strains (Young, 1983; Stewart *et al.*, 1985a). It has been previously described in this chapter that mutation and selection have been used to isolate derepressed mutants of brewing strains such that these strains possess the ability to metabolize maltose in the presence of glucose and thus have increased wort fermentation rates. Mutation and selection have also been used to induce auxotrophs and to select easily recognizable characteristics of brewing strains in order that such strains can be employed as spheroplast fusion partners and as recipients for transformation experiments. Spheroplast fusion has been employed to fuse strains constructed by hybridization with brewing strains in order to introduce the novel capabilities of the hybridized strains in the brewing strains whilst still maintaining all the characteristics of the latter (Stewart *et al.*, 1984). Finally, transformation is being employed in a number of brewing research laboratories to introduce genes from non-*Saccharomyces* yeast strains into brewing strains.

As previously discussed, the technique of rare mating has been employed to manipulate brewing yeast strains. When non-mating strains are mixed together at a high cell density, a few true hybrids with fused nuclei form which can usually be isolated selectively. An even more useful technique employs a yeast strain which harbours a specific nuclear gene mutation, designated *kar*. When this strain hybridizes with another strain, the nuclei will not fuse and this permits the formation of cell lines with mixed cytoplasmic contents (heteroplasmons). Rare mating has been employed to transfer zymocidal or killer factor from laboratory haploid strains to brewing yeast strains. Some strains of *Saccharomyces* spp. (and other yeast genera) secrete a proteinaceous toxin called a zymocide or killer toxin which is lethal to certain other strains of *Saccharomyces* (Young, 1983). Toxin producing strains are termed 'killers' and susceptible strains are termed 'sensitives'. However, there are strains that do not kill and are not themselves killed and these are called 'resistant'. The 'killer' factor has been renamed 'zymocide' to indicate that it is only lethal towards yeast and not bacteria or cells of higher organisms. Zymocidal yeasts have been recognized to be a serious problem in both batch and continuous fermentation systems. An infection of as little as 0.1% of the cell population can completely eliminate all the brewing yeast from the fermenter. Thus the brewer can protect the process from this occurrence in one of two ways: (i) maintain vigorous standards of hygiene to ensure that contamination with a wild yeast possessing zymocidal activity is prevented, or (ii) genetically modify the brewery yeast so that it is not susceptible to the zymocidal toxin. The first method is the one most brewers to date have been invoking to protect their process; however, genetic manipulation can also be used to produce a brewing strain that is less vulnerable to destruction by a zymocidal yeast infection.

The killer character of *Saccharomyces* spp. is determined by the presence to two types of cytoplasmically-located double stranded RNA (dsRNA) (for review see Gunge, 1983). M dsRNA (killer plasmid), which is killer strain specific, codes for killer toxin and also for a protein or proteins which render the host immune to the toxin. L dsRNA which is also present in many non-killer yeast strains, specifies a capsid protein which encapsulates both forms of dsRNA, thereby yielding virus-like particles (VLPs). Although the killer plasmid is contained within these VLPs, the killer genome is not naturally transmitted from cell to cell by any infection process. The killer plasmid behaves as a true cytoplasmic element, showing dominant non-Mendelian segregation, but it is dependent upon at least 29 different chromosomal genes (*mak*) for its maintenance in the cell. In addition, three other chromosomal genes (*kex1, kex2* and *rex*) are required for toxin production and resistance to toxin. Employing the technique of rare mating, brewing strains have been modified so that they are resistant to killing by a zymocidal yeast and secondly so that they will themselves be zymocidal and therefore eliminate any contaminating yeasts that are sensitive to the toxin. Rare mating has been successfully used to produce a number of brewing strains with zymocidal activity. In this laboratory (Stewart *et al.*, 1985a) and others, rare mating has been successfully used to produce a number of brewing strains with zymocidal activity. The zymocidal brewing strain isolated after rare mating of a brewing lager polyploid with a zymocidal haploid, possessed both L and M dsRNA. This lager strain was mixed at a concentration of 10% with a normal ale brewing strain. Within 10 hours the lager strain had almost totally eliminated the ale strain. The speed at which this occurs may well make the brewer apprehensive about using zymocidal yeast cultures in a brewery if the brewery normally employs several yeast strains for the production of different beers. An error on an operator's part could result in disaster. In a brewery where only one yeast strain is employed there would obviously be no cause for concern.

As well as the examples of genetic manipulation of brewer's yeast strains cited in this document, the potential for improving such strains is almost infinite. In the not too distant future, brewer's yeast strains will be 'tailor made' to a particular process, substrate requirement and beer flavour profile.

17.9 REFERENCES

Aalbers, V. J., B. Drost and P. van Eerde (1980). Practical methods for reduction of nitrosamine contents in malt. *Brauwelt*, **120**, 719.

Anderson, R. J. and B. H. Kirsop (1974). The control of volatile ester synthesis during the fermentation of wort of high specific gravity. *J. Inst. Brewing*, **80**, 48–55.

Anderson, R. J., B. H. Kirsop, H. Rennie and R. J. H. Wilson (1975). The practical use of oxygen during fermentation for the control of volatile acetate ester concentration in beer. In *Proceedings of the 15th Congress, European Brewery Convention*, pp. 243–253. Elsevier, Amsterdam.

Anness, B. J. (1981). The role of dimethyl sulphide in beer flavour. In *Monograph-VII, European Brewery Convention*, pp. 135–142. Verlag Hans Carl, Nürnberg.

Atkinson, B. (1983). Biotechnology in the brewing industry. In *Proceedings of the 19th Congress, European Brewery Convention*, pp. 339–351. IRL Press, Oxford.

Ayrapaa, T. and I. Lindstrom (1973). Influence of long-chain fatty acids on the formation of esters by brewer's yeast. In *Proceedings of the 14th Congress, European Brewery Convention*, pp. 271–283. Elsevier, Amsterdam.

Barker, R. L., D. E. F. Gracey, A. J. Irwin, P. Pipasts and E. Leiska (1983). Liberation of staling aldehydes during storage of beer. *J. Inst. Brewing*, **89**, 411–415.

Barnett, J. A., R. W. Payne and D. Yarrow (1983). *Yeasts—Characteristics and Identification*. Cambridge University Press, London.

Beavan, M. J., D. M. Belk, G. G. Stewart and A. H. Rose (1979). Changes in electrophoretic mobility and lytic enzyme ability associated with development of flocculating ability in *Saccharomyces cerevisiae*. *Can. J. Microbiol.*, **25**, 888–895.

Bengough, W. I. and G. Harris (1955). General composition of non-biological hazes of beers and some factors in their formation. Part I. *J. Inst. Brewing*, **61**, 134–145.

Calleja, G. B. (1984). *Microbial Aggregation*. CRC Press, Boca Raton, FL.

Chantrell, N. S. (1983). Practical and analytical results obtained from continuous high temperature wort boiling. In *Proceedings of the 19th Congress, European Brewery Convention*, pp. 89–96. IRL Press, Oxford.

Day, A. W., N. H. Poon and G. G. Stewart (1975). Fungal fimbriae. III. The effect on flocculation in *Saccharomyces*. *Can. J. Microbiol.*, **21**, 558–564.

Dougherty, J. J. (1977). Wort production. In *The Practical Brewer*, ed. H. M. Broderick, pp. 62–98. Master Brewers Assoc. of the Americas, Madison, WI.

Engan, S. (1981). Beer composition: volatile substances. In *Brewing Science*, ed. J. R. A. Pollock, vol. 2, pp. 93–165. Academic, London.

Evans, J. I. (1972). Control of beer foam. *Process Biochem.*, **16** (4), 29–31.

Gardner, R. J. and J. D. McGuinness (1977). Complex phenols in brewing—a critical survey. *MBAA Tech. Quart.*, **14**, 250–261.

Geiger, K. H. (1983). Brewing process. *Br. Pat.* 2 029 445.

Gilliland, R. B. (1951). The flocculation characteristics of brewing yeast during flocculation. In *Proceedings of the 3rd Congress, European Brewery Convention*, pp. 35–37. Elsevier, Amsterdam.

Godtfredsen, S. E., M. Ottesen, P. Sigsgaard, K. Erdal, T. Mathiasen and B. Ahrenst-Larsen (1983). Use of α-acetolactate decarboxylase for accelerated maturation of beer. In *Proceedings of the 19th Congress, European Brewery Convention*, pp. 161–168. IRL Press, Oxford.

Gracey, D. E. F. and R. L. Barker (1976). Studies on beer haze formation. *J. Inst. Brewing*, **82**, 72–84.

Gracey, D. E. F. and R. L. Barker (1981). A new perspective on the composition of beer hazes. In *Proceedings of the 18th Congress, European Brewery Convention*, pp. 471–478. IRL Press, Oxford.

Gramshaw, J. W. (1970). Beer polyphenols and the chemical basis of haze formation, Part II: Changes in polyphenols during the brewing and storage of beer—the composition of hazes. *MBAA Tech. Quart.*, **7**, 122–133.

Greenberg, D. M. (1975). *Metabolism of Sulphur Compounds*. Academic, London.

Gunge, N. (1983). Yeast DNA plasmids. *Annu. Rev. Microbiol.*, **37**, 253–276.

Gyllang, H. and E. Martinson (1976a). *Aspergillus fumigatus* and *Aspergillus amstelodami* as causes of gushing. *J. Inst. Brewing*, **82**, 182–183.

Gyllang, H. and E. Martinson (1976b). Studies on the mycoflora of malt. *J. Inst. Brewing*, **82**, 350–352.

Haikara, A. (1980). Gushing induced by fungi. In *Monograph-VI, European Brewery Convention*, pp. 251–259. Verlag Hans Carl, Nürnberg.

Hall, M. J., S. D. Dickinson, R. Pritchard and J. I. Evans (1973). Foams and foam control in fermentation processes. *Prog. Ind. Microbiol.*, **12**, 171–234.

Hargitt, R. and G. K. Buckee (1978). Carbohydrate balances and wort fermentability estimations. *J. Inst. Brewing*, **84**, 224–227.

Harold, F. V. and B. J. Clarke (1979). Liquid CO_2 hop extraction—the commercial reality. *Brewers Digest*, **54** (9), 45–52.

Hough, J. S. (1959). Flocculation characteristics of strains present in some typical British pitching yeasts. *J. Inst. Brewing*, **65**, 479–482.

Hough, J. S. (1976). Silica hydrogels for chill-proofing beer. *MBAA Tech. Quart.*, **13**, 34–39.

Hough, J. S. and Rudin, A. D. (1958). Experimental production of beer by continuous fermentation. *J. Inst. Brewing*, **64**, 404–410.

Hudson, J. L. (1979). Hops now and in the future. In *Proceedings of the 17th Congress, European Brewery Convention*, pp. 405–421. DSW, Dordrecht.

Hudson, J. L. (1983). Horace Brown Memorial Lecture—Expanding brewing technology. *J. Inst. Brewing*, **89**, 189–194.

Jayatissa, P. M. and A. H. Rose (1976). Role of wall phosphomannan in flocculation of *Saccharomyces cerevisiae*. *J. Gen. Microbiol.*, **96**, 165–174.

Johnston, J. R. and H. P. Reader (1983). Genetic control of flocculation. In *Yeast Genetics—Fundamental and Applied Aspects*, ed. J. F. T. Spencer, D. M. Spencer and A. R. W. Smith, pp. 205–224. Springer-Verlag, New York.

Jones, M. and J. S. Pierce (1967). The role of proline in the amino acid metabolism of germinating barley. *J. Inst. Brewing*, **73**, 577–583.

Kato, K., J. Haguja and M. Yonomura (1970). *Jpn. Pat.* 122 110.

Kirsop, B. H. (1974). Oxygen in brewery fermentation. *J. Inst. Brewing*, **80**, 252–259.

Kirsop, B. H. (1978). Fermentation: From wort to beer. In *Monograph V, European Brewery Convention*, pp. 3–16. Verlag Hans Carl, Nürnberg.

Krejer-van Rij, N. J. W. (1984). *The Yeasts—A Taxonomic Study*, 3rd edn. Elsevier, Amsterdam.

Laws, D. R. J. (1981). Hop extracts—a review. *J. Inst. Brewing*, **87**, 24–29.

Laws, D. R. J., N. A. Bath and J. A. Pickett (1977). Production of solvent-free isomerized extracts. *J. Am. Soc. Brew. Chemists.*, **35**, 187–190.

Lewis, M. J. (1968). Yeast and fermentation. *MBAA Tech. Quart.*, **5**, 103–113.

Lodder, J. (1970). *The Yeasts—A Taxonomic Study*, 2nd edn. North-Holland, Amsterdam.

MacDonald, J., P. T. V. Reeve, J. D. Ruddlesden and F. H. White (1984). Current approaches to brewery fermentations. In *Progress in Industrial Microbiology*, ed. M. E. Bushell, vol. 19, pp. 47–198. Elsevier, Amsterdam.

MacLeod, A. M. (1977). The impact of science on malting technology. In *Proceedings of the 16th Congress, European Brewery Convention*, pp. 63–76. DSW, Dordrecht.

Marshall, W. G. A., L. J. Denault, P. R. Glenister and J. Dower (1982). Enzymes in brewing. *Brewers Digest*, **57** (9), 14–22.

McFarlane, W. D. (1961). Determination of anthocyanogens. *J. Inst. Brewing*, **67**, 502–505.

McFarlane, W. D., E. Wye and H. L. Grant (1955). Studies on the flavonoid compounds in beer. In *Proceedings of the 5th Congress, European Brewery Convention*, pp. 298–310. Elsevier, Amsterdam.

McGuinness, J. D., R. Eastmond, D. R. J. Laws and R. J. Gardner (1975). The use of ^{14}C-labelled polyphenols to study haze formation in beer. *J. Inst. Brewing*, **81**, 287–292.

Murray, C. R., T. Barich, and D. Taylor (1984). The effect of yeast storage conditions on subsequent fermentation. *MBAA Tech. Quart.*, **21** (4), 189–194.

Nagami, K., T. Takahashi, K. Nakatani and J. Kumada (1980). Hydrogen sulfide in brewing—II. *MBAA Tech. Quart.*, **17**, 64–68.

Narziss, L. (1984). The German beer law. *J. Inst. Brewing*, **90**, 351–358.

Nielsen, E. B. (1971). Brewing with barley and enzymes—a review. In *Proceedings of the 13th Congress, European Brewery Convention*, pp. 149–170. Elsevier, Amsterdam.

Nordstrom, K. (1964). Formation of ethyl acetate in fermentation with brewer's yeast. V. Effects of some vitamins and mineral nutrients. *J. Inst. Brewing*, **70**, 209–221.

Palmer, G. H. (1969). Increased endosperm modification of abraded barley grain after gibberellic acid treatment. *J. Inst. Brewing*, **75**, 536–541.

Pasteur, M. L. (1876). *Etudes sur la Biere*, pp. 1–383. Gauthier-Villars, Paris.

Pechov, A. V., I. Ponomeresko and A. F. Prokopcuk (1968). *USSR Pat.* 167 798.

Pfisterer, E. A. (1971). The influence of various parameters on the accumulation of fermentable carbohydrates throughout the mash cycle. In *Proc. Am. Soc. Brew. Chemists*, 39–47.

Pfisterer, E. A. and G. G. Stewart (1975). Some aspects on the fermentation of high gravity worts. In *Proceedings of the 15th Congress, European Brewery Convention*, pp. 255–267. Elsevier, Amsterdam.

Pfisterer, E. A. and G. G. Stewart (1976). High gravity brewing. *Brewers Digest*, **51** (6), 34–42.

Pfisterer, E. A., I. F. Garrison and R. A. McKee (1978). Brewing with syrups. *MBAA Tech. Quart.*, **15**, 59–63.

Pollock, J. R. A., B. H. Kirsop and R. E. Essery (1955a). Studies in barley and malt. III. New and convenient germinative capacity test. *J. Inst. Brewing*, **61**, 295–300.

Pollock, J. R. A., B. H. Kirsop and R. E. Essery (1955b). Studies in barley and malt. IV. Experiments with dormant barley. *J. Inst. Brewing*, **61**, 301–307.

Portno, A. D. (1978). Continuous fermentation in the brewing industry—the future outlook. In *Monograph V, European Brewery Convention*, pp. 145–154. Verlag Hans Carl, Nürnberg.

Prentice, N. and W. Sloey (1960). Studies on barley microflora of possible importance to malting and brewing quality. I. The treatment of barley during malting with selected microorganisms. In *Proc. Am. Soc. Brew. Chemists*, 28–33.

Quain, D. E. and R. S. Tubb (1983). A rapid and simple method for the determination of glycogen in yeast. *J. Inst. Brewing*, **89**, 38–40.

Quittenton, R. C. (1969). Perspectives in brewing. *Wallerstein Lab. Commun.*, **32**, 77–79.

Ramos-Jeunehomme, C. L. and C. A. Masschelein (1977). Genetic control of the formation of vicinal diketones in *Saccharomyces cerevisiae*. In *Proceedings of the 16th Congress, European Brewing Convention*, pp. 267–283. DSW, Dordrecht.

Rennie, H. (1972). Is wort boiling necessary? *J. Inst. Brewing*, **78**, 162–164.

Roberts, R. T. (1977). Colloidal aspects of beer foam. *Brewers Digest*, **52** (6), 50–58.

Rose, A. H. (1984). Physiology of cell aggregation: Flocculation by *Saccharomyces cerevisiae* as a model system. In *Microbial Adhesion and Aggregation*, ed. K. C. Marshall, pp. 323–335. Springer-Verlag, Berlin.

Runkel, U.-D. (1975). Malt kilning and its influence on malt and beer quality. In *Monograph-II, European Brewery Convention*, pp. 222–235. Verlag Hans Carl, Nürnberg.

Russell, I., G. G. Stewart, H. P. Reader, J. R. Johnson and P. A. Martin (1980). Revised nomenclature of genes that control yeast flocculation. *J. Inst. Brewing*, **86**, 120–121.

Seaton, J. C., I. S. Forrest and A. Suggett (1981). High temperature wort boiling—consequence for beer flavour. In *Proceedings of the 18th Congress, European Brewery Convention*, pp. 161–168. IRL Press, Oxford.

Sills, A. M., I. Russell, and G. G. Stewart (1983). The production and use of yeast amylases in the brewing of low carbohydrate beer. In *Proceedings of the 19th Congress, European Brewery Convention*, pp. 377–384. IRL Press, Oxford.

Slater, E. C. (1977). Yeast, fermentation, fermenters and enzymes. In *Proceedings of the 16th Congress, European Brewery Convention*, pp. 1–11. DSW, Dordrecht.

Spiegelhalder, B., G. Eisenbrand and R. Preussmann (1979). Contamination of beer with trace quantities of N-nitrosodimethylamine. *Food Cosmet. Toxicol.*, **17**, 29–31.

Stewart, G. G. (1977). Fermentation—yesterday, today and tomorrow. *MBAA Tech. Quart.*, **14** (1), 1–15.

Stewart, G. G. and I. Russell (1981). Yeast flocculation. In *Brewing Science*, ed. J. R. A. Pollock, vol. 2, pp. 61–92. Academic, London.

Stewart, G. G., C. J. Panchal and I. Russell (1983). Current developments in the genetic manipulation of brewing yeast strains—a review. *J. Inst. Brewing*, **89**, 170–188.

Stewart, G. G., C. R. Murray, C. J. Panchal, I. Russell and A. M. Sills (1984). The selection and modification of brewer's yeast strains. *Food Microbiol.*, **1**, 289–302.

Stewart, G. G., C. A. Bilinski, C. J. Panchal, I. Russell and A. M. Sills (1985a). The genetic manipulation of brewer's yeast strains. *Microbiology 1985*, American Society for Microbiology (in press).

Stewart, G. G., R. M. Jones and I. Russell (1985b). The use of derepressed yeast mutants in the fermentation of brewery wort. In *Proceedings of the 20th Congress, European Brewery Convention*, in press. IRL Press, Oxford.

Takahashi, T. K., K. Nagami, K. Nakatani and J. Kumada (1980). Hydrogen sulfide in brewing—II. *MBAA Tech. Quart.*, **17**, 210–214.

Thorne, R. S. W. (1952). The genetics of flocculation in *Saccharomyces cerevisiae*. *C. R. Trav. Lab. Carlsberg*. **25**, 101–140.

Vernon, P. S. and A. H. Rose (1976). Adsorption of silicone antifoam by *Saccharomyces cerevisiae*. *J. Inst. Brewing*, **82**, 336–340.

Verzele, M. (1979). The chemistry of hops. In *Brewing Science*, ed. J. R. A. Pollock, vol. 1, pp. 280–322. Academic, London.

von Wettstein, D. (1983) Emil Christian Hansen Centennial Lecture: from pure yeast culture to genetic engineering of brewers yeast. In *Proceedings of the 19th Congress, European Brewery Convention*, pp. 97–119. IRL Press, Oxford.

von Wettstein, D., B. Jende-Strid, B. Ahrenst-Larsen and K. Erdal (1980). Proanthocyanidin-free barley prevents the formation of beer haze. *MBAA Tech. Quart.*, **17**, 16–23.

Wainwright, T. (1972). Sulfur tastes and smells in beer. *Brewers Digest*, **47** (7), 78–84.

Wainwright, T. (1973). Diacetyl—a review. *J. Inst. Brewing*, **70**, 451–470.

Wainwright, T. (1981). Nitrosodimethylamine: Formation and palliative measures. *J. Inst. Brewing*, **87**, 264–265.

Weig, A. J., J. Hollo and P. Varga (1969). Brewing beer with enzymes. *Process Biochem.*, **12** (5), 33–38.

Young, T. W. (1983). Brewing yeast with anti-contaminant properties. In *Proceedings of the 19th Congress, European Brewery Convention*, pp. 129–136. IRL Press, Oxford.

18
Whisky

M. SOBOLOV, D. M. BOOTH and R. G. ALDI
Hiram Walker and Sons Ltd, Windsor, Ontario, Canada

18.1 INTRODUCTION

The term 'whisky' is derived from the Gaelic phrase '*uisge beatha*' or 'water of life.' This alcoholic beverage has evolved to a refined flavourful beverage due primarily to advancements in technology. Excess grains were used to produce whiskies which were saleable at a higher price, while still leaving the residuals for animal feed.

In Scotland, malt whisky became a blended lighter 'Scotch' whisky with the advent of continuous distillation. This allowed whisky producers to produce a more acceptable beverage for European palates by blending the heavier malt whisky with lighter grain spirits. Although 'straight' malt whisky is still widely accepted as a good whisky, most people today have come to know 'Scotch' whisky as a blended product. Distillers are now able to produce a wide variety of products for virtually every palate.

A different type of whisky evolved in the United States. Bourbon, named after a county in Kentucky, used corn in conjunction with smaller grains. Modern day bourbon must consist of 51% corn, is not to be distilled over 160° proof (80% alcohol by volume), and must be barrelled in new oak cooperage.

Because corn was not a major crop in Canada for many years, Canadian whisky utilized wheat and rye as the grains of choice. For most distillers, rye and malt were used as the main whisky flavouring components with the lighter components being produced from wheat or rye. Today, the flavouring components remain the same; however, corn is mainly used for the lighter grain whisky.

Japanese whisky production is similar to other blended whiskies with a stronger emphasis on malt whiskies as the flavour components. Again, the use of grain whisky or other sources of alcohol is evident to produce the lighter whisky portion of the final product.

Even though whisky production is the emphasis here, it would be remiss not to mention other alcoholic products. These include vodka produced from such sources as molasses, grain and potatoes; tequila from the mescal plant; rum from sugar and molasses; as well as a host of other products specific to other areas and regions of the world.

The foregoing discussion is significant only as it points out the various grains used to impart particular flavours. In addition flavour is derived from the yeast during fermentation. Therefore particular operations call for specific yeasts.

The conditions of fermentation also play a significant role in congener production. Yeast will develop a variety of chemical entities other than alcohol. The side pathways of metabolism are controlled by pH, temperature and concentration. Further product control occurs at the distillation step. The amounts of congeners in the final product are dictated by a variety of operational parameters. Maturation in the oak barrels will also have an effect on congener content. Different processes occur in the barrel such as oxidation, reduction, atmospheric exchange and adsorption plus concentration.

The final controlling step is that of blending. The product may be left on its own, or may be blended to produce just the right combination of flavour components required for bottling.

Figure 1 is a simplified flow diagram of the continuous system using corn as the grain of choice. Values indicated are approximate using a basis of 50 tonnes of milled grain. Energy values in gigajoules are for steam usage at the indicated locations.

18.2 MILLING

Grains are conveyed to the distiller's mill house, where they are cleaned prior to processing.

The type of mill used by the distiller will depend upon the type of whisky that he is producing. In Scotland and Ireland, where clear wort mashing is practised, smooth roller mills are used for splitting the husk and crushing the starch-containing kernel endosperm. The undamaged husk is an essential component of the grain solids filtering bed when preparing clear wort from malted barley. With whole grain mashing, as is practised in North America, hammer mills are the mills of choice. However, fluted roll roller mills are sometimes used. The crushing or shearing action of the mill is intended to expose more endosperm starch surface area to the subsequent processes of cooking and conversion.

An ideal grain meal or grist will have an even particle size distribution, with coarse particles being preferred to flour. Hammer mills have a tendency to produce excess flour and are, therefore, unacceptable for clear wort mashing. Flour will plug the grain filter bed and impede throughputs. Fines must be minimized from both hammer mills and roller mills when stillage is dehydrated in an associated by-products recovery plant. Excessive fines can create evaporator fouling problems when concentrating stillage to 'distillers' solubles'.

18.3 MASHING

Mashing is the process of mixing grist with hot water to solubilize or gelatinize grain starches, followed by the employment of malt or other natural enzymes to convert soluble starch to fermentable sugar. Mashing must utilize optimum conditions of temperature, reaction time and pH in order to achieve the desired results. The physical and biological factors inherent in mashing necessitate strict adherence to these time, temperature and pH relationships.

18.3.1 Malted Grains Mashing

The mashing of malted grains, as in the manufacture of malt whisky, is rather unique in that starch solubilization, conversion and mash filtration to produce clear wort occur simultaneously. Grist and water at 66 °C are combined in a mixing conveyor known as a 'mashing machine.' From here the mash is propelled into a multi-purpose vessel which is either a mash tun or a lauter tun.

Both vessels are equipped with a false bottom, consisting of perforated plates, which permit the 'drawing off' or separation of the liquid malt extract from the spent grain. The malt extract solution (clear wort) contains trace amounts of suspended solids that harbour lipids, soluble malt-

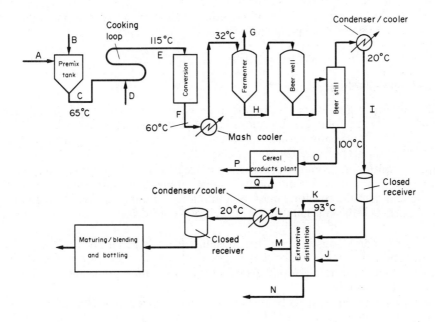

Mass and energy (steam) balance

Letter code	Description	Approximate values
A	Milled grain	50 t (basis)
B	Water	145 t
C	Mash	195 t
D	Steam	35 GJ
E	Cooked mash	210 t
F	Converted mash	189 t
G	Carbon dioxide	16 t
H	Fermented beer	173 t
I	High wines (80 % EtOH vol.)	20.6 t
J	Steam	172 GJ
K	Extraction water	260 t
L	Alcohol product (95 % EtOH vol.)	17.4 t
M	By – product (fusel oils, *etc.*)	23 kg
N	Extraction water	263 t
O	Stillage	152.4 t
P	Distiller's dried grains	21 t
Q	Steam	146 GJ

Figure 1 The continuous system for corn

ose sugar, soluble wort proteins, dextrins and many other components extracted from the grain that are utilized by yeasts in subsequent fermentation.

The malt whisky mashing process allows for sufficient holding time in the tun before 'drawing off' wort (usually one hour) to permit a partial pasteurization of the mash and for development of the grain filter bed. This is also an adequate time to effect a conversion of the majority of soluble starch to maltose. After drawing off the mashing liquid (first water), two subsequent hot water spargings of the grain bed are effected to extract further fermentable sugars and dextrins. The first hot wash at 78 °C removes most of the sugar and is pumped to the fermenter with the previous wort drawn from the tun. The second hot wash at 90 °C extracts some sugar and is recycled to be used for the mashing water of the following mash. This final wash is effective in sterilizing the vessel and associated piping in preparation for the following mash.

It is important in mashing and fermentation circuits to maintain relatively sterile conditions. If the distiller fails to do this, the resulting bacterial contamination will surely result in yield reductions and the production of 'off-flavours' attributed to bacterial growth. It is suggested by some distillers, however, that a modest amount of bacteria is necessary in a wort fermentation to create the desired mix of congeners to supplement those produced by yeast.

18.3.2 Large Grains Mashing

The mashing of large grains, such as corn and milo, necessitates a separate cooking stage prior to conversion, in order to solubilize the more complex starches characteristic of these grains. Large grains are usually pressure cooked to achieve temperatures ranging from 110 to 175 °C. Either batch or continuous pressure cooking systems are used.

In the popular continuous cooking process, corn meal is mixed with water in a small vessel called a premixer. Since it is a continuous system, grain meal and water are continuously delivered to the tank and the resulting mash is continuously pumped away. Since good mixing at this stage is important, efficient agitators with high-powered drives are employed for this purpose.

In the interest of energy conservation, the mashing water is generally a mixture of beer still condenser water and the supernatant of centrifuged beer column still bottoms (thin stillage). An ideal temperature to aim for here is 60 °C. Above this temperature, grain starch will begin to gel, restricting mixing and pumping.

Thin stillage has a pH value slightly below 4.0 and when utilized at 30% of total mashing water volume, a mash pH of between 4.8 and 5.0 is normally achieved. Mash pH should not be permitted to fall below 4.7 at the premixer, or amylases employed later for the conversion of gelatinized starch to sugar will not be functional.

The benefits of using stillage in corn grain mashes are many. The recycled stillage does not have to be evaporated in the feed recovery plant, which is a major energy saving. Secondly, the stillage provides a buffering capability to the mash, which maintains mash pH above 4.0 during the term of the fermentation, thus allowing dextrinases to continue their hydrolysis of dextrins, to release additional fermentable sugar to the medium. Thirdly, the stillage provides supplemented yeast nutrition as protein, which can otherwise be limiting when compared to protein levels found in small grains mashes.

From the premixer, the mash is pumped to a cooking loop where high pressure steam is injected into the mash as it enters the loop. The loop is basically a length of pipe through which the mash moves, over a retention time of two or three minutes, at a controlled temperature selected by the distiller. Temperatures ranging between 110 and 150 °C have been used in these loops.

At the end of the loop, the mash enters the top of a tall holding vessel at atmospheric pressure, where it flashes to 100 °C. A retention time of approximately 30 minutes in this vessel permits further gelatinization of starch. Some distillers have found that the holding tank can be eliminated by utilizing higher temperatures in the loop.

After cooking, barometric condensers are employed to flash cool the mash to a preselected temperature at which liquefying amylases are added to reduce mash viscosity by hydrolyzing complex starch polymers to lower molecular weight dextrins. Malt amylases and/or amylases of fungal or bacterial origin may be used. The variety of enzymes available for starch conversion all have their optimum operating conditions which include reaction temperature, reaction time and reaction pH. In most countries it is the option of the distiller to select the enzyme combinations that will best meet his needs.

The liquefaction process is completed in about 15 minutes in an appropriate vessel. Upon completion of liquefaction, the mash is cooled to fermentation temperature (20–30 °C) by pumping through water-cooled heat exchangers called 'mash coolers.' 'Spiral', 'plate and frame' and 'tube and shell' coolers are some common types used in practice.

The cooled mash is delivered to a fermenting vessel where yeast is added to perform the fermentation function. Most distillers also add to the fermenters, in addition to malt, saccharifying fungal amylase to improve yields. These enzymes may be manufactured in-house or purchased as concentrates from enzyme supply houses. These enzymes produce glucose instead of maltose and speed fermentation as well as increase the amount of sugar available for alcohol production.

Batch pressure cooking of corn usually employs temperatures as high as 175 °C after which similar batch conversion and cooling processes are used as have been discussed above for continuous pressure cooking.

18.3.3 Small Grains Mashing Other Than Malt

There are basically two approaches to mashing small grains, both of which are batch processes producing whole grain mashes.

One method is 'infusion mashing' whereby the complete grain bill, including rye, barley malt and perhaps rye malt, is simultaneously cooked and converted. In the process, all the grain meal is conveyed to the batch cooker–converter where it is mixed with water at 38 °C. The mixture is slowly raised to 55 °C, taking advantage of the proteolytic enzyme optimum of 46 °C on the way up, which provides essential yeast nutrients. At 55 °C there is a holding period of several minutes to encourage β-amylase activity which releases maltose sugar from the non-reducing ends of soluble starch molecules. A further increase to 65 °C, where a holding time of 45 minutes is common, encourages the activity of α-amylase, which randomly hydrolyzes starch to produce dextrins, maltose and some glucose. This holding time at 65 °C is also necessary to pasteurize the mash in order to be assured of contamination-free fermentation.

Mash coolers are used to cool the mash from 65 °C to set fermentation temperature (20–30 °C).

An alternative method of mashing small grains utilizes an atmospheric batch cooking stage for unmalted rye, employing temperatures ranging between 82 and 93 °C at the top of the cooking cycle. The rye cooks are cooled to 65 °C for malt enzyme conversion after the completion of the cook.

The cooking cycle begins with the mixing of rye meal and hot water in a batch cooker. The mash temperature is slowly raised to 82 °C, held for perhaps 15 minutes and then rapidly cooled to 68 °C by adding cold water.

Dry malt meal or cold malt slurry prepared with barley malt, and perhaps some rye malt for flavour considerations, is added to the cooked rye mash at 68 °C. The resulting temperature of the mash is now 65 °C, where it is held for perhaps 15 to 30 minutes to effect conversion and partial pasteurization of the malt.

The converted mash is pumped through a mash cooler to reduce the temperature to between 20 and 30 °C for fermentation.

18.4 DISTILLERY WHISKY FERMENTATIONS

Distillers' strains of *Saccharomyces cerevisiae* yeasts convert fermentable grain sugars to nearly equal parts of ethanol and carbon dioxide. Approximately 95% of the sugar is utilized in this reaction. Distillery yeasts have been selected on the basis of their adaptability to distillery environmental conditions and the types and quantities of congeners they produce.

The congeners present in flavouring distillates are of far more importance than ethanol yield. The yeast strain selected is largely but not entirely responsible for this. It is estimated that 5% of fermentable sugars are consumed in side reactions and the nature of these reactions is determined in part by yeast strain. The composition of the fermenting medium as well as the conditions of fermentation such as pH, temperature and bacterial flora are also important in influencing these side reactions. The products of these reactions influence the organoleptic character of the final whisky. Included in this group of secondary products are aldehydes, esters, fatty acids, fusel oils, sulfur compounds, phenols and many as yet unidentified compounds that contribute to the overall quality of the whisky.

The fermentation process begins when converted grain mashes or clear worts are inoculated with pure culture yeasts at a concentration of approximately 2×10^7 viable cells per ml of fermentation medium. Mashes containing glucamylase will ferment well using less than half the pitching rate generally recommended.

The temperature at which fermentations are set depends upon the degree of temperature control that is provided with the fermenter. Modern distilleries are equipped with fermenters capable of controlling the exothermic heat generated in the fermentation reaction, by having the capability of circulating mash through external fermenter coolers. Fermentations can be set at 33 °C and maintained there for the term of the fermentation with well-designed coolers.

Distilleries that do not have the capability to control fermentation temperature are required to set fermentations at 20 °C, permitting temperatures to rise naturally to 33 °C, at which point fermentation should be complete. Failure to commence fermentations at such a low temperature will result in uncontrollable heat generation prior to completion of fermentation that will ultimately kill the yeast and encourage bacterial growth. Poor yields and off-flavours are the result of such fermentations.

Generally, fermenters are set at lower temperatures, even though most fermenters are equipped with coolers. Lower temperatures are desirable during the initial few hours of fermentation when yeast numbers are increasing during the growth phase. Fermentations that are set at too high a temperature will encourage bacterial growth and retard yeast growth when yeast numbers should be increasing. Again, poor yields and undesirable flavours can be the outcome.

18.5 YEASTING

In Scotland, distillers' strains of yeasts are purchased in 'pressed' form for use in both malt and grain whisky fermentations. Brewers' yeasts are used in combination with the distillers' pressed yeast in the production of malt whisky. Brewers' yeasts impart a flavour to the distillate that is not achieved when distillers' yeasts are used alone.

In North America, yeast strains used in flavouring mash fermentations are generally produced in pure culture on the premises from slants maintained in the distillery laboratory.

Yeasts selected for neutral spirits fermentations are sometimes produced from pure culture slants on the premises. Today, however, more and more distillers are purchasing commercially produced strains of distillers' active dry yeast or distillers' compressed yeast for neutral spirits mash fermentations.

Yeast manufacturers often custom produce the distiller's own strain, to his specification, as an active dry yeast for use in flavouring or neutral spirits mashes. If, however, the distiller chooses to propagate pure culture yeasts in the distillery, several options are available to him, depending on his needs and/or wants.

The mash prepared in the plant for fermentation is ideally the base medium used for yeast propagation in the yeast department. This is certainly a convenient way to prepare plant yeast media and is entirely satisfactory, provided that the base medium is subjected to additional preparatory procedures prior to inoculation with laboratory scale pure culture strains.

In North America, malt-converted corn mash containing protein-rich backstillage is pumped to the yeast propagator where temperature is adjusted to 49 °C. Microbially produced enzymes (glucamylase) are added at the rate of 250 units per kg of grain in the propagator. The proteolytic and saccharifying enzymes in the glucamylase ferment are active at this temperature and release amino acids and sugars to the medium that will later support yeast growth. At this time, the distiller may elect to inoculate the yeast mash with a 'souring' organism (*Lactobacillus delbrucki*) which will ferment grain carbohydrate to lactic acid and reduce medium pH from 5.8–6.0 to 3.8–4.0 over a period of 8 to 12 hours. The purpose of souring is (a) to reduce mash viscosity and pH which facilitates the subsequent steam sterilization process, and (b) to reduce yeast mash pH to a level that will discourage the growth of certain low pH intolerant bacteria, indigenous to the distillery, that might otherwise contaminate the yeast during its growth and/or storage. The souring process is terminated by steam sterilization for two hours at 88–90 °C. After sterilization, the yeast mash is cooled to 26 °C and inoculated with the desired pure culture strain of yeast. When approximately half of the available sugar has been consumed (16–24 hours) the yeast is cooled for storage and is ready to be used for 'charging' or 'pitching' the distillery fermenters. Depending upon yeast cell concentrations achieved in the propagators (100–250 million cells ml^{-1}), pitching rates will range between 0.5% and 1.5% by volume. Such low pitching rates, common when corn-glucamylase yeast mash is used, are perhaps 20–50% of the pitching rates required when traditional yeast mashes are used that do not contain the stimulatory growth characteristics provided by glucamylase.

Some distillers elect not to use 'sour' yeast mash but rather use a 'sweet' yeast. Sweet yeasting does not include the *Lactobacillus* sp. souring stage in its manufacture. Consequently, to ensure proper sterilization of the more viscous and higher pH mash, a higher steam sterilization temperature over a longer period of time is suggested (*i.e.* 95 °C for 2.5 hours).

The corn-glucamylase yeast mash procedures that have been described are certainly the ideal way to go if the glucamylase ferment is available. They offer several advantages over traditional yeast mashes. Some of these advantages are as follows: (a) Lower pitching rates are required to achieve an equivalent fermentation performance. (b) With the corn-glucamylase medium, it is not necessary to shut down the distillery mashing process to prepare special yeast mash cooks containing the proper proportions of rye and malts. (c) It has been demonstrated that corn-glucamylase yeast preparations, when stored at reduced temperatures (15 °C), will maintain yeast viability for a much longer period of time than traditional yeast preparations. This is extremely convenient when it is necessary to store yeast over weekends or unscheduled shutdowns that might be required for many reasons, including maintenance, production rescheduling, *etc.*

The traditional method of propagating distillers' yeasts incorporates infusion batch booking of rye grain and varying percentages of barley malt and is still practised by some distillers today. Either 'sour' or 'sweet' yeasting is employed with this medium. Although this method is a proven one and certainly provides a satisfactory nutrient supply for yeast propagation, it does not have the same advantages as the corn-glucamylase system previously discussed.

18.6 DISTILLATION

Once the mash has been fermented, the 'beer' (or fermented mash) is distilled to produce the desired product. The type of beverage to be produced dictates the type of distillation or rectification required.

18.6.1 A Historical Look

Other than a few refinements, the methods and apparatus used by the original distillers of centuries ago are still in use today. The main refinement has been the advent of more and more efficiency, in terms of both quality and production. The two major parts of these original stills were a copper pot and a 'worm' condenser. Such an apparatus is known as a 'pot' still. Pot stills are still used exclusively in the distillation of Scotch malt whisky and most Irish malt whiskies, brandies, gins and some rums.

These pot stills are simple, but very uneconomical in terms of both time and materials. The product still contains considerable quantities of congeners (esters, aldehydes, higher alcohols, *etc.*). However, for certain products the presence of these congeners is to some extent desirable, as they impart characteristic tastes and odours. Figure 2 illustrates a fire-heated pot still. The neck was not as long (as that illustrated) in the earliest versions, but to prevent 'splash over' the longer neck was constructed.

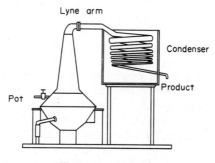

Figure 2 A pot still

Besides not being able to handle large volumes, pot stills provide little rectification. Rectification only occurred by repetitious pot distillation of product fractions until the desired result was achieved. Needless to say, this was very time consuming as well as costly.

With this in mind, various types of configurations were tried. These included discontinuous or Rayleigh rectifying stills and the Pistorius still. However, one of the first efficient continuous stills with plate columns (for rectification) was that designed by Aeneas Coffey in 1831. This type of still is still largely used to this day.

As can be seen in Figure 3, the distilling and rectifying employed two columns, one called the analyzer and the other the rectifier. The feed passes through a set of coils to preheat it before entering the analyzer still. The vapour from the analyzer still then enters the rectifier for further rectification, in which the majority of the water and congeners is removed. This configuration has formed the basis for most of the continuous stills in operation today.

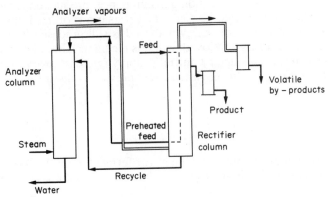

Figure 3 The Coffey still

18.6.2 Continuous Distillation

Continuous distillation is generally used for the processing of grain whisky. The end use of this spirit determines the extent of congener removal. For example, if the grain whisky is to be used with other whiskies, then congeners may be quite desirable; however, if vodka quality or 'neutral spirits' are required for cordial or gin production, then congeners are undesirable. The distillation process can be operated depending on the final product's destiny.

With this in mind, a closer look may be taken at the continuous distillation process. As was previously mentioned, upon completion of fermentation, the solids-bearing 'beer', ranging in alcohol concentration from 6 to 11% by volume, is sent to a beer column which essentially separates the solids while concentrating the alcohol anywhere from 65 to 80% by volume. The solids slurry flows from the base of the column. The alcohol–water mixture is either utilized for flavour in blending products (if used for whisky production) or sent for further distillation.

Various configurations can be utilized to concentrate the alcohol further depending on the end use. This is accomplished by controlling the amounts of congeners removed as well as the amount of energy used at each column. The types of configurations used may be direct fractionation and rectification or unique processes such as extractive distillation coupled with further rectification. Other concepts such as vacuum distillation are being actively investigated as possible means of more efficient operation.

Extractive distillation may be used when the differences in volatility between the species to be separated are so small as to necessitate a very large number of trays in a distillation operation. Such is the case for a majority of congeners associated with alcohol fermentations. In this case, the extractive agent is water as the ethyl alcohol is much more soluble in water than are these congeners. Because of the vast difference in volatility created by these phenomena, the alcohol–water mixture is separated and drawn off the column with relative ease. Rectification of this alcohol–water mixture follows. However, by adjustment of various heads and recycle streams, as well as adjusting extraction water rates and steam rates, the lightness or heaviness (degree of congeners) of the final product may be controlled. It is evident that the extractive distillation process has tremendous flexibility in terms of operational control.

18.6.3 Energy

As in all manufacturing, energy utilization is of major importance. Energy conservation steps which are commonly used are preheating the beer feed with beer column vapours, operational optimization, vapour feeding of columns, heat pumping and thermal (vapour or mechanical) recompression, column and pipe insulation, and waste heat scavenging.

18.7 BY-PRODUCT RECOVERY

This plays an important role, not only in terms of overall plant economics, but it may also be required for pollution control purposes.

One of these is the recovery of the de-alcoholized grain residues from the base of the beer column called stillage. This recovery involves a variety of dewatering operations, the first being centrifuging or screening. The liquid portion continues on to an evaporation step usually employing multi-stage or multi-effect evaporators. The solids content of this liquid portion is usually concentrated from approximately 4% to about 38–40% solids by weight. This evaporator 'syrup' is then combined with solid 'cake' from the centrifuges. The combined cake and syrup are then sent to dryers where the final distiller's dried grains with solubles, DDGS, are dried to a moisture content of approximately 12%. The final dried product is a valuable livestock feed as it is high in protein. Table 1 shows the approximate composition of this feed produced from corn.

Table 1 Composition of DDGS Produced from Corn

Component	Percentage
Moisture	12.0
Protein	27.0
Fat	8.0
Fibre	8.5
Ash	4.5

Other by-products, such as fusel oils (amyl alcohol, isoamyl alcohol and other higher alcohols), esters and aldehydes, are also produced during distillery fermentations. They are usually removed and stored during the extractive distillation process. These are used to some extent in the soap and perfume manufacturing industry and are sold whenever possible. If no markets exist, these by-products are usually burned in the power house boilers as an additional energy conservation step.

18.8 MATURING AND BLENDING

Maturation is the next step in the process of whisky making. This involves the storage of individual new whisky distillates or a combination of distillates in charred white oak casks.

The maturation of bourbon, by law, must take place in new charred white oak barrels. Others have the option of using new or used cooperage. The sources of used cooperage are predominantly once-used bourbon barrels, used American whisky barrels, used Canadian whisky barrels and sherry or brandy barrels.

As the whisky rests in barrels, housed in buildings known as maturing warehouses, a number of changes that influence the final outcome of the product, brought about by various interrelated factors, take place. Some of these factors are char thickness, barreling strength, storage time and atmospheric conditions — notably temperature and humidity.

The changes that occur in a maturing whisky are caused by a combination of physical and chemical reactions that occur simultaneously and are largely influenced by temperature. These include (a) the extraction from the wood, by the distillate, of wood components, (b) the oxidation of congeners in the original distillate and the oxidation of wood extracts, and (c) the reactions between congeners in the distillate or those extracted from the wood, to form new congeners, such as are produced in the slow organic chemical reactions of esterification and de-esterification. It is quite apparent that the maturation process is rather complex with many factors coming into play at a given time.

The rates of the reactions that take place during maturation are largely dependent upon temperature and the Q10 rule of chemistry which says that for every 10 °C rise in temperature, the rate of reaction will double. Since we are talking about slow organic reactions to begin with, the importance of temperature becomes very obvious in the maturation process. Since maturing warehouses are rarely heated, maturation reactions will naturally be more active in the warmer summer months than in the winter months. This phenomenon supports the significance of statements made by old time distillers who so often spoke of 'so many summers in the warehouse' when they discussed maturation times. The importance of whisky placement in the warehouse, and in particular the level in the warehouse, becomes obvious when the Q10 rule is appreciated.

During the maturation process, it has been observed that organic acids, esters and dissolved solids increase. Little is known about the interrelationships that occur between congeners of the volatile category. More is known of the mechanism of the extraction of tannins, colour and non-volatile acids.

The maturation time selected by the distiller or blender, for a particular whisky, will depend upon the time that is required to achieve the desired level of flavour and aroma in the whisky. The time required to accomplish the distiller's maturation objectives is in most cases well beyond the minimum legal maturation period.

It is quite remarkable that the flavour-producing components of whisky that we refer to as congeners amount to less than 1% of the total weight of the product. Evaporation losses that occur during maturation, and that ultimately concentrate the level of congeners in the barrel contents, are an important factor in the final organoleptic character of the product.

Apart from the impact that evaporation losses have on product quality, they are of major economic significance. The product losses attributed to evaporation account for a 2–3% loss of product per year of maturation, on the average.

Blending of matured whiskies is the final step in the creation of the desired end product. Lighter bodied grain whiskies are combined with heavier flavouring whiskies in formulations that are intended to produce a product acceptable to the consumer palate. The blending process will also even out any subtle flavour differences that can occur between various bonds of individual whiskies. Therefore, blending is also a quality control function which assures that a given whisky is always consistent with what the consumer is accustomed to purchasing.

Table 2 shows a comparison of the amount of various flavour-imparting components commonly found in final blended whiskies. These are the more prevalent species found. However, there are many others also present which are not shown.

Table 2   Flavour Imparting Components in Whiskies ($g\,l^{-1}$)

Spirit type	Acidity	Aldehyde	Ester	Tannin	Fusel oil
Light blended Canadian	0.0035	0.05	0.18	0.33	0.70
Heavy blended Canadian	0.78	0.09	0.35	0.65	1.30
Scotch blends	0.49	0.13	0.25	0.28	1.63
Bourbon	1.45	0.18	0.90	1.05	4.60

18.9 SUMMARY

A summary of the foregoing is illustrated in Table 3.

Table 3   A Summary of Whisky Production

	Bourbon	Blended US	Blended Scotch	Blended Canadian
Grain	(1) At least 51% corn, usually high malt and some rye. All one mash bill	(1) Rye mash bill (2) Bourbon mash bill (3) Corn and malt for grain whisky	(1) Malt for straight whisky (2) Corn and malt for grain whisky	(1) Rye and malt for straight whisky (2) Corn and malt for grain whisky
Fermentation yeast	Each type of product has its own yeast type to impart certain tastes and odours to the ferment			
Distillation				
High proof	Not done	For grain whisky or neutral spirits	(1) For grain whisky. Lower flavour	(1) For grain whisky. Lower flavour
Low proof	All done to retain flavour of above	Flavour whiskies Bourbon and rye	(2) Malt whisky — pot still	(2) Flavour — rye whisky (3) Some cases double distilled
Blending	Not blended	Minimum of 20% straight whiskies and rest can be grain neutral spirits or grain whisky	High percentage of malt whisky. Rest high proof grain whisky at age designated	All blended of different straight low proof whisky and high proof whiskies at age designated
Maturing	New charred oak barrels	Reused barrels for neutral spirits. Neutral spirits 1 to 6 months. Bourbon and rye new charred oak barrels	All in reused cooperage for varying times. All over 2 years to be called whisky	All reused cooperage for stated times
Final proof	Not under 40% alcohol for any — note that something like bourbon going from 50° to 90° proof would show a difference in the same whisky flavour			

18.10 BIBLIOGRAPHY

Daiches, D. (1976). *Scotch Whisky — Its Past and Present*. Collins, Glasgow.

Henley, E. J. and J. D. Seader (1981). *Equilibrium-Stage Separation Operations in Chemical Engineering*. Wiley, New York.

Murphy, B. (1979). *The World Book of Whisky*. Rand McNally, Chicago, Illinois.

Norman, W. S. (1961). *Absorption, Distillation and Cooling Towers*. Longman, London.

Packowski, G. W. (1963). Alcoholic beverages, distilled. In *Kirk–Othmer Encyclopedia of Chemical Technology*, 2nd edn., vol. 1. Wiley, New York.

Peppler, H. J. and D. Perlman (1979). *Microbial Technology — Fermentation Technology*, 2nd edn., vol. 2. Academic, New York.

Rose, A. H. (1977). *Economic Microbiology — Alcoholic Beverages*, 1st edn., vol. 1. Academic, New York.

Van Lanen, J. M., W. F. Maisch and M. B. Smith (1972). *An Improved Method For Propagating Distillers' Grain Yeast*. Presented before the Division of Microbial Chemistry and Technology, 164th Annual Meeting, American Chemical Society.

Young, S. (1922). *Distillation Principles and Processes*. Macmillan, London.

19

Traditional Fermented Soybean Foods

T. YOKOTSUKA

Kikkoman Corporation, Chiba, Japan

19.1   INTRODUCTION

19.1.1   Traditional Fermented Foods

Traditional fermented foodstuffs may be roughly divided into six categories as follows: (1) alcoholic beverages fermented with yeasts; (2) vinegars fermented with acetobacters; (3) fermented milk products treated with lactobacilli; (4) pickles fermented with lactobacilli in the presence of salt; (5) fermented fish or meat treated with enzymes together with lactobacilli in the presence of salt; and (6) fermented plant proteins treated with molds, with or without lactobacilli, and yeasts in the presence of salt.

From ancient times, nations all over the world have developed traditional processes for the production of alcoholic beverages. These beverages are prepared principally by converting the sugars present in the raw materials into alcohol by the action of yeasts. Vinegars have been made from almost all of these alcoholic beverages by converting the alcohol into acetic acid by the action of acetobacter. Thus, fruit wines have been made from sweet fruits such as grapes, apples and oranges, and the corresponding vinegars made from these wines.

To prepare wines from starchy raw materials such as wheat, barley, corn or potatoes, these raw materials must first be degraded into sugars, mainly glucose, to allow fermentation by yeasts; there are significant differences between the saccharification process of Western countries and that of the Orient. The amylolytic enzymes used for the saccharification in Western countries have been derived from sprouted barley or malt, but in the Orient, *Aspergillus* or *Rhizopus* molds have been utilized as the sources of amylolytic enzymes. Accordingly, in the preparation of beer, whisky, gin and vodka, the starch in the corresponding raw materials is first saccharified using malt, while in the preparation of alcoholic beverages from rice, wheat or sweet potatoes in the Orient, *Aspergillus* or *Rhizopus* molds are cultured on part of these raw materials to produce amylolytic enzymes. These cultured materials are called 'chu' in China and 'koji' in Japan; the koji is then mixed with the remainder of the raw material and water to make a mash. The mash is concurrently subjected to enzymatic saccharification, lactic fermentation and yeast fermentation.

Peoples throughout the world have developed different fermented foods and beverages in which sugars are principally converted into lactic acid by the action of lactobacilli. The milk of cows, goats or horses is sometimes fermented with lactobacilli or a mixture of lactobacilli and yeasts. Examples are yogurt in the Balkans, acidophilus milk in the USA, lactic beverages prepared from defatted cows' milk and sugars in Japan, kefir in the Caucasus, koumis in Central Asia, and many other similar products in Central Asia, the Middle East and Africa, such as idli, kishk, ogi and mahewu. The lactic fermentation in these cases protects the products from undesirable acetic fermentation or putrefaction. The lactic fermentation is preferable to the acetic fermentation in the preparation of pickles, and it is also important in the manufacture of alcoholic beverages in the Orient because it prevents undesirable acetic acid fermentation from occurring in the initial stage of manufacture before starting the yeast fermentation.

Procedures for enzymatically hydrolyzing certain protein foods into amino acids and lower peptides to make them more attractive and nourishing have been known for a long time. In the West, the flavor of cheese has been enriched by fermenting it with some *Penicillium* molds; in the Orient, the flavors of fish, poultry, meat, pulses and some cereals have been improved by fermenting them with proteolytic and amylolytic enzymes produced by *Aspergillus*, *Rhizopus* or *Mucor* molds or by *Bacillus*, sometimes accompanied by fermentation with lactobacilli and yeasts in the presence of high salt concentrations. These foods, formerly called chiang and shi in China, hishio or sho, and kuki in Japan, can still be found in every Asian country including Japan. They are believed to be the forerunners of foods such as shoyu and miso which are now in use; their records date back 3000 years in China and they are believed to have been introduced from China to Japan 1300 years ago or more.

19.1.2   Types of Proteinaceous Fermented Foods Produced in the Orient

The history of fermented proteinaceous foods in China, Japan and other East Asian countries may be summarized by dividing them into several types as follows. (References: *35 Years in the History of the Noda-shoyu Company*, 1955; *The History of Kikkoman*, 1977; Kinichiro Sakaguchi, 1979; Wang and Fuang, 1981; Bo Thi-An, 1982).

Type I: fish, poultry or meat was enzymatically degraded by the enzymes present in the animal's organs or in fermented grains such as rice, millet or wheat, in the presence of salt. Alcoholic beverages were sometimes added to the mixture to help prevent putrefaction of the raw materials. Typical modern examples are the fish sauces commonly produced in South-east Asia, for example nam-pla in Thailand, nuoc-mam in Vietnam, sambalikan in Singapore and Malaysia, patis in the Philippines, nam-pa in Lao, ketjapikan in Kalimantan, terosi in Indonesia, tuk-trey in Cambodia, and ngam-pya-ye in Burma. Shottsuru in Japan, garos in Greece, Pissala in Southern France and fessikh in the Sudan also belong to the same category. In the preparation of these fish products, the fish is kept in water, almost saturated with salt, for more than a year and then the liquid is drained off. In the case of shottsuru manufacture in Japan, fermented rice or rice koji is added to the mixture of fish and salt.

This type of fish product appeared for the first time in China under the name of chiang 3000 years ago. Fish which had been gutted, washed and dried was mixed in a pot with mold-cultured millet, salt and alcoholic beverages and sealed for 100 days. The salted fermented fish (or poultry or meat) was called shishi-bishio in Japan, a name which appeared in *Manyoshu*, an old book of songs (759–350 BC).

Type II: cooked soybeans, or other proteinaceous beans or pulses, were mixed with mold-covered grain such as rice-koji, salt and water, and stored. Alcoholic beverages were sometimes added to the starting mixture. One typical present day example is miso in Japan. Cooked soybeans, rice- or barley-koji, which is cooked rice or barley cultured with *Aspergillus* molds, and salt, with or without a small amount of water, are mixed and stored for fermentation. The earliest record of this type of foodstuff, under the name soybean-chiang, has been traced to China in a book named *Chi-Min-Yao-Shu* (532–549), that is 1000 years later than fish-chiang. The same product was first introduced into Japan some 1000 years ago directly from China and *via* Korea and was called mame-bishio, kama-bishio or miso.

Type III: cooked soybeans with a small amount of wheat or barley flour in a ratio of 3:2 or 5:1 were cultured with *Aspergillus* molds and the cultured materials were stored with salt and water to achieve enzymatic degradation of the raw materials. One typical modern example is soybean sauce, or soysauce, which is generally produced in Korea (under the name kanjang), Southern China (tou-yu, chiang-yu), Singapore, and Malaysia and Indonesia (kecap). It is also produced in Taiwan (in-yu), Thailand, the Philippines, and Japan in small amounts. The first reference to this type of process is found in *Ben-Chao-Gong-Mu*, written in 1590 in China, in which boiled soybeans are mixed with barley flour in a ratio of 3:2, and pressed into cakes which are left indoors until they are covered with yellow mold. The moldy cakes (chu or koji) are mixed with salt and water and aged in the sun. Nowadays, the same mixture of raw materials in granular form is cultured with yellow mold, and the salty mash is fermented. Fermentation is effected in the sun in South-east Asian countries, while in Japan and Korea the same mixture of raw materials is pressed into small balls or cakes, cultured with yellow mold, and the salty mash is fermented in porcelain jars in the sun or in wooden kegs indoors.

Type IV: Cooked wheat kernels, optionally along with a small amount of cooked or roasted soybeans, are cultured with yellow *Aspergillus* molds to make koji, which is mixed with cooked wheat kernels and salt water to make a mash and fermented. One typical example of this process is shiro-shoyu production in Japan and formation of a similar product in China. The first reference to this type of food is found in *Chi-Min-Yao-Shu* written in the sixth century AD in China.

Type V: A mixture of cooked soybeans or defatted soybean grits and roasted, crushed wheat kernels or dehulled barley in an approximate ratio of 1:1 or 2:3 is cultured with yellow *Aspergillus* molds to make koji. The koji is mixed with a volume of salt water 110–130% that of the raw materials to make a mash which is fermented with lactobacilli and yeasts. The final product is pasteurized at relatively high temperature to give a strong flavor and a deep reddish-brown color. A typical example of this type of product is koikuchi shoyu in Japan; this differs from the tao-yu or soybean sauce produced by a process of type III in several ways, namely the mixing ratio of the soybeans and wheat, the form of the wheat kernels (granules or flour), the koji molds culturing method (in granular form or on pressed cakes or small balls of raw materials), the degree of lactic and alcoholic fermentations of the salty mash, and the refining process, including the pasteurization of the final product.

Type VI: cooked soybeans are cultured with *Aspergillus*, *Rhizopus*, *Mucor* or *Bacillus*. The origin of this type of food seems to be the shi or tou-shi which appeared for the first time in Chinese history in *Shih-chi* written in the second century BC. It was the most popular seasoning, next to salt, sold at that time. The detailed method of preparing shi is found in the sixth century book, *Chi-Min-Yao-Shu*. Yellow or black soybeans were cooked, cooled, piled on a straw mat, and covered with straw. When the beans were covered with yellow mold, they were transferred to a pit and tightly covered with straw; after 10–15 days shi was prepared and sun dried. Salt was sometimes added to the mixture during preparation. Shi was extracted with boiling water to make shi-tche and used as seasoning. Shi or tou-shi was introduced into Japan around the time when soybean-chiang or mame-bishio was introduced, that is some 1000 years ago. It was called kuki in Japanese, written in the same Chinese character as that for shi.

A similar type of soybean food cultured with *Mucor racemosus* is found in Szechwan, China, and another bean product fermented with *Mucor* is known as la-pa-tou in China. The fermented mass of this product is sliced, sun dried and used as a flavoring or consumed fresh. Tempeh is another example of this type of fermented soybean product in Asia which probably has the same historical origin in China and is very popular in Indonesia today. In preparing tempeh, cooked soybeans are wrapped in banana leaves and cultured with *Rhizopus oligosporus* which is derived from the leaves. The mold-cultured materials are consumed fresh or after deep frying as a staple food. Shui-tou-shi in China is another soybean product fermented with *Bacillus*. Soaked and cooked soybeans are placed in a cloth bag and covered with straw and fermented for 1–2 days at 25–30 °C. The sticky beans cultured with *Bacillus* derived from the straw are mixed with minced ginger and salt and then tightly packed in jars and aged for one week before consumption. Natto is the same fermented soybean product and is very popular in Japan. Similar soybean foods cultured with *Bacillus* are also found in Thailand (tou-noa), and in North-east India (kinema; Saito *et al.*, 1983).

Generally speaking, yellow *Aspergillus* molds, such as *A. oryzae* or *A. sojae*, were the major molds involved in the fermentations, but *Rhizopus*, *Mucor* or *Bacillus* were also used. Some lactic fermentation is presumed to have occurred in the above-mentioned salt-containing mash, but almost no alcoholic fermentation by yeasts is recognized in these protein food fermentations apart from in production shoyu in Japan. The fermented salty mash was consumed as a seasoning and in many cases the clear liquid separated from the mash has been consumed. There were two ways of culturing molds on beans and grains: (a) beans or grains in granular form were cultured directly with molds, or (b) soybean or wheat flour, or a mixture of these flours, was made into cakes or small balls and cultured with molds. Soybeans were usually boiled or steamed, while crushed wheat or barley grains or flour were usually either roasted or used raw. The meat, poultry or gutted fish was usually dried in the sun before use, but whole fish were used raw.

Almost all these types of fermented proteinaceous foods are found nowadays in Asian countries along with many types of fermented alcoholic beverages. Two more modern types of fermented plant protein food can be added to the six types of traditional protein food.

Type VII: plant proteins such as defatted soybeans or soybean concentrate or isolate are hydrolyzed into amino acids and lower peptides by boiling with concentrated hydrochloric acid at more than 100 °C, followed by neutralization with alkali. This chemically-prepared protein hydrolyzate cannot be called 'fermented', but it is used as a substitute for or as an additive to fermented products. In more recent years, plant proteins have been hydrolyzed with dilute hydrochloric acid at less than 100 °C, neutralized with alkali and then mixed with shoyu-koji, which is a mixture of soybeans and wheat cultured with *Aspergillus* molds, and fermented with lactobacilli and yeasts. This semi-chemical or semi-fermentative method of soysauce production was invented in Japan some 40 years ago and is still very popular in some Asian countries, although it is not so popular in Japan as it was 20 years ago.

Type VIII: shoyu koji made of defatted soybeans and wheat, or more commonly wheat bran, is mixed with a small amount of diluted salt solution to make a hard, low salt mash containing 70–100% moisture and 6–8% salt. This mash is kept at 40–50 °C for 2–3 weeks, after which the aged mash is drained off, extracted with hot salt water and washed with hot water. This method is said to have been invented in China in 1955, and seems to be the standard method of soysauce manufacture in Peking, Shanghai and perhaps many other cities in China today.

19.1.3 Major Traditional Fermented Foods Produced in Japan

Nowadays the typical fermented foods produced in Japan are fermented soysauce or shoyu, fermented soybean paste or miso, rice wine or sake, distilled sake mash or shochu, rice vinegar, amasake, natto and pickles. With the exception of natto and pickles, there is a common method for the production of these fermented foods. The first step is to cultivate molds belonging to the *Aspergillus* group, such as *A. oryzae*, *A. sojae* or *A. niger*, on part or all of the raw material to produce plant tissue degrading enzymes such as amylases, proteases, lipases, cellulases and pectinases. The mold-cultured material is called koji in Japanese, and the koji is used to degrade raw materials mixed with water with or without salt.

According to old records, such molds as *Rhizopus* or *Mucor* were also used in China and are still used in other Oriental countries, for example Taiwan and Indonesia, but only *Aspergillus* molds have been used in Japan.

In Japan the annual production of shoyu is about 1.2 million kl. This means that the average yearly consumption per capita is about 11 l. Shoyu is the most popular liquid condiment used in Japanese cuisine as well as in that of other Oriental contries.

Miso or fermented soybean paste is used for cooking, mainly for making miso soup by diluting it with water and adding vegetables, chicken, fish or meat. The per capita consumption of miso is about one half of that of shoyu by weight. The production of miso in 1982 in Japan was estimated to be 578 000 tons, excluding the 100 000 tons which was estimated to be produced in the home.

Sake or rice wine is prepared by digesting a mixture of cooked rice and moldy rice or rice-koji in water, concurrently fermenting the mash with lactobacilli and yeasts. The annual Japanese consumption of sake is 1.45 million kl. Shochu is a type of wine containing 20 to 35% alcohol distilled from the mash, in which rice or dried sweet potato is saccharified with the enzymes from yellow or black *Aspergillus* molds and fermented with yeasts. Its annual production is about 500 000 kl (1984). Beer, whisky and grape wine are also popular in Japan. The consumption of shochu has been increasing in recent years, in contrast to consumption of sake and whisky. Notwithstanding, more than 50% of the Japanese intake of pure alcohol from alcoholic beverages in general is calculated to be from sake and shochu.

Mirin is a very sweet liquid containing about 35% glucose and 12% alcohol, which is mainly used for cooking. It is prepared by digesting a mixture of rice and rice-koji in water in the presence of 35% alcohol. The production of mirin in 1981 was about 67 000 kl. A fairly large amount of sake is also utilized for cooking, but both sake and mirin are subject to monopoly in Japan. The production of mirin-type and sake-type alcoholic seasonings containing 5% salt, which are not subject to monopoly, has been rapidly increasing in recent years and had reached 58 000 kl in 1981.

The annual production of vinegar in Japan is approximately 250 000 kl, of which about 10% is prepared from rice wine or sake by acetic acid fermentation. Amasake is a beverage prepared by saccharifying a mixture of cooked rice and rice-koji and is served after dilution with hot water. Per capita annual consumption of these fermented foods prepared from *Aspergillus* molds in Japan is summarized in Table 1.

Table 1 Fermented Foods and Beverages in Japan (1982)

Product	Total production/year (kl)	Per capita consumption/year (l)[c]
Shoyu[a]	1 190 500	10.0
Miso[a]	580 000[d]	4.89[e]
Sake[b]	1 544 560	13.0
Shochu[b]	325 970	2.75
Mirin[b]	66 675	0.56
Vinegar[c]	321 000	0.27

[a] Estimated by Bureau of Tax, Japan. [b] Estimated by Daily Economic Report Co. Ltd. [c] Japanese population (Oct. 1982): 118 690 000. [d] In tons. [e] In kg.

19.2 FERMENTED SOYBEAN FOODS PRODUCED IN THE ORIENT

Typical examples of fermented soybean foods now consumed in the Orient are shoyu, miso, tempeh, natto and sufu. Shoyu is the Japanese name for soysauce in general, which is a salty fermented seasoning produced from soybeans with or without wheat, barley and/or rice. Although most varieties are made from vegetable materials, fish soy is popular in South-east Asian countries and is even produced in Japan in small amounts. Chemical analyses of soysauce produced in the Orient and other amino acid containing seasonings produced world-wide are presented in Table 2. There are distinct differences in alcohol content, which indicate the amount of wheat used as raw material and the degree of alcoholic fermentation of the mash, and in the content of volatile flavor components.

Table 2 Analysis of Amino Acid Seasoning from Various Countries[a]

	Be^b	NaCl (g/100 ml)	TN (g/100 ml)[c]	RS (IS) (g/100 ml)[d]	Alc (ml/100 ml)[e]	Color intensity
Koikuchi shoyu (Japan)	23.6	17.0	1.70	5.07	2.50	++
Usukuchi shoyu (Japan)	22.2	18.0	1.18	4.00	2.00	+
Soysauce (Taiwan)	25.6	15.6	2.05	5.95	0.86	++
Soysauce (Korea)	21.9	17.3	1.50	2.10	0.39	++
Soysauce (Hong Kong)	28.5	26.2	1.54	4.22	0.00	+++
Soysauce (Philippines)	23.3	24.7	0.76	1.06	0.01	++
Soysauce (Singapore)	30.1	24.1	1.97	4.81	0.00	+++
Soysauce (Malaysia)	23.9	18.3	1.17	8.50	0.03	+++
Soysauce (Indonesia; kecap asin)	—	7.2	0.19	14.45	0.02	++
Soysauce (Indonesia; kecap manis)	—	5.9	0.19	11.1 (58.1)	0.09	+++
Soysauce (USA)	22.8	16.5	1.65	3.70	2.07	++
Chemical soysauce (USA)	23.8	19.7	1.51	0.82	0.01	++
H.V.P. (Europe)	30.6	21.4	4.75	0.00	0.06	+
Fish sauce (Thailand)	26.8	27.6	2.25	4.81	0.02	+

[a] Analyzed by the Kikkoman Corporation. [b] Be: Baume. [c] TN: total nitrogen. [d] RS (IS): reducing sugar (invert sugar). [e] Alc: alcohol.

Miso is fermented soybean paste and is popular in Japan, China, Korea and other countries. Cooked soybeans are mixed with mold-cultured rice or barley and salt and stored for fermentation. Miso is used for preparing miso soup and for seasoning in general.

Tempeh is very popular in Indonesia where it is a staple food. Cooked whole soybeans are cultured with *Rhizopus oligosporus* wrapped in banana leaves (*Pisau batu* is the best species) for about 40 hours at about 30 °C. The annual production of tempeh in Indonesia is about 80 000 tonnes and 14% of the soybeans produced is used for tempeh production. In Java the per capita daily consumption of fresh tempeh is in the order of 20 to 120 g, and the price of tempeh is the same as that of uncooked rice (Lee, 1981).

Natto is popular in Japan as a side dish to cooked rice and is served with shoyu, mustard and chopped Welsh onions. It is prepared by culturing autoclaved whole soybeans with *Bacillus natto*, identified as *B. subtilis*, at over 40 °C for 12–40 hours. It contains 0.1–0.8% viscous material composed of a mixture of glutamic polypeptide and a polymer of fructose which are produced by *B. natto*. Its annual production in Japan is about 150 000 tons.

Thuo-nao is the most common fermented soybean product in Northern Thailand and is similar to natto. The product is generally available as a dried paste. The microorganism used for production is *B. subtilis*. On average, the product is alkaline (pH 8.4), and contains approximately 62% moisture and about 5.2×10^9 organisms per gram. Because of the high protein (36.8%) and high fat (14.8%) contents of dried thuo-nao, the product is used to formulate high protein low cost foods. In preparing thuo-nao, whole soybeans are boiled for 3–4 hours and fermented at room temperature for 3–4 days. The raw thuo-nao is ground with or without flavoring agents, and then (1) steamed or roasted and packaged as cooked thuo-nao, or (2) pressed into thin chips and sun dried to make thuo-nao chips (Bhumiratana, 1980).

Sufu or soybean cheese is a fermented soybean curd, which has a very long history in China where it is known as tao-hu-yi, tohu-zu or tau-zu. Similar products are also found in Vietnam, the Philippines and the East Indies. There are wide variations in the production of sufu; the less common is yellowish in color and the more common is a reddish sufu produced in Thailand. Soybeans are soaked in water overnight, ground and centrifuged. Calcium sulfate is added to the supernatant which is then boiled and filtered, and then subjected to coagulation. The coagulant is pressed into a cake to make tofu. The tofu is cut into cubes 3 cm thick and inoculated with a mold culture, usually a *Mucor*, such as *Actinomucor elegans*, or a *Rhizopus*, and incubated at 12 to 30 °C for 2–7 days. The resultant cubes, known as pehtzes, are placed in a solution of 25–30% sodium chloride and about 10% ethanol and then aged for several months. One kg of soybeans is used to prepare 3 kg tofu, which will yield 1.9 kg sufu. The red color of the reddish type comes from *Monuscus purpureus* (Hesseltine and Wang, 1972; Lin, 1977).

19.2.1 Japanese Shoyu

Shoyu produced in Japan is classified in three different ways into three production methods, five types and three grades. The Japan Agricultural Standard (JAS) recognizes three production methods as follows: (1) honjozo or genuinely fermented; (2) shinshiki-jozo or semi-chemical in which fermented shoyu mash or shoyu is mixed with a chemical or enzymatic (*i.e.* using proteases) hydrolyzate of plant protein, in amounts of 50 and 30% respectively, on nitrogen basis, and fermented and aged for more than one month; and (3) amino acid solution mixed in which (1) or (2) is mixed with a chemical or enzymatic hydrolyzate of plant protein in amounts of less than 50 or 30%, respectively, on nitrogen basis. According to JAS, in the production of genuine fermented shoyu heat-treated raw materials, soybeans and wheat should be cultured with koji molds (*Aspergillus oryzae* or *A. Sojae*) to make koji, and the koji mixed with salt water to make a mash or moromi. Moromi is fermented with lactobacilli and yeasts and then aged. The total amount of shoyu checked by JAS in 1981 was about 1.076 million kl, of which 72.1% was genuine fermented, 24.2% was shinshiki-jozo, and 3.7% was amino acid solution mixed. The amounts of raw materials used for shoyu production in Japan in 1980 are indicated in Table 3.

Table 3 Raw Materials used for Shoyu Production in Japan (1980)

Raw material	Quantity (ton)
Wheat	181 637
Defatted soybeans	179 364
Soybeans	6 657
Salt	208 113
Chemical hydrolyzate of plant protein	87 206[a]

[a] In kl.

The JAS recognizes five types of shoyu, but fish sauce is not included in the JAS. The five types of shoyu and the amounts produced are shown in Table 4 and their chemical compositions are in Table 5.

Table 4 Production Quantities of Different Kinds of Shoyu in Japan (1982)[a]

Type of shoyu	Quantity (kl)[b]	Percentage of total
Koikuchi	902 862	84.3
Usukuchi	138 261	12.9
Tamari	20 885	2.0
Saishikomi	3 130	0.3
Shiro	5 042	0.5
Total	1 070 180	100.0
Total production	1 187 148	
Total sales	1 184 306	

[a] Figures for various types from Japan Shoyu Inspection Association. Figures for total production and sales from Bureau of Foods, Japan.
[b] Checked by Japanese Agricultural Standard.

Eighty-five percent of all shoyu consumed in Japan is of the koikuchi type; it is dark in color, and made from approximately equal parts of soybeans and wheat. The koikuchi mash is subjected to vigorous lactic and alcoholic fermentations and the finished product is pasteurized at a temperature of about 80 °C to give it a characteristic reddish-brown color and strong flavor.

Table 5 Typical Compositions of Different Types of Genuine Fermented Shoyu in Japan (May, 1982)[a,b]

Shoyu type	Baume	NaCl	TN	FN	RS	Alc	pH	Color[c]
Koikuchi	23.3	16.2	*1.94*	1.15	5.80	2.48	4.86	7
	22.7	16.9	1.68	1.00	4.75	2.37	4.74	10
	21.9	17.1	1.57	0.93	3.52	2.09	4.74	11
Koikuchi[d]	18.7	*13.2*	1.54	0.84	2.72	*3.20*	4.74	11
Koikuchi[e]	16.6	*8.91*	1.56	0.94	4.10	*3.24*	4.84	13
Usukuchi	22.5	18.0	*1.50*	0.91	4.98	2.79	4.79	28
	22.4	19.3	1.19	0.73	5.02	2.63	4.84	28
Tamari	23.2	17.6	*1.85*	1.03	3.45	2.87	4.90	3
Shiro	25.0	18.0	*0.49*	0.23	*17.9*	0.96	4.62	46
Saishikomi	29.6	13.1	*2.01*	0.94	*14.3*	*1.07*	4.78	2

[a] Abbreviations and units: TN, total nitrogen; FN, formol nitrogen; RS, reducing sugars (all % weight/volume); Alc, alcohol (% volume/volume). [b] Too high values of alcohol (italic) may be fortified as preservative. Cited from *J. Jpn. Soy Sauce Res. Inst.*, **8**, 195 (1982). [c] Shoyu color standard: the smaller the number, the darker the color. [d] Salt reduced by 20%. [e] Salt reduced by 50%.

These are the features that differentiate the Japanese koikuchi shoyu from other similar products in the Orient.

Usukuchi shoyu is made from a mixture containing less wheat and more soybeans than the koikuchi type. The saccharified rice-koji with water, which is called 'amasake', is sometimes added to usukuchi mash to improve the salty taste. The nitrogen content of the finished product does not exceed 1.2%. Usukuchi shoyu is used mainly for cooking when one wishes to preserve the original color and flavor of the foodstuff.

Tamari shoyu is made mostly from soybeans with only a small amount of wheat (ratio 10:1–2). Its nitrogen content is sometimes more than 2% and there is only a trace of alcohol.

Shiro shoyu is very light in color and is made from wheat with a small amount of soybeans with the ratio 10:1–2. Saishikomi shoyu is made by enzymatically degrading mold-cultured soybeans and wheat in shoyu instead of the usual salt water.

Good quality genuine fermented koikuchi shoyu is 1.5–1.8% (g/volume) total nitrogen, 3–5% reducing sugar (mainly glucose), 2–2.5% ethanol, 1–1.5% polyalcohol (primarily glycerol), 1–2% organic acid (predominantly lactic acid, pH 4.7–4.8), and 17–18% sodium chloride. In order to obtain a palatable taste, about one half of the nitrogenous compounds present must be free amino acids, and more than 10% must be free glutamic acid.

The JAS establishes three grades for each variety of shoyu: special, upper and standard. The grade is determined by organoleptic evaluation, total nitrogen content, soluble acids without sodium chloride content and alcohol content. Only high quality shoyu made by fermentation can qualify for the special grade. About 60% of Japanese shoyu was special grade in 1979. The JAS for the special grade of koikuchi shoyu is more than 1.5% total nitrogen, more than 16% extract and more than 0.8% alcohol.

Blending fermented shoyu with a chemical hydrolyzate of plant protein in an amount less than 50% on a nitrogen basis is permitted for upper and standard grades as long as the characteristic flavor of the fermented shoyu is not spoiled. The yearly consumption of shoyu per capita is about 10 l, 4.4 l of which is consumed at home and the remaining 5.6 l institutionally and industrially. The shoyu manufacturers in Japan are assumed to be less than 3200 in number. The five largest manufacturers produce 50% and the next 50 account for 25% of the total production. The largest producers are Kikkoman, Yamasa, Higashimaru, Higeta and Marukin, all of which produce genuine fermented koikuchi and usukuchi shoyu of the JAS special grade. Kikkoman is the largest, with an annual production of 360 000 to 400 000 kl in recent years.

19.2.2 Soysauce Produced in Oriental Countries Other than Japan

19.2.2.1 Korea

The amount of soysauce produced in Korea in 1971 was 223 000 kl, which included 107 000 kl of industrially produced soysauce and 116 000 kl produced domestically (Lee, 1981). The fermented soysauce produced industrially in Korea is of the Japanese koikuchi type. The home-made soysauce is prepared by a traditional method. Molds are grown on the surface and reside inside balls of cooked, dried soybeans, which are called meju. Dried meju is packed in a rice-straw bag to mature for a few months. The matured mejus are put into brine and kept in large china jars in the sunshine for several months to ferment. The ripened meju–brine mixture is separated into supernatant liquid and sediment. The liquid is soysauce and the residue is soybean paste. The residue is sometimes mixed with red pepper. In recent years, improved meju has been prepared industrially; cooked soybeans are inoculated with *A. oryzae* mold in a controlled fermentation room. The improved meju is then placed in brine to make soysauce and paste. Per capita daily consumptions of soybean sauce (kanjang), soybean paste (doenjang) and red pepper soybean sauce (gochoojang) were 23.3, 10.4 and 13.9 g, respectively, in 1971. The amount of soybean paste produced was 224 000 tonnes (including 56 000 tonnes of industrially-produced and 168 000 tonnes of home-made) and that of red pepper soybean paste was 112 000 tonnes (23 000 tonnes of industrially-produced and 89 000 tonnes of domestically-produced) in 1971 (Lee, 1981).

19.2.2.2 *Taiwan*

The production of soysauce in Taiwan in 1978 was estimated to be 160 000 kl, which is equivalent to 9 l per capita per year. About 80% is tou-yu, which is made from soybeans and wheat and originated in China. There were 433 soysauce plants in Taiwan in 1978; the eight largest producers supplied at least 45% of the local market. The soysauce products include genuine fermented soysauce (25%), chemically hydrolyzed soysauce (5%) and blended soysauce (75%; Su, 1980). Around 5 to 10% of Taiwan soysauce is estimated to be in-yu, which is made only from black soybeans and closely resembles Japanese tamari made from yellow soybeans. The black soybean-koji is washed with water before it is mixed with salt water to make a mash to remove the bitter taste of the product. There are three national standard grades of soysauce in Taiwan, and their total percentages of nitrogen were 1.4, 1.2 and 1.0 g/100 ml, in 1980.

19.2.2.3 *South-east Asian countries*

Fermented soybean sauce similar to in-yu and tamari is also produced in the southern part of China and it seems to be the prototype of the soysauce prepared using only soybeans. Tamari mash is usually fermented in wooden kegs in Japan, but the soybean sauce produced in Taiwan, Thailand, Singapore, Malaysia, Indonesia and Southern China is fermented in 150 l china jars.

There are about 10 major soysauce manufacturers in Singapore, each employing more than 50 workers. All are operated by the Chinese. The biggest is Yeo Hiap Seng Ltd. in which more than 10 000 kl of soybean sauce is prepared per year from soybeans and roasted wheat flour, used in a ratio of 5:1. The salty mash is fermented in 10 kl FRP tanks for 4 months in the sun. The total amount consumed by Singapore's population of 2 400 000 is estimated to be 14 000 kl, which is equivalent to about 6 l per capita per year and is made from 3158 tonnes of soybeans (Lian, 1981).

In Singapore, Malaysia and Indonesia, two types of soysauce are manufactured: a light soysauce (in Cantonese sung-show or in Hokkien chiuw-cheng), and a dark soysauce mixed with cane molasses (in Cantonese low-chow or in Hokkien tau-iu). In Indonesia soysauce is called ketjap or kecap; two types of kecap are popular: kecap-asin, which has a salty taste, and kecap-manis, which is very sweet and is blended with a large amount of cane or palm sugar. The former is popular in Sumatra and the latter in Java. The per capita daily consumption of kecap in Indonesia is 10–15 ml (Lee, 1981). The kecap mash is washed twice with salt water and the mixed extract is concentrated in the sun. The residue of the extraction is used for tauco or miso of an inferior grade, but genuine tauco is made from soybeans and wheat flour cultured with *A. oryzae*. The mold-cultured material or koji is dried in the sun, mixed with salt water, and fermented and concentrated in the sun for 3 to 7 weeks. Tauco is a slurry containing 10% protein; it is sold in glass bottles. It is different from Japanese miso, which is a paste containing 25% protein. In these countries the chemical acid hydrolyzate of soybeans and the semi-chemical or shinshiki soysauce are also popular, but the exact amounts produced are not known.

The major amino acid seasoning consumed in Thailand and the Philippines is fish sauce such as nam-pla and patis, and the amount of soybean sauce produced is small.

19.2.2.4 Peking and Shanghai

The soysauce consumed in Peking and Shanghai is different from those described above and is prepared as follows. The koji is prepared by large-scale cultivation of *A. oryzae* with a mixture of steamed soybeans and wheat or wheat bran (6:4). The koji is mixed with salt water to make a hard mash with a moisture content of about 80% and a salt concentration of about 6 to 8%. This hard, low-salt mash is kept at 45 to 50 °C for about 3 weeks to effect enzymatic digestion. The digested mash is extracted with hot salt water and then with hot water; the salt-free residue may be used for animal and poultry feed. There is no alcoholic fermentation or pressing of the mash as there is in the case of Japanese shoyu manufacture. The yield of soysauce on a nitrogen basis in 1979 was 75 to 80%, because the defatted soybeans used as raw material were cooked by the NK method. The highest government standard of soysauce is as follows: total nitrogen 1.6%, reducing sugar 4%, sodium chloride 19% or more. The chemical acid hydrolysis of plant protein for soysauce manufacture is illegal in China.

19.2.3 Miso

Miso is the Japanese name for a semi-solid salty food made by fermenting soybeans and rice, barley or rye, and salt. Miso is classified into three types on the basis of the raw materials used: kome-miso or rice-miso made from rice-koji, cooked soybeans and salt; mugi-miso made from barley- or rye-koji, cooked soybeans and salt; and soybean-miso made from soybean-koji and salt. These three types of miso are further classified into sweet, medium and salty, depending on their salt content, and white, light yellow and red, depending on their color. About 80% of the miso consumed in Japan is rice-miso in such varieties as white, edo, shinshu and sendai. The amounts of mugi- and soybean-miso consumed are equivalent to 12% and 8%, respectively, of the total. The constituents of some typical miso are indicated in Table 6.

Table 6 Constituents of Some Types of Miso[a]

Miso[b]	H₂O (%)	pH	NaCl (%)	Protein (%)	Fat (%)	RS (%)[c]
White miso	45	5.3	4.5	7.9	3.8	38
Edo sweet miso	49	5.4	5.8	11.0	4.5	15
Shinshu-miso	48	5.2	12.0	11.4	5.5	12
Sendai-miso	50	5.1	12.8	12.0	5.3	11
Mugi-miso (salty)	48	5.1	12.0	12.7	5.5	11
Mugi-miso (sweet)	47	5.2	9.5	10.0	4.9	17
Soybean-miso	46	5.0	11.5	20.0	10.5	4

[a] Ebine, 1980. [b] The ratio between soybeans and rice as raw materials in white miso is 10:20–25; edo-miso, 10:13; shinshu-miso, 10:6–10; and sendai-miso, 10:3–5. The ratio between soybeans and barley or rye as raw materials in mugi-miso is 10:10–20. The barley or rye is milled to make it 60–85% by weight.
[c] Reducing sugars.

The amount of miso produced in Japan was 568 000 tonnes in 1979, but it has been decreasing during the last decade. Per capita consumption of miso is about one half that of shoyu in Japan. There are about 2000 miso producers in Japan. The ten main producers supply 36% of the total amount produced, and the market share of the biggest producer among them was 4.7% in 1980. Miso is usually employed as an ingredient in miso soup and it is also used in cooking in general as a condiment.

The miso made from soybeans and a small amount of wheat flour is also very popular in Asian countries, but in most cases, the soybean-miso produced in countries other than Japan is a slurry instead of a paste and it is used mainly in general cooking rather than for making soups as in Japan.

19.3 CURRENT SHOYU AND MISO PRODUCTION PROCESSES

19.3.1 Raw Materials

The raw materials of shoyu and miso are soybeans, wheat, rice, barley, rye, salt and water.

19.3.1.1 Soybeans and defatted soybeans

The protein content of soybeans and the amino acid composition of that protein greatly influence the quality and yield of shoyu, which is usually calculated on the basis of nitrogen content. Soybeans of a high protein content or defatted soybeans are preferred for shoyu manufacture. The chemical compositions of whole and defatted soybeans are given in Table 7. Soybeans contain about 6% sucrose, 4% stachiose and less than 1% raffinose. Defatted soybeans, which are continuously extracted with *n*-hexane (BP 65–70 °C), are preferred to a defatting procedure involving pressing the soybeans through a nozzle, because the high temperature produced during pressing decreases the enzymatic digestibility of the proteins. Too high an NSI value (nitrogen solubility index: total soluble nitrogen/total nitrogen) of defatted soybeans (over 35%) makes soybeans undesirable as a raw material for shoyu because of their highly sticky nature which makes handling difficult during and after autoclaving. Defatted soybeans containing more than 7.8% protein, with an NSI value of 20–30%, and a granule size greater than 8 mesh are considered to be suitable for shoyu manufacture. For miso production, soybeans having a higher carbohydrate content and a larger granule size are preferred. Generally, soybeans contain about 20% crude oil, of which 94–97% is composed of glycerides of higher fatty acids (linoleic 10%; linolenic 55%; oleic 20%; stearic 10%; and palmitic 12%) and 2% of phospholipids. The soybean oil is degraded into higher fatty acids and glycerol by the action of lipase derived from koji molds mainly in the salty mash, and these degraded compounds remain totally in the miso. Only glycerol goes into shoyu. As a result of technological progress in recent years, shoyu is now mostly produced from defatted soybeans, but almost all miso is produced from whole soybeans because of the superior organoleptic qualities imparted, especially smoothness of texture of the cooked products.

Table 7 Chemical Analyses of Soybeans as the Raw Materials for Shoyu and Miso[a]

	H_2O (%)	Total nitrogen (%)	Invert sugar (%)	Crude fat (%)	Water total
Soybeans (USA)	10.03	5.99	17.7	19.1	—
Soybeans (China)	10.97	6.03	18.9	17.5	—
Soybeans (Japan)	10.91	5.98	20.8	15.2	—
Defatted dehulled soybeans	9.46	7.90	22.1	—	27.2
Defatted whole soybeans	10.18	7.60	21.6	—	26.2

[a] From Fukushima and Yokotsuka (1972).

19.3.1.2 Wheat, barley, rye and rice

The total nitrogen contained in koikuchi shoyu, which represents more than 85% of Japanese shoyu, is derived from soybeans (75%) and the remainder from wheat kernels. The ratio of soybeans and wheat as raw materials for koikuchi shoyu ranges from 6:4 to 4:6. The glutamic acid contents of soybeans and wheat are 20% and 30% of the total amino acids, respectively. Proteins present in wheat kernels are good sources of glutamic acid, which is an important taste ingredient of shoyu. The chemical analyses of wheat used for shoyu production are indicated in Table 8. Wheat bran is sometimes used instead of wheat kernels; this decreases the alcohol content of shoyu and makes the color darker and the color stability inferior because of the increased amount of pentoses in the shoyu.

Table 8 Chemical Analyses of Wheat used for Shoyu Production[a]

	H_2O (%)	Total nitrogen (%)	Starch (%)
Wheat (Japan)	11.0	2.0	65
Wheat (imported)	11.5	2.4	63
Wheat bran	11.5	2.4	45

[a] From Kikkoman Shoyu Co., Ltd. (1970).

Dehulled barley was commonly used as a raw material for shoyu in the past, but the use of large quantities of dehulled barley resulted in shoyu of poor organoleptic quality. As in miso making, dehulled or 15–40% refined barley or rye is commonly utilized nowadays. Milled rice is used for a wide variety of rice-misos in amounts ranging from 6 to 25% that of soybeans. The chemical compositions of these starchy raw materials are given in Table 9.

Table 9   Chemical Analyses of Starchy Raw Materials for Miso[a]

	H_2O (%)	Protein (%)	Carbohydrate (%)	Crude fat (%)
Rice (refined)[b]	15	6–6.5	77	0.8
Barley (refined)[c]	13	13–14	60	6.0
Rye (refined)[c]	13	13–14	62	4.0

[a] Ebine (1972). [b] Refined 92%. [c] Refined 82.5% or more.

19.3.1.3   Salt and water

Salt of purity greater than 95% is generally used; water low in Fe and Cu is preferred for both miso and shoyu, because these minerals reduce the color stability of the final products.

19.3.2   Koikuchi Shoyu

Production of Japanese fermented shoyu of the koikuchi type involves five processes: treatment of raw materials, koji making, mash controlling, pressing and refining. One example of the preparation of koikuchi shoyu is shown schematically in Figure 1.

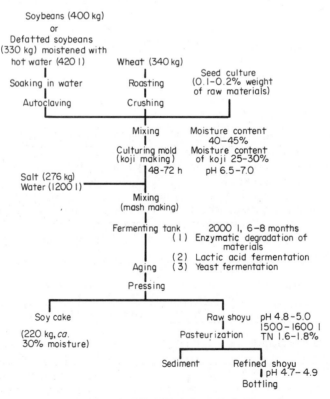

Figure 1   Koikuchi shoyu fermentation

19.3.2.1 Treatment of raw materials

Whole soybeans, or more commonly defatted soybean grits, are moistened and cooked with steam under pressure. This process greatly influences the digestibility of soybean protein. Wheat kernels are roasted at 160 to 180 °C for less than one minute, then coarsely crushed.

19.3.2.2 Koji production

The soybeans and wheat are inoculated with a small amount of seed mold or a pure culture of *A. oryzae* or *A. sojae*. This mixture is spread to a depth of 30 to 40 cm on a large perforated stainless steel rectangular plate 5 m wide and 12 m long or on a doughnut-shaped support of diameter 15 to 30 m. The heat-treated raw materials are aerated for 2 or 3 days with humidified air, which is fed through the bottom holes. The temperature is maintained at around 30 °C, and the 40 to 45% moisture content of the materials at the beginning of cultivation decreases to 25 to 35% after 2 or 3 days. This allows the mold to grow throughout the mass and provide the enzymes necessary to hydrolyze the protein, starch and other constituents of the raw materials. This cultured material is called koji.

19.3.2.3 Mash production and aging

In mash production, the koji is mixed with 120 to 130 volume percent water containing 22 to 23% salt. The mash or moromi is transferred to deep fermentation tanks. The 5–10 kl wooden kegs and 10–20 kl concrete tanks formerly used are now being replaced by resin-coated iron tanks of volume 50–300 kl. The moromi is kept for 4 to 8 months, depending upon its temperature, with occasional agitation with compressed air to mix the soluble components uniformly and to promote microbial growth. During the fermentation period, the enzymes from the koji mold hydrolyze most of the protein into amino acids and low molecular weight peptides. Around 20% of the starch is consumed by the mold during koji cultivation, but almost all of the remaining starch is converted into simple sugars; more than half of these are fermented to lactic acid and alcohol by lactobacilli and yeasts. The pH drops from an initial value of 6.5–7.0 to 4.7–4.9. The lactic acid fermentation is gradually replaced by yeast fermentations. Cultures of *Pediococcus halophylus* and *Saccharomyces rouxii* are sometimes added to the mash. The salt concentration of the mash reaches 17–18% (weight per volume) after one or two months. The high salt concentration of the mash limits the growth to a few desirable microorganisms.

19.3.2.4 Mash pressing

A matured mash is filtered at high pressure through cloth. Usually 12–13 l of shoyu mash are placed on a cloth sheet of area 1 m², the four sides of the cloth are folded into a square of area 70 × 70 cm and another smaller sheet of cloth, of area 65 × 65 cm, is placed on it so as to wrap up the mash. The next shoyu mash is put on top and wrapped in the same way. These operations are repeated in a wooden box, and 300 to 400 sheets of folded cloth containing the mash are piled and pressed for 2 or 3 days under hydraulic pressure. Sometimes a sheet of cloth twice as large as described above is used to wrap about 25 l of moromi. The pressure is increased in two or three steps, sometimes reaching 100 kg cm^{-2} in the final stage, which makes the moisture content of the cake less than 25%. A diaphragm-type pressing machine has recently been used for shoyu-mash filtration; the moisture content of the cake obtained is more than 30%. The residue from the pressing of the shoyu-mash, or shoyu-cake, is used as animal or poultry feed.

19.3.2.5 Refining

The liquid part of the mash obtained by pressing is stored in a tank and divided into three layers: sediment, a clear supernatant middle layer, and an oily layer floating on top. The middle layer is sometimes further clarified by filtration with Kieselgel as a filter aid in order to obtain raw shoyu. After adjusting the salt and nitrogen concentrations to commercial standards, the clarified raw shoyu is pasteurized at 70–80 °C and stored in a semi-closed tank. The clear middle layer is bottled, canned or spray-dried. The oily layer separated from the heated shoyu consists of free higher fatty acids and their ethyl esters derived from the yeast metabolism of soybean and wheat oils; it is sometimes mixed with paint as an antifreezing agent.

19.3.3 Tamari Shoyu

A mixture of cooked soybeans or defatted soybean grits and roasted wheat (20:3) is extruded to form granules 12–16 mm in diameter. These granulated soybeans are inoculated with a mixture of seed molds, *A. oryzae*, *A. sojae* or *A. tamarii*, and roasted powdered barley in amounts less than 1.5% of the total. These materials are incubated at 26–28 °C for about 45 hours to make koji. The moisture content of the koji is preferably about 35%, and its weight is about 120% that of the raw materials. This tamari-koji is sometimes dried so as to decrease the weight of koji by 7–8% before the mash making. The koji is mixed with 0.5 to 1.3 volumes of salt water to make the mash. The mash is usually too solid to stir, so the liquid part of the mash from the bottom is repeatedly spread over the top of the mash instead of agitating it with compressed air as is the case with koikuchi mash. Genuine tamari shoyu is rich in soluble solids and there is little contamination from film-forming yeasts, so the final product is not usually pasteurized.

19.3.4 Rice-miso and Barley-miso

Whole yellow soybeans are almost always used for the preparation of ordinary miso. Dehulled soybeans or soybean grits are sometimes employed for the production of white or pale yellow rice-miso. Soybeans are soaked in water until saturation and then cooked for 30 to 60 minutes at normal pressure, or cooked in four volumes of water for 20 to 30 minutes at a pressure of 0.5–0.7 $kg\,cm^{-2}$, or steamed for 20 minutes at a pressure of 0.7 $kg\,cm^{-2}$ (115 °C) either batchwise or continuously. Cooked soybean granules are preferably pressed using less than 0.5 $kg\,cm^{-2}$ of pressure. Milled rice or barley or rye is soaked in water and then steamed batchwise in an open cooker for 40 minutes or continuously on a net conveyor in a closed autoclave for 30 to 60 minutes. The koji cultivation on rice or barley or rye with *A. oryzae* is conducted at 35 to 38 °C, sometimes with an increase in temperature up to almost 40 °C in the final stage, for 40 to 48 hours. The finished koji is mixed with salt to stop further mold growth and to minimize the inactivation of enzymes. The amount of salt used is about 30% by weight of the koji. Finished koji is sometimes cooled instead of adding salt. Various types of koji fermenter are employed nowadays. Cooked soybeans are mixed with salted rice- or barley-koji, a small amount of water and an inoculum of cultured yeast and lactic acid bacteria, if necessary. It is important to mix these materials uniformly so that the variation in salt concentrations in the mash is less than 0.5%. The mixture is packed in a fermentation tank; it is moved from one tank to another at least twice during the fermentation period to mix the contents and to provide aerobic conditions suitable for microbial growth. Fermentation is conducted at around 30 °C for 1 to 3 months depending upon the type of miso. Well-ripened miso is then blended and mashed if necessary and pasteurized using a tube heater. About 2% alcohol is added to the product to stop the growth of yeasts. The production process for manufacturing salty rice-miso is shown in Figure 2, and the major differences between miso and shoyu are summarized in Table 10.

19.4 BIOCHEMISTRY AND PROCESSING OPERATIONS INVOLVED IN SHOYU AND MISO PRODUCTION IN JAPAN

19.4.1 Soybeans as Raw Materials

Only whole soybeans were used as raw materials for shoyu and miso before World War II, but defatted soybean grits are now widely used for shoyu production. Whole beans accounted for only 3.2% of the total soybeans used for the production of shoyu in 1978. In the case of miso, however, a large amount is still produced from whole beans.

In comparing the use of whole and defatted soybeans in shoyu manufacture the factors to be considered are cost, digestibility of proteins, fermentation period, difficulty of manufacture (especially in koji production and mash control), the quality of shoyu obtained in terms of chemical components such as glycerol, alcohol and lactic acid, organoleptic evaluation, and the stability of the product.

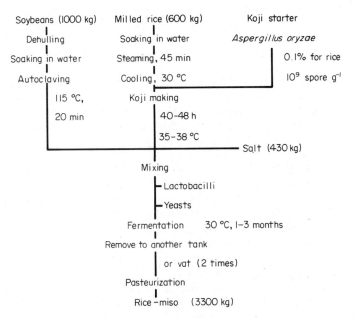

Figure 2 Rice-miso fermentation

Table 10 Major Differences between Miso and Shoyu

	Miso (kome- or mugi-miso)[a,c]	Shoyu (koikuchi shoyu)[b,c]
Raw materials	Soybeans, milled rice, dehulled barley, or rye	Soybeans (defatted soybeans), wheat grains
Mold	*A. oryzae*	*A. oryzae* or *sojae*
Koji is cultured on	Cooked rice, or dehulled barley, or rye	Total materials (cooked soybeans and roasted crushed wheat)
Mash making	Koji, cooked soybeans, salt, a small amount of water	Koji, a large amount of salt water
Period of mash fermentation	1–3 months	4–6 months
Lactobacilli in mash	10^5	10^{7-8}
Yeasts in mash	10^5	10^{6-7}
Pressing of mash	+	+
Product	Semi-solid or paste	Liquid
Total nitrogen	1.5–3.5%	1.0–2.5%
Amino nitrogen content of total nitrogen	20–30%	40–50%
Sodium chloride	6–12%	13–18%
Reducing sugars	10–40%	4–6%

[a] Kome means rice; mugi means barley or rye. [b] Koikuchi means dark in color.
[c] % means g per 100 g for miso, and g per 100 ml for shoyu.

Shoyu made from defatted beans, costing the same in terms of raw materials as whole beans, has a higher content of free amino acids including glutamic acid. The digestibility of defatted soybeans in the fermentative production of shoyu has been improved due to improvements in the preparation of defatted soybeans. An NSI value of about 20–30 for defatted soybeans is generally believed to be adequate for miso and shoyu production. The enzymatic digestibilities of the proteins contained in whole and defatted beans during shoyu production were reported to be 62 and 60%, respectively, in 1940 (Kawamori, 1940).

Whole soybeans were reported several years ago to have a slower rate of fermentation (about

15 months at room temperature for whole beans, 10 months for defatted beans; Yokotsuka, 1960), but the fermentation period for the shoyu mash is principally dependent upon the cooking conditions of the soybeans and the enzymatic activity of the koji. These two factors are considered to be associated with the manufacturing technology and not with the differences between whole and defatted soybeans. The average fermentation period of shoyu is about six months in both cases.

Because whole beans are larger than defatted beans, whole soybeans are more easily cooled by hand mixing, using wooden trays, during koji making. Recently it has become easy to prepare koji from defatted beans using equipment in which the temperature is easily controlled. Shoyu made from whole beans was reported to be lighter in color and to have better color stability, higher contents of alcohol and glycerol, lower contents of lactic acid and reducing sugars, and a better organoleptic evaluation compared to shoyu made from defatted beans (Okuhara and Yokotsuka, 1958; Moriguchi *et al.*, 1961). The glycerol contents of shoyu made from whole and defatted beans were reported to be 1–1.2% and 0.6–0.7%, respectively (Okuhara and Yokotsuka, 1962, 1963). The amount of glycerol derived from the degradation of soybean oil is calculated to be about 0.5%. Glycerol was also found to be produced by yeast fermentation of glucose in the presence of high salt concentrations. Shoyu mash is now subjected to much more vigorous yeast fermentation than previously, which results in a higher degree of glycerol formation, sometimes reaching 1.5–1.7%. Thus the difference in the glycerol contents of whole and defatted soybean shoyu has become much less than before. The lactic acid contents of shoyu are now freely adjusted by controlling the lactic acid fermentation in the mash regardless of the kind of soybean used. The differences between shoyus made from whole and defatted beans can be reduced by making the physical structure of whole beans similar to that of defatted beans by pressing (Okuhara and Yokotsuka, 1963).

Both the degree of color and the color stability of shoyu seem to be related to the degree of digestion of raw materials, and not to the type of soybeans. Whole soybean oil composed of glycerides of higher fatty acids is metabolized during shoyu-mash fermentation into shoyu oil which is composed of ethyl esters of these higher fatty acids and a large amount of free higher fatty acids. The shoyu oil is separated from raw shoyu as the upper layer. The shoyu oil is not edible and causes difficulties for shoyu producers.

From an environmental view point, the use of whole beans causes waste problems arising from the wastewater obtained by soaking whole beans and the sticky liquid drained from cooked soybeans.

19.4.2   Treatment of Raw Materials

19.4.2.1   Soybeans

The protein in raw soybeans is native and cannot be hydrolyzed by the proteases in koji, therefore it is necessary to denature the soybean protein so that it can be digested by the enzymes in the koji mold to make shoyu or miso. The soybeans used to be boiled in water at atmospheric pressure, but are now autoclaved under pressure. Soaked soybeans are cooked at a gauge pressure of 0.8 kg cm^{-2} for about one hour, and after discontinuing the supply of steam the cooked soybeans are kept in the vessel for several hours without breaking the seal. Kawano (1938) found that the highest enzymatic digestibility of the cooked soybeans and the highest free amino acid content of the shoyu prepared from cooked soybeans were obtained when the soybeans were cooked at a pressure of 0.5 kg cm^{-2} as compared with 0, 1.0, 1.5 and 2.0 kg cm^{-2}.

Tateno and Umeda (1955) invented the NK-method of soybean cooking, in which thoroughly soaked soybeans were cooked in a rotary cooker at a gauge pressure of 0.8 kg cm^{-2} for about one hour, followed by rapid cooling to below 40 °C by reducing the inside pressure with the aid of a jet condenser. The protein digestibility in shoyu manufacture, which is the ratio between the total nitrogen content of the shoyu and that of the raw materials, was increased from 69 to 73% by the NK-method as compared with conventional cooking methods (see Table 11).

Treatment of soybeans with water containing methanol, ethanol or propanol at its boiling point was found to give almost 90% enzymatic digestibility of protein (Yamaguchi, 1954; Fukushima and Mogi, 1955, 1957). These methods have not been used industrially mainly because of the difficulty of making koji and the inferior organoleptic quality of the final shoyu.

Yokotsuka *et al.* (1966a) and Yasuda *et al.* (1973a, 1973b) found it was possible to increase the enzymatic digestibility of cooked soybeans with a higher temperature and a shorter cooking time

Table 11 NK-Cooking Method for Soybeans as Compared
to the Conventional Method[a]

Cooking method	Digestibility of proteins in mash (salt 18%, room temperature 1 year)	Ratio of formol N to total N	Ratio of glutamic N to total N
Conventional[b]	68.7%	49.4%	5.5%
NK-Method[c]	73.1%	53.8%	7.3%
Increasing ratio	106.4%	108.8%*l*	135.4%

[a] Tateno and Umeda, 1955. Kikkoman Shoyu Co., Ltd. [b] Cooked at 0.8 kg cm^{-2} for 1 hour, soybeans left in the autoclave for additional 12 hours.
[c] Cooked at 0.8 kg cm^{-2} for 1 hour, soybeans taken out of the autoclave immediately.

(HTST) than the NK-method (see Table 12 and Figure 3). This method indicated the possibility of 92–93% protein digestion in shoyu production with better organoleptic qualities of the final product. It is important, especially in HTST cooking of soybeans, to avoid excessive heating because it causes over-denaturation of proteins and decreases their enzymatic digestibility. Similar results were reported by Harada and Kawaguchi (1968), in which defatted soybeans were cooked at a pressure of 4 kg cm^{-2} for 3 minutes.

Table 12 Effect of the Cooking Conditions of Soybeans
on the Enzymatic Digestibility of Protein[a]

Steam pressure (kg cm^{-2})	Cooking time (min)	Digestibility of protein in enzyme solution (%)[b]
0.9	45	86
1.2	10	91
1.8	8	91
2.0	5	92
3.0	3	93
4.0	2	94
5.0	1	95
6.0	$\frac{1}{2}$	95
7.0	$\frac{1}{4}$	95

[a] Yokotsuka *et al.*, 1966. [b] Salt 0%, 37 °C, 7 d.

Aonuma *et al.* (1970, 1971) reported a new cooking method for soybeans and wheat used for brewing, without soaking before cooking, using super-heated steam at a pressure of 4–8 kg cm^{-2} or at 200–289 °C for not less than 15 seconds, resulting in almost the same protein digestibility as that obtained by saturated steam. This method has the advantage that the heat-treated raw materials can be stored.

New HTST cooking methods for raw materials stimulated the development of several types of continuous cooker, as shown in Figure 4. At the same time, the NK-method was also greatly improved by the HTST method. Protein digestibility of 87.80% was achieved by cooking soybeans at 1.7 kg cm^{-2} for 8 minutes using an NK-cooker as compared with 81.80% obtained by the conventional NK-cooking conditions of 0.9 kg cm^{-2} for 40 minutes (Iijima *et al.*, 1973). The time for cooling of autoclaved soybeans was found to be related to their proteolytic digestibility (Yasuda *et al.*, 1973a). By enlarging the diameter of both the inlet and exhaust steam pipes of an NK-cooker in order to give rapid elevation of the temperature at the beginning of cooking and rapid cooling of the cooked soybeans, the protein digestibility increased by about 3% under the same conditions (Eguchi, 1977).

19.4.2.2 Wheat

Wheat kernels were originally roasted in an iron pan for shoyu production. This method was replaced at the beginning of the 1900s by roasting together with sand in a rotary roaster, in which

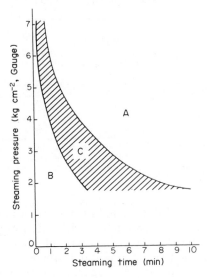

Figure 3   Denaturation of soybean protein by steaming with 130% moisture content: A over-denaturation region; B under-denaturation region; C acceptable denaturation region for shoyu production (Yokotsuka *et al.*, 1966)

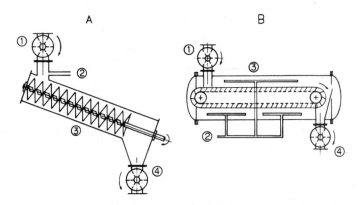

Figure 4   Continuous soybean cooker: A screw type; B net conveyor type. 1 rotary valve (charge); 2 steam; 3 cooker; 4 rotary valve (discharge)

only the heated sand is recycling in the oven. This method gives a relatively uniform roasting of the wheat and is still popular nowadays.

In roasting wheat kernels for shoyu production, there is a trend towards higher temperatures, resulting in a higher α-starch content, but lower digestibility of the protein (Figure 5). It is possible to increase the α-starch content of roasted wheat kernels without impairing the protein digestibility by increasing the moisture content of the wheat kernels to 15–25% before roasting. The same HTST cooking method as is used for soybeans, with the addition of 10% moisture before roasting, gives similar results to those obtained by roasting wheat kernels (Yokotsuka *et al.*, 1974).

According to Aiba (1982), good results were obtained by roasting wheat kernels at atmospheric pressure when the kernels contained more than 8% moisture and were treated with hot air at more than 150 °C for less than 45 s. The highest digestibility of starch in the roasted wheat kernels was 86%; the highest digestibility of protein was 97%. In an actual koikuchi shoyu production utilizing this roasted wheat, the improvements were a 2% increase in protein digestibility, a 0.3 to 0.4% increase in sugar content of the shoyu, and an increase in the amount of wheat kernels used from 20 to 36 kg (which can be roasted by 1 l of petroleum). One unit of the fluidized-bed roaster used is shown in Figure 6.

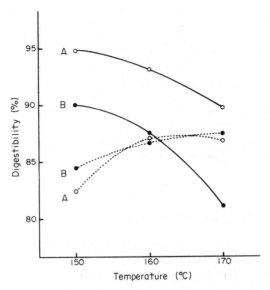

Figure 5 Enzymatic digestibility of wheat kernels roasted by the conventional method with sand; —— total nitrogen; --- total sugar (Aiba, 1982)

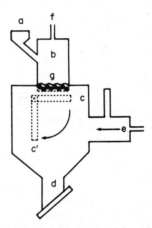

Figure 6 Wheat roaster with a fluidized bed; a wheat charger; b roasting tower; c perforated plates to regulate the stream of hot air; d outlet for roasted wheat kernels; e duct for hot air; f exhaust of hot air; g fluidized bed of wheat kernels. Wheat kernels are spread on g to a thickness of about 5 cm and roasted with a hot air stream (4–5 m s⁻¹ at about 300 °C for about 25 s); c moves to c′ and the roasted kernels drop to d. The wheat kernels are supplied to the three units of the roaster in succession so that the roasted wheat kernels come out almost continuously

19.4.3 Koji Molds

19.4.3.1 *Selection of koji molds*

When selecting the mold strains used for food fermentations the following points should be taken into account: the flavor and color of the final product; the spore forming ability necessary to prepare the seed starter; strong and rapid growth; high enzymatic activity especially of proteolytic and macerating enzymes in the case of shoyu-koji; strong activity of amylases in the case of rice-koji for miso manufacture; lower consumption of starch during growth; genetic stability; length of stalk (short-stalked strains are suitable for mechanical koji cultivation), and no toxin production.

Terada *et al.* (1981) isolated 128 strains of koji molds from the mold starters purchased in the Japanese market, which included 35 starters for shoyu production and 33 for miso production.

Sixty-eight percent of the starters for shoyu and 79% for miso were composed of more than two kinds of koji mold. The starters for shoyu tended to have short stalks and high spore productivity. The starters for miso production were good producers of amylase. Sixty-five strains of koji mold used for shoyu were composed of 80% *A. oryzae* and 20% *A. sojae*, while 63 strains used for miso were composed of 89% *A. oryzae* and 11% *A. sojae*.

The shoyu-koji cultured with *A. sojae* has the following characteristics compared to koji cultured with *A. oryzae* (Terada *et al.*, 1980, 1981; Hayashi *et al.*, 1981): (1) higher pH value due to the lower content of citric acid; (2) lower consumption of starch during koji making; (3) lower activity of α-amylase, acid protease, acid carboxypeptidase and higher end-polygalacturonase in the koji; (4) lower viscosity of mash or moromi; (5) lower enzymatic activity in the raw shoyu, which is related to the smaller amount of heat-coagulated products caused by pasteurization; and (6) lower content of reducing sugars, lactose and ammonia, and lower pH value of the raw shoyu.

19.4.3.2   Types and activities of enzymes produced by koji molds

As seen in Table 13, about 80% of the endopeptidases or proteinases produced by koji molds are alkaline proteinases having an optimum operating pH of 9–10. Three types of acid proteinases (optimum pH 3), two neutral proteinases (optimum pH 6–7), and one semi-alkaline proteinase have been isolated (Nakadai, 1977).

Table 13   Fractionation of Proteases Produced by
Aspergillus sojae through Sephadex G-100[a]

Protease	MW	Activity (units /g koji)
Acid I	39 000	41.1[b]
Acid II	100 000	10.0[b]
Acid III	31 000	4.6[b]
Neutral I	41 000	80.0[c]
Neutral II	19 300	8.7[c]
Semi-alkali	32 000	55.4[c]
Alkali	23 000	929.0[c]

[a] Nakadai (1977). [b] Activity on casein at pH 3.0.
[c] Activity on casein at pH 7.0.

The strong soybean-protein digesting abilities of the neutral proteinases I (optimum pH 7.0) and II (optimum pH 5.5–6.0, heat-resistant), especially of the former, have been pointed out (Sekine, 1972, 1976), but the percentage of neutral proteinases in the total proteinases of the *Aspergillus* molds checked was only 10 to 20% (Tagami and Sugawara, 1977). These exopeptidases or peptidases, especially leucine aminopeptidase (optimum pH 7–8), are strongly associated with the enzymatic formation of formol nitrogen and glutamic acid in the shoyu mash (Nakadai *et al.*, 1972).

The proteinases of koji molds degrade protein into peptides and not into individual amino acids, with the exception of acid proteinases (I and II), which degrade protein into amino acids and lower peptides especially at pH 6.0 rather than at pH 3.0. Glutamic acid and glutamine are separated from peptides by the action of carboxypeptidase and aminopeptidase. Glutamine is then converted into glutamic acid by the action of glutaminase in the koji mold, which has an optimum pH of 7 and temperature of 40–45 °C. The glutaminase in koji mold is not heat tolerant and is easily destroyed above 25 °C (Kuroshima *et al.*, 1969). The activity of glutaminase derived from koji mold is greatly reduced in the presence of high salt concentrations (Terada *et al.*, 1973). It was found that the amount of intercellular glutaminase in koji mold in the shoyu mash is 10 times that of the extracellular glutaminase, and the former was more resistant to heat and pH than the latter (Yomo *et al.*, 1978).

Ishii *et al.* (1972) searched for the *Aspergillus* mold strain which produces enzymes with strong plant tissue degrading activity, and isolated *A. sojae* 48. This strain of mold was found to produce strong pectin degrading enzymes, such as pectate transeliminase (*endo*-type hydrase, optimum pH 4.5), and pectinesterase (optimum pH 5.0–5.5), and hemicellulases such as xylanase (optimum pH 5.5) and arabinase (optimum pH 4–4.5). The strong macerating activity of this mold was attributed mostly to the pectinases produced by the mold, but, at the same time, this activity was found to be greatly reduced in salt solution. Nevertheless, the shoyu mash prepared from koji cultured with this strain of mold exhibited a greater digestibility of protein and carbohydrates and a lower viscosity as compared to those of a control mash prepared using ordinary koji mold. The amount of residue remaining after pressing this shoyu mash was smaller than the control. It is known that not only proteases, amylases, lipases and phosphatases, but also cellulases, hemicellulases, pectinases, β-galactosidases and pentosan-degrading enzymes are involved in the enzymatic degradation of plant tissues from soybeans and wheat kernels.

19.4.3.3 *Improvement of koji molds*

Improvements in the proteolytic activities of koji molds have been achieved by induced mutation, crossing or cell fusion. Iguchi (1955) obtained the induced mutant *A. sojae* X-816 by X-ray treatment, which increased both the total nitrogen and amino acid solubility of casein by a factor of two. In the industrial application of this mutant strain to shoyu fermentation, the total nitrogen digestibility of the raw materials increased by 2–3%. Ishitani and Sakaguchi (1956) reported natural heterocaryosis among koji molds and tried to improve the molds by crossing using anastomoses (Ishitani *et al.*, 1956). Uchida *et al.* (1958) succeeded in crossing *A. oryzae* and *A. sojae* by the above method. Oda and Iguchi (1963) tried to cross *A. sojae* and its induced mutant strain X-816, and observed that the diploid strains obtained from the crosses exhibited protease formation and spore forming ability intermediate between the two strains. A 2–6% increase in protein digestibility was reported in an experimental shoyu brewing by applying an induced mutant of *A. sojae*, with proteases increased by six times compared to the parent strain X-816 (Nasuno *et al.*, 1972).

Furuya *et al.* (1983) reported protoplast fusion between *A. oryzae* and its mutant induced by X-ray irradiation, which exhibits eight times the protease activity of the parent strain. Protoplasts were obtained from mycelia of two strains using snail gut juice. When these protoplasts were treated with a solution containing 20% polyethyleneglycerol, they fused effectively and formed heterocaryons with a fusion frequency of 1%. Two of the green strains derived from the heterocaryons showed high stability, fast growth and abundant sporulation, producing 2–3 times more protease than the original strain in koji culture. These strains were presumed to be heterocaryon diploids, because of the formation of recombinants.

19.4.3.4 *Safety of koji molds*

Many investigators have checked for aflatoxin production in koji molds, but so far all results suggest that koji molds do not produce aflatoxins (Hesseltine *et al.*, 1966; Aibara and Miyaki, 1965; Masuda *et al.*, 1965; Murakami *et al.*, 1967; Manabe *et al.*, 1968; and Kinoshita *et al.*, 1968). Yokotsuka *et al.* could not find aflatoxin producers among 200 strains of *Aspergillus* mold used in Japanese food fermentations (Yokotsuka *et al.*, 1966b, 1968a; Sasaki and Yokotsuka, 1972), In the course of this research, seven fluorescent pyrazines and one nontoxic pyrazine compound (lumichrome) were isolated along with other unknown compounds which are produced by molds in general and on thin layer chromatography exhibit *Rf* values similar to those of aflatoxin B or G depending on the solvent systems (Figure 7).

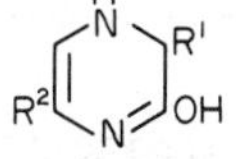

Figure 7 Fluorescent pyrazine compounds exhibiting similar *Rf* values as aflatoxin B1 on TLC: R^1, R^2 = isobutyl (**1**, flavacol); R^1 = isobutyl, R^2 = *s*-butyl (**2**); R^1 = isobutyl, R^2 = —C(OH)(Me)C$_2$H$_5$ (**3**); R^1 = isobutyl, R^2 = isopropyl (**4**); R^1 = isobutyl, R^2 = —C(OH)Me$_2$ (**5**); R^1, R^2 = *s*-butyl (**6**); R^1 = *s*-butyl, R^2 = —C(OH)(Me)C$_2$H$_5$ (**7**)

Twenty-nine out of 69 strains of *Aspergillus* mold were found to produce aspergillic acid, but the most prolific producer among them did not produce aspergillic acid on a solid substrate composed of soybeans and wheat within two days, which is the usual time period for koji cultivation (Yokotsuka *et al.*, 1969).

Aspergillus ochraceus, which is occasionally found in koji cultivation as a contaminant in the southern part of Japan, sometimes produces ochratoxin and some fluorescent toxic isocoumarin compounds.

Sasaki *et al.* (1970, 1980) and Yokotsuka (1977) were unable to detect aflatoxin, patulin, ochratoxin, sterigmatocystin, penicillic acid and cyclopiazonic acid in 33 strains of industrial *Aspergillus* mold.

19.4.4 Koji Production

The quality of koji is associated not only with the degree and rate of enzymatic degradation of the raw materials in the salty mash, but also with the chemical and organoleptic quality of the final product. It is necessary in making koji: (1) to obtain sufficient growth of mycelia; (2) to produce maximum amounts of the enzymes needed, such as proteases, amylases and other plant tissue degrading enzymes; (3) to maintain the activity of the enzymes once produced; (4) to minimize the consumption of starch caused by the growth of mold; and (5) to avoid bacterial and mold contamination.

19.4.4.1 *Starter mold*

The starter mold is cultured so as to induce spore formation. Wheat bran, crushed wheat, dehulled and pressed barley, rye, or crushed rice grains are used as raw materials for the cultivation of starter mold. These materials are moistened with water (45–55%) and autoclaved. Pure cultures of *Aspergillus* mold are inoculated and cultured at 25–30 °C for 75–100 h in a sterilized room.

19.4.4.2 *Temperature and humidity control*

The raw materials are mixed with 0.1–0.2% of starter mold, *A. oryzae* or *A. sojae*, for koji cultivation. The materials formerly were placed in flat wooden boxes or trays to a thickness of 3–5 cm and incubated in a warm room with temperature and humidity controlled using windows and charcoal fires. During cultivation, the materials were cooled, usually twice, by hand mixing. The major effort was directed towards obtaining good growth of mycelia and spores by culturing them at around 35 °C, but it was difficult to keep this temperature because of the absence of mechanical cooling; the materials sometimes became heated to over 40 °C. One example of temperature change during shoyu-koji cultivation in wooden trays is shown in Figure 8.

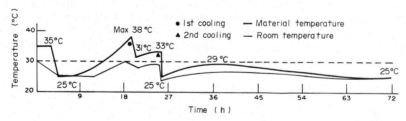

Figure 8   Temperature changes of materials during shoyu-koji cultivation by conventional methods using wooden trays

It was found that 30–36 °C was the preferred temperature for mycelium growth, around 25 °C for protease formation, but 35–40 °C for amylase formation. The rice or barley koji used for miso production, which is rich in amylase rather than proteases, is kept at 30–35 °C during the first half, and at about 40 °C during the second half of cultivation. In shoyu-koji cultivation, which requires protease rather than amylase, the temperature is kept at 30–35 °C in the early stage for mycelium growth and at less than 30 °C, preferably around 25 °C, in the latter stage for protease formation (Ohara *et al.*, 1959).

The through-flow system of koji cultivation or so called TF-method, in which the aerobic koji molds are cultured by passing temperature- and moisture-controlled air through the materials, was first utilized by miso producers in 1955, earlier than shoyu producers. This was partly because the calorific value or the generation of heat during koji cultivation and its maximum rate were much greater in shoyu-koji composed of soybeans and wheat than in miso-koji composed of rice or barley (Terui *et al.*, 1958). This is shown in Table 14.

Table 14 The Calorific Values of Miso and Shoyu-Koji Cultured with *Aspergillus oryzae*[a]

Type of koji	Maximum speed of heat generation (kJ kg^{-1} dry matter h^{-1})	Total calorific value (kJ kg^{-1})
Rice-koji for miso	38–54	500–835
Barley-koji for miso	50–67	500–920
Soybean- and wheat-koji for shoyu	130	1840–2090

[a] From Terui *et al.* (1958).

The development of mechanical controls for koji cultivation has made it possible to provide suitable temperature and humidity conditions; it has also made it possible to shorten the time required for koji cultivation from 72 to 48 h, to increase the enzymatic activities of koji, to reduce undesirable bacterial contamination in koji, and to reduce labor costs. By reducing the temperature of materials to below 35 °C after the second cooling in shoyu-koji cultivation, the yield of shoyu was improved by 7% compared to the conventional method (Harada, 1951), and when the koji was cultivated at a constant temperature of 25 °C, the protein digestibility increased by 7% as compared to the conventional wooden-tray method (Imai and Suzuki, 1967).

The progress of shoyu-koji culture is as follows. The spores of koji molds mixed with raw materials containing 40–45% moisture usually germinate after 8–10 h with a gradual increase in temperature. The temperature is initially kept at around 30 °C. When the temperature begins to increase and reaches around 35 °C, due to the vigorous growth of molds (usually after 16–18 h), the materials are mechanically mixed and cooled to below 30 °C. The mixture at this stage comprises a mass of mycelia, through which the passage of air is difficult. The temperature rises to over 30 °C again, usually 6–7 h after the first cooling, due to the continued vigorous growth of molds, and a second cooling stage is carried out. The temperature after the second cooling is controlled in different ways for miso or shoyu-koji cultivation. Proteases and amylases are formed in the latter stage of cultivation. The temperature of the materials is usually controlled by controlling the air flow, the temperature of which is kept at around 28–32 °C. The moisture content of the materials decreases due to the mechanical mixing. The preferred moisture content of materials in shoyu-koji cultivation is about 43% initially and about 30% in the finished koji. The relative humidity in the air is kept at almost saturation throughout the koji cultivation. Higher temperature and humidity increase the consumption of starch in the materials during the growth of the molds, and the amount of starch consumed during koji cultivation ranges from 15 to 30% of the total starch available. One example of temperature variation of materials during mechanical shoyu-koji cultivation with a through-flow system of aeration is shown in Figure 9.

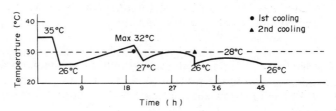

Figure 9 Temperature changes of materials during mechanical shoyu-koji cultivation with a through-flow system of aeration taking 3 days

19.4.4.3 Relationship between the raw materials and the enzyme composition of koji

The pH value and the C/N ratio of the raw materials influence the composition of the enzymes produced in koji. The pH values of the starting materials and those of the finished koji are between 6.0–7.0 in the case of shoyu koji composed of soybeans and wheat, and soybean koji, and around 5.5 in the case of milled rice or barley koji for miso. Higher amounts of ammonia are formed with smaller C/N ratios of the raw materials, and the ammonia shifts the pH value from acidic toward neutral or alkaline, increasing the alkaline protease content. More acidic protease is formed with neutral to acidic raw materials. This is illustrated in the comparison of enzyme compositions of rice koji, shoyu koji composed of soybeans and wheat, and wheat-bran koji shown in Table 15 (Nakadai *et al.*, 1971).

Table 15 Composition of Enzymes in Various Koji Cultures

Enzyme	Substrate for assay	pH for assay	Skoyu koji[a]	Rice koji[a]	Wheat-bran koji[a]	Submerged culture[b]
Total proteinase	Casein	7.0	1500	73	2540	665
Acid proteinase	Casein	3.0	295	417	3020	304
α-Amylase	Starch	5.0	3920	3100	11 700	257
Acid carboxypeptidase I	Cbz-Ala-Glu	4.0	0.456	2.153	5.650	0.665
Acid carboxypeptidase II, III, IV	Cbz-Glu-Tyr	3.0	0.708	1.491	7.430	0.842
Leucine aminopeptidase II, III	Leu-Gly-Gly	8.0	0.360	0.075	0.705	0.430
Leucine aminopeptidase I	Leu-β-NA	8.0	1.405	0.132	4.660	1.624
CM-cellulase	CM-cellulose	5.0	21.60	6.71	59.10	0.595
Pectin transeliminase	Pectin	5.5	12.33	1.84	12.88	0.036

[a] Units/g koji. [b] Units/ml culture filtrate.

19.4.4.4 Microbial contamination of koji

Although no health hazard due to bacterial contamination in koji has been reported in Japan to date, bacterial counts may reach 10^{8-9} g^{-1}. Bacterial contamination in koji not only reduces the proteolytic activities of koji but also impairs the quality of the shoyu and miso. The major contaminants in shoyu-koji cultivation in wooden trays with hand operation were *Bacillus subtilis* at high temperature and *Rhizopus nigricans* at low temperature. In the more recent mechanical koji cultivation by the TF-method, the major contaminant in koji tends to be *Micrococcus*, which is aerobic and of lower optimum temperature for growth than those of *Bacillus*. In addition *Leuconostoc*, *Lactobacillus* and *Saccharomyces rouxii* are sometimes found as contaminants in koji. It is possible to reduce the bacterial contamination in koji to 10^{6} g^{-1} or less by starting the koji cultivation with a bacterial count of 10^{2} g^{-1} or less in the raw materials. Reducing the moisture content of the surface of cooked soybeans by coating with finely crushed roasted wheat is effective. The optimal temperature to avoid bacterial contamination in koji making is considered to be below 34 °C (Ueda *et al.*, 1972).

19.4.4.5 Mechanical equipment

Mechanical equipment for koji cultivation can be classified as follows: (1) through-flow system of aeration; batch-type with a fixed rectangular perforated plate, batch-type with a moving circular perforated plate, and continuous-type with a moving circular perforated plate (Akao *et al.*, 1972); (2) rotary drum; (3) surface-flow system of aeration; the temperature and moisture controlled air flows over the materials placed in trays; and (4) liquid cultivation tanks. A rotary drum is sometimes used for rice koji cultivation. The third system is less popular nowadays. The application of liquid cultivation of koji in the protein food industry has not been successful, mainly because of the high cost of facilities, low enzyme activities, and poor flavor of the final products.

19.4.5 Mash Fermentation

19.4.5.1 Microbes present in the mash

The pH value of shoyu mash starts at 6.5–7.0, depending upon the pH value of the koji, which is greatly affected by the acid-forming bacterial contaminants such as *Micrococcus*. The water activity (*aw*) of 18% salt water is 0.88, and that of an average well-aged mash is 0.80. The types of microbe which can grow in the shoyu mash are limited by the 16–18% salt concentrations. The salt intolerant microbes derived from koji, such as wild yeasts, *Micrococcus* and *Bacillus*, are destroyed or stop growing at an early stage in the fermentation, and only salt-tolerant lactobacilli and yeasts can grow in shoyu mash. The major lactobacillus in shoyu mash has been identified as *Pediococcus halophilus*, which has an optimum pH of 5.5–9.0, *aw* of more than 0.81 and can grow in 24% salt solution at 20–42 °C.

The initial pH value of shoyu mash rapidly decreases due to enzymatic degradation of proteins and lactic acid fermentation. When the pH value changes to 5.5 or less, the growth of *S. rouxii* begins, replacing lactobacilli, and reaching a viable count of 10^6 to 10^7 ml^{-1}. The dominant strains of yeast found in shoyu mash are *S. rouxii*, but sometimes *Torulopsis* yeasts such as *T. versatilis* and *T. etchellsii* are found along with *S. rouxii* in shoyu mash fermentation. The *aw* values of *S. rouxii* and *Torulopsis* yeasts are 0.78–0.81 and 0.98–0.84, respectively, and both of them can grow in salt concentrations less than 24–26% (Yoshii, 1979). These yeasts can grow at pH 3–7 in salt-free media, but this range is narrowed to 4–5 in 18% salt solution (Ohnishi, 1963, 1979). The changes of microflora in shoyu mash are illustrated in Figure 10 (Tamagawa *et al.*, 1975).

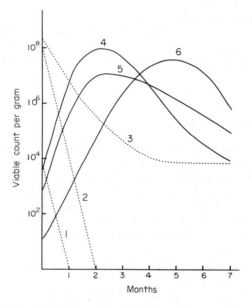

Figure 10 Microflora changes in shoyu mash fermentation: 1 wild yeast; 2 *Micrococcus*; 3 *Baccilus*; 4 *Lactobacilli*; 5 *Saccharomyces rouxii*; 6 *Torulopsis* yeasts (Tamagawa *et al.*, 1975)

19.4.5.2 Chemical changes in shoyu mash fermentation

The enzymatic degradation of proteins into lower peptides, free amino acids and ammonia almost stops after 2 or 3 months from the beginning of the fermentation, depending on the temperature. Carbohydrates are degraded into hexoses and pentoses, and these are metabolized partly into about 1% lactic acid and other organic acids by lactobacilli and partly into 2–3% ethanol and other minor compounds by yeasts, with 2–4% glucose and a trace amount of xylose usually remaining in the final mash. If the mash is adequately controlled, lactic and alcoholic fermentations are almost complete 3 to 4 months from the beginning, but an additional 3–4 months are necessary to finish the mash fermentation or for aging, which consists of 'browning reactions' such as Strecker degradations. The coloration of shoyu mash becomes about twice as strong in this latter stage of fermentation, and the final pH value is 4.8–5.0.

The diversity of lactobacilli in shoyu mash, which influences the aroma, pH and color of the shoyu (Fujimoto *et al.*, 1978, 1980), metabolism of organic acids (Terasawa *et al.*, 1979) sugars and some amino acids such as arginine, histidine, tyrosine, and aspartic acid (Uchida, 1978), has been pointed out. Uchida (1982) classified the lactobacilli in shoyu mash into 67 strains according to their metabolism of 10 different sugars including glucose, arabinose, lactose, melibiose, mannitol and sorbitose. It was found that some strains metabolize arginine into ornithine (Iizuka and Goan, 1973), or histidine into histamine, or tyrosine into tyramine, along with two moles of ammonia in each case, which increases the pH value of the shoyu mash. Some other strains metabolize aspartic acid into alanine and carbon dioxide and increase the pH value of the mash. Some of them degrade citric acid into acetic acid and malic acid, the latter of which is metabolized into lactic acid and acetic acid (Terasawa *et al.*, 1979).

19.4.5.3   Mash temperature

As a result of seasonal fluctuations in temperature, shoyu prepared during the summer has a lower nitrogen content, less amino nitrogen and glutamic acid and a greater content of organic acids compared to shoyu prepared during the winter. By reducing the temperature of a new mash, a 1–3% increase in protein digestibility is obtained, as the lower temperature prevents the rapid decrease in pH value caused by rapid lactic fermentation, which inactivates alkaline protease (Goan, 1957). Currently, koji is mixed with salt water at about 0 °C to keep the temperature of the new mash below 15 °C for several days and then gradually raise it to 28–30 °C after 20–30 days from the start of fermentation.

Glutaminase derived from koji molds is sensitive to heat and its activity rapidly decreases in a new mash (Kuroshima *et al.*, 1969). Adding salt and heat-tolerant glutaminase derived from yeasts to the new mash is effective in increasing the glutamic acid content of the final product by some 20%, regardless of the temperature of the new mash (Yokotsuka *et al.*, 1972; Iwasa *et al.*, 1972a, 1972b).

19.4.5.4   Mash fermentation period

It required 18 months for the glutamic acid content of the mash to reach a maximum in 1931, 10 months in 1953, and currently 3 months are required. The maximum glutamic acid content was taken as an indication of the completion of fermentation.

To accelerate the alcholic fermentation, pure cultured yeasts (*S. rouxii*) are sometimes added to the shoyu mash when its pH value reaches about 5.3, usually 3 to 4 weeks after making the mash (Watanabe *et al.*, 1970). The addition of *Torulopsis* yeasts along with *S. rouxii* is recommended as it gives a good flavor in the finished product (Suzuki *et al.*, 1972). The addition of a yeast starter, both of *Saccharomyces* and *Torulopsis*, becomes important when the mash is stored in resin-coated iron tanks, because the natural level of starter yeasts in these tanks is usually insufficient compared with that in wooden kegs or concrete tanks.

The factors which most hinder the activities of lactobacilli and yeasts in shoyu mash were found to be the salt content and the presence of ether-extractable compounds, such as guaiacol and vanillin, and of alcohol in the case of yeasts (Sakasai *et al.*, 1975a, 1975b; Noda *et al.*, 1976a, 1976b). Lactic acid fermentation is affected by the yeasts, including those derived from koji (Kusumoto *et al.*, 1977; Fujimoto *et al.*, 1978).

19.4.6   Refining

19.4.6.1   Mash pressing

The difficulties associated with pressing shoyu mash have been reduced as a result of the increase in protein digestibility in recent years. The activity of the plant tissue degrading enzymes in the koji is related to the viscosity and the amount of shoyu mash cake. Pectin transeliminase was claimed to be most important in this regard (Ishii *et al.*, 1972). The compounds which cause problems in the pressing of the shoyu mash are acidic polysaccharides derived from soybeans (Kikuchi, 1979; Kikuchi *et al.*, 1976).

19.4.6.2   Pasteurization

The filtrate of an aged mash is heated at 70 to 80 °C to stop the majority of microbial and enzymatic reactions. The formation of an agreeable brown color, the separation of heat-coagulated substances, increases in acidity, depth of color, clarity and anti-yeast potency, decreases in reducing sugar and amino acid contents, and the evaporation of volatile compounds are major changes which occur during this heating (Yokotsuka and Takimoto, 1956, 1958; Okuhara and Yokotsuka, 1962). In Japan, benzoic acid or butyl *p*-hydroxybenzoate are legally added to refined shoyu as a preservative but the trend seems to be toward either aseptic bottling of shoyu, or bottling shoyu which is fortified with ethanol as a preservative.

The coagulated substances produced by heating raw shoyu are equivalent to 10% by volume and 0.025–0.05% by weight of the shoyu, and they consist mainly of proteins derived from the koji mold. The proteins associate with each other through hydrophobic bonds due to the action of both heat and proteases. The acid proteinases (optimum pH 5.0) accelerate heat-induced coagulation. Adding proteases, in particular heat tolerant proteases, also accelerates this process (Hashimoto and Yokotsuka, 1976).

It has been claimed that alkaline protease accelerates heat-induced coagulation (Tomura *et al.*, 1982). Motai *et al.* (1983) reported that the alkaline protease accelerates the coagulation of raw shoyu; this coagulum consists of very small particles and sediment formation is difficult. This sedimentation in pasteurized shoyu is accelerated by the acid protease of koji, not by amylase. Alkaline protease hindered sediment formation of the coagulated substances. The volume of sediment derived from heat-coagulated substances in the bottom layer of pasteurized shoyu is related to the yield of refined shoyu. Noda *et al.* (1982) kept the aged shoyu mash at 40 °C for 120 h, at 45 °C for 72 h, or at 50 °C for 24 h in order to reduce the activity of the protease and amylase, thereby reducing heat-induced coagulation by 60%, 84%, or 99% respectively.

19.4.7   Quality of Koikuchi Shoyu

19.4.7.1   Color of shoyu

The color change in shoyu caused by an oxidative reaction after opening the package is an important problem, because flavor deterioration is observed together with darkening. The factors which influence the oxidative color change in shoyu are temperature, the initial color, the total nitrogen content and the amounts of sugars, carbonyl compounds, peptides, amino acids, 3-deoxyosones, Amadori compounds, reductones, organic acids, iron, copper, sodium chloride, *etc.* (Okuhara *et al.*, 1971, 1972, 1975; Hashiba, 1981).

19.4.7.2   Flavor evaluation of koikuchi shoyu

The aroma of fermented shoyu is found to be roughly proportional to its ethanol content. Tests indicated that of factors such as the partial correlation coefficients of 11 chemical components present, alcohol content, Baume, amount of sodium chloride, amount of reducing sugars, total nitrogen, formol nitrogen, titratable acidity and pH, the alcohol content had the most significant effect. Eleven types of odor were found to influence 96.5% of subjects in shoyu flavor tests. The smell of alcohol had a favorable influence, while the odor of chemical hydrolyzate of plant protein mixed with fermented shoyu, an oily smell, the smell of natto, an abnormal odor, the odor of butyric acid, the smell of shoyu mash fermented at too high a temperature, a cooked soybean smell, and a moldy smell, in descending order, all had unfavorable effects. Shoyu having a good flavor is produced by the totally fermentative method; it is free from the disagreeable odor derived from bacterial contamination. It must be well-fermented by yeasts, and its chemical components must be well balanced (Tanaka *et al.*, 1969a, 1969b, 1970). According to the results of multivariate analysis, an organoleptically preferable shoyu has balanced taste components and a good aroma (Mori, 1979).

19.4.7.3   Flavor components of shoyu

The importance of the volatile components of shoyu was indicated by the fact that the organoleptic ranking of two samples of shoyu could be reversed by exchanging their volatile fractions. Japanese investigators have identified nearly 300 volatile flavor compounds in the odor of koikuchi shoyu (Yokotsuka *et al.*, 1980). One example of the results of the quantitative analyses of the volatile flavor constituents of pasteurized koikuchi shoyu is given in Table 16 (Yokotsuka *et al.*, 1980).

Table 16   Results of Quantitative Analysis of Flavor Constituents in Koikuchi Shoyu[a]

Compound	Amount present (p.p.m.)	Compound	Amount present (p.p.m.)
Ethanol	31 501.10	Furfuryl alcohol	11.93
Lactic acid	14 346.57	Isopentyl alcohol	10.01
Glycerol	10 208.95	Acetoin	9.78
Acetic acid	2 107.74	*n*-Butyl alcohol	8.69
HMMF	256.36	HDMF	4.83
2,3-Butanediol	238.59	Acetaldehyde	4.63
Isovaleraldehyde	233.10	2-Phenylethanol	4.28
HEMF	239.04	*n*-Propyl alcohol	3.96
Methanol	62.37	Acetone	3.88
Acetol	24.60	Methionol	3.65
Ethyl lactate	24.29	2-Acetylpyrrole	2.86
2,6-Dimethoxyphenol	16.21	4-Ethylguaiacol	2.77
Ethyl acetate	15.13	Ethyl formate	2.63
Isobutyraldehyde	14.64	γ-Butyrolactone	2.02
Methyl acetate	13.84	4-Ethylphenol	Trace
Isobutyl alcohol	11.96		

[a] From Yokotsuka *et al.* (1980).

The most important part of the aroma characteristic to koikuchi shoyu seems to exist in its weakly acidic fraction, from which the following important flavor compounds have been isolated: (1) methionol (3-methylthio-1-propanol; Akabori and Kaneko, 1936); (2) 4-ethylguaiacol and *p*-ethylphenol (Yokotsuka, 1953; Asao and Yokotsuka, 1958); (3) esters of phenolic compounds such as 4-ethylguaiacol, vanillic acid and vanillin, with organic acids such as benzoic acid and acetic acid (Yokotsuka, 1953); (4) maltol (Nunomura *et al.*, 1976) and 5-hydroxymaltol (Nunomura *et al.*, 1980); (5) hydroxyfuranones (Nunomura *et al.*, 1976, 1978, 1980), namely HEMF (4-hydroxy-2(or 5)-ethyl-5(or 2)-methylfuran-3(2*H*)-one), HMF (4-hydroxy-5-methylfuran-3(2*H*)-one) and HDMF (4-hydroxy-2,5-dimethylfuran-3(2*H*)-one); (6) cyclothen (2-hydroxy-3-methylcyclopent-2-en-1-one); and (7) lactones such as 4-butanolide, 4-pentanolide, 2-methyl-4-butanolide (Nunomura *et al.*, 1980) and 4-hexanolide (Liardon and Philipposian, 1978).

About 25% of 50–70 samples of shoyu tested in 1964 contained 0.5–2.0 p.p.m. 4-ethylguaiacol (4EG). It was observed that 4EG also had a taste characteristic of fermented shoyu, and that it ameliorated the salty taste of shoyu. 4-Ethylguaiacol and *p*-ethylphenol are produced by *Torulopsis* yeasts, and not by *S. rouxii*, the predominant yeasts of shoyu mash (Yokotsuka *et al.*, 1967; Asao *et al.*, 1967). The yeast flora in 35 types of shoyu mash obtained in Hokkaido (northern most island of Japan) investigated in 1960, and it was found that the organoleptically good mashes were highly populated with *T. etchellsii* and *T. versatillis* (Sasaki and Yoshida, 1966; Yoshida, 1979). The 4EG contents of three popular brands of koikuchi shoyu were reported to be 1.0, 1.8 and 2.1 p.p.m., respectively, and those of three types of usukuchi shoyu to be 0.5, 1.3 and 0.3 p.p.m., respectively (Noda and Nakano, 1979).

Hydroxyfuranones such as HEMF, HMF and HDMF have a caramel-like flavor. HEMF seems to be the most important ingredient of fermented shoyu aroma as it is present in amounts of more than 100 p.p.m. and has a very low threshold value, 0.04 p.p.b. or less, in water (Nunomura *et al.*, 1976). Nunomura *et al.* (1978, 1980) identified 70 pyrazine compounds in shoyu. The presence of four pyrazines in major amounts and their significant increase in the course of pasteurization have been pointed out. The flavor differences between four major brands of shoyu on the Japanese market were explained by differences in the amounts (p.p.m.) of various volatile flavor components as follows: isobutyl alcohol 0.43–1.99, *n*-butyl alcohol 0.46–1.48, isopentyl alcohol 1.08–3.63, acetoin 0.74–1.51, ethyl lactate 4.04–13.15, furfuryl alcohol 2.02–3.55, methionol 1.47–3.39, 2-phenylethanol 1.61–3.67, 4-ethylguaiacol 0.31–1.31 (Sasaki *et al.*, 1970). According to Mori *et al.* (1982, 1983), the correlation coefficients of 27 odor components and the sensory evaluation of shoyu was 0.313 at highest, which suggested that it was difficult to predict the acceptability of shoyu on the basis of only one type of odor component. 4-Ethylguaiacol and methionol were found to have a synergistic effect, and this combination was revealed as the most significant variable which can influence the variation of sensory data. The optimum amounts of 4EG and methionol were found to be 0.8 p.p.m. and 3.9 p.p.m., respectively, by adding test amounts of both materials to shoyu.

19.4.7.4 *Nutritional value*

According to Government figures, the per capita daily consumptions in 1974 of miso and shoyu were 18.6 g and 30.0 ml, respectively. The former was equivalent to 122.1 kJ., 2.3 g protein, 0.6 g fat and 1.86 g salt, and the latter was equivalent to 58.5 kJ, 2.4 g protein, 0.2 g fat and 5.46 g salt. The salt contents were calculated on the basis of 10% for miso and 16% for shoyu. From these figures, it can be seen that in Japan the daily per capita salt intake from miso and shoyu is 7.32 g. The daily salt intake in Japan per person in 1972 was 14.5 g, while it decreased to 13.1 g in 1980, ranging from 10 g to over 15 g depending upon the district. A high incidence of high blood pressure, especially in the district of high salt consumption in the northern part of Japan, has been pointed out. Low-salt miso and shoyu, in which the salt contents are cut by 20% or 50%, have been gradually becoming more popular in the Japanese market.

19.4.7.5 *Safety problems*

The mutagenicity of the products resulting from heating some amino acids or proteins is generally recognised nowadays. Secliff and Mower (1977) reported that upon heating, soysauce produces mutagens from glucose, galactose and arabinose in shoyu. Lin *et al.* (1978) reported that when treated with nitrite at a level of 2000 p.p.m. soybean sauce produced a mutagenic substance, demonstrated by using the *Salmonella*/mammalian microsome mutagenicity test. Nagahori *et al.* (1980) reported a 60–80% suppression of *N*-nitrosodimethylamine formation by adding 5–7% fermented shoyu to a mixture of dimethylamine and nitrite at pH 3.6. The nitrosamine-inhibiting substances in shoyu were concluded to be amino acids, which more easily react with nitrite than dimethylamine by the Vanslyke reaction.

The long term effects of koikuchi shoyu (Kikkoman) on the gastric mucosa were assessed in intact rats and those having undergone a fundasectomy by MacDonald and Dueck in Canada (1976). The animals given shoyu for 29 months were smaller than the controls but healthier, more active and longer lived. Breast tumors developed in the control rats but none in those given shoyu. It was concluded that shoyu did not appear to be carcinogenic to the rats; its prolonged use impaired neither health nor longevity. Carcinogenicity was not observed in long term feeding tests of shoyu to mice and rats by the Kikkoman Co. (Ohshita *et al.*, 1977).

19.5 REFERENCES

Aiba, T. (1982). Modernization of roasting wheat for brewing. *Proceedings of the 14th Symposium on Brewing*, p. 20.

Aibara, K. and K. Miyaki (1965). Qualitative and quantitative analysis of aflatoxin. *J. Agric. Chem. Soc. Jpn.*, **39**, 86.

Akabori, S. and T. Kaneko (1936). On the flavor components of shoyu. II. Isolation and synthesis of a sulfur containing flavor compound from shoyu. *J. Chem. Soc. Jpn.*, **57**, 832–836.

Aonuma, A., A. Yasuda, T. Yuasa, A. Arai, K. Mogi and T. Yokotsuka (1970 and 1971). *Jpn. Pat.* 794 915 and *US Pat.* 3 764 708.

Akao, T., T. Sakasai, M. Sakai, A. Takano and T. Yokotsuka, (1972). A new plant for continuous production of Koji in the shoyu industry. *Proceedings of the First Pacific Chemical Engineering Congress, Kyoto, Japan*, p. 127.

Asao, Y. and T. Yokotsuka (1958). Studies on flavor substances in soy (17). Guaiacyl compounds in soy (1). *J. Agric. Chem. Soc. Jpn.*, **32**, 622–628.

Asao, Y., T. Sakasai and T. Yokotsuka (1967). Studies on flavor substances in shoyu (26). Flavor compounds produced by yeast fermentation (2). *J. Agric. Chem. Soc. Jpn.*, **41**, 434–441.

Bo, Thi-An (1982). Origin of chiang and chiang-yu and their technology of production. *J. Brew. Soc. Jpn.*, **77**, 365–371.

Bhumiratana, A. (1980). Traditional fermented foods in Thailand. *Proceedings of the Oriental Fermented Foods. Taipei, Taiwan, R.O.C.*, pp. 58–70.

Ebine, H. (1972). *Jozo-Seibun-Ichiran*, 3rd edn., p. 138. The Brewing Society of Japan, Takinogawa, Kitaku, Tokyo, Japan.

Ebine, H. (1980). *Sci. Foods*, **56**, 59.

Eguchi, U. (1977). HTST cooking of soybeans utilizing an NK-cooker. *J. Brew. Soc. Jpn.*, **72**, 250–255.

Fujimoto, H., M. Inao and M. Goan (1978). Studies on lactic acid bacteria in shoyu-mash. Part 1. Functional classification of lactic acid bacteria in shoyu-mash. *J. Jpn. Soy Sauce Res. Inst.*, **4**, 191–195.

Fujimoto, H., T. Aiba and M. Goan (1980). Studies on lactic acid bacteria in shoyu mash. Part 3. Effect of temperature on the growth and fermentation of lactic acid bacteria. *J. Jpn. Soy Sauce Res. Inst.*, **6**, 5–9.

Fukushima, D. and M. Mogi (1955). Treating method of soybean protein. *Jpn. Pat.* 236 368, 237 805.

Fukushima, D. and M. Mogi (1957). Treating method of soybean protein. *Jpn. Pat.* 248 103.

Fukushima, D. and T. Yokotsuka (1972). *Jozo-Seibun-Ichiran*, 3rd edn., p. 142. The Brewing Society of Japan, Takinogawa, Kitaku, Tokyo, Japan.

Furuya, T., M. Ishige, K. Uchida and H. Yoshino (1983). Koji-mold breeding by protoplast fusion for soy sauce production. *J. Agric. Chem. Soc. Jpn.*, **57**, 1–8.

Goan, M. (1957). Glutamic acid content in shoyu. *Seasoning Sci.*, **5** (1), 1–5.

Harada, Y. (1951). *Rep. Tatsuno Brew. Inst.*, **2**, 51.

Harada, Y. and S. Kawaguchi (1968). Dealing method of soybeans for soy sauce brewing. *Jpn. Pat.* 47 26 708.

Hashiba, H. (1981). Oxidative browning of soy sauce. Part III. Effect of organic acids, peptides and pentose. Part IV. Oxidative browning of Amadori compounds. Part V. Isolation and identification of Amadori compounds from soy sauce. *J. Jpn. Soy Sauce Res. Inst.*, **7**, 19–23, 116–120, 121–124.

Hashimoto, H. and T. Yokotsuka (1976). Mechanism of sediment formation during heating of raw shoyu. *J. Brew. Soc. Jpn.*, **71**, 496–499.

Hayashi, K., M. Terada and T. Mizunuma (1981). Some differences in characteristics of soy sauce fermentation between species of Koji-molds, *Aspergillus sojae* and *Aspergillus oryzae*. *J. Jpn. Soy Sauce Res. Inst.*, **7**, 166–172.

Hesseltine, C. W., O. L. Shotwell, J. J. Ellis and R. D. Stublefield (1966). Aflatoxin formation by *Aspergillus flavus*. *Bacteriol. Rev.*, **30**, 795–805.

Hesseltine, C. W. and H. L. Wang (1972). Fermented soybean food product. In *Soybeans: Chemistry and Technology*, pp. 389–419. Avi Publishing, Westport, CT.

Iguchi, N. (1955). Studies on Aspergilli. X. Changes of enzyme activities and induction of mutant having higher proteolytic activities in *Asp. sojae* by induced mutants. *J. Agric. Chem. Soc. Jpn.*, **29**, 73–78.

Iijima, K., M. Dejima, A. Tsuji, T. Watanabe and T. Kagami (1973). A new method for the cooking of soybean meal by a rotary NK-cooker. *Seasoning Sci.*, **20** (7), 13–19.

Imai, S. and K. Suzuki (1967). Studies on soy sauce Koji processed under automatic control (3). The properties of soy sauce Koji processed under the middle temperature in the early stage. *Seasoning Sci.*, **14** (2), 22–32.

Ishii, S., T. Kikuchi, T. Ogami and T. Yokotsuka, (1972). Effect of plant tissue degrading enzyme on shoyu production. *J. Agric. Chem. Soc. Jpn.*, **46**, 340–354.

Ishitani, C. and Sakaguchi Kinichiro (1956). Hereditary variation and recombination in koji-molds (*Aspergillus oryzae* and *Asp. sojae*). *J. Gen. Microbiol.*, **2**, 345–400.

Ishitani, C., Y. Ikeda, and Sakaguchi Kinichiro. (1956). Hereditary variation and genetic recombination in koji-molds (*Aspergillus oryzae* and *Asp. sojae*). VI. Genetic recombination in heterozygous diploids. *J. Gen. Microbiol.*, **2**, 401–430.

Iwasa, T., S. Fujii, T. Kakinuma and T. Yokotsuka (1972a). Glutaminase produced by *Cryptococcus albidus* ATCC 20293. Purification and properties of enzyme. Presented at the Annual Meeting of the Agricultural Chemical Society of Japan, Sendai, April 2.

Iwasa, T., S. Fujii and T. Yokotsuka (1972b). Microbial glutaminases and their properties. Presented at the 4th International Fermentation Symposium, Kyoto, March, 19–25.

Iizuka, K. and M. Goan (1973). Studies of L-arginine decomposition in soy-sauce mash. *Seasoning Sci.*, **20** (5), 17–24.

Kawamori, T. (1940). Brewing method for shoyu from the view point of food problems. *J. Brew.*, **16**, 860–873.

Kawano, Y. (1938). The ingredient change of soybeans during cooking. *J. Brew.*, **16**, 755–767.

Kikuchi, T., H. Sugimoto and T. Yokotsuka (1976). Polysaccharides in pressed cake and their effects on difficulty in press filtration of fermented soy sauce mash. *J. Agric. Chem. Soc. Jpn.*, **50**, 279–286.

Kikuchi, T. (1979). Relation between the difficulty of shoyu-mash pressing and the polysaccharides contained in soybeans. *J. Jpn. Soy Sauce Res. Inst.*, **5**, 71–77.

Kinoshita, R., T. Ishiko, S. Sugiyama, T. Seto, S. Igarashi and I. E. Goetz (1968). Mycotoxin in fermented foods. *Cancer Res.*, **28**, 2296–2311.

Kuroshima, E., Y. Oyama, T. Matsuo and T. Sugimori (1969). Biosynthesis and degradation of glutamic acid in microorganisms relating to soy sauce brewing. (III). Some factors affecting the glutamic acid and its related substances formation in soy sauce brewing. *J. Ferment. Technol.*, **47**, 693–700.

Kusumoto, E., K. Inamori, K. Uchida and H. Yoshino (1977). Studies on microorganisms isolated in fermentation of soy sauce mash (1). *J. Jpn. Soy Sauce Res. Inst.*, **3**, 283–289.

Lee, C. H. and J. Mogens (1981). The effect on Korean soysuce fermentation of the quality of soybean. Presented at the 8th ASKA meeting, Medan, Indonesia, February 9–15.

Lee, G. H. (1981). Nutritional aspects of fermented foods in Indonesia. Presented at the 8th ASKA meeting, Medan, Indonesia, February 9–15.

Lian, O. K. (1981). Some traditional foods in Singapore. Presented at the 8th ASKA meeting, Medan, Indonesia, February 9–15.

Liardon, R. and G. Philippossian (1978). Volatile components of fermented soya hydrolyzate. I. Identification of some lactones. *Z. Lebensm. Unters. Forsch.*, **167**, 180–185.

Lin, J. Y., H. I. Wang and Y. C. Yen (1978). Mutagenicity of soybean sauce. *Consumer Toxicol.*, **17**, 329–331.

Lin, L. P. (1977). Sufu. Presented at the 5th International Conference on Global Impacts of Applied Microbiology, Bangkok, December 21–26.

MacDonald, W. C. and J. W. Dueck (1976). Long term effect of shoyu on the gastric mucosa of the rat. *J. Natl. Cancer Inst.*, **56**, 1143–1147.

Manabe, M., S. Matsuura and M. Nakano (1968). Studies on the fluorescent compounds in fermented foods. Part 1. Chloroform soluble fluorescent compounds produced by Koji molds. *Nippon Shokuhin Kogyo Gakkaishi*, **15** (8), 7–12.

Masuda, Y., K. Mori and M. Kuratsune (1965). *Proceedings of the Annual Meeting of the Japanese Cancer Association*, Fukuoka, p.50.

Mori, S. (1979). Quality evaluation of shoyu and multivariate analysis. *J. Brew. Soc. Jpn.*, **74**, 526–531.

Mori, S., N. Nunomura and M. Sasaki (1982). A specific combination of flavor components influencing the odor preference of soy sauce assessed by sensory evaluation. *Proceedings of the Annual Meeting of the Agricultural Chemical Society of Japan*, p.36.

Mori, S., N. Nunomura and M. Sasaki (1983). The influence of 4-ethyl-2-methoxyphenol and 3-(methylthio)-1-propanol on the flavor of soy sauce. Presented at the Annual Meeting of the Agricultural Chemical Society of Japan, March 30.

Moriguchi, S., T. Kawaguchi, Y. Ishigami and R. Ueda (1961). Studies on the difference in shoyu made from raw and defatted soybeans. IV. The composition of organic acids and other components in shoyu moromi produced in every season. *Seasoning Sci.*, **9** (4), 34–38.

Motai, H., K. Hayashi, T. Ishiyama and T. Sonehara (1983). Role of the main enzyme proteins of koji-mold *Aspergillus* in sedimentation of coagula in soy sauce during pasteurization. *J. Agric. Chem. Soc. Jpn.*, **57**, 27–36.

Murakami, H., S. Takase and T. Ishii (1967). Non-productivity of aflatoxin by Japanese industrial strains of *Aspergillus*. I. Production of fluorescent substances in agar slant and shaking cultures. *J. Gen. Appl. Microbiol.*, **13**, 323–334.

Nagahori, T., H. Motai and A. Okuhara (1980). On the substances in shoyu to suppress the nitrosamine of dimethylamine. *Nutrition and Foods*, **33**, 151–160.

Nakadai, T., S. Nasuno and N. Iguchi (1971). Qualitative estimation of activities of peptidases in koji. *Seasoning Sci.*, **18**, 41–47.

Nakadai, T., S. Nasuno and N. Iguchi (1972a). The action of peptidases from *Aspergillus sojae* on soybean proteins *J. Agric. Biol. Chem.*, **36**, 1239–1243.

Nakadai, T., S. Nasuno and N. Iguchi (1972b). Quantitative estimation of activities of peptidase in koji. *Seasoning Sci.*, **18**, 41–47.

Nakadai, T. (1977). Estimation of composition of various enzymes produced by shoyu Koji culture of *Aspergillus sojae*. *J. Jpn. Soy Sauce Res. Inst.*, **3**, 99–104.

Nasuno, S., T. Ohara and N. Iguchi (1972). Improvement of Koji-mold for soy sauce production by induced mutation (I). Successive isolation of proteinase hyperproductive mutants of *Aspergillus sojae*. *Seasoning Sci.*, **19** (10), 27–31.

Noda, F., T. Sakasai and T. Yokotsuka (1976a). Studies on fermentation of soysauce mash (4). Identification of soysauce components which have an inhibitory effect on growth and an enhancing effect on fermentation activities of soysauce yeasts. *Nippon Shokuhin Kogyo Gakkaishi*, **23** (2), 53–58.

Noda, F., T. Sakasai and T. Yokotsuka (1976b). Studies on fermentation of soysauce mash (5). On the antibiotic activities of vanillin, guaiacol and isomers of vanillin. *Nippon Shokuhin Kogyo Gakkaishi*, **23** (2), 59–66.

Noda, Y. and M. Nakano (1979). Studies on the control of the fermentation of soy sauce mash. I. Determination of volatile phenols in soy sauce. *J. Jpn. Soy Sauce Res. Inst.*, **5**, 299–302.

Noda, Y., H. Mine and M. Nakano (1982). Effect of keeping raw shoyu or shoyu mash on the prevention of sediment produced by heating. *J. Jpn. Soy Sauce Res. Inst.*, **8**, 176–181.

Nunomura, N., M. Sasaki, Y. Asao and T. Yokotsuka (1976). Isolation and identification of 4-hydroxy-2(or 5)-ethyl-5(or 2)-methyl-3(2H)-furanone, as a flavor component in shoyu. *Agric. Biol. Chem.*, **40**, 491–495.

Nunomura, N., M. Sasaki, Y. Asao and T. Yokotsuka (1978). Shoyu (soy sauce) volatile flavor components: basic fraction. *Agric. Biol. Chem.*, **42**, 2123–2128.

Numomura, N., M. Sasaki and T. Yokotsuka (1980). Shoyu (soy sauce) flavor components: acidic fractions and the characteristic flavor component. *Agric. Biol. Chem.*, **44**, 339–351.

Oda, K. and N. Iguchi (1963). Genetic and biochemical studies on the formation of protease in *Aspergillus sojae*. *Agric. Biol. Chem.*, **27**, 758–766.

Ohara, H., S. Moriguchi and T. Nishiyama (1959). The effect of the temperature and moisture upon the cultivation of shoyu Koji. (1), (2). *Seasoning Sci.*, **7** (3), 25–29, 30–42.

Ohnishi, H. (1963). Shoyu yeasts. *J. Brew. Assoc. Jpn.*, **58**, 602–607.

Ohnishi, H. (1979). Shoyu brewing and microorganisms (I–III). *J. Jpn. Soy Sauce Res. Inst.*, **5**, 28–34, 78–82, 129–134.

Okuhara, A. and T. Yokotsuka (1958). Studies on the analytical method of shoyu (I). Analytical method of glycerol in shoyu. *J. Agric. Chem. Soc. Jpn.*, **32**, 138–143.

Okuhara, A. and T. Yokotsuka (1962). Studies on the analytical method of shoyu (III). Analytical method of total polyol in shoyu. *J. Agric. Chem. Soc. Jpn.*, **36**, 320–326.

Okuhara, A. and T. Yokotsuka (1963). Studies on the flavor substances in shoyu (22). Differences of shoyu made from whole and defatted soybean. *J. Agric. Chem. Soc. Jpn.*, **37**, 255–261.

Okuhara, A., N. Saito and T. Yokotsuka (1971). Color of soy sauce. VI. The effect of peptides on browning. *J. Agric. Chem. Soc. Jpn.*, **49**, 272–287.

Okuhara, A., N. Saito and T. Yokotsuka (1972). Color of soy sauce. VII. Oxidative browning. *J. Agric. Chem. Soc. Jpn.*, **50**, 264–272.

Okuhara, A., M. Saeki and T. Sasaki (1975). Browning of soy sauce. *J. Jpn. Soy Sauce Res. Inst.*, **1**, 185–189.

Oshita, K., Y. Nakashima, S. Sugiyama and R. Takahashi (1977). Safety evaluation of Shoyu. In Oral Presentation by T. Yokotsuka on 'Shoyu'. Select Committee of GRAS Substances Hearing, July 26, 1977, Bethesda, MD, USA.

Saito, K., T. Hasuo, M. Tadenuma and Y. Akiyama (1983). Fermented soybeans in India, Thailand and Japan. *J. Brew. Soc. Jpn.*, **78**, 69–72.

Sakaguchi Kinichiro (1979). Search for the route of shoyu. *Sekai*, **1**, 252–266.

Sakasai, T., F. Noda and T. Yokotsuka (1975a). Studies on fermentation of soy sauce mash (I). Effect of components in soy sauce on growth of microorganisms. *Nippon Shokuhin Kogyo Gakkaishi*, **22**, 474–480.

Sakasai, T., F. Noda and T. Yokotsuka (1975b). Studies on fermentation of soy sauce mash (II). Effect of components in soy sauce and physical environmental factors on growth of lactic acid bacteria and yeasts in soy sauce mash. *Nippon Shokuhin Kogyo Gakkaishi*, **22**, 481–487.

Sasaki, M., Y. Kaneko, K. Oshita, H. Takamatsu, Y. Asao and T. Yokotsuka (1970). Studies on compounds produced by molds. VII. Isolation of isocoumarin compounds. *Agric. Biol. Chem. Jpn.*, **34**, 1296–1300.

Sasaki, M. and T. Yokotsuka (1972). Fluorescent compounds produced by molds having *Rf* values resembling those of aflatoxins. *J. Ferment. Assoc.*, **30**, 205–216.

Sasaki, M. (1980). Unpublished results. Kikkoman Shoyu Co., Ltd., Japan.

Sasaki, Y. and T. Yoshida (1966). Microbial studies of soya mashes in Hokkaido (I). A taxonomic study and distribution of yeasts. *J. Ferment. Technol.*, **44**, 61–71.

Secliff, J. A. and H. F. Mower (1977). Mutagens in decomposition products of carbohydrate. *Federal Proc.*, **36**, 304.

Sekine, H. (1972). Neutral protease I and II of *Aspergillus sojae*. Isolation in homogeneous form, and some enzymatic properties. *Agric. Biol. Chem.*, **36**, 198–206, 207–216.

Sekine, H. (1976). Neutral protease I and II of *Aspergillus sojae*. Action on various substrates. *Agric. Biol. Chem.*, **40**, 703–709.

Su, Y. C. (1980). Traditional fermented foods in Taiwan. *Proceedings of the Oriental Fermented Foods*, Taipei, Republic of China, p.15–30.

Suzuki, T., N. Shibuya, Y. Watanabe and R. Tazaki (1972). The control of soy sauce fermentation (2). The additional effect of yeast. *Seasoning Sci.*, **19** (1), 30–35.

Tagami, H. and S. Sugawara (1977). Proteinase composition of various Koji molds. *J. Jpn. Soy Sauce Res. Inst.*, **3**, 111–115.

Tamagawa, Y., K. Yamada, K. Takinami, K. Kodama and T. Suga (1975). *Proceedings of the Annual Meeting on Fermentation Technology*, p.212.

Tamura, J., T. Aiba and K. Mogi (1982). Proteinaceous compounds associated with the heat coagulum formation of shoyu. *Proceedings of the Annual Meeting of the Agricultural Chemical Society of Japan*, p. 449.

Tanaka, T., N. Saito and T. Yokotsuka (1969a). Relation between chemical factors and preference of soy sauce. *Seasoning Sci.*, **16** (3), 21–26.

Tanaka, T., N. Saito, T. Nakajima and T. Yokotsuka (1969b). Studies on sensory evaluation of soy sauce (VI). Interrelation and harmony of flavor. *J. Ferment. Technol.*, **47**, 137–145.

Tanaka, T., N. Saito and T. Yokotsuka (1970). Studies on sensory evaluation of soy sauce (XI). Relation between odor patterns and chemical factors (2). *J. Ferment. Technol.*, **48**, 56–62.

Tateno, M. and I. Umeda (1955). Cooking method of soybeans and soybean cake as raw material for brewing. *Jpn. Pat.* 204 858.

Terada, M., K. Hayashi, T. Mizunuma and K. Mogi (1973). Effects of salt on the digestion of soybean proteins by the enzymes from *Aspergillus sojae* and on the digestion of proteinaceous compounds of shoyu-koji. *Seasoning Sci.*, **20** (2), 23–31.

Terada, M., K. Hayashi and T. Mizunuma (1980). Distinction between *Aspergillus oryzae* and *Aspergillus sojae* by the productivity of some hydrolytic enzymes. *J. Jpn. Soy Sauce Res. Inst.*, **6**, 75–81.

Terada, M., K. Hayashi and T. Mizunuma (1981). Enzyme productivity and identification of species of Koji-molds for soy sauce distributed in Japan. *J. Jpn. Soy Sauce Res. Inst.*, **7**, 158–172.

Terasawa, M., K. Kadowaki, H. Fujimoto and M. Goan (1979). Studies on lactic acid bacteria in shoyu mash. Part 2. Citric acid metabolism by lactic acid bacteria in shoyu mash. *J. Jpn. Soy Sauce Res. Inst.*, **5**, 15–20.

Terui, G., I. Shibasaki and T. Mochizuki (1958). Industrial culture of koji with through-flow system of aeration (II). Improved method. *J. Ferment. Technol.*, **36**, 109–116.

Uchida, K., C. Ishitani, Y. Ikeda and Sakaguchi Kinichiro. (1958). An attempt to produce interspecific hybrids between *Aspergillus oryzae* and *Asp. sojae*. *J. Gen. Appl. Microbiol.*, **4**, 31–38.

Uchida, K. (1978). The preferential fermentation of pentoses by *Pediococcus halophilus* X-160, lactobacilli in shoyu mash. *Proceedings of the Annual Meeting of the Agricultural Chemical Society of Japan*, p.458.

Uchida, K. (1982). Multiplicity in soy pediococci carbohydrate fermentation and its application for analysis of their flora. *J. Gen. Appl. Microbiol*, **28**, 215–223; *Proceedings of the 14th Symposium on Brewing*, pp. 67–69. Brewing Society of Japan, Tokyo.

Ueda, R., Y. Yamamoto, T. Okuno and Y. Ishigami (1972). Studies on changes in microflora and its metabolism during the process of making koji in soy sauce brewing. *Seasoning Sci.*, **19** (7), 30–48.

Wang, H. L. and S. F. Fuang (1981). *History of Chinese Fermented Foods*. USDA Miscellaneous Publication FL-MS-33. NRRL, US Dept. of Agric. Peoria, IL.

Watanabe, Y., M. Ishii and R. Tazaki (1970). The control of soy sauce fermentation (1). The addition of yeast and Lactobacteliaceae. *Seasoning Sci.*, **17** (1), 35–40.

Yamaguchi, S. (1954). Brewing method of shoyu. *Jpn. Pat.* 211 529. Kikkoman Shoyu. Co. Ltd.

Yasuda, A., K. Mogi and T. Yokotsuka (1973a). Studies on the cooking method of proteinaceous materials for soy sauce brewing. Part 1. High temperature and short time cooking method. *Seasoning Sci.*, **20** (7), 20–24.

Yasuda, A., A. Arai, N. Tsukada, K. Mogi, T. Aonuma and T. Yokotsuka (1973b). Studies on the cooking method of proteinaceous materials for soy sauce brewing. Part 2. Brewing test of soy sauce. *Seasoning Sci.*, **20** (7), 25–27.

Yokotsuka, T. (1953). Studies on flavor substances in soy (9). Isolation of phenolic flavor substances. *J. Agric. Chem. Soc. Jpn.*, **27**, 276–281.

Yokotsuka, T. and K. Takimoto (1956). Studies on flavor substances in soy (14). Flavor substances in heated soy (1). *J. Agric. Chem. Soc. Jpn.*, **30**, 66–71.

Yokotsuka, T. and K. Takimoto (1958). Studies on flavor substances in soy (15). Flavor substances in heated soy (2). *J. Agric. Chem. Soc. Jpn.* **32**, 23–26.

Yokotsuka, T. (1960). Aroma and flavor of Japanese soy sauce. *Advances in Food Research*, vol. 10, pp. 75–134. Academic, New York.

Yokotsuka, T., K. Mogi, D. Fukushima and A. Yasuda (1966a). Dealing method of proteinaceous raw materials for brewing. *Jpn. Pat.* 929 910. Kikkoman Shoyu Co., Ltd.

Yokotsuka, T., M. Sasaki, T. Kikuchi, Y. Asao and A. Nobuhara (1966b). Production of fluorescent compounds other than aflatoxins by Japanese industrial molds. In *Biochemistry of Some Foodborne Microbial Toxins*, ed. R. Mateles and G. N. Wogan, pp. 131–152. MIT Press, Cambridge, MA.

Yokotsuka, T., T. Sakasai and Y. Asao (1967). Studies on flavor substances in shoyu (25). Flavor components produced by yeast fermentation (1). *J. Agric. Chem. Soc. Jpn.*, **41**, 428.

Yokotsuka, T., Y. Asao, M. Sasaki and K. Oshita (1968a). Pyrazine compounds produced by molds. *Proceedings of the First US–Japan Conference*, Honolulu, Hawaii, p.133, UJNR Joint Panel on Toxic Microorganisms and US Dept. of the Interior.

Yokotsuka, T., K. Oshita, T. Kikuchi and M. Sasaki (1969). Studies on the compounds produced by molds. VI. Aspergillic acid, kojic acid, β-nitropropionic acid and oxalic acid in solid-Koji. *J. Agric. Chem. Soc. Jpn.*, **43**, 189–196.

Yokotsuka, T., T. Iwasa, S. Fujii and T. Kakinuma (1972). The role of glutaminase in shoyu brewing. Presented at the Annual Meeting of the Agricultural Chemical Society of Japan, Sendai, 1st April.

Yokotsuka, T. (1977). Oral Presentation. 'Shoyu'. Select Committee of GRAS Substances Hearing, July 26, Bethesda, MD, USA. Kikkoman Shoyu Co., Ltd.

Yokotsuka, T., M. Sasaki, N. Nunomura and Y. Asao (1980). Flavor of shoyu (1, 2). *J. Brew. Soc. Jpn.*, **75**, 516–522, 717–728.

Yomo, H., T. Yasui, U. Ishigami and K. Omori (1978). Studies on the glutaminase of shoyu Koji (Part 1). *J. Jpn. Soy Sauce Res. Inst.*, **4**, 48–52.

Yoshida, T. (1979). Studies on the yeasts in shoyu mash produced in Hokkaido Island. *Proceedings of the 11th Symposium on Brewing*, p. 62.

Yoshii, H. (1979). Fermented foods and water activity. *J. Ferment. Assoc.*, **74**, 213–218.

20

Production of Baker's Yeast

S. L. CHEN and M. CHIGER
Universal Foods Corporation, Milwaukee, WI, USA

20.1 INTRODUCTION

The first written record of the actual existence of bread dates to around 2600 BC in Babylonia. An extant clay tablet from the period of Urukagina, King of Lagash, revealed that a priest performing the burial rite would take for himself seven urns of beer, 420 pieces of bread, 60 quarts of grain, *etc.* By the time of Hammurabi, who ruled during the 12th century BC, baking had developed into a specialized craft. The discovery of leavened bread was generally attributed to ancient Egyptians. At the time of the early Pharaohs, the beginning of the third millennium BC, the peasant and the slave laborer received a daily ration of four loaves of bread and two jugs of beer as his compensation. Mixing of a fermenting beer with the wheat flour used in the palace household was probably done by a royal baker, leading to the development of a sour-dough process generally accepted by the Egyptian bakers (Jacob, 1944). The knowledge of making leavened bread spread from Egypt to other areas of the Mediterranean world in the 13th century BC. In ancient Jerusalem, for example, public bakeries were established which produced small breads similar to the present day bread rolls (Pyler, 1958). The practice of using beer yeast in sour-dough fermentation continued into the 19th century, when the commercial bakers obtained their yeast supplies from local breweries (Frey, 1930). Due to its bitter taste and variable fermentation activity, brewer's yeast was gradually replaced by distiller's yeast which in turn was replaced by baker's yeast.

The earliest production of pressed baker's yeast probably occurred around 1781 in Holland. With the so-called Dutch process, the yield of pressed yeast was equivalent to only 4–6% of the weight of the raw material used. In 1846, a Vienna process was developed by Mautner. In this process, yeast was recovered from the entire batch by continuously collecting the foam produced during fermentation. The yield of compressed yeast was increased to about 14%, plus a concurrent yield of 30% spirits. Aeration of the grain mash was introduced by Marquardt in 1879, resulting in the yield of compressed yeast gradually increasing to 50–60% and the amount of spirits falling to 20%. In 1919, a process was invented by Sak in Denmark and Hayduck in Germany in which sugar solution was fed to an aerated suspension of yeast instead of adding yeast to a diluted sugar solution. This process was known as 'Zulaufverfahren'. An incremental-feeding or fed-batch process was thus born. At about the same time, the traditional grain mash was replaced by molasses, because of the food shortage during World War I. These refinements gradually raised the yield of compressed yeast to the theoretical maximum of 50% by weight of the raw material used, with no concomitant spirit formation. Such accomplishments eventually led to the development of a baker's yeast industry independent of alcoholic beverage production (Pyke, 1957; Paturau, 1982). Several excellent books and reviews have been published on this subject matter in recent years (Burrows, 1970; Harrison, 1971; Peppler, 1979; Reed, 1982; Reed and Peppler, 1973; Rosen, 1977; White, 1954).

20.2 PHYSIOLOGY OF YEAST GROWTH

20.2.1 Species and Cultures of Baker's Yeast

Historically, beer yeast or brewer's yeast (*Saccharomyces uvarum*, sym. *S. carlsbergensis*) was used for baking purposes, but it was subsequently replaced by distiller's yeast and baker's yeast (*S. cerevisiae*). Attempts have been made periodically to evaluate the baking properties of other yeast strains and cultures. For example, Mitchell (1957) evaluated the baking properties of some 75 cultures including *Candida arborae*, *C. pseudotropicalis*, *C. tropicalis*, *Hansenula subpellicu-losa*, *Saccharomyces chevalieri*, *S. chodati*, *S. diastaticus*, *S. ellipsoideus thermophilus*, *S. fragilis*, *S. italicus*, *S. intermedius*, *S. logos*, *S. marxianus*, *S. osmophilus*, *S. oviformis*, *Schizosaccharo-myces pombe*, *Torula colliculosa*, *T. dattila*, *Zygosaccharomyces lactis*, *Z. drosophilae*, *etc.* Except possibly for the last one, none of these species and cultures has been found to be superior to the *S. cerevisiae*.

20.2.1.1 *Preservation of yeast cultures*

Maintenance of desirable yeast cultures is a major concern of the yeast industry. Several methods of preserving yeast cultures are in use at various organizations (Martin, 1964; Dalby, 1982; Hayner, *et al.* 1955; Beech and Davenport, 1971; Collins and Lyne, 1976).

(i) Periodic transfer

The storage of conventional agar slant or stab cultures at refrigerated temperature (*e.g.* 4–5 °C) has long been used as a means of reducing the metabolic activity of cultures and thereby increasing the duration for periodic transfer to 3–6 months. This method is commonly used for yeasts, algae, filamentous fungi, but seldom for bacteria. Several all purpose growth media, such as malt extract agar, yeast extract agar, beer wort agar, MYPD agar (malt yeast extract–peptone–dextrose), *etc.*, are being used for carrying stock cultures. Other media, such as potato-dextrose agar, yeast extract–peptone–dextrose agar, *etc.*, are commonly used for isolation and detection of yeasts and determination of yeast counts. Bacterial growth in these media is controlled by pH regulation or by addition of an antibiotic, such as tetracycline. Synthetic media, such as yeast morphology agar, yeast nitrogen base, yeast carbon base, vitamin-free yeast base, *etc.*, have been formulated and used for identification, characterization and classification of yeast (Wickerham, 1951). Most of these growth media are available commercially.

(ii) Storage under oil

The practice of covering actively growing slant or stab cultures with sterile mineral oil and subsequently storing them under refrigeration has frequently been used to increase the longevity of cultures and thereby lengthen the transfer intervals. Overlaying with oil prevents dehydration and slows down metabolic activity by reducing oxygen availability.

(iii) Storage in water

This method is extremely strain specific. Some bacteria have been reported to have survived suspension in distilled water for ten years at room temperature. Strains of *S. cerevisiae* stored in dilute buffer at 4 °C showed 2–19% survival after one year.

(iv) Storage in soil

Storage in sterile soil has been used quite commonly for extending the longevity of microorganisms, such as spore-forming bacteria, conidiating fungi, and some algae; however, it has been seldom used for yeasts. In its simplest form, this method consists of adding a suspension of cells or conidia to sterile soil, drying the mixture at room temperature, and then storing it in the refrigerator.

(v) Dehydration

Dehydration, without preliminary freezing, is a method which has taken many forms, from the simple drying of conventional slant cultures of sporulating molds, to some rather complex techniques of vacuum drying. Desiccation with silica gel or on cellulose material has also been used by some workers.

(vi) Freezing

With this method, a small volume of cell suspension is cooled slowly (*i.e* about 1 °C min^{-1}) to −20 °C and then as rapidly as possible until storage temperature (*i.e.* below −55 °C) is reached. Addition of a protective agent, such as glycerol (10–20%) or dimethyl sulfoxide (10%) to the cell suspension prior to freezing is recommended. Thawing should be accomplished as rapidly as possible. Use of liquid nitrogen for ultralow temperature (−130 °C) storage of cells has been successfully used for a variety of cells including fungi, bacteriophages, protozoa, algae, mammalian cells and bacteria (Norris and Ribbons, 1970).

(vii) Lyophilization or Freeze-drying

This is probably the most satisfactory method for long term preservation of yeast cultures. There are many advantages of this method: it requires no subculturing of cells; it causes no change in biochemical properties of the culture; the freeze-dried cells are genetically stable; there is no contamination problem once the ampoule is sealed; the finished ampoule can be shipped by mail; and the initial cost is low. This process consists of introducing a few drops of propagating yeast suspension (in a protective medium containing one part of fresh skim milk and one part of distilled water) into an ampoule, placing a piece of sterile cotton plug into the upper portion of the ampoule, freezing the yeast suspension to −35 °C and connecting the ampoule to a vacuum of 200 μm Hg for moisture removal by sublimation. After drying, the ampoule is sealed under vacuum and stored under refrigeration. During the revival of the yeast culture, the ampoule is opened under aseptic conditions, a small volume of sterile water is added to dissolve the pellet. After 30 minutes, the suspension is streaked in a growth agar medium. As an alternative, the pellet in the ampoule may be emptied into a sterilized flask, containing about 50 ml of a suitable culture medium such as malt extract.

20.2.2 Nutrition and Biosynthesis

The nutritional requirement for the growth of baker's yeast may be estimated from its elemental composition. Based upon the average values published by several authors (Frey *et al.*, 1936; White, 1954; Olbright, 1956; Harrison, 1967 and 1971; Wang *et al.*, 1977), baker's yeast contains (on a dry solids basis) about 46% carbon, 32% oxygen, 8.5% nitrogen, 6% hydrogen and 7.5% ash. On the assumption that 200 g of sucrose is required for the production of 100 g of yeast solids under efficient growth conditions, the following material balance equation may be established:

$$200 \text{ g sucrose} + 10.32 \text{ g NH}_3 + 100.44 \text{ g oxygen} + 7.5 \text{ g 'ash'} \rightarrow$$
$$100 \text{ g yeast solids} + 140.14 \text{ g CO}_2 + 78.12 \text{ g H}_2\text{O} \tag{1}$$

This equation shows that in addition to sucrose, 10.32 g of ammonia, 7.5 g of 'ash' and 100.44 g of oxygen (equivalent to 1 g of oxygen per g of yeast solids produced) are required. It also shows an overall respiratory quotient (RQ) of 1.02 during the course of efficient yeast propagation. The chemical constituents of the growth medium must supply all elemental and energy requirements. Assimilation of these nutrients is discussed in the following sections.

20.2.2.1 Carbon assimilation

The carbon source for baker's yeast propagation usually consists of assimilable sugars, such as glucose, fructose, mannose, galactose, sucrose, maltose and hydrolyzed lactose. Ethanol has also been used, at least partially, as a substrate for yeast production. In order to be assimilated, these compounds must be transported into the yeast cells. Monosaccharides are transported by carrier-mediated or facilitated diffusion. Three carrier systems have been identified: a specific carrier preferring pyranoses with an equatorial hydroxyl at C-4 (such as glucose), a non-specific one taking up all pyranose-type sugars, and an inducible one preferring pyranoses with an axial hydroxyl at C-4 (such as galactose). For disaccharides, sucrose is hydrolyzed in the periplasmic space by invertase before being transported into baker's yeast cells, while unhydrolyzed lactose is not taken up at all. Maltose enters baker's yeast cells either by facilitated diffusion or by active transport. Four transport systems are known. These systems may be induced by their respective substrates, such as maltose, trehalose, α-methyl-D-glucoside and glucose. Their half-lives are very short, of the order of 1 h (Alonso and Kotyk, 1978). Unlike sugars, ethanol enters yeast cells by simple diffusion (Barnett, 1976; Kotyk and Horak, 1981.)

Aerobic assimilation of glucose substrate is either by the glycolytic-TCA cycle pathway or by the pentose phospate pathway. By using C^{14}-labelled glucose substrates, Chen (1959a) demonstrated that aerobically growing *S. cerevisiae* cells catabolize 94% of the substrate through the glycolytic-TCA cycle pathway and only 6% through the pentose phosphate pathway. These results are in sharp contrast to those obtained with other oxidative yeasts, *e.g.* 35% through the pentose phosphate pathway in *Candida utilis* (Blumenthal *et al.*, 1954), 50% in *C. albicans* (Chattaway *et al.*, 1973) and 80–100% in *Rhodotorula gracilis* (Hofer *et al.*, 1971). The initial step in the assimilation of intracellular fructose and mannose is phosphorylation by the constitutive hexokinase. Fructose 6-phosphate is an intermediate for both pathways, while mannose 6-phosphate is either isomerized to fructose 6-phosphate or epimerized to glucose 6-phosphate (Slein, 1950; Noltmann and Bruns, 1958). The ability of baker's yeast to utilize galactose depends on the carbon source on which the yeast is grown. Transferring the yeast grown on D-galactose to a glucose medium leads to a loss of the D-galactose fermenting ability. Conversion of D-galactose to D-glucose 6-phosphate requires the participation of several enzymes. Galactokinase, hexosyl phosphate uridyl transferase, and UDP-D-glucose 4-empimerase are present in galactose-grown yeast, but not in glucose-grown baker's yeast (de Robichon-Szulmajster, 1958a, 1958b; Wilkinson, 1949).

Utilization of sucrose presents no problem, since it is hydrolyzed prior to being transported into the yeast cells as glucose and fructose. Unlike non-fermentable monosaccharides, lactose is not taken up by *S. cerevisiae* cells. However, hydrolyzed lactose has been used as a substrate for baker's yeast production (Anon, 1982; Stineman *et al.*, 1980). Assimilation of intracellular maltose involves the hydrolytic action of α-D-glucosidase, which is highly specific for the sugar moiety. It has no activity when the configuration of the sugar moiety is modified by (i) inversion at C-2 (α-

D-mannopyranosidyl), or C-4 (α-D-galactopyranosyl), (ii) substitution on the 6-hydroxyl group (raffinose), or (iii) other substitutions or replacement on C-2 to C-6. This enzyme also hydrolyzes maltotriose, which is known to be transported intact into baker's yeast (Yamamoto and Inone, 1961).

Oligo-(1,6)-D-glucosidase is also present in *S. cerevisiae* cells. It hydrolyzes isomaltose, methyl α-D-glucopyranoside but not maltose. Induction of these two α-D-glucosyl hydrolases has been investigated. Maltose induces both hydrolases, but α-D-methyl glucopyranoside induces only oligo-(1,6)-D-glucosidase, not α-D-glucosidase. Glucose often suppresses the synthesis of both enzymes. Another trisaccharide, raffinose, is known to be present in molasses. Baker's yeast can utilize only the fructose moiety of raffinose following its hydrolysis by invertase in the periplasmic space (Barnett, 1976).

Ethanol also has been used, at least partly, as a substrate for the production of *S. cerevisiae* yeast. The pathway for assimilating ethanol substrate into glycolytic intermediates and carbohydrates is *via* oxaloacetate (Haarasilta and Oura, 1975). The cell yield is reported to be 40–70 g of yeast solids per 100 g of ethanol (Mor and Fiechter, 1968; Suomalainen and Oura, 1978). In a mixed ethanol–glucose substrate (*i.e.* 15% ethanol–85% glucose), the yield obtained from ethanol is nearly 85 g yeast solids per 100 g of ethanol.

Ethanol carbon is as efficient as hexose carbon, *i.e.* 3 mol of ethanol are equivalent to 1 mol of hexose up to a mixture of 40% ethanol–60% molasses. If the proportion of ethanol is increased further, lower cell yields are obtained. Ridgeway *et al.* (1975) also reported on the use of ethanol for yeast production, but their results were obtained mainly with *C. utilis* yeast.

Yeasts are known to be capable of fixing CO_2. However, under efficient aerobic growth conditions reabsorption of metabolic CO_2 is insignificant, *e.g.* 1–3% (Chen, 1959a; Oura and Haarasilta, 1977).

20.2.2.2 Nitrogen assimilation

Assimilable nitrogen for commercial baker's yeast production is supplied in the form of aqueous ammonia, ammonium salts (such as phosphate, sulfate, chloride, bicarbonate, carbonate, *etc.*), and urea. Occasionally, amino acid mixtures, such as protein hydrolysates or autolysates, may be added. Nitrate and nitrite are not assimilated by *S. cerevisiae*. Ammonium ions are transported by the K^+-carrier system, while undissociate ammonium hydroxide and urea enter yeast cells by simple diffusion (Suomalainen and Oura, 1971). Ammonia is assimilated into glutamate and glutamine before being incorporated into other amino acids and nucleotides (Sims and Folkes, 1964; Witt *et al.*, 1964). Urea is hydrolyzed by urea amidolyase to ammonia, which is then assimilated (Roon and Levenberg 1968, 1972), and CO_2, which can be quantitatively recovered (Chen, 1959a).

The nitrogen of compounds such as allantoin and allantoic acid which can be converted to urea is available for yeast growth. However, the ureido nitrogen in purine bases is not available (DiCarlo *et al.*, 1953).

Amino acid uptake is mediated by ten active transport systems (Kotyk and Horak, 1981). The amino and amido groups of several L-amino acids (such as alanine, arginine, aspartic acid, asparagine, glutamic acid, leucine and valine) can be readily assimilated by baker's yeast, while those of other amino acids, such as cysteine, glycine, histidine, lysine and threonine, are not utilized (Schultz and Pomper, 1948). Mixtures of amino acids can be assimilated more rapidly than individual amino acids (Thorne, 1949). In the presence of assimilable sugars, the carbon skeletons of certain amino acids (such as aspartic acid, asparagine, proline and possibly glutamic acid) can also serve as carbon sources for the growth of *S. cerevisiae* yeast (Schultz *et al.*, 1949; White, 1954).

20.2.2.3 Inorganic elements

The total ash content of baker's yeast varies from 4.7 to 10.5%, which is made up of 1.9–5.5% of P_2O_5, 1.4–4.3% of K_2O, 0.1–0.7% MgO, less than 0.2% of CaO and SiO_2, and less than 0.1% of Al_2O_3, Fe_2O_3, SO_3, Cu and Cl (White, 1954). Trace elements, such as Ba, Cr, Au, Co, Mo, Ni, La, Pb, Mn, Rb, Pt, Ag, Tl, Sn and Zn, have been detected (Richard and Troutman, 1940). While some of these elements, such as P, K, Mg, Na, S, Fe, Cu and Zn must be supplied for proper yeast growth, other elements, such as B, Mn, Ca, Tl, Co, I and Sn have no effect on yeast growth (Olson and Johnson, 1949).

(i)   Phosphorus

In commercial yeast production, phosphorus is supplied in the form of phosphoric acid, ammonium phosphate, potassium phosphate, sodium phosphate, *etc*. Orthophosphate uptake in *S. cerevisiae* is mediated by at least two active transport systems. The first one has a relatively low affinity for its substrate and takes up two protons together with $H_2PO_4^-$ (a potassium ion leaves the cell to compensate for electric charge movement); this system is not affected by Na^+ (Cockburn *et al.*, 1975). The second carrier has a higher affinity for phosphate and is driven by a putative Na^+ gradient. Two Na^+ ions are bound per phosphate ion. Na^+ can be replaced by Li^+, but not by K^+ or Rb^+. This system is stimulated by Mg^{2+} and Ca^{2+} (Roomans *et al.*, 1977). Phosphorus is essential to the growth and survival of yeast cells, since it is involved in all phases of cellular metabolism. It is taken up by yeast cells only in the presence of an energy source such as glucose (Kotyk and Horak, 1981). Upon its absorption, ^{32}P-orthophosphate labels ten organic phosphate esters within 0.1 s (Miettinen, 1964). Orthophosphate is also stored as metaphosphate in volutin granules (Wiame, 1946, 1947).

Yoshida and Yamataka (1953) have established that metaphosphate could well be the store of phosphate energy in yeast cells. Trimeta- and tripoly-phosphate have also been isolated from baker's yeast (Kornberg, 1956). However, Suomalainen and Pfaffli (1961) believed that polyphosphate did not constitute any appreciable energy reserve in baker's yeast cells.

(ii)   Potassium and sodium

Potassium ions enter yeast cells through an active transport system. Absorption of K^+ occurs in the presence of a metabolizable substrate and is accompanied by an efflux of protons from the cells. Baker's yeast can take up K^+ from the medium against a concentration gradient of 1000:1. As the substrate becomes depleted, part of the intracellular potassium will be excreted in exchange for the H^+ in the medium. Uptake of sodium ions may involve the same carrier system (Foulkes, 1956) or two separate carrier systems (Conway *et al.*, 1954; Reilly, 1967). Unlike animal cells, active ion transport in yeast is not dependent on the energy derived from ATP, but is effected by means of the 'redox pump' (Conway, 1955). Potassium is necessary for yeast growth and for fermentation. When K^+ is absent from the medium, phosphate cannot be absorbed (Schmidt *et al.*, 1949). Attempts have been made to replace K^+ with sodium and ammonium ions. The 'sodium yeast' and the 'ammonia yeast' had lower growth rates and higher resting oxygen consumption than the reference 'potassium yeast' (Conway and Moore, 1954; Conway and Breen, 1945).

(iii)   Magnesium and calcium

Magnesium and calcium are taken up by an active transport system at the expense of ATP. Conway and his coworkers (Conway and Beary, 1956, 1958; Conway and Duggan, 1958) reported that Mg^{2+} was transferred into yeast cells by the same carrier as that for K^+ ions and that the uptake involved H^+ secretion. The activity of this carrier is dependent on the presence of oxygen. Magnesium is essential to yeast growth and cellular activities, since it serves as enzyme activator for many enzyme systems, but calcium is believed to be non-essential for yeast growth (Morris, 1958).

(iv)   Sulfur

Sulfur is usually supplied in the form of sulfate for yeast production. Its uptake is mediated by an active transport system and is enhanced by the presence of glucose in the medium. Breton and Surdin-Kerjan (1977) found that two permeases are involved: permease I has a high affinity and permease II has a low affinity for the substrate. Sulfate may be replaced by other inorganic compounds, such as sulfite and thiosulfate (Schultz and McManus, 1950), and organic compounds, such as methionine and glutathione (Maw, 1960). However, *S. cerevisiae* cannot utilize the sulfur in certain amino acids such as cysteine, or vitamins such as biotin and thiamine (Suomalainen and Oura, 1971). Assimilation of the sulfate ion in *S. cerevisiae* yeast involves its reduction to sulfite ion and H_2S through the formation of adenine monophosphate-SO_2 and phosphoadenosine phosphosulfate. Hydrogen sulfide is then incorporated into cysteine and homocysteine through sulfhydrylation of *O*-acetylserine and *O*-acetylhomoserine respectively. Methionine is formed by methylation of homocysteine (de Robichon-Szumajster and Surdin-Kerjan, 1971).

(v)   Trace elements

Yeast cells are relatively impermeable to bivalent cations, but elements such as Ba^{2+}, Zn^{2+}, Mn^{2+}, Sr^{2+}, Ni^{2+}, Co^{2+}, Fe^{2+} and Cu^{2+}, can equilibrate rapidly with the cell surface by combining with its phosphate and carboxyl groups. The binding of exogenous bivalent cations is rapid and reversible (Rothstein and Hayes, 1956; Van Steveninck and Booji, 1964). Like Mg^{2+} and Ca^{2+}, transport of these bivalent cations into yeast cells is mediated by an active transport system having the following affinities: $Mg^{2+}>Co^{2+}>Zn^{2+}>Mn^{2+}>Ni^{2+}>Ca^{2+}>Sr^{2+}$ (Fuhrman and Rothstein, 1968). Except for phosphate and sulfate, no data on the transport of other anions is available as yet (Kotyk and Horak, 1981).

Among these trace elements, *S. cerevisiae* requires 200 μg of Zn^{2+}, 75 μg of Fe^{2+}, and 12–15 μg of Cu^{2+} per liter of medium for optimum growth (Olsen and Johnson, 1949). Trace elements function, in general, as catalysts or activators in enzyme or vitamin systems. Zinc is known to be a constituent of alcohol dehydrogenase, carbonic anydrase, carboxypeptidase, glutamic dehydrogenase, lactic dehydrogenase, *etc*. Addition of Zn^{2+} to the growth medium results in a higher rate of fermentation (Frey *et al.*, 1967; Stone, 1965; Densky *et al.*, 1966) and protein autolysis (Maddox and Hough, 1970). Several metalloenzymes and proteins, such as cytochromes, cytochrome c reductase, catalase, peroxidase, *etc.*, contain iron, while others, such as ascorbic acid oxidase, tyrosinase, uricase, *etc.*, contain copper. On the other hand, many heavy metals are found to be inhibitory to yeast growth in trace quantities (White, 1954).

20.2.2.4   Vitamins

Due to the minute quantities of various vitamins in yeast cells, these compounds are ignored in material balance calculations. However, their contents in yeast cells have been well documented (Eddy, 1958). As many as six compounds in the vitamin B complex (*i.e.* biotin, pantothenic acid, inositol, thiamine, pyridoxine and niacin) have been found to be essential for the growth of certain yeast species and strains (Burkholder *et al.*, 1944). An attempt has been made to classify baker's yeast strains on the basis of their vitamin requirements (Schultz and Atkin, 1937). Among these vitamins, biotin (Bios 2B) is required by most species and strains. While its uptake into yeast cells is mediated by a carrier transport system, its efflux is probably by simple diffusion. Thiamine is transported actively by a system that includes a membrane-bound receptor protein, while pyridoxine and pyridoxal are transported by a constitutive system that is stimulated by K^+ and has a pH optimum of 3.5 (Kotyk and Horak, 1981). The uptake of both niacin and its amide is dependent upon the metabolic activity of yeast cells (Oura and Suomalainen, 1978).

Biotin requirements for yeast growth vary with growth conditions (Oura, 1978; Oura and Suomalainen, 1978). Increased aerobicity lowers the need for biotin. During anaerobic growth with glucose–urea substrate, three biotin-containing enzyme systems, *i.e* pyruvate carboxylase, acetyl-CoA carboxylase and urea amidolyase, are functioning; the biotin requirement is highest. When urea is replaced by an ammonium salt, only two enzymes, *i.e.* pyruvate carboxylase and acetyl-CoA carboxylase, are essential and the biotin requirement is lower. During aerobic growth on glucose, oxaloacetate formation through pyruvate carboxylation is partly supplemented by the glyoxylate cycle; the biotin requirement should be further decreased. The lowest biotin requirement occurs during aerobic growth on ethanol, when only acetyl-CoA carboxylase is needed. Biotin may be replaced by biocytin, biotine-D-sulfoxide, D-desthiobiotin and D-biotin methyl ester. According to Oura (1978), 100 μg of biotin/100 g sugar is required for aerobic yeast growth.

Pantothenic acid (Bios 2A) is required by many yeast species for growth. It is a component of coenzyme A, which participates in the transfer of the acyl group in carbohydrate and fatty acid metabolism. White (1954) found that the maximum yield of baker's yeast was achieved at 44 p.p.m. of pantothenic acid in the yeast cells. Only relatively few yeast species require inositol for growth. For these yeasts, deficiency of inositol (Bios 1) leads to morphological changes in the cell wall (Challinor and Power, 1964; Power and Challinor, 1969) and weakened glucose metabolism (Ridgeway and Douglas, 1958; Lewin, 1967). Ghosh and Bhattacharrya (1967) believed that phosphofrutose kinase activity is affected by inositol deficiency. Several yeast species, particularly the lactose-fermenting yeasts, require thiamine for growth (Rogosa, 1944). Baker's yeast can synthesize thiamine from thiazole and pyrimidine components (Schultz *et al.*, 1941). The thiamine added to the medium is taken up by the cells and esterified to thiamine pyrophosphate or cocarboxylase. It has been reported that thiamine stimulates the fermentation rate of certain baker's yeasts (Schultz *et al.*, 1937; Suomalainen and Axelson, 1956). Pyridoxine has been reported to promote growth of *S. cerevisiae* (Schultz *et al.*, 1939). Its requirement may be partly replaced by pyridoxal or pyridoxamine (Melnick *et al.*, 1945). Pyridoxal phosphate acts as a coenzyme of aminotransferases in amino acid metabolism. Only a few yeast species require niacin for growth. The ability of baker's yeast to synthesize niacin is limited under anaerobic conditions so it must be considered a necessary growth factor. Under aerobic conditions a considerable amount of niacin is secreted from the yeast cells. There may be a decreased need of NAD for yeast growth under aerobic conditions (Suomalainen *et al.*, 1965a, 1965b).

20.2.2.5 Oxygen requirement

On the basis of material balance calculations, 1 g of oxygen is required for the production of 1 g of yeast solids under efficient growth conditions (equation 1). This figure is in agreement with actual experimental values (Maxon and Johnson, 1953). Thus, oxygen, like other nutrients, must be supplied for efficient yeast growth. In commercial production of baker's yeast, oxygen is supplied by sparging an air stream through the fermenter broth. Under oxygen-limiting conditions, the amount of yeast which can be produced per unit fermenter volume and unit time is dependent upon the amount of oxygen which can be transferred from the gas phase to the liquid phase per unit volume and unit time. For example, for a volumetric productivity of 5 g of yeast solids l^{-1} h^{-1} the amount of oxygen required is 5 g or 156 mmol of $O_2\,l^{-1}\,h^{-1}$.

20.2.3 Kinetics of Yeast Growth

In a batch process, where adequate nutrients are present, the exponential growth of yeast may be described by the following equation:

$$dX/dt = \mu X \text{ or } dN/dt = \mu_n N \qquad (2)$$

where X = concentration of cell mass (g l^{-1}); N = number of cells l^{-1}; t = time; μ = specific growth rate for cell mass (h^{-1}); and μ_n = specific growth rate for cell number (h^{-1}).

The doubling time for cell mass is equal to 0.693/specific growth rate. The maximum specific growth rate for yeast has been reported to be 0.6 h^{-1} (Aiyar and Luedeking, 1966), equivalent to a doubling time or generation time of 1.16 h.

Since a fed-batch process, rather than a batch process, is normally used for commercial production of baker's yeast, application of these equations must be made with discretion. In the fed-batch process, yeast growth takes place under substrate-limiting conditions in order to achieve a maximum yield of biomass. The specific growth rate is not constant, but decreases progressively during fermentation. Thus, any calculations of the specific growth rate and/or doubling time merely reflect the instantaneous values for that moment. As such, they have little value in projecting and describing the course of a fed-batch fermentation. The ever-changing growth rate in the fed-batch fermentation may be estimated from the feeding schedule of the substrate. By assuming an appropriate yield factor for the substrate used, one may estimate the cell mass produced at various time intervals and thence the specific growth rates during the course of fed-batch fermentations.

The progressive decrease in the specific growth rate in the fed-batch process is probably due to the continuous increase of cell mass concentration and decrease of substrate concentration in the fermenter broth as dictated by the oxygen transfer capability of the fermenter system. As the cell mass increases and the yeast productivity approaches the limit of the oxygen transfer capability, the growth efficiency of biomass will begin to decrease, since aerobic growth will tend to shift gradually toward anaerobic growth. In order to maintain a desirable growth efficiency, the substrate concentration must be reduced to a proper level commensurate with the oxygen supply. The relationship between growth rate and substrate concentration has been found to be similar to the saturation kinetics exhibited by monomolecular absorption (Monod, 1949). For *S. cerevisiae*, K_s has been reported to be 25 mg glucose l^{-1}. Thus μ becomes a strong function of substrate concentration below $S = 10K_s$ or 250 mg glucose l^{-1} (Wang *et al.*, 1979).

In addition to substrate concentration other operating conditions, such as temperature, oxygen supply, pH, *etc.*, affect the growth rate μ. White (1954) has determined μ at different temperatures as follows: 20 °C, 0.149 h^{-1}; 24.5 °C, 0.207 h^{-1}; 30 °C, 0.311 h^{-1}; 32.5 °C, 0.324 h^{-1}, and 40 °C, 0.094 h^{-1}. Similar results have been reported by Keszler (1967). The critical dissolved oxygen concentration for yeast growth has been found to be 0.004 mM (Winzler, 1941).

20.2.4 Cellular Yields

While the information on the oxygen yield of yeast growth may provide useful data on the productivity of a fermenter system, the performance of a fermentation process as a whole must be evaluated on the basis of the efficiency of yeast growth, *i.e.* cell mass produced per unit weight of energy substrate, or substrate yield. Two methods have been proposed for calculating the theoretical substrate yields.

The first method is based on the 'available electrons' in the substrate. Mayberry *et al.* (1967) found the mean yield of cell mass per equivalent of available electrons to be 3.14 ± 0.11 g, regardless of the substrate or organism. Since four equivalents of electrons were required to reduce 1 mol of oxygen, complete oxidation of glucose required 24 electron equivalents. On the basis of their data, these authors predicted a substrate yield of 75.4 g cell mass of *Candida utilis* yeast per mole or 180 g of glucose, *i.e.* 41.9%. This predicted value is considerably lower than those reported in the literature. The low predicted value may be due to the fact that in their calculations the fraction of substrate being assimilated into biomass was not separated from the fraction of substrate actually oxidized.

The second method uses the amount of ATP produced from the substrate. Bauchop and Eldsden (1960) found that, under anaerobic conditions, the cell yield of *S. cerevisiae*, *Streptococcus fecalis* and *Pseudomonas lindneri* was proportional to the amount of ATP synthesized. The mean value of cell mass per mole of ATP (*i.e.* yield$_{ATP}$) was found to be 10.5 g (range: 8.3–12.6). An attempt has been made to estimate yield$_{ATP}$ under aerobic growth conditions by a material balance method. Assuming a P/O ratio of 3.0, the yield$_{ATP}$ was calculated to be 10.96 for *Candida utilis* and 6.21 for *S. cerevisiae* (Chen, 1964). After correcting the P/O ratio to 2.0 for *S. cerevisiae*, its yield$_{ATP}$ was recalculated to be 9.82 (Reed and Peppler, 1973, p. 69). These data are in general agreement with 7.0 for *C. utilis* at P/O = 3, 7.5 for *S. carlsbergensis* at P/O = 2 (Hernandez and Johnson, 1967), 8.7 for *S. cerevisiae* at P/O = 1.93 (Oura, 1972a) and 9.5–12 at P/O = 1.89 (Oura, 1973).

Maximum efficiency of yeast growth, whether it is expressed as cell mass per unit weight of substrate, or per mole of ATP synthesized, or per equivalent of available electron, can be achieved only under optimal conditions. Even in the presence of adequate nutrient supply, the efficiency of yeast growth is affected by many operating conditions, such as oxygen supply, substrate concentration, temperature, pH, as well as growth rate. The effect of oxygen supply on substrate yield is probably best understood. In the absence of oxygen, only 2 mol of ATP are synthesized per mole of glucose assimilated. Using Bauchop and Eldsden's (1960) data of 10.5 g cell mass per mole ATP for *S. cerevisiae*, a cell yield would be equal to: $(2 \times 10.5 \times 100)/180 = 11.65$ g cell mass per 100 g of glucose substrate. White (1954) determined a substrate yield of 10.5 under anaerobic conditions, while Reed (1982) reported a substrate yield of 7.5 g cell mass per 100 g of fermentable sugar. As oxygen supply increases, the growth efficiency increases according to Monod's saturation kinetics to about 50–54 g cell mass per 100 g of energy substrate (Reed, 1982).

Increased substrate concentration also leads to higher specific growth rate, but unlike oxygen supply, it lowers cell yield. This is due to the inhibition of yeast respiration and enhancement of alcoholic fermentation at higher substrate concentration even in the presence of oxygen. This phenomenon is known as the Crabtree effect (DeDeken, 1966; Rickard and Hogan, 1978). In order to achieve high efficiency, the fed-batch process, where calculated amount of substrate is fed continuously during the course of fermentation, is generally used in yeast production.

Cell yield is also a function of specific growth rate μ. Dellweg *et al.* (1977) reported that as μ exceeds 0.18 in a continuous fermenter, the substrate yield decreases sharply. This is due to increased substrate concentration at higher dilution rate, leading to higher alcoholic fermentation, as shown by greater RQ. von Meyenburg (1969) also reported lower substrate yield as μ exceeds 0.23. Thus, μ should not exceed 0.2 if a high yield of yeast is to be achieved.

The optimal temperature for maximum cell yields is considerably lower than that for maximal specific growth rate; μ is found to be highest at 32.5 °C, while the maximum growth efficiency occurs at 20 °C (White, 1954). Eroshin *et al.* (1976) reported the optimal temperature for growth efficiency at 28.5 °C. Lower cell yield at high temperatures is probably due to the greater maintenance energy requirement. Eroshin *et al.* (1976) also reported the optimal pH for growth efficiency to be 4.1 in their fermentation system.

In addition to various fermentation conditions, cell yield is also affected by the presence of metabolic by-products and other substances in the fermentation system. Chen and Gutmanis (1976) reported significantly lower yield as the CO_2 concentration exceeds 1.6×10^{-2} g l^{-1} in the fermenter broth. Molasses substrate used for yeast production usually contains several inhibitory substances, such as SO_2, aconitic acid, nitrite, heavy metals, *etc.* Lower substrate yields are obtained as the SO_2 concentration exceeds 100 p.p.m. in the fermenter broth. Similar results are obtained at high concentrations (*i.e.* 1–2%) of aconitic acid. Notkina *et al.* (1975) reported a loss in yield at 0.004–0.001% of nitrite.

20.3 COMMERCIAL PRACTICE OF BAKER'S YEAST PRODUCTION

20.3.1 General Description of Production Processes

Commercial baker's yeast is usually produced in a multiple-stage process. The early stages, one or more, are batch fermentations in that all ingredients are in the fermenters before yeast inoculum is added. The later stages are fed-batch fermentations, where the ingredients are added to the fermenters in a predetermined way before and during fermentation. The yeast cells produced in the later stages are separated by centrifugation; portions are used to inoculate the subsequent stage. Suomalainen (1963) reported an eight-stage process leading to a final production of 100 000 kg of compressed yeast in two weeks. Burrows (1970) reported a five-stage process with a final production of 125 000 kg of compressed yeast in 65 h, not counting the turn-around time between fermentations.

A variation of the multi-stage process is the DeLoffre process (DeLoffre, 1964), which normally involves a single batch fermentation stage followed by a three-stage fermentation, designated as A, B, C. Stage A is a highly alcoholic fermentation resembling a batch fermentation. Stage B is transitory to the highly aerated stage C, which corresponds to the final fermentation of the fed-batch process described above. Due to the constant adjustment of feeding rates based upon the yeast solids and ethanol contents in stages B and C, the results of the DeLoffre process can be quite variable. This process is not widely practiced.

Continuous fermentation processes have been used successfully in the production of *Candida* yeasts from spent sulfite liquor, hydrocarbons and ethanol substrates (Chen and Peppler, 1979), but its application to the production of baker's yeast from molasses substrate is limited (Olsen, 1960; Sher, 1960). The disadvantages of the continuous fermentation are as follows (Burrows, 1970): (1) inferior baking quality due to undesirable physiological changes, such as elongation of yeast cells, in continuous fermentation; (2) contamination problems, particularly with the fast-growing *Candida krusei* yeast; and (3) unfavorable economics because of costly equipment and lower productivity in the multi-vessel system.

20.3.2 Oxygen Supply and Fermenter Systems

It is generally recognized that the rate of transfer of atmospheric oxygen into the fermenter liquid is the limiting factor in yeast propagation. The major concern in fermenter design is the maximization of oxygen transfer capacity and economy (Finn, 1969).

The fermenters with mechanical agitation usually have higher oxygen transfer capacity, as measured by oxygen transfer rate or $K_L a$. An oxygen transfer rate close to 1 mol $O_2\ l^{-1}\ h^{-1}$ has been reported in a submerged turbine fermenter (Nyiri, 1974), even though the normal operating aeration capacity is lower. The $K_L a$ of this type of fermenter can be estimated from the power consumption per unit volume, air velocity, number and speed of impellers. A yeast concentration of 8–10% solids (Reed, 1982) or even 15% solids (Skiba, 1966) can be attained in this type of fermenter. Thus, the yeast productivity is expected to be very high in the submerged turbine fermenter, which has been used for commercial production of baker's yeast in many countries including the USA. The Vogelbusch fermenter with a rotating aeration wing also has been used in commercial-scale fermentations, particularly in European countries (Suomalainen, 1963; Hospodka *et al.*, 1962). The Waldhof fermenter with a rotating aeration wheel has been used commercially in the production of *Candida* yeast from spent sulfite liquor (Inskeep *et al.*, 1951), but seldom in the commercial production of baker's yeast from molasses substrate. The self-priming fermenter (Frings aerator), which sucks air through a hollow vertical shaft and aspirates through a rotating turbine, has been used for yeast production, but mostly on a pilot scale (Ebner *et al.*, 1967; Ebner and Enenkel, 1974). The Vogelbusch deep jet aeration system uses rotary pumps outside the fermenter to recirculate continuously the fermenter broth to the top of the fermenter where air is sucked into a hollow overflow shaft (Schreier, 1974). It has been used successfully for aerobic treatment of industrial and community waste water, but its application to baker's yeast production is not known with certainty.

Among the diffused-air fermenters, the gas-sparged fermenter with perforated horizontal pipes is used around the world for commercial production of baker's yeast. In this type of fermenter, the air is blown by a compresser–blower through a large number of small holes in the horizontal pipes near the bottom of the fermenter. Rosen (1977) reported a fermenter system of 150 m^3 which was aerated through a center pipe with 24 side tubes provided with 30 000 holes of 1.5 mm diameter. The aeration capacity, in terms of K_La, of a gas-sparged fermenter may be predicted from the dimension of the fermenter, airflow rate, air velocity at the orifice and the calculated air bubble size (Bhavaraju, *et al.*, 1978). Although the aeration capacity of the gas-sparged fermenter is lower than the submerged-turbine fermenter, its aeration economy as measured by lbs oxygen transferred per energy unit may be higher, particularly at low oxygen transfer rates (Hatch, 1975). The aeration tower system is similar in principle to the gas-sparged fermenter except for the dimension of the fermenter. An aeration capacity equivalent to a K_La value of 640 h^{-1} has been reported by Yoshida and Akita (1965) for this system. The air-lift fermenter has been used for semi-commercial production of single-cell protein from *n*-paraffins (Cooper *et al.*, 1975), but it is not known to be used for commercial production of baker's yeast.

The oxygen transfer rate required depends upon the fermentation process in use. If the maximum feed rate during the course of fermentation is 10 g of sugar l^{-1} h^{-1}, a yeast productivity of 5 g solids l^{-1} h^{-1} is expected. Thus, a minimum oxygen transfer rate of 5 g O_2 or 156 mol O_2 l^{-1} h^{-1} must be provided. A higher yeast productivity can not be achieved in a fermenter system which can not provide higher aeration capacities.

In evaluating competing aeration systems, the following factors must also be considered: (1) the power cost per unit volume of oxygen transferred; (2) the heat energy required per unit volume of oxygen transferred, since this will have an effect on the capital and operating costs of the system selected; (3) the ratio of the working volumes of various systems under consideration; (4) the effect of the air–liquid emulsion generated on the heat-transfer efficiency in the heat removal from the fermenter; and (5) the cost of installation.

There is considerable variation in the size of baker's yeast fermenters. The smaller ones have volumes in the neighborhood of 50 m^3 while the largest may be over 350 m^3. They have diameters in excess of 7.5 m and heights of up to 10 m. The fermenter size influences the type of aeration system selected. In fermenters with diameters up to about 4 m, mechanical or sparger aeration systems work well. In fermenters with diameters over 4 m, gas sparged systems are generally used because the distribution of air by a properly designed sparger is nonradial and therefore there is greater assurance of uniform air distribution and mixing throughout the fermenter. Fermenter size is an important factor in the selection of the fermenter cooling system and the method of heat removal should be compatible with the aeration system selected. Coils and/or plate heat exchangers are most commonly used for heat removal. The cooling area required for a given fermenter is a function of the yeast solids to be produced, the mechanical energy inputs, the temperature of the nutrient solutions and the temperature of the cooling water available. The quanitity of heat liberated during the production of baker's yeast is in the order of 20 × 10^3 kJ per kilogram of yeast solids (Fengl, 1969). The heat input from the fermenter mechanical systems, mainly aeration, and nutrient feed systems, mainly molasses, can amount to between 10 and 20% of the total heat load. Since natural cooling water temperature and availability are a function of location, the heat exchanger surface area required for a given fermenter will vary with location. The aeration system and heat removal surface are an integral part of any fermenter design. All other fermenter connected systems are auxiliary in that they can be by-passed in operation.

The aeration capacity and economy of various types of fermenters are tabulated in Table 1. For the high-price and small volume pharmaceuticals where productivity is a major consideration, fermenters with high aeration capacities, *e.g.* submerged turbine or similar mechanically-agitated fermenters, should be used. For the low-price and high volume products, such as baker's yeast, fermenters with high aeration economy, *e.g.* diffused-air fermenters, may have some advantages.

20.3.3 Preparation of Raw Materials

20.3.3.1 *Energy substrates*

Molasses has replaced grain mash as the primary energy substrate for yeast production since World War I. Beet and cane molasses are used singly or in combination. Decisions regarding the

Table 1 Aeration Capacity and Economy of Fermenter Systems[a]

Fermenter system	Normal operating O_2-transfer capacity (mmol $O_2\ l^{-1}\ h^{-1}$)	Aeration economy (1b $O_2\ HP^{-1}\ h^{-1}$)	Air utilization (%)
Mechanically agitated systems			
Submerged turbine	300–350	2.4	14–45
Waldhof aeration wheel	—	2.4–3.2	—
Phrix	—	1.7–2.5	14–22
Self-priming (Frings)	—	2.4	14
Vogelbusch aerating wing	140	2.3–2.7	14–19
Vogelbusch deep jet aerator	315–385	3.3	28
Diffused-air systems			
Gas-sparged fermenter	100–150	2.7	10–15
Air-lift fermenter	120	4.3–7.0	7–8

[a] Compiled from Cooney *et al.* (1977); Hatch (1973, 1975); Hospodka *et al.* (1962); Schreier (1974); Reed and Peppler (1973); Reed (1982).

type of molasses to be used are based on molasses availability and economics. When both types of molasses are available at comparable cost, higher proportions of beet molasses are used because it is easier to process and normally yields a lighter colored yeast.

Raw beet and cane molasses enter the factory at 80 to 90 °Brix and contain 80 to 85% solids of which up to 60% may be sugar. Dilution is necessary to facilitate pumping, clarification and sterilization. Dilution is accomplished in one of several ways. The simplest is to weigh a given quantity of molasses and add it to or add to it a given quantity of water. More complex systems use special pumps or blending systems, utilizing analog or digital instruments, to blend molasses and water to the desired Brix.

Factories using both cane and beet molasses may blend the two during the initial dilution, or at any stage of molasses treatment up to feeding to the fermentations, or the molasses may not be blended but fed to the fermentations in separate streams. The reasons why different manufacturers blend at different stages are based on experimentation which has indicated that benefits are to be derived from processing in a particular way, or on one method being the most consistent with plant processing conditions.

Cane and blends of cane and beet are usually diluted to between 30 and 40 °Brix for processing. Pure beet molasses may be processed at a lower dilution. The greater dilution of cane molasses helps clarification by providing a greater gradient between the specific gravities of the suspended solids and the liquid phase. The manufacture of sugar from cane starts with the crushing of the sugar cane to produce a sugar rich juice. The crushing and washing of the fibrous husk liberates inorganic and non-sugar organic compounds some of which are colloidal in nature. Beet sugar production starts with the slicing of the thin skinned beets and the resulting molasses contains less suspended solids.

The sludge and colloidal material in cane molasses cause problems in various stages of the manufacturing process and cane molasses is almost always clarified. Beet molasses on the other hand may or may not be clarified. Whether or not to clarify beet molasses is based on preference and perceived need on the part of each manufacturer. In general molasses is clarified to improve final yeast colors. In addition, yeast made from clarified molasses is easier to dewater and dry. Molasses clarification probably aids fermenter oxygen transfer and reduces the formation of fermentation foam.

Early clarification procedures involve precipitation of suspended matter by forming flocculents using alumina or phosphates and/or filtration with diatomaceous earth. Cane molasses is difficult to filter even after precipitation, and when necessary basket centrifuges were used to polish the decanted supernatant. Chemical clarification and filtration were later replaced by clarification in nozzle clarifiers equipped with sludge recycle. More recently, intermittent discharge centrifuges have come into use for molasses clarification. The movement to centrifugal clarification reduced losses, materials and labor, and improved overall process control.

Clarification can be done before or after sterilization depending on manufacturing philosophy. Clarifying after sterilization removes the precipitate formed in heating to sterilization temperatures but to some extent increases the possibility of recontaminating the sterile molasses in the clarifying equipment, while clarifying before sterilization reduces fouling of the sterilization equipment and reduces the risk of recontamination. Molasses sterilization is accomplished by direct steam injection on a batch or continuous basis. Batch sterilizations involve the heating of molasses in a tank, usually, but not always, at atmospheric pressure, to specified time and temperature conditions. Continuous sterilization is done in closed loops at higher temperatures with very short residence times, *e.g.* 1 s at 140–145 °C (Rosen, 1977). Continuous systems sometimes incorporate preheating in plate type heat exchangers, which are used in various configurations to save energy, and flash tanks that receive the heated molasses from the high pressure sterilization loop.

Recently, hydrolyzed cheese whey has been suggested as a substrate for baker's yeast production (Stineman *et al.*, 1980; Anon, 1982). The whey proteins and suspended solids are partially removed by centrifugation and ultrafiltration; the lactose in the permeate is hydrolyzed by lactase, either free or immobilized, and concentrated to a solids content of 65–70% or a fermentable sugar content of 50%. This substrate also has to be diluted, clarified and sterilized prior to use for baker's yeast production.

20.3.3.2 Nutrient supplementation

A comparison of the composition of molasses with that of baker's yeast shows that the molasses substrate is deficient in certain essential nutrients such as N, P, K, Mg, Zn and S (Harrison, 1971). These elements must be supplied for optimal yeast growth.

The nitrogen levels in baker's yeast are usually dictated by the leavening activity and the storage stability desired in the final yeast. Nitrogen normally makes up from 6.5 to 10% of the yeast solids. Beet molasses contains approximately 1.5% nitrogen one third of which is available to the baker's yeast. Cane molasses has a low nitrogen content and for practical purposes it contains no biologically available nitrogen for yeast growth. Baker's yeast preferentially takes up inorganic nitrogen. Aqueous ammonia, anhydrous ammonia, ammonium salts and urea are the usual sources of nitrogen in commercial fermentations. They are used either singly or in combination. The nitrogen source is selected on the basis of cost, availability and the effect of these compounds on fermentation pH control. In an efficient fermentation system, most of the ammonia fed is taken up by the yeast cells.

The P_2O_5 content in baker's yeast ranges from 1.5 to 3.5% of the yeast solids. While the yeast cells can take up larger quantities of phosphate, the commercial practice is to limit the P_2O_5 content to 25–30% of the nitrogen content. Small quantities of K, Mg and Zn are usually supplied in the form of chloride or sulfate, while S is supplied either as sulfuric acid or as sulfate, depending upon the fermentation process in use. There is no need to supply other trace elements, since molasses usually contains adequate quantities of these elements.

As pointed out in a previous section, baker's yeast requires some members of the vitamin B-complex, particularly biotin, for optimal growth. Since beet molasses is deficient in biotin (*i.e.* 0.01–0.13 μg g^{-1}), it is a common practice to blend beet molasses with blackstrap cane molasses which has a higher biotin content (*i.e.* 0.6–3.2 μg g^{-1}). Pure biotin may be used when cane molasses is not available. Oura (1978) showed that 100 μg biotin/100 g sugar is required for aerobic yeast growth. Addition of other vitamins, such as thiamine and pantothenic acid, is often practiced as well.

20.3.4 Multistage Fermentation

Peppler (1979) reported a six-stage fermentation process for the commercial production of baker's yeast. The first stage (F1) was a flask culture stage, the second (F2), third (F3) and fourth (F4) stages were batch fermentations, while the fifth (F5) and sixth (F6) stages were fed-batch fermentations. The procedure for the multi-stage process is described in the following sections.

20.3.4.1 Flask culture stage (F1)

Yeast cultures used for daily production of baker's yeast are maintained by periodic transfer on molasses or malt agar slants. Production cultures may be periodically renewed from stock cultures maintained under more stringent control procedures in a central quality control laboratory. These cultures are grown for 2–4 days through one or more flasks containing molasses or malt media with approximately 5% sugar for introduction to the factory; Erlenmeyer, Carlsberg or Pasteur flasks are commonly used. In some processes, the quantity of flask culture may be as much as 75 l and in others it may be as little as 1 l. Most processes use less than 5 l to inoculate the initial plant fermentation, the first pure culture stage.

20.3.4.2 Pure culture stages (F2, F3 and F4)

The plant production process begins with a pure culture fermentation (F2). The number of pure culture fermentations depends on the process in use and the quantity of contamination-free inoculum required. The pure culture fermentations are run in small fermenters which are easy to sterilize and the process is basically a continuation of the flask cultures except that the pure culture fermenters have provisions for sterile aeration and aseptic transfer to the next stage. The fermentations are batch in that all the nutrients are in the fermenters prior to inoculation with the flask culture or with the yeast from a previous pure culture stage. Rigorous sterilization of the fermentation media prior to inoculation is conducted by heating under pressure or by boiling at atmospheric pressure for extended periods. The critical factor in pure culture operations is sterility. Since *S. cerevisiae* has a long generation time compared to other microorganisms, a contaminating microorganism can easily outgrow the baker's yeast strain of interest.

Sterility and the exponential nature of yeast growth make it unnecessary to inoculate a pure culture fermentation with a large quantity of flask culture. If the quantity of yeast solids available from 1 l of flask culture is 5 g and the average doubling time is 2.5 h, this pure culture inoculum will produce, in the presence of adequate substrate and nutrients, 5.12 kg yeast solids in 25 h or 10 generations. Doubling the inoculant will reduce the fermentation time by 2.5 h, and a 100-fold increase in the inoculum will reduce the fermentation time only by a little over one half. The usual fermentation time for the first pure culture stage is 24 h and those for the subsequent stages are 9–11 h, according to Burrows (1970). Whether to use one or more pure culture fermentations depends on the quantity of yeast needed to inoculate the subsequent fed-batch fermentations. Generally, a single pure culture stage will be used for 5.0 kg of yeast solids or less, and a multivessel system will be used for greater quantities. In this case, the entire contents of a pure culture vessel is aseptically transferred to inoculate a subsequent fermentation (*i.e.* F3 or F4). The size relationship of one pure culture vessel to another varies with the manufacturer's scheme over a considerable range and depends to some extent on time considerations. The poor substrate yield and the practice of inoculating one fermentation with the entire contents of the previous fermentation limit the size of pure culture vessels to less than 30 m^3.

The need for process control in the pure culture operation is limited. Microbiological testing of the media before, during and after each fermentation is essential. The substrate concentration of the initial fermenter is often standardized to between 5 to 7.5% sugar. This is most easily done by manual measurement with a Brix hydrometer. The Brix measurement can also be used to monitor the progress of the fermentation, which is deemed complete when the Brix reading declines to a steady point. The pH may be monitored over the course of fermentation but any pH adjustment is made prior to sterilization of the media. Once the pure culture fermentation is started, the only controllable parameters are temperature and aeration. Control of aeration is not critical because of the excess sugar substrate. Temperature control is meaningful but control can be over a broad range.

20.3.4.3 Fed-batch fermentation (F5 and F6)

The main fermentations are of the fed-batch type and they are highly aerobic. The size of the fermenters and the auxiliary equipment required for incremental substrate supply, aeration and cooling make sterilization and the maintenance of sterility difficult. In practice, contamination limits the number of fed-batch fermentations after the pure culture stage. Some producers use the pure culture to inoculate a single fed-batch fermentation, while others may produce a seed yeast fermentation (F5) followed by the final fermentation (F6). Sometimes, as many as three to four successive seed yeast fermentations may be conducted prior to the final fermentation. One process calls for the drying of the seed yeast for long term storage. The dried seed yeast is then rehydrated as needed to inoculate three to four additional fermentations and the need for routine pure culture fermentations is eliminated.

The primary objective in the seed yeast fermentations, other than growing the biomass, is to limit contamination. This usually means greater attention to sterilization, operating at lower pH, shorter fermentation time and high growth rate at the expense of substrate yield. The final stage of commercial fermentation is conducted to maximize leavening activity, storage stability, substrate yield and the appearance of the final yeast product. Most of these objectives are achieved by manipulating various fermentation parameters including the pattern of substrate and nutrient addition.

The baker's yeast production process developed empirically, as an art, and while it is now possible to design *S. cerevisiae* fermentations with respect to the efficient production of a given amount of biomass, the development of fermentations to yield given physiological characteristics in the final yeast is still empirical. In practice, baker's yeast is not propagated with strictly exponential substrate additions nor is it propagated at the highest specific growth rate compatible with maximal substrate yield. Rather, the baker's yeast propagation is conducted to produce specific properties in the final yeast product. The substrate feed pattern and consequently the specific growth rate are a function of the conditions necessary to obtain the end results. Since the substrate feeding pattern in the fed-batch fermentations is developed empirically, it is expected to be different for different manufacturers. Some feeding patterns have been reported by Reed and Peppler (1973).

In order to achieve an acceptable substrate yield, the sugar concentration in the fermenter should be kept below certain levels. Wang *et al.* (1977) reported glucose repression of respiration at a glucose concentration of 0.13 g l^{-1} when the specific growth rate corresponded to $\mu = 0.25$ h^{-1} in a fed-batch fermentation. Dellweg *et al.* (1977) placed the critical concentration at 1.1 mmol l^{-1} (*i.e.* 0.2 g l^{-1}) in continuous fermentations with growth rates in the range of $\mu = 0.14$–0.18 h^{-1}. Exceeding the critical glucose concentration results in the production of ethanol and intermediate organic acids which tend to lower fermentation pH. These conditions have bacteriostatic effects and they are used by some manufacturers to limit contaminant growth in their seed yeast fermentations and in the early phase of their final commercial fermentations. The loss of substrate yield accompanying ethanol formation in the final fermentations can be minimized by allowing the yeast to metabolize the ethanol present late in the fermentation.

Aside from substrate concentration and feeding pattern, other fermentation parameters, such as pH and temperature, are also established empirically, not necessarily based upon their effects on specific growth rate and growth efficiency. Usually the pH lies between 4.0 to 6.5 and the temperature ranges between 28 to 32 °C. A properly designed final fed-batch fermentation should also permit the yeast cells to mature. This can be accomplished by stopping the feeding of nutrients at the end of fermentation but allowing a slight aeration to continue for an hour to 'ripen' the yeast, according to Oura *et al.* (1974). During this period, the unused substrate is assimilated; the cells with buds grow from two daughter cells and mature. These investigators also pointed out that the yeast samples taken at the middle of ripening period had the best leavening power and that the unripened yeast cells with buds were less stable in that they autolyzed easily.

The fermentation time for the final stage may vary from 10 to 20 h, during which the yeast may multiply six- to seven-fold (Burrows, 1970; Soumalainen, 1963). The yeast solids content at the end of fermentation may vary from 3 to 8% depending upon the aeration capacity of the fermenter system. In order to achieve a high productivity, the substrate concentration and the feeding pattern should be compatible with the aeration capacity of the fermenter.

20.3.5 Instrumental Control of Fermentation Process

In order to produce baker's yeast economically, a proper environment must be provided in the fermenter, so the efficiency and productivity of yeast growth can be maximized. Instrumental control of the physical environment in the fermenter, such as temperature, power input, gas flow, liquid flow, pressure, *etc.*, is relatively straightforward; adequate sensors are generally available. With the exception of pH and oxygen measurements, instrumental control of the chemical environment is in a less advanced state.

20.3.5.1 *Physical parameters*

(i) *Power input*

Two main systems have been used for measurement of shaft power: torsion dynamometer and strain gauge. The latter is preferred because of its greater accuracy. In this system, balancing strain gauges are mounted on the impeller shaft in the fermenter broth. Lead wires from the gauges are passed out of the vessel *via* an axial hole in the shaft. The electrical signal is picked up from the rotating shaft by an electrical slip ring.

(ii)   Impeller speed

This is routinely measured by a magnetic proximity sensor which counts revolutions of a gear or a magnetic spot on the shaft. The output of the sensor is electronically manipulated to give the r.p.m. or tip speed on a digital or meter readout.

(iii)   Liquid flow

Liquid flow may be effected by a centrifugal pump, a positive displacement pump, or a diaphragm metering pump, such as Interpace's Pulsafeeder. The pump is generally coupled with an electric actuator and a controller. The liquid feeding pattern is dictated by a curve follower. The actual flow rate may be monitored by a flow meter or by mounting the liquid reservoir on a load cell system, which weighs and records continuously the content in the reservoir.

(iv)   Pressure

Relatively simple diaphragm gauges can be used to measure the pressure in a fermenter. The resulting pneumatic signal can be either relayed directly or transduced through a simple device to an electronic signal.

20.3.5.2   Chemical parameters

(i)   pH

Sterilizable combination electrodes suitable for continuous operation are commercially available, *e.g.* Ingold's pressurized pH electrode (Buhler and Ingold, 1976). It may be coupled to a pH controller, such as New Brunswick's model 121 or 122, for monitoring and control of yeast fermentations.

(ii)   Dissolved oxygen

Two types of membrane probe are used: polarographic (or amperometric) and galvanic. The former requires an applied polarizing voltage of -0.7 to -0.9 V, while the latter is self-polarizing. The electrical signal from the oxygen controller is used to monitor, to record and to maintain the pre-set DO level by regulating the air-flow rate and/or impeller speed (Vincent, 1974).

(iii)   Oxygen in effluent air

The oxygen content in the exit gas may be monitored by a paramagnetic analyzer or by an oxygen probe. For the paramagnetic analyzer, the moisture in the exit gas must be removed, but this operation is not necessary for the oxygen probe. Satisfactory results have been obtained in the monitoring of yeast fermentations with a polarographic probe (Chen and Gutmanis, 1976). It is also possible to monitor effluent oxygen with a mass spectrometer, such as the Perkin-Elmer multiple gas analyzer MGA-1200.

(iv)   Residual glucose

There is no reliable method at the present time. Attempts have been made to monitor the residual glucose content in the fermenter broth by using an L and N's glucose analyzer coupled with a continuous sampling probe attachment (Chen and Gutmanis, 1976). Since this method is based upon the production of H_2O_2 by the immobilized glucose oxidase, any contamination by the catalase-containing yeast cells will seriously affect the accuracy of this method.

(v)   Other chemical constituents

Although several specific ion electrodes and enzyme electrodes have been developed in recent years, none of them are suitable for continuous monitoring operation at the present time.

20.3.5.3   Fermentation products

(i)   CO₂ in effluent gas

The CO_2 in the exit gas can be successfully monitored by an IR analyzer. Development of a membrane CO_2 probe suitable for monitoring dissolved CO_2 in the fermenter broth and the CO_2

in the exit gas has been reported by Shoda and Ishikawa (1981). The mass spectrometer may also be used to monitor the CO_2 in the exit gas.

(ii) Ethanol in exit gas

The ethanol content in the exit gas can be monitored by a long path IR analyzer, such as Miran II Analyzer, at 3.2 and 3.4 μm. Quantitative measurements were obtained by calibrating the instrument against ethanol solutions of known concentrations. The electrical signal from the IR analyzer can be used to control the substrate feeding rate in a fed-batch fermentation. For precise on-line measurement of ethanol in the exit gas, a mass spectrometer may be used. Monitoring of seven volatile compounds in the gas and liquid phases of fermenters by this method has been reported by Weaver *et al.* (1980). For approximate monitoring purposes a hydrocarbon analyzer may be adequate.

(iii) Biomass of baker's yeast

There is no satisfactory method for direct on-line monitoring of microbial biomass in the fermenter. Continuous measurement of turbidity with a turbidimeter or spectrophotometer has been attempted, but the results are not satisfactory. Indirect monitoring of microbial mass is based upon material balance (Cooney *et al.*, 1977), oxygen balance or the determination of certain cell constituents, such as ATP, NAD, *etc*. A constant chemical composition in the biomass is assumed in all indirect monitoring methods; the validity of this assumption remains to be established and confirmed.

20.3.5.4 Computer-coupled fermentation

In this application, a computer may perform the following functions: data acquisition and processing as well as process monitoring, control and optimization (Flynn, 1974). It becomes possible to obtain instantaneously many indirect measurements relating to the physiological state of the baker's yeast and to the fermentation conditions, such as the respiratory quotient (RQ), growth rate, cell density, $K_L a$, C/N ratio of the nutrient solution, *etc.* It appears that the crucial problem is to develop a control strategy for the computer-coupled fermentation, *i.e.* which parameters are to be selected for process control and optimization.

The respiratory quotient (RQ) is often used for the control of baker's yeast fermentation. As shown in equation (1), the RQ is calculated to be 1.02 under efficient growing conditions. Aiba *et al.* (1976) used a constant RQ of 1.0 to 1.2 to regulate the substrate feeding rate in a fed-batch culture of *S. cerevisiae*. A substrate yield of 0.55 g cell/g glucose and a specific growth rate of 0.24 h^{-1} were obtained.

If one accepts the fact that the substrate yield is determined mainly by the metabolic pathways of the microorganism and that the volumetric productivity is limited mainly by the aeration capacity of the fermenter, it becomes obvious that the computer aided fermentation will not be expected to make a significant impact on the efficiency and the productivity of baker's yeast production. The advantages of a computer-coupled system will lie in the greater uniformity of the product, better reproducibility of the fermentation process and the substantial reduction of manufacturing cost through more efficient utilization of raw material, fermenter and manpower.

20.3.6 Handling of Baker's Yeast

20.3.6.1 Separation, washing and cream storage

At the end of each fed-batch fermentation stage the yeast cells are centrifugally separated from the fermented media and subjected to one or more washing separations. Washing separations are run to reduce the non-yeast solids in the fermenter liquor which remain with the cells after the initial separation. This is necessary because the non-yeast solids hinder filtering and drying, darken the color of the final yeast, and affect the rehydration characteristics of active dry yeast. The number of washing separations depends on the fermentation stage and on the efficiency of the washing system. In practice the seed yeast fermentations usually receive fewer washings than the final commercial fermentations. Washing system efficiency is determined by the concentration of the yeast solids, the quantity of dilution water used and the solids content of the dilution water. The last factor operates when the washing water runs counter current to the yeast stream. This is usually done to reduce water usage.

If the final fermenter liquor contains 4% yeast solids and 4% non-yeast solids, and the yeast solids are 40% of the cell, then the 4% non-yeast solids will be contained in 90% of the fermenter volume. If the separator yeast discharge contains 20% yeast solids, the concentrate will consist of 50% yeast cells and 50% of the original fermenter liquor containing 4% non-yeast solids. Diluting the concentrate back to 110% of the original volume will result in the liquid phase solids being reduced to 0.4%. Reconcentration will yield a cream yeast containing 0.4% non-yeast solids in the liquid phase and each time the washing process is repeated a ten-fold reduction in the non-yeast solids is obtained. Washing obviously requires large volumes of water. Where water is scarce or effluent considerations limit its use, the supernatant discharge from the final washing separation can be used as the diluent in the previous stage. In this case the efficiency of the washing operation is reduced by the fraction of original solids running in the wash water when the system reaches equilibrium.

The separation process yields a light colored yeast cream containing up to 22% yeast solids. At this concentration the cells occupy 55% of the liquid volume and the cream appears to be highly viscous. Cream yeast is stored in agitated tanks at 2 to 4 °C for use in seeding additional fermentations or for additional dewatering if it is the final trade yeast. While cream yeast can be held for several weeks without appreciable deterioration it is seldom held for more than a few days. The pH of seed yeast creams is usually adjusted to between 2.5 and 3.5, and sodium bisulfite or ammonium persulfate may be used as a bacteriostat.

20.3.6.2 *Filtration*

While baker's yeast is occasionally sold in the cream form, the greatest portion is sold either as compressed yeast containing 27 to 30% yeast solids (Europe and America respectively) or as dry yeast containing from 92 to 96% solids (worldwide). There are many disadvantages to distributing cream yeast, primary among these are its greater water content, and attendant higher hauling costs, and the investment required to install a liquid yeast distribution system. A further 10% reduction in the water content of the cream yeast produces a yeast cake containing approximately 25% more yeast solids. The compressed yeast can be molded into blocks or bulked in bags for distribution.

Dewatering takes the cream yeast from 18–22% solids up to 28–33% solids. Filter press or rotary vacuum filters are used for this operation. The filtered yeast may or may not be mixed with emulsifiers prior to being extruded into yeast cakes or packaged in large multi-walled paper bags. Fresh baker's yeast may also be marketed in the form of free-flowing particles. This type of product is produced by adding certain hydrophobic SiO_2 and hydrophilic SiO_2, modified starch, micronized cellulose, *etc.* to the filtered yeast (Luca *et al.*, 1979; Pomper and Akerman, 1980). Yeast cakes may range in weights from approximately 10.5 g to 2.25 kg. Multi-walled bags vary in weight up to about 22.5 kg. The compressed yeast is then refrigerated at 2 to 4 °C until used.

20.4 COMMERCIAL PRACTICE OF ACTIVE DRY YEAST (ADY) PRODUCTION

Successful commercial production of active dry yeast started during World War II, even though many attempts to prepare this type of product had been patented in the 1920s. Frey (1957) reviewed the historically significant studies leading to the development of ADY; Thorn and Reed (1959) described the production and baking techniques for ADY.

20.4.1 General Description of Drying Processes

The production of ADY begins with the selection of baker's yeast strains which will yield the desired characteristics on drying. The selected strains are propagated using fermentation protocols known to condition the yeast cells for dehydration. The yeast so produced is filtered to as high a solids content as possible. The yeast cake is then extruded into strands or particles of various sizes, which are dried in an air stream under controlled conditions.

In order to produce an ADY product with acceptable leavening activity and storage stability, the following factors should be taken into consideration during dehydration: drying temperature, drying rate and the final moisture content in ADY. As a living microorganism, the vegetative yeast cell is rapidly killed at temperatures exceeding 50 °C (Reed and Peppler, 1973). Obviously, yeast should not be dried at such elevated temperatures. The equilibrium moisture content of ADY may be predicted from the predetermined desorption isotherms. A series of these curves has been published by Josic (1982).

20.4.2 Drying Methods and Systems

Although many methods have been explored, only a few are in use for the commercial production of ADY. Selection of a drying system is based upon the desired physical appearance and properties of the ADY as well as the cost of installation and operation. Some of these commercial drying systems are discussed below:

20.4.2.1 *Roto-Louvre dryer*

This dryer was originally developed in Sweden for drying of wood chips, paper pulp, *etc*. It was later used for dehydration of foods, chemicals, coal and cement (Erisman, 1938). The dryer consists of a hollow cylinder with radial louvre fin plates attached to the inside wall, dividing the cylinder into a number of small compartments. To these fin plates are attached the tangential louvres. During drying operation, the extruded yeast strands are fed into the cylinder, which rotates at 1 to 4 r.p.m. Heated air, at about 50–60 °C, is blown into the cylinder through the louvres and the tumbling yeast particles. The temperature of the yeast particles does not exceed 45 °C however. For batch operation, the total drying time may vary from 10–20 h (Reed and Peppler, 1973). The roto-louvre dryers range in size from 2.5 to 11.5 ft in diameter, and 8 to 35 ft in length. The largest unit is capable of evaporating 12 000 lb water h^{-1} (Marshall and Friedman, 1950). Operation of a roto-louvre dryer 4.85 m in length and 2.2 m in diameter for ADY production has been reported by Sysojewa and Gorochova (1965). It is possible to use a roto-louvre dryer for continuous operation.

20.4.2.2 *Through-circulation dryer*

This type of dryer may be used for continuous or batch operation. For continuous operation, the extruded yeast strands are spread as a layer, 1–6 in deep, on an endless perforated belt. It moves through several drying chambers, where heated air is blown alternately upward and downward through the yeast bed to avoid over- and under-drying. The size of each chamber may vary from 30 to 60 ft long and from 6 to 10 ft wide. Each chamber may be considered as an individual unit, complete with fan and heating coils, arranged in series to form a housing or tunnel through which the conveying belt travels. The air velocity, humidity and temperature in each chamber are individually controlled to achieve the desired drying rate and the final moisture content in the final ADY product. Belokon (1962) reported a drying time of 2 to 4 h for this type of dryer with inlet air temperatures of 42, 37, 32 and 28 °C, respectively, for four drying chambers.

In batch operation, the extruded yeast strands are placed on removable screen-bottom trays, suitably supported in the dryer. It is similar to a standard tray dryer except that the heated air passes through the yeast bed instead of across it. The pressure drop through the bed usually does not exceed 1 in of water (Marshall and Friedman, 1950). The direction of air flow may be reversed at regular time intervals. Again, the air velocity, humidity and temperature are controlled during the entire drying operation to achieve the desired properties in the ADY product. Chulina (1969) employed an inlet temperature of 50 °C at the beginning and 35 °C at the end of the drying period.

20.4.2.3 *Air-lift (fluidized-bed) dryer*

For batch operation, the extruded yeast strands are fed into a drying chamber with a metal screen or perforated plate at the bottom. Heated air is blown from the bottom through the yeast particles at velocities capable of suspending the yeast particles in a fluid bed. Use of this type of dryer for ADY production has been described by Simon (1976). Emulsifiers and swelling agents are often added to the yeast suspension prior to drying (Clement and Rossi, 1982). The drying time may vary from 10 min to 4 h. For rapid drying between 10 to 30 min, Langejan (1972, 1974) used an air temperature of 100–150 °C at the beginning of the drying period, while keeping the yeast temperature at 24–40 °C. The use of a multichamber air-lift dryer for continuous operation has been patented by Pressindustria (1971). The operating conditions are as follows: air velocity, 4000 m h^{-1}; airflow volume, 4000 m^3 h^{-1}; air temperature, 46, 36, 32 and 30 °C in four chambers; retention time, 3 h; yeast productivity, 160–350 kg h^{-1}. The final moisture content of the ADY was about 7%.

20.4.2.4 Spray dryer

With this process, the yeast suspension, containing 10–20% yeast solids, is atomized into a drying chamber where a stream of heated air is introduced. The moisture from the yeast suspension evaporates into the air stream; the dried yeast powder is separated from the air and collected while the moist cool air is exhausted. Use of this process for ADY production has been patented by Aizawa *et al.* (1968). The yeast suspension was dried at an inlet temperature of 100–120 °C and an outlet temperature of 65–67 °C. The moisture content of the powdery ADY was 6–7%. These investigators found that the addition of certain additives, such as alkali metal salts of inorganic and organic acids, polyhydric alcohols, non-fermentable sugars, urea, *etc.*, to the yeast suspension prior to spray drying resulted in higher fermentation activities of the final product. Presumably, these compounds imparted certain changes in the osmotic properties of the spray-dried ADY. Some other additives, including gelatin, carboxymethylcellulose, *etc.*, have been used by Wakamura *et al.* (1973). These authors used a drying temperature of 60–120 °C in their process; the moisture content of the powdery ADY was 5–6%.

20.4.2.5 Other drying systems

Johnston (1959) patented a process for ADY production by suspending yeast cells in a finely divided state in an edible oil. Water was removed from the yeast suspension by blowing a heated air stream (100 °F; 38 °C) from the bottom of the oil-bath chamber. At the end of drying, oil was removed and extracted from the yeast product. Fermentation activity was found to be quite high in the powdery ADY, which had a moisture content of 7–8%.

Grylls *et al.* (1978) patented a process and an apparatus for the production of powdery ADY by subjecting the yeast particles to disintegration forces while they are being dried in a fluidized-bed dryer. Vacuum drying of the yeast cream spread on an endless steel belt was reported by Hartmeier (1977).

Combination of two drying methods has also been used. Van 'Triet and De Bruijn (1981) patented a process of drying yeast initially in a fluidized-bed dryer at a temperature of 90–130 °C with an air velocity of 0.8–2.0 m s^{-1} and a loading of 100–1000 kg cm^{-2}. As the moisture content in the yeast reduced to 10–25%, the yeast particles were dried in a vacuum dryer at 25–45 °C and at a pressure of 1–10 mm Hg. The final moisture content may be as low as 2%.

20.4.3 Properties of Active Dry Yeast

The properties of active dry yeast are functions of its propagating conditions as well as its drying conditions. It was recognized at the early stage of ADY development that high protein yeast was not suitable for drying; activity loss was substantial and storage stability was poor (Frey, 1957). Until recently, ADY was prepared from yeast propagated under nitrogen limitation, resulting in a product with a low protein content of 40–45% and a relatively high level of carbohydrates. Since the fermentation activity of the yeast is related to its protein content, a low-protein yeast is less active, on a dry matter basis, than the compressed yeast grown to a 50–55% protein level.

The importance of high carbohydrate level in ADY, particularly trehalose, was also recognized at about the same time. Pollock and Holmstrom (1951) reported 16–18% of trehalose in ADY with good gassing activity. Simultaneous syntheses of trehalose and glycogen were observed in batch culture (Suomalainen and Pfaffli, 1961) and in non-proliferating culture (Grba *et al.*, 1975). A high incubation temperature of 45 °C favored trehalose formation, while a lower temperature of 30 °C was optimal for glycogen synthesis. In fed-batch fermentation, glycogen and trehalose were not formed in parallel, however. The former was formed during the initial exponential phase of growth, while the latter was synthesized at the late exponential growth phase (Grba *et al.*, 1979) or during the depletion of glucose substrate (Panek, 1975). Trehalose accumulated in baker's yeast during transfer from anaerobic to aerobic growth; its amount was highest at the final commercial fermentation stage (Suomalainen and Oura, 1956).

The physical appearance of ADY depends on the drying process. It is important with regard to the application and storage stability of the product. Pellet ADY is produced by roto-louvre dryers. This type of ADY has high storage stability due to the coating developed by the tumbling yeast particles during the drying process. It must be fully rehydrated prior to use. The granular ADY is produced by the through-circulation dryers. Since the yeast particles are not as dense as the pellet ADY, storage stability is somewhat inferior. The friability of the granular ADY may be increased by incorporating certain surfactants (Chen *et al.*, 1966) or by entrapping air (Carduck *et al.*, 1982) in the yeast strands prior to drying, or by controlling the rate of drying. Such friable yeast granules may be pulverized and blended directly with flour for baking (Cooper and Chen, 1966). Extremely fine yeast particles 0.2–3.0 mm long are produced in fluidized-bed dryers. Due to its uniform drying conditions, this drying process can be used to dry high protein yeast (Langejan, 1976, 1980). As a result, high activity dry yeast can be produced by this process (Clement and Hennette, 1982). The disadvantage of this type of product is its poor storage stability, due to its large specific surface area. Also, it should not be rehydrated in water directly due to its high leaching characteristics. The properties of powdery ADY produced in the spray dryers have not been fully investigated, because this product is not widely distributed.

20.5 QUALITY OF BAKER'S YEAST

While a good deal of information is available on the growth and production of baker's yeast biomass, the interrelationship between its propagating conditions and its baking quality is far from being understood. The baker's yeast biomass produced at the highest efficiency and productivity does not necessarily have the most desirable properties for baking. The primary functions of baker's yeast are two-fold: imparting an appetizing flavor and aroma to the baked products and leavening the baked products. Fulfillment of these requirements by baker's yeast will be discussed in this section.

20.5.1 Flavor and Aroma Development in Baked Products

There is little doubt that the predominant flavor and aroma of baked products originate from panary fermentation by baker's yeast. A loaf of bread with physical characteristics similar to yeast leavened bread can be produced by using proper chemical leavening agents such as glucono-Δ-lactone and $NaHCO_3$. Yet the resulting loaf is totally devoid of the flavor and aroma expected of yeast-leavened bread, in spite of the fact that identical ingredients and baking conditions were used. Similarly, the short fermentation continuous-mixing processes used for bread production in the United States during the 1960s are gradually being phased out, due to insufficient flavor and aroma in their products, even though higher levels of yeast are generally used in such processes. Many of the flavor compounds are produced during fermentation (Suomalainen and Lehtonen, 1978). Volatile and nonvolatile compounds, such as acids, alcohols, aldehydes, esters, ether derivatives, furan derivatives, hydrocarbons, ketones, lactone derivatives, pyrazines, pyrrole derivatives and sulfur compounds have been found in bread (Coffman, 1967; Maga, 1974). However, except for isolated cases, little information is available to relate the formation of these compounds to the origin and propagating conditions of baker's yeast. Sugihara *et al.* (1971) reported that *Saccharomyces exiguus* yeast was responsible for the leavening action in San Francisco sour-dough bread, but the bread's characteristic flavor was of bacterial origin (Kline and Sugihara, 1971). Chen and Peppler (1956a) reported the development of an off-odor in cinnamon breads due to the conversion of cinnamaldehyde to styrene by a mutant strain of baker's yeast.

20.5.2 Leavening of Baked Products

There are two facets of yeast leavening: CO_2 production during panary fermentation and CO_2 retention in dough systems.

20.5.2.1 *Carbon dioxide production during panary fermentation*

Yeast-leavened baked products are made from dough systems containing flour, water, yeast, salt and optional ingredients, such as sugar, shortening, emulsifier, milk solids, flavoring substances, yeast food, *etc*. The rate of CO_2 production (or fermentation activity) is dependent upon the intrinsic properties of yeast, concentration and composition of dough ingredients as well as environmental factors such as temperature, pH, *etc*. The effects of these factors will be examined separately.

(i) *Intrinsic properties of yeast*

Baker's yeast biomass is normally produced under aerobic conditions to achieve optimal growth efficiency and productivity, yet the panary fermentation in dough system is essentially anaerobic. It has been found that the fermentation activity of baker's yeast decreases when the yeast is transferred from anaerobic to aerobic growth conditions. This phenomenon is related to the decrease in the activities of several glycolytic enzymes in the aerobic yeasts, such as pyruvate decarboxylase (Suomalainen, 1963; Polakis and Bartley, 1965), hexokinase and alcohol dehydrogenase (Oura, 1972b; 1976). Besides aerobicity during yeast propagation, the glycolytic enzyme activities also are affected by the concentration and composition of growth substrates. In the presence of a high sugar concentration in the growth medium, yeast propagation shifts toward anaerobic growth, due to the Crabtree effect. As the glucose concentration increases from 0.6 to 20% in the growth medium, for example, the activities of several glycolytic enzymes, such as aldolase, triosephosphate isomerase, pyruvate kinase and pyruvate decarboxylase, increase 25 to 100 fold (Hommes, 1966). Oura (1972a, 1972b) reported that the highest fermentation activity and pyruvate decarboxylase activity were obtained from baker's yeast grown with glucose substrate, followed by those grown with ethanol and pyruvate substrates. Replacement of glucose substrate by acetate, ethanol, pyruvate or glycerol led to reduced enzyme levels for practically all glycolytic enzymes (Maitra and Lobo, 1971). Polakis and Bartley (1965) reported that the pyruvate decarboxylase activity of the glucose-grown cells was about five times that of the galactose-grown cells. Incubation of molasses-grown baker's yeast in a maltose solution resulted in higher contents of α-glucoside permease and α-glucosidase (Hautera and Lovgren, 1975b). To some extent, the level of glycolytic enzymes is reflected in the nitrogen content of yeast cells; high nitrogen content usually leads to higher fermentation activity. On the other hand, the baker's yeast grown at high temperature, high CO_2 tension (Chen and Gutmanis, 1976) or high aconitic acid content in the molasses substrate usually has lower fermentation activities.

All these data show that the fermentation activity of baker's yeast is related to its propagating conditions. For active dry yeast (ADY), other crucial factors, such as dehydration and rehydration conditions, must also be considered. During dehydration, the moisture content of the yeast cells is reduced to below 10%. It must be rehydrated to restore its full activity for baking. Due to the changes in the semi-permeability of the plasma membranes in the dehydrated cells (Harrison and Trevelyan, 1963), certain low molecular weight cell constituents leach out from the cells upon rehydration (Herrera *et al.*, 1956; Ramnietse *et al.*, 1978). The amount of cell constituents leached during rehydration is regulated by three factors: moisture content of ADY, rehydration temperature and rate of rehydration. The interrelationship of these factors is shown in Figure 1, based upon the data reported by Chen *et al.* (1966). This graph shows that (1) a greater amount of cell constituents leach out from a low-moisture (*i.e.* 5.3% H_2O) ADY than from a high moisture (*i.e.* 8.1%) ADY; (2) greater amount of cell constituents leach out at low rehydration temperature (*i.e.* 10 °C) than at higher rehydration temperature (*i.e.* 43 °C); and (3) addition of an emulsifier, such as sorbitan monostearate, can reduce the amount of leached substances from a low-moisture ADY by slowing down its rehydration rate (Mitchell and Enright, 1959). Blending of ADY with flour prior to rehydration can also slow down water absorption by yeast particles and thus minimize the leaching of their cell constituents (Cooper and Chen, 1966; Bruinsma and Finney, 1981). Even the properly rehydrated ADY has a longer lag phase than compressed yeast; it requires a longer recovering time to achieve its maximum fermentation activity, sugar consumption, oxygen consumption, specific growth rate as well as budding. Beker *et al.* (1974) believed that this was due to the structual damage incurred during dehydration; longer lag phase was required for the restoration of the polyphosphate and nucleic acids in rehydrated ADY. Generally speaking, the fermentation activity of the rehydrated ADY is inversely proportional to the amount of cell constituents leached during rehydration (Thorn and Reed, 1959; Ponte *et al.*, 1960; Kraus et al, 1981).

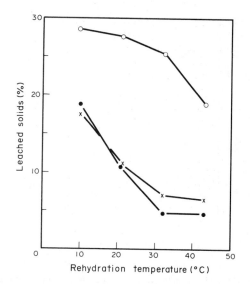

Figure 1 Effect of rehydration temperature, moisture content and surfactant treatment on the leaching characteristics of ADY: ●-●-●, 8.0% moisture content; o-o-o, 5.3% moisture content; x-x-x, 5.1% moisture content + 2% sorbitan monostearate

(ii) Dough ingredients and osmotic pressure

It is well recognized in the baking industry that yeast fermentation activity is substantially lower in dough systems containing high levels of sugar and/or salt. This phenomenon is interpreted as inhibition of yeast fermentation by the high osmotic pressure due to dough ingredients. The quantitative effect of osmotic pressure on yeast fermentation activity is shown in Figure 2. This graph shows that yeast fermentation activity decreases as the osmotic pressure in the dough system increases. Furthermore, D-xylose, a nonfermentable sugar, and NaCl have similar inhibitory effects at the same osmotic pressure, while glucose and sucrose have similar effects. These two lines run parallel to each other but with different ordinate intercepts, due to the additional CO_2 produced from glucose or sucrose supplementation. Thus, the inhibitory effects of fermentable sugars, nonfermentable sugar and salt all can be explained on a physicochemical basis.

The osmosensitivity of baker's yeast in dough systems is dependent on the yeast strain and its propagating conditions. Certain strains have greater tolerance to high osmotic pressure; these strains usually perform well in sweet dough systems containing 20–30% sugar. They tend to have lower invertase contents (Sato *et al.*, 1961; Reed and Peppler, 1973). The baker's yeast grown in a fed-batch process is less osmosensitive than that grown in a batch process; a slow-growing yeast is less osmosensitive than a fast-growing yeast (White, 1954). Pomper and Akerman (1970) propagated baker's yeast in the presence of an effective amount of non-nutritive salts, such as NaCl, Na_2SO_4, NaBr, Na_2CO_3, Na acetate and $SrCl_2$, to impart to the yeast superior leavening activity in sweet dough. Presumably, the yeast so produced adapts to the high osmotic pressure in dough systems. High salt concentration during propagation usually results in lower growth rate and substrate yield, due to a higher maintenance energy requirement (Watson, 1970).

In contrast to the inhibitory effect of high sugar concentration on yeast fermentation activity, the effect of sugar composition is difficult to demonstrate in panary fermentation. The formulas of various baked products may consist of no sugar supplementation in lean-dough systems to 30% sucrose or glucose in sweet-dough systems. The predominant fermentable sugar in the lean-dough system is maltose, even though small quantities of sucrose, glucose, fructose, *etc.* are present in the flour (Reed and Peppler, 1973). Many authors have reported low fermentation activity of commercial baker's yeast in synthetic maltose media (Atkin *et al.*, 1946; Seeley and Ziegler, 1962; Schultz, 1965). Correlation between maltose fermentation activity and α-glucoside per-

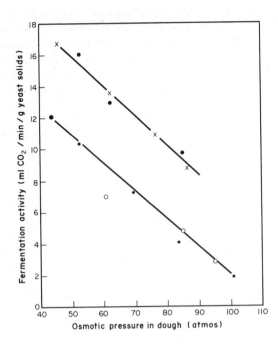

Figure 2 Yeast fermentation activity in relation to the osmotic pressure in the dough systems containing various amounts of ingredients: ●-●-●, sucrose; x-x-x, glucose; ····, D-xylose; o-o-o, NaCl

mease or α-glucosidase content has been demonstrated by Hautera and Lovgren (1975b), yet none of these criteria has any correlation to the leavening activity in lean-dough fermentation (Lovgren and Hautera, 1977; Suomalainen, 1975). These authors believed that the CO_2 produced in the early phase of lean-dough fermentation did not derive from maltose, but from other fermentable sugars in the flour. On the other hand, a high fermentation rate has been observed in the dough system containing no sugar supplementation (Figure 2), suggesting that the slow fermentation of maltose in the flour may have been counterbalanced by the favorable low-osmotic pressure in the lean-dough system.

With regard to other dough ingredients, a slight stimulating effect by nitrogen-containing yeast food on yeast fermentation was reported by Peppler (1960), while the inhibitory effect of mold inhibitors, such as calcium propionate and sodium diacetate, has been reported by Seeley and Ziegler (1962). The inhibition by the ethanol produced during panary fermentation has also been reported. Blish and Hughes (1932) reported alcohol contents of 2.79, 3.50 and 5.11% (w/w on flour weight) at the end of 24-hour fermentation in straight-doughs containing 0, 2.5 and 10% sugar (on flour weight), respectively. During the regular fermentation time of 3 h, the respective alcohol contents were 1.40, 1.38 and 1.25%. White (1954) reported about 13% inhibition of gassing rate toward the end of straight-dough fermentations, when 4 g of EtOH was added to 280 g flour (i.e. 1.42% w/w of ethanol on flour) and 156 ml of water. Inhibition of fermentation rate by ethanol in various liquid systems has been reported by many investigators (Brown et al., 1981; Franz, 1961; Gray and Sova, 1956; Rosá, 1963).

(iii) Environmental factors

The effect of dough temperature on the fermentation rate is well documented. Conventional sponge is generally set at 25–27 °C. After a four-hour fermentation and remixing, the dough temperature rises to about 30 °C; it is then proofed at 35–37 °C. Liquid preferment is usually set at 30 °C. After one to one and a half hour fermentation and mixing, the temperature of the dough rises to about 35–40 °C; it is proofed at about 45 °C. White (1954) reported an increased fermentation rate with temperature in a straight-dough system between 20 and 34 °C; the Q_{10} for gas production is calculated to be between 1.8 to 2.2. Harbrecht and Kautzmann's results (1967) on sponge fermentation between 27.5 and 35 °C showed a higher Q_{10} of 2.9 during the first 30 min. After a two-hour fermentation, the Q_{10} dropped to 2.4. Garver et al. (1966) reported increased CO_2 production rate at increased temperatures in a preferment system; it reached a maximum rate of 26 mmol CO_2 l^{-1} g^{-1} yeast solids at 38 °C, beyond which the rate began to decline.

With the exceptions of sour-dough bread and soda crackers, the conventional dough systems normally have a pH of 5.0–5.4 and they are well-buffered. Deviation from this range is seldom observed in actual baking practice. However, this is not true in preferments. Unless they are sufficiently buffered by the addition of inorganic buffering salts and/or milk solids, the pH of the preferment may drop to 3.0, due mainly to the dissolved CO_2 produced in fermentation. Garver *et al.* (1966) reported a significantly lower fermentation rate in a preferment system at pH 3.5; maximum fermentation rate was observed at pH 4.0–5.2. In a straight-dough system, the maximum fermentation rate was observed at pH 5.3, above which declining rates were noted (Seeley and Ziegler, 1962). Similar results were reported by Franz (1961) with molasses solutions.

20.5.2.2 *Carbon dioxide retention in dough systems*

In order to retain the CO_2 produced during panary fermentation in dough systems, the gluten in the flour must be properly hydrated and developed into a coherent and extensible film with desirable viscoelastic properties, otherwise much of the CO_2 will be lost from the fermented dough, resulting in low bread volume and inferior leavening of the baked products. To some extent, both baker's yeast and its fermentation action play important roles in the proper development of the flour constituents for gas retention.

It is well established that disulfide and hydrogen bonds are of primary importance for gluten structure, while electrostatic bonds and van der Waals forces are of minor significance (Bloksma, 1978). Yeast cells normally contain from 4–14 mg glutathione per gram of yeast solids, depending upon yeast strain and propagating conditions (Schultz and Swift, 1955). While little glutathione leaches out from the compressed baker's yeast during panary fermentation, a substantial portion of this tripeptide does leach out during rehydration of active dry yeast, varying according to the rehydration temperature and the moisture content of active dry yeast (Chen *et al.*, 1966). Glutathione has been found to participate in the disulfide–sulfhydryl interchange reaction in flour (Kuninori and Sullivan, 1968), leading to the cleavage of gluten molecules, lowering of relaxation constants in dough and reduction of development time during mixing (Ponte *et al.*, 1960). While such effects may be desirable in certain applications requiring prolonged mixing, such as pizza dough, bun dough and sweet dough, more precise control in dough mixing is required to avoid overdevelopment. The effect of yeast glutathione can be duplicated by the addition of crystalline glutathione to dough systems. Similarly, addition of reduced thiotic acid, a normal flour constituent, also results in weakened mixing behavior (Dahle and Hinz, 1966). The presence of thiotic acid reductase in yeast cells has been demonstrated (Black *et al.*, 1960).

Besides yeast cells *per se*, the substances produced during yeast fermentation also play an important role in modifying the physical properties of dough systems (Nagao *et al.*, 1981b). Incorporation of a prefermented dough piece into an unleavened dough led to marked improvements in gas retention and rheological properties, as measured by a Do-Corder instrument. Such an improvement was not likely to be due to the lowering of dough pH from 5.7 to 5.0 by the dissolved CO_2, since the addition of acetic acid to the same pH level had little beneficial effect. The function of CO_2 lies in the transformation of the developed gluten film into a vesicular structure by the expanding gas cells. In addition to CO_2, several organic acids, keto acids, glycerol, *etc.* are also produced in appreciable quantities in yeast fermentation (Cole *et al.*, 1962, 1966; Suomalainen and Keranen, 1967). Incorporation of several organic acids in the unleavened dough results in better rheological properties (Nagao *et al.*, 1981a). Decrease in sulfhydryl content has been observed during panary fermentation. This may also account for some improved physical properties of the fermented dough, as the addition of a fermented dough piece to a fresh sponge or dough produces a rheological change similar to the incorporation of potassium bromate, an oxidizing dough improver (Nagao *et al.*, 1981b). Little information is available on the effect of ethanol, a major fermentation product, on dough development. Blish and Hughes (1932) concluded on the basis of loaf volume that alcohol, up to 5.11% (w/w on flour) in the dough, produced no serious changes that might be regarded as gluten degradation. No rheological measurement was made, however.

20.5.3 Storage Stability of Baker's Yeast

Aside from flavor development and dough leavening, storage stability is also an important criterion for the quality of baker's yeast. A compressed baker's yeast usually contains 68–72%

water. As such, it is a perishable product requiring refrigeration during storage. The effect of storage temperature on yeast activity has been studied by many investigators. Hautera and Lovgren (1975a) reported little decrease in yeast activity after it was stored at 5 °C. for 28 days. At 23 °C, the fermentation activity remained reasonably constant for 16–18 days, while at 35 °C, yeast activity decreased linearly with storage time and approached zero activity after 7–9 days. Ginterova *et al.* (1966) found that the sucrose-fermenting activity of compressed yeast remained reasonably constant for 36 days at 7 °C, 5 days at 20 °C, and less than 3 days at 30 °C. However, the decline of maltose-fermenting activity occurred sooner in all cases.

Compressed yeast may also be stored in the frozen state in commerce. Thiessen (1942) found 2–6% yeast activity loss over 16 to 20 weeks in frozen storage, while Peppler (1960) reported little activity loss over 19 weeks at −30 °C. Although the freezing and thawing rates affected yeast viability (Mazur, 1970), Hsu *et al.* (1979) showed that the final freezing temperature was more important. At a freezing temperature of −10 to −20 °C, the prooftime of the frozen-dough was 71–72 min, while at −40 °C and −78 °C, the proof time increased to 132 and 360 min, respectively.

The effect of storage atmosphere on the fermentation activity of compressed yeast is not clear-cut. Edelman *et al.* (1978) reported lower leavening ability of commercial baker's yeast after it was stored under anaerobic conditions for 4 days at 30 °C. Similarly, the compressed yeast stored in 50 g pieces had higher leavening ability than that stored in 1 kg blocks (Edelman *et al.*, 1980). On the other hand, Takakuwa (1962) found that the liquefaction of compressed yeast occurred much later when it was stored in nitrogen than in oxygen or an air atmosphere. It was often observed that granulated compressed yeast was more difficult to store than the consolidated blocks (Burrows, 1970; Reed and Peppler, 1973), presumably due to higher endogenous carbohydrate breakdown under aerobic than under anaerobic conditions (Eaton, 1960).

In addition to endogenous consumption of reserve carbohydrates, such as glycogen and trehalose, the decrease of fermentation activity is also accompanied by an increase of dead cells (Suomalainen, 1975), proteolytic activity (Tohoyama and Takakuwa, 1971) and NADase activity (Takakuwa *et al.*, 1975) in the compressed yeast during storage. Its nucleic acid content shows a slight increase during the early phase of storage, followed by a rapid decrease in the late phase when autolysis is observed. Concurrently, the acid-soluble nucleotides increase slowly in the early phase and more rapidly in the late phase, when softening and autolysis of compressed yeast begin (Parkkinen *et al.*, 1973, 1974). These data suggest that when the reserve carbohydrates have been used up, the yeast cells start to attack other vital cellular constitutents, leading to rapid hydrolysis of proteins and nucleic acids, resulting in eventual autolysis.

Besides temperature and aerobicity, the storage stability of compressed yeast is also related to its propagating conditions: aerobically-grown yeast is more stable than anaerobically-grown yeast; low-protein yeast is more stable than high-protein yeast and a mature cell is more stable than a less mature yeast (Suomalainen, 1975). An attempt has been made to improve the storage stability of compressed yeast by post-harvest treatment with metabolic inhibitors (Takakuwa and Tohoyama, 1975); some degree of success has been reported with biuret treatment (Takakuwa, 1962).

The storage stability of active dry yeast (ADY) is significantly better than that of compressed yeast, since it contains less than 10% moisture. Its shelf life may be as short as 1 to 2 months to as long as 1 year, depending upon the storage temperature and atmosphere as well as its moisture content. Generally speaking, the storage stability is inversely proportional to storage temperature and moisture content. Dehydration of ADY to a lower moisture content (*i.e.* 4–6%) greatly improves its thermostability (Mitchell and Enright, 1957), but it does not eliminate the deteriorative effect of oxygen. Under an oxygen or air atmosphere, a pellet ADY with a rather impervious surface is more stable than a granular or finely particulated ADY. For improved shelf life, ADY is often packaged *in vacuo* or under an inert atmosphere such as nitrogen or CO_2. Incorporation of certain phenolic antioxidants, such as BHA, BHT or propyl gallate, also leads to improved storage stability in an air atmosphere (Chen and Cooper, 1962; Chen *et al.*, 1966). The biochemical changes for the deterioration of ADY are not fully understood. Chen and Peppler (1956b) showed an excellent correlation between pyruvate decarboxylation activity, anaerobic fermentation and baking activity of the ADY stored at 48 °C. The deterioration of pyruvate decarboxylation activity was found to be due to destruction of cocarboxylase, and not to denaturation of the apo-enzyme.

20.6 CONCLUSIONS

Due to the world-wide attention to single-cell protein and fuel alcohol, a good deal of effort has been devoted in recent years to the improvement of fermentation equipment and optimization of the fermentation process. Significant progress has been made toward the production of microbial biomass from various substrates (Mateles and Tannenbaum, 1968; Tannenbaum and Wang, 1975). However, in the production of baker's yeast, biomass formation is only a part of the picture. The quality of baker's yeast is equally, if not more, important. This is evidenced by the lack of acceptance of the continuous fermentation process in the baker's yeast industry. The current knowledge on the quality of baker's yeast is empirical and incomplete. For example, it is not known whether the panary fermentation activity of baker's yeast is determined mainly by its enzyme constitution, its sugar permeation rate, its osmotic sensitivity, or its drying tolerance. This basic information must be available before baker's yeast fermentation can be scientifically optimized.

ACKNOWLEDGEMENT

The authors are grateful to Messrs Gerald Reed, Henry Peppler, Richard Raymond, Mrs Aileen Mundstock and Mrs Betty Blue for their suggestions, comments and assistance during the preparation of this manuscript.

20.7 REFERENCES

Aiba, S., A. Humphrey and N. Mills (1973). *Biochemical Engineering*, 2nd edn. Academic, New York.
Aiba, S., S. Nagai and Y. Nishizawa (1976). Fed batch culture of *Saccharomyces cerevisiae*: A perspective of computer control to enhance the productivity in baker's yeast cultivation. *Biotechnol. Bioeng.*, **18**, 1001–1016.
Aiyar, A. S. and R. Luedeking (1966). A kinetic study of the alcoholic fermentation of glucose by *Saccharomyces cerevisiae. Chem. Eng. Prog., Symp. Ser.*, **62** (69), 55–59.
Aizawa, M., T. Matsuda, S. Kawabata, I. Omura, I. Amano and T. Nakamura (1968) Process for preparing an active dry powdery yeast. *US Pat.* 3 407 072.
Alonso, A. and A. Kotyk (1978). Apparent half-lives of sugar transport proteins in *S. cerevisiae. Folia Microbiol. (Prague)*, **23**, 118–125.
Anon (1982). Corning–Kroger combine technology to exploit lactose-hydrolyzed whey. *Food Dev.*, January, 34–35.
Armiger, W. B. and A. Humphrey (1979). Computer applications in fermentation technology. In *Microbial Technology*, ed. H. J. Peppler and D. Perlman, 2nd edn., vol. 2, chap. 15, pp. 375–401. Academic, New York.
Arnold, B. H. and R. Steel (1958). Oxygen supply and demand in aerobic fermentations. In *Biochemical Engineering*, ed. R. Steel, pp. 149–182. Heywood, London.
Atkin, L., A. S. Schultz and C. N. Frey (1946). Yeast fermentation. In *Enzymes and Their Role in Wheat Technology*, ed. J. A. Anderson. Interscience, New York.
Barnett, J. A. (1976). The utilization of sugars by yeasts. *Adv. Carbohydr. Chem. Biochem.*, **32**, 125–234.
Bauchop, T. and S. R. Eldsden (1960). The growth of micro-organisms in relation to their energy supply. *J. Gen. Microbiol.*, **23**, 457–469.
Beech, F. W. and R. P. Davenport (1971). Isolation, purification and maintenance of yeasts. In *Methods in Microbiology*, ed. B. Booth, vol. 4, chap. 5, pp. 153–182. Academic, New York.
Beker, M. E., B. E. Damberga, A. A. Upit, Y. E. Blumberg, I. Y. Krause, E. Y. Ventinya, I. K. Brivkalne and M. V. Popova (1974). Some peculiarities of the lag phase of the development of *Saccharomyces cerevisiae* after dehydration. *Mikrobiologia*, **43**, 1028–1033.
Belokon, V. N. (1962). Yeast drying on a belt drier. *Spirt. Promst.*, **1**, 40–42.
Bhavaraju, S. M., T. W. F. Russell and H. W. Blanch (1978). The design of gas sparged devices for viscous liquid systems. *AIChE J.*, **24**, 454–466.
Black, S., E. M. Harte, B. Hudson and L. Wartofsky (1960). A specific enzymic reduction of L(-)-methionine sulfoxide and related nonspecific reduction of disulfides. *J. Biol. Chem.*, **235**, 2910–2916.
Blish, M. J. and R. C. Hughes (1932). Some effects of varying sugar concentrations in bread dough on fermentation by-products and fermentation tolerance. *Cereal Chem.*, **9**, 331–356.
Bloksma, A. H. (1978). Rheology and chemistry of dough. In *Wheat Chemistry and Technology*, ed. Y. Pomeranz, 3rd edn. American Association for Cereal Chemistry, Inc., St. Paul, MN.
Blumenthal, H. J., K. F. Lewis and S. Weinhouse (1954). An estimation of pathways of glucose catabolism in yeast. *J. Am. Chem. Soc.*, **76**, 6063–6097.
Breton, A. and Y. Surdin-Kerjan (1977). Sulfate uptake in *Saccharomyces cerevisiae*: Biochemical and genetic study. *J. Bacteriol.*, **132**, 224–232.
Brown, S. W., S. G. Oliver, D. E. F. Harrison and R. C. Righelato (1981). Ethanol inhibition of yeast growth and fermentation: Difference in the magnitude and complexity of the effect. *Eur. J. Appl. Microbiol. Biotechnol.*, **11**, 151–155.
Bruinsma, B. L. and K. F. Kinney (1981). Functional (bread-making) properties of a new yeast. *Cereal Chem.*, **58**, 477–480.
Buhler, H. and W. Ingold (1976). Measuring pH and oxygen in fermentors. *Process Biochem.*, **11**, 19–24.
Burkholder, P. R., I. McVeigh and D. Moyer (1944). Studies on some growth factors of yeasts. *J. Bacteriol.*, **48**, 385–392.

Burrows, S. (1970). Baker's yeast. In *The Yeasts*, ed. A. H. Rose and J. S. Harrison, vol. 3. chap. 7, pp. 349–420. Academic, New York.

Carduck, F. J., D. Kloetzer and G. Veldman (1982). Preparation of porous active yeast granules. *US Pat.* 4 335 144.

Challinor, S. W. and D. M. Power (1964). Effect of inositol deficiency on yeast in particular reference to chemical composition of the cell and of the cell wall. *Nature (London)*, **203**, 250–251.

Chattaway, F. W., R. Bishop, M. R. Holmes, F. C. Odds, and A. J. E. Barlow (1973). Enzyme activities associated with carbohydrate synthesis and breakdown in the yeast and mycelial forms of *Candida albicans*. *J. Gen. Microbiol.*, **75**, 97–109.

Chen, S. L. (1959a). Carbohydrate assimilation in actively growing yeast, *S. cerevisiae*. I. Metabolic pathways for C^{14}-glucose utilization by yeast during aerobic fermentation. *Biochim. Biophys. Acta*, **32**, 470–479.

Chen, S. L. (1959b). Carbohydrate assimilation in actively growing yeast, *S. cerevisiae*. II. Synthesis of polysaccharides from C^{14}-glucose. *Biochim. Biophys. Acta*, **32**, 480–483.

Chen, S. L. (1964). Energy requirement for microbial growth. *Nature (London)*, **202**, 1135–1136.

Chen, S. L. and E. J. Cooper (1962). Production of active dry yeast. *US Pat.* 3 041 249.

Chen, S. L. and F. Gutmanis (1976). Carbon dioxide inhibition of yeast growth in biomass production. *Biotechnol. Bioeng.*, **18**, 1455–1462.

Chen, S. L. and H. J. Peppler (1956a). Conversion of cinnamaldehyde to styrene by a yeast mutant. *J. Biol. Chem.*, **221**, 101–106.

Chen, S. L. and H. J. Peppler (1956b). Destruction of cocarboxylase in active dry yeast. *Arch. Biochem. Biophys.*, **62**, 299–304.

Chen, S. L. and H. J. Peppler (1979). Single-cell proteins in food applications. *Dev. Ind. Microbiol.*, **19**, 79–94.

Chen, S. L., E. J. Cooper and F. Gutmanis (1966) Active dry yeast: Protection against oxidative deterioration during storage. *Food Technol.*, **20**, 79–83.

Chulina, E. P. (1969). The effect of drying temperature on the quality of yeast. *Khlebopek. Konditer., Promst.*, **13**(3), 27–28.

Clement, P. A. and A. Hennette (1982). Strains of yeast for breadmaking and novel strains of yeast thus prepared. *US Pat.* 4 318 930.

Clement, P. and J.-P. Rossi (1982). Active dried baker's yeast. *US Pat.* 4 328 250.

Cockburn, M., P. Earnshaw and A. A. Eddy (1975). The stoichiometry of the absorption of protons with phosphate and L-glutamate by yeast of the genus *Saccharomyces*. *Biochem. J.*, **146**, 705–712.

Coffman, J. R. (1967). Bread flavor. In *Chemistry and Physiology of Flavors*, ed. H. W. Schultz, E. A. Day and L. M. Libby. AVI Publishing Co., Westport, CT.

Cole, E. W., W. S. Hale and J. W. Pence (1962). The effect of processing variations on the alcohol, carbonyl, and organic acid contents of pre-ferments for bread baking. *Cereal Chem.*, **39**, 114–122.

Cole, E. W., V. Helmke and J. W. Pence (1966). Alpha-keto acids in bread preferments. *Cereal Chem.*, **43**, 357–361.

Collins, C. H. and P. M. Lyne (1976). *Microbiological Methods*, 4th edn. Butterworths, London.

Conway, E. J. (1955). Evidence for a redox pump in the active transport of cations. *Int. Rev. Cytol.*, **4**, 377–396.

Conway, E. J. and M. E. Beary (1956). Active transport of magnesium. *Nature (London)*, **178**, 1044.

Conway, E. J. and M. E. Beary (1958). Active transport of magnesium across yeast cell membrane. *Biochem. J.*, **69**, 275–280.

Conway, E. J. and J. Breen (1945). An 'ammonia' yeast and some of its properties. *Biochem. J.*, **39**, 368–371.

Conway, E. J. and P. F. Duggan (1958). A cation carrier in the yeast cell wall. *Biochem. J.*, **69**, 265–274.

Conway, E. J. and D. T. Moore (1954). A sodium yeast and some of its properties. *Biochem. J.*, **57**, 523–528.

Conway, E. J., H. Ryan and E. Carton (1954). Active transport of sodium ions from yeast cells. *Biochem. J.*, **58**, 158–167.

Cooney, C. L., H. Y. Wang and D. I. C. Wang (1977). Computer-aided material balancing for production of fermentation parameters. *Biotechnol. Bioeng.*, **19**, 55–67.

Cooper, E. J. and S. L. Chen (1966). Yeast leavened pre-mix for bakery products. *US Pat.* 3 231 389.

Cooper, E. J. and G. Reed (1968). Yeast fermentation. *Baker's Digest*, **42** (6), 22–24, 26, 28–29, 63.

Cooper, R. G., R. S. Silver and J. P. Boyle (1975). Semi-commercial studies of a petroprotein process based on *n*-paraffins. In *Single Cell Protein II*, ed. S. R. Tannenbaum and D. I. C. Wang, chap. 23, pp. 454–466. MIT Press, Cambridge, MA.

Dahle, L. K. and R. S. Hinz (1966). The weakening action of thioctic acid in unyeasted and yeasted doughs. *Cereal Chem.*, **43**, 682–688.

Dalby, D. K. (1982). Pure culture methods. In *Prescott and Dunn's Industrial Microbiology*, ed. G. Reed, 4th edn., chap. 3, pp. 44–62. AVI Publishing Co., Westport, CT.

DeDeken, R. H. (1966). The Crabtree effect: A regulatory system in yeast. *J. Gen. Microbiol.*, **44**, 149–156.

Dellweg, H., W. K. Bronn and W. Hartmeier (1977). Respiration rates of growing and fermenting yeast. *Kem.-Kemi*, **4**, 611–615.

DeLoffre, M. C. H. (1964). Improvements in multiple stage production of yeast. *US Pat.* 3 120 073.

Densky, H., P. J. Gray and A. Budat (1966). Further study on the determination of zinc and its effect on various yeasts. *Proc. Am. Soc. Brew. Chem.*, 93–100.

de Robichon-Szulmajster, H. (1958a). Uridine-diphosphogalactose-4-epimeriase, an adaptive enzyme in *Saccharomyces fragilis*. *Biochim. Biophys. Acta*, **29**, 270–272.

de Robichon-Szulmajster, H. (1958b). Induction of enzymes of the galactose pathway in mutants of *Saccharomyces cerevisiae*. *Science*, **127**, 28–29.

de Robichon-Szulmajster, H. and Y. Surdin-Kerjan (1971). Nucleic acid and protein synthesis in yeasts: regulation of synthesis and activity. In *The Yeasts*, ed. A. H. Rose and J. S. Harrison, vol. 2, chap. 9, pp. 336–418. Academic, New York.

DiCarlo, F. S., A. S. Schultz and A. M. Kent (1953). The mechanism of allantoin catabolism by yeast. *Arch. Biochem. Biophys.*, **44**, 468–474.

Eaton, N. R. (1960). Endogenous respiration of yeast. I. The endogenous substrate. *Arch. Biochem. Biophys.*, **88**, 17–25.

Ebner, H. and A. Enekel (1974). Device for aerating liquids. *US Pat.* 3 813 086.

Ebner, H., K. Pohl and A. Enekel (1967). Self-priming aerator and mechanical defoamer for microbiological processes. *Biotechnol. Bioeng.*, **9**, 357–364.

Eddy, A. A. (1958). Aspects of chemical composition of yeast. In *The Chemistry and Biology of Yeasts*, ed. A. H. Cook, chap. 5, pp. 157–249. Academic, New York.

Edelmann, K., P. Stelwagen and E. Oura (1978). Quality of baker's yeast stored under aerobic and anaerobic conditions. *6th International Specialist Symposium of Yeasts, Montpellier, France*, Abstr. SVII, pp. 15–16.

Edelmann, K., P. Stelwagen and E. Oura (1980). The influence of temperature and availability of oxygen on the carbohydrates of stored baker's yeast. In *Advances in Biotechnology*, ed. G. G. Stewart and I. Russell, vol. 4, Current Devel-opments in Yeast Research. Proceedings of the 5th International Yeast Symposium, London, Ontario, Canada. pp. 51–56. Pergamon, Toronto.

Erisman, J. L. (1938). Roto-Louvre dryer. *Ind. Eng. Chem.*, **30**, 996–997.

Eroshin, V. K., I. S. Utkin, S. V. Ladynichev, V. V. Samoylov, V. D. Kuvshinnikov and G. K. Skoyabin (1976). Influence of pH and temperature on substrate yield coefficient of yeast growth in chemostat. *Biotechnol. Bioeng.*, **18**, 289–295.

Fengl, Z. (1969). Production of microbial protein from carbon sources. In *Global Impacts of Applied Microbiology II*, ed. E. L. Gaden, Jr., pp. 63–70. Interscience, New York.

Finn, R. K. (1969). Energy cost of oxygen transfer. *Process Biochem.*, **4**, 17, 22.

Flynn, D. S. (1974). Computer control of fermentation processes. In *Advances in Microbial Engineering*, ed. A. Prokop and M. Novak, part 2, pp. 597–605. Wiley, New York.

Forsander, O. and H. Suomalainen (1957). Thiamine in baker's yeast. *Acta Chem. Scand.*, **11**, 1094.

Foulkes, E. C. (1956). Cation transport in yeast. *J. Gen. Physiol.*, **39**, 687–704.

Franz, B. (1961). Kinetics of the alcoholic fermentation during the propagation of baker's yeast. *Die Nahrung*, **5**, 458–481.

Frey, C. N. (1930). History and development of the modern yeast industry. *Ind. Eng. Chem.*, **22**, 1154–1162.

Frey, C. N. (1957). History of the development of active dry yeast. In *Yeast — Its Characteristics, Growth and Function in Baked Products*, ed. C. S. McWilliams and M. S. Peterson, pp. 7–33. Quartermaster Food and Container Institute for the Armed Force, Chicago.

Frey, C. N., G. W. Kirby and A. Schultz (1936). Yeast — physiology, manufacture and uses. *Ind. Eng. Chem.*, **28**, 879–884.

Frey, S. W., W. G. DeWitt and B. R. Bellamy (1967). The effect of several trace metals on fermentation. *Proc. Am. Soc. Brew. Chem.*, 199–205.

Fuhrmann, G. F. (1973). Dependence of divalent cation transport on ATP in yeast. *Experientia*, **29**, 742.

Fuhrmann, G. F. and A. Rothstein (1968). The transport of Zn^{2+}, Co^{2+} and Ni^{2+} into yeast cells. *Biochim. Biophys. Acta*, **163**, 325–330.

Garver, J. C., I. Navarine and A. M. Swanson (1966). Factors influencing the activation of baker's yeast. *Cereal Sci. Today*, **11**, 410–418.

Ghosh, A. and S. N. Bhattacharyya (1967). Changes in respiratory and glycolytic behaviour of *S. carlsbergensis* due to inositol deficiency. *Biochim. Biophys. Acta*, **136**, 19–26.

Ginterova, A., L. Mitterhauszerova and O. Janotkova (1966). Effect of temperature on the agglutination of stored baker's yeast. *Branntweinwirstschaft*, **106** (3), 57–59.

Gray, W. D. and C. Sova (1956). Relation of molecular size and structure to alcohol inhibition of glucose utilization by yeast. *J. Bacteriol.*, **72**, 349–356.

Grba, S., E. Oura and H. Suomalainen (1975). On the formation of glycogen and trehalose in baker's yeast. *Eur. J. Appl. Microbiol.*, **2** (1), 29–38.

Grba, S., E. Oura and H. Suomalainen (1979). Formation of trehalose and glycogen in growing baker's yeast. *Finn. Chem. Lett.*, 1979, 61–64.

Grylls, F. S. M., S. D. Rennie and M. Kelly (1978). Process and apparatus for producing active dried yeast. *US Pat.* 4 081 558.

Haarasilta, S. and E. Oura (1975). On the activity and regulation of anaplerotic and gluconeogenetic enzymes during the growth process of baker's yeast. *Eur. J. Biochem.*, **52**, 1–7.

Harbrecht, A. and R. Kautzmann (1967). Vergleichende Untersuchungren zur Triebkraftbestimmung von Backhefen. *Branntweinwirtschaft*, **107**, 21–23.

Harrison, J. S. (1967). Aspects of commercial yeast production. *Process Biochem.*, **2** (3), 41–45.

Harrison, J. S. (1971). Yeast production. *Prog. Ind. Microbiol.*, **10**, 129–177.

Harrison, J. S. and W. E. Trevelyan (1963). Phospholipid breakdown in baker's yeast during drying. *Nature (London)*, **200**, 1189–1190.

Hartmeier, W. (1977). Active dry yeast and method of its production. *Ger. Pat.* 2 515 029.

Hatch, R. T. (1973). Experimental and theoretical studies of oxygen transfer in the airlift fermentor. Ph.D. Thesis, Massachusetts Institute of Technology, Cambridge, MA.

Hatch, R. T. (1975). Fermentor design. In *Single-Cell Protein II*, ed. S. R. Tannenbaum and D. I. C. Wang. MIT Press, Cambridge, MA.

Hautera, P. and T. Lovgren (1975a). The fermentation activity of baker's yeast. *Baker's Digest*, **49**, 36–37, 49.

Hautera, P. and T. Lovgren (1975b). α-Glucosidase, α-glucoside permease, maltose fermentation and leavening ability of baker's yeast. *J. Inst. Brew.*, **81**, 309–311.

Hayduck, F. (1919). V. D. Spiritusfabrikanten in Deutschl., *Ger. Pat.* 300 662.

Hayner, W. C., L. J. Wickerham and C. W. Hesseltine (1955). Maintenance of cultures of industrially important microorganisms. *Appl. Microbiol.*, **3**, 361–368.

Hernandez, E. and M. J. Johnson (1967). Energy supply and cell yield in aerobically grown microorganisms. *J. Bacteriol.*, **94**, 996–1101.

Herrera, T., W. H. Peterson, E. J. Cooper and H. J. Peppler (1956). Loss of cell constituents on reconstitution of active dry yeast. *Arch. Biochem. Biophys.*, **63**, 131–143.

Hofer, M., K. Brand, K. Deckner and J. U. Becker (1971). Importance of the pentose phosphate pathway for D-glucose catabolism in the obligatory aerobic yeast *Rhodotorula gracilis*. *Biochem. J.*, **123**, 853–855.

Hommes, F. A. (1966). Effect of glucose on the level of glycolytic enzymes in yeast. *Arch. Biochem. Biophys.*, **114**, 221–233.

Hospodka, J., Z. Caslavsky, K. Beran and F. Stross (1962). The polarographic determination of oxygen uptake and transfer rate in aerobic steady-state yeast cultivation on laboratory plant scale. In *Continuous Cultivation of Microorganisms*, ed. I. Malke, K. Beran and J. Hospodka, pp. 353–368. Academic, New York.

Hsu, K. H., R. C. Hoseney and P. A. Seib (1979). Frozen dough. II. Effects of freezing and storing conditions on the stability of yeasted doughs. *Cereal Chem.*, **56**, 424–426.

Inskeep, G. C., A. J. Wiley, J. M. Holderby and L. P. Hughes (1951). Food yeast from sulfite liquor. *Ind. Eng. Chem.*, **43**, 1702–1711.

Irwin, R. (1954). Commercial yeast manufacture. In *Industrial Fermentation*, ed. L. A. Underkofler and R. J. Hickey, vol. 1, pp. 273–246. Chemical Publishing Co., New York.

Jacob, H. E. (1944). *Six Thousand Years of Bread*. Doubleday, Doran and Co., Garden City, NY.

Johnston, W. R. (1959). Active dry yeast product and processes for producing the same. *US Pat.* 2 919 194.

Jones, R. P. and P. F. Greenfield (1982). Effect of carbon dioxide on yeast growth and fermentation. *Enzym. Microbiol. Technol.*, **4**, 210–223.

Josic, D. (1982). Optimization of process conditions for the production of active dry yeast. *Lebensm.-Wiss. u.-Technol.*, **15**, 5–14.

Keszler, H. J. (1967). Some factors determining growth of yeast. *Zbet. Bakt. II*, **121**, 129–178.

Kline, L. and T. F. Sugihara (1971). Microorganisms of the San Francisco sour dough bread process. II. Isolation and characterization of undescribed bacterial species responsible for souring activity. *Appl. Microbiol.*, **21**, 459–465.

Kornberg, S. R. (1956). Tripolyphosphate and trimetaphosphate in yeast extract. *J. Biol. Chem.*, **218**, 23–31.

Kotyk, A. and J. Horak (1981). Transport processes in the plasma membrane. In *Yeast Cell Envelopes: Biochemistry, Biophysics and Ultrastructure*, ed. W. N. Arnold, vol. 1. chap. 4, pp. 49–64. CRC Press, Boca Raton, FL.

Kraus, J. K., R. Scopp and S. L. Chen (1981). Effect of rehydration on dry wine yeast activity. *Am. J. Enol. Vitic.*, **32**(2), 132–134.

Kuninori, T. and B. Sullivan (1968). Disulfide–sulfhydryl interchange studies of wheat flour. II. Reaction of glutathione. *Cereal Chem.*, **45**, 486–495.

Langejan, A. (1972). A novel type of active dry yeast baker's yeast. In *Fermentation Technology Today*, ed. G. Terui, pp. 669–671. Society of Fermentation Technology, Osaka, Japan.

Langejan, A. (1974). Preparation of active dried baker's yeast. *US Pat.* 3 843 800.

Langejan, A. (1976). High protein active dried baker's yeast. *US Pat.* 3 993 782.

Langejan, A. (1980). Active dried baker's yeast. *US Pat.* 4 217 420.

Lewin, L. M. (1967). Effect of *meso*-inositol deficiency on some important biological and chemical characteristics of yeasts. *J. Gen. Physiol.*, **41**, 215–244.

Lovgren, T. and P. Hautera (1977). Maltose fermentation and leavening ability of baker's yeast. *Eur. J. Appl. Microbiol.*, **4**, 37–43.

Luca, S. L., T. Thommel and W. K. Brown (1979). Freely-flowing powdered fresh baker's yeast preparation and method of producing it. *US Pat.* 4 160 040.

Maddox, I. S. and J. S. Hough (1970). Effect of zinc and cobalt on yeast growth and fermentation. *J. Inst. Brew.*, **76**, 262–264.

Maga, J. A. (1974). Bread flavor. *CRC Crit. Rev. Food Technol.*, **5**, 55.

Maitra, P. K. and Z. Lobo (1971). A kinetic study of glycolytic enzyme synthesis in yeast. *J. Biol. Chem.*, **246**, 475–488.

Marshall, W. R., Jr. and S. J. Friedman (1950). Drying. In *Chemical Engineers' Handbook*, ed. J. H. Perry, 3rd edn., pp. 799–884. McGraw-Hill, New York.

Martin, S. M. (1964). Conservation of microorganisms. *Annu. Rev. Microbiol.*, **18**, 1–16.

Mateles, R. I. and S. R. Tannenbaum (1968). *Single Cell Protein*. MIT Press, Cambridge, MA.

Maw, G. A. (1960). Utilization of sulphur compounds by a brewer's yeast. *J. Inst. Brew.*, **66**, 162–167.

Maxon, W. D. and M. J. Johnson (1953). Aeration studies on propagation of baker's yeast. *Ind. Eng. Chem.*, **45**, 2554–2560.

Mayberry, W. R., G. J. Prochazka and W. J. Payne (1967). Growth yields of bacteria on selected organic compounds. *Appl. Microbiol.*, **15**, 1332–1338.

Mazur, P. (1970). Cryobiology: The freezing of biological systems. *Science*, **168**, 939–949.

Melnick, D., M. Hockberg, H. M. Himes and B. L. Oser (1945). The multiple nature of vitamin B_6. Critique of methods for the determination of the complex and its components. *J. Biol. Chem.*, **160**, 1–14.

Meyenburg, von. H. K. (1969). Energetics of the budding cell of *S. cerevisiae* during glucose-limited aerobic growth. *Arch. Mikrobiol.*, **66**, 289–303.

Miettinen, J. K. (1964). Three fast sampling techniques for biokinetics experiments with radioisotopes. In *Rapid Mixing and Sampling Techniques in Biochemistry*, ed. B. Chances, *et al.*, pp. 303–310. Academic, New York.

Mitchell, J. H., Jr. (1957). Evaluation of the bread baking properties of various genera and species of yeasts. In *Yeasts — Its Characteristics, Growth, and Function in Baked Products*, ed. C. S. McWilliams and M. S. Peterson, pp. 67–69. Quartermaster Food and Container Institute for the Armed Force, Chicago.

Mitchell, J. H., Jr. and J. J. Enright (1957). Effect of low moisture levels on thermostability of active dry yeast. *Food Technol.*, **11**, 359–362.

Mitchell, J. H., Jr. and J. J. Enright (1959). Thermostable active dry yeast composition. *US Pat.* 2 894 842.

Monod, J. (1949). The growth of bacterial culture. *Annu. Rev. Microbiol.*, **3**, 371–394.

Mor, J. R. and A. Fiechter (1968). Continuous cultivation of *S. cerevisiae*. I. Growth on ethanol under steady-state conditions. *Biotechnol. Bioeng.*, **10**, 159–176.

Morris, E. O. (1958). Yeast growth. In *The Chemistry and Biology of Yeasts*, ed. A. H. Cook, chap. 6, 251–321. Academic, New York.

Morris, J. R. and D. W. Ribbons (1970). *Methods in Microbiology*, vol. 3A. Academic, New York.

Nagao, S., S. Endo, K. Takeya and K. Tanaka (1981a). The Do-Corder as a tool to evaluate the breadmaking properties of a dough. *Cereal Chem.*, **58**, 384–387.

Nagao, S., S. Endo and K. Tanaka (1981b). Effect of fermentation on the Do-Corder and bread-making properties of a dough. *Cereal Chem.*, **58**, 388–391.

Noltmann, E. and F. H. Bruns (1958). Phosphamannose isomerase. *Biochem. Z.*, **330**, 514–520.

Notkina, L. G., L. M. Balyberdina and L. D. Lavrenchuk (1975). The effect of nitrite in baker's yeast manufacture. *Khlebopek. Konditer. Promst.*, **19**(2), 28–31.

Nyiri, L. K. (1974). *Oxygen Transfer Rates in 270L Pilot-plant Fermentor*. Report of Fermentation Design, Inc. Bethlehem, PA.

Nyiri, L. K., G. M. Toth, C. S. Krishnaswami and D. V. Parmenter (1976). On-line analysis and control of fermentation processes. In *Workshop Computer Applications in Fermentation Technology*, ed. R. P. Jefferis, III, pp. 37–46. Verlag Chemie, Weinheim.

Olbright, H. (1956). *Die melasse*. Institut für Garungsgewerbe, Berlin.

Olbright, H. (1972). Baker's yeast production. In *International Molasses Report*, ed. F. O. Lights, vol. 9, no. 11, 12, 13, 15, 16.

Olsen, A. J. C. (1960). Manufacture of baker's yeast by continuous fermentation. I. Plant and process. *Soc. Chem. Ind. Monograph*, **12**, 81–93.

Olson, B. H. and M. J. Johnson (1949). Factors producing high yeast yields in synthetic media. *J. Bacteriol.*, **57**, 235–246.

Oura, E. (1972a). The production of energy during aerobic growth of baker's yeast on glucose and ethanol. *Proceedings of the 2nd Congress on Yugoslav Microbiology, Opatija*, pp. 237–246.

Oura, E. (1972b). The effect of aeration on the growth energetics and biochemical composition of baker's yeast. *Res. Lab. State Alcohol Monopoly, Helsinki*.

Oura, E. (1973). Energetics of yeast growth under different intensities of aeration. *Biotechnol. Bioeng. Symp. No. 4*, 117–127.

Oura, E. (1976). The effect of aeration intensity on the biochemical composition of baker's yeast: activities of enzymes of the glycolytic and pentose phosphate pathways. *Biotechnol. Bioeng.*, **18**, 415–420.

Oura, E. (1978). The effect of culture conditions on the requirement of biotin by baker's yeast. *6th International Specialist Symposium on Yeast, Montpellier, France*. Abstr. pp. SI, 1–2.

Oura, E. and S. Haarasilta (1977). Carbon dioxide fixation in yeast grown under gluconeogenic conditions. *Abstracts of the 11th FEBS Meeting, Copenhagen*.

Oura, E. and H. Suomalainen (1978). Biotin and the metabolism of baker's yeast. *J. Inst. Brew.*, **84**, 283–287.

Oura, E., H. Suomalainen and E. Parkkinen (1974). Changes in commercial baker's yeast during its ripening period. *Proceedings of the 4th International Symposium on Yeasts, Part I*, pp. 125–126.

Panek, A. D. (1975). Trehalose synthesis during starvation of baker's yeast. *Eur. J. Appl. Microbiol.*, **2**, 39–46.

Parkkinen, E., E. Oura and H. Suomalainen (1973). Changes in RNA fractions during storage of baker's yeast. *Proceedings of the 3rd International Specialist Symposium on Yeasts, Finland, Part 1*, abstr., pp. 62–63. Print OY, Helsinki.

Parkkinen, E., E. Oura and H. Suomalainen (1974). Effect of storage on the nucleic acid composition of baker's yeast. *J. Inst. Brew.*, **80**, 271–277.

Payen, R. (1949). Variations des teneurs en glycogene et trehalose pendant le sechage de la levure. *Can. J. Res.*, **27B**, 749–756.

Paturau, J. M. (1982). *By-Products of the Cane Sugar Industry*. Elsevier, New York.

Peppler, H. J. (1960). Yeast. In *Bakery Technology and Engineering*, ed. S. A. Matz. AVI Publishing Co., Westport, CT.

Peppler, H. J. (1979). Production of yeasts and yeast products. In *Microbial Technology*, ed. D. Perlman and H. J. Peppler, 2nd edn., vol. 1, pp. 157–185. Academic, New York.

Polakis, E. S. and W. Bartley (1965). Changes in the enzyme activities of *Saccharomyces cerevisiae* during aerobic growth on different carbon sources. *Biochem. J.*, **97**, 284–297.

Pollock, G. E. and C. D. Holmstrom (1951). The trehalose content and the quality of active dry yeast. *Cereal Chem.*, **28**, 498–505.

Pomper, S. and E. Akerman (1970). Propagation of yeast. *US Pat.* 3 617 306.

Pomper, S. and E. Akerman (1980). Preparation of free-flowing particulate yeast. *US Pat.* 4 232 045.

Ponte, J. G., R. L. Glass and W. F. Geddes (1960). Studies on the behaviour of active dry yeast in breadmaking. *Cereal Chem.*, **37**, 263–279.

Power, D. M. and S. W. Challinor (1969). The effects of inositol deficiency on the chemical composition of the yeast cell wall. *J. Gen. Microbiol.*, **55**, 169–176.

Pressindustria, SpA (1971). Process for drying bakery yeast. *Br. Pat.* 1 230 587.

Pyke, M. (1957). Bakers' yeast. In *Yeasts*, ed. W. Roman. Dr. W. Junk Publishers, The Hague.

Pyler, E. J. (1958). *Our Daily Bread*. Siebel Publishing Co., Chicago, IL.

Ramnietse, V. E., I. V. Skard and M. V. Popova (1978). Effect of glutamic acid on permeability of *Saccharomyces cerevisiae* cells during their rehydration after dehydration. *Mikrobiologia*, **47**, 430–435.

Reed, G. (1982). Production of bakers' yeast. In *Prescott and Dunn's Industrial Microbiology*, ed. G. Reed, 4th edn., chap. 14, pp. 593–633. AVI Publishing Co., Westport, CT.

Reed, G. and H. J. Peppler (1973). *Yeast Technology*. AVI Publishing Co., Westport, CT.

Reilly, C. (1967). Sodium and potassium transport in normal and mutant yeast: inhibition by *O*-phenanthroline. *Nature (London)*, **214**, 1330–1331.

Richard, O. W. and M. C. Troutman (1940). Spectroscopic analysis of the mineral content of yeast grown on synthetic and natural media. *J. Bacteriol.*, **39**, 739–746.

Richards, J. E. (1961). Studies in aeration and agitation. *Prog. Ind. Microbiol.*, **3**, 143–172.

Rickard, P. A. D. and C. B. J. Hogan (1978). Effect of glucose on the activity and synthesis of fermentative and respiratory pathways of *Saccharomyces* species. *Biotechnol. Bioeng.*, **20**, 1105–1110.

Ridgeway, G. J. and H. C. Douglas (1958). Unbalanced growth of yeast due to inositol deficiency. *J. Bacteriol.*, **76**, 163–166.

Ridgeway, J. A., T. A. Lappin, N. M. Benjamin, J. B. Corns and C. Akin (1975). Single-cell protein materials from ethanol. *US Pat.* 3 865 691.

Rogosa, M. (1944). Vitamin requirements of lactose-fermenting and certain other yeasts. *J. Bacteriol.*, **47**, 159–170.

Roomans, G. M., F. Blaso and G. Borst-Pauwels (1977). Cotransport of phosphate and sodium by yeast. *Biochim. Biophys. Acta*, **467**, 65–71.

Roomans, G. M., A. P. R. Theuvenet, T. Van DenBerg and G. Borst-Pauwels (1979). Kinetics of Ca^{2+} and Sr^{2+} uptake by yeast; effects of pH, cations, and phosphate. *Biochim. Biophys. Acta*, **551**, 187–196.

Roon, R. J. and B. Levenberg (1968). An adenosine triphosphate dependent, avidin sensitive enzymatic cleavage of urea in green algae. *J. Biol. Chem.*, **243**, 5213–5215.

Roon, R. J. and B. Levenberg (1972). Urea amidolyase. *J. Biol. Chem.*, **247**, 4107–4113.

Rosa, M. (1963). The fermentation activity of yeast during discontinuous and continuous alcoholic fermentations. *Die Nahrung*, **7**, 508–517.

Rosen, K. (1977). Production of baker's yeast. *Process Biochem.*, **12** (3), 10–12.

Rothstein, A. and A. D. Hayes (1956). The relationship of the cell surface to metabolism. XIII. The cation-binding properties of the yeast cell surface. *Arch. Biochem. Biophys.*, **63**, 87–99.

Sak, S. (1919). *Danish Pat.* 28 507.

Sato, T., N. Tsumura, Y. Tanaka, T. Okada and Y. Koyanagi (1961). Comparison of practical characters of baker's yeast of the world. *Nippon Jozo Kyokai*, **57** (1) 74–80; (2) 67–76.

Schmidt, G., L. Hecht and S. J. Thannhauser (1949). The effect of potassium ions on the absorption of orthophosphate and the formation of metaphosphate by baker's yeast. *J. Biol. Chem.*, **178**, 733–742.

Schreier, K. (1974). Bio-reactors: Stage of development and industrial application, especially with regard to systems for transfer of gas. *4th International Symposium on Yeasts*.

Schultz, A. (1965). Investigations of leavening activity of baker's yeasts. *Brot Gebaeck*, **19** (4), 61–65.

Schultz, A. S. and L. Atkin (1947). The utilization of Bios response in yeast classification and nomenclature. *Arch. Biochem.*, **14**, 369–380.

Schultz, A. S. and D. K. McManus (1950). Amino acids and inorganic sulfur as sulfur source for the growth of yeasts. *Arch. Biochem.*, **25**, 401–409.

Schultz, A. S. and S. Pomper (1948). Amino acids as nitrogen source for the growth of yeasts. *Arch. Biochem.*, **19**, 184–192.

Schultz, A. S. and F. R. Swift (1955). Manufacture of yeast. *US Pat.* 2 717 837.

Schultz, A. S., L. Atkin and C. N. Frey (1937). A fermentation test for vitamin B_1. *J. Am. Chem. Soc.*, **59**, 948.

Schultz, A. S., L. Atkin and C. N. Frey (1939). Vitamin B_6, a growth promoting factor for yeast. *J. Am. Chem. Soc.*, **61**, 1931.

Schultz, A. S., L. Atkin and C. N. Frey (1941). Synthesis of thiamin and product. *US Pat.* 2 262 735.

Schultz, A. S., D. K. McManus and S. Pomper (1949). Amino acids as carbon source for the growth of yeasts. *Arch. Biochem.*, **22**, 412–419.

Seeley, R. D. and H. F. Ziegler (1962). Yeast: Some aspects of its fermentative behaviours. *Baker's Digest*, **36** (4), 48–52.

Sher, H. N. (1960). Manufacture of baker's yeast by continuous fermentation. II. Instrumentation. *Soc. Chem. Ind. Monograph*, **12**, 94–115.

Shoda, M. and Y. Ishikawa (1981). Carbon dioxide sensor for fermentation systems. *Biotechnol. Bioeng.*, **23**, 461–466.

Sims, A. P. and B. F. Folkes (1964). A kinetic study of the assimilation of ^{15}N-ammonia and the synthesis of amino acids in an exponentially growing culture of *Candida utilis*. *Proc. R. Soc. London, Ser. B*, **159**, 479–502.

Simon, E. J. (1976). Drying of microorganisms in fluid bed driers under mild conditions. *Chemie-Tkechnik*, **7**, 277–280.

Skiba, M. (1966). *Abstracts of the 2nd International Congress on Food Science and Technology Warsaw*, p. 53. Cited by Burrows, S. (1970). *Baker's Yeast*, p. 363. Academic, New York.

Slein, M. W. (1950). Phosphomannose isomerase. *J. Biol. Chem.*, **186**, 753–761.

Stineman, T. L., J. D. Edwards and J. C. Grosskopf (1980). Production of baker's yeast from acid whey. *US Pat.* 4 192 918.

Stone, I. N. (1965). Treatment of wort with zinc. *US Pat.* 3 164 472.

Sugihara, T. F., L. Kline and M. W. Miller (1971). Microorganisms of the San Francisco sour dough bread process. I. Yeasts responsible for the leavening action. *Appl. Microbiol.*, **21**, 456–458.

Suomalainen, H. (1963). Changes in the cell constitutions of baker's yeast in changing growth conditions. *Pure Appl. Chem.*, **7**, 634–654.

Suomalainen, H. (1975). Some enzymological factors influencing the leavening capacity and keeping quality of baker's yeast. *Eur. J. Appl. Microbiol.*, **1**, 1–12.

Suomalainen, H. and E. Axelson (1956). Effect of thiamin on the rate of fermentation of zymohexose and of maltose by baker's yeast. *Biochim. Biophys. Acta*, **20**, 315–318.

Suomalainen, H. and A. J. A. Keranen (1967). Keto acids formed by baker's yeast. *J. Inst. Brew.*, **73**, 477–484.

Suomalainen, H. and M. Lehtonen (1978). Yeast as a producer of aroma compounds. *Proceedings of the 1st European Congress on Biotechnology, Interlaken, Monograph 82*, pp. 207–220.

Suomalainen, H. and E. Oura (1956). The fermentability of maltose by baker's yeast containing trehalose. *Biochim. Biophys. Acta*, **20**, 538–542.

Suomalainen, H. and E. Oura (1971). Yeast nutrition and solute uptake. In *The Yeasts*, ed. A. H. Rose and J. S. Harrison, vol. 2, chap. 2, pp. 3–74. Academic, New York.

Suomalainen, H. and E. Oura (1978). Ethanol as substrate for baker's yeast. *XII International Congress of Microbiology, München*, p. 60.

Suomalainen, H. and S. Pfaffli (1961). Changes in carbohydrate reserves of baker's yeast during growth and on standing. *J. Inst. Brew.*, **67**, 249–254.

Suomalainen, H., A. Boorklund, K. Vihervaara and E. Oura (1965a). Nicotinic acid and nicotinamide adenine dinucleotide contents of baker's yeast in changing culture conditions. *J. Inst. Brew.*, **71**, 221–226.

Suomalainen, H., T. Nurminen, K. Vihervaara and E. Oura (1965b). Effect of aeration on the syntheses of nicotinic acid and nicotinamide adenine dimucleotide by baker's yeast. *J. Inst. Brew.*, **71**, 227–231.

Sysojewa, J. L. and N. W. Gorochova (1965). Production of active dry yeast. *Khlebopek, Konditer. Promst.*, **9** (6), 37–41.

Takakuwa, M. (1962). Effect of biuret on preservability of pressed yeast and its mechanism. *Mem. Ehime Univ., Sect. VI*, **8**, 443–447.

Takakuwa, M. and H. Tohoyama (1975). Effect of metabolic inhibitors on the preservability of pressed baker's yeast and on the increase of intracellular proteinase activity during its storage. *J. Ferment. Technol.*, **53**, 895–897.

Takakuwa, M., K. Takeuchi and N. Yamasaki (1975). Increase of NADase activity during storage of pressed baker's yeast. *Agric. Biol. Chem.*, **39**, 2235–2236.

Tannen, L. P. and L. K. Nyiri (1979). Instrumentation of fermentation systems. In *Microbial Technology*, ed. H. J. Peppler and D. Perlman, 2nd edn., vol. 2, chap. 14, pp. 331–374. Academic, New York.

Tannenbaum, S. R. and D. I. C. Wang (1975). *Single Cell Protein II.* MIT Cambridge, MA.

Thiessen, E. J. (1942). The effect of temperature and viability and baking properties of dry and moist yeast stored for varied periods. *Cereal Chem.*, **19**, 773–784.

Thorn, J. A. and G. Reed (1959). Production and baking techniques for active dry yeast. *Cereal Sci. Today*, **4**, 198–201.

Thorne, R. S. W. (1949). Nitrogen metabolism of yeast. A consideration of the mode of assimilation of amino acids. *J. Inst. Brew.*, **55**, 201–222.

Tohoyama, H. and M. Takakuwa (1971). Increase in intracellular proteinase activity during storage of pressed baker's yeast. *J. Ferment Technol.*, **50**, 397–404.

Van Steveninck, J. and L. Booji (1964). The role of polyphosphates in the transport mechanism of glucose in yeast cells. *J. Gen. Physiol.*, **48**, 43–60.

Van 'Triet, J. and J. De Bruijn (1981). Process for the preparation of active dried microorganisms. *Int. Pat.* WO-81 01 415.

Vincent, A. (1974). Control of dissolved oxygen. *Process Biochem.*, **9**, 30–32.

Wakamura, A., Y. Yokoyama and K. Ueno (1973). A method to prepare powdery dry yeast for bread mix. *Jpn. Pat.* 48 35 475.

Walker, J. A. H. and A. B. Holdsworth (1958). Equipment design. In *Biochemical Engineering*, ed. R. Steel, pp. 223–274. Heywood, London.

Wang, H. Y., C. L. Cooney and D. I. C. Wang (1977). Computer aided baker's yeast fermentation. *Biotechnol. Bioeng.*, **19**, 69–86.

Wang, H. Y., C. L. Cooney and D. I. C. Wang (1979). Computer control of baker's yeast production. *Biotechnol. Bioeng.*, **21**, 975–996.

Watson, T. G. (1970). Effect of sodium chloride on steady state growth and metabolism of *Saccharomyces cerevisiae. J. Gen. Microbiol.*, **64**, 91–99.

Weaver, J. C., E. Pungor and C. L. Cooney (1980). Mass spectrometer monitoring of volatile compound in the gas and liquid phases of fermentors. *Am. Chem. Soc. Annu. Meeting*, Las Vegas.

White, J. (1954). *Yeast Technology.* Chapman and Hall, London.

Whitney, P. A. and T. G. Cooper (1970). Urea carboxylase and allophanate hydrolase: Two components of a multienzyme complex in *Saccharomyces cerevisiae. Biochem. Biophys. Res. Commun.*, **40**, 814–819.

Wiame, J. M. (1946). Basophile et metabolisme du phosphore chez la levure. *Bull. Soc. Chim. Biol.*, **28**, 552–556.

Wiame, J. M. (1947). Etude d'une substance polyphosphorée, basophile et meta-chromatique chez les levures. *Biochim. Biophys. Acta*, **1**, 234–255.

Wickerham, L. J. (1951). *Taxonomy of Yeasts.* US Dept. of Agriculture Tech. Bull. 1029.

Wilkinson, J. F. (1949). The pathway of the adaptive fermentation of galactose by yeast. *Biochem. J.*, **44**, 460–467.

Winter, R. (1956). Technical development problems in the production of yeast from sulfite liquor. *Chem. Technol.*, **8**, 144–157.

Winzler, R. J. (1941). The respiration of baker's yeast at low oxygen tension. *J. Cell. Comp. Physiol.*, **17**, 263–276.

Wise, W. S. (1951). The measurement of the aeration of culture media. *J. Gen. Microbiol.*, **5**, 167–177.

Witt, I., P. G. Weiler and H. Holzer (1964). Steigerung der CO_2 fixierung in glucose oxydierender hefe durch NH_4^+-salze. *Biochem. Z.*, **339**, 331–337.

Yamamoto, Y. and T. Inone (1961). Studies of the poor-attenuative yeast. I. Permeation of maltotriose through the cell wall of poor-attenuative yeast. *Rep. Res. Lab. Kiri Brew. Co.*, 49.

Yoshida, F. and K. Akita (1965). Performance of gas bubble columns: Volumetric liquid-phase mass transfer coefficient and gas hold-up. *AIChE J.*, **11**, 9–13.

Yoshida, A. and A. Yamataka (1953). On the metaphosphate of yeast. I. *J. Biochem.* **40**, 85–94.

21

Bacterial Biomass

J. H. LITCHFIELD
Battelle Memorial Institute, Columbus, OH, USA

21.1 INTRODUCTION

Why is bacterial biomass — the cell matter of bacteria — of interest as a source of single cell protein for human food or animal feed? To answer this question, in this chapter we will consider the microorganisms, processes, product quality, including food or feed, and economic and market aspects involved in producing bacterial biomass. In our definition of bacterial biomass we will also include the cells of actinomycetes, since these microorganisms have similar characteristics to those of bacteria used in single cell protein (SCP) production.

There are numerous published reviews and symposia on the subject of SCP production including bacterial biomass and the reader should consult these publications for further information (Cooney *et al.*, 1980; Davis, 1974; Litchfield, 1979, 1980, 1983a, 1983b; Rose, 1979; Solomons, 1983; Tannenbaum and Wang, 1975).

21.2 MICROORGANISMS

Numerous species of bacteria and some species of actinomycetes have been considered for use in SCP production. These organisms have a wide range of carbon and energy source utilization patterns and growth characteristics. In general, they have faster growth rates (20 min to 2 h generation times) than those of yeasts, molds and higher fungi (2 h to 16 h or more) although there is some overlap between the growth rates of slower-growing bacteria and those of the fastest growing yeasts.

Table 1 presents some of the photosynthetic bacteria and nonphotosynthetic bacteria and actinomycetes that have been investigated for SCP production from various substrates Photosynthetic bacteria of the genus *Rhodopseudomonas* have been grown on agricultural and industrial wastes (Kobayashi and Kurata, 1978; Shipman *et al.*, 1975) and on cow manure and anaerobic digester effluent (Vrati, 1984; Vrati and Verma, 1983). However, the requirements for light limit the applicability of photosynthetic bacteria to those regions where the available sunlight and temperatures permit open pond culture throughout the year. Consequently, nonphotosynthetic bacteria have greater utility for producing SCP and are emphasized in this chapter.

Some key criteria in selecting suitable strains are the following: (1) substrates utilized as carbon and energy sources, nitrogen sources and requirements for supplemental nutrients; (2) high specific growth rates, productivities and yields on a given substrate; (3) pH and temperature tolerance; (4) aeration requirements and foaming characteristics; (5) culture stability including freedom from bacteriophages; (6) nonpathogenicity to humans, animals or plants either by direct infection or by excretion of toxins; (7) absence of endotoxin; (8) lack of potential for mating with known pathogenic bacteria (*e.g.* members of the Enterobacteriaceae); and (9) ease of separation from the growth medium by agglomeration or flocculation.

Pure cultures are employed in most bacterial SCP processes. However, a two culture *Cellulomonas* sp.–*Alcaligenes faecalis* system was developed at Louisiana State University for cellulose utilization. The *Alcaligenes faecalis* culture scavenged soluble sugars produced by the *Cellulomonas* which had good cellulase but low β-glucosidase activity (Dunlap, 1975). For bacterial biomass production from methane, Shell Research, Ltd. in the United Kingdom found that a mixed culture system consisting of *Pseudomonas* sp., *Hyphomicrobium* sp., *Acinetobacter* sp. and *Flavobacterium* sp. gave a higher growth rate, higher yield coefficient, greater stability, greater assistance to contamination, and less foaming than pure cultures (Hamer, 1979).

21.3 CARBON AND ENERGY, NITROGEN AND SUPPLEMENTAL NUTRIENT SOURCES

21.3.1 Carbon and Energy Sources

Table 1 classifies carbon and energy sources for bacterial biomass according to their source, either from renewable resources such as carbohydrates or carbohydrate wastes, or proteinaceous materials, or from nonrenewable resources such as petroleum hydrocarbons, natural gas or chemicals derived from them. It also summarizes the extent of pretreatment required before a substrate can be utilized for bacterial biomass production.

21.3.1.1 Substrates from renewable resources

Substrates from renewable resources such as agricultural and forestry products include simple sugars (glucose, galactose, fructose, lactose, mannose and sucrose), starches, hemicellulose and cellulose. These substrates are often readily available at relatively low cost in a wide range of geographical regions. However, seasonal availability may be a problem in temperate zone regions for those substrates obtained from agricultural crops.

Many of the low cost agricultural or forestry waste materials shown in Table 1 contain complex carbohydrates in the form of starch or lignocellulose materials. Although some bacteria can utilize starch, in most cases, materials that contain starch must be hydrolyzed to simple sugars by acids or enzymes or a combination of both before they can be assimilated.

Lignocellulosic materials must be treated by physical or chemical methods or a combination of both to liberate cellulose and hemicellulose from lignin, since cellulose and hemicellulose in the lignin–hemicellulose–cellulose (LHC) complex are not accessible to enzymatic or acid hydrolysis (Bungay, 1982).

For example, sugar cane bagasse contains only 50% cellulose with the remaining portion being hemicellulose and lignin that are not useful as substrates for SCP production. Research at Louisiana State University showed that sugar cane bagasse pretreated with 0.01 to 0.2 g NaOH/g bagasse at 88–121 °C for 0.5–0.6 h gave the best growth of *Cellulomonas* sp. (Callihan and Clemmer, 1979). In other studies with sugar cane bagasse, treatment with a 10% NaOH solution for 1 h at 180 °C provided for best growth of *Cellulomonas* sp. (Enriques and Rodriguez, 1983) while treatment of bagasse pith with 1% NaOH for 24 h at 25 °C gave maximum protein production with a *Cellulomonas* sp.–*Bacillus subtilis* mixed culture (Molina *et al.*, 1984).

Table 1 Substrates and Microorganisms for Bacterial Biomass Production

Substrate	Microorganism	Utilizable constituents	Treatment required	References
From renewable resources, agricultural and forest products				
(a) Bagasse straw, paper mill wastes, sulfite waste liquor, mesquite wood	Cellulomonas sp. Thermomonospora fusca	Cellulose, glucose, pentoses	Physical or chemical treatment to separate lignin from hemicellulose and cellulose, hydrolysis of cellulose and hemicellulose except for cellulolytic organisms	Callihan and Clemmer (1979); Crawford et al. (1973); Camhi and Rogers (1976); Thayer et al. (1975)
(b) Manures (poultry, cattle, pig)	Pseudomonas fluorescens Rhodopseudomonas sp.	Urea, uric acid and other nonprotein nitrogen compounds, proteins	Slurrying in water	Shuler, et al. (1979); Vrati (1984); Vrati and Verna (1983)
Cane and beet sugars and molasses, fructose and glucose syrup, fruit processing wastes	Wide range	Sucrose, glucose, fructose	None	Litchfield (1979, 1983)
Cheese whey, whey permeate	Aeromonas hydrophila	Lactose	None for microorganisms that utilize lactose	Butany and Ingledew (1973)
Starches	Bacillus sp. and Lactobacillus sp.	Glucose, maltose	Sequential bacterial degradation	Busta et al. (1977)
Starch wastes, food wastes	Rhodopseudomonas capsulata		None	Kobayashi and Kurata (1978)
Wheat Bran	Rhodopseudomonas gelatinosa		None	Shipman et al. (1975)
Meat processing wastes	Bacillus megaterium	Collagen, nonprotein nitrogen compounds	Slurrying in water	Bough et al. (1972)
From nonrenewable resources				
Hydrocarbons	Achromobacter delvacvate	n-Alkanes	None	Ko et al. (1964)
Fuel oil	Pseudomonas 5401	n-Alkanes	None	Ko and Yu (1968)
Gas oil	Acinetobacter cerificans	n-Alkanes	None	Ertola et al. (1969)
n-Paraffins (liquid)	Corynebacterium hydrocarboclastus	n-Propane	None	Akiba et al. (1973)
n-Paraffins (gaseous)	Nocardia paraffinica	n-Butane	None	Sugimoto et al. (1972).
Natural gas	Methylococcus capsulatus, mixed cultures	Methane	None	Hamer (1979)
Chemicals derived from hydrocarbons				
Ethanol	Acinetobacter calcoaceticus	Ethanol	None	Laskin (1977)
Methanol	Methylophilus methylotrophus	Methanol	None	Margetts (1983)
	Methylomonas clara		None	Sittig (1983)
Petrochemicals	Pseudomonas fluorescens	Maleic acid, etc.	pH treatment	Edwards et al. (1972)

In general, the effectiveness of pretreatment of lignocellulosic materials depends upon their source and optimum conditions must be determined separately for each material. There is a need to evaluate some of the more recently developed explosive depressurization processes for facilitating the breakdown of the LHC complex, such as the Iotech and Stake process, as pretreatment methods for substrates for bacterial biomass production (Bungay, 1982). Pretreatment of lignocellulosic materials may contribute significantly to their cost and this additional cost must be considered in any evaluation of their economic suitability for use in single-cell protein production.

Livestock wastes (cattle, pig and poultry manures) may be available in relatively large quantities at large scale feed lots or laying hen, broiler chicken and turkey production facilities. However, the quantities of assimilable carbon and energy sources in manures is usually too low for their use as substrates unless an additional carbon source such as a sugar is added (Shuler *et al.*, 1979).

21.3.1.2 *Substrates from nonrenewable resources*

As shown in Table 1, substrates from nonrenewable sources include petroleum hydrocarbons, natural gas and chemicals derived from them, such as methanol, ethanol and chemical industry wastes. During the 1960s and early 1970s, these substrates had relatively low costs owing to the then low prevailing prices of petroleum and natural gas (Litchfield, 1979, 1980).

Extensive research by Shell Research Ltd. in the United Kingdom with methane (Harrison, 1976; Hamer, 1979) and in Japan with methane, ethane, propane, *n*- and iso-butane, propylene and butylene or mixtures of these compounds (Akiba *et al.*, 1973; Sugimoto *et al.*, 1972) demonstrated high productivities and yields with these substrates. However, capital costs are significantly higher, for example, in bacterial SCP production from gaseous hydrocarbons owing to potential explosive hazards.

Exxon Corporation, in a joint venture with Nestle Alimentana SA, investigated bacterial SCP production from purified *n*-alkanes and ethanol on a laboratory and small pilot plant scale (Guenther and Perkins, 1968; Laskin, 1977). Also, the Chinese Petroleum Corporation in Taiwan conducted pilot plant studies using diesel oil and fuel oil as substrates (Ko and Yu, 1968; Ko *et al.*, 1964). However, increases in the prices of petroleum and natural gas during the 1970s made these substrates economically unattractive.

Methanol has received attention as a result of its high solubility in water, lack of explosive hazards, freedom from undesirable impurities, and ease of removal from the cell product. Imperial Chemical Industries Ltd. (ICI) developed the most advanced bacterial SCP process based on methanol that has been operated on a large scale (70 000 tonnes per year capacity; Margetts, 1983; Stringer, 1983). This process and its SCP product will be discussed subsequently.

For batch processes, carbohydrate concentrations based on simple sugars such as glucose, maltose, sucrose and lactose are usually in the range of 1 to 10% in the production medium. For fed-batch or continuous processing, the rate of feed of the carbon and energy source (and nitrogen source) should be adjusted to provide for the growth requirements of the organism at each phase of the growth cycle without maintaining any excess contents in the medium.

In the case of hydrocarbon substrates including methane or *n*-alkanes, or methanol and ethanol, continuous processes are used for both economic and process control reasons. Substrate concentrations are maintained as close to zero as possible to prevent instability resulting from substrate inhibition of growth. It is particularly important to maintain substrate concentrations limiting for growth in methanol-based processes because of methanol toxicity.

21.3.2 Nitrogen and Supplemental Nutrient Sources

Suitable nitrogen sources for bacterial biomass production include anhydrous ammonia, ammonium salts and urea. Animal wastes contain urea, uric acid and nonprotein nitrogen while meat processing wastes contain collagen and other protein nitrogen sources. In general, the nitrogen source is added in sufficient quantity in the medium to meet the growth requirements of the microorganism. It is important to maintain the carbon to nitrogen ratio in the medium in the range of 10:1 or less. This ratio favors high cell protein contents and minimizes the accumulation of lipids or cell storage substances such as poly-β-hydroxybutyrate. In general, low cost, readily assimilable nitrogen sources should be selected.

Phosphorus requirements should be supplied as feed grade, rather than industrial grade phosphoric acid to prevent contamination of the cell product with arsenic or fluoride. Minerals such as iron, calcium, magnesium, manganese, potassium and sodium are usually present in adequate quantities in many natural water supplies. However, supplementation with some of these minerals may be necessary to compensate for deficiencies. To minimize corrosion problems, mineral salts should be added as sulfates or hydroxides rather than as chlorides.

21.4 PROCESS CONDITIONS

21.4.1 Temperature, pH and Oxygen Transfer

Table 2 summarizes the growth characteristics of bacteria and actinomycetes on various substrates. It is important to select strains having temperature tolerance in the range of 37–55 °C since the heat released during growth may result in significant cooling costs.

Holve (1976) calculated the heats of reaction of microbial growth on a variety of substrates. Wang *et al.* (1976) developed methods for calculating the heats of reaction from microbial growth on carbohydrates, hydrocarbons, methanol and ethanol. Their calculations showed that the heat generated decreases with increasing yield of cells (expressed in grams of cells per gram of substrate utilized). Cooney *et al.* (1969) developed an empirical factor of 0.46 kJ mmol^{-1} of oxygen consumed.

In a microcalorimetric study of aerobic growth of *Cellulomonas* sp. 21399, on glucose and cellulose, Dermoun *et al.* (1984) reported an experimental value of heat production of −1079 kJ mol^{-1} glucose equivalent as compared with the theoretical value of −1314 kJ mol^{-1}. This difference was attributed to the presence of an unutilized carbon compound in the medium.

The pH range for the bacterial processes shown in Table 2 is 6.0–7.2. Consequently, sterility must be maintained in these processes to prevent the growth of contaminating organisms including enteric pathogens that grow readily in this same range. Usually, pH is controlled by adjusting the range of addition of ammonia (or ammonium salts) and phosphoric acid to the production medium.

Oxygen transfer to the growing cells may become limiting in bacterial biomass production. Based on the elemental compositions of the carbon source and cells, molecular weight of the carbon source and yield based on carbon source consumed, Mateles (1971) developed an empirical equation for estimating oxygen requirements for cell production. As in the case of heat load, oxygen demand for bacterial cell production decreases with increasing cell yield.

21.4.2 Growth Rates and Yields

Table 2 presents growth rates, cell densities and yields for various bacterial biomass processes. Reported cell yields for processes based on carbohydrate or organic nitrogen substrates range from 0.25 to 0.61 g (dry weight)/g substrate utilized.

Recently, Dermoun *et al.* (1984) investigated growth rates and yields of *Cellulomonas* sp. 21399 from various carbohydrate substrates. The following equation describes cell synthesis from glucose and nitrogen by this organism.

$$C_6H_{12}O_6 + 0.96\,NH_4OH \rightarrow 6\,CH_{1.56}O_{0.54}N_{0.16} + 3.72\,H_2O$$

Growth rates (h^{-1}) and yields (g dry weight/g substrate utilized) from various carbohydrates, were, respectively: glucose, 0.321 and 0.73; cellobiose, 0.331 and 0.91; amorphous cellulose, 0.205 and 0.71; and crystalline cellulose, 0.073 and 0.72. They concluded that the structure of cellulose (amorphous or crystalline) markedly affected its degradation by this organism.

With hydrocarbons, methanol and ethanol, typical yields are 0.8–1.20, 0.4–0.6, and 0.75 g (dry weight)/g substrate utilized, respectively. According to Van Dijken and Harder (1975), the maximum theoretical bacterial cell yield from methane is 0.91 g (dry weight)/g substrate utilized assuming that a mixed function oxidase system is involved. They calculated a maximum theoretical yield of bacterial cells from methanol of 0.73 g (dry weight)/g substrate utilized through the ribulose monophosphate pathway assuming an ATP requirement (Y_{ATP}) for cell synthesis from 3-phosphoglycerate of 10.5 and a cell composition of $C_4H_8O_2N$. Studies of methanol utilization indicate that bacteria such as *Methanomonas methanolica* and *Pseudomonas* sp. using the ribulose monophosphate pathway have higher yields than methylotrophs such as *Methylosinus trichosporium* exhibiting the serine pathway (Goldberg *et al.*, 1976; Drozd and Wren, 1980).

Table 2 Growth Characteristics of Selected Bacteria and Actinomycetes on Various Substrates

Organism	Carbon and energy source	Scale Aeration Agitation	Temperature (°C)	pH	Specific growth rate or dilution rate (D; h^{-1})	Culture density (g/l; dry wt basis)	Yield (dry wt basis; g/g substrate used)	References
Achromobacter delvacvate	Diesel oil	6000 l fermenter 3000 l of medium l/v/v/min	35–36	7.0–7.2	—	10–15 (48 h)	—	Ko *et al.* (1964).
Acinetobacter calcoaceticus	Acetate ethanol	10 l fermenter 3 l of medium 0.3–1.0 v/v/min 2300–2800 r.p.m. continuous	33	7.2	0.22 0.96	6.85 1.80	0.40 0.75	DuPreez *et al.* (1981)
Acinetobacter (Micrococcus) cerificans	(a) Gas oil (b) *n*-Hexadecane	7.5 l fermenter 4.5 l of medium 1 v/v/min 500 r.p.m.	30	7.0	(a) 0.4–1.0 (b) 1.1–2.0	8–10 8–10	0.10–0.12 0.80–0.90	Ertola *et al.* (1969)
	(c) *n*-Hexadecane	7.5 l fermenter 3.5–7.0 mM O$_2$/l/min	30	6.8	1.33	—	1.20	Guenther and Perkins (1968)
Aeromonas hydrophila	Lactose-whey	9 l fermenter 15.7 l/min	30	6.7–6.8	—	1.1	0.598	Butany and Ingledew (1973)
Bacillus megaterium	Collagen meat packing waste	2.3 l working volume fermenter 3 v/v/min (continuous) 400 r.p.m.	34	7.0	0.25 (D)	2.3	—	Bough *et al.* (1972)
Brevibacterium sp.	Mesquite wood	14 l fermenter 1.0–1.5 v/v/min 1350–1500 r.p.m.	30–37	6.45–7.2	—	—	0.44	Fu and Thayer (1975) Thayer *et al.* (1975)
Cellulomonas sp.	Bagasse	5 and 40 l fermenters, 1 v/v/min	35	6.5	0.14–0.16	13.2	0.33–0.36	Enriques and Rodriguez (1983)
Cellulomonas sp.–Alcaligenes faecalis	Bagasse	7, 14, 530 l fermenters, batch and continuous	34	6.6–6.8	0.2–0.29 0.08–0.1 (D)	16 (batch) 10 (continuous)	0.44–0.50	Han *et al.* (1971)
Cellulomonas sp.– Candida utilis	Barley straw	5 l fermenter 1 v/v/min 700 r.p.m. continuous	33	6.2–6.5	0.12–0.14 (D)	0.25–1.06	0.32–0.61	Kristensen (1978)
Corynebacterium hydrocarbo- clastus	Propane	30 l tower fermenter 7 l of medium 0.15 l/min gas	—	6.8–7.0	0.046	0.9 (93 h)	0.30	Akiba *et al.* (1973)
Methylomonas sp.	Methanol	30 l fermenter 12 l of medium 2 v/v/min	40	7.0	—	19.2	0.48	Goto *et al.* (1978)
Methylococcus capsulatus	Methane	1200 r.p.m. continuous 2.8 l of medium, 6.7% CH$_4$, 17.1–19.4% O$_2$ v/v, 38.6–49.5 ml/min 1450 r.p.m.	37	6.9	0.14	0.4	1.00–1.03	Harwood and Pirt (1972)

Organism	Substrate	Fermenter	Temp	pH	D		Yield	Reference
Methylomonas clara	Methanol	10 l–4000 l fermenters	34–40	6.5–7.0	0.25–0.5	10–15	0.4–0.6	Sittig (1983); Faust et al. (1977)
Methylomonas methanolica	Methanol	4 l working fermenter 300–400 mM O_2/l/min 1.5 v/v/min 1200–1500 r.p.m. continuous	35	6.0	0.24 (D)	9.6	0.48	Dostalek et al. (1972); Dostalek and Molin (1975)
Methylophilus methylotrophus	Methanol	Pressure cycle airlift fermenter, continuous	35–40	6.0–7.0	0.38–0.5	30	0.5	MacLennan et al. (1973)
Mixed culture *Bacillus* sp. (4) *Cellulomonas* sp. (2) *Pseudomonas* sp. (1)	Rice hulls	50 l fermenter 35 l of medium 17.5 l/min 250 r.p.m.	37	7.0	—	4.29	0.25	Gow et al. (1975); Chang et al. (1980)
Nocardia sp. NBZ-23	n-Alkanes	7.5 l fermenter 6–8 l of medium 1500 r.p.m.	30	6.8	1.25	14.7	0.98	Wagner et al. (1969)
Nocardia paraffinica	(a) n-Propane (b) n-Butane	5 l fermenter 3 l of medium 3 l/min gas flow 600 r.p.m.	30 / 30	7.0 / 7.0	0.091 / 0.111	30 (112 h) / 22 (72 h)	1.36 / 0.95	Sugimoto et al. (1972)
Protaminobacter ruber	Methanol	1 l fermenter (medium volume) 500–600 r.p.m. fed batch DO control	30	7.0	—	85	—	Yano et al. (1978a, 1978b)
Pseudomonas sp. No. 5401	Fuel oil	6000 l fermenter (a) batch (b) continuous	36–38	7.0	0.16 / 0.12 (D) / 0.25 (D)	16 (24–26 h) / 10 / 8	1.00 / — / —	Ko and Yu (1968)
Pseudomonas sp. *Hyphomicrobium* sp. *Acinetobacter* sp. *Flavobacterium* sp. (mixed culture)	Methane 1 atm 470 r.p.m.	1.0 l fermenter 0.9 l of medium	32	5.7	0.06 (D)	0.8	0.99	Wilkinson and Harrison (1973)
Rhodopseudomonas gelatinosa	Bicarbonate, wheat brain	4 l working volume fermenter 75 W incandescent lamp 1100 r.p.m.	40	0.2	(a) Batch 0.31 (b) Continuous 0.028 (D)	4.33 / 3.15	—	Shipman et al. (1975)
Thermomonospora sp.	Cellulose	14, 70 l fermenters	55	7.2	0.48	2.3	0.44	Humphrey et al. (1977)
Thermomonospora fusca	Cellulose pulping fines	10 l fermenter 3 l/min 60 r.p.m.	55	7.4	—	—	0.35–0.40	Crawford et al. (1973)

Of the processes listed in Table 2, only the ICI Pruteen process for feed grade SCP production from methanol using *Methylophilus methylotrophus* has been operated on a full commercial scale. Campaigns of 6000–7000 tonnes production per month have been run on an intermittent basis. A novel 'pressure cycle' air lift fermenter design utilizes air or both agitation and aeration. Operating temperatures are in the range of 35–42 °C, which minimizes cooling costs (Gow *et al.*, 1975; MacLennan *et al.*, 1973, 1976; Stringer, 1983).

The following equation represents the conversion of methanol by the SCP product in the ICI Process (MacLennan *et al.*, 1973).

$$1.72\ CH_3OH + 0.23\ NH_3 + 1.51\ O_2 \rightarrow$$

$$1.0\ CH_{1.68}O_{0.36}\ N_{0.22} + 0.72\ CO_2 + 2.94\ H_2O$$

A relatively high specific growth rate ($0.50\ h^{-1}$), cell density [30 g (dry weight)/l] and yield [greater than 0.50 g (dry weight)/g substrate utilized] are obtained in this process.

A process for producing *Methylomonas clara* from methanol was developed by Hoechst-Uhde in West Germany. The goal in this case was a human food grade product (Probion) based on a 90% protein concentrate and nucleic acid by-products rather than an animal feed grade product as in the ICI process (Sittig, 1983). This process has been operated on a pilot plant scale (20 m³ fermenters), but has not been scaled-up to a commercial level. Puhar *et al.* (1982) found that the maximum cell density, yield and productivity obtainable for *M.clara* grown on methanol in a continuous draft tube bioreactor were 44.2 g (dry weight)/l; 0.5 g (dry weight)/g methanol utilized, and 11.72 g (dry weight)/l/h, respectively. Inhibitory by-products such as formaldehyde and formic acid limit growth and respiration.

Phillips Petroleum Company has conducted extensive studies on methanol utilization by both bacteria and yeasts for SCP production (Hitzman, 1982; Malick *et al.*, 1984). A proprietary fermenter design was developed enabling operation at high cell densities and recycle of the sterilized fermentation effluent as a portion of the fermentation medium. These processes have been operated on a pilot plant scale at Bartlesville, Oklahoma. Recent work has emphasized yeast rather than bacterial biomass production.

Other bacterial processes based on methanol include the Norprotein process of Norsk Hydro in Norway using *Methanomonas methanolica* on a 45 m³ scale, and in Japan, the Mitsubishi-Petrochemical Co. Ltd. process using *Pseudomonas* sp. on a 500 tonne per year scale (Litchfield, 1980) and the Sanraku-Ocean Co. Ltd. process using *Methylomonas* sp. on a 40 l scale (Goto *et al.*, 1978). None of these processes have been operated commercially.

The genetic modification of methylotrophic bacteria by introducing specific plasmids offers the potential opportunity to improve conversion of methanol to SCP. On the basis that ammonia assimilation through glutamate dehydrogenase (GDH) is more efficient than through the glutamate synthase, Windass *et al.* (1980) of ICI cloned the GDH gene of *Escherichia coli* into broad host range plasmids that were then introduced into *M. methylotrophus*. The gene was expressed and the mutants were stable without reversion. However, only a 3 to 5% increase in yield was obtained through this modification.

Stahl and Esser (1982) analyzed clones of *M. clara*, the organism used in the Hoechst-Uhde process, for this DNA content. They identified two plasmids, derived from each other in two separate strains. They concluded that the methylotrophy of *M. clara* was not coded by plasmids and could only be transferred by chromosomal DNA. Furthermore, *M. clara* could be used as a host cell for other genes for cloning.

It is apparent that it will be difficult to improve methanol conversion to SCP using methylotrophic bacteria to a sufficient extent by introducing plasmids governing nitrogen utilization pathways to justify the additional safety evaluations that will be required by regulatory agencies for these genetically modified strains.

During the 1960s, the National Aeronautics and Space Administration in the United States sponsored research on bioregenerative systems based on hydrogen-fixing bacteria for recycling expired carbon dioxide and urinary excretion products during extended space missions (Litchfield, 1979). *Alcaligenes eutrophus* and related species use hydrogen as an energy source and carbon dioxide as a carbon source in the presence of oxygen to yield bacterial biomass.

Miura *et al.* (1982) calculated that maximum cell productivities of *Alcaligenes hydrogenophilus* were obtained in continuous cultures as a dilution rate (D) of $0.21\ h^{-1}$ and a molar ratio of $H_2:O_2:CO_2$ of 67.4:29.2:2.7. There was good agreement between theoretical and experimental values of productivity and substrate utility expressed as substrate consumed as a percentage of substrate fed. In a study of the kinetics of growth of *A. eutrophus*, Siegel and Ollis (1984) reported maximum growth rates (μ_{max}) of 0.3 h^{-1} and 0.29 h^{-1} at 31 °C, and pH 6.5–6.7 under oxygen and hydrogen-limiting conditions. Yields in N$_2$- and O$_2$-limited cultures were Y_{H_2} = 4.1 g (dry weight)/mol H$_2$ and Y_{O_2} = 15.9 g (dry weight)/mole O$_2$. However, poly-β-hydroxybutryrate accumulated to the extent of 20% of cell dry weight under O$_2$-limitation.

21.5 PRODUCT RECOVERY

Bacterial cell recovery by conventional centrifugation is prohibitively expensive owing to their small size (1–2 μm), density close to that of water (1.003 g/cm^3) and large volumes of water that must be handled at cell densities in the range of 10–30 g (dry weight)/l. Vacuum filters are unsatisfactory because the small cell size and resulting packing density results in compaction on screens to the point that the void volume decreases to zero and filtration ceases (Labuza, 1975). Filter aids contaminate the cell product and make it unsuitable for food or feed uses. Consequently, there has been considerable interest in agglomeration and flocculation methods for concentrating bacterial biomass prior to final centrifugation without adding filter aids.

According to Gow *et al.* (1975), the ICI proprietary agglomeration process enables the concentration of *M. methylotrophus* grown on methanol without flocculating agents to yield a dewatered product having a concentration of 25 g (dry weight)/l or greater. This dewatered product can then be concentrated further by conventional decanter centrifuges prior to drying.

The Hoechst-Uhde process for producing *M. clara* from methanol utilizes an electrothermal flocculation process. The concentrate is then centrifuged and dried (Faust *et al.*, 1977).

Brevibacterium sp. was separated from a hydrocarbon-based fermentation by agitating the cells at 400–500 r.p.m. at pH 6–8 at the end of the exponential growth phase in the presence of the surfactant Span 20 (Shung and Ueda, 1976). *Pseudomonas* sp. grown on methanol could be separated from the growth medium by heating in a tank at 90 °C for 5–6 h at pH 8.4–9.2 followed by transfer to an aggregation tank where the cells aggregated and separated when the pH was held at 4.5–5.0 (Shung and Yang, 1983).

Phillips Petroleum Company has developed two processes for separating cells. The first process involves the use of a foam breaker inside the fermenter in which the cells are concentrated in the fluid in the foam breaker and withdrawn or the fluid is centrifuged to separate the organisms from the spent medium (Hitzman, 1982). In the second process, a fermentation broth containing bacterial cells is mixed with a second broth containing yeast cells. The mixed cell product is then recovered by centrifugation (Vanderveen *et al.*, 1983). Further details on the actual operation of these separation processes under pilot plant conditions are not yet available.

It is important to note that the spent growth medium and cell wash waters from recovery processes have high biochemical oxygen demands (BOD). Whenever possible, the spent medium should be purified, sterilized and recycled. In some cases, low concentrations of inhibitory compounds may be very costly to remove from the spent growth medium. In any event, the residues from cell separation must be treated before discharge.

21.6 PRODUCT QUALITY AND SAFETY

Bacterial biomass has three potential applications: (1) animal feed; (2) human food; and (3) functional protein concentrates and isolates for use in human foods. Nutritional characteristics are important in the first two applications while functional effects in food products are important in the third application. However, in all three cases, the products must be acceptable from a sensory (flavor and aroma) standpoint. Also, products for human food use must have low nucleic acid contents. Products for all three applications must be free from microbial toxins, pathogenic microorganisms and toxic heavy metals and chemical residues.

21.6.1 Nutritional and Functional Values

Table 3 presents proximate analyses and Table 4 amino acid analyses of bacteria and actinomycetes of interest in SCP production processes.

The crude protein values given are based on nitrogen multiplied by 6.25. Although this factor is used for calculating protein contents for use in feed formulations, it includes nonprotein nitrogen substances such as nucleic acids that have no nutritional value and are detrimental to humans.

Table 3   Proximate Analyses of Selected Bacteria and Actinomycetes Grown on Various Substrates

Organism	Substrate	Composition (g/100 g dry weight)							References
		Nitrogen	Protein	Fat	Total CHO	Crude Fiber	Ash	Energy (kJ/kg)	
Acinetobacter (Micrococcus) certificans	Hexadecane	11	72	—	—	—	—	—	Ertola et al. (1969)
Acinetobacter calcoaceticus	Ethanol	12.8	80	7	—	—	—	—	Laskin (1977)
Bacillus megaterium	Collagenous meat packing wastes	11.5	72	—	—	—	7.3	—	Bough et al. (1972)
Cellulomonas sp.	Bagasse	9.3	58	—	—	—	—	—	Enriques and Rodriguez (1983)
Cellulomonas sp.–	Bagasse	14	87	8	—	—	7	—	Han et al. (1971)
Alcaligenes *Cellulomonas* sp.–	Barley straw	9.3–10.6	58–66	—	—	—	9.4–11.4	—	Kristensen (1978)
Candida utilis	Propane	9.1	57	11.7	24.9	—	6.9	—	Akiba et al. (1973)
Corynebacterium hydrocarboclastus	Methanol	12.8–13.6	80–85	8–10	—	—	8–12	—	Faust et al. (1977)
Methylomonas clara	Methanol	13.1	82	—	—	—	—	—	Dostalek and Molin (1975)
Methylomonas methanolica *Methylophilus methylotrophus*	Methanol	13	83	7.4	—	<0.05	8.6	12 624	Gow et al. (1975)
MIXED CULTURE *Bacillus* sp. *Cellulomonas* sp. *Pseudomonas* sp.	Rice hulls	4.5	28	—	—	9.5	33.5	—	Chang et al. (1980)
Pseudomonas JM 127	Mesquite wood	3.5–10.4	22	6	—	—	—	—	Yang et al. (1977)
Pseudomonas denitrificans	Spent sulfite liquor	11.60	53	—	—	—	—	—	Camhi and Rogers (1976)
Pseudomonas fluorescens	(1) Poultry waste	7.2	45	7.5–11.3	—	—	12–19	13 443	Shuler et al. (1979)
	(2) Sodium maleate	10.7	67	16.7	—	—	—	—	Edwards et al. (1972)
Rhodopseudomonas capsulata	Biogas effluent	11.1	69	—	—	—	—	—	Vrati (1984)
Thermonomospora fusca	Pulping fines	4.8–5.6	30–35	—	—	—	—	—	Crawford et al. (1973)

Table 4 Amino Acid Analyses of Selected Bacteria and Actinomycetes Grown on Various Substrates[a]

Organism	Substrate	Ala	Arg	Asp	Cys	Glu	Gly	His	Ile	Leu	Lys	Met	Phe	Pro	Ser	Thr	Try	Tyr	Val	References
Acinetobacter calcoaceticus	Ethanol	—	—	—	—	—	—	—	—	—	6.7	2.8	—	—	—	—	—	—	—	Laskin (1977)
Acinetobacter (Micrococcus) cerificans	n-Alkanes	6.7	4.7	8.0	—	10.7	5.0	1.7	4.3	6.5	5.2	1.8	3.6	3.1	2.8	4.1	2.8	1.3	5.4	Ertola et al. (1969)
Bacillus megaterium	Collagenous meat wastes	8.4	3.9	7.4	0.4	16.9	5.8	1.8	3.4	5.4	6.3	3.2	3.0	3.8	2.4	3.2	0.5	2.5	4.2	Bough et al. (1972)
Cellulomonas sp.	Bagasse	—	7.8	—	—	—	—	4.0	2.9	6.7	4.2	0.6	3.3	—	—	3.9	—	3.1	4.9	Enriques and Rodriguez (1983)
Cellulomonas–Alcaligenes	Bagasse	—	6.5	—	—	—	—	7.8	5.4	7.4	7.6	2.0	4.7	—	—	5.5	—	—	7.1	Han et al. (1971)
Corynebacterium hydrocarboclastus	Propane	8.2	6.2	7.4	10.7	11.7	5.1	2.4	3.6	6.7	3.7	1.5	4.6	4.8	3.6	4.8	2.5	1.7	1.3	Akiba et al. (1973)
Methylococcus capsulatus	Methane	—	6.2	—	0.6	—	5.1	2.2	4.3	8.1	5.7	2.7	4.6	—	—	4.6	—	3.8	6.5	Hamer et al. (1975)
Methylomonas clara	Methanol	9.3	6.4	—	—	—	6.9	2.5	6.3	9.9	7.6	2.8	5.4	3.7	4.0	4.2	2.4	3.9	8.1	Faust et al. (1977); Sittig (1983)

[a] g/16 g N.

Table 5   Performance of Selected Bacterial Single Cell Protein Products in Animal Feeding Studies

Bacterial single cell protein and substrate	Animal	Treatment and percent in diet	Protein digestibility (%)	Protein efficiency ratio (PER)	Biological value (BV)	Feed conversion ratio (kg/kg weight gain)	References
Acinetobacter (*Micrococcus*) *cerificans* (*n*-hexadecane)	Rat	Dried	83.4	—	67	—	Ertola et al. (1969)
Bacillus megaterium (collagen waste)	Rat	Dried	—	1.88[a]	—	—	Bough et al. (1972)
Brevibacterium JM99B (mesquite wood)	Mouse	Lyophilized	—	—	—	0.638 (0.0569)[b]	Thayer et al. (1975)
Cellulomonas sp. (bagasse)	Rat	Dried	90	—	62	—	Enriques and Rodriguez (1983)
Methylophilus methylotrophus (methanol)	Broiler chicken	Flash-dried 15–43 g N/kg	—	—	—	0.741 (0.707)[b]	Felix D'Mello (1978)
		+0.091% methionine			—	0.742 (0.757)[b]	
		15–36 g N/kg			—	0.598 (0.628)[b]	
		+ 0.091% methionine			—	0.615 (0.678)[b]	
		+ 0.091% methionine, 1.09% lysine			—	0.561	
		+ 0.091% methionine, 1.09% lysine, 1.09% arginine			—	0.657	
Methylophilus methylotrophus (methanol)	Chick	Dried — 1.61% + methionine 0.15 g/kg 0.30 g/kg	—	—	—	0.614 (0.620)[b]	Abbey et al. (1980)
		0.60 g/kg				0.644 (0.647)[b]	
						0.709 (0.707)[b]	
		Dried — 4.68				—	
		+ lysine 1 g/kg				0.215 (0.227)[b]	
		2 g/kg				0.374 (0.378)[b]	
		4 g/kg				0.493 (0.490)[b]	

Methylophilus methylotrophus (methanol)	Pig	Dried — 3.34 + tryptophan 0.15 g/kg	—	—	—	—
		0.30 g/kg				0.440 (0.450)[b]
		0.60 g/kg				0.551 (0.551)[b]
						0.644 (0.640)[b]
		Dried — 20% (pair fed)	—	—	—	0.458 (0.450)[b]
		Individually fed				0.417 (0.416)[b]
						Whittemore *et al.* (1976)
Pseudomonas sp. (methanol)	Broiler chicken	Dried — 5	—	—	—	0.650
		10				0.639
		15				0.629 (0.670)[b]
						Waldroup and Payne (1974)

[a] Adjusted to PER of 2.5. [b] Control.

It was mentioned previously that lipid contents of bacterial cells are affected by the C/N ratio in the medium and mineral contents reflect the composition of the growth medium.

The amino acid values shown in Table 4 are compared with the Food and Agricultural Organization (FAO) of the United Nations reference protein. Some of the bacterial proteins have amino acid profiles, and, in particular, methionine contents that compare favorably with the FAO values. Yeast, fungal and soybean proteins tend to be deficient in methionine contents from the standpoint of human and animal nutrition.

The most important measure of nutritional value of bacterial SCP products is performance in feeding studies. Protein efficiency ratio (PER), biological value (BV) and net protein utilization (NPU) together with protein digestibility as measured in rats are measures of the nutritional value of bacterial SCP for human food applications. Metabolizable energy, protein digestibility and feed conversion ratio (weight of ration consumed/weight gain) have been used as indications of performance of bacterial SCP in rations for broiler chickens and swine.

Table 5 summarizes the performances of selected bacterial SCP products in animal feeding studies. The PER and BV data reported for the bacteria listed are lower than desirable for human nutrition (casein — PER, 2.5; BV, 75). The feed conversion ratios for *M. methylotrophus* (ICI's Pruteen) in broiler chickens (Abbey *et al.*, 1980; Felix d'Mello, 1978) and pigs (Whittemore *et al.*, 1976) were equivalent to controls. With *Pseudomonas* sp. grown on methanol (Phillips Petroleum Co.), up to 10% of this bacterial product in a pelleted ration gave an equivalent performance to pelleted rations at a 15% level in the diet but was poorer than that of the corn–soybean ration control (Waldroup and Payne, 1974).

In the case of *M. clara* (Hoechst-Uhde) grown on methanol, methionine was the first limiting amino acid in rats for both the crude bacterial biomass and the protein isolate from which nucleic acids were removed (Walz and Brune, 1984). The protein digestibility and biological value (BV) of the cell biomass were 85% and 0.70, respectively. Supplementation of the cell biomass with methionine gave equivalent results in piglets to those obtained with fish meal.

Griefe *et al.* (1984) fed *A. eutrophus* to growing pigs in place of soybean meal at levels equivalent to 30 and 60% of the protein content of the ration on an N × 6.25 basis. Nitrogen digestibility was 87.4% and 80% of the bacterial purines were excreted as allantoin and uric acid.

21.6.2 Functional Quality

Bacterial SCP protein concentrates and isolates have been prepared for evaluation of functional effectiveness in foods including water and fat binding properties, emulsion stability, dispersibility, gel formation, whippability, and thickening (McNairney, 1984). Table 6 presents a description of some of these products that include protein concentrates from which nucleic acids have been removed, and protein fibers for use in textured foods. In addition to these products, Hoechst-Uhde has developed a protein isolate (90% protein or greater) called 'Probion' by extraction of nucleic acids from *M. clara*; however, there is a lack of published information on the functional effectiveness of this product.

None of these functional protein products have been commercialized. It appears that the demonstrated effectiveness of these products may fall short of those of soy protein concentrates and isolates that are available for food uses and that consequently the cost of obtaining safety data to satisfy regulatory agency requirements is not justified.

21.6.3 Safety

Bacterial SCP products may have high nucleic acid contents up to 16% of cell dry weight. If not reduced, human consumption of more than 2 g nucleic acid equivalent per day could lead to kidney stone formation and gout. Bacterial SCP products having high nucleic acid contents are not a problem in ruminant feeding and are tolerated to a greater extent by other livestock than by humans. Also, human consumption of *A. eutrophus* has resulted in gastrointestinal disturbances, such as nausea and vomiting, while similar effects were not observed in animal feeding studies (Calloway, 1974).

Safety evaluations for SCP products for human and animal feeding are discussed by de Groot (1974) and Taylor *et al.* (1974). The Food and Drug Administration in the United States (1984) and the Protein Evaluation Group of the United Nations (1970a, 1970b, 1972, 1974) have developed guidelines for safety evaluations of SCP products in humans and domestic livestock.

Table 6 Preparation of bacterial SCP Concentrates or Isolates and Textured or Functional SCP Products

SCP	Product type	Description	References
Acinetobacter cerificans	Concentrate	Disrupted cells, treated with acid or alkali, isoelectric precipitation	Kalina (1973); Kalina and Nicholas (1974)
Bacillus megaterium	Concentrate	Cells disrupted in homogenizer, extracted with water and bicarbonate buffer	Tannenbaum *et al.* (1965)
Butane bacteria	Fibers	Alkaline bacterial cell slurry extruded into acetic acid, pH 4.5, to give fibers	Akin (1973)
	Translucent film	Alkaline bacterial cell slurry; boiled, centrifuged, residue resuspended, treated with carbonyl sulfide, centrifuged, dried	Akin (1974)
Cellulomonas sp.	Fibers	Disrupted cells slurried at 37 °C in 0.5–1.0 N NaOH, extruded into $HClO_4$	Daly and Ruiz (1974)
Methanol-oxidizing bacteria	Concentrate	Cells disrupted in homogenizer, pH adjusted to 12, centrifuged, adjusted to pH 7.5, heated at 80 °C to precipitate protein	Mogren *et al.* (1974)
Methane- or methanol-oxidising bacteria	Fibers	Cells disintegrated in freeze-press, dissolved in bicarbonate, centrifuged, eluate spun into fibers	Hedin *et al.* (1971)

21.7 ECONOMIC AND MARKET CONSIDERATIONS

Capital costs for bacterial biomass production will depend upon the equipment requirements for storing, processing and handling substrates, sterilization or cleaning operation, product separation, recovery and drying, and local land, site preparation and construction costs. Capital costs for SCP processes based on substrates that do not require pretreatment and clean rather than sterile operations will be lower than those of more complex processes. Bacterial SCP manufacturing costs are highly dependent upon the cost of the carbon and energy source to the producer of the product and may range from 13 to 57.5% depending upon the year of the estimate and the nature of the raw material (waste or purified substrate). Utilities costs including water, electricity and steam, labor costs, maintenance and depreciation, taxes and insurance, working capital, interest, and desired profit must also be taken into account.

The ICI Pruteen process plant in Billingham, England, constructed in 1976–79 is the most recent commercial scale bacterial SCP production facility. The estimated 1976 costs of $70 million for 50 000–75 000 tonnes per year capacity would be substantially greater today.

Current published information on SCP operating costs are not available, the most recent being made in 1975 for bacterial SCP production from methanol (Litchfield, 1979).

The most recent (1984) selling price for the only commercial bacterial SCP product, ICI's Pruteen, is approximately $600 per tonne for animal feed applications in Western European markets. The extent to which this product can compete with soybean meal or fish meal depends upon current delivered prices for these commodities in Western Europe. For example, in the US the 1984–1985 (12 month) price range for soybean meal (44% protein) was approximately $125–190 per tonne.

21.8 CONCLUSIONS

Bacteria and actinomycetes have high growth rates and favorable true protein contents and amino acid profiles for use in bacterial SCP production. Also, feeding performance in livestock of some bacterial SCP products is equivalent to that obtained with conventional protein feedstuffs.

Technological improvements, such as development of genetically engineered organisms having improved substrate and nitrogen source conversion efficiencies, processes enabling operation at very high cell densities (100 g/l) and productivities, and lower cost separation methods than centrifugation may result in decreased operating costs. Capital costs, however, will remain a major consideration in the development of novel processes.

It does not appear that food grade SCP products will enter human nutritional or functional food ingredient markets in the short term because of the relatively lower costs to the user of alternative proteins from plant and animal sources and the requirements of regulatory agencies for extensive safety studies. However, prices of competing proteins from conventional plant and animal sources will determine the future competitiveness of bacterial SCP products for feed applications.

21.9 REFERENCES

Abbey, B. W., K. N. Boorman and D. Lewis (1980). The availabilities of lysine, methionine and tryptophan on Pruteen by chick growth assay. *J. Sci. Food Agric.*, **31**, 421–431.

Akiba, T., S. Kajiyama and T. Fukimhara (1973). Cultivation of a strain of bacteria on propane in a tower-type fermenter with gas entrainment process. *J. Ferment. Technol.*, **51**, 343–347.

Akin, C. (1973). Process for texturizing microbial cells by alkali–acid treatment. *US Pat.* 3 781 264.

Akin, C. (1974). Process for preparing polycellular protein products. *US Pat.* 3 819 610.

Atkinson, B. and P. Sainter (1982). Development of downstream processing. *J. Chem. Technol. Biotechnol.*, **32**, 100–108.

Bough, W. A., W. L. Brown, J. D. Porsche and D. M. Doty (1972). Utilization of collagenous by-products from the meat-packing industry: production of single-cell protein by the continuous cultivation of *Bacillus megaterium*. *Appl. Microbiol.*, **24**, 226–235.

Bungay, H. R. (1982). Biomass refining. *Science*, **218**, 643–646.

Butany, G. and W. M. Ingledew (1973). Whey utilization by bacteria. *Can. Inst. Food Sci. Technol. J.*, **6**, 291–293.

Busta, F. F., B. E. Schmidt and L. L McKay (1977). Method of production and recovery of protein from food wastes. *US Pat.* 4 018 650.

Callihan, C. D. and J. E. Clemmer (1979). Biomass from cellulosic materials. In *Microbial Biomass — Economic Microbiology*, ed. A. H. Rose, vol. 4. pp. 271–288. Academic, New York.

Calloway, D. H. (1974). The place of SCP in man's diet. In *Single-Cell Protein*, ed. P. Davis, pp. 129–146. Academic, New York.

Camhi, J. D. and P. I. Rogers (1976). Continuous cultivation of bacteria (*Pseudomonas* sp.) on spent sulfite liquor. *J. Ferment. Technol.*, **54**, 450–458.

Chang, W. T. M., W.-H. Hsu, M. N. Lai and P. P. Chang (1980). Production of single-cell protein from rice hulls for animal feed. *Dev. Ind. Microbiol.*, **21**, 313–325.

Cooney, C. L., D. I. C. Wang and R. I. Mateles (1969). Measurement of heat evolution and correlation with oxygen consumption during microbial growth. *Biotechnol. Bioeng.*, **11**, 209–281.

Cooney, C. L., C. Rha and S. R. Tannenbaum (1980). Single-cell protein: engineering, economics and utilization in foods. *Adv. Food Res.*, **26**, 1–52.

Crawford, D. L., E. McCoy, J. M. Harkin and P. Jones (1973). Microbial protein from waste cellulose by *Thermomonosphera fusca*, a thermophilic actinomycete. *Biotechnol. Bioeng.*, **15**, 833–843.

Daly, W. H. and L. P. Ruiz (1974). Reduction of RNA in single-cell protein in conjunction with fiber formation. *Biotechnol Bioeng.*, **16**, 285–287.

Dasinger, B. L. and L. A. Nasland (1972). Detoxification of gram-negative bacteria grown in a fermentation process. *US Pat.* 3 644 175.

Davis, P. (ed.) (1974). *Single-Cell Protein*. Academic, New York.

de Groot, A. P. (1974). Minimal tests necessary to evaluate the nutritional qualities and the safety of SCP. In *Single-Cell Protein*, ed. P. Davis. Academic, New York.

Dermoun, Z., C. Gaudin and J. P. Belaich (1984). Microcalorimetric study of aerobic growth of *Cellulomonas* sp. 21399 on various carbohydrates. *Appl. Microbiol. Biotechnol.*, **19**, 281–287.

Dostalek, M. and N. Molin (1975). Studies of biomass production of methanol oxidizing bacteria. In *Single-Cell Protein II*, ed. S. R. Tannenbaum and D. I. C. Wang, pp. 385–401. MIT Press, Cambridge, MA.

Dostalek, M., L. Häggstrom and N. Molin (1972). Optimization of biomass production from methanol. *Fermentation Technology Today. Proceedings of the 4th International Fermentation Symposium (Kyoto)*, pp. 497–501. Society of Fermentation Technology, Osaka, Japan.

Drozd, J. W. and S. J. Wren (1980). Growth energetics in the production of bacterial single cell protein from methanol. *Biotechnol. Bioeng.*, **22**, 353–362.

Dunlap, C. E. (1975). Production of single-cell protein from insoluble agricultural wastes by mesophiles. In *Single-Cell Protein II*, ed. S. R. Tannenbaum and D. I. C. Wang, pp. 244–261. MIT Press, Cambridge, MA.

DuPreez, J. C., D. F. Toerien and P. M. Latagan (1981). Growth parameters of *Acinetobacter calcoaceticus* on acetate and ethanol. *Eur. J. Appl. Microbiol. Biotechnol.*, **13**, 45–53.

Edwards, V. H., J. E. Kinsella and D. B. Sholiton (1972). Continuous culture of *Pseudomonas fluorescens* with sodium maleate as a carbon source. *Biotechnol. Bioeng.*, **14**, 123–147.

Enriques, A. and M. Rodriguez (1983). High productivity and good nutritive values of cellulolytic bacteria grown on sugarcane bagasse. *Biotechnol. Bioeng.*, **25**, 877–880.

Ertola, R. J., L. A. Mazza, A. P. Balatti and J. Sanahuja (1969). Composition of cell material and biological value of cellular protein of a micrococcus strain grown on hydrocarbons. *Biotechnol. Bioeng.*, **11**, 409–416.

Faust, U., P. Prave and D. A. Sukatsch (1977). Continuous biomass production from methanol by *Methylomonas clara*. *J. Ferment. Technol.*, **55**, 609–614.

Felix D'Mello, J. P. (1978). Responses of young chicks to amino acid supplementation of methanol-grown dried microbial cells. *J. Sci. Food Agric.*, **29**, 453–460.

Food and Drug Administration (1984). *Guideline for New Animal Drugs and Food Additives Derived from a Fermentation: Human Food Safety Evaluation*. US Department of Health and Human Services, Washington, DC.

Fu, T. T. and D. W. Thayer, (1975). Comparison of batch and semicontinuous cultures for the production of protein from mesquite wood by *Brevibacterium* sp. J. M. 98A. *Biotechnol. Bioeng.*, **17**, 1749.

Goldberg, I., J. S. Ruck, A. Ben-Bassat and R. I. Mateles (1976). Bacterial yields on methanol, methylamine, formaldehyde and formate. *Biotechnol. Bioeng.*, **18**, 1657–1668.

Goto, S., M. Yamamoto, M. Tsuchiya, O. Kondo, R. Ukamoto and A. Takamatsu (1978). Isolation of methanol-assimilating bacteria and optimum medium composition. *J. Ferment. Technol.*, **56**, 516–523.

Gow, J. S., J. D. Littlehailes, S. R. L. Smith and R. B. Walter (1975). SCP production from methanol bacteria. In *Single-Cell Protein II*, ed S. R. Tannenbaum and D. I. C. Wang, pp. 370–378. MIT Press, Cambridge, MA.

Griefe, H. A., S. Moluar, T. Bos, M. Gossmann and K. D. Guenther (1984). Nitrogen metabolism in growing pigs receiving a bacterial protein supplement (*Alcaligenes eutrophus*) instead of soybean meal. *Arch. Tierenaehr.*, **34**, 179–190.

Guenther, K. R. and M. B. Perkins (1968). Process for producing high protein feed supplements from hydrocarbons. *US Pat.* 3 384 491.

Hamer, G. (1979). Biomass from natural gas. In *Microbial Biomass — Economic Microbiology*, ed. A. H. Rose, vol. 4, pp. 315–360. Academic, New York.

Hamer, G., D. E. F. Harrison, J. H. Harwood and H. H. Topiwala (1975). SCP from methane. In *Single-Cell Protein II*, ed. S. R. Tannenbaum and D. I. C. Wang. MIT Press, Cambridge, MA.

Han, Y. W., C. E Dunlap and C. D. Callihan (1971). Single-cell protein from cellulosic wastes. *Food Technol.*, **25**, 130–134, 154.

Harrison, D. E. F. (1976). Making protein from methane. *Chemtech*, **6**, 570–574.

Harwood, J. M. and S. J. Pirt (1972). Quantitative aspects of growth of the methane oxidizing bacterium *Methylococcus capsulatus* on methane in shake flask and continuous chemostat culture. *J. Appl. Bacteriol.*, **35, 597–607.**

Heden, C.-G., N. Molin, U. Olsson and A. Rupprecht (1971). Preliminary experiments on spinning bacterial proteins into fibers. *Biotechnol. Bioeng.*, **13**, 147–150.

Hedenskog, G. and H. Mogren (1973). Some methods for processing of single cell protein. *Biotechnol. Bioeng.*, **15**, 129.

Hitzman, D. O. (1982a). Fermentation of oxygenated hydrocarbons with thermophilic microorganisms and microorganisms therefrom. *US Reissue* 30 965.

Hitzman, D. O. (1982b). Fermentation process. *US Pat.* 4 340 677.

Holve, W. A. (1976). A general stoichiometric and thermochemical model for SCP yields. *Process Biochem.*, **14** (3), 2–4.

Humphrey, A. E., A. Moreira, W. Armiger and D. Zabriskie (1977). Production single cell protein from cellulose wastes. *Biotechnol. Bioeng. Symp.*, **7**, 45–64.

Kalina, V. (1973). Isolation of protein. *US Pat.* 3 718 541.

Kalina, V. and P. Nicholas (1974). Extraction of proteins from microorganisms. *US Pat.* 3 821 080.

Kato, N., K. Tsuji, Y. Tani and K. Ogata (1974). A methanol-utilizing actinomycete. *J. Ferment. Technol.*, **52**, 917–920.

Kristensen, T. P. (1978). Continuous single-cell protein production from *Cellulomonas* sp. and *Candida utilis* grown in mixture on barley straw. *Eur. J. Appl. Microbiol.*, **5**, 155–163.

Ko, P. C. and Y. Yu (1968). Production of SCP from hydrocarbons: Taiwan. In *Single-Cell Protein*, ed. R.I. Mateles and S. R. Tannenbaum, pp. 255–262. MIT Press, Cambridge, MA.

Ko, P. C., Y. Yu, and C.-S. Liu (1964). *Protein from Petroleum by Fermentation Process*. Chinese Solvent Works, Chinese Petroleum Corp., Taiwan.

Kobayashi, M. and S.-I. Kurata (1978). The mass culture and cell utilization of photosynthetic bacteria. *Process Biochem.*, **13** (9), 27–30.

Kobayashi, M., M. Kobayashi and H. Nakanishi (1971). Construction of a purification plant for polluted water using photosynthetic bacteria. *J. Ferment. Technol.*, **49**, 817–825.

Labuza, T. P. (1975). Cell collection recovery and drying for SCP manufacturing. In *Single Cell Protein II*, ed. S. R. Tannenbaum and D. I. C. Wang, pp. 69–104. MIT Press, Cambridge, MA.

Laskin, A. I. (1977). Ethanol as a substrate for single cell protein production. *Biotechnol. Bioeng. Symp.*, **7**, 91–103.

Litchfield, J. H. (1979). Production of single-cell protein for use in food or feed. In *Microbial Technology*, ed. H. J. Peppler and D. Perlman, 2nd edn. vol. 1. pp. 93–155. Academic, New York.

Litchfield, J. H. (1980). Microbial protein production. *Bioscience*, **30**, 387–395.

Litchfield, J. H. (1983a). Technical and economic prospects for industrial proteins in the coming decades. *International Symposium on Single-Cell Proteins* (Paris), pp. 9–33. Technique et Documentation, Paris, France.

Litchfield, J. H. (1983b). Single-cell proteins. *Science*, **219**, 740–746.

MacLennan, D. G., J. S. Gow and D. A. Stringer (1973). Methanol–bacterium process for SCP. *Process Biochem.*, **8** (6), 22–24.

MacLennan, D. G., J. C. Ousby, T. R. Owen and D. C. Steer (1976). Microbiological production of protein. *US Pat.* 3 989 594.

McNairney, J. (1984). Modification of a novel protein product. *J. Chem. Technol. Biotechnol.*, **34B**, 206–214.

Malick, E. A., J. W. Vanderveen, D. O. Hitzman and E. H. Wegner (1984). Production of single cell protein material. *US Pat.* 4 439 523.

Margetts, R. J. (1983). Economic and biotechnical problems facing large scale SCP commercialization. *International Symposium on Single-Cell Proteins* (Paris), pp. 165–169. Technique et Documentation, Paris, France.

Mateles, R. I. (1971). Calculation of the oxygen required for cell production. *Biotechnol. Bioeng.*, **13**, 581–582.

Minami, K., M. Yamamura, S. Shimizu, K. Ogawa and N. Sekine (1978). Methods and apparatus for methanol feeding with less growth-inhibition. *J. Ferment. Technol.*, **56**, 35–40.

Miura, Y., M. Okazaki, K. Ohi, T. Nishimura and S. Komemushi (1982). Optimization of biomass productivity and substrate utility of hydrogen bacterium *Alcaligenes hydrogenophilus*. *Biotechnol. Bioeng.*, **24**, 1173–1182.

Mogren, H. (1979). SCP from methanol — the Norprotein proces. *Process Biochem.*, **14** (3), 2–4, 7.

Mogren, H. L., G. O. Hedenskog and L. E. Enebo (1974). Process for extracting protein from microorganisms. *US Pat.* 3 848 812.

Molina, O. E., N. I. Perotti deGalvez, C. I. Frigerio and P. R. Cordoba (1984). Single-cell protein production from bagasse pith pretreated with sodium hydroxide at room temperature. *Appl. Microbiol. Biotechnol.*, **20**, 135–339.

Protein Advisory Group (1970a). *PAG Statement No. 4 on Single Cell Protein.* FAO/WHO/UNICEF, United Nations, New York.

Protein Advisory Group (1970b). *PAG Statement No. 6 for Pre-Clinical Testing of Novel Sources of Protein.* FAO/WHO/UNICEF, United Nations, New York.

Protein Advisory Group (1972). *PAG Statement No. 12 on the Production of Single Cell Protein for Human Consumption.* FAO/WHO/UNICEF, United Nations, New York.

Protein Advisory Group (1974). *PAG Guideline No. 15 on Nutritional and Safety Aspects of Novel Protein Sources for Animal Feeding.* FAO/WHO/UNICEF, United Nations, New York.

Puhar, E., I. Lorencez, L. M. Guerra and A. Fiechter (1982). Limit to cell mass production in continuous culture of *Methylomonas clara. J. Ferment. Technol.*, **60**, 321–326.

Puhar, E., I. Lorencez and A. Fiechter (1983). Influence of partial pressure of oxygen and carbon dioxide on *Methylomonas clara* in continuous culture. *Eur. J. Appl. Microbiol. Biotechnol.*, **18**, 131–134.

Rose, A. H. (ed.) (1979). *Microbial Biomass — Economic Microbiology*, vol. 4. Academic, New York.

Shipman, R. H., I. C. Kao and L. T. Fan (1975). Single-cell protein production by photosynthetic bacteria cultivation in agricultural by-products. *Biotechnol. Bioeng.*, **17**, 1561–1570.

Shuler, M. L., E. D. Roberts, D. W. Mitchell, F. Kargi, R. E. Austic, A. Henry, R. Vashon and H. W. Seeley, Jr. (1979). Process for the aerobic conversion of poultry manure into high-protein feedstuff. *Biotechnol. Bioeng.*, **21**, 19–38.

Shung, J.-T. and S. Ueda (1976). Separation of bacterial cells from fermentation broth by coagulation. *J. Ferment. Technol.*, **54**, 249–259.

Shung, J.-T. and C. C. Yang (1983). Separation of bacterial cells from fermentation broth — pilot test. *Hakkokogaku*, **61**, 63–68.

Siegel, R. S. and D. F. Ollis (1984). Kinetics of growth of the hydrogen oxidizing bacterium *Alcaligenes eutrophus. Biotechnol. Bioeng.*, **26**, 764–770.

Sittig, W. (1983). The economical aspects of protein production as consequence of technological decisions. *International Symposium on Single-Cell Proteins* (Paris), pp. 181–192. Technique et Documentation, Paris, France.

Solomons, G. L. (1983). Single-cell protein. *CRC Crit. Rev. Biotechnol.*, **1**, 21–58.

Stahl, U. and K. Esser (1982). Plasmids in *Methylomonas clara*, a methylotrophic producer of single-cell proteins. *Eur. J. Appl. Microbiol. Biotechnol.*, **15**, 223–226.

Stringer, D. A. (1983). Current views on the toxicological testing of SCP's. *International Symposium on Single-Cell Proteins* (Paris), pp. 170–180. Technique et Documentation, Paris, France.

Stringer, D. A. (1983). Process engineering of SCP-product safety considerations. *Biotech 83. Proceedings of the 1st International Conference on Commercial Applications and Implications of Biotechnology*, pp. 189–200. Online Publications, Ltd., Northwood, UK.

Srinivasan, V. R., M. B. Fleenor and K. J. Summers (1977). Gradient-feed method of growing high cell density cultures of *Cellulomonas* in a bench scale fermentor. *Biotechnol. Bioeng.*, **19**, 153–155.

Sugimoto, M., S. Yoko and O. Imada (1972). Biomass production from *n*-butane. *Fermentation Technology Today. Proceedings of the 4th International Fermentation Symposium*, pp. 503–507. Society of Fermentation Technology, Osaka, Japan.

Tanaka, K., K. Kimura and M. Yamamoto (1973). Microorganism production. *US Pat.* 3 751 337.

Tannenbaum, S. R. and D. I. C. Wang (eds.) (1967). *Single-Cell Protein II.* MIT Press, Cambridge, MA.

Tannenbaum, S. R., R. I. Mateles, and G. R. Capco (1965). Processing of bacteria for production of protein concentrates. *World Protein Resources*, pp. 254–260. American Chemical Society, Washington, DC.

Taylor, J. C., E. W. Lucas, D. A. Gable and G. Graber (1974). Evaluation of single cell protein for non-ruminants. In *Single-Cell Protein*, ed. P. Davis, pp. 179–186. Academic, New York.

Thayer, D. W., S. P. Yang, A. B. Key, H. H. Yang and J. W. Barker (1975). Production of cattle feed by the growth of bacteria on mesquite wood. *Dev. Ind. Microbiol.*, **16**, 465–474.

Vanderveen, J. W., D. O. Hitzman and E. H. Wegner (1983). Cellular product separation. *US Pat.* 4 399 223.

Van Dijken, J. P. and W. Harder (1975). Growth yields of microorganisms on methanol and methane. A theoretical study. *Biotechnol. Bioeng.*, **17**, 15–30.

Vrati, S. (1984). Single-cell protein production by photosynthetic bacteria grown on the clarified effluents of biogas plant. *Appl. Microbiol. Biotechnol.*, **19**, 199–202.

Vrati, S. and J. Verna (1983). Production of molecular hydrogen and single-cell protein by *Rhodopseudomonas capsulata* from cow dung. *J. Ferment. Technol.*, **61**, 157–162.

Wagner, F., Th. Kleeman and W. Zahn (1969). Microbial transformation of hydrocarbons. II. Growth constants and cell composition of microbial cells deprived from *n*-alkanes. *Biotechnol. Bioeng.*, **11**, 393–408.

Waldroup, P. W. and J. R. Payne (1974). Feeding value of methanol-derived single-cell protein from broiler chicks. *Poultry Sci.*, **53**, 1039–1042.

Walz, O. P. and H. Brune (1984). Protein utilization and effect of methionine supplementation on single-cell biomass of *Methylomonas clara* tested with rats and piglets. *Z. Tierphysiol. Tiernaehr. Futtermitteld.*, **51**, 236–249.

Wang, H. Y., D. G. Mou and J. R. Swartz (1976). Thermodynamic evaluation of microbial growth. *Biotechnol. Bioeng.*, **19**, 1811–1814.

Whitemore. C. T., I. W. Moffat and A. G. Taylor (1976). Evaluation by digestibility, growth, and daughter of microbial cells as a source of protein for young pigs. *J. Sci. Food Agric.*, **27**, 1163–1170.

Wilkinson, T. G. and D. E. F. Harrison (1973). The affinity for methane and methanol of mixed cultures grown on methane in continuous culture. *J. Appl. Bacteriol.*, **36**, 309–313.

Wilkinson, T. G., H. H. Topiwala and G. Hamer (1974). Interaction in a mixed bacterial population growing on methane in continuous culture. *Biotechnol. Bioeng.*, **16**, 41–59.

Windass, J. D., M. J. Worsey, E. M. Pioli, D. Pioli, P. T. Barth, K. T. Atherton, E. C. Dart, D. Byrom, K. Powell and P. J. Senior (1980). Improved conversion of methanol to single cell protein by *Methylophilus methylotrophus*. *Nature (London)*, **287**, 396–401.

Yamane, T., M. Kishimoto and F. Yoshida (1976). Semi-batch culture of methanol-assimilating bacteria with exponentially increased methanol feed. *J. Ferment. Technol.*, **54**, 229–240.

Yang, H. H., S. P. Yang and D. W. Thayer (1977). Evaluation of the protein quality of single-cell protein produced from mesquite. *J. Food Sci.*, **42**, 1247–1260.

Yano, T., T. Kobayashi and S. Shimizu (1978a). Silicone tubing sensor for detection of methanol. *J. Ferment. Technol.*, **56**, 421–427.

Yano, T., T. Kobayashi and S. Shimizu (1978b). Feed-batch culture of methanol-utilizing bacterium with DO-Stat. *J. Ferment. Technol.*, **56**, 416–420.

Yamamoto, M., Y. Serio, K. Kouno, R. Okamoto and T. Inui (1978). Isolation and characterization of marine methanol-utilizing bacteria. *J. Ferment. Technol.*, **56**, 451–458.

22

Production of Biomass by Filamentous Fungi

G. L. SOLOMONS
RHM Research Ltd, High Wycombe, Bucks, UK

22.1 INTRODUCTION

There is a long history of human consumption of the higher fungi and today the sporophores (fruiting bodies) of the basidiomycete *Agaricus bisporus* are cultivated on a wide scale and are sold as part of the normal food distribution system of Western countries.

Many other species are commonly consumed elsewhere in the world. The use of fungi as flavour enhancers in cheese production, *e.g.* Danish Blue, Roquefort, *etc.*, is well established, as is the use of solid substrate grown fungi to digest the protein of soyabean to a more acceptable form as in miso, tempeh, *etc.* (Hesseltine, 1965). Experiments in growing fungi on milk whey for supplementation of human diets in Germany during World War II showed some success, although it was carried out on a limited scale (Robinson, 1952). The submerged culture production of penicillin from 1944 onwards, also made available substantial amounts of fungal mycelium: the citric acid fermentations at that time, still being all surface culture, did not provide mycelium in such an easily utilizable form. Spent *Penicillium chrysogenum* was dried and used as an animal feed supplement, especially suitable for chickens. A detailed study of the composition and nutritional value of spent *Penicillium* mycelium was made by Fink *et al.* (1953). The problems of antibiotic residues and acceptability caused the abandonment of the use of the material in the West, but interest has continued elsewhere (Doctor and Kerue, 1968). With the further development of fermentation technology and the availability of reliable fermenter designs, several groups became involved in the production of protein by fungi (Reusser *et al.*, 1958; Falanghe *et al.*, 1964). It was however Gray and his coworkers in the United States who championed the use of fungi as a protein source (see Gray, 1970), but until the early 1960s virtually all of the research was confined to academic laboratories. The stimulus afforded to fermentation technology as a potential mass source of protein by British Petroleum's process for growing yeast on *n*-alkanes encouraged other industrial laboratories, as well as academic groups, to consider alternative organisms and substrates for use in protein production.

22.2 PRINCIPLES OF BIOMASS PRODUCTION

The consideration of a commercial process for the production of biomass, as of any other item, is dominated by the economics of the proposed process. Often, a company will have a preferred substrate available to it within its existing organization, *e.g.* oil companies are likely to have *n*-alkanes, chemical companies may have methanol, whilst food based organizations are likely to have or use substrates such as sugars or agricultural residues such as straw, *etc.* The company's interests are also likely to predetermine the market that the product is aimed at: animal feedstuffs or direct human consumption. Common to all processes is the need for the highest productivity, least capital and running costs, high yield factor on the limiting substrate, near complete utilization of the substrate, ease of recovery and final work-up of the material by drying or formulation into final products. Against this background will be considered the final overall process details: batch or continuous; aseptic or non-aseptic; single train or multi-unit operation; and scale of production in tonnes/annum. All of these considerations will then influence or even determine the choice of the type of organism that will optimize the overall process. This rationale has led over the past twenty years to large scale production facilities for single cell protein (SCP) for animal feeds to be based on yeasts or bacteria (Solomons, 1983); only in a minority of processes have fungi been chosen and the reasons for this choice are worth examination.

22.2.1 Substrates

The growth of bacteria and yeasts on a wide range of substrates has been studied with a view to producing biomass (Litchfield, 1979). Fungi can grow on some of the more unusual carbon sources, *Graphium* spp. on methane and ethane (Volesky and Zajic, 1971), *Trichoderma lignorum* on methanol (Tye and Willetts, 1977), but the majority of investigations into the use of fungi for protein production have concentrated on the use of substrates that are derived from agriculture or food processing by-products or wastes. Thus for fungi, starch, cellulosic wastes, milk whey, *etc.* feature prominently as the carbon sources (Rolz and Humphrey, 1982).

22.2.2 Economics

The economics of fermentation processes have been discussed by Bartholemew and Reisman (1979), whilst the economics of biomass production have been described by Litchfield (1977), Moo-Young (1977) and more recently by Solomons (1983). The total energy requirements for SCP production have been evaluated by Lewis (1976).

A particularly useful publication was that of Trilli (1977) who developed a model to predict costs in continuous fermentations. This model shows that at dilution rates higher than the critical, unit cost of growth-linked products, *e.g.* biomass, will decrease with run length. Using this model, the output of a biomass fermentation has been plotted (Figure 1). Also plotted are the effects of individual fermentation run times and it can be seen that above 400 hours there is little improvement in the cost per tonne. This suggests that continuous does not have to mean indefinite, as far as individual run time is concerned.

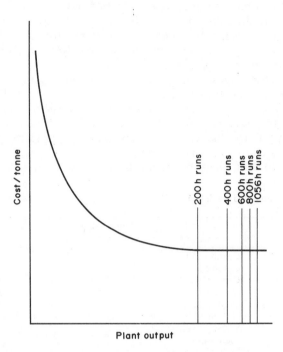

Figure 1 Production cost model of SCP fermentation run time and total output (Trilli, 1977)

22.3 THE CHOICE OF FUNGI AS A PROTEIN SOURCE: THE ADVANTAGES AND DISADVANTAGES

Table 1 lists the features, both for and against, that determine the choice of fungi as organisms to be used in a large scale industrial fermentation process to produce a final product that cannot be expected to command a high price in the market place. Consideration will be confined to the use of fungi capable of growing in a fermenter system rather than fungi suitable only for solid state fruiting body cultures, *i.e.* mushrooms.

Table 1 Some Supposed Advantages and Disadvantages of the Characteristics of Fungi as a Protein Source

Advantages
  (i) Good at breaking down a wide range of complex substrates, *e.g.* cellulose, hemicellulose, pectin, *etc.*
 (ii) Can tolerate low pH values, which helps in resisting infection
(iii) Few nutritional requirements for culture medium
 (iv) Ease of recovery of biomass by filtration
  (v) Ease of handling and of drying the biomass
 (vi) Structure conferred by hyphae can be used as a basis of food fabrication

Disadvantages
  (i) Growth rates are poor in comparison to yeasts and bacteria
 (ii) Tend to grow at lower temperatures than bacteria, which raises cooling requirements for production, and there are comparatively few thermotolerant strains to choose from
(iii) Protein content is unfavourable compared to yeasts or bacteria
 (iv) Fermentation broths are rheologically complex and difficult to aerate
  (v) Production of a range of undesirable metabolites, *e.g.* oxalic acid, mycotoxins
 (vi) Poor nutritional properties compared to yeast and bacteria
(vii) Genetically unstable

In practice, fungi vary in their growth and compositional characteristics as widely as do unicellular organisms and many of the points raised in Table 1 have proved to be unsubstantiated; this applies to supposed advantages as well as disadvantages.

The organism *Fusarium graminearum*, used in the RHM Myco-protein process for example, utilizes starch as a substrate, albeit rather slowly, and suffers a reduction in maximum growth rate as glucose chain length increases: glucose 0.28 h^{-1}; maltose 0.22 h^{-1}; maltotriose 0.18 h^{-1} (Anderson, *et al.*, 1975). It has an optimum growth rate in the pH range 4.5–7.0 with an operational pH of 6.0, which hardly constitutes low pH tolerance (Anderson and Solomons, 1983).

In sharp contrast, a new species, *Scytalidium acidophilum* (Sigler and Carmichael, 1974) has been shown to grow on hydrolysed waste paper at a pH <1. Not surprisingly, contaminants were never a problem (Ivarson and Morita, 1982).

Many of the listed disadvantages of fungi will be examined in detail later, but it is worth commenting here in general terms that although most fungi do grow more slowly than most bacteria or yeasts, the data in Table 2 shows that, for the practical consideration of biomass production, growth rates can be entirely adequate. As has been pointed out by Harrison (1976), growth rate is of considerably less importance than is productivity, and it is in this regard that fungi do show a real disadvantage compared to yeast or bacteria (Table 3). This is due to the difficulty of achieving high cell concentrations of fungi if they are growing in a mycelial form as this results in a

Table 2   Maximum Growth Rates of Filamentous Fungi Used for Biomass Production

Organism	Substrate	Temperature (°C)	μ_{max} (h^{-1})	Ref.
Aspergillus fumigatus I21A	Cassava	45	0.11	Reade and Gregory (1975)
Aspergillus niger M1	Carob extract	36	*ca.* 0.16	Imrie and Vlitos (1975)
Chaetomium cellulolyticum	Corn stover	37	>0.24	Moo-Young *et al.* (1979)
Fusarium graminearum	Glucose	30	0.28	Anderson *et al.* (1975)
Fusarium moniliforme	Carob extract	30	0.22	Macris and Kokke (1978)
Fusarium spp. M4	Not stated	35	0.30	Imrie and Righelato (1976)
Geotrichum candidum	Whisky distillery spent wash	22	0.385	Quinn and Marchant (1979)
Paecilomyces variotii	Sulfite liquor	38	0.31	Forss *et al.* (1974)
Penicillium cyclopium	Milk whey	28	0.20	Kim and Lebeault (1981)
Penicillium notatum-chrysogenum	Starch hydrolysate	30	0.20	Solomons and Spicer (1973)
Rhizopus oligosporus	Mung bean whey	32	0.16	Graham *et al.* (1976)
Trichoderma album	Not stated	28	*ca.* 0.46	Staron (1981a)
Trichoderma harzianum, Rifai	Coffee wastes	30	0.10	Aguirre *et al.* (1976)

Table 3   Comparison of Productivity (r_x) in Continuous Culture of Bacteria, Yeasts and Fungi

	Substrate	Y	x(kg m^{-3})	D(h^{-1})	r_x(kg m^{-3} h^{-1})	Ref.
Bacteria						
Methylomonas methanica	Methanol	0.5	20–25	0.4	8–10	Mogren (1979)
Methylomonas methylotrophus	Methanol	*ca.* 0.5	30	0.16–0.19	4.8–5.7	Waterworth (1981)
Yeasts						
Candida novellus (maltosa)	*n*-Alkane	1.1	15.2	0.21	3.2	Kanazawa (1975)
Candida utilis	Molasses	0.5	—	—	5.2	Moo-Young (1977)
Candida utilis	Sulfite liquor	0.6	10.7	0.33	3.6	Peppler (1970)
Candida intermedia	Whey	0.54	22.5	0.2	4.5	Meyrath and Bayer (1979)
Fungi						
Paecilomyces variotii	Sulfite liquor	0.55	13	0.17	2.7	Romantschuk and Lehtomäki (1978) Laine and Kuoppamäki (1979)
Fusarium moniliforme	Carob extract	0.71	3.3	0.21	0.69	Macris and Kokke (1978)
Penicillium cyclopium	Whey	0.68	12.8	0.16	2.0	Kim and Lebeault (1981)
Geotrichum candidum	Whisky distillery wash	0.57	18.0	0.125	2.24	Quinn and Marchant (1979)

highly viscous non-Newtonian culture broth which is very difficult to adequately aerate. Productivities of 3–4 kg m^{-3} h^{-1} are high for fungi, whereas values of perhaps 6–8 kg m^{-3} h^{-1} are obtainable using yeasts or bacteria. Fungi can grow as pellets, which eases the problem of liquid phase mass-transfer for oxygen, but results in an oxygen limited culture with a poor growth rate. Similarly, some fungi, such as *Mucor plumbeus*, can grow as apparent single-cells which look much the same as single cell yeast, which overcome all the problems of low productivity and oxygen limitation, but it then becomes difficult to see any real advantage over growing true yeasts.

The temperature at which a fermentation process is operated has important consequences on process economics, due to the cooling requirements it imposes; the higher the operating temperature, the less the cooling requirements. In a number of fermentations it is the restriction caused by cooling requirements in fermenter heat exchange surfaces which actually limit the size of the fermenter. In temperate climates, a fermentation temperature of around 40 °C is highly desirable but comparatively few fungi can tolerate these conditions (Table 2), notable exceptions being that range of organisms studied by the Canadian group (Reade and Gregory, 1975; Gregory *et al.*, 1977), which can grow in the temperature range of 45–55 °C. On the other hand, the genus *Fusarium*, which has been widely studied as a potential protein source (Solomons, 1983), tends to have an optimum growth temperature close to 30 °C.

The protein content of fungi is assumed to be low in comparison to yeast or bacteria. Whilst this is often true, those fungi selected for biomass production do have a true protein content not that inferior to yeasts, in the range 33–45% on dry weight and, whilst bacteria can reach values of 60–65% protein, this is naturally accompanied by much higher levels of RNA, which can reach 15–25% (Sinskey and Tannenbaum, 1975). The nutritional value of fungal protein has been shown to be very satisfactory (Duthie, 1975) and compares well with yeasts and bacteria.

Fungi can produce a wide range of undesirable metabolites, which give cause for concern as to their safety in use. Mycotoxins are a clear example of these very dangerous compounds and Litchfield (1979) advises against the use of species known to produce them, but they tend to be strain specific and invariably not produced under the conditions favourable for biomass production, that is fast growth at the highest temperature tolerated by the organism. An additional cause for concern is the potential of fungal cells to produce oxalates and other undesirable compounds. It is too often assumed that bacteria or yeasts have no such drawbacks, but the ingestion of *Hydrogenomonas* and *Aerobacter* in human volunteers was shown to cause violent reactions (Waslien *et al.*, 1969) and even yeasts have been shown to cause unpleasant effects on ingestion by humans (Scrimshaw, 1972). Of lesser consequence, *Methylophilus methylotrophus* grown on methanol contains a compound(s) which causes enlargement of chicken liver and which may require removal by treating the bacterial cells with 3% hydrogen peroxide (Talbot and Senior, 1980).

Fungi display genetic variability, but this is largely due to the ease with which changes are detected on agar culture plates, especially in the complex sporing apparatus of fungi. At the biochemical level of fungal mycelium, the changes that occur seem no more extensive than those experienced with other organisms. Fungi have been grown in continuous culture for many hundreds of generations and have proved to be as satisfactory in this regard as many unicellular organisms.

22.4 THE COMPOSITION OF FUNGI, NUTRITIONAL PROPERTIES AND SAFETY IN USE

22.4.1 Composition of Fungi and Nutritional Values

The compositions of some of the more important fungi that have been investigated for use in biomass production are shown in Table 4. The distinguishing feature of fungal composition lies in the distribution of the nitrogen content. Crude protein values based upon total nitrogen (TN) × 6.25 are really rather meaningless, even for unicellular SCP, because of the RNA content of these cells. In addition, fungi have a substantial amount of their nitrogen as *n*-acetylglucosamine in the form of chitin in the cell wall. The distribution of nitrogen in a strain of *Fusarium graminearum* grown in continuous culture is shown in Table 5. The true protein content of the cells accounts for approximately two-thirds of the total nitrogen, whilst RNA-N accounts for 15% and the chitin of the cell wall, 10%. The non-protein nitrogen values for 13 fungi grown on different media average 35% (Anderson *et al.*, 1975). Chemical composition is, of course, not a fixed entity; it depends greatly on limiting substrate, culture conditions, growth rate, temperature and pH (Mateles, 1979; Righelato, 1975).

Biomass is grown for its protein content and is therefore never operated under nitrogen limitation; in consequence, the lipid content of the cells is almost invariably low, since fungal cells tend to synthesize maximum lipid content only under nitrogen limitation. Lipid values depend

Table 4 Proximate Analysis, in % Dry Weight, of Fungi Examined for Biomass Production

Organism	Culture type[a]	Substrate	Temperature (°C)	pH	Crude protein (total N × 6.25)	True protein	Non-protein nitrogen	RNA	Lipid	Ash	Ref.
Aspergillus fumigatus I 21	b	Cassava extract	45	3.5	40	31.5	21.3	ND	12.2	ND	Khor *et al.* (1976)
Aspergillus fumigatus I 21 A	b	Cassava	45	3.5	37	27	27	ND	ND	ND	Reade and Gregory (1975)
Aspergillus oryzae CMI 44242	b	Ground barley	30	4.5	39.4	30.2	23.4	2.5	ND	ND	Smith *et al.* (1975)
Aspergillus oryzae NRRL 3483	b	Hydrolysed potato	30	6.0	40	22	45	ND	2	9	Smith *et al.* (1975)
Aspergillus oryzae NRRL 3484	b	Hydrolysed potato	30	6.0	39	25	35.9	ND	2	9	Smith *et al.* (1975)
Cephalosporium eichhorniae	b	Cassava starch	45	3.5	49.5	37.8	23.6	ND	ND	ND	Gregory *et al.* (1977)
Chaetomium cellulolyticum	c	Crop residues	37	5.5	45	ND	—	5	10	5	Moo-Young *et al.* (1979)
Fusarium graminearum CMI 145425	c	Glucose	30	6.0	60	42	30	10	13	6	Anderson and Solomons (1983)
Fusarium moniliforme	c	Carob extract	30	4.5	43	30	30	8	5	ND	Macris and Kokke (1978)
Fusarium semitectum CMI 135410	ND	Glucose	ND	ND	48	34.5	28.1	3.2	ND	ND	Smith (1975)
Paecilomyces variotii	c	Sulfite liquor	37	4.5	55	ND	—	10	1.3	6	Romantschuk and Lehtomäki (1978)
Penicillium notatum-chrysogenum CMI 138291	b	Hydrolysed potato	30	5.5	43	36	16.3	ND	1.6	5	Solomons and Spicer (1973)
Penicillium cyclopium	c	Milk whey	28	3.5	54	38	29.6	9	ND	ND	Kim and Lebeault (1981)
Rhizopus chinensis	b	Cassava extract	45	3.5	49	37	24.5	ND	ND	ND	Gregory *et al.* (1977)
Scytalidium acidophilum	b	Waste paper	a.22	<1.0	45	36	20	6.2	2.6	3.5	Ivarson and Morita (1982)
Sporotrichum pulverulentum	b	Fibre board waste water	38	4.8	43	30	16.2	ND	10	6	Thomke *et al.* (1980)
Sporotrichum thermophile	b	Cassava extract	45	3.5	37	26	29.7	ND	6.6	ND	Khor *et al.* (1976)
Trichoderma album	b	ND	28	3.7	64	54	16	4-6	6-12	6-9	Staron (1981b)

[a] b, batch; c, continuous; ND, no data.

Table 5 The Distribution of Nitrogen in *Fusarium
graminearum* Grown in Continuous Culture at a Growth
Rate of 0.1 h^{-1a}

Nitrogen	Nitrogen (%)	Nitrogen (%) of total
Free amino acid	0.71	7.35
Protein	6.34	65.63
Nucleotide	0.15	1.55
RNA	1.46	15.11
N-Acetylglucosamine	1.00	10.35
Total	9.66	
Total N (by Kjeldhal)	9.72	

[a] Anderson *et al.*, 1975.

upon the analytical methods used; with *Fusarium*, petroleum ether extraction gives a value of 2–3%, acid hydrolysis followed by hexane extraction increases the value by approximately three-fold, whilst extraction with 3:1 chloroform:methanol provides values of 13–14%. The increase is largely due to the extraction of phospholipids by the more polar solvent system.

Clearly, the most important component of the cell in a biomass fermentation is the protein or amino acid composition. Table 6 shows the amino acid analysis of a number of fungi grown in the course of biomass studies, together with the nutritional data from animal feeding studies. Amino acid analysis by itself provides little indication of the true nutritional usefulness of a product. Digestibility of the protein is equally important, as is the absence of toxic factors which harm the animals. For example, *Mucor plumbeus* possesses an adequate amino acid profile, but repeated rat assays showed it to possess poor nutritional qualities, whilst *Alternaria tenuis* actually killed the animals.

22.4.2 Safety in Use

Only two SCP animal feed products have gained regulatory clearance in a number of countries, British Petroleum's 'Toprina' and ICI's 'Pruteen', produced by a yeast on *n*-alkane and a bacterium grown on methanol, respectively. Both of these products were tested by very extensive nutritional and toxicological programmes costing several million dollars. The figure of $6 million has been estimated for the ICI product (Anon, 1980). For other details on toxicological testing, see Solomons (1983). Only one biomass project using fungi has yet achieved a commercial scale of production, the Finnish Pekilo process, and although it has received clearance for animal feed in that country (Romantschuk, 1975) it is not known if clearance has been requested or obtained elsewhere. Certainly, the scale of testing will not have approached that of BP or ICI, but the use of sulfite liquor as substrate undoubtedly gives it an advantage with regard to acceptability over materials produced from a hydrocarbon based substrate. A similar situation applies to the Waterloo SCP process based on *Chaetomium cellulolyticum* growing on agricultural residues such as straw. After a ten year programme of toxicological studies, RHM were given clearance by the UK Government in 1980, to market their Myco-protein product for human foods.

The safety of a product also has to include safety during production, that is to the people exposed to it, and also to the environment around the production plant. These considerations are described in detail by Sargeant and Evans (1979).

22.5 GROWTH FORMS, STRAIN SELECTION AND STABILITY

The three growth forms that fungi can adopt in submerged culture have been described as single-cell, pelleted and filamentous (Solomons, 1975). Almost all work on biomass production has been concerned with using filamentous growth, since single cells are only produced by a limited number of fungi and pelleted growth does not maximize growth rate or yield and may lead to secondary metabolite formation caused by anaerobiosis inside the pellet. For detailed discussion on pellet formation see Metz (1976) and Van Suijdam (1980). The general factors which affect population dominance in culture medium have recently been discussed (Solomons, 1983). However the organism or strain is selected, it is important that it remains stable in continuous

Table 6   Amino Acid Analysis of Fungi and Nutritional Evaluation[a]

AMINO ACID	Aspergillus fumigatus I 21[b]	Aspergillus fumigatus I 21A[b,c]	Aspergillus niger	Aspergillus oryzae CMI 44242[d]	Aspergillus oryzae NRRL 3483	Aspergillus oryzae NRRL 3484	Chaetomium cellulolyticum[e]	Fusarium graminearum CMI 145425	Fusarium moniliforme[g]	Fusarium semitectum CMI 135410[d]	Paecilomyces varioti[h]	Penicillium notatum-chrysogenum CMI 13829I	Rhizopus arrhizus[d]	Scytalidium acidophilum[h]	Sporotrichum pulverulentum[i]	Sporotrichum thermophile[h]	Trichoderma album[k]	Trichoderma viride[d]
Aspartic acid	6.0	4.3	5.9	6.9	7.0	7.5	—	8.4	7.0	6.7	—	8.2	6.8	8.1	6.6	4.7	9.9	8.3
Threonine	3.7	3.7	3.1	3.5	4.8	3.7	6.1	4.3	4.3	3.8	4.8	6.1	3.3	5.2	3.6	3.4	4.8	4.2
Serine	3.9	3.7	3.4	3.6	3.4	3.8	—	4.5	3.9	3.3	—	4.4	3.2	4.6	4.6	3.1	4.7	4.1
Glutamic acid	7.8	11.4	7.7	12.4	6.2	8.5	—	11.9	10.8	10.7	—	13.1	8.7	11.1	9.6	8.9	13.7	9.8
Proline	3.4	7.0	1.6	5.2	2.8	3.2	—	4.5	3.5	3.0	—	3.8	4.6	3.9	3.8	3.2	4.7	3.8
Glycine	5.0	4.3	3.0	3.7	3.0	3.8	—	4.3	4.1	3.6	—	5.3	3.5	4.5	3.9	3.8	3.6	4.5
Alanine	7.2	5.4	4.2	4.6	3.4	3.8	—	6.3	6.2	5.0	—	4.8	4.4	5.8	6.0	7.8	5.7	6.2
Valine	5.4	4.6	3.4	4.6	3.0	3.3	5.8	4.9	4.2	3.8	5.0	4.9	4.0	4.9	4.5	4.2	5.7	5.0
Cysteine	0.7	0.3	0.6	1.0	0.7	0.7	2.6	0.5	—	1.0	—	0.7	1.5	1.7	1.8	0.6	2.2	1.1
Methionine	1.4	1.3	1.0	1.3	1.9	2.5	—	1.6	0.9	1.4	1.6	1.3	1.6	1.4	1.8	1.1	2.2	1.7
Isoleucine	4.5	3.6	2.5	3.5	2.4	3.4	4.7	3.5	3.3	3.2	4.6	4.3	3.2	3.9	3.4	3.5	5.1	4.4
Leucine	7.1	6.0	5.0	5.8	4.6	5.1	7.5	4.6	5.4	4.6	7.1	7.2	5.2	6.1	6.1	5.5	9.6	7.0
Tyrosine	2.8	2.6	2.1	5.0	1.5	2.0	3.3	3.1	7.3	3.8	4.0	3.0	3.1	3.3	2.5	2.4	3.8	5.2
Phenylalanine	4.8	3.5	2.9	3.8	1.9	2.4	3.8	3.6	3.2	3.9	4.2	3.4	2.7	3.4	2.9	3.7	3.8	4.8
Lysine	7.0	5.3	3.0	4.2	4.9	5.4	6.8	6.1	8.1	5.0	6.5	9.3	5.3	5.4	4.4	5.6	8.1	5.4
Histidine	1.3	1.5	1.4	1.9	1.1	1.2	—	2.1	1.5	1.8	—	2.0	2.7	2.2	1.9	1.2	2.4	2.1
Arginine	5.8	5.0	2.8	4.4	3.2	3.4	—	5.4	4.9	6.0	—	4.3	4.0	5.4	4.0	5.4	4.5	4.9
Tryptophan	1.0	—	—	1.4	—	—	—	1.8	1.6	1.1	—	—	1.2	—	1.2	1.2	2.0	1.1
TOTAL AMINO ACIDS	78.8	73.5	53.6	76.8	55.8	63.7	—	81.4	80.2	71.7	—	80.1	69.0	80.9	70.7	69.3	96.5	83.6
α-Amino nitrogen (%)	4.4	4.3	3.6	4.3	3.0	3.5	—	7.0	4.3	4.8	—	5.3	2.8	5.2	3.8	3.6	8.7	5.3
Total nitrogen (%)	6.4	5.2	7.7	6.4	6.4	6.2	7.2	9.7	6.1	7.7	8.8	7.0	4.4	7.3	6.1	5.9	10.3	7.2
NPU (based on TN)	—	—	41.0	41.0	50.0	58.0	—	65–73[f]	42.0	42.0	—	55.0	44.0	—	28.0	—	—	40.0
NPU (based on α-amino nitrogen)	—	—	69.0	—	92.0	86.0	—	—	—	—	—	—	—	—	—	—	—	—
PER (normalized, casein = 2.5)	2.1	2.2	—	—	—	—	—	1.15	—	—	—	—	—	—	1.7	—	2.5	—
Digestibility	—	—	—	49.0	—	—	—	78,88[f]	—	83.0	—	—	69.0	—	77.5	—	—	83.0
Biological value	—	—	59.0	59.0	—	—	—	84,84[f]	—	50.0	—	—	64.0	—	36.0	—	—	48.0

[a] All figures in g/16 g N.   [b] Khor *et al.* (1976).   [c] Gregory *et al.* (1977).   [d] Smith *et al.* (1975).   [e] Moo-Young *et al.* (1979).   [f] Data on humans, before and after chitin N excluded.   [g] Drouliscos *et al.* (1976).   [h] Romantschuk and Lehtomäki (1978).   [i] Ivarson and Morita (1982).   [j] Thomke *et al.* (1980).   [k] Staron (1979, 1981b).

culture for a sufficient time to allow economic operation. Moreover, for production reasons, changes in composition, yield or productivity need to be avoided. The requirement for stability is made more demanding because regulatory clearance for either animal or human use usually insists upon the minimum variation in cell characteristics, since it is usually argued that an obvious change could accompany an unobvious change which could lead to the formation of undesirable compounds that could present a toxicological hazard. However unlikely such an occurrence, regulatory authorities usually possess the means by which to enforce adherence to their required standards.

A specific factor of fungi is that wild types tend to grow with long hyphae, with few branches, and this is the type of growth form often seen in Petri-dish culture as well as growth in submerged culture. The fact that branch number has an important consequence for maximum growth rate can, in submerged culture conditions, lead to selection pressures so that a variant which branches more frequently will possess a higher growth rate and hence be able, in a continuous culture, to take over the population (Righelato, 1975). Just such a phenomenon was described by Forss *et al.* (1974) who state 'When these microorganisms have been cultivated continuously for some time, usually for a week, a significant change becomes apparent in the properties of the microorganism. The long, rather sparingly branched mycelium is transformed into a short, thick and abundantly branched modification, retained on further cultivation of the mycelium.' Times for these changes to occur in five fungi are shown in Table 7. The change in morphology was accompanied by an increase in growth rate and protein content and an apparent decrease in K_s value. Moreover, the culture broth shows a marked reduction in viscosity, with a consequent favourable effect upon the mixing and aeration in the fermenter. We have observed similar changes in the morphology of *Fusarium graminearum* grown in a glucose/salts medium, but that change does not occur until the culture is 1000–1200 hours old. We have not found similar changes with *Penicillium notatum-chrysogenum* when in culture for 1500–2000 hours. Changes in the morphology of *Penicillium chrysogenum* grown in continuous culture were first described by Pirt and Callow (1959), but these changes were pH dependent. The changes described above are probably brought about by the selection of an aneuploid or diploid strain. These exhibit restricted colony morphology and grow on surface culture as cerebriform colonies, which appear to be slower growing than the less branched wild types. In submerged culture, however, they exhibit increased growth rate. Treatment of these strains with *p*-fluorophenylanine often produces sectored reversion to haploid forms (Lhoas, 1961).

Table 7 Time for Selection of Faster Growing Variants of Fungi Grown in Continuous Culture

Organism	Substrate	Time (h)
Paecilomyces variotii[a]	Sulfite waste liquor (SWL)	100
	Glucose/salts	150–200
Paecilomyces puntonii[a]	SWL	150
Gliocladium virens[a]	SWL	150
Trichoderma viride[a]	SWL	300
Byssochlamys nivea[a]	SWL	200
Fusarium graminearum[b]	Glucose/salts	1000–1200

[a] Forss *et al.* (1974).   [b] Solomons and Scammell (1976).

A change in growth rate brought about by differentiation has been shown to exist for *Geotrichum candidum*. With low levels of glucose (0.1%), the fungus grows with a single exponential growth rate in batch culture of 0.39 h^{-1}. At higher initial levels of glucose (0.5%), the initial fast growth rate is followed by a second slow phase of 0.14 h^{-1}, once the glucose level falls to around 15% of its initial value. The reason for this change is the differentiation of the cells into spores (Kier *et al.*, 1976). Fungal strains used for biomass production have been either wild type (Imrie and Vlitos, 1975; Solomons and Scammell, 1976), or strains isolated from continuous cultures (Forss *et al.*, 1974). A strain development programme of substantial complexity was used by Staron (1981a) to improve a wild type *Trichoderma viride* to isolate a strain which has been named as *Trichoderma album*. This strain was white and lost all ability to form gliotoxins or trichodermine as well as pigment. The screening involved sequential incubation and UV irradiation, employing in all eleven different stages and the cycle of stages then repeated. After 1400 such cycles, the new strain was isolated and its characteristics compared with the original and intermediate isolates as shown in Table 8. These considerable benefits are however only brought about at the cost of a low maximum growth temperature of 28 °C which clearly imposes a problem of fermenter cooling requirements and a sensitivity to shear. This inability to grow in a conventional stirred fermenter has necessitated the development of a complex fermenter which would have little prospect of economical scale-up (Staron, 1981b), but it would be interesting to examine the organism for growth in an air-lift system which would impose little mechanical shear on the cells (Barber and Worgan, 1981).

Table 8 Characteristics of *Trichoderma album* Isolate[a]

Number of selection cycles	Doubling time (h^{-1})	Protein (% DW)	Glucosamine (% in 16 g N)	Antibiotic peptides in culture filtrate $(g\,l^{-1})$
Wild type	6.0	40	18	2–5
100	5.0	45	15	1–2
300	3.5	47	9.5	0.5
1000	2.3	60	2.2	0
1400	1–2	65–70	0.7	0

[a] Staron (1981b).

Other changes can be brought about by growth of fungi in continuous culture. These changes are not expressed in the fermenter vessel, since growth is in the mycelial form, but is evidenced when samples are plated on to agar cultures. Using an exceptionally stable strain of *Penicillium notatum-chrysogenum*, Solomons and Spicer (1973) found that in carbohydrate limited continuous culture medium, at a dilution rate of $0.1\ h^{-1}$, sporulation capacity of the organism on agar medium was suddenly reduced after 300 hours cultivation and that after 900 hours a further loss resulted in the appearance of colonies which had all but lost the ability to form spores. This loss of sporulation was repeated in many experiments and the time taken for the two stages was remarkably consistent. Subsequent experience with the culture tended to suggest that the loss of sporulation was due to lack of gene expression, rather than the loss of the necessary genetic information. Although there appeared to be little, if any, change in the morphology of the organism in the fermenter, the two isolates appeared to increase their nitrogen content from the parent strain's value of 6.1% to 6.4 and 7.1% of dry weight.

22.6 INOCULUM AND BATCH PHASE GROWTH

Conventional batch fermentation technology usually makes elaborate provision for fermenter inoculum (Hockenhull, 1980). For penicillin production, it would be common to find inoculum prepared so as to provide a 10% inoculum level to all of the fermentation stages. This technique is helpful in reducing the organism's ability to revert to a more stable but less productive strain. Moreover, in batch culture, the grow-up period to maximize cell growth is non-productive, since secondary metabolites are more usually formed in the stationary phase of growth and time spent in large production vessels not producing product is uneconomic. By contrast, in the continuous culture of biomass, inoculum is provided at extremely low levels. Even in the 3000 m^3 ICI fermenter (admittedly growing bacteria, not fungi) the inoculum is provided from the laboratory as 10–20 l of culture (Anon, 1982a). Since most organisms used in biomass production have fast maximum growth rates ($>0.2\ h^{-1}$) cell concentrations of 15–20 kg m^{-3} are achievable in around 50 hours. Moreover, since the strains used for biomass production are usually stable wild types, unlike the highly unstable, mutated strains used in many secondary metabolite fermentations, there is little risk of strain degeneration. Another added advantage of small inoculums is that the high capital costs involved with serial inoculum preparations are avoided, especially since they would only be required for a very few times in a year.

The use of specialist culture media to produce fungal spores in submerged culture has greatly simplified the preparation of a satisfactory inoculum (Capellini and Petersen, 1965; Foster *et al.*, 1945; Vézina and Singh, 1975). Some examples of the culture media used are given in Table 9. A schematic design of the stages used to produce an inoculum is shown in Figure 2.

The influence of different inoculum levels (0.5–5%) on the batch growth of *Trichoderma reesei* showed that when adequate aeration was available, growth rates were substantially unaffected, but time to peak growth was reduced (Brown and Zainudeen, 1978). The use of very low inoculum levels in biomass fermentations requires strict control of aeration levels in the batch growth phase. As a rule, air will be supplied to the fermenter to maintain a dissolved oxygen level of from 20–25% saturation. Excess aeration often leads to poor growth, perhaps due to a too rapid removal of carbon dioxide (Berry, 1975). If a stirrer is fitted with variable speed control it is customary to stir at low speeds at the start of the batch phase, gradually increasing stirrer speed and aeration in order to maintain the dissolved oxygen at a predetermined level.

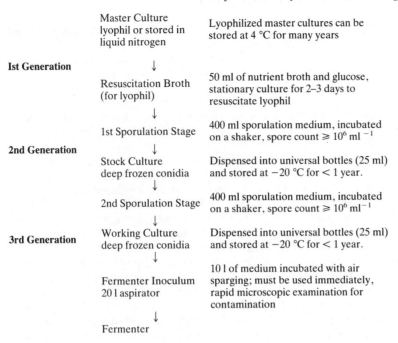

Master Culture lyophil or stored in liquid nitrogen	Lyophilized master cultures can be stored at 4 °C for many years
1st Generation ↓	
Resuscitation Broth (for lyophil)	50 ml of nutrient broth and glucose, stationary culture for 2–3 days to resuscitate lyophil
↓	
1st Sporulation Stage	400 ml sporulation medium, incubated on a shaker, spore count $\geq 10^6$ ml^{-1}
2nd Generation ↓	
Stock Culture deep frozen conidia	Dispensed into universal bottles (25 ml) and stored at -20 °C for < 1 year.
↓	
2nd Sporulation Stage	400 ml sporulation medium, incubated on a shaker, spore count $\geq 10^6$ ml^{-1}
3rd Generation ↓	
Working Culture deep frozen conidia	Dispensed into universal bottles (25 ml) and stored at -20 °C for < 1 year.
↓	
Fermenter Inoculum 20 l aspirator	10 l of medium incubated with air sparging; must be used immediately, rapid microscopic examination for contamination
↓	
Fermenter	

Figure 2 Schematic design of inoculum preparation

Table 9 Culture Medium Used for the Production of Fungal Spores in Submerged Culture[a]

Penicillium notatum[b]		Aspergillus ochraceus[c]		Gibberella zeae[d]	
Sucrose	20	Glucose	25	Carboxymethylcellulose (7MP) (Hercules Powder Co.)	15
NaNO$_3$	6	NaCl	25	NH$_4$NO$_3$	1
KH$_2$PO$_4$	1.5	Corn-steep liquor	5	KH$_2$PO$_4$	1
MgSO$_4$·7H$_2$O	0.5	Molasses (Blackstrap)	50	MgSO$_4$·7 H$_2$O	0.5
CaCl$_2$	25			Yeast extract	1

[a] All figures g l.$^{-1}$ [b] Foster *et al.* (1945).   [c] Vézina *et al.* (1965).   [d] Capellini and Peterson (1965).

22.7 FERMENTER DESIGN

Production economics dictate that the growth of biomass should be carried out in the minimum number of growing units (*i.e.* fermenters), which in turn means that for a substantial scale of operation the fermenters must be very large. For the production of 100 000 tonnes per annum of yeast, the British Petroleum–ANIC partnership constructed three 1800 m^3 vessels whilst ICI have built and operated a 3000 m^3 vessel to produce 50–60 000 tonnes per annum of bacterial protein (Solomons, 1983). However, vessels in excess of 300 m^3 have been employed for some time for antibiotic fermentation (Casida, 1964) whilst the Pekilo process has made use of 2×360 m^3 continuous fermenters (Romantschuk and Lehtomäki, 1978). The performance of these two vessels, which are of different design, provides important information as to criteria that can lead to economic production of fungal biomass. If we assess the performance of a fermenter in terms of the productivity it can support as kg m^{-3} h^{-1}, assuming sterility and the mechanical factors are adequate, then two parameters become overriding in their importance: oxygen transfer to the cells and heat removal from the culture broth. Actually, the two are very interlinked in that high oxygen transfer promotes high respiration rates which, in turn, promote increased heat production, which necessitates rapid heat removal. The problems of oxygen transfer and heat removal require solutions that are both economically and technically sound.

It may be assumed that the mass balance for the Pekilo process is similar to that of other organisms used in biomass production and can be expressed by:

$$\underset{2.0}{C_6H_{12}O_6} + \underset{0.7}{O_2} + \underset{0.1}{N.P.K.Mg.S.} \rightarrow \underset{1}{Biomass} + \underset{1.1}{CO_2} + \underset{0.7}{H_2O}$$

There is a stoichometric relationship between oxygen utilization and heat evolution (Cooney *et al.*, 1969) which approximates to 3 kcal (13 kJ) g^{-1} cell mass or 4 kcal (17 kJ) g^{-1} oxygen utilized (Forage and Righelato, 1979). For a productivity of 2.8 kg m^{-3} h^{-1} an oxygen transfer rate of approximately 2.0 kg m^{-3} h^{-1} would be necessary and this in turn would generate heat at a rate of 8400 kcal (35.2 MJ) m^{-3} h^{-1}. In addition to the heat generated metabolically, heat derived from the energy input *via* the agitator would have to be removed; at 220 kW input in 280 m^3, this would amount to 0.79 kW (2.83 MJ) m^{-3} h^{-1}, (Laine and Kuoppamäki, 1979). Total heat removed then becomes 38 MJ m^{-3} h^{-1} and although the total heat exchange surface is not stated in the literature, we can expect the heat transfer coefficient, U, to vary from 80–100 at surfaces which have sufficient turbulence to perhaps 10–20 in areas of stagnant liquid films.

The two Pekilo fermenters are described as: (i) central tube and (ii) propeller agitated. Regrettably, there are some conflicting statements as to the detailed design of the central tube fermenter, since Romantschuk and Lehtomäki (1978) state that this vessel is fitted with a bottom drive turbine, whilst Laine and Kuoppamäki (1979) make no mention of an agitator fitted to this fermenter. Using tracer experiments, Laine and Kuoppmäki (1979) showed that gas channelling was contributing to poor mixing in the vessel (Figure 3), and that liquid flow away from the central tube was poor. Moreover, there was little or no horizontal mixing, exchange being concurrent flow only. The time constants measured in the gas stream showed values for the gas–liquid phase of 9.5 s as opposed to the 20 s calculated and a reverse situation in the gas fermenter head space, with a value of 20 s compared to an expected value of 48 s. By rearrangement of the sparger and other internal configurations, a 40% increase in fermenter productivity was obtained, with increase of feed rate from 21 to 39 m^3 h^{-1}, mycelium concentration of 11 to 13 kg m^{-3} and productivity from 290 to 510 kg h^{-1}. The average long period residence time in the vessel was reduced from 9.5 to 4.7 h. The propeller agitated vessel was fitted with an unusual design of impeller, which was clearly intended to provide opposing vertical flow of the culture, but the results of the tracer experiments made it clear that it did not do so effectively. Whilst the gas-phase was satisfactorily mixed, the liquid-phase showed horizontal areas of stagnation (Figure 4). The fluid circulation rate was 500 m^3 h^{-1}, which when compared to an operating volume of around 280 m^3 was very inadequate. A new design of agitator has increased productivity by 20–30% with the same power input and with a reduction in air flow.

The problems experienced with these vessels highlight the areas of fermenter design which are most necessary in achieving high levels of productivity. Continuous culture vessels require for their maximum productivity to be homogenous or, at least, homogenous within the critical time constants for the growth of the microorganism, in particular, the availability of sufficient dissolved oxygen to prevent respiratory limitation.

Volumes of unagitated culture broth cause not merely a reduction in productivity, by not supporting growth, but can lead to the formation of undesirable anaerobic or semi-aerobic metabolites which can produce undesirable compounds which contribute to colour, smell, taste, *etc.* and, in any case, lower the yield of cells from the substrate consumed. The use of conventional Rushton turbines alone is unsatisfactory in very large vessels, due to insufficient mixing. In order to optimize mass-transfer and mass-flow, Solomons and LeGrys (1981) devised a design (Figure 5) which separates these functions with two impellers, each driven by a separate shaft and run at optimum speeds. By this means, the organism should be returned to the volume of high dissolved oxygen before limitation sets in (Anderson *et al.*, 1982). The concept of using all air agitated vessels thus avoiding the complexity and hence expense of mechanical agitation has obvious attractions and is clearly effective with true single-celled organisms (Hamer, 1979). With fungi growing in a pelleted form and intended for secondary metabolite formation, there is much to recommend this approach (Atkinson and Lewis, 1980; Greenshields and Smith, 1974). Some fungi which are particularly sensitive to mechanical shear, *e.g.* some *Mucor* spp., may only grow in a filamentous form in an air sparged system. There is also a complex relationship between shear and cell morphology, such that low shear induces long (>1 cm) unbranched mycelium, which has few growing tips and hence low growth rate. The long mycelium, even at low cell densities (<10 g dm^{-3}) tends however to high pseudoplastic viscosities. Some of the chemical engineering studies carried out (Calderbank *et al.*, 1965; LeGrys, 1978) indicate that for viscous fluids, air sparging would not

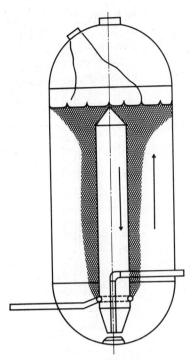

Figure 3 Air distribution in central tube fermenter (Laine and Kuoppomäki, 1979); hatched area is region of high gas velocity

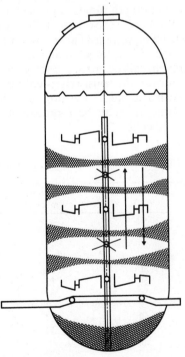

Figure 4 Liquid phase distribution in a propeller agitated fermenter (Laine and Kuoppomäki, 1979); hatched areas are regions of passive flow, low turbulence and mixing

provide an effective oxygen transfer system. An attempt to compare small mechanically stirred and air-lift fermenters was made by Barber and Worgan (1981).

22.8 RNA REDUCTION PROCESSES

In order to meet the requirements of FAO/WHO PAG (1976) to limit the ingestion of RNA from non-conventional food sources to 2 g day^{-1}, various methods of RNA reduction have been

Dual impeller
fermenter

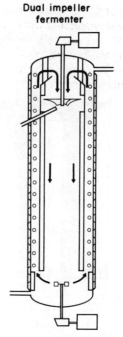

Figure 5 Design for fermenter for viscous culture broths (Solomons and LeGrys, 1981)

investigated, but little work has been applied to fungi. In their review of RNA removal, Sinskey and Tannenbaum (1975) made no mention of fungi, but working with *Paecilomyces variotii* from the Pekilo process, Viikari and Linko (1977) showed that they could, using alkali extraction, successfully lower the RNA levels of the mycelium (Table 10). Treatment of proteins with alkalies can lead to the formation of the dipeptide lysinoalanine which is undesirable in food materials, and care has to be exercised to prevent this occurring (Shetty and Kinsella, 1980). It was claimed that the use of alkali extraction improved the consistency, colour and odour of Pekilo protein biomass if the alkali was neutralized with acid before washing. Care had to be taken to ensure the pH did not fall below 6.0, as this caused RNA to be reprecipitated onto the biomass and hence increased the RNA level of recovered cells.

Table 10 Reduction of RNA Content of *Paecilomyces variotii* Using Alkali Extraction[a]

Type of alkali	Concentration (N)	Temperature (°C)	Treatment time (min)	Nucleic acid (% DW)	Protein content (% DW)
NaOH	0.075	50	30	2.8	50
NaOH	0.100	50	30	1.7	49
NaOH	0.125	20	30	2.8	50
NaOH	0.125	50	30	1.4	49
NaOH	0.125	80	30	1.8	46
NaOH	0.125	50	10	2.5	50
NaOH	0.125	50	20	2.3	49
NaOH	0.125	50	60	1.0	45
NaOH	0.125	50	120	0.9	37
NH$_4$OH	0.2	50	30	6.3	51
NH$_4$OH	0.5	50	30	4.8	51
NH$_4$OH	1.0	50	30	4.5	51
NH$_4$OH	2.0	50	30	4.2	50
NH$_4$OH	1.0	60	30	3.7	50
NH$_4$OH	1.0	70	30	3.4	50
NH$_4$OH	1.0	80	30	2.9	51
NH$_4$OH	1.0	50	60	4.2	50
NH$_4$OH	1.0	50	120	3.8	48

[a] Viikari and Linko (1977).

Endogenous enzymic hydrolysis was also examined by Viikari and Linko (1977) by separating the biomass from spent medium, washing and resuspending in water adjusted to pH 7.5 – 8.5 with ammonium hydroxide and incubating at 65 °C. After four hours, the RNA levels were reduced from 9% to less than 2%. A similar heat shock method developed by Towersey *et al.* (1977) used a single temperature treatment of 64 °C, but the cells were not separated from the culture broth. The method relies upon rapidly heating the culture to 64 °C which inactivates the fungal proteases and allows the endogenous RNases to hydrolyse the disrupted ribosomal RNA. Products, mostly 5′ nucleotides, diffuse through the cell wall into the culture broth. The time course of this reaction is shown in Figure 6, 20–30 minutes residence time in a CSTR being sufficient with *Fusarium graminearum* to reduce RNA levels to below 2%.

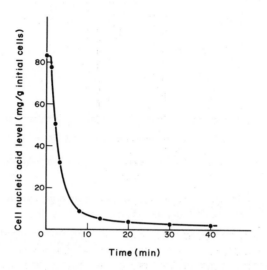

Figure 6 Time course for RNase reduction in *Fusarium graminearum* (Anderson and Solomons, 1983)

The double reciprocal plot of rate of removal *versus* the square of the RNA (Figure 7) shows that the rate of nucleic acid removal can be expressed as:

$$V = \frac{V_m C^2}{K + C^2}$$

where V is rate of nucleic acid removal mg g^{-1} ml^{-1}; V_m is maximum rate; K is a constant; and C is nucleic acid, mg g^{-1}, initial level in cells. This expression suggests that the enzyme reaction is activated by the substrate (Anderson and Solomons, 1982). The RNA reduction process leads to substantial changes in the cell composition (Table 11).

22.9 BIOMASS RECOVERY

22.9.1 Recovery of Biomass from Culture Broths

One of the major advantages possessed by the filamentous fungi over single-celled organisms is the ease with which the former can be separated from the culture medium. On a small scale, filtration using filter paper and Buchner funnels is usually adequate. For larger volumes, a low-speed, perforated bowl centrifuge gives good results. On a large scale, rotary vacuum filters are the method of choice, nylon filter cloths of suitable retentivity can normally recover 99.9% + of biomass mycelium and provision can be made for spray washing as part of the filtration; removal is by scraper blade or string discharge (Imrie and Vlitos, 1975). With a vacuum of 60–65 cm Hg, filtration rates of around 70–80 kg m^{-2} h^{-1} at 20% total solids are achievable. In order to reduce drying costs, various de-watering equipment has been tried to further reduce the 80% water content of filter cake. Continuous screw expellers of the type used in the brewing industry for de-watering spent grains have also been tried. High volumetric throughput necessitates continuous equipment rather than batch processes, which are very effective but too labour intensive.

The Pekilo process makes use of mechanical de-watering to produce a material of 35–45% total solids (Romantschuk, 1975).

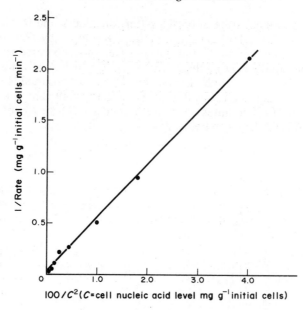

Figure 7　Reaction kinetics of RNA removal in *Fusarium graminearum* (Anderson and Solomons, 1983)

Table 11　The Composition of *Fusarium graminearum*[a]
(% DW) Before and After RNA Reduction

	RNA reduction (% DW)	
	Before	*After*
Protein (α-amino nitrogen $\times$ 6.22)	42.4	44.3
RNA	9.7	1.1
Lipid	13.0	13.8
Ash	6.6	3.1
Dietary fibre	16.3	25.1
Carbohydrate (by difference)	12.0	12.6
Nitrogen distribution		
Free amino N		0
Protein N		7.1
Nucleotide N		0
RNA N		0.1
N-acetylglucosamine N		1.5
Total N		8.7
TN Kjeldahl analysis		8.8

[a] Anderson and Solomons (1983).

22.9.2　Drying

Whilst economic considerations are very important, the drying of biomass must be carried out in a manner that is not detrimental to the nutritional value of the dried product. Fungal biomass is in fact rather easy to dry since its structure does not tend to collapse and lead to case hardening, as do single cell organisms. Using a continuous band drier with single pass warm air down-flow, it was found that an air temperature of 75 °C was optimal for drying *Penicillium* mycelium ex-vacuum filter at 20% solids; a residence time of 20–30 minutes produced a product of 8–10% moisture. Heating at too high a temperature reduced the nutritional value of the product (Table 12; Spicer, 1971), due to alteration of lysine availability. Other forms of simple drier such as rotary drum driers are also applicable.

22.10　END USE APPLICATIONS

Most of the fungal biomass processes have had animal feed as their end use, but because of the low monetary value this provides, there have been attempts to use the products for human con-

Table 12 The effect of Heating on the Available Lysine Content of a Fungal Protein[a]

Drying temperature (°C)	g available lysine/16 g N
70	6.1
80	5.0
90	2.7
100	3.2

[a] Spicer (1971).

sumption. Because of the high RNA content or the crude nature of the substrate imparting colour and odour to the product, this approach has not had much success. Attempts to use Pekilo protein in bread making were unsuccessful due to the distinctive off-flavours and adverse effect on baking characteristics (Koivurinta *et al.*, 1980).

One material that was intended for human consumption is RHM's 'Myco-protein' now sold in limited market trials in the UK. Based on food grade materials and RNA reduced, it is formulated into a variety of foods, special emphasis having been placed on meat-like products, because the filamentous nature of microfungi lends itself to this type of texture (Anon, 1981)

22.11 FUNGAL BIOMASS PROCESSES

22.11.1 Pekilo

The process for growing *Paecilomyces variotii* on stripped sulfite waste liquor was developed in Finland and has resulted in two 10 000 tonne per annum plants being erected, both in that country. A process flow diagram is shown in Figure 8 (Romantschuk and Lehtomäki, 1978). The substrate is a complex mixture (Table 13) from which the monosaccharides and acetic acid, some of the aldonic acids and oligosaccharides are utilized. Two 360 m^3 fermenters are used at pH 4–5 and at a temperature of around 38 °C. Cell concentrations of approximately 13 kg m^{-3} are normal and dilution rates of 0.14–0.20 h^{-1}, giving a biomass productivity of 2.7–2.8 kg m^{-3} h^{-1} (Laine and Kuoppamäki, 1979).

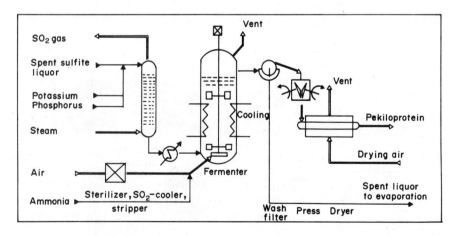

Figure 8 The Pekilo process: a flow diagram (Romantschuk and Lehtomäki, 1978)

Considerable claims are made for the economy of the process (Romantschuk, 1975) but a 10 000 tonne per annum plant is not usually considered to be of an economic scale (Humphrey, 1975) and inflation in manufacturing costs *versus* agricultural commodities has not been favourable since 1975.

Pekilo protein has been cleared by the Finnish authorities for use in animal feedstuffs and its composition is given in Tables 4 and 6.

Table 13   Composition of Spent Spruce Sulfite Waste
Liquor[a]

Compound	% DW
Lignosulfonic acids	43
Hemilignin compounds	12
Incompletely hydrolysed hemicellulose compounds and uronic acids	7
Monosaccharides	22
D-Glucose	(2.6)
D-Xylose	(4.6)
D-Mannose	(11.0)
D-Galactose	(2.6)
L-Arabinose	(0.9)
Acetic acid	6
Aldonic acids and other substances	10

[a] Romantschuk (1975).

22.11.2   RHM Myco-protein

Research into this process commenced in 1963 and from 1969 onwards has made use of *Fusarium graminearum* ATCC 20334. The distinctive feature of this programme is that it has been aimed, from its inception, at human foodstuffs so that the process utilizes all food-grade components of the culture medium and is run in either a strictly aseptic mode for fermentation and RNA reduction, or food-hygiene mode, for recovery and product storage. After a major safety evaluation programme, this product was cleared in 1980 by the UK regulatory authorities for the sale of Myco-protein to the public, the first biomass product to be so approved (Gellender, 1981). A process flow diagram is shown in Figure 9; the substrate is food grade glucose syrup which, with salts and biotin, can support an operating dilution rate of up to 0.2 h^{-1}. The operating temperature is 30 °C and the pH of 6 is controlled by addition of gaseous ammonia fed into the inlet air stream. Cell concentrations are in the range 15–20 kg m^{-3}. The fermenter output is separated from the spent air by means of a cyclone and the culture passes to a CSTR operating at 64 °C and with a residence time of 20–30 minutes. During this holding period, the RNA level falls from around 9% to <2%. Cells are then recovered by vacuum filtration. The product, 30% total solids, can then be used to fabricate a range of products, the most important of which are meat analogues. White meat and pork analogues are difficult to tell from actual meats (Anon, 1981).

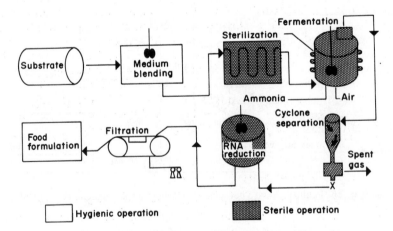

Figure 9   The RHM Myco-protein process: a flow diagram (Anderson and Solomons, 1983)

The current capacity is a pilot plant equipped with 1.3 m^3 continuous fermenters. The process is clearly more expensive to operate than an animal feed grade plant, but the products will of course have to command a realistic price in the market place. Consumer reaction to the products now under test has been very encouraging. The composition of Myco-protein is given in Tables 4, 5, 6 and 11.

22.11.3 Heurtey

This French process was developed as a method of upgrading milk whey, which is produced in a substantial amount as a by-product of cheese manufacture. Existing processes were based on a yeast strain that could utilize lactose, the sugar present in whey, but Heurtey isolated a strain of *Penicillium cyclopium*, which they claim has a better amino acid composition and of course provides much easier recovery and drying (Kosman, 1978).

A block diagram of the process is shown in Figure 10. The aseptic process operates at 28 °C and pH 3.5; the organism produces a high yield factor of 0.68 on whey, since it probably utilizes lactic acid or lactate as well as the sugar. Its growth characteristics have been reported by Kim and Lebeault (1981) and its composition is shown in Tables 4 and 6.

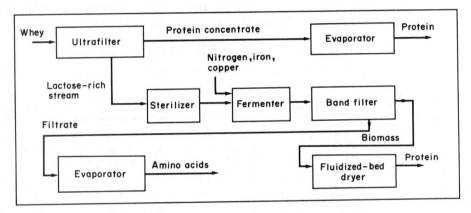

Figure 10 The Heurtey process: a flow diagram (Kosman, 1978)

22.11.4 Waterloo Process

This Canadian development uses a new cellulotytic organism *Chaetomium cellulolyticum* in conjunction with thermal or chemically pretreated cellulosic material, in a solid-substrate fermentation system. The product, which is intended for animal feed, has been produced in a 1 m³ pilot plant (Moo-Young *et al.*, 1979). The process is aimed at farm operatives and other organizations that have access to suitable substrate supplies. It is reported that a commercial pilot plant costing $1.5 million is to be built in British Columbia, using this process (Anon, 1982b). An outline of this process is shown in Figure 11 and can make use of a new tubular fermenter which has been designed for continuous semi-solid fermentation (Moo-Young *et al.*, 1979).

The process, run for example on corn stover, operates at 37 °C and pH 5.5 with a dilution rate of 0.24 h^{-1}. The composition of the product is shown in Tables 4 and 6. The minimum economic plant size, using Kraft pulp-mill sludge, is in the range 10–20 tonnes day^{-1}.

22.11.5 Miscellaneous Fungal Biomass Developments

The following are processes which have been evaluated on pilot scale and represent alternative uses and/or methodologies for utilizing filamentous fungi for human food or animal feeds.

22.11.5.1 *Protein enriched fermented feeds (PEFF)*

This novel French development makes use of *Aspergillus niger* to ferment a starchy substrate such as banana, cassava or potato in batch culture using semi-solid media at 55–75% moisture

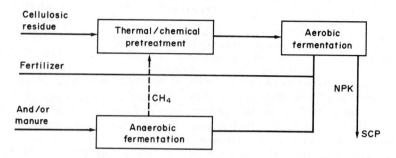

Figure 11 The Waterloo process: a flow diagram (Moo-Young *et al.*, 1979)

levels (Deschamps, 1979). The non-aseptic process operates at 38 °C and pH 4. A 1.2 m³ container can produce 150–200 kg of product, with a substantial protein content, in 30 hours (Table 14). The process is intended for rural communities at the village or farm level and experimental units are said to be operating in several tropical countries (Senez *et al.*, 1980)

Table 14 The Protein Enrichment of Starchy Substrates Using *Aspergillus niger* in the PEFF Process[a]

Substrate	Initial composition (%)		Final composition (%)	
	Protein	Sugars	Protein	Sugars
Cassava	2.5	90	18	30
Banana	6.4	80	20	25
Potato	5	90	20	35

[a] Deschamps (1979).

22.11.5.2 Use of fungi for effluent treatment

A pilot scheme for the treatment of corn and pea canning wastes was described by Church *et al.* (1972, 1973). Several designs of lagoon were examined including a 50 000 gallon open, polyethylene lined concrete tank, in which effluent was aerated by low pressure blowers and with a residence time of 18–22 hours. Both *Trichoderma viride* and *Gliocladium deliquescens* were used for inoculation, but according to the type of waste and the season, other species could predominate; the pea waste quickly becomes dominated by *Fusarium* spp. Nevertheless, cells could be grown and recovered and used in animal feedstuffs. There was also a substantial reduction of the BOD.

Although not tried on a large pilot scale, the use of fungi as fixed films (Anderson and Blain, 1980) would appear to hold promise as a means of dealing with dilute effluent streams.

A great deal of development work has been carried out at ICAITI in Guatemala into using fungi to treat coffee wastes. Aguirre *et al.* (1976) described the use of 2 × 5000 gallon non-aseptic fermenters for growing a number of fungi, but particularly *Trichoderma harzianum*. Batch culture was found to be preferable to continuous culture, since yeasts and bacteria overgrew the fungus after 100 hours.

Finally, as an example of a new approach to an old technology, semi-solid fermentations for foods are well established in the Orient (Hesseltine, 1965). By partially gelatinizing the starch in a material such as barley and ensuring that all the particles are less than 3 mm diameter, MacLennan and Lawson (1981) were able by using an amylotic strain of *Rhizopus oligosporus* to produce a product which they claim has a similar appearance to meat.

22.12 REFERENCES

Aguirre, F., O. Maldonado, C. Rolz, J. F. Menchú, R. Espinosa and S. de Cabrera (1976). Protein from waste: Growing fungi on coffee waste. *Chemtech.*, **6**, 636–642.
Anderson, C., G. A. LeGrys and G. L. Solomons (1982). Concepts in the design of large-scale fermenters for viscous culture broths. *Chem. Eng. (London)*, Feb., 43–49.

Anderson, C., J. Longton, C. Maddix, G. W. Scammell and G. L. Solomons (1975). The growth of microfungi on carbohydrates. In *Single-Cell Protein II*, ed. S.R. Tannenbaum and D. I. C. Wang, pp. 314–329. MIT Press, Cambridge, MA.

Anderson, C. and G. L. Solomons (1983). Primary metabolism and biomass production from *Fusarium*. In *The Applied Mycology of Fusarium*, pp. 231–250. Cambridge University Press, Cambridge.

Anderson, J. G. and J. A. Blain (1980). Novel developments in microbial film reactors. In *Fungal Biotechnology*, ed. J. E. Smith, D. R. Berry and B. Kristiansen, pp. 125–152. Academic, New York.

Anon (1980). Britain puts bugs into production. *New Scientist*, Dec. 18/25, p. 780.

Anon (1981). Food from a fermenter looks and tastes like meat. *Food Eng.*, May, 117–118.

Anon (1982a). Solving the problems of biotechnology scale-up. *ECN Contractors Rev.* (Achema issue), June 7, 16–18 and 23.

Anon (1982b). Canadian plant will test single cell protein process. *Feedstuffs*, **54** (9), 6.

Atkinson, B. and P. J. S. Lewis (1980). The development of immobilised fungal particles and their use in fluidised bed fermenters. In *Fungal Biotechnology*, ed. J. E. Smith, D. R. Berry and B. Kristiansen, pp. 153–173. Academic, New York.

Barber, T. W. and J. T. Worgan (1981). The application of air-lift fermenters to the cultivation of filamentous fungi. *Eur. J. Appl. Microbiol. Biotechnol.*, **13**, 77–83.

Bartholomew, W. H. and H. B. Reisman (1979). In *Microbial Technology: Fermentation Technology*, ed. H. J. Peppler and D. Perlman, vol. 2., 2nd edn., pp. 463–496. Academic, New York.

Berry, D. R. (1975). The microbial control of the physiology of filamentous fungi. In *The Filamentous Fungi*, ed. J. E. Smith and D. R. Berry, vol. 1, pp. 16–32. Edward Arnold, London.

Brown, D. E. and M. A. Zainudeen (1978). Effect of inoculum size on the aeration pattern of batch culture of a fungal micro-organism. *Biotechnol. Bioeng.*, **20**, 1045–1061.

Calderbank, P. H., M. Moo-Young and R. Bibby (1965). Coalescence in bubble reactors and absorbers. *Chem. Eng. Sci.*, **20**, 91–111.

Capellini, R. A. and J. L. Petersen (1965). Macroconidia formation in submerged culture by a non-sporulating strain of *Gibberella zeae. Mycologia*, **57**, 962–66.

Casida, L. E., Jr. (1964). In *Industry Microbiology*, pp. 28–31. Wiley, New York.

Church, B. D., E. E. Erickson and C. M. Widmer (1972). Fungal digestion of food processing wastes at a pilot level. *AIChE 72nd National Meeting*, May 21–26, pp. 1–16.

Church, B. D., E. E. Erickson and C. M. Widmer (1973). Fungal digestion of food processing wastes. *Food Technol.*, **27** (2), 36–42.

Cooney, C., D. I. C. Wang and R. I. Mateles (1969). Measurement of heat evolution and correlation with oxygen consumption during microbial growth. *Biotechnol. Bioeng.*, **11**, 269–281.

Deschamps, F. (1979). Production de proteines d'organismes unicellulaires a partir d'amidon. *Ann. Congr. Int. (Paris)*, **1**, 49–63.

Doctor, V. M. and L. Kerue (1968). *Penicillium* mycelium waste as protein supplement in animals. *Appl. Microbiol.*, **16**, 1723–1726.

Drouliscos, N. J., B. J. Macris and R. Kokke (1976). Growth of *Fusarium moniliforme* on carob aqueous extract and nutritional evaluation of its biomass. *Appl. Environ. Microbiol.*, **31**, 691–694.

Duthie, I. F. (1975). Animal feeding trials with a microfungal protein. In *Single Cell Protein II.*, ed. S. R. Tannenbaum and D. I. C. Wang, pp. 505–544. MIT Press, Cambridge, MA.

Falanghe, H., A. K. Smith and J. J. Rackis (1964). Production of fungal mycelial protein in submerged culture of soybean whey. *Appl. Microbiol.*, **12**, 330–334.

Fink, H., I. Schlie and U. Ruge (1953). Über die Zusammensetzung, die Verdanlichkeit und die Eiweiss-Qualität des Mycels von *Penicillium notatum. Hoppe-Seyler's Z. Physiol. Chem.*, **292**, 251–263.

Forage, A. J. and R. C. Righelato (1979). Biomass from carbohydrates. In *Microbiol Biomass*, ed. A. H. Rose, pp. 289–313. Academic, New York.

Forss, K. G., G. O. Gadd, R. O. Lundell and H. W. Williamson (1974). Process for the manufacture of protein-containing substances for fodder, foodstuffs and technical application. *US Pat.* 3 809 614.

Foster, J. W., L. W. McDaniel, H. B. Woodruff and J. L. Stokes (1945). Microbial aspects of penicillin. Production of conidia in submerged culture of *Penicillium notatum. J. Bacteriol.*, **50**, 365–381.

Gellender, M. (1981). Microbes upgrade starch to high-protein foods. *Chem. Int.*, No. 1 (February), 21–25.

Graham, D. C. W., K. H. Stenikraus and L. R. Hackler (1976). Factors affecting production of mold mycelium and protein in synthetic media. *Appl. Environ. Microbiol.*, **32**, 381–387.

Gray, W. D. (1970). The use of fungi as food and in food processing. *CRC Crit. Rev. Food Technol.*, **1**, 225–329.

Greenshields, R. N. and E. L. Smith (1974). The tubular reactor in fermentation. *Process Biochem.*, **9** (3), 11–13, 15, 17 and 28.

Gregory, K. F., A. E. Reade, J. Santos-Nunez, J. C. Alexander, R. E. Smith and S. J. Maclean (1977). Further thermotolerant fungi for the conversion of cassava starch to protein. *Ann. Feed Sci. Technol.*, **2**, 7–19.

Hamer, G. (1979). Biomass from natural gas. In *Microbial Biomass*, ed. A. H. Rose, vol. 4, pp. 315–360. Academic, New York.

Harrison, D. E. F. (1976). Making protein from methane. *Chem. Technol.*, **6**, 570–574.

Hesseltine, C. W. (1965). A millenium of fungi, food and fermentation. *Mycologia*, **57**, 149–197.

Hockenhull, D. J. D. (1980). Inoculum development with particular reference to *Aspergillus* and *Penicillium*. In *Fungal Biotechnology*, ed. J. E. Smith, D. R. Berry and B. Kristiansen, pp. 1–24. Academic, New York.

Humphrey, A. E. (1975). Product outlook and technical feasibility of SCP. In *Single Cell Protein II*, ed. S. R. Tannenbaum, and D. I. C. Wang, pp. 1–23. MIT Press, Cambridge, MA.

Imrie, F. K. E. and R. C. Righelato (1976). Production of microbial protein from carbohydrate wastes in developing countries. In *Food from Waste*, ed. G. G. Birch, K. J. Parker and J. T. Worgan, pp. 79–93. Applied Science Publishers, London.

Imrie, F. K. E. and A. J. Vlitos (1975). Production of fungal protein from carob (*Ceratonia siliqua* L.). In *Single Cell Protein II*, ed. S. R. Tannenbaum and D. I. C. Wang, pp. 223–243. MIT Press, Cambridge, MA.

Ivarson, K. C. and H. Morita (1982). Single-cell protein production by the acid-tolerant fungus *Scytalidium acidophilum* from acid hydrolysates of waste paper. *Appl. Environ. Microbiol.*, **43**, 643–647.

Kanazawa, M. (1975). The production of yeast from *n*-paraffins. In *Single-Cell Protein II*, ed. S. R. Tannenbaum and D. I. C. Wang, pp. 438–453. MIT Press, Cambridge, MA.

Khor, G. L., J. C. Alexander, J. Santos-Nunez, A. E. Reade and K. F. Gregory (1976). Nutritive value of thermotolerant fungi grown on cassava. *J. Inst. Can. Sci. Technol. Aliment.*, **9**, 139–143.

Kier, I., K. Allerman, F. Floto, J. Olsen and O. Sortkjaer (1976). Changes of exponential growth rates in relation to differentation of *Geotrichum candidum* in submerged culture. *Physiol. Plants*, **38**, 6–12.

Kim, J. H. and J. M. Lebeault (1981). Protein production from whey using *Penicillium cyclopum*; growth parameters and cellular composition. *Eur. J. Appl. Microbiol. Biotechnol.*, **13**, 151–154.

Kosman, W. (1978). A better way to make protein from whey? *Chem. Eng.*, March 13th, 36 C and D.

Koivurinta, J., R. Kurkela and D. Koivistoinen (1980). Bread making properties of Pekilo, a microfungus biomass from *Paecilomyces variotii*. *Nahrung*, **24**, 597–606.

Laine, J. and R. Kuoppamäki (1979). Development of the design of large-scale fermenters. *Ind. Eng. Chem. Process. Des. Dev.*, **18**, 501–506.

LeGrys, G. A. (1978). Power demand and mass transfer capability of mechanically agitated gas–liquid contactors and their relationship to air lift fermenters. *Chem. Eng. Sci.*, **33**, 83–86.

Lewis, C. W. (1976). Energy requirements for single cell protein production. *J. Appl. Chem. Biotechnol.*, **26**, 568–575.

Lhoas, P. (1961). Mitotic haploidization by treatment of *Aspergillus niger* diploids with para-fluorophenylalanine. *Nature*, *(London)*, **190**, 744.

Litchfield, J. H. (1977). Comparative technical and economical aspects of single-cell protein processes. *Adv. Appl. Microbiol.*, **22**, 267–305.

Litchfield, J. H. (1979). Production of single-cell protein for use in food or feed. In *Microbial Technology*, ed. H. J. Peppler and D. Perlman, vol. 1, 2nd edn., pp. 93–155. Academic, New York.

MacLennan, M. and M. Lawson (1981). Protein containing food material. *US Pat.* 4 265 915.

Macris, B. J. and R. Kokke (1978). Continuous fermentation to produce fungal protein. Effect of growth rate on the biomass yield and chemical composition of *Fusarium moniliforme*. *Biotechnol. Bioeng.*, **20**, 1027–1035.

Mateles, R. I. (1979). The physiology of single-cell protein (SCP) production. In *Microbial Technology: Current State, Future Prospects* (SGM Symp. 29), ed. A. T. Bull, D. C. Ellwood and C. Ratledge, pp. 29–52. Cambridge University Press, London.

Metz, B. (1976). From pulp to pellet. *Ph.D thesis*. Delft Univ. of Technology.

Meyrath, J. and K. Bayer (1979). Biomass from whey. In *Microbial Biomass*, ed. A. H. Rose, vol. 4, pp. 207–269. Academic, New York.

Mogren, H. (1979). SCP from methanol—The Norprotein process. *Process Biochem.*, **14** (1), 2–4 and 7.

Moo-Young, M. (1977). Economics of SCP production. *Process Biochem.*, **12** (4), 6–10.

Moo-Young, M., A. J. Daugulis, D. S. Chahal and D. C. Macdonald (1979). The Waterloo process for SCP production from waste biomass. *Process Biochem.*, **14** (10), 38–40.

Moo-Young, M., G. van Dedem and A. Binder (1979). Design of scraped tubular fermenters. *Biotechnol. Bioeng.*, **21**, 593–607.

PAG ad hoc (1975). Working group meeting on clinical evaluation and acceptable nucleic acid levels of SCP for human consumption. *PAG Bulletin*, **5** (3), 17–26.

Peppler, H. J. (1970). Food yeasts. In *The Yeasts*, ed. A. H. Rose and J. S. Harrison, vol. 3, pp. 421–462. Academic, New York.

Pirt, S. J. and D. S. Callow (1959). Continuous-flow culture of the filamentous mould *Penicillium chrysogenum* and the control of its morphology. *Nature (London)*, **184**, 307–310.

Quinn, J. P. and R. Marchant (1979). The growth of *Geotrichum candidum* on whiskey distillery spent wash. *Eur. J. Appl. Microbiol. Biotechnol.*, **6**, 251–261.

Reade, A. E. and K. F. Gregory (1975). High temperature production of protein-enriched feed from cassava by fungi. *Appl. Microbiol.*, **30**, 897–904.

Reusser, F., J. F. T. Spencer and H. R. Sallans (1958). *Tricholoma nudum* as a source of microbiological protein. *Appl. Microbiol.*, **6**, 5–8.

Righelato, R. C. (1975). Growth kinetics of mycelial fungi. In *The Filamentous Fungi*, ed. J. E. Smith and D. R. Berry, vol. 1, pp. 79–103. Edward Arnold, London.

Robinson, R. F. (1952). Food production by fungi. *Sci. Monthly*, **75**, 149–154.

Rolz, C. and A. E. Humphrey (1982). Microbial biomass from renewables: Review of alternatives. *Adv. Biochem. Eng.*, **21**, 1–54.

Romantschuk, H. (1975). The Pekilo process: protein from spent sulfite liquor. In *Single-Cell Protein II*, ed. S. R. Tannenbaum and D. I. C. Wang, pp. 344–356. MIT Press, Cambridge, MA.

Romantschuk, H. and M. Lehtomäki (1978). Operational experiences of first full scale Pekilo SCP-mill application. *Process Biochem.*, **13** (3), 16, 17, 29.

Sargeant, K. and C. G. T. Evans (1979). Hazards involved in the industrial use of micro-organisms. *Commission of the European Communities*, EUR 6349 EN.

Scrimshaw, N. (1972). The future outlook for feeding the human race. The PAG's recommendations Nos. 6 and 7. In *Protein from Hydrocarbons*, ed. G. de Pontanel, pp. 189–201. Academic, New York.

Senez, J. C., M. Raimboult and F. Deschamps (1980). Protein enriched fermented feeds (P.E.F.F.) *VIth International Fermentation Symposium*, London, Ontario, July 20th–25th.

Shetty, J. K. and J. E. Kinsella (1980). Lysinoalanine formation in yeast proteins isolated by alkaline methods. *J. Agric. Food. Chem.*, **28**, 798–800.

Sigler, L. and J. W. Carmichael (1974). A new acidophilic *Scytalidium*. *Can. J. Microbiol.*, **20**, 267–268.

Sinskey, A. J. and S. R. Tannenbaum (1975). Removal of nucleic acids in SCP. In *Single-Cell Protein II*, ed. S. R. Tannenbaum and D. I. C.Wang, pp. 158–178. MIT Press, Cambridge, MA.

Smith, R. H., R. Palmer and A. E. Reade (1975). A chemical and biological assessment of *Aspergillus oryzae* and other filamentous fungi as protein sources for single stomached animals. *J. Sci. Food. Agric.*, **26**, 785–795.

Solomons, G. L. (1975). Submerged culture production of mycelial biomass. In *The Filamentous Fungi*, ed. J. E. Smith and D. R. Berry, vol. 1, pp. 249–264. Edward Arnold, London.

Solomons, G. L. (1983). Single cell protein. *CRC Crit. Rev. Biotechnol.*, **1**, 21–58.

Solomons, G. L. and G. A. LeGrys (1981). Reactor system such as a fermenter system. *US Pat.* 4 256 839.

Solomons, G. L. and G. W. Scammell (1976). Production of edible protein substances. *US Pat.* 3 937 654.

Solomons, G. L. and A. Spicer (1973). Improvements in the production of edible protein substances. *Br. Pat.* 1 331 471.

Spicer, A. (1971). Synthetic proteins for human and animal consumption. *Vet. Rec.*, **89**, 482–486.

Staron, T. J. (1979). La culture des moisissures. *Ann. Congr. Int.* (*Paris*), **1**, 125–166.

Staron, T. J. (1981a). Selection process for obtaining a fungal micro-organism of the genus *Trichoderma* having advantageous characteristics. *Br. Pat.* 1 604 782.

Staron, T. J. (1981b). Production of proteins by a mycological process. *Br. Pat.* 1 604 781.

Talbot, C. J. and P. J. Senior (1980). Single cell protein and its production. *Eur. Pat.* 15 082.

Thomke, S., M. Rundgren and S. Eriksson (1980). Nutritional evaluation of the white-rot fungus *Sporotrichum pulverulentum* as a feedstuff to rats, pigs and sheep. *Biotechnol. Bioeng.*, **22**, 2285–2303.

Towersey, P. J., J. Longton and G. N. Cockram (1977). Production of edible protein containing substances. *US Pat.* 4 041 189.

Trilli, A. (1977). Prediction of costs in continuous fermentations. *J. Appl. Chem. Biotechnol.*, **27**, 251–259.

Tye, R. and A. Willets (1977). Fungal growth on C_1 compounds: quantitive aspects of growth of a methanol utilising strain of *Trichoderma lignorum* in batch culture. *Appl. Environ. Microbiol.*, **33**, 758–761.

Van Suijdam, J. C. (1980). Mycelial pellet suspension: Biotechnological aspects. *Ph.D. thesis*, Delft Univ. of Technology.

Vézina, C. and K. Singh (1975). Transformation of organic compounds by fungal spores. In *The Filamentous Fungi*, ed. J. E. Smith and D. F. Berry, vol. 1, pp. 158–192. Edward Arnold, London.

Vézina, C., K. Singh and S. N. Sehgal (1965). Sporulation of filamentous fungi in submerged culture. *Mycologia*, **57**, 722–736.

Viikari, L. and M. Linko (1977). Reduction of nucleic acid content of SCP. *Process Biochem.*, **12** (4), 17–19, 35.

Volesky, B. and J. E. Zajic (1971). Batch production of protein from ethane and ethane–methane mixtures. *Appl. Microbiol.*, **21**, 614–622.

Waslien, C. I., D. H. Calloway and S. Margen (1969). Human intolerance to bacteria as food. *Nature* (*London*), **221**, 84–85.

Waterworth, D. G. (1981). Single cell protein. *New Scientist*, Dec. 4th, 403–408.

23

Cheese Starters

R. J. HALL
University of New South Wales, Sydney, NSW, Australia
and
P. A. FRANKS
Mauri Dairy Laboratories, Sydney, NSW, Australia

23.1 INTRODUCTION

Cheese starters have been the subject of a number of recent reviews. This article will only summarize those aspects which are satisfactorily covered by these reviews and deal in slightly more detail with those more recent aspects which are not so covered. The interested reader is referred to the following reviews for more detail: general reviews (Reiter and Møller-Madsen, 1963; Lawrence *et al.*, 1976; Law, 1982a), reviews on bacterial classification taxonomy, metabolism and ecology (Sandine *et al.*, 1972; Law and Sharpe, 1978), energy metabolism, pathways and regulation (McKay, 1982), genetics (Davies and Gasson, 1981; Gasson 1983), proteolysis and provision of nitrogen for growth (Thomas and Mills, 1981; Law, 1982b; Law and Kolstad, 1983), and production of starter (Cox and Stanley, 1978).

The action of starter bacteria is central to the production of cheese. Their primary purpose is the production of the required lactic acid from lactose to produce the desired pH at all stages throughout the process. The processes of curd formation, whey expulsion, final curd fusion and texture development are all significantly influenced by the pH. Equally important, starter bacteria are responsible for discharge of any residual carbohydrate, and this, in association with the lowered moisture content, is responsible for the microbial stability of cheese. Finally, proteolysis and residual metabolism of the starter bacteria are responsible for, or at least essential to, the development of flavour during maturation.

In pre-industrial cheese manufacture the starter cultures were undefined mixed cultures selected because of largely fortuitous desirable properties. The day's production was frequently simply inoculated with whey from the previous day's make. That this system worked at all is probably due to the inhibition of undesirable species by the lactic acid and other metabolites produced by the lactic acid bacteria (Sandine *et al.*, 1972). These cultures were dominated by the group N streptococci, *Streptococcus lactis*, *Streptococcus cremoris* and *Streptococcus diacetylactis* (now referred to as *S. lactis* subsp. *diacetylactis*), but also including *Leuconostoc cremoris* in cheese types made at temperatures not exceeding about 40 °C, as well as the more thermophilic strains including *Lactobacilli* and *Streptococcus thermophilus* in cooked cheeses (Sandine *et al.*, 1972; Law and Sharpe, 1978).

As cheese making progressed from its cottage industry origin towards the modern large throughput mechanized factory which currently processes several hundred thousand litres of milk per day into cheese, substantial improvements in the reliability of these starter cultures were required.

While few cheese makers will openly admit to starter problems, it remains true that despite considerable advances in cheese starter technology, mechanization and automation of the cheese industry have raced ahead of the development of reliable quantification and control of starter performance. Even in the largest most modern cheese factories some degree of art remains and, more seriously, substantial economic losses directly attributable to starter failure or variability continue to occur.

The impact of biotechnology on the cheese industry is only comparatively recent, and as yet extends significantly only to the large throughput products, such as Cheddar cheese and some closely related types. Fine cheeses and other smaller throughput products are benefitting from some 'spin off' from this application of biotechnology, but manufacturing procedures are still essentially only streamlined versions of the original cottage industry production procedures. This article will direct attention therefore primarily to the technology of Cheddar and related cheese starters.

23.2 STARTER SYSTEMS

During the earlier part of this century the major defect in matured cheese was open texture due to gas production by bacteria that were present in the mixed strain starter cultures. Whitehead in the early 1930s isolated pure strains of non-gas producing lactic acid bacteria (strains of *S. lactis* and *S. cremoris*) from the mixed cultures and successfully used them as single strain starters to manufacture Cheddar cheese with a closed texture. The use of single strain starters was instrumental in defining the factors causing variability in starter performance and in identifying the dependence of cheese quality on strain characteristics. Today both undefined mixed cultures and defined multiples of single strains are used. These cultures may either be grown within the cheese factory in bulk starter units or purchased in concentrated form for direct addition to the cheese vat.

Traditionally, starter cultures were maintained by what was essentially a continuing subculturing in whey by the cheese maker himself. The evolution of microbiological awareness and technique has seen this activity become more specialized, both within the cheese factories and in specialist commercial suppliers. A whole support industry has now grown out of the need to produce better and more reliable starters, although many large factories continue to develop and propagate their own cultures by the conventional methods.

23.2.1 Conventional Starter Production

The current 'in-house' cheese inoculum preparation involves three sequential pH limited fermentations in sterilized milk. The first step is the inoculation of 20–30 ml milk with a freeze dried or frozen ampoule followed by overnight incubation at 22 °C. The second step is a further overnight incubation of several litres of milk and the third is the overnight (generally pH uncontrolled) incubation of several thousand litres in a bulk starter unit. In this conventional system the contents of the bulk starter unit may be used to inoculate many vats over the course of one day. The inoculum is generally 1–2% depending on the factory but metering of the inoculum is inaccurate and generally no attempt is made to compensate for any loss of activity, as indeed no activity test is made on the inoculum. The maintenance and propagation of well characterized cheese strains is an expensive and labour intensive activity and a growing number of larger companies are turning to commercial suppliers.

23.2.2 Commercial Starters

Commercial starters are available which can: (a) provide the seed for step one in the conventional process; (b) replace step two (the seed for the bulk starter unit); or (c) replace step three (the inoculum into the vat). The seed cultures for step one are generally either frozen or freeze dried ampoules of several millilitres. The bulk set (bulk starter inocula) replacements are concentrated cultures, either frozen or freeze dried. Finally, direct vat inocula of frozen concentrated starter are available which eliminate any need for culture propagation within the factory. Direct-to-the vat frozen concentrated starter is poured into the vat during filling at about 1% of the rate of conventional bulk starter inocula, on a mass basis. The frozen concentrate is activity tested by the concentrate manufacturer and added at the level desired by the cheese maker.

23.3 STRAIN SYSTEMS

As mentioned earlier the performance and flavour characteristics of the starter can determine the success or failure of the final cheese. Generally no single strain satisfies all the necessary requirements for acid production, temperature resistance, salt tolerance, reliability and flavour development. Single strains are therefore combined in a variety of systems to ensure that these criteria are met.

The major concern of the cheese maker in modern high volume cheese factories is reliability of make time and starter performance. Despite recent advances in bacteriophage (phage) control the greatest single factor affecting reliability remains phage attack and it is primarily to combat this problem that the many strain systems in current use have been devised. No sensible discussion of strain systems can therefore be made without first considering the major characteristics of bacteriophage propagation and virulence in a cheese factory.

Bacteriophages are submicroscopic viral bodies which infect bacterial cells and result in their eventual lysis and death. There are several identifiable stages in bacteriophage attack (for reviews see Keogh, 1973; Lawrence *et al.*, 1976). The bacteriophages first adsorb to the cell and then inject their DNA into the cytoplasm. Bacteriophage DNA takes over anabolism and hundreds of new phage particles are produced within the cell. Finally, a lysin is produced which ruptures the cell releasing the bacteriophage into the medium for subsequent infection of sensitive cells. The virulence of a phage on its host in a defined system (the rate at which the host bacteria are attacked) is determined by the latent period (between adsorption and DNA injection), the rise period (between injection and lysis) and the burst size. For lactic streptococcal bacteriophages the replication time (from adsorption to lysis) is characteristically of the order of 30–60 minutes (Keogh, 1973) with a burst size of about 2–105. When this growth rate is compared to the relatively long generation time of typical starters of 50–60 minutes at best (see, for example, Thompson *et al.*, 1978) it can be clearly seen that bacteriophage attack on a single strain can be rapid and catastrophic. Addition of a high titre bacteriophage inoculum to a growing culture can bring acid production to a halt in 30 minutes in milk.

So far we have dealt only with bacteriophage proliferation in a well mixed controlled environment. The actual virulence in a cheese factory is manifested as the ability to disturb acid production and interfere with cheese production. It is determined by the intrinsic properties of the individual phage and bacteria species as described above but also by extrinsic factors associated with the factory situation and with the various stresses and non-idealities within the cheese vat and subsequent stages. The rate of bacteriophage proliferation depends on the specific growth rate of the bacteria and as will be described later, this varies throughout the cheese making process.

The crucial concept underlying many of the current strain systems is that the starter can operate even in the presence of bacteriophages for its components, so long as it achieves its acid production objectives before significant bacteriophage proliferation has occurred. The continued success of a starter system is therefore based on a favourable dynamic balance being established between starter and bacteriophage growth.

To understand the nature of this dynamic balance it is necessary to consider the quantitative and qualitative sources of phage in the cheese vat. Whereas starter is added to each vat at essentially the same population level each time, there is enormous scope for variability in the phage type and inoculum concentration, depending on factory hygiene, plant layout and piping design. Each phage is capable of attacking a characteristic set of starter strains (termed its host range) but the virulence varies widely with the strain. Strains sensitive to the same bacteriophage are said to be phage related even if the virulence varies widely, because of the possibility of host mediated phage modification. One line of cell resistance to phage attack is the modification restriction (M/R) system which either destroys or modifies foreign DNA (Boussemaer *et al.*, 1980). A given bacteriophage can be modified by the host strain's M/R system to become more virulent on that host and less virulent on its original host. This can be expected to change the host range of the bacteriophage as well as its virulence and is probably important as both a qualitative and quantitative source of bacteriophage. Other qualitative sources of bacteriophage are held variously to be wild bacteriophage and organisms in silo milk and the variety of strains themselves (Thunell *et al.*, 1981). It has been proposed that almost all strains carry prophages (lysogenic or dormant phages) which are lytic at only low rates on their host but which may be highly virulent on other 'indicator' strains in the culture. The main quantitative sources of bacteriophage are whey residues in cheese vats and pipes and air borne contamination (Hull, 1978; Lawrence, 1978).

The various strain systems currently available have been designed implicitly to withstand bacteriophage related failure. Systems used range from pairs of known single strains (New Zealand, Australia, and to a lesser extent the UK), through triples of known strains (New Zealand, Australia, the USA) to multiples of four to six known strains. Larger undefined mixtures containing many different classes of organisms are still used widely in Europe and the UK.

Using phage stability as a criterion, the move to defined mixtures of phage unrelated strains is easy to understand, the theory being that a catastrophic phage attack on one strain will only destroy a fraction of the total starter. It may be tempting to extrapolate to the extreme and propose that the greater the number of strains in a culture the more stable will be the culture, but the converse has been found by experience to be true. True bacteriophage resistance, based on resistance to phage adsorption, is very rare and M/R based immunity is more common. Experience has shown that as the number of strains present in a cheese factory increases, so too does the likelihood of failure (Lawrence, 1978). Each strain introduced into the factory increases the likelihood of a prophage that can indicate strongly on the other strains in the mixture. Furthermore, there appears to be only a small number of M/R systems (Boussemaer *et al.*, 1980), so that if there are a large number of strains relying on M/R systems for phage immunity some strains will have the same M/R system and are therefore potentially phage related. If a phage appears for one of these strains and the M/R defence is breached, the other strains will also collapse.

Much of the work on characterization of phage/host relationships and the evaluation of alternative strain systems has been performed in New Zealand by Lawrence and his coworkers (see, for example, Heap and Lawrence, 1976; Thunell *et al.*, 1981). From their original multiple of six in the early 1970s New Zealand has moved towards smaller multiples and has successfully run several seasons on the same pair of strains. Similar success has been enjoyed in Australia in several large factories which have employed Hull's factory deriving technique (Hull, 1977) and now operate on pairs or triples. Commercial suppliers generally offer a range of starter cultures each containing between two and six different strains. Another system of strain management employed by cheese makers is rotation. Starter rotation is the technique of replacing one starter before a large population of phage builds up for it. The rotation concept is applied to both undefined mixed cultures and defined multiples. Its success depends on whether the decreased risk of bacteriophage build up outweighs the increased risk of modified and new virulent bacteriophages being introduced by the large number of strains.

The ultimate criterion for success of a culture system in the industry is that it survives long enough to do its job. The facts are that a variety of systems are currently employed and that all are relatively successful. The degree of success is best judged by the length of time the system functions reliably before the need for replacement. Also each system can withstand a certain degree of phage stress in the factory and those factories with very low phage stress can survive with a less robust system than their high stress counterparts. Evolution of these strain systems has substantially reduced phage related problems but phages remain a significant cause of unreliable starter performance. Current starter systems require that the primary strain characteristic, phage tolerance, should be followed by activity (rate of acid production) with flavour production relegated to third place.

Research into the development of totally phage resistant strains is being pursued throughout the world. It is likely that as these appear the complexity of strain systems will be reduced and flavour, quality and make time will become the main criteria for constructing strain systems.

23.4 METABOLISM OF THE LACTIC STREPTOCOCCI

Recognition of two independent factors has resulted in a proliferation of fundamental work in this area over the last five to seven years. The first is the commercial significance of the lactic streptococci and the second is the relative simplicity of these organisms (Mason *et al.*, 1981; Otto *et al.*, 1983). Fundamental interest in these organisms has been stimulated by this simplicity coupled with the fact that while the overall essential functionality must be similar, the regulation of the metabolism of these and similar organisms is different in at least one basic and essential aspect. This essential difference is that in lactic streptococci the only endogenous source of energy essential to the recovery from starvation is a pool of phosphoenolpyruvate (PEP) accumulated during the transition to starvation (Thompson and Thomas, 1977). Consequently the ATP pool is not regulated to maintain a constant energy charge [(ATP + 0.5 ADP/(ATP + ADP + AMP)], and the regulation of catabolism does not involve cAMP as a signal of restricted energy supply.

Much of the basic biochemical details of these organisms has been summarized in a number of reviews (Lawrence *et al.*, 1976; Law *et al.*, 1976; Law and Sharpe, 1978; Thomas and Mills, 1981; McKay, 1982). This discussion will therefore only summarize this work, while covering the more recent work in a little more detail.

The lactic streptococci are normally homolactic, converting around 95% of incoming carbohydrate to lactic acid (Thomas, 1976), and are nutritionally fastidious, requiring the exogenous supply of a substantial range of amino acids. Intermediary metabolism is consequently limited. The amounts of amino acids and directly assimilable peptides in milk are inadequate for normal growth and in such media growth is dependent on active proteases and peptidases. The metabolism of the lactic streptococci will be discussed under the three headings: energy metabolism, provision of nitrogen, and genetics.

23.4.1 Energy Metabolism

The lactic streptococci contain no cytochromes and consequently energy production depends wholly on substrate level phosphorylation, normally exclusively from carbohydrate.

23.4.1.1 *Carbohydrate uptake*

The lactic streptococci possess two mechanisms for transport of sugars (Law and Sharpe, 1978; McKay, 1982; Konings and Otto, 1983), a phosphoenolpyruvate dependent phosphotransferase system (PTS) and a permease linked to the proton motive force (PMF) generated by an ATP dependent membrane bound ATPase. PTS carriers have been characterized for glucose (Thompson, 1978), lactose (Thompson, 1978, 1979) and sucrose (Thompson and Chassy, 1981). The glucose PTS appears similar to the less specific PTS carrier for glucose characterized in *Escherichia coli* and *Salmonella typhimurium* and is referred to as the mannose PTS (Curtis and Epstein, 1975); it is constitutive (Thompson and Chassy, 1981). The lactose PTS transports both lactose and galactose and is inducible by either lactose or galactose. In this aspect the lactose PTS in the lactic streptococci is similar to that in other streptococci and Gram-positive organisms such as *Staphylococcus aureus* (Morse *et al.*, 1968), but differs from that in *Lactobacillis casei* where separate PTS carriers for lactose and galactose have been characterized (Chassy and Thompson, 1983a, 1983b).

The enzymes additional to the constitutive glycolytic enzymes necessary for the catabolism of the transported sugars are coordinately produced with the relevant inducible PTS (Davies and Gasson, 1981; McKay, 1982) with the exception of the mannofructokinase necessary to phosphorylate the fructose from sucrose to fructose 6-phosphate. The lac enzymes are coordinately produced and are coded for on a large plasmid in most strains of lactic streptococci examined

(Davies and Gasson, 1981; McKay, 1982). Enzymatic control of the PTS carriers is central to the control of catabolism (Collins and Thomas, 1974; Thomas, 1976; Thompson and Thomas, 1977; Thompson 1978, 1979; Thompson and Chassy, 1981). The identity of the biological effector is, as in most such cases, difficult to determine, but regulation of the distribution of energy supply to the PTS and to ATP production is clearly achieved by control of the PEP pool by pyruvate kinase (Thompson and Thomas, 1977; Thompson, 1978; Mason *et al.*, 1981) and has also been implicated in the control of the rate of carbohydrate uptake.

It has been demonstrated that during energy starvation, inhibition of pyruvate kinase by increased inorganic phosphate levels (Mason *et al.*, 1981) in the absence of activation provided by the early glycolytic intermediates (Collins and Thomas 1974; Thomas, 1976) results in an order of magnitude increase (to a concentration of about 40 mM) in the pool of phosphoenolpyruvate intermediates (actually the sum of 2-phosphoglycerate, 3-phosphoglycerate and phosphoenolpyruvate, hereafter simply referred to as the PEP pool). This pool represents the only endogenous energy supply available in the cell and it is reserved for transport and subsequent phosphorylation of PTS sugars during recovery from starvation. ATP levels, and consequently membrane energization by the PMF, are allowed to run down resulting in, presumably, direct limitation of energy consuming reactions (*cf.* the maintenance of energy charge in *E. coli* and the resulting requirement for an indirect limitation of energy consuming reactions, Atkinson, 1968).

It has further been demonstrated that in metabolizing cells a clear correlation exists between the PEP pool and PTS activity (Thompson, 1978). Consequently, it has been suggested that the PEP pool controls both carbohydrate uptake and disposition of energy between carbohydrate (PTS) transport at ATP production (Thompson, 1978). While the latter function seems highly likely it is the authors' opinion that simultaneous control of overall rate of carbohydrate uptake is dynamically unlikely.

This conclusion is suggested as follows. In the event of a transient excessive uptake of carbohydrate, accumulation of additional PEP (and hence further stimulation of carbohydrate uptake) can be eliminated only if the additional PEP is converted to pyruvate with the associated phosphorylation of ADP to ATP. It would seem that the pool size of ADP is too small to prevent such a system from becoming unstable. Unfortunately, direct feedback control of the PTS, the most likely alternative, has not been demonstrated, but examination of the data by Mason *et al.* (1981) does suggest a correlation with the hexose diphosphate concentration (compare rate of accumulation of lactate with fructose 1,6-diphosphate concentration in Figure 2 in Mason *et al.*, 1981).

Despite this difficulty in identification of the biological effector, strict coordination of PTS activity, glycolysis and energy demand is clearly established and while, as always, cause and effect are difficult to establish, it now appears clear that growth in the presence of an active PTS is not limited by substrate uptake, but rather PTS activity is controlled to match glycolysis or, more likely, energy demand. Kinetic study of the PTS (Thompson, 1978), growth and resting catabolic rates on different sugars and in the presence of more than one simultaneously assimilated sugar (Thompson *et al.*, 1978), 'turn up' and 'turn down' ratios by artificial variation in energy demand (Mason *et al.*, 1981), the similarity of growth rates for mutants constitutively overproducing the lac enzymes (Schifsky and McKay, 1975), and finally the significantly different internal metabolic balance observed during growth on poorly assimilable sugars (*e.g.* galactose, Thomas *et al.*, 1980; or PTS defective strains, Thompson *et al.*, 1978) all combine to attest to this fact. The only report suggesting a direct effect of a PTS on growth rate involved a report by Schifsky and McKay (1975) which suggested a reduction in lag phase for mutants constitutively overproducing the lac enzymes. In view of the multiplicity of causes of lag phase and the fact that lag phases similar to that reported do not seem to be normally encountered, it must now be concluded that except under abnormal circumstances the PTS and associated enzymes are present in excess and their activity controlled to meet demand.

Substantially less investigation of the PMF dependent permease carriers has been reported. Possession of a functional PTS appears to be a prerequisite for rapid growth and homolactic fermentation in the group N streptococci (Thompson *et al.*, 1978), since transport rates supported by permease alone are low. Consequently permease carriers are only significant for growth on galactose (for which the PTS carrier is normally less active) or PTS deficient strains (Thompson *et al.*, 1978; Thomas *et al.*, 1980). In the case of growth on galactose, growth rates supported by both the PTS and permease are similar and the relative importance of each is strain dependent. What little data is available suggests that maximum rates of PTS galactose uptake are somewhat faster than the maximum rate supported by the galactose permease but the affinity for galactose by the permease is much higher (K_m for PTS 1.0 mM compared with 0.13 mM for permease in *S. lactis* ML3, Thompson, 1980). It should be noted that since transport by a permease will be associated with the cotransport of one proton at least, and that most likely one proton is equivalent to 0.5 ATP, and further that sugars must be phosphorylated as normal, the net energy yield per hexose is reduced from 2 to 1.5 ATP.

These organisms do not contain cAMP, and competition between sugars appears to follow directly from the above described facts with the additional fact that an active PTS appears to inhibit permease operation (Thomas *et al.*, 1980). Glucose uptake is constitutive, whereas uptake of other sugars is dependent on induction (Thompson and Chassy, 1981). Sugars transported by separate induced PTS are assimilated simultaneously at rates presumably in proportion to their PTS activity, whereas substrates cotransported by the same PTS are subject to competitive inhibition (Thomas *et al.*, 1980). Thus cells suitably induced will simultaneously utilize glucose, lactose and sucrose with glucose uptake being most rapid, but will not utilize galactose until these are almost fully utilized.

23.4.1.2 Glycolytic pathways

The glycolytic pathways used are summarized in a number of reviews and elsewhere (see, for example, Lawrence *et al.*, 1976; Law and Sharpe, 1978; Kandler, 1983). Hexoses, with the exception of galactose, are catabolized by the Embden–Meyerhof pathway. Galactose transported by the PTS (distinguished because it is phosphorylated during transport) is assimilated using the D-tagatose 6-phosphate pathway and galactose transported by the permease by the Leloir pathway (Thomas *et al.*, 1980).

Control of these pathways is essentially similar. The pathways can be divided into a number of pools which can be considered to be essentially in equilibrium. The first is the hexose 6-phosphate pool including the hexose 6-phosphate intermediates down to, but not including, the hexose 1,6-diphosphate (*i.e.* either fructose 1,6-diphosphate or tagatose 1,6-diphosphate). The second is the hexose 1,6-diphosphate pool containing hexose 1,6-diphosphate, glyceraldehyde 3-phosphate and dihydroxyacetone phosphate, and finally there is the PEP pool containing 3-phosphoglycerate, 2-phosphoglycerate and PEP (Mason *et al.*, 1981).

Control is achieved by limiting PTS uptake to demand, either *via* the PEP pool or some other unidentified feed-back effector, as discussed above. Control of the PEP pool controls energy to PTS substrate uptake or ATP production. The concentration of the hexose diphosphate, fructose 1,6-diphosphate or tagatose 1,6-diphosphate, is approximately proportional to the incoming carbohydrate supply rate (Thomas *et al.*, 1979) and appears to be the primary signal of energy source availability. Under all conditions of reduced energy substrate availability examined (carbon limitations in chemostat, Thomas *et al.*, 1979; growth on galactose, Thomas *et al.*, 1980; growth with a defective PTS, Thompson *et al.*, 1978) the concentration of this intermediate is reduced, and in response to this signal the lactic streptococci induce alternative pathways for energy production, as follows: both *S. lactis* and *S. cremoris* switch from the usual homolactic to heterolactic fermentation, resulting in the production of only small amounts (as low as about 5% incoming carbon when fully induced) of lactic acid, the major products being formate, acetate and ethanol. The change results in an additional ATP per acetate produced, and when fully induced increases yield from 36 to 44 g dry wt/mol of hexose (Thomas *et al.*, 1979).

S. lactis under the same conditions induces the arginine deaminase pathway (Crow and Thomas, 1982) which results in the production of 1 ATP from carbamyl phosphate produced when arginine is converted to ornithine. *S. lactis* subsp. *diacetylactis* is also capable of fermenting citrate to produce acetate and CO_2 as well as diacetyl. It is the production of diacetyl which imparts the characteristic aroma to cottage cheese and fermented creams (Law and Sharpe, 1978).

23.4.1.3 Membrane processes and lactate efflux

It has recently been realized that the absence of endogenous energy sources, the ability of starved cells to accumulate and retain PEP after the ATP pool has been discharged, and the ability to produce ATP from arginine without production of PEP all combine to make the lactic streptococci almost ideal test organisms for the investigation of membrane processes. A recent review of membrane processes in the lactic acid bacteria is now available and this present work will only briefly summarize this area (Kandler, 1983).

The lactic streptococci contain a typical Ca^{2+}/Mg^{2+} activated membrane bound ATPase (Maloney, 1982; Maloney and Hansen, 1982; Kandler, 1983) which maintains a PMF coupled to the ATP pool and hence isolated from the PEP pool. The PMF, Δp, consists of a transmembrane potential ($\Delta\psi$, negative inside) and a transmembrane pH gradient (ΔpH, alkaline inside). Expressed in millivolts we have:

$$\Delta p = \Delta\psi - z\Delta pH$$

where $z = 2.303RT/F = 59$ mV per pH unit at 25 °C and R, T and F have the usual meanings. The total PMF measured as a function of growth rate is in the range 130 to 170 mV (Otto *et al.*, 1983; Kashket *et al.*, 1980) when produced by hydrolysis of ATP.

The PMF is coupled to a potassium ion gradient (Kashket *et al.*, 1980) and most energy requiring membrane processes other than PTS are apparently coupled to this combined gradient (Otto *et al.*, 1983; Kandler, 1983). The comparatively high energy substrate excess, non-growth associated catabolic rate typical of these organisms (see, for example, Kanasaki *et al.*, 1975) appears directly coupled to this gradient (Mason *et al.*, 1981), and does not appear to be simply a failure to control catabolism as has been considered (Forrest, 1967). The nature of a leakage flux of this magnitude is not at all clear.

Manipulation of the contribution of the potential gradient and the pH gradient at constant overall PMF by varying the contribution of the potassium ion gradient limits the change in internal pH as a consequence of changes in external pH (Kashket *et al.*, 1980; Otto *et al.*, 1983). At an external pH of 7.0 the contribution from the pH gradient is small and the internal pH is also essentially 7.0. As the pH is reduced the contribution of the pH gradient increases at the expense of the potential gradient so that at an external pH 5.7 the internal pH is 6.7 (Otto *et al.*, 1983).

Lactate efflux is carrier mediated and at essentially neutral pH and, provided the external lactate concentration is not high, is associated with a proton symport resulting in some energy recycling in association with lactate excretion (ten Brink and Konings, 1982). Significant increases in yield at neutral pH (*ca.* 25%) in batch growth were observed as a result of this process. The increase in yield decreases to zero at pH 6.0 (ten Brink and Konings, 1982). In energy limited chemostats loss of this source of energy recycling by increase in external lactate concentration resulted in a similar loss of yield (Otto *et al.*, 1980).

The importance of membrane processes has only relatively recently been recognized and one looks to this area for interesting and significant advances.

23.4.2   Supply of Nitrogen for Growth

The lactic acid bacteria are all nutritionally fastidious. Nitrogen requirements and the supply of that nitrogen from milk and other sources have been the subject of a substantial amount of work which is discussed in a number of recent reviews (Thomas and Mills, 1981; Law, 1982; Law and Kolstad, 1983). Nitrogen sources in milk include free amino acids and short peptides (referred to as non-protein nitrogen), soluble (whey) proteins and insoluble casein.

23.4.2.1   *Nitrogen requirement and uptake*

Nitrogen requirements of the group N streptococci have been studied in detail (Law *et al.*, 1976; Thomas and Mills, 1981; Law and Kolstad, 1983). Actual requirements are somewhat strain dependent. All *S. lactis* strains examined required glutamine, valine, methionine, leucine, isoleucine and histidine, and some strains also required phenylalanine and arginine. *S. cremoris* strains were found to have a wider range of requirements. All strains required in addition to the requirements found for *S. lactis*, proline and phenylalanine, and some strains tested required serine, glycine, alanine, tyrosine, lysine and tryptophan. *S. thermophilus* strains appear to have amino acid requirements similar to those of *S. lactis* except that methionine is not required, while cystine and tryptophan are usually essential.

While all these amino acids are available in milk in a directly assimilable form either as free amino acids or low molecular weight peptides (*ca.* 5 residues or less), they are not present in sufficient amounts to support high cell densities (cell densities of 8–16% of that found in coagulated milk, Thomas and Mills, 1981). Examination of uptake indicates that these available sources of amino acids are utilized preferentially initially but subsequent growth depends increasingly on other sources (Thomas and Mills, 1981). The addition of amino acids to milk has little effect on the growth rate of the lactic streptococci, while the addition of low molecular weight peptides results in a small increase in growth rate (Reiter and Møller-Madsen, 1963).

Uptake of both amino acids and peptides are normally energy dependent but some passive transport of amino acids has been reported (Rice *et al.*, 1978). Transport is carrier mediated (permease) and is linked to the PMF in most cases but is in some cases driven directly by ATP (Konings and Otto, 1983).

Separate carriers appear to exist for small groups of structurally related amino acids (Rice *et al.*, 1978). Consequently, competitive inhibition of amino acid uptake by structurally related amino acids occurs (Reiter and Møller-Madsen, 1963). Separate carriers for dipeptides, tripeptides and oligopeptides exist (Rice *et al.*, 1978; Law and Kolstad, 1983), although not all of these necessarily exist in all strains (Law, 1978). Early reports suggest that competitive inhibition of peptide uptake did not occur (Reiter and Møller-Madsen, 1963) but more recent work has demonstrated the existence of competitive inhibition between peptides transported by the same carrier (Law, 1978; Rice *et al.*, 1978). Presumably the early failure to detect this inhibition was caused by the then unrecognized existence of a number of different carriers specific for peptides containing different numbers of residues and further, in *S. lactis* strains, by the existence of membrane associated peptidases allowing transport of residues supplied in peptides as individual amino acids (Law and Kolstad, 1983).

23.4.2.2 *Proteolytic enzymes of the lactic acid bacteria*

Despite their dependence on proteolysis, the lactic acid bacteria are only weakly proteolytic compared with other groups of bacteria such as the *Bacillus*, *Proteus*, *Pseudomonas* and coliforms. Nevertheless, detailed study of the proteolytic enzymes of the lactic acid bacteria over the past decades has revealed a complex array of enzymes involved in proteolysis in these bacteria. This complexity relates both to the number and location of these enzymes (Law and Kolstad, 1983), and is likely to be a major source of variation between otherwise closely related strains.

The hydrolysis and uptake of proteins involves the following three identifiable steps: (a) proteolysis of native proteins to peptides; (b) further hydrolysis by peptidases eventually to individual amino acids; and (c) transport into the cell. As is clear from the above discussion, amino acids can be transported as short chain peptides requiring intracellular peptidases or as individual amino acids after hydrolysis by extracellular or membrane bound peptidases. The classification of enzymes into proteinases or peptidases can lead to confusion in some cases. Law and Kolstad (1983) have suggested that distinction based on the above functionality results if enzymes detected only by their action on native proteins are referred to as proteinases and enzymes detected with peptides or peptide derivatives, irrespective of their activity on native proteins, as peptidases.

Proteinases in the lactic acid bacteria occur both intracellularly, on the exterior of the cell wall, and extracellularly (Law and Kolstad, 1983). Intracellular proteinases do not appear to be related to the provision of nitrogen for growth but provide for protein turnover and the hydrolysis of denatured or defective proteins apparently essential to the normal growth of probably all organisms (Law and Kolstad, 1983). Little work has been devoted to the characterization of these proteinases, but it seems highly likely that they are both functionally and genetically unrelated to the proteinases involved in the supply of nitrogen. These intracellular proteinases are involved in the proteolysis essential to maturation of cheese (for example Law and Kolstad, 1983).

Extracellular proteinases in the group N lactic streptococci now appear to be an artifact of lysis or destabilization and it would seem certain now that these enzymes are naturally located on the exterior of the cell wall (Thomas and Mills, 1981; Law and Kolstad, 1983). At least three separate wall bound proteinases have been characterized in the group N streptococci based on pH and temperature optima (Exterkate, 1976). The distribution of these proteinases is strain dependent, with ML1 apparently containing none of these (although not being proteinase negative), and other strains containing one, two or all three types.

All the group N streptococci readily lose the ability to produce proteinases (Prt$^-$) and it seems essentially certain that the genetic determinants for these enzymes are on one or more of the large, low copy number plasmids present in the lactic streptococci (McKay, 1982; Davies and Gasson, 1981). Prt$^-$ isolates grow in milk at the normal growth rate when supplied with a source of hydrolysed protein and in unenriched milk grow at the normal growth rate to 10 to 25% of the cell density of Prt$^+$ organisms found in coagulated milk (see, for example, Thomas and Mills, 1981). The products of the cell wall bound proteinases in Prt$^+$ lactic streptococci are released into the medium and consequently natural populations of these species normally contain substantial proportions of Prt⁻ individuals. Overall growth rates of mixtures of Prt⁻ and Prt⁺ isolates of *S. cremoris* NZ346 were not significantly less than the parent strain until the Prt⁻ population reached 80% of the population (Hall, unpublished data).

Peptidase activity has been located in the cytoplasmic membrane in some species. Law and Kolstad (1983) report that membrane bound peptidases would seem to be present in *S. lactis* but not *S. cremoris*, however Exterkate (1981) reports evidence of membrane associated peptidases in *S. cremoris*. Whatever the location, a complex range of peptidases adequate for the final hydrolysis of proteins to amino acids clearly exists and is essential to the growth of these organisms (Law, 1978).

23.4.3   Genetics of the Lactic Acid Bacteria

The genetics of the lactic streptococci were largely ignored until comparatively recently, but a number of developments have stimulated more recent interest. It has been recognized, for example, that enzymes essential to growth in milk (specifically the lac enzymes and the proteinases) are unstable and that instability appears to be related to the unusually large amount of extrachromosomal plasmid DNA present in these organisms (McKay, 1982; Davies and Gasson, 1981; Gasson, 1983).

The group N streptococci would appear to be genetically consistent with the hypothesis that, with the assistance of genes provided by plasmid DNA, they have only comparatively recently invaded milk as a growth environment and are still undergoing relatively rapid genetic reorganization of the necessary additional enzymes. These organisms are therefore regarded as highly variable and their cultures potentially inhomogeneous even after normal strain purification procedures. This reputation is clearly enhanced by the use of phage sensitivities as the definitive determinant for strain identification. It is possible that phage/host relationships are hypervariable and use of this property as a strain determinant may result in the proliferation of strains differing only in their phage sensitivities, as the use of sera type did for *Salmonella typhimurium*. While distinctions based on phage/host relationships are clearly of commercial importance, unless correlated with other genetic determinants, strain distinction based on this property may well be less than ideal and potentially confusing. As suggested below, plasmid profiles which are now readily obtained may provide a more generally useful alternative determinant.

Most lactic streptococci contain an unusually large complement of plasmid DNA (Davies and Gasson, 1981). Plasmids of between 1 and 40 MD or greater can commonly be found and up to 14 plasmid bands may be resolved for a single strain. Plasmid profiles have been shown to be strain specific and while individual plasmids may be lost (particularly the larger, low copy number plasmids) the overall profiles remain sufficiently unaltered to be clearly recognizable (Davies and Gasson, 1981).

Although the lactic streptococci generally have large numbers of plasmids, most appear cryptic and only a few have been shown to encode for known functions. The two most important such functions are lactose metabolism (specifically the two inducible components of the lac PTS, phospho-β-galactosidase and the enzymes of the tagatose 6-phosphate pathway, Crow *et al.*, 1983) and the cell wall bound proteinases. Other functions associated with plasmids include citrate fermentation (specifically citrate permease in *S. lactis* subsp. *diacetylactis*) inorganic salt resistance, and nisin production (Davies and Gasson, 1981).

The relationship between specific plasmids and phenotypic trait has been generally implied from the association between loss of the function and the plasmid loss. Conclusions drawn on this basis have resulted in confusing and often contradictory evidence for the location of the relevant gene (McKay, 1982; Davies and Gasson, 1981; Gasson, 1983). The frequency of loss of the larger plasmids is such that unrelated loss of plasmid and function is possible. Further, some reports in the literature (see for example McKay, 1982) and some initial work carried out in our laboratories are difficult to explain unless rearrangement of DNA between various replicons is assumed to occur. In any event it has been accepted for some time that more specific evidence must be sought in order to confirm that a specific plasmid carries genes for a particular property.

Unequivocal evidence for association of lactose metabolism (McKay, 1982) and wall bound proteinases (Gasson, 1983) with specific plasmids in some strains has now been provided. Techniques developed for directed genetic modification of the group N lactic streptococci are summarized in Davies and Gasson (1981) and Gasson (1983).

23.5 CHEESE STARTERS and FLAVOUR DEVELOPMENT

Besides its obvious roles of acid production and lactose removal, cheese starter affects cheese flavour either directly or indirectly by the nature and quantity of its end products. The effect may be either beneficial, resulting in 'cheesy' flavours and 'mature' taste (Grieve and Dulley, 1983) or detrimental, resulting in off flavours such as bitterness (Lowrie and Lawrence, 1972) or fruitiness (Bills *et al.*, 1965).

The field of flavour research is complex and characterized by a lack of reproducibility. Furthermore the agents responsible for flavour are present in minute concentrations and their modes of action are still largely unknown despite some 30 years of ongoing research in cheese flavour. Nevertheless, some heavily qualified progress has been made, and the reader is referred to a recent review on Cheddar cheese flavour for further detail (Aston and Dulley, 1982). This section will concern itself only with a summary of perceived contribution of the starter bacteria to cheese flavour.

The primary causative agent in flavour development produced by the starter appears to be the enzymes released following lysis, which affect the degradation of curd components during maturation. The major classes of degradation products deemed to be important in flavour development are (a) free amino acids, (b) carbonyls, (c) fatty acids, and (d) sulfur compounds (Lawrence *et al.*, 1976; Aston and Dulley, 1982). The free amino acids appear to be responsible for the 'cheesy' flavour but do not contribute to odour (Harper and Swanson, 1949), and the intensity of this flavour increases with amino acid concentration. Recent studies have also shown that flavour (maturity) correlated well with phosphotungstic acid (PTA) soluble amino nitrogen levels (Jarrett *et al.*, 1982). The level of intracellular proteinases and peptidases in the cheese determine the rate of release of amino acids and therefore the intensity of the taste at a given time (Grieve and Dulley, 1983; Aston *et al.*, 1983).

Carbonyl compounds such as acetaldehyde, propionaldehyde, butyraldehyde and acetone have been detected in Cheddar cheese but their role in flavour is uncertain. They have been implicated in the formation of esters (Aston and Dulley, 1982) but their true role remains unresolved. They are produced as end products (*e.g.* diacetyl and acetone) by heterofermentative strains such as *Streptococcus diacetylactis* (Lawrence *et al.*, 1976).

Fatty acids can arise from either lipolysis of milk fat (the principal source of C_4 and greater free fatty acids, Aston and Dulley, 1983) or from the metabolism of carbohydrates and amino acids by bacteria. The starter bacteria may also be the source of lipases which can catalyse milk fat hydrolysis during maturation. It has been proposed that amino acid metabolism by lactic streptococci is the principle source of volatile fatty acids in cheese (Lawrence *et al.*, 1976), and that the formation of esters from these C_2 and C_4 acids and ethanol may explain the relatively high incidence of fruitiness in Cheddar cheese made with *S. diacetylactis* strains. If the fatty acid concentration is too high a 'rancid' flavour develops.

Sulfur compounds are normally found only at very low levels in cheese but recent technological developments have allowed detection and quantification of their presence. The main compounds studied have been hydrogen sulfide, methional, methanethiol and dimethyl sulfide. No unequivocal evidence for the role of H_2S or methional has yet been discovered, and it is unclear whether they are essential to Cheddar flavour or simply contribute to them. Methanethiol has been correlated highly with flavour intensity and has been shown not to develop in the absence of starter (Manning, 1974; Manning *et al.*, 1976).

Law and Sharpe (1977) concluded that neither the starter bacteria nor their enzymes are directly responsible for typical Cheddar flavour and that they are only present to produce the correct conditions for ripening reactions to occur and for the product to be stable *viz.*: (1) supply of flavour precursors; (2) low pH to suppress spoilage organisms; and (3) low redox potential. This last factor is important to maintain sulfur compounds such as methanethiol in their reduced form.

Insofar as the rates of many flavour development reactions depend on the concentration of proteases and other enzymes released by cell lysis, the starter population clearly affects the rate of flavour development during maturation. Lawrence and coworkers observed (1976) that growth of bacteriophages during cheese making to levels which restrict starter numbers without markedly increasing the time for sufficient acid to be produced has a striking if indirect effect on cheese flavour. They proposed that many anomalous results in factory trials of flavour production could be explained by this theory. In particular they claimed that comparisons between open and closed vats may be invalid if significant air borne bacteriophage infection of the open vats occurs during production.

It has been found that cheese made with some *S. cremoris* strains is always of good strong flavour whilst other strains always give little Cheddar flavour under normal cheese making conditions. The manufacturing conditions themselves can, however, significantly affect the cheese flavour and more specifically control defects such as bitterness.

The bitter defect in cheese is generally attributed to the presence of low molecular weight bitter peptides. These peptides are the result of proteolysis of curd proteins by cell lysates. The mechanism by which these bitter peptides remain undegraded in bitter cheeses remains a point of some contention but it appears to involve a balance of proteinase and peptidase activities in the cheese and the presence of rennet enzymes.

Grieve and Dulley (1983) reported that bitterness produced in control cheese using strain E8 was absent in test cheese containing five fold extra lac⁻, Prt⁻ cells. The authors attributed this to the probable surfeit of intracellular peptidases (and the absence of membrane bound proteinases) in the mutant cells. Lowrie and Lawrence (1972) proposed that the bitter defect arose simply because of high populations of starter in the curd following cooking and they claimed that reducing the cooking temperature enabled normally non-bitter starters to produce bitter cheese by reaching higher populations in the curd. A more complicated explanation was proposed by Visser and coworkers (1977). They suggest that some strains produce more bitter peptide degrading peptidases and hence do not produce bitterness at high cell numbers. It is not known whether the peptidases are different or simply present in greater numbers.

Whichever theory proves to be true the process manipulations during the initial stages of cheese production (*e.g.* ripening time, cooking temperature, make time) clearly have been shown to affect subsequent flavour development during maturation. As mentioned earlier, subjective assessment of flavour and maturity is intrinsically very variable and recent studies have gone to considerable trouble to control and quantify flavour assessments (Aston *et al.*, 1983). Such disciplines are essential for meaningful work in flavour assessment and will undoubtedly be adopted by workers in the future.

23.6 STARTER GROWTH AND ACID PRODUCTION DURING CHEESE MANUFACTURE

The translation of traditional cheese production into large scale industrial manufacture has placed much greater demands on the regularity and control of starter performance. Traditional control of starter performance evolved from the recognition that over-fast acid production in the vat produced bleached, over acid cheese with texture and flavour defects, but that quality cheese could be made by relying on slower acid production (within limits) provided the curd was held in the vat long enough to allow sufficient acid development. Consequently 'time in the whey' was a primary control variable in traditional cheese manufacture. The realities of large scale industrial manufacture, and the fact that a multiple batch vat stage is used to feed continuous curd handling (cheddaring) equipment, demand control of the vat stage to achieve fixed vat times. The cheese fermentation is a complex interactive process involving cell growth and acid production in a two phase (curd and whey) system, with expulsion of whey from the curd phase. The quantitative understanding of the interactions involved, required to develop a control philosophy to replace the traditional procedures, demands a kinetic analysis of these processes and their interactions. Such work faces considerable measurement problems in the quantification of cell growth and acid production in both milk and curd and in the quantification of the process of syneresis. While at the time of writing some progress towards this end has been made (see, for example, Franks *et al.*, 1980) these attempts represent only a beginning. It is anticipated that the demands of large scale industrial production and the stimulation provided by the fundamental advances described above will result in substantial progress in this area over the next few years. It is therefore desirable at this stage to discuss in general terms the nature of the interactions involved in the quantification and manipulation of growth and acid production rather than to discuss the specifics of this embryonic work.

Growth and acid production during manufacture involve three clear stages: starter inoculum production, growth and acid production during the vat stage, and acidification of the curd during cheddaring (or its equivalent in other products).

23.6.1 Starter Production

Much of the starter produced today is simply grown in a pH uncontrolled batch fermentation in skim milk or skim milk based media. In such a fermentation culture growth is limited by pH inhibition. Growth ceases before acid production is arrested and the final acidification is achieved by uncoupled or non-growth associated acid production and is, of course, not accompanied by cell mass increase. Strain to strain or day to day variations in final cell density may not therefore be detected by measurement of the final pH alone.

Improvements in starter production have involved attempts to improve productivity by neutralization of the acid produced (see, for example, Cogan *et al.*, 1971) combined with medium enrichment (see, for example, Peebles *et al.*, 1969; also reviewed in Cox and Stanley, 1978). Assuming adequate nutrient enrichment, growth in neutralized starter medium must eventually be limited by an end product, *e.g.* lactate. Increases in cell counts of the order of ten fold are possible by neutralization (Cox and Stanley, 1978), however without substantial manipulation of the medium, and relying on sodium hydroxide for neutralization, activity tends to be little more than doubled, with larger increases being achieved using ammonia for neutralization.

Growth rate during starter production (as opposed to cheese production) is of only secondary importance, while sensitivity to the stresses of growth limitation assume greater importance (pH, end product, starvation sensitivities). Investigation of the importance of these sensitivities will necessitate a kinetic analysis of the growth response to transient stresses. The old cell counts and empirical activity tests are inadequate for such kinetic analysis because they provide a very poor description of the complex cellular processes described in earlier sections of this chapter. Any sensible description will probably involve measurement of cell mass and recognition of subsections of the cell metabolism, that is a structured growth model (see, for example, Franks *et al.*, 1980). Reliable independent estimation of cell mass in milk has proved an elusive goal but effort in this area must continue if specific activity is to be quantified.

Empirical activity tests have been developed (see, for example, Cox and Lewis, 1972) but the results of such tests correlate only moderately well with performance (Reiter and Møller-Madsen, 1963). Furthermore they take too long to be useful as a means of production monitoring and control.

Centralized production of pretested and stable starter (for example frozen concentrated starter) has obvious technical advantages, and this has undoubtedly contributed to the success of the product. Full kinetic characterization of each strain in the starter system can be undertaken by the supplier of frozen concentrated starter and each batch assayed for activity. The starter is then supplied in predefined units and the cheesemaker can with no further technology add easily metered amounts to the cheese vat.

23.6.2 Vat Stage

Cell densities required for cheese production are substantially less than are achievable in milk. Acid production in the vat stage is determined by a combination of inoculum size and temperature stress. Response to temperature stress is far from simple and involves significant transients (Franks, 1981; Franks *et al.*, 1980). Recognition of the transient nature of growth response to milk temperature stress is not yet general and despite the fact that this was first reported some time ago (Shaw, 1967) steady state analyses of the response to temperature stress are still reported (see, for example, Ratkowsky *et al.*, 1983).

Growth rate is more rapidly affected by mild temperature stress than is acid production (Franks, 1981; Franks *et al.*, 1980). In fact over the time span occurring in the vat, temperature stress imposed does not result in significant decay in specific acid production rate (that is acid production rate per unit cell mass). Consequently acid production in the vat is controlled indirectly by limiting the cell mass. This may be achieved by a combination of coarse control *via* inoculum size and fine tuning *via* variation of the vat (cooking) temperature.

23.6.3 Cheddaring

After whey off, starter bacteria are required to reduce the pH finally to *ca.* pH 5.2 (depending on the cheese type), and simultaneously discharge all lactose. Growth during cheddaring is particularly difficult to measure and the analytical problems in quantification of pH and lactate production are by no means simple (see, for example, Czulak *et al.*, 1969). Growth in cheddaring curd presumably stops as moisture levels reduce, but potential for conversion of lactose to lactic acid continues for some time.

Control of acidification during cheddaring is complicated by the fact that there is little that can be done to manipulate conditions in the mass of cheddaring curd. For all practical purposes then, control of acid production during cheddaring must be achieved by manipulation of initial conditions (primarily cell mass) in the curd at whey off.

The causes of variability in starter performance have been canvassed in a number of reviews (see, for example, Reiter and Møller-Madsen, 1963; Lawrence *et al.*, 1976). Major causes are variability in starter activity, milk composition, phage background, inhibition naturally occurring in milk and inhibition accidentally included in milk (antibiotics, *etc.*). Starter activity is presumably controllable, but the other factors have to be compensated for. The commercial advantage of a starter addition test using today's starter in today's milk, and capable of yielding a meaningful culture activity in time to allow calculation of the inoculum desired for the first vat, is obvious.

The significance of the amount of acid production occurring in the vat can be appreciated from the following. Firstly pH development in the vat must be sufficient to stimulate syneresis, and the curd population at whey off must be such that again adequate pH fall occurs during cheddaring to stimulate the textural changes required in that stage of cheese making. On the other hand overproduction of acid during the vat stage results in the loss of additional calcium and phosphate. This occurs because the solubility of the colloidal calcium phosphate present in the curd increases as the pH falls. This colloidal calcium phosphate dissolves, largely into the whey remaining in the curd phase only, and does so to a concentration set by the pH. At whey off, a rush of moisture from the curd occurs carrying with it any dissolved calcium phosphate. Therefore if the pH in the vat at whey off is too low, excessive calcium phosphate is removed with this rush of whey. Calcium phosphate has a direct effect on curd texture and makes a significant contribution to the curd buffer capacity. Low calcium phosphate therefore results in low pH (over acid flavour) and poor texture cheese. This simplified description explains in principle at least part of the basis for traditional control of acid production.

In conclusion, it can be seen that control of acid production is indirect and to a substantial extent feed forward in nature, that is control of the vat by manipulation of the inoculum and control of acidification during cheddaring by control of the composition of the curd at the end of the vat. Translation of such a process into a technology based industrial process will require a detailed quantitative understanding of the essential interactions involved. Further optimization of the process, which will depend on an understanding of the reasons for the traditional 'set points' of pH, moisture and temperature, *etc.*, will require a deeper understanding of the physicochemical processes involved in cheese production. A discussion of this complex area is beyond the scope of this work.

While this section has dealt in detail with Cheddar cheese production the principles relevant to the biotechnologist are common to all cheeses. The growth kinetics of the starter cannot be considered in isolation and must be superimposed on the kinetics of syneresis and the chemical buffering and maturation reactions. Finally the physical and chemical goals in each stage of cheese making have to be identified before these kinetics can all be combined to determine ideal control procedures.

Cheese starter research is continuing at an accelerating pace. Increased effort has been directed recently into genetic characterization of starter strains with a view to genetic engineering for desired properties. This and the many other fields of research mentioned here are being pursued in rather disappointing isolation from each other. Despite the many studies characterizing the biochemical nature of starter growth, there is a dearth of information on the nature and importance of the various cheese making set points, and yet these are clearly essential to the definition of 'desirable' starter properties, for which the geneticist will strive. An holistic approach to cheese starter performance must be pursued in future, in which the cell metabolism, growth kinetics, environmental stresses, and mass transfer (syneresis, diffusion) are drawn together.

23.7 REFERENCES

Aston, J. W. and J. R. Dulley (1982). Cheddar cheese flavour. *Aust. J. Dairy Technol.*, **37** (2), 59–66.
Aston, J. W., P. A. Grieve, I. G. Durwood and J. R. Dulley (1983). Proteolysis and flavour development in Cheddar cheese subjected to accelerated ripening treatments. *Aust. J. Dairy Technol.*, **38**, June, 59–65.
Atkinson, D. E. (1968). The energy charge of the adenylate pool as a regulatory parameter. Interaction with feedback modifiers. *Biochemistry*, **7**, 430.
Bills, D. D., M. E. Morgan, L. M. Libbey and E. A. Day (1965). Identification of components responsible for the fruity flavour defect of some experimental cheeses. *J. Dairy Sci.*, **48**, 765.

Boussemaer, J. P., P.P. Schrauwen, J. L. Sourrouille and P. Guy (1980). Multiple modification restriction systems in lactic streptococci and their significance in defining a phage typing system. *J. Dairy Res.*, **47**, 401.

Chassy, B. M. and J. Thompson (1983a). Regulation of lactose–phosphoenolpyruvate-dependent phosphotransferase system and β-D-phosphogalactoside galactohydrolase activities in *Lactobacillus casei. J. Bacteriol.*, **154**, 1195–1203.

Chassy, B. M. and J. Thompson (1983b). Regulation and characterization of the galactose–phosphoenolpyruvate-dependent phosphotransferase system in *Lactobacillus casei. J. Bacteriol.*, **154**, 1204–1214.

Cogan, T.M., D. J. Buckley and S. Condon (1971). Optimum growth parameters of lactic streptococci used for the production of concentrated starter cultures. *J. Appl. Bacteriol.*, **34**, 403–409.

Collins, L. B. and T. D. Thomas (1974). Pyruvate kinase of *Streptococcus lactis. J. Bacteriol.* **120**, 52–58.

Cox, W. A. and J. E. Lewis (1972). Methods of handling and testing starter cultures. In *Safety in Microbiology*, ed. D. A. Shapton and R. G. Board, The Society for Applied Bacteriology Technical Services N. G. Academic, New York.

Cox, W. A. and G. Stanley (1978). Starters: purpose, production and problems. In *Streptococci, The Society of Applied Bacteriology Symposium Series 7*, ed. F. A. Skinner and L. B. Quesnel, pp. 279–296. Academic, London.

Crow, V. L. and T. D. Thomas (1982). Arginine metabolism in lactic streptococci. *J. Bacteriol.*, **150**, 1024–1032.

Crow, V. L., G. P. Davey, L. E. Pearce and T. D. Thomas (1983). Plasmid linkage of the D-tagatose 6-phosphate pathway in *Streptococcus lactis*: Effect on lactose and galactose metabolism. *J. Bacteriol.*, **153**, 76–83.

Curtis, S. J. and W. Epstein (1975). Phosphorylation of D-glucose in *Escherichia coli* mutants defective in glucosephosphotransferase, mannosephosphotransferase and glucokinase. *J. Bacteriol.*, **122**, 1189–1199.

Czulak, J., J. Conochie, B. J. Sutherland and H. J. M. van Leeuwen (1969). Lactose, lactic acid and mineral equilibria in Cheddar cheese manufacture. *J. Dairy Res.*, **36**, 93–101.

Davies, F. L. and M. J. Gasson (1981). Genetics of lactic acid bacteria. *J. Dairy Res.*, **48**, 363–376.

Exterkate, F. A. (1976). Comparison of strains of *Streptococcus cremoris* for proteolytic activities associated with the cell wall. *Neth. Milk Dairy J.*, **30**, 95–105.

Exterkate, F. A. (1981). Membrane-bound peptidases *Streptococcus cremoris. Neth. Milk Dairy J.*, **35**, 328–332.

Forrest, W. W. (1967). Energies of activation and uncoupled growth of *Streptococcus faecalis* and *Zymomonas mobilis. J. Bacteriol*, **94**, 1459–1463.

Franks, P. A. (1981). *The Effect of Temperature on the Growth and Acid Production of Cheese Starters.* Thesis, University of NSW, Australia.

Franks, P. A., R. J. Hall and P. M. Linklater (1980). Mechanistic model of the growth of *Streptococcus cremoris* HP at superoptimal temperatures. *Biotechnol. Bioeng.*, **22**, 1465–1487.

Franks, P. A., R. J. Hall and P. M. Linklater (1981). The development of control techniques in Cheddar cheese production. In *Advances in Biotechnology*, ed. M. Moo-Young and C. W. Robinson, vol. II, p. 479. Pergamon, Oxford.

Gasson, M. J. (1983). Genetic transfer systems in lactic acid bacteria. *Antonie van Leeuwenhoek*, **49**, 275–282.

Grieve, P. A. and J. R. Dulley (1983). Use of *Streptococcus lactis* lac⁻ mutants for accelerating Cheddar cheese ripening. 2. Their effect on the rate of proteolysis and flavour development. *Aust. J. Dairy Technol.*, **38**, June, 49–54.

Harper, W. J. and A. M. Swanson (1949). Determination of the amino acids in Cheddar cheese and their relation to the development of flavour. *XII Int. Dairy Cong.*, **2**, 147.

Heap, H. A. and R. C. Lawrence (1976). The selection of starter strains for cheesemaking. *N. Z. J. Dairy Sci. Technol.*, **11**, 16.

Hull, R. R. (1977). Control of bacteriophage in cheese factories. *Aust. J. Dairy Technol.* **32**, June, 65.

Hull, R. R. (1978) (ed.). Factory derived cheese starters. *Proceedings of a Meeting organized by the Dairy Research Laboratory, Division of Food Research CSIRO and the Australian Dairy Corporation, Highett*, CSIRO, Highett, Victoria.

Jarrett, W. D., J. W. Aston and J. R. Dulley (1982). A simple method for estimating free amino acids in Cheddar cheese. *Aust. J. Dairy Technol.*, **37**, 55–58.

Kanasaki, M., S. Breheny, A. J. Hillier and G. R. Jago (1975). Effect of temperature on the growth and acid production of lactic acid bacteria. *Aust. J. Dairy Technol.*, **30**, 142–152.

Kandler, O. (1983). Carbohydrate metabolism in lactic bacteria. *Antonie van Leeuwenhoek*, **49**, 209–224.

Kashket, E. R., A. G. Blanchard and W. C. Metzger (1980). Proton motive force during growth of *Streptococcus lactis* cells. *J. Bacteriol.*, **143**, 128–134.

Keogh, B. P. (1973). Adsorption, latent period and burst size of phages of some strains of lactic streptococci. *J. Dairy Res.*, **40**, 303–309.

Konings, W. N. and R. Otto (1983). Energy transduction and solute transport in streptococci. *Antonie van Leeuwenhoek*, **49**, 247–257.

Law, B. A. (1978). Peptide utilization by group N streptococci. *J. Gen. Microbiol.*, **105**, 113–118.

Law, B. A. (1982a). Cheeses. In *Economic Microbiology*, ed. A. H. Rose, vol. 7. Academic, London.

Law, B. A. (1982b). Microbial proteolysis of milk proteins. In *Food Proteins, Kellogg Foundation International Symposium on Food Proteins*, ed. P. F. Fox, pp. 307–328. Applied Science, London.

Law, B. A. and M. E. Sharpe (1977). The influence of the microflora of Cheddar cheese on flavour development. *Dairy Ind. Int.*, **42**, 10–11.

Law, B. A. and M. E. Sharpe (1978). Streptococci in dairy industry. In *Streptococci, The Society of Applied Bacteriology Symposium Series 7*, ed. F. A. Skinner and L. B. Quesnel. pp. 263–297. Academic, London.

Law, B. A., E. Sezgin and M. E. Sharpe (1976). Amino acid nutrition of some commercial cheese starters in relation to their growth in peptone-supplemented whey media. *J. Dairy Res.*, **43**, 291–300.

Law, B. A. and J. Kolstad (1983). Proteolytic systems in lactic acid bacteria. *Antonie van Leeuwenhoek*, **49**, 225–245.

Lawrence, R. D. (1978). Action of bacteriophage on lactic acid bacteria: Consequences and protection. *N. Z. J. Dairy Sci. Technol.*, **13**, 129.

Lawrence, R. C., T. D. Thomas and B. E. Terzaghi (1976). Cheese starters. *J. Dairy Res.*, **43**, 141–193.

Lowrie, R. J. and R. C. Lawrence (1972). Cheddar cheese flavour. IV. A new hypothesis to account for the development of bitterness. *N. Z. J. Dairy Sci. Technol.*, **7**, 51–53.

Maloney, P. C. (1982). Energy coupling to ATP synthesis by the proton-translocating ATPase *J. Membr. Biol.*, **67**, 1–12.

Maloney, P. C. and F. C. Hansen, III (1982). Stoichiometry of proton movements coupled to ATP synthesis driven by a pH gradient in *Streptococcus lactis. J. Membr. Biol.*, **66**, 63–75.

Manning, D. J. (1974). Flavour analysis of Cheddar cheese. *XIX Int. Dairy Cong.*, **1E**, 291.

Manning, D. J., H. R. Chapman and Z. D. Hosking (1976). The production of sulphur compounds in Cheddar cheese and their significance in flavour development. *J. Dairy Res.*, **43**, 313–320.

Martley, F. G. (1972). The effect of cell numbers in streptococcal chains on plate counting. *N. Z. J. Dairy Sci. Technol.*, **7**, 7–11.

Mason, P. W., D. P. Carbone, R. A. Cushman and A. S. Waggoner (1981). The importance of inorganic phosphate in regulation of energy metabolism of *Streptococcus lactis. J. Biol. Chem.*, **25**, 1861–1866.

McKay, L. L. (1982). Regulation of lactose metabolism in dairy streptococci. In *Developments in Food Microbiology*, ed. R. Davies, pp. 153–182. Applied Science, London.

Morse, M. L., K. L. Hill, J. B. Egan and W. Hengstenberg (1968). Metabolism of lactose by *Staphylococcus aureus* and its genetic basis. *J. Bacteriol.*, **95**, 2270–2274.

Otto, R., A. S. M. Sonnenberg, H. Veldkamp and W. N. Konings (1980). Generation of an electrochemical proton gradient in *Streptococcus cremoris* by lactate efflux. *Proc. Natl. Acad. Sci. USA*, **77**, 5502–5506.

Otto, R., B. ten Brink, H. Veldkamp and W. N. Konings (1983). The relation between growth rate and electrochemical proton gradient of *Streptococcus cremoris. FEMS Microbiol Lett.*, **16**, 69–74.

Peebles, M. M., S. E. Gilliland and M. L. Speck (1969). Preparation of concentrated lactic streptococci starters. *Appl. Microbiol.*, **17**, 805–810.

Ratkowsky, D. A., R. K. Lowry, T. A. McMeekin, A. N. Stokes and R. E. Chandler (1983). Model for bacterial culture growth rate throughout the entire biokinetic temperature range. *J. Bacteriol.*, **154**, 1222–1226.

Reiter, B. and A. Møller-Madsen (1963). Reviews of the progress of dairy science; cheese and butter starters. *J. Dairy Res.*, **30**, 419–456.

Rice, G. H., F. H. C. Stewart, A. J. Hillier and G. R. Jago (1978). The uptake of amino acids and peptides by *Streptococcus lactis. J. Dairy Res.*, **45**, 93–107.

Saier, M. H. and S. Roseman (1976). Sugar transport inducer exclusion and regulation of the melibiose, maltose, glycerol and lactose transport systems by the phosphoenolpyruvate sugar phosphotransferase system. *J. Biol. Chem.*, **251**, 6606–6615.

Sandine, W. E., P. C. Radich and P. R. Elliker (1972). Ecology of the lactic streptococci. A review. *J. Milk Food Technol.*, **35**, 176–184.

Schifsky, R. F. and L. L. McKay (1975). Isolation of *Streptococcus lactis* C2 mutants with high phospho-β-galactosidase activity. *J. Dairy Sci.*, **58**, 482–493.

Shaw, M. K. (1967). Effect of abrupt temperature shift on the growth of mesophilic and psychrophilic yeasts. *J. Bacteriol.*, **93**, 1332–1336.

Slade, H. D. and W. C. Slamp (1956). Sonic oscillation as an aid in the counting of group A streptococci by the pour-plate method. *J. Bacteriol.*, **71**, 624–625.

ten Brink, B. and W. N. Konings (1982). Electrochemical proton gradient and lactate concentration gradient in *Streptococcus cremoris* cells grown in batch culture. *J. Bacteriol.*, **152**, 682–686.

Thomas, T. D. (1976). Regulation of lactose fermentation in group N streptococci, 1976. *Appl. Environ. Microbiol.*, **32**, 474–478.

Thomas, T. D., D. C. Ellwood and V. M. Longyear (1979). Change from homo- to heterolactic fermentation by *Streptococcus lactis* resulting from glucose limitation in anaerobic chemostat cultures. *J. Bacteriol.*, **138**, 109–117.

Thomas, T. D., K. W. Turner and V. L. Crow (1980). Galactose fermentation by *Streptococcus lactis* and *Streptococcus cremoris*: Pathways, products, and regulation. *J. Bacteriol.*, **144**, 672–682.

Thomas, T. D. and O. E. Mills (1981). Proteolytic enzymes of starter bacteria. *Neth. Milk Dairy J.*, **35**, 255–273.

Thompson, J. (1978). *In vivo* regulation of glycolysis and characterization of sugar: Phosphotransferase systems in *Streptococcus lactis. J. Bacteriol.*, **136**, 465–476.

Thompson, J. (1979). Lactose metabolism in *Streptococcus lactis*: Phosphorylation of galactose and glucose moieties *in vivo. J. Bacteriol.*, **140**, 774–785.

Thompson, J. (1980). Galactose transport systems in *Streptococcus lactis. J. Bacteriol.*, **144**, 683–691.

Thompson, J. and T. D. Thomas (1977). Phosphoenolpyruvate and 2-phosphoglycerate: Endogenous energy source(s) for sugar accumulation by starved cells of *Streptococcus lactis. J. Bacteriol.*, **130**, 583–595.

Thompson, J., K. W. Turner and T. D. Thomas (1978). Catabolite inhibition and sequential metabolism of sugars by *Streptococcus lactis. J. Bacteriol.*, **133**, 1163–1174.

Thompson, J. and B. M. Chassy (1981). Uptake and metabolism of sucrose by *Streptococcus lactis. J. Bacteriol.*, **147**, 543–551.

Thunell, R. K., W. E. Sandine and F. W. Bodyfelt (1981). Phage insensitive multiple-strain starter approach to Cheddar cheese making. *J. Dairy Sci.*, **64**, 2270.

Visser, F. M. W. (1977). Contribution of enzymes from rennet, starter bacteria and milk to proteolysis and flavour development of Gouda cheese. 2. Development of bitterness and cheese flavour. *Neth. Milk Dairy J.*, **31**, 188–209.

24

Cheese Technology

D. M. IRVINE and A. R. HILL
University of Guelph, Ontario, Canada

24.1   INTRODUCTION

Cheese has been a food for thousands of years. The origins of cheesemaking are unknown, but ancient records refer to the domestication of the cow (genus *Bos*) and the use of milk and production of cheese between the Tigress and Euphrates rivers in what is now Iraq (6000–7000 BC; Scott, 1981). Cheese was made and consumed in biblical times and many references to cheese are found in the Bible. David was carrying cheese when he slew Goliath. Homer, *ca.* 1184 BC, made reference to cheese. Cheese sustained the armies of Ghengis Khan, *ca.* 1162, and enhanced the banquet tables of Caesar. The art of cheesemaking was brought from Asia to Europe and cheese was made in many parts of the Roman Empire. Cheesemaking was introduced, and improved, by the monks in the monasteries of Europe. The settlers to North America brought the art of cheesemaking with them.

Until the middle of the 19th century cheese was made on farms or monasteries, but in the 1850s, cheesemaking became localized in factories. Since this time, there has been a great centralization of the cheese industry with factories daily receiving over 2 million kg of milk to be made into cheese.

Cheese represents a method of preserving the nutritionally high value protein which cannot be readily stored.

There are over 800 named cheeses, many of which are similar, differing only in shape, size, degree of ripening, type of milk, condiments, packaging and locality of manufacturing.

The classification of cheese is complicated. The FAO/WHO (1972) jointly defined a code of quality standards for cheese which is continually being reviewed and updated (Codex Alimentarius Commission, 1973). Lists of cheese have been compiled by committees of the International Dairy Federation (Burkhalter, 1971). For the purposes of the following discussion, cheese will be classified according to the basic method of manufacture as follows: hard cheese — Cheddar, Swiss, Romano, *etc.* (high acid development, high cooking temperatures); soft and semi soft cheese — brick, Gouda, Feta, *etc.* (acid develops slowly, control of lactose by washing, minimal cooking temperatures); fresh cheese — cottage, cream, quarg (high acid development by bacteria); processed cheese — high heating of cheese to stop ripening. A schematic flow chart of the cheesemaking processes is found in Figure 1.

24.2 COMPOSITION OF MILK

24.2.1 Milk Quality

The quality of milk has a profound effect on the quality of the finished cheese. The milk must be of high quality and free from absorbed flavours or odours and from sediment or extraneous matter. Milk that has the following defects should be rejected for cheesemaking purposes. (1) Poor-flavoured milk. Milk which may have taken on the flavour of feeds or weeds, such as turnips and leeks or may have a cowy–barny odour from a poorly ventilated barn is undesirable. Rancid milk caused either by cows in late lactation or excessive agitation is also unacceptable for cheesemaking. Misuse of chemicals, sanitizers, gasoline and sprays can result in poor-flavoured milk. (2) Mastitic milk. Milk from cows suffering from mastitis is not acceptable for cheesemaking as it may have a high bacteria count, be abnormal in composition or may inhibit the growth of culture organisms. (3) Milk containing antibiotics. Antibiotics are commonly used to treat cows for mastitis. Milk containing antibiotics is not suitable for cheesemaking. (4) Milk with a high bacterial count (Section 24.2.3). Milk that is improperly cooled and/or handled in unsanitary equipment has high numbers of bacteria. These bacteria may cause undesirable flavours in milk. (5) Milk containing sediment. It is important that any foreign material be excluded from the milk supply. Such material may contain large numbers of bacteria and will contaminate the milk.

The most important tests of milk quality are: (1) organoleptic evaluation by competent grader (smell and taste); (2) bacterial plate counts (see Section 24.2.3) (ICMSF, 1978; Milk Industry Foundation, 1964); (3) inhibitory substances test (see Section 24.2.3) (ICMSF, 1978); and (4) fermentation and curd tests (Irvine 1982).

24.2.2 Chemical

Cheese has been made from the milk of many mammals, notably the cow, goat, sheep, reindeer, buffalo, camel, llama, zebu and the yak. Most cheese is made from cows' milk but milk from the goat, sheep and buffalo are used in some areas of the world. The composition of cows' milk is presented in Table 1 and that of other mammals may be found in the literature (Davis, 1955; Johnson, 1972).

Milk is a complex biological system which varies in composition depending on the breed of cow, stage of lactation, age of lactating cow, season of the year, feed, health and nutritional level of the animal, gestation and oestrium. Milk from cows suffering from mastitis will also have an abnormal composition. These factors may exert a large influence on the composition of milk from a cow or even a herd.

The stage of lactation has a marked influence on the composition of milk. The fat and protein contents of milk are at a minimum 75 days after calving and continue to increase until the end of lactation. The volume of milk produced increases dramatically from the onset of lactation, doubling in 45 days from which level it decreases until the end of lactation. The lactose content of milk is at a maximum at the beginning of lactation, decreasing slowly until the end of lactation (Barnum *et al.*, 1966).

There are major seasonal changes in the milk supply. In the northern hemisphere the fat content drops to a low in August and rises rapidly to a high in December. The protein content is generally correlated to the fat content, but low protein/fat ratios are obtained at the end of May and a high *P/F* ratio occurs in August–October. The lactose content is fairly uniform throughout the year. Factors affecting compositional variations of milk are adequately described by Johnson

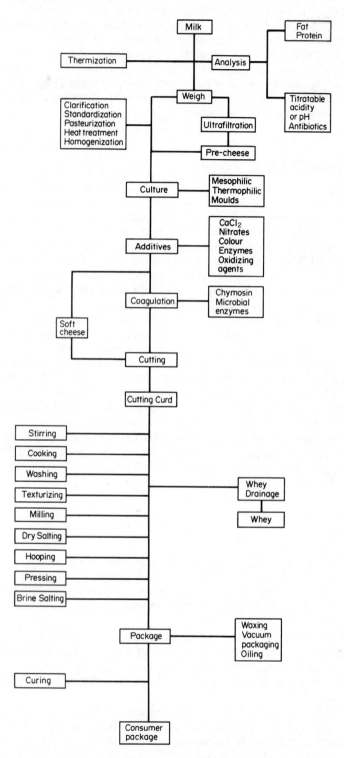

Figure 1 Flow diagram of the cheesemaking process

(1972). The seasonal variations will be compounded when all cows 'freshen' about the same time (*e.g.* New Zealand).

The modern cheese factory may receive 2 million kg of milk daily transported in large tankers. The milk is mixed or pooled and hence the variation between cows and herds are minimized. For example, the Ontario data indicate bi-monthly ranges of fat, protein and lactose of 3.62–3.91, 3.19–3.44 and 4.99–5.12, respectively. These data represent the herd bi-monthly analyses over a

5 year period of all Ontario milk producers (12 000) analyzed by the Infra Red Milk Analyzer at the Central Milk Testing Laboratory, Ontario. These data indicated that in a large pooled milk supply there were no significant variations on a daily basis.

Table 1 Composition of Milk[a]

Class	% in milk	Constituents (% of class unless stated)	
Water	87.1		
Lipids[b]	3.8	True fats	
		triglycerides	95–96%
		di- and mono-glycerides	1.452%
		Phospholipids	
		lecithins, cephalin, sphingomyelin	0.6–0.8%
		Sterols	0.3% or 120 p.p.m.
Proteins[b]	3.31		% total N
		Caseins (α, α_{S1}, β, $\varkappa$, γ)	76–86%
		Whey protein	18%
		β-lactoglobulin 8%	
		α-lactalbumin 4%	
		proteose peptone 3.5%	
		immunoglobulin 1.6%	
		serum albumin 0.9%	
		Non-protein nitrogen	6%
Lactose[b]	5.06	Glucose	7 mg 100 l^{-1}
		Galactose	2 mg 100 l^{-1}
Salts	0.73	Bases	
		K_2O	0.175%
		Na_2O	0.070%
		CaO	0.070%
		MgO	0.017%
		Fe	0.3–0.77 p.p.m.
		Cu	0.05–0.14 p.p.m.
		Zn	2.0–5.0 p.p.m.
		Mn	0.03 p.p.m.
		Traces of Al, Co, Ti, V, Rb, Li and Sr	
		Acids	
		P_2O_5	0.195%
		Cl	0.100%
		SO_3	0.027%
		citric acid	0.200%
		I	Trace–0.276 p.p.m.
		F	0.138 p.p.m.
		Li	2.0 p.p.m.

Pigments
 Carotene and the carotenoids, xanthophyll and riboflavin
Vitamins
 Fat soluble
  Vitamin A and carotenoids
  Vitamin D (cholecalciferol–ergocalciferol)
  Vitamin E (α-tocopherol)
  Vitamin K
 Water soluble
  Vitamin C (L-ascorbic acid and dehydro-L-ascorbic acid)
  B vitamins
   Thiamine B_1
   Riboflavin B_2
   Niacin or nicotinic acid
   Pantothenic acid
   Pyridoxine B_6
   Vitamin B_{12}
   Biotin
   Inositol
   Choline
   p-Amino benzoic acid
   Folocin
   Grass juice factor

Table 1 Composition of Milk[a] (*continued*)

Class
Enzymes
Aldolase, amylase, lipase, esterase (A, B and C), catalase, peroxidase (lactoperoxidase), protease, alkaline phosphatase, acid phosphatase (reductase), xanthine oxidase, ribonuclease, lysozyme, carbonic anhydrase, salolase, rhodonase, lactase
Miscellaneous constituents
Gases — CO_2, N_2 and O_2
Nitrogenous (non protein N) 28 mg/100 ml — Urea, creatine, creatinine, uric acid, ammonia N, alpha amino N unaccounted
Cellular constituents — fragments of secreting cells, nuclear material, leucocytes, bacteria
Miscellaneous factors — flavour depending upon feeds consumed

[a] Johnson, 1972. [b] The fat, protein and lactose are the means of bi-monthly samples for 4 years of 12–15 000 Ontario producers. The samples are analyzed by infrared analyses at the Central Milk Testing laboratory of the Ontario Ministry of Agriculture and Food.

Water may be added to milk inadvertently by poor pipeline installation, careless handling, or deliberately. Milk is regularly examined for added water by cryoscopic techniques (Richardson, 1981). Deviations in the lactose content of herd milk can be used to screen the milk samples for added water, prior to positive identification by the cryoscope (Richardson, 1981).

The chemical composition of milk for payment to the farmers is analyzed by infrared instrumentation which gives data on fat, protein and lactose (Biggs, 1967; Richardson, 1981). This modern instrumentation is available for in-line determination of the chemical constituents of milk, and with the proper controls the desired fat–protein contents (Richardson, 1981) of cheese milk can be automatically delivered to the cheese vats.

24.2.3 Bacteriological

Milk is an excellent growth medium for many different kinds of organisms and, whether milked by hand or machine, contains bacteria. It is a desirable medium for microbiological growth and it is essential for it to be cooled as quickly as possible to keep bacterial numbers to a minimum. In modern dairy farms the milk is extracted from the teat of the cow by a pulsator, which draws a vacuum of 50 kPa, after which the milk is conveyed in a glass, stainless steel or plastic pipeline to a refrigerated bulk tank. The warm milk (39 °C) is rapidly chilled to 5 °C where it is held for four milkings or a maximum of two days. These temperatures inhibit bacterial growth and it is normal to obtain milk with less than 5000 bacteria per cm^3. The psychrophiles grow slowly until they increase rapidly in numbers after 96 h. The psychrophiles are reputed to cause flavour problems (Hylmar, 1980; Law, 1979; Marshall, R. J., 1979, 1982) and protein digestion, which results in loss of cheese yield (Hicks *et al.*, 1982; IDF, 1980a; Reimerdes, 1982). These organisms find their way to the milk from the udder of the cow, hair of the cow, air and dust in barns, and unclean milking equipment. Many species of streptococci, micrococci, staphylococci, pseudomonads, coliforms, corynebacteria, lactobacilli, clostridia, and yeasts and moulds have been reported in milk (Scott, 1981).

Aside from the organisms that may cause flavour and other defects, there are a number of pathogens which are of vital concern to the cheesemaker. These include pathogenic organisms of the following type: *Mycobacterium tuberculosis*, *Coxiella burnetti*, *Staphylococcus aureus*, *Streptococcus pyogenes*, *Cl. botulinum*, *Bacillus anthrax*, *Salmonella* spp., *Shigella* and *E. coli*, *Campylobacter* spp., *Yersinia enterocolitica*, *B. cereus*, and *Cl. perfringens* (IDF, 1980b; Snyder *et al.*, 1978). All of these except the spore formers and the enterococci are destroyed by pasteurization (Chapman and Sharpe, 1981).

In cheesemaking operations, all phases of the operation must be monitored. The raw milk should be sampled for total bacterial count, the presence of coliforms, thermodurics and psychrotrophs, and should be screened for abnormal milk such as that encountered with severe mastitis. Mastitic milk will exhibit high leucocyte counts which may be enumerated by microscopic, viscometric and electronic means (Marth, 1978). Tests for bacteria are carried out by standard plate counts, dye reduction tests and other specific tests for gas forming clostridia, coliform and staphylococci (ICMSF, 1978; Marth, 1978).

Cultures used for cheesemaking are very sensitive to antibiotics such as penicillin, aureomycin and tetracycline, which are used to control mastitis in dairy cattle. The antibiotic residues may be shed into the milk supply causing widespread damage as a tenth of a unit of penicillin will inhibit a lactic culture (Sellars and Babel, 1978). Microbiological and chemical tests are commonly used to select and reject antibiotic contaminated milk for cheesemaking such as the rapid Delvotest (Marth, 1978), the Charm Test (Thorogood *et al.*, 1983) and by the *Bacillus subtilis* Disc Assay which requires at least 5–7 h (Messer *et al.*, 1978).

There are countries where it is not practical to cool milk and where raw milk with high bacteriological counts (usually lactics) are made into cheese. Indeed, the uncooled milk containing indigenous lactic bacteria may be made into cheese without any heat or other treatment and without the use of cultures.

24.3 TREATMENT OF MILK FOR CHEESEMAKING

24.3.1 Standardization

Standardization is the term used when the composition of milk is adjusted. Milk is usually priced on the basis of the fat content and it is in the best interests of the cheesemaker to make as many kilograms of cheese per kilogram of fat in compliance with the regulations. Standardization of milk for cheesemaking is desirable in order to regulate the correct proportions of fat and protein in the finished cheese. Government regulations usually define minimum fat and maximum moisture. The cheesemaker should standardize the milk, at least during a portion of the year when the protein/fat ratio is low. The most efficient way to make cheese is to have a definite protein/fat ratio in the milk depending on the desired cheese composition. This ratio decreases as the percentage of fat in the milk increases and may be adjusted by: (1) removing fat in a standardizing centrifuge or a cream separator; (2) the addition of skim milk powder to the milk; or (3) the addition of skim milk. Tables indicating the amount of fat to be removed, skim milk powder or skim milk to be added to the milk of varying composition are available (Irvine, 1982).

Some of the costs of standardizing include costs of skim milk powder, time, labour and equipment to add and mix the powder, costs of analyzing milk and cream for fat or protein and/or costs of cleaning a separator. The cheesemaker must compare the costs of standardizing the milk with the extra yield of cheese or cream that will be realized from the standardization process. The addition of skim milk to lower the fat content is not usually practised unless there is an excess of skim milk and excess vat capacity.

Many cheeses are made from milk of reduced fat content. The fat in the dry matter (*F/DM*) varies from 48–52% for whole milk cheese to a low of 32% for Grana cheese and 31% *F/DM* for part-skim Mozzarella.

24.3.1.1 *Method for standardizing milk*

(i) Determine the fat and protein content of cheese milk

If instrumentation for determining protein is not available, it may be estimated from the fat content (equation 1) or from tables (Irvine, 1982).

$$\% \text{ protein} = 1.53 + 0.449 \,(\% \text{ fat}) \tag{1}$$

(ii) Determine the required P/F ratio of the milk

The *P/F* ratio depends on the desired *F/DM* of the cheese, the ratio of protein to casein, the moisture and salt content of the cheese, and if the cheese curd is washed. For *F/DM* values of 50–51, 47–49, 44–47, 38–42 and 29–32, *P/F* ratios of 0.96, 1.07, 1.19, 1.47 and 2.3, respectively, are required.

(iii) Calculate amount of fat which must be removed

For example, for milk of 3% fat and 2.88% protein, the required adjusted fat content for the manufacture of low fat Mozzarella cheese is (equation 2):

$$\text{Adjusted fat content} = \frac{\text{Milk protein}}{P/F \text{ ratio}} = \frac{2.88}{1.34} = 2.14\% \tag{2}$$

The amount of fat to be removed is $3 - 2.14 = 0.86\%$.

The alternative is to calculate the amount of protein which must be added as skim milk powder or skim milk.

24.3.1.2   Guidelines for standardizing milk

Accurately determine the composition of cheese milk daily, as well as metering the milk in the vats. Use only high quality, antibiotic-free powder for standardizing. Determine the composition of the standardized cheese and if necessary adjust the proportions of fat and protein in the cheese milk on succeeding days. If the bulk starter is being made from skim milk powder, then reduce the amount of skim milk powder added to the cheese vat accordingly. If skim milk powder is being added to the cheese milk then compensate for titratable acidity (titration increases 0.06% for each 1% increase in solids).

24.3.2   Filtration

Milk may be filtered to remove large particles of sediment. The milk is warmed to 30 °C through a preheater and filtered through cotton flannel bags. Cheesecloth strainers are poor and cold milk filters are slow. Filtering removes gross sediment and is of value only from an aesthetic standpoint.

24.3.3   Clarification (Sediment Removal)

Centrifuging milk through a clarifier is much more efficient than cloth filters. Milk is warmed to 32–38 °C and pumped through a clarifier which may vary in size from 7000 to 50 000 l/h at speeds of 6000 r.p.m. and above (Davis, 1965). The clarifier removes sediment and other particulate matter in milk that has a specific gravity greater than 1.032 (specific gravity of milk). The clarifier also removes bacterial cells, breaks up clumps of bacteria and disrupts fat globules. It is essential that milk be pasteurized immediately after clarification, otherwise the lipase enzymes will act on the disrupted fat globules causing rancidity. The eye formation of Swiss cheese will be improved due to the absence of particulate matter around which eyes develop (Reinbold, 1972).

24.3.4   Bactofugation (Bacterial Removal)

The bactofuge is a high speed centrifuge that operates at a force of 980 000 kPa and at 54 °C. It may remove up to 99% of the bacteria in milk but under practical conditions usually removes 80–90%. It usually removes 95% of the spores in the milk. Bactofugation usually employs two bactofuges which operate at 5000–10 000 l/h. The need to use this equipment is largely confined to Europe where problems arise from spore-forming anaerobes such as *Cl. butyricum* that are not destroyed by pasteurization.

Bactofugation removes a portion of the milk, casein and leucocytes, as well as the bacteria, which may amount to 1–2% of the total quantity of the milk. The bactofugate (sludge) may contain 12–16% dry matter. In order to avoid losses, this bactofugate is sterilized at 130–140 °C for 3–4 s and re-introduced back into the original milk (Kessler, 1981).

24.3.5   Thermization

Many cheese factories operate on a five day working week and store milk in large silo tanks. This may result in some milk being held for 3–4 days before use. Such milk even when held at 4 °C supports the growth of psychrophiles. These organisms have been shown to cause flavour defects and protein degradation (Chapman and Sharpe, 1981; Hicks *et al.*, 1982; Johnston *et al.*, 1983; Reimerdes, 1982).

Thermization is a sub-pasteurization heat treatment in which the milk is heated to 63–65 °C followed by cooling to 4–8 °C for storage. Reimerdes (1982) states that thermization at 60 °C for 30 min results in a protein content in the milk serum that re-establishes the original cheesemaking properties. Thermized milk is phosphatase positive and must be followed by proper pasteurization. Milk is thermized immediately when it is brought to the plant and when the milk is to be held in storage for some period of time (see heat treatments, Section 24.3.8).

24.3.6 Ultrafiltration

Milk for certain types of soft cheese like Feta, quarg, Camembert, Havarti, cream and St. Paulin may be fractionated by ultrafiltration (UF) (Mann, 1982). The milk has been concentrated at 650 kN/m^2 to 12–14% protein. Pressures of 200–280 kN/m^2 and 40 °C have been reported (Mahaut and Maubois, 1978; Maubois and Mocquot, 1975). It may be necessary to reduce the pH prior to UF to pH 5.9–6.0 in order to dissociate the colloidal calcium and phosphate which interfere with culture growth (Narasimhan and Ernstrom, 1977) and may cause bitterness. Milk may be concentrated to a pre-cheese with the desired composition of the final cheese to avoid whey drainage, or it may be partially fractionated, followed by syneresis. Increased yields of up to 30% have been reported due to the incorporation of the whey proteins in the pre-cheese (Maubois and Mocquot, 1975). This method has been successful for some of the soft cheese but has not been successful with hard cheese.

24.3.7 Homogenization

Homogenization of milk is not usual for cheesemaking. Homogenization subjects the milk to pressures of 3400 to 20 000 kN/m^2 during which the fat globules are reduced in size from 3–4 μm to less than 1 μm. The number of globules is greatly increased and the fat surface area is increased 15 fold (Mulder and Walstra, 1974). Homogenization disrupts the protective fat globule membrane and the exposed fat is subject to rapid deterioration by lipase unless immediately pasteurized.

Homogenized milk exhibits a soft, weak curd at renneting due to the disruptive effect of the increased number of globules in the curd structure or the attraction of the casein particles to the new surface area of the negatively charged fat globule (Mulder and Walstra, 1974). If milk is to be homogenized it is essential that pressures less than 3400 kN/m^2 be used or that the milk be fortified with extra solids (Maxcy *et al.*, 1955). Low pressure homogenization may be used to form an emulsion with butter or other oils when incorporated with reconstituted skim milk.

In the manufacture of blue mould cheese from cow's milk, the milk may be separated and the cream homogenized at 14 000 kN/m^2 for the first stage and 3450 kN/m^2 for the second stage at temperatures of 37–43 °C. The purpose is to increase the surface area of the fat so that the lipolytic action of the mould (*P. roqueforti*) may be stimulated to accelerate flavour development (Morris, H. A., 1981). Cream (10–15% fat) is homogenized at 7000 kN/m^2 to 10 000 kN/m^2 pressure and at a temperature of 62 °C during the manufacture of cream cheese. The result is to cause the adsorption of the protein on the new fat globule surface which results in a more viscous body (Mulder and Walstra, 1974). Skim milk for cottage cheese may be homogenized in order to inactivate the agglutenins formed by some lactic bacteria (Emmons and Elliot, 1967). Processed cheese spreads may also be homogenized at 7000 to 14 000 kN/m^2 to make the spread smoother and more homogeneous.

24.3.8 Heat Treatments — Pasteurization

The purpose of heating milk is to destroy microorganisms and inactivate enzymes which would affect the healthiness or palatability of the cheese. However, pasteurized milk cheese never develops the 'full' flavour of raw milk cheese. Pasteurization is that temperature and time combination (63 °C, 30 min or 72 °C, 16 s) above which the alkaline phosphatase enzyme is destroyed. If the time of holding or temperature of pasteurization is increased to over 74 °C or held over 20 s, a very weak curd will result. It is important that the cheese milk be pasteurized in H.T.S.T. equipment with proper controls, a positive pump and a flow diversion valve. Proper pasteurization will destroy the pathogens (except spores) and the coliform organisms.

Many cheesemakers use a sub-pasteurization heat treatment to inhibit but not necessarily destroy the enzymes or bacteria in order to control the making schedule and to develop more flavour than a completely pasteurized cheese. This heat treatment may be 72 °C for a few seconds or 60–68 °C for 10–15 s.

24.4   ADDITIVES TO CHEESE MILK

24.4.1   Calcium Chloride

Calcium chloride may be added to pasteurized milk at a level of 0.02%. Some of the calcium of the milk is heat precipitated as tricalcium phosphate and is not available. Calcium is required for rennet coagulation and the addition of calcium may restore the curd forming properties and reduce the amount of rennet required.

24.4.2   Nitrates

In some types of cheese like Edam, Gouda and Swiss, it has been found necessary to add sodium or potassium nitrate to the milk at 200 p.p.m. to control the undesirable effects of the butyric acid bacteria *Clostridium butyricum*, or *Cl. tyrobutyricum*. These organisms cause late gas, which results in blowing of the cheese and the development of an off flavour.

Nitrates do not affect the growth of lactic organisms or coliforms and may even be stimulatory. During curing the nitrates change to nitrites which under mild acid conditions may react with various amines, amino acids and proteins to form nitrosamines. Nitrosamines are potent carcinogens. Although the levels of nitrosamines found in cheese has been absent or in very low levels (p.p.b.), there is concern about the use of nitrates (World Health Organization, 1978). Indeed, work in our laboratory has indicated that nitrosamines are formed in cheese by the complexing of nitrites and amines (Aurora, unpublished data).

24.4.3   Colour

The colour of milk and cheese has consumer appeal. It has been traditional to colour or decolour cheese milk. An orange colour is desired in many types of cheese like Cheddar, whereas cheese made from goats' or sheeps' milk is flat white in colour. Milk contains carotene and carotenoids which impart the yellow colour to milk, cream, butter and cheese. Riboflavin imparts a yellowish-green colour but this is of little consequence in cheese as it is lost in the whey. The carotenoids in cows' milk are increased with fresh forage and are higher in milk from the Jersey and Guernsey breeds. The milk normally varies in colour seasonally. In order to standardize the colour throughout the year, colour may be added. The carotenoids are absent in goat and sheep milk.

The colours that are added are: annatto, a yellow-red obtained from a shrub, *Bixa orellana*; β-carotene, yellow-red obtained from carrots or manufactured synthetically; and paprika. The colour is usually added at the rate of 88 g per 1000 kg. The colour extract is concentrated and should be diluted ten times with sterile cold water. The carotenoids in milk for blue cheese may be decolourized with benzoyl peroxides in order that the veins of blue mould will be blue and not green. Cows' milk may be bleached to simulate sheep or goat milk cheese.

24.4.4   Hydrogen Peroxide

In some countries hydrogen peroxide may be added to cheese milk at levels of 0.04–0.08% of the weight of milk. This is an alternative treatment to full pasteurization or where pasteurization equipment is not available. Hydrogen peroxide treatment is at 50–54 °C for 30 min in a vat pasteurizer or at 60 °C for 16 s in an H.T.S.T. pasteurizer (Morris and Jezeski, 1964). The excess H_2O_2 is removed by the addition of catalase, otherwise the excess H_2O_2 hinders the lactic cultures. This treatment, which reduces the numbers of bacteria, is somewhat selective in destroying organisms, *e.g. S. lactis* is readily destroyed but spore formers (*Bacillus cereus* or *subtilis*) are not (Scott, 1981), and does not guarantee freedom from pathogens or coliforms. This treatment is not common as the time and space required to hold milk at 50–54 °C for 30 min is a bottleneck in large modern factories. The temperature at which the H_2O_2 treatment is effective is reasonably close to pasteurizing temperatures and it is more practical to pasteurize.

24.4.5 Lipases

Lipolytic enzymes which liberate fatty acids by hydrolysis are important flavour enhancers for such cheeses as Feta, Romano and Parmesan.

Lipases are normally present in all raw milk, probably bound to the casein complex. Lipolysis may be induced by warming cold milk to 29.4 °C and cooling, or subjecting raw milk to violent agitation, pumping, foaming or homogenization. Lipases are normally inactivated at pasteurization temperatures but there are some microbial lipases which withstand pasteurization and may affect the flavour of the finished cheese. Lipases are completely inactivated by heating at 80 °C for 20 s (Nilsson and Willart, 1961). The optimum temperature of activity is 37 °C. Some of the flavour profile of a raw milk Cheddar cheese may be attributed to normal milk lipases (Ernstrom and Wong, 1974).

Traditional rennet paste which originated from comminuted fourth stomachs of calves, kids and lambs contained not only rennet but also lipases which accounted for the rancidity in some Italian cheese. Farnham (1950) discovered that glandular lipases from the mouths of suckling mammals secreted lipases. He extracted and standardized these lipases which are commonly used today for Italian cheese. They are used at the rate of 125–187 g per 1000 kg depending upon the degree of lipolysis required. These lipases are also used when cows' milk is being used to replace goat or sheep milk to reproduce the traditional flavour profile in cheeses like Feta and Kefalotyri which are normally made from sheep or goat milk.

It must also be noted that these lipases which produce rancid milk may adversely affect the growth of lactic bacteria (Costilow and Speck, 1951).

24.5 CULTURES (STARTERS)

The fermentation of lactose by lactic acid bacteria of the genera *Streptococcus* and *Lactobacillus* provides the basis of cheesemaking. The purpose of the culture is to produce lactic acid which contributes to the flavour, texture and keeping quality of the cheese. Lactic acid is essential to cheesemaking to cause syneresis of the curd which results in the expulsion of the whey. Some of the by-products of the lactic fermentation provide the necessary growth factors for other microorganisms that are responsible for the desired flavour and body breakdown of ripened cheese. The function of the culture or starter is to produce acid, to assist in the coagulation of milk, to help form the curd, to produce volatile components, to produce proteolytic or lipolytic enzymes and to decrease the pH to the desired level for keeping quality and maturation (McKay *et al.*, 1971).

24.5.1 Mesophilic Starter Cultures

24.5.1.1 *Streptococcus*

Bacteria of the genus *Streptococcus* are most commonly used for the production of cheese such as Cheddar, brick, blue, Gouda, cottage, Feta, *etc.*; some examples are *S. lactis*, *S. lactis* ssp. *diacetylactis*, *S. cremoris* and *S. thermophilus* (Buchanan and Gibbons, 1974). These bacteria are homofermentative, producing only lactic acid. *S. thermophilus* is a thermoduric organism but *S. lactis* and *S. cremoris* are mesophilic. Lactic acid bacteria vary in size from 0.5–1.0 μm in diameter. If enough nutrient is available, these organisms will grow and produce acid at temperatures of 5–40 °C but the optimum growth is at 30 °C. *S. lactis* ssp. *diacetylactis* is used because of its ability to form flavour compounds. It grows under similar conditions to *S. lactis* but is different in that it rapidly ferments citrate to citric acid, and produces large amounts of diacetyl and carbon dioxide. It produces a cheese with open texture and a tendency for slit eyes and may be used for Gouda cheese.

24.5.1.2 *Leuconostoc*

This genus contains *L. cremoris* (*L. citrovorus*) and *L. dextranicum* which ferment citric acid and produce lactic acid, carbon dioxide and aroma compounds (Lawrence *et al.*, 1976). These organisms are used for cultured products but not usually for cheese because of the undesirable production of carbon dioxide which causes gas holes in the cheese.

24.5.1.3 *Lactobacillus*

In this genus are *L. bulgaricus*, *L. lactis*, *L. acidophilus*, *L. casei*, *L. helveticus*, *L. plantarum* (Sharpe 1978, 1979) which are usually thermophiles and are used for the manufacture of Grana and Swiss cheese.

24.5.2 Thermophilic Starter Cultures

L. acidophilus and *L. bulgaricus* are rod-shaped organisms that have their optimum growth at 37 °C. They will grow at 22–49 °C and survive temperatures as high as 55 °C, but acid production is seriously inhibited at temperatures over 42 °C. These organisms are used for Italian cheese (Romano, Provoloni, Parmesan), Swiss cheese and yoghurt. *L. bulgaricus* is a short rod-shaped organism that is incubated for 8–16 h at 39– 47 °C and may produce 2–4% lactic acid. These organisms utilize lactose for growth but also require nitrogenous compounds such as peptides and nucleic acid compounds. They grow well in sterilized milk. *L. lactis* is very similar to *L. bulgaricus* while *L. helveticus* is a mutant of *L. jugurti*. *L. helveticus* and *L. lactis* are used for cheese-making in Europe. *L. acidophilus* is used for Romano and Swiss cheese. The culture is incubated at 37 °C and is usually added at the rate of 0.03–0.1% for Swiss cheese and develops 1–1.1% lactic acid. *S. thermophilus* is a thermoduric Gram-positive organism that reduces litmus milk. It appears as spherical or ovoid cells (0.7–0.9 μm in diameter) in pairs or long chains. It does not grow below 20 °C but grows rapidly at 45 °C. It does not grow well at salt levels above 2%. It ferments fructose, glucose, lactose and sucrose.

24.5.3 Other Cheese Cultures

Brevibacterium linens, surface ripened brick; *Propionibacterium freudenreichii* spp. *shermanii*, eye former for Swiss; *moulds*, *Penicillium camembertii*, Camembert cheese; *Penicillium caseiocolum*, Camembert cheese, surface mould; *Penicillium candidum*, Camembert cheese, surface mould; *Penicillium roqueforti*, blue mould cheese; *Mucor rasmusen*, Norwegian skim milk cheese.

24.5.4 New Culture Developments

There has been a great deal of interest in the genetics of lactic bacteria in the hope of modifying cheese cultures for more efficient lactose utilization and the development of proteinase deficient variants (Smiley, 1982). The existence of plasmids in both lactic streptococci and lactobacilli is now well established. Davies and Gasson (1981) reviewed extraction and examination of plasmid DNA, plasmid components of lactic bacteria, and properties encoded by plasmids such as lactose metabolism, proteinase production, citrate metabolism, inorganic salt resistance and nisin production. The genes may be transferred by conjugation, transduction or by protoplast fusion (Davies and Gasson, 1981; Gasson, 1980; Gasson and Davies, 1980a,b; Walsh and McKay, 1982) but transduction and transformation were ruled out as mechanisms of genetic transfer (Snook and McKay, 1981). Plasmids play a fundamental role in the proteinase system and the lactose utilization system (Refstrup and Vogensen, 1980). Proteinase deficient variants have been used in cultures, which resulted in less bitter cheese. McKay and Baldwin (1978) and Walsh and McKay (1981) have shown that Lac$^+$ transconjugants were able to transfer lactose fermenting ability at a frequency higher than 10^{-1} per donor on milk agar plates. The study of plasmids and gene transfer should lead to great improvements of lactic cultures.

24.5.5 Commercial Culture Practice

Lactic cultures are available from commercial suppliers in liquid, dried or frozen forms. The liquid cultures are not very satisfactory as the activity of the culture decreases rapidly at a low pH unless the culture is buffered to pH 6 or above. If the time between the log phase of growth and the time of use is minimal, then they can be successfully used.

Spray dried or lyophilized culture may be held without deterioration for up to 6 months at 21 °C and longer if held at lower temperatures (Porubcan and Sellers, 1975). The culture contains

1–2% viable cells and usually requires two to three transfers to attain maximum activity. The method of carrying or adding cultures is shown in Figure 2.

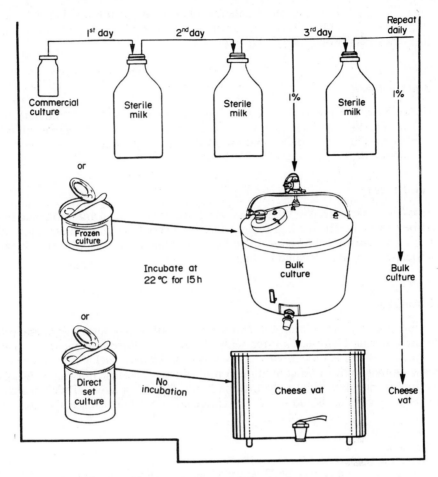

Figure 2 Culture handling techniques

24.5.5.1 *Frozen cultures*

Cultures are preserved by freezing concentrated cell suspensions. The cells are placed in buffered media, concentrated in a centrifuge and neutralized (pH 6). The concentrated culture with up to 10^{11} cfu ml^{-1} (Osborne, 1977) is fast frozen and stored. Liquid nitrogen ($-196\,°C$) is the best method to freeze and maintain the culture (Gilliland and Speck, 1974). Cultures so frozen may be stored in a deep freeze cabinet at $-44\,°C$, without deterioration, for several months. Frozen cultures can be used to prepare bulk culture or added directly to the cheese vat without any pre-incubation.

24.5.5.2 *Media for culture growth*

Fresh milk from healthy cows may be used if it contains enough milk solids and lactose (>8.5 S.N.F.) to support bacterial growth and has a low bacterial count as a high bacterial count milk may contain antibiotic material (nisin). Milk from mastitic cows must not be used as it may interfere with normal culture growth. Milk must not contain antibiotics, sulfonamides or other therapeutic agents. Rancid milk also interferes with bacterial growth.

The production of pooled milk makes it difficult to ascertain the source of the milk. It is recommended that milk be replaced with a pretested, antibiotic-free, low temperature skim milk powder reconstituted to 10–12% solids for culture growth. Whey has also been used for many years as a suitable culture medium for Swiss and Italian cheese. A new phage-inhibitory whey-based medium in which the pH is maintained at pH 6.0–6.2 by the addition of anhydrous ammonia is claimed to be more economical (Jonas *et al.*, 1977). The fresh sweet whey is diluted to 3–3.75% lactose, fortified with yeast autolysate and phosphates, heat treated at 90 °C for 45 min, incubated at 27 °C for 14–20 h and used to inoculate cheese milk at 10–40% of regular volume (Richardson, 1978; Richardson *et al.*, 1977, 1979).

24.5.5.3 Heat treatment

Milk for starter growth should be heated to a temperature of at least 88 °C and held at this temperature for 1 h. The heat treatment destroys antagonistic bacteria and some inhibitory substances, and releases minor constituents that improve bacterial growth.

24.5.5.4 Culture room

If a culture is being transferred in the factory, a culture room with an anteroom and self-closing doors should be provided. The air ingress into the room should be filtered and kept at positive pressure. It is important that the room, including floors and drains, can be sterilized. Culture tanks and autoclaves should be located in the room.

24.5.5.5 Inoculation

Use sterile pipettes for inoculating the culture and sterilize laboratory equipment with 200 p.p.m. chlorine. Inoculate sterile cultures with 1% active culture and incubate. Take every precaution to ensure there is no contamination. Many specialized techniques have been developed for ensuring that there is no contamination (Tamime, 1982) such as the Lewis system in which the culture is inoculated in a polyethylene bottle through a hypodermic needle. Some of the culture vats have water seals and sterile air is injected into the deadspace above the culture.

24.5.5.6 Culturing

Incubate mesophilic lactic culture at 22 °C for 15 h and thermophilic cultures at 39 °C for 8–12 h. When mesophilic cultures have attained the desired acidity (pH 4.5), they should be cooled to 5 °C until used. Thermophilic cultures should not be cooled below 20 °C prior to use.

The cultures develop different degrees of acidity as indicated below: *S. lactis* 0.64–0.75%, *L. bulgaricus* 2–4%, *L. acidophilus* 1–1.1% and *S. thermophilus* 0.7–0.8%.

24.5.6 Culture Problems

24.5.6.1 Bacteriophage (phage)

Phages are bacterial viruses which multiply on and at the expense of lactic bacteria. Lactic phages resemble sperm cells in shape. The phage is 0.8–0.28 μm long with a head of 0.7 μm. The core is made up of nucleic acid (DNA) or ribonucleic acid (RNA) which is protected by a protein layer (Sandine, 1979). The phages of lactic cultures are less than a micron in size and contain a single chromophore (double strand of DNA). The phage is attracted to the bacterial cell wall where it injects protein DNA into the cell through its tail. The viruses multiply and after a time the cell bursts (lysis) releasing 40–130 new phage to destroy other lactic cells. A lysed cell is destroyed and is not effective as an acid producer. Phages multiply so rapidly that a factory may produce 10^{16} phages daily and acid production may stop rapidly after a phage attack. By the time phages are detected, the culture is inactive resulting in financial losses to the cheesemaker. Phages are not destroyed by pasteurization or at 55 °C for 6 h but are destroyed by sterilization (Sozzi *et al.*, 1982).

The DNA material injected into the cell replicates the phages which lyse the cell. Sandine

(1979) classifies cultures on the basis of sensitivity: (1) phage insensitive, (2) phage carrier, or (3) phage sensitive. Phage–host interaction may be categorized as virulent, temperate and carrier. In virulent phages there may be lysis or partial lysis of the cell and the survivors are phage resistant. In temperate phages there are different types of lactic cell–phage interactions, including release of residing pro-phages, cell lysis transduction and special transduction (Sandine, 1979) in carrier phages in which a small number of free phages are always present.

Phage variants occur due to mutation or genetic recombination. All lactic bacteria develop phages. If a new strain is brought into a factory, a phage will eventually attack the strain. Strains resistant to certain phage variants are sensitive to others. The problem of phage resistant strains is that they are slow acid producers. Cultures resistant to phage may have charged cell walls that do not attract the phage and may develop different permeabilities and metabolic activities which make them unsuitable as cheese cultures. Bacteria may develop resistance to a phage and become a phage carrier. This phage may remain dormant in the cell for many cell generations without any harmful effect on the cell. This phage may impart immunity to the host bacteria against attack by another phage (Gasson and Davies, 1980b). Bacteria may become multiple carriers and resistant to more than one type of phage. The dormant pro-phage may reappear as phage particles called a lysogeny state. If the cheesemaker uses lysogenic cultures, he contaminates his milk (Davies and Gasson, 1981). Host specificity to lytic phages may be altered in mutant strains of *S. cremoris* (Sinha, 1980).

Phages are strain specific, that is, a phage will attack one strain of lactic culture. After a culture failure or a slowdown of acid development due to phage, it is desirable to change the strain of culture. If acid development commences normally and stops, then phages are very likely to be responsible and a new strain of lactic bacteria that is not susceptible to that phage must be used. This new strain may be used for a day or more until it slows down. Culture suppliers can usually recommend a rotational plan for their cultures. Some suggest that a different culture be used daily with as many as 10 different cultures on a systematic rotation. Culture rotation is an excellent method of preventing costly losses due to phage.

Culture media prepared from calcium-reduced skim milk and containing phosphates have been effective in reducing phage attacks. This material is effective for mother and bulk cultures because phage cannot reproduce due to lack of calcium. Cheese milk contains a large percentage of calcium and phage can grow in the cheese vat, but by using phage resistant media, a large inoculum of phage is not added to the cheese milk from the bulk culture.

Thus, culture control measures for preventing infection are: (i) practise aseptic techniques for carrying cultures in a proper culture room; (ii) rotate cultures daily; (iii) proper sanitation of all equipment; (iv) use a phage resistant medium; (v) use a direct-to-vat culture which eliminates transferring cultures; (vi) fog the make room with 200 p.p.m. chlorine or iodine; (vii) use a mixed strain culture of two closely related strains and not a single strain; and (viii) remove and dispose of whey daily.

24.6 COAGULATION OF MILK

Milk is a lacteal secretion containing fat, colloidal casein, dissolved protein, lactose and salts. Physical treatments such as cooling, pasteurization and concentration disturb the colloidal equilibrium. The caseins (phosphoproteins) of molecular weight 20 000–30 000, constitute about 76% of the true protein (Szijarto *et al.*, 1973) and are composed of α_{s1}, β and $\varkappa$, and the serum proteins constitute the remainder (Table 1).

Destabilization of the milk system is due to the destabilization of the casein micelles. The protein may be agglomerated by enzymes, acid and/or heat, singly or in combination. Caseins associate to form aggregates in the presence of calcium. Larger aggregates are formed in combination with calcium phosphate. It is this property of the caseins that results in the formation of cheese and yoghurt. Most of the calcium of milk is located in the casein micelle, usually bound or complexed with the phosphate or citrate (Rose and Colvin, 1966). The amino acid profiles and chemical characteristics of the caseins, serum proteins and the effects of pH, concentration, salt balance and temperature on curd agglomeration have been reported in some excellent papers (Cheeseman, 1981; Dalgleish, 1982a; Farrell and Thompson, 1974; McKenzie, 1971a, 1971b; Parry, 1974). The kinetics of the clotting reaction are described by Dalgleish (1982a). The factors affecting protein denaturation relative to cheesemaking parameters will be discussed in Sections 24.6.1, 24.6.2 and 24.6.3.

24.6.1 Heat Coagulation

Serum (whey) proteins constitute 0.4% of milk or 18% of total milk nitrogen, made up of 46% β-lactoglobulin, 21% lactalbumin, 19% proteose–peptone, some casein and the remainder globulins and enyzmes. The effect of heat on proteins is a two stage process in which the secondary and tertiary structures are altered causing denaturation, followed by aggregation of the proteins which results in coagulation. Excellent reviews of heat denaturation have been published by McKenzie (1971), Dalgleish (1982b) and Sawyer (1969). About 80% of serum protein is denatured at high temperatures, but the proteose–peptone fraction cannot be recovered by heat precipitation. Some of the serum proteins will complex with caseins at temperatures greater than 70 °C and produce what are called coprecipitates (Sawyer, 1969). This high heat treatment has been used to increase the yields of fresh cheese (Muller, 1971). A heat treatment of 85 °C for 5 min is used to form a heat acid (pH 5.3) precipitated curd called Queso Blanco (Parnell-Clunies *et al.*, 1982; Torres and Chandan 1981).

24.6.2 Acid Coagulation

Casein is agglomerated by acid commencing at pH 5.3–5.2 and titratable acidity greater than 0.52%. The maximum coagulation is at the isoelectric point of casein which is pH 4.6. This coagulation is dependent upon temperature and the salt balance, namely the ratio of insoluble to soluble calcium phosphate. Milk may be abnormal in this balance if the cow is suffering from mastitis.

Cottage, bakers' cheese and quarg coagulate when a low pH of 4.8–4.5 is attained. In cottage cheese making, a small amount of chymosin may be added to the milk (2 cm³/1000 kg) to make the curd more elastic and less subject to breakage.

24.6.3 Enzyme Coagulation

Many proteolytic enzymes cause milk to clot or form a coagulum. These enzymes are acid proteinases (aspartate proteinases) which are active in the presence of aspartate residues (Foltmann *et al.*, 1978). $\varkappa$-Casein has a stabilizing effect on the casein micelle thus preventing coagulation. The development of micellar instability results from the splitting or hydrolysis of the highly sensitive phenylalanine–methionine (105–106) peptide bond of $\varkappa$-casein. Splitting of this bond causes the formation of *para*-$\varkappa$-casein and a highly hydrophilic glycomacropeptide. As a result of this action on the micelles, the stabilizing effect of the $\varkappa$-casein is negated and the milk coagulates or clots in the presence of calcium (McKenzie, 1971). The desired enzyme has a high ratio of clotting to proteolytic activity. Traditionally, chymosin (rennet) has been used for this purpose, because it has a high clotting power with relatively little proteolysis. Chymosin has more specificity for attacking the phenylalaine–methionine bond than other coagulants. Excessive proteolysis is undesirable as it may result in bitter peptides and texture defects.

The enzymes increase in activity as the pH is lowered below the pH of fresh milk (6.7–6.8). Most milk-clotting enzymes are irreversibly denatured above this pH. The optimum temperature for coagulation is 40 °C while below 27 °C the curd is soft and at 30 °C the curd is firm and elastic. At temperatures below 20 °C and above 50 °C, clotting activity is very low. Most enzyme extracts are standardized so that 200 cm³ coagulate 1000 kg of milk or 1 part in 5000.

24.6.3.1 Milk treatments for coagulation

Previous heat treatment of milk may seriously affect the calcium phosphate balance and the colloidal calcium phosphate–caseinate complex. This shift increases the chymosin clotting time which is inversely related to the ratio of colloidal to soluble phosphate and to the total calcium content of the milk (Dalgleish, 1982a). Soluble calcium salts may be added to milk to restore normal clotting. Milk will coagulate instantaneously at 80 °C and a pH of 5.3 due to casein agglomeration. The casein is agglomerated by low pH but the serum protein is unaffected by low pH.

Preheating milk to temperatures > 70 °C reduces the susceptibility of the $\varkappa$-casein to attack by chymosin. Fewer peptides are released when previously heated milk is exposed to the enzyme. Inhibition appears to result from the complexing of β-lactoglobulin and $\varkappa$-casein.

Cold storage changes the calcium phosphate–citrate equilibrium in milk and soluble caseins increase. Loss of cheese yield may be overcome by pre-ripening with lactic cultures, the addition of calcium chloride and by adjusting the pH. Heat treatment, like thermization of milk, is a useful way to cause re-adsorption of the solubilized casein on the micelle and to re-establish the original cheesemaking properties of milk (Ali *et al.*, 1980).

From a public health standpoint, it is necessary to heat treat or pasteurize milk to destroy pathogenic organisms. Heat treatments in excess of 64 °C for 30 min or 75 °C for 16 s complex some $\varkappa$-casein and seriously affect enzyme hydrolysis, which results in a soft or weak curd.

24.6.3.2 Enzymes

Chymosin (rennin) is a saline extract from the abomasum of milk fed calves. Chymosin is the major protease in the young calf (88–94%) but is replaced by pepsin as the calf matures. Chymosin extracts contain some pepsin depending upon the age of the calf. Worldwide cheese production has increased more rapidly than the availability of chymosin which has resulted in chymosin shortages and subsequent higher prices.

Pepsins, obtained from the stomach mucosa of adult bovines, porcines and chickens can be used as chymosin substitutes, but slow coagulation may occur if the pH is too high. Between pH 6.3–6.8, the milk clotting activity of the pepsin decreases more rapidly than that of chymosin. A mixture called 50/50 is used commercially which represents 50% of pepsin and 50% of chymosin. Chymosin substitutes can also be extracted from the moulds *Mucor pusillus*, *Mucor miehei* and *Endothia parasiticus* as well as from bacteria *Bacillus subtilis* and *Bacillus polymyxa* (Mann, 1967). The difference between the microbial rennets, pepsins and chymosin is in the coagulation–proteolysis ratio. Chymosin has the highest ratio followed by *Mucor* rennet, *Endothia* rennet and bovine pepsin in the order 1.40, 0.52, 0.15, 0.04, respectively (Law, 1982).

The various rennets have proprietary names and each has its own characteristics. The *Endothia parasiticus* rennet may be used for cheese with high cooking temperatures which destroy the proteolytic activity of the enzyme. The *Mucor* rennets are heat stable in whey and precaution must be taken to avoid continued proteolysis.

24.6.4 New Developments

Automated continuous cheesemaking techniques have encouraged attempts to accelerate the operation and reduce costs. It is possible to immobilize enzymes by covalent binding on an insoluble carrier for continuous and controlled proteolysis of milk. Degradation of $\varkappa$-casein follows Michaelis–Menton kinetics and the binding of the enzyme is not greatly affected by immobilization (Dalgleish, 1979). Such a system is difficult to use as the resultant curd clogs the column. When milk is renneted below 15 °C, the enzyme degrades the $\varkappa$-casein without coagulation. Clotting can then be induced by heating the rennet-activated milk to 25–30 °C. Pepsin is the most suitable coagulant when attached to porous glass (Dalgleish, 1982a). Bound chymosin is active but possesses only a fraction of normal chymosin activity. Immobilization inhibits enzyme diffusion and the rate of proteolysis depends on casein micellar diffusion which is slower due to the size of the casein. Immobilized proteases produce *para*-$\varkappa$-casein and glycomacropeptide and release NPN into solution, but there is increased proteolysis of other casein compounds. Chymosin and pepsin appear to be more successful than mucor rennets (Dalgleish, 1982a).

It has been claimed that the use of immobilized enzymes produces a satisfactory curd for cheesemaking (Ohmiya, 1981; Ohmiya *et al.*, 1981), but Cheeseman (1981) suggests that although a satisfactory curd is formed, there is the necessity for a portion of the enzyme to be released for subsequent ripening. He also suggested that the process is not likely to be successful until better process control for curd forming and ripening is attained. Dalgleish (1982a) comments that the low activity of the bound enzymes and the removal of inhibitory characteristics of the milk system have prevented their commercial application.

Molecular cloning and nucleotide sequence of cDNA coding for calf preprochymosin is an exciting new development. A biologically active specific mRNA was isolated from the digestive tract of the calf and was a suitable template for the generation of specific cDNA for prorennin mRNA sequences (Uchiyama *et al.*, 1980). Double stranded cDNA was prepared from prorennin-specific mRNA by sequential actions of reverse transcriptase, DNA polymerase and S_1 nuclease, and inserted into the site. Transformation of *E. coli* by the hybrid plasmid yielded transformants containing prorennin cDNA (Nishimori *et al.*, 1981, 1982a,b). Clones were identified which contained a complete copy of prochymosin in which the nucleotide sequence is in substantial agreement with the reported amino acid sequence of prochymosin (Harris *et al.*, 1982). Expression of preprorennin in a host cell is obtained by generating a DNA sequence that codes for preprorennin. This sequence has attached to it a transcriptional promoter and a ribosomal binding site. The DNA is then transformed into the host cells. The host cells are cloned and those that have a high level of expression of preprorennin are selected. Any species of bacteria which are considered safe, such as *E.coli*, *Bacillus subtilis* and various *Lactobacillus* and *Micrococcus* species, may be used for recombinant DNA. This genetic material is often contained in the form of a plasmid which is capable of replicating in the host cell and has inserted therein genetic material from a donor cell (Alford *et al.*, 1982). It is possible that this technology may yield a stable chymosin that is economical with high coagulating and low proteolytic activities for cheesemaking.

24.7 CURD TREATMENT

24.7.1 Ripening the Milk

The milk for cheesemaking at the optimum bacterial growth temperature is inoculated with the correct proportion of lactic culture which may vary from 0.1 to 5% of the weight of milk. A 1% culture is the normal amount for most types of cheese. Culture may be weighed or metered into the cheese milk. Inoculated milk may be held for periods of up to 2 h for the desired degree of acid development to be attained. Long ripening periods in open vats expose the milk and culture to airborne contamination, notably bacteriophage. One hour of ripening is common for some hard cheeses like Cheddar, during which time the acidity increases 0.01% which is an indication to the cheesemaker that bacterial activity is progressing. When the cheesemaker has determined that the bacteria are active and growing and have developed acidity, he proceeds to set the milk (rennet addition).

24.7.2 Setting the Vat (Addition of Coagulant)

The addition of rennet to the vat is called setting. Rennet is usually added at the rate of 20 cm^3 liquid extract or 1 g powder per 100 kg milk. The correct amount of rennet mixture is diluted with 20 times its volume of sterile cold water in order to obtain a uniform distribution. The rennet is added ahead of the agitator and stirred for 3–5 min. The vat is covered and left undisturbed until coagulation occurs. The milk should be kept warm during this period. The first signs of coagulation will be noticed in 10–14 min after setting. If a spatula is inserted in the milk, it will show flakes of curd on the blade. The curd will be firm enough for cutting in twice the time noted for the first signs of coagulation. Various cheesemakers desire different degrees of firmness at cutting depending upon the type of cheese. If the curd is cut too soon, when soft and mushy, there will be large protein and fat losses in the whey, whereas if the curd is too firm at cutting the cubes of curd are brittle and will shatter, resulting in higher than normal losses. An acid curd (fresh cheese) at low pH (4.6–5.0) is flocculent and is not cohesive. A rennet curd at higher pH (6.7–6.0) is cohesive, elastic and shrinkable. Some of the factors that affect the firmness of the curd are: (i) amount and strength of coagulating enzyme; (ii) too low or too high setting temperature; (iii) too low calcium and protein contents of the milk; (iv) previous heat treatment (over pasteurized); (v) acidity of the milk; and (vi) colostrum milk or milk of abnormal composition.

24.7.3 Cutting the Curd

Mechanical devices have been developed for the determination of the correct cutting firmness, but these have not been adopted and reliance is placed on the organoleptic assessment by the cheesemaker. The curd is ready to cut when it breaks clean after a spatula is inserted at a 45° angle to the surface and slowly raised. The firm curd divides into two portions without breaking.

The curd may be cut into various sized cubes depending on the type of cheese being made. Curd is normally cut with wire curd knives, horizontal and vertical, which are spaced from 3 mm for Swiss to 1.2 cm apart for cottage. Traditionally, horizontal knives are used to cut the curd into layers. These layers are cut with vertical knives in two directions so that the resulting curd particles are uniformly sized cubes. It is important that all cubes be identical in size so that the expulsion of whey on heating will be at a uniform rate.

Harps are used in some plants for cutting curd in round bottomed vats as in Swiss and Romano. In mechanized cheesemaking operations the curd is cut with mechanical cutters attached to the vat agitator. With some types of high solids milk (sheep), the curd is too firm and must be broken by stirring devices.

In the production of fresh cheese, the curd may not be cut but ladled directly into cloth lined or perforated moulds or placed in cotton bags. For some fresh cheeses like quarg and cream, the curd may be pumped into a sludge separator which removes the whey quickly. In this case the soft curd will be ejected from the centrifuge and packaged directly.

The curd after cutting is quite fragile and must be handled gently to avoid curd damage. Agitation can become more rapid when enough whey is released so that the curd may be stirred without breakage. The stirring operation may require from 15 to 60 min. Stirring maintains the curd in discrete particles allowing the uniform release of whey. If stirring is inadequate, the curd particles may fuse together preventing the free release of whey.

24.7.4 Cooking (Heating the Curd)

When the curd is firm enough for handling, the curd–whey mixture may be heated to the desired temperature for the type of cheese by turning on the steam in the jacketed vat.

Cooking causes the syneresis of the curd. As the temperature is increased there is a rearrangement and shrinkage of the gel structure with resulting expulsion of whey. The heating may favour further degradation of $\varkappa$-casein and provide for increased crosslinking in the gel which results in a firmer more rubbery curd (Dalgleish, 1982a). As the firmness of the coagulum increases, the water holding capacity decreases. The shrinkage of the curd is also accelerated by the decrease in pH from pH 6.7–5.0 due to the action of the lactic bacteria (Beeby *et al.*, 1971). Cooking relates to both the time and temperature of heating; the higher the temperature and the longer the time of cooking, the less the moisture content of the cheese.

The starter bacteria are 95% enmeshed in the curd at renneting and are retained in the curd particle. These ferment lactose in the curd and the development of acid takes place more rapidly in and around the curd. The acid development assists in the expulsion of moisture. The number of lactic bacteria may have increased from 2×10^7 cfu ml^{-1} at ripening to 4×10^8 cfu g^{-1} at cooking (Chapman and Sharpe, 1981).

Hard cheeses like Cheddar are usually cooked to 38 °C and Swiss is cooked to 52 °C depending upon the heat-tolerance of the lactic cultures, while high moisture, soft and semi soft cheeses may not be cooked. The curd is held in the whey until the desired degree of firmness (loss of moisture) or pH is attained. The pH of separating the curd and the whey is very critical for some types of cheese. In hard cheeses, the pH of the curd should have decreased to at least 6.2 and preferably 6.1. The cheesemaker determines the end of cooking by organoleptically testing the firmness of the curd.

In all cheesemaking techniques, it is important to control the calcium phosphate content of the curd. This is achieved by the decrease in pH. The insoluble salts are rendered soluble by the acid and are largely lost in the whey. One fourth of the phosphorus is held in organic combination and the calcium becomes soluble more rapidly than the phosphorus and other mineral components. The Ca/P ratio and the percentage of calcium are low in high acid cheeses. The mineral content of the curd will be further reduced if the curd is washed. About 60–65% of the calcium and 50–60% of the phosphorus are retained in Cheddar cheese while only 20% and 37% of the calcium and phosphorus respectively are found in fresh cheeses such as cottage (Ernstrom and Wong, 1974).

One of the most important aspects of cheesemaking is to control the amount of lactose in the curd and in the resultant cheese. Lactose may be removed by fermentation, syneresis or washing. It is the desire of the cheesemaker to reduce the lactose to a level that when left in cheese will bring fermentation to a final pH of 5.1 at 3–4 days of age. Residual lactose remaining in the curd will eventually ferment and reduce the pH. A low pH (< 5.0) results in cheese which is acid, sour and bitter while a cheese with pH > 5.3 will be fruity and fermented. If acidity fails to develop properly, there will be too little syneresis and the curd will be high in moisture. This curd containing high moisture and lactose will eventually ferment resulting in a sour, acid cheese.

The control of lactose may be the result of fermentation which causes low pH and shrinks the curd, releasing lactose containing whey. The whey draining removes most of the lactose.

In some of the soft cheeses where acid development is purposely slow, the curd retains a large percentage of whey and lactose. The milk may be diluted or the curd may be washed with water to leach some of the developed acid and lactose from the curd. Careful control of the amount of water at washing determines the amount of lactose left in the curd. This residual lactose will ferment and reduce the pH to the desired level of 5.1–5.2. It is common in some types of cheese to remove one half to two thirds of the whey and replace it with water at the same temperature. The lactose equilibrates with the water and it is important that the water is removed within 15–20 min.

24.7.5 Curd Handling

When the curd has attained the proper degree of firmness and/or pH, the whey will be removed from the curd. In some cases the curd is allowed to settle and the whey removed from the top or the curd–whey mixture can be pumped or drained by gravity to whey draining vats. In some types of cheese the curd–whey mixture may be placed in perforated moulds for draining, forming and shaping. A battery of moulds may be placed on conveyors so that they can be filled mechanically. For some cheeses the curd–whey mixture is placed in press vats in which plates are placed over the curd to help expel the whey. The large blocks of curd are automatically moved forward and cut into small blocks that are put in the moulds. Whey composition, processing and utilization are discussed in Section 24.11.

There is much proprietary equipment designed for separating curd from whey and for handling, forming and shaping the curd. The reader is advised to contact the equipment manufacturers relative to the type of cheese being produced.

Most cheeses are placed in forms when the curd is warm and the rheological properties of the curd allow it to fuse and form the desired shape. The curd fuses in the absence of whey and is much more plastic in the absence of salt. The cheese so formed is usually allowed to remain in the forms or in the presses over night, during which time the pH decreases. After 12 h in the forms the cheese is salted.

In types of cheese like Cheddar and Colby, the cheese is kept warm for 2 h, in a vat, curd sink, cheddaring belts or towers until the correct pH develops (5.35–5.45) and the proper texture develops. The curd is then comminuted and mixed with salt. After thorough mixing and penetration of the salt, the curd is hooped and pressed.

24.7.6 Pressing

Pressure is exerted on cheese in the moulds to press the cheese into the desired shape and to assist in expelling whey. The pressure may be little or none for soft cheese and up to 172 kPa for firm Cheddar cheese. The warmer the curd, usually the less pressure required. Cheese may be pressed individually or in long continuous hydraulic presses or in large block presses. In order to close up the texture of the cheese (reduce openings), a vacuum of up to 101 kPa may be drawn on the blocks of curd either before, during or after pressing. These blocks may be from 200 g to 300 kg.

24.7.7 Salting

The purpose of salting is to aid in whey removal, shrink the curd, slow down acid development and check undesirable forms of bacteria. There are three methods of salting: (1) rubbing dry salt on the surface of the cheese; (2) placing in a brine bath at a concentration of 16–25% NaCl; and (3) addition of salt to the dry curd. The salt content of cheese should be between 1.5–2%, although some cheese may be salted as high as 4%. Cheddar and Colby and similar cheeses are salted in the vat prior to pressing and hooping but most cheeses are now brine salted, although a few are dry salted. The amount of salt is important as it stops the lactic fermentation and controls the rate of ripening. Too much salt will result in slow curing and inadequate salt leads to off flavours.

The cheese is often immersed in the brine tank in mechanical racks and left until adequate salt has penetrated into the cheese. The brine tank is located in the refrigerated storage to assist in cooling the curd quickly. The length of time in the brine depends on the size of the cheese.

Cheeses of 100–150 kg, such as Swiss and Parmesan, require 3–5 days, 2–5 kg cheeses require 24 h, while cheeses of 250–350 g may only require 1–2 h.

24.7.8 Cheese Ripening

Ripening refers to the storage of cheese under controlled conditions of time, temperature and humidity during which the desired body, flavour and texture develop. Ripening embodies the physical and chemical changes that occur in the curd which results in the characteristic flavour and body. Fresh curd is often tough and rubbery but during aging it changes to a smooth plastic or soft consistency. During this time the ripening agents bring about degradation of the lactose, proteins and fats which result in the desired flavour profile of that cheese.

The ripening agents in the cheese are the bacteria and enzymes of the milk, lactic culture, rennet, lipases, added moulds or yeasts and environmental contaminants. The cheesemaker controls these agents by manipulating temperature, time, moisture, pH and salt.

24.7.8.1 *Temperatures*

Cheese may be cured at temperatures of 4–15 °C for periods of a few weeks to two years. The higher the temperature of curing, the more rapidly flavour will develop, but higher temperatures are often associated with flavour and texture defects. If the cheese is properly made from milk of good quality, it may be cured at temperatures as high as 10 °C which increases the rate of proteolysis by 40–100% (Scott, 1981; Wong, 1974). Some types of cheese require special periods of controlled temperature and humidity which favour the growth of selected microorganisms such as *Propionibacterium shermanii* and *Brevibacterium linens*, in Swiss and brick, respectively.

24.7.8.2 *Moisture of curd*

The cheesemaker controls the moisture content of the curd by manipulating pH and temperature during the making process. The moisture of the curd is bound to the casein and the fat–protein matrix and is stabilized when maximum acidity is attained. This moisture contains the salt-soluble fractions and the enzymes and microorganisms and, therefore, the amount of moisture controls the rate of ripening. A high moisture (> 50%) cheese may ripen in a few weeks while a low moisture Parmesan (< 32%) may require two years to attain the desired flavour level. Added salt (normally 1.7%) which is dissolved in the water phase controls the growth of some undesirable organisms (Chapman and Sharpe, 1981). A 1.7% salt in a Cheddar cheese (36% moisture) is equivalent to $1.7/36 = 4.7\%$ salt in the moisture of the cheese.

The water activity (a_W) is the ratio of the vapour pressure of the food to that of pure water at the same temperature. The a_W for fresh cheese is 0.98, for processed cheese and semi soft types between 0.98 and 0.93 a_W, while aged Cheddar is below 0.93 and some aged cheese may be below 0.85 a_W (ICMSF, 1980a).

A high salt content on the surface of surface-ripened cheese may suppress some undesirable moulds (but permit the growth of desirable ripening organisms as in the case of *B. linens* on a Port Salut cheese. Work in our laboratory has shown that the plasticity of the curd is inversely related to the salt content. In brined cheese, high salt at the surface diffuses through the cheese with time. The eye formation and the growth of *Propionibacterium shermanii* takes place in areas of low salt in the centre of the cheese. A high salt will also inhibit the production of gas by *E. coli*.

The pH of the cheese controls the growth of microorganisms and greatly affects the plasticity and other rheological properties of the finished cheese. For instance, a low pH curd (< 5.0) will usually result in a short, brittle, crumbly cheese with an acid, sour taste.

24.7.8.3 *Lactose degradation*

Changes in lactose occur largely during the making process and during the first stages of ripening. Most of the lactose disappears at about 3 days, when cheese attains its minimum pH. Conversion of lactose to lactic acid involves about 14 enzymatic steps (Wong, 1974). Lactic acid inhibits undesirable organisms such as the *coli-aerogenes* group and allows normal ripening and keeping quality. The different types of cheese are subjected to different curing parameters and as a result a great number of degradation products are formed, *e.g. S. diacetylactis* produces diacetyl; *L. bulgaricus*, acetaldehyde; yeasts, ethanol; *Propionibacterium shermanii*, propionic acid, *etc*. The intermediate and end products of fermentation which produce the distinctive character of a cheese are reported in some excellent reviews (Marth, 1974; Ney, 1981; Wong, 1974). According to the following metabolic pathway, lactose is changed to pyruvic acid → lactic acid → diacetyl → acetaldehyde → ethanol and acetic acid and carbon dioxide. The lactic organisms may cause a slight degree of proteolysis and some may result in the formation of bitter peptides.

24.7.8.4 *Protein degradation*

The degree of proteolysis of cheese may be measured by the increase in the water soluble nitrogen. Vakaleris and Price (1959) found that the degree of ripening could be assessed by a measure of the tyrosine content. Other measurements of the degree of ripening have been suggested (Ney, 1981; Wong, 1974). Proteolysis of casein and whey proteins forms peptones and peptides, which are further broken down to amino acids. The formation of free amino acids commences during the making process and continues during ripening. Transaminase and decarboxylase enzymes are elaborated by microorganisms and released into the cheese. Eighteen different free amino acids have been reported (Wong, 1974). A simplistic degradation of the proteins is as follows: proteins → peptones → peptides → smaller peptides → amino acids. Some of the end products that have been reported are as follows: tyramine, serine, proline, agmatine, cadavarine, histamine, putrescine, tryptamine, α-aminobutyric acid, glutamine, glutamic acid, proline, aldehydes, tryptophan, carbon dioxide, hydrogen sulfide and ammonia. These are some of the products which have been found in different cheeses in various concentrations and which in varying proportions give a cheese its specific flavour.

24.7.8.5 *Lipid degradation*

The lipids contribute more to flavour than other constituents. Milk lipases from raw milk, glandular lipases and microbial lipases may all produce free fatty acids. Some lipolytic organisms from silage or low grade milk may also produce rancidity. The principal ripening of blue-veined cheese is caused by the lipolytic and proteolytic activity of *P. roqueforti*. The lipases release the short chain fatty acids — butyric, caproic, caprylic and capric which may be hydrolyzed into methyl ketones such as heptanone 2; the mechanism is by oxidation of the fatty acids to the β-keto acid and by decarboxylation (Schwartz and Parks, 1974). Some Italian cheeses (Parmesan, Romano, Provolone) owe their 'piquant' flavour to the release of the free fatty acids (Nelson *et al.*, 1977). Romano cheese flavour is due to free butyric and glutamic acids (Wong, 1974).

24.7.8.6 *Cheese flavour development*

Flavour of cheese is hard to define and to determine. Many excellent papers have been prepared on the subject of cheese flavour (Fryer, 1969; Harper, 1959; Mabbit, 1955; Marth, 1963; Wong, 1974). New techniques of extraction, isolation and characterization have elucidated the complexities of flavour. Flavour is produced from breakdown products of the fat, protein and lactose due to slow release of the ripening enzymes. There is no one enzyme or bacterium, or group of enzymes or bacteria that are responsible for a specific flavour. Flavour is the end result of the release of many compounds, all of which in the correct proportion produce the desired ripened cheese flavour.

There has been considerable interest in accelerating flavour development as a means of eliminating curing room costs. Harper and Wang (1980), Harper *et al.* (1979), Kristofferson *et al.* (1967) and Law *et al.* (1979) have investigated cheese slurries as a means of producing flavour more efficiently. A water–cheese slurry is formed and held under carefully controlled conditions at 30 °C for 7 days (Chapman and Sharpe, 1981). Such slurries have been added to curd and resulted in more rapid flavour development often associated with undesirable off flavours. There has also been much interest in accelerating cheese ripening in the traditional sizes and shapes (IDF, 1980c). This group has suggested some of the following for acceleration of flavour — use of modified starter (doubling number of viable cells), the addition of exogenous enzymes (Sood and Kosikowski, 1979) and the development of more flavour in cheese of low fat content. The addition of encapsulated enzymes to cheese milk has been recommended in which the capsules slowly digest as the cheese matures and release the ripening enzymes (Magee and Olson, 1981).

24.7.9 Packaging

Packaging prevents desiccation, mould growth, growth of surface proteolytic organisms and ingress of oxygen, and protects the cheese from insects and vermin. The cheese is usually packaged after it is made but in some cases, like Camembert, the cheese cannot be packaged until after the mould or smear has developed and then only in an oxygen permeable wrapper.

Fresh unripened cheese may be packaged directly in rigid containers of polystyrene or poly(vinyl chloride) with heat-sealed or snap-on closures.

Soft cheese may be wrapped in grease proof paper or foil-laminated paper which are not heat sealed as the surface organisms require oxygen. Laminated films of polyester, aluminum and polyethylene may also be used.

Some cheeses (Scamorze, Provolone) have a smooth moulded surface and are allowed to dry and form a hard rind. The surface may be oiled with butter or olive oil to prevent cracking of the rind and desiccation. Waxing is a common method of protecting cheese. The cheese, when the surface is dry, is immersed in a vat of flexible microcrystalline wax at 105–121 °C for 5 s. Wax slows down the loss of moisture and allows gases to permeate from the cheese. Wax, a by-product of the oil industry, is expensive and in some traditional cheeses the wax is being replaced with plastic. Plastic films provide effective barriers against oxygen, moisture loss and insect pests.

Moisture proof cellophane, polyethylene, saran, poly(vinylidene chloride) (PVDC)–vinyl chloride copolymers and nylon 6 are commonly used for wrapping cheese. Many of these materials are copolymerized or laminated to provide the desired degree of protection.

Hard cheeses are often cured in laminates of nylon and polyethylene and vinylidene chloride–vinyl chloride copolymers. The film may vary in thickness from 10–40 μm. For consumer packages a shrink wrap of PVDC is commonly used. Some consumer packages are CO_2 or N_2 flushed and others are vacuum packed.

Growth of specific mould species is desirable for mould ripened cheese, *e.g.* blue and Camembert, but is undesirable in other types of cheese. Moulds may impart flavour and produce mycotoxins (aflotoxins). Mould growth may increase the pH, even above 7.0, which favours the growth of proteolytic organisms and staphylococci (Duitschaever and Irvine, 1971). These proteolytic organisms may produce an undesirable flavour in the cheese. Mould growth can be controlled by flushing with an inert gas or pulling a vacuum to eliminate oxygen. Antimycotic agents may also be used for coating the surface or by impregnating in the wrapping material. Sorbic acid, propionates and pimarcin are effective antimycotic agents. Pimarcin is a non-toxic antibiotic produced by *Streptomyces* which suppresses the growth of yeasts and moulds.

24.8 CHEESE YIELD

The main factors that determine the yield of cheese are: composition of the milk, amount of constituents lost in the whey, amount of salt added and the amount of water retained in the cheese. Two methods used for calculating yield are: (1) kg of cheese made from 100 kg of milk, or (2) kg of cheese made per kg of fat. There have been many formulae devised for various types of cheese (Davis, 1965; Scott, 1981) but a simple formula will approximate the yield of cheese. Table 2, which is based on Ontario milk, contains yield values for various fat–protein levels and various moisture contents of cheese. A general formula for yield of cheese is fat plus protein multiplied by a factor which represents the percentage of moisture in the cheese, *i.e.* yield of cheese = (fat + protein) x factor. The moisture levels and the x values are, respectively: 30–1.29, 31–1.31, 32–1.33, 33–1.35, 34–1.37, 35–1.40, 36–1.42, 37–1.44, 38–1.46, 39–1.48, 40–1.51, 41–1.53, 42–1.56, 43–1.59, 44–1.62 and 45–1.65. For example, if a 37% moisture cheese is being made from milk of 3.5% fat and 3.10% protein, then yield of cheese per 100 kg milk is (3.5 + 3.1) 1.44 = 9.5 kg.

Some of the factors that cause poor yield are: inferior milk with high bacterial count, milk with low solids, poor coagulation, careless cutting of the curd, cooking too fast, salting too soon, high temperatures during pressing, proteolytic culture and proteolytic coagulating enzyme.

Table 2 Cheese Yields (kg/100kg)[a]

Milk		Moisture content of cheese (%)													
Fat (%)	Protein[b] (%)	32	33	34	35	36	37	38	39	40	41	42	43	44	45
2.8	2.78	7.55	7.66	7.78	7.90	8.02	8.15	8.28	8.42	8.56	8.70	8.85	9.01	9.17	9.3
2.9	2.83	7.73	7.85	7.97	8.09	8.22	8.35	8.48	8.62	8.76	8.91	9.07	9.23	9.39	9.5
3.0	2.88	7.91	8.03	8.15	8.28	8.41	8.54	8.68	8.82	8.97	9.12	9.28	9.44	9.61	9.7
3.1	2.92	8.10	8.22	8.34	8.47	8.60	8.74	8.88	9.03	9.17	9.33	9.49	9.66	9.83	10.0
3.2	2.97	8.28	8.40	8.53	8.66	8.80	8.93	9.08	9.23	9.38	9.54	9.71	9.88	10.05	10.2
3.3	3.02	8.46	8.59	8.72	8.85	8.99	9.13	9.28	9.43	9.59	9.75	9.92	10.09	10.27	10.4
3.4	3.06	8.63	8.76	8.89	9.03	9.17	9.32	9.47	9.62	9.78	9.94	10.12	10.30	10.48	10.6
3.5	3.10	8.76	8.93	9.07	9.21	9.35	9.50	9.66	9.81	9.98	10.15	10.32	10.50	10.69	10.8
3.6	3.15	8.98	9.12	9.26	9.40	9.55	9.70	9.85	10.01	10.18	10.36	10.53	10.72	10.91	11.1
3.7	3.19	9.17	9.30	9.44	9.59	9.74	9.89	10.05	10.22	10.39	10.56	10.75	10.93	11.13	11.3
3.8	3.24	9.35	9.49	9.63	9.78	9.93	10.09	10.25	10.42	10.59	10.77	10.96	11.15	11.35	11.5
3.9	3.28	9.52	9.66	9.81	9.96	10.12	10.28	10.44	10.61	10.79	10.97	11.16	11.36	11.56	11.7
4.0	3.33	9.70	9.85	10.00	10.15	10.31	10.47	10.64	10.82	11.00	11.18	11.37	11.57	11.78	12.0
4.1	3.37	9.88	10.03	10.18	10.34	10.50	10.67	10.84	11.02	11.20	11.39	11.58	11.79	12.00	12.2
4.2	3.42	10.06	10.22	10.37	10.53	10.69	10.86	11.04	11.22	11.40	11.60	11.79	12.01	12.22	12.4
4.3	3.46	10.24	10.39	10.55	10.71	10.88	11.05	11.23	11.41	11.60	11.80	12.00	12.21	12.43	12.6
4.4	3.51	10.42	10.57	10.73	10.90	11.07	11.25	11.43	11.61	11.80	12.01	12.21	12.43	12.65	12.8
4.5	3.55	10.59	10.75	10.91	11.08	11.25	11.43	11.62	11.81	12.00	12.21	12.42	12.64	12.86	13.0

[a] Irvine, 1982. [b] Based on weighted averages of Ontario industrial milk, 1973 and 1974, from the Central Milk Testing Laboratory, Ontario Ministry of Agriculture and Food.

24.9 CHEESE DEFECTS

The factors that control the keeping quality of cheese are the moisture, pH, lactose, salt, oxygen content and the presence of undesirable microorganisms. The cheesemaker must control all of these parameters very closely in order to produce a first grade cheese.

The flavour of the cheese may be affected by undesirable organisms such as *E. coli* and *Cl. tyrobutyricum* both of which produce gas and flavour defects. The proper development of the culture is important because it establishes the pH at which the desirable ripening organisms flourish. For instance in Cheddar cheese if the pH is above 5.3 or below 5.0 serious flavour defects will develop.

The body of the cheese may have defects due to poor manufacturing practices or poor control of the culture. Crumbly, short, brittle, weak and pasty are common body defects.

Texture relates to the openings in the cheese. Mechanical openings are irregular and non-uniform and are due to the inclusion of whey in the curd at pressing. Gas holes are large round shiny eyes due to gas-forming organisms. Control of the size, shape and location of these openings as in Swiss cheese determine the grade of the cheese, but gas holes in Cheddar are undesirable.

The colour of cheese may also influence the grade of the cheese. The intensity of colour is not a defect but the colour must be uniform. Variations in colours such as mottled, streaky and acid cut are manufacturing defects.

The finish is the outward appearance of the cheese including the packaging. Defective finish may exhibit mould growth, cracking of the rind, misshapen or poor waxing, all due to careless manufacturing practices.

Grading of cheese is usually performed by expert plant or government graders. Many of the defects of cheese are interrelated, *e.g.* an acid flavoured cheese may have a weak pasty body and mechanical openings. The scores for ideal cheese are defined by most governments and the seriousness of the defects is spelled out in the regulations. The reader is advised to consult government regulations and grade standards (Chapman and Sharpe, 1981; Nelson and Trout, 1964; Scott, 1981).

Pathogenic bacteria may be found in cheese made from raw or improperly pasteurized milk or from post pasteurization contamination of the milk or curd. The only organisms that survive pasteurization are spore formers and enterococci. In many countries, raw milk or sub-pasteurized milk cheese must be held for 60 days at a curing temperature of > 2 °C. Most pathogenic bacteria will die out in properly made cheese at correct levels of temperature, moisture, salt and pH, due to competition from other bacteria. For more detailed information on pathogenic organisms, the reader is advised to refer to the International Dairy Federation Report (IDF, 1980b), Chapman and Sharpe (1981) and ICMSF (1980b).

24.10 PROCESS CHEESE

Process cheese is made from hard or semi soft cheese to which may be added fat (butter, cream or whey cream), skim milk powder, whey powder, emulsifying agents, water, salt, pH adjusting agents, colour and condiments in the correct proportion to meet government specifications of fat and moisture. The cheese blend is comminuted along with the dry ingredients. If a hard cheese is being used it may be necessary to grind it through rollers.

The cheese should be selected on the basis of moisture, age, colour, flavour and body. If the cheese is firm, corky or young, it imparts firmness to the finished cheese; but if the cheese is soft and weak in body it imparts this characteristic to the finished product. A young cheese may be difficult to melt down while an old cheese may impart graininess to the product. Usually, a three month blend will yield a desirable product. Cheddar-type stirred curd which has been fast-cured in polyethylene lined barrels has been used as a base, UF milk may also be used as an ingredient. The cheese should be trimmed and cleaned prior to processing.

Fat and moisture content of the cheese are determined and the correct proportion of the ingredients are calculated to meet the standards for fat, moisture and solids-not-fat. A process cheese is composed of cheese, fat from a dairy source for balancing the formula and emulsifier, whereas a cheese food may have added milk solids in the form of whey powder or skim milk powder. A cheese spread will have more moisture than a process cheese, 55–60% and 40–45%, respectively.

The 'emulsifier' is usually sodium citrate or disodium phosphate. Tri- and tetra-sodium phosphates have also been used in small amounts. The term 'emulsifier' is commonly used in the industry but these salts do not directly promote emulsification of the fat and water phases and the term 'sequestering agents' would be more appropriate. The cheese protein is in the form of a calcium *para*-caseinate. When the cheese mixture is heated, the calcium is sequestered and ion exchange takes place resulting in the formation of sodium caseinate and calcium citrate or phosphate. The sodium caseinate has greater water-holding capacity than the calcium *para*-caseinate at the same pH due to its electrostatic properties. Sodium caseinate also holds the fat in suspension and results in a firmer body and smoother texture. The more sequestering agents that are added, the firmer the cheese. Sequestering salts (emulsifiers) are usually added at the rate of 1.5 and 3% for process cheese and cheese spreads, respectively. These salts are usually dissolved in water and added to the process kettle.

The pH of the process cheese should be approximately 5.6. The added salts raise the pH and it may be necessary to add organic acids (citric) to lower the pH. The body of the cheese will be smoother at a higher pH but there is the risk of clostridia growth above 5.6.

The process kettle may be steam jacketed which may lead to slow cooking, over cooking and 'burn on'. The kettle should be equipped for direct steam injection which gives faster cooking and assists in homogenizing the mixture. Large plants have lay-down cookers where suitable worm screw agitators and direct steam injectors heat the curd mixture quickly and efficiently. It is important to have culinary steam for direct steam injection cooking.

Some process kettles are designed to cook the cheese under pressure or in a vacuum. The pressure may be increased to raise the cooking temperature of the mixture to 110 °C (Kessler, 1981). Cooking in a nonpressurized kettle is at 80–85 °C for 5–10 min, usually 80 °C for process cheese and 85 °C for cheese spread. The hot mixture, while still liquid, is poured or pumped to packaging machines.

High salt content, low pH and heat treatment give process cheese excellent keeping qualities. The only species that survive the heat treatment are spore formers such as clostridia. Heat treatment stimulates germination of these spores which originate from the cheese and may cause blowing and off flavours.

24.11 WHEY TECHNOLOGY

Cheese whey is the liquid by-product of cheese manufacture which is yellowish green in colour and has a salty flavour. The following discussion considers whey to be the product prior to salting. Salty whey occurs in press drippings from cheese salted before pressing (*e.g.* Cheddar) or where cheese is washed in a whey brine. Salty whey is difficult to process and should not be used for animal or human consumption. Like most dairy products, whey is perishable and must be handled in a sanitary way. Whey is 94–96% water which makes processing and handling expensive. Annual world production is about 65 million tonnes of whey containing 180 000 tonnes of protein and 1.1 million tonnes of lactose (Kosaric and Asher, 1982).

Whey has traditionally been dumped into surface water or fed to livestock. However, current environmental regulations and levies are forcing cheesemakers to treat whey before disposal to reduce its biological oxygen demand (BOD) from 40 000–60 000 p.p.m. to less than 500 p.p.m., and because of factory centralization the cost of transporting whole whey for feed use has become prohibitive. Therefore, whey has become a liability and a great amount of research has been focused on converting this liability to an asset. The composition, processing and utilization of cheese whey and its valuable components are discussed below.

24.11.1 Whey Composition

Cheese whey contains more than 50% of the original milk solids and constitutes 85–90% of the original milk volume (Kosikowski, 1979). Whey contains about 14, 22, 74 and 98% of the original milk fat, protein, ash and lactose, respectively (Table 3). Compositions of fluid and dried wheys are given in Table 4. Cerbulis *et al.* (1972) report the average composition of blended whey powders as 9.7% protein (protein nitrogen × 6.4), 71.7% carbohydrate, 1.2% lipid and 8.2% ash.

Table 3 Distribution of the Major Milk Components During Cheese Manufacture[a]

	Fat	Protein	Carbohydrate	Ash
100 kg milk contains (kg)	3.8	3.3	5.0	0.73
10 kg cheese contains (kg)	3.3	2.6	0.2[c]	0.19
% recovery[b]	86	78	4.0	26
90 kg whey contains (kg)	0.5	0.7	4.8[d]	0.54
% recovery[b]	14	22	96	74.0

[a] This table is calculated assuming an average yield of 10 kg of cheese from each 100 kg of milk. The actual yields vary with moisture content of the cheese but fat and protein recoveries are independent of final cheese moisture.   [b] % recovery is weight of component in cheese or whey/weight of component in milk.   [c] This value was estimated assuming cheese moisture of 40% and a carbohydrate content of 5.0% in the water phase of the cheese. Residual lactose in cheese is quickly fermented to lactic acid which reacts with Ca to form calcium lactate.   [d] Most carbohydrate in whey exists as lactose with a small amount of lactic acid.

Table 4 Gross Composition of Liquid and Dried Whey

Component	Sweet whey			Acid whey		
	Fluid[a]	Dried[a]	Dried[b]	Fluid[a]	Dried[a]	Dried[b]
Total solids	6.35	.96.5	96.3	6.5	96.0	95.4
Protein (N × 6.38)	0.80	13.1	13.0	0.75	12.5	11.7
Lactose	4.85	75.0	69.4	5.0	67.4	63.2
Fat[c]	0.50	0.8	1.0	0.04	0.6	0.48
Ash	0.50	7.3	8.3	0.80	11.8	10.6
Lactic acid	0.05	0.2	—	0.40	4.2	—

[a] Kosikowski, 1979.   [b] Glass and Hedrick, 1977a.   [c] Normally whey is centrifuged to <0.1% fat.

Whey contains small curd particles ('fines') which may be profitably recovered by filtration on a vibrating screen (fines saver) and included in the cheese. Because whey is normally defatted by centrifugation before any further processing, the remainder of this discussion will relate to defatted whey with less than 0.1% milk fat. The recovered fat or whey cream can be used in place of regular cream.

The carbohydrate content of whey is mainly lactose but also includes about 0.05–0.4% lactic acid depending on the type of whey (Table 4). Assuming the carbohydrate content in the water phase of cheese is similar to that of whey (4.8–5.0%), the carbohydrate content of cheese is about 2%, mostly in the form of calcium lactate (Table 3).

The principal whey protein fractions are β-lactoglobulin, α-lactalbumin, serum albumin, proteose–peptone and immunoglobulin. The reported amounts of the fractions depend on the method of fractionation (Harper, 1979). Estimates of the proportions of β-lactoglobulin range from 33 to 62% and average 46% of the true protein. The proportions of α-lactalbumin, serum albumin, proteose–peptone and immunoglobulin fractions are about 21, 5, 19 and 9%, respectively. About 25% of whey nitrogen is non-protein nitrogen (NPN).

The principal mineral components, calcium, phosphorus, sodium and potassium, constitute about 60% of whey ash content (Table 5). Whey also contains significant amounts of thiamine, riboflavin, pantothenic acid and folic acid (Gillies, 1974) and ascorbic acid (Mavropoulou and Kosikowski, 1973). Phospholipids, mainly phosphatidyl ethanolamine and phosphatidyl choline, constitute about 0.2% of dry whey (calculated from data of Mavropoulou and Kosikowski, 1973). Nutritional evaluations of whey have been reported by Forsum and Hambraeus (1977), and Glass and Hedrick (1977a, 1977b).

Table 5   Mineral Contents of Sweet and Acid Whey Powders

Component	Sweet		Acid	
	a	b	c	b
P^d	0.66	1.10	1.78	1.59
Ca^d	0.77	0.88	1.12	2.40
K^d	2.54	1.86	2.64	1.92
Mg^d	0.12	0.18	0.15	0.22
Na^d	0.83	1.29	0.76	1.09
Zn	9.0	2.1	41.7	81.0
Mn	24.0	—	24.0	—
Fe	6.0	9.0	12.1	13.0
Cu	2.0	2.8	2.3	5.3
B	4.3	—	5.3	—
Al	21.3	—	23.3	—
Pb	—	1.15	—	1.68
I	—	6.8	—	8.6
Hg	—	0.02	—	0.03
Cd	—	0.11	—	0.14
As	—	0.77	—	0.59
Se	—	0.06	—	0.03

[a] Average of 3 samples (Mavropoulou and Kosikowski, 1973). [b] Average of 12–15 commercial samples analyzed each quarter for one year (Glass and Hedrick, 1977a). [c] Average of 8 samples (Mavropoulou and Koskikowski, 1973). [d] %.

24.11.1.1   *Factors affecting whey composition*

The composition of whey varies with cheese manufacturing procedures, where the most important determining factor is the pH at which the curd is separated from the whey, *i.e.* the dipping pH. Sweet whey, pH 5.9–6.5, is a by-product of the manufacture of hard and semi soft cheese (*e.g.* Cheddar, Swiss, Edam, Mozzarella) and acid whey, pH 4.4–4.8, is obtained from the manufacture of fresh cheese and casein. Acid whey contains more lactic acid and ash and less protein than sweet whey (Table 4).

Difference in ash content is mainly due to higher levels of calcium and phosphorus in acid whey. Calcium phosphate is bound to the casein and removed with it during coagulation. Acid development during cheese manufacture releases calcium to the whey as illustrated in the work of van den Berg and de Vries (1975). Robinson *et al.* (1976) compared the pH and calcium content of a variety of wheys (Table 6). Effects of cooking temperature and dipping pH on whey composition were studied by Hill *et al.* (1984). They reported that levels of calcium, phosphorus, magnesium and NPN in whey increased with decreasing dipping pH, levels of sodium and potassium were unaffected by dipping pH, and there was no effect of cooking temperature on whey composition.

Table 6   Natural pH and Calcium Content of a
Variety of Wheys[a]

Type of whey	pH	Calcium (mg l^{-1})
Rennet casein	6.5	500
Gouda	6.3	550
Cheddar	5.9	640
Hydrochloric	4.5	1400
Lactic	4.5	1500

[a] Robinson *et al.*, 1976.

24.11.2   Concentration and Fractionation of Whey

Concentration of whey by evaporation and spray drying is well established. Concentration by reverse osmosis (RO) and fractionation by UF, electrodialysis, ion exchange and gel filtration has facilitated the development of numerous whey products in the past 20 years as well as increasing the process efficiency of traditional products.

24.11.2.1   Reverse osmosis and ultrafiltration

RO and UF are based on the same principle, where solute separation is governed by both the chemical nature and porous structure of the membrane (Sourirajan, 1977). The distinction between RO and UF is, therefore, arbitrary. With respect to dairy processing, RO normally refers to 'tight' or 'concentration' membranes which have a salt rejection of 75–99% and allow only water and certain small chemical species to pass into the permeate. UF normally refers to fractionation processes with a molecular weight cut off of 3000–15 000 which retain proteins and fats but allow lactose and soluble salts to pass into the permeate. Usually, RO membranes are cellulose acetate and UF membranes are cellulose acetate or preferably polysulfone (Modler, 1982a). Because solute rejection depends on temperature, flow rate, pressure, turbulence and feed characteristics (Peri and Dunkley, 1971a,b), membrane rejection characteristics can only be defined for specific conditions and 'molecular weight cut off' and 'salt rejection' values as specified by the manufacturer are useful as general guides only. For example, the same membrane under the same operating conditions will recover a greater proportion of small nitrogenous compounds from acid whey than from sweet whey. Generally, calcium and phosphorus do not readily pass through the membranes and cause membrane fouling (Hiddink *et al.*, 1978). Potassium, sodium and chloride ions and lactic acid are more readily lost in the permeate (Peri and Dunkley, 1971a).

Concentration by RO offers many advantages to cheese manufacturers: (1) RO minimizes pollution. Concentration membranes reduce BOD from 45 000–55 000 p.p.m. in the whey to 80–600 p.p.m. in the permeate. The permeate is suitable for use as rinse water. (2) Reduced transportation costs. Whey volume can readily be reduced by 50–70% to facilitate transport to feed lots, land spraying sites or central processing plants. (3) Preconcentration by RO facilitates further processing of whey to numerous end products. (4) RO is less costly than evaporation up to a maximum concentration of 12–28% depending on the type of membrane, membrane configuration, size of plant, the type of process (*e.g.* batch or continuous) and other conditions related to individual plants.

The costs of RO and evaporation of whey have been compared (de Boer *et al.*, 1977). The cost of energy is much less for RO than for evaporation, but other operating costs are higher. Other conditions also need to be considered. For example, it may be more economical to concentrate beyond the optimum maximum concentration by RO than to increase evaporation or boiler capacity. Generally, the practical maximum level of whey concentration by RO is about 18% (3×).

Batch processing is more efficient with respect to membrane area required than continuous processing but allows more microbial growth due to longer residence time and is more difficult to integrate with other processes. Pepper and Orchard (1982) compared three types of continuous

RO processes, a once through tapered plant, a once through with partial recycle plant, and a new multi-stage recycle (MSR) plant. They reported that the average membrane flux in concentrating whey from 6 to 28% solids with the MSR was equal to that achieved in a once through tapered plant concentrating only from 6 to 12% solids. The principle of the MSR design has traditionally been applied in UF plants but was impractical for RO until recently when high pressure centrifugal pumps became available.

For most practical purposes the composition of RO concentrated wheys can be calculated from the original whey composition assuming no losses in the permeate (Roualeyn *et al.*, 1971).

The main purpose of UF in whey processing is to produce protein powders ranging from 35 to 75% whey protein (typically 35, 50 and 75). A 35% protein powder can be prepared by direct UF of whey to about 10% total solids followed by evaporation to 40% solids and spray drying. Each 100 kg of whey yields 18 kg of UF retentate of 10% solids or 1.8 kg of 35% protein powder (Goldsmith, 1981). For concentrates with more than 50% protein, diafiltration is normally employed, although Goldsmith (1981) reports that spiral-wound modules are capable of achieving 65% protein purity (25% total solids) without diafiltration. Greater processing efficiency can be achieved by various pretreatments before UF (Hickey *et al.*, 1980; Muller and Harper, 1979) and by preconcentration with RO to about 20% solids (de Boer *et al.*, 1977).

Typical compositions of UF whey retentate and permeate are given in Table 7. Because the fractionation process is less than ideal, some concentration of lactose, minerals, NPN and certain ions occurs. Hiddink *et al.* (1978) demonstrated that the retention of ions was governed by the 'Donnan effect', *i.e.* at pH 6.6 anions such as Cl^- and NO_3^- are preferentially removed while at pH 3.2, below the isoelectric point of the whey proteins, cations such as Na^+, K^+ and Ca^{2+} are preferentially removed. They reported that low ash protein concentrates can be produced by direct UF of whey or neutralized whey to produce a sweet permeate, followed by diafiltration (dilution to 300% of rententate volume) at pH 3.6.

Table 7 Composition of Ultrafiltered Whey Retentate and Permeate[a]

Component (%)	Feed	Concentrate[b]	Permeate
True protein	0.60	3.27	0.02
NPN	0.20	0.30	0.18
Lactose	4.80	5.46	4.66
Ash	0.55	0.77	0.50
Fat	0.05	0.28	0.00
Total solids	6.2	10.1	5.4

[a] Goldsmith, 1981. [b] Total nitrogen/total solids = 35%.

24.11.2.2 Demineralization

Demineralization of whey has been reviewed (Delaney, 1976; Houldsworth, 1980). Although several techniques are available, only electrodialysis and ion exchange are commercially significant.

Sodium, potassium and chloride ions are easily removed by electrodialysis and the residual ash is mainly calcium, magnesium and phosphorus. The main advantage of electrodialysis is low cost waste disposal. Where high percentage ash removal is required, as in infant formulations, resin ion exchange columns are more efficient and economical than electrodialysis with respect to fixed and operating costs, including waste management.

24.11.3 Separation and Fractionation of Whey Proteins

Numerous methods are available to separate or fractionate whey proteins to produce protein concentrates with particular nutritional or functional properties. An important process for separating proteins from whey, UF, has been discussed in Section 24.11.2.1. Protein concentrates can also be prepared by crystallizing lactose from concentrated whey (Section 24.11.4.1) followed by demineralization (Section 24.11.2.2). Other protein separation processes include heat–acid precipitation, coprecipitation, chemical precipitation, ion exchange and gel filtration.

24.11.3.1   Heat–acid precipitation

The principles of heat–acid precipitation of whey proteins have been discussed (Hill *et al.*, 1982a). Optimum procedures for heat–acid precipitation are difficult to define because of numerous factors which act and interact on whey systems. The recovery of protein and the composition of the protein isolate depend on pH adjustments, temperature and calcium addition, and the optimum levels of these factors depend on the type of whey and the type of solids separation employed. Generally, the principles governing heat–acid precipitation of whey proteins are as follows. Maximum protein recovery is obtained by denaturing whey proteins at pH in the range 6.0–7.0 and temperatures greater than 90 °C for 10–30 min, followed by precipitation at pH 4.4–5.5. (The term 'denaturation' is used loosely here. It is probable that heating at neutral pH induces primary aggregation through S—S bridges which must precede isoelectric precipitation for maximum protein recovery.) The pH dependence of whey protein aggregation is decreased in the presence of calcium and, therefore, the pH of precipitation of proteins from acid whey is less critical than from sweet whey. The ash content of the protein isolate is minimized by precipitation at low pH (3.5–4.6). Heating and precipitation at low pH maximize the solubility of the protein isolate.

The theoretical maximum recovery of crude protein (total nitrogen × 6.4) from whey is 55–65% (not including recovery of nonprecipitable nitrogen in the water phase of the isolate) because the heat stable proteose–peptone fraction and NPN constitute 35–45% of whey nitrogen. Commercially feasible processes should recover at least 50% of the crude protein.

Recovery of heat-precipitated whey proteins is normally best accomplished by a centrifuge or decanter. For small operations, recovery of heat-precipitated proteins from very sweet whey (dipping pH 6.3–6.5) by filtration may be feasible (Hill *et al.*, 1982b).

24.11.3.2   Coprecipitates

Whey proteins may be coprecipitated with casein or with other proteins such as blood, egg and soya proteins, but only casein–whey protein coprecipitates appear to be of practical interest. The principles (Sawyer, 1969) and processing (Southward and Goldman, 1975) of casein–whey protein coprecipitates have been reviewed.

Casein is insensitive to heat denaturation but is very sensitive to isoelectric precipitation. When whey proteins are heated in the presence of casein they form complexes (Section 24.6.1) which can be efficiently recovered by isoelectric precipitation (acidification), followed by centrifugation or filtration.

24.11.3.3   Chemical precipitants, ion exchange and gel filtration

Chemical precipitants such as polyphosphates, carboxymethylcellulose (CMC), polyacrylic acid, bentonite and alcohols have been suggested for precipitation of whey proteins (Hill *et al.*, 1982a). CMC has been used commercially to recover proteins from whey. Anionic hydrocolloids, such as CMC, form ionic complexes with β-lactoglobulin which can be isolated from solution by isoelectric precipitation, *i.e.* by acidification to pH 3.2 which is the isoelectric point of CMC.

Separation of whey proteins by ion exchange is being done on a commercial scale in France (Morris, 1981). One operation processes 100 000 l of sweet whey day^{-1} and recovers 450 kg of protein day^{-1}. The main advantage of the process is the purity of the concentrates, 90% protein by ion exchange but 80% maximum by UF. The process is more economical than UF for products of more than 35% protein. Ion exchange columns are also available for separation of proteins from acid whey and for the fractionation of whey proteins but are not in commercial use.

Gel filtration is an important technique in the purification of enzymes. Commercial application of gel filtration to the separation of whey proteins has been attempted (Richert, 1975), but proved to be uneconomical. Large scale fractionation of whey protein concentrates by gel filtration for the production of β-lactoglobulin rich concentrates and α-lactalbumin rich concentrates was proposed in 1974 (Forsum *et al.*). Although research is still continuing, the technology has not yet been applied commercially.

24.11.4 Lactose Processing

Lactose processing will be discussed here under three headings: manufacture of lactose, lactose hydrolysis and chemical modification of lactose. Utilization of lactose as a substrate for fermentation processes will be discussed in Section 24.11.5.

24.11.4.1 *Manufacture of lactose*

Conventional lactose manufacture produces α-lactose monohydrate by crystallization from concentrated, deproteinated whey. Deproteination is carried out by neutralization and heat precipitation followed by centrifugation. Deproteinated whey is then evaporated to 55–65% total solids (recently preconcentration by RO has been employed), and subjected to crystallization. In the case of lactic acid whey, lactic acid must be removed by yeast fermentation or electrodialysis before crystalization (Short, 1978). Depending on required purity, crude lactose may be decolorized and recrystallized.

The by-product of conventional lactose manufacture is a high ash protein isolate which is unfit for human use. Two alternatives are available to avoid this problem: (1) decrease the ash content of the protein isolate from 24% (dry basis) to 1–3% by washing the isolate at pH 3.5 (Harwalker and Emmons, 1969); or (2) deproteinate whey by UF or heat–acid precipitation and then demineralize by ion exchange or electrodialysis.

The production of crystalline β-lactose can be accomplished by roller or drum drying deproteinated and demineralized whey at temperatures above 93 °C (lactose in solution at 93 °C crystallizes in the beta form). Spray drying produces β-lactose in the glass state which is extremely hygroscopic and rapidly forms α-lactose monohydrate crystals on exposure to the atmosphere. The same reaction causes clumping of spray dried skim milk powder.

24.11.4.2 *Lactose hydrolysis*

Procedures of lactose hydrolysis for the manufacture of glucose/galactose syrups include acid hydrolysis, enzymatic hydrolysis and cation exchange. Acid hydrolysis is unattractive due to formation of reversion products and dark colour (Short, 1978).

The use of free enzymes has been uneconomical because the enzyme is not recovered. Enzyme recovery is possible by UF but again is too costly (Short, 1978). Economics of using free enzymes may be improved by substituting lactase producing yeasts for expensive enzyme extracts (Briscoe, 1983, personal communication). Lactase producing yeasts can be used to produce alcohol from whey; the yeast cells can be recovered by centrifugation and subjected to freezing or alcohol contact to deactivate the cells without deactivating the lactase. The cell suspension containing lactase is then used to hydrolyze another batch of whey which is subsequently concentrated to produce a sweet whey syrup. An integrated system, then, could produce alcohol and sweet syrup with the only direct inputs being whey and yeast inoculum.

Commercial lactose hydrolysis plants using immobilized enzymes are operating in France, New Zealand, England and the US. Greenberg and Mahoney (1981) have reviewed the various lactases and immobilization systems which could be used for lactose hydrolysis. Enzyme sources include *Aspergillus niger*, *Escherichia coli*, *Kluyveromyces fragilis* and *Kluyveromyces lactis*. Lactases have been immobilized by adsorption, covalent linkages and entrapment, and numerous materials have been used for supports. For example, an enzyme system developed by Corning Glass Company uses lactase from *Aspergillus niger* which is covalently linked to glass beads by silane–gluteraldehyde linkage.

Lactose hydrolysis can be accomplished using ion exchange by passing deproteinated–desalinated whey over a cation exchange (H^+) column (Ralph, 1982). This treatment lowers the pH to 1.2–1.5 and with subsequent heat treatment effects acid hydrolysis more efficiently than direct acidification with mineral acids.

Coton (1980) compared the economics of various hydrolysis systems and concluded that lactose hydrolysis by ion exchange or immobilized enzyme was economically feasible and that free lactase was impractical.

24.11.4.3 *Chemical modification of lactose*

The principal derivatives of lactose are lactulose, lactitol, lactobionic acid and lactosyl urea (Figure 3). Lactulose is an isomer of lactose in which the glucose moiety is converted to fructose in alkaline solution. Lactitol is analogous to sorbitol and can be produced by high pressure hydrogenation or electrolytic reduction. Lactobionic acid is produced by oxidation of lactose, and lactosyl urea is formed by blending lactose and urea in alcoholic solution (Thelwall, 1980).

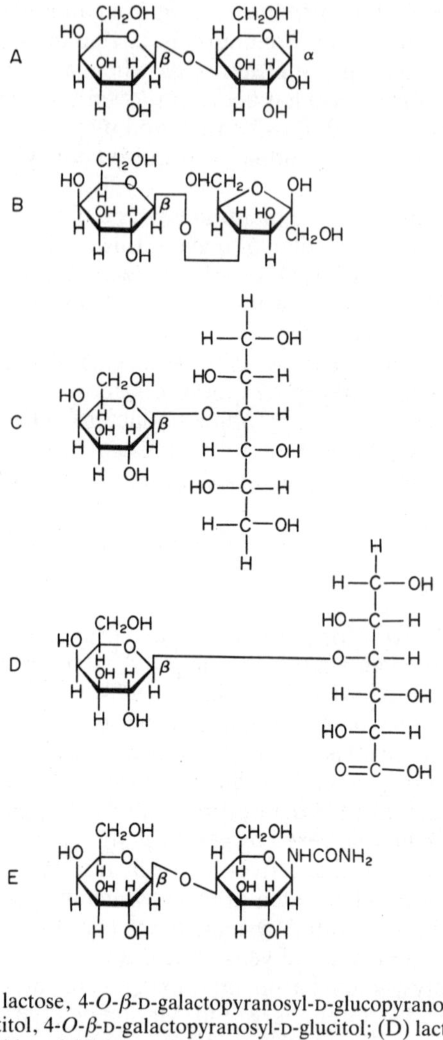

Figure 3 Lactose derivatives: (A) lactose, 4-*O*-β-D-galactopyranosyl-D-glucopyranose; (B) lactulose, 4-*O*-β-D-galactopyranosyl-D-fructofuranose; (C) lactitol, 4-*O*-β-D-galactopyranosyl-D-glucitol; (D) lactobionic acid, 4-*O*-β-D-galactopyranosyl-D-gluconic acid; and (E) lactosyl urea, 4-*O*-β-D-galactopyranosyl-D-glucosyl urea

The use of whey powder as a cheap source of polyol for the manufacture of polyurethane foams was reported in 1970 (Hustad *et al.*). The process was economical and produced high quality foams, but has not been utilized because on-site formation of foams containing lactose (*e.g.* for insulation) caused excessive wear of high pressure nozzles. Recently, Richardson (personal communication, 1983) produced polyether polyols by propoxylation of lactose for use in polyurethane foams to avoid the problem of abrasion.

The formation and application of surfactants derived from disaccharides have been reviewed (Hurford, 1980). High yields of surfactants can be obtained by esterification of lactitol with fatty acids at 200–230 °C in the presence of a basic catalyst.

24.11.5 Whey Fermentation

Whey fermentation processes are available to produce single cell protein (SCP), ethanol, fermented beverages, organic acids, vitamins, antibiotics and lactic cultures. Generally unmodified whey is the best substrate for fermentation but may not optimize returns from individual whey components or produce effluent with minimum BOD. Deproteinated whey, UF milk permeate and concentrated whey have been used as substrates. Whey products have been used as supplementary carbon sources in the production of grain alcohol and beer. This section will discuss the principal whey fermentation processes, excepting fermented beverages which will be considered in Section 24.11.6.

24.11.5.1 Single cell protein

The technology of producing yeast SCP is well established and has been extensively reviewed (Meyrath and Bayer, 1979). Whey or deproteinated whey is fermented aerobically with yeasts of the *Kluyveromyces*, *Candida*, *Torula* or *Saccharomyces* species. A nitrogen source and sometimes yeast extract and/or corn steep liquor are added. In the case of deproteinated whey and milk permeate, fermentation time and foaming can be reduced by adding about 10% of whole whey. Maximum yields of biomass of about 1.8% by weight of whey are obtained in 6–10 h.

Biomass is normally separated by centrifugation resulting in a high BOD supernatant. In the case of feed grade biomass the entire fermentation liquor can be concentrated and dried to eliminate waste effluent (Coton, 1980). Yeast biomass contains about 45–50% protein and 12–16% ash on a dry basis (fermentation solubles not included) (Bernstein and Plantz, 1977).

Yeasts are selected for their ability to ferment lactose rapidly and to produce maximum yields of biomass. Most commonly used are strains of *S. fragilis*, now considered to be *K. fragilis*. In practice, continuous fermentation or the recycling of yeast cells results in a balanced flora. Moulin *et al.* (1983) studied the nature of a balanced yeast in a whey processing plant in France after 20 years of continuous fermentation. They reported an association of *K. fragilis*, *T. spherica* and *T. bovina*. *T. bovina* was entirely dependent on the alcohol produced by *K. fragilis* which utilized lactose while *T. spherica* used only lactic acid. In effect, a flora evolved which utilized all carbon substrates present or formed during the process.

The use of concentrated whey or deproteinated whey offers increased efficiency with respect to materials handling, but has an adverse effect on fermentation rate and yield of biomass. At high concentrations, yeasts produce more alcohol and less cells (Simpson, 1980). One solution is to recover both alcohol and biomass (Anonymous, 1978). This also permits production of more or less biomass and alcohol depending on market conditions. Giec and Kosikowski (1982) reported accelerated fermentation and increased biomass from concentrated whey with reduced ash and high aeration.

Filamentous fungi are also used to produce biomass which is easily separated from the fermentation broth and appears to be nutritionally superior to yeast biomass. SCP from filamentous fungi such as *P. cylopium* has low nucleic acid content and a better distribution of essential amino acids than yeast (Short, 1978). The main disadvantages of mycelial biomass are longer fermentation time and a general concern (which may not be justified) about potential production of aflatoxin.

24.11.5.2 Ethanol

Commercial whey to alcohol conversion plants are operating in most, if not all, major milk producing countries. The basic technology is similar to that described for yeast SCP except that fermentation is anaerobic and culture selection is based on alcohol tolerance as well as the ability to utilize lactose. Alcohol yield should be 80% of the theoretical value according to equation (3).

$$C_{12}H_{22}O_{11} + H_2O \rightarrow 4C_2H_5OH + 4CO_2 \tag{3}$$

This corresponds to 1 l of alcohol per 42 l of whey containing 4.4% lactose (Reesen, 1978). Substrate may be whey or deproteinated whey and the by-product of the process is a denatured yeast–whey or yeast protein concentrate which is recovered by centrifugation following distillation. The remaining supernatant can profitably be converted to methane by anaerobic digestion (Reesen and Strube, 1978).

There are two major difficulties which prevent more whey utilization for alcohol production.

Firstly, yeast strains presently in use are unable to tolerate high levels of alcohol and the process is, therefore, restricted to low lactose substrates. For example, one of the most successful whey to alcohol conversion plants, the Carberry plant in Ireland, uses deproteinated whey containing only 4.7% lactose giving an alcohol content of 2.59% in the feed to the still (Barry, 1982). One solution is to hydrolyze the lactose, concentrate, and ferment with alcohol tolerant industrial strains of *S. cerevisiae*, but this approach results in low yields and slow fermentation, due to poor assimilation of galactose, and also involves the costly hydrolysis step. A potentially more practical approach is to develop mutant strains or select strains of yeasts which have both high alcohol tolerance and the ability to utilize lactose. Bailey *et al.* (1982) successfully produced mutant strains of *S. cerevisiae* which completely utilized equimolar mixtures of glucose and galactose but were still dependent on the hydrolysis step. Moulin and Galzy (1980) reported the selection of a strain of *K. fragilis* which produced 12% alcohol using a 20% lactose substrate. Lactose conversion was 86% using the 20% lactose substrate *versus* 90% using 5 or 10% lactose substrate.

Secondly, large quantities of whey are required for whey conversion plants to compete with molasses and grain conversion plants or synthetic alcohol. This problem can be solved without long distance transport of whey by combining whey and grain fermentation (Friend and Shahani, 1979). Normal grain fermentation can proceed if lactase is added to make lactose available to *S. cerevisiae*. Alternatively, whey and grain fermentations can be carried on separately in parallel in the same plant. The economics of combined and parallel whey/corn alcohol conversion were tested for Southwestern New York and both systems were considered profitable (New York State Energy Research and Development Authority, 1981).

24.11.5.3 *Other fermented products*

Production of lactic acid from whey and deproteinated whey is established technology. Whey or deproteinated whey is supplemented with yeast extract and/or corn steep liquor and fermented with lactic bacteria such as *Lactobacillus bulgaricus*. The pH of the fermentation liquor is maintained at 5.5 by adding calcium hydroxide or calcium carbonate. Nearly complete conversion of lactose to calcium lactate can be achieved in 14–24 h (Friend and Shahani, 1979). After concentration of the liquor, calcium lactate is recovered by crystallization and converted to lactic acid with sulfuric acid. The by-product is calcium sulfate.

The cost of lactic acid production is mainly associated with the recovery process. Short (1978) has reviewed possible alternative methods of recovery including anion exchange, solvent extraction and RO. Lactic acid was separated from lactose at pH near 5.5 by recycling the liquor through an RO plant during the fermentation (Smith *et al.*, 1977). This prevented product inhibition and recovered lactic acid in a single step. However, subsequent demineralization was necessary for technical grade lactic acid and flux rates were decreased in the presence of corn steep liquor.

Neutralization with ammonia or ammonium salts during lactic fermentation yields ammonium lactate which has value as a feed for ruminants.

Acetic acid vinegar is being produced commercially in France, Switzerland and the US (Short, 1978). Following lactose conversion to alcohol and yeast recovery, *Acetobacter* sp. are added to convert alcohol to acetic acid.

Whey is also the substrate for commercial production of penicillin and has been suggested for the production of gums (Stauffer and Leeder, 1978). The technology of whey based lactic starter culture media was discussed in Section 24.5.5.2. Moon and Hammond (1978) evaluated the feasibility of converting lactose to oil and SCP. Fermentation of deproteinated whey with strains of *Candida curvata* and *Trichosporon cutaneum* for 72 h produced 4.0–15.6 g oil and 2.2–3.2 g SCP per litre of permeate.

24.11.6 Whey Beverages

The development of whey beverages from the first patented whey soft drink in 1898 has been studied in an extensive review by Holsinger *et al.* (1974) who considered four main types of beverages, namely, beverages from whole whey, nonalcoholic beverages from deproteinated whey, alcoholic beverages from whey and protein beverages. A total of about 70 processes for the production of whey containing beverages was described. This section is a general discussion of whey beverage processing and only a few successful commercial processes will be described in detail.

24.11.6.1 Nonfermented whey/fruit drinks

The simplest type of process is to blend whey, concentrated whey or dried whey with fruit juices or concentrates. The blends may optionally be carbonated, fortified, concentrated and packaged aseptically. Generally, whey flavour is most compatible with citrus juices. The only successful whey beverage of this type that has been reported in the literature is 'Freshi' which is distributed in Switzerland. Freshi is 50% whey and contains natural orange, lemon and grapefruit flavours, sugar and water. The mixture is heated at 90 °C, packaged aseptically and has a shelf life of 6 months without refrigeration. A similar product 'Lactofruit' is made from hydrolyzed whey (Lang and Lang, 1979). Hydrolysis avoids the problem of lactose intolerance and permits reduced levels of sucrose. Sparkling fruit flavoured beverages can be made from deproteinated whey, although some authors reported that further clarification by filtration through diatomaceous earth was necessary (Holsinger *et al.*, 1974). A mixture of fruit juice, whey protein and hydrolyzed lactose has been introduced in the US (Elliot, 1984).

24.11.6.2 Nonalcoholic fermentation

A second type of process involves fermentation of whey or deproteinized whey. Three successful whey beverages are made commercially by nonalcoholic fermentation. Rivella, first marketed in Switzerland in 1952, is prepared by fermenting deproteinized whey with lactic acid bacteria, filtering, condensing to a 7:1 concentrate, adding sugar and herbs, refiltering, diluting and carbonating. Rivella contains 9.7% total solids, 0.125% nitrogen and the pH is about 3.7 (Holsinger *et al.*, 1974). It is clear and sparkling and is marketed in most of Western Europe. Two fermented nonalcoholic whey beverages are being produced on a small scale in Poland, namely, whey champagne and whey kwas. Whey champagne is a sparkling beverage made by fermentation of deproteinated whey containing 7% sucrose with baker's yeast, followed by colouring with caramel and flavouring. Whey kwas is prepared from sweet whey by lactic acid fermentation followed by addition of yeast and colour, and bottling. After 40 h, the kwas is ready for consumption (Holsinger *et al.*, 1974). Other whey beverages made by lactic acid fermentation of acid whey were recently developed in the Soviet Union (Lang and Lang, 1979).

24.11.6.3 Whey wines

Although optimism about the manufacture of whey wines has been expressed in the literature (Friend and Shahani, 1979; Lang and Lang, 1979), none of the various suggested processes has been commercially successful. In order to obtain 10–12% alcohol, whey or deproteinated whey must be fortified with 10–16% sucrose (Yoo and Mattick, 1969) or with lactose (22.5% total lactose) followed by fermentation in the presence of lactase (Lang and Lang, 1979). Clarity and flavour of whey wine were improved by fining with bentonite and partial demineralization, but expert and consumer panels indicated that even the best whey wine was unpalatable (Ip, 1981). In another study, encouraging taste panel results were obtained for grape wines which had been supplemented with hydrolyzed lactose syrup (Roland and Alm, 1975). There is also potential for the utilization of whey in 'pop' wines (Holsinger *et al.*, 1974) because they contain lower levels of alcohol and the whey taint is masked by fruit flavours.

24.11.6.4 Whey beer

Holsinger *et al.* (1974) noted the properties of whey which make it compatible with beer manufacture. Like beer, whey has a great capacity for binding carbonic acid and has a high salt content. Lactose in whey reacts with other whey components during cooking to form a caramel colour and flavour and does not alter the sweetness of the finished beer. Lactic acid is compatible with the acids of wort and beer (Anonymous, 1981) although the pH of whey is crucial to the proper development of acidity in the mash (Holsinger *et al.*, 1974).

Several beer-like beverages have been developed and one, namely, Bodrost, is produced commercially in Russia. Bodrost is a mixture of deproteinated whey, sucrose, raisins, flavouring and caramel colouring which is fermented with Kefir starter (Kefir starter is a mixture of yeasts and lactic bacteria). However, the greatest potential for utilization in beer is to extend the normal adjuncts (nonmalt sources of energy) in the manufacture of established beers. A successful commercial brew was made by replacing part of the normal adjunct, corn, with whey where 12.2% of the total extract was derived from whey (Anonymous, 1981). Chemical and flavour analysis of the whey-containing beer and the control beer were the same. Sediment and flavour stability were also normal. Part of the whey-containing beer was mixed with an equal part of regular beer and marketed normally without comment.

24.11.7 Whey Utilization

The foregoing has described the principal whey processes. The purpose of this section is to discuss the present and potential uses for whey as summarized in Figure 4.

24.11.7.1 *Utilization of whole whey*

Utilization of whole whey as powder or animal feed has the obvious advantages of zero processing wastes and simplicity. The principal use of whey powder is in feeds, especially as a binder in pelleted feeds and in milk replacers for calves. Important food uses include imitation chocolate, ice cream, sherbets, process cheese, confections, salad dressing and baby products. Generally, it is used to replace or extend skim milk powder. At present the market is static, production is high and returns to the cheese plant are low or negative.

The use of liquid whey as an animal feed, especially for small animals, has been emphasized (Modler *et al.*, 1980). About 34% of the whey produced in the European Economic Community in 1979 was fed to animals (mostly pigs) in the liquid form (Kuipers, 1980). Lactose modified whole whey (lactosyl urea) can be profitably fed to ruminants if it is handled in liquid form to avoid the cost of drying (Coton, 1980). Whole whey can also be concentrated, crystallized and fed to cattle as 'lick blocks' but the economics do not appear to be attractive.

Beverages are a significant outlet for whole whey in Western Europe but have been unsuccessful in North America. Whey beverages are more expensive than traditional soft drinks and the general public is not yet willing to pay for nutrition. Whey beverages have been used in US food aid programs (Holsinger, 1978).

Mysost is a Norwegian cheese which shows some market potential in North America (Jelen and Buchheim, 1976). It completely utilizes whey solids and is made by evaporating a blend of whey, cream and milk. Hard Mysost contains about 18% water, 28% fat, 11% protein and 37% lactose. A higher moisture, low fat spread is produced by a similar process (Jelen and Buchheim, 1976).

24.11.7.2 *Utilization of whey proteins*

Heat–acid precipitation procedures yield denatured whey protein isolates with reduced functionality. Fresh, heat-precipitated whey protein has good water-holding capacity and can be used to replace or extend regular cheese in the manufacture of process cheese spreads (Irvine *et al.*, 1982). Dried, heat-precipitated protein such as 'insoluble lactalbumin', which is a product of New Zealand, has little water-holding capacity and is used for nonfunctional purposes such as protein fortification of pasta and breakfast cereals. Ricotta cheese is a whey protein isolate or a casein–whey protein coprecipitate made by heat-precipitation from sweet whey or sweet whey plus milk (Hill *et al.*, 1982a). It is used in Italian dishes such as lasagna.

A process known as the Centri-whey process incorporates heat-precipitated whey proteins in semi-soft cheese (Marshall, K. R., 1982). Up to 93% of the true protein of whey is heat-precipitated, recovered by centrifugation and blended with cheese milk giving a yield increase of about 15%. This technique has not been successfully applied to hard cheese or soft cheese, and for semi-soft cheese it is being replaced by UF (Section 24.3.6).

Whey proteins which retain native properties (solubility and the ability to gel, whip and emulsify) can be obtained by chemical precipitation, ion exchange and membrane processing (Morr, 1978). Utilization of chemical precipitants is limited due to high contents of ash and residual precipitant in the protein isolate. CMC-precipitated whey protein has been used in sandwich spreads, dips and whipped products. Functional whey protein concentrates prepared by ion exchange or membrane processing have many applications: as skim milk replacer in bread, margarine, confections, ice cream, process cheese and yoghurt; as egg replacer in baked and whipped

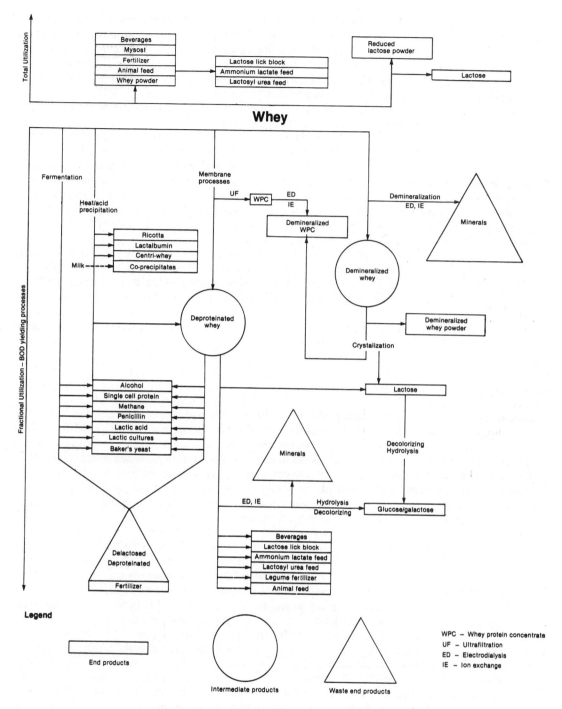

Figure 4 Whey utilization (Adapted from Modler, 1982b)

products; as binders and extenders in processed meats; in baby foods, especially β-lactalbumin rich concentrates; and for high quality protein fortification of soft drinks and snack foods.

24.11.7.3 *Utilization of lactose*

Traditional uses of lactose have been reviewed (Staff, 1956). A few of the possible uses will be mentioned here. Lactose is a flavour enhancer and is used to extend flavours such as monosodium glutamate and protein hydrolysates and to sharpen fruit flavours in soft drinks. Its participation in the Maillard reaction and its lack of sweetness make lactose ideal to improve colour of fish and meat batters, potato chips and baked products. It is used to reduce sweetness in icings and cured meats, to seed sweetened condensed milk and high solids ice cream mixes, to improve free-flowing properties of dried foods such as powdered soups and cream, to maintain clarity in beers and ales, to control crystallization and sweetness in sweet confections, to fortify infant formulae, as a coating for pills and as a carrier for vitamins, oils and flavours. Its numerous applications in confections (Hugunin and Nishikawa, 1978) have been reviewed.

In spite of its many applications, the market for unmodified lactose is static and increasing production has forced prices close to or below production costs.

An attractive alternative is the manufacture of glucose/galactose syrups. Coton (1980) studied the applications and economics of glucose/galactose syrups and concluded that this alternative could be a profitable solution to the problem of using lactose. Hydrolyzed lactose syrups can be used to partially replace sucrose and/or corn syrup in beer production, ice cream and sugar confectioneries. Coton (1980) calculated that if all available lactose was converted to glucose/galactose syrups, it would represent only 3.8% of world sucrose production. Therefore, if hydrolyzed lactose syrups can be produced at significantly less cost than sucrose, all available lactose could be utilized without seriously threatening the sucrose market.

Conversion of lactose to alcohol is a profitable alternative depending on the economies of scale which can be achieved and the cost of competing products. In Brazil, whey alcohol is profitably used in gasahol. In Ireland potable alcohol is being successfully produced from whey. In North America, whey alcohol faces heavy price competition from grain alcohol and inexpensive liquid fuels.

Chemical modification of lactose may have important specialized applications but is unlikely to become a significant outlet for surplus lactose.

24.12 SUMMARY

Cheese manufacture is a process of concentrating and separating proteins and fat from milk. Milk contains proteins, fat, lactose, minerals, vitamins, pigments and enzymes. Milk proteins consist of about 80% casein and 19% whey proteins. Caseins are sensitive to isoelectric precipitation by acidification or addition of salts, and to enzymic precipitation. Whey proteins are precipitated by heat but are not sensitive to isoelectric or enzymic coagulation. Most of the fat is occluded with the protein while most of the soluble components (lactose, salts, whey proteins) are excluded in the water phase or the whey which remains after the coagulated protein is removed. Ripened cheese such as Cheddar is casein coagulated with rennet (chymosin) or rennet substitutes, while fresh cheese is made by lactic fermentation resulting in acid coagulation. Protein coagulation is affected by milk quality, types and numbers of bacteria, acidity, antibiotics, milk composition, total solids, protein/fat ratio, calcium phosphate equilibria and heat treatment. Pasteurization or subpasteurization of milk is used to destroy pathogens but overheating will render casein insensitive to rennet. Fresh cheese milk may be subjected to high heat treatment before fermentation to coprecipitate whey proteins with the casein. In some soft cheeses, whey proteins can be recovered by using ultrafiltration to separate the proteins from milk solubles before coagulation. Selected lactic acid producing bacteria are required to develop acidity, control cheese moisture and body, and to develop characteristic cheese flavour, except some heat precipitated cheese such as Queso Blanco. Lactic cultures are usually grown in skim milk or whey based media which may be buffered to avoid product inhibition. Culture rotation is essential to control the growth of bacteriophage. After protein coagulation, the curd is cut and cooked (39–52 °C) to expel moisture, whey is removed, and the curd is textured, placed in moulds, salted and packaged. Excepting fresh cheese, curd is ripened at various temperatures for periods of two weeks to two years with microbiological agents until the characteristic flavour profile for the particular cheese is achieved. Process cheese is ripened cheese which has been blended with sequestering agents and cooked.

Cheese whey constitutes about 90% of the original milk volume and contains more than half of the original milk solids. Whey is about 6% solids of which about 10, 72, 1, 8.2 and 2% are protein, carbohydrate, lipids, ash and nonprotein nitrogen, respectively. Whey contains 40 000–60 000 p.p.m. of BOD and because of environmental regulations and levies has become a heavy liability. The protein fraction can be recovered and/or purified by heat precipitation, chemical precipitation, ultrafiltration, ion exchange and electrodialysis for protein fortification of cereals and beverages, meat extenders and incorporation into cheese and process cheese. Lactose is recovered by crystallization from concentrated whey and deproteinated whey and is useful as a non-sweet sugar to promote browning of batters, to impart free flowing properties to powdered foods, control crystallization in sweet confections and high solids dairy products, fortify infant formulations, and enhance and extend flavours. Hydrolyzed lactose syrups are made using immobilized lactase and represent a large potential outlet for excess lactose. Fermentation processes are available to convert whey and deproteinated whey to single cell protein (SCP) and/or alcohol. The technology of whey–yeast SCP is well established but is generally not profitable. Commercial whey to alcohol conversion plants are operating successfully in all major cheese producing countries. The economics of alcohol production from whey could be greatly improved by the selection or development of a high alcohol tolerant yeast. Whey beverages such as 'Rivella' are significant outlets for whey in Western Europe but have not been successful in North America. Whey powder and direct feeding to animals are the most important outlets for whey with respect to the amounts utilized.

24.13 REFERENCES

Alford, B. L., J. I. Mao and D. T. Moir (1982). Recombinant DNA. *Br. Pat.* 2 091 272.

Ali, A. E., A. T. Andrews and G. C. Cheeseman (1980). Influence of storage of milk on casein distribution between the micellar and soluble phases and its relationship to cheesemaking parameters. *J. Dairy Res.*, **47**, 371–382.

Anonymous (1978). New use for liquid whey. *Food Eng.*, May, 100–102.

Anonymous (1981). Whey: a prospective brewing ingredient? *Food Eng.*, May, 84–85.

Bailey, R. B., T. Benitez and A. Woodward (1982). *Saccharomyces cerevisiae* mutants resistant to catabolite repression: use in cheese whey hydrolysate fermentation. *Appl. Environ. Microbiol.*, **44**, 631–639.

Barnum, D. A., D. A. Biggs, D. M. Irvine and J. C. Rennie (1966). A summary of the study of milk composition in Ontario 1961–1965. *Bull. Ont. Dept. Agric. Food*, 1966. Toronto, Ontario.

Barry, J. A. (1982). Alcohol production from cheese whey. *Dairy Ind. Int.*, October, 19–21.

Beeby, R., R. D. Hill and N. S. Snow (1971). Milk protein research and milk technology. In *Milk Protein Chemistry and Molecular Biology*, ed. H. A. McKenzie, vol. 2, pp. 421–459. Academic, New York.

Bernstein, S. and P. E. Plantz (1977). Ferments whey into yeast. *Food Eng.*, November, 74–75.

Biggs, D. A. (1967). Milk analyses with the infrared milk analyzer. *J. Dairy Sci.*, **50**, 799–803.

Buchanan, R. E. and N. E. Gibbons (eds.) (1974). *Bergey's Manual of Determinative Bacteriology*, 8th edn. William and Wilkins, Baltimore.

Burkhalter, G. (1971). Preliminary catalogue of cheese varieties. *Int. Dairy Fed. Annu. Bull. (4)*, I.D.F. Doc 62.

Cerbulis, J., J. H. Woychick and M. V. Wondolowski (1972). Composition of commercial wheys. *J. Agric. Food Chem.*, **20**, 1057–1059.

Chapman, H. R. and M. E. Sharpe (1981). Microbiology of cheese. In *Dairy Microbiology*, ed. R. K. Robinson, vol. 2, pp. 157–243. Applied Science, London.

Cheeseman, G. C. (1981). Rennet and cheesemaking. In *Enzymes and Food Processing*, ed. G. C. Birch, N. Blakebrough and K. J. Parker, pp. 195–211. Applied Science, London.

Codex Alimentarius Commission (1973). *Joint FAO/WHO Food Standards Program*, 7th edn. Food and Agriculture Organization of the United Nations World Health Organization.

Costilow, R. N. and M. L. Speck (1951). Inhibitory effect of rancid milk on certain bacteria. *J. Dairy Sci.*, **34**, 1119–1127.

Coton, S. G. (1980). The utilization of permeates from the ultrafiltration of whey and skim milk. *J. Soc. Dairy Technol.*, **33**, 89–94.

Dalgleish, D. G. (1979). Proteolysis and aggregation of casein micelles treated with immobilized or soluble chymosin. *J. Dairy Res.*, **46**, 653–661.

Dalgleish, D. G. (1982a). The enzymatic coagulation of milk. In *Developments in Dairy Chemistry*, ed. P. F. Fox, vol. 1, pp. 157–188. Applied Science, London.

Dalgleish, D. G. (1982b). Milk proteins. In *Chemistry and Physics in Food Proteins*, ed. P. F. Fox and J. J. Condon, pp. 155–178. Applied Science, London.

Davies, L. F. and M. J. Gasson (1981). Reviews of the progress of dairy science: Genetics of lactic acid bacteria. *J. Dairy Res.*, **48**, 363–376.

Davis, J. G. (1955). *A Dictionary of Dairying*, 2nd edn. Leonard Hill, London.

Davis, J. G. (1965). *Cheese*. Elsevier, New York.

de Boer, R., J. N. de Wit and J. Hiddink (1977). Processing of whey by means of membranes and some applications of whey protein concentrate. *J. Soc. Dairy Technol.*, **30**, 112–120.

Delaney, R. A. M. (1976). Demineralization of whey. *Aust. J. Dairy Technol.*, **31**, 12–17.

Duitschaever, C. L. and D. M. Irvine (1971). A case study: effect of mold on growth of coagulase-positive staphylococci in Cheddar cheese. *J. Food Technol.*, **34**, 593.

Elliot, R. (1984). Swiss Valley has eye for innovation. *Dairy Field*, June, 38–42.

Emmons, D. B. and J. A. Elliot (1967). Effect of homogenization of skim milk on rate of acid development, sediment formation and quality of Cottage cheese made with agglutinating cultures. *J. Dairy Sci.*, **6**, Abstr. 957.

Ernstrom, C. A. and N. P. Wong (1974). Milk clotting enzymes and cheese chemistry. In *Fundamentals of Dairy Chemistry*, ed. B. H. Webb, A. H. Johnson and J. A. Alford, pp. 662–718. AVI, Westport, CT.

FAO/WHO (1972). *Recommended International Standards for Cheese and Government Acceptances*. FAO/WHO Food Standards Program, FAO, Rome.

Farnham, M. (1950). Animal lipolytic and proteolytic extract. *US Pat.* 2 531 329.

Farrell, H. M., Jr. and M. P. Thompson (1974). Physical equilibria. In *Fundamentals of Dairy Chemistry*, ed. B. H. Webb, A. H. Johnson and J. A. Alford, pp. 442–470. AVI, Westport. CT.

Foltmann, B., A. M. Faerch and V. B. Pedersen (1978). Recent studies on structure–function relationships of milk clotting enzymes. *XX Int. Dairy Congress*. 33ST. MIR, Moscow.

Forsum, E. and L. Hambraeus (1977). Nutritional and biochemical studies of whey products. *J. Dairy Sci.*, **60**, 370–377.

Forsum, E., L. Hambraeus and I. H. Siddigi (1974). Large-scale fractionation of whey protein concentrates. *J. Dairy Sci.*, **57**, 659–664.

Friend, B. A. and K. M. Shahani (1979). Review: whey fermentation. *J. Dairy Sci. Technol.*, **14**, 143–152.

Fryer, T. F. (1969). Microflora of Cheddar cheese and its influence on cheese flavour. *Dairy Sci. Abstr.*, **31**, 471–490.

Gasson, M. J. (1980). Production, regeneration and fusion of protoplasts in lactic streptococci. *Microbiol. Lett.*, **9**, 99–102.

Gasson, M. J. and F. L. Davies (1980a). Conjugal transfer of the drug-resistant plasma $\varrho AM\beta$ in the lactic streptococci. *FEMS Microbiol. Lett.*, **7**, 51–53.

Gasson, M. J. and F. L. Davies (1980b). Prophage-cured derivatives of *Streptococcus lactis* and *Streptococcus cremoris*. *Appl. Environ. Microbiol.*, **40**, 964–966.

Giec, A. and F. V. Kosikowski (1982). Ability of lactose fermenting yeasts in producing biomass from concentrated whey permeates. *J. Food Sci.*, **47**, 1892–1893.

Gillies, M. T. (1974). *Whey Processing and Utilization*. Noyes Data Corp., Park Ridge, NJ.

Gilliland, S. E. and M. L. Speck (1974). Frozen concentrated cultures of lactic acid bacteria: a review. *J. Milk Food Technol.*, **37**, 107–111.

Glass, L. and T. I. Hedrick (1977a). Nutritional composition of sweet and acid type wheys: I. Major factors including amino acids. *J. Dairy Sci.*, **60**, 185–189.

Glass, L. and T. I. Hedrick (1977b). Nutritional composition of sweet and acid type wheys: II. Vitamin, mineral, and calorie contents. *J. Dairy Sci.*, **60**, 190–196.

Goldsmith, R. L. (1981). Ultrafiltration production of whey protein concentrates. *Dairy Field*, August, **88**, 92–95.

Greenberg, N. A. and R. R. Mahoney (1981). Immobilisation of lactase (β-galactosidase) for use in dairy processing: A review. *Process Biochem.*, Feb./March, 3–8.

Harwalker, V. R. and D. B. Emmons (1969). Low-ash lactalbumin as a by-product of lactose production. *Can. Inst. Food Technol. J.*, **2**, 9–11.

Harper, W. J. (1959). Chemistry of cheese flavour. *J. Dairy Sci.*, **42**, 207–213.

Harper, W. J. (1979). Analytical procedures for whey and whey products. *NZ J. Dairy Sci. Technol.*, **14**, 156–169.

Harper, W. J. and J. Y. Wang (1980). Amino acid catabolism in cheddar cheese slurries. II. Evaluation of transamination. *Milchwissenschaft*, **35**, 598–599.

Harper, W. J., J. Y. Wang and T. Kristoffersen (1979). Free fatty acids and amino acids in fat modified cheese. *Milchwissenschaft*, **34**, 525–527.

Harris, T. J. R., P. A. Lowe, A. Lyons, P. G. Thomas, M. A. W. Eaton, T. A. Millicon, T. P. Patel, C. C. Bose, N. H. Carey and M. T. Doel (1982). Molecular cloning and nucleotide sequence of cDNA cloning for calf preprochymosin. *Nucleic Acid Res.*, **10**, 2177–2187.

Hickey, M. W., R. D. Hill and B. R. Smith (1980). Investigations into the ultra-filtration and reverse osmosis of wheys: I. The effects of certain pretreatments. *NZ J. Dairy Sci. Technol.*, **15**, 109–121.

Hicks, C. L., M. Allauddin, B. E. Langlois and J. O'Leary (1982). Psychotrophic bacteria reduce cheese yields. *J. Food Protection*, **45**, 331–334.

Hiddink, J., R. de Boer and D J. Romijin (1978). Removal of milk salts during ultrafiltration of whey and buttermilk. *Neth. Milk Dairy J.*, **32**, 80–93.

Hill, A. R., D. M. Irvine and D. H. Bullock (1982a). Precipitation and recovery of whey proteins: a review. *Can. Inst. Food Sci. Technol. J.*, **15**, 155–160.

Hill, A. R., D. H. Bullock and D. M. Irvine (1982b). Recovery of whey proteins from concentrated sweet whey. *Can. Inst. Food Sci. Technol. J.*, **15**, 180–184.

Hill, A. R., D. H. Bullock and D. M. Irvine (1985). Composition of cheese whey: effect of pH and temperature of dipping. *Can. Inst. Food Sci. Technol. J.*, **18**, 53–57.

Holsinger, V. H. (1978). Lactose-modified milk and whey. *Food Technol.*, March, 102–105.

Holsinger, V. H., L. P. Posati and E. D. de Vilbiss (1974). Whey beverages: a review. *J. Dairy Sci.*, **57**, 849–859.

Houldsworth, D. W. (1980). Demineralization of whey by means of ion exchange and electrodialysis. *J. Soc. Dairy Technol.*, **33**, 45–51.

Hugunin, A. G. and R. K. Nishikawa (1978). Milk-derived ingredients lend flavour. *Food Prod. Dev.*, February, 46–48.

Hurford, J. R. (1980). Surface active agents derived from some selected disaccharides. In *Developments in Food Carbohydrates*, ed. C. K. Lee, pp. 327–350. Applied Science, London.

Hustad, G. O., T. Richardson and C. H. Amundson (1970). Polyurethane foams from dried whey. *J. Dairy Sci.*, **53**, 18–23.

Hylmar, B. (1980). Effects of psychrotrophic microflora in raw milk in dairy processing. *Prumysl Potravin*, **31**, 688–690.

ICMSF of the International Association of Microbiological Societies (1978). *Microorganisms in Foods. 1. Their Significance and Methods of Enumeration*. University of Toronto Press, Toronto, Canada.

ICMSF (1980a). Factors affecting life and death of organisms. In *Microbial Ecology of Foods*, vol. 1, pp. 70–91. Academic, New York.

ICMSF (1980b). Food commodities. In *Microbial Ecology of Foods*, vol. 2, pp. 499–513. Academic, New York.

IDF (1980a). *Factors influencing the bacteriological quality of raw milk*. International Dairy Federation, Document 120.

IDF (1980b). *Behaviour of pathogens in cheese*. International Dairy Federation, Document 122.

IDF (1980c). *Accelerated cheese ripening*. International Dairy Federation, Document 79.

Irvine, D. M. (1982). *Cheddar cheese. Bull. Ont. Dept. Agr. of Food*, Toronto, Ontario.

Irvine, D. M., D. H. Bullock, A. R. Hill (1982). Utilization of sweet whey lactalbumin in cheese spread. *XXI Int. Dairy Congress*, vol. I, book I, p. 491. MIR, Moscow.

Ip, E. (1981). The utilization of ultrafiltration whey permeate for the production of whey wine. MSc. Thesis, University of Guelph. pp. 132–134.

Jelen, P. and W. Buchheim (1976). Characteristics of Norwegian whey cheese. *Food Technol.*, November, 62–74.

Johnson, A. H. (1972). The composition of milk. In *Fundamentals of Dairy Chemistry*, ed. B. H. Webb, A. H. Johnson and J. A. Alford, pp. 4–21. AVI, Westport, CT.

Johnston, D. E., R. J. Murphy and N. R. Whittaker (1983). Effects of thermization and cold storage on some properties of rennet-induced gels. *J. Dairy Res.*, **50**, 231–236.

Jonas, L. L., C. A. Ernstrom and G. H. Richardson (1977). Major savings realized in the production of bulk lactic culture with whey based phage inhibitory medium. *Cultured Prod. J.*, **12** (2), 12–14.

Kessler, H. G. (1981). *Dairy Engineering and Food Technology*, ed. V. A. Kessler. F.R., Germany.

Kosaric, N. and Y. Asher (1982). Cheese whey and its utilization. *Conservation and Recycling*, **5** (1), 23–32.

Kosikowski, F. V. (1979). Whey utilization and whey products. *J. Dairy Sci.*, **62**, 1149–1160.

Kristoffersen, T., E. M. Mikolajak and I. A. Gould (1967). Cheddar cheese flavour: IV. Directed and accelerated ripening process. *J. Dairy Sci.*, **50**, 292–297.

Kuipers, A. (1980). Update: whey utilization in animal feeds — EEC. Proceedings 1980 Whey Products Conference. *Dairy Sci. Abstr.* 1982, No. 5253, 587.

Lang, F. and A. Lang (1979). Whey for the production of soft drinks and alcoholic beverages. *Milk Ind.*, **81** (11), 30–31.

Law, B. A. (1979). Reviews of the progress of dairy science: enzymes of psychroytrophic bacteria and their effects on milk and milk products. *J. Dairy Res.*, **46**, 573–588.

Law, B. A. (1982). Microbial proteolysis of milk protein. In *Food Protein*, ed. P. F. Fox and J. J. Condon, p. 307. Applied Science, London.

Law, B. A., Z. D. Hosking and H. R. Chapman (1979). The effect of some manufacturing conditions on the development of flavour in Cheddar cheese. *J. Soc. Dairy Technol.*, **32** (2), 87–90.

Lawrence, R. C., T. D. Thomas and Terzaghi (1976). Reviews of the progress of dairy science. *J. Dairy Sci.*, **43**, 141–193.

Mabbitt, L. A. (1955). Quantitative estimation of the amino acids in Cheddar cheese and their importance in flavour. *J. Dairy Res.*, **22**, 224–231.

McKay, L. L. and K. A. Baldwin (1978). Stabilization of lactose metabolism in *Streptococcus lactis* C_2. *Appl. Environ. Microbiol.*, **36**, 360–367.

McKay, L. L., W. E. Sandine and P. R. Elliker (1971). Lactose utilization by lactic acid bacteria: A review. *Dairy Sci. Abstr.*, **33**, 493–499.

McKenzie, H. A. (1971a). *Milk Proteins: Chemistry and Molecular Biology*, vol. 2, pp. 75–77. Academic, New York.

McKenzie, H. A. (1971b). *Milk Proteins: Chemistry and Molecular Biology*, vol. 2, pp. 316–318.

Magee, E. L., Jr. and N. F. Olson (1981). Microencapsulation of cheese ripening systems: formation of microcapsules. *J. Dairy Sci.*, **64**, 600–610.

Mahaut, M. and J. L. Maubois (1978). Application of the MMV process to the making of Blue cheeses. *XX Int. Dairy Cong.*, vol. E., p. 793. MIR, Moscow.

Mann, E. J. (1967). Rennet substitutes. *J. Dairy Ind.*, **32**, 761–62.

Mann, E. J. (1982). Ultrafiltration of milk for cheesemaking. *Dairy Ind. Int.*, **47** (12), 11–12.

Marshall, K. R. (1982). Industrial isolation of milk proteins: Whey proteins. In *Developments in Dairy Chemistry*, ed. P. F. Fox, vol. 1, pp. 339–367. Applied Science, London.

Marshall, R. J. (1979). Psychotrophic bacteria — their relationship to raw milk quality and keeping quality of Cottage cheese. *Proceedings from the first biennial Marschall International Cheese Conference*. Miles Laboratories Inc., Madison, WI.

Marshall, R. J. (1982). Relationship between the bacteriological quality of raw milk and the final products. A review of basic information and practical aspects. *Milchwirtschaftliche Forschungsberichte*, **34** (1), 149–157.

Marth, E. H. (1963). Microbiological and chemical aspects of Cheddar cheese ripening: a review. *J. Dairy Sci.*, **46**, 869–890.

Marth, E. H. (1974). Fermentations. In *Fundamentals of Dairy Chemistry*, ed. B. H. Webb, A. H. Johnson and J. A. Alford, pp. 772–872. AVI, Westport, CT.

Marth, E. H. (1978). *Standard Methods for the Examination of Dairy Products*, 14th edn., pp. 338–344. American Public Health Association, Washington, DC.

Maubois, L. and G. Mocquot (1975). Application of membrane ultrafiltration to preparation of various types of cheeses. *J. Dairy Sci.*, **58**, 1001–1007.

Mavropoulou, I. P. and F. V. Kosikowski (1973). Composition, solubility, and stability of whey powders. *J. Dairy Sci.*, **56**, 1128–1134.

Maxcy, R. B., W. V. Price and D. M. Irvine (1955). Improving curd-forming properties of homogenized milk. *J. Dairy Sci.*, **38**, 80–84.

Meyrath, J. and K. Bayer (1979). Biomass from whey. *Economic Microbiol.*, **4**, 207–269.

Messer, J. W., L. L. Claypool, G. A. Houghtby, E. M. McKolajcik and E. L. Sing (1978). Detection of antibiotic residues in milk and dairy products. In *Standard Method for the Examination of Dairy Products*, ed. E. H. Marth, pp. 141–150. American Public Health Association, Washington, DC.

Milk Industry Foundation (1964). *Laboratory manual: Methods of Analysis of Milk and its Products*. Milk Industry Foundation, Washington, DC.

Modler, H. W. (1982a). Exploring latest ultrafiltration hardware. *Dairy Field*, August, 45–48.

Modler, H. W. (1982b). Wiping out our whey woes. *Cultured Dairy Prod. J.*, May, 11–14.

Modler, H. W., P. G. Muller, J. A. Elliot and D. B. Emmons (1980). Economic and technical aspects of feeding whey to livestock. *J. Dairy Sci.*, **63**, 838–855.

Moon, N. J. and E. G. Hammond (1978). Conversion of cheese whey and whey permeate to oil and single-cell protein. *J. Dairy Sci.*, **61**, 1537–1547.

Morr, C. V. (1978). Functionality of whey protein products. *NZ J. Dairy Sci. Technol.*, **14**, 185–194.

Morris, C. E. (1981). Recovers whey proteins of 90% purity. *Food Eng.*, April, 92–95.

Morris, H. A. (1981). Blue-veined cheeses. *Pfizer Cheese Monographs*, vol. VII, p. 9. Pfizer, New York.

Morris, H. A. and J. J. Jezeski (1964). Influence of the hydrogen peroxide–catalase milk treatment on Cheddar cheese hardness. *J. Diary Sci.*, **47**, 681 (Abstr.).

Moulin, G. and P. Galzy (1980). Alcohol production from whey. In *Advances in Biotechnology*, ed. M. Moo-Young, vol. II, pp. 181–187. Pergamon, Toronto.

Moulin, G., B. Malige and P. Galzy (1983). Balanced flora of an industrial fermenter: production of yeast from whey. *J. Dairy Sci.*, **66**, 21–28.

Mulder, H. and P. Walstra (1974). *The Milk Fat Globule*. Commonwealth Agricultural Bureau, Farnham Royal, England.

Muller, L. L. (1971). Manufacture and uses of casein and co-precipitate. *Dairy Sci. Abstr.*, **33**, 659–674.

Muller, L. L. and W. J. Harper (1979). Effects on membrane processing of pretreatments of whey. *J. Agric. Food Chem.*, **27**, 662–664.

Narasimhan, R. and C. A. Ernstrom (1977). Phosphate inhibition of lactic cultures in ultrafiltered skim milk concentrates. *J. Dairy Sci.*, **60**, Abstr. 36.

Nelson, J. A. and G. M. Trout (1964). *Judging Dairy Products*, 4th edn., pp. 174–252. AVI, Westport, CT.

Nelson, J. H., R. G. Jensen and R. E. Pitas (1977). Pregastric esterase and other oral lipases: a review. *J. Dairy Sci.*, **60**, 327–362.

New York State Energy Research and Development Authority (1981). *Ethanol production in southwestern New York: Technical and economic feasibility*, Report 81–3, pp. xix–xxxvii. Albany, New York.

Ney, K. H. (1981). Recent advances in cheese flavour research. In *The Quality of Foods and Beverages*, ed. G. Charalambous and G. Inglett, vol. 1, pp. 389–435. Academic, New York.

Nilsson R. and S. Willart (1961). Lipolytic activity in milk: II. The heat inactivation of the fat splitting in milk. *Rep. Milk Dairy Res., Alnarp, Sweden*, **64**, 10.

Nishimori, K., Y. Kawaguchi, M. Hidaka, T. Uozumi and T. Beppu (1981). Cloning in *Escherichia coli* of the structural gene of prorennin, the precursor of calf milk-clotting enzyme rennin. *J. Biochem*, **90**, 901–904.

Nishimori, K., Y. Kawaguchi, M. Hidaka, T. Uozumi and T. Beppu (1982a). Nucleotide sequence of calf prorennin cDNA cloned in *Escherichia coli. J. Biochem.*, **91**, 1085–1088.

Nishimori, K., Y. Kawaguchi, M. Hidaka, T. Uozumi and T. Beppu (1982b). Expression of cloned calf prochymosin gene sequence in *Escherichia coli. Gene*, **19**, 337–344.

Ohmiya, K., S. E. Yun, T. Kobayashi and S. Shimizu (1981). Application of immobilized proteases to milk curdling. *Neth. Milk Dairy J.*, **35**, 318–322.

Ohmiya, K. (1981). Immobilization and application of milk clotting enzyme. *Jpn. J. Dairy Food Sci.*, **30** (1), 11–12.

Osborne, R. J. W. (1977). The manufacture of starters: III. Production of frozen concentrated cheese starter by diffusion culture. *J. Soc. Dairy Technol.*, **30** (1), 40–44.

Parnell-Clunies, E. M., D. H. Bullock and D. M. Irvine (1982). Improved procedure for large scale manufacture of Queso Blanco using acid coagulated solutions of low concentration. *J. Dairy Sci.*, **65** (Suppl. 1), Abstr. 67.

Parry, R. M. (1974). Milk Coagulation. In *Fundamentals of Dairy Chemistry*, ed. B. H. Webb, A. H. Johnson and J. A. Alford, pp. 603–661. AVI, Westport, CT.

Pepper, D. and A. C. J. Orchard (1982). Improvements in the concentration of whey and milk by reverse osmosis. *J. Soc. Dairy Technol.*, **35**, 49–53.

Peri, C. and W. L. Dunkley (1971a). Reverse osmosis of Cottage cheese whey: I. Influence of composition of feed. *J. Food Sci.*, **36**, 25–30.

Peri, C. and W. L. Dunkley (1971b). Reverse osmosis of Cottage cheese whey: II. Influence of flow conditions. *J. Food Sci.*, **36**, 395–396.

Porubcan, R. S. and R. L. Sellers (1975). Spray drying of yoghurt and related cultures. *J. Dairy Sci.*, **58**, 787 (Abstr.).

Ralph, W. (1982). Profits in whey. *Rural Res.*, **116**, 22–27.

Reesen, L. (1978). Alcohol production of whey. *Dairy Ind. Int.*, **43** (1), 9–16.

Reesen, L. and R. Strube (1978). Complete utilization of whey for alcohol and methane production. *Process Biochem.*, November, 21–24.

Refstrup, E. and F. K. Vogensen (1980). Plasmids in starter bacteria — occurrence and significance. *Maelkeritidende*, **93**, 612–615.

Reinbold, G. W. (1972). Swiss cheese varieties. *Pfizer Cheese Monographs*, vol. V. Pfizer, New York.

Reimerdes, E. H. (1982). Changes in the proteins of raw milk during storage. In *Developments in Dairy Chemistry*, ed. P. F. Fox, vol. 1, pp. 280–281. Applied Science, London.

Richardson, G. H. (1978). Solving bulk-culture problems through the use of active lactic concentrates. *Dairy Ice Cream Field*, **161** (9), 80A–80D.

Richardson, G. H. (1981). Automated testing of milk. *J. Dairy Sci.*, **64**, 1087–1095.

Richardson, G. H., C. T. Cheng and R. Young (1977). Lactic bulk culture system utilizing a whey-based bacteriophage inhibitory medium and pH control: I. Application to American style cheese, II. Reduction of phosphate requirements under pH control. *J. Dairy Sci.*, **60**, 1245–1251.

Richardson, G. H., G. L. Hong and C. A. Ernstrom (1979). USU lactic culture system. *Utah Sci.*, **40** (4), 94, 96–99.

Richert, S. H. (1975). Current milk protein manufacturing processes. *J. Dairy Sci.*, **58**, 985–993.

Robinson, B. P., J. L. Short and K. R. Marshall (1976). Traditional lactalbumin — manufacture, properties and uses. *NZ J. Dairy Sci. Technol.*, **11**, 114–126.

Roland, J. F. and W. L. Alm (1975). Wine fermentations using membrane processed hydrolyzed whey. *Biotechnol. Bioeng.*, **17**, 1443–1453.

Rose, D. and J. R. Colvin (1966). Internal structure of casein micelles from bovine milk. *J. Dairy Sci.*, **49**, 351–355.

Roualeyn, I. F., C. G. Hill and C. H. Amundson (1971). Use of ultrafiltration/reverse osmosis systems for the concentration and fractionation of whey. *J. Food Sci.*, **36**, 14–21.

Sandine, W. E. (1979). In lactic starter culture technology. *Pfizer Cheese Monographs*, vol. VI, pp. 24–30. Pfizer, New York.

Sawyer, W. H. (1969). Complex between β-lactoglobulin and ϰ-casein: a review. *J. Dairy Sci.*, **52**, 1347–1355.

Schwartz, D. P. and O. W. Parks (1974). The lipids of milk: Deterioration. In *Fundamentals of Dairy Chemistry*, ed. B. H. Webb, A. H. Johnson and J. A. Alford, pp. 220–239. AVI, Westport, CT.

Scott, R. (1981). *Cheesemaking Practice*. Applied Science, London.

Sellars, R. L. and F. J. Babel (1978). *Cultures for the Manufacture of Dairy Products*. Chr. Hansen's Laboratory Inc., Milwaukee, WI.

Sharpe, M. E. (1978). Lactic acid bacteria (including the Leuconostocs). *XX Int. Dairy Congress*. MIR, Moscow.

Sharpe, M. E. (1979). Lactic acid bacteria in the dairy industry. *J. Soc. Dairy Technol.*, **32** (1), 9–18.

Short, J. L. (1978). Prospects for the utilization of deproteinated whey in New Zealand: A review. *NZ J. Dairy Sci. Technol.*, **13**, 181–194.

Simpson, D. S. (1980). *Production of Yeast–Whey Biomass for Single Cell Protein*. MSc Thesis, University of Guelph.

Sinha, R. P. (1980). Alteration of host specificity to lytic bacteriophages in *Streptococcus cremoris*. *Appl. Environ. Microbiol.*, **40**, 326–332.

Smiley, M. B. (1982). Bacterial genetics, molecular biology and dairy cultures. In *Nestle Research News 1980/81*, pp. 39–51. Nestle Products Technical Assistance Co., Switzerland.

Smith, B. R., R. D. MacBean and G. C. Cox (1977). Separation of lactic acid from lactose fermentation liquors by reverse osmosis. *Aust. J. Dairy Technol.*, **32**, 23–26.

Snook, R. J. and L. L. McKay (1981). Conjugal transfer of lactose-fermenting ability among *Streptococcus cremoris* and *Streptococcus lactis* strains. *Appl. Environ. Microbiol.*, **42**, 904–911.

Snyder, I. S., W. Johnson and E. A. Zottola (1978). Significant pathogens in dairy products. In *Standard Methods for the Examination of Dairy Products*, ed. E. H. Marth, pp. 11–32. American Public Health Assoc., Washington, DC.

Sood, V. K. and F. Kosikowski (1979). Ripening changes and flavour development in microbial enzyme treated Cheddar cheese slurries. *J. Food Sci.*, **44**, 1690–1694.

Sourirajan, S. (1977). Reverse osmosis — a general separation technique. In *Reverse Osmosis and Synthetic Membranes*, p. 3. National Research Council of Canada, Ottawa, Canada.

Sozzi, T., J. M. Poulin and R. Maret (1982). Effect of incubation temperature on the development of lactic acid bacteria and their phages. In *Nestle Research News 1980/81*. Nestle Products Technical Assistance Co., Switzerland.

Southward, C. R. and A. Goldman (1975). Co-precipitates: A review. *NZ J. Dairy Sci. Technol.*, **10**, 101–112.

Szijarto, L., D. A. Biggs and D. M. Irvine (1973). Variability of casein, serum protein and nonprotein nitrogen in plant milk supplies in Ontario. *J. Dairy Sci.*, **56**, 45–51.

Staff, F. (1956). New twists in lactose uses. *Food Eng.*, **28** (10), 56–57.

Stauffer, K. R. and J. G. Leeder (1978). Extracellular microbial polysaccharide production by fermentation on whey or hydrolyzed whey. *J. Food Sci.*, **43**, 756–758.

Tamime, A. Y. (1982). Microbiology of starter cultures. In *Dairy Microbiology*, ed. R. K. Robinson, vol. 2, pp. 113–151. Applied Science, London.

Thelwall, L. A. W. (1980). Lactose. In *Developments in Food Carbohydrates*, ed. C. K. Lee. Applied Science, London.

Thorogood, S. A., P. D. P. Wood and G. A. Prentice (1983). An evaluation of the Charm Test—a rapid method for the detection of penicillin in milk. *J. Dairy Res.*, **50**, 185–191.

Torres, N. and R. C. Chandan (1981). Latin American white cheese: A review. *J. Dairy Sci.*, **64**, 552–557.

Uchiyama, H., T. Uozumi, T. Beppu and K. Arima (1980). Purification of prorennin mRNA and its translation *in vitro*. *Agric. Biol. Chem.*, **44**, 1373–1381.

Vakaleris, D. G. and W. V. Price (1959). A rapid spectrophotometric method for measuring cheese ripening. *J. Dairy Sci.*, **42**, 264–276.

van den Berg, G. and E. de Vries (1975). Whey composition during the course of cheese manufacture as affected by the amount of starter and curd washing water. *Neth. Milk Dairy J.*, **29**, 181–197.

Walsh, P. M. and L. L. McKay (1981). Recombinant plasmid associated with cell aggregation and high frequency conjugation of *Streptococcus lactis*. *J. Bacteriol.* **146**, 937–944.

Walsh, P. M. and L. L. McKay (1982). Restriction endonuclease analysis of the lactose plasmid in *Streptococcus lactis* ML3 and two recombinant lactose plasmids. *Appl. Environ. Microbiol.*, **43**, 1006–1010.

Wong, N. P. (1974). Cheese chemistry. In *Fundamentals of Dairy Chemistry*, ed. B. H. Webb, A. H. Johnson and J. A. Alford, 2nd edn., pp. 719–753. AVI, Westport, CT.

World Health Organization (1978). *Environmental Health Criteria 5. Nitrates, Nitrites and N-Nitroso Compounds*. World Health Organization, Geneva.

Yoo, B. W. and J. F. Mattick (1969). Utilization of acid and sweet whey in wine production. *J. Dairy Sci.*, **52**, 900.

25

Fermented Dairy Products

B. A. FRIEND
Ralston Purina Company, St Louis, MO, USA

and

K. M. SHAHANI
University of Nebraska, Lincoln, NE, USA

25.1 INTRODUCTION

The fermentation of milk with lactic acid bacteria is one of the oldest methods of food processing and food preservation still used by mankind. Although the art of preparing fermented dairy products has been practiced for ages, recent scientific and technological advances in starter culture management and process control have yielded a wide variety of products with improved chemical, physical, nutritional and sanitary qualities.

Nearly every civilization has consumed fermented milk products of one type or another and these products have been, and still are, of extreme importance in human nutrition throughout the world. The total consumption of yogurt, sour cream and dips, buttermilk and cottage cheese remained constant at just under 6 kg per person per year during the past decade (Milk Industry Foundation, 1982). There have been, however, significant changes in the popularity of the individual products (Figure 1). In 1970, buttermilk was the most popular product at 2.6 kg per person and 42% of the market. Cottage cheese followed with 2.3 kg and 40% of the market. Sour cream and dips at 0.5 kg and yogurt at 0.4 kg occupied only 9% and 7% of the market, respectively. By 1980, buttermilk consumption had fallen to 1.9 kg and only 32% of the market. Although cottage cheese consumption dropped to 2.0 kg, it remained the market leader at 34%. The consumption of sour cream and dips increased to 0.83 kg and 14% of the market. Yogurt consumption increased to 1.19 kg and 20% of the market, which was slightly above that of the sour cream and dips. The growing popularity of yogurt has been ascribed to its image as a natural, convenient, low calorie food. The potential for additional growth in the United States appears bright, especially if our 1.19 kg is compared to the 16 to 34 kg of yogurt consumed each year by persons in northern Europe (Shahani and Friend, 1983).

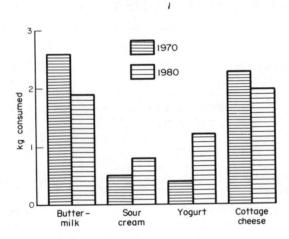

Figure 1　Per capita consumption of yogurt, sour cream and dips, buttermilk and cottage cheese in 1970 and 1980 in the United States (From Milk Industry Foundation, 1982)

This chapter begins with a brief overview of commercial lactic starter cultures used for the fermentation of milk. The remainder of the chapter outlines general principles and specific procedures of practical importance in the manufacture of fermented dairy products (other than hard cheeses) and, finally, the nutritional properties of these products.

25.2　LACTIC STARTER CULTURES

Milk is an ideal culture medium containing many factors required for the growth of the fastidious lactic acid bacteria. Thousands of years ago, milk set aside for later use was fermented spontaneously by the natural milk flora. At some time it was recognized that a small portion of soured milk could be added to fresh milk (back-slop) to prepare additional sour milk. The early fermentations were often slow and unpredictable, making the preparation of fermented milk products an art which was passed on from generation to generation. Sellars (1967) noted that the isolation and identification of pure starter cultures for milk fermentation did not begin until early in the twentieth century. During the past 25 years, research has covered the biochemistry of lactic cultures, including culture strain dominance and interactions, factors affecting the activity of lactic cultures and the control of bacteriophage infection of starter cultures with specialized media. The latter years have resulted in much research into the development of hybrid and/or genetically cloned cultures (Sellars, 1981). Today, seven major commercial suppliers provide a variety of single-strain and multiple- or mixed-strain lactic starter cultures with standardized performance potentials and characteristics.

25.2.1 Physiology and Biochemistry of Starters

The major lactic starter organisms and their physiological characteristics are shown in Table 1. Generally, *Streptococcus cremoris*, *Streptococcus lactis*, *Streptococcus thermophilus* and *Lactobacillus acidophilus* are used for acid production, *Streptococcus diacetylactis* and *Lactobacillus bulgaricus* for acid and flavor production and *Leuconostoc cremoris* and *Leuconostoc dextranicum* for flavor production. The combination of different species and strains of these cultures allows a high degree of versatility in acid and flavor production.

Table 1 Characteristics of Starter Organisms Used in Manufacturing Fermented Dairy Products[a]

Culture	Incubation temperature	Growth 10 °C	45 °C	Acid production (% in milk)	Diacetyl production	Litmus reduction	Salt tolerance (% max)
Streptococcus cremoris	22–30	+	−	0.8–1.0	−	++	4.0
Streptococcus lactis	21–30	+	−	0.8–1.0	−	++	4.0–6.5
Streptococcus thermophilus	40–45	−	+	0.8–1.0	±	+	2.0
Streptococcus diacetylactis	22–28	+	−	0.8–1.0	+	++	4.0–6.5
Lactobacillus acidophilus	37	−	+	0.3–2.0	−	±	6.5
Lactobacillus bulgaricus	42	−	+	1.5–4.0	−	++	2.0
Leuconostoc cremoris	20–25	+	−	0.1–0.3	+	±	6.5
Leuconostoc dextranicum	20–25	+	−	0.1–0.3	+	±	4.0–6.5

[a] From Chandan (1982) and Kosikowski (1977).

In the United States, most commercial starters are sold as mixed strain concentrates, the individual cultures of which have been selected on the basis of strain or species compatibility, resistance to antibiotics and phage infection, development of uniform acidity and flavor, production of carbon dioxide and synthesis and accumulation of compounds which contribute to acceptable flavor, color and body of the final product (Sellars, 1967).

The effect of starter culture compatibility on acid production and flavor in yogurt is illustrated in Table 2. Various commercial strains of *Streptococcus thermophilus* and *Lactobacillus bulgaricus* were combined in a 1:1 ratio and incubated at either 30 °C (long incubation) or 42 °C (short incubation) until the pH reached 4.2. The times required to reach the final pH and flavor score data for four representative starters are shown. Starter 1 produced acid rapidly at 30 °C and 42 °C, reaching the final pH within the expected time limit. Starters 2, 3 and 4 were slow starters at both temperatures. Starter 1 produced a yogurt with an acceptable flavor (score of 3.0 or above) when incubated at either 30 °C or 42 °C. Starters 3 and 4 produced an acceptable yogurt only at 42 °C. Starter 2 did not produce an acceptable yogurt at either 30 °C or 42 °C.

Table 2 Performance of Four Different Yogurt Starters Incubated at 30 °C and 42 °C[a]

Parameter	Yogurt starter 1	2	3	4
Incubation at 30 °C				
Time (h)	13	19	20	25
pH	4.0	4.1	4.1	4.2
Flavor score	3.6	2.3	2.0	2.0
Incubation at 42 °C				
Time (h)	3.5	6.5	7.0	5.5
pH	4.2	4.2	4.2	4.2
Flavor score	4.0	2.7	3.0	3.0

[a] From Friend *et al.* (1983).

The basic biochemical reactions of lactic starter cultures in milk are illustrated in Figure 2. Homofermentative streptococci and lactobacilli split lactose to glucose and galactose. The glucose is then converted to pyruvate and then to lactic acid with trace amounts of acetic acid and carbon dioxide (Collins, 1977). Certain species and strains of homofermentative organisms may produce aroma and flavor-enhancing intermediates in addition to lactic acid. Moreover, under highly aerobic conditions, the homofermentative lactobacilli will produce acetic acid rather than lactic acid (Collins, 1977).

The citric acid fermenting starters, *Streptococcus diacetylactis*, *Leuconostoc cremoris* and

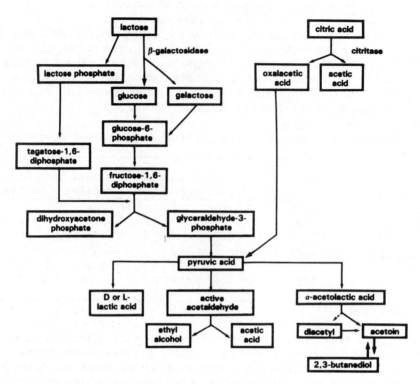

Figure 2 Biochemical reactions of dairy starter cultures (From Chandan, 1982 and Vedamuthu, 1979)

Leuconostoc dextranicum, convert citric acid to pyruvate and then to acetoin, diacetyl and carbon dioxide. Diacetyl imparts the characteristic 'buttery' flavor and aroma to buttermilk and sour cream. Diacetyl has been thought to be formed by condensation of active aldehyde and acetyl-CoA or formed *via* the intermediate α-acetolactic acid (Vedamuthu, 1979). Under appropriate conditions, acetoin and diacetyl may be reduced to the flavorless and odorless compound, 2,3-butanediol (Figure 2).

Acetaldehyde, the major flavor component of yogurt, is produced in significant amounts through the symbiotic action of *Lactobacillus bulgaricus* and *Streptococcus thermophilus* (Tamime and Deeth, 1980). *Streptococcus diacetylactis*, a common buttermilk and sour cream starter, will produce acetaldehyde as well as acetoin (Collins, 1977). While the accumulation of acetaldehyde is important in yogurt flavor, it will result in undesirable green, harsh or yogurt flavors in buttermilk. These defects may be controlled by the inclusion of *Leuconostoc cremoris* in the starter culture, since this organism can convert the excess acetaldehyde to ethanol (Collins, 1977).

The chemical compounds produced during incubation of a commercial buttermilk starter are shown in Table 3. During fermentation only a small portion of the lactose is converted to lactic acid, while nearly all of the citric acid is utilized. However, the flavor compounds (acetoin, diacetyl and volatile acids) do not accumulate until after the pH drops to 5.0. If the incubation period is extended beyond 16 hours, there will be loss in flavor due to the conversion of acetoin and diacetyl to 2,3-butanediol (Babel, 1976). Diacetyl production may be enhanced by adding 0.15% citric acid to milk prior to inoculation, monitoring the incubation time to allow sufficient acid production and synthesis of flavor compounds, aerating gently, chilling rapidly to avoid reduction to acetoin and 2,3-butanediol and by selecting starter strains which produce diacetyl in preference to acetoin (Vedamuthu, 1979; Chandan, 1982).

25.2.2 Propagation and Management of Starters

Before the widespread availability and acceptance of commercial starter culture concentrates, cultures were maintained, propagated and evaluated at the plant level. This required specially-trained personnel, separate culture handling facilities, extensive culture collections and time-consuming successive transfers to ensure the availability of sufficient active culture for inoculation. Today, standardized culture concentrates are purchased on a regular basis from commercial

Table 3 Compounds Produced in Milk Inoculated with a Mixed Starter of Lactic Acid
Fermenting and Citric Acid Fermenting Organisms[a]

Incubation time (h)	Titratable acidity (%)	pH	Lactose (%)	Citric acid (%)	Acetoin (p.p.m.)	Diacetyl (p.p.m.)	Volatile acid (%)
0	0.18	6.48	5.1	0.15	1.2	0.1	0.16
4	0.20	6.29	5.1	0.14	1.8	0.1	0.20
8	0.37	5.52	5.0	0.12	3.6	0.4	0.59
12	0.71	4.80	4.8	0.08	32.5	1.3	1.25
16	0.90	4.46	4.4	0.03	99.0	2.7	2.82

[a] From Babel (1976).

suppliers in three major forms: cryogenically frozen preparations designed for direct inoculation into product mixes (direct sets), cryogenic concentrates for preparation of bulk starter (bulk sets) and lyophilized powders. Information regarding the preparation and utilization of culture concentrates has been reviewed by Chandan (1982), Gilliland (1977) and Sellars and his associates (Porubcan and Sellars, 1977; Sellars and Babel, 1978).

Porubcan and Sellars (1977) outlined the steps required in a typical batch-type fermentation process used for large-scale preparation of concentrated starter cultures. Specialized media containing complex nutrients, simple nutrients, buffers and defoamers are formulated and sterilized by ultra-high temperature (UHT) processes (140–150 °C, 4–8 s). The mixture is transferred aseptically into sterile fermenters and inoculated with 1–2% mother culture. Throughout fermentation, the temperature, pH, redox potential and agitation level are strictly controlled to ensure maximum populations of organisms. The organisms are then concentrated by centrifugation or ultrafiltration. For preparation of frozen concentrates, the concentrate is packaged and then flash-frozen in liquid nitrogen at -196 °C. For preparation of lyophilized powders, the concentrate is frozen in thin films, freeze dried and then packaged.

Frozen concentrates are shipped either in liquid nitrogen or under dry ice. Although storage at -196 °C is preferable, cultures may be stored at -40 to -45 °C for up to six weeks without loss in activity (Sellars and Babel, 1978). Direct set cultures are packed in 130–500 ml units containing $50 \times 10^9 - 400 \times 10^9$ cfu ml^{-1}. These cultures are sufficient for inoculating up to 2600 ml of milk at a concentration equivalent to 1% inoculum of bulk culture at 4×10^9 cfu ml^{-1} (Porubcan and Sellars, 1977). Direct set cultures eliminate the need for culture propagation, eliminate preparation of intermediate or bulk culture and allow culture rotation on a vat-to-vat basis.

Propagation techniques for commercial bulk set cultures and lyophilized powders for use in buttermilk, cream cheese, cultured cream and cottage cheese are illustrated in Figure 3. Media are formulated at 10–12% solids using antibiotic-free skim milk, special phage inhibitory media or a combination of both. The media are then heat treated to destroy microbial contaminants and to stimulate the growth of starter cultures through partial hydrolysis of casein, release of sulfhydryl groups and conversion of lactose to formate (Chandan, 1982). Sellars and Babel (1978) suggest that media for mother cultures be held at 90 °C for 1 h or autoclaved at 120 °C for 15–20 min, while media for intermediate and bulk cultures be held at 85–96 °C for 1 h. Autoclaving is not used for intermediate or bulk cultures since undesirable caramelized color and flavor may carry over into the final product (Chandan, 1982).

Feldstein and Westhoff (1979) reviewed the effect of heat treatments on the activity of lactic starter cultures and noted a wide range of activity for each heat treatment. While UHT processed milk appeared to be suitable for the growth of starter cultures, these authors suggested that quality factors as well as growth factors be examined before milk treated in this manner is utilized commercially for the manufacture of cultured products (Feldstein and Westhoff, 1979).

Bulk sets are sold in 70 to 130 ml units which contain from $5 \times 10^9 - 50 \times 10^9$ cfu ml^{-1} and are sufficient for inoculating from 379–1893 l of substrate (Porubcan and Sellars, 1977). The use of bulk set concentrates eliminates the need for propagation of cultures and preparation of intermediate cultures.

Lyophilized powders may be shipped and stored for short periods of time without refrigeration. One package of dried culture may be used for the preparation of 750 ml of mother culture or up to 37.8 l (10 gal) of intermediate or bulk culture, without preparation of mother culture (Porubcan and Sellars, 1977). The performance of lyophilized powders may be limited, however, if they are used directly without one or two intermediate propagations (Porubcan and Sellars, 1977).

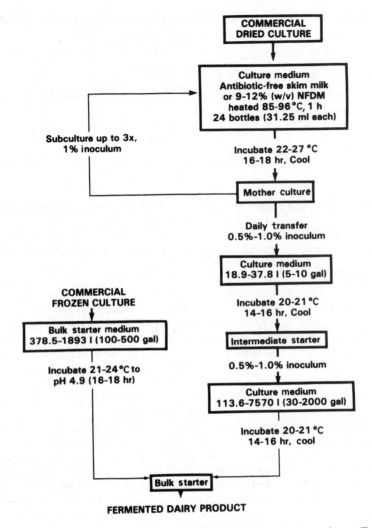

Figure 3   Preparation of bulk cultures for use in buttermilk, cultured cream and cottage cheese (From Chandan, 1982 and Porubcan and Sellars, 1977)

Generally, the mixed strain starter cultures used for buttermilk, sour cream and cottage cheese are incubated at 20 to 21 °C to maintain a proper balance between lactic streptococci and citric acid fermenting organisms. In cases where rapid acid production is desired, the incubation temperature may be increased to approximately 30 °C (Sellars, 1967).

To assure maximum culture activity, media should be at the proper temperature prior to inoculation, inoculated media should not be agitated or aerated and cultures should not be held at incubation temperature for extended periods of time after coagulation (Sellars and Babel, 1978).

By increasing the incubation temperature to 40–45 °C, the procedures outlined in Figure 3 may also be used to prepare yogurt starters (Sellars and Babel, 1978). Yogurt starter contains *Streptococcus thermophilus* and *Lactobacillus bulgaricus* growing in a symbiotic relationship. Incubation at 40 °C for a shorter time period will favor the growth of *Streptococcus thermophilus*, while incubation at 50 °C for a longer time will favor the growth of *Lactobacillus bulgaricus*. The level of inoculum, incubation temperature and incubation time may be varied in order to maintain the two organisms in the desired 1:1 ratio during successive transfers.

25.2.3 Control of Starter Culture Inhibition

Inhibition of dairy starter culture growth may be noted by a decrease in the production of lactic acid (slow starter), by a complete cessation in the production of acid (dead starter) or by a decrease in the production of flavor compounds from citric acid. The primary causes of starter culture inhibition and suggested methods of control are summarized in Table 4.

Table 4 Common Inhibitors of Dairy Starter Cultures[a]

Substance	Suggested controls
Bacteriophage	Improve plant sanitation
	Sterilize with chlorine compounds
	Use multiple strain cultures, culture rotation and/or phage inhibitory media
Milk factors	
Antibiotics	Test milk supply and reject contaminated milk
Cleaning compounds	Monitor sanitation practices
Composition	Standardize composition
	Pasteurize to destroy natural inhibitors
Mastitis	Screen for mastitis
Rancidity	Avoid agitation of raw milk
	Pasteurize before homogenization

[a] From Sellars and Babel (1978).

25.2.3.1 *Bacteriophage infection*

Bacteriophage infection is the major cause of slow acid production by the lactic cultures. The bacteriophage adsorbs to the surface of sensitive host cells, penetrates the host with DNA, replicates and assembles the replicated phage particles, lyses the cells and releases up to 200 new phage particles which can attack additional phage-sensitive organisms (Kosikowski, 1977). Since phage particles require living cells to multiply, any material in which lactic cultures have grown becomes a potential source of phage contamination. Proper disposal of cottage cheese whey and careful cleaning and sanitation of equipment and rooms with chlorine compounds (200–300 p.p.m. solution for equipment and 500–1000 p.p.m. fog for rooms) are important for controlling bacteriophage infection (Chandan, 1982).

If cultures are propagated in the plant, a separate culture preparation room must be provided to avoid contact with phage-containing materials. It is especially important to avoid contamination of the mother culture to ensure that phage multiplication does not occur during subsequent transfers to intermediate and bulk cultures. The use of specialized commercial media, which inhibit phage growth, reduces phage problems during culture propagation. These media are highly buffered and contain demineralized whey, non-fat dry milk, phosphate, citrate and yeast extract (Chandan, 1982). The phosphate binds calcium and prevents calcium-dependent phage growth. Citrate contributes to buffering capacity and provides substrate for diacetyl production. The yeast extract provides growth factors for the lactic culture. Alternatively, in-plant propagation problems may be avoided altogether by using direct set cultures.

The final means of phage control used in the United States is the use of commercial multi-strain cultures which are rotated regularly on the basis of their phage sensitivity. The combination of good sanitation, phage inhibitory media and proper rotation of multi-strain cultures will reduce the probability of bacteriophage problems.

25.2.3.2 *Milk factors*

Pharmaceutical antibiotics may be present in the milk supply if milk is not withheld from the market for at least 96 h after treatment of cows for mastitis or other diseases (Sellars, 1967). As shown in Table 5, as little as 0.05 IU of penicillin or 0.03 μg of aureomycin will completely inhibit the growth of lactic starter cultures. Thus, all milk samples should be tested for antibiotics before use. Charm (1979) developed a rapid antibiotic assay which can detect as little as 0.05 IU ml^{-1} penicillin with a simplified on-farm test, or 0.005 IU ml^{-1} penicillin with an in-plant test.

In a recent study, Wulf and Sandine (1983) were able to isolate and characterize fast acid-producing streptomycin-resistant mutants of lactic streptococci. Although antibiotic-resistant isolates had been isolated previously, they were slow acid producers and thus unsuitable for starter cultures.

Many sanitizers used routinely on the farm and in the plant may be present as milk contaminants. 20–200 p.p.m. of chlorine compounds will inhibit acid production by most lactic starters, although individual strains vary in their sensitivity (Sellars, 1967). Similarly, 50–100 p.p.m. iodo-

Table 5   Minimum Levels of Antibiotic Required to Inhibit
Growth of Dairy Starter Cultures[a]

Starter culture	Penicillin $(IU\ ml^{-1})$	Aureomycin $(\mu g\ ml^{-1})$
Streptococcus lactis 9	0.05	0.05
Streptococcus thermophilus H	0.05	0.03
Lactobacillus bulgaricus 444	0.10	5.0
Lactobacillus bulgaricus 488	0.10	3.0
Lactobacillus bulgaricus V12	0.05	0.3

[a] From Shahani and Harper (1958).

phores or 25–75 p.p.m. quaternary ammonium compounds will reduce or completely inhibit acid production (Sellars, 1967). In addition, as little as 0.5 p.p.m. quaternary ammonium compounds will inhibit flavor production by the citric acid fermenting cultures (Sellars, 1967).

Seasonal variations in the composition of milk can also affect culture growth. Milk containing a higher level of solids-non-fat (SNF) will favor growth and acid production by lactic cultures (Sellars, 1967). Variations in the composition of milk may also influence the ratio between leuconostocs and streptococci (Sellars, 1967). Raw milk also contains several natural inhibitory compounds including lactenins, lactoperoxidase, agglutenins and lysozyme (Chandan, 1982). These compounds are heat labile and will be destroyed during HTST or UHT pasteurization treatments (Chandan, 1982).

Milk from mastitis-infected cows contains less lactose, less unhydrolyzed protein, more chloride and a higher pH than normal milk and, therefore, is considered to be a poor microbial growth substrate. Consequently, when such milk is used for the manufacture of cultured products, it does not support adequate acid development (Sellars, 1967). The high concentrations of leukocytes associated with mastitis may also inhibit starter growth by phagocytic action, although this can be controlled by adequate heat treatment of the milk (Sellars, 1967).

Rancid milk results from the release of fatty acids from milk lipids through the action of inherent milk lipases or through the growth of lipolytic contaminants such as psychrotrophs. The presence of short chain fatty acids not only gives an objectionable flavor to the milk, but may also inhibit the growth of lactic starter cultures (Chandan, 1982). Although both milk and microbial lipases are destroyed by pasteurization, problems may occur in raw milk which has undergone vigorous agitation or homogenization, temperature activation of lipase or extended storage (Sellars and Babel, 1978). Senyk *et al.* (1982) have shown that rancidity problems due to inherent lipase of the growth of psychrotrophic contaminants can be controlled by routine, on-farm subpasteurization treatments of milk. Moreover, significantly higher cottage cheese yields have been noted for milk which has been subjected to a subpasteurization heat treatment and storage (Djurec and Zall, 1982). Alternatively, White and Shilotri (1979) have suggested that raw milk in farm bulk tanks be routinely inoculated with citric acid fermenting organisms to prevent the growth and activity of psychrotrophic contaminants.

25.3   GENERAL MANUFACTURING PRINCIPLES

Fermented dairy products can be divided into three classes: liquid products which include acidophilus milk, buttermilk, kefir and koumiss; semi-solid products which include cultured cream and yogurt; and the unripened soft cheeses which include bakers' cheese, cottage cheese, cream cheese and quarg (Table 6). Buttermilk, cultured cream and cream cheese are prepared from starters containing lactic streptococci (*Streptococcus lactis* and *Streptococcus cremoris*) and citric acid fermenting organism(s) (*Streptococcus diacetylactis* and/or *Leuconostoc cremoris*). Bakers' cheese, cottage cheese and quarg utilize only the lactic streptococci. Lactobacilli are used for the preparation of acidophilus milk (*Lactobacillus acidophilus*), kefir (*Lactobacillus causasicus* plus lactic streptococci, *Leuconostoc* spp. and yeasts), koumiss (*Lactobacillus bulgaricus* plus Torula yeast) and yogurt (*Lactobacillus bulgaricus* and *Streptococcus thermophilus*).

25.3.1   Processing Steps and Ingredients

Chandan (1982) described the general design of plants used for the manufacture of cultured dairy products. Each plant is equipped with a separate receiving room to receive, weigh and store

Table 6 Major Fermented Dairy Products Other Than Hard Cheeses[a]

Product	Process conditions	Starter cultures
Liquid products		
Acidophilus milk	37–40 °C, 16–18 h	*Lactobacillus acidophilus*
Buttermilk	22 °C, 18 h	*Streptococcus cremoris* or *S. lactis*, *S. diacetylactis* or *Leuconostoc cremoris*
Kefir	18–22 °C, 12 h or 10 °C, 1–3 d	Kefir grains (*Streptococci* spp., *Lactobacillus caucasicus*, *Leuconostoc* spp., yeasts)
Koumiss	28 °C, 2 h	*Lactobacillus bulgaricus* *Torulopsis holmii*
Semi-solid products		
Cultured cream	22 °C, 18 h	*Streptococcus cremoris* or *S. lactis*, *S. diacetylactis* or *Leuconostoc cremoris*
Yogurt	43–45 °C, 3 h	*Streptococcus thermophilus*, *Lactobacillus bulgaricus*
Unripened soft cheeses		
Bakers' cheese	31 °C, 5 h	*Streptococcus cremoris* or *S. lactis*
Cottage cheese	22 °C, 18 h or 35 °C, 5 h	*Streptococcus cremoris*, *S. lactis*
Cream cheese	22 °C, 18 h	*Streptococcus cremoris* or *S. lactis*, *S. diacetylactis* or *Leuconostoc cremoris*
Quarg	31 °C, 5 h	*Streptococcus cremoris* or *S. lactis*

[a] From Kosikowski (1977).

milk and other raw materials. The plant also contains an ambient-temperature dry storage area, a refrigerated storage area, a processing room, a packaging room, a utility room, a culture propagation room and a quality control laboratory. Dry storage is used for heat-stable raw materials and packaging supplies, while the refrigerated storage area is used for fruits, other heat-labile raw materials and finished products. The processing room contains the equipment and controls for the separation, standardization, pasteurization and homogenization of milk and the formulation, inoculation and incubation of product. Plants for the manufacture of cottage cheese contain cheese vats with mechanical agitation devices, cutting knives, and systems for whey draining, curd washing and creaming. The utility room is used for plant maintenance and engineering services. A separate culture preparation room is required only if cultures are maintained and propagated in the plant. Finally, a quality control program (established and administered by laboratory personnel) is needed to monitor culture preparation, the quality of raw materials, process control, product composition and shelf life to ensure compliance with regulatory and company standards.

Heldman and Seiberling (1976) discussed the importance of environmental sanitation in the production, processing and distribution of milk and milk products. These authors define environmental sanitation as the 'maintenance of all substances, surfaces and materials which contact the product directly or indirectly in a state which (a) is free from disease-producing bacteria and toxic-substances, (b) is free from foreign materials, (c) has low bacterial count, (d) does not influence product safety or keeping quality, and (e) promotes the high nutritive value of the product'. To achieve this, Heldman and Seiberling (1976) suggest consideration be given first to the layout of rooms and the ventilation system to exclude outside contamination, as well as to prevent cross contamination between raw materials and finished product within the plant. Secondly, the floors, walls and ceilings should be constructed of smooth, impervious materials. Floors must be sloped to ensure rapid drainage of water, whey and other liquid waste. Thirdly, the design and installation of equipment and pipelines and the materials used in their construction are of critical importance since the product is in direct contact with their surfaces during the entire production process. Stainless steel and glass are used routinely for equipment and pipelines since these materials provide surfaces which are smooth, impervious, corrosion-resistant, non-toxic and easily cleaned. Fourthly, all surfaces which come in contact with the product must be cleaned immediately after use to remove all evidence of soil, and then should be sanitized with a bactericidal agent. Other plant surfaces such as the floors, the walls and the ceilings should be given less rigorous, though similar, attention.

The general process steps used to manufacture fermented dairy products are outlined in Table 7. While milk from cow, water buffalo, goat, mare, sheep and sow are used for cultured products in many parts of the world, only cow milk is used in the US. All milk must be of high quality and free from inhibitors to ensure optimum culture growth. After reception at the plant, milk may be stored for up to 72 h at 10 °C in large vertical silos holding up to 100 000 1 (Chandan, 1982). Extraneous matter (including leukocytes, some bacteria, dust and dirt) is removed from the milk in a high speed centrifuge known as a clarifier (Jones and Harper, 1976). Milk is then centrifuged using a separator or standardizing clarifier to separate the cream and skim milk for standardization of fat levels.

Table 7 Process Steps for the Manufacture of Fermented Dairy Products[a]

Process step	Remarks
1. Procurement of milk from the farm	Antibiotic-free milk from healthy cows should be cooled in refrigerated bulk tanks to 10 °C in 1 h and below 5 °C in 2 h to ensure microbiological control. Milk pick-up should be at 48 h intervals in insulated tanks. Avoid unnecessary agitation to prevent lipolytic rancidity problems in milk
2. Reception and storage of milk in plant	Quality of milk checked prior to reception and storage at 10 °C for not more that 72 h
3. Centrifugal clarification and separation	Leucocytes and sediment are removed. Milk separated into cream and skim milk, or standardized to a specific fat level
4. Mix preparation	Ingredients are blended together to a desired formulation at 50 °C in a mix tank
5. Heat treatment	Mix is heated above pasteurization (85–95 °C, 10–40 min). Heat treatment destroys contaminating microorganisms, releases growth factors, generates microaerophilic conditions and improves body and texture in fermented product
6. Homogenization	Mix is homogenized in two stage process to prevent aggregation of fat globules during incubation and storage of products. Aids in uniform dispersion of stabilizer and other mix components
7. Inoculation and incubation	The mix is cooled to incubation temperature, inoculated with 0.5–5% starter culture and incubated quiescently until desired pH level is reached
8. Cooling, incorporation of fruit and flavoring and packaging	The coagulated mix is cooled to 5–22 °C, depending on the product. Fruit, fruit flavors, other flavoring or seasoning ingredients are incorporated before packaging
9. Storage and distribution	Low temperature storage ensures adequate shelf life by decreasing the rate of physical, chemical and microbiological degradation

[a] From Jones and Harper (1976) and Chandan (1982).

Raw whole milk (3.7% butterfat) may be separated into full cream (37% fat) and skim milk (0.1% fat) (Figure 4). By adjusting the cream flow valve, the relative quantity of butterfat can be standardized to that of low fat milk (2% fat), light cream (18% fat) or other levels. Non-fat dry milk (NFDM) is prepared from condensed skim milk which has undergone either spray or roller drying. These ingredients are blended together in a mix tank at 50 °C at a predetermined level of fat and solids-non-fat (Table 7).

Milk is pasteurized by vat, HTST or UHT processes (Chandan, 1982). Vat pasteurization is conducted at 63 °C (66 °C if sweeteners are present in the mix) for a minimum of 30 min. For HTST, the equivalent temperature is 73 °C (75 °C with sweeteners) for 15 s. UHT systems are used at 90–148 °C for 2 s. Generally, standardized mixes are heated to above pasteurization temperatures in order to destroy contaminating microorganisms, to release growth factors, to generate microaerophilic conditions required for starter culture growth and to achieve the proper body and texture in the fermented product. One exception is cottage cheese manufacture, where skim milk is only heated at 72 °C for 16–17 s (Chandan, 1982).

After heating, the mix is homogenized in a two-stage treatment (Chandan, 1982). The first stage (6.8–13.7 kPa) reduces the average size of the milk fat globules from a 4 μm diameter to 1 μm. The second stage (3.4 kPa) breaks apart fat globule clusters to prevent creaming. Stabilizer may also be added at this time to improve the body and texture of the product.

The mix is cooled and transferred to a jacketed, culturing vat equipped with an agitation system. The mix is inoculated either with a vat set culture concentrate or with a bulk starter, and is incubated until the desired pH is achieved. Rennet may also be added to increase the body of the product. The fermented mix is then cooled quickly to 5–22 °C, depending on type of product to be manufactured. High quality fruits, fruit flavors and other seasoning and flavoring ingredients are incorporated prior to packaging. Chandan (1982) noted that most plants attempt to synchronize packaging lines with termination of the incubation period. Proper storage of the product at low temperature will ensure an adequate shelf life by decreasing the rate of physical, chemical and microbiological degradation.

The manufacture of cottage cheese curd requires several modifications to the general scheme. Since skim milk is used rather than a mix, cottage cheese requires a less severe heat treatment, and no homogenization. After inoculation and coagulation, the curd is cut, cooked, drained and

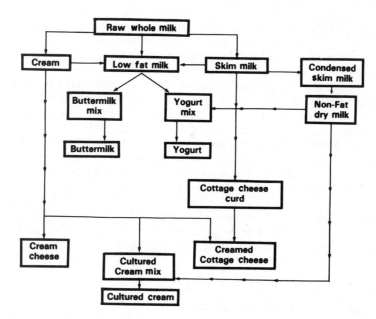

Figure 4 Milk ingredients used for the manufacture of fermented dairy products

rinsed. A special cream 'dressing' (cream, salt and stabilizer mixture) is then mixed with the dry curd before packaging and storage.

25.3.2 Extension of Shelf Life

Lactic starter cultures synthesize a number of antimicrobial compounds including lactic acid, acetic acid, benzoic acid and hydrogen peroxide. In addition, natural antibiotics are produced by several species of lactic cultures including nisin from *Streptococcus lactis*, diplococcin from *Streptococcus cremoris*, bulgarican from *Lactobacillus bulgaricus* and acidolin, acidophilin and lactocidin from *Lactobacillus acidophilus*. Shahani and Friend (1983) reviewed the production of acidophilin and bulgarican by *Lactobacillus acidophilus* and *Lactobacillus bulgaricus*, respectively and noted that the level of antibiotic was dependent on the culture strain, type of culture medium and incubation conditions. Both acidophilin and bulgarican possess a wide spectrum of activity, inhibiting many pathogenic and non-pathogenic organisms found in foods.

The contribution of antimicrobial compounds to shelf life extension and inhibition of foodborne pathogens in fermented dairy products has been extensively reviewed (Babel, 1977; Chandan, 1982; Gilliland and Speck, 1972, 1977; Shahani and Chandan, 1979). One commercial application has been the addition of *Leuconostoc cremoris* to cottage cheese creaming mixtures to increase shelf life by preventing the growth of coliforms and slime-forming *Pseudomonas* spp. (Babel, 1977).

A post-incubation heat treatment may also be employed to stabilize the fermented product during storage and to increase shelf life. This has been used most often in the preparation of long-life cultured creams, sour cream dips and pasteurized yogurts.

25.3.3 Use of Direct Acidification

Direct acidification has been used as an alternative to fermentation in the production of sour milk and imitation cultured milk products. Sellars (1981) noted that sour cream was the first dairy product to be prepared commercially by direct acidification. Commercial processes for buttermilk and cottage cheese have also been introduced.

The base for these products consists of dairy ingredients or an emulsified vegetable oil stabilized with sodium caseinate (Chandan, 1982). Food-grade acidulant replaces the starter culture used in fermented products. Citric, phosphoric, gluconic and lactic acids are used, although the high cost of lactic acid limits its application. Alternatively, gluconic acid may be formed in milk by the hydrolysis of glucono-δ-lactone. Culture distillates containing diacetyl are added for flavor. The body and texture of the fermented product are simulated by the addition of emulsifiers and stabilizers including vegetable gums, starch, carrageenan, gelatin, partial glycerides, caseinates and sodium phosphate (Chandan, 1982). Specific processing steps used for the manufacture of directly acidified sour cream and cottage cheese will be given later.

Chandan (1982) noted that direct acidification has been claimed: (a) to be easier to control and less expensive than fermentation since cultures and expensive culturing vessels are not needed; (b) to be quicker and more economical in terms of manpower; and (c) to yield products with uniform, reproducible flavor and texture. Since culture-related problems with inhibitors or contaminants are eliminated with direct acidification, product failure is less likely to occur. Studies in our laboratory showed that the appearance and texture of direct acidified sour cream and cultured cream were identical, but the flavor of the acidified product was inferior (Kwan *et al.*, 1982). Consequently, acidified sour cream is used extensively as a base in seasoned dips, dressings and other food products where superior body and texture, rather than flavor, are the most important attributes (Sellars, 1981). Because of increased demand for these products in the United States, the majority of the sour cream manufactured today is thought to be prepared by direct acidification rather than by fermentation (Sellars, 1981).

25.4 MANUFACTURE OF LIQUID PRODUCTS

25.4.1 Acidophilus and Sweet Acidophilus Milks

Certain strains of *Lactobacillus acidophilus*, a normal inhabitant of the human intestine, can be transplanted to the intestine and provide therapeutic benefits by suppressing the growth of undesirable microorganisms (Gilliland, 1979; Gilliland and Speck, 1977; Sandine, 1979; Shahani and Ayebo, 1980). Recent research has centered on the selection and propagation of *Lactobacillus acidophilus* strains known to be compatible with the human intestine and the manufacture of products containing large numbers of viable cells from compatible strains (Gilliland, 1979; Klaenhammer, 1982; Speck, 1980).

The general preparation of acidophilus milk has been discussed by Kosikowski (1977) and Chandan (1982). According to Kosikowski (1977), low fat or skim milk should be sterilized at 120 °C for 20 min (15 p.s.i.), cooled to 38 °C, inoculated with 5% *Lactobacillus acidophilus* and incubated for 18 to 24 h until a curd forms with 1% acidity. In contrast, Chandan (1982) recommended ultraheating whole or skim milk at 98 °C for 30 min or 145 °C for 2 to 3 s rather than autoclaving to increase the growth rate of the organism. He also noted the culture should be incubated at 37 °C only until the pH reached 4.7 (0.65% acidity), since increased acid production decreases cell viability (Chandan, 1982). While fresh acidophilus milk contains more than 500 million viable cells per ml, viability decreases rapidly during storage (Chandan, 1982). In addition, the strong acid flavor of this product has little appeal in the United States.

A sweet acidophilus milk, in which *Lactobacillus acidophilus* cell concentrates are added to fresh low fat milk, has been marketed in the United States to overcome the problems of culture viability and consumer appeal. Sweet acidophilus milk contains more than 2 million viable cells per ml, yet retains its sweet, unfermented flavor if held at refrigeration temperatures during distribution and storage (Speck, 1980).

Another alternative to improving the flavor of *Lactobacillus acidophilus* containing products is to use a mixed culture fermentation. Products prepared in this manner include bioghurt fermented by *Lactobacillus acidophilus* and *Streptococcus lactis*, and acidophilus yogurt fermented by *Lactobacillus acidophilus*, *Streptococcus thermophilus* and, in some cases, also *Lactobacillus bulgaricus*. *Lactobacillus acidophilus* concentrates have also been blended with fruit juices and/or cultured buttermilk to improve their flavor.

The Japanese consume a therapeutic beverage called yakult, prepared with *Lactobacillus casei* strain Shirota (formerly *Lactobacillus acidophilus* strain Shirota). Shirota (1971) noted that this strain can also be implanted in the intestine where it inhibits the growth of pathogenic organisms and synthesizes B vitamins.

25.4.2 Cultured Buttermilk

Cultured buttermilk is prepared from skim or low fat milk fermented with *Streptococcus lactis* or *Streptococcus cremoris* for acid production and *Leuconostoc cremoris*, *Leuconostoc dextrani-*

cum or *Streptococcus diacetylactis* for flavor production (Figure 5). Although milk fat levels from 0.5–1.8% have been used (Chandan, 1982), the proposed federal buttermilk standards require a maximum fat content of 0.5% and a minimum milk solids-non-fat (MSNF) of 8.25% (Anon, 1977). The product must also possess a minimum titratable acidity of 0.5% calculated as lactic acid. In addition to traditional dairy ingredients, any milk-derived ingredient may be used to standardize the solids content of the mix provided there is no decrease in either the ratio of protein to MSNF or the protein efficiency ratio. Other legal ingredients include characteristic flavorings (starter distillate, butter flavor, diacetyl), nutritive carbohydrate sweeteners, stabilizers and colorings. Vitamin fortification with a minimum of 2000 IU of vitamin A and 400 IU of vitamin D per 0.95 l is also allowed. Milk which is low in citric acid is normally supplemented with 0.20–0.25% sodium citrate for flavor enhancement. In certain parts of the United States, butter granules or flakes are added to the finished buttermilk to give a heavier product with smoother taste. Butterflakes are prepared by churning 18–20% fat cream, or by spraying melted butter oil at 71.1 °C onto the chilled buttermilk.

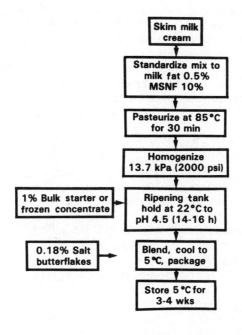

CULTURED BUTTERMILK

Figure 5 Flowchart for the manufacture of cultured buttermilk (From Chandan and Shahani, 1982)

The causes of flavor and texture defects in buttermilk and their prevention have been reviewed by Chandan (1982), Custer (1982) and Kosikowski (1977). In general, a good buttermilk flavor and texture will be produced by using high quality milk ingredients formulated at the proper solids level, heating at 85 °C for 30 min, using active starter, incubating at 22 °C to maintain the balance between the starter organisms, breaking the curd at appropriate acid level, cooling rapidly with gentle agitation to avoid destruction of diacetyl and avoiding contamination problems through proper sanitation. The curd should be broken at pH 4.65–4.7 which is equivalent to a titratable acidity (TA) of 0.85% for mix containing 9% MSNF and a TA of 0.90% for 10% MSNF or a TA of 0.95% for mix containing 11% MSNF (Custer, 1982).

The use of 'return milk', poor quality ingredients or the presence of microbial contaminants may cause bitter, stale, cheesy, yeasty, putrid or unclean flavors. Also, poor flavor development may result from low acidity and/or diacetyl production due to poor culture balance, poor starter activity or improper handling. On the other hand, excessive production of acid or diacetyl may cause harsh or bitter flavors, although harshness may be cut by the addition of 0.1–0.2% salt. Acetaldehyde accumulation gives a green or yogurt flavor. This can be prevented by using cultures free of contamination with *Streptococcus diacetylactis*, or by using direct vat inoculation of each batch of buttermilk.

A weak bodied, thin buttermilk is produced if the mix contains less than 9% milk solids, improper heat treatment is used, the culture is inactive, acid development is low or there is excessive agitation. A thick, viscous product will occur if the mix contains high solids, excess acid, insufficient agitation, slime forming organisms or other microbial contamination. A grainy or lumpy buttermilk will occur if excess acid is produced, the ingredients are not dissolved completely, the heat treatment is too low, the incubation temperature is too high, there is insufficient agitation during breaking or poor quality milk is used.

25.4.3 Kefir and Koumiss

Kefir and koumiss are carbonated, fermented milks prepared in Eastern Europe and Western Asia through the action of a lactic acid bacteria and yeasts. Kefir is produced by fermenting the milk of the cow, sheep or goat with kefir grains. According to Kosikowski (1977) these grains are white to yellow, gelatinous granules which contain a mixed microflora of lactic streptococci, *Lactobacillus caucasicus*, *Leuconostoc* spp. and *Saccharomyces kefir* and *Torula kefir*.

Commercial preparations of freeze-dried grains are rehydrated and activated by transferring twice into sterile 0.9% NaCl (Kosikowski, 1977). The grains are then transferred to pasteurized skim milk (95 °C for 30 min) and incubated for 24 h at 25 °C. Kefir is prepared by pasteurizing whole milk at 85 °C for 30 min, cooling to 22 °C, seeding with active kefir grains and then holding for 12–18 h until curdled. After removal of the grains by sieving, the curdled milk is allowed to ripen for an additional 1–3 days at 10 °C. The recovered grains may be stored in water at 4 °C for up to 8–10 days, or dried at room temperature for 36–48 h and stored up to 12–18 months. Kefir contains 0.8% lactic acid, 1.0% alcohol, carbon dioxide and a ratio of diacetyl to acetaldehyde of 3.1 (Kosikowski, 1977).

Although traditional koumiss is prepared from mare's milk, the shortage of this product has led to the development of procedures using cow's milk as well. Kosikowski (1977) noted that bulk starter for koumiss is prepared by inoculating two flasks of pasteurized (70 °C for 30 min) skimmed cow's milk with *Torulopsis holmii* and incubating at 30 °C for 15 h. Another flask of milk is inoculated with *Lactobacillus bulgaricus* and incubated at 37 °C for 7 h. The three starter flasks are mixed with one flask of fresh mare's milk and then incubated at 28 °C. During the next three days, the acidity is maintained at 0.65 to 70% through the addition of mare's milk. After that time, the acidity of the starter is allowed to reach 1.4%.

Koumiss is prepared by inoculating fresh mare's milk with 30% bulk starter, incubating for 2 h under agitation at 28 °C, bottling and capping the mixture, reincubating for 2 h at 20 °C and then storing the fermented mixture at 4 °C (Kosikowski, 1977). Koumiss varies in composition from 0.7% lactic acid and 1.0% alcohol (weak koumiss) to 1.8% lactic acid and 2.5% alcohol (strong koumiss). A product similar to koumiss has been prepared in the United States from skim milk and a mixed culture of *Lactobacillus acidophilus*, *Lactobacillus bulgaricus*, *Lactobacillus caucasicus* and yeast (Kosikowski, 1977).

25.5 MANUFACTURE OF SEMI-SOLID PRODUCTS

25.5.1 Cultured Cream

Cultured cream or sour cream is prepared by fermenting a minimum of 18% fat cream to a titratable acidity of 0.5% (US Department of Agriculture, 1977) using the same lactic starter as buttermilk. Optional ingredients include stabilizers, rennet, salt, nutritive sweeteners, flavorings and colors. Up to 0.1% sodium citrate may also be added for flavor enhancement.

The procedures for preparing cultured cream, long life cultured cream, sour cream dip and acidifed sour cream are shown in Figure 6. Raw cream should be fresh and of high bacterial quality. Heavy, full-bodied products are obtained by formulating mixes with at least 6.8% MSNF, using HTST pasteurization (74 °C for 16 s) rather than vat pasteurization, homogenizing in two stages at 72 °C and 17.2 kPa, adding rennet, mixing without incorporation of air and allowing sufficient acid production (Kosikowski, 1977; Chandan, 1982).

Long-life or hot-pack cultured cream is prepared by adding 0.20–0.5% stabilizer to normal cultured cream and then heating, homogenizing and hot-packing the mixture into glass containers or metal cans. Heating destroys proteolytic enzymes of the lactic acid streptococci which can cause bitterness during storage (Kosikowski, 1977). In addition, the combination of the heat treatment

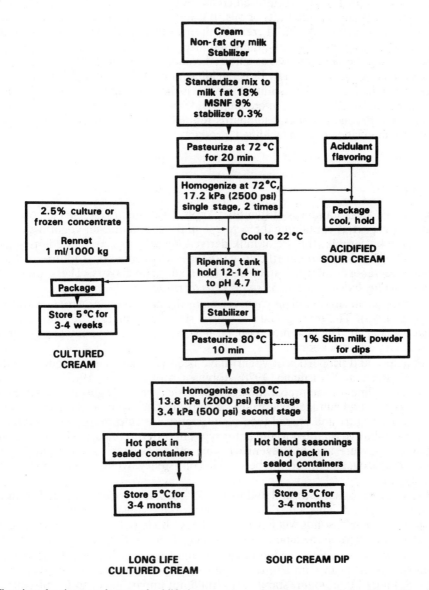

Figure 6 Flowchart for the manufacture of acidified sour cream, cultured creams and dips (From Chandan, 1982 and Chandan and Shahani, 1982)

and natural vacuum which forms after cooling prevents the growth of molds and yeasts and minimizes oxidative deterioration of the fat (Kosikowski, 1977; Chandan, 1982). Kosikowski (1977) noted that hot-pack cultured cream may retain its quality for as long as 4–12 months before oxidation occurs.

Sour cream dips are prepared by adding appropriate seasonings to cultured cream (16–18% fat) or long-life cultured cream. Cheese condiments are introduced at the rate of 20% for blue cheese and 100% for cheddar cheese; concentrated flavorings such as horseradish are added at 1.5% (Kosikowski, 1977). To improve the body and stability of dips prepared by the hot-pack process, Chandan (1982) suggests the incorporation of 1–2% non-fat dry milk and 0.8–1.0% stabilizer at 80 °C prior to homogenization. If the viscosity of the product is too heavy after ingredient addition, a lower fat mix or a longer agitation time should be used (Kosikowski, 1977).

Cultured half and half has an advantage over cultured cream in terms of lower price and lower caloric content, but has a lower viscosity which makes it subject to wheying off (Kosikowski, 1977). It is manufactured using the same process as cultured cream except that the standardized mix contains 10.5–18% fat and 10–12% MSNF (Chandan, 1982). Sour half and half has also been used as a base for low-calorie creamy salad dressing since it contains 10.5% fat *versus* the 30–80% fat in regular dressing (Chandan, 1982).

Chandan (1982) defines artificial cultured cream as cultured cream in which all or part of the milk fat has been replaced by other fats or oils. It is less expensive to produce and contains fewer calories than cultured cream. Artificial cream is prepared by emulsifying fat in skim milk, casein suspensions or soybean products and adding stabilizer, flavoring materials and colors. The mixture is cultured using the same process as cultured cream (Chandan, 1982).

Sour cream can also be prepared by direct acidification. Acidulant and flavoring are added to the homogenized mix at 22 °C. The acidified product is then packaged, cooled to 4 °C and held for 24 h before distribution. Acidified sour cream provides an excellent, low-cost cream base which is used in many food products. Federal standards require, however, that sour cream prepared by direct acidification be labeled as an 'acidified' product.

25.5.2 Yogurt

The proposed Federal Standards of Identity (Anon, 1977) identify yogurt as containing 3.25% milk fat, low-fat yogurt as containing 0.5% to 2% milk fat and non-fat yogurt as containing less than 0.5%. As in buttermilk, the use of milk-derived ingredients is permitted provided the ratio of protein to MSNF and the protein efficiency ratio are not decreased. Additives include nutritive carbohydrate sweeteners, coloring, stabilizer, fruits and fruit flavors. The starter culture must contain *Lactobacillus bulgaricus* and *Streptococcus thermophilus*.

According to a recent survey, plain yogurt remains the most popular variety with 18.3% of the sales (Przybyla, 1983). The most popular flavor is strawberry at 13.7%, followed by raspberry and peach at 9.6% each. The next most popular flavors are blueberry, pineapple, lemon, cherry, apple, boysenberry and blackberry.

The manufacture and properties of yogurt have been extensively reviewed (Chambers, 1979; Chandan, 1982; Rasic and Kurmann, 1978; Robinson and Tamime, 1975; Tamime and Deeth, 1980). The basic outline for the preparation of four basic types of yogurt is given in Figure 7. The set-type plain yogurt and sundae-style fruited yogurt are packaged before incubation, while the Swiss-style fruited yogurt and stirred-type plain yogurt are cultured in vats before packaging. From 0.5–0.7% stabilizer is added to the latter types in order to impart gel structure, to ensure a smooth body and texture and to prevent wheying off or syneresis after packaging (Chandan, 1982). Yogurt that is overstabilized will be solid in consistency, while yogurt which is under-stabilized will be liquid.

Generally, yogurt mixes are formulated to 10–12% MSNF. Yogurt prepared from mixes containing 12–15% solids will have a medium firm, custard-like consistency while yogurt from mixes containing more than 15% solids will have a firm, heavy body (Chambers, 1979). The actual consistency is also dependent on the amount of stabilizer, if any, and the heat treatment of the mix.

Gelatin, seaweed gums (algin, sodium alginate and carrageenan), vegetable gums (carboxymethylcellulose, locust bean gum and carob gum) and pectin are used as stabilizers in yogurt. Chandan (1982) noted that yogurt stabilizers should not impart flavor to the product, should be active at low pH and should be dispersible at the normal temperature of the yogurt. He also noted that 0.3–0.5% gelatin gives a smooth, shiny appearance to refrigerated yogurt, and a pudding-like consistency to frozen yogurt. The activity of gelatin is temperature dependent and will degrade at high temperatures. In contrast, the seaweed gums, guar gum and carboxymethylcellulose are heat-stable.

A heat treatment of 85 °C for 30 min is needed (a) to destroy microbial contaminants, (b) to provide a microaerophilic environment and release peptone, sulfur groups and formate required for starter culture growth and (c) to denature and coagulate milk albumins and globulins and thus increase the viscosity and improve the body and texture of the final product (Chandan, 1982). Temperatures above 90 °C for over 30 min will decrease water binding capacity of the whey proteins, weaken the yogurt curd and increase syneresis in the product (Chambers, 1979). Proper homogenization of the mixture improves the texture and decreases surface creaming and wheying off (Chandan, 1982).

Chandan (1982) noted three methods of preparing fruited yogurts: the fruit-on-bottom or Eastern sundae-style yogurt, Western sundae-style yogurt and stirred or Swiss-style yogurt. In the Eastern sundae-style, 59 ml (2 oz) of fruit preserves are layered in the bottom of the container prior to the addition of 177 ml (6 oz) of unflavored, unsweetened yogurt mix which has been inoculated with starter culture. The containers are incubated until their contents reach pH 4.2–4.4, and then placed in refrigerated rooms for cooling. In contrast, Western sundae style

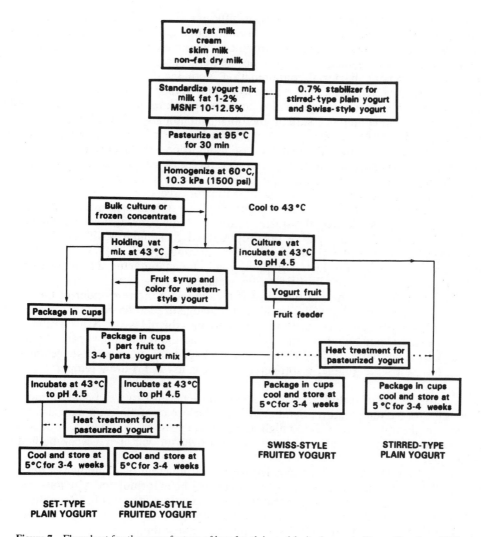

Figure 7 Flowchart for the manufacture of low-fat plain and fruited yogurts (From Chandan, 1982)

yogurt has fruit preserves or fruit preparations on the bottom and yogurt containing sweeteners, coloring and flavoring (2–4%) on the top. Identity standards require that fruit preserves contain 55% sugar and a minimum of 45% fruit, usually frozen fruits or juices, which are cooked to a final soluble solids content of 65–68% (Chandan, 1982). Fruit preserves also contain approximately 0.5% pectin and sufficient citric acid or other food grade acidulant to adjust the pH to 3.0–3.5%.

In Swiss-style yogurt the fruit is added after fermentation. To increase smoothness, the mixture may be pumped through a valve or stainless steel screen prior to packaging (Chandan, 1982). The packaged mixture is stored under refrigeration for 48 h to allow for thickening. The special fruit bases used in stirred yogurt contain 17–41% fruit, 22–40% sugar, 10–24% corn syrup solids, 3.5–5.0% modified food starch, 0.1% natural or 1.25% artificial flavor, 0.01% color, 0.1% potassium sorbate and sufficient citric acid to adjust the pH to 3.7–4.2 (Chandan, 1982). In addition, calcium chloride and food-grade phosphates may be used in specialized preparations. Fruit bases contain 60–65% soluble solids, a standardized viscosity and a total microbial count of less than 500 cfu per gram with coliform, yeast and mold counts of less than 10 per gram.

Connolly (1982) reported on the defects found in plain yogurt and Swiss-style and sundae-style strawberry yogurts. Flavor problems included: (a) unnatural flavoring due to poor quality fruit and/or flavoring materials; (b) low flavor and/or low acid development due to insufficient heat treatment of mix, poor culture activity, high incubation temperature, excessive sweetener or improper flavoring; (c) high acid flavor due to culture imbalance, increased incubation temperature and time, insufficient cooling or inadequate refrigeration during storage; (d) old ingredient flavor from aged, stale or otherwise poor quality ingredients; (e) over-sweetened flavor due to excess sweetener, excess flavoring materials, insufficient acid production or insufficient flavor production; and (f) bitterness due to poor or contaminated cultures, poor quality dairy ingredients, poor quality fruit or flavoring, psychrotrophic contamination or excessive preservatives.

Body and texture defects included: (a) lumpy texture due to improper or excess stabilizer, uneven acid development, bacterial contamination, poor incorporation of fruit or inadequate cooling after packaging; (b) weak curd due to low solids and/or protein in mix, improper use of stabilizer, excessive heat treatment, high incubation temperature, insufficient acid, excessive agitation, excessive line and pump pressure or microbial contamination; (c) gel-like or firm curd due to high solids in mix, excess stabilizer or dehydration because of poor packaging materials; (d) grainy texture due to high acid milk, unstable casein, improper stabilizer selection, poor cultures or improper culture balance, high incubation temperature, fast acid development or homogenization at too high a temperature; and (e) ropy texture due to overpasteurization, improper choice of stabilizer, culture selection, microbial contamination or excessive sweetener.

While post-production pasteurization of yogurt at 60–65 °C will increase the shelf life to 6–8 weeks at 12 °C, it will also destroy the starter culture and any benefits associated with the presence of a viable culture. For this reason, the proposed federal standards will require any thermal processed yogurt to be labeled as 'heat treated after culturing' (Anon, 1977).

Frozen yogurt and novelty products containing yogurt are manufactured from mixes containing 1.0–1.5% milk fat, 13–15% MSNF, 0.15–0.20% gelatin, 7–10% sucrose and 4–5% corn syrup solids (Chandan, 1982). One-half of the sugar is blended with the remainder of the ingredients and the mixture is pasteurized at 88 °C for 40 min, homogenized at 58–63 °C at 10.3 kPa, cooled to 44 °C, inoculated with starter culture and incubated until the pH is 3.9. After fermentation, the mix is cooled to 25 °C and blended with 15–20% fruit and the remaining sugar. The mix is then frozen in an ice cream freezer at 50–60% overrun, packaged and hardened like ice cream (Chandan, 1982). For soft-serve yogurt, a soft-serve freezer is used at a draw temperature of −8 °C (Chandan, 1982).

Defects in frozen yogurt include a coarse or icy texture because of ice crystal formation during fluctuations in storage temperature. Sandiness due to lactose crystallization may also occur if the level of milk solids or whey solids is too high. Generally, excess MSNF or excess sugar will give a soggy or gummy product. Insufficient solids or high overrun will give a weak-bodied product (Chandan, 1982).

The increased popularity of yogurt has led to the introduction of a number of new products (Przybyla, 1983). For example, a custard-style yogurt has been formulated with milk, sugar, nonfat dry milk, fruit puree, corn starch, natural flavors, gelatin and coloring. Recently, breakfast yogurt has appeared on the market, containing wheat grains, walnuts, raisins and other berries, citrus fruits, orchard fruits, tropical fruits or apples and cinnamon.

Tamime and Deeth (1980) have reviewed the preparation of yogurt from milk pre-treated with β-galactosidase to hydrolyze lactose. This product is sweeter and milder in flavor than conventional yogurt and thus require less sucrose to achieve an equivalent degree of sweetness in flavored preparations. In addition, it contains less lactose and thus may be of benefit to lactose intolerant persons.

Acidophilus yogurt has been prepared by replacing all or part of the *Lactobacillus bulgaricus* in the starter with *Lactobacillus acidophilus*. Total replacement of *Lactobacillus bulgaricus* in the starter, however, will decrease the flavor and creamy character of the final product (Kosikowski, 1977). Acidophilus yogurt is prepared by incorporating concentrates of *Lactobacillus acidophilus* along with the yogurt starter prior to incubation at 36 °C, or by incorporating concentrates into finished yogurt prior to storage (Kosikowski, 1977). The viability of the *Lactobacillus acidophilus* may be limited under these conditions (Chandan, 1982).

Several liquid products such as yogurt drink (3.5% butterfat, 8.25–10.0% MSNF), low fat yogurt drink (0.5–2.0% fat) and non-fat (less than 0.5% fat) yogurt drink have also been formulated to meet the proposed Federal Regulatory Standards (Anon, 1977). Stabilizer (0.2–0.3%) and 4.0% sucrose are also added. After pasteurization and homogenization, the product is incubated at 45 °C for 8–9 h until the pH drops to 4.3. The mixture is then cooled to 32 °C with agitation. Sugar syrup and flavoring are added and the mixture is cooled to 4 °C and packaged. The product is then held overnight at 4 °C to allow development of body, texture and mouth feel. Carbonation has also been added to yogurt beverages (Tamine and Deeth, 1980).

Kosikowski (1977) described the manufacture of yogurt cheese. Pasteurized mix is inoculated with 2.5% yogurt starter at 35 °C, incubated for one hour and then inoculated with a 1% commercial starter containing *Streptococcus lactis* or *Streptococcus cremoris* along with rennet and annatto coloring. The milk is held for 2–4 h until the pH reaches 4.7. The curd may be cut, or may be dipped and drained using the same procedure as bakers' cheese. The final product resembles a cream cheese with a smooth grained texture and mild nutty flavor (Kosikowski, 1977).

25.6 MANUFACTURE OF UNRIPENED SOFT CHEESES

25.6.1 Cottage Cheese

Cottage cheese is prepared in the United States by a short set method which uses 5–7% starter and an incubation of 4.5–5.5 h at 31–32 °C, by a medium set method which uses 2–4% starter and an incubation of 8–10 h at 27–28 °C and by a long set method which uses 0.1% starter and an incubation of 12–16 h at 21–22 °C (Kosikowski, 1977; Chandan, 1982).

The short set method is illustrated in Figure 8. Raw skim milk containing at least 9% solids is pasteurized, cooled to 32 °C and inoculated with starter culture containing lactic streptococci and citric acid fermenting organisms. The titratable acidity of the mix is determined and the mix held for 1.5 h, stirring every 30 min. At this point the titratable acidity should be 0.05–0.07% greater than the initial reading. Chandan (1982) suggests that if acidity increase is less than 0.05%, an additional 1% of culture should be added for each 0.01% increment below 0.05%. Rennet is added at the rate of 1 ml per 454 g for large and medium curd cottage cheese, and 0.3–0.5 ml per 454 g for small curd. The vat is covered and allowed to incubate for an additional 2 h before the curd formation is checked.

Cottage cheese curd is ready to be cut when the titratable acidity of the whey is 0.34 to 0.36% higher than the initial value, when the curd pH is 4.6–4.7 or when the end point of the acid coagulation test is reached (Chandan, 1982). The curd is first cut lengthwise with a horizontal knife. It is then cut lengthwise with a vertical knife and finally cut crosswise with a vertical knife (Chandan, 1982). The size of the curd varies from 0.95 cm for small curd to 1.9 cm for large curd cottage cheese.

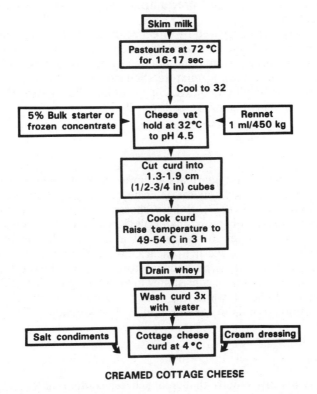

Figure 8 Flowchart for the manufacture of cottage cheese (From Chandan and Shahani, 1982)

The curd is held without agitation for 10 min to allow the curd surfaces to firm. The curd is pushed to the center of the vat and cooked by raising the temperature to 49–54 °C within 3 h. The curd is then washed three times with water which has been charcoal-filtered to remove off-flavors, acidified with citric or phosphoric acid to pH 4.0–5.0 and chlorinated to kill spoilage organisms. After the final wash, the curd is trenched and allowed to drain for 30–60 min before creaming.

Cottage cheese dressing normally contains 12.5% fat, 8.5% MSNF, 2.7% salt and 0.25% stabilizer, while low fat dressing contains 3.0% fat, 15.0% MSNF, 2.7% salt and 0.25% stabilizer (Chandan, 1982). Dry ingredients are mixed separately and added to a mixing vat by centrifugal pump. After mixing, the dressing is vat pasteurized at 75–77 °C for 30 min, homogenized at 57 °C and 13.8 kPA and cooled to 4 °C. One part dressing is then mixed with two parts dried cheese curd. Federal standards require creamed cottage cheese to contain not more than 80% moisture and not less than 4% fat *versus* 0.5–2.0% fat in low-fat cottage cheese (US Department of Agriculture, 1977). Other ingredients include silicone antifoams, sorbate preservatives and flavorings. In addition, citric acid fermenting organisms may be incorporated into the cream dressing to enhance the flavor and increase shelf life (Chandan, 1982).

Defects found in cottage cheese have been discussed by Chandan (1982), Kosikowski (1977) and White (1982). Flavor defects include high acid, bitter, unclean, coarse (high diacetyl), fermented, fruity, flat, stale, oxidized, rancid and whey flavors. Body, texture and appearance defects include mealy, grainy, firm, rubbery, pasty, weak, soft, matted or shattered curds, slick texture, lack of uniformity and appearance of free cream. The use of poor quality milk can result in stale, oxidized or rancid flavors. Excess acid production will give a high acid flavor and a mealy, grainy, soft or shattered curd. Too little acid will result in a firm, rubbery or matted curd and a free cream appearance. Improper cooking temperatures will result in mealy, rubbery, weak or pasty curd. Microbial contaminants, particularly psychrotrophs, may cause bitter, unclean, fruity or flat flavors and a gelatinous curd.

The methods used to produce cottage cheese by direct acidification have been reviewed (Chandan, 1982; Kosikowsi, 1977; Sellars, 1981). Sellars (1981) noted that two processes have been commercialized: the Ernstrom-CP process which involves the addition of food grade hydrochloric acid to 14% solids milk at 5 °C, and the Vitex process which involves the addition of glucono-δ-lactone to skim milk at various temperatures. To further improve the body and texture of the product, Ernstrom and Kale (1975) suggest preculturing the milk with lactic starter to pH 5.5–5.7 prior to direct acidification.

Sellars (1981) noted that curd prepared by direct acidification had the same body and texture, but a milder and more bland flavor than curd prepared by culturing. After creaming, the curd reportedly cannot be distinguished from that obtained by culturing, especially if starter distillates or cultures are added to the creaming mixture (Sellars, 1981). Federal Standards of Identity require packages containing acidified cottage cheese to be labeled 'directly set' or 'curd set by direct acidification' (US Department of Agriculture, 1977).

25.6.2 Bakers' Cheese

Bakers' cheese, cream cheese and quarg are all unripened soft cheeses, which differ from cottage cheese in the manner of whey separation, whey drainage, individual curd structure and cream incorporation (Kosikowski, 1977). Bakers' cheese is utilized widely for the preparation of pastries, cheese cakes and cake decorating materials. Kosikowski (1977) noted bakers' cheese is prepared by adding 5% lactic starter and rennet to skim milk and then incubating the mixture at 31.1 °C for approximately 5 h until the pH reaches 4.4. The mixture is chilled, placed in mesh bags overnight for whey drainage, bulk packed and stored frozen. The product should have a dry, pliable, smooth curd which is similar in consistency to bread dough. High acid formation gives the cheese an undesirable harsh flavor, while low acid production gives the cheese a grainy texture and decreases its absorption of milk in baked products (Kosikowski, 1977).

25.6.3 Cream Cheese

Federal Standards of Identity require that cream cheese contains at least 33% fat and not more than 55% moisture (US Department of Agriculture, 1977). To produce cheese of legal composition, the mix must contain a minimum of 11.5% fat and 7.8% MSNF (Kosikowski, 1977). According to Kosikowski (1977), standardized cream cheese mix is pasteurized at 71.1 °C for 30 min, homogenized at 17.2 kPa, cooled to 31.1 °C and inoculated with starter culture (5%) and rennet (1 ml 454 kg^{-1}). The mixture is held for approximately 5 h until the pH reaches 4.6. The curd is agitated, heated to 54 °C, cooled to 32.2 °C, diluted with water to facilitate whey drainage and then further cooled to 7.2 °C. The whey is drained by mechanical centrifugation or by transferring to mesh bags.

For the cold pack process, dried curd is blended with 0.5% stabilizer and 1% salt before packaging and storage at 4.4 °C. While this product possesses an excellent flavor and texture, it is subject to yeast or mold spoilage within 1 to 2 weeks (Kosikowski, 1977). For the hot pack process, dried curd is mixed with 1.0% salt, heated to 73.9 °C and then standardized for fat and total solids using a special mixture containing 15% SNF and 0.35% gum at pH 4.6. The standardized mix is pasteurized by holding at 73.9 °C for 30 min, homogenized and packaged hot. Hot pack cream cheese is free of coliform, yeasts and molds, but is susceptible to chemical deterioration after 2 to 3 months storage (Kosikowski, 1977).

If desired, condiments such as pimentos, relish, chives, scallions, olives, cherries, pineapples or nuts can be added to the hot mixture immediately after homogenization (Kosikowski, 1977). Since these additions reduce fat levels, the minimum legal fat requirement for cream cheese containing condiments is 27% rather than 33% fat.

Neufchatel cheese, containing from 20 to 33% fat and not more than 65% water, has the same flavor as cream cheese but a slightly more grainy texture (Kosikowski, 1977). It is prepared from a standardized mix containing at least 5% fat and 8.4% SNF using the same procedures outlined for cream cheese (Kosikowski, 1977)..

25.6.4 Quarg

Quarg or quark is an unripened fresh cheese similar to bakers' cheese. According to Kosikowski (1977), quarg may be made from skim milk, reconstituted skim milk powder or standardized mixes containing up to 12% added fat. Its smooth texture and mild acid flavor increase with increasing levels of fat (Kosikowski, 1977). Quarg is prepared by inoculating milk with lactic starter culture and rennet, incubating to pH 4.6 and separating the whey by bag drainage or mechanical centrifugation (Kosikowski, 1977). Kosikowski (1977) noted that quarg may be packaged plain or with salt, fruit preserves, spices and other condiments. It is used as a garnish for soups or salads, or served with fruit for dessert or with cereals for breakfast.

25.7 NUTRITIONAL PROPERTIES OF FERMENTED DAIRY PRODUCTS

Customarily, claims regarding the nutritional and health-giving properties of fermented dairy products were considered folklore with no scientific basis. It was assumed that during the manufacture of fermented dairy products, the major effects of the starter culture were the production of lactic acid and, in some cases, the synthesis of other flavor compounds. Recent reviews (Deeth and Tamime, 1981; Shahani and Chandan, 1979; Shahani and Friend, 1983; Speck and Katz, 1980) indicate that selected starter cultures augment a number of changes which enhance the nutritional value and possibly also the health value of milk and milk products. These include the prehydrolysis of milk components and the production of antibiotics, anticholesteremic factor(s), anticarcinogenic factor(s), B vitamins and enzymes.

25.7.1 Nutrient Composition

The nutrient content of skim milk, cream and selected fermented dairy products is given in Table 8. All dairy products are good dietary sources of high quality protein, calcium, phosphorus, potassium and riboflavin. While many lactic organisms require B vitamins for growth, several species are capable of synthesizing vitamins. The extent of biosynthesis depends on the culture species and strain, the incubation conditions and other processing parameters (Reif *et al.*, 1976; Friend *et al.*, 1983).

Reif *et al.* (1976) observed that cottage cheese starter actively synthesized vitamin B_{12} and folacin during the setting period. *Lactobacillus casei* used in the preparation of yakult has been shown to synthesize thiamin, riboflavin, B_6 and B_{12} (Shirota, 1971).

Friend *et al.* (1983) monitored the synthesis and utilization of B vitamins by three strains each of *Streptococcus thermophilus* and *Lactobacillus bulgaricus* incubated for 24 h in skim milk. *Streptococcus thermophilus* strains 1 and 3 synthesized significant amounts of folacin. None of the

Table 8   Nutrient Composition of 100 g Portions of Skim Milk, Cream and Selected Fermented Dairy Products[a]

Nutrient	Skim milk	Cream (light)	Buttermilk	Cultured cream	Plain yogurt (low fat)	Cottage cheese (low fat)
Energy (kJ)	146	818	169	987	265	303
Carbohydrate (g)	4.8	3.7	4.8	4.3	7.0	2.7
Fat (g)	0.2	19.3	0.9	21.0	1.6	1.0
Protein (g)	3.4	2.7	3.3	3.2	5.2	12.4
Calcium (mg)	123	96	116	116	183	61
Iron (μg)	40	40	50	60	80	140
Magnesium (mg)	11	9	11	11	17	5
Phosphorus (mg)	101	80	89	85	144	134
Potassium (mg)	166	122	151	144	234	86
Sodium (mg)	52	40	105	53	70	406
Zinc (μg)	400	270	420	270	890	380
Ascorbic acid (mg)	1.0	0.8	1.0	0.9	0.8	trace
Thiamine (μg)	36	32	34	35	44	21
Riboflavin (μg)	140	148	154	149	214	165
Niacin (μg)	88	57	58	67	114	128
Panthothenic acid (μg)	329	276	275	360	591	215
Vitamin B_6 (μg)	40	32	34	16	49	68
Folacin (μg)	5	2	trace	11	11	12
Vitamin B_{12} (μg)	0.38	0.22	0.22	0.30	0.56	0.63
Vitamin A(IU)	204	720	33	790	66	37

[a] From US Department of Agriculture Handbook 8 (1976).

strains had a significant effect on the synthesis of biotin, B_{12} or niacin. *Lactobacillus bulgaricus* strains 1 and 2 utilized significant amounts of biotin; strains 2 and 3 utilized folacin; strain 3 utilized niacin. None of the strains of *Lactobacillus bulgaricus* synthesized B vitamins. A survey of the literature by Friend *et al.* (1983) revealed that biotin, B_{12} and niacin decrease during the preparation of yogurt while folacin increases. Cultured cream, yogurt and cottage cheese contain more than twice the level of folacin than the milk or cream from which they are made (Table 8).

25.7.2   Prehydrolysis of Milk Components

The nutritional value of a food is dependent not only on its nutritional content, but also on the availability, digestibility and assimilability of its nutrients. Thus, fermentation may increase the nutritional value of milk through the enzymatic hydrolysis of carbohydrates, fats and proteins by the lactic starter cultures.

The milk carbohydrate lactose may not be digested properly in persons who are deficient in the intestinal enzyme β-galactosidase (β-gal). Infants, children and adolescents require the calcium in milk for proper bone growth and development and a restriction in their intake of dairy products because of problems with lactose intolerance may have serious nutritional consequences. Up to 50% of the lactose in yogurt, 48% of the lactose in acidophilus milk and 26% of the lactose in buttermilk may be hydrolyzed during manufacture and storage (Alm, 1982). In addition, yogurt starter culture produces significant quantities of β-gal which may enhance the digestion of lactose when yogurt is consumed by lactose intolerant individuals (Goodenough and Kleyn, 1976; Kilara, 1976). Alm (1982) reported that the consumption of acidophilus milk or yogurt caused no intestinal discomfort in a test group of lactose intolerant individuals. Recently, lactose intolerant persons consuming a non-fermented sweet acidophilus milk also demonstrated improved lactose utilization as monitored by the breath hydrogen test (Kim and Gilliland, 1983).

Chandan (1982) summarized previous studies on the lipolytic and proteolytic activities of various lactic cultures and noted that the streptococci and leuconostocs play an important role in the prehydrolysis of milk fat, while the lactobacilli are important in the prehydrolysis of the milk protein.

25.7.3   Digestibility and Protein Quality of Yogurt

Yogurt proteins have been reported to possess a higher *in vitro* digestibility (Breslaw and Kleyn, 1973), and higher protein quality as estimated by the biological value (Rasic *et al.*, 1971) and gross protein value (Simhaee and Keshavarz, 1974). In addition, rats fed yogurt had better weight gains and higher food efficiency than rats fed milk, other fermented milks or acidified milks (Hargrove and Alford, 1978).

Studies in our laboratory have examined the effects of various processing parameters on the nutritional value of yogurt protein (Table 9). The level of predigested protein, the *in vivo* digestibility and protein quality of yogurt mix were monitored after heating and after heating plus fermentation. Only 1–2% of the protein was predigested in the unfermented mix or in the heated mix. 6.3–6.4% of the yogurt protein was predigested after fermentation to pH 4.2. Fermentation also increased the apparent *in vivo* digestibility of the product from 84.34 to 86.07%.

Table 9 Effect of Heating and Fermentation on the Level of Predigestion, *In Vivo* Digestibility and Protein Quality of Yogurt Mix

Product	Predigestion (%)	Apparent digestibility (%)	NPR (Adjusted)	C-PER (Adjusted)
Unheated yogurt mix	2.31	84.34	4.79 ± 0.39	2.07
Heated yogurt mix	1.39	84.56	5.08 ± 0.24	2.70
Cultured yogurt	6.33	86.07	5.26 ± 0.23[a]	2.45
Casein standard	—	90.03	4.50 ± 0.29	2.50

[a] Significantly different (P<0.05) from unheated mix.

The protein quality of the products was estimated *in vitro*, using the rapid C-PER method (calculated protein efficiency ratio) based on the amino acid profile and digestibility, and *in vivo*, using the classical NPR method (net protein ratio) based on feeding trials with weanling rats. The protein quality, as estimated by the C-PER, was increased from 2.07 to 2.70 after heating and to 2.45 after heating and fermentation. However, in the rat, although both heating and fermentation increased the NPR, only the combination of the two treatments resulted in a significant change.

25.8 SUMMARY

A cursory look at the immediate past performance of fermented dairy products suggests that their prospects are indeed bright. The consumption of traditional fermented products such as yogurt should continue to rise. The widespread availability and use of mixed strain culture concentrates will ensure the manufacture of uniformly high quality fermented products. In addition, the appearance of simulated and modified products seems to hold promise in the marketplace, particularly for satisfying the needs of a specific clientele. While these products simulate the desirable organoleptic and textural qualities of the traditional fermented products, they do not possess the same nutritional and therapeutic properties.

With the exception of lactic acid or other acidulants, products prepared by direct acidification possess essentially the same composition as milk (Table 10). In contrast, fermented products contain starter culture, lactic acid and hydrolyzed milk fat, lactose and protein as well as specific metabolites associated with the growth of lactic starter cultures including anticholesteremic factor(s), anticarcinogenic factor(s), antibiotics and enzymes. Thus, while products prepared by acidification do possess a better keeping quality than milk because of their low pH, they lack other nutritional or healthful factors found in similar products prepared by fermentation.

It is important to note, however, that since the activity of certain enzymes or the ability to synthesize specific metabolites is often strain specific as well as species specific, not every metabolite listed in Table 10 will be present to the same degree in every fermented product. For example, while many species of lactobacilli produce natural antibiotics, screening studies in our laboratory showed that *Lactobacillus acidophilus* DDS 1 produced the highest level of the antibiotic acidophilin and *Lactobacillus bulgaricus* DDS 14 produced the highest level of the antibiotic bulgarican. Other lactobacilli species or strains produced much lower quantities of the antibiotic metabolites (Shahani and Friend, 1983). Similarly, only yogurt has been reported to stimulate the growth of rats. Other fermented or acidified products have no effect (Hargrove and Alford, 1978).

The effects of strain selection on the synthesis of metabolites in yogurt are shown in Table 11.

Table 10   Major Differences between Fermented and Acidified Dairy Products

Factor	Fermented	Acidified
Sensory properties		
Flavor	Natural, slight variations	Added, uniform
Texture	Excellent	Excellent
Gross composition		
Compounds plus lactic acid	Present	Absent or low levels
Fat	Partly digested	Unchanged
Lactic starter	High levels	Absent or low levels
Lactose	Partly digested	Unchanged
Lactic acid	Present	Present
Protein	Partly digested	Unchanged
Specific metabolites		
Antibiotics	Present	Absent
Anticholesteremic factor(s)	Present	Absent
Anticarcinogenic factor(s)	Present	Absent
B vitamins	Increased folic acid, slight increases or decreases in others	Unchanged
Enzymes	Present	Absent

Starter 1 produced yogurt with the highest flavor score, highest β-gal activity, lowest niacin content and median antimicrobial activity. In contrast, Starter 3 produced yogurt with the highest niacin content, the second highest β-gal activity, the lowest antimicrobial activity and a median flavor score.

Table 11   Effect of Starter Culture Selection on the Characteristics of Yogurt[a]

Yogurt starter	Flavor score	pH	Zone of inhibition (mm)	Viable organisms (log cfu ml^{-1})	β-Gal activity (μmol min^{-1}g^{-1})	Niacin content (μg)
1	4.0	4.2	16.2	8.94	7.8	117
2	2.9	4.2	15.8	8.68	3.9	125
3	3.1	4.3	15.2	7.65	5.0	132
4	2.7	4.2	16.6	9.0	4.0	131
5	3.3	4.2	17.0	8.56	4.1	130

[a] From Friend *et al.* (1983).

Finally, lactic starter cultures are rather fastidious in their growth and metabolic requirements. Therefore, strains which produce specific beneficial metabolites in products prepared under standardized conditions (composition of growth medium, length and temperature of growth, presence of stabilizing agents, conditions of processing, freezing, drying and storage) may not necessarily produce the same metabolites under different conditions.

25.9   REFERENCES

Alm, L. (1982). Effect of fermentation on lactose, glucose, and galactose content in milk and suitability of fermented milk products for lactose intolerant individuals. *J. Dairy Sci.*, **65**, 346–352.

Anon (1977). Cultured and acidified buttermilk, yogurts, cultured and acidified milks, and eggnog. Proposal to establish new identity standards. *Fed. Reg.*, **42**, 29 919–29 925.

Babel, F. J. (1976). Technology of dairy products manufactured with selected microorganisms. In *Dairy Technology and Engineering* ed. W. J. Harper and C. W. Hall, pp. 213–271. AVI Publishing Co., Westport, CT.

Babel, F. J. (1977). Antibiosis by lactic acid bacteria. *J. Dairy Sci.*, **60**, 815–821.

Breslaw, E. S. and D. H. Kleyn (1973). *In vitro* digestibility of yogurt at various stages of processing. *J. Food Sci.*, **38**, 1016–1021.

Chambers, J. V. (1979). Culture and processing techniques important to the manufacture of good quality yogurt. *Cultured Dairy Prod. J.*, **14** (2), 28–33.

Chandan, R. C. (1982). Other fermented dairy products. In *Prescott and Dunn's Industrial Microbiology*, ed. G. Reed, 4th edn., pp. 113–184 AVI Publishing Co., Westport, CT.

Chandan, R. C. and K. M. Shahani (1982). Cultured milk products. In *Handbook of Processing and Utilization in Agriculture*, ed. I. A. Wolff, vol. I, pp. 365–378. CRC Press, Boca Raton, FL.

Charm, S. E. (1979). A 15-minute assay for penicillin and other antibiotics. *Cultured Dairy Prod. J.*, **14** (2), 24–27.

Collins, E. B. (1977). Influence of medium and temperature on end products and growth. *J. Dairy Sci.*, **60**, 779–804.

Connolly, E. J. (1982). Yogurt. In *Cultured Dairy Foods. Quality Improvement Manual*. American Cultured Dairy Products Institute, Washington, DC.

Custer, E. W. (1982). Buttermilk. In *Cultured Dairy Foods. Quality Improvement Manual*. American Cultured Dairy Products Institute, Washington, DC.

Deeth, H. C. and A. Y. Tamime (1981). Yogurt: nutritive and therapeutic aspects. *J. Food Protect.*, **44**, 78–86.

Djurec, D. J. and R. R. Zall (1982). Effect of on-farm heating and storage of milk on cottage cheese yield. *J. Dairy Sci.*, **65**, 2296–2300.

Ernstrom, C. A. and C. G. Kale (1975). Continuous manufacture of Cottage and other uncured cheese varieties. *J. Dairy Sci.*, **58**, 1008–1014.

Feldstein, F. J. and D. C. Westhoff (1979). The influence of heat treatment of milk on starter activity. What about UHT? *Cultured Dairy Prod. J.*, **14** (2), 11–15.

Friend, B. A., J. M. Fiedler and K. M. Shahani (1983). Influence of culture selection on the flavor, antimicrobial activity, β-galactosidase and B-vitamins of yogurt. *Milchwissenschaft*, **38**, 133–136.

Gilliland, S. E. (1977). Preparation and storage of concentrated cultures of lactic streptococci. *J. Dairy Sci.*, **60**, 805–809.

Gilliland, S. E. (1979). Beneficial interrelationships between certain microorganisms and humans: candidate microorganisms for use as dietary adjuncts. *J. Food Protect.*, **42**, 164–167.

Gilliland, S. E. and M. L. Speck (1972). Interactions of food starter cultures and food-borne pathogens: lactic streptococci *versus* staphylococci and salmonellae. *J. Milk Food Technol.*, **35**, 307–310.

Gilliland, S. E. and M. L. Speck (1977). Antagonistic action of *Lactobacillus acidophilus* toward intestinal and food borne pathogens in associative cultures. *J. Food Protect.*, **40**, 820–823.

Goodenough, E. R. and D. H. Kleyn (1976). Influence of viable yogurt microflora on digestion of lactose by the rat. *J. Dairy Sci.*, **59**, 601–606.

Hargrove, R. E. and J. A. Alford (1978). Growth rate and feed efficiency of rats fed yogurt and other fermented milks. *J. Dairy Sci.*, **61**, 11–19.

Heldman, D. R. and D. A. Seiberling (1976). Environmental sanitation. In *Dairy Technology and Engineering*, ed. W. J. Harper and C. W. Hall, pp. 272–321. AVI Publishing Co., Westport, CT.

Jones, V. A. and W. J. Harper (1976). General processes for fluid milks. In *Dairy Technology and Engineering*, ed. W. J. Harper and C. W. Hall, pp. 141–184. AVI Publishing Co., Westport, CT.

Kilara, A. and K. M. Shahani (1976). Lactase activity of cultured and acidified dairy products. *J. Dairy Sci.*, **59**, 2031–2035.

Kim, H. S. and S. E. Gilliland (1983). *Lactobacillus acidophilus* as a dietary adjunct for milk to aid lactose digestion in humans. *J. Dairy Sci.*, **66**, 959–966.

Klaenhammer, T. R. (1982). Microbiological considerations in selection and preparation of lactobacillus strains for use as dietary adjuncts. *J. Dairy Sci.*, **65**, 1339–1349.

Kosikowski, F. V. (1977). *Cheese and Fermented Milk Foods*, 2nd edn. Edwards Bros., Ann Arbor, MI.

Kwan, A. J., A. Kilara, B. A. Friend and K. M. Shahani (1982). Comparative B-vitamin content and organoleptic qualities of cultured and acidified sour cream. *J. Dairy Sci.*, **65**, 697–701.

Milk Industry Foundation (1982). *Milk Facts*, pp. 12–15. Milk Industry Foundation, Washington, DC.

Porubcan, R. S. and R. L. Sellars (1977). Lactic starter concentrates. In *Microbial Technology*, ed. H. J. Peppler and D. Pearlman, vol. 1, 2nd edn., pp. 59–92. Academic, New York.

Przybyla, A. (1983). Yogurt bandwagon creates spectrum of new products. *Prepared Foods*, March, 124–125.

Rasic, J. L. and J. A. Kurmann (1978). *Yogurt*. Technical Dairy Publishing House, Jyllingevez 39.

Rasic, J., T. Stojsauljevic and R. Curcic (1971). A study of the amino acids of yogurt. II. Amino acid content and biological value of protein of different kinds of yogurt. *Milchwissenschaft*, **26**, 219–224.

Reif, G. D., K. M. Shahani, J. R. Vakil and L. K. Crowe (1976). Factors affecting the B-vitamin content of cottage cheese. *J. Dairy Sci.*, **59**, 410–415.

Robinson, R. K. and A. Y. Tamime (1975). Yogurt — a review of the product and its manufacture. *J. Soc. Dairy Technol.*, **28** (3), 149–163.

Sandine, W. E. (1979). Role of lactobacillus in the intestinal tract. *J. Food Protect.*, **42**, 259–262.

Sellars, R. L. (1967). Bacterial starter cultures. In *Microbial Technology*, ed. H. J. Peppler, pp. 34–75. Reinhold, New York.

Sellars, R. L. (1981). Fermented dairy foods. *J. Dairy Sci.*, **64**, 1070–1076.

Sellars, R. L. and F. J. Babel (1978). *Cultures for the Manufacture of Dairy Products*. Chr. Hansen's Laboratory, Milwaukee, WI.

Senyk, G. F., R. R. Zall and W. F. Shipe (1982). Subpasteurization heat treatment to inactivate lipase and control bacterial growth in raw milk. *J. Food Protect.*, **45**, 513–515.

Shahani, K. M. and A. D. Ayebo (1980). Role of dietary lactobacilli in gastrointestinal microecology. *Am. J. Clin. Nutr.*, **33**, 2448–2457.

Shahani, K. M. and B. A. Friend (1983). Properties of and prospects for cultured dairy foods. In *Food Microbiology*, ed. F. A. Skinner and T. A. Roberts, pp. 257–269. Academic, London.

Shahani, K. M. and R. C. Chandan (1979). Nutritional and healthful aspects of cultured and culture-containing dairy foods. *J. Dairy Sci.*, **62**, 1685–1694.

Shahani, K. M. and W. J. Harper (1958). The development of antibiotic resistance in cheese starter cultures. *Milk Prod. J.*, **49**, 15–16, 53–54.

Shirota, M. (1971). Report on yakult-strain. In *The Summary of Reports on Yakult*, No. 1, ed. S. Kotani, pp. 10–12. Yakult Honsha Co. Ltd., Tokyo.

Simhaee, I. and K. Keshavarz (1974). Comparison of gross protein value and metabolizable energy of dried skim milk and dried yogurt. *Poultry Sci.*, **53**, 184–191.

Speck, M. L. (1980). Preparation of lactobacilli for dietary uses. *J. Food Protect.*, **42**, 65–67.

Speck, M. L. and R. S. Katz (1980). Nutritive and health values of cultured dairy foods. ACDPI Status Paper. *Cultured Dairy Prod. J.*, **15** (4), 10–11.

Tamime, A. Y. and H. C. Deeth (1980). Yogurt: technology and biochemistry. *J. Food Protect.*, **43**, 939–977.

US Department of Agriculture (1976). *Composition of Foods*. US Dept. Agric., Agric. Handbook 8.

US Department of Agriculture (1977). *Federal and State Standards for the Composition of Milk Products (and Certain Non-Milk Fat Products)*. US Dept. Agric., Agric. Handbook 51.

Vedamuthu, E. R. (1979). Microbiologically induced desirable flavors in the fermented foods of the west. *Dev. Ind. Microbiol.*, **20**, 187–202.

White, C. H. (1982). Cottage cheese. In *Cultured Dairy Foods. Quality Improvement Manual.* American Cultured Dairy Products Institute, Washington, DC.

White, C. H. and S. G. Shilotri (1979). Inoculation of citric acid-fermenting bacteria into raw milk in farm bulk tanks. *J. Food Protect.*, **42**, 51–54.

Wulf, J. J. and W. E. Sandine (1983). Isolation and characterization of fast acid-producing antibiotic-resistant mutants of lactic streptococci. *J. Dairy Sci.*, **66**, 1835–1842.

26

L-Glutamic Acid Fermentation

Y. HIROSE, H. ENEI AND H. SHIBAI
Ajinomoto Co. Inc., Kawasaki, Japan

26.1 INTRODUCTION

The flavor enhancing property of konbu, a kelp-like seaweed traditionally used as a seasoning source in Japan, was identified as being due to L-glutamic acid (Ikeda, 1908). This discovery led to the industrial production of monosodium L-glutamate by the Ajinomoto Co. In those days, L-glutamic acid was produced by the acid hydrolysis of wheat gluten or soybean protein. Over a period of 50 years, L-glutamic acid-producing microorganisms were isolated (Kinoshita *et al.*, 1957) and subsequent research resulted in an economical fermentative process for the production of L-glutamic acid.

The establishment of L-glutamic acid fermentation provided a significant impetus to the development of microbial production of primary metabolites. Essential metabolites, *e.g.* L-glutamic acid, were never considered to be sufficiently concentrated in microbial cultures owing to the regulatory mechanisms in cells. Encouraged by the establishment of the L-glutamic acid fermentation, various research projects have been carried out in the attempt to isolate wild strains or derive genetic mutants producing various kinds of amino acids. As a result, almost all of the amino acids are now commercially produced by fermentation.

The annual production of monosodium L-glutamate, produced exclusively by fermentation, exceeds 370 000 tonnes. L-Glutamic acid is a seasoning in widespread use throughout the world. It is also used as a starting material for the synthesis of various kinds of speciality chemicals. *N*-Acylglutamate is commercially available as a biodegradable surfactant with low skin irritation properties which is valued as an additive in cosmetics, soaps and shampoos. Oxopyrrolidinecarboxylic acid, another derivative of L-glutamic acid, is used as a natural moisturizing factor in cosmetics, playing an important role in maintaining water in the cornified layer, acting synergistically with glycine, threonine, alanine, aspartic acid, glutamic acid and serine. Amides of acylglutamate are utilized as gelatinization agents. They render a wide variety of hydrocarbon and vegetable oils jelly-like, and have applications in oil dispersion for marine antipollution purposes.

Three basic aspects of L-glutamic acid fermentation will be discussed: production of L-glutamic acid from sugars, microbial physiology of the fermentation and large scale performance of the fermentation with particular emphasis on industrial aspects.

26.2 PRODUCTION OF L-GLUTAMIC ACID

26.2.1 Microbial Strains

A number of wild strains that have been isolated as L-glutamic acid-producing bacteria are shown in Table 1 (Yamada *et al.*, 1972). Most of these L-glutamic acid-producing bacteria are Gram-positive, non spore-forming, non-motile and require biotin for growth. Among these strains, bacteria belonging to the genera *Corynebacterium* and *Brevibacterium* are in widespread use along with an oleic acid-requiring auxotrophic mutant, which was derived from biotin-requiring *Brevibacterium thiogenitalis*.

Table 1 Microbial Strains Producing L-Glutamic Acid

Genus	Species
Corynebacterium	*C. glutamicum, C. lilium, C. callunae, C. herculis*
Brevibacterium	*B. divaricatum, B. aminogenes, B. flavum, B. lactofermentum, B. saccharolyticum, B. roseum, B. immariophilum, B. alanicum, B. ammoniagenes, B. thiogenitalis*
Microbacterium	*M. salicinovolum, M. ammoniaphilum, M. flavum* var. *glutamicum*
Arthrobacter	*A. globiformis, A. aminofaciens*

26.2.2 Culture Conditions

26.2.2.1 Carbon source

L-Glutamic acid-producing bacteria can utilize various carbon sources, such as glucose, fructose, sucrose, maltose, ribose or xylose, as the substrate for cell growth and L-glutamic acid biosynthesis. For industrial production, molasses and starch hydrolyzates are generally employed as the carbon source.

In order to obtain high yields of L-glutamic acid, the biotin concentration in the medium must be strictly controlled at a suboptimum level for the maximum cell growth. Therefore, biotin-rich raw materials, *e.g.* beet molasses and cane molasses, could not be used until the discovery of the biotin mediating effects of penicillins and C_{16}–C_{18} saturated fatty acids. Oleic acid-requiring mutants accumulate L-glutamic acid in biotin-rich media only when the oleic acid concentration is controlled at the suboptimal level for maximum growth.

26.2.2.2 Nitrogen source and pH control

The ample supply of a suitable nitrogen source is essential for L-glutamic acid fermentation, since the molecule contains 9.5% nitrogen. Ammonium salts such as ammonium chloride or ammonium sulfate are assimilable. L-Glutamic acid-producing bacteria also have a strong urease activity, so urea is also utilizable as a nitrogen source. The ammonium ion is detrimental to both cell growth and product formation, and its concentration in the medium must be maintained at a low level. The pH of the culture medium is very apt to become acidic as ammonium ions are assimilated and L-glutamic acid is excreted. Gaseous ammonia has a great advantage over aqueous bases in maintaining the pH at 7.0–8.0, the optimum for L-glutamic acid accumulation. It serves as a pH-controlling agent and as a nitrogen source, and solves various technological problems.

Automatic addition of gaseous ammonia makes precise pH control possible, and avoids the harmful effects of ammonia and the undesirable dilution of the fermentation liquid.

26.2.2.3 Growth factors

L-Glutamic acid-producing bacteria require biotin for growth and its concentration must be strictly controlled for the maximum yield of product. The effect of biotin on L-glutamic acid fermentation, extensively investigated in relation to the permeability of L-glutamic acid to the cell membrane, is described in Section 26.4.1.

26.2.2.4 Oxygen supply

The biosynthesis of L-glutamic acid is an aerobic process requiring oxygen throughout the fermentation. For maximum L-glutamic acid production, control of dissolved oxygen at its optimum level is essential. Actively respiring cells will consume all the oxygen in a saturated broth (7 p.p.m.) within a few seconds, therefore oxygen must be supplied continuously to maintain the optimum dissolved oxygen concentration. For effective aeration, both physiological and engineering factors need to be considered. Equation (1) is a simple expression describing oxygen transfer during fermentation:

$$R_{ab} = K_d(P_b - P_L) = Kr_m \ (P_L > P_{L_{crit}}) \atop < Kr_m \ (P_L < P_{L_{crit}}) \tag{1}$$

where R_{ab} is the rate of cell respiration (mol O_2 ml^{-1} min^{-1}); K_d is the oxygen absorption coefficient (mol O_2 ml^{-1} min^{-1} atm^{-1}); P_L is the dissolved oxygen tension (atm); Kr_m is the maximum rate of cell respiration (mol O_2 ml^{-1} min^{-1}); and $P_{L_{crit}}$ is the critical level of the dissolved oxygen for cell respiration (atm). Rate of cell respiration (R_{ab}) equals Kr_m at the P_L levels above $P_{L_{crit}}$.

Values of $P_{L_{crit}}$ are usually lower than 0.01 atm, too low to be determined with conventional oxygen electrodes. As a result, a new method was developed in which both the P_L and the redox potentials of the culture medium were measured simultaneously. The value of P_L measured in this way was found to be 0.0002 atm for the L-glutamic acid-producing bacterium *Brevibacterium flavum* (Akashi *et al.*, 1978). For maximum production of L-glutamic acid, satisfaction of the cell's respiratory requirements ($R_{ab}/Kr_m = 1.0$) is necessary; oxygen supply must be strictly controlled to maintain the P_L above $P_{L_{crit}}$. Practically, the degree of agitation and aeration in L-glutamic acid fermentation is controlled to maintain the P_L level slightly above 0.01 atm by using the conventional oxygen electrodes (Hirose *et al.*, 1966). The lack of quantitative analysis on oxygen transfer sometimes led to erroneous conclusions concerning cell metabolism. For example, L-glutamic acid-producing bacteria had been supposed to excrete lactic acid in biotin rich media, but it was found that the excretion was due to oxygen deficiency not due to excess biotin. In a biotin-rich medium, cell density became high resulting in increased oxygen demand of culture liquids, and consequently oxygen shortage occurred. Cells do not accumulate lactic acid in a biotin rich medium when sufficient oxygen is supplied at the R_{ab}/Kr_m value of 1.0 (Hirose *et al.*, 1968).

26.3 ACCUMULATION OF OTHER PRODUCTS IN RELATION TO CHANGE IN CULTURE CONDITIONS

26.3.1 Lactic Acid and Succinic Acid

L-Glutamic acid-producing *Brevibacterium flavum* accumulate lactic acid and succinic acid when cultured under limited oxygen supply. As the rate of oxygen supply decreases from the condition of complete saturation to various degrees of satisfaction of oxygen requirement, the main product changes from L-glutamic acid to succinic acid, and then to lactic acid. More than 30 g l^{-1} of succinic acid or 45 g l^{-1} of lactic acid can accumulate in 72 h at optimum conditions (Okada *et al.*, 1961).

26.3.2 α-Ketoglutaric Acid

It was demonstrated that the absence of ammonium ions, but with sufficient oxygen supply, resulted in the accumulation of α-ketoglutaric acid in place of L-glutamic acid. When the pH controlling agent was switched from NH_4OH to $NaOH$ at the end of the growth phase, 18 g l^{-1} of α-ketoglutaric acid accumulated at a yield of 0.20 g g^{-1} substrate in 72 h cultivation (Tanaka *et al.*, 1960).

26.3.3 L-Giutamine

L-Glutamic acid is converted into L-glutamine when the culture is performed in the presence of excess ammonium chloride at a weakly acidic pH in the presence of zinc ions. In a medium containing 40 g l^{-1} of ammonium chloride and 10 mg l^{-1} of zinc sulfate, the cells accumulated more than 40 g l^{-1} of L-glutamine at 0.30 g g^{-1} carbon source. A high concentration of ammonium ions in weakly acidic conditions results in the production of *N*-acetyl-L-glutamine. Zinc ions are effective in decreasing the excretion of *N*-acetyl-L-glutamine in favor of L-glutamine accumulation (Nakanishi *et al.*, 1975).

26.4 MICROBIAL PHYSIOLOGY OF L-GLUTAMIC ACID FERMENTATION

26.4.1 Permeability of Cell Membrane to L-Glutamic Acid in Relation to Biotin Concentration

A key compound controlling L-glutamic acid fermentation is biotin. The accumulation of L-glutamic acid is at a maximum when the biotin concentration is suboptimal for maximum growth. Excess biotin supports abundant cell growth but seriously decreases the L-glutamic acid accumulation. The critical biotin content of the cells for the accumulation of L-glutamic acid is $0.5 \text{ }\mu\text{g g}^{-1}$ of dry cells.

However, in the presence of excess biotin, the addition of penicillin, which is known to inhibit the formation of cross-links in the peptidoglycan of bacteria at the growth phase, permits cells to accumulate a large amount of L-glutamic acid. Other antibiotics, such as cephalosporin C, which also inhibits cell wall synthesis, can replace penicillin. The addition of C_{16}–C_{18} saturated fatty acids or their esters with hydrophilic polyalcohols during the growth phase also permits cells to accumulate L-glutamic acid in a biotin-rich medium. The use of these antibiotics and C_{16}–C_{18} saturated fatty acids allowed industrial utilization of biotin-rich raw materials such as cane and beet molasses.

The accumulation of L-glutamic acid is governed not by its biosynthesis but by its excretion. The excretion of L-glutamic acid is closely related to the cell permeability, which is associated with both chemical and physical constituents of the cell membrane. L-Glutamic acid-producing cells grown with limited biotin or grown with excess biotin and treated with either penicillin or Tween-60 excreted intracellular L-glutamic acid when washed with phosphate buffer. Cells grown with excess biotin without the treatment of penicillin or Tween-60, however, do not. This phenomenon has also been observed for L-aspartic acid. Other amino acids were washed out of the cells even when grown under biotin-rich condition. L-Glutamic acid-excreting cells, notwithstanding biotin limitation, oleic acid requirements or C_{16}–C_{18} saturated fatty acid treatment, have a low content of phospholipids in the cell membrane. On the other hand, the cells with a poor ability to accumulate L-glutamic acid in a biotin-rich medium have a much higher concentration of membrane phospholipids.

As shown in Figure 1, biotin is a cofactor of acetyl-CoA carboxylase, the first enzyme in the biosynthesis of oleic acid, and C_{16}–C_{18} saturated fatty acids inhibit the biosynthesis of oleic acid by repressing acetyl-CoA carboxylase (Izumi *et al.*, 1973; Kamiryo *et al.*, 1976). Limited amounts of biotin or C_{16}–C_{18} saturated fatty acids cause incomplete biosynthesis of oleic acid resulting in a decrease in phospholipid concentration. Consequently, phospholipids such as cardiolipin and phosphatidylinositol dimannoside were thought to be involved in the regulation of the permeability of cells to L-glutamic acid. The effect of penicillin on L-glutamic acid permeability is not explained by the phospholipid content in the cell membrane. The permeability of penicillin-treated cells was sensitive to changes in osmotic pressure. Under reduced osmotic pressure, penicillin promoted the excretion of L-glutamic acid in a biotin-rich medium and microscopic study revealed that penicillin induced elongated and swollen cells. C_{16}–C_{18} saturated fatty acids, on the other hand, enhanced the excretion of L-glutamic acid in a biotin-rich medium independently of the osmotic pressure. From these findings, it is plausible to consider that penicillin has a secondary effect on the membrane function. Primarily, it inhibits the cell wall synthesis, leaving the cell membrane unprotected, and the permeability barrier of the cell membrane is thus liable to damage.

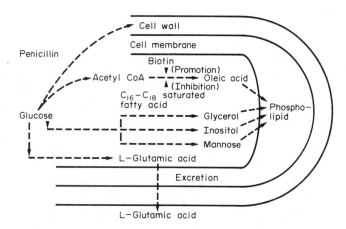

Figure 1 Cell permeability to L-glutamic acid in relation to phospholipid content in the membrane

26.4.2 Regulatory Mechanisms of L-Glutamic Acid Biosynthesis

Although the permeability barrier is more important, the regulatory mechanisms of L-glutamic acid biosynthesis have been studied in order to obtain mutants with increased productivity.

In the production of L-glutamic acid, two enzymes have been shown to play important roles; phosphoenolpyruvate carboxylase (PC), which catalyzes carboxylation of phosphoenolpyruvate to form oxaloacetate, and α-ketoglutarate dehydrogenase (KD), which converts α-ketoglutarate to succinyl-CoA (Shiio and Ujigawa, 1980). The efficiency of carbon dioxide fixation to give oxaloacetic acid depends on PC activity. Aspartic acid showed both inhibition and repression of the enzyme, and the inhibition was enhanced by α-ketoglutaric acid. Therefore, the endogeneous pool of aspartic acid and α-ketoglutaric acid must be minimized if the production of L-glutamic acid is to be maximized. KD is essential for complete oxidation of glucose to carbon dioxide. This enzyme is strongly inhibited by *cis*-aconitate, succinyl-CoA, NADH, NADPH, pyruvate and oxaloacetate, while being stimulated by acetyl-CoA. The properties of KD of L-glutamic acid-producing bacteria are favorable for the preferential synthesis of L-glutamic acid from α-ketoglutaric acid, preventing the further oxidation of α-ketoglutaric acid to carbon dioxide and H_2O *via* succinyl-CoA. The K_m value of KD for α-ketoglutaric acid was shown to be about one seventieth of that of L-glutamic acid dehydrogenase, which catalyzes the formation of L-glutamic acid from α-ketoglutarate. On the other hand, V_{max} of L-glutamic acid dehydrogenase was about 150 times larger than that of KD. Consequently, the endogeneous concentration of α-ketoglutaric acid, which controls the relative metabolic flow of α-ketoglutarate leading to L-glutamic acid biosynthesis or further oxidation, was shown to be high enough to overproduce L-glutamic acid preferentially (Shiio and Ujigawa, 1980).

26.4.3 Genetic Improvement of L-Glutamic Acid-producing Microorganisms

Initial overproduction of L-glutamic acid was performed with wild strains in which the permeability barrier was modified, but the productivity was further increased by microbial breeding. In one example, the cell's permeability barrier to L-glutamic acid was modified by mutation: a temperature sensitive mutant, which showed normal growth at 30 °C but little or no growth at 37 °C, produced a large amount of L-glutamic acid even in a medium with excess biotin when the culture temperature was shifted from 30 °C to 40 °C during the cultivation. The membrane synthesis of this mutant was considered to be insufficient at 37–40 °C, thus permitting L-glutamic acid excretion. No chemical control by penicillin or C_{16}–C_{18} saturated fatty acids was needed for the overproduction of L-glutamic acid in biotin rich media (Momose and Takagi, 1978).

Another attempt to improve the production yield involved increasing the carbon dioxide fixation. L-Glutamic acid is biosynthesized through the glyoxylate cycle as an oxaloacetate generating system without carbon dioxide fixation (Figure 2a), and through phosphoenolpyruvate to form oxaloacetate with carbon dioxide fixation (Figure 2b). The increase in carbon dioxide fixation might improve the production yield.

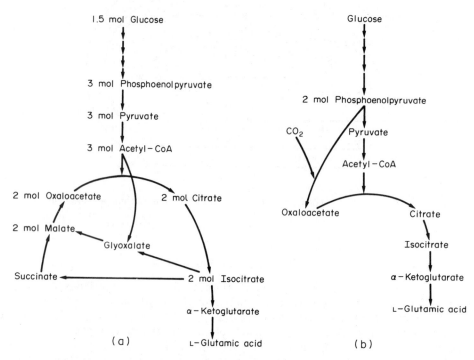

Figure 2 (a) Biosynthetic pathway of L-glutamic acid through the glyoxylate cycle as oxaloacetate generating system without carbon dioxide formation. (b) Biosynthetic pathway of L-glutamic acid through phosphoenolpyruvate as oxaloacetate generating system with carbon dioxide fixation

Several of the monofluoroacetate resistant mutants derived from *Brevibacterium lactofermentum* showed improved productivity of L-glutamic acid in parallel with the increased PC activity (Yamada *et al.*, 1978). One of these mutants with decreased isocitrate lyase activity also accumulated increased amounts of L-glutamic acid (Tosaka *et al.*, 1981). The extent of carbon dioxide fixation was increased in these improved mutants.

A pyruvate dehydrogenase leaky mutant derived from *Brevibacterium lactofermentum* utilized acetic acid and glucose simultaneously to give the higher yield, in which acetic acid was considered to be assimilated as a substrate of acetyl-CoA and glucose as a substrate of oxaloacetate (Ono *et al.*, 1980).

Application of DNA recombination techniques to the improvement of L-glutamic acid-producing bacteria is a promising new route. Several kinds of plasmids of *Brevibacterium* (Kaneko *et al.*, 1979; Momose *et al.*, 1976; Shapiro, 1976; Tsuchida *et al.*, 1981) and a plasmid of *Corynebacterium* relating to spectinomycin resistance were found to be suitable as a possible vector system. Construction of a chimera plasmid involving a gene associated with L-glutamic acid biosynthesis was performed with *Brevibacterium lactofermentum* (Tsuchida *et al.*, 1981).

26.4.4   Performance of Large Scale L-Glutamic Acid Fermentations

Figure 3 shows a typical large scale fermentation system equipped with automatic control. Continuous rather than batchwise sterilization is more successful in eliminating undesirable foreign microbes in the large volume of the media. The advantages are (1) energy saving, (2) better quality control and (3) improved productivity. Air filters packed with glass wool are usually employed for air sterilization.

In L-glutamic acid fermentation, less power input is needed for agitation than in antibiotic fer-

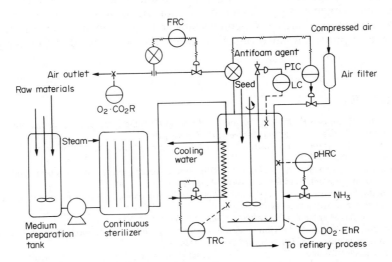

Figure 3 An example of an automatically controlled fermentation process: LC level control; PIC pressure control; pHRC pH recording and control; FRC flow rate recording and control; TRC temperature recording and control; EhR redox potential recording

mentation as the bacterial culture fluids have a lower viscosity than mycelial culture fluids. However, it should be noted that oxygen requirement and heat evolution per unit time and unit volume of the culture are higher because of the higher rate of sugar assimilation and cell respiration.

For a successful fermentation operation, dissolved oxygen tension, temperature and pH must be optimized throughout the fermentation, the dissolved oxygen maintained above 0.01 atm by changing air flow rate, temperature controlled through a cooling device, and culture pH maintained at a constant level by gaseous ammonia. These controls can be performed by computer-aided systems (Hikuma *et al.*, 1981). In addition, the sequential control of many operations, for example sterilizing the system, continuous medium sterilization, introduction of the medium to the fermenter, feeding the concentrated sugar solution into the fermenter, and then washing the fermenter with water, can be easily programmed to occur synchronously (Kuratani and Goto, 1980).

26.4.5 Commercial Aspects of L-Glutamic Acid Fermentation

Annual production of L-glutamic acid amounts to 370 000 tonnes world-wide. It is produced in Japan, Korea, Taiwan, Thailand, Malaysia, Indonesia, the Philippines, France, Italy, Spain, Brazil, Peru and the US. Among these, Japan is the biggest producer with the Ajinomoto Co., Asahi-Kasei Co., Kyowa-Hakko Co. and Takeda-Yakuhin Co. manufacturing 107 000 tonnes between them.

Cane molasses or tapioca starch have been exclusively employed as the main raw materials of L-glutamic acid. The cost is currently about $95 ton^{-1} for cane molasses (sugar content 60%) and about $360 ton^{-1} for tapioca starch. The international price of L-glutamic acid is currently approximately $2 kg^{-1}.

26.5 CONCLUSIONS

Research and development of L-glutamic acid fermentation changed the commercial production method of monosodium L-glutamate from a protein hydrolysis process to a microbial production process. The protein hydrolysis method was more costly, using expensive wheat gluten or soybean protein as raw materials, and resulted in a large amount of byproducts such as starch or amino acid mixtures. L-Glutamic acid fermentation, on the other hand, does not yield any particular byproducts, and has now completely replaced the protein hydrolysis method.

Moreover, recent technological innovation, such as DNA recombination, cell fusion and bioreactor development, are now being applied for further improvement of L-glutamic acid fermentation. DNA recombination and cell fusion techniques might be useful for the genetic construction of microorganisms with higher production yields or with the capability to assimilate less expensive raw materials such as C_1-compounds and cellulosic materials. Bioreactors packed with L-glutamic acid-producing microorganisms are being investigated in an attempt to improve the productivity.

26.6 REFERENCES

Akashi, K., I. Shigeho, H. Shibai, K. Kobayashi and Y. Hirose (1978). Determination of redox potential levels critical for cell respiration for L-leucine production. *Biotechnol. Bioeng.*, **20**, 27.

Hikuma, M., T. Yasuda, I. Karube and S. Suzuki (1981). Applications of microbial sensors to the fermentation process. *Ann. N.Y. Acad. Sci.*, **369**, 307–319.

Hirose, Y., H. Sonoda, K. Kinoshita and H. Okada (1966). Studies on oxygen transfer in submerged fermentations *versus* effects of aeration on glutamic acid fermentation. *Agric. Biol. Chem.*, **30**, 585.

Hirose, Y., H. Sonoda, K. Kinoshita and H. Okada (1968). Studies on oxygen transfer in submerged fermentations and oxygen demand in glutamic acid fermentation. *Agric. Biol. Chem.*, **32**, 855.

Ikeda, K. (1908). A new flavor enhancer. *J. Tokyo Chem. Soc.*, **30**, 820.

Izumi, Y., Y. Tani and K. Ogata (1973). Conversion of bisnorbiotin and bisnordethiobiotin to biotin and dethiobiotin, respectively, by microorganisms. *Biochim. Biophys. Acta*, **326**, 485.

Kamiryo, T., S. Parthasarathy and S. Numa (1976). Evidence that acyl-CoA synthetase activity is required for the repression of yeast acetyl-CoA carboxylase. *Proc. Natl. Acad. Sci. USA*, **73**, 386–390.

Kaneko, H., T. Tanaka and K. Sakaguchi (1979). Isolation and characterization of a plasmid from *Brevibacterium lactofermentum*. *Agric. Biol. Chem.*, **43**, 867–868.

Kinoshita, S., S. Udaka and M. Shimono (1957). Amino acid fermentation. I. Production of L-glutamic acid by various microorganisms. *J. Gen. Appl. Microbiol.*, **3**, 193–205.

Kuratani, T. and Y. Goto (1980). Computer system for fermentation process. *Chem. Ind. (Jpn.)*, **31**, 900–904.

Momose, H., S. Miyashiro and M. Oba (1976). Transducing phages in glutamic acid producing bacteria. *J. Gen. Appl. Microbiol.*, **22**, 119–129.

Momose, H. and T. Takagi (1978). Glutamic acid production in biotin-rich media by temperature-sensitive mutants of *Brevibacterium lactofermentum*, a novel fermentation process. *Agric. Biol. Chem.*, **42**, 1911–1917.

Nakanishi, T., J. Nakaljima and K. Kanda (1975). Conditions for conversion of L-glutamic acid fermentation by *Corynebacterium glutamicum* into L-glutamine production. *J. Ferment. Technol.*, **53**, 543–550.

Okada, H., I. Kameyama, S. Okumura and T. Tsunoda (1961). L-Glutamic acid and succinic acid metabolism by *Brevibacterium flavin* 1996. *J. Gen. Appl. Microbiol.*, **7**, 177–199.

Ono, E., O. Tosaka and K. Takimami (1980). Fermentative production of L-glutamic acid. *Jpn. Pat.* 55 21 762.

Shapiro, J. A. (1976). Observations on lysogeny in glutamic acid bacteria. *Appl. Environ. Microbiol.*, **32**, 179–182.

Shiio, I. and K. Ujigawa (1980). Presence and regulation of α-ketoglutarate dehydrogenase complex in a glutamate-producing bacterium, *Brevibacterium flavum*. *Agric. Biol. Chem.*, **42**, 1897–1904.

Tanka, K., S. Akita, K. Kimira and S. Kinishita (1960). Studies on L-glutamic acid fermentation. VI. The role of biotin. *J. Agric. Chem. Soc. Jpn.*, **34**, 600.

Tosaka, O., Y. Murakami, K. Akashi, S. Ikada and H. Yoshii (1981). L-Glutamic acid by fermentation with mutants. *Jpn. Pat.* 56 92 795.

Tsuchida, T., K. Miwa, S. Nakamori and H. Momose (1981). L-Glutamic acid by fermentation with microorganisms obtained by genetic transformations. *Jpn. Pat.* 56 148 295.

Yamada K., S. Kinoshita, T. Tsunoda and K. Aida (1972). Taxonomic classification of L-glutamic acid producing bacteria. In *The Microbial Production of Amino Acids*, pp. 5. Kodansha, Tokyo.

Yamada, Y., T. Takagi and K. Takinami (1978). Fermentative production of L-glutamic acid. *Jpn. Pat.* 53 32 193.

27
Phenylalanine

H. ENEI and Y. HIROSE
Ajinomoto Co. Inc., Kawasaki, Japan

27.1 INTRODUCTION

Phenylalanine was until recently only a minor part of the amino acid market. Production was approximately 100 tonnes per annum, mainly for incorporation into intravenous nutrient solutions. The use of L-aspartylphenylalanine as a sweetener has caused an increase in the market share of phenylalanine, as well as for aspartic acid. Various methods of phenylalanine production have been investigated, including chemical synthesis and three types of microbial method: direct fermentation from glucose (Coates and Nester, 1974; Goto *et al.*, 1981; Hagino and Nakayama, 1974a), precursor addition (Asai *et al.*, 1960) and enzymatic methods (Chibata, 1981; Sano *et al.*, 1978). All three of these methods are reviewed in this chapter but only chemical synthesis and direct fermentation from sugars have been applied in an industrial situation.

27.2 CONTROL MECHANISMS OF PHENYLALANINE BIOSYNTHESIS

Maximal production of phenylalanine by direct fermentation depends on the circumvention of the complex metabolic regulation caused by the three common products of phenylalanine synthesis: tyrosine, tryptophan and phenylalanine itself. Fundamental studies of metabolic regulation have provided effective guiding principles for the development of high yielding mutants, and the selection of phenylalanine producers of *Bacillus* (Gibson and Pittard, 1968; Umbarger, 1978), *Brevibacterium* (Shiio, 1982) and *Corynebacterium* (Hagino and Nakayama, 1974a) has been carried out in parallel with the investigation on the regulatory mechanism of phenylalanine biosynthesis.

The principal regulatory mechanisms for the biosynthesis of phenylalanine in *Bacillus subtilis* are summarized in Figure 1. DAHP synthetase, the first enzyme in the phenylalanine–tyrosine–tryptophan pathway, is inhibited by chorismate and prephenate, and repressed by tyrosine alone and synergistically with phenylalanine. The activity of shikimate kinase is also inhibited by both chorismate and prephenate *via* feedback control. Prephenate dehydratase, the first enzyme of the phenylalanine specific pathway, is inhibited by phenylalanine itself (Nester and Jensen, 1966; Rebello and Jensen, 1970).

Figure 2 summarizes the other types of regulatory mechanism in *Brevibacterium flavum* (Shiio, 1982). DAHP synthetase in the common pathway is controlled by phenylalanine and tyrosine *via*

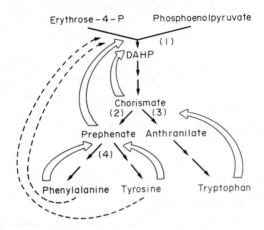

Figure 1   Regulatory mechanisms for L-phenylalanine biosynthesis in *Bacillus subtilis*. ◁▭ , feedback inhibition; ←−−−− feedback repression. (1) DAHP synthetase; (2) chorismate mutase; (3) anthranilate synthetase; (4) prephenate dehydratase

negative feedback. Anthranilate synthetase, in tryptophan synthesis, is strongly inhibited by tryptophan, whereas chorismate mutase is free from regulation by phenylalanine and tyrosine in the synthesis of both phenylalanine and tyrosine. Prephenate aminotransferase, specific for tyrosine synthesis, is not regulated by tyrosine, while prephenate dehydratase for phenylalanine synthesis is regulated by phenylalanine (Shiio *et al.*, 1974; Sugimoto and Shiio, 1974, 1980; Shiio and Sugimoto, 1976, 1981a, 1981b).

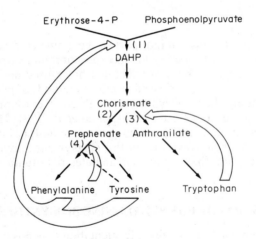

Figure 2   Regulatory mechanisms for L-phenylalanine biosynthesis in *Brevibacterium flavum*. ◁▭ , feedback inhibition; ←−−−− removal of inhibition. (1) DAHP synthetase; (2) chorismate mutase; (3) anthranilate synthetase; (4) prephenate dehydratase

In *Corynebacterium glutamicum*, DAHP synthetase is inhibited synergistically by tyrosine, phenylalanine and tryptophan, the effects of single amino acids being only very slight. The activity of prephenate dehydratase is inhibited by phenylalanine and tryptophan, although stimulated by tyrosine (Hagino and Nakayama, 1974a).

27.3   DIRECT FERMENTATIVE PRODUCTION OF PHENYLALANINE FROM SUGARS

Maximal production of phenylalanine in *Bacillus subtilis* depends on the desensitization of prephenate dehydratase to negative feedback by phenylalanine. A β-thienylalanine resistant mutant, whose prephenate dehydratase was desensitized and thus resistant to the feedback control, excreted about 4 g l^{-1} of L-phenylalanine (Coates and Nester, 1974). Similarly, other phenylalanine producers resistant to such phenylalanine analogs as *p*-fluorophenylalanine (PFP), *m*-fluorophenylalanine (MFP), *p*-aminophenylalanine (PAP) and 5-fluorotryptophan (5FT) were derived; a mutant resistant to 5FT produced 6 g l^{-1} of L-phenylalanine and 4 g l^{-1} of L-tryptophan (Shiio *et al.*, 1973).

In *Brevibacterium flavum*, the prime goal was to desensitize the DAHP synthetase and/or prephenate dehydratase to feedback inhibition by phenylalanine. MFP resistant mutants accumulated 2 g l^{-1} of L-phenylalanine (Shiio and Sugimoto, 1976; Sugimoto *et al.*, 1973). Either DAHP synthetase or prephenate dehydratase was free from feedback inhibition by phenylalanine in these mutants. The development of tyrosine auxotrophic mutants is also an effective means by which to avoid feedback inhibition of DAHP synthetase by tyrosine. The productivity of tyrosine auxotrophic mutants was about 1.6 g l^{-1}, lower than that of MFP resistant mutants (Shiio and Sugimoto, 1976). However, better producers of phenylalanine were obtained by the combination of both auxotrophic and regulatory mutants. The production of 25 g l^{-1} of L-phenylalanine was reported by a tyrosine auxotrophic, PFP and 5MT resistant mutant derived from *Brevibacterium lactofermentum* (Goto *et al.*, 1981).

A tyrosine auxotrophic mutant resistant to PFP and PAP, derived from *Corynebacterium glutamicum*, also produced 9.5 g l^{-1} of L-phenylalanine in a molasses medium (Hagino and Nakayama, 1974a). Phenylalanine production in this mutant was stimulated by tryptophan but inhibited by tyrosine. The prephenate dehydratase of this mutant was resistant to inhibition by phenylalanine and tryptophan, although its DAHP synthetase still remained sensitive (Table 1).

Table 1   Production of L-Phenylalanine by Direct Fermentation

Strain	Marker[a]	Substrate	L-Phe formed (g l^{-1})	Yield[b] (%)
Bacillus subtilis	5FT^r	Glucose	6.0	7.5
Brevibacterium lactofermentum	5MT^r, PFP^r, Dec^r, Tyr$^-$, Met$^-$	Glucose	25.0	19.0
Corynebacterium glutamicum	PFP^r, PAP^r, Tyr$^-$	Cane molasses	9.5	9.5

[a] Abbreviations: 5FT, 5-fluorotryptophan; 5MT, 5-methyltryptophan; PFP, *p*-fluorophenylalanine; PAP, *p*-aminophenylalanine; Dec, decoyinine. [b] g phenylalanine/g substrate consumed × 100.

27.4 PRODUCTION WITH PHENYLPYRUVIC ACID AS A PRECURSOR

The use of intermediates as substrates in phenylalanine synthesis avoids the inhibition by metabolites. Phenylpyruvic acid, an intermediate precursor in the biosynthesis of phenylalanine, is converted into L-phenylalanine by *Alcaligenes*, *Pseudomonas* and *Escherichia* (Asai *et al.*, 1960). For example, dried cells of *Alcaligenes faecalis* B141-1 converted 10.9 g l^{-1} of phenylpyruvic acid into L-phenylalanine with a conversion efficiency of 78%. In this reaction, L-aspartic acid, L-leucine and L-glutamic acid were used as amino donors, and better results were obtained when these amino acids were used in combination as shown in Table 2. This probably means that this strain had several L-phenylalanine aminotransferases specific to each amino donor.

Table 2   Performance of Certain Amino Acids as Amino Donors in L-Phenylalanine Synthesis[a]

Amino acid	Concentration (μmol)	L-Phe formed (mg tube^{-1})	Yield (%)
L-Aspartate	200	9.00	54.5
L-Glutamate	200	7.67	46.5
L-Leucine	200	8.25	50.0
L-Aspartate	100		
L-Glutamate	100	9.40	56.9
L-Aspartate	70		
L-Glutamate	70	11.65	70.6
L-Leucine	70		

[a] Reaction mixture: dried cells, 30 mg; phenylpyruvic acid, 100 μmol; M/15 phosphate buffer (pH 7.6) in a total volume of 1.5 ml; 37 °C; 20 h.

27.5 ENZYMATIC METHODS

Phenylalanine ammonia-lyase and hydantoinase have both been used independently for the production of L-phenylalanine. L-Phenylalanine ammonia-lyase of *Sporobomyces roseus* generally catalyzes the breakdown of L-phenylalanine to phenylsuccinate and ammonia. However, this process for L-phenylalanine production utilizes the lyase to synthesize L-phenylalanine from phenylsuccinate and uses excess ammonium ions to reverse the reaction. The concentration of L-phenylalanine obtained using this method was 28 g l^{-1} and the production yield was 80% on a molar basis (Chibata, 1981).

L-Phenylalanine has also been enzymatically synthesized from DL-5-benzylhydantoin. A wild strain of *Flavobacter aminogenes*, which hydrolyzes DL-5-indolylmethylhydantoin to L-tryptophan, was used for the formation of L-phenylalanine from DL-5-benzylhydantoin. The hydantoin-hydrolyzing enzyme of this strain was induced by DL-5-indolylmethylhydantoin. Fifty mg ml^{-1} of wet cells converted 10 mg ml^{-1} of DL-5-benzylhydantoin to L-phenylalanine with a yield of 100% on a molar basis. The fact that the D-form of hydantoin was racemized to its L-form by the intact cells of *Flavobacter aminogenes* was advantageous in obtaining such high reaction yields (Sano *et al.*, 1977).

27.6 ISOMERIZATION OF D-PHENYLALANINE TO L-PHENYLALANINE

Chemical synthesis of L-phenylalanine, a method currently still in commercial use, involves isomerization of DL-phenylalanine. The asymmetric hydrolysis of acyl-DL-amino acids by microbial aminoacylase was found to be better than any other enzymatic method, and was further improved by immobilization. The reaction can be carried out continuously in columns packed with amino-acylase immobilized on a DEAE Sephadex support (Izumi *et al.*, 1978).

27.7 REFERENCES

Asai, T., K. Aida and K. Oishi (1960). Fermentative production of tryptophan, phenylalanine and tyrosine. *Amino Acid*, **2**, 114.
Chibata, I. (1981). Production of phenylalanine. *Jpn. Pat.* 56 26 197.
Coates, J. H. and E. W. Nester (1967). Regulation reversal mutation: Characterization of end-product activated mutants of *Bacillus subtilis*. *J. Biol. Chem.*, **242**, 4948.
Gibson, F. and J. Pittard (1968). Pathways of biosynthesis of aromatic amino acids and vitamins and their control in microorganisms. *Bacteriol. Rev.*, **32**, 465.
Goto, I., M. Ishihara, S. Sakurai, H. Enei and K. Takinami (1981). Production of phenylalanine. *Jpn. Pat.* 59 64 793.
Hagino, H. and K. Nakayama (1974a). Production of aromatic amino acids by microorganisms. 4. L-Phenylalanine production by analog-resistant mutants of *Corynebacterium glutamicum*. *Agric. Biol. Chem.*, **38**, 157.
Hagino, H. and K. Nakayama (1974b). Production of aromatic amino acids by microorganisms. 6. Regulatory properties of prephenate dehydrogenase and prephenate dehydratase from *Corynebacterium glutamicum*. *Agric. Biol. Chem.*, **38**, 2367.
Izumi, Y., I. Chibata and T. Itoh (1978). Production and utilization of amino acids. *Angew. Chem.*, **17**, 176–183.
Nester, E. W. and R. A. Jensen (1966). Control of aromatic amino acid biosynthesis in *Bacillus subtilis* sequential feedback inhibition. *J. Bacteriol.*, **91**, 1594.
Rebello, J. L. and R. A. Jensen (1970). The multi-metabolite control of prephenate dehydratase activity in *Bacillus subtilis*. *J. Biol. Chem.*, **245**, 3738.
Sano, K., K. Yokozeki, C. Eguchi, T. Kagawa, I. Noda and K. Mitsugi (1977). Enzymatic production of L-tryptophan from L-5-indolylmethylhydantoin and DL-5-indolylmethylhydantoin by newly isolated bacterium. *Agric. Biol. Chem.*, **41**, 819.
Shiio, I. (1982). Fermentative production of tryptophan, phenylalanine and tyrosine. *Ferment. Ind.*, **40**, 211.
Shiio, I. and S. Sugimoto (1976). Altered prephenate dehydratase in phenylalanine-excreting mutants of *Brevibacterium flavum*. *J. Biochem.*, **79**, 173.
Shiio, I. and S. Sugimoto (1981a). Effect of enzyme concentration on regulation of dissociable chorismate mutase in *Brevibacterium flavum*. *J. Biochem.*, **89**, 1483.
Shiio, I. and S. Sugimoto (1981b). Regulation at metabolic branch points of aromatic amino acid biosynthesis in *Brevibacterium flavum*. *Agric. Biol. Chem.*, **45**, 2197.
Shiio, I., K. Ishii and K. Yokozeki (1973). Production of L-tryptophan by 5-fluorotryptophan resistant mutant of *Bacillus subtilis*. *Agric. Biol. Chem.*, **37**, 1991.
Shiio, I., S. Sugimoto and R. Mivajima (1974). Regulation of 3-deoxy-D-arabinoheptulosonate-7-phosphate synthetase in *Brevibacterium flavum*. *J. Biochem.*, **75**, 987.
Sugimoto, S. and I. Shiio (1974). Regulation of prephenate dehydratase in *Brevibacterium flavum*. *J. Biochem.*, **76**, 1103–1111.
Sugimoto, S. and I. Shiio (1980). Purification and properties of bifunctional 3-deoxy-D-arabinoheptulosanate-7-phosphate synthetase chorismate mutase component A from *Brevibacterium flavum*. *J. Biochem.*, **87**, 881–890.

Sugimoto, S., M. Nakagawa, T. Tsuchida and I. Shiio (1973). Regulation of aromatic amino acid synthesis and production of tyrosine and phenylalanine in *Brevibacterium flavum. Agric. Biol. Chem.*, **37**, 2327.

Umbarger, H. E. (1978). Amino acid biosynthesis and its regulation — a review. *Annu. Rev. Biochem.*, **47**, 533.

28
Lysine

K. NAKAYAMA
Bior Inc., Kanagawa, Japan

28.1 INTRODUCTION

Lack of the essential amino acid L-lysine in cereals triggered studies on L-lysine production by both chemical synthesis and microbial methods. Production of L-lysine on a large scale started in 1958, when a fermentation process using an auxotrophic mutant of *Corynebacterium glutamicum* was established by the Kyowa Hakko Kogyo Company. Recently another process using a regulatory mutant of *Brevibacterium flavum* was developed by the Ajinomoto Company. Both processes have been further improved and are used for industrial production, which has reached 40 000 ton yr^{-1}. This amount is subject to change depending on the relative costs of production of soybean meals and fish meal. More recently, a third industrial process was developed by the Toray Company. In the process, racemic α-aminocaprolactam synthesized from cyclohexane is converted into L-lysine by the combined action of L-aminocaprolactam hydrolase and aminocaprolactam racemase.

28.2 BIOSYNTHETIC PATHWAY OF LYSINE

Biosynthetically L-lysine is a member of the aspartate family (except in fungi) and the regulation of lysine synthesis has a close relationship with that of the other amino acids in the aspartate family. The aspartate family consists of aspartate, asparagine, methionine, threonine, lysine and isoleucine. The carbon skeletons of lysine and isoleucine are derived in part from pyruvate but are still considered part of the aspartate family. Isoleucine biosynthesis is best considered along with the pyruvate family of amino acids since four of its biosynthetic enzymes are also needed in the valine pathway. Diaminopimelate, a lysine precursor, is needed for cell wall synthesis in bacteria but not for protein synthesis. It can therefore be considered another member of the aspartate family. The methyl carbon of methionine is derived to a major extent from the β-carbon of serine and in most organisms its sulfur is from cysteine.

Since the aspartate family of amino acids is formed *via* a highly branched pathway, there are numerous branch points that are potential sites of regulation of carbon flow. Several patterns of control over this branching pathway have evolved in various organisms (Umbarger, 1978).

The obvious way for aspartate biosynthesis to occur is *via* a glutamate aspartate transaminase, which catalyzes the formation of aspartate from oxaloacetate, that has long been known. It has only recently become known, however, which enzyme did in fact perform this reaction in *Escherichia coli*, essentially the only organism that has been studied for its variety of transaminases. The tyrosine–glutamate transaminase appears to have a minor role in aspartate formation, but transaminase A (the *aspC* gene product) plays the major role. The common pathway for the aspartate family of amino acids consists of three reactions that convert aspartate to the two branch point intermediates, β-aspartyl semialdehyde and homoserine.

In *E. coli*, the formation of β-aspartyl phosphate (reaction 2, Figure 1) is catalyzed by three different aspartokinases, which have specifically evolved to fulfill the function of threonine (and isoleucine) formation, methionine formation and lysine formation, respectively. Although each aspartokinase would appear to be adapted for synthesis of β-aspartyl phosphate in an amount required for the specific end product, there appears to be no channelling of product and function in such a way. Rather, there appears to be a common pool of β-aspartyl phosphate from which all products are derived, as is demonstrated by the fact that the predominant aspartokinase varies from one strain to another. Furthermore, the loss of one or even two of the three aspartokinase activities does not lead to auxotrophy, *i.e.* a deficiency in activity by any one aspartokinase is compensated for by derepression of one or both of the remaining isozymes.

Aspartokinase I activity is associated with a second activity, homoserine dehydrogenase I, car-

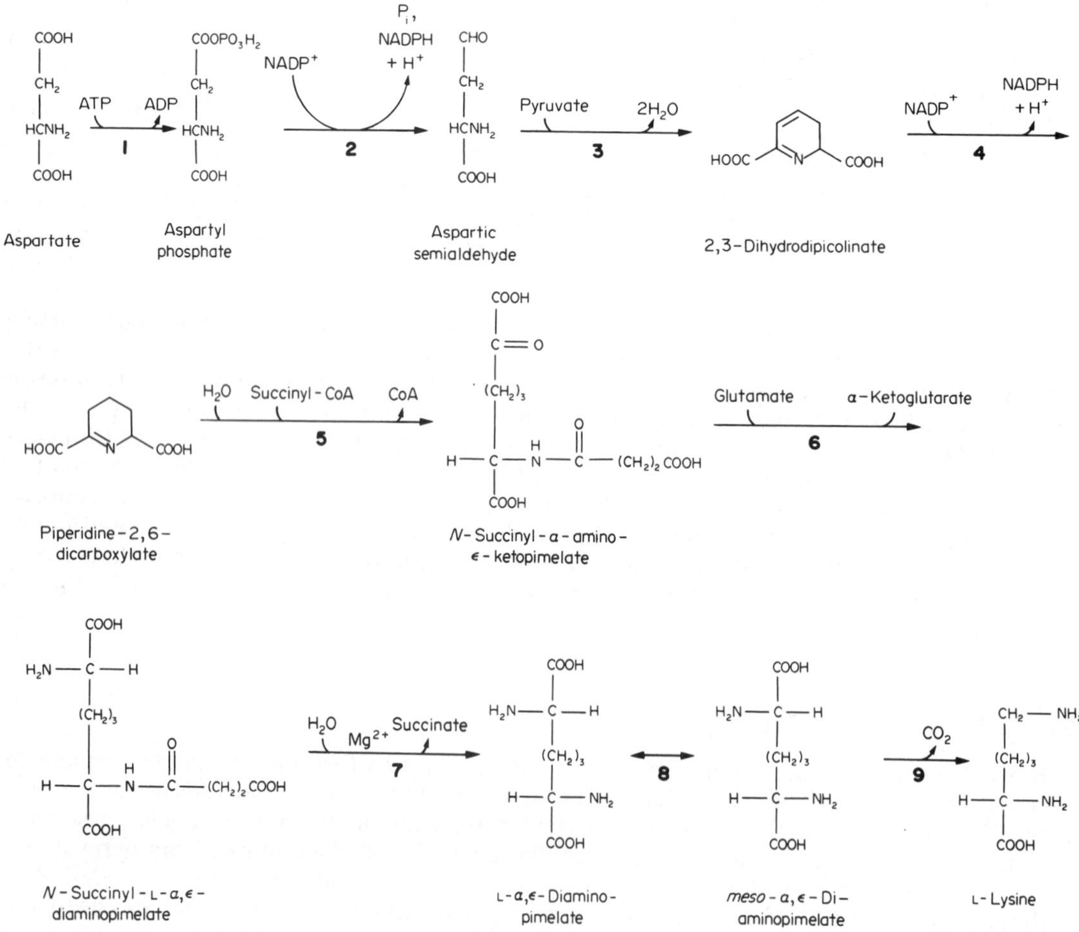

Figure 1 Pathway of lysine biosynthesis in bacteria (diaminopimelic acid pathway). Trivial names of enzymes: (1) β-aspartokinase, (2) aspartic β-semialdehyde dehydrogenase, (3) dihydrodipicolinate synthetase, (4) dihydrodipicolinate reductase, (5) N-succinyl-ε-keto-L-α-aminopimelic acid synthetase, (6) N-succinyldiaminopimelate transaminase, (7) N-succinyldiaminopimelate deacylase, (8) diaminopimelate epimerase, (9) diaminopimelate decarboxylase

ried on a bifunctional enzyme. Both activities are inhibited by threonine, but to different extents. The inhibition of aspartokinase activity is competitive with both substrates, whereas that of homoserine dehydrogenase activity is not. In *E. coli* K12 and presumably *Salmonella typhimurium*, aspartokinase II is also associated with homoserine dehydrogenase activity in a bifunctional protein. The third aspartokinase in the Enterobacteriaceae appears not to be associated with any additional activities.

The pattern of multiple enzymes found in the Enterobacteriaceae for the multifunctional step, β-aspartyl phosphate formation, is not the only way the reaction is controlled. In *Bacillus polymyxa*, *Rhodopseudomonas capsulata* and the pseudomonads, there are only single aspartokinases, and their regulation is achieved by either a multivalent or a synergistic inhibition involving both lysine and threonine. In those organisms in which a single aspartokinase is found, homoserine dehydrogenase activity is found on a separate protein. In general, homoserine dehydrogenase is inhibited by threonine but to various extents, so that there may be among these organisms some that contain two homoserine dehydrogenases. The remaining enzyme in the common pathway, β-aspartic semialdehyde dehydrogenase, catalyzes the NADPH-dependent reduction of β-aspartyl phosphate (reaction 2, Figure 1).

With the exception of fungi in which lysine is formed by the pathway through α-aminoadipic acid, the biosynthesis of lysine occurs as a branch of the aspartate family of amino acids. The condensation of aspartic semialdehyde with pyruvate (reaction 3, Figure 1) yields a compound that spontaneously cyclizes and is then reduced in an NADPH-linked reaction (reaction 4, Figure 1). Ring opening for subsequent reactions is achieved by 'trapping' the open chain form by succinylation as in *E. coli* or by acetylation as in bacilli (reaction 5, Figure 1). After a transamination reaction (reaction 6), the acyl group is removed to yield L,L-diaminopimelate (reaction 7). Either this form or the *meso* form obtained by a specific recemase (reaction 8) or both (depending upon the organism) is used as a constituent of bacterial cell wall synthesis. The *meso* form is the substrate for the decarboxylase that catalyzes the final step in lysine synthesis (reaction 9).

28.3 LYSINE PRODUCTION WITH HOMOSERINE AUXOTROPHS

Microbial production of L-lysine was reviewed by Nakayama (1972, 1978, 1982). A microbial process for L-lysine production was first developed by a combination of diaminopimelate production by a lysine-requiring auxotroph or a lysine–histidine-requiring double auxotroph of *E. coli* and decarboxylation of the compound by *Aerobacter aerogenes* or wild-type *E. coli*. The process was developed by workers of E. I. de Pont de Nemours and Co. Inc. The yield of diaminopimelic acid reached 24 g l^{-1}. Direct production of L-lysine from carbohydrate was developed first with a homoserine- or threonine- plus methionine-requiring auxotroph of *C. glutamicum* by the present author. The same type of process was reported with a homoserine-requiring auxotroph of *B. flavum*. The leaky homoserine-requiring auxotroph was recognized as a threonine-sensitive mutant because growth was inhibited by excessive threonine and the inhibition was released by addition of methionine. This phenomenon is due to feedback inhibition of residual homoserine dehydrogenase by threonine. Homeserine- (or threonine plus methionine)- requiring auxotrophs of other bacteria were also found to produce L-lysine, but the yields were lower than that from the homosereine-requiring auxotroph of coryneform bacteria.

Threonine auxotrophs and leucine auxotrophs of *C. glutamicum* produce fairly large amounts of L-lysine, but they are inferior to the homoserine auxotroph. Other auxotrophs of *C. glutamicum* and other bacteria were also inferior to the homoserine auxotroph of *C. glutamicum*. Double auxotrophs, which require, in addition to homoserine, at least one of the amino acids, threonine, isoleucine or methionine for growth, have been found to be highly stabilized, showing little tendency to revert to homoserine independence. It is possible not only to prevent reversion of the cultures to a wild-type state, but also to produce lysine in higher yields since many of the microorganisms are double mutants in the homoserine pathway.

Cane molasses is now generally used as a carbon source in the industrial production of lysine, though other carbohydrate materials, acetic acid and ethanol can be used. The pH value of the medium is maintained near neutrality during the fermentation by feeding ammonia or urea. Ammonia and ammonium salts are generally good nitrogen sources, and urea can be used for organisms having urease activity. An example of a fermentation using cane molasses is as follows. The medium for first seed culture contained 2% glucose, 1% peptone, 0.5% meat extract and 0.25% NaCl in tap water. For the second seed culture, the medium contained 5% cane molasses, 2% $(NH_4)_2SO_4$, 5% corn-steep liquor and 1% $CaCO_3$ in tap water. The fermentation medium contained 20% cane molasses as glucose and 1.8% soybean meal hydrolyzate (as weight of meal before hydrolysis with 6 N H_2SO_4 and neutralization with ammonia water) in tap water. The fermentation was carried out at 28 °C. Figure 2 shows the time course of the fermentation using *C. glutamicum* No. 901 (a homoserine-requiring auxotroph), which produced 44 g of L-lysine per liter in 60 h. (The amount of L-lysine was expressed as the weight of L-lysine monohydrochloride because the hydrochloride is the common form for use.) Foaming in the aerated culture can be repressed by addition of proper antifoaming agents. The amount of the growth factors (homoserine or threonine and methionine) should be appropriate for the production of L-lysine. These are supplied in limited amounts and are suboptimal for the growth. The biotin concentration in the medium must generally be greater than 30 $\mu g\,l^{-1}$. Cane molasses usually supplies enough biotin, but when beet molasses or starch hydrolysate is used, biotin must be added. Yields of L-lysine as the monohydrochloride reach 30–40% in relation to the initial sugar concentration.

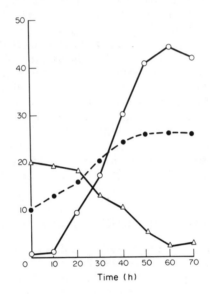

Figure 2　Time course of lysine fermentation; $\bigcirc$, L-lysine $(g\,l^{-1})$; $\triangle$, residual sugar (%); $\bullet$, dry cell weight $(g\,l^{-1})$

Coryneform glutamic acid-producing bacteria can utilize acetic acid as a carbon source for growth and lysine production. L-Lysine production from acetic acid by a homoserine-leaky (threonine-sensitive) threonine auxotroph mutant of *Brevibacterium flavum* reached 75 g l^{-1} (as monohydrochloride) or 29% on the basis of acetic acid and glucose supplied. The medium contained 0.7% acetic acid, 0.2% KH_2PO_4, 0.04% $MgSO_4\cdot7H_2O$, 0.001% $FeSO_4\cdot7H_2O$, 0.001% $MnSO_4\cdot H_2O$, 3.5% hydrolyzate of soybean protein, 3.0% glucose, 50 $\mu g\,l^{-1}$ biotin, and 40 $\mu g\,l^{-1}$ thiamin HCl (pH 6.0). Fermentation was carried out at 33 °C with feeding of a solution of acetic acid. The feeding solution contained 60% acetic acid composed of a mixture of acetic acid and ammonium acetate having a molar ratio of 4:1, and 3% glucose. Feeding was controlled automatically until the end of the fermentation, keeping the pH value of the medium at 7.4. Some patents have been issued for a process to produce L-lysine from *n*-paraffins.

L-Lysine in fermentation broth is recovered by adsorption on a cation exchange resin, and elution of the adsorbed lysine with dilute alkali. The eluate is neutralized with hydrochloric acid and crystals of L-lysine monohydrochloride are obtained by concentration of the eluate.

28.4　DEREGULATION IN LYSINE OVERPRODUCING MUTANTS

The mechanism of regulation of lysine biosynthesis in *C. glutamicum* is shown in Figure 3. A similar regulatory pattern was also observed in *B. flavum*. Generally the reaction yielding the initial product on a biosynthetic pathway is subject to end product inhibition. Lysine inhibition of the aspartic semialdehyde–pyruvate condensation, the branch point to lysine biosynthesis, has in fact been demonstrated in *E. coli*, but in *C. glutamicum* neither lysine nor any other amino acid exerted an inhibitory effect even at concentrations of 10^{-2} M. Similar insensitivity of dihydrodipicolinate synthetase to lysine has been reported in *B. flavum*, *Bacillus cereus*, *Bacillus subtilis* and *Streptococcus faecalis*. Though homoserine dehydrogenase was inhibited by threonine and repressed by methionine in *C. glutamicum* and *B. flavum*, aspartic-β-semialdehyde dehydrogenase was not inhibited or repressed by any amino acid tested. Thus, the overproduction of lysine by homoserine or threonine auxotrophs can be accounted for by the release of their aspartokinases from concerted feedback inhibition, due to inability to produce threonine, and by the absence of any other major regulatory mechanism on the lysine biosynthetic pathway (Figure 3). The blocking of homoserine synthesis at homoserine dehydrogenase results in the release of the concerted feedback inhibition by threonine and lysine on aspartokinase, and the aspartic semialdehyde produced proceeds to lysine through the lysine synthetic pathway on which no feedback inhibition is found, a situation which differs from that in *E. coli*. Lysine production by homoserine or threonine auxotrophs of *B. subtilis* may involve a similar mechanism, because similar metabolic control has been reported in *Bacillus* species.

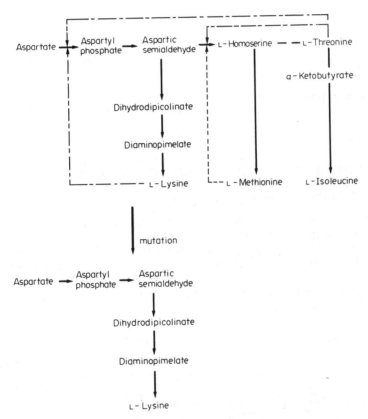

Figure 3 Deregulation of lysine biosynthesis in homoserine auxotroph of *Corynebacterium glutamicum*; ---→, feedback inhibition; −−→, repression

No significant degradation or decarboxylation of lysine occurs in *C. glutamicum*, favoring lysine accumulation. In *E. coli* however, lysine can be decarboxylated to yield cadaverine. Excretion is also an important factor in lysine accumulation, and the situation is in contrast to that observed in yeasts. When penicillin was added to the homoserine auxotrophs of *C. glutamicum* and *Brevibacterium* No. 22 during an early stage of the fermentation, glutamic acid was produced instead of lysine.

A mutant of *E. coli* with aspartokinase insensitive to feedback inhibition by the end product but with normal catalytic activity has been isolated. Amino acid analogs have been used to select such regulatory, amino acid hyperproducing mutants in *E. coli*. The principle of this selective technique is simple: the analog acts as a pseudofeedback inhibitor, inhibiting the synthesis of the end product. The microorganism is unable to grow because it cannot replace the end product nutritionally. Deregulatory mutants are obtained as they are resistant to the analog and able to grow in its presence. A mutant of *B. flavum* resistant to 5-(β-aminoethyl)-L-cysteine (AEC), a lysine analog, produced fairly large amounts of L-lysine. Resistance to AEC brought about by the desensitization of aspartokinase also releases the concerted feedback inhibition. The conversion of aspartic semialdehyde to threonine is feedback inhibited by L-threonine. Thus the overproduced aspartic semialdehyde is channelled into L-lysine production (Figure 4).

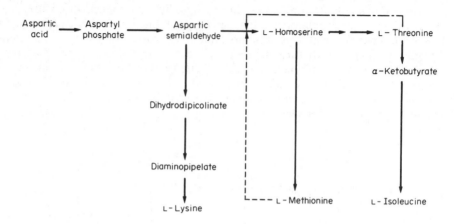

Figure 4   Deregulation of lysine biosynthesis in *S*-(*β*-aminoethyl)cysteine (AEC)-resistant mutant of *Brevibacterium flavum*; —·→, feedback inhibition; – – –→, repression

28.5   PRODUCTION WITH MULTIPLY IMPROVED MUTANTS

28.5.1   *Corynebacterium glutamicum*

An increase in lysine yield (more than 10%) was obtained using a mutant of *C. glutamicum* which requires homoserine and leucine and is resistant to AEC. It produced 39.5 g L-lysine l^{-1} in a medium containing 10% reducing sugars, expressed as invert (as cane molasses), while the homoserine plus leucine auxotroph produced 34.5 g L-lysine l^{-1} (Nakayama and Araki, 1973). As described earlier, leucine auxotrophs of *C. glutamicum* produced a fairly large amount of lysine. The results shown in Table 1 demonstrated the contribution of leucine auxotrophy and resistance to lysine, threonine and isoleucine analogs.

Table 1   L-Lysine Production by some Auxotrophic Regulatory Mutants of
Corynebacterium glutamicum

Strain	Genetic markers[a]	L-*Lysine yield*[b] (l^{-1})
ATCC 21526	Homoserine$^-$, leucine$^-$, AEC^r	39.4
ATCC 21523	Homoserine$^-$, leucine$^-$, AHV^r	38.2
ATCC 21544	Homoserine$^-$, leucine$^-$, AMTB^r	38.1
ATCC 21253	Homoserine$^-$, leucine$^-$,	34.5
ATCC 13032	Homoserine$^-$	28–30

[a] AEC = *S*-(*β*-aminoethyl)cysteine, AHB = *α*-amino-*β*-hydroxyvaleric acid, AMTB = 2-amino-3-methylthiobutyric acid.   [b] Medium containing 100 g l^{-1} reducing sugars.

C. glutamicum mutant, which requires homoserine, leucine and pantothenic acid and is resistant to AEC, produced 42 g L-lysine l^{-1} in a cane molasses medium containing 10% reducing sugars expressed as invert (Nakayama and Araki, 1981).

28.5.2   *Brevibacterium flavum*

The prototrophic revertant 15-8 was derived from a citrate synthetase (CS) defective glutamate auxotroph of *B. flavum*. It showed low citrate synthetase activity and overproduced L-aspartic acid. The maximum production was 10.6 g l^{-1} (about a 30% yield) when the strain was cultured in medium containing 36 g l^{-1} of glucose for 48 h. Aspartate production by these mutants markedly depended on biotin concentration of the medium, as did the glutamate production by the original wild strain. When the revertant strain was cultured with excess biotin, acetic acid accumulated instead of aspartic acid. Therefore, the mechanism for the aspartate overproduction seems to be explained in the same way as that for the glutamate production. Limited concentration of biotin in the medium will induce the formation of defective cell membrane with enhanced permeability to the amino acid, lowering its intracellular concentration and reducing the feedback control exerted over its biosynthesis. The amount of aspartate produced by the wild strain was markedly less than that of glutamate production. This seems to be due to preferential synthesis of glutamate at the biosynthetic branch point for the two amino acids. On the other hand, the predominant production of aspartate by mutant 15-8 seems to be a result of preferential synthesis of aspartate due to the low level of CS, the enzyme at the branch point for glutamate biosynthesis.

A mutant resistant to AEC plus threonine derived from the revertant strain overproduced lysine with yields of 36% (Shiio *et al.*, 1982). This value is clearly higher than that obtained with an AEC-resistant strain which was directly derived from the wild strain. A homoserine auxotroph derived from the revertant strain produced 33 g L-lysine l^{-1} (33% yield), which was almost the same amount as that derived directly from the original wild strain.

28.5.3 *Brevibacterium lactofermentum*

As in *C. glutamicum*, leucine auxotrophy increased the lysine productivity of an AEC-resistant mutant of *B. lactofermentum*. A leucine auxotroph derived from AEC resistant mutants produced 41 g L-lysine l^{-1} while the parental strain produced about 18 g L-lysine l^{-1} (Tosaka *et al.*, 1978a). A kind of metabolic interlock was found in *B. lactofermentum*. Dihydrodipicolinate synthetase, an enzyme for the initial step of lysine biosynthesis, was repressed by L-lysine. This amino acid activated α-isopropylmalic acid synthetase, an initial enzyme for leucine biosynthesis, and also reversed the inhibition of the enzyme by leucine (Figure 5) (Tosaka *et al.*, 1978b). These facts explain the effect of leucine auxotrophy in a lysine overproducing mutant.

During the study of the effect of leucine on the accumulation of amino acids, it was noted that

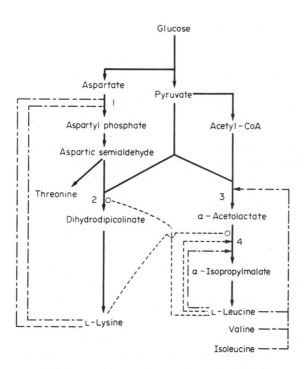

Figure 5 Mutual regulation in biosynthesis of lysine and leucine in *Brevibacterium lactofermentum*; trivial names of enzymes: (1) aspartokinase, (2) dihydrodipicolinate synthetase, (3) α-acetolactate synthetase, (4) α-isopropylmalic acid synthetase; – – →, feedback inhibition; – – →, repression; – – – –○, activation

the accumulation of alanine increased as that of L-lysine decreased. Alanine can be formed from pyruvate by transamination, and directly from aspartate by aspartate β-decarboxylase (Figure 6). But the activity of the aspartate β-decarboxylase was found to be approximately one tenth to one fiftieth of the transaminase activity. An alanine auxotroph was obtained by mutation from strain AJ 3445 (AEC^r). The productivity of L-lysine was inversely proportional to the level of pyruvate-L-amino acid transaminase. The best L-lysine producer, AJ 3799, yielded 39 g L-lysine l^{-1} and lacked pyruvate-L-amino acid transaminase (Tosaka *et al.*, 1978c).

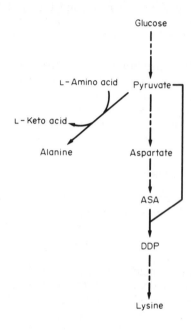

Figure 6   Pathway of lysine biosynthesis and alanine transaminase. ASA, aspartic β-semialdehyde; DDP, dihydrodipicolinate

B. lactofermentum AJ 3799 (AEC^r, Ala$^-$) accumulated 34 g lysine l^{-1} in the culture medium. This increase in L-lysine production was observed with various kinds of lysine producers and also with resting cells grown in biotin-poor medium (Tosaka *et al.*, 1979a). This effect was observed only when glucose or pyruvate was used as the sole carbon source, and accompanied with the specific incorporation of $^{13}CO_2$ into the γ-CH$_2$ group of L-lysine. *B. lactofermentum* AJ 3445 (AEC^r) could grow on pyruvate medium supplemented with biotin at more than 200 μg l^{-1}, while the same growth was observed with the addition of acids of the TCA cycle or glutamate to pyruvate medium.

The anaplerotic synthesis of oxaloacetate from three-carbon precursors is usually accomplished in bacteria through the action of one of the two CO$_2$-fixing enzymes, the biotin-dependent pyruvate carboxylase, which catalyzes the reaction in equation (1):

$$\text{Pyruvate} + \text{ATP} + \text{HCO}_3 \rightarrow \text{oxaloacetate} + \text{ADP} + \text{P} \tag{1}$$

or the biotin-independent phosphoenolpyruvate (PEP) carboxylase, which catalyzes that shown in equation (2).

$$\text{Phosphoenolpyruvate} + \text{HCO}_3 \rightarrow \text{oxaloacetate} + \text{P} \tag{2}$$

Since both pyruvate and phosphoenolpyruvate carboxylases are supposed to fulfill the same function in cell metabolism, it was thought that they were not likely to be simultaneously present in the same cell. In *B. flavum*, PEP carboxylase was found at significant levels in cell-free extracts. On the other hand, pyruvate carboxylase has never been detected. *B. lactofermentum* contained both pyruvate carboxylase and PEP carboxylase, as has been previously reported for *Pseudomonas citronellolis*, *P. fluorescens* and *Azetobacter vinelandii*.

A PEP carboxylase-deficient mutant derived from AJ 3445 could not grow on glucose as the sole carbon source, but grew on glucose when the medium contained 200 μg l^{-1} of biotin. AJ 3445 grown on lactate medium with 500 μg l^{-1} of biotin and KHCO$_3$ contained the biotin-dependent pyruvate carboxylase (Tosaka *et al.*, 1979).

Pyruvate carboxylase was insensitive to acetyl-CoA and L-aspartate, while PEP carboxylase was sensitive to L-aspartate. This regulatory difference suggests that the two carboxylases may have different functions in the living cells. At high concentrations of biotin in the culture medium, aspartate cannot permeate through the cell membrane and consequently intracellular levels of aspartate may be increased. Therefore, PEP carboxylase activity is strongly inhibited by aspartate at concentrations less than 1 mM. These circumstances are unfavorable for the synthesis of cell constituents. On the other hand, the apparent affinity of PEP for PEP carboxylase is about one tenth of that for pyruvate kinase. PEP is converted to pyruvate preferentially by pyruvate kinase and pyruvate may induce pyruvate carboxylase. A higher concentration of biotin is required for pyruvate carboxylase activity. This sequence involves a net conversion of pyruvate to oxaloacetate and permits growth on three-carbon sources.

Thus the promotion of L-lysine production in *B. lactofermentum* by high concentrations of biotin is explained by the stimulation of pyruvate carboxylase by biotin which consequently leads to the increase of aspartate formation through the increase of oxaloacetate formation (Figure 7).

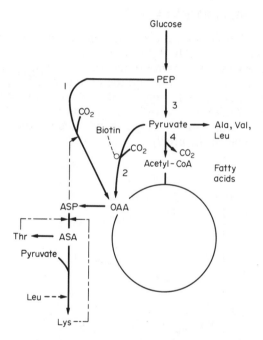

Figure 7 Pathway and regulation of lysine biosynthesis from glucose in *Brevibacterium lactofermentum*; $--\rightarrow$, feedback inhibition; $---\rightarrow$, repression; $---\!\!-\!\!\bigcirc$, activation; trivial names of enzymes: (1) PEP carboxylase, (2) pyruvate carboxylase, (3) pyruvate kinase, (4) pyruvate dehydrogenase

Fluoropyruvate inhibited pyruvate dehydrogenase and increased lysine production of *B. lactofermentum* AJ 3799 (AEC^r, Ala$^-$) in the presence of high concentrations of biotin. Furthermore, the mutant selected as fluoropyruvate-sensitive after mutagenization of AJ 3799 gave increased lysine production compared with the parental strain (Tosaka *et al.*, 1981). The activity of its pyruvate dehydrogenase was found to be $\sim$10–15% of the parental strain.

Aspartokinase of *B. lactofermentum* is subject to concerted feedback inhibition by lysine plus threonine as in *C. glutamicum* and *B. flavum*. A mutant resistant to AEC that grows in the presence of AEC plus threonine has been isolated. But α-chlorocaprolactam (CCL) or γ-methyllysine (ML) strongly inhibited the growth of *B. lactofermentum* by single addition, and the inhibition was reversed by lysine. The mutant isolated as a resistant to CCL or ML after mutagenic treatment of AJ 3424 (AEC^r, Ala$^-$) produced 43 g L-lysine l^{-1} (Kubota *et al.*, 1976).

As a result of the breeding described above, *B. lactofermentum* AJ 11204 has been selected and produces 48 g L-lysine l^{-1} in a medium containing 10% glucose. The genealogy of the strain is shown in Figure 8 (Tosaka and Takinami, 1982).

Brevibacterium lactofermentum	*Lysine productivity* (g l^{-1})
AJ 1511 (wild)	0
↓	
AJ 3445 (AEC^r)	16
↓	
AJ 3424 (AEC^r, Ala$^-$)	33
↓	
AJ 3796 (AEC^r, Ala$^-$, CCL^r)	39
↓	
AJ 3991 (AEC^r, Ala$^-$, CCL^r, ML^r)	43
↓	
AJ 11204 (AEC^r, Ala$^-$, CCL^r, ML^r, FP^s)	48

Figure 8   Genealogy of lysine-producing mutants in *Brevibacterium lactofermentum*; AEC, *S*-(β-aminoethyl)-L-cysteine; CCL, α-chlorocaprolactam; ML, γ-methyllysine; FP, β-fluoropyruvate

28.6   AERATION IN LYSINE FERMENTATION

The effect of oxygen tension (P_L) on the production of L-lysine was studied employing the mutant of *B. lactofermentum* (AEC^r, Ala$^-$). Sufficient supply of oxygen to satisfy the cells' oxygen demand was essential for the maximum production of L-lysine. The dissolved oxygen level must be controlled at greater than 0.01 atm, and the optimum redox potential of culture media should be above −170 mV. An extremely oxygen deficient condition, when the degree of oxygen satisfaction represented by *rab*/*KrM* (*rab* = cells' respiration rate, ml O$_2$ ml^{-1}; *KrM* = maximum oxygen demand of cells, ml O$_2$ ml^{-1} min^{-1}) is less than 0.3, led to the production of lactic acid at the expense of the lysine produced (Figure 9).

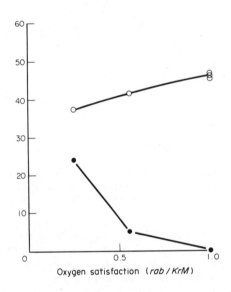

Figure 9   Relationship between L-lysine production and degree of oxygen satisfaction (from Akashi *et al.* 1979); ○——○, lysine (g l^{-1}); ●——●, lactate (g l^{-1}); *KrM*=11.6×10^{-7} mol O$_2$ ml^{-1} min^{-1}

The production of L-lysine, L-threonine and L-isoleucine, which are biosynthesized *via* L-aspartic acid, decreased only slightly under insufficient oxygen supply.

An L-glutamic acid producing strain of *Brevibacterium* is generally able to convert α-ketoglutaric acid into succinic acid, and this lack of production is considered to favor the overproduction of glutamic acid. Two kinds of routes are considered for the biosynthesis of glutamic acid. One involves a glyoxylate cycle in which glutamic acid is biosynthesized by way of pyruvate, citrate and α-ketoglutarate. The other involves a PEP carboxylation system in which glutamic acid is formed by way of PEP, oxaloacetic acid, isocitric acid and α-ketoglutaric acid. In glutamic acid fermentation, it was experimentally shown that the product was accumulated according to equation (3).

$$\text{Glucose} \rightarrow 0.82 \text{ glutamic acid} + 1.94CO_2 \tag{3}$$

Judging from the amount of carbon dioxide evolved during the cultivation, the amino acid was produced from glucose employing both the glyoxylate cycle and the PEP carboxylation system. Biosynthesis of glutamic acid generates $NAD(P)H_2$ whether it is accomplished through the glyoxylate cycle or the PEP carboxylation system. This characteristic of glutamate biosynthesis was considered to cause marked inhibition of the product formation in oxygen-limited culture, because oxygen is required in amino acid fermentation mainly to reoxidize $NADH_2$ generated in the process of amino acid biosynthesis. Aspartic acid is also made through the glyoxylate cycle or the PEP carboxylation system. Biosynthesis of amino acid through the glyoxylate cycle is an $NAD(P)H_2$ generation process in marked contrast to that through the PEP carboxylation system which is an $NAD(P)H_2$ consuming one (Figure 10).

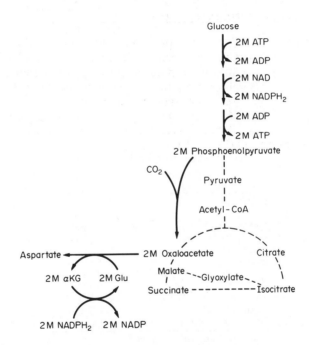

Figure 10 Metabolic pathway of aspartate biosynthesis involving carbon dioxide fixation (from Akashi *et al.*, 1979)

In lysine fermentation, the degree of decrease in product formation in oxygen limitation was shown to be less significant than that in amino acid fermentations of the glutamic acid family. Biosynthesis of these amino acids of the aspartic acid family might be performed by way of PEP and oxaloacetate including carbon dioxide fixation in oxygen-limited culture (Akashi *et al.*, 1979).

A typical example of fermentation kinetics is shown in Figure 11 (Akashi *et al.*, 1979). Sufficient oxygen was supplied to maintain the P_L level between 0.01 and 0.10 atm throughout the fermentation, and the cells' oxygen demand was satisfied. The medium used for fermentation is shown in Table 2.

28.7 ENZYMATIC METHOD

L-Lysine production from DL-α-aminocaprolactam was first studied using a fungus, *Aspergillus ustus*, which hydrolyzed only the L-form of α-aminocaprolactam. The remaining D-form was recycled after racemization. However, the yield was low. Later, a more efficient process was

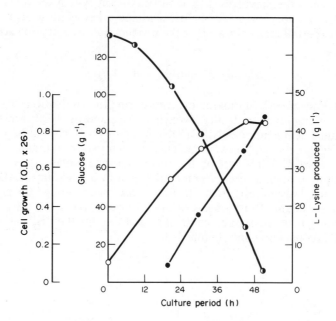

Figure 11 Time course of L-lysine fermentation in a jar fermenter (from Akashi *et al.* 1979). Conditions of oxygen supply: agitation speed, 1200 r.p.m.; air flow rate, 0.5 vvm —O—, cell growth (O.D. 26); —O—, glucose (g l^{-1}); —●—, L-lysine (g l^{-1})

Table 2 Production Medium for Lysine
Fermentation

Glucose	130 g l^{-1}
KH$_2$PO$_4$	1 g l^{-1}
MgSO$_4$·7H$_2$O	0.4 g l^{-1}
FeSO$_4$·7H$_2$O	0.01 g l^{-1}
MnSO$_4$·4H$_2$O	0.01 g l^{-1}
(NH$_4$)$_2$SO$_4$	25 g l^{-1}
DL-Alanine	0.35 g l^{-1}
Biotin	50 μg l^{-1}
Thiamine hydrochloride	200 μg l^{-1}
Nicotinamide	0.5 mg l^{-1}
Soybean hydrolysate[a]	15 mg l^{-1}

[a] 64g/liter of nitrogen is contained.

developed for the conversion (Fukumura, 1976a, 1976b). Incubation of a mixture of 100 ml 10% DL-α-aminocaprolactam (adjusted to pH 8.0 with HCl), 0.1 g acetone-dried cells of *Cryptococcus laurentii* and 0.1 g acetone-dried cells of *Achromobacter obae* nov. sp. with gentle shaking at 40 °C for 24 h resulted in the conversion of DL-α-aminocaprolactam to L-lysine with 99.8% yield.

$$\text{D-}\alpha\text{-Aminocaprolactam} \underset{}{\overset{\text{racemization}}{\rightleftharpoons}} \text{L-aminocaprolactam} \overset{\text{hydrolysis}}{\longrightarrow} \text{L-lysine}$$

C. laurentii produces L-aminocaprolactam hydrolase inductively in a medium containing L-α-aminocaprolactam, glucose and other ingredients. *A. obae* produces aminocaprolactam racemase using both D- and L-α-aminocaprolactam as an inducer (Sato *et al.*, 1974). A similar optimal pH value of both enzymes allows the efficient conversion in what appears to be a single step.

Lysine is produced by this enzymatic route by the Toray Company, Japan. Present capacity is 7500 ton yr^{-1}, increasing from 3000 (1980) to 4000 ton yr^{-1} (1981). This process uses resting cells, and high grade L-lysine is obtained by carbon treatment and crystallization. The Toray process uses DL-aminocaprolactam, with an approximate conversion cost of 140–160 yen kg^{-1}, including purification. A schematic illustration of the enzymatic process is shown in Figure 12.

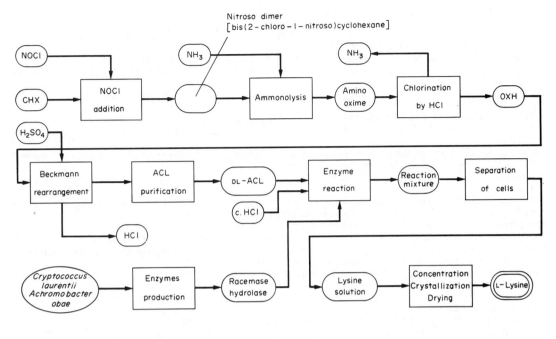

Figure 12 Process flow diagram for enzymatic L-lysine production

CHX : cyclohexane
OXH : α-aminocyclohexane oxime
ACL : aminocaprolactam

28.8 SUMMARY

At present a fairly high yield has been attained in lysine production by fermentation using *C. glutamicum*, *B. flavum* or *B. lactofermentum*. Conversion yields of 48% on weight basis from carbohydrate have been attained in the laboratory. However, there still remains room for improvement, even when the expense of carbohydrate for making cells is considered. The theoretical conversion yields calculated for the main pathways of lysine biosynthesis led to the conclusion that the most efficient formation of lysine is *via* oxaloacetate formed by carboxylation of phosphoenolpyruvate (PEP). Therefore, a strain with high PEP carboxylase activity is desirable (Shvinka *et al.*, 1980).

Further genetic improvement of microorganisms based on the knowledge of microbial physiology should give higher yields. Application of recombinant DNA techniques to this effect has been already tried in *E. coli* and *Saccharomyces cerevisiae* (Chenais *et al.*, 1981). Furthermore, application of continuous fermentation and immobilized cells must be studied in detail, especially for lowering the expense of carbon and energy sources for making the cells.

The major producers of L-lysine and historical production levels are shown in Table 3.

Table 3 Major Producers of L-Lysine and the Amounts
Produced (tons)

Supplier	1972	1977	1982
Kyowa (Japan)	7000	7300 (2300)	7000 (1500)
Ajinomoto (Japan)	2000–3000	8500 (1500)	10 000 (1000)
Toray (Japan)	0	0	4000 (500)
Miwon (Korea)	0	2700	5000
Eurolysine (France)	0	6000	10 000
Fermex (Mexico)	0	500	4000
Total	9000–10 000	25 000 (3800)	40 000 (3000)
Price (f.o.b. Japan) ($ kg^{-1})	3.35	3.40	3.30

[a] ( )= Japanese market.

28.9 REFERENCES

Akashi, K., H. Shibai and Y. Hirose (1979). Effect of oxygen supply on L-lysine, L-threonine and L-isoleucine fermentations. *Agric. Biol. Chem.*, **43**, 2087–2092.

Chenais, J., C. Richaud, J. Ronceray, H. Cherest, Y. Surdin-Kerjan and J.-C. Patte (1981). Construction of hybrid plasmids containing *lyse A* gene of *Escherichia coli*: studies of expression in *Escherichia coli* and *Saccharomyces cerevisiae*. *Mol. Gen. Genet.*, **182**, 456–461.

Fukumura, T. (1976a). Screening, classification and distribution of α-amino-ε-caprolactam-hydrolyzing yeasts. *Agric. Biol. Chem.*, **40**, 1687–1693.

Fukumura, T. (1976b). Hydrolysis of L-α-amino-ε-caprolactam by yeasts. *Agric. Biol. Chem.*, **40**, 1695–1698.

Kubota, K., O. Tosaka, Y. Yoshihara, H. Hirakawa and Y. Hirose (1976). Fermentative production of L-lysine. *Jpn. Kokai* 76 19 186.

Nakayama, K. (1972). Lysine and diaminopimelic acid. In *The Microbial Production of Amino Acids*, ed. K. Yamada, S. Kinoshita, T. Tsunoda and K. Aida, pp. 369–397. Kodansha, Tokyo.

Nakayama, K. (1978). Amino acids. In *Economic Microbiology*, vol. 2, ed. A. H. Rose, pp. 209–261. Academic, London.

Nakayama, K. (1982). Amino acids. In *Prescott and Dunn's Industrial Microbiology*, ed. G. Reed, 4th edn. AVI Publishing Co., pp. 748–801. AVI Publishing Co., Westport, Connecticut.

Nakayama, K. and K. Araki (1973). Process for producing L-lysine. *US Pat.* 3 708 395.

Nakayama, K. and K. Araki (1981). Process for producing L-lysine by fermentation. *Jpn. Kokai* 81 8692.

Sato, E., Y. Kawahata and T. Fukumura (1974). Production of L-aminocaprolactam hydrolase and aminocaprolactam racemase. *Abstr. Symp. Amino Acid Nucleic Acid Assoc.*, *23rd, Tokyo*, p. 2.

Shiio, I., H. Ozaki and K. Ujigawa-Takeda (1982). Production of aspartic acid and lysine by citrate synthetase mutants of *Brevibacterium flavum*. *Agric. Biol. Chem.*, **46**, 101–107.

Shvinka, J., U. Viesturs and M. Ruklisha (1980). Yield regulation of lysine biosynthesis in *Brevibacterium flavum*. *Biotechnol. Bioeng.*, **22**, 897–912.

Tosaka, O. and K. Takinami (1982). L-Lysine fermentation. *Hakko To Kogyo*, **40**, 110–127.

Tosaka, O., K. Takinami and Y. Hirose (1978a). Production of L-lysine by leucine auxotrophs derived from AEC resistant mutant of *Brevibacterium lactofermentum*. *Agric. Biol. Chem.*, **42**, 1181–1186.

Tosaka, O., H. Hirakawa, K. Takinami and Y. Hirose (1978b). Regulation of lysine biosynthesis by leucine in *Brevibacterium lactofermentum*. *Agric. Biol. Chem.*, **42**, 1501–1506.

Tosaka, O., H. Hirakawa, Y. Yoshimura, K. Takinami and Y. Hirose (1978c). Production of L-lysine by alanine auxotrophs derived from AEC resistant mutant of *Brevibacterium lactofermentum*. *Agric. Biol. Chem.*, **42**, 1773–1778.

Tosaka, O., H. Hirakawa and K. Takinami (1979a). Effect of biotin levels on L-lysine formation in *Brevibacterium lactofermentum*. *Agric. Biol. Chem.*, **43**, 491–495.

Tosaka, O., H. Morioka and K. Takinami (1979b). The role of biotin-dependent pyruvate carboxylase in L-lysine production. *Agric. Biol. Chem.*, **43**, 1513–1519.

Tosaka, O., E. Ono, M. Ishihara, H. Morioka and K. Takinami (1981). Method for the production of L-lysine. *US Pat.* 4 275 157.

Umbarger, H. E. (1978). Amino acid biosynthesis and its regulation. *Annu. Rev. Biochem.*, **47**, 533–606.

29

Tryptophan

K. NAKAYAMA
Bior Inc., Kanagawa, Japan

29.1 INTRODUCTION

L-Tryptophan was first isolated in 1902 by Hopkins and Cole. It was the first amino acid that was proved to be essential for human and animal nutrition, though it was later found that the requirement for it is only one half to one fifth as much as that for other essential amino acids. Although D-tryptophan is partly converted to the L-form *in vivo*, its nutritional value is much lower than that of the L-form.

At present, L-tryptophan is mainly used therapeutically, especially as a component of solutions for transfusion, but because it is an essential nutrient for humans and animals and various cereals such as corn are deficient in it, there is potential for its use as a food- and feed-supplement.

Two types of processes are being studied for the production of L-tryptophan. One is the conversion of precursors such as indole or anthranilic acid into L-tryptophan using suitable microorganisms or an enzyme preparation. In this type of process fermentative conversion and conversion with immobilized enzymes or immobilized cells are employed. Genetic improvement of the microorganisms using recombinant DNA techniques is being examined. The other process is a direct fermentation method, namely, production of L-tryptophan from carbon and nitrogen sources such as glucose and ammonia. For this purpose, mutant strains have been bred from bacteria such as *Corynebacterium glutamicum* and *Brevibacterium flavum*.

29.2 BIOSYNTHESIS OF TRYPTOPHAN AND ITS REGULATION

A general outline of the pathways consists of a 'common pathway' leading through shikimate to chorismate, after which there is branching to the specific pathways: to the three amino acids, to *p*-aminobenzoate, to menadione, to ubiquinone and to enterochelin (Figure 1).

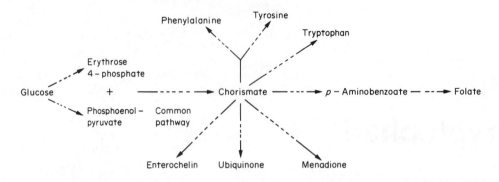

Figure 1   General outline of pathways for the formation of aromatic amino acids and vitamins in *Escherichia coli*

29.2.1   The Common Aromatic Pathway

Aromatic biosynthesis (Gibson *et al.*, 1968; Umbarger, 1978) begins with the condensation of phosphoenolpyruvate and erythrose 4-phosphate to yield 3-deoxy-D-arabinoheptulosonate 7-phosphate (DAHP; reaction i, Figure 2). DAHP is cyclized upon the removal of phosphate to yield 5-dehydroquinate (reaction ii). The enzyme catalyzing this reaction requires NAD as a cofactor. Although there is no net oxidation or reduction in the overall reaction, there is an internal redox change; what had been carbon 7 of DAHP is reduced and what had been carbon 6 is oxidized (carbons 6 and 5 of 5-dehydroquinate). Removal of water and an NADPH-dependent reduction yield shikimate (reactions iii and iv). Shikimate is phosphorylated and condensed with another molecule of phosphoenolpyruvate (reactions v and vi). A second double bond is generated upon removal of the ring phosphate to yield the branch point compound, chorismate (reactions v to vii).

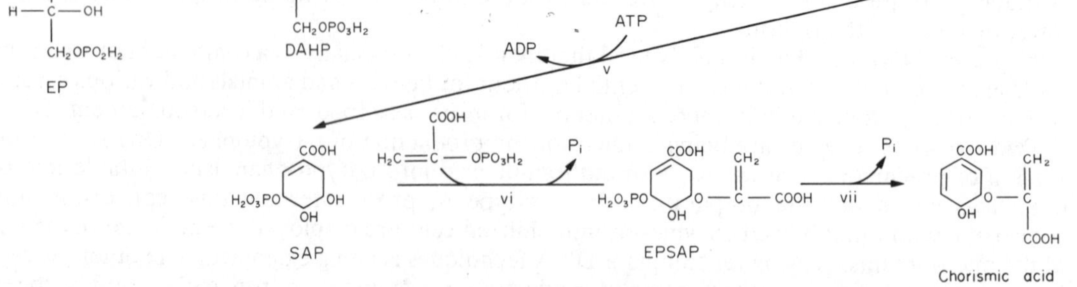

Figure 2   Intermediates in the common pathway of aromatic biosynthesis. Abbreviations: PEP, phosphoenolpyruvate; EP, erythrose 4-phosphate; DAHP, 3-deoxy-D-arabinoheptulosonic acid 7-phosphate; DHQ, 5-dehydroquinic acid; DHS, 5-dehydroshikimic acid; SAP, shikimic acid 5-phosphate; EPSAP, 3-enolpyruvylshikimic acid 5-phosphate. Trivial names of enzymes: i, 3-deoxy-D-arabinoheptulosonate-7-phosphate synthetase (DAHP synthetase); ii, 5-dehydroquinate synthetase; iii, dehydroquinase; iv, dehydroshikimate reductase; v, shikimate kinase; vi, 3-enolpyruvylshikimate-5-phosphate synthetase; vii, chorismate synthetase

In *Escherichia coli* and *Salmonella typhimurium* there are three DAHP synthetases: one is inhibited by phenylalanine, one by tyrosine and one by tryptophan. The same isozymic pattern has also been found in *Neurospora crassa*. In contrast to this isozymic pattern in *Bacillus subtilis* there is only a single DAHP synthetase activity and it is carried on the same protein that carries chorismate mutase. This protein (subunit A) forms a complex with another protein (subunit B) that has almost no activity by itself but which with subunit A exhibits shikimate kinase activity. Both shikimate kinase and DAHP synthetase activities are inhibited by chorismate and prephenate. That the inhibition may be due to binding of chorismate or prephenate to the active site of the mutase is suggested by the fact that proteolytic cleavage of the chorismate mutase fragment from the complex was accompanied by a loss of feedback sensitivity of DAHP synthetase.

In *Corynebacterium glutamicum* there is a strong synergistic inhibition of DAHP synthetase by phenylalanine and tyrosine. This inhibition was increased still further by the presence of exogenous tryptophan (nearly 90%). Still another pattern of inhibition found in organisms with single DAHP synthetase is that in which the enzyme is inhibited by only a single aromatic amino acid. In *Streptomyces aureofaciens* the enzyme is inhibited by tryptophan alone. The single enzyme in *Pseudomonas aeruginosa* is somewhat similar in that tyrosine is the most effective inhibitor, although phenylpyruvate and tryptophan are also inhibitors.

Repression of enzyme formation in the common aromatic pathway has again been studied primarily in the Enterobacteriaceae. In *E. coli*, in addition to the tryptophan-specific repression of the tryptophan-sensitive DAHP synthetase and the tyrosine-specific repression of the tyrosine-sensitive enzyme, there was a multiple repression of the phenylalanine-sensitive enzyme by phenylalanine plus tryptophan. The repression of the tryptophan and tyrosine-sensitive isozymes was dependent upon the *trpR* and *tyrR* loci, respectively. The *tyrR* gene specifies a repressor for the *tyr* regulon. The repression of the phenylalanine plus tryptophan-repressible isozyme also involves the participation of *tyrR* but not *trpR*. No *pheR* gene has been found in *E. coli*, but in *S. typhimurium*, in which there is a *pheR* gene, it does not affect the multiply repressed, phenylalanine-sensitive isozyme. At least five of the other enzymes catalyzing the remaining steps of the common aromatic pathway are not end-product controlled.

29.2.2 The Tryptophan Pathway

Tryptophan is formed *via* the same sequence of reactions in all organisms, but the distribution of activities as separate proteins is variable. The pathway is initiated by the conversion of chorismate to anthranilate in a glutamine-dependent reaction (reaction i, Figure 3). The phosphoribosyl moiety of phosphoribosyl pyrophosphate is transferred to anthranilate (reaction ii). The indole ring is formed in two steps involving first an isomerization converting the ribose group to a ribulose and then a cyclization reaction to yield indoleglycerol phosphate (reactions iii and iv). The final reaction in the pathway is always catalyzed by a single enzyme which may contain either one or two types of subunit and consists of the cleavage of indoleglyceraldehyde 3-phosphate and condensation of the indole group with serine (reaction v).

The control of metabolite flow in the tryptophan pathway occurs by the inhibition of anthranilate synthetase by tryptophan. The enzyme has been extensively studied in *E. coli* and *S. typhimurium* with respect to both the structure of the protein and the interaction with substrates and inhibitors. The anthranilate synthetase of both organisms consists of two components. One, anthranilate synthetase component I (CoI), can by itself catalyze the formation of anthranilate with ammonia as substrate. For the more efficient glutamine-dependent reaction, anthranilate synthetase component II (CoII) activity is also needed. CoII activity (amido transferase activity) resides in a portion of anthranilate phosphoribosyltransferase, the enzyme catalyzing reaction ii (Figure 3). Thus, anthranilate is a complex of the two kinds of proteins.

The CoII activity in some other organisms is associated with other proteins. For example, in *N. crassa*, it is carried on the same protein that converts phosphoribosylanthranilate in two steps to indoleglycerol phosphate. In *B. subtilis*, *Serratia marcescens* and *Pseudomonas putida*, the amido transferase is carried on a separate protein apparently exhibiting no other activity. The anthranilate synthetase from *B. subtilis* could use only ammonia for the formation of anthranilate. Thus it appeared that anthranilate synthetase could actually function *in vivo* with ammonia.

Genetic studies on tryptophan biosynthesis in both *E. coli* and *S. typhimurium* have demonstrated a regulatory gene (*trpR*) that was thought to specify the repressor for the *trp* operon, and a *trpO* gene was thought to be the site of the repressor interaction. There was nearly a coordinate control over the levels of the five enzymes except under extreme tryptophan starvation conditions and under conditions of full repression. A deviation from coordinate repression under repressing conditions also occurs and is attributed in both *E. coli* and *S. typhimurium* to the presence of a

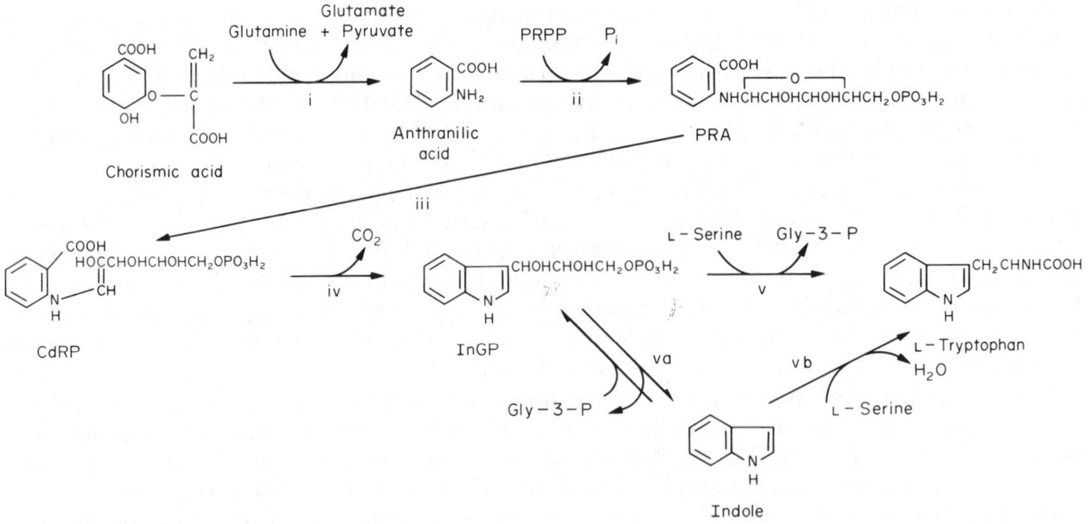

Figure 3 Intermediates in the tryptophan pathway. Abbreviations: PRPP, phosphoribosyl pyrophosphate; PRA, *N*-5'-phosphoribosylanthranilate; CdRP, 1-(*O*-carboxyphenylamino)-1-deoxyribulose 5-phosphate; InGP, indoleglycerol phosphate; Gly-3-P, glyceraldehyde 3-phosphate. Trivial names of enzymes: i, anthranilate synthetase; ii, anthranilate phosphoribosylpyrophosphate phosphoribosyltransferase; iii, *N*-5'-phosphoribosylanthranilate isomerase; iv, indole-3-glycerol-phosphate synthetase; v (a or b), tryptophan synthetase (A or B)

'low level' promoter within the *trpD* gene of *E. coli* (*trpB* gene of *S. typhimurium*) that allows the unregulated expression of the last three genes in the operon.

Relative to what has been known concerning the regulation of formation of the tryptophan biosynthetic enzymes in *E. coli*, little is known in other organisms. In view of the differences between the gene–enzyme relationships in various organisms, some differences between mechanisms of enzyme regulation should be expected. For example, in *P. putida*, phosphoribosylanthranilate isomerase and indoleglycerol-phosphate synthetase are specified by separate genes. The gene for the isomerase, *trpF*, is alone and separate from the *trp* gene cluster, and is apparently constitutive. The two genes for tryptophan synthetase are also separate from the *trp* gene cluster and are induced by indole or indoleglycerol phosphate. The regulatory molecule for this induction is the *trpA* product itself. The regulation of tryptophan synthetase is thus autogenous. On the other hand, the regulation of the cluster of four genes is under the influence of a regulatory gene, *trpR*, but their expression is not coordinate.

It has been proposed that in *B. subtilis*, the entire tryptophan pathway is induced by chorismate. A single regulatory locus, *mtr*, has been described, which upon mutation leads to constitutive enzyme formation and excretion of tryptophan.

The regulation of tryptophan biosynthesis in *N. crassa* has been shown to be achieved at least in part by a kind of general amino acid control. Upon derepression of the tryptophan biosynthetic pathway with limiting tryptophan, there was a concurrent derepression of the histidine and arginine biosynthetic enzymes.

29.2.3 Regulation of Tryptophan Biosynthesis in *Brevibacterium flavum*

Regulation of aromatic biosynthesis in *Brevibacterium flavum* is similar to that of *Corynebacterium glutamicum*. Because of the industrial importance of these bacteria and of the more detailed studies of the former, the results with *B. flavum* will be described here.

In *B. flavum* DAHP synthetase of the common pathway is feedback-inhibited by only two of the three end products, phenylalanine and tyrosine (Shiio and Sugimoto, 1981). Other amino acids and intermediate metabolites, including tryptophan, chorismate and prephenate, were not inhibitory. Formation of DAHP synthetase was affected neither by tyrosine, phenylalanine and tryptophan alone nor by a combination of the three amino acids (Shiio *et al.*, 1974). In a phenylalanine auxotroph under phenylalanine-limiting conditions, overproduction of tyrosine strongly repressed the enzyme, while in a tyrosine auxotroph under tyrosine-limiting conditions, enzyme production was slightly derepressed. Tryptophan markedly reversed the repressive action of tyrosine.

The formation of DAHP synthetase was strongly repressed by overproduced tyrosine inside the cells under phenylalanine-limiting conditions, but formation of DAHP was only weakly repressed by addition of excess tyrosine to the culture medium, probably due to low permeability of the cells to tyrosine. It was also considerably derepressed under tyrosine-limiting conditions, therefore it seems that the DAHP synthetase of *B. flavum* is repressed by tyrosine but not by phenylalanine. Tyrosine at a normal level in wild-type cells seems to repress the enzyme only partially unless the tryptophan level is markedly reduced. Furthermore, the strong repressive action of overproduced tyrosine was also reversed by the addition of tryptophan. Thus, tryptophan does not repress enzyme formation but reverses the repressive action of tyrosine in *B. flavum*.

Repression of DAHP synthetase by tyrosine does not seem to be a significant mechanism of control of biosynthesis of aromatic amino acids under normal conditions *in vivo* but may be useful for reducing tyrosine overproduction under phenylalanine-limiting conditions, or in combination with the tryptophan effect, to improve the balance of the phenylalanine and tyrosine levels in the cells, corresponding to various tryptophan levels which may be established under special conditions.

Chorismate mutase is composed of two inactive components (A, B) which are separated from each other by gel-filtration but associate to form an active complex in the presence of the substrate (Shiio and Sugimoto, 1979). Component A is a bifunctional enzyme and acts as DAHP synthetase (Sugimoto and Shiio, 1980).

Anthranilate synthetase from *Brevibacterium flavum* required chorismate, glutamine and Mg^{2+} for its activity (Shiio *et al.*, 1972). The enzyme was strongly and specifically inhibited by tryptophan, the metabolic end product. The inhibition by tryptophan was competitive with respect to chorismate and uncompetitive with respect to glutamine. In the presence of tryptophan, homotropic cooperativity of chorismate was observed. D-Tryptophan, phenylalanine, tyrosine, histidine and indole scarcely affected the enzyme activity.

Among the six tryptophan enzymes, two dissociable enzyme complexes were observed; one included tryptophan synthetase A (TS-A) and tryptophan synthetase B (TS-B), the other *N*-5′-phosphoribosylanthranilate isomerase (reaction iii; Figure 3) and indole-3-glycerol-phosphate synthetase (reaction iv; Figure 3; Sugimoto and Shiio, 1977). When tryptophan auxotrophs derived from *B. flavum* which lacked TS-A, TS-B or anthranilate phosphoribosylpyrophosphate phosphoribosyltransferase (reaction ii; Figure 3) were cultured in media containing limiting or excess tryptophan, the specific activities of all six tryptophan enzymes were much higher under conditions of limiting tryptophan than under conditions of excess tryptophan. This indicates that the formation of all the tryptophan enzymes was repressed by tryptophan. However, the ratio of specific activities under the two conditions was not constant among these enzymes. When the rates of synthesis of the enzymes during tryptophan starvation were compared, those of anthranilate synthetase (reaction i; Figure 3), *N*-5′-phosphoribosylanthranilate isomerase (reaction iii; Figure 3), TS-A and TS-B were coordinate, while the specific activity of indole-3-glycerol-phosphate synthetase (reaction iv; Figure 3) did not increase significantly.

In the biosynthetic pathway of aromatic amino acids of *B. flavum*, the ratios of each biosynthetic flow at the chorismate branch-point were calculated from the reaction velocities of anthranilate synthetase for tryptophan, and chorismate mutase for phenylalanine and tyrosine at steady state concentrations of chorismate. When these aromatic amino acids were absent, the ratio was 61, showing an extremely preferential synthesis of tryptophan (Shiio and Sugimoto, 1981).

Since tryptophan does not take part in the feedback control of DAHP synthetase, the key enzyme of the common pathway for aromatic amino acid biosynthesis in *B. flavum*, the results are consistent with a general regulatory mechanism that includes preferential synthesis. The preferential synthesis of tryptophan is also consistent with the evidence that, *in vivo*, mutants lacking regulation in the tryptophan-specific branch overproduce tryptophan and that their overproduction is depressed by the addition of phenylalanine and tyrosine.

Prephenate aminotransferase, the first enzyme of the tyrosine-specific branch, was not regulated at all by tyrosine nor by the other aromatic amino acids, while prephenate dehydratase, the enzyme for phenylalanine synthesis, was inhibited, but not repressed, by phenylalanine and activated by tyrosine. Moreover, this inhibition by phenylalanine was competitively recovered by the presence of tyrosine. The ratio of each biosynthetic flow for tyrosine and phenylalanine at the prephenate branch-point was calculated from the kinetic equations of prephenate aminotransferase and prephenate dehydratase, the first enzyme in the phenylalanine-specific branch. It showed that tyrosine was synthesized in preference to phenylalanine when phenylalanine and tyrosine were absent. Furthermore, this preferential synthesis was diverted to a balanced synthesis of phenylalanine and tyrosine through activation of prephenate dehydratase by the tyrosine thus synthesized. The feedback inhibition of prephenate dehydratase by phenylalanine was proposed to play a role in maintaining a balanced synthesis when supply of prephenate was decreased by feedback inhibition of DAHP synthetase, the common key enzyme. Overproduction of the end products in various regulatory mutants was also explained by these results. Figure 4 summarizes the major regulatory mechanisms for aromatic amino acid biosynthesis in *B. flavum* (Shiio and Sugimoto, 1981).

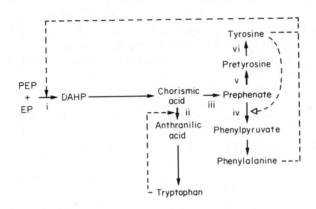

Figure 4   Regulation of aromatic amino acid biosynthesis in *Brevibacterium flavum*. Trivial names of enzymes and symbols: i, DAHP synthetase; ii, anthranilate synthetase; iii, chorismate mutase; iv, prephenate dehydratase; v, prephenate aminotransferase; vi, pretyrosine dehydrogenase; ————▶, feedback inhibition; ————▷, activation

29.3   TRYPTOPHAN PRODUCTION BY FERMENTATION

29.3.1   *Corynebacterium glutamicum*

Figure 5 shows the genealogy of L-tryptophan-producing mutants of *Corynebacterium glutamicum*. Mutants producing a large amount of L-tryptophan were derived from a phenylalanine and tyrosine double auxotroph of *C. glutamicum* KY 9456 which produced only a trace amount of L-tryptophan and anthranilate (Nakayama *et al.*, 1976). A mutant (4MT-11), which, in a stepwise manner, acquired resistance to 5-methyltryptophan (5MT), tryptophan hydroxamate (TrpHx), 6-fluorotryptophan (6FT) and 4-methyltryptophan (4MT), produced L-tryptophan to a concentration of 4.9 g l^{-1} in a cane molasses medium containing 10% reducing sugar as invert. L-Tryptophan production with this mutant was inhibited by L-phenylalanine and L-tyrosine. Accordingly, mutants resistant to phenylalanine and tyrosine analogs, such as *p*-fluorophenylalanine (PFP), *p*-aminophenylalanine (PAP), tyrosine hydroxamate (TyrHx) and phenylalanine hydroxamate (PheHx), were derived from this mutant. One of the mutants thus obtained (Px-115-97) produced 12 g l^{-1} of L-tryptophan in the molasses medium. The medium used had the following composition: 10% reducing sugars as invert (as cane molasses), 0.05% KH_2PO_4, 0.05% K_2HPO_4, 0.025% $MgSO_4 \cdot 7H_2O$, 2% $(NH_4)_2SO_4$, 1% corn-steep liquor and 2% $CaCO_3$ (pH 7.2). Production of L-tryptophan with the mutant was still sensitive to L-phenylalanine and L-tyrosine. Hence, further genetic improvement of the strain may be possible.

29.3.2   Mechanism of Overproduction of Aromatic Amino Acids

Regulatory properties of the enzyme involved in aromatic amino acid biosynthesis in *C. glutamicum* wild and mutant strains were investigated (Nakayama *et al.*, 1976). The overall control pattern (Figure 6) is a new addition to the list of control patterns in aromatic amino acid biosynthesis in microorganisms. A phenylalanine and tyrosine double auxotrophic L-tryptophan producer, Px-115-97, has anthranilate synthetase partially released from the inhibition by L-tryptophan and DAHP synthetase of a wild type. L-Tryptophan production by the mutant

	L-Tryptophan produced (g l^{-1})
KY9456 Phe$^-$, Tyr$^-$	0.15
↓ 5MT^r, TrpHx^r, 6FT^r, 4MT^r	
4MT-11	4.9
↓ PFP^r	
PFP-2-32	5.7
↓ PAP^r	
PAP-126-50	7.1
↓ TyrHx^r	
Tx-49	10.0
↓ PheHx^r	
Px-115-97	12.0

Figure 5 Genealogy of L-tryptophan-producing mutants of *Corynebacterium glutamicum* and their L-tryptophan productivity. Medium used for production: cane molasses containing 10% reducing sugar as invert. Abbreviations: 5MT, 5-methyltryptophan; TrpHx, tryptophan hydroxamate; 6FT, 6-fluorotryptophan; 4MT, 4-methyltryptophan; PFP, *p*-fluorophenylalanine; PAP, *p*-aminophenylalanine; TyrHx, tyrosine hydroxamate; PheHx, phenylalanine hydroxamate

appeared to be caused by the release from the feedback inhibition of anthranilate synthetase by L-tryptophan and blockage of chorismate mutase.

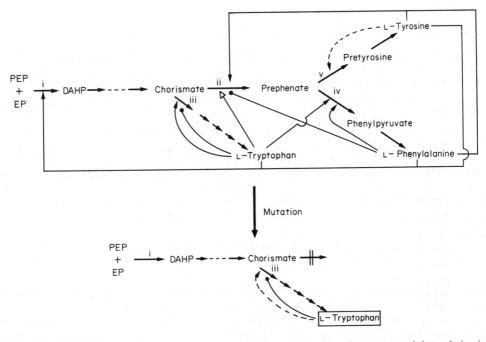

Figure 6 Regulation in aromatic amino acid biosynthesis in *Corynebacterium glutamicum* and deregulation in tryptophan producing mutant. Abbreviations: PEP, phosphoenolpyruvate; EP, erythrose 4-phosphate; DAHP, 3-deoxy-D-arabinoheptulosonic acid 7-phosphate. Trivial names of enzymes and symbols: i, DAHP synthetase; ii, chorismate mutase; iii, anthranilate synthetase; iv, prephenate dehydratase; v, pretyrosine dehydrogenase; ──▶, feedback inhibition; ----▶, partial inhibition; ──▷, activation; ▭, overproduced metabolite; ─╫▶, blocked reaction; ──●, repression

29.3.3 *Brevibacterium flavum*

A 5-fluorotryptophan (5FT) resistant histidine auxotrophic mutant of *Brevibacterium flavum* produced 2.4 g l^{-1} of L-tryptophan. Production was increased to 3.8 g l^{-1} using a *m*-fluorophenylalanine (MFP) resistant mutant derived from the 5FT resistant one. A phenylalanine auxotrophic mutant from the latter mutant produced 6.2 g l^{-1} of L-tryptophan (Shiio *et al.*, 1975).

A 5-FT-resistant mutant, No. 187, derived from a tyrosine auxotrophic *p*-fluorophenylalanine (PFP) resistant mutant produced 8.0 g l^{-1} of L-tryptophan. Addition of resistance to a substrate analog of anthranilate synthetase, azaserine, gave higher tryptophan productivity. Thus an azaserine resistant mutant A-100 produced 10.3 g l^{-1} of L-tryptophan (Shiio *et al.*, 1982). The production increased to 11.4 g l^{-1} when L-serine was added. In the mutant anthranilate synthetase activity increased to a level twice that in the parent strain though it was inhibited in a similar manner to that of the parent. Furthermore, DAHP synthetase increased three-fold and was less sensitive to feedback inhibition by phenylalanine and tyrosine.

29.3.4 *Bacillus subtilis* and *Enterobacter cloacae*

A 5FT-resistant mutant derived from *B. subtilis* produced 4 g l^{-1} of each of L-tryptophan and L-phenylalanine. A leucine auxotroph derived from the mutant produced 6.15 g l^{-1} of L-tryptophan in a medium containing 300 μg ml^{-1} of L-leucine (Shiio *et al.*, 1973). Suppressor prototrophic revertants from the No. 149 *rec*$^{-}$ histidine auxotroph of *B. subtilis* produced L-tryptophan in the medium. Elimination of *rec*$^{-}$ mutation from the gene using genetic transformation resulted in an increase in tryptophan productivity by 60%. This strain produced 5–6 g l^{-1} of L-tryptophan (Alikhanian, 1972).

A mutant of *Enterobacter cloacae* TA 599 which is resistant to 5-methyltryptophan, β-2-thienylalanine (an analog of phenylalanine) and streptomycin and requires tyrosine produced 9.8 g l^{-1} of L-tryptophan in a medium containing 12% glucose (Arima *et al.*, 1976).

29.4 PRODUCTION BY MICROBIAL CONVERSION

Terui *et al.* selected *Hansenula anomala* for microbial production of L-tryptophan using precursors such as anthranilic acid or indole. With *Hansenula anomala*, L-tryptophan production from anthranilic acid reached 5.7 g l^{-1} (Terui, 1972).

Afterwards, various microorganisms including auxotrophic and regulatory mutants were selected by different workers. They include *Candida fumicola*, *Corynebacterium glutamicum*, *Bacillus subtilis*, *Escherichia coli* and other species. By feeding anthranilic acid to a derepressed anthranilic acid auxotroph of *Bacillus subtilis*, 5.5 g l^{-1} of L-tryptophan were produced from 5 g l^{-1} of anthranilic acid (Arima *et al.*, 1971). *Candida utilis* (synonym *Torulopsis utilis*) 295-t produced 6.4 g l^{-1} of L-tryptophan from 4.2 g l^{-1} of anthranilic acid in 36 h (Bekers *et al.*, 1971). By feeding indole, a 5-methyltryptophan-resistant mutant of *B. subtilis* ATCC 21336 produced 10.4 g l^{-1} of L-tryptophan in 96 h with a medium containing 7% glucose (Thieman and Pagani, 1972). A strain carrying an *F Try* episome in addition to the chromosomal *try* operon was obtained by sexduction from a feedback-resistant and derepressed mutant of *E. coli* K12 which is resistant to 5-methyltryptophan (5MT) and 5-fluorotryptophan (5FT). This strain produced 5 g l^{-1} of L-tryptophan as a result of feeding with indole (1.5 g l^{-1}) and L-serine (7 g l^{-1}; Sahm and Zähner, 1971).

A *Bacillus subtilis* mutant resistant to 5-fluorotryptophan produced 9.6 g l^{-1} of L-tryptophan in a medium containing 1% glucose and 5% soluble starch with continuous feeding of anthranilic acid (Fukui *et al.*, 1974). A further improved mutant which is resistant to 5-fluorotryptophan and 8-azaguanine produced 15.6 g l^{-1} of L-tryptophan in a medium containing 10% glucose with feeding of anthranilate solution. Yield of L-tryptophan compared with consumed glucose was 17.4% and conversion of anthranilate into L-tryptophan was 99% (Nakayama *et al.*, 1982).

Recombinant DNA techniques were applied for enhancement of the tryptophan productivity of *E. coli*. A multiple mutant (*trpR*, *traA*) of *E. coli* was transformed with a recombinant plasmid containing a feedback resistant *trp* operon. The transformed strain produced 5.5 g l^{-1} of L-tryptophan by culturing for 24 h in a medium containing 5% glucose with feeding of anthranilic acid (Aiba *et al.*, 1980).

29.5 PRODUCTION BY ENZYMATIC METHODS

Tryptophanase, which catalyzes synthesis of L-tryptophan by reversal of the α,β-elimination reaction at rates similar to the forward reaction, was utilized for production of L-tryptophan and related compounds such as 5-hydroxytryptophan (Nakazawa *et al.*, 1972a). The culture broth of *Proteus rettgeri* (AJ 2770) was used as the enzyme for the reaction. For the synthesis of L-tryptophan, a reaction mixture contained 6.0 g of indole in 10 ml of methanol, 8.0 g of sodium pyruvate, 8.0 g of ammonium acetate, 0.001 g of pyridoxal phosphate, 0.1 g of Na$_2$SO$_4$ and 100 ml of the cultured broth in a total volume of 120 ml. After the pH value of the mixture was adjusted to 8.8 with 6 N KOH, it was incubated at 34 °C for 48 h. Under these conditions, 7.5 g of L-tryptophan were synthesized. Similarly, 5-hydroxy-L-tryptophan was synthesized from 5-hydroxyindole, pyruvate and ammonia.

$$\text{indole} + CH_3COCOOH + NH_3 \rightarrow \text{L-tryptophan} + H_2O$$

When the produced L-tryptophan was removed from the reaction system by precipitation with inosine, 83.3 g l^{-1} of L-tryptophan were produced. The molar conversion yield was 96% (Nakazawa *et al.*, 1972b). More recently *Achromobacter liquidum* was used for L-tryptophan production from indole and L-serine (Ujimaru *et al.*, 1981). High enzyme activity was obtained by growing in a medium containing 0.2% L-tryptophan and 0.5% L-glutamate in addition to other ingredients. The cells thus obtained produced 96.6 g l$^-$ of L-tryptophan after 3 days' reaction with a mixture containing indole 60 g l^{-1}, L-serine 60 g l^{-1}, pyridoxal phosphate 0.5 mM and potassium phosphate buffer 0.1 M (pH 9.0). Crystals of L-tryptophan separated out.

Tryptophan synthetase of *E.coli* was utilized for L-tryptophan production from indole and L-serine or DL-serine (Bang *et al.*, 1978; Asai *et al.*, 1980). When the cells of *Pseudomonas putida* were added to the reaction mixture to racemize serine, 23.5 g l^{-1} of L-tryptophan were produced and conversion of DL-serine into L-tryptophan reached 81% (Asai *et al.*, 1980).

Flavobacterum aminogenes nov. sp. has an intracellular enzyme system which degrades aromatic amino acid hydantoins into corresponding L-amino acids. With intact cells, 50 g l^{-1} as wet base, 10 g l^{-1} of DL-5-indolylmethylhydantoin were consumed and 7.4 g l^{-1} of L-tryptophan (a molar yield of 82%) were produced after 35 h incubation (Sano *et al.*, 1977). Molar conversion of DL-tryptophanhydantoin into L-tryptophan was 100% when 5% each of DL-tryptophanhydantoin and inosine were treated at 40 °C for 100 h with cells of a mutant of *F. aminogenes* in which the activity to break down tryptophan had been reduced (Yokozeki *et al.*, 1976). In this system inosine removes L-tryptophan from the reaction system by forming a complex with it. Spontaneous racemization of the substrate allowed the conversion of the D-form of hydantoin into L-tryptophan. The DL-5-indolylmethylhydantoin-hydrolyzing enzymes were found to be inducible and intracellular (Figure 7).

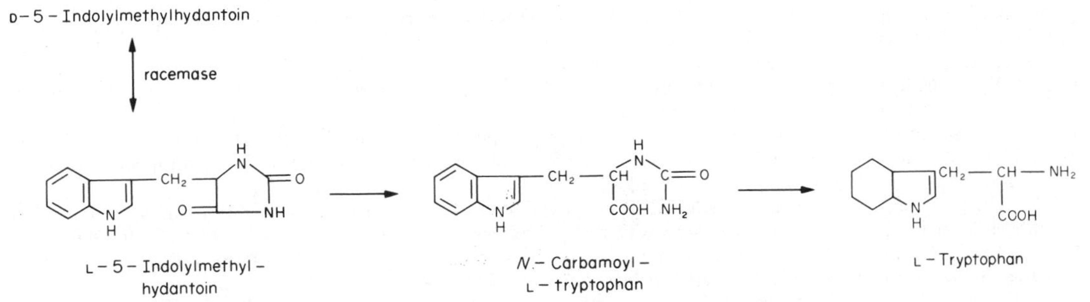

Figure 7 Possible scheme of L-tryptophan production from indolylmethylhydantoin

29.6 RECOVERY

L-Tryptophan in the culture broth is recovered conveniently using ion exchange resin. For example, strongly acidic cation exchange resins (H$^+$ form) adsorb the L-tryptophan in the culture supernatant. After washing, the resin is subjected to elution with 0.5 N aqueous ammonia and then the resulting eluate is concentrated to obtain crude crystals of L-tryptophan. The crude crystals are dissolved in a small amount of hot 50% aqueous ethanol. The resulting solution is decolorized with active carbon and cooled, whereby L-tryptophan is recrystallized.

Non-ionic synthetic polymers like Permtit DR can adsorb aromatic amino acids such as tryptophan, phenylalanine and tyrosine. These characteristics are applied for the selective adsorption of aromatic amino acids from mixtures of amino acids.

29.7 SUMMARY

Various processes including fermentation, microbial conversion of indole or anthranilic acid, and enzymatic methods are now being studied. In enzymatic methods, those using immobilized enzymes or immobilized cells have already been examined (Bang *et al.*, 1978; Decottignies-Le Mareohal *et al.*, 1979; Azerad *et al.*, 1980). Increasing the production from microbial conversion or enzymatic methods is being studied using recombinant DNA techniques. But in these processes, the costs of precursor compounds such as indole, anthranilic acid, pyruvate and serine are generally high. Solving this problem is important. Direct production of L-tryptophan from carbohydrate by fermentation has been developed to a certain level, and production is being carried out to supply products for medical use. To satisfy the expected large market for feed-supplements the present processes must be further improved. Current production levels are given in Table 1. Precise control of biosynthesis by genetic manipulation and culture based on the physiology of the microorganisms will be required to reduce costs sufficiently to allow the use of tryptophan in the feed supplement market.

Table 1 L-Tryptophan Market (Pharmaceutical Use)

	Amount (tons)		
Supplier	*1972*	*1977*	*1982*
Kyowa (Japan)	—	—	50
Ajinomoto (Japan)	—	—	100
Tanabe (Japan)	—	—	50
Total	Not available	80[a]	200[b]

[a] Price 25 yen/kg. [b] Price 26 yen/kg.

29.8 REFERENCES

Aiba, S., T. Imanaka and H. Tsunekawa (1980). Enhancement of tryptophan productivity of *Escherichia coli* as an application of genetic engineering. *Biotechnol. Lett.*, **2**, 525–530.
Alikhanian, S. I. (1972). Genetic methods for the improvement of microbiological fermentation. In *Fermentation Technology Today*, ed. G. Terui, pp. 233–237. Society of Fermentation Technology, Osaka, Japan.
Arima, K., I. Nogami. and M. Yoneda (1971). L-Tryptophan production from anthranilic acid by bacteria. *Abstr. Meet. Agric. Chem. Soc. Jpn., Tokyo*, 153.
Arima, K., I. Nogami and M. Yoneda (1976). Production of L-tryptophan. *Jpn. Pat.* 76 57 888.
Azerad, R., R. Calderon-Seguin and P. Decottignies-Le Mareohal (1980). Production of L-tryptophan by immobilized bacteria. *Bull. Soc. Chim. Fr.*, (1–2, Pr.2), 83–86.
Asai, Y., M. Shimada and K. Soda (1980). Production of L-tryptophan by enzymes. *Jpn. Pat.* 80 148 095.
Bang, W.-G., S. Lang, H. Sahm and F. Wagner (1978). Production of L-tryptophan with free and entrapped *Escherichia coli* cells. *Prepr.-Eur. Congr. Biotechnol., 1st*, 186–189.
Bekers, M., T. Abolins, U. Viesturs, S. Selga, L. Ramina and S. Ozolins (1971). Effect of aeration on L-tryptophan biosynthesis from anthranilic acid by *Candida utilis* (*Torulopsis utilis*) 295-t. *Prik. Biokhim. Mikrobiol.*, **7**, 103–106.
Decottignies-Le Mareohal, P., R. Calderon-Seguin, J. P. Vandecasteele and R. Azerad (1979). Synthesis of L-tryptophan by immobilized *Escherichia coli* cells. *Eur. J. Appl. Microbiol.*, **7**, 33–34.
Fukui, K., Y. Torigoe and T. Akashiba (1974). Production of L-tryptophan by fermentation. *Jpn. Pat.* 74 20 391.
Gibson, F. and J. Pittard (1968). Pathways of biosynthesis of aromatic amino acids and vitamins and their control in microorganisms. *Bacteriol. Rev.*, **32**, 465–492.
Nakayama, K., K. Araki, H. Hagino, H. Kase and H. Yoshida (1976). Amino acid fermentation using regulatory mutants of *Corynebacterium glutamicum*. In *Genetics of Industrial Microorganisms*, ed. K. D. MacDonald, pp. 437–449. Academic, London.
Nakayama, A., A. Murata and T. Akashiba (1982). Production of L-tryptophan. *Jpn. Pat.* 82 5694.
Nakazawa, H., H. Enei, S. Okumura, H. Yoshida and H. Yamada (1972a). Enzymatic production of L-tryptophan and 5-hydroxy-L-tryptophan. *FEBS Lett.*, **25**, 43–45.
Nakazawa, H., H. Enei, S. Okumura and H. Yamada (1972b). Biosynthesis of L-tryptophan and its derivatives. II. Induction of enzyme and reaction condition. *21st Symposium of The Association of Amino Acids and Nucleic Acids*. Abstract, p. 4.
Sahm. H. and H. Zähner (1971). Metabolic products of microorganisms 90. Studies on the formation of tryptophan by *Escherichia coli* K12. *Arch. Microbiol.*, **76**, 223–251.
Sano, K., K. Yokozeki, C. Eguchi, T. Kagawa, I. Noda and K. Mitsugi (1977). Enzymatic production of L-tryptophan from L-5-indolylmethylhydantoin and DL-5-indolylmethylhydantoin by newly isolated bacterium. *Agric. Biol. Chem.*, **41**, 819–825.
Shiio, I. and S. Sugimoto (1979). Two components of chorismate mutase in *Brevibacterium flavum*. *J. Biochem.*, **86**. 17–25.
Shiio, I. and S. Sugimoto (1981). Regulation at metabolic branch points of aromatic amino acid biosynthesis in *Brevibacterium flavum*. *Agric. Biol. Chem.*, **45**, 2197–2207.

Shiio. I., R. Miyajima and M. Nakagawa (1972). Regulation of aromatic amino acid biosynthesis in *Brevibacterium flavum*. I. Regulation of anthranilate synthetase. *J. Biochem.*, **72**, 1447–1455.

Shiio, I., K. Ishii and K. Yokozeki (1973). Production of L-tryptophan by 5-fluorotryptophan resistant mutants of *Bacillus subtilis*. *Agric. Biol. Chem.*, **37**, 1991–2000.

Shiio. I., S. Sugimoto and R. Miyajima (1974). Regulation of 3-deoxy-D-arabinoheptulosonate-7-phosphate synthetase in *Brevibacterium flavum*. *J. Biochem.*, **75**, 987–997.

Shiio, I., S. Sugimoto and M. Nakagawa (1975). Production of L-tryptophan by mutants of *Brevibacterium flavum* resistant to both tryptophan and phenylalanine analogues. *Agric. Biol. Chem.*, **39**, 627–635.

Shiio, I., S. Sugimoto and K. Kawamura (1982). Production of L-tryptophan by azaserine-resistant mutants of *Brevibacterium flavum*. *Agric. Biol. Chem.*, **46**, 1849–1854.

Sugimoto, S. and I. Shiio (1977). Enzymes for the tryptophan synthetic pathway in *Brevibacterium flavum*. *J. Biochem.*, **81**, 823–833.

Sugimoto, S. and I. Shiio (1980). Purification and properties of bifunctional 3-deoxy-D-arabinoheptulosonate-7-phosphate synthetase–chorismate mutase component A from *Brevibacterium flavum*. *J. Biochem.*, **87**, 881–890.

Terui, G. (1972). Tryptophan. In *The Microbial Production of Amino Acids*, ed. K. Yamada, S. Kinoshita, T. Tsunoda and K. Aida, pp. 515–531. Kodansha, Tokyo.

Thieman, J. E. and H. Pagani (1972). Production of L-tryptophan by fermentation. *US Pat.* 3 700 558.

Ujimaru, T., T. Kakimoto and I. Chibata (1981). Production of L-tryptophan using tryptophanase of *Achromobacter liquidum*. *Annual Meeting of the Society of Fermentation Technologists, Japan*, Abstract, p. 61.

Umbarger, H. E. (1978). Amino acid biosynthesis and its regulation. *Annu. Rev. Biochem.*, **47**, 533–606.

Yokozeki, K., K. Sano, A. Eguchi, N. Yasuda, I. Noda and K. Mitsugi (1976). L-Amino acid production from the corresponding hydantoin of aromatic amino acid by enzymic method. *Abstr. Meet. Agric. Chem. Soc. Jpn.*, *Kyoto*, 238.

30
Aspartic Acid

I. CHIBATA, T. TOSA and T. SATO
Tanabe Seiyaku Co. Ltd., Osaka, Japan

30.1 INTRODUCTION

Apartic acid was first obtained by hydrolysis of asparagine isolated from the juice of asparagus, and its chemical structure was confirmed by chemical synthesis in 1887.

The amino acid has received considerable attention as a medicine since the report on its physiological and therapeutic importance (Laborit *et al.*, 1958a). At present, it is widely used, not only in medicines but also as a food additive.

Recently Aspartame®, a dipeptide of L-aspartic acid and L-phenylalanine methyl ester, was commercialized as a low calorie sweetener. Thus, demand for L-aspartic acid is expected to increase rapidly as it is a raw material for the synthesis of the dipeptide.

Initially, L-aspartic acid was prepared by the hydrolysis of asparagine isolated from plant material such as ethiolated seedlings of lupines, by its purification from protein hydrolyzates, and by optical resolution of chemically synthesized DL-aspartic acid.

Quastel and Woolf (1926) discovered that deamination of aspartic acid to fumaric acid took place in the presence of resting cells of *Escherichia coli*. The enzyme responsible for this reaction was subsequently designated as 'aspartase' (Woolf, 1929). This enzyme can also catalyze the reverse reaction. Virtanen and Tarnanen (1932) obtained a cell-free aspartase preparation from dried cells of *Pseudomonas fluorescens*. This was the first enzyme isolated from microbial cells and the first which could catalyze the synthesis of an amino acid. Recently, the enzyme has been purified from extracts of several microorganisms by different workers (Sekijo *et al.*, 1965; Rudolph and Fromm, 1971; Suzuki *et al.*, 1973) and its enzymatic and physicochemical properties have been investigated in detail (Watanabe *et al.*, 1981). Since 1958, industrial production of L-aspartic acid has been carried out by fermentative or enzymatic methods from fumaric acid and ammonia. Thus, the supply of large quantities of L-aspartic acid became possible. The enzymatic reaction is shown in equation (1).

$$HO_2CCH=CHCO_2H + NH_3 \underset{aspartase}{\rightleftharpoons} L\text{-}HO_2CCH_2CH(NH_2)CO_2H \qquad (1)$$

Fumaric acid   L-Aspartic acid

In 1973, continuous industrial production of L-aspartic acid from fumaric acid and ammonia using immobilized microbial cells with high aspartase activity was reported by the Tanabe Seiyaku Co.

In this chapter, the conventional fermentative and enzymatic productions of L-aspartic acid, the continuous production of L-aspartic acid from ammonium fumarate using immobilized enzyme and microbial cells, and the utilization of L-aspartic acid are reviewed.

## 30.2	PRODUCTION OF L-ASPARTIC ACID

The methods for industrial production of L-aspartic acid can be classified into two categories: fermentative and enzymatic.

### 30.2.1	Fermentative Production

Many papers have been published on the aspartic acid fermentation using glucose or fumaric acid as main carbon sources. In practice, industrial production of L-aspartic acid has been carried out by the fermentation method using fumaric acid.

#### 30.2.1.1	*Use of glucose as a main carbon source*

One of the several patents in this area describes a glutamate auxotroph strain derived from *Brevibacterium flavum* (Uchino *et al.*, 1976). When this strain was cultured in a nutrient medium containing 3.6% glucose and 0.5% L-glutamic acid at 30 °C for 4 days, the final concentration of L-aspartic acid in the medium was 4 g l^{-1}.

A prototrophic revertant derived from a citrate synthase-defective glutamate auxotroph of *Brevibacterium flavum* has been employed for the overproduction of L-aspartic acid (Shiio *et al.*, 1982). The revertant showed low citrate synthase activity and overproduced L-aspartic acid. When the revertant was cultured in a nutrient medium containing 36 g l^{-1} of glucose at 30 °C for 48 h, the maximum production was 10.6 g l^{-1} corresponding to a yield of approximately 30% from glucose.

Conversion of fumaric acid to aspartic acid by fermentation has been investigated in mixed cultures of *Rhizopus* sp., producing fumaric acid from glucose, and *Proteus vulgaris* having high aspartase activity (Takao and Hotta, 1972; Hotta and Takao, 1973). In the fermentation, the yield of L-aspartic acid was found to depend on the amount of fumaric acid accumulated in the initial fermentation process.

As described above, however, fermentation using glucose as a main carbon source is not efficient for industrial production of L-aspartic acid. Consequently, this fermentation procedure has not been commercialized.

#### 30.2.1.2	*Use of fumaric acid as a main carbon source*

In the course of screening for amino acid-producing microorganisms, Kinoshita *et al.* (1958) found that a number of them belonging to *Bacillus* sp. had the ability to form L-aspartic acid from fumaric acid. Among the microorganisms tested, a strain of *Bacillus megaterium* was chosen as the most suitable for production of L-aspartic acid, and its cultural conditions for the formation of L-aspartic acid from fumaric acid were investigated. The strain was precultured in a medium containing 0.5% fumaric acid for 24 h and a fumaric solution neutralized with ammonia was then added to the fermentative broth. When the broth was further incubated for 72 h, L-aspartic acid was accumulated at the level of 10.6 g l^{-1} and obtained at near 80% yield from fumaric acid.

Kisumi *et al.* (1960) screened a number of microorganisms accumulating L-aspartic acid from fumaric acid and ammonium salts, and selected *Pseudomonas fluorescens* 6009-2 and a mutant, Ki-1023, derived from *Escherichia coli* K-12 as suitable strains for production of L-aspartic acid. These were investigated further and their productivities compared. The results are shown in Figure 1. In the case of *P. fluorescens*, fumaric acid was completely consumed in 3 days, and conversion of fumaric acid to L-aspartic acid was greater than 95% on the partially stationary culture, *i.e.* stationary culture after shaking for one day. In the case of *E. coli*, fumaric acid was completely consumed in 3 days and the maximum conversion of fumaric acid to L-aspartic acid was attained in 4 days. The maximum yield of L-aspartic acid almost reached the theoretical value of 56 g l^{-1} under partially stationary culture. This method was improved by using the mutant

Ki-1023 derived from *E. coli* K-12 and by feeding fumaric acid and ammonia during the cultivation of these bacteria, and from 1960 to 1973 L-aspartic acid was industrially produced by this improved method at the Tanabe Seiyaku Co. in Japan.

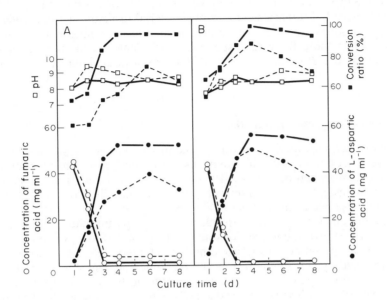

Figure 1 Time course of fermentation by *Pseudomonas fluorescens* (A) and *Escherichia coli* K-12, Ki-1023 (B). Fermentation was carried out in a nutrient medium containing 6.5% fumaric acid at 37 °C (——— stationary, – – – shaken)

30.2.2 Enzymatic Production

Enzymatic production of L-aspartic acid from fumaric acid and ammonia by the action of aspartase can be classified as either batch or continuous processes.

30.2.2.1 Batchwise enzyme reaction

The batchwise reaction is essentially similar to the above-mentioned fermentative method using fumaric acid as a main carbon source, but in this case there is no accompanying growth of microorganisms.

Sumiki (1928) reported the preparation of L-aspartic acid from sodium fumarate and ammonium chloride using resting yeast cells. Since 1958, studies on enzymatic production of L-aspartic acid on an industrial scale have become very active.

Production of L-aspartic acid from ammonium fumarate using dried cells of *E. coli* No. 2 having high aspartase activity was investigated (Kitahara *et al.*, 1960). In this case, 0.5 g of the dried cells were added to 100 ml of 20% ammonium fumarate solution (pH 7.2–7.4), the mixture was incubated at 37 °C for 18 h and L-aspartic acid was obtained with a yield of 88.1% from fumaric acid. In this experiment, in which the fermentative broth was used as an enzyme source, L-aspartic acid was released efficiently by the addition of a surfactant such as cetylpyridinium chloride.

The effect of chemical compounds on the preparation of L-aspartic acid was also investigated in detail using the fermentative broth of *Brevibacterium ammoniagenes* (Watanabe *et al.*, 1964). Some surfactants such as cetyltrimethylammonium bromide and cetylpyridinium chloride, higher fatty acids such as oleic acid and linoleic acid, and penicillin were found to enhance L-aspartic acid production.

On the other hand, the fermentative broth of *Pseudomonas trefollii* could produce high levels of L-aspartic acid without addition of surfactant, and under optimum conditions 400 g of L-aspartic acid were formed per liter of the reaction mixture (Takahashi *et al.*, 1963).

Production of L-aspartic acid from ammonium fumarate was also investigated using a fumaric acid-assimilating thermophilic bacteria, *Bacillus stearothermophilus* (Suzuki *et al.*, 1980) and an α-amino-*n*-butyric acid-resistant strain of *Brevibacterium* sp. (Yukawa *et al.*, 1981). However, the aspartase activity and the stability of these microorganisms were not satisfactory for industrial purposes.

In addition to the procedures described above, production of L-aspartic acid was investigated using maleic acid instead of fumaric acid as a raw material (Takamura *et al.*, 1966). *Alcaligenes faecalis* isolated from soil was cultured in a nutrient medium containing maleic acid, and the cells were harvested. When the cells were suspended in 1% maleic acid solution neutralized with ammonia and incubated at 25 °C for 48 h, the yield of L-aspartic acid reached 63% of the theoretical amount.

30.2.2.2   *Continuous enzymatic production*

Fermentative or enzymatic methods for industrial production of L-aspartic acid can be carried out by batchwise reaction. In order to isolate L-aspartic acid from the reaction mixture, it is necessary to remove microbial cells and enzyme proteins by pH and/or heat treatment. Even if enzyme activity remains in the reaction mixture, the microbial cells have to be discarded, resulting in uneconomical use of enzyme or microbial cells.

To develop a more efficient continuous production of L-aspartic acid using immobilized aspartase, Tosa *et al.*, (1973) investigated various immobilization methods for the enzyme extracted from *E. coli*. Among the immobilization methods attempted, relatively active immobilization aspartase was obtained by entrapping it in polyacrylamide gel. The stability of the immobilized enzyme column was investigated by operating it continuously for a long period. As shown in Figure 2, the activity of the column decreased to 50% of the initial value after operation for about 30 days at 37 °C. For industrial application of this method, the enzyme has to be extracted from microbial cells. Furthermore, intracellular enzymes are unstable when extracted from the cells, and the activity yield and the stability of the immobilized enzyme are not satisfactory for industrial purposes.

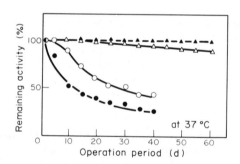

Figure 2   Stability of various aspartase preparations: ● intact *E. coli* cells, half-life = 10 days; ○ immobilized aspartase (polyacrylamide), half-life = 30 days; △ immobilized *E. coli* (polyacrylamide), half-life = 120 days; ▲ immobilized *E. coli* (carrageenan + hexamethylenediamine + glutaraldehyde), half-life = 680 days

By immobilizing whole cells these disadvantages can be overcome, and Chibita *et al.* (1974) examined various methods for immobilization of *E. coli* cells having high enzyme activity. Among the methods tested, the most active immobilized *E. coli* cells were obtained by entrapping the cells in a polyacrylamide gel lattice.

An interesting phenomenon was observed with the immobilized cells. When freshly prepared immobilized *E. coli* cells were suspended at 37 °C for 24–48 h in a substrate solution, their activity increased about 10 times. As this phenomenon is advantageous for continuous production of L-aspartic acid, the activation mechanism was investigated in detail. The apparent enzyme activity was found to be elevated by an increase of membrane permeability for substrate and/or product due to autolysis of the cells in the gel lattice. This phenomenon was similar to the case of the batchwise reaction using the fermentative broth described above, *i.e.* the production of L-aspartic acid was efficiently carried out by the addition of a surfactant to the reaction mixture.

By using a column packed with immobilized *E. coli* cells, the operational stability was investigated. As shown in Figure 2, the immobilized cells column was very stable and its half-life was 120 days at 37 °C (Tosa *et al.*, 1974; Sato *et al.*, 1975).

Contaminants such as microbial cells, proteins and so on are not present in the effluent from the column. Therefore, L-aspartic acid of high purity can be obtained in high yield without recrystallization by a very simple procedure such as adjusting the pH of the effluent to the isoelectric point of the acid. As shown in Figure 3, the effluent of appropriate volume is adjusted to pH 2.8 with 60% H_2SO_4 at 90 °C and then cooled to 15 °C. The L-aspartic acid crystallizes out and is collected by centrifugation or filtration and washed with water. The yield is 90–95% of theoretical.

Immobilized *E. coli* column

      passed through 1.2 M ammonium fumarate
      (1mM Mg^{2+}, pH 8.5) at a flow rate of SV = 0.8
      at 37 °C

Effluent

      adjusted to pH 2.8 with 60% H_2SO_4
      cooled to 15 °C
      collected by centrifugation
      washed with water

Crystalline L-aspartic acid
$[\alpha]_D^{20}$ = + 25.5 (C = 8, 6 M HCl)
Yield : 90-95%

Figure 3 Continuous production of L-aspartic acid from ammonium fumarate by immobilized *E. coli*

For industrial application of this technique, Sato *et al.* (1975) carried out an analysis of continuous enzyme reaction using a column packed with immobilized *E. coli* cells, and the aspartase reactor system was designed. As this aspartase reaction is exothermic, the column reactor used for industrial production of L-aspartic acid was designed as a multi-stage system with cooling.

This immobilized cell system has been operated industrially since 1973 by the Tanabe Seiyaku Co. Ltd., Japan. A comparison of the costs for production of L-aspartic acid by the conventional batch process using intact cells and the continuous process using immobilized cells is shown in Figure 4. The overall production cost by this system was reduced to about 60% of that of the conventional batchwise reaction using intact cells due to the marked increase of productivity of L-aspartic acid per unit of cells, reduction of labor costs due to automation, and an increase in the yield of L-aspartic acid (Chibata *et al.*, 1976). Furthermore, the procedure employing immobilized cells is advantageous from the standpoint of waste treatment. This was the first industrial application of immobilized microbial cells as a solid catalyst.

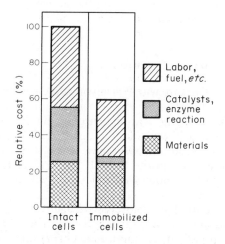

Figure 4 Comparison of cost for production of L-aspartic acid

To improve this process further, an immobilization technique using ϰ-carrageenan, a polysaccharide isolated from seaweed, was examined (Nishida *et al.*, 1979; Sato *et al.*, 1979). This polysaccharide is composed of β-D-galactose sulfate and 3,6-anhydro-α-D-galactose, which are non-toxic and widely used as food additives.

The aspartase activity and the operational stability of *E. coli* cells immobilized with ϰ-carrageenan were compared with those immobilized with polyacrylamide. As shown in Figure 2, the enzyme activity of immobilized cells prepared with ϰ-carrageenan was much higher, and the operational stability was increased by hardening treatment with glutaraldehyde and hexamethylenediamine. Its half-life was 680 days, almost two years. L-Aspartic acid productivities of *E. coli* cells immobilized with polyacrylamide and with carrageenan are compared in Table 1. The productivity of cells immobilized with carrageenan and hardened with glutaraldehyde and hexamethylenediamine is about 15 times that of the polyacrylamide-immobilized system. The carrageenan method replaced the polyacrylamide method for industrial production of L-aspartic acid by the Tanabe Seiyaku Co. in 1978. With a 1000 liter column, the theoretical yield of L-aspartic acid is 3.4 ton d^{-1}.

Table 1 Comparison of L-Aspartic Acid Productivities of *Escherichia coli* Immobilized with Polyacrylamide and with Carrageenan

Immobilization method[a]	Aspartase activity (units per g cells)	Stability at 37 °C (half-life, d)	Relative productivity[b]
Polyacrylamide	18 850	120	100
Carrageenan	56 340	70	174
Carrageenan (GA)	37 460	240	397
Carrageenan (GA + HMDA)	49 400	680	1498

[a] GA = glutaraldehyde, HMDA = hexamethylenediamine. [b] Productivity = $\int_0^t E_o \exp(-k_d t) \mathrm{d}t$ where E_o = initial activity, k_d = decay constant, t = operational period.

The continuous production of L-aspartic acid from ammonium fumarate has been investigated in the laboratory scale using *E. coli* cells immobilized with polyurethane (Fusee *et al.*, 1981). Yokote *et al.* (1978) immobilized aspartase, extracted from *E. coli* cells in the presence of the substrate, to a weakly basic anion exchange resin, Duolite A-7, by ionic binding and investigated the continuous production of L-aspartic acid. The half-life of the immobilized enzyme was found to be 18 days at 37 °C and when 2 M ammonium fumarate was passed through the column packed with the immobilized enzyme with a residence time of less than 0.75 h and at a temperature of 37 °C, the percentage conversion of the substrate solution could be maintained at more than 99% for 3 months. L-Aspartic acid has been industrially produced by this immobilized enzyme system by the Kyowa Hakko Kogyo Co. Ltd. since 1974.

In addition to the processes described above, other methods of immobilization of aspartase have been carried out. Aspartase extracted from *Escherichia intermedia* was entrapped in cellulose triacetate porous fibers (Pittalis *et al.*, 1979), and the enzyme from a thermophilic bacterium, *Bacillus aminogenes*, was immobilized ionically on the above-mentioned Duolite A-7 and then treated with glutaraldehyde (Kimura *et al.*, 1981).

30.3 UTILIZATION OF L-ASPARTIC ACID

L-Aspartic acid and its derivatives are widely used in medicines, foods and cosmetics. Clarification of the specific physiological role of L-aspartic acid has brought about its therapeutic application. The importance of K^+ and Mg^{2+} in myocardial cells function has been the subject of increased interest, and L-aspartic acid was found to be an ideal 'carrier substance' for these ions. Thus, potassium magnesium aspartate is widely used to assist recovery from fatigue, heart failure, and liver disease of diabetes (Laborit *et al.*, 1958b, 1958c, 1959). In addition, potassium aspartate is used for alleviating leg cramp syndrome and liver disease, and ferrous aspartate is used for treating anemia.

Some L-aspartic acid salts are used as surface active agents (Kalopissis *et al.*, 1966, 1967) and for the treatment of keratodermatitis (Suzue *et al.*, 1968). The addition of L-aspartic acid or its derivatives, together with vitamin B_6, to cosmetic bases has been suggested to be effective in preventing ageing of the skin or revitalizing old skin (Taniguchi, 1965).

In food processing, although L-aspartic acid is acidic, its neutralized solution is tasty. Sodium aspartate is used as a seasoning in orange juice. It is known that there is a synergistic action in terms of taste between amino acid and mononucleotides. The taste of sodium inosinate, which was identified as a taste substance of 'Katsuobushi' (dried bonito), is not strong in itself, but is enhanced by the addition of monosodium L-aspartate (Kuninaka, 1960).

In the course of studies on the synthesis of gastrin, a hormone-stimulating gastric juice, Mazur *et al.* (1969, 1970) found that an intermediate of gastrin synthesis, L-aspartyl-L-phenylalanine methyl ester, showed very strong sweetness. The dipeptide Aspartame® was found to be about 200 times sweeter than 4% sucrose in an aqueous solution and to be low in calories. Aspartame is now being developed by several companies and will become one of the important materials in the market for artificial sweeteners.

30.4 REFERENCES

Chibata, I., T. Tosa and T. Sato (1974). Immobilized aspartase-containing microbial cells: preparation and enzymatic properties. *Appl. Microbiol.*, **27**, 878–885.

Chibata, I., T. Tosa and T. Sato (1976). Production of L-aspartic acid by immobilized microbial cells. *Methods Enzymol.*, **44**, 739–746.

Fusee, M. C., W. E. Swann and G. J. Calton (1981). Immobilization of *Escherichia coli* cells containing aspartase activity with polyurethane and its application for L-aspartic acid production. *Appl. Environ. Microbiol.*, **42**, 672–676.

Hotta, K. and S. Takao (1973). Conversion of fumaric acid fermentation to aspartic acid fermentation by the association of *Rhizopus* and bacteria. II. Production of aspartic acid by the combination of *Rhizopus* and *Proteus vulgaris*. *J. Ferment. Technol.*, **51**, 12–18.

Kalopissis, G., G. Vanlerverghe and A. Viout (1966). Surface-active agents. Amphoteric derivatives of asparagine and their utilization in cosmetics. *Arch. Biochim. Cosmetol.*, **9**, 206–213.

Kalopissis, G., G. Vanlerverghe and A. Viout (1967). New amphoteric surface-active agents derived from asparagine and their use in cosmetics. *Parfum. Cosmet. Savons*, **10**, 293–297.

Kimura, K., K. Takamaya, Y. Ado, T. Kawamoto and I. Masunaga (1981). Preparation of thermophilic aspartase. *Jpn. Pat.* 81 75 097.

Kinoshita, S., K. Nakamaya and S. Kitada (1958). Production of aspartic acid from fumaric acid by microorganisms. *Hakko Kyokaishi*, **16**, 517–520.

Kisumi, M., Y. Ashikaga and I. Chibata (1960). Studies on the fermentative preparation of L-aspartic acid from fumaric acid. *Bull. Agric. Chem. Soc. Jpn.*, **24**, 296–305.

Kitahara, K., S. Fukui and M. Misawa. (1960). Preparation of L-aspartic acid by bacterial aspartase. *Nogeikagaku Kaishi*, **34**, 44–48 (see also *J. Gen. Appl. Microbiol.*, **5**, 74–77 (1959)).

Kuninaka, A. (1960). Studies on taste of ribonucleic acid derivatives. *Nogeikagaku Kaishi*, **34**, 489–492.

Laborit, H., R. Moynier, R. Coirault, J. Thiebault, G. Guiot, P. Niaussat, B. Weber, J. M. Jouany and C. Baron (1958a). The place of certain salts of DL-aspartic acid in the mechanism of the maintenance of reactivity to the environment. Summary of experimental and clinical study. *Presse Méd.*, **66**, 1307–1309.

Laborit, H., B. Weber, J. M. Jouany, P. Niaussat and C. Baron (1958b). The metabolism of ammonia and its disturbances: therapeutic value of salts of aspartic acid. *Presse Méd.*, **66**, 2125–2128.

Laborit, H., B. Weber, P. Niaussat, J. M. Jouany, G. Guiot, J. Zawadoski and C. Baron (1958c). Effect of several salts of DL-aspartic acid. *Anesth. Analg. Reanim.*, **15**, 480–494.

Laborit, H., B. Weber, J. M. Jouany, P. Niaussat, B. Broussolle, M. Reynier and C. Baron (1959). New concepts concerning the metabolism of ammonia and its disturbances. *Anesth. Analg. Reanim.*, **16**, 379–392.

Mazur, R. H., J. M. Schlatter and A. H. Goldkamp (1969). Structure–taste relationships of some dipeptides. *J. Am. Chem. Soc.*, **91**, 2684–2691.

Mazur, R. H., A. H. Goldkamp, P. A. James and J. M. Schlatter (1979). Structure–taste relationships of aspartic acid amides. *J. Med. Chem.*, **13**, 1217–1221.

Nishida, Y., T. Sato, T. Tosa and I. Chibata (1979). Immobilization of *Escherichia coli* cells having aspartase activity with carrageenan and locust bean gum. *Enzyme Microb. Technol.*, **1**, 95–99.

Pittalis, F., F. Bartoli and F. Morisi (1979). Properties of fiber-entrapped aspartase. *Enzyme Microb. Technol.*, **1**, 189–192.

Quastel, J. H. and B. Woolf (1926). Equilibrium between L-aspartic acid, fumaric acid and ammonia in presence of resting bacteria. *Biochem. J.*, **20**, 545–555.

Rudolph, F. B. and H. J. Fromm (1971). Purification and properties of aspartase from *Escherichia coli*. *Arch. Biochem. Biophys.*, **147**, 92–98.

Sato, T., T. Mori, T. Tosa, I. Chibata, M. Furui, K. Yamashita and A. Sumi (1975). Engineering analysis of continuous production of L-aspartic acid by immobilized *Escherichia coli* cells in fixed beds. *Biotechnol. Bioeng.*, **17**, 1797–1804.

Sato, T., Y. Nishida, T. Tosa and I. Chibata (1979). Immobilization of *Escherichia coli* cells containing aspartase activity with κ-carrageenan. Enzymic properties and application for L-aspartic acid production. *Biochim. Biophys. Acta*, **570**, 179–186.

Sekijo, C., N. Sunahara and S. Iwakumo (1965). Studies on aspartase from *Escherichia freundii*. *Amino Acid Nucleic Acid* (*Tokyo*), **11**, 139–146.

Shiio, I., H. Ozaki and K. Ujigawa-Takeda (1982). Production of aspartic acid and lysine by citrate synthesis mutants of *Brevibacterium flavum*. *Agric. Biol. Chem.*, **46**, 101–107.

Sumiki, Y. (1928). Synthesis of optically active compounds by means of yeast. I. Synthesis of L-aspartic acid from fumaric acid. *Bull. Jpn. Soc. Ferment.*, **23**, 33–41.

Suzue, K., T. Yamashiki, T. Banno and S. Yasui (1968). Photosensitive aspartate dyes. *Jpn. Pat.* 65 3884.

Suzuki, S., J. Yamaguchi and M. Tokushige (1973). Studies on aspartase. I. Purification and molecular properties of aspartase from *Escherichia coli*. *Biochem. Biophys. Acta*, **321**, 369–381.

Suzuki, Y., T. Yasui, Y. Mino and S. Abe (1980). Production of L-aspartic acid from fumaric acid by a fumaric acid-assimilating obligate thermophile, *Bacillus stearothermophilus* KP 1041. *Eur. J. Appl. Microbiol. Biotechnol.*, **11**, 23–27.

Takahashi, M., T. Hino, S. Okamura, Y. Miura and M. Ishida (1963). Studies on L-aspartic acid fermentation. *Amino Acid Nucleic Acid* (*Tokyo*), **7**, 23–28.

Takamura, Y., I. Kitamura, M. Iikura, K. Kono and A. Ozaki (1966). Studies on the enzymatic production of L-aspartic acid from maleic acid. *Agric. Biol. Chem.*, **30**, 338–344.

Takao, S. and K. Hotta (1972). Conversion of fumaric acid fermentation to aspartic acid fermentation by the association of *Rhizopus* and bacteria. I. Screening of bacterial strains having a high aspartase activity and some conditions for the conversion. *J. Ferment. Technol.*, **50**, 751–757.

Taniguchi, T. (1965). Cosmetics. *Jpn. Pat.* 65 24 640.

Tosa, T., T. Sato, T. Mori, Y. Matsuo and I. Chibata (1973). Continuous production of L-aspartic acid by immobilized aspartase. *Biotechnol. Bioeng.*, **15**, 69–84.

Tosa, T., T. Sato, T. Mori and I. Chibata (1974). Basic studies for continuous production of L-aspartic acid by immobilized *Escherichia coli* cells. *Appl. Microbiol.*, **27**, 886–889.

Uchino, R., K. Kubota, I. Maeyashiki and Y. Hirose (1976). Fermentative production of L-aspartic acid. *Jpn. Pat.* 76 24 592.

Virtanen, A. I. and J. Tarnanen (1932). Enzymic hydrolysis and synthesis of aspartic acid. *Biochem. Z.*, **250**, 193–211.

Watanabe, S., T. Osawa and T. Nobukuni (1964). Studies on L-aspartic acid fermentation. *Amino Acid Nucleic Acid* (*Tokyo*), **7**, 97–104.

Watanabe, Y., M. Iwakura, M. Tokushige and G. Eguchi (1981). Studies on aspartase. VII. Subunit arrangement of *Escherichia coli* aspartase. *Biochim. Biophys. Acta*, **661**, 261–266.

Woolf, B. (1929). Some enzymes in *E. coli* communities which act on fumaric acid. *Biochem. J.*, **23**, 472–482.

Yokote, Y., S. Maeda, H. Yabushita, S. Noguchi, K. Kimura and H. Samejima (1978). Production of L-aspartic acid by *E. coli* aspartase immobilized on phenol-formaldehyde resin. *J. Solid-Phase Biochem.*, **3**, 247–261.

Yukawa, H., T. Nara and Y. Takayama (1981). Production of L-aspartic acid. *Jpn. Kokai* 81 26 196.

31
Threonine

K. SHIMURA
Tohoku University, Sendai, Japan

31.1 INTRODUCTION

Threonine, or α-amino-β-hydroxy-n-butyric acid, was first isolated from an acid hydrolyzate of oat glutelin by Schryver and Buston in 1925, but no nutritional significance was attributed to the compound at the time, nor was there any concept of its optical configuration. Rose and his collaborators (1931–1932) predicted the existence of an unknown essential amino acid in protein hydrolyzates from their experiments involving the feeding of animals with synthetic mixtures of highly purified amino acids. In 1935 Rose *et al.* identified it as α-amino-β-hydroxy-n-butyric acid which was subsequently named threonine on the basis of its configurational similarity to D-threose.

As threonine contains two asymmetric centers, there are four steroisomers of the molecule. Of these isomers, only L-threonine is effective in nutrition. A deficiency of L-threonine produces stunted growth followed by metabolic disorders throughout the body, such as decrease in body-weight and anaemia, and may induce functional disorders of the liver.

The pathway of biosynthesis of threonine was established by isotopic and enzymic studies. Teas *et al.* (1948, 1950) first obtained evidence with a mutant strain of *Neurospora crassa* that aspartate is converted through homoserine to threonine. Further progress in the elucidation of the biosynthetic route to threonine was made by Cohen and Hirsh (1954), Black and Wright (1955) and Watanabe *et al.* (1957). Three new intermediates in the aspartate–threonine pathway, namely, β-aspartyl phosphate, aspartate β-semialdehyde and O-phosphohomoserine, were found and the intermediate reactions were elucidated with partially purified enzyme preparations from yeast. These studies led to the establishment of the biosynthetic pathway of threonine shown in Figure 1 for yeast, *Neurospora*, *E. coli* and other bacteria and higher plants.

Threonine is deaminated by threonine hydratase to α-ketobutyrate, which is the starting compound for the biosynthesis of isoleucine. The biosynthetic pathways of lysine and methionine

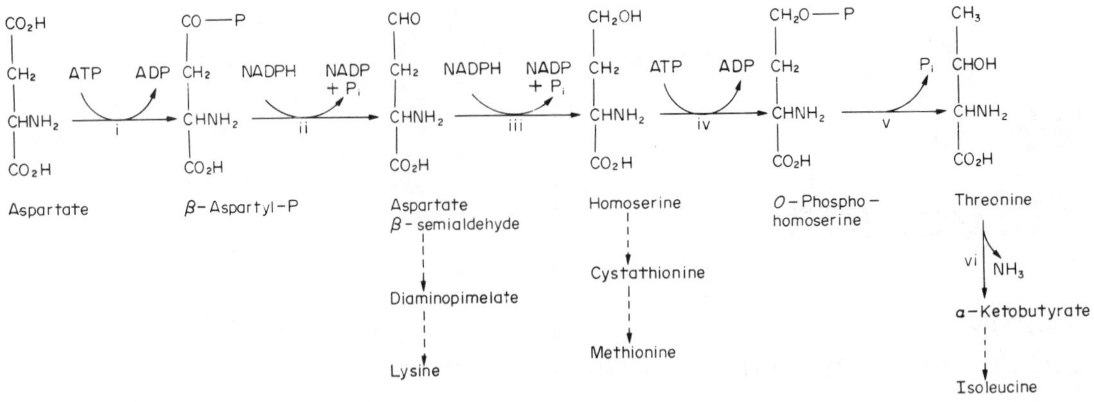

Figure 1 Biosynthetic pathway of the aspartate family of amino acids: i, aspartate kinase; ii, aspartate-semialdehyde dehydrogenase; iii, homoserine dehydrogenase; iv, homoserine kinase; v, threonine synthase; and vi, threonine dehydratase (deaminase)

branch at aspartate β-semialdehyde and homoserine, respectively. Feedback inhibition of enzyme activity and repression of enzyme formation by product amino acids are known to operate on various steps in these pathways. This will be described in detail in the following section. Thus, because of control by metabolic regulation, it is not expected that threonine would accumulate to more than a limited level in normal microorganisms. To increase threonine production, the effects of feedback inhibition and repression must be circumvented and the flow of threonine and its precursor into the branch pathways leading to lysine, methionine and isoleucine synthesis must be prevented.

In the early stages of the studies on the production of L-threonine by fermentative processes, two approaches for over production were examined (Shimura, 1972). One method is fermentation conducted in a culture medium containing homoserine as a precursor of threonine, and the other is a direct fermentative method, starting from sugars and ammonium salts, using appropriate mutants of microorganisms. Recently, however, it has been generally accepted that the direct method is more promising than precursor feeding. Therefore, the production of threonine mutants will be discussed in this chapter as this is the method used in industrial production of the amino acid.

31.2 REGULATORY MECHANISMS OF THREONINE BIOSYNTHESIS IN SEVERAL MICROORGANISMS

As mentioned above, it is necessary to circumvent the regulatory mechanisms operating in the threonine biosynthetic pathway in order to accumulate a large amount of threonine in the fermentation broth. The mode of regulation in the threonine synthetic pathway seems to vary in different microorganisms. The regulatory mechanisms of threonine biosynthesis in two typical threonine producing bacteria strains of *E. coli* K-12 and *Brevibacterium flavum*, are shown in Figure 2.

In *E. coli* K-12, three aspartate kinases are known, one competitively inhibited and repressed by L-threonine, the second one not inhibited by L-methionine but repressed by it, and the third one non-competitively inhibited and repressed by L-lysine (Patte *et al.*, 1967). In contrast to this great complexity of the aspartate kinase system in *E. coli*, in *Brevibacterium flavum* only one aspartate kinase is known, and its activity is regulated by a feedback inhibition with L-threonine and L-lysine, but not by end-product repression (Shiio *et al.*, 1970, Shiio and Nakamori 1970). In addition, no evidence showing the operation of repressions by L-threonine, L-isoleucine or L-lysine in the synthetic pathway of the aspartate family of amino acids has been obtained so far. This fact suggests that regulatory mechanisms in threonine biosynthesis may vary in different microorganisms. Therefore, it is a prerequisite to elucidate the regulatory mechanism of threonine biosynthesis in the individual microorganism which is selected as the threonine producer.

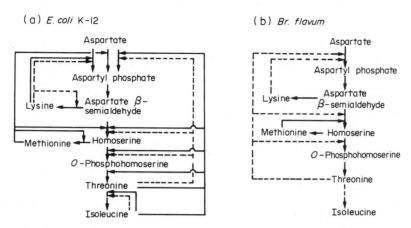

Figure 2 Summary of feedback regulation in the biosynthesis of the aspartate family amino acids in (a) *E. coli* K-12 and (b) *Brevibacterium flavum*: feedback inhibition – – –→, repression →

31.3 SELECTION OF THREONINE-PRODUCING BACTERIA

31.3.1 *Escherichia coli*

Cohen and Patte (1963) first reported that an *E. coli* mutant resistant to α-amino-β-hydroxyvaleric acid (AHV), an analogue of threonine, excreted L-threonine into media, and that homoserine dehydrogenase activity of the mutant was less sensitive to feedback inhibition of threonine than that of the parental strain. Based on this observation, an attempt to isolate threonine-producing *E. coli* mutants resistant to AHV was made by Shiio and Nakamori (1969). AHV resistant mutants, both spontaneous and induced by *N*-methyl-*N'*-nitro-*N*-nitrosoguanidine (NG) treatment of *E. coli* PB-8, were obtained and the resistant colonies were isolated on a medium containing 1 mg ml^{-1} of AHV after 48 h incubation at 37 °C. Among many AHV resistant mutants obtained, one of the isoleucine auxotrophs, strain βI-67, produced about 4.7 mg ml^{-1} of threonine, and one of the methionine auxotrophs, strain βIM-4, derived from strain βI-67, produced about 6.1 mg ml^{-1} of threonine, when cultured in a medium containing 3% glucose supplemented with 100 μg ml^{-1} of L-isoleucine and L-methionine.

The threonine production by auxotrophs derived from AHV sensitive and resistant mutant strains is shown in Table 1. It is of interest that simple methionine auxotrophs accumulated threonine, but simple isoleucine or lysine auxotrophs did not. These results suggest that the alteration of the sensitivity of homoserine dehydrogenase to the threonine inhibition as is the case with AHV resistant mutants, or formation of another threonine insensitive homoserine dehydrogenase and aspartokinase as occurs with methionine auxotrophs, would be essential for threonine production in *E. coli*. As seen in the cases of strains PBMI-13 and PBMI-14, the addition of isoleucine requirement to methionine auxotrophs enhanced their threonine production, although simple isoleucine auxotrophs did not accumulate threonine. This result suggests that isoleucine-repressible enzymes in the pathway, other than homoserine dehydrogenase, are rate limiting for threonine production in these methionine auxotrophs.

Kase *et al.* (1971) treated various bacteria with UV light or nitrosoguanidine (NG) to derive auxotrophic mutants, which were screened for their ability to produce threonine. Among the auxotrophs obtained, triple auxotrophs [DAP$^-$ (α,ε-diaminopimelic acid), Met$^-$, Ile$^-$] of *E. coli* and their isoleucine revertants were screened for their threonine productivity. One of the isoleucine revertants, KY 8280, was further studied to determine optimal culture conditions to attain a maximal accumulation of threonine.

Among various carbon sources tested, fructose was found to give the best production of threonine. Threonine accumulation during the fermentation in a 5 l jar fermenter with a medium containing 7.5% fructose, DL-methionine (50 μg ml^{-1}), DAP (100 μg ml^{-1}) and isoleucine (25 μg ml^{-1}) was examined. Cell concentration reached a maximum in about 48 h, after which the concentration of L-threonine increased rapidly, with a concurrent consumption of fructose and attained a level of 13.8 mg ml^{-1} in 120 h. Huang (1961) described threonine production by DAP

Table 1 Threonine Accumulation by Auxotrophs derived from AHV-sensitive and -resistant *E. coli* Strains[a]

Strain	Nutritional requirement	AHV resistance[b]	Growth (O.D.$_{562}$)	Final pH	L-Thr (g l^{-1})
PB-8	—	−	0.362	4.8	0.01
PBI-6	Ile	−	0.251	4.8	0.01
PBI-11	Ile	−	0.241	5.0	0.01
PBL-1	Lys	−	0.134	5.2	0.09
PBM-1	Met	−	0.369	4.9	1.74
PBM-28	Met	−	0.432	4.8	1.34
PBMI-13	Met, Ile	−	0.348	4.8	2.99
PBMI-14	Met, Ile	−	0.310	4.8	2.67
β-101	—	+	0.424	4.8	1.87
β-133	Met, Ile	+	0.393	4.8	1.20
βI-67	Ile	+	0.392	4.8	4.69
βM-7	Met	+	0.375	5.2	3.78
βM-9	Met	+	0.430	5.0	3.59
βL-1	Lys	+	0.221	5.2	0.11
βIM-4	Ile, Met	+	0.350	5.0	6.10
βIML-6	Ile, Met, Lys	+	0.215	5.2	3.40

[a] Shiio and Nakamori, 1969. [b] − and + represent AHV sensitive and resistant respectively. Fermentation medium was supplemented with 100 mg l^{-1} of required L-amino acids.

and DAP plus methionine auxotrophs of *E. coli*. However, it was difficult to increase the yield of threonine with these auxotrophs. It is noteworthy that the imposition of a nutritional requirement for isoleucine appreciably increased threonine accumulation, and that some isoleucine revertants of these mutants produced more threonine than parent triple auxotrophs.

Hirakawa *et al.* (1973) also selected threonine producing auxotrophs of *E. coli* C-6 by treating them with UV radiation or NG. One of the methionine–valine auxotrophs, strain No. 234, was found to produce a large amount of L-threonine in a culture medium supplemented with DL-methionine (60 μg ml^{-1} and L-valine (400 μg ml^{-1}). The yield of threonine was greatly dependent on the temperature of incubation. Higher yields of threonine were obtained when the temperature was kept at a relatively high level (38–40 °C). In addition, it appears that there is a close relationship between threonine production and the cell yield. When the yield of cells was repressed, the accumulation of threonine increased with increasing temperature (Hirakawa, 1973). The yield of L-threonine in the broth of the glucose medium was also enhanced by the temperature shift-down technique, in which the culture temperature was shifted from 38 to 28 °C at 17 h after the start of cultivation. This procedure was particularly effective in maintaining a high level of threonine production throughout the fermentation (Hirakawa and Watanabe, 1974).

The addition of an antibiotic, borrelidin, together with L-aspartic acid to the culture medium of *E. coli* strain No. 234 increased threonine accumulation to about 15 mg ml^{-1} (Hirakawa *et al.*, 1974). There was more than five-fold derepression of the threonine biosynthetic enzymes when *E. coli* was grown in the presence of trace amounts of borrelidin, therefore the stimulative effect of borrelidin on the threonine accumulation may possibly be caused by an increase in the activities of the enzymes responsible for threonine biosynthesis.

31.3.2 *Serratia marcescens*

Komatsubara *et al.* (1978a) attempted to isolate threonine producers from the mutant strains of *Serratia marcescens* which had been used by them for isoleucine production studies. These mutants were known to lack both feedback inhibition of threonine deaminase and repression of isoleucine biosynthetic enzymes (Komatsubara, 1978b), and, therefore, produce a large amount of isoleucine in a medium containing glucose and urea. These facts suggest that a mutant strain of *S. marcescens* might be attractive for threonine production, if the absence of threonine degrading enzymes could be included in the mutant strains. Thus, AHV-resistant mutants were selected from cells of the isoleucine auxotroph of *S. marcescens* strain D-60 by treating it with NG.

The selection was carried out by spreading the mutated cells of strain D-60 on a minimal agar plate containing glycerol as a carbon source and high concentrations of AHV, L-isoleucine, L-methionine and L-lysine. These three natural amino acids were added to select mutants lacking feedback controls to a greater extent. Among the AHV-resistant mutants obtained, three representative mutants were tested for threonine production. As shown in Table 2, the parent strain

D-60 produced minimal levels of threonine in a medium containing a limiting amount or an excess of isoleucine. Strain NHr21 produced approximately 11 mg ml^{-1} of threonine on addition of a limiting amount of isoleucine. This production was decreased with a higher concentration of isoleucine, owing to its repressive effect on threonine synthetic enzymes. Strain HNr59 also produced threonine to a lesser extent than that of strain HNr21, owing to the feedback inhibition of aspartate kinase. Strain HNr53 produced a trace amount of threonine, a result of feedback inhibition of aspartate kinase and homoserine dehydrogenase. Strain E-84, a DAP auxotroph derived from strain HNr59, produced an increased amount of threonine (13 mg ml^{-1}) with the addition of appropriate amounts of DAP and lysine. In this auxotroph the lysine-sensitive aspartate kinase might be physiologically released from feedback controls by limiting the lysine source. This indicates that a genotype lacking feedback control of the lysine-sensitive aspartate kinase would increase the threonine productivity of *S. marcescens* strains.

Table 2 Threonine Production by AHV-resistant Mutants of *S. marcescens*[a]

Strain	Addition of L-isoleucine to the medium (mg l^{-1})	Growth (g dry wt l^{-1}) at:				L-Threonine produced (g l^{-1}) at:			
		48 h	72 h	96 h	120 h	48 h	72 h	96 h	120 h
D-60	2	5.8	7.4	12.8	23.8	0.1	0.1	0.3	0.3
	10	17.2	27.3	25.2	19.4	0.1	0.1	0.1	0.1
HNr21	2	7.0	11.2	19.0	18.1	1.9	7.4	8.9	10.9
	10	16.8	22.7	24.2	20.0	0.9	3.6	4.2	3.7
HNr53	2	1.6	7.3	11.3	18.3	0.1	0.3	0.3	0.3
	10	12.8	18.3	19.4	18.3	0.3	0.3	0.3	0.3
HNr59	2	3.2	13.3	26.1	28.9	1.2	2.9	4.8	5.0
	10	8.4	25.3	27.4	26.0	1.0	3.0	4.6	5.1

[a] Komatsubara *et al.*, 1978a.

Thus, Komatsubara *et al.* (1979a) have tried to isolate regulatory mutants of *S. marcescens* for lysine-sensitive aspartate kinase using *S*-2-aminoethylcysteine (AES) as a lysine antagonist. Strains AECr174 and AECr301, derived from strains HNr31 and HNr53, respectively, lacked both feedback inhibition and repression of lysine-sensitive aspartate kinase. These strains produced about 7 mg ml^{-1} of threonine in the medium containing glucose and urea.

31.3.3 *Brevibacterium*

Shiio and Nakamori (1970) observed that threonine and homoserine auxotrophs derived from *Brevibacterium flavum* No. 2247, a glutamate producing bacterium, accumulated L-lysine in the broth to a level high enough to be used for industrial scale production. Since homoserine dehydrogenase of this strain was inhibited strongly by L-threonine as in the case of *E. coli*, it was expected that AHV-resistant mutants of glutamate producing bacteria would produce as much threonine as lysine.

Shiio and Nakamori (1970) selected threonine producers from AHV-resistant mutants of *Br. flavum* No. 2247. The best producer, strain B-183, was isolated from a plate containing 5 mg ml^{-1} of AHV and accumulated 10.5 mg ml^{-1} of L-threonine. The culture medium used in the threonine production contained per liter 100 g glucose, 30 g ammonium sulfate, 15 g KH$_2$PO$_4$, 0.4 g MgSO$_4$·7H$_2$O, 2 p.p.m. Fe^{2+}, 2 p.p.m. Mg^{2+}, 200 μg D-biotin, 300 μg thiamine·HC1, 4 ml Mieki (a mixture of amino acids) and 50 g CaCO$_3$ (pH 7.2).

In these experiments, it appeared that mutants resistant to higher concentrations of AHV accumulated larger amounts of threonine than those less resistant to AHV. Thus, further selection of the AHV mutants derived from strain B-183 by treating it with NG was carried out. Among the mutants tested, about twenty strains produced more threonine than did the parent strain. Out of these mutants, strains BB-82, BB-69 and BB-24 accumulated 13.5, 12.9 and 11.9 mg ml^{-1} of threonine with the culture medium described above.

The time courses of growth, pH and threonine production using BB-82 strain are given in Figure 3. Maximal levels of threonine accumulation (13.5 mg ml^{-1}) were obtained between 60 and 70 h, just after reaching a plateau of the growth curve. In the broth of the strains resistant to AHV, in addition to L-threonine, a significant amount of L-homoserine (2.7 mg ml^{-1}) and lower amounts of other amino acids such as L-valine, L-isoleucine, L-leucine, DL-alanine and glycine also accumulated. It is, of course, desirable to reduce the accumulation of any other amino acids for industrial production of threonine.

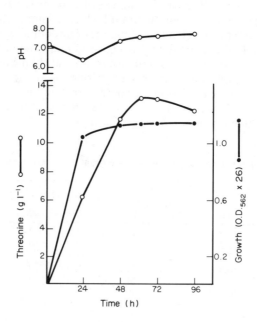

Figure 3   Time course of threonine production in *Br. flavum* (Shiio and Nakamori, 1970)

To increase threonine production by *Br. flavum*, Nakamori and Shiio (1972) further attempted to get methionine and lysine auxotrophs from strains BB-69 and BB-2, threonine producers derived from *Br. flavum* No. 2247 mentioned above. Five lysine auxotrophs and 76 methionine auxotrophs were isolated. None of the lysine auxotrophs produced more threonine than the parental strain; one methionine auxotroph, strain BBM-21, accumulated more threonine than the parent strain when cultured in a medium containing 200 μg ml^{-1} of L-methionine. The maximum production of threonine (17.5 mg ml^{-1}), which was 50% more than with the parental strain, was obtained when the initial concentration of methionine was 750–1000 μg ml^{-1}; further addition of methionine decreased the level of threonine. The maximum cell concentration was attained at more than 250 μg ml^{-1} of methionine. It is noteworthy that the addition of methionine markedly reduced homoserine accumulation.

It has been shown that the homoserine dehydrogenase of threonine producers selected as AHV resistant mutants, and the aspartate kinase of lysine producers selected as AEC resistant mutants, derived from *Br. flavum* No. 2247, are genetically desensitized to feedback inhibition by their end products, L-threonine and L-threonine plus L-lysine, respectively. As these two enzymes are involved in the synthesis of threonine, it seems useful for threonine overproduction to derive mutants in which both aspartate kinase and homoserine dehydrogenase are desensitized to feedback inhibition by L-threonine plus L-lysine and L-threonine, respectively.

Nakamori and Shiio (1973) have selected AHV-resistant mutants from the AEC resistant strains of *Br. flavum* FA-1-30 and FA-3-115, which had been isolated as lysine producers. Many resistant mutants isolated from a plate containing 2 mg ml^{-1} of AHV accumulated lysine as well as threonine, while none of the mutants from plates containing 3 or 5 mg ml^{-1} of AHV accumulated lysine. The best producer, FAB-44, isolated from NG-treated FA-1-30 accumulated 15 mg ml^{-1} of threonine in a medium containing 5 mg ml^{-1} of AHV.

31.3.4   Other Microorganisms

In addition to the threonine-producing bacteria mentioned above, auxotrophic mutants of various bacteria have been examined for their ability to accumulate threonine by Kase *et al.* (1971). Fifteen strains of bacteria representing seven families were mutagenized to derive auxotrophs of amino acids directly related to threonine biosynthesis, *i.e.* methionine, isoleucine, lysine and α,ε-

diaminopimelic acid. Among the isolated auxotrophs, those which produced appreciable amounts of L-threonine were found in *Aerobacter aerogenes*, *S. marcescens* and *E. coli*, all of which belong to the family Enterobacteriaceae.

Auxotrophic mutants of *Arthrobacter paraffineus* were also selected as threonine producers. Among them, five mutants produced above 12 mg ml^{-1} of threonine with the medium containing 10% *n*-paraffins (C_{12}–C_{14} rich) and 2 mg ml^{-1} of DL-methionine (Kase and Nakayama, 1973).

31.4 THREONINE PRODUCERS OBTAINED BY GENETIC MANIPULATION

Microbial strains having a high productivity of a specified amino acid require multiple mutations leading to (i) release of feedback controls of the amino acid biosynthetic pathway, (ii) defects in enzymes degrading the amino acid, (iii) defects in enzymes involved in the flow of biosynthetic precursors to side pathways, (iv) enhancement of the formation of the precursors, (v) prevention of the accumulation of by-products, and (vi) genetic stabilization of the amino acid producing strain. As described in the preceding sections, most of the work intended to obtain such strains has been done by mutagenizing the microbes sequentially to add the above mutations to a single strain. However, this method does not always result in a multiple mutant that has all of the required mutations. When a microorganism has complex regulatory mechanisms, sequential mutagenesis does not provide multiple mutants that satisfactorily circumvent individual regulatory mechanisms.

Experimental procedures to obtain better strains for amino acid production by using genetic techniques such as transduction, transformation and conjugation have been developed. Komatsubara *et al.* (1979b) have applied a transductional technique for construction of threonine producing strains of *S. marcescens*. This bacterium has the advantages that the regulatory mechanisms operating in the biosynthetic pathway of threonine is very similar to that of *E. coli* and that bacteriophage BP-20 is known to be an effective transducing phage in *S. marcescens*. In the threonine synthesis pathway, aspartate kinase and homoserine dehydrogenase are rate-limiting in *S. marcescens*, which has three aspartate kinases and two homoserine dehydrogenases. All these enzymes are feedback-controlled by threonine, isoleucine, methionine and lysine, singly or in combination (Figure 4). With phage PS-20-mediated transduction, Komatsubara *et al.* (1979b) transferred the *thrA$_1$1* and *thrA$_2$1* mutations (*thrA$_1$1*: lack of feedback inhibition of threonine-sensitive aspartate kinase; *thrA$_2$1* and *thrA$_2$2*: lack of feedback inhibition of homoserine dehydrogenase I and II, respectively) into strain E-60 lacking repression by the threonine operon (*hnr-1*, lack of repression of threonine) in the initial stages of the study. One of the transductants, strain T-570, produced about 8 mg ml^{-1} of L-threonine in a medium containing sucrose and urea. This strain was assumed to be a recombinant carrying *thrA$_1$*, *thrA$_2$* and *hnr-1*.

Subsequently, an additional regulatory mutation lacking both feedback inhibition and repression of lysine-sensitive aspartate kinase was introduced into strain T-570 using strain AECr174 as donor. The strain AECr174 lacks both feedback inhibition and repression of lysine-sensitive aspartate kinase, *lysC1* mutation (lack of both feedback inhibition and repression of lysine-sensitive aspartate kinase). Among the transductants obtained, a representative strain T-693 exhibited a higher activity of aspartate kinase than that of strain T-570 or AECr174. Aspartate kinase was not inhibited by either lysine or threonine or a combination of these. Its homoserine dehydrogenase was derepressed and insensitive to threonine-mediated feedback inhibition. These data indicate that strain T-693 carries the expected genotype: *lysC1*, *thrA$_1$2*, *thrA$_2$1* and *hnr-1*. Strain T-693, carrying the four regulatory mutations for threonine biosynthesis, accumulated the largest amounts of L-threonine (about 25 mg ml^{-1}) among the nine strains tested. The other strain, lacking more than one of the four mutations, accumulated less than 14 mg ml^{-1} of threonine. These results indicate that each regulatory mutation contributes to the high threonine production in this strain.

A further improvement in the transductional construction of threonine-producing strains of *S. marcescens* has been achieved by Kisumi (1982). The six regulatory mutations were independently selected by isolating mutants resistant to analogs of threonine, methionine and lysine. The genealogy of the main strains used in the transductional experiments is given in Figure 5. Strain Mu-910 was isolated from strain 8000 (wild-type) and lacks threonine dehydrogenase. Strain D-60 was derived from strain Mu-910 as an isoleucine auxotroph, lacking both threonine dehydro-

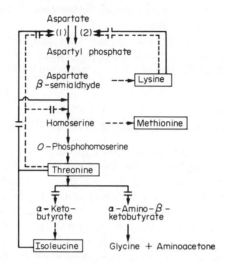

Figure 4 Regulation of biosynthesis and degradation of threonine in *S. marcescens*: feedback inhibition: _ _ _→, repression: →

genase and threonine deaminase. The strains carrying $thrA_11$ and $thrA_21$ mutations lack feedback inhibition of aspartate kinase I and homoserine dehydrogenase I. Mutations $hnrA1$ and $hnrB2$ lead to derepression of these two enzymes. Mutation $hnrA1$ is linked to the *thr* locus, but mutation $hnrB2$ is not. A mutation causing lack of both feedback inhibition and repression of aspartate kinase III is denoted as $lysC1$. Mutation $etr-1$ causes derepression of aspartate kinase II and homoserine dehydrogenase II.

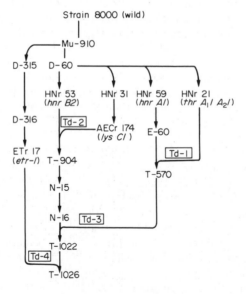

Figure 5 Genealogy of main strains used in the transductional construction of threonine-producers of *S. marcescens*: Td, transduction (Kisumi, 1982)

By four transductional crosses as indicated in Figure 5 (Td-1–4), the mutations described above were combined in a single strain D-60 defective in threonine-degrading enzymes. Of the transductants selected, strain T-1026 was found to be relieved of both feedback inhibition and repression of three aspartate kinases and two homoserine dehydrogenases owing to the six regulatory mutations. This strain produced about 40 mg ml^{-1} of threonine, whereas the other strains, lacking more than one of the six mutations, produced less than 30 mg ml^{-1} of threonine (Table 3). Strain T-1026 was further improved to strain P-200, which had both isoleucine and methionine bradytrophies. Strain P-200 accumulated more than 40 mg ml^{-1} of threonine in a simple fermentation medium containing sucrose and urea. This strain was quite stable and the cultured medium contained no detectable amounts of amino acids other than threonine.

Table 3 Threonine Production by Transductionally Constructed Strains of *S. marcescens*[a]

Strain	lys C	thr A₁	thr A₂	hnr A	hnr B	etr	L-Thr produced (g l⁻¹)[b]
D-60	+	+	+	+	+	+	<0.1
HNr21	+	1	1	+	+	+	4.2
HNr53	+	+	+	+	2	+	0.7
HNr59	+	+	2	1	+	+	4.3
AECr174	1	+	+	+	+	+	7.4
Etr17	+	+	−	+	+	1	7.5
T-570	+	1	1	1	+	+	8.8
T-904	1	+	+	+	2	+	10.2
T-1021	1	+	+	+	2	1	28.9
T-1025	1	1	1	1	2	+	25.8
T-1026	1	1	1	1	2	1	40.3

[a] Kisumi, 1982. [b] Medium: 15% sucrose, 1.5% urea, 0.05% $(NH_4)_2SO_4$, 0.1% K_2HPO_4, 0.1% $MgSO_4 \cdot 7 H_2O$, 0.2% C.S.L., 10mM Ile, 10mM Met and 1% $CaCO_3$, 96–120 h.

Recently Debanov (1982) presented a paper reporting that he has obtained *E. coli* K-12 strains capable of threonine overproduction by using methods of selection for resistance to a structural analog. In such strains, the threonine operon (the entire operon or part of it) is combined with the pBR322 plasmid and the hybrid plasmids are transformed into appropriate recipients. In addition, the genes involved in saccharose utilization have been introduced into the strains obtained. Special mutations in *E. coli* chromosomes ensure the stability of strains carrying hybrid plasmids. The optimization of fermentation conditions enabled producers yielding up to 55 mg ml⁻¹ of threonine when saccharose conversion into amino acids exceeds 40% to be obtained.

31.5 DISCUSSION

Among the enzymes concerned with the synthesis of threonine in bacteria, aspartate kinase and homoserine dehydrogenase are considered to play a dominant role in feedback controls of threonine synthesis (Umbarger, 1978). Recently, it was demonstrated that aspartate kinase I activity in *E. coli* is associated with a second activity, homoserine dehydrogenase I, carried on a bifunctional enzyme, which is a tetramer containing a single kind of subunit (Starnes *et al*, 1972). The polypeptide chain contains two distinct functional regions: one carried by the amino-terminal portion yields a threonine-sensitive aspartate kinase devoid of homoserine dehydrogenase activity and the other carried by the carboxy-terminal portion exhibits only homoserine dehydrogenase activity. The bifunctional enzyme, aspartate kinase I–homoserine dehydrogenase I, is specified by the *thrA* gene in *E. coli* and multivalently repressed by threonine plus isoleucine. When selecting threonine producers, therefore, it is absolutely necessary to obtain mutants in which aspartate kinase I–homoserine dehydrogenase I has become insensitive to threonine. The feedback controls by isoleucine could be avoided by introducing isoleucine auxotrophy into the mutant. As shown in Table 1, one of the *E. coli* AHV resistant mutants, strain β-101, accumulated about 1.9 mg ml⁻¹ of L-threonine, and an isoleucine auxotroph, βI-67, derived from β-101 produced about 4.7 mg ml⁻¹.

In *E. coli* K-12, aspartate kinase II is also associated with a homoserine dehydrogenase activity in a bifunctional protein. Neither activity is inhibited by methionine or any other amino acid of the aspartate family. The enzyme is, however, repressed by methionine. Aspartate kinase II–homoserine dehydrogenase II is a dimer of identical subunits. Thus, we can anticipate further enhancement of threonine production by using a methionine auxotroph derived from the βI-67 strain described above. In fact, an isoleucine and methionine double auxotroph, strain βIM-4, was selected and shown to accumulate threonine more than the parent strain βI-67.

Aspartate kinase III in *E. coli* appears not to be associated with any additional activities. It is a dimer of identical polypeptide chains and is both repressed and inhibited by lysine (Funkhouser *et al.*, 1974; Richaud *et al.*, 1974). The addition of lysine requirement to isoleucine plus methionine auxotrophs, however, was rather inhibitory for threonine production (see strain βIML-6 in Table 1). A possible explanation of this negative effect could be that the lysine limiting conditions may relieve the enzyme of the first step in the lysine biosynthetic pathway from feedback inhibition, causing the consumption of much larger amounts of the common intermediate to the lysine pathway than occurs in the parent strains, and may thus decrease threonine production. To avoid the negative effect of lysine limitation on threonine production, it is necessary to obtain a lysine auxotroph lacking the first step enzyme in the lysine pathway.

Most attempts to obtain better threonine-producing strains have been performed by adding sequentially required mutations to a single strain. However, this method does not always provide multiple mutants that would have all of the required mutations. It is difficult to obtain mutants that completely lack individual regulatory mechanisms because of cross-resistance to analogs or the reversion of prior mutations. These problems have been solved by applying a transduction technique. Using *S. marcescens* individual mutations to relieve feedback controls by threonine, isoleucine, methionine and lysine can be independently selected, and combined by cotransducing with appropriate selecting markers. Thus, the constructed strain of *S. marcescens* T-1026 exhibited the highest productivity of threonine (Table 3).

Although some undesirable problems may arise, such as the instability of the gene introduced and the variability of the microorganisms constructed when these newly constructed strains are used on an industrial scale, genetic manipulation including transduction, transformation or cell fusion will become increasingly the preferred routes for future microbial production of not only threonine but other amino acids, the biosyntheses of which are under strict feedback controls in wild strains.

31.6 SUMMARY

Microbial production of L-threonine by various mutant strains relieved of feedback controls of the aspartate family of amino acids is reviewed. Among the microorganisms examined, *Escherichia coli*, *Serratia marcescens* and *Brevibacterium flavum* seem to be promising microbes for the production of threonine.

To enhance the threonine productivity of these bacteria, various mutants resistant to α-amino-β-hydroxyvaleric acid and/or S-(2-aminoethyl)-L-cysteine, which are analogs of methionine, were selected. In addition, isoleucine, methionine or lysine auxotrophs were derived from the resistant strains. The best producers obtained in *E. coli*, *S. marcescens* and *Br. flavum* accumulated about 14 mg ml^{-1}, 15 mg ml^{-1} and 17.5 mg ml^{-1} of threonine, respectively.

Recently, attempts to construct a threonine-producing strain of *Serratia marcescens* have been successful. One of the transductants accumulated about 40 mg ml^{-1} of threonine, indicating that new genetic techniques will provide the most useful method for the production of not only threonine but other amino acids.

31.7 REFERENCES

Black, S. and N. G. Wright (1955). β-Aspartokinase and β-aspartyl phosphate. *J. Biol. Chem.*, **213**, 27–38.
Cohen, G. N. and M. L. Hirsh (1954). Threonine synthease: A system synthesizing L-threonine from L-homoserine. *J. Bacteriol.*, **67**, 182–190.
Cohen, G. N. and J. C. Patte (1963). Some aspects of the regulation of amino acid biosynthesis in a branched pathway. *Cold Spring Harbor Symp. Quant. Biol..*, **28**, 513–516.
Debabov, V. (1982). Construction of strains producing L-threonine. *Proceedings of the Fourth International Symposium on Genetics of Industrial Microorganisms*, pp. 254–258. Kodansha, Tokyo.
Funkhouser, J. D., A. Abraham, A. V. Smith and W. G. Smith (1974). Kinetic and molecular properties of lysine-sensitive aspartokinase. *J. Biol. Chem.*, **249**, 5478–5484.
Hirakawa, T. (1973). Effect of temperature on threonine production. *Agric. Biol. Chem.*, **37**, 243–250.
Hirakawa, T., T. Tanaka and K. Watanabe (1973). L-Threonine production by auxotrophs of *E. coli*. *Agric. Biol. Chem.*, **37**, 123–130.
Hirakawa, T. and K. Watanabe (1974). Mechanism of L-threonine production in *E. coli* auxotrophs. *Agric. Biol Chem.*, **38**, 77–84.
Hirakawa, T., H. Morinaga and K. Watanabe (1974). Effect of the antibiotic, borrelidin, on the production of L-threonine by *E. coli* auxotrophs. *Agric. Biol. Chem.*, **38**, 85–89.
Huang, H. T. (1961). Production of L-threonine by auxotrophic mutants of *E. coli*. *Appl. Microbiol.*, **9**, 419–424.
Kase, H. and K. Nakayama (1973). L-Threonine production by a mutant of *Arthrobacter paraffinens*. *Agric. Biol. Chem.*, **37**, 1643–1649.
Kase, H., H. Tanaka and K. Nakayama (1971). Studies on L-threonine fermentation Part I. Production of L-threonine by auxotrophic mutants of various bacteria. *Agric. Biol. Chem.*, **35**, 2089–2096.

Kisumi, M. (1982). Transductional construction of amino acid producing strains of *Serratia marcescens*. *Proceedings of the Fourth International Symposium on Genetics of Industrial Microorganisms*, pp. 247–253. Kodansha, Tokyo.

Komatsubara, S., M. Kisumi, K. Murata and I. Chibata (1978a). Threonine production by regulatory mutants of *Serratia marcescens*. *Appl. Environ. Microbiol.*, **35**, 834–840.

Komatsubara, S., K. Murata, M. Kisumi and I. Chibata (1978b). Threonine degradation by *Serratia marcescens*. *J. Bacteriol.*, **135**, 318–323.

Komatsubara, S., M. Kisumi and I. Chibata (1979a). Participation of lysine-sensitive aspartokinase in threonine production by S-2-aminoethylcysteine-resistant mutants of *Serratia marcescens*. *Appl. Environ Microbiol.*, **38**, 777–782.

Komatsubara, S., M. Kisumi and I. Chibata (1979b). Transductional construction of a threonine-producing strain of *Serratia marcescens*. *Appl. Environ. Microbiol.*, **38**, 1045–1051.

Nakamori, S. and I. Shiio (1972). Microbial production of L-threonine Part III. Production by methionine and lysine auxotrophs derived from α-amino-β-hydroxyvaleric acid-resistant mutants of *Brevibacterium flavum*. *Agric. Biol. Chem.*, **36**, 1209–1216.

Nakamori, S. and I. Shiio (1973). Production of L-threonine by mutants resistant to both α-amino-β-hydroxyvaleric acid and S-(2-aminoethyl)-L-cysteine derived from *Brevibacterium flavum*. *Agric. Biol. Chem.*, **37**, 653–659.

Patte, J. C., G. L. Bras and G. N. Cohen (1967). Regulation by methionine of the synthesis of a third aspartokinase and of a second homoserine dehydrogenase in *Escherichia coli* K-12. *Biochem. Biophys. Acta*, **136**, 245–257.

Richaud, C., J. P. Mazat, C. C. Gros and J.-C. Patte (1974). Subunit structure of aspartokinase III of *Escherichia coli* K-12. *Eur. J. Biochem.*, **40**, 619–629.

Rose, W. C. (1931–1932). Feeding experiments with mixtures of highly purified amino acids. I. The inadequacy of diets containing nineteen amino acids. *J. Biol. Chem.*, **94**, 155.

Rose, W. C., R. H. McCoy, C. E. Meyer, H. E. Carter, M. Womack and E. T. Mertz (1935). Isolation of the 'unkown essential' present in proteins. *J. Biol. Chem.*, **109**, 77.

Schryver, S. B. and H. W. Buston (1925–1926). The isolation of some hitherto undescribed products of hydrolysis of proteins. Part II. *Proc. R. Soc. London, Ser. B*, **99**, 476–487.

Shiio, I. and S. Nakamori (1969). Microbial production of L-threonine. Part I. Production by *Escherichia coli* mutant resistant to α-amino-β-hydroxyvaleric acid. *Agric. Biol. Chem.*, **33**, 1152–1160.

Shiio, I. and S. Nakamori (1970). Microbial production of L-threonine. Part II. Production by α-amino-β-hydroxyvaleric acid resistant mutants of glutamate producing bacteria. *Agric. Biol. Chem.*, **34**, 448–456.

Shiio, I., R. Miyajima and S. Nakamori (1970). Homoserine dehydrogenase genetically desensitized to the feedback inhibition in *Brevibacterium flavum*. *J. Biochem.*, **68**, 859–866.

Shimura, K. (1972). Threonine. In *The Microbial Production of Amino Acids*, ed. K. Yamada, chap. 17, pp. 453–472. Kodansha, Tokyo.

Starnes, W. L., P. Munk, S. B. Maul, G. N. Cunningham, D. J. Cox and W. Shive (1972). Threonine-sensitive aspartokinase–homoserine dehydrogenase complex, amino acid composition, molecular weight, and subunit composition of the complex. *Biochemistry*, **11**, 677–687.

Teas, H. J., N. H. Horowitz and M. Fling (1948). Homoserine as a precursor of threonine and methionine in *Neurospora*. *J. Biol. Chem.*, **172**, 651–658.

Teas, H. J. (1950). Mutants of *Bacillus subtilis* that require threonine plus methionine. *J. Bacteriol.*, **59**, 93–104.

Umbarger, H. E. (1978). Amino acid biosynthesis and its regulation. *Annu. Rev. Biochem.*, **47**, 533–606.

Watanabe, Y., S. Konishi and K. Shimura (1957). Biosynthesis of threonine from homoserine. VI. Homoserine kinase. *J. Biochem.*, **44**, 299–307.

32

5'-Guanosine Monophosphate

H. ENEI, H. SHIBAI and Y. HIROSE
Ajinomoto Co. Inc., Kawasaki, Japan

32.1 INTRODUCTION

In Japan, konbu (kelp-like seaweed), katsubushi (dried bonito) and shiitake (a type of mushroom) have been habitually used as flavor-enhancing materials. The flavor enhancing effects of konbu and katsubushi were found to be due to monosodium glutamate and the histidine salt of 5'-inosine monophosphate (IMP), respectively. In 1960 Kuninaka *et al.* found that 5'-guanosine monophosphate (GMP) was also a flavor enhancing component. Its action is more powerful than that of IMP and it is now produced in Japan at the rate of 1000 tons per annum. The methods for the production of GMP are as follows: (1) enzymatic hydrolysis of yeast ribonucleic acid (RNA); (2) chemical synthesis from 5-amino-4-imidazolecarboxamide riboside (AICAR); (3) chemical phosphorylation of guanosine; (4) enzymatic conversion of 5'-xanthosine monophosphate (XMP) to GMP; and (5) direct fermentation from glucose. The research and development of these production processes was mainly undertaken in Japan by applying recent techniques of microorganism mutation and optimization of fermentation conditions.

32.2 REGULATION OF GMP BIOSYNTHESIS

Fundamental studies of metabolic regulation have provided an effective guide for the breeding of mutants with enhanced purine nucleotide and nucleoside productivity. The biosynthetic pathways of purine nucleotides were elucidated by Buchanan, employing a homogenate of pigeon liver (Figure 1; Buchanan, 1972) and have been confirmed to operate in microorganisms belonging to the genera *Enterobacter* and *Bacillus*.

Negative feedback inhibition and repression were reported to be important in the regulation of purine nucleotide biosynthesis. In *Bacillus subtilis* (Shiio, 1972), phosphoribosylpyrophosphate amidotransferase, the first enzyme of the purine nucleotide biosynthetic pathway, was competitively inhibited by adenine derivatives such as AMP and ADP, while guanine derivatives showed only weak inhibition. IMP dehydrogenase, the first enzyme of the GMP-specific pathway, was strongly inhibited by GMP and XMP, but only negligibly by AMP and ADP. The inhibition by GMP was competitive with IMP. GMP synthetase, the last enzyme of the pathway, was slightly inhibited by GMP and AMP.

Three enzymes of the common pathway of AMP and GMP synthesis, phosphoribosylpyro-

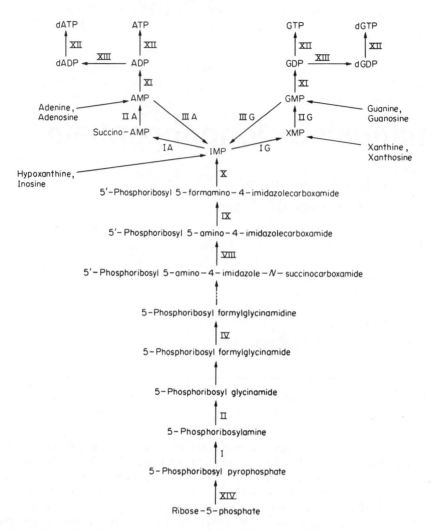

Figure 1 Biosynthetic pathway of purine nucleotides. I, 5-phosphoribosylpyrophosphate amidotransferase; IA, adeny-losuccinate synthetase; IG, IMP aminase; II, 5-phosphoribosylglycinamide synthetase; IIA, VIII, adenylosuccinate lyase; IIG, XMP reductase; IIIA, AMP deaminase; IIIG, GMP reductase; IV, 5-phosphoribosylformylglycinamide amidotransferase; IX, 5'-phosphoribosyl-5-amino-4-imidazolecarboxamide transformylase; X, IMP cyclohydrolase; XI, nucleotide kinase; XII, nucleoside diphosphate kinase; XIII, ribonucleotide reductase; XIV, 5-phosphoribosylpyrophosphate synthetase

phosphate amidotransferase, adenylosuccinate lyase and IMP transformylase, are repressed by end products such as adenosine and guanosine, while the intermediate derivatives such as inosine and xanthosine show only partial repression. IMP dehydrogenase, the enzyme specific for GMP synthesis, is repressed by guanosine, although xanthosine and inosine have no repressive effect.

Among the enzymes that are involved in the biosynthesis of nucleotides, nucleosides and bases, 5'-nucleotidase and nucleosidase have been shown to play important roles in nucleotide and nucleoside fermentation (Figure 2).

32.3 PRODUCTION OF GMP

32.3.1 Enzymatic Hydrolysis of Yeast RNA

The production of 5'-nucleotides from yeast RNA involves the production of an RNA-hydro-lyzing enzyme, hydrolysis of RNA by the enzyme, and separation and purification of 5'-nucleotides from the hydrolyzate.

A microbial enzyme which hydrolyzes yeast RNA to 5'-nucleotides was first obtained from a water extract of a solid culture of *Penicillium citrinum* on wheat bran by Kuninaka *et al.* (1959,

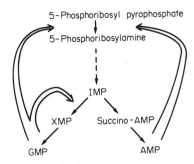

Figure 2 Regulation of GMP biosynthesis in *Bacillus subtilis*: ◁▭, inhibition and repression

1961). Ogata *et al.* (1963), after extensive studies, found that similar enzymes are distributed in the culture broth of a variety of microorganisms belonging to the Ascomycetes, Fungi Imperfecti, actinomycetes and bacteria. Strain improvement was carried out to enhance the production of endonuclease, exonuclease and AMP deaminase, and to reduce that of 5'-nucleotidase and alkaline phosphatase. The enzymes now used industrially are obtained from cultures of *Penicillium citrinum* and *Streptomyces aureus*.

Yeast cells, the source of RNA, are cultivated in a medium composed of waste from the sulfite pulp process. The RNA content in the yeasts varies between 10–15% of the dry cell weight. RNA is extracted from the cells with dilute alkaline NaCl solution and precipitated by acidification or the addition of an organic solvent to the extracted solution. With heat-treated yeast cells, the process of RNA extraction is omitted, and the cells are subjected to the enzyme reaction.

Enzyme solutions extracted from solid cultures, or the culture filtrate, contain many enzymes that hydrolyze RNA, nucleotides and nucleosides. Kuninaka (1961) increased the yield of 5'-nucleotides by up to 90% of the original RNA content by hydrolyzing the RNA at 65 °C and pH 5.0. The heat-labile phosphomonoesterase was shown to be inactivated.

32.3.2 Chemical Synthesis from AICAR

AICAR is produced by a fermentative process and separated from culture liquid by adsorption on a cation exchange resin, followed by elution, concentration and drying. The process for synthesizing GMP from AICAR is shown in Figure 3. AICAR is allowed to react with sodium ethylxanthate, giving 2-mercaptoinosine. The 2-mercapto derivative is then oxidized to give inosine-2-sulfonic acid and then aminated by ammonia to produce guanosine. Guanosine is phosphorylated with phosphoryl chloride to give 5'-GMP (Yamazaki *et al.*, 1967).

To produce sufficient quantities of AICAR the inhibitors of regulation of purine synthesis must be overcome. The main requirements for obtaining AICAR-producing mutants are as follows: (1) purine synthesizing capacity of the mutants should be high; (2) the mutants should lack AICARP formyltransferase, which is responsible for the conversion of AICARP to FAICARP; (3) the mutants should be free from feedback inhibition and repression of biosynthetic enzymes, especially of PRPP amidotransferase by intracellular purine nucleotides; and (4) the mutants should be deficient in AICAR hydrolyzing activity. An excellent producer of AICAR, *Bacillus megaterium* No. 335, was obtained by Kinoshita *et al.*; this strain satisfied all of the above conditions and excreted 16 g l^{-1} of AICAR from 80 g l^{-1} of glucose (Kinoshita *et al.*, 1969).

In AICAR fermentation using a spore-forming *Bacillus megaterium*, the yield is markedly influenced by the degree of spore formation. Since only vegetative cells form AICAR, it is essential to restrict the spore formation for sufficient production of AICAR. The vegetative cells do not form spores when cell respiration is restricted ($R_{ab}/Kr_M < 1.00$); however, AICAR accumulation is seriously depressed in oxygen-starved culture medium. On the other hand, when oxygen is in excess ($R_{ab}/Kr_M = 1.0$), a large amount of spores appear and AICAR production is substantially reduced. Therefore, two incompatible factors must be satisfied in a culture: one is sufficient oxygen supply to ensure AICAR biosynthesis, and the other is insufficient oxygen supply to restrict spore formation. It is only during a period 8–12 h after the beginning of the culture that spore formation is absolutely influenced by oxygen. When cell respiration is satisfied during this period, a large number of spores are formed during the stationary phase of cell growth. In contrast, when cell respiration is not satisfied at this time, spores are not formed at the stationary

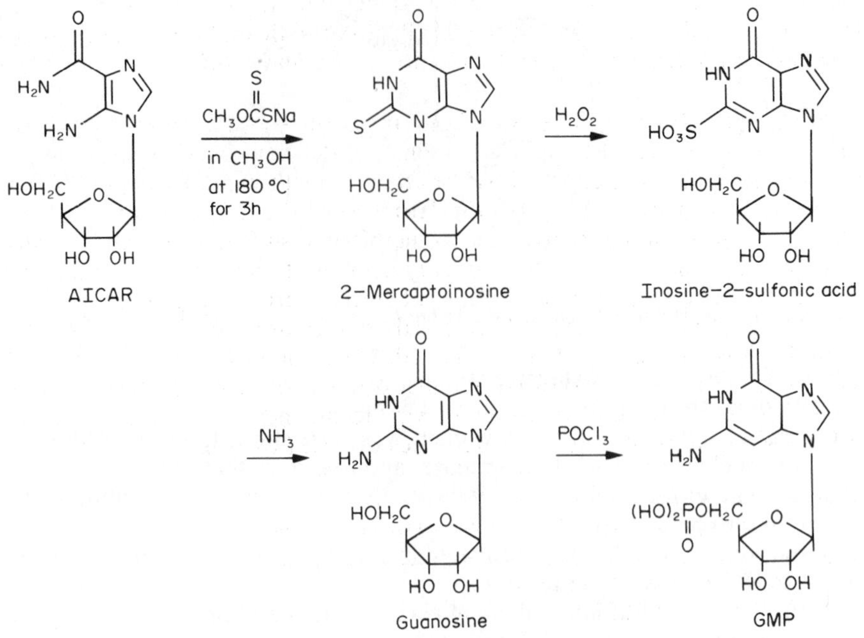

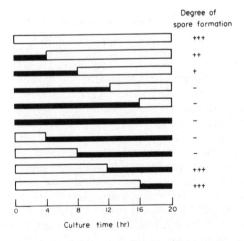

Figure 3 Synthesis of GMP from AICAR

phase. Therefore, oxygen supply must be controlled at a level between 0.85 and 1.0 R_{ab}/Kr_M during the critical period, and 1.0 throughout the other periods of fermentation. Under these conditions, satisfactory accumulation of AICAR is observed (Figure 4; Kinoshita *et al.*, 1969).

Figure 4 Effect of oxygen tension on spore formation: □, sufficient oxygen supply (dissolved oxygen above critical value for respiration); ■, insufficient oxygen supply (dissolved oxygen recorded as zero by the membrane coated oxygen electrode)

32.3.3 Phosphorylation of Guanosine

Guanosine, an intermediate in the chemical synthesis of 5'-GMP from AICAR (Figure 3), was produced from glucose by mutants of *Bacillus subtilis*, and chemically phosphorylated to form 5'-GMP for industrial production.

In order to derive guanosine-producing strains, it is necessary to satisfy the following conditions based on the regulatory mechanisms in purine nucleotide biosynthesis: (1) adenylosuccinate synthetase deficiency; (2) GMP reductase deficiency; (3) deficiency in purine nucleotide hydrolyzing activity; and (4) enzymes of the GMP biosynthetic pathway, especially phosphoribosylpyrophosphate amidotransferase, IMP dehydrogenase and GMP synthetase being free from regulation.

The best producer, MG-1, was derived from an inosine- and guanosine-producing strain of *Bacillus subtilis*, which was a histidine auxotroph, and deficient in adenylosuccinate synthetase, GMP reductase and nucleosidase. MG-1 was resistant to methionine sulfoxide, psicofuranine and decoynine, and produced 16 g l^{-1} of guanosine from 80 g l^{-1} of glucose (Enei, 1976; Matsi *et al.*, 1978, 1979a, b, c). In the mutants resistant to methionine sulfoxide, which is a glutamine analogue, the activity of 5'-nucleotidase was greatly decreased and conversely the activity of IMP dehydrogenase was larger than that of the parent strain. Furthermore, in the psicofuranine and decoynine resistant mutants, GMP synthetase activity increased remarkably, and the repression and inhibition by GMP were completely removed. Consequently, guanosine was overproduced by GMP synthetic enzymes free from the feedback effects of GMP, and accumulated at a high yield in preference to inosine and xanthosine owing to the increased enzyme level of IMP dehydrogenase and GMP synthetase rather than 5'-nucleotidase (Table 1).

Table 1 Derivation of Guanosine Producers

Strain	Phenotype	Productivity (g l^{-1}) IR	XR	GR
1411	ade$^-$, his$^-$, GMP reductase$^-$	11.0	0	5.5
14119	ade$^-$, his$^-$, GMP reductase$^-$, methionine sulfoxide^r	4.8	0	9.6
AG169	ade$^-$, his$^-$, GMP reductase$^-$, methionine sulfoxide^{rr}	0	6.0	8.0
GP-1	ade$^-$, his$^-$, GMP reductase$^-$, methionine sulfoxide^{rr}, psicofuranine^r	0	3.4	10.6
MG-1	ade$^-$, his$^-$, GMP reductase$^-$, methionine sulfoxide^{rr}, psicofuranine^r, decoyinine^r	0	0	16.0

32.3.4 Microbial Conversion of XMP to GMP

To produce GMP, a mixed culture of two different mutants of *Brevibacterium ammoniagenes* was examined, one of which produces XMP from glucose, the other converting XMP into GMP. A maximum production of 2.5 g l^{-1} of GMP was obtained from 100 g l^{-1} of glucose by selecting suitable inoculum proportions of the two strains and by feeding glucose and urea during the culture (Furuya, 1973).

32.3.5 Fermentative Accumulation of GMP from Glucose

Genetic breeding of strains producing GMP from glucose are summarized in Table 2 (Okumura *et al.*, 1964; Momose, 1968; Abe and Udagawa, 1966a, 1966b). The permeability barrier of the cell membrane to GMP was the major problem. GMP, though accumulated within the cells as a result of enhanced biosynthesis by mutation, did not permeate through the cell membrane, but is excreted in the form of guanosine.

Table 2 Fermentative Production of GMP

Microorganism	Yield (g l^{-1})	(%)	Ref.
Bacillus subtilis (ade$^-$, amino acid$^-$)	2.2	3	Okumura *et al.*, 1967
Escherichia coli (ade$^-$, nucleotidase^{weak})	0.9	1.8	Momose, 1968
Micrococcus glutamicus (ade$^-$)	5.1	5.1	Abe and Udagawa, 1966a
Brevibacterium ammoniagenes (ade$^-$)	4.6	4.6	Abe and Udagawa, 1966b

658 *Food and Beverage Products*

32.4 REFERENCES

Abe, S. and K. Udagawa (1966a). Fermentative production of 5'-guanosine monophosphate. *Jpn. Pat.* 66 12 690.
Abe, S. and K. Udagawa (1966b). Fermentative production of 5'-guanosine monophosphate. *Jpn. Pat.* 66 12 796.
Buchanan, M. J. (1972). In *Nucleic Acids*, ed. E. Chargaff and N. J. Davidson, vol. III, p. 303. Academic, New York.
Enei, H. (1976). *Proc. Ann. Meet. Agric. Chem. Soc. Jpn.*, 119.
Furuya, A., R. Okachi, K. Takayama and S. Abe (1973). Accumulation of 5-guanine nucleotides by mutants of *Brevibacterium ammoniagenes*. *Biotechnol. Bioeng.*, **15**, 795.
Kinoshita, K., H. Tsuri, S. Sakai, H. Yasunaga, H. Okada and T. Shiro (1969a). Fermentative production of 4-amino-5-imidazole carboxamide riboside (AICA-riboside). III. Identification of gluconate as a metabolic intermediate. *J. Agric. Chem. Soc. Jpn.*, **43**, 400–403.
Kinoshita, K., S. Sakai, M. Yasunaga, H. Sasaki and T. Shiro (1969b). Fermentative production of 4-amino-5-imidazole carboxamide riboside (AICA-riboside). V. Mechanism of the antisporulating effect of butyric acid. *J. Agric. Chem. Soc. Jpn.*, **43**, 473–477.
Kuninaka, A., S. Otsuka, Y. Kobayashi and K. Sakaguchi (1959). Studies on 5'-phosphodiesterases in microorganisms. I. Formation of nucleoside-5'-monophosphate from yeast ribonucleic acid by *Penicillium citrinum*. *Agric. Biol. Chem.*, **23**, 239.
Kuninaka, A., M. Kibi, H. Yoshino and K. Sakaguchi (1961). Studies on 5'-phosphodiesterase in microorganisms. II. Properties and application of *Penicillium citrinum* 5'-phosphodiesterase. *Agric. Biol. Chem.*, **25**, 693.
Matsui, H., K. Sato, H. Enei and Y. Hirose (1978). Production of guanosine by sulfaguanidine resistant mutants of *Bacillus subtilis*. *Agric. Biol. Chem.*, **42**, 637.
Matsui, H., K. Sato, H. Enei and Y. Hirose (1979a). A guanosine-producing mutant of *Bacillus subtilis* with high productivity. *Agric. Biol. Chem.*, **43**, 393.
Matsui, H., K. Sato, H. Enei and Y. Hirose (1979b). Guanosine production and purine nucleotide biosynthetic enzymes in guanosine producing mutants of *Bacillus subtilis*. *Agric. Biol. Chem.*, **43**, 1317.
Matsui, H., K. Sato, H. Enei, H. Shibai and Y. Hirose (1979c). Medium components essential for guanosine production in *Bacillus subtilis*. *Agric. Biol. Chem.*, **43**, 1325.
Momose, H. (1968). Production of 5'-guanosine-monophosphate by microorganisms. *Jpn. Pat.* 68 11 760.
Ogata, K., Y. Nakao, S. Igarashi, E. Omura, Y. Sugino, M. Yoneda and I. Suhara (1963). Degradation of nucleic acids and their related compounds by microbial enzymes. I. Distribution of extracellular enzymes capable of degrading ribonucleic acid into 5'-mononucleotides in microorganisms. *Agric. Biol. Chem.*, **27**, 110.
Okumura, S., R. Aoki, K. Komogata, T. Shiro, N. Matsumura, T. Tsunoda, M. Takahaski and S. Motozaki (1967). Production of purine nucleoside by microorganisms. *Jpn. Pat.* 67 16 347.
Shiio, I. (1972). Regulation of nucleotide biosynthesis. *Seikagaku*, **44**, 7.
Yamazaki, A., I. Kumashiro and T. Takenishi (1967). Synthesis of guanosine and its derivatives from 5-amino-1-beta-D-ribofuranosyl-4-imidazolecarboxamide. I. Ring closure with benzoyl isothiocyanate. *J. Org. Chem.*, **32**, 1825.

INDUSTRIAL CHEMICALS, BIOCHEMICALS AND FUELS

33
Introduction

H. W. BLANCH
University of California, Berkeley, CA, USA

33.1 INTRODUCTION

The use of microorganisms to produce chemicals and fuels in the broadest sense dates back to the production of ethanol and acetic acid (as vinegar) in prehistoric times. The earliest recorded production of beverage ethanol is reported around 2000 BC in Assyria. This chapter reviews the most significant chemicals and fuels currently or formerly produced by biological routes. In some cases this includes single or multiple enzyme reaction steps, as is the case with the production of fructose and glucose from starch-containing materials. A number of speciality chemicals, such as steroids, are produced by enzymatic or whole cell routes and these are considered in other sections in this series. Such specialized activities as enzymatic resolution of enantiomers are covered in Volume 2 of the series. The market for each of the major chemicals is described in each section and where possible the companies involved in production are indicated.

The acetone–butanol fermentation is illustrative of the history and use of biological methods for the production of chemicals. Although Pasteur observed in the mid 19th century that butanol could be formed as a by-product in the butyric acid fermentation, commercial production of acetone and butanol, one of the earliest commercial processes, did not occur until the discovery of the organism *Clostridium acetobutylicum* around 1914. This organism was able to ferment starchy materials, such as corn, to acetone, butanol and ethanol. The first plants were built in Canada and then later in the United States (at Terre Haute, Indiana) during World War I to provide acetone, which was an important raw material in the manufacture of cordite, used as a propellant for heavy artillery and in small arms munition. With the development of nitrocellulose lacquers for the automobile industry by Du Pont, butyl acetate was used as the solvent of choice and the acetone–butanol fermentation was started as a private venture in the United States.

The manufacture of commodity chemicals by biological means was made economically unattractive to a large extent by the growth of the petroleum-based chemical industry following

World War I and it has not been till recently that the cost of petroleum has risen to a point where the reintroduction of renewable biological raw materials has initiated a re-evaluation of biological routes to commodity and some intermediate chemicals.

This section contains chapters on the key chemicals in these areas; some of the small volume speciality chemicals which are now produced from synthetic routes are not included, but are reviewed in the following sections.

33.2   ORGANIC ACIDS

33.2.1   Oxygluconic Acids

The production of 2-oxygluconic acid was once a commercial undertaking, the acid being used for the production of erythorbic acid and its esters, which are water or fat soluble antioxidants used in canned vegetables, fruits and meats. The fermentation of glucose by *Gluconobacter suboxydans* produces a mixture of 2-oxygluconate and 5-oxygluconate *via* the gluconolactone intermediate. *Pseudomonas* species may yield 2-oxygluconate in weight yields in excess of 90%. Details of the fermentation, which is similar to the gluconic acid production process, may be found in Lockwood (1979).

33.2.2   Fumaric Acid

Fumaric acid was formerly produced by fermentation, but is now produced by hydrocarbon oxidation. Many *Rhizopus* strains produce fumaric acid in large amounts from glucose. It arises in the citric acid cycle from succinic acid by the action of succinic dehydrogenase, although it has been reported that ethanol can serve as a primary carbon source with 50 to 80% being converted to fumaric acid. Fumaric acid can also be formed from the glyoxylic acid cycle, *via* acetate. Fumaric acid is used in the manufacture of polyester fabrics and for acidification of beverages. It is finding increased use in the manufacture of the amino acid aspartic acid, by use of fumarase in a variety of immobilized forms. Aspartic acid is a component of the dipeptide sweetener aspartame. Details of the fermentation process for the production of fumaric acid can be found in Rehm (1967a).

33.2.3   Kojic Acid

Kojic acid is produced in rather small amounts *via* a variety of *Aspergillus* species. It can be produced from glucose, xylose and other pentoses. Kojic acid can also be chemically synthesized from glucose and consists of a six-membered ring structure, with one atom of oxygen in the ring. The fermentation operates at a pH of 2 to 5, with lower pH favoring the production of the acid. Ammonium sulfate usually serves as the nitrogen source. The fermentation runs for 9 to 20 days, with a yield of kojic acid of 0.5 to 0.6 g acid/g glucose fermented. Kojic acid is recovered by the addition of $ZnSO_4$ or related salts which form a metal chelate with kojic acid; the chelate can be filtered from the fermentation broth.

Kojic acid finds use as an antibiotic and has properties as an insecticide. It is also used as a reagent in the determination of iron concentration.

33.2.4   Gallic Acid

Gallic acid was produced in the United States around 1916 from tannase produced by *Aspergillus wentii* grown on extracts of gallotannins. Gallic acid is a 3,4,5-trihydroxybenzoic acid. Usual fermentation time is 48 hours with aeration and agitation discontinued at some point in the fermentation.

Gallic acid finds use in printing and in the pharmaceutical industry. It is the raw material for the production of alizarin brown; condensation reactions with sulfuric acid yield hexahydroxyanthraquinone. Gallic acid is also used in the production of other dyes and pigments. It is currently produced by hydrolysis of tannins so that the fermentation route is no longer employed. Further details can be obtained in Rehm (1967b).

33.2.5 Pyruvic Acid

Although pyruvic acid is not currently produced commercially, it may find use as a raw material in the production of the amino acids tryptophan, tyrosine and L-DOPA *via* enzymatic transformation. It can be produced by fermentation of glucose with several strains of *E. coli* and attractive yields have been reported (El-Sayed, 1980; Asai *et al.*, 1952).

33.2.6 Tartaric Acid

Tartaric acid is currently produced from the precipitation of potassium tartrate during the manufacture of wine and finds use in the food industry. The annual production of tartaric acid is approximately 40 000 to 45 000 tons, with the supply and demand of the market being fairly well balanced. However, there has been some research on the biological production of tartaric acid by fermentation using *Acetobacter* and *Gluconobacter* species. The conversion of glucose to 5-oxogluconic acid is the starting point for the fermentative production. Continued fermentation with certain strains results in the conversion of 5-oxogluconic acid to tartrate. Final concentrations in the process are low, as are the yields (0.05 g g^{-1}), so the process is economically unattractive. Epoxysuccinic acid has also been used as a starting material with microbial transformation to tartrate (Kamatani *et al.*, 1976). The fermentation route is described in Yamada *et al.* (1969).

33.2.7 Poly(hydroxybutyric Acid)

Many microorganisms have been known to store poly(hydroxybutyric acid) as an energy source. It has received attention recently as a possible polymeric material to replace organic polymers of specific functionality. The difficulty in the past with the fermentation has been due to the expensive recovery process involved in extracting the intracellular material. The use of large quantities of solvent was required and costly and complex precipitation steps were involved. A recent patent describes the use of a liquid cyclic carbonic acid ester as a solvent for poly(hydroxybutyric acid) which avoids some of these difficulties (Lafferty and Heinzle, 1978).

33.3 GIBBERELLINS

Gibberellins are diterpenoids containing four cyclic carbon rings. The gibberellins produced by the fungi have hydroxy and carboxylic acid functionality. The producing organisms are related to the strain *Gibberella fujikuroi*, which was first isolated from infected rice plants. Gibberellins were produced in surface culture originally, with glucose, saccharose or glycerine as the carbon source and a fermentation time of 15 to 30 days. Typical concentrations of 40 to 60 mg l^{-1} were reported. In submerged culture, a 170 to 350 hour fermentation results in concentrations up to 880 mg l^{-1}. The fermentation proceeds in four stages and the product is recovered on a cation exchanger. The main use of gibberellins is in the production of malt, where they accelerate the germination of the barley and increase the activities of α- and β-amylase. Details of the fermentation can be found in Rehm (1967c).

33.4 CAROTENOIDS

Carotenoids are liposoluble tetraterpenes, usually red or yellow in color, and are one of the most important families of natural pigments. They are precursors of vitamin A and are thus used as food supplements. They are used as food additives for modifying the color of fats, oils, cheese and beverages and are also used in animal feeds. The overall market for carotenes is around 30 tons/year, with approximately half of the market being within the United States. Most are produced by extraction of plants. A variety of microorganisms have been examined for the fermentative production of carotenes, including the *Mucorales* group of fungi and the green algae. The most widely examined fungus is *Blakeslea trispora*, which can be grown on vegetable or animal oils or on linolenic acid-containing materials. The carotenes are extracted from the mycelium with methylene chloride, washed with acetone and crystallized. The fermentation process and general information on related carotenes, including lycopene, xanthophylls and other pigments, are described in Ninet and Renault (1979).

33.5 OXIDATIVE AND REDUCTIVE REACTIONS

33.5.1 Sorbose

Sorbose is the key starting material for the production of vitamin C. It is produced by the action of *Acetobacter suboxydans*. Sorbitol is the substrate for the oxidation; it is produced by catalytic or electrolytic reduction of glucose. The submerged fermentation process involves a 20% concentration of sorbitol (higher concentrations are inhibitory to the organism), with the addition of yeast extract or corn-steep liquor. Intense aeration and agitation are required for maximal yields and higher sorbitol amounts can be fed by use of a fed-batch approach. The recovery procedure is relatively simple, as there is little contamination in the fermentation broth, so that evaporative crystallization of the sorbose is possible (Muller, 1960).

33.5.2 Dihydroxyacetone

Dihydroxyacetone is formed biologically from the dehydration of glycerine. A variety of organisms can carry out the reaction, including *Acetobacter suboxydans* and *Bacterium orleanse*. Early fermentation processes required long batch times of the order of 20 days for complete conversion of a 6% glycerine solution. A recent patent (Charney, 1978) describes the use of *A. suboxydans* (ATCC 621) with 9 to 12 wt% glycerol, yeast hydrolysate and a pH of 3.3 to 4.3. The fermentation is complete in 24 to 48 hours, with the product being recovered by filtration of the whole broth, removal of inorganic cations and anions by use of ion exchange, concentration of the eluant and crystallization of dihydroxyacetone from the concentrated eluant.

33.6 OTHER PRODUCTS

In addition to the chemicals described above, there are a number of examples of the use of enzymes for resolution of optical isomers and production of speciality chemicals that are described in other sections of this series. Details of some of these minor activities can be found in Duffy (1980).

33.7 REFERENCES

Asai, T., K. Aida and Y. Ueno (1952). Studies on the oxidative bacteria. Part 5. *J. Agric. Chem. Soc. Jpn.*, **26**, 625.
Charney, W. (1978). *US Pat.* 4 076 589.
Duffy, I. J. (1980). *Chemicals by Enzymatic and Microbial Processes—Recent Advances*. Noyes Data Corporation, New Jersey.
El-Sayed, E. (1980). Process for the production of pyruvic acid and citric acid. *US Pat.* 4 326 030.
Kamatani, Y., H. Okazali, K. Imai, N. Fujita, Y. Yamakazi and K. Ogino (1976). Process for the production of tartaric acid. *Ger. Pat.* 2 600 589.
Lafferty, R. M. and E. Heinzle (1978). *US Pat.* 4 101 533.
Lockwood, L. B. (1979). Production of organic acids by fermentation. In *Microbial Technology*, ed. H. J. Peppler and D. Perlman, vol. 2, chap. 11. Academic, New York.
Muller, J. (1960). *Zentral Bakteriol. Paristenk. Abt II*, **120**, 349.
Ninet, L. and J. Renault (1979). In *Microbial Technology*, ed. H. J. Pepler and D. Perlman, vol. 1, chap. 17, p. 529. Academic, New York.
Rehm, H. J. (1967a). *Industrielle Mikrobiologie*, p. 375. Springer-Verlag, Berlin.
Rehm H. J. (1967b). *Industrielle Mikrobiologie*, p. 383. Springer-Verlag, Berlin.
Rehm H. J. (1967c). *Industrielle Mikrobiologie*, p. 357. Springer-Verlag, Berlin.
Yamada, K., Y. Minoda, T. Kodama and U. Kotera (1969). In *Fermentation Advances*, ed. D. Perlman, pp. 541–560. Academic, New York.

34
Citric Acid

P. E. MILSOM and J. L. MEERS
John & E. Sturge Limited, Selby, North Yorkshire, UK

34.1 INTRODUCTION

Citric acid or 2-hydroxy-1,2,3-propanetricarboxylic acid (77-92-9) was first isolated from lemon juice by Scheele (1784). Although it occurs in rather high concentrations in citrus fruits, citric acid is ubiquitous in nature forming as it does an intermediate in the citric acid (Krebs) cycle whereby carbohydrates are oxidized to carbon dioxide. The widespread presence of citric acid in the animal and plant kingdoms is an assurance of its non-toxic nature and it has long been used as an acidulent in the manufacture of soft drinks, as an aid to the setting of jams and in other ways in the confectionery industry.

It was first produced commercially by John & Edmund Sturge in England in increasing amounts from 1826 onwards. It was made from Italian produced calcium citrate which derived from lemon juice and this method of manufacture was established in the US and in France and Germany well before the end of the century.

However, the Italians commenced to make citric acid themselves and during the first two decades of the twentieth century established what was essentially a monopoly in the product with concomitant high prices. Other means of citric acid production were sought.

In 1880 citric acid had been synthesized from glycerol (Grimoux and Adam, 1880) and since that time a number of syntheses from other raw materials by different routes have been published. This was especially the case around 1970 when the possibility of the widespread use of sodium citrate in detergents as a substitute for polyphosphates was being considered. All these synthetic methods proved to be unsuitable because of expensive or hazardous raw materials, or an excessive number of reaction steps leading to low yields.

The possibility of a fermentation route for citric acid manufacture was indicated by the discovery (Wehmer, 1893) that certain species of *Penicillium* (termed by Wehmer *Citromyces*) were able to accumulate significant quantities of citric acid when grown on solutions containing sugar. Although Wehmer attempted to use this finding to establish a commercial citric acid process he was unsuccessful. However, somewhat later Currie (1917) found that a strain of *Aspergillus niger* was able to produce better yields of citric acid. Currie subsequently joined Chas. Pfizer & Co. Inc. with the result that a citric acid plant using the new method opened in the US in 1923.

Subsequently fermentation processes were started in England, Belgium, Czechoslovakia and Germany. All these plants used the so-called surface process in which *A. niger* grows on static medium held in trays housed in ventilated rooms. At first only media prepared from sucrose and inorganic salts were employed (Doelger and Prescott, 1934) but soon processes based on the cheaper beet molasses were introduced.

Following the Second World War submerged fermentation processes for the production of citric acid using *A. niger* and media based on either purified glucose syrups, or beet or cane molasses were developed (Perlman, 1949).

From 1965 onwards there was progress towards the introduction of processes in which certain yeasts were used in the production of citric acid, first of all from carbohydrates, and then from *n*-alkanes (Stottmeister *et al.*, 1982). At the time the processes using hydrocarbons as substrate were being developed, petroleum products were still cheap and economic advantages were envisaged. However, the economics have remained in favour of carbohydrates and a plant built by Chas. Pfizer Inc. which is able to utilize either carbohydrates or *n*-alkanes is reported to have been switched back to carbohydrates after a period of use with hydrocarbons. Likewise a plant built by Liquichimica in southern Italy to produce 50 000 tonnes of sodium citrate per annum from *n*-alkanes has closed after only a brief trial period of operation.

34.2   USES AND PROPERTIES OF CITRIC ACID

Citric acid is produced either in the anhydrous form or as the monohydrate. The transition temperature between the two forms is 36.6 °C. Thus the anhydrous form is obtained by crystallization from hot aqueous solutions whereas the monohydrate is obtained by crystallization at temperatures below 36.6 °C. Both forms are utilized in commerce.

Sales of citric acid worldwide are divided amongst the principal fields of use approximately as follows: food, confectionery and beverages (75%), pharmaceutical (10%) and industrial (15%).

In food, sugar confectionery and beverages, citric acid is the most versatile and widely used food acidulent. The use of citric acid as a food acidulent depends in part on its strength as an acid. (At 18 °C, $K_1 = 8.2 \times 10^{-4}$, $K_2 = 1.77 \times 10^{-5}$ and $K_3 = 10^{-7}$; Grayson and Eckroth, 1979.) However, its pleasant taste and its property of enhancing existing flavours have ensured its dominant position in this market.

Citric acid is able to complex heavy metals such as iron and copper. This property has led to its increasing use as a stabilizer of oils and fats where it greatly reduces oxidation catalysed by these metals. The ability to complex metals combined with its low degree of attack on special steels allows the use of solutions of citric acid in the cleaning of power station boilers and similar installations.

The sequestering action of citric acid is also used in the pharmaceutical field, for example in the stabilization of ascorbic acid. Another large pharmaceutical use is based on the effervescent effect it produces when combined with carbonates and bicarbonates, for example in antacid and soluble aspirin preparations. Citric acid is often used as the anion in pharmaceutical preparations employing basic substances as the active agent.

Citric acid forms a wide range of metallic salts, many of which are articles of commerce. In terms of volume, trisodium and tripotassium citrates are probably the most important. Trisodium citrate is widely used as a blood preservative, where it prevents clotting by complexing calcium. It is also employed as an aid to emulsification in the manufacture of processed foodstuffs, *e.g.* cheese.

In the areas where there are restrictions on phosphates in detergents trisodium citrate is replacing phosphates in speciality cleaners and heavy-duty liquids.

Ferric ammonium citrate is still used in the treatment of anaemia although other iron salts are increasingly preferred.

Mixtures of citric acid and its salts have good buffering capacity and are extensively used for this purpose in the pharmaceutical, toiletry and food industries.

In a process for the removal of sulfur dioxide from the flue gases of power stations and metal smelters proposed by the US Bureau of Mines, a buffer solution containing principally H_2Cit^- is used as a scrubbing agent. A complex ion, $H_2CitHSO_3^{2-}$, is formed which in a second stage reacts with H_2S to produce elemental sulfur. The sulfur is readily separated from the citrate solution which may be recycled. The advantage of using citrate solution in this process is that certain side reactions yielding sulfate and thiosulfate, both giving rise to disposal problems, are substantially inhibited by the citrate (Rosenbaum *et al.*, 1973).

Citric acid esters of a wide range of alcohols are known. In particular the triethyl, tributyl and acetyltributyl esters are employed as non-toxic plasticizers in plastic films used to protect foodstuffs. Monostearyl citrate can be used instead of citric acid as an antioxidant in oils and fats. It is more easily incorporated than the free acid, and an improved method of preparation has been devised in the authors' laboratories.

The properties and uses of citric acid and its derivatives have been reviewed (Grayson and Eckroth, 1979; Schulz and Rauch, 1975; Fricke and Jensen, 1975).

34.3 PRODUCTION PROCESSES—FERMENTATION

From among the historically used processes for the production of citric acid the following are still important:

(a) A. niger
(i) Surface fermentation using beet molasses; (ii) submerged fermentation using beet or cane molasses or glucose syrup. Submerged processes using sucrose as carbohydrate source are also believed to be running in areas where sugar is cheap.

(b) Yeast
(i) Submerged fermentation using beet molasses or glucose syrup.

34.3.1 Media

The use of cane molasses is not practicable in the surface mode because very low yields are obtained. Glucose syrup as a carbohydrate source is at present not economic in Europe for fiscal and other reasons and its use is confined to the U.S.A. and other areas where cheap corn starch is available.

Both beet and cane molasses are very variable in quality both from season to season and from refinery to refinery. In spite of many investigations no clear reason for this has emerged, probably because the composition of molasses is so complex (Tressl *et al.*, 1976). It is necessary, therefore, to make a selection of available molasses on the basis of performance.

In order to obtain good yields of citric acid, particularly when *A. niger* is used, it is especially important to keep available levels of heavy metals, including iron and manganese, below certain critical levels. This is done in molasses media by making additions of sodium or potassium ferrocyanide (Clark *et al.*, 1965; Hustede and Rudy, 1976). Other inorganic nutrients are supplemented where necessary but most inorganic nutrients are already present in molasses.

Where glucose syrup is employed as carbohydrate source heavy metals are removed by ion exchange (Swarthout, 1966). Pretreatment of sucrose-based media by additions of ferrocyanide have also been proposed (Jungbunzlauer, 1974). In processes where pure glucose or sucrose is used as the substrate, additions of nitrogen, phosphate and other essential nutrients are made (for example Shu and Johnson, 1948; Batti, 1967; Kristiansen and Sinclair, 1978).

34.3.2 Yields

The yield of citric acid in the fermentation is expressed as kg citric acid monohydrate per 100 kg carbohydrate supplied. Yields in the range 70–90% on this basis have been reported. It should be mentioned that the theoretical yield of citric acid monohydrate from sucrose assuming no carbon is diverted to biomass, carbon dioxide or other by-products is 123% and that from anhydrous glucose is 117%. Thus up to about three quarters of the supplied carbon is converted to citric acid in a good fermentation.

34.3.3 Surface Process with *A. niger*

The surface fermentation using *A. niger* with beet molasses as raw material is still extensively employed by major manufacturers. Although somewhat labour intensive, the power requirements are less than in the submerged fermentation. Because citric acid manufacturers keep their methods secret, little authorative material has been published.

Beet molasses is diluted with water to a suitable sugar concentration, *e.g.* 150 kg m^{-3}, and the pH adjusted. An initial pH of 5 to 7 is usually employed because *A. niger* will not germinate at higher hydrogen ion concentrations. This effect was unknown when media based on sucrose were being used and starting pH values as low as 2 were in some cases employed. The lack of germination in molasses at low pH is ascribed to the presence of acetic acid, which is a normal constituent of molasses (Fencl and Leopold, 1957, 1959). It appears that unionized acetic acid is the species that prevents the germination, acetate being harmless in this respect. Hence the effect of pH.

Additional nutrients and alkali ferrocyanide are then added, and the whole boiled or otherwise sterilized. After cooling, the prepared medium is run down into a series of trays supported on racks in a ventilated chamber. The trays, which are usually made of very high purity aluminum, are filled to a depth of between 0.05 and 0.20 m.

Spores of *A. niger* are obtained by growing a selected strain on a sporulation medium. The spores are collected and distributed over the surface of the medium in the trays. Sterile air is supplied to the fermentation chamber. The air performs the dual function of supplying oxygen and carrying away fermentation heat and the rate of flow of the air is regulated accordingly. A temperature in the region of 30 °C is often employed. The mycelium forms a coherent felt on the surface of the liquid becoming progressively more convoluted. The removal of the heavy metals by the ferrocyanide severely restricts sporulation. After a period of 7 to 15 days the trays are emptied, the mycelium being at the same time separated from the fermented liquor. The liquors are pumped forward to the recovery section.

Unwanted by-products of the process are gluconic and oxalic acids. In many processes oxalic acid production is minimized by careful strain selection.

34.3.4 Submerged Fermentations

As indicated above there are a number of variants of this process.

34.3.4.1 Cultivation of *A. niger* in carbohydrate media

Figure 1 is a flow diagram of the fermentation layout of a possible citric acid plant which consists of a medium preparation section, a fermenter section and a section for separating the fermented liquor from the organism.

The medium preparation (suitable in this case for molasses) shows an in-line sterilization step, but sterilization of the medium in the fermenters is a possible alternative. Where in-line sterilization is used the fermenters are sterilized separately using steam at not less than 1 bar.

In the plant shown, a vegetative inoculum stage is used where spores of *A. niger* are allowed to germinate in an inoculum medium before being transferred to the main fermentation medium in a larger vessel. In some processes the spores are introduced directly into the main fermentation but in all cases the inoculum stage, or the initial growth stage in the main fermenter, is of the utmost importance to the success of the fermentation. The morphology of the mycelium at this point is crucial according to many reports, not only in relation to the shape of the hyphae themselves, but also in the aggregation of the growth into small spherical pellets. Thus the hyphae should be abnormally short, stubby, forked and bulbous (Snell and Schweiger, 1951; Kisser *et al.*, 1980; Hustede and Rudy, 1976a). This state of affairs is brought about by a deficiency of manganese in the medium (Kisser *et al.*, 1980) or the obviously related additions of ferrocyanide ion (Hustede and Rudy, 1976a). The mycelial pellets should be small (0.2 to 0.5 mm) with a hard,

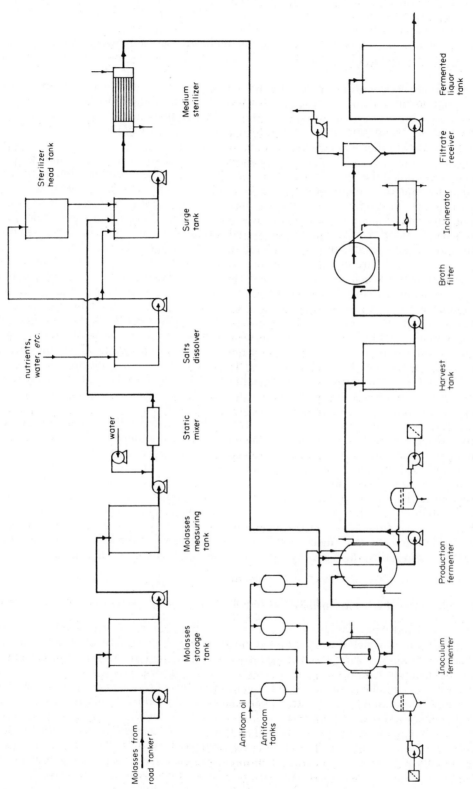

Figure 1 A possible plant layout for citric acid production by submerged fermentation from molasses

smooth surface (Clark, 1962). Factors leading to the production of such pellets are correct ferro-cyanide level (Clark, 1964; Hustede and Rudy, 1976b), an iron concentration of less than 1 p.p.m. (Snell and Schweiger, 1951), adjustment of pH (Fried and Sandza, 1959), adjustment of aeration and agitation (Svenska Sockerfabriks AB, 1964), concentration of manganese (Clark *et al.*, 1966) and amount of spore inoculum (Berry *et al.*, 1977). Whether the aggregation of the deformed hyphae into pellets is really necessary is doubtful, especially when stirred fermenters are being used (Clark and Lentz, 1963), but the pellet form does give a broth that is more readily mixed.

Where a separate inoculum stage is employed, a suspension of spores of *A. niger*, usually grown on a solid medium, is introduced into sterilized medium in the inoculum fermenter. The medium is aerated and, in some processes, agitated and the mould allowed to grow at a tempera-ture of about 30 °C for a period of from 18 to 30 hours as judged by pH level reached (Hustede and Rudy, 1976b) or in other ways (Martin and Waters, 1952).

The fermentation medium is prepared and transferred to the main fermenter and the grown inoculum incorporated at the rate of about 1 m^3 inoculum to 10 m^3 fermentation medium.

Where the inoculum or fermentation medium is based on molasses the initial pH is normally in the range 5 to 7. As mentioned above, most *A. niger* strains do not germinate or grow at lower pH values in this type of medium. On the other hand lower initial pH values can be tolerated in glucose- or sucrose-based media, and are often used with advantage. There is a lesser chance of infection by adventitious organisms. The fermentation is conducted at about 30 °C.

Two types of fermenter are in common use, namely the stirred, aerated, baffled tank and the aerated tower fermenter which has a much higher aspect ratio than the former and often contains an internal draught tube to promote circulation. Both types are constructed of high grade stain-less steel and have provision for cooling. Both designs are sparged from the base with sterile air, although extra oxygen is sometimes used in the tower type (Svenska Sockerfabriks AB, 1964; Martin and Waters, 1952). Sometimes superatmospheric pressure in the fermenter is used to increase the oxygen solution rate. The effect of dissolved oxygen tension (DOT) on the citric fer-mentation has been studied (Kubicek *et al.*, 1980), and DOT measuring electrodes are usually provided. The kinetics of the citric acid fermentation with *A. niger* have been reported by Röhr *et al.* (1981). A growth phase is followed by a citric acid producing phase during which only a small amount of growth occurs.

When the rate of increase in citric acid concentration has reached a point where it is unecono-mic to proceed, the fermentation is discontinued and the broth pumped to a filter or centrifuge where the mould is separated from the liquor which is passed forward to the recovery section. Siebert and Hustede (1982) have described in some detail the control of a commercial citric acid fermentation by means of a computer.

A continuous fermentation with a mould of the *A. niger* group has been studied on the labora-tory scale (Kristiansen and Sinclair, 1979) but no commercial production by this type of process is known. Because the accumulation of citric acid is only partly growth associated, and because for economic reasons the substrate sugar must be almost completely utilized, it is necessary to use several fermenters in series which greatly reduces the potential savings of this method.

34.3.4.2   *Cultivation of yeast in carbohydrate media*

In 1965 Tabuchi and Abe (1968) filed a patent for the manufacture of citric acid from technical glucose and molasses in which the organisms claimed were eight genera of yeasts including *Can-dida*. The fermentation was carried out at a neutral pH maintained by incorporating calcium car-bonate in the medium. A further patent (Iizuka, 1971) reports yields of about 65% from blackstrap molasses containing calcium carbonate using strains of *Candida oleophila*. However, it was soon found that a limitation on citric acid yield was the production of quantities of L-(+)-isocitric acid (Kyowa Hakko Kogyo Ltd., 1972). Efforts were made to select yeast strains not having this defect. Kimura *et al.* (1974) selected a subspecies of *Candida guilliermondii* producing only small amounts of isocitrate. Takeda Yakuhin Kogyo K.K. (1974) claimed a method of obtaining mutants of yeasts giving negligible quantities of isocitric acid wherein the mutants were selected for an inability to grow on media containing a halocitric acid or a precursor thereof. A mutant of *Debaromyces claussenii* producing a 73% yield from glucose was obtained. Takayama *et al.* (1976) claimed the use of mutants of species of *Candida* (*e.g. C. zeylanoides*) requiring a higher iron content in the medium than the parent. Chas. Pfizer & Co. Inc. (1970) claimed the use of strains of *Candida, Endomycopsis, Torulopsis, Hansenula* and *Pichia* grown in a molasses medium containing α-chloro- or α-fluoro-substituted lower alkanoic mono- or di-carboxylic acids or their water soluble salts or amides such as fluoroacetate or fluoroacetamide. Fried (1972) proposed to cultivate species of *Candida* in media containing glucose or blackstrap molasses to which 1.5 kg m^{-3} lead acetate had been added, obtaining thereby higher yields of citric acid. Additions of lead acetate, together with either *n*-hexadecyl citric acid or *trans*-aconitic acid have also been proposed (Pfizer Inc., 1972). By using strains of osmophilic yeasts (Pfizer Inc., 1974), initial concentrations of sugar as high as 280 kg m^{-3} may be employed. In this particular process no pretreatment of medium prepared from crude carbohydrates to remove metal ions is apparently necessary. Calcium carbonate, lead acetate and fluoroacetamide are also added as before.

A continuous fermentation process for the production of citric acid from blackstrap molasses using *Candida guilliermondii* has been described (Miall and Parker, 1975).

The advantages of using a yeast, rather than *A. niger*, are the possibility of using very high initial sugar concentrations, together with a much faster fermentation. This combination gives a high productivity per run, to which must be added the reported insensitivity of the fermentation to variations in the heavy metal content of the crude carbohydrates that are often used for economic reasons.

Fermenters of the tower type previously described are suitable, but more effective cooling is required because of the rapid rate of heat output during the shorter fermentation.

An inoculum culture is prepared in a smaller fermenter by introducing cells of the selected yeast strain previously grown in slope culture. When the inoculum is sufficiently grown it is transferred to the fermentation medium in the main fermenter. The fermentation is conducted at a temperature between 25 and 37 °C, depending on the organism. The initial pH must not be too low or yeast growth will be impaired, this presumably being the reason for adding the calcium carbonate to the medium. The pH can subsequently be allowed to fall. (In the continuous process (Miall and Parker, 1975) the pH is controlled at 3.5 by means of aqueous ammonia.)

The broth is harvested when citric acid accumulation has become uneconomically low.

34.3.4.3 *Cultivation of yeasts and some bacteria in medium containing* n-*alkanes*

At the present time this process is of academic interest only, for reasons given above. Since plants designed to use it have actually been constructed however, it is worthwhile to give it some consideration. The subject has been reviewed in some detail by Stottmeister *et al.* (1982).

That some microorganisms are able to grow on hydrocarbons as substrate has long been known (Bushnell and Haas, 1941; Foster, 1962). Johnson (1964) noted that organisms preferred straight chain alkanes to branched compounds and that the preferred chain length differed depending on strain.

The production of citric acid by yeasts from *n*-alkanes was first reported in the patent literature (Takeda Yakuhin Kogyo K.K., 1970). The patent claims the use of species of *Candida* (and in particular *C. lipolytica*, *C. tropicalis*, *C. intermedia*, *C. parapsilosis* and *C. guilliermondii*). It has a priority date in Japan of 7 June, 1967 and seems to be a master patent where the use of *Candida* is concerned. It specifies a culture medium containing at least one of the normal alkanes with from 9 to 20 carbon atoms. An aqueous medium for use with *C. lipolytica* contains (kg m^{-3}) NH$_4$Cl (4), KH$_2$PO$_4$ (0.5), MgSO$_4$·7H$_2$O (0.5), yeast extract (1) and CaCO$_3$ (30). *n*-Decane (10% by weight) was incorporated and the whole inoculated with a culture of *C. lipolytica*. Incubation was for seven days at 26 °C. A weight yield based on hydrocarbon supplied was 107% as anhydrous citric acid (117% as monohydrate). Higher yields have been reported from species of *Candida* where mutants have been selected for an inability to grow on citric acid (Fukuda *et al.*, 1974). For instance a mutant of *Candida* sp. IFO1461 so selected gave a yield of 138% with negligible L-(+)-isocitric acid after incubation for three days at 28 °C. It has been claimed advantageous to select mutants of *Candida* for a higher iron requirement than the parent (Takayama *et al.*, 1976), so as to eliminate the production of isocitric acid. For this purpose mutants selected for their inability to grow on media containing a halocitric acid or its precursors (Takeda Yakuhin Kogyo K.K., 1974) are also useful.

It should be mentioned that, starting from a highly reduced form of carbon such as *n*-alkanes, very high weight yields of citric acid are possible. The 'theoretical' yield is about 250% but allowing for biomass and carbon dioxide production a reasonable target yield would be 175%.

It has been proposed to control the pH of the fermentation between 5 and 7 using sodium hydroxide (Hitachi Chemical Co. Ltd., 1975). A continuous fermentation has been described (Charpentier *et al.*, 1976).

Organisms other than yeasts have been proposed. Thus, species of *Arthrobacter*, especially *A. paraffineus* (Kyowa Hakko Kogyo Co. Ltd., 1966), and *Corynebacterium* (Takeda Chemical Industries Ltd., 1971) may be cultivated on media containing *n*-alkanes to yield citric acid.

n-Alkanes are, of course, insoluble in aqueous solutions and it is essential during fermentations of this type to maintain a fine dispersion of the hydrocarbon in the aqueous phase. To aid this dispersion the addition of from 20 to 200 p.p.m. of polyoxypropylene glycol diether has been claimed to be advantageous (Kanegafuchi Chemical Industries Co. Ltd., 1976).

Little detailed information is available about the plants able to ferment *n*-alkanes to citric acid. It is thought that both tower fermenters with draught tubes and stirred tank fermenters may be used. It is clear that because of the greater degree of oxidation required to convert the hydrocarbon to citric acid, more heat will be evolved than in a carbohydrate-based fermentation. Thus fermenter cooling arrangements will be necessarily more elaborate.

A short description of the Liquichimica plant at Saline, Reggio Calabria, Italy has been published (Ferrara *et al.*, 1977). The fermentation is a scale-up of a pilot plant process developed by Takeda Chemical Industries using a mutant of *C. lipolytica* selected for low aconitase activity (Takeda Yakuhin Kogyo K.K., 1974). A batch process is used in which the main fermenters are aerated agitated tanks of 400 m³ capacity operating on a 72 hour cycle. The conversion of hydrocarbon to citric acid is said to exceed 130% by weight. The process exhibits three phases. In the first, yeast growth occurs and in the second a very fast conversion of *n*-alkane to citric acid takes place. The last phase is a slower exhaustion of the substrate. During the fermentation pH is controlled by addition of, for example, sodium hydroxide. At the end of the fermentation the yeast is removed by centrifugation, leaving an exceptionally pure fermented liquor.

34.4 BIOCHEMISTRY OF THE CITRIC ACID FERMENTATIONS

As has been described above, citric acid is accumulated in very large quantities by *A. niger*, certain yeasts and some bacteria. This is not a normal phenomenon but is brought about by growing the organisms under certain constraints, helped by selection of strains and the obtaining of mutants whose metabolism is altered in some way which predisposes them to citrate accumulation. A large amount of work has been published on the citric acid fermentation, mostly on *A. niger*. This work has by and large concentrated on nutritional requirements and the way nutritional constraints affect the amount and activity of critical enzymes at various stages in the fermentation. Some ^{14}C tracer investigations have been published and there has been some work on the activity and properties of enzymes isolated from mutants which differ in their ability to accumulate citrate. However, very little, if any, truly genetic work has been done on the system.

An examination of the voluminous published work on this subject will reveal many contradictory conclusions which may puzzle those not familiar with the field. There are several reasons for this. One is that different strains have been used in the work and the media have in many cases also been different. There is a certain amount of evidence to suggest that there may be more than one set of conditions which can lead to citric acid accumulation. Another factor which is at least as important as those mentioned above is the level of yield obtained in the experiments. Citric acid manufacturers have long kept secret their know-how as well as their strains and this leads to statements being made in the literature that good yields have been obtained when in fact they are mediocre. At mediocre yield levels, side reactions occur which may nullify the conclusions drawn, particularly if they are extrapolated to commercial processes. The literature on the biochemistry of the citric acid fermentation will be examined with these reservations in mind. The greater part of the review will refer to *A. niger* with some supplementary comments on the situation in yeast fermentations.

34.4.1 Nutritional Considerations

The growth of *A. niger* requires, as would be expected, in addition to a source of carbon, supplies of nitrogen, phosphate, potassium, magnesium and sulfur. In addition, small quantities of zinc, iron, copper and manganese are necessary, as well as molybdenum if nitrate is to be metabolized. When a sufficiency of these nutrients is available the mould grows to its full extent finally entering a sporulation phase. In order that citric acid accumulation can occur to the extent required by a commercial process, neither full growth nor sporulation must take place. This implies a limitation of nutrients. Kristiansen and Sinclair (1979) grew a mould of the *A. niger* group in continuous culture. They found a sharp maximum of citric acid production at $0.8 \, \text{kg m}^{-3}$ NH_4NO_3 and concluded that nitrogen limitation was an essential requirement for citric acid production. However, Kubicek and Röhr (1977) demonstrated that cultures otherwise conditioned to citrate accumulation moved out of the growth phase when phosphate was exhausted even when nitrogen was not limiting. Probably either nitrogen or phosphate limitation is effective in inducing citric acid production depending upon strain and other conditions. The positive effect of phosphate limitation would explain the successful use of beet molasses which contains relatively large amounts of metabolizable nitrogen.

The importance to the citric acid fermentation of the correct levels in the medium of the trace elements zinc, iron, copper and manganese has long been appreciated (Perlman *et al.*, 1946; Shu and Johnson, 1948). Optimum additions of iron and manganese were found to be necessary for best citric acid yields and the optima varied from strain to strain. Using a purified glucose medium and the well-known *A. niger* Wisconsin 72-4 additions of 0.3 p.p.m. zinc and 1.3 p.p.m. iron were optimal. Additions of manganese at any level reduced the yield. Shu and Johnson (1947) found that manganese carried over in spores, when these were produced on a medium containing manganese, was sufficient to reduce citric acid yields.

The importance of the zinc, iron and manganese concentrations in the medium thus explains the necessity for the ion-exchange purification of glucose solutions or the treatment of molasses with ferrocyanide ions. Clark *et al.* (1965) have examined the effect of temperature and pH on the precipitation of iron, copper and manganese from molasses solutions. Clark *et al.* (1966) found that additions of as little as 1 p.p.b. manganese to ferrocyanide-treated beet molasses reduced the yield of citric acid by 10%. Martin (1955) concluded that ferrocyanide ion directly inhibited mould growth as well as removing unwanted heavy metals from molasses. Heyer and Schwartz (1964) found that a small excess of free ferrocyanide in the medium after removal of metals led to increased yields and reached the same conclusion.

34.4.2 Metabolic Pathways Leading to Citric Acid Formation in *A. niger*

Figure 2 shows a slightly simplified version of the various enzymatic reactions relating to the synthesis and breakdown of citric acid. Early ^{14}C tracer experiments under conditions giving only moderate yields of citric acid had led to the conclusion that 40% of the acid was produced by a C_2,C_2 condensation, and 60% *via* a C_1,C_3 condensation. Moreover, nearly 40% was produced from recycled C_4 acids (Shu *et al.*, 1954). However, a major contribution to the solving of the problem was made by Cleland and Johnson (1954). Using glucose-3,4-^{14}C under good yielding conditions they showed that a near quantitative recovery of the labelling was obtained in carbon 4 of the citric acid. Carbon 6 had a specific activity of about 15% of the original. None of the other carbons was labelled to any significant degree. It was concluded that there was no recycling of C_4 acids and that the citric acid was formed by an initial symmetrical 3:3 split of the glucose and that decarboxylation of one fragment produced a two-carbon fragment. The other three-carbon fragment was carboxylated to give a four-carbon compound and finally citric acid was produced by condensation of the two- and four-carbon moieties. Carbon introduced by carboxylation was carbon 6, the specific activity having been diluted by the carbon dioxide pool in the hyphal cells (see Figure 2).

Verhoff and Spradlin (1976) examined mass balances for the system in which the input was pyruvate giving rise to citric acid, oxalic acid and carbon dioxide. Their analysis gave rise to two metabolic schemes, one of which was identical with that of Cleland and Johnson (1954). The other involved the carboxylation of two molecules of pyruvate giving two molecules of oxaloacetate. One of these was hydrolysed to oxalic acid and acetate by oxaloacetic hydrolase, thus providing the acetate moiety for a condensation with the other molecule of oxaloacetate to give citric acid. According to this proposal, the oxalate is then reduced to glyoxylate which is condensed with succinate by isocitrate lyase operating in reverse to give isocitrate. The latter is then metabolized by the normal citric acid cycle enzymes back to succinate, losing two molecules of carbon dioxide on the way which are used to carboxylate the two molecules of pyruvate at the beginning of the scheme. It is claimed by Verhoff and Spradlin that the results of Cleland and Johnson can be used to support either interpretation.

Verhoff and Spradlin also adduce as evidence in favour of their proposal certain enzyme activities measured during the accumulation of citric acid. They state that work carried out at Miles Laboratories Inc. and at other laboratories (Wongchai and Jefferson, 1974; Joshi and Ramakrishnan, 1959) indicates that pyruvate carboxylase, isocitrate lyase and oxaloacetic hydrolase are

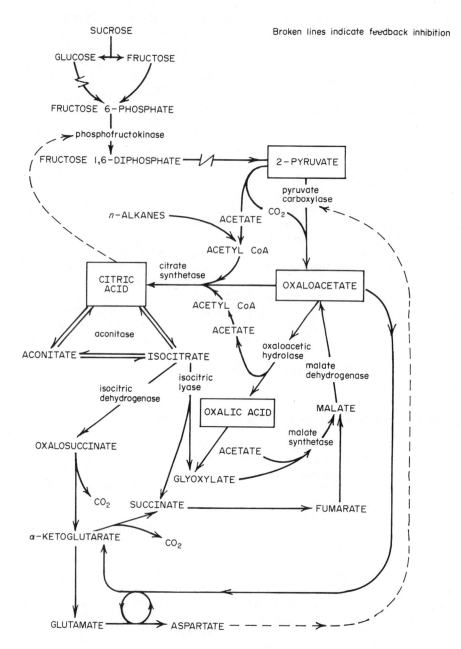

Figure 2 A simplified metabolic scheme showing citric acid production from carbohydrates and *n*-alkanes

more active during citric acid and oxalate production. Joshi and Ramakrishnan in fact found that oxaloacetic hydrolase activity disappeared when citric acid accumulation began and Müller (1967) found no evidence for the presence of oxaloacetic hydrolase at all. Kubicek and Röhr (1977) were unable to detect citrate lyase during citrate accumulation. Nor were they able to detect α-ketoglutarate dehydrogenase under these conditions.

Thus there is a conflict of evidence over the exact metabolic pathway leading to citric acid formation, the balance probably being in favour of the simpler scheme of Cleland and Johnson (1954).

34.4.3 Enzymatic Considerations

The previous section has considered tracer evidence and other matters in relation to citric acid accumulation by *A. niger* but some enzymatic evidence was inevitably adduced. Further evidence of enzymatic activity during the citric acid fermentation together with aspects of the regulation of these activities under conditions of high citric acid yield will now be considered.

34.4.3.1 *The Embden–Meyerhof pathway*

The enzymes of this pathway are well attested in *A. niger* (Damodaran *et al.*, 1955).

The aldolase of *A. niger* requires zinc for activity (Jagannanthan and Singh, 1954) and this may explain the often expressed need for zinc in the fermentation medium. Citrate is a known feedback inhibitor of phosphofructokinase (PFK) and mitigation of this inhibition is obviously required for good citric acid accumulation. Bowes (1980) has studied the inhibition of PFK by citrate in a high yielding and a low yielding strain of *A. niger*. She found that the enzyme from the low yielding strain was greatly inhibited by addition of 5.5 mM citrate, whereas the enzyme from the high yielding strain was slightly activated under the same conditions. Kubicek and Röhr (1977) and Habison *et al.* (1979) have shown that under manganese deficiency, an accumulation of NH_4^+ ions occurs in the cell during citric acid production and that this condition has the effect of antagonizing the inhibition of PFK by citrate.

34.4.3.2 *Citric acid formation from pyruvate*

Citric acid is formed by the condensation of acetyl coenzyme A and oxaloacetic acid under the influence of citrate synthase. Normally the oxaloacetate is provided by the operation of the citric acid cycle. When citric acid accumulates, the cycle is to a greater or lesser degree blocked and another, so-called anaplerotic, reaction is needed to supply the oxaloacetate. This anaplerotic reaction is the carboxylation of pyruvate by pyruvate carboxylase.

It has been shown (Woronick and Johnson, 1959) that CO_2 fixation does indeed occur and pyruvate carboxylase has been isolated from *A. niger* (Feir and Suzuki, 1969). The enzyme required no acetyl coenzyme A but was inhibited by aspartate. Since α-ketoglutarate dehydrogenase is practically absent during citric acid accumulation (Kubicek and Röhr, 1977) aspartate concentrations must be low and inhibition minimal.

Citric synthase (Figure 2) has been studied by Kubicek and Röhr (1980), who found that it was poorly regulated and that citric acid formation was related only to the oxaloacetic acid available. Since both citrate synthase and pyruvate carboxylase are not effectively regulated, the rate of citric acid formation is related to the speed at which carbohydrate is converted to pyruvate.

34.4.3.3 *Why the normal citric acid catabolism does not occur*

Citric acid is normally catabolized *via* the citric acid cycle. The first step in this series of reactions, conversion to a mixture of aconitic and isocitric acids, is mediated by aconitase (aconitate hydratase) (ACH). Neilson (1956) using the high yielding strain Wisconsin 72-4 was unable to detect ACH under citric acid accumulating conditions. Additions of manganese reduced the citric acid yield and caused the appearance of ACH (there was some evidence that two aconitases were present). Ramakrishnan *et al.* (1955) using a ferrocyanide treated beet molasses medium observed the disappearance of ACH during citric acid accumulation. Szczodrak (1981) in a beet molasses fermentation with production strains found that whilst ACH activity greatly decreased during citric acid production, it did not disappear altogether. This was in contrast with citrate synthase which maintained its level almost throughout. La Nauze (1966) demonstrated the activation of *A. niger* ACH by iron. Twenty times the optimal level doubled the ACH activity in a mutant of Wisconsin 72-4 but reduced the yield only by a factor of 0.75. The citric yields she obtained were, however, not good and her conclusions on the relation between ACH levels and citric acid production should be treated with caution. Overall, the effects of Mn and of Fe on the presence and activity of ACH accord well with experience of the fermentation.

Isocitric dehydrogenase (ICDH) and isocitric lyase (ICL) break down isocitric acid. ICL has been shown to be absent from *A. niger* when citric acid is accumulating (Kubicek and Röhr, 1977). The case of ICDH is more complex, there being two enzymes with NADH and NADPH specificities (Ramakrishnan and Martin, 1955). The NADH enzyme is cytoplasmic whereas the NADPH enzyme is located in the mitochondria and is inhibited by physiological concentrations of citrate (Mattey, 1977). Mattey considers the possibility that the inhibition of mitochondrial

ICDH by citrate causes isocitrate to accumulate and since isocitrate is in equilibrium with a much higher concentration of citrate under the influence of ACH, citrate will also accumulate. However, since only low or zero ACH activities have been detected during citric acid accumulation it is difficult to agree with this conclusion. Moreover, although there is an equilibrium between citrate, isocitrate and aconitate in the presence of ACH, isocitric acid (or aconitic acid) is not found as an impurity of citric acid made with *A. niger*. The situation when yeasts are used is of course very different. It is possible that the permeability of the mitochondria in *A. niger* is much greater for citrate than isocitrate.

The exact mechanism of citric acid accumulation is still not clear. The most likely picture is of high activity in the glycolytic enzymes, pyruvate carboxylase and citrate synthase providing an accumulation of citric acid which the greatly reduced activities of the citric acid cycle enzymes are unable to deal with. This hypothesis is supported by the observed catabolism of citric acid once the carbohydrate substrate reaches a low level in the medium.

34.4.3.4 *Citric acid production by yeasts*

Whether glucose or *n*-alkanes are used as substrate the fermentation pattern shows a growth phase followed by an acid accumulation phase, the transition being brought about by nitrogen limitation. When *n*-alkanes are used as substrate they are converted by β-oxidation to acetyl coenzyme A which is combined with oxaloacetic acid in the usual way (Figure 2). It is thought that when glucose is used as substrate pyruvate carboxylase provides the anaplerotic reaction as in *A. niger*. However, when *n*-alkanes are the substrate, anaplerosis is provided by the glyoxylate cycle, and this is reflected by a higher level of ICL (Tabuchi and Igoshi, 1978). At the same time the levels of ACH are higher than when glucose is being utilized. A problem with the yeast process is the very high ratios of isocitric acid to citric acid which are produced in some cases. These ratios are much higher than the equilibrium values mediated by ACH. It has been suggested (Marchal *et al.*, 1980) that a selective transport of isocitric acid from the mitochondrion to the cytoplasm occurs.

34.5 PRODUCTION PROCESSES—PRODUCT RECOVERY

Removal of the fermenting organism from the final broth leaves an aqueous solution of citric acid contaminated by various organics and inorganics depending on the initial carbon source.

34.5.1 The Classical Citric Acid Recovery Process

The classical citric acid recovery process, which is particularly suitable for use with the very impure liquors derived from molasses, is to heat the fermented liquor and add lime. The insoluble calcium citrate tetrahydrate is precipitated (Figure 3). The washed precipitate is treated in aqueous suspension with H_2SO_4 yielding an aqueous solution of citric acid and a precipitate of by-product $CaSO_4$ (gypsum). This set of operations has the effect of removing most of the impurities, either those derived from the substrate or those generated in the fermentation. Figure 3 shows a sequence of steps leading to packed saleable citric acid. As mentioned above, the conditions of the concentration/crystallization steps can be varied to produce either the anhydrous acid or the monohydrate.

34.5.2 Solvent Extraction

Solvent extraction is a possible alternative to the classical method but, because the available solvents tend to extract some of the impurities contained in molasses-derived liquors, it is easier to apply to the products from glucose or alkane-based substrates. The advantage of the solvent extraction method is that is avoids the use of lime and H_2SO_4 and the concomitant problem of gypsum disposal.

Les Usines de Melle (1961) proposed the use of butan-2-ol as an extractant. Later it was

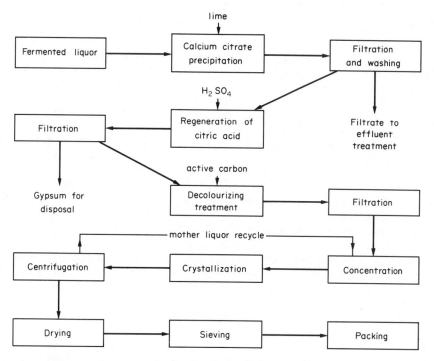

Figure 3 A diagram showing the recovery of citric acid by the classical process

claimed (Les Usines de Melle, 1963) that tributyl phosphate diluted with a minor amount of kerosene could be used. In this process a better recovery could be realized by extracting the citric acid into the solvent at a low temperature, subsequently stripping the solvent with hot water. Further details of a process of this type which can be carried out in mixer-settlers or centrifugal extractors are given by Wennerston (1980).

Another variant of the solvent extraction process is the ion pair extraction system (Baniel *et al.*, 1976) in which the extractant consists of secondary or tertiary amines having in total at least 20 carbon atoms dissolved in a water immiscible solvent. Again the extraction of the citric acid from the fermented liquor is done at a lower temperature ($\not< 20\,°C$) and the stripping stage at a higher ($\not< 80\,°C$). It is believed that this process has been licensed to Miles Laboratories Inc. who have secured FDA approval for the use of a mixture of tridodecylamine, octyl alcohol and isoalkanes. A further development of this process is the use of *N*-substituted alkyl amides as extractants (Alter and Blumberg, 1981).

In fermentations with *n*-alkanes and in particular where sodium hydroxide is used for pH control, monosodium or trisodium citrate can be directly crystallized from the clarified fermented liquor (Ferrara *et al.*, 1977). In this case citric acid may be produced from the sodium citrate by electrodialysis.

34.6 EFFLUENT DISPOSAL

Reference has already been made to the production of gypsum in the calcium citrate precipitation method of citric acid recovery. The disposal of this solid waste can pose a problem. A more serious problem is the disposal of the filtrate from the calcium citrate precipitation, especially where molasses is used as the starting material. This waste, although non-toxic, has a very high oxygen demand which makes it unacceptable in rivers without treatment. Braun *et al.* (1979) have proposed the cultivation of yeasts on the effluent producing a material suitable for animal feed. Another possibility is to evaporate the effluent to produce a concentrated molasses-like material usually called condensed molasses solubles (CMS). This material can be used in feedstuff formulations. Another method of treatment is anaerobic digestion which has the advantage of producing a fuel gas as a by-product—see De Zeeue and Lettinga (1980), Frostell (1981) and Binot *et al.* (1982).

34.7 FUTURE DEVELOPMENTS

The current world market for citric acid (and its derivatives) is upwards of 300 000 tonnes per annum and it may be regarded as a mature product, although an upward trend in its usage may be expected to continue. This trend will include additional sales as an acidulent for food use as well as an increasing quantity for uses where its complexing properties are required. Its use as a biodegradable substitute for polyphosphates in detergents has not developed except in a few areas with special problems. In part this is due to price considerations. A decrease in fermentation recovery costs could change the situation, and further developments in the newer methods such as solvent extraction in its widest sense can be expected.

In the case of the fermentation stage, although most manufacturers already possess high yielding strains, scope for some further improvement probably remains, particularly in relation to the shortening of fermentation times. Mutagenic treatment of the organism followed by selection is still the normal mode of new strain development, although the selection process is made less cumbersome by use of replica plate cultures and selection pressures (Takeda Yakuhin Kogyo K.K., 1974). Whether the methods of gene transfer will be applicable to the citric acid fermentation in the near future is doubtful. In the present state of the art, gene transfer techniques are very powerful where the desired products are single peptides or proteins. In a 10-stage conversion such as that from glucose to citric acid, progress could only be made by identifying limiting steps and improving the flow of metabolites through these. Unfortunately, at the present time, the knowledge of the metabolic steps and of the genetics of the organisms involved is somewhat wanting in detail.

34.8 SUMMARY

For the last 60 years citric acid has been produced by fermentation of carbohydrates. At first the surface process utilizing the mould *Aspergillus niger* was used, beet molasses eventually replacing pure sucrose as carbon source. Latterly, submerged fermentations of beet or cane molasses or glucose syrups by *A. niger* have been introduced. More recently still, *A. niger* has been replaced by strains of yeast which exhibit higher productivity and less sensitivity to variations in the crude carbohydrate media.

The biochemistry of citric acid accumulation by *A. niger* has been extensively studied and the key roles of the levels of zinc, iron and particularly manganese in the medium have been identified. The restriction in the levels of iron and manganese greatly reduces the activities of aconitate hydratase and isocitric dehydrogenase whilst at the same time the inhibition of phosphofructokinase by citrate is removed. Anaplerosis is provided by pyruvate carboxylase and neither the latter enzyme nor citrate synthase are regulated to any extent during citric acid accumulation.

The variety of processes and plants is indicated and the essentials of the layout and operation of a submerged fermentation plant together with the classical citric acid recovery are described. Other possible methods of recovery are outlined and some solutions to the effluent problem discussed.

34.9 REFERENCES

Aktiengesellschaft Jungbunzlauer Spiritus und Chemische Fabrik (1974). *Br. Pat.* 1 342 311 (*Chem. Abstr.*, **76**, 44 691z).
Alter, J. E. and R. Blumberg (Miles Laboratories Inc.) (1981). Extraction of citric acid. *US Pat.* 4 251 671 (*Chem. Abstr.*, **95**, 61 507d).
Baniel, A. M., R. Blumberg and K. Hajdu (IMI (TAMI) Institute for Research and Development) (1976). Recovery of acids from aqueous solutions. *Br. Pat.* 1 426 018 (*Chem. Abstr.*, **80**, 132 807e).
Batti, M. A. (Miles Laboratories Inc.) (1967). *US Pat.* 3 335 067 (*Chem. Abstr.*, **67**, 89 808y).
Berry, D. R., A. Chmiel and Z. Al Obaidi (1977). Citric acid production by *A. niger*. In *Genetics and Physiology of Aspergillus*, ed. J. E. Smith and J. A. Pateman, pp. 405–426. Academic, London.
Binot, R. A., T. Bol, H.-P. Naveau and E.-J. Nyns (1982). Biomethanation by immobilised fluidised cells. *Preprint IAWPR Specialised Seminar*, June 1982, Copenhagen, Denmark.
Bowes, I. (1980). Citric acid production by *A. niger*. Ph.D. Thesis, University of Strathclyde, Glasgow.
Braun, R., J. Meyrath, W. Stuparek and G. Zerlauth (1979). Feed yeast production from citric acid waste. *Process Biochem.*, **14** (1), 16–20.
Bushnell, L. D. and H. E. Haas (1941). The utilisation of certain hydrocarbons by microorganisms. *J. Bacteriol.*, **41** (5), 653–675.
Charpentier, J.-M., G. Glikmans and P. Maldonado (Institut Français du Petrole) (1976). Fermentative production of citric acid. *Br. Pat.* 1 428 440 (*Chem. Abstr.*, **82**, 15 242x).
Clark, D. S. (1962). Submerged citric acid fermentation of ferrocyanide-treated beet molasses: morphology of pellets of *A. niger. Can. J. Microbiol.*, **8**, 133–136.

Clark, D. S. (National Research Council of Canada) (1964). Citric acid production. *US Pat.* 3 118 821 (*Chem. Abstr.*, **61**, 4926d).

Clark, D. S. and C. P. Lentz (1963). Submerged citric acid fermentation of beet molasses in tank type fermenters. *Biotechnol. Bioeng.*, **5**, 193–199.

Clark, D. S., K. Ito and P. Tymchuk (1965). Effect of potassium ferrocyanide addition on the chemical composition of molasses mash used in the citric acid fermentation. *Biotechnol. Bioeng.*, **7**, 269–278.

Clark, D. S., K. Ito and H. Horitsu (1966). Effect of manganese and other heavy metals on submerged citric acid fermentation of molasses. *Biotechnol. Bioeng.*, **8**, 465–471.

Cleland, W. W. and M. J. Johnson (1954). Tracer experiments on the mechanism of citric acid formation by *A. niger*. *J. Biol. Chem.*, **208**, 679–689.

Currie, J. N. (1917). The citric acid fermentation of *A. niger*. *J. Biol. Chem.*, **31**, 15–37.

Damodaran, M., V. Jagannathan and K. Singh (1955). Carbohydrate metabolism in citric acid fermentation. *Enzymologia*, **17**, 199–206.

De Zeeuw, W. J. and G. Lettinga (1980). Use of anaerobic digestion for waste water treatment. *Antonie van Leeuwenhoek*, **46** (1), 110–112.

Doelger, W. P. and S. C. Prescott (1934). Citric acid fermentation. *Ind. Eng. Chem.*, **26**, 1142–1149.

Feir, H. A. and I. Suzuki (1969). Pyruvate carboxylase of *A. niger*. *Can. J. Biochem.*, **47**, 697–710.

Fencl, Z. and J. Leopold (1957). Mechanism of inhibition by acetic acid of the germination of spores of *A. niger*. *Nature (London)*, **179**, 922.

Fencl, Z. and J. Leopold (1959). Inhibition of spore germination of the mould *A. niger* by acetic acid. *Folia Microbiol.*, **4**, 7–15.

Ferrara, L., L. De Cesari and E. Salvini (1977). *n*-Paraffine: una nuova fonte di materia prima per la produzione di acido citrico e derivate. *Chim. Ind. (Milan)*, **59** (3), 202–206.

Foster, J. W. (1962). Hydrocarbons as substrates for microorganisms. *Antonie van Leeuwenhoek*, **28** (3), 241–262.

Fricke, H. and S. B. Jensen (1975). Citric acid—a versatile food additive. *Food Process. Ind.*, **44** (528), 37, 39, 41, 44.

Fried, J. H. (Chas. Pfizer & Co. Inc.) (1972). Method of producing citric acid by fermentation. *Br. Pat.* 1 264 578 (*Chem. Abstr.*, **76**, 139 056k).

Fried, J. H. and J. G. Sandza (Stauffer Chemical Company) (1959). Production of citric acid. *US Pat.* 2 910 409 (*Chem. Abstr.*, **54**, 7063b).

Frostell, B. (1981). Anamet® anaerobic–aerobic treatment of concentrated waste waters. *Proc. Ind. Waste Conf. 36th*. Purdue University, Indiana.

Fukuda, H., T. Suzuki, S. Akiyama and Y. Sumino (Takeda Chemical Industries Ltd.) (1974). Method for producing citric acid. *US Pat.* 3 799 840 (*Chem. Abstr.*, **74**, 30 744r).

Grayson, M. and D. Eckroth (eds.) (1979). *Kirk–Othmer Encyclopedia of Chemical Technology*, 3rd edn., vol. 6, pp. 150–178. Wiley, New York.

Grimoux, E. and P. Adam (1880). Synthèse de l'acide citrique. *C. R. Hebd. Seances Acad. Sci.*, **90**, 1252–1255.

Habison, A., C. P. Kubicek and M. Röhr (1979). Phosphofructokinase as a regulatory enzyme in citric acid producing *A. niger*. *FEMS Microbiol. Lett.*, **5**, 39–42.

Heyer, J. and W. Schwartz (1964). Wirkung von Hexacyanoferrate auf die Citronensäuregärung von *A. niger* im Melassemedium. *Z. Allg. Mikrobiol.*, **4** (3), 199–224.

Hitachi Chemical Co. Ltd. (1975). Process for producing sodium citrate and isocitrate by fermentation. *Br. Pat.* 1 381 047 (*Chem. Abstr.*, **78**, 27 899x).

Hustede, H. and H. Rudy (Joh. A. Benckiser GmbH) (1976a). Manufacture of citric acid by submerged fermentation. *US Pat.* 3 941 656 (*Chem. Abstr.*, **84**, 178 222w).

Hustede, H. and H. Rudy (Joh. A. Benckiser GmbH) (1976b). Manufacture of citric acid by submerged fermentation. *US Pat.* 3 940 315 (*Chem. Abstr.*, **85**, 3869f).

Iizuka, H. (Shibaura Seito K.K.) (1971). Preparation of citric acid by a fermentation process. *Br. Pat.* 1 218 476 (*Chem. Abstr.*, **71**, 122 382n).

Jagannathan, V. and K. Singh (1954). Metal activation of *A. niger* aldolase. *Biochim. Biophys. Acta*, **15**, 138.

Johnson, M. J. (1964). Utilisation of hydrocarbons by microorganisms. *Chem. Ind. (London)*, 1532–1537.

Joshi, A. P. and C. V. Ramakrishnan (1959). Mechanism of formation and accumulation of citric acid in *A. niger*. Part I. Citric acid formation and oxaloacetic hydrolase in a citric acid producing strain of *A. niger*. *Enzymologia*, **21**, 43–51.

Kanegafuchi Chemical Industries Co. Ltd. (1976). Cultivation of microorganisms. *Br. Pat.* 1 434 606.

Kimura, K., K. Takayama and T. Nakanishi (Kyowa Hakko Kogyo Ltd.) (1974). Citric acid production. *Br. Pat.* 1 366 526, example 1 (*Chem. Abstr.*, **77**, 60 092f).

Kisser, M., C. P. Kubicek and M. Röhr (1980). Influence of manganese on morphology and cell wall composition of *A. niger* during citric acid fermentation. *Arch. Microbiol.*, **128**, 26–33.

Kristiansen, B. and C. G. Sinclair (1978). Production of citric acid in batch culture. *Biotechnol. Bioeng.*, **20**, 1711–1722.

Kristiansen, B. and C. G. Sinclair (1979). Production of citric acid in continuous culture. *Biotechnol. Bioeng.*, **21**, 297–315.

Kubicek, C. P. and M. Röhr (1977). Influence of manganese on enzyme synthesis and citric acid accumulation in *A. niger*. *Eur. J. Appl. Microbiol.*, **4**, 167–175.

Kubicek, C. P. and M. Röhr (1980). Regulation of citrate synthase from the citric acid accumulating fungus, *A. niger*. *Biochim. Biophys. Acta*, **615**, 449–457.

Kubicek, C. P., O. Zehentgruber, H. El-Kalak and M. Röhr (1980). Regulation of citric acid production by oxygen. *Eur. J. Appl. Microbiol. Biotechnol.*, **9**, 101–115.

Kyowa Hakko Kogyo Ltd. (1972). Process for the production of citric acid and isocitric acid . *Br. Pat.* 1 278 013, example 3 (*Chem. Abstr.*, **77**, 112 491g).

La Nauze, J. M. (1966). Aconitase and isocitric dehydrogenases of *A. niger* in relation to citric acid production. *J. Gen. Microbiol.*, **44**, 73–81.

Les Usines de Melle (1961). Process for extracting citric acid from aqueous solutions thereof. *Br. Pat.* 874 030 (*Chem. Abstr.*, **55**, 18 010c).

Les Usines de Melle (1963). Process for extracting carboxylic acids produced by fermentation. *Br. Pat.* 936 339 (*Chem. Abstr.*, **57**, 15 620a).

Marchal, R., M. Metche and J.-P. Vandecasteele (1980). Intracellular concentrations of citric and isocitric acids in cultures of the citric acid-excreting yeast *Saccharomycopsis lipolytica* grown on alkanes. *J. Gen. Microbiol.*, **116**, 535–538.

Martin, S. M. (1955). Effect of ferrocyanide on growth and acid production of *A. niger*. *Can. J. Microbiol.*, **1**, 644–652.

Martin, S. M. and W. R. Waters (1952). Production of citric acid by submerged fermentation. *Ind. Eng. Chem.*, **44**, 2229–2240.

Mattey, M. (1977). Citrate regulation of citric acid production in *A. niger*. *FEMS Microbiol. Lett.*, **2**, 71–74.

Miall, L. M. and G. F. Parker (Pfizer Ltd.) (1975). Citric acid production. *Br. Pat.* 1 418 561 (*Chem. Abstr.*, **83**, 7119t).

Müller, H.-M. (1967). Säureanhäufung und Enzymactivatät bei *A. niger*. *Ber. Deutsch. Bot. Ges.*, **80** (2), 109–110.

Neilson, N. E. (1956). Presence of aconitase and 'aconitic hydrase' in *A. niger*. *J. Bacteriol.*, **71**, 356–361.

Perlman, D. (1949). Mycological production of citric acid—the submerged culture method. *Economic Botany*, **3** (4), 360–374.

Perlman, D., W. W. Dorrell and M. J. Johnson (1946). Effect of metallic ions on the production of citric acid by *A. niger*. *Arch. Biochem.*, **11**, 131–143.

Pfizer, Chas. & Co. Inc. (1970). Fermentation process for the production of citric acid. *Br. Pat.* 1 182 983 (*Chem. Abstr.*, **71**, 90 055q).

Pfizer Inc. (1972). Stimulatory effect of organic acids in citric acid fermentation. *Br. Pat.* 1 293 786 (*Chem. Abstr.*, **77**, 124 768r).

Pfizer Inc. (1974). Fermentation process for the production of citric acid. *Br. Pat.* 1 364 094 (*Chem. Abstr.*, **78**, 70 192y).

Ramakrishnan, C. V. and S. M. Martin (1955). Isocitric dehydrogenase in *A. niger*. *Arch. Biochem. Biophys.*, **55** (2), 403–407.

Ramakrishnan, C. V., R. Steel and C. P. Lentz (1955). Mechanism of citric acid formation and accumulation in *A. niger*. *Arch. Biochem. Biophys.*, **55** (1), 270–273.

Röhr, M., O. Zehentgruber and C. P. Kubicek (1981). Kinetics of biomass formation and citric acid production by *A. niger* on a pilot plant scale. *Biotechnol. Bioeng.*, **23**, 2433–2445.

Rosenbaum, J. B., W. A. McKinney, H. R. Beard, L. Crocker and W. I. Nissen (1973). Sulfur dioxide emission control by hydrogen sulfide reaction in aqueous solution. The citrate system. US Bureau of Mines, Report 1774.

Schulz, G. and J. Rauch (1975). Citronensäure. In *Ullmans Encyklopaedie der technischen Chemie*, 4th edn., vol. 9, pp. 624–636. Verlag Chemie, Weinheim.

Shu, P. and M. J. Johnson (1947). Effect of the composition of the sporulation medium on citric acid production by *A. niger* in submerged culture. *J. Bacteriol.*, **54**, 161–167.

Shu, P. and M. J. Johnson (1948). The interdependence of medium constituents in citric acid production by submerged fermentation, *J. Bacteriol.*, **56**, 577–585.

Shu, P., A. Funk and A. C. Neish (1954). Mechanism of citric acid formation from glucose by *A. niger*. *Can. J. Biochem. Physiol.*, **32**, 68–80.

Siebert, D. and H. Hustede (1982). Citronensäure-Fermentation—biotechnologische Probleme und Möglichkeiten der Rechnersteuerung. *Chem.-Ing.-Tech.*, **54**, 659–669.

Snell, R. L. and L. B. Schweiger (Miles Laboratories Inc.) (1951). Citric acid by fermentation. *Br. Pat.* 653 808 (*Chem. Abstr.*, **45**, 8719a).

Stottmeister, U., U. Behrens, E. Weissbrodt, G. Barth, D. Franke-Rinker and E. Schulze (1982). Nutzung von Paraffinen und anderen Nichtkohlenhydrate-Kohlenstoffquellen zur mikrobiellen Citronensäuresynthese. *Z. Allg. Mikrobiol.*, **22** (6), 399–424.

Svenska Sockerfabriks AB (1964). A method of producing citric acid. *Br. Pat.* 951 629 (*Chem. Abstr.*, **60**, 2304a).

Swarthout, E. J. (Miles Laboratories Inc.) (1966). Citric acid production by fungal fermentation of sugar. *US Pat.* 3 285 831 (*Chem. Abstr.*, **64**, 5714a).

Szczodrak, J. (1981). Biosynthesis of citric acid in relation to the activity of selected enzymes of the Krebs cycle in *A. niger* mycelium. *Eur. J. Appl. Microbiol. Biotechnol.*, **13**, 107–112.

Tabuchi, T. and M. Abe (Takeda Yakuhin Kogyo K.K.) (1968). Process for the production of citric acid. *Jpn. Pat.* 68 20 707.

Tabuchi, T. and K. Igoshi (1978). Regulation of enzyme synthesis of the glyoxylate, the citric acid and the methylcitric acid cycles in *Candida lipolytica*. *Agric. Biol. Chem.*, **42** (12), 2381–2386.

Takayama, K., T. Adachi, M. Kohata, K. Hattori and T. Tomiyama (Kyowa Hakko Kogyo Ltd.) (1976). Process for the production of citric acid. *Br. Pat.* 1 455 486, example 8 (*Chem. Abstr.*, **82**, 15 276m).

Takeda Yakuhin Kogyo K.K. (1970). Method of producing citric acid by fermentation. *Br. Pat.* 1 211 246 (*Chem. Abstr.*, **72**, 77 462w).

Takeda Yakuhin Kogyo K.K. (1974). Method of producing citric acid. *Br. Pat.* 1 353 480, example 3 (*Chem. Abstr.*, **76**, 44 687c).

Tressl, Von R., R. Jakob, T. Kossa and W. K. Brown (1976). Gaschromatographisch-massenspektrometrische Untersuchung flüchtiger Inhaltsstoffe von Melasse. *Branntweinwirtschaft*, **116** (8), 117–119.

Verhoff, F. H. and J. E. Spradlin (1976). Mass and energy balance analysis of metabolic pathways applied to citric acid production by *A. niger*. *Biotechnol. Bioeng.*, **18**, 425–432.

Wehmer, C. (1893). Note sur la fermentation citrique. *Bull. Soc. Chim. Fr.*, **9**, 728–730.

Wennerston, R. (1980). A new method for the purification of citric acid by liquid–liquid extraction. In *Proc. Int. Solvent Extr. Conf.*, vol. 2, paper 80–63. Assoc. Ing. Univ. Liège, Liège.

Wongchai, V. and W. E. Jefferson (1974). Pyruvate carboxylase of *A. niger*: partial purification and some properties. *Fed. Proc.*, **33**, 1378.

Woronick, C. L. and M. J. Johnson (1959). Carbon dioxide fixation by cell-free extracts of *A. niger*. *J. Biol. Chem.*, **235**, 9–15.

35

Gluconic and Itaconic Acids

P. E. MILSOM and J. L. MEERS
John and E. Sturge Ltd., Selby, North Yorkshire, UK

35.1 INTRODUCTION AND PROPERTIES OF GLUCONIC ACID AND ITS LACTONES

D-Gluconic acid (pentahydroxycaproic acid; Figure 1; 526-95-4) is an oxidation product of D-glucose and has the same stereochemical configuration, having a specific rotation $\alpha_D^{20} = -6.7°$ (Prescott *et al.*, 1953) $\alpha_D^{25} = -5.40°$ (Sawyer and Bagger, 1959). Removal of two hydrogen atoms from D-glucopyranose yields D-glucono-δ-lactone (90-80-2) which in aqueous solution is in chemical equilibrium with D-gluconic acid and with D-glucono-γ-lactone (1198-69-2). D-Gluconic acid ionizes in water to give the gluconate ion and a proton. An aqueous solution of 'gluconic acid' is thus a somewhat complex system. Because of the various equilibria existing in such a solution, the apparent acid dissociation constant of gluconic acid must be corrected for lactone formation. Sawyer and Bagger (1959) gave a value for $K_A = 1.99 \times 10^{-4}$ equivalent to $pK_a = 3.70$. Skibsted and Kilde (1971) report $pK_a = 3.62$ at zero ionic strength. Both these sets of authors calculate values for the equilibrium gluconic acid/lactone but both fail to distinguish between the δ- and γ-lactones.

The individual products, *i.e.* gluconic acid and the δ- and γ-lactones, may be separated as crystalline solids from a 'gluconic acid' solution by obtaining supersaturated solutions at defined tem-

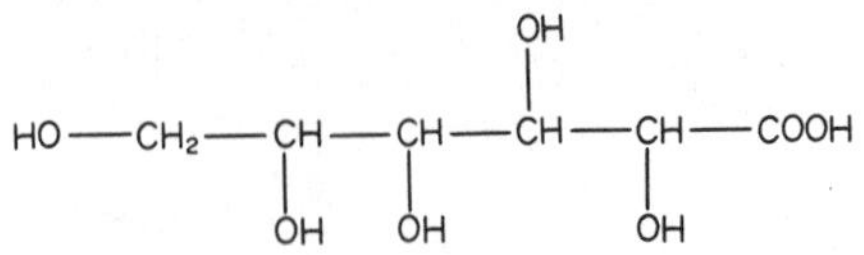

$$HO—CH_2—CH—CH—CH—CH—COOH$$

Figure 1 D-Gluconic acid

peratures and seeding them with crystals of the desired compound. Thus a supersaturated solution below 30 °C yields crystalline gluconic acid. Between 30 and 70 °C, or better between 36 and 57 °C (Yamauchi and Shimizu, 1975), the δ-lactone is obtained and above 70 °C the γ-lactone crystallizes (Pasternak and Giles, 1934). The equilibrium between gluconic acid or δ-gluconolactone and the γ-lactone is attained only slowly and many hours must be allowed for this process when the γ-lactone is required. Isbell (1934) describes a continuous process for making crystalline gluconic acid whereby an aqueous solution containing 50–75% of dry substance is seeded at a temperature not exceeding 60 °C, with crystals of gluconic acid and fresh gluconic acid solution added at the same rate as the gluconic acid crystallizes. Noury and Van der Lande (1962) state that anhydrous gluconic acid crystallizes as fine needles, which are difficult to free from mother liquor, and claim a process giving the monohydrate which has a more handlable crystalline form. The transition temperature between the two species is 23 °C and crystallization should take place below this temperature and preferably in the range 0–3 °C to obtain the monohydrate.

35.2 USES OF GLUCONIC ACID AND ITS DERIVATIVES

D-Gluconic acid is not at the present time marketed as the crystalline solid, but is available as a 50% solution. Crystalline D-glucono-δ-lactone is commercially available in large quantities, but the γ-lactone is made only in small quantities as a speciality.

Gluconic acid and its lactones are non-toxic and find uses in food. This is particularly the case with the δ-lactone which is extensively used as a latent acid in baking powders for use in dry cake mixes and in instant chemically leavened bread mixes. Both gluconic acid and especially its δ-lactone are used in the preparation of gluconate salts. Although many metallic salts of gluconic acid have been prepared and studied (May *et al.*, 1929; Prescott *et al.*, 1953; Sawyer, 1964), commercially the most important are sodium, calcium and iron(II) gluconates.

The sodium salt, which is available either as a solid or a solution, finds a major use in the sequestering of calcium and iron. The sequestering action of gluconate for calcium is rather poor even up to pH 14 but in the presence of free sodium hydroxide it becomes outstanding (Prescott *et al.*, 1953). This improvement is attributed to the ionization of the hydroxyl groups allowing the formation of chelate rings (Sawyer, 1964). This property allows the use of sodium gluconate in the washing of glass bottles, a process which is normally carried out with the aid of hot sodium hydroxide solution of up to 5% concentration. In hard water districts especially, the formation of calcium and magnesium hydroxides forms a scum not easily rinsed away. If sodium gluconate is added to the alkali at the rate of about one part gluconate to ten parts NaOH, the alkaline earth metals are sequestered and the problem does not arise (Dvorkovitz and Hawley, 1952a, 1952b).

Sodium gluconate has a very high sequestering action for iron over a wide pH range and this ability is not diminished by the presence of up to 4% NaOH (Prescott *et al.*, 1953). This property is exploited in the use of sodium hydroxide solutions containing sodium gluconate in alkaline derusting of ferrous metals. Sodium gluconate is also widely used as an additive in cement mixes where it modifies the setting and other properties in advantageous ways (Ward, 1967; Lockwood, 1979).

Calcium gluconate is widely used to treat diseases caused by a deficiency of calcium in the body. In this application small quantities of lactate, saccharate or heptonate are added to the gluconate to increase its solubility and obtain a stable non-crystallizing, sterilizable solution of high calcium content.

Iron(II) gluconate is often used to supply iron in cases of anaemia. Gluconate salts of iron and other trace metals are also used in foliar feed formulations in horticulture.

Gluconate is employed as the anion in pharmaceutical preparations containing nitrogen bases. A similar application is in the disinfectant chlorhexidine gluconate. There are, in addition, many minor outlets for gluconic acid and its salts. These are described in reviews (Prescott *et al.*, 1953; Wengraf, 1952; Ward, 1967; Anon, 1980).

35.3 PRODUCTION PROCESSES—FERMENTATION

The conversion of glucose to D-glucono-δ-lactone is a simple process and chemical methods have been used in manufacturing processes. One method is to employ electrochemical oxidation in the presence of bromide ions (Isbell *et al.*, 1932). Another is to use air or oxygen in the presence of a catalyst (Pfizer, 1957; Acres and Budd, 1970; de Wilt, 1972). However, the fermentation method is fully competitive with chemical techniques and is at present the method of choice. A combination of the fermentation and chemical techniques would be the use of glucose oxidase as a catalyst in the oxidation of glucose by air. This may prove to be the method of the future (Baker, 1953; CPC International Inc., 1972; Boehringer Mannheim G.m.b.H., 1972; Richter and Heinecker, 1979; Hartmeier and Tegge, 1979).

The existence of gluconic acid and methods for its preparation by biochemical means have long been known. Gluconic acid was first identified by Hlasiwetz and Habermann (1870). Boutroux (1880) isolated the calcium salt from a fermentation of glucose in the presence of calcium carbonate using a strain of *Mycoderma aceti*. Subsequently Molliard (1922) showed that gluconic acid was produced by a strain of *Sterigmatocystis nigra* (probably *Aspergillus niger*) when grown on a sucrose medium deficient in nitrogen. Bernhauer (1924), investigating the citric acid fermentation with *A. niger*, identified gluconic acid as a by-product. Bernhauer (1928) selected a strain of *A. niger* which under specified conditions produced only gluconic acid. May *et al.* of the US Department of Agriculture (1927) screened many types of fungi and selected strains in the *Penicillium luteum purpurogenum* group capable of producing gluconic acid from commercial dextrose without significant by-products and optimized the conditions for a surface fermentation (Herrick and May, 1928, 1929). These authors and co-workers (May *et al.*, 1929) studied the *P. luteum purpurogenum* fermentation on the pilot plant scale using high quality aluminum pans and obtained a 57% yield of gluconic acid in 11 days. Moyer *et al.* (1936) looked for a better organism and selected a strain of *P. chrysogenum* out of 50 *Penicillia* tested. The same authors grew the new strain under submerged conditions in sparged glass bottles and found that additions of calcium carbonate and increased air pressure improved its performance.

As a continuation of this work the use of a rotating drum fermenter was explored (Wells *et al.*, 1937). In order to obtain sufficient spores for inoculation the use of *P. chrysogenum* was discontinued and a strain of *A. niger* (NRRL 67) employed instead. The improved aeration in the rotating drum equipment led to much higher yields in considerably shorter times. The process was optimized (Moyer *et al.*, 1937) and larger, pilot scale drums used (Gastrock *et al.*, 1938) giving over 95% yields of gluconic acid in 24 hours from a medium containing glucose (150 kg m^{-3}). The submerged *A. niger* mycelium could be collected and re-used in semicontinuous production (Porges *et al.*, 1940). Finally this group of workers reported sodium gluconate production using a stirred aerated tank fermenter in which the gluconic acid was continuously neutralized with sodium hydroxide (Blom *et al.*, 1952). Yields approaching 100% were obtained in 19 hours. It is interesting to note that what must surely have been the first use of a stirred aerated fermenter was patented by Currie *et al.* (1931) more than 20 years earlier. Aeration was of the vortex type. A 90% yield of gluconic acid as the calcium salt was obtained from a glucose medium (200 kg m^{-3}) in 48 to 60 hours using *A. niger*.

Whilst the *A. niger* process described by Blom *et al.* (1952) is essentially that used by most manufacturers today, methods using other organisms have been reported. The organism described by Boutroux (1880) is likely to have been a species of *Acetobacter*. The genus *Acetobacter* has been more recently divided into those organisms which oxidize ethanol well and those which oxidize glucose well (Asai, 1971). The latter are now classified as *Gluconobacter*, whereas the former remain in the genus *Acetobacter*. Thus some strains have changed their name, *e.g. Acetobacter suboxydans* NRRL B-52 now often, but not always, appears as *Gluconobacter suboxydans* NRRL B-52. This can be a source of confusion. In this chapter the nomenclature of the original author will be followed.

It is a feature of the metabolism of *Gluconobacter* (*Acetobacter*) that gluconate is readily oxidized further to 2- or 5-ketogluconate (Stubbs *et al.*, 1940; Kheshghi *et al.*, 1954). However, the oxidation of the gluconate seems to be inhibited by glucose, so that gluconate can be obtained by stopping the fermentation at the correct time.

Currie and Carter (1930) used a modification of the so called quick vinegar process to produce gluconic acid by organisms from the *Acetobacter* group. Verhave (1932) used a two phase aeration scheme (low aeration for growth, high aeration for acid production) in the production of gluconic acid by *Acetobacter suboxydans*. More recently Nyeste *et al.* (1980) have modelled and optimized the gluconic acid fermentation using *Acetobacter suboxydans* ATCC 621 and Ooster-huis *et al.* (1983) have investigated the effect of physical parameters on the fermentation with *Gluconobacter oxydans* ATCC 621H. Huchette and Devos (1971) describe the production of gluconic acid by *Acetobacter suboxydans* as an adjunct to the production of sorbitol from starch.

Strains of *Pseudomonas ovalis* normally oxidize glucose through gluconic acid to 2-ketogluconic acid. Lockwood *et al.* (1941) identified a non-ketogenic strain which produced only gluconic acid. Production of gluconic acid by this organism has been studied as a measure of aeration in stirred tank reactors (Tsao and Kempe, 1960; Bennett and Kempe, 1964; Humphrey and Reilly, 1965).

Another study has used the *Ps. ovalis* fermentation as a demonstration of modelling by the continuous maximum principle (Constantinides and Rai, 1974). The kinetics of the fermentation have been studied (Bull and Kempe, 1970; Ghose and Ghosh, 1976; Ghose and Mukhopadhyay, 1976).

As well as the above mentioned organisms, *Pseudomonas fluorescens* and species of *Morexella*, *Tetracoccus*, *Pullularia*, *Micrococcus*, *Enterobacter*, *Scopulariopsis*, *Gonatobotrys* and *Endomycopsis* have been reported as producers of gluconic acid. However, the processes in use today employ either *A. niger* or *Acetobacter* (*Gluconobacter*) *suboxydans* in submerged culture.

35.3.1 *A. niger* Process—Media

The carbohydrate source for gluconate production is glucose either in the form of glucose monohydrate crystals (Blom *et al.*, 1952) or dextrose syrup (Hatcher, 1972). In addition, sources of nitrogen (ammonium salts, urea, corn steep liquor), phosphate, potassium and magnesium must be provided for the growth of the mould. Too much nitrogen will give rise to excessive growth and decreased gluconate production. The use of crude nitrogen and phosphate sources such as corn steep liquor can be avoided (Quadeer *et al.*, 1975) but in that case traces of iron, copper and zinc must be incorporated in the medium.

A separate growth of vegetative inoculum is usually carried out. Such inoculum media normally contain less glucose but most contain inorganic nutrients sufficient for growth. If adequate growth is secured in the inoculum stage a production medium containing little more than glucose can be used (Hatcher, 1972).

35.3.2 *A. niger* Process—Yields

The 'theoretical' yield of gluconic acid from anhydrous glucose, assuming no production of mycelium or CO_2, is 109% (kg gluconic acid per 100 kg glucose supplied). If the starting material is glucose monohydrate the corresponding figure is 99%. In practice, yields of gluconic acid exceed 90% in a good fermentation.

35.3.3 *A. niger* Process—Process Details

Figure 2 shows the possible layout of a plant suitable for sodium gluconate or gluconic acid production.

The prepared medium is shown as being sterilized by a continuous method using a plate heat exchanger but sterilization in the fermenter is a possible variant, although excessive darkening of the medium can occur. Where the medium is sterilized outside the fermenter the latter must be steam sterilized separately.

The first fermentation stage is the growth of a vegetative inoculum. To this end a spore suspension of a selected strain of *A. niger* (*e.g.* NRRL 3) prepared from a culture grown on a solid sporulation medium is introduced into the inoculum fermenter. The initial pH of the inoculum medium is adjusted to about 6.5 with NaOH. Both the inoculum and production fermenters are stirred, baffled, stainless steel tanks, sparged with air from the base and provided with a means of cooling.

After a period for germination of the spores and subsequent growth of mycelium at 30 to 33 °C,

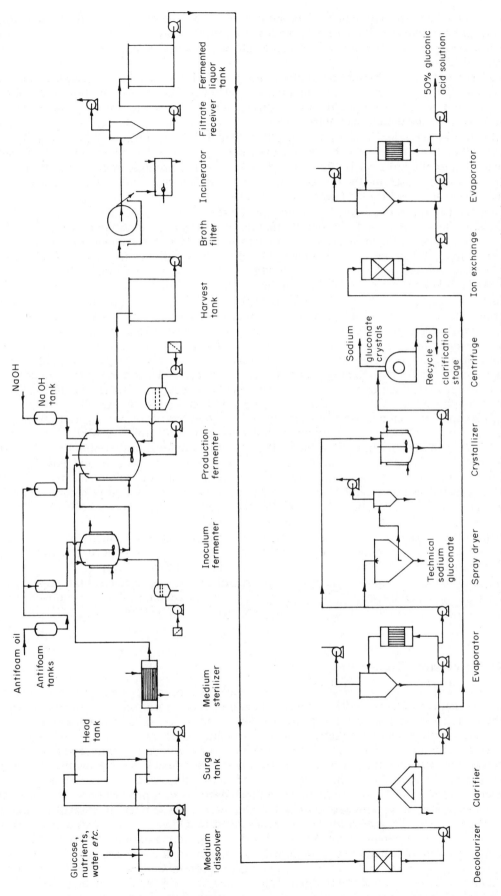

Figure 2 A possible layout for the production of sodium gluconate or gluconic acid by submerged fermentation using *A. niger*

the inoculum is transferred to the production fermenter at the rate of about 1 part inoculum to 10 parts production medium. The correct time for this transfer may be judged by the amount of mycelial growth or by the rate of increase of glucose oxidase activity in the mycelial cells (Hatcher, 1972).

In order to promote the formation of gluconate, the production medium, which may contain, for example, 220 kg glucose m^{-3}, is maintained at a pH of between 6 and 7, by addition of sodium hydroxide solution. The alkali addition is controlled automatically with the aid of a sterilizable pH electrode installed in the fermenter.

During the fermentation, which is conducted at 30 to 33 °C, the broth is agitated and sparged with air at the rate of up to 1.5 volumes of air per medium volume per minute. A back pressure on the fermenter of up to 2 bar is said to be advantageous. The progress of gluconate production may be followed by observing the rate of addition of the sodium hydroxide solution. The fermentation can be as short as 19 hours (Blom *et al.*, 1952).

As a variant of this process, the mycelium may be re-used up to twice more. In this case the mycelium is allowed to settle for about 30 minutes after stopping aeration and agitation, after which it is forced out of the fermenter together with about 20% of the original liquid volume and transferred to a fresh fermenter containing glucose solution without other nutrients (Hatcher, 1972). Because no lag time for mycelial growth is required, even less time is needed for the second and third cycles.

A further variant is described by Ziffer *et al.* (1971). It is desirable that the broth at the end of the fermentation should have as high a product concentration as possible to save on evaporation costs. However, high glucose concentrations, *e.g.* in excess of 600 kg m^{-3}, are not readily fermented. Furthermore, final concentrations of sodium gluconate in the region of 600 kg m^{-3} are difficult to handle in recovery because the saturation concentration is exceeded except at temperatures above 70 °C. Great problems are encountered with crystallization in pipe work and plant items. In the process of Ziffer *et al.* the glucose is added in stages. For instance the initial glucose concentration might be 270 kg m^{-3} and subsequent additions of 300, 80 and 90 kg m^{-3} may be made. It is apparently not necessary to sterilize the later additions of glucose. Typically 97 to 99% yields of product expressed as gluconic acid are obtained in 60 to 70 hours. Another feature of the Ziffer process is that neutralization of the medium is carried out only until optimum cell growth and glucose oxidase content are achieved. Subsequently the pH is allowed to fall below 5, values of 3.2 to 3.5 being reached. The final broth thus contains a mixture of sodium gluconate and gluconic acid which does not crystallize and problems are avoided. An alternative method of avoiding premature product crystallization, which involves adding sodium borate (Moyer, 1944), is effective but seems likely to give rise to more problems than it solves.

In yet another variant, $CaCO_3$ is used to neutralize the gluconic acid formed, leading to a fermented liquor containing calcium gluconate solution. In this process it is important to sterilize the $CaCO_3$ separately from the glucose to avoid decomposition of the carbohydrate.

At the end of the fermentation the broth is filtered or centrifuged to remove the cells and the clarified liquor pumped forward to the recovery stage. The mycelium of *A. niger* removed in this step may either be incinerated or used as a source of glucose oxidase (Lockwood, 1975).

35.3.4 *Acetobacter (Gluconobacter)* process

In contrast with the *A. niger* process far fewer details have been reported for the *Acetobacter* process. Huchette and Devos (1971) propose the use of a substrate consisting of starch hydrolysed by bacterial α-amylase to a DE (dextrose equivalent) of 20 to 25. Nutrients such as corn steep liquor and potassium phosphate are added. A culture of *A. suboxydans* is added and the broth aerated at 1 volume of air per volume of medium with vigorous agitation. The temperature is maintained at 30 °C and the pH controlled initially at 6, although it is subsequently allowed to fall to 3.2. After 24 hours the DE has fallen to 10 and all free glucose has been consumed.

Oosterhuis *et al.* (1983) have calculated the time constants of a production scale reactor by studying the fermentation with *Gluconobacter suboxydans* ATCC 621H in a small laboratory reactor. The fermentation conditions were as described by Olijve (1978). Nyeste *et al.* (1980) have determined an optimum temperature profile for the fermentation with *A. suboxydans* ATCC 621 by means of a modelling procedure. The temperature should initially be 30 °C rising in a logarithmic type curve to 36 °C at 5 hours and remaining constant at that level for the rest of the fermentation.

35.4 BIOCHEMISTRY OF THE GLUCONIC FERMENTATIONS

The steps by which *A. niger*, *A. suboxydans* or *Ps. ovalis* produce gluconic acid or gluconates are relatively few in number, consisting of the removal of two hydrogen atoms from β-D-gluco-pyranose to yield D-glucono-δ-lactone and the hydrolysis of the latter to gluconic acid or a gluconate. These steps are shown in the upper part of Figure 3.

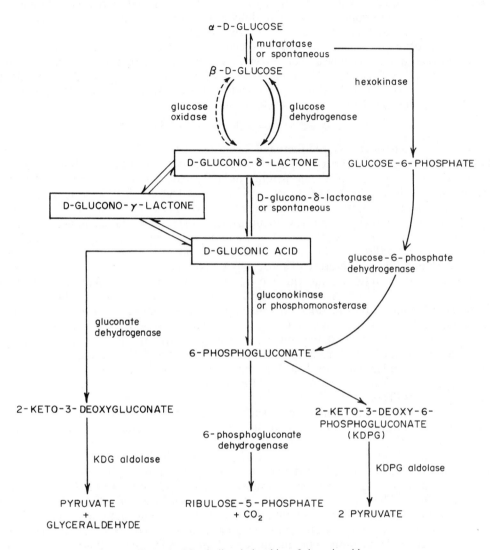

Figure 3 Metabolic relationships of gluconic acid

The first point to note is that it is the β-form of glucose which is dehydrogenated whether the enzyme is glucose oxidase or glucose dehydrogenase. In practice, although crystalline glucose monohydrate is in the α-form, this causes little difficulty. Not only does the α-form convert to the β-form spontaneously in solution but the enzyme mutarotase is present, at least in *A. niger*, and serves to accelerate the reaction.

The figure shows a pathway not involving direct dehydrogenation which has been observed in *A. niger*, *i.e.* via glucose 6-phosphate and 6-phosphogluconate (Müller, 1977). However, no evidence that this path is important in the gluconic fermentation has been found with *A. niger*. Although a glucose dehydrogenase utilizing either NAD or NADP having a pH optimum of 6.5 has been identified in *A. niger* (Müller, 1977), it is thought that glucose oxidase is by far the most important enzyme for the production of glucono-δ-lactone by this organism.

On the other hand in *Acetobacter* (*Gluconobacter*) *suboxydans* ATCC 621H, two glucose dehydrogenases are present (Olijve, 1978). One, a particulate enzyme, is largely inactive in the presence of glucose. The other, a soluble NADP specific enzyme, occurs in large amounts in the presence of glucose. In *A. suboxydans* the alternative pathway *via* 6-phosphogluconate is repressed at high glucose concentrations.

The oxidizing enzyme in *Ps. ovalis* is variously stated to be an NAD specific dehydrogenase (Humphrey and Reilly, 1965; Bull and Kempe, 1970) or a glucose oxidase (Constantinides and Rai, 1974; Ghose and Ghosh, 1976). However, neither of these statements is based on a study of the enzyme isolated from *Ps. ovalis* or even from other *Pseudomonads*. The evidence from *Ps. fluorescens* (Wood and Schwerdt, 1953; Jermyn, 1960) and other *Pseudomonads* (Hauge, 1966) is that species of this genus contain particulate dehydrogenases requiring neither NAD or NADP. It is likely, therefore, that the enzyme of *Ps. ovalis* is of this type.

Although the principal enzyme in *A. niger* is commonly called glucose oxidase, its true description is glucose aerodehydrogenase. The enzyme, a flavoprotein, removes two hydrogens from glucose becoming itself reduced. The reduced form of enzyme is then re-oxidized by molecular oxygen, giving hydrogen peroxide as a by-product (Bentley and Neuberger, 1949). In *A. niger* a catalase decomposes the H_2O_2 giving a net reaction:

$$\text{D-glucose} + 0.5\,O_2 \rightarrow \text{D-glucono-}\delta\text{-lactone} + H_2O$$

It has been demonstrated (Van Dijken and Veenhuis, 1980) that in *A. niger* both the glucose oxidase and the catalase are located in peroxisomes. The containment of the two enzymes within an organelle effectively prevents the occurrence of H_2O_2 toxicity during gluconate production.

The final step in the production of gluconic acid is the hydrolysis of the lactone which can occur spontaneously or can be assisted by the enzyme glucono-δ-lactonase. This enzyme is present in *A. niger* (Dumontier and Hanss, 1974; Cho and Bailey, 1977, 1978). Gluconolactonase has been demonstrated in *Ps. fluorescens* (Jermyn, 1960) and has been reported present by implication in *Ps. ovalis* (Constantinides and Rai, 1974). Whether or not it is present in *Ps. ovalis* seems to depend on fermentation temperature (Tanner and Yunker, 1977).

Accumulation of the intermediate glucono-δ-lactone has a negative effect on the rate of glucose oxidation. Therefore removal of the lactone by hydrolysis is important for the overall rate of production of gluconic acid or its salts. At neutral pH values, such as are brought about in gluconic fermentation by neutralization with $CaCO_3$ or NaOH, the spontaneous hydrolysis of the lactone is quite rapid. However at the lower pH values used in the latter part of some of the processes described above, the presence of the lactonase can be more important.

The effect of pH on lactone hydrolysis is not the only influence this factor has on gluconic acid production. It seems that higher values of pH predispose *A. niger* to produce gluconate rather than other products (Tessi *et al.*, 1967).

The fact that gluconic acid accumulates in such large quantities in cultures of *A. niger* and other organisms under suitable conditions implies that it is not being metabolized by the organism to any significant extent.

The lower half of Figure 3 shows some pathways by which gluconate could be metabolized. Lakshminarayana *et al.* (1969) adapted a strain of *A. niger* to grow on gluconate and studied its behaviour. It was found that growth on this substrate occurred only at low pH and was maximal in the range 1 to 3. The biochemistry of the strain was not investigated. Elzainy *et al.* (1973) demonstrated the non-phosphorylated path *via* 2-keto-3-deoxygluconate in a strain of *A. niger* when grown on gluconate as substrate. Ichikawa and Imanaka (1960) showed the existence of the pathway *via* 2-keto-3-deoxy-6-phosphogluconate in *A. niger* ATCC 4417. It is therefore clear that strains of mould differ in the pathways they use to metabolize gluconate and that the enzymes of these pathways are probably induced by the presence of gluconate. In a gluconate accumulating medium either the prevailing pH or excess glucose may inhibit gluconate catabolism. Strains of the organism may also differ considerably in their ability to dissimilate gluconate.

35.5 PRODUCTION PROCESSES—RECOVERY

The various fermentation processes described above lead to fermentation broths containing calcium gluconate solution, sodium gluconate solution and in the case of the process of Ziffer *et al.* (1971) a solution of partly neutralized gluconic acid. The latter mixture may be readily converted to sodium gluconate solution, if required, by addition of extra NaOH.

Figure 2 shows three ways of treating a fermented liquor containing only sodium gluconate. The clarified and decolourized broth (previously freed from cells of the organism and containing minimal residual glucose) is concentrated by evaporation. A choice can now be made. If a technical grade of sodium gluconate, *e.g.* of 98% purity is required, the liquor may be dried directly. A method of spray drying sodium gluconate solutions is described by Gillenwater (1975). Two of the advantages of the Ziffer process become apparent at this point. Because the product concentration is so high, little, if any, evaporation will be necessary before spray or drum drying. Further more because much of the glucose will have been added unsterilized the fermented liquor will be much lighter in colour. A better technical grade will therefore be obtained with less processing.

If a pure grade of sodium gluconate is required, the liquor after concentration may be transferred to a crystallizer and the crystals so obtained separated in a centrifuge. The mother liquor may be recirculated to an earlier stage of the process.

If a 50% gluconic acid solution is required the evaporated liquor may be passed through a cation exchanger in the hydrogen form to remove the sodium ions. Again the process of Ziffer *et al.* exhibits an advantage in that the fermentation broth as produced contains less sodium ion per unit weight of gluconic acid and a smaller cation exchange column will be necessary. The methods used to produce D-glucono-δ-lactone or D-glucono-γ-lactone from gluconic acid solution have already been described on pages 681 and 682.

A variant for recovering gluconic acid from a clarified and decolourized broth containing sodium gluconate is described by Huchette and Devos (1971) whereby the solution is passed through a strong anion exchanger such as Amberlite IRA 400 where the gluconic acid is retained.

When the fermentation broth has been kept neutral by addition of $CaCO_3$ either calcium gluconate or gluconic acid or another salt may be recovered. In order to recover calcium gluconate, the clarified and decolourized liquor is concentrated and cooled to below 10 °C with or without addition of alcohol when the calcium gluconate crystallizes and may be recovered and further purified. Gluconic acid may be obtained from a solution of the calcium salt either by addition of H_2SO_4 to remove the calcium as $CaSO_4$ or by passing the solution through a column containing a strongly acid cation exchanger to absorb the calcium ions.

35.6 FUTURE DEVELOPMENTS

The market for gluconic acid has been developing for some 40 years and now amounts to upwards of 50 000 tonnes per annum worldwide. There is evidence to suggest that the market is still growing in some areas of usage.

Great improvements in manufacture by fermentation seem unlikely. There are many strains, particularly of *A. niger*, which under known conditions give yields close to the theoretical and the reported fermentation times are exceedingly short. Re-use of mycelium reduces the initial lag time needed for growth and glucose feeding leads to very high concentrations of product in the final broth. By-product formation is minimal. A possible development may be in the use of immobilized cells or immobilized glucose oxidase. Work has been done on both these techniques but neither has yet shown the efficiency exhibited by the fermentation. The use of immobilized biocatalysts in fixed beds is difficult because of the need to supply gaseous or dissolved oxygen. Hydrogen peroxide has been proposed as an oxygen source, but not only does H_2O_2 tend to inactivate glucose oxidase, it is also expensive. Perhaps the most likely development is the use of immobilized cells of, for example, *A. niger* in a stirred tank reactor.

35.7 ITACONIC ACID—INTRODUCTION

Itaconic acid (methylene succinic acid, methylenebutanedioic acid; Figure 4; 97-65-4), was discovered by Baup (1836) who obtained it by destructive distillation of citric acid. K. Kinoshita (1931, 1932) found that it was produced by strains of an osmophilic species of *Aspergillus* which he named *A. itaconicus*. A few years later Calam *et al.* (1939) examined six strains of *Aspergillus terreus* Thom and identified itaconic acid as a product formed by one of them. S. Kinoshita and Tanaka (1961) obtained a patent claiming the use of *A. itaconicus* for the production of itaconic acid from untreated molasses and Kobayashi claimed process details for strains of the same species using a variety of media (Kobayashi and Tabuchi, 1957a, 1957b; Kobayashi, 1960). Despite these developments, industrial processes have used strains of *A. terreus*, this organism being preferred to the slow growing *A. itaconicus*.

The main impetus for the use of *A. terreus* came from work at the Northern Regional Research Laboratory of the US Department of Agriculture. A screening of wild type isolations led to the finding of *A. terreus* NRRL 265 (Moyer and Coghill, 1945) which was studied in surface culture.

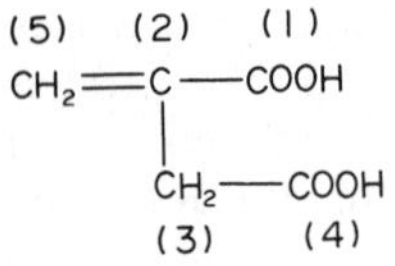

Figure 4 Itaconic acid

A further more extensive screening of new isolations in surface culture led to the discovery of NRRL 1960, an isolate from Texas soil, which gave an outstanding performance both in flask surface culture (Lockwood and Reeves, 1945) and in semipilot scale pan culture (Lockwood and Ward, 1945). Strain NRRL 1960 was also studied in submerged culture both in shake flasks (Lockwood and Nelson, 1946) and in 20 l stirred fermenter cultures (Nelson *et al.*, 1952), leading to the development of a pilot scale process on the 2.2 m^3 scale (Pfeifer *et al.*, 1952). Unfortunately the process was optimized at an initial glucose concentration of only 60 kg m^{-3}, which is too low to be economic. Nevertheless, the strain NRRL 1960 and mutants thereof have been the basis for other, more successful, developments.

In more recent years it has been discovered that certain yeasts can accumulate itaconic acid. Thus Kawamura *et al.* (1981) selected three strains of yeast producing this acid and a patent was applied for naming a species of *Rhodotorula* (Shizuoka Prefecture, 1981). Tabuchi *et al.* (1981) isolated a species thought to belong to the genus *Candida* and obtained a mutant from it giving 35% yield of itaconic acid. Again a patent was applied for (Mitsubishi Chemical Industries Co. Ltd., 1980). The production of itaconic acid by *Ustilago zeae*, *Penicillium charlesii*, a yellow-green *Aspergillus*, and *Helicobasidium mompa* has also been reported.

Fermentation processes are fully competitive with synthetic methods despite the fact that a number of chemical routes have been investigated. For example great improvements have been made in the destructive distillation of citric acid. In addition condensation of formaldehyde with succinic acid derivatives, the use of the cyanhydrin reaction with ethyl acetoacetate, as well as other methods, have been proposed. High raw material costs and/or relatively low yields have militated against the success of these syntheses.

35.8 PROPERTIES AND USES OF ITACONIC ACID AND ITS DERIVATIVES

Itaconic acid is a crystalline substance mp 167–168 °C. In water it dissolves to the extent of about 7 kg m^{-3} at 20 °C and about 60 kg m^{-3} at 80 °C, making it quite easy to purify by crystallization. It is relatively non-toxic (Finkelstein *et al.*, 1947).

Itaconic acid forms a range of metallic salts. The calcium salt has a higher solubility than calcium citrate and cannot readily be used, as is the latter, in a precipitation step in the purification of the acid.

Diesters of itaconic acid are easily prepared and dimethyl itaconate and di-*n*-butyl itaconate are available in commercial quantities. Itaconic anhydride can be made by any of the normal methods and may be used for the preparation of mono-esters including monomethyl itaconate which has been offered for sale.

The key to the properties of itaconic acid which make it a uniquely valuable intermediate is the conjunction of its two carboxyl groups and its methylene group. The methylene group is able to take part in addition polymerization giving polymers with many free carboxyl groups which confer advantageous properties on the resulting polymer. Itaconic acid itself polymerizes only slowly to give rather low molecular weight products and it is best used in copolymers. Fibres made with an acrylonitrile/itaconic acid copolymer have improved dyeing properties. Styrene–butadiene latices incorporating itaconic acid display superior adhesion and are extensively used in carpet backings and in paper coatings. Itaconic acid is also used in emulsion paints where it improves the adhesion of the acrylic polymer. Other uses of itaconic acid in synthetic polymers have been reviewed (Billington, 1969; Luskin, 1974).

The dimethyl and di-*n*-butyl esters of itaconic acid can also be used in copolymers, *e.g.* for adhesives, and esters with long chain alcohols have been proposed as plasticizers.

Another important outlet for itaconic acid arises from its reaction with amines to yield *N*-substituted pyrrolidinones (pyrrolidones) as shown in Figure 5. Pyrrolidones produced in this way from aromatic diamines have been proposed as thickeners for greases (Gordon and Coupland, 1980). Other pyrrolidones made from itaconic acid and a wide range of amines have actual or potential uses in detergents, shampoos, pharmaceuticals and herbicides.

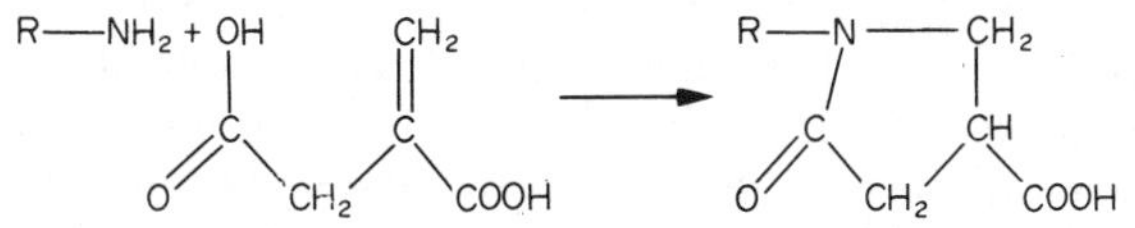

Figure 5 Reaction of itaconic acid with amines to yield *N*-substituted pyrrolidinones

A condensate of lauric acid and aminoethylethanolamine reacts with itaconic acid to give, it is claimed, an imidazoline derivative useful as an active ingredient in shampoos (Christiansen, 1980).

35.9 PRODUCTION PROCESSES—FERMENTATION

It is thought that all current processes utilize strains of *A. terreus*, either in the submerged or surface mode. Pfizer, the largest manufacturer, is said to use a submerged process in the US. There is production by Iwata in Japan and in Riga in the Soviet Union probably by a surface method (Karklin' and Agafonova, 1969).

35.9.1 Media

Various carbohydrate sources have been reported as suitable for itaconic acid production. These include glucose (Pfeiffer *et al.*, 1952), sucrose (Buendia and Garrido, 1958, 1959), decationized Hitest molasses (Batti and Schweiger, 1963) and beet and cane molasses (Nubel and Ratajak, 1962). Production from wood hydrolysate has also been studied (Kobayashi, 1978; Vitola *et al.*, 1969).

It is said that sucrose is superior to glucose as a carbohydrate source (Tandon and Mehrotra, 1970; Elnaghy and Megalla, 1975).

Beet molasses is a very variable raw material for itaconic acid production (Karklin' and Agafanova, 1969). The effect of colouring matter in beet molasses on acid production was investigated by Nowakowska-Waszczuk and Zakowska (1971) who concluded that one fraction of the melanoidins was toxic to *A. terreus*. The same authors identified the volatile aliphatic carboxylic acids contained in beet molasses media as factors preventing growth of the mould at low pH. Valeric and caproic acids were found to be particularly harmful. Nakamura *et al.* (1975) state that organic acids, *e.g.* formic and acetic acids, must be removed from blackstrap molasses before the itaconic fermentation. Nubel and Ratajak (1962) recommend supplying 10 to 30% of the total carbohydrate as beet molasses, for instance by using beet molasses to grow a vegetative inoculum.

Except where beet molasses is used as a substrate it is essential to supply a suitable source of nitrogen. Alkali metal nitrates are considered to be poor nitrogen sources for the fermentation, giving abundant growth with little or no itaconic acid production (Moyer and Coghill, 1945; Nowakowska-Waszczuk *et al.*, 1969, 1971; Tandon and Mehrotra, 1970). Ammonium sulfate and ammonium nitrate are good nitrogen sources for itaconic acid production. The unfavourable effect of the alkali nitrates may be ascribed either to the effect of the alkali metal ions or to pH effects.

Many authors recommend the use of an organic nitrogen source such as corn steep liquor in addition to inorganic nitrogen compounds. However, Nowakowska-Waszczuk and Zakowska (1969) found that the positive effect of corn steep liquor was ascribable to its mineral content.

In addition to a source of phosphate it is important to add relatively large amounts of magnesium (Lockwood and Reeves, 1945) and calcium (Batti and Schweiger, 1963). Nakamura *et al.* (1975) found that the potassium content of a blackstrap molasses medium should not exceed 0.14 kg m^{-3} for good yields of itaconic acid with *A. terreus* K26.

As in the citric acid fermentation, trace metals play an important role in itaconic acid accumulation. Lockwood and Reeves (1945) using *A. terreus* NRRL 1960 in surface culture with a medium containing corn steep liquor found that small additions of iron increased yields of itaconic acid and that this effect was more marked at low initial pH. In the presence of 1 p.p.m. zinc, however, additions of 10 p.p.m. iron reduced itaconic accumulation. Additions of zinc up to 1 p.p.m. also increased acid production but not in the presence of 10 p.p.m. iron. Batti and Schweiger (1963) using the same strain of mould in a decationized Hitest molasses medium containing inorganic salts, found that as little as 1 p.p.m. iron greatly reduced itaconic production but that this effect could be offset by additions of copper and zinc. These authors showed that optimum zinc and copper additions were dependent on the calcium concentration in the medium. Von Fries (1966) claimed the use of specified levels of ferrocyanide ion in the growth and acid forming stages when beet molasses or technical glucose were fermented with *A. terreus*, NRRL 1960.

35.9.2 Yields

Initial carbohydrate concentrations are reported to be within the range 100 to 180 kg m^{-3}. Yields of itaconic acid are expressed as kg acid per 100 kg carbohydrate supplied and are said to be between 55 and 65%. The theoretical yield assuming no diversion of carbon to mycelium, CO_2 or other products is 76% from sucrose and 72% from anhydrous glucose. In making this calculation it is assumed that one hexose gives one molecule of itaconic acid, the remaining carbon atom being lost as CO_2.

35.9.3 Process Details

Figure 6 shows the layout of a plant designed to produce itaconic acid from sucrose or glucose. If molasses were to be used, the plant would differ to some degree. After dissolving the carbohydrate and inorganic salts in a dissolver, the medium is pumped *via* a sterilizing heat exchanger either to one, or more, inoculum vessels or to the main fermenter as required. Medium intended for the preparation of a vegetative inoculum would normally contain less carbohydrate. All fermenters are of very high grade stainless steel and are aerated stirred baffled tanks with provision for temperature control, antifoam addition and pH measurement. The sizing of the inoculum fermenter(s) depends on the volume of vegetative inoculum to be transferred to the next stage. Reported ratios of inoculum to main fermentation volume vary widely from 1 to 20%.

The pH of the medium is of great importance in the itaconic acid fermentation although there is some disagreement in the literature about the best pH profile to employ. It seems that a low pH must be reached before the production of itaconic acid commences. Some authors have found that values as low as 1.8 to 2.0 are required (Larsen and Eimhjellen, 1955) but others report that pH values just above 3 are optimal. Growth of the organisms, however, is better at higher pH values and the growth of the vegetative inoculum may commence in the range 5 to 7. Batti (1964) proposed to carry out the fermentation stage starting at about 4.3, continuing until the itaconic concentration reached 4.6 to 5.0 kg m^{-3}. At this point the pH had reached about 2 and part of the acid was then neutralized using NH_4OH, $Ca(OH)_2$ or $CaCO_3$ so that the pH rose to a value between 3.1 and 3.9. The fermentation was then allowed to proceed to its conclusion. By thus increasing the pH, higher yields of itaconic acid were obtained and the formation of by-products such as itatartaric acid were avoided. Nubel and Ratajak (1962) proposed a similar pH regime.

A. terreus is capable of growing at quite high temperature. Pfeiffer *et al.* (1952) recommended 35 °C. Nubel and Ratajak (1962) proposed the use of temperatures between 35 and 42 °C. Pfeiffer *et al.* using a glucose concentration of 60 kg m^{-3} found that the fermentation was complete in about 65 hours. Nubel and Ratajak, using an initial carbohydrate concentration of about 150 kg m^{-3}, reported fermentation times of about 72 hours.

The itaconic acid fermentation with *A. terreus* is highly aerobic. Aeration rates of between 0.25 and 0.5 volumes of air per volume of medium are quoted. Back pressure on the fermenter of 1 bar over atmospheric has been recommended (Pfeiffer *et al.*, 1952). What is more important, however, is that aeration and agitation must be continuous, as even very short intermissions cause a complete cessation of itaconic acid production which is only very slowly regained, if at all, on resumption. A continuous process for itaconic acid production has been proposed (Kobayashi, 1967). The fermentation is continued until the rate of itaconic acid production

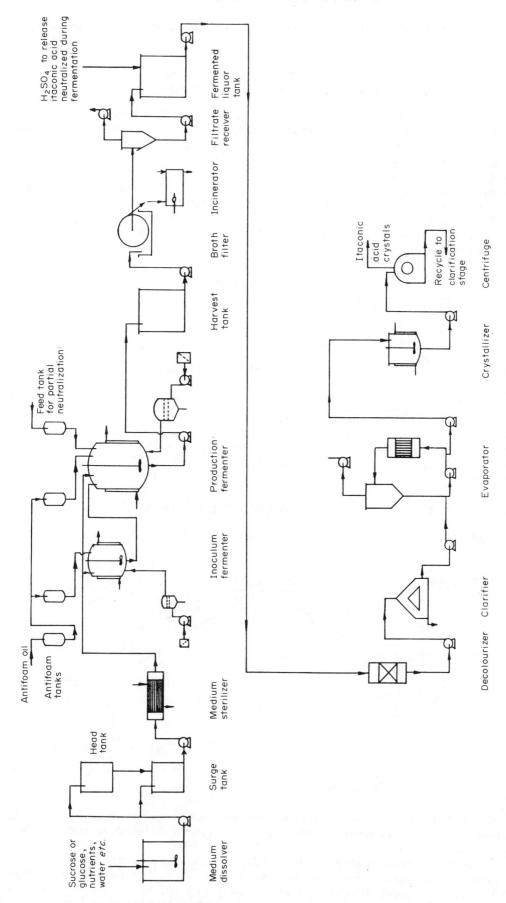

Figure 6 Possible layout of a plant designed to produce itaconic acid from sucrose or glucose

becomes uneconomically slow. At this point the mycelium is removed by filtration and the liquor transferred to the recovery stage.

35.10   BIOCHEMISTRY OF ITACONIC ACID ACCUMULATION

The mechanism of itaconic acid formation in *A. terreus* is not yet clear. Recorded instances of itaconic acid accumulation are not widespread (see above) but do cover a number of genera of the fungi including yeasts. Metabolism of itaconic acid to pyruvate and acetyl coenzyme A by *Pseudomonas B₂aba* has been reported (Cooper and Kornberg, 1962), and a similar scheme was proposed for liver mitochondria by Adler *et al.*, (1957).

The formation of itaconic acid *via* citric and *cis*-aconitic acids was first proposed by Kinoshita (1931) and a scheme which embodies this route of formation with the path of degradation proposed by Cooper and Kornberg (1962) is shown in Figure 7. Pyruvate is produced from hexose *via* the Embden–Meyerhof pathway, and, as with the citric fermentation, anaplerosis is by carboxylation of pyruvate.

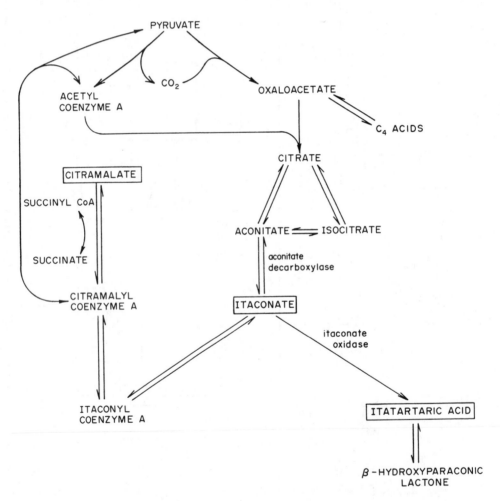

Figure 7   Metabolic relationships of itaconic acid

The presence of the enzymes catabolizing itaconate in *A. terreus* can be inferred from the formation from itaconic acid of itatartrate (2-hydroxy-2-hydroxymethylbutanedioic acid [2957-09-7]) and L(−)-citramalate (2-hydroxy-2-methylbutanedioic acid [2306-22-1]) in replacement surface cultures fed with itaconate (Jakubowska *et al.*, 1974). Experiments with cell free extracts of *A. terreus* showed the enzymic reactions leading from itaconic acid to citramalate were reversible (Jakubowska and Metodiewa, 1974).

Arpai (1959) isolated a mutant which produced itatartaric acid and identified the properties of an itaconic oxidase producing this substance from itaconic acid. Jakubowska *et al.* (1974) were unable to confirm the presence of itaconic oxidase and further found that if arsenite was present in the itaconic production medium, itatartaric acid appeared as the sole product. They therefore disputed the statement that itatartaric acid was a product of itaconic acid oxidation and suggested that itatartrate was an intermediate in itaconate formation, *e.g.* from citramalate.

Eimhjellen and Larsen (1955) found that *A. terreus* NRRL 1960 grown on glucose in shake flasks at pH 2.1 would convert citric acid to itaconic acid. Cell free extracts of this mould converted citric acid to itaconic acid (Jenssen *et al.*, 1956) but, in contrast to the finding of the previous authors, itaconic acid was produced much more effectively from aconitic acid. Aconitase was detected in the extract. Bentley and Thiessen (1955, 1956) also found aconitase in cell free extracts of *A. terreus* grown under itaconic acid producing conditions. In addition they established the presence of an active *cis*-aconitic decarboxylase, so that the extracts converted citric acid to itaconic acid. Citrate, α-ketoglutarate, succinate and malate were detected in the medium of itaconic acid accumulating cultures (Bentley and Thiessen, 1957b).

^{14}C from 1-^{14}C-acetate was incorporated into the two carboxyls of itaconic acid equally, whilst when 2-^{14}C-acetate was used 60% of the radioactivity was found in carbon 5 (Bentley and Thiessen, 1957a). In addition, it was demonstrated by similar methods that the skeleton of succinate was incorporated in its entirety as carbons 1, 2, 3 and 4 of itaconate, confirming the conclusion of Corzo and Tatum (1953) that it was the acetate carboxyl which was lost in the decarboxylation of *cis*-aconitic acid. Bentley and Thiessen (1957c) showed that the decarboxylation of *cis*-aconitic acid took place at the primary carboxyl and that allylic rearrangement followed. This mechanism was recently confirmed by Ranzi *et al.* (1981) using a ^{2}H NMR method. These findings are consistent with the usual mode of action of aconitase (Bentley and Thiessen, 1957c).

Nowakowska-Waszczuk (1973) and Smith *et al.* (1974) studied intact and disrupted mitochondria from *A. terreus* 1960 and two mutants thereof and found them incapable of oxidizing citric acid cycle intermediates whether the cultures used were accumulating itaconic acid or not. It was concluded that the citric acid cycle enzymes played no part in itaconic acid accumulation.

Nowakowska-Waszczuk (1973) and Smith *et al.* (1974) support a suggestion of Shimi and El Dein (1962) that pyruvate is cleaved to acetyl phosphate, three molecules of which are progressively condensed to 1,2,3-tricarboxypropane. The latter is then converted to *cis*-aconitate, which is subsequently decarboxylated to itaconate. This scheme is said to be consistent with the results of the published radiotracer studies outlined above.

The literature is thus very contradictory. As with citric acid production, it is necessary to be wary of results obtained in conditions other than under those giving maximum yield. Unless these conditions are used, various side reactions occur giving rise to erroneous conclusions. Another criterion to be applied is that of mass balance. A calculation shows that the pathway *via* 1,2,3-tricarboxypropane, in which one carbon is lost for every pyruvate molecule cleaved, cannot be consistent with 55 to 65% yields unless a powerful mechanism for carbon dioxide fixation is postulated. Similarly, the published yields cannot be obtained if the oxidative segment of the citric acid cycle is active to more than a minor degree. In this sense it is easy to agree that the citric acid cycle plays no part (except for the first step) in the accumulation of itaconic acid. It is, on the other hand, difficult to explain why no oxidation of citric acid cycle intermediates occurred with mitochondria from a strain accumulating no itaconic acid. The detection of aconitase in cell free extracts and the identification of citric acid as a minor component of the fermentation medium seem, however, to be important indications of the metabolic path. That citric acid was found in some cases to be a poor precursor of itaconic acid in experiments with cell free extracts could be attributed to the extreme instability of fungal aconitase *in vitro*.

Taking all the evidence into account it would seem most likely that citric acid is an intermediate in the itaconic fermentation and this conclusion is strongly supported by the recent work of Winskill (1983).

35.11 PRODUCTION PROCESSES—RECOVERY

When the fermentation medium is essentially composed of sucrose or glucose and salts, and particularly when the residual carbohydrate at the end of the fermentation is very low, a direct crystallization of itaconic acid from clarified, decolourized broth is possible (Pfeiffer *et al.*, 1952). A suitable plant layout is illustrated in Figure 5. The primary crop of crystals is of a good commercial grade suitable for ester preparation and copolymerization. Recrystallization from water gives a pure grade of material.

When crude carbohydrates are used as fermentation substrate, the itaconic acid may be precipitated as the lead salt by addition of basic lead carbonate (Kobayashi and Nakamura, 1971). The itaconic acid is redissolved from the precipitate by treatment with an alkali metal (or ammonium) carbonate solution. After removal of the regenerated lead carbonate, the alkali metal cations are absorbed on a cation exchanger giving a purified solution of itaconic acid. Recovery of itaconic acid by solvent extraction has also been employed. For instance, Les Usines de Melle (1962) have used tributyl phosphate mixed with one or more diluents. The filtered broth is extracted with the solvent at 10 to 30 °C and the acid laden solvent stripped with water at 70 to 95 ° C.

Ion exchange absorption of itaconic acid from fermented broth has been investigated (Nakagawa and Kobayashi, 1968; Korol'kov *et al.*, 1969; Nakagawa *et al.*, 1975). Weakly basic exchangers such as Amberlite IR-45 have been tried with some success both in fixed and fluidized bed reactors. Adams *et al.* (1970) have proposed reverse osmosis using specially selected membranes which allow substantial amounts of itaconic acid and water to pass while retaining substantial amounts of organic and inorganic impurities. In one example, while the itaconic concentration in the permeate was similar to that of the feed, the ash and colour had decreased to 10% and 5% of the feed liquor values respectively. Kobayashi *et al.* (1972) have designed a complete continuous fermentation process starting from Hitest molasses in which recovery is by continuous electrodialysis.

35.12 FUTURE DEVELOPMENTS FOR ITACONIC ACID

Itaconic acid has been made by fermentation for some 30 years and was produced in small quantities by a chemical method before that. The current production is measured in thousands of tonnes. New uses for itaconic acid both in the polymer field and in *N*-substituted pyrrolidones are continually being proposed and it is expected that the market will increase overall.

Reported yields of 65% on sugar supplied are very nearly maximal, given the loss of one molecule of CO_2 per molecule of *cis*-aconitic acid converted to itaconic acid, unless a way could be found to re-incorporate this carbon in the metabolic flow. Although the fermentation with *A. terreus* is quite fast there might be some scope for shortening the duration of the fermentation by the use of yeasts and a little work has already been done in this direction.

There appears to be little immediate scope for the use of gene transfer techniques because of the multistep nature of itaconic acid production from sugars and because the steps of the pathway have not, as yet, been properly defined.

35.13 SUMMARY

Gluconic acid, its δ-lactone, and sodium and calcium gluconates have been produced for 40 years by fermentation processes using strains of *Aspergillus niger* and *Acetobacter suboxydans* acting on glucose. The physical and chemical factors affecting these processes have been described and the necessary items of plant required outlined. The relatively simple dehydrogenation of D-glucopyranose to the intermediate D-glucono-δ-lactone by either glucose oxidase (*A. niger*) or glucose dehydrogenase (*A. suboxydans*) has been discussed and the chemical procedures used for producing 50% gluconic acid solution, gluconate salts, or glucono-δ-lactone, defined. The fields of usage of gluconic acid and its derivatives, which range from food to bottle washing, have been indicated.

Although itaconic acid is produced by several different microorganisms, manufacturing processes have so far used strains of *Aspergillus terreus* acting on media based on either pure sugars or crude carbohydrates. The critical role of pH as well as the importance of the nitrogen source and the presence in the medium of the correct concentrations of calcium, magnesium, iron, zinc and copper have been discussed. The conflicting results of investigations into the biochemical mechanisms of itaconic acid accumulation have been presented and the view has been taken that the most likely pathway involves the dehydration of citric acid and the decarboxylation of the *cis*-aconitic acid so produced.

The layout of a plant suitable for a fermentation starting from a sucrose or glucose/salts medium and a recovery by direct crystallization of itaconic acid from the filtered, clarified and the decolourized broth has been outlined. An account has been given of the many uses of itaconic acid both as a monomer yielding valuable copolymers and as an intermediate in the chemical synthesis of pyrrolidones.

35.14 REFERENCES

Acres, G. J. K. and A. E. R. Budd (Johnson Matthey and Co. Ltd.) (1970). Improvements in and relating to the catalytic oxidation of glucose. *Br. Pat.* 1 208 101 (*Chem. Abstr.*, **74**, 14 347h).

Adams, F., L. F. Rice and R. J. Taylor (Chas. Pfizer and Co. Inc.) (1970). Itaconic acid purification by reverse osmosis. *US Pat.* 3 544 455 (*Chem. Abstr.*, **74**, 124 834x).

Adler, J., S.-F. Wang and H. A. Lardy (1957). The metabolism of itaconic acid by liver mitochondria. *J. Biol. Chem.*, **229**, 865–879.

Anon. (India) (1980). Sodium gluconate as a speciality chemical in industry. *Indian Chem. Manuf.*, **18** (3), 27–32.

Arpai, J. (1959). Ultraviolet induced mutational changes in enzyme activity of *A. terreus. J. Bacteriol.*, **78**, 153–158.

Asai, T. (1971). The classification of acetic acid bacteria. In *Biochemical and Industrial Aspects of Fermentation*, ed. K. Sakaguchi, T. Uemura and S. Kinoshita, pp. 201–232. Kodansha Ltd., Tokyo.

Baker, D. L. (B. L. Sarett) (1953). Enzymatic process for producing gluconic acid. *US Pat.* 2 651 592 (*Chem. Abstr.*, **48**, 334h).

Batti, M. and L. B. Schweiger (Miles Laboratories Inc.) (1963). Process for the production of itaconic acid. *US Pat.* 3 078 217 (*Chem. Abstr.*, **58**, 14 668a).

Batti, M. A. (Miles Laboratories Inc.) (1964). Process for the production of itaconic acid. *US Pat.* 3 162 582 (*Chem. Abstr.*, **62**, 5857gh).

Baup, S. (1836). Ueber eine neue Pyrogen-Citronensäure u.s.w. *Annalen*, **19**, 29–38.

Bennett, G. F. and L. L. Kempe (1964). Oxygen transfer mechanisms in the gluconic acid fermentation by *Pseudomonas ovalis. Biotechnol. Bioeng.*, **6**, 347–360.

Bentley, R. and A. Neuberger (1949). The mechanism of action of notatin. *Biochem. J.*, **45**, 584–590.

Bentley, R. and C. P. Thiessen (1955). *cis*-Aconitic decarboxylase. *Science*, **122**, 330.

Bentley, R. and C. P. Thiessen (1956). Mechanism of action of aconitase from *A. terreus. Fed. Proc.*, **15**, 219.

Bentley, R. and C. P. Thiessen (1957a). Biosynthesis of itaconic acid in *A. terreus*, 1. Tracer studies with ^{14}C labelled substrates. *J. Biol. Chem.*, **226**, 673–687.

Bentley, R. and C. P. Thiessen (1957b). Biosynthesis of itaconic acid in *A. terreus*, 2. Early stages in glucose dissimilation and the role of citrate. *J. Biol. Chem.*, **226**, 689–701.

Bentley, R. and C. P. Thiessen (1957c). Biosynthesis of itaconic acid in *A. terreus*, 3. The properties and reaction mechanism of *cis*-aconitic decarboxylase. *J. Biol. Chem.*, **226**, 703–720.

Bernhauer, K. (1924). Zum Problem der Säurebildung durch *A. niger. Biochem. Z.*, **153**, 517–521.

Bernhauer, K. (1928). Ueber die Characterisierung der Stämme von *A. niger* auf Grund ihres biochemischen Verhaltens. *Biochem. Z.*, **197**, 278–287.

Billington, R. H. (1969). Versatile itaconic acid and its derivatives. *Chem. Processing (London)*, **15** (9), 8–11.

Blom, R. H., V. F. Pfeiffer, A. J. Moyer, D. H. Traufler, H. F. Conway, C. K. Crocker, R. E. Farison and D. V. Hannibal (1952). Sodium gluconate production—fermentation with *A. niger. Ind. Eng. Chem.*, **44**, 435–440.

Boehringer Mannheim G.m.b.H. (1974). Process for the conversion of glucose into gluconic acid. *Br. Pat.* 1 373 562 (*Chem. Abstr.*, **80**, 37 430t).

Boutroux, L. (1880). Sur une fermentation nouvelle du glucose. *C. R. Hebd. Seances Acad. Sci.*, **91**, 236–238.

Buendia, M. and J. M. Garrido (1958). Estudios sobre la produccion de acido itaconico. *Revista de Ciencia Aplicade (Madrid)*, 481–495.

Buendia, M. and J. M. Garrido (1959). Estudios sobre la produccion de acido itaconico. *Revista de Ciencia Aplicade (Madrid)*, 23–33 and 130–140.

Bull, D. N. and L. L. Kempe (1970). Kinetics of the conversion of glucose to gluconic acid by *Pseudomonas ovalis. Biotechnol. Bioeng.*, **12**, 273–290.

Calam, C. T., A. E. Oxford and H. Raistrick (1939). Studies in the biochemistry of microorganisms, XLIII, itaconic acid, a metabolic product of *A. terreus* Thom. *Biochem. J.*, **33**, 1488–1495.

Cho, Y. K. and J. E. Bailey (1977). Glucoamylase and glucose oxidase preparations and their combined application for the conversion of maltose to gluconic acid. *Biotechnol. Bioeng.*, **19**, 185–198.

Cho, Y. K. and J. E. Bailey (1978). Immobilisation of enzymes on active carbon: Properties of immobilised glucoamylase, glucose oxidase and gluconolactonase. *Biotechnol. Bioeng.*, **20**, 1651–1665.

Christiansen, A. (Miranol Chemical Company Inc.) (1980). Surface active amides and imidazolines. *Br. Pat.* 1 574 916 (*Chem. Abstr.*, **94**, 109 083w).

Constantinides, A. and C. R. Rai (1974). Application of the continuous maximum principle to fermentation processes. *Biotechnol. Bioeng. Symp. Ser.*, **4**, 663–680.

Cooper, R. A. and H. L. Kornberg (1962). Identification of enzymes involved in the formation of pyruvate from itaconyl-coenzyme A. *Biochem. Biophys. Acta*, **62**, 438–440.

Corzo, R. H. and E. L. Tatum (1953). Biosynthesis of itaconic acid. *Fed. Proc.*, **12**, 470.

CPC International Inc. (1972). Enzymatic production of gluconic acid. *Brit. Pat.* 1 276 245 (*Chem. Abstr.*, **73**, 108 274p).

Currie, J. N. and R. H. Carter (Chas. Pfizer and Co.) (1930). *US Pat.* 1 896 811 (*Chem. Abstr.*, **27**, 2757).

Currie, J. N., J. H. Kane and A. Finlay (Chas. Pfizer and Co.) (1931). Process for producing gluconic acid by fungi. *US Pat.* 1 893 819 (*Chem. Abstr.*, **27**, 2249).

de Wilt, H. G. J. (1972). Oxidation of glucose to gluconic acid. *Ind. Eng. Chem. Prod. Res. Dev.*, **11**, 370–378.

Dumontier, M. and M. Hanss (1974). Mesure conductimétrique d'une activité gluconolactonasique. *Biochimie*, **56**, 1291–1292.

Dvorkovitz, V. and T. G. Hawley (Diversey Corp.) (1952a). Washing composition. *US Pat.* 2 584 017 (*Chem. Abstr.*, **46**, 4258h).

Dvorkovitz, V. and T. G. Hawley (Diversey Corp.) (1952b). Washing composition. *US Pat.* 2 615 846 (*Chem. Abstr.*, 47, 1958a).

Eimhjellen, K. E. and H. Larsen (1955). The mechanism of itaconic acid formation by *A. terreus. Biochem. J.*, 60, 139–147.

Elnaghy, M. A. and S. E. Megalla (1975). Itaconic acid production by a local strain of *A. terreus. Eur. J. Appl. Microbiol.*, 2, 159–172.

Elzainy, T. A., M. M. Hassan and A. L. Allam (1973). Occurrence of the non-phosphorylated pathway for gluconate degradation in different fungi. *Biochem. System. Ecol.*, 1, 127–128.

Finkelstein, M., H. Gold and C. A. Paterno (1947). Pharmacology of itaconic acid and its sodium, magnesium and calcium salts. *J. Am. Pharm. Assoc.*, 36 (6), 173–179.

Gastrock, E. A., N. Porges, P. A. Wells and A. J. Moyer (1938). Gluconic acid production on pilot plant scale—effect of variables on production by submerged mold growths. *Ind. Eng. Chem.*, 30, 782–789.

Ghose, T. K. and P. Ghosh (1976). Kinetic analysis of gluconic production by *Pseudomonas ovalis. J. Appl. Chem. Biotechnol.*, 26, 768–777.

Ghose, T. K. and Mukhopadhyay (1976). Kinetic studies of gluconic acid fermentation in horizontal rotary fermenter by *Pseudomonas ovalis. J. Ferment. Technol.*, 54, 738–750.

Gillenwater, D. L. (Grain Processing Corp.) (1975). Verfahren zur Herstellung von Alkalimetallgluconaten. *Ger. Pat.* 2 437 848 (*Chem. Abstr.*, 83, 41 507c).

Gordon, A. A. and K. Coupland (Exxon Research and Engineering Co.) (1980). Mehrzweckschmiermittel. *Ger. Pat.* 3 001 000 (*Chem. Abstr.*, 94, 106 212b).

Hartmeier, W. and G. Tegge (1979). Versuche zur Glucoseoxidation in Glucose–Fructose-Gemischen mittels fixierter Glucoseoxidase und Katalase. *Starch/Staerke*, 31, 348–353.

Hatcher, J. H. (Economics Laboratory Inc.) (1972). Gluconic acid production. *US Pat.* 3 669 840 (*Chem. Abstr.*, 77, 99 657u).

Hauge, J. G. (1966). Glucose dehydrogenases—particulate. In *Methods in Enzymology*, ed. W. A. Wood, 1st edn., vol. 9, pp. 92–98. Academic, New York.

Herrick, H. T. and O. E. May (1928). The production of gluconic acid by the *Penicillium luteum purpurogenum* group, II. Some optimal conditions for acid formation. *J. Biol. Chem.*, 77, 185–195.

Herrick, H. T. and O. E. May (Government and People of the US) (1929). Process for the manufacture of gluconic acid. *US Pat.* 1 726 067 (*Chem. Abstr.*, 23, 5004).

Hlasiwetz, H. and J. Habermann (1870). Zur Kenntnis einiger Zuckerarten (Glucose, Rohrzucker, Levulose, Sorbin, Phloroglucin). *Ann. Chem. Pharm.*, 155, 128–144.

Huchette, M. and F. Devos (Roquette Frères) (1971). Procédé de traitement des hydrolysats d'amidon. *Fr. Pat.* 2 054 829 (*Chem. Abstr.*, 74, 127 947x).

Humphrey, A. E. and P. J. Reilly. (1965) Kinetic studies of gluconic acid fermentations. *Biotechnol. Bioeng.*, 7, 229–243.

Ichikawa, Y. and Imanaka (1960). Decomposition of gluconic acid by *A. niger. Nippon Nogei Kagaku Kaishi*, 34, 966–971.

Isbell, H. S. (Secretary of Commerce, US Government) (1934). Process for the production of crystalline gluconic acid. *US Pat.* 1 985 255 (*Chem. Abstr.*, 29, 1102).

Isbell, H. S., H. L. Frush and F. J. Bates (1932). Manufacture of calcium gluconate by electrolytic oxidation of glucose. *Ind. Eng. Chem.*, 24, 375–378.

Jakubowska, J. and D. Metodiewa (1974). Studies on the metabolic pathway for itatartaric acid formation by *A. terreus. Acta Microbiol. Pol.*, Ser. B, 6, 51–61.

Jakubowska, J., D. Metodiewa and Z. Zakowska (1974). Studies on the metabolic pathway for itatartaric acid formation by *A. terreus. Acta Microbiol. Pol.*, Ser. B, 6, 43–50.

Jenssen, E. B., H. Larsen and J. G. Ormerod (1956). Formation of itaconic acid from the Krebs cycle tricarboxylic acids by extracts of *A. terreus. Acta Chem. Scand.*, 10, 1047.

Jermyn, M. A. (1960). Studies on the glucono-δ-lactonase of *Ps. fluorescens. Biochim. Biophys. Acta*, 37, 78–92.

Karklin', R. Ya. and V. F. Agafonova (1969). Production of itaconic acid from molasses by *A. terreus* moulds. *Kul'tirovanie Mikroorganismov* (*Riga*), 1969, 141–155.

Kawamura, D., O. Saito, H. Matsui and K. Morita (1981). Production of itaconic acid by yeasts. 1. Screening of yeasts producing itaconic acid. *Shizuoka-ken Kogyo Shikenjo Hokoku*, 25, 65–68 (*Chem. Abstr.*, 96, 197 868a).

Kheshghi, S., H. R. Roberts and W. Bucek (1954). Studies on the production of 5-ketogluconic acid by *Acetobacter suboxydans. Appl. Microbiol.*, 2, 183–190.

Kinoshita, K. (1931). Über eine neue *Aspergillus* Art., *A. itaconicus. Bot. Mag.* (*Tokyo*), 45, 45–61.

Kinoshita, K. (1932). Über die Produktion von Itaconsäure und Mannit durch einen neuen Schimmelpilz, *Aspergillus itaconicus. Acta Phytochimica*, 5, 271–287.

Kinoshita, S. and R. Tanaka (Kyowa Hakko Kogyo K.K.) (1961). Process for the production of itaconic acid by fermentation. *Br. Pat.* 878 152 (*Chem. Abstr.*, 55, 20 324).

Kobayashi, T. (1960). Itaconic acid manufacture by fermentation. *Jpn. Pat.* 60 147 (*Chem. Abstr.*, 54, 18 880).

Kobayashi, T. (1967). Itaconic acid fermentation. *Process Biochem.*, 2, 61–65.

Kobasyashi, T. (1978). Production of itaconic acid from wood waste. *Process Biochem.*, 5, 15–22.

Kobayashi, T., I. Nakamura and M. Nakagawa (1972). Process design for itaconic acid fermentation. In *Fermentation Technology Today*, ed. G. Terrui, *Proc. 4th International Fermentation Symp.*, pp. 215–221. Society of Fermentation Technology, Japan.

Kobayashi, T. and I. Nakamura (1971). Process for recovering itaconic acid and salts thereof from fermented broth. *US Pat.* 3 621 053 (*Chem. Abstr.*, 74, 75 205g).

Kobayashi, T. and B. Tabuchi (1957a). Itaconic acid. *Jpn. Pat.* 57 1100 (*Chem. Abstr.*, 52, 15 647i).

Kobayashi, T. and T. Tabuchi (1957b). Treatment of waste molasses for itaconic acid fermentation. *Jpn. Pat.* 57 9394 (*Chem. Abstr.*, 52, 14 076g).

Korol'kov, N. M., I. I. Brod and K. N. Lossakaya (1969). Dynamics of itaconic acid sorption on AV-17 and AN-2F anion exchangers. *Massoobmennye Protsessy Khim. Teckhnol.*, 4, 132–3, (*Chem. Abstr.*, 73, 29 312w).

Lakshminarayana, K., V. V. Modi and V. K. Shah (1969). Studies on gluconate metabolism in *A. niger. Arch. Mikrobiol.*, **66**, 389–395.

Larsen, H. and K. E. Eimhjellen (1955). The mechanism of itaconic acid formation by *A. terreus*. 1. The effect of acidity. *Biochem. J.*, **60**, 135–147.

Les Usines de Melle (1962). Procédé d'extraction d'acides organiques de fermentation. *Fr. Pat.* 1 300 250 (*Chem. Abstr.*, **57**, 15 620ab).

Lockwood, L. B. (1975). Organic acid production. In *Filamentous Fungi*, ed. J. E. Smith and D. R. Berry, vol. 1, pp. 140–157. Wiley, New York.

Lockwood, L. B. (1979). Production of organic acids by fermentation. In *Microbial Technology*, ed. H. J. Peppler and D. Perlman, 2nd edn., vol. 1, pp. 376–383. Academic, New York.

Lockwood, L. B. and G. E. Nelson (1946). Some factors affecting the production of itaconic acid by *A. terreus* in agitated cultures. *Arch. Biochem.*, **10**, 365–374.

Lockwood, L. B. and M. D. Reeves (1945). Some factors affecting the production of itaconic acid by *A. terreus. Arch. Biochem.*, **6**, 455–469.

Lockwood, L. B. and G. E. Ward (1945). Fermentation processes for itaconic acid. *Ind. Eng. Chem.*, **37**, 405–406.

Lockwood, L. B., B. Tabenkin and G. E. Ward (1941). The production of gluconic acid and 2-ketogluconic acid from glucose by species of *Pseudomonas* and *Phytomonas. J. Bacteriol.*, **42**, 51–61.

Luskin, L. S. (1974). Acidic monomers. III. Itaconic acid. In *Functional Monomers, Their Preparation, Polymerisation, and Application*, ed. R. H. Yocum and E. B. Nyquist, 1st edn., vol. 2, pp. 465–501. Marcel Dekker, New York.

May, O. E., H. T. Herrick, C. Thom and N. B. Church (1927). The production of gluconic acid by *Penicillium luteum purpurogenum* Group I. *J. Biol. Chem.*, **75**, 417–422.

May, O. E., H. T. Herrick, A. J. Moyer and R. Hellbach (1929). Semi-plant scale production of gluconic acid by mold fermentation. *Ind. Eng. Chem.*, **21**, 1198–1203.

May, O. E., H. T. Herrick, A. J. Moyer and P. A. Wells (1934). Gluconic acid — production by submerged mold growths under increased air pressure. *Ind. Eng. Chem.*, **26**, 575–578.

May, O. E., S. M. Weisberg and H. T. Herrick (1929). Some physical constants of D-gluconic acid and several of its salts. *J. Washington Acad. Sci.*, **19**, 443–447.

Mitsubishi Chemical Industries Co. Ltd. (1980). Itaconic acid. *Jpn. Pat.* 80 34 017 (*Chem. Abstr.*, **93**, 43 962c).

Molliard, M. (1922). Sur une nouvelle fermentation acide produite par le *Sterigmatocystis nigra. C. R. Hebd. Seances Acad. Sci.*, **174**, 881–883.

Moyer, A. J. (US Secretary of Agriculture) (1944). Process for gluconic acid production. *US Pat.* 2 351 500 (*Chem. Abstr.*, **38**, 5360).

Moyer, A. J. and R. D. Coghill (1945). Laboratory scale production of itaconic acid by *A. terreus. Arch. Biochem.*, **7**, 167–183.

Moyer, A. J., O. E. May and H. T. Herrick (1936). Production of gluconic acid by *P. chrysogenum. Zentr. Bakteriol. Parasitenk. Abt. II*, **95**, 311–324.

Moyer, A. J., P. A. Wells, J. J. Stubbs, H. T. Herrick and O. E. May (1937). Gluconic acid production—development of inoculum and composition of fermentation solution for gluconic acid production by submerged mold growths under increased air pressure. *Ind. Eng. Chem.*, **29**, 777–781.

Müller, H.-M. (1977). Gluconic acid forming enzymes in *A. niger. Zbl. Bakt. Abt. II*, **132**, 14–24.

Nakagawa, M. and T. Kobayashi (1968). Concentrating itaconic acid from fermented liquor with ion exchangers. *J. Ferment. Technol.*, **46**, 158–168.

Nakamura, I., M. Nakagawa and T. Kobayashi (1975). Effects of organic acids and metal ions in molasses on the itaconic acid fermentation with *A. terreus* K26. *Hakko Kogaku Zasshi*, **53**, 435–442 (*Chem. Abstr.*, **83**, 129 946f).

Nakagawa, M., I. Nakamura and T. Kobayashi (1975). Product separation from fermented liquor. 4. Processes for concentrating and purifying itaconic acid from fermented liquor using ion exchangers. *Hakko Kogaku Zasshi*, **53**, 127–134 (*Chem. Abstr.*, **83**, 7042n).

Nelson, G. E., D. H. Traufler, S. E. Kelley and L. B. Lockwood (1952). Production of itaconic acid by *A. terreus* in 20 litre fermenters. *Ind Eng. Chem.* **44**, 1166–1168.

Noury and Van der Lande (1962). Process for the preparation of gluconic acid monohydrate. *Br. Pat.* 902 609 (*Chem. Abstr.*, **57**, 13 020g).

Nowakowska-Waszczuk, A. (1970). Assimilatory and dissimilatory utilisation of nitrate by *A. terreus* mutants. *Acta Microbiol. Pol., Ser. B.* **2**, 83–94.

Nowakowska-Waszczuk, A. (1973). Utilisation of some tricarboxylic-acid-cycle intermediates by mitochondria and growing mycelium of *A. terreus. J. Gen. Microbiol.*, **79**, 19–29.

Nowakowska-Waszczuk, A. and Z. Zakowska (1971). Influence of volatile acid and colouring substances of beet molasses on the production of itaconic acid. *Rocz. Technol. Chem. Zywn.*, **21**, 39–49.

Nowakowska-Waszczuk, A. and Z. Zakowska (1969). Corn steep liquor value for itaconic fermentation. *Acta Microbiol. Pol., Ser. B.* **1**, 111–116.

Nowakowska-Waszczuk, A., Z. Zakowska and B. Sobocka (1969). The effect of nitrogen sources on acid production by *A. terreus. Acta Microbiol. Pol., Ser. B.*, **1**, 105–110.

Nubel, R. C. and E. J. Ratajak (Chas. Pfizer & Co., Inc.) (1962). Process for producing itaconic acid. *US Pat.* 3 044 941 (*Chem. Abstr.*, **57**, 11 676i).

Nyeste, L., B. Savella, L. Szigeti, A. Szöke and J. Hollo (1980). Modelling and off line optimisation of batch gluconic acid fermentation. *Eur. J. Appl. Microbiol. Biotechnol.*, **10**, 87–94.

Olijve, A. (1978). Glucose metabolism in *Gluconobacter suboxydans*. Ph.D.Thesis, State University, Groningen, The Netherlands.

Oosterhuis, N. M. G., N. M. Groesbeek, A. P. C. Olivier and N. W. F. Kossen (1983). Scale down aspects of the gluconic acid fermentation. *Biotechnol. Lett.*, **5**, 141–146.

Pasternak, R. and W. Giles (Chas. Pfizer and Co.) (1934). Process for the preparation of gluconic acid and its lactones. *US Pat.* 1 942 660 (*Chem. Abstr.*, **28**, 1719)

Pfeiffer, V. F., C. Vojnovich and E. N. Heger (1952). Itaconic acid by fermentation with *A. terreus. Ind. Eng. Chem.*, **44**, 2975–2980.

Pfizer, Chas. and Co. Inc. (1957). Improvements in or relating to the preparation of metal gluconates. *Br. Pat.* 786 288 (*Chem. Abstr.*, **52**, 8190c).

Porges, N., T. F. Clark and E. A. Gastrock (1940). Gluconic acid production—repeated use of submerged *A. niger* for semicontinuous production. *Ind. Eng. Chem.*, **32**, 107–111.

Prescott, F. J., J. K. Shaw, J. P. Bilello and G. O. Cragwall (1953). Gluconic acid and its derivatives. *Ind. Eng. Chem.*, **45**, 338–342.

Quadeer, M. A., M. A. Baig and O. Yunus (1975). Production of calcium gluconate by *A. niger* in 50-l fermenter. *Pakistan J. Sci. Ind. Res.*, **18**, 227–228.

Ranzi, B. M., F. Ronchetti, G. Russo and L. Toma (1981). Stereochemistry of the introduction of the hydrogen atom at C-2 of *cis*-aconitic acid during its transformation into itaconic acid in *A. terreus*: a ^{2}H N.M.R. approach. *J. Chem. Soc., Chem. Commun.*, 1050–1051.

Richter, G. and H. Heinecker (1979). Conversion of glucose into gluconic acid by means of immobilised glucose oxidase. *Starch/Staerke*, **31**, 418–422.

Sawyer, D. T. (1964). Metal gluconate complexes. *Chem. Rev.*, **64**, 633–643.

Sawyer, D. T. and J. B. Bagger (1959). The lactone–acid–salt equilibria for D-glucono-δ-lactone and the hydrolysis kinetics for this lactone. *J. Am. Chem. Soc.*, **81**, 5302–5306.

Shimi, I. R. and M. S. Nour El Dein (1962). Biosynthesis of itaconic acid by *A. terreus*. *Arch. Microbiol.*, **44**, 181–188.

Shizuoka Prefecture Banda Kagaku Kogyo K.K. (1981). Production of itaconic acid by fermentation. *Jpn. Pat.* 81 137 893 (*Chem. Abstr.*, **96**, 50 711k).

Skibsted, L. H. and G. Kilde (1971). Strength and lactone formation of gluconic acid. *Dansk Tidsskr. Farm.*, **45**, 320–324.

Smith, J. E., A. Nowakowska-Waszcznuk and J. G. Anderson (1974). Organic acid production by mycelial fungi. *Ind. Aspects Biochem.*, **30**, (1), 297–317 (*Chem. Abstr.*, **82**, 123 230a).

Stubbs, J. J., L. B. Lockwood, E. T. Roe, B. Tabenkin and G. E. Ward (1940). Ketogluconic acids from glucose. *Ind. Eng. Chem.*, **32**, 1626–1630.

Tabuchi, T., T. Sugisawa, T. Ishidori, T. Nakahara and J. Sugiyama (1981). Itaconic acid fermentation by a yeast belonging to the genus *Candida. Agric. Biol. Chem.*, **45**, 475–479.

Tandon, T. G. and B. S. Mehrotra (1970). Mycological production of itaconic acid. II. Suitability of some carbon and nitrogen sources. *Hindustan Antibiot. Bull.*, **12** (4), 156–163.

Tanner, R. D. and J. M. Yunker (1977). Temperature dependent hysteresis oscillations in the gluconic acid fermentation. *J. Ferment. Technol.*, **55**, 143–150.

Tessi, M. A., B. G. Riera and E. Emiliani (1967). Sobre el quimismo de la asimilacion del acido gluconico por el *A. niger. Rev. Fac. Ing. Quim. Univ. Nac. Litoral*, **35**, 171–178.

Tsao, G. T. and L. L. Kempe (1960). Oxygen transfer in fermentation systems. I. Use of gluconic acid fermentation for determination of instantaneous oxygen transfer rates. *J. Biochem. Microbiol. Tech. Eng.*, **2** (2), 129–142.

Van Dijken, J. P. and M. Veenhuis (1980). Cytochemical localisation of glucose oxidase in peroxisomes of *A. niger. Eur. J. Appl. Microbiol. Biotechnol.*, **9**, 275–283.

Verhave, T. H. (1932). Verfahren zur bakteriellen Oxydation organischer Verbindungen zwecks Herstellung von Oxydationsprodukten, wie Dioxyaceton usw. *Ger. Pat.* 563 758 (*Chem. Abstr.*, **27**, 1085).

Vitola, G. A., B. T. Luka, R. Ya. Karklin', A. K. Proboks and B. Ya. Kestere (1969). The use of wood waste to obtain itaconic acid. *Latvijas PSR Zinatnu Akademijas Vestis*, **268**, (11), 87–91.

Von Fries, H. (1966). Verfahren zur fermentation Herstellung von Itaconsäure durch submers-aerobe Schimmelpilzgarung. *Ger. Pat.* 1 219 430 (*Chem. Abstr.*, **65**, 9698ef).

Ward, G. E. (1967). Production of gluconic acid, glucose oxidase, *etc.* In *Microbial Technology*, ed. H. J. Peppler, pp. 200–221. Van Nostrand-Reinhold, Princeton, NJ.

Wells, P. A., A. J. Moyer, J. J. Stubbs, H. T. Herrick and O. E. May (1937). Gluconic acid production—effect of pressure, air flow, and agitation on gluconic acid production by submerged mold growths. *Ind. Eng. Chem.*, **29**, 653–656.

Wengraf, P. (1952). Glukonsäure und ihre technische Verwendung. *Textil-Rundschau*, **7**, 319–323.

Winskill, N. (1983). Tricarboxylic acid activity in relation to itaconic acid biosynthesis by *Aspergillus terreus. J. Gen. Microbiol.*, **129**, 2877–2883.

Wood, W. A. and R. F. Schwerdt (1953). Carbohydrate oxidation by *Ps. fluorescens. J. Biol. Chem.*, **201**, 501–511.

Yamauchi, T. and K. Shimizu (Daiichi Kogyo Seiyaku Co. Ltd.) (1975). Glucono-δ-lactone crystallisation. *Jpn. Pat.* 75 123 674 (*Chem. Abstr.*, **85**, 6008k).

Ziffer, J., A. S. Gaffney, S. Rothenberg and T. J. Cairney (Pabst Brewing Company) (1971). Aldonic acid and aldonate compositions and production thereof. *Br. Pat.* 1 249 347 (*Chem. Abstr.*, **79**, 113 995c).

36
Acetic Acid

T. K. GHOSE and A. BHADRA
Indian Institute of Technology, New Delhi, India

36.1 INTRODUCTION

Vinegar, that is aqueous solutions of acetic acid used as a condiment and in pickling, has been known and consumed for as long as the art of wine making has been practised and therefore dates back to at least 10 000 BC (Nickol, 1979). Apparently the earliest records which refer to vinegar are the Old and New Testaments; vinegar is said to have been offered as a drink to Jesus (Conner and Allgeier, 1976b). The Bible describes vinegar as a popular nostrum and one of the important compounds of alchemists (Vaughn, 1954). The word vinegar is derived from two French words, *viz. vin* (wine) and *aigre* (sour); the Latin word *acetum* means literally sour or sharp wine. So, it can be assumed that the first vinegar used by ancient peoples was simply spoiled wine.

Early methods obtained acetic acid from natural carbohydrates by biochemical oxidation of alcohol and destructive distillation of wood. Vinegar may be produced from a wide variety of raw materials, *e.g.* apples, malt, grapes, grain, molasses, pears, peaches, oranges, whey, pineapples, and other fruits, the main requirement being a satisfactory, economic source of alcohol and accessory flavouring constituents; the term vinegar is applied to the product of acetous fermentation of ethanol from any of the above mentioned sources.

Today acetic acid is an important industrial organic chemical with approximately 2.45 million ton/year being produced and consumed world wide, half of it in the USA (Helsel, 1977). It ranks 34th amongst all chemicals produced and its use is considered to be the fastest growing amongst the important aliphatic intermediates. The industrial importance of acetic acid can be understood from Figure 1 which illustrates the market for it.

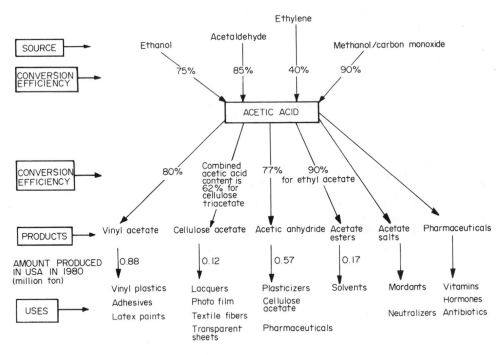

Figure 1   Market for acetic acid (Grayson and Eckroth, 1979; Anon., 1982a)

Although from 1960 to 1973 acetic acid production increased three-fold in the USA and West Germany and seven-fold in Japan (Patankar, 1982), the recent statistics show a downward trend. The production was lower by 9% in the USA and 13% in Japan in the year 1980–81 compared to the previous year (Anon., 1982b). The average price increase of acetic acid has been 10% annually. Ethylene, a raw material for acetic acid production, which was sold in the 1970s at 6¢ per kg, was selling at 30¢ per kg in 1980.

The most likely step to cope with the predicted shortage of acetic acid appears to be the use of microbial methods of production to supply a significant proportion of the acetic acid consumed industrially. Brazil provides an interesting example. It achieved an increase in acetic acid production of almost 500% during the last five years (Filho and Sattamin, 1981). The acetic acid was derived solely from alcohol which, in turn, was produced mainly from biomass (especially sugar cane and cassava). This was the result of the establishment of the 'National Alcohol Programme' which has increased alcohol production in Brazil from 0.73 billion l in 1975 to 4.0 billion l in 1981. A target of 10.7 billion l is expected to be reached in 1985.

Since 1950 synthetic methods have provided the major part of the world's acetic acid supply. Because it was possible for a time to hold the price of organic chemicals derived from fossil fuels within certain limits, despite a steep rise in the cost of crude oil, the need to find alternative methods of production, such as by biological routes, did not arise. The current shortfall in supply appears to be an incentive for more serious consideration of the revival of biological routes.

36.2 PRODUCTION TECHNOLOGY

36.2.1 Thermodynamics and Kinetics

In an aerobic system, 1 mol of glucose produces 2 mol of ethanol; the free energy of formation values show that the reaction proceeds spontaneously towards ethanol formation with an adequate decrease in free energy. When ethanol is oxidized to acetic acid, a further decrease in free energy occurs. The free energy changes and the heats of reaction are presented below (equations 1–3; Perry and Chilton, 1973; Hougen *et al.*, 1967).

$$C_6H_{12}O_6 \rightarrow 2\,C_2H_5OH + 2\,CO_2 \tag{1}$$
$$\Delta G° = -\,218.78 \text{ kJ mol}^{-1}; \Delta H° = -\,81.93 \text{ kJ mol}^{-1}$$

$$2\,C_2H_5OH \xrightarrow{(O_2)} 2\,CH_3CO_2H + 2\,H_2O \tag{2}$$
$$\Delta G° = -(2 \times 453.53) = -907.06 \text{ kJ mol}^{-1}; \Delta H° = -(2 \times 497.42) = -994.84 \text{ kJ mol}^{-1}$$

Overall

$$C_6H_{12}O_6 \xrightarrow{(O_2)} 2\,CH_3CO_2H + 2\,CO_2 + 2\,H_2O \tag{3}$$
$$\Delta G° = -1125.67 \text{ kJ mol}^{-1}; \Delta H° = -1076.77 \text{ kJ mol}^{-1}$$

There is a large decrease in free energy due to the spontaneous production of CO_2 and H_2O, resulting in a substantial loss of carbon and hydrogen. In an anaerobic process, reduction of carbon dioxide occurs to form an additional mole of acetic acid. The free energy change for the reduction of CO_2 to acetic acid is given below (equation 4).

$$2\,CO_2 + 4\,H_2 \rightarrow CH_3CO_2H + 2\,H_2O \tag{4}$$
$$\Delta G° = -77.0 \text{ kJ mol}^{-1}; \Delta H° -271.7 \text{ kJ mol}^{-1}$$

The mechanism of this reaction will be discussed later.

In an aerobic process four carbon atoms of the glucose are transferred to acetic and two carbon atoms are lost, whereas theoretically all the carbon atoms of glucose may be obtained as acetic acid in anaerobic acidogenesis.

36.2.1.1 *Aerobic acetic acid fermentations*

The kinetics of acetic acid fermentation in submerged culture has been studied by Mori and Terui (1972), Vera and Wang (1977), Perdin and Kafol (1978), Levonenmunoz and Cabezudo (1981), and Mori *et al.* (1972).

Employing a strain of *Acetobacter rancen*, the specific growth rate (μ) and cell mass (X) under the influence of product inhibition are related by equation (5), which approximates to the experimental results (Figure 2; Mori and Terui, 1972).

$$\mu = \frac{k_3}{1 + k_4(P_o + k_1X + k_2X^2)^2} \tag{5}$$

Here P_o is the acetic acid concentration in the inflowing medium and k_1, k_2, k_3 and k_4 are the system constants.

The specific rate of product formation follows a parabolic growth relationship (Figure 3; equation 6) with μ even if the product inhibition is substantial.

$$v = \frac{1}{X}\frac{dP}{dt} = a\mu - b\mu^2 \tag{6}$$

In equation (6) *a* and *b* are constants. Yeast extract in the culture medium has a considerable effect on both μ and v (Mori *et al.*, 1972). μ follows a Monod relationship with yeast extract concentration; the value of the saturation constant is 2.15 g l^{-1} and μ_m is 0.4 h^{-1}.

When acetic acid is added at the beginning of the fermentation (>17.6 g l^{-1}), an increase in the value of μ is perceived, which is regarded as being as a result of adaptation of cells to acetic acid on being exposed for a longer time. In the presence of yeast extract, μ is expressed by a mixed product inhibition equation (7).

$$\mu = \frac{\mu_m S_Y}{K_{SY}\,(1 + P^2/K_P^2) + S_Y(1 + P^2/\alpha K_P^2)} \tag{7}$$

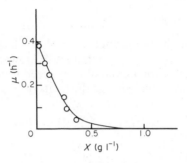

Figure 2 Fit of equation (5) with experimental data: μ, specific rate growth rate; X, cell mass

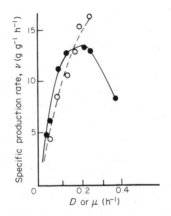

Figure 3 Relation of μ to v in batch and continuous culture: ○——○ batch; ●——● continuous

When $P = 0$, equation (7) becomes equation (8).

$$\mu = \frac{\mu_m S_Y}{K_{SY} + S_Y} \tag{8}$$

Addition of yeast extract results in an increase in the initial growth rate when $P \approx 0$; when $S_Y \to \infty$ the equation becomes

$$\mu = \frac{\mu_m}{(1 + P^2/\alpha K_P^2)} \tag{9}$$

i.e. large amounts of yeast extract reduce the inhibitory effects of acetic acid.

A hypothetical model for the mechanism of mixed inhibition by acetic acid is given by Mori *et al.* (1972; Figure 4). Ethanol, up to about 5% (v/v) concentration, shows no significant inhibitory effect on acetic acid production; generally there is a tendency for acetic accumulation to be proportional to the initial ethanol concentration provided it is lower than 50.6 ml 1^{-1} (Figure 5). The maximum cell yield is obtained at a lower ethanol concentration, but the presence of yeast extract results in a greater cell yield at a higher initial ethanol concentration.

The highest ethanol tolerance which allows the formation of acetic acid varies with the species and ranges between 5 and 11% (v/v). The growth of *Acetobacter* exhibits combined inhibition by ethanol (S_A), as expressed by equation (10) (Mori and Terui, 1972).

$$\mu = \frac{1}{K_1 + K_2 S_A + K_3/S_A} - K_4 S_A^2 \tag{10}$$

In equation (10) K_1, K_2, K_3 and K_4 are combined constants consisting of μ_m, saturation constants and yeast extract concentration terms.

Vera and Wang (1977) claimed that the relationship between growth and production is a mixed growth associated process. With *Acetobacter aceti*, Levonenmunoz and Cabezudo (1981) also

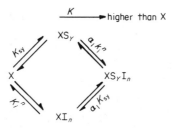

Figure 4 Hypothetical scheme for the mechanism of inhibition by acetic acid: X, component of cell mass which is μ-limiting and duplicates as the cell duplicates and which combines with a component of yeast extract (YE); α_1, competition coefficient; I, inhibiting substance (acetic acid); K, dissociation constant or saturation constant

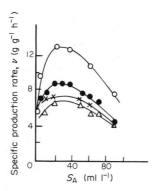

Figure 5 Change of v with ethanol concentration (S_A) in the presence of varying amounts of yeast extract (YE): ○——○ 2.5 g l^{-1}; ●——● 5.3 g l^{-1}; ×——× 10.6 g l^{-1}; △——△ 21.3 g l^{-1}

developed a mixed growth associated model given by equation (11) but the growth associated term in this case has a negative impact.

$$dP/dt = a - b\mu X \tag{11}$$

According to Perdih and Kafol (1978) acetic acid is produced by maintenance metabolism.

36.2.1.2 Anaerobic acidogenesis

The kinetic behaviour of acidogenesis depends on the source of microflora and the type of substrate used. Very little information is available on the kinetic behaviour of acid-formers in the absence of methanogens. Most studies are known to have been carried out with the simultaneous production of methane from acids produced by acid-formers. The kinetic characteristics of acid-formers have been studied by Lozano *et al.* (1980), Ghose and Bhadra (1981) using pure substrates, and Ghosh *et al.* (1975) using wastewater solids. The parameters which are important for the design and operation of this process are (a) rate of product formation; (b) composition of the acid produced; and (c) rate of substrate degradation. It is generally observed that acetic acid constitutes the largest part of the volatile fatty acids produced. Propionic and butyric acids are also produced in significant amounts with small amounts of valeric and isobutyric acids. Theoretically the rate of formation of acid increases with the organic loading but loading above 65 kg m^{-3} d^{-1}, corresponding to a feed total solid concentration of 9%, may result in serious transport problems.

Using activated sludge as substrate the values of the maximum growth rate (μ_m) and the saturation constant (K_S) have been reported as 0.16 h^{-1} and 26 g l^{-1} as volatile solids, respectively, which indicate a very low affinity of acid-formers towards the substrate. Low μ and high K_S values are probably caused by a slow hydrolysis step (Ghosh *et al.*, 1975). For a given temperature, pH and mixing regime, the efficiency of the process is governed by two important parameters; the loading and the detention time. Figure 6 shows that little additional product formation is achieved after 24 h while the rate of acid production increases with loading.

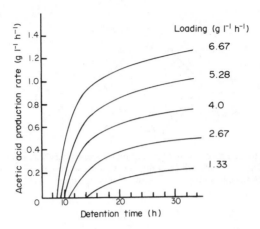

Figure 6 Simulated plots for volatile acid production: $\mu_m = 0.16 \text{ h}^{-1}$, $K_s = 25 \text{ g l}^{-1}$, $m = 0.033 \text{ h}^{-1}$, $U_p = 1.12$, $U_e = 1.35$, $\alpha_{HAC} = 0.28$

The effect of temperature, pH and initial substrate concentration on acid production and yield have been studied by Lozano *et al.* (1980) in a continuous multireactor system with cell recycle (Figure 7) using sucrose as substrate and inoculum isolated from a municipal sludge digester.

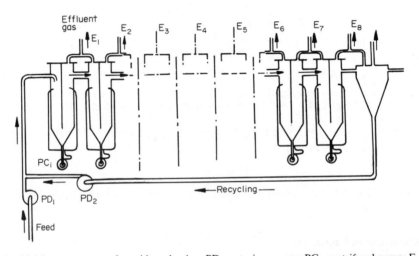

Figure 7 Multi-reactor system for acid production: PD, metering pumps; PC_i, centrifugal pump; E, stages

Control of pH at 6 results in a significant increase in acid production compared to no control of the pH system (Figure 8). The effect of temperature on acid composition is shown in Table 1. The concentration of propionic acid decreases when the temperature increases and at 40 °C there is no propionic acid. Since at this temperature the toxic effect of organic acids decreases, it appears that propionic acid exercises increased toxic effects relative to acetic and butyric acids. The amount of butyric acid increases at higher temperatures.

The effects of S_0 and residence time on acid production are shown in Figure 9. The maximum acid concentration attained was 61 g l^{-1} after 96 hours with $S_0 = 150$ g l^{-1} sucrose.

36.2.2 Microbiology

36.2.2.1 *Aerobic process (acetic acid bacteria)*

The empirical approach to the manufacture of vinegar from an alcoholic mash has been known for a long time. Vinegar had been produced by natural fermentation in the Orleans region of France long before the discovery of acetic acid bacteria. The biological nature of 'mother of vinegar' was first suggested by Boerhaave (1732) and in 1822 Persoon first reported a bacterial study

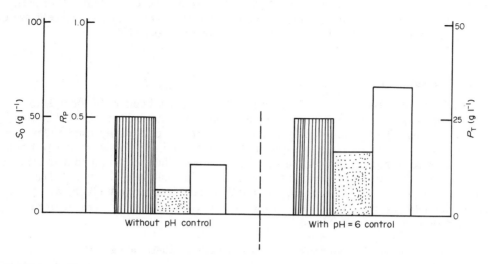

Figure 8 Effect of pH regulation on acid production: ⊞ initial sucrose concentration, S_0; ▨ total acid concentration expressed as acetic acid, P_T; ☐ ratio of product formation to substrate consumption, R_P

Table 1 The Effect of Temperature on Acid Composition

Temperature (°C)	Composition (%, w/w)		
	Acetic	Propionic	Butyric
20	55.6	18.0	24.5
30	64.7	5.5	29.7
40	49.8	0	52.0

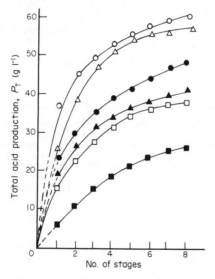

Figure 9 Organic acid production at different values of initial sucrose concentration (S_0, g l^{-1}) and reaction times (RT, h): ■ S_0 50, RT 5; ☐ S_0 50, RT 12; ▲ S_0, 75, RT 5; △ S_0 75, RT 12; ● S_0 150, RT 5; ○ S_0 150, RT 12; pH = 6; temperature = 30 °C

of 'mother of vinegar' and named the film appearing on the liquid layer 'Mycoderma'. Kutzing (1837) recognized that minute organisms of the mother liquor were responsible for acetification of alcohol. Pasteur (1868) confirmed Kutzing's findings and reported that the conversion of wine to vinegar was brought about by one species of bacteria which he named '*Mycoderma aceti*'.

The acetic acid bacteria belong to the family Pseudomonadaceae. The cells are rod-shaped, but elongated, swollen or branched forms may occur. They may be motile or non-motile; they do not form endospores. The species which do not convert ethanol to acetic acid are either peritrichous flagellate (*Acetobacter aceti*) or non-flagellate (*Acetobacter rancens*). The young cells are Gram-negative in character while the older ones are Gram-positive (Asai, 1968).

36.2.2.2 Classification

The details of the classification of acetic acid bacteria has been well documented by Asai (1968). Acetic acid bacteria have been given a number of generic names since Persoon named them 'Mycoderma'. The generic designation *Acetobacter* is universally used today. The first attempt to classify acetic acid bacteria was made by Hansen in 1894; the species were differentiated on the basis of morphological characteristics of the pellicle in fluid media and their iodine reaction. In 1925, Vissert Hooft divided the *Acetobacter* into catalase-negative and catalase-positive groups. This system was developed into another classification by Vaughn, which is used by Bergey's Manual (7th edn., 1957). According to this classification, species of *Acetobacter* are divided into two classes (Table 2).

Table 2 Classification of Acetic Acid Bacteria according to Vaughn[a]

Class I	Class II
Oxidize acetic acid to carbon dioxide and water	Do not oxidize acetic acid
(a) Utilize ammonium salts as the sole source of nitrogen, e.g. A. aceti	(a) Form pigments in glucose media (i) Dark brown to blackish pigment, e.g. A. melanogenus (ii) Pink to rose pigment, e.g. A. roseus
(b) Do not utilize ammonium salts as the sole source of nitrogen (i) Form a thick, zoogloeal cellulose membrane on the surface of liquid media, e.g. A. xylinum (ii) Do not form a thick zoogloeal cellulose membrane on the surface of liquid media, e.g. A. racens	(b) Do not form pigments (i) Optimum temperature 30–35 °C, e.g. A. suboxydans (ii) Optimum temperature 18–21 °C, e.g. A. oxydans

[a] Asai (1968).

Frateur (1950) proposed a new system with biochemical properties being taken as the sole criteria for acetic acid production. He classified the *Acetobacter* into four groups *viz. peroxydans, oxydans, mesoxydans* and *suboxydans*. This classification was useful because it was based on relatively simple reactions, but after acceptance of the genus *Gluconobacter*, a redefinition of the genus *Acetobacter* was necessary and a modified Frateur classification is shown in Table 3 (Bergey's Manual, 8th edn., 1974).

Table 3 Characteristics Differentiating the Species and Subspecies of Genus *Acetobacter*

	Acetobacter aceti; sub sp. aceti, orleanensis xylinum, liquefaciens	Acetobacter pasteurianus; sub sp. pasteurianus lovaniensis, rancen ascendens, paradoxus	Acetobacter peroxydans
Catalase	+	+ (− for paradoxus)	−
Ketogenesis in glycerol	+	−	−
Formation of			
5-ketogluconate	+	−	−
2-Ketogluconate	+	−	−
Gluconate	+	+ or −	−
Growth on ethanol	+ or −	+ or −	+
Produces cellulose	Except xylinum	−	−

36.2.3 Earlier Processes

36.2.3.1 The 'slow' process

Vinegar used to be produced by keeping wine in open, partially-filled containers (Allgeier, 1960). Later it was found that addition of some vinegar to the partially filled casks speeded up the reaction. This concept was already incorporated into the process known as the 'Orleans process' (1670), the 'French method' or the 'slow process'.

In this process, wooden barrels are partially filled with good quality vinegar which acts as a source of inoculum and wine is added at weekly intervals for four weeks. After five weeks, a portion of vinegar is withdrawn and the same amount of wine is introduced and the process is repeated resulting in a very slow continuous process. Air is admitted through holes at a height just above the surface of the liquid. The acetic acid bacteria grow on the surface of the liquid and a gelatinous zoogloeal mat known as the 'mother of vinegar' is formed; this contains a large number of bacteria.

This process was modified by immobilization of the cells on a floating light wooden grating on the surface of the medium so that addition of alcoholic solution did not break the mat. The costly 'slow process' was used extensively for a long time and was eventually replaced by the 'quick process'.

36.2.3.2 The 'quick' process

The idea that vinegar can be produced rapidly by trickling wine through packed pumice was discovered by Boerheave in the early part of the nineteenth century. It was improved by Schüzenbach (1823) to make it the 'quick process', also known as the 'German process', which is the basis for modern methods of manufacture of vinegar using a generator. The generator consists of a wooden or metal-coated tank packed with beech wood shavings on which cells are allowed to grow. The feed trickles from the top through the wood shavings. A large volume of air is sparged into the tank through perforations in the bottom. Employing 12% (v/v) alcohol, 98% conversion into acetic acid is attained in 5 days by this process (Conner and Allgeier, 1970). This process has been used for the commercial production of vinegar for about a century.

36.2.3.3 Frings process

Major improvements in the quick process took place in 1929 when forced aeration and temperature control were introduced and the trickling generator widely used today emerged. A cross-section of the Frings generator is shown in Figure 10. The significant advantages of this process include the following: (1) the cost is low, it is relatively simple and easy to control; (2) higher acetic acid concentrations are obtained; (3) the tank occupies less space; and (4) evaporation losses are low (Prescott and Dunn, 1959).

36.2.3.4 The submerged culture process

The application of submerged fermentation to the oxidation of ethanol to acetic acid by Hromatka and Ebner (1949) was the next technical advance in the commercial production of vinegar. They noted that *Acetobacter* species in submerged conditions were very sensitive to oxygen-deficiency and that the fermentation stopped when the level of oxygen in the gas phase was less than 5% (Hromatka and Ebner, 1950). The success of this process depends largely on the efficiency of the aeration of the broth. The acetic acid bacteria during submerged exponential fermentation have an average oxygen uptake rate of 7.75 l O_2 per gram cell per hour (Vaughn, 1954).

The advantages of submerged cultivation (Figure 11) over the trickling generator are: (1) the submerged cultivation permits 30 times faster oxidation of alcohol; (2) a smaller reactor volume is needed (about 16% of the trickling generator) to produce an equivalent amount of vinegar; (3) greater efficiency is achieved; yields are 5–8% higher and more than 90% of the theoretical yield is obtained; (4) the process can be highly automated; and (5) increased economy owing to the elimination of clogging by shavings, interruptions, *etc*. The ratio of productivity to capital investment is much higher in the case of submerged cultivation.

After the discovery of submerged acetification, a number of processes were developed (Nickol, 1979), *viz*. (i) Yeomans cavitator (1967), in use in the US and Japan; (ii) the Bourgeois process used in Spain and Italy; and (iii) the Fardon process, used mainly in Africa for making malt vinegar. These processes were not used extensively on a commercial scale.

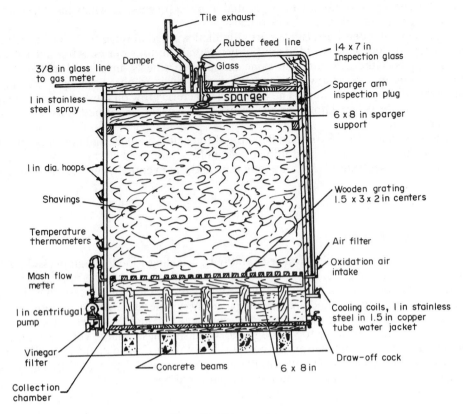

Figure 10 Frings acetator

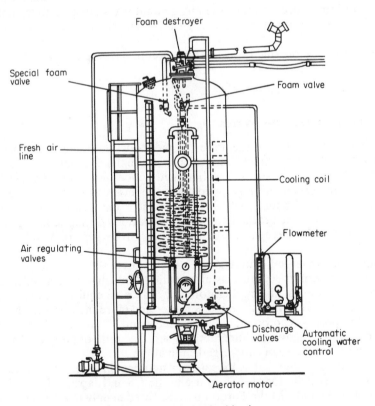

Figure 11 Submerged cultivation

36.2.4 Recent Developments in Acetic Acid Fermentation

The application of immobilized cell techniques for the production of acetic acid dates back to the early part of the nineteenth century, but little has been done to improve the process into an efficient technology. Only recently has considerable emphasis been placed on the immobilization of whole cells.

A variety of immobilization techniques and carriers have been reported but there is no ideal or universal method. Factors to be considered include the organism, the product(s) formed and the diffusion characteristics of the system. The general methods of immobilization are adsorption on a carrier; covalent attachment to a carrier; encapsulation within a confinement or gel entrapment; and cross-linking of cells within the carrier or between cells in the absence of a carrier.

Live immobilized microbial cells have been used for multienzyme reactions such as alcohol production (Folin and Ciocatten, 1927; Ghose and Bandyopadhyay, 1980; Tyagi and Ghose, 1982) but aerobic reactions involving large oxygen requirements like acetic acid production pose problems of mass transfer.

Production of vinegar by live cells of *Acetobacter aceti* immobilized on fibrous and other carriers has been studied (Chattopadhyay, 1980). Both adsorption and covalent binding techniques were employed for the cell immobilization. The performances of the carriers in terms of cell loading and productivity in batch experiments are summarized in Table 4. In terms of cell loading, PVC, bagasse and rice husks appear to be better, but the rate of acetic acid production is higher in the case of inorganic materials. Substrate concentrations above a level of 10% are inhibitory to product formation. The effect of initial ethanol concentration on productivity is shown in Table 5.

Table 4 Comparative Performances of Different Carriers

Carrier	Cell immobilization technique	Cell retention (g cells/g carrier)	Acetic acid production rate (g/l/h)	Specific acetic acid production rate (g/l/h/g cells immobilized)
PVC	Adsorption	0.163	0.23	1.36
Pumice stone	Covalent binding[a]	0.021	0.10	4.53
Concrete	Adsorption	0.034	0.10	2.82
Concrete	Covalent binding[a]	0.020	0.07	3.64
Bagasse	Adsorption	0.130	0.23	1.78
Bagasse	Covalent binding[b]	0.119	0.15	1.30
Wool	Adsorption	0.061	0.12	2.03
Wool	Covalent binding	0.037	0.10	2.59
Viscose rayon	Covalent binding[b]	0.062	0.12	1.92
Rice husk	Covalent binding[a]	0.175	0.312	1.78

[a] Covalent treatment: 0.5% glutaraldehyde + 10% silane. [b] Covalent treatment: carbodiimide. [c] Covalent treatment: 0.5% glutaraldehyde.

Table 5 Effect of Increased Ethanol Concentration on Productivity and Percentage Conversion in PFR[a]

Dilution rate (h^{-1})	$Y_{P/S}$ (g acid/g ethanol)[b]				Productivity (g l^{-1} h^{-1})[b]				Conversion (%)[b]			
	A	B	C	D	A	B	C	D	A	B	C	D
0.032	1.223	1.2	1.24	1.18	1.48	1.74	2.08	1.70	74	70	69	45
0.037	1.235	1.186	1.27	1.15	1.72	1.94	2.25	1.31	74	67	64	37
0.046	1.22	1.20	1.26	1.13	2.02	2.1	2.53	1.24	69	58	58	23
0.055	1.20	1.15	1.20	1.08	1.81	2.0	2.70	1.21	51	45	50	18
0.069	0.84	1.18	1.21	1.00	0.81	1.61	2.50	1.14	20	29	39	14
0.091	0.934	1.0	0.9	0.96	0.60	1.0	1.52	1.10	11	14	21	10

[a] Air flow rate 50 ml min^{-1}. [b] A: 6.5% (v/v) ethanol in feed; B: 8.0% (v/v) ethanol in feed; C: 10.0% (v/v) ethanol in feed; D: 12.1% (v/v) ethanol in feed.

Comparison of the growth of cells in free and immobilized states shows considerable differences. The μ_m values for free and immobilized cells were estimated as 0.25 h^{-1} and 0.05 h^{-1}, respectively, and the corresponding cell yields were 0.104 and 0.008, respectively. Both oxygen uptake and carbon dioxide evolution rates are higher in free cells than in immobilized cells but the respiratory quotients for both the systems remain almost the same. According to Kennedy *et al.* (1976) in the case of *Acetobacter aceti* immobilized on an adsorbed Ti/Zn hydroxide bed (supposedly having partial covalent binding) the oxygen consumption takes place at a rate 30% of that used by free cells. An activity decay profile of the immobilized *Acetobacter* follows first order kinetics and the half-life of these cells is computed as 378 h.

Recently, a modification of the tubular reactor (tower fermenter) has been applied to the acetification of alcoholic mash (Brown, 1963), and a volumetric efficiency of up to 1.0 fermenter volume output per day has been achieved; this is 100% higher than the performance of the Frings process.

In a highly aerobic acetic acid fermentation, the mass transfer and two-phase flow phenomena are the dominant features of the fluidized bed tower. Generally the overall rate is limited by the rate of oxygen transfer from the air bubbles into the liquid phase. The use of a tower bioreactor reduces loss of alcohol due to evaporation. It has been reported (Akiba and Fukimbara, 1973) that at an aeration rate of 0.5 vvm in a stirred tank reactor, the loss of alcohol in 48 h using 5% alcohol is nearly 28%, whereas the loss becomes about 50% at an aeration rate of 1 vvm. On the other hand in a tower fermenter very little loss due to evaporation is observed in 48 h. The loss of alcohol ultimately lowers the yield in a stirred tank by about 27% compared to a tower reactor.

According to Kennedy *et al.* (1980) *Acetobacter* species which produce extracellular polysaccharides aggregate in the presence of hydrous titanium(IV) oxide thereby enabling a higher medium flow rate and a subsequent increased acetic acid output to be achieved. Strains of *Acetobacter* which cannot produce polysaccharides do not give this effect with hydrous Ti(IV) oxide but produce more acetic acid when a Ti(IV)–cellulose chelate is added to the system. The mechanism which results in a better performance by extracellular polysaccharide-forming cells is believed to be as described below.

Water insoluble Ti compounds can react with extracellular cellulose by a cross-linking process; this encourages the formation of aggregates, thus increasing the cell density and making the cells more resistant to washout by the increased flow rate at which they form a sediment in the reactor. This gives a greater cell mass per unit volume, which leads to an increase in the rate of conversion of ethanol in the reactor. The authors reported that in a tower fermenter (2.6 l) with an aspect ratio of 8:1, the daily production of acetic acid could be increased from 87 g d^{-1} to 263 g d^{-1} by using hydrous Ti(IV) oxide in the case of aggregating bacteria; with non-aggregating bacteria, addition of a Ti(IV)–cellulose chelate produced only a 20% rise in the daily production of acetic acid.

In the oxidative conversion of ethanol to acetic acid, delivery of oxygen from the gas phase to the bacteria can be described as occurring in three stages. First, oxygen (from air) has to dissolve in the liquid medium; the saturation concentration (9 p.p.m.) being very low, the dissolution of oxygen can obviously become limited by the rate of supply of air to the fermenter. Second, the dissolved oxygen has to diffuse through the liquid to the bacteria, and finally the oxygen has to enter the bacterium and participate in the oxidation reaction. Using *Acetobacter aceti* the overall mass transfer coefficient, $K_L a$, was found to be 4200 h^{-1} in a pulsed reactor (Ghommidh *et al.*, 1982); the high value of $K_L a$ reflects the high oxygen transfer efficiency of the pulsed flow, also observed by Serieys *et al.* (1978).

Ghommidh *et al.* (1981, 1982) studied the performance of a pulsed reactor using cylindrical monoliths of cordierite, a porous ceramic which has good mechanical properties and is chemically inert. Cell adsorption was irreversible and followed first-order kinetics, maximum cell loading was 2.6 mg g^{-1} of cordierite. The effect of dilution rate on product formation in this system is shown in Table 6. At low D (<0.1 h^{-1}), even in the presence of residual alcohol, the acetic acid concentration is almost constant (30–35 g l^{-1}); this is attributed to the inhibitory effects of acetic acid. At high dilution rates the acetic acid production may be limited by the partial pressure of oxygen. Using pure oxygen a maximum productivity of 10.4 g l^{-1} h^{-1} was observed which is about 2.3 times that obtained using air in the pulsed reactor system (Table 7).

Much work has been done to increase acid concentration in the reactor but very little effort has been made to improve the design and performance of bioreactors used for this purpose. Muller (1978) has developed a modern bioreactor for vinegar production, called a 'vinegator', with application of modern instrumentation and process control. Here reliable measurement and precise control of DO, which are essential for acetic acid formation, are possible. In this reactor, air is drawn into the vinegator by a self priming agitator turbine; the DO is measured by a polarographic oxygen electrode. The signal is amplified and controls a pneumatic valve in the aeration circuit. At the beginning of the process, when O_2 consumption is very low, the major part of the process air is recycled back into the bioreactor, which reduces loss of ethanol by evaporation and economizes on air utilization. As the reaction becomes faster, the oxygen controller increases the aeration rate automatically. The vinegator has all the other features of a modern fermenter.

Table 6 Steady-state Values of Various Parameters for Different Dilution Rates

| Dilution rate D (h^{-1}) | Effluent concentration | | | Productivity |
	Cell concentration X $(mg\,l^{-1})$	Acetic acid P $(g\,l^{-1})$	Ethanol S $(g\,l^{-1})$	P × D $(g\,l^{-1}\,h^{-1})$
0.024	16	34	8	0.82
0.057	25	35	10.5	1.05
0.074	23	32	10	1.55
0.09	25	35	11	3.15
0.153	32	30	15	4.60
0.207	25	22	21.2	4.55
0.247	27	18.5	24	4.55
0.444	34	10.5	30.5	4.65
0.551	24	9	32	4.95

Table 7 Effect of Oxygen Partial Pressure on Acetic Acid Productivity[a]

Partial pressure P_{O_2} (atm)	Oxygen uptake rate calculated from equation (22) Q_{O_2} $(g\,l^{-1}\,h^{-1})$	Productivity P × D $(g\,l^{-1}\,h^{-1})$
0.115	1.44	2.7
0.21	2.45	4.6
0.34	3.09	5.8
0.49	4.11	7.7
0.64	4.51	8.45
1.0	5.55	10.4

[a] $D = 0.513\,h^{-1}$

Diagrams of a pulsed immobilized reactor, a tower reactor and chemostat culvation are presented in Figure 12.

36.3 THE BIOLOGY OF ACETIC ACID PRODUCERS

36.3.1 Aerobic Acetic Acid Fermentations

36.3.1.1 Biochemical mechanism of acetic acid formation

The oxidation of ethanol by *Acetobacter* is a two-step process: the first is a partial oxidation of ethanol to acetaldehyde and the second completes the oxidation of acetaldehyde to acetic acid. Under anaerobic conditions 1 mol of each of ethanol and acetic acid are produced from 2 mol of acetaldehyde by the Cannizaro dismutation reaction; 1 mol of ethanol is formed by hydrogenation of acetaldehyde and one of acetic acid by oxidation of acetaldehyde (equation 12).

$$2\,CH_3CHO + H_2O \rightarrow C_2H_5OH + CH_3CO_2H \tag{12}$$

Neuberg and Molinari (1926) postulated that under aerobic conditions 1 mol of acetaldehyde is directly oxidized to acetic acid and the other mole is dismutated according to equation (12). With adequate aeration the oxidation and the dismutation proceed side by side converting all the acetaldehyde to acetic acid:

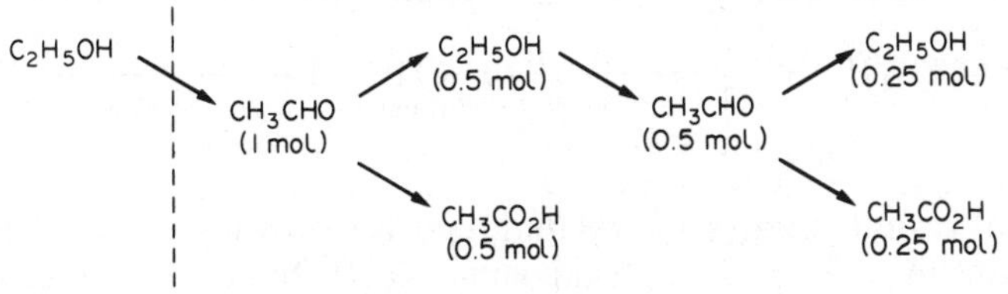

Since it is reported (Wieland, 1913) that acetic acid is also formed in the absence of oxygen but in the presence of hydrogen acceptors like methylene blue and benzoquinone, it was concluded that acetic acid was formed by the action of dehydrogenases. It is assumed that hydrated

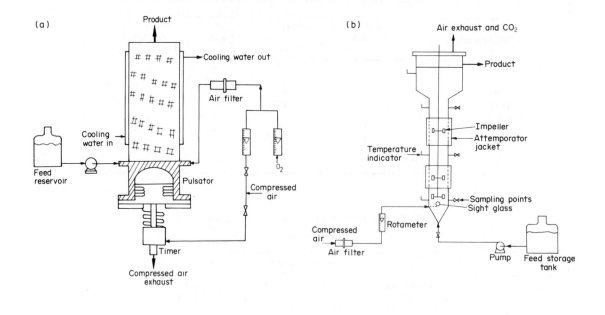

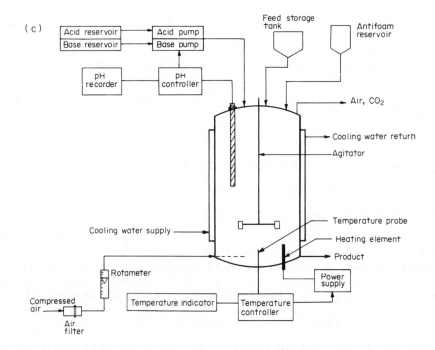

Figure 12   (a) Pulsed immobilized cell reactor; (b) tower bioreactor; (c) acetic acid production in chemostat

acetaldehyde is formed as an intermediate. In the first step alcohol dehydrogenase and in the second acetaldehyde dehydrogenase are involved as shown below (equation 13).

$$C_2H_5OH \xrightarrow[\substack{\text{alcohol} \\ \text{dehydrogenase}}]{} CH_3CHO + H_2O \rightleftharpoons \underset{\substack{\text{(hydrated} \\ \text{acetaldehyde)}}}{CH_3CH(OH)_2} \xrightarrow[\substack{\text{acetaldehyde} \\ \text{dehydrogenase}}]{} CH_3CO_2H + H_2O \quad (13)$$

From the results of various studies and related reports (Lutwak-Mann, 1938; King and Cheldelin, 1952, 1956; Atkinson, 1956a, 1956b; Nakayama, 1959), Nakayama (1961a, 1961b) postulated the scheme shown in Figure 13 for oxidation of ethanol by *Acetobacter* species. Ethanol is first oxidized to acetaldehyde by E_1, which is reduced by the liberated electrons; the acetaldehyde thus formed is further oxidized either by E_2, when liberated electrons are delivered to E_1, or by

E_3 when NADP is reduced to $NADPH_2$. Reduced E_1 is then oxidized by a cytochrome oxidase. On the basis of inhibition effects by cyanide and PCMB, it is presumed that the cytochrome system is driven in the ethanol and acetaldehyde oxidation by E_1 and E_2, whereas the $NADPH_2$ produced by E_3 inhibits further oxidation of acetic acid through the TCA cycle by changing the equilibrium $NADPH_2 \rightleftharpoons NADH_2$.

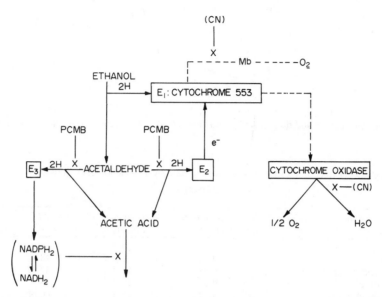

Figure 13 Scheme of ethanol oxidation by *Acetobacter* species: E_1 alcohol-cytochrome-553 reductase; E_2 coenzyme-independent aldehyde dehydrogenase; E_3 NADP-dependent aldehyde dehydrogenase; PCMB *p*-chloromercuriobenzoate; X inhibition (Asai, 1968).

36.3.1.2 The role of metabolites

The importance of the addition of yeast extract to the medium with regard to higher ethanol tolerance and higher productivity has already been discussed. It has another important effect, *viz.* shortening of the lag phase in acetic acid fermentation (Hiyikata *et al.*, 1972a; Figure 14). Lactic acid, pyruvic acid, glycerol, glycenic acid, α-glycerophosphate, alanine, succinic acid and fumaric acid are reported to exercise such effects; decreased lag effect is accompanied by an increase in the ratio of acid produced to cell growth (Table 8) implying that the flow of metabolism in cells is changed in favour of acetic acid production rather than cell growth. A remarkable acceleration in acetic acid fermentation has been observed by supplying low-priced malt sprouts to the medium; an increase in the *P/X* ratio of about 50% is obtained using malt sprouts compared to that using yeast extract (Hiyikata *et al.*, 1972b).

Tetracoccus soyae, a homofermentative lactic acid bacterium, can utilize glycerol as an energy source only under aerobic conditions and both growing and resting cells produce acetic acid as the end product from glycerol (Kawasaki *et al.*, 1975). The acid yield is about 70% and its concentration reaches 12 g l^{-1}. Glycerol can be completely dissimilated to acetic acid; lactic acid is produced initially but it readily undergoes oxidative degradation to acetic acid. Glycerol dissimilation (Figure 15; Rush *et al.*, 1957; Kawasaki *et al.*, 1975) is catalyzed by inducible enzymes and cells possess high levels of catalase and cytochromes. H_2O_2 produced in the oxidation of L-glycerol 3-phosphate is decomposed by catalase.

36.3.1.3 The effect of oxygen

It has long been recognized that oxidative microbial production of acetic acid is a highly aerobic process and the effect of the oxygen level has been studied by many workers (Hromatka and Ebner, 1949; Mori *et al.*, 1970; Greenshields and Smith, 1974; Levonenmunoz and Cabezudo, 1981). Values of oxygen transfer coefficients for *Acetobacter aceti* are presented in Table 9; these

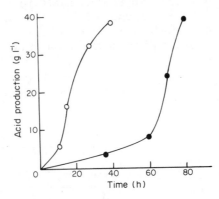

Figure 14　Effect of yeast extract addition on acid production: ○——○ with yeast extract; ●——● without yeast extract

Table 8　Effect of Various Metabolites on the Acid Production compared with Yeast Extract

Metabolite (2×10^{-3} M)	Cell mass X (mg ml^{-1})	Acid production P (mg ml^{-1})	P/X (mg acid/mg cell)
Yeast extract (0.5%)	0.6	20.4	34.0
Lactic acid	0.43	21.0	48.84
Pyruvic acid	0.38	18.0	47.37
Glycerol	0.35	16.8	48.0
Glyceric acid	0.32	15.0	46.88
Succinic acid	0.3	18.6	62.0
Oxaloacetic acid	0.3	17.4	58.0
Fumaric acid	0.26	18.6	70.8
α-Glycerophosphate	0.18	15.0	83.33
Alanine (5×10^{-3} M)	0.34	16.8	49.41

show that an increase in $K_{\mathrm{L}}a$ gives rise to better growth and production rates (Levonenmunoz and Cabezudo, 1981).

Hromatka *et al.* (1951) observed that complete stoppage of aeration for only 30 s caused a rapid decline in acid production. The extent of the damage to acid production due to suspension of aeration and/or increased agitation is a function of the acetic acid concentration. Muraoka *et al.* (1982) reported that in *Acetobacter aceti* submerged fermentation, when the acid concentration reached 6%, if aeration and agitation were interrupted for 10 s, complete inhibition of the subsequent acid production occurred, whereas there was no damage even under conditions of oxygen deficiency for 720 s at an acid concentration of 4%. The impairment of acid production in submerged culture by oxygen deficiency was found to be caused by damage to bacterial cells. This was revealed by observation of morphological changes in acetic acid-producing bacteria at various phases of growth. So far there have been few morphological studies of oxygen deficiency on acetic acid bacteria and consequent decrease in acid production. Phase contrast microscopic observation revealed that at a lower acid concentration (≤4%), when aeration and agitation are adequate, cells collected from the exponential to stationary phases appear non-refractile, large and regularly shaped; when acid concentrations reach levels above 8% and aeration and agitation are interrupted, the contrast between cytoplasm and the cell envelope becomes stronger and cells become irregular in shape.

Both respiration of cells in the culture medium and acid production are inhibited by acetic acid in the medium; acetate ions play a role in the higher pH range and both H^{+} and undissociated acetic acid in the lower pH range (Muraoka *et al.*, 1982; Mori and Terui, 1972).

It is presumed that there are physiological differences, especially with regard to respiratory behaviour between surface grown and submerged grown cells. Mori *et al.* (1970) reported that acetic acid bacteria grown under submerged conditions exhibit characteristics of the cytochrome system having K_{m} values for oxygen of the order of 10^{-5} to 10^{-7} M. In the case of surface grown acetic acid bacteria this value was found to be 2.6×10^{-4} M, which appears to indicate that flavin enzyme is the oxygen terminal entity. An inhibition study with Na$_3$N, which is a specific inhibitor for cytochrome (Dolin, 1961), showed 93% inhibition, so involvement of cytochrome oxidase can

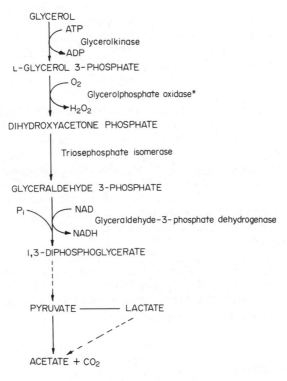

Figure 15 Pathway for glycerol dissimilation in *Tetracoccus soyae*. *This enzyme is NAD-independent flavin linked to oxygen (Kawasaki *et al.*, 1975)

Table 9 Effect of Different Oxygen Transfer Rates on Rates of Growth and Production[a]

Rotation rate (r.p.m.)	Q_{O_2} [(mmol O_2) 1^{-1} h^{-1}]	K_La (h^{-1})	$1/X \cdot dX/dt$ (h^{-1})	$1/X \cdot dP/dt$ (g $g^{-1}h^{-1}$)
400	48	219	0.048	8.5
450	65	299	0.08	18.5
500	93	425	0.10	22.0
550	124	566	0.172	30.0

[a] Aeration rate 0.75 vvm.

also be presumed. Thus, in the case of surface grown acetic acid-producing cells, no concrete conclusion can be made with regard to the oxygen terminal entity.

36.3.2 Anaerobic Conversions

36.3.2.1 Microbiology

The mixed flora popularly known as acid-formers, which are responsible for the conversion of cellulose to volatile fatty acids, comprise two distinct groups of bacterial species: cellulolytic and non-cellulolytic. The interaction between these two groups in acidogenesis is very complex. Biosynthesis and extracellular cellulase activities are constrained by non-cellulolytic bacteria as competition exists between these two groups for the soluble products of hydrolysis. It is likely that the non-cellulolytic organisms provide some nutritional inputs such as vitamins, growth factors and branched chain fatty acids necessary for cellulolytic species (Bryant, 1973).

The first step in acidogenesis is the solubilization of cellulose and hemicellulose. The most active cellulolytic species have been shown to be Gram-negative, short rods (*Bacteroides* sp.) and cocci (*Ruminococcus* sp.; Bryant, 1973; Hobson, 1965; Hobson, 1973). The anaerobic cellulase activity of *Ruminococcus* is partially inhibited by oxygen (Smith *et al.*, 1973) and the optimum pH is 6.2. Cellobiose and glucose inhibit enzyme activity but their effect on cellulase biosynthesis is not yet resolved (Scharer and Moo-Young, 1979). Some authors have reported its constitutive nature (Smith *et al.*, 1973; Hammerstrom *et al.*, 1955), while according to others anaerobic bacterial cellulases undergo catabolic repression (Leatherwood, 1965). Whatever may be the case, this factor is of little importance in anaerobic acidogenesis since no significant amount of cellobiose or glucose is ever observed in the fermenting media.

In spite of their critical role, cellulolytic bacteria constitute a small fraction of the total acid formers (Hobson, 1973) and the distribution of population depends on the type of waste on which they grow. In digesting piggery waste 76% of the total population (10^7 ml^{-1}) is *Streptococcus* (Hobson and Shaw, 1974). They are facultative Gram-positive, oxidase- and catalase-negative and they do not contain any hydrolytic enzymes. Generally *Clostridium butyricum* is found in large numbers and it produces acetic and butyric acids from glucose. These are anaerobic, motile, Gram-positive rods with subterminal spores. The cellulolytic bacterial count is less than 1% of the total population. These are strict anaerobes, mostly Gram-negative, rod shaped or coccobacillus. Formic, acetic, butyric, propionic, lactic, succinic and isobutyric acids are produced from the degradation of cellulose.

In the case of wastewater sludge digestion the predominant non-cellulolytic bacteria other than streptococci are Enterobacteriaceae which are Gram-negative, facultative, anaerobic, motile oxidase-negative, coccobacillus. Non-cellulolytic anaerobic bacteria, such as clostridium, propiobacterium and bacteroids are generally found to be present in most digesting systems. Usually obligate anaerobes outnumber facultative flora.

36.3.2.2 *Biochemical mechanisms*

The primary metabolic products of cellulolytic bacteria include aliphatic acids (formic, acetic, propionic, butyric, valeric), lactic acid, succinic acid, ethanol, carbon dioxide and hydrogen (Hungate, 1947). The product distribution and energy yields suggest that the Embden–Meyerhof pathway represents the principal route of carbohydrate metabolism (Horvath, 1979). In some species the hexose monophosphate pathway plays a supplementary but important role. Acetic acid is often produced as the major product independent of source. Ethanol formation is commonly observed but normally at lower concentration. In mixed culture some of the products of cellulolytic bacteria, *viz.* lactic acid, succinic acid and ethanol, are rapidy metabolized by non-cellulolytic bacteria (Scharer and Moo-Young, 1979). The oxidation of ethanol to acetic acid regularly occurs in acidogenesis (Chynoweth and Mah, 1971). The combined activity of cellulolytic and non-cellulolytic mixed flora results in rapid and efficient hydrolysis of cellulose to solubilized saccharides and their conversion to volatile fatty acids (mainly acetic acid), carbon dioxide and hydrogen.

The production of acetic acid from carbon dioxide and hydrogen is known to be catalyzed by a number of bacteria in pure cultures (Schoberth, 1977; Ohwaki and Hungate, 1977) *via* a common pathway. This concept developed when it was discovered that *Clostridium thermoaceticum* reduces carbon dioxide to acetate during fermentation of sugars (Fontaine *et al.*, 1942; Poston *et al.*, 1966). *Cl. thermoaceticum* lacks hydrogenase activity and is unable to utilize hydrogen gas, instead using carbon dioxide as an electron sink (Ljungdahl and Andreesen, 1976). It has been reported (Ljungdahl and Wood, 1969) that formation of acetate from carbon dioxide occurs *via* formate, tetrahydrofolate (THF) and corrinoid proteins as shown in Figure 16. A number of microorganisms are known to incorporate carbon dioxide into monocarboxylic acid during fermentation in combination with other substrates. These microorganisms include: (i) *Clostridia* (*e.g. Cl. aceticum, Cl. formicoaceticum*), (ii) *Butyribacterium rettgeri*, and (iii) *Diplococcus glycinophilus*.

Cl. thermocellum ferments glucose and fructose almost stoichiometrially to acetate according to the following steps:

$$C_6H_{12}O_6 + H_2O \rightarrow CH_3COCO_2H + CH_3CO_2H + CO_2 + 6\,H \tag{14}$$

$$CH_3COCO_2H + CO_2 + 6\,H \rightarrow 2\,CH_3CO_2H + H_2O \tag{15}$$

Overall reaction: $C_6H_{12}O_6 \rightarrow 3\,CH_3CO_2H$ $\tag{16}$

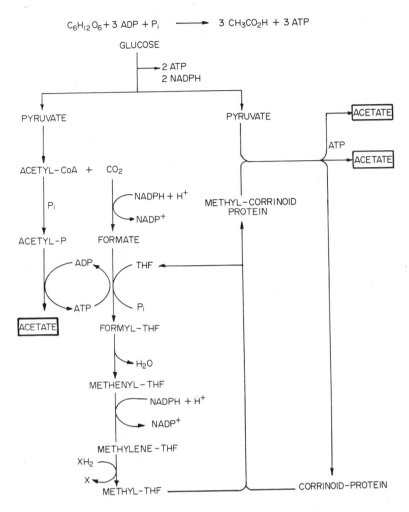

Figure 16 Pathway for conversion of hexose and carbon dioxide to acetate by *Clostridium thermoaceticum* (Ljungdahl and Wood, 1969)

In the first reaction the hexose is fermented *via* the fructose diphosphate pathway to 2 mol of pyruvate of which one is further metabolized to acetate and carbon dioxide; six equivalents of electrons are formed during this process. In the second step, these electrons are used to reduce carbon dioxide to the methyl group which yields 2 mol of acetate in a transcarboxylation reaction involving the carboxyl group of pyruvate formed in the first step.

36.3.2.3 *Substrates*

Dynatech workers (Sanderson *et al.*, 1979) have successfully fermented aquatic weeds to produce organic acids in high yield in a fixed packed bed bioreactor. The primary types of macro algae are brown, green and red-coloured. Instead of containing an appreciable quantity of glucose, they contain carboxylated or sulfonated galactose and/or mannose chains, which have been shown to be rapidly and thoroughly converted to organic acids. Conversion of carbon in marine algal biomass to acids has been demonstrated to be essentially complete and the maximum conversion of the algal species is about 43–45%; 40% to 60% of the total acid is found to be acetic acid but up to 80% acetic acid could be obtained at low conversion and by thermophillic species (55 °C).

Figure 17 shows batch and continuous fixed bed bioreactors for the production of acids. The liquid is circulated through the loop of the packed bed. Toxic compounds may be removed from the aqueous phase by passing through an activated charcoal cleaner. The acid is removed in an external subsystem by solvent extraction. Fixed packed bed fermenters can improve operating parameters and can remove a number of constraints, like mass transfer, generally faced in conventional digestion systems.

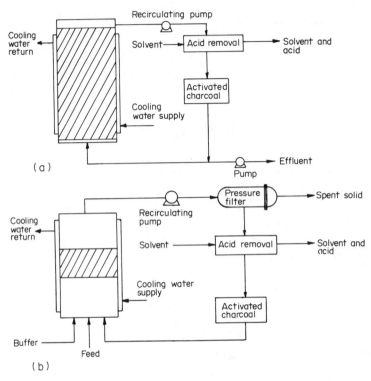

Figure 17   Possible arrangements for packed bed fermenters: (a) batch system; (b) continuous flow

(i)   Effects of pretreatment of substrate on acid production

Sodium hydroxide pretreatment to increase biodegradability of lignocellulosics by ruminants is well known. Several studies have elucidated the physicochemical mechanisms for increased digestibility (Donefer *et al.*, 1969; Tarkow and Feist, 1969; Stone *et al.*, 1969). The important chemical reaction with dilute alkali is saponification of esters of uronic acid associated with xylan chains, breaking the cross-links of polymeric units. Consequently, swelling allows diffusion of hydrolytic enzymes to the substrate and provides improved enzyme–substrate interaction even though no delignification occurs. Organic acid yields of around 0.55 g acetic acid/g dry ash free stover have been reported. The volatile fatty acid profile for the fermentation of 5% (w/v) corn stover pretreated for 48 hours with 0.5% $Ca(OH)_2$ + 0.663% Na_2CO_3 is shown in Figure 18. Initially the acetic acid yield is very high but butyric acid accumulation increased rapidly after 3 days.

The effect of alkali pretreatment of corn stover has been studied by Dutta (1981a). Mild pretreatment with 1% NaOH solution (16 ml per g stover) produces a two-fold increase in conversion (Figure 19). Lignin degradation in anaerobic digestion has been questioned by some and there is experimental evidence that lignin is not degraded in anaerobic digestion (Ghose and Bhadra, 1981). But Boruff and Buswell (1930) and Dutta (1981b) have, however, reported degradation of some portion of lignin. Akin (1980) has isolated a filamentous, facultative anaerobic rumen microorganism that attacks lignin and can grow on xylan as well as phenolic acid. The biochemistry of anaerobic degradation of lignin to organic acids is still unknown.

36.4   RECOVERY AND PURIFICATION

The separation of acetic acid from water has been industrially important for many years. Goering suggested ethyl acetate as an extractant for acetic acid in 1833 (Jones, 1967). The physical methods which have been considered include: (i) fractional distillation; (ii) azeotropic dehy-

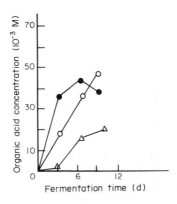

Figure 18 Mixed acid production in batch: substrate 5% (w/v) corn stover [Ca(OH)$_2$ + Na$_2$CO$_3$ treated]; temperature 25 °C. ●——● acetic acid; ○——○ butyric acid; △——△ propionic acid

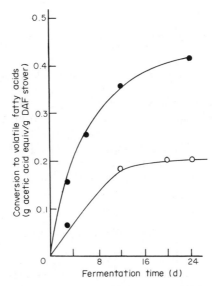

Figure 19 Effect of NaOH pretreatment on corn stover fermentation: temperature 25 °C; ○——○ untreated; ●——● 1% NaOH treated, 48 h; slurry concentration 1% w/v

dration distillation; (iii) solvent extraction; (iv) combination of the above methods (Othmer, 1958); (v) extractive distillation; and (vi) carbon adsorption. Concentration of acetic acid by rectification is impractical because although water/acetic acid do not form an azeotrope, a large number of equilibrium stages and a very high reflux ratio are required to obtain glacial acetic acid by distillation. At high acid concentrations (>50%), simple fractionation can be considered.

In extractive distillation, counter-current washing of mixed vapors in a distillation column takes place *via* a descending stream of high boiling point liquid, which is a preferential solvent for one of the components. This method was first developed by Suida in Austria (Othmer and Robert, 1942) for the removal of acetic acid from pyroligneous acid. This technique requires more expensive equipment and uses greater amounts of steam than other methods.

As an alternative, to reduce energy consumption azeotropic dehydration, first proposed by Othmer (1941), can be employed. In this method a water insoluble 'withdrawing' liquid called the entrainer is added, which decreases the effective boiling point of water relative to that of acetic acid by the formation of a low boiling point azeotrope. Low molecular weight esters such as butyl acetate are generally used as entrainers. A flow diagram of azeotropic distillation is shown in Figure 20.

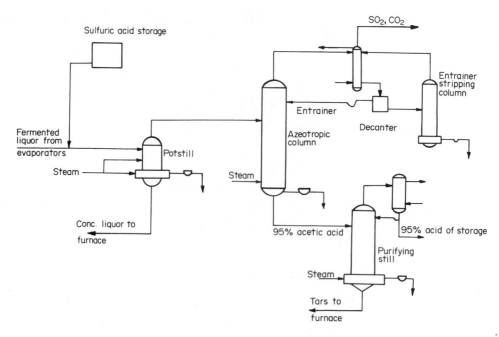

Figure 20   Othmer azeotropic distillation recovery process (Othmer, 1958)

For intermediate concentrations (10 to 50%), liquid–liquid extraction is employed and usually is followed by azeotropic distillation. Low molecular weight solvents (esters, ethers and ketones) have sufficiently high distribution coefficients for acetic acid at low concentrations (Helsel, 1977). Ethyl acetate along with diethyl ether are still considered to be the most efficient extractants (Brown, 1963).

The energy requirements for different processes employed for acetic acid recovery are presented in Table 10. As the concentration of acetic acid gets lower, the related energy costs rapidly increase along with the capital cost associated with the distillation equipment needed (Helsel, 1977).

Table 10   Steam Requirements for Acid Recovery by Various Methods[a]

Recovery method	Tons of steam required/tons of glacial acetic acid produced	
	6% Feed	25% Feed
Entrainment distillation		
Othmer entrainer	32.3	6.3
Methyl propyl ketone	—	9.6
Isopropyl acetate	—	12.0
Paraldehyde	—	5.7
Extraction processes		
Methyl acetate	16.4	5.8
Methyl propyl ketone	45.6	9.1
Cyclohexanone	15.5	6.9
Methyl cyclohexanone	11.9	3.5
75% Ethyl acetate–25% benzene mixture	25.4	—
Straight rectification	67.0	—

[a] Eaglesfield *et al.* (1953).

Othmer (1958) suggested a combination of extraction, azeotropic distillation and possibly extractive distillation for the production of glacial acetic acid. Amberg *et al.* (1969) modified this proposition and recommended the following steps for the recovery of volatile fatty acids from fermentation liquors: (i) evaporation of the fermented liquor to 40–50% solids, (ii) solvent extraction using methyl ethyl ether as solvent; (iii) dehydration and recovery of solvent by azeotropic distillation; and (iv) refining 95% acetic acid to the glacial state. A schematic diagram is shown in Figure 21.

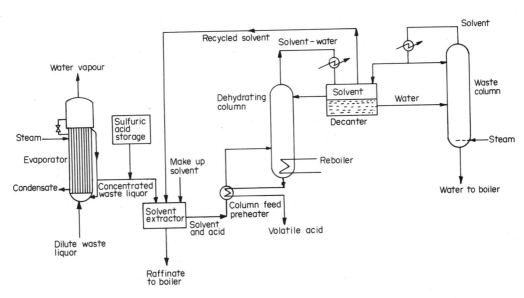

Figure 21 Flow diagram for recovery of acetic acid by a combination of evaporation, solvent extraction and azeotropic distillation

An economically attractive scheme for the recovery of acetic acid from a dilute aqueous stream by extraction with a hydrocarbon followed by distillation has been developed by Helsel (1977). The extractant used is a high boiling (bp 460 °C) trioctylphosphine oxide (TOPO). It is soluble in many polar and non-polar solvents but has a very low water solubility (1 p.p.m.). The four operations of the system are shown in Figure 22. The inherent advantages of this new technology are: (i) a high distribution coefficient of acetic acid in very dilute aqueous solution allows a small solvent (TOPO) volume to be used; (ii) good phase separation reduces the size of the extraction equipment; (iii) the stability and high boiling point of the solvent enable a small volume of acid to be recovered from a much larger solvent flow; and (iv) the low solubility of the solvent in water permits higher selectivity of the solvent and obviates the need for raffinate stripping. These advantages represent significant energy and capital savings compared with other recovery processes, especially when the acetic acid concentration is less than 5%. For 2% acetic acid solution the energy required for recovery is about 15 GJ/ton of glacial acetic acid.

To avoid product inhibition in the reactor, a concurrent extraction process is required which will remove acetic acid continuously to keep its level below the critical point in the fermenter. The extraction methods suggested are ion-exchange resins, solvent extraction and membrane separation.

An anion-exchange resin with quarternary ammonium salts was examined by the Dynatech Co. (Sanderson *et al.*, 1979) but it was found to be unsuitable for several reasons: (i) replacement of acetate increases the pH of the fermenter; (ii) PO_4^{3-} and other nutrient anions are removed; and (iii) generation of acid from the resin requires large amounts of mineral acids and alkali.

Solvent extraction by TOPO may be very useful in removing the acids preferentially for concentrations from a low level (1–3%) to levels of 20 or 25%.

Recently a membrane system has been developed which appears to have potential on a commercial scale. It is a hollow fiber membrane device consisting of silicone rubber tubing swollen with Freon TF and soaked in a 20% TOPO in Freon solution. This system is sufficiently resistant to biological attack and does not suffer mechanical fatigue (Sanderson *et al.*, 1979).

The final step in acetic acid recovery is rectification, after application of any of the above mentioned methods either singly or in combination, to give 95% acetic acid. For the final purification the acid is distilled in the presence of $KMnO_4$, $K_2Cr_2O_7$ or similar oxidants. $KMnO_4$ is more expensive but oxidizes a wider range of impurities. Glacial acetic acid is obtained at the top. The bottom product is passed through a solvent extraction system containing toluene or butyl acetate.

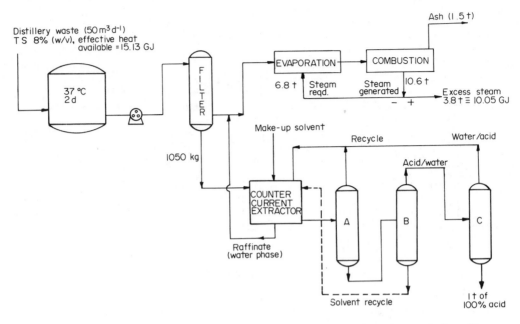

Figure 22 Flow sheet for the production of acetic acid from distillery waste: A first solvent stripping still to separate majority of water; B second solvent stripping still to separate majority of acid and the remaining water; C distillation column to remove remaining water

The oxidized organic impurities are removed with the solvent layer while the MnO_4^- stays in the water layer and can be reused.

36.5　COMPARATIVE MERITS OF AEROBIC AND ANAEROBIC PROCESSES

36.5.1　Advantages of Acidogenic Fermentations over Acetic Acid Fermentations

Acidogenic fermentations have several advantages. Since acidogenesis is a non-sterile fermentation, fermentation costs are less, operation is relatively simple and contamination problems are avoided. Aeration costs are avoided and there is a reduction in equipment design constraints. As it is a mixed flora system, it can utilize cellulose, hemicellulose and other carbohydrates, proteins and lipids to produce volatile fatty acids. A maximum of 3 mol of acetic acid can be obtained from 1 mol of glucose, whereas in the case of an aerobic process only 2 mol of ethanol can be obtained from 1 mol of glucose; two carbon atoms of the glucose molecule are lost as CO_2, so the theoretical yield based on glucose is 1.0 in the case of acidogenesis, whereas for aerobic processes it is 0.66. Conversion of carbon sources to cellular material is minimized. The primary products are volatile fatty acids, a single class of compounds which can be decarboxylated to produce volatile liquid fuel and/or chemical feedstock. The energy requirement for the production of acids by anaerobiosis is less than that required in conventional chemical processes, so this can be an effective means to supply volatile fatty acids as industrial chemical feedstock. Wastes can be treated to reduce pollution problems.

36.5.2　Advantages of Aerobic Acetic Acid Fermentations over Acidogenesis

Aerobic fermentations have the following advantages over acidogenesis. In acidogenesis it is necessary to suppress methanogenesis so that organic acids are the principal products. Acidogenesis is slower than the aerobic process. Vinegar is produced only by the aerobic process because in an anaerobic process with acetic acid, propionic and butyric acids are also produced in significant quantities. To comply with the definition of vinegar it is necessary to adopt the aerobic process for vinegar production. Further processing of the mixture of volatile fatty acids is necessary to make useful volatile combustible products. A higher concentration of acetic acid (up to 11%) is obtained in an aerobic process.

36.5.3 Comments

The yield of acid can be increased in both aerobic and anaerobic processes by control of pH, temperature and residence time. In the case of an aerobic process, the most critical parameter is oxygenation, so this should be controlled very carefully, and loss of ethanol due to evaporation should be minimized. In acidogenesis, the main problem lies with the methanogens that are associated with the acid-formers in the mixed culture. When acids are the desired products, any conversion of acid to methane and carbon dioxide is a loss, so methanogens must be suppressed or washed out of the system. Since methanogens grow much more slowly than acid-formers, these are gradually outnumbered by the acid-formers. Use of this enriched culture of acid-formers in the digester at pH<6 stops subsequent conversion of acid to methane. The washout of methanogens is achieved by controlling the dilution rate in a continuous system. A D value above $0.3 \, d^{-1}$ excludes methanogens from the reactor.

Another bottleneck in acidogenesis is the low acid tolerance of acid-formers. This can be improved by injecting acid into the reactor to adapt the crude culture to higher acid concentrations. It has been found that addition of acetic acid at a level of 0.2% increases release by 30% (Ghose and Bhadra, 1981).

36.6 ENERGY ANALYSIS OF ACETIC ACID PRODUCTION

One of the earliest methods for preparing acetic acid was by destructive distillation of wood to give pyroligneous acid, which contained 10% acetic acid. The acid used to be recovered by solvent extraction and distillation. This method was later replaced by a synthetic chemical process in which energy requirement is about 45% of that required in the wood distillation process (Chilton, 1960).

The energy requirements for the production of acetic acid by the following routes can be compared: (i) the chemical route (from acetylene); and (ii) the biochemical route *via* (a) the aerobic process, (b) the anaerobic process.

36.6.1 Chemical Route

$$C_2H_2 \xrightarrow{\; H_2O \;} CH_3CHO \xrightarrow{\; 0.5 \, O_2 \;} CH_3CO_2H$$

26 tons acetylene produce 60 tons acetic acid. The energy diagram for the production of 1 ton of acetic acid is shown below (Chilton, 1960).

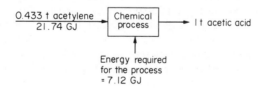

36.6.2 Aerobic Process

Basis: 1 ton of acetic acid; 10% acetic acid is produced in the broth. The energy content of the raw material is calculated on the basis of the net energy available by incineration. For this, the following basis has been taken (Alfa-Laval; steam at 4 bar):

Energy required for evaporation in five-stage evaporator (*i.e.* to bring the slurry up to 60% TS) $= 0.227 \dfrac{\text{tons steam}}{\text{tons water evaporated}}$

Energy obtained by combustion of concentrated slurry (60% TS) $= 3.85 \dfrac{\text{tons steam generated}}{\text{tons of dry solid}}$

Although the raw material has a calorific value of 23.55 GJ, the net energy available through incineration is 11.11 GJ. The temperature of fermentation is about 30 °C, so the energy required for heating can be neglected. The results are summarized in Table 11.

Table 11 Energy Analysis of Different Routes for the Production of 1 ton of Acetic Acid

| | | Biochemical route | |
| | Chemical | Anaerobic | Aerobic |
Parameter	route	process	process
Energy input as raw material (GJ)	21.74	5.08	11.11
Energy required (GJ) for			
Aeration	—	0.0	6.45
Agitation	—	0.6	9.41
Heating	—	1.5	0.0
Recovery of acid	—	14.0	10.0
Total process required (GJ)	7.12	16.1	25.86
Net energy requirement (GJ)	28.86	21.53	36.97

Since a higher acetic acid concentration is obtained in the aerobic system than in the anaerobic process, the energy required for the acid recovery is less.

36.6.3 Anaerobic Acidogenesis

Distillery waste has been found to produce acetic acid anaerobically in an appreciable quantity in a very short time (2 days). Basis for calculation: 1 ton acetic acid; distillery waste containing 8% TS (BOD 45 000 p.p.m.). In this case no energy is required for aeration but acid concentration is low (about 2%), so much energy is required for separation. The energy requirements for heating, recovery of acid, *etc.* are shown in Table 11. A schematic diagram of anaerobic acid production and separation is shown in Figure 22.

36.6.4 Comments

The low energy input in acidogenesis is due to the fact that, even after acidogenesis, the effluent (TS≃5.5%) can be incinerated to produce some energy which has been deducted from the raw material energy (effective heats available from raw material and from the effluent are 15.13 GJ and 10.05 GJ respectively).

Biological processes produce about 1.5 tons of ash per ton of acetic acid produced. This ash is very rich in potash (about 40%).

The energy requirement in the aerobic process is very high, so it appears that the aerobic process is not suitable for industrial acetic acid production, but for vinegar production it is probably the only suitable method. However, it would be interesting to investigate the conversion of sugar extracts of different fruits directly to acetic acid by the anaerobic process because the theoretical availability of acetic acid from 1 mol of glucose is 3 mol in the anaerobic process whereas it is 2 mol in the aerobic process.

In the chemical process the process energy requirement is less, but energy input as raw material is tremendously high; the high cost of the petrobased raw materials should also be kept in mind.

36.7 CONCLUSIONS

The choice of process to be used, *i.e.* aerobic or anaerobic, depends on the product required; for vinegar, the aerobic process is the method of choice, but to produce acetic acid as a chemical feedstock, the energy requirement is less in the case of anaerobic acidogenesis if the acid concentration in the digester is at least 2%. Sulfite waste liquor and distillery waste are potential substrates for acidogenesis, but solid wastes like cow dung or agricultural residues are not very promising for acid production since the maximum acid concentration never exceeds 1% with these substrates. These substrates can be better digested for the production of fuel and methane. In acidogenesis adaptation of the mixed flora, a higher acid concentration is necessary. Acetic acid injection to the digester up to a certain level may improve acid tolerance of the flora and enhance the acid productivity.

In both processes, acetic acid has an inhibitory effect on the rate of acid production, so simultaneous removal of acid from the system should allow more conversion of substrate to acids. This can be done by attaching a subsystem outside the fermenter; this may be a solvent extractor or a membrane separator.

36.8 TERMS AND SYMBOLS

A	fraction of substrate unconverted, $(S_0 - S)/S_0$
D	dilution rate, h^{-1}
$\Delta G°$	standard free energy change, $kJ\,mol^{-1}$
$\Delta H°$	heat of reaction, $kJ\,mol^{-1}$
I	inhibiting substance (acetic acid)
K_P	inhibition constant for product, $g\,l^{-1}$
K_{SY}	saturation constant for yeast extract, $g\,l^{-1}$
K_S	saturation constant for substrate, $g\,l^{-1}$
$K_L a$	oxygen transfer coefficient, h^{-1}
K'	true product yield constant
m	maintenance coefficient, g substrate $(g\,cell)^{-1}\,h^{-1}$
P	product (acetic acid) concentration, $g\,l^{-1}$
P_o	acetic acid concentration in the inflowing medium, $g\,l^{-1}$
dP/dt	rate of product formation, $g\,l^{-1}\,h^{-1}$
Q_{O_2max}	maximum oxygen uptake rate, $g\,l^{-1}\,h^{-1}$
r_P	rate of product formation, dP/dt, $g\,l^{-1}\,h^{-1}$
r_S	rate of substrate consumption, dS/dt, $g\,l^{-1}\,h^{-1}$
r_X	rate of growth, dX/dt, $g\,l^{-1}\,h^{-1}$
R_X	r_X/r_S
R_P	r_P/r_S
S	substrate concentration, $g\,l^{-1}$
S_0	initial substrate concentration, $g\,l^{-1}$
S_A	ethanol concentration, $g\,l^{-1}$
S_Y	yeast extract concentration, $g\,l^{-1}$
X	cell mass, $g\,l^{-1}$
α	internal mass transfer coefficient, $g^{0.5}\,l^{-0.5}\,h^{-1}$
α_1	competition coefficient
v	specific rate of product formation, $(g\,product)\,(g\,cell)^{-1}\,h^{-1}$
μ	specific growth rate, h^{-1}
μ_m	maximum specific growth rate, h^{-1}

36.9 REFERENCES

Akiba, T. and T. Fukimbara (1973). Tower-type fermentor with a gas entrainment process. 6. Fermentation of a volatile substrate in a tower-type fermentor with a gas entrainment process. *Ferment. Technol. J.*, **51**, 134–141.

Akin, D. E. (1980). Attack on lignified grass cell walls by a facultatively anaerobic bacterium. *Appl. Environ. Microbiol.*, **40**, 809–820.

Allgeier, R. J. and F. M. Hildebrandt (1960). Developments in vinegar manufacture. *Adv. Appl. Microbiol.*, **8**, 163.

Amberg, H. R., T. R. Aspitarte and J. F. Cormack (1969). Fermentation of spent sulfite liquor for production of volatile acids. *J. Water Pollut. Control Fed.*, **41**, 419.

Anon. (1982a). *Chem. Eng. News*, **60** (24), 30.

Anon. (1982b). *Chemical Weekly Annual*.

Asai, T. (1968). *Acetic Acid Bacteria*. University of Tokyo Press, Tokyo.

Atkinson, D. E. (1956a). Hydrogen metabolism in *Actinobacter peroxydans*. *Bacteriol. J.*, **72**, 189.

Atkinson, D. E. (1956b). Oxidation of ethanol and tricarboxylic acid cycle intermediates by *Actinobacter peroxydans*. *Bacteriol. J.*, **72**, 195.

Boerhaave, H. (1732). *Elementa Chemiae, Luduni Balavorum*, **2**, 179, 207.

Boruff, C. S. and A. M. Buswell (1930). Fermentation products from cornstalks. *Ind. Eng. Chem.*, **22**, 931–933.

Brown, W. V. (1963). *Chem. Eng. Progr.*, **59**, 10.

Bryant, M. P. (1973). Nutritional requirements of predominant rumen cellulytic bacteria. *Fed. Proc., Fed. Am. Soc. Exp. Biol.*, **32**, 1809–1813.

Chattapodhyay, P. K. (1980). M. Tech Thesis, BERC, IIT Delhi, India.

Chilton, C. H. (1960). *Cost Engineering in the Process Industries*. McGraw-Hill, New York.

Chynoweth, D. P. and R. A. Mah (1971). Volatile acid formation in a sludge digestion. *Adv. Chem. Ser.*, **105**, 41.

Conner, H. A. and R. J. Allgeier (1976a). *Adv. Appl. Microbiol.*, **20**, 81–133.

Conner, H. A. and R. J. Allgeier (1976b). *Adv. Appl. Microbiol.*, **20**, 166.

Dolin, M. I. (1961). In *The Bacteria*, ed. I. C. Gunsalus and R. Y. Stanier, vol. 2, p. 319. Academic, New York.

Donefer, E., I. Adeleye and T. Jones (1969). Effect of urea supplementation on the nutritive value of NaOH-treated oat straw. *Adv. Chem. Ser.*, **95**, 328.

Dutta, R. (1981a). *Biotechnol. Bioeng.*, **23**, 61.

Dutta, R. (1981b). *Biotechnol. Bioeng.*, **23**, 2167.

Eaglesfield P., B. K. Kelly and J. F. Short (1953). Recovery of acetic acid from dilute solutions by liquid–liquid extractions. *Ind. Chem. Chem. Manuf.*, **29**, 147–151.

Filho, A. P. daSilva and L. C. Sattamin (1981). *Proceedings of the 2nd World Congress of Chemical Engineers, Montreal, Canada*, vol. 1, p. 39.

Folin, O. and V. Ciocatten (1927). Tyrosine and tryptophan determinations in proteins. *Biochem. J.*, **73**, 627.

Fontaine, F., W. H. Peterson, E. McCoy and M. J. Johnson (1942). A new type of glucose fermentation by *Clostridium thermoaceticum* n. sp. *Bacteriol. J.*, **43**, 701.

Frateur, J. (1950). *La Cellule*, **53** (3), 288.

Ghommidh, C., J. M. Navarro and G. Durand (1981). Acetic acid production by immobilized actinobacter cells. *Biotechnol. Lett.*, **3**, 93–98.

Ghommidh, C., J. M. Navarro and G. Durand (1982). A study of acetic acid production by immobilized acetobacter cells—oxygen transfer. *Biotechnol. Bioeng.*, **24**, 605–617.

Ghosh, S., J. R. Conrad and D. L. Klass (1975). Anaerobic acidogenesis of wastewater sludge. *WPCF, J.*, **47**, 30–45.

Ghose, T. K. and K. K. Bandyopadhyay (1980). Rapid ethanol fermentation in an immobilized yeast-cell reactor. *Biotechnol. Bioeng.*, **22**, 1489–1496.

Ghose, T. K. and A. Bhadra (1981). Maximization of energy recovery in the biomethanation process. 1. Use of cow dung as substrate in a multi-reactor system. *Process Biochem.*, **16** (6), 23–25.

Grayson, M. and D. Eckroth (ed.) (1979). *Kirk–Othmer Encyclopedia of Chemical Technology*, vol. 5, 3rd edn. Wiley, New York.

Grayson, M. and D. Eckroth (ed.) (1980). *Kirk–Othmer Encyclopedia of Chemical Technology*, vol. 9, 3rd edn. Wiley, New York.

Greenshields, R. N. and R. N. Smith (1974). *Process Biochem.*, **9** (3), 11.

Hammerstrom, R. A., K. D. Claus, J. W. Coghlan and R. H. McBee (1955). The constitutive nature of bacterial cellulases. *Arch. Biochem. Biophys.*, **56**, 123–129.

Helsel, R. W. (1977). Removing carboxylic acids from aqueous wastes. *Chem. Eng. Prog.*, **73** (5), 55–59.

Hijikata, Y., H. Okumura and G. Terui (1972a). Acceleration of acetic acid fermentation by malt sprouts. *Ferment. Technol. J.*, **49**, 577.

Hijikata, Y., H. Okumura and G. Terui (1972b). Studies on factors to promote acetic acid formation by *Actinobacter rancens*. 2. Effect of some simple metabolites. *Ferment. Technol. J.*, **50**, 7.

Hobson, P. N. (1965). Continuous culture of some anaerobic and facultatively anaerobic rumen bacteria. *Gen. Microbiol. J.*, **38**, 167.

Hobson, P. N. (1973). Bacteriology of anaerobic sewage digestion. *Process Biochem.*, **8**, 19.

Hobson, P. N. and B. G. Shaw (1974). Bacterial population of piggery-waste anaerobic digesters. *Water Res.*, **8**, 507.

Horvath, R. S. (1974). Evolution of anaerobic energy-yielding metabolic pathways of prokaryotes. *Theor. Biol. J.*, **48**, 361.

Hougen, O. A., K. M. Watson and R. A. Rogatz (1967). *Chemical Process Principles, Part I*. Asia Publishing House.

Hromatka, O. and H. Ebner (1949). Investigations on vinegar fermentation. 1. Generator vinegar fermentation and aeration procedures. *Enzymologia*, **13**, 369.

Hromatka, O. and H. Ebner (1950). Acetic fermentation. 2. Lack of oxygen in generator fermentation. *Enzymologia*, **14**, 96.

Hromatka, O., G. Kastner and H. Ebner (1951). Investigations on vinegar fermentation. 5. Influence of temperature and total concentration on the submerged fermentation. *Enzymologia*, **15**, 337.

Hungate, R. E. (1947). Studies on cellulose. 3. The culture and isolation of cellulose-decomposing bacteria from the rumen of cattle. *Bacteriol. J.*, **53**, 631.

Jones, E. L. (1967). Economic saving through the use of solvent extraction. *Chem. Ind. (London)*, 1590.

Kawasaki, H., S. Okano and S. Omata (1975). Oxidative dissimilation of glycerol in *Tetracoccus soyae*. *Ferment. Technol. J.*, **53**, 207–217.

Kennedy, J. F., S. A. Barker and J. D. Humphrey (1976). Microbial cells living immobilized on metal hydroxides. *Nature (London)*, **261**, 242–244.

Kennedy, J. F., J. D. Humphrey, S. A. Barker and R. N. Greenshields (1980). Application of living immobilized cells to the acceleration of continuous conversions of ethanol to acetic acid—hydrous titanium(IV) oxide-immobilized acetobacter species. *Enzyme Microb. Technol.*, **2**, 209–216.

King, T. E. and V. H. Cheldelin (1952). Oxidative dissimilation in *Acetobacter suboxydans*. *Biol. Chem. J.*, **198**, 127.

King, T. E. and V. H. Cheldelin (1956). Oxidation of acetaldehyde by *Acetobacter suboxydans*. *Biol. Chem. J.*, **220**, 177.

Kutzing, F. T. (1837). *Prakt Chem. J.*, **11**, 385.

Leatherwood, J. M. (1965). Cellulase from *Ruminococcus albus* and mixed rumen microorganisms. *Appl. Microbiol.*, **13**, 771.

Ley, J. D. and J. Frateur (1971). In *Bergey's Manual of Determinative Bacteriology*, 8th edn. William and Wilkins, Baltimore, MD.

Ljungdahl, L. G. and J. R. Andreesen (1976). In *Microbial Production and Utilization of Gases*, ed. H. G. Schlegel, Gottingen.

Ljungdahl, L. G. and H. G. Wood (1969). Total synthesis of acetate from carbon dioxide by heterotrophic bacteria. *Annu. Rev. Microbiol.*, **23**, 515.

Lozano, I. de la T., I. Cernok and G. Goma (1980). *Proceedings of the 2nd Symposium on Bioconversion and Biotechnology, New Delhi*, vol. 2, p. 113.

Lutwak-Mann, C. (1938). Alcohol dehydrogenase of animal tissues. *Biochem. J.*, **32**, 1364.

Mori, A. and G. Terui (1972). Kinetic studies on submerged acetic acid fermentation. 2. Process kinetics. *Ferment. Technol. J.*, **50**, 776.

Mori, A., N. Konno and G. Terui (1970). Kinetic studies on submerged acetic acid fermentation. 1. Behaviors of *Actinobacter rancens* cells toward dissolved oxygen. *Ferment. Technol. J.*, **48**, 203.

Mori, A., H. Yoshikawa and G. Terui (1972). Kinetic studies on submerged acetic acid fermentation. 4. Product inhibition and transient adaptation of cells to product. *Ferment. Technol. J.*, **50**, 518.

Levonenmunoz, E. and M. D. Cabezudo (1981). Influence of oxygen-transfer rate on vinegar production by *Acetobacter aceti* in submerged fermentation. *Biotechnol. Lett.*, **3**, 27–32.

Muller, F. (1978). Modern bioreactor for vinegar production. *Process Biochem.*, **13**, 10–11.

Muraoka, H., Y. Watabe and N. Ogasawara (1982). Effect of oxygen deficiency on acid production and morphology of bacterial cells in submerged acetic fermentation by *Acetobacter aceti*. *Ferment. Technol. J.*, **60**, 171–180.

Nakayama, T. (1959). Studies on acetic bacteria. 1. Biochemical studies on ethanol oxidation. *Biochem. J.*, **46**, 1217–1225.

Nakayama, T. (1961a). Studies on acetic acid bacteria. 3. Purification and properties of a co-enzyme-independent aldehyde dehydrogenase. *Biochem. J.*, **49**, 158.

Nakayama, T. (1961b). Studies on acetic acid bacteria. 4. Purification and properties of a new type of alcohol dehydrogenase, alcohol-cytochrome-553 reductase. *Biochem. J.*, **49**, 240.

Neuberg, C. and E. Molinari (1926). The mechanism of acetic acid fermentation. *Naturwissenschaften*, **14**, 758.

Nickol, G. B. (1979). In *Microbial Technology*, ed. H. J. Peppler and D. Perlman, vol. 2, p. 155. Academic, New York.

Ohwaki, K. and R. E. Hungate (1977). Hydrogen utilization by clostridia in sewage sludge. *Appl. Environ. Microbiol.*, **33**, 1270–1274.

Othmer, D. F. (1941). Azeotropic distillation for dehydrating acetic acid. *Chem. Met. Eng.*, **48** (6), 91–94.

Othmer, D. F. (1958). Acetic acid recovery methods. *Chem. Eng. Progr.*, **54** (7), 48–52.

Othmer, D. F. and H. R. Robert (1942). *Ind. Eng. Chem.*, **34**, 274.

Pasteur, L. (1868). *Etudes Sur le Vinaigre*. Paris.

Patankar, A. D. (1982). *Chemical Weekly Annual*, 177.

Perdih, A. and P. Kafol (1978). 1st Eur. Congr. Biotechnol., Interlaken, Part-II.

Perry, R. H. and C. H. Chilton (1973). *Chemical Engineers' Handbook*. McGraw-Hill, Kogakusha, Tokyo.

Persoon, C. H. (1822). *Mycologia enropaea*, **1**, 96.

Poston, J. M., K. Kuratomi and E. R. Stadtman (1966). Conversion of carbon dioxide to acetate. 1. The use of cobalt methylcobalamine as a source of methyl groups for synthesis of acetate by cell-free extracts of *Clostridium thermoaceticum*. *Biol. Chem. J.*, **241**, 4209.

Prescott, S. C. and C. G. Dunn (1959). *Industrial Microbiology*, pp. 428. McGraw-Hill, Kogakusha, Tokyo.

Rush, D., D. Karibian, M. L. Karnofsky and B. Magasanik (1957). Pathways of glycerol dissimilation in two strains of *Aerobacter aerogenes*—enzymatic and tracer studies. *Biol. Chem. J.*, **226**, 891–899.

Sanderson, J. E., D. L. Wise and D. C. Augenstein (1979). *Biotechnol. Bioeng. Symp.*, **8**, 131.

Scharer, J. M. and M. Moo-Young (1979). In *Advances in Biochemical Engineering*, ed. T. K. Ghose *et al.*, vol. 2, p. 85. Springer-Verlag, Berlin.

Schoberth, S. (1977). Acetic acid from H_2 and CO_2—formation of acetate by cell extracts of *Acetobacterium woodii*. *Arch. Microbiol.*, **114**, 143–148.

Serieys, M., G. Goma and G. Durand (1978). Design and oxygen transfer potential of a pulsed continuous tubular fermentor. *Biotechnol. Bioeng.*, **20**, 1393–1406.

Smith, W. R., I. Yu and R. E. Hungate (1973). Factors affecting cellulolysis by *Ruminococcus albus*. *Bacteriol. J.*, **114**, 729–737.

Standen, A. *et al.* (ed.) (1970). *Kirk–Othmer Encyclopedia of Chemical Technology*, vol. 21, 2nd edn. Wiley, New York.

Stone, J. E., A. M. Scallan, E. Donefer and E. Ahlgren (1969). Digestibility as a simple function of a molecule of similar size to a cellulase enzyme. *Adv. Chem. Ser.*, **95**, 219.

Tarkow, H. and W. C. Feist (1969). A mechanism for improving digestibility of lignocellulosic materials with dilute alkali and liquid ammonia. *Adv. Chem. Ser.*, **95**, 197.

Tyagi, R. D. and T. K. Ghose (1982). Studies on immobilized *Saccharomyces cerevisiae*. 1. Analysis of continuous rapid ethanol fermentation in an immobilized cell reactor. *Biotechnol. Bioeng.*, **24**, 781–795.

Vaughn, R. H. (1954). In *Industrial Fermentations*, ed. L. A. Underkofler and R. J. Hickey, vol. 1, p. 498. Chemical Publishing Co., New York.

Vera, F. M. and D. I. C. Wang (1977). 174th ACS National Meeting, Chicago.

Wieland, H. (1913). Mechanism of oxidation processes. *Ber. Chem. Ges.*, **46**, 3327–3342.

37
Propionic and Butyric Acids

M. J. PLAYNE
CSIRO, Clayton, Victoria, Australia

37.1 INTRODUCTION

Commercial production of propionic and butyric acids is entirely by petrochemical routes, although since early this century a number of pilot plants have been designed and built to produce these acids by fermentation processes. Fermentation processes have not been used commercially, primarily because separation of the product acids from the fermentation medium and concentration of the acids have proved too expensive. The acids are present in relatively low concentrations (20–30 g l^{-1}) in the fermentation media, and the differences in volatility between the acids and water are small. Furthermore, the expense of producing pure propionic or butyric acid is increased because these acids are generally produced concurrently with other acids such as acetic and lactic acids.

Consequently, there has been an interest in producing mixtures of ketones (Lefranc, 1943; Playne, 1980), paraffins and olefins (Sanderson *et al.*, 1978) and esters (Datta, 1981b) from mixed acid fermentations, and in using these derived compounds as liquid fuels. In these cases, the boiling range and other properties are more important than the production of a pure chemical compound. Whilst petrochemical routes are preferred at present, routes employing renewable feedstocks would be desirable when supplies and costs of oil make existing processes uneconomic. Therefore, the purpose of this chapter is to summarize research and development on the production of propionic and butyric acids by fermentation processes, to indicate the problems and limitations of such processes, and to suggest areas where more research is needed.

Relevant physical data on propionic and butyric acids are given in Table 1. Other useful data (*e.g.* vapour pressure) may be found in Lurie (1964) and Wocasek (1968).

37.2 MICROBIOLOGY OF THE FERMENTATIONS

37.2.1 History of the Fermentation Route to Propionic Acid

Propionic acid derived from fermentation was first observed by Strecker (1854) and subsequently by Pasteur (1879) and Fitz (1879) on various sugar alcohol and organic acid substrates. These observations resulted in the formulation of the Fitz equation (propionic acid yield of 54.8% (w/w)):

$$3 \text{ lactic acid} \rightarrow 2 \text{ propionic acid} + 1 \text{ acetic acid} + 1 \text{ CO}_2 + 1 \text{ H}_2\text{O} \tag{1}$$

Our knowledge of *Propionibacterium*, which stems from their role in cheese manufacture, dates back to Freudenreich and Orla Jensen (1906), Van Niel (1928), and Werkman and Kendall (1931). These authors named some 11 species of *Propionibacterium*. Most species will utilize glucose and maltose, some species will utilize the pentose sugars, arabinose and xylose, and others, starch. Pectin, inulin and cellulose are not recorded as being attacked by any *Propionibacterium* species. *Propionibacterium* species are described as non-motile, short non-sporulating rods, and are all Gram-positive. They are mesophilic (30 °C), anaerobic species, which assume various involutions in acid media or in aerobic conditions.

Several other genera of anaerobic bacteria produce propionic acid as a major product. These are listed in Table 2. The major genera are *Veillonella*, *Selenomonas* and the species *Clostridium propionicum*. These bacteria are commonly found in the gut of herbivores. Further reference to them is made in Section 37.2.3 which is concerned with the metabolic pathways of fermentation to propionic acid.

37.2.2 History of the Fermentation Route to Butyric Acid

The history of this fermentation route has been reviewed by Péaud-Lenoël (1952). The butyric acid-producing fermentation was discovered in 1861 by Pasteur who suggested the name 'anaerobic' to describe the organism since it was killed by air. Fitz, Gruber and Grimbert made further studies of the fermentation in the 1880s but did not differentiate between butyric acid- and butanol-producing species. Winogradsky and Beijerinck did make this differentiation (Péaud-Lenoël, 1952). All these workers studied various *Clostridium* species. Butanol- and acetone-producing organisms were isolated by Fernbach and Weizmann before 1915. The work of Weizmann led to the well-known commercial process for the production of butanol and acetone by *Clostridium acetobutylicum*. This process is still used in some countries (Spivey, 1978). Currently, there is an active interest in developing a continuous butanol fermentation process by immobilizing the bacteria on a support, by enhancing yields of butanol through addition of butyric acid and by finding ways to overcome product inhibition of the fermentation. In the traditional batch process, butyric acid is produced as an intermediate in concentrations of about 5 g l^{-1} in the early phases of the fermentation.

Although many species of *Clostridium* produce butyric acid as a major product, several non spore-forming bacteria also produce butyric acid. These are listed in Table 2.

Table 1 Some Physical Data on Propionic and Butyric Acids

Common name	IUPAC name	Formula	Molecular weight	Melting point (°C)	Boiling point (°C)	Density d_4^{20}	Heat of combustion (kJ mol^{-1})	pK_a 25 °C
Propionic acid	Propanoic acid	CH_3CH_2COOH	74.08	−22.4	141.1	0.992	1536	4.88
n-Butyric acid	Butanoic acid	$CH_3CH_2CH_2COOH$	88.11	−7.9	163.5	0.959	2194	4.83

Table 2 Bacteria Capable of Producing Either Propionic or Butyric Acid as a
Major End-product

Propionic acid-producers	Butyric acid-producers
Species of major importance	
Propionibacterium (9/9)[a]	*Clostridium* spp. (37/8)[a]
e.g. *P. shermanii*	e.g. *C. butyricum*
	C. kluyveri
Veillonella parvula	*C. pasteurianium*
Veillonella alcalescens	
Selenomonas ruminantium (pH 5)	*Butyrivibrio fibrosolvens* (pH 5.6–7)
	Sarcina maxima (pH 1–9)
Selenomonas sputigena (pH 5)	*Eubacterium* spp. (17/25)[a]
Clostridium (5/78)[a]	(includes *Butyribacterium* spp.)
e.g. *C. propionicum* (25 °C)[b]	e.g. *E. limosum*
C. novyi (45 °C)[b]	(= *B. rettgeri*)[d]
	E. multiforme (25–40 °C)
Megasphaera elsdenii (pH 4–8)[b]	*Fusobacterium* spp. (13/13)[a]
Bacteriodes spp. (17/30)[a]	e.g. *F. nucleatum*
e.g. *B. fragilis*	*Megasphaera elsdenii*
B. ruminicola[b]	
Fusobacterium necrophorum	
Species of minor importance	
Arachnia propionica	*Bacteriodes melaninogenicus*
(*Anaerovibrio lipolytica*)[d]	*Treponema phagedenis*
	Acidaminococcus fermentans[c]
	Peptococcus asaccharolyticus[c]
	(*Coprococcus* spp.)[d]
	(*Gemmiger* spp.)[d]
	(*Peptococcus prevotii*)[c,d]
	(*Gaffkya anaerobia*)[c,d]

[a] Proportion of species in the genus which produce the acid as a major product. [b] Acrylic acid pathway operates in this species. [c] Produced mainly from amino acids only. [d] Species is not included in the Bergey manual (8th edition), but is in VPI manual (4th edition).

37.2.3 Metabolic Pathways to Propionate and Butyrate of Anaerobic Bacteria

The generalized pathways from glucose to the major fermentation products formed are shown in Figure 1.

The two major branch points where regulation occurs for the types of products formed are pyruvate and acetyl-SCoA. Propionate is mostly formed *via* the dicarboxylic acid pathway, but some species (*Megasphaera elsdenii, Clostridium propionicum, Bacteriodes ruminicola*) form propionate from lactate *via* the acrylate pathway. Butyrate is formed by only one basic pathway from acetyl-SCoA.

Propionate and butyrate are scarcely ever formed as sole products. Formation of either is usually accompanied by formation of acetate. This occurs for stoichiometric reasons, and to maintain hydrogen and redox balance. In addition, product ratios are controlled for thermodynamic reasons, such as management of adenosine 5′-triphosphate (ATP) production and entropy generation (Thauer *et al.*, 1977). For example, 4 mol of ATP are produced when 1 mol of glucose is converted to acetate, but only 3 mol of ATP are formed when 1 mol of glucose is converted to butyrate. ATP production is related to bacterial cell yield. The products formed are also regulated by biotin, heme and vitamin B_{12}. Without vitamin B_{12} and biotin, propionate is not produced by the dicarboxylic acid pathway. Without heme, succinate is not formed (Chen and Wolin, 1981). Similarly, *Propionibacterium* species produce lactate rather than propionate from glucose if the pH is allowed to drop (Phelps *et al.*, 1939).

37.2.3.1 Dicarboxylic acid pathway of propionic acid formation

This is the most common pathway for the formation of propionic acid. Lactate is used preferentially to glucose as a substrate by most propionic acid-producing bacteria. Propionate may be

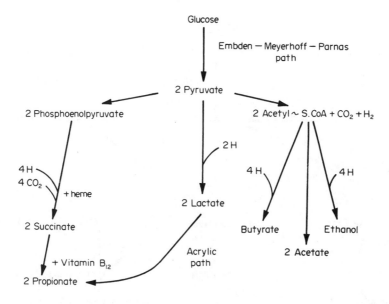

Figure 1 Generalized pathways to major products of anaerobic bacteria

formed from lactate by either the dicarboxylic acid pathway or the acrylic pathway. The route taken depends on the bacterial species. The dicarboxylic acid pathway from glucose and from lactate is shown in Figure 2.

The simplified summary equations for this path are:

$$\text{glucose} + 2\,H_2O \rightarrow 2\,\text{acetic acid} + 2\,CO_2 + 8\,H \tag{2}$$

$$\underline{2\,\text{glucose} + 8\,H \rightarrow 4\,\text{propionic acid} + 4\,H_2O} \tag{3}$$

$$\text{Sum: } 3\,\text{glucose} \rightarrow 4\,\text{propionic acid} + 2\,\text{acetic acid} + 2\,CO_2 + 2\,H_2O + 12\,\text{ATP} \tag{4}$$

Theoretical maximum yields from sugar are, for equation (3), 82.2% (w/w) as propionic acid, and for the balanced equation (4), 54.8% (w/w) as propionic acid and 77.0% (w/w) as total acids.

The enzyme, *S*-methylmalonyl-*S*CoA:pyruvate transcarboxylase, is a key to the cyclic nature of this pathway, since it enables a carboxyl group to be transferred from *S*-methylmalonyl-*S*CoA to pyruvate to form oxaloacetate and propionyl-*S*CoA. The enzyme requires biotin to act as a 'carrier' of the carboxyl group from *S*-methylmalonyl-CoA to oxaloacetate. In *Bacteriodes fragilis* and in *Veillonella* species, the transcarboxylase is replaced by methylmalonyl-CoA decarboxylase. Otherwise, the pathways are essentially the same as in *Propionibacterium*.

37.2.3.2 Acrylic acid pathway of propionic acid formation

This pathway is restricted to a few species of bacteria, and has been described most fully for *Clostridium propionicum* (Johns, 1952) and *Megasphaera elsdenii* (Ladd and Walker, 1965). The pathway is shown in Figure 3. Sinskey and others (1981) have investigated this pathway for production of acrylic acid from either lactic or propionic acids. Their work suggests the importance of β-alanine in the operation of the pathway.

The simplified summary equations for this pathway are:

$$\text{lactic acid} + H_2O \rightarrow \text{acetic acid} + CO_2 + 4\,H \tag{5}$$

$$\underline{2\,\text{lactic acid} + 4\,H \rightarrow 2\,\text{propionic acid} + 2\,H_2O} \tag{6}$$

$$\text{Sum: } 3\,\text{lactic acid} \rightarrow 2\,\text{propionic acid} + 1\,\text{acetic acid} + 1\,H_2O + 1\,CO_2 + 1\,\text{ATP} \tag{7}$$

The theoretical weight yields from lactic acid to propionic acid are the same as those for the dicarboxylic acid pathway.

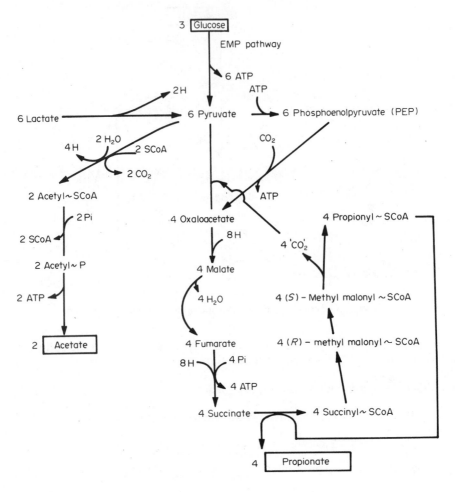

Figure 2 The dicarboxylic pathway to propionate

37.2.3.3 Butyric acid pathway

The butyric acid pathway is essentially the same for all bacteria. Figure 4 shows minor variations in the way butyrate is formed from butyryl-SCoA, and their relationship to the formation of acetone and butanol by species such as *Clostridium acetobutylicum*.

Butyrate is rarely formed as the sole acid salt, since the concurrent formation of acetate is usually needed for thermodynamic reasons. Acetate from glucose yields 4 mol ATP, while butyrate fermentation yields 3 mol ATP. Acetyl-SCoA and reduced nicotinamide adenine dinucleotide (NADH) regulate the system for control of ATP and entropy generation. This is further discussed by Thauer *et al.* (1977). Miller and Jenesel (1979) have examined the pathways and enzymology of butyrate formation in *Butyrivibrio fibrosolvens* in detail, and have found that the butyryl phosphate alternative pathway exists at least in that species.

Butyrate formation from glucose yields 3 mol ATP per mol of glucose, whereas propionate formation yielded 4 mol ATP per mol of glucose.

Typical simplified summary equations are:

$$2 \text{ glucose} + 4 \text{ H}_2\text{O} \rightarrow 4 \text{ acetic acid} + 4 \text{ CO}_2 + 16 \text{ H} \tag{8}$$

$$\underline{2 \text{ glucose} + 16 \text{ H} \rightarrow 3 \text{ butyric acid} + 6 \text{ H}_2\text{O} + 2 \text{ H}_2} \tag{9}$$

$$\text{Sum: } 4 \text{ glucose} \rightarrow 3 \text{ butyric acid} + 4 \text{ acetic acid} + 4 \text{ CO}_2 + 2 \text{ H}_2\text{O} + 2 \text{ H}_2 \tag{10}$$

Theoretical weight yields from glucose are, for equation (9), 73.3% (w/w) as butyric acid; and for equation (10), 36.7% (w/w) as butyric acid and 70.0% (w/w) as total acids. Alternatively, a

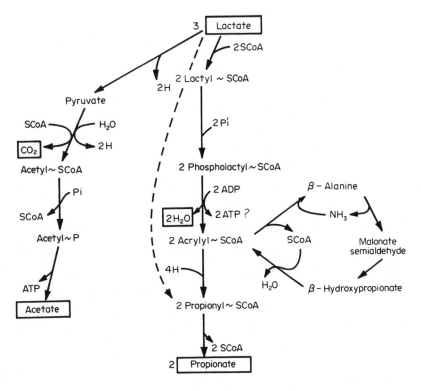

Figure 3 The acrylic acid pathway to propionate

stoichiometric balance can also be obtained by the following equation, which would give a weight yield of 48.9% (w/w) as butyric acid.

$$\text{glucose} \rightarrow \text{butyric acid} + 2\,CO_2 + 2\,H_2 + 3\,\text{ATP} \tag{11}$$

From the viewpoint of engineering design, theoretical weight yields and energy of microbial conversions are important considerations. These are summarized in Table 3 for propionic and butyric acid, and compared with conversions to lactic acid, ethanol, acetone and butanol. The carbon loss is low in all these anaerobic fermentations relative to losses in aerobic fermentations and in thermal methods of conversion of biomass or coal to useful chemicals and fuels. Furthermore, energy recovery (based on heats of combustion of product and substrate) is high. The energy recovery is higher for propionic and butyric acids than for acetic acid. The aceto–propionic fermentation is particularly efficient if the value of both acetic and propionic acid is included in the calculations. The free energy of the reactions provides an indication of the maximum ATP production possible. Approximately 1 mole ATP will allow formation of 10 g microbial cell-dry matter.

37.2.4 The Bacteria

Bacteria capable of producing either propionic or butyric acid as a major fermentation product are listed in Table 2. None of these bacteria will produce from glucose the acid required as the sole product. Production of acetic acid usually accompanies the formation of either propionic or butyric acid. The reasons for this have been given earlier (see Section 37.2.3). Apparently, it is possible to obtain propionic acid exclusively from lactic acid using *Clostridium propionicum* (Johns, 1952) and from glycerol by *Propionibacterium* (Werkman *et al.*, 1929). These claims have not been verified by recent work. Table 2 was derived mainly from data available in Bergey's Manual (Buchanan and Gibbons, 1974) and in the VPI Manual (Holdeman *et al.*, 1977). Where many species of a genus produce the required acid, an indication of the proportion of species doing so is given in Table 2. Only species whose acid production has been well characterized and studied are listed by species name, unless the particular species is of possible industrial value because of a useful characteristic. For example, data in the VPI manual suggest that *Eubacterium multiforme* produces an unusually high ratio of butyric to acetic acid and exhibits excess

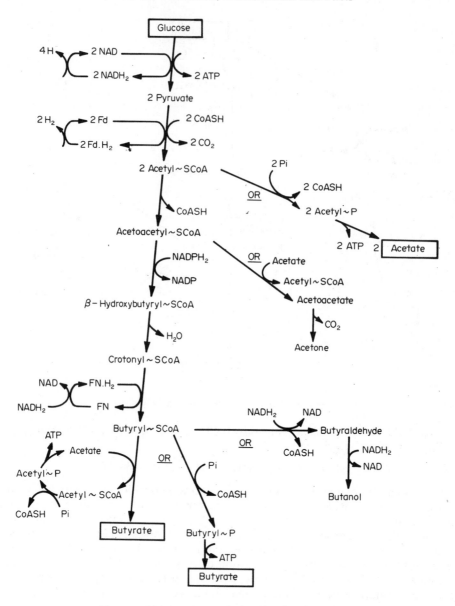

Figure 4 Pathways of butyrate formation from glucose

hydrogen gas production. Temperature optima for growth of the species are given where these were available and differ from the usual 30–40 °C optimum, as this is an important characteristic for industrial fermentations. Likewise, when optimum pH for growth differed from the pH range 6–8, this is indicated.

Information on substrates used by the bacteria can be obtained in Bergey's Manual and the VPI Manual. Such information is incomplete, and strains of a species may show a considerable variation (especially *Butyrivibrio*). Only *Butyrivibrio fibrosolvens* and *Clostridium polysaccharolyticum* are able to utilize cellulose (van Gylswyk *et al.*, 1980), although several of the species listed in Table 2 are able to use cellobiose. Nearly all can use a wide range of hexose sugars, but pentose sugars are not able to be used by several species. *Veillonella parvula* differs in that it cannot use carbohydrates, but uses lactate, pyruvate and succinate.

The growth rate of species is an important characteristic for fermenter design. *Propionibacterium* are known to have a slow growth rate. This would suggest that immobilized cell systems would be advisable for this species. Species able to grow at low pH are important in acidogenic fermentation processes, because separation of acids from the fermentation solution requires that the acids be in the unionized state. Such species include *Selenomonas*, *Megasphaera elsdenii*, *Butyrivibrio fibrosolvens* and *Sarcina maxima*. A butyric acid-producing spore-forming bacterium

Table 3 Stoichiometric Yield of Dry Matter (g) and Energy (kJ), and Free Energy of the Reactions ($\Delta G^{o'}$) From Glucose or Lactate to Various Fermentation Products, and Expected and Maximum ATP Yields.

	Recovery (%) as acids		$\Delta G^{o'}$ (kJ mol^{-1} substrate)[a]	*ATP produced*[b] (as mol mol^{-1} substrate)	
	wt. basis	*energy basis*		*expected*	*maximum*
1. Glucose + 4 H$_2$O → 2 acetate$^-$ + 2 HCO$_3^-$ + 4 H$^+$ + 4 H$_2$	67	62	−206.3	4	6.4
2. Glucose → 4/3 propionate$^-$ + 2/3 acetate$^-$ + 2/3 HCO$_3^-$ + 8/3 H$^+$	55, 77[c]	73, 93[c]	−308.0	4	9.6
3. Glucose + 0.6 H$_2$O → 0.7 butyrate$^-$ + 0.6 acetate$^-$ + 1.3 H$^+$ + 2 CO$_2$ + 2.6 H	34, 74[c]	53, 73[c]	−249.5	3.3	7.7
4. Glucose + 2H$_2$O → butyrate$^-$ + 2 HCO$_3^-$ + 2 H$_2$ + 3 H$^+$	49	78	−254.8	3	7.9
5. Lactate → 2/3 propionate$^-$ + 1/3 acetate$^-$ + 1/3 H$_2$O + 1/3 CO$_2$	55, 77[c]	75, 96[c]	−56.6	1?	1.7
6. Glucose → 2 lactate	100	97	−198.3	2	6.1
7. Glucose → 2 ethanol + 2 CO$_2$	51	97	−226.0	2	7.0
8. Glucose + H$_2$O → acetone + 4 H$_2$ + 3 CO$_2$	32	63	−189.8	2	5.9
9. Glucose → butanol + 2 CO$_2$ + H$_2$O	41	95	−280.5	2	8.7

[a] $\Delta G^{o'}$ of the reactions were calculated at pH 7 in aqueous solutions (except for CO$_2$ and H$_2$) at 1 mol kg^{-1}. ATP formed was not included in the calculations. Data were obtained from Thauer *et al.* (1977). [b] The theoretical maximum was based on a $\Delta G^{o'}$ of 32.18 kJ being required for the formation of 1 mol of ATP. [c] The first value refers to the primary acid, and the second value to the sum of the two acids, as products recovered.

able to grow at low pH (5.0–5.9) has been recorded in the early patent literature. The species was named *Clostridium saccharo-aceto-propionic-butyricum* (Beesch and Legg, 1951).

Some of the genera listed in Table 2 include strains which may prove pathogenic to man, and which may require special growth factors. Thus, their use may not be safe or practical in industrial fermentations. For example, *Treponema phagadenis* is a non-pathogenic species in a genus containing many pathogens. Furthermore, it requires animal serum for growth.

37.3 PROCESS DEVELOPMENT

Industrial production of propionic or butyric acid involves not simply their formation (at 20 to 30 g l^{-1}) from a single water-soluble carbohydrate substrate (*i.e.* glucose) in a fermenter, but rather a series of process steps. These steps include the preparation (*e.g.* hydrolysis) of the chosen substrate for fermentation, the fermentation process conditions, the separation and concentration of the mixture of acids formed in a dilute state in the fermentation solution, and their purification or further conversion to other chemicals. The several steps in the production process interact with each other. Thus, it is of limited value to examine only one step in isolation. Unfortunately, much research has been restricted to examining a single step in isolation.

Acidogenic fermentations are typically most productive if the pH is maintained between pH 6 and 7. The concentration of total acids in the fermenter is usually less than 50 g l^{-1}. Product inhibition of microbial growth occurs. Usually a mixture of acids is produced, even when pure microbial cultures are employed. The acids are hydrophilic and when they are in dilute solutions, the relative volatility of water to the acids is very low. For these reasons, it is difficult to separate the acids from fermentation solutions cheaply or with energy efficiency.

The above limitations of acidogenic fermentations are some reasons why no commercial process has yet been developed for the production of propionic or butyric acids.

37.3.1 Fermentation Processes

The purpose of this section is to review past fermentation processes developed for the production of propionic and butyric acids, to draw conclusions about the difficulties encountered in these processes, to assess current research, and to indicate possible fruitful avenues for new fermentation research. None of the processes developed has gone beyond the pilot plant stage. Thus, definitive estimates of cost of production of the acids, or of steam or electricity requirements are not provided.

37.3.1.1 *Propionic acid*

Sherman and Shaw (1923a) appear to have been the first workers to advocate industrial production of propionic acid by fermentation. They and their co-workers established some important characteristics of fermentation using *Propionibacterium* which was a slow growing species. This could be accelerated by using mixed inoculum of a lactic-acid producing organism (*e.g. Lactobacillus casei*; at about 5–10% of the inoculum) with the chosen *Propionibacterium* species (Sherman and Shaw, 1921; 1923b). Yields of 48% (w/w) propionic acid or 68% (w/w) as propionic and acetic acids were obtained from lactose using the above mixed culture (Whittier and Sherman, 1923; Sherman and Shaw, 1923c). These authors also suggested that the acids could be converted to ketones by pyrolysis, rather than being refined to pure acids. Various carbohydrate substrates were examined in subsequent work (Whittier *et al.*, 1924; Wilson *et al.*, 1930). In all cases, fermentation was slow and took about 10 days to reach completion. The yield of acids was low (45–59%) compared with the theoretical maximum yield of about 75%. A search was made for suitable cheap sources of nitrogen and vitamins for the fermentation. Wilson *et al.* (1930) and Wilson (1933) suggested using residues from acetone–butanol fermentations. Steep water (from maize grain processing) and maize gluten were found to be satisfactory N and vitamin sources which could replace yeast extract (Wood and Werkman, 1934). These two workers achieved yields of up to 74.8% (w/w) total acids, a 96% utilization of the carbohydrate source, total acid concentrations of 21.6 g l^{-1}, and a propionic to acetic ratio of 4:1 (w/w). They used calcium carbonate to maintain pH during batch fermentation. Nevertheless, fermentation still took 14 days to reach completion. Fromageot and Bost (1938) made an extensive study on maize juice and potatoes as substrates, but did not achieve satisfactory propionic acid production until they added malt to the fermentation. They also studied various processes for the separation of the acids produced from the fermentation solution. Antila and Hietaranta (1953) made quite an extensive study of inhibition of microbial growth, and showed sodium to be more inhibitory than calcium salts, and that growth of *Propionibacterium* ceased at pH 5.6.

This period of research between 1920 and 1953 led to some 17 patents on propionic acid fermentation methods. These are listed in the references (Section 37.7) and Table 6. Most of these were taken out in the US by J. M. Sherman, R. H. Shaw, J. C. Woodruff, P. W. Wilson, H. R. Stiles and co-workers before 1936 and most were assigned to the Commercial Solvents Corporation (now owned by International Minerals and Chemical Corporation) or to the Wilbur White Chemical Company, both located in the US. Essentially, these patents described methods to accelerate the fermentation, to obtain high yields of propionic acid and to define cheap and suitable sources of carbohydrate substrate, nitrogen and vitamin sources.

Sherman (1932) claimed that a mixture of strains and species of *Propionibacterium* increased the rate of production and yield of propionic acid. A 30 g l^{-1} solution of propionic acid could be obtained within 14 days. He also claimed (Sherman, 1933a) that the redox potential of the fermentation solution could be adjusted by aeration of the solution twice daily for 0.5 h. This resulted in propionic acid concentrations some 30–40% higher than those without aeration. Sherman (1933b) also developed methods to increase fermentation rate by increasing microbial biomass. These proposals were based on the retention of cells by the addition of porous inert substances, or by fermentation in a tower, column or bed, or by the use of a filter to retain cells. A fed-batch system was proposed. Stiles and Wilson (1933) obtained some of the highest acid concentrations ever recorded in work where they compared the use of molasses and starch hydrolysate as substrates. Molasses fermented slowly and yields of propionic acid from it were low. Using a two-step fermentation, consisting of a lactic acid stage and a propionic acid stage, they obtained up to 40.9 g l^{-1} propionic plus acetic acids from starch hydrolysate (supplied at 50 g l^{-1} sugars in the fermenter). A one-stage system, containing both *Lactobacillus* and *Propionibacterium*, only gave 32.7 g l^{-1} during the fermentation period of 7.5 days.

Fermentation of fructose mashes was examined by Stiles (1934). He obtained increased yields and a more rapid fermentation if mannitol-forming bacteria were added to *Propionibacterium* inoculum. This accelerated the fermentation to propionic acid in a similar manner to the accelerating effect of lactic acid bacteria on propionic acid formation from glucose mentioned earlier. Fermentations were satisfactory at pH 5.4–6.0 and at 30 °C.

Retention of bacteria in the fermenter by various systems of immobilization was proposed in several patents (*e.g.* Stiles, 1935), in addition to that of Sherman (1933b) mentioned above. Werkman *et al.* (1935) described a semi-continuous system where a sugar syrup was drawn through strata of crushed rock, sand and limestone, fermented and then drawn off again leaving the microbial biomass adhering to the strata. The limestone also acted as a pH buffer. A similar proposal was made later by Wayman *et al.* (1962), and is discussed in detail below.

After World War II, propionic acid processes were developed to utilize sulfite waste liquors from wood pulp processing (McCarthy 1947; Fortress and White, 1954; Wayman *et al.*, 1962). The examples given in the patent of Fortress and White incorporated most of the ideas developed previously, and included a semi-continuous fermentation process, and a solvent extraction separation process. Wayman *et al.* (1962) further developed a continuous process which included immobilization of *Propionibacteria* on limestone particles. The process of Wayman *et al.* (1962) was developed on a small pilot plant scale at the Columbia Cellulose Co. Ltd, Prince Rupert, British Columbia, Canada (see Martin *et al.*, 1961).

The Engineering Department of the University of British Columbia prepared preliminary design estimates for treating waste sulfite liquor from a 500 tonne d^{-1} mill based on the Wayman proposal for the Ferguson Point Pulp and Paper Company in 1966 (Engineering Dept, 1966). These proposals did not eventuate in a commercial plant. The concepts of Wayman, Martin and Graf were further investigated (Nishikawa *et al.*, 1970) and the co-production of vitamin B_{12} with propionic and acetic acids was evaluated.

Fortress and White (1954) gave four examples of batch fermentation techniques to produce propionic acid from waste sulfite liquor. Excess SO_2 was removed either by steam stripping or by making the liquor alkaline (pH 8.5–9.5) with lime and filtering off the precipitate of sulfites and sulfates before adjusting the pH of the filtrate to 7 using carbon dioxide. The strength of the liquor was adjusted so that it contained between 22.5 and 45.0 g l^{-1} sugars. Yeast extract (10 g l^{-1}) and either urea or diammonium phosphate (10 g l^{-1}) were added. Inocula of several *Propionibacterium* species were tested together with *Lactobacillus casei*. Fermentations ran for 12–16 days at 30 °C. The pH was maintained at pH 6.5–7.5 with lime or NaOH. The fermentation was terminated with H_2SO_4 so that the organic acids were in the unionized state. Products were removed from solution by solvent extraction. Yields (w/w) of acids ranged from 55–70% of the sugar substrate (67–85% of theoretical maximum). A continuous process was also developed which used three fermentation vessels connected sequentially. Retention time was 3.2 days, and yields (w/w) ranged from 57 to 62% (68–74%) of theoretical maximum.

The Wayman process (Wayman *et al.*, 1962) was a continuous process based on waste sulfite liquor in which *Propionibacterium arabinosum* was immobilized on limestone pebbles (>3 mm) in a series of vertical columns. The liquor was recycled which helped to reduce mould growth, assisted with buffering, and improved acid yield. The pH had to be maintained above pH 5.7. The fermentation was conducted at 35–38 °C. The acids were separated, after acidification with H_2SO_4, by steam distillation or by solvent extraction. Excess sulfurous acid in the feed was removed by boiling waste liquor under aeration for 50–60 min. This raised the liquor pH from pH 1.4 to about pH 2.7. The residual acid was neutralized by the limestone and the recycling fermentation liquid. Wayman *et al.* (1962) described the design of a plant using about 4000 l min^{-1} of hot waste sulfite liquor. This plant would produce approximately 50 tonnes of propionic and acetic acids each day. Separation of the acids involved extraction of them from the fermentation solution with ethyl acetate–benzene (70:30). Martin *et al.* (1961) described a small pilot plant based on the same concepts. With a retention time of 54 h, acid yields (w/w) of 65% of the original sugar were obtained (theoretical maximum yield 77%). 83% of sugars were converted, and acid concentrations of 230 mM were obtained. It appears that these workers were able to achieve good acid yields in a fermentation operating at a pH of 5.7–6.2, and that even pH values below pH 5.7 were tolerated by the immobilized non-growing bacteria. Requirements for yeast extract and vitamins were much reduced (except for the preparation of the initial inoculum) because immobilized non-growing bacteria were used in the process.

The design estimate for treating waste sulfite liquor prepared by the Engineering Department of the University of British Columbia (1966) concluded that for a capital investment of CAN $4 024 000 (at 1966 values) and an operating cost of CAN $1 883 000 per year, a revenue from sale of acetic and propionic acids of CAN $6 820 000 could be obtained. These calculations assumed a no-cost feedstock at 4.0 kl min^{-1} (or 86 tonnes d^{-1} of sugar equivalent), a 75% conversion of the sugars to propionic and acetic acids in a 2:1 ratio, an 80% recovery of propionic acid during distillation, no land or service costs, and no cooling water charges. Yields each day of about 10.58 tonnes of acetic acid and 14.40 tonnes of propionic acid were expected. Each day, an input of 40 tonnes of limestone would be required. The reasons why commercial plants have not been built by pulp mills is not known, in view of the apparently favourable return on investment.

Nishikawa *et al.* (1970) investigated the co-production of vitamin B_{12} and propionic and acetic acids using *Propionibacterium freudenreichii* instead of *P. arabinosum*. Strains of the first organism can produce vitamin B_{12} if cobalt ions are present. In this preliminary investigation, difficulties were encountered with inhibitors in the waste liquor, and after 7 days fermentation, yields of acids were only about one-quarter of those obtained by Martin *et al.* (1961). Up to 2 mg l^{-1} of vitamin B_{12} was found—about two thirds of this was cell-associated and one third in the medium.

In industrial production of vitamin B_{12} from propionibacteria, much higher concentrations of the vitamin are obtained (at least 25 to 40 mg l^{-1}; see review by Florent and Ninet, 1979).

Clausen and Gaddy (1981) recently reported the fermentation of acid hydrolysates of orchard grass by *P. acidi-propionici*. Conversion of glucose sugar was ten times faster than that of xylose. Overall, 80% conversion of sugars was obtained in continuous fermentation with a retention time of about 75 h. Fermentation was not pH-controlled. Initial pH was 6.8 and this was allowed to drop to pH 4.2. Temperature was 35 °C. Preliminary designs for a plant using 200 tonnes day^{-1} of feed were made and costs calculated. It was estimated that 32 million kg of acids could be produced annually at a concentration of 20 g l^{-1}. Cost of production was 10.8 ¢ kg^{-1} (supplied as dilute acids, 20 g l^{-1}). Separation methods such as solvent extraction were calculated to increase total costs to 30 ¢ kg^{-1}. This total refined cost for the acids was compared with market prices of 46 ¢ kg^{-1} for acetic acid and 54 ¢ kg^{-1} for propionic acid.

Humphrey (1977) advocated further process development of propionic acid fermentation because weight yields are greater than those for ethanol fermentation, and because the conversion of propionic acid by hydrogenation and dehydration to propylene gives one-third greater weight yields than the conversion of ethanol to ethylene. Research and development workers have apparently not yet responded to his call.

37.3.1.2 Butyric acid

The first and only major effort to produce butyric acid commercially by fermentation was that of Lefranc and his colleagues. His first patent (Lefranc, 1914) described the fermentation of a range of carbohydrate substrates by a mixture of two bacteria. Calcium carbonate was used to maintain pH. In 1922, Lefranc's company examined acid hydrolysis of waste cellulosic materials and their conversion to butyric acid by fermentation. The use of bacteria derived from the gut of herbivora and from compost is mentioned (Lefranc et Cie, 1923; Lefranc, 1927).

The following procedure was proposed: wood waste was milled to reduce particle size, and leached with hot or cold water to remove tannins. It was then hydrolyzed by adding two to five times its weight of an aqueous solution of 2 to 5% H_2SO_4. The paste was agitated, and steamed at 170 °C for 30 min (pressure 7 to 7.55% kg cm^{-2}). The mass was neutralized with milky $CaCO_3$ or with $Ca(OH)_2$. A sugar diffusion unit was used to extract solubilized sugars. The furfurals and resins present with the sugars were removed before fermentation by making the wort alkaline with $Ca(OH)_2$ and then removing the precipitate. Alternatively, charcoal could be used. The wort containing 80–120 g l^{-1} sugar was fermented at 40 °C for 6 to 15 days. The butyric acid organism used gave quite pure butyric acid, but yields were only 70–75% of the maximum possible. A second organism used in symbiotic relation with the first increased yields but more acetic acid was formed. Thus, 100 kg wood yielded 25 kg sugars which produced 8–9 kg butyric acid using the single organism. Using two organisms, 9–9.5 kg of butyric acid and 2 kg of acetic acid were produced.

The acid salts were concentrated by evaporating the solution under vacuum, and removing $CaSO_4$ precipitates as they formed. Sulfuric or hydrochloric acids or preferably $NaHSO_4$ (since it is a waste from nitric acid manufacture) was used to release unionized butyric and acetic acids. These were then recovered by fractional distillation. The waste cellulose was used to provide steam. The fermentation gases, CO_2 and H_2, were also collected and used in the process.

In 1923, the Société Lefranc et Cie (1925) described the conversion of calcium butyrate mixed with sand or clay to dipropyl ketone by pyrolysis at 300–400 °C. It was proposed to use the ketone as a motor fuel. Lefranc and his company, Société des Brevets Étrangers Lefrance et Cie, took out some 23 patents in France, Britain, Germany, Canada and the USA between 1914 and 1934. Many of these were concerned with ketone fuel production. Those concerned with fermentation are listed in Section 37.7 and Table 6. They are concerned with the further development of the processes described above, and their adaptation to utilization of other substrates such as whey (lactose). For example, the Société des Brevets Étrangers Lefranc et Cie (1930) obtained about 45 kg of butyric acid and 8 kg of acetic acid from 100 kg of lactose. This work emphasized the use of practical industrial substrates, cheap neutralizing chemicals, and cheap separation processes. Processes were designed to fit in with existing sugar mills and refineries and to use their waste products. The development of the Lefranc process has been described by Depasse (1926). He stated that several hundred kilograms per day of calcium butyrate crystals were being produced at the Ris Orangis factory near Paris, and that a factory was being planned to produce 15–18 tonnes day^{-1} of butyrate from an input of 80 tonnes day^{-1} of sawdust. It is not known if this factory was built. During World War II, the Lefranc process was described in several publications (Depasse, 1943; Lefranc, 1943; Depasse, 1945). Depasse (1945) included a detailed description for a process to produce 5000 l day^{-1} of ketones from beet molasses *via* the acidogenic fermentation process. The conversion of acid salts to ketones is discussed later in this chapter (see Section 37.5.1).

Except for the developmental work of Lefranc and co-workers there have been few attempts to develop butyric acid fermentation processes. Dupont (1921) described the production of butyric acid (40 kg), acetic acid (80 kg) and iodine (1.5 kg) from fermentation of 1000 kg fresh marine algae. Langwell (1932) summarized his investigations on the thermophilic fermentation of cellulosic substrates by mixed bacterial cultures to produce ethanol, acetic and butyric acids. These studies were mostly performed with pilot plant scale fermenters. He claimed to obtain yields (w/w) of 33% for acetic plus butyric acids from so-called 'cellulose'. Acetic acid yields were at least four times larger than butyric acid yields. He found that added $CaCO_3$ inhibited fermentation and attributed this inhibition to precipitation of insoluble calcium phosphates. Thus, the use of $NaHCO_3$ or of NH_3 was recommended for neutralization during fermentation. He registered about 10 patents on this work. Christensen continued Langwell's work in the USA and registered four patents (*e.g.* Christensen and McCutchan, 1932).

Yields of butyric acid from the Langwell process were too low for an economically viable process. However, his work apparently created great interest, probably because of his widespread use of quite large fermentation vats, rather than for the scientific merit of the work.

Legg and Stiles (1933) found that butyric acid fermentations could be accelerated if *Lactobacillus*, *Proteus* or *Alcaligenes* species were added to the inoculum containing the butyric acid-producing organism. This paralleled the similar finding (Sherman and Shaw, 1921) for propionic acid production, but has less biochemical logic.

Maister (1934) produced 255 g butyric acid from 1400 g molasses distillery slops plus 200 g molasses. Concentrations of 32.6 g l^{-1} butyric acid, 1.8 g l^{-1} acetic acid and 8.5 g l^{-1} lactic acid were achieved. The substrates were fermented first with yeast at pH 4.9 and at 30 °C to generate essential nutrients (*e.g.* vitamins) for the acidogenic fermentation. The pH was then raised to pH 7 with lime, the solution inoculated with butyric acid bacteria, and the temperature gradually raised to 45 °C.

Arroyo (1939) described a strain of *Clostridium saccharobutyricum* able to use a wide range of carbohydrate substrates, and produce 46% (w/w) yields of acid, 99% of which was butyric acid. Concentrations of 30 g l^{-1} butyric acid were obtained.

Beesch and Legg (1951) also described a new *Clostridium* species which gave about 50% (w/w) yields of total acids and which fermented faster than previously used butyric acid organisms. Fermentation pH was exceptionally low (pH 5.0–5.9). Molasses was used as the substrate. Acid composition was butyric acid 43%, propionic acid 12% and acetic acid 45%.

Peldán (1938) gave clear evidence that calcium was necessary to obtain good yields of butyric acid. This finding is supported by most workers, except Langwell (1932), who found $CaCO_3$ reduced acid production from straw when using thermophilic organisms.

The only comprehensive review of butyric acid fermentations and of attempts to develop an industrial process is that of Péaud-Lenoël (1952). It is surprising that there have been no definitive studies aimed at maximizing the production of butyric acid by pure cultures of species such as *Clostridium butyricum*. Recently, Sharpell and Stegmann (1982) reported preliminary findings of this type.

Arroyo (1934) appropriately summarized the patent literature on butyric acid fermentations by finding it hard to understand how patents could be obtained 'on such vague, indefinite and entirely unscientific data'. Unfortunately, he was not able to provide the data so sorely needed.

37.3.1.3 Mixed acids

There have been a number of attempts to produce mixtures of acetic, propionic and butyric acids, as opposed to production of either propionic or butyric acid. For example, Weizmann

(1945) described *Clostridium butylo-butyricum* which he claimed was able to utilize lignocellulose as well as sugars and to produce the acids in about equal proportions. The Distillers Company (1930) described a cellulolytic thermophile which produced acetic and butyric acids. Langwell (1932) also fermented lignocellulose and produced a mixture of acids, of which acetic acid was predominant. Beesch and Legg (1951) found a *Clostridium* species which fermented molasses at pH 5.0 to 5.9 and yielded 37.6 g l^{-1} total acids in the following proportions: acetic, 45%; propionic, 12%; and butyric, 43%.

Much of the recent research is also aimed at producing mixtures of acids. One reason is that this work is aimed at producing liquid fuels using derivatives of fatty acids or fatty acid salts (*e.g.* paraffins, olefins, ketones and esters). Secondly, mixed acidogenic fermentations of a wide range of waste carbohydrates are being examined by research groups primarily interested in anaerobic digestion to methane or in rumen fermentations. Good examples of this are the excellent studies of the acidogenic phase of methane production by Cohen (1982) and Zoetemeyer (1982) at the University of Amsterdam. Thirdly, many workers (Sanderson *et al.*, 1978; Playne, 1980; Datta, 1981) are using mixed cultures of bacteria to produce acids to avoid costs of sterilization of substrates, to simplify fermentation procedures and to take advantage of the ability of mixed bacterial populations to adjust their species dominance to the heterogeneous substrate supplied. Waste substrates, particularly lignocellulosic wastes, are typically very heterogeneous in their chemical and physical nature. Research groups currently working on acidogenic fermentations are listed in Table 4. Research aimed at producing acetic acid using the efficient homoacetic organism, *Clostridium thermoaceticum*, is also included. Schwartz and Keller (1982), working with this organism, have very clearly specified the requirements that have to be met for an industrial process to be viable. These requirements are largely related to making the product recovery process economically viable (*i.e.* high concentrations of unionized acid in the fermentation liquor). Their targets are a fermentation pH of 4.5 or less; productivity, 5 g l^{-1} h^{-1}; dilution rate, 0.1 h^{-1}; and an acetic acid concentration of 50 g l^{-1}. Similar values would be necessary for processes producing mixtures of acids, or propionic or butyric acid.

A description of current research and development by the Dynatech, CSIRO, Exxon, Toulouse and University of Amsterdam workers follows. The Dynatech Biorefining Project is to develop a process for the production of liquid alkane fuels, and consists of three steps. Carboxylic acids are produced from the carbohydrate substrate by non-sterile anaerobic fermentation. The acids formed are separated and concentrated by liquid–liquid extraction and are then converted by electrolytic oxidation (Kolbé electrolysis) to alkanes. Because of the long retention time employed in the fermentation, it is necessary to suppress methane and this has been done using the specific inhibitor 2-bromoethanesulfonic acid. The fermentation is conducted in a 300 l fixed bed fermenter constructed of 300 mm diameter glass tubing, 4.5 m high. It is maintained at 37–38 °C. The fermentation broth is circulated through the bed at a flow rate of about 12 l h^{-1}. Solids content of the fixed bed is initially 10% (w/w). Removal of the product acids from the fermenter is done by liquid–liquid extraction using kerosene, which extracts butyric and longer-chain acids. Trioctyl phosphine oxide (TOPO) (20% solution in kerosene) is used to extract acetic and propionic acids. The extracted acids are transferred from the kerosene into dilute NaOH solutions in a second extractor, giving a final concentration of acid salts of up to 1 M.

The acid salts are electrolytically reduced to aliphatic hydrocarbons. NaOH is concurrently reformed for recycle. The composition of products formed depends on current density, pH, concentrations and proportions of acid salts, and temperature. Olefins are produced as well as paraffins. Hydrogen is also released and could be used to produce electricity. About one-third of the electricity requirements could be met from the hydrogen produced. Research is aimed at minimizing the voltage required for electrolysis since the products formed are directly related to the current density. A design for, and an economic analysis of, a 1000 tonne d^{-1} plant are being examined. A tapered auger device has been designed for adding substrate and removing residue from the packed bed. Current cost (1982 prices) based on utility financing, including a 'reasonable' return of investment is US \$5.17 GJ^{-1} which is competitive with the cost of processes for producing liquid fuels from renewable resources.

Retention time of over 17 days for solids in the fermenter necessitates the use of a fermenter with large capacity. The large amount of platinum electrodes required is another disadvantage of the process. In addition to the publications of Sanderson *et al.* (1978) and Levy *et al.* (1981), a detailed report (Dynatech R/D Co., 1980) contains useful details of the development of the process.

The Australian CSIRO Process is being developed to produce short-chain fatty acids as chemical feedstocks and mixed ketone derivatives as liquid fuels primarily from lignocellulosic agricul-

Table 4 Current Research on Acidogenic Fermentations

Research Group and Address	Ref.	Process details	Products
Dynatech R/D Company, 99 Erie St, Cambridge, MA 02139, USA.	Sanderson *et al.* (1978), Levy *et al.* (1981)	Conversion of seaweeds by mixed bacterial culture in packed bed. Extraction of higher VFA with kerosene, and acetic and propionic with kerosene + TOPO. Kolbé electrolysis of acids to olefins + paraffins.	Paraffins, olefins, esters
CSIRO Div. Chem. Wood Tech., Private Bag 10, Clayton, Vic. 3168, Australia.	Playne (1980), Playne and Smith (1982)	Lignocellulose to acetate, propionate and butyrate by mixed bacterial cultures. Extraction of acids by coupled transport membranes. Conversion of acids to ketones by pyrolysis.	Acetic, propionic butyric acids or mixed ketones
Dept. Chem. Eng., Univ. Arkansas, Fayetteville, AR 72701, USA.	Clausen and Gaddy (1981)	Acid hydrolysate of lignocellulose fermented by *Propionibacterium* liquid–liquid extraction of acids.	Propionic, acetic acids
Lab de Genie Biochimique, Institut National des Sciences Appliquees, Ave. de Rangueil F31077, Toulouse, France.	Torre and Goma (1981), Torre *et al.* (1981)	Kinetics of acid formation by mixed cultures from soluble organic wastes and sugars.	Acetic, propionic, butyric acids
Exxon Res. Eng. Co., PO Box 45, Linden, NJ 07036, USA.	Datta (1981a), Datta (1981b)	Lignocellulose pretreated with Na_2CO_3/$Ca(OH)_2$ fermented by mixed cultures to VFA. Separation of VFA by forming exters in various ways is proposed.	Ethyl esters of acetic, propionic and butyric acids
Givauden Corp., Clifton, NJ 07014, USA.	Sharpell and Stegmann (1981)	Butyric acid from glucose by *Clostridium butyricum*.	Butyric acid
Mass. Inst. of Tech., Dept. Nutr. and Food Science, Cambridge, MA 02139, USA.	Sinskey *et al.* (1981)	Acrylic acid from propionate by *Cl. propionicum*	Acrylic acid
Mass. Inst. Tech., Dept. Nutr. Food Sci., USA	Goldberg and Cooney (1981)	Conversion of H_2 and CO_2 to VFA by mixed cultures.	Acetic, propionic, butyric acids
Rowett Research Inst., Bucksburn, Scotland, UK.	Hobson *et al.* (personal communication)	Not available yet	
Dept. Chem. Eng., Univ. of Queensland, St Lucia, Brisbane, Australia.	Sheehan (1982), Henry *et al.* (1976)	Mixed cultures fermenting pig wastes to VFA and lactic acid. Utilization of the acids by the yeast, *Candida ingens*.	Feed yeast
Div. Biological Sciences N.R.C., Canada, Ottawa, Canada.	Saddler and Khan (1979)	Conversion of cellulose to acetic acid by *Acetovibrio* sp.	Acetic acid
Mass. Inst. Tech., Dept. Nutr. Food Sci., USA	Wang *et al.* (1978)	Glucose to acetic acid by *Clostridium thermoaceticum*	Acetic acid
Union Carbide Corp, R/D Dept., South Charleston, WV 25303, USA	Schwartz and Keller (1982)	Glucose to acetic acid by *Clostridium thermoaceticum*	Acetic acid
E.I. DuPont de Nemours & Co., Inc., Central Res. and Develop. Dept., Exptl. Sta., Wilmington, DE 19898, USA	Busche *et al.* (1982)	Glucose to acetic acid by *Clostridium thermoaceticum*	Acetic acid
Cawthron Institute, Nelson, New Zealand	Kaspar and Mountfort (personal communication)	Not available yet	Acetic acid

tural feedstocks. A principal aim of the project is to produce a predominance of propionic and higher fatty acids (Playne, 1980). In addition, this process should operate even more efficiently on soluble carbohydrate waste streams.

The process has four basic steps which are: (i) pretreatment of lignocellulose to make it more fermentable; (ii) fermentation by mixed anaerobic bacterial cultures manipulated to produce predominantly propionic acid; (iii) continuous extraction of acids by a coupled-transport membrane separation system; and (iv) separation of the acids or pyrolysis of calcium salts of the acids to ketone mixtures.

An envisaged process flowsheet is shown schematically in Figure 5. Research on the process up to November 1981 has been summarized by Playne and Smith (1982a). On economic grounds, calcium hydroxide is favoured for pretreatment of lignocellulose and has been shown to increase digestibility of bagasse from 20% to 65%. Steam explosion has also been examined as a pretreatment, and although it can increase digestibility to 80%, the process may be too expensive to use to produce relatively low value products such as acetic, propionic and butyric acids, and ketone fuels.

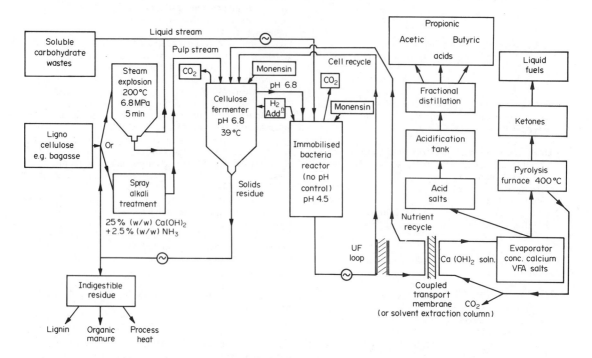

Figure 5 Preliminary process flowsheet of the CSIRO ketone fuel process (1982)

Aims of the fermentation research are to improve the cellulolytic ability of the cultures (Matei and Playne, 1984), to obtain an efficient fermentation at a lower pH (pH 5.5 instead of pH 6.8), to overcome product inhibition of microbial growth (Playne, 1981), and to increase the proportion of propionic acid produced. Separation of the acids from the fermentation solution is by coupled-transport membranes (see Section 37.3.2.7), which selectively pass unionized acids using a proton gradient generated by having a base solution of calcium hydroxide on the far side of the membrane. To increase separation efficiency, the concentration of unionized acids in the fermentation solution should be high. Thus, the aims of the fermentation step are dictated by this requirement. The calcium salts of the acids which are obtained can be converted to free acids by acidification and fractional distillation, or to ketones by pyrolysis (Lefranc, 1943; Depasse, 1926). Cost of production of ketones from sugar cane bagasse has been estimated at US $0.50 l^{-1} (1979 prices) but this could be as low as US $0.23 l^{-1} if less conservative assumptions are made about costs of membrane separation (Smith and Playne, 1982).

In the Exxon Process, lignocellulosic substrates, such as corn stover, are pretreated with alkali, using mixtures of Na_2CO_3 and $Ca(OH)_2$, and fermented with mixed cultures of bacteria. Yields greater than 50–55% (as acetic acid) have been obtained. The toxicity of various extractants for the acids have been tested, with a view to developing a liquid–liquid extraction system for recovery of the acids produced (Datta, 1981a). Datta (1981b) has proposed formation of ethyl esters of organic acids as a means to aid removal of acids from fermentation solutions. Esterification is to be performed by either chemical or biological methods yet to be developed.

At the Institut National Scientifique Appliquée, Toulouse, Torre *et al.* (1981) using sucrose as the substrate in a plug flow reactor obtained volatile fatty acid (VFA) production up to 1.7 g l^{-1} h^{-1}, a yield of 0.75 g acid/g sucrose, and a VFA concentration of 42 g l^{-1}. The plug flow design was chosen to minimize product inhibition of microbial growth. Fermentations were conducted at pH 6, and at temperatures between 30 and 40 °C. At higher temperatures (*e.g.* >40 °C) propionic acid was not produced and butyric acid became dominant (52% of total VFA). A VFA yield of 60 g l^{-1} was obtained from 150 g l^{-1} sucrose. Product inhibition was less at the higher temperature.

At the University of Amsterdam, Netherlands, Cohen and coworkers (1979, 1980, 1982) have examined the stoichiometry of acid fermentation, and manipulated fermentations to produce either propionate- or butyrate-dominant fermentations from soluble carbohydrate wastewaters using mixed bacterial cultures. Zoetemeyer and co-workers (1982a, 1982b, 1982c) have examined the effects of pH, temperature and product inhibition on the fermentations. This work is particularly relevant to the formation of butyric acid from soluble carbohydrates using mixed cultures, and provides extensive information on means to manipulate the proportions of acetic, propionic and butyric acids produced.

37.3.1.4 *Manipulation of mixed cultures*

In nature, mixed cultures of anaerobic bacteria (*e.g.* in the rumen of herbivores) efficiently convert heterogeneous cellulosic substrates to fatty acids. Acetic acid is the dominant acid formed. Production of propionic acid has been shown earlier to be advantageous in terms of yields and energy conversion (Table 3). Thus, means to increase the proportions of propionic acid produced, and to a lesser extent, butyric acid are relevant in mixed culture fermentations.

The proportions of acids can be varied in a variety of ways (*e.g.* pH, temperature, dilution rate, chemical additives and inhibitors). Fermentation control aims to either alter the dominant bacterial species present, or block or change the metabolic pathways operating in a bacterium. Propionate is a more reduced product than acetate. Thus, addition of H_2 gas to the fermenter, or deletion of alternative hydrogen 'sinks' such as methane production are obvious ways to increase the proportion of propionate produced. Van Nevel *et al.* (1974) have unequivocally shown that H_2 gas increases propionate levels, and secondly that there is an inverse relation between methane and propionate in the rumen. Henderson (1980) found that addition of H_2 gas enhanced propionate production in cultures of three propionate-producing species (*Anaerovibrio lipolytica*, *Selenomonas ruminantium* and *Megasphaera elsdenii*).

Whilst significant production of methane is unlikely in acidogenic fermentations where dilution rate is greater than $0.02\ h^{-1}$ or where pH is below 6, significant amounts are found in the Dynatech fermentation (Levy *et al.*, 1981) and in rumen fermentations (van Nevel *et al.*, 1974). Methanogenesis has been eliminated by addition of halogenated methane analogues and other polyhalogenated compounds (Chalupa, 1980), or by addition of the coenzyme M analogue, bromoethanesulfonic acid (BESA) (Levy *et al.*, 1981). The effects of many other additives on methane inhibition and propionate production have been reviewed thoroughly by Prins (1978). The antibiotic, monensin, which is an ionophore, has been used widely in rumen studies to inhibit methane and enhance propionate production. It probably acts by modulating the movement of cations across bacterial cell membranes, and by selectively inhibiting bacterial species in a mixed population (Chalupa, 1980). Selective inhibition of species has been examined by Chen and Wolin (1979), Henderson *et al.* (1981) and Dennis *et al.* (1981). Cellulolytic bacteria, except for *Bacteriodes succinogenes*, and lactate- and butyrate-producers tend to be inhibited by monensin. Hydrogen regulation is central to the control of product ratios (Demeyer and Van Nevel, 1975). This has been discussed earlier in Section 37.2.3 in relation to metabolic pathways.

37.3.2 Product Recovery and Purification of Acids

Energy budget studies and economic analyses have indicated that product separation is the most energy-intensive and expensive stage in fermentation processes for the manufacture of bulk chemicals and liquid fuels. The only exception to this generalization is the production of methane by anaerobic digestion of wastes, where the insolubility of the product causes spontaneous separation. With products such as ethanol, acetone and butanol, distillation can be used because there is sufficient difference in volatility between the product and water. This is not the case with acetic, propionic and butyric acids. The optimum pH range for most acidogenic fermentations is usually pH 6.0–7.5, which means that the acids are in the ionized state, and hence are completely non-volatile. Reduction of pH to around the pK_a values of the acids (pH 4.8) is necessary before sufficient concentrations of free unionized acids are generated. Fermentation media, especially ones which involve lignocellulosic substrates, contain microbial biomass, inorganic salts, colloids, dissolved ammonia and carbon dioxide, and other organic non-electrolytes, in addition to fatty acids. The latter are usually present in concentrations less than $50\ g\ l^{-1}$. Ideally, a process is required which separates the acids only and recycles the remaining material to earlier stages in the process. Good outlines of the problems of product recovery are given by Wang *et al.* (1981) and, for acidogenic fermentations, by Smith and Playne (1982).

37.3.2.1 *Acidification of the fermentation liquor*

The first requirement for acid recovery is to design a process where the pH of the fermentation liquor is reduced to pH 4.8 or less from the optimum fermentation pH which may frequently be near pH 7. This requirement can be met in several ways. The bacteria used can be selected for good growth at low pH (*e.g.* Schwartz and Keller, 1982), or immobilized bacteria can be used in which case growth is not as important as maintenance of metabolic rate. A two-stage fermentation system can be designed where the first stage is maintained at pH 6–7 to allow for optimal conversion of a cellulosic substrate, and a second fermentation stage which is not neutralized by alkali but allowed to self-acidify. Acidogenic fermentations acidify quite rapidly to pH 4.8–5.2 if alkalis are not added. Traditionally, addition of H_2SO_4 has been practised (*e.g.* Lefranc, 1923), but precipitates of $CaSO_4$ create disposal problems. Woodruff *et al.* (1934) advocated the use of CO_2 under pressure to acidify the fermentation solution. This idea has recently been revived (Yates, 1981; Busche *et al.*, 1982). Certainly, CO_2 is an attractive choice because the gas is readily available from nearly all fermentation processes and does not cause any disposal or immediate environmental problems. Alternatively, acidification could be achieved by addition of cation-exchange resins (in H^+ form). Regeneration of the resin could be performed externally with acids, and the resin recycled.

Acidification may be avoided if an anion-exchange system could be developed for direct removal of the carboxylate anions. Unfortunately, the capacity of anion-exchange resins is reduced rapidly by colloidal particles present in fermentation liquor. Other anions (HCO_3^-, Cl^-, SO_4^{2-}, S^{2-}, NO_3^-, HPO_4^{2-}) present are also adsorbed on to the resin. This would lead to depletion of essential nutrients for microbial growth, and competition with acid anions for exchange sites. Kawabata *et al.* (1981) have recently reported work on the use of cross-linked poly(4-vinylpyridine) as an adsorbent for butyric, propionic and acetic acids, and have made comparisons with the Amberlite resins IRA 400 and IRA 45, and the adsorbents XAD 2 and XAD 4. Anion exchange homogeneous membranes are a possible approach to overcome problems of colloid interference but would require considerable research and development.

37.3.2.2 *Solvent extraction and distillation*

Early methods for recovery of acids included evaporation of the acid salts in the fermentation liquor and their subsequent decomposition to free acids, distillation with an entrainer (such as butanol, ketones, ethyl acetate, *etc.*), or solvent extraction followed by entrainment distillation. An entrainer is included in distillation methods to avoid refluxing large volumes of water and to alter the composition of the azeotropic mixture.

Solvent extraction and distillation is only advantageous over distillation with an entrainer if the ratio of acid to water in the solvent is higher than in the aqueous feed. Solvent extraction can also act as a purification step if the solvent chosen does not absorb impurities. The solvent should also act as an entrainer in the subsequent distillation step. The above processes have been reviewed in detail by Shah and Tiwari (1978) and Eaglesfield, Kelly and Short (1953) who list many of the patents taken out in this area. Groves (1931) and Stone (1933) list some of the solvent characteristics required for solvent extraction–distillation systems.

Many of the solvent extraction–distillation schemes described in the literature only operate effectively with relatively high concentrations (20% or more) of acids from aqueous solutions and are not really applicable to concentrations of acids found in fermentation liquors. The major reason is usually that the partition coefficient diminishes as the acid concentration in the aqueous layer becomes lower. Guinot and Chassaing (1948) claim that tetrahydrofuran and its derivatives act in the reverse way and are thus potentially useful additives to extraction solvents such as benzene and toluene. Fortress *et al.* (1951) overcame the same problem in their work on extraction of propionic acid by using a solvent mixture of 70% trialkyl phosphate and 30% hydrophobic diluent (such as benzene or dichloromethane). The hydrophobic solvent and water boiled off first, the propionic acid second and the trialkyl phosphate last. Trialkyl phosphate increased the partition coefficient, but did not have to be distilled. This system enabled a larger volume of solvent to be used without increasing distillation costs.

The choice of solvent systems is very critical and has to be tailored to the acid being extracted. It is a compromise between a high partition coefficient for the acid and solubility of water in the solvent. On these grounds, dichloromethane has been claimed to be a good solvent for acid extraction. Conventional solvent extraction followed by distillation has not proved to be an economic system for acid extraction from solutions of less than 50 g l^{-1} concentration. Even complicated and modern versions such as that devised by Othmer (1958) result in costs of separation of around US $0.37 kg^{-1} (1977 prices) for 10 g l^{-1} acetic acid feeds (Helsel, 1977).

37.3.2.3 *Precipitation of acids*

Precipitation of insoluble acid salts is not commonly practised because most acid salts are soluble in water (silver salts are the exception). However, McDermott and Glasgow (1922) described a separation process for butyric acid, which involved the addition of iron(II) sulfate to precipitate proteins and microbial cells, filtration, and then addition of cupric chloride in solid form to precipitate cupric butyrate.

37.3.2.4 *'Salting out' procedures*

'Salting out' has been proposed as a separation procedure for propionic and butyric acids. Martin and Krchma (1933) described a method for separating butyric acid from acetic acid produced in the Langwell fermentation procedure. The acid salts are concentrated by evaporation to a 30% strength. The free acids are liberated using H_2SO_4. Butyric acid is salted out using a mixture of an aqueous solution (saturated with sodium sulfate) and benzene. The acetic acid remains in the aqueous phase. Calcium chloride or sodium chloride can be used instead of sulfate.

37.3.2.5 *Esterification of acids*

Esterification of the acids with ethanol or butanol has been proposed as a means to separate product acids. This process takes advantage of the lower water solubility of esters compared with that of their respective acids and their lower boiling points (see Table 5). A detailed procedure has been described by Krchma and Bloomer (1933) in which butyl esters were formed at 75 °C from butanol and acetic and butyric acids produced in a Langwell fermentation. More recently, Playne (1978) examined the thermodynamics of ester formation from ethanol and various short chain fatty acids, with a view to developing enzyme systems to carry out the esterification. The reaction thermodynamics were unfavourable unless C_5 or C_6 acids were used. These are not likely to be available in large quantities from fermentations. Datta (1981b) has proposed three approaches to ester formation: (i) adsorption of organic acids with simultaneous catalytic conversion to esters by alcohol vapours; (ii) extraction of the acids into an organic solvent phase, followed by chemical esterification in that phase with added alcohol; and (iii) enzymic esterification in dilute solutions by suitable microorganisms. These approaches have not yet been tested experimentally. Sasson *et al.* (1981) have examined the formation of butyl esters using *n*-butyl bromide by a process of phase transfer catalysis which permitted selective esterification of carboxylic acid salts.

Table 5 Boiling Points and Solubilities in Water of Acids and Their Esters

Ester	Boiling Point (°C)	Solubility in water (g l^{-1})	Acid	Boiling Point (°C)	Solubility in water (g l^{-1})
Ethyl acetate	77.2	85.0	Acetic	118	∞
Ethyl propionate	99.1	24.0	Propionic	141	∞
Ethyl butyrate	121.6	6.8	Butyric	164	∞

37.3.2.6 *Modern solvent extraction techniques*

Most of the solvents used earlier had relatively weak chemical attractions for acetic acid (*e.g.* hydrogen bonding) and small partition coefficients (≈ 1.0). Recently, solvent systems have been developed in which chemical reaction with the acid is much stronger, and association complexes are formed. Examples are tributyl phosphate (Fortress *et al.*, 1951), trioctyl phosphine oxide (TOPO; Helsel, 1977) and high molecular weight amines (Baniel *et al.*, 1976; Ricker *et al.*, 1979). In general, these compounds are solids or viscous liquids, and need to be dissolved in a suitable hydrophobic diluent. For the long-chain amines, 2-ethylhexanol seems a good choice (Ricker and King, 1978). These authors have made some preliminary cost estimates for removal of acetic acid (15 g l^{-1}) from waste streams in the USA using either amines or TOPO and have arrived at costs ranging between US $0.07 and US $0.13 kg^{-1} acetic acid (1977) prices). Helsel (1977) has calculated a cost of about US $0.14 kg^{-1} for acetic acid recovery using TOPO from 20 gl^{-1} acetic acid feeds. Thus, costs are less than half those for traditional solvent extraction. However, there is little advantage in using TOPO or amines when the concentration of the acid feed exceeds about 100 g l^{-1}.

Sanderson *et al.* (1978) have used kerosene to extract fermentation fatty acids in low concentrations. They claimed efficient extraction of butyric and longer chain acids. However, they recommended the use of TOPO–kerosene mixtures if extraction of acetic and propionic acids was required. In their system, they back-extracted the acids from kerosene using a small volume of an alkaline solution. This procedure resulted in concentration of the acids to 60–100 g l^{-1} from fermentation concentrations of 10–20 g l^{-1}. Most of the above work has been tested using relatively simple waste streams. With fermentation solutions, emulsification can be a major problem which can be overcome by ultrafiltration before solvent extraction. Special solvent extraction column designs may be required (Karr *et al.*, 1980). Recently, Urbas (1984) proposed recovery of propionic and butyric acids from dilute aqueous solutions of the acids by converting them to their calcium salts, and adding a molar equivalent of a water-soluble tertiary amine carbonate (tributylamine is preferred and the carbonate is formed *in situ* using CO_2). Trialkylammonium salts of the acids are formed, and extracted with a suitable organic solvent such as chloroform. The extracted trialkylammonium salt is heated finally to liberate the amine and acids. No examples of the efficiency of removal of propionic or butyric acid from fermentation broths were given in this patent.

37.3.2.7 *Membrane techniques*

Emulsification problems during solvent extraction of acids from fermentation solutions have led to an interest in the use of membranes for separating acids from water. The use of ultrafiltration membranes to reduce fouling and emulsification problems in solvent extraction of acids has already been mentioned. Jeffries *et al.* (1978) have suggested that ultrafiltration membranes of the sulfonic acid type with a molecular weight cut-off of about 10 000 daltons would be satisfactory and would maintain an adequate flux. This approach has been pursued by Omstead *et al.* (1980).

Membranes selective to short-chain fatty acids are being developed. These function either as water-splitting, electrodialysis membranes (Jeffries *et al.*, 1978) or as carrier-mediated (or coupled-transport) hydrophobic membranes (Smith, 1979; Smith and Playne, 1982). The advantage of the electrodialysis system is that alkali (NaOH) is generated and can be recycled into the fermenter to control pH. Secondly, since these are essentially ion-exchange membranes, the feed can consist of the sodium salts of the acids. Disadvantages of the system are that nutrient cations and anions have to be separated and recycled to the fermenter, and electrical current densities of 20–50 mA cm^{-2} have to be applied across the membrane system.

Carrier-mediated transport membranes are analogous to solvent extraction in their principle of operation. Microporous membranes were found by Smith (1979) not to be durable. Subsequently, he has worked with homogeneous membranes, which possess a semi-solid structure, and as a result have lower permeability coefficients than microporous membranes. This problem can be solved by fabricating thin membranes and if necessary providing a porous support for them. Most of Smith's work (1979) has been with poly(vinyl chloride) membranes in which acid-selective carriers such as TOPO and long-chain amines have made incorporated. The best result so far achieved has been a flux of 0.80 kg acetic acid per m^2 of membrane per day from a solution of 60 g l^{-1} acetic acid. Plasticizers (*e.g.* tributyl phosphate) to make membranes more flexible have also been investigated. However, TOPO is able to act both as a carrier and a plasticizer. A schematic diagram (Figure 6) shows the principle of operation of a carrier-mediated transport membrane. A proton gradient is maintained across the membrane by an alkaline solution (*e.g.* $Ca(OH)_2$) facilitating diffusion of acids across the membrane.

Unfortunately, there is little practical experimental data yet available for either the electrodialysis membrane or the carrier-mediated transport membrane systems. A preliminary costing of membrane separation systems (Smith, 1979, unpublished) showed that separation costs may exceed 50% of the total production cost of producing acids or derived ketones from lignocellulose. Insufficient information exists on membrane costs, acid transfer rates across membranes and the functional life of membranes.

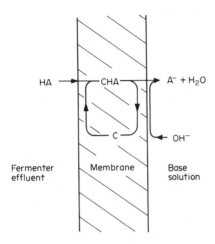

Figure 6 Carrier mediated transport membrane system. C=carrier (*e.g.* TOPO), HA=unionized acid

37.3.2.8 Toxicities of carriers and solvents

To avoid product inhibition of microbial growth, continuous removal of acids from fermenters is desirable. Economic processes to achieve continuous removal are likely to be solvent extraction systems or membrane systems. Since return of the stripped effluent to the fermenter would be desirable, the toxicity to bacteria of the extraction reagents used in these systems needs to be known. A limited amount of work has been done on toxicity by Datta (1981a) and Playne and Smith (1983). The latter authors examined 30 chemicals. Paraffins, phthalate esters, organophosphorus compounds (including TOPO) and tertiary long-chain amines were not toxic. Alcohols, ketones, aromatics, several esters and primary and secondary long-chain amines were toxic to varying degrees. These results were obtained from batch fermentation bioassays, and some adaptation by bacteria to toxic chemicals may occur in continuous fermentation systems. Toxicity effects must be considered when suitable solvent and carrier systems are being selected.

37.3.2.9 Other separation techniques

Supercritical fluid extraction with CO_2 has been examined for the separation of ethanol from fermentation solutions, and possibilities exist for separation of acetic, propionic and butyric acids by such methods (Snedekar, 1956; De Fillipi and Moses, 1982).

Adsorption on silicalite, a zeolite analogue, has been used to separate ethanol and butanol from dilute aqueous solutions (Milestone and Bibby, 1981), and there are indications that short-chain fatty acids could be separated this way. Desorption, at low energy cost, of the product from silicalite is a difficulty.

37.4 CURRENT MANUFACTURE AND USES

37.4.1 Manufacturing Processes

Most propionic and butyric acid is manufactured either by the oxo process or by liquid-phase oxidation of propane or butane. In the oxo process, an olefin is reacted with carbon monoxide and hydrogen in the presence of a catalyst. In the older, high-pressure oxo process, cobalt carbonyl was used as a catalyst at 20–30 MPa and 140–180 °C. More recently, low pressure (<2 MPa) oxo processes have been introduced using a rhodium–phosphine complex as catalyst. Such a process has been constructed by the Union Carbide Corporation in the USA and can produce 67 000 tonnes of propionic acid annually from ethylene *via* propionaldehyde. Butyric acid can be produced by a similar process from propylene through isobutyraldehyde. Liquid phase air oxidation of lower paraffin hydrocarbons at 170–208 °C and 65 atm yields mainly acetic acid, but some 8% of the yield can be propionic acid. Power costs are high, with air compression being the major cost. The process was developed by the Celanese Chemical Company, and now accounts for some 30% of propionic acid production. Different versions of it have been developed by the Celanese Chemical Company (USA), Chemische Werke Hüls (W. Germany), the Distillers Company (UK), Union Carbide Corporation (USA) and in Moscow (USSR). The Chemische Werke Hüls process has been described by Höferman (1964). Butyric acid is also prepared by oxidation of *n*-butanol.

Lurie (1964) stated that butyric acid has been produced industrially in conjunction with lactic acid by fermentation of starch and molasses. However, no details of this have been found in the literature.

37.4.2 Market Size and Prices

Reliable statistics on production of the acids, plant capacities and price trends outside the USA are not available. Production volume of butyric acid is approximately one-fifth to one-tenth that of propionic acid, but no detailed information on butyric acid is available.

In the USA, production of propionic acid rose from about 10 000 tonnes in 1958 to a current level around 50 000 tonnes (1981) annually. Plant capacities in the USA have risen from 27 000 tonnes in 1973 to a capacity in 1982 of 107 000 tonnes. Major manufacturers in that country are Union Carbide, Eastman and Celanese. Annual consumption of butyric acid in the USA reached 20 000 tonnes, but has declined as other types of thermoplastic moulding compounds have replaced cellulose acetate butyrate (Bogan *et al.*, 1979).

Current price (May 1982) in the USA for propionic acid was US $0.73 kg^{-1}, and for butyric acid, $US 1.50 kg^{-1}. Some fluctuation in the propionic acid market is predicted because the extent of agricultural applications, such as grain preservation, will vary with climatic conditions. Price quotations for the acids, their salts and for cellulose esters can be found in the Chemical Marketing Reporter (Schnell Publishing Co., New York).

37.4.3 Uses

The major uses in the US in 1981 for propionic acid were; calcium and sodium propionates (28%), grain preservation (27%), cellulose plastics (20%), herbicides (20%) and miscellaneous uses (5%) (Chem. Marketing Reporter, 1982). Calcium and sodium propionates are mainly used as antifungal agents in bread and other foods. A major shift in use pattern has occurred with grain preservation increasing from only 5% of the market in 1977. This increase has also led to some instability in the market due to climatic factors affecting grain yields and grain moisture content. In addition to these major uses, propionic acid and derivatives have been used for manufacture of antiarthritic drugs, perfumes and flavours (particularly citronellyl propionate and geranyl propionate), plasticizers (*e.g.* glycerol tripropionate, phenyl propionate), mould preventatives in silage and hays (Lacey *et al.*, 1981) and as solvents (pentyl propionate as a solvent for nitrocellulose).

The major use for butyric acid has been in the production of cellulose acetate butyrate plastics. These esters are produced with varying degrees of butyryl substitution up to about 60% substitution. These thermoplastics find a myriad of uses, especially where resistance to high temperature and sunlight is required. Esters of butyric acid find a wide use as flavours especially in the soft drink and chewing gum industries. Glycerol tributyrate and other esters are used as plasticizers. Calcium butyrate is used in leather tanning. Butyric acid derivatives are used to produce antithyroid and vasoconstrictor drugs, and in anaesthetics. A derivative has also been used as an antioxidant and stabilizer in olefin polymers.

37.5 PRODUCT DEVELOPMENT

37.5.1 Derived Liquid Fuels

New and revived uses for propionic and butyric acids will, perhaps, be influenced by the need for fossil fuel alternatives in the petrochemical and liquid fuel industries over the next 30 years.

Production of ketone fuels from fatty acid dates back to the proposals of Lefranc and colleagues in 1923 (Depasse, 1926). In France, many engine tests were conducted by several car manufacturers on ketone fuels or ketone-blended fuels (Lefranc, 1943). The patent of Baronnet *et al.* (1978) shows the continuing interest in France in the fuel properties of ketones. Other derivatives from the acids include paraffins and olefins (Sanderson *et al.*, 1978) and esters (Datta, 1981b). The success of using acidogenic fermentation to produce a liquid fuel will depend primarily on the ability of the process to utilize cheap and abundant substrates, such as lignocellulose, and on the development of an effective on-line process to extract acids from dilute solutions of fermentation liquid. Preliminary cost estimates by Sanderson *et al.* (1978), Smith and Playne (1982) and Clausen and Gaddy (1981) suggest that acidogenic fermentation processes are competitive with other proposals to produce liquid fuels from renewable resources, and indeed with many of the fossil fuel alternatives being investigated.

37.5.2 Chemical Industry Feedstocks

Propionic and butyric acids can be readily integrated into existing petrochemical industry processes as feedstocks. Humphrey (1977) has indicated the potential of propionic acid in this regard. Recently, it has been predicted that there will be shortfall in C_4 feedstocks (Anon., 1981), in which case, butyric acid would be in demand.

Recently, Japanese workers have isolated mutants of *Candida rugosa* which are able to produce β-hydroxycarboxylic acids from the corresponding carboxylic acids (Hasegawa *et al.*, 1983). This development may lead to the use of propionic and butyric acids as substrates for the production of specialized hydroxy acids.

37.5.3 Plastics

Cellulose acetate, cellulose propionate, cellulose butyrates, cellulose acetate propionates and cellulose acetate butyrates are currently used in textiles, filters, reserve osmosis membranes, plastic sheeting, film products, lacquer formulations and moulding plastics. Cellulose acetate ester has a limited compatibility with plasticizers and other resins which restricts its use to filter and textile applications. However, the higher esters, containing propionate and butyrate, are thermoplastic and this property widens the range of possible applications. However, between 1960 and 1975, there has been little expansion in the market for these higher esters (Bogan *et al.*, 1979).

Thermoplastic biodegradable polymers produced from microbial poly (β-hydroxybutyrate) are being developed and marketed as 'Biopol' by Marlborough Biopolymers, an associated company of ICI Ltd. Holmes *et al.* (1983) have patented a process to modify the crystallinity and other properties of this polymer. A copolymer is made by incorporating β-hydroxyvalerate units into the β-hydroxybutyrate chains. This can be achieved by adding propionic acid as a substrate for the microorganism (*Alcaligenes eutrophus*) which is used to form poly (β-hydroxybutyrate). Thus, propionic acid may have a future role as a feedstock for the microbial production of specialty polymers.

37.5.4 Butanol Production

Butanol production from carbohydrates by *Clostridium acetobutylicum* is enhanced by the addition of butyric acid (Gottschal and Morris, 1981; Soni *et al.*, 1982). Thus, dilute butyric acid or calcium butyrate streams may be required if the acetone–butanol fermentation industry revives.

37.5.5 Other Products

Creuze *et al.* (1941) proposed the manufacture of long chain condensed compounds ($C_7H_{14}O_5$) from the reaction of butyric acid with lactic acid, both these acids being obtained by fermentation. The condensed product would have uses as a lubricant or soap or as a fatty oil with resistance to high temperatures. The Battelle Columbus Laboratories suggested that the production of hydrogen by photoelectrolysis could be economic if the formation of oxygen in the process was prevented by the coproduction of acetic or butyric acid made from lignocellulose.

Acrylic acid may be produced from propionic acid by biochemical pathways (Sinskey *et al.*, 1981).

37.6 SUMMARY

Neither propionic nor butyric acid is currently produced commercially by fermentation. Propionic acid production by *Propionibacterium* species has been examined extensively. Yields of propionic acid generally exceed 40% (w/w) of the sugar utilized. However, for stoichiometric reasons, acetic acid is a coproduct, and yields of total acids from sugar usually range from 50–75% (w/w). Growth of *Propionibacterium* is slow. Attempts to accelerate the fermentation have included coculture of *Propionibacterium* species with *Lactobacillus* species, since lactic acid is a favoured substrate for *Propionibacterium*. Immobilization of propionibacteria on supports has also been successful. Pilot processes were developed and showed economic promise for the production of propionic acid from waste sulfite liquors. No information exists on attempts to produce propionic acid industrially from other propionic acid-producing species, such as *Veillonella* and *Selenomonas* species.

Production of butyric acid was developed by Lefranc and colleagues to a large pilot plant scale between 1918 and 1943. The main emphasis was on the conversion of the acids to ketones for use as liquid fuels. Yields of butyric acid from sugars of around 50% (w/w) are usually achieved. *Clostridium*-dominant mixed cultures have usually been used for butyric acid production. Neutralization of the fermentation has been achieved by using calcium carbonate for both propionic and butyric acid production, and maintenance of pH above pH 6 has been shown to be necessary in most cases for microbial growth. There is some evidence that calcium ions enhance butyric acid production.

Product recovery has proved expensive, but recent developments in solvent extraction and in membrane technology show promise for cheaper and more energy-efficient recovery of acids from fermentation solutions.

Mixed culture fermentations have been extensively studied in anaerobic digestion to methane and in ruminant nutrition. This knowledge is being applied in several projects aimed at producing acetic, propionic and butyric acids, and derived liquid fuels such as ketones, esters, olefins and paraffins. Manipulation of mixed culture fermentations can result in increased concentrations of propionic acid by the use of hydrogen gas and the antibiotic, monensin.

Research aimed at maximizing butyric acid production from sugars by pure cultures of bacteria is lacking, even for *Clostridium butyricum*. Good candidate species for further pure culture experiments are ones which grow well at low pH as the low pH aids product recovery. Spore-formers such as *Clostridium* species are especially useful as they can be adapted successfully to cell immobilization techniques.

ACKNOWLEDGMENTS

I thank M. Dua and J. A. Davies for assistance with literature searches; B. R. Smith for helpful discussions; C. C. Hobbs of Celanese Chemical Co. and M. C. Bachman of I. M. C. Corp. for information on commercial processes; H. Van of Chemical Marketing Reporter for information on markets and prices; R. Branion of the University of British Columbia and M. Wayman of the University of Toronto for information on pilot propionic acid processes; and C. A. Brown for typing the manuscript.

37.7 REFERENCES

Anon. (1981). The changing picture for C_4 feedstocks. *Chemical Week*, Dec. 16, 30–34.
Antila, M. and M. Hietaranta (1953). Wachstumshemmung der Propionsäure-bakterien durch Propionat. *Finnish J. Dairy Sci.*, **15**, 3–10 (*Chem. Abstr.*, **49**, 7640).
Arroyo, R. (1934). The utilization of waste molasses in the production of (I) acetone and butanol, (II) normal butyric acid. *Puerto Rico Univ. J. Agric.*, **18**, 463–479.
Arroyo, R. (1939). Fermentation process for producing butyric acid. *US Pat.* 2 181 310 (*Chem. Abstr.*, **34**, 2132).

Baniel, A. M., R. Blumberg and K. Hajdu (1976). Recovery of acids from aqueous solutions. *Br. Pat.* 1 426 018 (*Chem. Abstr.*, **80**, 132 807).

Baronnet, F., M. Niclausa, A. Ahmed, R. Vichnievsky and L. Charpenet (1978). Carburant à fort pouvoir antidetonant. *Fr. Pat.* 2 367 110 (*Chem. Abstr.*, **90**, 171 295).

Beesch, S. C. and D. A. Legg (1951). Process for production of lower aliphatic acids by fermentation. *US Pat.* 2 549 765 (*Chem. Abstr.*, **45**, 58756).

Bogan, R. T., C. M. Kuo and R. J. Brewer (1979). Cellulose derivatives, esters. In *Kirk-Othmer Encyclopaedia of Chemical Technology*, ed. M. Grayson and D. Eckroth, 3rd edn., vol. 5, pp. 118–143. Wiley, New York.

Buchanan, R. E. and N. E. Gibbons (eds.) (1974). *Bergey's Manual of Determinative Bacteriology*, 8th edn. Williams and Wilkins, Baltimore.

Busche, R. M., E. J. Shimshick and R. A. Yates (1982). Recovery of acetic acid from dilute acetate solution. *Biotechnol. Bioeng. Symp. Ser.*, **12**, 249–262.

Chalupa, W. (1980). Chemical control of rumen microbial metabolism. In *Digestive Physiology and Metabolism in Ruminants*, ed. Y. Ruckebusch and P. Thivend, pp. 325–347. MTP Press, London.

Chemical Marketing Reporter (1982). Chemical Profile: Propionic Acid. January 25 (Schnell Publishing Co. New York).

Chen, M. and M. J. Wolin (1979). Effect of monesin and lasalocid-sodium on the growth of methanogenic and rumen saccharolytic bacteria. *Appl. Environ. Microbiol.*, **38**, 72–77.

Chen, M. and M. J. Wolin (1981). Influence of heme and vitamin B_{12} on growth and fermentations of *Bacteriodes* species. *J. Bacteriol.*, **145**, 466–471.

Christensen, L. M. and W. N. McCutchan (1932). Process of producing acids by fermentation of cellulosic materials. *US Pat.* 1 875 368 (*Chem. Abstr.*, **27**, 161).

Clausen, E. C. and J. L. Gaddy (1981). Fermentation of biomass into acetic and propionic acids with *Propionibacterium acidi-propionici*. In *Advances in Biotechnology*, ed. M. Moo-Young and C. W. Robinson, vol. II, pp. 63–69. Pergamon, Toronto.

Cohen, A. (1982). Optimization of anaerobic digestion of soluble carbohydrate containing wastewaters by phase separation. Doctoral Thesis, University of Amsterdam.

Cohen, A., R. J. Zoetemeyer, A. Van Deursen and J. G. Van Andel (1979). Anaerobic digestion of glucose with separated acid production and methane formation. *Water Res.*, **13**, 571–580.

Cohen, A., A. M. Breure, J. G. Van Andel and A. Van Deursen (1980). Influence of phase separation on the anaerobic digestion of glucose. 1. Maximum COD-turnover rate during continuous operation. *Water Res.*, **14**, 1439–1448.

Cohen, A., A. M. Breure, J. G. Van Andel and A. Van Deursen (1982). Influence of phase separation on the anaerobic digestion of glucose. II. Stability and kinetic responses to shock loadings. *Water Res.*, **16**, 449–455.

Creuze, P., P.-J. Malvezin and L. Rignault (1941). Procédé d'obtention d'acides gras, saponifiés ou non, à partir de substances cellulosiques saccharifiables. *Fr. Pat.* 974 067 (*Chem. Abstr.*, **52**, 14 200).

Datta, R. (1981a). Acidogenic fermentation of corn stover. *Biotechnol. Bioeng.*, **23**, 61–77.

Datta, R. (1981b). Production of organic acid esters from biomass-novel processes and concepts. *Biotechnol. Bioeng. Symp. Ser.*, **11**, 521–532.

De Fillipi, R. P. and J. M. Moses (1982). Extraction of organics from aqueous solutions using critical fluid carbon dioxide. *Biotechnol. Bioeng. Symp. Ser.*, **12**, 205–219.

Demeyer, D. I. and C. J. Van Nevel (1975). Methanogenesis, an integrated part of carbohydrate fermentation, and its control. In *Digestion and Metabolism in the Ruminant*, ed. I. W. McDonald and A. C. I. Warner, pp. 366–382. University of New England Publishing Unit, Armidale, Australia.

Dennis, S. M., T. G. Nagaraja and E. E. Bartley (1981). Effects of lasalocid or monensin on lactate-producing or -using rumen bacteria. *J. Anim. Sci.*, **52**, 418–426.

Depasse, E. (1926). Sur un nouveau carburant: le Ketol. *Bull. Assoc. Chim. Sucr. Distill. Fr.*, **43**, 409–414 (*Chem. Abstr.*, **21**, 296).

Depasse, E. (1943). Les cétones-carburants. *Bull. Assoc. Chim. Sucr. Distill. Fr.*, **60**, 400–409 (*Chem. Abstr.*, **40**, 2603).

Depasse, E. (1945). Vue d'ensemble d'une production industrielle de cétones. *Bull. Assoc. Chim. Sucr. Distill. Fr.*, **62**, 317–339 (*Chem. Abstr.*, **40**, 6747).

Distillers Co., Ltd. (1930). Improvements in or relating to the production of aliphatic acids by fermentation. *Br. Pat.* 337 153 (*Chem. Abstr.*, **25**, 1945).

Dupont, L. (1921). Process for the utilization of marine algae for the manufacture of acetic and butyric acids. *US Pat.* 1 371 611 (*Chem. Abstr.*, **15**, 1779).

Dynatech R/D Co. (1980). Liquid fuels production from biomass. Final Report, 30 June 1980. Contract ACO2-77ET 20050 (issued by US Nat. Tech. Inf. Serv. Report No. DOE/ET/20050-T4).

Eaglesfield, P., B. K. Kelly and J. F. Short (1953). Recovery of acetic acid from dilute aqueous solutions by liquid-liquid extraction. Parts 1 and 2. *The Industrial Chemist.*, April 1953, 147–151; June 1953, 243–250.

Engineering Department, Univ. Brit. Columbia (1966). Preliminary design estimate for the proposed waste sulfite liquor sugar fermentation process (prepared by Chemical Engineering 453 Class) Report.

Fitz, A. (1879). Üeber spalt pilzgärung. *Ber.*, **12**, 474–481.

Florent, J. and L. Ninet (1979). Vitamin B_{12}. In *Microbial Technology*, ed. H. J. Peppler and D. Perlman, 2nd edn., vol. 1, pp. 497–519. Academic, New York.

Fortress, F., A. J. Rosenthal and B. B. White (1951). Concentration of organic acids. *US Pat.* 2 572 128 (*Chem. Abstr.*, **46**, 3557).

Fortress, F. and B. B. White (1954). Propionic acid from wood pulp waste liquor. *US Pat.* 2 689 817 (*Chem. Abstr.*, **48**, 14 110).

Fromageot, C. and G. Bost (1938). Production d'acide propionique par fermentation de mouts a base de mais. *Annales des Fermentations*, **4**, 463–488 (*Chem. Abstr.*, **33**, 7954).

Freudenreich, E. von and Orla Jensen (1906). Über die im Emmentalerkäse stattfindende Propionsäuregärung. *Zentralblatt fur Bakteriologie, Parasitenkunde, Infektionskrankheiten und Hygiene Abt II*, **17**, 529–546.

Goldberg, I. and C. L. Cooney (1981). Formation of short-chain fatty acids from H_2 and CO_2 by a mixed culture of bacteria. *Appl. Environ. Microbiol.*, **41**, 148–154.

Gottschal, J. C. and J. G. Morris (1981). Induction of acetone and butanol production in cultures of *Clostridium acetobutylicum* by elevated concentrations of acetate and butyrate. *FEMS Microbiol. Lett.*, **12**, 385–390.

Groves, W. W. (1931). Process of concentrating dilute aliphatic acids. *Br. Pat.* 371 554 (*Chem. Abstr.*, **27**, 2964).

Guinot, H. M. and P. Chassaing (1948). Extraction of aliphatic acids. *US Pat.* 2 437 519 (*Chem. Abstr.*, **42**, 4603).

Hasegawa, J., M. Ogura, H. Kanema, H. Kawaharada and K. Watanabe (1983). Production of D-β-hydroxycarboxylic acids from the corresponding carboxylic acids by a mutant of *Candida rugosa*. *J. Ferment. Technol.*, **61**, 37–42.

Helsel, R. W. (1977). Waste recovery: removing carboxylic acids from aqueous wastes. *Chem. Eng. Progr.*, **73** (5), 55–59.

Henderson, C. (1980). The influence of extracellular hydrogen on the metabolism of *Bacteriodes ruminicola*, *Anaerovilbrio lipolytica* and *Selenomones ruminantium*. *J. Gen. Microbiol.*, **119**, 485–491.

Henderson, C., C. S. Stewart and F. V. Nekrep (1981). The effect of monensin on pure and mixed cultures of rumen bacteria. *J. Appl. Bacteriol.*, **51**, 159–169.

Henry, D. P., R. H. Thomson, D. J. Sizemore and J. A. O'Leary (1976). Study of *Candida ingens* grown on the supernatant derived from the anaerobic fermentation of monogastric animal wastes. *Appl. Environ. Microbiol.*, **31**, 813–818.

Höferman, H. (1964). Butane oxidation for the production of aliphatic acids and solvents. *Chem. Ing. Technol.*, **36**, 422–429 (in German).

Holdeman, L. V., E. P. Cato and W. E. C. Moore (1977). *Anaerobe Laboratory Manual*, 4th edn. VPI Anaerobe Laboratory, Blacksburg, USA.

Holmes, P. A., L. F. Wright and S. H. Collins (1983). 3-Hydroxybutyrate polymers. *Eur Pat.* 069 497 A2 (*Chem. Abstr.*, **98**, 141 883).

Humphrey, A. E. (1977). Fermentation technology. *Chem. Eng. Progr.*, **73** (5), 85–91.

Jeffries, T. W., D. R. Omstead, R. R. Cardenas and H. P. Gregor (1978). Membrane-controlled digestion: effect of ultrafiltration on anaerobic digestion of glucose. *Biotechnol. Bioeng. Symp. Ser.*, **8**, 37–49.

Johns, A. T. (1952). The mechanism of propionic acid formation by *Clostridium propionicum*. *J. Gen. Microbiol.*, **6**, 123–127.

Karr, A. E., W. Gebert and M. Wang (1980). Extraction of whole fermentation broth with Karr reciprocating plate extraction column. *Can. J. Chem. Eng.*, **58**, 249–252.

Kawabata, N., J.-I. Yoshida and Y. Tanigawa (1981). Removal and recovery of organic pollutants from aquatic environments. 4. Separation of carboxylic acids from aqueous solution using crosslinked poly (4-vinylpyridine). *Ind. Eng. Chem. Prod. Res. Dev.*, **20**, 386–390.

Krchma, I. J. and W. J. Bloomer (1933). Recovery of organic acids. *US Pat.* 1 908 708 (*Chem. Abstr.*, **27**, 3723).

Lacey, J., K. A. Lord and G. R. Cayley (1981). Chemicals for preventing moulding in damp hay. *Anim. Feed Sci. Technol.*, **6**, 323–336.

Ladd, J. N. and D. J. Walker (1965). Fermentation of lactic acid by the rumen microorganism, *Peptostreptococcus elsdenii*. *Ann. N.Y. Acad. Sci.*, **119**, 1038–1047.

Langwell, H. (1932). Cellulose fermentation. *Chem. Ind.* (*London*), **51**, 988–994.

Lefranc, L. (1914). Improvements in the process for the commercial manufacture of acids of the fatty series and of butyric acid in particular. *Br. Pat.* 17 776 (*Chem. Abstr.*, **13**, 1595).

Lefranc, L. (1927). Manufacture of butyric acids and other aliphatic acids. *US Pat.* 1 625 732 (*Chem. Abstr.*, **21**, 2045).

Lefranc, L. et Cie (1923). A process for the manufacture of butyric acid and other fatty acids with recovery of the gases of fermentation. *Br. Pat.* 186 572 (*Chem. Abstr.*, **17**, 324).

Lefranc. J. (1943). Les cetones, carburant à grand pouvoir indetonant. Leur fabrication dans le cadre des matières premières nationales et 'impériales'. Conference of 28 Oct 1941, Conservatoire National des Arts et Métiers, pp. 3–27. *Actualités Scientifiques et industrielles*, **936**. Hermann et Cie, Paris (*Chem. Abstr.*, **39**, 3296).

Legg, D. A. and H. R. Stiles (1933). Process of accelerating the production of butyric acid by fermentation. *US Pat.* 1 927 813 (*Chem. Abstr.*, **27**, 5888).

Levy, P. F., J. E. Sanderson, R. G. Kispert and D. L. Wise (1981). Biorefining of biomass to liquid fuels and organic chemicals. *Enzyme Microb. Technol.*, **3**, 207–215.

Lurie, A. P. (1964). Butyric acid and butyric anhydride. In *Kirk-Othmer Encyclopaedia of Chemical Technology*, ed. A. Stander, 2nd edn., vol. 3, pp. 878–883. Wiley, New York.

McCarthy, J. L. (1947). Production of useful products by microorganisms acting upon prepared sulfite waste liquor. *US Pat.* 2 430 355 (*Chem. Abstr.*, **42**, 761).

McDermott, F. R. and R. Glasgow (1922). Manufacture of butyric acid. *US Pat.* 1 405 055 (*Chem. Abstr.*, **16**, 1635).

Maister, H. G. (1934). Process of producing calcium butyrate. *US Pat.* 1 951 250 (*Chem. Abstr.*, **28**, 3520).

Martin, J. and I. J. Krchma (1933). Method of separating and recovering acetic and butyric acids. *US Pat.* 1 917 660 (*Chem. Abstr.*, **27**, 4545).

Martin, M. E., M. Wayman and G. Graf (1961). Fermentation of sulfite waste liquor to produce organic acids. *Can. J. Microbiol.*, **7**, 341–346.

Matei, C. H. and M. J. Playne (1984). Production of volatile fatty acids from bagasse by rumen bacteria. *Appl. Microbiol. Biotechnol.*, **20**, 170–175.

Milestone, N. B. and D. M. Bibby (1981). Concentration of alcohols by adsorption on silicalite. *J. Chem. Technol. Biotechnol.*, **31**, 732–736.

Miller, T. L. and S. E. Jenesel (1979). Enzymology of butyrate formation by *Butyrivibrio fibrosolvens*. *J. Bacteriol.*, **138**, 99–104.

Nishikawa, M., R. M. R. Branion, K. L. Pinder and G. A. Strasdine (1970). Fermentation of spent sulfite liquor to produce acetic acid, propionic acid, and vitamin B_{12}. *Pulp and Paper Magazine of Canada*, **71** (3), T59–T64.

Omstead, D. R., T. W. Jeffries, R. Naughton and H. P. Gregor (1980). Membrane-controlled digestion: anaerobic production of methane and organic acids. *Biotechnol. Bioeng. Symp. Ser.*, **10**, 247–258.

Othmer, D. F. (1958). Acetic acid recovery methods. *Chem. Eng. Progr.*, **54**, (7), 48.

Pasteur, L. (1879). *Studies on Fermentation*. London.

Péaud-Lenoël, C. (1952). La Production d'acides gras volatils par fermentation. *Industries agricoles et alimentaires*, **69**, 211–220 (*Chem. Abstr.*, **47**, 6597).

Peldán, H. (1938). Glukosegärung mit Buttersäurebacillen. *Suomen Kemi.*, **11B**, 5 (*Chem. Abstr.*, **32**, 3547).

Phelps, A. S., M. J. Johnson and W. H. Peterson (1939). CXCVII. The production and utilization of lactic acid by certain propionic acid bacteria. *Biochem. J.*, **33**, 1606–1610.

Playne, M. J. (1978). The thermodynamics of ester formation from solutions of ethanol and various volatile fatty acids. CSIRO Div. Chem. Tech (Australia) Internal Report EB5.

Playne, M. J. (1980). Microbial conversion of cereal straw and bran into volatile fatty acids—key intermediates in the production of liquid fuels. *Food Technol. Aust.*, **32**, 451–456.

Playne, M. J. (1981). Volatile fatty acid production by anaerobic fermentation of ligno-cellulosic substrates. In *Advances in Biotechnology*, ed. M. Moo-Young and C. W. Robinson, vol. II, pp. 85–90. Pergamon, Toronto.

Playne, M. J. and B. R. Smith (1982). Acidogenic fermentations of wastes to produce chemicals and liquid fuels. *Proceedings of the 1st ASEAN Workshop on Fermentation Technology (Kuala Lumpur, 1982)*, pp. 474–509. (ASEAN Committee on Science and Technology).

Playne, M. J. and B. R. Smith (1983). Toxicity of organic extraction reagents to anaerobic bacteria. *Biotechnol. Bioeng.*, **25**, 1251–1265.

Prins, R. A. (1978). Nutritional impact of intestinal drug–microbe interactions. In *Nutrition and Drug Interrelations*, pp. 189–251. Academic, New York.

Ricker, N. L. and C. J. King (1978). Solvent extraction for treatment of wastewaters from acetic acid manufacture. *Water AIChE Symp. Ser. No. 178*, **74**, 204–209.

Ricker, N. L., J. N. Michaels and C. J. King (1979). Solvent properties of organic bases for extraction of acetic acid from water. *J. Sep. Proc. Technol.*, **1**, 36–41.

Saddler, J. N. and A. W. Khan (1979). Cellulose degradation by a new isolate from sewage sludge, a member of the *Bacteriodaceae* family. *Can. J. Microbiol.*, **25**, 1427–1431.

Sanderson, J. E., D. L. Wise and D. C. Augenstein (1978). Organic chemicals and liquid fuels from algal biomass. *Biotechnol. Bioeng. Symp. Ser.*, **8**, 131–151.

Sasson, Y., M. Yonovich-Weiss, and E. Grushka (1981). Catalytic extraction-phase transfer catalysis. Selective esterification of carboxylic acid salts. *Sep. Sci. Technol.*, **16**, 195–199.

Schwartz, R. D. and F. A. Keller, Jr. (1982). Isolation of a strain of *Clostridium thermoaceticum* capable of growth and acetic acid production at pH 4.5. *Appl. Environ. Microbiol.*, **43**, 117–123.

Shah, D. H. and K. K. Tiwari (1978). Recovery of acetic acid from effluents by liquid–liquid extraction. *AIChE. Symp. Ser.*, **54**, 257–266.

Sharpell, F. and C. Stegmann (1981). Development of fermentation media for the production of butyric acid. In *Advances in Biotechnology*, ed. M. Moo-Young and C. W. Robinson, vol II, pp. 71–77. Pergamon, Toronto.

Sheehan, G. J. (1982). Kinetics of heterogeneous acidogenic fermentations. PhD thesis, University of Queensland.

Sherman, J. M. (1932). Propionic acid fermentation by the use of mixed strains of propionic bacteria. *US Pat.* 1 865 146 (*Chem. Abstr.*, **26**, 4410).

Sherman, J. M. (1933a). Method of accelerating the propionic acid fermentation. *US Pat.* 1 910 130. (*Chem. Abstr.*, **27**, 4022).

Sherman, J. M. (1933b). Method of accelerating propionic fermentation. *US Pat.* 1 937 672 (*Chem. Abstr.*, **28**, 1135).

Sherman, J. M. and R. H. Shaw (1921). Associative bacterial action in the propionic acid fermentation. *J. Gen. Physiol.*, **3**, 657–658.

Sherman, J. M. and R. H. Shaw (1923a). The propionic fermentation of lactose. *J. Biol. Chem.*, **56**, 695–700.

Sherman, J. M. and R. H. Shaw (1923b). Process for the acceleration of propionic fermentation. *US Pat.* 1 459 959 (*Chem. Abstr.*, **17**, 3072).

Sherman, J. M. and R. H. Shaw (1923c). Process for the production of propionates and propionic acid. *US Pat.* 1 470 885 (*Chem. Abstr.*, **18**, 146).

Sinskey, A. J., M. Akedo and C. L. Cooney (1981). Acrylate fermentations. In *Trends in the Biology of Fermentations*, ed. A. Hollaender *et al.*, pp. 473–492. Plenum, New York.

Smith, B. R. (1979). The separation of volatile fatty acids from fermenter effluent. *Research Review 1979 (CSIRO Div. Chem. Tech., Australia)*, pp. 73–80.

Smith, B. R. and M. J. Playne (1982). Recovery of volatile fatty acids from fermenter effluents and their conversion to liquid fuels. *Proceedings of the 1st ASEAN Workshop on Fermentation Technology (Kuala Lumpur, 1982)*, pp. 510–524 (ASEAN Committee on Science and Technology).

Snedekar, R. A. (1956). Phase equilibrium in systems with supercritical carbon dioxide. PhD Thesis, Princeton University.

Société des Brevets Étrangers Lefranc et Cie (1930). Procédé de fabrication de l'acide butyrique et de ses homologues à partir des résidus lactosés de laiterie. *Fr. Pat.* 717 769 (*Chem. Abstr.*, **26**, 2821).

Société Lefranc et Cie (1925). An improved process for the manufacture of dipropylketone. *Br. Pat.* 216 120 (*Chem. Abstr.*, **19**, 77).

Soni, B. K., K. Das and T. K. Ghose (1982). Bioconversion of agro-wastes into acetone butanol. *Biotechnol. Lett.*, **4**, 19–22.

Spivey, M. J. (1978) The acetone/butanol/ethanol fermentation. *Process Biochem.*, **13** (11), 2–4, 25.

Stiles, H. R. (1934). Propionic acid fermentation of fructose-containing mashes. *US Pat.* 1 946 447 (*Chem. Abstr.*, **28**, 2462).

Stiles, H. R. (1935). Propionic acid fermentation process. *US Pat.* 2 020 251 (*Chem. Abstr.*, **30**, 567).

Stiles, H. R. and P. W. Wilson (1933). Production of propionic acid. *US Pat.* 1 932 755 (*Chem. Abstr.*, **28**, 565).

Stone, H. G. (1933). Process of dehydrating aqueous solutions containing propionic acid. *US Pat* 1 939 237 (*Chem. Abstr.*, **28**, 1363).

Strecker, A. (1854). *Ann. Chem.*, xcii, 80.

Thauer, R. K., K. Jungermann and K. Decker (1977). Energy conservation in chemotrophic anaerobic bacteria. *Bacteriol. Rev.*, **41**, 100–180.

Torre, I. de la and G. Goma (1981). Characterization of anaerobic microbial culture with high acidogenic activity. *Biotechnol. Bioeng.*, **23**, 185–199.

Torre Lozano, I. de la, I. Cernok and G. Goma (1981). Kinetic study of the organic acid production with mixed culture isolated from municipal sludge digester. *Proceedings of the 2nd Symposium on Bioconversions and Biochemical Engin-eering (New Delhi, March 1980)*, ed. T. K. Ghose, vol. II, pp. 113–131. Indian Institute of Technology, New Delhi.

Urbas, B. (1984). Recovery of organic acids from a fermentation broth. *US Pat.* 4 444 881 (*Chem. Abstr.*, **101**, 35 826).

Van Gylswyk, N. O., E. J. Morris and H. J. Els (1980). Sporulation and cell wall structure of *Clostridium polysaccharoly-ticum* comb. nov. (formerly *Fusobacterium polysaccharolyticum*). *J. Gen. Microbiol.*, **121**, 491–493.

Van Nevel, C. J., R. A. Prins and D. I. Demeyer (1974). On the inverse relationship between methane and propionate in the rumen. *Zeitschrift für Tierphysiologie, Tiernährung und Futtermittelkunde*, **33**, 121–125.

Van Niel, C. B. (1928). The propionic acid bacteria. N. V. Uitgeversaak, J. W. Boissevan and Co., Haarlem (Thesis, Delft).

Wang, D. I. C., R. J. Fleischaker, and G. Y. Wang (1978). A novel route to the production of acetic acid by fermen-tation. *Biochemical Engineering: Renewable Sources. AIChE Symp. Ser. No. 181*, **74**, 105–110.

Wang, H. Y., L. A. Kominek and J. L. Jost (1981). On-line extraction fermentation processes. In *Advances in Biotechno-logy*, ed. M. Moo-Young and C. W. Robinson, vol. 1, pp. 601–607. Pergamon, Toronto.

Wayman, M., M. E. Martin and G. Graf (1962). Propionic acid fermentation. *US Pat.* 3 067 107 (*Chem. Abstr.*, **58**, 7337).

Weizmann, C. (1945). Fermentation processes. *US Pat.* 2 386 374 (*Chem. Abstr.*, **40**, 166).

Werkman, C. H. and S. E. Kendall (1931). The propionic acid bacteria. I. Classification and nomenclature. *Iowa State Coll. J. Sci.*, **6**, 17–32.

Werkman, C. H., R. M. Hixon, E. I. Fulmer and C. H. Rayburn (1929). The production of propionic acid from pentoses by *Propionibacterium pentosaceum*. *Iowa Acad. Sci. Proc.*, **36**, 111–112.

Werkman, C. H., C. H. Rayburn and R. M. Hixon (1935). Process for producing products of fermentation. *US Pat.* 1 991 993 (*Chem. Abstr.*, **29**, 2298).

Whittier, E. O. and J. M. Sherman (1923). Propionic acid and ketones from whey. *Ind. Eng. Chem.*, **15**, 729–731.

Whittier, E. O., J. M. Sherman and W. R. Albus (1924). The rates of fermentation of sugars by the propionic organism. *Ind. Eng. Chem.*, **16**, 122.

Wilson, P. W. (1933). Propionic acid fermentation. *US Pat* 1 898 329 (*Chem. Abstr.*, **27**, 2757).

Wilson, P. W., E. B. Fred and W. H. Peterson (1930). Bildung und Identifizierung der von verschiedenen Stämmen von propionsäurebakterien gebildeten säuren. *Biochem. Ztschr.*, **229**, 271–280 (*Chem. Abstr.*, **25**, 1627).

Wocasek, J. J. (1968). Propionic acid. In *Kirk-Othmer Encyclopaedia of Chemical Technology*, ed. A. Standen, 2nd edn., vol. 16, pp. 554–558. Wiley, New York.

Wood, H. G. and C. H. Werkman (1934). The utilization of agricultural byproducts in the production of propionic acid. *J. Agric. Res.*, **49**, 1017–1024.

Woodruff, J. C., G. Bloomfield and I. J. Krchma (1934). Process for recovery of organic acids. *US Pat.* 1 946 419 (*Chem. Abstr.*, **28**, 2369).

Yates, R. A. (1981). Removal and concentration of lower molecular weight organic acids from dilute solutions. *US Pat.* 4 282 323 (*Chem. Abstr.*, **95**, 167 167).

Zoetemeyer, R. J. (1982). Acidogenesis of soluble carbohydrate-containing wastewaters. Doctoral thesis, University of Amsterdam.

Zoetemeyer, R. J., J. C. Van den Heuval and A. Cohen (1982a). pH influence on acidogenic dissimilation of glucose in an anaerobic digestor. *Water Res.*, **16**, 303–311.

Zoetemeyer, R. J., P. Arnoldy, A. Cohen and C. Boelhouwer (1982b). Influence of temperature on the anaerobic acidifi-cation of glucose in a mixed culture forming part of a two-stage digestion process. *Water Res.*, **16**, 313–321.

Zoetemeyer, R. J., A. J. C. M. Matthijsen, A. Cohen and C. Boelhouwer (1982c). Product inhibition in the acid forming stage of the anaerobic digestion process. *Water Res.*, **16**, 633–639.

ADDENDUM

A summary of additional patents on the production of propionic and butyric acids is given in the following table.

Table 6   Additional Patents on Production of Propionic and Butyric Acids by Fermentation

Inventor	Year	Patent Number	Chem. Abstr. ref.
Christensen, L.M.	1931	*Can. Pat.* 317 238	**26**, 1386
		Can. Pat. 317 239	not located
	1932	*US Pat.* 1 857 429	**26**, 3870
	1933	*US Pat.* 1 875 688	**27**, 161
		US Pat. 1 911 172	**27**, 4022
De Vecchis, I.	1933	*Fr. Pat.* 779 677	**29**, 5593
Langwell, H. (with others)	1919	*Br. Pat.* 134 265	**14**, 796
		Br. Pat. 161 294	not located
	1924	*Br. Pat.* 248 795	**21**, 796
	1925	*Br. Pat.* 271 254	**22**, 1650
	1929	*Br. Pat.* 334 900	**25**, 1329
	1923	*US Pat.* 1 443 881	**17**, 1298

Table 6 (*continued*)

Inventor	Year	Patent Number	Chem. Abstr. ref.
	1926	*US Pat.* 1 602 306	**20**, 3771
	1932	*US Pat.* 1 864 839	**26**, 1386
	1932	*US Pat.* 1 875 689	**27**, 161
	1928	*Can. Pat.* 282 039	**22**, 3485
Lefranc, J.G.A.	1934	*Fr. Pat.* 759 495	**28**, 3520
Lefrance et Cie	1922	*Ger. Pat.* 478 116	**23**, 4230
Legg, D.A. and L.M. Christensen	1932	*US Pat.* 1 864 746	**26**, 4410
Le Ketol (E. Desparmet)	1928	*Fr. Pat.* 662 321	**24**, 461
Naldi, P.	1943	*Fr. Pat.* 888 022	**47**, 3517
Société des Brevets Étrangers Lefranc et Cie	1926	*Br. Pat.* 276 617	**22**, 2378
		Fr. Pat. 633 446	**22**, 3417
	1930	*Fr. Pat.* 710 047	**26**, 1061
		Fr. Pat. 717 726	**26**, 2821
		Fr. Pat. 717 769	**26**, 2821
		Fr. Pat. 717 798	**26**, 2821
Stiles, H. R.	1931	*Can. Pat.* 317 240	**26**, 1300
	1933	*US Pat.* 1 913 346	**27**, 4342
		Br. Pat. 390 769	**27**, 4874
		Br. Pat. 396 968	**28**, 852
Synthèse et Fermentation	1942	*Fr. Pat.* 978 453	**47**, 6090
Weizmann, C. (with others)	1918	*Br. Pat.* 164 366	**16**, 460
	1938	*Br. Pat.* 489 170	**33**, 310
	1945	*Br. Pat.* 573 930	**43**, 4810
		US Pat. 2 386 374	**40**, 166
Wilson, P.W.	1930	*Can. Pat.* 302 531	**24**, 4354
Woodruff, J.C. and P.W. Wilson	1930	*Can. Pat.* 302 532	**24**, 4354
	1932	*US Pat.* 1 875 401	**27**, 161

38
Lactic Acid

T. B. VICKROY
University of California, Berkeley, CA, USA

38.1 INTRODUCTION

38.1.1 Historical Perspective

Lactic acid (2-hydroxypropanoic acid, 2-hydroxypropionic acid) is an organic hydroxy acid whose occurrence in nature is widespread. It was first produced commercially by Charles E. Avery at Littleton, Massachusetts, USA in 1881. The venture was unsuccessful in its attempts to market calcium lactate as a substitute for cream of tartar in baking powder. The first successful uses in the leather and textile industries began about 1894 (Garrett, 1930) and production levels were about 5000 kg y^{-1} on a 100% basis (Inskeep *et al.*, 1952). In 1942, about half of the 2.7×10^6 kg y^{-1} produced in the US was used by the leather industry, and an emerging use in food products consumed about 20% (Filachione, 1952). US production peaked at 4.1×10^6 kg y^{-1} during World War II and then leveled off to about 2.3×10^6 kg y^{-1}. A 90×10^6 kg y^{-1} (Needle and Aries, 1949) market for lactic acid in the plastics industry was predicted in the late 1940s and early 1950s which encouraged a large, but unsuccessful, research effort to reduce costs and increase purity. A decade later, the need for heat stable lactic acid to produce stearoyl-2-lactylates for the baking industry opened the way for a synthetic route to lactic acid (Anon., 1963). Presently, the 1982 world-wide production of lactic acid is $24–28 \times 10^6$ kg y^{-1}. More than 50% of the lactic acid produced is used in food as an acidulent and a preservative. The production of stearoyl-2-lactylates consumes another 20%. The rest of the lactic acid is used by the pharmaceutical industry or is used in numerous industrial applications. Fermentation is presently used to make about half of the world's total production.

38.1.2   Physical and Chemical Properties

Lactic acid was first isolated from sour milk by Scheele in 1780 (Lockwood *et al.*, 1965). The chemical and physical properties of lactic acid have been extensively reviewed by Holten *et al.* (1971). Lactic acid exists in two optically active isomeric forms shown in (**1**) and (**2**).

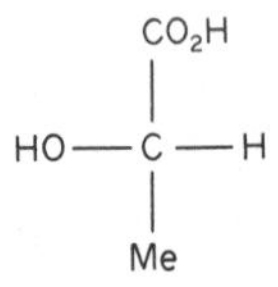

$$
\begin{array}{cc}
\mathrm{CO_2H} & \mathrm{CO_2H} \\
| & | \\
\mathrm{HO-C-H} & \mathrm{H-C-OH} \\
| & | \\
\mathrm{Me} & \mathrm{Me}
\end{array}
$$

(**1**) L(+) – lactic acid 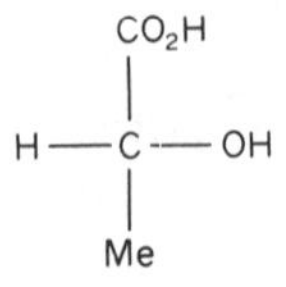   (**2**) D(−) – lactic acid

Lockwood *et al.* (1965) state that although the L(+) form appears to be dextrorotatory, it may actually be levorotatory as are its salts and esters. The apparent reversal in optical rotation may be due to the 'formation of an ethylene oxide bridge between carbon atoms 1 and 2 by tautomeric shift of the hydroxyl group on carbon atom 2 to the carbonyl group of the carboxyl radical'. Salts and esters of L(+)-lactic acid cannot form this epoxide ring and are levorotatory. The L(+) isomer (sarcolactic acid, paralactic acid) is present in humans, although both the L(+) and D(−) isomers are found in biological systems. Some properties of general interest are given in Table 1.

Table 1   Physical Properties of Lactic Acid[a]

Molecular weight	90.08
Melting point D(−) or L(+)	52.8–54 °C
DL (varies with composition)	16.8–33 °C
Boiling point DL	82 °C at 0.5 mmHg
	122 °C at 14 mmHg
Dissociation constant (K_a at 25 °C)	1.37×10^{-4}
Heat of combustion (ΔH_c)	1361 kJ mol^{-1}
Specific heat (C_p at 20 °C)	190 J mol^{-1} °C^{-1}

[a] From Holten *et al.* (1971) and Lockwood *et al.* (1965).

Optically active, high-purity lactic acid can form colorless monoclinic crystals (Lockwood *et al.*, 1965). Lactic acid is soluble in all proportions with water and exhibits a low volatility. In solutions with roughly 20% or more lactic acid, self-esterification occurs because of the hydroxyl and carboxyl functional groups. Lactic acid may form a cyclic dimer (lactide) or form linear polymers with the general formula H[OCH(CH$_3$)CO]$_n$OH. Figure 1 shows the equilibrium relationship between water, lactic acid and polymers of lactic acid. Filachione and Fisher (1944) have reviewed the preparation, properties and chemistry of these polymers.

Lactic acid may react as an organic acid as well as an organic alcohol and can participate in numerous types of chemical reactions. Holten *et al.* (1971) provide an excellent review of the versatile chemistry of lactic acid.

The analysis of lactic acid in aqueous solution has been simplified by the commercial availability of stereospecific NAD$^+$ linked lactate dehydrogenases. A procedure for this enzymatic, colorimetric assay is described by Drewes (1974). Drewes (1974) and Holten *et al.* (1971) have reviewed the techniques used for lactic acid determination. A non-enzymatic, non-stereospecific, colorimetric determination is described by Pryce (1969). Another method, which involves carefully performing a partial oxidation of lactic acid to acetaldehyde, is described by Freidmann and Graeser (1933). The acetaldehyde may then be titrated iodometrically or analyzed by gas chromatography. A simple titration of lactic acid with strong base may be used when other components in a sample will not interfere. Gas chromatography can be used; however, self-esterification generally makes the method unsatisfactory for quantitative purposes (Van Ness,

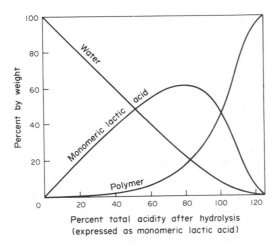

Figure 1 Water, lactic acid and poly(lactic acid) equilibrium data. Reprinted with permission from Filachione and Fischer (1944)

1981). Liquid chromatography can be used for the quantitative analysis of lactic acid (Lunder and Messori, 1979). Variations of the HPLC techniques used by Pirkle *et al.* (1981) can be used to separate the enantioners of lactic acid and lactic acid esters.

38.2 PRODUCTION TECHNOLOGY

38.2.1 Microorganisms and Raw Materials

38.2.1.1 Microorganisms

Garrett (1930) has summarized the early history of lactic acid fermentations. Pasteur discovered in 1857 that the souring of milk was caused by microorganisms. In the 1860s and 1870s the presence of lactic acid bacteria in distilleries was noted and their optimum growth temperature was investigated. Leichmann isolated a pure culture of *B. delbreuckii* from soured yeast in 1878 (Garrett, 1930). Strains of *Lactobacillus delbreuckii* which probably are quite similar to the strain that was isolated by Leichmann have often been used for the commercial production of lactic acid. Today, the strains used in industry are proprietary; however, it is believed that most of the organisms used belong to the genus *Lactobacillus*. Prescott and Dunn (1959) have given a summary of the vast number of bacteria that produce lactic acid in large quantities. The bacteria may be classified as homofermentative, producing lactic acid, cells and little else, or heterofermentative, producing lactic acid, cells and other by-products such as acetic acid, carbon dioxide, ethanol and glycerol. Only the homofermentative organisms are of industrial importance for lactic acid manufacture. The homofermentative lactic acid bacteria are from the genera *Lactobacillus*, *Streptococcus* and *Pediococcus* (Stanier *et al.*, 1976). The industrially important organisms grow optimally at temperatures above 40 °C and at a pH of 5–7. The organisms are facultative anaerobes but do not use respiration to generate ATP. Because of the high temperature, low oxygen concentration, high lactate concentration and low pH, contamination is usually not a severe problem.

The homofermentative lactic acid bacteria catabolize glucose *via* the Embden–Meyerhof pathway. Two lactic acid molecules are produced from each molecule of glucose, typically with a yield of better than 90 g per 100 g glucose. Pentose sugars are also metabolized by some homofermentative species and acetic acid and lactic acid are the products of this metabolism. Organisms may produce D(−)-, L(+)- or DL-lactic acid. The formation of the racemic mixture could arise from the action of two stereospecific lactate dehydrogenases or from one stereospecific dehydrogenase and a racemase (Gasser, 1970). The classification of many strains of lactic acid bacteria is still in much confusion at the present. Consequently, it is best to consult the literature on a specific strain or perform an analysis to determine the stereospecificity of the lactic acid it produces. The lactic acid bacteria are extremely limited in their synthetic capabilities. They always require B vitamins and almost without exception require a large number of amino acids (Stanier *et al.*, 1976). In addition, there are many growth promoting factors that have considerable effect on the fermentation rate. Koser (1968) has reviewed the nutritional requirements of lactobacilli. Ledesma *et al.* (1977) have proposed a synthetic medium for comparative nutritional studies of lactobacilli. Rees and Pirt (1979) have examined the stability of lactic acid production by non-growing cells of *L. delbreuckii*. Maintenance of the glycolytic activity in non-growing immobilized cells might lead to an increase in lactic acid yields, and reduce the amount of cell mass to be disposed.

The selection of an organism depends primarily on the carbohydrate to be fermented. Gasser (1970) has tabulated the ability of several lactobacilli to grow on different sugars. *L. bulgaricus*, *L. casei* or *S. lactis* are used to ferment lactose. Adapted strains of *L. delbreuckii* and *L. leichmannii* are typically used to ferment glucose. *L. pentosus* has been used to ferment sulfite waste liquor (Leonard *et al.*, 1948). Nakamura and Crowell (1979) have isolated a homofermentative strain called *L. amylophilus* which is capable of fermenting starch to L(+)-lactic acid with 90 wt% yields. Mixtures of strains as well as pure cultures have been used for the commercial manufacture of lactic acid (Childs and Welsby, 1966; Viniegra-Gonzalez and Gomez, 1983).

Some fungi of the species *Rhizopus*, particularly *R. oryzae*, can be used to produce L(+)-lactic acid. This organism has less complex nutritional requirements than the lactic acid bacteria. Yields and fermentation rates on glucose are cited by Snell and Lowery (1964) in their patent and are comparable to those of *Lactobacilli*. *Rhizopus* species can also utilize starch feedstocks. The process, however, never achieved lasting commercial success. The use of organisms such as *Rhizopus* with less exacting nutritional requirements could possibly reduce the feed costs and simplify the recovery process.

In general, the desirable characteristics of industrial organisms are the ability to rapidly and completely ferment cheap feedstocks with minimal amounts of nitrogenous substances. The organism is preferred to give high yields of stereospecific lactic acid under conditions of low pH and high temperature, with the production of low amounts of cell mass and negligible amounts of other by-products.

38.2.1.2 Raw materials

A large number of carbohydrate materials have been used, tested or proposed for the manufacture of lactic acid by fermentation. It is useful to compare feedstocks on the basis of the following desirable qualities: (1) low cost, (2) low levels of contaminants, (3) fast fermentation rate, (4) high lactic acid yields, (5) little or no by-product formation, (6) ability to be fermented with little or no pretreatment, and (7) year-round availability. Crude feedstocks have historically been avoided because high levels of extraneous materials can cause troublesome separation problems in the recovery stage. Use of pentose sugars results in the production of acetic acid which will incur extra process equipment for separation and could potentially become a nuisance from a marketing standpoint.

Sucrose from cane and beet sugar, whey containing lactose, and maltose and dextrose from hydrolyzed starch are presently used commercially. Refined sucrose, although expensive, is the most commonly used substrate, followed by dextrose. Not surprisingly, some lactic acid manufacturers are connected with the beet or cane sugar business. Dextrose from corn starch was the most commonly used feedstock in the late 1950s (Machell, 1959). Concentrated whey has been used without any other pretreatment by the Sheffield Product Co. of Norwich, NY, USA in a process that has remained substantially unchanged since 1936. Olive (1936) and Burton (1937) discuss some advantages and disadvantages of using whey in this process. Cordon *et al.* (1950) employed fungal amylases to hydrolyze potato starch for lactic acid production. They state that potato starch was used in Germany on an industrial scale. Cellulosic materials such as corn cobs, corn stalks, cottonseed hulls, and straw (Schopmeyer, 1954) and sulfite waste liquor (Leonard *et al.*, 1948) have been used. Acetic acid is produced as a by-product from cellulose derived sugars. Sulfite liquor also requires some pretreatment for rapid microbial growth (Schopmeyer, 1954; Leonard *et al.*, 1948). Residual lignin and the recovery of lactic acid from the crude broths present additional problems for the conversion of cellulosic materials (Leonard *et al.*, 1948). Needle and Aries (1949) have described a process utilizing molasses. Although molasses was used commercially, its complex nature introduces difficulties in the recovery stages (Machell, 1959). Schopmeyer (1954) reports that cull grapefruit and Jerusalem artichokes have also been examined as potential feedstocks. The direct fermentation of starch materials by *L. thermophillus* was reviewed by Schopmeyer (1954). Nakamura and Crowell (1979) report the fermentation of corn starch in high yields by *L. amylophilus*. Their results appear encouraging; however, a complete process has not been developed. The use of inexpensive and relatively pure corn starch could reduce the raw material cost by roughly $0.20 kg^{-1} if it were used to replace sucrose or dextrose.

Nitrogenous sources such as malt sprouts, malt extract, corn-steep liquor, barley, yeast extract or undenatured milk must supplement most carbohydrate sources to give fast and heavy growth. Some growth promoting substances in these nitrogen sources are sensitive to heat. In commercial practice, minimal amounts of the substances are used in order to simplify the recovery process.

Additional minerals are occasionally required when the carbohydrate and nitrogenous sources lack sufficient quantities. Calcium carbonate and calcium hydroxide are typically used to neutralize the acid that is formed.

38.2.2 Fermentation Processes

Batch fermentation has been the method used industrially. Fermenters have been constructed of wood or 316 stainless steel, and are equipped with heat transfer coils for temperature control. Minimal agitation is provided by top or side mounted stirrers in order to keep the contents mixed. Fermenters are typically steamed, heated with water to boiling (Inskeep *et al.*, 1952) or chemically sterilized (Burton, 1937) before filling with a pasteurized medium. Often, fermenter covers are loose fitting. Contamination is not a large problem: the most serious contamination problems are due to the growth of butyric acid bacteria at the end of the fermentation. Final product concentrations are less than 12–15% depending on the other fermentation conditions to prevent precipitation of calcium lactate (Peckham, 1944). Fermentation conditions are different for each industrial producer but are typically in the range of 45–60 °C with a pH of 5.0–6.5 for *L. delbreuckii* (Inskeep *et al.*, 1952; Peckham, 1944); 43 °C and a pH of 6–7 for *L. bulgaricus* (Burton, 1937); and 30–50 °C and a pH below 6 for *Rhizopus* (Snell and Lowery, 1964). The inoculum size is usually 5–10% of the liquid volume in the fermenter. The inoculum can be propagated in seed tanks or taken from a completed large scale fermentation. The acid formed is neutralized by calcium hydroxide or calcium carbonate. The neutralizing agent can be added in excess as a slurry at the beginning of the fermentation or added intermittently during the fermentation on the basis of pH or acid titration measurements. The fermentation time is 1–2 days for a 5% sugar source such as whey and 2–6 days for a 15% sugar source such as glucose or sucrose. Reactor productivities are in the range of 1–3 kg m^{-3} h^{-1}. Under optimal laboratory conditions the fermentation takes one to two days. The yield of lactic acid after the fermentation stage is 90–95 wt% based on the initial sugar or starch concentration. The residual sugar concentration is typically less than 0.1%. Cell mass yields can be as large as 30 wt% but generally are less than 15 wt% based on the initial sugar concentration. The yield of cell mass depends heavily on the amount of nitrogenous nutrients used. The fermentation rate depends primarily on the temperature, pH, concentration of nitrogenous nutrients, and the lactic acid concentration. A controlled pH batch fermentation will proceed quickly at first. Minimum cell mass doubling times are about one hour, but these rates are not achieved under industrial operating conditions where the amounts of nitrogenous nutrients are suboptimal. It has been noted that mixtures of strains may have a symbiotic relationship that produces faster fermentation rates than the pure isolated strains (Childs and Welsby, 1966; Viniegra-Gonzalez and Gomez, 1983). As the fermentation proceeds, the rate begins to slow because of the depletion of non-essential but stimulatory growth substances and the accumulation of lactic acid. Tsao and Hanson (1975) have modeled the effect of these stimulatory growth substances. The undissociated, electroneutral form of lactic acid rather than lactate appears to be the species which inhibits the fermentation (Viniegra-Gonzalez and Gomez, 1983; Blanch *et al.*, 1984). Mathematical models for the lactic acid fermentation have been proposed by Leudeking and Piret (1959a), Hanson and Tsao (1972), Tsao and Hanson (1975), Keller and Gerhardt (1975), Aborhey and Williamson (1977) and Samuel *et al.* (1980). These models are all based on laboratory studies in which large amounts of nitrogenous nutrients were used.

The commercial fermentation of pasteurized milk described by Burton (1937) and Olive (1936) has remained substantially unchanged except a pure culture of *L. bulgaricus* is used at the present time. The fermentation of dextrose from corn is detailed by Inskeep *et al.* (1952) and Peckham (1944). The fermentation of glucose by *Rhizopus* is partially described in a patent by Snell and Lowery (1964). Cordon *et al.* (1950) describe the fermentation of potato hydrolysate and Leonard *et al.* (1948) describe the fermentation of sulfite waste liquor. The fermentation of a crude sorghum extract by *L. plantarum* was studied by Samuel *et al.* (1980). Data from batch laboratory studies of the fermentation of *L. delbreuckii* on glucose are given by Kempe *et al.* (1950), Finn *et al.* (1950), Leudeking and Piret (1959a), Hanson and Tsao (1972) and Tsao and Hanson (1975).

38.2.2.1 *Continuous, high cell density and immobilized cell reactors*

The continuous fermentation of sweet whey has been carried out in the laboratory and at a 2 m^3 pilot plant by Whittier and Rogers (1931). The fermenter was clean but not necessarily sterile. A residence time of one day gave a 90% yield of lactic acid based on the original lactose concentration. The reactor's productivity was 2–2.5 kg lactic acid m^{-3} h^{-1}. The pH was kept at 5.0–5.8 and no contamination was noted in a two-week trial run. Olive (1936) states, however, that evaluations of the process made by the Sheffield Products Co. indicated that 'side reactions' made the process unattractive. These 'side reactions' may have been caused by contaminating organisms and might be avoided by using sterile procedures. Childs and Welsby (1966) of Bowman's Chemicals examined the continuous fermentation of a corn starch hydrolysate using *L. delbreuckii*. Based on laboratory results, they speculated that a large scale, two or three stage continuous fermentation could reduce the fermentation time to 40% of that required for a batch process. They stress that it is difficult to relate commercial practice to a number of other laboratory studies which use large amounts of nitrogenous nutrients and leave unacceptably high levels of residual sugar. Leudeking and Piret (1959b) describe some laboratory scale continuous fermentations of 5% glucose containing ample nitrogenous sources using *L. delbreuckii*. A residence time of 5.7 h resulted in a residual glucose level of less than 0.1% and a productivity of 6.7 kg lactic acid m^{-3} h^{-1}. Hanson and Tsao (1972) used this same organism for continuous studies but were unable to achieve consistent yields of lactic acid and reproducible steady states. Keller and Gerhardt (1975) performed a two-stage laboratory scale non-aseptic continuous fermentation of whey. They were able to achieve greater than 98% conversion at pH 5.5 with a total retention time of 31 h.

A batch dialysis system was used by Freidman and Gaden (1970) in which a 60% increase in the fermentation rate was achieved by reducing the lactic acid concentration. The experiments were performed at pH 5.8 using *L. delbreuckii* on a nutrient rich glucose medium. Stieber *et al.* (1977) non-aseptically fermented whey for 94 days without contamination at pH 5.3 in a continuous dialysis fermenter. A lactose consumption rate of greater than 11 kg m^{-3} h^{-1} with 97% conversion of lactose to all products was achieved. Sortland and Wilke (1969) used a rotating filter fermenter to produce lactic acid using *Streptococcus faecalus*. The reactor productivity was 15 kg m^{-3} h^{-1} with a 0.7 wt% glucose feed. Vick Roy *et al.* (1983) fermented glucose to lactic acid using *L. delbreuckii* in a continuous stirred tank reactor with cell recycle. A volumetric productivity of 76 kg m^{-3} h^{-1} was obtained with an effluent lactic acid concentration of 3.5 wt% and residual glucose levels of less than 0.002 wt%.

Lactic acid bacteria have been immobilized in gel supports by several investigators. Compere and Griffith (1976) used a 3 m high, 5 cm wide column with 0.64 cm berl saddles upon which a mixed culture of yeast and lactobacilli were immobilized in a glutaraldehyde cross-linked gelatin coating. Sour whey was fermented from 1.4% lactic acid to 2.1% lactic acid with a superficial residence time of 10–20 h. Stenroos *et al.* (1982) have immobilized *L. delbreuckii* in calcium alginate beads and used them in continuous flow column reactors. A maximum yield of 97% lactic acid from 4.8% glucose was obtained with a residence time of 18 hours. Solid calcium carbonate was used as a buffer and the effluent pH was 5.5–5.7. Vick Roy *et al.* (1982) have immobilized *L. delbreuckii* in a hollow fiber fermenter. Reactor productivities were as high as 100 kg lactic acid m^{-3} h^{-1}. Excessive growth of the organisms reduced the long term operation of the reactor system.

38.2.3 Recovery Processes

Lactic acid is sold in three major grades: technical, food (FCC) and pharmaceutical (USP). The grades are listed in order of increasing purity and more elaborate recovery processes are needed to produce the higher quality material. In addition, heat stable lactic acid, which does not discolor significantly upon heating to about 200 °C for a few hours, has a large market. The recovery of lactic acid or lactate salts from the fermentation broth is a large part of the total cost of manufacture. Synthetically made lactic acid may be purified with less effort and thus in the past has been preferred for uses where heat stability is needed.

Materials of construction for fermentation and recovery equipment are limited by the very corrosive nature of lactic acid, and contribute significantly to the product's final cost. Information on the performance of some materials is given by Peckham (1944), Inskeep *et al.* (1952), Schopmeyer (1954), Thorne (1969) and Holten *et al.* (1971). Iron, copper, copper alloys, steel, chrome steel and monel are unsatisfactory. Inconel and nickel are better but not recommended. Low iron alloys with large amounts of nickel and chromium corrode at an even slower rate. High molybdenum stainless steels such as 316 SS are satisfactory but still encounter problems, especially at improperly annealed welds and at gas/liquid interfaces where oxygen is present. Silver and tantalum are suitable, but too expensive for general use. In addition to equipment failure, corrosion increases the number of metal ions in the product which must be removed for some end uses. Wood, especially cypress and pitch pine, are satisfactory for dilute solutions but become dried out when exposed to concentrated solutions. Rubber is suitable for low temperature applications. Glass and ceramics are resistant, but their brittleness and poor heat transfer properties limit their usefulness. Some plastics are softened by warm concentrated lactic acid; however, heresite-lined, saran-lined, teflon-lined and polyester materials have been used. Plasticizers and other additives in plastic and rubber materials may be extracted or decomposed by lactic acid under some conditions (Holten *et al.*, 1971). Advances in plastic, rubber, ceramic, composite materials and metal alloys may provide some new materials choices with more attractive cost and acid resistance.

The first step in all recovery processes is to raise the fermentation liquor's temperature to 80–100 °C and increase the pH to 10 or 11. This procedure kills the organisms, coagulates the proteins, solubilizes the calcium lactate, and degrades some of the residual sugars. The liquid is then decanted or filtered. For some purposes, acidification of this liquor yields a usable product; however, for most applications further processing by one of the following methods is required. It should also be noted that use of cheap but impure raw materials must be weighed against higher purification costs.

38.2.3.1 *Filtration, carbon treatment and evaporation*

One of the methods for commercially producing lactic acid relies on the fermentation of relatively pure sugars with minimal amounts of nitrogenous nutrients. Thus, by using a pure feedstock the recovery process is simplified. Figure 2 shows the process that was used by the American Maize-Products Co. to produce lactic acid from glucose. Further details of this process are given by Inskeep *et al.* (1952) and a similar process that was used by the Clinton Company is described by Peckham (1944). The process may be used to produce technical or food grade acid. After the fermenter broth is filtered, activated vegetable carbon is used to bleach the calcium lactate for production of food grade acid. No carbon treatment is used for the technical grade. Next, the calcium lactate is evaporated to a 37% concentration at 70 °C and 0.57 atm. The concentrated lactate is then acidified with 63% sulfuric acid, and the calcium sulfate precipitate is removed by a continuous filter and sent back to the first filter which treats the fermenter liquor. The filtered acid is then treated with activated carbon from the filter cakes of the first, third and fourth carbon treatments. The carbon from this step is discarded. The lactic acid is then evaporated from 8 to 52% or 82% in 316 stainless steel evaporators. Technical grade acid is then diluted to 50% or 80% and treated with sodium sulfide to remove heavy metals if needed. Edible grade acid is diluted to 50 or 80%, bleached with activated carbon for a third time, and treated with sodium sulfide to remove heavy metals. It is then bleached a fourth time with carbon before packaging. Other manufacturers have used fewer carbon treatments. Heavy metals could be removed by ion exchange which may also remove some of the amino acids present (Machell, 1959). Heavy metals can also be removed by the stoichiometric addition of calcium or sodium ferrocyanide to form insoluble ferrocyanide salts of the heavy metals. Whether or not this process is presently used is not widely known.

38.2.3.2 *Calcium lactate crystallization*

Lactic acid may be recovered from fermentations utilizing cruder raw materials such as whey or molasses. Details and flowsheets of a process used by the Sheffield Products Co. are given by Burton (1937) and summarized by Prescott and Dunn (1959). Several grades of calcium lactate and lactic acid were produced from whey. The filtered liquor from the fermenters was treated with carbon first under slightly alkaline and then under slightly acidic conditions. The crude calcium lactate liquor was then evaporated under vacuum to a density of about 1.12 kg m^{-3}. Technical grade acid was made from this liquor after evaporation, acidification, filtration of the precipitated calcium sulfate, carbon treatment and heavy metals precipitation. To make higher grades of product the liquor was cooled, crystallized and washed. The mother liquor and wash water were also cooled, crystallized and washed. The crystals were redissolved and similarly recrystallized as in earlier steps to create purer grades. Acids of different purity were made from

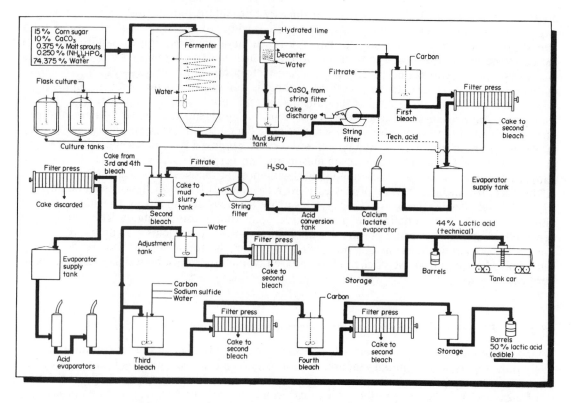

Figure 2   Lactic acid process used by the American Maize-Products Co. Reprinted with permission from Inskeep *et al.* (1952)

the different grades of crystals by dissolution in water, acidification, calcium sulfate precipitation, filtration, evaporation, carbon treatment and heavy metals precipitation. Sheffield is currently making only calcium lactate. Crystallization is presently performed in a single unit and a stainless steel double effect evaporator is used. Impurities are removed by filtration and carbon treatments. The hydration of the final product is controlled in the drying stage. C.V. Chemie Combinatie Amsterdam also uses a recovery process based on calcium lactate crystallization to recover part of their production.

Peckham (1944) describes a process for the purification of lactic acid by calcium lactate precipitation. The fermenter liquor is filtered and evaporated to 25% lactic acid. The calcium lactate is then crystallized and separated from the mother liquor. The mother liquor can be used for technical acid. The crystals are made into acid by a series of treatments similar to those used by the Sheffield Products Co.

These methods provide a product that is low in unfermented carbohydrates but may contain some ash which is mainly calcium sulfate (Schopmeyer, 1954). The crystals tend to form clusters and can be difficult to wash. The wash water and mother liquor contain high amounts of calcium lactate due to its high solubility, and must be recycled (Machell, 1959). Important costs in the purification process are related to the energy for water removal, losses in yield, and labor.

38.2.3.3   Liquid–liquid extraction

The extraction of lactic acid into an immiscible solvent phase has been researched by many investigators. Lactic acid can be purified in this way from fermentations using crude raw materials. In all such processes the acid must first be extracted from the crude liquor by the solvent, and then recovered from the solvent by some means such as back extraction into water or distillation of the solvent–lactic acid mixture.

The extraction solvent should have a low water solubility, a high distribution coefficient for lactic acid, and a low distribution coefficient for impurities such as the residual sugars. King (1980) discusses some other desirable solvent properties. The distribution coefficient is defined as the concentration of lactic acid in the solvent phase divided by the concentration of lactic acid in the water phase. Extraction data for solvents with lactic acid and water and some data for solvents with crude lactic acid fermentation liquors are given by Leonard *et al.* (1948), Ratchford *et al.* (1951), Weiser and Geankoplis (1955) and Holten *et al.* (1971). The effect of adding inorganic salts to the aqueous phase and the distribution coefficients of sucrose and lactose with several solvents was examined by Weiser and Geankoplis (1955), who advocated the use of isopentyl alcohol as a solvent. They included a flowsheet for their proposed process in which the lactic acid is recovered from the solvent by steam stripping.

Croda-Bowmans Chemicals Ltd. (England) has used a countercurrent extraction with isopropyl ether as the solvent. A description and block flow diagram of the process was published when the operation began (Anon., 1959). The fermentation liquor is filtered and then acidified with sulfuric acid; the calcium sulfate is filtered off. Next, the crude lactic acid is decolorized with activated carbon and then heavy metals, calcium and amino acids are removed by ion exchange. The acid is then evaporated under vacuum before it enters the countercurrent extraction columns. The acid is recovered from the solvent by countercurrent extraction into water. Next, the acid is given additional activated carbon and ion exchange treatments as needed. Lastly, the acid is evaporated to its final concentration. A similar process was patented by Jenemann in 1933 (Peckham, 1944). The solubility of isopropyl ether in water is low and the loss of solvent is tolerable. Croda presently uses the same process except that another solvent may now be used. Luis Ayuso S.A. (Spain) has also used a similar liquid–liquid extraction process.

Molini (1956) investigated the use of liquid–liquid extraction to remove lactic acid from a continuous fermenter, thereby increasing the reactor's productivity. The alcohols used, *n*-butyl and isopentyl, were found to be toxic to the organisms. A non-toxic solvent would simplify the proposed process.

Lactic acid refined by liquid–liquid extraction is substantially free from ash, but contains other impurities from the raw materials and needs additional treatment by activated carbon, oxidation and other means (Schopmeyer, 1954).

38.2.3.4 Distillation of lactate esters

High quality lactic acid, substantially free from residual sugars and other impurities, can be prepared by the esterification of lactic acid with a low molecular weight alcohol, distillation of the lactic ester, hydrolysis of the distilled lactate ester to yield the alcohol and lactic acid, and distillation of the alcohol from the regenerated lactic acid. A process involving methyl lactate was proposed by Hillringhaus and Heilman in 1905 and by Byk in 1912 (Schopmeyer, 1954). Smith and Claborn (1939) presented a modification of the process. Sulfuric acid was used to catalyze the reaction in an excess of methanol. An overall yield of 85% was reported. Filachione and Fisher (1946) studied the effects of the concentrations of methanol, lactic acid, and the sulfuric acid catalyst on the 'volatilization' of lactic acid. Childs and Welsby (1966) state that self-esterification and impurities from the crude fermentation liquor can pose significant problems. Dietz *et al.* (1947) esterified lactic acid to form an alkyl lactate and then extracted the ester into a solvent such as 1,2-dichloroethane. Filachione and Costello (1952) give details for a procedure to make lactic acid esters directly from ammonium lactate which is the crude fermentation product if ammonia or one of its salts is used to neutralize the acid during the fermentation. Schopmeyer (1954) discusses the operation of a commercial unit that was used to continuously refine lactic acid by the distillation of methyl lactate. The product obtained was ash free and low in all other impurities. Corrosion of the stainless steel columns contaminated the product with iron. Ceramic equipment was found to be unsuitable because of frequent temperature changes and the strong acid. Problems were also encountered with gasket materials.

A heat stable fermentation derived lactic acid is produced by C.V. Chemie Combinatie Amsterdam using a process in which a lactate ester is formed. This enables the use of fermentation derived lactic acid in the manufacture of stearoyl-2-lactylates. Synthetic lactic acid previously was the only lactic acid suitable for this large end use. The further development of this purification route could have a considerable effect on the competitive position of fermentation derived lactic acid *versus* synthetically derived lactic acid.

38.2.3.5 Other recovery processes

Lactic acid may be recovered by the adsorption of lactic acid on solid adsorbents or by the adsorption of lactate on ion exchange resins. Luis Ayuso S.A. (Spain) might be using an ion exchange or liquid–liquid extraction process to commercially recover lactic acid, but no details of the process are known. Sugimoto *et al.* (1976) patented a process for the production of L(+)-lactic acid in which strongly acidic and alkaline ion exchange resins were used to separate the acid from the broth.

Several other methods for the purification of lactic acid have been proposed but are not in use commercially. Smith and Claborn (1939) and Molini (1956) discuss several of these methods. Zinc or magnesium lactate may be recrystallized and dissolved in water. The zinc may be precipitated with hydrogen sulfide and magnesium may be precipitated as magnesium sulfate. An edible grade can be produced by partial oxidation of the crude liquor containing the free acid or lactate salts. Various distillation schemes utilizing steam, hot air, inert gases and vacuum have been unsuccessfully tried. Krumphanzel and Dyr (1964) have examined the use of electrodialysis for the continuous removal of lactic acid from fermentation. Other mobile ionic species in the fermentation broth pose a problem for this method.

38.2.4 Synthetic Manufacture

The synthetic manufacture of lactic acid on a commercial scale began around 1963 in Japan and in the United States. Today these two countries produce roughly 50% of the world's lactic acid. Synthetic lactic acid production is based on the hydrolysis of lactonitrile by a strong acid such as HCl:

$$MeCH(OH)CN + 2 H_2O + HCl \rightarrow MeCH(OH)CO_2H + NH_4Cl$$

An ammonium salt is formed as a by-product of this reaction. Lactonitrile was obtained along with acetaldehyde as a by-product of acetylene based acrylonitrile synthesis, but it is presently made from hydrogen cyanide and acetaldehyde:

$$HCN + MeCHO \rightarrow MeCH(OH)CN$$

Hydrogen cyanide may be obtained as a by-product from the propylene ammoxidation route to acrylonitrile (Thorne, 1969), or from the reaction of ammonia and light hydrocarbons such as methane. Synthetic lactic acid made substantial gains when it was introduced to the market place because the process used by-products from other synthetic routes, and perhaps more importantly, the production of stearoyl-2-lactylates required a high purity, heat-stable lactic acid. The synthetic lactic acid contains no residual sugars and does not discolor significantly upon heating. Although by-products were used in synthetic manufacture in the past, lactic acid is now made starting from acetaldehyde and hydrogen cyanide. Monsanto (USA) starts from acetaldehyde and hydrogen cyanide, whereas Mushashino (Japan) purchases the lactonitrile from another manufacturer.

The lactic acid is purified by forming methyl lactate, distilling the ester, and then hydrolyzing the methyl lactate. Methanol, hydrogen cyanide and other impurities are then removed by a combinaton of steaming, carbon treatment and ion exchange (Van Ness, 1981). The acid can also be recovered by solvent extraction (Thorne, 1969).

Many methods for the synthetic preparation of lactic acid are summarized by Holten *et al.* (1971) and Van Ness (1981). Among these methods are the alkaline degradation of sugars; the synthesis from carbon monoxide, acetaldehyde and water; the hydrolysis of chloropropionic acid; and the nitric acid oxidation of propylene. Synthetic means of preparation form a racemic DL mixture of the stereoisomers. Rhone Poulenc (France) attempted large-scale production of lactic acid *via* the oxidation of propylene with nitric acid. It is generally believed, however, that this route is no longer being pursued because of safety considerations.

38.3 ECONOMICS

38.3.1 Market Size, Manufacturers, Prices

Economic information about lactic acid is not widely available. The complicated flow of lactic acid from the manufacterers through importers, distributors and sub-distributors to the final users makes it difficult to gather precise figures and information about the uses of lactic acid. The information that follows is a combination of the knowledge shared by firms in the business (Bergman,

1983; Brown, 1983; Davis, 1983; Holstein, 1983; Marnett, 1983; Merrill, 1983; Reed, 1983; Rusch, 1983; and Wilke, 1983).

Estimates of the manufacturing capacity, production levels, and consumption of lactic acid in the major producing countries are given in Table 2. The estimates provided include lactic acid salts and derivatives as an equivalent amount of 100% lactic acid when possible. Some of the estimates obtained may not include these salts and derivatives and would therefore be underestimated. On the other hand, the figures tend to be overestimated because lactic acid is sold primarily in 88, 80 and 50% concentrations. Capacity figures are also complicated by differences in definition. For example, capacity may refer to the production possible at a given moment or to the production possible with a month or more of debottlenecking and modifications. Production figures are often proprietary and therefore many figures represent the range of estimates made by competitors. Consumption figures are probably the least reliable since it is difficult to follow the distribution of the product. Information on lactic acid production outside the free world is not well known. There is some production in communist China, and there appears to be some produced in the USSR, Czechoslovakia and perhaps some other socialist countries.

Table 2 Lactic Acid and Lactic Acid Derivative Market Estimates[a]

Country	Capacity	Production	Consumption
United States	7.7–8.5	6.3–7.7	8.6–10.0
Japan	6.6–7.5	5.0–6.1	2.3–3.6
Europe	11.5–13.5	9.3–11.1	8.2–9.5
Holland	≈6.8	≈5.7	?
England	2.0–2.7	1.3–1.8	?
Spain	2.7–4.0	2.3–3.6	?
Brazil	≈4.1	≈3.3	?
WORLD	30–33	24–28	24–28

[a] 10^6 kg y^{-1}; 1982 figures on a 100% lactic acid equivalent basis.

Table 3 lists the major manufacturers of lactic acid. Some other small producers do exist in Spain and Japan and perhaps elsewhere, but were too small to be considered here. Capacities and production levels for the major manufacturers are the figures shown in Table 2 for their respective countries. The Sheffield Products Co. is the smallest producer listed. They probably produce less than approximately 0.3×10^6 kg y^{-1} of calcium lactate from whey, although their exact production rate is not available. The largest producer is C.C.A., which controls operations in both Holland and Brazil. In addition to lactic acid, C.C.A. produces lactate salts, ester derivatives and some stearoyl-2-lactylates. Two fermentation plants, Archer Daniels Midland in Clinton, Iowa, and Cabisa Castelo Bioquimica S.A. in Brazil were shut down in the early 1980s. Fermentation presently provides 48–56% of the world's lactic acid.

Table 3 Major Lactic Acid and Lactic Acid Derivative Producers

Synthetic
  Monsanto Co., Texas City, Texas, USA
  Mushashino Chemical Laboratory Ltd., Isohara, Japan

Fermentation
  Luis Ayuso S.A., Barcelona, Spain
  Croda-Bowmans Chemicals Ltd., Cheshire, England
  C.V. Chemie Combinatie Amsterdam C.C.A., Gorinchem, Holland
  Industria Quimica de Sintesis y Fermentacoes S.A., Campos, Brazil
  Sheffield Products Co., Norwich, NY, USA

Lactic acid prices are highly variable. Table 4 shows the 1983 price range of lactic acid and some major lactic acid derivatives. List prices are always at the top of this range; however, actual selling prices are often lower.

Fermentation acid generally sells for 0.20–0.40 US $ kg^{-1} less than synthetic acid. Its price appears to be linked to the price of synthetic acid, although in recent years the price of synthetic acid has risen faster than that of fermentation derived acid.

Historically, lactic acid production in the US grew from 180 metric tons y^{-1} in 1900 to 1360 metric tons in 1940. Production peaked at 3600–4500 metric tons y^{-1} during World War II, when potassium and sodium lactate were used as a coolant for armored tanks and as a substitute for

Table 4 Lactic Acid and Lactic Acid Derivative
Prices[a]

Product	Price
Lactic acid (USP) (as 100% acid)	1.65–2.65
(FCC) (,, ,, ,,)	1.65–2.65
(tech) (,, ,, ,,)	1.50–2.40
Sodium and calcium lactate	4.40–6.20
Stearoyl-2-lactylates	≈2.20
Lactate esters	≈3.30

[a] All costs in US $ kg^{-1}; 1983 prices.

glycerol (Peckham and Filachione, 1967; Schopmeyer, 1954). Production leveled off at about 2300 metric tons y^{-1} from 1948 to 1963. With the introduction of synthetic lactic acid, and a new use in stearoyl-2-lactylates, production grew again reaching its present level around 1980. Growth has been slow in recent years, and is expected to be less than 6% for the next few years. Growth of the major end uses for lactic acid, for stearoyl-2-lactylate manufacture and as a food acidulent, is expected to be low (Anon., 1983). The price of lactic acid fell during the introduction of the synthetic route. Prices have increased steadily from 0.84 US $ kg^{-1} in 1970 to 2.42 US $ kg^{-1} in 1982 for food grade acid. The market is very competitive and volatile at present and price fluctuations make the prediction of future trends difficult.

38.3.2 Uses and Applications

An exhaustive list of the myriad uses for lactic acid is not available, and care must be taken in distinguishing former uses or potential applications as present day uses. The uses of lactic acid can be broken down by grade, and by major lactic acid derivatives.

Lactic acid is sold in technical, food and pharmaceutical grades, although most lactic acid meets the food and pharmaceutical requirements. The most common concentrations are 88 and 50% lactic acid. Lactic acid concentrations above 90% are difficult to pump and handle. The higher quality grades have lower concentrations of contaminants such as sugar, metals, chloride, sulfate and ash. Fermentation grade acid generally contains some residual sugars and other impurities from the carbohydrates and nitrogenous nutrients used in the fermentation. Fermentation lactic acid generally has a yellow color which is darkest for the concentrated technical grade and a pale yellow for the food grade. A 'colorless', heat stable, pharmaceutical grade from fermentation derived acid is made by C.C.A. The corrosiveness and liquid form of lactic acid may present a handling problem for some users. Thus, some manufacturers supply a powdered form consisting of lactic acid on a calcium lactate base for uses that require a solid form. Aqueous solutions of lactic acid are sold in plastic-lined tank cars and drums, and in plastic carboys.

Lactic acid finds medical applications as an intermediate for pharmaceutical manufacture, for adjusting the pH of preparations, and in topical wart medications. Biodegradable plastic made of poly(lactic acid) is used for sutures that do not need to be removed surgically, and has been evaluated for use as a biodegradable implant for the repair of fractures and other injuries (Kulkarni *et al.*, 1971; Wehrenberg, 1981). The largest single use of high quality 'heat stable' food or pharmaceutical grade lactic acid is for the production of stearoyl-2-lactylates. Approximately 3.6–4.0 ×10^6 kg y^{-1} and 5.0–6.0 × 10^6 kg y^{-1} of lactic acid are used for this purpose in the US and the world, respectively. Most of this acid is produced synthetically, because residual sugar from the fermentation causes a caramelization color, odor and flavor during the manufacture of the stearoyl-2-lactylates. Calcium stearoyl-2-lactylate (CSL) is used mostly in baking. CSL acts as a 'dough conditioner' by combining with the gluten in the dough, making it more tolerant to mixing and processing conditions as well as allowing a wider variation of bread ingredients. It also acts as a 'crumb softener' by complexing the starch in the flour which produces baked products with a softer texture. Sodium stearoyl-2-lactylate (SSL) behaves similarly to CSL and also acts as an emulsifier as well. Both CSL and SSL help extend the shelf life of baked products. Stearoyl-2-lactylates are also used as starch conditioners in other food products such as dehydrated potatoes, and as emulsifiers in cosmetics and food products. Stearoyl-2-lactic acid is hard to handle and the lactylate moiety depolymerizes easily; however, a small amount is used in prepared food mixes. Lactylated fatty acid esters of mono- and di-glycerides such as glycerol lactopalmitate and glycerol lactostearate are used as emulsifiers in cake mixes, bakery products, liquid shortenings and cosmetics. Stearoyl-2-lactylates and lactylated fatty esters compete with other emulsifiers such as ethoxylated mono- and di-glycerides and succinylated monoglycerides. The principal competition in the US for stearoyl-2-lactylates for baking purposes is a combination of monoglycerides and ethoxylated monoglycerides. This combination is cheaper to use, but gives an inferior performance. In the rest of the world, diacetyltartaric acid esters appear to be used in preference to stearoyl-2-lactylates. Growth for these products has been projected to be about 2.5% (Anon., 1983).

Lactic acid is mainly used directly as a food ingredient. Perhaps more than 50% of all lactic acid is used for this purpose. Both synthetic and fermentation derived acid are used for this purpose. Lactic acid is used as a food acidulent because it naturally occurs in many foodstuffs, has a mild acid taste, and has no strong flavors or odors of its own. Lactic acid is also used as a preservative, sometimes in combination with other food acids such as propanoic and acetic acid. As a food acidulent, lactic acid experiences competition primarily from citric, acetic and phosphoric acid and to a lesser extent from malic, adipic, fumaric, propanoic, formic and tartaric acids. Lactic acid is generally more expensive to use than other food acids, but it is sometimes preferred because it adds less of its own flavor to the food. Lactic acid is used in brines for processing and packaging foods such as olives, pickles and sauerkraut. Many cheese products such as cheese spreads, cold pack cheese, process cheese and cheese food contain lactic acid. Creamy salad dressings in both liquid and powder form often contain lactic acid. Some powder mixes for preparation of dips, sour cream and cheese cake contain lactic acid. A few meat products such as salami contain lactic acid. Lactic acid's use in soft drinks has been largely replaced by citric acid, phosphoric acid and other food acids in the US. It is still found in a few formulations such as pepper-type soft drinks. Lactic acid is used directly in the production of some rye and sourdough breads. It is also used in a few bakery products for its qualities as a preservative. The use of lactic acid in jams, jellies, pie fillings and packaged pectin powders has been largely replaced by citric, malic and fumaric acid in the US, although it is still used to some extent in Europe. Lactic acid competes with phosphoric acid for use in adjusting the pH of the water used for beer brewing. It is not used significantly for this purpose in the US, but is used to some extent in Canada, Europe and South America. Lactic acid buffered with sodium lactate is used in the production of confectionery products. Some animal feeds contain lactic acid. The use of crude ammonium lactate from fermentation as a potential animal feed supplement has attracted attention in recent years (Keller and Gerhardt, 1975; Samuel *et al.*, 1980). Lactic acid is sometimes used in the production of wine. The market growth of food acidulents in general has been flat for several years and is expected to remain that way in the near future (Anon., 1983).

The technical uses for lactic acid comprise a relatively small portion of the world's production. It is used in the manufacture of cellophane to control the pH in the film coating bath. It finds some uses in plastics for the production of phenol–formaldehyde resins, and can be reacted with alcohols and acids to make polyesters which are useful as plasticizers. It is used in treating metal surfaces, the manufacture of rubber products, electrostatic painting, textile and paper printing, the 'brightening' of silk and rayon, and textile dyeing. It was once widely used for the deliming of hides and in other parts of leather manufacture; however, it appears its use has been largely replaced by sulfuric and formic acids. It has potential for use in combination with other co-polymers for the production of biodegradable plastics (Lipinsky, 1981; Wehrenberg, 1981). Its use is likely to be limited to high-value, low-volume speciality plastics such as medical applications and the controlled release of pesticides unless the price of lactic acid is considerably lowered. Lactic acid is also used for the manufacture of some herbicides, fungicides and pesticides. Stereospecificity may have some advantages here as well as in some pharmaceutical products.

The calcium salts of lactic acid are produced in a granular and powdered form. Calcium lactate trihydrate is used in pharmaceuticals primarily as a dietary calcium source and also as a blood coagulant for use in the treatment of hemorrhages and to inhibit bleeding during dental operations. Calcium lactate pentahydrate is used primarily to firm potatoes and other vegetables and fruits for packaging. Calcium lactate is also used to a small extent in baking powders. Sodium lactate is very hygroscopic and is typically sold as a 60% solution. It is used primarily as an electrolyte replenisher in intravenous feeding formulations. It is also used in the production of some antibiotics and to buffer some pharmaceutical preparations. As a humectant, it is used for the production of cosmetics, personal care products, paper and tobacco. The world production of lactic acid salts is of the order of 10^6 kg y^{-1}. Most of this is sodium lactate, followed by calcium lactate. Potassium lactate and metal salts of lactic acid are produced in much smaller quantities. Van Ness (1981) cites some applications for the metal salts of lactic acid.

The alkyl esters of lactic acid find uses in pharmaceuticals, foods and industry. The ethyl and butyl lactates are used as flavor ingredients and for solvents. Ethyl lactate is also used as a tablet lubricant, in cryogenic greases, in the manufacture of pesticides and herbicides and in the manufacture of some anti-inflamatory drugs. Isopropyl lactate is also used for pesticides and herbicides. Alkyl lactates are also used in brushing and stripping lacquers, in printing inks and as flavors.

38.4 SUMMARY

The commercial production of lactic acid *via* fermentation began in 1881 and continues today. The outlook for the production of lactic acid by fermentation has been previously viewed as unfavorable (Lockwood, 1979). Today, however, because of the rising cost of petrochemical feedstocks and the recent establishment of better commercial scale recovery methods for fermentation acid, the outlook for the process looks more favorable. In addition, many opportunities still exist to further improve the process with alternative raw materials; improved fermentation organisms, conditions and equipment; and improved recovery processes.

Historically, lactic acid started as an industrial chemical. The leather and textile industry consumed 80–90% of the 1930 US production of lactic acid. In 1942, 53% of the production was used by the leather industry, 28% was used in other technical applications and 18% was used in food products (Filachione, 1952). During World War II, lactate salts were used for armored tank coolants and as a substitute for glycerol. After World War II, uses in food grew considerably. In the early 1960s a new use in the manufacture of stearoyl-2-lactylates was formed. Today, more than half the world's production goes to food uses, about one-fifth is used for stearoyl-2-lactylate manufacture and the rest is used by the pharmaceutical industry or in technical applications.

38.5 REFERENCES

Aborhey, S. and D. Williamson (1977). Modeling of lactic acid production by *Streptococcus cremoris* HP. *J. Gen. Appl. Microbiol.*, **23**, 7–21.
Anon. (1959). Cheaper lactic acid ahead? *Chem. Eng. News*, **37**, June 15, 77.
Anon. (1963). Monsanto starts new lactic acid plant. *Chem. Eng. News*, **41** (49), 44–46.
Anon. (1983). Food additives shake-out seen coming in the 1980's as competition strengthens. *Chem. Mark. Rep.*, Feb. 14, pp. 7 and 34.
Bergman, G. (1983). Pettibone World Trade, Chicago, IL, personal communication.
Blanch, H. W., T. B. Vickroy and C. R. Wilke (1984). Growth of prokaryotic cells in hollow-fiber reactors. *Ann. N.Y. Acad. Sci.*, **434**, 373–381.
Brown, R. (1983). Ramsay Brown and Co., Montclair, CA, personal communication.
Burton, L. V. (1937). By-products of milk. *Food. Ind.*, **9**, 571–575, 617–618, 634–636.
Childs, C. G. and B. Welsby (1966). Continuous lactic fermentation. *Process Biochem.*, **1**, 441–444.
Compere, A. L. and W. L. Griffith (1976). Fermentation of waste materials to produce industrial intermediates. *Dev. Ind. Microbiol.*, **18**, 135–143.
Cordon, T. C., R. H. Treadway, M. D. Walsh and M. F. Osborne (1950). Lactic acid from potatoes. *Ind. Eng. Chem.*, **42**, 1833–1836.
Davis, R. (1983). Monsanto Co., St. Louis, MO, personal communication.
Dietz, A. A., E. D. Degering and H. H. Schopmeyer (1947). Recovery of lactic acid from dilute solutions. *Ind. Eng. Chem.*, **39**, 82–85.
Drewes, P. A. (1974). Lactic acid. In *Clinical Chemistry: Principles and Techniques*, ed. R. J. Henry, D. C. Cannona and J. W. Winkelman, 2nd edn., chap. 26, pp. 1328–1334. Harper and Row, New York.
Filachione, E. M. (1952). In *Encyclopedia of Chemical Technology*, ed. R. E. Kirk and D. F. Othmer, 1st edn., vol. 8, pp. 167–180. Wiley, New York.
Filachione, E. M. and E. J. Costello (1952). Lactic esters by reaction of ammonium lactate with alcohols. *Ind. Eng. Chem.*, **44**, 2189–2191.
Filachione, E. M. and C. H. Fischer (1944). Lactic acid condensation polymers. *Ind. Eng. Chem.*, **36**, 223–228.
Filachione, E. M. and C. H. Fischer (1946). Purification of lactic acid. *Ind. Eng. Chem.*, **38**, 228–232.
Finn, R. K., H. O. Halvorson and E. L. Piret (1950). Lactic acid fermentation rate. *Ind. Eng. Chem.*, **42**, 1857–1861.
Freidman, M. R. and E. L. Gaden (1970). Growth and acid production by *Lactobacillus* (L.) *delbreuckii* in a dialysis culture system. *Biotechnol. Bioeng.*, **12**, 961–974.
Freidmann, T. E. and J. B. Graeser (1933). The determination of lactic acid. *J. Biol. Chem.*, **100**, 291–308.
Garrett, J. F. (1930). Lactic acid. *Ind. Eng. Chem.*, **22**, 1153–1154.
Gasser, F. (1970). Electrophoretic characterization of lactate dehydrogenases in the genus *Lactobacillus*. *J. Gen. Microbiol.*, **62**, 223–239.
Hanson, T. P. and G. T. Tsao (1972). Kinetic studies of the lactic acid fermentation in batch and continuous cultures. *Biotechnol. Bioeng.*, **14**, 233–252.
Holstein, A. G. (1983). Phanstiehl Laboratories Inc., Walkegan, IL, personal communication.
Holten, C. M. (ed.) with A. Muller and D. Rehbinder (1971). *Lactic Acid.* Verlag Chemie, Weinheim/Bergstr.
Inskeep, G. C., G. G. Taylor and W. C. Breitzke (1952). Lactic acid from corn sugar. *Ind. Eng. Chem.*, **44**, 1955–1966.
Keller, A. K. and P. Gerhardt (1975). Continuous lactic acid fermentation of whey to produce a ruminant feed supplement high in crude protein. *Biotechnol. Bioeng.*, **17**, 997–1018.

Kempe, L. L., H. O. Halvorson and E. L. Piret (1950). Effect of continuously controlled pH on lactic acid fermentation. *Ind. Eng. Chem.*, **42**, 1852–1857.

King, C. J. (1980). *Separation Processes*, p. 757. McGraw-Hill, New York.

Koser, S. A. (1968). *Vitamin Requirements of Bacteria and Yeasts*, chap. 21, *Lactobacillus*, pp. 340–365. Charles E. Thomas, Springfield, IL.

Krumphanzel, V. and J. Dyr (1964). Continuous fermentation and isolation of lactic acid. In *Continuous Culture of Microorganisms, Proc. 2nd Int. Symp. Continuous Culture of Organisms, Prague, June 1962*, ed. I. Malek, K. Beran and J. Hospodka, pp. 235–244. Academic, New York.

Kulkarni, R. K., E. G. Moore, A. F. Hegyell and F. Leonard (1971). Biodegradable poly(lactic acid) polymers. *J. Biomed. Mater. Res.*, **5**, 169–181.

Ledesma, O. V., A. P. DeRuiz Holgardo, G. Oliver, G. S. DeGiori, P. Raibaud and J. V. Galpin (1977). A synthetic medium for comparative nutritional studies of *Lactobacilli*. *J. Appl. Bacteriol.*, **42**, 123–133.

Leonard, R. H., W. H. Peterson and M. J. Johnson (1948). Lactic acid from fermentation of sulfite waste liquor. *Ind. Eng. Chem.*, **40**, 57–67.

Leudeking, R. and E. L. Piret (1959a). A kinetic study of the lactic acid fermentation. *J. Biochem. Microb. Technol. Eng.*, **1**, 393–412.

Leudeking, R. and E. L. Piret (1959b). Transient and steady states in continuous fermentation, theory and experiment. *J. Biochem. Microb. Tech. Eng.*, **1**, 431–459.

Lipinsky, E. S. (1981). Chemicals from biomass: petrochemical substitution options. *Science*, **212**, 1465–1471.

Lockwood, L. B. (1979). Production of organic acids by fermentation. In *Microbial Technology*, ed. H. J. Peppler and D. Perlman, 2nd edn., vol. 1, pp. 373–376, 385. Academic, New York.

Lockwood, L. B., D. E. Yoder and M. Zienty (1965). Lactic acid. *Ann. N.Y. Acad. Sci.*, **119**, 854–867.

Lunder, T. L. and E. Messori (1979). Determination of ten organic acids by low pressure liquid chromatography. *Chromatographia*, **12**, 716–719.

Machell, G. (1959). Production and applications of lactic acid. *Ind. Eng. Chem.*, **35**, 283–290.

Marnett, L. F. (1983). C. J. Patterson Co., Kansas City, MO, personal communication.

Merrill, T. (1983). Sheffield Products Co., Norwich, NY, personal communication.

Molini, A. E. (1956). *Proposed New Process for the Manufacture of Lactic Acid*. Ph.D. Thesis, University of Michigan.

Nakamura, L. K. and C. D. Crowell (1979). *Lactobacillus amylophilus*, a new starch hydrolysing species from swine waste-corn fermentation. *Dev. Ind. Microbiol.*, **20**, 531–540.

Needle, M. C. and R. S. Aries (1949). Lactic acid and lactates. *Sugar*, **44**, 32–36.

Olive, T. R. (1936). Waste lactose, a new material for a new lactic acid process. *Chem. Met. Eng.*, **43**, 480–483.

Peckham, G. T. (1944). The commercial manufacture of lactic acid. *Chem. Eng. News*, **22**, 440.

Peckham, G. T. and E. M. Filachione (1967). In *Encyclopedia of Chemical Technology*, ed. H. F. Mark, J. J. McKetta and D. F. Othmer, 2nd ed., vol. 12, pp. 170–188. Wiley, New York.

Pirkle, W. H., J. M. Finn, J. L. Schreiner and B. C. Hamper (1981). A widely useful chiral stationary phase for the high performance liquid chromatography separation of enantiomers. *J. Am. Chem. Soc.*, **103**, 3964–3966.

Prescott, S. C. and C. G. Dunn (1959). *Industrial Microbiology*. McGraw-Hill, New York.

Pryce, J. D. (1969). A modification of the Barker–Summerson method for the determination of lactic acid. *Analyst*, **94**, 1151–1152.

Ratchford, W. P., E. H. Harris Jr., C. H. Fisher and C. O. Willtis (1951). Extraction of lactic acid from water solution. *Ind. Eng. Chem.*, **43**, 778–781.

Reed, T. (1983). Croda-Bowmans Chemicals Ltd., Cheshire, England, personal communication.

Rees, J. F. and S. J. Pirt (1979). The stability of lactic acid production in resting suspensions of *Lactobacillus delbreuckii*. *J. Chem. Technol. Biotechnol.*, **29**, 591–602.

Rusch, D. (1983). C. J. Paterson Co., Kansas City, MO, personal communication.

Samuel, W. A., Y. Y. Lee and W. B. Anthony (1980). Lactic acid fermentation of crude sorghum extract. *Biotechnol. Bioeng.*, **22**, 757–777.

Schopmeyer, H. H. (1954). Lactic acid. In *Industrial Fermentations*, ed. L. A. Underkofler and R. J. Hickey, pp. 391–419. Chemical Publishing Co., New York.

Smith, L. T. and H. V. Claborn (1939). The production of pure lactic acid. *Ind. Eng. Chem. News Ed.*, **17**, 641.

Snell, R. L. and C. E. Lowery (1964). Calcium (L+) lactic acid production. *US Pat.* 3 125 494 (March 17, 1964).

Sortland, L. D. and C. R. Wilke (1969). Growth of *Streptococcus faecalis* in dense culture. *Biotechnol. Bioeng.*, **11**, 805–841.

Stanier, R. Y., E. A. Adleburg and J. L. Ingraham (1976). *The Microbial World*, pp. 678–684. Prentice Hall, Englewood Cliffs, NJ.

Stenroos, S. L., Y. Y. Linko and P. Linko (1982). Production of L-lactic acid with immobilized *Lactobacillus delbreuckii*. *Biotechnol. Lett.*, **4**, 159–164.

Stieber, R. W., G. A. Coulman and P. Gerhardt (1977). Dialysis continuous process for ammonium-lactate fermentation of whey: experimental tests. *Appl. Environ. Microbiol.*, **34**, 733–739.

Sugimoto, S., M. Ikeda. A. Aoki and T. Nakamura (1976). *Jpn. Kokai.* 76 12 990.

Thorne, J. G. M. (1969). Synthetic lactic acid. *Chem. Process.*, **15**, 8–9.

Tsao, G. T. and T. P. Hanson (1975). Extended Monod equation for batch cultures with multiple exponential phases. *Biotechnol. Bioeng.*, **17**, 1591–1598.

Van Ness, J. H. (1981). Hydroxycarboxylic acids, lactic acid. In *Encyclopedia of Chemical Technology*, ed. H. F. Mark, D. E. Othmer, C. G. Overberger and G. T. Seaborg, 3rd edn., vol. 13, pp. 80–103. Wiley, New York.

Vick Roy, T. B., H. W. Blanch and C. R. Wilke (1982). Lactic acid production by *Lactobacillus delbreuckii* in a hollow fiber fermenter. *Biotechnol. Lett.*, **4**, 483–488.

Vick Roy, T. B., D. K. Mandel, D. K. Dea, H. W. Blanch and C. R. Wilke (1983). Lactic acid production in a CSTR with cell recycle. *Biotechnol. Lett.*, **5**, 665–670.

Viniegra-Gonzalez, G. and J. Gomez (1984). Lactic acid production by pure and mixed bacterial cultures. In *Bioconversion Systems*, ed. D. Wise, pp. 17–39. CRC Press, Boca Raton, FL.

Wehrenberg. R. H. II (1981). Lactic acid polymers: strong, degradable thermoplastics. *Mater. Eng.*, **94**, 63–66.

Weiser, R. B. and C. J. Geankoplis (1955). Lactic acid purification by extraction. *Ind. Eng. Chem.*, **47**, 858–863.

Whittier, E. O and L. A. Rogers (1931). Continuous fermentation in the production of lactic acid. *Ind. Eng. Chem.*, **23**, 532–534.

Wilke, W. (1983). Wilke International, Overland Park, KS, personal communication.

39

Starch Conversion Processes

L. E. COKER

The Hubinger Company, Subsidiary of H. J. Heinz Company, Pittsburgh, PA, USA

and

K. VENKATASUBRAMANIAN

H. J. Heinz Company, Pittsburgh, PA and Rutgers University, New Brunswick, NJ, USA

39.1 INTRODUCTION

The conversion of starch to sugars has registered a very impressive growth from its modest beginning in the laboratory of a Russian chemist nearly 175 years ago. In 1982, shipments of all refinery products by the corn refining industry amounted to nearly 15 billion pounds (*Corn Annual*, 1983). The sweetener consumption pattern in the US is being reshaped by new and sweeter corn syrups, particularly high fructose corn syrup. In 1984, sales of high fructose corn syrup exceeded 10 billion pounds or about 34% of the US sweetener market.

This chapter reviews the major technological developments of the corn sweetener industry from conventional acid converted corn syrup to the modern all enzyme manufacturing processes for high fructose corn syrup. Syrups of virtually any combination of viscosity and sweetness as well as other functional properties can be produced from corn starch. Economic and marketing aspects of corn sweeteners are also discussed briefly.

39.2 HISTORICAL PERSPECTIVES

The evolution of technology for the production of sweeteners from corn is an interesting history that encompasses some of the most significant events in chemistry and biotechnology. In some respects, the development of high fructose corn syrup—a mixture of dextrose and fructose—completes a circle back to wild honey, the oldest sweetener known to man. This circle also encompasses sucrose and the wide variety of corn syrups manufactured from corn starch.

It is believed that cane sugar dates back at least 8000 years to the South Pacific. The sweetness of cane sugar approximates that of wild honey. It was not until 1744 that a German chemist found that the sugar isolated from sugar beets was identical to the sugar derived from sugar cane. Sucrose, manufactured from cane or sugar beets, was very much more abundant than honey and, as such, became the sweetener of commerce. This position was unchallenged until the development of high fructose corn syrup.

The development of corn sweeteners can be traced back to 1811 when the Russian chemist Kirchoff reported his discovery that starch yielded a sweet substance when heated with acid (Kirchoff, 1811). In 1815, de Saussure identified acid hydrolysis as the underlying reaction of Kirchoff's observation. He found that the end product of the hydrolytic reaction was dextrose (de Saussure, 1815).

The early experiments established the basis for the commercial production of syrups and crude sugars from starch. As the demand for starch-derived syrups increased, corn emerged as the preferred source of starch because it was low in cost, plentiful, and easily stored. The result was the beginning of the present day corn wet milling industry. Significantly, this development opened up the corn growing temperate agricultural zones of the world to the production of raw materials for sweeteners with profound and continued impact on national economics.

During the next 150 years, starch syrup technology improved considerably. In 1935, only one type of corn syrup was available. This product was the 42 DE (dextrose equivalent) acid-converted corn syrup. Although this product was useful in many applications, it lacked the sweetness of sucrose and could not compete as a sweetener. When corn syrup was acid-converted to higher degrees of hydrolysis, the resulting products were often highly discolored and contained bitter by-products from the acid-mediated reaction.

At the scientific level, investigators continued to follow the lead of Kirchoff and de Saussure in elucidating the chemical nature and fine structure of starch. At the same time, the application of the fundamentals of enzyme chemistry and modern biochemistry began to blossom in the 1930s.

The first major technological breakthrough since Kirchoff occurred in 1940 when Dale and Langlois patented the use of commercially available enzymes and hydrolyzed starch (Langlois and Dale, 1940). While their discovery resulted almost immediately in the manufacture of commercial corn syrups of unique properties, it took many years for the real significance of their discovery to be fully appreciated. Although the new enzyme syrups found many commercial applications, their new technology was not used to produce large quantities of corn syrup until the 1950s.

The discovery, isolation and application of various carbohydrase enzymes resulted in the development of many new corn syrups. These enzymes could be used singly, in sequence, or in combination on starch to produce a variety of saccharile components and a broad range of syrup compositions. This, in turn, resulted in a wide variety of syrup property differences. As these property differences were exploited in many applications, the term corn sweetener became an important part of the language of food technology.

In the early 1960s, the first commercially significant quantities of crystalline dextrose were being marketed. It was produced from corn starch by an all enzyme process using bacterial α-amylase for starch liquefaction and dextrinization and fungal glucoamylase for saccharification of the hydrolysate to produce approximately 95% dextrose levels. The significant features of this new technology were patented by Langois and Dale (1940), and Wallerstein (1950).

The commercial development of glucose isomerase, which converts glucose to its sweeter isomer, fructose, was a major milestone in the corn sweetener industry. A historical perspective of this exciting development is presented by Casey (1977).

The enzymatic transformation of glucose to fructose was first introduced to corn sweetener production in 1967. The first high fructose corn syrup, commonly referred to as HFCS, contained 15% fructose. The manufacturing process, known as isomerization, originally involved the direct addition of an isomerase enzyme to a dextrose substrate in a batch reactor. Further process improvements resulted in the HFCS products containing 42 and 55% fructose.

In 1972, batch production was replaced by a continuous process utilizing isomerase enzyme immobilized in an insoluble carrier. This continuous process for the production of high fructose corn syrup involved the first large scale use of an immobilized enzyme system in the world. The continuous process resulted in a significant reduction in production costs. The design and operation of the continuous immobilized isomerase reaction for HFCS has been discussed extensively (Oestergaard and Knudson, 1976; Hupkes and van Telburg, 1976; Venkatasubramanian, 1978, 1979).

39.3 FUNCTIONAL PROPERTIES OF CORN SYRUP

The term 'corn sweetener' has been the generic term for a wide variety of nutritive carbohydrates prepared by hydrolysis of corn starch. These products are differentiated principally by the degree of hydrolysis. The functional properties of a corn sweetener are a consequence of the particular type and mixture of hydrolysis products.

The hydrolysis products of starch can be classified conveniently on the basis of the 'reducing sugars' they contain. Since starch is a homopolymer of anhydroglucose, its depolymerization through hydrolysis results in a mixture of glucose oligomers, ranging typically from the monomer, glucose, to glucose-multiples containing many units of glucose linked together. The reducing sugars contained in these mixtures can be assayed and expressed as they relate to the reducing power of dextrose, in terms of a 'dextrose equivalent' (DE). The greater the degree of hydrolysis, the higher is the DE value.

Different corn sweeteners are categorized according to their DE values (Table 1). High fructose corn syrups (HFCS) are not classified according to DE values since they contain substantial amounts of fructose. Indeed, HFCS syrups are differentiated on the basis of their fructose content.

Table 1 Types of Corn Sweeteners

Product	DE value	Product	DE value
Maltodextrins	<20	Type III	58–73
Corn syrups			
Type I	20–37	Type IV	>73
Type II	38–57	Dextrose	>99.5

The actual carbohydrate composition of different hydrolysis products plays a key role in determining its functional properties. Typical compositions of different products are shown in Table 2. Maltodextrins and Type I corn syrups contain little glucose while Type III and Type IV corn syrups are characterized by a large content of mono- and di-saccharides.

The important physical and chemical properties of different starch-derived products and their significance in various applications are outlined in Tables 3 and 4.

A more detailed analysis of the composition–properties–applications matrix can be found elsewhere (Horn, 1981).

Maltodextrin and various corn syrups are versatile ingredients which find numerous applications in the food processing and related industries. In fact, there is hardly any sector of the food industry in which some starch-derived product is not used. These applications are based on a wide variety of functional properties, employed either individually or in combination. An extensive range of applications can be found in many manufacturers' catalogs and a convenient handbook (Corn Refiners Association, 1979).

39.4 HIGH FRUCTOSE CORN SYRUP PRODUCTS

Three high fructose corn syrup products differentiated by fructose content are commercially available. These products contain 42, 55 or 90% fructose. The typical carbohydrate compositions of these products are shown in Table 5.

In most applications, HFCS syrups replace sucrose as the sweetener. High fructose corn syrups containing 42% fructose can be used in most food products that make use of a liquid sweetener. The development of 'second generation' syrups with higher fructose levels greatly extended the use of HFCS in many applications. This is particularly true with respect to the use of HFCS with 55% fructose as a replacement for sucrose in soft drinks. Fructose corn syrup containing 90% fructose has found wide application in many new and speciality food products. It is an ideal sweetener for reduced calorie foods such as jams and jellies. Today, there are several companies in the

Table 2 Compositional Data[a]

Sample	Dextrose equivalent	Saccharides (%) carbohydrate basis						
		DP_1	DP_2	DP_3	DP_4	DP_5	DP_6	DP_7
Maltodextrin	12	1	3	4	3	3	6	80
Corn syrup AC	27	9	9	8	7	7	6	54
Corn syrup AC	36	14	12	10	9	8	7	40
Corn syrup AC	42	20	14	12	9	8	7	30
Corn syrup AC	55	31	18	12	10	7	5	17
Corn syrup HM, DC	43	8	40	15	7	2	2	26
Corn syrup HM, DC	49	9	52	15	1	2	2	19
Corn syrup DC	65	39	31	7	5	4	3	11
Corn syrup DC	70	47	27	5	5	4	3	9
Corn syrup DC	95	92	4	1	1		2^b	
High fructose corn syrup	—	94^c	4			2^d		

[a] Corn Refiner's Association Handbook. [b] Sum of $DP_{5,6,7}$. [c] Dextrose plus fructose. [d] Sum of $DP_{3,4,5,6,7}$

Table 3 Properties of Corn Sweeteners

Property	Significance
Solubility	Control of fermentation rates in baking
Hygroscopicity	Moisture conditioning; food, pasticizing, crystallization inhibition
Textural properties	Body; chewiness
Osmotic pressure	Inhibition of microbial spoilage
Freezing point depression	Ice cream and frozen deserts
Viscosity	Handling characteristics

Table 4 Chemical Properties of Corn Sweeteners

Property	Significance
Sweetness	Determine application; cost
Fermentability	Baking and brewing industries
Browning reactions	Non-enzymatic browning
Reducing properties	Decreases oxidative degradation in foods; color

Table 5 Typical Analysis of Carbohydrate Composition of High Fructose Corn Syrups

Carbohydrate	42 HFCS[a]	55 HFCS[a]	90 VEFCS[a]
Fructose	42	55	90
Dextrose	52	42	7
Monosaccharides	94	97	97
Higher saccharides	6	3	3

[a] Percent dry basis.

US and abroad manufacturing high fructose corn syrup. These US companies, together with their location, plant size and HFCS production capacities, are listed in Table 6.

Table 6   US HFCS Producers

Producer	Plant location	Plant size (Bu/day)	HFCS capacity[a] (lb C.B./year)
Archer Daniels Midland	Cedar Rapids, IA	220 000	1 264 000
	Decatur, IL	200 000	1 714 000
American Maize Products	Decatur, IL	35 000	500 000
Amstar	Dimmitt, TX	30 000	336 000
Cargill	Dayton, OH	150 000	696 000
	Memphis, TN	90 000	696 000
Corn Products Company	Argo, IL	145 000	551 000
	Stockton, CA	32 000	352 000
	Winston-Salem, NC	32 000	351 000
Clinton Corn Processing	Clinton, IA	110 000	948 000
	Montezuma, NY	35 000	423 000
Hubinger	Keokuk, IA	83 000	632 000
A. E. Staley	Decatur, IL	150 000	600 000
	Morrisville, PA	85 000	700 000
	Lafayette, IN	155 000	1 300 000
	Loudon, TN	70 000	600 000
		Total =	11 663 000

[a] 42 HFCS and 55 EFCS.

39.5   MODERN PROCESS TECHNOLOGY

A simplified flow diagram for the manufacture of 42% high fructose corn syrup from starch is shown schematically in Figure 1. This diagram also outlines the production of 55% and 90% fructose syrups from 42% HFCS. The manufacturing process may be divided into about 18 separate steps in five principal phases of operation, each of which has a major objective. These phases include dextrose production, primary refining and chemical treatment to produce dextrose feedstock, isomerization of dextrose feedstock to produce 42% fructose, secondary refining of 42% fructose, and fractionation of 42% fructose for the production of 55% fructose.

The heart of the entire process is the enzymatic isomerization step utilizing immobilized glucose isomerase. The economic and technical success of the isomerization reaction depends upon the production and delivery of high quality dextrose feedstock to the isomerization reactors.

39.5.1   The Manufacture of Dextrose from Starch

The manufacture of dextrose feedstock from starch is a multistage process involving thermostable α-amylase and glucoamylase in successive enzymatic steps. These enzymes catalyze the hydrolysis of the starch polymer to dextrose monomer.

In the first step, starch slurry is gelatinized by cooking at high temperature. The gelatinized starch is then liquefied and dextrinized by thermostable α-amylase in a continuous two stage reaction. The product of this reaction is a soluble dextrin hydrolyzate with a DE of 10–15, suitable for the following saccharification step. Typical conditions for the liquefaction–dextrinization reaction are summarized in Table 7.

During the liquefaction–dextrinization reaction, the major process variables of starch quality and concentration, α-amylase dose, temperature, pH, starch flow and time must be maintained within relatively narrow limits to insure that the starch is properly cooked and dextrinized. If any one of these variables is significantly out of process specifications, the finished liquor may contain undercooked and unconverted starch. This leads to poor conversion and filtration problems, or the production of anomalous saccharides such as maltulose which can result in flavor problems in the finished high fructose corn syrup.

The total soluble protein levels in the starch slurry should be maintained below about 0.3% total protein and 0.03% soluble protein. This is necessary to minimize color formation as a result of the Maillard reaction between amino acids and sugars under conditions of high temperature

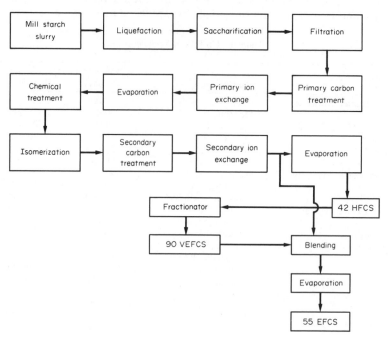

Figure 1 Schematic diagram showing the process used to manufacture high fructose corn syrup

Table 7 Liquefaction–Dextrinization Reaction Conditions

	Condition
Starch slurry	30–35% w/w d.b.
Total protein	≤0.3%
Soluble protein	≤0.03%
pH	6.0–6.5
Thermostable α-amylase	0.05–0.10% d.b. starch
First stage liquefaction	
Temperature	104–107 °C
Hold	5–8 min
DE	0.5–1.5
Second stage liquefaction	
Temperature	94–97 °C
Hold	90–120 min
DE	10–15

and pH. If the starch slurry is high in protein, the syrup produced has a greater potential to develop color exceeding the capacity of the carbon and ion-exchange refining systems.

Following liquefaction–dextrinization, the pH and temperature of the 10–15 DE hydrolyzate is adjusted for saccharification. During saccharification, the hydrolyzate is further converted to dextrose by the action of glucoamylase enzyme. Although saccharification can be carried out as a batch reaction, a continuous saccharification is carried out in most modern plants. In the continuous saccharification reaction, glucoamylase is added to 10–15 DE hydrolyzate following pH and temperature adjustment. The saccharifying liquor is then pumped through a number of large reactors in series. Typical saccharification reaction conditions are summarized in Table 8. The total time for saccharification is particularly sensitive to the amount of glucoamylase. Commonly, a reaction time of 65–75 hours is used to obtain a hydrolyzate containing 94–96% dextrose.

During saccharification, the important variables monitored are the initial DE, pH, temperature, dry substance, glucoamylase dose, the time course of conversion, and the presence of unconverted starch in the liquor during this reaction. These variables must again be controlled within relatively narrow limits to insure that a minimum 94% dextrose level is reached at the end of the reaction. This minimum level of 94% dextrose is necessary to meet the specifications for the carbohydrate composition of 42% HFCS in which the higher saccharides level cannot exceed

Table 8 Saccharification Reaction
Conditions

	Condition
Dry substance	30–35%
pH	4.0–4.4
Temperature	60–62 °C
Glucoamylase	1 l/ton d.s. starch
Reaction time	65–75 h

6%. The carbohydrate composition of a typical high dextrose saccharification liquor is summarized in Table 9. The high dextrose hydrolyzate is then refined to produce dextrose feedstock for the isomerization reaction.

Table 9 Typical Analysis of the
Product of Dextrose Saccharification

Component	d.b. (%)
Dextrose	94–96
Maltose	2–3
Maltotriose	0.3–0.5
Higher saccharides	1–2
Dry substance	35–37

39.5.2 Preparation of Dextrose Feedstock for Isomerization

Preparation of very high quality dextrose feedstock for isomerization is first necessary because of the very low color and ash specification for high fructose corn syrup. A high purity feedstock is also required for efficient utilization of the immobilized isomerase enzyme column.

Immobilized isomerase enzyme columns are used continuously over a period of several months. During this period very large volumes of dextrose feedstock pass through the columns. Extremely low levels of impurities such as ash, metal ions, or protein in the feedstock can accumulate and lead to decreased productivity of the enzyme. For these reasons, dextrose feedstock is refined to a color of 0.1 (CRA × 100) and a conductivity of 5–10 μmho.

In the refining process, the high dextrose conversion liquor is filtered on rotary precoat filters to remove 'mud' or coagulated protein and oil. This filtered liquor is then passed through a series of check filters and polish filters to remove traces of particulate matter. Color in the hydrolyzate is first removed in a series of granulated carbon columns. These columns are renewed by the removal of spent carbon and the addition of regenerated carbon on a regular basis. Carbon treated liquor is filtered again and passed through an ion-exchange system where it is deionized by an ion-exchange resin. This resin is usually arranged in a dual pass cation–anion–cation–anion system. Ion exchange is designed to remove all heavy metals and ash that might be detrimental to the immobilized isomerase enzyme. Ion-exchange resin also removes significant amounts of color bodies that are not removed by the carbon columns.

The carbon-treated and deionized high dextrose liquor is evaporated to the proper solids level for isomerization. In addition, the feedstock is chemically treated by the addition of magnesium ions, which not only activate the immobilized isomerase, but also competitively inhibit the action of any residual calcium ions which are potent inhibitors of isomerase.

39.5.3 Isomerization Process

The isomerization reaction is carried out under the conditions summarized in Table 10. The design and operation of a commercial immobilized glucose isomerase reactor system has been described (Venkatasubramanian, 1978; Venkatasubramanian and Harrow, 1979). Careful integration of design and operation are necessary to minimize production fluctuations with respect to capacity and conversion level.

The conversion of glucose to fructose is a reversible reaction with an equilibrium constant of

Table 10   Isomerization Feedstock and
Reaction Conditions

	Conditions
Dry substance	40–45%
Dextrose	94–96%
Higher saccharides	4–6%
pH	7.5–8.2 at 25 °C
pH drop	0.2–0.4
Temperature	55–65 °C
Activator	0.004 M Mg^{2+}
Residence time	0.5–4 h

about 1.0 at 60 °C. Thus, one would expect to obtain a fructose level of about 47–48% at equilibrium, starting from a feedstock containing 94–96% dextrose. However, the reaction rate near the equilibrium value is so slow that it is prudent to terminate the reaction at a conversion level of 42% fructose to achieve practical reaction residence times.

In a given isocolumn (immobilized isomerase column), the rate of conversion of dextrose to fructose is proportional to the enzyme activity of the immobilized isomerase. This activity decays in a nearly exponential manner as a function of time. When the column is new and the activity is high, the flow of feedstock through the column is relatively high since a shorter residence time is required to achieve the 42% fructose level. The flow through the column must be reduced proportionately to give a longer residence time compensating for the lowered activity in order to achieve a constant conversion level as the column ages.

Since the activity of the bound enzyme decreases continuously, it is necessary to have a number of isocolumns to minimize production fluctuations with respect to capacity and conversion level. Both series and parallel operation of the isocolumn are possible. In practice, the parallel operation of at least six isocolumns offers the greatest operational flexibility. In this arrangement, each isocolumn can be operated essentially independent of the others. Each isocolumn can be brought into and taken out of service very readily.

It is necessary to operate the isocolumn battery to obtain a finished product of uniform quality and constant average fructose content of 42%. The variation in total flow of the isocolumns must be maintained within relatively narrow limits because of the requirements of evaporation and other finishing operations. In practice, flow cannot be precisely controlled at all times to obtain 42% fructose, but this can easily be achieved on an averaged basis.

A number of important process control strategies can be used to achieve an approximate steady state operation. These include modest variations in operating temperature, conversion level, and feedstock pH control. Final uniformity of product quality and fructose level is usually accomplished by automatic back-blending operations controlled by an in-line polarimeter.

The effect of temperature on overall productivity can be dramatic. A feed temperature of 60 °C is considered as an optimum. Higher temperatures result in faster flow rates but also cause accelerated decay rates. A feed temperature as low as 55 °C can be used to extend enzyme life at the expense of slower flow rates. Some risk of microbial contamination does exist at the lower operation temperature.

One of the most critical operating variables is the internal isocolumn pH. The operating pH is usually a compromise between the pH of maximum activity, typically around pH 8, and the pH of maximum stability, typically pH 7.0–7.5. This is complicated by the fact that dextrose feedstock is not stable to pH at temperatures around 60 °C. Some decomposition to produce acidic by-products results in a pH drop across the isocolumn during operation. Since it is difficult to control pH directly in the column, pH adjustment is made to feedstock to maintain a constant pH in the column effluent.

The operational stability of the immobilized isomerase system can be characterized by the half life, or the time span during which the enzyme activity is reduced by one-half. Typically, a single isocolumn is operated for a period of at least two half-lives, after which the enzyme is discarded and replaced with a fresh batch. The operating life of a column is determined by such considerations as the number of isocolumns in the battery, the average decay rate of the individual columns, the maximum allowable variance in individual isocolumn flow, minimum total flow, and total required production capacity. Enzyme half-lives of 70–120 days are common resulting in column replacement 1.5–2 times per year.

Isomerization enzyme cost is a significant fraction of total operating cost. The enzyme can be purchased commercially or manufactured by the user. In the former case, most enzyme is supplied on a performance guarantee with minimum stated activity. The quality of immobilized isomerase has improved with respect to longer half-life and higher initial activity with a net higher productivity. In general, isomerization costs have decreased significantly from the original cost of 50–70 cents per cwt. HFCS dry basis. The overall process economics is improved by every percent increase in enzyme productivity.

The principal producers of immobilized glucose isomerase are listed in Table 11. Most of these preparations consist of fixed whole cells containing glucose isomerase activity. The products of Novo and UOP are believed to be manufactured from isolated and purified glucose isomerase.

Table 11 Principal Producers of Immobilized
Isomerase

Company	Trade name of product
Gist-Brocades	Maxazyme GI Immob.
Miles Laboratories, Inc.	Takasweet
Novo Industries	Sweetzyme Q
UOP Process Division	Ketomax GI-100

An objective comparison of the performance of different immobilized enzymes is essential from a user's point of view. Since the enzyme supplier's data are generally provided on different bases, laboratory and pilot plant evaluations are generally required to compare competing systems. The biochemical, mechanical and hydraulic characteristics of importance in evaluating the suitability of an immobilized enzyme for a particular reactor design have been reviewed by Venkatasubramanian (1982).

39.5.4 Secondary Refining of Isomerized Feedstock

Following isomerization, the manufacturing process involves the secondary or polish refining of the 42% HFCS product. Some additional color is picked up during the chemical treatment and isomerization when the feedstock is held at higher pH and temperature for a period of time. The product also contains some additional ash from the chemicals added for isomerization. This color and ash is removed by the secondary carbon and ion-exchange systems. The refined 42% HFCS is then evaporated to 71% solids for shipment.

39.5.5 Manufacture of Enriched Fructose Products

The HFCS product from the isomerization reaction typically contains 42% fructose, 52% unconverted dextrose and about 6% oligosaccharides. For reasons previously discussed, this product represents the practical maximum level of fructose attainable. In order to obtain products with higher levels of fructose, it is necessary to selectively concentrate the fructose. Many common separation techniques are not applicable for this purpose, since they do not readily discriminate between the two isomers of essentially the same molecular size. However, fructose preferentially forms a complex with different cations such as calcium. This difference has been exploited to develop commercial processes. There are basically two different commercial processes available today for the large scale purification of fructose. In both instances, resins in the preferred cationic form are used in packed bed systems. One process employs an inorganic resin leading to a selective molecular adsorption of fructose (Jensen, 1978). Chromotographic fractionation using organic resins is the basis for the second commercial separation process (Venkatasubramanian, 1982). When an aqueous solution of dextrose and fructose such as 42% HFCS is fed to a fractionating column, fructose is selectively held by the resin to a greater degree than dextrose. Deionized and deoxygenated water is used as the eluant. Typically, the separation is achieved in a column packed with a bed of low cross-linked fine mesh polystyrene sulfonate cation-exchange resin using calcium as the preferred salt form. The enriched fructose, containing approximately 90% fructose, is referred to as very enriched fructose corn syrup (VEFCS). This VEFCS fraction can be blended back with the 42 HFCS feed material to obtain products having fructose contents between 42% and 90%. The most typical of these products is 55% enriched fructose corn syrup, called 55 EFCS. The fractionation and blending process is schematically outlined as a portion of Figure 1.

The treatment of other raffinate streams in the fractionation process is an important consideration. In general, the dextrose rich raffinate stream is recycled to the dextrose feed of the iso-column system for further conversion to 42 HFCS. The raffinate stream containing dextrose with fructose levels higher than the feed level is generally recycled through the fractionator to maintain a high solids level and reduce water usage. The raffinate stream rich in oligosaccharides is recycled back to the saccharification system.

Since water is used as the elution medium, it has a great impact on the overall evaporation load on the system. Very low solids concentrations contribute to the risk of microbiological problems within the system. Thus, the most important design parameter dictating overall process economics is the maximization of solids yield at acceptable purity while reducing the dilution effect of the eluant rinse to a minimum. The efficiency of feed and water usage must be maximized for optimal yield. The yield is important to reduce the cost of re-isomerization.

Procedures available for achieving these goals include recycling techniques, higher equalization of the resin phase with proper redistribution in a packed column, and the addition of multiple entry/exit points in the column. These approaches to increase the purity and the yield are sometimes referred to as feed enrichment. A small apparent increase in the purity of feed to the column, *i.e.* higher fructose levels, results in a much larger gain in production through increased yield at a given product purity. In practice, this translates into maximization of the ratio of sugar volume fed per volume of resin per cycle, minimization of the ratio of water volume required per volume of resin per cycle, and provision for careful fluid distribution to the columns.

39.6 MARKETING AND ECONOMIC CONSIDERATIONS

Sucrose has long been the dominant caloric sweetener in the United States. However, since 1972 it has been losing ground to HFCS. The remarkable growth of the HFCS industry in the last 15 years was dramatically triggered by high sugar prices in the early 1970s. In 1971, the wholesale price of sugar (Northeast) was 11.5 cents per pound. By 1976, the price rose to 19.2 cents per pound and in 1980 a record price of 41.0 cents per pound was reached. In comparison, in 1976 high fructose corn syrup (Illinois) sold for 14.0 cents per pound or 73% of sugar cost. In 1980, during the record rise in sugar prices, HFCS was selling for 23.6 cents per pound or only 58% of the sugar cost. In 1981, sugar dropped by about 25% to 30.6 cents per pound; HFCS prices dropped by only about 9% to 21.5 cents per pound or 70% of the sugar cost (Venkatasubramanian, 1982).

The price differential between HFCS and sugar offered strong incentive for the substitution of HFCS for sugar in soft drinks, canned fruits, ice cream and certain bakery products. As a result, sugar usage by industrial users has been declining relative to total sugar and particularly relative to total sweeteners because of the inroads of high fructose corn syrup. This decline in sugar usage is dramatically illustrated by the beverage industry, the largest industrial users of sugar. In 1978, annual sugar usage peaked at 5.12 billion pounds. By 1981, this usage had declined by 28.5% to 3.66 billion pounds. The total industrial sugar usage declined by about 18%, from 13.9 billion pounds in 1978 to 11.4 billion pounds in 1981. In late 1984, all major soft drinks companies approved the total substitution of sucrose by HFCS. The trend of HFCS substitution for sugar in canned fruits, ice cream and certain bakery items is expected to slow down as HFCS saturates these markets by 1985. Caloric sweetener consumption in the United States reached a peak of 129.3 pounds per capita in 1977, after growing at a rate of about one pound per person per year since 1960. By 1982, this had declined to about 122.4 pounds per person. An annual consumption of about 130 pounds per capita has been considered a safe plateau. Future growth in total caloric sweetener consumption is expected to occur only in direct proportion to population growth. It is expected that sugar will continue to be replaced by HFCS particularly in the processed foods sector. Consequently, the sale of HFCS is projected to grow, but at a decreasing rate.

39.7 REFERENCES

Beverage World (1982). August, 26–28.
Casey, S. P. (1977). *Die Stark*, **29**, 196–204.
Corn Annual (1983). Corn Refiners Association.

Corn Refiners Association (1979). *Nutritive Sweeteners from Corn*, 2nd edn.

de Saussure, T. (1815). *Ann. Physik*, **48**, 129.

Horn, H. E. (1981). *Cereal Foods World*, **26**, 219.

Hupkes, J. V. and R. van Telburg (1976). *Die Starke*, **28**, 356.

Jensen, R. H. (1978). *Abstr. Inst. Chem. Eng. 85th Natl. Mtg.*, Philadelphia.

Kirchoff, G. S. C. (1811). *Sead. Emp. Sci., St. Petersburg, Mem.*, **4**, 27.

Langois, D. P. and J. K. Dale (1940). *US Pat.* 2 201 609.

Oestergaard, J. and S. L. Knudson (1976). *Die Stark*, **28**, 350.

Rentshler, D., D. P. Langois, R. F. Larson and L. H. Alverson (1962). *US Pat.* 3 039 935.

Venkatasubramanian, K. (1978). In *Enzyme, The Interface Between Technology and Economics*, ed. J. P. Dansky and B.
 Wolnak, p. 35. Marcel Dekker, New York.

Venkatasubramanian, K. (1982). *Enzyme Eng.*, **6**, 37.

Venkatasubramanian, K. and L. S. Harrow (1979). *Ann. N.Y. Acad. Sci.*, **326**, 141.

Wallerstein, L. (1950). *US Pat.* 2 531 999.

40
Proteolytic Enzymes

O. P. WARD
National Institute for Higher Education, Dublin, Ireland

40.1 INTRODUCTION

Proteolytic enzymes account for nearly 60% of the industrial enzyme market. Proteases of commercial importance are produced from microbial, animal and plant sources. The proteases constitute a very large and complex group of enzymes which differ in properties such as substrate specificity, active site and catalytic mechanism, pH and temperature activity and stability profiles. Commercial proteases have application in a range of processes which take advantage of the unique physical and catalytic properties of individual proteolytic enzyme types.

Milk clotting enzymes have been used to transform milk into products such as cheese since about 5000 BC, when it was observed that milk carried in calf stomachs tended to curdle. Pancreatic proteases were used for dehairing and bating of hides and as presoak detergents since about 1910. Major interest in commercial detergent proteases developed in the 1950s and enzyme detergents captured a substantial share of the detergent market during the 1960s. In 1969 the industry met with unfavourable publicity when some detergent workers developed allergies due to the dusting effect of the enzymes. The problem was overcome by the introduction of dust-free preparations for use in detergents. Alkaline detergent proteases now account for 25% of the total industrial enzyme sales. Use of animal and microbial rennets in cheesemaking and proteases of *Aspergillus oryzae* in baking are the other two predominant applications of microbial proteases.

Proteases have a wide range of functions in nature. Extracellular microbial proteinases contribute to the nutritional well-being of the producing organism by hydrolyzing large polypeptide substrates into smaller molecules that the cell can absorb. Mammalian pancreatic proteases and intestinal and stomach peptidases generally perform a similar nutritional role in the digestion and absorption processes of these species.

Proteolytic enzymes are also involved in the regulation of biological metabolic processes such as spore formation, spore germination, protein maturation in viral assembly, activation of certain viruses of importance for pathogenicity, various stages of the mammalian fertilization process, blood coagulation, fibrinolysis, complement activation, phagocytosis and blood pressure control. Some of the processes, such as the coagulation and complement systems, involve cascades of proteolytic events, where a number of inactive enzyme precursors or zymogens are converted to active proteases which in turn activate other proteases. These cascades allow control and amplification of the processes involved. In all cell systems there is a balance between metabolic processes involving protein synthesis and breakdown and intracellular proteinases play a vital role in these protein turnover processes. In growing bacteria, 1–3% of the cell proteins are degraded to amino acids per hour and similar values are reported for plant cells in culture. Protein turnover is essential for the adaptation of cells to new environmental conditions, particularly in environments lacking in amino acids, as it generates an amino acid pool for synthesis of newly required enzymes and other proteins. Proteolytic activity is also important during cell differentiation.

There is evidence that proteases are involved in the modulation of gene expression, and in enzyme modification and secretion. The possible role of proteases in the regulation of translation by modification of ribosomal proteins has been suggested. Proteases convert inactive enzymes and other biologically inactive protein molecules into their active forms. In addition, the removal of *N*-formylmethionine or methionine from nascent peptide chains and the cleavage of the translation product of monocistronic mRNA, the coding for several distinct peptide chains, is mediated by proteases. Finally, during extracellular enzyme secretion, the hydrophobic peptide extension which facilitates the passage of the enzyme through the cell membrane is cleaved by proteolytic action, thus releasing the extracellular enzyme.

This chapter describes the production, properties and industrial applications of proteolytic enzymes. The reader is also referred to earlier reviews of various aspects of this topic (Keay, 1971; Yamamoto, 1975; Aunstrup, 1974, 1980; Whitaker and Puigserver, 1982; Ward, 1983).

40.2 PRODUCTION AND EXTRACTION OF PROTEASES

Microbial proteases are produced from high yielding strains by fermentation under controlled conditions in surface or submerged culture. The enzymes are produced extracellularly and recovery involves separation of the cell free liquor by filtration or centrifugation. Depending on the product and the degree of purity required, further purification might involve steps such as concentration, precipitation and stabilization. Production of enzymes from plant or animal sources usually involves grinding or mincing of the material followed by a series of extraction and purification steps to achieve a product of stable and standardized activity.

Regardless of source, enzyme products are supplied as liquid or solid preparations. In liquid preparations, the enzyme solution has to be stabilized against chemical and microbial denaturation. High concentrations of salts, and chemicals such as glycerol, often increase general enzyme stability and product shelf life. The pH of the liquid should be adjusted to optimize stability. Other chemicals, such as divalent cations, oxidizing or reducing agents, may be used to stabilize specific enzymes and their actions are often directed towards the active site of the enzyme. While some of these additives may also contribute to microbial preservation of the enzyme liquid, suitable microbial preservatives may be used in addition. Solid enzyme preparations are usually supplied as dry powders or granules of standardized activity. Most enzymes exhibit good stability characteristics in powdered form. Guidelines on suitable stabilizers and preservatives and levels of use are available (General Standards for Enzyme Regulations, 1980). A general flow diagram for production/extraction of enzymes is given in Figure 1.

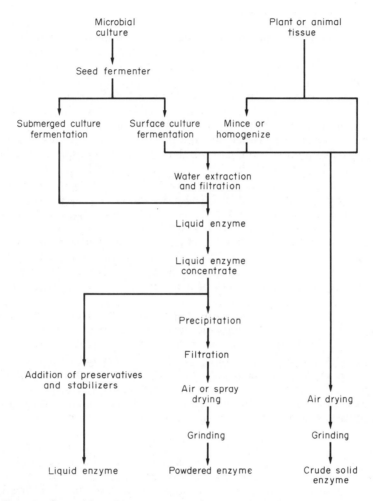

Figure 1 General flow diagram for production and extraction of industrial enzymes

40.2.1 Microbial Protease Production

The first stage of the development of an industrial fermentation process is to isolate a strain with the potential capability of producing the particular product in commercial yields. This involves designing screening procedures which can test a large number of strains easily for production capability. With extracellular microbial products the usual practice is to grow large numbers of organisms on agar plate media and to relate each organism's production capability to the radius of the product's zone of diffusion around the colony. Particularly with proteases, which may be inducible or repressible, care must be taken that the agar medium is designed to optimize induction and minimize the chances of repression. It has been reported that *B. licheniformis* produces very narrow zones of hydrolysis on casein-agar despite giving very good protease production in submerged cultures (Aunstrup, 1974). Nasuno and Onara (1972) observed a relationship between colony morphology and protease production. Proteinase formation by *A. oryzae* was associated with production of smooth conidia whereas enzyme producing *A. sojae* strains had echinulate or tuberculate conidia. The actual screening approach used will vary with the particular enzyme and the producing organism and should be aimed at effectively testing as many organisms as possible. Bu'Lock *et al.* (1982) discuss general approaches to strain isolation.

Having isolated the best overproducing strain, it may be necessary to further increase its production capability using mutation selection programmes and other genetic manipulation techniques appropriate to industrial microorganisms (Sebek and Laskin, 1979).

40.2.1.1 Control of enzyme production

Most published media for production of proteases contain nitrogen sources, such as soybean meal, casein, corn-steep powder or liquor, distillers' solubles, brewers' yeast and carbohydrate sources such as starch, ground barley or lactose. Details of some of these media are given by Aunstrup (1974, 1980). As there are many examples where high carbohydrate concentrations repress enzyme production, it is often the practice to add carbohydrate continuously or at timed intervals throughout the fermentation. This approach also reduces the viscosity of the medium and hence the fermenter's power requirements. It has also been observed that free amino acids often repress protease formation. Isoleucine and proline repress protease synthesis in *Bacillus species*. Amino acids also repress proteinase production by *Sarcina* spp. and *Arthrobacter* (Hofsten and Tjeder, 1965). Proteases of *Micrococcus caseolyticus*, *B. megaterium* and *Vibrio parahemolyticus*, on the other hand, appear to be insensitive to amino acid repression. Proteases of *Pseudomonas maltophilia* (Boethling, 1975) and *Aspergillus nidulans* are constitutive. Priest (1983) has described aspects of induction, derepression and secretion of the alkaline protease of *Neurospora crassa* and a model for regulation of synthesis of the enzyme.

Peptides or proteins have been reported to induce proteinase synthesis in a number of microorganisms (Murakami *et al.*, 1969; Lasure, 1980). A range of individual amino acids used as sole nitrogen source induced protease production by *A. terricola* (Imshenetskii *et al.*, 1971), while other workers demonstrated stimulation of protease production in *A. niger* by casein or casein hydrolysates.

It is difficult to draw general conclusions from the above observations on the factors affecting protease production. Conclusions on cultivation experiments should be made with extreme caution unless parameters (such as pH) are controlled, since changes in media constituents may affect these parameters, which may in turn be responsible for the altered enzyme production. Chemostats have been used to overcome these problems (Wiersma and Harder, 1978). However, since most commercially important extracellular enzymes are produced during the idiophase (Glen, 1976), this technique does not provide a real solution. Under most growth conditions *Bacillus* spp. produce extracellular proteinases during the post-exponential phase (Ward, 1983), although some examples of growth associated protease production are also cited. An inverse relationship has been observed between growth rate and the rate of exoenzyme production by *Bacillus*. In addition, the best catabolite repressors of exoenzyme synthesis support the highest growth rates, suggesting that there is a relationship between catabolite repression and growth rate (Debabov, 1982). These correlations apply to protease synthesis by *B. licheniformis* (Laishley and Bernlohr, 1968), *B. subtilis* and *B. amyloliquefaciens* (Heineken and O'Connor, 1972). Alkaline protease from *B. subtilis* which is involved in the sporulation process could be eliminated by preparation of non-sporulating mutants. For spore-forming organisms, which produce enzymes in the post-exponential phase, non-sporulating mutants may be used in fermentations to reduce the possibility of enzyme production being terminated by sporulation. Alkaline protease positive non-sporulating mutants have also been found with yields which were increased five-fold (Aunstrup, 1974). Debabov (1982) discussed the relationships between extracellular serine protease and other exoenzymes and sporulation of *Bacillus* spp. and concluded that the synthesis of most of these enzymes (excluding α-amylase) requires triggering mechanisms similar to those responsible for the initiation of sporulation.

The most accepted hypothesis for excretion of bacterial extracellular enzymes is based on concepts developed for the secretory proteins of eukaryotes, *i.e.* the signal hypothesis (Blobel and Dobberstein, 1975). According to this hypothesis, synthesis proceeds on ribosomes located on membranes. The primary precursors of these proteins contain a hydrophobic amino acid sequence at the *N*-terminus which is recognized by and becomes associated with receptor membrane proteins, after which secretion of the growing peptide is facilitated. The signal sequence is cleaved proteolytically on the external side of the membrane and the protein folds and acquires a stable conformation. Such precursors have been described for the amylase and protease of *B. amyloliquefaciens* (Nagata *et al.*, 1974; Sanders and May, 1975).

Some fungal enzymes, including proteases from *Aspergillus* species and *Mucor pusillus*, are often produced in higher yields in semi-solid culture, incubated at 30–40 °C in media such as wheat bran, containing 50% moisture. In certain cases the nature of the fermentation, semi-solid or submerged, influences the amount of carbohydrate associated with the enzyme. Acid protein-ase of *A. oryzae* contains less carbohydrate when produced in semi-solid rather than submerged culture. Microbial rennet from *M. pusillus*, produced in semi-solid culture, contains no carbohy-drate while the *M. miehei* enzyme in submerged culture is a glycoprotein. There is a much greater tendency for moulds grown in surface culture to produce spores and fruiting bodies and this con-trast in physiology obviously influences enzyme production. General aspects of production of mould enzymes by solid-substrate fermentations are discussed by Aidoo and others (1982). They describe three methods, a drum method, a pot method and a tray method, the drum method alone having agitation. These methods have been found to be particularly suitable for strains of *Aspergillus, Mucor, Penicillium* and *Rhizopus* and have proved very successful industrially for *Aspergillus oryzae* fermentations.

Meyrath and Volavsek (1975) discuss various aspects of microbial protease production includ-ing strain selection procedures, environmental factors and production technology.

40.2.2 Plant and Animal Protease Extraction

40.2.2.1 Animal proteases

The enzyme containing organs usually come from the slaughterhouses and this requires an organized and cost-effective collection system. Organs are preserved until ready for extraction by drying, freezing or salting.

A number of processes exist for production of rennet. Dried calf stomachs or vells may be cut into strips, blended or ground, and extracted with 5–10% salt. Alternatively, salted stomachs are washed and dried, prior to extraction. In some countries, such as New Zealand, stomachs are cleaned and frozen before extraction. Single strength rennet can be produced by such extraction processes. Double or triple strength requires use of ultrafiltration equipment.

Pancreatin, which contains a mixture of enzymes including protease, is extracted from the pan-creas. Crude material is produced by defatting, mincing and drying. Purification may involve removal of insoluble material and activation of proteases. While these preparations contain chy-motrypsin and trypsin, these enzymes are generally produced industrially as by-products of insu-lin production. Pepsin is prepared from the fundus portion of hog stomachs, by acid extraction and filtration.

40.2.2.2 Plant proteases

Papain and ficin are prepared by water extraction of crude material from *Carica papaya* and *Ficus carica* respectively. Further purification involves filtration, solvent precipitation and drying (Aunstrup, 1974). Bromelain is usually obtained from the stems of the pineapple plant by extrac-tion and fractional solvent precipitation. Kling *et al.* (1982) recently described the purification of a protease from agave juice, a by-product of Brazil agave fibre production. They predict that this enzyme has real commercial potential.

Where high purity proteolytic enzymes are required, a range of protein separation techniques are used which purify particular protein fractions. These techniques exploit the characteristic dif-ferences in properties of different proteins such as charge, size, solubility and biological affinity, in order to isolate the required fraction. These techniques are discussed by Wang *et al.* (1979).

40.3 PROPERTIES OF PROTEINASES

The specificity of proteolytic enzymes is primarily related to the nature of the amino acid side-chains and other groups close to the bond being hydrolyzed. In addition to this, these enzymes can be separated into two major groups based on their ability to hydrolyze bonds of amino acids which are at the outside (exopeptidases) or in the middle (endopeptidases) of the peptide chain. An earlier view that proteolytic specificity is related to substrate molecular size has no scientific basis.

Recent studies on protease specificity have concentrated on assessing the relative activities of an enzyme on synthetic di- and tri-peptides and larger natural and synthetic peptides where sequences are known. Combining the results of this approach with observations on hydrolysis of pure large natural proteins, the following general comments on specificity can be made. The presence or absence of charged groups in particular positions relative to the susceptible bond can be the basis for the distinction between endopeptidases and exopeptidases. Uncharged and nonpolar groups can contribute to substrate specificity. Aromatic groups or the pyrrolidine rings of imino acids are of importance in particular cases. Indeed, if the sensitive bond is represented by the structure RCO—X (where X = NHR, in the case of peptide bonds), then provided the structure of R is appropriate and the CO—X bond is activated, it is not always essential that CO—X is a peptide bond. The group X is only important inasmuch as it influences the chemical stability of the CO—X bond, and peptidases which hydrolyze esters and amides have been observed also to catalyze transpeptidations of the acyl type (Brenner *et al.*, 1950; Gibson, 1953).

Godfrey and Reichelt (1983) have reported on the comparative reactivity of various commercially important proteinases on different substrates when each is assayed at its optimum pH. This demonstrates that the relative reactivity of these enzymes varies with substrate. Another aspect of the complexity of protease catalysis is the observation that the optimum pH for activity may vary with substrate. *Aspergillus* acid protease has a pH optimum of 8.0 with casein substrate but 4–5 with haemoglobin as substrate. Fujimaki *et al.* (1977) have reviewed factors which influence degradation of proteins. These include enzyme specificity, substrate and enzyme concentration, extent of protein denaturation, pH, temperature, ionic strength, inhibitors and activators. Because of their compact conformation, native proteins are generally not very susceptible to degradation by proteases (Robinson and Jencks, 1965). Peptide bonds are more exposed and susceptible to proteolytic attack when proteins are unfolded due to denaturation.

All of the proteinases of significant industrial importance are endo- rather than exo-peptidases. An up-to-date classification of some of these endopeptidases is given in Table 1. This is only a partial list. Excluded from it are endopeptidases from human tissues, blood, snake venoms and remote animal and plant species. The EC numbers and enzyme names are those recommended by the Nomenclature Committee of the International Union of Biochemistry (Enzyme Nomenclature, 1978). Proteinases belong to sub-group 4 of group 3 which classifies the hydrolases (3.4). Endopeptidases are further divided into sub-sub-groups based on catalytic mechanisms, and individual enzymes in each sub-sub-group are distinguished on the basis of substrate specificity (see also Figure 2).

In the following sections, some of the properties of the different groups of endopeptidases will be discussed with emphasis being placed on the diverse catalytic nature and industrial importance of this group of enzymes.

40.3.1 Serine Proteinases

This group of enzymes, which has both serine and histidine at the active site, has been subdivided by Morihara (1974) into four sub-groups: trypsin-like proteinases, alkaline proteinases, *Myxobacter* α-lytic proteinase and *Staphylococcal* proteinase. The latter two are not commercially significant.

40.3.1.1 Trypsin-like proteinases

Apart from trypsin itself and chymotrypsin, these enzymes are produced by a number of *Streptomyces* species, *e.g. S. griseus* (Juresek *et al.*, 1969), *S. fradiae* (Morihara, 1974) and *S. erythreus*. They all show similar catalytic specificity towards the insulin B-chain, hydrolyzing only two bonds, namely Arg-Gly (22–23) and Lys-Ala (29–30). Trypsin and the microbial enzymes have pH optima at 8.0 and are inhibited by diisopropyl fluorophosphate (DFP), soybean trypsin inhibitor and tosyl-L-lysine chloromethyl ketone (TLCK). Considerable sequence homology exists between bovine trypsin and the *Streptomyces* enzymes (Juresek *et al.*, 1969), and they also exhibit similar esterase activity. The *Streptomyces* enzymes have much higher amidase activity than trypsin.

Trypsin, which is important in nature as a digestive enzyme, is formed in the intestinal tract from an inactive precursor, trypsinogen, which is secreted by the pancreas. Trypsin catalyzes the conversion of trypsinogen to trypsin by hydrolysis of a single peptide bond. Another digestive serine proteinase, chymotrypsin, the precursor of which is also produced in the pancreas, has been used successfully to remove the bitter taste of peptic hydrolysates (Tanimoto *et al.*, 1972).

The *Streptomyces* enzyme is a component of pronase, which is of little commercial importance.

Trypsin is a component of pancreatin, an extract from porcine and bovine pancreas which has industrial applications in the food processing, leather and pharmaceutical industry. Despite the rather high pH optimum of trypsin, the enzyme is active over a broad pH range (4–11) and its temperature optimum for activity is around 50 °C.

40.3.1.2 *Alkaline proteinases*

The serine alkaline proteinases are produced by a range of bacteria, yeasts and moulds and are also present in mammalian tissues. Commercially this is an extremely important group of enzymes, containing as it does the subtilisins, which are the major enzyme constituents of alkaline enzyme detergents. Other industrial enzymes included in this group are chymotrypsin, which with trypsin is a component of pancreatin, and the fungal serine alkaline proteinase of *Aspergillus orzyae*, which is a component of commercial *Aspergillus* proteinase preparations.

Table 1 Classification of Endopeptidases (Sub-sub-groups 3.4.21–3.4.24)

EC number	Recommended name	Source/comments
3.4.21	*Serine proteinases*	
3.4.21.1	Chymotrypsin	Pancreas
3.4.21.2	Chymotrypsin C	Pancreas
3.4.21.4	Trypsin	Pancreas
3.4.21.11	Elastase	Pancreas, *Pseudomonas*
3.4.21.12	*Myxobacter* α-lytic proteinase	*Mycobacterium sorangium*
3.4.21.14	Subtilisin	*Bacillus subtilis*
	E. coli periplasmic alkaline proteinase	
	Aspergillus alkaline proteinase	
	Tritirachium alkaline proteinase	*Tritirachium album*
	Arthrobacter serine proteinase	
	Pseudomonas serine proteinase	*Pseudomonas a ruginosa*
	Thermomycolin	*Malbranchea pulchella*
	Thermophilic *Streptomyces* serine proteinase	*Streptomyces rectus*
	Candida lipolytica serine proteinase	
3.4.21.19	*Staphylococcal* serine proteinase	*Staphylococcus aureus*
3.4.21.31	Urokinase	Kidney
3.4.22	*Thiol proteinases*	
3.4.22.2	Papain	*Papaya* latex
3.4.22.3	Ficin	*Ficus* latex
3.4.22.4	Bromelain	*Ananas comosus*
3.4.22.6	Chymopapain	*Papaya* latex
3.4.22.8	Clostripain	*Clostridium histolyticum*
3.4.22.9	Yeast proteinase B	
3.4.22.10	*Streptococcal* proteinase	
3.4.22.13	*Staphylococcal* thiol proteinase	
3.4.23	*Carboxyl (acid) proteinases*	
3.4.23.1	Pepsin A	Gastric juice
3.4.23.2	Pepsin B	Gastric juice
3.4.23.3	Pepsin C	Gastric juice
3.4.23.4	Chymosin	Gastric juice of young animals
3.4.23.6	Microbial carboxyl proteinases:	
	Aspergillus oryzae carboxyl proteinase	
	Aspergillus saitoi carboxyl proteinase	
	Aspergillus niger var *macrosporus* carboxyl proteinase	
	Penicillium janthinellum carboxyl proteinase	
	Rhizopus carboxyl proteinase	*Rhizopus chinensis*
	Endothia carboxyl proteinase	*Endothia parasitica*
	Mucor pusillus carboxyl proteinase	*M. pusillus, M. miehei*
	Candida albicans carboxl proteinase	
	Paecilomyces varioti carboxyl proteinase	
	Saccharomyces carboxyl proteinase	Baker's yeast
	Rhodotorula carboxyl proteinase	*Rhodotorula glutinis*
	Physarum carboxyl proteinase	*Physarum polycephalum*
	Tetrahymena carboxyl proteinase	*Tetrahymena pyriformis*
	Plasmodium carboxyl proteinase	*Plasmodium berghei*

Table 1 Classification of Endopeptidases (Sub-sub-groups 3.4.21–3.4.24)—*cont.*

EC number	Recommended name	Source/comments
3.4.24	*Metalloproteinases*	
3.4.24.3	*Clostridium histolyticum* collagenase	Similar in many bacteria
3.4.24.4	Microbial metalloproteinases:	
	Streptomyces thermophilus intracellular proteinase	Also *S. diacetilactis*
	Sarcina neutral proteinase	
	Micrococcus caseolyticus neutral proteinase	
	Staphylococcus aureus neutral proteinase	
	Bacillus subtilis neutral proteinase	
	Aeromonas proteolytica neutral proteinase	
	Pseudomonas aeruginosa neutral proteinase	
	Pseudomonas aeruginosa alkaline proteinase	
	Escherichia freundii proteinase	
	Bacillus thermoproteolyticus neutral proteinase	
	Penicillium roqueforti neutral proteinase	
	Streptomyces griseus neutral proteinase	A component of pronase
	Aspergillus oryzae neutral proteinase	
	Myxobacter β-lytic proteinase	
	Serratia marcescens extracellular proteinase	
3.4.24.7	Vertebrate collagenase	Animal tissues
3.4.24.8	*Achromobacter iophagus* collagenase	
3.4.24.9	*Trichophyton mentagrophytes* keratinase	

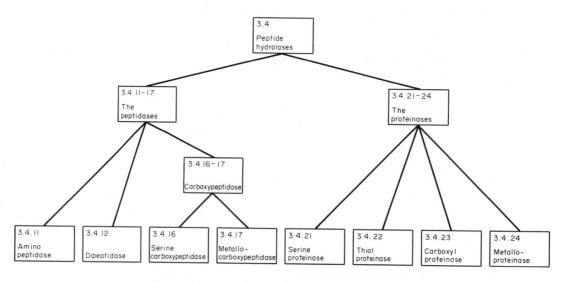

Figure 2 Classification scheme for the peptide hydrolases

These enzymes are most active at pH values of about pH 10. They are sensitive to DFP and a potato inhibitor but not to TLCK or tosyl-L-phenylalanine chloromethyl ketone (TPCK). While their specificities towards the B-chain of insulin vary somewhat with enzyme source, as a group they have much broader specificity towards this substrate than the trypsin-like enzymes. They are all specific against aromatic or hydrophobic amino acid residues at the carboxyl side of the splitting point (Johansen *et al.*, 1968).

(i) Subtilisins

Subtilisin Carlsberg, produced by *B. licheniformis*, was first prepared and crystallized by Guntelberg and Ottesen (1952). Originally it was observed to have one component but other workers separated the enzyme into several components (Hall *et al.* 1966; Zuidweg *et al.*, 1972). Verbruggen (1975) reported two antigenically different components, each of which underwent autodigestion. The major component was identified as subtilisin Carlsberg, which is the most important detergent proteinase.

Subtilisin BPN derived its name from a commercial enzyme preparation, Bacterial Protease Nagarase, from which it was purified. Smith *et al.* (1968) reported that another enzyme, subtilisin

Novo, isolated by Ottesen and Spector (1960), had an amino acid sequence identical with that of subtilisin BPN. While this enzyme is also reported to be used in detergents (Barfoed, 1981) it is of much less commercial significance than subtilisin Carlsberg.

Some of the properties of subtilisin Carlsberg and subtilisin BPN are compared in Table 2. It should be noted from this table that these enzymes have broad specificities. About 30–40% of the peptide bonds in casein are hydrolyzed by subtilisin Carlsberg. The Carlsberg enzyme is active over a broader pH range than the BPN enzyme, and while calcium improves the stability of both enzymes at high temperature or extremes of pH, the BPN enzyme is more dependent on calcium for this stability. All amino acid differences between Carlsberg and BPN proteinases with one exception (position 31, Ile in BPN but Leu in Carlsberg) are located on exterior chain segments of the molecule, which implies that most of the substitutions in amino acid content of the two sub-tilisins are conservative (Richards and Vithayathil, 1959). Alkaline variants of the subtilisin Carlsberg enzyme have been produced which exhibit higher pH–activity profiles than the original enzyme.

Table 2 Some Properties of Subtilisin Carlsberg and Subtilisin BPN

Property	Carlsberg	BPN
Number of peptide chains	1	1
Number of amino acids	274	275
Number of homologous amino acids	217	217
Number of cysteine residues	0	0
Isoelectric point	9.4	9.1
Dependence on calcium for stabilization at high temperatures or pH extremes	Less	More
pH optimum	8–9	8–9
% of maximum activity at pH 7.0	40–70	20
Inhibitors	DFP, PMSF	DFP, PMSF
	Cereal and legume inhibitors	Cereal and legume inhibitors
	Organophosphorus compounds	
Relative activities on BTEE[a]	3–4	1
Number of bonds hydrolyzed in insulin B chain	7	7

[a] BTEE = benzyl-L-tyrosine ethyl ester.

(ii) Other microbial alkaline serine proteinases

A range of *Aspergilli* including *A. flavus*, *A. oryzae*, *A. sojae* and *A. sydowi* (Nakagawa, 1970) have been shown to produce alkaline serine proteinases. Differences in specificity have been observed by Nakadai *et al.* (1973). With detergent applications in mind, thermostable fungal enzymes from this group have been isolated from *Tritirachium album* (Ebeling *et al.*, 1971), *Malbranchea pulchella* (Ong and Gaucher, 1976), *Acremonium kiliense*, *Fusarium* and *Gibberella* spp. (Isono *et al.*, 1972). The *Tritirachium* enzyme had a broad pH optimum, pH 7–12, and a temperature optimum of 65–70 °C. None, however, appears to have achieved any commercial success. Another interesting serine alkaline proteinase is produced by *Streptomyces rimosus* under industrial fermentation conditions for oxytetracycline production (Pokorny and Vitale, 1980).

40.3.2 Thiol Proteinases

Enzymes in this group have cysteine at their active site. In general they are activated by reducing compounds and inhibited by oxidizing agents. They are susceptible to sulfhydryl (SH) reagents such as pCMB and are activated by reducing agents such as cysteine or hydrogen cyanide. pH optima for these enzymes are in the neutral range. The important commercial thiol proteinases are the plant proteinases, papain, ficin and bromelain. Papain is obtained together with chymopapain and lysozyme from the latex of *Papaya*. Bromelain is present in both the fruit juice and stems of pineapple (Heinicke, 1953). Unlike papain and ficin, bromelain is a glycoprotein (Yasuda *et al.*, 1970). Ficin is extracted from the latex of different species of the genus *Ficus*. Of these, papain has by far the major market share while ficin is the least significant. Some of the general properties of these enzymes are given in Table 3. It should be noted that papain is the most thermostable of these enzymes. It is widely used as a broad spectrum protease and rarely produces bitterness in food protein hydrolysates (Godfrey and Reichelt, 1983).

Table 3   Properties of Commercial Plant Thiol Proteinases

Enzyme	Papain	Bromelain	Ficin
EC number	3.4.22.2	3.4.22.4	3.4.22.3
Source	*Carica papaya*	*Bromeliaceae* spp.	*Ficus carica*
pH optimum	4.5–7.0	5.0–8.0	5.0–8.0
Temperature optimum	60–75 °C	50 °C	60 °C
Amino acid specificity	Broad	Preference for basic and aromatic	Preference for aromatic
Inhibitors	All are inhibited by oxidizing agents, sulfhydryl reagents and heavy metals		
Activators	Reducing agents, thiol compounds, EDTA		
Industrial importance	Major	Limited	Very limited
Industrial applications	Brewing, meat tenderizing, fish processing, leather, textiles, animal feeds, pharmaceuticals, baking, general protein hydrolysis		

Cysteine 25, at the active site of papain, is essential for its activity and its oxidation inhibits activity. Activity can be restored by reducing agents or EDTA. Activation by cyanide or thiol compounds is due to regeneration of an SH group from a S—S bond. Papain is very active in the hydrolysis of esters and amides and is strongly inhibited by oligopeptides that possess phenylalanine as the second amino acid from the carboxyl terminus (Schechter and Berger, 1968). Chymopapain, which can be separated into two components, has many similarities with papain. Papain and chymopapain show different relative activities in haemoglobin and milk clotting assays and chymopapain is somewhat more stable at 75 °C, pH 7.2 and at pH 2.0 (Kunimatsu and Yasunobu, 1970).

Papain, ficin and bromelain resemble each other in the amino acid sequence around the reactive cysteine (Takahashi *et al.*, 1973).

The best known microbial thiol proteinases are clostripain and *Streptococcal* proteinase, but these are not of direct commercial interest. Clostripain shows strong specificity against basic amino acids at the carboxyl side of the splitting point and cleaves the insulin B chain at the same two points attacked by bovine trypsin, although compared with trypsin it has low esterase activity (Morihara, 1974). The specificity of *Streptococcal* proteinase as assessed by hydrolysis of the insulin B chain is much broader and similar to that of papain.

40.3.3   Acid Proteinases

Acid proteinases are widely distributed in animal cells, moulds and yeasts but are seldom found in bacteria. Many of them contain aspartate residues at their active site (Kovaleva *et al.*, 1972) and they show specificity towards aromatic or bulky amino acid residues at both sides of the splitting point. Their pH optima are in the range 3–4 and they are insensitive to inhibitors such as DFP, pCMB and EDTA.

These enzymes may be classified into two groups based on whether their catalytic activity resembles rennin or pepsin (Matsubara and Feder, 1971). Renin-like enzymes such as calf rennet or microbial rennets are characterized by their ability to clot milk and have major commercial importance in cheese manufacture. The group of pepsin-like acid proteinases includes enzymes such as bovine and porcine pepsin and commercial microbial acid proteinases mainly of *Aspergilli*. The microbial acid proteinases have application in the hydrolysis of soybean protein in soy sauce manufacture, improvement of baking properties of flour and digestive aids. Bovine and porcine pepsins are used commercially as 50:50 mixtures with calf rennet as suitable formulations for cheese making.

40.3.3.1   *Rennin and rennin-like proteinases*

These enzymes are highly specific enzymes which act on the κ-casein fraction of milk. Casein exists in fresh milk and fresh skim milk as micelles of diameter 400–2800 Å which are in equilibrium with soluble casein. The major components of these micelles are α-S, β- and κ-casein (McKenzie, 1971; Jolles, 1975). The soluble calcium salt of κ-casein combines with α-casein, which forms an insoluble salt, giving stability to the resultant micelle. Rennet acts on the κ-casein component of the micelle, releasing a soluble peptide and leaving an insoluble fraction p-κ-casein which causes the micelle to be destabilized and the milk to clot (Waugh, 1971). An essential feature of these enzymes is their proteolytic specificity towards κ-casein, without attacking the α-S or β-casein fractions with a resultant production of bitter peptides.

(i) Animal rennet

Calf rennet is still the most prevalent cheesemaking enzyme. It is extracted from the fourth stomach of the unweaned calf and contains 88–94% chymosin and 6–12% pepsin (Burgess and Shaw, 1983). It should be noted that the percentages in an older animal are quite different, with pepsin amounting to 90–94% of the total enzyme and chymosin only constituting 6–10%. It follows from this that the ratio of the two activities varies with the age of the calf at slaughter. Rennin is formed from the inactive precursor, prorennin, mainly by partial hydrolysis by autocatalysis. Rennin rapidly hydrolyzes a specific phenylalanylmethionine peptide bond in κ-casein (Jolles *et al.*, 1968). It only attacks two bonds in the insulin B chain, a Glu-Ala and Leu-Val bond. The same two bonds are hydrolyzed by the rennet of *Mucor miehei*. In contrast, pepsin hydrolyzes five bonds and this reflects also its broader spectrum of activity on casein. Rennin has a very narrow pH and temperature range for activity and calcium is required for the reaction to proceed to coagulation; the optimum pH is 6.4–6.6.

(ii) Microbial rennet

The most important microbial cheesemaking enzymes are the rennin-like enzymes from *Mucor miehei* and *Mucor pusillus* and these have been used with considerable success since they were introduced a dozen years ago. For reviews on microbial rennets the reader is referred to Sardinas (1972), Arima *et al.* (1970), Ottesen and Rickert (1970) and Whitaker (1970). It has been demonstrated that the microbial rennets of *M. miehei* and *M. pusillus* produce a pattern of hydrolysis products in milk similar to that produced by calf rennet (Vanderpoorten and Weckx, 1972). The rate of liberation of non-protein nitrogen from whole casein and from the α-S and β-fractions were higher for the microbial rennets than for calf rennet. All the coagulating enzymes tested liberated large quantities of non-protein nitrogen from κ-casein. Comparative trials have shown that *M. miehei* rennet compared favourably with rennet in cheesemaking trials with respect to curd yield and firmness. Organoleptic tests have also been satisfactory. Reference has been made to the presence of contaminating lipases and proteinases (Green, 1977) and purification procedures are used to eliminate these contaminants from commercial products. For example, the presence of undesirable unspecific protease has been observed as a side activity in *M. pusillus* culture broths and this can be removed by silicate absorption. Lipase may be removed from crude preparations of *Mucor* enzyme by heating, salt or organic solvent treatment or pH inactivation.

Martens and Naudts (1976) drew the following conclusions in a technological evaluation of microbial rennets. *M. miehei* enzyme produces good quality cheese, although in some cases of slightly lower quality than calf rennet. Bitter flavours were seldom noticed. *M. pusillus* rennet may be used for quite a number of cheeses without encountering any major problem as far as organoleptic quality and cheese yield are concerned. The age of the cheese and the method of production are sometimes important.

Another microbial rennet substitute, *Endothia parasitica* proteinase, catalyzes more extensive hydrolysis of casein and κ-casein than rennet or pepsin. While good quality cheese can sometimes be produced, particularly when high heating temperatures are applied, in most cases the enzyme cannot be used.

40.3.3.2 Pepsin and pepsin-like acid proteinases

Pepsin acts in the pH range 1.5–4.0 and has a temperature optimum of 60 °C. The enzyme is stable between pH 2–4. Consequently it can hydrolyze proteins denatured by acid precipitation and the hydrolysis is often characterized by release of bitter peptides. It is formed from a precursor, pepsinogen, which is found in the stomach mucosa of animals (man, beef, swine, chicken). Pepsinogen is converted to pepsin by pepsin itself at pH values of less than 5.0. Haard and coworkers (1982) recently discussed the properties of pepsins, isolated from the stomachs of three marine fish, which had low temperature optima (15–20 °C) and evaluated their possible application as food processing aids.

Microbial pepsin-like acid proteases are mainly produced by strains of *Aspergillus* and *Rhizopus* spp. Yoshida (1956) isolated the enzyme described as Aspergillopeptidase A and this enzyme had its main application in digestive aid preparations, having a pH optimum at 2.5–3.0. The acid proteinase of *Aspergillus oryzae* has a pH optimum at 4–4.5. Two components are produced in semi-solid culture, one which is carbohydrate free and the other containing 50% carbohydrate. Tsujita and Endo (1977) suggest that the carbohydrate appears to confer better heat stability on these enzymes. All of the enzyme isolated from submerged culture is glycoprotein. The fungal acid protease of *Rhizopus* spp. (Godfrey and Reichelt, 1983) has a pH optimum between 5–6 and a temperature optimum of 50–55 °C. It has broad specificity and will clot milk. These authors also describe two *Aspergillus* fungal acid proteases. One of these, effective in the range pH 3–6 but of relatively low temperature tolerance, is produced by *A. niger* var. *macrosporus*. A more substrate specific variant enzyme with respect to pH, which has a pH optimum of 5.0 on haemoglobin and 7.5–8.0 on casein, is produced by another *Aspergillus* strain. This enzyme has greater temperature tolerance and performs well at 60 °C at pH 4.0.

40.3.4 Metallo-proteinases

There are two groups of metallo-proteinases which are of commercial interest: the neutral and alkaline metallo-proteinases. All have a metal ion involved in the catalytic mechanism and are consequently not surprisingly sensitive to chelating agents such as EDTA. They are insensitive to sulfhydryl agents such as pCMB and inhibitors such as DFP.

40.3.4.1 Neutral proteinases

These enzymes contain an essential metal atom, usually zinc, and have pH optima near neutral. They are widely distributed among microorganisms and have specificity towards hydrophobic amino acids on the amino side of the peptide linkage being attacked. They exhibit high activity with the synthetic substrate FAGLA (furylacrolylglycylleucinamide), which is not cleaved by most serine proteinases. These enzymes are produced with particular abundance by *Bacillus* spp. (Ward, 1983) and also by *Aspergillus* spp.

(i)   Bacterial neutral proteinases

Bacillus subtilis neutral proteinase has been reported (Godfrey and Reichelt, 1983) to have quite a narrow specificity of action with a preference for bonds linking aromatic amino acids. This report appears to be at variance with the work of Morihara *et al.* (1968) who observed, using synthetic substrates, that *Bacillus* proteinases were more specific to aliphatic than aromatic residues while the reverse was the case with the enzymes from *P. aeruginosa*, *S. griseus* and *A. oryzae*. Differences here have been observed between the specificities of the neutral proteinases of *B. subtilis* and *B. stearothermophilus* which is called thermolysin (Feder and Shuck, 1970; Morihara and Tsuzuki, 1970), based on their action on synthetic peptides. The mechanism of binding and hydrolysis of substrate by thermolysin has recently been evaluated and the catalytic properties of this enzyme were compared with carboxypeptidase A (Lipscombe, 1983). The *B. subtilis* enzyme is stable in the pH range 5–8 and has a temperature optimum for activity of 55 °C. Of significance to its application in brewing is that this enzyme does not appear to be sensitive to inhibitors of alkaline proteinases that are present in raw barley. It has been reported that the ratio of neutral proteinase to subtilisin is between 1:1 and 4:1 (Millet, 1970; Yoneda and Maruo, 1975). There are also reports that subtilisin degrades neutral proteinase and that PMSF protects neutral proteinase from such degradation. Calcium also stabilizes the enzyme. Other *Bacillus* species which produce this enzyme include *B. cereus*, *B. megaterium*, *B. stearothermophilus* (Keay, 1972), *B. thuringiensis* (Li and Yousten, 1975), *B. pumilus* (Tran Chau and Urbanek, 1974), *B. polymyxa* (Griffin and Fogarty, 1973) and *B. amyloliquefaciens* which is closely related to *B. subtilis*.

Structural studies on thermolysin, produced by *B. stearothermophilus*, indicate that this single chain peptide enzyme, of molecular weight 34 000, is folded to form two lobes separated by a deep cleft, in the bottom of which the essential zinc atom is located. Two histidine residues and glutamic acid bind the ion and only slight conformational changes occur when it is removed. It has therefore been suggested that zinc is not essential for stability (Weaver *et al.*, 1976). This enzyme has high thermostability, retaining 50% of its activity after 1 h at 80 °C (Endo, 1962), and four calcium atoms, which the enzymes also contains, contribute to this heat stability. Although there is considerable homology and structural similarity between *B. amyloliquefaciens* neutral proteinase and thermolysin, it is not at all as thermostable as thermolysin, losing 50% of its activity at 59 °C after 15 min. It has been suggested that the lower percentage of hydrophobic residues in the *B. amyloliquefaciens* enzyme, and the fact that it contains only two calcium atoms, may account for its lower temperature stability.

(ii) Fungal neutral proteinases

This enzyme is the most important protease component of commercial fungal protease preparations which have applications in baking, food processing, protein modification and in the leather, animal feeds and pharmaceutical industries. *Aspergillus oryzae* is the predominant source of the enzyme. Its affinity for hydrophobic amino acid residues is an advantage in minimizing bitterness in protein hydrolysates. This enzyme has a pH optimum near 7.0 and a temperature optimum of 45 °C, implying that it is not very thermostable. Commercial preparations containing it also usually contain acid and alkaline proteinases and a considerable amount of peptidase activity. The proteolytic mixture thus exhibits a wide pH–activity spectrum of broad specificity which can extensively degrade proteins.

Nakadai *et al.* (1972, 1973) identified two metallo-proteinase components in *A. oryzae* enzyme preparations. One had similar pH and temperature activity properties to those described above, while the second had a pH optimum of 5.5–6.0 and higher temperature stability. Sekine (1972) likewise reported isolation of two metallo-proteinase components from *A. sojae*, each having one zinc ion per molecule but having different structures, pH properties and substrate specificities. EDTA sensitive proteinases have also been identified from other *Aspergillus* species (Kishida and Yoshimura, 1964; Dhar and Bose, 1964).

40.3.4.2 Alkaline metallo-proteinases

These enzymes have pH optima in the region pH 7–9. They appear to be less sensitive to EDTA than neutral proteinases, requiring high concentrations of 10^{-2} M for inactivation, whereas neutral proteinases are sensitive below 10^{-3} M (Hampson *et al.*, 1963). The enzymes from *Pseudomonas aeruginosa* and *Serratia* spp. have been characterized.

Alkaline variants of neutral metallo-proteinases have been produced by *B. subtilis* (Godfrey and Reichelt, 1983). Full activity of one of these enzymes can be maintained at 60 °C and pH 9.0 for at least 1 h, making it suitable for use in the detergent industry. In addition, sensitivity to EDTA and phosphates is low and tolerance to oxidizing agents such as perborates is good.

40.4 COMMERCIAL APPLICATIONS OF MICROBIAL PROTEASES

Estimated world sales of industrial enzymes in 1981 indicate that proteases account for nearly 60% of the total industrial enzyme market (Table 4). It is predicted that the total value of the market will increase from $400 to $600 million between 1981 and 1985 (Godfrey and Reichelt, 1983). It is clear from the previous sections that proteases constitute a complex range of enzymes of widely varying properties with regard to specificity and pH/temperature activity and stability properties. Proteases with major industrial applications are distributed among all the sub-sub-groups and also come from animal, plant and microbial sources (Table 5).

40.4.1 Effect of Proteases on Functional and Taste Properties of Proteins

Protein hydrolysis can cause changes in the solubility and functional properties of proteins and foods. When protein concentrates are partially hydrolyzed there is a decrease in the viscosity and gelation properties of the solution. Protein foam volume is increased and protein foam stability is decreased as a result of limited proteolysis of egg albumin and whey protein concentrates. Meat tenderization occurs on limited proteolysis of the myofibrillar and stromal proteins of muscle. Dough viscosity is reduced and the structural properties of bread are altered as a result of partial hydrolysis of dough proteins during baking. Specific proteolytic attack on the ϰ-casein fraction of milk causes destabilization of the casein micelle, resulting in clot formation, which is an essential part of the cheesemaking process. Limited hydrolysis of proteins in wort or beer leads to the prevention of chill haze formation due to protein–phenolic interactions. These examples serve to

Table 4 Estimated Market Distribution of Industrial Proteases and Other
Enzymes in 1981

Enzyme	Type	World market ($ million)	Market share (%)
Proteases	Alkaline (detergent)	100	25.0
	Other alkaline	24	6.0
	Neutral	48	12.0
	Animal rennet	26	6.5
	Microbial rennet	14	3.5
	Trypsins	12	3.0
	Other acid proteases	12	3.0
α-Amylases		20	5.0
β-Amylases		52	13.0
Glucose isomerase		24	6.0
Pectinase		12	3.0
Lipase		12	3.0
All others		44	11.0

Table 5 Catalytic Nature and Sources of Major Industrial Proteases

Catalytic nature	Animal	Source Plant	Microbial
Serine proteinases	Trypsins		Alkaline proteases (*Bacillus* spp.)
Thiol proteinases		Papain Bromelain Ficin	
Carboxyl (acid) proteinases	Rennets		Microbial rennets (*Mucor* spp.)
	Pepsins		Acid proteases (*Aspergillus* and *Rhizopus* spp.)
Metallo-proteinases			Neutral proteases (*Bacillus* spp. and *Aspergillus* spp.)

illustrate some of the important functional properties contributed by proteins to foodstuffs and the applications of enzymes in food processing.

Hydrolysis of proteins can confer a bitter taste to the product. It is well known that the higher the content of hydrophobic amino acids, in the particular protein being hydrolyzed, the more pronounced is the tendency to form bitter tasting hydrolysates (Ney, 1971). There are three methods of controlling bitterness (Cowan, 1983) which may be used singly or in combinations, namely masking, removal or prevention. Polyphosphates mask the bitterness of casein hydrolysates. Gelatin has also been shown to mask the bitterness of protein hydrolysates, an effect which may be due to its high content of glycine (Schwille *et al.*, 1976; Stanley, 1981). Bitter peptides may be removed with activated carbon and glass fibre filters (Helbig *et al.*, 1980). They may also be masked and reduced by acid precipitation at pH 4.5 (Adler-Nissen and Olsen, 1982). Prevention of the production of bitter peptides may be achieved by limiting the proteolytic reaction time and by using proteases which minimize the production of hydrophobic peptides. As there can be a correlation between yield of soluble hydrolysate and the degree of hydrolysis, it is often necessary in practice to establish an acceptable compromise between the degree of hydrolysis and the level of bitter tasting peptides present.

Because of the highly nutritious nature of proteins it is important to guard against microbial infection during the protein hydrolysis process. This is of particular importance where cruder proteinaceous materials, for example fish or meat, are being used which may already contain microorganisms. In such cases consideration should be given to using enzymes which are active at higher temperatures or pH extremes and the incubation time should be reduced as far as possible. Mackie (1982) achieved satisfactory hydrolysis of fish protein using papain (65 °C, pH 6.5), alkaline protease (55 °C, pH 8.5) and neutral protease (50 °C, pH 7.0). In each case incubation time was limited to 30 min.

40.4.2 Protein Hydrolysis

Because of the high cost of animal protein relative to vegetable protein there has been a lot of interest in recent years in developing methods for processing and recovering protein hydrolysates from vegetable materials and wastes. In the development of commercial products from these materials, emphasis has to be placed on achieving a consistent product in high yields. In protein hydrolysis procedures this means tightly controlling the process so that the protein yield, solubility and degree of hydrolysis is constant from batch to batch.

Yield and solubility measurements can be made using conventional nitrogen estimation and centrifugation methods respectively. The degree of hydrolysis (DH), that is, the percentage of total peptide bonds hydrolyzed by the enzyme, can be regulated and standardized by using a pH-stat method developed by Olsen and Adler-Nissen (1979). For a given enzyme substrate system, five independent parameters can be defined: S, E/S, pH, temperature and time (S = substrate concentration, E = enzyme concentration). The advantage of controlling the process based on the DH concept is that all of the parameters except pH can be substituted by DH (Adler-Nissen, 1981). When the hydrolysis is carried out at a pH above neutrality, at a constant pH using a pH stat, the degree of hydrolysis is proportional to the consumption of alkali. Consequently, the end point of the reaction can be simply related to the amount of alkali used and the reaction terminated.

40.4.2.1 Soy protein hydrolysis

The potential applications of soy protein hydrolysates are now widely appreciated. Defatted soybean in particular, which is a residue obtained when oil is extracted from the soybean, is widely used as a foodstuff, because it has a high protein content and its amino acid composition is nutritionally excellent. The functional properties of unmodified soy proteins are not always appropriate and hence the need for protein hydrolysis. There are two classes of soy protein hydrolysates, highly functional hydrolysates and highly soluble hydrolysates. Highly functional hydrolysates (DH = 3) are normally incorporated into products at a low rate of inclusion and consequently the impact of bitterness is not important. Highly soluble soy hydrolysates (DH = 10) are used at high inclusion rates, for example in protein fortified soft drinks and special dietetic foods, and consequently strict control on bitterness is important.

A summary of two methods for production of highly functional hydrolysates using alkaline protease from *B. licheniformis* is given in Table 6. It should be noted that, although the same proteolytic conditions (enzyme dose rate, pH, temperature) were applied in both cases, the differences in other parts of the process result in formation of products which have very different whipping expansion properties. The differences in isoelectric solubilities and emulsification capacities were less significant (Cowan, 1983). The procedure for production of highly soluble hydrolysate is similar to process 1 of Table 6, except that the proteolysis is allowed to continue until a DH of 9–10 is reached. The reaction is stopped by reducing the pH to 4.0, inactivating the enzyme and precipitating the unconverted protein material which can be separated. The soluble protein fraction accounts for 65% of the starting material (Adler-Nissen, 1978; Olsen and Adler-Nissen, 1979).

Processes have also been described for production of soy hydrolysates with good whipping characteristics using pepsin in combination with acid or alkaline hydrolysis (Pour-El and Swenson, 1976; Gunther, 1974) or papain (Sawada *et al.*, 1977).

40.4.2.2 Production of soy sauce

Soy sauce manufacture involves the fermentation of steeped, dehulled, cooled soybeans initially with strains of *Aspergillus oryzae* for three days at 30 °C and later, after addition of brine, by other microorganisms such as yeasts and lactic acid bacteria. A flow diagram of the overall process is given in Figure 3. Wood (1982) has illustrated protease enzyme production and nitrogen levels for 100 hours after inoculation with *A. oryzae* and observed that maximum enzyme activity and acid soluble nitrogenous compounds are observed 40–45 hours after inoculation while maximum amino nitrogen content is achieved after about 75 hours. Acid, neutral and alkaline proteases are probably involved. Other enzymes associated with the process include lipase,

Table 6 Summary of Processes for Production of Highly Functional Soy Protein Hydrolysate

	Process 1	*Process 2*
Raw material	Soy white flakes	Soy meal
Pretreatment	Stepwise extraction (centrifugation, washing) at the protein's isoelectric point, pH 4.5	Extraction at pH 8.0 followed by centrifugation, sludge removal and ultrafiltration/concentration
Properties of pre-hydrolysis material		
Isoelectric solubility	5%	20%
Emulsification capacity	100 ml g^{-1}	130 ml g^{-1}
Whipping expansion	20%	380%
Proteolytic conditions	pH-stat hydrolysis, pH 8.0, 50 °C Alkaline protease, 12 000 Anson units per ton protein	pH stat hydrolysis, pH 8.0, 50 °C Alkaline protease, 12 000 Anson units per ton protein
Properties of post-hydrolysis material		
DH	3	3
Isoelectric solubility	42%	44%
Emulsification capacity	280 ml g^{-1}	220 ml g^{-1}
Whipping expansion	200%	1100%

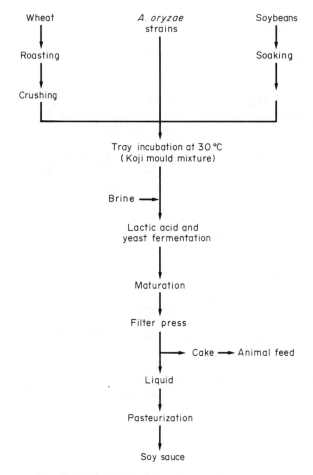

Figure 3 General flow diagram for soy sauce manufacture

phosphatase, cellulase, amylase and amyloglucosidase. Wood (1982) has provided detailed statistics on world trade in soy sauce. In 1978, Japan produced 1 434 000 tonnes of soy sauce, and other figures which the author has obtained allow him to support the view that fermented soy products come third after alcohol and milk proucts in a world 'league table' of fermented foods.

40.4.2.3 Gelatine hydrolysis

Gelatine hydrolysates have low levels of some amino acids (*e.g.* tryptophan) and hence they are of low nutritional value. They are incorporated into foods in combination with other proteins for dietetic purposes. They may also be used to give body to low-calorie drinks where they have an advantage over unhydrolyzed gelatine in that problems of gelation are avoided. Other applications are found in the cosmetic industry as a component of shampoos and ointments. In a typical process a 30% solution of gelatine is incubated with a mixture of alkaline and neutral proteases for up to 4 h, depending on the properties required, followed by pasteurization at 90 °C.

40.4.2.4 Casein and whey proteins

Tryptic and peptic casein hydrolysates have application in dietetic products for treatment of cystic fibrosis, where enzyme secretion by the pancreas is impaired. They suffer the disadvantage of having a bitter taste. Alkaline and neutral proteinase formulations can also be used to produce stable suspension calf milk replacers containing soy protein, whey powder, sodium caseinate and minerals (Van Krenemburg, 1979). Proteolytic modification of whey proteins using microbial proteinases and trypsin has been shown to improve the solubility and functionality of these proteins.

Mixtures of gelatine and whey may be used to produce mixed hydrolysates by treatment with pancreatic proteases. The hydrolysate has good nutritional properties, in contrast to the low essential amino acid composition of pure gelatine hydrolysates, and other properties of the product such as wettability, water solubility, clarity and taste are excellent (Schwille *et al.*, 1976).

40.4.2.5 Meat protein recovery

Proteases may be used to aid recovery of meat from bones, which can amount to 5% of bone weight. This enables production of a meat extract product which can be incorporated into canned meats and soups. Cowan (1983) describes a process whereby bones are crushed, slurried with an equal volume of water and heated in the presence of alkaline protease to 60 °C and held for 3–4 hours. After separation of solids the liquid is pasteurized at 98 °C for 15 seconds. Other processes are described by Denten *et al.* (1963), Criswell *et al.* (1964) and Connelly *et al.* (1966). Formerly acid hydrolysis was used for conversion of such meat scraps but this results in denaturation of tryptophan.

40.4.2.6 Fish protein hydrolysis

Proteases may be used for modification of fish protein to recover fish waste or to speed up processing of inedible fish to produce oil, fish solids and fish solubles. Use of added enzymes allows the process to be controlled so as to produce soluble fish hydrolysate of high nutritional value and to prevent the development of bitter flavours. Problems of rancidity are avoided by using fish with oil levels of less than 1%. With higher oil levels, anti-oxidants should be incorporated. It is desirable that enzyme incubations should be carried out at or above 60 °C to reduce microbial infection. Use of proteases also helps reduce viscosity of fish extracts, thus facilitating evaporation. The stages involved in this process are illustrated in Figure 4. Mackie (1982) has recently reviewed the various aspects of fish protein hydrolysis.

40.4.2.7 Meat tenderization

Conventional methods for tenderization of meat involving 'aging processes' are described by Bernholdt (1975). In these procedures the meat's endogenous proteinases, especially catheptic enzymes and a neutral protease, produce a tenderizing effect at cool temperatures over a period of 10 days to 4 weeks. Undesirable features were problems of meat shrinkage and microbial spoilage. With the development of vaccum packaging and controlled refrigeration processes, these latter problems have essentially been eliminated. Enzymes have played a prominent role in meat tenderization since the 1940s, particularly the plant proteases, papain, bromelain and ficin. In 1962, the process was revolutionized by Swift and Co. in the United States, who patented a pre-

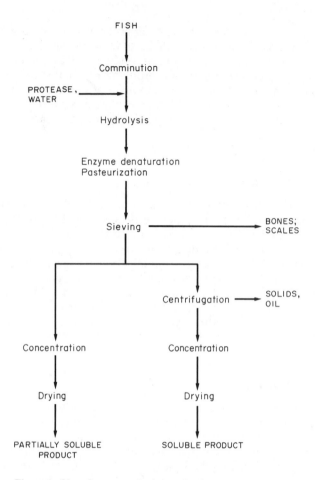

Figure 4   Use of proteases in the production of fish hydrolysates

tendering process (Beuk *et al.*, 1959) whereby papain is introduced into the animal's blood stream prior to slaughter, ensuring rapid and uniform distribution throughout the tissues. The following advantages have been claimed. (1) The pretendering process does not affect flavour, appearance or keeping quality of meat and aging is not necessary. (2) Because of uniform enzyme distribution all meat cuts are tenderized. (3) The product can be marketed fresh or frozen and greater carcass utilization is achieved.

Other proteolytic tenderization processes involve sprinkling the enzyme powder on the meat, dipping the meat in the enzyme solution, spraying the enzyme on to the meat or injecting the meat after slaughter. The disadvantages associated with these procedures relate to the difficulties of achieving uniform enzyme distribution and consequently product consistency. Differences have been observed in the tenderizing actions of plant and microbial proteinases and advantage has been taken of these differences in formulating meat tenderizers containing combinations of plant and microbial enzymes. Plant proteinases appear to act on the connective tissues, mainly collagen and elastin, and exhibit some action also on muscle fibre proteins. Bacterial and fungal preparations show only slight action on collagen fibres but have considerable effect on muscle fibres.

40.4.3   Protein Synthesis

Enzymatic peptide synthesis, involving ordered condensation of amino acids or peptides into polypeptides of defined sequences, represents one of the major potential applications of proteinases (Glass, 1981). A recent very significant example of this application is the Novo process for conversion of porcine to human insulin. The equilibrium of the enzymic proteolytic reaction

$$R^1\text{—CO—NHR}^2 + H_2O \rightleftharpoons R^1\text{—CO—OH} + R^2NH_2$$

lies far to the right under normal aqueous conditions. In principle the reaction is reversible and the condensation reaction is favoured by any scavenging mechanism which can keep the condensation product below its equilibrium concentration (Saltman *et al.*, 1977). This can be achieved if the condensation product is an insoluble peptide and this is the basis of the plastein reaction whereby concentrated peptide solutions are converted into insoluble protein by proteolytic enzymes (Isowa, 1978). The equilibrium of the reaction may also be shifted in favour of condensation by adding organic solvents such as 1,4-butanediol or glycerol, which suppress ionization of the carboxyl groups (Parr *et al.*, 1978). Where one of the condensation reactants is of high value and consequently limited concentration, a high concentration of a second expendable or recoverable reactant may be used to shift the equilibrium in favour of synthesis, provided the catalytic enzyme becomes saturated with the limiting substrate at low concentration. The semi-synthetic conversion of porcine insulin to human insulin is based on this principle (Morihara *et al.*, 1979). Enzymatic peptide bond resynthesis is favoured by the following conditions. (1) Substrates should be low in molecular weight and preferably be prepared by enzymatic degradation. (2) The substrate concentration should be in the range 20–40%. (3) Most pH optima for synthesis lie in the range pH 4–7 and proteinases may have different pH optima for peptide bond degradation and resynthesis.

Observations on the formation of plasteins were made since the beginning of this century and Fujimaki and coworkers (1977) have discussed possible applications of this reaction including improvement of the solubility and/or nutritional value of various proteins, preparation of special dietetic foods and removal of unwanted protein components such as bitter peptides. Watanabe and Arai (1982) have recently discussed the role of papain in the covalent modification of proteins, such as fish protein concentrate, soy protein isolate, casein, ovalbumin and gelatine, with resultant changes in their functional properties.

40.4.4 Detergent Enzymes

The market for enzyme detergents has grown rapidly in the past two decades, despite a temporary set back in the early 1970s when detergent workers developed allergies. This enzyme market is in excess of $100 million. Use of enzymes has gained widespread acceptance in a range of detergent types and the market share of such enzyme detergents is between 30 and 60% in most industrialized countries. Formulations for a number of enzyme containing detergents are reported by Barfoed (1981). Some of the major constituents are described in Table 7.

Table 7 Examples of Proteolytic Enzyme Detergent Formulations

Constituent	Pre-wash	Detergent type Liquid	Heavy duty
Anionics	5–15%	5–10%	5–15%
Nonionics	1–3%	10–40%	1–6%
Sodium tripolyphosphate	25–50%		20–40%
Sodium silicate	0–5%		5–10%
Sodium perborate			15–30%
Sodium carboxymethylcellulose	0.5–1%		0.5–1%
Sequestering agents		0–15%	
Optical brighteners/perfume	0.1–0.2%	0.1–0.5%	0.5–1%
Surfactants		5–15%	2–5%
Water		35–50%	
Sodium sulfate	35–60%		0–40%
pH	8–9.55	7–9.5	9.5–10.5
Alkaline protease	0.5–1%	0.4–0.8%	0.4–0.8%

Close examination of these formulations, together with an appreciation of the types of temperature washing programmes used, enables one to establish criteria related to the suitability of an enzyme preparation for use in detergents. Detergent enzymes must be stable during the normal shelf life of the detergent. The enzyme must also have adequate stability and activity in the detergent wash solution and consequently must be able to withstand high pH and temperature levels, sequestering agents and oxidizing agents. The enzyme should ideally have broad specificity and should be present in detergent powder in a dust-free form. When the properties of a number of industrial proteases are assessed in this way the unchallenged suitability of microbial alkaline proteases becomes clear (Ward, 1983). Unlike neutral metallo-proteases, they are not inhibited by chelating agents. Unlike thiol proteinases, they are not sensitive to oxidizing agents. Alkaline proteases have also high pH activity/stability properties and good thermostability.

While the alkaline serine protease of *B. licheniformis* is the predominant commercial detergent protease, there are reports of other sources of alkaline proteases suitable for detergents. The enzyme produced by *B. firmis* has a pH optimum of 11.0 and a temperature optimum of 60 °C (Aunstrup and Andresen, 1972).

Laboratory model washing systems have been developed to screen, develop or evaluate enzyme detergents and these are designed to simulate standard washing processes. It is possible to test efficiently a range of parameters such as enzyme type and dose rate, physical and chemical properties of the washing solution and types of stains/fabrics which can be treated.

Following the allergy problems encountered in 1981, methods were developed to reduce the dusting properties of powdered enzymes in detergents involving enzyme encapsulation or granulation. These processes also appear to improve the stability of the enzyme in the detergent powder. Non-ionic surfactants or sodium chloride/polyethylene glycol mixtures are used to encapsulate the enzymes. One process involves mixing the enzymes with melted non-ionic detergent and then spraying into a cold chamber so as to solidify the droplets as granules. It has been reported that cellulose aids the granulation process and produces particles with improved resistance to mechanical damage.

40.4.5 Dairy Industry

The single major application of proteases in the dairy industry is in cheesemaking, which accounts for 10% of the total industrial enzyme market. The market is divided between animal rennets and microbial rennets in the ratio 65:35. Calf rennet is the principal commercial animal enzyme, which is extracted from the abomasum or fourth stomach of suckling calves. The predominant enzyme of standard rennet preparations is the milk coagulating enzyme, chymosin or rennin, but extracts also contain varying amounts of pepsin (6–12%) depending on the age of the animal. Pepsin also has milk coagulating ability, although less than rennin, and greater proteolysis occurs. Nevertheless, encouraging results have been obtained using 50:50 mixtures of standard rennet and porcine or bovine pepsin. The requirement to find rennet extenders, such as pepsin or microbial rennet substitutes, relates both to the increasing scarcity of calf rennet, particularly in the United States, and to a lesser extent to religious and ethnic regulations against animal derived enzymes. Since 1960 there has been a worldwide increase in milk production and yield per cow and cheese manufacture has increased in step with this. Calf slaughter rate, on the other hand, has declined, thus creating a shortage in availability of rennet. Total calf slaughter rates in the US dropped from 8 million to 4 million per annum between 1960 and 1970, while in the same period cheese manufacture in the US increased from 1.5 to 2.2 billion pounds per annum and these trends created a demand for microbial rennet substitutes. A range of commercially available rennets are listed in Table 8.

The standard process for manufacture of Cheddar cheese involves adding rennet to the milk at 31 °C. With rates of approximately 200 ml per 1000 litres of milk, initial clotting occurs within 15 minutes and the curd is suitable for cutting after about 45 minutes. A higher temperature, about 35 °C, is used in manufacture of Swiss cheese. After addition of the enzyme, up to five minutes is allowed for mixing to disperse the enzyme. Then the milk is allowed to stand to allow the development of a compact curd. Important process parameters which vary from cheese to cheese are milk temperature, milk acidity and rennet dose rate. At lower temperatures, 20–25 °C, a soft cheese curd is produced. At temperatures of 30–32 °C, curds will be firm and of good cutting quality. Coagulation at higher temperatures produces tough, rubbery curds. The normal pH range for coagulation of milk is 6.4–6.6 and control of this is important as lower pH values down to 5.8 increase curd firmness. Below this pH, firmness decreases. Rennet dose rates can range from 30–50 ml per 1000 litres of milk, for production of cottage cheeses, where a curd acid precipitation step is involved, to 100–450 ml per 1000 litres of milk for other cheeses, including Cheddar.

During cheesemaking, milk coagulating enzymes act in three phases: the enzymic phase, the clotting phase and the proteolytic phase. The enzymic and clotting phases may be accelerated by increasing the temperature up to 45 °C. At temperatures below 15 °C the enzymic phase proceeds but clotting does not occur until the milk is heated and advantage can be taken of this in continuous cheese-making processes where the cheese is first cold-renneted and later coagulated continuously in special devices (Kosikowski, 1975). In the proteolytic phase, further degradation by

Table 8 Commercial Rennet Preparations

Product name	Manufacturer	Source
Caft rennet	Hansens	Calf
Calf rennet	Dairyland Food Laboratories	Calf
Beef rennet	Dairyland Food Laboratories	Cow
Cabo	Hansens	Calf/ox
Quickset	Dairyland Food Laboratories	Calf/pig
Stabo	Hansens	Ox
50:50	Hansens	Ox/pig
Regalase	Dairyland Food Laboratories	Microbial/pig
Hannilase	Hansens	*Mucor miehei*
Fromase	Gist Brocades	*Mucor miehei*
Marzyme I	Miles Laboratories	*Mucor miehei*
Morcural	Pfizer Inc.	*Mucor miehei*
Rennilase	Novo	*Mucor miehei*
Modilase[a]	Hansens	*Mucor miehei*
Marzyme II[a]	Miles Laboratories	*Mucor miehei*
Rennilase TL[a]	Novo	*Mucor miehei*
Emporase	Dairyland Food Laboratories	*Mucor pusillus*

[a] These are second generation, thermolabile microbial rennets.

rennet of the casein fractions is thought to occur. Processes for continuous cheese production have been described by Berridge (1970), Nikolic (1970) and Stenne (1975).

Burgess and Shaw (1983) have provided details of costings (sterling) of various commercial rennets as in October 1981. Standard rennet costs £4.05 per litre. Other mammalian rennets and rennet mixtures cost £2.85–3.35. Most microbial rennets are produced as triple stength material but the equivalent cost of single strength material would be £2.50–2.70 per litre. 10 000 litres of milk produces approximately 1 tonne of cheese and requires 2.5 litres of standard strength rennet when applied at an average rate of 0.25% v/v milk. It was estimated that the enzyme costs only amounted to 0.5% or 0.76% of the raw material milk costs for microbial or calf rennet respectively. It is clear from these details that there is very little cost advantage in using rennet substitutes and they are not commercially acceptable if even the slightest reduction in yield or quality is observed.

Residual coagulating activity in whey produced from milk treated with microbial rennet can cause problems when used in dairy foods, baby foods or dietary aids. This problem has been overcome by the introduction of a number of second generation thermolabile microbial rennets (see Table 8) which are more easily denatured by heat treatment during processing of whey.

Proteases other than rennet hydrolyze the curd proteins during the cheese ripening stage. Evidence has been provided which would indicate that proteases associated with *Penicillium caseicolum*, *P. roqueforti* and *Streptococci* all contribute to degradation of polypeptides during cheese ripening. Law (1982) has discussed the role of proteases from various sources on cheese ripening. Native milk proteases appear to have little effect. Mesophilic starter bacteria play a significant proteolytic role, contributing both to flavour and texture development. Thermophilic starters also contribute to proteolysis but their action is less well understood. Because of the capital equipment costs associated with cheese processing due to the length of the ripening stage, methods for accelerating the ripening of cheese by use of added enzymes are being investigated. Law and Wigmore (1982) tested a range of proteases for their ability to accelerate cheese ripening and demonstrated that, while acid and alkaline proteases produced off-flavours, *B. subtilis* neutral protease significantly enhanced flavour intensity without producing bitterness. Proteases in combination with lipases can be used to produce strongly flavoured cheese by adding the enzymes to scalded curds and then curing at 10–25 °C for 1–2 months (Godfrey, 1983).

An appreciation of the contribution of proteases to cheese flavour production has led to the development of enzyme modified cheeses (EMC). These are cheese flavour additives, produced by addition of specific proteases and lipases to a slurry of a natural cheese variety and incubated for several weeks under controlled conditions. EMC pastes of Cheddar, Edam, Swiss, Provolone, Romano, Parmesan, Mozzarella, Cream and Blue are available. These products are incorporated into processed cheeses, cheese spreads, cheese dips and other products. Meat flavours may be produced also by proteolysis of milk (Jaeggi *et al.*, 1977).

A method for production of processed cheese from a combination of natural cheese and rework process cheese is described by Kichline and Sharpf (1972). This involves the incubation of protease from *Aspergillus flavus-oryzae* with a mixture of the two cheese batches for a short period during the pasteurization step, thereby providing the processed cheese with improved properties.

40.4.6 Brewing

Proteases have two major applications in brewing. They may be used during the cereal mashing process to increase the yield of extract and level of α-amino nitrogen of the wort produced. Proteases are also used during the finishing stages of beer production to remove chill haze (chillproofing) from beer. The overall brewing process is illustrated in Figure 5.

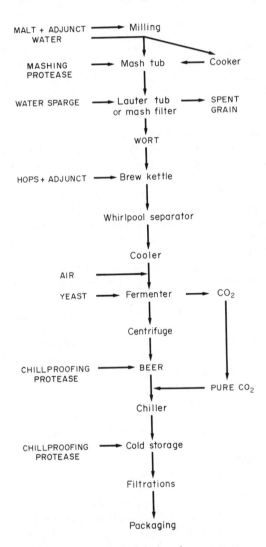

Figure 5   Summary of the brewing process

In an all-malt mash, four enzymes contribute to the production of high yields of fermentable extract: α- and β-amylases, which liquefy and saccharify the malt starch, producing a carbohydrate mixture of mainly glucose and maltose which is 80% fermentable; β-glucanase, which hydrolyzes the β-glucans of barley and malt, reducing the viscosity of the wort and improving its yield and filterability; and peptidases, which hyrolyze proteins in malt, thereby increasing the level of α-amino nitrogen and possibly also extract yield. Mashing temperature programmes are designed to have temperature holds to allow these various enzymes to act. A typical all-malt infusion mash programme is illustrated in Figure 6. The 30 minute hold near 50 °C is primarily a proteolytic stand, as the malt proteolytic enzymes are denatured at temperatures of 55–58 °C.

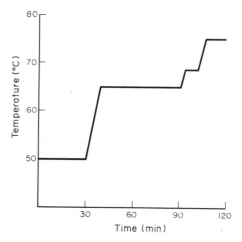

Figure 6 Temperature programme for an all malt infusion mash

Germinating barley contains a complex range of enzymes including three carboxypeptides, four neutral aminopeptidases, leucine aminopeptidase and dipeptidase (Enari and Mikola, 1972). Mikola *et al.* (1972) studied the inactivation of these enzymes during mashing. At pH 5.7–5.8 all the enzymes were stable at 45 °C. The aminopeptidases were inactivated during a 30 minute incubation at 55 °C, while the carboxypeptidases retained 70–100% of their activity under these conditions. A strong correlation was found between leucine aminopeptidase activity of malt and the yield of low-molecular-weight nitrogen on mashing. These peptidases have very low activity in barley. During germination, carboxypeptidase activity increases 10–20 times. Synthesis of proteases during germination is controlled by gibberellic acid.

The introduction of commercial enzymes, including proteases, to the brewing process has contributed to the overall economics and flexibility of the brewing operation. The cost differential between the price of malt and barley or other unmalted cereal or adjunct has encouraged the brewer to consider the inclusion of a proportion of unmalted material in the mash. As these adjuncts differ from malt, in that they lack the necessary mashing enzymes, added enzymes have to be used to compensate. A level of 140–180 mg of α-amino nitrogen per litre of wort is generally required for a normal fermentation and this level is produced during the malting process and by the action of malt proteinases at the proteolytic stand of a malt mash. Brewers' yeast uses the amino acids of wort during the fermentation in a preferential sequence in which glutamate, aspartate, asparagine, glutamine, serine, threonine, lysine and arginine are most rapidly absorbed while proline is scarcely utilized and predominates in finished beer (Gilliland, 1981). During the malting process there is substantial proteolytic conversion of the major barley protein fractions to peptides and amino acids (Wainwright, 1971) so that there is a much greater proportion of insoluble protein in barley, particularly hordein, than in malt. The necessary addition of protease to malt–barley mixed mashes, containing a substantial proportion of barley, relates both to the higher level of insoluble protein in the barley and the overall lower levels of malt proteinases present in the mash. Microbial proteinases from *Bacillus subtilis*, *Aspergillus oryzae* and plant proteases can be used effectively to enhance yield of extract and α-amino nitrogen in such cases. In worts produced for brewing it is important not to use too high a dose rate of added proteinases as this can lead to production of excess α-amino nitrogen in the wort which is not utilized by the fementing yeast. Residual amino acids in finished beer can have an adverse effect on the stability characteristics, quality and shelf life of the product. Over-hydrolysis of the wort proteins can also affect beer head retention so that careful process control is necessary.

Mashing processes involving enzymes are also used to produce malt, barley or other cereal extracts for general use in the food industry and the primary role of added proteases in such cases is to increase yield of extract. The cereal raw materials used vary from country to country and from plant to plant depending on local raw material availability, processing plant design and the properties required in the product. These factors will all contribute to determining the nature and dose rate of the protease to be applied.

The plant protease papain is used to counteract the problem of chill haze formation in beer. These hazes appear to result from formation of complexes between polyphenols, carbohydrate fragments and proteins. Papain is added at the cold maturation stage of beer finishing and causes chill haze proteins to be either hydrolyzed or precipitated. Ficin, bromelain and pepsin have also been used as chillproofing enzymes but these are now of little commercial importance (Bass and Cayle, 1975). Ferns (1977) reported a procedure whereby protease from *Streptomyces fradiae* was used successfully for beer clarification. The microbial rennet produced by *Mucor pusillus* was also effective (Nelson and Witt, 1973).

40.4.7 Baking

The unique and variable rheological properties of wheat flour doughs are largely due to the insoluble gluten proteins they contain. This protein confers on doughs their special viscoelastic and other mechanical properties. The 'strength' of dough is related to the quantity and quality of the gluten present and the number of disulfide bonds which form when dough is mixed. Proteases and peptidases are normally present at very low levels in wheat but are increased by malting. However, wheat gluten is not susceptible to attack by flour proteinases and protease supplementation is usually essential.

Addition of proteases to doughs improves their handling properties and increases loaf volume, thereby producing a more symmetrical loaf with good crumb and crust texture. Excess protease produces a sticky loaf with poor volume and crumb properties. While enzyme treatment has to be carefully controlled, it allows for increased automation of the process and the operation of new accelerated methods with highly standardized and optimized mixing and dough handling procedures. Proteolytic treatment also enables the baker to adapt flours of different qualities to a standard process. Fungal proteases from *Aspergillus oryzae* are normally used in bread production due to their lower temperature stability and also their amylase content, which also contributes to the baking process (Reichelt, 1983). It is estimated that two-thirds of the white bread baked in the US is treated with *A. oryzae* protease.

A. oryzae fungal protease contains both endo- and exo-peptidase. Endopeptidase modifies the viscoelastic properties of the dough by hydrolyzing interior gluten peptide bonds. Exopeptidase releases amino acids from gluten which can react with sugars through the Maillard reaction during baking, thereby contributing to flavour and crust colour. Improvements in the aroma of bread due to fungal protease have been observed (El-Dash and Johnson, 1967; Barrett, 1975).

Both fungal and bacterial proteases have application in the manufacture of cracker and biscuit dough. Bacterial neutral protease improves the extensibility of cracker dough, allowing it to be rolled very thinly without tearing and reducing bubbling during baking. This enzyme is also used in high-protein flour doughs for cookie, pizza and biscuit production. Papain may be used to weaken the gluten network in unbaked dough in a manner which reduces crust shrinkage.

40.4.8 Leather

The major steps involved in the processing of leather are summarized in Table 9. These are aimed at making it soft and resistant to decay or tough and resilient. Proteolytic enzymes may be used in the process at the soaking, dehairing and bathing stages.

Table 9 Steps Involved in the Processing of Leather

Curing	Steeping in brine bath and drying
Soaking	Rehydration and washing at tannery in the presence of sulfides or chlorites, surfactants and alkali
Dehairing and dewooling	Soaking of swollen skins in strong alkali and treatment with sulfides to solubilize root hair proteins and release hair
Bathing	Deswelling of skins and partial protein hydrolysis to make skins soft and easy to dye
Tanning	Acid pretreatment and chrome tanning or direct treatment with chemical stains and preservatives.

Alkaline proteases have been used very effectively at the soaking stage to aid the general washing of dirt and excess fats from the skins. The recommended enzyme dose rate is 70–150 Anson units per 100 kg hides or skins and this aids the uptake of water by the skin. Bovine or procine pancreatic trypsins are also effective and benefit from the additional presence of lipase and amylase in such preparations.

While chemical processes for dehairing are cost-effective, they suffer the disadvantages of creating both a safety hazard, due to the build-up of hydrogen sulfide, and a waste disposal problem. Treatment with alkaline protease, however, reduces the requirement for lime by 50%, reduces the dehairing process time from days to about 6 hours, and lowers the waste treatment costs. In addition, skin area is increased, more even dyeing is observed, and more hair is recovered undamaged. The overall process is easier to control and produces very clean hides. Coarse-wooled skins may be dewooled by surface application of alkaline protease to the flesh side of the skins followed by incubation overnight at ambient temperatures. For treatment of fine-wooled skins, neutral bacterial protease formulations are used.

Bating enzymic preparations may contain any of a range of enzymes: bacterial alkaline or neutral protease, fungal acid protease, trypsin or plant proteases. Criteria for selection of a particular enzyme or mixture relate to their pattern of action on the major leather proteins. In general, plant proteases exhibit greater relative activity towards collagen and elastin whereas microbial proteases are more active towards muscle proteins. Hides pretreated with alkali are much more sensitive to protease treatment than acid treated hides.

The applications of proteases to the removal of wool from skin and in hide and skin processing are discussed by Lewis (1981) and Haines (1981) respectively.

40.4.9 Other Applications of Proteinases

It is evident from the sections above that the proteases have a wide range of industrial uses. In addition to the applications already described, proteases are used to a lesser extent in a large number of specialist processes and in medicine.

Papain, bromelain and microbial proteases are incorporated into animal feeds to improve their nutritional value. Papain and other neutral proteases are used in the manufacture of yeast extract and production of other single-cell proteins. Proteases are also used to aid extraction of flavour and colour compounds from plants.

Pepsin, trypsin and papain have application in the manufacture of microbiological culture media while trypsin and alkaline protease are used in animal cell culture procedures. Neutral and alkaline proteases are used in membrane and equipment cleaning processes. Neutral proteases are also used to hydrolyze gelatin on photographic film to aid silver recovery.

Papain, bromelain, pepsin and *Aspergillus* proteases have been used as digestive aids in gastrointestinal disorders such as dyspepsia.

Urokinase, extracted from urine or produced from kidney cells in tissue culture, is used for treatment of clotting disorders. Brinase, a plasmin-like protease produced by *A. oryzae*, which appears to be a pepsin-like acid protease, hydrolyzes fibrin and fibrinogen. Patients on chronic haemodialysis with clotted arteriovenous cannulae showed restored vessel function several minutes after brinase treatment. The enzyme exhibits toxic side effects and is also inhibited by serum inhibitors, which consequently have to be neutralized by saturation with the enzyme prior to it becoming effective.

Collagenase, which hydrolyzes native collagen, has been used for debridement of dermal ulcers and burns and may also have application in lysis of diseased invertebral disks. Lysostaphin, a bacterial enzyme, may have application against coagulase positive *Staphylococcus aureus*. It has been administered to patients only topically for reduction of the *Staphylococcus* carrier rate in the nose and throat where it is both effective and non-toxic. It may have application as a bacteriocidal agent in instances of endocarditis, where rapid reduction of viable bacterial counts is necessary, and also against methicillin resistant *S. aureus* strains.

40.4.10 Protease Inhibitors

Proteases can be extremely destructive outside their normal environment in mammalian tissues and natural human plasma inhibitors inhibit most proteases that escape. An imbalance in protease–protease inhibitor activities can result in disease symptoms. Much effort has concentrated on investigating the role of proteases in disease and the evaluation of protease inhibitors in therapy (Powers, 1982).

In the disease pulmonary emphysema, the protease elastase attacks the major elastic protein of lung, elastin. As the disease progresses, with loss of lung elasticity, respiration becomes difficult and death often results. The serum α_1-protease inhibitor inhibits elastase and this protects the lung from proteases such as elastase and cathepsin G which are present in granules of human leucocytes. α_1-Proteinase inhibitor is inactivated by oxidants in cigarette smoke and the chronic obstructive pulmonary disease to which cigarette smokers are predisposed appears to be related at least in part to the removal of α_1-proteinase inhibitor. Emphysema also occurs in PiZ individuals who produce an altered α_1-protease inhibitor which accumulates in the liver and is not exported to the plasma.

In humans suffering from arthritis, collagenase produced by leukocytes and other phagocytic cells infiltrates the synovial fluid of arthritic joints and degrades collagen, the body's major connective tissue protein. The enzyme is also involved in idiopathic pulmonary fibrosis, where lung collagen is degraded. In both these diseases a protease–protease inhibitor imbalance is thought to exist.

In muscular dystrophy, extensive degradation of contractile proteins occurs. Lysosomal cysteine proteases such as cathepsin B may be involved. Proteolytic degradation of myelin in multiple sclerosis may be related to muscular dystrophy and emphysema but little is known of the proteases involved or their natural inhibitors.

A number of invasive tumours have been shown to secrete collagenase, which gives the malignant cells the ability to invade other cells and tissues by dissolution of connective tissue. The plasminogen activator protease increases in many neoplastic cells, which results in high levels of plasmin being produced near transformed cells. Protease inhibitors have been shown to block tumour promotion in mice and the low level of breast and bladder cancer in countries where diet is high in plant proteins may be due to the plant protease inhibitors present in these foods.

Bacterial and viral proteases appear to contribute to the infectivity of a number of diseases. For example, hemorrhagic pneumonia, corneal ulcers and cystic fibrosis may result from *Pseudomonas aeruginosa* infection. Most strains of *P. aeruginosa* produce elastase and another neutral protease, and elastase positive strains are much more pathogenic than elastase negative strains. *P. aeruginosa* elastase inactivates α_1-protease inhibitor.

Considerable research has been devoted to the extraction or synthesis of protease inhibitors for possible therapeutic treatment of diseases such as those mentioned above. Protein–protease inhibitors, isolated from animal, plant or microbial sources, have the disadvantage of immunogenicity when injected. Low molecular weight, non-protein protease inhibitors, produced by fermentation or synthesis, do not have this disadvantage and might also be taken orally. They may be directed to the active sites of proteases and designed to bind the enzyme reversibly or irreversibly.

A number of protease inhibitors are currently being used or evaluated for disease treatment. Pancreatic trypsin inhibitor is being used for treatment of pancreatitis, where pancreatic proteases leak into plasma. A reversible inhibitor of the angiotensin converting enzyme is being tested for hypertension treatment. Synthetic elastase inhibitors were used effectively to treat emphysema in laboratory animals. It is expected that selective synthetic protease inhibitors will be used in treatment of a range of conditions from the common cold to chronic diseases such as cancer and emphysema.

40.5 SUMMARY

Proteases constitute a very large and complex group of enzymes which play important nutritional and regulatory roles in nature. Commercially they are the most important industrial enzymes, accounting for 60% of the total enzyme market. Industrial proteases are produced by fermentation using overproducing strains of *Bacilli*, *Aspergilli*, *Rhizopus*, *Mucor* and other organisms of lesser importance or by extraction of protease containing animal glands or plant tissues.

All of the major industrial proteases are endo- rather than exo-cleaving enzymes. They can be subdivided, based on catalytic mechanism into four groups.

1. The serine proteinases have both serine and histidine at their active site. Commercial enzymes within this group include the pancreatic digestive proteases, trypsin and chymotrypsin, and the alkaline proteinases of *Bacilli* and *Aspergilli*. Subtilisin Carlsberg, produced by *Bacillus licheniformis*, is the predominant detergent enzyme, having a market amounting to $100 million.

2. The thiol proteinases have cysteine at their active site and are inhibited by sulfhydryl reagents and oxidizing agents. Papain, the plant protease, is the most important industrial

enzyme in this group. While its best known applications are in beer chillproofing and meat tenderization, it has a variety of other applications, mainly in food processing.

3. The carboxyl or acid proteases generally have pH optima in the range 3–4. This group contains the animal and microbial rennets, which have narrow specificity and cause coagulation of milk by hydrolysis of a specific bond in *x*-casein. Other important acid proteases are pepsin and the pepsin-like enzymes from *Aspergillus* and *Rhizopus*.

4. The metallo-proteinases all have a metal ion involved in their catalytic mechanism and are sensitive to chelating agents such as EDTA. The neutral proteases of *Bacillus subtilis* and *Aspergillus oryzae*, which belong to this group, have a wide range of applications in food processing, and in the leather, animal feed and pharmaceutical industries.

The major applications of proteases in food processing are in cheese making at the coagulation and maturation stage; in baking, to reduce dough viscosity and aid process automation; in tenderization of meat; in brewing, to increase α-amino nitrogen levels; and in the general manufacture of protein hydrolysates. Applications of proteases in protein synthesis have also been discussed. Proteases are used widely in detergents for hydrolysis of protein stains and the high pH optimum, stability properties, and resistance to oxidizing and chelating agents are all essential features of the detergent enzyme, subtilisin Carlsberg.

Proteases in mammals can be extremely destructive outside their normal environment and natural plasma protease inhibitors protect the tissues from such proteases. Imbalances in protease–protease inhibitor activities can cause disease and it is predicted that synthetic protease inhibitors will be used widely in therapy.

40.6 REFERENCES

Abiose, S. H., M. C. Allan and B. J. B. Wood (1982). Microbiology and biochemistry of miso (soy paste) fermentation. *Adv. Appl. Microbiol.*, **28**, 239–265.

Adler-Nissen, J. (1978). Enzymatic hydrolysis of soy protein for nutritional fortification of low pH food. *Ann. Nutr. Aliment.*, **32**, 205–216.

Adler-Nissen, J. (1981). Limited enzymic degradation of proteins. A new approach in the industrial application of hydrolases. *Abstr., 2nd European Congress of Biotechnology*, p. 39.

Adler-Nissen, J. and H. S. Olsen (1982). Taste and taste evaluation of soy protein hydrolysates. In *Chemistry of Foods and Beverages: Recent Developments*, ed. G. Charelambous and G. Inglett, pp. 149–169. Academic, New York.

Aidoo, K. E., R. Hendry and B. J. B. Wood (1982). Solid substrate fermentations. *Adv. Appl. Microbiol.*, **28**, 201–237.

Arima, K., J. Yu and S. Iwasaki (1970). Milk-clotting enzyme from *Mucor pusillus* var. *Lindt. Methods Enzymol.*, **19**, 446–459.

Aunstrup, K. (1974). Industrial production of proteolytic enzymes. In *Industrial Aspects of Biochemistry*, ed. B. Spencer, pp. 23–46. Federation of European Biochemical Societies.

Aunstrup, K. (1980). Proteinases. In *Microbial Enzymes and Bioconversions*, *Economic Microbiology*, ed. A. H. Rose, vol. 5, pp. 50–114. Academic, New York.

Aunstrup, K. and O. Andresen (1972). Proteolytic enzymes—useful in washing agents and as depilatory agents. *US Pat.* 3 827 938.

Barfoed, H. C. (1981). Detergents. In *Industrial Enzymology*, ed. T. Godfrey and J. Reichelt, pp. 284–293. Macmillan, London.

Barrett, F. F. (1975). Enzyme uses in the milling and baking industries. In *Enzymes in Food Processing*, ed. G. Reed, pp. 301–330. Academic, New York.

Bass, E. J. and T. Cayle (1975). Beer. In *Enzymes in Food Processing*, ed. G. Reed, pp. 455–471. Academic, New York.

Bernholdt, H. F. (1975). Meat and other proteinaceous foods. In *Enzymes in Food Processing*, ed. G. Reed, pp. 473–492. Academic, New York.

Berridge, N. J. (1970). Continuous production of curd. *US Pat.* 3 520 697.

Beuk, J. F., A. L. Savich, P. A. Goeser and J. M. Hogan (1959). *US Pat.* 2 903 362.

Blobel, G. and B. Dobberstein (1975). Transfer of proteins across membranes. *J. Cell Biol.*, **67**, 835–851.

Boethling, R. S. (1975). Regulation of extracellular protease secretion in *Pseudomonas maltophilia*. *J. Bacteriol.*, **123**, 954–961.

Brenner, M., H. R. Mullen and R. W. Pfister (1950). A new enzymic peptide synthesis. *Helv. Chim. Acta*, **33**, 568–591.

Bu'Lock, J. D., K. J. Nisbet and D. J. Winstanley (1982). *Bioactive Microbial Products: Search and Discovery*. Academic, London.

Burgess, K. and M. Shaw. (1983). Dairy. In *Industrial Enzymology*, ed. T. Godfrey and J. Reichelt, pp. 260–283. Macmillan, London.

Connelly, J. J., V. G. Vely, W. H. Mink, J. F. Sachsel and J. H. Litchfield (1966). *Food Technol.*, **20**, 829.

Cowan, D. (1983). Proteins. In *Industrial Enzymology*, ed. T. Godfrey and K. Reichelt, pp. 352–374. Macmillan, London.

Criswell, L. G., R. W. Schatz, J. H. Litchfield, V. G. Vely and J. F. Sachsel (1964). *Food Technol.*, **18**, 1489.

Debabov, V. G. (1982). The industrial uses of bacilli. In *The Molecular Biology of the Bacilli*, vol. 1, *Bacillus subtilis*, ed. D. A. Dubnau, pp. 331–370. Academic, New York.

Denton, A. E., J. F. Beuk, J. M. Hogan and W. J. Bradley (1963). *US Pat.* 3 098 014.

Dhar, S. C. and S. M. Bose (1964). The specificity of crystalline protease of *Aspergillus parasiticus* on the B-chain of oxidized insulin and bovine serum albumin. *Enzymologia*, **28** (2), 89–99.

Ebeling, W., H. Metz, H. D. Orth and M. Klochow (1971). Neues proteolytisches Enzyme und Verfahren zu seiner Herstellung. *Ger. Offen.*, 1 965 281.

El-Dash, A. A. and J. A. Johnson (1967). Protease enzymes: effect on bread flavour. *Cereal Sci. Today*, **12**, 282–288.

Enari, T. M. and J. Mikola (1972). Peptidases in germinating barley grain: properties, localization and possible functions. *Ciba Found. Symp.*, **50**, 335–352.

Endo, S. (1962). The protease produced by thermophilic bacteria. *Hakko Kogaku Zasshi*, **40**, 346–353.

Enzyme Nomenclature (1978). *Recommendations of the International Union of Biochemistry*. Academic, New York.

Feder, J. and J. M. Schuck (1970). Studies on the *Bacillus subtilis* neutral protease and *Bacillus thermoproteolyticus* thermolysin-catalyzed hydrolysis of dipeptide substrates. *Biochemistry*, **9**, 2784–2791.

Ferns, R. S. (1977). Clearing or prevention of clouding of aqueous liquids, *e.g.* drinks, by treatment with *Streptomyces* protease. *US Pat.* 4 038 419.

Fujimaki, M., S. Arai and M. Yamashita (1977). In *Enzymatic Modification of Food Proteins*, ed. R. E. Feeney and J. R. Whitaker, *Adv. Chem. Ser.*, No. 160, pp. 156–183. American Chemical Society, Washington.

General Standards for Enzyme Regulations (1980). In *Report of Association of Microbial Food Enzyme Producers*, Brussels.

Gibson, K. D. (1953). True and apparent activation energies of enzymic reactions. *Biochim. Biophys. Acta*, **10**, 221–229.

Gilliland, R. B. (1981). Brewing yeast. In *Brewing Science*, ed. J. R. A. Pollock, vol. 2, pp. 1–60. Academic, New York.

Glass, J. D. (1981). Enzymes as reagents in the synthesis of peptides. *Enzyme Microb. Technol.*, **3**, 2–18.

Glen, A. R. (1976). Production of extracellular proteins by bacteria. *Annu. Rev. Microbiol.*, **30**, 41–62.

Godfrey, T. (1983). Flavouring and colouring. In *Industrial Enzymology*, ed. T. Godfrey and J. Reichelt, pp. 305–314. Macmillan, London.

Godfrey, T. and J. Reichelt (1983). *Industrial Enzymology*. Macmillan, London.

Green, M. L. (1977). Review of the progress of dairy science: milk coagulants. *J. Dairy Res.*, **44**, 159–188.

Griffin, P. J. and W. M. Fogarty (1973). Physicochemical properties of the native, zinc- and manganese-prepared metalloprotease of *Bacillus polymyxa*. *Appl. Microbiol.*, **26**, 191–195.

Guntelberg, A. V. and M. Ottesen (1952). Preparation of crystals containing the plakalbumin-forming enzyme from *Bacillus subtilis*. *Nature (London)*, **170**, 802.

Gunther, R. C. (1974). Soy protein aerating agents—obtained by hydrolyzing the protein in acid or alkali before pepsin modification. *US Pat.* 3 814 816.

Haard, N. F., L. A. W. Feltham, N. Elbig and E. J. Squires (1982). Modification of proteins with proteolytic enzymes from the marine environment. In *Modification of Proteins: Food, Nutritional and Pharmacological Aspects*, ed. R. E. Feeney and J. R. Whitaker, *Adv. Chem. Ser.*, No. 198, pp. 223–244. American Chemical Society, Washington.

Haines, B. M. (1981). Hides and skins. In *Microbial Biodeterioration, Economic Microbiology*, ed. A. H. Rose, vol. 6, pp. 131–149. Academic, New York.

Hall, F. F., H. O. Kunkel and J. M. Prescott (1966). Multiple proteolytic enzymes of *Bacillus lichenformis*. *Arch. Biochem. Biophys.*, **114**, 145–153.

Hampson, S. E., G. L. Mills and T. Spencer (1963). Observations on a protease from *Proteus mirabilis*. *Biochim. Biophys. Acta*, **73**, 476–487.

Heineken, F. G. and R. J. O'Connor (1972). Continuous culture studies on the biosynthesis of alkaline protease, neutral protease and alpha-amylase by *Bacillus subtilis* NRRL B-3411. *J. Gen. Microbiol.*, **73**, 35–44.

Heinicke, R. M. (1953). Complementary enzyme actions in the clotting of milk. *Science*, **118**, 753–754.

Helbig, N. B., L. Ho, G. E. Christy and S. Nakai (1980). Debittering of skim milk hydrolysates by adsorption for incorporation into acidic beverages. *J. Food Sci.*, **45**, 331–335.

Hofsten, B. V. and C. Tjeder (1965). An extracellular proteolytic enzyme from a strain of *Arthrobacter*. Pt. 1. Formation of the enzyyme and isolation of mutant strains without proteolytic activity. *Biochim. Biophys. Acta*, **110**, 576–584.

Imshenetskii, A. A., I. D. Kasatkina and E. T. Zheltova (1971). Repression of protease synthesis in *Aspergillus terricola* by exogenic amino acid. *Mikrobiologiya*, **40**, 382–386.

Isono, M., K. Tomoda, K. Miyata, K. Maejima and K. Tsubaki (1972). Alkali protease. *US Pat.* 3 652 399.

Isowa, Y. (1978). Enzymic peptide synthesis. *Yuki Gosei Kagaku Kyokaishi*, **36**, 195–205.

Jaeggi, K., P. Krasnobajew, P. Weber and K. Wild (1977). Flavourant production from milk products by proteolysis, lactic acid fermentation and heating with cystein. *US Pat.* 4 001 437.

Johansen, J. T., M. Otteson, I. Svendsen and G. Wybrant (1968). Degradation of the B-chain of oxidized insulin by two subtilisins and their succinylated and *N*-carbamylated derivatives. *Carlsberg Res. Commun.*, **36**, 365–384.

Jolles, J., C. Alais and P. Jolles (1968). The tryptic peptide with the rennin-sensitive linkage of cow's kappa-casein. *Biochim. Biophys. Acta*, **168**, 591–593.

Jolles, P. (1975). Structural aspects of the milk clotting process. Comparative features with the blood clotting process. *Mol. Cell. Biochem.*, **7**, 73–85.

Juresek, L., D. Facie and L. B. Smillie (1969). Remarkable homology about the disulfide bridges of a trypsin-like enzyme from *Streptomyces griseus*. *Biochem. Biophys. Res. Commun.*, **37**, 99–105.

Keay, L. (1971). Microbial proteases. *Process Biochem.*, August, 17–21.

Keay, L. (1972). Proteases of the genus *Bacillus*. In *Fermentation Technology Today: Proceedings of the Fourth International Fermentation Symposium, Kyoto*, ed. G. Terilli, pp. 289–298. Society of Fermentation Technology, Japan.

Kichline, T. P. and L. G. Sharpf (1972). Process cheese, composition containing emulsifying agent and proteolytic enzyme. *US Pat.* 3 692 630.

Kishida, T. and S. Yoshimura (1964). Mode of action of *Aspergillus ochraceus* proteinase on oxidized insulin. *J. Biochem. (Tokyo)*, **55**, 95–101.

Kling, S. H., J. S. Araujo Neto and J. C. Perrone (1982). Proteases from by-products of agave production: its quantitative evaluation, concentration and purification. *Process Biochem.*, **17**, 29–31.

Kosikowski, F. V. (1975). Potential of enzymes in continuous cheesemaking. *J. Dairy Sci.*, **58**, 994–1000.

Kovaleva, G. G., M. P. Shimanskaya and V. M. Stepanov (1972). Site of diazoacetyl inhibitor attachment to acid proteinase of *Aspergillus awamori*, an analog of penicillopepsin and pepsin. *Biochem. Biophys. Res. Commun.*, **49**, 1075–1081.

Kunimatsu, D. K. and K. T. Yasunobu (1970). Chymopapain B. *Methods Enzymol.*, **19**, 244–252.

Laishley, E. and R. W. Bernlohr (1968). Regulation of arginine and proline catabolism in *Bacillus licheniformis*. *J. Bacteriol.*, **96**, 322–329.

Lasure, L. L. (1980). Regulation of extracellular acid protease in *Mucor miehei*. *Mycologia*, **72**, 483–493.

Law, B. A. (1982). Cheeses. In *Fermented Foods, Economic Microbiology*, ed. A. H. Rose, vol. 7, pp. 147–198. Academic, New York.

Law, B. A. and A. S. Wigmore (1982). Microbial proteinases as agents for accelerated cheese ripening. *J. Soc. Dairy Technol.*, **35**, 75–76.

Lewis, J. (1981). Wool. In *Microbial Biodeterioration, Economic Microbiology*, ed. A. H. Rose, vol. 6, pp. 81–130. Academic, New York.

Li, E. and A. A. Yousten (1975). Metalloprotease from *Bacillus thuringiensis*. *Appl. Microbiol.*, **30**, 354–361.

Lipscombe, W. N. (1983). Structure and catalysis of enzymes. *Annu. Rev. Biochem.*, **52**, 17–34.

Mackie, I. M. (1982). Fish protein hydrolysates. *Process Biochem.*, **17**, 26–28, 31.

Martens, R. and M. Naudts (1976). Technological suitability of rennet substitutes. *Annu. Bull. Int. Dairy Fed.*, 1–12.

Matsubara, H. and J. Feder (1971). Other bacterial, mold and yeast proteases. In *The Enzymes*, ed. P. D. Boyer, vol. 3, pp. 721–795. Academic, New York.

McKenzie, H. A. (1971). An account of the isolation, properties and zone electrophoresis of whole casein. In *Milk Proteins, Chemistry and Molecular Biology*, ed. H. A. McKenzie, vol. 2, pp. 87–116. Academic, New York.

Meyrath, J. and G. Volavsek (1975). Production of microbial enzymes. In *Enzymes in Food Processing*, ed. G. Reed, pp. 255–300. Academic, New York.

Mikola, J., K. Pietila and T.-M. Enari (1972). Activities of various peptidases in barley, green malt and malt. *J. Inst. Brewing.* **78**, 388–391.

Millet, J. (1970). Characterization of proteinases excreted by *Bacillus subtilis* Marburg strain during sporulation. *J. Appl. Bacteriol.*, **33**, 207–219.

Morihara, K. (1974). Comparative specificity of microbial proteinases. *Adv. Enzymol.*, **41**, 179–243.

Morihara, K. and H. Tsuzuki (1970). Thermolysin: kinetic study with oligopeptides. *Eur. J. Biochem.*, **15**, 374–380.

Morihara, K., H. Tsuzuki and T. Oka (1968). Comparison of the specificities of various neutral proteinases from microorganisms. *Arch. Biochem. Biophys.*, **123**, 572–588.

Morihara, K., T. Oka, H. Tsuzuki, K. Inouye and S. Sakakibara (1979). Semisynthesis of human insulin: trypsin catalyzed replacement of alanine B30 by threonine in porcine insulin. In *Peptide Structure and Biological Function*, ed. E. Gross and J. Meienhofer, *Proc. 6th Am. Peptide Symp.*, pp. 617–620. Pierce Chemical Company, Rockford, IL, USA.

Murakami, M., K. Fukunaga, M. Matsuhashi and M. Ono (1969). Stimulation effect of proteins on protease formation by *Serratia* spp. *Biochim. Biophys. Acta*, **192**, 378–380.

Nagata, Y., K. Yamaguchi and B. Maruo (1974). Genetic and biochemical studies on cell-bound alpha-amylase in *Bacillus subtilis*. *J. Bacteriol.*, **119**, 425–430.

Nakadai, T., S. Nasuno and N. Iguchi (1972). The action of peptidases from *Aspergillus oryzae* in digestion of soybean proteins. *Agric. Biol. Chem.*, **36**, 261–268.

Nakadai, T., S. Nasuno and N. Iguchi (1973). Purification and properties of alkaline proteinase from *Aspergillus oryzae*. *Agric. Biol. Chem.*, **37**, 2685–2694.

Nakagawa, Y. (1970). Alkaline proteinases from *Aspergillus*. *Methods Enzymol.*, **19**, 581–591.

Nasuno, S. and T. Onara (1972). Purification of alkaline proteinase from *Aspergillus candidus*. *Agric. Biol. Chem.*, **36**, 1791–1796.

Nelson, J. H. and P. R. Witt (1973). *Mucor pusillus* Lindt enzyme for chillproofing beer. *US Pat.* 3 740 233.

Ney, K. H. (1971). Prediction of bitterness of peptides from their amino acid composition. *Z. Lebensm. Unters. Forsch.*, **147** (2), 64–68.

Nikolic, V. (1970). Continuous production of Telemes cheese. *US Pat.* 3 518 094.

Olsen, H. S. and J. Adler-Nissen (1979). Industrial production and applications of a soluble enzymatic hydrolysate of soya protein. *Process Biochem.*, **14** (7), 6–11.

Ong, P. S. and M. Gaucher (1976). Production, purification and characterisation of thermomycolase, the extracellular serine protease of the thermophilic fungus *Malbranchea pulchella* var. *sulfurea*. *Can. J. Microbiol.*, **22**, 165–176.

Ottesen, M. and W. Rickert (1970). The acid protease of *Mucor miehei*. *Method Enzymol.*, **19**, 459.

Ottesen, M. and A. Spector (1960). A comparison of two proteinases from *Bacillus subtilis*. *C. R. Trav. Lab. Carlsberg*, **32**, 63–74.

Parr, G. R., R. R. Hantgan and H. Taniuchi (1978). Formation of two alternative complementing structures from a cytochrome *c* heme fragment (residues 1 to 38) and the apoprotein. *J. Biol. Chem.*, **253**, 5381–5388.

Pokorny, M. and L. Vitale (1980). Enzymes as by-products during biosynthesis of antibiotics. *Proc. FEBS Meet.*, **61**, 13–25.

Pour-El, A. and T. C. Swenson (1976). Enzymatic gellable and whippable protein production, useful as an egg white substitute in food toppings. *US Pat.* 3 932 672.

Powers, J. C. (1982). Proteolytic enzymes and their active site specific inhibitors: role in the treatment of disease. In *Modification of Proteins: Food, Nutritional and Pharmacologial Aspects*, ed. R. E. Feeney and J. R. Whitaker, *Adv. Chem. Ser.*, No. 198, pp. 347–367. American Chemical Society, Washington.

Priest, F. G. (1983). Enzyme synthesis by microorganisms. In *Microbial Enzymes and Biotechnology*, ed. W. M. Fogarty, pp. 319–366, Applied Science, London.

Reichelt, J. R. (1983). Baking. In *Industrial Enzymology*, ed. T. Godfrey and J. Reichelt, pp. 210–220. Macmillan, London.

Richards, F. M. and P. J. Vithayathil (1959). Preparation of subtilisin modified ribonuclease and the separation of the peptide and protein components. *J. Biol. Chem.*, **234**, 1459–1465.

Robinson, D. R. and W. P. Jencks (1965). The effect of compounds of the urea–guanidinium class on the activity coefficient of acetyltetraglycine ethyl ester and related compounds. *J. Am. Chem. Soc.*, **87**, 2462–2470.

Saltman, R., D. Vlach and P. Luigi Luisi (1977). Co-oligopeptides of aromatic amino acids and glycine with variable distance between the aromatic residues. VII. Enzymic synthesis of *N*-protected peptide amides. *Biopolymers*, **16**, 631–638.

Sanders, R. L. and B. K. May (1975). Evidence for extrusion of unfolded extracellular enzyme peptide chains through membranes of *Bacillus amyloliquefaciens*. *J. Bacteriol.*, **123**, 806–814.

Sardinas, J. L. (1972). Microbial rennets. *Adv. Appl. Microbiol.*, **15**, 39–67.

Sawada, K., M. Kajikawa and K. Kotani (1977). Foamable soya production for foodstuffs. Use made by partial hydrolysis with enzymes and heat treatment without removing soya whey. *US Pat.* 4 015 019.

Schechter, I. and A. Berger (1968). On the active site of proteases. Part II. Mapping the active site of papain; specific peptide inhibitors of papain. *Biochem. Biophys. Res. Commun.*, **32**, 898–902.

Schwille, D., H. Seiz, E. Song and U. Sommer (1976). Water soluble protein materials production by treating natural protein in aqueous suspension with proteolytic enzyme. *US Pat.* 3 974 294.

Sebek, O. K. and A. I. Laskin (1979). *Genetics of Industrial Microorganisms*. American Society for Microbiology, Washington.

Sekine, H. (1972). Neutral proteinases I and II of *Aspergillus sojae*; some enzymatic properties. *Agric. Biol. Chem.*, **36**, 207–216.

Smith, E. L., R. J. DeLange, W. H. Evans, M. Landon and F. S. Markland (1968). Subtilisin Carlsberg. V. The complete sequence: comparison with subtilisin BPN; evolutionary relationship. *J. Biol. Chem.*, **243**, 2184–2191.

Stanley, D. W. (1981). Non-bitter protein hydrolysates. *Can. Inst. Food Sci. Technol. J.*, **14**, 49–52.

Stenne, P. (1975). Continuous milk curd production—using pulsed flow through the coagulator to prevent build-up of deposits. *U.S. Pat.* 3 899 595.

Takahashi, N., Y. Yasuda, K. Goto, T. Miyake and T. Murachi (1973). Multiple molecular forms of stem bromelain. Isolation and characterization of two closely related components. *J. Biochem. (Tokyo)*, **74**, 355–373.

Tanimoto, S., M. Yamashita, A. Arai and M. Fujimaki (1972). Probes for catalytic action of α-chymotrypsin in plastein synthesis. *Agric. Biol. Chem.*, **36**, 1595–1602.

Tran Chau, P. and H. Urbanek (1974). Serine neutral proteinase from *Bacillus pumilus* as metalloenzyme. *Acta Microbiol. Pol., Ser B*, **6** (23), 21–25.

Tsujita, Y. and A. Endo (1977). Extracellular acid protease of *Aspergillus oryzae* grown on liquid media: multiple forms due to association with heterogeneous polysaccharides. *J. Bacteriol.*, **130**, 48–56.

Vanderpoorten, R. and M. Weckx (1972). Breakdown of casein by rennet and microbial milk-clotting enzymes. *Neth. Milk Dairy J.*, **26**, 47–59.

Van Krenemburg, S. (1979). Stable liquid animal feed, milk substitute for calves prepared using casein hydrolyzed by a proteolytic enzyme to give peptide of molecular weight below 10 000. *US Pat.* 4 259 357.

Verbruggen, R. (1975). Subtilopeptidase A isoenzyme system. *Biochem. J.*, **151**, 149–155.

Wainwright, T. (1971). Biochemistry of brewing. In *Modern Brewing Technology*, ed. W. P. K. Findlay, pp. 129–164. Macmillan, London.

Wang, D. I. C., C. L. Cooney, A. L. Demain, P. Dunnill, A. E. Humphrey and M. D. Lilly (1979). *Fermentation and Enzyme Technology*, pp. 238–310. Wiley, New York.

Ward, O. P. (1983). Proteinases. In *Microbial Enzymes and Biotechnology*, ed. W. M. Fogarty, pp. 251–317. Applied Science, London.

Watanabe, M. and S. Arai (1982). Proteinaceous surfactants prepared by covalent attachment of L-leucine *N*-alkyl esters to food proteins by modification with papain. In *Modification of Proteins: Food, Nutritional and Pharmacological Aspects*, ed. R. E. Feeney and J. R. Whitaker, *Adv. Chem. Ser.*, No. 198, pp. 199–220. American Chemical Society, Washington.

Waugh, D. F. (1971). The formation and structure of casein micelles. In *Milk Proteins. Chemical and Molecular Biology*, ed. H. A. McKenzie, vol. 2, pp. 3–85. Academic, New York.

Weaver, L. H., W. R. Kester, L. F. Ten Eyck and B. W. Matthews (1976). The structure and stability of thermolysin. *Experientia, Suppl.*, **26**, 31–39.

Whitaker, J. R. (1970). Proteases of *Endothia parasitica*. *Methods Enzymol.*, **19**, 436–445.

Whitaker, J. R. and A. J. Puigserver (1982). Fundamentals and applications of enzymatic modification of proteins: an overview. In *Modification of Proteins: Nutritional and Pharmacological Aspects*, ed. R. E. Feeney and J. R. Whitaker, *Adv. Chem. Ser.*, No. 198, pp. 57–87. American Chemical Society, Washington.

Wiersma, M. and W. Harder (1978). A continuous culture study of the regulation of extracellular protease production in *Vibrio* SA 1. *Antonie van Leeuwenhoek*, **44** (2), 141–155.

Wood, B. J. (1982). Soy sauce and miso. In *Fermented Foods*, *Economic Microbiology*, ed. A. H. Rose, vol. 7, pp. 39–86. Academic, New York.

Yamamoto, A. (1975). Proteolytic enzymes. In *Enzymes in Food Processing*, ed. G. Reed, pp. 123–179. Academic, New York.

Yasuda, Y., N. Takahashi and T. Murachi (1970). Composition and structure of carbohydrate moiety of stem bromelain. *Biochemistry*, **9**, 25–32.

Yoneda, Y. and B. Maruo (1975). Mutation of *Bacillus subtilis* causing hyperproduction of α-amylase and protease and its synergistic effect. *J. Bacteriol.*, **124**, 48–54.

Yoshida, F. (1956). Proteolytic enzymes of black *Aspergillus*. I. Strains of black *Aspergillus* producing high yields of superior proteinase and the crystallization of proteolytic enzyme from *Aspergillus saitoi*. *Bull. Agric. Chem. Soc. Jpn.*, **20**, 252–256.

Zuidweg, M. H. J., C. J. K. Bos and H. Van Welzen (1972). Proteolytic components of alkaline protease of *Bacillus* strains. Zymograms and electrophoretic isolation. *Biotechnol. Bioeng.*, **14**, 685–714.

41

Hydrolytic Enzymes

O. P. WARD
National Institute for Higher Education, Dublin, Ireland

41.1 INTRODUCTION

In the preceding chapters, the most important commercial hydrolytic enzymes, namely the starch degrading enzymes and the proteases, have been discussed. This chapter deals with other hydrolytic industrial enzymes. While these other enzymes as a group command a much smaller share of the total industrial enzyme market than either the starch hydrolytic enzymes or proteases, they perform important functions in particular processes, especially in the food processing area.

The cellulases, other β-glucanases, hemicellulases, xylanases and dextranase are all depolymerizing enzymes which act by hydrolysis of glycosidic linkages. Lactase is a disaccharidase which hydrolyzes the disaccharide lactose to glucose and galactose. The pectinases contain two types of depolymerizing enzymes, galacturonases and transeliminases, which cleave the α-1,4-galacturonosyl linkages by hydrolysis or elimination respectively. While the transeliminases are not classified as hydrolytic enzymes, they are discussed here because they act in combination with other pectinases in commercial enzyme preparations. Pectinases also contain pectin methyl esterase which hydrolyzes the methanol groups esterified to the galacturonic acid carboxyl residues. Lipase acts also by hydrolysis of ester linkages. In this case glycerol is the alcohol component and long chain fatty acids provide the carboxyl groups of the ester linkage.

With the exception of lactase and lipase, the major applications of these enzymes are in the processing of plant materials such as fruits, vegetables or cereals. The discussion on cellulase applications is confined to current uses and does not deal with the potential applications of this enzyme in large scale lignocellulose conversions. The major application of lactase is in the dairy industry. Lipase is used as a flavour producing enzyme in the dairy and food industry and has other applications in the chemical and pharmaceutical industries.

41.2 PECTINASES

41.2.1 Pectic Substances and Pectinolytic Enzymes

Pectic substances are complex acidic polysaccharides which occur in varying amounts in the middle lamella or intercellular spaces of all higher plants. They consist of α-1,4-D-polygalacturonide polymers, often complexed with two other polymers, a highly branched L-araban and a β-1,4-D-galactan (Whistler and Smart,1953). The carboxyl groups of the galacturonic acid units are partially esterified with methanol. The methoxyl content varies with source and in the case of fruits varies with fruit development and ripening. Reported molecular weights of pectic polysaccharides extracted from fruits range from 25 000 to 360 000 (Fogarty and Kelly, 1983). Vegetable and fruit texture is greatly influenced by the quantity and nature of the pectic substances present. Most pectin in unripened fruit is present as protopectin (water insoluble pectic substances). During maturation and ripening this protopectin is transformed into soluble pectin, leading to fruit softening (Dilley, 1970). Soluble pectins in fruit or vegetable extracts increase the viscosity and gelling properties of these extracts. Protopectin and partially soluble pectic substances confer cloudy or haze properties on such extracts. Commercial pectinases are used to increase yields of extract or to clarify cloudy pectin solutions.

A range of different pectinolytic enzymes occur in nature, produced by plant and microbial species. The most important enzymes involved and the reactions they catalyze are illustrated in Figure 1. Pectin esterase is produced by higher plants, moulds, yeasts and bacteria. The enzymes from tomato, *Fusarium oxysporum* and *Clostridium multifermentans* (Miller and Macmillan, 1971) have been well characterized. Exopolygalacturonases produced by fungi, including *Aspergillus niger*, have pH optima from 4.0 to 6.0 whereas the bacterial enzyme produced by *Erwinia aroideae* has a pH optimum of 7.2 (Hatanaka and Ozawa, 1969). Oligogalacturonases produced by *Aspergillus niger* and *Bacillus* spp. (Hasegawa and Nagel, 1968) have been purified and characterized. Endopolygalacturonases are produced by a wide range of fungi including *Aspergillus* spp., *Fusarium* spp. and *Rhizopus* spp. They all have pH optima in the range 3.5–5.0.

The other important group of pectic depolymerizing enzymes are the lyases or transeliminases. Although they are not in fact hydrolytic enzymes, they will be discussed here because they act in combination with the true hydrolytic pectic depolymerizing enzymes and pectin esterase in commercial enzyme preparations. They cause depolymerization with cleavage of the α-1,4-galacturonosyl linkages by elimination of Δ-4,5-D-galacturonate residues at the non-reducing side of the bond being split. The reaction may be followed spectrophotometrically at 235 nm due to maximum absorption at this wavelength by the double bond formed. Endopolygalacturonate lyases are produced by a wide range of bacteria, including *Bacillus* spp. (Ward, 1973; Ward and Fogarty, 1972), *Erwinia* spp., *Clostridium* spp., *Streptomyces* spp. and *Fusarium* spp. and these enzymes also have alkaline pH optima and a requirement for calcium for activity. Endopolymethylgalacturonate lyases are produced mainly by species such as *Aspergillus* and *Penicillium*. pH optima vary with source and reports indicate enzymes with pH optima from 5.5 to 8.6 Van Houdenhoven (1975) has investigated the enzyme produced by *Aspergillus niger*.

Commercial pectinases are produced from strains of *Aspergillus niger*. The major enzyme components of these preparations are pectin esterase, endopolymethylgalacturonate lyase and polygalacturonase. Pectinases are produced commercially using both surface and submerged cultivation techniques. Different activity ratios of the three major enzyme components and of other contaminating enzymes such as cellulases and hemicellulases are observed in surface and submerged pectinase fermentations. Surface culture is generally considered more suitable, as high yields of endopolymethylgalacturonate lyase are produced under these conditions but not in submerged fermentations.

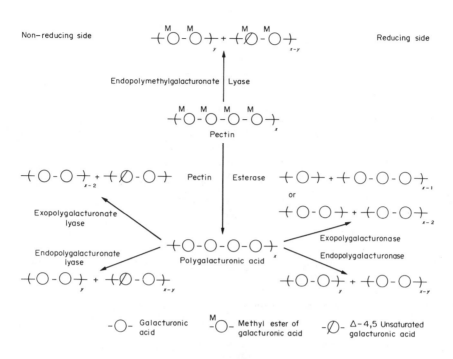

Figure 1 Pectinases and their mode of action

41.2.2 Applications of Pectinases

41.2.2.1 Fruit juice extraction

When fruit tissue is crushed during processing, some of the pectin appears in either soluble or insoluble form in the juice and this can affect the quality or yield of the juice which is recovered. Pectinases are used to modify this pectin material with a view to improving both the yield of juice extracted and its consistency or quality. The addition of pectinase results in a rapid reduction of the viscosity of the juice and results in a 'break' in the turbidity, *i.e.* cloudy particles coagulate and settle out. Flocculating agents such as gelatin and bentonite are used to aid this clarification process. Cellulases may be used in fruit juice extraction where optimum colour extraction (usually associated with the skin) is desirable or where processing involves complete maceration of the tissue. Amylase and amyloglucosidase may be used to aid juice extraction from immature apples or pears to hydrolyze gelatinized starch which can cause filtration or haze problems. A generalized process summary for fruit juice production from core fruits, stone fruits, berries and grapes is illustrated in Figure 2. The actual process varies from fruit to fruit and also differs depending on the final form of the product required.

With core fruits, enzyme incubations are usually carried out at 20–25 °C. Higher temperatures tend to destroy the physical structure of core fruits which is essential for effective pressing. With berries and stone fruits, including red grapes, where colour extraction is optimized, incubations are carried out at 50 °C with a cellulase supplement or at 60–65 °C where cell wall and membrane permeability and colour extraction are aided by plasmolysis. Intermediate temperature holds are best avoided in order to reduce the growth of yeasts, *Lactobacilli* and other microorganisms. In the case of red grape juice extraction, the mash may be heated to 80 °C prior to pectinase treatment to denature oxidases which cause loss of colour. In addition commercial enzyme preparations used for extraction and clarification of red grape juice should not contain anthocyanase (a β-glucosidase) which hydrolyzes coloured anthocyanins to anthocyanidins and glucose, after which transformation of the aglucone to a colourless derivative occurs (Neubeck, 1975).

With apple juice production, pectinase preparations must be capable of rapidly degrading

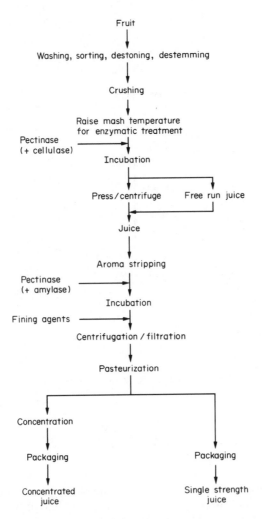

Figure 2 Summary of fruit juice production process

highly esterified apple pectin. Pectin esterase and endopolygalacturonase are likely to be the most important enzymes involved. Their pattern of action has been discussed by Rombouts and Pilnik (1980). While apple juice clarification is also possible with pectin lyase, this enzyme does not give good clarification where the pectin has a lower degree of esterification (45–60%). Arabinofuranosidase, often present in commercial preparations, can cause haze problems (Whitaker, 1984).

41.2.2.2 *Processing of citrus fruits*

In citrus processing, pectinases are used mainly for the production of by-products. A general flow diagram of the citrus fruit processing operation is given in Figure 3. The fruit is initially separated by pressing into two fractions, peel and juice/seeds/internal membranes/pulp. The latter fraction is separated into two components, the juice and coarse pulp. The juice is pasteurized and packaged as single strength material or treated with pectinase to lower juice viscosity, where a concentration step is involved. The coarse pulp still contains substantial amounts of juice after pressing and this is recovered by incubation with pectinase, followed by a three to five step countercurrent extraction of the pulp with water. During this process, pectin degradation is limited to soluble material, so that the insoluble pectin fraction which maintains cloud stability is not affected. The pectinase treatment enables the pulp wash to be concentrated to 60° Brix without gel formation.

The main waste products from the above processes, pulp, rags and outer peel, may be used for

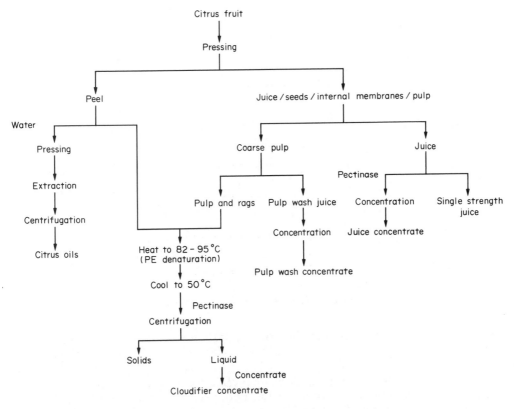

Figure 3 Summary of citrus fruit processing operation

preparation of natural cloudifier, which has application in the beverage industry. In this process the raw materials are ground, mixed with water, heated to 90 °C, cooled to about 50 °C and incubated with pectinase. The enzymes macerate the material releasing various cloudy substances. The cloudy material is recovered and concentrated.

Citrus oils recovery involves expressing the oils in the presence of water. Pectinases hydrolyze the pectin gel associated with the extract, substantially reduce the volume of water required for extraction and improve the yield of recoverable oils.

In the processing of the pulp/rags/peel mash, the 90 °C heating step is incorporated to denature fruit pectin esterase which causes cloud coagulation problems. Pectin esterase is located mainly in the juice sacs and is released on extraction of the juice. In the absence of heat inactivation, this native pectin esterase demethoxylates the soluble pectin in the juice to produce low methoxyl pectin, which reacts with polyvalent cations to form insoluble pectate salts. Occlusion by pectate of the cloud particles then occurs, leading to sedimentation. Commercial pectinase preparations used for production of citrus juices and cloudifier are produced from *Aspergillus* strains which have been selected for low pectin esterase content. Pectin esterase deficient pectinases are not suitable for juice clarification.

Fungal commercial pectinases are usually active in the pH range 3.0–8.0 and hence have not traditionally been used in processes for clarification of lemon juice, which has a pH of about 2.5. Recently a commercial pectinase has been produced from a variant strain of *Aspergillus* which has a pH optimum in the range 2.0–3.0. This enzyme may be used in combination with non-enzymatic coagulants to speed up lemon juice clarification procedures.

41.2.2.3 Vegetable and fruit maceration

Maceration results from the selective hydrolysis of the middle lamella pectin, producing suspensions of loose cells from fruit or vegetable tissue. The process may be used to produce comminuted fruit juices or nectars, pulpy drinks obtained from apples, apricots, peaches, pears and other fruits. The general production process is summarized in Figure 4. The process is also useful for preparation of finely dispersed baby food vegetable preparations.

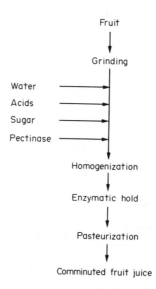

Figure 4 Production of comminuted fruit juices

With fleshy fruits, pectinase used at the homogenization stage improves cloud stability. In addition, the resultant reduction in viscosity of the product makes it possible to prepare concentrated nectar preparations.

Maceration may be achieved using high endopolygalacturonase activity preparations having low or negligible pectin esterase or pectin lyase. Pure pectin lyase has also been shown to be effective. Enzymatic treatment of pulp with more complex mixtures and hemicellulases can result in almost complete liquefaction of a fruit, so that waste material can be reduced with almost all of the raw material being converted to soluble product. There is now substantial interest in commercial enzyme preparations having this capability.

41.2.2.4 Wine

Pectinases are used both to increase juice yield during crushing and for clarification of the juice or wine.

Where fermentation of red wine is carried out on the skins, pectinase treatment of the crushed grapes speeds up juice and colour extraction and the fermentation process. Total yield of wine is increased and subsequent filtration and clarification are more rapid. The fermentation time saving ranges from 30–50%, thereby increasing the overall production capacity of the plant. Continuous extraction processes are now widely used. Pectinases can improve colour extraction provided the preparation can act on the highly esterified insoluble pectins associated with the skins. It has also been observed that pectinases reduce tannic acid extraction and speed up wine maturation. For effective enzyme treatment, where fermentations are carried out on the skins, thorough mash mixing is required initially and at intervals throughout the fermentation. The desired colour intensity is achieved in the wine after about 60 hours of the fermentation, at which stage the liquid is drained from the mash.

In the alternative method of wine production the juice is first extracted using pectinase. As discussed in general for juices, higher temperatures (50–65 °C) cause plasmolysis of the cells, thereby increasing their permeability so that an increased rate of extraction of colour and juice material is observed. Enzymic treatment is usually terminated after about four hours, when the desired colour intensity is reached. The high temperature extraction is not necessary for white grapes. However, pectinase treatment does improve yield, filterability and wine quality.

Where pectinases are not used for juice extraction, wines may contain considerable amounts of pectin which causes clarification and filtration processing problems. In order to overcome this, pectinase should be added as early as possible during the fermentation. Enzyme action can benefit from the relatively higher temperatures occurring during the active fermentation. Neubeck

(1975) has suggested that alcohol inhibits pectinase activity. Consequently higher enzyme dose rates are needed to clarify fermented wine.

41.3 CELLULASES

There are three main types of cellulase produced by fungi: endoglucanase, cellobiohydrolase and β-glucosidase. Their actions on cellulose are summarized in Table 1.

Table 1 Properties of the Major Cellulases of Fungi

E.C. number	3.2.1.4	3.2.1.91	3.2.1.21
Enzyme name	Endoglucanase 1,4β-D-Glucanohydrolase	Cellobiohydrolase 1,4β-D-Glucan cellobiohydrolase	β-Glucosidase β-D-Glucoside glucanohydrolase
Reaction catalyzed	$(G)_n$—$(G)_m \rightarrow (G)_n + (G)_m$	$(G)_2$—$(G)_n$-reducing end $\rightarrow (G)_2 + (G)_n$-reducing end	$(G)_2 \rightarrow G + G$ Also small cellodextrins
Substrates:			
Crystalline cellulose	−	+	−
CMC	+	−	−
Amorphous cellulose	+	+	−
Cellotetraose	+	+	+
Cellobiose	−	−	+
Other substrates	Substituted celluloses attacked	Substituted celluloses not attacked	
Commercial producing organisms	*Trichoderma reesei* *Aspergillus niger* *Penicillium funiculosum* *Rhizopus* spp.	*Trichoderma reesei*	*Trichoderma reesei* *Aspergillus niger* *Penicillium funiculosum*

The cellulases currently used commercially are produced by strains of *Aspergillus niger*, *Trichoderma reesei*, *Penicillium funiculosum*, and *Rhizopus* sp. The *Trichoderma* enzyme preparation is most effective on native or crystalline cellulose. Furthermore, this cellulase complex is resistant to chemical inhibitors and is stable in stirred tank reactors at pH 4.8 and 50 °C, for 48 h (Ryu and Mandels, 1980). The major reaction product is cellobiose which inhibits further cellulose hydrolysis. Lignin is not digested and cellobiase (β-glucosidase) activity is low. Nevertheless, the *Trichoderma* enzyme complex is widely recognized as the best cellulase available. In a saccharification reaction glucose does accumulate. Cellobiase is inhibited by glucose, so that supplemental cellobiase has to be added to give effective cellobiose hydrolysis. Efforts to increase cellobiase activity in *Trichoderma* have met with limited success. Cellobiase produced by *Aspergillus niger* is particularly suitable as a supplement to *Trichoderma* cellulase, for cellulose saccharification, as it is less susceptible to end product inhibition than the *Trichoderma* enzyme. The *A. niger* cellulase does not attack crystalline cellulose effectively but its high cellobiase or β-glucosidase enables glucose to be produced as end product, thereby removing the inhibitory effect which cellobiose has on the depolymerizing enzymes. Good cellobiase activity has also been reported with other strains of *Aspergillus*, including *A. awamori* (Enari *et al.*, 1975) and *A. phoenicis* (Sternberg *et al.*, 1977). The endoglucanase of *Penicillium funiculosum* is less sensitive to inhibition by cellobiose than the *Trichoderma* enzyme.

41.3.1 Cellulase Production

Cellulase is an inducible enzyme system and cellulose is generally considered to be the best inducer of the complete enzyme complex. Induction by cellulose, cellobiose and lactose requires high substrate concentrations whereas another inducer, sophorose, is effective at very low concentrations. Glucose causes catabolite repression of cellulase synthesis at concentrations greater than 0.1 mg ml^{-1} (Gallo *et al.*, 1979). Catabolite repression of biosynthesis occurs at the translation level. There is also evidence that enzyme biosynthesis is closely related to amino acid assimilation and metabolic regulation. Cellulase induction, synthesis and secretion in *Trichoderma* are closely associated or are concurrent events occurring at the cell surface (Sternberg and Mandels,

1979). In general, strains of *Trichoderma*, mutated for increased cellulase production, have similar patterns of cellulase enzymes when grown in the same medium and respond similarly to inducers and fermentation conditions.

With *Trichoderma reesei*, mutant QM 9414, final enzyme yields in the fermenter were found to be proportional to initial cellulose concentration, up to 8% cellulose as sole carbon source (Sternbert and Dorval, 1979). *T. reesei* grows rapidly on simple sugars such as glucose and complex nitrogen sources. It grows more slowly with a lag phase on lactose or cellulose and ammonium salt. Complex nitrogen sources appear to stimulate growth and enzyme production. The pH optimum for growth is 4.0 while enzyme production is highest at pH 3.0. Cellulase synthesis and release occur mainly in the stationary phase. During enzyme production, the level of oxygen must be sufficient to supply the cells' maintenance energy requirements. Otherwise cellulase production stops.

The intracellular β-glucosidases of *T. reesei*, *A. awamori*, *A. phoenicis* and *A. niger* are all thought to be constitutive and relatively insensitive to catabolite repression. The extracellular β-glucosidases are repressed by glucose and induced by β-glucosides such as methyl-β-D-glucoside, amygdalin, salicin and cellulose. Cellobiose is not an inducer. Suitable media for production of β-glucosidase by *Aspergillus* spp. contain starch syrups or starch-containing cereals. Enzyme production occurs in the stationary phase of growth (Enari, 1983).

41.3.2 Current Applications of Cellulases

The potential applications of cellulases in large scale production of glucose and alcohol from cellulose and lignocellulose are discussed elsewhere in these volumes. Cellulases are currently used in cereal processing, brewing, alcohol production, plant extraction, fruit processing, wine manufacture and waste treatment. In many of these processes cellulase supplements the effects of other enzymes.

In brewing and cereal extraction, cellulase speeds up mash filtration and can increase extract yield. It may supplement the action of β-glucanase. In alcoholic fermentations, depending on the nature of the raw material, cellulase can increase the yield of alcohol produced, when used as a supplement to starch-degrading enzymes. Alcohol-producing yeasts in general do not ferment cellobiose, so it is essential that the cellulase used degrades cellobiose. Enzymatic cellulose breakdown aids polysaccharide hydrolysis during cassava root liquefaction and saccharification and is reported to increase alcohol yield (Lyons, 1983).

Cellulases, used to supplement pectinases, speed up colour extraction from the skins of fruits and also have an important application as part of an enzyme complex in total liquefaction or maceration of vegetables or fruits. Cellulases and β-glucanases are effective on skins and pomaces for production of pectin. The skins and pomaces are incubated with the enzymes for 2–6 hours at 20–45 °C with agitation. Cellulases are also used for extraction or refining of cereal proteins and for recovery of alginates from seaweed. Cellulase, combined with pectinase and hemicellulase, is used to hydrolyze mucilage during coffee extraction. Use of *Trichoderma* cellulase in silage processes speeds up the process by more rapid release of fermentable sugars and nitrogen.

41.4 OTHER β-GLUCANASES

41.4.1 β-1,3/β-1,4-Glucanases

The major application of these endohydrolases is in hydrolysis of β-glucans of barley and other cereals in brewing and cereal extraction. Barley contains substantial levels of β-glucan which is partly solubilized during mashing, making worts viscous and difficult to filter. Malted barley contains less β-glucan and active β-1,3/β-1,4-glucanases. Barley contains *endo*- and *exo*-β-glucanases and pentosanases (Moll, 1979). During malting gibberellic acid promotes synthesis of *endo*-β-glucanases (Macleod, 1979). Manners and Marshall (1969) showed that there is a substantial rise in *endo*-β-1,3-glucanase immediately after steeping but no significant increase in *endo*-β-1,4-glucanase. Barley *endo*-β-1,3/β-1,4-glucanase activity also increases dramatically after steeping (Preece and Hoggan, 1956).

In mashes containing a substantial proportion of malt, the malt β-glucanases hydrolyze the soluble β-glucans, thereby reducing wort viscosity and improving filtration and extract yield. Where mashes contain higher proportions of adjuncts such as barley, containing β-glucans, exogenous commercial β-glucanases are added to the mash. Without added enzymes when good malt is used it is possible to supplement the malt with up to 30% barley without encountering major lautering problems. β-Glucanase addition, however, usually improves the filtration rate and increases extract yields under these conditions. With poor malt or with higher proportions of barley, it is essential to use β-glucanase in the mash. Use of β-glucanase in combination with α-amylase and protease makes it possible to obtain high yields of extract from barley mashes containing as little as 5–20% malt, although longer mashing times are usually necessary.

Malt β-glucanases are most active in the pH range 4.0–5.5 and have temperature optima of 43–45 °C. They are inactivated at 60 °C. The β-glucanase of *Bacillus subtilis* has a specificity identical to that of malt β-glucanase. Both have no effect on β-1,3- and β-1,4-homopolyglucosides (Bass and Cayle, 1975). Both specificities involve an affinity for that part of the β-glucan molecule where β-1,3- and β-1,4-linkages alternate. The β-glucanase of *Penicillium emersonii* also hydrolyzes these linkages at a high rate and has broader specificity towards other bonds than the *B. subtilis* enzyme. β-Glucanase produced by *Aspergillus niger*, while also showing broad specificity towards a range of bonds, does not have substrate specificity properties identical to the *P. emersonii* enzyme. Table 2 shows a comparison of the temperature and pH optima of the most important commercial β-glucanases.

Table 2 pH and Temperature Optima for Activity of Malt and Microbial Barley β-Glucanases

Enzyme source	pH optimum	Temperature optimum (°C)
Malt	4.0–5.5	43–45
Bacillus subtilis	5.0–7.0	58–60
Pencillium emersonii	3.5–6.0	68–70
Aspergillus niger	4.5–5.5	60–63

The higher thermal stability of microbial β-glucanases, compared with malt, enables the brewer to adjust the mashing process, perhaps using higher temperatures and mashing programmes of shorter duration. In addition, wort filtration is often the rate limiting step in the brewing process and the availability of a range of commercial β-glucanases insures that this stage can be speeded up and problem free, thereby increasing overall throughput.

β-Glucanases may also be added at the fermenter or maturation stage to treat glucan hazes which may give problems during the final beer filtration steps or show up in the packaged product. The various stages of brewing are illustrated in Figure 5 of the preceding chapter (Proteolytic Enzymes).

β-Glucanase is produced by *Bacillus amyloliquefaciens* concomitantly with α-amylase and protease. The fermentation conditions are optimized for glucanase content in commercial production processes. *Aspergillus niger* β-glucanase is often formed as a side activity in pectinase production. Again, the fermentation is adjusted to produce high β-glucanase yields.

41.4.2 β-1,3/β-1,6-Glucanases

Grapes infected by *Botrytis cinerea* present serious clarification and filtration problems during winemaking. The basic problem relates to the presence in the grapes of a glucan polysaccharide, produced by the fungus, which has a β-1,3-glucan backbone and β-1,6-glucan side chains. The polysaccharide is not hydrolyzed by any of the β-glucanases described above. Recently, Dubourdieu *et al.* (1981) reported trials with a β-1,3/β-1,6-glucanase which effected a 10–20-fold improvement in filterability and improved wine clarification. The enzyme was most active in the pH range 3.0–4.5 and had a temperature optimum of 30 °C. In addition, the presence of the *Botrytis* polysaccharide in the fermentation prevents normal sedimentation of yeasts and bacteria and higher levels of sulfur are required to prevent microbial infection. Application of the effective enzyme improves filtration, reduces filter aid costs and filtration time, reduces the risk of infection and sulfur dioxide requirement and produces a wine of considerably improved quality.

41.5 HEMICELLULASES and XYLANASES

The hemicellulases and xylanases cover a complex group of enzymes, many of which are poorly characterized, which hydrolyze arabans, galactans, mannans and xylans. These polysaccharides and gums are found in various forms (arabinogalactans, galactomannans, *etc.*) in plant tissues. Hemicellulases are usually present as contaminating enzymes in commercial pectinase and cellulase preparations and contribute to plant/fruit processing where pectinases or cellulases are used, particularly for complete tissue maceration and colour extraction. In addition pentosanases, present in hemicellulase preparations, confer anti-staling characteristics on bread made from wheat or rye which contains 5–10% pentosans (Reichelt, 1983; Barrett, 1975). Pentosanase application to brewing mashes containing wheat improves extract yield and filtration.

Hemicellulases are also effective in silage production. Galactomannanase reduces the viscosity of gums present in coffee extract and reduces problems at the evaporation and spray drying stages of coffee processing. Arunga (1982) has discussed the use of commercial enzymes containing mixtures of hemicellulase with pectinases and cellulases in coffee making. They are applied by mixing in the fermentation tanks where they supplement the enzymatic activity of organisms such as *Bacillus*, *Erwinia*, *Aspergillus*, *Penicillium* and *Fusarium* involved in the coffee fermentation, in removing mucilage material. Hemicellulases are also used to hydrolyze gum-containing oil drilling fluids.

Hemicellulase may be added to methane fermentation processes utilizing plant waste materials as energy feeds. *Apergillus niger* hemicellulase addition to a methane fermentation, using soy bean seed coat as raw material, doubled the fermentation rate and reduced the chemical oxygen demand of the post-fermentation residue.

Aspergillus spp. produce a range of hemicellulase and pentosanase activities which hydrolyze mannans, arabans and xylans. *Aspergillus niger* is the predominant source of commercial hemicellulases.

41.6 DEXTRANASE

Dextran is a branched polymer of glucose linked by α-1,6-glucosidic bonds but containing also some α-1,3-glucosidic linkages. The polysaccharide is produced in substantial quantities by *Leuconostoc mesenteroides* and related species. Commercial dextranases (*endo-α-1,6-glucanase* EC 3.2.1.11) are produced from strains of *Penicillium lilacinum* and *Penicillium funiculosum*. They are most active in the pH range 5–7 and at temperatures up to 60 °C. The major end products of the enzymatic hydrolysis of dextran are isomaltose and isomaltotriose.

Dextrans are produced by *Leuconostoc* spp. in damaged sugar cane and sugar cane juices. These dextrans increase the viscosity of the juice and cause problems at the juice clarification and evaporation stages and in the concentration of molasses. Treatment of dextran-containing raw juices with dextranase between extraction and clarification by inserting an enzymatic hold of 15 min at 50–55 °C overcomes these problems. Scott (1975) cited an example where the dextranase-treated juice had a specific viscosity of 0.024 prior to heating and liming whereas the control had a specific viscosity of 0.11. The processing rate increased from 256 to 273 tons of cane per hour under these conditions. Treatment may also be carried out at the raw sugar syrup stage. The degradation products do not appear to crystallize with the sugar and consequently remain in the molasses. A summary of the process for manufacture of cane sugar is illustrated in Figure 5. The process is discussed in detail by Abram and Ramage (1979). Dextranases may also be used during processing of sugar beet when *Leuconostoc* infection of beet, damaged by freezing and thawing, occurs. Enzyme preparations for use in the processing of sugar should not contain invertase activity.

Dextranases have been advocated for use to remove dental plaque, by incorporation into tooth paste. Enzymes produced by *Penicillium* spp. are of little value as the dextrans of dental plaque usually contain α-1,3-linkages which are not hydrolyzed by the *Penicillium* enzymes, which only attack α-1,6-linkages (Aunstrup, 1979).

41.7 LACTASE

Lactase or β-galactosidase (EC 3.2.1.23) catalyzes the conversion of lactose to glucose and galactose (Figure 6). During hydrolysis small amounts of di- and tri-saccharides are produced as a

Cane → Crushing → Extracted → Centrifugation →
harvesting   juice

Boiling under → Raw → Affination[a] → Clarification[b] →
vacuum   sugar

Decolourization → Crystallization[c] → Recovery

[a] Washing process to remove the bulk of total impurities. It involves heating with an affination syrup and low centrifugation. [b] Involves filtration, phosphation or carbonation. [c] Involves evaporation, pan boiling, separation, conditioning, drying and cooling.

Figure 5 Process for manufacture of cane sugar

result of transgalactosidation reactions, particularly at high substrate concentrations. Lactase is very specific with regard to the glycone component of the substrate (galactose), whereas the aglycone can consist of other glycosides, alkyl or aryl alcohols (Kulp, 1975). The enzyme occurs in plants, such as apples, peaches and apricots, animals and in a wide range of bacteria, fungi and yeasts. Lactases produced by bacteria and yeasts are intracellular, whereas most fungi produce extracellular lactases. *Aspergillus* lactases are usually produced by semi-solid cultivation in acidified wheat bran at 30 °C (Aunstrup, 1979). Milk does not contain the enzyme. Commercial lactases are produced by *Kluyveromyces lactis*, *Kluyveromyces fragilis*, *Aspergillus oryzae* and *Bacillus* sp. The properties of these enzymes are summarized in Table 3. It should be noted that a major difference between yeast and fungal enzymes is their pH optima. Fungal β-galactosidases have pH optima near 5.0, while yeast enzymes show maximum activity in the range pH 6–7. Fungal lactases are more suitable for hydrolysis of acid wheys, produced from manufacture of acid casein and cottage cheese, whereas yeast lactases are more suitable for hydrolysis of milk lactose or neutral sweet cheese whey.

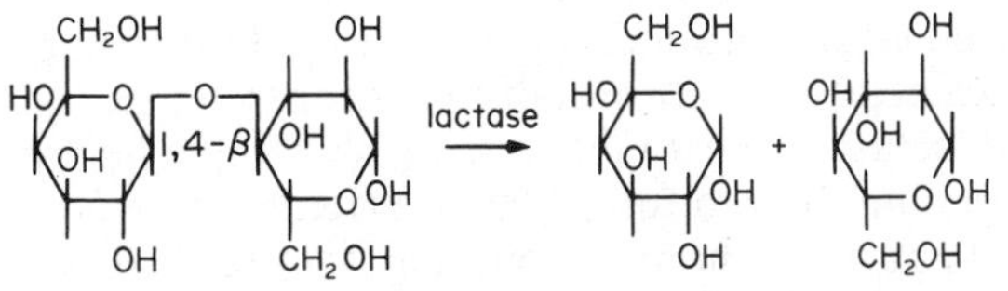

Figure 6 Hydrolysis of lactose

Table 3 Properties of Commercial Lactases

Source	Form	pH optimum	Temperature optimum[a] (°C)	Other properties
K. lactis	Soluble	6.5	40–50	
K. fragilis	Soluble	6.7	35–45	
A. oryzae	Soluble	5.0	55	Variants have pH optima near 6.0
Bacillus sp.	Immobilized	6.5	60	Optimum pH for long term stability is 7.5

[a] Temperature optima can vary markedly depending on pH.
Activators: Potassium, magnesium and manganese. Phosphates (complexes with calcium which inhibits yeast lactase). Reducing compounds, cysteine, sodium sulfide, sodium sulfite (overcome effect of metal inhibitors).
Inhibitors: Heavy metals (copper, zinc, mercury), galactose. Yeast lactase also inhibited by calcium.

41.7.1 Applications of Lactases

The most important applications of lactase relate to hydrolysis of lactose in whey and milk. These applications are summarized in Table 4.

Lactose enzymatic hydrolysis is generally carried out using soluble lactose in batch procedures.

Table 4   Major Applications of Lactase

Raw material	Product	Comments/advantages
Whey	Animal feeds	Allows more whey to be incorporated into animal foods. Prevents crystallization of lactose in whey concentrate
	Lactose-hydrolyzed whey syrup	Used as a food ingredient in bakery, confectionary and ice cream products
Deproteinized whey	Lactose-hydrolyzed permeate syrup	Properties similar to glucose syrups of medium dextrose equivalent
Milk	Lactose-hydrolyzed milk	Improves digestibility in lactose intolerance. Increases sweetness. Prevents crystallization in milk concentrates

A summary of the processes involved is given in Figure 7. In designing a lactose hydrolysis process the following considerations should be borne in mind. The lactase to be used should have a pH optimum as close as possible to the pH of the lactose containing raw materials so that the need for pH adjustment is avoided or minimized. As sodium inhibits yeast lactases, whereas potassium is a lactase activator, potassium hydroxide should be used to raise the pH of wheys that are too acidic. Other inhibitors of lactose should be avoided. Yeast β-galactosidases are inhibited by calcium ions. The inhibitory effect of calcium can be reduced by addition of phosphates followed by heat treatment to precipitate the calcium as calcium phosphate. Where whey is demineralized, prior to treatment with yeast lactase, potassium should be added to the demineralized whey at a concentration of 10^{-1}–10^{-2} M to activate the enzyme. Demineralization has the beneficial effect of reducing the levels of lactase inhibitors such as sodium and calcium. The holding time at 40 °C for yeast lactase treatment of milk has to be short to prevent appreciable microbial growth taking place. Alternatively, efficient lactose hydrolysis can be carried out at 5–10 °C if the incubation period is extended to 16–24 h.

A number of immobilized lactase products have been produced using supports such as cellulose acetate fibres, porous silica beads and absorbent resins for treatment of milk and whey products. Use of yeast-immobilized lactase to treat milk is seriously limited by problems of infection. Infection is less of a problem in immobilized fungal lactase treatment of acid whey or whey permeate, provided the immobilized enzyme column is routinely sanitized by back-flushing with a suitable agent such as dilute acetic acid. The more acidic conditions prevailing in this situation inhibit microbial growth. In the Corning Glass continuous process a fixed bed with downward flow is used. Demineralized whey is pumped through the column at pH 3.5, at a rate of 10 litres per kg immobilized enzyme per hour, and an initial temperature of 32 °C. There is a gradual loss of enzyme activity and the temperature is adjusted upwards eventually to 50 °C to compensate for this and maintain a standard degree of hydrolysis.

Immobilized lactase has not yet made a substantial commercial impact because of the relatively high associated capital equipment costs, the problems of microbial infection and the relatively low hydrolysis costs of the soluble enzyme. Burgess and Shaw (1983) have compared the advantages and disadvantages of soluble and immobilized lactase processes.

41.8   LIPASES

Lipases (EC 3.1.1.3) hydrolyze triglycerides to free fatty acid, partial glycerides and glycerol. Their natural substrates are triglycerides of long-chain fatty acids which are insoluble in water (Shahani, 1975). Lipases hydrolyze the ester bonds at the interface between the aqueous phase, in which the enzyme is soluble, and the insoluble substrate phase. Their relative activity towards water soluble fatty acyl esters is low. It is this ability to hydrolyze insoluble fatty acyl esters which distinguishes lipases from esterases.

Commercial lipases are produced from porcine and bovine pancreas, from the yeast *Candida* and from *Aspergillus*, *Rhizopus* and *Mucor* spp. Betzing and Lakim (1975) described a method for extraction of lipase from pancreatic tissue. The pancreas is partially degreased and dehydrated and mixed with 9 parts per volume of chloroform and one part butanol and left standing at 0–4 °C for 24–96 h. After further degreasing with acetone, the dehydrated glands are extracted with 5% aqueous ethanol. The enzyme, contained in the extract, is precipitated with acetone.

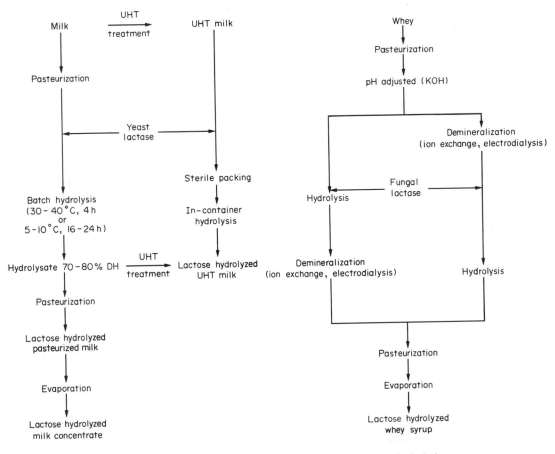

Figure 7 Summary of batch processes for enzymatic lactose hydrolysis

The preferred method for production of microbial lipase is by semi-solid culture (Aunstrup, 1979). Usually it is not necessary to have lipids in the production medium and it has been suggested that, in some instances, they repress enzyme synthesis. Lipase may be recovered as a by-product of the production of microbial rennet by *Mucor miehei* by adsorption on clay minerals. Commercial phospholipase is also extracted from porcine pancreas. The pH and temperature optima for activity of these enzymes and other properties are summarized in Table 5.

Table 5 Properties of Commercial Lipases

Enzyme source	*pH optimum*	*Temperature optimum* (°C)	*Positional specificity*	*Other properties*
Pancreas	8.0	50	1,3-Specific	Preference for long chain fatty acids (>12 carbons). Reaction rate TG > DG >MG. Activated by sodium ions. Porcine enzyme shows enhanced thermostability with calcium
Candida cylindracae	7.5	50	Non-specific	
Rhizopus arrhizus	7.0	40	1,3-Specific	Hydrolyzes long chain fatty acids and tributyrin at similar rates
Aspergillus niger	7.0	45	1,3-Specific	Preference for short chain fatty acids (<12 carbons)
Mucor lipolyticus	7.5	50	1,3-Specific	Contains two components. Lower molecular weight component prefers medium and long chain TGs. Higher molecular weight component hydrolyzes tributyrin as well as medium and long chain TGs.
Pancreatic lecithinase				Produces a lysophosphatide as product. Requires calcium

Lipases can be divided into two groups based on their positional specificity, *i.e.* the ester linkages of the triglyceride that they preferentially cleave. Non-specific lipases release fatty acids from all three positions of the glycerol molecule. 1,3-Specific lipases release fatty acids only from the 1- and 3-positions. They convert triglycerides to 1,2-(2,3-)diglycerides, which in turn are converted to 2-monoglycerides. Since these intermediates are chemically unstable and undergo acyl migration to give 1,3-diglyceride and 1-monoglyceride respectively, prolonged incubation results in complete hydrolysis of the triglyceride. These reactions are summarized in Figure 8.

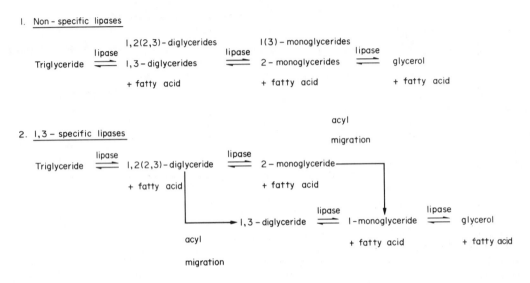

Figure 8 Specificity of lipases

41.8.1 Industrial Applications of Lipases

Lipases have a number of existing or potential industrial applications:

1. They are incorporated into digestive aids by the pharmaceutical industry. Suitable enzymes should be able to withstand the low pH values encountered in the gut.

2. *Candida cylindracae* lipase may be used to hydrolyze oils for soap manufacture, to provide products of better odour and colour characteristics than those obtained by the conventional chemical process.

3. Lipase may be used to generate fatty acids from unstable oils such as those containing conjugated or highly fatty acids (Sonntag, 1979).

4. Enzymes may be used instead of chemicals to promote interesterification of oils and fats. As lipase reactions are reversible, hydrolysis and resynthesis of glycerides occur when lipases are incubated with fats and oils, thereby resulting in acyl migration. The interesterification reaction is favoured over the hydrolysis reaction when the amount of water is restricted, as with peptide synthesis (see Section 40.4.3). The enzymatic reaction is manipulated to produce different product mixtures, by judicious choice of lipase, in respect of positional specificity or non-specificity and fatty acid chain length specificity. The stages involved are summarized in Figure 9. These processes are discussed in detail by Macrae (1983) and Wisdom *et al.* (1984).

5. Lipase may be used in formation of fatty acyl esters. The reaction is driven by using a large molar excess of the alcohol reactant (Tsujisaka *et al.*, 1977).

6. Animal pregastric esterases from lamb, calf and kid and lipase from *Mucor miehei* are used to enhance flavour formation, particularly in Italian hard cheese, blue cheese and processed cheese, or to accelerate cheese ripening (Law, 1982; Richardson, 1975). If the esterases are to be added to the milk prior to renneting, they need to be free from coagulating proteases. Excessive lipase treatment results in the development of methyl ketones in cheese.

7. Lipases may be used for manufacture of cheese and butter flavours. Lipase-modified butterfat is a major source of flavour for butter (Arnold *et al.*, 1975). Enzymatic modification of butterfat releases butyric, caproic, caprylic, capric and long chain fatty acids, and their flavour can be standardized. Butterfat flavours have applications in confectionery (toffees and caramels), as cof-

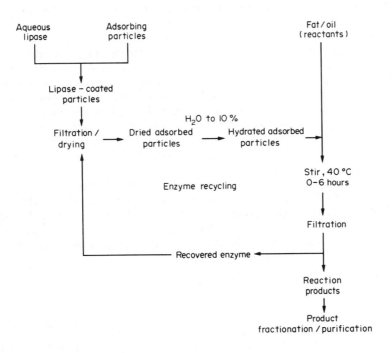

Figure 9 Stages involved in enzymatic interesterification of fatty acids

fee whiteners to give a creamy flavour, in margarines to give a buttery flavour as well as in cheese products.

8. There is renewed interest in lipase-containing detergents, since fat stains are difficult to remove from fabrics using standard laundering processes. Better presoak preparations may be produced by using lipase in combination with synthetic activators such as naphthalene sulfonates.

9. Lipase may be incorporated into fat-containing animal pet foods to improve their flavour. The fat may be emulsified with a proteinaceous emulsifying agent and then treated with a mixture of enzymes including lipase and protease.

41.9 SUMMARY

Commercial pectinase preparations are produced by strains of *Aspergillus niger*. The major enzyme components are pectin esterase, endopolymethylgalacturonate lyase and polygalacturonase. In fruit juice extraction and wine making processes pectinases increase juice yield, reduce viscosity and aid clarification. Used in combination with cellulase/hemicellulase preparations, pectinase improves colour extraction from skins of fruit and is also used to completely macerate fruit or vegetable tissues in preparation of comminuted fruit juices and dispersed baby food preparations.

The major sources of commercial cellulases are strains of *Trichoderma reesei* and *Aspergillus niger*. The *Trichoderma* enzyme complex attacks crystalline cellulose but contains low cellobiase activity. *A. niger* cellulase in contrast does not attack crystalline cellulose but has high cellobiase activity. *A. niger* cellobiase is also less susceptible to end product inhibition by glucose than *Trichoderma* cellobiase and as such is particularly suitable for application in cellulose saccharification. *T. reesei* and *A. niger* cellulases are produced in the stationary phase. Cellulose appears to be the best inducer of the *Trichoderma* enzyme complex. In addition to applications mentioned above, cellulases are used in cereal processing, brewing, alcohol production and waste treatment.

Microbial commercial β-1,3/β-1,4-glucanases are produced by strains of *Bacillus amyloliquefaciens*, *Aspergillus niger* and *Penicillium emersonii*. Used in brewing mashes containing barley or poor malt, these enzymes improve wort filtration and extract yield. They allow a greater proportion of unmalted barley to be incorporated into the mash. β-Glucanases may also be used to treat glucan hazes in beer. β-1,3/β-1,6-Glucanase may be used to hydrolyze the polysaccharide produced by *Botrytis cinerea* in infected grapes, which presents serious clarification and filtration problems during winemaking.

Hemicellulases and xylanases are used, often in combination with other enzymes, in tissue maceration and fruit colour extraction, in baking doughs and brewing mashes containing wheat or rye, in coffee processing, silage production, hydrolysis of gum-containing oil drilling fluids and waste treatment. *A. niger* is the predominant commercial source of this enzyme.

Commercial dextranases, produced by strains of *Penicillium* spp., may be used to hydrolyze dextrans produced by *Leuconostoc* in damaged sugar cane juices thereby eliminating problems caused by the viscous polysaccharide during the juice clarification and evaporation stages.

Lactases are used to hydrolyze lactose in whey, deproteinized whey and milk, preventing lactose crystallization in concentrates, increasing product sweetness and improving product digestibility. Commercial yeast lactase has a pH optimum around 7.0 and is used for lactose hydrolysis in milk. *Aspergillus* lactase has a pH optimum near 5.0 and is more suitable for enzymatic lactose hydrolysis of acid wheys.

Lipases are produced from animal pancreas and by strains of fungi and yeasts. They vary in specificity with respect to the ester positions on the glycerol molecule they attack and in relation to the fatty acid chain length. They have applications as digestive aids in the pharmaceutical industry or in ester synthesis of fats and oils in the chemical industry. Lipases are used in the dairy and food industries to produce specific flavours and have other applications in detergents and waste treatment.

41.10 REFERENCES

Abram, J. C. and J. T. Ramage (1979). Sugar refining: present technology and future developments. In *Sugar Science and Technology*, ed. G. G. Birch and K. J. Parker, pp. 49–95. Applied Science Publishers, London.

Arnold, R. G., K. M. Shahani and B. K. Dwivedi (1975). Application of lipolytic enzymes to flavor development in dairy products. *J. Dairy Sci.*, **58**, 1127–1143.

Arunga, R. O. (1982). Coffee. In *Fermented Foods, Economic Microbiology*, ed. A. H. Rose, vol. 7, pp. 259–292. Academic, London.

Aunstrup, K. (1979). Production of extracellular enzymes. In *Applied Biochemistry and Bioengineering*, ed. L. B. Wingard, Jr., E. Katchalski-Katzin and L. Golstein, vol. 2, pp. 27–69. Academic, New York.

Barrett, F. R., (1975). Milling and baking industries. In *Enzymes in Food Processing*, ed. G. Reed, pp. 301–330. Academic, New York.

Bass, E. J. and T. Cayle (1975). Beer. In *Enzymes in Food Processing*, ed. G. Reed, pp. 455–471. Academic, New York.

Betzing, H. and D. Lakim (1975). Highly active lipase from pancreas. *US Pat.* 3 925 158.

Burgess, K. and M. Shaw (1983). Dairy. In *Industrial Enzymology*, ed. T. Godfrey and J. Reichelt, pp. 260–283. Macmillan, London.

Dilley, D. R. (1970). Enzymes. In *The Biochemistry of Fruits and Their Products*, ed. A. C. Hulme, vol. 1, pp. 179–207. Academic Press, London.

Dubourdieu, D., J.-C. Villettaz, C. Desplaques and P. Ribereau-Gayon (1981). Enzymatic degradation of glucan from *Botrytis cinerea*: clarification of wines of botrytized grapes. *Connaissance de la Vigne et du Vin*, **15** (3), 161–177.

Enari, T.-M. (1983). Microbial cellulases. In *Microbial Enzymes and Biotechnology*, ed. W. M. Fogarty, pp. 183–223. Applied Science Publishers, London.

Enari, T.-M., P. Markkanen and E. Korhonen (1975). In *Symposium on Enzymatic Hydrolysis of Cellulose*, ed. M. Bailey, T.-M. Enari and M. Linko, pp. 171–180. SITRA, Helsinki.

Fogarty, W. M. and C. T. Kelly (1983). Pectic enzymes. In *Microbial Enzymes and Biotechnology*, ed. W. M. Fogarty, pp. 131–182. Applied Science Publishers, London.

Gallo, B. J., R. Andreotti, C. Roche, D. Ryu and M. Mandels (1979). *Biotechnol. Bioeng. Symp.*, **8**, pp. 89–101.

Hasegawa, S. and C. W. Nagel (1968). Isolation of an oligogalacturonate hydronase from a bacillus specie. *Arch. Biochem. Biophys.*, **124**, 513–528.

Hatanaka, C. and J. Ozawa (1969). Isolation of a new exopolygalacturonase producing digalacturonic acid from pectic acid. *Agric. Biol. Chem.*, **33**, 116–121.

Kulp, K. (1975). Carbohydrases. In *Enzymes in Food Processing*, ed. G. Reed, pp. 53–122. Academic, New York.

Law, B. A. (1982). Cheeses. In *Fermented Foods, Economic Microbiology*, ed. A. H. Rose, vol. 7, pp. 147–198. Academic, London.

Lyons, T. P. (1983). Alcohol-power/fuel. In *Industrial Enzymology*, ed. T. Godfrey and J. Reichelt, pp. 179–193. Macmillan, London.

Macleod, A. M. (1979). The physiology of malting. In *Brewing Science*, vol. 1, pp. 145–232. Academic, London.

Macrae, A. R. (1983). Lipase catalyzed interesterification of oils and fats. *J. Am. Oil Chem. Soc.*, **60**, 291–294.

Manners, D. J. and J. J. Marshall (1969). Carbohydrate-metabolising enzymes. XXII. The β-glucanase system of malted barley. *J. Inst. Brewing*, **75**, 550–561.

Miller, L. and J. D. Macmillan (1971). Purification and pattern of action of pectinesterase of *Fusarium oxysporum* f. *vasinfectum*. *Biochemistry*, **10**, 570–576.

Moll, M. (1979). Analysis and composition of barley and malt. In *Brewing Science*, ed. J. R. A. Pollock, vol. 1, pp. 1–143. Academic, London.

Moran, F., S. Nasuno and M. P. Starr (1968). Extracellular and intracellular polygalacturonic acid transeliminases of *Erwinia caratovora*. *Arch. Biochem. Biophys.*, **123**, 298–306.

Neubeck, C. E. (1975). Fruits, fruit products and wines. In *Enzymes in Food Processing*, ed. G. Reed, pp. 397–442. Academic, New York.

Preece, I. A. and J. Hoggan (1956). *J. Inst. Brewing*, **62**, 486.

Reichelt, J. R. (1983). The application of enzymes in industry. In *Industrial Enzymology*, ed. T. Godfrey and J. Reichelt, pp. 210–220. Nature Press, New York.

Richardson, G. H. (1975). Dairy industry. In *Enzymes in Food Processing*, ed. G. Reed, pp. 361–395. Academic, New York.

Rombouts, F. M. and W. Pilnik (1980). In *Microbial Enzymes and Bioconversions, Economic Microbiology*, ed. A. H. Rose, vol. 5, pp. 227–282. Academic, London.

Rombouts, F. M., C. H. Spaansen, C. Versteeg and W. Pilnik (1979). Thermostability and orange juice cloud destabilizing properties of multiple pectinesterases from orange. *J. Food Sci.*, **45**, 969.

Ryu, D. Y. and M. Mandels (1980). Cellulases: biosynthesis and applications. *Enzyme Microb. Technol.*, **2** (2), 91–102.

Scott, D. (1975). Miscellaneous applications of enzymes. In *Enzymes in Food Processing*, ed. G. Reed, pp. 493–517. Academic, New York.

Shahani, K. M. (1975). Lipases and esterases. In *Enzymes in Food Processing*, ed. G. Reed, pp. 181–217. Academic, New York.

Sonntag, N. O. V. (1979). Fat splitting. *J. Am. Oil Chem. Soc.*, **56**, 729A–732A.

Sternberg, D. and S. Dorval (1979).Cellulase production and ammonia metabolism in *Trichoderma reesei* on high levels of cellulose. *Biotechnol. Bioeng.*, **21**, 181–191.

Sternberg, D. and G. R. Mandels (1979). Induction of cellulolytic enzymes in *Trichoderma reesei* by sophorose. *J. Bacteriol.*, **139**, 761–769.

Sternberg, D., P. Vijayakumar and E.T. Reese (1977). β-Glucosidase: microbial production and effect on enzymatic hydrolysis of cellulose. *Can. J. Microbiol.*, **23**, 139–147.

Tsujisaka, Y. S. Okumura and M. Iwai (1977). Glyceride synthesis by four kinds of microbial lipase. *Biochim. Biophys. Acta*, **489**, 415–422.

Van Houdenhoven, F. E. A. (1975). *Studies on Pectin Lyase*. Agricultural University, Wageningen, The Netherlands.

Ward, O. P. (1973). *A Study of the Bacterial Extracellular Enzymes Associated with Increase in Permeability of Water Stored Sitka Spruce (Picea sitchensis)*. Ph.D. Thesis, National University of Ireland.

Ward, O. P. and W. M. Fogarty (1972). Polygalacturonate lyase of a *Bacillus* species associated with increase in permeability of sitka spruce (*Picea sitchensis*). *J. Gen. Microbiol.*, **73**, 439–446.

Whistler, R. L. and C. L. Smart (1953). *Polysaccharide Chemistry*, p.161. Academic, New York.

Whitaker, J. R. (1984). Pectic substances, pectic enzymes and haze formation in fruit juices. *Enzyme Microb. Technol.*, **6**, 337–384.

Wisdom, R. A., P. Dunnill and M. D. Lilly (1984). Enzymic interesterification of fats: factors influencing the choice of support for immobilized lipase. *Enzyme Microb. Technol.*, **6**, 443–446.

42

Glucose Isomerase

F. H. VERHOFF, G. BOGUSLAWSKI, O. J. LANTERO, S. T. SCHLAGER
and Y. C. JAO

Miles Laboratories, Elkhart, IN, USA

42.1 INTRODUCTION

Prior to the 1970s, the main source of sweeteners for food was sugar cane and sugar beets. For some time it had been known that the enzyme glucose isomerase (GI) could convert glucose from various sources into fructose, a much sweeter substance. However, this conversion was not economical because the cost of the enzyme was prohibitive when used in solution. The high cost could be overcome if the enzyme was immobilized because the enzyme could then be utilized to convert much more glucose per unit of enzyme activity.

In the United States the principal source of glucose is corn. For years the corn milling industry has produced a glucose syrup by treating the corn starch with the enzymes α-amylase and amyloglucosidase in solution. The glucose syrup need only be treated further with glucose isomerase to produce a corn syrup with a significant fructose content and hence a greater sweetening capacity. The process for the conversion of corn starch to a glucose–fructose syrup is now well established and is a third major source of food sweetener along with cane sugar and beet sugar.

This glucose–fructose syrup is produced in a variety of glucose–fructose proportions. The conversion of glucose to fructose is limited by equilibrium to approximately equal concentrations of each. Since in practice the conversion is not carried to equilibrium, the effluent from the GI reactor has a somewhat higher glucose concentration. The glucose–fructose mixture with the equivalent sweetening power of sucrose contains approximately 55% fructose. Many producers of high fructose corn syrup concentrate the fructose, usually using a chromatographic technique, and supply the concentrated material as a sweetener. Also, a market has developed for a sweetener with 90% fructose and this product is commercially available.

The markets for high fructose corn sweetener were initially established in the packing industry. As plant capacity for producing high fructose corn syrup came on stream, the markets expanded and now include other sweetener uses and the large soft drink industry. Initially, the high fructose corn syrup was priced in relationship to sugar of a corresponding sweetening power. However, it is expected that the market for high fructose corn syrup will eventually seek its own price which probably will be somewhat independent of sugar.

In this article we will discuss the technology used in the manufacture of the glucose isomerase enzyme used for the glucose to fructose conversion. The review will primarily focus on the processes used in the production of Miles glucose isomerase. During the period that Miles has been producing this enzyme, two different organisms have been used. The processes used for both are similar and therefore the discussions may pertain to one or the other.

42.2 BIOCHEMISTRY AND REGULATION OF GLUCOSE ISOMERASE

42.2.1 Background

The enzyme, which is primarily D-xylose isomerase, catalyzes the aldose–ketose isomerization of a number of sugars. The unexpected discovery of Marshall and Kooi (1957) that D-glucose can be isomerized to D-fructose without prior phosphorylation generated considerable interest in the exploitation of the enzyme for industrial purposes. This seems to have occurred for the wrong reason as the glucose isomerase described by Marshall and Kooi was soon found to be a phosphoglucose isomerase and could isomerize the unphosphorylated sugar only in the presence of arsenate (Natake, 1966; Natake and Yoshimura, 1964; Natake, 1968). The situation was quickly clarified, however, and the existence of a true glucose/xylose isomerizing enzyme has been demonstrated in a variety of bacteria. The reader is referred to the recent excellent reviews for further historical details and the data on general properties and distribution of glucose isomerases (Bucke, 1977; Antrim *et al.*, 1979; Hemmingsen, 1979). Here we shall concentrate on an example of glucose isomerase from a single bacterial species, *Flavobacterium arborescens*, and shall present some of our findings concerning the regulation of the enzyme synthesis in this species.

42.2.2 Characterization of Glucose Isomerase

42.2.2.1 Purification

The yellow soil bacterium *F. arborescens* grows very well on a variety of media, both complex and defined, with the generation time of under 50 min under optimal conditions. The final density in a rich growth medium may reach 10^{11} cells ml^{-1}. Thus, it is a simple matter to obtain large quantities of cells from which glucose isomerase can be readily isolated. We have purified the enzyme to an apparent homogeneity (Verhoff *et al.*, 1983). The molecular weight of native glucose isomerase has been estimated as 180 000, and the enzyme is composed of 4 subunits of 44 700 daltons each. As such, the *F. arborescens* isomerase resembles the isomerases from other sources (Chen and Anderson, 1979; Kasumi *et al.*, 1981; Yamanaka, 1975). The enzyme is thermostable and withstands prolonged heating at 70 °C.

42.2.2.2 Kinetic data

The enzyme's affinities for its two major substrates, D-glucose and D-xylose, reflect strong preference for D-xylose (measured Michaelis–Menten constants are 0.110 M for glucose and 0.51 mM for xylose). The isomerization of either sugar is linear with respect to time and enzyme concentration over a relatively broad range (data not shown).

42.2.2.3 Metal cofactors

All known glucose isomerases are metalloenzymes, and the *F. arborescens* isomerase conforms to this rule. The enzyme in the solubilized form requires cobalt (0.1 mM) and magnesium (0.1 M) ions for full activity and, as Table 1 shows, these metals act in a synergistic fashion. Important for the purpose of commercial production of the enzyme is the fact that the activity of glucose isomerase in whole cells is almost completely independent of cobalt (Table 2). This presumably means that the cells sequester sufficient amounts of cobalt from the medium, an ability which is not shared by all bacteria. Some other metals tested had little effect except for ferric ions which were strongly inhibitory (Table 3).

Table 1 Synergistic Effect of Cobalt and Magnesium Ions on Glucose Isomerase Activity

Cation	Activity (% maximum)
None	0.0
Co^{2+} only	23.8
Mg^{2+} only	21.2
Both	100.0[a]

[a] 35.3 units mg^{-1}.

Table 2 Effect of Cobalt Chloride on the Apparent Activity of Glucose Isomerase

Strain	Enzyme fraction	Activity (units l^{-1}) $-CoCl_2$	$+CoCl_2$
F. arborescens	Whole cells	11 021	12 843
	Purified enzyme	2 331	10 993
B. licheniformis	Whole cells	1220	6150

Table 3 Effect of Selected Metal Cations on Glucose Isomerase Activity

Cation (mM)		Activity (% of control)[a]
None		100.0
Ca^{2+}	0.01	97.3
	0.10	93.6
	1.00	99.3
Fe^{3+}	0.01	93.6
	0.10	31.1
	1.00	0.0
Mn^{2+}	0.01	99.3
	0.10	97.3
	1.00	53.7

[a] 100 mM $MgSO_4$ and 0.1 mM $CoCl_2$ were present throughout.

42.2.2.4 Regulation

Very little is known about genetic factors responsible for the synthesis of glucose isomerase in bacteria. An early communication by David and Wiesmeyer (1970) dealt with the xylose utilization pathway in *Escherichia coli*. These authors established the coordinate control of xylose catabolic enzymes by D-xylose and showed that glucose repressed the enzyme synthesis. This work was followed or paralleled by other reports on D-xylose being the inducer of glucose isomerase synthesis (Danno, 1970; Sanchez and Quinto, 1975; Sanchez and Smiley, 1975). To date, the most detailed study of D-xylose metabolism in bacteria is that of Shamanna and Sanderson (1979a, 1979b). In a series of elegant experiments they showed that in *Salmonella typhimurium* D-xylose utilization is controlled by an operon-like cluster of loci consisting of three structural genes (*xyl T*, *xyl B*, and *xyl A*, responsible for xylose transport, D-xylulokinase, and xylose (glucose) isomerase, respectively). Mutants deficient in glucose isomerase cannot utilize xylose for growth. These authors also obtained genetic, but not yet biochemical, evidence that the three structural genes are under the control of a regulatory gene, *xyl R*. In the absence of the xylose inducer, the *xyl R* gene product is postulated to repress the operon. When xylose is added, it combines with *xyl R* gene product, and the resulting complex activates the transcription of the xylose operon. These regulatory relationships are summarized in Figure 1.

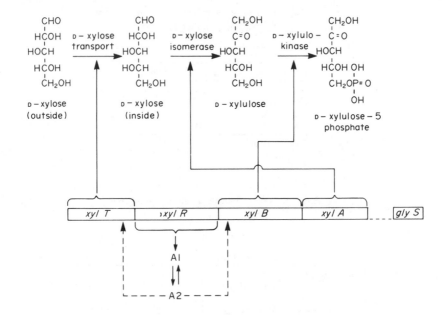

Figure 1 The pathway of xylose catabolism and the organization of the relevant gene cluster in *Salmonella typhimurium*. The product of the regulatory gene *xyl R*, A1, becomes an activator, A2, after it interacts with the inducer, D-xylose (reprinted from Shamanna and Sanderson, 1979b, with permission)

No detailed studies of this type have been carried out on bacteria used in commercial scale production of glucose isomerase. The main concern there has been to obtain mutants constitutive for the enzyme synthesis so that the expensive xylose inducer would not have to be included in the fermentation broth. A selection scheme for isolation of such mutants has been described (Sanchez and Quinto, 1975). Our own laboratory used 2-deoxyglucose to select for constitutive mutants in *Bacillus licheniformis* (Boguslawski and Rynski, 1982).

42.2.2.5 Inhibitor studies

Although 2-deoxyglucose has been used successfully for mutant selection, its effect on the *F. arborescens* glucose isomerase is quite weak (5% inhibition at 0.05 M, and 47% at 0.5 M concentration). We obtained several mutants resistant to 2–10% (w/v) 2-deoxyglucose but they were all unchanged in the enzyme content and regulation. Therefore, we concentrated our efforts on another glucose analogue, β-D-glucose oxime (Figure 2). This compound was initially shown by P. Finch to be a good competitive inhibitor of the *Streptomyces albus* glucose isomerase (unpublished data). We confirmed this and have found that in the case of *F. arborescens* the inhibitory constant, K_i, for the oxime is 150 times lower (0.7 mM) than the affinity of the enzyme for glucose substrate (0.110 M). Under our assay conditions (Boguslawski and Bertch, 1980), glucose oxime binds to glucose isomerase (Figure 3). The binding does not take place at 0 °C. Presumably, the hydroxylamine group of the inhibitor is reactive with the protein only at the elevated temperature (60–70 °C).

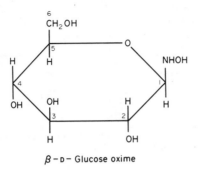

β–D– Glucose oxime

Figure 2 The structure of β-D-glucose oxime. The compound was synthesized according to the method of Finch and Merchant (1975) by M. P. Kotick of Miles Laboratories, Inc.

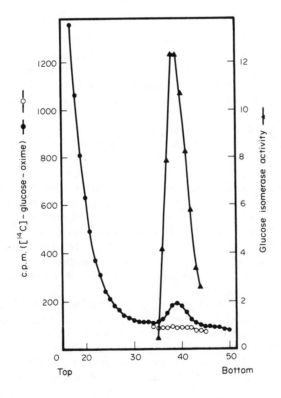

Figure 3 Binding of [^{14}C]-labeled glucose oxime to purified glucose oxime. The enzyme sample (350 μl) was mixed with 0.193 μmole (0.1 μCi) of glucose oxime and heated at 60 °C for 45 min (-●-). Another sample was mixed with glucose oxime and kept on ice for 45 min (-○-). After incubation, the samples were layered on top of a 10–40% sucrose gradient (11 ml) and centrifuged for 40 hours in a Beckman SW41Ti ultracentrifuge rotor. Fractions (0.2 ml) were collected and radioactivity distribution determined by scintillation counting. A parallel gradient received an untreated enzyme sample, and glucose isomerase activity (▲) was determined in the usual manner (Boguslawski and Bertch, 1980)

Glucose oxime effects *in vivo* are complex. The compound inhibits cell growth even when carbon sources other than xylose are used. Therefore, the inhibition cannot be due only to the inactivation of glucose isomerase. Surprisingly, we have found that the presence of glucose oxime in the growth medium results in a two-fold increase in the specific activity of at least two unrelated enzymes, glucose isomerase and β-galactosidase (Table 4). Our data suggest that glucose oxime interferes with an aspect of catabolite repression and thereby allows the increased synthesis of enzymes involved in sugar metabolism (not shown). We are attempting to isolate mutants resistant to glucose oxime in order to understand better the mechanism of its action *in vivo* and *in vitro*.

Table 4 Effect of Glucose Oxime on the Level of Activity of
Glucose Isomerase and β-Galactosidase in Whole Cells of
F. arborescens

Enzyme tested	Glucose oxime	Activity	Relative increase
Glucose isomerase	—	15.3	1.0
	0.1%	34.2	2.2
β-Galactosidase[a]	—	0.47	1.0
	0.1%	0.88	1.9

[a] Assayed as described by Miller (1972) and activity expressed as increase in A_{420} nm due to liberation of *o*-nitrophenol.

42.3 FERMENTATION AND RECOVERY

42.3.1 Background

Glucose isomerase (GI; EC 5.3.1.18) is surprisingly ubiquitous, Chen (1980a, 1980b) reporting over 65 microorganisms as producers of the enzyme. This may reflect not only the remarkable ability of microbes to supply man with products of interest, but the intense interest of man in this particular product. It is also of interest that almost all of the industrial enzymes are in the form of immobilized whole cells rather than free enzyme bound to an artificial support. This is because most of the enzymes are intracellular and it was economically efficient to utilize the cell structure itself as the support for the immobilized system.

Bucke (1977) reviews the history of the enzyme from 1957 to 1977. He points out the major problems with the early processes and hence, the main objectives of fermentation research became yield improvement, replacement of xylose as an enzyme inducer by cheaper media, and elimination of the need for cobalt ions in the fermentation media. Another objective was the elimination of corn steep liquor in the media to reduce fermentation variability.

Yield improvement or cost reduction is essential to the enzyme manufacturer for a competitive product such as GI. The search for new efficient enzyme-producing organisms continues. Demnerova *et al.* (1979) compared 11 strains of *Streptomyces* to two reference strains of *S. phaeochromogenes*. They found the highest GI producer to be *S. nigrificans* 3014 producing almost 80% as much as the highest reference strain. Once a good organism is found, mutation is used to enhance production even further (Demnerova *et al.*, 1982).

The reduction or elimination of xylose in the fermentation was needed to reduce costs. Drazic and coworkers (1980) produced GI within the cells of *Streptomyces bambergiensis* grown in a medium containing xylose as an ingredient. They found that other carbon sources such as starch, glucose, sorbitol or glycerol could be substituted for part of the xylose as long as the xylose concentration was at least 25% of the total substrate concentration. CPC International (1982) working with *S. olivochromogenes*, induced GI production utilizing ratios of xylose to total assimilable carbon of 20 to 66%.

Generally, researchers have found constitutive mutants which eliminate the need for an enzyme inducer. Outtrup (1976a,b), working with *Bacillus coagulans*, isolated an atypical, asporogenic variant which would produce GI without xylose in the medium. Miles mutated *S. olivaceous* to abolish the xylose requirement (unpublished data), and Lee (1981) described mutants of *Flavobacterium arborescens* ATCC 4358 which produce more GI than the parent on lactose instead of xylose as a carbon source.

Cobalt ions in the substrate feed to an immobilized enzyme reactor are found to be activators of the enzyme. However, their use is restricted because of health problems related to the high fructose corn syrup and the fermentation waste disposal problems of the spent media. Kowser and Joseph (1982) found cobalt chloride was a necessary medium component for GI enzyme production by *Streptomyces fradiae* SCF5. Since inclusion of high levels of peptone in the medium caused low enzyme production, they inferred that the peptone affected enzyme production by chelating cobalt ions and making them unavailable at some critical period in the life cycle of the organism.

Lilly (1979) noted the advantages of using simplified media for the production of intracellular enzymes so that control of enzyme synthesis may be applied. He further notes that defined media may, unfortunately, lead to slower growth rates and lower cell mass yields. Corn steep liquor was found to be an excellent and cheap nitrogen source for many GI producing organisms. However, due to its great variability, both seasonal and from batch to batch, it proves to be an unreliable and non-reproducible factor in the fermentation media. Various pretreatments including pH adjustment (Shieh, 1976), centrifugation and filtration have been used to remove sludge, including phytates, from the corn steep liquor. This reduces problems but substitutes for CSL are still being evaluated, even at higher cost. Shieh (1977) showed that *Actinoplanes missouriensis* produced 50% higher enzyme yields growing on a beet molasses–soy flour medium in 72 hours compared to a beet molasses–corn steep liquor medium. The *B. coagulans* variant described by Outtrup (1976a,b) was able to produce enzyme on a medium devoid of an organic nitrogen source.

The use of continuous culture has been used by Diers (1976) to study the influence of growth conditions on GI enzyme production by *B. coagulans*. Severe repression of enzyme synthesis occurred when inorganic compounds were growth limiting and glucose was in excess. Carbon limited growth was beneficial for enzyme production and dual carbon/oxygen limitation increased productivity even more, but with lower specific enzyme activity. He also reported no problems with respect to the development of lower yielding mutants (or overproducers either), and no more contamination problems than with batch culture. Meers (1981) described carbon-limited growth conditions enhancing the specific enzyme yield and carbon conversion efficiency for GI produced by both an *Anthrobacter nov.* sp. and *Mycobacterium smegmatis* in continuous culture. The use of continuous culture can reduce product cost by providing higher equipment productivity and minimizing product variability. However, optimization of continuous fermentation process conditions for a new medium or mutants can be difficult since the operating conditions to be optimized (*e.g.* dilution rate, cell mass, aeration) are all a function of the new parameters. Skot (1982) used batch culture experiments to obtain operating parameters for *B. coagulans* which were then verified in continuous culture experiments. The batch experiments took 48 hours, while to obtain the same data in continuous culture experiments would take more than 2000 hours.

42.3.2 Commercial Processes

Sweigart (1979) listed six companies with immobilized glucose isomerase enzyme or technology including Car-Mi (now Miles), Clinton Corn Processing, Corning Glass Works, Gist-Biocades, ICI, Novo and Snamprogette. Others are Anheuser-Busch, CPC International, Miles-Kali Chemie, R. J. Reynolds and UOP. Process details are not available for most of the fermentations so a flow sheet and process description for the Miles process utilizing *S. olivaceous* (Figure 4) will be given as a representative example.

Growth (*S. olivaceous* NRRL 3916) is initiated from a lyophil onto a plate. A typical colony is picked and used to inoculate several sporulation agar slants containing 1% 2-deoxyglucose (2-DG) to check on mutant reversion. These are incubated at 30 °C for 7 to 9 days and are then stored at 4 °C for use as master slants. Working slants are made from the masters, again utilizing 2-DG and are stored at 4 °C (new working slants may be prepared from these but should never be more than two transfers from a master slant). Spores from the working slant are used to inoculate a shake flask which is incubated for 24–48 hours at 30 °C before transfer to the pre-seed tank. The pre-seed tank medium contains soy protein, yeast extract, glucose, phosphate and soft water (calcium is removed for all fermenter makeup).

Seed tank medium contains corn steep liquor, yeast extract, corn-starch and silicon based antifoam (used throughout). Fermenter medium is the same but with added dextrose. Inoculum build-up is two-step with about 5% inoculum into the seed tank and then into the fermenter. Appropriate purity and quality checks are performed at each step. Fermentation medium is assembled in a charge tank utilizing corn steep liquor which can be pretreated to remove unwanted solids. The medium is subjected to a high temperature–short time sterilization (150 °C,

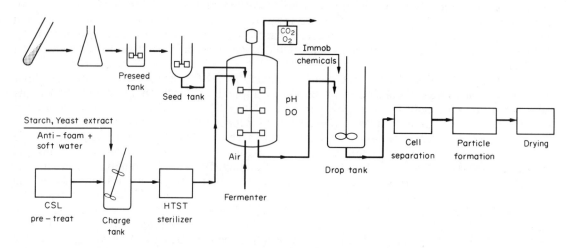

Figure 4 Process for the production of glucose isomerase from the organism slant and nutrient media to the final product

1–2 min) with heat reclaiming, and transferred to a presterilized fermenter. After inoculation, pH, dissolved oxygen (DO), and carbon dioxide and oxygen in the fermenter off-gas are continuously monitored on-line. Periodic samples are checked for carbohydrate concentration, dissolved solids, cell-mass, enzyme activity, *etc.*

At the end of the fermentation (24–48 hours), the fermenter is cooled and the beer sent to recovery. For this organism, immobilization then takes place, followed by solid–liquid separation, particle formation and drying. Other organisms would require preliminary cell separation.

42.4 IMMOBILIZATION OF GLUCOSE ISOMERASE

42.4.1 Introduction

The reports by Bucholz (1979), Mosbach (1976), Pitcher (1980) and Wingard *et al.* (1976) describe many of the methods to immobilize enzymes. More specifically, the development of immobilized glucose isomerase and its application for the production of high fructose corn syrup (HFCS) has been the subject of several pertinent reviews by Antrim *et al.* (1979), Bucke (1977), Hamilton *et al.* (1974), Hemmingsen (1979) and MacAllister (1980). Many methods have been reported for the immobilization of glucose isomerase. However, only a few have been used by enzyme suppliers to produce economically immobilized glucose isomerases that have the necessary properties for the commercial production of HFCS. This section will be limited to the discussion of immobilization processes used in preparing commercially available immobilized glucose isomerase.

A selected list of commercially available immobilized glucose isomerases is given in Table 5. More are given in the reviews of Antrim *et al.* (1979) and van Tilburg (1983). The commercial immobilization processes are of two general types, *i.e.* immobilization of glucose isomerase associated with the microorganism and immobilization of the cell-free enzyme.

Table 5 Commercially Available Immobilized Glucose Isomerase

Company	Enzyme source	Method of immobilization	Enzyme form
Gist-Brocades	*Actinoplanes missouriensis*	Whole cells entrapped in gelatin treated with glutaraldehyde	Hydrated spherical particles
Kali-Chemie	*Streptomyces rubiginosus*	Enzyme absorbed to inorganic support (silicate)	Hydrated
Miles Laboratories, Inc.	*Flavobacterium arborescens*	Flocculated whole cells treated with glutaraldehyde	Dry elliptical sphere particles
Novo Industri	*Bacillus coagulans*	Glutaraldehyde treated lysed cells	Dry cylindrical particles
Universal Oil Products, Inc.	*Actinoplanes missouriensis*	Enzyme bound to inorganic support (alumina)	Hydrated?

42.4.2   Cell-free Immobilization

All the known glucose isomerase producing microorganisms retain the enzyme intracellularly. To immobilize the cell-free enzyme, the enzyme must be solubilized. The enzyme can be solubilized by treating the cells with a detergent or with enzymes, or by homogenization. The free enzyme is separated from the cell debris, and then added to the desired support. The binding of the enzyme to the support can be covalent or adsorptive, or a combination of both. Many examples of binding cell-free glucose isomerase to insoluble carriers have been described in the reviews by Antrim *et al.* (1979) and MacAllister (1980). Both inorganic and organic supports have been used for binding soluble glucose isomerase. Inorganic supports include porous glass beads, and various types of silicates and aluminas. Common organic supports include various types of cellulose, collagen, chitosan, chitin and synthetic polymers, such as ion exchange resins and polyurethanes.

A distinct advantage of binding soluble enzymes to insoluble supports is that the support itself has good structural integrity. Therefore, by careful selection of the support matrix to immobilize enzymes, little emphasis has to be placed on improving particle integrity. It has been observed by one of the authors that soluble glucose isomerase bound to a support appears to be more susceptible to inactivation by minor perturbation in the substrate than whole cell immobilized enzymes.

Kali-Chemie offers, under the trade name Optisweet 22, a soluble glucose isomerase bound to fine SiO_2 beads (Weidenbach *et al.*, 1983). The method of immobilization is rather simple. The glucose isomerase is extracted from *Streptomyces rubiginosus* cells and then adsorbed to silicate beads. It is recommended by the manufacturer that a special precolumn material be used to treat the glucose feed prior to being converted to fructose by the immobilized enzyme for better stability of the enzyme.

UOP, Inc. has been granted several patents on immobilizing glucose isomerase extracted from *Actinoplanes missouriensis* on a derivatized inorganic support. Winans (1981) describes a method in which γ-alumina particles (60–80 mesh) are coated with polyethylenimine prior to treating with glutaraldehyde. The enzyme is then added to the alumina–polyethylenimine–glutaraldehyde complex. In this manner, the amino groups of polyethylenimine bind the enzyme *via* the bifunctional reagent glutaraldehyde. Rohrbach and Lester (1981) similarly immobilized soluble glucose isomerase to alumina particles that are first treated with polystyrene, then nitrated with nitric acid, and finally reduced to obtain an amino–polystyrene–alumina complex. This complex is then treated with glutaraldehyde prior to adding the enzyme. UOP offers a commercial immobilized glucose isomerase which is attached to derivatized alumina particles.

42.4.3   Non Cell-free Immobilization

Non cell-free immobilization represents the technique used for the majority of the commercially available immobilized glucose isomerases. This type of immobilization encompasses a wide variety of immobilization methods, which in a broad sense can be divided into two groups. One group of methods involves the immobilization of the intact microorganism. By this method the glucose isomerase remains associated with the cell. The second method is characterized by procedures that involve releasing glucose isomerase from the cell prior to immobilization. This second method differs from the cell-free immobilization in that no attempts are made to remove the cell debris prior to immobilization. Often the release of glucose isomerase from the cell is not complete but only partial. The excellent reviews by Antrim *et al.* (1979) and van Tilburg (1983) give examples of each type of immobilization. One of the main concerns when immobilizing glucose isomerase without the use of rigid supports is forming an immobilized enzyme particle sufficiently rigid to hold up in large industrial column reactors.

Hupkes and van Tilburg (1976) describe a method of preparing a form of immobilized glucose isomerase suitable for large column use which is commercially available under the trade name of Maxazyme® from Gist-Brocades N.V. The glucose isomerase containing mycelium from *Actinoplanes missouriensis* is mixed with gelatin, and shaped into spherical particles by transferring the gelatin–mycelium mixture into a water-immiscible solvent, *e.g.* butyl acetate. The particles are washed and then treated with glutaraldehyde. Particle integrity is obtained essentially by the glutaraldehyde cross-linking with gelatin. The hydrated particles obtained by the process are about 1 mm in diameter. Since these particles are hydrated, a preservative, *e.g.* propylene glycol, must be added to prevent spoilage.

The immobilization process used by Novo Industri may be considered a non-cell-free type of immobilization, since the cells are disrupted during the immobilization process (Hemmingsen, 1979). More specifically, the glucose isomerase-producing cells of *Bacillus coagulans* are separated from the culture broth by centrifugation. The resulting cell slurry is homogenized and then cross-linked with glutaraldehyde. During homogenization a certain number of cells are disrupted exposing functional groups capable of reacting with glutaraldehyde. The homogenization also releases some of the glucose isomerase from the cell membrane. However, during the glutaraldehyde treatment, enough cross-linking takes place so that the soluble glucose isomerase becomes bound into the cell mixture. The cross-linked cell mixture is dewatered, extruded and shaped into the desired particle form. The particles are then dried, resulting in cylindrically shaped particles with a diameter less than 1 mm. This immobilized glucose isomerase is available from Novo Industri under the trade name of Sweetzyme®.

Recently Miles Laboratories has introduced an immobilized whole cell glucose isomerase (Lantero, 1982). The essential steps in the immobilization are given in Figure 5. The glucose isomerase producing *Flavobacterium arborescens* cells are isolated from the culture broth by centrifugation. The resulting cell slurry is then subjected to a pretreatment which includes heating. The heating may 'fix' the intracellular glucose isomerase to the cell membrane or wall, as was demonstrated by Takasaki *et al.* (1969) who immobilized the glucose isomerase of *Streptomyces albus* cells by heating. Heating the cells can also increase the permeability of the cellular membrane. Van Kuelen *et al.* (1981) reported that permeability of the cell membrane of *Arthrobacter* was increased by heating. Increasing the permeability of the cell would tend to reduce the diffusional resistance, and therefore increase the efficiency of the intracellularly bound enzyme.

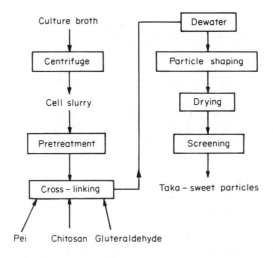

Figure 5   Method for immobilizing the glucose isomerase

The cross-linking step required considerable investigation so that the small *F. arborescens* cells would be recovered conveniently and be shaped into a sufficiently rigid particle. In our previous commercially available immobilized whole cell glucose isomerase, the mycelium of *Streptomyces olivaceus* was aggregated by use of a cationic flocculating agent (Borglum, 1980). The aggregate was cross-linked with glutaraldehyde, recovered by a filtration process and dried. The smaller *Flavobacterium* cells did not aggregate as easily as the *Streptomyces* cells. It was found that several reagents were required to aggregate the *Flavobacterium* cells to facilitate the recovery from the reaction mixture and the shaping into a desirable particle.

The cross-linking is carried out by adding a high molecular weight polyimine, *i.e.* polyethylenimine (PEI), to the heated cell slurry. The large PEI molecule is believed to be adsorbed to the cell surface. According to Treweek and Morgan (1977) this should cause cell flocculation. However, the addition of PEI to the *Flavobacterium* cell slurry causes only a very slight amount of cell aggregation. Previously, Lantero (1982) immobilized whole cells of the glucose isomerase producing microorganisms *Streptomyces olivaceous* and *Bacillus licheniformis* by a process using PEI.

The next step is to add chitosan which increases the cell aggregation slightly. Chitosan has been used in other immobilizations as well, and further information of the use of chitosan on immobilization of enzymes is given in the review by Muzzarelli (1980). One interesting property of chitosan is its ability to bind metal ions, due mainly to its cationic properties. Muzzarelli (1980) reported that the metal ion inhibition of a chitosan-bound glucose isomerase was considerably reduced.

The last step in the cross-linking process is the addition of glutaraldehyde. Complete cell flocculation occurs immediately upon addition of glutaraldehyde. Presumably the glutaraldehyde reacts with the amino groups of the PEI adsorbed to the cell surface and with amino groups of chitosan to cause cell flocculation.

The cell mass is then collected and dewatered by filtration to a point where the cell mass can be shaped. During the shaping process elliptical spheres are obtained and then dried. Drying is a critical step for the development of particle structural integrity. This form of immobilized glucose isomerase is commercially available from Miles Laboratories under the trade name of Taka-Sweet FM®.

42.5 IMMOBILIZED GLUCOSE ISOMERASE PARTICLE FORMATION

42.5.1 Introduction

The major commercially available immobilized glucose isomerase (IGI) products are represented by Sweetzyme® (*Bacillus coagulans*) of Novo Industri A/S, Maxazyme® GI-Immob (*Actinoplanes missouriensis*) of Gist-Brocades N.V., Taka-Sweet® (*Streptomyces olivaceous*) (no longer available) and Taka-Sweet® FM (*Flavobacterium arborescens*) of Miles Laboratories, Inc., Ketozyme® (*Actinoplanes missouriensis*) of Universal Oil Products, Inc. and Optisweet 22 (*Streptomyces rubiginosus*) of Kali-Chemie. The methodology used for manufacturing of each product is quite different. The first three products are derived from whole cell or cell debris immobilization coupled with a particle formation technique (Amotz *et al.*, 1976; Outtrup, 1976; Van Velzen, 1974; Chen and Jao, 1981), while the others are derived from refined soluble glucose isomerase which is covalently bound to chemically modified matrices such as inorganic oxides (Jackson and Yoshiba, 1980; Teague and Huebner, 1982). Particle formation of the first three products will be discussed.

42.5.2 Particle Formation

42.5.2.1 Novo Industri A/S process

The simplified manufacturing scheme of Sweetzyme® presented by Amotz *et al.* (1976) indicated that a fermented culture broth was centrifuged to a biomass slurry followed by either flocculation or homogenization, and then cross-linked with glutaraldehyde. The coherent gel was dewatered and shaped into a particulate form with a final moisture of about 12%. Size distribution of commercially available Sweetzyme® type Q is in the range of 0.25 to 1.10 mm with the majority being from 0.5 to 0.90 mm. The bulk density is about 66 g 100 ml^{-1}. Morphological observation indicated that the particle extrusion was apparently followed by either downsizing on the die face before drying, or by grinding after drying (Hemmingsen, 1979). It is also possible that the extrudates are downsized before drying by a brief spheronization process (Bucke, 1977).

42.5.2.2 Gist-Brocades N.V. process

Immobilization and particle formation of Maxazyme® GI-Immob have been reported quite extensively (Van Velzen, 1974; Hupkes and Van Tilburg, 1976; Hupkes, 1978; Proefschrift and Van Tilburg, 1983). The process is as follows. Fermented beer is concentrated to a crude enzyme-containing mycelium and mixed with a gelatin–water mixture above 40 °C to a final gelatin concentration of 8% (w/v). The mixture is spray-chilled into a cold, water-immiscible solvent such as *n*-butanol or butyl acetate for particle formation. Particles so collected are washed repeatedly with a water-miscible solvent such as ethanol. The solvent is then removed by filtration or suction. The particles so obtained are then suspended in water and cross-linked with glutaraldehyde at a concentration of 2.5% (v/v) to develop the desirable consistency and hardness. The excess glutaraldehyde and soluble impurities are washed out with tap water. The product is then transferred to a 25% (v/v) aqueous solution of propylene glycol, *p*-hydroxybenzoic acid or formaldehyde, and drained before shipment. The bulk density of the product is about 70 g 100 ml^{-1}. Particle size falls in a wide range with an average of 1.5 mm.

42.5.2.3 *Miles Laboratories, Inc. process*

Since 1974, Miles has introduced Taka-Sweet® and Taka-Sweet® FM to the wet milling industry. A process scheme diagram is presented in Figure 6 covering the outlines for both products.

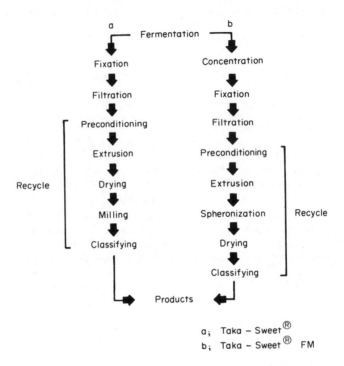

Figure 6 Two processes for particle formation in glucose isomerase

(i) *Taka-Sweet*®

The bacterial cell aggregates used as the starting material in making particles can be produced by contacting the fermented broth with the cross-linking reagents mixture of glutaraldehyde and epihalohydrin polyamine copolymer at pH 8 to 9 at room temperature for 0.5 to 1.5 hours. The resulting bacterial cell aggregate slurry is processed through a rotary vacuum filter resulting in a filter cake with a moisture content of 70 to 75%. The moist cake is then extruded through a die with an opening diameter of 3.18 mm followed by drying, milling and sieving. The physical toughness of the particles is improved by recycling down-sized dry fines to the filter cake prior to extrusion (Chen and Jao, 1981). The final product size ranges from 0.71 to 1.00 mm with a bulk density of 56 g 100 ml^{-1}.

(ii) *Taka-Sweet*® FM

Biomass from a fermented broth is concentrated and fixed with a cross-linking system consisting of polyethyleneimine, chitosan and glutaraldehyde. The coherent gel slurry is then filtered to a specific moisture content of about 72%, followed by preconditioning and extruding the filter cake through openings with diameters approximately that of the spherically shaped cell aggregates to be produced, onto the rotating plate of a spheronizing machine. The discharged spheres are dried in a fluidized bed dryer and classified. Compared with Taka-Sweet®, Taka-Sweet® FM shows a better size and morphological uniformity, better physical toughness, and less friability. Product particle size is ranged from 0.30 mm to 0.85 mm. Bulk density of the product is about 70 g 100 ml^{-1}.

42.5.3 Physical Property Evaluation

Various types of bioreactors are available for immobilized glucose isomerase to convert dextrose syrup into a high fructose syrup. However, the down flow fixed bed reactor has been overwhelmingly utilized through the industry. Mathematical modeling has been extensively employed to translate laboratory and/or pilot plant data into a production-scale operation (Hupkes, 1978; Proefschrift and Van Tilburg, 1983; Verhoff and Furjanic, 1983; Bucholz and Godelmann, 1979). Evaluation of physical parameters on the immobilized enzyme became essential to validate the theoretical approach. The following parameters of Taka-Sweet® were evaluated with the collaboration of the Soil Mechanics Department of Ohio State University (Goldstein *et al.*, 1979). Void volume (ε): 42–49%; static friction against stainless steel (μ_s): 0.44; sliding friction (μ'): 0.16–0.17; coefficient of internal friction (K'): 0.5–0.6; and compressibility (α): 0.28~0.30. A major enzyme particle testing method is presented in the following section.

42.5.3.1 *Mechanical toughness test*

This testing procedure was described in a great detail by Chen and Jao (1981). Immobilized glucose isomerase is hydrated using 40% (w/w) dextrose solution at pH 8.0 and room temperature for one hour before testing. The apparatus depicted in Figure 7 shows an Instron Universal Tester Model 1102 with a stainless steel cylindrical plunger of 4.3 cm diameter and 13.6 cm length attached to a 50 kg load cell for an axial compression. The lower part of the apparatus is a sample cell with an inside diameter of 4.37 cm and a height of 21.75 cm. A spinnerette with 14 500 holes of 0.2 mm opening diameter is located on the bottom of the cell to support the sample and to drain the syrup during compression. Hydrated enzyme sample is charged to the container to a height of 11 cm. The plunger is manually moved down to slightly touch the sample surface, followed by an automatic downward movement for 1 in and withdrawal. This movement is repeated for a second compression. The resistant force and plunger travel distance of the second compression is then correlated to a quadratic equation and further integrated from zero to 1 in for a work value (kg in) as the sample toughness.

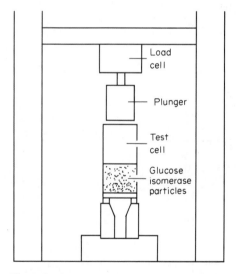

Figure 7 Apparatus for compression testing of immobilized glucose isomerase particles

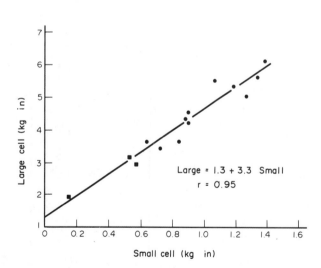

Figure 8 Linear relationship between small and large scale compression tests

A modified system, using a plunger of 1.58 cm diameter with a sample cell of 1.75 cm inside diameter and 3.5 cm height, requires much less sample with only one uniaxial compression. Table 6 presents a computer analysis on data obtained from a Taka-Sweet® FM sample. From the small cell testing system, Sweetzyme® type Q shows a toughness of 2.56 kg in. A linear relationship between the toughness of Taka-Sweet® obtained from the large cell and the small cell is shown in Figure 8.

Table 6   Instron Testing of GI Mechanical Strength, Instron
Crosshead Speed 0.5 in/min

Crosshead travel distance x (in)	Resistant force Y (kg)
0.10	0.78
0.20	2.87
0.30	6.60
0.40	12.55
0.50	21.50
Equation fit for quadratic	

Example: $Y = ax^2 + bx + c$; $a = 114$; $b = -17.2$; $c = 1.49$; $\gamma^2 = 0.9993$;
Work $= \int_0^{0.5} Y dx = 3.342$ kg in

42.5.3.2   Hydraulic test

Applications of fluid dynamics on compressible packed beds to an immobilized glucose isomerase reactor were extensively studied and reported (Zittan *et al.*, 1979; Verhoff and Furjanic, 1983). A research vehicle in the laboratory was designed to measure parameters including pressure across the enzyme bed, enzyme bed height and syrup flow rate through the bed. The correlations of these parameters were used to predict the potential hydraulic behavior of the enzyme. Zittan *et al.* (1979) presented a device which can be used to examine the pressure effect of both fluid and dead load on the packed enzyme bed. Verhoff and Furjanic (1983) presented a bench device used for testing Taka-Sweet®. A pilot plant 18 in enzyme column for studying the hydrodynamic behavior of the enzyme bed is shown in Figure 9 (Goldstein *et al.*, 1979).

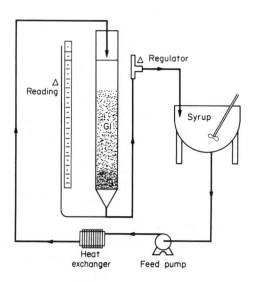

Figure 9   Pilot scale flow test apparatus

Hydraulic behavior of an enzyme bed reactor is expressed as a function of previously mentioned physical parameters as well as of viscosity and specific gravity of the substrate system (Verhoff and Furjanic, 1983). Pilot scale enzyme bed hydrodynamics and enzyme productivity evaluation are critical steps for introducing the product to the market. The relationship of pressure drop and superficial velocity from an 18 in Taka-Sweet® FM iso-column is shown in Figure 10. This hydraulic test result was collected after the iso-column had produced high-fructose corn syrup for 28 days. Data from the figure revealed that for a syrup flowing with a superficial velocity of 23 cm min⁻¹ the pressure drop through the bed is 3.4 p.s.i. Further stepwise velocity changes followed the same pressure drop profile. This hydraulic phenomenon indicates that the enzyme bed was compressed initially and then remained unchanged by the pressure impact resulting from flow rate variation.

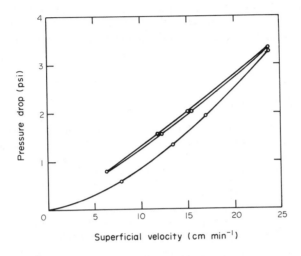

Figure 10   Pressure drop in glucose isomerase bed as a function of flow rate

42.6   MATHEMATICAL MODELING OF REACTOR

Many aspects of the reactor are amenable to design and mathematical modeling. These range from the mathematical analysis of flow through in the reactor to various analyses of the enzyme activity decay, to the optimization of the reactor in some sense. For examples of optimization one is referred to papers by Havewala and Pitcher (1974) and Kim *et al.* (1978). Since the authors' own work has been in the area of flow and enzyme decay, these two areas will be discussed.

42.6.1   Flow Through Compressible GI Reactor

Although there are many types of supports available for glucose isomerase, *e.g.* porous glass, alumina, dextrans and fixed cellular material, the most economical support is the enzyme-containing microorganism. Thus, many glucose isomerase products used commercially are compressible materials. When the dextrose syrup is passed through a column of such immobilized cells, the pressure drop of the fluid causes a solids pressure to develop in the packing material. The compressible packing material then reduces its porosity and limits the flow rate through the packed bed.

The phenomena to be included in the model are the flow through porous media, the compressibility of the solids, the wall friction support of the solids and the interaction between the solids pressure and the fluid pressure. First, the model is used to investigate the influence of various parameters such as the particle size of the packing, the wall friction coefficient and the compressibility coefficient of the enzyme. The results are then compared with experimental data obtained from glucose isomerase columns. The model is used for the scale-up of glucose isomerase columns and for developing experimental compressibility tests for different enzyme formulations. Finally, it should be noted that the theory of fluid flow in compressible beds is not only applicable to glucose isomerase reactors, but also to many other compressible packed beds such as gel permeation columns, ion-exchange columns and deep bed filters.

42.6.1.1 Previous work

Johanson (1979) discusses the processes involved in compressible beds from the bulk solid mechanics point of view. He qualitatively discusses the development of large pressure drops near the bottom of compressible packed beds. Further evidence of the importance of pressure drop in glucose isomerase packed beds is given by two recent publications. Bucholz and Godelmann (1978) describe the pressure drop along enzyme compressible beds. They found that there appears to be a maximum flow rate which can be achieved through such beds. Norsker *et al.* (1970) performed many experiments on the pressure drop in glucose isomerase packed beds. They developed empirical methods for evaluating the pressure drop properties of such packed beds. It appears that the compressible packed bed enzyme literature does not contain a comprehensive approach to the analysis of the fluid dynamics and solid mechanics involved in the reactors.

42.6.1.2 Development of the mathematical model

To analyze fully the forces acting on a packed bed of deformable or mobile particles, all forces as shown in Figure 11 must be taken into account. These include the weight of the particles and support from the walls of the container as well as drag on the particles due to flow of a viscous fluid. Except during settling there is no acceleration, so the sum of the forces is zero and the force balance is:

$$dF = dF_g + dF_F + dF_W \tag{1}$$

Force F_g is due to the weight of the particles, and its contribution is:

$$dF_g = AP_g = R\pi(\Delta\varrho)(1-\varepsilon)dZ \tag{2}$$

where:

$$\Delta\varrho = (\varrho_s - \varrho_L)\frac{g}{g_c}$$

Downward drag on the bed due to viscous flow follows the Kozeny–Carman form:

$$dF_F = AdP_F = \pi R^2 f\frac{(1-\varepsilon)^2}{\varepsilon^3}dZ \tag{3}$$

where:

$$f = \frac{150\mu v_0}{\phi^2 D_P^2}$$

from which follows:

$$\frac{\Delta P_F}{Z_L} = \frac{1}{Z_L}\int_0^{Z_L}\frac{K\mu v_0}{D_P^2}\frac{(1-\varepsilon)^2}{\varepsilon^3}dZ \tag{4}$$

Support from the container walls follows the treatment of pressures of granular solids in silos by McCabe and Smith (1976). In their nomenclature, K' is the ratio of the lateral pressure to the vertical pressure and μ' is the coefficient of friction, giving:

$$dF_W = SdP_W = \pi R^2 \omega P_s dZ \tag{5}$$

where $\omega = 2\mu' K'/R$. Combining the above equations, ignoring turbulence, and dividing through by the volume, AdZ, of the differential element yields:

$$\frac{dP_s}{dZ} = \Delta\varrho(1-\varepsilon) + f\frac{(1-\varepsilon)^2}{\varepsilon^3} - \omega P_s \tag{6}$$

An important extension is needed for a complete description of the phenomena we are attempting to analyze, and that is the physical properties of the compressible enzyme bed. By compressible we mean that the bed as a whole responds to compressive forces by deforming into a more densely packed configuration. In doing so, interparticle void volume is lost. Thus, the proper measure of compressibility of the enzyme is a description of the void volume as a function of pressure applied. Void volume may be lost by two mechanisms: rearrangement of the particles from a loose, random packing to a more dense packing, and deformation of individual particles into the void space. Since we were unable to discover a satisfactory theoretical description of this phenomenon, empirical models were fitted to the data obtained from the compression cell described later. The most promising of these were:

$$\varepsilon = \varepsilon_0/(1 + P_s) \tag{7}$$

and:

$$\varepsilon = \varepsilon_0 e^{-\beta P_s} \tag{8}$$

These have the advantage that the compressibility of the bed is lumped into a single parameter. Critical comparison of the shape of the curves as well as the correlation coefficient of curve fitting led to the selection of equation (7) as the function of choice for this study. Completion of the model is accomplished by substitution of equation (7) into (6), but little is to be gained by rewriting the full form here.

42.6.1.3 *Behavior of the modeled system*

There are three types of solution for the model. This is most clearly seen by expressing the flow term as a quadratic in P_s and solving the resulting Riccati equation. It may also be demonstrated by proper choice of the parameters in (6), as is shown in Figure 11. The middle curve corresponds to incompressible material with no wall support and, like static fluid, it exhibits a linear relationship. The lower curve is for dominant wall effect and it corresponds to the situation described by McCabe and Smith (1976) for storage of granular solids in silos. The upper curve is for the case where compressibility dominates; that is, the wall support is not sufficient to overcome the forces generated. This is the case that causes the most difficulty in reactor operations.

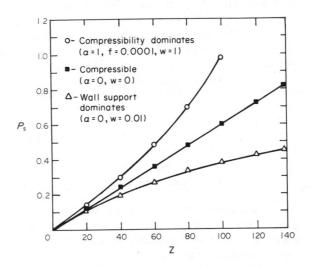

Figure 11 Three characteristic types of solids pressure as a function of flow rate through a packed bed

42.6.2 Decay of Activity

Immobilized enzymes and microorganisms are being considered for the production of various bulk chemicals. In particular, immobilized glucose isomerase has been commercialized and is now used to produce a sweetener (high fructose corn syrup) suitable for applications formerly requiring sucrose. This technology employs a packed bed of immobilized enzyme through which is pumped a dextrose solution that has been obtained by enzymatic hydrolysis of corn starch. As the glucose in the solution moves through the column, it diffuses to and into the porous immobilized enzyme beads and reacts to form fructose. Thus, the glucose concentration decreases and the fructose concentration increases as the solution moves through the column. At the exit of the reactor, the glucose concentration has decreased so that it is nearly in equilibrium with the fructose concentration.

The activity of the glucose isomerase, as well as of most other enzymes, declines with time. Hence, the production rate of the reactor decreases. At some time, the reactor must be stopped and recharged with new enzyme. The factor of economic importance is the amount of product produced by a given amount of immobilized enzyme or microbial particle (specific activity). Two parameters influence this factor: the initial effective enzyme activity and the deactivation rate of the immobilized enzyme. In this review, the prime variable of importance will be the deactivation rate. However, the implications of the initial enzyme activity as related to deactivation rate will be discussed.

There have been several investigations into the influences of diffusion restrictions on the apparent deactivation rate of enzymes. Ollis (1972) pointed out that the apparent deactivation rate was reduced by one-half under strong intraparticle diffusion limitations assuming that the deactivation occurred at an exponential rate and that the reaction rate was first order. Korus and O'Driscoll (1975, 1976) extended this analysis from the first-order reaction case to that of the usual Michaelis–Menten kinetics. They again concluded that the possible increase in the apparent decay half-life was to double the half-life. The implication of this analysis for productivity is somewhat obscure. It is easy to show that the increasing intraparticle resistance (*e.g.* increasing the diameter) decreases the productivity of the enzyme although the apparent half-life of the enzyme has been increased. It follows from this simple analysis that one would want to make the enzyme particle as small as is practical to permit liquid flow through the packed bed without excessive pressure drop.

The above analysis assumes that the decay rate of the enzyme is the same whether the enzyme site is vacant or occupied by substrate. For many enzymes, the different forms of the enzyme are known to have different decay rate constants. Lee *et al.* (1978) have considered the case for which the substrate binds with the vacant enzyme to form an enzyme–substrate complex. They define the first-order deactivation rate constant for the vacant enzyme to be k_d, and the deactivation rate for the complex to be δk_d. Thus, when $\delta < 1$, they describe the enzyme as substrate protected, and when $\delta > 1$, the enzyme is poisoned by the substrate. Generally, it is concluded that diffusional resistance in the immobilized enzyme pellet is very advantageous when the enzyme is susceptible to substrate poisoning. They also investigated the ratio of the half saturation constant to the substrate concentration in the bulk solution. When this ratio is less than one, the effects of substrate protection and/or substrate poisoning are enhanced.

There have been other studies investigating the effect of diffusional resistances on the performance of enzymes. Naik and Karanth (1978) analyzed the influence of the external diffusional resistance as well as the internal resistance. Engasser and Coulet (1977) developed a graphical technique for extracting the intrinsic stability of bound and free enzymes from the influences of diffusion. Roels and van Tilburg (1979) relate the effect of temperature and diffusional resistance to the performance of the enzyme pellet. In particular, they show how the apparent activation energy for the reaction and for the deactivation of the enzyme are influenced by diffusion and temperature.

Essentially, the apparent deactivation of enzymes is affected by diffusion in two ways. The first will be called the changing Thiele modulus phenomenon. As deactivation occurs in the enzyme pellet, the Thiele modulus, the ratio of the internal reaction rate to the diffusion rate, decreases which yields a higher effectiveness factor for the enzyme pellet. Since the apparent reaction rate in the pellet is the product of the effectiveness factor and the internal reaction rate, the apparent reaction rate does not decrease as fast as the actual reaction rate. This yields a decreased apparent deactivation rate in comparison to the internal deactivation rate. The article by Ollis (1972) and many subsequent articles (*e.g.* Naik and Karanth, 1978) deal with this phenomenon.

The second way in which diffusion reactions influence the apparent deactivation rate of enzymes is by creating concentration gradients within a packed-bed reactor. Lee *et al.* (1978) have investigated this phenomenon for the simple Michaelis–Menten model of enzyme reactions. Basically, the diffusion limitation causes gradients in the two enzyme forms, vacant site enzyme and enzyme–substrate complex. Since these two enzyme forms do not deactivate at the same rate, the degree of diffusional resistance alters the relative ratio of the two enzyme forms and hence alters the apparent deactivation rate of the enzyme. Of course, the two phenomena by which diffusion can alter the deactivation rate could be operating simultaneously.

Herein we investigate the influence of spatial concentration variations on the deactivation of enzymes in packed bed reactors. These concentration gradients can be caused by diffusion limitations within the enzyme pellet. Gradients can also be caused by the changing concentrations in the direction of flow as the solution proceeds through the column. It will be assumed that the enzyme in its different forms can decay at different rates. Since the reaction for the formation of fructose from glucose is of interest, the enzyme, glucose isomerase, will be considered in three forms; vacant enzyme, enzyme complexed with glucose and enzyme complexed with fructose.

Of course, the decay rate of the glucose isomerase depends upon many factors. Commercially, temperature, pH and cations are principal factors. These factors affect various binding and kinetic rate constants used in the models contained herein. Therefore, this work then also serves to determine how these parameters influence the diffusion limitation and the apparent deactivation rate of the enzyme.

The mechanism of the reaction plays a central role in the influence of concentration gradients on the stability of the enzyme. In order for these gradients to influence stability, at least two different forms of a single enzyme or two different enzymes must be involved in the mechanisms of the reaction. Lee *et al.* (1978) have treated the irreversible reaction with two different forms of a single enzyme. Herein, the reversible reaction with three different forms of the same enzyme will be presented because it has application to glucose isomerase. It should be obvious that many other enzyme combinations exist which would cause a variation in stability with concentration gradients (*e.g.* metalloenzyme complexes). Carrying out the calculations, one obtains two different equations.

The enzyme activity decays by first-order kinetics, according to the following equations, assuming that deactivation is slow compared to the interconversion between different enzyme forms.

$$\frac{dC_E}{dt} = -k_d D_E; \quad \frac{dC_{EG}}{dt} = -k_d \alpha C_{EG}; \quad \frac{dC_{EF}}{dt} = -k_d \beta C_{EF} \tag{9}$$

where t is the time; k_d is the first-order deactivation rate of vacant enzyme; α is the ratio of glucose complex to vacant enzyme deactivation rate; and β is the ratio of fructose complex to vacant enzyme deactivation rate.

Summing these three equations and using equation (3) gives the following expression for deactivation of total enzyme concentration. This equation does not involve an assumption about slow deactivation.

$$dC_T/dt = -k_d(C_E + \alpha C_{EG} + \beta C_{EF}) \tag{10}$$

Using equations (2)–(4), the expression for the deactivation of the enzyme becomes the following:

$$\frac{dC_T}{dt} = \frac{-k_d C_T[1 + \beta C_S/K_F + (\alpha/K_G - \beta/K_F)C_G]}{[1 + C_S/K_F + (1/K_G - 1/K_F)C_G]} \tag{11}$$

A mass balance on an infinitesimal distance in the enzyme pellet gives the following differential equation.

$$D_G \frac{1}{r^2} \frac{\partial}{\partial r} \frac{(r^2 \partial C_G)}{\partial r} - \frac{\partial C_G}{\partial t}$$
$$= \frac{[(k_f/K_G + k_r/K_F)C_G - k_r/K_F C_S]C_T}{1 + C_S/K_F + (1/K_G - 1/K_F)C_G} \tag{12}$$

where D_G is the diffusivity of glucose (and fructose), and r is the distance measurement in the pellet.

In addition to the differential equations describing the diffusion and reaction in the pellet, the boundary conditions must be stated. The two dependent variables, the glucose concentration and the total enzyme site concentration, must be specified, *i.e.* for $r = R$, $C_G = wC_S$; for $r = 0$, $\partial C_G/\partial r = 0$; for $t = 0$, $C_G = wC_S$; for $t = 0$, $C_T = C_{T_0}$; where R is the radius of spherical pellet; w is the glucose fraction of total monomer (fructose and glucose); and C_T is the initial concentration of active enzyme sites.

An analytical solution of these equations is not apparent and hence numerical solutions are required. The solution is simplified somewhat by converting to dimensionless variables as defined below.

$$\Psi = C_G/C_S; \quad \phi = C_T/C_{T_0}; \quad \varepsilon = r/R; \quad \tau = k_d t$$

These dimensionless variables yield the following eight dimensionless groups.

$$Q_1 = \frac{k_f R^2 C_{T_0}}{D_G C_S}; \quad Q_2 = \frac{k_d C_S}{k_f C_{T_0}} \tag{13}$$
$$Q_3 = C_S/K_G; \quad Q_4 = C_S/K_F; \quad Q_5 = k_r/k_f$$
$$\alpha, \beta, w$$

For the reversible reaction of commercial importance, it is useful to discuss the relationship of productivity to enzyme decay rate. As has been shown by prior authors, the enzyme decay rate decreases as the internal mass transfer resistance increases (increasing Thiele modulus). However, increasing mass transfer resistance yields decreasing productivity of the enzyme bed.

The usual method for altering the decay rate would be to change the size of the enzyme pellet. Reducing the pellet diameter yields higher productivity, but increased apparent decay rate. Also, the pressure drop through the reactor and physical stability of the enzyme bed must be considered as particle size is reduced.

The particle size radius R appears in only one dimensionless parameter Q_1. If the particle radius is doubled, the parameter Q_1 is quadrupled. Figure 12 contains a plot of the dimensionless enzyme productivity as a function of time for two different values of the parameter Q_1. As can be seen from Figure 12, the productivity of the smaller particle ($Q_1=20$) is significantly better than that of the larger particle although the decay rate of the smaller particle is greater than that of the larger particle. Generally, it is true for enzyme reactions as well as other reactions that increased internal mass transfer resistance yields lower productivity; exceptions exist however, such as inverse-order reactions or nonuniform decay of enzymes.

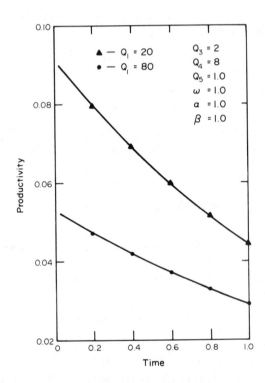

Figure 12 Thiele modulus effect on enzyme activity as a function of time

As an example of the types of different decay rates, Figure 13 contains a plot of the enzyme activity as a function of pellet radius for specific times and values of α and β. It can be seen that, if the value of β is greater than the value of α, the enzyme activity profiles will be concave upward, *i.e.* the activity will be greater at the surface of the pellet as compared to the center of the pellet. This is so because the decay rate of the glucose form is slower than the fructose form and the glucose concentration is greater at the surface and the fructose concentration is greater in the center of the pellet. The reverse is true if α is greater than β in which case the profiles are concave downward. Also, it should be noted that the magnitude of the values of these decay ratios, α and β, has an effect upon the degree of concavity because the enzyme without fructose or glucose also decays. When the decay ratios are low, the main loss of enzyme activity occurs in the unbound enzyme form. For the case of α and β equal but less than one, a concentration gradient of enzyme activity still can exist since the binding equilibrium constants for glucose and fructose could differ significantly and cause a different overall decay rate in the center of the pellet as compared to the surface of the pellet. Only if the decay rate of all three enzyme forms is all the same, *i.e.* α and β

equal to one, will the enzyme activity necessarily be uniform across the pellet, *i.e.* a horizontal line on Figure 13.

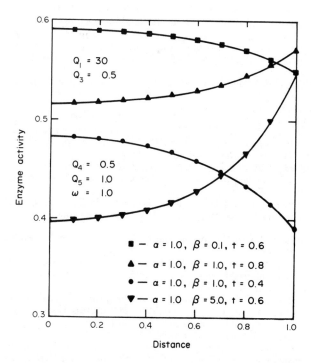

Figure 13   Concentration gradients in an enzyme pellet under different decay conditions

42.7 REFERENCES

Amotz, S., T. K. Nielsen and N. O. Thiesen (1976). Immobilization of glucose isomerase. *US Pat.* 3 980 521.
Antrim, R. L., W. Colilla and B. J. Schnyder (1979). In *Applied Biochemistry and Biotechnology*, ed. L. B. Wingard, Jr., E. Katchalski-Katzir and L. Goldstein, vol. 2, pp. 97–155. Academic, New York.
Boguslawski, G. and S. W. Bertch (1980). *J. Appl. Biochem.*, **2**, 367–372.
Boguslawski, G. and M. J. Rynski (1982). *US Pat.* 4 355 103.
Borglum, G. B. (1980). Production of bacterial cell aggregate. *US Pat.* 4 212 943.
Bucholz, K. (1979). Characterization of immobilized biocatalysts. *Dechema*, **84**.
Bucholz, K. and B. B. Godelmann (1978). *J. Enzyme Eng.*, **4**, 89.
Bucholz, K. and B. B. Godelmann (1979). Characterization of immobilized biocatalysts. *Dechema*, **84**, 127–135.
Bucke, C. (1977). In *Topics in Enzyme and Fermentation Technology*, ed. A. Wiseman, vol. 1, pp. 147–171. Halsted Press, London.
Chen, W. (1980a). Glucose isomerase (a review). *Process Biochemistry*, **15** (June–July), 30–35.
Chen, W. (1980b). Glucose isomerase (a review). *Process Biochemistry*, **15** (Aug.–Sept.), 36–41.
Chen, A. H. and Y. C. Jao (1981). Preparation of a bacterial cell aggregate. *US Pat.* 4 251 632.
Chen, W. P. and A. W. Anderson (1979). *Appl. Environ. Microbiol.*, **38**, 1111–1119.
CPC International (1982). A process for glucose isomerase production. *Havant, Eng., Industrial Opportunities*, **218**, 221.
Danno, G. (1970). *Agric. Biol. Chem.*, **34**, 1658–1667.
David, J. D. and H. Wiesmeyer (1970). *Biochem. Biophys. Acta*, **201**, 497–499.
Demnerova, K., S. Pospiel, O. Valentova and J. Kas (1979). Production of glucose isomerase by different strains of streptomyces. *Biotechnol. Lett.*, **1**, 293–298.
Demnerova, K., I. Safarik and B. Kralova (1982). Glucose isomerase extraction from *S. nigrificans*: a comparison of methods. *Biotechnol. Lett.*, **4**, 431–435.
Diers, I. (1976). Glucose isomerase in *Bacillus coagulans*. In *Continuous Culture 6: Applications in New Fields*, ed. A. C. R. Dean, pp. 208–225. Ellis Horwood Ltd., Chichester.
Drazic, M., Z. Golubic and S. Cizmek (1980). Isomerization of glucose to fructose using microbial enzymes. *Periodicum Biologorum*, **82**, 481–484.
Engasser, J. M. and P. R. Coulet (1977). *Biochim. Biophys. Acta*, **485**, 2936.
Finch, P. and Z. Merchant (1975). *J. Chem. Soc., Perkin Trans. 1*, 1682–1686.
Goldstein, W. E., A. H. Chen, Y. C. Jao, O. J. Lantero, J. Furjanic, and J. J. Fish (1979). Process considerations, kinetic and stability, hydraulic property testing and scale-up studies for Taka-Sweet® research & development. Biotechnology Group, Miles Laboratories, Inc.
Hamilton, B. K., C. K. Carlton and C. L. Cooneg (1974). Glucose isomerase — a case study of enzyme-catalyzed process

technology. In *Immobilized Enzymes in Food and Microbiology Processes*, ed. A. C. Olson and C. L. Cooney, pp. 85–131. Plenum, New York.

Havewala, N. B. and W. H. Pitcher, Jr. (1974). In *Enzyme Engineering II*, ed. E. K. Pye and L. B. Wingard. Plenum, New York.

Hemmingsen, S. H. (1979). Development of an immobilized glucose isomerase for industrial application. In *Applied Biochemistry and Bioengineering*, ed. L. B. Wingard, Jr., E. Katchalski-Katzir and L. Goldstein, vol. 2, pp. 157–183. Academic, New York.

Hupkes, J. V. (1978). Practical process conditions for the use of immobilized glucose isomerase. *Starch/Stärke*, **30**, 24–28.

Hupkes, J. V. and R. Van Tilburg (1976). Production and properties of an immobilized glucose isomerase. *Starch/Stärke*, **28**, 356–360.

Jackson, D. M. and T. Yoshiba (1980). Method for purifying glucose isomerase and a composition for storing same. *US Pat.* 4 237 231.

Johanson, J. R. (1979). *Chem. Eng.*, **86**, 77.

Kasumi, T., N. Hayashi and T. Takagi (1981). *Agric. Biol. Chem.*, **45**, 1087–1095.

Kim, S. M., S. B. Lee and D. Y. Rya (1978). Effect of design parameter on the performance of immobilized glucose isomerase reactor system. In *Enzyme Engineering IV*, ed. G. B. Brown *et al.*, p. 101. Plenum, New York.

Korus, R. A. and K. F. O'Driscoll (1975). *Biotechnol. Bioeng.*, **17**, 441.

Korus, R. A. and K. F. O'Driscoll (1976). *Biotechnol. Bioeng.*, **18**, 1656.

Kowser, N. and R. Joseph (1982). Influence of peptone and cobalt levels on glucose isomerase production by *Streptomyces fradiee*. *Starch/Stärke*, **34**, 172–175.

Lantero, O. J., Jr. (1977). Properties of a cell-bound glucose isomerase. In *Enzyme Engineering*, ed. G. H. Brown, vol. IV. From 4th Enzyme Engineering Conference, sponsored by Engineering Foundation, New York, and DECHEMA. Plenum, New York.

Lantero, O. J., Jr. (1982). Gluteraldehyde/polyethylenimine immobilization of whole microbial cells. *US Pat.* 4 355 105.

Lee, C. K. (1981). *US Pat.* 4 283 496.

Lee, D. D., G. K. Lee, P. J. Reilly and Y. Y. Lee (1978). *Abstr. Pap. Am. Chem. Soc.*, **176**, 55.

Lilly, M. D. (1979). Production of intracellular microbial enzymes. In *Applied Biochemistry and Bioengineering*, ed. L. E. Wingard, Jr., vol. 2, pp. 1–26. Academic, New York.

MacAllister, R. V. (1980). Manufacture of high fructose corn syrup using immobilized glucose isomerase. In *Immobilized Enzymes for Food Processing*, ed. W. E. Pitcher, Jr., pp. 81–111. CRC Press, Boca Raton, FL.

McCabe, W. L. and J. M. Smith (1976). *Unit Operations of Chemical Engineering*. McGraw-Hill, New York.

Marshall, R. O. and E. R. Kooi (1957). *Science*, **125**, 648–649.

Meers, J. L. (1981). The continuous production of glucose isomerase. *Biotechnol. Lett.*, **3**, 136–141.

Miller, J. H. (1972). *Experiments in Molecular Genetics*, pp. 352–355. Cold Spring Harbor Laboratory Press, New York.

Mosbach, K. (ed.) (1976). *Methods Enzymol.*, **54**.

Muzzarelli, R. A. A. (1980). Immobilization of enzyme on chitin and chitosan. *Enzyme Microb. Technol.*, **2**, 177–184.

Naik, S. S. and N. G. Karanth (1978). *J. Appl. Chem. Biotechnol.*, **28**, 569.

Natake, M. (1966). *Agric. Biol. Chem.*, **30**, 887–895.

Natake, M. (1968). *Agric. Biol. Chem.*, **32**, 303–312.

Natake, M. and S. Yoshimura (1964). *Agric. Biol. Chem.*, **28**, 510–516.

Norsker, O., K. Gibson and L. Zittan (1970). *Starch/Stärke*, **18**, 13.

Ollis, D. F. (1972). *Biotechnol. Bioeng.*, **14**, 871.

Outtrup, H. (1976a). *US Pat.* 3 979 261.

Outtrup, H. (1976b). *Br. Pat.* 1 455 993.

Outtrup, H. (1976c). Production of glucose isomerase by *Bacillus coagulans*. *US Pat.* 3 979 261.

Pitcher, W. H., Jr. (ed.) (1980). *Immobilized Enzymes for Food Processing*. CRC Press, Boca Raton, FL.

Proefschrift and R. Van Tilburg (1983). *Engineering Aspects of Biocatalysts in Industrial Starch Conversion Technology*. Delftse Universitaire Press, Delft, Netherlands.

Roels, J. A. and R. Van Tilburg (1979). *Starch/Stärke*, **31**, 1724.

Rohrbach, R. P. and G. W. Lester (1981). Preparation of support matrices for immobilized enzymes. *US Pat.* 4 250 080.

Sanchez, S. and C. M. Quinto (1975). *Appl. Microbiol.*, **30**, 750–754.

Sanchez, S. and K. L. Smiley (1975). *Appl. Microbiol.*, **29**, 745–750.

Shamanna, D. K. and K. E. Sanderson (1979a). *J. Bacteriol.*, **139**, 64–70.

Shamanna, D. K. and K. E. Sanderson (1979b). *J. Bacteriol.*, **139**, 71–79.

Shieh, K. K. (1976). *US Pat.* 3 992 262.

Shieh, K. K. (1977). *US Pat.* 4 003 793.

Skot, G. (1982). Batch culture experiments in the optimization of continuous fermentation of an intracellular enzyme. In *Enzyme Engineering*, ed. G. B. Brown, G. Manecke and L. B. Wingard, Jr., vol. 4, pp. 33–39. Plenum, New York.

Sweigart, R. D. (1979). Industrial applications of immobilized enzymes: a commercial overview. In *Applied Biochemistry and Bioengineering*, ed. L. B. Wingard, Jr. vol. 2, pp. 209–218. Academic, New York.

Takasaki, Y., Y. Kosugi and A. Kanbayashi (1969). *Streptomyces* glucose isomerase. In *Fermentation Advances*, ed. D. Perlman, pp. 561–589. Academic, New York.

Teague, J. R. and A. L. Huebner (1982). Enhanced immobilization of a glucose isomerase. *US Pat.* 4 337 172.

Treweek, G. P. and J. J. Morgan (1977). Polymer flocculation of bacteria—the mechanism of *E. coli* aggregation by polyethylenimine. *J. Colloid Interface Sci.*, **60**, 258–273.

Van Kuelen, M. A., K. Vellenga and G. E. H. Joosten (1981). Kinetics of the isomerization of D-glucose into D-fructose catalyzed by glucose isomerase containing *Arithrobacter* cells in immobilized and non-immobilized form. *Biotechnol. Bioeng.*, **23**, 1437–1448.

van Tilburg, R. (ed.) (1983). Engineering aspects of biocatalysts. In *Industrial Starch Conversion Technology*. Delftse Universitaire Press, Delft, Netherlands.

Van Velzen, A. G. (1974). Water-insoluble enzyme composition. *US Pat.* 3 838 007.

Verhoff, F. H. and J. J. Furjanic (1983). Compressible packed bed fluid dynamics with application on a glucose isomerase reactor. *I & EC Process Design and Development*, **22**, 192–198.

Verhoff, F. H., G. Boguslawski, O. J. Lantero, S. T. Schlager and Y. C. Jao (1983). The purification and characterization of glucose isomerase from *Flavobacterium arborescens*. *J. Appl. Biochem.*, **5**, 186–196.

Weidenbach, G., D. Bonse and G. Ritchie (1983). Isomerization with a newly developed, high-productivity immobilized glucose isomerase—Optisweet 22. Poster at Biotech 83, London.

Winans, V. (1981). Method of purification of thermally stable enzymes. *US Pat.* 4 250 263.

Wingard, L. B., Jr., E. Katchalski-Katzir and L. Goldstein (eds.) (1976). *Applied Biochemistry and Biotechnology*, vol. 1. Academic, New York.

Yamanaka, K. (1975). *Methods Enzymol.*, **4**, 466–471.

Zittan, L., Norsker, K., and Gibson, K. (1979). *Starch/Stärke*, **31**, 13–16.

43
Ethanol

B. L. MAIORELLA
University of California, Berkeley, CA, USA

43.1 INTRODUCTION

43.1.1 Historical Perspective

The preparation of distilled alcohol spirits was first described in the 12th century version of *Mappae Clavicula* (Key to Painting) (Underwood, 1935; Forbes, 1970), with the first reported (nonbeverage) uses of spirits as an incendiary, a solvent and later medicinally. The preparation of absolute alcohol (by repeated distillation over potassium carbonate) was first reported in 1796.

Extensive industrial use of ethyl alcohol began in the late 1800s with the growth of the synthetic chemical industry and expanded rapidly in the United States after 1906 when tax regulations were amended to allow the tax-free preparation of denatured alcohol for industrial use (US Denatured Alcohol Act, 1906). Following the repeal of prohibition, regulations were further tightened to separate beverage and industrial alcohol production, thus establishing the industrial ethanol fermentation industry (US Department of Treasury, 1942).

Industrial alcohol has been valuable as a solvent, germicide, antifreeze, fuel and chemical raw material, and use in the US grew to 564 million liters per year in 1941 (77% by fermentation). During World War II, US industrial alcohol production was increased 3.8 fold (BATF, 1984) on an emergency basis to provide alcohol for synthetic rubber (*via* butadiene) and smokeless powder production.

The availability after World War II of cheap ethylene (as a by-product of natural gas recovery and of gasoline production) led to the rapid growth of the synthetic ethanol industry based on the esterification–hydrolysis of ethylene in concentrated sulfuric acid–water solution (Sherman and Kavasmaneck, 1978). Pressure from this source combined with the excess capacity left after the war emergency led to the mothballing or dismantling of many fermentative ethanol plants. The adoption in the late 1960s and 1970s of the less capital intensive catalytic vapor-phase ethylene hydration process (Sherman and Kavasmaneck, 1978) further reduced ethanol production cost, supplanting the earlier sulfuric acid process.

43.1.2 Current Status and Future Outlook

The 1973 oil embargo and accompanying increase in oil price from \$2.59 to \$11.65 per barrel led to worldwide interest in the development of alternative renewable fuels. The most ambitious program—in Brazil—called for an increase in fermentative ethanol production from 4.14 Ml y^{-1} in 1978 to 5 Gl y^{-1} in 1985, corresponding to 2% of Brazil's total primary energy consumption (Yand and Trindade, 1979). Actual 1980 production reached 3.8 Gl. In the US, funding for research into improved ethanol production and recovery techniques was sponsored by the Department of Energy. The 1980 Energy Security Act and companion legislation established tax incentives, investment loan guarantees and federal purchase guarantees with the goal of increasing US fermentative ethanol production to 1.9 Gl y^{-1} in 1981 and as much as 41.6 Gl y^{-1} in 1990 (Vervalin, 1980b). The oil glut of 1981–1982 has slowed ethanol fuel production growth.

Tax incentives allow continued ethanol sale as a gasoline fuel extender and octane enhancer (Staff, Chem. Eng. News, 1982). Current petroleum costs (\$41 per barrel—OPEC) do allow direct competition of fermentative ethanol with synthetic ethanol (without incentives) as long as grain or sugar feedstock prices remain moderate (Staff, Chem. Mark. Rep., 1982a). Ethanol is currently the largest volume (liquid) industrial fermentation product worldwide. Continued growth of fermentative alcohol production will depend on process improvements allowing reduced fermentative alcohol cost, and on continued increases in petroleum costs to reduce competition from synthetic ethanol for chemical uses and increase demand for fermentative ethanol as a renewable fuel.

43.2 RAW MATERIALS SELECTION AND PREPARATION

43.2.1 Raw Materials Alternatives

Table 1 summarizes historical trends in raw material use for 190 proof and greater rectified alcohol production (BATF, 1984). These figures somewhat overstate industrial alcohol production, as a portion of rectified spirits is blended back for beverage use.

Table 1 Raw Materials for Production of 190 Proof and Higher Alcohol

| | *Alcohol production* (proof gallons)[a] | | | | | | |
	1947	*1955*	*1960*	*1965*	*1970*	*1975*	*1980*
Ethyl sulfate	133.3	274.6	390.6	447.6	305.1	30.6	0
Ethylene gas	0	52.7	80.6	117.6	279.3	395.7	374.6
Grain[b]	39.7	8.6	7.9	80.2	84.8	111.7	251.0
From redistillation	49.5	34.2	13.9	33.1	45.3	25.4	7.0
Molasses	54.2	66.9	30.7	4.2	2.8	12.9	1.9
Fruit	0.2	0.6	0.4	21.7	25.9	25.9	25.1
Sulfite liquor	4.4	5.6	7.6	7.1	7.1	7.9	9.8
Crude alcohol mixture	0.7	0.7	0.7	0.4	0	0	0
Cellulose pulp and chemical mixtures	1.7	1.5	1.8	1.6	0	0	0
Potatoes	15.1	0	0	0.02	0	0	0
Whey	0	0.6	0	0.3	0.4	0.5	0.0
Total	298.8	446.1	534.2	713.9	750.8	610.5	669.4

[a] Proof gallon is the alcoholic equivalent of one US gallon containing 50% ethanol by volume (1.496 kg ethanol). [b] Some grain and fruit alcohol is blended back into beverages.

Raw materials for fermentative ethanol production can be divided into three classifications by carbohydrate type: saccharine materials, starchy materials and cellulosics. Yeast (the primary organism used for ethanolic fermentation) will convert simple hexose sugar monomers (*e.g.* glucose and fructose) and some disaccharides (*e.g.* maltose and sucrose) to ethanol. The choice of raw material is critical as raw material costs typically make up 55–75% of the final alcohol selling price. Saccharine materials—with sugars available in fermentable form—require the least extensive preparation, but are generally the most expensive to obtain. Starch bearing materials are often cheaper but require processing to solubilize and convert the starch to fermentable sugars.

Cellulosics can be available as cheap waste residue, but require the most extensive and costly preparation.

43.2.2 Saccharine Materials

Major saccharine raw materials are sugar cane or sugar beet juice, high test molasses, blackstrap molasses, fruit pulp and juice wastes, cane sorghum and whey. Blackstrap molasses, fruit wastes, sorghum and whey have alternate use as cattle feed components, and their prices are set primarily based upon feed value and shipping costs. Sugar cane and sugar beets are used for edible sugar production and this sets their market value. Prices for all of these commodities are subject to large fluctuations. The price of blackstrap molasses in the US, for instance, has ranged from \$40 to \$124 per metric ton between 1975 and 1980, with fluctuations of as much as \$36 per ton in a single year (Emery, 1981).

43.2.2.1 Sugar cane juice

Owing to transportation cost, the direct use of sugar cane is limited to alcohol plants located within farming districts. In Brazil, 43% of alcohol plants process cane directly as raw material (Yand and Trindade, 1979).

Sugar cane juice is recovered by a milling train consisting of knife cutters followed by a pounding mill to rupture the cell walls and 3–5 high-pressure roller presses to express the juice. Crushed cane is hot water rinsed to aid sugar extraction between pressing cycles. Total water used is 25% of the weight of the cane and 85–90% of the fermentable sugars are extracted. The resulting juice is 12–16% sucrose sugar. The yield of sugar from cane is typically 125 kg per metric ton (250 lbs per US ton) (Lipinski *et al.*, 1976; Spencer, 1963). A mascerated solid bagasse residue (50% moisture) remains after sugar expression, and this can be used as a boiler fuel, resulting in a considerable economy in distillation and stillage evaporation fuel. Two kg of bagasse (dry weight), with a heating value of 19 700 kJ kg^{-1} (8500 Btu lb^{-1}), are produced for every kg of sugar (Curtis, 1950). Sugar cane wax and aconitic acid can also be recovered as by-products (Jacobs, 1950). Cane crushers are high-pressure steam driven using from 1.0–1.2 kg steam per kg sugar recovered. Low-pressure exhaust steam from the crushers is reused for distillation (Lipinski *et al.*, 1976).

43.2.2.2 Sugar beet juice

Direct use of sugar beets requires location of the alcohol plant within the farming district. Unlike sugar cane, however, sugar beet production is not limited to tropical and subtropical climates. A large sugar beet alcohol industry existed in France in the 1940s.

Beet sugar is expressed by first slicing beets into V-shaped cossette wedges, followed by sugar extraction in continuous counter-current hot-water (85 °C) diffusers to produce a 10–15 wt% sugar solution (McGinnis, 1951). The cellulosic beet pulp residue is generally used as a low-grade cattle feed roughage valued at \$130–220 per metric ton (Staff, Feedstuffs, 1982). Each metric ton of beets expressed produces 140–190 kg of sugar and 70–100 kg of pulp (dry weight).

43.2.2.3 High test molasses

High test molasses (Hodge and Hildebrandt, 1954) is a concentrated sugar solution, allowing easy shipping and long-term storage. Sugar cane or beet juice is first partially hydrolyzed with dilute acid to convert sucrose to noncrystallizable invert sugar monomers (glucose and fructose). The juice is then concentrated to 70–78 wt% sugar. Table 2 gives a typical composition of high test molasses. Owing to thermal decomposition of sugars during evaporation, and to the presence of small concentrations of yeast inhibitors, high test molasses sugars are generally only 95% fermentable.

Up to 1280 Ml y^{-1} of high test molasses, exclusively for alcohol production, was imported into the US between 1935 and 1942, when a sugar quota restricted sugar imports for human consumption. Except for such special circumstances, food use generally limits the availability of high test molasses for industrial purposes.

43.2.2.4 *Blackstrap molasses*

Blackstrap molasses (Hodge and Hildebrandt, 1954) is the noncrystallizable residue by-product of table sugar manufacture. The sugar juice is first treated with lime (0.75 kg per metric ton cane) at 100 °C to neutralize organic acids. On cooling, salts, coagulated albumen, fats and gums are precipitated. The clarified juice is concentrated to 60 wt% solids by multieffect evaporation and sucrose is precipitated by further evaporation in vacuum pans. After three or four evaporations, high viscosity prevents further crystallization. The sugars are primarily sucrose and noncrystallizable invert mixed with the now concentrated inorganic salts and soluble nonsugar organic residues retained from the juice or generated during evaporation (Table 2). Approximately 27 kg of blackstrap are obtained from every metric ton of cane processed, and 2.5 l of molasses (3.5 kg) is required to produce one liter of 95 wt% ethanol (Hodge and Hildebrandt, 1954).

Table 2 Composition of High Test and Blackstrap Molasses

	High test molasses	Blackstrap molasses
Total solids	80–85%	83–85%
Invert sugar	40–60%	12–18%
Sucrose	35–15%	40–30%
Fermentable sugar	70–78%	50–55%
Organic nonsugars	4–8%	20–25%
Ash	2–4%	7–10%

Up to 17% of blackstrap molasses is nonfermentable reducing compounds resulting from high-temperature destruction of sugars during evaporation. Fructose undergoes dehydration and reduction to 1,3-fructopyranose, and both glucose and fructose undergo salt-catalyzed condensation with cane amino acids, resulting in dark nonfermentable caramel residues. Sugars are also decomposed to volatile hydroxymethylfurfural, acetoin and formic and levulinic acids (Zerban and Sattler, 1942; Sattler and Zerban, 1945, 1949). Vapor-recompression vacuum sugar evaporation reduces evaporation temperatures and can thus reduce sugar decomposition and substantially increase the yield of ethanol from blackstrap molasses. Flash cooling of molasses after high-temperature evaporation can also reduce sugar decomposition and increase the ethanol yield.

In blackstrap molasses, components which are inhibitory to yeast (primarily calcium salts from the lime treatment and organic decomposition compounds) are quite concentrated due to the repeated crystallizations. These reduce final alcohol yield and typically only 90% of untreated blackstrap molasses sugar is utilizable. In the Arroyo process (Arroyo, 1949) ammonium sulfate and calcium superphosphate are added to concentrated molasses, which is heated to 80 °C with the pH adjusted to 4.5–5.2. A sludge, containing calcium sulfate and organic inhibitors in colloidal form, settles from the solution and the clarified molasses is then used in fermentation. In the Reich process (Reich, 1942), concentrated molasses is heated to 70–90 °C and sulfuric acid is added to adjust the pH to 3–4. Again, a sludge is removed and the clarified solution used for fermentation.

Unfortunately, both of these methods also result in the precipitation of natural yeast growth factors present in the raw molasses, and additional nutrients must be added to clarified molasses for rapid yeast growth. Some sugar decomposition also takes place at the elevated clarification temperature. Blackstrap molasses clarification is not necessary for simple batch fermentation, but is essential if yeast or stillage is to be recycled. In addition, molasses clarification can be beneficial in reducing scale formation (from calcium salt deposition) in distillation and stillage evaporation heat-transfer equipment.

Inhibitor concentrations are much lower in high test molasses and cane juices, but these may also be pretreated to further increase yield and eliminate scaling problems.

43.2.2.5 *Sweet sorghum*

Sweet sorghum is a tall grass with high sucrose content in the stems. It is widely grown for cattle forage. In 1942, a USDA test project successfully used 23 250 US tons of sweet sorghum for alcohol production (Jacobs, 1950). Sweet sorghum has received much attention because it can be readily cultivated under a wide variety of growth conditions and has a yield of sugar per hectare slightly higher than that found for sugar beets. After sugar expression, the cellulosic sorghum residue can be used as a cattle feed roughage source.

43.2.2.6 *Fruits and juices*

Fruit cannery wastes can be used as a substrate for alcohol production. After orange, apple and pineapple juices have been squeezed, additional sugars can be extracted from the pulp by hot water diffusion. Pineapple canning yields excess juice, and in Hawaii up to 1.25 million liters per year of alcohol has been produced from this waste with 15–25 liters of juice required per liter of alcohol product (Jacobs, 1950). Ion exchange processes now allow the recovery of sugar from very dilute waste streams, and dilute cannery wastes could be used as an alcohol source while reducing the BOD of cannery effluents. Production from this source will, of course, be limited, but does well illustrate the potential to integrate alcohol production into a waste disposal process, thereby producing a valuable product at very low raw material cost.

43.2.2.7 *Whey*

Whey, produced as a by-product of cheese manufacture, is high (4.5–5.0%) in lactose. Lactose is a disaccharide composed of a glucose and a galactose unit and is not fermentable by most yeasts. The strains *Torula cremoris* and *Candida pseudotropicalis* readily ferment lactose to ethanol and this process has been used at many small installations since the 1940s (Jacobs, 1950; Maiorella and Castillo, 1984). Ethanol production from whey is limited and must compete with use for whey in cattle feed and antibiotic production.

43.2.3 Starchy Materials

Starchy raw materials include cereal grains, starch root plants and some cacti. The grains and root plants are all used both for human consumption and animal feed. Prices are set based on these uses and fluctuate widely based on annual crop yields and international demand. The price of corn in the US, for instance, has ranged from $65 to $131 per metric ton for the period 1975–1980, with fluctuations of as much as $35 per ton in a single year (Emery, 1981).

The carbohydrate in the starchy plants is not directly fermentable by yeast and these materials must be pretreated to hydrolyze the starch to simple sugars. Starch is first hydrated and gelatinized by milling and cooking, and then broken down to fermentable sugars by diastatic enzymes or weak acids. Capital investment for starchy material feed pretreatment can typically make up 15–20% of the total alcohol plant cost.

The cereals contain valuable secondary products, and considerable further preprocessing may be employed to recover these prior to the fermentation to offset the high cost of cereal materials.

43.2.3.1 *Cereal grains*

Important cereal grains are corn, wheat, rice, barley and grain sorghum. Cereal grains are generally 50–65% starch and as high as 80% for rice. Corn is the major grain used for alcoholic fermentation in the US.

Grain processing (Wilkie and Prochaska, 1943; Stark, 1954) has traditionally been by batch methods, but these are now often replaced by fully continuous processes. The grain is first air classified to remove dirt and cracked hulls. It is then ground (typically to 20–60 mesh) to allow easy wetting and then gelatinized by cooking.

Traditional atmospheric pressure processes have been almost entirely replaced by pressure cooking. The higher temperature cooking allows reduced cooking time with less sugar degradation and 3–5% higher final yields. A coarser grind is allowable. Pressure cooking sterilizes the mash. Batch pressure cooking is by steam injection at typically 690 kPa (100 p.s.i.), into horizontal pressure vessels of up to 40 000 liter capacity with horizontal rake agitators. The mash, with a grain concentration of roughly 0.27 kg per liter of water and pH adjusted to 5.5, is heated to 135–150 °C and held for 10–30 minutes. Rapid cooling by pressure blowdown and then vacuum evaporation reduces the temperature to below 65 °C for the addition of enzymes. Blowdown steam may be recovered for other uses.

In the continuous cooking process (Unger *et al.*, 1944), milled grain is first slurried with water in a stirred tank with a 1–5 min residence time. The slurry, again at roughly 0.27 kg grain per liter of water and pH 5–6, is continuously metered through a steam-jet heater into a cooking tube with a plug flow residence time of less than 5 min and at up to 180 °C cooking temperature. Flash cooling is used to reduce rapidly the mash temperature and halt sugar breakdown. The continuous cooking process gives more uniform cooking at lower steam consumption and capital investment than the batch process.

Traditional starch hydrolysis has been conducted using barley malt (Hudson, 1972). Sprouted barley contains a high diastase activity. Ground dried barley malt is reactivated by slurrying with warm water (0.19 kg l^{-1}). After 5 min the malt solution is mixed with cooled mash to achieve a ratio of typically 1 part malt to 10 parts other grain at 63 °C. Grain starch is composed of straight-chained amylose and branched amylopectin. The amylose is readily hydrolyzed to maltose. Amylopectin is broken down partly to maltose, but largely to branched nonfermentable dextrins. After 40 min, 75–80% of grain starch is converted to maltose, the remainder to limit dextrins. The mash is then fermented, with starch hydrolysis continuing. The conversion of maltose to ethanol during fermentation favors further breakdown of dextrins to maltose. Fermentation pH must be kept above 4.1 to maintain enzyme activity. For conventional grain fermentations this final dextrin conversion, not the ethanol fermentation itself, becomes the rate limiting step in fermentation.

The continuous flow hydrolysis process (Gallagher *et al.*, 1942; Norman, 1978) is used in combination with continuous cooking (Figure 1). A reactivated malt solution is continuously metered into the stream of cooked, cooled mash at 63 °C in a plug flow pipe reactor with a 2 min residence time. The mash is then further cooled to fermentation temperature. The conversion of starch to maltose is reduced to 70%, but the abbreviated time results in reduced denaturation of the diastatic enzymes. Conversion of residual dextrins in the fermenter is faster and more complete with a 2% final alcohol yield increase typical.

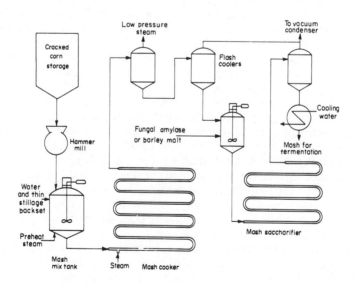

Figure 1 Continuous flow grain cooking and saccharification

An alternative to barley malt is mold bran (Underkofler *et al.*, 1946; Ziffer and Iosit, 1982). *Aspergillus oryzae* is grown on moist sprouting bran and produces amylase enzymes. The moldy bran can then be treated exactly as barley malt in carrying out the hydrolysis. Under some conditions, mold bran can be less expensive than barley malt and 2% final alcohol yield increases are typical.

Barley or bran can at no time be heated above 63 °C or the malt enzymes will be destroyed.

Thus sterilization is impossible, and a major disadvantage of malt and mold bran hydrolysis is the increased occurrence of fermenter contamination. Lactic acid and other bacteria introduced in the malt can reduce alcohol yield by 20%.

The use of sterilely produced microbial culture amylases overcomes the contamination problems inherent in barley malt starch hydrolysis. Microbial amylase systems have been developed which are also of higher activity than barley malt.

The original commercial, submerged culture, microbial starch saccharification method was the Amylo process (Owen, 1933), applied widely in European distilleries. After cooking and cooling, the mash is inoculated in the fermenter with a culture of the mold strain *Rhizopus delemar*, *Mucor rouxii* or *M. boulard*. In vigorously aerated culture at 40 °C the mold grows and produces extracellular amylase enzymes which hydrolyze the starch. After 24 hours, yeast is added with hydrolysis continuing during the fermentation. Sterility problems are overcome and the high barley malt cost is eliminated. The 24 hour increase in fermenter residence time for the mold growth period requires considerable added vessel capacity and some starch is consumed for the mold growth.

To overcome these problems, sterile enzyme solutions can be separately prepared and then used in batch or continuous hydrolysis (USDA, 1950). Thermostable bacterial (*Bacillus licheniformis*) α-amylase can be used in high temperature (100 °C) continuous liquefaction. Submerged culture fungal amylase production (Norman, 1978) provides the highest activity glucoamylase solutions. *Rhizopus*, *Endomycopsis bispona*, *Aspergillus oryzae*, *A. batatae*, *A. awamori* and *A. niger* strains with high amylase activity have been selected and used in submerged culture amylase production. Many strains can be grown aerobically on fermentation stillage waste, resulting in further economies (Yarovenko, 1975). Amylases from two or more strains with differing straight- and branched-chain amylase activity can be combined or used sequentially to obtain the best overall conversion of amylose and amylopectine. Temperature-sensitive glucoamylase enzymes of high activity toward dextrins can be produced separately and added to the alcohol fermentation to promote dextrin conversion. Submerged culture amylases thus provide the greatest flexibility in starch conversion and also allow completely aseptic operation. Economies are possible at large distilleries producing their own amylases on fermentation stillage. For smaller distilleries which cannot justify the added capital investment, concentrated microbial amylases are now available commercially at moderate cost (Novo Industrias, 1976). Final alcohol yields using high-activity microbial amylases are typically 3–7% higher than for barley malt hydrolysis. In the US, microbial amylase has almost completely supplanted malt for non-beverage alcohol.

Katzen (1979) gives energy requirements for a 190 Ml y^{-1} corn grain alcohol plant. Grain milling requires 507 kW (12 W h kg^{-1} sugar) of electricity. Mashing, cooking and saccharification uses an additional 216 kW of electricity and 27 800 kg h^{-1} (0.66 kg kg^{-1} sugar) of steam. Fungal amylase production requires 1696 kW (40 W h kg^{-1} sugar), mainly for air compression and agitation to maintain the highly aerobic conditions for rapid fungal growth.

A major disadvantage of most enzymatic grain starch conversion processes is the protracted fermentation time required as residual dextrins are slowly hydrolyzed to simple sugars. This problem is overcome by the dilute acid hydrolysis process (Ruf *et al.*, 1948). After grinding, grain is cooked in 0.15–0.20 N sulfuric acid solution (2.5 l water kg^{-1} grain) for 8 min at 160 °C. An acid-resistant continuous flow pressure vessel with direct steam injection is used. The cooked mash is neutralized with a 10% calcium carbonate slurry and flash cooled. Precipitated calcium sulfate is settled free and the mash is then diluted for fermentation. This method does result in high starch conversion but the acid hydrolysis destroys naturally occurring yeast nutrients, and costly supplementation with corn-steep liquor, yeast autolysate or bran cultures is required to maintain high yields. Capital costs for acid-resistant hydrolysis vessels are higher. Some sugar is also lost in the calcium sulfate precipitate slurry.

High-value by-products can be recovered from most grains prior to fermentation. Corn, wheat and rye can be treated to recover bran and germ by careful dry milling followed by mechanical separation. Alternatively, the grain can be steeped in dilute sulfurous acid. After 48 h the corn is rough ground and the bran and germ are floated free of the starch (Kerr, 1950). Solvent extraction of the germ can recover valuable vegetable oil (Parekh, 1964). After starch hydrolysis, fiber and gluten can be separated to produce a high-protein meal. For each metric ton of corn processed, 100 kg of germ, 85 kg of bran, 22 kg of oil and 98 kg of protein meal (50% protein) can be recovered (Process Engineering Co., 1980). The protein and hull content of grain sorghum and rice is too low to justify pretreatment.

If secondary products are not recovered from corn, wheat or rye grains, then these products pass through the fermentation relatively unaffected and contribute to the nutrient value of stillage for use as a mixed cattle feed supplement.

43.2.3.2 Roots, tubers and cacti

Roots, tubers and cacti can all be very high in carbohydrate content, giving final alcohol yields per hectare higher than the cereal grains. High water content makes storage difficult and seasonal operation is required without costly drying to prevent spoilage. The protein content of these raw materials is low so that by-product recovery is not feasible and substantial by-product credits are not available to offset the raw materials costs. Materials used for alcohol production include potatoes, sweet potatoes, Jerusaleum artichoke, manioc and sotol.

Potatoes may be 15–20% starch with alcohol yields of 4700 liters per hectare (476 gal per acre) (USDOE, 1979). Sweet potatoes, with a starch content as high as 30%, can have yields of up to 7200 l ha^{-1}. Potatoes were used extensively for industrial alcohol production in pre-World War II Germany under government price supports. 184 000 metric tons were used in the US in 1947 during a grain shortage (Jacobs, 1950). Except for such special circumstances, potato costs are generally prohibitive for ethanol manufacture.

Jerusalem artichoke (wild sunflower) tubers and manioc (tapioca, cassava) roots are high in inulin (a plant starch). Jerusalem artichokes have been used for industrial alcohol production in France (Jacobs, 1950). Manioc growth is limited to tropical climates, but yields as high as 19 000 l ha^{-1} have been achieved (USDOE, 1979). 9% of Brazilian alcohol manufacture is now from manioc (Yand and Trindade, 1979). Prussic acid (a yeast inhibitor) content is high in manioc, but this can be broken down by sun drying the kibbled root on the fields.

Dasylirion cactus (sotol) is a wild desert plant of interest as it would be harvested from otherwise unutilized lands. Dasylirion was harvested in Texas in 1944 for alcohol manufacture with a yield of up to 127 l t^{-1} (Jacobs, 1950).

Processing for the roots, tubers and cacti are all similar. After water rinsing, the materials are cut or chipped and then hammer milled to a mash. Processing is then similar to the steps for grain cooking and hydrolysis. An exception is that special inulase enzymes are used for the hydrolysis of Jerusalem artichoke or manioc inulin to fructose. For soft materials like potatoes and sweet potatoes, mashing and cooking may be partly combined in the Henze cooker (Foth, 1935). Whole or chopped substrate is batch loaded into a vertical pressure cooker. After cooking, the mash is blown down through the conical vessel bottom into a lower-pressure drop tub with violent boiling resulting. This 'steam explosion' disintegrates the substrate, thus eliminating the need for milling. Conventional hydrolysis processes are then used.

43.2.4 Cellulosic Raw Materials

Compositions of several proposed cellulosic feedstock are summarized in Table 3 (Blanch and Sciamanna, 1980). These materials are soft woods or woody agricultural residues and are made up of cellulose, hemicellulose and lignin bound into a structure which gives a plant its rigidity and support. Cellulose polymer can be hydrolyzed to its glucose monomer units for fermentation to ethanol.

Table 3 Composition of Cellulosic Raw Materials for Ethanol Production

	Corn stover	Wheat straw	Rice straw	Rice hulls	Bagasse fiber	Cotton gin trash	Newsprint	Populus tristis	Douglas fir
Carbohydrate (% sugar equivalent)									
Glucose	39.0	36.6	41.0	36.1	38.1	20.0	64.4	40.0	50.0
Mannose	0.3	0.8	1.8	3.0	NA	2.1	16.6	8.0	12.0
Galactose	0.8	2.4	0.4	0.1	1.1	0.1	NA	NA	1.3
Xylose	14.8	19.2	14.8	14.0	23.3	4.6	4.6	13.0	3.4
Arabinose	3.2	2.4	4.5	2.6	2.5	2.3	0.5	2.0	1.1
Non-carbohydrate (%)									
Lignin	15.1	14.5	9.9	19.4	18.4	17.6	21.0	20.0	28.3
Ash	4.3	9.6	12.4	20.1	2.8	14.8	0.4	1.0	0.2
Protein	4.0	3.0	NA	NA	3.0	3.0	NA	NA	NA

For every 1 kg of grain harvested, 1–1.5 kg of straw, cobs, stover or other residue are generated. Only one-third of tree biomass is recovered as finished lumber. Cellulosics have received great attention as possible feedstocks on the assumption that raw material costs for agricultural and lumber waste materials should be negligible. This unfortunately neglects the high collection cost for agricultural residues—typically \$33–50 t^{-1} (Vervalin, 1980a). Special cases do exist where the residues are collected as part of harvesting, but in these cases residues are already put to valuable use. Corn cobs and beet pulp, for instance, are used as animal feed. Sugar cane bagasse, sawdust and wood chips are used for process fuel or as fiberboard components.

Cellulosic paper wastes can be separated from municipal garbage by air flotation, but in this case the careless disposal of a pesticide or other toxic chemical by a single household could contaminate the material and force a plant shutdown. Intensive tree farming has been proposed and studies suggest that fast growing poplar varieties could be grown with a delivered dry wood price of only \$16.5 t^{-1} (Bungay, 1981), but this has never been substantiated by an industrial trial.

Cellulose is far more chemically stable than starch and harsher acid hydrolysis conditions or more potent enzyme solutions are required for hydrolysis. Extensive pretreatment is often required to increase the accessibility of the cellulose which is otherwise well protected in its natural lignocellulosic structure (Chang *et al.*, 1981). The cost of cellulose conversion to sugar is far higher than the cost of starch conversion. Yields are typically also much lower, requiring more raw materials and resulting in substantial waste streams requiring costly disposal.

A special case is sulfite waste liquor (McCarthy, 1954), which is the residue of partial wood hydrolysis from paper pulp production. In this case a partial acid hydrolysis is carried out to prepare a valuable bleached cellulose pulp and a dilute sugar solution is produced as an unwanted by-product. After pretreatment to neutralize yeast toxins, this waste can be fermented to produce alcohol. This process must compete, however, with aerobic yeast feed production from the sulfite liquor which requires less capital investment and is more effective in reducing the BOD of the waste liquor in preparation for discharge.

Few by-products of cellulose hydrolysis have proven high value, but there are promising possibilities. Lignin with a heating value of 29 500 kJ kg^{-1} (12 700 Btu lb^{-1}) (Falkehag, 1975) is readily utilized as a boiler fuel. Lignin is a complex three-dimensional polymer of phenolic origin, and potentially higher value chemical applications for lignin have been reviewed (Pearl, 1958; Drew, 1978). Lignin can be used as a source of vanillin and syringic aldehydes, as an extender in phenolic plastics and as a lignin–formaldehyde binder for particle board, replacing phenol–formaldehyde. Reactive chemical lignin is valued at \$120 t^{-1}. Further development of high-value applications for lignin and other by-products could go a long way toward offsetting the high cost of cellulose hydrolysis.

Most hydrolysis processes also break down the hemicellulose fraction of woody materials, producing xylose (Gong *et al.*, 1981). Five-carbon sugars are not directly fermentable by ordinary yeast, but can be fermented by select bacteria to mixtures of either ethanol or butanol and organic acids. The further development of such methods to utilize xylose would greatly increase the yield of valuable products from woody materials, again improving the economics of their use.

43.2.4.1 *Slow acid hydrolysis*

Twenty-one alcohol plants utilizing wood were operated in Europe and two in the US prior to or during World War II (Jacobs, 1950). These plants used either the Scholler dilute (0.2–1.0%) sulfuric acid process or the Bergius concentrated (40–45%) hydrochloric acid process for wood hydrolysis. Yields for both processes were similar with up to 500 kg of sugar (70% fermentable glucose) produced per metric ton of coniferous sawdust or bark-free chips. The Scholler process was less capital intensive and was therefore preferred. The US Forest Products Madison Laboratories optimized operating conditions for the dilute sulfuric acid hydrolysis and a 200 metric ton dry wood per day plant was built based on this improved Madison process (Harris and Beglinger, 1946). Wood chips were charged into Herculoy lined, acid-resistant pressure vessels (14 ton wood capacity). Dilute acid of 0.5–0.6% concentration was percolated down through the packed bed with continuous removal of sugar solution at the base. Reactors were heated by direct steam injection beginning at 350 kPa (50 p.s.i.) and increasing to 1135 kPa to hydrolyze the more resistant materials remaining at the end of a run. Total hydrolysis time was 2.5–3.5 hours. After removal from the vessel, the sugar solution was immediately neutralized with lime and flash cooled to minimize sugar decomposition caused by the harsh acid conditions. The resulting sugar solution is only 5–6 wt% sugars, 80% fermentable by yeast, and fermentation produces dilute alcohol solutions that are costly to distil. Despite the careful control of temperature and rapid flash cooling of product solutions, 40% of total sugars and 35% of fermentable product glucose are not recovered or are decomposed further by the acid to unwanted by-products such as furfural, methanol and sugar acids (especially levulinic). In addition to the loss of yield, these by-products are inhibitory to the yeast fermentation. After hydrolysis, the reactors were blown down and the lignin residue was filter pressed to 50% moisture and burned to help provide the 500 000 kg d^{-1} (5 kg kg^{-1} sugar) steam requirement of the plant. Chemical requirements for the plant were 0.65 kg sulfuric acid and 0.50 kg lime per kg of fermentable sugar produced.

While no longer in industrial use in the US and Western Europe, the dilute slow acid hydrolysis process has been revived in a small pilot plant by the New Zealand Forest Research Institute. Dilute acid hydrolysis for ethanol manufacture continues as an industrial process in Russia and the Russian technology is being applied in a new Brazilian plant (Worthy, 1981).

43.2.4.2 Rapid acid hydrolysis

Careful kinetic studies by Graethlein at Dartmouth have shown that high conversions of cellulose to sugar can be achieved and sugar degradation minimized by reacting at high temperature and pressure for very short periods (Knappert *et al.*, 1980). Based on these results, Rugg at New York University has developed and tested, at one ton cellulose per day pilot-plant scale, a continuous fast acid hydrolysis reactor (Staff, Chem. Eng. News, 1979). A twin corotating screw extruder (of the type used in plastics compounding and extrusion) is used. Hydropulped newspaper or sawdust is cram fed into the extruder which shears the material into a slurry and compresses it to a solid plug at 3450 kPa pressure, dewatering it to a cellulose concentration of 50%. High-pressure superheated steam is injected, raising the temperature to 240 °C. Near the extruder barrel outlet, 0.5% sulfuric acid is injected. The hydrolysis takes place with up to 60% sugar conversion in 20 seconds before the plug is expelled from the screw through a high-pressure valve. The brown paste slug, containing up to 30% glucose, is flash cooled to prevent decomposition and the sugars are extracted in a two-stage countercurrent washer. For each kg of sugar produced, 2.1 kg of steam for heating and 0.75 kWh of electricity to drive the screw reactor are consumed. The solid residue can be burned to provide all required process steam.

Production of concentrated sugar solutions overcomes the problem of high distillation and evaporation costs associated with the slow acid hydrolysis process. Experiments are necessary to determine whether further treatment of the sugar solutions is necessary to remove inhibitors before fermentation. Plans to build a 50 ton per day demonstration plant have not been carried through.

43.2.4.3 Enzymatic hydrolysis of cellulosics

Enzymatic hydrolysis has the potential to overcome many of the drawbacks of acid hydrolysis. The conversion is carried out at ambient temperature and pressure, thus greatly reducing the cost of hydrolysis equipment. Sugar decomposition is eliminated, thus eliminating this cause for loss in yield and producing clean sugar streams for further processing. Costly neutralization and purification equipment is unnecessary, and hard to dispose of waste streams from acid neutralization are eliminated. Balancing these potential savings, extensive pretreatment to break down lignin and increase cellulose accessibility is required to achieve good yields and the cost of high activity cellulolytic enzyme solutions is at present very high. Extensive research to overcome these problems is under way (Bungay, 1981).

No enzymatic cellulose hydrolysis process has ever been operated commercially, but a large industrial pilot test has been completed.

Mutation and selection methods have been used to develop and isolate *Fusorium*, *Phanerochaete* and *Trichoderma* fungal strains of high cellulolytic activity. Three classes of enzymes—all necessary for the complete hydrolysis of natural cellulose to glucose—are produced. Endoglucanases cleave long cellulose polymer chains at random to produce shorter more accessible chains. Exoglucanases cleave cellobiose (glucose dimer) exclusively from the nonreducing ends of chains. β-Glucosidase hydrolyzes cellobiose to fermentable glucose. The highest activity enzyme solutions thus far prepared are those of *Trichodema reesei* Rut C-30, isolated at Rutgers University (Montenecourt and Eveleigh, 1977).

For untreated substrates, cellulose is often present in a crystalline form which limits accessibility and defies enzymatic hydrolysis except for slow end cleavage. The cellulose is also protected by a lignin sheath inert to cellulolytic enzymes. Without pretreatment, sugar conversion from natural feedstocks is generally limited to less than 25%. Pretreatment processes are necessary to convert cellulose to an accessible amorphous form for rapid hydrolysis (Chang *et al.*, 1981).

Ball or hammer milling reduces the degree of polymerization of both cellulose and lignin and reduces cellulose crystallinity. Sugar conversion can be increased to 60–70% by milling to −200 mesh, but milling is very energy and capital intensive. Partial delignification can be achieved by extraction with hot alcohol solutions (butanol and higher) and the lignin recovered in a thermoplastic form. Cellulose crystallinity can be reduced by chemically swelling the polymer with alkali, ammonia or nitric oxide to break interstrand hydrogen bonds. These treatments can also remove a portion of the lignin. Treatment with ozone or cadoxen reduces the degree of polymerization of cellulose, decreasing crystallinity and increasing solubility. All of these chemical processes are costly, and extensive chemical recovery or waste disposal provisions are necessary just as in the direct acid hydrolysis processes. One particularly promising pretreatment is steam explosion, developed by the Masonite Co. and later refined by IOTECH (Jurasek, 1978). Wood chips or other cellulosic wastes are pressure cooked for 5–10 min with direct steam injection at 3895–4240 kPa during which time steam fully impregnates the chips. The material is then blown down to atmospheric pressure. The explosive decompression breaks the substrate apart into wool-like amorphous fibers which are readily hydrolyzable with up to 80% sugar conversion. Low-pressure steam is recovered and can be used elsewhere in the plant. A reactive thermoplastic lignin can be washed from the exploded substrate.

Several alternative enzymatic hydrolysis processes have been proposed and were thoroughly reviewed by Wilke *et al.* (1980).

A process involving simultaneous saccharification and fermentation is the only one to have been tested at pilot scale over a substantial period. The Gulf Jayhawk pilot plant (Emert *et al.*, 1980a,b) converted 1 ton per day of cellulosics. Various raw materials (including paper mill pulp digester rejects and air classified municipal waste paper) and pretreatments were tested. When chemical pretreatments were not used, material size was first reduced in an attritor mill. A 6% solids slurry was then prepared in a 75 HP hydropulper and substrate was further degraded with a paper pulp disc refiner. A dewatering press then concentrated the slurry which was finally sterilized before use. Enzyme solutions were prepared by sterile semi-continuous aerobic culture of *Trichoderma reesei* on a portion of the refined substrate in 1100 liter enzyme reactors for 48 h. The remaining substrate was then simultaneously treated with the enzyme solution and yeast in two 1100 l vessels. Saccharification and fermentation took place simultaneously. As with the final (in fermenter) limit dextrin hydrolysis of starch, the consumption of sugars for ethanol production favors more complete hydrolysis. In semi-continuous flow mode at 40 °C and with a residence time of 24–48 h, an 8% cellulose feed was hydrolyzed to glucose which was fermented to produce up to a 3.6% ethanol beer slurried with solids; 60% conversions were reported but it is not clear for what substrate and pretreatment conditions. After receiving much attention, plans for a 50 ton per day pilot plant and 2000 ton per day municipal refuse conversion plant were cancelled—apparently for economic reasons.

43.2.4.4 Sulfite waste liquor

Alcohol is currently produced from sulfite pulping wastes at several large mills in Europe and the US. The sulfite pulping process is a treatment to remove hemicellulose and delignify wood to produce a clean cellulose fiber pulp for paper or rayon manufacture (McCarthy, 1954). Wood chips are pressure cooked at 140 °C with an aqueous solution of sulfurous acid containing calcium, ammonium or magnesium cation to provide 5–7% free sulfur dioxide. The dilute acid hydrolyzes hemicellulose almost completely to its monomer pentose sugars and renders lignin soluble by sulfonation. The cations neutralize the strong lignosulfuric acids to prevent excessive degradation of the cellulose. Amorphous cellulose, however, is hydrolyzed to glucose which can be fermented. After blowing down to atmospheric pressure, the liquor is screened or pressed from the cellulose fibers and the fibers are then washed in countercurrent flow. The resulting sulfite waste liquors contain 8–12% dissolved solids—organic decomposition products equal to almost half the weight of the original wood, in addition to the inorganic pulping chemicals. Sugars make up typically 20–30% of the dissolved solids. The proportion of fermentable hexose sugars is highest for soft woods and harsh pulping conditions and can reach up to 70%. For each metric ton of finished pulp produced, 4000–6000 l of waste liquor is generated.

Free and loosely combined sulfur dioxide is present at 5–15 g l^{-1}. Sulfur dioxide in low concentrations reduces ethanol yield by increasing acetaldehyde production. At the high concentrations present in raw waste liquor, it is completely toxic to yeast (Hagglund, 1925). Early plants (1907–1945) used direct addition of lime to neutralize sulfurous acid and simultaneously precipitate a portion of the lignosulfonic acids to detoxify the liquor for fermentation. Modern plants steam strip the liquor with 0.06 kg steam per l of liquor in a roughly 20-stage column (Walker and Morgen, 1946). 10 kg of sulfur for recycle to the pulp digesters is recovered per metric ton of liquor and this offsets the steam cost.

A final lime neutralization (with approximately 0.4 kg lime per 1000 l liquor) is used to precipitate organics and adjust the pH.

A major drawback to the use of sulfite waste liquor is the low concentration of fermentable sugars—only 2–3 wt%. Evaporation prior to fermentation is impractical due to salt deposition problems in heat transfer equipment. Sugar bearing waste liquor can be recycled to the digesters to increase the sugar concentration of the subsequent liquor recovered. Up to 55 g l^{-1} of sugar have been obtained in small scale experiments, but sugar degradation to sugar sulfonic acids and bisulfite addition compounds is increased. Molasses has been added to sulfite liquor in one US plant to increase the final alcohol product concentration and simplify distillation. If none of these methods are employed, then very dilute 1–1.5 wt% ethanol solutions are obtained from fermentation and special distillation methods must be employed to prevent extremely high separation energy requirements. Distillation is also complicated by the presence of increased levels of fermentation by-products—fusel oils, borneol, limonene, camphene and organic acids (McCarthy, 1954). Depending on the initial pulping conditions, 65–105 liters of ethanol can be produced per metric ton (15–25 gal per US ton) of pulp manufactured.

An important aspect of ethanol production from sulfite waste liquor is its role in waste disposal. Alcohol production reduces the liquor biological oxygen demand by 45–50%. (Tyler, 1947). Growth of cattle feed yeast competes with use of the sulfite liquor for ethanol manufacture. Aerobically grown *Torula* and *Candida* feed yeast can consume acetic acid and pentose sugars as well as hexose sugars; 95% of all reducing sugars are typically consumed with a 60% overall reduction in BOD (Wiley, 1954).

43.2.5 Feedstock Comparison

Raw materials costs make up 55–75% of the final alcohol selling price. By-product credits from the raw materials differ and possible credits must be considered in establishing a net raw material cost for comparison. For conventional feedstocks, prices, conversion efficiencies and possible by-product credits are established. For cellulosics, conversion efficiencies are uncertain, by-product markets unproven, and even the raw material costs are disputed. Available substrate cost and by-product credit information is summarized in Table 4. Pretreatment costs also differ between the various raw materials. Chemical, steam and electric power requirements of various pretreatment processes were included in the process descriptions.

Additional cost differences arise due to differences in stillage handling and distillation processes as is discussed later. At present in the US, grain sorghum and corn compete as economical feedstocks subject to the variations in their market prices.

43.3 FERMENTATION

Depending on design, fermentation equipment makes up 10–25% of the total fixed capital cost of an ethanol plant. Further, the fermentation process, by dictating feed sugar concentration and product cell and ethanol concentrations, specifies the major flows and concentrations throughout the remainder of the plant. The fermentation process, thus, is central to the overall plant design.

43.3.1 Organisms for Ethanol Fermentation

43.3.1.1 Organism alternatives

Yeasts are the only organisms currently used for large-scale industrial ethanol production. Yeasts produce ethanol with very high selectivity (only traces of by-products), are very hardy and are large compared with bacteria (allowing simplified handling).

Clostridium thermosaccharolyticum, *Thermoanaerobacter ethanolicus* and other thermophilic

Table 4 Net Raw Materials Costs for Alternative Ethanol Production Feedstocks

	Saccharine materials			Starchy materials				Cellulosics	
	Sugar cane	Sugar beet	Blackstrap molasses	Corn	Wheat	Grain sorghum	Potatoes	Corn stover	Douglas fir
Raw material cost[a] ($ t^{-1})	32.1	36.9	84.5	98.1	149.1	79.4	170.4	38.0	17.0
Conversion (l ethanol t^{-1})	66.0	84.0	275.0	340.0	360.0	370.0	105.0	136.0[b]	215.0[b]
Unit cost (¢ l^{-1} ethanol)	49.0	44.0	31.0	29.0	41.0	21.0	162.0	28.0	8.0
By-products (kg l^{-1} ethanol)	Bagasse (3.50) Stillage yeast (0.18)	Pulp (1.10) Stillage yeast (0.18)	Condensed solubles (1.30) Stillage yeast (1.30)	Dark grains (0.78)	Dark grains (0.96)	Stillage yeast (0.18)	Pulp (0.31) Stillage yeast (0.18)	Lignin (0.95) Stillage yeast (0.18)	Lignin (1.13) Stillage yeast (0.18)
Unit by-product credit (¢ l^{-1} ethanol)	18.0	26.0	12.0	15.0	18.0	8.0	9.0	12.0	13.0
Net raw material cost (¢ l^{-1} ethanol)	31.0	18.0	19.0	14.0	23.0	13.0	153.0	16.0	5.0

[a] Average for 1978–1980.   [b] Assuming 60% fermentable sugar conversion.

bacteria as well as *Pachysolen tannophilus* yeast (Staff, Chem. Eng. News, 1981) are under intensive study for use in fermenting pentose sugars which are nonfermentable by ordinary yeast. These bacteria also convert hexose sugars and have been considered as an alternative to yeast since very high-temperature reactions would allow simple continuous stripping of ethanol product from the active fermenting mixture, thus eliminating end-product inhibition effects. Organisms so far studied, however, produce excessive quantities of undesirable by-products or are limited to producing only dilute ethanol beers. The bacteria also require strict anaerobic conditions which would be difficult to maintain on an industrial scale. Thermophilic bacteria for ethanol production have recently been reviewed by Gong *et al.* (1981) and by Rosenberg (1980).

The bacterium *Zymomonas mobilis* is also under intensive study (Rogers *et al.*, 1979, 1980). This organism ferments glucose to ethanol with a typical yield 5–10% higher than for most yeasts, but is less ethanol tolerant than industrial yeast strains. The small bacterium is also difficult to centrifuge. An important possibility for the future is the development of an organism especially tuned to rapid ethanol production. *Zymomonas* is a simple procaryote and, hence, is more amenable than yeast to genetic modification. Attempts are under way to increase the ethanol tolerance of *Zymomonas*, to allow utilization of pentose sugars and to impart flocculent characteristics for improved centrifugability (Dunphy, 1980; Lee *et al.*, 1982).

43.3.1.2 Yeast strain selection

Yeast strains are generally chosen from among *Saccharomyces cerevisiae*, *S. ellypsoideus*, *S. carlsbergensis*, *S. fragilis* and *Schizosaccharomyces pombe*. For whey fermentation, *Torula cremoris* or *Candida pseudotropicalis* is used (Castillo *et al.*, 1982).

Yeasts are carefully selected for: (1) high growth and fermentation rate; (2) high ethanol yield; (3) ethanol and glucose tolerance; (4) osmotolerance; (5) low pH fermentation optimum; (6) high temperature fermentation optimum; (7) general hardiness under physical and chemical stress. High growth and fermentation rate allows the use of smaller fermentation equipment. Ethanol and glucose tolerance allows the conversion of concentrated feeds to concentrated products, reducing energy requirements for distillation and stillage handling. Osmotolerance allows the handling of relatively 'dirty' raw materials such as blackstrap molasses with its high salt content. Osmotolerance also allows the recycle of a large portion of stillage liquids, thus reducing stillage handling costs. Low pH fermentation combats contamination by competing organisms. High temperature tolerance simplifies fermenter cooling. General hardiness allows yeast to survive both the ordinary stress of handling (such as centrifugation) as well as the stresses arising from a plant upset.

The years of careful selection by industrial use have led to yeast strains with these desirable characteristics. Many of the best strains are proprietary, but others are available from the culture collections (Davis and Jung, 1974).

43.3.2 Fermentation Kinetics

43.3.2.1 Yeast metabolic pathways

In the anaerobic pathway, glucose is converted to ethanol and carbon dioxide *via* glycolysis. The overall reaction (equation 1) produces two moles of ethanol and carbon dioxide for every mole of glucose consumed, with the reaction energy stored as two moles of ATP for use in biosynthesis or maintenance.

$$C_6H_{12}O_6 \rightarrow 2\ C_2H_5OH + 2\ CO_2 + \text{energy (stored as ATP)} \tag{1}$$

Via this pathway, every gram of glucose converted will yield 0.511 g of ethanol. Secondary reactions consume a small portion of the glucose feed, however, to produce biomass and secondary products and Pasteur found that the actual yield of ethanol from fermentation by yeast is reduced to 95% of the theoretical maximum (Table 5) (Hodge and Hildebrandt, 1954). When complex substrates, typical of industrial practice, are used, further by-products are generated and the ethanol yield is reduced typically to only 90% of the theoretical ($0.46\ \mathrm{g\ g^{-1}}$ glucose) (Venkiteswaran, 1964a).

Table 5   Optimum Yields from
Anaerobic Fermentation by Yeast

Product	g per 100 g glucose
Ethanol	48.4
Carbon dioxide	46.6
Glycerol	3.3
Succinic acid	0.6
Cell mass	1.2

Via aerobic metabolism, sugar is converted completely to carbon dioxide, cell mass and by-products, with no ethanol formed, and aerobic metabolism must be avoided.

43.3.2.2   *Effect of sugar concentration*

Hexose sugar (glucose, fructose, galactose or maltose) is the primary reactant in the yeast metabolism. Under fermentative conditions, the rate of ethanol production is related to the available sugar concentrations by a Monod type equation:

$$v = v_{max}C_s/(K_s + C_s) \tag{2}$$

where v is the specific ethanol productivity (g ethanol g^{-1} cells h^{-1}), C_s is the sugar substrate concentration (g l^{-1}) and K_s is a saturation constant having a very low value, typically 0.2–0.4 g l^{-1}.

At very low substrate concentrations (below about 3 g l^{-1}), the yeast is starved and productivity decreases (Levenspiel, 1980). At higher concentrations a saturation limit is reached so that the rate of ethanol production per cell is essentially at its maximum up to 150 g l^{-1} sugar concentration. Above 150 g l^{-1}, catabolite (sugar) inhibition of enzymes in the fermentative pathway becomes important, and the conversion rate is slowed (Holzer, 1968; Wang *et al.*, 1979).

An important secondary effect of sugar is catabolite repression of the oxidative pathways (the Crabtree effect). At above 3–30 g l^{-1} sugar concentration (depending on yeast strain), the production of oxidative enzymes is inhibited (Moss *et al.*, 1971; DeDeken, 1966), thus forcing fermentative metabolism. This catabolite repression is not found in all yeasts and is a desirable property which is selected in industrial strains.

43.3.2.3   *Effect of ethanol*

Ethanol is toxic to yeast and high ethanol tolerance is a desirable trait selected for in industrial strains. As illustrated in Figure 2 (Bazua and Wilke, 1977), the inhibitory effect of ethanol is generally negligible at low alcohol concentrations (less than 20 g l^{-1}) but increases rapidly at higher concentrations. For most strains, ethanol production and cell growth are halted completely at above 110 g ethanol l^{-1} although some very slow fermenting sake yeast (*Saccharomyces sake*) can tolerate ethanol concentrations as high as 160 g l^{-1} at low temperatures (Hayashida and Ohta, 1981). Ethanol inhibition is directly related to the inhibition and denaturation of important glycolytic enzymes, as well as to modification of the cell membrane (Millar *et al.*, 1982; Rose and Beavan, 1981).

43.3.2.4   *Effect of oxygen*

It is important to avoid a high degree of aerobic metabolism which utilizes sugar substrate but produces no ethanol. It has been found, however, that trace amounts of oxygen may greatly stimulate yeast fermentation. Oxygen is required for yeast growth as a building block for the biosynthesis of polyunsaturated fats and lipids required in mitochondria and the plasma membrane (Haukeli and Lie, 1971). High sugar concentration is adequate to repress aerobic sugar consumption in yeasts which show the Crabtree effect. For other yeasts or at low sugar concentrations, the oxygen supply should be limited. Trace amounts (0.7 mmHg oxygen tension) of oxygen are adequate and do not promote aerobic metabolism (Cysewski and Wilke, 1976a).

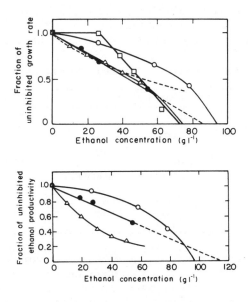

Figure 2 Effect of ethanol on cell growth and ethanol production (various strains of *Saccharomyces*)

43.3.2.5 Effect of pH

Fermentation rate is sensitive to pH, but most distiller's yeasts show a broad pH optimum from at least pH 4–6 (Cysewski and Wilke, 1976b). This range is lower than that for typical bacteria. Further, most yeasts can tolerate exposure to acid solutions of pH as low as 2 without permanent damage (Hodge and Hildebrandt, 1954).

43.3.2.6 Effect of temperature

High temperature tolerance is a desirable characteristic selected for in distillery yeasts and most distillery yeasts have a temperature growth optimum of 30–35 °C (Gray *et al.*, 1942). The optimum fermentation temperature at low alcohol concentrations is often slightly higher (up to 38 °C), but alcohol tolerance is improved at reduced temperatures (Hayashida and Ohta, 1981). Exposure to temperatures above the optimum results in excessive enzyme degradation and loss of yeast viability. Yeast metabolism liberates 11.7 kcal of heat for each kilogram of substrate consumed (Lamprecht and Meggers, 1969).

Yeasts can be stored inactive at low temperatures (above 0 °C) and are readily revived (Stark, 1954).

43.3.2.7 Additional nutrient requirements

In addition to providing a sugar source for ethanol production, fermentation mash must also provide the secondary nutrients necessary for cell maintenance and growth (Jones *et al.*, 1981; Grylls, 1961). In laboratory tests, very rapid cell growth and ethanol production and high yields are achieved with a glucose medium supplemented with NH_4Cl, $MgSO_4$, $CaCl_2$ and yeast extract (Cysewski and Wilke, 1976a,b). Ammonium ions provide nitrogen for protein and nucleic acid synthesis. Yeast extract is the water-soluble extract of autolyzed yeast and contains all necessary yeast growth factors: amino acids, purines, pyrimidines and vitamins as well as minerals. Phosphorus, potassium (from the yeast extract), magnesium and calcium are incorporated into cell mass and are also cofactors activating several enzymes. The wide variation in media compositions used is apparent from Oura's (1974) comparison of 12 of the best known culture media.

To establish better the yeast nutritional requirements, Wong (1983) developed a synthetic, totally defined medium (Table 6) which (after adaptation of the yeast in continuous culture) gives high productivity and yield. The minerals, $(NH_4)_2SO_4$ and KCl, and the vitamins, biotin and pantothenate, are stimulatory to growth, while thiamine and pyridoxine increase specific ethanol productivity. Thiamine may also increase ethanol tolerance (Rahn, 1952).

Table 6   Synthetic Medium for Yeast Fermentation

Nutrient	Concentration	Nutrient	Concentration
Glucose	100 g l^{-1}	$MnSO_4$	1.3 mg l^{-1}
$(NH_4)_2SO_4$	3.54 g l^{-1}	KI	0.5 mg l^{-1}
KCl	0.625 g l^{-1}	$FeSO_4$	0.5 mg l^{-1}
H_3PO_4	0.411 g l^{-1}	$CoSO_4$	0.3 mg l^{-1}
$MgSO_4$	0.061 g l^{-1}	Pyridoxine	3.33 mg l^{-1}
$CaCl_2$	0.022 g l^{-1}	Pantothenate	2.00 mg l^{-1}
H_3BO_3	5.0 mg l^{-1}	Thiamine	1.34 mg l^{-1}
$ZnSO_4$	2.8 mg l^{-1}	Inositol	1.26 mg l^{-1}
$Al_2(SO_4)$	1.5 mg l^{-1}	Biotin	$5.25 \text{ }\mu\text{g l}^{-1}$
$CuSO_4$	1.3 mg l^{-1}		

Industrial substrates contain many of the necessary yeast nutrients (Solomons, 1969). For industrial fermentation, blackstrap molasses is normally supplemented with 0.06–0.40 kg ammonium sulfate per 1000 liters of mash and adjusted to pH 4–5 with sulfuric acid (Hodge and Hildebrandt, 1954). High test molasses and cane or beet juice are supplemented with 0.40–0.80 kg ammonium sulfate and 0.30–0.60 kg sodium phosphate per 1000 liters of mash (Hodge and Hildebrandt, 1954). For grain hydrolyzate fermentation, most necessary nutrients are available from the malt or amylase culture liquid (Stark, 1954). If concentrated fungal amylase preparations are used, then supplementation with papain and urea or ammonium sulfate may be necessary. Sulfite liquor is neutralized with lime and supplemented with ammonium hydroxide or urea at 0.01–0.06 kg per 1000 liters of mash (McCarthy, 1954).

Cell growth and fermentation efficiency are both greatly improved by employing backset, where a portion of the dealcoholized stillage from previous production is recycled to the new fermenter mash. The stillage contains concentrated soluble nutrients from previous fermentation, as well as extracts from the autolysis of yeast residue in the stillage stripper. The stillage also has excellent buffer capacity and reduces requirements for pH adjustment.

For high cell yields in the production of a yeast inoculum, semiaerobic or aerobic growth may be used with much higher levels of nutrient supplementation (Unger *et al.*, 1942). Complex media supplements such as corn-steep liquor, yeast extract, malt extract or casein hydrolyzate (Solomons, 1969) may be added but these procedures vary widely between plants.

43.3.2.8 Secondary component inhibition

Yeast growth and ethanol production can be inhibited by fermentation by-products or by nonmetabolized feed components. These components become concentrated when backsetting is used and this limits the fraction of distiller's residue which can be recycled.

Acetate and lactate are the most important inhibitory fermentation by-products (Maiorella *et al.*, 1983a). Lactic acid and, to a lesser extent, acetic acid are partially removed in stripping and fermentation by-products do not normally establish the limit on backsetting.

Specific feeds may be high in certain inhibitors. Sulfite waste liquor may be high in sulfurous acid and furfural. Blackstrap molasses may contain high concentrations of calcium salts. High-temperature sugar concentration and sterilization in the presence of salts (especially phosphates) and proteins can produce yeast toxins (Bridson and Brecker, 1970; Mehta and Khanna, 1964).

When important individual inhibitors are not present, a combination of inhibitors or generalized osmotic pressure effects will be limiting. High salt concentrations also encourage the production of undesirable by-products such as glycerol (Watson, 1970). A 16–20 wt% nonfermentable dissolved solids content sets the practical upper limit for most yeasts in the absence of specific toxic inhibitors (Goggin and Thorsson, 1982; Kujala, 1979).

In industrial practice, 10–20% of the mash volume is recycled for blackstrap molasses, and up to 50% for high-test molasses or cane juice (Hodge and Hildebrandt, 1954); 20–25% of the mash volume is recycled for fermentation of hydrolyzed grain (Stark, 1954). Backsetting is not normally practiced in waste sulfite liquor plants (Wiley, 1954). Laboratory scale tests on grain mash indicate that the degree of backset may be increased for selected yeast strains, with up to 85% of the mash volume recycled up to ten times without adverse effect (Staff, Chem. Eng., 1981; Ronkainen, 1978).

43.3.3 Fermentation Processes

43.3.3.1 Conventional batch fermentation

Batch fermentation begins with the production of an active yeast inoculum. This can be either by the conventional serial growth method or by the rapid semiaerobic method. Aseptic techniques are used throughout. In the serial growth method (Hodge and Hildebrandt, 1954) a pure culture inoculum from an agar slant is used to seed a laboratory shake flask. At the peak of growth (12–24 h) this culture is used to seed a succeeding culture 30–50 times larger. This is repeated, generally through three laboratory stages and two or three plant semiworks stages, to produce the final 2–5 vol% inoculum for the primary fermentation. The inoculum is grown on a medium similar to the final fermentation mash to minimize acclimatization time in the final fermenter, but higher levels of yeast growth nutrients may be used to produce a high cell density (typically 150 billion cells l^{-1}).

An inoculum 3–4 times more concentrated in yeast can be produced by the rapid semiaerobic method (Unger *et al.*, 1942; Aries, 1947). Yeasts are grown in an aerated and agitated semiwork fermenter operated in fed batch mode. A large portion (20–25%) of the previous batch is retained to provide an inoculum. A high nutrient medium is added and pH and temperature are controlled. Sterile air is sparged at a rate of one-eighth volume of air per fermenter volume per minute. Aerobic metabolism is stimulated and a cell density of 500 billion cells l^{-1} is reached in 5 h. This high cell density allows the use of a proportionately smaller inoculum to the final fermenter, and a smaller propagating fermenter can be used. However, aerobically grown yeast may require additional time to acclimatize to anaerobic fermentation conditions, resulting in an increased lag period between inoculation and rapid fermentation.

Cylindro-conical Nathan vessels (Hudson, 1972) are preferred for fermentation as these promote better circulation and allow thorough drainage for cleaning. For very large plants, sloped-bottom cone-roof tanks of up to 1 million liters volume are used and these large vessels are often agitated only by carbon dioxide evolution during fermentation.

After emptying and rinsing from the previous batch, mash at 13–17 wt% sugar is pumped to the fermenters. Once 20% full, the inoculum is added to allow growth during the remainder of the filling cycle, which can last 4–6 h.

Fermentation temperature is regulated by circulating cooling water through submerged coils, circulating the mash through external heat exchanges, or simply spraying the vessel walls with cool water (adequate for small fermenters only). The feed is generally introduced at 25–30 °C, and the temperature allowed to gradually rise as heat is evolved. The temperature thus varies from 30–35 °C during the initial period (which is optimal for yeast growth). Cooling is then used to prevent the temperature from exceeding 35–38 °C, which is optimal for ethanol production. These temperatures may be modified depending on the yeast strain used.

Stillage backset provides excellent buffering. The pH is set initially at from 4.5–5.5 and decreases only slowly, generally holding at pH 4.0 or above. This is especially important for fermentation of grain mashes with simultaneous dextrin hydrolysis, as many amylase enzymes are rapidly denatured at lower pH (Stark, 1954).

The time course of a typical batch fermentation of molasses is presented in Figure 3. The ethanol production rate is the product of specific (per cell) productivity and the concentration of cells. Initially, the rate of alcohol production is quite low, but as the number of yeast cells increases the overall rate increases, and with rapid carbon dioxide evolution the beer appears to boil. After 20 hours a maximum in ethanol productivity is reached. The effects of reduced sugar concentration and ethanol inhibition then become important. The fermentation continues at a decreasing rate until, at 36 hours, 94% of the sugar is utilized and a final ethanol content of 69 g l^{-1} is achieved. The average volumetric ethanol productivity over the course of the fermentation is 1.9 g l^{-1} h^{-1}.

Fermentation time will vary depending on yeast strain and substrate. Hawaiian and Cuban blackstrap molasses fermentation usually requires 36 h. Molasses from Java may require as long as 72 h. These times are reduced when molasses clarification is used. Grain fermentation requires 40–50 h to allow complete residual dextrin conversion.

After fermentation, the beer is pumped to a beer well to provide a continuous feed to distillation. The fermenters are then cleaned and prepared for another cycle.

The fermentation rate can be increased 30–40% by improved agitation and temperature regu-

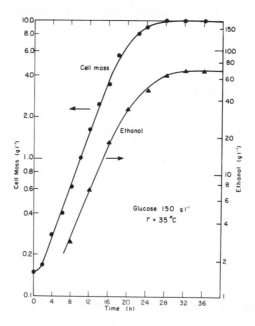

Figure 3   Time course of a batch fermentation

lation (Aries, 1947). Turbine impellers have been used in smaller fermenters (100 000 l or less). For large fermenters, improved agitation and temperature control are achieved by rapidly circulating the beer through an external heat exchanger. The cooled beer is pumped back into the head of the fermenter tangentially and at a high velocity. The greatest portion of heat generation occurs during the 'boiling' fermentation period, creating a high peak cooling demand. Investment in cooling equipment can be reduced by teaming three or four fermenters to share a single bank of exchangers. The fermentations are then carefully scheduled to give a uniform heat load (Katzen, 1979). With improved agitation, continuous pH control can be used. Ammonium hydroxide or other base is added to the stirred fermenter or mixed with returning cycled beer. The pH can thus also be held at an optimum setpoint throughout the reaction period.

Contamination by lactic acid bacteria is occasionally a problem and alcohol yield can be reduced by as much as 20% (Stark, 1954). Such contamination is more likely when stillage backset is used, allowing contaminating organisms to accumulate and acclimatize to the fermentation conditions. Aseptic operation, with complete sterilization of the very large mash volume, was considered impractical until recently. Growth of organisms other than the seed yeast is generally restricted by the adverse conditions of low pH and high sugar or alcohol concentration, and the rapid growth of the yeast compared with contaminants was relied upon in place of aseptic techniques. Highly efficient continuous media sterilization now makes aseptic operation quite practical (Pfeifer and Vojnovich, 1952). The medium is heated by steam injection to 135–140 °C and held in plug flow for 1–2 min, resulting in essentially complete sterilization. Cooling is by flashing to regenerate steam or by heat exchange to preheat incoming feed. The steam requirement is only 3.5 kg steam per 100 kg mash. Small spherical head fermenter vessels can be sterilized by pressurizing with steam. For very large fermenters this is impractical and antiseptic solutions such as ammonium bifluoride (Hodge and Hildebrandt, 1954), iodine (Katzen, 1979), sodium hypochlorite or formalin are automatically sprayed to rinse the vessel walls before filling. Heat exchangers and transfer lines can be steamed.

A batch fermentation layout incorporating teamed heat exchange and chemical sterilization systems is shown in Figure 4.

43.3.3.2   The Mellé–Boinot fermentation process

The Mellé–Boinot process (Lagomasino, 1949) achieves a reduced fermentation time and increased yield by recycling yeast. Cell density at the beginning of a conventional fermentation is quite low (5–10 billion cells l^{-1} *vs.* 100 billion cells l^{-1} at completion). Ethanol production rate is proportional to cell density and the initial growth phase of a conventional fermentation (lasting

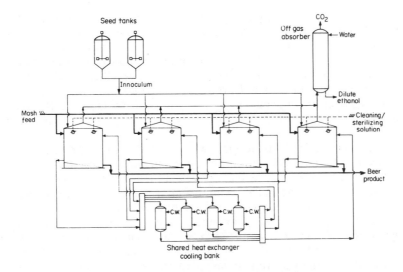

Figure 4 Batch fermentation equipment layout incorporating teamed heat exchange and chemical sterilization systems.

up to 15 h) is thus relatively unproductive. In the Mellé–Boinot process, yeast cells from the previous fermentation are recovered by centrifugation and up to 80% are recycled. The initial cell density of a batch is thus as high as 80 billion cells l^{-1} and very rapid ('boiling') fermentation begins almost immediately. With the long growth phase eliminated, the overall fermentation cycle time is reduced by one-half to two-thirds, increasing volumetric productivity to typically 6 g $l^{-1} h^{-1}$.

Yeast reuse also results in a decrease in new cell growth with more sugar available for conversion to ethanol and a corresponding increase of ethanol yield by 2–7% (Espinosa *et al.*, 1978). The reduction in yeast growth has been attributed to achieving a cellular saturation point at about 8 g l^{-1} (200 billion cells l^{-1}), beyond which cells will not grow (Hodge and Hildebrandt, 1954). This explanation is contradicted, however, by results of continuous laboratory scale fermentations with cell recycle where rapid growth was maintained at cell densities as high as 83 g l^{-1} (the limit of the cell recycling equipment) (Cysewski and Wilke, 1977). Environmental factors are a more likely explanation for the reduced growth in industrial batch yeast recycle. Upon introduction to fresh medium, a lag phase is required before biosynthetic pathways can be reactivated and rapid growth begun. Rapid fermentation can, however, begin almost immediately. Heat is evolved and the fermentation temperature rises much more rapidly than in conventional batch fermentation, so that the lower temperature period which favors yeast growth is shortened. Sugar is consumed and inhibitory ethanol is produced more rapidly. Ideal growth conditions may thus be already passed before the yeast can reacclimatize to the fresh medium and grow effectively.

Contamination can be a problem with the continual recycle of microorganisms. This is mitigated by the yeast recycle technique. Centrifugation conditions are adjusted to recover most yeast but not lighter bacterial contaminants. The centrifuged yeast is then held at pH 2 in the presence of carbon dioxide for 4 h, sufficient to kill nonsporulated lactic acid bacteria. Sterile feeds and aseptic fermentation techniques are preferred for the recycle technique.

The Mellé–Boinot process was generally developed for and is widely used in sulfite waste liquor fermentation, where the low sugar concentrations require maximum yield and cell recycle achieves much higher cell densities and reduced fermentation time (McCarthy, 1954). The process is widely used in Europe for molasses fermentation (Hodge and Hildebrandt, 1954).

Sugar clarification is essential, though, for good centrifuge performance and to maintain high yeast viability (Samaniego and Srivastas, 1971). Yeast recycle will reduce the fermentation time for grain fermentations only if the starch is preconverted to glucose by acid hydrolysis or high activity glucoamylase enzymes. Otherwise dextrin hydrolysis in the fermentation will still be limiting and little advantage gained.

43.3.3.3 Simple continuous flow fermentation

Continuous flow fermentation processes have been used in industrial sulfite waste liquor fermentation since the 1930s (McCarthy, 1954). The antiseptic qualities of sulfite liquor minimize the possibility of adverse contamination and allow long continuous runs without shutdowns for cleaning. Early attempts at continuous fermentation of molasses and grain hydrolyzates on an industrial scale were unsuccessful due to contamination problems and these plants were retrofit for batch operation (Hough *et al.*, 1977). With continuous media sterilization and aseptic plant techniques, the contamination problem has been overcome, as is illustrated by the success of continuous molasses fermentation plants in Europe (Rosen, 1978) and Japan (Karaki *et al.*, 1972), and many continuous beer brewing plants especially in New Zealand and Britain (Hough *et al.*, 1977). Continuous culture techniques have been reviewed extensively (Yarovenko, 1978; Dawson, 1976).

Detailed results of pilot scale (1800 l fermenter) tests of molasses fermentation in a simple continuous stirred flow reactor have been published (Borzani and Aquarone, 1957). Feed is pumped continuously into the fermenter, displacing beer which then overflows from the vessel. The product beer composition is the same as the composition in the fermenter vessel and thus, if a high ethanol product concentration is desired, the fermentation will be relatively slow as the entire course of the fermentation must take place under inhibitory high ethanol concentration conditions. The throughput rate must be adjusted to be slow enough to allow growth of new yeast in the fermenter to replace yeast washed out in the overflow and to allow essentially complete utilization of the sugar. Agitation by stirring or gas sparging was found to be especially important for successful continuous flow fermentation. With agitation, blackstrap molasses at 140 g l^{-1} concentration could be 95% utilized in a 21 h residence time compared with the batch fermentation time of 40 h. Without agitation, the continuous flow fermentation residence time for complete sugar utilization was 55 h.

In laboratory tests with carefully optimized conditions of temperature, pH, agitation and flow rate, much higher productivities have been achieved. Molasses at 130 g sugar l^{-1} has been fermented to completion in a 7 h residence time (Bilford *et al.*, 1942), corresponding to a volumetric productivity of 8.3 g l^{-1} h^{-1}. A nutrient supplemented glucose medium of 130 g sugar l^{-1} could be completely utilized in 9.5 h residence time (Cysewski and Wilke, 1976a). This fermentation was continued for 60 days without decline in productivity.

43.3.3.4 *Continuous fermentation with yeast recycle*

The productivity of continuous fermentations can be greatly enhanced by yeast cell recycle (DelRosario *et al.*, 1979). A centrifuge is used to separate yeast from the product flow and the concentrated yeast cream (up to 150 g yeast l^{-1}) is recycled to the fermentation vessel. A small bleed of cells is required to maintain a viable culture (Pirt and Kurowski, 1970). In laboratory tests, a simple chilled settler was used to recycle cells, increasing the cell density four-fold to 50 g l^{-1} (Cysewski and Wilke, 1977; Ghose and Tyagei, 1979a). The residence time for complete conversion of a 100 g glucose l^{-1} feed was reduced to 1.6 h with a corresponding productivity of 30 g ethanol l^{-1} h^{-1}. Over a two-week total period the viable yeast fraction (96%) did not decrease and the specific ethanol productivity (per cell) was the same as in nonrecycle experiments.

Capital costs for centrifuges are high and sealed sterilizable centrifuges have become available only recently. This has slowed the application of cell recycling in industrial fermentation. Attempts have been made to develop cheaper alternative cell recycle methods. Simple cell settling systems have been developed wherein the cells are thermally shocked (to temporarily halt CO_2 evolution) and allowed to gravity settle. Very large settling vessels are required, however (Walsh and Bungay, 1979). Whirlpool separators have been used industrially in beer manufacture (Hudson, 1972). The fermenter broth is pumped tangentially into a verticle cylindrical vessel and flocculent yeast cells are deposited in a central cone for recycle. A very simple partial recycle fermenter has been developed and tested at pilot scale (Hough *et al.*, 1962). The overflow is taken from a vertical pipe rising through the fermenter base and jacketed by a baffled sleeve. The region between overflow pipe and sleeve escapes agitation, allowing yeast to settle back and separate from rising beer. The cell density in the fermenter can be increased 2.5 times using flocculent rapid settling yeast strains with essentially no added equipment cost.

43.3.3.5 *Series arranged, continuous flow fermentation*

Fermentation throughput can be increased when very complete sugar utilization or high ethanol product concentration is required by using continuous flow fermenters arranged in series.

In laboratory tests the residence time for complete sugar utilization (to 0.3% residual sugar) in a single fermenter increased from 9.5 h for a 100 g l^{-1} glucose feed to 25 h for a 160 g l^{-1} feed (Cysewski and Wilke, 1976a).

With two fermenters arranged in series the residence time can be chosen so that the sugar is only partially utilized in the first with fermentation completed in the second. Ethanol inhibition is reduced in the first vessel, allowing a faster throughput. The second, lower productivity fermenter must now convert less sugar than if operated alone and it too can be operated at an increased throughput.

For a high product concentration (80 g ethanol l^{-1}) the productivity of a two-stage system has been as much as 2.3 times higher than a single stage (Ghose and Tyagei, 1979b). In other tests, by adding equal volume fermentation stages at constant molasses feed rate, feed sugar concentration could be increased. Alcohol product concentration at essentially complete sugar utilization was increased from 52 g l^{-1} for one stage to 73 g l^{-1} for two and 99 g l^{-1} for three stages (Chu *et al.*, 1970).

In Denmark, a two-stage process has been used industrially (170 000 liter volume vessels), producing a 66 g l^{-1} ethanol product in 21 h residence time (Rosen, 1978). In the Netherlands, beet molasses is fermented in a series of seven vessels of 70 000 liters each to produce 86 g l^{-1} ethanol beer in 8 h (Chen, 1980). A Japanese company uses six fermenters in series (total volume 100 000 liters) to produce 95 g l^{-1} beer in 8.5 h, corresponding to a volumetric ethanol productivity of 11.2 g l^{-1} h^{-1} (Karaki *et al.*, 1972).

When multiple vessels are used in series, conditions in each can be controlled independently for optimal performance. In the Japanese system, for instance, the first vessel is kept at 27 °C and aerated; conditions to optimize cell growth to start the fermentation. The remaining vessels are not aerated and are kept at 30 °C for higher specific ethanol productivity.

For fermentation of grain, dextrin conversion would ordinarily be limiting and no advantage would be found in the series continuous flow process. In New Zealand, however, it has been found economical to hydrolyze the grain completely and then conduct a continuous fermentation. Two fermenters in series are generally used with the first aerated to promote yeast growth. Slow fermenting brewer's yeast and low temperatures (15 °C) are used so that the fermentation time is typically 28 h. This is a considerable improvement, however, over the typical 2–4 day batch beer fermentation. In the US, acid hydrolyzed grain has been fermented in a two-stage continuous flow system at pilot plant (4000 l fermenter) scale to produce palatable beer in a 12 h residence time (Ruf *et al.*, 1948). Stable operation was maintained for 6 months. A Russian industrial process combines simultaneous *Aspergillus* enzymatic dextrin conversion and yeast fermentation in a four-vessel continuous series flow system with yeast recycle. Conversion of glucose to ethanol apparently increases the dextrin conversion rate and the fermentation time is reduced from 68 h to only 17 (Yarovenko, 1978).

Series flow arrangements are especially advantageous in sulfite liquor processing, where the sugar concentration is low and a high degree of utilization results in yeast starvation effects in a single vessel. An American plant built in the 1940s used eight 300 000 l fermenters in series to produce 9 g l^{-1} alcohol in 12 h (McCarthy, 1954); 75% of the sugar was converted in the first two fermenters while all eight were required for a 95% conversion, leaving less than 1 g l^{-1} residual fermentable sugar.

43.3.3.6 *The Biostil process*

The Biostil process (Figure 5) (Goggin and Thorsson, 1982; Alfa-Laval, 1981, 1982) is a modification of the continuous fermentation process with recycle, whereby the fermentation and distillation are closely coupled and a very high stillage backsetting rate is used. Fermenter beer is continually cycled (through a centrifuge for yeast recycle) to a small rectifying column where ethanol is removed. The majority of the ethanol depleted beer (with residual sugars and nonfermentables) is then recycled to the fermenters. The yeast cell density is maintained at 500 billion cells per liter. The fermenter ethanol concentration can be maintained at any desired non-inhibitory level by adjustment of the beer cycle rate. The large liquid recycle provides an internal dilution so that very concentrated feeds can be processed. Liquid flows external to the recycle loop are greatly reduced. Less water is consumed and a more concentrated stillage is produced. The flow rate capacities of most auxiliary equipment can be substantially reduced.

Extensive heat exchange is incorporated into the beer cycle loop to maintain energy efficiency.

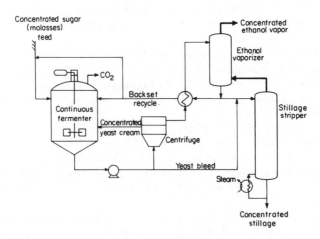

Figure 5 Biostil fermentation process

As bacteria are not well separated by the centrifuge, the cycling through a hot distillation stage also provides a continuous pasteurization and infection risk is reduced. As with the Mellé–Boinot process, ethanol yield is increased by cell recycling.

With ethanol inhibition overcome, the Biostil process is limited instead by the build-up of toxic nonfermentable feed components and fermentation by-products. In pilot plant studies the stillage flow could be reduced by a factor of 20 for concentrated cane syrup, but only a factor of 3 for blackstrap molasses with its high nonfermentable content (Goggin and Thorsson, 1982). The process can also be applied to prehydrolyzed starch feeds.

A 37 Ml y^{-1} Biostil alcohol plant was commissioned in Brazil in 1982. The increase in stillage concentration and reduction in stillage volume, compared with conventional processes with low backset rates, results in substantial energy and cost savings for stillage processing.

43.3.3.7 Tower fermenter

An interesting fermenter arrangement which has been suggested for rapid industrial alcohol production but not yet used commercially is the APV tower fermenter (Hough *et al.*, 1977; Ault *et al.*, 1969; Greenshields and Smith, 1971). This system has been used commercially since 1965 for the production of beverage beers.

The fermenter consists of a cylindrical tower typically 2 m in diameter and 15 m tall and is topped by a larger diameter settling zone fitted with baffles. A 'sticky' flocculent yeast with very high settling rate is used. The prehydrolyzed sugar wort is pumped into the base of the tower and reaction proceeds progressively as the beer rises upward through a dense yeast plug. The yeast tends to settle back against the flow and very high cell densities (50–80 g l^{-1}) are achieved without the need for centrifuge cell recycle. Since volumetric productivity increases in proportion to active yeast cell density, very high productivities are achieved. The residence time for complete sugar utilization to produce a 6.5 wt% beer is 4 h. Yeast strains used have been limited to slower fermenting varieties which have the necessary flocculence properties and will produce beer of the desired flavor characteristics. Fermentation temperature is regulated by cooling jackets at ordinary low brewing temperatures (16–23 °C) for best flavor and this also reduces the specific ethanol production rate.

A major drawback of the APV system is the long time required for start-up. Two to three weeks are required to build up the desired high cell density and achieve stable operation. Continuous run times of 12 months and greater have been achieved and compensate for the slow start-ups.

Process modifications would be necessary for large-scale industrial ethanol production. Larger diameter columns with higher throughput would be advantageous. Higher product concentration would also be desirable. Increased fermentation rates have recently been demonstrated in a laboratory scale tower operating at higher fermentation temperature and without the constraint of product palatability. An ethanol productivity of 27 g l^{-1} h^{-1} was achieved with *S. cerevisiae* and 99 g l^{-1} h^{-1} with *Zymomonas mobilis* (Prince and Barford, 1982a,b).

43.3.3.8 *Experimental fermentation processes*

Considerable attention has been given to the development of improved higher-rate ethanol fermentation processes, with as much new research published in the past 15 years as in the previous 50. Immobilized cell and dialysis membrane reactors have been used to increase yeast cell density, and methods have been developed for the continuous removal of alcohol from an actively fermenting beer to eliminate end-product inhibition. Volumetric ethanol productivities as high as $83 \text{ g l}^{-1} \text{ h}^{-1}$ have been achieved in laboratory tests (Cysewski and Wilke, 1977).

Fermentation process research has been comprehensively reviewed and new processes evaluated for commercial application (Maiorella *et al.*, 1981, 1984b). Further development is required before these new methods can be used industrially.

43.3.3.9 *Industrial fermentation process comparison*

Zuid-Nederlandse Spiritusfabriek currently operates both Mellé–Boinot batch and series vessel continuous molasses alcohol plants as well as having experience with conventional batch fermentation. Descriptions for these plants have been published along with an evaluation by the company's fermentation director (Chen, 1980). A direct comparison of these processes under comparable industrial conditions is thus possible.

Table 7 compares fermentation cycle times and alcohol product concentrations for these processes. The Mellé–Boinot process results in a 66% reduction in required fermenter capacity compared with batch fermentation. Two 80 000 l fermenter vessels are eliminated while one added 250 l min^{-1} capacity centrifuge is required. A net reduction in purchased equipment cost results. The continuous process gives an 80% reduction in required fermenter capacity for an even greater reduction in capital investment.

Table 7 Industrial Fermentation Process Comparison

	Cycle time			Product concentration (g l^{-1})	Volumetric productivity $(\text{g l}^{-1} \text{ h}^{-1})$
Process	Fermentation (h)	Emptying (h)	Total (h)		
Simple batch	32	4	36	79	2.2
Mellé–Boinot	8	4	12	79	6.6
Series continuous	8	0	8	86	10.8

Additional advantages cited for the continuous process are: (1) stable steady state fermentation characteristics; (2) improved automation and better control based on real time sampling of the product; (3) a slight increase in ethanol yield; (4) reduced labor; (5) reduced distillation and stillage drying energy requirement for the higher concentration beer and (6) reduced cleaning expenses.

Continuous fermentation is also preferred for waste sulfite liquor fermentation (McCarthy, 1954). The series continuous process can be used in the simultaneous microbial enzyme saccharification and yeast fermentation of starch with a 70% reduction in fermentation time (Yarovenko, 1978). Fermentation savings comparable to those achieved with molasses are possible with continuous fermentation of starch which has been completely prehydrolyzed with acid or high activity glucoamylase. The further adoption of continuous fermentation for starches should be expected.

The Biostil process allows substantial savings in stillage processing costs and should become more widely accepted as further industrial experience is accumulated.

43.4 ALCOHOL RECOVERY

43.4.1 Introduction

Alcohol product recovery is energy intensive, typically accounting for more than 50% of the total fermentative ethanol plant energy consumption (Eakin *et al.*, 1981). When heat from burning of raw material residues (such as bagasse) is not available, this constitutes a significant operating cost. Depending on recovery system design, recovery equipment cost generally makes up 6–12% of the plant total capital investment (Katzen, 1979; Lipinski *et al.*, 1976).

Industrial alcohol is produced in various grades. The majority is 190 proof (95 vol% or 92.4 wt%, minimum) alcohol used for solvent, pharmaceutical, cosmetic and chemical applications. Technical grade alcohol (containing up to 5% volatile organic aldehyde, esters and sometimes methanol) is used for industrial solvents and some chemical syntheses. A high-purity 200 proof anhydrous alcohol product (99.85 wt%) is produced for specialized chemical applications. For fuel use in mixtures with gasoline (gasohol), a nearly anhydrous (99.2 wt%) alcohol, but with higher allowable levels of organic impurities, is used (Katzen, 1979).

Today, distillation is used almost exclusively as the means for ethanol recovery and purification and various designs are used to produce the different product grades. Ethanol distillation technology was highly refined during the 1940s to reduce energy consumption to approximately 2.5 kg of steam per liter of anhydrous ethanol produced (Aries, 1947). Recent further refinements in distillation technique make possible marginal improvements with increased capital investment. Alternatives to distillation are under study to further reduce costs (Parkinson, 1981).

43.4.2   The Separation Problem

43.4.2.1   Ethanol/water equilibrium

The ethanol/water vapor/liquid equilibrium is presented graphically in Figure 6 (Otsuki and Williams, 1953). Dilute alcohol is readily concentrated by distillation as the volatility of ethanol in dilute solution is much higher than the volatility of water. Ethanol and water form an azeotrope at 89 mol% ethanol (95.7 wt%) (Perry and Chilton, 1973). The volatilities of ethanol and water are then identical. Continued boiling produces a vapor of the same composition as the liquid and no further enrichment can be achieved by simple distillation. To produce anhydrous ethanol, the distillation equilibrium must be altered (generally by the addition of a third chemical component) or other separation procedures must be employed.

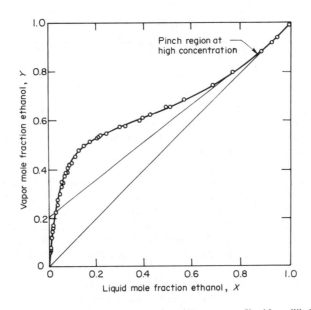

Figure 6   Atmospheric pressure ethanol/water vapor/liquid equilibrium

Distillation to produce 190 proof alcohol can be thought of as composed of two steps: (1) stripping, to recover alcohol from the dilute beer; and (2) rectification, to concentrate the stripped vapor and produce the relatively water-free product. In a single column, both operations are accomplished in series by the same steam energy input and the step with the larger energy requirement sets the overall steam consumption.

In rectification of concentrated feeds, a distillation pinch associated with the azeotrope sets the energy requirement (Maiorella, 1985b). As the azeotropic composition is approached from below, the relative volatility of ethanol over water becomes increasingly smaller, requiring many trays and a high steam input for the separation. For a 14 wt% feed the minimum theoretical recti-fying steam requirement is increased 64% for a 95 wt% product (near the azeotrope) compared with a 93 wt% ethanol product. In a single column and for feeds of greater than about 7 wt% ethanol, the rectification energy requirement as fixed by the pinch establishes the overall distilla-tion energy consumption to produce 95 wt% alcohol (Maiorella, 1985a). No energy reduction (except for small savings in preheating the lower volume of feed) results from further increases in the feed beer concentration (Gundu Rao, 1964; Panicker, 1964).

For a 93 wt% product the pinch becomes limiting for a feed of above 13.5 wt% alcohol (Maior-ella, 1985a; Eakin, 1981). At lower feed concentrations the stripping energy requirement is the controlling factor, and the distillation energy rises as feed concentration decreases. The minimum energy requirement for simple one-column distillation of a 1 wt% ethanol feed is five times the energy for a typical 6 wt% feed (Maiorella, 1985a). Special multicolumn distillation system arrangements are used to overcome these limitations.

43.4.2.2 By-product effects

Refinement of industrial alcohol from fermenter beer is further complicated by the presence of fermentation by-products and nonmetabolized feed components which must be removed.

Yeast cells, nutrient salts, residual sugars, organic acids and glycerol are of low volatility and are recovered in the stripper stillage. These components do have a small effect on the ethanol/water equilibrium (Maiorella, 1984). More important, though, are physical effects, especially promotion of foaming on the stripper trays and fouling and corroding of trays and heat transfer equipment (Katzen, 1979; Ramana Rao, 1964).

Acetaldehyde (and smaller amounts of other aldehydes and volatile esters) are produced at a rate typically of 1 l per 1000 l of ethanol for molasses or grain fermentation (Suomalainen and Ronkainen, 1968) and at higher rates for the fermentation of sulfite waste liquor (Hagglund, 1925). In sulfite liquor fermentation, methanol is also present (McCarthy, 1954). Acetaldehyde and methanol are very volatile and can be distilled from 95 wt% alcohol product. With a boiling point of 21 °C, however, condensation of pure acetaldehyde to provide a reflux is impractical and, instead, a portion of the alcohol product must be recovered as a separate 'heads' technical grade to carry away the aldehyde. A supplemental aldehyde distillation at elevated pressure or with a refrigerated condenser can then be performed.

Fusel oils are a fermentation by-product produced at a rate of 1–5 l per 1000 l of alcohol depending on the substrate (Aries, 1947; Mahlingham, 1977). Fusel oil is composed of a mixture of amyl (pentyl) and propyl alcohol isomers (Aries, 1947; Harrison and Graham, 1970). Because of its unusual equilibrium behavior, fusel oil concentrates on the lower plates of a rectifying col-umn. Dry fusel oils boil in the range of 128–137 °C (Weast, 1971) and hence do not distill over-head with the more volatile alcohol product. In the presence of water, however, the volatility of fusel oils is greatly enhanced with only 2.4% amyl alcohol in water exerting the same partial pressure as 90.7% amyl alcohol (Lecat, 1949). Fusel oils, thus, are forced into the vapor phase in the stripping section and cannot be recovered as a bottoms product. Fusel oil miscibility with eth-anol and water decreases at reduced temperatures and fusel oils are therefore separated by cool-ing and decanting fusel oil laden liquid bled from the rectifier lower plates.

43.4.3 Azeotropic Alcohol Production—Conventional Practice

43.4.3.1 Simple two-column system

Simple one- or two-column systems with only a stripping and rectifying section are used to pro-duce lower quality industrial alcohol and azeotropic alcohol for further dehydration to fuel grade (Aries, 1947; Paturau, 1982; Humphrey and Nolan, 1979). Beer at typically 6–8 wt% alcohol con-centration is fed to the top of the beer still (or stripping section in a single column unit). A reboiler, or direct steam injection at the base of the still, provides the vapor flow to strip alcohol, aldehydes and fusel oil from the descending beer. The alcohol depleted stillage is recovered as the bottom product. Vapor from the stripper flows into the base of the rectifier and is concen-trated by contact with the descending reflux.

Aldehydes, as the most volatile components, concentrate at the head of the column and a technical grade heads alcohol is condensed to provide reflux with enough (2–5% of the total product) withdrawn to purge the aldehydes. The main alcohol product is taken as a liquid side-draw a few plates below the head. A liquid bleed from the lower plates of the rectifying section is cooled to decant fusel oil, which is washed by countercurrent flow with cool water to recover dilute ethanol for recycle to the stripper.

About 2.4–3.0 kg of steam is required per liter of 94 wt% alcohol product with this scheme. Uses of the alcohol are limited as it is contaminated with some aldehydes.

43.4.3.2 Three-column Barbet system

For higher quality alcohol, the three-column Barbet system was commonly employed (Figure 7) (Aries, 1947, Paturau, 1982; Barron, 1944; Webb, 1964). In this system the stripped beer is first purged of most aldehydes in a purifying column which then feeds a relatively dilute alcohol stream to a stripper–rectifier. The alcohol product is taken as a side draw a few plates below the head of the rectifier. A small purge of additional aldehydes from the head of the rectifier is recycled to the purifying column. Fusel oils are tapped from the lower plates of the rectifier, and in some cases an additional fusel oil concentrating column was employed. The energy consumption for the four column system is 4.1 kg of steam per liter of 94 wt% alcohol and the product is still contaminated with traces of aldehydes (Aries, 1947).

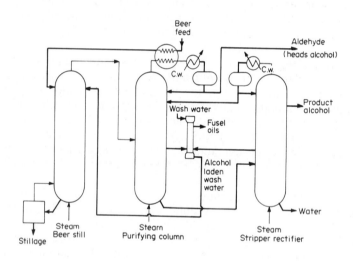

Figure 7   Three-column Barbet distillation system for 190 proof ethanol

43.4.3.3 Three-column Othmer system

In the 1930s an alternative three-column Othmer system incorporating vapor reuse methods to reduce steam consumption was developed (Figure 8) (Aries, 1947; Othmer, 1936). A third aldehyde stripping column is added to the basic two-column design. The product alcohol side-draw from the rectifier is fed to the top plate of the aldehyde column and, in descending, is stripped of volatiles to produce an essentially pure azeotropic alcohol bottoms product. Aldehyde carrying vapors from the head of the aldehyde stripper are returned to the rectifying column and a technical grade heads alcohol product is tapped from the rectifier reflux to purge the aldehydes as in the simple two-column design. The main product is virtually free of aldehydes and is of higher quality than the Barbet system product.

Heat for alcohol vapor boil-up in the aldehyde stripper reboiler is provided by condensation of a portion of the stripper vapor product, which is then returned to the stripper as reflux to reduce the stripper tray requirement. This reuse of heat increases energy efficiency and the steam requirement is not increased over the simple two-column design (2.4–3.0 kg per liter of 94 wt% product) (Aries, 1947).

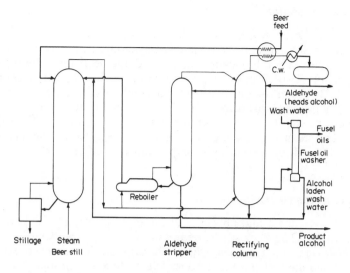

Figure 8   Three-column Othmer distillation system for 190 proof ethanol

43.4.3.4   *Vacuum rectification*

Vacuum rectification can be used to reduce further the distillation energy requirement (Maiorella, 1985a; Othmer, 1936). The ethanol/water equilibrium is pressure sensitive, and the relative volatility of ethanol over water at high ethanol concentration is increased at low pressure (Beebe *et al.*, 1942). As the pressure is reduced from one atmosphere the azeotropic composition increases, the azeotrope disappearing at below 86 mmHg (11.5 kPa) (Pemberton and Mash, 1978). Anhydrous alcohol can thus be produced by vacuum distillation, but the energy requirement for such a 'one-step' distillation is high and the process is uneconomical (Eakin *et al.*, 1981; Black, 1980). Vacuum rectification to produce azeotropic alcohol is, however, very efficient with the required reflux ratio as set by the rectification head pinch reduced by 45% for distillation at 100 mmHg (13.3 kPa) compared with one atmosphere. A concentrated feed to the rectifier is necessary, though, to achieve these savings.

An efficient arrangement, then, combines an atmospheric pressure beer still with a vacuum stripper rectifier. The beer still produces a partially concentrated vapor. This hot beer vapor is then condensed in the vacuum column and aldehyde side-stripper reboilers to reclaim heat for boil-up in these columns. The concentrated condensed liquid is then fed to the stripper/rectifier for efficient vacuum distillation to the 94 wt% product. The complete vapor heat reuse is also possible because of the vacuum operation which reduces the boiling temperatures in the reboilers and facilitates heat transfer.

The total steam consumption for this process is 1.8–2.2 kg of steam per liter 94 wt% product (Aries, 1947; Maiorella, 1985a). A further advantage is that a large portion of the water is now recovered as a relatively clean product from the base of the vacuum stripper rectifier, producing a more concentrated stillage from the beer still for stillage by-product recovery. A drawback of this process is the requirement of lower temperature cooling water as the vacuum rectifier condenser temperature is reduced to 45 °C at 170 mmHg (22.6 kPa) (Maiorella, 1985a).

43.4.3.5   *Vapor recompression*

In some new commercial installations, vapor recompression heating is used as an alternative method for vapor heat reuse (Process Engineering Co., 1980). The overhead vapors are compressed to increase the saturation temperature. The heated vapors can then be condensed in the reboiler of the same column to provide heat for boil-up with a substantial energy savings (Eakin *et al.*, 1981; Parkinson, 1981). The cooling water requirement for product condensation is also reduced. Capital investment for compressors is, however, high.

43.4.3.6 Multieffect distillation

Special measures have been developed for the processing of very dilute feeds such as the beer from sulfite waste liquor fermentation. One industrial method employed a multieffect distillation (Ericson, 1947; McCarthy, 1954; Panicker, 1964). For dilute beers the stripping heat requirement is the controlling factor. Two strippers are operated in parallel, each processing approximately half the dilute feed. By operating one of the strippers at elevated pressure, the head vapors are sufficiently hot to generate low-pressure steam for operation of the other stripper and the rectifier. By this method the steam consumption for concentration of a 1 wt% feed is reduced from 13.8 to 8.3 kg l^{-1}. The steam consumption can be further reduced by the inclusion of additional stripping effects, trading off increased capital expenditure for steam cost savings.

43.4.3.7 Six-column reagent alcohol system

For ultra high purity reagent and pharmaceutical applications, a six-column system has been developed (Aries, 1947). The beer still is operated under vacuum to reduce temperatures and minimize the breakdown of stillage yeast, which otherwise produces trace organics contaminating the alcohol product and imparting a slight odor. This system also produces an almost pure aldehyde product, eliminating the technical head product and increasing recovery of alcohol in the main product to over 99%. The capital investment and energy consumption are very high, however.

43.4.4 Anhydrous Ethanol Production

43.4.4.1 Azeotropic distillation

Azeotropic distillation is today the prevalent method for ethanol dehydration to produce anhydrous product (David *et al.*, 1978; Paturau, 1982). The benzene azeotropic distillation is exemplary (Figure 9) (Young and Fortey, 1902; Guinot and Clark, 1938; Robinson and Gilliland, 1950). Water is only slightly soluble in nonpolar benzene, while ethanol is freely soluble. The vapor pressure of water over a benzene/ethanol/water liquid mixture is thus greatly enhanced relative to the ethanol (Barbaudy, 1927). In the distillation the ethanol/water mixture is fed at a midpoint in the column while an organic-rich phase (75 mol% benzene decanted from the cooled distillate) is used as reflux at the column head. Benzene concentrates in the liquid phase on the plates above the feed, enhancing the volatility of water and forcing it into the rising vapor phase. Ethanol concentrates with the benzene and is carried downward. The water-concentrated benzene/ethanol/water ternary azeotrope (24% ethanol, 54% benzene, 22% water) is taken as the column head product. Below the feed tray, benzene is stripped from the ethanol and anhydrous ethanol is recovered from the azeotropic column reboiler. The distillate of the azeotropic column, when cooled and decanted, separates into two phases: the organic-rich phase used for reflux and a water-rich phase (35% ethanol, 4% benzene, 61% water) which is treated in the benzene recovery column. This column produces the benzene concentrated azeotrope as head product for recycle to the decanter and dilute alcohol as the bottom product. The dilute alcohol can be reconcentrated in an additional column or recycled to the primary alcohol recovery column. The distillation columns are costly as the azeotropic column alone requires 50 plates and is of similar diameter to the primary stripper rectifier (Aries, 1947; Katzen, 1979).

In the US, anhydrous alcohol is generally refined from 94 wt% azeotropic alcohol. The additional steam requirement for anhydrous alcohol production with benzene as the entrainer is then 1.0 kg of steam per liter of product (Aries, 1947).

In Europe, a modification (the Mellé 5th method) has been widely used (Figure 10) (Paturau, 1982; Jumentier, 1931). In this process a concentrating column, with the decanted benzene-rich phase as reflux, serves as a combined primary rectifier and benzene recovery column so that a dilute alcohol beer can be processed directly. The beer is fed to a stripper for removal of nonvolatile stillage. The stripper is also equipped with a small diameter rectifier for aldehyde heads impurities removal. Dilute vapor (20% ethanol) from the top stripper plate is passed to the base of the concentrating column with a portion condensed in the azeotropic dehydration column reboiler for heat recovery. The concentrating column is a rectifier, and vapors become concentrated in alcohol as they rise. Fusel oils are tapped from a lower rectifying plate. The decanted water-rich phase is fed higher up in the rectifier so that benzene is also stripped in the lower sec-

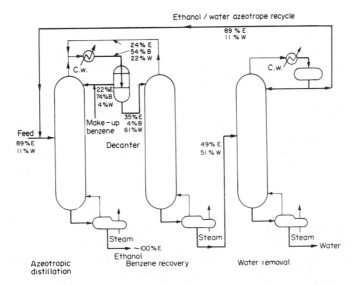

Figure 9 Benzene azeotropic distillation for anhydrous ethanol

tion to produce a dilute, benzene free, liquid bottom product for recycle to the stripping column. Above the water-rich benzene phase addition plate the liquid phase becomes more and more concentrated in benzene and alcohol. A portion of this alcohol-rich liquid is tapped from the concentrator column and fed to the azeotropic dehydrating column, which produces anhydrous alcohol as the bottom product. Both the concentrating and dehydrating columns produce the ternary azeotrope as head products and these are condensed, separated into the organic and water rich phases, and recycled.

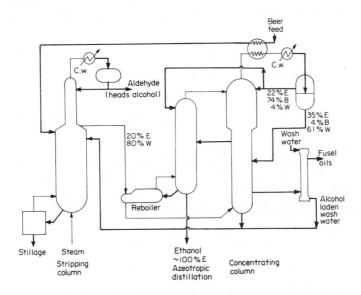

Figure 10 Mellé distillation system for anhydrous ethanol

The steam requirement for this process (3.0 kg l⁻¹ starting with a 6.5 wt% feed) (Paturau, 1982) is slightly lower than for separate production of 94 wt% ethanol and subsequent dehydration to anhydrous (3.4 kg l⁻¹). With efficient dual pressure vapor reuse methods, however, the steam can be largely reutilized if 94% and anhydrous alcohol production are carried out in a continuous process and the steam requirements of the two methods are then comparable (Katzen, 1979).

The ACR process (Hartline, 1979; Black, 1980) is a simple variant of the benzene azeotropic distillation. Gasoline is used in place of benzene as the extractant. The stripping section of the azeotropic column can be reduced as it is no longer necessary to remove all of the extractant (gasoline) from the bottom ethanol product.

Benzene is the most common entrainer if the end product is for fuel use (Paturau, 1982). Cyclohexane is most common when the alcohol is to be used in perfumes or as chemical reagent (Paturau, 1982; Aries, 1947). Trichloroethylene is used in the German Drawinol process (Fritzweiler and Dietrich, 1932; Colburn and Phillips, 1944) and *n*-pentane (at 440 kPa, 4.4 atm) in the Shell Oil process (Black *et al.*, 1972).

The *n*-pentane process requires smaller columns than the others and has a slightly reduced steam requirement, but it has been applied only at pilot plant scale and problems, including excessive foaming, were reported.

Diethyl ether is used as entrainer in the Othmer and Wentworth process (Wentworth and Othmer, 1940; Othmer and Wentworth, 1940; Wentworth *et al.*, 1943). Diethyl ether offers an advantage as entrainer in that a ternary ethanol/water/ether azeotrope is not formed and the azeotropic column head product is the alcohol-free ether/water binary azeotrope. All the alcohol fed to the azeotropic column is recovered as anhydrous product from the reboiler, thus eliminating the energy inefficient recycling of alcohol/water mixtures.

Water is only slightly soluble in ether so a very high ether boil-up rate is required to carry the water overhead. The water solubility is increased by operating the azeotropic column under pressure (1000 kPa, 10 atm), but the head product is still only 4.4 wt% water.

When the binary azeotrope is cooled (to 20 °C) the organic phase for recycle still contains 1.2 wt% water and, thus, 28 kg of ether must be distilled to remove 1 kg of water.

The heat of vaporization of ether is very low (380 kJ kg^{-1}) and under these design conditions the minimum theoretical steam requirement for the ether removal is only 0.24 kg steam l^{-1} alcohol. The actual steam consumption in a European industrial distillery is 0.6 kg steam l^{-1} (Chen, 1980).

An integrated vapor reuse process employing the ether azeotropic distillation was operated at the USDA Peoria alcohol pilot plant (Jacobs, 1950). In this design the ether column is operated at very high reflux with a steam consumption as high as 2.4 kg steam l^{-1}. However, the high condenser temperature allows the condenser to serve simultaneously as the reboiler for the primary stripper. With this efficient heat reuse, the tray requirement in the azeotropic column is minimized and the total steam requirement for the process is only 2.5 kg l^{-1} anhydrous alcohol produced from 6.5 wt% beer.

43.4.5 Alternative Ethanol Recovery Methods

Extractive distillation using ethylene glycol, glycerine or a molten eutectic mixture of potassium and sodium acetate to depress the volatility of water and allow the distillation of anhydrous ethanol was widely practiced during World War II (Paturau, 1982; Jacobs, 1950).

New nondistillative processes including dehydration by vapor phase water adsorption on to solids, molecular sieve drying, solvent extraction, extraction with supercritical fluids and membrane separations are all currently under study. Special processes for the production of gasohol using gasoline as an ethanol extractant are also under development.

These processes are critically appraised in recent reviews (Parkinson, 1981; Eakin *et al.*, 1981; Hartline, 1979; Maiorella, 1983b).

43.4.6 Ethanol Recovery Process Comparison

Table 8 compares steam consumption and equipment requirements for several of the conventional distillation processes for ethanol recovery and purification (Aries, 1947; Paturau, 1982; Black *et al.*, 1972). These processes are quite efficient, utilizing typically less than one-third the fuel value of the ethanol product in producing anhydrous ethanol from relatively dilute (6 wt%) fermentation beers. These processes also provide for the separate recovery of aldehyde, fusel oil and stillage by-products. Table 9 presents recent (1980) estimates for the operating cost of various conventional distillation processes to produce anhydrous alcohol (Black, 1980). The pentane azeotropic distillation was found to be the least costly at 3.8¢ per liter but costs for all of the processes studied are similar. The designs evaluated did not incorporate vapor heat reuse methods. These would further reduce operating costs and bring the costs of the various processes even closer together.

Table 8a Conventional Ethanol Recovery Process Comparison:
Azeotropic Ethanol Production

Process	Steam consumption (kg l⁻¹ product)	Equipment
Simple one or two column distillation to 94 wt% crude ethanol from 6 wt% beer	2.4–3.0	20 tray beer still 30 tray beer rectifier
Four column Barbet system for high quality 94 wt% neutral spirits from 6 wt% beer	4.0–4.2	20 tray beer still 30 tray purifying column 54 tray stripper rectifier small fusel oil column
Three column Othmer system for high quality 94 wt% neutral spirits from 6 wt% beer	2.4–3.0	20 tray beer still 45 tray aldehyde column 30 tray rectifier
Three column vacuum distillation for high quality 94 wt% neutral spirits from 6 wt% beer	1.8–2.2	20 tray beer still 45 tray aldehyde column 54 tray stripper/rectifier
Dual stripper, multieffect distillation for high quality 94 wt% neutral spirits from 1 wt% beer	8.3 (*vs.* 13.8 for a single stripper effect and 5.5 for three stripper effects)	Two 35 tray beer stills 45 tray aldehyde column 30 tray rectifier

Table 8b Conventional Ethanol Recovery Process Comparison:
Anhydrous Ethanol Production

Process	Steam consumption (kg l⁻¹ product)	Equipment
Benzene dehydration from 94–99.9 wt% ethanol	1.0	50 tray dehydrating column 30 tray benzene recovery column Supplementary rectifier
Pentane dehydration from 94–99.9 wt% ethanol	0.5	23 tray dehydrating column 18 tray pentane recovery column Supplementary rectifier
Ether dehydration from 94–99.9 wt% ethanol	0.6	60 tray dehydrating column 20 tray ether recovery column
Ethylene glycol extractive distillation from 94–99.9 wt%	1.1	50 tray dehydrating column 10 tray glycol recovery column
Mellé benzene process for 99.9 wt% ethanol from 6 wt% beer	3.1	Ethanol stripper/rectifier Concentrating column Dehydrating column
Combined Othmer three column and high-pressure ether distillation with vapor reuse for 99.9 wt% ethanol from 6 wt% beer	2.5	20 tray beer still 45 tray aldehyde column 54 tray stripper/rectifier 30 tray dehydrator 20 tray ether recovery column
Mariller glycerol extractive distillation process for 99.9 wt% ethanol from 6 wt% beer	3.4	Ethanol stripper/rectifier Extractive distillation column Glycerol recovery

Table 9 Operating Costs for Various Distillation
Methods Producing Anhydrous Alcohol

Process	Operating cost (¢ l⁻¹)
Simple distillation followed by:	
Benzene azeotropic distillation	4.24
Pentane azeotropic distillation	3.78
Diethyl ether azeotropic distillation	4.72
Ethylene glycol extractive distillation	7.08

The sales price of anhydrous ethanol is 4¢ per liter higher than for 190 proof alcohol (Staff, Chem. Mark. Rep., 1982d). Concentration of ethanol from 190 proof to anhydrous results, however, in a 6% volume reduction so that the apparent large premium for anhydrous alcohol is largely a result of the increase in delivered alcohol, not the distillation cost.

43.5 FERMENTATION WASTE TREATMENT AND BY-PRODUCT RECOVERY

43.5.1 Introduction

Waste treatment and by-product recovery are extremely important factors in the economics of ethanol production (Noguchi, 1981). Secondary products recoverable from the feed during sugar preparation and prior to fermentation were considered under the feedstock alternatives sections (43.2.1–43.2.4). Here, postfermentation wastes and by-products are reviewed. The ethanol-depleted spent beer stillage is the most important waste stream. Without backset, typically 10–15 l of stillage are produced for every liter of alcohol product. Corn stillage has a waste biological oxygen demand of 15 000–25 000 p.p.m. (Rudolfs, 1953) and a single 100 Ml y^{-1} plant thus generates a pollution load equivalent to a city of 1.4 million inhabitants. Balancing this, stillage can be treated to become a valuable by-product. For corn, the treated stillage can be sold for a credit equal to 52% of the cost of the original corn (see Table 4).

Fusel oil, aldehydes and carbon dioxide can also be recovered as valuable by-products.

43.5.2 Stillage Treatment and By-products

43.5.2.1 Stillage comparison

This section follows closely the recent review by Maiorella *et al.* (1983c).

Table 10 compares typical properties of stillage from fermentation of blackstrap molasses (Chen, 1980), corn (Stark, 1954; Solomons, 1969) and sulfite waste liquor (McCarthy, 1954; Wiley, 1954). The biological oxygen demand (pollution load) varies from 60 000 for blackstrap molasses to 15 000 for corn stillage.

Table 10 Properties of Ethanol Fermentation Stillages

	Blackstrap molasses	Sulfite waste liquor	Corn
Biological oxygen demand (p.p.m.)	50–60 000	40–50 000	15–25 000
Solids (wt%)	8.5	10	7.5
Ash (wt%)	2.5	2.1	1.5
Sugar (wt%)	0.9	0.8	0.5
Protein (wt%)	0.8	0.1	2.3
Vitamins	High	Low	High

Nutrient (protein and vitamin) content is highest for corn stillage containing corn protein residue from the germ and bran which are largely unaffected by the fermentation. Blackstrap molasses stillage is second highest in nutrient value with less protein than corn stillage, but high vitamin content. Mineral (ash) content is also high. Cane and beet juice stillages are similar in nutrient make-up to blackstrap stillage, but with lower nutrient concentrations. Potato stillage and cornstarch stillage (from distilleries with complete corn germ and bran removal prior to fermentation) are low in nutrient value. Sulfite liquor and wood hydrolyzate stillage are low in protein and vitamin nutrients but high in BOD, primarily as pentose sugars.

43.5.2.2 Stillage drying for cattle feed production

Figure 11 shows the process for cattle feed production from corn stillage (Stark, 1954; Katzen, 1979). Bulk solids are removed by centrifugation or by screening followed by dewatering in a rotary press. These solids then are dried to less than 5% water in rotary driers to produce the distiller's light grains product. The pressed liquid thin stillage is concentrated to 35% solids in forced convection vertical tube evaporators and then dried to a powder using drum or spray driers (Sastry and Hohanrao, 1964), yielding distiller's solubles. Alternatively, the concentrated liquid solubles can be blended back with the pressed solids and these rotary dried together to produce distiller's dark grains. About 700–900 g of stillage feed can be recovered per liter of alcohol produced.

The feed value of corn stillage is similar to that of soybean (Table 11) (Aries, 1947; Sastry and Hohanrao, 1964) and corn stillage is valued at approximately 95% of soybean cost for use in

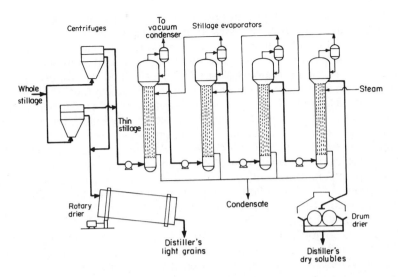

Figure 11 Distiller's grains recovery process

cattle feed (Staff, Feedstuffs, 1982). 454 700 metric tons of distiller's dried corn stillage (primarily from beverage distilleries) was used in the US in 1981 (Potter, 1982) and this consumption is readily expansible with corn stillage displacing more costly feeds.

Table 11 Nutrient Value of Grain Distiller's By-product

	Distiller's solubles	Dark grains
Moisture (%)	5	5
Protein (%)	27	29
Fat (%)	7	9
Fiber (%)	2	7
Ash (%)	8	4
Carbohydrate (%)	51	46
Vitamins (p.p.m.)		
Thiamine (B_1)	8	4
Riboflavin (B_2)	22	8
Pantothenic acid	29	11
Niacin	125	65
Pyridoxine (B_6)	9	—
Biotin	0.5	0.2
Choline	6500	4500
Carotene	0.8	1
p-Aminobenzoic acid	10	—
Folic acid	4	—
Zeaxanthin	8	8
Cryptoxanthin	4	5
Price ($ t^{-1}) (1982)	185	191

The corn stillage cattle feed process recovers all the stillage solids, leaving no waste stream for further processing. The evaporation energy load to achieve this is, however, very high and a conventional plant may use as much steam for stillage drying as for alcohol distillation (Scheller and Mohr, 1977a). To reduce steam consumption, vapor recompression heating of the evaporator can be used (Renshaw *et al.*, 1982). Steam consumption for the final grains drying has also been reduced by using carefully filtered hot boiler stack gases for rotary drier heating (Katzen, 1979). The total steam consumption for cattle feed drying has thus been reduced to only 11% of the distillation steam consumption. An additional 111 kWh of electric power per metric ton of dried grains is consumed, primarily to drive the centrifuges and evaporator thick syrup recirculation pumps. This plant recycles 10% of the thin stillage as backset.

The importance of backset (as discussed in Section 43.3.2.8) to steam economy is now apparent. The larger the fraction of stillage which is recycled, the less new water is added with the feed, and the higher the stillage solids content. The evaporative load for stillage drying is proportionately decreased (Ronkainen, 1978).

Molasses stillage (vinasse) is lower in suspended solids content than corn stillage, and the yeast may be recovered as a valuable by-product by simple centrifugation. The stillage is high in vitamin content. For use as cattle feed, molasses stillage is evaporated to typically 50–65 wt% solids (sufficiently concentrated to prevent spoilage) and marketed as condensed molasses solubles or 'vacatone' (Bu'Lock, 1979; Reich, 1945). Feeding tests show that the concentrate has roughly 65% of the nutritive feed value of molasses (Lewicki, 1978) and it is widely used as a molasses feed substitute in Europe. Based on protein and carbohydrate content, the value of condensed molasses solubles at 65% solids has been estimated at as much as $99 per metric ton (Chen, 1980). Unfortunately the high mineral (especially potassium) content of molasses stillage imparts a laxative effect which limits molasses stillage use in ruminants to less than 10% of the diet and in pigs to less than 2% (Lewicki, 1978; Rudolfs, 1953). Consequently, the actual sale price of condensed molasses solubles (in Europe) is only $25–48 per metric ton (Lewicki, 1978; Chen, 1980). In some European plants, molasses stillage is refined *via* an ammonia precipitation and neutralization process to produce a depotassified, high protein content (44%) cattle feed which can compete with soybean meal. The economics of this upgrading are favorable though only when soybean is in short supply (Lewicki, 1978).

Condensed molasses solubles are viscous, hygroscopic and sticky, making feed compounding difficult. Addition of coagulants such as ammonium phosphate or phosphoric acid improves handling properties (Gillies, 1978). The high viscosity of untreated stillage also results in poor heat transfer rates. Heat transfer coefficients for natural circulation and forced convection evaporators, respectively, are 9400 and 13 600 kJ m^{-2} h^{-1} K^{-1} at 20 wt% solids and 2100 and 5900 at 60 wt% solids. Large costly evaporators are therefore required (Chakrabarty, 1964). These problems are overcome in the Carver–Greenshield process whereby the stillage is mixed with a vegetable oil prior to evaporation. The oil is an effective heat transfer medium and solids carrier, allowing almost complete water evaporation. A belt filter press then expresses the oil from the dried evaporator slurry for reuse, producing a blendable powdered stillage. This process should also be applicable to corn stillage drying and might reduce processing costs.

Cane and beet juice stillages may be treated similarly to molasses stillage. Alternatively these stillages may be mixed with bagasse fiber or beet pulp residue and dried to produce a nutrient enriched roughage feed.

The soluble nutrient content of corn starch, potato and sulfite waste liquor stillage is not sufficient to justify the high cost of evaporation (David *et al.*, 1978). For these stillages the yeast may be recovered by centrifugation and then spray or drum dried. Secondary treatment is required to reduce the stillage pollution load.

Distiller's yeast is very valuable as a cattle feed, selling for 45¢ kg^{-1} (Staff, Feedstuffs, 1982). The nutritive value of yeast from molasses distillery sludge is summarized by Rastogi and Krishna Murti (1964).

43.5.2.3 Secondary (aerobic) yeast production

Secondary aerobic yeast production is an effective means to reduce BOD and produce a valuable by-product. Sulfite liquor stillage and wood hydrolysis stillage are high in nonfermentable sugars from hemicellulose breakdown. Molasses stillage also has sufficient sugars, largely derived from hydrolysis of nonfermentables during distillation (Hodge and Hildebrandt, 1954), to make secondary fermentation economically attractive.

Torula yeast (*Candida utilis*) is grown with fine bubble aeration (10 m^3 air kg^{-1} yeast produced) in large agitated vessels (Schmidt, 1947). Nutrient supplementation is sometimes required as 5.6 kg ammonia, 1.8 kg phosphorus and 0.4 kg magnesium are required per 100 kg of dry yeast (Wiley, 1954). The optimum growth temperature is normally approximately 35 °C and cooling must be provided as the aerobic growth is exothermic (17 000 kJ kg^{-1} yeast product) (Schmidt, 1947). Approximately 51 kg of yeast are produced per 100 kg of aerobically consumed sugar. Other simple carbon compounds such as residual ethanol, glycerol and acetic acid can also be consumed. Secondary fermentation of waste sulfite liquor thus yields a new crop of typically 5–6 kg of additional yeast per metric ton of stillage (420 g l^{-1} alcohol), while reducing the BOD by approximately 10 000 p.p.m.

Molasses stillage secondary fermentation yields 16 kg of additional yeast per metric ton of stillage (52 g l^{-1} alcohol) (Chang and Yang, 1973). In addition, the yeast consumes otherwise inhibitory components and the fraction of blackstrap molasses stillage backset can be increased, thus reducing stillage evaporation costs. Residual BOD is reduced to 10 000–15 000 p.p.m.

Stillage has also been used as fermentation medium for other important products (Sastry and Hohanrao, 1964). Grain stillage will support mold growth for amylase enzyme production. Supplemented grain and molasses stillage has been used for growth of *Penicillium notatum*. *Ashbya gossypii* growth to increase B vitamin content of stillage for cattle feed has also been practiced.

43.5.2.4 Anaerobic digestion

For stillage with low feed value but high BOD (*i.e.* corn starch, potato, sulfite waste liquor), anaerobic digestion can be cost effective (Skogman, 1979). Low-grade distillery waste heat can be used to maintain high temperatures for rapid thermophilic (50 °C) digestion. Distillery waste is fed to large slowly agitated vessels and digested by a mixed culture sludge blanket (the initial inoculum usually taken from a municipal sewage treatment facility (Sundaram and Pachaiyappan, 1975). Acid forming bacteria convert sugar to acetic and butyric acids. Amino acids are broken down, releasing ammonium and sulfide ions. Methanogenic bacteria then convert the organic acids to methane and carbon dioxide; 580–720 l of gas, 65% methane, are produced per kg of BOD consumed (Jackman, 1977). In an industrial, heated, agitated two-stage system with sludge recycle, yeast waste with a BOD of up to 15 000 p.p.m. is treated in a two-day residence time with 90% BOD reduction (Rudolfs and Trubnick, 1950). High BOD molasses distillery wastes are treated in 7–10 days (Sastry and Hohanrao, 1964). New reactor designs, such as the upflow reactor, are being evaluated at pilot plant scale and may greatly reduce the treatment time (Pipyn and Verstraete, 1979).

In some plants, ammonia recovery is also practiced (Skogman, 1979). The effluent liquid pH is increased to release ammonium ions as free ammonia gas. The liquid is stripped with air which is then counter current contacted with acid solution in an absorber to produce concentrated ammonium salts for recycle as a fermentation nutrient or for sale as a fertilizer; 90% nitrogen reduction in the stillage is typical.

For final BOD reduction an aerobic polishing stage is required (Jackson and Lines, 1972). An aerobic activated biosludge reactor or a trickling bed reactor can be used. Loadings of up to 4.8 kg BOD m^{-3} of stones in a trickling bed bioreactor can be handled (Rudolfs and Trubnick, 1950). An overall BOD reduction for the complete system of 98–99% can be maintained (Skogman, 1979). In the commercial ANOX process (Savage, 1979) this effluent is further treated with ozone over a fixed bed metal catalyst for essentially complete waste oxidation. The resulting water is of drinking quality.

Typical biogas production rates for molasses, sulfite waste liquor and corn starch stillages, respectively, are 290, 230 and 80 l gas l^{-1} alcohol produced. This gas (65% methane), with a net heating value of approximately 25 000 kJ m^{-3}, can be burned for steam or electric generation. For biogas from digestion of molasses stillage wastes with an initial BOD of 50 000 p.p.m., 2.9 kg of steam can be generated per liter of alcohol.

43.5.2.5 Stillage disposal as agricultural irrigant and fertilizer

Molasses stillage is high in minerals with as much as 1.1% potassium and 3.1% total ash (Chen, 1980; Jackman, 1977). Repeated growing cycles deplete the soil of necessary minerals. The production of 100 tons of cane, for instance, is accompanied by a loss of 250 kg of K$_2$O (Humbert, 1971). Direct disposal of untreated molasses stillage back to the crop lands is widely practiced.

After removal of valuable yeast, the stillage is trucked to cane plantations. In Brazil, 650–1000 m^3 of raw stillage is spread per hectare, preferably 1–4 weeks prior to cane planting (Monteiro, 1975). In France, much lower levels (2.5–30 m^3 ha^{-1}) are sprayed on the fields after harvesting (Lewicki, 1978). Advantages cited (Chen, 1980) are: (1) disposal of the effluent; (2) replenishment of depleted minerals; (3) initial increase in soil pH; (4) improved soil physical properties; (5) increased water and salts retention capacity; (6) increased soil microflora population. These positive factors have combined in Brazil to increase crop yield (Monteiro, 1975).

Strong odor and insect problems are associated with spreading of untreated stillage. Also, European experience has shown that while an advantage may be gained initially, stillage use must be severely reduced after several years due to problems of soil acidity increase, salt leaching and putrescity (Jackman, 1977). The high concentration of calcium and magnesium in stillage initially acts to buffer the soil and beneficially increase pH (Monteiro, 1975). Repeated applications, however, lead to a build-up of sulfates. These are reduced in the soil to hydrogen sulfide (giving a pungent odor), which is then reoxidized by sulfur bacteria to sulfuric acid (Sastry and Hohanrao, 1964). The problem of sulfite build-up rules out untreated sulfite waste liquor or wood acid hydrolysis stillages from direct applications to crop lands.

Stillage is initially also quite corrosive, requiring specialized storage, trucking and spreading equipment. The cost to spread stillage in Brazil was reported as 18 cruzeros (US$1) per 1000 liters of stillage (Jackman, 1977). This corresponds to a cost of 1.15¢ l^{-1} alcohol product.

43.5.2.6 Stillage incineration

Stillage incineration can be attractive as a means to recover the mineral content of stillage with total consumption of the organic content (Chakrabarty, 1964). For stillages of high organic content (such as molasses, waste sulfite liquor and wood acid hydrolysis wastes) a positive energy return is possible. This process has been used commercially at European waste sulfite liquor (Chen, 1980) and American and Brazilian molasses distilleries (Sastry and Hohanrao, 1964; Monteiro, 1975).

Stillage is concentrated typically to 60 wt% solids before incineration. The heat of combustion of molasses or cane juice stillage solids is 12 500–15 000 kJ kg^{-1} and combustion generates sufficient heat for the evaporation step when a four- or five-effect evaporator is used (Jackman, 1977). The heat generated by sulfite waste stillage combustion is even greater.

Ash recovery is important to the economics of incineration processes. Molasses or cane juice ash is typically 30–40% potash (K_2O) and 2–3% P_2O_5 (Chakrabarty, 1964). After dissolving in water and neutralization with sulfuric acid, a high-value potassium fertilizer is produced with 25–35 kg recovered per 1000 m^3 of stillage incinerated (Chakrabarty, 1964). The potassium sulfate salt is contaminated with typically 16% potassium chloride and 7% potassium carbonate, reducing the sale value from $212 per metric ton for refined potassium sulfate (Huffman, 1978) to $120 per metric ton for the mixed salts (Paturau, 1982). Additional ammonium sulfate can be recovered by absorption of sulfur dioxide from the boiler gas (Sastry and Hohanrao, 1964). Sodium sulfite ash is recovered from waste sulfite stillage incineration and recycled for use in the paper manufacturing process (Chen, 1980).

Special boiler designs are required to handle the high ash content of distillery wastes. This is especially problematic for molasses stillage incineration as the potassium ash fusion temperature is only approximately 700 °C. Conventional stoked boilers must be operated at reduced hearth temperatures to prevent molten ash incrustations. A fluid bed boiler has been tested (Kujala *et al.*, 1976) and a high temperature (1100 °C) atomized mist boiler is operating commercially with excellent success producing high-pressure steam (Chen, 1980).

43.5.2.7 Industrial stillage handling process comparison

Table 12 summarizes capital and operating costs for the four major commercial stillage handling processes, treating 1000 tons per day of molasses stillage (equivalent to approximately 30 Ml y^{-1} alcohol). Data are compiled from several sources (Paturau, 1982; Chen, 1980; Cysewski and Wilke, 1976a; Skogman, 1979; Noguchi, 1981) scaled to the uniform 1000 metric ton per day capacity and updated to 1980 dollars. Economic assumptions used in the various sources are not necessarily identical and this tabulation should be considered only for a rough initial comparison.

Because of the high sales value of yeast, the secondary aerobic yeast growth process, with a net credit of $1 045 000 y^{-1} (3.4¢ l^{-1} alcohol), appears most attractive for molasses stillage treatment. However, this process results in only a 76% reduction in the stillage BOD and secondary treatment is required. The residue can be dried and incinerated for potash recovery, but additional steam is required for evaporation of the lower organic content residue. This secondary treatment will cost approximately $743 000 y^{-1} (assuming the full potash credit), thus reducing the net profit to $302 000 y^{-1} (1.01¢ l^{-1} alcohol).

For grain stillage handling, the high stillage feed value ($191 t^{-1} dark grains *vs.* only $35 t^{-1} for molasses stillage) makes evaporation for cattle feed production most attractive, with a net credit of $3 100 000 y^{-1} (10.3¢ l^{-1} alcohol).

Anaerobic digestion is most favorable for corn-starch and potato stillage as these have negligible cattle feed or fertilizer value and insufficient combustion heat to drive their own evaporation.

With high residual pentose sugar concentration and high combustion energy from stillage

Table 12 Stillage Handling Process Comparison

	Evaporation to cattle feed	Incineration to recover potash	Anaerobic digestion and methane production	Aerobic torula yeast growth
Capital Investment ($)	3 000 000	6 400 000	3 500 000	2 400 000
Utilities:				
Steam (kg m^{-3} stillage)	180	—	—	30
Electricity (kWh m^{-3})	5	6	3	12
Chemicals	—	H_2SO_4	—	$(NH_4)_2SO_4$, $MgSO_4$, H_3PO_4
Products:				
60% Solids concentrate (kg m^{-3})	120	—	—	—
Torula yeast (kg m^{-3})	—	—	—	16
Potash (kg m^{-3})	—	30	—	—
Biogas (m^3 kg^{-1} BOD)	—	—	0.65	—
Sludge (kg m^{-3})	—	—	2.5	—
Excess steam (kg m^{-3})	—	80	280 (from biogas combustion)	—
Annual Cost ($)				
Steam ($7 t^{-1})	378 000	—	—	63 000
Electricity (7¢ kWh^{-1})	105 000	126 000	63 000	252 000
Fixed costs (10% of C.I.)	300 000	640 000	350 000	240 000
Maintenance (6% of C.I.)	180 000	384 000	210 000	144 000
Operation (6% of C.I.)	180 000	384 000	210 000	144 000
Chemicals	—	42 000	—	272 000
Gross annual costs ($)	1 143 000	1 576 000	833 000	1 115 000
By-product credits ($)				
Concentrate ($35 t^{-1})	1 260 000	—	—	—
Yeast (45¢ kg^{-1})	—	—	—	2 160 000
Potash ($120 t^{-1})	—	1 080 000	—	—
Steam	—	168 000	588 000	—
Net operating revenue ($ y^{-1})	117 000	(328 000)	(245 000)	1 045 000
Residual wastes	none	none	750 t y^{-1} sludge 3000 p.p.m. residual BOD in stillage	12 000 p.p.m. residual BOD in stillage

solids, sulfite waste liquor and wood acid hydrolysis stillages are most economically treated by aerobic yeast growth or stillage incineration.

43.5.3 Distillation By-products

Recovery of acetaldehyde and fusel oils in the distillation of ethanol was described in Section 43.4.2.2.

Typically only 1 l of acetaldehyde (Suomalainen and Ronkainen, 1968) and 5 l of fusel oil (Aries, 1947) are produced per 1000 l of alcohol. Glycerol is produced at a rate of 41 l per 1000 l of alcohol (Neish and Blackwood, 1951) but its recovery from the stillage is generally not economical.

43.5.4 Carbon Dioxide

Carbon dioxide is produced at a rate of 770 g l^{-1} ethanol during fermentation. The carbon dioxide vapor is saturated with water and carries traces of ethanol, acetaldehyde, organic acids, fusel oils and compressor lubricating oils (CO_2 from sulfite liquor fermentation also contains methanol and hydrogen sulfide). After countercurrent absorption with water for alcohol recovery, further impurities are removed by adsorption over activated carbon (Backus process) or by oxidation with potassium dichromate solution followed by sulfuric acid scrubbing for dehydration and dichromate removal (Reich process) (Ballou, 1978). 80% of the total fermentation CO_2 produced can be recovered at very high purity, and purification costs are lower than for CO_2 recovered from flue gases.

The fixed capital investment for a 24 t d^{-1} liquid CO_2 plant built annexed to an alcohol refinery in the Philippines in 1978 was $6 550 000 (Isidro, 1980). Carbon dioxide price is $71 per metric ton (Staff, Chem. Mark. Rep., 1982d).

43.5.5 Electricity

Cogeneration can greatly improve the usable energy return in alcohol production (Ziminski, 1981). Alcohol production typically requires 3.0–5.0 kg steam per liter of alcohol depending on the process.

In a cogeneration scheme, fuel is burnt to produce high-pressure superheated steam which drives turbines for electric power generation. The low-pressure turbine exhaust steam is then reused to provide the process heat for distillery operation. Using a 40 atmosphere boiler, 0.17 kWh of electricity can be generated for every kg of process steam. A 100 Ml y^{-1} alcohol plant boiler can thus also power an 8 MW electric generator. The favorable economic return for cogeneration is well proven, and in Hawaii 16% of all power is produced by cogeneration in sugar mills using bagasse fuel.

43.6 INTEGRATED ALCOHOL PLANT DESIGN AND ECONOMICS

Several good engineering designs for complete alcohol production plants are presented in the literature (Katzen, 1979; Lipinski *et al.*, 1976; Cysewski and Wilke, 1978). A design prepared by Raphael Katzen Associates (Katzen, 1979) for the US Department of Energy will be reviewed here as exemplary of current practice in the United States.

43.6.1 Alcohol Plant Design Basis

Plant flows are summarized in Figure 12. 194 Ml y^{-1} (50 million gallons y^{-1}) of alcohol are produced from 495 million kg (19.4 million bushels) of corn.

The plant receives shelled corn which is hammer milled and then saccharified in a continuous tubular cooker/saccharifier using *Aspergillus niger* amylase (whole cell suspension) produced on site by batch fermentation. Batch ethanol fermentation is in 16 vessels, each of 950 000 l. Fermenters are inoculated with 140 kg of yeast press cake (purchased from a central yeast producer). 7.1 wt% alcohol beer is produced in a 48 hour fermentation cycle time (including emptying, cleaning and refilling time).

Distillation is by an efficient two-pressure vapor heat reuse method. A combined stripper/rectifier produces 94 wt% alcohol as its main product along with a heads (aldehyde) and a fusel oils draw. Fuel grade 99.2 wt% alcohol is produced by azeotropic distillation with a hydrocarbon entrainer. The stripper rectifier is operated at elevated pressure (440 kPa, 4.4 atm) so that heat can be recovered by condensing the vapor products in the azeotropic column reboiler. The final distillation to produce anhydrous alcohol from the azeotrope is thus operated with no additional steam input.

Stillage is centrifuged. A portion of the thin stillage is recycled to the saccharification section to make up 10% of the new mash. Vapor recompression evaporation is used for syrup concentration. A distiller's dark grains product of less than 10% moisture is then produced from the syrup and light grain solids by drying in rotary drum driers. Steam is generated on site with a 4250 kPa (42 atm) boiler burning coal. Ammonium sulfate is produced in the flue gas scrubbing process. Feed inputs and plant products are summarized in Table 13.

43.6.2 Plant Energy Requirements

Total plant energy consumption is 91 000 kg h^{-1} of steam and 8314 kW as summarized in Table 14. Electricity cogeneration is not used, but shaft work is recovered by expansion of high-pressure (42 atm) boiler steam through back pressure turbine motors to drive the largest energy users. The medium-pressure (11 atm) turbine exhaust steam is then used for process heating. In this way, one third of the plant power requirement is obtained from steam turbines. In the distillation section, low-grade heat (such as from the hot stillage and 2 atm flash vapors from the mash cooler) is used for feed preheating. The 11 atm steam is condensed in the stripper/rectifier reboiler. The rectifier product heat is then recovered: 11% condensed for additional feed preheating, 82% to heat the azeotropic dehydration column and 7% to heat the hydrocarbon stripper for entrainer regeneration. Thus the high-pressure steam is used in up to three successive operations with very high heat use efficiency. Efficiency is further improved by recovering flue gas waste heat for grains drying.

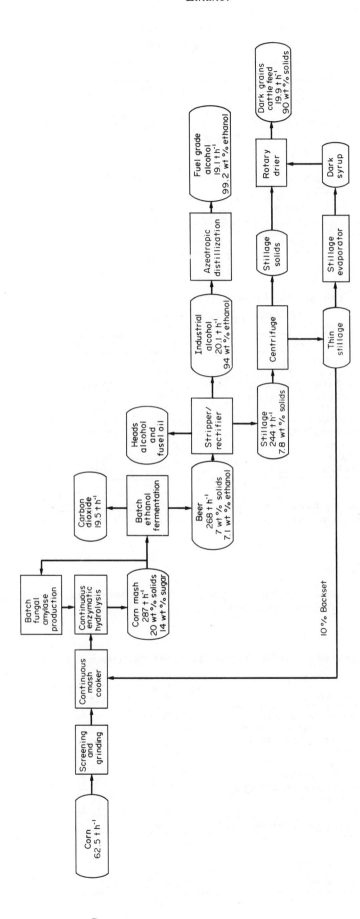

Figure 12 Material flows: 190 million l y⁻¹ fermentative ethanol plant (after Katzen, 1979)

Table 13   190 Million Liter Per Year Corn
Alcohol Plant Material Flows

Inputs	Rate
Corn	1500 t d^{-1}
Coal	270 t d^{-1}
Yeast	1090 kg d^{-1}
Denaturant	5700 l d^{-1}
Ammonia (for coal flue gas purification)	8.4 t d^{-1}
Products	
199 Proof motor fuel	573 500 l d^{-1}
By-products	
1. Distiller's dark grains	478 t d^{-1}
2. $(NH_4)_2SO_4$ from coal flue gas purification	29 t d^{-1}
3. Fusel oil	2600 l d^{-1}

Table 14   Process Steam and Electrial Power Consumption
190 Million Liter Per Year Alcohol Plant

Process section	Process steam	Electrical
Total plant requirement	90 900 kg h^{-1}	8314 kW
Receiving, storage and milling	0.0%	6.1%
Mash cooking and saccharification	30.5%	2.6%
Fungal amylase production	0.7%	20.4%
Fermentation	0.2%	4.0%
Distillation	58.5%	1.6%
Dark grain recovery	6.4%	27.1%
Storage	0.0%	0.7%
Utilities	2.7%	37.0%
Buildings	1.0%	0.5%

The distillation steam requirement is only 2.57 kg l^{-1} alcohol of which 0.34 kg is low-pressure flash vapor recovered from mashing, and the total steam usage is 3.81 kg l^{-1} alcohol.

43.6.3   Plant Capital Investment

Capital investment for the 190 million liter per year fuel alcohol plant is summarized by plant section in Table 15. The total investment (1980 basis) is $70 120 700.

Table 15   190 Million Liter per Year Alcohol Plant Investment

Process section	Capital investment ($, 1980)	Percent of investment
Receiving, storage, milling	2 523 300	3.96
Cooking and saccharification	3 415 100	5.36
Fungal amylase production	4 215 100	6.61
Fermentation	5 073 200	7.96
Distillation	6 195 600	9.72
Dark grain recovery	15 741 500	24.69
Alcohol storage, denaturing and by-product storage	5 320 200	8.34
Utilities	18 246 400	28.62
Building, general services and land	3 015 700	4.73
	63 746 100	100.00
Add 10% contingency	6 374 600	
Total plant cost	70 120 700	

43.6.4 Alcohol Manufacturing Costs

Manufacturing costs (1980 basis) are summarized in Table 16. The largest cost component is for corn raw material ($28.8¢\,l^{-1}$). This cost is partly offset by the distiller's dark grain credit ($15.9¢\,l^{-1}$). After including fixed charges, labor and utilities, a net production cost of $24.6¢\,l^{-1}$ results. The alcohol selling price (including miscellaneous and marketing expenses) required for a 15% after tax return on investment is then $38.3¢\,l^{-1}$.

Table 16 190 Million Liter Per Year Alcohol Plant Operating Cost Summary

		Cost ($¢\,l^{-1}$)
Raw materials		
Corn ($110 t^{-1})		28.8
Coal ($38 t^{-1})		1.8
Active yeast ($1.01 kg^{-1})		0.2
Ammonia ($153 t^{-1})		0.2
Miscellaneous chemicals		0.2
	Raw material subtotal	31.2
Utilities		
Electric power ($7¢\,kWh^{-1}$)		2.4
Steam (from plant)		0.0
Cooling water (from plant)		0.0
	Utilities subtotal	2.4
Labor		
Management		0.1
Supervisors/operators		0.3
Office and laborers	(total 26 people/shift)	1.7
	Labor subtotal	2.1
	Direct production cost	35.6
Fixed charges		
Depreciation (18 years)		2.1
Local tax (3% F.C.I.)		1.1
Insurance (0.7% F.C.I.)		0.2
Maintenance (4% F.C.I.)		1.5
	Fixed charges subtotal	4.9
	Total production cost	40.6

43.7 ETHANOL MARKET HISTORY

43.7.1 Industrial Ethanol End Uses

43.7.1.1 Introduction

Industrial ethanol end uses may be categorized into three classes: solvent, chemical intermediate and fuel. In each of these applications, fermentation ethanol must compete with petroleum derived ethanol as well as with alternative commodities. The extent of ethanol use and the mix of fermentative *vs.* petroleum derived ethanol varies substantially throughout the world.

The fermentation route is the more prevalent in less industrialized nations. Alcohol production can begin in relatively small capacity plants which can then serve as a first step toward a diversified chemical industry (Paturau, 1982; Venkiteswaran, 1964b). The Union Carbide Corporation in the US established a large chemical business in the 1930s, producing over 70 chemicals based on ethanol (Kochar, 1982). Today, both India and Brazil are expanding their chemical industries based on fermentation ethanol and Brazil is now actively marketing its ethanol technology to other less industrialized nations (Kislin, 1980). Fermentative ethanol has remained important in

Belgium, Italy, France and the Netherlands, and is growing in Japan (Sherman and Kavasmaneck, 1978).

In the industrialized oil refining nations the advantages of scale and stable raw material supply favored synthetic over fermentative ethanol production through the 1960s and early 1970s (Sherman and Kavasmaneck, 1978). Fermentative ethanol made up only 6% of US industrial alcohol production in 1975 (Johnston, 1976).

The redevelopment of ethanol fuel use in response to petroleum price increases is the most significant current factor influencing the world ethanol market. Ethanol fuel use has generally been instituted under government regulation (Remirez *et al.*, 1980; Anderson, 1982). World fuel use of fermentative ethanol has risen to at least 4.4 billion liters, primarily in Brazil (Kislin, 1980), the US (Staff, Fuel Alcohol, 1982) and West Germany (Muensterer, 1976). Subsidies are needed to allow fermentative ethanol to compete with petroleum as a fuel (Remirez *et al.*, 1980). Under favorable raw material price conditions, however, alcohol from fuel plants can compete in the high-priced industrial ethanol chemicals market (Staff, Chem. Mark. Rep., 1982a).

43.7.1.2 Product specifications

The specifications of ethanol product quality set by the US Pharmacopeia (USP, 1980) and American Chemical Society (ACS Committee on Analytical Reagents, 1974) are generally accepted. 190 proof alcohol must have a minimum 95 vol% ethanol (92.42 wt%) purity with a specific gravity of 0.816 or less and less than 0.0001 g nonvolatile matter per liter. 200 proof alcohol must have a minimum 99.9 vol% (99.85 wt%) purity with a specific gravity of 0.7905 or less. Mehlenbocher provides a survey of analytical methods (Mehlenbocher, 1953).

For industrial use, denaturants are generally added to make the alcohol nonpotable, thus allowing its sale without payment of the beverage tax. Various formulations are allowed and must be selected to not interfere with the intended end use (Anderson, 1977).

43.7.1.3 Ethanol solvent use

Ethanol ranks second only to water as an industrial solvent (Rose, 1967). Solvent applications include resins, pharmaceuticals, cosmetics, household cleaning products and industrial solvents; 50–55% of US industrial (nonfuel) alcohol use is for solvent applications, as summarized in Table 17 for 1975 (Sherman and Kavasmaneck, 1978). Solvent usage is mature with few major new applications emerging and a slow growth rate of 2–3% per year is expected (Staff, Chem. Eng. News, 1980).

Table 17 United States Ethanol Solvent
Consumption (1975)

Use	Volume (Ml)
Resins and lacquers	12.9
Pharmaceuticals and cosmetics	142.7
Cleaning preparations	80.2
Industrial solvent	47.3
Proprietary solvent formulation	91.2
Total	385.7

In Brazil, by contrast, the chemical industry is expanding and fermentative ethanol use in solvents is expected to grow from 210 million liters in 1975 to 350 million liters in 1985 (Yand and Trindade, 1979).

43.7.1.4 Ethanol use as chemical intermediate

Figure 13 summarizes several of the more important chemicals that can be derived from ethanol (Paturau, 1982; Sherman and Kavasmaneck, 1978). More exhaustive listings along with reaction details are presented in several sources (Hatch, 1962; Thampy, 1964). A new process allows the production of a hydrocarbon mixture similar to high-octane gasoline using a shape-selective zeolite catalyst (Maiorella, 1982). Thus there is the potential to produce practically all petroleum derived chemicals from ethanol.

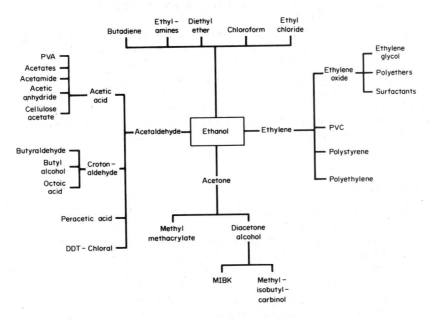

Figure 13 Chemical products derivable from ethanol.

In the US, butadiene production for synthetic rubber led to the tremendous growth in the ethanol fermentation industry during World War II (Jacobs, 1950). After the war, petroleum replaced ethanol as the raw material for butadiene. Acetaldehyde became the major intermediate based on ethanol with a maximum production of 568 million liters in 1963 (Sherman and Kavasmaneck, 1978). A direct route to acetaldehyde from ethylene was then developed and acetaldehyde production from ethanol was reduced to 87 million liters in 1975. Table 18 compares US ethanol use in chemicals for 1945 and 1975 (Aries, 1947; Sherman and Kavasmaneck, 1978). The trend to replace ethanol as a chemical intermediate has continued and in 1981 the major chemical products from ethanol were esters, amines and ethers (consuming 68 million, 34 million and 34 million liters of ethanol, respectively) (Greek, 1981). Table 19 compares production costs of various chemicals from ethanol and petroleum feedstocks (Paturau, 1982). Substitution of ethylene for ethanol resulted in a 4.5% y^{-1} average decrease in (nonfuel) use of ethanol from 1132 Ml y^{-1} in 1970 to 681 Ml y^{-1} in 1981 (Greek, 1981). This is only 67% of installed US synthetic ethanol capacity. Chemical use is expected to remain constant with the exception of vinegar production which may grow at a rate of 6% per year (Staff, Chem. Eng. News, 1980). Similar trends are expected in Europe (Marchal, 1976).

Table 18 United States Ethanol Consumption as a Chemical Intermediate

| | | | | Ethanol used in manufacture (Ml) | | | | | |
Year	Synthetic rubber	Acetaldehyde	Acetic acid	Vinegar	Ethyl acetate	Ethyl chloride	Ethylene dibromide	Others	Total
1945	1195.8	210.9	18.9	27.2	38.4	56.3	10.8	51.1	1609.4
1955	—	439.1	20.8	31.4	22.7	25.0	18.1	50.3	607.5
1965	—	548.5	19.3	43.5	25.0	24.6	0.4	133.2	794.5
1975	—	87.1	14.8	74.2	14.0	1.1	—	174.9	366.0

The replacement of ethanol by ethylene is reversed in the less industrialized nations. In Brazil and India, ethylene and its chemical derivatives are produced by catalytic dehydration of fermentative ethanol (Kochar, 1982). Current Brazilian chemical production capacity from ethanol is

Table 19 Production Cost of Various Chemicals
Using Ethanol as Feedstock

| | Production cost (¢ kg^{-1}) | |
| | From petroleum | From ethanol |
Chemical	feedstock	(at 40¢ l^{-1})
Acetaldehyde	60	66
Acetic acid	50	63
Butadiene	64	145
Ethylene	44	95
2-Ethylhexanol	61	166

summarized by Anderson (1982). Production is expected to grow, with 1 billion liters to be consumed for chemical synthesis in 1985 *vs.* 56 million liters in 1975 (Yand and Trindade, 1979).

43.7.1.5 Ethanol fuel use

Fuel use in internal combustion engines is the fastest growing application for fermentative ethanol. The original Otto engine was developed using anhydrous ethanol fuel. Ethanol use in blends of up to 20 vol% with gasoline was widespread during World War II induced petroleum shortages (Pleeth, 1949; API, 1976). As early as 1931, Brazil legislated the use of a 5 vol% alcohol blend with gasoline as a means to utilize sugar refinery molasses waste and to stabilize sugar prices (Yand and Trindade, 1979).

Petroleum imports account for roughly 43% of Brazil's primary energy consumption. With price rises following the 1973 oil shortages, this resulted in a cost of $4 billion in 1977 (31% of the total import expenditure) (Yand and Trindade, 1979). To combat the resulting balance of payments problems, the Brazilian National Gasohol Program was established in 1975 with a goal to increase the ethanol production from 414 Ml to 10 Gl in 1985 (Anderson, 1982). Alcohol is purchased at a subsidized price based on current sugar price and then resold at a discounted price for fuel (65% of gasoline price) or chemical feedstock (35% of ethylene cost) (Anderson, 1982). New plant capacity financing is also subsidized. In 1978, Brazilian fuel alcohol content reached 12 vol% on average (Yand and Trindade, 1979). With a gasoline price of 74.5¢ l^{-1}, pure alcohol cars were introduced by General Motors, Vokswagen, Ford, Chrysler and Fiat, and 250 000 (one quarter of all sales) were sold in 1980. Total 1980 ethanol production was 4.1 Gl with 3.0 Gl used in gasohol and 500 Ml as neat fuel (Anderson, 1982).

In the US, the 1980 Energy Security Act and associated legislation established goals of 1.9 Gl of fermentative ethanol production in 1981 and 5.7 Gl in 1985 (Remirez *et al.*, 1980). Total nationwide adoption of gasohol, by contrast, would use 38 Gl of alcohol (requiring an unrealistic 50% of current US corn production). The major incentive established was a 1.06¢ l^{-1} fuel tax reduction on gasohol containing 10 vol% ethanol, corresponding to a 10.6¢ l^{-1} ethanol credit. A 1.3¢ l^{-1} ethanol entitlement from the Department of Energy and loan guarantees of $1.2 billion through DOE, USDA and FHA were also provided, totalling a subsidy of roughly 13.2¢ l^{-1} ethanol from the federal government. State gasohol fuel tax reductions of 0.5–1.8¢ l^{-1} were also granted in 30 states so that total subsidies could be as high as 31.2¢ l^{-1} ethanol (AGA Policy Eval. Group, 1980). With these incentives, permits were sought in 1980 for 20 new plants of over 2 Ml y^{-1} capacity each and 8200 additional smaller plants (USBATF, 1980).

With the increase in world oil supply in 1981 and following the change in administrations in the US, loan guarantees were withdrawn and plans for most smaller plants were cancelled (Maize, 1981). Fermentative alcohol capacity did rise to 1.59 Gl in 1982 (Staff, Chem. Mark. Rep., 1982c).

Addition of 10 vol % ethanol to gasoline results in a three point increase in average octane rating. Tetraethyllead is being withdrawn in the US as an octane booster to meet new pollution standards. Methyl *t*-butyl ether and toluene can each be substituted and are 26% and 16% cheaper on a per octane increase basis than ethanol (Staff, Chem. Eng. News., 1982). Ethanol can compete effectively with these other organic octane boosters, however, as long as the 1.05¢ l^{-1} federal gasohol tax credit is maintained. This product is successfully marketed as a premium fuel, 'super unleaded with ethanol', and is expected to provide a sustained outlet for the new fermentative ethanol capacity.

43.7.2 Ethanol Sales Volume and Price History

43.7.2.1 World ethanol production

Total world ethanol rectified spirits production was 8.9 billion liters in 1975, as summarized in Table 20 (Marchal, 1976). This amount has grown by approximately 4 billion liters per year for 1981 due to expanded fuel use. A large portion of rectified spirit production was used in beverage blending, but it is estimated that 4.3 billion liters of alcohol are used in nonfuel industrial applications worldwide (Scheller and Mohr, 1977a). In 1975, industrial alcohol consumption accounted for 752 million out of 11 700 million liters in the United States, 143 Ml out of 199 Ml in Britain, 188 Ml out of 254 Ml in West Germany, 214 Ml out of 316 Ml in France and 118 Ml out of 379 Ml in Japan (Marchal, 1976; Suzuki, 1976).

Table 20 World Rectified Spirits Production
(1975)

Region	Ethanol production (Ml 100% ethanol)
Western Europe	1606
USSR and Eastern Europe	3270
North America	1265
Latin America	1478
Asia	949
Africa	145
Oceania	60
A.C.P. Countries	82
Total	8855

The US accounts for 25% of world (nonfuel) industrial demand and 55% of synthetic based ethanol production (Johnston, 1976). Fermentation provides 85% of non-US production (Scheller and Mohr, 1977a) and fermentative production for fuel use is expanding rapidly. New synthetic ethanol capacity is, however, also under construction (Staff, Chem. Mark. Rep., 1982c).

Molasses is the most important fermentative ethanol raw material worldwide (OECD, 1976) and much production in the industrialized nations is based on imported molasses. Japan, for instance, imported 440 thousand tons of molasses to produce 103 thousand tons of ethanol in 1975 (Suzuki, 1976). The high penalty for shipping molasses *vs.* ethanol, however, is leading now to a shift in alcohol production back to the sugar cane growing nations. Rectified alcohol production from farm wastes, such as from excess grapes in the French wine country, is also increasing (Staff, Chem. Mark. Rep., 1982c).

Import duties intended to protect home production are interfering with world trade. Tariffs have been long standing in Europe (Brown, 1976) and the US imposed a levy in 1982 against imported ethanol to slow Brazilian competition (Staff, Chem. Mark. Rep., 1982c). Nonetheless, the trend to shift alcohol production to the sugar producing countries is expected to continue.

43.7.2.2 United States ethanol production

US rectified spirits production was 1.57 billion liters in 1981, 55% by fermentation (Staff, Chem. Mark. Rep., 1982a): 775 million liters were used in industrial solvent and chemical applications; 269 million liters were used as fuel and 38 million liters were exported (Greek, 1981). Table 21 presents the historic trends in US ethanol consumption and price. Table 22 lists major US producers and their capacities (Staff, Chem. Mark. Rep., 1982b). Total synthetic capacity is 1.03 billion liters. Total fermentative capacity is 1.59 billion liters and this will expand by 760 million liters in 1983 based on plants under construction (Staff, Chem. Mark. Rep., 1982c). The industrial alcohol list price (January, 1982) is 52.0¢ l^{-1} for 190 proof neutral spirits and 56.0¢ l^{-1} for 200 proof alcohol. The list price for fuel grade alcohol, by contrast, varies from 44.9 to 50.2¢ l^{-1}, remaining relatively low due to good crop years with corn selling for $2.50 per bushel compared with $3.20 per bushel in 1974 (Staff, Fuel Alcohol, 1982). Fuel grade alcohol can replace synthetic alcohol in most solvent and chemical applications and (with improved distillation systems) can even be used in cosmetics and perfumes (Staff, Chem. Mark. Rep., 1982a). As a result,

113 million liters of fuel alcohol entered the industrial alcohol market in 1981. This competition has resulted in price declines for synthetic ethanol (Publicker dropped its 200 proof list price to 48.1¢ l^{-1} in February 1982) with further reductions as manufacturers allow 1.5¢ l^{-1} discounts below the list price. Thus, the outlook for ethanol in the US depends heavily on the continuation of the alcohol fuels program. Synthetic ethanol plants are already operating at only 67% of capacity. Without the 1.05¢ l^{-1} gasohol tax incentive, fermentative ethanol cannot compete with gasoline as a fuel or even with MTBE as an octane booster, and withdrawal of the present tax credit could result in a total alcohol capacity of over 2.62 billion liters per year for a 1.26 billion liters per year industrial plus beverage blending market.

Table 21   United States Ethanol Consumption and Price History

| | Alcohol consumption | | | |
| | Industrial | | Fuel | |
Year	Volume (Ml)	Price (¢ l^{-1} 200 proof)	Volume (Ml)	Price (¢ l^{-1})
1950	676	11.6	—	—
1955	824	11.9	—	—
1960	1071	15.6	—	—
1965	1144	15.6	—	—
1970	1092	16.0	—	—
1975	752	28.2	—	—
1981	775	55.5	269	46.9

Table 22   United States Ethanol Production Capacity (1982)

Producer	Capacity (Ml y^{-1})	Producer	Capacity (Ml y^{-1})
Synthetic		Fermentation	
Union Carbide	454	Archer Daniels	549
U.S.I.	250	Grain Processing	227
Publicker	227	Pekin Energy	227
Eastman	95	Publicker	227
		Midwest Solvents	114
Total Synthetic	1026	Clinton Corn Processing	102
		Amber Labs	38
		National Distillers	30
		Georgia Pacific (waste sulfite)	19
		Others	60
		Total Fermentation	1593

43.8   CONCLUSION—THE OUTLOOK FOR FERMENTATIVE ETHANOL

Current ethanol production and the distribution between petroleum and fermentation sources is highly influenced by government regulation. The costs of ethanol produced from ethylene and molasses based on conventional technology and adjusted to a 1982 basis are given approximately by the equations (Brown, 1976; Moundlic, 1976):

Fermentation:
$$\text{alcohol cost (\$ m}^{-3}) = 3.6 \times \text{molasses cost (\$ t}^{-1}) + 176$$

Synthetic:
$$\text{alcohol cost (\$ m}^{-3}) = 0.53 \times \text{ethylene cost (\$ t}^{-1}) + 143$$

Capital related costs for both processes are nearly identical. The total nonvariable component for fermentative alcohol is higher due to higher costs for purification and waste treatment as well as added nutrients and other materials.

Under current conditions in the US (with ethylene at \$593 t^{-1} and molasses at \$93.5 t^{-1}), ethanol production from ethylene (or corn) is more favorable. In tropical sugar producing countries such as Brazil, however, the preference is reversed and fermentative ethanol production may be expected to continue to grow in these countries (Brown, 1976). Further, the price of petroleum is expected in the long term to continue to rise. As a rough rule, a \$1 rise in the per barrel petroleum price leads to a 2.2¢ kg^{-1} increase in ethylene price (Johnston, 1976). The price of sugar crops is affected by oil prices to a much lesser extent (Scheller and Mohr, 1977a) and fermentative ethanol should continue to become more competitive both as an industrial chemical and a fuel.

43.9 REFERENCES

ACS Committee on Analytical Reagents (1974). *Reagent Chemicals—ACS Specifications*, 5th edn. American Chemical Society, Washington.
AGA Policy Evaluation Group (1980). Preliminary comparison of methane *vs.* ethanol. In *Energy Analysis*, Report no. 1980–2, p. 1. American Gas Association, Arlington, Va.
Alfa-Laval (1981). The Biostil technique. Tech. Rept. IB-8104-03, Alfa-Laval, Tumba, Sweden.
Alfa-Laval (1982). Biostil. New concept for continuous alcohol production. Report PB-40661E-8205, Alfa-Laval, Tumba, Sweden.
Anderson, E. V. (1977). Plummeting exports hurt U.S. ethanol makers. *Chem. Eng. News*, **55** (2), 12.
Anderson, E. (1982). Brazil sets lofty goals for ethanol. *Chem. Eng. News*, **60** (31), 15.
API (1976). *Alcohols: A Technical Assessment of Their Application as Fuels*. Pub. No. 4261, American Petroleum Institute, Washington.
Aries, R. S. (1947). Industrial alcohol. In *Encyclopedia of Chemical Technology*, ed. R. E. Kirk and D. F. Othmer, Vol. 1, p. 252. Interscience, New York.
Arroyo, R. (1949). The Arroyo fermentation process for alcohol and light rum from blackstrap molasses. *Sugar J.*, **11** (8), 5.
Ault, R. G., A. N. Hampton, R. Newton and R. H. Roberts (1969). Biological and biochemical aspects of tower fermentation. *J. Inst. Brew.*, **75**, 260.
Ballou, W. R. (1978). Carbon dioxide. In *Kirk-Othmer Encyclopedia of Chemical Technology*, 3rd edn., ed. M. Grayson, p. 725. Wiley, New York.
Barbaudy, J. (1927). The distillation of heterogeneous ternary mixtures. *J. Chim. Phys.*, **24**, 1.
Barron, H. (1944). *Distillation of Alcohol*. Div. of Education, Joseph E. Seagram and Sons, Louisville, Ky.
BATF (1984). *Summary Statistics*. US Dept. of the Treasury, Bureau of Alcohol, Tobacco and Firearms, Washington.
Bazua, C. D. and C. R. Wilke (1977). Ethanol effects on the kinetics of a continuous fermentation with *Saccharomyces cerevisiae*. *Biotechnol. Bioeng. Symp.*, **7**, 105.
Beebe, A. H., K. E. Coulter, R. A. Lindsay and E. M. Baker (1942). Equilibrium in ethanol–water system at pressures less than atmospheric. *Ind. Eng. Chem.*, **34**, 1501.
Bilford, H. R., R. E. Scalf, W. H. Stark and P. J. Kolachov (1942). Alcoholic fermentation of molasses. *Ind. Eng. Chem.*, **34**, 1406.
Black, C. (1980). Distillation modeling of ethanol recovery and dehydration processes for ethanol and gasohol. *Chem. Eng. Prog.*, **76** (9), 78.
Black, C., R. A. Golding and D. E. Ditsler (1972). Azeotropic distillation results from automatic computer calculations. *Adv. Chem. Ser.*, **115**, 64.
Blanch, H. W. and A. Sciamanna (1980). Composition of cellulosic raw materials for sugar production. *Alcohol Fuels Process R/D Newsletter*, Winter Quarter, p. 69. USDOE Solar Energy Research Institute, Golden, Co.
Borzani, W. and E. Aquarone (1957). Continuous fermentation. *Agric. Food Chem.*, **5**, 610.
Bridson, E. Y. and A. Brecker (1970). Design of media. In *Methods in Microbiology*, ed. J. Norris and D. Ribbins, Vol. 3A, p. 229. Academic, New York.
Brooks, B. T. (1952). Recovery of glycerol from fermented molasses. *US Pat.* 2 614 964.
Brown, O. M. R. (1976). Potential for fermentation ethanol as a feed stock. In *Molasses and Industrial Alcohol*, p. 137. Dev. Centre, OECD, Paris.
Bu'Lock, J. D. (1979). Industrial alcohol. In *Microbial Technology: Current State, Future Prospects*, ed. A. T. Bull, D. C. Ellwood, and C. Rutledge, Vol. 29, p. 310. Cambridge Univ. Press, Cambridge.
Bungay, H. R. (1981). Biochemical engineering for fuel production in the United States. *Adv. Biochem. Eng.*, **20**, 1.
Castillo, F. J., M. E. Izaguirre, V. Michelena and B. Moreno (1982). Optimization of fermentation conditions for ethanol production from whey. *Biotechnol. Lett.*, **4**, 567.
Chakrabarty, R. N. (1964). Potash recovery—a method of disposal of distillery wastes. In *Ethyl Alcohol Production Technique*, p. 93. Noyes Development Corporation, Pearl River, New York.
Chang, C. T. and W. C. Yang (1973). Study on feed yeast production from molasses distillery stillage. *Taiwan Sugar*, **20**, 200.
Chang, M. M., T. Y. C. Chou and G. T. Tsao (1981). Structure, pretreatment and hydrolysis of cellulose. *Adv. Biochem. Eng.*, **20**, 15.
Chen, C. S. (1980). *Hawaii Ethanol From Molasses Project—Final Report*. HNEI-80-03. Hawaii Natural Energy Institute, University of Hawaii at Manoa.
Chu, C. J., C. J. Chiou, A. K. Ng, T. C. Lin, E. C. Hwang and C. H. Rao (1970). Multi-stage continuous alcohol fermentation. *Chung Kuo Nung Yeh Hua Hsueh Hui Chih*, **8**, 84 (*Chem. Abstr.*, 1971, **75**, 3981).
Colburn, A. P. and J. C. Phillips (1944). Experimental study of azeotropic distillation—use of trichloroethylene in dehydration of ethanol. *Trans. Am. Inst. Chem. Eng.*, **40**, 333.
Curtis, H. A. (1950). Solid fuels. In *Chemical Engineers' Handbook*, 2nd edn., ed. J. H. Perry, McGraw-Hill, New York.
Cysewski, G. R. and C. R. Wilke (1976a). Utilization of cellulosic materials through enzymatic hydrolysis. *Biotechnol. Bioeng.*, **18**, 1297.
Cysewski, G. R. and C. R. Wilke (1976b). *Fermentation Kinetics and Process Economics for the Production of Ethanol*. Report LBL-4480, Lawrence Berkeley Laboratory, Berkeley, Ca.

Cysewski, G. R. and C. R. Wilke (1977). Rapid ethanol fermentation using vacuum and cell recycle. *Biotechnol. Bioeng.*, **19**, 1125.

Cysewski, G. R. and C. R. Wilke (1978). Process design and economic studies of alternative fermentation methods for the production of ethanol. *Biotechnol. Bioeng.*, **20**, 1421.

David, M. L., G. S. Hammaker, R. J. Buzenberg and J. P. Wagner (1978). *Gasohol Economic Feasibility Study*. Development Planning and Research Associates, Manhattan, Ks.

Davis, E. E. and S. C. Jung (1974). *The American Type Culture Collection*. ATCC, Rockville, Md.

Dawson, P. S. S. (1976). Continuous fermentations. *Annu. Rep. Ferment. Processes*, **1**, 74.

DeDeken, R. H. (1966). The Crabtree effect: a regulatory system in yeast. *J. Gen. Microbiol.*, **44**, 149.

DelRosario, E. J., K. J. Lee and P. L. Rogers (1979). Kinetics of alcohol fermentation at high yeast levels. *Biotechnol. Bioeng.*, **21**, 1477.

Drew, S. W. (1978). Chemical feedstocks and fuel from lignin. *AIChE Symp. Series No. 181*, Vol. 74, p. 21. American Institute of Chemical Engineers, New York.

Dunphy, J. (1980). Alcohol from glucose without using yeast. *Chem. Week*, **127** (25), 19.

Eakin, D. E., J. M. Donovan, G. R. Cysewski, S. E. Petty and J. V. Maxham (1981). *Preliminary Evaluation of Alternative Ethanol/Water Separation Processes*. Battelle Pacific Northwest Laboratory, US National Technical Information Service, No. PNL-3823.

Emert, G. H. and R. Katzen (1980a). Gulf's cellulose-to-ethanol process. *Chemtech*, **10**, 610.

Emert, G. H., R. Katzen, R. E. Fredrickson and K. F. Kaupish (1980b). Economic update of the Gulf cellulose alcohol process. *Chem. Eng. Prog.*, **76** (9), 47.

Emery, W. L. (1981). *Commodity Year Book*. Commodity Research Bureau, New York.

Ericson, E. O. (1947). Alcohol from sulfite waste liquor. *Chem. Eng. Prog.*, **43**, 165.

Espinosa, R., V. Cojulun and F. Marroquin (1978). Alternatives for energy savings at plant level for the production of alcohol for use as automotive fuel. *Biotechnol. Bioeng. Symp.*, **8**, 69.

Falkehag, S. I. (1975). Lignin in materials. *J. Appl. Polym. Sci. (Appl. Polymer Symp. 28)*, 247.

Forbes, R. J. (1970). *A Short History of the Art of Distillation*. Brill, Leiden, Netherlands.

Foth, G. (1935). *Die Praxis des Brennereibetriebes auf Wissenschaftlicher Grundlage*. Parey, Berlin.

Fritzweiler, R. and K. R. Dietrich (1932). Azeotropism and its application in a new method for dehydration of ethyl alcohol. *Angew. Chem.*, **45**, 605.

Gallagher, F. H., H. R. Bilford, W. H. Stark and P. Kolachov (1942). Fast conversion of distillery mash for use in a continuous process. *Ind. Eng. Chem.*, **34**, 1395.

Ghose, T. K. and R. D. Tyagi (1979a). Rapid ethanol fermentation of cellulose hydrolysate, I. *Biotechnol. Bioeng.*, **21**, 1387.

Ghose, T. K. and R. D. Tyagi (1979b). Rapid ethanol fermentation of cellulose hydrolysate, II. *Biotechnol. Bioeng.*, **21**, 1401.

Gillies, M. T. (1978). Animal feeds from waste materials. *Food Technol. Rev.*, **46**, 346.

Goggin, B. and G. Thorsson (1982). *Operating Experience with Biostil in a Commercial Distillery*. Alfa-Laval, Tumba, Sweden.

Gong, C. S., L. F. Chen, G. T. Tsao and M. C. Flickinger (1981). Conversion of hemicellulose carbohydrates. *Adv. Biochem. Eng.*, **20**, 93.

Gray, W. D., W. H. Stark and P. Kolachov (1942). *J. Bacteriol.*, **43**, 270.

Greek, B. (1981). Key chemicals—ethanol. *Chem. Eng. News*, **59** (45), 14.

Greenshields, R. N. and E. L. Smith (1971). Tower fermentation systems and their applications. *Chem. Eng. (London)*, **249**, 182.

Grylls, F. S. M. (1961). The chemical composition of yeasts. In *Biochemist's Handbook*, ed. C. Long, p. 1050. Spon, London.

Guinot, H. M. and F. W. Clark (1938). Azeotropic distillation in industry. *Trans. Inst. Chem. Eng. (London)*, **16**, 189.

Gundu Rao, G. (1964). The overall steam requirement in an industrial alcohol distillery. In *Ethyl Alcohol Production Technique*, p. 32. Noyes Development Corp., Pearl River, New York.

Hagglund, E. (1925). *Paper Trade J.*, **80** (26), 43.

Harris, E. E. and E. Beglinger (1946). Madison wood sugar process. *Ind. Eng. Chem.*, **38**, 890.

Harrison, J. and J. Graham (1970). Yeast in distillery practice. In *The Yeasts*, ed. A. H. Rose and J. S. Harrison, Vol. 13. Academic, New York.

Hartline, F. H. (1979). Lowering the cost of alcohol. *Science*, **206**, 41.

Hatch, L. F. (1962). *Ethyl Alcohol*. Enjay Chemical Co., a division of Humble Oil and Refining Co., New York.

Haukeli, A. D. and S. Lie (1971). Controlled supply of trace amounts of oxygen in laboratory scale fermentations. *Biotechnol. Bioeng.*, **13**, 619.

Hayashida, S. and K. Ohta (1981). Formation of high concentrations of alcohol by various yeasts. *J. Inst. Brew.*, **87**, 42.

Hodge, H. M. and F. M. Hildebrandt (1954). Alcoholic fermentation of molasses. In *Industrial Fermentations*, ed. L. A. Underkofler and R. J. Hickey, Vol. 1, p. 73. Chemical Pub. Co., New York.

Holzer, H. (1968). Biochemistry of adaptation of yeast. In *Aspects of Yeast Metabolism*, ed. A. Mills and S. Krebbs, p. 155. Blackwell, Oxford.

Hough, J. S., P. E. Gough and A. D. Davis (1962). Continuous fermentation on the pilot scale. *J. Inst. Brew.*, **68**, 478.

Hough, J. S., C. W. Keevil, V. Maric, G. Philliskirk and T. W. Young (1977). Continuous culture in brewing. In *Continuous Culture: Applications and New Fields*, ed. A. Dean and D. Ellwood, p. 226. Holsted Press, London.

Hudson, J. R. (1972). Recent changes in brewing technology. In *Fermentation Technology Today: Proc. 4th Int. Ferment. Symp.*, ed. G. Terni, p. 629. Society of Fermentation Technology, Japan.

Huffman, E. O. (1978). Fertilizers. In *Kirk-Othmer Encyclopedia of Chemical Technology*, 3rd edn., ed. M. Grayson, Vol. 1, p. 31. Wiley, New York.

Humbert, R. P. (1979). The growing of sugarcane for energy. In *Alcohol Fuels Technology*, Third International Symposium, Asilomar, Ca. US Department of Energy, Washington.

Humphrey, A. E. and E. J. Nolan (1979). *Biological Production of Liquid Fuels and Chemical Feedstocks*. Report to the Congress of the United States, Office of Technology Assessment, Washington.

Isidro, R. G. (1980). CO$_2$ recovery plant of central Don Pedro. In *Proc. 17th Congress ISSCT, Philippines.*

Jackman, E. A. (1977). Distillery effluent treatment in the Brazilian national alcohol programme. *Chem. Eng. (London)*, **319**, 239.

Jackson, L. V. and G. T. Lines (1972). Measures against water pollution in the fermentation industries. *Pure Appl. Chem.*, **29**, 381.

Jacobs, P. B. (1950). *Industrial Alcohol*, Misc. Pub. No. 695, US Department of Agriculture, Washington.

Johnston, P. J. (1976). The industrial market for ethanol in the United States. In *Molasses and Industrial Alcohol*, p. 81. Dev. Centre, OECD, Paris.

Jones, R. P., N. Pamment and P. F. Greenfield (1981). Alcohol fermentation by yeasts. *Process Biochem.*, **16**, 42.

Jumentier, R. (1931). Fabrication de l'alcool absolu par les procedes des usines de Mellé. *Bull. Assoc. Chim. Sucr.*, **48**, 396.

Jurasek, L. (1978). Enzymatic hydrolysis of pretreated aspen wood. *Dev. Ind. Microbiol.*, **20**, 177.

Karaki, I., M. Konishi, K. Amakai and F. Ishikawa (1972). Alcohol production by continuous fermentation of molasses. *Hakko Kyokaishi*, **30**, 106.

Katzen, R. (1979). *Grain Motor Fuel Alcohol, Technical and Economic Assessment Study.* Raphael Katzen Assoc., Cincinnati, Ohio.

Kerr, R. W. (ed.) (1950). *Chemistry and Industry of Starch*, 2nd edn. Academic, New York.

Kislin, A. H. (1980). In Brazil, there's a run on the all-alcohol car. *Chem. Week*, **127** (26), 26.

Knappert, D., H. Grethlein and A. Converse (1980). Partial acid hydrolysis of cellulose materials as a pretreatment for enzymatic hydrolysis. *Biotechnol. Bioeng.*, **22**, 1449.

Kochar, K. (1982). Vast new horizons opening for ethanol. *J. Commerce*, March 29, p. 1c.

Kujala, P. (1979). Distillery fuel savings by efficient molasses processing and stillage utilization. *Sugar y Azucar*, **74** (10), 13.

Kujala, P., R. Hall, F. Engstrom and E. A. Jackman (1976). Alcohol from molasses as a possible fuel and the economics of distillery effluent treatment. *Sugar y Azucar*, **71** (3), 28.

Lagomasino, J. M. (1949). The Mellé–Boinot alcoholic fermentation method. *Int. Sugar J.*, **51**, 338.

Lamprecht, I. and C. Meggers (1969). Microcalorimetric study on the effect of stirring on the growth of yeast. *Z. Naturforsch., Teil B*, **24**, 1205.

Lecat, M. (1949). *Tables Azeotropiques.* Uccle, Bruxelles.

Lee, J. H., M. L. Skotnicki and P. L. Rogers (1982). Kinetic studies on a flocculent strain of *Zymomonas mobilis*. *Biotechnol. Lett.*, **4**, 615.

LeMense E. H., J. Corman, J. M. Van Lanen and A. F. Langlykke (1947). Production of mold amylases in submerged culture, *J. Bacteriol.*, **54**, 149.

Levenspiel, O. (1980). The Monod equation: a revisit and a generalization to product inhibition situations. *Biotechnol. Bioeng.*, **22**, 1671.

Lewicki, W. (1978). Production, application and marketing of concentrated molasses-fermentation-effluent. *Process Biochem.*, **13** (6), 12.

Lipinski, E. S., R. A. Nathan, W. J. Sheppard and J. L. Otis (1976). *Final Report on Systems Study of Fuels from Sugarcane, Sweet Sorghum, and Sugar Beets*, Vol. III. Battelle Laboratories, Columbus, Ohio.

Mahlingham, P. R. (1977). Manufacture of industrial alcohol. *Chem. Age India*, **28**, 303.

Maiorella, B. L., C. R. Wilke and H. W. Blanch (1981). Alcohol production and recovery. *Adv. Biochem. Eng.*, **20**, 43.

Maiorella, B. (1982). Fermentation alcohol: better to convert to fuel. *Hydrocarbon Process.*, **61** (8), 95.

Maiorella, B. L., H. W. Blanch and C. R. Wilke (1983a). By-product inhibition effects on ethanolic fermentation by *Saccharomyces cerevisiae*. *Biotechnol. Bioeng.*, **25**, 103.

Maiorella, B. L. (1983b). *Fermentative Ethanol Production and Recovery.* Doctoral thesis. Dept. Chem. Eng., Univ. California, Berkeley, Ca. University Microfilms International, Ann Arbor, Mi.

Maiorella, B. L., H. W. Blanch and C. R. Wilke (1983c). Ethanol stillage treatment, *Process Biochem.*, **18** (4), 5.

Maiorella, B. L. (1985a). Ethanol distillation technology. In *Ethanol Separation Handbook*, Report no. SP–2037, ed. L. Douglas. US Department of Energy, Solar Energy Research Institute, Golden, Co.

Maiorella, B. L. (1985b). Ethanol/water, physical/chemical properties for separation process design. In *Ethanol Separation Handbook*, Report no. SP–2037, ed. L. Douglass. US Department of Energy, Solar Energy Research Institute, Golden, Co.

Maiorella, B. L. and F. J. Castillo (1984). Ethanol biomass and enzyme production for whey taste abatement. *Process Biochem.*, **19** (4), 157.

Maiorella, B. L., H. W. Blanch and C. R. Wilke (1984a). Ethanol volatility in fermentation systems; salt and metabolite effects. *IEC Fund.*, **23**, 332.

Maiorella, B. L., H. W. Blanch and C. R. Wilke (1984b). Economic evaluation of alternative fermentation processes. *Biotechnol. Bioeng.*, **26**, 1003.

Maize, K. P. (1981). Alcohol fuel politics. *BioCycle*, Jan/Feb., 58.

Marchal, E. (1976). The world market of alcohol. In *Molasses and Industrial Alcohol*, p. 67. Dev. Centre, OECD, Paris.

McCarthy, J. L. (1954). Alcoholic fermentation of sulfite waste liquor. In *Industrial Fermentations*, ed. L. A. Underkofler and R. J. Hickey, Vol. 1, p. 95. Chem. Pub. Co., New York.

McGinnis, R. A. (1951). *Beet Sugar Technology.* Reinhold, New York.

Mehlenbocher, V. C. (1953). Determination of alcohols. In *Organic Analysis*, ed. J. Mitchell, Jr. Vol. 1, p. 1. Interscience, New York.

Mehta, N. N. and R. S. Khanna (1964). Nutrients as accelerator of alcoholic fermentation of carbonation molasses. In *Ethyl Alcohol Production Technique*, p. 67. Noyes Dev. Corp., Pearl River, New York.

Millar, D. G., K. Griffiths-Smith, E. Algar and R. K. Scopes (1982). Activity and stability of glycolytic enzymes in the presence of ethanol. *Biotechnol. Lett.*, **4**, 601.

Monteiro, C. E. (1975). Brazilian experience with the disposal of wastewater from the cane and alcohol industry. *Process Biochem.*, **10** (9), 33.

Montenecourt, B. S. and D. E. Eveleigh (1977). Preparation of mutants of *Trichoderma reesei* with enhanced cellulase production. *Appl. Environ. Microbiol.*, **34**, 777.

Moss, F. J., P. A. D. Rickard, F. E. Bush and P. Caiger (1971). The response by microorganisms to steady-state growth in controlled concentrations of oxygen and glucose. *Biotechnol. Bioeng.*, **13**, 63.

Moundlic J. (1976). Costs of manufacturing alcohol from molasses and by synthetic process. In *Molasses and Industrial Alcohol*, p. 135. Dev. Centre, OECD, Paris.

Muensterer, L. (1976). The industrial market for ethyl alcohol in the Federal Republic of Germany. In *Molasses and Industrial Alcohol*, p. 101. Dev. Centre, OECD, Paris.

Neish, A. C. and A. C. Blackwood (1951). Dissimilation of glucose by yeast at poised hydrogen-ion concentrations, *Can. J. Technol.*, **29**, 123.

Noguchi, S. (1981). Technology of ethanol manufacturing. Use of byproducts and wastewater treatment technology. *Hakko To Kogyo*, **39**, 700.

Norman, B. E. (1978). The application of polysaccharide degrading enzymes in the starch industry. In *Microbial Polysaccharides and Polysaccharases*, ed. R. C. W. Berkley, G. W. Gooday and D. C. Ellwood, p. 339. Academic, London.

Novo Industrias (1976). Novo enzymes for alcohol production. Novo Industrias, Bagsvaerd, Denmark.

OECD Staff (1976). Summary of proceedings of the meeting of experts on molasses and industrial alcohol. In *Molasses and Industrial Alcohol*, p. 13. Dev. Centre, OECD, Paris.

Othmer, D. F. (1936). Vapor reuse method, separation of mixtures of volatile liquids. *Ind. Eng. Chem.*, **28**, 1435.

Othmer, D. F. and T. O. Wentworth (1940). Absolute alcohol; an economical method for its manufacture. *Ind. Eng. Chem.*, **32**, 1588.

Otsuki, H. and F. Williams (1953). Effect of pressure on vapor–liquid equilibria for the system ethyl alcohol–water. *Chem. Eng. Prog. Symp. Ser. 49*, **6**, 55.

Oura, E. (1974). Effect of aeration on the biochemical composition of baker's yeast. *Biotechnol. Bioeng.*, **16**, 1197.

Owen, W. L. (1933). Production of industrial alcohol from grain by Amylo process. *Ind. Eng. Chem.*, **25**, 87.

Panicker, P. K. N. (1964). Vapor re-compression system for steam economy in alcohol distillation plants. In *Ethyl Alcohol Production Technique*, p. 37. Noyes Dev. Corp., Pearl River, New York.

Parekh, H. V. (1964). *Solvent Extraction of Vegetable Oils*, 2nd edn. India Central Oilseed Committee, Hyberabad.

Parkinson, G. (1981). Battelle maps ways to pare ethanol costs. *Chem. Eng. (NY)*, **88** (11), 29.

Paturau, J. M. (1982). *By-Products of the Cane Sugar Industry*, 2nd edn. Elsevier, New York.

Pearl, I. A. (1958). Lignin as a raw material for the production of pure chemicals. *J. Chem. Educ.*, **35**, 502.

Pemberton, R. C. and C. J. Mash (1978). Thermodynamic properties of aqueous non-electrolyte mixtures. *J. Chem. Thermodyn.*, **10**, 867.

Perez, J. P. (1981). *Enzymatic Hydrolysis of Corn Stover. Process, Development and Evaluation*. Report LBL-14223, Lawrence Berkeley Laboratory, Berkeley, Ca.

Perry, R. H. and C. H. Chilton (1973). *Chemical Engineers' Handbook*, 5th edn. McGraw-Hill, New York.

Pfeifer, V. F. and C. Vojnovich (1952). Continuous sterilization of media in biochemical processes. *Ind. Eng. Chem.*, **44**, 1940.

Pipyn, P. and W. Verstraete (1979). A pilot scale anaerobic upflow reactor treating distillery wastewaters. *Biotechnol. Lett.*, **1**, 495.

Pirt, S. J. and W. Kurowski (1970). Extension of the theory of chemostat with feedback of organisms. *J. Gen. Microbiol.*, **63**, 357.

Pleeth, S. J. W. (1949). *Alcohol: A fuel for Internal Combustion Engines*. Chapman and Hall, London.

Potter, F. (1982). Fuel alcohol trade statistics. *Fuel Alcohol USA*, **4** (2), 35.

Prince, I. G. and J. P. Barford (1982a). Tower fermentation using *Zymomonas mobilis* for ethanol production. *Biotechnol. Lett.*, **4**, 525.

Prince, I. G. and J. P. Barford (1982b). Induced flocculation of yeasts for use in the tower fermenter. *Biotechnol. Lett.*, **4**, 621.

Process Engineering Co. (1980). *Power Alcohol*. Process Engineering Co., Mannedorf, Switzerland.

Rahn, O. (1952). Acid and alcohol tolerance imparted by thiamine, *Growth*, **16**, 59.

Ramana Rao, B. V. (1964). Scaling in wash columns. In *Ethyl Alcohol Production Technique*, p. 55. Noyes Dev. Corp., Pearl River, New York.

Rastogi, M. K. and C. R. Krishna Murti (1964). Preparation and properties of a protein digest made from distillery sludge. In *Ethyl Alcohol Production Technique*, p. 98. Noyes Dev. Corp., Pearl River, New York.

Reich, G. T. (1942). Molasses elaboration. *Trans. Am. Inst. Chem. Eng.*, **38**, 1049.

Reich, G. T. (1945). Molasses stillage. *Ind. Eng. Chem.*, **37**, 534.

Remirez, R., R. Grover and L. Marion (1980). The ethanol race: waiting for the government plan. *Chem. Eng. (NY)*, **87** (5), 80.

Renshaw, T. A., S. F. Sapakie and M. C. Hanson (1982). Concentration economics in the food industry. *Chem. Eng. Prog.*, **78**, (5), 33.

Ricard, E. and H. Guinot (1928). Process of dehydration of alcohol for carburants. US Pat. 1 659 958.

Robinson, C. S. and E. R. Gilliland (1950). *Elements of Fractional Distillation*, 4th edn., Chap. 10. McGraw-Hill, New York.

Rogers, P. L., K. J. Lee and D. E. Tribe (1979). Kinetics of alcohol production by *Zymomonas mobilis* at high sugar concentrations. *Biotechnol. Lett.*, **1**, 165.

Rogers, P. L., D. Phil, K. J. Lee, M. E. Tribe and D. E. Tribe (1980). High productivity ethanol fermentations with *Zymomonas mobilis*. *Process Biochem.*, **15** (6), 7.

Ronkainen, P. (1978). The use of stillage water in the mashing of grain as a means of energy conservation. *J. Inst. Brew.*, **84**, 115.

Rose, A. H. (1967). Alcoholic fermentation. In *The Encyclopedia of Biochemistry*, ed. R. J. Williams and E. M. Lansford, Jr., p. 25. Reinhold, New York.

Rose, A. H. and M. J. Beavan (1981). End-product tolerance of ethanol. In *Trends in the Biology of Fermentations for Fuels and Chemicals*, ed. A. Hollaender, p. 513. Plenum, New York.

Rosen, K. (1978). Continuous production of alcohol. *Process Biochem.*, **13** (5), 25.

Rosenberg, S. (1980). Fermentations of pentose sugars to ethanol and other neutral products by microorganisms. *Enzyme Microb. Technol.*, **2**, 185.

Rudolfs, W. (1953). *Industrial Wastes*. Reinhold, New York.

Rudolfs, W. and E. H. Trubnick (1950). Treatment of compressed yeast wastes. *Ind. Eng. Chem.*, **42**, 612.

Ruf, E. W., W. H. Stark, L. A. Smith and E. E. Allen (1948). Alcoholic fermentation of acid-hydrolyzed grain mashes. *Ind. Eng. Chem.*, **40**, 1154.

Samaniego, R. and R. L. Srivastas (1971). Effect of pretreatment of molasses and recycling of yeast on ethanol fermentation. *Sugar News*, **47**, 301.

Sastry, C. A. and G. J. Hohanrao (1964). Treatment and disposal of distillery wastes. In *Ethyl Alcohol Production Technique*, p. 88. Noyes Dev. Corp., Pearl River, New York.

Sattler, L. and F. W. Zerban (1945). Unfermentable reducing substances in molasses. *Ind. Eng. Chem.*, **37**, 1133.

Sattler, L. and F. W. Zerban (1949). Unfermentable reducing substances in molasses. *Ind. Eng. Chem.*, **41**, 1401.

Savage, P. R. (1979). Waste disposal with an energy bonus. *Chem. Eng. (NY)*, **86** (11), 116.

Scheller, W. A. and B. J. Mohr (1977a). Gasoline does, too, mix with alcohol. *Chemtech*, **7**, 616.

Scheller, W. A. and B. J. Mohr (1977b). *Net Energy Analysis of Ethanol Production*. Dept. Chem. Eng., Univ. Nebraska, Omaha, Ne.

Schmidt, E. (1947). Protein and fat production from sulfite liquor by means of yeast. *Angew. Chem.*, **A19**, 16.

Sherman, P. D. and P. R. Kavasmaneck (1978). Ethanol. In *Kirk-Othmer Encyclopedia of Chemical Technology*, 3rd edn., ed. M. Grayson, Vol. 9, p. 338. Wiley, New York.

Skogman, H. (1979). Effluent treatment of molasses based fermentation wastes. *Process Biochem.*, **14** (1), 5.

Solomons, G. L. (1969). Constituents of fermentation culture medium. In *Materials and Methods in Fermentation*, ed. G. Solomons, p. 115. Academic, New York.

Spatz, D. D. (1977). Reclamation via reverse osmosis. *Chemtech*, **7**, 696.

Spencer, G. L. (1963). *Cane Sugar Handbook*, 9th edn. Wiley, New York.

Staff, Chem. Eng. (1981). Chementator. *Chem. Eng. (NY)*, **88** (16), 17.

Staff, Chem. Eng. News (1979). Continuous cellulose to glucose process. *Chem. Eng. News*, **57** (41), 19.

Staff, Chem. Eng. News (1981). USDA process converts xylose to ethanol. *Chem. Eng. News*, **59** (25), 55.

Staff, Chem. Eng. News (1980). Gasohol holds key to ethanol outlooks. *Chem. Eng. News*, **58** (20), 21.

Staff, Chem. Eng. News (1982). Growth slows for alcohols as octane boosters. *Chem. Eng. News*, **60** (14), 21.

Staff, Chem. Mark. Rep. (1982a). Ethanol makers drop prices as grain product gains edge on producers of synthetic. *Chem. Mark. Rep.*, **221** (8), 7.

Staff, Chem. Mark. Rep. (1982b). Chemical profile. *Chem. Mark. Rep.*, **221** (21), 58.

Staff, Chem. Mark. Rep. (1982c). Ethanol capacity to rise by 200 million gallons in the next eight months. *Chem. Mark. Rep.*, **221** (21), 5.

Staff, Chem. Mark. Rep. (1982d). Weekly chemical commodity price quotes. *Chem. Mark. Rep.*, **221** (1), 42.

Staff, Feedstuffs (1982). Ingredient market. *Feedstuffs*, **54** (36), 36.

Staff, Fuel Alcohol (1982). Fuel market update. *Fuel Alcohol USA*, **4** (7), 41.

Stark, W. H. (1954). Alcoholic fermentation of grain. In *Industrial Fermentations*, ed. L. A. Underkofler and R. J. Hickey, Vol. 1, p. 17. Chem. Pub. Co., New York.

Stoddart, W. E. (1962). Continuous malting. *Food Eng.*, **34** (4), 55.

Sundaram, S. and V. Pachaiyappan (1975). Distillery wastes—disposal and by-product recovery. *Chem. Age India*, **26** (2), 97.

Suomalainen, H. and P. Ronkainen (1968). *Tech. Q. Master Brew. Assoc. Am.*, **5**, 119.

Suzuki, M. (1976). The Japanese alcohol market. In *Molasses and Industrial Alcohol*, p. 98. Dev. Centre, OECD, Paris.

Thampy, R. T. (1964). Chemicals from alcohol. In *Ethyl Alcohol Production Technique*, p. 132. Noyes Dev. Corp., Pearl River, New York.

Tyler, R. G. (1947). Effect of alcoholic fermentation and growing fodder yeast on B.O.D. of sulfite waste liquor. *Sewage Works J.*, **19** (1), 70.

Underkofler, L. A., G. M. Severson and K. J. Goering (1946). Saccharification of grain mashes for alcoholic fermentation. *Ind. Eng. Chem.*, **38**, 980.

Underwood, A. J. V. (1935). The historical development of distilling plant. *Trans. Inst. Chem. Eng. (London)*, **13**, 34.

Unger, E. D., W. H. Stark, R. E. Scalf and P. J. Kolachov (1942). Continuous aerobic process for distiller's yeast. *Ind. Eng. Chem.*, **34**, 1402.

Unger, E. D., H. F. Wilkie and H. C. Blankmeyer (1944). The development and design of a continuous cooking and mashing system for cereal grains. *Trans. Am. Inst. Chem. Eng.*, **40**, 421.

USBATF (1980). *Monthly Statistical Release—Distilled Spirits*. October. US Bureau of Alcohol, Tobacco and Firearms, Washington.

USDA (1950). Methods and costs of producing alcohol from grain by the fungal amylase process on a commercial scale. *Tech. Bull.*, No. 1024. US Dept. Agriculture, Washington.

US Department of Treasury (1942). *Regulations 3*. US Dept. Treasury, Bureau of Industrial Alcohol, Washington.

USDOE (June 1979). Report PE0012. US Department of Energy, Washington.

USP (1980). *USPXX—NFXV*. Mack Pub. Co., Easton, Pa.

Venkiteswaran, S. L. (1964a). Operational efficiency in the ethyl alcohol industry in India. In *Ethyl Alcohol Production Technique*, p. 49. Noyes Dev. Corp., Pearl River, New York.

Venkiteswaran, S. L. (1964b). Ethyl alcohol *vs.* petrochemical raw materials for organic chemical industries in India. In *Ethyl Alcohol Production Technique*, p. 111. Noyes Dev. Corp., Pearl River, New York.

Vervalin, C. H. (1980a). Biomass fuels may account for 10% of US energy in year 2020. *Hydrocarbon Process.*, **59** (8), 15.

Vervalin, C. H. (1980b). Gasohol, alcohol consumption forecast in F. & S. report. *Hydrocarbon Process.*, **59** (8), 15.

Walker, R. D. and R. A. Morgen (1946). Protein feed from sulfite waste liquor. *Paper Trade J.*, **123** (6), 43.

Walsh, T. J. and H. R. Bungay (1979). Shallow depth sedimentation of yeast cells. *Biotechnol. Bioeng.*, **21**, 1081.

Wang, D. I. C., C. L. Cooney, A. L. Demain, P. Dunnil, A. E. Humphrey and M. D. Lilly (1979). *Fermentation and Enzyme Technology*. Wiley, New York.

Watson, G. T. (1970). Effects of sodium chloride on steady-state growth and metabolism of *Saccharomyces cerevisiae*. *J. Gen. Microbiol.*, **64**, 91.

Weast, R. (1971). *Handbook of Chemistry and Physics*, 55th edn. Chemical Rubber Co., Cleveland, Oh.

Webb, F. C. (1964). Industrial alcohol. In *Biochemical Engineering*, ed. F. C. Webb, p. 630. Van Nostrand, London.

Wentworth, T. O. and D. F. Othmer (1940). Absolute alcohol, an economical method for its manufacture. *Trans. Am. Inst. Chem. Eng.*, **36**, 785.

Wentworth, T. O., D. F. Othmer and G. M. Pohler (1943). Absolute alcohol, an economical method for its manufacture. Vol. II, plant data. *Trans. Am. Inst. Chem. Eng.*, **39**, 565.

Wiley, A. J. (1954). Food and feed yeast. In *Industrial Fermentations*, ed. L. A. Underkofler and R. J. Hickey, Vol. 1, p. 307. Chemical Pub. Co., New York.

Wilke, C. R., B. Maiorella, A. Sciamanna, K. Tangnu, D. Wiley and H. Wong (1983). *Enzymatic Hydrolysis of Cellulose: Theory and Applications*. Haigh and Hochland, Manchester.

Wilkie, H. F. and J. A. Prochaska (1943). *Fundamentals of Distillery Practice*. Joseph E. Seagram and Sons, Louisville, Ky.

Wong, H. (1983). *Ph.D. Thesis*. University of California, Berkeley, Ca.

Worthy, W. (1981). Cellulose to ethanol projects losing momentum. *Chem. Eng. News*, **59** (49), 35.

Yand, V. and S. Trindade (1979). Brazil's gasohol program. *Chem. Eng. Prog.*, **75** (4), 11.

Yarovenko, V. L. (1975). O factore sterilnosti pri neprenyvnom kultivirovanii microorganismov. *Microbiol. Prom.*, **6** (126), 4.

Yarovenko, V. L. (1978). Theory and practice of continuous cultivation of microorganisms in industrial alcohol processes. *Adv. Biochem. Eng.*, **9**, 1.

Young, S. and Fortey (1902). *J. Chem. Soc.*, **81**, 739.

Zerban, F. W. and L. Sattler (1942). Unfermentable reducing substances in molasses. *Ind. Eng. Chem.*, **34**, 1180.

Ziffer, J. and M. C. Iosit (1982). High ethanol yields using *Aspergillus oryzae* koji and corn media. *Biotechnol. Lett.*, **4**, 573.

Ziminski, R. D. (1981). Conversion of cellulose waste to low cost ethanol. Presented at the 1981 Annual Meeting of Am. Inst. Chem. Eng., New Orleans.

44

Acetone and Butanol

J. C. LINDEN, A. R. MOREIRA and T. G. LENZ
Colorado State University, Fort Collins, CO, USA

44.1 INTRODUCTION

44.1.1 History

The acetone–butanol fermentation has a long history as a successful industrial fermentation process. The earliest work on this fermentation was performed by Pasteur in 1862, who investigated the production of butanol from lactic acid and calcium lactate (Compere and Griffith, 1979). Prazmowski proposed the name of 'clostridium' for these anaerobic spore formers that produced butanol, resulting in the genus *Clostridium* (Buchanan and Gibbons, 1974).

The fact that acetone was also a fermentation product was not known until Schardinger made the discovery in 1905 (Underkofler and Hickey, 1954). The interest in commercializing the process began in 1909 in England primarily as a means to obtain butadiene as raw material for synthetic rubber. Strange and Graham Ltd., obtained the services of Perkin, Weizmann, Fernback, Schoen and others to study the problems associated with this process (Prescott and Dunn, 1959). After considerable investigation, it was decided that butadiene was a most desirable raw material for synthetic rubber, and this diene could be produced from butanol. In 1912, Weizmann broke from the group, but continued to study the fermentation independently. After about two years, he found an organism which he called *Clostridium acetobutylicum* which successfully fermented starchy grains to produce acetone, butanol and ethanol. He applied for patents on the process and British Patent 4845 (1915) and US Patent 1 315 585 (1919) were issued (Shreve, 1967).

With the outbreak of World War I in 1914, the production of acetone was of great interest to the UK for the manufacture of cordite, the explosive used in naval warfare. Strange and Graham Ltd. was contracted to produce acetone, but their process was unsuccessful until Weizmann was placed in charge of the plant and installed his process. Because of the shortage of corn in Britain the process was transferred in August 1916 to Canada where it operated until 1918 (Prescott and Dunn, 1959). Shortly after the United States entered the war, the Weizmann process was operated at Terre Haute, Indiana, for acetone production. Considerable quantities of acetone were produced, along with twice as much butanol. The storage and disposal of the butanol presented a problem for some time. It was shortly after the war that E. I. DuPont de Nemours and Company developed nitrocellulose lacquers for the automobile industry and it was found that butyl acetate, derived from butanol, was the solvent of choice for this coating system (Steel, 1958). During 1940, over 45×10^6 and 90×10^6 kg of butanol and acetone, respectively, were produced (Shreve, 1967). However, changes in the availability of carbohydrate feedstocks, problems which developed with phage infestation, and the availability of inexpensive petrochemicals in the 1950s, caused a decrease in the production of acetone and butanol by fermentation. It was estimated in 1965 that only 5% of the acetone and 10% of the butanol produced in the world were obtained by fermentation (Faith *et al.*, 1965). Former producers by fermentation in the United States that ceased operation during the 1950s included Commercial Solvents Corp., Terre Haute, IN; Publiker Industries, Inc., Philadelphia, PA; US Industrial Chemicals Co., New York, NY; and Western Condensing Co., Adell, WI. Production by fermentation was abandoned during the 1960s in Japan (Peppler, 1967).

Abstracts reporting activities in the acetone–butanol fermentation of beet molasses in the Soviet Union continued to appear throughout the 1970s. Commercialized production is still carried out by the fermentation of cane molasses in South Africa (Spivey, 1978).

44.1.2 Current Production and Uses

Currently, butanol is manufactured from ethylene and triethylaluminum and is used primarily in the manufacture of lacquers, rayon, detergents, brake fluids and amines for gasoline additives. It can also be used as a solvent for fats, waxes, resins, shellac and varnish. Acetone is a by-product from the manufacture of phenol from cumene, a by-product of oxidation cracking of propane, or can be manufactured by chemical reduction of isopropanol. Acetone is used mostly as a solvent for fats, oils, waxes, resins, rubber plastics, lacquers, varnishes and rubber cements.

44.1.3 Proposed Uses

The acetone–butanol fermentation currently has potential because butanol has many characteristics which make it a better liquid fuel extender than ethanol, now used in the formulation of gasohol. Three of the more important characteristics which make butanol a better liquid fuel extender are: (1) its low vapor pressure, (2) its low miscibility with water and (3) butanol, unlike ethanol, is completely miscible with diesel fuel even at low temperatures (see Table 1).

Although economically attractive when based on waste-type materials such as cheese whey (Lenz and Moreira, 1980), the acetone–butanol fermentation has a number of drawbacks which must be addressed before any attempt for commercial production is made. The major one is the very low levels of acetone and *n*-butanol obtained in the final fermentation broth. With current technology, such levels are only 0.7 and 1.4% (w/v), respectively. This results in the need for large vessels for fermentation and an energy-intensive distillation recovery of the solvents. Sterilization of large volumes of media is also highly energy intensive. Additional difficulties with this fermentation include the need for strict anaerobic conditions and delicate culture maintenance and propagation.

Table 1 Characteristics of Chemically Pure Fuels

Fuel	Molecular weight	Specific gravity	Boiling point (°C)	Vapor pressure at 37.7 °C (p.s.i.)	Combustion energy (kJ kg^{-1})	Latent heat (kJ kg^{-1})	Solubility (parts in 100 parts H$_2$O)	Stoichiometric air–fuel ratio
Methanol	32	0.79	65	4.6	23 864.8	1170.0	∞	6.5
Ethanol	46	0.79	78	2.2	30 610.6	921.1	∞	9
Butanol	74	0.81	117	0.3	36 681.0	432.6	9	11.2
Octane	114	0.70	210	1.72	48 264.5	360.5	insoluble	15.2
Hexadecane	240	0.79	287	3.46	47 264.3	—	insoluble	15

44.2 FERMENTATIVE PRODUCTION

The organism *Clostridium acetobutylicum* will ferment a variety of carbohydrates such as lactose, glucose, xylose, fructose, arabinose, galactose, maltose, mannose, starch and sucrose (Buchanan and Gibbons, 1974), and produce a variety of organic solvents including butyric and acetic acids, butanol, acetone, ethanol, carbon dioxide, hydrogen, isopropanol, formic acid, acetone (acetylmethylcarbinol or 3-hydroxy-2-butanone) and a yellow oil, which is a complex mixture of higher alcohols, higher acids and esters (Stanier *et al.*, 1976). Butanol, acetone and ethanol are normally considered to be the principal products of this fermentation.

44.2.1 Raw Materials

44.2.1.1 Starch

The production of acetone and butanol by the Weizmann process utilized starch as substrate (Beech, 1953). *C. acetobutylicum* (Weizmann) possessed amylolytic and saccharolytic enzyme activities required to hydrolyze gelatinized starch to glucose and maltose. Concentrations of corn of 8 to 10% (5 to 6.5% starch) were readily utilized; average solvent yields of 38.0%, based on sugar fermented were reported. Wheat, milo and rye served equally well. Potatoes could not be handled at high concentrations because of viscosity problems. Normally, the germ and bran were removed from the grains prior to milling in order to recover these valuable by-products. The grain mash was first gelatinized for 20 min at 65 °C and then sterilized for 60 min at 102 °C. This cooked mash was cooled by means of double-pipe heat exchangers to 35 °C as it was pumped through steam-sterilized lines to fermenters of 250 000–2 000 000 l capacity. The fermenter was inoculated either during or after the filling process with 3% (v/v) inoculum from a 24 hour culture tank. In some cases, 30 to 40% of the total aqueous volume of the mash was provided by stillage from which neutral fermentation solvents from previous fermentations had been stripped. This so-called slopping-back procedure added certain nutrients to the medium which increased yield, aided in foam control and reduced the amount of stillage which required evaporation and drying for feed supplement preparations.

The fermentations which lasted 2 to 2.5 days, passed through three phases. The first was characterized by rapid growth, production of acetic and butyric acids, and evolution of carbon dioxide and hydrogen, primarily the latter. The pH, initially 6.0 to 6.5, decreased during this phase to a constant value determined by the particular buffering capacity of the substrate and medium. The 'titratable acidity' however increased to a maximum which defined the end of the first phase. The second phase, called the acid break, began with a sharp decrease in the 'titratable acidity' after 12 to 14 hours. The organisms had begun to convert the acidic products to neutral solvents. Gas production increased, a greater proportion being carbon dioxide than in the earlier phase of the fermentation. During the third phase, the rates of gas and solvent production decreased until the cells autolyzed. The corn substrate fermentation ended at between pH 4.2 and 4.4.

44.2.1.2 Molasses

Beech (1952) described the use of a variety of other *Clostridium* species used for fermentation of sugar in molasses. The fermentation process was similar to that described above for starch except that the mash could be sterilized at a lower temperature, the fermentation was conducted at 31 to 32 °C, cleaning of tanks and lines was easier, higher concentrations of sugars could be used and a greater ratio of butanol to acetone plus ethanol was produced. It was necessary to add nitrogen and phosphate nutrients to molasses fermentations (*e.g.* 1.0 to 1.4% ammonia based on sugar present). The ammonia was usually added to maintain the pH at 5.6 after the 'titratable acidity' dropped the pH from an initial value of 6.5 during the first 16 hours of fermentation. The molasses production medium contained 6% sugar calculated as sucrose; average solvent yields of 30.0% were reported (Beech, 1952). Shin *et al.* (1982) have reported successful fermentation of sorghum molasses to produce acetone and butanol.

44.2.1.3   Cheese whey

Since *C. acetobutylicum* had the capability of metabolizing lactose, it was possible to utilize cheese whey as a feedstock for the acetone–butanol fermentations. Cheese whey, the fluid which remains following separation of the curd when converting milk into cheese, contains about 6% lactose by weight and in the order of 1% protein, depending on the type of cheese manufacture. The level of sugar in cheese whey coincided with that required in the medium for the acetone–butanol fermentation. Cheese whey preparations from which approximately 75% of the protein was removed by simple heat denaturation, acidification or ultrafiltration have proven to be suitable fermentation substrates.

The fermentation of lactose in cheese whey produced some interesting modifications in the product distribution and yield (Lindberg and Moreira, 1982a). The classical solvent ratio produced by *C. acetobutylicum* from corn mash was 2 to 1 butanol to acetone. With pure glucose as substrate, a ratio of 2.7 to 1 was obtained. Under the same conditions, substitution of lactose for glucose resulted in a product stream in which the relative proportion of butanol to acetone was 2.9 to 1. The fermentation of cheese whey resulted in a strikingly different set of product concentration profiles. For instance, the normal intermediary build-up of acetic and butyric acid was not observed and the final ratio of butanol to acetone was 9.7 to 1. Maddox (1980) also reported a 10 to 1 butanol to acetone ratio in fermentation of an ultrafiltrate of cheese whey. However, Lindberg and Moreira (1982b) reported that fermentations on cheese whey from which 80% of the protein was removed by heat precipitation and ultrafiltration yielded butanol to acetone ratios of 12.3 to 1 at 37 °C and 20.6 to 1 at 33 °C. Compere and Griffith (1979) also obtained 10 to 1 ratios of butanol to acetone with ultrafiltered cheese whey, using one particular strain of *C. acetobutylicum*, but that result was not consistent under all experimental conditions. The dependence of product distribution on micronutrient components will be discussed below in Section 44.4.1.1.

44.2.1.4   Jerusalem artichoke

An assortment of raw materials such as beet molasses, wheat, rice, horse chestnuts, potatoes and Jerusalem artichokes, have also been used in this fermentation. Wendland, Fulmer and Underkofler (1941) studied the acetone–butanol fermentation with Jerusalem artichoke. The French government was recently making a concerted effort to produce quantities of butanol for emergency fuel purposes from Jerusalem artichokes. The potential for utilizing not only fructose from tuber-derived inulin, but also the structural carbohydrates in the vegetative stalk of the plant has been investigated (Linden *et al.*, 1982). Until anthesis, the inulin content of the stalk and leaves is greater than that of the tubers. The protein content of the leaves is great enough to offer substantial by-product credit from extracted leaf protein. Butanol to acetone ratios of 2.75 to 1 (w/w) from acid hydrolyzed inulin from Jerusalem artichoke tubers have been obtained.

44.2.1.5   Lignocellulosic hydrolysates

C. acetobutylicum can anaerobically ferment different carbohydrates into a spectrum of products including *n*-butanol, acetone, ethanol, butyric acid, acetic acid and acetoin. Hydrolysates of lignocellulosics, *i.e.* wood, paper, crop residues, *etc.*, primarily contain glucose and cellobiose as well as galactose, mannose and the pentose sugars such as D-xylose and L-arabinose, from the hemicellulose fraction. Glucose, cellobiose, mannose and arabinose have been shown to be fermentable (Yu and Saddler, 1983). Galactose and xylose were utilized poorly and primarily yielded acids. Mes-Hartree and Saddler (1982) also noted that under the conditions of their study, better solvent production was obtained from sugars arising from the *erythro* configuration (glucose, mannose, arabinose) as compared to the *threo* configuration (xylose, galactose).

Compere and Griffiths (1979) have shown that widely varying product yields were obtained when different substrate concentrations of xylose and arabinose were used. Yu and Saddler (1983) reported a significant increase in acetone and butanol production occurred when *C. acetobutylicum* was grown on D-xylose in the presence of 1 g l^{-1} acetic acid or 3 g l^{-1} butyric acid added prior to inoculations. Saddler *et al.* (1983) reported very low conversion rates of the pentose fraction from steam exploded aspen wood by *C. acetobutylicum*, but this may have been due to the extreme sensitivity of the organism to furfurals (Langlykke *et al.*, 1948). Langlykke *et al.* (1948) reported an average of 30.6% yield, based on sugar fermented, of solvents with normal product distribution when fermenting corn cob saccharification liquors, composed primarily of xylose, with *C. butylicum* (NRRL 594). Wiley *et al.* (1941) also reported similar product ratios and yield using *C. butylicum* for the production of acetone and butanol from the wood sugars in waste sulfite liquor which is primarily xylose, following lignin precipitation with calcium ions. Xylose utilization by *C. acetobutylicum* was also shown to be improved with concentrations of up to 10 g l^{-1} calcium carbonate by Mes-Hartree and Saddler (1982).

44.2.2 Current and Potential Processes

44.2.2.1 South African operation

The commercialized production of acetone and butanol is still carried out by the fermentation of cane molasses with the organism *C. acetobutylicum* in South Africa by National Chemical Products Ltd., Germiston, South Africa. The fermenters used there are 90 000 l working volume and considerable detail has been given (Spivey 1978) about the operation of the plant. Distillation is used to recover solvent from the acetone–butanol fermentation broth, which contains approximately 2% (w/v) solvents (Beech, 1952). The fermentation broth is fed at a constant rate to the top of a beer still containing 30 perforated plates, from which an approximately 40/60 (wt %) solvent/water stream was removed overhead from slops (water and stillage).

The mixed solvents are then separated by batch fractionation. Acetone, butanol and a mixture of ethanol and isopropyl alcohol (formed by reduction of acetone in the beer well) are obtained as separate fractions. The butanol is dried by removal of the distillate through a decanter. A high boiling fraction containing higher alcohols, esters and organic acids is also obtained. The slops contained bacterial cells rich in riboflavin and B vitamins which are concentrated, dried and shipped around the world as animal feed. Rumumco Ltd. of Burton-on-Trent in Great Britain has handled the distribution of the dried products from the South African NCP operation.

44.2.2.2 Cheese whey

Lenz and Moreira (1980) presented preliminary design criteria for an acetone–butanol fermentation facility based on high-quality molasses feedstock and found the economics to be unattractive. The total production costs were found to be slightly higher than the total annual income even when credit was taken for all of the fermentation by-products. This was mainly due to the high cost of the molasses feedstock. Indeed, even though the molasses price has been under government control, the South African operation described above has been intermittent recently for this reason.

The economics for the acetone–butanol fermentation using a liquid whey feed are strikingly different from those for molasses feed. Lindberg and Moreira (1982b) updated the previous preliminary design (Lenz and Moreira, 1980) used as a base-case model to reflect 1982 prices and the substantially more attractive butanol yields obtained from whey as compared to molasses. Lenz and Moreira (1980) obtained information relative to size and location of typical cheese manufacturing facilities from the Wisconsin Department of Agriculture. This and other information regarding waste-treatment and trucking costs for liquid cheese whey were integrated with base-case economics to give the whey-based economics for a 45×10^6 kg of solvents year^{-1} plant. This analysis differs from the former in that the capital and energy intensive ultrafiltration which was considered by Lenz and Moreira (1980) was replaced by protein recovery through heat denaturation and rotary drum filtration. There was little likelihood for sale of carbon dioxide in Wisconsin at prices paid for oil recovery operation, so gaseous products were eliminated from the income summary. An anaerobic waste treatment stage which would generate methane was considered in the whey analysis by Lenz and Moreira (1980) and was retained in this analysis for partial supply of energy requirements by utilization of minor fermentation product streams such as acetic acid, butyric acid, acetoin and other organics. Since separation and compression costs for hydrogen were not considered in the original equipment and operating analysis, allowance for the anaerobic conversion to methane of 25% of the carbon dioxide and 100% of the hydrogen produced in the *C. acetobutylicum* fermentations by methanogenic bacteria in the waste treatment stages would replace approximately 10% of the product recovery costs. It was assumed that the liquid whey was available for the cost of trucking an average 160 km from cheese plants to a central fermentation facility at current trucking rates.

This economic study provided a preliminary estimate ($\pm$ 25%) and was based on the flow sheet from the Lenz–Moreira (1980) acetone–butanol fermentation plant design in Figure 1. The following characteristics were used: (a) about 45×10^6 kg per year solvent; (b) location in the state of Wisconsin; (c) 300 day per year operation, with 6 hour staging of fermentations; (d) costs reported are end 1982 values; and (e) protein is recovered through heat denaturation and rotary-drum filtration.

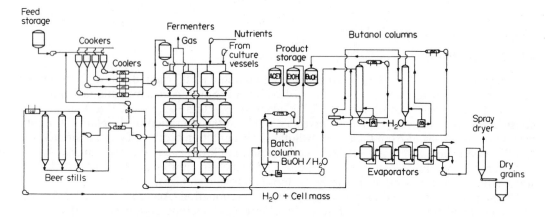

Figure 1   The process flow-sheet for the acetone–butanol fermentation (Lenz and Moreira, 1980)

The equipment costs were utilized to generate fixed capital and total capital investment costs (Peters and Timmerhaus, 1980). The results obtained are shown in Tables 2 and 3. These capital cost figures were then used to generate the manufacturing and production cost estimates shown in Table 4. Finally, the annual sales revenues were estimated as shown in Table 5; the solvent values used in this table are those given in the Chemical Marketing Reporter at the end of 1982.

The economics of protein recovery and the production of acetone and butanol *via* fermentation have been examined for a 45×10^6 kg solvents year^{-1} plant in Table 6. This production facility required a total capital investment of $27.7 million and produced an after tax income of about $9 million, yielding a net return on investment of 32.5%. Considering the preliminary nature of the estimate ($\pm$ 25%), the economics for the fermentation production of acetone and butanol would be marginally attractive. Maddox *et al.* (1981) considered the economics of *n*-butanol production from whey ultrafiltrate on a smaller scale, *i.e.* 1.0×10^6 kg solvents year^{-1}, which could be supplied with whey from a medium-sized cheese plant. The analysis showed that *n*-butanol from such a facility would cost about twice current market value to produce. However, the authors demonstrated how greater through-put would reduce production costs to near the breakeven point.

44.2.2.3   Lignocellulose hydrolysates

A process for production of butanol from cellulose waste has been given by Gibbs (1983), and includes a discussion of the impact of genetic manipulation of *C. acetobutylicum*. The yields predicted were thus greater than currently feasible and the fermentation temperature was twice the current optimum.

Once subjected to various types of pretreatment, the hemicellulose and cellulose fractions of

Table 2 Equipment Costs for a 45×10^6 kg Solvents per Year Acetone–
Butanol Fermentation Facility[a]

Item	Size	Cost ($\$ \times 10^3$)
Whey storage tank	13 249 m³ (3.5×10^6 gal)	398
Cookers (4 units)	18.9 m³ (5.0×10^3 gal)	72.6
Rotary drum filter	84 m³ (900 ft²)	65
Coolers (4 units)	279 m² (3.0×10^3 ft²)	102
Holding tank	2800 m³ (7.4×10^5 gal)	113.2
Fermenters (16 units)	1514 m³ (4.0×10^5 gal)	2487.5
Beer still preheater	100 m² (1077 ft²)	14.2
Beer still (3 units)	2.44 m (8 ft) diameter	534.8
Beer still condenser	80.7 m² (869 ft²)	13.9
Batch column	1.83 (6 ft) diameter	190.3
Batch column reboiler	8.4 m² (90 ft²)	11.6
Acetone condenser	31.4 m² (337 ft²)	6
Ethanol cooler	1.23 m² (13.2 ft²)	0.6
Butanol columns (2 units)	1.38 m (4.5 ft) diameter	130
Butanol condensers	134 m² (1447 ft²)	22.2
	192 m² (2079 ft²)	27.2
Surge tank preevaporators	1514 m³ (40×10^5 gal)	155.5
Evaporators (5 units)	300 m² (3200 ft²)	1318
Spray dryer	5.9 m (19.5 ft) diameter	360.7
Butanol storage tank	5110 m³ (13.5×10^5 gal)	425
Acetone storage tank	870 m³ (2.3×10^5 gal)	72
Ethanol storage tank	568 m³ (1.5×10^5 gal)	47.3
Culture inoculum tank	37.9 m³ (10×10^4 gal)	12.1
Purchased and installed equipment cost total		6581.5

[a] Adapted with permission from Lenz and Moreira, 1980.

Table 3 Fixed Capital and Total Capital Investment Summary for a 45×10^6
kg Solvents per Year Acetone–Butanol Fermentation Facility[a]

	Based on fluid processing plant ($)
Direct costs	
Purchased equipment (delivered and installed)	6 581 500
Instrumentation and controls (installed)	802 943
Piping (installed)	2 961 675
Electrical (installed)	493 612
Buildings (including service)	802 943
Yard improvements	460 705
Service facilities (installed)	3 132 794
Land purchased	269 841
Total direct plant cost	15 506 013
Indirect costs	
Engineering and supervision	1 488 577
Construction expenses	1 837 418
Total direct and indirect plant costs	18 832 008
Contractor's fee	941 600
Contingency	1 883 200
Fixed capital investment (FCI)	21 656 808
Working capital	3 854 912
Total capital investment	25 511 720

[a] Adapted with permission from Lenz and Moreira, 1980.

wood residues would undergo enzymatic hydrolysis and would yield the monomeric form of the corresponding sugars, primarily xylose and glucose. Both pentose and hexose sugar solutions could be fed to a fermentation vessel where *C. acetobutylicum* would ferment those sugars to ace-

Table 4 Estimated Annual Manufacturing and Total Production Costs for a 45 × 10⁶ kg Solvents per Year Acetone–Butanol Fermentation Facility[a]

Direct production costs	
Cheese whey (trucking expenses @ $0.57/100 kg)	$17 100 000
Operating labor (10% TPC)	3 665 061
Direct supervisory and clerical labor (1.0% TPC)	366 506
Utilities (18% TPC)	6 597 110
Maintenance and repairs (6% FCI)	1 299 408
Operating supplies (0.75% FCI)	162 426
Laboratory charges (1% TPC)	366 506
Total	$29 557 017
Fixed charges	
Depreciation	$ 2 185 754
Local taxes (2.5% FCI)	541 420
Insurance (0.7% FCI)	151 598
Total	$ 2 878 772
Plant overhead costs (5% TPC)	$ 1 832 530
Total manufacturing cost	$34 268 319
General expenses	
Administrative costs (2% TPC)	733 012
Distribution and selling costs (4.5% TPC)	1 649 277
Total productions costs (TPC)	$36 650 608

[a] Adapted with permission from Lenz and Moreira, 1980.

Table 5 Income Summary for a 45 × 10⁶ kg Solvents per Year Acetone–Butanol Fermentation Facility[a]

Item	Annual quantity [kg (lb)]	Value [$/kg ($/lb)]	Income ($/y)
n-Butanol	40.4 × 10⁶ (88.8 × 10⁶)	0.75 (0.34)	$29 784 000
Acetone	3.4 × 10⁶ (7.5 × 10⁶)	0.71 (0.32)	2 392 000
Ethanol	3.3 × 10⁶ (7.3 × 10⁶)	0.57 (0.26)	1 900 000
Dry slops	64.3 × 10⁶ (142 × 10⁶)	0.094 (0.042)	6 000 000
Protein	21.0 × 10⁶ (46.2 × 10⁶)	0.60 (0.27)	12 600 000
		Total	$52 676 000

[a] Adapted with permission from Lenz and Moreira, 1980.

Table 6 Summary of the Economics of a 45 × 10⁶ kg Solvents per Year Acetone–Butanol Fermentation Facility Using Liquid Whey Waste as Substrate[a]

Total capital investment	$25 511 720
Wastewater treatment	2 238 731
New TCI total	27 750 451
Total production cost	36 650 608
Less 10% of utilities (CH₄ credit)	659 711
	35 990 897
Annual income	52 676 000
Gross profit	
$52 676 000 less $35 990 897	16 685 103
After-tax income (0.54) (16 685 103)	9 009 955
Net return on investment (NROI)	
$\frac{9\ 009\ 955}{27\ 750\ 450} = 32.47\%$	

[a] Adapted with permission from Lenz and Moreira, 1980.

tone, butanol and ethanol. Hydrogen and carbon dioxide would also be products of the fermentation.

Based on sawdust availability of 250 ton day^{-1} and an average composition of 25% lignin and 65% carbohydrate (dry basis), a plant operating 300 days per year would produce 15 000 ton per year of lignin and 13.6×10^6 l per year mixed solvents (acetone + butanol + ethanol). This estimate for solvents production was based on available solvent yield data for oak hydrolysates (Leonard *et al.*, 1947). Utilizing as an estimate the hydrogen and carbon dioxide data for blackstrap molasses (Beech, 1952) and the current selling prices for the relevant commodities, an income estimate per ton of residue was prepared and is shown in Table 7. The total estimated income from the sale of the products derived from 1 ton of residue would be $181.50; even in the worst situation of non-crediting hydrogen, carbon dioxide and fermentation slops, the estimated income would be $139.00. Considering a raw material cost of $20 ton^{-1}, there would seem to be considerable room for processing costs within an economically feasible operation.

Table 7 Estimated Income for Products Derived from 1 ton of Wood Biomass

Product	Value (¢/lb)	Quantity (lb)	Income ($)
Lignin	10.0	400	40.0
Mixed solvents	31.4	316	99.0
Hydrogen	13.0	15	2.0
Carbon dioxide	5.0	600	30.0
Slops	6.0	175	10.5
		Total	181.5

44.3 IMPACT OF ADDED FERMENTATIVE PRODUCTION

44.3.1 Basis: Available Wood Residues

According to a recent report by Arthur D. Little, Inc. (1979), the supply of wood residues available for conversion into fuels and chemicals in the US was estimated at 179 million dry tons in 1976 with the following relative distribution: 163 million tons from commercial logging residues and 16 million tons from primary wood manufacturing residues, including bark. Although the composition of these residues is species specific, the following average figures (on a dry basis) will be assumed in our projections for reasons of simplicity (Cote, 1977; Millett *et al.*, 1976): lignin, 25%; carbohydrate (cellulose + hemicellulose), 65%. Based on the previous figures and assuming reasonable product yields in the proposed conversion scheme, the data shown in Table 8 were then calculated. This table also includes the 1981 US production figures for those chemicals expected to be obtained in the proposed process (Anon., 1982).

Table 8 Estimated Potential Chemicals Production from Wood Biomass

Chemical	Potential quantity (10^9 lb)	1981 US production (10^9 lb)	
Lignin	72	2.5	(phenol)
Butanol	38	0.8	
Acetone	18	2.2	
Ethanol	6	1.1	(synthetic)

In Table 8, the lignin production capacity is compared to the US annual phenol production in view of the intended application of lignin in this process as a phenol–formaldehyde resin extender. It can be seen from this table that if all the biomass were to be used in such a conversion process, the amount of chemicals produced would be one to two orders of magnitude higher than the current production levels.

44.3.2 Petrochemical Displacement

The possibility of achieving total replacement of the petrochemicals used for phenol, butanol, acetone and ethanol manufacture and still stimulate the creation of new domestic markets, or the development of new export markets, certainly points to a promising future for wood as a source of chemicals. Alternatively, if the butanol produced from wood biomass is envisioned as being utilized as a gasoline extender (gasohol production), the 3.8 billion pounds show in Table 8 could replace 5% of the US current annual gasoline consumption. If used in the same way, the ethanol produced could add almost another 1% to this consideration.

While the production potentials discussed above and listed in Table 8 may represent valid maxima today, economically collectable quantities of wood residues would probably be limited to those from primary manufacturing, within limitations of current technology and economy. Thus, the practical potentials would be an order of magnitude lower than those of Table 8, and in rather good balance with current production (demand). It is appropriate, then, to examine the effect of reasonable displacements upon demands for petrochemical feedstocks.

44.3.2.1 *Phenol*

Suitable lignin products could reasonably satisfy 25% of the phenol requirements of the one billion pounds of phenolic resins produced annually. Such displacement would represent about 175 million pounds of phenol. Production of such an amount of phenol requires about 210 million pounds of benzene and 113 million pounds of propylene. These latter two chemicals are primarily derived from petroleum. Note also that production of phenol and its precursors consumes considerable amounts of energy, which are unfortunately not very readily quantifiable.

44.3.2.2 *Butanol*

If some 18% of the butanol derivable from primary wood residues were to be used to displace the butanol currently produced by the petrochemical industry, this would relieve propylene demand by about 420 million pounds per year. Thus, this propylene, now consumed in butanol production, would be available for other feedstocks or for use in gasoline, as an octane improver. (While propylene is one of the principal petrochemical building blocks, 55% of US production goes into gasoline). Energy expended in the butanol manufacturing scheme would be saved in addition.

44.3.2.3 *Acetone*

Similarly, the 1.8 billion pounds of acetone from primary residues could displace 50% of the petrochemical acetone, manufactured by isopropanol dehydrogenation (the other 50% is a by-product of phenol production from cumene hydroperoxide and is not truly displaceable), and the ethanol could approximately halve the current demands of synthetic ethanol upon supplies of ethylene, our most important petrochemical building block. Again, the energy now consumed in manufacture of the displaceable acetone and ethanol (and their precursors) would go on the credit side of the ledger for the hypothetical process.

The preceding analysis, although simplistic in nature, indicates that the butanol, acetone, ethanol and phenol quantities potentially derivable annually from primary manufacturing wood residues are equivalent to a total of about 6 billion pounds of petrochemicals. Although information on the energy required to produce these chemicals from petroleum is not readily available, it can be safely stated that at least a like quantity of crude petroleum will be replaced and can be re-routed to other uses. The energy requirements for the proposed alternative process (wood → chemicals) are also rather speculative at this time; very preliminary calculations show a total energy consumption of about 150 000 BTU to derive one gallon of mixed solvents and 10 pounds of lignin from the wood residue. It seems reasonable to assume at this stage that the energy requirements for both processes will be of the same order of magnitude and consequently the alternate process spares at least as much petroleum as the 6 billion pounds of chemicals produced. Since wood residues can be burned simultaneously to supply energy to the proposed bioconversion scheme (as opposed to what happens, for instance, in a petroleum refinery), the potential petroleum displacement of this project can ultimately be significantly enhanced relative to the 6 billion pounds figure cited earlier.

44.3.2.4 *Hydrogen and carbon dioxide*

The total weight of the gases formed during a fermentation exceeds that of the solvents. In fact, only 38% yields of neutral solvents based on substrate consumption can be expected from the fermentation; the remainder of the fermentable sugars end up as carbon dioxide and hydrogen. Since the most important factor in the economics of the process is substrate cost, utilization of these by-products should be considered in an integrated process. The gases can be separated by selective absorption of carbon dioxide in a liquid amine or by using a membrane separation. Pure hydrogen can be used as a fuel, in fuel cell application to generate electricity or as a reductant for chemical reactions. Examples of the latter are methane formation by methanogenic bacteria, synthesis of ammonia by catalytic reaction with nitrogen at high pressure and temperature, and methanol production by reaction of either CO_2 or CO with hydrogen. Prior to the close of the acetone–butanol fermentation industry in the United States, Commercial Solvents produced ammonia and methanol from fermenter off-gases. Using current technology, the latter has been shown to yield an incremental rate of return of 25 to 30%. Carbon dioxide could be used for dry ice production or for oil recovery operations where a market existed.

44.4 PROCESS DEVELOPMENT

44.4.1 Regulation of the Fermentation

The regulation of the changeover from acid production to the production of neutral solvents appears to be more complex than can be explained by dropping the pH of the medium to 5.0 as once thought (Peterson and Fred, 1932). The exact reason for the shift to production of neutral solvents is not clearly understood, but recent research has uncovered interesting hypotheses. Because butanol currently has a higher commercial value (5.5¢ mol^{-1}) than acetone (4.0¢ mol^{-1}), it would be desirable to direct the fermentation toward maximal butanol production. Hence, the possibilities for regulation of product distribution will be briefly covered in this section.

44.4.1.1 *Phosphate limitation*

Bahl *et al.* (1982a) have described batch fermentations of *C. acetobutylicum* which demonstrate the regulation of butanol and butyric acid production. The biochemical reason for the observed regulation appears to lie in the involvement of butyryl phosphate in the energy metabolism of the cell. This was first appreciated by Gottschalk and Bahl (1981). During the acid phase (while sufficient phosphate is present), butyrate is primarily formed by reaction of two enzymes which occur in *C. acetobutylicum* (Gottschalk and Andreesen, 1979):

(i) phosphotransbutyrylase:

$$\text{butyryl-CoA} + P_i \xrightarrow{P_i} \text{butyryl phosphate} + \text{CoASH}$$

(ii) butyrate kinase:

$$\text{butyryl phosphate} + \text{ADP} \rightarrow \text{butyrate} + \text{ATP}$$

Phosphate limitation does not allow the formation of butyryl phosphate. With the resulting accumulation of butyryl-CoA, butyraldehyde dehydrogenase is induced or activated by an unknown mechanism and catalyzes the formation of butyraldehyde which is further reduced to butanol.

Addition of acetic and butyric acids in continuous cultures by Bahl *et al.* (1982b) and in batch cultures by Yu and Saddler (1983) improved butanol productivities. Concentrations of butyric acid in the nondissociated form in the fermenter have been correlated with the acid shift by Monot *et al.* (1983). Components which accompany the fermentable carbohydrate from various practical sources influence the distribution of fermentation products. For instance, calcium is a low molecular weight component of cheese whey, which is capable of complexing phosphate and which has been historically used to improve butanol yields in the fermentation of grains, molasses and wood wastes.

44.4.1.2 Agitation control

The two phases of the *C. acetobutylicum* fermentation described above are manifestations of the different ways in which the organism regenerates its biochemical reducing potential (Petit-demange *et al.*, 1976). During the acid phase, this is done by production of hydrogen gas. The reaction is catalyzed by a ferredoxin-linked hydrogenase, which also has the ability to shunt the reducing power to the cofactor, NAD. Excess reducing power in this form can be disposed of by reducing butyryl-CoA to butanol in two steps in the second phase of the fermentation. The equilibrium between H_2 and the reduced $NADH_2$ has an important influence on solvent production. Since the hydrogenase catalyzes a reversible reaction, an increase in the concentration of dissolved hydrogen would prohibit the production of more hydrogen and encourage the formation of neutral solvents.

Agitation and gas head space pressure are two means of controlling dissolved gas concentrations. Su *et al.* (1981) found that stirring had an effect on dissolved hydrogen concentrations in the fermentation broth of *C. thermocellum*. Doremus (1983) has shown that greater agitation rates favored the simultaneous production of hydrogen and butyric acid in *C. acetobutylicum* fermentations. Inversely, hydrogen supersaturation in the medium, favored at lower agitation rates, *e.g.* 25 r.p.m., enhanced butanol formation. Butanol and ethanol productivity were also increased under pressurized conditions (15.2 p.s.i.g.). Since the butyric acid productivity peaked earlier in highly agitated fermentations than at lower agitation rates, overall fermenter productivity would be improved by a combination of moderate to high agitation (*e.g.* 300 r.p.m.) during the acid phase followed by low agitation during the solvent phase.

44.4.1.3 Extractive fermentation

Butanol recovery by extraction has represented an area of considerable research. Not only will extractive fermentation increase fermenter productivity, the toxicity of butanol to *C. acetobutylicum* may be controlled. One type of extractive fermentation which has been studied is one in which the extraction involves contacting a recycle stream of fermentation broth outside the fermenter (Pitt *et al.*, 1983). Alternatively, approaches which have shown partial success are *in situ* extraction systems. Corn oil, paraffin oil, kerosene and dibutylphthalate have been examined and found not to affect cell growth (Wang *et al.*, 1979). Corn oil did not affect the productivities and conversion yields of acetone and butanol in an *in situ* fermentation; the total concentrations of acetone and butanol in both phases were 10.6 and 19.6 g l^{-1}, respectively. Activated carbon has also been used to increase the overall solvent concentrations during *in situ* fermentations.

Using *n*-butyl *n*-butyrate, which could be formed directly from two of the fermentation products, Sierks (1982) showed the ester extractant exhibited the following distribution coefficients with five fermentation products: *n*-butanol, 3.50; butyric acid, 2.15; ethanol, 1.20; acetone, 0.76; and acetic acid, 0.04. In a controlled fermentation in which a 4 to 1 volumetric ratio of aqueous phase of ester was used, 19 g l^{-1} butanol and a normal distribution of other fermentation products were obtained.

Experiments reported with polypropylene glycol extractant produced changes in water activity which inhibited cellular functionality (Mattiason, 1983).

The principles developed for the application of extractive fermentation to ethanol production by Wang (1982), Minier and Goma (1982) and Murphy *et al.* (1982) have importance in development of modern acetone–butanol fermentation technology.

44.4.1.4 Continuous fermentation

Fermentations operated continuously or semi-continuously could improve reactor productivity. Leung and Wang (1981) reported that volumetric butanol productivity could be increased three-fold in continuous culture over that obtained in batch culture (2.5 g l^{-1} h^{-1} *versus*

$0.8\,\mathrm{g\,l^{-1}\,h^{-1}}$). It was also shown that specific productivities of acetone and butanol increased with dilution rate to maximum values of 0.3 and 0.2 $\mathrm{g\,g^{-1}}$ cells $\mathrm{h^{-1}}$, respectively, at the dilution rate of $0.22\,\mathrm{h^{-1}}$; above that dilution rate, the fermentation favored butyric acid production.

The use of continuous culture has also been employed for studies of nutritional and environmental factors which affect solvent production. Monot and Engasser (1983) studied the fermentation under nitrogen limitation; Bahl and coworkers (1982a) studied the fermentation under phosphate limitation. The latter two groups have also used continuous culture to study the effect of pH and of butyrate concentration on solvent production (Monot *et al.*, 1983; Bahl *et al.*, 1982b). Near theoretical conversion of glucose into solvents has been obtained using continuous culture, but as in batch culture, butanol toxicity limits high product concentrations and volumetric productivities.

44.4.1.5 Immobilization

The continuous production of butanol has been reported from immobilized *C. acetobutylicum* (Haeggstroem and Molin, 1979) and *C. butylicum* (Krouwel *et al.*, 1980). Spores and vegetative cells were immobilized in calcium alginate gel and studied under what were termed non-growth conditions. The productivity of these cells was found to be reasonable ($1.0\,\mathrm{g}$ butanol $\mathrm{l^{-1}\,h^{-1}}$), but because butanol toxicity rapidly reduced cell activity within the immobilization matrix, extensive investigation will be required to establish economical working conditions for such a process.

44.4.2 Biochemical and Genetic Developments

Spivey (1978) and Peterson and Fred (1932) describe variations in the morphology of *C. acetobutylicum* during different stages of the fermentation process depending upon the strain of organism used and the formulation of the medium. An encapsulated form predominates during neutral solvent production (Jones *et al.*, 1982). Cho and Doy (1973) have described some of the ultrastructure of *C. acetobutylicum*. Fundamental information derived from such studies has been used to improve butanol tolerance in *C. acetobutylicum*.

44.4.2.1 Toxicity of products

For each fermentation product, there exists a threshold concentration below which no growth inhibition occurs, and above which a linear decrease in growth rate is observed (Costa and Moreira, 1983). These values are given in Table 9. The fermentation end products are in dilute solution primarily because of butanol toxicity. Research has been aimed at gaining a fundamental understanding of alcohol toxicity in fermentations and at investigating potential means of overcoming the problem (Linden and Moreira, 1983). Aliphatic alcohols were found to inhibit cell growth and cytoplasmic membrane functionality in *C. acetobutylicum*. Concentrations on the order of 1 M ethanol, 0.1 M butanol and 0.01 M hexanol were found to inhibit by 50% the cell growth rate, the active nutrient transport process and the membrane-bound ATPase activity. The intensity of the alcohol-induced effects was proportional to the hydrophobicity of the aliphatic alcohols at any given concentration.

Table 9 Concentrations of Fermentation Products Added to Active Cultures which were Inhibitory to *C. acetobutylicum* (ATCC 824)

Product	Concentration below which growth was not inhibited (M)	$(\mathrm{g\,l^{-1}})$	Concentration at which growth was inhibited 50% (M)	$(\mathrm{g\,l^{-1}})$
Butyric acid	0.02	1.7	0.07	6.0
Butanol	0.05	3.7	0.15	11.0
Acetic acid	0.04	2.5	0.13	8.0
Ethanol	0.25	11.6	1.10	51.0
Acetone	0.36	20.9	1.45	84.0

Alteration of phospholipid composition of the cytoplasmic membrane of *C. acetobutylicum* by selective incorporation of specified fatty acids was accomplished by blocking fatty acid synthesis in biotin deficient media. The cell membrane was enriched in oleic acid or elaidic acid when supplemented into respective media at 10 mg l^{-1}. Cell growth and membrane functionality (ATPase activity) were inhibited by *n*-butanol to a lower degree in these cells than in cells grown on normal soluble medium. Addition of 10 mg l^{-1} of oleic acid (18:1 *cis*) or elaidic (18:1 *trans*) acid to a biotin deficient culture medium has also shown the potential for obtaining greater butanol levels in the fermentation.

The economic impact of increasing the butanol concentration in the fermentation broth to 20 g l^{-1} would be significant. Phillips and Humphrey (1983) have presented curves from McCabe–Thiele calculations for the butanol dehydration column operating at 1 atm. Figure 2 shows a diagram for the energy requirements for recovery of butanol as a function of feed composition at two reflux ratios. The calculations were based on recovery of the azeotrope as the overhead product from feed of saturated liquid at the given fermenter concentrations. The calculations showed that to increase fermenter concentration from the typical 1.2 to 1.9% (w/w) would halve the energy consumption for distillation. Concentrations greater than the latter would have diminishing impact on distillation costs.

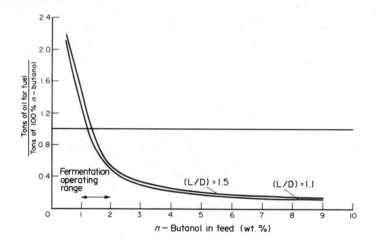

Figure 2   Energy requirements for *n*-butanol recovery by distillation (Phillips and Humphrey, 1983)

44.4.2.2   Genetic developments

The production of solvents during the acetone–butanol fermentation is not only affected by product inhibition, but may also be inhibited by the production of an autolysin (Webster *et al.*, 1981). During the fermentation of a molasses medium, it was found that an autolysin was released toward the end of the exponential growth phase and accompanied the lysis of the culture. The exact role of the autolysin in the fermentation process is open to speculation; it may be a factor accounting for the low level of solvents achieved.

Recently Allcock *et al.* (1981) have isolated an autolysin-deficient mutant to determine if such a microorganism can enhance the concentrations of solvents obtained during a fermentation. The mutant produced 70% less autolysin and had a cell wall which was more resistant to the wild type autolysin unless treated with penicillin. The mutant cells could tolerate up to 31 g l^{-1} butanol without cell death. The mutant *C. acetobutylicum* grew in up to 16 g l^{-1} of solvents; however, the highest solvent production obtained was 20 g l^{-1} of solvents. Hence, the level of solvent produced was not improved over that from the wild type culture (Westhuizen *et al.*, 1982; Long *et al.*, 1983).

Anaerobic fermentations which operate at high temperatures are of particular interest for the production of fuels and chemicals, because process heat can help in recovering volatile products, contamination risks are reduced and oxygen transfer limitations are not critical. The utility of *C. acetobutylicum* would be greatly improved if thermophilic characteristics were incorporated by genetic or *in vitro* recombinant DNA technologies. Several thermophilic clostridia are particularly promising for the production of chemicals from biomass: (1) *C. thermoaceticum* for the production of acetic acid; (2) *C. thermosacchrolyticum* for butyric acid production; (3) *C. thermocellum* for its ability to degrade cellulose; and (4) *C. thermohydrosulfuricum* for its high yields of ethanol (Linden and Moreira, 1983). Genetic manipulation of *C. acetobutylicum* and of the thermophilic clostridia would be possible were a suitable vector for transfer of recombinant DNA available. The capability for transformation (Reid *et al.*, 1983) and regeneration (Allcock *et al.*, 1982) of *C. acetobutylicum* protoplasts and the ability to enumerate *C. acetobutylicum* using conventional microbiological techniques for aerobes have been demonstrated. The latter has been made possible by incorporation of membrane particles from facultative anaerobes into the *C. acetobutylicum* growth medium (Adler and Crow, 1981). These developments are certainly encouraging in this regard.

44.5 SUMMARY

Butanol and acetone may be produced by fermentation processes. Although subsidized commercial production by fermentation of molasses is currently used in South Africa, analyses presented above show that only processes based on waste materials, such as cheese whey and lignocellulosic wastes, would compete economically with petrochemical production. However, Cooney (1983) has pointed out several problems in the development of biotechnology for the chemical process industry. In the case of acetone and butanol, the following must be carefully considered: unproven pricing and processing of biomass wastes, low volumetric productivity, capital intensive processes, high water requirement and coproduct separation processes. These problems and the paramount difficulty, butanol toxicity, may be solved by one or a combination of biochemical, genetic or biochemical engineering approaches. Some of the new developments arising from recent research with *C. acetobutylicum* may provide avenues around some problems.

44.6 REFERENCES

Adler, H. I. and W. D. Crow (1981). A novel approach to the growth of anaerobic microorganisms. *Biotechnol. Bioeng. Symp.*, **11**, 533–540.

Allcock, E. R., S. J. Reid, D. T. Jones and D. W. Woods (1981). Autolytic activity and an autolysis-deficient mutant of *Clostridium acetobutylicum*. *Appl. Environ. Microbiol.*, **42**, 929–935.

Allcock, E. R., S. J. Reid, D. T. Jones and D. R. Woods (1982). Protoplast formation and regeneration in *Clostridium acetobutylicum*. *Appl. Environ. Microbiol.*, **43**, 719–721.

Anon. (1982). Production by the U.S. chemical industry. *Chem. Eng. News*, June 14, 33.

Arthur D. Little, Inc. (1979). Use of wood residues. Report to U.S.D.O.E. Division of Industrial Energy Conservation under Contract No. EC–77–C–03–1692 INDUS 006, February.

Bahl, H., W. Andersch and G. Gottschalk (1982a). Continuous production of acetone and butanol by *Clostridium acetobutylicum* in a two-stage phosphate limited chemostat. *Eur. J. Appl. Microbiol. Biotechnol.*, **15**, 201–205.

Bahl, H., W. Andersch, K. Braun and G. Gottschalk (1982b). Effect of pH and butyrate concentration on the production of acetone and butanol by *Clostridium acetobutylicum* grown in continuous culture. *Eur. J. Appl. Microbiol. Biotechnol.*, **14**, 17–20.

Beech, S. C. (1952). Acetone–butanol fermentation of sugars. *Ind. Eng. Chem.*, **44**, 1677–1682.

Beech, S. C. (1953). Acetone–butanol fermentation of starches. *Appl. Microbiol.*, **2**, 85–95.

Buchanan, R. E. and N. E. Gibbons (ed.) (1974). *Bergey's Manual of Determinative Bacteriology*, 8th edn. pp. 556–560. Williams and Wilkins, Baltimore, MD.

Cho, K. Y. and C. H. Doy (1973). Ultrastructure of the obligately anaerobic bacteria *Clostridium kluvveri* and *Cl. acetobutylicum*. *Aust. J. Biol. Sci.*, **26**, 547–558.

Compere, A. L. and W. L. Griffith (1979). Evaluation of substrates for butanol production. *Dev. Ind. Microbiol.*, **20**, 509–517.

Cooney, C. L. (1983). Prospects for chemicals and fuels production by fermentation. In *Biological Basis for New Developments in Biotechnology*, ed. A. Hollaender *et al.*, pp. 307–316. Plenum, New York.

Costa, J. M. and A. R. Moreira (1983). Growth kinetics of the acetone–butanol fermentation. In *Foundations of Biochemical Engineering, Kinetics and Thermodynamics in Biological Systems*, ed. H. W. Blanch, E. T. Papoutsakis and G. Stephanopoulos. *ACS Symposium Series No. 207*, pp. 501–504. American Chemical Society, Washington, DC.

Cote, W. G. (1977). Wood ultrastructure in relation to chemical composition. In *Recent Advances in Phytochemistry*, ed. F. A. Loewus and V. C. Runechles, vol. 11, pp. 1–44. Plenum, New York.

Doremus, M. G. (1983). Agitation and pressure effects on the acetone–butanol fermentation. MS Thesis, Colorado State University, Colorado.

Faith, W. L., D. B. Keyes and R. L. Clark (1965). *Industrial Chemicals*, 3rd edn., pp. 181–188. Wiley, New York.

Gibbs, D. F. (1983). The rise and fall (...and rise?) of acetone/butanol fermentations. *Trends Biotechnol.*, **1**, 12–15.

Gottschalk, G. and J. R. Andreesen (1979). Energy metabolism in anaerobes. In *International Review of Biochemistry: Microbial Biochemistry*, ed. J. R. Quayle, vol. 21, pp. 85–115. University Park Press, Baltimore.

Gottschalk, G. and H. Bahl (1981). Feasible improvements of the butanol production by *Clostridium acetobutylicum*. In *Trends in the Biology of Fermentations for Fuels and Chemicals*, ed. A. Hollaender, vol. 18, pp. 463–471. Plenum, New York.

Haeggstroem, L. and N. Molin (1980). Calcium alginate immobilized cells of *Clostridium acetobutylicum* for solvent production. *Biotechnol. Lett.*, **2**, 242–246.

Jones, D. T., A. van der Westhuizen, S. Long, E. R. Allcock, S. J. Reid and D. R. Woods (1982). Solvent production and morphological changes in *Clostridium acetobutylicum*. *Appl. Environ. Microbiol.*, **43**, 1434–1439.

Krouwel, P. G., W. F. M. van der Laan and N. W. F. Kossen (1980). Continuous production of *n*-butanol and isopropanol by immobilized, growing *Clostridium butylicum* cells. *Biotechnol. Lett.*, **2**, 253–258.

Langlykke, A. F., J. M. Van Lanen and D. R. Fraser (1948). Butyl alcohol from xylose saccharification liquors from corn cobs. *Ind. Eng. Chem.*, **40**, 1716–1719.

Lenz, T. G. and A. R. Moreira (1980). Economic evaluation of acetone–butanol fermentation. *Ind. Eng. Chem. Prod. Res. Dev.*, **19**, 478–483.

Leonard, R. H., W. H. Peterson and G. J. Ritter (1947). Butanol–acetone fermentation of wood sugar. *Ind. Eng. Chem.*, **11**, 1443–1445.

Leung, J. C. Y. and D. I. C. Wang (1981). Production of acetone and butanol by *Clostridium acetobutylicum* in continuous culture using free cells and immobilized cells. *Proceedings of the 2nd World Congress of Chemical Engineering and World Chemical Exposition*, ed. C. W. Robinson, vol. 1, pp. 348–352. Canadian Society for Chemical Engineering, Ottawa.

Lindberg, S. L. and A. R. Moreira (1982a). Acetone–butanol fermentation of cheese whey. Presented at the AIChE Winter Meeting, Orlando, FL, February 28–March 3.

Lindberg, S. L. and A. R. Moreira (1982b). Production of neutral solvents from cheese whey. Presented at the ACS National Meeting, Kansas City, MO, September 12–17.

Linden, J. C., V. G. Murphy and D. H. Smith (1982). Pretreatment of crop cellulosics. In *International Symposium on Ethanol from Biomass*, ed. H. E. Duckworth and E. A. Thompson, pp. 391–414. Royal Society of Canada, Ottawa.

Linden, J. C. and A. R. Moreira (1983). Anaerobic production of chemicals. In *Biological Basis for New Developments in Biotechnology*, ed. A. Hollaender *et al.*, pp. 377–404. Plenum, New York.

Long, S., D. T. Jones and D. R. Woods (1983). Sporulation of *Clostridium acetobutylicum* P262 in a defined medium. *Appl. Environ. Microbiol.*, **45**, 1389–1393.

Maddox, I. S., J. R. Gapes and V. G. Larsen (1981). Production of *n*-butanol from whey ultrafiltrate. *Proceedings of the 9th Australian Conference on Chemical Engineering, Christchurch, New Zealand*, pp. 535–542. The Institution of Chemical Engineers, Sydney.

Maddox, I. S. (1980). Production of *n*-butanol from whey filtrate using *Clostridium acetobutylicum* N.C.T.B. 2951. *Biotechnol. Lett.*, **2**, 493–498.

Mattiason, B. (1983). Applications of aqueous two-phase systems in biotechnology. *Trends Biotechnol.*, **1**, 16–20.

Mes-Hartree, M. and J. N. Saddler (1982). Butanol production of *Clostridium acetobutylicum* grown on sugars found in hemicellulose hydrolysates. *Biotechnol. Lett.*, **4**, 247–252.

Millett, M. A., A. J. Baker and L. D. Satter (1976). Physical and chemical pretreatments for enhancing cellulose saccharification. *Biotechnol. Bioeng. Symp.*, **6**, 125–153.

Minier, M. and G. Goma (1982). Ethanol production by extractive fermentation. *Biotechnol. Bioeng.*, **24**, 1565–1579.

Monot, F. and J. M. Engasser (1983). Production of acetone and butanol by batch and continuous culture of *Clostridium acetobutylicum* under nitrogen limitation. *Biotechnol. Lett.*, **5**, 213–218.

Monot, F., J. M. Engasser and H. Petitdemange (1983). Regulation of acetone butanol production in batch and continuous cultures of *Clostridium acetobutylicum*. *Biotechnol. Bioeng. Symp.*, **13**, 207–216.

Moreira, A. R., B. E. Dale and M. G. Doremus (1982). Utilization of the fermentor off-gases from an acetone–butanol fermentation. *Biotechnol. Bioeng. Symp.*, **12**, 261–277.

Murphy, T. K., H. W. Blanch and C. R. Wilke (1982). Water recycling in extractive fermentation. *Process Biochem.*, (November-December), 6, 7, 8, 9, 40.

Peppler, H. J. (1967). *Microbial Technology*, pp. 403–416. Rheinhold, New York.

Peters, M. S. and K. D. Timmerhaus (1980). *Plant Design and Economics for Chemical Engineers*, 3rd edn., pp. 147–224. McGraw-Hill, New York.

Peterson, W. H. and E. B. Fred (1932). Butyl-acetone fermentation of corn meal; interrelations of substrates and products. *Ind. Eng. Chem.*, **24**, 237–242.

Petitdemange, H., C. Cherrier, G. Raval and R. Gay (1976). Regulation of the NADH and NADPH-ferredoxin oxidoreductases in clostridia of the butyric group. *Biochim. Biophys. Acta*, **421**, 334–347.

Phillips, J. A. and A. E. Humphrey (1983). Process technology for the biological conversion of lignocellulosic materials to fermentables and alcohols. In *Wood and Agricultural Residues: Research on Use for Feed, Fuels and Chemicals*, ed. E. J. Soltes, Academic, New York.

Pitt, W. W. Jr., G. L. Haag and D. D. Lee (1983). Recovery of ethanol from fermentation broths using selective sorption–desorption. *Biotechnol. Bioeng.*, **25**, 123–131.

Prescott, S. C. and C. G. Dunn (1959). *Industrial Microbiology*, 3rd edn., pp. 250–281. McGraw-Hill, New York.

Reid, S. J., E. R. Allcock, D. T. Jones and D. R. Woods (1983). Transformation of *Clostridium acetobutylicum* protoplasts with bacteriophage DNA. *Appl. Environ. Microbiol.*, **45**, 305–307.

Saddler, J. N., E. K. C. Yu, M. Mes-Hartree, N. Levitin and H. H. Brownell (1983). Utilization of enzymatically hydrolyzed wood hemicelluloses by microorganisms for production of liquid fuels. *Appl. Environ. Microbiol.*, **45**, 153–160.

Sierks, M. (1982). Extractive butanol fermentation. In *Proceedings of the 12th Biochemical Engineering Symposium*, ed. L. E. Erickson and L. T. Fan, pp. 49–55. Kansas State University, Manhattan, KS.

Shin, K. C., B. Hong and L. T. Fan (1982). Fermentative production of butanol from sorghum molasses. Paper No. 52 in the Cellulose, Paper and Textile Division presented at the 184th ACS National Meeting, Kansas City, MO, September 12–17.

Shreve, R. N. (1967). *Chemical Process Industries*, 3rd edn., pp. 591–670. McGraw-Hill, New York.

Spivey, M. J. (1978). The acetone/butanol/ethanol fermentation. *Process Biochem.*, **13**, 2–4, 25.

Stanier, R. Y., E. A. Adelberg and J. L. Ingrahm (1976). *Microbial World*, 4th edn., pp. 846–847. Prentice-Hall, NJ.

Steel, R. (1958). *Biochemical Engineering: Unit Processes in Fermentation*, pp. 125–148. Macmillan, New York.

Su, T. M., R. Lamed and J. H. Lobos (1981). Effect of stirring and H_2 on ethanol production by thermophilic fermentation. *Proceedings of the 2nd World Congress of Chemical Engineering and World Chemical Exposition*, ed. C. W. Robinson, vol. 1, pp. 353–356. Canadian Society for Chemical Engineering, Ottawa.

Underkofler, L. A. and R. J. Hickey (1954). *Industrial Fermentations*, vol. 1, pp. 347–388. Chemical Publishing Co., New York.

Wang, D. I. C., C. L. Cooney, A. L. Demain, R. F. Gomez and A. J. Sinskey (1979). Degradation of cellulosic biomass and its subsequent utilization for the production of chemical feedstocks. MIT Quarterly Report to the US DOE under Contract COG-4198-9, February 28.

Wang, H. Y. (1982). Interaction between fermentation and subsequent product recovery processes. Paper presented at the Biochemical Engineering III. Symposium Engr. Foundation, Santa Barbara, California, September 19–24.

Webster, J. R., S. J. Reid, D. T. Jones, and D. R. Woods (1981). Purification and characterization of an autolysin from *Clostridium acetobutylicum*. *Appl. Environ. Microbiol.*, **41**, 371–374.

Wendland, R. T., E. I. Fulmer and L. A. Underkofler (1941). Butyl-acetonic fermentation of Jerusalem artichokes. *Ind. Eng. Chem.*, **33**, 1078–1081.

Westhuizen, A. van der, D. T. Jones and D. R. Woods (1982). Autolytic activity and butanol tolerance of *Clostridium acetobutylicum*. *Appl. Environ. Microbiol.*, **44**, 1277–1281.

Wiley, A. J., M. J. Johnson, E. McCoy and W. H. Peterson (1941). Acetone–butyl alcohol fermentation of waste sulfite liquor. *Ind. Eng. Chem.*, **33**, 606–610.

Yu, E. K. C. and J. N. Saddler (1983). Enhanced acetone–butanol fermentation by *Clostridium acetobutylicum* grown on D-xylose in the presence of acetic or butyric acid. *FEMS Microbiol. Lett.*, **18**, 103–109.

45

2,3-Butanediol

M. VOLOCH, N. B. JANSEN, M. R. LADISCH, G. T. TSAO, R. NARAYAN
and V. W. RODWELL
Purdue University, West Lafayette, IN, USA

45.1 INTRODUCTION

Biomass conversion gives both pentoses and hexoses as products. While the hexoses (primarily glucose) are readily fermented, routes for pentose fermentation are still being developed. Hence, pentoses represent a potentially significant source of sugars with xylose being the major product.

Fermentation of xylose as well as glucose by *Klebsiella oxytoca*, ATCC 8724 (formerly known as *Klebsiella pneumoniae* and *Aerobacter aerogenes*) yields 2,3-butanediol as the major product. Other microorganisms capable of producing 2,3-butanediol (abbreviated 2,3-BD) include *Bacillus subtilis* (Ford strain), *Aeromonas hydrophilia* and several species of *Serratia* (Ledingham and Neish, 1954). Secondary products formed include acetoin, ethanol, lactic acid and glycerol. While *K. oxytoca* is able to yield high concentrations of 2,3-BD as mixtures of stereoisomers from monosaccharides, it is unable to utilize polysaccharides (Ledingham and Neish, 1954). In comparison, *B. polymyxa* is able to ferment starch directly giving L-2,3-butanediol and ethanol in almost equal amounts (Long and Patrick, 1963; Prescott and Dunn, 1959). However, *B. polymyxa* is unstable and difficult to maintain (Long and Patrick, 1963).

Both *K. oxytoca* and *B. polymyxa* have been used in pilot scale fermentation (Ledingham and Neish, 1954; Blackwood *et al.*, 1949), especially during World War II, as a possible means of producing 2,3-BD and subsequently 1,3-butadiene, an organic intermediate for rubber production. In the 1940s, process development was carried out through the pilot plant stage at the National

Research Laboratories in Ottawa, Canada. A 90% fermentation efficiency was attained on a 750-gallon scale for sugars obtained from whole wheat (Blackwood *et al.*, 1949). Process evaluation with barley as a feedstock indicated a 2,3-BD cost of 13 to 18 cents per pound (Tomkins *et al.*, 1948). Development was discontinued because less expensive routes for chemically producing 1,3-butadiene from petroleum became available. In recent times, the long-term prospects of rising petroleum prices have revived significant interest in producing alcohols, including 2,3-BD, from biomass.

45.2 PROPERTIES

45.2.1 Stereochemical Configurations

There are three isomeric forms of 2,3-BD: D-(−), L-(+) and *meso* (Figure 1). Both the *meso* and (±) forms exist, to a considerable extent, in the conformation in which the hydroxy groups are *gauche* to each other (Figure 2). This conformation is favored because of the energy gained in the formation of the hydrogen bond. As a result, the methyl groups are *gauche* in the *meso* form and *anti* in the optically active form. Hence, the optically active form of the isomer is more stable than the *meso* form.

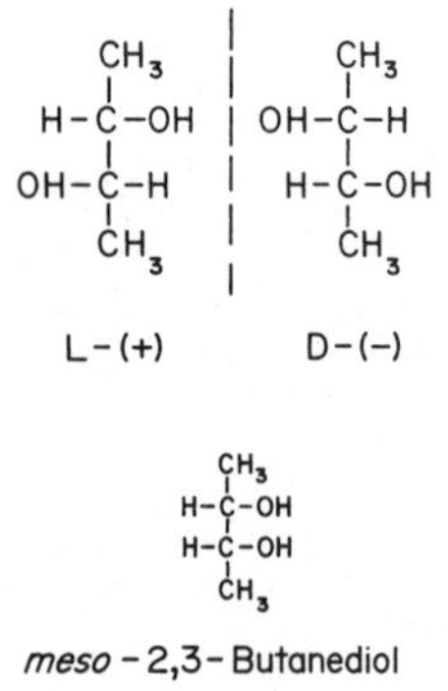

Figure 1   Three stereoisomers of 2,3-butanediol

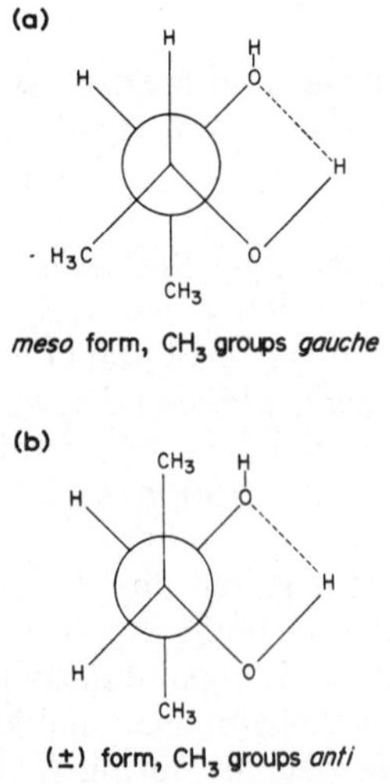

Figure 2   *Anti* and *gauche* forms of 2,3-butanediol

45.2.2 Physical Properties

Vapor–liquid equilibrium data for 2,3-butanediol are given by Othmer *et al.* (1945b) and other physical properties by Ledingham and Neish (1954). Briefly summarized, the water/butanediol equilibrium data show no azeotrope. The distillation of 2,3-butanediol removed the water overhead with 2,3-butanediol being the bottom product. While the boiling points of *meso* (181–182 °C), D (179–180 °C) and racemic (177 °C) 2,3-BD are slightly different, they all have boiling points much higher than water. Hence, the recovery of 2,3-BD from fermentation broth requires a large quantity of water to be evaporated. In practice, it was found that an excess of lime must be added to filtered fermentation liquor if 90% recovery were to be obtained through steam distillation. If the liquor was not filtered and lime was added, recovery was only 50% due to decomposition (Othmer *et al.*, 1945b). Process economics would dictate that essentially complete recovery be attained. An alternate approach suggested is liquid–liquid extraction, with *n*-butanol being a preferred extractant. Butanediol is then recovered as a bottom product in a subsequent distillation step with butanol being recycled (Othmer *et al.*, 1945a).

A major challenge in the economic production of 2,3-butanediol would still appear to be in separation of 2,3-BD from water (or fermentation broth) in an efficient manner. Hence, further research to carry forward the excellent work of these early pioneers seems needed.

45.3 FERMENTATION

45.3.1 Substrates

The single major cost in most biomass conversion processes appears to be the substrate cost (Ladisch *et al.*, 1983). Hence, the availability of an inexpensive carbohydrate material is essential for developing an economical fermentation process for production of 2,3-BD. Substrates suitable for the *K. oxytoca* fermentation include molasses (Long and Patrick, 1963), enzymatically hydrolyzed cereal mashes (Rose, 1961), acid hydrolyzed starch (Ward *et al.*, 1945) and wheat (Olson and Johnson, 1948), wood hydrolysates (Perlman, 1944) and sulfite waste liquor (Murphy and Stranks, 1951). *B. polymyxa* secretes amylolytic enzymes and hence is able to utilize cornstarch (Kooi *et al.*, 1948) and whole grain mashes of wheat (Blackwood *et al.*, 1949). The high post-fermentation solids content of whole grains impairs the recovery of 2,3-BD from the fermentation broth, and hence renders the use of such mashes to be less than optimum.

K. oxytoca and *B. polymyxa* are both able to utilize pentoses as well as hexoses (Ledingham and Neish, 1954). This is of considerable practical importance since hydrolysate from biomass materials can have pentose:glucose ratios of 1:1.5 (Tsao *et al.*, 1982). As a result, almost all of the sugar present in hemicellulose and cellulose hydrolysates can be converted to 2,3-BD (Flickinger and Tsao, 1979; Yu and Saddler, 1982).

45.3.2 Fermentation Conditions

The optimum pH for butanediol production by *K. oxytoca* is in the range pH 5.0–6.0 (Neish and Ledingham, 1949; Pirt and Callow, 1958; Jansen, 1984). The specific substrate utilization rate is maximum at pH 5.5 (Pirt and Callow, 1958). Above pH 6 the activity of one of the key enzymes in the butanediol pathway decreases sharply (Stormer, 1968).

The optimum temperature for growth, sugar uptake and butanediol production is 37 °C (Pirt and Callow, 1958; Esener *et al.*, 1981a). It is interesting to note that while 37 °C appears to be the best temperature for growth and fermentation, the highest butanediol concentration reported in the literature was achieved at 30 °C (Olson and Johnson, 1948).

The most important variable affecting the butanediol yield and the fermentation rate is the availability of oxygen. Although 2,3-BD is a product of anaerobic metabolism, aeration has been shown to enhance its production (Ledingham and Neish, 1954; Long and Patrick, 1963). Pirt and Callow (1959) suggested that aeration increases the butanediol productivity by increasing the cell concentration. However, too much aeration can decrease the yield of 2,3-BD. *K. oxytoca* is a facultative anaerobe which is able to obtain the energy it needs for growth by two different pathways: respiration and 'fermentation' (Figure 3). During oxygen limited growth (DOT < 5 mmHg as reported by Harrison and Loveless, 1971), both energy producing pathways are active simultaneously, and the yield of butanediol depends on the relative activities of each of the three pathways depicted in Figure 3. The butanediol yield can be maximized by minimizing the oxygen availability because this limits respiration. However, with a small oxygen supply, little cell mass is produced and, therefore, the conversion rates are slow. The butanediol production rate can be maximized by increasing the oxygen supply rate because this leads to a higher cell density (Jansen, 1984).

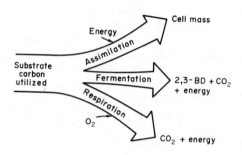

Figure 3 Pathways of substrate utilization for *K. oxytoca*

The oxygen supply rate is also important because the ratio of oxygen demand to oxygen supply can control the proportions of various metabolites produced (Vollbrecht, 1982). Fermentation products excreted by *K. oxytoca*, other than 2,3-BD, include acetoin, ethanol, acetate and others (Neish and Ledingham, 1949; Vollbrecht, 1982). In the absence of oxygen, ethanol is produced in approximately equimolar amounts with 2,3-BD (Neish and Ledingham, 1949). The presence of some oxygen appears to inhibit ethanol production. Increasing the oxygen supply rate toward the value of the potential oxygen demand results in an increase in the acetoin:butanediol ratio (Pirt and Callow, 1958). If the oxygen supply rate exceeds the microbial oxygen demand, oxygen is no longer limiting and the only products of sugar metabolism are cell mass and carbon dioxide (Pirt, 1957). Therefore, in order to maximize butanediol production, a limited but non-zero supply of oxygen is required.

Substrate concentrations used for this fermentation are generally in the range of 5–10% (Long and Patrick, 1963). Substrates commonly used in industrial-scale fermentations are usually diluted to even lower sugar concentrations. In experiments employing D-xylose as the sole carbon source, metabolic rates decreased significantly at xylose concentrations greater than 20 g l^{-1}. Indeed, when the xylose concentration exceeded 160 g l^{-1}, no growth at all was observed (Jansen, 1984). This apparent substrate inhibition may be explained by the dependence of the microbial growth rate on the water activity. When any solute decreases the water activity of the fermentation broth, the growth rate of *Klebsiella* also decreases (Esener *et al.*, 1981b). The low osmotic tolerance of *K. oxytoca* may be an important factor to consider in developing processes utilizing natural carbohydrate sources.

45.3.3 Fermentation Yields and Rates

The theoretical maximum yield of 2,3-butanediol from glucose is 0.50 g g^{-1}. The yield is the same from pentoses, which are converted to glyceraldehyde 3-phosphate by the pentose phosphate pathway (Jansen and Tsao, 1983). Actual butanediol yields obtained using *K. oxytoca* can exceed 80–90% of theory (Pirt and Callow, 1958).

The efficiency of the 2,3-butanediol fermentation can be judged by the product yield from sugar, the final butanediol concentration and the volumetric butanediol production rate. Table 1 lists values for these performance indexes that have been reported for various types of reactor configurations. These results were obtained using glucose or sucrose as the carbon source. When substrates of industrial interest are used, the butanediol yield is usually slightly lower (Blackwood *et al.*, 1949; Tomkins *et al.*, 1948).

The fed-batch reactor has the capability of producing a high final butanediol concentration while minimizing the effects of initial substrate inhibition and final product inhibition. With continuous reactors, much higher rates are possible; however, product inhibition and incomplete substrate utilization are problems. The two-stage continuous culture system devised by Pirt and Callow (1959) is outstanding in its ability to rapidly produce a high butanediol concentration with a good yield from sugar. Another promising system employs immobilized cells in an attempt to

Table 1   Comparison of Yields for Different Types of Fermentation

Reactor type	Butanediol yield $(g\,g^{-1})$	Butanediol concentration $(g\,l^{-1})$	Butanediol productivity $(g\,l^{-1}\,h^{-1})$	Ref.
Batch	0.43	65	1.6	Freeman and Morrison (1947)
Fed-batch	0.37	99	0.9	Olson and Johnson (1948)
Continuous	0.32	30	3.0	Pirt and Callow (1958)
Two-stage continuous	0.46	67	2.7	Pirt and Callow (1959)
Immobilized cells	0.25	3.4	1.3	Chua *et al.* (1980)

increase the conversion efficiency. The initial work of Chua *et al.* (1980) was carried out at pH 6.5 which may have caused the relatively poor performance. Future advances in the development of the immobilized cell reactor may result in a system that rivals or surpasses the two-stage continuous system in terms of overall performance.

45.4 BIOCHEMISTRY

The major intermediates in the conversion of a pentose or hexose to 2,3-butanediol are shown in Figure 4. The last step in the biological pathway of the fermentation involves the reduction of acetoin (2-hydroxy-2-butanone) to 2,3-butanediol. While 2,3-BD has two assymmetrical centers, acetoin has only one, and hence two stereoisomeric forms, D-(−) and L-(+). Hence, the reduction of acetoin to 2,3-BD may involve as many as two substrates and three products.

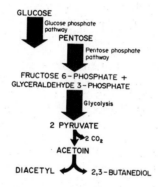

Figure 4   Major intermediates in conversion of a pentose or a hexose to 2,3-butanediol. Heavy arrows represent reactions of the pentose phosphate and glycolytic pathway. Light arrows indicate individual reactions

Juni (1952) reported that *K. oxytoca* forms acetoin from pyruvate by the action of two enzymes. An acetolactate-forming enzyme catalyzes the condensation of two pyruvate molecules combined with a single decarboxylation to yield acetolactate and CO_2. The decarboxylase is specific for the dextrorotatory isomer and the product is the levorotatory isomer of acetoin (D-(−)-acetoin). Both the decarboxylase and acetolactate-forming enzymes have been partially purified and characterized (Stormer, 1967; Stormer, 1968; Loken and Stormer, 1970; Malthe-Sorenssen and Stormer, 1970).

The acetoin can be oxidized to 2,3-butanedione (diacetyl) by O_2 present in the fermentation medium, or enzymatically reduced (with NADH as a cofactor) to 2,3-BD. Three stereoisomers are possible; indeed, for 40 years they have been known to exist in the fermentation broth. The isomeric composition varies with the microorganism used in the fermentation.

Walpole (1911) reported that *K. oxytoca* produced a mixture of *meso* and L-(+)-2,3-BD. *Aeromonas hydrophilia* and *Aerobacillus polymyxa* produce only D-(−)-2,3-BD (Stanier and Adams, 1944; Adams and Stanier, 1945; Neish, 1945). Bacterial oxidation of 2,3-BD is a function of the stereo configuration as summarized in Table 2 (Stanier and Fratkin, 1944; Sebek and Randles, 1952).

A mechanism for the formation of 2,3-BD stereoisomers by microorganisms was proposed by

Table 2   Bacterial Oxidation of 2,3-Butanediol by Various Microorganisms

Microorganism	Observation	Ref.
K. oxytoca	*meso* and L−(+) are oxidized	Stanier and Fratkin (1944)
A. polymyxa	D-(−) oxidized faster than *meso*	Stanier and Fratkin (1944)
A. hydrophila	Only *meso*-2,3-BD is oxidized	Stanier and Fratkin (1944)
Pseudomonas fluorescens	All three isomers of 2,3-BD oxidized	Sebek and Randles (1952)

Ledingham and Neish (1954). These investigators postulated the existence of two 2,3-BD dehydrogenases: (1) a dehydrogenase catalyzing reduction of D-(−)-acetoin to *meso*-2,3-BD, and (2) a dehydrogenase catalyzing reduction of D-(−)-acetoin to D-(−)-2,3-BD. The existence of an acetoin racemase was mentioned as a possibility. These investigators did not consider the formation of L-(+)-2,3-BD. Experimental data were not provided for the model.

In 1960, Taylor and Juni (1960) proposed a model for the formation of 2,3-BD stereoisomers from acetoin (Figure 5), based on the observed optical rotation of acetoin produced from pyruvate, the composition of 2,3-BD stereoisomers formed in the fermentation, and rates of oxidation of 2,3-BD stereoisomers. The model proposed the existence of three enzymes: an acetoin racemase, L-(+)-2,3-BD dehydrogenase and D-(−)-2,3-BD dehydrogenase. The dehydrogenases were said to be non-specific with respect to acetoin stereoisomers. That is, they would accept either acetoin isomer as substrate, but the reaction product would still be dependent on the acetoin isomer reduced. For example, the L-(+)-dehydrogenase would reduce L-(+)-acetoin to L-(+)-2,3-BD and D-(−)-acetoin to *meso*-2,3-BD.

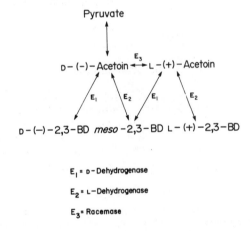

Figure 5   Mechanism for the formation of 2,3-butanediol stereoisomers by bacteria as proposed by Taylor and Juni (1960). For *K. oxytoca*, the presence of an acetoin racemase and L-(+)-2,3-butanediol dehydrogenase was proposed.

For *K. oxytoca* Taylor and Juni (1960) proposed the presence of acetoin racemase and the L-(+)-2,3-BD dehydrogenase. While they were unable to document the presence of acetoin racemase in acetone-dried preparations of *K. oxytoca*, they speculated that cell extracts prepared by other procedures would be shown to contain an acetoin racemase activity.

45.4.1   Acetoin Reductase

Bryn *et al.* (1971) reported the purification and characterization of diacetyl (acetoin) reductase from *K. oxytoca*, which catalyzed the reduction of acetoin to 2,3-BD and the reduction of diacetyl to acetoin (Scheme 1).

Acetoin reductase is a tetramer with a molecular weight of 100 000 (Hetland *et al.*, 1970). Upon isoelectric focusing of an apparently homogeneous preparation, at least 12 species all possessing enzymatic activity with respect to acetoin were detected (Hetland *et al.*, 1971). Kinetic constants were determined using commercially available acetoin and 2,3-BD and are summarized in Table 3.

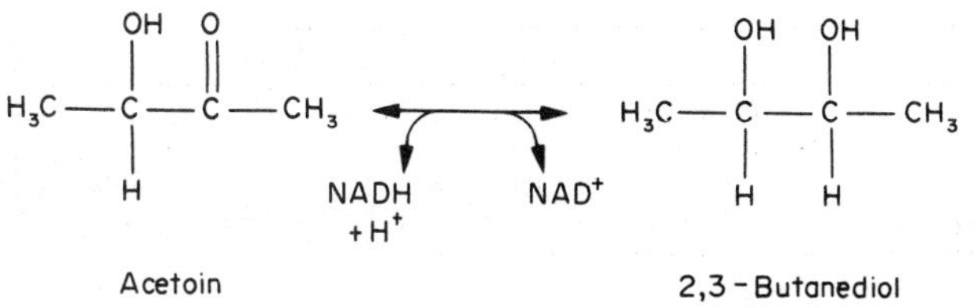

Scheme 1

Table 3 Kinetic Constants for Acetoin (Diacetyl) Reductase[a]

Constant	Value (μM)
$K_m(\text{NADH})$	9
$K_m(\text{NAD}^+)$	180
$K_m(\text{acetoin})$	530
$K_m(\text{2,3-BD})$	11 300
$K_i(\text{NADH})$	11
$K_i(\text{NAD}^+)$	140

[a] Phosphate buffer, pH 7.0, 35 °C (Larsen and Stormer, 1973).

The reduction of acetoin followed an ordered sequential Bi–Bi mechanism (Larsen and Stormer, 1973).

45.4.2 Kinetics of 2,3-Butanediol Formation from Acetoin

The results of Juni and co-workers and Ledingham and Neish (1954) have been extended by Voloch *et al.* (1983) to formulate a kinetic model based on studies using cell-free extracts from *K. oxytoca* grown on D-xylose, as well as acetoin reductases (E_1 and E_2) isolated from the extracts. The cell-free extract was found to have acetoin reductase activity. Based on their activities with respect to D-($-$) and racemic acetoin, E_1 and E_2 were found to be *meso*-2,3-butanediol:NAD$^+$ oxidoreductase (D-($-$)-acetoin forming) and L-($+$)-2,3-butanediol:NAD$^+$ oxidoreductase (L-($+$)-acetoin forming), respectively.

The kinetic model which describes the activities of these enzymes is shown in Figure 6. The major difference between this model and the one of Taylor and Juni (Figure 5) lies in the stereospecificity of acetoin reductases.

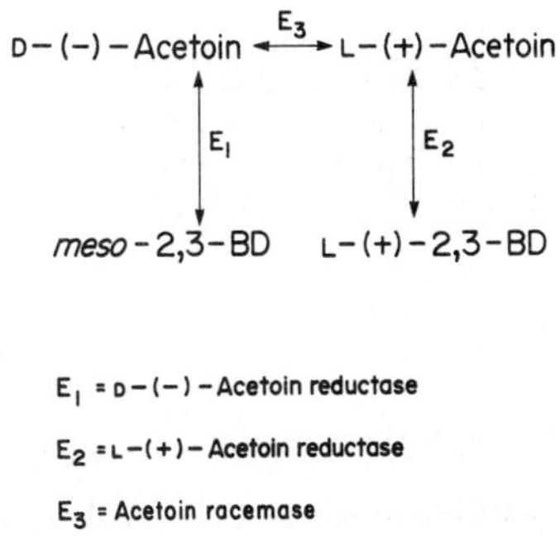

Figure 6 Modified mechanism for formation of 2,3-butanediol stereoisomers based on studies with enzyme activities isolated from *K. oxytoca* enzyme preparation

The properties and kinetics of D-($-$)- and L-($+$)-acetoin reductases have been determined by Voloch (1981). These reductases exhibit an ordered Bi–Bi mechanism.

The reaction equation for D-(−)-acetoin reductase is:

$$v = \frac{V_f V_r([A][B] - ([P][Q]/K_{app}))}{DEN} \tag{1}$$

where

$$DEN = V_r K_{iA} K_{mB} + V_r K_{mB}[A] + V_r K_{mA}[B] + \frac{V_f K_{mQ}[P]}{K_{app}}$$

$$+ \frac{V_f K_{mP}[Q]}{K_{app}} + V_r[B][A] + \frac{V_f K_{mQ}[A][P]}{K_{app} K_{iA}} + \frac{V_f[P][Q]}{K_{app}}$$

$$+ \frac{V_r K_{mA}[B][Q]}{K_{iQ}} + \frac{V_f K_{mB}[P][Q]}{K_{iB} K_{app}} + \frac{V_r[B][A][P]}{K_{iP}} \tag{2}$$

$[A]$ = [NADH]
$[B]$ = [D-(−)-acetoin]
$[P]$ = [*meso*-2,3-BD]
$[Q]$ = [NAD$^+$]
v = reaction velocity
$K_{app} = \left(\frac{[P][Q]}{[A][B]}\right)_{eq}$

The constants are defined in Table 4.

In the absence of products, equation (1) reduces to:

$$v = \frac{V_f[A][B]}{K_{iA} K_{mB} + K_{mB}[A] + K_{mA}[B] + [A][B]} \tag{3}$$

where v is the reaction velocity and the other parameters are defined in Table 4. This equation represents the rate which would be expected during the initial period of reaction when product accumulation is small.

Table 4 Definition of Constants for the Ordered Bi–Bi Mechanism (Segal, 1975)

K_{mA}	$k_3 k_4/k_1(k_3 + k_4)$
K_{mB}	$k_4(k_{-2} + k_3)/k_2(k_3 + k_4)$
K_{iA}	k_{-1}/k_1
K_{iB}	$(k_{-1} + k_{-2})/k_2$
K_{mP}	$k_{-1}(k_{-2} + k_3)/k_{-3}(k_{-1} + k_{-2})$
K_{mQ}	$k_{-1}k_{-2}/k_{-4}(k_{-1} + k_{-2})$
K_{iP}	$(k_3 + k_4)/k_{-3}$
K_{iQ}	k_4/k_{-4}
V_f	$k_3 k_4[E]_t/(k_3 + k_4)$
V_r	$k_{-1}k_{-2}[E]_t/(k_{-1} + k_{-2})$

Constants refer to reaction sequence given in the general reaction sequences:

$$E + NADH \underset{k_{-1}}{\overset{k_1}{\rightleftharpoons}} E \cdot NADH$$

$$E \cdot NADH + acetoin \underset{k_{-2}}{\overset{k_2}{\rightleftharpoons}} E \cdot NADH \cdot acetoin$$

$$E \cdot NADH \cdot acetoin \underset{k_{-p}}{\overset{k_p}{\rightleftharpoons}} E \cdot NAD \cdot 2,3\text{-}BD$$

$$E \cdot NAD \cdot 2,3\text{-}BD \underset{k_{-3}}{\overset{k_3}{\rightleftharpoons}} E \cdot NAD + 2,3\text{-}BD$$

$$E \cdot NAD \underset{k_{-4}}{\overset{k_4}{\rightleftharpoons}} E + NAD$$

where the k's are kinetic constants, E_t = total enzyme, and stereo configurations given in Figure 5.

Values of these constants based on initial rate studies for E_1 are given in Table 5. The apparent equilibrium constant may be calculated by using a Haldane relationship (Segal, 1975):

$$K_{app} = \frac{V_f K_{iQ} K_{mP}}{V_r K_{iA} K_{mB}} \tag{4}$$

Table 5 Kinetic Constants for Acetoin Reductases

Constant[a]	E_1 D-(−)-Acetoin reductase value	E_2 L-(+)-Acetoin reductase value
K_{mA}	7.4 μM	—
K_{mB}	460 μM	—
K_{iA}	10 μM	17 μM
K_{iB}	3900 μM	—
K_{mP}	2200 μM	—
K_{mQ}	56 μM	—
K_{iQ}	150 μM	20 μM
K_{iP}	29 000 μM	—
V_f	0.59 IU μg^{-1} protein	0.016 IU μg^{-1} protein
V_r	0.41 IU μg^{-1} protein	—
$K_{iB}{}^{b}$	—	6600 μM

[a] A = NADH, B = D-(−)-acetoin, P = *meso*-2,3-BD, Q = NAD$^+$. 1 IU = formation of 1 μmol of product at 30 °C, pH 7.5. [b] B = L-(+)-acetoin.

Substituting the values for the constants given in Table 5 in equation (4) yields K_{app} = 103. The K_{app} determined experimentally has a similar value of 113. Hence, this shows that the kinetic constants obtained are consistent with the experimental data.

The data indicated that the enzymatic reduction of D-(−)-acetoin is essentially irreversible (*i.e.* $k_{-2} = k_{-p} = k_{-3} \simeq 0$ in Table 4). A similar observation was made for L-(+)-acetoin. Hence, an integrated rate approach was used to determine the key kinetic constants for L-(+)-acetoin reductase. These values are also given in Table 5 (see E_2).

The magnitudes of the constants are similar to those reported by Larsen and Stormer in 1973 (see Table 3) although their model was different.

45.4.3 Analysis of Butanediol by Liquid Chromatography

Modeling of the reduction of the stereoisomers of acetoin to the stereoisomers of 2,3-BD requires an analytical tool which allows the separation of at least some of the stereoisomers. A liquid chromatography technique has been reported (Voloch *et al.*, 1981) which resolves *meso*- from L- and/or D-2,3-BD, from diacetyl or acetoin, and from their precursor sugar, xylose. This technique permits precise quantification without prior sample work-up and is suitable for preparation scale procedures.

The technique consists of injecting a 10 to 100 μl sample and eluting it with water through a 6 mm ID × 60 cm length column packed with Aminex 50W × 4 (Biorad Laboratories, Richmond, CA) at 85 °C. A typical chromatogram is shown in Figure 7. Details on the involved procedures for positively identifying the 2,3-butanediol peaks are given by Voloch *et al.* (1981).

45.5 CHEMISTRY

There are several interesting chemical reactions of 2,3-butanediol. These include dehydration to methyl ethyl ketone (industrial solvent), reaction with acetone to produce a 'tetramethyl' compound (a possible gasoline blending agent), and formation of butene and butadiene.

45.5.1 Preparation of Methyl Ethyl Ketone (MEK)

MEK can be produced by dehydration of 2,3-butanediol. The dehydration can be carried out using catalysts such as alumina or by direct reaction with sulfuric acid (Emerson *et al.*, 1982). The reaction mechanism involves a hydride shift (Scheme 2).

MEK is an industrial solvent and may find use as a liquid fuel additive.

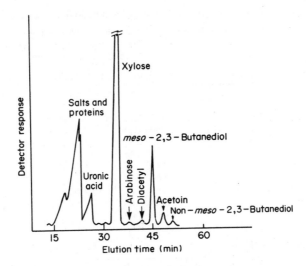

Figure 7   Liquid chromatogram of a 3 l fermentation broth of D-xylose by *K. oxytoca*, after 8 h. Fermentation conditions given by Jansen (1984)

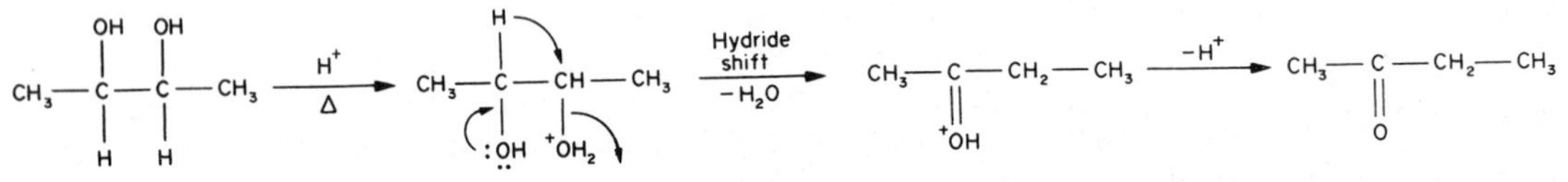

Scheme 2

45.5.2 Tetramethyl Compound

The tetramethyl ether may find use as a blending agent for gasoline, similar to MTBE (methyl *t*-butyl ether). In fact, it has been pointed out that synthesis of MTBE and other new alkyl ether blending agents from sources other than petroleum stocks is essential if they are to be of much real benefit in extending gasoline supply (Stinson, 1979). Both acetone and 2,3-BD are fermentation products.

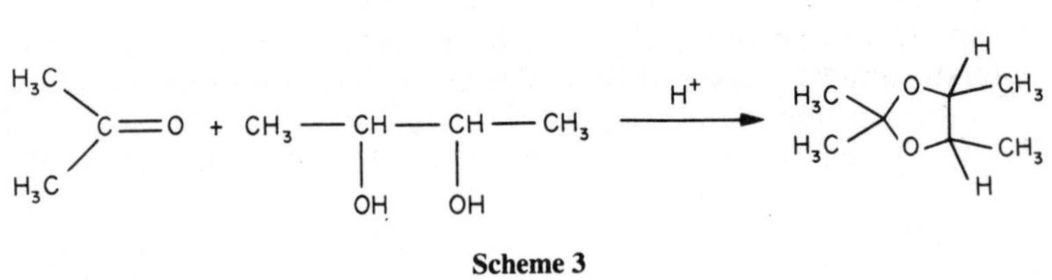

Scheme 3

45.5.3 Preparation of 2-Butene and 1,3-Butadiene

Several reductive elimination reactions have been described in the literature (Corey and Winter, 1963, 1965; Tipson and Cohen, 1965; Josan and Eastwood, 1968) which allow the conversion of a 1,2-diol into the corresponding alkene by the breakdown of the intermediate 1,3-dioxolane according to Scheme 4. These reactions proceed with a high degree of *syn* stereospecificity and can be readily applied to the preparation of *cis* and *trans* isomers of but-2-ene from 2,3-butanediol.

Treatment of the diol with PBr$_3$/HBr, followed by Zn powder, should also result in the formation of 2-butenes (Scheme 5). The debrominations also proceed with a high degree of *anti* stereospecificity (House and Ro, 1958; Gordon and Hay, 1968), the *meso* isomer giving the *trans*-butene, and the (±) isomer the *cis* butene.

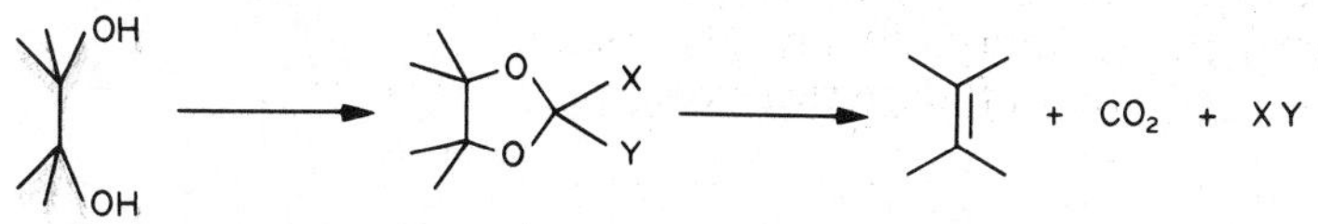

Scheme 4

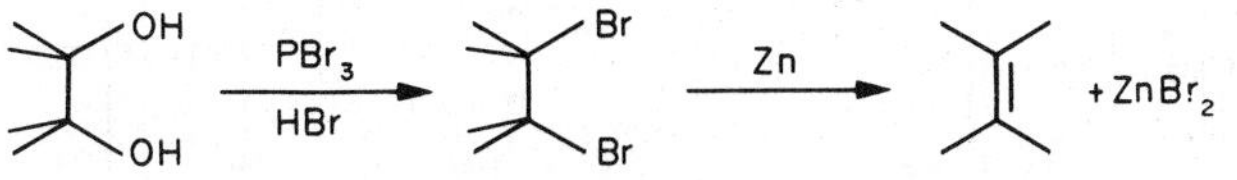

Scheme 5

The butenes can be catalytically dehydrogenated to 1,3-butadiene in the presence of super-heated steam as a diluent and a heating medium (Kearby, 1955).

Butadiene can also be obtained in good yield by the direct dehydration of 2,3-butanediol over thoria catalyst, although most other dehydration catalysts give methyl ethyl ketone as the main product (Winfield, 1945).

Earlier work reported the esterification of 2,3-butanediol with acetic acid (Schlecter *et al.*, 1945b) followed by pyrolysis of the diacetate to butadiene (Scheme 6) (Schlecter *et al.*, 1945a).

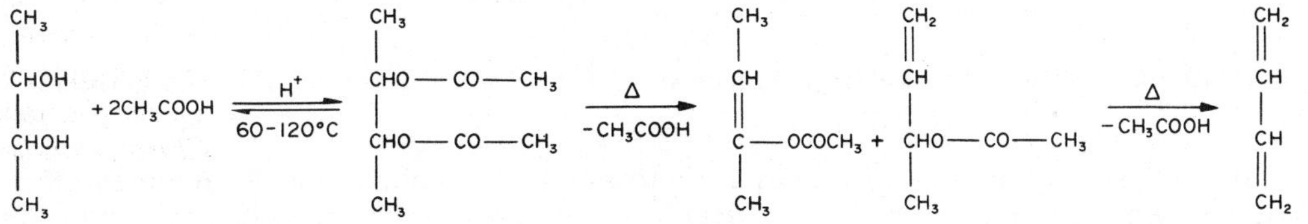

Scheme 6

Butene and butadiene are important industrial chemicals and are currently obtained from cracked petroleum.

45.5.4 Plasticizers

The esters of butanediol and suitable monobasic acids could find use as effective plasticizers for thermoplastic polymers such as cellulose nitrate, cellulose triacetate, cellulose acetate butyrate, polyvinyl chloride, polyvinyl esters, polyacrylates and polymethylacrylates. The diesters can be prepared by the usual esterification reactions with monobasic acids or their functional equivalents (Scheme 7).

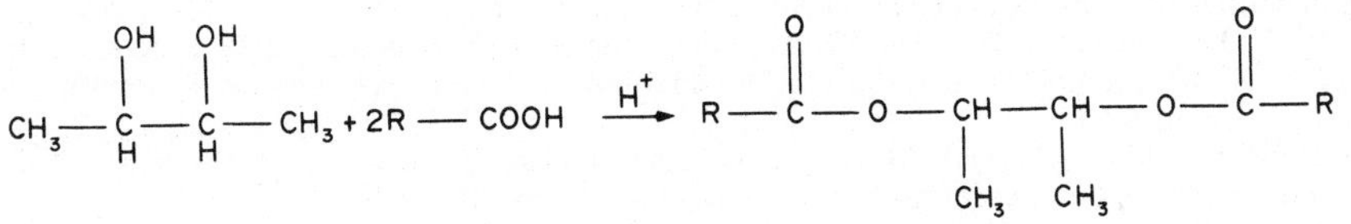

Scheme 7

45.6 CONCLUSIONS

2,3-Butanediol is an example of a potential bulk chemical which can be produced by fermentation (Palsson *et al.*, 1981). While a process (Figures 8 and 9) for 2,3-BD recovery has been piloted (Wheat *et al.*, 1948), enhanced efficiency, both in energy consumption and product recovery, will aid the scale-up of the laboratory fermentations.

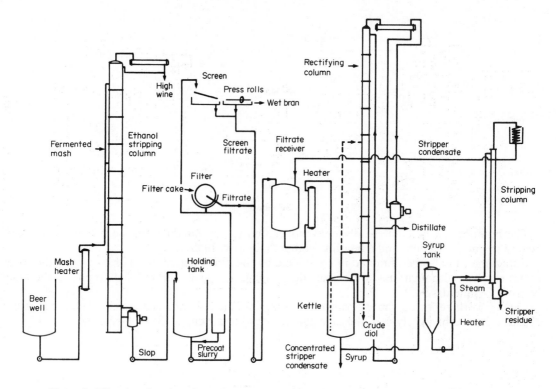

Figure 8   Pilot plant flowsheet (from Wheat *et al.*, 1948) for 2,3-BD recovery from fermentation beer

Meanwhile, further laboratory studies are warranted, for example, the development of strains of bacteria which are able to produce 2,3-BD from solutions containing high total solute concentration (derived from hydrolysis of biomass) at the same rate as from laboratory-prepared sugar solutions, and the development of more efficient bioreactors which will maximize conversion rate while minimizing the net production of cell mass. There is also need for further understanding of the kinetic behavior of the different enzymes involved in the biochemical pathway. The use of new analytical techniques has opened the way to re-evaluation of models existent in the literature.

The improvement of our knowledge of the 2,3-BD fermentation will not only enhance the chance of its commercial use but will also add to our general knowledge of bacterial fermentation.

45.7 REFERENCES

Adams, G. A. and R. Y. Stainer (1945). Production and properties of 2,3-butanediol. III. Studies on the biochemistry of carbohydrate fermentation by *Aerobacillus polymyxa*. *Can. J. Res.*, **23B**, 1–9.
Blackwood, A. C., J. A. Wheat, J. D. Leslie, G. A. Ledingham and F. T. Simpson (1949). Production and properties of 2,3-butanediol. XXXI. Pilot plant studies on the fermentation of wheat by *Aerobacillus polymyxa*. *Can. J. Res.*, **27F**, 199–210.
Bryn, K., O. Hetland and F. C. Stormer (1971). The reduction of diacetyl and acetoin in *Aerobacter aerogenes*. Evidence for one enzyme catalyzing both reactions. *Eur. J. Biochem.*, **18**, 116–119.
Corey, E. J. and R. A. E. Winger (1963). A new, stereospecific olefin synthesis from 1,2-diols. *J. Am. Chem. Soc.*, **85**, 2677–2678.
Corey, E. J., F. A. Carey and R. A. E. Winter (1965). Stereospecific synthesis of olefins from 1,2-thionocarbonates and 1,2-tri-thiocarbonates, *J. Am. Chem. Soc.*, **87**, 934–935.
Chua, J. W., A. Erarslan, S. Kinoshita and H. Taguchi (1980). 2,3-Butanediol production of immobilized *Enterobacter aerogenes* IAM1133 with ×-carrageenan. *J. Ferment. Technol.*, **58**, 123–127.
Emerson, R. R., M. C. Flickinger and G. T. Tsao (1982). Kinetics of dehydration of aqueous 2,3-butanediol to methyl ethyl ketone. *Ind. Eng. Chem. Prod. Res. Dev.*, **21**, 473–477.
Esener, A. A., J. A. Roels and N. W. F. Kossen (1981a). The influence of temperature on the maximum specific growth rate of *Klebsiella pneumoniae*. *Biotechnol. Bioeng.*, **23**, 1401–1405.
Esener, A. A., G. Bol, N. W. F. Kossen and J. A. Roels (1981b). Effect of water activity on microbial growth. In *Advances in Biotechnology*, ed. M. Moo-Young, C. W. Robinson and C. Vezina, vol. 1, pp. 339–344. Pergamon, Toronto.

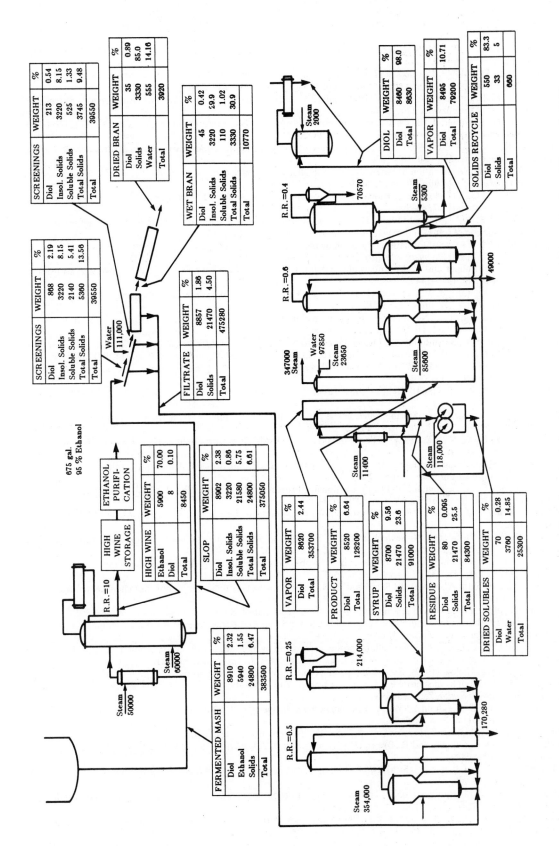

Figure 9 Material balance for commercial flowsheet for the production of (−)-2,3-BD based on 1000 bu. of wheat per day (from Wheat *et al.*, 1948)

Flickinger, M. C. and G. T. Tsao (1979). Growth of *Klebsiella pneumoniae* on a hemicellulose hydrolysate. 178th Annual ACS Meeting, MICR Division, Paper No. 44, Washington, DC.

Freeman, G. G. and R. I. Morrison (1947). Production of 2,3-butylene glycol by fermentation of molasses. *J. Soc. Chem. Ind., London*, 66, 216–221.

Gordon, M. and J. V. Hay (1968). Stereochemistry of vapor phase dehalogenation of *meso* and DL-2,3-dibromobutane with zinc. *J. Org. Chem.*, 33, 427.

Harrison, D. E. F. and J. E. Loveless (1971). The effect of growth conditions on respiratory activity and growth efficiency in facultative anaerobes grown in chemostat culture. *J. Gen. Microbiol.*, 68, 35–43.

Hetland, O., K. Bryn and F. C. Stormer (1971). Diacetyl (acetoin) reductase from *Aerobacter aerogenes*: evidence for multiple forms of the enzyme. *Eur. J. Biochem.*, 20, 206–208.

Hetland, O., G. R. Olsen, T. B. Christensen and F. C. Stormer (1971). Diacetyl (acetoin) reductase from *Aerobacter aerogenes*: structural properties. *Eur. J. Biochem.*, 20, 200–205.

House, H. O. and R. S. Ro (1958). The stereochemistry of elimination reactions involving halohydrin derivatives and metals. *J. Am. Chem. Soc.*, 80, 182–187.

Jansen, N. B. (1984). Application of bioenergetics to modelling the batch fermentation of D-xylose to 2,3-butanediol by *Klebsiella oxytoca. Biotechnol. Bioeng.*, 26, 362–369.

Jansen, N. B. and G. T. Tsao (1983). Bioconversion of pentoses to 2,3-butanediol by *Klebsiella pneumoniae. Adv. Biochem. Eng. Biotechnol.*, 27, 85–99.

Josan, J. S. and F. W. Eastwood (1968). Derivatives of orthoacids. IV: Acid catalyzed thermal elimination and best induced elimination of some 2-ethoxy-1,2-dioxolans. *Aust. J. Chem.*, 21, 2013–2020.

Juni, E. (1952). Mechanism of formation of acetoin by bacteria. *J. Biol. Chem.*, 195, 715–726.

Kearby, K. (1955). In *The chemistry of petroleum hydrocarbons*, ed. B. T. Brooks *et al.*, vol. 2. Reinhold, New York.

Kooi, E. R., E. I. Fulmer and L. A. Underkofler (1948). Production of 2,3–butanediol by fermentation of cornstarch. *Ind. Eng. Chem.*, 40, 1440–1445.

Ladisch, M., K. W. Lin, M. Voloch and G. T. Tsao (1983). Process considerations in the enzymatic hydrolysis of biomass. *Enzyme Microb. Technol.*, 5, 82–102.

Ledingham, G. A. and A. C. Neish (1954). Fermentative production of 2,3-butanediol. In *Industrial Fermentations*, ed. L. A. Underkofler and R. J. Hickey, vol. 2, pp. 27–93. Chemical Publishing Co., New York.

Larsen, S. H. and F. C. Stormer (1973). Diacetyl (acetoin) reductase from *Aerobacter aerogenes*: kinetic mechanism and regulation by acetate of the reversible reduction of acetoin to 2,3-butanediol. *Eur. J. Biochem.*, 34, 100–106.

Loken, J. A. and F. C. Stormer (1970). Acetolactate decarboxylase from *Aerobacter aerogenes*. Purification and properties. *Eur. J. Biochem.*, 14, 133–137.

Long, S. K. and R. Patrick (1963). The present status of the 2,3-butylene glycol fermentation. *Adv. Appl. Microbiol.*, 5, 135–155.

Malthe-Sorenssen, D. and F. C. Stormer (1970). The pH 6 acetolactate-forming enzyme from *Serratia marcesans*: purification and properties. *Eur. J. Biochem.*, 14, 127–132.

Murphy, D. and D. W. Stranks (1951). The production of 2,3-butanediol from sulphite waste liquor. *Can. J. Technol.*, 29, 413–420.

Neish, A. C. (1945). Production and properties of 2,3-butanediol. IV. Purity of the laboratory 2,3-butanediol produced by *Aerobacillus polymyxa. Can. J. Res.*, 23B, 10–16.

Neish, A. C. and G. A. Ledingham (1949). Production and properties of 2,3-butanediol. XXXII. Fermentations at poised hydrogen ion concentrations. *Can. J. Res.*, 27B, 694–704.

Olson, B. H. and M. J. Johnson (1948). The production of 2,3-butylene glycol by *Aerobacter aerogenes* 199. *J. Bacteriol.*, 55, 209–222.

Othmer, D. F., W. B. Sergen, N. Schlechter and P. F. Bruins (1945a). Liquid–liquid extraction data, systems used in butadiene manufacture from butylene glycol. *Ind. Eng. Chem.*, 37 (9), 890–894.

Othmer, D. F., N. Schlechter and W. A. Koszalka (1945b). Composition of vapors from boiling binary solutions. *Ind. Eng. Chem.*, 37 (9), 895–900.

Palsson, B. D., S. Fathi-Afshar, D. F. Rudd and E. N. Lightfoot (1981). Biomass as a source of chemical feedstocks: an economic evaluation. *Science*, 213, 513–517.

Perlman, D. (1944). Production of 2,3-butylene glycol from wood hydrolyzates. *Ind. Eng. Chem.*, 36, 803–804.

Pirt, S. J. (1957). The oxygen requirement of growing cultures of an *Aerobacter* species determined by means of the continuous culture techniques. *J. Gen. Microbiol.*, 16, 59–75.

Pirt, S. J. and D. S. Callow (1958). Exocellular product formation by microorganisms in continuous culture. I. Production of 2,3-butanediol by *Aerobacter aerogenes* in a single stage process. *J. Appl. Bacteriol.*, 21, 188–205.

Pirt, S. J. and D. S. Callow (1959). Exocellular product formation by microorganisms in continuous culture. II. Production of 2,3-butanediol from sucrose by *Aerobacter aerogenes* in a two-stage process. *Selected Scientific Papers from the Istituto Superiore di Sanità*, 2, 292–313.

Prescott, S. C. and C. G. Dunn (1959). The production and properties of 2,3-butanediol. *Industrial Microbiology*, pp. 399–427. McGraw-Hill, New York.

Rose, A. H. (1961). *Industrial Microbiology*, pp. 255–259. Butterworths, Washington, DC.

Schlechter, N., D. F. Othmer and R. Brand (1945a). Pyrolysis of 2,3-butylene glycol diacetate to butadiene. *Ind. Eng. Chem.*, 37 (9), 905–908.

Schlechter, N., D. F. Othmer and S. Marshak (1945b). Esterification of 2,3-butylene glycol with acetic acid. *Ind. Eng. Chem.*, 37 (9), 900–905.

Sebek, O. K. and C. I. Randles (1952). The oxidation of stereoisomeric 2,3-butanediols by *Pseudomonas. Arch. Biochim. Biophys.*, 40, 373–379.

Segal, I. H. (1975). *Enzyme Kinetics*, pp. 560–590. Wiley, New York.

Stanier, R. Y. and G. A. Adams (1944). The nature of the *Aeromonas* fermentation. *Biochem J.*, 38, 168–171.

Stanier, R. Y. and S. B. Fratkin (1944). Studies on the bacterial oxidation of 2,3-butanediol and related compounds. *Can. J. Res.*, 22B, 140–153.

Stinson, S. C. (1979). New plants processes set for octane booster. *Chem. Eng. News*, 57 (26), June 25, 35–36.

Stormer, F. C. (1967). Isolation of crystalline pH 6 acetolactate-forming enzyme from *Aerobacter aerogenes*. *J. Biol. Chem.*, **242**, 1756–1759.

Stormer, F. C. (1968). The pH 6 acetolactate-forming enzyme from *Aerobacter aerogenes*. I. Kinetic studies. *J. Biol. Chem.*, **243**, 3735–3739.

Taylor, M. B. and E. Juni (1960). Steroisomeric specificities of 2,3-butanediol dehydrogenases. *Biochim. Biophys. Acta*, **39**, 448–460.

Tipson, R. S. and A. Cohen (1965). Action of zinc dust and sodium iodide in *N,N*-dimethyl formamide on contiguous secondary sulfonyloxy groups: a simple method for introducing non-terminal unsaturation. *Carbohydr. Res.*, **1**, 338–340.

Tomkins, R. V., S. D. Scott and F. J. Simpson (1948). Production and properties of 2,3-butanediol. XXIX. Pilot plant studies on fermentation of barley by *Aerobacillus polymyxa* and recovery of the products. *Can. J. Res.*, **26F**, 497–502.

Tsao, G. T., M. R. Ladisch, M. Voloch and P. Bienkowski (1982). Production of ethanol and chemicals from cellulosic materials. *Process Biochem.*, **17** (5), 34–38.

Voloch, M. (1981). Reduction of acetoin to 2,3-butanediol in *Klebsiella pneumoniae*. A stereochemical model and kinetics. Ph.D. Thesis, Purdue University, West Lafayette, IN.

Voloch, M., M. R. Ladisch, V. W. Rodwell and G. T. Tsao (1981). Separation of *meso* and racemic 2,3-butanediol by aqueous liquid chromatography. *Biotechnol. Bioeng.*, **23**, 1289–1296.

Voloch, M., M. R. Ladisch, V. Rodwell and G. T. Tsao (1983). Reduction of acetoin to 2,3-butanediol in *Klebsiella pneumoniae*: a new model. *Biotechnol. Bioeng.*, **25**, 173–183.

Vollbrecht, D. (1982). Restricted oxygen supply and excretion of metabolites. II. *Escherichia coli* K12, *Enterobacter aerogenes* and *Brevibacterium lactofermentum*. *Eur. J. Appl. Microbiol. Biotechnol.*, **15**, 111–116.

Walpole, G. C. (1911). The action of *Bacillus lactis aerogenes* on glucose and mannitol. III. The investigation of the 2,3-butanediol and the acetylmethylcarbinol formed. *Proc. R. Soc. London, Ser. B*, **83**, 272–286.

Ward, G. F., O. G. Pettijohn and R. D. Coghill (1945). Production of 2,3-butanediol from acid-hydrolyzed starch. *Ind. Eng. Chem.*, **37**, 1189–1194.

Wheat, J. A., J. D. Leslie, R. V. Tomkins, H. E. Mitton, D. S. Scott and G. A. Ledingham (1948). Production and properties of 2,3-butanediol. XXVIII. Pilot plant recovery of levo-2-3-butanediol from whole wheat mashes fermented by *Aerobacillus polymyxa*. *Can. J. Res.*, **26F**, 469–496.

Winfield, M. E. (1945). The catalytic dehydration of 2,3-butanediol to 1,3-butadiene. *J. Council Sci. Ind. Res.*, **18**, 412–423 (*Chem. Abstr.*, **40**, 3719).

Yu, E. K. C. and J. N. Saddler (1982). Power solvent production by *Klebsiella pneumoniae* grown on sugars present in wood hemicellulose. *Biotechnol. Lett.*, **4**, 121–126.

46
Microbial Insecticides

H. STOCKDALE
Shell Research Ltd., Sittingbourne, Kent, UK

46.1 INTRODUCTION AND HISTORICAL BACKGROUND

The creatures with which humanity is in conflict, either as a competitor, hunter or victim, are themselves held in abeyance by other species living at their expense. This system of checks and balances may be stable in general, but is not so in detail, and fluctations of feast and famine can be seen within an overall equilibrium of population and resources.

Until recent historical times virtually all control of mankind's antagonists could be said to be of this 'biological' type, and there was no recourse to an alternative when it proved inadequate. However, the 20th century has seen the widespread usage of chemical pesticides, inventions of human wit for human need. Nearly all agricultural production and a large part of public health projects require their use. Chemical pesticides have been eminently successful, but the drawbacks attendant on their use are now manifest. The two commonest criticisms are that a chemical agent may kill non-target organisms and that the target itself may become resistant to the agent. There is a continuing search for effective and specific pesticides and those of biological origin are among the candidates being considered by industry (Ayers *et al.*, 1977). The theoretical aspects of biological control agents (Huffaker and Messenger, 1976) and many instances of practical use (Burges, 1981a) are well documented. Target pests include insects and mites (Burges and Hall,

1982), weeds (Hasan, 1979), nematodes (Stirling and Wachtel, 1980) and plant diseases (Schroth and Hancock, 1981).

If we confine our attention to the control of insect pests it is seen that the enemies that can be marshalled against them fall into two broad classes, first predators and parasites, and secondly pathogenic microorganisms. A dramatic increase in insect numbers in a field or forest is often followed by an increase in insect enemies, macroscopic or microscopic, in turn causing the collapse of such an outbreak population. Observation of this process gives cold comfort to those whose crops have been destroyed: successful practice of biological control depends on application of the agent at such a time or in such amounts that economic damage is minimized.

Conscious attempts to control insect pests by biological means have been made over an extended period of our history. As early as 900 AD ants were sold in China for the protection of citrus trees (Coppel and Mertins, 1977). Recognition and attempted use of microbes is obviously a more recent development. Ordish (1967) quotes John Evelyn's book 'Sylva', published in 1664, as recommending that bushes should be sprinkled with a decoction of caterpillars so as to protect them from further attack. Although Evelyn does not appear to give a reason, it is likely that some of the larvae were diseased. The 19th century saw an increasing awareness of the potential of microbial methods for crop protection which was fostered by the elucidation of the nature of pathogens, both in vertebrates and invertebrates. The period has been summarized entertainingly by Steinhaus (1956). Highlights include the observations of Agostino Bassi, published in 1834, on the disease of silkworms, 'mal del segno', caused by a fungus later to be known as *Beauveria bassiana*. Apparently this is the first recorded instance of a microorganism being recognized as the causative transmittable agent of an animal disease. Pasteur's later studies of the silkworm diseases 'pébrine' and 'flacherie' provided direct practical benefit to the stricken French silk industry. In 1878 Elie Metchnikoff began a study of diseases of the grain beetle *Anisoplia austriaca* which resulted in the discovery of the fungus *Metarhizium anisopliae*. As far as can be determined, Metchnikoff attempted to propagate the fungus on artificial media and actually undertook some field experiments. Since those days the number of insecticidal microorganisms available for research has increased tremendously, and records of their field trials are now legion.

46.2 OVERALL VIEW OF PRESENT USE

While the use of predators and parasites for insect control provides a fascinating topic, it falls outside the scope of the present article. Methods of production and application are radically different from those of pathogens, which in this regard have several characteristics in common with chemical insecticides.

Insect pathogens have representatives among the bacteria, fungi, viruses, rickettsiae, protozoa and nematodes. Over a thousand pathogens have been recorded, and many of these could be put to a practical use (Ignoffo, 1975). Indeed, at some time all classes of pathogens, with the exception of the rickettsiae, have been used to control some pest insect. The use of insect pathogens, while not carried out on a massive scale, is extremely widespread. They are employed in the areas of forestry, crop protection, disease vector control and the preservation of stored products. Geographically, they are employed across both hemispheres, from the coniferous forests to the tropics. In parts of South America peasants have been known to collect virus-infected lepidopteran larvae for later use on crops, and the same may be true in South East Asia. Across the industrialized North a range of pathogens are produced on a factory scale. Among the Western countries *Bacillus thuringiensis* (B.t.) is the major product, while several viruses are used for forest and crop protection. In Eastern Europe and Russia B.t. is employed in forests. The fungi *B. bassiana* and *M. anisopliae* are produced for use against the Colorado beetle *Leptinotarsa decemlineata*. Two viruses are produced for forest use, a nuclear polyhedrosis virus (NPV) of the Siberian silkworm *Dendrolimus sibiricus* and an entomopoxvirus (EPV) for use against the winter moth *Operophthera brumata*. In China B.t., *B. bassiana* and *M. anisopliae* are produced in communes. In Japan a cytoplasmic polyhedrosis virus (CPV) is produced by Chugai-Kumiai under the name 'Matsukemin' for control of *Dendrolimus spectabilis*, while Sumitomo produces a preparation of *Bacillus moritai* for use against fly larvae (Ignoffo, 1973). Table 1 lists the major insect pathogens in use and their producers. By far the greater part is made up of bacteria and viruses, with some fungi. A search of the patents filed by producing organizations shows a similar pattern (Table 2).

Future developments are difficult to predict, but work with a relatively recent strain of B.t., namely var. *israelensis* (H14), may well prove efficacious against medically-important dipteran larvae (Goldberg and Margalit, 1977). It is doubtful if protozoans will ever assume a major role in

Table 1 Major Microbial Insecticides in Production

Organism	Target	Characteristic of production	
Bacteria			
Bacillus thuringiensis	Lepidopteran larvae	By fermentation by several firms	Commercial
Bacillus thuringiensis israelensis	Dipteran larvae	By fermentation for WHO	Non-commercial
Bacillus popilliae	Japanese beetle	In collected larvae by Fairfax Biologicals	Commercial
Bacillus sphaericus	Mosquito larvae	By fermentation by Government agencies	Non-commercial
Fungi			
Beauveria bassiana	Numerous	By fermentation in factories in USSR	Non-commercial
Beauveria tenella	Numerous	By fermentation in factories in USSR	Non-commercial
Metarhizium anisopliae	Numerous	By fermentation in factories in USSR	Non-commercial
Hirsutella thompsonii	Citrus mites	By fermentation by Abbott Laboratories	Commercial
Verticillium lecanii	Glasshouse aphids	By fermentation by Tate and Lyle	Commercial
Viruses (forest)			
Sawfly NPV	Sawflies	In collected larvae by Government agencies	Non-commercial
Tussock moth NPV	*Hemerocampa* spp.	In mass-reared larvae by Government agencies	Non-commercial
Gypsy moth NPV	*Porthetria dispar*	In mass-reared larvae by Government agencies	Non-commercial
Spruce budworm NPV	*Choristoneura fumiferana*	In mass-reared larvae by Government agencies	Non-commercial
Siberian silkworm NPV	*Dendrolimus sibiricus*	Presumably in mass-reared insects in USSR	Non-commercial
Winter moth	*Operophthera brumata*	Presumably in mass-reared insects in USSR	Non-commercial
Viruses (agricultural)			
Codling moth GV	*Cydia pomonella*	In mass-reared larvae by Government agencies	Non-commercial
Potato tuber moth GV	*Phthorimaea operculella*	In mass-reared larvae by Government agencies	Non-commercial
Cotton bollworm NPV	*Heliothis* spp.	In mass-reared larvae by Sandoz	Commercial
Cabbage looper NPV	*Trichoplusia ni*	In mass-reared larvae by Government agencies	Non-commercial
Alfalfa looper NPV	*Plusia* spp.	In mass-reared larvae by Government agencies	Non-commercial
Cotton leafworm NPV	*Spodoptera littoralis*	In mass-reared larvae by Government agencies	Non-commercial

insect pest control. However, *Nosema* spp. have been reported effective against grasshoppers (Henry, 1971). McLaughlin (1974) reported some success against the boll weevil using *Glugea* (*Nosema*) *gasti* in a bait, although at an uneconomic price. Nematodes attack a wide range of insects, but their efficacy in the field will depend on formulations which can maintain the agents' viability (Nickle, 1980). In general it can be said that while one can confidently expect production of new insecticidal microbes, an immediate requirement is for methods of improving the effectiveness of those agents now in common use.

46.3 PRODUCTION METHODS

The technology used in the production of microbial insecticides varies from that of a cottage industry to that equal in sophistication to the pharmaceutical industry. Present methods can be divided into three classes demanding progressively less labour and more costly equipment. They are: collection and infection, mass rearing and fermentation methods.

46.3.1 Collection and Infection

In cases where a pathogen is highly infectious, or alternatively where protection of an expensive commodity justifies the effort, a control agent may be produced from insects collected from the field. Several sawfly species are economically-important pests in coniferous forests. Their NPV are often highly infectious and enough material can be obtained from field-collected larvae for use in trials or control measures. In the USA, subterranean larvae of the imported beetle *Popillia japonica* constitute a threat to amenity grasslands. Two commercial enterprises, Fairfax Biological Laboratories and Reuter Laboratories, produce microbial control agents for this pest. Field-collected larvae are infected by injection with *Bacillus popilliae* or *Bacillus lentimorbus*. The resultant spore-laden cadavers are formulated as a wettable powder for protection of lawns, field crops and market gardens.

Table 2   Insect Pathogens Concerning which Patents have been Filed

Organism	Target	Organization	Patent No.
Bacteria			
Bacillus thuringiensis	Lepidopteran larvae	29 Organizations	47 Patents
Bacillus cereus	Lepidopteran larvae	M.G. Gandman	RU 144 579
Bacillus insectus	Lepidopteran larvae	A.B. Gukasyan	RU 343 674
Bacillus popillae	Japanese beetle	US Sec. of Agriculture	US 3 308 038
Bacillus popillae	Japanese beetle	Research Corporation	US 3 503 851
Bacillus popillae	Japanese beetle	Amstar	US 3 950 225
Bacillus sphaericus	Mosquito larvae	IMC	US 3 420 933
Bacillus sphaericus	Mosquito larvae	Sumitomo	J 53 145 915
Bacillus moritai	Dipteran larvae	S. Morita	UK 1 113 319
Fungi			
Beauveria bassiana	Fleas	Rostov Pest. Res. Inst.	RU 367 835
Beauveria bassiana	Colorado beetle	Forschzentrum Graz	DT 2 617 892
Coelomomyces iliensis	Culex larvae	Kazakhstan Zool. Inst.	RU 425 938
Viruses			
Production *in vitro*	Lepidopteran larvae	MGK	UK 1 300 447
Production *in vitro*	Lepidopteran larvae	Shell	BE *e.g.* 806 708
Microencapsulation	Lepidopteran larvae	NCR	US 3 541 203
Microencapsulation	Lepidopteran larvae	Battelle Dev. Co.	BE 870 561
Dendrolimus NPV	*Dendrolimus sibiricus*	Katukara Ind. Co.	J 71 21 755
Dendrolimus CPV	*Dendrolimus sibiricus*	Chugai Pharmaceuticals	J 47 23 527
Lymantria NPV	*Lymantria dispar*	Katakura Ind. Co.	J 71 17 924
Spodoptera NPV	*Spodoptera frugiperda*	MGK	US 3 716 634
Heliothis NPV	*Heliothis* spp.	IMC/Sandoz	5 Patents
Nematodes			
Heterorhabditis bacteriophora	Numerous targets	CSIRO	DT 2 632 990 Aus PC 2930
Neoaplectana carpocapsae	Numerous targets	CSIRO	J 100 735

46.3.2   Mass Rearing

Mass rearing techniques have been most notably employed in the production and release of radiation-sterilized adults as a means of reducing progeny populations. A striking example is that of the screw-worm fly *Cochliomyia hominivorax* where $50–100 \times 10^6$ individuals a week were released over the Southwestern USA and Northern Mexico. Some lepidopteran pests are also susceptible to control by this method. The USDA produces 1.3×10^6 sterile adults of the pink bollworm *Pectinophora gossypiella* daily (Waterhouse *et al.*, 1976). Mass rearing programmes for pathogen production have been directed commonly, but not exclusively, to lepidopteran viruses. Suitable agents are to be found among the so-called occluded group of viruses in which the infectious units or virions are enclosed, either singly or as multiples, in bodies of a dense crystalline protein matrix usually $0.5–2.0$ μm in diameter. These inclusion bodies (IB) confer stability on the virions until such time as they are ingested by a larva when the matrix protein dissolves in the normally alkaline gut juice to release infectious virions. Occluded viruses are of several types. In two groups polyhedral inclusion bodies (PIB) are formed, either in the cytoplasm or nuclei of the host cells. The PIB of cytoplasmic polyhedrosis viruses (CPV) enclose icosahedral virions. PIB of nuclear polyhedrosis viruses (NPV) may contain numerous rod-shaped virions, either singly (SNPV), or in bundles or multiples (MNPV). In the granulosis viruses (GV) each PIB contains a single rod-shaped virion. NPV and GV are collectively termed baculoviruses. PIB of entomopoxviruses (EPV) contain numerous large, complex, slab-shaped virions. A succinct account of insect virus morphology has been given by Payne and Kelly (1981).

In the USA, forests are subject to attack by the gypsy moth *Lymantria dispar* and the tussock moths *Hemerocampa pseudotsugata* and *H. leucostigma*, and mass rearing programmes for the production of their respective NPV are carried out under the auspices of the USDA. In Japan the CPV of *D. spectabilis* is produced by mass rearing. The cotton bollworm or corn earworm *Heliothis zea* and the tobacco budworm *H. virescens* are major crop pests, notably in the USA, but also worldwide. They are both susceptible to *Heliothis* SNPV, and Sandoz now has a commercial plant in Florida producing the virus under the trade name of 'Elcar'.

Successful production of lepidopteran larvae depends on the availability of a synthetic diet as larvae reared on imported foliage are susceptible to epizoötics. Antimicrobials incorporated into diet provide some safeguard against this. Larvae are raised on an agar-based diet, either singly or severally, depending on their cannibalistic proclivities. The diet is inoculated with the requisite pathogen and after it has multiplied in the larvae, cadavers and moribund larvae are collected for processing and formulation. Probably the most complete account of such a procedure is given for the *Heliothis* SNPV virus by Ignoffo (1973a). Productivity depends on judicious choice of the time of infecting and harvesting larvae and the dose of pathogen given. Although the amount of product obviously varies greatly with the host and pathogen, some degree of standardization has been introduced, for example the number of PIB of NPV recovered from a larva of the cabbage looper *Trichoplusia ni* is commonly considered as 6×10^9. This is known as a larval equivalent (LE). Mass rearing has two main drawbacks. It is labour intensive, and it is inflexible in that a programme cannot be adapted rapidly to production of another species. Some steps can be taken to alleviate the first problem by automation where possible, *e.g.* in diet dispensation. The second appears intractable.

At present, attention is focussed on the performance of 'Elcar', but several other viruses can be produced for crop protection by mass rearing. Among likely candidates are the MNPV of *T. ni*, the MNPV of the alfalfa looper *Autographa californica*, and the GV of the codling moth *Cydia pomonella*. Only one nematode pesticide has been produced commercially by mass rearing techniques. This was a preparation of *Romanomermis culicivorax* obtained from *Culex pipiens* larvae and marketed for mosquito control under the name of 'Skeeterdoom'. Production has now been suspended.

46.3.3 Fermentation Techniques

In general, fermentation offers the best route for the industrialization of insect pathogens. It is possible to produce a diversity of organisms *in vitro* although most efforts have been confined to small-scale laboratory studies. In many cases the physicochemical and nutritional requirements for production of insect pathogens of optimal quality and concentration are poorly understood, and research into these conditions is inhibited by the relatively small markets the products are expected to command. However, some notable exceptions exist.

46.3.3.1 Nematodes

Members of seven families of nematodes have insecticidal properties and interest in their use is widespread although none has been successfully industrialized. Some non-entomogenous nematodes have been reared in artificial media and so may be amenable to fermenter production (Hansen and Cryan, 1966). It is not yet known whether any entomogenous nematodes can be induced to reach maturity in deep liquid culture.

Neoaplectana carpocapsae, which attacks about 200 species of insect, can be grown on a variety of solid nutrient media. It is usually produced with its associated bacterium *Achromobacter nematophilus* which apparently supplies nutrients to the nematode, although Poinar (1979) mentions axenic growth of this organism. In the USA the Nutrilite Corporation has investigated the DD-136 strain of *N. carpocapsae* and mass produced it on dog food.

Bedding (1976) working at CSIRO in Tasmania produced batches of *N. carpocapsae* in culture vessels of up to 80 l capacity. A solid substrate such as foam rubber or wood wool was required to allow the adults to mate. The nematodes were cultured in the presence of their bacteria. Complex meat-based media such as homogenized chicken hearts and pig kidneys with beef fat were used. Bedding (1981) extended his studies with this type of culture to the production of three *Neoaplectana* and three *Heterorhabditis* species. *Deladenus siricidola* proved highly effective in controlling the woodwasp *Sirex* in Tasmanian forests (Bedding, 1974). The nematode has two life cycles; it parasitizes the woodwasp, but can also feed on a fungus associated with the woodwasp and then give rise to insect-infective individuals. Bedding (1974) was able to culture the fungus on a wheat and water medium and then, on this, rear and harvest the nematode for *Sirex* control.

Mermithid nematodes are not associated with specific microorganisms but obtain nutrients from the haemolymph of the host by absorption through the cuticle (Peterson, 1982). Recently Fassuliotis and Creighton (1982) have reared the mermithid *Filipjevimeris leipsandra* to pre-adult stage in an insect tissue culture medium, Schneider's *Drosophila* medium, with foetal bovine serum. Although a solid substrate was required for the final moult to the adult stage this system appears to be the most promising yet for mermithid development *in vitro*

46.3.3.2 Fungi

Few entomogenous fungi have been produced on an industrial scale. Those that have include the genera *Beauveria, Metarhizium, Verticillium, Hirsutella* and *Aschersonia* (Ignoffo and Anderson, 1979). This should not be taken as evidence of limited usefulness, rather of the difficulties connected with the development of fungal insecticides. In many respects fungi form the most diverse group of entomopathogens in their life cycles, nutrition and targets, and new candidates for insect control are frequently reported (Soares *et al.*, 1979; McInnis and Zattau, 1982).

Production of entomogenous fungi *in vitro* is a complex subject. It has recently been treated in detail by Hall and Papierok (1982), but several broad points can be made here. In some fungi the infectious units are conidia (conidiospores) which are not produced readily in submerged culture. Consequently these agents must be produced on solid or semi-solid media. Infectious spores of *Nomurea rileyi* (Ignoffo, 1981) and *Hirsutella thompsonii* (McCoy and Kanavel, 1969) have been produced on a series of agar media of varying nutritional quality, but media based on sterilized cereal grains, corn mash or bran are often used for industrial production. The mechanisms of production vary with choice or with the facilities available. Cultures have been produced in open trays or in sterilized bottles or plastic bags. Horizontally-rotating fermenters have also been used. Hall and Papierok (1982) point out that such methods permit exposure to light, which can increase productivity and resistance of spores to field conditions.

Other fungi will form conidia in submerged culture, and infective spore preparations of *B. bassiana, Culicinomyces clavosporus* and a single strain of *H. thompsonii* have been produced in this manner (Hall and Papierok, 1982). McCoy *et al.* (1975) produced mycelia of *H. thompsonii* in submerged culture in 12 l batches, in the expectation that mycelial fragments would produce infectious conidia when applied in the field. Where it is feasible, submerged culture can produce high spore concentrations coupled with a more complete and hence economic use of nutrients. It is then possible to determine limiting nutrients and to control the physicochemical parameters of the fermentation more closely.

Samsinakova *et al.* (1981) have described a method for producing spores of *Beauveria bassiana* in still liquid culture with a final concentration of 3.5×10^{11} spores per gram of mycelium.

The nutrition of insect-pathogenic fungi is complex, important in its effect on activity, and as yet poorly understood. Some fungi have not been cultured satisfactorily *in vitro*. Smith and Grula (1981) reported the ability of *B. bassiana* to utilize a wide variety of substrates for germination and growth. However, more subtle events may be greatly influenced by substrates. Kučera (1981) found that production of the toxic protease of *M. anisopliae* varied according to which one of 29 nitrogen sources was supplied. A homogenate of larvae of the wax moth *Galleria mellonella* proved most effective. It would appear that many opportunities exist for research into the production and optimization of fungal insecticides.

46.3.3.3 Bacteria

Insect-pathogenic bacteria have been classified as obligate or facultative pathogens (Falcon, 1971). The so-called obligate pathogens cannot be cultivated successfully *in vitro*, while of the facultative pathogens only those that produce a resistant spore or stable toxic moiety are worth cultivating. Candidates are therefore restricted to the crystalliferous spore former *Bacillus thuringiensis* (B.t.) and its recently discovered variant *B. thuringiensis* var. *israelensis* (H-14) (Goldberg and Margalit, 1977) or to the non-crystal producer *B. sphaericus* in which the toxin is in the cell wall. B.t. is a lepidopteran pathogen producing a proteinaceous crystalline toxin (δ-endotoxin) which causes paralysis of the larval gut and a thermostable β-exotoxin capable of killing flies and other insects. *B. sphaericus* is a promising pathogen of mosquito larvae (Kellen *et al.*, 1965), as is *B. thuringiensis* H-14, which has additional activity against blackfly (*Simulidae*) larvae.

The production of B.t. is far in excess of that of any other commercially-produced microbial insecticide. Burges (1982) estimated that production in Western countries is about 2000 tons per annum active ingredient although it is also produced and used widely in Eastern Europe. Judicious strain selection has led to a 100–600 fold improvement in performance since it was first produced commercially (Burges, 1982). The characteristics of δ-endotoxin toxicity to lepidopteran larvae are complex. A particular strain of B.t. can be active against one species but not against another. There is still much scope for isolation of more effective strains for specific targets such as that reported by Salama and Foda (1982) for the cotton leafworm *Spodoptera littoralis*. Use of β-exotoxin-producing strains is not encouraged in North America and Western Europe, but β-exotoxin-containing preparations have been used against the Colorado beetle *L. decemlineata* (Cantwell and Cantelo, 1981) and one has been registered for suppression of flies in pig houses in Finland (Burges, 1982).

Media used for B.t. production are complex and are based on such natural products as fishmeal, starch, tryptone and cottonseed flour (Pendleton, 1969; Dulmage, 1970; Bulla and Yousten, 1979). Efficacy of the product can vary greatly according to the medium used (Dulmage, 1970). Fermenters of up to 12 000 gal capacity have been used. Described conditions favour an optimum temperature of 30 °C with aeration. The starting pH is 7.2–7.6. The process is a batch one, being completed in 14–72 h depending on how the chosen medium influences sporulation and concomitant crystal toxin production (Couch and Ross, 1980). Refinements of this basic procedure have been reported. Goldberg *et al.* (1980) describe the optimization of a medium for high $(4 \times 10^9$ spores ml^{-1}) spore crystal production in a chemostat with pH maintenance at 7.0. Holmberg *et al.* (1980) report the determination of optimum conditions of pH, temperature, aeration, agitation and limiting nutrient (total sugar) concentration for β-exotoxin production in a batch process.

The nutritional requirements for production of toxin of *B. thuringiensis* H-14 are not substantially different from those of B.t. It has been produced in a medium previously used for the culture of B.t. subsp. *kurstaki* (Tyrell *et al.*, 1979).

B. sphaericus is a potentially useful agent for mosquito control in that it is active against several different genera of larvae and persists in the environment. It has been produced on a pilot scale (Burges, 1981). Its nutrition appears to be undemanding in that its toxicity is similar when it is grown on synthetic, complex or hay infusion media although certain strains may be exceptional (Singer, 1980).

B. popilliae is a useful agent for the control of the Japanese beetle *P. japonica*. Infection is effected by ingestion of the spores, but infectious spores cannot as yet be produced *in vitro* (Sharpe *et al.*, 1970). However, Sharpe and Bulla (1978), basing their experimental approach on the observation that cultures which sporulated after growth in the haemocoel of *P. japonica* had very long population doubling times (*ca.* 19 h), grew the bacillus at progressively slower rates in a chemostat. They observed about 1% sporulation after 7 d at slow growth rates (dilution rates below 0.1 h^{-1}). It was not stated whether or not these spores were infectious *per se*. It may well be that successful production of spores depends on a critical combination of nutritional and physicochemical factors affecting growth rate.

46.3.3.4 *Viruses*

Insect viruses can only be produced in an appreciable quantity by infecting susceptible insects (*in vivo*) or established cell lines (*in vitro*). It is debatable if the *in vitro* process will ever compete commercially with *in vivo* methods. Nevertheless, a large body of knowledge regarding *in vitro* techniques now exists, and the production of some insect viruses by such means is feasible if not economic.

The state of the art was reviewed recently by Stockdale and Priston (1981) since when there has been little change in the overall picture. The history of the *in vitro* process can be recalled briefly by reference to several key events. In 1962 Grace established four continuously-dividing cell lines of the silkmoth *Antheraea eucalypti*. Establishment of lines from other lepidoptera and other orders of insects followed rapidly. Hink (1976) listed 121 invertebrate cell lines of which the majority are insect, predominantly lepidopteran and dipteran.

Bellet and Mercer (1964) infected *A. eucalypti* cells with the non-occluded iridescent virus of the beetle *Sericesthis* (SIV). *In vitro* production of occluded viruses took a little longer to realize. Goodwin *et al.* (1970) established a cell line from the fall armyworm *Spodoptera frugiperda* and infected it with its homologous NPV. Most occluded viruses established in cell culture are NPV although a CPV (Granados *et al.*, 1974) and an EPV (Granados and Naughton, 1975) have also been reported. Production of complete PIB of GV has now been achieved (Naser *et al.*, 1984).

When Vaughn (1968) adapted cells of *A. eucalypti* to growth in suspension culture he provided the basic technique for producing insect viruses in large batch cultures. A great deal of subsequent effort has been expended in optimizing cell growth and virus replication, providing cheaper media and scaling up the process. Nearly all published work refers to production of NPV in lepidopteran cell lines, but viruses from other insect orders might prove equally amenable to *in vitro* production.

There have been several differing approaches to large-scale insect virus production. Weiss *et al.* (1981) describe an optimized roller-bottle technique for production of *A. californica* multiply-embedded NPV (ACMNPV) in *S. frugiperda* cells. Bottles with a surface area of 490 cm^2 containing 100 ml medium and seeded with 1.25×10^5 cells ml^{-1} of medium were incubated for 6 d at 26–27.5 °C while rotating at 1 rev. 8.5 min^{-1}. At that stage the attached cells had grown to 85% confluency on the bottle surface and were infected with a multiplicity of infection (MOI) of 0.75 virus particles per cell. A further 8 d incubation produced 9×10^9 PIB per bottle. This gave 1.5 larval equivalents (LE = 6×10^9 PIB) per bottle. Weiss *et al.* (1981) mention present production costs as being uneconomic. It may be that further economies can be made by reducing medium costs.

Röder, working at Hoechst AG, noted that PIB production of ACMNPV and *T. ni* (TNMNPV) in lines of *S. littoralis* and cabbage armyworm *Mamestra brassicae* cells could be obtained with equal ease in monolayer or suspension culture and that production in suspension culture was not adversely affected by scale-up (Röder, 1978). With ACMNPV he obtained concentrations of 1×10^6 PIB ml^{-1} with *M. brassicae* cells and 2×10^7 PIB ml^{-1} with *S. littoralis* cells compared to 9×10^7 PIB ml^{-1} with *S. frugiperda* reported by Weiss *et al.* (1981). Röder and Gröner (1982) described the growth of cells of *S. frugiperda*, *S. littoralis* and *M. brassicae* in Biostat V (Braun Melsungen) and Microferm (New Brunswick) fermenters of up to 10 l capacity for periods of up to 200 d with periodic withdrawal and replenishment of the cell suspension with new medium. In the Biostat fermenters they obviated the problems of sparging by passing air through a silicone rubber tube coiled inside the vessel. In these studies they employed a medium in which the (normally) most expensive component, foetal bovine serum, was replaced by egg yolk emulsion (Röder, 1982).

At present the only publication which deals specifically with the problems of large-scale production of insect cells and viruses on a theoretical basis is that of Pollard and Khosrovi (1978). Animal cells in culture are very susceptible to mechanical damage, especially to shear forces, while such characteristics of their growth as oxygen demand, heat generation and productivity are significantly lower than those of bacterial fermentations. Laboratory-scale batch processes for producing animal cells, *e.g.* roller bottles, may not be suitable models for scale-up as capital costs may be prohibitive. With these considerations in mind Pollard and Khosrovi examined the factors causing disruption of fragile cells, and the ways in which this could be avoided. They concluded that the most suitable vessel design would be that of a vertical tubular flow reactor where cells are kept in suspension by a medium current upstream of the inoculation point. During continuous operation inoculum is provided by the recycling of a small proportion of product cells. Virus production can be effected by infecting the culture downstream of this point. Several methods of oxygenation were considered as alternatives to the disruptive method of sparging, including saturation of inlet medium, electrolysis and oxygenating compounds. It was concluded that the most practicable method was by transfer across a silicone membrane. The maximum shear stress in a reactor of this design is several orders of magnitude lower than that encountered in a mechanically-agitated reactor. Pollard and Khosrovi give data for the operation of a 185 m^3 capacity vessel and point out that production on this scale could only be equalled by 400 separate stirred-tank fermenters if shear forces are not to become damaging, or alternatively by 1500 roller vessels 5 m long and 2 m in diameter. Using such a system, variable costs would form a substantial percentage of total costs (68.1% for materials and 1.4% for utilities) for an annual output of 7.5 tonnes of material containing 20% (m/m) PIB. Fixed costs would be 21.4% and capital costs, of which the reactor vessel constitutes 47%, would be 9.1% over a 10 year depreciation.

46.3.3.5 Protozoa

Very little work has been carried out on the production of insect-pathogenic protozoa. Several studies have been made on the multiplication of the obligate parasites, the microsporidia, in both insect and mammalian cell culture in the laboratory, but such is the cost of the starting materials that production for control purposes by this method is not feasible. The encysting ciliates, *e.g.* the genera *Lambornella* and *Tetrahymena*, have potential for insect control and can be cultured *in vitro* but have not been investigated in depth (Canning, 1982).

46.4 FORMULATION AND APPLICATION

Depending on the manner in which they can be handled, insect pathogens can be divided into two groups. Vegetative cells and fungal mycelia often lost viability rapidly. Indeed this may also be true of structures normally thought of as 'resting' or 'resistant', such as some fungal spores. In such cases a normal marketing and distribution system cannot be used. The pathogen must be applied close to its place of manufacture and within a short time.

However, several classes of insect pathogens form relatively durable resting bodies which at least offer the hope of a product having a practicable shelf life of 18 months to 2 years. Among these are bacterial and fungal spores, protozoan cysts and the inclusion bodies of insect viruses. Within certain limits such agents may be formulated in a manner similar to that of chemical insecticides. Suspensions or powders can be produced, and fillers, spreader-stickers and protectants incorporated.

Formulation has, as has application, a critical effect on the efficacy of microbial insecticides. The subject of formulation is too fragmentary for comprehensive treatment in a short article and several reviews exist in the literature, one of the most recent being that of Couch and Ignoffo (1981). However, several general points can be made. Because insect pathogens have high activities a large amount of filler is often included so as to ensure an efficient spread of the agent, for instance 'Elcar' contains 0.4% (m/m) PIB, the remainder being mainly filler. An alternative method of ensuring efficient spread is to use an ultra low volume (ULV) spray for maximal droplet dispersal. Shelf stability of a microbial insecticide is often adversely affected by temperature, *e.g.* as with NPV (Ignoffo and Hostetter, 1977), and little seems possible in the way of preventing inactivation by formulation. High humidity can exacerbate the effect of temperature (Couch and Ignoffo, 1981).

In the field the factor most likely to inactivate a microbial insecticide is sunlight. Consequently sunlight protectants such as dyes, carbon black or titanium dioxide are added. Ignoffo and Batzer (1971) used protectants in conjunction with microencapsulation with *Heliothis* SNPV. At present the evidence gathered does not point to a marked or unequivocal benefit conferred by sunlight protectants.

Insect pathogens can be formulated with, or applied at the same time as, some chemical insecticides. Morris (1977) found that B.t. was compatible with Orthene, Dylox, Sevin, Zectran, Dimilin and Lannate and was incompatible with several others out of a total of 27 insecticides tested. A general effect of chemical–biological combinations is that lower doses of either agent are required for the same result although cases of independent action and antagonism also occur. The subject has been reviewed in detail by Benz (1971).

Much work remains to be done on improving the formulation of microbial insecticides. For instance Nickle (1980) pointed out that an effective method of preventing water loss from nematode preparations would greatly enhance their usefulness. Ignoffo *et al.* (1976) reported enhancement of the performance of 'Elcar' when a gustatory stimulant was added. Yamamoto and Tanada (1980) found that infectivity of a virus is increased when it is ingested along with acylamines. It is not yet known if this could improve efficacy in the field. Shapiro and Bell (1982) recorded enhanced effectiveness of *L. dispar* NPV when it was formulated with boric acid.

Several different strategies have been used in applying insect pathogens in the field. To some extent the strategy depends on the degree of control required and the time necessary to achieve it. Pathogens have been released at points to set up foci of infection that will eventually spread. Where the pest is highly mobile, infected insects may be released to infect their fellows or their progeny. Virus-infected rhinoceros beetles have been released for this purpose (Bedford, 1981). Adult *Heliothis* have been lured to traps where they were contaminated with an NPV-containing dust, and escaped to produce infected egg-batches.

However, widespread infestation of cash crops usually dictates a 'blanket' application of the control agent. Often this is done when the pest becomes apparent. A 'prophylactic' approach, in which low levels of pathogens are used to control young and emergent larvae, would work in principle provided that the techniques of scouting and forecasting are adequate (McClelland and Collins, 1978). Because of their generally restricted host range microbial insecticides are ideally suited to integrated pest control programmes. Arthropod predators and parasites of the pest are left intact to control survivors. This bonus can mean that two or three applications of a pathogen can replace a dozen applications of a chemical spread over the earlier parts of a growing season. Should the system be seen to be failing in the latter part of the season, the pest, along with its enemies, can be attacked with chemical insecticides.

Various methods of application are under study. Microbial pesticides are amenable to conventional ways of dispersal, either as sprays or powders. Pathogens can be delivered as ULV sprays with droplet sizes ranging from 5–30 μm. The effective life of a pathogen can be prolonged by application to the underside of foliage (Milstead *et al.*, 1980).

46.5 FIELD PERFORMANCE

The efficacy of microbial control is affected, as are other forms of control to a certain extent, by a multiplicity of factors. Prominent among these are the habitat in which control is to be attempted and the degree of control required, the age and habits of the target, and the weather. The areas in which insect pathogens have been used may be listed as: ponds and streams; forests; stored products; and field crops (subdivided into broad-leaved crops, and topfruit, stems and tubers).

46.5.1 Ponds and Streams

Problems in this area are of either a medical or amenity nature and are usually caused by mosquitos, midges or blackflies. Larvae of these groups are susceptible to several pathogens. The World Health Organization has recommended that several pathogens should be investigated as potential candidates for mosquito control. These include the bacteria *B. sphaericus* and *B. thuringiensis* H-14, the fungi *Coelomomyces* spp. and *Culiciniomyces clavosporus*, the protozoa *Nosema algerae* and *Vavraia culicis*, and the nematode *Romanomermis culiciovorax* (Anon, 1979). The effectiveness of *B. sphaericus* strain 1593 was confirmed by several field studies, and *B. thuringiensis* H-14 is being developed rapidly for use both against mosquito and blackfly (*Simulidae*) larvae. *Coelomomyces* is an effective antimosquito agent as the motile zoospore actively seeks out its host, but the complexity of its life cycle precludes easy manufacture. In contrast the Australian Army Medical Corps has mass-produced *C. clavosporus* in a simple medium and demonstrated its efficacy against larvae of mosquitos and some midges (*Chironomidae* and *Ceratopogonidae*).

46.5.2 Forests

In North America large stretches of forest are treated with B.t. and viruses. Major pests are the sawflies *Neodiprion sertifer* and *N. leconti* and the spruce budworm *Choristoneura fumiferana* in coniferous forests, and the tussock moths *Hemerocampa* spp., the gypsy moth *Lymantria dispar* and the forest tent caterpillar *Malacasoma disstria* in deciduous forests. In the USSR the Siberian silkmoth *D. sibiricus* and the winter moth *O. brumata* assume importance. From the numerous instances of the use of pathogens in forests some general conclusions can be drawn. As forests can tolerate some defoliation, as opposed to tree killing, control can be less vigorous than that required for cash crops. Viral control of lepidoptera is generally more difficult than that of sawflies, but where this is not feasible B.t. may be used. Both B.t. and viruses have been used in combination with chemical insecticides (Morris, 1977a, 1977b).

46.5.3 Stored Products

Those occluded viruses which have been found in stored product insects provide a potentially effective means for control. They can be applied at the time of storage, stable conditions aid reproducible performance and the absence of light prolongs activity. Viruses have been used against the almond moth *Cadra cautella*, the Indian meal moth *Plodia interpunctella* and the Mediterranean flour moth *Ephistia kuehniella* (Hunter, 1974). Protozoans have been used against beetles of the genera *Tenebrio* and *Tribolium* (McLaughlin, 1971).

46.5.4 Field Crops

Field crops provide the major market for insecticides. From the point of view of application they can be divided into two categories: broad-leaved crops; and top fruit, stems and tubers.

Most work in this area has been done with B.t. and viruses, with a smaller amount of effort being expended on protozoa, nematodes and fungi. The degree of success reported by authors varies from superiority to chemicals, through comparability with chemicals, to inactivity. Variability in results can be ascribed to a multiplicity of factors which include the intrinsic activity of the agent and how it is handled and applied. In at least one publication an insect pathogen has been deemed inactive when it was used against the wrong pest (Harrison, 1974).

Broad-leaved crops can be protected with relative ease both by chemical and microbial insecticides. When they are applied correctly microbials can exert a degree of control equal or superior to that obtained with chemicals. Jaques (1977) concluded that ACMNPV was as effective against *T. ni* on cabbage as Lannate (methomyl), while Vail *et al.* (1980) found that ACMNPV with reduced application rates of chemicals gave as good a control of *T. ni* as did chemicals at normal rates.

Burrowing pests of fruits, stems and tubers are difficult to control by either biological or chemical means as there is only a limited period when the insect is exposed. Normally the pest must be treated at an early age in order to achieve adequate control (Falcon, 1976). Recently Payne (1982) obtained good protection of apples with the GV of the codling moth *Cydia pomonella*. Bedding and Miller (1981) found that larvae of the borer *Synanthedon tipuliformis* ensconced in blackcurrant stems could be killed by a nematode which travelled to the pest.

As in other areas, B.t. and viruses have been found to be compatible with chemicals when used on field crops. In some cases a synergistic effect may be seen, with lower doses of both agents being required to obtain the same degree of control (Creighton *et al.*, 1970). Infection with a pathogen, such as a subacute polyhedrosis, can increase the susceptibility of an insect to chemical insecticides (Girardeau and Mitchell, 1968).

46.6 SAFETY CONSIDERATIONS

In broad terms, authorities charged with environmental protection look favourably on biological methods of pest control. They are regarded as having the twin advantages of lowering the burden of chemical residues on the environment and, in some cases, being effective where the development of resistance has defeated chemical pesticides (Enlund, 1980).

Regulatory authorities do not require safety data to allow the use of predators and parasites for biological control; these are required only in the case of pathogens. Nematodes present an interesting borderline case. In the USA the Environmental Protection Agency (EPA) has ruled that *N. carpocapsae* and its attendant bacterium *Xenorhabdus nematophilus* do not require registration (Kaya, 1982).

The specificity of insect pathogens has a direct bearing on the manner in which their safety is investigated. Their toxicology differs from that of chemicals in two respects. Pathogens act by infection, therefore selected non-target organisms must be examined to demonstrate their immunity to attack, rather than to establish an LD_{50}. In general, where infection does not occur, no lethal dosage can be measured, unless the agent carries a non-specific toxin. In cases where no activity has been shown with non-target organisms, the inference has been drawn that residues of insect pathogens on crops are innocuous. This can lead to some agents being granted exemptions from tolerance as in the case of B.t. spores (Anon, 1973) or the *Heliothis* NPV 'Elcar' in 1975 (Engler and Rogoff, 1980).

The EPA is committed to reduce the use of chemical pesticides and to promote integrated pest management schemes (Engler and Rogoff, 1980). To this end it has drawn up interim guidelines for the registration of microbial pesticides and hopes to finalize them shortly. Permits for the experimental use of microbials can be issued after the submission of preliminary safety data. Tests for potential hazards which might occur with certain products are described, *e.g.* inhalation tests with fungi and allergenicity tests for materials containing insect remains. The guidelines have a several-tier approach. Maximum hazard tests (MHT) are proposed in the first tier. MHT are expected to take into account the worst case that could be encountered, either in size of dosage or in the sensitivity of the route of infection. Tests in subsequent tiers are only required if a potential hazard has been indicated in the first. Similar steps are being taken in the UK and guidelines have been drawn up for inclusion in the Pesticides Safety Precautions Scheme (PSPS). The EEC may eventually produce unified guidelines for registration of biological pesticides.

Several microbial insecticides have now been cleared for use in North America or Western Europe. B.t. has been employed commercially for many years and 'Elcar' has been used in cotton over the last seven. Tate and Lyle registered the fungus *Verticillium lecanii* in the UK in 1980, primarily for use against chrysanthemum aphids in glasshouses. Several viruses are in widespread non-commercial use in the USA. The USDA has registered two NPV preparations for use against the Douglas fir tussock moth, 'Biocontrol-1' (1976), and gypsy moth, 'Gyp-check' (1978). The USA and Canadian forestry authorities plan to register NPV for control of the sawflies *Neodiprion sertifer* and *N. leconti*, and for the spruce budworm, *Choristoneura fumiferana*. In the USA Sandoz is testing and developing *T. ni* MNPV for use on cotton and cole crops, and the broad-spectrum ACMNPV against a complex of pests (*Spodoptera*, *Trichoplusia*, *Estigmene* and *Plutella*) on a variety of crops (vegetables, cotton, soybeans and sugar beet). Sandoz is also cooperating with several universities in the development of *C. pomonella* GV for top fruit protection.

Because of their ability to infect non-insect hosts, especially warm-blooded animals, the idea of using rickettsiae for insect control is not seriously entertained (Krieg, 1971). By the same token the aflatoxin-producing insect pathogens *Aspergillus flavus* and *A. parasiticus* are unlikely to be widely employed. However, it is generally recognized that fungal spore preparations may induce allergic reactions or be carried deep into the respiratory tract, and these facts should be taken into account during production and application (Austwick, 1980). It appears that some insect-pathogenic fungi may be able to infect the respiratory tracts of reptiles (Austwick and Keymer, 1981).

In some cases the judicious use of a pathogen can actually reduce the final concentration of that agent on a crop. In the Eastern USA *T. ni* on cole crops is subject to an almost-annual epizootic of NPV which builds up during late summer. Larval cadavers deposit a residue of PIB which may amount to as much as 10^6 g^{-1} of washed late-season supermarket cole crops (Heimpel *et al.*, 1973). Early suppression of the pest with NPV prevents such an accumulation of virus.

The selective advantages of pathogens can best be shown with non-target insects. So far little work has been done on the predators and parasites of pest insects, although there are some examples of certain broad-spectrum microsporidians affecting the hymenopteran and dipteran parasites of the host (Smith, 1973). However, such is the specificity of NPV and GV that enemies of lepidoptera are most unlikely to be harmed by them. Lehnert and Cantwell (1978) have collated data from several studies of insect pathogens on honey bees. No pathogenicity was detected with eight NPV, three GV, five protozoans, four bacilli and a single fungus, *Hirsutella*.

In summary, it would appear that insect pathogens are not attended by deleterious effects which are overt and widespread, and that this is being recognized by the regulatory authorities to an increasing extent. It is probable that the registration of microbial insecticides will become increasingly easier, the process being facilitated by usage as more agents are submitted for registration.

46.7 ECONOMICS

Commercial organizations may produce microbial insecticides for several reasons. They may do so for simple profit, by direct selling or by contract to another firm, but other factors may play a part. Microbials may be produced in spare plant capacity or in order to diversify. From the point of view of the large pesticide manufacturer, microbial insecticides have one major drawback, real or imaginary. Specificity limits market size, resulting in relatively small volume production more comparable to the fermentation than the petrochemical industry. Set against this is the ease with which microbials can now be registered (Engler and Rogoff, 1980) and their low development costs. Ignoffo (1975) estimated this cost as 1.8×10^6. Lisansky (1982) assessed it at $2.0–4.0 \times 10^6$.

Manufacturing costs of microbial insecticides are but poorly documented. Ignoffo (1975) calculated that the cost of producing an entomopathogen by fermentation was 15–45 ¢ per gallon of culture, regardless of activity. Bacterial fermentations normally produce 10–20, and on rare occasions 60, mg dry weight ml^{-1}. These values give the cost of 1 kg of material as $1.65–$3.30 (at 15 ¢ gal^{-1}) to $4.95–$9.90 (at 45 ¢ gal^{-1}). Extremely high bacterial concentrations could lower costs to 56 ¢ – $1.67 kg^{-1}. Ignoffo (1975) could not foresee any dramatic reduction in manufacturing costs. These figures are in agreement with the 1975 end-user cost for B.t. of $11.30 per kg of active ingredient (Ayers *et al.*, 1977). Obligate insect pathogens are produced in living hosts, frequently lepidopteran larvae. Ignoffo (1975) estimated rearing costs, the major costs of virus production pogrammes, as 0.1–4.0 ¢ per larva. It is generally accepted that a medium-sized larva, such as *H. zea*, produces about 6×10^9 PIB (1 LE) of an NPV. Application rates of viruses are usually 50–300 LE acre^{-1}. A manufacturing cost of 4 ¢ per larva is equivalent to protection at $2.0–$12.0 acre^{-1}. 'Elcar' for example has a recommended application rate of 150–300 g ha^{-1}

(100–200 LE ha^{-1} or 42–84 LE acre^{-1}). A manufacturing cost of 4 ¢ per larva suggests that protection could be conferred at \$1.66–\$3.32 acre^{-1}. This, however, does not take into account further processing costs and sales and distribution costs which can account for 60% of retail costs (Lisansky, 1982).

A good microbial insecticide is as cost-effective as a (good) chemical. The potential performance of a microbial is optimal when the crop is valuable, the environment predictable, the pest susceptible and the delivery system adequate. These conditions are optimized by the glasshouse environment (Burges, 1981) where the cost of the crop and increasing pest resistance to chemicals have stimulated the application of integrated control programmes. This does not mean that performance of microbials under less-regulated conditions is inadequate. Ayers *et al.* (1977) ranked B.t. in cost-effectiveness with middle-cost chemicals for the protection of cole crops while Vail *et al.* (1980) showed that NPV of *T. ni* and *A. californica* were as cost-effective as acephate and methomyl for the protection of lettuce.

46.8 THE FUTURE

At present microbial insecticides command but a small proportion of the insecticide market, about 1% of the total in monetary terms. However, they will probably play an increasingly important role in intregrated control schemes and in niches where chemical insecticides have lost efficacy or never were effective. Non-commercial organizations and international bodies will continue to play a key part in their development. Although only a small proportion of agents discovered will reach development, microorganisms with novel insecticidal properties are likely to be reported at a steady if not increasing rate.

The cloning and expression of *Bacillus thuringiensis* toxin genes in *Escherichia coli* and *Bacillus subtilis* (Held *et al.*, 1982; Klier *et al.*, 1982) provide a method of powerful potential for the improvement of this microbial insecticide, by increasing the rate of toxin production or its concentration in the product, or in the spectrum of its targets. The possibility of externalizing a hitherto intracellular product suggests a route for further increasing toxin production many fold (Priest, 1982).

In many respects the production and use of microbial insecticides is still an infant subject and numerous disciplines can contribute to its growth. There is a pressing need for improvements in formulation and application methods. There is no reason why entomopathogens other than B.t. should not be equally amenable to improvement by genetic manipulation.

ACKNOWLEDGEMENTS

I am very grateful to Miss G.Y. Pugh for a search of the patent literature and to Dr. B. N. Herbert for his helpful suggestions.

46.9 REFERENCES

Anon (1973). Viable spores of the microorganism *Bacillus thuringiensis*. Exemption from the requirement of tolerance. *Federal Register*, **38**, 5337.
Anon (1979). Biological control of vectors. *Third Annual Report*, chap. 9, pp. 151–167. UNDP/World Bank/WHO.
Austwick, P. K. C. (1980). The pathogenic aspects of the use of fungi: the need for risk analysis and the registration of fungi. In *Environmental Protection and Biological Forms of Control of Pest Organisms*, ed. B. Lundholm and M. Stakerud, *Ecol. Bull.* (*Stockholm*), **31**, 91–102.
Austwick, P. K. C. and I. F. Keymer (1981). Fungi and actinomycetes. In *Diseases of the Reptilia*, ed. J. E. Cooper and O. F. Jackson, chap. 7, pp. 193–231. Academic, New York.
Ayers, J. H., T. A. Blue, R. R. Bramhall, T. J. Braunstein, E. A. Daris, J. I. DeGraw, T. E. Elward, R. E. Inman, O. H. Johnson, E. B. Leaf, F. L. Offsend and P. D. Stent (1977). New, innovative pesticides: an evaluation of incentives and disincentives for commercial development by industry. *Stanford Research Institute Report*, prepared for EPA.
Bedding, R. A. (1974). Five new species of *Deladenus* (Neotylenchidae) entomophagous–mycetophagous nematodes parasitic in siricid woodwasps. *Nematologica*, **20**, 204–225.
Bedding, R. A. (1976). New methods increase the feasibility of using *Neoaplectana* spp. (Nematoda) for the control of insect pests. *Proc. 1st Int. Coll. Invertebr. Pathol, Kingston, Canada*, 250–254.
Bedding, R. A. (1981). Low cost *in vitro* mass production of *Neoaplectana* and *Heterorhabditis* species for field control of insect pests. *Nematologica*, **27**, 109–114.
Bedding, R. A. and L. A. Miller (1981). Disinfesting blackcurrant cuttings of *Synanthedon tipuliformis* using the insect-parasitic nematode *Neoaplectana bibionis*. *Environ. Entomol.*, **10**, 499–457.

Bedford, G. O. (1981). Control of the rhinoceros beetle by baculovirus. In *Microbial Control of Pests and Plant Diseases 1970–1980*, ed. H. D. Burges, chap. 20, pp. 409–426. Academic, London.

Bellet, A. J. D. and E. H. Mercer (1964). The multiplication of *Sericesthis* iridescent virus in cell cultures from *Antheraea encalypti* Scott. I Qualitative experiment. *Virology*, **24**, 645–653.

Benz, G. (1971). Synergism of microorganisms and chemical insecticides. In *Microbial Control of Insects and Mites*, ed. H. D. Burges and N. W. Hussey, chap. 14, pp. 327–355. Academic, New York.

Bulla, L. A. and A. A. Yousten (1979). Bacterial insecticides. In *Economic Microbiology*, ed. A. H. Rose, vol. 4, chap. 4, pp. 91–114. Academic, New York.

Burges, H. D. (ed.) (1981a). *Microbial Control of Pests and Plant Diseases 1970–1980*. Academic, New York.

Burges, H. D. (1981b). Progress in the microbial control of pests 1970–1980. In *Microbial Control of Pests and Plant Diseases 1970–1980*, ed. H. D. Burges, chap. 1, pp. 1–6. Academic, New York.

Burges, H. D. (1982). Control of insects by bacteria. *Parasitology*, **84**, 79–117.

Burges, H. D. and R. A. Hall (1982). Bacteria and fungi as insecticides. *Outlook on Agriculture*, **11**, 79–86.

Canning, E. U. (1982). An evaluation of protozoal characteristics in relation to biological control of pests. *Parasitology*, **84**, 119–149.

Cantwell, G. E. and W. W. Cantelo (1981). *Bacillus thuringiensis*, a potential control agent for the Colorado potato beetle. *Am. Potato J.*, **58**, 457–468.

Coppel, H. C. and J. W. Mertins (1977). *Biological Insect Pest Suppression*. Springer-Verlag, Berlin.

Couch, T. L. and C. M. Ignoffo (1981). Formulation of insect pathogens. In *Microbial Control of Pests and Plant Diseases 1970–1980*, ed. H. Burges, chap. 34, pp. 621–634. Academic, New York.

Couch, T. L. and D. A. Ross (1980). Production and utilization of *Bacillus thuringiensis*. *Biotechnol. Bioeng.*, **22**, 1297–1304.

Creighton, C. S., T. McFadden and J. V. Bell (1970). Pathogens and chemicals tested against caterpillars on cabbage. *USDA Prod. Res. Rep.*, **114**, 1–10.

Dulmage, H. T. (1970). Production of the spore-δ-endotoxin complex by variants of *Bacillus thuringiensis* in two fermentation media. *J. Invertebr. Pathol.*, **16**, 385–389.

Engler, R. and M. H. Rogoff (1980). Registration and regulation of microbial pesticides. *Biotechnol. Bioeng.*, **22**, 1441–1448.

Enlund, E. (1980). Opening address. In *Environmental Protection and Biological Forms of Control of Pest Organisms*. ed. B. Lundholm and M. Stakerud, *Ecol. Bull. (Stockholm)*, **31**, 9–10.

Falcon, L. A. (1971). Use of bacteria for microbial control of insects. In *Microbial Control of Insects and Mites*, ed. H. D. Burges and N. W. Hussey, chap. 3, pp. 67–95. Academic, New York.

Falcon, L. A. (1976). Problems associated with the use of arthropod viruses in pest control. *Annu. Rev. Entomol.*, **21**, 305–324.

Fassuliotis, G. and C. S. Creighton (1982). *In vitro* cultivation of the entomogenous nematode *Filipjevimermis leipsandra*. *J. Nematol.*, **14**, 126–131.

Girardeau, J. H. and E. R. Mitchell (1968). The influence of a sub-acute infection of polyhedrosis virus in the cabbage looper on susceptibility to chemical insecticides. *J. Econ. Entomol.*, **61**, 312–313.

Goldberg, I., B. Sneh, E. Battat and D. Klein (1980). Optimisation of a medium for a high yield production of spore-crystal preparation of *Bacillus thuringiensis* effective against the cotton leafworm *Spodoptera littoralis* Boisd. *Biotechnol. Lett.*, **2**, 419–426.

Goldberg, L. J. and J. Margalit (1977). A bacterial spore demonstrating rapid larvicidal activity against *Anopheles sergentii, Uranotaenia unguiculata, Culex univittatus, Aedes aegypti* and *Culex pipiens*. *Mosquito News*, **37**, 355–358.

Goodwin, R. H., J. L. Vaughn, J. R. Adams and S. J. Louloudes (1970). Replication of a nuclear polyhedrosis virus in an established insect cell line. *J. Invertebr. Pathol.*, **16**, 284–288.

Grace, T. D. C. (1962). Establishment of four strains of cells from insect tissues grown *in vitro*. *Nature (London)*, **195**, 788–789.

Granados, R. R., W. J. McCarthy and M. Naughton (1974). Replication of a cytoplasmic polyhedrosis virus in an established cell line of *Trichoplusia ni* cells. *Virology*, **59**, 584–586.

Granados, R. R. and M. Naughton (1975). Development of *Amsacta moorei* entomopoxvirus in ovarian and hemocyte cultures from *Estigmene acrea* larvae. *Intervirology*, **5**, 62–68.

Hall. R. A. and B. Papierok (1982). Fungi as biological control agents of arthropods of agricultural and medical importance. *Parasitology*, **84**, 205–240.

Hansen, E. L. and W. S. Cryan (1966). Continuous axenic culture of free-living nematodes. *Nematologica*, **12**, 138–142.

Harrison, E. P. (1974). Chemical control of ear-infesting insects of sweet corn. *J. Econ. Entomol.*, **67**, 548–550.

Hasan, S. (1979). Current research on plant pathogens as biocontrol agents for weeds of Mediterranean origin. *Proc. EWRS Symp.*, 333–341.

Heimpel, A. M., E. D. Thomas, J. R. Adams and L. J. Smith (1973). The presence of nuclear polyhedrosis viruses of *Trichoplusia ni* on cabbage from the market shelf. *Environ. Entomol.*, **2**, 72–75.

Held, G. A., L. A. Bulla, Jr., E. Ferrari, J. Hoch, A. I. Aronson and S. A. Minnich (1982). Cloning and localization of the lepidopteran protoxin gene of *Bacillus thuringiensis* subsp. *kurstaki*. *Proc. Natl. Acad. Sci. USA*, **79**, 6065–6069.

Henry, J. E. (1971). Experimental application of *Nosema locustae* for control of grasshoppers. *J. Invertebr. Pathol.*, **18**, 389–394.

Hink, W. F. (1976). A compilation of invertebrate cell lines and culture media. In *Invertebrate Tissue Culture: Research Applications*, ed. K. Maramorosch, chap. 17, pp. 319–369. Academic, New York.

Holmberg, A., R. Sienvänen and G. Carlberg (1980). Fermentation of *Bacillus thuringiensis* for exotoxin production: process analysis study. *Biotechnol. Bioeng.*, **22**, 1707–1724.

Huffaker, C. B. and P.S. Messenger (1976). *Theory and Practice of Biological Control*. Academic, New York.

Hunter, D. K. (1974). Viruses of stored product insects and their potential as control agents. *Misc. Publ. Entomol. Soc. Am.*, **9**, 62–65.

Ignoffo, C. M. (1973a). Development of a viral insecticide: concept to commercialization. *Environ. Parasitol.*, **33**, 380–406.

Ignoffo, C. M. (1973b). Effects of entomopathogens on invertebrate. *Ann. N.Y. Acad. Sci.*, **217**, 141–172.

Ignoffo, C. M. (1975). Entomopathogens as insecticides. *Environ. Lett.*, **8**, 23–40.

Ignoffo, C. M. (1981). The fungus *Nomuraea rileyi* as a microbial insecticide. In *Microbial Control of Pests and Plant Diseases 1970–1980*, ed. H. D. Burges, chap. 27, pp. 513–538. Academic, New York.

Ignoffo, C. M. and R. F. Anderson (1979). Bioinsecticides. In *Microbial Technology*, ed. H. J. Peppler and D. Perlman, 2nd edn., vol. 1, chap 1, pp. 1–28. Academic, New York.

Ignoffo, C. M. and O. F. Batzer (1971). Microencapsulation and ultraviolet protectants to increase sunlight stability of an insect virus. *J. Econ. Entomol.*, **64**, 850–853.

Ignoffo, C. M. and D. L. Hostetter (1977). Environmental stability of microbial insecticides. *Misc. Publ. Entomol. Soc. Am.*, **10**, 117–119.

Ignoffo, C. M., D. L. Hostetter and D. B. Smith (1976). Gustatory stimulant, sunlight protectant, evaporation retardant: three characteristics of a microbial insecticide adjuvant. *J. Econ. Entomol.*, **69**, 207–210.

Jaques, R. P. (1977). Field efficacy of virus infections to the cabbage looper and imported cabbageworm on late cabbage. *J. Econ. Entomol.*, **70**, 111–118.

Kaya, H. K. (1982). The nematodes *Neoaplectana carpocapsae* and *Heterorhabditis* spp. in biological control of insect pests. *Proc. 3rd Int. Coll. Invertebr. Pathol., Brighton, UK*, 107–112.

Kellen, W. R., T. B. Clark, J. E. Lindegren, B. C. Ho, M. H. Rogoff and S. Singer (1965). *Bacillus sphaericus* Neide as a pathogen of mosquitoes. *J. Invertebr. Pathol.*, **7**, 442–448.

Klier, A., F. Fargette, J. Ribier and G. Rapoport (1982). Cloning and expression of the crystal protein genes from *Bacillus thuringiensis* strain *berliner* 1715. *EMBO J.*, **1**, 791–799.

Krieg, A. (1971). Possible use of rickettsiae for microbial control of insects. In *Microbial Control of Insects and Mites*, ed. H. D. Burges and N. W. Hussey, chap. 7, pp. 173–179. Academic, New York.

Kučera, M. (1981). The production of toxic protease by the entomopathogenous fungus *Metarhizium anisopliae* in submerged culture. *J. Invertebr. Pathol.*, **38**, 33–38.

Lehnert, T. and G. E. Cantwell (1978). The effects of microbial insecticides on the honey bee: a review. *Am. Bee J.*, **118**, 674–675.

Lisanksy, S. G. (1982). Commercial aspects of the development of entomopathogens. *Proc. 3rd Int. Coll. Invertebr. Pathol., Brighton, UK*, Abstracts of offered papers, p. 251 (Poster Session).

McClelland, A. J. and P. Collins (1978). UK investigates virus insecticides. *Nature (London)*, **276**, 548–549.

McCoy, C. W., A. J. Hill and R. F. Kanavel (1975). Large-scale production of the fungal pathogen *Hirsutella thompsonii* in submerged culture and its formulation for application in the field. *Entomophaga*, **20**, 229–240.

McCoy, C. W. and R. F. Kanavel (1969). Isolation of *Hirsutella thompsonii* from the citrus fruit mite *Phyllocptruta oleivora* and its cultivation on various synthetic media. *J. Invertebr. Pathol.*. **14**, 386–390.

McInnis, R., Jr. and W. C. Zattau (1982). Experimental infection of mosquito larvae by a species of the aquatic fungus *Leptolegnia*. *J. Invertebr. Pathol.*, **39**, 98–104.

McLaughlin, R. E. (1971). Use of protozoans for microbial control of insects. In *Microbial Control of Insects and Mites*, ed. H. D. Burges and N. W. Hussey, chap. 6, pp. 151–172. Academic, New York.

McLaughlin, R. E. (1974). Protozoa as microbial control agents. *Misc. Publ. Entomol. Soc. Am.*, **9**, 95–98.

Milstead, J. E., D. Odom and M. Kirby (1980). Control of early larval stages of the California oakworm by low concentrations of *Bacillus thuringiensis* applied to lower leaf surfaces. *J. Econ. Entomol.*, **73**, 344–345.

Morris, O. N. (1977a). Long term effects of aerial applications of virus–fenitrothion combinations against the spruce budworm *Choristoneura fumiferana* (Lepidoptera:Tortricidae). *Can. Entomol.*, **109**, 9–14.

Morris, O. N. (1977b). Compatibility of 27 chemical insecticides with *Bacillus thuringiensis* var. *Kurstaki*. *Can. Entomol.*, **109**, 855–864.

Naser, W. L., H. G. Miltenburger, J. P. Harvey, J. Huber and A. M. Huger (1984). *In vitro* replication of the *Cydia pomonella* (codling moth) granulosis virus. *FEMS Microbiol. Lett.*, **24**, 117–121.

Nickle, W. R. (1980). Possible commercial formulations of insect-parasitic nematodes. *Biotechnol. Bioeng.*, **22**, 1407–1414.

Ordish, G. (1967). *Biological Methods in Crop Pest Control*. Constable, London.

Payne, C. C. (1982). Insect viruses as control agents. *Parasitology*, **84**, 35–77.

Payne, C. C. and D. C. Kelly (1981). Identification of insect and mite viruses. In *Microbial Control of Pest and Plant Diseases 1970–1980*, ed. H. D. Burges, chap. 5, pp. 61–91. Academic, New York.

Pendleton, I. R. (1969). Insecticides of crystal-forming bacteria. *Process Biochem.*, **4**, 29–32.

Peterson, J. J. (1982). Current status of nematodes for the biological control of insects. *Parasitology*, **84**, 177–204.

Poinar, G. O., Jr. (1979). *Nematodes for Biological Control of Insects*, p. 148. CRC Press, Boca Raton, FL.

Pollard, R. and B. Khosrovi (1978). Reactor design for fermentation of fragile tissue cells. *Process Biochem.*, **13**, 31–37.

Priest, F. G. (1982). Strategies for the improvement of industrial bacteria. *Proc. 3rd Int. Coll. Invertebr. Pathol. Brighton, UK*, 6–10.

Roberts, D. W. and W. G. Yendol (1971). Use of fungi for microbial control of insects. In *Microbial Control of Insects and Mites*, ed. H. D. Burges and N. W. Hussey, chap. 5, pp. 125–149. Academic, London.

Röder, A. (1978). Vermehrung von kernpolyederviren in insektenzellinien. *Mitt. Dtsch. Ges. Allg. Angew. Ent.*, **1**, 116–122.

Röder, A. (1982). Development of a serum-free medium for cultivation of insect cells. *Naturwissenschaften*, **69**, 92.

Röder, A. and A. Gröner (1982). Growth conditions of insect cells and NPV in suspension. *Proc. 3rd Int. Coll. Invertebr. Pathol. Brighton, UK*, offered papers, p. 166.

Salama, H. S. and M. S. Foda (1982). A strain of *Bacillus thuringiensis* var. *entomocidus* with high potential activity on *Spodoptera littoralis*. *J Invertebr. Pathol.*, **39**, 110–111.

Samsinakova, A., S. Kalalova, V. Vleck and J. Kybal (1981). Mass production of *Beauveria bassiana* for regulation of *Leptinotarsa decemlineata* populations. *J. Invertebr. Pathol.*, **38**, 169–174.

Schroth, M. N. and J. G. Hancock (1981). Selected topics in biological control. *Annu. Rev. Microbiol.*, **35**, 453–476.

Shapiro, M. and R. A. Bell (1982). Enhanced effectiveness of *Lymantria dispar* (Lepidoptera:Lymantriidae) nucleopolyhedrosis virus formulated with boric acid. *Ann. Entomol. Soc. Am.*, **75**, 346–349.

Sharpe, E. S. and L. A. Bulla (1978). Characteristics of the constituent substrains of *Bacillus popilliae* growing in batch and continuous cultures. *Appl. Environ. Microbiol.*, **35**, 601–609.

Sharpe, E. S., G. St. Julian and C. Crowell (1970). Characteristics of a new strain of *Bacillus popilliae* sporogenic *in vitro*. *Appl. Microbiol.*, **19**, 681–688.

Singer, S. (1980). *Bacillus sphaericus* for the control of mosquitos. *Biotechnol. Bioeng.*, **22**, 1335–1355.

Smith, R. F. (1973). Considerations on the safety of certain biological agents for arthropod control. *Bull. WHO*, **48**, 685–698.

Smith, R. J. and E. A. Grula (1981). Nutritional requirements for conidial germination and hyphal growth of *Beauveria bassiana*. *J. Invertebr. Pathol.*, **37**, 222–230.

Soares, G. G., D. E. Pinnock and R. A. Samson (1979). *Tolypocladium*, a new fungal pathogen of mosquito larvae with promise for use in microbial control. *Calif. Mosq. Vector Control Assoc.*, **47**, 51–54.

Steinhaus, E. A. (1956). Microbial control — the emergence of an idea. A brief history of insect pathology through the nineteenth century. *Hilgardia*, **26**, 107–160.

Stirling, G. R. and M. F. Wachtel (1980). Mass production of *Bacillus penetrans* for the biological control of root-knot nematodes *Meloidogyne*. *Nematologica*, **26**, 308–312.

Stockdale, H. and R. A. J. Priston (1981). Production of insect viruses in cell culture. In *Microbial Control for Pest and Plant Diseases 1970–1980*, ed. H. D. Burges, chap. 16, pp. 318–328. Academic, New York.

Tyrell, D. J., L. I. Davidson, L. A. J. R. Bulla and W. A. Ramoska (1979). Toxicity of parasporal crystals of *Bacillus thuringiensis* subsp. *israelensis* to mosquitos. *Appl. Environ. Microbiol.*, **38**, 656–658.

Vail, P. V., R. E. Seay and J. DeBolt (1980). Microbial and chemical control of the cabbage looper and fall lettuce. *J. Econ. Entomol.*, **73**, 72–75.

Vaughn, J. L. (1968). Growth of insect cell lines in suspension culture. *Proc. Int. Coll. Invertebr. Tissue Culture 2nd, 1967*, 119–125.

Waterhouse, D. F., L. E. LaChance and M. J. Whitten (1976). Use of autocidal methods. In *Theory and Practice of Biological Control*, ed. C. B. Huffaker and P. S. Messenger, chap. 26, pp. 637–659. Academic, New York.

Weiss, S. A., G. C. Smith, S. S. Kalter, J. L. Vaughn and E. Dougherty (1981). Improved replication of *Autographa californica* nuclear polyhedrosis virus in roller bottles. Characterization of the progeny virus. *Intervirology*, **15**, 213–222.

Yamamoto, T. and Y. Tanada (1980). Acylamines enhance the infection of a baculovirus of the armyworm *Pseudaletia unipuncta* (Noctuidae:Lepidoptera). *J. Invertebr. Pathol.*, **35**, 265–268.

47

Microbial Flavors and Fragrances

F. H. SHARPELL, JR.
Givaudan Corporation, Clifton, NJ, USA

47.1 INTRODUCTION

The production of specific flavor and aroma chemicals by microorganisms has been recognized for over 60 years. Stärkle in 1924 noted that molds were responsible for the methyl ketones in cheese. Eight years later, Heublyum *et al.* (1932) determined that diacetyl, a chemical with a butter-like flavor and aroma, was microbial in origin. Subsequently, numerous other components of flavors and aromas were shown to be the results of microbial growth.

Many microbially produced compounds may be chemically synthesized. Many other non-microbial individual components of flavors and fragrances may also be chemically synthesized. This has enabled the flavorist and the aroma chemist to reconstitute products, imparting desired taste and odor nuances. Most flavors and aromas, however, are complex mixtures of major and minor components reacting synergistically whose total effect is hard to duplicate. In addition, today's consumer prefers 'natural' products, and the labeling of a product as natural rather than as artificial is a strong selling point. As a result, companies have begun to examine the use of microorganisms for naturally synthesizing both individual chemicals and complex flavor and fragrance mixtures.

The industrial success of a microbial product is dependent either upon a high yield of that product in a reasonable period of time from a cheap, available substrate or upon the manufacture of a unique product which is in demand but difficult or impossible to obtain by other methods. Com-

965

plementing this is the economical recovery of the product in usable form. Chemicals produced microbially have, in many instances, this desired uniqueness.

Although many papers and patents have been published about microbial flavors and aromas, little is available about the commercialization of these processes and products. Nevertheless, it must be assumed that the major flavor and aroma corporations are actively engaged in research and development if not in actual production.

47.2 FLAVOR AND FRAGRANCE CHEMICALS

Due to the close relationship between flavors and aromas, *e.g.* citrus notes find application in both flavors and fragrances, no attempt has been made to categorize various chemicals. Discussion of microbially produced chemicals and products will include methods of potential manufacture as well as theoretical possibilities.

47.2.1 Methyl Ketones

Of all the flavors in demand, there is no doubt that chemicals that contribute to cheese flavors are most important. Stärkle (1924) first demonstrated the importance of methyl ketones, $RCOCH_3$ (**1**; R = C_3H_7, C_5H_{11}, *etc.*), and their formation in mold ripened cheese. Numerous other investigators have attributed the odor and taste of mold ripened cheese to the presence of methyl ketones, particularly methyl *n*-pentyl ketone (2-heptanone) as well as other short chain ketones such as 2-pentanone and 2-nonanone. Low concentrations of methyl ketones may also contribute fruity-spicy notes to fragrances (Arctander, 1969).

The formation of methyl ketones from fatty acids was first attributed to *Penicillium roqueforti* mold spores and not to the mycelium (Gehrig and Knight, 1958). It was subsequently shown by Lawrence and Hawke (1968) that the *P. roqueforti* mycelium was capable of converting fatty acids with less than 14 carbon atoms, $RCH_2CH_2CO_2H$ (**2**; R = C_3H_7, C_5H_{11}, *etc.*), to methyl ketones. In every case, the acids were oxidized to methyl ketones with one less carbon atom than the original acid. This has been confirmed in the laboratories of the Givaudan Corporation (unpublished data) using a variety of fungi. It is also possible to convert vegetable oil and triglycerides to methyl ketones.

The formation of methyl ketones from fatty acids proceeds *via* interrupted β-oxidation. The various steps have been discussed by Fonken and Johnson (1972). The reaction proceeds according to the scheme shown in Figure 1 (Naipawer, personal communication).

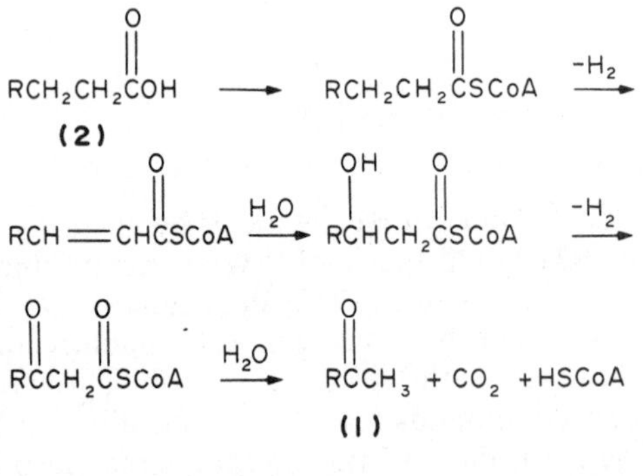

Figure 1   The formation of methyl ketones (**1**) from fatty acids (**2**; R = C_3H_7, C_5H_{11}, *etc.*); CoA = coenzyme A

The commercial production of methyl ketones presents several problems, including a relatively high cost. Fortunately, the uniqueness of a naturally produced methyl ketone for use in fortifying cheese and other flavors may offset the manufacturing expense. Among the production difficulties encountered are the inherent toxicity of the methyl ketones towards fungi and the volatility of these chemicals. Normal fermentation techniques must be supplemented with enzymatic conversions. Since aeration during the fermentation process removes the volatile products, provisions must be made to condense vapors before they escape into the atmosphere. Residual methyl ketones may then be easily recovered *via* solvent extraction or steam distillation. A typical process for producing 2-heptanone might proceed as shown in Figure 2.

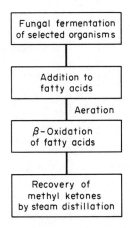

Figure 2 Flow chart for the production of methyl ketones

47.2.2 Diacetyl (Biacetyl; 2,3-Butadione)

Diacetyl, $CH_3COCOCH_3$ (**3**), is a naturally occurring chemical characterized by a powerful and diffusive odor resembling butter when dilute (Arctander, 1969). It is extensively used in imitation butter and other dairy flavors and in numerous flavors where butter notes are desirable. Diacetyl also finds limited use in perfumes, primarily in reconstituting essential oils. Closely related to diacetyl is acetoin (3-hydroxy-2-butanone; $CH_3COCH(OH)CH_3$, **4**). Acetoin is frequently found with diacetyl but probably contributes little or no flavor by itself. For many years it had been assumed that diacetyl was produced from acetoin by microbiological oxidation. Numerous recent studies have shown that this is not the case. Collins (1972) summarized the following: (1) diacetyl is not produced from acetoin by either bacteria or yeast; (2) the mechanisms by which bacteria and yeast produce diacetyl are the same, but differ from the mechanisms used to produce acetoin; and (3) diacetyl cannot be detected in several microorganisms which produce acetoin.

It has been accepted that citric acid is a precursor for both diacetyl and acetoin. The metabolism of citric acid has been thoroughly investigated by Collins and his associates at the University of California at Davis, by Seitz *et al.* (1963) and by Stadhouders (1974). More recently, Jönsson and Pettersson (1977) have contributed to our knowledge of the metabolic pathways of diacetyl production. In addition to the precursor citrate, the production of diacetyl is enhanced by a pH below 5.5, low temperature and aeration. A pH below 5.5 favors citric acid permease activity and restricts diacetyl reductase activity. Aeration promotes both the formation and accumulation of diacetyl by increasing the oxidation–reduction potential of the culture. This results in enzymatic stimulation and the spontaneous oxidative decarboxylation of α-acetolactic acid to diacetyl.

The various theoretical pathways to diacetyl synthesis have been summarized by Kempler and McKay (1981). The routes are shown in Figure 3.

Suomalainen and Jännes (1946) demonstrated that diacetyl could be obtained from aerated bakers' and brewers' yeasts acting on a saccharose (sucrose) substrate. It was not indicated whether the yeast first hydrolyzed saccharose *via* the enzyme invertase to glucose and fructose. Today, the organisms of commercial significance which produce diacetyl are *Streptococcus lactis* ssp. *diacetilactis* and several *Leuconostoc* species. Troller (1981) has patented a method for increasing the diacetyl production of bacteria such as *S. diacetilactis*, *S. cremoris* and *S. lactis*. The use of a humectant, such as glycerol or sucrose, which lowers the A_w value (water activity) of the medium results in greater diacetyl production.

Gupta *et al.* (1978) described conditions for laboratory scale production of acetoin plus diacetyl using *Enterobacter cloacae*. Sucrose was converted in 70% yield to a mixture of diacetyl and acetoin. Further oxidation of acetoin by iron(III) sulfate and iron(II) chloride gave an overall recovery of 60% diacetyl on the basis of sugar utilized.

Natural diacetyl may also be obtained by distillation of starter distillate, a by-product from the

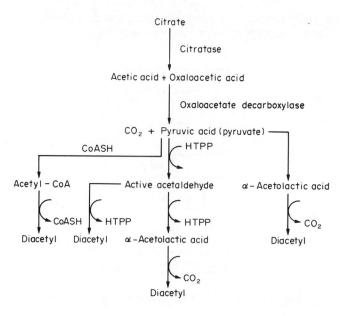

Figure 3 Pathways in diacetyl synthesis: HTPP = α-hydroxyethylthiamine pyrophosphate (active acetaldehyde)

manufacture of starter cultures. Diacetyl is sold commercially in the United States as a natural product imported from France. Although exact details of the manufacturing procedure are not available, it must be assumed that this imported diacetyl is bacterial in origin.

47.2.3 Lactones

Lactones are internal (cyclic) esters of primarily γ- and δ-hydroxy acids (**5** and **6**). Lactones are ubiquitous in food, contributing taste and flavor nuances. Numerous odor and taste characteristics have been attributed to lactones. Among these are oily–peachy, creamy, fruity, nut-like, coconut, honey, and so on. Maga (1976) summarized the types of lactones identified in fruits and vegetables. Many of these have been chemically synthesized, finding use in a variety of artificial flavors. The use, however, of microbially produced natural lactones in flavors would find wide acceptance among consumers.

$$\underset{(5)}{\overset{RCHCH_2CH_2}{\underset{O-C}{\bigg|}\overset{\bigg|}{}}\diagdown_O}\qquad\underset{(6)}{\overset{RCHCH_2CH_2CH_2}{\underset{O-C}{\bigg|}\overset{\bigg|}{}}\diagdown_O}$$

Labows *et al.* (1979) discussed the production of lactones by yeast of the genus *Pityrosporum* and applied for a patent for the 'Production of gamma-lactone rich flavor additives by *Pityrosporum* species cultured on lipid rich substrates'. The substrates include triolein, sebum, lecithin, oleic acid and Tween 80. Among the lactones claimed are the γ-hexa-, γ-hepta-, γ-nona-, γ-deca-, γ-undeca- and γ-dodeca-lactones. Lanza *et al.* (1976) also noted the volatile components of cultures of *Ceratocystis moniliformis* as having 'fruity–banana', 'peach–pear', and 'citrus' aromas. Identification of these by gas chromatographic head space analysis revealed a number of compounds including γ- and δ-decalactone. The above lactones are known for their peach-like aroma.

A coconut aroma is highly desired by flavorists. γ-Octalactone and γ-nonalactone possess this aroma. Another lactone having a coconut odor is 6-pentyl-2-pyrone (**7**). This chemical was found by Collins and Halim (1972) to be the major volatile constituent of the fungus *Trichoderma viride*. It is also claimed that the spores rather than the mycelia are responsible for the aroma.

Further work is required to confirm this. A natural microbial synthesis of coconut chemicals would have considerable impact upon the market.

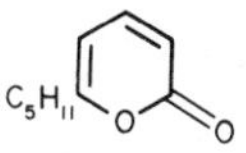

(7)

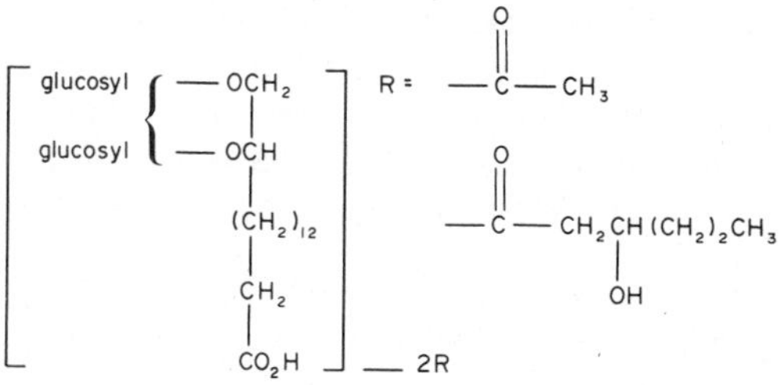

(8) 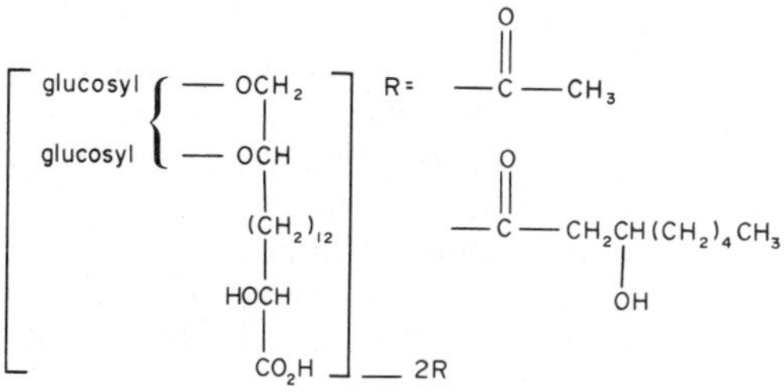   (9)

Roxburgh *et al.* (1954) and other workers at the Prairie Regional Laboratory, Saskatoon, Canada have demonstrated that *Ustilago zeae*, a corn smut fungus, may be used to produce ustilagic acids A (8) and B (9), potentially useful as precursors for the synthesis of lactonic macrocyclic musks. Macrocyclic musks are extensively used in the perfume industry.

The production of ustilagic acid has been scaled up in 200 gal fermenters (Roxburgh *et al.*, 1954). Yields of 22 g l^{-1} of fermentation broth have been obtained. Ustilagic acid, a mixture of monoacidic D-glucolipids, may be converted into its respective ustilic acids. Ustilic acid A (15D,16-dihydroxyhexadecanoic acid; 10; Figure 4) can be converted to 15-hydroxypentadecanoic acid (11) which in turn can be transformed into the macrocyclic musk 15-pentadecanolactone (12), known commercially as Exaltolide®. Usitilic acid A may also be converted to cyclopentadecanone (Exaltone®).

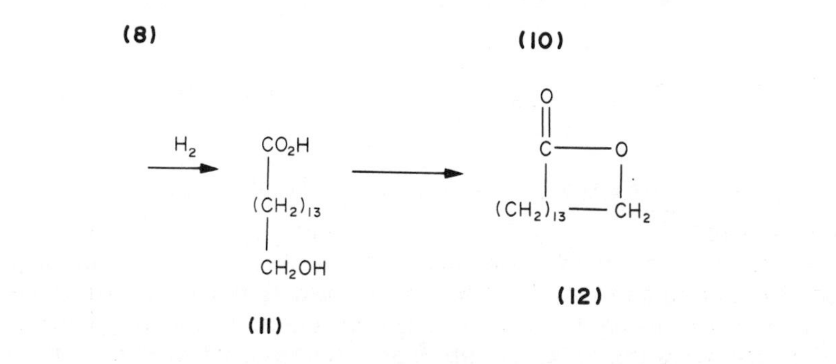

(8)   (10)

(11)   (12)

Figure 4 Production of Exaltolide® (15-pentadecanolactone; 12) from ustilagic acid A (8; R = —C(O)CH$_3$, —C(O)CH$_2$CH(OH)(CH$_2$)$_2$CH$_3$)

Two patents (Haskins, 1955; Lemieux, 1960) have been issued covering the processes described.

In addition to the above, Goldman and Perret (1965) described a fermentation process for 15- and 16-hydroxypalmitic acids. These products are intermediates for the valuable macrocyclic musks 15- and 16-hexadecanolides. The procedure utilizes *Torulopsis magnoliae* to hydroxylate palmitic acid. The hydroxypalmitic acids formed are then cyclized to produce the lactones.

47.2.4 Butyric Acid

Butyric acid, $CH_3CH_2CH_2CO_2H$ (13), at low concentrations is used to supply butter-like notes to flavors. It finds particular application in natural cheese flavors. The esters of butyric acid also may contribute to the flavor of various products. Pentyl butyrate provides a strong, ethereal, fruity odor reminiscent of apricot, banana and pineapple; isobutyl butyrate supplies an ethereal, fruity, somewhat pungent odor suggestive of pear, pineapple and banana (Arctander, 1969).

Although natural butyric acid as an ester may be found at a concentration of 2–4% in butter, its isolation is an expensive and difficult process. As a result, the fermentative production of natural butyric acid is a valuable alternative.

Butyric acid is primarily produced by obligate anaerobes of the genera *Clostridium*, *Butyrivibrio*, *Eubacterium* and *Fusarium*. The clostridia, particularly *C. acetylbutyricum*, have been studied in detail. Their ability to produce organic solvents such as acetone and butanol has led to commercial processes which may be modified and adapted to produce butyric acid.

The mechanism for butyric acid production has been summarized by Gottschalk (1979; Figure 5).

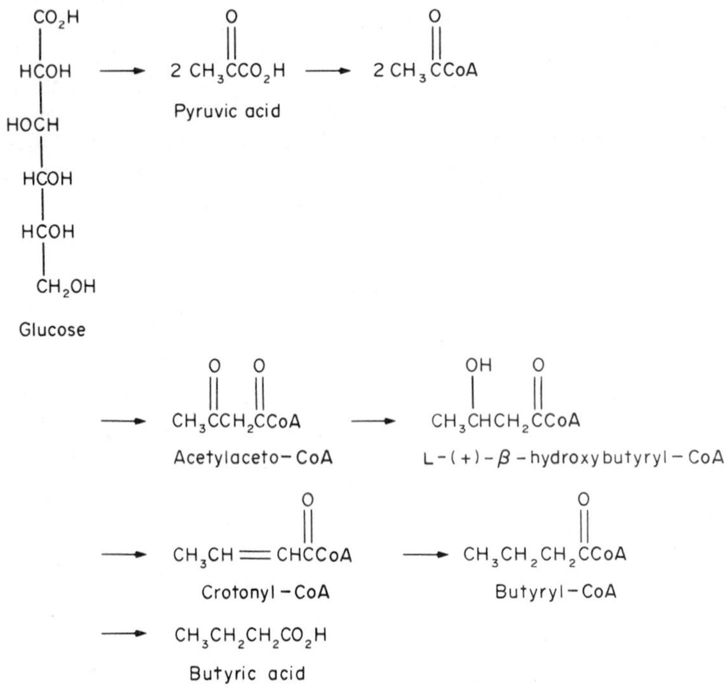

Figure 5 Mechanism of butyric acid production

Besides proper selection of the microorganisms, it is necessary to maintain the pH above 5.0 in order to direct the fermentation away from solvent formation and towards butyric acid formation. Sharpell and Stegmann (1979) have investigated conditions for the optimum production of butyric acid. They showed that calcium carbonate may be used to control the pH above 5.0. When other methods of controlling pH were attempted, butyric acid yields were low, indicating that calcium carbonate plays another role, possibly serving as a point of attachment for the microorganisms. Using a simple medium consisting of cerelose (commercial dextrose), dried yeast and calcium carbonate, yields of 1% butyric acid were obtained.

Additional work by Sharpell and his coworkers showed that increased yields of butyric acid could be obtained with further modifications of the medium and operating conditions. The toxic nature of butyric acid towards *Clostridium* species, however, appears to be a limiting factor.

47.2.5 Isovaleric Acid

Isovaleric acid, $(CH_3)_2CHCH_2CO_2H$ (14), is undoubtedly one of the most offensive odors encountered in the flavor/fragrance industry. Not only does it possess an acid–acrid odor commonly described as 'locker room' or 'dirty feet', but isovaleric acid has a tenacious affinity for the skin. Despite this, in extremely dilute concentrations isovaleric acid becomes agreeable and herbaceous. Furthermore, its esters find widespread use in the flavor industry. Prominent among these are ethyl isovalerate, which possesses a powerful apple–fruity odor and finds application in numerous fruit flavors as well as in candies and chewing gums; isopentyl isovalerate, which has a fruity–apple–raspberry odor and finds use in apple flavors and as a modifier in numerous fruit and nut flavors; and isobutyl isovalerate, also apple–raspberry-like and used not only in fruit flavors, but also in perfumes for lipsticks.

Isovaleric acid may be produced synthetically by the oxidation of isopentyl alcohol. Subsequent direct esterification leads to the various esters. If natural isovaleric acid is desired, this process is not suitable. Two potential methods exist for the microbial production of natural isovaleric acid. The first, the microbial oxidation of isopentyl alcohol (*ex* fusel oil from yeast fermentations) is theoretically possible since the oxidation of terminal alcohol groups to their corresponding carboxyl groups is well known. This procedure is limited both by the toxicity and the low solubility of isopentyl alcohol. Furthermore, no microorganisms are known at present that will oxidize isopentyl alcohol. Nevertheless, the method remains an interesting challenge. The second possibility involves the conversion of leucine to isovaleric acid. Numerous investigators have demonstrated that this is possible *via* the Stickland reaction. Their work has been reviewed by Nisman (1954). The reaction employs amino acid utilizating anaerobes to facilitate coupled oxidation–reduction between pairs of amino acids. One amino acid is oxidatively deaminated and decarboxylated; the other is reductively deaminated. The Stickland reaction is summarized in Figure 6.

$$
\begin{array}{c}
\underset{\displaystyle\overset{|}{NH_2}}{RCHCO_2H} + 2\ \underset{\displaystyle\overset{|}{NH_2}}{R'CHCO_2H} + 2\ H_2O \\[2em]
\longrightarrow RCO_2H + CO_2 + 3\ NH_3 + 2\ R'CH_2CO_2H
\end{array}
$$

Figure 6 The Stickland reaction

Britz and Wilkinson (1982), using cell suspensions, demonstrated that several species of the genus *Clostridium* and *Peptostreptococcus anaerobius* convert leucine to both isovaleric acid and isocaproic acid in a mechanism compatible with the Stickland reaction (Figure 7). In this case, leucine serves as both the proton donor and acceptor within one reaction. If however, the clostridial species does not produce isocaproic acid, leucine functions only as a proton donor, and proton acceptors must be supplied.

$$
3\ \underset{\text{Leucine}}{(CH_3)_2CHCH_2\overset{\displaystyle\overset{H_2N}{|}}{C}HCO_2H} + 2\ H_2O \longrightarrow
$$

$$
\underset{\text{Isovaleric acid}}{(CH_3)_2CHCH_2CO_2H} + CO_2 + 3\ NH_3 + 2\ \underset{\text{Isocaproic acid}}{(CH_3)_2CHCH_2CH_2CO_2H}
$$

Figure 7 Microbial conversion of leucine to isovaleric and isocaproic acids

Sharpell and Stegmann (1979) showed that by proper selection of the medium and clostridial species, good conversions of leucine to isovaleric acid may be obtained.

Thijsse and van der Linden (1963) described an interesting method of obtaining isovaleric acid. Cultures of *Pseudomonas aeruginosa* were initially grown on heptane. A small amount of chloramphenicol was then added with the substrate, 2-methylhexane, resulting in the conversion of the substrate to 30–40% isovaleric acid. It is assumed that chloramphenicol prevents the metabolism of intermediate products so that further oxidation is blocked.

47.2.6 Terpenes and Terpene Transformations

Terpenes are natural products whose basic structure consists of isoprene units linked head to tail (Figure 8).

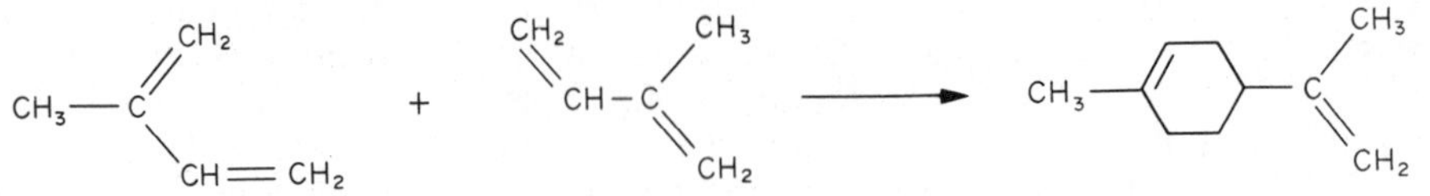

Figure 8 Production of the monoterpene (±)-limonene by head to tail linkage of two isoprene units

Terpenes are the basic components of essential oils and find widespread use in flavors and aromas. Other terpenes, including the carotenes and gibberellins, are discussed elsewhere.

The general structural formula for terpenes is $(C_5H_8)_n$. For the monoterpenes, $n=2$, for the sesquiterpenes, $n=3$, and so on. Also found in the plant world are the closely related cyclic alcohols and ketones such as menthol ($C_{10}H_{20}O$) and menthone ($C_{10}H_{18}O$). Complete discussions of the chemistry of terpenes may be found in the books listed in Section 47.6.

Microorganisms have the unique ability not only to degrade or transform terpenes, but also to synthesize them. In addition, microorganisms may be used to racemize terpenes.

Following the initial studies of Collins and Halim (1970) involving the synthesis of various monoterpenes by the ascomycete *Ceratocystis variospora*, numerous other microorganisms were shown to synthesize terpenes. Table 1, compiled by Schindler and Schmid (1982), lists the terpenes and some other products produced directly by fungi.

Table 1 Fragrance Compounds from Microorganisms

Microorganism	Fragrance	Chemical
Ascoidea hylacoeti	Fruity, rose	β-Phenylethanol, furan-2-carboxylate
Ceratocystis moniliformis	Fruity, banana, peach, pear, rose	3-Methylbutyl acetate, δ- and γ-decalactone, geraniol, citronellol, nerol, linalool, α-terpineol
Ceratocystis variospora	Fragrant, geranium	Citronellol, citronellyl acetate, geranial, neral, geraniol, linalool, geranyl acetate
Ceratocystis virescens	Rose, fruity	6-Methyl-5-hepten-2-ol acetate, citronellol, linalool, geraniol, geranyl acetate
Inocybe corydalina	Fruity, jasmine	Cinnamic acid methyl ester
Kluyveromyces lactis	Fruity, rose	Citronellol, linalool, geraniol
Mycoacia uda	Fruity, grassy, almond	p-Methylacetophenone, p-tolyl-1-ethanol, p-tolylaldehyde
Penicillium decumbens	Pine, rose, apple, mushroom	Thujopsene, 3-octanone, 1-octen-3-ol, nerolidol, β-phenylethanol
Phellinus species	Fruity, rose, wintergreen	Methyl benzoate, methyl salicylate, β-phenylethanol, γ-Decalactone
Sporobolomyces odorus	Peach	
Streptomyces odorifer	Earthy, camphor	*trans*-1,10 Dimethyl-*trans*-9-decalol, 2-exohydroxy-2-methylbornane
Trametes odorata	Honey, rose, fruity, anise	Methyl phenylacetate, geraniol, nerol, citronellol
Trichoderma viride	Coconut	6-Pentyl-2-pyrone

Undoubtedly, there will be numerous additions to this list in the future. Microbiological production of terpenes, however, cannot compete economically at present with the recovery of terpenes from essential oils. Nevertheless, improvements in yield combined with a diminished supply of natural plant terpenes may make the microbiological production of terpenes an attractive possibility. Already, Schindler and Bruns (1980) have demonstrated with *C. variospora* that terpene yields approaching 2 g l^{-1} could be obtained by fermentation provided that toxic end products were removed using ion exchange resins. Further improvements would increase the practicality.

Microbial transformations of terpenes have been reviewed by numerous authors (Ciegler, 1969; Voishillo *et al.*, 1970) and more recently by Johnson (1978). Krasnobajew (1983) reviewed the microbial transformations of terpenoid chemicals, considering their practical applications. The next section lists representative transformations discussed in his review.

47.2.6.1 Monoterpenoids

Microbial transformations of monoterpenoid compounds are summarized in Table 2.

Table 2 Microbial Transformations of Monoterpenoids

Substrate	Organism	Major product(s)
	Acyclic monoterpenoids	
Citronellal	Yeast	(+)-Citronellol
	Candida reukaufii	(−)-β-Citronellol
	Pseudomonas aeruginosa	Citronellic acid
	Ps. digitatum	Menthol
Citral	*Ps. convexa*	Geranic acid
Linalool	*Ps. pseudomallei*	Camphor
	Ps. incognita	Various cyclic compounds and acids
	Monocyclic monoterpenoids	
Limonene	Pseudomonad	Carveol, carvone, dihydrocarvone, perillyl alcohol *etc.*
(+)-Limonene	*Penicillium digitatum*	Carvone
	P. italicum	Carvone
	Cladisporium sp. T-7	Limonene-1,2-diol
	Cladisporium sp. T-12	(+)-α-Terpineol
(±)-Menthol (acetates)	*Trichoderma viride*	1-Menthol
	Rhodotorula mucilaginosa	1-Menthol
	Arginomonas non-fermentans	1-Menthol
(±)-Menthol (lactic acid esters)	*Bacillus subtilis*	1-Menthol
	Bicyclic monoterpenoids	
α-Pinene	*Aspergillus niger*	(+)-Verbenone (+)-*cis*-Verbenol (+)-*trans*-Sobrerol
α-Pinene	*Ps. maltophila*	2-(4-Methyl-3-cyclohexenylidene)propionic acid
β-Pinene	*A. niger*	Pinocarveol, pinocarvone, myrtenol
1,8-Cineole (ecualyptol)	*Ps. flava*	2-Oxocineole, isomers of 2-hydroxycineole, 5,5-dimethyl-4-(3′-oxobutyl)-4,5-dihydrofuran 2-(3*H*)-one

47.2.6.2 Ionones

Transformation of β-ionone (**15b**) has been studied by Mikami and his associates (1981) and by Krasnobajew and Helmlinger (1982). Using *Aspergillus niger*, Mikami demonstrated that β-ionone could be transformed to a complex mixture resembling an essential oil with an odor similar to that of tobacco. He suggested that these products could be used as tobacco flavoring compounds. Among the transformation products were various hydroxy- and oxo-ionone derivatives. Krasnobajew showed that *Lasiodiplodia theobromae* could also transform β-ionone into a variety of essential oil type products with a tobacco like odor. In this case, the molecule is degraded by a Baeyer–Villiger type oxidation. The main product is β-cyclogeraniol.

α-Ionone (15a)   β-Ionone (15b)   γ-Ionone (15c)

Krasnobajew (1982) demonstrated that α-ionone (**15a**) could also be transformed by *L. theobromae*, although not as readily as β-ionone. Similarly, essential oil type products with a tobacco odor were obtained. These compounds had previously been reported as flavor components of burley tobacco.

47.2.6.3 *β-Damascone*

The microbiological transformation of β-damascone (16) has been reported by Helmlinger and Krasnobajew (unpublished data). Using fungi of the genera *Aspergillus*, *Botryosphaeria* and *Lasiodiplodia*, β-damascone was transformed to 4-hydroxy- and 2-hydroxy-β-damascone. The metabolic products were found to be useful as tobacco flavors.

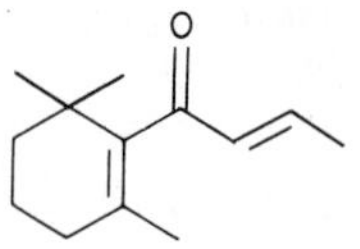

(16)

47.2.6.4 *Sesquiterpenoids*

Sesquiterpenes are found in essential oils along with monoterpenoids. Although there are at least 2000 sesquiterpenes found in nature (Ohloff, 1978), reports on their microbial transformation are relatively rare. Of interest is the transformation of valencene (17) to nootkatone (18; Figure 9).

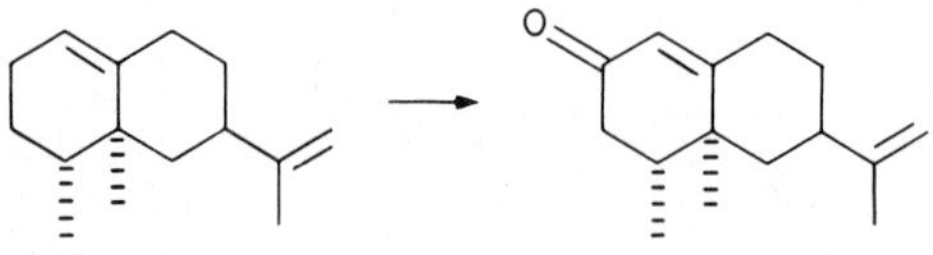

(17)   (18)

Figure 9   Conversion of valencene (17) to nootkatone (18)

Valencene, readily available from orange oil, is in itself of little commercial use. Dhavlikar and Albroscheit (1973) showed that both a bacterium isolated from Dutch soil and a bacterium isolated from infected beer were capable of transforming valencene to nootkatone, a main flavoring ingredient of grapefruit. The commercialization of this may be of considerable potential.

47.2.6.5 *Patchouli alcohol*

Patchouli alcohol or patchoulol (19) is an important ingredient of the patchouli oil used in perfumery. The odor carrier in patchouli oil is thought to be norpatchoulenol (20; Teisseire *et al.*, 1974). 10-Hydroxypatchoulol can be chemically converted to norpatchoulenol. Extensive screening of microorganisms by Becher *et al.* (1981) revealed that many would hydroxylate the C-10 methyl group (Figure 10). Numerous other hydroxylation products were also found.

Curvularia lunata was found to produce 8-hydroxypatchoulol, a precursor for patchoulion, an important trace component of patchouli oil. Although yields of valuable transformation products of patchoulol are low, optimization of the fermentation and further strain selection might yield practical amounts of 8- or 10-hydroxypatchoulol.

47.2.7   Geosmin

Geosmin (*trans*-1,10-dimethyl-*trans*-9-decalol; 21), first isolated by Gerber and Lechevalier (1965), is an earthy-smelling chemical produced by blue-green algae, myxomycetes, actinomy-

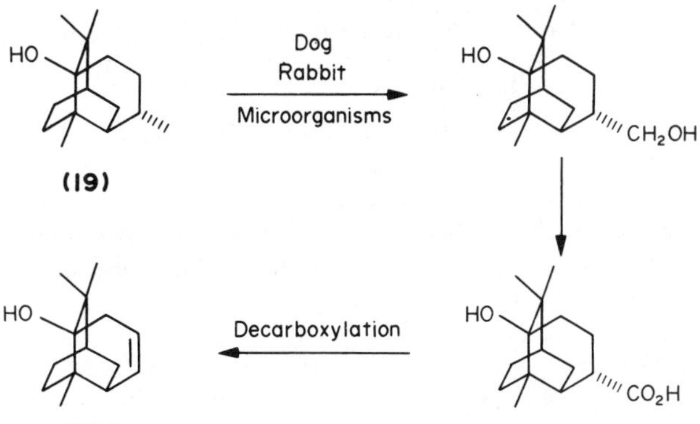

Figure 10 Conversion of patchoulol (**19**) to norpatchoulenol (**20**)

cetes and other microorganisms. Geosmin has frequently been found contaminating water supplies. Not only does this result in unpalatable water, but also fish and animals drinking the water become unacceptable for consumption. Geosmin has two unique properties. Its odor is detectable in water at a concentration of 0.2 μg l^{-1} (0.2 p.p.b.), and it has the ability to 'fatigue' the nose rapidly, making organoleptic evaluation difficult. The earthy notes of geosmin are useful in imparting or modifying amber notes in perfume (Escher and Morris, 1981). At a concentration of 1–100 p.p.m., geosmin is described as being useful for the reconstitution of natural essential oils.

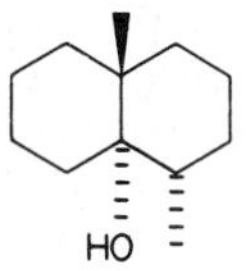

(21)

Gerber and Lechevalier (1977) produced geosmin in fermenters utilizing *Streptomyces griseus* LP-16 in a soybean meal–peptone–salt–glucose medium. Yields were 6 mg l^{-1} after three days fermentation. Geosmin was recovered by extracting a steam distillate with methylene chloride and obtaining the geosmin directly by preparative gas chromatography. Since the chemical synthesis of geosmin is complicated (Marshall and Hochstetler, 1968), its production by fermentation might be a practical commercial process.

47.3 COMPLEX FLAVOR AND AROMA PRODUCTS

In the preceding section, we dealt with individual chemicals produced by microorganisms, chemicals which might find application in the flavor and/or fragrance industry. In this section, we will discuss the production of complex mixtures of chemicals giving rise to various flavors and fragrances associated with natural products. Where pertinent, we will also examine enhancement of flavors *via* the application of microbial enzymes and the acceleration of natural aging. In many cases, the creation of natural flavors is elusive but economically rewarding.

47.3.1 Dairy Flavors

Dairy flavors, primarily cheese, are widely used in the food industry. Cheese flavors find application in snacks, sauces, baked goods and numerous other products. To a lesser extent, yogurt and buttermilk flavors are also useful. Research has been directed towards shortening the aging time and enhancing the flavor potential of natural cheese. In the United States, cheddar cheese and its milder varieties (process, American cheese) dominate the market.

Kristoffersen (1973) has discussed the biogenesis of cheese flavors. The flavor results from the action of microorganisms and enzymes on the proteins, fats and carbohydrates of the milk and curd. The numerous breakdown products such as short chain fatty acids, acetic and lactic acids, alcohols, aldehydes, ketones, esters, ammonia, amines, sulfides and mercaptans all contribute to the flavor. The characteristic flavor of various cheeses is determined by the concentration and ratio of these compounds.

Law (1981) summarized the various methods used to accelerate the flavor ripening process. Raising the storage temperature speeds up flavor formation, but off flavors are promoted and microbial spoilage occurs. If the flavor of blue cheese is required, commercial *Aspergillus* enzyme preparations may be used to accelerate fatty acid and δ-lactone formation. Incubating high-moisture curd slurries with *Penicillium* at 30 °C results in a flavor suitable for dips or other foods. More subtle flavors like cheddar require a balanced selection of enzymes and many investigators have attempted to do this by adding microbial proteinases. If this process is not carefully controlled, however, bitter-tasting peptides may mask the desired cheddar flavor. The neutral proteinase of *Bacillus subtilis* gives good organoleptic results, achieving an overall cheddar flavor in approximately half the time as natural ripening. Recently, commercial preparations of β-galactosidase (lactase) have been used to convert lactose to glucose and galactose in milk and whey.

When applied to cheesemaking, glucose supposedly stimulates the growth and production of starter bacteria, resulting in a 50 to 70% reduction in ripening time. On the other hand, it has been suggested that a contaminating, heat-resistant proteinase may be responsible for the accelerated cheese ripening, since glucose itself stimulates starter cultures but does not accelerate ripening.

A liquid or powdered product having an intense cheese background flavor suitable for flavoring food products has been decribed by Kasik (1973). Non-toxic lactic acid-producing organisms of the genus *Streptococcus* and non-toxic *Acetobacter* were grown in a reconstituted milk medium with mixing and aeration. After 24 hours fermentation, the mixture was pasteurized and then either used directly or spray dried and incorporated into food products. Further modification of this process (Kasik and Luksas, 1973) utilized non-toxic organisms of the genera *Bacillus* and *Streptococcus* to ferment milk. Naturally produced cheddar cheese was added and, after an additional day of fermentation, the mixture was pasteurized and spray dried. The resulting powder had a complete cheddar cheese taste and desirable texture. Johnson and Southworth (1973) utlized an esterase (lipolytic enzymes which hydrolyze esters of a C_{10} chain length or lower) to hydrolyze a pasteurized cheese and butter mixture. The mixture was fermented with a lactic starter culture, and then buttermilk, acid whey and sweet whey were added. Following pasteurization, the mixture was spray dried to yield a product with a strong, cheese-like flavor.

Knight (1963) claimed that *Penicillium roqueforti* could be grown rapidly in milk and that the resulting products without further aging could be used directly for flavoring foodstuffs with a blue cheese type flavor. Luksas (1973) modified this process by substituting sodium caseinate and butterfat for milk, avoiding some of the flavor defects in the milk process. Dwivedi and Kinsella (1974) described a process for the continuous production of a blue-type cheese flavor by submerged fermentation of *P. roqueforti*. Mycelium was used to convert lipolyzed milk fat to a mixture that included lipids, free fatty acids and various carbonyls. Sensory evaluation of the fermented flavor in a dip was equivalent to that of an authentic blue cheese dip. Jolly and Kosikowski (1975) used a commercial *P. roqueforti* spore powder and microbial lipases acting on a substrate of reconstituted acid whey and either milk fat or coconut fat. The resulting typical blue cheese flavor was considered excellent. Enzymatic techniques have been used to produce a variety of commercial products. These products can be used to intensify cheese flavor in a processed cheese formula or to replace part of the cheese in a food formulation, thereby significantly reducing ingredient costs. Among the companies producing enzyme modified cheese (EMC) are Dairyland, Mid-America, Borden's, International Flavors and Fragrances, and Miles Laboratories, Marschall Division. Although the exact details of manufacturing EMC are confidential, all companies use controlled multiple enzyme curing processes which enable them to maintain consistent flavor character and balance. Most companies provide a full line of cheese flavors including cheddar, Swiss, provolone, romano, parmesan, mozzarella and others. In addition, since no artificial flavors are used, EMC complies with federal regulations as a natural product.

47.3.2 Bread Flavors

The flavor of white bread is dependent upon four factors (Jackel, 1969). These are (1) the ingredients; (2) yeast and bacterial fermentation products; (3) mechanical and/or biochemical degradations; and (4) thermal reaction products. There are at least 100 volatile flavor chemicals in bread (Maga, 1974), many of which are yeast or bacterial fermentation products. These chemicals, when heated, give bread its final flavor and its appealing character.

Yeast leavened dough (sponge), requires a time consuming fermentation to assure full development of the bread flavor. Shortening this time by supplementing the flavor would be desirable in bakeries. In addition, new types of dough products rely upon chemical leavening agents. The omission of yeast from these modern-day breads results in a flavor deficient product. To overcome this problem, considerable interest exists in supplementing the dough with a bread flavor, either natural, artificial, or a combination of both.

Thrusts at enhancing bread flavor have taken two directions. In the first, attempts have been made to develop a preferment flavor. Swortfiguer (1956) described a 'stable ferment process' designed to replace the more time consuming sponge method currently in use. A mixture of yeast, water, yeast food, sugar, salt and non-fat dried milk solids was allowed to ferment for six hours. The mixture was then added directly to dough and the dough processed, omitting the sponge step. Robinson *et al.* (1958) suggested the use of selected microorganisms, particularly *Lactobacillus bulgaricus* plus a buttermilk culture, in preferments or in the sponge to enhance bread flavor. Bundus and Luksas (1969a) grew yeast or yeast–cocci in a whey culture, claiming that a synergistic effect occurs among yeast, enterococci of serological group D, streptococci of group N, and lactobacilli with respect to bread flavor. After fermentation, the mixture was concentrated, spray dried and added to dough. In a later patent, Bundas and Luksas (1969b) elaborated on this, claiming that *Saccharomyces* yeast may be grown in a whey containing lactic acid. The lactic acid may be added directly or developed from lactic acid-producing microorganisms. Lendvay (1970) fermented yeast in a sugar and maltol mixture. The yeast was allowed to autolyze after fermentation. The resulting liquid could either be added directly to chemically leavened dough or dried and incorporated into the product.

In the second, various enzyme preparations including microbial amylases and proteases have been recommended for improving the manufacture of bread. Colby (1965) described a method for treating flour to substantially reduce development time during mixing to form the dough. Numerous commercial bacterial and fungal enzymes are available for use in the baking industry from such companies as Novo, Gist-Brocade, Genencor and Miles Laboratories. The main function of enzymes, however, is to improve dough characteristics rather than improve flavor.

47.3.3 Mushroom Flavors

Commercially-important mushrooms belong to the orders Ascomycetes and Basidiomycetes. The Ascomycetes are represented by the truffles (*Tuber* sp.) and the morels (*Morchella* sp.). The Basidiomycetes are represented primarily by *Agaricus bisporus* and *A. bitorquis*, the Shiitake (*Lentinus edodes*), the Padi straw mushroom (*Volvariella volvacea*), *Pleurotus* sp., *Coprinus fimetarius* and *Flammulina velutipes*.

Numerous studies (Dijkstra, 1976; Pyysalo and Honkanen, 1976; Card and Avisse, 1977) have confirmed that 1-octen-3-ol, $H_2C=CHC(OH)(CH_2)_4 CH_3$ (**22**), is the main chemical responsible for the mushroom aroma, although numerous other chemicals including glutamic acid and 5'-guanylic acid modify the flavor giving each mushroom species its distinctive odor. The search for a mushroom flavor has not, however, concentrated on producing these chemicals or combinations of them, but rather on growing the mushroom mycelium in submerged culture and then utilizing the dried mycelium as a flavorant.

Almost all mushrooms are capable of growing in submerged culture. Sugihara and Humfield (1954) grew 20 different varieties in a chemically defined medium. Samples of *Agaricus campestris* mycelium were cooked and submitted to a panel of tasters. Although the panelists on the whole found the taste of the mycelium pleasant, they preferred samples of natural mushrooms. Of the other species, only *Lipiota rachodes* had a pleasant flavor. Mycelium grown on a solid medium often has a typical mushroom odor not detectable in submerged growth. The fermentation of mushroom mycelium possessing a distinct mushroom flavor is a prerequisite to commercial development. It is doubtful, however, that such a process has been developed which permits the consistent production of mycelium with the true flavor nuances of fleshy mushrooms (Margalith, 1981). Martin (1982) grew *Morchella esculenta* in sphagnum peat hydrolysates as a potential food or feed supplement, but no mention of aroma or taste was made. Castle and Cooke Foods promote a mushroom concentrate from fresh mushrooms. The flavor intensity is roughly 18 times that of whole fresh mushrooms and about 12 times that of canned mushroom slices. Enhancement of flavor is also claimed. Whether the concentrate is produced by submerged fermentation or by another process is not known.

47.4 PRODUCTION OF FLAVOR AND AROMA CHEMICALS BY PLANT TISSUE CULTURE

Various natural flavor and aroma chemicals, *e.g.* vanilla and peppermint, may eventually become scarce due to economic or political situations. It is obvious that plant tissue culture should be exploited for the production of these compounds. Lee and Scott (1979) stated that there are three ways to utilize plant tissue culture for the production of useful products: (1) direct extraction of compounds from the cells or the medium; (2) biotransformation; and (3) enzymatic synthesis. To this may be added a fourth: utilization of the entire cellular mass for further processing. By far, most studies have been involved with the direct isolation of compounds and, of these, most have been aimed at producing chemicals of medicinal value.

Table 3 lists aromatic flavor substances produced by plant cell cultures (Drawert and Berger, 1981).

Table 3 Biosynthetic Products of Plant Cell Cultures

Substance	Species	Culture type[a]
Cinnamic acid	*Nicotiana tabacum*	C
4-Hydroxy-3-methoxybenzoic acid	*Linum usitatissimum*	C
Caryophyllene	*Lindera strychnifolia*	C
2-Undecanone, 2-undecanyl acetate	*Ruta graveolens*	C
Stevioside	*Stevia rebaudiana*	C
Limonene, linalool	*Perilla fructescens*	C
Anethol	*Foeniculum vulgare*	C
'Ess. oil'	*Pimpinella anisum*	C
Diallyl disulfide	*Allium cepa*	C
Farnesol	*Andrographis paniculata*	S
2-Phenylethylglycoside	*Tropeaolum majus*	S
Glycyrrhizin	*Glycyrrhiza glabra*	S
'Apple aroma'	*Malus silvestris*	S
L-Glutamine	*Symphytum officinale*	S

[a] C = callus culture; S = suspension culture.

Townsley and his associates at the University of British Columbia have studied both coffee and chocolate flavor production by plant cell culture and have applied for patents (Townsley, 1974). Other areas of investigation are terpenes (Reinhard and Alfermann, 1980), onion (Turnbull *et al.*, 1980), peppermint (Rodov and Reznikova, 1982), and apple (Ambid and Fallot, 1980).

All the techniques applicable to microbial fermentations may be utilized for the production of flavor and aroma chemicals by plant tissue culture. This includes the selection of overproducing cells, the use of precursors, medium improvement and hormone application. The commercialization of such processes depends, however, on economic considerations and it is this more than anything else that will limit the application of plant tissue culture in industry. While Japan and Germany appear to be in the forefront of research, the United States remains largely uncommitted.

47.5 EVALUATION OF PRODUCTS

Whether one is producing a pure chemical or a melange of aromas and sensations as in a cheese flavor, it is necessary to submit the product for organoleptic evaluation. Initially, the product is screened by a flavor expert who is experienced in organoleptic evaluation and whose nose is trained to characterize the various nuances present. The product, either in its pure state or in use concentration, may then be submitted for panel evaluation. Favorable samples proceed up the normal chain of marketing and production. The details of sensory evaluation have been published by the American Society for Testing and Materials in a series of monographs. It is suggested the reader consult the bibliography for further information in this area.

A necessary adjunct to any evaluation program is the monitoring and analysis of product and product formation. Identifying new or sought after chemicals is almost routine and involves the whole realm of analytical chemistry. Among the newer techniques are nuclear magnetic resonance, mass spectrometry, and gas–liquid chromatography (GLC). Of all the instrumentation available, GLC is undoubtedly the most useful for separating chemicals and detecting trace impurities in manufactured products. GLC has also been used in conjunction with mass spectrometry to run such tests as head space analysis. Several publications outlining analytical methods are listed in the bibliography. An excellent compilation of flavor chemicals may be found in 'Fenaroli's Handbook of Flavor Ingredients'.

The quality control of microbially produced flavors and aromas differs somewhat from the quality control of standard food products. By the nature of their production from microorganisms, these flavors and aromas are susceptible to biological variability. Frequently, the sole judge of a process is the organoleptic evaluator. Regardless of what analytical controls are established, the final line is 'How does this compare with what we've already made regarding taste and odor?'.

47.6 GENERAL REFERENCE BOOKS

Aroma and Flavor Ingredients
> *Fenaroli's Handbook of Flavor Ingredients*, 2nd edn. (1975). Ed. T. E. Furia and N. Bellanca. CRC Press, Cleveland, OH.
> *Flavor of Foods and Beverages—Chemistry and Technology* (1978). Ed. G. Charalambous and G. Inglett. Academic, New York.
> *Perfume and Flavor Chemicals* (1969). S. Arctander. Steffen Arctander, Montclair, NJ.

Terpenes
> *The Biosynthesis of Steroids, Terpenes and Acetogens* (1964). J. H. Richards and J. B. Hendrickson. Benjamin Co. Inc., New York.
> *Chemistry of Terpenes and Terpenoids* (1972). Ed. A. A. Newman. Academic, New York.
> *Handbook of Terpenoids*, vol. I and II (1982). Ed. S. Dev. CRC Press, Boca Raton, FL.
> *The Molecules of Nature* (1965). J. B. Hendrickson. Benjamin Co. Inc., New York.

Sensory and Analytical Evaluation
> American Society for Testing and Materials, Philadelphia, PA.
>> *Basic Principles of Sensory Evaluation*. STP 433 (1968).
>> *Manual on Sensory Testing Methods*. STP 434 (1968).
>> *Correlation of Subjective–Objective Methods in the Study of Odors and Tastes*. STP 440 (1968).
>> *Manual on Consumer Sensory Evaluation*. STP 682 (1979).
> *Principles of Sensory Evaluation of Foods* (1965). A. M. Amerine, R. M. Pangborn and E. B. Roessler. Academic, New York.
> *Food Analysis: Theory and Practice Rev. Ed.* (1978). Y. Pomeranz and C. E. Meloan. Avi, Westport, CT.

47.7 REFERENCES

Ambid, C. and J. Fallot (1980). Bioconversion of fatty acids and aldehydes by apple cells cultured *in vitro. Bull. Soc. Chim. Fr.*, **11**, 104–107.
Arctander, S. (1969). *Perfume and Flavor Chemicals*. Steffen Arctander, Montclair, NJ.
Becher, E., W. Schüep, P. K. Matzinger, P. J. Teisseire, C. Ehret, H. Maruyama, Y. Suhara, S. Ito, M. Ogawa, K. Yokose, T. Sawada, A. Fujiwara, M. Fujiwara, M. Tazoe and Y. Shlomi (1981). Process for the preparation of patchoulol derivatives. *Br. Pat.* 1 586 859.
Britz, M. L. and R. G. Wilkinson (1982). Leucine dissimilation to isovaleric and isocaproic acids by cell suspensions of amino acid fermenting anaerobes: The Stickland reaction revisited. *Can. J. Microbiol.*, **28**, 291–300.
Bundas, R. H. and A. J. Luksas (1969a). Bread flavor. *US Pat.* 3 466 174.
Bundas, R. H. and A. J. Luksas (1969b). Dough flavor concentrate. *US Pat.* 3 485 641.
Card, A. and C. Avisse (1977). Comparative study of the aroma of raw and cooked mushrooms (*Agaricus bisporus L.*). *Ann. Technol. Agric.*, **27**, 287–293.
Ciegler, A. (1969). Microbiological transformation of terpenes. In *Fermentation Advances*, ed. D. Perlman, pp. 689–714. Academic, New York.
Colby, E. E. (1965). Slurry treatment of flours with proteolytic enzymes and dry mixes utilizing the resulting flour. *US Pat.* 3 167 432.
Collins, E. B. (1972). Biosynthesis of flavor compounds by microorganisms. *J. Dairy Sci.*, **55**, 1022–1028.
Collins, R. P. and A. F. Halim (1970). Production of monoterpenes by the filamentous fungus *Ceratocystis variospora*. *Lloydia*, **33**, 481–482.
Collins, R. P. and A. F. Halim (1972). Characterization of the major aroma constituents of the fungus *Trichoderma viride* (Pers.). *J. Agric. Food Chem.*, **20**, 437–438.

Dhavlikar, R. S. and G. Albroscheit (1973) Microbial transformation of terpenoids: valencene. *Dragoco Rep.*, **12**, 251–258.

Dijkstra, F. Y. (1976). Studies on mushroom flavors. 3. Some flavor compounds in fresh, canned and dried edible mushrooms. *Z. Lebensm. Unters. -Forsch.*, **160**, 401–405.

Drawert, F. and R. Berger (1981). Possibilities of the biotechnological production of aroma substances by plant tissue culture. In *Flavour '81*, ed. P. Schreier, pp. 509–527. de Gruyter and Co., Berlin.

Dwivedi, B. K. and J. E. Kinsella (1974). Continuous production of blue-type cheese flavor by submerged fermentation of *Penicillium roqueforti*. *J. Food Sci.*, **39**, 620–622.

Escher, S. D. and A. F. Morris (1981). Use of a hydroxylic bicyclic derivative as perfuming ingredient. *US Pat.* 4 248 742.

Fonken, G. S. and R. A. Johnson (1972). *Chemical Oxidations with Microorganisms*. Dekker, New York.

Gehrig, R. F. and S. G. Knight (1958). Formation of ketones from fatty acids by spores of *Penicillium roqueforti*. *Nature (London)*, **182**, 1237.

Gerber, N. N. and H. A. Lechevalier (1965). Geosmin, an earthy-smelling substance isolated from *Actinomycetes*. *Appl. Microbiol.*, **13**, 935–938.

Gerber, N. N. and H. A. Lechevalier (1977). Production of geosmin in fermentors and extraction with an ion exchange resin. *Appl. Environ. Microbiol.*, **34**, 857–858.

Goldman, I. M. and M. C. Perret (1965). Process for the preparation of the lactones of hydroxypalmitic acid for use in preparing perfumes. *Fr. Pat.* 1 406 122.

Gottschalk, G. (1979). *Bacterial Metabolism*. Springer-Verlag, New York.

Gupta, K. G., N. K. Yadov and S. Dhawan (1978). Laboratory-scale production of acetoin plus diacetyl by *Enterobacter cloacae* ATCC 27613. *Biotechnol. Bioeng.*, **20**, 1895–1901.

Haskins, R. H. (1955). Ustilagic acid. *US Pat.* 2 698 843.

Heublyum, R., M. Barlas, M. Klyachko and A. Zevalina-Blokh (1932). The preparation of biacetyl by fermentation. *Maslob. Zhir. Deb.*, **4–5**, 54–60.

Jackel, S. S. (1969). Fermentation flavors of white bread. *Baker's Dig.*, **43**, 24–28, 64.

Johnson, J. D. and D. L. Southworth (1973). Method for imparting cheese-like flavor to proteinaceous materials. *US Pat.* 3 780 182.

Johnson, R. A. (1978). In *Oxidation in Organic Chemistry, Part C*, ed. W. S. Trahanovsky, pp. 131–210. Academic, New York.

Jolly, R. and F. V. Kosikowski (1975). Blue cheese flavor by microbial lipases and mold spores utilizing whey powder, butter and coconut fats. *J. Food Sci.*, **40**, 285–287.

Jönsson, H. and H.-E. Pettersson (1977). Studies on the citric acid fermentation in lactic starter cultures with special interest in α-acetolactic acid. 2. Metabolic studies. *Milchwissenschaft*, **32**, 587–594.

Kasik, R. L. (1973). Cheese flavor. *US Pat.* 3 729 326.

Kasik, R. L. and A. J. Luksas (1973). Natural cheese culture. *US Pat.* 3 765 905.

Kempler, G. M. and L. L. McKay (1981). Biochemistry and genetics of citrate utilization in *Streptococcus lactis* ssp. *diacetylactis*. *J. Dairy Sci.*, **64**, 1527–1539.

Knight, G. S. (1963). Process for preparing flavor compositions. *US Pat.* 3 100 153.

Krasnobajew, V. (1982). Microbiological transformation of ionone compounds. *US Pat.* 4 311 860.

Krasnobajew, V. (1983). In *Biotechnology*, ed. H. Rehm and G. Reed, vol. 6. Verlag Chemie, Weinheim.

Krasnobajew, V. and D. Helmlinger (1982). Fermentation of fragrances: biotransformation of β-ionone by *Lasiodiplodia theobromae*. *Helv. Chim. Acta*, **65**, 1590–1601.

Kristoffersen, T. (1973). The biogenesis of cheese flavors. *J. Agric. Food. Chem.*, **21**, 573–575.

Labows, J. N., K. J. McGinley, J. J. Leyden and G. F. Webster (1979). Characteristic γ-lactone odor production of the genus *Pityrosporum*. *Appl. Environ. Microbiol.*, **38**, 412–415.

Lanza, E., K. H. Ko and J. K. Palmer (1976). Aroma production by cultures of *Ceratocystis moniliformis*. *J. Agric. Food Chem.*, **24**, 1247–1250.

Law, B. (1981). Short cuts to faster flavor. *Food, Flav., Ingred., Process., Packag.*, **3**, 17–19.

Lawrence, R. C. and J. C. Hawke (1968). The oxidation of fatty acids by mycelium of *Penicillium roqueforti*. *J. Gen. Microbiol.*, **51**, 289.

Lee, S. L. and A. I. Scott (1979). In *Developments in Industrial Microbiology*, ed. L. A. Underkofler, vol. 20, chap. 34, pp. 381–391. Impressions Ltd., Gaithersburg, MD.

Lemieux, R. V. (1960). Ustilic acids. *Can. Pat.* 600 121.

Lendvay, A. T. (1970). Improving the flavor of baked goods. *US Pat.* 3 499 765.

Luksas, A. J. (1973). Preparation of bleu cheese flavored composition. *US Pat.* 3 720 520.

Maga, J. A. (1974). Bread flavor. *CRC Rev. Food Technol.*, **5**, 55–141.

Maga, J. A. (1976). Lactones in foods. *CRC Crit. Rev. Food Sci. Nutr.*, **8**, 1–56.

Margalith, P. Z. (1981). *Flavor Microbiology*. Charles C. Thomas, Springfield, IL.

Marshall, J. A. and A. R. Hochstetler (1968). The synthesis of (±)–geosmin and the other 1,10-dimethyl-9-decalol-isomers. *J. Org. Chem.*, **33**, 2593–2595.

Martin, A. M. (1982). Submerged growth of *Morchella esculenta* in peat hydrolysates. *Biotechnol. Lett.*, **4**, 13–18.

Mikami, Y., Y. Fukunaga, M. Arita and T. Kisaki (1981). Microbial transformation of β-ionone and β-methylionone. *Appl. Environ. Microbiol.*, **3**, 610–617.

Nisman, B. (1954). The Stickland reaction. *Bacteriol. Rev.*, **18**, 16–42.

Ohloff, G. (1978). In *Progress in the Chemistry of Organic Natural Products*, ed. W. Herz, H. Grisebach and G. Kirby, pp. 431–527. Springer-Verlag, Wien.

Pyysalo, H. and E. Honkanen (1976). The aroma of fresh mushrooms. *4th Nordic Symposium on the Sensory Properties of Foods, Skövde, Sweden*, pp. 159–162.

Reinhard, E. and A. W. Alfermann (1980). Biotransformation by plant cell cultures. *Adv. Biochem. Eng.*, **16**, 49–83.

Robinson, R. J., T. H. Lord, J. A. Johnson and B. S. Miller (1958). The aerobic microbiological population of preferments and the use of selected bacteria for flavor production. *Cereal Chem.*, **35**, 295–305.

Rodov, V. A. and S. A. Reznikova (1982). Production and characteristics of peppermint (*Mentha piperita*) cell suspension culture in relation to terpenoid biosynthesis. *Fiziol. Rast.*, **29**, 644–648.

Roxburgh, J. M., J. F. T. Spencer and H. R. Sallans (1954). Factors affecting the production of ustilagic acid by *Ustilago zeae. J. Agric. Food Chem.*, **2**, 1121–1124.

Schindler, J. and K. Bruns (1980). Process for producing monoterpene containing aromas by fermentation. *Ger. Pat.* 2 840 143.

Schindler, J. and R. D. Schmid (1982). Fragrance or aroma chemicals—microbial synthesis and enzymatic transformation — a review. *Process Biochem.*, **17**, 2–8.

Seitz, E. W., W. F. Sandine, P. R. Elliker and E. A. Day (1963). Studies on diacetyl synthesis by *Streptococcus diacetilactis. Can. J. Microbiol.*, **9**, 431–441.

Sharpell, F. and C. Stegmann (1979). Development of fermentation media for the production of butyric acid. In *Advances in Biotechnology*, ed. M. Moo-Young, vol. II, pp. 71–77. Pergamon, Toronto.

Stadhouders, J. (1974). Dairy starter cultures. *Milchwissenschaft*, **29**, 329–337.

Stärkle, M. (1924). Methyl ketones in the oxidative degradation of triglycerides (fatty acids for example) by mold regarding the particular rancidity of coconut fat. *Biochem. Z.*, **151**, 371.

Sugihara, T. F. and H. Humfield (1954). Submerged culture of the mycelium of various species of mushroom. *Appl. Microbiol.*, **2**, 170–172.

Suomalainen, H. and I. Jännes (1946). Fermentative formation of diacetyl. *Nature (London)*, **157**, 336–337.

Swortfiguer, M. J. (1956). ADMI stable ferment for the retail baker. *American Independent Baker*, March 1.

Teisseire P., P. Maupetit, B. Corbier and P. C. Rouiller (1974). Contribution to the knowledge of patchouli oil. *Recherches*, **19**, 8.

Thijsse, G. J. E. and A. C. van der Linden (1963). Pathways of hydrocarbon dissimilation by a *Pseudomonas* as revealed by chloramphenicol. *Antonie van Leeuwenhoek*, **29**, 89–100.

Townsley, P. M. (1974). Chocolate aroma from plant cells. *J. Inst. Can. Sci. Technol. Aliment.*, **7**, 76–78.

Troller, J. A. (1981). Method for increasing the diacetyl production of diacetyl-producing bacteria. *US Pat.* 4 304 862.

Turnbull, A., I. J. Galpin and H. A. Collin (1980). Flavor production in tissue cultures of onion. In *Plant Cell Culture: Results and Perspectives*, ed. F. Sala, B. Parisi, K. Cella and O. Ciferri, pp. 359–362. Elsevier, New York.

Voishillo, N. E., A. A. Akhrem and Y. A. Titov (1970). Metabolism of terpenes by microorganisms. *Prikl. Biokhim. Microbiol.*, **6**, 491.

48

Fats and Oils

C. RATLEDGE and C. A. BOULTON*
University of Hull, UK

48.1 COMMERCIAL OILS AND FATS

The market for oils and fats is a slowly expanding one, growing probably at a rate slightly faster than the increase in population. The demand for oils and fats is met largely from plant sources with animal fats and marine oils contributing less than 25% of the total production. The expansion of trade naturally puts pressure on the commodity and in the first instance the increased demand can be met by the simple expedient of growing more crops. The crop which contributes most significantly is the soybean which, although having only a 16% oil content, nevertheless contributes over 15 million tonnes of oil out of a total annual production of 60 million tonnes from all sources. There is, though, a natural limit to the extent to which the soybean can be grown. The two countries which are principally involved are the USA, which contributes 65% of the world's production, and Brazil, which contributes about 11% of the total. It has been suggested (Bartholemew, 1980) that further expansion of the soybean crop is likely to be limited because of availability of suitable land. Similar limitations to the expansion of palm oil production in Malaysia probably also exist. Altogether an expansion in production of some 9 million tonnes of oil can be anticipated in this present decade. Good though this figure may appear it is not enough to meet the expected increase in demand and a shortfall between production and demand is likely to appear by the mid-1980s which, by the end of the decade, will have widened to some 9 million tonnes.

*Present address: Bass Brewing Plc, Burton-on-Trent, UK.

983

This anticipated shortfall will put pressure on prices and, if the predictions are correct, increased world prices in this sector of the commodity market can be anticipated (Bartholomew, 1980).

Production of oils and fats (see Table 1) is dominated by seven major plant crops: soybean, groundnuts, cottonseed, rapeseed, palm, coconut and sunflower. The USA is a major world producer of three of these: soybeans, sunflower and cottonseed. In Europe, only rapeseed is grown as an oil crop to any significant extent: the leading producers (1982) are France (27% of Europe's total), UK (13.5%), West Germany (12%), Poland (10%) and Denmark (8%). Over the past four years (1979–1982), of the European countries the UK alone has had a vigorous programme of expansion of rapeseed cultivation. This means that this crop now satisfies about 15–18% of the national annual demand for oils and fats and has been responsible for the slight decline in the importation of oil and fat commodities into the UK over this period. Nevertheless a country such as the UK still imports well over a million tonnes of oils and fats each year at an estimated cost of over £400 million.

The prices of oils and fats (Table 2) have remained relatively stable over the past 13 years apart from a period of volatility between 1974 and 1975 when the commodity market as a whole was in a state of flux following the sharp increase in petroleum prices in 1973 precipitated by the OPEC consortium. The price index for all fats and oils for 1982 is not markedly different from that of 1973. 1982, as a whole, saw a decrease in the world price index for all parts of the fats and oil market. This, however, was no more than had been experienced the previous year, and indeed since 1979 prices of all sectors of this market have been declining steadily. As already discussed, the prediction is that these prices will begin to rise, perhaps more steeply towards the end of this decade, when increased demand begins to outstrip the ability to continue the expansion of production.

Other factors which can affect prices of oils and fats include the changes in demand for individual commodities, such as the polyunsaturated products derived from sunflower oil. Prices can be affected by the producing countries beginning to supply oil *per se* rather than the oilseed for crushing elsewhere. But possibly the biggest, but unknown, factor is the weather. A disastrous harvest for any one of the major world commodities would have repercussions throughout the market. Add to these factors the political considerations of a country deciding to export, or not to export, its oil and we enter a period of price instability.

A useful series of papers discussing various commercial aspects of oils and fats have recently been published (Leysen, 1982; Hughes, 1982; Franke, 1982; Vedeler, 1982; Hancock, 1982; Chapman, 1982; Lysons, 1982; Airey, 1982) and these might be consulted for further and more detailed information.

The uses of fats and oils are presented in Table 3. Approximately 30% of the total consumption goes into industrial products. The fatty acid composition of the major oils and fats is given in Table 4. Current uses and predictions for future trends for the uses of conventional oils and fats have been reviewed by Pryde (1979, 1981). Princen (1979) has considered the possibilities of new crops being developed to produce oils and fats specifically for the industrial market. Such commodities include oils with hydroxy fatty acids (to replace castor oil), polyhydroxy and hydroxy-unsaturated fatty acids, epoxy fatty acids (for use in plastics) and fatty acids with conjugated unsaturation (to replace tung oil). Waxes and oils, either with short chain fatty acids or having a composition that differs from the common triacylglycerol structure, have all been identified as producible by various plant crops. It would be interesting to contemplate if similar materials could be found amongst microorganisms.

48.2 OLEAGINOUS MICROORGANISMS

48.2.1 Background and Definitions

Microorganisms have long been known to produce lipids and therefore to be potentially useful for the production of oils and fats. Such organisms may be termed oleaginous in keeping with the terminology used for oil-bearing plant seeds. The historical background has been reviewed by Woodbine (1959) and more recently by Ratledge (1976, 1982). Principal interest has centred upon yeasts and moulds though both bacteria and algae may be considered as of potential interest. A general review of the biosynthesis of fatty acids and lipids has been written for Volume 1 (Chapter 24) of this work. This could be consulted for information regarding the general distribution of the various lipid types in microorganisms as well as for details of their biosyntheses.

For the most part, the oils produced by the oleaginous strains of eukaryotic microorganisms approximate to the oil produced by plants; that is they contain mainly C_{16} and C_{18} fatty acids esterified in the form of triacylglycerols. The opportunities for a microbial oil displacing a

Table 1 Production of Oils and Fats (1980–1982): World Production and Major Producers (from *FAO Monthly Bulletin of Statistics*, 1983, **6**, no. 1)

Commodity	Production (thousand tonnes)			Oil equivalent (thousand tonnes)		
	1980	1981	1982	1980	1981	1982
Castor beans						
World	809	814	804	405	407	402
Brazil	281	278	212	141	139	106
India	227	210	250	114	105	125
Copra (dry coconut)						
World	4680	4908	4876	(calculated at 69% oil content)		
Philippines	1853	2090	2120			
Indonesia	1301	1254	1260			
Cottonseed						
World	26 578	28 779	27 539	(calculated at 15.5% oil content)		
China	5414	6000	6258			
USSR	6077	5879	5700			
USA	4056	5803	4307			
Olives						
World	11 252	8238	10 671	2217	1659	2140
Italy	3491	2990	3200	735	625	679
Greece	1550	1350	1600	336	280	340
Spain	2255	1553	2540	497	329	557
Groundnuts (in shell)						
World	17 123	19 891	19 482	(calculated at 30% oil content)		
India	5020	6200	5300			
China	3686	3908	3985			
USA	1044	1809	1557			
Linseed						
World	2113	2262	2805	(calculated at 34% oil content)		
Canada	465	467	714			
Argentina	585	598	720			
India	270	428	475			
Palm kernels						
World	1829	1875	2239	(calculated at 46% oil content)		
Malaysia	557	589	900			
Nigeria	345	350	350			
Brazil	266	275	280			
Palm oil						
World				5080	5389	6350
Nigeria				675	675	700
Malaysia				2576	2822	3600
Rapeseed						
World	10 591	12 321	14 319	(calculated at 40% oil content)		
India	1428	2247	2700			
China	2386	4067	4702			
Canada	2483	1837	2073			
UK	300	325	580			
Europe	3737	3646	4295			
Sesame seed						
World	1782	2034	2075	(calculated at 50% oil content)		
China	260	511	481			
India	437	500	475			
Sudan	200	200	200			
Soybeans						
World	81 026	88 551	96 103	(calculated at 16% oil content)		
USA	48 772	54 436	62 584			
Brazil	15 156	14 978	12 810			
China	7906	9341	10 017			
Sunflower seed						
World	13 577	14 232	15 966	(calculated at 25% oil content)		
USA	1748	2098	2547			
USSR	4652	4600	5200			
Argentina	1650	1280	1780			
Tung oil						
World				100	104	106
Lard						
World				4685	4659	4640
Tallow						
World				5654	5549	5677

Not included in the list are corn oil, cocoa butter and safflower oil which contribute less than 1.5 million tonnes of oil. Also not included are fish and whale oils which contribute just over 1 million tonnes.

Table 2 FAO Indices of International Market Prices for Fats and Oils and Oilcakes and Meals
(from *FAO Monthly Bulletin of Statistics* 1983, **6**, no. 1)

	Fats and oils (excluding butter)						
	Edible/soap fats and oils					Technical oils	All fats and oils
	Soft oils		Lauric oils	Others	All		
Years and months	Olive oil	Others					
	1975–1977 = 100						
ANNUAL AVERAGES							
1970	49	54	70	56	56	38	55
1971	50	58	64	54	57	32	55
1972	64	48	42	47	48	40	47
1973	91	79	105	87	85	91	85
1974	134	152	218	135	155	122	152
1975	123	108	84	97	102	85	101
1976	87	85	90	92	88	88	88
1977	91	108	126	112	110	126	112
1978	100	117	149	125	122	108	121
1979	119	123	212	142	139	115	137
1980	129	110	144	125	119	118	119
1981	115	102	122	120	110	114	110
1982	177	82	99	101	91	104	92
1982 January	122	87	116	112	99	113	100
February	120	87	113	113	99	111	100
March	120	85	104	110	96	106	96
April	121	90	108	111	99	108	100
May	120	93	105	113	101	114	102
June	118	86	105	108	96	112	97
July	117	84	97	98	90	105	91
August	116	79	86	91	85	103	86
September	116	78	86	91	84	99	85
October	113	75	84	88	81	94	82
November	113	73	87	89	81	94	82
December	108	72	91	91	81	90	82

Table 3 Uses of Fats and Oils

Edible	
Margarine	Soybean oil, groundnut oil, cottonseed oil, sunflower oil, rapeseed oil, sesame oil, palm oil, some fish oils, olive oil, castor oil, lard and tallow
Cooking fat	
Cooking oils	
Salad oils/mayonnaise/table oils	
Ice cream	
Confectionery	
Cosmetics, toiletries and pharmaceuticals	Coconut oil, palm kernel oil, castor oil

Non-edible	
Detergents and surfactants	Palm kernel, coconut oil
Soaps, metallic soaps, synthetic waxes	Palm oil
Paints and coatings	Linseed oil, tung oil, soybean oil, sunflower oil
Varnishes and lacquers	Linseed oil, tung oil
Inks	Various, mainly castor oil
Plastics and additives	Various, mainly soybean oil
Lubricants and cutting oils	Castor oil, coconut oil
Wood dressings, polishes	Tung oil
Leather dressing	Fish oils
Metal industry	Palm oil and tallow
Agrichemicals, long chain quaternary compounds as herbicides, insecticides and fungicides	Various, mainly soybean oil
Evaporation retardants	Fatty alcohols from any source
Fabric softeners	Tallow

Table 4 Fatty Acyl Composition of Commercial Plant Seed Oils and Animal Fats (from Procter and Gamble, 1979)

Commodity	12:0 and below	14:0	16:0	16:1	18:0	18:1	18:2	18:3	HO–18:1	20:0 and over
Castor			2	—	1	6	3	—	87.5	—
Coconut	63	18	8.5	—	3	6	1	0.5	—	—
Corn			12	—	2.5	29	56	0.5	—	—
Cottonseed		1	24	1	2	17	55	0.3	—	—
Groundnut	—	—	115	—	2	52	27	—	—	7.5
Linseed			6	0.5	3.5	19	14	57	—	—
Olive	—		12	2	2	70	13	0.5	—	0.5
Palm		1	45	—	3.8	40	10	0.2	—	—
Palm kernel	56	18	8	—	2	14	2	—	—	—
Rapeseed[a]	—	—	5	—	2	63	20	9	—	1
Soybean	0.5	0.5	12	—	4	25	52	6	—	1
Sunflower	—	—	8	—	3	20	67.8	0.5	—	0.7
Tung	—	—	4	—	1	8	4	83[b]	—	—
Lard	—	1.5	27	3	14	43.5	10.5	0.5	—	—
Tallow	—	3	25	3.5	21	43	4	0.5	—	—

[a] Canadian (Canola or Canbra) rape (zero erucic). [b] 3 parts = linoleic acid, 80 parts = eleostearic acid.

conventional oil, such as soybean oil or palm oil, must however be considered remote. The economies of large-scale fermentations involving high technology would not seem to be able to compete against the low technology of agriculture. Thus costs of microbial oils must be considered to be several-fold more than plant oils. There could though be opportunities for the production of higher value-added commodities as well as for producing oils and fats from waste materials. The latter could be more a cost-effective process than producing single cell protein to be sold as animal feed (see also Section 48.4). There are also opportunities to develop microbial lipids in those countries which may have a surfeit of cheap fermentable substrates but are unable to grow the requisite plants and face difficulties with the necessary balance of payments with which to effect the necessary importation of these commodities.

A definition for what constitutes an oleaginous microorganism poses some difficulty. A pragmatic definition would suggest that a microorganism containing more than 20–25% oil could be deemed a suitable candidate for commercial consideration. However, this should not be taken so as to deny that organisms with less lipid may not also be useful but these organisms could hardly be classed as oil-bearing if their oil content was much less than, say, 15%.

A biochemical definition for yeasts (and probably for moulds and eukaryotic algae) can be offered, though this will not hold for bacteria. A correlation has been observed (Boulton and Ratledge, 1981) between the possession of the enzyme ATP:citrate lyase and the ability of a yeast to accumulate more than 20% of its biomass as lipid. The significance of the enzyme is that it serves to produce the substrate for fatty acid biosynthesis, acetyl-CoA, from citrate:

$$\text{citrate} + \text{ATP} + \text{CoA} \rightarrow \text{acetyl-CoA} + \text{oxaloacetate} + \text{ADP} + \text{P}_i \tag{1}$$

Acetyl-CoA cannot be produced in the cytoplasm from pyruvate (this reaction proceeds in the mitochondria). Oleaginous yeasts and, as suggested above, probably other oleaginous eukaryotic microorganisms accumulate citrate in the mitochondria which is then transported into the cytoplasm and there cleaved by ATP:citrate lyase. Non-oleaginous organisms do not possess the citrate-cleaving enzyme and must rely on less effective means of producing acetyl-CoA in the cytoplasm. The biochemistry of lipid accumulation has been reviewed in some detail elsewhere (Ratledge *et al.*, 1984).

As prokaryotic microorganisms do not have the compartmentalization of the mitochondrion to separate acetyl-CoA formation from acetyl-CoA utilization for fatty acid biosynthesis, there is no need for ATP:citrate lyase. Thus the absence of this enzyme in oleaginous bacteria, such as there are, has no biochemical significance.

48.2.2 Accumulation of Lipid

The patterns of lipid accumulation were reviewed in Chapter 24 of Volume 1. The salient points may be usefully repeated here.

As lipid represents a reserve storage material, it is not unexpected to find that lipid accumulation is favoured by oleaginous microorganisms growing in a medium with a high carbon to nitrogen ratio. Usually a C:N ratio of 50:1 is employed. In a batch culture, the organism grows until the nitrogen is consumed but thereafter it continues to take up the excess carbon and convert this to lipid. Thus a biphasic growth pattern can be envisaged.

With some of the slower growing moulds, the rate of lipid accumulation appears to coincide with the growth rate. Although this is probably fortuitous, the result is that the lipid content of the cells increases at the same rate as growth proceeds.

In continuous culture, lipid accumulation is achieved by growing oleaginous microorganisms under nitrogen-limiting conditions at a dilution rate (= specific growth rate) of about 30% of the maximum. The build-up of lipid is dependent upon the correct balance being achieved between growth rate and the specific rate of lipid biosynthesis so that the optimum amount of carbon can be diverted into lipid and the minimum into other cell components.

The efficiency of lipid accumulation in continuous culture is often the same as or better than in batch cultures where the same organism has been studied under both conditions. With *Candida* sp. 107, *Rhodotorula glutinis*, *R. gracilis* and *C. curvata* (Gill *et al.*, 1977; Hall and Ratledge, 1977; Yoon *et al.*, 1982; Choi *et al.*, 1982; Evans and Ratledge, 1983), lipid yields of 17–22% have been obtained under both conditions of growth. A conversion of carbohydrate to lipid of 20% would appear near to a possible practical limit as the theoretical maximum is about 33 g triacylglycerol from 100 g glucose (Ratledge, 1982) assuming that all the carbon of the medium is converted into lipid without synthesis of any other cell component.

48.2.3 Bacteria

48.2.3.1 *Triacylglycerols*

Only a few species are known which produce appreciable amounts of extractable glycerol lipids. The mycobacteria-nocardia group of organisms are well known for their high lipid contents but these lipids are complex structures often in a bound form as part of the cell envelope structure. Some of these species do contain triacylglycerols but exploitation of them is not sensible as the coextraction of toxic or allergenic substances from the mycobacteria is highly likely.

The only bacterium which has been reported as producing significant amounts of triacylglycerol is *Arthrobacter* AK19 (McLee *et al.*, 1972; Wayman *et al.*, 1984). This organism is unlike any other bacterium in that it can contain up to 80% of its biomass as lipid; this lipid, moreover, is predominantly composed of triacylglycerols and would thus seem an excellent candidate for commercial exploitation. The composition of the fatty acids of this organism is given in Table 5. The only drawback with the organism is its slow growth rate. However, Wayman and Whiteley (1979) have considered it possible that the bacterium could be efficiently cultivated as a symbiont along with an algal culture. The provision of an external carbon source, other than CO_2, would then be obviated.

48.2.3.2 *Poly-β-hydroxybutyrate*

Many bacterial species produce the polymeric poly-β-hydroxybutyrate (PHB) as a reserve storage polymer (Dawes and Senior, 1973). Although PHB is not a fatty acid-containing lipid (*cf.* Figure 1), it is nevertheless classified as a lipid in view of its solubility in chloroform and similar solvents. Like the lipids of the eukaryotic microorganisms, PHB is produced in increased quantities when nitrogen is exhausted from the medium. However, its synthesis also responds to the concentration of O_2. At low partial pressures of oxygen the ability of the bacteria to reoxidize NADH becomes limiting. By producing PHB, the organism is able to achieve reoxidation of NADH, in much the same way as ethanol and butanol are produced as reduced fermentation products by other organisms. The structure of PHB is given in Figure 1; its monomer, β-hydroxybutyrate, is synthesized from two acetyl-CoA units. The degree of polymerization is between 4000 and 10 000 with the average molecular weight being 5×10^5.

PHB is currently being considered for commercial exploitation by ICI Ltd. at Billingham, UK (King, 1982; Howells, 1982). It has been given the trade name of Biopol and is classed as a biodegradable thermoplastic. Its applications vary from acting as a substitute for plastics in roles where biodegradability would be an important attribute, to applications in the manufacture of hi-fi

Table 5 Fatty Acyl Composition of *Arthrobacter*
AK19 (from Wayman *et al.*, 1984)

	Time of growth (days)		
	3	8	13
Total lipids, (% cell dry wt)	16.0	39.5	78.3
Fatty acyl composition of triacylglycerols (% w/w)			
14:0	2.7	3.2	3.1
15:0	2.8	2.2	1.9
16:0	36.1	23.9	30.1
16:1	12.0	13.2	14.4
br17:0	≤ 0.3	≤ 0.3	0.6
17:0	4.4	1.4	2.5
17:1	7.3	8.4	7.0
br18:0	≤ 0.3	≤ 0.3	0.4
18:0	7.8	5.1	5.4
18:1	23.9	35.0	29.8
br19:0	1.4	1.7	2.7
Other	1.0	0.3	2.1

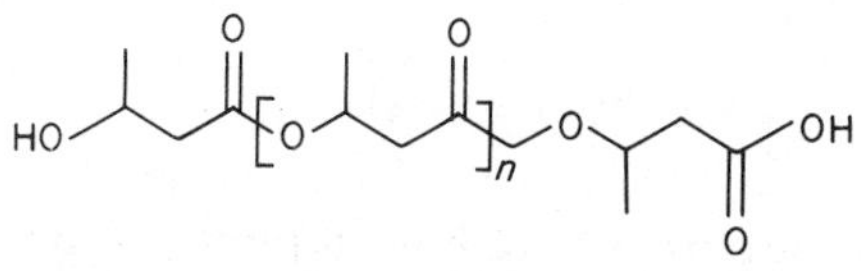

Figure 1 Structure of poly-β-hydroxybutyrate

equipment where its piezoelectric properties would be useful. In the former case, the manufacture of surgical pins and sutures would be an important application.

The organism of choice for PHB production is *Alcaligenes eutrophus* which produces between 70% (Howells, 1982) and 80% (King, 1982) of its biomass as the polymer. The substrate currently used is glucose, though the possibilities either of being able to use another organism or of genetically engineering *A. eutrophus* into being able to utilize CO_2 would significantly reduce costs and thus open up new horizons for its applications.

48.2.3.3 Waxes

Waxes, although unusual amongst microorganisms, appear to be a common, though not major, lipid constituent of the *Acinetobacter*. Their composition is of a simple ester of a fatty acid with a fatty alcohol, $CH_3(CH_2)_xC(=O)OCH_2(CH_2)_yCH_3$ where x and y are usually either 14 or 16, though shorter chain alcohols with $y = 1$ to 3 have been reported (Gallagher, 1971). Unsaturation in both the alcohol and fatty acid components has been reported (Gallagher, 1971; Russell, 1974; Thorne *et al.*, 1973).

The amount of wax produced is usually small when these bacteria are grown on succinate or acetate (the bacteria do not grow on glucose). Typical figures indicate that the waxes constitute about 10–25% of the lipid, but this only represents about 2–3% of the cell biomass. However, growth on hydrocarbons and fatty alcohols increases the yield many-fold, with up to 15% of the cell biomass as waxes being recorded for *Acinetobacter* sp. HO1-N after growth on hexadecanol (Finnerty, 1984).

Although the early studies with this strain did not reveal the occurrence of unsaturated waxes, these have subsequently been identified by capillary gas chromatography (Dewitt *et al.*, 1982; Ervin *et al.*, 1984). These waxes are thus similar to the commercially important sperm whale and jojoba oils (Table 6) though the percentage of diunsaturated wax is only about 10% of the total wax when hexadecane is used as substrate. Growth on eicosane (C_{20}) at 17 °C, however, produces 6% of the biomass as waxes with 71% being diunsaturated (Ervin *et al.*, 1984).

Table 6 Comparison of the Wax Esters Produced by *Acinetobacter* sp. HO1-N Grown on Hydrocarbons and Those Found in Sperm Whale and Jojoba Oils (from Dewitt *et al.*, 1982; Ervin *et al.*, 1984)

	Sperm whale oil	Microbially produced wax esters	Jojoba oil
Carbon number of intact wax esters	28–40	32–40[a]	36–44
Acyl segments			
Carbon number	14–22	16–20[a]	16–24
Number of unsaturations	0.1	0.1–0.7	1
Alkoxy segments			
Carbon number	16–20	16–20[a]	18–24
Number of unsaturations	0.1	0.1–0.7	1
Predominant sites of unsaturation	$\omega 7$, $\omega 9$ and $\omega 11$	$\omega 7$ and $\omega 9$	$\omega 9$
Yield of waxes (% cell dry wt)	—	2.6–8.5	—

[a] Carbon numbers dependent on *n*-alkane used as substrate.

48.2.3.4 Other lipids

Glycolipids with surfactant properties have been reported to be produced by a number of bacteria. Some of these compounds are more correctly classed as non-lipids for, although they do contain some lipid moiety, they are not soluble to any extent in organic solvents. Most of these glycolipids are produced in significantly increased amounts when the organisms are grown on hydrocarbons. A brief account of the structures of some of the more important surfactants is given in Chapter 24 of Volume 1 of this work.

48.2.4 Algae

The oil content of microalgae has been reported to be as high as 70% with *Chlorella pyrenoidosa* (*cf.* Ratledge, 1982). Although even higher claims have been made, more reasonable ones would put this maximum no higher than 50% (Dubinsky *et al.*, 1978; Shifrin and Chisholm, 1980; Shifrin, 1984). The major constraints to development of algae are availability of land, water, sunlight and a warm ambient temperature.

Shifrin (1984) has calculated that microalgae cultures should reasonably be expected to be able to achieve productivities of 15–25 g m^{-2} d^{-1}, which is less than 10% of the theoretical maximum. Given an oil content of 50% of the biomass, yields of oil could be 25 tonne ha^{-1} y^{-1} which is significantly higher than any plant oilseed crop. Dubinsky *et al.* (1978) calculated similarly that an annual oil yield per hectare by algal culture should realize 22.8 tonnes. However both Shifrin and Dubinsky and colleagues conclude that this yield is not high enough to be economically viable in view of the large capital investment costs that would be required. The suggested solution would be to couple algal culture to sewage treatment which would then make the entire process economically feasible (Dubinsky *et al.*, 1978). This could also apply to algal culture in more temperate climes.

Although there are many algae which will produce more than 20% of their biomass as lipid (Berner *et al.*, 1982) few appear to exceed 40%. From the survey of Dubinsky *et al.* (1978) these species are *Botryococcus braunii* (53% lipid), *Dunaliella salina* (47%) and *Radiosphaera negevensis* (43%). Shifrin (1984) identified *Chlorella vulgaris* (~40%), *Nannochloris* sp. (49%) *Ourococcus* sp. (50%), *Scenedesmus obliquus* (~49%), *Nitzschia palea* (~40%), *Navicula pelliculosa* (~45%), *Monalanthus salina* (~70%) and *Biddulphia aurita* (~40%), as well as *Botryococcus braunii* (~70%), as good lipid producers. *Chlorella pyrenoidosa*, at about 37% lipid, would also seem to be potentially useful.

Botryococcus braunii has an established reputation as a lipid producer having been implicated in the formation of oily droplets of the type known as Boghead Coal where is was found to contain up to 86% of its mass as hydrocarbon (Brown *et al.*, 1969). This would appear to be an exceptionally high figure and 30% would appear more reasonable under laboratory conditions. Unfortunately this species grows very slowly with doubling times of up to 6 weeks. Nevertheless, the organism can be grown in laboratory conditions to produce over 50% of its biomass as lipid with this lipid containing 20% hydrocarbon. Hillen *et al.* (1982) have shown that hydrocracking of the hydrocarbons, whose principal component is botryococcene (Figure 2), gives 67% gasoline, 15% aviation turbine fuel, 15% diesel fuel and 3% residual oil.

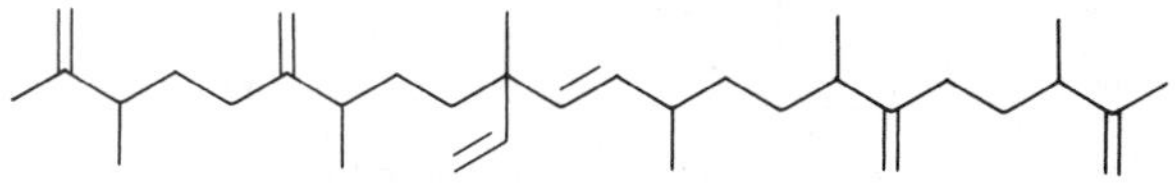

Figure 2 Structure of botryococcene, a major lipid component from *Botryococcus braunii* (from Hillen *et al.*, 1982)

The future exploitation of this alga would therefore appear to lie as an alternative means of producing fuel oil. An added advantage with this species is that, although classed as a freshwater alga, it can be cultured on media with osmotic potentials up to those of seawater with no adverse effects (Dubinsky *et al.*, 1978).

Of the more conventional algae, species of *Chlorella* (either *C. pyrenoidosa* or *C. vulgaris*) would appear to be worth investigating further. Shifrin (1984) has shown that a copper-tolerant clone of *C. vulgaris* produces 40% lipid content in about 9 days culture whereas the copper-sensitive strain produces less than 30% lipid. A recent report from the Solar Energy Research Institute, California (Anon., 1983) has also indicated *Chlorella* to be an organism of choice for the production of oils to act as vegetable oil substitutes.

With the possibilities of genetic improvement or of strain selection and improvement, perhaps by utilizing copper tolerance as indicative of a high lipid content as in Shifrin's work, the way seems to be open for striking developments in this field. The opportunities for large scale algal cultivation in such countries as Israel would appear to offer excellent possibilities of deriving acceptable replacements for existing plant seed oils.

The fatty acids produced by potentially useful algae are given in Table 7. As will be seen, algal lipids are marked by their exceptionally high proportions of polyunsaturated fatty acids. Unfortunately as these are the type which are also found in fish oil, the frequent complaint against algal lipids is their unpleasant 'fishy' odour. The desirability of including polyunsaturated acids in the diet might suggest that a proportion of algal oils could be mixed with a more saturated or monounsaturated plant oil, *viz.* palm oil or rapeseed oil (see Table 4) to give a nutritionally acceptable blend.

48.2.5 Yeasts and Moulds

48.2.5.1 Organisms

The number of oleaginous organisms is not extensive. For the yeasts, the list is given in Table 8. Some 16 classified species have been reported as producing better than 25% lipid. In addition there are two species identified only to the genus level (*Candida* sp. 107 and *Lipomyces* sp. 33). It should be pointed out that many of the designated species contain numerous strains, some of which readily attain high lipid contents whereas others bearing the same name do not.

New oleaginous species are continually being added to the list. It is possible that other known yeasts could prove to be oleaginous if they were grown under the appropriate conditions. The prerequisite for lipid accumulation is the possession of ATP:citrate lyase (*cf.* Section 48.2.1) and it is this activity which should be determined in order to ascertain if a given yeast might be capable of lipid accumulation. The enzyme has been found in all oleaginous yeasts so far examined (Boulton and Ratledge, 1981). It is not present in non-oleaginous species, nor even in non-oleaginous strains of organisms such as *Lipomyces starkeyi* and *L. lipofer*. It is therefore an extremely powerful determinant for lipid production.

The number of oleaginous moulds is considerably greater than yeasts. However, to give a list for moulds similar to that given in Table 8 for yeasts could be misleading as much of the infor-

Table 7 Fatty Acid Composition of Selected Oleaginous Algae

Alga	14:0	16:0	16:1	16:2	16:3	16:4	18:0	18:1	18:2	18:3 (α)	18:3 (γ)	18:4	20:U	Ref.
						Relative % (w/w)								
Spirulina platensis	1	43	3	5	—	—	1	6	22	0.3	18	—	—	Capella et al. (1977)
Chlorella pyrenoidosa	1	21–34	2–8	8–16	3–13	—	0.3–3	4–17	19–30	8–26	—	—	—	Shifrin (1984)
Chlorella vulgaris	2	26	8	7	2	20	2	2	34	20	—	—	—	Wood (1974)
Scenedesmus actus	1	15	1	tr	4	6	tr	8	20	30	0.4	—	—	Capella et al. (1977)
Dunaliella primolecta	5	11	10	8	7	—	—	6	6	10	2	7	14	Wood (1974)
Navicula pelliculosa	3	9	31	3	18	—	—	6	4	2	—	—	19	Shifrin (1984)
Nitzschia palea	6	23	45	4	2	—	—	3	—	—	—	—	18	Shifrin (1984)

tr = trace.

Table 8 Oleaginous Yeasts [compiled from data of Ratledge (1982) and Suzuki (1980)]

Yeast	Lipid (%)
Candida curvata (spp. R and D)	51–58
Candida guilliermondii	25
Candida (*Saccharomycopsis*, now *Yarrowia*) *lipolytica*	36
Candida paralipolytica	32
Candida sp. 107 (NCYC 911; CBS 329.80)	42
Cryptococcus terricolus (syn. *C. albidus*)	55–65
Endomyces vernalis ⎫ *Endomycopsis vernalis* ⎬ = *Trichosporon pullans* (q.v.)	
Hansenula saturnus	28
Lipomyces lipofer (= *lipofera*, *lipoferus*)	64
Lipomyces starkeyi	63
Lipomyces tetrasporus (sp. 5011F)	64
Lipomyces sp. no. 33 (= *L. lipofer?*)	67
Rhodosporidium toruloides	66
Rhodotorula glutinis (syn. *R. gracilis* and *R. suganii*)	74
Rhodotorula graminis	41
Rhodotorula mucilaginosa	28
Trichosporon cutaneum	45
Trichosporon pullulans (syn. *Geotrichum candidum*)	65
Trigonopsis variabilis	40

mation on oil production has come from work in which the moulds have been grown as 'felts' in static culture, often for two to three weeks. In the authors' experience, many of the claims from early work (*cf.* Woodbine, 1959) cannot be repeated when the moulds are regrown in a stirred tank reactor. A list of some oleaginous moulds is given in Table 9 but the veracity of this cannot be verified. Whether the presence of ATP:citrate lyase could be used to indicate lipid-accumulating ability, as with yeasts, is completely unknown as no systematic survey has so far been carried out with moulds to examine this property.

48.2.5.2 *Triacylglycerols and fatty acids*

The major accumulating lipid of yeasts and fungi is the triacylglycerol fraction which can account for up to 92% of the total lipid of a cell. Where positional analysis of the fatty acyl residues has been carried out, the *sn*-2 position of the glycerol appears to be almost entirely occupied by unsaturated fatty acids (see Ratledge, 1982). This, therefore, is the same type of distribution which occurs in plant seed oils but differs from that in animal fat where saturated fatty acids occur on the *sn*-2 position.

In their detailed analysis of the oil from *Cunninghamella blakesleeanus*, DeBell and Jack (1975) distinguished between the *sn*-1 and *sn*-3 positions of the glycerol. These two positions were not acylated by the same fatty acids: the *sn*-1 position contained 65% of its acyl groups as palmitic acid (16:0) and oleic acid (18:1) whereas the *sn*-3 position contained only 39% palmitate and 10% oleic acid, the remainder being accounted for by linoleic (18:2) and linolenic (18:3) acids.

Haley and Jack (1974) carried out a similar stereospecific analysis of the triacylglycerol from *Lipomyces lipofer* and reported that the *sn*-1 position was occupied 61% by oleic acid whereas there was only 37% of this acid on the *sn*-3 position. The *sn*-2 position was occupied to 98.8% by unsaturated acids of which oleic acid contributed 88%.

The fatty acids of yeasts are usually in approximate order of abundance: oleic, palmitic, linoleic and stearic acids. Linolenic acid (18:3) and palmitoleic acid (16:1) may be found a few per cent in some cases. A list of the fatty acids of the major oleaginous species is given in Table 10. It should be appreciated that these values are only a guide as variations, sometimes considerable, can be achieved by cultivating the organism of choice under different conditions. Such conditions would include variation in oxygen tension (Choi *et al.*, 1982), choice of growth substrate (Suzuki, 1980; Yoon *et al.*, 1982; Evans and Ratledge, 1983), nitrogen source (Yoon *et al.*, 1982), growth temperature (Ratledge and Hall, 1979) as well as the growth rate of the organism itself (Gill *et al.*, 1977; Choi *et al.*, 1982). These factors have been discussed in detail elsewhere (Rattray *et al.*, 1975; Ratledge, 1982; see also Volume 1, Chapter 24).

Table 9   Some Moulds Which Have Been Claimed to be Oleaginous (see text)

Mould	Lipid (%)	Ref.
Entomophthorales		
Entomophthora conica	38	Popova *et al.* (1980)
Entomophthora coronata	47	Popova *et al.* (1980)
Entomophthora obscura	34	See Ratledge (1982)
Entomophthora thaxteriana	26	Popova *et al.* (1980)
Mucorales		
Absidia corymbifera	27	Suzuki (1980)
Absidia spinosa	28	See Ratledge (1982)
Blakeslea trispora	37	Bekhtereva and Yakovleva (1980)
Cunninghamella elegans	44	Bekhtereva and Yakovleva (1980)
Cunninghamella homothallica	38	Bekhtereva and Yakovleva (1980)
Mortierella isabellia	63	Suzuki (1980)
Mortierella vinacea	66	See Ratledge (1982)
Mucor circinelloides	65	See Ratledge (1982)
Mucor ramannianus	56	See Ratledge (1982)
Mucor spinosus	47	See Ratledge (1982)
Rhizopus arrhizus	49	See Ratledge (1982)
Rhizopus oryzea	32	Suzuki (1980)
Ascomycetes		
Aspergillus fischeri	53	See Ratledge (1982)
Aspergillus nidulans	25	See Ratledge (1982)
Aspergillus ochraceus	48	See Ratledge (1982)
Aspergillus terreus	57	See Ratledge (1982)
Fusarium sp. N-11	39	See Ratledge (1982)
Fusarium oxysporum	51	Naim and Saad (1978)
Gibberella fujikuroi	48	See Ratledge (1982)
Penicillium lilacinum	51	Suzuki (1980)
Penicillium spinulosum	64	See Ratledge (1982)
Hyphomycetes		
Cladosporium herbarum	49	Suzuki (1980)
Pellicularia		
Pellicularia practicola	39	Suzuki (1980)
Basidomycete		
Ustilago zeae	59	See Ratledge (1982)

Table 10   Fatty Acyl Groups of Lipids from Various Oleaginous Yeasts

Yeast[a]		Relative %						Ref.
	14:0	16:0	16:1	18:0	18:1	18:2	18:3	
Candida curvata D	tr	36	—	14	40	6	—	Evans and Ratledge (1983)
Candida guilliermondii		21	6	5	61	5	1	Suzuki (1980)
Candida 107	3	44	5	8	31	9	1	See Ratledge (1982)
Hansenula saturnus	1	24	3	4	30	25	12	See Ratledge (1982)
Lipomyces lipofer	2	16	3	3	62	9	1	See Ratledge (1982)
Lipomyces starkeyi	—	40	6	5	44	4	—	See Ratledge (1982)
Rhodotorula glutinis	2	16	—	7	39	24	1	See Ratledge (1982)
Rhodotorula gracilis	—	25	—	12	51	11	1	Choi *et al.* (1982)
Trichosporon cutaneum	—	30	—	13	46	11	—	See Ratledge (1982)

[a] All yeasts were grown on glucose except *T. cutaneum* which was grown on lactose.

The fatty acids of moulds show a greater range and diversity than yeasts. Members of the Entomophthoraceae are characterized by the presence, often in substantial amounts, of shorter chain fatty acids (C_{10} to C_{14}) as well as still containing C_{18} polyunsaturated acids. Branched chain acids have also been reported in these organisms. The fatty acids of three oleaginous species are given in Table 11. Fatty acids of other fungi are given in Table 12. As with yeasts, considerable variations in the fatty acyl composition may occur under different growth conditions or with different substrates and nutrients being used (Weete, 1980; Ratledge, 1982; see also Volume 1, Chapter 24).

Hydroxy fatty acids, including ricinoleic acid (12-hydroxystearic acid), occur up to 62% of the total lipid in *Claviceps* spp. Their synthesis, though, only occurs in the sclerotial form of the mould which is a form difficult to propagate in a stable manner in the laboratory. Although *Claviceps* lipid may be thought of as an attractive alternative to castor oil, the main value of any large-

Table 11 Relative % of Fatty Acids in Lipids from Three Oleaginous Species of *Entomophthora* Grown on Glucose in Shake Culture for 120 h (from Popova *et al.*, 1980)

	10:0	12:0	13:0	14:0	16:0	16:1	16:2	18:0	18:1[a]	18:2[a]	18:3[a]	19:0	20:4
E. conica	—	1.7	0.6	11.7	18.0	28.0	1.0	2.9	20.1	2.0	1.4	—	8.8
E. coronata	2.2	17.9	2.7	29.5	12.9	0.6	—	1.9	12.3	9.5	tr	5.2	2.1
E. thaxteriana	1.8	0.4	0.5	8.7	25.5	17.3	0.6	4.6	19.7	3.8	1.5	—	7.0

All samples also contained traces of 15:0, 17:0, 20:1, 20:2 and 20:3; 11:0, 17:1 and 17:2 were also present in *E. coronata*, with traces of 22:0 and 24:0 in *E. conica*

[a] Mixture of positional isomers: for 18:1, two isomers; 18:2, three isomers; 18:3, two isomers.

Table 12 Fatty Acids of the Lipid from Various Oleaginous Moulds

	16:0	16:1	18:0	18:1	18:2	18:3	20+	Ref.
Absidia corymbifera	23	1	20	22	21	3[a]		Suzuki (1980)
Blakeslea trispora	18	4	2	10	62	tr[a]	3	Bekhtereva and Yakovleva (1980)
Cunninghamella elegans	20	2	5	40	19	11[a]	2	Bekhtereva and Yakovleva (1980)
Mortierella isabellia	35	6	3	42	8	5[a]	—	Suzuki (1980)
Rhizopus oryzea	20	2	8	48	11	6[a]	—	Suzuki (1980)
Aspergillus nidulans	20	3	12	39	13	9[b]	—	Farag *et al.* (1981)
Aspergillus terreus	23	0.1	0.3	14	39	21	—	Singh and Sood (1973)
Cladosporium herbarum	31	tr	13	35	18	0.6	—	Suzuki (1980)
Fusarium N₁	21	1	←——	76[c]	——→	—	—	Abraham and Srinivasan (1979)
Fusarium oxysporum	17	0.4	8	20	46	5	—	Suzuki (1980)
Pellicularia practicola	12	0.3	7	20	47	10	—	Suzuki (1980)
Penicillium lilacinum	28	0.3	13	37	20	0.1	—	Suzuki (1980)
Penicillium spinulosum	15	1	7	42	31	1	—	Suzuki (1980)

In all species there are traces of 12:0 and 14:0 fatty acids

[a] 6,9,12–18:3 (γ–linolenic acid); in all other cases this is the 9,12,15–18:3 isomer (α-linolenic acid). [b] Said by authors to be 20:0 but probably confused with α-linolenic acid. [c] 18:0, 18:1 and 18:2 not distinguished.

scale cultivation of this mould would be that it also produces ergot alkaloids when in its sclerotial form. The commercial value of the culture would then be in the high value minor component rather than in its oil. Although it would not seem impossible to extract the ricinoleic acid from such cultures, and thus produce a valuable by-product, the volume likely to be obtained would be so insignificant (in terms of the quantities produced by the castor oil industry) that its recovery would probably not be worthwhile.

48.2.5.3 Steroids

Sterols can be produced in some abundance by eukaryotic microorganisms and have been obtained in commercial quantities by extraction of spent fungal mycelium recovered from various fermentation processes. As far as we are aware, few of these processes are still carried out though ergosterol is produced in one or two instances by extraction of *Saccharomyces cerevisiae*, though it is not certain whether the organism is specifically grown for this purpose or whether extraction is only carried out on spent brewers' yeast.

Ergosterol (see Figure 3) is a common constituent of most yeasts and moulds (see Rattray *et al.*, 1975; Weete, 1980). In a detailed study of 558 yeast cultures, including 240 species of *S. cerevisiae*, Dulaney *et al.* (1954) found that species of *Saccharomyces* were the only ones which consistently produced more than 0.1% of their biomass as ergosterol. They identified eight cultures of *S. cerevisiae* which produced ergosterol from 7 to 10% of the dry cell biomass. The best of these yeasts could be grown to produce cell dry weights in excess of 30 g l⁻¹ giving an ergosterol yield of between 3 and 4 g per litre of culture. Some commercial production of ergosterol (but not necessarily by Merck and Co.) appears to have been carried out using the same or similar high yielding strains. Owing to industrial secrecy, it is not known whether this production is still carried out today.

El-Refai and El-Kady (1968a) in a smaller survey than that of Dulaney *et al.* identified *Saccharomyces carlsbergensis* and *S. fermentati*, besides *S. cerevisiae*, as being of potential value in ergosterol production with the best strains giving yields up to 4.4% of the biomass. By judicious selection of the culture conditions they were able to achieve over 14% of the dry weight of *S. fermentati* as sterols (El-Refai and El-Kady, 1968b). The highest yield of ergosterol was 9.6% of the biomass. Conditions for the optimum productivity (g per litre of fermenter volume) were not however evaluated.

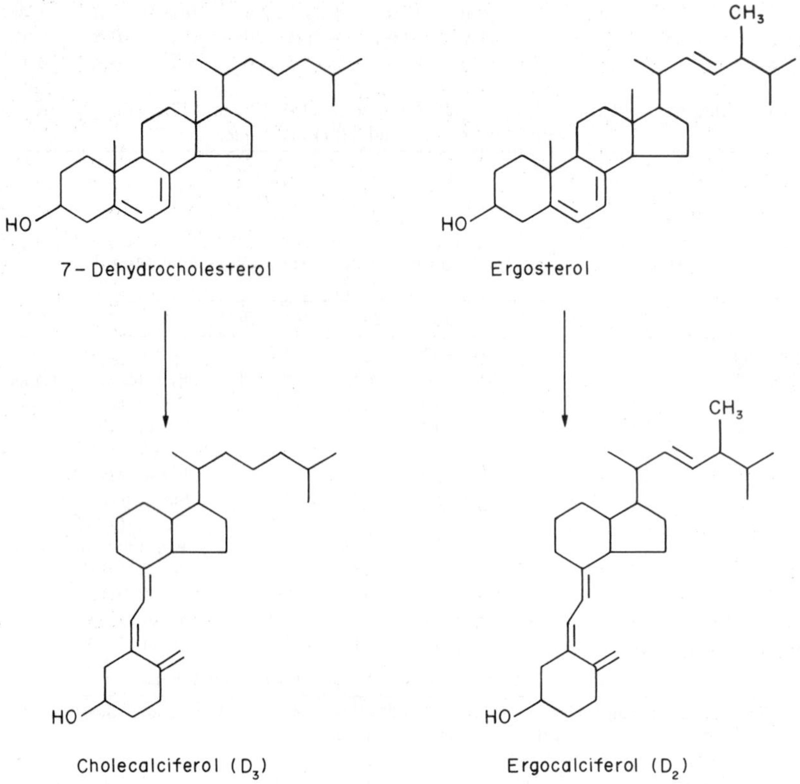

Figure 3 Relationship of ergosterol and 7-dehydrocholesterol to vitamins D_2 and D_3 (from Parks *et al.*, 1984)

The main application of ergosterol is its use as an analogue for cholecalciferol, vitamin D_3. Ergocalciferol—vitamin D_2—(see Figure 3) is formed from ergosterol by the action of ultraviolet light in the same way as vitamin D_3 is formed from 7-dehydrocholesterol. However it has a decreased effectiveness in some animals; in chickens, for example, it is only about as 10% as good. The real need therefore is for 7-dehydrocholesterol rather than ergosterol but an inexpensive source of this pro-vitamin D_3 material is not readily available. Recent work by Parks *et al.* (1984) would seem to indicate that this problem may be resolved by using yeast mutants. As yet there is still some way to go before this goal can be realized though other potentially valuable sterols have been identified.

Ergosterol is also a common constituent in most moulds. It has been reported as accounting for 92% of the total sterols in *Aspergillus fumigatus* which can produce up to 5% of its cell dry weight as sterols (Osman *et al.*, 1969). A short but detailed account of the wide variety of sterols to be found in fungi has been given by Weete (1980). As stated at the beginning of this section, the natural occurrence of relatively high levels (at 2 to 3% of the biomass) of sterols in various moulds has led to some extractions being carried out on mycelium recovered from antibiotic and other fermentation processes. However, the problems of coextraction of residual antibiotics, or of there being potentially deleterious materials, would now necessitate adequate toxicological tests being carried out before the fungal sterols may be permitted back into the food chain.

48.2.5.4 *Other lipids*

A wide variety of lipids, besides triacylglycerols and fatty acids, can be obtained from eukaryotic microorganisms. Of these, probably the greatest commercial attention has been paid to the carotenoids, of which a wide variety of molecular species are produced (see Volume 1, Chapter 24). The principal carotenoid of interest is β-carotene; not only is it an important red colorant but it also acts as a pro-vitamin A compound. Whilst various patents have been taken out for caro-tene production, particularly in *Blakeslea trispora* where yields may be up to 1 g l^{-1} [*i.e.* about 2% of the cell biomass (Ninet and Renaut, 1979)], no sustained commercial process appears to be in existence.

Phospholipids occur in all living systems. They may reach up to about 5% of the cell biomass in bacteria, especially in those grown on methane or methanol where extensive phospholipid mem-branes occur as part of the mechanism for the organism being able to deal with these C$_1$ com-pounds. Phospholipids can provide useful chemical properties with the combined presence of a large polar group such as choline, —OCH$_2$CH$_2$N$^+$(CH$_3$)$_3$, attached to a large non-polar group— the diacylglycerol moiety. Such materials are potential emulsificants and surfactants. Some com-mercial production of yeast lipids, which appear to be particularly rich in phospholipids, occurs in the USSR (Voigt *et al.*, 1979). The process of lipid extraction operates as an adjunct to the pro-duction of fodder yeast [probably *Yarrowia* (*Candida*) *lipolytica*] from hydrocarbons. Without the defatting, the yeast would probably not be acceptable as an animal feedstuff because of the inclusion of residual alkanes. The extracted lipid has been given the name Biolipid and is des-cribed as a dark brown, slightly viscous, combustible liquid with a characteristic odour. It has a high content of alkanes (45 to 55%), which reinforces the view of the unacceptable nature of the original whole yeast, contains phospholipids from 20% to 30%, some triacylglycerols (10–20%), free fatty acids (up to 10%) with small amounts (1%) of sterol and ubiquinone. Thus, apart from the inert residual hydrocarbons, which will be of the gas oil or kerosene variety, the major com-ponent of this material is the phospholipid which then gives the lipid useful boundary surface properties. A variety of technical applications for the whole oil have been suggested: an additive to fuel oil to improve dispersion and flame temperature; a mould-releasing agent in the casting of concrete parts in the building industry; a plant protective agent; an additive to bitumen to improve its adhesion characteristics; and a protective layer or conditioning layer over deliques-cent chemicals such as granular fertilizers and potash salts. The individual components have been separated and some applications for each have been suggested.

48.3 SUBSTRATES

48.3.1 Hydrocarbons

The use of hydrocarbons as a means of producing yeast either for food or for fodder has fallen into disfavour, though perhaps only temporarily. Thus if the production of lipids were to be con-templated from hydrocarbons, they would have to be considered for technical uses only (see end of Section 48.2.5.3).

Hydrocarbons, and alkanes in particular, have the advantage over other substrates in that they can predetermine the chain length of the ensuing fatty acids found in the extracted lipids. This may be of considerable advantage if a lipid with particular fatty acid substituents should be wanted for any reason. Hydrocarbons, in general, also lead to the greater production of lipid, as a percentage of the cell biomass, than do carbohydrates. This may again be of advantage where a product such as a wax (*cf.* Section 48.2.3.3) may be wanted but is normally only found as a small percentage of the total biomass.

The main routes of alkane oxidation and incorporation into lipid are given in Figure 4. Attack is usually at one of the terminal methyl groups though sub-terminal oxidation is not unknown. A brief account of alkane metabolism was given in Volume 1 (Chapter 24) and has also been given in greater detail elsewhere (Boulton and Ratledge, 1984). Several reviews dealing specifically with the various microbial lipids which can be derived from hydrocarbons have also been written (Ratledge, 1980, 1984; Finnerty, 1984).

Besides being useful for the production of specific fatty acids which are then recoverable as triacylglycerols or phospholipids, hydrocarbons can lead to the production of both ω and $\omega-1$ hydroxy fatty acids and dicarboxylic acids. These arise by microbial attack of the alkane chain occurring at both ends of the molecule. The hydroxy fatty acids which are formed with *Torulopsis bombicola* and related species (Spencer *et al.*, 1979) are recovered as esterified to a disaccharide, sophorose. These sophorosides can be produced in some quantity and may be thought of as a bio-logical Tween. Various applications of these compounds have been considered but none has seemingly warranted commercial production; the cheapness of producing similar molecules by chemical means, *e.g.* fatty acyl sucroses, has overridden all possibility of being able to produce such materials competitively by microbial means.

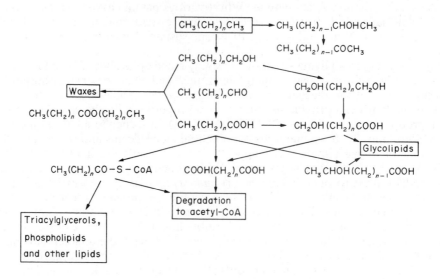

$$n = 8 \text{ to } 16, \text{ especially } 11 \text{ to } 14$$

Figure 4 Pathways of alkane oxidation and assimilation in microorganisms

ω and $\omega-1$ Hydroxy fatty acids, isolated from *T. bombicola* have been used to produce polyester material of high molecular weight (Howe and Jones, 1969):

$$-\text{OCHR}(\text{CH}_2)_n\text{COOCHR}(\text{CH}_2)_n\text{CO}-$$

where, if $R = CH_3$, $n = 15$; if $R = H$, $n = 16$. However the intrinsic properties of the polyester were apparently not substantially different from those of synthetic polyesters to warrant further consideration as a commercially viable product.

Dioic acid production from alkanes has been well documented (see Ratledge, 1980; Boulton and Ratledge, 1984). The route of biosynthesis is usually considered to be *via* the ω-hydroxy fatty acid as given in Figure 4 though Yi and Rehm (1982) have indicated the dihydroxy acid may be an intermediate. The process has been the subject of several patents (see Ratledge, 1980) which detail how the dioic acid is produced extracellularly from C_{10} to C_{16} alkanes by various *Candida* yeasts in yields of 70% (based on the original alkane) and up to 56 g l^{-1}. Shorter chain dicarboxylic acids, from glutaric acid to azelaic acid, can also be produced by *Candida* spp. though here yields appear to be fairly low at about 70 mg l^{-1}.

48.3.2 Fatty Acids, Soapstocks and Oils

Hydrocarbons have severe limitations when it comes to producing lipids intended for incorporation into human foodstuffs. Embargoes, arising either indirectly by public opinion or by government legislation, on the use of alkanes as fermentation feedstocks do not however apply to fatty acids which can metabolically produce the same alterations in lipid compositions as alkanes. Various patents have been taken out which have sought to demonstrate that desirable lipids can be produced by cultivating appropriate yeasts, usually of the genus *Candida*, *Torulopsis* and *Trichosporon*, though *Saccharomyces cerevisiae* and *C. utilis* have also been used, on a mixed carbon source which includes a fatty acid or material containing a fatty acid (Fuji Oil Co., 1979; CPC International Inc., 1979, 1982a, 1982b). The fatty acids or oils may be up to 20 g l^{-1} in the growth medium and, like alkanes, these then lead to high lipid contents: up to 65% and 67% have been reported. Equally importantly, high relative percentages of stearic acid (if stearic acid had been included in the medium) may also be achieved in the yeast oil. This, it is claimed, can then lead to a yeast lipid which has some of the characteristics of cocoa butter. A typical example of this type of work is given in Table 13.

Oils themselves can be presented to yeasts, again with the object of producing triacylglycerols with altered acyl substituents. The possibilities of upgrading the cheap vegetable oils, such as

Table 13 Fatty Acid and Triacylglycerol Composition of a Yeast Grown on Fatty Acids

		A	B	C
Lipid % in cells		51.5	34.8	52.8
Yield of fats and oil (per consumed material)		38.8%	23.7%	18.7%
Major fatty acids[a]	16:0	6.6	10.4	4.2
	18:0	35.6	56.6	40.2
	18:1	51.7	29.5	47.9
	18:2	3.9	1.7	4.8
Triacylglycerols	SSS[b]	3.9	21.4	6.8
	SUS	31.5	55.0	28.8
	SSU	5.2	6.2	9.1
	SUU	43.0	15.9	39.0
	USU	1.8	0.4	3.1
	UUU	14.6	1.1	13.2

Torulopsis sp. ATCC 20507 was grown in a 150 litre fermenter with 100 litres of medium for 28 h containing (A) methyl stearate (18 g) and methyl palmitate (2 g); (B) stearyl alcohol (17 g) and methyl palmitate (3 g); or (C) methyl stearate (20 g) (Fuji Oil Co., 1979).

[a] Traces of 14:0 and 18:3 also present. [b] S = saturated acid (16:0 or 18:0), U = unsaturated acid (18:1 or 18:2)

palm oil, to more expensive materials have been considered (Glatz *et al.*, 1984). Such a process relies upon the various lipases of the organism carrying out the initial hydrolysis; the ensuing fatty acids are then incorporated directly into new triacylglycerols in the same manner as occurs when alkanes are used as substrate. Lipases though may be isolated from microorganisms and, as immobilized preparations, then used to carry out transesterification reactions either between two different oils or between an oil and a fatty acid (Macrae, 1984). Such processes are under active consideration by several companies to produce high value-added commodities such as cocoa butter.

48.3.3 Other Substrates

A wide variety of substrates have from time to time been considered for the production of oils and fats (Ratledge, 1982). These include various starchy crops and wastes, molasses, whey, peat hydrolysates and ethanol. The only substrate which is abundant and is not too expensive but which cannot be used by oleaginous microorganisms is methanol. Although methanol-utilizing yeasts are well known, none has been described which accumulates more than 15% lipid, though even here the triacylglycerols account for 72% of the total lipid (Suzuki *et al.*, 1979). The fatty acids of methanol-utilizing microorganisms are not exceptional and were reported in Volume 1 (Chapter 24). The metabolic reason for methanol-utilizing yeasts, which include various *Candida* and *Rhodoturula* spp., being unable to accumulate lipid is unknown though it may be presumed that these yeasts lack the necessary ATP:citrate lyase with which to produce acetyl-CoA in the cytosol (see Section 48.2.1). Whether the genetic introduction of this enzyme into a methanol-utilizing yeast would be sufficient to promote lipid accumulation is, of course, unknown.

Cellulolytic oleaginous microorganisms have not yet been reported though there would appear to be no good reason why they might not be found. Cellulose degradation should not be incompatible with lipid accumulation. The use of hydrolysed cellulose, which would include a wide range of materials including peat, would of course be no different from using carbohydrate. However, concerted programmes for cellulose utilization are needed if large scale production of bulk materials, including lipids, is to be contemplated.

The utilization of pentoses arising from the hydrolysis of hemicelluloses is not detrimental for the formation of lipids (Evans and Ratledge, 1983) and thus these materials could be considered as a convenient cheap substrate. Again, a concerted programme for the total utilization of woody materials would be needed to provide sufficient hemicelluloses to develop fermentation industries using pentoses and hexoses. Obviously lipid production would be only one of many bulk processes which could be provided by such a programme.

The emphasis of all work with fermentation substrates focuses primarily on availability and cheapness. For the production of those lipids which do not command high prices, the need is for very inexpensive, or even negative cost, substrates which would be available throughout the year.

This then restricts the possibilities to a small number of choices. Such a substrate in several countries could be whey (Ratledge, 1977). Lactose-utilizing oleaginous yeasts are well known and efficient conversion of whey to oils has been achieved (Moon *et al.*, 1977; Iowa University, 1980). Such a process has the added advantage of being able to produce a spent fermentation broth of relatively low COD thus indicating that the fermentation removes not only the lactose but also most of the other organic solids within the whey. This process would obviously be worthy of consideration where whey disposal was a problem. The organism of choice for this work would appear to be *Candida curvata* (see Tables 8 and 10).

48.4 FUTURE PROSPECTS

As far as we are aware there is no commercial process in existence which has been specifically designed for the production of any microbial lipid. Microbial lipids are, however, produced as part of a defatting process of yeasts grown on hydrocarbons (Section 48.3.1) and although these lipids may have interesting properties, there would seem little likelihood of similar materials being deliberately produced elsewhere for reasons of cost.

The future prospects for microbial oils might be seen to lie in three possible areas: (1) as substitutes for high value plant oils; (2) as novel materials unavailable from other sources; (3) as a saleable end-product from waste processing.

With the first prospect, the higher the value of the oil to be replaced, the smaller tends to be its market. The highest priced oils have only small markets, perhaps of the order of only a few tonnes per annum. The production of small tonnages of microbial oils is of course much more expensive than production in bulk and it requires a perceptive understanding of the market to identify a potentially useful target. With respect to oils which may be produced on a much larger scale, it has to be said that few, if any, of the major oils and fats companies have much insight or commitment into fermentation technology. They therefore usually remain sceptical of the concept of a microbial oil being a satisfactory alternative to a plant oil. Moreover they are extremely reluctant to enter an entirely new field of high technology activity and begin production of microbial oils for themselves. Thus, if microbial oils were to be produced they would have to be produced by a company which was not in the oils and fats industry. Not unnaturally, fermentation companies are reluctant to produce materials for another highly competitive industry. They will certainly not undertake production of oils for marketing themselves as this is an area of commercial activity in which they have no expertise but which they know to be very traditional in outlook. There is thus something of a Catch 22 situation: those that can use the oils cannot and will not produce them; those that could produce them will not do so as they do not have the necessary market know-how.

Microbial lipids which are novel materials in their own right do exist but materials similar to them can often be produced chemically, and this usually means cheaper. However, the lipids of many microorganisms still remain to be critically evaluated; for example, it is surprising that little systematic work has been carried out with the lipids of many oleaginous moulds. The oils of moulds have been examined for the most part in only a cursory manner and yet it would not be unreasonable to suppose that some novel, and perhaps useful, lipid could be found. Closer liaison between oil chemists and microbiologists may help to identify potentially useful materials.

Single cell protein (SCP) has long been produced as a valuable adjunct to waste processing. The SCP, though, is only suitable for animal feed and as such commands little more than the price of soybean meal. It has been argued elsewhere (Ratledge, 1982) that SCO (single cell oil) could be a more valuable product than SCP as the oil; at worst, it would compete with the lowest priced oils which are already some 3 to 3.5 times higher in value than animal feed protein. Credit would also be obtained for the defatted microbial biomass which would, like its oilseed meal counterpart, be sold as animal fodder. The argument that SCP production is uneconomical no matter what substrate is used may be true when the substrate has to be purchased but when SCP is a saleable end-product arising from waste which would otherwise be costly to dispose of, then the argument no longer holds. Single cell oil has the advantage over SCP in that it can be used for technical purposes and therefore expensive toxicological trials are unnecessary. If the waste in question is such that the end-product could not be returned into the food chain, then clearly SCP is not a sensible product, but SCO could be.

There are few materials which a range of microorganisms can each produce and which can find a range of technological applications. With fermentation products such as ethanol or butanol one is restricted to using a small number of organisms. Moreover the substrates which need to be degraded in a particular waste-processing system may not be suitable for these particular organisms to convert into ethanol or butanol. Methane production is the resort of the unimaginative. With the range of oleaginous organisms which are now available, it would be surprising if one could not be found which could usefully attack almost any waste organic material. Microbial oils therefore could be an important addition to the biotechnologist's armoury.

48.5 SUMMARY

A survey of the current world scene in oils and fats is given as background to the material which follows. Oil-accumulating (oleaginous) organisms are defined and the process of oil accumulation is described. Oleaginous bacteria, algae, yeasts and moulds are listed together with their major products: triacylglycerols, fatty acids, sterols, waxes, poly-β-hydroxybutyrates, *etc.* The use of various feedstocks is described in relation to oil production. Hydrocarbons, especially alkanes, may be particularly useful for the production of certain lipids. The use of fatty acids and cheap vegetable oils to improve the quality of microbial oils is mentioned. The chapter concludes with a discussion of the future prospects for microbial oils which may lie in three possible areas: (i) as substitutes for existing plant oils; (ii) as novel materials not available from elsewhere; and (iii) as an alternative to SCP production in waste-processing systems.

48.6 REFERENCES

Abraham, M. J. and R. A. Srinivasan (1979). Utilization of whey for the production of microbial protein and lipid. *J. Food Sci. Technol.*, **16**, 11–15.
Airey, D. (1982). The role of the U.K. seed crushing and oil processing industry and the issues we face. *Chem. Ind.* (*London*), 453–459.
Anon. (1983). New source for vegetable oils: algae. *Chem. Eng.* [*Int. Ed.*], **90**, 9–10.
Bartholomew, D. M. (1980). Oilseed demand never saturated. *J. Am. Oil Chem. Soc.*, **57**, 340A–344A.
Bekhtereva, M. N. and M. B. Yakovleva (1980). The fatty acid composition of phospholipids in some fungal species belonging to the family *Choanephoraceae*. *Mikrobiologiya*, **49**, 827–829.
Berner, T., A. Dubinsky and S. Aaronson (1982). The total lipid content of microalgae and its use as a potential lipid source for industry. *Environ. Sci. Res.*, **23**, 439–442.
Boulton, C. A. and C. Ratledge (1981). Correlation of lipid accumulation in yeasts with possession and ATP:citrate lyase. *J. Gen. Microbiol.*, **127**, 169–176.
Boulton, C. A. and C. Ratledge (1984). *Adv. Enzyme Ferment. Technol.*, **9**, 11–17.
Brown, A. C., B. A. Knights and E. Conway (1969). Hydrocarbon content of *Botryococcus braunii*. *Phytochemistry*, **8**, 543–547.
Capella, P., C. Paoletti and R. Materassi (1977). Sulla composizione dei lipidi algali in confronto a quelle dell 'olio di oliva con particolare riguardo all' in saponificabile. *La Rivista delle Sostanze Grasse*, **54**, 271–273.
Chapman, B. (1982). The procurement of oils and fats. *Chem. Ind.* (*London*), 446–449.
Choi, S. Y., D. D. Y. Ryu and J. S. Rhee (1982). Production of microbial lipid: effects of growth rate and oxygen on lipid synthesis and fatty acid composition of *Rhodotorula gracilis*. *Biotechnol. Bioeng.*, **14**, 1165–1172.
CPC International Inc. (1979). Process for the microbiological production of oil. *Eur. Pat.* 0 005 277.
CPC International Inc. (1982a). Multistage process for the preparation of fats and oils. *Br. Pat.* 2 091 285 A.
CPC International Inc. (1982b). Preparation of fats and oils. *Br. Pat.*, 2 091 286 A.
Dawes, E. A. and P. J. Senior (1973). The role and regulation of energy reserve polymers in microorganisms. *Adv. Microb. Physiol.*, **10**, 136–266.
DeBell, R. M. and R. C. Jack (1975). Stereospecific analysis of major glycerolipids of *Phycomyces blakesleeanus* sporangiophores and mycelium. *J. Bacteriol.*, **124**, 220–224.
Dewitt, S., J. L. Ervin, D. Howe-Orchison, D. Dalietos and S. L. Neidleman (1982). Saturated and unsaturated wax esters produced by *Acinetobacter* sp. HO1-N grown on C_{16}–C_{20} *n*-alkanes. *J. Am. Oil Chem. Soc.*, **59**, 69–74.
Dubinsky, Z., T. Berner and S. Aaronson (1978). Potential of large-scale algal culture for biomass and lipid production in arid lands. *Biotechnol. Bioeng. Symp.*, **8**, 51–68.
Dulaney, E. L., E. O. Stapley and K. Simpf (1954). Studies on ergosterol production by yeasts. *Appl. Microbiol.*, **2**, 371–379.
El-Refai, A. H. and I. A. El-Kady (1968a). Sterol production of yeast strains *Z. Allg. Mikrobiol.*, **8**, 355–360.
El-Refai, A. H. and I. A. El-Kady (1968b). Sterol biosynthesis in *Saccharomyces fermentati*. *Z. Allg. Mikrobiol.*, **8**, 361–366.
Ervin, J. L., J. Geigert, S. L. Neidleman and J. Wadsworth (1984). In *Biotechnology for the Oils and Fats Industry*, ed. C. Ratledge, J. B. M. Rattray and P. S. S. Dawson, *Am. Oil Chem. Soc. Monograph*, no. 11, pp. 217–222.
Evans, C. T. and C. Ratledge (1983). A comparison of the oleaginous yeast, *Candida curvata*, grown on different carbon sources in continuous and batch culture. *Lipids* (in press).
Farag, R. S., A. M. Youssef, F. A. Khalil and R. A. Taha (1981). The lipids of various fungi grown on an artificial medium. *J. Am. Oil Chem. Soc.*, **58**, 765–768.
Finnerty, W. R. (1984). In *Biotechnology for the Oils and Fats Industry*, ed. C. Ratledge, J. B. M. Rattray and P. S. S. Dawson, *Am. Oil Chem. Soc. Monograph*, no. 11, pp. 199–215.
Franke, A. (1982). Commercial aspects of oils and fats. *Chem. Ind.* (*London*), 433–435.

Fuji Oil Co. Ltd (1979). Method for producing cacao butter substitute. *Br. Pat.* 1 555 000.

Gallagher, I. H. C. (1971). Occurrence of waxes in *Acinetobacter. J. Gen. Microbiol.*, **68**, 245–247.

Gill, C. O., M. J. Hall and C. Ratledge (1977). Lipid accumulation in an oleaginous yeast, *Candida* 107, growing on glucose in a single-stage continuous culture. *Appl. Environ. Microbiol.*, **33**, 231–239.

Glatz, B. A., E. G. Hammond, K. H. Hsu, L. Baehman, N. Bati, W. Bednarski, D. Brown and M. Floetenmeyer (1984). In *Biotechnology for the Oils and Fats Industry*, ed. C. Ratledge, J. B. M. Rattray and P. S. S. Dawson, *Amer. Oil Chem. Soc. Monograph*, no. 11, pp. 163–176.

Haley, J. E. and Jack, R. C. (1974). Stereospecific analysis of triacylglycerols and major phosphoglycerides of *Lipomyces lipoferus. Lipids*, **9**, 679–681.

Hall, M. J. and C. Ratledge (1977). Lipid accumulation in an oleaginous yeast (*Candida* 107) growing on glucose under various conditions in a one- and two-stage continuous culture. *Appl. Environ. Microbiol.*, **33**, 577–584.

Hancock, R. F. (1982). The changing role of developing countries in the world economy of fats and oils; effects on the EEC. *Chem. Ind. (London)*, 439–445.

Hillen, L. W., G. Pollard, L. V. Wake and N. White (1982). Hydrocracking of the oils of *Botryococcus braunii* to transport fuels. *Biotechnol. Bioeng.*, **24**, 193–205.

Howe, R. and D. F. Jones (1969). Conference report: Hydrocarbon fermentations. *Chem. Ind. (London)*, 1181.

Howells, E. R. (1982). Single-cell protein and related technology. *Chem. Ind. (London)*, 508–511.

Hughes, B. (1982). Procurement of oils and fats—a manufacturer's view. *Chem. Ind. (London)*, 431–433.

Iowa University (1980). Process for converting whey permeate to oil-containing yeast. *US Pat.* 4 235 933.

King, P. P. (1982). Biotechnology. An industrial view. *J. Chem. Technol. Biotechnol.*, **32**, 2–8.

Leysen, R. (1982). Soya oil consumption trends in the EEC. *Chem. Ind. (London)*, 428–431.

Lysons, A. (1982). Long term developments in the world oils and fats market. *Chem. Ind. (London)*, 449–453.

McLee, A. G., A. C. Kormendy and M. Wayman (1972). Isolation and characterization of *n*-butanol-utilizing microorganisms. *Can. J. Microbiol.*, **18**, 1191–1195.

Macrae, A. R. (1984). In *Biotechnology for the Oils and Fats Industry*, ed. C. Ratledge, J. B. M. Rattray and P. S. S. Dawson, *Am. Oil Chem. Soc. Monograph*, no. 11, 189–198.

Moon, N. J., E. G. Hammond and B. A. Glatz (1977). Conversion of cheese whey and whey permeate to oil and single cell protein. *J. Dairy Sci.*, **61**, 1537–1547.

Naim, N. and R. R. Saad (1978). Production of lipids and sterols by *Fusarium oxysporum* (Schlecht). Part 1. *Egypt. J. Bot.*, **21**, 197–205.

Ninet, L. and J. Renaut (1979). In *Microbial Technology*, 2nd edn., ed. H. J. Peppler and D. Perlman, vol. 1, pp. 529–544. Academic, New York.

Osman, H. G., M. A. Mostafa and A. H. El-Refai (1969). Production of lipids and sterols by *Aspergillus fumigatus*. Part I. *J. Chem. UAR*, **12**, 185–197.

Parks, L. W., R. J. Rodriguez and M. T. McCammon (1984). In *Biotechnology for the Oils and Fats Industry*, ed. C. Ratledge, J. B. M. Rattray and P. S. S. Dawson, *Am. Oil Chem. Soc. Monograph*, no. 11, pp. 177–187.

Perry, J. J. (1979). Microbial co-oxidations involving hydrocarbons. *Microbiol. Rev.*, **43**, 59–72.

Popova, N. I., M. N. Bekhtereva, L. A. Galanina, L. Boikova and F. A. Medvedev (1980). Age change in the fatty acid composition of the lipids of certain cultures of the family Entomophthoraceae. *Mikrobiologiya*, **49**, 588–594; *Microbiol. (Engl. Transl.)*, 511–516.

Princen, L. H. (1979). New crop developments for industrial oils. *J. Am. Oil Chem. Soc.*, **56**, 845–848.

Procter and Gamble (1979). Industrial Chemical Division: *Procter and Gamble Fatty Chemicals*.

Pryde, E. H. (1979). Fats and oils as chemical intermediates: present and future uses. *J. Am. Oil Chem. Soc.*, **56**, 849–854.

Pryde, E. H. and H. O. Doty (1981). World fats and oil situation. In *New Sources of Fats and Oils*, ed. E. H. Pryde, L. H. Princen and K. D. Mukherjee, *Am. Oil Chem. Soc. Monograph*, no. 9, pp. 3–14.

Ratledge, C. (1976). Lipids and fatty acids. In *Economic Microbiology*, ed. A. H. Rose, vol. 2, pp. 263–302. Academic, London.

Ratledge, C. (1977). Fermentation substrates. *Annu. Rep. Ferment. Proc.*, **1**, 49–71.

Ratledge, C. (1980). In *Hydrocarbons in Biotechnology*, ed. D. E. F. Harrison, I. J. Higgins and R. J. Watkinson, pp. 133–153. Heyden, London.

Ratledge, C. (1982). Microbial oils and fats: an assessment of their commercial potential. *Prog. Indust. Microbiol.*, **16**, 119–206.

Ratledge, C. (1984). Microbial conversions of alkanes and fatty acids. *J. Am. Oil Chem. Soc.*, **61**, 447–453.

Ratledge, C. and M. J. Hall (1979). Accumulation of lipid by *Rhodotorula glutinis* in continuous culture. *Biotechnol. Lett.*, **1**, 115–120.

Ratledge, C., C. A. Boulton and C. T. Evans (1984). In *Continuous Culture of Microorganisms*, ed. A. C. R. Dean, D. C. Ellwood and C. G. T. Evans, vol. 8, pp. 272–291. Ellis Horwood, Chichester.

Rattray, J. B. M., A. Schibeci and D. K. Kidby (1975). Lipids of yeasts. *Bacteriol. Rev.*, **39**, 197–231.

Russell, N. J. (1974). Lipid composition of *Micrococcus cryophilus. J. Gen. Microbiol.*, **80**, 217–225.

Shifrin, N. (1984). Oils from microalgae. In *Biotechnology for the Oils and Fats Industry*, ed. C. Ratledge, J. B. M. Rattray and P. S. S. Dawson, *Am. Oil Chem. Soc. Monograph*, no. 11, pp. 145–162.

Shifrin, N. S. and S. W. Chisholm (1980). In *Algal Biomass: Production and Uses*, ed. G. Shelef and C. J. Soeder, pp. 627–645. Elsevier/North Holland, Amsterdam.

Singh, J. and M. G. Sood (1973). Component fatty acids of *Aspergillus terreus* fat. *J. Am. Oil Chem. Soc.*, **50**, 485–486.

Spencer, J. F. T., D. M. Spencer and A. P. Tulloch (1979). In *Economic Microbiology*, ed. A. H. Rose, vol. 2, pp. 523–540. Academic, London.

Suzuki, O., Y. Jigami and S. Nakasato (1979). Changes in lipid composition of methanol-grown *Candida guilliermondii. Agric. Biol. Chem.*, **43**, 1343–1345.

Suzuki, O. (1980). Production of lipids. *Kagiken Nyusu Kagaku Kogyo*, **15**, 58–64.

Thorne, K. J. I., M. J. Thornley and A. M. Gauert (1973). Analysis of outer layers of cell envelopes of *Acinetobacter. J. Bacteriol.*, **116**, 410–417.

Vedeler, B. C. (1982). Fish oil production and marketing. *Chem. Ind. (London)*, 435–439.

Voigt, B., H. Seidel, H. Müller, D. Beck, M. Ringpfeil, M. Riedel, J. Bauch, H. Gentzsch and D. Bohlman (1979). Bio-lipidextract—ein neuer Rohstoff aus der Produktion von Fermosin (TM)-Futterhefe auf Basis Erdöldestillat. *Chem. Tech. (Leipzig)*, **31**, 409–411.

Wayman, M. and M. Whiteley (1979). Liquid fuels from carbonates by a microbial system. *Am. Chem. Soc. Symp. Ser.*, **90**, 120–156.

Wayman, M., A. D. Jenkins and A. C. Kormendy (1984). In *Biotechnology for the Oils and Fats Industry*, ed. C. Ratledge, J. B. M. Rattray and P. S. S. Dawson, *Am. Oil Chem. Soc. Monograph*, no. 11, pp. 129–143.

Weete, J. D. (1980). *Lipid Biochemistry of Fungi and Other Organisms*. Plenum Press, New York.

Wood, B. J. B. (1974). Fatty acids and saponifiable lipids. In *Algal Physiology and Biochemistry*. ed. W. D. P. Stewart, pp. 236–265. Blackwell Scientific Publications, Oxford.

Woodbine, M. (1959). Microbial fat: microorganisms as potential fat producers. *Prog. Ind. Microbiol.*, **1**, 179–245.

Yi, Z. H. and H. J. Rehm (1982). Metabolic formation of dodecanoic acid from *n*-dodecane by a mutant of *Candida tropicalis*. *Eur. J. Appl. Microbiol. Biotechnol.*, **14**, 254–258.

Yoon, S. H., J. W. Rhim, S. Y. Choi, D. D. Y. Ryu and J. S. Rhee (1982). Effect of carbon and nitrogen sources on lipid production of *Rhodotorula gracilis*. *J. Ferment. Technol.*, **60**, 243–246.

49

Microbial Polysaccharides

A. MARGARITIS
The University of Western Ontario, London, Ontario, Canada
and
G. W. PACE
Biotechnology Australia Pty Ltd, East Roseville, New South Wales, Australia

49.1 INTRODUCTION

Exopolysaccharides produced by a wide variety of microorganisms are water soluble gums which have novel and unique physical properties. Because of their wide diversity in structure and physical properties microbial exopolysaccharides have found a wide range of applications in the food, pharmaceutical and other industries. Some of these applications include their use as emulsifiers, stabilizers, binders, gelling agents, coagulants, lubricants, film formers, thickening and suspending agents. These biopolymers are rapidly emerging as a new and industrially important source of polymeric materials which are gradually becoming economically competitive with natural gums produced from marine algae and other plants. The potential use of genetically modified microorganisms under controlled fermentation conditions may result in the production of new exopolysaccharides having novel superior properties which will open up new areas of industrial applications and thus increase their demand.

In this chapter we present an updated overview of the status of exopolysaccharide production by microbial systems. This chapter includes the uses and properties of these biopolymers, microbial kinetics and biochemical aspects of exopolysaccharide fermentation, process development, product recovery and economic feasibility aspects of exopolysaccharide production.

49.2 SOURCES AND APPLICATIONS OF EXOPOLYSACCHARIDES

Microbial polysaccharides, which serve different functions in a microbial cell, may be distinguished into three main types: (a) intracellular polysaccharides which may provide mechanisms for storing carbon or energy for the cell; (b) structural polysaccharides which are components of cell structures such as lipopolysaccharides and teichoic acids present as integral components of cell walls; and (c) extra cellular polysaccharides referred to in this chapter as exopolysaccharides. Depending on the microbial system, some exopolysaccharides form capsules around the cell thus becoming part of the cell wall, while others form slimes which accumulate outside the cell wall and have the ability to diffuse away into the liquid phase during the course of fermentation. As a result of exopolyaccharide production the viscosity and rheology of the fermentation broth may undergo profound changes, starting out as a low-viscosity Newtonian fluid and ending up as a highly viscous non-Newtonian fluid. Those microorganisms that produce large amounts of polysaccharide slimes have the greatest potential for commercialization since these exopolysaccharides may be recovered easily from the fermentation broth.

49.2.1 Sources of Microbial Exopolysaccharides

Production of exopolysaccharides is found in many species of Gram-positive and Gram-negative bacteria, some algae and many fungi (Margaritis and Zajic, 1978; Lawson, 1977; Cottrell, 1980). Table 1 summarizes the production of exopolysaccharides by various microorganisms using a wide variety of substrates, such as glucose, fructose, sucrose, lactose, hydrolyzed starch, methanol and different hydrocarbons (Slodki and Cadmus, 1978). Exopolysaccharide production from C_9, C_{10} and C_{12} *n*-alkanes as principal carbon sources and mineral salts media has been reported for a large number of mycobacterial isolates by Grechushkina and Rozanova (1971). These authors found that the best exopolysaccharide producer was *Mycobacterium lacticolun* grown on *n*-dodecane. Yamada and Furukawa (1973) reported the production of exopolysaccharides *Corynebacterium viscosus* in a medium containing a C_{13}–C_{16} *n*-alkane mixture under aerobic conditions.

49.2.2 Applications of Microbial Exopolysaccharides

Microbial exopolysaccharides have found a wide range of applications in the chemical, food and pharmaceutical industries. In view of their unique and novel chemical and physical proper-

Table 1 Production of Exopolysaccharides by Various Microorganisms

Product	Substrate	Microorganism	Yield	Ref.
Alginate	Sucrose	*Azotobacter vinelandii* NCIB 9068	5% batch	Bucke *et al.*, 1975
Polymer of: D-glucose, D-mannose, D-ribose, 6-deoxy-L-mannose	2% glucose	*Xanthomonas fuscans*	25% batch, 45% continuous, 35%	Deavin *et al.*, 1977, Deavin *et al.*, 1977, Konicek *et al.*, 1977
Curdlan succinoglucan	Glucose	*Alcaligenes faecalis* var. *myxogenes* 10C3		Amemura *et al.*, 1977
Curdlan types	5% glucose	*Alcaligenes faecalis* var. *myxogenes* 10C3 LFO 13140	50%	Harada, 1977
Erwina gum (Zanflo)	Lactose, hydrolyzed starch	*Erwina tahitica*	n.a.	Kang *et al.*, 1977
Polymer of: D-glucose 81.9%, L-rhamnose 14%, L-glucose 0.7%, D-mannose 1.9%, D-galactose 1.5%	1% w/v methanol	*Methylocytis parvus* OBBP	62% fed	Hou *et al.*, 1978
Polysaccharide	4.55% methanol	*Methylomonas mucose* NRRL B-5696	45.2%	Tam and Finn 1977
Polysaccharide	0.3% v/v methanol	*Methylomonas methanolica* M13V mutant		Haggstrom, 1977
Polymer of: galactose, glucose, mannose, glucuronic acid	1% methanol	*Pseudomonas* sp. *viscogena* TS-1004	21%	Misaki *et al.*, 1979
Polymer of: glucose, galactose, rhamnose, mannose, acetatepyruvate	2% glucose	*Pseudomonas* NCIB 11264	n.a.	Williams and Wimpenny, 1978
PS-60 gum: glucose 41%, rhamnose 30%, uronic acid 29%		*Pseudomonas* sp.	n.a.	Cottrell, 1980
Levan	2% sucrose	*Zymomonas mobilis* NCIB 8938	< 2%	Dawes *et al.*, 1966; Ribbons *et al.* 1962
Scleroglucan	3% glucose	*Sclerotium rolfsii* ATCC 15206	1.5–2.2% w/v	Griffith and Compere, 1978
Scleroglucan	5% w/v starch	*Sclerotium delphinii*, *S. glucanicum*	1.4% w/v, 1.8% w/v	Compere and Griffith, 1978
PS-7 gum: glucose 73%, rhamnose 16%, glucuronic acid 11%	3% glucose	*Beijerinckia indica* var. *mysogenes*	n.a.	Cottrell, 1980; Kang and McNeely, 1977
Pullulan	5% sucrose	*Aureobasidium pullulans* S-1	50–60%	Ono, Yasada and Ueda, 1977
Phosphomannan	Hydrolyzed whey, 4.4% sugars	*Hansenula holstii* NRRL y-2448	20.0%	Stauffer and Leeder, 1978
Levan	6% lactose	*Alcaligenes viscosus* NRRL B-182	2.5%	Stauffer and Leeder, 1978
Xanthan gum	6% lactose	*Xanthomonas campestris* NRRL B-1459	38.3%	Stauffer and Leeder, 1978
Galactoglucan	6% lactose	*Zoogloa ramigera* NRRL B-3669	55.6%	Stauffer and Leeder, 1978
Polymer of: galactose, glucose, glucuronic acid	6% lactose	*Arthrobacter viscosus* MRRL B-1973	0.7%	Stauffer and Leeder, 1978
O-acetylated acid polysaccharide: glucose, mannose, galactose	10% v/v C_{12}–C_{17} n-parrafin mixture	*Corynebacterium* sp. *Brevibacterium* sp.	n.a.	Kanamaru and Yamatodani, 1969
Gellan gum (S-60 polysaccharide)	Carbohydrate	*Pseudomonas elodea* ATCC 31461		Kang and Veeder, 1982; Kang and Veeder, 1983

ties, microbial exopolysaccharides are being used as gelling agents, emulsifiers, stabilizers, binders, coagulants, lubricants, film formers, thickening and suspending agents. Typical applications of an exopolysaccharide, xantham gum, are summarized in Table 2 (Baird *et al.*, 1983).

Table 3 is based on published literature data and shows the estimated 1975 US consumption and approximate price for different types of commercial polysaccharides used in food and other industrial applications (Cottrell and Kang, 1978; te Bokkel, 1983). These data show that the majority of industrial polysaccharides used in the USA are starch based, comprising about 41% of the total market value, while chemically modified cellulose polysaccharides account for about 28% of the total market. Plant-derived polysaccharides, such as gum arabic, guar gum, constitute about 6.6% of the total market, while algae-derived alginates account for about 6.7% of the total market value. Xanthan gum is a microbial exopolysaccharide derived from *Xanthomonas campestris* fermentation and accounts for 4% of the total market value. Although the xanthan gum production has increased over the last few years, the relative market value contribution considering all other industrial polysaccharides did not vary significantly from the estimated data shown in Table 3. These data also show that on the same weight basis microbially-derived xanthan gum is less competitive than cornstarch and cellulose-derived polysaccharides. However, the competitive advantage of microbial exopolysaccharides will improve with the development of more efficient fermentation processes coupled with the discovery of new microorganisms which produce exopolysaccharides in high yields and having unique physical and chemical properties. In the case of alginate, considerable commercial interest is underway for its production using *Azotobacter vinelandii* strains, to gradually substitute alginates derived from algae. Other interesting developments are the production of exopolysaccharides pullulan from *Aureobasidium pullulans*, scleroglucan using various species of *Sclerotium*, and curdlan from *Alcaligenes faecalis*. A new development of considerable interest and commercial potential is the production of Gellan exopolysaccharide by *Pseudomonas elodea* (Baird *et al.*, 1983). This new polysaccharide (Gelrite) has unique characteristics and excellent gelling properties and has the potential of substituting the seaweed-derived agar for its use in bacteriological media (Shungu *et al.*, 1983).

Table 2 Major Food and Industrial Applications of Xanthan Gum[a]

Food Applications
Dressings (high oil, low oil, no oil);
Relishes and sauces; Syrups and toppings;
Starch based products (canned desserts, sauces, fillings, retort pouches);
Dry mix products (desserts, gravies, beverages, sauces, dressings);
Farinaceous foods (cakes);
Beverages; Dairy products (ice cream, shakes, processed cheese spread, cottage cheese); Confectionery.
Industrial Chemical Applications
Flowable pesticides; Liquid feed supplements;
Cleaners, abrasives, and polishes;
Metal working; Ceramics, Foundry coatings; Texturized coatings;
Slurry explosives; Dye and pigment suspensions.
Oil Field Applications
Drilling fluids (muds); Workover and completion fluids; Enhanced oil recovery (polymer flooding).

---

[a] Adapted from Baird *et al.*, 1983.

Xanthan gum is the only microbial exopolysaccharide which constitutes an important component of the total world polysaccharide market. The estimated 1980 world production of xanthan gum was about 8000 tonnes (Pace and Righelato, 1980). Xanthan gum, because of its excellent rheological characteristics over a wide range of temperatures and pH, has a great potential for its use to recover an estimated 43.3 billion barrels of oil in the USA by tertiary oil recovery methods. Each barrel of oil produced requires about 0.5 kg of a high viscosity polymer such as xanthan gum (Compere and Griffith, 1978). Therefore, in order to produce 100 000 barrels of oil per day by the tertiary oil recovery technique, about 50 000 kg of xanthan gum per day will be required, which represents about 2.1 times the 1980 daily world production. The potential of microbial exopolysaccharides for capturing a substantial portion of the world polysaccharide market is excellent, as more research and development work will undoubtedly result in lower production costs and wider versatility in industrial applications.

49.3 STRUCTURE AND COMPOSITION OF EXOPOLYSACCHARIDES

Exopolysaccharides produced by a wide variety of microorganisms are important biopolymers which have unique physical and chemical properties. Their diversity in structure and composition makes these biopolymers very versatile in a wide range of industrial applications. Of particular

Table 3 Estimated 1975 Consumption and Price of Industrial Polysaccharides in the US[a]

Industrial polysaccharide	Food usage (tonnes)	Industrial usage (tonnes)	Total usage (tonnes)	Price ($ kg^{-1})	Approx. total value (million $)	% of total value
Cornstarch	203 000	1 013 000	1 216 000	0.20	243.2	40.77
Carboxymethyl cellulose	6 100	40 000	46 100	2.00	92.2	15.46
Methylcellulose	820	21 500	22 320	3.30	73.7	12.36
Alginate	3700	3600	7300	5.50	40.2	6.74
Pectin	4900	0	4900	4.85	23.8	3.99
Xanthan	950	2500	3450	6.90	23.8	3.99
Gum arabic	9340	2850	12 190	1.65	20.1	3.36
Guar gum	6070	14 180	20 250	0.96	19.4	3.20
Carrageenan	3700	90	3790	4.40	16.7	2.80
Tragacanth	526	81	607	26.50	16.1	2.77
Locust bean gum	3650	1620	5270	2.00	10.5	1.76
Karaya	410	2900	3310	2.10	6.9	1.15
Ghatti	4050	410	4460	1.15	5.1	0.85
Agar	125	165	290	16.6	4.8	0.80
Total	247 341	1 102 896	1 350 237		596.5	100.00

[a] Adapted from Cottrell and Kang, 1978; te Bokkel, 1983.

importance is the rheological characteristics of exopolysaccharides, which are influenced by their structure and composition. The structure and composition of microbial exopolysaccharides depends on many different factors, such as microbial species, nature of substrate and other fermentation conditions. In this section we describe the structure and composition of a few selected microbial exopolysaccharides as an example to illustrate the nature and diversity of these biopolymers.

49.3.1 A Brief Classification of Polysaccharides

Microbial polysaccharides result from the condensation of monosaccharide units by eliminating water between the C_1 hydroxyl group of one unit and an available hydroxyl group of another monosaccharide. Polysaccharides may be classified into two types, namely, homopolysaccharides, which contain only one type of sugar moiety, and heteropolysaccharides, which contain two or more sugar moieties. Heteropolysaccharides are usually produced by microorganisms from any utilizable carbon source and complex enzyme systems are usually involved, whereas homopolysaccharide synthesis involves a single or simple enzyme system. Examples of homopolysaccharides include pullulan and dextrans; xanthan gum is a typical example of a heteropolysaccharide. Depending on the structural arrangement of the monomeric units, polysaccharides are either linear or branched. Another important distinction of polysaccharides is based on their charge properties and they may be classified as naturally anionic, neutral or cationic. Some anionic microbial exopolysaccharides include the following: xanthan gum, phosphomannan and alginate. Neutral exopolysaccharide examples include: levan, scleroglucan, pullulan, dextran and curdlan. Some polysaccharides have anionic properties and they contain acidic groups, such as carboxyl, phosphate or sulfate. Other polysaccharides in their natural state may have some free amino groups and as a result these biopolymers possess cationic properties.

49.3.2 Xanthan Gum

Xanthan gum produced by *Xanthomonas campestris* is a branched anionic heteropolysaccharide. As shown in Figure 1, the xanthan gum biopolymer has a repeat unit of five sugars and pending on the microbial source and fermentation conditions the molecular weight may be greater than 10^6. Xanthan contains D-glucose, D-mannose, D-glucuronate and variable amounts of acetate and pyruvate (McNeely and Kang, 1973; Lawson and Sutherland, 1977). The macromolecule has a cellulose backbone (β-(1→4) linked glucose) with a trisaccharide sidechain composed of two mannose and one glucuronate residues (Figure 1). The acetate and pyruvate substituents are attached to the mannose residues (Figure 1). In addition to the different sugar units acetyl and pyruvate acetal groups, xanthan gum also contains monovalent cations as shown in Figure 1. There is some debate as a result of X-ray diffraction studies where xanthan gum exists as a single or double helix (Moorhouse *et al.*, 1977). Xanthan gum in aqueous solutions has many interesting properties resulting in the use of this biopolymer in a wide variety of industrial applications (see Table 2). The viscosity of aqueous solutions of xanthan gum is almost entirely independent of temperature over a range 10–70 °C, and is also fairly constant for pH between 6 and 9 (McNeely and Kang, 1973). Xanthan gum aqueous solutions have very good compatibility with high concentrations of salts and exposure to temperatures up to 80 °C for a long period of time has little effect on physical properties.

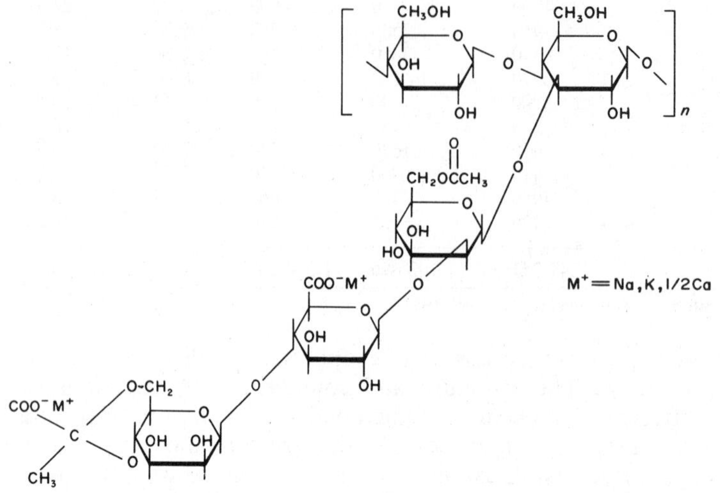

Figure 1 Polymeric structure of xanthan gum

49.3.3 Dextran

Dextrans are primarily produced commercially by two microorganisms *Leuconostoc mesenteroides* and *Leuconostoc dextranicum*. Depending on the microbial species employed, the type of substrate and fermentation conditions, a wide variety of dextrans may be produced with structures which are slightly or highly branched. Dextran is also synthesized from sucrose using cell-free culture filtrates of *Leuconostoc mesenteroides* NRRL B-512(F) strain. The culture filtrate contains the extracellular enzyme dextransucrase and the resulting dextran biopolymer has a molecular weight between $30–9 \times 10^6$. Further fractionation by acid hydrolysis yields a series of lower molecular weight dextrans which are used in different pharmaceutical and other industrial applications. Dextran is a branched neutral homopolysaccharide which is composed exclusively of α-D-glucopyranosidic residues, and 95% of these residues are linked through carbons 1 or 1 and 6. The remaining 5% of the α-D-glucopyranosidic residues in the macromolecule carry branches of about one or two glucose units attached at carbon 3 positions. Figure 2 shows the structure of dextran. One of the most successful applications of dextrans is in the manufacture of a series of important molecular sieves. The degree of cross-linking of dextran with different epoxy compounds and sodium hydroxide, determines the pore size and water regain value of these molecular sieves and thus their molecular exclusion characteristics.

49.3.4 Alginate

Although most of the commercial alginate produced today is derived primarily from the sea kelp *Macrocystis pyrifera*, microbially derived alginates are under development and their future is very promising (Lawson, 1977; Pace and Righelato, 1980). The wide variety of products obtained from *Azotobacter vinelandii* have been found to possess physical properties similar to the alginates derived from marine algae. The biopolymers derived from *Azotobacter vinelandii* have a wide range of molecular weights and it has been postulated that the extracellular enzyme alginate lyase may play an important role in the molecular weight of alginates. Alginate is a general term used to describe the salts of alginic acid, the most notable of which is sodium alginate. Alginic acid

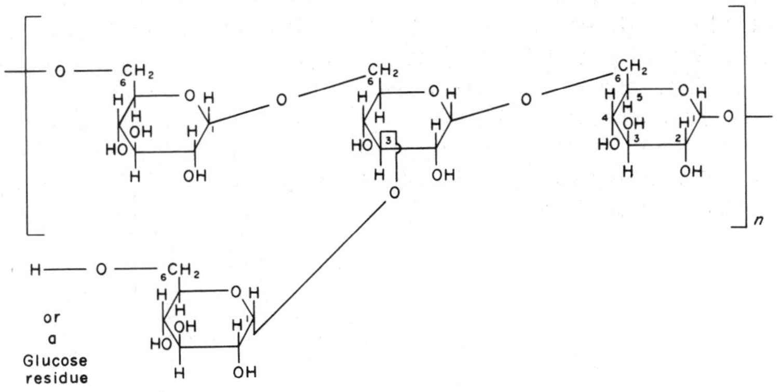

Figure 2 Structure of dextran exopolysaccharide

is a weak organic acid which readily forms salts with different bases. Alginic acids are linear polysaccharides composed of varying proportions of β-(1→4) linked D-mannopyranosyluronic acid and α-(1→4) linked L-guluronic residues in block and alternating sequences in the linear chain (Linker and Evans, 1976). The presence of L-guluronic acid in the alginic acid macromolecule is important because increasing its content improves the gelling characteristics of alginate in the presence of calcium ions. Sodium alginate is used widely in research as a gelling agent to immobilise a wide variety of cells, such as microbial cells, plant and mammalian cells (Margaritis and Merchant, 1984). Alginates have ion-exchange properties similar to ion-exchange resins. The relative affinities of divalent metal ions depend on the relative amounts of D-mannuronic and L-guluronic acid units which are present in the macromolecule.

49.3.5 Pullulan

Pullulan is a neutral linear homopolysaccharide which is composed of glucose units polymerized into repeating maltotriose units (Figure 3). The repeating maltotriose units are linked by α-(1→6) glucosidic bonds, and within each maltotriose unit the glucopyranose units are linked by α-(1→4) glucosidic bonds (Catley and Whelan, 1971). Because of its structure pullulan has been used as a source of maltotriose, and also as a substrate to measure the activity of the enzyme α-(1→6)-glucanohydrolase (pullulanase), an important enzyme used to depolymerize amylopectin. Depending on the substrate used and fermentation conditions, the yield and molecular weight of pullulan may vary widely, with reported values of 5×10^4–4×10^6. Kato and Shiosaka (1975) report that for different substrates, such as starch syrup, glucose and sucrose, initial pH values of 5–6 resulted in the production of high molecular weight pullulan, 1.5–4×10^6, in seven days. For shorter fermentation times and initial pH values of 7–8, low molecular weight pullulan was produced in the range of 5–10×10^4. The unique structure and physical properties of pullulan form the basis for a wide range of industrial applications such as adhesives, fibers, molded articles, coatings, films and esterified derivative applications (Yuen, 1974). Starch hydrolysates, at a concentraton of 10%, were used to produce pullulan with high yields of 75%. One notable physical property of pullulan is its low oxygen permeability when cross-linked with other materials to form thin polymeric film, which is used as food packaging material thus minimizing food oxidation. Pullulan is produced commercially in Japan by Hayashibara Corp. using *Azotobacter pullulans*.

49.3.6 Scleroglucan

Scleroglucan is a highly branched neutral homopolysaccharide composed of glucose units. This exopolysaccharide is produced from different carbohydrate substrates using *Sclerotium glucanicum* NRRL 3006 and other related *Sclerotium* fungal species (Bluhm *et al.*, 1982). The scleroglucan macromolecule is highly branched and has a backbone chain of about 90 β-(1→3) linked glucopyranosyl residues and every third or fourth backbone residue is linked to a single D-glucopyranosyl side group by β-(1→6) glucosidic bond. For details regarding the structure of scleroglucan see Figure 8, in Sutherland (1983).

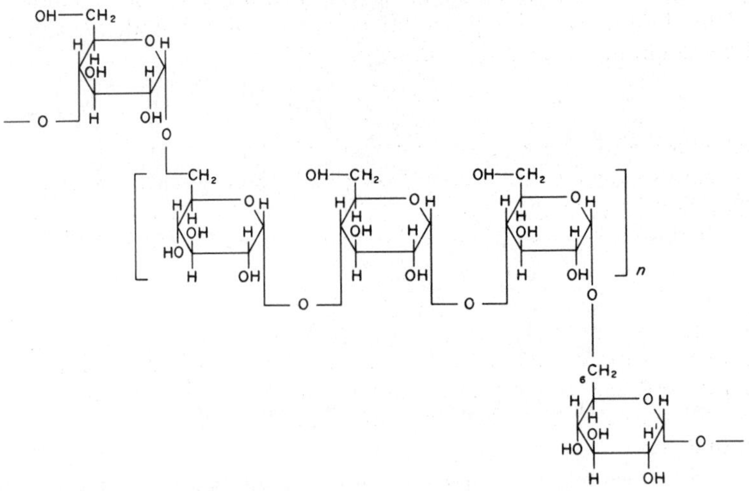

Figure 3 Structure of pullulan exopolysaccharide

49.4 RHEOLOGICAL CHARACTERISTICS OF EXOPOLYSACCHARIDES

The rheological properties of aqueous solutions of exopolysaccharides are of paramount importance to the mixing, oxygen mass transfer and heat transfer characteristics of polysaccharide fermentations (see Section 49.7). In this section we present some basic information on the rheological classification of fluids, and some recent experimental rheological data on different exopolysaccharide aqueous solutions.

49.4.1 Rheological Classification of Fluids

The generalized relationship between shear stress and shear rate for a fluid is given by equation (1):

$$\tau = \tau_0 + K(\dot{\gamma})^n \tag{1}$$

where τ is the shear stress applied on the fluid (FL^{-2}), τ_0 is the yield stress (FL^{-2}), K is the consistency index $(FL^{-2}T^n)$, n is the dimensionless flow index, and $\dot{\gamma}$ is the fluid velocity gradient, also referred to as the shear rate, (T^{-1}). Depending on the numerical values of τ_0 and K the following rheological classification of fluids is possible:

(a) Newtonian fluids for which $\tau_0 = 0$ and $n = 1$ in which case equation (1) is reduced to equation (2):

$$\tau = K\dot{\gamma} \tag{2}$$

Newtonian fluids have a linear relationship between shear stress τ and shear rate $\dot{\gamma}$ and the proportionality constant is referred to as viscosity μ $(K = \mu)$, which is uniquely defined and is independent of shear rate. Therefore, a plot of τ *versus* $\dot{\gamma}$ yields a straight line going through the origin with slope μ $(FL^{-2}T)$.

Non-Newtonian fluids have a more complex relationship between τ and $\dot{\gamma}$ and $n \neq 1$. Depending on the mathematical model one can use to fit experimental rheological data, non-Newtonian fluids may be further classified into pseudoplastic, dilatant, Bingham plastic and Casson.

(b) For pseudoplastic fluids $\tau_0 = 0$ and $n < 1$ and equation (3) applies.

$$\tau = K(\dot{\gamma})^n \tag{3}$$

(c) For dilatant fluids $\tau_0 = 0$ and $n > 1$ and equation (3) also applies. For both pseudoplastic and dilatant fluids, the apparent viscosity $\mu_a (FL^{-2}T)$ is not constant but depends on the fluid shear rate $\dot{\gamma}$, the consistency index K and flow index n according to equation (4).

$$\mu_a = K(\dot{\gamma})^{n-1} \tag{4}$$

In the case of power law fluids (pseudoplastic and dilatant) obeying equation (3), one can plot experimental data of $\log \tau$ versus $\log \dot{\gamma}$, and obtain the consistency index K, and flow index n from the slope of the straight line according to equation (5).

$$\log \tau = \log K + n \log \dot{\gamma} \tag{5}$$

The consistency index K is evaluated at a shear rate $\dot{\gamma} = 1\ s^{-1}$.

(d) Bingham plastic fluids are characterized by their finite shear yield stress, $\tau_o > 0$, which is needed to initiate flow. For Bingham plastic fluids $n = 1$, which means that they behave like Newtonian fluids after the yield stress τ_o has been applied. Equation (6) applies to Bingham plastic fluids.

$$\tau = \tau_0 + K\dot{\gamma} \tag{6}$$

Therefore, a plot of τ versus $\dot{\gamma}$ gives a straight line with slope K and intercept τ_0. The consistency index K is also referred to as the pseudoplasticity term.

(e) A number of non-Newtonian fluids have been described by the Casson mathematical model given by equation (7)

$$\tau^{0.5} = \tau_0^{0.5} + K_c (\dot{\gamma})^{0.5} \tag{7}$$

where K_c is a constant.

Equations (3) to (7) are the mathematical models most commonly used to describe the rheological behaviour of non-Newtonian fluids. Depending on the required accuracy of fitting given experimental shear stress *versus* shear rate data, the proper rheological classification of exopolysaccharide solutions may be at times somewhat difficult. This is particularly true when one tries to fit experimental rheological data of a given exopolysaccharide solution system over a wide range of shear rates. It has been reported (Chang and Ollis, 1982; Margaritis and te Bokkel, 1984) that for a given exopolysaccharide system one rheological model may apply at the low and medium shear rate range, while a different model may apply for the same system at the high shear rate region. Chang and Ollis (1982) reviewed the different mathematical models used in the literature to correlate rheological experimental data of exopolysaccharide solutions and presented a generalized power law model for the low and medium shear rate regions (equation 8)

$$\mu_a = \mu_0 [1 + (\dot{\gamma}/\dot{\gamma}_0)]^{n-1} \tag{8}$$

where μ_a is the apparent viscosity of the non-Newtonian exopolysaccharide solution, μ_0 is the low shear limit viscosity, and $\dot{\gamma}_0$ is the characteristic shear rate dividing the low shear rate region with the medium shear rate power law region. It was found that depending on the type of exopolysaccharide system used, the values of μ_0 and $\dot{\gamma}_0$ may vary appreciably.

49.4.2 Rheological Behaviour of Some Exopolysaccharide Aqueous Solutions

Representative rheological experimental data for a wide variety of microbial exopolysaccharide systems have been reported by several investigators (Baird *et al.*, 1983; Jeanes, 1974; Lawson, 1977; Margaritis and Zajic, 1978; Pace and Righelato, 1980; te Bokkel, 1983; Whitcomb *et al.*, 1977).

In this section we present some recent experimental rheological data on xanthan gum and dextran aqueous solutions of different concentrations (Margaritis and te Bokkel, 1984). Figure 4 shows a logarithmic plot of shear stress *versus* shear rate for aqueous xanthan gum solutions having four different concentrations, ranging from 0.375–4.762% w/w. The molecular weight of the xanthan gum used was about $1.4–3.6 \times 10^6$ as reported previously by Dintzis *et al.* (1970). As shown in Figure 4, the experimental data fit very well according to the power law equation (5). A regression analysis of the data was performed and the correlation coefficient was found to be higher than 0.999. The slope of the straight line n, and intercept K were found by regression analysis of the data shown in Figure 4. It was found that both the flow index n and consistency index K varied with the concentration of xanthan gum over the shear rate ranges studied. It must be emphasized that for the very low shear and very high shear rate regions more sophisticated mathematical rheological models may be applicable, *i.e.* other than the simple power law model given by equation (5). Figure 5 shows a rheogram for a low molecular weight dextran (MW = 176 800) at four different concentrations varying from 13.04–44.44% w/w. The experimental data fit the power law equation (5) very well. Figure 6 shows the rheological data for a high molecular

weight dextran aqueous solution (MW = 5–40 × 10⁶) at four different concentrations from
2.44–16.67% w/w. For the shear rate ranges studied an excellent power law fit of the experimental data was found. In order to find the effect of molecular weight of dextran on the rheological characteristics of aqueous solutions, the flow index, n, and consistency index, K, are compared for low and high molecular weight solutions (MW = 176 800 and 5–40 × 10⁶). A systematic comparison of the two systems with corresponding shear rate ranges is shown in Table 4. These data show that the flow index n of the high molecular weight dextran solutions was much less than the flow index for the low molecular weight solutions. In fact, the high molecular weight dextran solutions showed a much higher degree of pseudoplasticity than the lower molecular weight form, giving values of n close to 1.000 and behaving almost like Newtonian systems. As shown in Table 4, the consistency index K of the high molecular weight dextran solutions was found to be higher than the K values of the low molecular weight solutions for the same dextran concentrations. Figure 4 also shows the high degree of pseudoplasticity or shear thinning of xanthan gum solutions, since the apparent viscosity μ_a decreases drastically as the shear rate $\dot{\gamma}$ reaches high values.

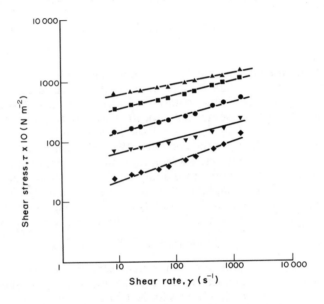

Figure 4 Rheogram of xanthan gum aqueous solutions (MW = 1.4–3.6 × 10⁶). ◆, 0.375% w/w; ▼, 0.744% w/w; ●, 1.478% w/w; ■, 2.913% w/w; ▲, 4.762% w/w. From Margaritis and te Bokkel, 1984

By comparison, as Figure 5 shows, the low molecular weight dextran solutions show almost no pseudoplasticity at all since there is almost no change as shear rate $\dot{\gamma}$ changes from low to high values. Figure 6 also shows the pseudoplastic behaviour of the high molecular weight dextran solutions; however, their degree of pseudoplasticity is less than that found in the xanthan gum solutions (Figure 4). Systematic comparisons of the flow index, n, and consistency index, K, data for the xanthan gum and high molecular weight dextran solutions are shown in Figures 7 and 8 respectively. Figure 7 is a semi-logarithmic plot of n *versus* the concentration of polysaccharide aqueous solutions of xanthan gum and high molecular weight dextran. For both systems n decreases with increasing biopolymer concentration. These data also show clearly that for the same biopolymer concentration, the values n for the xanthan gum solutions are much lower than those for the dextran solutions, indicating a much lower degree of pseudoplasticity for the latter. The data for the xantham gum solutions shown in Figure 7 agree very well with earlier data on xanthan gum presented by Whitcomb *et al.* (1977). Figure 8 is a semi-logarithmic plot of K *versus* biopolymer concentration of aqueous solutions of xanthan gum and high molecular weight dextran. These data show that the consistency index increases with biopolymer concentration and then levels off with further concentration increase. For the same biopolymer, concentration K is higher for the xanthan gum than the dextran aqueous solutions. The higher K values for xanthan gum compared with those of dextran solutions may be explained on the basis of their different molecular structures as shown in Figures 1 and 2 respectively. Xanthan gum is a highly branched anionic heteropolysaccharide which tends to form viscous solutions, whereas dextran is not highly branched and it is a neutral homopolysaccharide.

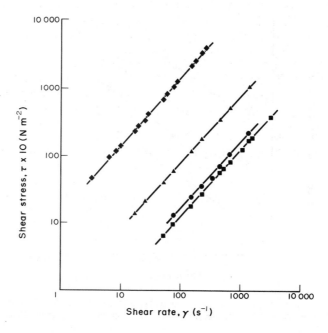

Figure 5 Rheogram of low molecular weight dextran solutions (MW = 176 800. ■, 13.043% w/w; ●, 16.667% w/w; ▲, 28.571% w/w; ◆, 44.444% w/w. From Margaritis and te Bokkel, 1984

Table 4 Rheological Characteristics of Low Molecular Weight (MW = 176 800) and High Molecular Weight (MW = 5–40 × 10⁶) Aqueous Solutions of Dextran[a]

Dextran conc. (% w/w)	Shear rate range, $\dot{\gamma}$ (s^{-1})		Consistency index K (dyne s^n cm^{-2})		Flow index n (dimensionless)	
	low MW	high MW[b]	low MW	high MW[b]	low MW	high MW[b]
2.44	115–5000	8–2000	0.034	1.568	0.964	0.697
4.76	100–5000	8–2000	0.041	7.349	0.984	0.668
9.09	25–5000	8–1000	0.187	25.960	0.841	0.633
13.04	20–5000	n.d.	0.137	n.d.	0.982	n.d.
16.67	8–2000	8–200	0.165	110.300	0.988	0.635
28.57	3–2000	n.d.	0.816	n.d.	0.989	n.d.
37.50	3–800	n.d.	3.876	n.d.	1.053	n.d.
47.37	2–450	n.d.	23.84	n.d.	0.997	n.d.
50.00	1.5–150	n.d.	37.24	n.d.	1.002	n.d.

[a] Margaritis and te Bokkel, 1984. [b] n.d. Not measured.

49.5 MICROBIAL BIOSYNTHESIS OF EXOPOLYSACCHARIDES

A systematic perusal of the literature revealed that, despite the commercial importance of exopolysaccharide producing microorganisms, very little is known about the biochemical pathways involved during the biosynthesis of these different exopolysaccharides. A good understanding of the pathways involved in exopolysaccharide biosynthesis is important because this information can be used to control and optimize rates and yields of fermentation, as well as the physicochemical characteristics of exopolysaccharides.

The majority of microbial exopolysaccharides are assumed to be synthesized within the cell in an analogous mechanism to that involved in cell wall synthesis. Very few exopolysaccharides have been reported to be synthesized outside the cell (Pace and Righelato, 1980). In this section an attempt is made to review the basic information available in the literature on the biosynthesis of some exopolysaccharides both by whole cell and cell-free systems. A comprehensive review on bacterial exopolysaccharide biosynthesis is given by Sutherland (1972).

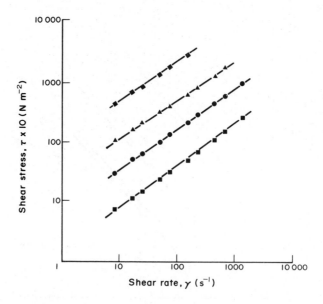

Figure 6 Rheogram of high molecular weight dextran aqueous solutions (MW = 5 × 10⁶ to 40 × 10⁶). ■, 2.439% w/w; ●, 4.761% w/w; ▲, 9.901% w/w; ◆, 16.667% w/w. From Margaritis and te Bokkel, 1984

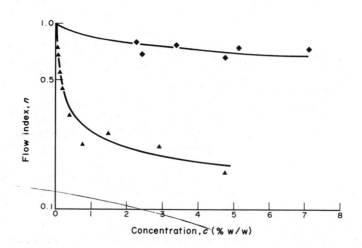

Figure 7 Flow index, *n, versus* concentration for aqueous polysaccharide solutions. ▲, xanthan gum (MW = 1.4 × 10⁶ to 3.6 × 10⁶); ◆, dextran (MW = 5 × 10⁶ to 40 × 10⁶). From Margaritis and te Bokkel, 1984

49.5.1 Cell Biosynthesis of Exopolysaccharides

According to Sutherland (1972), the enzymes involved in exopolysaccharide biosynthesis may be classified into four types: Group I, those enzymes which are involved in the initial metabolism of the substrate, such as hexokinases; Group II, enzymes responsible for the synthesis and interconversion of sugar nucleotides (examples include the enzymes UDP-glucose pyrophosphorylase and UDP-glucose dehydrogenase); Group III, transferases, which are enzymes responsible for the formation of the repeating monosaccharide unit attached to the carrier lipid; and Group IV, translocases or polymerases which form the exopolysaccharide biopolymer molecule. It has been postulated that each group of these enzymes is located in different regions of the microbial cell, thus having specific functions in the overall biosynthesis of the exopolysaccharide macromolecule. A model for exopolysaccharide biosynthesis is presented based on known biochemical reactions in *Klebsiella aerogenes* and related species (see Figure 15 in Sutherland, 1972). This model,

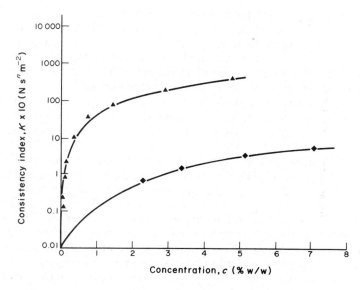

Figure 8 Consistency index, K, *versus* concentration for aqueous polysaccharide solutions. ▲, xanthan gum (MW = 1.4 × 10^6 to 3.6 × 10^6); ◆, dextran (MW = 5 × 10^6 to 40 × 10^6). From Margaritis and te Bokkel, 1984

which is described briefly below, may be used as an important guide to make useful inferences regarding possible biochemical reactions involved in other microbial species which produce exopolysaccharides. Control strategies for exopolysaccharide biosynthesis are possible at any of the four different group levels of enzymes described earlier.

The first group of enzymes are found intracellularly and are also involved in many other cell metabolic processes. The first of these enzymes, hexokinase, is involved in the phosphorylation of glucose (Glc) to glucose 6-phosphate (Glc-6-P) and then the second enzyme phosphoglucomutase converts Glc-6-P to Glc-1-P.

The second group of enzymes are believed to be intracellular. One of these enzymes, uridine diphosphate-glucose pyrophosphorylase (UDP-glucose pyrophosphorylase), catalyzes the conversion of Glc-1-P to uridine diphosphate-glucose (UDP-Glc), which is a key sugar nucleotide and of great importance because UDP-Glc is the main precursor during exopolysaccharide biosynthesis. At this point it is important to emphasize the dual role sugar nucleotides such as UDP-Glc play in exopolysaccharide biosynthesis. The first role of UDP-Glc is its interconversion to sugar nucleotide UDP-galactose (UDP-Gal) with the enzyme UDP-galactose-4-epimerase, and/or its interconversion to the sugar nucleotide UDP-glucoronic acid (UDP-Glc UA) with the enzyme UDP-glucose dehydrogenase. The second role sugar nucleotides play is that they are donors of monosaccharide residues during the biosynthesis of the exopolysaccharide polymer molecule (Lehninger, 1970). The coenzyme uridine diphosphate (UDP) is an essential carrier of monosaccharide residues which become activated and transferred to suitable acceptors during polymer biosynthesis. It has been reported, for example, that the free energy of hydrolysis of the sugar nucleotide UDP-glucose is −31920 J, with similar values for other sugar nucleotides. It is believed that the only homopolysaccharides which apparently are not formed from activated sugar nucleotides are levans and dextrans. During their biosynthesis, sucrose or other related oligosaccharides are required from which only fructose or glucose is polymerized into levans and dextrans respectively. However, all other homopolysaccharides require sugar nucleotides as donors of activated monosaccharide residues (Sutherland, 1972).

The third group of enzymes are located in the cell periplasmic membrane and are referred to as glycosyl transferases. They are involved in the transfer of sugar nucleotides, UDP-glucose or UDP-galactose and/or UDP-glucoronic acid to form the repeating unit attached to glycosyl carrier lipid present in the cell membrane. During this step the UDP is released while the sugar moiety is attached to the carrier lipid. Troy *et al.* (1971), using a strain of *Aerobacter aerogenes*, found a sequential transfer of monosaccharide residues from the sugar nucleotide donors to the glycosyl carrier lipid. The carrier lipid was identified as an isoprenoid alcohol, and its terminal alcohol group was attached to a monosaccharide residue through a pyrophosphate bridge. Not much information is available on the biochemistry of glycosyl transferase enzymes involved in exopolysaccharide biosynthesis.

The fourth group of enzymes are the translocases or polymerases which are believed to be located on the outside part of the cell membrane and the cell wall. Although not much is known about these enzymes, it is presumed that they are involved in the polymerization of the macromolecule and then the exopolysaccharide is extruded from the cell surface to form a loose slime or a well attached polysaccharide capsule surrounding the cell. The exact mechanism of this 'extrusion' process of the exopolysaccharide macromolecules is not known.

An interesting finding has been reported by Haug and Larsen (1971), who showed that an extracellular enzyme, polymannuronic acid C_5-epimerase, was able to modify the exopolysaccharide after its biosynthesis and extrusion from the cells of *Azotobacter vinelandii*. This epimerase enzyme was responsible for converting D-mannuronic acid residues present in the polysaccharide macromolecule to L-guluronic acid, and the enzyme activity depended on the presence of Ca^{2+}. Garegg *et al.* (1971), reported the acylation of some exopolysaccharides and they postulated that these exopolysaccharides are formed with a basic acyl group, such as pyruvate, which could then be modified by other enzymes present in the periplasmic region or outside the cell. Sutherland (1972) reported preliminary experimental results with *Klebsiella aerogenes* exopolysaccharide, which contains acetate and pyruvate in addition to the other monosaccharides, glucose, galactose and mannose. It was observed that addition of potential acyl donors, such as acetyl-CoA or acetyl phosphate and phosphoenolpyruvate, did not have a stimulatory effect on the transfer of nucleotide sugars to the exopolysaccharide polymer or the carrier lipid.

Ginsburg (1964) reported that the nucleotide phosphorylases were the main controlling step in exopolysaccharide production. These enzymes were shown to control the transfer of monosaccharides into the polysaccharide molecule, possibly through negative feedback mechanisms on these enzymes. Bernstein and Robbins (1965) also studied the control aspects of UDP-glucose and thymidine diphosphate (TDP)-glucose synthesis by microbial enzymes and came to the same conclusion as Ginsburg (1964) regarding the controlling nature of nucleotide phosphorylases.

Detailed studies on the proposed biochemical pathways involved in the biosynthesis of alginic acid by *Azotobacter vinelandii*, and xanthan gum by *Xanthomonas campestris*, have been reported (see Figures 1 and 2 in Pace and Righelato, 1980).

49.5.2 Cell-Free Biosynthesis of Exopolysaccharides

The synthesis of the homopolymers, dextrans and levans, is catalyzed by extracellular enzymes acting on specific sugars, and therefore many of the constraints associated with a fermentation no longer apply to the design of a process for making these polymers. For example, the synthesis of dextran using immobilized dextransucrase has been demonstrated (Kaboli and Reilly, 1980) and the use of cell free native dextransucrase is the basis of some commercial production (Jeanes, 1974). The synthesis of dextran by dextransucrase is well understood, for example controlled synthesis of dextrans of various molecular weights can be achieved by altering enzyme and sucrose concentrations, temperature and the addition of small oligosaccharides or fractions of dextrans which act as primers (Hellman *et al.*, 1955; Ebert and Schenk, 1968).

In contrast to the homopolysaccharides, the synthesis of heteropolysaccharides involved several enzymes and cofactors thus making the commercial production of these polymers by cell free synthesis or immobilized cells less likely. However, some data in the literature does indicate its technical feasibility and more work in this area may be fruitful. For example, the synthesis of an exopolymer by washed resting cells of *Pseudomonas* NCIB11264 (Williams and Wimpenny, 1980) and the production of pullulan by protoplasts of *Aureobasidium pullulans* (Finkelman and Vardanis, 1982) have been reported. Further support is provided by Yuen (1974) who reported the synthesis of pullulan by cell free extracts of *Aureobasidium pullulans*, although no supporting data was provided, and Church (1980) described the synthesis of xanthan by lysed cells of *Xanthomonas campestris*.

49.6 PROCESSES FOR MICROBIAL EXOPOLYSACCHARIDE PRODUCTION

Processes for the production of microbial exopolysaccharides are characterized by the extreme rheology of the fermentation broth, the low product concentrations at which this occurs and the diversity of subtle structural and conformational changes which can occur throughout the entire process and the marked effect of these changes on the product's end application performance. In the following section a description is given of how these general characteristics and more specific features of polysaccharides affect the design and operation of processes for their production.

49.6.1 Fermentation of Exopolysaccharides

The successful design of the fermentation stage of the process relies on producing a product to a set specification while achieving a product concentration, yield and productivity set from economic targets. These goals can best be reached with minimum risk by establishing how the microorganism's performance is controlled by its environment and, in turn, how this relates to equipment design operation (Figure 9). The specific controls, which exist at these two levels, (*i.e.* environmental and equipment design and operation) and are characteristic of microbial exopolysaccharide fermentations, are described below.

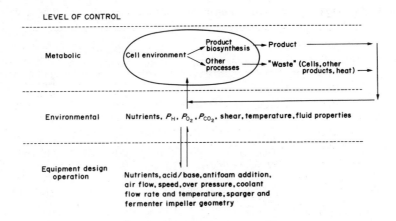

Figure 9 Levels of control governing microbial processes

49.6.1.1 Environmental control

Control of the composition of the fermentation medium, its addition and other environmental parameters are critical in achieving the desired rates of synthesis and yields of microbial polysaccharides, and the complete utilization of the full heat and mass transfer capabilities, or energy input, of existing fermenters. These parameters are also prime determinants of the purity of the product, its chemical composition and molecular weight, which in turn determine its end use performance. Also, the strategy adopted in the running of the fermentation, *e.g.* batch, draw fill, continuous, controlled feeding, *etc.*, is also determined by the relative kinetics of cell and polymer production. In the following section these key parameters which affect the design of the fermentation stage of exopolysaccharide production are discussed.

49.6.1.2 Fermentation medium

The growth media which are suited to the production of different exopolysaccharides by microorganisms vary widely and this probably reflects the differing role of each polysaccharide in nature. For example, both xanthan produced by *Xanthomonas campestris* (Ellwood *et al.*, 1978) and alginate produced by *Azotobacter vinelandii* (Jarman *et al.*, 1978) are made under carbon limited conditions whereas polymers produced by *Pseudomonas NCIB11264* (Williams and Wimpenny, 1978) and *Klebsiella aerogenes* (Neijssel and Tempest, 1975) are not made under these conditions. Thus, for example, it can be argued that xanthan and alginate do not represent methods by which the cells store excess carbon, whereas the synthesis of the polymers by *Pseudomonas NCIB11264* and *Klebsiella aerogenes* may be for storage. Although it is difficult to draw general conclusions, it is instructive to consider the effects on polymer biosynthesis rates, yields and compositions of varying growth media, as these may have wider application.

49.6.1.3   Effect of medium on specific rate of exopolysaccharide synthesis

Continuous culture studies on the effect of the growth limiting substrate on the synthesis of exopolysaccharides, by several different types of microorganisms, clearly demonstrate that the composition of the growth medium can dramatically affect the specific rate of polymer synthesis. Such effects are illustrated by the summary of some published data given in Table 5, and it is probable that in all the examples shown, the bacteria use the activated sugar nucleotide pathway for polymer synthesis. It can be seen that for a given microorganism, although the polymer is produced under a variety of conditions, specific rates of synthesis can easily change by a factor of two or even cease under certain limitations. As expected, the absolute values of the specific rates appear to vary between species and strains, and also there appears to be no particular limitation which gives a common highest relative rate of production between organisms.

Table 5   The Effect of Growth Limiting on the Specific Rate of Exopolysaccharide Production

Microorganism	Polymer	Limiting nutrient	Specific rate of polymer production[a] $(g\,g^{-1}\,cells\,h^{-1})$	Ref.
X. campestris	Xanthan	Glucose	0.12	Davidson, 1978 (calculated
		NH_4^+	0.22	from data given)
		PO_4^{3-}	0.09	
A. vinelandii	Alginate	Sucrose	0.25	Deavin *et al.*, 1977
		N_2	0.22	
		PO_4^{3-}	0.28	
Ps. aeruginosa	Alginate	Glucose	0.19	Mian *et al.*, 1979
		N (yeast extract)	0.27	
		NH_4^+	0.34	
Rhizobium trifolli	Not identified	Mannitol–asparagine	0.0017	de Hollander *et al.*, 1979
		SO_4^{2-}	0.0049	
Pseudomonas NCIB11264	Acetylated polymer containing glucose, galactose, rhamnose and mannose	Glucose	0	Williams and Wimpenny, 1978
		NH_4^+	Produced	
K. aerogenes	Not identified	Glucose	0	Neijssel and Tempest, 1975
		NH_4^+	Produced	

[a] Each set of results are from experiments conducted at constant dilution rate.

49.6.1.4   Effect of medium on exopolysaccharide yield

The amount of (carbon) substrate converted by the cell to polymer depends on the composition of the growth medium and, under certain conditions, the product may not be made at all. Generally, media containing a high carbon to limiting nutrient ratio, often nitrogen, are favoured for polysaccharide production (Wilkinson, 1958; Sutherland, 1979). Conversions of 70–80% of utilized glucose into crude polymer are commonly found in high yielding polysaccharide fermentations. Care must, however, be taken in the interpretation of such yields as the crude product will often contain cells, other organic material and inorganic salts which are coprecipitated with the polymer when it is recovered from the fermentation broth. For example when corrected for the presence of contaminants, yields of 50–60% xanthan from glucose consumed can be obtained in well run processes, and this compares with a theoretical yield of about 0.85, with no cell growth, depending on the degree of pyruvulation of the polymer and the energy efficiency (P/O ratio) of the cell (Table 6; Jarman and Pace, 1984). Some examples of the effect of medium composition on the yield of cell free polymer are given in Table 7. The effects of a variety of carbohydrate and nitrogen sources on the xanthan fermentation are also described by Souw and Demain (1979).

The composition of the growth medium can also indirectly affect polymer yield, for example by governing the pH change which can occur during fermentations without pH control. In addition, high media concentrations, resulting in high polymer concentrations, may lead to oxygen limitations or heterogeneity due to increased viscosity within the fermenter before exhaustion of the carbon source, and this might indirectly affect yield as well as the overall production rate.

Table 6 Maximum Theoretical Yields
of Xanthan[a] at Different P/O Ratios

P/O ratio	$\dfrac{Xanthan\ (g)}{Glucose\ (g)}$	$\dfrac{Xanthan\ (g)}{Oxygen\ (g)}$
1	0.81	5.9
2	0.86	10.7
3	0.87	12.5

[a] Acid form of xanthan containing 1 mole of
acetate and 0.5 mole pyruvate per repeating
subunit.

Table 7 The Effect of Growth Limiting Nutrient on Yield of
Exopolysaccharide

Microorganism	Polymer	Limiting nutrient	Yield[a]	Ref.
X. campestris	Xanthan	Glucose	0.54	Davidson, 1978
		NH_4^+	0.60	(calculated from
		SO_4^+	0.53	data given)
		Mg	0.55	
		K^+	0.42	
		PO_4^{3-}	0.31	
Ps. aerugenosa	Alginate	Glucose	0.33	Mian et al., 1978
		N (yeast extract)	0.61	
		NH_4^+	0.53	

[a] Each set of results are from experiments conducted at constant dilution rate, and the value
given is the amount of cell free polymer produced per unit of glucose consumed.

49.6.1.5 Effect of medium on exopolysaccharide composition

Manipulation of exopolysaccharide composition by changing the growth medium is one possible mechanism by which the properties of the polymer can be tailored to the end application.
Experimental work on this aspect appears limited, but does indicate that although it may not be
possible to alter the structure or composition of the base repeating unit of the polysaccharide, it is
possible to change the degree of substitution of the repeating unit by various groups and the
degree of polymerization (molecular weight).

Numerous studies on the chemical structure of xanthan produced by *Xanthomonas campestris*
grown under differing fermentation conditions show that the structure of its base repeating unit is
invariant with growth conditions. In one systematic study, Davidson (1978) found that the ratios
of the sugars present in the repeating unit of xanthan remained constant when the product was
made in continuous culture under six different growth limitations. Although the content of guluronic and mannuronic acids of alginate produced by *Azotobacter vinelandii* and *Pseudomonas aeruginosa* varies with culture conditions, the primary form of the polymer produced in the cell
membrane/wall area does not and is composed of pure mannuronic acid (Jarman, 1978). An
extracellular enzyme, alginate epimerase, then catalyzes the isomerization of parts of the polymannuronic acid to guluronic acid resulting in the alginate which is recovered. The activity of the
epimerase is dependent upon the growth medium, with high levels of calcium ions stimulating
activity and hence a high guluronic acid content in the polymer (Haug and Larsen, 1971). Alginates with high guluronic to mannuronic ratios are preferred in gelling applications as they form
stronger gels owing to the cross-linking of guluronic acid residues on adjacent chains by calcium
ions. In another study on polymer composition Williams and Wimpenny (1978) found that the
formation of an exopolysaccharide by *Pseudomonas* NCIB11264 was constant under a variety of
growth conditions.

It appears from some examples in the literature that where a polymer contains groups, such as
acetate, pyruvate, formate, succinate or phosphate, attached to its base unit, it may be possible to
alter the properties of the product by changing growth conditions to vary the content of the
groups. The pyruvate content of xanthan can be varied virtually from 0% to its theoretical maximum of 7.5% by the use of certain growth media or strain (Sutherland, 1979; Davidson, 1978;
Cadmus *et al.*, 1978; Jarman and Pace, 1981). The combination of the general properties

exhibited by solutions of xanthan is very unusual and includes high viscosity and pseudoplasticity at low concentrations; yield stress; synergistic increases in viscosity and sometimes gelation when mixed with galactomannans; high stability at extremes of pH, temperature and shear; and unusual viscous behaviour with changes in pH and ionic strength. These properties result from its chemical structure (see Figure 1), together with the various macromolecular forms of xanthan that can occur in solution (Figure 10). Variation in the pyruvate content of the polymer tends not to change the polymer's gross characteristic properties, but it does affect the incremental changes in rheology which occur when the concentration of the polymer or the solution's ionic strength, pH or temperature are altered. These effects appear to be primarily associated with a promotion of the aggregated form of xanthan by pyruvate residues, whereas products which have a low pyruvate content tend to exist in disaggregated subunits (Symes, 1980; Smith *et al.*, 1981). Solutions in which the aggregated form dominates exhibit high apparent viscosity, pseuodplasticity and yield stress. Thus control of the pyruvate content of xanthan, for example, through manipulation of the growth medium, represents a very useful way of fine tuning the properties of a polysaccharide.

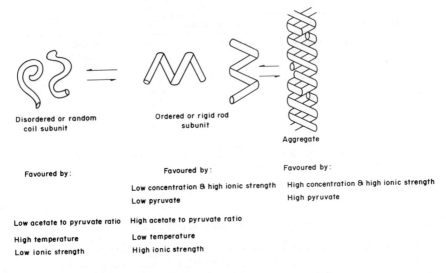

Figure 10 Conformational forms of xanthan in solutions

Similarly, it has been shown that the acetate content of alginate is variable and appears associated only with mannuronic acid residues, and it has been proposed that acetylated mannuronic acid residues are not epimerased to guluronic acid (Jarman, 1978). However, environmental methods for controlling the acetate content of alginate and its influence on rheological properties are not well understood.

The reduction of the phosphate concentration to growth limiting levels in the culture medium for certain yeasts results in a shift from the production of type 11 phosphomannans, in which the terminal non-reducing end group is phosphorylated, to a structurally related mannan and it has been proposed that there is a common pathway for the two polymers in these yeasts (Slodki *et al.*, 1972). The concentration of phosphate in the polymer would be expected to markedly affect its rheological properties, through polyelectrolytic effects on the polymer's conformation.

Theoretical calculations show that the net energy required for polymer synthesis can vary from a positive to negative value in exopolymers containing oxidized groups, such as xanthan and alginate, as the content of these groups increases (Figure 11). These calculations account for the ATP used in polymerization and the ATP generated from the glucose used in polymer synthesis when metabolized *via* the operating pathways, *i.e.* the Entner–Doudoroff pathway, the TCA pathway and oxidative phosphorylation of the net reduced pyridine nucleotide produced. The results suggest that although these moieties are not essential for growth and do not affect the gross characteristics of the polymer, they may be important to the energy metabolism of the cell providing either a source or sink for energy when incorporated into the exopolysaccharide. Thus successful strategies for manipulating the content of oxidized groups may centre on affecting the energy metabolism of the cell, through, for example, composition of the growth medium (Jarman and Pace, 1984).

The molecular weight of a polymer has a direct influence on its rheological behaviour, with larger molecules having high viscosity. Although it is known that the molecular weight of xanthan

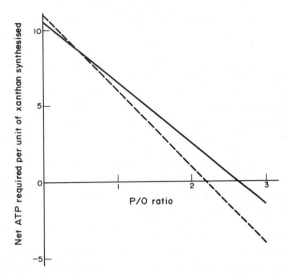

Figure 11 Net ATP requirements for synthesis of xanthan with various substituents. ---------, 1 acetate + 1 pyruvate; ——————, 1 acetate + 0 pyruvate

and other microbial exopolysaccharides can vary, little is known about chain termination and how the growth medium can be altered to control it. In some fermentations exocellular enzymes are secreted which degrade their own polymer. Suppression of the activity of the lytic enzyme, alginate lyase, in the *Azotobacter vinelandii* fermentation remains one of the major problems in economical production of high molecular weight microbial alginates. However, recently a possible solution has been suggested, being illustrated by the increase in viscosity that followed the addition of protease to a *Pseudomonas mendocina* and presumably resulted from proteolytic degradation of the alginate lyase (Hacking *et al.*, 1983). A similar problem is encountered in the pullulan fermentation in which there is a decline in the culture viscosity and molecular weight towards the latter stages of growth (Catley, 1978; Leduy *et al.*, 1974).

In addition to the effects of medium composition on fermentation performance, as discussed above, other considerations must be taken into account. The composition of the medium can affect the nature of the recovery processing or the purity of the final product. For example, insoluble media solids either have to be removed prior to polysaccharide precipitation or they will be coprecipitated with the polymer. Also, high concentrations of inorganic salts in the medium can result in coprecipitates being formed when the polymer is isolated, giving the final product an increased ash content.

The choice of growth medium can affect the stability of the microbial culture. In the early work on xanthan fermentation, the formation of nonpolysaccharide producing mutants was reported (Silman and Rogovin, 1972) and much effort was put into culture maintenance (Cadmus *et al.*, 1976; Kidby *et al.*, 1977) and the selection of stable strains. However, latter work has shown that by the choice of certain media and growth under limitations, nonxanthan producing mutants are not formed (Evans *et al.*, 1978; Manchee *et al.*, 1979). This approach to maintaining culture stability has been demonstrated with other systems (Righelato, 1976), and undoubtedly will be applicable to other polysaccharide fermentations. As with other fermentations, care must be taken in the selection of raw materials to ensure they are within economic constraints and they do not contain impurities that will adversely affect fermentation performance. Also, it is important to ensure correct sterilization procedures are used which minimize reactions that are deleterious to growth or product formation such as non-enzymatic browning or essential metal precipitation.

49.6.1.6 *Other environmental parameters affecting exopolysaccharide fermentation*

The growth and production of exopolysaccharides by microorganisms are determined by a wide range of environmental parameters, in addition to the effects of the culture medium described in the previous section. However, the influence on the fermentation of variables such as temperature, pH, dissolved oxygen and carbon dioxide, shear, *etc.*, appears to have received less attention in published studies.

One early study reported the optimum temperature for xanthan production, as measured by culture viscosity, as 28 °C and for growth 24 to 30 °C (Moraine and Rogovin, 1966). In a more detailed study, Williams and Wimpenny (1978) studied the effect of temperature on the production of exopolysaccharide by *Pseudomonas* NCIB11264 under controlled conditions in continuous culture. They found that in a steady state under nitrogen limited conditions, the cell concentration remained constant between 20 to 37.5 °C, whereas the conversion efficiency of glucose to polymer, and culture viscosity, varied sharply with temperature showing a maximum at 30 °C. Interestingly, the polymer composition appeared unaffected by the fermentation temperature. A number of other studies also demonstrate the importance of temperature in determining the specific rate of synthesis and yield of polysaccharide and the variability which exists between species and strains (Wilkinson, 1958; Norval and Sutherland, 1969; Goto *et al.*, 1971; Evans and Linker, 1973; Piggott *et al.*, 1982; Finkelman and Vardanis, 1982).

Early work on the xanthan fermentation demonstrated the beneficial effects of pH control on the production of charged polymers. Lack of pH control and poor medium buffering results in a rapid fall in the pH of the xanthan fermentation with a cessation of growth and production of the polymer at about pH 5.5. A pH optimum of 7 has been reported for the production of an acidic exopolysaccharide by *Xanthomonas campestris* (Moraine and Rogovin, 1973), *Pseudomonas* NCIB11264 (Williams and Wimpenny, 1978) and other organisms (Wilkinson, 1958). In the case of polymer production by *Pseudomonas* NCIB11264 it was also shown that cell growth was less sensitive to pH than product synthesis and that the composition of the gum was unaffected by pH. The optimal pH for production of the neutral glucan, pullulan, by *Aureobasidium pullulans* was found to be about 5.3 and again was much more sensitive to pH than cell growth (Finkelman and Vardanis, 1982).

Oxygen is usually required in both the synthesis of components of the polymer (*e.g.* sugar acid) or indirectly in the oxidation of reduced pyridine nucleotides generated. Theoretical calculations for the synthesis of xanthan illustrate the amount of oxygen required and its dependence on the energy efficiency of the cell (Table 6). However, little information is available on the effect of dissolved oxygen tension and oxygen availability on growth and product formation in polysaccharide fermentations. Experiments and their interpretations are made difficult by the poor mixing often encountered in laboratory scale polysaccharide fermentations and the failure to independently separate the variables of oxygen tension and supply from shear and vessel homogenicity. In studies on alginate production by the nitrogen fixing obligate aerobe *Azotobacter vinelandii* (Jarman *et al.*, 1978; Horan *et al.*, 1983) it has been found that the conversion efficiency of sucrose to alginate increased with decreasing specific respiration rate in response to changes in oxygen supply. Under conditions of high dissolved oxygen tension the organism increases its respiration rate in an apparent attempt to protect its oxygen sensitive nitrogenase, with the resultant effect of diverting sucrose from polymer production to carbon dioxide.

In an experiment designed to investigate the effect of shear on the xanthan fermentation, Moraine and Rogovin (1973) found that when the dissolved oxygen tension, measured by a single probe in the fermenter, was maintained over 90% and the agitator speed was increased, the specific rate of polymer production increased. Based on this observation they proposed a shear sensitive slime layer existing around the cell that, by acting as a diffusion barrier to nutrients, affected polymer synthesis. However, in the absence of oxygen uptake measurements it is possible that the effect was due to improved oxygen transfer due to better mixing. In poorly mixed fermenters high oxygen tensions can be obtained in the well mixed region around the impeller while the remainder of the vessel remains poorly mixed and at a very low tension. Similarly Tanzer's (Tanzer *et al.*, 1969) conclusion, that in the dextran fermentation dextran forms a diffusion barrier around clumps of cells and limits growth, could be due to oxygen limitation because of poor mass transfer at high dextran concentrations.

49.6.2 Kinetics of Microbial Exopolysaccharide Synthesis

The relationships between growth and polysaccharide formation vary between microorganisms and for a particular fermentation they change with growth conditions and microbial strain. An interesting example is the specific rate of alginate synthesis in continuous culture, which is relatively independent of growth rate when produced by *Azotobacter vinelandii* (Jarman *et al.*, 1978) but is strongly dependent upon growth rate when produced by *Pseudomonas aeruginosa* (Mian *et al.*, 1978). Other examples are listed in Table 8.

In growth associated fermentations, a high rate continuous process is best suited to high rates

Table 8 Kinetics of Polymer Formation by Microorganisms Under Various Conditions

Microorganism/polymer	Growth conditions	Kinetics of polymer formation	Ref.
Acinetobacter calcoaceticus/ rhamnose–glucose exopolymer	Batch	Mixed	Kaplan and Rosenberg (1982)
Zoogloea ramigera/ exopolymer	Batch, N limited	Non-growth associated	Norberg and Enfors (1982)
Alcaligenes faecalis/ curdlan	Batch, N limited	Non-growth associated	Railton et al. (1981)
Pseudomonas aeruginosa/ alginate	Batch and continuous N limited	Growth associated	Mian et al. (1978)
Alcaligenes vinelandii/ alginate (parent strain) (mutant)	Batch, P limitation continuous, P limitation	Growth associated Non-growth associated[a]	Deavin et al. (1977) Jarman et al. (1978)
Rhizobium trifolii/ exopolymer	Batch, P limitation continuous, mannitol, asparagine limited	Mixed Growth associated	Horan et al. (1981) de Hollander et al. (1979)
Pseudomonas NCIB11264/ exopolymer	Batch and continuous, N limited	Non-growth associated[a]	Williams and Wimpenny (1977, 1978)
Xanthomonas campestris/ xanthan	Batch and continuous, N limited	Mixed	Moraine and Rogovin (1973)
X. juglandis/xanthan	Continuous, S limited	Non-growth associated[a]	Evans et al. (1979)

[a] Non-growth associated production in continuous culture occurs when the specific rate of polymer formation is independent of growth rate.

of polysaccharide production. Whereas with non-growth associated products, the best process will be a batch process in which there is a rapid period of cell growth followed by a stationary phase (or by greatly reduced growth achieved by controlled feeding of a limiting nutrient), during which time the product is generated. Draw fill techniques may also be applied to the batch process to increase overall fermenter productivity.

Ollis and co-workers (Weiss and Ollis, 1980; Klimek and Ollis, 1980) found the Leudeking–Piret equation:

$$\frac{dP}{dt} = nX + m\frac{dX}{dt}$$

(9)

where dP/dt and dX/dt are respectively the rates of product and cell formation, X is the cell concentration and n and m are constants, adequately described the production of polysaccharide in batch cultures by *Xanthomonas campestris*, *Pseudomonas* NCIB11264, *Azotobacter vinelandii* and *Aureobasidium pullulans*. However, as the authors also point out, the constants derived for the cases considered should not be expected to hold for a variety of conditions or strains. This is illustrated by the examination of the results of a series of papers on the production of alginate by *Azotobacter vinelandii*. Klimek and Ollis (1980) found that a wholly growth associated form of the Leudeking–Piret equation reasonably described alginate synthesis by the parent strain of *Azotobacter vinelandii* in batch culture(s) under phosphate limiting conditions. However, growth of the same strain using similar growth conditions in continuous culture (Deavin et al., 1977; Jarman et al., 1978) indicated that the specific production rate was relatively independent of growth rate, i.e. the contant, m, in the Leudeking–Piret equation is zero. These differences were most likely due to oxygen limitation which probably occurred in the late stages of growth in the batch culture but, through control, was avoided in continuous culture experiments (Jarman, 1983). Increases in cellular poly(hydroxy butyrate) under oxygen limited conditions in the batch culture may have further affected the nature of the Leudeking–Piret model. In a later paper (Horan et al., 1981) the synthesis of alginate by mutant strain, derived from the parent in batch culture using phosphate limited conditions, is described and, in marked contrast to the parent alginate synthesis, continued at a high rate after growth had ceased. The Leudeking–Piret equation derived from the data in the paper indicates mixed kinetics of polymer production by the mutant.

Although the production of pullulan by *Aureobasidium pullulans* can be described as wholly growth associated, detailed observations during the fermentation suggest that it is associated with blastospore formation which occurs throughout the growth phase and is not produced by the parent mycelium (Catley, 1980) and therefore it can be argued that the product is not strictly wholly growth associated.

49.7 FERMENTER DESIGN AND OPERATION

An understanding of the relationships between the environment in a fermentation and the equipment and operating variables is the key to the design and scale up of equipment for economic production. In microbial exopolysaccharide fermentations, the extreme rheology encountered has the major influence on the nature of these relationships compared to other low viscosity fermentations, such as bacterial and yeast cultures. The picture is further complicated in that changes in heat, mass and momentum transfer caused by increases in culture rheology can feed back to further effect either the polymer's rheological behaviour or the ability of the microorganism to produce the polymer.

The major contributor to the rheology of exopolysaccharide culture fluids is the polymer dissolved in the continuous phase, which contrasts with filamentous fungal fermentations in which the high viscosity is caused by mycelium or the discontinuous phase (Pace, 1980). This can lead to differences in the heat and mass transfer achieved in experiments conducted on mycelial and polysaccharide fermentations at equivalent measured rheologies on the same equipment (Pace, 1980; Blakebrough *et al.*, 1978).

The rheology exhibited by microbial polysaccharide culture fluids is complex and includes pseudoplastic flow behaviour (which is commonly observed), thixotropy, yield stress and viscoelasticity. These terms are described in Table 9 and some examples are provided (also see Section 49.4).

Table 9 Rheological Properties of Some Polysaccharide Fermentations

Rheological property	Description	Examples of fermentations showing relations
Pseudoplastic flow or shear thinning	Apparent viscosity of the fluid decreases with increasing shear rate	Pullulan (Leduy *et al.*, 1974), xanthan (Jeanes, 1974; Pace and Righelato, 1980), zoogloea ramiger (Norberg and Enfors, 1982), mannans and phosphomannans (Slodki *et al.*, 1974), alginate (Lawson and Sutherland, 1978)
Yield stress	A given stress has to be applied to the fluid before movement occurs	Xanthan (Jeanes, 1974; Solomon, 1980)
Thixotropy	Apparent viscosity of a pseudoplastic fluid decreases with time at a constant shear rate	Alginate fermentations containing high levels of calcium.
Viscoelasticity	Combinations of viscous and elastic behaviour. This can result in such effects as the fluid having a 'memory' (a tendency to regain its original state on relase of a stress); normal stress which results in the stirrer climbing or Weissenberg effect	Xanthan (Solomon, 1980)

In the following sections a brief review is given of how fluid rheology affects power consumption, mass and heat transfer and mixing and how these combine to affect the design of fermenters for use in microbial exopolysaccharide fermentations; further information on this topic can be found elsewhere (Skelland, 1967; Blanch and Bhavaraju, 1976; Pace, 1978; Margaritis and Zajic, 1978; Moo-Young and Blanch, 1981; Schugerl, 1981).

49.7.1 Power Requirements

The power drawn by the agitator of a fermenter is prime in determining its heat and mass transfer and mixing capabilities. In the turbulent regime the power, P, drawn by a given impeller in an ungassed fluid is insensitive to changes in fluid rheology and is given by equation (10),

$$P \propto \varrho N^3 D^5 \tag{10}$$

where ϱ is the fluid's density, N the impeller speed and D its diameter.

In the laminar and transition regions the power drawn by a given impeller is dependent on the solution's rheology. It is well established in fluids containing inelastic pseudoplastic polymers that the power drawn by an agitator can be correlated with a modified Reynolds number (Metzner and Otto, 1957; Foresti and Liu, 1959; Calderbank and Moo-Young, 1959; Calderbank and Moo-Young, 1961; Metzner *et al.*, 1961), for example, containing an estimate of the apparent viscosity, μ_a, in the immediate vicinity of the impeller. To determine this viscosity it is assumed that the average fluid shear rate, $\dot{\gamma}_{av}$, caused by the impeller is given by equation (11):

$$\dot{\gamma}_{av} = kN \tag{11}$$

and thus for a pseudophastic power law fluid (Consistency index K; flow index n), the apparent viscosity is given by equation (12):

$$\mu_a = K(kN)^{n-1} \tag{12}$$

thus giving a modified Reynolds number (N_{Re}):

$$N_{Re} = \frac{\varrho N^{2-n} D^2}{K(k)^{n-1}} \tag{13}$$

In the laminar regime the power drawn is inversely proportional to this modified Reynolds number. The value of k is characteristic of the impeller and for standard disc turbines is about 11.5.

As shown in Figure 12, using the modified Reynolds number a plot of the power number (N_{P_0}) curve for unaerated fluids of complex rheology yields several interesting features (Metzner and Otto, 1957; Metzner *et al.*, 1961; Kale *et al.*, 1975; Ranade and Ulbrecht, 1977; Solomon, 1980; Solomon *et al.*, 1981c). Firstly, in the laminar flow regime for highly viscoelastic fluids the N_{P_0} is increased above the value for Newtonian and inelastic pseudoplastic fluids. In the transition region (say $50 < N_{Re} < 1000$) the N_{P_0} for both inelastic and viscoelastic fluids is considerably less than that for a Newtonian fluid, and the curve for the inelastic fluid shows a characteristic sharp minimum which is not shown by highly elastic fluids. At higher values of Reynolds number the curves flatten out illustrating the increasing dominance of inertial over viscous and elastic forces. Values of impeller Reynolds number, in large vessels, of less than 1000 are likely in high viscous pseudoplastic fermentations (Solomon *et al.*, 1981c).

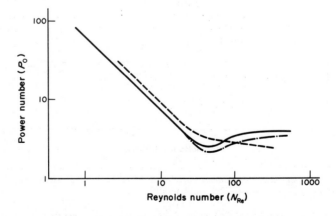

Figure 12 Power number–Reynolds number correlation for a flat bladed disc turbine agitating unaerated inelastic fluids (————) and fluids of high viscoelasticity (---------, 1.4% CMC and 4.5% xanthan) and low viscoelasticity (—·—·—, 0.3%, 0.17% carbopol and 2% xanthan)

As with low viscosity fermentations, aeration of polysaccharide cultures results in decreased power consumption due to the formation of air cavities on the trailing side of the agitator blade. The size and shape of the cavities formed appears to be a function of the fluid's rheology (Bruijn *et al.*, 1974; Ranade and Ulbrecht, 1977; Solomon, 1980; Solomon *et al.*, 1981a) and in highly viscous systems the cavities can remain adhered to the blade even when the air is turned off and the stirrer speed reduced, and may lead to values of the ratio of ungassed (P_o) to gassed (P_g) power of less than 1 under unaerated conditions (Solomon, 1980). The point at which the ratio of gassed to ungassed power changes in highly viscous systems occurs at a Froude number of about 0.1 and flow visualization studies show this to be the point at which large stable cavities are formed or grow if they were already present (Solomon, 1980; Solomon *et al.*, 1981c). Similarly, van't Riet (1975) found that only when the Froude number exceeded 0.1 in water did stable cavities form on the impeller and the gassed to ungassed power ratio fall.

A number of examples exist in the literature on the effects of aeration of highly viscous fluids on power consumption (Edney and Edwards, 1976; Paca and Gregr, 1976; Ranade and Ulbrecht, 1977; Kipke, 1978; Schugerl, 1981) and it appears that the conventional aeration number correlation does not apply and the effects of fluid rheology, gassing rate and impeller speed and geometry need to be treated separately. In a recent series of papers these effects have been considered (Solomon *et al.*, 1981a, 1981b, 1981c; Nienow *et al.*, 1981; Solomon, 1980) and the following generalization can be made.

In the laminar region the gassed and ungassed power numbers are approximately equal. However, if the solution under test has a high yield stress and has been previously aerated at higher speeds, air cavities may remain adhered to the blades and give lower values of gassed power. Such an effect is only likely to occur on laboratory scale where the absolute size of the retained gas cavity is small.

For a given highly viscous fluid and agitator, in the region of $10 < N_{Re} < 900$ the gassed power number (N_{Pg}) appears independent of gas flow rate and shows a minimum value around N_{Re} of 250 to 450 depending on the fluid rheology (Figure 13). The occurrence of this minimum and the independence of gas flow rate is thought to be due to a number of effects and not simple flooding of the impeller or changes in Froude number. Two important factors contributing to these effects are believed to be: (1) as the agitator speed is increased, independent of gassing rate, the cavities grow in size until they fill the space between the impeller blades and then near the minimum value of N_{Pg}, small gas bubbles begin to form and the cavity is thought to decrease in size; and (2) this minimum value is partly correlated with the first normal stress difference of the fluid and probably reflects the influence of viscoelasticity in decreased power consumption in the transition regime. In addition an underlying factor may be the existence, in aerated highly pseudoplastic fluids and those exhibiting a yield stress, of a well mixed region (cavern) in the vicinity of the impeller which tends to hold small gas bubbles and grows in size as the impeller speed is increased.

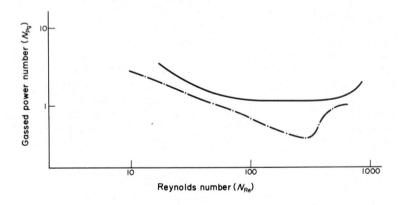

Figure 13 Gassed power number–Reynolds number correlation for a flat bladed disc turbine agitating highly viscous fluids at different gas flow rates (——————, 2% xanthan at 0.5 and 1.0 vvm; —·——·—, 4.5% xanthan at 0.5 and 1.0 vvm)

At higher Reynolds number (>900) the gassed power number for a given impeller becomes dependent on gas flow rate, provided the impeller is not flooded, and shows variations similar to those observed with water.

The dependence of gassed power on variations in impeller diameter and geometry appears related to the rheology of the fluid being mixed and examples are given by Solomon (1980).

49.7.2 Mass Transfer Characteristics

Some empirical correlations relating the overall mass transfer coefficient ($k_L a$) to equipment and operating variables have been published (Hattori *et al.*, 1972; Perez and Sandall, 1974; Yagi and Yoshida, 1975; Paca *et al.*, 1976; Ranade and Ulbrecht, 1978; Kipke, 1978; Henzler, 1981). Schugerl (1981) points out that the majority of these equations fall into two types: one in which a modified Sherwood number is correlated with Reynolds, Froude and Schmidt numbers and groups accounting for gas flow and rheology, or where $k_L a$ is correlated with dimensionless specific power input with respect to gas flow, Schmidt number and gas flow per unit volume.

Some examples of these two types of correlation are: (1) The Yagi and Yoshida (1975) correlation given by equation (14)

$$\frac{k_L a}{D_L} = 0.06\left(\frac{D^2 N \varrho}{\mu a}\right)^{1.5}\left(\frac{DN^2}{g}\right)^{0.19}\left(\frac{\mu a}{\varrho D_L}\right)^{0.5}\left(\frac{\mu a V_s}{\sigma}\right)^{0.6}\left(\frac{ND}{V_s}\right)^{0.32}(1 + 2(\lambda N)^{0.5})^{-0.67} \tag{14}$$

where D_L is liquid diffusivity, V_s is superficial gas velocity, σ surface tension, and λ is derived from a plot of normalized apparent viscosity *versus* shear rate; other symbols as defined elsewhere in the text.

(2) The Henzler (1981) correlation is given by equation (15)

$$(k_L a)\left(\frac{V}{q}\right) = 0.082\left[\frac{\mu a}{\varrho D_L}\right]^{-0.3}\left[\frac{P_g}{(q\varrho^{0.32})(g\mu a)^{0.67}}\right]^{0.6} \tag{15}$$

where V is the liquid volume and q is the volumetric gas flow rate.

In general these equations only take account of viscosity changes and do not take account of other rheological effects such as yield stress and viscoelasticity (note that Ranade and Ulbrecht's (1977) expression is an exception and accounts for a fluid's normal stress). The possible importance of these parameters is shown by more fundamental studies on bubble behaviour, for example the rise velocity of bubbles moving through given fluids under certain conditions has been shown to be dependent on the degree of viscoelasticity and the yield stress of the fluid (Astarita and Apuzzo, 1965). Another example is given by Moo-Young and Hirose (1972) who showed that mass transfer from bubbles with mobile interfaces moving in creeping flow conditions was enhanced by the viscoelastic properties of fluids. Also, the coalescence of bubbles is affected by various rheological properties of fluids (Acharya and Ulbrecht, 1978; McBride *et al.*, 1981). Additional caution is also warranted in the use of empirical expressions such as those given above, as they are based on experiments conducted with model systems on a small scale and their applicability to larger scale and real fermentations has not been demonstrated in the literature.

In well mixed polysaccharide fermentations it is likely that swarms of small, spherical, rigid interfaced bubbles exist, whereas in poorly mixed conditions as may exist at the end of fermentations, large bubbles, which are probably spherical cap in shape with mobile interfaces, are often seen. In a range of rheological complex fluids, Solomon *et al.* (1981a) found that small bubbles were only generated by agitators when a modified Reynolds number, using an apparent viscosity based on the shear rate at the impeller tip, exceeded about 300. Further data is necessary to demonstrate whether this concept is applicable to large scales. Some information exists in the literature on the separate prediction of mass transfer coefficient (k_L) and the interfacial area (a) for such bubbles in model systems and a more detailed review is given by Moo-Young and Blanch (1981).

49.7.3 Heat Transfer Characteristics

Studies on the heat transfer from viscous pseudoplastic polymer solutions indicate that for a given vessel, equations of the following form can be used to predict the heat transfer coefficient through the Nusselt number (N_{Nu}) under unaerated conditions:

$$N_{Nu} = A(N_{Re})^a(N_{Pr})^b\left(\frac{\mu_a}{\mu_{aw}}\right)^c[\text{dimensionless geometric factors raised to some power}] \tag{16}$$

where μ_{aw} is the apparent viscosity at the mean temperature of the heat transfer surface, and an apparent viscosity (μ_a) related to impeller speed (see equation 12) is used in the Reynolds (N_{Re}) and Prandtl (N_{Pr}) numbers (Edwards and Wilkinson 1972; Edney and Edwards, 1976). At given Reynolds numbers greater than 10 the heat transfer coefficient for viscous pseudoplastic fluids appears relatively unaffected by changes in aeration (Edney and Edwards, 1976; Schugerl, 1981); however, at lower Reynolds numbers, presumably under poorly mixed conditions, aeration has a greater effect (Schugerl, 1981).

There is considerable scope for published studies to quantify the effect of aeration, scale, reactor and impeller geometry and other non-Newtonian properties on the heat transfer coefficient for microbial polysaccharide solutions.

49.7.4 Mixing Characteristics

The flow patterns and mixing in agitated vessels are markedly affected by the fluid's rheology. Highly viscous pseudoplastic fluids show good movement in the region of the impeller where shear rates are high (Figure 14). However, away from the impeller the movement decreases due to the apparent increase in viscosity resulting from the pseudoplastic nature of the fluid (Metzner and Taylor, 1960). High viscosity also decreases the pumping capacity of the impeller (Norwood and Metzner, 1960) which further compounds the problem. Thus these effects promote gas funnelling through the centre of the vessel and inhomogeneity, particularly in areas away from the impeller near the vessel wall, and virtually static areas may form around baffles and in small vessels around probes. The tendency to form dead zones is further enhanced if the fluid possesses a yield stress (Solomon *et al.*, 1981a). As shown in Figure 14, in highly viscous pseudoplastic fluids, particularly if they have a yield stress (*e.g.* xanthan), a well mixed region, termed a cavern, can exist around the impeller with no movement in the surrounding fluid (Wichterle and Wien, 1979; Solomon *et al.*, 1981b). The size of this cavern for both aerated and unaerated conditions has been measured and a model for predicting its size developed (Solomon *et al.*, 1981b), and such an approach may be helpful in predicting the conditions necessary for good mixing. Einsele and Finn (1980) have also found that the blending time in vessels, particularly those containing highly viscous fluids, increased with gas hold up. For impeller Reynolds number in the transition regime viscoelastic fluids show reversed flow patterns compared to elastic fluids (Geisekus, 1965; Ide and White, 1974). Although flow reversal is unlikely to occur in the impeller region in a polysaccharide fermentation, such effects may be important in determining local mixing at some distance from the impeller (Figure 14).

49.7.5 Fermenter Design

It is generally agreed that a mechanically agitated fermenter is required to achieve good heat and mass transfer at high viscosities (Margaritis and Zajic, 1978; Pace and Righelato, 1980; Sittig, 1982) and is supported by the data of Bhavaraju *et al.* (1978) and Henzler (1981).

At extreme viscosities, the pumping capacity of the standard flat bladed turbine impellers falls dramatically and results in a poorly mixed fermenter capable only of low rates of heat and mass transfer. On the other hand, in low viscosity fermentations this type of impeller appears to give adequate mixing and in most cases this does not limit heat and mass transfer. Thus the design of the agitation system for use in polysaccharide fermentations requires special attention to give the correct distribution of power to ensure good culture homogeneity and turbulence, to minimize bubble coalescence, to promote small bubble formation and to achieve adequate fluid movement at heat transfer surfaces, and will inevitably require non-geometric scale up. The effect of impeller geometry on the mixing patterns of viscous systems is described by Margaritis and Zajic (1978).

Improvements over the flat bladed turbine include the use of backward swept turbines for intermediate viscosity fluids or the use of two different types of impeller, one for mass transfer (small bubble generation) and one type for mixing (promotion of bulk flow) at higher viscosities. Through the use of flow visualization techniques, Solomon *et al.* (1981b) has shown that the power required to mix fully a small vessel is markedly decreased when a combination of a flat bladed turbine, and a 45° axial flow impeller *vs.* a two flat bladed impeller of the same diameter, are used to agitate high viscous pseudoplastic fluids. In order to minimize the power input to achieve the desired rates of heat, mass and momentum transfer into a fermenter used for viscous mycelial fermentation, Anderson *et al.* (1982) used two separate drives for the gassing dispersing turbine impeller and the slower moving bulk flow promoting axial flow propeller.

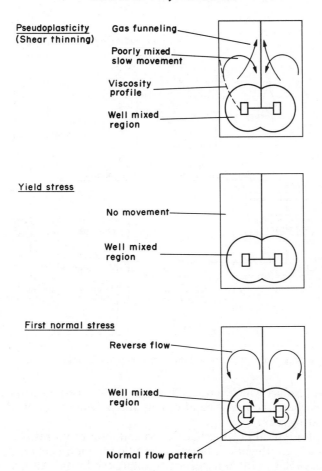

Figure 14 The effects of various forms of rheological behaviour on flow patterns in agitated vessels

49.8 RECOVERY OF THE EXOPOLYSACCHARIDE PRODUCT

The cost of recovery of microbial polysaccharides, including concentration, isolation and purification, is a significant part of the total production cost (see Section 49.9) and is due to the dilute nature of the stream leaving the fermenter (about 15 to 30 kg m^{-3}), the presence of contaminating solids (*e.g.* cells) and solutes in the stream and its high viscosity. The objectives of the recovery processing are (Smith and Pace, 1981): (i) concentration of the fermentation broth or extract to a form, usually a solid which is microbiologically stable, easy to handle, transport and store and can be readily redissolved or diluted for use in a particular application; (ii) purification to reduce the level of non-polymer solids, such as cells or salts, and to improve the functional performance, colour, odour or taste of the product; (iii) deactivation of undesirable contaminating enzymes, such as cellulases, pectinases, *etc.*, and (iv) modification of the chemical properties of the polymer to alter either the functional performance, the solid's handling properties of the dried product or the dispersion and solution rate characteristics.

Two examples of recovery processes which have been suggested for the production of microbial alginate and xanthan are shown in Figure 15, and in general the processing steps are similar to those used in the harvesting of plant and algal gums (Pace and Righelato, 1981). In the following sections the commonly described operations are discussed and further information can be found elsewhere (Pace and Righelato, 1981; Smith and Pace, 1982).

49.8.1 Cell Removal

The presence of cells in the final product can directly affects its end application performance as well as diluting the amount of polymer present. Cell removal may be essential, for example, if the

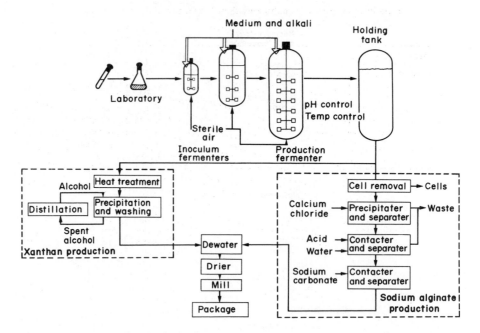

Figure 15 Processes for the production of the microbial polysaccharides xanthan and alginate

gum is to be used in a product which is optically clear, such as certain cosmetics or foods, or in enhanced oil recovery where the presence of cells in the polymer flood fluid causes plugging of the pores of the oil bearing rock.

The high viscosity of a typical polysaccharide fermentation broth results in conventional cell separation steps such as centrifugation, filtration and flocculation being expensive, and to give practical rates of processing it is often necessary to dilute the broth. Also, some advantage may be derived during processing by heating the broth to lower its viscosity, however care must be taken not to degrade the product and, in the case of xanthan, the fluid must be heated to more than 100 °C before the viscosity falls (Pace and Righelato, 1981). Several patents exist which cover these physical methods of cell separation (for example, O'Connell, 1967; Corley and Richmon, 1979; Walker and Dunlop, 1979; Naslund and Laskin, 1980).

An alternative approach to cell removal is the degradation of cells by chemical or enzymatic methods (for example, Colgrove, 1970; Colgrove, 1976; Stokke, 1979; Holding and Pace, 1981), however, the resultant product may contain cell fragments and although it may be suited to, say, enhanced oil recovery applications, it may not form optically clear solutions.

49.8.2 Isolation of Exopolysaccharide

The isolation of a polysaccharide is achieved by lowering of the solubility of the polymer to result in either a precipitate or a phase consisting of a concentrated solution, a coacervate, by either the addition of a water miscible solvent, *e.g.* methanol, ethanol, isopropranol, acetone (note that in strict terms it is a non-solvent for the polymer), or salt or acid. The molecular mechanisms behind these methods are described in detail by Smith and Pace (1982), but broadly rely on a decrease in the polymer–solvent (water) affinity by either changing the net charge on the polymer (addition of salt or an acid) and/or through competition with the polymer for water (addition of large amounts of salt), or by changing the hydrophilicity of the solvent (addition of alcohol). The method of choice for isolation of the polysaccharide is determined by economics, practicability and the final specification of the product, with alcohol precipitation being widely favoured. Some examples of alcohol, salt and acid precipitation are given in Table 10 and by Pace and Righelato (1981) and Smith and Pace (1982).

The volume of alcohol required for precipitation of xanthan is relatively independent of polymer concentration but is affected by the concentration of certain salts. For example, addition of potassium chloride to xanthan broths decreases the amount of isopropanol required for precipitation by about 30% (Pace and Righelato, 1980). Thus increasing the concentration of product

Table 10 Examples of Methods used for the Precipitation of Microbial Polysaccharides

Method	Example	Comments
Alcohol precipitation	Isopropanol	45–60% (w/w) of isopropanol required for xanthan cultures; precipitation conditions and washing can affect subsequent processing and final product (Bouniot, 1976; Roche, 1981)
Salt precipitation	Calcium hydroxide	Insoluble salt of alginate or xanthan formed; required acid titration to form soluble salt of polymers (Mehltretter, 1965; see Figure 7). Suitable for most polyanionic polysaccharides
	Quaternary ammonium compounds	Precipitate washed with methanol to give soluble salt (Albrecht *et al.*, 1965); unlikely to be acceptable for a food grade process
Acid precipitation	Lower pH with hydrochloric acid	Alginic acid precipitated at pH<4; used in some commercial processes for extraction of algal alginate

during fermentation and increasing the broth's salt content prior to precipitation will decrease the amount of alcohol processed per unit weight of polymer recovered. Such improvements can significantly affect production economics (see Section 49.9). The coprecipitation of cells, certain inorganic salts and colour pigments with the polymer leads to decreased product purity and, hence, rheological performance or appearance and washing with alcohol/water will only partly remove such impurities. Also, adjustments in the fermentation medium and conditions can also be used to minimize coprecipitation and improve final product purity.

49.8.3 Dewatering and Drying

The final purity of the product and the cost of drying can be improved if the wet precipitate is dewatered prior to drying, for example by pressing or centrifugation. Exposure of the product to excessive heat or mechanical stresses during dewatering can lead to product degradation.

Forced air or vacuum continuous or batch driers are normally used to dry the dewatered precipitate. The main aim of drying is to lower the water content of the product to a level where it is microbiologically stable and in an easily handled form. Drying conditions can also affect other properties of the polymer including its colour, solubility, rheological properties and solvent content. High drying temperatures can lead to poor colour and solubility, and product degradation.

An alternative approach to the recovery of the product is drying of the whole culture fluid (Albrecht and Sohns, 1965). However, the product obtained contains a high level of impurities and colour and as such is not commercially desirable.

49.8.4 Milling and Packaging

The particle size of the dried product is controlled to ensure a proper balance between dispersability, *i.e.* the ability of the polymer to be dispersed in water without clumping, and dissolution rate, *i.e.* the rate at which it dissolves into solution. For example, large particles tend to be more readily dispersed but slower dissolving compared to smaller particles, which tend to enter solution quickly but are more difficult to disperse. Control of these properties is important to the end user as it determines the method and type of equipment required for making up solutions. Again excessive heating or mechanical stresses during milling can lower product quality.

Owing to the hydroscopic nature of the dried polysaccharide it is important that the polymer be packaged in containers with a low permeability to water. Moisture pick-up can lead to clumping and subsequent problems in solution make-up, and in some cases hydrolytic degradation of the product.

49.8.5 Additional Processing

Gums may be physically or chemically treated or modified during recovery to affect their purity, cosmetic appearance, handling characteristics, rheological or other physical properties and the processing method used for a given polymer depends on its physical structure and chemical reactivity. Some specific examples of the types of changes that can be carried out on a microbial gum, xanthan, are given below.

Xanthan is an unusual polymer in that when solutions of high concentration containing salt are heated the viscosity remains approximately constant to temperatures in excess of 100 °C. In addition to this phenomenon, controlled heating of xanthan fermentation broth in the region of 100 °C for a few minutes results in a final dried product with enhanced low concentration and shear rheology owing to conformational changes or cross-linking of polymer chains (Kang and Burnett, 1977; Pace and Smith, 1982). Heating of most other polymers at this temperature results in degradation, reflected in the low viscosity of the finished product, and only when xanthan is heated above about 150 °C does it degrade rapidly. The dispersability of xanthan can be improved by reacting it with a dialdehyde such as glyoxal (Cottrell and Hartneck, 1976). When the dried glyoxal xanthan complex is added to water its low solubility aids its dispersion and as the glyoxal–xanthan bonds hydrolyse the polymer enters the solution.

Microbial exopolysaccharides may contain extraneous enzymes which restrict the range of materials with which the gum is compatible. For example, xanthan directly precipitated from culture fluid contains cellulase which is very active against other cellulosic gums, such as carboxymethylcellulose (CMC), and thus cannot be used together with CMC in an end application. However, heating or treatment of xanthan with an oxidizing agent such as propylene oxide may be used to inactivate the cellulase and render it compatible with CMC (Empey and Pettitt, 1978).

Polysaccharides can also be reacted with various chemicals to modify their structure and hence their rheological properties. Xanthan can be deacetylated by heating the gum under alkaline conditions, and the viscosities of solutions of the resultant product are claimed to be enhanced by the addition of salt (Patton, 1976). Cross-linking xanthan with formaldehyde gives a stable, water soluble form with enhanced viscosity (Patton, 1962). A graft polymer between polyacrylamide and xanthan can be formed and is claimed to have superior properties compared to xanthan when used as an additive in polymer floods in enhanced recovery (Stratton, 1974).

49.9 PROCESS ECONOMICS OF MICROBIAL EXOPOLYSACCHARIDE PRODUCTION

To illustrate the likely approximate costs involved in the production of microbial exopolysaccharides and the effect of certain process variables on them, an economic analysis of a 'greenfields' sited plant producing 2000 tons per annum of product is presented in this section. The key assumptions used in the costing are listed in Table 11 and, although hypothetical, they are in line with information in the literature defining the xanthan production process. A process flow identical to that shown in Figure 15 for xanthan production was assumed. A discounted cash flow return on investment calculation was used in the analysis which accounted for the total project costs, its life and the time value of money. As the selling price of gums varies with the type, grade and the quantity sold, an arbitrary 20% after tax discounted return on total investment was chosen, and for the different cases examined the average product selling price, which must be achieved to obtain that return, was determined. The estimations of capital and production costs were based on published literature (Anson, 1977: Allen, 1980; Peters and Timmerhaus, 1968), manufacturers information and internal models. The absolute values of the selling prices can only be used as a guide as they will be affected by inaccuracies in the costings and the varied approaches of companies, based on their own internal cost structure, to estimating selling price.

The various cost elements derived in determining the selling price for the product from the base case plant are summarized in Tables 12, 13, 14, and 15. The calculated selling price yielding

Table 11 Key Assumptions Used in the Economic Analysis of a 2000 tonnes/year Microbial Polysaccharide Production Process

---

Project assumptions

Greenfields site

Research and Development Costs $2.0 m; expenditure rate $0.5 m year 0, $1.0 m year 1, $0.5 m year 2

Plant construction takes 2.5 years and started after 2.5 years from project commencement; plant capital expenditure rate 10%, year 2; 30%, year 3; 60%, year 4

Plant life 15 years

Project profitability 20% after tax discounted cash flow return on total investment (*i.e.* fixed capital plus working capital and R and D costs)

Annual production rate 2000 tonnes/year of product at 5% moisture

Process assumptions

Batch process: total cycle time = 50 h
turnround time = 10 h
fermentation time = 40 h

Final product concentration = 25 kg m^{-3} of dried product

Yield = 0.7 kg dried product kg^{-1} of glucose

Fermentation medium — defined glucose salts (Souw and Demain, 1979)

Total product losses = 5% of dried product

Moisture content of finished product = 5%

Plant availability = 8000 ha^{-1}

Fermentation temperature = 30 °C

Distillation and solvent processing losses = 0.35 kg alcohol kg^{-1} product

---

a 20% after tax discounted rate of return for the finished product from the 2000 tons per annum capacity plant was $13.20 kg^{-1}. Although this price of about $13 kg^{-1} is only indicative, it does illustrate that microbial gums are expensive when compared to other polymers such as starch, modified starches and some cellulose derivatives (see Table 3). Thus, microbial gums are most likely to be saleable when they fulfil a market niche which other products cannot, and where the market will pay a relatively high price. For these reasons, most industrial research into microbial polysaccharides appears to be aimed at high price speciality markets. One exception may be the use of polymers in enhanced oil recovery, where the gum would be produced in large quantities near the oil field and the fermentation broth used directly in the make-up of the polymer flooding solution. For example, if a greenfields sited plant producing 20 000 tons per annum of fermentation broth at 25 kg m^{-3} is considered, the cost of the product falls to approximately $4.60 kg^{-1} of polymer solids.

Table 12 Estimated Total Capital Costs for a 2000 tonnes/year Microbial Polysaccharide Plant

Element		Cost ($ × 10³)
Purchased equipment cost		7138
Purchased equipment installation		2859
Piping (installed)		3560
Instrumentation (installed)		1100
Electrical (installed)		1059
Buildings/civil and site improvements		2781
Land		428
Engineering/supervision/commissioning		2284
Construction expenses		2427
	Sub-total	23 636
Contractors' fees (8%)		1891
Contingency (20%)		5105
Total fixed capital		30 632
Working capital		3223
Total capital		33 855

A variety of factors affect the selling price of the product and the previous example demonstrated the influence of the scale of operation and the lack of a product recovery stage in the process on the price. In the following, the relative effects on the gum's selling price of variations of certain parameters from the base values are examined. The variables, which are of prime concern

Table 13   Estimated Purchased Equipment Cost for 2000
tonnes/year Microbial Polysaccharide Plant

Plant section	Purchased equipment cost ($ \times 10^3$)
Raw material storage and media make-up	230
Sterilization and fermentation	3209
Culture fluid storage and heat treatment	244
Product isolation, drying and packaging	1080
Solvent distillation and storage	706
Utilities	1671
Total	7140

Table 14   Estimated Production Cost for 2000^{-1} tonnes/year Microbial
Polysaccharide Plant

Element			Cost ($ \times 10^3$/year)
Raw materials:	Fermentation	glucose	1000
		others	334
	Recovery	isopropanol	448
	Packaging		80
Utilities			2775
Salaries and wages			767
Overheads and general expenses			640
Maintenance materials and labour[a]			1532
Selling and distribution[b]			1337
Total			8913

[a] Assumed 5% of total fixed capital. [b] Assumed 15% of total operating cost.

Table 15   Estimated Utilities Usage and Costs for 2000 tonnes/year
Microbial Polysaccharide Plant

Utility	Usage (year^{-1})	Cost ($ \times 10^3$)	Principal use
Gas	2.72×10^5 GJ	1142	Steam generation for distillation
	1.9×10^4 GJ	80	Hot air generation for product drier
Electricity	5.03×10^7 kwh	1509	Fermenter agitation aeration and cooling
Water	80 000 m^3	20	Process
Effluent	80 000 m^3	24	Process
Total		2775	

to biotechnologists, group together as those affecting either the capital and/or operating costs of the plant. The sensitivity of the selling price to variations in these is given in Figure 16 and shows the higher sensitivity to changes in capital than operating costs. The main components of the capital and production costs are listed in Tables 12, 13, 14 and 15.

Changes in the conversion efficiency, kg product kg^{-1} glucose consumed; productivity, kg product m^{-3} of fermenter h^{-1}; or product concentration, kg product m^{-3} of culture fluid, of the base case fermentation affects the selling price of the gum by altering specific elements in the capital and production costs. Examples of the effects of these parameters and the cost elements which they alter significantly are given in Figures 17, 18 and 19 and Table 16.

Figure 17 demonstrates the high sensitivity of the gum's selling price to the final product concentration in the culture fluid. If the product concentration is varied, the volume of culture fluid required to be processed to produce 2000 tons per annum of polysaccharide changes proportion-

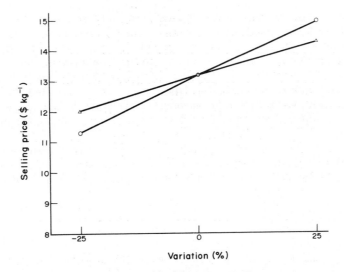

Figure 16 The effect of variations in product and capital cost on product selling price for a 2000 tonnes/year microbial polysaccharide plant. △, production cost variation; ○, capital cost variation

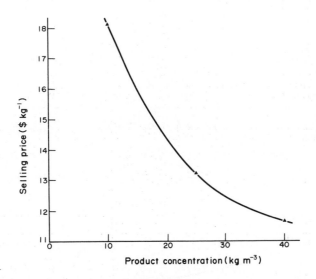

Figure 17 The effect of variations in product concentration on product selling price for a 2000 tonnes/year microbial polysaccharide plant

Table 16 Main Capital and Production Cost Elements Affected by Changes in Fermentation Conversion Efficiency, Productivity and Product Concentration

Parameter varied	Main cost elements affected
Conversion Efficiency	Raw material (glucose) usage
Productivity[a]	Fermentation capital cost[b]
Product Concentration	Solvent losses
	Steam usage (distillation)
	Fermentation, precipitation and
	distillation capital costs[b]

[a] The method examined for increasing fermentation productivity used the drawfill technique which decreases the ratio of turnaround time to the overall cycle time. [b] The maintenance charge also was varied as this is factored from the total capital cost (Table 14).

ately, and thus the size or capital cost of the plant processing the culture fluid, *i.e.* fermentation, precipitation and distillation, is affected. As the charge for maintenance materials and labour is taken as a proportion of the total capital cost, it will also vary. Furthermore, the volume of solvent required for precipitation is almost directly proportional to the volume of fermentation broth to be processed and relatively independent of product concentration, hence process solvent losses and the steam requirements for distillation are markedly affected. Other operating costs which are changed, but to a lesser extent, include the fermentation heat and mass transfer related utilities which are controlled by the culture fluid rheology or product concentration and characteristics.

Changes in fermentation conversion efficiency and productivity result mainly in varied raw material (glucose) usage and fermenter size, however, as these represent a low proportion of the total costs (Tables 13 and 14) their effect on selling price is slight (Figures 18 and 19).

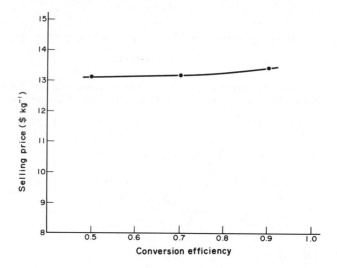

Figure 18 The effect of variations in conversion efficiency on product selling price for a 20 000 tonnes/year microbial polysaccharide plant

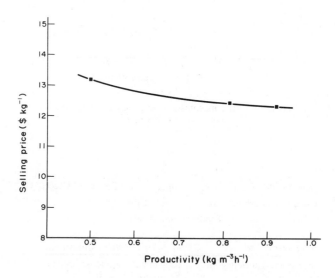

Figure 19 The effect of variations in productivity on plant selling price for a 20 000 tonnes/year microbial polysaccharide plant

All examples considered up to this point assume a plant built on a greenfields site, thus if an existing plant is available and can be readily adapted for polysaccharide production, the selling price to give the desired 20% return will be less due to likely lower capital and research and development charges and faster entry to the market place. A hypothetical example to demonstrate these combined effects is given in Table 17 and clearly shows the benefits of minimizing expenditure in the early stages of the project and shortening the time taken to get the product onto the market.

Table 17 The Effect of Decreasing Research and Development and Capital Costs and Expenditure Period on Project Profitability[a]

	R and D Costs ($ × 10⁶)	Expenditure period (year)	Capital cost ($ × 10⁶)	Expenditure period (year)	Plant on stream (year)	Estimated selling price ($ kg⁻¹)
Base case	2.0	3	33.8	2.5	5	13.2
Existing plant	1.0	2	10.0	1.5	3	7.6

[a] From the point of first sales the rate of market penetration was assumed to be the same for both cases.

49.10 CONCLUSIONS

Microbial exopolysaccharides are water soluble biopolymers produced by a wide variety of microorganisms grown on different carbohydrate and other substrates under submerged culture conditions. In view of their unique physical and chemical properties, exopolysaccharides have found a wide range of important applications in the pharmaceutical, food, chemical and other industries. The rheological characteristics of exopolysaccharides are of particular importance, and this resulted in their use as emulsifiers, gelling agents, stabilizers, coagulants, binders, film formers, suspending agents and lubricants.

The structure and composition of microbial exopolysaccharides depends on many different factors, which may include medium composition, type of carbon and energy source, microbial system used, and other fermentation conditions such as pH, temperature and oxygen concentration. Despite the commercial importance of microorganisms which produce exopolysaccharides, very little is known about the biochemistry of these microbial systems. A good understanding of the biochemical pathways is important because this information can be used to optimize and control the biosynthesis of these exopolysaccharides during fermentation. More research is needed to elucidate the various biochemical pathways involved in different exopolysaccharide systems.

Depending on the microbial system used and fermentation conditions, the kinetics of exopolysaccharide production was found to be cell growth or non-growth associated or mixed. This diversity in the kinetics of different systems necessitates the choice of the best suited modes of fermentation, *i.e.* continuous, batch or semi-batch, in order to maximize the rate of exopolysaccharide production. During the course of fermentation, the broth starts as a low viscosity Newtonian system and ends up as a highly viscous non-Newtonian system as the exopolysaccharide concentration increases with time. These extreme rheological changes have profound effects on the mixing, mass transfer and heat transfer characteristics of exopolysaccharide fermentations. Optimal fermenter design and operation must take into account these rheological changes. A lot of work remains to be done on mixing, mass transfer and especially heat transfer characteristics of exopolysaccharide fermentations.

A process economic feasibility study revealed that the cost of downstream recovery and purification of the exopolysaccharide product represents a significant part of the total production cost. Very little information is available in the research literature on the optimization of recovery and purification of exopolysaccharides. Microbial exopolysaccharides represent only a small portion of the total polysaccharide market, most notable of which are xanthan gum and dextrans. At the present time the production cost of microbial polysaccharides is higher than the cost of other traditional polysaccharides, such as cornstarch and cellulose-derived products, which dominate the market. However, the competitive advantage of microbial polysaccharides will improve as more research and development work will undoubtedly result in lower production costs, and new industrial speciality applications will create higher demand for these products.

ACKNOWLEDGEMENTS

The authors wish to thank Mr. Bruce R. Wallace, Mr. Wayne Vollick and Mr. Derek te Bokkel for their assistance with the drawings in this chapter.

49.11 REFERENCES

Acharya, A. and J. J. Ulbrecht (1978). Note on the influence of viscoelasticity on the coalescence rate of bubbles and drops. *AIChE J.*, **24**, 348–351.

Albrecht, W. J., V. E. Sohns and S. P. Rogovin (1963). Pilot-plant process for the isolation of a microbial polysaccharide with a quaternary ammonium compound. *Biotechnol. Bioeng.*, **5**, 91–99.

Allen, D. H. (1980). *A Guide to the Economic Evaluation of Projects*. The Institution of Chemical Engineers, Rugby, England.

Amemura, A., M. Hisamatsu and T. Harada (1977). Spontaneous mutation of polysaccharide production in *Alcaligenes faecalis* var. *myxogenes* 10C3. *Appl. Environ. Microbiol.*, **34**, 617–620.

Anderson, C., G. A. LeGrys and G. L. Solomons (1982). Concepts in the design of large-scale fermenters for viscous culture broths. *Chem. Eng.* (London), February, 43–49.

Anson, H. H. (1977). *A New Guide to Capital Cost Estimating*. The Institution of Chemical Engineers, Rugby, England.

Astarita, G. and G. Apuzzo (1965). Motion of gas bubbles in non-Newtonian liquids. *AIChE J.*, **11**, 815–820.

Baird, J. K., P. A. Sandford and I. W. Cottrell (1983). Industrial applications of some new microbial polysaccharides. *Bio/Technology*, **1**, 778–783.

Bernstein, R. L. and P. W. Robbins (1965). Control aspects of uridine 5'-diphosphate glucose and thymidine 5'-diphosphate (TDP) glucose synthesis by microbial enzymes. *J. Biol. Chem.*, **240**, 391–397.

Bhavaraju, S. M., T. W. F. Russell and H. W. Blanch (1978). The design of gas sparged devices for viscous liquid systems, *AIChE J.*, **24**, 454–460.

Blakebrough, N., W. J. McManamey and K. R. Tart (1978). Heat transfer to fermentation systems in an air-lift fermenter, *Trans. I. Chem. E.*, **56**, 127–135.

Blanch, H. W. and S. M. Bhavaraju (1976). Non-Newtonian fermentation broths: rheology and mass transfer. *Biotechnol. Bioeng.*, **17**, 745–790.

Bluhm, T. L., Y. Deslandes, R. H. Marchessault, S. Perez and M. Rinaudo (1982). Solid state and solution conformation of scleroglucan. *Carbohydr. Res.*, **100**, 117–130.

Bouniot, A. (1976). Process for the production of fermentation polysaccharides having a non-fibrous structure. *US Pat.* 3 988 313.

Bruijn, W., K. van't Riet, and J. M. Smith (1974). Power consumption with aerated rushton turbines. *Trans. Inst. Chem. Eng.*, **52**, 88–104.

Bucke, C., L. Deavin, C. J. Lawson and D. F. Pindar (1975). The production of industrially important bacterial polysaccharides. *Biochem. Soc. Trans.*, **3**, 844–847.

Cadmus, M. C., C. A. Knutson, A. A. Lagoda, J. E. Pittsley and K. A. Burton (1978). Synthetic media for production of quality xanthan gum in 20 liter fermentors. *Biotechnol. Bioeng.*, **32**, 1103–1014.

Cadmus, M. C., S. P. Rogovin, K. A. Burton, J. E. Pittsley, C. A. Knutson and A. Jeanes (1976). Colonial variation in *Xanthomonas campestris* NRRL B-1459 and characterization of the polysaccharide from a variant strain. *Can. J. Microbiol.*, **22**, 942–948.

Calderbank, P. H. and M. B. Moo-Young (1959). The prediction of power consumption in the agitation of non-Newtonian fluids. *Trans. I. Chem. E.*, **37**, 26–33.

Calderbank, P. H. and M. B. Moo-Young (1961). The power characteristics of agitators for the mixing of Newtonian and non-Newtonian fluids. *Trans. I. Chem. E.*, **39**, 337–347.

Catley, B. J. (1971). Role of pH and nitrogen limitation in the elaboration of the extracellular polysaccharide pullulan by *Pullularia pullulans. Appl. Microbiol.*, **22**, 650–654.

Catley, B. J. and W. J. Whelan (1971). Observations on the structure of pullulan. *Arch. Biochem. Biophys.*, **143**, 138–142.

Catley, B. J. (1978). Pullulan synthesis by *Aureobasidium pullulans*. In *Microbial Polysaccharides and Polysaccharases*, ed. R. C. W. Berkeley, G. W. Gooday and D. C. Ellwood. Academic, London.

Catley, B. J. (1980). The extracellular polysaccharide, pullulan, produced by *Aureobasidium pullulans*: a relationship between elaboration rate and morphology. *J. Gen. Microbiol.*, **120**, 265–268.

Church, B. D. (1980). Two staged continuous fermentation process for production of heteropolysaccharide. *US Pat.* 4 218 538.

Colegrove, G. T. (1970). Treatment of *Xanthomonas* hydrophillic colloid and resulting product, *US Pat.* 3 516 983.

Colegrove, G. T. (1976). Clarification of Xanthan gum, *US Pat.* 3 966 618.

Compere, A. L. and W. L. Griffith (1978). Production of high viscosity glucans from hydrolyzed cellulosics. *Dev. Ind. Microbiol.*, **19**, 601–607.

Corley, F. E. and J. B. Richmon (1979). Treatment of xanthan gum to improve clarity. *US Pat.* 4 135 979.

Cottrell, I. W. and K. S. Kang (1978). Xanthan gum, a unique bacterial polysaccharide for food applications. *Dev. Ind. Microbiol.*, **19**, 117–131.

Cottrell, I. W. and H. G. Hartnek (1979). Improving polysaccharide dispersability, *Br. Pat.*, 1 547 030.

Cottrell, I. W. (1980). Industrial potential of fungal and bacterial polysaccharides. *ACS Symp. Ser.* **126**, 251–270.

Davidson, I. W. (1978). Production of polysaccharide by *Xanthomonas campestris* in continuous culture. *FEMS Microbiol. Lett.* **3**, 347–349.

Dawes, E. A., D. W. Ribbons and D. A. Rees (1966). Sucrose utilization by *Zymomonas mobilis*: formation of a levan. *Biochem. J.*, **98**, 804.

Deavin, L., T. R. Jarman, C. J. Lawson, R. C. Righelato and S. Slocombe (1977). The production of alginic acid by *Azotobacter vinelandii* in batch and continuous culture. *ACS Symp. Ser.*, **45**.

De Hollander, J. A., C. W. Bettenhaussen and A. H. Stouthamer (1979). Growth yields, polysaccharide production and energy conservation in chemostat cultures of *Rhizobium trifolii*. *Antonie van Leeuwenhoek*, **45**, 401–415.

Dintzis, E. R., G. E. Babcock and R. Tobin (1970). Studies on dilute solutions and dispersions of the polysaccharide from *Xanthomonas campestris* NRRL B-1459. *Carbohydr. Res.*, **13**, 257–267.

Ebert, K. H. and G. Schenk (1968). Dextran formation. *Adv. Enzymol.*, **30**, 190–202.

Edney, H. G. S. and M. F. Edwards (1976). Heat transfer to non-Newtonian and aerated fluids in stirred tanks. *Trans. Inst. Chem. Eng.*, **54**, 160–166.

Edward, M. F. and W. L. Wilkinson (1972). Heat transfer in agitated vessels, Part II—non-newtonian fluids. *Chem. Eng.* (London), September, 328–335.

Einsel, A. and R. K. Finn (1980). Influence of gas flow rates and gas holdup on blending efficiency in stirred tanks. *Ind. Eng. Chem. Process Des. Dev.*, **19**, 600–603.

Empey, R. A. and D. J. Pettitt (1978). Cellulase-free xanthan gum and process for producing same. *US Pat.* 4 070 535.

Ellwood, D. C., R. C. Righelato and C. E. Smith (1978). Process for the production of xanthan gum. *Br. Pat.* 2 008 138.

Evans, L. R. and A. Linker (1973). Production and characterization of the slime polysaccharide of *Pseudomonas aeruginosa. J. Bacteriol.*, **116**, 915–924.

Evans, C. G. T., R. G. Yeo and D. C. Ellwood (1978). Continuous culture studies on the production of extracellular polysaccharides by *Xanthomonas juglandis*. In *Microbial Polysaccharides and Polysaccharases*, ed. R. C. W. Berkeley, G. W. Gooday and D. C. Ellwood. Academic, London.

Finkelman, M. A. J. and A. Vardanis (1982). Simplified microassay for pullulan synthesis. *Appl. Environ. Microbiol.*, **43**, 483–485.

Finkelman, M. A. J. and A. Vardanis (1982). Pullulan elaboration by *Aureobasidium pullulans* protoplasts. *Appl. Environ. Microbiol.*, **44**, 121–127.

Foresti, Jr., R. and T. Liu (1959). Agitation of non-Newtonian liquids. *Ind. Eng. Chem.*, **51**, 860–864.

Garegg, P. J., B. Lindberg, T. Onn and I. W. Sutherland (1971). Comparative structural studies on the M-antigen from *Salmonella typhimurium, Escherichia coli* and *Aerobacter cloacae. Acta Chem. Scand.*, **25**, 2103–2108.

Giesekus, V. H. (1965). Secondary currents in visco-elastic fluids during stationary and period movement. *Rheol. Acta*, **4**, 85–101.

Ginsburg, V. (1964). Sugar nucleotides and the synthesis of carbohydrates. *Adv. Enzymol.* **26**, 35–88.

Goto, S., S. Enomoto, Y. Takahashi and R. Motomatsu (1971). Slime production by *Pseudomonas aeruginosa. Jpn. J. Microbiol.*, **15**, 317–324.

Grechushkina, N. N. and L. I. Rozanova (1971). *Mikrobiologiya*, **40**, 820.

Griffith, W. L. and A. L. Compere (1978). Production of a high viscosity glucan by *Sclerotium rolfsii* ATCC 15206. *Dev. Ind. Microbiol.*, **19**, 609–616.

Haggström, L. (1977). Mutant of *Methylomonas methonolica* and its characterization with respect to biomass production from methanol. *Appl. Environ. Microbiol.*, **33**, 567–576.

Hacking, A. J., I. W. F. Taylor and T. R. Jarman (1983). Alginate biosynthesis by *Pseudomonas mendocina*. To be published.

Harada, T. (1977). Production, properties and application of Curdlan. *ACS Symp. Ser.*, **45**, 265–283.

Hattori, K., K. Yokoo and O. Imada (1972). Oxygen transfer in a highly viscous solution. *J. Ferment. Technol.*, **50**, 737–741.

Haug, A. and B. Larsen (1971). Biosynthesis of alginate. II Polymannuronic acid C_5 epimerase from *Azotobacter vinelandii. Carbohydr. Res.*, **17**, 297–308.

Hellman, N. N., H. M. Tsuchiya, S. P. Rogovin, B. L Lamberts, R. Tobin, C. A. Glass, C. S. Stringer, R. W. Jackson and F. R. Senti (1955). Controlled enzymatic synthesis of dextran. *Ind. Eng. Chem. Prod. Proc. Dev.*, **47**, 1593–1598.

Henzler, H. J. (1981). Gas–liquid contacting in highly viscous liquids. In *Advances in Biotechnology*, ed. M. Moo-Young and C. W. Robinson. **1**, pp. 497–502. Pergamon, Toronto.

Holding, T. J. and G. W. Pace (1981). Enzymatic clarification of polysaccharide solutions. *Br. Pat.* 2 065 689.

Horan, N. J., T. R. Jarman and E. A. Dawes (1983). Studies on some enzymes of alginic acid biosynthesis in *Azotobacter vinelandii* grown in continuous culture. To be published.

Horan, N. J., T. R. Jarman and E. A. Dawes (1981). Effects of carbon source and inorganic phosphate concentration on the production of alginic acid by a mutant of *Azotobacter vinelandii* and on the enzymes involved in its biosynthesis. *J. Gen. Microbiol.*, **127**, 185–191.

Hou, C. T., A. I. Laskin, and R. N. Patel (1978). Growth and polysaccharide production by *Methylocystis parvus* OBBP on methanol. *Appl. Environ. Microbiol.*, **37**, 800–804.

Ide, Y., and J. L. White (1974). Rheological phenomena in polymerization reactors: rheological properties and flow patterns around agitators in polystyrene–styrene solutions. *J. Appl. Polymer Sci.*, **18**, 2997–3018.

Jarman, T. R. and G. W. Pace (1984). Energy requirements for microbial exopolysaccharide synthesis. *Arch. Microbiol.*, **37**, 231–235.

Jarman, T. R. (1978). Bacterial alginate synthesis. In *Microbial Polysaccharides and Polysaccharases*, ed. R. C. W. Berkeley, G. W. Gooday and D. C. Ellwood, Academic, London.

Jarman, T. R., L. Deavin, S. Slocombe and R. C. Righelato (1978). Investigation of the effect of environmental conditions on the rate of exopolysaccharide synthesis in *Azotobacter vinelandii. J. Gen. Microbiol.*, **107**, 59–64.

Jarman, T. R., and G. W. Pace (1981). Production of xanthan having a low pyruvate content. *Br. Pat.* 2 115 854.

Jeanes, A. (1974). New hydrocolloids of interest to the food industry, *Food Technol.*, May, 34–40.

Kale, D. D., R. A. Mashelkar and J. Ulbrecht (1975). Rotational viscoelastic laminar boundary layer flow around a rotating disc. *Rheol. Acta*, **14**, 631–640.

Kanamaru, T. and S. Yamatodani (1969). *Agric. Biol. Chem.*, **33**, 1521.

Kang, K. S., G. T. Veeder and D. D. Richey (1977). Zanflo— A novel bacterial heteropolysaccharide. *ACS Symp. Ser.*, **45**, 211–219.

Kang, K. S. and W. H. McNeely (1977). PS-7—a new bacterial heteropolysaccharide. *ACS. Symp. Ser.*, **45**, 220–230.

Kang, K. S. and D. B. Burnett (1977). Xanthan gum. *Br. Pat.* 1 488 645.

Kang, K. S. and G. T. Veeder (1982). Polysaccharide S-60 and Bacterial process for its preparation. *US Pat.* 4 326 053.

Kang, K. S. and G. T. Veeder (1983). Fermentation process for its preparation of polysaccharide S-60. *US Pat.* 4 377 636.

Kaplan, N. and E. Rosenberg (1982). Exopolysaccharide distribution of and bioemulsifier production by *Acinetobacter calcoaceticus* BD4 and BD413. *Appl. Environ. Microbiol.*, **44**, 1335–1441.

Kato, K. and M. Shiosaka (1975). *US Pat.* 3 912 591.

Kirby, D., P. Sandford, A. Herman and M. Cadmus (1977). Maintenance procedures for the curtailment of genetic instability: *Xanthomonas campestris* NRRL B-1459. *Appl. Environ. Microbiol.*, **33**, 840–845.

Kipke, K. (1978). Gas dispersion in non-Newtonian liquids. *International Symposium on Mixing*. Faculte Polytechnique de Mons, C5-1-22.

Klimek, J. and D. F. Ollis (1980). Extracellular microbial polysaccharides: kinetics of *Pseudomonas* sp., *Azotobacter vinelandii*, and *Aureobasidium pullulans* batch fermentations. *Biotechnol. Bioeng.*, **22**, 2321–2342.

Koboli, H. and P. J. Reilly (1980). Immobilization and properties of *Leuconostoc mesenteroids* dextransucrase. *Biotechnol. Bioeng.*, **22**, 1055–1069.

Konicek, J., J. Lasik and M. Wurst (1977). Production and characteristics of the exocellular polysaccharide in mutant strains of *Xanthomonas fuscans*. *Folia Microbiol.*, **22**, 12–18.

Lawson, C. J. (1977). Production of industrially important gums with particular reference to Xanthan gum and microbial alginate. *ACS Symp. Ser.*, **41**, 282–296.

Lawson, C. J. and I. W. Sutherland (1978). Polysaccharides. In *Economic Microbiology*, ed. A. H. Rose, vol. 2. Academic, London.

Leduy, A., A. A. Marsan and B. Coupal (1974). A study of the rheological properties of a non-Newtonian fermentation broth. *Biotechnol. Bioeng.*, **16**, 61–76.

Lehninger, A. L. (1970). *Biochemistry*. Worth, N.Y.

Linker, A. and L. R. Evans (1976). Unusual properties of glycuronans. *Carbohydr. Res.*, **47**, 179–187.

Manchee, R. J., C. G. T. Evans and A. Robinson (1980). Strain variation in continuous culture. In *The Stability of Industrial Organisms*. ed. B. E. Kirsop. Commonwealth Mycological Institute, England.

Margaritis, A. and J. E. Zajic (1978). Mixing, mass transfer, and scale-up of polysaccharide fermentations. *Biotechnol. Bioeng.*, **20**, 939–1001.

Margaritis, A. and D. te Bokkel (1984). Rheology and air bubble terminal velocity measurements in non-Newtonian polysaccharide solutions. Paper presented at the 187th Annual Meeting, American Chemical Society. Philadelphia, August 26–31.

Margaritis, A. and F. Merchant (1984). Advances in ethanol production using immobilized cell systems. *CRC Crit. Rev. Biotechnol.*, **1**, 339–393.

McBride, C., J. Walter, H. W. Blanch and T. W. F. Russel (1981). Bubble coalescence and break-up in fermentations. In *Advances in Biotechnology*, ed. M. Moo-Young and C. W. Robinson, vol. 1, pp. 489–496. Pergamon, Toronto.

McNeely, W. H. and K. S. King (1973). Xanthan and other biosynthetic gums. In *Industrial Gums*, ed. R. L. Whistler and J. N. BeMiller, pp. 473–497. Academic, N.Y.

Mehltretter, C. L. (1965). Isolation of bacterial polysaccharide B-1459 through calcium hydroxide complex formation. *Biotechnol. Bioeng.*, **7**, 171–175.

Metzner, A. B., and R. E. Otto (1957). Agitation of non-Newtonian fluids. *AIChE J.*, **3**, 3–10.

Metzner, A. B., R. H. Feehs, H. L. Ramos, R. E. Otto and J. D. Tuthill (1961). Agitation of viscous Newtonian and non-Newtonian fluids. *AIChE J.*, **7**, 3–9.

Metzner, A. B. and J. S. Taylor (1960). Flow patterns in agitated vessels. *AIChE J.*, **6**, 109–114.

Mian, F. A., T. R. Jarman and R. C. Righelato (1978). Biosynthesis of exopolysaccharide by *Pseudomonas aeruginosa*. *J. Bacteriol.*, **134**, 418–422.

Misaki, A., Y. Tsumuraya, M. Kakuta, H. Takemoto and T. Igarashi (1979). D-Allose-containing polysaccharide synthesized from methanol by *Pseudomonas* sp. *Carbohydr. Res.*, **75**, C8–C10.

Moorhouse, R., M. D. Walkinshaw and S. Arnott (1977). Xanthan gum-molecular conformation and interactions. *ACS Symp. Ser.*, **45**, 90–102.

Moo-Young, M. and H. W. Blanch (1981). Design of biochemical reactors, mass transfer criteria for simple and complex systems. *Adv. Biochem. Eng.*, **19**, 1–69.

Moo-Young, M. and T. Hirose (1972). Bubble mass transfer in creeping flow viscoelastic fluids. *Can. J. Chem. Eng.*, **50**, 128–130.

Moraine, R. A. and P. Rogovin (1966). Kinetics of polysaccharide B-1459 fermentation. *Biotechnol. Bioeng.*, **8**, 511–524.

Moraine, R. A. and P. Rogovin (1973). Kinetics of the xanthan fermentation. *Biotechnol. Bioeng.*, **15**, 225–237.

Naslund, L. A. and A. I. Laskin (1980). Modified heteropolysaccharides and their preparation, *US Pat.* 4 182 860.

Neijssel, O. M. and D. W. Tempest (1975). The regulation of carbohydrate metabolism in *Klebsiella aerogenes* NCTC 418 organisms, growing in chemostat culture. *Arch. Microbiol.*, **106**, 251–258.

Nienow, A. W., D. J. Wisdom, J. Solomon, V. Machon and J. Vlcek (1981). The effect of rheology on power consumption in an aerated, agitated vessel, paper A.2.3., *Chisa Congress*, Prague.

Norberg, A. B. and S. Enfors (1982). Production of extracellular polysaccharide by *Zoogloea ramigera*. *Appl. Environ. Microbiol.*, **44**, 1231–1237.

Norval, M. and I. W. Sutherland (1969). A group of *Klebsiella* mutants showing temperature-dependent polysaccharides synthesis. *J. Gen. Microbiol.*, **57**, 369–377.

Norwood, K. W. and A. B. Metzner (1960). Flow patterns and mixing rates in agitated vessel. *AIChE J.*, **6**, 432–437.

O'Connell, J. J. (1967). Treatment of Xanthomonas hydrophilic colloid and resulting product. *US Pat.* 3 355 447.

Ono, K., N. Yasuda and S. Ueda (1977). Effect of pH on pullulan elaboration by *Aureobasidium pullulans* S-1. *Agric. Biol. Chem.*, **41**, 2113–2118.

Paca, J., P. Ettler and V. Gregr (1976). Hydrodynamic behaviour and oxygen transfer rate in a pilot plant fermenter. *J. Appl. Chem. Biotechnol.*, **26**, 309–317.

Pace, G. W. (1978). Mixing of highly viscous fermentation broths. *Chem. Eng.(London)* November, 833–837.

Pace, G. W. (1980). Rheology of mycelial fermentation broths. In *Fungal Biotechnology*, ed. J. E. Smith, D. R. Berry and B. Kristiansen. pp. 95–110. Academic, New York.

Pace, G. W. and R. C. Righelato (1980). Production of extracellular microbial polysaccharides. *Adv. Biochem. Eng.* **15**, 41–70.

Parsons, A. B. and P. R. Dugan (1971). Production of extracellular polysaccharide matrix by *Zoogloea ramigera*. *Appl. Microbiol.*, **21**, 657–661.

Patton, J. T. (1962). Thickening agent and process for producing same. *US Pat.* 3 020 207.

Patton, J. T. (1976). Modified heteropolysaccharides, *US Pat.* 2 964 972.

Perez, J. F. and O. C. Sandall (1974). Gas absorption by non-Newtonian fluids in agitated vessels. *AIChE J.*, **20**, 770–775.

Peters, M. S. and K. D. Timmerhaus (1968). *Plant Design and Economics for Chemical Engineers*. McGraw-Hill, London.

Piggott, N. H., I. W. Sutherland and T. R. Jarman (1982). Alginate synthesis by mucoid strains of *Pseudomonas aeruginosa* PAO. *Eur. J. Appl. Microbiol. Biotechnol.*, **16**, 131–135.

Railton, K., D. Farago, C. R. MacKenzie, G. R. Lawford, J. Pik and H. G. Lawford (1981). Gel-forming exopolysaccharide production by *Alcaligenes faecalis* growing in nitrogen-limited continuous culture. In *Advances in Biotechnology*, ed. M. Moo-Young, C. W. Robinson, vol. 1, pp. 243–248. Pergamon, Toronto.

Ranade, V. R. and J. Ulbrecht (1977). Gas dispersion in agitated viscous inelastic and viscoelastic liquids. *Second European Conference on Mixing, Cambridge*, Paper F6, pp. 83–100.

Ranade, V. R. and J. J. Ulbrecht (1978). Influence of polymer additives on the gas-liquid mass transfer in stirred tanks. *AIChE Journal*, **24**, 796–803.

Righelato, R. C. (1976). Selection of strains of *Penicillium chrysogenum* with reduced penicillin yield in continuous cultures. *J. Appl. Chem. Biotechnol.*, **26**, 153–159.

Roche, R. E. (1981). Recovery of xanthan gum. *Br. Pat.* 2 051 104.

Rogovin, P., W. Albrecht and V. Sohns (1965). Production of industrial grade polysaccharide B-1459. *Biotechnol. Bioeng.*, **7**, 161–169.

Schugerl, K. (1981). Oxygen transfer into highly viscous media. *Adv. Biochem. Eng.*, **19**, 71–174.

Shungu, D., M. Valiant, V. Tutlane, E. Weinberg, B. Weissberger, L. Koupal, H. Gadebusch and E. Stapley (1983). Gelrite as an agar substitute in bacteriological media. *Appl. Environ. Microbiol.*, **46** (4), 840–845.

Silman, R. W. and P. Rogovin (1972). Continuous fermentation to produce xanthan biopolymer: effect of dilution rate. *Biotechnol. Bioeng.*, **14**, 23–31.

Sittig, W. (1982). The present state of fermentation reactors. *J. Chem. Technol. Biotechnol.*, **32**, 47–58.

Skelland, A. H. P. (1967). *Non-Newtonian flow and heat transfer*. Wiley, New York.

Slodki, M. E., R. M. Ward and M. C. Cadmus (1972). Extracellular mannans from yeast. *Dev. Ind. Microbiol.*, **13**, 428–435.

Slodki, M. E., R. M. Ward, J. A. Boundy, and M. C. Cadmus (1972). Extracellular mannans and phosphomannans: structural and ibosynthetic relationships. *Ferment. Technol. Today*, 597–601.

Slodki, M. E. and M. C. Cadmus (1978). Production of Microbial Polysaccharides. *Adv. Appl. Microbiol.*, **23**, 19–54.

Smith, I. H., K. C. Symes, C. J. Lawson and E. R. Morris (1981). Influence of the pyruvate content of xanthan on macromolecular association in solution. *Int. J. Biol. Macromol.*, **3**, 129–133.

Smith, I. H. and G. W. Pace (1982). Recovery of microbial polysaccharides. *J. Chem. Technol. Biotechnol.*, **32**, 119–129.

Stauffer, K. R. and J. G. Leeder (1978). Extracellular microbial polysaccharide production by fermentation on whey or hydrolyzed whey. *J. Food Sci.*, **43**, 756–758.

Sutherland, I. W. (1983). Microbial exopolysaccharides. Their role in microbial adhesion in aqueous systems. *CRC Crit. Rev. Microbiol.*, **10**, 173–201.

Solomon, J. (1980). Mixing, Aeration and Rheology of Highly Viscous Solutions. PhD Thesis, University of London.

Solomon, J., A. W. Nienow and G. W. Pace (1981b). Flow patterns in agitated plastic and pseudoplastic viscoelastic fluids. *Inst. Chem. Eng. Symp. Ser.*, **64**, A1–A13.

Solomon, J., T. P. Elson, A. W. Nienow and G. W. Pace (1981a). Cavern sizes in agitated fluids with a yield stress, *Chem. Eng. Commun.*, **11**, 143–164.

Solomon, J., A. W. Nienow and G. W. Pace (1981c). Power and hold-up in the mixing of aerated highly viscous non-Newtonian fluids. In *Advances in Biotechnology*, ed. M. Moo-Young and C. W. Robinson, vol. 1, pp. 503–509. Pergamon, Toronto.

Souw, P. and A. L. Demain (1979). Nutritional studies on xanthan production by *Xanthomonas campestris* NRRL B1459. *Appl. Environ. Microbiol.*, 1186–1192.

Stokke, D. M. (1979). Biopolymer filterability improvement by caustic-enzyme treatment. *US Pat.* 4 165 257.

Stratton, C. (1974). Method of controlling subterranean formation permeability employing graft polymerized bacterial gum polysaccharide compositions. *US Pat.* 3 844 348.

Sutherland, I. W. (1979). Microbial exopolysaccharides: control of synthesis and acylation. In *Microbial Polysaccharides and Polysaccharases*, ed. R. C. W. Berkeley, G. W. Gooday and D. C. Ellwood. Academic, London.

Sutherland, I. W. (1979). Enhancement of polysaccharide viscosity of mutagenesis. *J. Appl. Biochem.*, **1**, 60–70.

Sutherland, I. W., and G. A. Keen (1981). Alginases from *Beneckea pelegia* and *Pseudomonas* spp. *J. Appl. Biochem.*, **3**, 48–57.

Symes, K. C. (1980). The relationship between the covalent structure of the *Xanthomonas* polysaccharide (xanthan) and its function as a thickening, suspending and gelling agent. *Food Chem.*, **6**, 63–76.

Tam, K. T., and R. K. Finn (1977). Polysaccharide formation by a *Methylomonas*. *ACS Symp. Ser.*, **45**, 58–80.

Tanzer, J. M., W. I. Wood and M. I. Krichevsky (1970). Linear growth kinetics of plaque-forming streptococci in the presence of sucrose. *J. Gen. Microbiol.*, **58**, 125–133.

te Bokkel, D. W. (1983). Study of terminal velocities of air bubbles in non-Newtonian polysaccharide solutions using laser techniques. M.E.Sc. Thesis, Biochemical Engineering, University of Western Ontario, London, Ontario, Canada.

Troy, F. A., F. E. Frerman and E. C. Heath (1971). Biosynthesis of capsular polysaccharide in *Aerobacter aerogenes*. *J. Biol. Chem.*, **246**, 118–133.

Unz, R. F. and S. R. Farrah (1976). Exopolymer production and flocculation by *Zooglaea* MP6. *Appl. Environ. Microbiol.*, **31**, 623–626.

Van't Riet, K. (1975). Turbine Agitator Hydrodynamics and Dispersion Performance. PhD Thesis, Delft University of Technology.

Walker, J. and E. H. Dunlop (1979). Flocculation and removal of organic/inorganic matter from suspensions. *Br. Pat.* 2 019 378.

Weiss, R. M. and D. F. Ollis (1980). Extracellular microbial polysaccharides. I. Substrate, biomass, and product kinetic equations for batch xanthan gum fermentation. *Biotechnol. Bioeng.*, **22**, 859–873.

Whitcomb, P. J., B. J. Ek and C. W. Macosko (1977). Rheology of xanthan gum solutions. *ACS Symp. Ser.*, **45**, 160–173.

Wichterle, K. and Wein (1979). Onset of mixing in non-Newtonian liquids. *Chemicky Prumsyl*, **29**, 113–115.

Williams, A. G. and J. W. T. Wimpenny (1978). Exopolysaccharide production by *Pseudomonas* NCIB11264 grown in continuous culture. *J. Gen. Microbiol.*, **104**, 47–57.

Yamada, K. and T. Furukawa (1973). *Jpn. Pat.* 73/33,396.

Yuen, S. (1974). Pullulan and its applications. *Process Biochem.*, **9**, 7–9.

50
Enzymes in Food Technology

A. KILARA
Pennsylvania State University, University Park, PA, USA
and
K. M. SHAHANI
University of Nebraska–Lincoln, Lincoln, NE, USA

50.1 INTRODUCTION

Enzymes are organic biocatalysts which govern, initiate and control biological reactions important for life processes. They are produced by living cells but can act independently of the cell if appropriate environmental conditions are created. All known enzymes are proteins and some contain nonprotein moieties termed prosthetic groups that are essential for the manifestation of catalytic activity. Enzymes, therefore, are colloidal, thermolabile macromolecules exhibiting high degrees of stereochemical and substrate specificities. The importance of enzymes in sustaining and promoting growth and development of cells is a recognized fact (Dixon and Webb, 1979).

The use of enzymes in food preservation and processing predates modern civilization. Fermentations of common substrates such as fruits, vegetables, meat and milk provide a diverse array of foods in the human diet, *e.g.* beer, wine, sauerkraut, pickles, sausages, salami, ham, yogurt cheese and buttermilk. Whatever the origin of these fermented products, they can be said to result from the enzymatic modification of constituents in the substrate (Reed, 1975). Grapes grown in relatively warm climates around the world contain sucrose as the main carbohydrate and this is fermented to ethanol. Therefore, wine can be manufactured with relative ease in such climates. The decline of the Roman Empire has been speculated to be due to consumption of large amounts of lead leached into wine storage utensils! This is an interesting speculation. In the colder climates of the world grapes cannot be grown with ease and cereal grains such as barley and corn predominate. These grains contain carbohydrate polymers that cannot be readily converted by yeast to ethanol. Germinated grains produce malt with 'saccharolytic' enzymes capable of providing simple sugars to support yeast growth and fermentation. The beverage so produced is beer. In the tropics and equatorial regions where neither grapes nor barley and corn are grown, rice is the staple cereal crop. Mold, rather than yeast or bacteria, predominates and mold growth in turn can produce sake (rice wine) (Rose, 1981). These microorganisms and plants can be considered to be 'pre-packaged' enzymes with the package being an intact living cell. Enzymes isolated from these living cells are also used in a variety of enterprises (Demain and Solomon, 1981; Phaff, 1981).

It has been estimated that in 1981 about 65 000 tonnes of enzymes valued at \$400 million were available on a worldwide basis. In 1985 this market has been estimated to be \$600 million representing about 75 000 tonnes of enzymes. Production of enzymes in the U.S.S.R., People's Republic of China and Eastern European nations is thought to be significant. No production figures are, however, available. In the Western nations, Denmark alone accounts for 50% of the enzymes produced followed by the Netherlands (20%) and the U.S. (12%). The remaining 18% is attributable to Japan, West Germany, France, Switzerland and the U.K. (Godfrey and Reichelt, 1983).

Several enzyme classifications have been proposed or used. Most enzymes are broadly classified according to the substrates they act upon. For example, carbohydrases act upon carbohydrates, proteases upon proteins, lipases on lipids and so forth. The classification proposed by the International Union of Biochemistry and its Commission on Enzymes is the scientifically accepted method of nomenclature. This categorizes enzymes into six classes, namely oxidoreductases, transferases, hydrolases, lyases, isomerases and ligases. Oxidorectuctases catalyze oxidation or reduction reactions; transferases are involved in specific transfer of chemical moieties; hydrolases hydrolyze substrates with the uptake of water; lyases remove or add specific moieties to substrates; isomerases catalyze isomerization and ligases catalyze the synthesis or bonding together of substrate units (Schwimmer, 1981).

Hydrolases are the most commonly used enzymes accounting for nearly 80% of all commercially produced enzymes. Approximately 59% of all enzymes used by industrial concerns are proteases. Proteases are used in the detergent, dairy, meat and leather industries. About 28% of all enzymes produced are carbohydrases which find applications in baking, brewing, distilling, starch and textile industries. Malt enzymes alone are estimated to be produced at a level of 10 000 tonnes per annum. The remainder of the enzymes are accounted for by lipolytic enzymes (3%), oxidoreductases and enzymes used for analytical, pharmaceutical and developmental purposes (10%) (Godfrey and Reichelt, 1983).

Enzymes are derived from plant, animal and microbial sources. From an economic standpoint microbial enzymes are desirable since they can be readily grown and extracted in all parts of the world. Plants can grow only in certain climatological zones and are therefore not universally available. Further land requirements for growing plants are such that they may compete with other cash crops also cultivated in the same zone. Animals as sources of enzymes are subject to economic competition from the pharmaceutical industry. For example, the pancreas from beef cattle or swine serves as a source of trypsin and chymotrypsin. Additionally the pancreas is also a source of insulin for use in the therapy of diabetes. The manufacturers of insulin can pay a much higher price for this endocrine gland than can the enzyme manufacturers. Techniques pioneered by genetic engineering may improve the availability of many desirable enzymes from microorganisms. Enzymes of interest to the food industry are often extracellular rather than intracellular (Anstrup, 1977; Beckhorn *et al.*, 1965; Underkofler, 1966). In this chapter an overview of the enzymes inherently present in foods, technological aspects of enzymes, immobilized enzymes and lactose hydrolysis will be considered.

50.2 ENZYMES INHERENT IN FOODS

Food and food products are derived from living systems of plant or animal origin. Since life processes are intimately dependent upon enzymes, raw materials used as foods or further processed into food products contain a diversity of enzymes. Control of the activities of these enzymes leads to the production of good quality food (Schwimmer, 1981). For example, meat contains various enzymes including proteases. Control of the activity of these proteases leads to a resolution of *rigor mortis* and to the tenderization of the meat (Dransfield and Ethrington, 1981). Further, low molecular weight peptides released during proteolysis are responsible for the desirable brown color of cooked meat and for improved flavor of the cooked product. If the meat undergoes extensive proteolysis a loss of desirable texture is observed. Fruits contain a number of different enzymes. If the external surface of the fruit is damaged, polyphenol oxidase is activated leading to the formation of melanins that are black or dark brown in color. The melanins create an appearance defect leading to the loss of 'eye appeal' of a product (Schwimmer, 1981). Milk is the lacteal product secreted by normal healthy udders of cows, buffaloes, goats, sheep and other mammals. This fluid contains numerous enzymes as well and the control of these enzymes plays a crucial role in determining the quality of milk (Shahani *et al.*, 1974). Milk, for instance, contains lipase which can create off-flavors and rancidity due to the breakdown of triglycerides. Lipolytic off-flavors are considered to be a flavor defect. Controlled lipolysis of milk fat by the same enzyme can enhance the flavor of milk chocolate and create flavor concentrates such as enzyme-modified butter oil and enzyme modified cheese (Shahani *et al.*, 1976). Depending upon the food system and the type of enzyme, changes induced by enzymes inherent in the raw product can be desirable or undesirable.

50.2.1 Enzymes Inherent in Meats

Visual appearance of meat in the retail stores may influence consumer preference but another unrelated trait that influences meat purchase is tenderness of the meat. Tenderness and appearance are unrelated quality characteristics and the tenderness is very variable. Some ante-mortem characteristics such as breed, age, feed and health are all considered to influence tenderness (Briskey *et al.*, 1966, 1967). Post-mortem factors such as the method of slaughter, time and temperature of holding carcasses, the cut of meat, *etc.* are all related to the eating characteristics of meat. Enzymes inherently present in meat affect glycolysis and proteolysis and the overall changes are manifested in the palatability of meat.

Several hours after exsanguination of the animal, the oxygen supply to muscle is depleted. Adenosine triphosphate (ATP) is consumed to maintain temperature and cellular integrity. Regeneration of ATP occurs *via* anaerobic glycolysis leading to the production of lactic acid and a decrease in pH. When the pH of the cells reaches 6.0, loss of physiological integrity of the muscle is observed wherein actin and myosin combine to form actinomyosin, a condition called *rigor mortis* (Huxley, 1965). The rate and extent of glycolysis are important determinants of the tenderness of cooked meat. Glycolysis, of course, is a series of enzyme reactions leading to the formation of pyruvate, lactate and CO_2 from glycogen. During 'conditioning' of meat tenderness is enhanced through the action of muscle proteinases which include lysosomal cathepsins B, D, L and N (Table 1) and the nonlysosomal calcium-activated neutral proteinase. Levels of individual enzymes, their respective activities and their relative distribution in muscle and nonmuscle cells depend on age, breed, sex and plane of nutrition. In post-mortem, muscle temperature and pH are key determinants of enzyme activity. Additionally activities of muscle proteinases can be affected by the presence of activators and inhibitors (Dransfield and Ethrington, 1981).

Table 1 Proteinases of Muscle[a]

Enzyme		pH range for activity
Cathepsin	B	3–6
	D	2.5–4.5
	L	3.0–6.5
	N	3.0–6.0
Calcium-activated neutral protease		6.0–8.5

[a] Dransfield and Ethrington (1981).

50.2.2 Enzymes Inherent in Milk

Milk is a dynamic fluid containing about 30 or so enzymes. The presence of enzymes in milk has not been shown to be of any biological significance to the nurturing of young of that species; the enzymes are considered to be residuals of the milk secretion process. The origin of enzymes in milk therefore is due to blood, secretory cytoplasm and the apical membrane of the secretory cell (Shahani *et al.*, 1974). These enzymes do, however, have a profound influence on the keeping quality and the technological properties of the milk. Some of these are shown in Table 2. The biochemical properties of these enzymes have been fairly well understood. Catalase in milk, for example, has a molecular weight of 200 000, isoelectric point of pH 5.5, pH stability between 5 and 10, is strongly inhibited by heavy metal ions, cyanide and nitrate ions and can be completely destroyed at 70 °C for 60 min. Catalase levels vary with feed, stage of lactation and udder health; catalase acts as a prooxidant for lipids due to the presence of heme iron. Similarly, milk proteinases prefer β- and α_s2-casein as substrates, and hydrolysis products of β-casein have electrophoretic mobilities similar to caseins. Milk proteinase is thought to be very similar to blood plasmin. The enzyme plays an important role in hydrolysis of β-casein during cold storage of milk which in turn affects ability to form curds and thermal stability of milk. It can cause age gelation of UHT milk. Technological significance of other enzymes listed in Table 2 can be elaborated upon very easily (Fox and Morrissey, 1981).

Table 2 Some Enzymes Inherent in Milk and their Technological Importance[a]

Enzyme	Property
Lipase	Rancidity development, flavor generation in cheeses
Proteinase	Reduced stability of UHT milk, development of cheddar flavor
Catalase	Promoter of oxidative reactions
Lysozyme	Antibacterial
Xanthine oxidase	Promoter of oxidative reactions
Sulfhydryl oxidase	Reduction of cooked flavor in milk
Superoxide dismutase	Antioxidant
Lactoperoxidase	Antibacterial, promoter of oxidation
Acid phosphatase	Reduced heat stability of milk, cheese ripening

[a] Fox and Morrissey (1981).

50.2.3 Enzymes Inherent in Higher Plants

The main enzymes of concern are those acting on structural polysaccharides of fruits and vegetables and polyphenol oxidase, the enzyme that initiates enzymatic browning reactions. Pectic substances are the structural cement in middle lamellae and primary cell walls of higher plants (Doseburg, 1965; Pilnik and Voragen, 1970). The pectic substances are thought to be responsible for texture of fruits and vegetables. Action of native pectic enzymes results in post-harvest textural changes observed in fruits and vegetables. The native pectic enzymes of importance are pectinesterase and polygalacturonase. Pectinesterase EC 3.1.1.11 deesterifies pectins giving rise to low ester pectins and pectic acid. Pectinesterases are highly specific for methyl esters of polygalacturonic acid and methyl esters of polymannuronic acid are not attacked. Plant pectinesterases are thought to attack either at the reducing end or next to a free carboxyl group and then to proceed along the molecule by a single chain mechanism creating regions of free polygalacturonic acid which are extremely sensitive to calcium. Polygalacturonases depolymerize pectic acid. Pectic acid is polygalacturonic acid. The enzymes can be *exo* (EC 3.2.1.15) or *endo* (EC 3.2.1.67) in their attack on pectic acid meaning thereby that *exo* enzymes initiate depolymerization from the outer end of the polymer, whereas *endo* enzymes hydrolyze pectic acid from within the polymer. Properties of some of these enzymes are shown in Table 3 (Rombouts and Pilnik, 1978; MacMillan and Sheiman, 1974; Fogarty and Ward, 1972; Kulp, 1975). Pectinesterases occurring in conjunction with polygalacturonases, as in tomatoes, result in active pectin depolymerizing enzyme systems. Tomato juices are called hot break or cold break depending on whether the native pectinesterase and polygalacturonase are immediately heat inactivated or only after a holding time. Hot break juices are pulpy and viscous and suitable for beverage use, while cold break juices are thin and suitable for use in concentration to manufacture tomato paste. Enzymatic deesterification of pectin in juices leads to the coagulation of calcium pectate which is a loss of cloudiness in turbid juices and therefore undesirable or in the case of lemon and lime juices a desirable clarification (Fogarty and Kelly, 1983). The coagulation phenomenon in ground citrus pulp facilitates mechanical removal of water and uses less fuel to dry the peel and pulp sold as cattle feed! In several fruits and vegetables native pectinesterase is used for protecting the texture during processing. Low temperature–long time blanching activates esterases causing a partial deesterification of the pectins which then can react with added calcium ions and promote intercellular cohesion. Pectinesterases should be heat inactivated prior to fermenting fruit to minimize the production of methanol (Pilnik and Rombouts, 1981; Baumann, 1981).

Table 3 Selected Properties of Polygalacturonase and Pectinesterase from Plants[a]

Enzyme	Source		Molecular weight	Isoelectric point	Optimum pH	K_m value for pectate or pectin (mg ml^{-1})
Endopolygalacturonase	Tomato		52 000	—	4.5	2.7
		I	84 000	—	4.5	—
		II	44 000	—	5.0	—
Pectinesterase	Banana	I	30 000	8.9	6.0	—
		II	30 000	9.4	6.0	—
	Orange	I	36 200	10.0	7.6	0.083
		II	36 200	11.0	8.0	0.0046
	Tomato		27 500	—	6–9	0.74
			26 300	8.4	—	—
			—	—	8.0	2.40
			27 800	—	—	—

[a] Pilnik and Rombouts (1981).

Therefore this brief discussion points out the importance of indigenous or inherent enzymes in color, flavor and textural quality of various food products. Careful control of these inherent enzymes can lead to beneficial alterations in sensory and tactual attributes of foods while mismanagement of enzymes can lead to undesirable quality deterioration.

50.3 TECHNOLOGICAL ASPECTS

The main objectives of using enzymes in industrial processes are to produce an end product in greater amounts at lower costs and to do so at faster rates than nonenzymatic reactions can. To achieve these objectives it becomes necessary to know the quantity of enzyme needed, the process times necessary to achieve optimal amounts of the end product and the process conditions of pH, ionic strength and temperatures required to be maintained for optimal productivity. Additionally, it becomes necessary to know what conditions lead to the 'poisoning' of the catalysts, *e.g.* heavy metal ions. To fully understand these process variables a knowledge of enzyme kinetics is essential. Indeed an even more basic consideration will be the choice of an appropriate enzyme. This choice must take into consideration such factors as specificity, pH, temperature, activator and inhibitor requirements, availability, cost, technical service support from supplier, toxicity, legality and ease of control (Reed, 1975). With a wide variety of enzyme suppliers providing diverse enzymes capable of catalyzing similar reactions, the choice becomes a difficult task. One criterion often used is to take into consideration the activity of the enzyme. 'Activity' provides a quantitative description of the catalytic effect of an enzyme. Activity is expressed in units per unit weight or volume. There are many different ways in which these activity units are determined. Biochemists prefer to define one unit of activity as that mount of enzyme which will catalyze the transformation of one micromole of substrate per minute under defined conditions (Dixon and Webb, 1979).

The biochemists modified the original definition and redefined one SI unit (a katal) as the amount of enzyme which catalyzes the transformation of one millimole of substrate per second under specified conditions. These definitions are useful if purified or pure substrates are used for assays (Dixon and Webb, 1979). In practice, say for starch hydrolysis, one ton of corn is slurried and the starch gelatinized and modified by amylases. In this instance, what is the molecular weight of starch? If this is not known then how can one unit of katal be calculated? Therefore, alternative definitions of activity have been used for such complex substrates. Depending upon the manufacturer this alternative definition can vary considerably and great caution has to be exercised in ensuring that activity units when compared were determined under similar conditions and are defined similarly. Often the wording used is similar causing further confusion. To a person not aware, the two statements 'one unit of activity is the amount of enzyme that will convert 10 mg of starch per minute at 25 °C and pH 7.0' and 'one unit of activity is that amount of enzyme necessary to convert 10 mg of starch per hour at 25 °C and pH 7.0' may seem very similar but the latter enzyme has only 1/60th the activity if the former definition is applied! Neither definition would be satisfactory if this enzyme was to be used for, say, maltose production because the units mentioned conversion of a quantity of starch per unit time which is not the same as specifying that the quantity of starch was converted to maltose per unit time. Therefore, if productivity of the enzyme is the important consideration this should be simply stated as weight of product per unit weight of enzyme under operating conditions specified. Most enzymes used commercially are impure and are standardized by the addition of diluents and carriers. Further, the physical state of the enzyme can also vary, *e.g.* powder, liquid or immobilized (Fullbrook, 1983a).

Productivity of an enzyme is therefore important but not the sole determining factor in the choice. Reaction times are also critical and enzyme reactions can take place during a few minutes to several days. It may be desirable for the reaction to be complete in the shortest possible time. The cost of the enzyme and the process may become prohibitive if both high productivity and shortest duration are adopted. In such instances a suitable compromise has to be arrived at.

50.3.1 Nature of Enzyme Reactions

Two key elements in an enzyme reaction are the substrate and the enzyme, both contained in a suitable medium providing the necessary environment needed for 'reactivity'. In some enzyme reactions cofactors or coenzymes may also be necessary. Assuming then that all these components are present, the reaction can be divided into three phases. The first phase of the reaction is the initiation of the reaction and involves the binding of the substrate by the enzyme. This first phase is achieved very rapidly (fractions of seconds to seconds) and is transient. The moment the first molecule of substrate is bound by the enzyme the catalytic action begins and soon the product will be produced (Dixon and Webb, 1979). Depending on the amount of enzyme and substrate present, the second phase is initiated and involves the attainment of a steady state, *i.e.* an increasing amount of substrate being consumed and an increasing amount of product accumulating in the reaction *milieu*. Eventually the third phase called the nonlinear phase is attained wherein the initial velocity in the second phase decreases. The reaction is in this phase until either no more substrate remains to be converted, or the enzyme is inhibited by high product concentration or destroyed or denatured due to a variety of reasons. The time course for the reaction (phase I and II) is shown in Figure 1, whereas the formation of an enzyme substrate complex is (phase I) shown in Figure 2 and for all three phases in Figure 3.

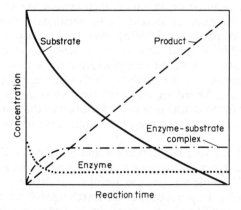

Figure 1   Time–concentration relationship in an enzyme-catalyzed reaction

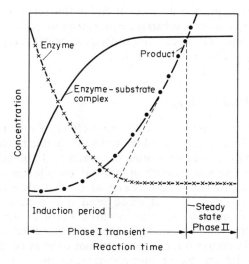

Figure 2 Time–concentration relationship during the formation of the enzyme–substrate complex

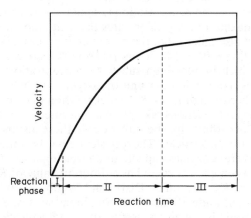

Figure 3 Velocity–time relationship for an ideal enzyme reaction

50.3.2 Effect of Substrate Concentration

If enzyme concentration is kept constant in a reaction and the enzyme is not exposed to harsh conditions, then the reaction will appreciably slow down when the substrate is depleted. Further, at high substrate concentrations maximal reaction velocities are attained: a parameter called V_{max} (the maximal velocity) and at low concentrations the reaction velocity is directly proportional to the substrate concentration (Dixon and Webb, 1979). Neither of these conditions is helpful to estimate for industrial scale use of enzymes. Rather, the velocity at intermediate substrate concentrations (phase II of reaction) provides a key to the behavior of enzymes in commercial scale operations. Kinetic analysis of this phase of the reaction yields a term called K_m or the Michaelis–Menten constant which is the substrate concentration needed to achieve half maximal velocity (K_m = substrate concentration at which velocity = $0.5V_{max}$). A word of caution is warranted. The velocity–substrate concentration curve is not linear in phase III and therefore V_{max} is attained asymptotically. Therefore, a substrate concentration of $2 \times K_m$ will not yield V_{max}, nor will $10 \times K_m$. The K_m value of commercially used enzymes lies in the range 10^{-1}–10^{-5} M under normal conditions. The K_m value varies with the type of substrate, and because of this property K_m is sometimes used as an indicator of the affinity of the enzyme for a substrate in question. A low K_m value means a high affinity and a high K_m value represents a low affinity. The practical use of K_m values lies in the choice of substrates and substrate concentrations or for the choice of an appropriate enzyme for a given substrate. If K_m is the substrate concentration needed to achieve half maximal velocity and a substrate concentration of $10K_m$ is used, 91% of V_{max} is achieved. If on the other hand only a substrate concentration of $0.1K_m$ can be used then the reaction proceeds at 9% of V_{max} (Schwimmer, 1981).

50.3.3 Effect of Enzyme Concentration

In the previous condition in Section 50.3.2. where effect of substrate concentration was discussed, enzyme concentrations were always kept constant. In the industrial context choice of substrate concentration suited to an enzyme is often not practical. Manufacturers of enzymes provide the conditions for the entire process and often this may include the use of highest possible substrate concentrations. If the process, for one reason or another, does not meet the required reaction velocities the net effect is a loss of throughput in the plant. To meet the prescribed throughput, adjustments are made by altering enzyme concentration. If substrate concentrations are kept constant and enzyme amounts are varied several conclusions can be drawn.

First, reaction velocity is always directly proportional to enzyme concentration and the constant of proportionality is related to the rate constant for the formation of the product (K_3). Second, at high substrate concentrations K_3 is the turnover number while at intermediate substrate concentrations the value of K_3 is influenced by the value of K_m and at low substrate concentrations enzyme is predominantly unbound to the substrate and is influenced by the substrate concentration and values of K_m and K_3. Last, unlike K_3, V_{max} is not a constant for the enzyme since it depends on the free enzyme concentration (Bender and Brubacher, 1973).

If the achievement of maximum conversion rates per unit amount of enzyme is the commercially desirable goal then it also means that the best overall process results with the least possible contamination of the reaction *milieu*. Maximum reaction rates occur when the enzyme is saturated with the substrate which implies high substrate concentration. If high substrate concentrations are used the fastest conversion per unit of enzyme will be given by those enzymes that have a high turnover number or high K_3 or K_{cat} values. Often in commercial conditions this utopian choice is not available and reactions take place at intermediate substrate concentrations and with enzymes that have a low affinity for the substrate. In these instances K_3, K_m and operational substrate concentrations are all critical. The problem now becomes the choice of the most efficient enzyme, given several enzymes capable of converting a substrate to a required product under defined conditions when each enzyme has a different value of K_m and K_3. Enzyme kineticists evolved the concept of $K_{cat}:K_m$ ratio (Dixon and Webb, 1979). The higher this ratio, the more substrate converted, but under commercial conditions substrate concentrations exceed K_m and therefore the $K_{cat}:K_m$ ratio is not a very useful criterion for the choice of an enzyme. Instead a more suitable approach is to evaluate these enzymes by taking the ratio $V_{max}:K_m$. The V_{max} term then incorporates both K_{cat} and enzyme concentration terms and provides a practical solution to the dilemma mentioned above concerning the choice of an appropriate enzyme.

Enzymes, like all catalysts, do not alter the equilibrium of a reaction but rather speed up the attainment of the equilibrium by lowering the activation energy for the reactions. Therefore as this equilibrium point is attained or neared, reaction velocity decreases and the reverse reaction begins to take place. For example, hydrolysis of liquefied starch by amyloglucosidase (EC 3.2.1.3) at a concentration of >40% dry matter is less than at 35% dry matter. At the higher substrate concentration less glucose and more maltose was produced than at the lower substrate concentration. For a successful commercial process maximum product yields are desired, which implies that the reaction needs to be run to near completion or to the point just prior to the initiation of the reverse reaction. From a practical standpoint this is controlled by the enzyme concentration and the reaction time (Fullbrook, 1983b).

Using the elementary equation derived from Michaelis–Menten assumptions under steady state conditions, the velocity of a reaction is obtained by

$$v = \frac{K_3 E_t A}{A + K_m} \tag{1}$$

where v = velocity of the reaction, K_3 is the rate constant for the undirectional breakdown of the enzyme substrate complex to free enzyme and product (EA → E + P), E_t is the total enzyme concentration, A is the substrate concentration, and K_m is the Michaelis constant, $(K_2 + K_3)/K_1$. The velocity of an enzyme reaction can also be defined by the change of substrate concentration (dA) as a function of time (dt) or $v = dA/dt$.

Therefore

$$v = \frac{d_A}{dt} = \frac{K_3 E_t A}{A + K_m} \tag{2}$$

or

$$k_3 E_t dt = (A + K_m)dA = \left(1 + \frac{K_m}{A}\right)dA \tag{3}$$

Integrating equation (3) under the limits $t = t_0$, $t = t$ and $A = A_0$, $A = A$, *i.e.* A_0 is the substrate concentration at time t_0 and A is the substrate concentration at time t, we obtain

$$k_3 E_t \int_{t_0}^{t} dt = K_m \int_{A_0}^{A} \frac{1}{A} dA + \int_{A_0}^{A} dA$$

$$k_3 E_t t = (A_0 - A) + K_m \ln(A_0/A)$$

or

$$t = \frac{(A_0 - A) + K_m \ln(A_0/A)}{K_3 E_t} \tag{4}$$

Equation (4) implies that the time needed to change the substrate concentration from A_0 to A at an enzyme concentration of E_t is dependent on two constants, K_3 and K_m, provided activity of the enzyme remains unchanged at t_0 and t. Product yield in many instances is calculated as conversion (x) and $x = (A_0 - A)/A_0$. In equation (4) the conversion equation can be substituted to obtain:

$$t = \frac{A_0 x + K_m \ln[1/(1 - x)]}{E_t K_3} \tag{5}$$

If the assumption that enzyme activity remains unchanged throughout the reaction is valid, then and only then is equation (5) useful in calculating the minimum usage of enzyme required to convert substrate within a finite time. This can be one method by which industrial enzyme efficiency can be judged, *i.e.* the enzyme that achieves the reaction in the shortest time implies faster throughput and therefore higher overall product yields.

Another useful concept for the comparative evaluation of enzymes is half-life. The half-life of an enzyme $(t_{0.5})$ is defined as the time required for an enzyme to lose 50% of its activity. If equation (4) were derived by integrating within the limits $(A, A_{0.5})$ and (t, t_0) and $V_{max} = K_3 E_t$ then the result would be

$$t_{0.5} = \ln 2 \frac{K_m}{V_{max}} + \frac{A}{2V_{max}} \tag{6}$$

K_m and V_{max} may not be clearly definable and obtainable entities in commercial enzymes. Because of impurities, kinetic analyses may hinder the determination of K_m and V_{max}. If it is assumed that $A \ll K_m$ then $A/2V_{max} \simeq 0$ and

$$t_{0.5} = \frac{0.7 K_m}{E_t K_3} \tag{7}$$

Another definition of $t_{0.5}$ can be the time required to reduce the substrate concentration by 50% which would be useful as well if the concentration of enzyme was kept constant and carefully defined. In this context $t_{0.5}$ or K_m [rather than $t_{0.5}(V_{max}/K_m)$] suggesting therefore that a lower $t_{0.5}$ would translate to increased efficiency of the overall enzyme-catalyzed reaction.

Therefore evaluation of the effect of enzyme concentration is not as straightforward because the concentration of the enzyme is closely allied with substrate concentration as well.

50.3.4 Effect of pH

Enzymes are active only within a narrow pH range. The pH optimum, however, is dependent on other factors such as time of reaction, temperature, type of substrate, substrate concentration, chemical properties of the medium in which the reaction is being carried out, ionic strength, type and source of the enzyme and the purity of the enzyme. Additionally, pH optimum should be considered for reactivity and for storage stability. Indeed temperature alone could affect ionization, dissociation, solubility and position of equilibrium, and the combined effect of these alterations could be manifested as altered reaction rates. The reactant species being affected in the reaction may be the enzyme, the substrate or the product (Schwimmer, 1981).

The pH optimum for activity can result from a true, reversible effect in V_{max}, K_m or both. Many reactions of commercial interest are not run at fixed pH values and during the course of the reaction the pH of the bulk solution can slowly drift. In some reactions, particularly those involving proteinaceous substrates, the pH is more or less fixed due to the buffering capacity of the substrates (Whitaker, 1972).

Enzymes can catalyze reactions in either direction, *e.g.* trypsin can hydrolyze proteins and also synthesize peptide bonds (transpeptidation). The transpeptidation is however favored at a pH level different from that for breakdown of peptide bonds.

The pH stability optimum may be considerably different from the pH optimum for activity. Change in pH can cause irreversible denaturation in some enzymes. The proximity of certain amino acid residues in a fixed manner in an enzyme is believed to be essential for the active catalytic site. If this configuration is altered in manner, such as the electrostatic repulsion caused by pH-induced changes in charges, then the geometry of the active site is altered as well, leading to a loss of catalytic activity. The pH stability of enzymes is dependent on a number of factors such as incubation conditions, presence or absence of cofactors and other small molecules, buffer type and concentration, presence or absence of substrate, ionic strength and dielectric constant. Mechanistically pH can affect ionization of substrate, enzyme, enzyme–substrate complex, enzyme–product complex and/or products, all of which have a direct bearing on reaction rate and direction (Dixon and Webb, 1979).

50.3.5 Effect of Temperature

Temperature plays a very important role in all reactions. Enzyme-catalyzed reactions are also included in this general rule. Temperature affects stability of enzymes, stability of the substrate, availability of substrates and cofactors, affinity of enzyme for substrate, activators and inhibitors, formation of by-products, velocity for the breakdown of enzyme–substrate complex, heat of activation of reaction, pH functions of reacting components, transfer of rate-limiting functions in multi-enzyme systems, changes in solubility of gases and competing reactions. Temperature effects are subdivided into temperature optimum for reactivity and temperature optimum for storage stability (Whitaker, 1972).

Enzymes are produced to catalyze physiologically significant reactions and therefore the life of the enzyme parallels the life of the cell from which the enzyme was isolated. During the process of isolation and purification conditions are often drastically different than *in vivo* and stability of enzymes can be enhanced by removal of moisture in many instances.

Increase in temperature results in faster enzymatic reactions. Generally for every 10 °C rise in temperature (Q_{10}) the reaction rates increase by a factor of two to three. Concomitant to the increase in reaction rates, denaturation rates also increase. The net effect is faster reaction rates for shorter periods of time or decreased product yields beyond certain limits. This dichotomy therefore makes it hard to define a temperature 'optimum' for reactivity unless this concept can be inalienably linked to specified reaction time and other reaction parameters (Dixon and Webb, 1979).

Temperature stability of enzymes is a function not only of temperature but also of pH, ionic strength, nature of buffer, presence or absence of substrate, concentration of enzyme, time of incubation and presence or absence of activators and inhibitors. Further conditions for maximum stability of crude enzyme preparations may be quite different than for purified enzymes. In general, the lower the temperature the greater is the stability of the enzyme. Similarly, enzymes of smaller molecular weight ($12–50 \times 10^3$) composed of single polypeptide chains with disulfide bonds are resistant to high temperature, whereas enzymes with molecular weights $>50 \times 10^3$ have more complex tertiary and quaternary structures and are more susceptible to higher temperatures (Fullbrook, 1983b).

A practical comparison of stability of enzymes is to determine the half-life of an enzyme ($t_{0.5}$), defined as the reaction time for the enzyme activity to drop to exactly half the initial value under specified conditions. The concept finds much greater use in immobilized enzyme catalyzed reactions rather than in soluble enzyme catalyzed reactions. Temperature is a very critical parameter in determining $t_{0.5}$, *e.g.* immobilized glucoamylase at 60 °C had an operational $t_{0.5}$ of 15 days but at 40 °C the same enzyme preparation had a $t_{0.5}$ of 900 days. Owing to this increase in $t_{0.5}$ the disadvantage of lowered reaction rates can be offset. Mathematically, the enzyme stability measured in terms of $t_{0.5}$ can be expressed as

$$\ln\left[\frac{\Delta DE}{\Delta t}/f(DE)\right] = \ln(aE_0) - t\ln(2/t_{0.5}) \tag{8}$$

where DE = dextrose equivalent, E_0 = initial enzyme concentration, a = activity of enzyme, $f(DE)$ = correction factor (1.0 for DE = 12), t = reaction time and in $2/t_{0.5}$ is the decay constant for constant temperature and pH (Vieth *et al.*, 1976).

Enzymes are produced and standardized to specified levels of activity which are usually 10–15% higher than the value stated on the label. Activity can be lost during storage in the factory, shipment to client and/or storage in clients' facilities. Therefore storage stability becomes of prime concern to enzyme manufacturers. The most stable form of enzyme is the crystalline or dry powder form. Industrially, the handling of dry enzymes poses potential health hazards and therefore it is customary to sell enzymes in stabilized liquid form. In such instances temperature effects on storage stability have to be studied in detail by the manufacturer to ensure an enzyme solution of predictable activity.

50.3.6 Effect of Ionic Strength

Every enzymologist acknowledges the importance of pH, temperature and ionic strength in affecting enzyme-catalyzed reactions but while reporting data these same enzymologists neglect to specify ionic strength or to study effects of varying ionic strength on enzyme activity. Ionic strength may assume even greater importance in immobilized enzyme reactions. The reaction rate constant for two charged species X and Y can be obtained by Debye–Hückel theory as being

$$\log K = \log K_0 + 1.02Z_X Z_Y \mu^{1/2} \tag{9}$$

where K = reaction rate constant, K_0 = rate constant at zero ionic strength, Z_X and Z_Y = electrostatic charges on reacting species X and Y, respectively, and μ = ionic strength of the solution in which the reaction is occurring (Dale and White, 1983).

Enzymes immobilized on charged supports may be subject to significant electrostatic effects. It has been suggested that high ionic strength media can effectively negate these electrostatic effects which may negate changes in the microenvironment but may at the same time alter the overall kinetics of the reaction. It has been suggested that the use of high ionic strength is as inappropriate in studies with immobilized enzymes as the use of high temperatures or high pH in such studies. Changing activity coefficients with changes in ionic strength is a concept that should not be neglected but routinely is. The danger of such a practice is to overestimate the activity of immobilized enzymes at normal ionic strength values when extrapolating from data obtained at high ionic strength. Unfortunately, due to the lack of literature on this subject it is difficult to give suggestions of practical interest (Vieth *et al.*, 1976).

50.3.7 Effect of Pressure

Increase in pressure translates to a reduction in volume and therefore alters the equilibrium of reactions. In applying this general rule it must be borne in mind that enzymes are specific catalysts and that catalysis occurs in a homogeneous phase. Therefore in most instances pressure does not affect the enzyme-catalyzed reaction, and in instances where it does, the effect is to reduce enzyme activity presumably by altering the molecular architecture (tertiary and quaternary structure) of enzymes. The latter occurs while expressing enzymes from cells using a French press. Increase in operational pressures is also detrimental to the activity and stability of immobilized enzymes, and an attendant and secondary effect is to reduce rates due to compaction of the enzyme reactor (Pitcher, 1980).

50.3.8 Overall Considerations

The preceding discussion should have made it clear that enzymes are active only within narrow ranges of temperature, pH, ionic strength, substrate concentration and enzyme concentration. Each range when being specified takes into account all the other factors; hence the disclaimers 'under conditions of the assay' or 'under specified condition.' Some conditions such as pH optimum for reactivity and stability can be clearly differentiated because their effects are on the overall reaction. The same is not true of temperature optimum for reactivity and stability because of the difficulty in separating the individual effects of temperature on both these parameters. Be that as it may, regions where optima overlap can be effectively utilized to increase the productivity of the enzyme, and imposing conditions away from this overlap provides for effective control of the enzyme reaction. This is critical where only limited hydrolysis of starches and proteins is desired to achieve specific functional ingredients. In other hydrolysis reactions where a maximum amount of product is desired, careful manipulation of the reaction conditions is useful. In many other instances it is absolutely essential to have no residual enzyme activity in the product. Imagine the consequences of residual amylase on the rheological and microbiological stability of products using starch as an ingredient in soups, gravies, sauces, *etc.*! In all the foregoing problems it becomes critical to understand the regions in which temperature or pH can be used as a tool to achieve the desired results.

50.4   CHEMICAL CONTROL OF ENZYME REACTIONS

Chemical control of enzyme reactions is much more precise and specific than physical control. An understanding of these effects of chemicals on enzyme-catalyzed reactions can be helpful in explaining the lack of performance of some enzymes on complex substrates (*i.e.* lower than expected performance) and can also provide a powerful tool to exercise control on an enzyme-catalyzed reaction at the molecular level. A better understanding of use of chemicals to control enzymatic reactions can be gained by studying mechanisms of regulation *in vivo*.

50.4.1   *In Vivo* Control of Enzyme Activity

In vivo the end product of one enzyme reaction serves as the substrate for one or more subsequent reactions. Therefore the regulation of enzymes *in vivo* can occur by one of many mechanisms (Koshland, 1970; Stadtman, 1970; Atkinson, 1970).

Regulation of enzyme concentration is one such mode. This mode includes the induction of enzyme synthesis by the substrate, catabolite repression, feedback repression of enzyme synthesis and balance between enzyme synthesis and degradation.

Feedback inhibition is another method by which enzyme activity can be controlled *in vivo*. Allosteric regulation proposed by Monod *et al.* (1965) and cooperative effects are mechanisms operational in this mode of control. Another mode of control of enzyme activity *in vivo* is the covalent modification of regulatory enzymes, *e.g.* phosphorylation of glycogen phosphorylase, as are the proteolysis of enzymes, compartmentalization of enzymes and the undirectional orientation of otherwise reversible reactions. Many of these modes may not have a significant bearing on the technological use of isolated and purified enzymes because they no longer produce substrates for subsequent reactions in a metabolic pathway.

50.4.2   Activation

Certain compounds can enhance or promote the expression of enzyme activity and are called activators whereas other compounds prevent the expression of maximal activity and are called deactivators. The difference between deactivators and inhibitors is important and subtle but the end result is indistinguishable. Deactivators prevent action of activators by binding the activators. Inhibitors on the other hand prevent the catalytic action of enzymes through many different mechanisms which will be discussed later. Both activation and deactivation provide a means of exerting crude chemical control of enzyme reactions (Mildvan, 1970).

It has been estimated that nearly 75% of all known enzymes require the presence of metal ions to express their full catalytic action. Metal ions can be either monovalent (Na^+, K^+) or divalent (Ca^{2+}, Mg^{2+}, Co^{2+}, Mn^{2+}, Zn^{2+}), and these ions are required at very low levels. Generally, food substrates have enough of these ions to ensure maximal activity. In fact, sometimes metal ions have to be removed from substrates as is the case with the glucose syrups subjected to glucose isomerase actions. If starch is hydrolyzed by α-amylase derived from the bacteria of genus *Bacillus* then Ca^{2+} is necessary to activate this enzyme. For xylose isomerase (glucose isomerase), however, Ca^{2+} serves as a deactivator. Therefore glucose syrups are subjected to ion exchange chromatography prior to the isomerization step. Another interesting method of activation is to use chelating agents to remove metal ions that may otherwise have deactivated the enzyme. Addition of chelating agents is commonly practiced with serine and sulfhydryl proteases (Bucke, 1983).

Cofactor activation is important in oxidative enzymes where molecular oxygen serves as the final electron acceptor. Removal of the cofactor deactivates the enzyme. Glucose oxidase (EC 1.1.3.4) contains flavin adenine dinucleotide (FAD) as a cofactor. Removal of FAD by dialysis, reverse osmosis or ultrafiltration causes this enzyme to be deactivated. Addition of FAD restores activity (Scott, 1975).

50.4.3 Promoters and Inhibitors

Promoters enhance activity of already activated enzymes while inhibitors decrease or suppress enzyme activity. Promoters are sometimes called stabilizers. Stabilizers have additional functions of protecting the enzymes from becoming inactive. Substrate *per se* can act as a stabilizer and the implication is that the enzyme will be active and stable at higher temperatures in the presence of a high substrate concentration when compared with a low substrate concentration. The qualitative use of high and low rather than precise quantitative units is purposeful since different enzymes have differing substrate concentrations that meet the 'high' and 'low' qualifications (Wiseman, 1978).

Similarly, enzymes can be stabilized by the presence of the products as well and this observation may impact on the ability to destroy enzyme activity after a desired reaction has reached completion. In commerce, amylases are stabilized by starch hydrolyzates, sugars or their derivatives and proteases by peptides in conjunction with salt, antifungal agents, antioxidants and other preservatives. Metal ions, in certain instances, can serve as stabilizers as well. α-Amylases derived from *Bacillus licheniformis* and *B. amyloliquifaciens* are both stabilized by calcium ions but the enzyme from the latter organisms requires much more calcium to be stabilized. The stabilized α-amylase hydrolyzes amylose to glucose monomers and if these monomers are to be isomerized to fructose by the enzyme xylose/glucose isomerase then the calcium ions needed for stability of α-amylase have to be removed prior to the syrup being isomerized. Calcium ions inhibit xylose/glucose isomerase. The end result in a commercial process would be that the use of *B. licheniformis* enzyme would require much less calcium ion addition as a stabilizer and therefore much less ion would have to be removed from the glucose syrup prior to isomerization than with syrups obtained from *B. amyloliquifaciens* amylase (Fullbrook, 1983b).

Use of inhibitors as elements to control enzymatic activity offers exciting possibilities in biotechnology. Inhibitors can interact with an enzyme either irreversibly or reversibly. The latter type of inhibition can be classified as (a) competitive, (b) uncompetitive, (c) noncompetitive, (d) mixed, (e) partial, (f) substrate, (g) product and (h) allosteric. Kinetic models for each type of inhibition are available in the literature (Dixon and Webb, 1979) and the application of these principles in industrial enzymology is not yet very clear.

Irreversible inhibitors or dead end inhibitors combine with the enzyme in a manner such that the enzyme–inhibitor complex is not dissociable. Kinetic methods to distinguish this mode of inhibition have been proposed by Cleland (1970). In a practical sense the mechanism of inhibition and kinetics may not be as important as resolving the problem and restoring throughput. Solution to inhibition may be as simple as destroying an inhibitor by heat treatment, *e.g.* soybean trypsin inhibitors, or more difficult as shifting the equilibrium of the enzyme–inhibitor complex formation to a state where dissociation of this complex is favored. The latter can be achieved by processes such as ultrafiltration or increasing the substrate concentration so as to favor the formation of the enzyme–substrate complex rather than the enzyme–inhibitor complex (Fullbrook, 1983a).

50.4.4 Interrelationships of Parameters

A thorough understanding of enzyme kinetics leading to an eventual delineation of the interactions between enzyme and substrate, enzyme–substrate cofactor, enzyme–product and enzyme–product-modified cofactor has limited applications in applied enzymology. Instead the aim in applied enzymology is to obtain maximum amount of product with minimum use of enzyme. Since industrial enzymologists are often dealing with complex and impure substrates and the enzymes *per se* are also seldom pure, optimization of processes is geared towards (a) maximum enzyme productivity or (b) absolute destruction of residual enzyme activity in the product.

To achieve maximum enzyme productivity, physical conditions are critical and therefore these are controlled and monitored. Very often physical changes can translate into activity and stability of an enzyme. These parameters, activity and stability, are also related to chemical changes in the enzyme molecules. This is the case when an enzyme is immobilized. Enzyme productivity can be defined as the product of the enzyme activity and enzyme stability. Activity is expressed in terms of weight of substrate converted or weight of product obtained per unit weight or volume of enzyme while stability of an enzyme is expressed as half-life. When dealing with soluble enzymes the concept of half-life is not applicable because a given quantity of enzyme cannot be reused, nor can it be easily separated from the reaction *milieu*. Half-life is a useful parameter for immobilized enzyme systems. For example, if the productivity of an enzyme was given as 600 000 for 2 half-lives and the half-life was stated as 28 days, this would mean that 1 kg of enzyme would yield 600 000 kg product during 56 days of working life. Temperature of operation, a physical factor, has a great bearing on operational half-life of enzymes. Immobilized glucoamylase with starch as the substrate had half-lives of 15, 30 and 900 days at 60, 50 and 40 °C, respectively. While the half-life increased at 40 °C the activity was only 25% of that observed at 60 °C. In this particular instance the increased half-life would likely compensate for the lowered reaction rates. In fact 1 kg of immobilized enzyme had 3×10^6 units, and 1 unit of activity produced 0.0138 g glucose per h at 60 °C. At 60 °C then, 41.4 kg dextrose h^{-1} could be produced per day by 1 kg of the immobilized enzyme or 993.6 kg^{-1} or 14 904 kg per half-life. At 40 °C the activity was only 25% as much so 1 kg of immobilized enzyme had 7.5×10^5 units capable of producing 10.35 kg glucose h^{-1} or 257 kg d^{-1}, but with a 900 day half-life 231 384 kg per half-life or 15.5 times the amount of product could be produced per kg of immobilized enzyme preparation at 40 °C when compared to 60 °C operational temperatures. What should be optimized? Should it be maximum product per unit time or maximum product over the life of the enzyme? This compromise is dependent on various considerations including capital investments, enzyme availability and cost, immobilization cost, system performance, clean-up cost, marketing costs, *etc.* Another important consideration with all food substrates is the potential for microbial contamination of product leading to economic loss of product or lack of safety of product due to growth of toxigenic microorganisms. From this consideration, 40 °C may support the growth of mesophilic and thermoduric organisms while at 60 °C very few thermophilic organisms may be able to grow. At least the potential for growth of toxigenic microorganisms can be eliminated at 60 °C and economic spoilage of product can be controlled with rigorous sanitation practices. At an operational temperature of 50 °C, however, even rigorous sanitation practices cannot ensure the prevention of contamination and the resulting growth of toxigenic microorganisms. In such cases hazard analysis and critical control point evaluations of the process have to be implemented (Roland, 1980).

The absolute destruction of residual enzyme activity in products implies that the enzymatic reaction has to be maximal and then some condition in the process has to be altered leading to the destruction of catalytic properties of the enzyme. The condition that is most frequently altered in enzymatic processes is temperature. pH can also be altered in some instances or a combination of temperature and pH in yet other processes. In all these instances, due to a change in activity and stability of the enzyme, productivity is altered. During the enzymatic saccharification of carbohydrates pH is altered to control productivity. In this process a final temperature of 80 °C is achieved when activity declines and stability diminishes. The last portion of the conversion is actually taking place as the activity of the enzyme is rapidly approaching zero. Figure 4 shows the effect of reaction time and reaction temperature on the activity of *Bacillus subtilis* protease during hydrolysis of whole milled barley (20% w/w) at pH 5.7 and temperatures of 50–63 °C in the presence of 0.01 M $CaCl_2$ (Fullbrook, 1983b). In this process enzyme productivity was being altered constantly and the problem arose of how to calculate the amount of enzyme to obtain a desired product yield within a defined reaction time. Unscientific as it may sound, empirical approach yielded the answer to the above problem and classical enzyme kinetic parameters were not considered at all. Such a program is successful as long as the magnitude and time sequences of changes in physical parameters are faithfully duplicated from the bench scale process to the commercial production scale.

50.5 IMMOBILIZED ENZYMES

Immobilization of an enzyme refers to the localization or confinement of enzymes during a process which allows the enzyme to be readily separated from substrate and product for reuse. The advantages of immobilizing enzymes are as follows: (a) multiple or repetitive use of a given batch

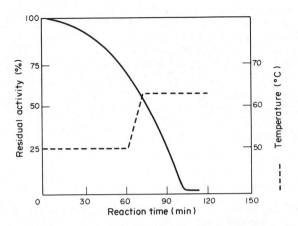

Figure 4 Whole milled barley (20% w/w) was hydrolyzed by *Bacillus subtilis* protease at pH 5.7, in the presence of 0.01 M CaCl₂ and temperature was varied between 50 and 63 °C. The graph shows the effect of reaction time and temperature on enzyme activity (Fullbrook, 1983)

of enzyme, (b) better process control, (c) enhanced stability, (d) enzyme-free products, (e) long half-lives and predictable decay rates and (f) good models for the study of *in vivo* kinetics of enzymes.

A variety of physical and chemical methods are available to immobilize enzymes. These methods are reviewed by Kilara and Shahani (1979). The choice of a carrier for immobilizing enzymes depends upon such factors as mechanical strength, microbial resistance, thermal stability, chemical durability, chemical functionality, low cost, hydrophilicity, regeneration and high capacity for enzyme. Since economics is the key determinant of eventual success of a commercial process, items such as cost of immobilization, price of support, cost of enzyme, cost of specialized equipment, *etc.* should be borne in mind. Factors affecting enzyme productivity discussed in earlier sections should not be overlooked either.

50.5.1 Immobilized Lactase Systems

During the manufacture of cheese, casein and whey proteins an aqueous solution of lactose and minerals is obtained. This solution called whey imposes a high biochemical oxygen demand on waste water treatment facilities. Of the nearly two billion pounds of whey solids produced annually in the United States, one billion pounds is not reclaimed and thus wasted. The problem of whey disposal is appropriately viewed as a problem of lactose disposal since lactose is the most abundant constituent in whey solids and the $\beta 1 \rightarrow 4$ galactosidic linkage is not readily broken by sewage microflora. Cheese whey, for example, contains 63–69% lactose. Lactose is not as sweet as glucose and the sweetness of lactose has been increased by isomerization to lactulose or hydrolysis to its constituent monosaccharides glucose and galactose (Wierzbicki and Kosikowski, 1973). Other applications of lactase have been suggested by Bouvy (1975), Moore (1980) and Finnocchiaro *et al.* (1980). Development of use of lactase in foods, especially using immobilized enzyme systems, where diffusional limitations exist, was hampered by severe end product inhibition. These limitations have been overcome to an extent that commercial feasibility has been discussed (Hinberg *et al.*, 1974; Hasselburger *et al.*, 1974; Shukla, 1975; Wierzbicki *et al.*, 1974; Coughlin, 1977; Woychik *et al.*, 1974; Pastore *et al.*, 1974; Portelle and Thonart, 1975; Kilara *et al.*, 1977; Coughlin and Charles, 1980). In these reports, a variety of methods to immobilize lactase have been utilized. Some examples are covalent attachment of glass, stainless steel, collagen, Sepharose and glutaraldehyde; entrapment in hollow fibers, polyacrylamide gel particles, porous cellulose sheets and cellulose triacetate; copolymerization with acrylate and methacrylate and ionizing radiation-induced polymerization and binding to magnetic supports.

Kay *et al.* (1968) immobilized lactase on cellulose sheets using chloro-*s*-triazinyl linking agent. Although this derivative was stable for up to several months, a majority of the activity was lost upon storage for three years (Lilly, 1971). Further, when the porous sheets were freeze-dried, the enzyme activity was severely decreased.

From this beginning GenenCor, a joint venture of GenenTech and Corning Biosystems, has emerged. In 1974 Corning developed a commercial process for using immobilized lactase in which the cost of lactose hydrolysis was estimated to be $1–4 per lb of lactose hydrolyzed. The enzyme was a fungal lactase derived from *Aspergillus niger* and was covalently immobilized on controlled pore glass or titania. Whey demineralization costs were $2–5 per lb, and concentration costs of $2–5 per lb lactose were not included in these estimates (Charles and Coughlin, 1980). A fruition of these efforts is the joint venture with Corning and Kroger under the name Nutritek to produce hydrolyzed lactose from cottage cheese whey, the whey coming from Kroger's cheese plants and the hydrolyzed lactose whey being used by Kroger's bakery operations. The immobilized enzymes are provided by GenenCor. These developments are encouraging and may lead to the full scale commercialization of immobilized lactase technology.

50.5.2 Other Immobilized Enzyme Systems

Commercially, the glucose isomerase and amino acylase systems have been successes. Progress in the glucose isomerase area has been recently reviewed by Bucke (1983). Some of the companies using these systems are listed along with properties of the enzymes in Table 4. Similarly, the developments in the use of aminoacylase have been reviewed by Chibata and Tosa (1976) and a summary is shown in Table 5.

Immobilized multienzymes or the immobilization of more than one enzyme on the same support can (a) enhance substrate transfer efficiencies, (b) facilitate, enhance or inhibit reactions by aggregation, (c) stabilize intermediates, (d) share cofactors, (e) provide undirectional control of reactions, (f) establish hydrogen ion gradient or a redox potential, and (g) provide hydrophobic areas of reactions. In nature such enzyme systems carry out sequential reactions where the product of one enzyme reaction serves as a substrate for the next reaction and so on. From a commercial point of view very few studies have concentrated on developing immobilized multiple enzyme systems. Glucose oxidase and catalase were immobilized on controlled pore titania by Messing (1974) who observed the system to be very stable. Catalase acted as both a stabilizer and an activator for glucose oxidase. Hultin (1974) reported the immobilization of glucose oxidase and catalase on 5% nickel-impregnated silica alumina support. The rate of hydrogen peroxide utilization was measured as an index of efficiency of the system. It was reported that the efficiency of the coimmobilized system was greater than one observed with a corresponding soluble system. The activities immobilized were of greater significance than the ratio of glucose oxidase to catalase offered for immobilization. Invoking the principle of proximity effects, it was explained that when greater activities of each enzyme were immobilized, hydrogen peroxide generated by glucose oxidase had a decreased tendency to escape into the bulk medium than if the enzyme molecules were separated by considerable distances. When molecules of glucose oxidase and catalase are packed near one another the probability of hydrogen peroxide being utilized immediately by a high turnover rate enzyme, such as catalase, would be greater.

Kilara *et al.* (1977) immobilized lactase, papain and lipase concomitantly as derivatized Sepharose-4B. Equal molar concentrations of each enzyme were allowed to react with activated Sepharose and a preferential binding of lipase and papain over lactase was observed. This preferential binding was explained on the basis of differences in diffusion rates of enzymes resulting in greater availability of binding sites within the beads to certain enzyme species. The effects of pH on assay, temperature and substrate concentration on each individual enzyme were determined and compared to the corresponding soluble enzymes. Some differences in pH and temperature optima for immobilized enzymes were observed; however the affinities of the enzymes for their substrates were substantially the same. Such a system could possibly be used for the simultaneous hydrolysis of lactose, protein and fat in cheese whey.

50.6 SUMMARY

Enzyme immobilization reported as early as 1916 by Nelson and Griffin was not exploited until the past 15 years. It holds potential for increased usage in the food industry. High fructose corn syrup manufacture utilizing glucose isomerase leads the list of immobilized enzyme processes using this technology; a significant share of the market for sweeteners is captured by fructose sweeteners. Such spectacular success in other food and pharmaceutical areas has not yet materialized with any other product.

Klibanov *et al.* (1977) suggest using enzymatic processes in a water:water-immiscible organic

Table 4 Summary of Glucose Isomerase Processes to Produce High Fructose Corn Syrup

Company	Enzyme source	Immobilization procedure	Enzyme form	Productivity (kg 42% fructose syrup per kg enzyme)	Reactor type	Patent numbers (country)	Commercial name
Clinton Corn Processing Co.	*Streptomyces*	Adsorption on DEAE cellulose	Fibrous	7200–9000	Shallow bed	3 788 945 3 623 953 (US)	Isomerose
Novo Industri	*Bacillus coagulans*	Lysed cells cross-linked with glutaraldehyde	Granular	1000–1600	Plug flow	1 381 387 1 362 365 (UK)	—[a]
Gist Brocades	*Actinoplanes*	Gelatin-entrapped cells cross-linked with glutaraldehyde	Granular	1778	Varied	3 834 848 3 838 007 (US)	—
ICI Americas	*Arthrobacter*	Whole cell flocs	Granular	2000	Plug flow	3 645 848 3 821 086 3 935 068 (US)	—
Miles Laboratories	*Streptomyces*	Whole cells cross-linked with glutaraldehyde	Amorphous	1000	Plug flow	3 625 828 3 654 081 3 779 869 (US)	
CPC International	*Streptomyces*	Adsorption on inorganic carrier	Granular	—	—	—	Invertose
American Maize	—	—	—	—	—	—	Trusweet
Amstar Corp.	—	—	—	—	—	—	Amerose
ADM Corn Sweeteners	—	—	—	—	—	—	Corn Sweet
Cargill Inc.	—	—	—	—	—	—	Isoclear
Hubinger Co.	—	—	—	—	—	—	Hisweet
A.E. Staley Mfg. Co.	—	—	—	—	—	—	Isosweet

[a] Data not available.

Table 5 Summary of Aminoacylase System using Acetyl-DL-methionine as Substrate at 37 °C, pH 7.0[a]

Properties	Soluble enzyme	Immobilized enzyme preparations using		
		DEAE Sephadex (ionic)	Iodoacetyl cellulose (covalent)	Polyacrylamide entrapment
Optimum pH	7.5–8.0	7.0	7.5–8.0	7.0
Optimum temperature (°C)	60	72	55	65
Activation energy (kJ mol^{-1})	6.9	7.0	3.9	5.3
K_m (mM)	5.7	8.7	6.7	5.0
v_m (mol h^{-1})	1.52	3.33	4.65	2.33
Half life (days at °C)	—	65 at 50 °C	—	48 at 37 °C
Ease of preparation		Easy	Difficult	Medium
Enzyme activity		High	High	High
Cost of immobilization		Low	High	Moderate
Binding force		Medium	Strong	Strong
Operational stability		High	—	Medium
Regeneration	Impossible	Possible	Impossible	Impossible
Optimum Co^{2+} (mM)	0.5	0.5	0.5	0.5

[a] Chibata and Tosa (1979).

solvent system as a means of improving productivity of enzymes. The content of water in such a system can be extremely low shifting the equilibrium of reactions forming water as a product toward the product side of the equation. Such would be the case in the synthesis of amides and esters, polymerization of amino acids and sugars and dehydration reactions. Such biphasic systems provide free energies of transfer of a reagent from one phase to another as a driving force for shifts in equilibria. Overall this concept is exciting and can help transform the practical applications of enzymology in a very significant manner.

Enzymology has been called a solution in search of a problem. In the food area descaling of fish, modification of beer wort, beverage clarification, production of hydrolyzate-based beverages for infants, geriatrics and invalids and the removal of antinutritional factors from foods serve as examples of problems that may someday be solved with enzymes. For the time being, however, economic constraints prevent the routine use of enzymes for such processes.

50.7 REFERENCES

Anstrup, K. (1977). Industrial approach to enzyme production. In *Biotechnological Application of Proteins and Enzymes*, ed. Z. Bohak and N. Sharon, pp. 39–49. Academic, New York.

Atkinson, D. E. (1970). Enzymes as control elements in metabolic regulation. In *The Enzymes*, ed. P. D. Boyer, vol. 1, pp. 461–490. Academic, New York.

Baumann, J. W. (1981). Application of enzymes in fruit juice technology. In *Enzymes and Food Processing*, ed. G. G. Birch, N. Blakebrough and K. J. Parker, pp. 129–148. Applied Science, London.

Beckhorn, E. J., M. D. Labbee and L. A. Underkofler (1965). Production and use of microbial enzymes for food processing. *J. Agric. Food Chem.*, **13**, 30–34.

Bender, M. L. and L. H. Brubacher (1973). *Catalysis and Enzyme Action*. McGraw-Hill, New York.

Bouvy, F. A. M. (1975). Applications for lactase treated whey. *Food Prod. Dev.*, **9** (2), 10–13.

Briskey, E. J., R. G. Casens and J. C. Trautmann (1966). *The Physiology and Biochemistry of Muscle as Food*, volume I. University of Wisconsin Press, Madison.

Briskey, E. J., R. G. Cassens and J. C. Trautmann (1967). *The Physiology and Biochemistry of Muscle as Food*, volume II. University of Wisconsin Press, Madison.

Bucke, C. (1983). Glucose transforming enzymes. In *Microbial Enzymes and Biotechnology*, ed. W. M. Fogarty, pp. 93–130. Applied Science, London.

Chibata, I. and T. Tosa (1976). Industrial application to immobilized enzymes and immobilized microbial cells. In *Applied Biochemistry and Bioengineering*, ed. L. B. Wingard, Jr., E. Katchalski-Katzir and L. Goldstein, vol. 1, pp. 329–358. Academic, New York.

Cleland, W. W. (1970). Steady state kinetics. In *The Enzymes*, ed. P. D. Boyer, vol. 2, pp. 1–65. Academic, New York.

Coughlin, R. W. (1977). Immobilized lactase technology. Recent developments and applications. 37th Annual Meeting of Institute of Food Technologists, Philadelphia, PA.

Coughlin, R. W. and M. Charles (1980). Applications of lactose and immobilized lactase. In *Immobilized Enzymes for Food Processing*, ed. W. H. Pitcher, Jr., pp. 153–174. CRC Press, Boca Raton, FL.

Dale, B. E. and D. H. White (1983). Ionic strength: a neglected variable in enzyme technology. *Enzyme Microb. Technol.*, **5**, 227–229.

Demain, A. L. and N. A. Solomon (1981). *Ind. Microbiol. Sci. Am.*, **245** (3), 66–75.

Dixon, M. and E. C. Webb (1979). *Enzymes*, 3rd edn. Academic, New York.

Doseburg, J. J. (1965). *Pectic Substances in Fresh and Preserved Fruits and Vegetables* (IBVT–Commun. No. 25). Sprenger Institute, Wageningen, The Netherlands.

Dransfield, E. and D. Ethrington (1981). Enzymes in the tenderization of meat. In *Enzymes and Food Processing*, ed. G. G. Birch, N. Blakenbrough and K. J. Parker, pp. 177–194. Applied Science, London.

Finnochiaro, T. T., T. Richardson and N. F. Olson (1980). Lactase immobilized on alumina. *J. Dairy Sci.*, **63**, 215–222.

Fogarty, W. M. and C. T. Kelly (1983). Pectic enzymes. In *Microbial Enzymes and Biotechnology*, ed. W. M. Fogarty, pp. 131–182. Applied Science, London.

Fogarty, W. M. and O. P. Ward (1972). Pectic substances and pectolytic enzymes. *Process Biochem.*, **7** (8), 13–15.

Fox, P. F. and P. A. Morrissey (1981). Indigenous enzymes in bovine milk. In *Enzymes and Food Processing*, ed. G. G. Birch, N. Blakenbrough and K. H. Parker. Applied Science, London.

Fullbrook, P. D. (1983a). Practical applied kinetics. In *Industrial Enzymology: The Application of Enzymes in Industry*, ed. T. Godfrey and J. Reichelt, pp. 8–40. Nature Press, New York.

Fullbrook, P. D. (1983b). Practical limits and prospects. In *Industrial Enzymology: The Applications of Enzymes in Industry*, ed. T. Godfrey and J. Reichelt, pp. 41–110. Nature Press, New York.

Godfrey, T. and J. R. Reichelt (1983). Introduction to industrial enzymology. In *Industrial Enzymology: The Application of Enzymes in Industry*, ed. T. Godfrey and J. Reichelt, pp 1–7. Nature Press, New York.

Hasselburger, F. X., B. Allen, M. C. Paruchuri and R. W. Coughlin (1974). Immobilized enzymes: lactase bonded to stainless steel and other dense carriers for use in fluidized bed reactors. *Biochem.Biophys. Res. Commun.*, **57**, 1054–1062.

Hinberg, I., R. Korus and K. F. O'Driscoll (1974). Gel entrapped enzymes: kinetic studies of immobilized β-galactosidase. *Biotechnol. Bioeng.*, **16**, 943–963.

Hultin, H. O. (1974). Characteristics of immobilized multienzymic systems. *J. Food Sci.*, **39**, 647–653.

Huxley, H. E. (1965). The mechanisms of muscle contraction. *Sci. Am.*, **213** (6), 18–27.

Kay, G., M. D. Lilly, A. K. Sharp and R. J. H. Wilson (1960). Preparation and use of porous sheets with enzyme action. *Nature (London)*, **217**, 641–643.

Kilara, A. and K. M. Shahani (1979). Immobilized enzymes in the food industry: a review. *CRC Crit. Rev. Food Sci. Nutr.*, **12**, 161–198.

Kilara, A., K. M. Shahani and F. W. Wagner (1977a). Preparation and characterization of immobilized lactase. *Lebensm. Wiss. Technol.*, **10**, 84–88.

Kilara, A., K. M. Shahani and F. W. Wagner (1977b). The kinetic properties of lactase, papain and lipase immobilized on a single support. *J. Food Biochem.*, **1**, 261–273.

Klibanov, A. M., G. P. Samokhin, K. Martinek and I. W. Berezin (1977). A new approach to preparative enzymatic synthesis. *Biotechnol. Bioeng.*, **19**, 1351–1361.

Koshland, D. E., Jr. (1970). The molecular basis for enzyme regulation. In *The Enzymes*, ed. P. D. Boyer, vol. 1, pp. 342–397. Academic, New York.

Kulp, K. (1975). Carbohydrases. In *Enzymes in Food Processing*, ed. G. Reed, pp. 54–122. Academic, New York.

Lilly, M. D. (1971). Stability of immobilized β-galactosidase on prolonged storage. *Biotechnol. Bioeng.*, **13**, 589–596.

MacMillan, J. D. and M. I. Sheiman (1974). Pectic enzymes. In *Food Related Enzymes*, ed. J. R. Whitaker, pp. 101–130. American Chemical Society, Washington, DC.

Messing. R. A. (1974). Simultaneously immobilized glucose oxidase and catalase in controlled pore titania. *Biotechnol. Bioeng.*, **16**, 897–901.

Mildvan, A. S. (1970). Metals in enzyme catalysis. In *The Enzymes*, ed. P. D. Boyer, vol. 2, pp. 446–536. Academic, New York.

Monod, J., J. Wyman and J. P. Changeux (1965). On the nature of allosteric transitions: a plausible model. *J. Mol. Biol.*, **12**, 88–118.

Moore, K. (1980). Immobilized enzyme technology commercially hydrolyzes lactose. *Food Prod. Dev.*, **14** (1), 50–51.

Nelson, J. M. and E. G. Griffin (1916). Adsorption of invertase. *J. Am. Chem. Soc.*, **38**, 1109–1111.

Pastore, M., F. Morisi and A. Viglia (1974). Reductions of lactose in milk by entrapped β-galactosidase II. Conditions for an industrial continuous process. *J. Dairy Sci.*, **57**, 269–272.

Phaff, H. J. (1981). Industrial microorganisms. *Sci. Am.*, **245** (3), 76–89.

Pilnik, W. and F. M. Rombouts (1981). Pectic enzymes. In *Enzymes and Food Processing*, ed. G. G. Birch, N. Blakenbrough and K. J. Parker, pp. 105–128. Applied Science, London.

Pilnik, W. and A. G. J. Voragen (1970). Pectic substances and other uronides. In *The Biochemistry of Fruits and Their Products*, ed. A. C. Hulme, vol. 1, pp. 53–87. Academic, New York.

Pitcher, W. H., Jr. (1980). Immobilized enzyme engineering. In *Immobilized Enzymes for Food Processing*, ed. W. H. Pitcher, Jr., pp. 15–54. CRC Press, Boca Raton, FL.

Portelle, D. and P. H. Thonart (1975). Immobilized β-galactosidase on insoluble carrier. II. Technological applications of immobilized β-galactosidase. *Lebensm. Wiss. Technol.*, **8**, 274–277.

Reed, G. (1975). *Enzymes in Food Processing*. Academic, New York.

Roland, J. F. (1980). Requirements unique to the food and beverage industry. In *Immobilized Enzymes for Food Processing*, ed. W. H. Pitcher, Jr., pp. 55–80. CRC Press, Boca Raton, FL.

Rombouts, F. M. and W. Pilnik (1978). Enzymes in fruit and vegetable technology. *Process Biochem.*, **13** (8), 9–13.

Rose, A. H. (1981). The microbiological production of food and drink. *Sci. Am.*, **245** (3), 126–139.

Schwimmer, S. (1969). Trends and perspectives in the enzymology of foods. *Lebensm. Wiss. Technol.*, **2**, 97–103.

Schwimmer, S. (1981). *Source Book of Food Enzymology*. AVI Publishing Co., Westport, CT.

Scott, D. (1975). Oxidoreductases. In *Enzymes in Food Processing*, ed. G. Reed, pp. 222–230. Academic, New York.

Shahani, K. M., R. G. Arnold, A. Kilara and B. K. Dwivedi (1976). Role of microbial enzymes in flavor development in foods. *Biotechnol. Bioeng.*, **18**, 891–917.

Shahani, K. M., W. J. Harper, R. G. Jensen, R. M. Parry, Jr. and C. A. Zittle (1974). Enzymes in bovine milk. A review. *J. Dairy Sci.*, **56**, 531–543.

Shukla, T. P. (1975). Beta galactosidase technology: a solution to the lactose problem. *CRC Crit. Rev. Food Technol.*, **5**, 325–356.

Stadtman, E. R. (1970). Mechanisms of enzyme regulation in metabolism. In *The Enzymes*, ed. P. D. Boyer, vol. 1, pp. 398–460. Academic, New York.

Underkofler, L. A. (1966). Manufacture and use of industrial microbial enzymes. *Chem. Eng. Prog. Symp. Ser.*, **62** (69), 11–20.

Vieth, W. R., K. Venkatsubramanian, A. Constantinides and B. Davidson (1976). Design and analysis of immobilized reactors. In *Applied Biochemistry and Bioengineering*, ed. L. B. Wingard, Jr., E. Katchalski-Katzir and L. Goldstein, vol. 1, pp. 222–328. Academic, New York.

Whitaker, J. R. (1972). *Principles of Enzymology for the Food Sciences*. Dekker, New York.

Wierzbicki, L. E. and F. V. Kosikowski (1973). Kinetics of lactose hydrolysis in acid whey by β-galactosidase from *Aspergillus niger*. *J. Dairy Sci.*, **56**, 1396–1399.

Wierzbicki, L. E., V. H. Edwards and F. V. Kosikowski (1974). Hydrolysis of lactose in acid whey by lactase bound to porous glass particles in tubular reactors. *J. Food Sci.*, **39**, 374–378.

Wiseman, D. (1978). Stabilization of enzymes. In *Topics Enzyme and Fermentation Biotechnology*, ed. A. Wiseman, vol. 2, pp. 280–303. Ellis Horwood, Chichester.

Woychik, J. H., M. V. Wonolowski and K. J. Dahl (1974). Preparation and applications of immobilized beta galactosidase. In *Immobilized Enzymes in Food and Microbial Processes*, ed. A. C. Olson and C. L. Cooney, pp. 41–52. Plenum Press, New York.

Appendix 1: Glossary of Terms

Because of the broad multidisciplinary nature of biotechnology, both beginners and specialists in this field may find this glossary useful. It covers terms often used in the relevant areas of the biological, chemical and engineering sciences. It is not intended to be exhaustive. The material was generated primarily from four sources: 'Commercial Biotechnology', Office of Technology Assessment, Congress of the US (1984); 'Advances in Biotechnology', M Moo-Young *et al.* (1981); *Pure and Applied Chemistry*, Vol. 54, No. 9, pp. 1743–1749 (1982); and 'Dictionary of Biochemistry', J. Stenish (1975).

Acclimatization: The biological process whereby an organism adapts to a new environment. For example, it describes the process of developing microorganisms that degrade toxic wastes in the environment.

Activated sludge: Biological growth that occurs in aerobic, organic-containing systems. These growths develop into suspensions that settle and possess clarification and oxidative properties.

Activation energy: The difference in energy between that of the activated complex and that of the reactants; the energy that must be supplied to the reactants before they can undergo transformation to products.

Active immunity: Disease resistance in a person or animal due to antibody production after exposure to a microbial antigen following disease, inapparent infection or inoculation. Active immunity is usually long-lasting.

Active transport: The movement of a solute across a biological membrane such that the movement is directed against the concentration gradient and requires the expenditure of energy.

Activity: A measure of the effective concentration of an enzyme, drug, hormone or some other substance. It is also the product of the molar concentration of an ionic solute and its activity coefficient.

Adsorption: The taking up of molecules of gases, dissolved substances or liquids at the surfaces of solids or liquids with which they are in contact.

Aerobic: Living or acting only in the presence of free-form oxygen, as in air.

Affinity chromatography: The use of compounds, such as antibodies, bound to an immobile matrix to 'capture' other compounds as a highly specific means of separation and purification.

Airlift fermenter: Vessel in which a bioconversion process takes place; the sparged gas is the only source of agitation. The presence of a draft tube inside a fermenter distinguishes this type of fermenter from a bubble column.

Alga (pl. **algae**): A chlorophyll-containing, photosynthetic protist; algae are unicellular or multicellular, generally aquatic and either eukaryotic or prokaryotic.

Amino acids: The building blocks of proteins. There are 20 common amino acids.

Amino acid sequence: The linear order of amino acids in a protein.

Amylase: An enzyme that catalyzes the hydrolysis of starch.

Anabolism: The phase of intermediary metabolism that encompasses the biosynthetic and energy-requiring reactions whereby cell components are produced. Also, the cellular assimilation of macromolecules and complex substances from low-molecular weight precursors (*cf.* Catabolism).

Anaerobic: Living or acting in the absence of free-form oxygen.

Anaerobic digestion: The energy-yielding metabolic breakdown of organic compounds by microorganisms that generally proceeds under anaerobic conditions and with the evolution of gas. The term is most often used to describe the reduction of waste sludges to less solid mass and for methane gas production.

Antibiotic: A specific type of chemical substance that is administered to fight infections, usually bacterial infections, in humans or animals. Many antibiotics are produced by using microorganisms; others are produced synthetically.

Antibody: A protein (immunoglobulin) produced by humans or higher animals in response to

1065

exposure to a specific antigen and characterized by specific reactivity with its complementary antigen (see also Monoclonal antibodies).

Antigen: A substance, usually a protein or carbohydrate, which, when introduced into the body of a human or higher animal, stimulates the production of an antibody that will react specifically with it.

Antiserum: Blood serum containing antibodies from animals that have been inoculated with an antigen. When administered to other animals or humans, antiserum produces passive immunity.

Aromatic compound: A compound containing a benzene ring. Many speciality and commodity chemicals are aromatic compounds.

Ascites: Liquid accumulations in the peritoneal cavity. Used as a method for producing monoclonal antibodies.

Asepsis: The prevention of access of microorganisms causing disease, decay or putrefaction to the site of a potential infection.

Aseptic: Of, or pertaining to, asepsis. In directed fermentation processes, the exclusion of unwanted (contaminating) organisms.

Assay: A technique that measures a biological response.

Attenuated vaccine: Whole, pathogenic organisms that are treated with chemical, radioactive, or other means to render them incapable of producing infection. Attenuated vaccines are injected into the body, which then produces protective antibodies against the pathogen to protect against disease.

Autolysis: The self-destruction of a cell as a result of the action of its own hydrolytic enzymes.

Autotrophic: Capable of self-nourishment (opposed to heterotrophic).

Axial dispersion: Mixing along the flow path of fluids during processing as in a bioreactor.

Bacteria: Any of a large group of microscopic organisms having round, rodlike, spiral or filamentous unicellular or noncellular bodies that are often aggregated into colonies, are enclosed by a cell wall or membrane, and lack fully differentiated nuclei. Bacteria may exist as free-living organisms in soil, water, organic matter, or as parasites in the live bodies of plants and animals.

Bacteriophage (or **phage**)/**bacterial virus**: A virus that multiplies in bacteria. Bacteriophage lambda is commonly used as a vector in rDNA experiments.

Batch processing: A method of processing in which a bioreactor, for example, is loaded with raw materials and microorganisms, and the process is run to completion, at which time products are removed (*cf.* Continuous processing).

Binary fission: Asexual division in which a cell divides into two, approximately equal, parts, as in the growth method of some single cells.

Biocatalyst: An enzyme, in cell-free or whole-cell forms, that plays a fundamental role in living organisms or industrially by activating or accelerating a process.

Biochemical: Characterized by, produced by, or involving chemical reactions in living organisms; a product produced by chemical reactions in living organisms.

Biochip: An electronic device that uses biological molecules as the framework for molecules which act as semiconductors and functions as an integrated circuit.

Bioconversion: A chemical conversion using a biocatalyst.

Biodegradation: The breakdown of substances by biological agents, especially microbes.

Biological oxygen demand (BOD): The oxygen used in meeting the metabolic needs of aerobic organisms in water containing organic compounds.

Biological response modifier: Generic term for hormones, neuroactive compounds and immunoactive compounds that act at the cellular level; many are possible targets for production with biotechnology.

Biologics: Vaccines, therapeutic serums, toxoids, antitoxins and analogous biological products used to induce immunity to infectious diseases or harmful substances of biological origin.

Biomass: Organic matter of biological origin such as microbial and plant material.

Biooxidation: Oxidation (the loss of electrons) catalyzed by a biocatalyst.

Biopolymers: Naturally occurring macromolecules that include proteins, nucleic acids and polysaccharides.

Bioprocess: Any process that uses complete living cells or their components (*e.g.* enzymes, chloroplasts) to effect desired physical or chemical changes.

Bioreactor: Vessel in which a bioprocess takes place; examples include fermenter, enzyme reactor.

Biosensor: A device, usually electronic, that uses biological molecules to detect specific compounds.

Biosurfactant: A compound produced by living organisms that helps solubilize compounds such as organic molecules (*e.g.* oil and tar) by reducing surface tension between the compound and liquid.

Biosynthesis: Production, by synthesis or degradation, of a chemical compound by a living organism.

Biotechnology: Use of biological agents or materials to produce goods or services for industry, trade and commerce; a multidisciplinary field.

Bubble column: A gas–liquid contacting vessel in which sparged gas is the only source of agitation.

Bubbly flow: Type of two-phase flow in which the gas phase is distributed in the liquid in the form of bubbles whose dimensions are small compared to the characteristic dimension of the flow cross-section, as in certain designs of bioreactor geometrics.

Budding: A form of asexual reproduction typical of yeast, in which a new cell is formed as an outgrowth from the parent cell.

Buffer: A solution containing a mixture of a weak acid and its conjugate weak base that is capable of resisting substantial changes in pH upon the addition of small amounts of acidic or basic substances, as used for some fermentation media.

Bulking: Increase in volumetric solids as in certain waste treatment bioreactors, which limits the weight concentration that a clarifier can handle.

Callus: An undifferentiated cluster of plant cells that is a first step in regeneration of plants from tissue culture.

Carbohydrate: An aldehyde or a ketone derivative of a polyhydroxy alcohol that is synthesized by living cells. Carbohydrates may be classified either on the basis of their size into mono-, oligo- and poly-saccharides, or on the basis of their functional group into aldehyde or ketone derivatives.

Carboxylation: The addition of an organic acid group (COOH) to a molecule.

Catabolism: The phase of intermediary metabolism that encompasses the degradative and energy-yielding reactions whereby nutrients are metabolized. Also, the cellular breakdown of complex substances and macromolecules to low-molecular weight compounds (*cf.* Anabolism).

Catalysis: A modification, especially an increase, in the rate of a chemical reaction induced by a material (*e.g.* enzyme) that is chemically unchanged at the end of the reaction.

Catalyst: A substance that induces catalysis; an agent that enables a chemical reaction to proceed under milder conditions (*e.g.* at a lower temperature) than otherwise possible. Biological catalysts are enzymes; some nonbiological catalysts include metallic complexes.

Cell: The smallest structural unit of living matter capable of functioning independently; a microscopic mass of protoplasm surrounded by a semipermeable membrane, usually including one or more nuclei and various nonliving products, capable alone, or interacting with other cells, of performing all the fundamental functions of life.

Cell culture: The *in vitro* growth of cells usually isolated from a mixture of organisms. These cells are usually of one type.

Cell differentiation: The process whereby descendants of a common parental cell achieve and maintain specialization of structure and function.

Cell fusion: Formation of a single hybrid cell with nuclei and cytoplasm from different cells (as in 'cell fusion technology'; see also Hybridoma.)

Cell line: Cells that acquire the ability to multiply indefinitely *in vitro* (especially in plant cell tissues).

Cellulase: The enzyme that digests cellulose to sugars.

Cellulose: A polymer of six carbon sugars found in all plant matter; the most abundant biological compound on earth.

Chemical clarification: Characterization of a wastewater process involving distinct operations: coagulation, flocculation and sedimentation.

Chemostat: An apparatus for maintaining microorganisms in the (exponential) phase of growth over prolonged periods of time. This is achieved by the continuous addition of fresh medium and the continuous removal of effluent, so that the volume of the growing culture remains constant.

Chemostat selection: Screening process used to identify microorganisms with desired properties, such as microorganisms that degrade toxic chemicals (see Acclimatization).

Chemotherapeutic agent: A chemical that interferes with the growth of either microorganisms or cancer cells at concentrations at which it is tolerated by the host cells.

Chemotherapy: The treatment of a disease by means of chemotherapeutic agents.

Chitin: A homopolysaccharide of *N*-acetyl-D-glucosamine that is a major constituent of the hard, horny exoskeleton of insects and crustaceans.

Chlorophyll: The green pigment that occurs in plants and functions in photosynthesis by absorbing and utilizing the radiant energy of the sun.

Chloroplasts: Cellular organelles where photosynthesis occurs.

Chromosome: A structure in the nucleus of eukaryotic cells that consists of one or more large double-helical DNA molecules that are associated with RNA and histones; the DNA of the chromosome contains the genes and functions in the storage and in the transmission of the genetic information of the organism.

Chromatography: A process of separating gases, liquids or solids in a mixture or solution by adsorption as the mixture or solution flows over the adsorbent medium, often in a column. The substances are separated because of their differing chemical interaction with the adsorbent medium.

Chromosomes: The rodlike structures of a cell's nucleus that store and transmit genetic information; the physical structures that contain genes. Chromosomes are composed mostly of DNA and protein and contain most of the cell's DNA. Each species has a characteristic number of chromosomes.

Clinical trial: One of the final stages in the collection of data for drug approval where the drug is tested in humans.

Clone: A group of genetically identical cells or organisms produced asexually from a common ancestor.

Cloning: The amplification of segments of DNA, usually genes.

Coagulation: The process whereby chemicals are added to wastewater resulting in a reduction of the forces tending to keep suspended particles apart.

Coagulation–flocculation aids: Materials used in relatively small concentrations which are added either to the coagulation and/or flocculation basins and may be classified as: oxidants, such as chlorine or ozone; weighting agents, such as bentomite clay; activated silica; and polyelectrolytes.

Coding sequence: The region of a gene (DNA) that encodes the amino acid sequence of a protein.

Codon: The sequence of three adjacent nucleotides that occurs in messenger RNA (mRNA) and that functions as a coding unit for a specific amino acid in protein synthesis. The codon determines which amino acid will be incorporated into the protein at a particular position in the polypeptide chain.

Coefficient of thermal conductivity: A physical parameter characterizing intensity of heat conduction in a substance; it is numerically equal to the conductive heat flux density due to a temperature gradient of unity.

Coenzyme: The organic molecule that functions as a cofactor of an enzyme.

Cofactor: The nonprotein component that is required by an enzyme for its activity. The cofactor may be either a metal ion (activator) or an organic molecule (coenzyme) and it may be attached either loosely or tightly to the enzyme; a tightly attached cofactor is known as a prosthetic group.

Colony: A group of contiguous cells that grow in or on a solid medium and are derived from a single cell.

Complementary DNA (cDNA): DNA that is complementary to messenger RNA; used for cloning or as a probe in DNA hybridization studies.

Compulsory licensing: Laws that require the licensing of patents, presumably to ensure early application of a technology and to diffuse control over a technology.

Constitutive enzyme: An enzyme that is present in a given cell in nearly constant amounts regardless of the composition of either the tissue or the medium in which the cell is contained.

Conjugation: The covalent or noncovalent combination of a large molecule, such as a protein or a bile acid, with another molecule. Also, the alternating sequence of single and double bonds in a molecule. Also, the genetic recombination in bacteria and in other unicellular organisms that resemble sexual reproduction and that entails a transfer of DNA between two cells of opposite mating type which are associated side by side.

Continuous processing: Method of processing in which raw materials are supplied and products are removed continuously, at volumetrically equal rates (*cf*. Batch processing).

Continuum: A medium whose discrete heterogeneous structure can be neglected, as in certain fluid flow problems.

Convective mass transfer: Mass transfer produced by simultaneous convection and molecular diffusion. The term is usually used to describe mass transfer associated with fluid flow and involves the mass transfer between a moving fluid and a boundary surface or between two immiscible moving fluids.

Convective transfer: The transfer of mass, heat or momentum in a medium with a nonhomogeneous distribution of velocity, temperature, or concentration; it is accompanied by the displacement of macroscopic elements through the medium.

Cosmid: A DNA cloning vector consisting of plasmid and phage sequences.

Corporate venture capital: Capital provided by major corporations exclusively for high-risk investments.

Crabtree effect: The inhibition of oxygen consumption in cellular respiration that is produced by increasing concentrations of glucose (see also Pasteur effect).

Critical dilution rate: Dilution rate at which wash-out conditions of cells in a bioreactor occurs (see Chemostat).

Cross flow filtration: Method of operating a filtration device whereby the processed material prevents undue build-up of filtered material on filter.

Crystalloid: A noncolloidal low-molecular weight substance.

Culture medium: Any nutrient system for the artificial cultivation of bacteria or other cells; usually a complex mixture of organic and inorganic materials.

Cyclic batch culture: Method of operating a bioreactor whereby a fill-and-dump approach retains enough biocatalyst to avoid need for re-inoculation. In cell cultures, relatively insignificant growth between fill and dump stages occurs.

Cytoplasm: The 'liquid' portion of a cell outside and surrounding the nucleus.

Cytotoxic: Damaging to cells.

Declining phase (or **Death phase**): The phase of growth of a culture of cells that follows the stationary phase and during which there is a decrease in the number (or the mass) of the cells.

Denitrification: The formation of molecular nitrogen from nitrate by way of nitrite.

Deoxyribonucleic acid (DNA): A linear polymer, made up of deoxyribonucleotide repeating units, that is the carrier of genetic information; present in chromosomes and chromosomal material of cell organelles such as mitochondria and chloroplasts, and also present in some viruses. The genetic material found in all living organisms. Every inherited characteristic has its origin somewhere in the code of each individual's DNA.

Diagnostic products: Products that recognize molecules associated with disease or other biological conditions and are used to diagnose these conditions.

Dialysis: The separation of macromolecules from ions and low-molecular weight compounds by means of a semipermeable membrane that is impermeable to (colloidal) macromolecules but is freely permeable to crystalloids and liquid medium.

Dicots (dicotyledons): Plants with two first embryonic leaves and nonparallel veined mature leaves. Examples are soybean and most flowering plants.

Diffusion boundary layer: Characterized by a transverse concentration gradient of a given component in a mixture; the effect of the gradient produces a transverse (mass transfer) of this component.

Diffusion coefficient: A physical parameter that appears as a proportionality coefficient with the gradient of concentration of a specified component in the equation which establishes the dependence of the mass diffusion flux density of the given component on the concentration gradients of all the components in the mixture. 'Self-diffusion coefficient' denotes a physical parameter which characterizes the diffusion of some molecules in the same medium with respect to others for a single-component medium.

Diffusional mass flux: Mass flux due to molecular diffusion.

Dilution rate: Reciprocal of the residence time of a culture in a bioreactor; given by the flow rate divided by bioreactor volume.

Dimensional analysis: Method of determining the number and structure of dimensionless groups consisting of variables essential to a given process on the basis of a comparison of the dimensions of these variables.

Diploidy (or **Diploid state**): The chromosome state in which each of the various chromosomes, except the sex chromosome, is represented twice.

Disclosure requirements: A patent requirement for adequate public disclosure of an invention that enables other people to build and use the invention without 'undue' experimentation.

Distal: Remote from a particular location or from a point of attachment.

DNA: Deoxyribonucleic acid (see above).

DNA base pair: A pair of DNA nucleotide bases. Nucleotide bases pair across the double helix in a very specific way: adenine can only pair with thymine; cytosine can only pair with guanine.

DNA probe: A sequence of DNA that is used to detect the presence of a particular nucleotide sequence.

DNA sequence: The order of nucleotide bases in the DNA helix; the DNA sequence is essential to the storage of genetic information.

DNA synthesis: The synthesis of DNA in the laboratory by the sequential addition of nucleotide bases.

Doubling time: The observed time required for a cell population to double in either the number of cells or the cell mass; it is equal to the generation time only if all the cells in the population are capable of doubling, have the same generation time, and do not undergo lysis (see Generation time).

Downstream processing: After bioconversion of materials in a bioreactor, the separation and purification of the product(s).

Drug: Any chemical compound that may be administered to humans or animals as an aid in the treatment of disease.

Dry weight: The weight of a sample from which liquid has been removed, usually by drying of filtered material.

Eddy mass diffusivity: A quantity characterizing the intensity of turbulent mass transfer of a particular component, as in intensely mixed bioreactors.

Electrophoresis: The movement of charged particles through a stationary liquid under the influence of an electric field. Electrophoresis is a tool for the separation of particles in both preparative and analytical studies of macromolecules. Separation is achieved primarily on the basis of the charge on the particles and to a lesser extent on the basis of the size and shape of the particles. Potentially useful in downstream processing.

Elution: The removal of adsorbed material from an adsorbent, such as the removal of a product from an enzyme bound in a chromatography column.

Emulsification: The process of making lipids, oils, fats, more soluble in water.

Endoenzyme: An enzyme that acts at random in cleaving molecules of substrate.

Endorphins: Opiate-like, naturally occurring peptides with a variety of analgesic effects throughout the endocrine and nervous systems.

Enrichment culture: A culture used for the selection of specific strains of an organism from among a mixture; such a culture favors the growth of the desired strain under the conditions used.

Enzyme: Any of a group of catalytic proteins that are produced by living cells and that mediate and promote the chemical processes of life without themselves being altered or destroyed.

Enzyme induction: The process whereby an inducible enzyme is synthesized in response to an inducer. The inducer combines with a repressor and thereby prevents the blocking of an operator by the repressor.

***Escherichia coli* (*E. coli*)**: A species of bacteria that inhabits the intestinal tract of most vertebrates. Some strains are pathogenic to humans and animals. Many nonpathogenic strains are used experimentally as hosts for rDNA.

Eukaryote: A cell or organism with membrane-bound, structurally discrete nuclei and well-developed cell organelles. Eukaryotes include all organisms except viruses, bacteria and blue-green algae (*cf.* Prokaryote).

Exoenzyme: An enzyme that acts by cleaving the ends of molecular chains in a substrate.

Exponential growth: The growth of cells in which the number of cells (or the cell mass) increases exponentially.

Export controls: Laws that restrict technology transfer and trade for reasons of national security, foreign policy or economic policy.

Facultative: Capable of living under more than one set of conditions, usually with respect to aerobic or anaerobic conditions (*e.g.* a facultative anaerobe is an organism or a cell that can grow either in the absence, or in the presence, of molecular oxygen).

Fatty acids: Organic acids with long carbon chains. Fatty acids are abundant in cell membranes and are widely used as industrial emulsifiers.

Fed-batch culture: As in 'cyclic batch culture' except that significant changes in the medium (*e.g.* cell growth) occur during the addition and/or removal of materials from bioreactor.

Feedback inhibition: A negative feedback mechanism in which a product of an enzymatic reaction inhibits the activity of an enzyme that functions in the synthesis of this product.

Feedstocks: Raw materials used for the production of chemicals.

Fermentation: A bioprocess. Fermentation is carried out in bioreactors and is used in various industrial processes for the manufacture of products such as antibiotics, alcohols, acids and vaccines by the action of living organisms (strictly speaking, anaerobically).

Film boiling: Boiling in which a continuous film of vapor that collapses periodically into the bulk of the liquid is formed on the heated surface, as in certain evaporation processes.

Flagellum: (pl. **flagella**): A threadlike, cellular extension that functions in the locomotion of bacterial cells and of unicellular eukaryotic organisms.

Flavin adenine dinucleotide (FAD): The flavin nucleotide, riboflavin adenosine diphosphate, which is a coenzyme form of the vitamin riboflavin, and which functions in dehydrogenation reactions catalyzed by flavoproteins.

Flocculating agent: A reagent added to a dispersion of solids in a liquid to bring together the fine particles into larger masses.

Flocculation: The agglomeration of suspended material to form particles that will settle by gravity, as in the 'tertiary' treatment of waste materials.

Food additive (or **Food ingredient**): A substance that becomes a component of food or affects the characteristics of food and, as such, is regulated, *e.g.* by the US Food and Drug Administration.

Forced convection: Motion of fluid elements induced by external forces, *e.g.* in a bioreactor by a mechanical stirrer.

Free convection: Motion of fluid elements induced by 'natural' forces, *e.g.* by density differences caused by concentration or temperature gradients.

Free-living organism: An organism that does not depend on other organisms for survival.

Fruiting body: A mass of vegetative cells which swarm together at the same stage of growth.

Fungus: Any of a major group of saprophytic and parasitic plants that lack chlorophyll, including molds, rusts, mildews, smuts and mushrooms.

Gamma globulin (GG): A protein component of blood that contains antibodies and confers passive immunity.

Gene: The basic unit of heredity; an ordered sequence of nucleotide bases, comprising a segment of DNA. A gene contains the sequence of DNA that encodes one polypeptide chain (*via* RNA).

Gene amplification: In biotechnology, an increase in gene number for a certain protein so that the protein is produced at elevated levels.

Gene expression: The mechanism whereby the genetic directions in any particular cell are decoded and processed into the final functioning product, usually a protein (see also Transcription and Translation).

Gene transfer: The use of genetic or physical manipulation to introduce foreign genes into host cells to achieve desired characteristics in progeny.

Generation time: The time required by a cell for the completion of one cycle of growth (see also Doubling time).

Genetic engineering: Loose term used to describe any gene manipulative technique, especially recombinant DNA techniques.

Genome: The genetic endowment of an organism or individual.

Genus: A taxonomic category that includes groups of closely related species.

Germ cell: The male and female reproductive cells; egg and sperm.

Germplasm: The total genetic variability available to a species.

Glycoproteins: Proteins with attached sugar groups.

Glycoside: A mixed acetal (or ketal) derived from the cyclic hemiacetal (or hemiketal) form of an aldose (or a ketose); a compound formed by replacing the hydrogen or the hydroxyl group of the anomeric carbon of the carbohydrate with an alkyl or aryl group.

Glycosylation: The attachment of sugar groups to a molecule, such as a protein.

Gram negative: Designating a bacterium that does not retain the initial Gram stain but retains the counterstain. Gram-negative bacteria possess a relatively thin cell wall that is not readily digested by the enzyme lysozyme, and in which the peptidoglycan layer is covered with lipopolysaccharide.

Gram positive: Designating a bacterium that retains the initial Gram stain and is not stained by the counterstain. Gram-positive bacteria generally possess a relatively thick and rigid cell wall that is readily digested by the enzyme lysozyme, and that consists of a layer of peptidoglycan.

Gram stain: A set of two stains (chemicals) that are used to stain bacteria; the staining depends on the composition and the structure of the bacterial cell wall.

Growth hormone (GH): A group of peptides involved in regulating growth in higher animals.

Heat flux: The quantity of heat that passes through an arbitrary surface per unit time.

Heat flux density: Heat flux per unit area.

Heat transfer: Spontaneous irreversible process of heat transmission in a space with a nonisothermal temperature field, as in the cooling or heating of bioreactors and ancilliary equipment.

Hemicellulose: A polymer of D-xylose that contains side chains of other sugars and that serves to cement plant cellulose fibers together.

Herbicide: An agent (*e.g.* a chemical) used to destroy or inhibit plant growth; specifically, a selective weed killer that is not injurious to crop plants.

Heterofermentative lactic acid bacteria: Lactic acid bacteria that produce in fermentation less than 1.8 moles of lactic acid per mole of glucose; in addition to lactic acid, these organisms produce ethanol, acetate, glycerol, mannitol and carbon dioxide (see also Homofermentative lactic acid bacteria).

Heterotroph: A cell or organism that requires a variety of carbon-containing compounds from animals and plants as its source of carbon, and that synthesizes all of its carbon-containing biomolecules from these compounds and from small inorganic molecules.

Heterotrophic: Pertaining to a regulatory enzyme in which the effector is a metabolite other than the substance of the enzyme.

High performance liquid chromatography (HPLC): A recently developed type of chromatography that is potentially important in downstream processing of bioreactor products.

Histone: A basic, globular, and simple protein that is characterized by its high content of arginine and lysine. Histones are found in association with nucleic acids in the nuclei of many eukaryotic cells.

Homofermentative lactic acid bacteria: Lactic acid bacteria that produce 1.8–2.0 moles of lactic acid per mole of glucose during fermentation.

Hormone: A chemical messenger found in the circulation of higher organisms that transmits regulatory messages to cells.

Host: A cell whose metabolism is used for growth and reproduction of a virus, plasmid or other form of foreign DNA.

Host–vector system: Compatible combinations of host (*e.g.* bacterium) and vector (*e.g.* plasmid) that allow stable introduction of foreign DNA into cells.

Hybrid: The offspring of two genetically dissimilar parents (*e.g.* a new variety of plant or animal that results from cross-breeding two different existing varieties; a cell derived from two different cultured cell lines that have fused).

Hybridization: The act or process of producing hybrids.

Hybridoma: Product of fusion between myeloma cell (which divides continuously in culture and is 'immortal') and lymphocyte (antibody-producing cell); the resulting cell grows in culture and produces monoclonal antibodies.

Hybridoma technology: See Monoclonal antibody technology.

Hydrolysis: Chemical reaction involving addition of water to break bonds.

Hydroxylation: Chemical reaction involving the addition of hydroxyl (OH) groups to chemical compounds.

Hypha (pl. **hyphae**): The filamentous and branched tube that forms the network which contains the cytoplasm of the mycelium of a fungus.

Immobilized enzyme or cell techniques: Techniques used for the fixation of enzymes or cells on to solid supports. Immobilized cells and enzymes are used in continuous bioprocessing in bioreactors and upstream or downstream processing of materials.

Immune response: The reaction of an organism to invasion by a foreign substance. Immune responses are often complex, and may involve the production of antibodies from special cells (lymphocytes), as well as the removal of the foreign substance by other cells.

Immunization: The administration of an antigen to an animal organism to stimulate the production of antibodies by that organism. Also, the administration of antigens, antibodies or lymphocytes to an animal organism to produce the corresponding active, passive or adoptive immunity.

Immunoassay: The use of antibodies to identify and quantify substances. The binding of antibodies to antigen, the substance being measured, is often followed by tracers such as radioisotopes.

Immunogenic: Capable of causing an immune response (see also Antigen).

Immunotoxin: A molecule attached to an antibody capable of killing cells that display the antigen to which the antibody binds.

Inducible enzyme: An enzyme that is normally either absent from a cell or present in very small amounts, but that is synthesized in appreciable amounts in response to an inducer in the process medium.

Interface: The boundary between two phases.

Interferons (Ifns): A class of glycoproteins (proteins with sugar groups attached at specific locations) important in immune function and thought to inhibit viral infections.

In vitro: Literally, in glass; pertaining to a biological reaction taking place in an artificial apparatus; sometimes used to include the growth of cells from multicellular organisms under cell culture conditions. *In vitro* diagnostic products are products used to diagnose disease outside of the body after a sample has been taken from the body.

In vivo: Literally, in life; pertaining to a biological reaction taking place in a living cell or organism. *In vivo* products are products used within the body.

Ionic strength: A measure of the ionic concentration of a solution.

Laminar flow: Fluid motion in which the existence of steady fluid particle trajectories can exist; in processing equipment, it represents relatively low levels of mixing intensities.

Isoelectric pH (isoelectric point): The pH at which a molecule has a net zero charge; the pH at which the molecule has an equal number of positive and negative charges, which includes those due to any ions bound by the molecule.

Isoelectrophoretic pH (isoelectrophoretic point): The pH at which the electrophoretic mobility of a protein is zero; this pH may coincide with the theoretical isoelectric pH of the protein, depending on the surface structure of the protein, the ionic strength, and the nature of the ionic double layer around the protein.

Lag phase: That phase of growth of a cell that precedes the exponential phase and during which there is only little or no growth.

Leaching: The removal of a soluble compound such as an ore from a solid mixture by washing or percolating.

Lignocellulose: The composition of woody biomass, including lignin and cellulose.

Lignolytic: Pertaining to the breakdown of lignin.

Lime: Various natural forms of the chemical compound CaO, as in hydrated lime and dolomitic lime.

Linker: A small fragment of synthetic DNA that has a restriction site useful for gene cloning, which is used for joining DNA strands together.

Lipase: An enzyme that catalyzes the hydrolysis of fats.

Lipids: A large, varied class of water-insoluble fat-based organic molecules; includes steroids, fatty acids, prostaglandins, terpenes and waxes.

Lipopolysaccharide: A water-soluble lipid–polysaccharide complex.

Liposome transfer: The process of enclosing biological compounds inside a lipid membrane and allowing the complex to be taken up by a cell.

Lymphocytes: Specialized white blood cells involved in the immune response; B lymphocytes produce antibodies.

Lymphokines: Proteins that mediate interactions among lymphocytes and are vital to proper immune function.

Lyophilization: The removal of water under vacuum from a frozen sample; a relatively gentle process in which water sublimes.

Lysis: The rupture and dissolution of cells.

Mass exchange: Mass transfer across an interface or a permeable wall (membrane) between two phases.

Mass flux: The mass of a given mixture component passing per unit time across any surface.

Mass flux density: Mass flux per unit area of surface.

Mass transfer: Spontaneous irreversible process of transfer of mass of a given component in a space with a nonhomogenous field of the chemical potential of the component. In the simplest case, the driving force is the difference in concentration (in liquids) or partial pressure (in gases) of the component. Other physical quantities, *e.g.* temperature difference (thermal diffusion), can also induce mass transfer.

Mass transfer coefficient: A quantity characterizing the intensity of mass transfer; it is numerically equal to the ratio of the mass flux to the difference of its mass fractions. For the case of mass transfer between a liquid medium and a gas, the mass fraction of a given component in the liquid is determined by phase equilibrium parameters (distribution coefficient) with allowance, if necessary, for resistance to transfer at the phase boundary *per se*.

Mass velocity: Mass flow rate across a unit area perpendicular to the direction of the velocity vector.

Mesophile: An organism that grows at moderate temperatures in the range 20–45 °C, and that has an optimum growth temperature in the range 30–39 °C.

Mesophilic: Of, or pertaining to, mesophiles.

Messenger RNA (mRNA): RNA that serves as the template for protein synthesis; it carries the transcribed genetic code from the DNA to the protein synthesizing complex to direct protein synthesis.

Metabolism: The physical and chemical processes by which chemical components are synthesized into complex elements, complex substances are transformed into simpler ones, and energy is made available for use by an organism.

Metabolite: Any reactant, intermediate or product in the reactions of metabolism.

Metallothioneins: Proteins, found in higher organisms, that have a high affinity for heavy metals.

Methanogens: Bacteria that produce methane as a metabolic product.

Microorganisms: Microscopic living entities; microorganisms can be viruses, prokaryotes (*e.g.* bacteria) or eukaryotes (*e.g.* fungi). Also referred to as microbes.

Microencapsulation: The process of surrounding cells with a permeable membrane.

Mitochondrion (pl. **mitochondria**): A subcellular organelle in aerobic eukaryotic cells that is the site of cellular respiration and that carries out the reactions of the citric acid cycle, electron transport, and oxidative phosphorylation. Mitochondria contain DNA and ribosomes, carry out protein synthesis, and are capable of self-replication.

Mixed culture: Culture containing two or more types of microorganisms.

Molecular diffusion: Mass transfer resulting from thermal motion. Concentration diffusion refers to molecular diffusion resulting from a nonhomogenous distribution of concentrations of components of a mixture.

Monoclonal antibodies (MAbs): Homogeneous antibodies derived from a single clone of cells; MAbs recognize only one chemical structure. MAbs are useful in a variety of industrial and medical capacities since they are easily produced in large quantities and have remarkable specificity.

Monoclonal antibody technology: The use of hybridomas that produce monoclonal antibodies for a variety of purposes. Hybridomas are maintained in cell culture or, on a larger scale, as tumors (ascites) in mice. Also referred to as 'hybridoma' technology.

Monocots (monocotyledons): Plants with single first embryonic leaves, parallel-veined leaves, and simple stems and roots. Examples are cereal grains such as corn, wheat, rye, barley and rice.

Monosaccharide: A polyhydroxy alcohol containing either an aldehyde or a ketone group; a simple sugar.

Multigenic: A trait specialized by several genes.

Multi-phase medium: A medium consisting of two or more single-phase portions with physical properties changing discontinuously (stepwise) at the boundaries of the medium, as in a gas–liquid dispersion used in aerobic fermentations.

Mutagenesis: The induction of mutation in the genetic material of an organism; researchers may use physical or chemical means to cause mutations that improve the production of capabilities of organisms.

Mutagen: An agent that causes mutation.

Mutant: An organism with one or more DNA mutations, making its genetic function or structure different from that of a corresponding wild-type organism.

Mutation: A permanent change in a DNA sequence.

Mycelium (pl. **mycelia**): The vegetative structure of a fungus that consists of a multinucleate mass of cytoplasm, enclosed within a branched network of filamentous tubes known as hyphae.

Myeloma: Antibody-producing tumor cells.

Myeloma cell line: Myeloma cells established in culture.

Natural convection: Free motion due to gravitational forces in a system with a non-homogeneous density distribution (see also Free convection).

Neurotransmitters: Small molecules found at nerve junctions that transmit signals across those junctions.

Newtonian fluid: A fluid, the viscosity of which is independent of the rate and/or duration of shear.

NIH Guidelines: Guidelines, established by the US National Institutes of Health, on the safety of research involving recombinant DNA.

Nitrogen fixation: The conversion of atmospheric nitrogen gas to a chemically combined form, ammonia (NH_3), which is essential to growth. Only a limited number of microorganisms can fix nitrogen.

Nodule: The anatomical part of a plant root in which nitrogen-fixing bacteria are maintained in a symbiotic relationship with the plant.

Nodulins: Proteins, possibly enzymes, present in nodules; function unknown.

Non-Newtonian fluid: A fluid, the viscosity of which depends on the rate and/or duration of shear.

Nucleate boiling: Boiling in which vapor is generated in the form of periodically produced and growing discrete bubbles.

Nucleic acids: Macromolecules composed of sequences of nucleotide bases. There are two kinds of nucleic acids: DNA, which contains the sugar deoxyribose, and RNA, which contains the sugar ribose.

Nucleoside: A glycoside composed of D-ribose or 2-deoxy-D-ribose and either a purine or a pyrimidine.

Nucleotide base: A structural unit of nucleic acid. The bases present in DNA are adenine, cytosine, guanine and thymine. In RNA, uracil substitutes for thymine.

Nucleus: In the biological sciences, a relatively large spherical body inside a cell that contains the chromosomes.

Oligomer: A molecule that consists of two or more monomers linked together, covalently or non-covalently.

Oligonucleotides: Short segments of DNA or RNA.

Optical density (absorbance): A measure of the light absorbed by a solution.

Organelle: A specialized part of a cell that conducts certain functions. Examples are nuclei, chloroplasts and mitochondria, which contain most of the genetic material, conduct photosynthesis and provide energy, respectively.

Organic compounds: Molecules that contain carbon.

Organic micropollutant: Low molecular weight organic compounds considered hazardous to humans or the environment.

Osmosis: The movement of water or another solvent across a semipermeable membrane from a region of low solute concentration to one of a higher solute concentration.

Osmotic pressure: The pressure that causes water or another solvent to move in osmosis from a solution having a low solute concentration to one having a high solute concentration; it is equal to the hydrostatic pressure that has to be applied to the more concentrated solution to prevent the movement of water (solvent) into it.

Oxidation ponds: Quiescent earthen basins that provide sufficient hydraulic hold-up time for the natural processes to effect removal and stabilization of organic matter.

Oxygen transfer rate (OTR): Mass transfer for oxygen solute as in fermentation medium.

Parasite: An organism that lives in or upon another organism from which it derives some or all of its nutrients.

Passive immunity: Disease resistance in a person or animal due to the injection of antibodies from another person or animal. Passive immunity is usually short-lasting (*cf.* Active immunity).

Pasteur effect: The inhibition of glycolysis and the decrease of lactic acid accumulation that is produced by increasing concentrations of oxygen.

Patent: A limiting property right granted to inventors by a government allowing the inventor of a new invention the right to exclude all others from making, using or selling the invention unless specifically approved by the inventor, for a specified time period in return for full disclosure by the inventor about the invention.

Pathogen: A disease-producing agent, usually restricted to a living agent such as a bacterium or virus.

Pectin: A polysaccharide that occurs in fruits and that consists of a form of pectic acid in which many of its carboxyl groups have been methylated.

Peptide: A linear polymer of amino acids. A polymer of numerous amino acids is called a *polypeptide*. Polypeptides may be grouped by function, such as 'neuroactive' polypeptides.

Permease: An enzyme that is instrumental in transporting material across a biological membrane or within a biological fluid.

Pharmaceuticals: Products intended for use in humans, as well as *in vitro* applications to humans, including drugs, vaccines, diagnostics and biological response modifiers.

Photorespiration: Reaction in plants that competes with the photosynthetic process. Instead of fixing CO_2, RuBPCase can utilize oxygen, which results in a net loss of fixed CO_2.

Photosynthesis: The reaction carried out by plants where carbon dioxide from the atmosphere is fixed into sugars in the presence of sunlight; the transformation of solar energy into biological energy.

Plasma: The liquid (noncellular) fraction of blood. In vertebrates, it contains many important proteins (*e.g.* fibrinogen, responsible for clotting).

Plasmid: An extrachromosomal, self-replicating, circular segment of DNA; plasmids (and some viruses) are used as 'vectors' for cloning DNA in bacterial 'host' cells.

Plug flow: Flow of materials in which there is no mixing in the direction of flow (see Axial dispersion).

Polymer: A linear or branched molecule of repeating subunits.

Polypeptide: A long peptide, which consists of amino acids.

Polysaccharide: A polymer of sugars.

Pool boiling: Boiling with convective (free) motion in a liquid volume whose dimensions in all directions are large compared to the breakaway diameter of the bubble.

Primary metabolite: Metabolite that is required for the function of the organism's life support system.

Proinsulin: A precursor protein of insulin.

Prokaryote: A cell or organism lacking membrane-bound, structurally discrete nuclei and organelles. Prokaryotes include bacteria and the blue-green algae. (*cf.* eukaryote).

Promoter: A DNA sequence in front of a gene that controls the initiation of 'transcription' (see below).

Prophylaxis: Prevention of disease.

Protease: Protein-digesting enzyme.

Protein: A polypeptide consisting of amino acids. In their biologically active states, proteins function as catalysts in metabolism and, to some extent, as structural elements of cells and tissues.

Protist: A unicellular or multicellular organism that lacks the tissue differentiation and the elaborate organization that is characteristic of plants and animals.

Protoplast fusion: The joining of two cells in the laboratory to achieve desired results, such as increased viability of antibiotic-producing cells.

Protozoa: Diverse forms of eukaryotic microorganisms; structure varies from simple single cells to colonial forms; some protozoa are pathogenic.

Psychrophile: An organism that grows at low temperatures in the range of 0–25 °C, and that has an optimum growth temperature in the range 20–25 °C.

Pure culture: A culture containing only microorganisms from one species.

Pyrogenicity: The tendency for some bacterial cells or parts of cells to cause inflammatory reactions in the body, which may detract from their usefulness as pharmaceutical products.

Recarbonation: Unit water treatment process in which carbon dioxide is added to a lime-treated water. Basic purpose is the downward adjustment of the pH of the water.

Recombinant DNA (rDNA): The hybrid DNA produced by joining pieces of DNA from different organisms together *in vitro* (*i.e.* in an artificial apparatus).

Recombinant DNA technology: The use of recombinant DNA for a specific purpose, such as the formation of a product or the study of a gene.

Recombination: Formation of a new association of genes or DNA sequences from different parental origins.

Reducing sugar: A sugar that will reduce certain inorganic ions in solution, such as the copper(II) ions of Fehling's or Benedict's reagent; the reducing property of the sugar is due to its aldehyde or potential aldehyde group.

Regeneration: In biological sciences, the laboratory process of growing a whole plant from a single cell or small clump of cells.

Regulatory sequence: A DNA sequence involved in regulating the expression of a gene.

Repressible enzyme: An enzyme, the synthesis of which is decreased when the intracellular concentration of specific metabolites reaches a certain level.

Resistance gene: Gene that provides resistance to an environmental stress such as an antibiotic or other chemical compound.

Respiration: The cellular oxidative reactions of metabolism, particularly the terminal steps, by which nutrients are broken down; the reactions which require oxygen as the terminal electron acceptor, produce carbon dioxide as a waste product, and yield utilizable energy.

Restriction enzymes: Enzymes that cut DNA at specific DNA sequences.

Ribosome: One of a large number of subcellular, nucleoprotein particles that are composed of approximately equal amounts of RNA and protein and that are the sites of protein synthesis in the cell.

Ribosomal RNA: The RNA that is linked noncovalently to the ribosomal proteins in the two ribosomal subunits and that constitutes about 80% of the total cellular RNA.

RNA: Ribonucleic acid (see also Messenger RNA).

RuBPCase (ribulosebisphosphate carboxylase): An enzyme that catalyzes the critical step of the photosynthetic CO_2 cycle.

Salting out: The decrease in the solubility of a protein that is produced in solutions of high ionic strength by an increase of the concentrations of neutral salts.

Saccharification: The degradation of polysaccharides to sugars.

Scale-up: The transition of a process from research laboratory bench scale to engineering pilot plant or industrial scale.

Secondary metabolite: Metabolite that is not required by the producing organism for its life-support system.

Semiconductor: A material such as silicon or germanium with electrical conductivities intermediate between good conductors such as copper wire and insulators such as glass.

Shake flask: A laboratory flask for culturing microorganisms in a shaker–incubator which provides mixing and aeration.

Shear rate: The variation in velocity within a flowing material, as in a bioreactor.

Single cell protein (SCP): Cells, or protein extracts, of microorganisms grown in large quantities for use as human or animal protein supplements. A misnomer for multicellular SCP products.

Single-phase medium: Continuous single- or multi-component medium whose properties in space can vary in a continuous manner with no phase boundaries, *e.g.* a gas or a liquid.

Slaking: Process of adding water to quicklime, or recalcined lime, to produce a slurry of hydrated lime.

Slant culture: A culture grown in a tube that contains a solid nutrient medium which was solidified while the tube was kept in a slanted position.

Slimes: Aggregations of microbial cells that pose environmental and industrial problems; may be amenable to biological control.

Sludge: Precipitated solid matter produced by water and sewage treatment or industrial problems; may be amenable to biological control.

Slug flow: Type of two-phase flow in which the gas phase flows in the form of large bubbles whose transverse dimensions are commensurate with the characteristic dimension of the flow cross section as in some pipeline operations.

Somaclonal variation: Genetic variation produced from the culture of plant cells from a pure breeding strain; the source of the variation is not known.

Species: A taxonomic subdivision of a genus. A group of closely related, morphologically similar individuals which actually or potentially interbreed.

Specific growth rate: The rate of growth of a population of microorganisms, per unit mass of cells.

Spectrometer: An instrument used for analyzing the structure of compounds on the basis of their light-absorbing properties.

Spheroplast: A bacterial cell that is largely, but not entirely, freed of its cell wall.

Spore: A dormant cellular form, derived from a bacterial or a fungal cell, that is devoid of metabolic activity and that can give rise to a vegetative cell upon germination; it is dehydrated and can survive for prolonged periods of time under drastic environmental conditions.

Starch: The major form of storage carbohydrates in plants. It is a homopolysaccharide, composed of D-glucose units, that occurs in two forms: amylose, which consists of straight chains, and in which the glucose residues are linked by means of alpha(1–4) glycosidic bonds; and amylopectin, which consists of branched chains, and in which the glucose residues are linked by means of both alpha(1–4) and alpha(1–6) glycosidic bonds.

Stationary phase: The phase of growth of a culture of microorganisms that follows the exponential phase and in which there is little or no growth.

Sterile: Free from viable microorganisms.

Sterilization: The complete destruction of all viable microorganisms in a material by physical and/or chemical means.

Steroid: A group of organic compounds, some of which act as hormones to stimulate cell growth in higher animals and humans.

Stirred tank bioreactor: Agitated vessel in which a bioprocess takes place; mixing is provided by the mechanical action of an impeller/agitator.

Storage protein genes: Genes coding for the major proteins found in plant seeds.

Strain: A group of organisms of the same species having distinctive characteristics but not usually considered a separate breed or variety. A genetically homogeneous population of organisms at a subspecies level that can be differentiated by a biochemical, pathogenic or other taxonomic feature.

Substrate: A substance acted upon, for example, by an enzyme.

Subunit vaccine: A vaccine that contains only portions of a surface molecule of a pathogen. Subunit vaccines can be prepared by using rDNA technology to produce all or part of the surface protein molecule or by artificial (chemical) synthesis of short peptides.

Surfactant: A substance that alters the surface tension of a liquid, generally lowering it; detergents and soaps are typical examples.

Symbiont: An organism living in symbiosis, usually the smaller member of a symbiotic pair of dissimilar size.

Symbiosis: In the biological sciences, the living together of two dissimilar organisms in mutually beneficial relationships.

Synchronous growth: Growth in which all of the cells are at the same stage in cell division at any given time.

Tangential flow filtration: See Cross-flow filtration.

Taxis: The movement of an organism in response to a stimulus.

Taxonomy: The scientific classification of plants and animals that is based on their natural relationships; includes the systematic grouping, ordering and naming of the organisms.

Therapeutics: Pharmaceutical products used in the treatment of disease.

Thermal diffusivity: Numerically equal to the ratio of the coefficient of thermal conductivity to the volumetric specific heat of a substance (see also Heat transfer).

Thermophile: An organism that grows at high temperatures in the range 45–70 °C (or higher temperatures) and that has an optimum growth temperature in the range 50–55 °C.

Thermophilic: Heat loving. Usually refers to microorganisms that are capable of surviving at elevated temperatures; this capability may make them more compatible with industrial biotechnology schemes.

Thermotolerant: Capable of withstanding relatively high temperatures (45–70 °C).

Thrombolytic enzymes: Enzymes such as streptokinase and urokinase that initiate the dissolution of blood clots.

Ti plasmid: Plasmid from *Agrobacterium tumefacciens*, used as a plant vector.

Tissue plasminogen activator (TPA): A hormone that selectively dissolves blood clots that cause heart attacks and strokes.

Totipotency: The capacity of a higher organism cell to differentiate into an entire organism. A totipotent cell contains all the genetic information necessary for complete development.

Toxicity: The ability of a substance to produce a harmful effect on an organism by physical contact, ingestion or inhalation.

Toxin: A substance, produced in some cases by disease-causing microorganisms, which is toxic to other living organisms.

Toxoid: Detoxified toxin, but with antigenic properties intact.

Transcription: The synthesis of messenger RNA on a DNA template; the resulting RNA sequence is complementary to the DNA sequence. This is the first step in gene expression (see also Translation).

Transfer RNA (tRNA): A low-molecular weight RNA molecule, containing about 70–80 nucleotides, that binds an amino acid and transfers it to the ribosomes for incorporation into a polypeptide chain during translation.

Transformation: In the biological sciences, the introduction of new genetic information into a cell using naked DNA.

Transistor: An active component of an electrical circuit consisting of semiconductor material to which at least three electrical contacts are made so that it acts as an amplifier, detector or switch.

Translation: In the biological sciences, the process in which the genetic code contained in the nucleotide base sequence of messenger RNA directs the synthesis of a specific order of amino acids to produce a protein. This is the second step in gene expression (see also Transcription).

Transposable element: Segment of DNA which moves from one location to another among or within chromosomes in possibly a predetermined fashion, causing genetic change; may be useful as a vector for manipulating DNA.

Turbulent flow: In the engineering sciences, fluid motion with particle trajectories varying chaotically (randomly) with time; irregular fluctuations of velocity, pressure and other parameters, non-uniformly distributed in the flow; indicates relatively high levels of mixing (*cf.* Laminar flow).

Vaccine: A suspension of attenuated or killed bacteria or viruses, or portions thereof, injected to produce active immunity (see also Subunit vaccine).

Vector: DNA molecule used to introduce foreign DNA into host cells. Vectors include plasmids, bacteriophages (virus) and other forms of DNA. A vector must be capable of replicating autonomously and must have cloning sites for the introduction of foreign DNA.

Viable: Describing a cell or an organism that is alive and capable of reproduction.

Viable count: The number of viable cells in a culture of microorganisms.

Virus: Any of a large group of submicroscopic agents infecting plants, animals and bacteria and unable to reproduce outside the tissues of the host. A fully formed virus consists of nucleic acid (DNA or RNA) surrounded by a protein or protein and lipid coat.

Viscosity: A measure of a liquid's resistance to flow.

Volatile fatty acids (VFAs): Mixture of acids, primarily acetic and propionic, produced by acidogenic microorganisms during anaerobic digestion.

Volatile organic compounds (VOCs): Group of toxic compounds found in ground water and that pose environmental hazards; their destruction during water purification may be done biologically.

Wash out: Condition at which the critical dilution rate is exceeded in a chemostat.

Wild-type: The most frequently encountered phenotype in natural breeding populations.

Yeast: A fungus of the family Saccharomycetacea that is used especially in the making of alcoholic liquors and fodder yeast for animal feeds, and as leavening in baking. Yeasts are also commonly used in bioprocesses.

Yield: For a general chemical reaction: the weight of product obtained divided by the theoretical amount expected. Also, for the isolation of an enzyme: the total activity at a given step in the isolation divided by the total activity at a reference step.

Appendix 2:
Nomenclature Guidelines

Provisional list of symbols and units recommended for use in biotechnology by the IUPAC Commission on Biotechnology as reported in *Pure Appl. Chem.*, **54**, 1743–1749 (1982). This list is intended for use in conjunction with other recommendations on symbols, in particular *Manual of Symbols and Terminology for Physicochemical Quantities and Units* (Pergamon, 1979) and 'Letter Symbols for Chemical Engineering', *Chem. Eng. Prog.*, 73–80 (1978).

A2.1 GENERAL CONCEPTS

	Symbol	SI units	Other units
Activation energy	E	J mol^{-1}	cal mol^{-1}
for growth	E_g		
for death	E_d		
Area dimensions			
area per unit volume	a	m^{-1}	cm^{-1}
Linear dimension			
impeller diameter	D_i	m	m
tank diameter	D_t	m	m
liquid depth	D_l	m	m
width of baffle	D_b	m	m
Pressure	p	Pa	atm, bar
denote partial pressure with appropriate subscript, *e.g.* P_{O_2} for partial pressure of oxygen			
Ratio, in general	R		
for stoichiometric mass ratio, *e.g.* mass of substrate A consumed per mass of substrate B consumed	$R_{A/B}$		
for stoichiometric molar ratio, *e.g.* mole of substrate A consumed per mole of B consumed	$R_{MA/B}$		
Temperature			
absolute	T	K	K
general	t, T	°C	°C
Time	t	s	min, h
identify specific time periods by appropriate subscripts, *e.g.* t_d for			

	Symbol	SI units	Other units
doubling time, t_1 for lag time, and t_r for replacement or mean residence time			
Volume dimensions			
volume	V	m^3, L	L
identify by subscript, *e.g.* V_1 for volume of stage 1, *etc.*			
Yield, general mass ratio expressing output over input	Y		
without further definition, Y refers to the mass conversion ratio in terms of g dry weight biomass per g mass of substrate used. It should be further defined by subscripts to denote other ratios, *e.g.* $Y_{P/S}$ and $Y_{P/X}$ for g mass of product per g mass of substrate and per g dry weight of biomass, respectively			
Yield, growth mass ratio corrected for maintenance, where $$\frac{1}{Y} = \frac{1}{Y_G} + \frac{m}{\mu}$$ or $$q_s = \frac{\mu}{Y_G} + m$$	Y_G		
Yield, molar growth	Y_{GM}	$kg\ mol^{-1}$	$g\ mol^{-1}$
kg biomass formed per mole of mass used, or further defined as above to denote other molar yields			

A2.2 CONCENTRATIONS AND AMOUNTS

Concentration

Biomass*

	Symbol	SI units	Other units
total mass (dry wt. basis)	x	kg	g
mass concentration (dry wt. basis)	X	$kg\ m^{-3}$	$g\ L^{-1}$
volume fraction	ϕ		
total number	N		
number concentration	n	m^{-3}	
Substrate concentration mass or moles per unit volume	C_S	$kg\ m^{-3}$, $kmol\ m^{-3}$	$mg\ L^{-1}$, $mmol\ L^{-1}$
Product concentration mass or moles per unit volume	C_P	$kg\ m^{-3}$, $kmol\ m^{-3}$	$mg\ L^{-1}$, $mmol\ L^{-1}$
Gas hold-up volume of gas per volume of dispersion	ε_G		

* Note: because of the difficulty in expressing biomass (cells) in molar terms, a separate symbol (other than C) is recommended.

	Symbol	*SI units*	*Other units*
Inhibitor concentration mass or moles per unit volume	C_i	$\mathrm{kg\,m^{-3}}$, $\mathrm{kmol\,m^{-3}}$	$\mathrm{mg\,L^{-1}}$, $\mathrm{mmol\,L^{-1}}$
Inhibitor constant dissociation constant of inhibitor–biomass complex	K_i	$\mathrm{kg\,m^{-3}}$	$\mathrm{g\,L^{-1}}$
Saturation constant as in the growth rate expression $\mu = \mu_m C_S/(K_s + C_S)$	K_s	$\mathrm{kmol\,m^{-3}}$	$\mathrm{g\,L^{-1}}$, $\mathrm{mmol\,L^{-1}}$
Total amount, *e.g.* mass or moles	C	kg, kmol	g, mol

A2.3   INTENSIVE PROPERTIES

	Symbol	*SI units*	*Other units*
Density, mass	ϱ	$\mathrm{kg\,m^{-3}}$	$\mathrm{g\,L^{-1}}$
Diffusivity, molecular, volumetric	D_v	$\mathrm{m^2\,s^{-1}}$	$\mathrm{cm^2\,s^{-1}}$
Enthalpy, mass, of growth	H_X	$\mathrm{J\,kg^{-1}}$	$\mathrm{J\,g^{-1}}$
heat produced per unit of dry weight biomass formed			
Enthalpy, molar, of substrate consumption or of product formation	H_S, H_P	$\mathrm{J\,mol^{-1}}$	$\mathrm{J\,mol^{-1}}$
Vapor pressure denote with appropriate subscript, *e.g.* $p_i^* = $ vapor pressure of material i	p^*	Pa	atm, bar
Viscosity, absolute	μ	Pa s	poise
Viscosity, kinematic	ν	$\mathrm{m^2\,s^{-1}}$	$\mathrm{cm^2\,s^{-1}}$

A2.4   RATE CONCEPTS

	Symbol	*SI units*	*Other units*
Death rate, specific $\delta = -(\mathrm{d}n/\mathrm{d}t)/n$	δ	$\mathrm{s^{-1}}$	$\mathrm{s^{-1}}$, $\mathrm{h^{-1}}$
Dilution rate volume flow rate/culture volume	D	$\mathrm{s^{-1}}$	$\mathrm{h^{-1}}$, $\mathrm{d^{-1}}$
Dilution rate, critical value at which biomass washout occurs in continuous flow culture	D_c	$\mathrm{s^{-1}}$	$\mathrm{h^{-1}}$, $\mathrm{d^{-1}}$
Doubling time, biomass $t_d = (\ln 2)/\mu$	t_d	s	min, h
Flow rate, volumetric identify stream by appropriate subscript, *e.g.* A for air, G for gas, L for liquid, *etc.*	F	$\mathrm{m^3\,s^{-1}}$	$\mathrm{L\,h^{-1}}$
Growth rate, colony radial rate of extension of biomass colony on a surface	K_r	$\mathrm{m\,s^{-1}}$	$\mathrm{\mu m\,h^{-1}}$
Growth rate, maximum specific	μ_m	$\mathrm{s^{-1}}$	$\mathrm{h^{-1}}$, $\mathrm{d^{-1}}$
Growth rate, specific $\mu = (\mathrm{d}x/\mathrm{d}t)/x$	μ	$\mathrm{s^{-1}}$	$\mathrm{h^{-1}}$, $\mathrm{d^{-1}}$

	Symbol	SI units	Other units
Heat transfer coefficient			
individual	h	$\text{W m}^{-2}\,\text{K}^{-1}$	$\text{cal h}^{-1}\,\text{cm}^{-2}\,{}^{\circ}\text{C}^{-1}$
overall	U	$\text{W m}^{-2}\,\text{K}^{-1}$	$\text{cal h}^{-1}\,\text{cm}^{-2}\,{}^{\circ}\text{C}^{-1}$
Maintenance coefficient, substrate or non-growth term associated with substrate consumption as defined in yield relationship (see yield term)	m	s^{-1}	h^{-1}
Mass transfer coefficient (molar basis)			
Individual, area basis	k	$\text{kmol m}^{-2}\,\text{s}^{-1}$ (driving force)$^{-1}$	$\text{gmol h}^{-1}\,\text{cm}^{-2}$ (driving force)$^{-1}$
gas film	k_{G}	$\text{kmol m}^{-2}\,\text{s}^{-1}\,\text{kPa}^{-1}$	"
liquid film	k_{L}	m s^{-1}	"
Overall, area basis	K	"	"
gas film	K_{G}	"	"
liquid film	K_{L}	"	"
Individual, volumetric basis	ka	$\text{kmol m}^{-3}\,\text{s}^{-1}\,\text{kPa}^{-1}$	h^{-1}
gas film	$k_{\text{G}}a$	"	h^{-1}
liquid film	$k_{\text{L}}a$	s^{-1}	h^{-1}
Metabolic rate, maximum specific	q_{m}	s^{-1}	h^{-1}
Metabolic rate, specific	q	s^{-1}	h^{-1}

$q = (\mathrm{d}C/\mathrm{d}t)/X$
where C may be a substrate or product mass concentration. Subscripts may further define the rates, *e.g.* q_{S}, q_{P}, q_{O_2}, which are substrate utilization, product formation, and oxygen uptake rates, respectively

	Symbol	SI units	Other units
Mutation rate	w	s^{-1}	h^{-1}
Power	P	W	W
Productivity, mass concentration rate basis, use appropriate subscripts, *e.g.* r_{X} for biomass productivity and r_{P} for product productivity	r	$\text{kg m}^{-3}\,\text{s}^{-1}$	$\text{kg m}^{-3}\,\text{h}^{-1}$
Revolutions per unit time or stirring speed	N	s^{-1}	s^{-1}
Velocity	V	m s^{-1}	cm s^{-1}

V_{s} for superficial gas velocity = $F_{\text{G}}/\pi D_{\text{t}}^{2}$

V_{i} for impeller tip velocity = $\pi N D_{\text{i}}$

Subject Index